Springer Collected Works in Mathematics

Goro Shimura

Collected Papers I

1954 – 1966

Reprint of the 2002 Edition

 Springer

Goro Shimura
Mathematics Department
Princeton University
Princeton, NJ 08544-1000
USA

ISSN 2194-9875
ISBN 978-1-4939-1809-6   (Softcover)
       978-0-387-95406-6   (Hardcover)
DOI 10.1007/978-1-4612-2074-9
Springer New York Heidelberg Dordrecht London

Library of Congress Control Number: 2012954381

Printed on acid-free paper

Springer is part of Springer Science+Business Media (www.springer.com)

Contents

Preface

A natural question that confronts anyone who wishes to publish a collection of one's own papers is whether it is really worth doing it. There are of course many kinds of justification. For example, later papers rely on the results proved in earlier ones, and each cited result was proved just once, and so it is meaningful to collect the papers in one sequence. However, what about the papers which have been completely superseded by later investigations? Even in that case, I have recognized the following fact: Mathematical contents of proved facts may have been superseded, but methods of proof, perspective, and readability are different, and the earlier papers have some good points in that respect in which their successors are not necessarily superior. Still, one must carefully examine the suitability of each article.

In the following pages the reader will find a list of my mathematical articles and an explanation about which of them are included here. At the end of each volume I have added *Notes* to the articles. They are, for the most part, corrections, clarifications, the proof of some facts which were stated but not proved in the original articles, and answers to some natural questions related to the subject treated in them. A few of them may be called new theorems. Some of my recollections are included with the hope that they may help the reader have a better perspective. I have also mentioned the results in my later articles which supersede or are related to those in the article at issue. However, I decided not to mention the results of other later investigators, mainly in order to make my task easier.

It gives me great pleasure to acknowledge my gratitude to Koji Doi, Alice Silverberg, and Hiroyuki Yoshida, who kindly helped me in various ways in this project.

Princeton
June, 2001

Goro Shimura

The Articles Included in This Collection

The list starting on the next page consists of all published articles of the author related to mathematics and four unpublished articles. One of them, [68c], was circulated only in preprint form; [64e] was distributed in mimeographed form to the participants of the conference noted in the title of the article, but never published otherwise. The last two articles [01b] and [01c] were submitted to a well-known journal in October, 2000, and accepted in April and May, 2001, but a chain of events forced me to publish them only in this collection, not in that journal. There are some more notes and one-page abstracts of my lectures at other conferences similar to [64e], but they are not listed.

All the articles on the list, excluding those belonging to the following two categories, are reproduced in these volumes. The first category consists of those published in book form; they are [57b], [61a], [68b], [71d], [97a], [98], and [00]. Those belonging to the second category are [52], [58c], [60b], [61c], [73b], [73c], [81e], [90c], [93a], and [01a]. Most of them are of expository nature, and their contents are covered by some articles reproduced here. For this and other reasons I did not think their reproduction worthwhile.

Most of the articles are photocopied from the original volumes of publication, but some are newly typeset for various reasons; they are [57a], [58b], [63e], [64e], [67c], [68c], [71c], [78c], [96b], [97b], [99c], [99d], [99e], [01b], and [01c]. In all cases except [64e], [68c], [01b], and [01c] the volume of publication, the page numbers, and the date of receipt, if any, are mentioned under the title. However, we put an asterisk at the end of the line to indicate that the article was newly typeset.

There is one article which was published as a joint paper with S. Chowla:

On the representation of zero by a linear combination of k-th powers, Det Kongelige Norske Videnskabers Selskabs Forhandlinger, 36 (1963), 169–176.

This is completely due to Chowla, who included my name as a coauthor without my knowledge or consent, for reasons only he knew. I spent July and August of 1963 in Boulder, Colorado, and at a conference held there, I gave a lecture whose text is [63e].

A few weeks before the conference, Chowla, who was at the University of Colorado at that time, showed me a manuscript, and I made some comments on it, whose nature I don't remember. Then, sometime in 1964 I received a package of reprints, on which I found my name as a coauthor. Anyway, this paper is not included in the list, since I have no responsibility for it.

List of Articles

The numbers in parentheses are the articles not included in this collection.

Volume I

[(52)] On a certain ideal of the center of a Frobeniusean algebra, Scientific Papers of the College of General Education, University of Tokyo, 2 (1952), 117–124.

[54] A note on the normalization-theorem of an integral domain, Scientific Papers of the College of General Education, University of Tokyo, 4 (1954), 1–8.

[55] Reduction of algebraic varieties with respect to a discrete valuation of the basic field, American Journal of Mathematics, 77 (1955), 134–176.

[56] On complex multiplications, Proceedings of the International Symposium on Algebraic Number Theory, Tokyo-Nikko, 1955, Science Council of Japan, Tokyo (1956), 23–30.

[57a] La fonction ζ du corps des fonctions modulaires elliptiques, Comptes rendus des séances de l'Académie des Sciences, 244 (1957), 2127–2130.

[(57b)] Kindai-teki Seisu-ron (Modern Number Theory, in Japanese, with Yutaka Taniyama), Kyoritsu Shuppan, 1957.

[58a] Correspondances modulaires et les fonctions zeta de courbes algébriques, Journal of the Mathematical Society of Japan, 10 (1958), 1–28.

[58b] Modules des variétés abéliennes polarisées et fonctions modulaires, Séminaire Henri Cartan, École Normale Supérieure, 1957/58, *Fonctions Automorphes*, Exposé 18–20 (1958).

[(58c)] Fonctions automorphes et variétés abéliennes, Séminaire Bourbaki, 1957/58, Exposé 167 (1958).

[59a] Fonctions automorphes et correspondances modulaires, Proceedings of the International Congress of Mathematicians, Edinburgh, 1958 (1959), 330–338.

[59b] On the theory of automorphic functions, Annals of Mathematics, 70 (1959), 101–144.

[59c] Sur les intégrales attachées aux formes automorphes, Journal of the Mathematical Society of Japan, 11 (1959), 291–311.

[59d] On specializations of abelian varieties (with Shoji Koizumi), Scientific Papers of the College of General Education, University of Tokyo, 9 (1959), 187–211.

[60a] On vector differential forms attached to automorphic forms (with Michio Kuga), Journal of the Mathematical Society of Japan, 12 (1960), 258–270.

[(60b)] Automorphic functions and number theory: I (in Japanese), Sugaku, 11 (1960), 193–205.

[(61a)] Complex multiplication of abelian varieties and its applications to number theory (with Yutaka Taniyama), Publications of the Mathematical Society of Japan, No. 6 (1961).

[61b] On the zeta functions of the algebraic curves uniformized by certain automorphic functions, Journal of the Mathematical Society of Japan, 13 (1961), 275–331.

[(61c)] Automorphic functions and number theory: II (in Japanese), Sugaku, 13 (1961), 65–80.

[62a] On Dirichlet series and abelian varieties attached to automorphic forms, Annals of Mathematics, 76 (1962), 237–294.

[62b] On the class-fields obtained by complex multiplication of abelian varieties, Osaka Mathematical Journal, 14 (1962), 33-44.

[63a] Arithmetic of alternating forms and quaternion hermitian forms, Journal of the Mathematical Society of Japan, 15 (1963), 33–65.

[63b] On analytic families of polarized abelian varieties and automorphic functions, Annals of Mathematics, 78 (1963), 149–192.

[63c] On the cohomology groups attached to certain vector valued differential forms on the product of the upper half planes (with Yozo Matsushima), Annals of Mathematics, 78 (1963), 417–449.

[63d] On modular correspondences for $Sp(n, \mathbf{Z})$ and their congruence relations, Proceedings of the National Academy of Sciences, 49 (1963), 824–828.

[63e] On the fields of definition for fields of automorphic functions, Proceedings of the 1963 Number Theory Conference, Boulder, Colorado, August 5–24, 1963, 26–32.

[64a] Arithmetic of unitary groups, Annals of Mathematics, 79 (1964), 369–409.

[64b] On the field of definition for a field of automorphic functions, Annals of Mathematics, 80 (1964), 160–189.

[64c] Class-fields and automorphic functions, Annals of Mathematics, 80 (1964), 444–463.

[64d] On purely transcendental fields of automorphic functions of several variables, Osaka Journal of Mathematics, 1 (1964), 1–14.

[64e] The zeta function of an algebraic variety and automorphic functions, Summer Research Institute on Algebraic Geometry, Woods Hole, Massachusetts, July 6–31, 1964 (mimeographed notes of lectures).

[65a] On the field of definition for a field of automorphic functions: II, Annals of Mathematics, 81 (1965), 124–165.

[65b] On the zeta function of a fibre variety whose fibres are abelian varieties (with Michio Kuga), Annals of Mathematics, 82 (1965), 478–539.

[66a] A reciprocity law in non-solvable extensions, Journal für die reine und angewandte Mathematik, 221 (1966), 209–220.

[66b] Moduli and fibre systems of abelian varieties, Annals of Mathematics, 83 (1966), 294–338.

[66c] On the field of definition for a field of automorphic functions: III, Annals of Mathematics, 83 (1966), 377–385.

[66d] Moduli of abelian varieties and number theory, Proceedings of the Symposium on Pure Mathematics, 9, *Algebraic groups and discontinuous subgroups*, 1965, American Mathematical Society, (1966), 312–332.

Volume II

[67a] Discontinuous groups and abelian varieties, Mathematische Annalen, 168 (1967), 171–199.

[67b] Construction of class fields and zeta functions of algebraic curves, Annals of Mathematics, 85 (1967), 58–159.

[67c] Number fields and zeta functions associated with discontinuous groups and algebraic varieties, Proceedings of the International Congress of Mathematicians, Moscow, 1966, 100–107 (1967).

[67d] Algebraic number fields and symplectic discontinuous groups, Annals of Mathematics, 86 (1967), 503–592.

[68a] Algebraic varieties without deformation and the Chow variety, Journal of the Mathematical Society of Japan, 20 (1968), 336–341.

[(68b)] Automorphic functions and number theory, Lecture notes in mathematics 54, Springer, 1968.

[68c] An ℓ-adic method in the theory of automorphic forms, the text of a lecture at the conference *Automorphic functions for arithmetically defined groups*, Oberwolfach, Germany, July 28–August 3, 1968.

[69] Local representations of Galois groups, Annals of Mathematics, 89 (1969), 99–124.

[70a] On canonical models of arithmetic quotients of bounded symmetric domains, Annals of Mathematics, 91 (1970), 144–222.

[70b] On canonical models of arithmetic quotients of bounded symmetric domains: II, Annals of Mathematics, 92 (1970), 528–549.

[71a] On arithmetic automorphic functions, Proceedings of the International Congress of Mathematicians, Nice, 1970, vol. 2, 343–348 (1971).

[71b] On the zeta-function of an abelian variety with complex multiplication, Annals of Mathematics, 94 (1971), 504–533.

[71c] Class fields over real quadratic fields in the theory of modular functions, in *Several Complex Variables* II, Maryland 1970, Lecture notes in mathematics 185 (1971), 169–188.

[(71d)] Introduction to the arithmetic theory of automorphic functions, Publications of the Mathematical Society of Japan, No. 11, Iwanami Shoten and Princeton University Press, 1971.

[71e] On elliptic curves with complex multiplication as factors of the Jacobians of modular function fields, Nagoya Mathematical Journal, 43 (1971), 199–208.

[72a] On the field of rationality for an abelian variety, Nagoya Mathematical Journal, 45 (1972), 167–178.

[72b] Class fields over real quadratic fields and Hecke operators, Annals of Mathematics, 95 (1972), 130–190.

[73a] On modular forms of half integral weight, Annals of Mathematics, 97 (1973), 440–481.

[(73b)] Complex multiplication, Proceedings of the International Summer School of Modular Functions of One Variable, Antwerp, 1972, Lecure notes in mathematics, 320 (1973), 37–56.

[(73c)] Modular forms of half integral weight, Proceedings of the International Summer School of Modular Functions of One Variable, Antwerp, 1972, Lecure notes in mathematics, 320 (1973), 57–74.

[73d] On the factors of the jacobian variety of a modular function field, Journal of the Mathematical Society of Japan, 25 (1973), 523–544.

[74] On the trace formula for Hecke operators, Acta mathematica, 132 (1974), 245–281.

[75a] On the holomorphy of certain Dirichlet series, Proceedings of the London Mathematical Society, 3rd ser. 31 (1975), 79–98.

[75b] On the real points of an arithmetic quotient of a bounded symmetric domain, Mathematische Annalen, 215 (1975), 135–164.

Volume III

[81d] Arithmetic of differential operators on symmetric domains, Duke Mathematical Journal 48 (1981), 813–843.

[(81e)] Corrections to "The special values of the zeta functions associated with Hilbert modular forms," vol 45 (1978), 637–679, Duke Mathematical Journal 48 (1981), 697.

[82a] Models of an abelian variety with complex multiplication over small fields, Journal of Number Theory, 15 (1982), 25–35.

[82b] The periods of certain automorphic forms of arithmetic type, Journal of the Faculty of Science, University of Tokyo, Sec. IA, 28 (1982), 605–632.

[82c] Confluent hypergeometric functions on tube domains, Mathematische Annalen, 260 (1982), 269–302.

[83a] Algebraic relations between critical values of zeta functions and inner products, American Journal of Mathematics, 105 (1983), 253–285.

[83b] On Eisenstein series, Duke Mathematical Journal, 50 (1983), 417–476.

[84a] Differential operators and the singular values of Eisenstein series, Duke Mathematical Journal, 51 (1984), 261–329.

[84b] On differential operators attached to certain representations of classical groups, Inventiones mathematicae, 77 (1984), 463–488.

[85a] On Eisenstein series of half-integral weight, Duke Mathematical Journal, 52 (1985), 281–314.

[85b] On the Eisenstein series of Hilbert modular groups, Revista Matemática Iberoamericana, 1 (1985), 1–42.

[86] On a class of nearly holomorphic automorphic forms, Annals of Mathematics, 123 (1986), 347–406.

[87a] Nearly holomorphic functions on hermitian symmetric spaces, Mathematische Annalen, 278 (1987), 1–28.

[87b] On Hilbert modular forms of half-integral weight, Duke Mathematical Journal, 55 (1987), 765–838.

[88] On the critical values of certain Dirichlet series and the periods of automorphic forms, Inventiones mathematicae, 94 (1988), 245–305.

Volume IV

[89a] Yutaka Taniyama and his time, Bulletin of the London Mathematical Society, 21 (1989), 186–196.

[89b] *L*-functions and eigenvalue problems, *Algebraic analysis, geometry, and number theory*, Proceedings of the JAMI Inaugural Conference 1988, Supplement to the American Journal of Mathematics, 1989, 341–396.

[90a] Invariant differential operators on hermitian symmetric spaces, Annals of Mathematics, 132 (1990), 237–272.

[90b] On the fundamental periods of automorphic forms of arithmetic type, Inventiones mathematicae, 102 (1990), 399–428.

[(90c)] Some old and recent arithmetical results concerning modular forms and related zeta functions, University of Istanbul, Faculty of Science, Journal of Mathematics, 49 (1990), 45–56.

[91] The critical values of certain Dirichlet series attached to Hilbert modular forms, Duke Mathematical Journal, 63 (1991), 557–613.

[(93a)] Arithmeticity of the special values of various zeta functions and the periods of abelian integrals (in Japanese), Sugaku, 45 (1993), 111–127; English translation by Toshitsune Miyake, Sugaku expositions, 8 (1995), 17–38.

[93b] On the transformation formulas of theta series, American Journal of Mathematics, 115 (1993), 1011–1052.

[93c] On the Fourier coefficients of Hilbert modular forms of half-integral weight, Duke Mathematical Journal, 71 (1993), 501–557.

[94a] Fractional and trigonometric expressions for matrices, American Mathematical Monthly, 101 (1994), 744–758.

[94b] Euler products and Fourier coefficients of automorphic forms on symplectic groups, Inventiones mathematicae, 116 (1994), 531–576.

[94c] Differential operators, holomorphic projection, and singular forms, Duke Mathematical Journal, 76 (1994), 141–173.

[95a] Eisenstein series and zeta functions on symplectic groups, Inventiones mathematicae, 119 (1995), 539–584.

[95b] Zeta functions and Eisenstein series on metaplectic groups, Inventiones mathematicae, 121 (1995), 21–60.

[96a] Convergence of zeta functions on symplectic and metaplectic groups, Duke Mathematical Journal, 82 (1996), 327–347.

[96b] Response, Notices of the American Mathematical Society, vol. 43, No. 11 (November 1996), 1344–1347.

[(97a)] Euler Products and Eisenstein series, CBMS Regional Conference Series in Mathematics, No. 93, American Mathematical Society, 1997.

[97b] Zeta functions and Eisenstein series on classical groups, Proceedings of the National Academy of Sciences, 94, 11133–11137 (1997).

[(98)] Abelian varieties with complex multiplication and modular functions, Princeton University Press, 1998.

[99a] An exact mass formula for orthogonal groups, Duke Mathematical Journal, 97 (1999), 1–66.

[99b] The number of representations of an integer by a quadratic form, Duke Mathematical Journal, 100 (1999), 59–92.

[99c] Generalized Bessel functions on symmetric spaces, Journal für die reine und angewandte Mathematik, 509 (1999), 35–66.

[99d] Some exact formulas on quaternion unitary groups, Journal für die reine und angewandte Mathematik, 509 (1999), 67–102.

[99e] André Weil as I knew him, Notices of the American Mathematical Society, vol. 46, No. 4 (April 1999), 428–433

[(00)] Arithmeticity in the theory of automorphic forms, Mathematical Surveys and Monographs, vol. 82, American Mathematical Society, 2000.

[(01a)] Letter to the editor, Notices of the American Mathematical Society, vol. 48, No. 7 (August 2001), 678.

[01b] Arithmeticity of Dirichlet series and automorphic forms on unitary groups, unpublished.

[01c] The relative regulator of an algebraic number field, unpublished.

Acknowledgment

The author and Springer-Verlag acknowledge that all the previously published articles in this collection have been reprinted, when required, with the permission of the publishers or editors of the journals and proceedings in which they originally appeared. Here is an alphabetical list of the journals and publishers, and the article numbers, except for one item, [76b], for which such acknowledgement appears on the first page of the article. In some cases the author can reprint without specific permission, but the following list includes even those cases. The papers published by Springer-Verlag are not included.

Academiae Scientiarum Fennica
[79b].

Acta Mathematica (Institut Mittag-Leffler)
[74], [78a].

American Journal of Mathematics (The Johns Hopkins University Press)
[55], [83a], [89b], [93b].

American Mathematical Monthly (Mathematical Association of America)
[94a].

American Mathematical Society
[66b].

Annals of Mathematics (Priceton University Press)
[59b], [62a], [63b], [63c], [64a], [64b], [64c], [65a], [65b], [66b], [66c], [67b], [67d], [69], [70a], [70b], [71b], [72b], [73a], [75c], [78b], [80], [81b], [81c], [86], [90a].

Bulletin of the London Mathematical Society
[89a].

Cambridge University Press
[59a].

Duke Mathematical Journal (Duke University Press)
   [76a], [77c], [78c], [81d], [83b], [84a], [85a], [87b], [91], [93c], [94c], [96a], [99a], [99b].

Journal für die reine und angewandte Mathematik (Walter de Gruyter GmbH & Co. KG)
   [66a], [99c], [99d].

Journal of Number Theory (Academic Press)
   [82a].

Journal of the Mathematical Society of Japan
   [58a], [59c], [60a], [61b], [63a], [68a], [73d], [79a], [81a].

Nachrichten der Akademie der Wissenschaften in Göttingen
   [75d].

Nagoya Mathematical Journal (Nagoya University)
   [71e], [72a].

Osaka Mathematical Journal, Osaka Journal of Mathematics (Osaka University)
   [62b], [64d].

Proceedings of the London Mathematical Society
   [75a], [77a].

Proceedings of the National Academy of Sciences
   [63d], [97b].

Revista Matemática Iberoamericana (Departamento de Matemáticas, Universidad Autonoma de Madrid)
   [85b].

Scientific Papers of the College of General Education, University of Tokyo
   [54], [59d], [82b].

A note on the normalization-theorem of an integral domain

Scientific Papers of the College of General Education, University of Tokyo, 4 (1954), 1-8

(Received May 15, 1954)

Let K be a field and $K(x)=K(x_1, \cdots\cdots, x_n)$ an extension of K, of dimension r. Then there exist r elements $y_1, \cdots\cdots, y_r$ in the ring $K[x]$ such that the x_j are all integral over $K[y]$. This fact is what is called the normalization-theorem.[1] If we take an integral domain I having K as its quotient field, in place of the basic field K, can we find r elements $y_1, \cdots\cdots, y_r$ in $I[x]$ such that the x_j are all integral over $I[y]$? In general, this is impossible. In this note we shall deal with this problem in case where the set (x) is homogeneous, namely, the ideal determined by the set (x) over K is a homogeneous ideal. In this case, we shall prove the existence of such a set (y) when I is a discrete valuation ring or the ring of algebraic integers in an algebraic number-field. We shall prove this in constructing r hypersurfaces in the projective space in such a way that, for every prime divisor $\mathfrak{p}$ of I, the intersection of those hypersurfaces and the locus of (x) is empty in the reduced space with respect to $\mathfrak{p}$. For that purpose we shall make use of some facts and terminologies from the theory on "reduction modulo $\mathfrak{p}$" of algebraic varieties which we have studied in a previous paper.[2]

1. Let k be a field with a discrete valuation and $\mathfrak{o}$ be its valuation ring. We denote by $\mathfrak{p}$ the maximal ideal of $\mathfrak{o}$ and by κ the residue field $\mathfrak{o}/\mathfrak{p}$. In this section we shall use these notations k, $\mathfrak{o}$, $\mathfrak{p}$ and κ always in this sense.

Let $(x)=(x_0, \cdots\cdots, x_n)$ be a set of elements in an extension field of k. We say that (x) is homogeneous over k if the set of all polynomials $F(X)$ in $k[X]=k[X_0, \cdots\cdots, X_n]$, such that $F(x)=0$, is a homogeneous ideal.

LEMMA 1. *Let $(x)=(x_0, \cdots\cdots, x_n)$ be a set of elements which is homogeneous over k and $(y)=(y_0, \cdots\cdots, y_r)$ a set of homogeneous forms in (x) with coefficients in $\mathfrak{o}$, of the same degree. Then, the x_j are all integral over $\mathfrak{o}[y]$*

---

1) For a proof of this theorem, see O. Zariski, Foundations of a general theory of birational correspondences, Trans. Amer. Math. Soc., vol. 53 (1943).

2) G. Shimura, Reduction of algebraic varieties with respect to a discrete valuation of the basic field, to appear in Amer. Journ. Math. soon. This paper will be referred to as [RA].

if and only if (x) has $(0, \cdots\cdots, 0)$ as the only finite specialization over

$$(y_0, \cdots\cdots, y_r) \,[\text{ref } \mathfrak{o}] \to (0, \cdots\cdots, 0).$$

Proof. The 'only if' part is almost obvious; so we shall only prove the 'if' part. Put $z_j = x_j^d$ for $0 \leq j \leq n$ where d is the degree of the y_i. Let $(\zeta_0, \cdots\cdots, \zeta_n)$ be a specialization of $(z_0, \cdots\cdots, z_n)$ over

$$(y, \cdots\cdots, y_r)\,[\text{ref } \mathfrak{o}] \to (0, \cdots\cdots, 0).$$

Suppose that $(\zeta_0, \cdots\cdots, \zeta_n)$ is not $(0, \cdots\cdots, 0)$; then we can find one index j, say j_0, such that, if we put $v_i = y_i/z_{j_0}$ and $w_j = z_j/z_{j_0}$ for $0 \leq i \leq r$ and $0 \leq j \leq n$, then $(v_0, \cdots\cdots, v_r, w_0, \cdots\cdots, w_n)$ has a finite specialization $(0. \cdots\cdots, 0, \omega_0, \cdots\cdots, \omega_n)$ over

$$(y_0, \cdots\cdots, y_r, z_0, \cdots\cdots, z_n)\,[\text{ref } \mathfrak{o}] \to (0, \cdots\cdots, 0, \zeta_0, \cdots\cdots, \zeta_n).^{3)}$$

Since the set (z, y) is homogeneous over k, z_{j_0} is an independent variable over $k(v, w)$; so that we have

$$(v_0, \cdots\cdots, v_r, w_0, \cdots\cdots, w_n, z_{j_0})\,[\text{ref } \mathfrak{o}] \to (0, \cdots\cdots, 0, \omega_0, \cdots\cdots, \omega_n, 1).$$

Then, by the relations $z_j = z_{j_0} w_j$ and $y_i = z_{j_0} v_i$, we see that

$$(y_0, \cdots\cdots, y_r, z_0, \cdots\cdots, z_n)\,[\text{ref } \mathfrak{o}] \to (0, \cdots\cdots, 0, \omega_0, \cdots\cdots, \omega_n).$$

By our assumption on (x), we have $(\omega_0, \cdots\cdots, \omega_n) = (0, \cdots\cdots, 0)$. This is a contradiction since $\omega_{j_0} = 1$; so that $(\zeta_0, \cdots\cdots, \zeta_n)$ must be $(0, \cdots\cdots, 0)$. Hence, for every j, $(0, \cdots\cdots, 0, 1)$ is not a specialization of $(y_0, \cdots\cdots, y_r, z_j)$ over $\mathfrak{o}$; so there exists a homogeneous polynomial $F(Z_j, Y_0, \cdots\cdots, Y_r)$ in $\mathfrak{o}[Z_j, Y_0, \cdots\cdots, Y_r]$ such that

$$F(z_i, y_0, \cdots\cdots, y_r) = 0, \quad \bar{F}(1, 0, \cdots\cdots, 0) \neq 0,$$

where $\bar{F}$ denotes the class of the polynomial F modulo $\mathfrak{p}$. Expressing $F(Z_j, Y)$ in a form

$$F(Z_j, Y) = \sum_{\nu=0}^{N} F_\nu(Y) Z_j^{N-\nu}$$

where, for each ν, $F_\nu(Y)$ is a homogeneous polynomial in $\mathfrak{o}[Y_0, \cdots\cdots, Y_r]$ of degree ν, we see that F_0 is a unit in $\mathfrak{o}$. This shows that z_j is integral over $\mathfrak{o}[y]$ for every j; so that the x_j are all integral over $\mathfrak{o}[y]$.

Lemma 2. *Let V be a bunch in the affine n-space S^n, which is normally*

---

3) This is proved by the fact that every specialization ring can be extended to a valuation ring.

algebraic over k, and denote by $\bar{V}$ the bunch obtained from V by the reduction with respect to $\mathfrak{o}^{4)}$; let (α) be a point in $\bar{V}$ which is algebraic over κ. Then there exists a point (a) in V which is algebraic over k such that

$$(a) \, [\text{ref } \mathfrak{o}] \to (\alpha).$$

PROOF. We may assume that V is a prime rational cycle over k. Let r be the dimension of V. We can find a linear variety

$$\mathfrak{L}^{n-r} = \left(\sum_{j=1}^{n} \gamma_{ij} X_j - \gamma_i = 0 \qquad (1 \leqq i \leqq r) \right)$$

such that the γ_{ij} and the γ_i are algebraic over κ and that (α) is a component of the intersection $\mathfrak{L} \frown \bar{V}$. By prop. 9 of [RA] §1, there exist an algebraic extension k_1 of k, a prolongation $\mathfrak{o}_1$ of $\mathfrak{o}$ in k_1 and a set (c_{ij}, c_i) in $\mathfrak{o}_1$ such that

$$(c_{ij}, c_i) \, [\text{ref } \mathfrak{o}] \to (\gamma_{ij}, \gamma_i).$$

Denote by L^{n-r} the linear variety

$$\left(\sum_{j=1}^{n} c_{ij} X_j - c_i = 0 \qquad (1 \leqq i \leqq r) \right).$$

Then, by th.9 of [RA] §3, there exists a component (a) in $V \frown L$ such that $(a) \, [\text{ref } \mathfrak{o}_1] \to (\alpha)$. Since L is algebraic over k and V is rational over k, (a) is algebraic over k. This proves our lemma.

2. Let $\mathfrak{o}$, $\mathfrak{p}$, k and κ be the same as in 1. Then we have

THEOREM 1. *Let $(x) = (x_0, \cdots\cdots, x_n)$ be a set of $n+1$ elements of dimension $r+1$ over k. If (x) is homogeneous over k, then there exist $r+1$ homogeneous elements $y_0, \cdots\cdots, y_r$ in $\mathfrak{o}[x]$ such that the x_j are all integral over $\mathfrak{o}[y]$.*

PROOF. Let V^r be the locus of $(x_0, \cdots\cdots, x_n)$ over k in the projective n-space L^n and denote by $\bar{V}$ the bunch obtained from V by the reduction with respect to $\mathfrak{p}$. Then we can easily find $r+1$ hypersurfaces $\mathfrak{H}_0^{n-1}, \cdots\cdots, \mathfrak{H}_r^{n-1}$ which are rational over κ such that

$$\bar{V} \frown \mathfrak{H}_0 \frown \cdots\cdots \frown \mathfrak{H}_r = \phi.$$

Let, for each i, $\bar{F}_i(X) = 0$ be a defining equation for $\mathfrak{H}_i$ and $F_i(X)$ a polynomial in $\mathfrak{o}[X]$ such that $\bar{F}_i$ is the class of F_i modulo $\mathfrak{p}$. Put

$$y_i = F_i(x)^{d/d_i} \qquad (0 \leqq i \leqq r)$$

where d_i is the degree of F_i and $d = \prod_{i=0}^{r} d_i$; then since $\bar{V} \frown \mathfrak{H}_0 \frown \cdots\cdots \frown \mathfrak{H}_r = \phi$,

---

4) See [RA] §3.

(x) has (0) as the only finite specialization over

$$(y_0, \cdots\cdots, y_r)\, [\text{ref } \mathfrak{o}] \to (0, \cdots\cdots, 0).$$

By lemma 1, the x_j are all integral over $\mathfrak{o}[y]$.

3. Now we shall deal with the case where the basic ring I is the ring of algebraic integers in an algebraic number-field.

Lemma 3. *Let the* a_{ij}, *for* $1 \leq i \leq h$ *and* $1 \leq j \leq n$, *be* hn *algebraic integers. Suppose that, for each* i, *the ideal* $(a_{i1}, \cdots\cdots, a_{in})$ *is equal to the unit ideal* (1). *Then there exists a homogeneous polynomial* $F(X_1, \cdots\cdots, X_n)$ *whose coefficients are rational integers such that* $F(a_{i1}, \cdots\cdots, a_{in})$ *is a unit for every* i.

Proof.[5] Let K be an algebraic number-field of finite degree which contains all the a_{ij}. We shall first prove, by induction on h, that there exists a homogeneous polynomial $H(X)$ whose coefficients are algebraic integers in K such that $H(a_{i1}, \cdots\cdots, a_{in})$ is a unit for every i. If $h=1$, this is obvious since $(a_{11}, \cdots\cdots, a_{1n})=1$. Suppose that there exists a homogeneous polynomial $H_0(X)$ whose coefficients are algebraic integers in K such that $H_0(a_{i1}, \cdots\cdots, a_{in})$ are units for $1 \leq i \leq h-1$. Put

$$u_i = H_0(a_{i1}, \cdots\cdots, a_{in}) \qquad (1 \leq i \leq h-1),$$

$$G_j(X) = \prod_{i=1}^{h-1} [u_i X_j^d - a_{ij}^d H_0(X)] \qquad (1 \leq j \leq n)$$

where d is the degree of $H_0(X)$; then the degree of the $G_j(X)$ is $d(h-1)$; and we have $G_j(a_{i1}, \cdots\cdots, a_{in})=0$ for $1 \leq i \leq h-1$. Put

$$b_j = G_j(a_{h1}, \cdots\cdots, a_{hn}) \qquad (1 \leq j \leq n),$$

$$a = H(a_{h1}, \cdots\cdots, a_{hn});$$

then we have

$$b_j \equiv \left(\prod_{i=1}^{h-1} u_i \right) a_{hj}^{d(h-1)} \pmod{a} \qquad (1 \leq j \leq n).$$

By our assumption $(a_{h1}, \cdots\cdots, a_{hn})=1$, we see that $(a, b_1, \cdots\cdots, b_n)=1$; so we can find a positive integer m and a unit v such that

$$a^m \equiv v \pmod{(b_1, \cdots\cdots, b_n)}.[6]$$

Put $c = a^{h-1}$, then we have

---

5) For this proof, the author is much indebted to Mr. Nagata.

6) If $(b_1, \cdots\cdots, b_n) \neq (0)$, we can take 1 as v; and if $(b_1, \cdots\cdots, b_n)=(0)$ we take a itself as v.

$$a^{m(h-1)}=c^m\equiv v^{h-1} \qquad (\mathrm{mod}\ (b_1,\ \cdots\cdots,\ b_n)).$$

Since $(c, b_1,\ \cdots\cdots,\ b_n)=1$, we have

$$(b_1,\ \cdots\cdots,\ b_n)=(b_1 c^{m-1},\ \cdots\cdots,\ b_n c^{m-1},\ b_1{}^m,\ \cdots\cdots,\ b_n^m);$$

so we have an expression

$$c^m-v^{h-1}=\sum_{j=1}^{n} d_j b_j c^{m-1}+\sum_{j=1}^{n} e_j b_j^m$$

where the d_j and the e_j are algebraic integers in K. Put

$$H(X)=H_0(X)^{m(h-1)}-\sum_{j=1}^{n} d_j G_j(X) H_0(X)^{(m-1)(h-1)}-\sum_{j=1}^{n} e_j G_j(X)^m,$$

then $H(X)$ is a homogeneous polynomial of degree $dm(h-1)$ whose coefficients are algebraic integers in K; and by our construction we have

$$H(a_{i1},\ \cdots\cdots,\ a_{in})=u_i{}^{m(h-1)} \qquad (1\leqq i\leqq h-1),$$

$$H(a_{h1},\ \cdots\cdots,\ a_{hn})=v^{h-1}.$$

Thus we have proved the existence of the homogeneous polynomial $H(X)$ whose coefficients are algebraic integers in K such that the $H(a_{i1},\ \cdots\cdots,\ a_{in})$ are units for $1\leqq i\leqq h$. Now let K_0 be a Galois extension of the field of rational numbers which contains the a_{ij}, and $\mathfrak{G}=\{\sigma\}$ the Galois group of this extension. By what is proved above, we can find a homogeneous polynomial $H(X)$ whose coefficients are algebraic integers in K_0 such that for every i and every σ in $\mathfrak{G}$, $H(a_{i1}^{\sigma},\ \cdots\cdots,\ a_{in}^{\sigma})$ is a unit. Put

$$F(X)=\prod_{\sigma\in\mathfrak{G}} H^{\sigma}(X);$$

then $F(X)$ is a homogeneous polynomial whose coefficients are rational integers. Since

$$H^{\sigma}(a_{i1},\ \cdots\cdots,\ a_{in})=H(a_{i1}^{\sigma^{-1}},\ \cdots\cdots,\ a_{in}^{\sigma^{-1}})^{\sigma},$$

$F(a_{i1},\ \cdots\cdots,\ a_{in})$ are units for $1\leqq i\leqq h$. This proves our lemma.

LEMMA 4. *Let the a_{ij}, for $1\leqq i\leqq h$ and $1\leqq j\leqq n$, be hn algebraic integers and denote by the $\mathfrak{a}_i$ the ideals $(a_{i1},\ \cdots\cdots,\ a_{in})$ for $1\leqq i\leqq h$. Then there exists a homogeneous polynomial $F(X_1,\ \cdots\cdots,\ X_n)$ whose coefficients are rational integers such that*

$$(F(a_{i1},\ \cdots\cdots,\ a_{in}))=\mathfrak{a}_i{}^d \qquad (1\leqq i\leqq h)$$

where d is the degree of $F(X)$.

PROOF. We can find a positive integer l such that $\mathfrak{a}_i{}^l$ is a principal ideal (b_i) for every i. Put $c_i = \sqrt[l]{b_i}$ for $1 \leq i \leq h$; then we have $\mathfrak{a}_i = (c_i)$ for $1 \leq i \leq h$. By lemma 3, there exists a homogeneous polynomial $F(X)$ whose coefficients are rational integers such that $F(c_i^{-1}a_{i1}, \cdots\cdots, c_i^{-1}a_{in})$ are units for $1 \leq i \leq h$. Then we have $(F(a_{i1}, \cdots\cdots, a_{in})) = (c_i{}^d)$ for $1 \leq i \leq h$ where d is the degree of $F(X)$. This proves our lemma.

LEMMA 5. *Let K be an algebraic number-field, L^n the projective n-space defined over K and $P = (a_0, \cdots\cdots, a_n)$ be a point in L^n with the a_j in K; and denote by $\mathfrak{a}$ the ideal $(a_0, \cdots\cdots, a_n)$ in K. Let $F(X_0, \cdots\cdots, X_n)$ be a homogeneous polynomial, whose coefficients are algebraic integers in K having no common divisor; and let H be the hypersurface in L^n defined by the equation $F(X) = 0$. Then the bunch $\bar{H}^{(\mathfrak{p})}$ obtained from H by the reduction modulo $\mathfrak{p}$ does not contain the reduced point $\bar{P}^{(\mathfrak{p})}$ for every prime divisor $\mathfrak{p}$ of K, if and only if $(F(a_0, \cdots\cdots, a_n)) = \mathfrak{a}^d$ where d is the degree of $F(X)$.*

This is easily proved by the definition of the reduced bunch ([RA] §3).

LEMMA 6. *Let K be an algebraic number-field of finite degree and V a pure r-dimensional bunch in the projective n-space L^n which is normally algebraic over K. Then there exists a hypersurface H^{n-1} in L^n which is rational over the field of rational numbers such that, for every prime divisor $\mathfrak{p}$ of K, $\bar{V}^{(\mathfrak{p})} \frown \bar{H}^{(\mathfrak{p})}$ is of dimension $r-1$, where $\bar{V}^{(\mathfrak{p})}$ and $\bar{H}^{(\mathfrak{p})}$ denote the bunches obtained from V and H by the reduction modulo $\mathfrak{p}$, respectively.*

PROOF. Let $V_1, \cdots\cdots, V_t$ be the components of V and K_0 a finite extension of K such that all the V_j are defined over K_0. By th.26 of [RA] §6, there exists a finite set of prime divisors $\{\mathfrak{p}_1, \cdots\cdots, \mathfrak{p}_s\}$ of K_0 such that if $\mathfrak{p} \neq \mathfrak{p}_\lambda$ $(1 \leq \lambda \leq s)$, then the bunch $\bar{V}_j^{(\mathfrak{p})}$ consists of only one component for each j. Let, for each j, P_j be an absolutely algebraic point[7] in V_j. Denote by $\mathfrak{B}_\nu^{(\lambda)}$ $(1 \leq \lambda \leq s, 1 \leq \nu \leq l_\lambda)$ the components of the $\bar{V}^{(\mathfrak{p}_\lambda)}$, and let $\bar{P}_{\lambda\nu}$ be an absolutely algebraic point in $\mathfrak{B}_\nu^{(\lambda)}$ for each λ and ν. By lemma 2, for each λ and ν, there exists an absolutely algebraic point $P_{\lambda\nu}$ such that $\bar{P}_{\lambda\nu}$ is a specialization of $P_{\lambda\nu}$ with respect to $\mathfrak{p}_\lambda$. Denote by $(a_{i0}, \cdots\cdots, a_{in})$ for $1 \leq i \leq h$ the coordinates of $h = t + \sum_{\lambda=1}^{s} l_\lambda$ points P_j and $P_{\lambda\nu}$; where we may assume that the a_{ij} are algebraic integers. Then, applying lemma 4 to these sets $(a_{i0}, \cdots, a_{in})$, we obtain a homogeneous polynomial $F(X)$ satisfying the conditions as in lemma 4. Denote by H^{n-1} the hypersurface defined by the equation $F(X) = 0$.

---

7) We understand by an absolutely algebraic point a point whose non-homogeneous coordinates are algebraic over a prime field.

Then, by lemma 5, we can easily verify that $\bar{H}^{(\mathfrak{p})}$ does not contain any component of $\bar{V}^{(\mathfrak{p})}$ for every prime divisor of K_0. This proves our lemma.

THEOREM 2. *Let K be an algebraic number-field of finite degree and I the ring of algebraic integers in K; let $(x)=(x_0, \cdots\cdots, x_n)$ be a set of $n+1$ elements of dimension $r+1$ over K. If (x) is homogeneous over K, then there exists a set of homogeneous forms $(y_0, \cdots\cdots, y_r)$ in (x) with coefficients in I such that the x_j are all integral over $I[y]$.*

PROOF. Let V^r be the locus of (x) over K in the projective n-space L^n. Then by lemma 6, we can find $r+1$ hypersurfaces $H_0^{n-1}, \cdots\cdots, H_r^{n-1}$ which are rational over the field of rational numbers such that, for every prime divisor $\mathfrak{p}$ of K, we have

$$\bar{V}^{(\mathfrak{p})} \frown \bar{H}_0^{(\mathfrak{p})} \frown \cdots\cdots \frown \bar{H}_r^{(\mathfrak{p})} = \phi$$

where $\bar{V}^{(\mathfrak{p})}$ and the $\bar{H}_i^{(\mathfrak{p})}$ denote the bunches obtained from V and the H_i by the reduction modulo $\mathfrak{p}$, respectively. Let, for each i, $F_i(X)=0$ be a defining equation for H_i where we may assume that the coefficients of F_i are rational integers without common divisor. Put

$$y_i = F_i(x)^{d/d_i} \qquad (0 \leq i \leq r)$$

where d_i is the degree of F_i and $d = \prod_{i=0}^{r} d_i$. Then, by lemma 1, the x_j are integral over $\mathfrak{o}_\mathfrak{p}[y]$ for every prime divisor $\mathfrak{p}$ of K where $\mathfrak{o}_\mathfrak{p}$ denotes the valuation ring corresponding to a prime divisor $\mathfrak{p}$, for each $\mathfrak{p}$. Therefore, the x_j are integral over $I[y]$.

4. We remark that the conclusion of th.2 does not necessarily hold, if we take a more general integral domain in place of the ring of algebraic integers, even in case where the basic ring is a polynomial domain of one variable over a field. Let C be the field of complex numbers, t a variable over C and denote by I the polynomial domain $C[t]$. We can easily find a finite algebraic extension K of $C(t)$ and an ideal $\mathfrak{a}$ of the integral closure I^* of I in K such that, for any positive integer m, $\mathfrak{a}^m$ is not a principal ideal. Let $(a_0, \cdots\cdots, a_n)$ be a basis of this ideal $\mathfrak{a}$; and u a variable over K; and put $x_j = u a_j$ for $0 \leq i \leq n$. Then $(x_0, \cdots\cdots, x_n)$ is a homogeneous set over $C(t)$; but we can not find a homogeneous element y in $I[x]$ such that the x_j are all integral over $I[y]$. For, suppose that there exists such an element y in $I[x]$. Then by lemma 1, $(0, \cdots, 0)$ is the only finite specialization of $(x_0, \cdots, x_n)$ over $y[\text{ref } \mathfrak{o}_\mathfrak{p}] \to 0$ for every prime divisor $\mathfrak{p}$ of K. Expressing y in a

form $y=F(x)$ where $F(X)$ is a homogeneous polynomial in $I[X_0, \cdots, X_n]$, we see that the coefficients of $F(X)$ have no common divisor. Then by lemma 5, the ideal $(F(a_0, \cdots, a_n))$ must coincide with the ideal $\mathfrak{a}^d$ where d is the degree of $F(X)$. This is a contradiction since $\mathfrak{a}^d$ is not a principal ideal.

Reduction of algebraic varieties with respect to a discrete valuation of the basic field

American Journal of Mathematics, 77 (1955), 134-176

In the arithmetic theory of algebraic varieties it is useful to have a general theory on "reduction modulo $\mathfrak{p}$" of varieties. Here "reduction" means, roughly spoken, the following process: let V be an algebraic variety defined over a field k with a discrete valuation, and denote by $\mathfrak{p}$ the maximal ideal of that valuation; then we have an algebraic variety $\bar{V}$ over the residue field κ of $\mathfrak{p}$ defined by the equations for V modulo $\mathfrak{p}$; such a variety $\bar{V}$ may be called the variety obtained from the variety V by the reduction modulo $\mathfrak{p}$. When one tries to construct the general theory of reduction in this sense, the first thing to do will be to investigate how the algebraico-geometric properties of V behave in the reduction process. The object of this paper is concerned with that problem.

We shall give a precise definition of the reduction of varieties and the reduction of cycles on a variety. In the case when the ambient varieties V and $\bar{V}$ are both absolutely irreducible, we shall show that the reduction of cycles on V defines a linear mapping from the group of rational V-cycles over k into the group of rational $\bar{V}$-cycles over κ of the same dimension; and shall show that the operations on cycles (the intersection-product, the direct product and the algebraic projection) are all preserved by that mapping (Section 4). If we consider reduction as specialization, this result may be regarded as a specialization-theory of cycles on algebraic varieties, in a certain sense. In fact, from this reduction-theory of cycles we can easily obtain a more general specialization-theory of cycles (Section 5). The latter will be a generalization of the specialization-theory of cycles which has been given by T. Matsusaka [4] and P. Samuel [6]. Our theory may be also called a "local" theory of varieties at a divisor $\mathfrak{p}$, whereas $\mathfrak{p}$ is not a usual divisor on varieties. We shall also show the following fact: let K be a field with a set of infinitely many valuations which satisfies a certain condition (condition (I) in Section 6); if V is an absolutely irreducible variety defined over K, then for almost all those valuations, it remains absolutely irreducible by the reduction process.

---

* Received February 15, 1954.

M. Deuring has given a reduction-theory of an algebraic function-field of one variable (Deuring [1]) and applied it to his theory of complex multiplication (Deuring [2]). Our results will supply an algebraico-geometric method for these theories.

We proceed in following the method of A. Weil in his book "Foundations of algebraic geometry."[1] We consider specializations over local domains,[2] and by extending Weil's theory to our case, we obtain our results without difficulty. As in Weil's book we shall first deal with affine varieties, and then proceed to abstract and projective varieties.

1. Specialization over an S-domain. We shall call a commutative ring with the identity element an S-*ring* if the set of all non-units forms an ideal. An S-ring which has no zero-divisor will be called an S-*domain*. We shall call a Noetherian S-ring (or S-domain) a *local ring* (or *local domain*). If $\Re$ is a local ring and $\mathfrak{m}$ is the maximal ideal of $\Re$, then we have a topology on $\Re$ by the powers of $\mathfrak{m}$. With respect to this topology, $\Re$ has a *completion* $\Re^*$, which is a local ring containing $\Re$ as subring and subspace, and in which $\Re$ is dense. A valuation ring is an S-domain; and it is a local domain if it is discrete or is a field.

Let $\Re$ be an S-*domain* with the quotient field K and the maximal ideal $\mathfrak{m}$, and $\mathrm{K} = \Re/\mathfrak{m}$ its residue field. For convenience we consider two "*universal domains*" K and $\Re$ which are algebraically closed fields of infinite degree of transcendency over K and K, respectively. We call an element of K (or $\Re$) a *quantity of* K (or $\Re$).[3] Let $(x) = (x_1, \cdots, x_n)$ be a set of n quantities in K and $(\xi) = (\xi_1, \cdots, \xi_n)$ a set of n quantities in $\Re$. We say that (ξ) is a *specialization of* (x) *over* $\Re$ and denote $(x) \xrightarrow{\Re} (\xi)$ if the natural homomorphism of $\Re$ to $\Re/\mathfrak{m}$ can be extended to a homomorphism of $\Re[x]$ onto $\mathrm{K}[\xi]$ which maps (x) on (ξ). It is easy to see that (ξ) is a specialization of (x) over $\Re$ if and only if for every polynomial $F(X)$ in $\Re[X]$ such that $F(x) = 0$ we have $\bar{F}(\xi) = 0$ where $\bar{F}(X)$ is the class of $F(X)$ modulo $\mathfrak{m}$. As in [WF] we add an *infinite* element ∞ both to K and $\Re$ with operations $\infty^{-1} = 0$ and $0^{-1} = \infty$, and by a *generalized quantity* of K (or $\Re$) we understand either a quantity of K (or $\Re$) or ∞. So we say that a set of generalized quantities (x) in K (or (ξ) in $\Re$) is *finite* or *infinite*

---

[1] Hereafter we shall use the same notations and terminologies as in this book. We shall quote this book as [WF].

[2] Such a specialization has been defined in D. G. Northcott [5] and applied to prove O. Zariski's main theorem on birational correspondences.

[3] The elements of K will be denoted by Latin, those of $\Re$ by Greek letters.

according as none of the x_j (or ξ_j) is ∞ or otherwise. Let (x) be a set of n generalized quantities in K and (ξ) a set of n generalized quantities in $\Re$; we call (ξ) a *specialization of* (x) *over* $\Re$, if there exists a set of n integers $\epsilon_j = \pm 1$, such that $(x') = (x_1^{\epsilon_1}, \cdots, x_n^{\epsilon_n})$ and $(\xi') = (\xi_1^{\epsilon_1}, \cdots, \xi_n^{\epsilon_n})$ are both finite and (ξ') is a specialization of (x') over $\Re$; and also denote by $(x) \xrightarrow{\Re} (\xi)$. We can easily show that if we have $(x) \xrightarrow{K} (x')$ and $(x') \xrightarrow{\Re} (\xi)$ then we have $(x) \xrightarrow{\Re} (\xi)$ and that if we have $(x) \xrightarrow{\Re} (\xi)$ and $(\xi) \xrightarrow{K} (\xi')$ then we have $(x) \xrightarrow{\Re} (\xi')$.

The following theorem is well-known. A proof is given in Weil [7] (Theorem 1.) for a more general case.

THEOREM 1. *Let (x) and (y) be two sets of generalized quantities in K. Then every specialization $(x) \xrightarrow{\Re} (\xi)$ can be extended to a specialization $(x, y) \xrightarrow{\Re} (\xi, \eta)$.*

This theorem asserts that, for every S-domain $\Re$ contained in a field K_1, there exists a valuation ring $\Re_1$ with the quotient field K_1 such that $\Re_1 \supset \Re$, $\mathfrak{m}_1 \cap \Re = \mathfrak{m}$ where $\mathfrak{m}$ and $\mathfrak{m}_1$ are respectively the maximal ideals of $\Re$ and $\Re_1$.

We say that (η) is a *specialization of* (y) *over* $(x) \xrightarrow{\Re} (\xi)$ if (ξ, η) is a specialization of (x, y) over $\Re$.

Let (ξ) be a finite specialization of (x) over $\Re$. We call the set $\{F(x)/G(x) \mid \bar{G}(\xi) \neq 0\}$ the *specialization ring of* (ξ) *in* $K(x)$, where $F(X)$ and $G(X)$ are polynomials in $\Re[X]$, and $\bar{G}(X)$ is the class of $G(X)$ modulo $\mathfrak{m}$.[3'] This set is also an S-domain and will be denoted by $[(x) \xrightarrow{\Re} (\xi)]$. If $\Re$ is a local domain, the specialization ring $[(x) \xrightarrow{\Re} (\xi)]$ is also a local domain.

Let (x) be a set of quantities in K; we say that (x) is *finite over* $\Re$ if every specialization (ξ) of (x) over $\Re$ is finite. The following proposition is also well-known (Weil [7], Theorem 2.).

PROPOSITION 1. *A set of quantities* $(x) = (x_1, \cdots, x_n)$ *is finite over $\Re$ if and only if every one of the x_j is integral over $\Re$.*

The following two propositions are easily proved.

PROPOSITION 2. *Let (ξ) be a specialization of (x) over $\Re$. If $\Re$ is a valuation ring, then $\dim_K (\xi) \leqq \dim_K (x)$.*

PROPOSITION 3. *Let (x) be a set of n independent variables over K and (ξ) a set of n independent variables over K. If $\Re$ is a (discrete) valuation ring, then $[(x) \xrightarrow{\Re} (\xi)]$ is also a (discrete) valuation ring.*

[3'] Hereafter we use the same notation for the class of a polynomial modulo the maximal ideal. Also we shall use the notation $\bar{k}$ for the algebraic closure of a field k.

LEMMA 1. *Let A be an integral domain and $(t_1, \cdots, t_n)$ a set of n independent variables over A. If A is integrally closed, then the polynomial domain $A[t]$ is also integrally closed.*

Proof. It is sufficient to prove our lemma in the case $n = 1$. It is well-known that an integral domain is integrally closed if and only if it is an intersection of valuation rings. (This follows from Theorem 1 and Proposition 1.) By this fact, we have a representation of A as an intersection $\cap \, \mathfrak{R}_\omega$ of valuation rings $\mathfrak{R}_\omega$. Denote by K the quotient field of A and by $\mathfrak{R}'_\omega$ the specialization ring $[t \xrightarrow{\;\mathfrak{R}_\omega\;} \tau_\omega]$ where τ_ω is an independent variable over the residue field of $\mathfrak{R}_\omega$. Then, by Proposition 3, $\mathfrak{R}'_\omega$ is also a valuation ring for every ω. It can be easily verified that

$$A[t] = \bigcap_\omega \mathfrak{R}_\omega \cap K[t].$$

As is well-known $K[t]$ is integrally closed; so that $A[t]$ is integrally closed.

PROPOSITION 4. *Let $(t_1, \cdots, t_n)$ be a set of n independent variables over K and $(\tau_1, \cdots, \tau_n)$ a set of n quantities in $\mathfrak{R}$. If $\mathfrak{R}$ is an integrally closed S-domain, then the specialization ring $[(t) \xrightarrow{\;\mathfrak{R}\;} (\tau)]$ is also integrally closed.*

This follows immediately from Lemma 1.

PROPOSITION 5. *Let (ξ) be a finite specialization of (x) over $\mathfrak{R}$ such that the specialization ring $\mathfrak{S} = [(x) \xrightarrow{\;\mathfrak{R}\;} (\xi)]$ is integrally closed. If $\mathfrak{R}$ is a (discrete) valuation ring, and if $\dim_K (x) = \dim_K (\xi)$, then $\mathfrak{S}$ is also a (discrete) valuation ring.*

Proof. Let y be a quantity in $K(x)$ which is not in $\mathfrak{S}$. Then, by Proposition 1 and by our assumption, we have $y \xrightarrow{\;\mathfrak{S}\;} \infty$. If η is any finite specialization of y over $\mathfrak{S}$, then, by Proposition 2, we have

$$\dim_K (\xi, \eta) \leqq \dim_K (x, y) = \dim_K (x) = \dim_K (\xi).$$

This shows that η is algebraic over $K(\xi)$. Let t be an independent variable over $K(x)$ and τ an independent variable over $K(\xi)$. Put $\mathfrak{T} = [t \xrightarrow{\;\mathfrak{S}\;} \tau]$ and $z = (t + y)^{-1}$. By Proposition 4, $\mathfrak{T}$ is integrally closed; and it is easy to see that z is finite over $\mathfrak{T}$; so that z is contained in $\mathfrak{T}$, and has a uniquely determined specialization over $t \xrightarrow{\;\mathfrak{S}\;} \tau$. This specialization of z must be 0, since $y \xrightarrow{\;\mathfrak{S}\;} \infty$; this shows that y has the only specialization ∞ over $\mathfrak{S}$. Hence y^{-1} is finite over $\mathfrak{S}$, so that y^{-1} is contained in $\mathfrak{S}$, by Proposition 1. Thus we have proved that $\mathfrak{S}$ is a valuation ring. If $\mathfrak{R}$ is discrete, then, by the assumption $\dim_K (x) = \dim_K (\xi)$, $\mathfrak{S}$ is also discrete.

Let K' be any extension of K and let $\Re'$ be an S-domain with the quotient field K' and the maximal ideal $\mathfrak{m}'$, such that $\mathfrak{m}' \cap \Re = \mathfrak{m}$, $\Re' \supset \Re$. When we fix a homomorphism of $\Re'$ into $\Re$ which is an extension of the natural mapping of $\Re$ onto K, we shall say that $\Re'$ is a *prolongation of $\Re$ in K'*, and call the image of $\Re'$ by that homomorphism the *residue field* of $\Re'$. Thus, when we speak of a prolongation of an S-domain or a valuation ring, it should be remembered that " prolongation " is a combined conception of ring and homomorphism. Also we regard the specialization ring $[(x) \xrightarrow{\ \Re\ } (\xi)]$ with its natural mapping $(x) \longrightarrow (\xi)$ as a prolongation of $\Re$.

PROPOSITION 6. *Let $\Re$ be a discrete valuation ring and $\Re'$ a prolongation of $\Re$ in an extension K' of K. If (ξ) is a specialization of (x) over $\Re$, then there exists a generic specialization (x') of (x) over K, such that (ξ) is a specialization of (x') over $\Re'$.*

Proof. We may assume that (x) and (ξ) are both finite. Let $\mathfrak{A}$ be the ideal determined by (x) over the field K. We have a representation of the ideal $K'\mathfrak{A}$ of $K'[X]$ as an intersection of primary ideals $\mathfrak{Q}_i$, belonging respectively to prime ideals $\mathfrak{P}_i$. For every i, we have a set of quantities $(x^{(i)})$ such that $\mathfrak{P}_i$ is the ideal determined by $(x^{(i)})$ over K'. Then we have $(x) \xrightarrow{\ \ \ } (x^{(i)})$. It is easy to prove that there exists, for every i, a generic specialization $(x'^{(i)})$ of (x) over K such that $(x^{(i)})$ is a specialization of $(x'^{(i)})$ over K'. Hence our proposition is proved if we show that for at least one value of i, $(x^{(i)})$ has (ξ) as a specialization over $\Re'$. Assume that this is impossible. Then there exists, for every i, a polynomial $F_i(X)$ in $\Re'[X]$ such that $F_i(x^{(i)}) = 0$ and that $\bar{F}_i(\xi) \neq 0$. Then $F_i(X) \, \varepsilon \, \mathfrak{P}_i$ for every i; so we can find an integer ρ such that the polynomial $F(X) = \prod_i F_i(X)^\rho$ is contained in $K'\mathfrak{A}$. As $F(X)$ is contained in $\Re'[X]$, we obtain an expression

$$F(X) = \sum_{\nu=1}^{s} b_\nu G_\nu(X)$$

where $b_\nu \, \varepsilon \, \Re'$ and $G_\nu(X) \, \varepsilon \, \Re[X]$. Since $\Re$ is a discrete valuation ring, the finite $\Re$-module $\Re b_1 + \cdots + \Re b_s$ has an $\Re$-basis $(c_1, \cdots, c_r)$ such that $c_1, \cdots, c_r$ are linearly independent over K. Expressing b_ν in a form $b_\nu = \sum_\lambda u_{\nu\lambda} c_\lambda$ with $u_{\nu\lambda} \, \varepsilon \, \Re$, we have $F(X) = \sum_\nu b_\nu G_\nu(X)$ $= \sum_\lambda c_\lambda H_\lambda(X)$, where $H_\lambda(X) = \sum_\nu u_{\nu\lambda} G_\nu(X) \, \varepsilon \, \Re[X]$. Then it follows from $F(X) \, \varepsilon \, K'\mathfrak{A}$ that $H_\lambda(X) \, \varepsilon \, \mathfrak{A}$ for every λ. Then we have $\bar{F}(\xi) = \sum_\lambda \bar{c}_\lambda \bar{H}_\lambda(\xi)$ $= 0$; this is a contradiction.

PROPOSITION 7. *Let $\Re$ be a discrete valuation ring and $\Re'$ a prolongation of $\Re$ in an extension K' of K. Let (x) be a set of quantities such that*

$K(x)$ and K' are linearly disjoint over K, and (ξ) a finite specialization of (x) over $\mathfrak{R}$. Then (ξ) is a specialization of (x) over $\mathfrak{R}'$; and we have $[(x) \xrightarrow{\mathfrak{R}} (\xi)] = K(x) \cap [(x) \xrightarrow{\mathfrak{R}'} (\xi)]$.

Proof. Let $F(X)$ be a polynomial in $\mathfrak{R}'[X]$ such that $F(x) = 0$. We have an expression $F(X) = \sum_{\lambda=1}^{r} c_\lambda H_\lambda(X)$ where $c_\lambda \, \varepsilon \, \mathfrak{R}'$, $H_\lambda(X) \, \varepsilon \, \mathfrak{R}[X]$ for every λ. By the same procedure as in the proof of Proposition 6, we may assume that $c_1, \, \cdots, c_r$ are linearly independent over K. Then, by our assumption that K' and $K(x)$ are linearly disjoint over K, the relation $\sum_\lambda c_\lambda H_\lambda(x) = F(x) = 0$ implies that $H_\lambda(x) = 0$ for every λ. Since (ξ) is a specialization of (x) over $\mathfrak{R}$, we have $\bar{F}(\xi) = \sum_\lambda \bar{c}_\lambda \bar{H}_\lambda(\xi) = 0$; this proves the first assertion. Now let y be a quantity in $K(x) \cap [(x) \xrightarrow{\mathfrak{R}'} (\xi)]$; then we have an expression $y = P(x)/Q(x)$ with $P(X)$ and $Q(X)$ in $\mathfrak{R}'[X]$ and $\bar{Q}(\xi) \neq 0$. Let $(b_1, \cdots, b_s)$ be the set of all coefficients of P and Q; as in the proof of Proposition 6, we have an $\mathfrak{R}$-basis $d_1, \cdots, d_t$ of $\mathfrak{R}b_1 + \cdots + \mathfrak{R}b_s$ which are linearly independent over K. Then we have expressions $P(X) = \sum_\nu d_\nu F_\nu(X)$, $Q(X) = \sum_\nu d_\nu G_\nu(X)$ where the $F_\nu(X)$ and the $G_\nu(X)$ are in $\mathfrak{R}[X]$. By our assumption, from the relation $\sum_\nu d_\nu (F_\nu(x) - yG_\nu(x)) = 0$ follows that $F_\nu(x) = yG_\nu(x)$ for every ν. Since $\bar{Q}(\xi) \neq 0$, we have $\bar{G}_\nu(\xi) \neq 0$ for some ν; for that ν we have $y = F_\nu(x)/G_\nu(x)$; this proves the second assertion.

PROPOSITION 8. *Let $K(x)$ and $K(y)$ be two extensions of K which are linearly disjoint over K. If $\mathfrak{R}$ is a discrete valuation ring, then (ξ, η) is a specialization of (x, y) over $\mathfrak{R}$, if and only if (ξ) is a specialization of (x) over $\mathfrak{R}$ and (η) is a specialization of (y) over $\mathfrak{R}$.*

Proof. We shall prove only the direct part since the converse is obvious. We may assume that (ξ) and (η) are both finite. Put $K' = K(y)$ and $\mathfrak{R}' = [(y) \xrightarrow{\mathfrak{R}} (\eta)]$. We may consider $\mathfrak{R}'$ is a prolongation of $\mathfrak{R}$ in K'. Then our proposition is an immediate consequence of Proposition 7.

PROPOSITION 9. *Let $\mathfrak{R}$ be a discrete valuation ring. For any finitely generated extension $\mathbf{K}'$ of $\mathbf{K}$, there exist a finitely generated extension K' of K and a prolongation $\mathfrak{R}'$ of $\mathfrak{R}$ in K' which is a discrete valuation ring, such that $\mathbf{K}'$ is the residue field of $\mathfrak{R}'$.*

Proof. Our proposition follows from Proposition 3 if K' is purely transcendental over K; so that we have only to prove our proposition in case where K' is a simple algebraic extension $K(\xi)$ of K. Let $\bar{F}(X) = 0$ be an irreducible equation for ξ over K; where we may assume that $\bar{F}$ is the class of a polynomial F in $\mathfrak{R}[X]$ of the same degree as $\bar{F}$, modulo $\mathfrak{m}$. Let x be a root of the equation $F(X) = 0$ in $\bar{K}$, and put $K' = K(x)$. We have a discrete valuation ring $\mathfrak{R}'$ in K' with the maximal ideal $\mathfrak{m}'$, such that $\mathfrak{R}' \supset \mathfrak{R}$, $\mathfrak{m}' \cap \mathfrak{R} = \mathfrak{m}$. Since the class of x modulo $\mathfrak{m}'$ satisfies the equation $\bar{F}(X) = 0$, we have $[K':K] \leqq [\mathfrak{R}'/\mathfrak{m}':\mathfrak{R}/\mathfrak{m}]$. It is well-known that $[\mathfrak{R}'/\mathfrak{m}':\mathfrak{R}/\mathfrak{m}] \leqq [K':K]$; so, by our definition of K', we have $[K':K] = [\mathfrak{R}'/\mathfrak{m}':\mathfrak{R}/\mathfrak{m}] = [K':K]$. This shows that $\mathfrak{R}'/\mathfrak{m}'$ is isomorphic to K' over K; so our proposition is proved.

2. Multiplicity of a proper specialization. As before, let $\mathfrak{R}$ be an S-domain with the quotient field K and the residue field K. We say that (η) is an *isolated specialization of* (y) *over* $\mathfrak{R}$, if whenever we have $(y) \xrightarrow{\mathfrak{R}} (\eta')$ and $(\eta') \xrightarrow{K} (\eta)$, we also have $(\eta) \xrightarrow{K} (\eta')$. A finite specialization of (y) over $\mathfrak{R}$ is called a *proper specialization of* (y) *over* $\mathfrak{R}$ if it is isolated and algebraic over K.

Let (y) be a set of algebraic quantities over K and let $(y^{(1)}, \cdots, y^{(n)})$ be a complete set of conjugates of (y) over K. Let (η) be a specialization of (y) over $\mathfrak{R}$. If there exists a number μ, such that for every specialization $(\eta^{(1)}, \cdots, \eta^{(n)})$ of $(y^{(1)}, \cdots, y^{(n)})$ over $\mathfrak{R}$, the set (η) occurs exactly μ times among $(\eta^{(\nu)})$, then we call the number μ the *multiplicity of* (η) *as a specialization of* (y) *over* $\mathfrak{R}$. The main object of this section is to show that a proper specialization over a certain local domain has a well-defined multiplicity.

PROPOSITION 10. *Let $\mathfrak{R}$ be an integrally closed S-domain with the quotient field K, and (y) a set of algebraic quantities over K. If (y) is finite over $\mathfrak{R}$, then every specialization of (y) over $\mathfrak{R}$ is proper and has a multiplicity defined above.*

Proof. By Proposition 1 of Section 1, (y) is integral over $\mathfrak{R}$; so that every specialization of (y) over $\mathfrak{R}$ is finite algebraic over K. From this follows that every specialization of (y) over $\mathfrak{R}$ is proper. As for the assertion concerned with the multiplicity, first we shall prove it in case where the set (y) consists of a single quantity y. Let $F(Y) = 0$ be the irreducible equation for y over K with the leading coefficient 1. Then $F(Y)$ is a polynomial contained in $\mathfrak{R}[Y]$ since $\mathfrak{R}$ is integrally closed. It can be easily verified that every specialization of the complete set of conjugates of y coincides with the set of all roots of the equation $\bar{F}(Y) = 0$. This proves our proposition in the case when (y) consists of one quantity. To prove the general case we put $z = \sum_i t_i y_i$ where (t) is a set of independent variables over K. Let (τ) be a set of independent variables over $\mathbf{K}$, and denote by $\mathfrak{S}$ the specialization ring $[(t) \xrightarrow{\mathfrak{R}} (\tau)]$. Then, by Proposition 4, $\mathfrak{S}$ is also an integrally closed S-domain; and it is easy to see that z is finite over $\mathfrak{S}$. If we denote by $(y^{(1)}, \cdots, y^{(n)})$ the complete set of conjugates of (y) over K and put $z_\nu = \sum_i t_i y_i^{(\nu)}$, then the set $(z_1, \cdots, z_n)$ consists of a repetition in a certain number of times of the complete set of conjugates of z over $K(t)$. So the general case is reduced to the case of one quantity.

Let $\mathfrak{R}$ be a local domain such that the completion $\mathfrak{R}^*$ of $\mathfrak{R}$ has no zero-divisor. We identify the quotient field K of $\mathfrak{R}$ with a subfield of the quotient field Ω of $\mathfrak{R}^*$ in the usual manner. The following two propositions are generalizations of Theorem 1 and Theorem 3 of [WF] Chapter III, Section 3. Proofs by the same method as in [WF] have been stated in Northcott [5] in a general form.

PROPOSITION 11. *Let $\mathfrak{R}^*$ be a complete local domain with the quotient field Ω and (y) a set of elements in some field containing Ω. If (y) has a proper specialization over $\mathfrak{R}^*$, then (y) is integral over $\mathfrak{R}^*$; and any two specializations of (y) over $\mathfrak{R}^*$ are conjugates of each other over the residue field of $\mathfrak{R}^*$.*

PROPOSITION 12. *Let $\mathfrak{R}$ be a local domain such that the completion $\mathfrak{R}^*$ of $\mathfrak{R}$ has no zero-divisor, (y) a set of generalized quantities which is algebraic over K, and (η) a specialization of (y) over $\mathfrak{R}$. Let Ω be the quotient field of $\mathfrak{R}^*$ and $\bar{\Omega}$ the algebraic closure of Ω. Then there exists an isomorphism of $K(y)$ over K into $\bar{\Omega}$ such that if (y') is the image of (y) we have $(y') \xrightarrow{\mathfrak{R}^*} (\eta)$.*

Now we consider specializations over a special type of local domains. Let $\mathfrak{o}$ be a discrete valuation ring with the quotient field k and the maximal ideal $\mathfrak{p}$, and let $\kappa = \mathfrak{o}/\mathfrak{p}$ be its residue field. Let $(x_1, \cdots, x_n)$ be a set of independent variables over k, and denote by $\mathfrak{R}$ the specialization ring $[(x_1, \cdots, x_n) \xrightarrow{\mathfrak{o}} (0, \cdots, 0)]$ and by $\mathfrak{R}^*$ its completion. Then $\mathfrak{R}^*$ is the ring of the formal power-series $\sum a_{(i)}{}^* x_1^{i_1} \cdots x_n^{i_n}$ with $a_{(i)}{}^*$ in the completion $\mathfrak{o}^*$ of $\mathfrak{o}$. The complete local domain $\mathfrak{R}^*$ is a regular local ring in Krull's sense [4] and is integrally closed.

PROPOSITION 13. *Let $\mathfrak{R}^*$ be the completion of the specialization ring $[(x_1, \cdots, x_n) \xrightarrow{\mathfrak{o}} (0, \cdots, 0)]$ in the above notations; let Ω be the quotient*

---

[4] Krull [3].

*field of $\Re^*$ and (y) a set of algebraic elements over Ω. If (η) is a proper
specialization of (y) over $\Re^*$, then (η) has a multiplicity which is a multiple
of $[\kappa(\eta):\kappa]_i$.*

Proof. By Proposition 11, (y) is integral over $\Re^*$; hence, by Proposition 10, every specialization (η') of (y) over $\Re^*$ has a multiplicity. Let $(y^{(1)}, \cdots, y^{(d)})$ be a complete set of conjugates of (y) over Ω and $(\eta^{(1)}, \cdots, \eta^{(d)})$ a specialization of $(y^{(1)}, \cdots, y^{(d)})$ over $\Re^*$. By Proposition 11, the set $(\eta^{(1)}, \cdots, \eta^{(d)})$ consists of conjugates of (η) over κ. Everyone of conjugates of (η) over κ is a specialization of (y) over $\Re^*$; so that it has a positive multiplicity as a specialization of (y) over $\Re^*$. Thus $(\eta^{(1)}, \cdots, \eta^{(d)})$ contains all the conjugates of (η) over κ; and it is easy to see that all the conjugates of (η) occur in $(\eta^{(\nu)})$ in the same multiplicity. Hence our proposition is proved when we show that $d = [\Omega(y):\Omega]$ is a multiple of $[\kappa(\eta):\kappa]$. Let (t) be a set of n independent variables over k and (τ) a set of n independent variables over κ. We denote by $\Re_1$ the specialization ring $[(t) \xrightarrow{\;o\;} (\tau)]$ and by $\Re_1^*$ its completion. By Proposition 3 of Section 1, $\Re_1$ is a discrete valuation ring. We may regard the completion o^* of o as a subring of $\Re_1^*$. Now we shall define an isomorphic mapping f of $\Re^*$ over o^* into $\Re_1^*$. For an element

$$A = \sum a_{(i)}^* x_1^{i_1} \cdots x_n^{i_n} \, \varepsilon \, \Re^*,$$

we define

$$f(A) = \sum a_{(i)}^* \pi^{i_1 + \cdots + i_n} t_1^{i_1} \cdots t_n^{i_n}$$

where π is a prime element in o. It is easy to see that this mapping f is actually an isomorphic mapping into $\Re_1^*$. Then f is extended to an isomorphism of Ω into the quotient field Ω_1 of $\Re_1^*$, and is also extended to an isomorphism g of $\bar{\Omega}$ into $\bar{\Omega}_1$. We denote by (y') the image of (y) by g and by $(y'^{(1)}, \cdots, y'^{(d)})$ the image of $(y^{(1)}, \cdots, y^{(d)})$. If (ζ) is a specialization of $(y'^{(\nu)})$ over $\Re_1^*$, then (ζ) is a specialization of (y) over $\Re^*$; hence, by Proposition 11, (ζ) is a conjugate of (η) over κ. Now it is well-known that $[\Omega_1(y'^{(\nu)}):\Omega_1]$ is a multiple of $[\kappa(\tau, \zeta):\kappa(\tau)]$ since $\Re_1^*$ is a complete discrete valuation ring; the latter number is clearly equal to $[\kappa(\eta):\kappa]$, while $[\Omega(y):\Omega]$ is a sum of several $[\Omega_1(y'^{(\nu)}):\Omega_1]$; this completes our proof.

The following theorem is an extension of Theorem 4 of [WF] Chapter III, Section 4, and is basic for our whole theory.

THEOREM 2. *Let (x) be a set of n independent variables over k and (y) a set of algebraic quantities over $k(x)$. Let (ξ) be a finite specialization of (x) over o and denote by $\Re$ the specialization ring $[(x) \xrightarrow{\;o\;} (\xi)]$. If (η)*

is a proper specialization of (y) *over* $\mathfrak{R}$, *then it has a multiplicity which is a multiple of* $[\kappa(\xi,\eta):\kappa(\xi)]_i$.[5]

Proof. We first assume that (ξ) is contained in κ. Then we have a set of quantities (a) in $\mathfrak{o}$ such that $(a) \xrightarrow{\ \mathfrak{o}\ } (\xi)$. Replacing the set (x_j) by $(x_j - a_j)$, we may assume that $(\xi) = (0)$. In that special case, we can obtain our theorem by Proposition 12 and Proposition 13, in the same procedure as in the proof of Proposition 7 of [WF] Chapter III, Section 3. Now we shall prove the general case. Let s be the dimension of (ξ) over κ; if s is not 0, we may assume that $\xi_1, \cdots, \xi_s$ are independent variables over κ, and that (ξ) is algebraic over $\kappa(\xi_1, \cdots, \xi_s)$. By Proposition 3 of Section 1, the specialization ring $[(x_1, \cdots, x_s) \xrightarrow{\ \mathfrak{o}\ } (\xi_1, \cdots, \xi_s)]$ is a discrete valuation ring; if we denote it by $\mathfrak{o}_1$, we have $\mathfrak{R} = [(x_{s+1}, \cdots, x_n) \xrightarrow{\ \mathfrak{o}_1\ } (\xi_{s+1}, \cdots, \xi_n)]$. Hence, it is sufficient to prove our theorem in case where (ξ) is algebraic over κ. Suppose that (ξ) is algebraic over κ. By Proposition 9 of Section 1, we can find an algebraic extension k' of k and a valuation ring $\mathfrak{o}'$ which is a prolongation of $\mathfrak{o}$ in k', such that the residue field κ' of $\mathfrak{o}'$ is $\kappa(\xi)$. Let $(y^{(1)}, \cdots, y^{(d)})$ be the complete set of conjugates of (y) over $k(x)$ and $(\eta^{(1)}, \cdots, \eta^{(d)})$ a specialization of $(y^{(1)}, \cdots, y^{(d)})$ over $\mathfrak{R}$. Then by Proposition 6 of Section 1, there exists a generic specialization $(x', y'^{(1)}, \cdots, y'^{(d)})$ of $(x, y^{(1)}, \cdots, y^{(d)})$ such that $(x', y'^{(1)}, \cdots, y'^{(d)}) \xrightarrow{\ \mathfrak{o}'\ } (\xi, \eta^{(1)}, \cdots, \eta^{(d)})$. Since k' is algebraic over k, (x) is a generic specialization of (x') over k', and is extended to a generic specialization $(x, y''^{(1)}, \cdots, y''^{(d)})$ of $(x', y'^{(1)}, \cdots, y'^{(d)})$ over k'. Then $(x, y''^{(1)}, \cdots, y''^{(d)})$ is a generic specialization of $(x, y^{(1)}, \cdots, y^{(d)})$ over k and has the specialization $(\xi, \eta^{(1)}, \cdots, \eta^{(d)})$ over $\mathfrak{o}'$. The set $(y''^{(1)}, \cdots, y''^{(d)})$ is a permutation of $(y^{(1)}, \cdots, y^{(d)})$ since $(y^{(1)}, \cdots, y^{(d)})$ is the complete set of conjugates of (y) over $k(x)$. Therefore, it is sufficient to prove that for every specialization $(\eta^{(1)}, \cdots, \eta^{(d)})$ of $(y^{(1)}, \cdots, y^{(d)})$ over $(x) \xrightarrow{\ \mathfrak{o}'\ } (\xi)$ the set (η) occurs in $(\eta^{(1)}, \cdots, \eta^{(d)})$ in the same times μ, and that μ is a multiple of $[\kappa(\xi,\eta):\kappa(\xi)]_i$. Now the set $(y^{(1)}, \cdots, y^{(d)})$ consists of several complete sets of conjugates of sets $(y^{(\nu)})$ over $k'(x)$. Since $\kappa' = \kappa(\xi)$, we have $[\kappa(\xi,\eta):\kappa(\xi)]_i = [\kappa'(\eta):\kappa']_i$. Thus what we have to prove is reduced to the case in which our theorem is already proved.

The following three theorems are generalizations of Proposition 10, Proposition 11 and Theorem 5 in [WF] Chapter III, Section 4, and can be proved by the same arguments as there.

---

[5] This theorem holds for any regular local ring $\mathfrak{R}$. It can be easily verified that the specialization ring $[(x) \xrightarrow{\ \mathfrak{o}\ } (\xi)]$ in our theorem is a regular local ring.

THEOREM 3. *Let (x) be a set of independent variables over k, (y) and (z) two sets of algebraic quantities over $k(x)$, and (ξ, η) a finite specialization of (x, y) over $\mathfrak{o}$. Assume that (η) is a proper specialization, of multiplicity μ, of (y) over $(x) \xrightarrow{\mathfrak{o}} (\xi)$, and that (z) is finite over $(x, y) \xrightarrow{\mathfrak{o}} (\xi, \eta)$. Then the specializations $(\zeta^{(\sigma)})$ of (z) over $(x, y) \xrightarrow{\mathfrak{o}} (\xi, \eta)$ are in a finite number; for every σ, $(\eta, \zeta^{(\sigma)})$ is a proper specialization of (y, z) over $(x) \xrightarrow{\mathfrak{o}} (\xi)$; and if μ_σ is the multiplicity of that specialization, we have the relation*

$$\sum_\sigma \mu_\sigma = \mu[k(x, y, z) : k(x, y)].$$

THEOREM 4. *Let (x) and (z) be two sets of independent variables over k; let (y) be a set of algebraic quantities over $k(x)$ and (w) a set of algebraic quantities over $k(z)$. Assume that $k(x, y)$ and $k(z, w)$ are linearly disjoint over k. Let (ξ, ζ) be a finite specialization of (x, z) over $\mathfrak{o}$; let (η) be a proper specialization of (y), of multiplicity μ, over $(x) \xrightarrow{\mathfrak{o}} (\xi)$, and let (ω) be a proper specialization of (w), of multiplicity ν, over $(z) \xrightarrow{\mathfrak{o}} (\zeta)$. Then (η, ω) is a proper specialization of (y, w) over $(x, z) \xrightarrow{\mathfrak{o}} (\xi, \zeta)$; and the multiplicity of this specialization is equal to $\mu\nu$.*

THEOREM 5. *Let (x) be a set of independent variables over k and (y) a set of m algebraic quantities over $k(x)$. Let (ξ, η) be a finite specialization of (x, y) over $\mathfrak{o}$, and assume that there are m polynomials $F_j(X, Y)$ in $\mathfrak{o}[X, Y]$, such that $F_j(x, y) = 0$ for $1 \leqq j \leqq m$ and that the determinant $|\partial F_j/\partial Y_h(\xi, \eta)|$ is not 0. Then (η) is a proper specialization of (y) over $(x) \xrightarrow{\mathfrak{o}} (\xi)$; and its multiplicity is 1.*

PROPOSITION 14. *Let (x) be a set of independent variables over k and (y) a set of algebraic quantities over $k(x)$. Let (ξ) be a finite specialization of (x) over $\mathfrak{o}$ and (η) a proper specialization of (y) over $(x) \xrightarrow{\mathfrak{o}} (\xi)$. Then the following two conclusions hold.*

(I) *Let (x') be a finite specialization of (x) over k such that (ξ) is a specialization of (x') over $\mathfrak{o}$. Then there exists a specialization (x', y') of (x, y) over k, which has (ξ, η) as a specialization over $\mathfrak{o}$.*

(II) *Let (ξ') be a finite specialization of (x) over $\mathfrak{o}$ such that (ξ) is a specialization of (ξ') over κ. Then there exists a specialization (ξ', η') of (x, y) over $\mathfrak{o}$, which has (ξ, η) as a specialization over κ.*

This is also proved as in Proposition 12 of [WF] Chapter III, Section 4.

PROPOSITION 15. *Let (x) be a set of n quantities of dimension r over k. Let the t_{ij} $(1 \leqq i \leqq r, 1 \leqq j \leqq n)$ be rn independent variables over $k(x)$*

and the τ_{ij} $(1\leq i\leq r, 1\leq j\leq n)$ rn independent variables over κ. Put $\mathfrak{o}'=[(t)\xrightarrow{\mathfrak{o}}(\tau)]$, and $y_i=\sum_j t_{ij}x_j$ for $1\leq i\leq r$. If (ξ,η) is a finite specialization of (x,y) over $\mathfrak{o}'$, then (ξ) is a proper specialization of (x) over $(y)\xrightarrow{\mathfrak{o}'}(\eta)$.

Proof. It is enough to prove that if (ξ,η) is a finite specialization of (x,y) over $\mathfrak{o}'$ then (ξ) is algebraic over $\kappa(\tau,\eta)$. Let (ξ,η) be a finite specialization of (x,y) over $\mathfrak{o}'$ and s the dimension of (ξ) over κ. Then by Proposition 2 of Section 1, we have $s\leq r$. Let (ξ') be a generic specialization of (ξ) over κ such that (ξ') and (τ) are independent over κ. Put $\eta_i'=\sum_j \tau_{ij}\xi_j'$ for $1\leq i\leq r$. Since $(\xi',\tau)\xrightarrow{\kappa}(\xi,\tau)$, we have $(\xi',\tau,\eta')\xrightarrow{\kappa}(\xi,\tau,\eta)$. Now by Proposition 24 of [WF] Chapter II, Section 5, (ξ') is finite over every finite specialization of $(\eta_1',\cdots,\eta_s')$ with reference to $\kappa(\tau_{11},\cdots,\tau_{sn})$. Hence (ξ) is algebraic over $\kappa(\tau,\eta)$.

PROPOSITION 16. *Let (x), (y), (t), (τ) and $\mathfrak{o}'$ be the same as in Proposition 15. Let the s_j $(1\leq j\leq n)$ be n independent variables over $k(t,x)$ and the σ_j $(1\leq j\leq n)$ n independent variables over $\kappa(\tau)$. Put $\mathfrak{o}''=[(s)\xrightarrow{\mathfrak{o}'}(\sigma)]$ and $z=\sum_j s_jx_j$. Let (η,ζ) be a finite specialization of (y,z) over $\mathfrak{o}''$. Then ζ is a proper specialization of z over $(y)\xrightarrow{\mathfrak{o}''}(\eta)$; and (x) is finite over $(y,z)\xrightarrow{\mathfrak{o}''}(\eta,\zeta)$. Let (ξ) be a specialization of (x) over $(y,z)\xrightarrow{\mathfrak{o}''}(\eta,\zeta)$ such that (ξ) and (τ,σ) are independent over κ. If $k(x)$ is separably generated over k, then we have $k(t,s,x)=k(t,s,y,z)$; and the multiplicity of (ξ) as a specialization of (x) over $(y)\xrightarrow{\mathfrak{o}'}(\eta)$ is equal to the multiplicity of ζ as a specialization of z over $(y)\xrightarrow{\mathfrak{o}''}(\eta)$.*

Proof. By the same method as in the proof of Proposition 24 of [WF] Chapter II, Section 5, we can prove that (x) is finite over $(y,z)\xrightarrow{\mathfrak{o}''}(\eta,\zeta)$. Let ξ be a specialization of (x) over $(y,z)\xrightarrow{\mathfrak{o}''}(\eta,\zeta)$. Then, by Proposition 15, (ξ) is algebraic over $\kappa(\tau,\eta)$; so that $\zeta=\sum_j \sigma_j\xi_j$ is algebraic over $\kappa(\tau,\sigma,\eta)$. This shows that every finite specialization of z over $(y)\xrightarrow{\mathfrak{o}''}(\eta)$ is algebraic over $\kappa(\tau,\sigma,\eta)$. Hence ζ is a proper specialization of z over $(y)\xrightarrow{\mathfrak{o}''}(\eta)$. Suppose that $k(x)$ is separably generated over k. Then $k(t,x)$ is separably algebraic over $k(t,y)$. Denote by $(x^{(1)},\cdots,x^{(d)})$ the complete set of conjugates of (x) over $k(t,y)$, and put $z^{(\nu)}=\sum_j s_jx_j^{(\nu)}$ for $1\leq\nu\leq d$; then it is easy to see that $(z^{(1)},\cdots,z^{(d)})$ is the complete set of conjugates of z over $k(t,s,y)$. This shows $[k(t,s,y,z):k(t,s,y)]=d=[k(t,s,x):k(t,s,y)]$; so we have $k(t,s,x)=k(t,s,y,z)$. Let (ξ) be a specialization of (x)

over $(y,z) \xrightarrow{\quad\sigma''\quad} (\eta,\zeta)$ such that (ξ) and (τ,σ) are independent over κ. Let $(\xi^{(1)}, \cdots, \xi^{(d)})$ be a specialization of $(x^{(1)}, \cdots, x^{(d)})$ over $(y) \xrightarrow{\quad\sigma'\quad} (\eta)$. We extend this specialization to a specialization

$$(x^{(1)}, \cdots, x^{(d)}, y, z^{(1)}, \cdots, z^{(d)}) \xrightarrow{\quad\sigma''\quad} (\xi^{(1)}, \cdots, \xi^{(d)}, \eta, \zeta^{(1)}, \cdots, \zeta^{(d)}).$$

If $(\xi^{(\nu)}) = (\xi)$, then we have $\zeta^{(\nu)} = \sum_j \sigma_j \xi_j^{(\nu)} = \sum_j \sigma_j \xi_j = \zeta$. Conversely, if $\zeta^{(\nu)} = \zeta$, then we have $(x,y,z,t,s) \xrightarrow{\quad k\quad} (x^{(\nu)}, y, z^{(\nu)}, t, s) \xrightarrow{\quad\sigma\quad} (\xi^{(\nu)}, \eta, \zeta, \tau, \sigma)$. Since (x) is finite over $(y,z) \xrightarrow{\quad\sigma''\quad} (\eta,\zeta)$, $(\xi^{(\nu)})$ is finite; so we have $\zeta = \sum_j \sigma_j \xi_j^{(\nu)}$. By Proposition 15, $(\xi^{(\nu)})$ is algebraic over $\kappa(\tau,\eta)$; since $\eta_i = \sum_j \tau_{ij} \xi_j$, $(\xi^{(\nu)})$ is also algebraic over $\kappa(\tau,\xi)$. By our assumption, (σ) is a set of independent variables over $\kappa(\tau,\xi)$; so that from $\sum_j \sigma_j \xi_j^{(\nu)} = \sum_j \sigma_j \xi_j$ follows $\xi_j^{(\nu)} = \xi_j$ $(1 \leq j \leq n)$. Thus we have proved that $\zeta^{(\nu)} = \zeta$ if and only if $(\xi^{(\nu)}) = (\xi)$. This proves the final assertion of our proposition.

By Proposition 14 and by Proposition 15, the following theorem can be proved in the same procedure as in the proof of Proposition 13 in [WF] Chapter III, Section 4.

THEOREM 6. *Let (x) and (y) be two sets of quantities, and (ξ, η) a finite specialization of (x,y) over $\mathfrak{o}$. Then if (η) is an isolated specialization of (y) over $(x) \xrightarrow{\quad\mathfrak{o}\quad} (\xi)$, the dimension of (η) over $\kappa(\xi)$ is at least equal to that of (y) over $k(x)$.*

PROPOSITION 17. *Let (y) be a set of independent variables over k and z an algebraic quantity over $k(y)$. Let $F(Y,Z) = 0$ be an irreducible equation for (y,z) over k and assume that F is a primitive polynomial in $\mathfrak{o}[Y,Z]$. Let ζ be a proper specialization of z, of multiplicity 1, over a finite specialization $(y) \xrightarrow{\quad\mathfrak{o}\quad} (\eta)$. Then we have $\partial F/\partial Z(\eta,\zeta) \neq 0$; and $[(y,z) \xrightarrow{\quad\mathfrak{o}\quad} (\eta,\zeta)]$ is integrally closed.*

This proposition is a translation of Proposition 19 in [WF] Chapter V, Section 3 to our case and can be proved by the same argument as there.

3. Reduction of affine varieties. Let k be a field with a discrete valuation, and $\mathfrak{o}$ be its valuation ring. We denote by $\mathfrak{p}$ the maximal ideal of $\mathfrak{o}$ and by κ the residue field $\mathfrak{o}/\mathfrak{p}$. We fix this field k and the valuation ring $\mathfrak{o}$; in the following we shall use these notations k, $\mathfrak{o}$, $\mathfrak{p}$ and κ always in this sense; and whenever we speak of a prolongation $\mathfrak{o}'$ of $\mathfrak{o}$ we presuppose that $\mathfrak{o}'$ is also a discrete valuation ring.

Let S^n and $\mathfrak{S}^n$ be the affine n-spaces defined with respect to the field k and κ, respectively. First we define the reduction of a variety as a point set.

Let U be a bunch [6] in S^n which is normally algebraic over k. We call a bunch attached to a set $\{(\alpha) \mid (\alpha) \; \varepsilon \; \mathfrak{S}^n, (a) \overset{o}{\longrightarrow} (\alpha)$ for some $(a) \; \varepsilon \; U\}$ the *bunch obtained from U by the reduction with respect to $\mathfrak{p}$ (or o), or the reduced bunch,* and denote it by $\bar{U}$. It may happen that $\bar{U}$ is an empty set even if U is not empty.

THEOREM 7. *Let U be a bunch in S^n which is normally algebraic over k, and let $\mathfrak{a}$ be the ideal $\{F(X) \mid F(X) \; \varepsilon \; o[X], F(a) = 0$ for every $(a) \; \varepsilon \; U\}$ in $o[X]$. Then, we have $\bar{U} = \{(\alpha) \mid (\alpha) \; \varepsilon \; \mathfrak{S}^n, \bar{F}(\alpha) = 0$ for every $F(X) \; \varepsilon \; \mathfrak{a}\}$; so that $\bar{U}$ is a bunch in $\mathfrak{S}^n$ which is normally algebraic over κ. Furthermore if k' is any extension of k and o' is a prolongation of o in k', then we have $\bar{U} = \{(\alpha) \mid (\alpha) \; \varepsilon \; \mathfrak{S}^n, (a) \overset{o'}{\longrightarrow} (\alpha)$ for some $(a) \; \varepsilon \; U\}$.*

Proof. By our definition of $\bar{U}$, it is obvious that if $(\alpha) \; \varepsilon \; \bar{U}$ then $\bar{F}(\alpha) = 0$ for every $F(X) \; \varepsilon \; \mathfrak{a}$. Now we express U as the union of prime rational cycles U_i over k: $U = U_1 \cup \cdots \cup U_h$. Let $(x^{(i)})$ be a generic point of U_i over k for every i; then we have $\mathfrak{a} = \mathfrak{a}_1 \cap \cdots \cap \mathfrak{a}_h$ where

$$\mathfrak{a}_i = \{F(X) \mid F(X) \; \varepsilon \; o[X], F(x^{(i)}) = 0\}$$

for every i. Suppose that, for every i, (α) is not a specialization of $(x^{(i)})$ over o. Then there exists, for every i, a polynomial $F_i(X)$ in $\mathfrak{a}_i$ such that $\bar{F}_i(\alpha) \neq 0$. If we put $F(X) = \prod_i F_i(X)$ we have $F(X) \; \varepsilon \; \mathfrak{a}$ and $\bar{F}(\alpha) \neq 0$. This proves our first assertion. The second assertion follows from our definition of $\bar{U}$ and Proposition 6 of Section 1.

By this theorem the reduced bunch $\bar{U}$ is invariant under the extension of the basic field k and the prolongation of o in that extension. This property of reduction also holds for reduction of cycles which is to be defined later.

PROPOSITION 18. *Let U and V be bunches which are normally algebraic over k. Then we have*

　　i) *If U and V are in the same ambient space S^n, then we have $\overline{U \cup V} = \bar{U} \cup \bar{V}$, $\overline{U \cap V} \subset \bar{U} \cap \bar{V}$.*

　　ii) *If U is in S^n and V is in S^m, then we have $\overline{U \times V} = \bar{U} \times \bar{V}$.*

Proof. The assertion i) is an immediate consequence of our definition. Now we shall prove ii). Let k' be a finite algebraic extension of k such that

---

[6] From now on we shall identify a bunch with the point set attached to that bunch. Also we shall use the term " prime rational cycle " for the bunch whose components are components of that cycle.

all the components of U and V are defined over k'. We have representations $U = \bigcup_i U_i$ and $V = \bigcup_j V_j$ where U_i and V_j are varieties defined over k'. Let $\mathfrak{o}'$ be a prolongation of $\mathfrak{o}$ in k', we consider the reduction with respect to $\mathfrak{o}'$. Since $\overline{U \times V} = \bigcup_{i,j} \overline{U_i \times V_j}$, it is sufficient to prove that $\overline{U_i \times V_j} = \overline{U_i} \times \overline{V_j}$ for every i and j. This is proved by Proposition 8 of Section 1 and by our definition.

PROPOSITION 19. *Let U be a bunch in S^n which is normally algebraic over k and assume that the components of U are all of dimension r. If $\bar{U}$ is not empty, then the components of $\bar{U}$ are all of dimension r.*

This is an immediate consequence of Theorem 6 of Section 2.

Let A be a prime rational S^n-cycle over k. Now we shall define a rational $\mathfrak{S}^n$-cycle $\rho(A)$ over κ which will be called the cycle obtained from A by the reduction with respect to $\mathfrak{p}$ (or $\mathfrak{o}$).

LEMMA 2. *Let A^r be a prime rational S^n-cycle over a field K. Let the $F_i(X) = \sum_j c_{ij} X_j$ $(1 \leq i \leq r)$ be r linearly independent linear forms in $K[X]$, and (v) a set of r independent variables over K; and let L^{n-r} be the variety defined by the set of equations $F_i(X) - v_i = 0$ $(1 \leq i \leq r)$. Then, if there is a point P in $A \cap L$, this point is algebraic over $K(v)$; and is a generic point of A over K; moreover the intersection product $U \cdot L$ is a prime rational cycle over $K(v)$ having P as a generic point over $K(v)$.*

This lemma is a generalization of Theorem 3 of [WF] Chapter V, Section 1, and follows immediately from that theorem.

Let A^r be a prime rational S^n-cycle over k. Suppose that $\bar{A}$ is not empty, and let $\mathfrak{A}_{11}$ be a component of $\bar{A}$. By Proposition 19, $\mathfrak{A}_{11}$ has the same dimension r as A. We shall define a certain multiplicity which will be denoted by $\mu(A, \mathfrak{A}_{11})$. Let

$$L^{n-r} = \left(\sum_j t_{ij} X_j - t_i = 0 \quad (1 \leq i \leq r) \right)$$

be a generic linear variety [7] over k in S^n; and let

$$\mathfrak{L}^{n-r} = \left(\sum \tau_{ij} X_j - \tau_i = 0 \quad (1 \leq i \leq r) \right)$$

be a generic linear variety over κ, in $\mathfrak{S}^n$. Then we have a point (x) in $L \cap A$, which is a generic point of A over $k(t_{ij})$, and we have a point (ξ) in

---

[7] By a linear variety $(\sum c_{ij} X_j - c_i = 0 \ (1 \leq i \leq r))$ we understand the linear variety defined by the equations $\sum c_{ij} X_j - c_i = 0$ $(1 \leq i \leq r)$. We call it a generic linear variety over K if (c_i, c_{ij}) is a set of independent variables over K.

$\mathfrak{L} \cap \mathfrak{A}_{11}$ which is a generic point of $\mathfrak{A}_{11}$ over $\bar{\kappa}(\tau_{ij})$. The point (x) is algebraic over $k(t_i, t_{ij})$, and (ξ) is algebraic over $\kappa(\tau_i, \tau_{ij})$. Since $\mathfrak{A}_{11}$ is a component of $\bar{A}$, (ξ) is an isolated specialization of (x) over $\mathfrak{o}$. Hence (ξ) is a proper specialization of (x) over $(t_i, t_{ij}) \xrightarrow{\mathfrak{o}} (\tau_i, \tau_{ij})$; so that by Theorem 2 of Section 2, it has a multiplicity μ defined in Section 2. It is easy to see that this multiplicity μ depends only upon the cycle A and the component $\mathfrak{A}_{11}$ of A, and not upon the choices of (t), (τ), (x), and (ξ). We shall denote it by $\mu(A, \mathfrak{A}_{11})$. By Theorem 2, $\mu(A, \mathfrak{A}_{11})$, is a multiple of $[\kappa(\tau_i, \tau_{ij}, \xi) : \kappa(\tau_i, \tau_{ij})]_i$. It can be easily verified that the latter number is equal to the order of inseparability of $\mathfrak{A}_{11}$ over κ.[8] Let $\mathfrak{A}_1$ be the prime rational $\mathfrak{S}^n$-cycle over κ with the component $\mathfrak{A}_{11}$. We denote by $\mu(A, \mathfrak{A}_1)$ the integer $\mu(A, \mathfrak{A}_{11})/[\mathfrak{A}_{11} : \kappa]_i$. It is easy to see that $\mu(A, \mathfrak{A}_1)$ is actually determined only by A and $\mathfrak{A}_1$. We define a $\mathfrak{S}^n$-cycle $\rho(A)$ by $\rho(A) = \sum_{\nu} \mu(A, \mathfrak{A}_\nu)\mathfrak{A}_\nu$

where the sum is taken over all the prime rational $\mathfrak{S}^n$-cycles $\mathfrak{A}_\nu$ over κ whose components are components of $\bar{A}$. If $\bar{A}$ is empty, we put $\rho(A) = 0$. It is obvious that

$$\rho(A) = \sum_{\nu, i} \mu(A, \mathfrak{A}_{\nu i})\mathfrak{A}_{\nu i}$$

where the sum is taken over all the components $\mathfrak{A}_{\nu i}$ of $\bar{A}$. Let X^r be a rational S^n-cycle over k; and let $X = \sum a_\alpha A_\alpha$ be a representation of X as a linear combination of prime rational S^n-cycles A_α over k. We put $\rho(X) = \sum a_\alpha \rho(A_\alpha)$ and call it the *cycle obtained from X by the reduction with respect to $\mathfrak{p}$ (or $\mathfrak{o}$)*. Then, ρ is an additive mapping of the group of rational S^n-cycles over k into the group of rational $\mathfrak{S}^n$-cycles over κ of the same dimension. We shall define in Section 4 such a mapping ρ also for cycles on an abstract variety.

PROPOSITION 20. *Let k' be an extension of k and $\mathfrak{o}'$ a prolongation of $\mathfrak{o}$ in k'. Let X^r be a rational S^n-cycle over k. If we define a cycle $\rho'(X)$ by the same procedure as above, with respect to $\mathfrak{o}'$, then $\rho'(X)$ coincides with $\rho(X)$.*

Proof. By linearity it is sufficient to prove our proposition in case where X is a prime rational cycle A over k. Let $A = \sum a_\alpha A_\alpha$ be a representation of A as a linear combination of prime rational cycles A over k'. Let

$$L^{n-r} = (\sum t_{ij}X_j - t_i = 0 \quad (1 \leq i \leq r))$$

be a generic linear variety over k' and

$$\mathfrak{L}^{n-r} = (\sum \tau_{ij}X_j - \tau_i = 0 \quad (1 \leq i \leq r))$$

---

[8] We shall denote by $[V : K]_i$ the order of inseparability of the variety V over a field K. If A is a prime rational cycle over K with the component V then we denote it also by $[A : K]_i$.

be a generic linear variety over the residue field κ' of $\mathfrak{o}'$. Let (x) be a point in $L \cap A$ and (ξ) a point in $\mathfrak{L} \cap \mathfrak{A}_{11}$, where $\mathfrak{A}_{11}$ is a component of $\bar{A}$. By Lemma 2, and by the equation $L \cdot A = \sum a_\alpha L \cdot A_\alpha$, the complete set of conjugates of (x) over $k(t_i, t_{ij})$ consists of the complete set of conjugates of $(x^{(\alpha)})$ over $k'(t_i, t_{ij})$, each set being repeated a_α times, where $(x^{(\alpha)})$ is a point in $L \cap A_\alpha$. By counting the number of (ξ) in a specialization of the complete set of conjugates of (x) over $k(t_i, t_{ij})$, over $(t_i, t_{ij}) \xrightarrow{\mathfrak{o}'} (\tau_i, \tau_{ij})$, we have $\mu(A, \mathfrak{A}_{11}) = \sum a_\alpha \mu'(A_\alpha, \mathfrak{A}_{11})$ where μ' is the multiplicity defined with respect to $\mathfrak{o}'$. By Theorem 7, a variety $\mathfrak{A}^r$ in $\mathfrak{S}^n$ is a component of $\rho(A)$ if and only if it is a component of $\rho'(A)$; so our proposition is proved.

PROPOSITION 21. *Let A^r be a prime rational S^n-cycle over k, and $\mathfrak{A}^r$ a prime rational $\mathfrak{S}^n$-cycle over κ with a component in $\bar{A}$. Let*

$$L^{n-r} = (\sum d_{ij}X_j - e_i = 0 \quad (1 \leqq i \leqq r))$$

be a linear variety in S^n, such that all the elements d_{ij} which are not in $\mathfrak{o}$ form a set of independent variables over k and that (e) is a set of independent variables over $k(d)$. Let

$$\mathfrak{L}^{n-r} = (\sum \delta_{ij}X_j - \epsilon_i = 0 \quad (1 \leqq i \leqq r))$$

be a linear variety in $\mathfrak{S}^n$, such that $\kappa(\delta)$ is regular over κ and that (ϵ) is a set of independent variables over $\kappa(\delta)$. Suppose that (δ, ϵ) is a specialization of (d, e) over $\mathfrak{o}$ and that there are a point (x) in $L \cap A$ and a point (ξ) in $\mathfrak{L} \cap \mathfrak{A}$. Then (ξ) is a proper specialization of (x) over $(d, e) \xrightarrow{\mathfrak{o}} (\delta, \epsilon)$, of multiplicity $\mu(A, \mathfrak{A}) \cdot [\kappa(\delta, \epsilon, \xi) : \kappa(\delta, \epsilon)]_i$.

Proof. First we remark that A is prime rational over $k(d)$ and that $\mathfrak{A}$ is prime rational over $\kappa(\delta)$. By Lemma 2, (x) is a generic point of A over $k(d)$ and (ξ) is a generic point of $\mathfrak{A}$ over $\kappa(\delta)$. Now it is obvious that (ξ) is a proper specialization of (x) over $(d, e) \xrightarrow{\mathfrak{o}} (\delta, \epsilon)$, since $(x) \longrightarrow (\xi)$ is isolated over $\mathfrak{o}$. By our assumption for the set (d) and by Theorem 2 of Section 2, (ξ) has a multiplicity μ as a specialization of (x) over $(d, e) \xrightarrow{\mathfrak{o}} (\delta, \epsilon)$. Let

$$M^{n-r} = (\sum t_{ij}X_j - t_i = 0 \quad (1 \leqq i \leqq r))$$

be a generic linear variety over $k(d, e)$ and

$$\mathfrak{M}^{n-r} = (\sum \tau_{ij}X_j - \tau_i = 0 \quad (1 \leqq i \leqq r))$$

a generic linear variety over $\kappa(\delta, \epsilon)$. Then we have a point (y) in $M \cap A$ and a point (η) in $\mathfrak{M} \cap \mathfrak{A}$. Obviously, (η) is also a proper specialization of

(y) over $(t_{ij}, t_i) \overset{\mathfrak{o}}{\longrightarrow} (\tau_{ij}, \tau_i)$; so that it has a multiplicity μ_1. We shall show $\mu = \mu_1$. Let $(y^{(1)}, \cdots, y^{(m)})$ be the complete set of conjugates of (y) over $k(t)$ and $(x^{(1)}, \cdots, x^{(m)})$ a specialization of $(y^{(1)}, \cdots, y^{(m)})$ over $(t_{ij}, t_i) \overset{k}{\longrightarrow} (d_{ij}, e_i)$. Then, by Lemma 2 and the definition of intersection-multiplicity,[8'] $(x^{(1)}, \cdots, x^{(m)})$ consists of the complete set of conjugates of (x) over $k(d, e)$ and pseudo-points which may arise or may not arise. Let $(\xi^{(1)}, \cdots, \xi^{(m)})$ be a specialization of $(x^{(1)}, \cdots, x^{(m)})$ over $(d, e) \overset{\mathfrak{o}}{\longrightarrow} (\delta, \epsilon)$; then we see that (ξ) occurs in $(\xi^{(1)}, \cdots, \xi^{(m)})$ exactly μ times. Thus we have shown $\mu = \mu_1$. Let $(\eta^{(1)}, \cdots, \eta^{(m)})$ be a specialization of $(y^{(1)}, \cdots, y^{(m)})$ over $(t_{ij}, t_i) \overset{\mathfrak{o}}{\longrightarrow} (\tau_{ij}, \tau_i)$ and $(\xi'^{(1)}, \cdots, \xi'^{(m)})$, a specialization of $(\eta^{(1)}, \cdots, \eta^{(m)})$ over $(\tau_{ij}, \tau_i) \overset{\kappa}{\longrightarrow} (\delta_{ij}, \epsilon_i)$. By definition of μ_1, we have (ξ) in $(\xi'^{(1)}, \cdots, \xi'^{(m)})$ exactly μ_1 times. If $(\xi'^{(\nu)}) = (\xi)$, then we have

$$(x) \overset{k}{\longrightarrow} (y^{(\nu)}) \overset{\mathfrak{o}}{\longrightarrow} (\eta^{(\nu)}) \overset{\kappa}{\longrightarrow} (\xi).$$

By the isolatedness of $(x) \overset{\mathfrak{o}}{\longrightarrow} (\xi)$, we have $(\xi) \overset{\kappa}{\longrightarrow} (\eta^{(\nu)})$, this shows that $(\eta^{(\nu)})$ is in $\mathfrak{A}$. Since $(y^{(\nu)}, t_{ij}, t_i) \overset{\mathfrak{o}}{\longrightarrow} (\eta^{(\nu)}, \tau_{ij}, \tau_i)$ and since $(y^{(\nu)}) \, \varepsilon \, M$, we see that $(\eta^{(\nu)})$ is in $\mathfrak{M} \cap \mathfrak{A}$, so that $(\eta^{(\nu)})$ is a conjugate of (η) over $\kappa(\tau)$. Now by our definition of $\mu(A, \mathfrak{A})$, the complete set of conjugates of (η) occurs in $(\eta^{(1)}, \cdots, \eta^{(m)})$ exactly $\mu(A, \mathfrak{A})$ times. Then, by Lemma 2 and by the definition of intersection-multiplicity, we have $\mu(A, \mathfrak{A}) [\kappa(\delta, \epsilon, \xi) : \kappa(\delta, \epsilon)]_i$ = the number of (ξ) which occurs in $(\xi'^{(1)}, \cdots, \xi'^{(m)})$. This completes our proof.

THEOREM 8. *Let A^r be a rational S^m-cycle over k and B^s a rational S^n-cycle over k. Then we have $\rho(A \times B) = \rho(A) \times \rho(B)$.*

Proof. By linearity and by Proposition 20, it is sufficient to prove our theorem in case where A and B are varieties defined over k. Furthermore, we may assume that all the components of $\bar{A}$ and $\bar{B}$ are defined over κ, because for every extension κ' of κ, there exists, by Proposition 9 of Section 1, a prolongation $\mathfrak{o}'$ of $\mathfrak{o}$ such that κ' is the residue field of $\mathfrak{o}'$. By Proposition 18, a variety $\mathfrak{C}$ in $\mathfrak{S}^n \times \mathfrak{S}^m$ is a component of $\overline{A \times B}$ if and only if it is a product $\mathfrak{A} \times \mathfrak{B}$ where $\mathfrak{A}$ is a component of $\bar{A}$ and $\mathfrak{B}$ is a component of $\bar{B}$. Hence it suffices to prove $\mu(A \times B, \mathfrak{A} \times \mathfrak{B}) = \mu(A, \mathfrak{A}) \cdot \mu(B, \mathfrak{B})$. Let $M^{m-r} = (\sum t_{ij}X_j - t_i = 0 \ (1 \leq i \leq r))$ be a generic linear variety over k in S^m and $N^{n-s} = (\sum z_{ij}X_j - z_i = 0 \ (1 \leq i \leq s))$ a generic linear variety over $k(t)$ in S^n. Correspondingly, let $\mathfrak{M}^{m-r} = (\sum \tau_{ij}X_j - \tau_i = 0 \ (1 \leq i \leq r))$ be a generic linear variety over κ in $\mathfrak{S}^m$ and $\mathfrak{N}^{n-s} = (\sum \zeta_{ij}X_j - \zeta_i = 0 \ (1 \leq i \leq s))$ a generic linear variety over $\kappa(\tau)$ in $\mathfrak{S}^n$. We have points (x),

---

[8'] See [WF] Chapter V, Section 1.

(y), (ξ) and (η) respectively in $M \cap A$, $N \cap B$, $\mathfrak{M} \cap \mathfrak{A}$ and $\mathfrak{N} \cap \mathfrak{B}$. Applying Proposition 21 to $A \times B$, $\mathfrak{A} \times \mathfrak{B}$, $M \times N$ and $\mathfrak{M} \times \mathfrak{N}$, we see that the multiplicity of (ξ, η) as a specialization of (x, y) over $(t_{ii}, z_{ij}, t_i, z_i) \overset{0}{\longrightarrow} (\tau_{ij}, \zeta_{ij}, \tau_i, \zeta_i)$ is equal to $\mu(A \times B, \mathfrak{A} \times \mathfrak{B})$. By Theorem 4 of Section 2 and by our definition of $\mu(A, \mathfrak{A})$, this multiplicity is equal to $\mu(A, \mathfrak{A}) \cdot \mu(B, \mathfrak{B})$.

Now we consider the relation between our multiplicity $\mu(A, \mathfrak{A})$ and the intersection-multiplicity.

Let U^n be a variety defined over a field K; let A^r and B^s be prime rational U-cycles over K. If there is a component C_1 of the bunch $A \cap B$, which is simple on U and has dimension $r + s - n$, then every conjugate of C_1 over K is also a simple component of $A \cap B$ on U with the dimension $r + s - n$. Consider the number

$$[A:K]_i [B:K]_i [C_1:K]_i^{-1} \sum_{p,q} i(A_p \cdot B_q, C_1; U)$$

where the sum is taken over all the A_p and all the B_q which are respectively components of A and B, and which contains C_1. It is easy to see that this number is invariant when we replace C_1 with any conjugate of C_1 over K. We denote this number by $i_K(A \cdot B, C; U)$ where C is a prime rational U-cycle over K with the component C_1. We shall say that C is a *prime component of $A \cap B$ on U*, over K, if U, A, B and C are in the above situation. We shall also use the notation $j_K(V \cdot L, W)$ for an intersection of a prime rational cycle and a linear variety.

THEOREM 9. *Let V^r be a variety defined over k in S^n. Let*

$$L^{n-s} = (\sum c_{ij} X_j - c_i = 0 \quad (1 \leqq i \leqq s))$$

be a linear variety in S^n where $c_{ij} \, \varepsilon \, \mathfrak{o}$, $c_i \, \varepsilon \, \mathfrak{o}$; and let

$$\mathfrak{L}^{n-s} = (\sum \gamma_{ij} X_j - \gamma_i = 0 \quad (1 \leqq i \leqq s))$$

be a linear variety in $\mathfrak{S}^n$ such that $(c_{ij}, c_i) \overset{0}{\longrightarrow} (\gamma_{ij}, \gamma_i)$. If there is a component $\mathfrak{W}_1$ of dimension $r - s$ in $\bar{V} \cap \bar{L}$, then there exists a prime rational cycle W^{r-s} over k with a component in $V \cap L$ such that $\mathfrak{W}_1$ is a component of $\bar{W}$. If we denote by the W_ν such cycles, then for every ν, W_ν is a prime component of $V \cap L$ over k; and we have

$$\sum_\nu j_k(V \cdot L, W_\nu) \mu(W_\nu, \mathfrak{B}) = \sum_\nu \mu(V, \mathfrak{B}_\nu) j_\kappa(\mathfrak{B}_\nu \cdot \mathfrak{L}, \mathfrak{B}),$$

where $\mathfrak{B}$ is the prime rational cycle over κ with the component $\mathfrak{W}_1$, and the $\mathfrak{B}_\nu$ are all the prime components of $\rho(V)$[9] over κ which contain $\mathfrak{W}_1$.

---

[*] We understand by a prime component over K, of a rational cycle over K, a prime rational cycle over K with a component which appears in the reduced expression of that rational cycle.

Proof. Let $M^{n-s} = (\sum t_{ij}X_j - t_i = 0 \ (1 \leq i \leq s))$ be a generic linear variety over k and $N^{n-r+s} = (\sum z_{ij}X_j - z_i = 0 \ (1 \leq i \leq r-s))$ a generic linear variety over $k(t)$. Correspondingly, let $\mathfrak{M}^{n-s} = (\sum \tau_{ij}X_j - \tau_i = 0 \ (1 \leq i \leq s))$ be a generic linear variety over κ and $\mathfrak{N}^{n-r+s} = (\sum \zeta_{ij}X_j - \zeta_i = 0 \ (1 \leq i \leq r-s))$ a generic linear variety over $\kappa(\tau)$. Then we have a point (x) in $V \cap M \cap N$ and a point (η) in $\mathfrak{W}_1 \cap \mathfrak{N}$. By Lemma 2, (x) is a generic point of V over $k(t_{ij}, z_{ij})$; and (η) is a generic point of $\mathfrak{W}_1$ over $\bar{\kappa}(\zeta_{ij})$. It is easy to see that (η) is a specialization of (x) over $(t_{ij}, t_i, z_{ij}, z_i) \xrightarrow{\mathfrak{o}} (\gamma_{ij}, \gamma_i, \zeta_{ij}, \zeta_i)$. Moreover (η) is an isolated specialization of (x) over $(t_{ij}, t_i) \xrightarrow{\mathfrak{o}} (\gamma_{ij}, \gamma_i)$. For, if $(x, t) \xrightarrow{\mathfrak{o}} (\eta', \gamma) \xrightarrow{\kappa} (\eta, \gamma)$, then (η') is in $\bar{V} \cap \bar{L}$ and has (η) as a specialization over κ. Since (η) is a generic point of a component $\mathfrak{W}_1$ of $\bar{V} \cap \bar{L}$, we have $(\eta) \xrightarrow{\kappa} (\eta')$. Hence (η) is a proper specialization of (x) over $(t, z) \xrightarrow{\mathfrak{o}} (\gamma, \zeta)$, and by Theorem 2 of Section 2, has a multiplicity μ. Now consider a series of specializations $(t, z) \xrightarrow{k} (c, z) \xrightarrow{\mathfrak{o}} (\gamma, \zeta)$. By (I) of Proposition 14 of Section 2, there exists a set (y) such that $(x, t, z) \xrightarrow{k} (y, c, z) \xrightarrow{\mathfrak{o}} (\eta, \gamma, \zeta)$. Since $\sum t_{ij}x_j = t_i \ (1 \leq i \leq s)$, we have $\sum c_{ij}y_j = c_i \ (1 \leq i \leq s)$; so that (y) is contained in $V \cap L$. Let W_0 be the locus of (y) over k, and W_0' the component of $V \cap L$ which contains W_0. Then we have $\bar{W}' \supset \mathfrak{W}_1$ where W' is the prime rational cycle over k, with the component W_0'. Since $\mathfrak{W}_1$ is a component of $\bar{V} \cap \bar{L}$, $\mathfrak{W}_1$ must be a component of $\bar{W}'$; so that we have $\dim \mathfrak{W}_1 = \dim W' \geq \dim W_0 \geq \dim \mathfrak{W}_1$. From this follows that W_0 is a component of $V \cap L$ and that $\dim W_0 = \dim \mathfrak{W}_1$. This proves the existence of a prime rational cycle W over k in our theorem. Let the W_ν be such cycles; then by the same argument as above we have $\dim W_\nu = \dim \mathfrak{W}_1 = r + s - n$; so that the W_ν are prime components of $V \cap L$ over k. Let k' be an extension of k and $\mathfrak{o}'$ a prolongation of $\mathfrak{o}$ in k' such that all the components of $V \cap L$ are defined over k' and that all the components of $\bar{V}$ and $\bar{V} \cap \bar{L}$ are defined over the residue field κ' of $\mathfrak{o}'$.[9'] We denote by $\mu'(A, \mathfrak{A})$ the multiplicity with respect to $\mathfrak{o}'$. Denote by the $W_{\nu\lambda}$ the components of W_ν for every ν and by the $\mathfrak{W}_{\nu\lambda}$ the components of $\mathfrak{W}_\nu$ for every ν. By Proposition 20, we have

$$[\mathfrak{W} : k]_i \sum_\nu j_k(V \cdot L, W_\nu)\mu(W_\nu, \mathfrak{W})$$

$$= \sum_\nu j(V \cdot L, W_{\nu 1})[W_\nu : k]_i^{-1} \sum_\lambda [W_\nu : k]_i \mu'(W_{\nu\lambda}, \mathfrak{W}_1)$$

$$= \sum_{\nu, \lambda} j(V \cdot L, W_{\nu\lambda})\mu'(W_{\nu\lambda}, \mathfrak{W}_1).$$

[9'] The existences of such an extension k' and a prolongation $\mathfrak{o}'$ are assured by Proposition 9 in Section 1.

Similarly we have

$$[\mathfrak{W}:k]_i \sum_{\nu} \mu(V, \mathfrak{B}_\nu) j_\kappa(\mathfrak{B}_\nu \cdot \mathfrak{L}, \mathfrak{W}) = \sum_{\nu, \lambda} \mu'(V, \mathfrak{B}_{\nu\lambda}) j(\mathfrak{B}_{\nu\lambda} \cdot \mathfrak{L}, \mathfrak{W}_1).$$

By these equations it suffices to prove the formula in our theorem in case where all the components of $V \cap L$ are defined over k and where all the components of $\bar{V}$ and $\bar{V} \cap \bar{L}$ are defined over κ. Assume that we are in this situation. Let (y') be a point in $W_\nu \cap N$, then (y') is a generic point of W_ν over k; and we have $(x, t, z) \xrightarrow{k} (y', c, z) \xrightarrow{o} (\eta, \gamma, \zeta)$. Conversely if we have $(x, t, z) \xrightarrow{k} (y', c, z) \xrightarrow{o} (\eta, \gamma, \zeta)$, then (y') is a generic point of some W_ν and is contained in $W_\nu \cap N$. This is already proved. Any such two points (y') and (y'') are conjugates of each other over $k(z)$ if and only if the loci of them are the same. Now let $(x^{(1)}, \cdots, x^{(d)})$ be the complete set of conjugates of (x) over $k(t, z)$. We extend the specializations $(t, z) \xrightarrow{k} (c, z) \xrightarrow{o} (\gamma, \zeta)$ to specializations

$$(x^{(1)}, \cdots, x^{(d)}, t, z) \xrightarrow{k} (y^{(1)}, \cdots, y^{(d)}, c, z) \xrightarrow{o} (\eta^{(1)}, \cdots, \eta^{(d)}, \gamma, \zeta).$$

If $(\eta^{(\lambda)}) = (\eta)$, then $(y^{(\lambda)})$ is in $W_\nu \cap N$ for some W_ν. By the definition of $j(V \cdot L, W_\nu)$ ([WF] Chapter V, Section 2), we have the complete set of conjugates of $(y^{(\lambda)})$ over $k(z)$ in $(y^{(1)}, \cdots, y^{(d)})$, exactly $j(V \cdot L, W_\nu)$ times. Hence we have $\sum_{\nu} j(V \cdot L, W_\nu) \mu(W_\nu, \mathfrak{W})$ as the number μ of (η) which occurs in $(\eta^{(1)}, \cdots, \eta^{(d)})$. We consider another series of specializations $(t, z) \xrightarrow{o} (\tau, \zeta) \xrightarrow{\kappa} (\gamma, \zeta)$. Then a point (ξ) is a generic point of one of the $\mathfrak{B}_\nu$ and is contained in $\mathfrak{M} \cap \mathfrak{N}$ if and only if

$$(x, t, z) \xrightarrow{o} (\xi, \tau, \zeta) \xrightarrow{\kappa} (\eta, \gamma, \zeta).$$

The loci of any such two points (ξ') and (ξ'') are the same if and only if they are conjugates of each other over $\kappa(\tau, \zeta)$. Hence we have as above, $\mu = \sum_{\nu} \mu(V, \mathfrak{B}_\nu) j(\mathfrak{B}_\nu \cdot \mathfrak{L}, \mathfrak{W})$. This completes our proof.

We denote by $j_\kappa(V \cdot \mathfrak{L}, \mathfrak{W})$ the number $\sum_{\nu} \mu(V, \mathfrak{B}_\nu) j_\kappa(\mathfrak{B}_\nu \cdot \mathfrak{L}, \mathfrak{W})$ in the above theorem. It is easily verified that the number $j_\kappa(V \cdot \mathfrak{L}, \mathfrak{W}) [\mathfrak{W}:\kappa]_i$ is equal to the multiplicity of (η) as a specialization of (x) over $(t, z) \xrightarrow{n} (\gamma, \zeta)$ where the notations are the same as in the above proof. We shall also denote by $j(V \cdot \mathfrak{L}, \mathfrak{W}_1)$ this number $j_\kappa(V \cdot \mathfrak{L}, \mathfrak{W}) [\mathfrak{W}:\kappa]_i$ where $\mathfrak{W}_1$ is a component of $\mathfrak{W}$. Using this j, we have $\mu(V, \mathfrak{B}) = j(V \cdot \mathfrak{S}^n, \mathfrak{B})$ for every prime component $\mathfrak{B}$ of $\rho(V)$.

Before we deal with intersections in the general case, we extend the conception of the simple point. Let V^r be a variety defined over k in S^n, and denote by $\mathfrak{a}$ the ideal $\{F(X)\,|\,F(X)\,\varepsilon\,\mathfrak{o}[X],\,F(a)=0$ for every $(a)\,\varepsilon\,V\}$ in $\mathfrak{o}[X]$. We say that a point (α) in $\bar{V}$ is *simple on* V when there exists a set of polynomials $F_i(X)$ $(1\leq i\leq n-r)$ in $\mathfrak{a}$ such that rank $\|\partial F_i/\partial X_j(\alpha)\|$ is equal to $n-r$. We call the linear variety $(\sum\partial F_i/\partial X_j(\alpha)(X_j-\alpha_j)=0$ $(1\leq i\leq n-r))$ in $\mathfrak{S}^n$ the *tangent linear variety to* V *at* (α). Also we say that a variety $\mathfrak{W}$ contained in $\bar{V}$ is *simple on* V if some point on $\mathfrak{W}$ is simple on V.

The following proposition is proved as in Theorem 5 of [WF] Chapter IV, Section 3.

PROPOSITION 22. *Let V and W be varieties defined over k, respectively in S^n and S^m. Let (α) be a point in $\bar{V}$ and (β) a point in $\bar{W}$. Then the point $(\alpha)\times(\beta)$ is simple on $V\times W$ if and only if (α) is simple on V and (β) is simple on W.*

Let $\mathfrak{A}$ be a variety in $\bar{V}$ and suppose that $\mathfrak{A}$ is simple on V. We shall use the term "*uniformizing set of linear forms for V along $\mathfrak{A}$.*" We understand, by this term, a set of r linear forms $\Phi_i(X)=\sum\gamma_{ij}X_j$ $(1\leq i\leq r)$ such that the linear variety $(\Phi_i(X)=0$ $(1\leq i\leq r))$ in $\mathfrak{S}^n$ is of dimension $n-r$ and is transversal to the tangent linear variety to V at some point of $\mathfrak{A}$. This is an analogy of that for the ordinary case in [WF] Chapter IV, Section 6.

THEOREM 10. *Let V^r be a variety in S^n defined over k and (α) a point in $\bar{V}$ which is simple on V. Let $\mathfrak{L}^{n-r}=(\sum\gamma_{ij}X_j-\gamma_i=0$ $(1\leq i\leq r))$ be a linear variety in $\mathfrak{S}^n$ which is transversal to the tangent linear variety $\mathfrak{X}^r$ to V at (α). Suppose that (α) is rational over κ and that (γ_{ij},γ_i) is in κ. Then (α) is a component of $\bar{V}\cap\mathfrak{L}$; and we have $j(V\cdot\mathfrak{L},(\alpha))=1$. In particular, this shows that (α) is contained in only one component $\mathfrak{B}$ of $\bar{V}$. Moreover (α) is a simple point on $\mathfrak{B}$; and $\mathfrak{X}$ is the tangent linear variety to $\mathfrak{B}$ at (α).*

Proof. By Theorem 5 of Section 2, and by the same argument as in the proof of Proposition 7 of [WF] Chapter V, Section 1, it can be proved that $j(V\cdot\mathfrak{L},(\alpha))=1$. Then if we denote by $\mathfrak{B}_1,\cdots,\mathfrak{B}_h$ all the components of V which contain (α), we have

$$1=\sum_{\nu=1}^{h}\mu(V,\mathfrak{B}_\nu)j(\mathfrak{B}_\nu\cdot\mathfrak{L},(\alpha)).$$

Hence h must be equal to 1. The remaining part of our theorem is obvious.

COROLLARY. *Let V be a variety defined over k and $\mathfrak{B}$ a component of* by the above theorem and the proof of it. From this follows $[\mathfrak{B}:\kappa]_i=1$.

Proof. By our definition, there exists a point (α) on $\mathfrak{B}$ which is simple on V. By Proposition 9 of Section 1, there exist an extension k' of k and a prolongation $\mathfrak{o}'$ of $\mathfrak{o}$ in k' such that the residue field of $\mathfrak{o}'$ is $\kappa(\alpha)$. Since $\mu(V, \mathfrak{B})$ is invariant when considered with respect to $\mathfrak{o}'$, we have $\mu(V, \mathfrak{B}) = 1$ by the above theorem and the proof of it. From this follows $[\mathfrak{B} : \kappa]_i = 1$.

THEOREM 11. *Let U^n be a variety defined over k; let A^r and B^r be two prime rational U-cycles over k. Suppose that there exists a component $\mathfrak{C}_1{}^{r+s-n}$ of $\bar{A} \cap \bar{B}$ which is simple on U. Then, there exists a prime rational cycle C over k with a component in $A \cap B$ such that $\mathfrak{C}_1$ is a component of $\bar{C}$. If we denote by the C_ν such cycles, then every C_ν is a prime component of $A \cap B$ on U over k. Let the $\mathfrak{A}_p{}^r$ and the $\mathfrak{B}_q{}^s$ be those prime rational cycles over κ which contain $\mathfrak{C}_1$, and of which components are components of $\bar{A}$ and $\bar{B}$, respectively; and let $\mathfrak{U}$ be the component of U which contains $\mathfrak{C}_1$ (by Theorem 10, such $\mathfrak{U}$ is uniquely determined), and $\mathfrak{C}$ the prime rational cycle over κ with the component $\mathfrak{C}_1$. If $\mathfrak{U}$ is defined over κ, then, the $\mathfrak{A}_p$ and the $\mathfrak{B}_q$ are all contained in $\mathfrak{U}$; and $\mathfrak{C}$ is a prime component of $\mathfrak{A}_p \cap \mathfrak{B}_q$ on $\mathfrak{U}$ over κ, for every p and q; and we have the following formula:*

$$\sum_\nu i_k(A \cdot B, C_\nu; U)\mu(C_\nu, \mathfrak{C}) = \sum_{p,q} \mu(A, \mathfrak{A}_p)\mu(B, \mathfrak{B}_q)i_\kappa(\mathfrak{A}_p \cdot \mathfrak{B}_q, \mathfrak{C}; \mathfrak{U}).$$

Proof. Let k' be an extension of k and $\mathfrak{o}'$ a prolongation of $\mathfrak{o}$ in k' such that all the components of A, B and $A \cap B$ are defined over k', and that all the components of $\bar{A}$, $\bar{B}$ and $\bar{A} \cap \bar{B}$ are defined over the residue field κ' of $\mathfrak{o}'$. We shall denote by $\mu'(A, \mathfrak{A})$ our multiplicity with respect to $\mathfrak{o}'$. Suppose that there exists a component C_0 of $A \cap B$ such that $\bar{C}_0 \supset \mathfrak{C}_1$, and denote by C the prime rational cycle over k with the component C_0. Then, similarly as in the proof of Theorem 9, we can prove that $\mathfrak{C}_1$ is a component of $\bar{C}$ and that C is a prime component of $A \cap B$ on U over k. If we define the C_ν, the $\mathfrak{A}_p$, the $\mathfrak{B}_q$ and $\mathfrak{U}$ as in our theorem, then it is obvious that the $\mathfrak{A}_p$ and the $\mathfrak{B}_q$ are all contained in $\mathfrak{U}$ and that $\mathfrak{C}$ is a prime component of $\mathfrak{A}_p \cap \mathfrak{B}_q$ on $\mathfrak{U}$ over κ. Denote the components of A, B, C_ν, $\mathfrak{A}_p$ and $\mathfrak{B}_q$ respectively by A_α, B_β, $C_{\nu\lambda}$, $\mathfrak{A}_{p\mu}$ and $\mathfrak{B}_{qv}$. By Proposition 20 and by our definition of $i_k(A \cdot B, C_\nu; U)$, we have

$$[A:k]_i^{-1}[B:k]_i^{-1}[\mathfrak{C}:\kappa]_i \sum_\nu i_k(A \cdot B, C_\nu; U)\mu(C_\nu, \mathfrak{C})$$
$$= \sum_{\alpha,\beta,\nu,\lambda} i(A_\alpha \cdot B_\beta, C_{\nu\lambda}; U)\mu'(C_{\nu\lambda}, \mathfrak{C}_1)$$

and

$$[A:k]_i^{-1}[B:k]_i^{-1}[\mathfrak{C}:\kappa]_i \sum_{p,q} \mu(A, \mathfrak{A}_p)\mu(B, \mathfrak{B}_q)i_\kappa(\mathfrak{A}_p \cdot \mathfrak{B}_q, \mathfrak{C}; \mathfrak{U})$$
$$= \sum_{\alpha,\beta,p,q,u,v} \mu'(A_\alpha, \mathfrak{A}_{pu})\mu'(B_\beta, \mathfrak{B}_{qv})i(\mathfrak{A}_{pu} \cdot \mathfrak{B}_{qv}, \mathfrak{C}_1; \mathfrak{U}).$$

By these equations, it is sufficient to prove our theorem in case where A, B, and the components of $A \cap B$ are all defined over k and where all the components of $\bar{A}$, B and $\bar{A} \cap \bar{B}$ are defined over κ. Assume that we are in this case. Let S^N and $\mathfrak{S}^N$ be respectively the ambient spaces for U and $\mathfrak{U}$. Let $\bar{F}_i(X)$ $(1 \leq i \leq n)$ be a uniformizing set of linear forms for U along $\mathfrak{C}$. We may assume that for every i, $\bar{F}_i(X)$ is the class of a linear form $F_i(X)$ in $\mathfrak{o}[X_i]$ modulo $\mathfrak{p}$. We denote by Λ^{2N-n} and $\bar{\Lambda}^{2N-n}$ respectively the linear variety $(F_i(X - X') = 0 \; (1 \leq i \leq n))$ in $S^N \times S^N$ and the linear variety $(\bar{F}_i(X - X') = 0 \; (1 \leq i \leq n)$ in $\mathfrak{S}^N \times \mathfrak{S}^N$. Let (γ) be a generic point of $\mathfrak{C}$ over κ and $\Delta_{\mathfrak{C}}$ the locus of (γ, γ) over κ. Then, $\Delta_{\mathfrak{C}}$ is a proper component of $\bar{\Lambda} \cap (\mathfrak{A}_p \times \mathfrak{B}_q)$; and by definition of i ([WF] Chapter VI, Section 1), we have $i(\mathfrak{A}_p \cdot \mathfrak{B}_q, \mathfrak{C}; \mathfrak{U}) = j[(\mathfrak{A}_p \times \mathfrak{B}_q) \cdot \bar{\Lambda}, \Delta_{\mathfrak{C}}]$. By Theorem 9, there exists a component D of $\Lambda \cap (A \times B)$ such that $\bar{D} \supset \Delta_{\mathfrak{C}}$. (We may assume that the components of $\Lambda \cap (A \times B)$ are all defined over k.) If we denote by $D_1, \cdots, D_h$ such components of $\Lambda \cap (A \times B)$, then by the same theorem, we have

$$\sum_\nu j[(A \times B) \cdot \Lambda, D_\nu]\mu(D_\nu, \Delta_{\mathfrak{C}})$$
$$= \sum_{p,q} \mu(A \times B, \mathfrak{A}_p \times \mathfrak{B}_q)j[(\mathfrak{A}_p \times \mathfrak{B}_q) \cdot \bar{\Lambda}, \Delta_{\mathfrak{C}}].$$

Let (c, c') be a generic point of D over k. Put $k_1 = k(c, c')$. As it is well-known, we can extend the specialization ring $[(c, c') \xrightarrow{\mathfrak{o}} (\gamma, \gamma)]$ to a valuation ring $\mathfrak{o}_1$ in k_1.[9'''] Since $\dim_k (c, c') = \dim_\kappa (\gamma, \gamma)$, $\mathfrak{o}_1$ is discrete. Put

$$L^{N-n} = (F_i(X - c') = 0 \; (1 \leq i \leq n)), \quad \mathfrak{L}^{N-n} = (\bar{F}_i(X - \gamma) = 0 \; (1 \leq i \leq n)).$$

Then, since $\mathfrak{L}$ is transversal to the tangent linear variety to U at (γ), we have, by Theorem 10, $j(U \cdot \mathfrak{L}, (\gamma)) = 1$. If we denote by $(c^{(1)}), \cdots, (c^{(m)})$ the component of $U \cap L$ such that $\overline{(c^{(i)})} = (\gamma)$ (with respect to a prolongation of $\mathfrak{o}_1$ in an extension of k_1), we have, by Theorem 9

$$1 = \sum_{i=1}^m j(U \cdot L, (c^{(i)}))\mu((c^{(i)}), (\gamma));$$

so we have $m = 1$ and $j(U \cdot L, (c^{(1)})) = 1$. As the point (c, c') is in Λ, (c) and (c') are both contained in L; the coordinates of (c) and (c') are all in $\mathfrak{o}_1$; and $\overline{(c)} = \overline{(c')} = (\gamma)$. So we must have $(c) = (c')$; this shows that D is the diagonal Δ_C of a variety C which is the locus of (c) over k. It can be easily verified that C is a proper component of $A \cap B$ on U. For every D_ν, we have a variety C_ν such that $D_\nu = \Delta_{C_\nu}$. Conversely, if we have a component C' of $A \cap B$ such that $\overline{C'} \supset \mathfrak{C}$, then it is easily proved that $\Delta_{C'}$ is

---

[9'''] See Theorem 1 and the remark below that theorem.

a component of $\Lambda \cap (A \times B)$; so C' coincides with one of the C_ν. For every ν, the set of linear forms $F_i(X)$ $(1 \leqq i \leqq n)$ is a uniformizing set of linear forms for U along C_ν, since $\bar{C}_\nu \supset \mathfrak{C}$ and since the set $\bar{F}_i(X)$ $(1 \leqq i \leqq n)$ is that for U along $\mathfrak{C}$. So we have $i(A \cdot B, C_\nu; U) = j[(A \times B) \cdot \Lambda, \Delta_{C_\nu}]$. By Theorem 8, we have $\mu(A \times B, \mathfrak{A}_p \times \mathfrak{B}_q) = \mu(A, \mathfrak{A}_p)\mu(B, \mathfrak{B}_q)$. Therefore our theorem will be proved if we show that $\mu(\Delta_{C_\nu}, \Delta_{\mathfrak{C}}) = \mu(C_\nu, \mathfrak{C})$. But this is easily proved by means of Proposition 21.[10]

Now, we shall prove the birational invariance of our multiplicity. Let U be a variety in $S^n \times S^m$ defined over k and U' the projection of U on S^n. Let $(x) \times (y)$ be a generic point of U over k and (α) a point in $\overline{U'}$. We say that U is *finite over* (α) if (y) is finite over $(x) \overset{\mathfrak{o}}{\longrightarrow} (\alpha)$. Also we say that the projection from U to U' is *regular at* (α) if all the y_j are contained in the specialization ring $[(x) \overset{\mathfrak{o}}{\longrightarrow} (\alpha)]$. Let k' be an extension of k and $\mathfrak{o}'$ a prolongation of $\mathfrak{o}$ in k'. Then by Proposition 6 of Section 1, U is finite over (α) if and only if U is finite over (α) with respect to $\mathfrak{o}'$. Similarly the projection from U to U' is regular at (α) if and only if that projection is regular at (α) with respect to $\mathfrak{o}'$. Let (β) be a specialization of (α) over κ; then the point (β) is a point in U'. It is easy to see that if U is finite over (β), U is also finite over (α), and that if the projection from U to U' is regular at (β), it is also regular at (α). So we say that U is finite over a variety in $\overline{U'}$ if U is finite over some point in that variety, and also we say that the projection from U to U' is regular along a variety in $\overline{U'}$ if that projection is regular at some point on that variety.

THEOREM 12. *Let U be a variety in $S^n \times S^m$ defined over k, having the same dimension r as its projection U' on S^n. Suppose that there exists a component $\mathfrak{U}'$ of $\overline{U'}$ such that U is finite over $\mathfrak{U}'$. Then there exists a component of $\bar{U}$ whose projection on $\mathfrak{S}^n$ is $\mathfrak{U}'$. If we denote by the $\mathfrak{U}_\nu$ all such components of $\bar{U}$, then we have*

$$[U : U']\mu(U', \mathfrak{U}') = \sum_\nu \mu(U, \mathfrak{U}_\nu)[\mathfrak{U}_\nu : \mathfrak{U}']$$

Proof. Let $L^{n-r} = (\sum t_{ij}X_j - t_i = 0 \; (1 \leqq i \leqq r))$ be a generic linear variety in S^n over k, and $\mathfrak{L}^{n-r} = (\sum \tau_{ij}X_j - \tau_i = 0 \; (1 \leqq i \leqq r))$ a generic linear variety in $\mathfrak{S}^n$ over κ. Then we have a point (x) in $U' \cap L$ and a point (ξ) in $\mathfrak{U}' \cap \mathfrak{L}$. There exists a point (y) in S^m such that (x, y) is a generic point of U over k. By our assumption, (y) is finite over $(x) \overset{\mathfrak{o}}{\longrightarrow} (\xi)$. Let (η) be a specialization of (y) over $(x) \overset{\mathfrak{o}}{\longrightarrow} (\xi)$. Then the locus $\mathfrak{U}$ of

---

[10] Let $M = (\sum t_{ij}X_j - t_i = 0 \; (1 \leqq i \leqq r + s - n))$ be a generic linear variety in S^N over k and $\mathfrak{M} = (\sum \tau_{ij}X_j - \tau_i = 0 \; (1 \leqq i \leqq r + s - n))$ a generic linear variety in $\mathfrak{S}^N$ over κ. Apply Proposition 21 to Δ_{C_ν}, $\Delta_{\mathfrak{C}}$, $M \times S^N$ and $\mathfrak{M} \times \mathfrak{S}^N$.

(ξ, η) over $\bar{\kappa}$ is contained in U and has the projection on $\mathfrak{S}^n$. If we denote by $(\eta^{(\nu)})$ all the specializations of (η) over $(x) \xrightarrow{\;o\;} (\xi)$, then, by Theorem 3 of Section 2, we have

$$\sum \mu_\nu = \mu(U', \mathfrak{U}')\,[k(t, x, y) : k(t, x)],$$

where μ_ν is the multiplicity of $(\xi, \eta^{(\nu)})$ as a specialization of (x, y) over $(t) \xrightarrow{\;o\;} (\tau)$. It is obvious that $[U : U'] = [k(t, x, y) : k(t, x)]$. Applying Proposition 21 to U, $\mathfrak{U}_\nu$, $L \times S^m$ and $\mathfrak{L} \times \mathfrak{S}^m$, we have

$$\mu_\nu = \mu(U, \mathfrak{U}_\nu)\,[\kappa(\tau, \xi, \eta^{(\nu)}) : \kappa(\tau)]_i\,[\mathfrak{U}_\nu : \kappa]_i^{-1},$$

if we denote by $\mathfrak{U}_\nu$ the locus of $(\xi, \eta^{(\nu)})$ over $\bar{\kappa}$. We have $\mathfrak{U}_\nu = \mathfrak{U}_\lambda$ if and only if $(\eta^{(\nu)})$ and $(\eta^{(\lambda)})$ are conjugates of each other over $\bar{\kappa}(\xi)$. Hence we have

$$[U : U']\mu(U', \mathfrak{U}')$$
$$= \sum{}' \mu(U, \mathfrak{U}_\nu)\,[\bar{\kappa}(\xi, \eta^{(\nu)}) : \bar{\kappa}(\xi)]_s\,[\kappa(\tau, \xi, \eta^{(\nu)}) : \kappa(\tau)]_i\,[\mathfrak{U}_\nu : \kappa]_i^{-1}.$$

where the sum is extended over all the components $\mathfrak{U}_\nu$ of $\bar{U}$ having $\mathfrak{U}'$ as their projections on $\mathfrak{S}^n$. This proves our theorem.

The following theorem is a translation of Theorem 15 of [WF] Chapter IV, Section 7 to our case, and can be proved similarly.

THEOREM 13. *Let U be a variety in $S^n \times S^m$ defined over k with the projection U' on S^n. Let $\mathfrak{Z}'$ be a variety in $\overline{U'}$ such that the projection from U to U' is regular along $\mathfrak{Z}'$. Then there is one and only one variety $\mathfrak{Z}$ in $\bar{U}$ with the projection $\mathfrak{Z}'$ on $\mathfrak{S}^n$; $\mathfrak{Z}$ is simple on U if and only if $\mathfrak{Z}'$ is simple on U; and we have $[\mathfrak{Z} : \mathfrak{Z}'] = 1$.*

Let T be a birational correspondence between two varieties V and W, defined over k. Let $\mathfrak{Z}_V$ be a variety in V such that the projection from T to V is regular along $\mathfrak{Z}_V$. Then, by Theorem 13, there is one and only one variety $\mathfrak{Z}$ in $\bar{T}$ with the projection $\mathfrak{Z}_V$ on $\bar{V}$. If the projection from T to W is also regular along the projection $\mathfrak{Z}_W$ of $\mathfrak{Z}$ on W, then we say that T is *biregular along* $\mathfrak{Z}_V$ and *along* $\mathfrak{Z}_W$, and that $\mathfrak{Z}_V$ and $\mathfrak{Z}_W$ are *regularly corresponding varieties in $\bar{V}$ and in $\bar{W}$ by T.*

THEOREM 14. *Let T be a birational correspondence between two varieties V and V', defined over k. Let A^r be a prime rational cycle over k, which is contained in V, and $\mathfrak{A}$ a prime component of $\rho(A)$ over κ. Suppose that T is biregular along a component of $\mathfrak{A}$ and denote by A' and $\mathfrak{A}'$ the cycles respectively corresponding to A and $\mathfrak{A}$ by T. Then $\mathfrak{A}'$ is a prime component of $\rho(A')$ over κ; and we have $\mu(A, \mathfrak{A}) = \mu(A', \mathfrak{A}')$.*

This follows immediately from Theorem 12, Theorem 13 and from Proposition 20.

Now we shall prove the converse of Theorem 10.

THEOREM 15. *Let V^r be a variety in S^n defined over k and (ξ) a point in $\bar{V}$. Then (ξ) is simple on V if and only if (ξ) is contained in only one component $\mathfrak{B}$ of $\bar{V}$, $\mu(V, \mathfrak{B}) = 1$ and (ξ) is simple on $\mathfrak{B}$. Moreover, if that is so, and if (x) is a generic point of V over k, then the specialization ring $[(x) \xrightarrow{\mathfrak{o}} (\xi)]$ is integrally closed.*

Proof. We have only to prove the direct part. Let the t_{ij} and the s_j $(1 \leq i \leq r, 1 \leq j \leq n)$ be $(r+1)n$ independent variables over $k(x)$, let the τ_{ij} and the σ_{ij} be $(r+1)n$ independent variables over $\kappa(\xi)$. Put $y_i = \sum t_{ij}x_j$ and $\eta_i = \sum \tau_{ij}\xi_j$ for $1 \leq i \leq r$, $z = \sum s_j x_j$ and $\zeta = \sum \sigma_j \xi_j$. Denote by $\mathfrak{L}$ the linear variety $(\sum \tau_{ij}X_j - \eta_i = 0 \ (1 \leq i \leq r))$. Since $\mathfrak{L}$ is transversal to the tangent linear variety to $\mathfrak{B}$ at (ξ), we have $j(\mathfrak{B} \cdot \mathfrak{L}, (\xi)) = 1$. By Proposition 9 of Section 1, we can find an extension k_1 of k and a prolongation $\mathfrak{o}_1$ of $\mathfrak{o}$ in k_1 such that the residue field of $\mathfrak{o}_1$ is $\kappa(\xi, \tau)$. We may assume that $k(x, t)$ and k_1 are independent over k. With respect to $\mathfrak{o}_1$, our assumption gives

$$j(V \cdot \mathfrak{L}, (\xi)) = \mu(V, \mathfrak{B}) j(\mathfrak{B} \cdot \mathfrak{L}, (\xi)) = 1,$$

so that (ξ) is a proper specialization of (x) over $(t, y) \xrightarrow{\mathfrak{o}_1} (\tau, \eta)$ of multiplicity 1.[10'] Since $k(x, t)$ and k_1 are linearly disjoint over k, we see that (ξ) is a proper specialization of (x) over $(t, y) \xrightarrow{\mathfrak{o}} (\tau, \eta)$ of multiplicity 1. By Proposition 16 of Section 2, ζ is a proper specialization of z over $(t, s, y) \xrightarrow{\mathfrak{o}} (\tau, \sigma, \eta)$, of multiplicity 1. By the same proposition, we have $k(t, s, y, z) = k(t, s, x)$. Put $k' = k(t, s)$ and $\mathfrak{o}' = [(t, s) \xrightarrow{\mathfrak{o}} (\tau, \sigma)]$. By Proposition 17 of Section 2, the specialization ring $[(y, z) \xrightarrow{\mathfrak{o}'} (\eta, \zeta)]$ is integrally closed, so it contains all the x_j since (x) is finite over this specialization ring by Proposition 16 of Section 2. Let Z be the locus of (x, y, z) over k' and W the locus of (y, z) over k'. Then Z is a birational correspondence between V and W; and the above results shows that (ξ) and (η, ζ) are regularly corresponding points by Z. By Proposition 17 of Section 2, (η, ζ) is simple on W; from this and Theorem 13, follows that (ξ) is simple on V. Moreover we have $[(x) \xrightarrow{\mathfrak{o}} (\xi)] = [(y, z) \xrightarrow{\mathfrak{o}'} (\eta, \zeta)]$. By this and Proposition 7 of Section 1, $[(x) \xrightarrow{\mathfrak{o}} (\xi)]$ is integrally closed.

COROLLARY 1. *Let V be a variety defined over k and $\mathfrak{B}$ a component of $\bar{V}$. Then $\mathfrak{B}$ is simple on V if and only if $\mu(V, \mathfrak{B}) = 1$.*

[10'] Note that the linear variety $(\sum t_{ij}X_j - y_i = 0 \ (1 \leq i \leq r))$ is generic over k and see the remark below the proof of Theorem 9.

This is an immediate consequence of the above theorem and the corollary of Theorem 10.

COROLLARY 2. *Let U be a variety defined over k and $\mathfrak{U}$ a component of $\bar{U}$; let (x) and (ξ) be respectively a generic point of U over k and a generic point of $\mathfrak{U}$ over $\bar{\kappa}$; and denote by $\mathfrak{R}$ the specialization ring $[(x) \xrightarrow{\mathfrak{o}} (\xi)]$. If $\mathfrak{U}$ is simple on U, then the maximal ideal of $\mathfrak{R}$ is the ideal $\mathfrak{p}\mathfrak{R}$.*

Proof. Let the t_{ij} $(1 \leqq i \leqq r+1, 1 \leqq j \leqq n)$ be $(r+1)n$ independent variables over $k(x)$ and the τ_{ij} $(1 \leqq i \leqq r+1, 1 \leqq j \leqq n)$ be $(r+1)n$ independent variables over $\kappa(\xi)$. By Proposition 9 of Section 1, we can find a finite algebraic extension k_1 of k and a prolongation $\mathfrak{o}_1$ of $\mathfrak{o}$ in k_1 such that $\mathfrak{U}$ is defined over the residue field κ_1 of $\mathfrak{o}_1$. We may assume that the maximal ideal $\mathfrak{p}_1$ of $\mathfrak{o}_1$ is the ideal $\mathfrak{p}\mathfrak{o}_1$.[10''] Put $\mathfrak{o}' = [(t) \xrightarrow{\mathfrak{o}_1} (\tau)]$, $\mathfrak{p}' = \mathfrak{p}_1\mathfrak{o}'$, $\kappa' = \kappa_1(\tau)$, $y_i = \sum t_{ij}x_j$ for $1 \leqq i \leqq r+1$ and $\eta_i = \sum \tau_{ij}\xi_j$ for $1 \leqq i \leqq r+1$. Then as in the proof of Theorem 15, we have $[(x) \xrightarrow{\mathfrak{o}'} (\xi)] = [(y) \xrightarrow{\mathfrak{o}'} (\eta)]$. Moreover, if we denote by V the locus of (y) over $k(t)$, then (η) is a point in $\bar{V}$ and it is simple on V. Let $F(Y) = 0$ be an irreducible equation for (y) over $k(t)$ and $\bar{F}_0(Y) = 0$ an irreducible equation for (η) over κ'. We may assume that $F(Y)$ is a primitive polynomial in $\mathfrak{o}'[Y]$ and that $\bar{F}_0(Y)$ is the class of a ploynomial $F_0(Y)$ in $\mathfrak{o}'[Y]$ modulo $\mathfrak{p}'$. We have $\bar{F}(Y) = \bar{F}_0(Y)\bar{F}_1(Y)$ where $\bar{F}_1(Y)$ is a polynomial in $\kappa'[Y]$. Since (η) is simple on V, we have $\partial \bar{F}/\partial Y_j(\eta) \neq 0$ for some j; so that we have $\bar{F}_1(\eta) \neq 0$. Let z be a quantity in $k(x)$ contained in the maximal ideal of $[(x) \xrightarrow{\mathfrak{o}} (\xi)]$. Then we have an expression $z = P(y)/Q(y)$ where $P(Y)$ and $Q(Y)$ are polynomials in $\mathfrak{o}'[Y]$, $\bar{P}(\eta) = 0$ and $\bar{Q}(\eta) \neq 0$. Since $\bar{F}_0(Y) = 0$ is an irreducible equation for (η) over κ', we have $\bar{P}(Y) = \bar{F}_0(Y)\bar{G}(Y)$ where $\bar{G}(Y)$ is a polynomial in $\kappa'[Y]$. Then we have $\bar{P}(Y)\bar{F}_1(Y) = \bar{F}(Y)\bar{G}(Y)$. Let $F_1(Y)$ and $G(Y)$ be polynomials in $\mathfrak{o}'[Y]$ such that $\bar{F}_1(Y)$ and $\bar{G}(Y)$ are respectively the class of $F_1(Y)$ and the class of $G(Y)$ modulo $\mathfrak{p}'$. Then we have

$$P(Y)F_1(Y) = F(Y)G(Y) + \pi H(Y),$$

where $H(Y)$ is a polynomial in $\mathfrak{o}'[Y]$ and π is a prime element of $\mathfrak{o}$. Substituting (y) for (Y), we have $P(y)F_1(y) = \pi H(y)$, so we have $z/\pi = H(y)/[Q(y)F_1(y)]$. Since $\bar{Q}(\eta)\bar{F}_1(\eta) \neq 0$, z/π is contained in $[(y) \xrightarrow{\mathfrak{o}'} (\eta)]$, so in $[(x) \xrightarrow{\mathfrak{o}'} (\xi)]$. Then by Proposition 7 of Section 1, z/π is contained in $[(x) \xrightarrow{\mathfrak{o}} (\xi)]$. This proves our corollary.

[10''] This is assured since we can take k_1 and $\mathfrak{o}_1$ in such a way that $[k_1: k] = [\kappa_1: \kappa]$ as in the proof of Proposition 9.

4. Reduction of abstract varieties. Let $[V_\alpha; F_\alpha; T_{\beta\alpha}]$ be an abstract variety in Weil's sense, defined over k; let, for each α, $\mathfrak{F}_\alpha$ be a bunch in $\bar{V}_\alpha$ which is normally algebraic over κ and which contains $\bar{F}_\alpha$. We call the system $[V_\alpha; F_\alpha; \mathfrak{F}_\alpha; T_{\beta\alpha}]$ *a $\mathfrak{p}$-variety defined over k having* $[V_\alpha; F_\alpha; T_{\beta\alpha}]$ *as its underlying abstract variety* if the following condition is satisfied.

Whenever, for any α and β, $\bar{P}_\alpha$ is a point in $\bar{V}_\alpha - \mathfrak{F}_\alpha$ and $\bar{P}_\beta$ a point in $\bar{V}_\beta - \mathfrak{F}_\beta$, such that $(\bar{P}_\alpha, \bar{P}_\beta)$ is in $T_{\beta\alpha}$, then $\bar{P}_\alpha$ and $\bar{P}_\beta$ are regularly corresponding points by $T_{\beta\alpha}$.

We can easily prove the following facts.

i)   Let $[V_\alpha; F_\alpha; \mathfrak{F}_\alpha; T_{\beta\alpha}]$ be a $\mathfrak{p}$-variety defined over k and

$$[U_{\alpha_\lambda}; F_{\alpha_\lambda} \cap U_{\alpha_\lambda}; R_{\mu\lambda}]$$

a subvariety of the abstract variety $[V_\alpha; F_\alpha; T_{\beta\alpha}]$. Then the system

$$[U_{\alpha_\lambda}; F_{\alpha_\lambda} \cap U_{\alpha_\lambda}; \mathfrak{F}_{\alpha_\lambda} \cap \bar{U}_{\alpha_\lambda}; R_{\mu\lambda}]$$

defines a $\mathfrak{p}$-variety which will be called a *subvariety* of the $\mathfrak{p}$-variety

$$[V_\alpha; F_\alpha; \mathfrak{F}_\alpha; T_{\beta\alpha}].$$

ii)   Let $[V_\alpha; F_\alpha; \mathfrak{F}_\alpha; T_{\beta\alpha}]$ and $[W_\lambda; G_\lambda; \mathfrak{G}_\lambda; S_{\mu\lambda}]$ be two $\mathfrak{p}$-varieties defined over k and $[V_\alpha \times W_\lambda; H_{\alpha\lambda}; U_{\beta\mu,\alpha\lambda}]$ the product-variety of the abstract varieties $[V_\alpha; F_\alpha; T_{\beta\alpha}]$ and $[W_\lambda; G_\lambda; S_{\mu\lambda}]$. Put

$$\mathfrak{H}_{\alpha\lambda} = (\mathfrak{F}_\alpha \times \bar{W}_\lambda) \cup (\bar{V}_\alpha \times \mathfrak{G}_\lambda).$$

Then the system $[V_\alpha \times W_\lambda; H_{\alpha\lambda}; \mathfrak{H}_{\alpha\lambda}; U_{\beta\mu,\alpha\lambda}]$ defines a $\mathfrak{p}$-variety which will be called the *product-variety* of $[V_\alpha; F_\alpha; \mathfrak{g}_\alpha; T_{\beta\alpha}]$ and $[W_\lambda; G_\lambda; \mathfrak{G}_\lambda; S_{\mu\lambda}]$.

iii)   The projective space of any dimension defines a $\mathfrak{p}$-variety with the empty $\mathfrak{F}_\alpha$; so that every projective variety defines a uniquely determined $\mathfrak{p}$-variety with the empty $\mathfrak{F}_\alpha$.

Let $[V] = [V_\alpha; F_\alpha; \mathfrak{F}_\alpha; T_{\beta\alpha}]$ be a $\mathfrak{p}$-variety defined over k. Let the M_α be corresponding generic points of the V_α over k by the $T_{\beta\alpha}$ and $(\bar{P}_1, \cdots, \bar{P}_h)$ be any specialization of $(M_1, \cdots, M_h)$ over $\mathfrak{o}$. Similarly as in [WF] p. 168, by a *full set of representatives attached to* $[\bar{V}]$, we understand the set $(\bar{P}_{\alpha_1}, \cdots, \bar{P}_{\alpha_l})$ of all the $\bar{P}_\alpha$ which is finite and not in $\mathfrak{F}_\alpha$. Also we say that $[V]$ is *$\mathfrak{p}$-complete* if no full set of representatives attached to $[\bar{V}]$ is empty. The following facts are easily proved; i) the underlying abstract variety of a $\mathfrak{p}$-complete $\mathfrak{p}$-variety is complete; ii) every subvariety of a $\mathfrak{p}$-complete $\mathfrak{p}$-variety is $\mathfrak{p}$-complete; iii) the product of $\mathfrak{p}$-complete $\mathfrak{p}$-varieties is $\mathfrak{p}$-complete; iv) every projective variety defines a $\mathfrak{p}$-complete $\mathfrak{p}$-variety.

Let $[V]$ be the same as above, and let $[\mathfrak{W}] = [\mathfrak{W}_\lambda; \mathfrak{G}_\lambda; \mathfrak{S}_{\mu\lambda}]$ be an abstract variety defined over an extension field κ' of κ. We say that $[\mathfrak{W}]$ is a *variety in* $[\bar{V}]$ if the following conditions are satisfied: i) there exists a full set of representatives $(\bar{P}_{a_1}, \cdots, \bar{P}_{a_l})$ attached to $[\bar{V}]$ such that, for every λ, $\mathfrak{W}_\lambda$ is the locus of $\bar{P}_{a_\lambda}$ over κ'; ii) $\mathfrak{G}_\lambda = \mathfrak{W}_\lambda \cap \mathfrak{F}_{a_\lambda}$; iii) $\mathfrak{S}_{\mu\lambda}$ is the variety in $\bar{T}_{a_\mu a_\lambda}$ with the projections $\mathfrak{W}_\lambda$ on $\bar{V}_{a_\lambda}$ and $\mathfrak{W}_\mu$ on $\bar{V}_{a_\mu}$. We call $\mathfrak{W}_\lambda$ the *representative of* $[\mathfrak{W}]$ *in* V_{a_λ}. We can easily prove that for any variety $\mathfrak{W}_a$ in $\bar{V}_a$ which is not contained in $\mathfrak{F}_a$ there is one and only one variety $[\mathfrak{W}]$ in $[\bar{V}]$ such that $\mathfrak{W}_a$ is a representative of $[\mathfrak{W}]$ in V_a. By a *point in* $[\bar{V}]$ we understand a zero-dimensional variety in $[\bar{V}]$. It is obvious that all the representatives of a point in $[\bar{V}]$ forms a full set of representatives attached to $[\bar{V}]$ and conversely. Let $[P]$ be a point in $[V]$ and $[\mathfrak{P}]$ a point in $[\bar{V}]$. We say that $[\mathfrak{P}]$ is a *specialization of* $[P]$ *over* $\mathfrak{o}$ (on $[V]$) if $[P]$ and $[\mathfrak{P}]$ have in some representative V_a of $[V]$, representatives P_a and $\mathfrak{P}_a$ such that $\mathfrak{P}_a$ is a specialization of P_a over $\mathfrak{o}$. We say that a variety $[\mathfrak{W}]$ in $[\bar{V}]$ is *simple on* $[\bar{V}]$ if a representative $\mathfrak{W}_a$ of $[\mathfrak{W}]$ is simple on V_a. We call a finite set of varieties in $[\bar{V}]$ a *bunch in* $[\bar{V}]$; we may identify a bunch with the point-set in $[\bar{V}]$ attached to that bunch. Also, we say that a variety $[\mathfrak{W}]$ is a *component of* a bunch in $[\bar{V}]$ if $[\mathfrak{W}]$ is maximal in that bunch. These definitions are analogies of the definitions in $[\text{WF}]$ Chapter VII.

Now our theory of the reduction with respect to $\mathfrak{p}$ can be extended to abstract varieties on which $\mathfrak{p}$-varieties are defined.

Let $[V]$ be a $\mathfrak{p}$-variety defined over k. Let $[A]^r$ be a prime rational $[V]$-cycle over k (which means a prime rational cycle on the underlying abstract variety of $[V]$). We denote by $[\bar{A}]$ the set of all the points in $[\bar{V}]$ which are specializations of some points in $[A]$, over $\mathfrak{o}$. Now we may consider the notation $[\bar{V}]$ in this sense. It is easy to see that the set $[\bar{A}]$ defines a bunch in $[\bar{V}]$. If $[\bar{A}]$ is not empty, then, by Proposition 19 of Section 3, every component of $[\bar{A}]$ is of dimension r. Let $[\mathfrak{A}]_1$ be a component of $[\bar{A}]$. By our definition of $[\bar{A}]$, there exists a representative V_a in which $[A]$ and $[\mathfrak{A}]_1$ have representatives A_a and $\mathfrak{A}_{1a}$.[10] By Theorem 14, of Section 3, the multiplicity $\mu(A_a, \mathfrak{A}_{1a})$ is independent of the choice of representation V_a, A_a, $\mathfrak{A}_{1a}$ for $[V]$, $[A]$ and $[\mathfrak{A}]_1$. We shall denote it by $\mu([A], [\mathfrak{A}]_1)$. Now the results in Section 3 are all translated to the corresponding results on abstract varieties. We shall state them in the following theorems for varieties with a property which we shall call $\mathfrak{p}$-simplicity. We

---

[10] We shall use the term "representative" also for a prime rational cycle.

shall say that a $\mathfrak{p}$-variety $[V]$ is $\mathfrak{p}$-*simple* if there exists only one component $[\mathfrak{B}]$ in $[\bar{V}]$ and if $\mu([V],[\mathfrak{B}]) = 1$; and we shall call this $[\mathfrak{B}]$ the *abstract variety obtained from* $[V]$ *by the reduction with respect to* $\mathfrak{p}$.

THEOREM 16. *Let* $[V]$ *be a* $\mathfrak{p}$-*simple* $\mathfrak{p}$-*variety defined over* k, *then the abstract variety* $[\mathfrak{B}]$ *obtained from* $[V]$ *by the reduction with respect to* $\mathfrak{p}$ *is defined over* κ; *and a point on* $[\mathfrak{B}]$ *is simple on* $[\mathfrak{B}]$ *if and only if it is simple on* $[V]$.

Proof. As for the first assertion, it is sufficient to prove that every representative of $[\mathfrak{B}]$ is defined over κ. Let V_α be a representative of $[V]$ and $\mathfrak{B}_\alpha$ a representative of $[\mathfrak{B}]$ in V_α. By Corollary 1 of Theorem 15 of Section 3 and by corollary of Theorem 10 of Section 3, we have $[\mathfrak{B}_\alpha : \kappa]_i = 1$. This shows $[\kappa' : \kappa]_i = 1$ if we denote by κ' the smallest field of definition for $\mathfrak{B}_\alpha$ which contains κ. Since the bunch $\mathfrak{F}_\alpha$ in $\bar{V}_\alpha$ is normally algebraic over κ, every conjugate of $\mathfrak{B}_\alpha$ over κ is not contained in $\mathfrak{F}_\alpha$; so that, by our assumption, there is no conjugate of $\mathfrak{B}_\alpha$ over κ other than $\mathfrak{B}_\alpha$ itself; so we have $[\kappa' : \kappa]_s = 1$. Hence we have $\kappa' = \kappa$. The remaining part of the theorem follows immediately from Theorem 15 of Section 3.

PROPOSITION 23. *If* $[U]$ *and* $[V]$ *are* $\mathfrak{p}$-*simple* $\mathfrak{p}$-*varieties defined over* k, *then the product-variety* $[U] \times [V]$ *is also* $\mathfrak{p}$-*simple*.

This is easily proved since the result corresponding to Proposition 5 of [WF] Chapter VII, Section 3 also holds in our case.

Let $[V]$ be a $\mathfrak{p}$-simple variety defined over k and $[\mathfrak{B}]$ the abstract variety obtained from $[V]$ by the reduction with respect to $\mathfrak{p}$. Then every " variety in $[\bar{V}]$ " is a subvariety of $[\mathfrak{B}]$. Thus we are able to define the reduction of $[V]$-cycles. Let $[A]^r$ be a prime rational $[V]$-cycle over k and $[\mathfrak{A}]_1{}^r$ a prime rational $[\mathfrak{B}]$-cycle over κ with a component $[\mathfrak{A}]_{11}$ which is a component of $[\bar{A}]$. We denote by $\mu([A],[\mathfrak{A}]_1)$ the number

$$\mu([A],[\mathfrak{A}]_{11})/[[\mathfrak{A}]_{11}:\kappa]_i,$$

and define the $[\mathfrak{B}]$-cycle $\rho([A])$ as follows:

$$\rho([A]) = \sum_\nu \mu([A],[\mathfrak{A}]_\nu)[\mathfrak{A}]_\nu,$$

where the sum is taken over all the prime rational $[\mathfrak{B}]$-cycles $[\mathfrak{A}]_\nu$ over κ with the components which are components of $[\bar{A}]$. Even if $[\bar{A}]$ is not empty the cycle $\rho([A])$ vanishes when every component of $[\bar{A}]$ is not simple on $[\mathfrak{B}]$. Let $[X]$ be a rational $[V]$-cycle over k; we have an expression $[X] = \sum_\lambda a_\lambda [A]_\lambda$ where, for every λ, $[A]_\lambda$ is a prime rational $[V]$-cycle over k. We put

$\rho([X]) = \sum_{\lambda} a_{\lambda} \rho([A]_{\lambda})$ and call it the *cycle obtained from* $[X]$ *by the reduction with respect to* $\mathfrak{p}$. It is obvious that ρ is a linear mapping of the group of rational $[V]$-cycles over k into the group of rational $[\mathfrak{B}]$-cycles over κ. Similarly as in the affine space, $\rho([X])$ is invariant under the extension of the basic field k and the prolongation of $\mathfrak{o}$. It should be mentioned that the mapping ρ is defined with reference to the ambient variety $[V]$. But we shall use the same notation ρ for mappings of cycles on different ambient varieties, since there will be no confusion.

THEOREM 17. *Let* $[V]$ *be a* $\mathfrak{p}$-*simple* $\mathfrak{p}$-*variety defined over* k. *Let* $[X]^r$ *and* $[Y]^s$ *be positive rational* $[V]$-*cycles over* k *such that the intersection-products* $[X] \cdot [Y]$ *and* $\rho([X]) \cdot \rho([Y])$ *are both defined. Then we have* $\rho([X] \cdot [Y]) = \rho([X]) \cdot \rho([Y])$.

Proof. By linearity it is sufficient to prove our theorem in case where $[X]$ and $[Y]$ are prime rational $[V]$-cycles $[A]^r$ and $[B]^s$ over k, respectively. Moreover we may assume that $[A]$ and $[B]$ are both varieties defined over k. For, we may take a suitable extension of k and a prolongation of $\mathfrak{o}$ instead of k and $\mathfrak{o}$, since ρ is invariant under such a change of the basic field and ring. We may also assume that every component of $[A] \cap [B]$ is defined over k. Now denote by $[\mathfrak{B}]$ the abstract variety obtained from $[V]$ by reduction. Let $[C]$ be a proper component of $[A] \cap [B]$ on $[V]$ and $[\mathfrak{C}]$ a component of $[\bar{C}]$. If $[\mathfrak{C}]$ is simple on $[\mathfrak{B}]$, it is a component of $\rho([A]) \cdot \rho([B])$. To see this let $[\mathfrak{C}]'$ be a component of $[\bar{A}] \cap [\bar{B}]$ containing $[\mathfrak{C}]$; then $[\mathfrak{C}]'$ is simple on $[\mathfrak{B}]$. By our assumption that $\rho([A]) \cdot \rho([B])$ is defined on $[\mathfrak{B}]$, we have dim $[\mathfrak{C}]' = r + s - n$ where n is the dimension of $[V]$ and of $[\mathfrak{B}]$; so that we have dim $[\mathfrak{C}]' = $ dim $[C]$ $=$ dim $[\mathfrak{C}]$. Hence $[\mathfrak{C}]$ is a component of $\rho([A]) \cdot \rho([B])$. Conversely, let $[\mathfrak{C}]$ be a component of $\rho([A]) \cdot \rho([B])$. Then it is a component of $[\bar{A}] \cap [\bar{B}]$ of dimension $r + s - n$ and is simple on $[\mathfrak{B}]$. By Theorem 16, $[\mathfrak{C}]$ is also simple on $[V]$; then, by Theorem 11 of Section 3, there exists a component $[C]$ of $[A] \cap [B]$ such that $[\mathfrak{C}]$ is a component of $[\bar{C}]$. Hence we see that $[\mathfrak{C}]$ is a component of $\rho([A] \cdot [B])$. Thus we have proved that a variety on $[\mathfrak{B}]$ is a component of $\rho([A] \cdot [B])$ if and only if it is a component of $\rho([A]) \cdot \rho([B])$. Now our theorem is an immediate consequence of the formula in Theorem 11 of Section 3.

THEOREM 18. *Let* $[U]$ *and* $[V]$ *be* $\mathfrak{p}$-*simple* $\mathfrak{p}$-*varieties defined over* k. *If* $[X]$ *is a rational* $[U]$-*cycle over* k *and* $[Y]$ *is a rational* $[V]$-*cycle over* k, *then we have* $\rho([X] \times [Y]) = \rho([X]) \times \rho([Y])$.

This follows immediately from Theorem 8 of Section 3 and from our definition of ρ.

THEOREM 19. *Let $[U]$ and $[V]$ be $\mathfrak{p}$-simple $\mathfrak{p}$-varieties defined over k; and let $[\mathfrak{U}]$ and $[\mathfrak{V}]$ be respectively the abstract varieties obtained from $[U]$ and $[V]$ by the reduction with respect to $\mathfrak{p}$. Suppose that $[V]$ is $\mathfrak{p}$-complete and that $[\mathfrak{V}]$ has no multiple point. If $[X]$ is a rational $[U] \times [V]$-cycle over k, then we have $\rho\{\mathrm{pr}_{[U]}([X])\} = \mathrm{pr}_{[\mathfrak{U}]}\{\rho([X])\}$.*

Proof. By linearity, it is enough to prove our theorem in case where $[X]$ is a prime rational cycle $[A]$. Moreover, as in the proof of Theorem 17, we may assume that $[A]$ is a variety defined over k. Let $[A]'$ be the projection of $[A]$ on $[U]$, and $[P] \times [Q]$ a generic point of $[A]$ over k. Our theorem is obvious when $\dim [A]' < \dim [A]$; so we assume that $[A]'$ has the same dimension r as $[A]$ and put $[[A]:[A]'] = d$. Let $[\mathfrak{A}]'$ be a component of $\rho([A]')$ and $[\bar{P}]$ a generic point of $[\mathfrak{A}]'$ over $\bar{\kappa}$. Since $[V]$ is $\mathfrak{p}$-complete, there exists a point $[\bar{Q}]$ on $[\mathfrak{V}]$ which is a specialization of $[Q]$ over $[P] \xrightarrow{\mathrm{o}} [\bar{P}]$. If we denote by $[\mathfrak{A}]$ the locus of $[\bar{P}] \times [\bar{Q}]$ over $\bar{\kappa}$, then $[\mathfrak{A}]$ is a component of $\rho([A])$ with the projection $[\mathfrak{A}]'$ on $[\mathfrak{U}]$. Thus every component of $\rho\{\mathrm{pr}_{[U]}([A])\}$ is a component of $\mathrm{pr}_{[\mathfrak{U}]}\{\rho([A])\}$. It is easy to see that the converse is also true. Now we may assume that the components of $\rho([A])$ and $\rho([A]')$ are all defined over κ. Our theorem is proved when we show that, for every component $[\mathfrak{A}]'$ of $\rho([A]')$,

$$d \cdot \mu([A]', [\mathfrak{A}]') = \sum_{\sigma} \mu([A], [\mathfrak{A}]_\sigma) [[\mathfrak{A}]_\sigma : [\mathfrak{A}]'],$$

where the sum is taken over every component $[\mathfrak{A}]_\sigma$ of $\rho([A])$ with the projection $[\mathfrak{A}]'$ on $[\mathfrak{U}]$. Let U, $\mathfrak{U}$, A', $\mathfrak{A}'$, P, $\bar{P}$ be representations of $[U]$, $[\mathfrak{U}]$, $[A]'$ $[\mathfrak{A}]'$, $[P]$ and $[\bar{P}]$, respectively; and let S^N and $\mathfrak{S}^N$ be respectively the ambient spaces for U and for $\mathfrak{U}$. We may assume that $[U] = U$ and $[\mathfrak{U}] = \mathfrak{U}$. Let $L^{N-r} = (\sum t_{ij}X_j - t_i = 0 \ \ (1 \leq i \leq r))$ be a generic linear variety in S^N over k and $\mathfrak{L}^{N-r} = (\sum \tau_{ij}X_j - \tau_i = 0 \ \ (1 \leq i \leq r))$ a generic linear variety in $\mathfrak{S}^N$ over κ. We may assume that the point P is in L and that $\bar{P}$ is in $\mathfrak{L}$. Denote by $(P_1, \cdots, P_\delta)$ the complete set of conjugates of P over $k(t)$ and denote by $([Q]_1, \cdots, [Q]_d)$ the complete set of conjugates of $[Q]$ over $k(P)$; then we have the set of δd points $P_\lambda \times [Q]_{\lambda\nu}$ $(1 \leq \lambda \leq \delta, 1 \leq \nu \leq d)$ as the complete set of conjugates of $P \times [Q]$ over $k(t)$, where $([Q]_{\lambda 1}, \cdots, [Q]_{\lambda d})$ is the transform of $([Q]_1, \cdots, [Q]_d)$ by some automorphism over $k(t)$ of $\overline{k(t)}$. Let $(\bar{P}_\lambda \times [\bar{Q}]_{\lambda\nu})$ be a specialization of $(P_\lambda \times [Q]_{\lambda\nu})$ over $(t) \xrightarrow{\mathrm{o}} (\tau)$. By our assumption that $[V]$ is $\mathfrak{p}$-complete, $[\bar{Q}]_{\lambda\nu}$ is a point of $[\mathfrak{V}]$ for every (λ, ν). If $\bar{P}_\lambda$ is a conjugate of $\bar{P}$

over $\kappa(\tau)$, then $\bar{P}_\lambda \times [\bar{Q}]_{\lambda\nu}$ is a generic point of one of the $[\mathfrak{A}]_\sigma$, say $[\mathfrak{A}]_1$, over κ. If $[\mathfrak{A}]_1$ has a representative in $U \times V_\alpha$ then every generic point of $[\mathfrak{A}]_1$ over κ has a representative in $U \times V_\alpha$. Hence we may consider that we are in the affine space; so that by Proposition 21 of Section 3,[11] the number of (λ, ν) such that $\bar{P}_\lambda \times [\bar{Q}]_{\lambda\nu}$ is a generic point of $[\mathfrak{A}]_1$ over k is equal to $\mu([A], [\mathfrak{A}]_1)[\kappa(\bar{P}_\lambda \times [\bar{Q}]_{\lambda\nu}, \tau) : \kappa(\tau)]$. On the other hand, the number of $\bar{P}_\lambda \times [\bar{Q}]_{\lambda\nu}$ such that $\bar{P}_\lambda$ is a conjugate of $\bar{P}$ over $\kappa(\tau)$ is equal to

$$d \cdot \mu([A]', [\mathfrak{A}]')[\kappa(\bar{P}, \tau) : \kappa(\tau)].$$

Hence we have

$$d \cdot \mu([A]', [\mathfrak{A}]')[\kappa(\bar{P}, \tau) : \kappa(\tau)]$$
$$= \sum_\sigma \mu([A], [\mathfrak{A}]_\sigma)[[\mathfrak{A}]_\sigma : [\mathfrak{A}]'][\kappa(\bar{P}, \tau) : \kappa(\tau)].$$

This completes our proof.

PROPOSITION 24. *Let $[V]$ be a $\mathfrak{p}$-simple $\mathfrak{p}$-variety defined over k and $[\mathfrak{B}]$ the abstract variety obtained from $[V]$ by the reduction with respect to $\mathfrak{p}$. Suppose that $[V]$ is $\mathfrak{p}$-complete and $[\mathfrak{B}]$ has no multiple point. Then for every rational $[V]$-cycle $[X]$ over k, of dimension 0, we have*

$$\deg([X]) = \deg\{\rho([X])\}.$$

Proof. As in the proof of Theorem 17, we may assume that every component of $[X]$ is rational over k. Since every rational point over k has a uniquely determined specialization over $\mathfrak{o}$, our proposition is an immediate consequence of our definition of ρ.

PROPOSITION 25. *Let $[L]^n$ be the projective space defined over k and $[X]$ a rational $[L]$-cycle over k. Consider the abstract variety obtained from $[L]$ by the reduction as the projective space $[\mathfrak{L}]^n$ defined over κ. Then we have $\deg([X]) = \deg\{\rho([X])\}$.*

This follows immediately from Theorem 17 and Proposition 24.

Let $[V]$ be a $\mathfrak{p}$-simple $\mathfrak{p}$-variety defined over k and $[\mathfrak{B}]$ the abstract variety obtained from $[V]$ by the reduction with respect to $\mathfrak{p}$. We denote by $\boldsymbol{F}$ the field of algebraic functions on $[V]$ defined over k and by Φ the field of algebraic functions on $[\mathfrak{B}]$ defined over κ. Let $[P]$ be a generic point of $[V]$ over k and $[\bar{P}]$ a generic point of $[\mathfrak{B}]$ over κ. Let ϕ be an element of $\boldsymbol{F}$ which is defined by $z = \phi([P])$. By our assumption and by Theorem 15 of Section 3, the specialization ring $[[P] \xrightarrow{\mathfrak{o}} [\bar{P}]]$ is integrally

---

closed, so that by Proposition 5 of Section 1, it is a discrete valuation ring. Hence z has a uniquely determined specialization ζ over $[P] \xrightarrow{\text{o}} [\bar{P}]$; and ζ is a generalized quantity in $\kappa([\bar{P}])$. Denote by $\bar{\phi}$ the generalized function on $[\mathfrak{W}]$ defined by $\zeta = \bar{\phi}([\bar{P}])$ over κ. Thus we obtain a mapping $\phi \to \bar{\phi}$ from F to Φ; it is easy to see that this mapping does not depend on the choice of $[P]$ and $[\bar{P}]$. Therefore, we may call the field Φ the *field obtained from* F *by the reduction with respect to* $\mathfrak{p}$; of course, the field Φ depends upon the variety $[V]$ which is a "model" of F.

THEOREM 20. *Let* $[V]$ *be a* $\mathfrak{p}$*-simple* $\mathfrak{p}$*-variety defined over* k *and* $[\mathfrak{W}]$ *the abstract variety obtained from* $[V]$ *by the reduction with respect to* $\mathfrak{p}$. *Let* ϕ *be a function on* $[V]$ *defined over* k *such that* $\bar{\phi}$ *is a function on* $[\mathfrak{W}]$ *other than the constant* 0. *Then* $\rho((\phi)) = (\bar{\phi})$.

Proof. Let $[D]$ be the projective straight line defined over k and $[\mathfrak{D}]$ be the projective straight line defined over κ. We may consider $[\mathfrak{D}]$ as the abstract variety obtained from $[D]$ by reduction. Let Γ_ϕ be the graph of ϕ and $\Gamma_{\bar{\phi}}$ the graph of $\bar{\phi}$; then $\Gamma_{\bar{\phi}}$ is a component of $\bar{\Gamma}_\phi$. By Theorem 19, we have

$$\mathrm{pr}_{[\mathfrak{W}]}\{\rho(\Gamma_\phi)\} = \rho\{\mathrm{pr}_{[V]}(\Gamma_\phi)\} = \rho([V]) = [\mathfrak{W}].$$

By this equation we have $\rho(\Gamma_\phi) = \Gamma_{\bar{\phi}} + [\mathfrak{X}] \times [\mathfrak{D}]$, where $[\mathfrak{X}]$ is a $[\mathfrak{W}]$-divisor. Now by the definition of $(\phi)_0$, Theorem 17, Theorem 18, and Theorem 19, we have

$$\rho((\phi)_0) = \rho\{\mathrm{pr}_{[V]}(\Gamma_\phi \cdot [V]_0)\} = \mathrm{pr}_{[\mathfrak{W}]}\{(\Gamma_{\bar{\phi}} + [\mathfrak{X}] \times [\mathfrak{D}]) \cdot ([\mathfrak{W}] \times (0))\}$$

$$= \mathrm{pr}_{[\mathfrak{W}]}\{\Gamma_{\bar{\phi}} \cdot [\mathfrak{W}]_0 + [\mathfrak{X}] \times (0)\} = (\bar{\phi})_0 + [\mathfrak{X}].$$

Similarly, we have $\rho((\phi)_\infty) = (\bar{\phi})_\infty + [\mathfrak{X}]$; so we have $\rho((\phi)) = (\bar{\phi})$.

THEOREM 21. *Let* A^{n-1} *be a prime rational* S^n*-divisor over* k *and* (x) *a generic point of* A *over* k. *Let* $F(X) = 0$ *be an irreducible equation for* (x) *over* k *and suppose that* $F(X)$ *is a primitive polynomial in* $\mathfrak{o}[X]$. *Let* $\bar{F}(X) = \prod_{\alpha} \bar{G}_\alpha(X)^{\mu_\alpha}$ *be an expression of* $\bar{F}(X)$ *as a product of irreducible polynomials in* $\kappa[X]$. *Then, we have* $\rho(A) = \sum_{\alpha} \mu_\alpha \mathfrak{A}_\alpha$ *where, for each* α, $\mathfrak{A}_\alpha$ *is the prime rational* $\mathfrak{S}^n$*-divisor over* κ *which has a generic point* $(\xi^{(\alpha)})$ *over* κ *such that* $\bar{G}_\alpha(\xi^{(\alpha)}) = 0$.

Proof. Let $P = (t)$ and $\bar{P} = (\tau)$ be respectively a generic point of S^n over k and a generic point of $\mathfrak{S}^n$ over κ. Let ϕ be the function on S^n, defined over k by $F(t) = \phi(P)$. Then the function $\bar{\phi}$ on $\mathfrak{S}^n$ is defined over κ by $\bar{F}(\tau) = \bar{\phi}(\bar{P})$. By our assumption on $F(X)$, the function $\bar{\phi}$ is not the constant 0; so that by Theorem 8 of [WF] Chapter VIII, Section 3 and by Theorem 20, we have $\rho(A) = \rho((\phi)) = (\bar{\phi})$. Let $\bar{\psi}_\alpha$ be the function on $\mathfrak{S}^n$ defined over κ by $\bar{G}_\alpha(\tau) = \bar{\psi}_\alpha(\bar{P})$, for each α. Then we have $\bar{\phi} = \prod_\alpha \bar{\psi}_\alpha{}^{\mu_\alpha}$; so we have by Theorem 6 and Theorem 8 of [WF] Chapter VIII, $(\bar{\phi}) = \sum_\alpha \mu_\alpha(\bar{\psi}_\alpha) = \sum_\alpha \mu_\alpha \mathfrak{A}_\alpha$. This proves our theorem.

5. Specialization-theory of cycles. When we consider the reduction of cycles with respect to $\mathfrak{p}$ as the specialization of cycles, our theory in Section 4 is concerned only with the specialization of the cycles which are rational over the basic field. In this section, we shall define a more general concept of the specialization of cycles on abstract varieties.

Let $[V]_1, \cdots, [V]_n$ be $\mathfrak{p}$-simple $\mathfrak{p}$-varieties defined over k and $[\mathfrak{B}]_1, \cdots, [\mathfrak{B}]_n$ be the abstract varieties obtained from $[V]_1, \cdots, [V]_n$ by the reduction with respect to $\mathfrak{p}$, respectively. Let $[X]_1, \cdots, [X]_n$ be respectively cycles on $[V]_1, \cdots, [V]_n$; and let $[\mathfrak{X}]_1, \cdots, [\mathfrak{X}]_n$ be respectively cycles on $[\mathfrak{B}]_1, \cdots, [\mathfrak{B}]_n$. We say that $([\mathfrak{X}]_1, \cdots, [\mathfrak{X}]_n)$ is a *specialization* of $([X]_1, \cdots, [X]_n)$ *over* $\mathfrak{o}$ if there exist an extension k' of k and a prolongation $\mathfrak{o}'$ of $\mathfrak{o}$ in k' such that $[X]_1, \cdots, [X]_n$ are all rational over k' and $[\mathfrak{X}]_1, \cdots, [\mathfrak{X}]_n$ are respectively the cycles obtained from $[X]_1, \cdots, [X]_n$ by the reduction with respect to $\mathfrak{o}'$. We shall denote by
$$([X]_1, \cdots, [X]_n) \xrightarrow{\mathfrak{o}} ([\mathfrak{X}]_1, \cdots, [\mathfrak{X}]_n).$$

PROPOSITION 26. *Let (x) be a set of n quantities in $\mathbf{K}$ and (ξ) a specialization of (x) over $\mathfrak{o}$. Then there exists a prolongation $\mathfrak{o}'$ of $\mathfrak{o}$ in $k(x)$ such that (ξ) is a specialization of (x) over $\mathfrak{o}'$.*

Proof. By Proposition 9 of Section 1, there exist an extension k_1 of k and a prolongation $\mathfrak{o}_1$ of $\mathfrak{o}$ in k_1 such that the residue field κ_1 of $\mathfrak{o}_1$ is $\kappa(\xi)$. Let (b) be a set of quantities in $\mathfrak{o}_1$ such that $(b) \xrightarrow{\mathfrak{o}_1} (\xi)$. We may assume that $k_1 = k(b)$. Let r be the dimension of (x) over k. Let the t_{ij} and the t_i for $1 \leq i \leq r$, $1 \leq j \leq n$ be $r(n+1)$ independent variables over k_1; and let the τ_{ij} and the τ_i for $1 \leq i \leq r$, $1 \leq j \leq n$ be $r(n+1)$ independent variables over κ_1. Put $k_2 = k_1(t_i, t_{ij})$ and $\mathfrak{o}_2 = [(t_i, t_{ij}) \xrightarrow{\mathfrak{o}_1} (\tau_i, \tau_{ij})]$; then $\mathfrak{o}_2$ is a prolongation of $\mathfrak{o}_1$ in k_2. Denote by L^{n-r} the linear variety

$$(\sum t_{ij}(X_j - b_j) - \pi t_i = 0 \quad (1 \leq i \leq r))$$

in S^n where π is a prime element in $\mathfrak{o}$ and by $\mathfrak{L}^{n-r}$ the linear variety

$$(\sum \tau_{ij}(X_j - \xi_j) = 0 \quad (1 \leq i \leq r))$$

in $\mathfrak{S}^n$; denote by A^r the prime rational cycle in S^n with the generic point (x) over k. Now we consider the reduction with respect to $\mathfrak{o}_2$. Then we have $L = \mathfrak{L}$. The point (ξ) is contained in $\bar{A} \cap \mathfrak{L}$; moreover, since (τ) is a set of independent variables over κ_1, it can be easily proved that (ξ) is a component of $\bar{A} \cap \mathfrak{L}$. By Theorem 9 of Section 2, there exists a point (y) in $A \cap L$ such that (ξ) is a specialization of (y) over $\mathfrak{o}_2$. We extend the specialization ring $[(y) \xrightarrow{\mathfrak{o}_2} (\xi)]$ to a valuation ring $\mathfrak{o}_3$ in $k_3 = k_2(y)$; this valuation ring $\mathfrak{o}_3$ is also discrete as (y) is algebraic over k. Since L is a generic linear variety over k, we have $(x) \xleftrightarrow{k} (y)$. We extend this generic specialization to a generic specialization $(x, a, s) \xleftrightarrow{k} (y, b, t)$. Put $k_4 = k(x, a, s)$ and denote by $\mathfrak{o}_4$ the prolongation of $\mathfrak{o}$ in k_4 which corresponds to the prolongation $\mathfrak{o}_3$ in k_3 by the specialization $(x, a, s) \xleftrightarrow{k} (y, b, t)$. Put $\mathfrak{o}' = k(x) \cap \mathfrak{o}_4$; then $\mathfrak{o}'$ is a discrete valuation ring in $k(x)$ and it defines a prolongation of $\mathfrak{o}$ in $k(x)$ such that $(x) \xrightarrow{\mathfrak{o}} (\xi)$; so our proposition is proved.

By this proposition, our definition of the specialization of cycles does not conflict with the definition of the specialization of points. But it should be remarked that singular points or a pseudo-point are considered as the cycle 0.

PROPOSITION 27. *Let* $[U]_1, \cdots, [U]_n$, $[V]_1, \cdots, [V]_m$ *be* $\mathfrak{p}$-*simple* $\mathfrak{p}$-*varieties defined over* k; *and let* $[X]_1, \cdots, [X]_n$, $[Y]_1, \cdots, [Y]_m$ *be respectively cycles on* $[U]_1, \cdots, [U]_n$, $[V]_1, \cdots, [V]_m$. *Every specialization of* $([X]_1, \cdots, [X]_n)$ *over* $\mathfrak{o}$ *can be extended to a specialization of* $([X]_1, \cdots, [X]_n, [Y]_1, \cdots, [Y]_m)$ *over* $\mathfrak{o}$.

Proof. Let $([\mathfrak{X}]_1, \cdots, [\mathfrak{X}]_n)$ be a specialization of $([X]_1, \cdots, [X]_n)$ over $\mathfrak{o}$. Then, by our definition, there exist an extension k' of k and a prolongation $\mathfrak{o}'$ of $\mathfrak{o}$ in k' such that all the $[X]_i$ are rational over k' and that $\rho'([X]_i) = [\mathfrak{X}]_i$ for $1 \leq i \leq n$ where ρ' denotes the reduction-mapping with respect to $\mathfrak{o}'$. Let k'' be an extension of k' such that $[Y]_1, \cdots, [Y]_m$ are all rational over k'' and $\mathfrak{o}''$ a prolongation of $\mathfrak{o}'$ in k''. Put $[\mathfrak{Y}]_i = \rho''([Y]_i)$ for $1 \leq i \leq m$ where ρ'' denotes the reduction-mapping with respect to $\mathfrak{o}''$. Since $\rho''([X]_i) = \rho'([X]_i)$, we have

$$([X]_1, \cdots, [X]_n, [Y]_1, \cdots, [Y]_m) \xrightarrow{\mathfrak{o}} ([\mathfrak{X}]_1, \cdots, [\mathfrak{X}]_n, [\mathfrak{Y}]_1, \cdots, [\mathfrak{Y}]_m).$$

This proves our proposition.

Now we can easily prove that the specialization of cycles defined above preserves the operations on cycles.

THEOREM 22. *Let* $[V]$ *be a* $\mathfrak{p}$-*simple* $\mathfrak{p}$-*variety defined over* k, $[X]$ *and* $[Y]$ *two* $[V]$-*cycles of the same dimension and* $([\mathfrak{X}], [\mathfrak{Y}])$ *a specialization of* $([X], [Y])$ *over* $\mathfrak{o}$. *Then* $[\mathfrak{X}] + [\mathfrak{Y}]$ *is a uniquely determined specialization of* $[X] + [Y]$ *over* $([X], [Y]) \overset{\mathfrak{o}}{\longrightarrow} ([\mathfrak{X}], [\mathfrak{Y}])$.

THEOREM 23. *Let* $[V]$ *be a* $\mathfrak{p}$-*simple* $\mathfrak{p}$-*variety defined over* k, $[X]^r$ *and* $[Y]^s$ *two positive* $[V]$-*cycles and* $([\mathfrak{X}], [\mathfrak{Y}])$ *a specialization of* $([X], [Y])$ *over* $\mathfrak{o}$. *If* $[X] \cdot [Y]$ *and* $[\mathfrak{X}] \cdot [\mathfrak{Y}]$ *are both defined, then* $[\mathfrak{X}] \cdot [\mathfrak{Y}]$ *is a uniquely determined specialization of* $[X] \cdot [Y]$ *over*

$$([X], [Y]) \overset{\mathfrak{o}}{\longrightarrow} ([\mathfrak{X}], [\mathfrak{Y}]).$$

THEOREM 24. *Let* $[U]$ *and* $[V]$ *be two* $\mathfrak{p}$-*simple* $\mathfrak{p}$-*varieties defined over* k, $[X] \times [Y]$ *a* $[U] \times [V]$-*cycle and* $([\mathfrak{X}], [\mathfrak{Y}])$ *a specialization of* $([X], [Y])$ *over* $\mathfrak{o}$. *Then,* $[\mathfrak{X}] \times [\mathfrak{Y}]$ *is a uniquely determined specialization of* $[X] \times [Y]$ *over* $([X], [Y]) \overset{\mathfrak{o}}{\longrightarrow} ([\mathfrak{X}], [\mathfrak{Y}])$.

THEOREM 25. *Let* $[U]$ *and* $[V]$ *be two* $\mathfrak{p}$-*simple* $\mathfrak{p}$-*varieties defined over* k *and denote by* $[\mathfrak{U}]$ *and* $[\mathfrak{W}]$ *the varieties obtained from* $[U]$ *and* $[V]$ *by the reduction with respect to* $\mathfrak{p}$, *respectively. Suppose that* $[V]$ *is* $\mathfrak{p}$-*complete and that* $[\mathfrak{W}]$ *has no multiple point. Let* $[X]$ *be a* $[U] \times [V]$-*cycle and* $[\mathfrak{X}]$ *a specialization of* $[X]$ *over* $\mathfrak{o}$. *Then* $\mathrm{pr}_{[\mathfrak{U}]}([\mathfrak{X}])$ *is a uniquely determined specialization of* $\mathrm{pr}_{[U]}([X])$ *over* $[X] \overset{\mathfrak{o}}{\longrightarrow} [\mathfrak{X}]$.

We shall only prove Theorem 23 since the others are proved in the same way.

Proof of Theorem 23. Let $([\mathfrak{X}], [\mathfrak{Y}], [\mathfrak{Z}])$ be a specialization of $([X], [Y], [X] \cdot [Y])$ over $\mathfrak{o}$. By our definition, there exist an extension k' of k and a prolongation $\mathfrak{o}'$ of $\mathfrak{o}$ in k' such that $[X], [Y]$ are rational over k' and $\rho'([X]) = [\mathfrak{X}]$, $\rho'(Y) = [\mathfrak{Y}]$, $\rho'([X] \cdot [Y]) = [\mathfrak{Z}]$, where ρ' denotes the reduction-mapping defined with respect to $\mathfrak{o}'$. Since $[\mathfrak{X}] \cdot [\mathfrak{Y}]$ is defined by our assumption, we have by Theorem 18 of Section 4, $\rho'([X] \cdot [Y]) = [\mathfrak{X}] \cdot [\mathfrak{Y}]$. This proves our theorem.

In case where $k = \mathfrak{o}$, we have a specialization-theory of cycles over a field. In this case, we have $[\mathfrak{U}] = [U]$ (or $[\mathfrak{U}] = [U]^\sigma$ where σ is an isomorphism of k) for an ambient variety $[U]$.

We have to prove that the specialization of cycles thus defined is transitive since this is not so obvious.

PROPOSITION 28. *Let* $[V]$ *be a* $\mathfrak{p}$-*simple* $\mathfrak{p}$-*variety defined over* k, $[X]'$ *a* V-*cycle and* $[\mathfrak{X}]$ *a specialization of* $[X]'$ *over* $\mathfrak{o}$.

i) *If* $[X]'$ *is a specialization of a* $\cdot [V]$-*cycle* $[X]$ *over* k, *then* $[\mathfrak{X}]$ *is a specialization of* $[X]$ *over* $\mathfrak{o}$.

ii) *Every specialization of* $[\mathfrak{X}]$ *over* κ *is also a specialization of* $[X]'$ *over* $\mathfrak{o}$.

Proof. Suppose that $[X] \xrightarrow{h} [X]' \xrightarrow{o} [\mathfrak{X}]$. Then, by our definition, there exist an extension k_1 of k and a discrete valuation ring o_1 in k_1 which contains k such that $[X]$ is rational over k_1 and $\rho_1([X]) = [X]'$ where ρ_1 is the reduction-mapping with respect to o_1; and there exist an extension k_2 of k and a prolongation o_2 of o in k_2 such that $[X]'$ is rational over k_2 and $\rho_2([X]') = [\mathfrak{X}]$ where ρ_2 is the reduction-mapping with respect to o_2. By Proposition 9 of Section 1, there exist an extension k_3 of k_1 and a prolongation o_3 of o_1 in k_3 such that the residue field k_4 of o_3 contains k_2. Let o_4 be a prolongation of o_2 in k_4. Then we have $\rho_3([X]) = [X]'$, $\rho_4([X]') = [\mathfrak{X}]$ where ρ_3 and ρ_4 are the reduction-mappings with respect to o_3 and o_4, respectively. Let $\mathfrak{R}$ be the inverse image of o_4 by the natural homomorphism of o_3 onto k_4. Then $\mathfrak{R}$ is a (non-discrete) valuation ring with the quotient field k_3; we may consider the residue field of o_4 as the residue field of $\mathfrak{R}$. Denote by the $[A]_\lambda$ all the components of $[X]$. We may assume that the $[A]_\lambda$ are all defined over k_3. We fix our attention to one component $[A]_\lambda$ and one of its representative $A_{\lambda\alpha}$. Let S^n be the ambient space for $A_{\lambda\alpha}$, (x) a generic point of $A_{\lambda\alpha}$ over k_3 and the t_{ij} for $1 \leq i \leq r$, $1 \leq j \leq n$ rn independent variables over k where r is the dimension of $A_{\lambda\alpha}$. Put $y_i = \sum t_{ij} x_j$ for $1 \leq i \leq r$. Then we have an irreducible equation $F_{\lambda\alpha}(Y, T) = 0$ for (y_i, t_{ij}) over k_3. For each $[A]_\lambda$ and for each representative $A_{\lambda\alpha}$ of $[A]_\lambda$, we obtain such an equation. We may assume that the polynomials $F_{\lambda\alpha}$ are all primitive polynomials in $\mathfrak{R}[Y, T]$. Denote by c_ν all the coefficients of the $F_{\lambda\alpha}$. Let (γ) be the uniquely determined specialization of (c) over $\mathfrak{R}$. Then obviously we have $(c) \xrightarrow{o} (\gamma)$. By Proposition 26, there exists a prolongation o' of o in $k(c)$ such that $(c) \xrightarrow{o'} (\gamma)$. By Proposition 16 of Section 2 and by our definition of the reduction, we can easily prove that $[X]$ is rational over $k(c)$ and $\rho'([X]) = [\mathfrak{X}]$ where ρ' is the reduction-mapping with respect to o'. This proves the assertion i). The second assertion can be proved similarly.

6. Reduction with respect to infinitely many $\mathfrak{p}$. Let K be a field with a set of infinitely many discrete valuations $\{\omega_\lambda\}$. We denote by o_λ, $\mathfrak{p}_\lambda$ and κ_λ the valuation ring, the valuation ideal and the residue field of ω_λ, for every λ; denote by σ that set of valuations. In this section, we shall use these notations always in this sense. We assume that the set σ satisfies the following condition:

(I) *Every element of K other than zero is a $\mathfrak{p}_\lambda$-unit for almost all* [11] *the $\mathfrak{p}_\lambda$.*

---

[11] We shall use the term " almost all " in the sense of " all but a finite number of."

In the following we shall use the notations $\sigma_1, \sigma_2, \cdots$ for subsets of σ which contain almost all the $\mathfrak{p}_\lambda$.

Let V be a variety defined over K. We denote by $\bar{V}^{(\lambda)}$ the bunch obtained from V by the reduction with respect to $\mathfrak{p}_\lambda$, for every λ. Similarly, we use the notations $\rho_\lambda(X)$, $[\bar{V}]^{(\lambda)}$ and $\rho_\lambda([X])$ in the sense of the reduction with respect to $\mathfrak{p}_\lambda$.

PROPOSITION 29. *Let U be a variety in $S^n \times S^m$ defined over K with the projection V on S^n. If $[U : V] = 1$, then the projection from U to V is regular along every component of $\bar{V}^{(\lambda)}$ for almost all the $\mathfrak{p}_\lambda$.*

Proof. Let (x, y) be a generic point of U over K; then we have expressions $y_j = F_j(x)/F(x)$ where $F(X)$ and $F_j(X)$ are polynomials in $K[X]$. By condition (I), $F(X)$ and the $F_j(X)$ are all contained in $\mathfrak{o}_\lambda[X]$ for almost all the $\mathfrak{p}_\lambda$. Let the t_{ij} $(1 \leqq i \leqq r, 1 \leqq j \leqq n)$ be rn independent variables over $K(x)$ and put $t_i = \sum t_{ij} x_j$ for $1 \leqq i \leqq r$. Let the $\tau_i^{(\lambda)}$ and the $\tau_{ij}^{(\lambda)}$ $(1 \leqq i \leqq r, 1 \leqq j \leqq n)$ be $r(n+1)$ independent variables over κ_λ, for every λ. Put $K' = K(t_i, t_{ij})$, $\mathfrak{o}_\lambda' = [(t_i, t_{ij}) \xrightarrow{\mathfrak{o}_\lambda} (\tau_i^{(\lambda)}, \tau_{ij}^{(\lambda)})]$; then, by Proposition 3 of Section 1, for every λ, $\mathfrak{o}_\lambda'$ is a discrete valuation ring with the quotient field K'. It is easy to see that the set σ' of all the valuations with the valuation rings $\mathfrak{o}_\lambda'$ also satisfies the condition (I). Since $K'(x)$ is finite algebraic over K', the set σ'' of all the valuations which are prolongations on $K'(x)$ of valuations in σ' also satisfies the condition (I); in particular, $F(x)$ is a unit for almost all valuations in σ''. Let $\mathfrak{B}_\lambda$ be a component of $\bar{V}^{(\lambda)}$ and $\mathfrak{L}_\lambda$ the linear variety $(\sum \tau_{ij}^{(\lambda)} X_j - \tau_i^{(\lambda)} = 0$ $(1 \leqq i \leqq r))$. Then we have a point $(\xi^{(\lambda)})$ in $\mathfrak{B}_\lambda \cap \mathfrak{L}_\lambda$. The valuation obtained from the specialization $(x) \xrightarrow{\mathfrak{o}_\lambda'} (\xi^{(\lambda)})$ is clearly in σ'', so that $\bar{F}^{(\lambda)}(\xi^{(\lambda)}) = 0$ only for a finite number of the $\mathfrak{p}_\lambda$, where $\bar{F}^{(\lambda)}$ is the class of F modulo $\mathfrak{p}_\lambda$. This proves our proposition.

LEMMA 3. *Let $F(X)$ be an absolutely irreducible polynomial in $K[X]$. Suppose that $F(X)$ is in $\mathfrak{o}_\lambda[X]$ for every λ, and denote by $\bar{F}^{(\lambda)}(X)$ the class of $F(X)$ modulo $\mathfrak{p}_\lambda$. Then $\bar{F}^{(\lambda)}(X)$ is absolutely irreducible for almost all the $\mathfrak{p}_\lambda$.*

Proof. We may assume that the degree of $F(X)$ is greater than 1 and that $\bar{F}^{(\lambda)}(X)$ has the same degree as $F(X)$ for every λ. Let the $M_\alpha^{(r)}$ be all the monomials in X whose degrees are less than or equal to r. Let d be the degree of F; and let r and s be a pair of positive integers such that $d = r + s$. Let $G_1(X) = \sum x_\alpha M_\alpha^{(r)}$ and $G_2(X) = \sum y_\beta M_\beta^{(s)}$ be two polynomials such that (x_α, y_β) is a set of independent variables over K, and put

$$H(X) = \sum z_\gamma M_\gamma^{(d)} = G_1(X) G_2(X).$$

We consider (x) and (y) as homogeneous coordinates of generic points of the projective spaces $[L]^{(r)}$ and $[L]^{(s)}$ over K, respectively. Also we consider (z) as homogeneous coordinates of a generic point of some projective variety $[V]$ over K. It is easy to see that (z) has actually a locus over K. By considering the "generic polynomials" over κ_λ, in the same way as above, we have points $(\xi^{(\lambda)})$, $(\eta^{(\lambda)})$ and $(\zeta^{(\lambda)})$ with loci $[\mathfrak{L}]_\lambda^{(r)}$, $[\mathfrak{L}]_\lambda^{(s)}$ and $[\mathfrak{V}]_\lambda$ over κ_λ, for every λ. It is easy to see that $(\xi^{(\lambda)}) \times (\eta^{(\lambda)}) \times (\zeta^{(\lambda)})$ is a specialization of $(x) \times (y) \times (z)$ over $\mathfrak{o}_\lambda$; so that $[\mathfrak{V}]_\lambda$ is contained in $[\bar{V}]_\lambda$. Now a polynomial $\sum c_\gamma M_\gamma^{(d)}$ in $K[X]$ of degree d has a factor of degree r in an extension of K if and only if (c) is a point of $[V]$; and a similar fact holds for a polynomial in $\kappa_\lambda[X]$ and the variety $[\mathfrak{V}]_\lambda$. In fact, the converse part is obvious; so we have only to prove the direct part. If (c) is a point in $[V]$, then (c) is a specialization of (z) over K. This specialization can be extended to a specialization $(a) \times (b) \times (c)$ of $(x) \times (y) \times (z)$ where $(a) \times (b)$ is a point in $[\mathfrak{L}]^{(r)} \times [\mathfrak{L}]^{(s)}$. By the equation $(\sum x_\alpha M_\alpha^{(r)})(\sum y_\beta M_\beta^{(s)}) = \sum z_\gamma M_\gamma^{(d)}$, we can easily verify (if necessary, we consider affine representatives) $(\sum a_\alpha M_\alpha^{(r)})(\sum b_\beta M_\beta^{(s)}) = h \sum c_\gamma M_\gamma^{(d)}$ where h is a quantity. Thus we have proved the above fact. Let $\mathfrak{a}$ be the ideal in $K[Z]$ determined by the set (z). Put $F(X) = \sum c_\gamma M_\gamma^{(d)}$; then by our assumption that F is absolutely irreducible, (c) is not a point in $[V]$; so that there exists a homogeneous polynomial $P(Z)$ in $\mathfrak{a}$ such that $P(c) \neq 0$. By our condition (I), the coefficients of $P(Z)$ and $P(c)$ are all $\mathfrak{p}_\lambda$-units for every $\mathfrak{p}_\lambda$ in some $\sigma_1 \subset \sigma$. Put $\bar{F}^{(\lambda)}(X) = \sum \bar{c}_\gamma^{(\lambda)} M_\gamma^{(d)}$ for every $\mathfrak{p}_\lambda$ in σ_1. If $\bar{F}^{(\lambda)}$ has a factor of degree r, then $(\bar{c}^{(\lambda)})$ is a specialization of $(\zeta^{(\lambda)})$ over κ_λ; so that it is a specialization of (z) over $\mathfrak{o}_\lambda$. Then from $P(z) = 0$ follows $\bar{P}^{(\lambda)}(\bar{c}^{(\lambda)}) = 0$; this is a contradiction if $\mathfrak{p}_\lambda \varepsilon \sigma_1$. Now if $\bar{F}^{(\lambda)}(X)$ is reducible, the degree r of one of the factors must satisfy $1 \leqq r \leqq [d/2]$; so our lemma is proved.

PROPOSITION 30. *Let V^r be a variety in S^{r+1} defined over K. Then V is $\mathfrak{p}_\lambda$-simple for almost all the $\mathfrak{p}_\lambda$.*

Proof. By Proposition 2 of [WF] Chapter IV, Section 1, V is defined by an absolutely irreducible equation $F(X) = 0$ with coefficients in K. By our condition (I), the coefficients of $F(X)$ are all $\mathfrak{p}_\lambda$-units for every $\mathfrak{p}_\lambda$ in some $\sigma_1 \subset \sigma$. Denote by $\bar{F}^{(\lambda)}(X)$ the class of $F(X)$ modulo $\mathfrak{p}_\lambda$ for every $\mathfrak{p}_\lambda$ in σ_1. Then by Lemma 3, $\bar{F}^{(\lambda)}$ is absolutely irreducible for every $\mathfrak{p}_\lambda$ in some $\sigma_2 \subset \sigma_1$. By Theorem 21 of Section 4, V is $\mathfrak{p}_\lambda$-simple for every $\mathfrak{p}_\lambda$ in σ_2; this proves our proposition.

THEOREM 26. *Let $[V]$ be an abstract variety defined over K; let $[V]_\lambda$ be a $\mathfrak{p}_\lambda$-variety with the underlying abstract variety $[V]$, for every λ. Suppose that $[\bar{V}]_\lambda$ is not empty for almost all the $\mathfrak{p}_\lambda$. Then $[V]_\lambda$ is $\mathfrak{p}_\lambda$-simple for almost all the $\mathfrak{p}_\lambda$.*[12]

Therefore, in this case, for almost all the $\mathfrak{p}_\lambda$, we can obtain the abstract variety $[\mathfrak{W}]_\lambda$ from $[V]_\lambda$ by the reduction with respect to $\mathfrak{p}_\lambda$; and $[\mathfrak{W}]_\lambda$ is defined over κ_λ.

Proof. We shall first prove our theorem in case where $[V]$ is a variety V^r in the affine space S^n. Let (x) be a generic point of V over K. Let the t_{ij} $(0 \le i \le r, 1 \le j \le n)$ be $(r+1)n$ independent variables over $K(x)$; and put $K' = K(t)$ and $y_i = \sum t_{ij}x_j$ for $0 \le i \le r$. Then by Proposition 16 of Section 2, we have $K'(x) = K'(y)$. Let W be the locus of (y) over K' in S^{r+1} and T the birational correspondence between V and W with the corresponding generic points (x) and (y) over K'. Let the $\tau_{ij}^{(\lambda)}$ $(0 \le i \le r, 1 \le j \le n)$ be $(r+1)n$ independent variables over κ_λ and put $\mathfrak{o}_\lambda' = [(t) \xrightarrow{\mathfrak{o}_\lambda} (\tau^{(\lambda)})]$. Then by Proposition 3 of Section 1, for every λ, $\mathfrak{o}_\lambda'$ is a discrete valuation ring with the quotient field K'. It is easy to see that the set σ' of all the valuations whose valuation rings are $\mathfrak{o}_\lambda'$ also satisfies (I). By Proposition 30, W is $\mathfrak{p}_\lambda'$-simple for every $\mathfrak{p}_\lambda'$ in some $\sigma_1' \subset \sigma'$. Denote by σ_1 the subset of σ corresponding to the subset σ_1' of σ'. Let $(\xi^{(\lambda)})$ be a generic point of a component $\mathfrak{W}_{\lambda 1}$ of $\bar{V}^{(\lambda)}$ over $\bar{\kappa}_\lambda(\tau^{(\lambda)})$, for every $\mathfrak{p}_\lambda$ in σ_1. Put $\eta_i^{(\lambda)} = \sum \tau_{ij}^{(\lambda)}\xi_j^{(\lambda)}$ for $0 \le i \le r$; then $(\xi^{(\lambda)}, \eta^{(\lambda)})$ is a specialization of (x, y) over $\mathfrak{o}_\lambda'$; and by Proposition 16 of Section 2, (x) is finite over $(y) \xrightarrow{\mathfrak{o}_\lambda'} (\eta^{(\lambda)})$. Let $\mathfrak{W}_\lambda$ be the variety obtained from W by the reduction with respect to $\mathfrak{p}_\lambda'$, for every $\mathfrak{p}_\lambda'$ in σ_1'. Since $\eta_i^{(\lambda)} = \sum \tau_{ij}^{(\lambda)}\xi_j^{(\lambda)}$, the dimension of $(\eta^{(\lambda)})$ over $\kappa_\lambda(\tau^{(\lambda)})$ is r; so that $(\eta^{(\lambda)})$ is a generic point of $\mathfrak{W}_\lambda$ over $\kappa_\lambda(\tau^{(\lambda)})$. By Theorem 15 of Section 3, $[(y) \xrightarrow{\mathfrak{o}_\lambda'} (\eta^{(\lambda)})]$ is integrally closed, so contains all the x_j since (x) is finite over $(y) \xrightarrow{\mathfrak{o}_\lambda'} (\eta^{(\lambda)})$. This shows that $(\xi^{(\lambda)})$ and $(\eta^{(\lambda)})$ are biregularly corresponding points by T. Then by Theorem 14 of Section 3, we have $\mu(V, \mathfrak{W}_{\lambda 1}) = \mu(W, \mathfrak{W}_\lambda) = 1$. Moreover if $\mathfrak{W}_{\lambda 2}$ is another component of $\bar{V}^{(\lambda)}$ then $\mathfrak{W}_{\lambda 2}$ also biregularly corresponds to $\mathfrak{W}_\lambda$, so it must coincide with $\mathfrak{W}_{\lambda 1}$. Thus our theorem is proved for an affine variety. Now we shall prove our theorem for an abstract variety $[V] = [V_\alpha; F_\alpha; T_{\beta\alpha}]$ on which a $\mathfrak{p}_\lambda$-variety $[V]_\lambda = [V_\alpha; F_\alpha; \mathfrak{F}_\alpha^{(\lambda)}; T_{\beta\alpha}]$ is

---

[12] This theorem may be considered as a generalization of a theorem of Bertini for pencils. (See O. Zariski, " Pencils on an algebraic variety and a new proof of a theorem of Bertini," *Transactions of the American Mathematical Society*, vol. 50 (1941), pp. 48-70.)

defined for every λ. As our theorem is proved for affine varieties, the varieties V_α and $T_{\beta\alpha}$ are all $\mathfrak{p}_\lambda$-simple for every $\mathfrak{p}_\lambda$ in some $\sigma_1 \subset \sigma$. Denote by $\mathfrak{V}_\alpha{}^{(\lambda)}$ the variety obtained from V_α by the reduction with respect to $\mathfrak{p}_\lambda$, for every α and every $\mathfrak{p}_\lambda$ in σ_1. By Proposition 29, for every α and β, $T_{\beta\alpha}$ is regular along $\mathfrak{V}_\alpha{}^{(\lambda)}$ and $\mathfrak{V}_\beta{}^{(\lambda)}$ for every $\mathfrak{p}_\lambda$ in some $\sigma_2 \subset \sigma_1$. Then by Theorem 13 of Section 3, there exists a component $\mathfrak{T}_{\beta\alpha}{}^{(\lambda)}$ of $\bar{T}_{\beta\alpha}{}^{(\lambda)}$ having the projection $\mathfrak{V}_\alpha{}^{(\lambda)}$. Since $T_{\beta\alpha}$ is $\mathfrak{p}_\lambda$-simple if $\mathfrak{p}_\lambda \, \varepsilon \, \sigma_1$, $\mathfrak{T}_{\beta\alpha}{}^{(\lambda)}$ is the only component of $\bar{T}_{\beta\alpha}{}^{(\lambda)}$ if $\mathfrak{p}_\lambda \, \varepsilon \, \sigma_2$; so that, by the same reason, the projection of $\mathfrak{T}_{\beta\alpha}{}^{(\lambda)}$ on $\bar{V}_\beta{}^{(\lambda)}$ is equal to $\mathfrak{V}_\beta{}^{(\lambda)}$. This shows that $\mathfrak{V}_\alpha{}^{(\lambda)}$ and $\mathfrak{V}_\beta{}^{(\lambda)}$ are biregularly corresponding varieties by $T_{\beta\alpha}$ for every $\mathfrak{p}_\lambda$ in σ_2. From this follows that, for every $\mathfrak{p}_\lambda$ in σ_2, $[\bar{V}]_\lambda$ has only one component $[\mathfrak{V}]_\lambda$. Furthermore, by our definition we have $\mu([V]_\lambda, [\mathfrak{V}]_\lambda) = \mu(V_\alpha, \mathfrak{V}_\alpha{}^{(\lambda)}) = 1$, if $[\mathfrak{V}]_\lambda$ has a representative in V_α. This proves our theorem.

UNIVERSITY OF TOKYO.

REFERENCES.

[1] M. Deuring, "Reduktion algebraischer Funktionenkörper nach Primdivisoren des Konstantenkörpers," *Mathematische Zeitschrift*, vol. 47 (1942), pp. 643-654.

[2] ———, "Die Struktur der elliptischen Funktionenkörper und die Klassenkörper der imaginären quadratischen Zahlkörper," *Mathematische Annalen*, vol. 124 (1952), pp. 393-426.

[3] W. Krull, "Dimensionstheorie in Stellenringen," *Journal für die Reine und Angewandte Mathematik*, vol. 179 (1938), pp. 204-226.

[4] T. Matsusaka, "Specialization of cycles on a projective model," *Memoirs of the College of Science, University of Kyoto*, vol. 26 (1950), pp. 167-173.

[5] D. G. Northcott, "Specialization over a local domain," *Proceedings of the London Mathematical Society*, (III), vol. 1 (1951), pp. 129-137.

[6] P. Samuel, "La notion de multiplicité en algèbre et en géométrie algébrique," *Journal de Mathématique Pures et Appliquées*, (IX), vol. 30 (1951), pp. 159-274.

[7] A. Weil, *Foundations of algebraic geometry*, New York, 1946.

[8] ———, "Arithmetic on algebraic varieties," *Annals of Mathematics*, vol. 53 (1951), pp. 412-444.

On complex multiplications

Proceedings of the International Symposium on Algebraic Number Theory,
Tokyo-Nikko, 1955, Science Council of Japan, Tokyo (1956), 23-30

It is well known that the theory of class-fields over imaginary
quadratic fields can be described in terms of the complex multiplication
of elliptic functions. In the classical treatment of this theory as well
as in the purely algebraic treatment by M. Deuring ([1]), the congruence relation on elliptic functions, which was first given by L.
Kronecker, has played a central role. If one obtains some similar
relation on abelian functions of higher dimension, then one can study
the arithmetic of such abelian functions particularly in connection
with class-field theory. In the present paper we shall give some
results in this direction. Using the notion of reduction modulo $\mathfrak{p}$ of
algebraic varieties, we obtain a certain congruence relation for an
abelian variety A of dimension n whose endomorphism-ring contains
a subring isomorphic to the ring of integers of an algebraic number-
field K of degree $2n$. By means of this congruence relation we can
prove that a field of definition for such an abelian variety A always
contains a certain class-field over K_0 and that the fields generated
by division points on A contain class-fields over K_0, corresponding to
the ideal-groups determined by some relations which we can write
down, where K_0 denotes a certain algebraic number-field determined
by K and some isomorphisms of K.

1. Let A be an abelian variety defined over a field k. We
shall denote by $\mathcal{A}(A)$ the ring of all endomorphisms of A. Let K
be an algebraic number-field of degree $2n$ and R the ring of all
integers in K. By an *abelian variety having R as operator-domain*,
we shall understand a pair (A, ι) of an abelian variety A of dimension
n and an isomorphism ι of R into $\mathcal{A}(A)$. We shall denote simply by
A such a pair (A, ι) if there is no fear of misunderstanding. Let
(A', ι') be another abelian variety having R as operator-domain. We
shall understand by an *R-homomorphism of A into A'* a homomorphism
λ of A into A' such that $\lambda\iota(\mu)x = \iota'(\mu)\lambda x$ for every $\mu \in R$ and every $x \in A$.
We shall say that (A, ι) is defined over a field k if k is a field of
definition for A and for every element of $\iota(R)$.

Now let $\mathfrak{a}$ be an ideal of R other than the zero-ideal. We shall denote by $\mathfrak{g}(\mathfrak{a}, A)$ the set of all points t of A such that $\iota(\mu)t = 0$ for every $\mu \in \mathfrak{a}$. We shall call an element t in $\mathfrak{g}(\mathfrak{a}, A)$ a *primitive element in* $\mathfrak{g}(\mathfrak{a}, A)$ if $\iota(\mu)t = 0$ implies $\mu \in \mathfrak{a}$. Such an element exists if $\mathfrak{a}$ is prime to the characteristic of the ground field k. Let (A', ι') be an abelian variety having R as operator-domain and λ an R-homomorphism of (A, ι) into (A', ι'), both defined over k. We shall call λ an $\mathfrak{a}$-*multiplication of A onto A'* if the following condition is satisfied: if x is a generic point of A over k, then the field $k(\lambda x)$ is the composite of all the fields $k(\iota(\mu)x)$ such that $\mu \in \mathfrak{a}$. We shall call (A', ι') an $\mathfrak{a}$-*transform of A* if there exists an $\mathfrak{a}$-multiplication of A onto A'. For every ideal $\mathfrak{a}$ of R, there exist an $\mathfrak{a}$-transform A' of A and an $\mathfrak{a}$-multiplication λ of A onto A'; the kernel of λ is equal to $\mathfrak{g}(\mathfrak{a}, A)$. One may consider an $\mathfrak{a}$-multiplication an "ideal number" which makes $\mathfrak{a}$ principal. Let $\mathfrak{a}$ and $\mathfrak{b}$ be two ideals of R. Then, an $\mathfrak{a}$-transform of A and a $\mathfrak{b}$-transform of A are R-isomorphic to each other if and only if $\mathfrak{a}$ and $\mathfrak{b}$ belong to the same ideal-class. Let c be an ideal-class of K. We shall call A' a *c-transform of A* if A' is an $\mathfrak{a}$-transform of A for an ideal $\mathfrak{a}$ in c. We can easily prove the following proposition.

PROPOSITION 1. *Let c be an ideal-class of K and A' a c-transform of A. Let $\mathfrak{m}$ be an ideal of R and t a primitive element in* $\mathfrak{g}(\mathfrak{m}, A)$. *If $t' \in \mathfrak{g}(\mathfrak{m}, A')$, then there exist an ideal $\mathfrak{a}$ in c and an $\mathfrak{a}$-multiplication $\lambda_{\mathfrak{a}}$ of A onto A' such that $t' = \lambda_{\mathfrak{a}}t$; t' is primitive in* $\mathfrak{g}(\mathfrak{m}, A')$ *if and only if $\mathfrak{a}$ is prime to $\mathfrak{m}$. If t' is primitive in* $\mathfrak{g}(\mathfrak{m}, A')$, *the ideal-class of $\mathfrak{a}$ modulo $\mathfrak{m}$ is uniquely determined by t'.*

Let σ be an isomorphism of the ground field k onto a field k^σ. Then we have an abelian variety A^σ defined over k^σ, the transform of A by σ. Let μ be an element of R and W the graph of $\iota(\mu)$. The transform W^σ of W by σ is a subvariety of $A^\sigma \times A^\sigma$; there exists an endomorphism of A^σ having W^σ as its graph; denote by $\iota^\sigma(\mu)$ that endomorphism. Then the mapping $\mu \to \iota^\sigma(\mu)$ gives an isomorphism of R into $\mathcal{A}(A^\sigma)$; thus we obtain an abelian variety having R as operator-domain (A^σ, ι^σ) defined over k^σ. When k has a prime characteristic p, we obtain an isomorphism σ of k, for any power $q = p^f$ of p, defined by $z^\sigma = z^q$ for every z in k. In this situation, we shall denote (A^σ, ι^σ) simply by A^q.

2. Let k be a field with a discrete valuation and $\mathfrak{o}$ be its valuation ring; denote by $\mathfrak{p}$ the maximal ideal of $\mathfrak{o}$ and by κ the residue field $\mathfrak{o}/\mathfrak{p}$. Let V be a variety in the projective N-space P^N, defined over k and $\mathfrak{A}$ the set of all polynomials $F(X)$ in $\mathfrak{o}[X_0, \cdots, X_N]$ such that $F(x) = 0$ for every (x) on V. Let $\bar{P}^N$ be the projective

N-space defined over κ and denote by $\overline{V}$ the algebraic set in $\overline{P}^N$ defined by the equations $\overline{F}(X)=0$ for $F\in\mathfrak{A}$ where $\overline{F}$ denotes the class of F modulo $\mathfrak{p}$. If $\overline{V}$ has the only one component and that component has the multiplicity one ([3] p. 148), we say that V is $\mathfrak{p}$-*simple* and call $\overline{V}$ *the variety obtained from V by reduction modulo* $\mathfrak{p}$. In [3], we have given a theory of reduction modulo $\mathfrak{p}$ of algebraic varieties thus defined. As shown there, we can define reduction of an abstract variety. We shall use the same terminologies and notations as in [3].

Let V and W be two $\mathfrak{p}$-simple varieties defined over k: denote by $\overline{V}$ and by $\overline{W}$ the varieties obtained from V and W by reduction modulo $\mathfrak{p}$, respectively. Then $\overline{V}$ and $\overline{W}$ are abstract varieties defined over κ. Let f be a rational mapping of V into W, defined over k; denote by T the graph of f. Let ξ be a point on $\overline{V}$. We shall say that f *is defined at* ξ if there exists a point η on $\overline{W}$ such that, for some representatives $V_\alpha, W_\lambda, T_{\alpha\lambda}, \xi_\alpha$ and η_λ of V, W, T, ξ and η, we have $\xi_\alpha\times\eta_\lambda\in\overline{T}_{\alpha\lambda}$ and the projection from $T_{\alpha\lambda}$ to V_α is regular at ξ_α (namely, if $x_\alpha\times y_\lambda$ is generic on $T_{\alpha\lambda}$ over k, the coordinates of y_λ are all contained in the specialization-ring $[x_\alpha\to\xi_\alpha]$).

Now let A be an abelian variety defined over k. Denote by f the rational mapping of $A\times A$ into A such that $f(x,y)=x+y$ for $x\in A$, $y\in A$ and by g the rational mapping of A into A such that $g(x)=-x$ for $x\in A$. We shall say that A *has no defect for* $\mathfrak{p}$ if the following conditions (1–4) are satisfied. (1) A *is* $\mathfrak{p}$-*simple*. Denote by $\overline{A}$ the variety obtained from A by reduction modulo $\mathfrak{p}$. (2) A *is* $\mathfrak{p}$-*complete*. (If A is a subvariety of a projective space, this is always satisfied.) (3) f *is everywhere defined on* $\overline{A}\times\overline{A}$. (4) g *is everywhere defined on* $\overline{A}$. If A has no defect for $\mathfrak{p}$, the variety $\overline{A}$ obtained from A by reduction modulo $\mathfrak{p}$ becomes an abelian variety defined over κ, in a natural manner. We shall call $\overline{A}$ *the abelian variety obtained from A by reduction modulo* $\mathfrak{p}$.

Let A and B be two abelian varieties defined over k, having no defect for $\mathfrak{p}$; denote by $\overline{A}$ and by $\overline{B}$ the abelian varieties obtained from A and from B by reduction modulo $\mathfrak{p}$, respectively. Denote by $\mathcal{H}(A,B;k)$ the set of all homomorphisms of A into B, defined over k, and by $\mathcal{H}(\overline{A},\overline{B};\kappa)$ the set of all homomorphisms of $\overline{A}$ into $\overline{B}$, defined over κ. Then, for every $\lambda\in\mathcal{H}(A,B;k)$, there exists a uniquely determined element $\overline{\lambda}\in\mathcal{H}(\overline{A},\overline{B};\kappa)$ such that the graph of λ is the variety obtained from the graph of λ by reduction modulo $\mathfrak{p}$. The correspondence $\lambda \to \bar{\lambda}$ defines an isomorphism of the additive group $\mathscr{H}(A, B; k)$ into the additive group $\mathscr{H}(\bar{A}, \bar{B}; \kappa)$. If $A = B$, this isomorphism is a ring-isomorphism.

Now as in 1, let R be the ring of integers in an algebraic number-field K and (A, ι) an abelian variety having R as operator-domain, defined over k. Suppose that A has no defect for $\mathfrak{p}$; denote by $\bar{A}$ the abelian variety obtained from A by reduction modulo $\mathfrak{p}$. Let μ be an element of R; put $\iota(\mu) = \mu_A$ and denote by $\bar{\mu}_A$ the corresponding endomorphism of $\bar{A}$. Then the mapping $\bar{\iota}$ of R into $\mathscr{A}(\bar{A})$ defined by $\bar{\iota}(\mu) = \bar{\mu}_A$ is an isomorphism. Hence we obtain an abelian variety having R as operator-domain $(\bar{A}, \bar{\iota})$; we denote it also by $\bar{A}$ and call *the abelian variety having R as operator-domain obtained from (A, ι) by reduction modulo $\mathfrak{p}$*. Let c be an ideal-class of K and A' a c-transform of A, defined over k; let $\mathfrak{a}$ be an ideal in c and $\lambda_{\mathfrak{a}}$ an $\mathfrak{a}$-multiplication of A onto A' defined over k. Suppose that A' has no defect for $\mathfrak{p}$. Then $\bar{A}'$ is a c-transform of $\bar{A}$ and $\bar{\lambda}_{\mathfrak{a}}$ is an $\mathfrak{a}$-multiplication of $\bar{A}$ onto $\bar{A}'$. Let $\mathfrak{m}$ be an ideal of R which is prime to the characteristic of the residue field κ. Let t be a primitive element in $\mathfrak{g}(\mathfrak{m}, A)$ and suppose that t is rational over k. Then the point $\bar{t}$ obtained from t by reduction modulo $\mathfrak{p}$ is a primitive element in $\mathfrak{g}(\mathfrak{m}, \bar{A})$.

If A is an abelian variety defined over an algebraic number-field, then A has no defect for all but a finite number of prime divisors of that field.

Let V be a $\mathfrak{p}$-simple variety defined over k and denote by $\bar{V}$ the variety obtained from V by reduction modulo $\mathfrak{p}$. Let f be a function on V defined over k and denote by $\bar{f}$ the generalized function on $\bar{V}$ obtained from f by reduction modulo $\mathfrak{p}$ ([3] pp. 167–168). We shall say that f is $\mathfrak{p}$-*finite* if $\bar{f} \neq \infty$. Let ω be a differential form on V defined over k. We shall say tat ω is $\mathfrak{p}$-*finite* if ω is written in a form $\omega = \sum_{(i)} f_{(i)} dg_{i_1} \cdots dg_{i_r}$ where the $f_{(i)}$ and the g_i are $\mathfrak{p}$-finite functions on V defined over k. We can prove that the differential form $\bar{\omega} = \sum_{(i)} \bar{f}_{(i)} d\bar{g}_{i_1} \cdots d\bar{g}_{i_r}$ on $\bar{V}$ does not depend upon the choice of the $f_{(i)}$ and the g_i. We shall call $\bar{\omega}$ *the differential form obtained from ω by reduction modulo $\mathfrak{p}$*. If ω is of the first kind, and if $\bar{V}$ is a complete non-singular variety then $\bar{\omega}$ is of the first kind.

3. Let E be a complete non-singular curve of genus one, defined over a field k, having a rational point over k. Then E becomes an abelian variety of dimension one, defined over k. We shall understand by an *elliptic curve defined over a field k*, an abelian variety of dimension one, defined over k.

Let Φ be an imaginary quadratic field, R the ring of integers in Φ. Then there exists an elliptic curve E defined over an algebraic number-field k such that $\mathcal{A}(E)$ is isomorphic to R. Let ω be a differential form of the first kind on E. If $\mu_0 \in \mathcal{A}(E)$, there exists a number μ such that $\delta\mu_0\omega = \mu\omega$, where $\delta\mu_0$ denotes the differential of the rational mapping μ_0. The correspondence $\mu \to \mu_0$ is an isomorphism of R onto $\mathcal{A}(E)$; this isomorphism does not depend upon the choice of ω; denote it by ι. Thus we obtain an abelian variety having R as operator-domain (E, ι); we denote it also by E.

THEOREM 1. *Notations being as above, suppose that k contains Φ. Let $\mathfrak{p}$ be a prime ideal of R and $\mathfrak{P}$ a prime divisor of $\mathfrak{p}$ in k. Suppose that E has no defect for $\mathfrak{P}$ and denote by $\overline{E}$ the elliptic curve having R as operator-domain, obtained from E by reduction modulo $\mathfrak{P}$. Denote by π the rational mapping of $\overline{E}$ onto $\overline{E}^{N\mathfrak{p}}$ such that $\pi\bar{t} = \bar{t}^{N\mathfrak{p}}$ for every $\bar{t} \in \overline{E}$. Then $\overline{E}^{N\mathfrak{p}}$ is a $\mathfrak{p}$-transform of $\overline{E}$; and π is a $\mathfrak{p}$-multiplication of $\overline{E}$ onto $\overline{E}^{N\mathfrak{p}}$.*

This theorem is an algebro-geometric formulation of Kronecker's congruence relation on elliptic functions ([2] XI, § 14), though our theorem is concerned only with a singular modulus. From this we can derive the law of reciprocity for Strahl-class-fields over imaginary quadratic fields, with no use of the general class-field theory. Moreover we can determine the ramification in case where the conductor is prime to 2.

4. As in 1, let K be an algebraic number-field of degree $2n$ and R the ring of all integers in K. Let (A, ι) be an abelian variety having R as operator-domain defined over the field of complex numbers. Let K^* be the smallest normal extension of the rational number field Q containing K; denote by G the Galois group of the extension K^*/Q and by H the subgroup of G corresponding to K. Denote by $D(A)$ the set of all invariant differential forms on A of degree one. Let μ be an element of R and denote by $\delta\mu$ the differential of the rational mapping $\iota(\mu)$. Then $\delta\mu$ defines a linear endomorphism of the linear space $D(A)$. We can find n elements $\sigma_1, \cdots, \sigma_n$ in G such that $\mu^{\sigma_1}, \cdots, \mu^{\sigma_n}$ are the characteristic roots of the linear endomorphism $\delta\mu$ for every $\mu \in R$. We shall say that A has the *type* $(K, \sigma_1, \cdots, \sigma_n)$ if the situation is as above. We can prove that if there exists an abelian variety of the type $(K, \sigma_1, \cdots, \sigma_n)$, there exists one defined over an algebraic number-field of a finite degree. Now, notations being as above, denote by H_0 the subgroup of G consisting of all the elements σ such that $\bigcup_{j=1}^{n} H\sigma_j\sigma = \bigcup_{j=1}^{n} H\sigma_j$. Then we can find elements $\tau_1, \cdots, \tau_s$ in G such that $\bigcup_{j=1}^{n}\sigma_j^{-1}H = \bigcup_{i=1}^{s} H_0\tau_i$ and $n[H:1] = s[H_0:1]$. Denote by K_0 the subfield of K^* corresponding to H_0; then we have $[K_0:Q]=2s$. For this field K_0, the following proposition holds.

PROPOSITION 2. *Notations being as above, let (A, ι) be an abelian variety having R as operator-domain, of the type $(K, \sigma_1, \cdots, \sigma_n)$ defined over an algebraic number-field k of a finite degree. Then k contains K_0. Furthermore, let σ be an isomorphism of k. Then (A^σ, ι^σ) is a c-transform of (A, ι) for some ideal-class c of K if and only if σ fixes every element of K_0.*

Now we have a congruence relation on the abelian variety A.

THEOREM 2. *Notations being as in Proposition 2, let $\mathfrak{p}$ be a prime ideal of K_0 which is of the absolute degree one and $\mathfrak{P}$ a prime divisor of $\mathfrak{p}$ in k; put $N\mathfrak{p}=p$. Suppose that p is unramified in K^* and A has no defect for $\mathfrak{P}$. Denote by $\bar{A}$ the abelian variety having R as operator-domain obtained from A by reduction modulo $\mathfrak{P}$. Denote by π the rational mapping of $\bar{A}$ onto $\bar{A}^p$ such that $\pi\bar{t}=\bar{t}^p$ for every $\bar{t} \in \bar{A}$. Then $\bar{A}^p$ is a $\mathfrak{p}^{\tau_1}\cdots\mathfrak{p}^{\tau_s}$-transform of $\bar{A}$ and π is a $\mathfrak{p}^{\tau_1}\cdots\mathfrak{p}^{\tau_s}$-multiplication of $\bar{A}$ onto $\bar{A}^p$.*

By this theorem we obtain the following result.

THEOREM 3. *Let the notations be as in Proposition 2. Then,*

1) k contains the class-field $K_{(1)}$ over K_0 which corresponds to the ideal-group $H_{(1)}$ consisting of all ideals $\mathfrak{a}$ in K_0 such that $\mathfrak{a}^{\tau_1}\cdots\mathfrak{a}^{\tau_s}$ is principal in K;

2) let $\mathfrak{m}$ be an ideal of R and t a primitive element in $g(\mathfrak{m}, A)$; then the field $k(t)$ contains the class-field $K_\mathfrak{m}$ over K_0 which corresponds to the ideal-group $H_\mathfrak{m}$ consisting of all ideals $\mathfrak{a}$ in K_0 such that $\mathfrak{a}^{\tau_1}\cdots\mathfrak{a}^{\tau_s}$ belongs to the Strahl modulo $\mathfrak{m}$ in K.

The field $K_{(1)}$ and $K_\mathfrak{m}$ are represented by means of Chow-points as follows. Supposing that A is a variety in a projective space P^N, let $\{A_1, \cdots, A_r\}$ be the set of all the conjugates A' of A over K_0 such that A' is R-isomorphic to A. Denote by (a) the Chow-point of the cycle $A_1 + \cdots + A_r$ in P^N. Then we have $K_{(1)}=K_0(a)$. Let $\{t_1, \cdots, t_u\}$ be the set of all the conjugates t' of t over K_0 such that there exists an isomorphism σ of $k(t)$ for which we have $A^\sigma = A_i$ for some i and $t^\sigma = t' = \lambda_\mathfrak{a} t$ where $\mathfrak{a}$ is an ideal belonging to the Strahl modulo $\mathfrak{m}$ and $\lambda_\mathfrak{a}$ is an $\mathfrak{a}$-multiplication of A onto A^σ. Denote by (m) the Chow-point of the cycle $t_1 + \cdots + t_u$ in P^N. Then we have $K_\mathfrak{m} = K_0(a, m)$.

Now we shall sketch the proofs of Theorem 2 and Theorem 3. Let $\mathfrak{p}$ and $\mathfrak{P}$ be as in Theorem 2. It is easy to see that $\mathfrak{p}^{\tau_1} \cdots \mathfrak{p}^{\tau_s}$ is really an ideal of K and $N_{K/Q}(\mathfrak{p}^{\tau_1} \cdots \mathfrak{p}^{\tau_s}) = p^n$. We can find n invariant differential forms $\omega_1, \cdots, \omega_n$ on A such that $\delta \mu \omega_i = \mu^{\sigma_i} \omega_i$ $(1 \leq i \leq n)$ for every $\mu \in R$. Moreover we can take the ω_i in such a way that they are $\mathfrak{P}'$-finite and the forms $\bar{\omega}_i$ obtained from the ω_i by reduction modulo $\mathfrak{P}'$ form a basis of the linear space of linear differential forms on $\bar{A}$, where $\mathfrak{P}'$ denotes a prime divisor of $\mathfrak{P}$ in a field of definition for the ω_i. If μ is contained in $\mathfrak{p}^{\tau_1} \cdots \mathfrak{p}^{\tau_s}$, then μ^{τ_i} is divisible by $\mathfrak{P}$ for every i. Hence we have $\delta \bar{\mu} \bar{\omega}_i = \bar{\mu}^{\sigma_i} \bar{\omega} = 0$ $(1 \leq i \leq n)$ for every $\mu \in \mathfrak{p}^{\tau_1} \cdots \mathfrak{p}^{\tau_s}$ where a letter with a bar denotes an object obtained from the corresponding one by reduction modulo $\mathfrak{P}'$. This shows that $\delta \bar{\mu} = 0$ for every $\mu \in \mathfrak{p}^{\tau_1} \cdots \mathfrak{p}^{\tau_s}$. Therefore, if $\bar{x}$ is a generic point of A over the residue field κ of $\mathfrak{P}$, we have $\kappa(\bar{\mu}\bar{x}) \subset \kappa(\bar{x}^p)$ for every μ in $\mathfrak{p}^{\tau_1} \cdots \mathfrak{p}^{\tau_s}$. From this we obtain Theorem 2.

Let $\mathfrak{m}$ and t be as in Theorem 3. Let k^* be a finite normal extension of K_0 such that $k^* \supset kK^*(t)$ and that every homomorphism of A into any conjugates of A over K_0 is defined over k^*. Denote by G^* the Galois group of the extension k^*/K_0. If σ is an element of G^*, then t^σ is a primitive element in $\mathfrak{g}(\mathfrak{m}, A^\sigma)$. By Proposition 2, A^σ is a c-transform of A for some ideal-class c of K. Then, by Proposition 1, there exist an ideal $\mathfrak{a}$ in c and an $\mathfrak{a}$-multiplication $\lambda_\mathfrak{a}$ of A onto A^σ such that $t^\sigma = \lambda_\mathfrak{a} t$. The correspondence $\sigma \to \mathfrak{a}$ defines a homomorphism of G^* into the group of ideal-classes modulo $\mathfrak{m}$ in K. Denote by N^* the kernel of that homomorphism and by $K_\mathfrak{m}$ the subfield of k^* corresponding to N^*. Then $K_\mathfrak{m}$ is an abelian extension of K_0. If an element σ of G^* fixes every element of $k(t)$, then σ is contained in N^* as we have $A^\sigma = A$, $t^\sigma = t$. This shows that $K_\mathfrak{m}$ is a subfield of $k(t)$. Now let $\mathfrak{p}$ be a prime ideal of K_0 satisfying the following conditions: i) $\mathfrak{p}$ is of the absolute degree one, ii) $N\mathfrak{p} = p$ is unramified in k^*, iii) the A^σ for all $\sigma \in G^*$ have no defect for every prime divisor of $\mathfrak{p}$ in k^*. Let $\mathfrak{P}$ be a prime divisor of $\mathfrak{p}$ in k^* and σ a Frobenius substitution of $\mathfrak{P}$; then σ induces $\left(\dfrac{K_\mathfrak{m}/K_0}{\mathfrak{p}} \right)$ in $K_\mathfrak{m}$. For this σ, we obtain an ideal $\mathfrak{a}$ and an $\mathfrak{a}$-multiplication $\lambda_\mathfrak{a}$ of A onto

A^σ such that $t^\sigma = \lambda_\mathfrak{a} t$. Denoting by bars over letters the objects obtained by reduction modulo $\mathfrak{P}$, we have $\dot{\bar{A}}^\sigma = \bar{A}^p$ and $\bar{t}^\sigma = \bar{\lambda}_\mathfrak{a}\bar{t}$. On the other hand, by Theorem 2, we have $\bar{t}^\sigma = \bar{t}^p = \pi\bar{t}$ where π denotes the $\mathfrak{p}^{\tau_1}\cdots\mathfrak{p}^{\tau_s}$-multiplication of $\bar{A}$ onto $\bar{A}^p$ as in that theorem. Now if $\mathfrak{p}$ is prime to $N\mathfrak{m}$, $\bar{t}$ is a primitive element in $\mathfrak{g}(\mathfrak{m}, \bar{A})$. By Proposition 1, the relation $\bar{\lambda}_\mathfrak{a}\bar{t} = \pi\bar{t}$ implies that $\mathfrak{a}$ and $\mathfrak{p}^{\tau_1}\cdots\mathfrak{p}^{\tau_s}$ belong to the same ideal-class modulo $\mathfrak{m}$. Thus we have proved that a prime ideal $\mathfrak{p}$ of the absolute degree one is contained in $H_\mathfrak{m}$ if and only if $\left(\dfrac{K_\mathfrak{m}/K_0}{\mathfrak{p}}\right)$ is equal to the identity with a finite number of exceptions. By a result of class-field theory, we have 2) of Theorem 3.

UNIVERSITY OF TOKYO

REFERENCES

[1] M. Deuring, Die Struktur der elliptischen Funktionenkörper und die Klassenkörper der imaginären quadratischen Zahlkörper, Math. Ann., **124** (1952), pp. 393–426.

[2] L. Kronecker, Zur Theorie der elliptischen Funktionen, Werke IV.

[3] G. Shimura, Reduction of algebraic varieties with respect to a discrete valuation of the basic field, Amer. Journ. of Math., **77** (1955), pp. 134–176.

La fonction ζ du corps des fonctions modulaires elliptiques

Comptes rendus des séances de l'Académie des Sciences, 244 (1957), 2127-2130*

Soit M le corps complet des fonctions modulaires elliptiques d'espèce («Stufe») n; M est un corps de fonctions algébriques d'une variable sur le corps des nombres complexes $\mathbf{C}$. Il y a un sous-corps K ayant le corps des nombres rationnels $\mathbf{Q}$ comme corps de constantes, qui engendre M sur $\mathbf{C}$. Dans cette Note nous allons démontrer que la fonction ζ d'un tel corps K, choisi convenablement, est représentée par la foncion ζ de Riemann et un produit d'Euler introduit par E. Hecke (^{1}) et que les valeurs absolues des racines caractéristiques de l'opérateur T_p de Hecke pour les formes paraboliques de degré 2 ne dépassent pas $2\sqrt{p}$. C'est une extension des résultats obtenus par M. Eichler pour un certain sous-corps de M qui a une relation particulière avec les formes quadratiques (^{2}).

Soient γ une variables sur $\mathbf{Q}$ et E la courbe $X_0 X_2^2 = 4X_1^3 - \gamma X_0^2 X_1 - \gamma X_0^3$ dans le plan projectif; E est une courbe elliptique dont l'invariant est $\gamma/(\gamma - 27)$ que nous désignons par j. En prenant le point $(X_0, X_1, X_2) = (0, 0, 1)$ pour zéro, nous considérons E comme une variété abélienne définie sur le corps $\mathbf{Q}(j)$. Soient $w(n, E)$ l'ensemble $\{t ; t \in E, nt = 0\}$, h la fonction X_1/X_0 sur E et K_n le corps $\mathbf{Q}(j, h(t); t \in w(n, E))$; K_n ist isomorphe au corps engendré sur $\mathbf{Q}$ par les fonctions modulaires

$$j(\tau) = \frac{g_2^3}{g_2^3 - 27g_3^2}, \qquad f_{\alpha\beta}(\tau) = \frac{g_2}{g_3}\wp\left(\frac{\alpha\omega_1 + \beta\omega_2}{n}; \omega_1, \omega_2\right), \qquad \left(\tau = \frac{\omega_1}{\omega_2}\right),$$

en faisant correspondre j, $h(\alpha t_1 + \beta t_2)$ aux $j(\tau)$, $f_{\alpha\beta}(\tau)$ où t_1, t_2 sont deux générateurs de $w(n, E)$. Nous identifions les deux corps par cet isomorphisme; K_n engendre M sur $\mathbf{C}$; il est galoisien sur $\mathbf{Q}(j)$ et le groupe de Galois de $K_n/\mathbf{Q}(j)$ est isomorphe au groupe $G_n = GL(2, \mathbf{Z}/(n))/\{\pm 1\}$, où $\mathbf{Z}$ désigne l'anneau des entiers rationnels, de telle façon que pour chaque $\begin{pmatrix} a & b \\ c & d \end{pmatrix} \in G_n$, l'application

$$f_{\alpha\beta} \to f_{\alpha'\beta'} : (\alpha'\ \beta') = (\alpha\ \beta)\begin{pmatrix} a & b \\ c & d \end{pmatrix}$$ donne un automorphisme de $K_n/\mathbf{Q}(j)$.

Si $ad - bc = 1$, on a $f_{\alpha'\beta'}(\tau) = f_{\alpha\beta}((a\tau + b)/(c\tau + d))$. Le sous-corps de K_n correspondant au sous-groupe $S_n = SL(2, \mathbf{Z}/(n))/\{\pm 1\}$ de G_n est le corps $\mathbf{Q}(j, \zeta_n)$ où ζ_n désigne $e^{2\pi i/n}$; $\mathbf{Q}(\zeta_n)$ est algébriquement fermé dans K_n.

Soient p un nombre premier et w_ν ($1 \leq \nu \leq p + 1$) les sous-groupes d'ordre p de E. Pour chaque ν, il y a une courbe elliptique E_ν et un homomorphism λ_ν de E sur E_ν tel que le noyau de λ_ν soit w_ν. Nous prenons l'équation de

1

E_ν dans la forme $X_0 X_2^2 = 4X_1^3 - \gamma_\nu X_0^2 X_1 - \gamma_\nu X_0^3$; l'invariant de E_ν est alors $\gamma_\nu/(\gamma_\nu - 27)$. Comme γ_ν est aussi une variable sur $\mathbf{Q}$, il y a un isomorphisme σ_ν de $\mathbf{Q}(j) = \mathbf{Q}(\gamma)$ sur $\mathbf{Q}(\gamma_\nu)$ tel que $\gamma^{\sigma_\nu} = \gamma_\nu$ et, par conséquent, $E^{\sigma_\nu} = E_\nu$. Posons $j_\nu = j^{\sigma_\nu}$, $h_\nu = h^{\sigma_\nu}$. Si $n \not\equiv 0 \pmod{p}$, il y a un isomorphisme de K_n prolongeant σ_ν, que nous désignons aussi par σ_ν, tel que $h(t)^{\sigma_\nu} = h_\nu(\lambda_\nu t)$ pour $t \in w(n, E)$. Soit P un diviseur premier de p dans une clôture algébrique de $\mathbf{Q}(j)$ tel que j modulo P soit transcendant sur le corps premier. Nous indiquerons par la barre la réduction modulo P. La réduction modulo P donne un homomorphisme de $w(p, E)$ sur $w(p, \overline{E})$. Comme $w(p, \overline{E})$ est d'ordre p, le noyau de cet homomorphisme est un des groupes w_ν, mettons w_1. Alors, $\overline{\lambda}_1$ est inséparable et les $\overline{\lambda}_\nu$ $(\nu > 1)$ sont séparable. Désignant par μ_ν l'homomorphisme de E_ν sur E tel que $\mu_\nu \lambda_\nu = p\delta_E$, $\overline{\mu}_1$ est séparable et les $\overline{\mu}_\nu$ $(\nu > 1)$ sont inséparable. On a donc $\overline{j} = \overline{j}_1^{1/p} = \overline{j}_\nu^p$, $\overline{E} = \overline{E}_1^{1/p} = \overline{E}_\nu^p$ $(\nu > 1)$, d'où résultent les formules

$$(1) \qquad \overline{h_1(\lambda_1 t)} = \overline{h(t)}^p, \quad \overline{h_\nu(\lambda_1 t)}^p = \overline{h(pt)} \quad (\nu > 1) \quad \text{pour} \quad t \in w(n, E).$$

Maintenant soit K le sous-corps de K_n correspondant au sous-groupe $\left\{ \begin{pmatrix} a & 0 \\ 0 & \pm 1 \end{pmatrix}; \ (a, n) = 1 \right\} \Big/ \{\pm 1\}$ de G_n; alors on a $K_n = K(\zeta_n)$, $K = \mathbf{Q}\big(j(\tau),$ $j(\tau/n), f_{01}(\tau)\big)$; K engendre M sur $\mathbf{C}$. Comme $\mathbf{Q}$ est algébriquement fermé dans K, il y a une courbe complète Γ sans point multiple avec un point générique x par rapport à $\mathbf{Q}$ tel que $\mathbf{Q}(x) = K$. Soient ρ_p et ψ les isomorphismes de K donnés par $\begin{pmatrix} p & 0 \\ 0 & p \end{pmatrix}$ et par $\begin{pmatrix} 0 & 1 \\ -1 & 0 \end{pmatrix}$. Soient X_p, Y_p et A respectivement les lieux de $x \times x^{\sigma_1}$, de $x \times x^{\rho_p}$ et de $x \times x^\psi$ par rapport à $\mathbf{Q}(\zeta_n)$. Il s'ensuit de la définition des σ_ν que $X_p(x) = x^{\sigma_1} + \cdots + x^{\sigma_{p+1}}$. D'après les formules (1), on a $\overline{X}_p(\overline{x}) = \overline{x}^p + p\overline{Y}_p(\overline{x})^{1/p}$. Désignant par ξ_p et par η_p les classes de la correspondance X_p et de Y_p, on a donc pour presque tous les nombres premiers p,

$$(2) \qquad \overline{\xi}_p = \pi + \pi^* \overline{\eta}_p,$$

où π est la classe de la correspondance $\overline{x} \to \overline{x}^p$ sur $\overline{\Gamma}$ et $*$ désigne l'antiautomorphisme de Rosati. De plus, désignant par α la classe de la correspondance A, on a

$$(3) \qquad \pi^* \overline{\eta}_p = \overline{\alpha}^{-1} \pi^* \overline{\alpha},$$

Cette formule est démontrée par le fait que $\begin{pmatrix} a & b \\ c & d \end{pmatrix} \in G_n$ donne l'automorphisme de Frobenius pour p sur $\mathbf{Q}(\zeta_n)/\mathbf{Q}$ si $ad - bc \equiv p \pmod{n}$.

Soit g le genre de Γ. D'après les formules (2), (3) et $\pi\pi^* = p\delta$, on a

$$\det(I_{2g} - M_l(\overline{\xi}_p)p^{-s} + M_l(\overline{\eta}_p)p^{1-2s}) = \det(I_{2g} - M_l(\pi)p^{-s})^2,$$

où I_m désigne la matrice unité de m lignes et M_l désigne la représentation l-adique. D'autre part, on peut mettre $M_l(\xi_p) = M_l(\overline{\xi}_p)$, $M_l(\eta_p) = M_l(\overline{\eta}_p)$.

Comme Γ, ξ_p, η_p sont définis sur $\mathbf{Q}$, on a $M_l(\xi_p) \sim M^d(\xi_p) \oplus M^d(\xi_p)$, $M_l(\eta_p) \sim M^d(\eta_p) \oplus M^d(\eta_p)$ où M^d désigne la représentation par les formes différentielles de première espèce. Alors la fonction ζ de la courbe $\overline{\Gamma}$ a l'expression

$$\zeta(s,\, \Gamma,\, p) = \frac{\det\left(I_{2g} - M_l(\pi)p^{-s}\right)}{(1 - p^{-s})(1 - p^{1-s})} = \frac{\det\left(I_g - M^d(\xi_p)p^{-s} + M^d(\eta_p)p^{1-2s}\right)}{(1 - p^{-s})(1 - p^{1-s})}.$$

L'opérateur de Hecke T_p pour les formes paraboliques de degré 2 et d'espèce n peut être regardé comme la différentielles de ξ_p, si l'on identifie ces formes avec les formes différentielles de première espèce sur Γ; la différentielles de η_p représente l'opérateur $R_p = \begin{bmatrix} p^{-1} & 0 \\ 0 & p \end{bmatrix}$ dans les traveux de Hecke. Alors $M^d(\xi_p)$, $M^d(\eta_p)$ sont les représentations de T_p et de R_p par les formes paraboliques de degré 2.

Nous avons ainsi pour la fonction ζ de Γ,

$$\zeta(s,\, \Gamma) = \prod_p \zeta(s,\, \Gamma,\, p) = f(s)\zeta(s)\zeta(s - 1)\Phi(s)^{-1},$$

$$\Phi(s) = \prod_{t,\varepsilon} \det F_{t,\varepsilon}(s),$$

où $\zeta(s)$ est la fonction ζ de Riemann, $f(s)$ est une fonction rationnelle de p^{-s} et $F_{t,\varepsilon}(s)$ désigne le produit d'Euler introduit par Hecke, attaché aux formes paraboliques de degré 2, de diviseur t et de caractère $\varepsilon(p)$. La fonction $\zeta(s,\, \Gamma)$ est donc méromorphe sur tout le s-plan.

Soient H un sous-groupe du groupe multiplicatif des entiers modulo n, et K_H le sous-corps de K_n correspondant au sous-groupe

$$\left\{ \begin{pmatrix} a & b \\ 0 & d \end{pmatrix};\ (a,\, n) = 1,\, d \in H,\, b \in \mathbf{Z} \right\} \Big/ \{\pm 1\} \quad \text{de } G_n.$$

On peut obtenir des résultats analogues aussi pour la fonction ζ du corps K_H; le cas où H a l'index 1 ou 2 a été traité par Eichler.

D'après A. Weil (^{3}) les racines caractéristiques de π ont la valeur absolue $\sqrt{p}$. Il s'ensuit donc de la formule (2) que les valeurs absolues des racines caractéristiques de T_p pour les formes paraboliques de degré 2 ne dépassent pas $2\sqrt{p}$ pour presque tous les nombres premiers p.

(^{1}) E. HECKE, *Math. Ann.* **114**, 1937, p. 1–28 et p. 316–351.

(^{2}) M. EICHLER, *Arch. d. Math.*, **5**, 1954, p. 355–366.

(^{3}) A. WEIL, *Act. Sc. et Ind.*, n° **1041**, Hermann et Cie, Paris, 1948.

58a

Correspondances modulaires et les fonctions zeta de courbes algébriques

Journal of the Mathematical Society of Japan, 10 (1958), 1-28

(Reçu le 5 Nov., 1957)

M. Eichler [3] a découvert qu'il y a une relation étroite entre l'opérateur T_p défini par E. Hecke et les fonctions ζ de certains corps de fonctions modulaires elliptiques qui sont en rapport avec les formes quadratiques. Dans ce travail nous allons généraliser cette relation.

Soient N un entier positif et $\mathfrak{M}_N$ le corps complet des fonctions modulaires elliptiques d'espèce («Stufe») N; $\mathfrak{M}_N$ est un corps de fonctions algébriques d'une variable sur le corps des nombres complexes C. Nous pouvons montrer qu'il y a un sous-corps $\Re$ ayant le corps des nombres rationnels Q comme corps de constantes, qui engendre $\mathfrak{M}_N$ sur C. Nous allons démontrer que les fonctions ζ d'un tel corps $\Re$, choisi convenablement, et de certains sous-corps de $\Re$ sont représentées explicitement par la fonction ζ de Riemann et les produits d'Euler introduits par E. Hecke et que les valeurs absolues des racines caractéristiques de l'opérateur T_p de Hecke pour les formes paraboliques («Spitzenformen») de degré 2 ne dépassent pas $2\sqrt{p}$ pour presque tous les nombres premiers p.

Nous démontrerons ces résultats en établissant deux formules de congruence ((I) et (II) dans §3) pour les correspondances modulaires. Nous représentons d'abord le corps $\mathfrak{M}_N$ par les coordonnées des points t tels que $Nt=0$ sur une courbe elliptique E dont l'invariant est transcendant sur Q. L'opérateur T_p de Hecke peut être regardé comme la différentielle d'une correspondance algébrique que nous pouvons définir d'une manière algébro-géométrique par un homomorphisme λ de E tel que $\nu(\lambda)=p$ et que nous appelons la correspondance modulaire de degré p. Dans la théorie de la multiplication complexe ([6], [7]), la relation de congruence ou la décomposition de l'homomorphisme π a été obtenue par la réduction d'homomorphismes d'une variété abélienne modulo un diviseur premier du corps de base. Une méthode analogue est employée ici pour démontrer la formule (I); nous considérerons la réduction de l'homomorphisme λ modulo un diviseur premier. Seulement nous nous occupons au cas présent d'une courbe elliptique sans multiplication complexe, mais on pourrait considérer λ comme une «multiplication générique» de la courbe elliptique. La formule complémentaire (II) est démontrée au moyen du fait qui s'exprime dans la prop. 7 (§2). Le résultat pour les racines caractéristiques de l'opérateur T_p est une conséquence de la formule (I).

§1. Courbes elliptiques.

1. Nous désignerons par Z, Q et C respectivement l'anneau des entiers rationnels, le corps des nombres rationnels et le corps des nombres complexes.

Soit A une variété abélienne[1]. Nous désignerons par $\mathcal{A}(A)$ l'anneau des endomorphismes de A, par $\mathcal{A}_0(A)$ le produit tensoriel $\mathcal{A}(A) \times Q$ et par δ_A l'élément unité de $\mathcal{A}(A)$. Soient B une autre variété abélienne de même dimension que A et λ un homomorphisme de A sur B. Soient k un corps de définition pour A, B et λ; x un point générique de A par rapport à k. Nous poserons

$$\nu(\lambda)=[k(x): k(\lambda(x))], \qquad \nu_s(\lambda)=[k(x): k(\lambda x)]_s,$$

$$\nu_i(\lambda)=[k(x): k(\lambda x)]_i.$$

Ces entiers ne dépendent que de A, B et λ et non du choix de k et de x. Nous désignerons par $\mathfrak{g}(\lambda)$ le noyau de λ et par $\mathfrak{g}(n, A)$ le noyau de $n\delta_A$ pour chaque entier n. Le groupe $\mathfrak{g}(\lambda)$ est d'ordre $\nu_s(\lambda)$. Si A est de dimension d, on a $\nu(n\delta_A)=n^{2d}$; par suite $\mathfrak{g}(n, A)$ est d'ordre n^{2d} si n n'est pas multiple de la caractéristique de k ([11]). Pour le cas où n est la caractéristique de k, on a le lemme suivant dont la démonstration est donnée dans [8].

Lemme 1. *Soient k un corps de caractéristique $p \neq 0$ et A une variété abélienne de dimension d définie par rapport à k. On a alors*

$$\nu_i(p\delta_A) \geqq p^d, \qquad \nu_s(p\delta_A) \leqq p^d.$$

Soient k un corps de caractéristique $p \neq 0$, $q=p^f$ une puissance de p, où f est un entier positif ou négatif. Nous désignerons par k^q le corps des q-ièmes puissances des éléments de k. En faisant correspondre $z \in k$ à $z^q \in k^q$ on obtient un isomorphisme σ de k sur k^q. Soient V une variété définie par rapport à k et x un point de V. Nous désignerons par V^q et par x^q la variété V^σ transformée par σ et le point de V^q dont les coordonnées sont les q-ièmes puissances de celles de x. Soient h une application rationnelle de V dans une variété W, définie par rapport à k, et H le graphe de h. Nous désignerons par h^q l'application de V^q dans W^q dont le graphe est

---

1) Il sera constamment fait usage des définitions et des résultats de [9], [10], [11].

H^q. Soient A une variété abélienne définie par rapport à k et O l'élément neutre de A. Nous entendrons par la notation A^q toujours la variété abélienne dont l'élément neutre est O^q. Si λ est un homomorphisme de A dans une variété abélienne, défini par rapport à k, λ^q est un homomorphisme de A^q. Soit x un point générique de A par rapport à k; x^q est alors un point générique de A^q par rapport à k. Comme on a $k(x) \supset k(x^q)$ on obtient une application rationnelle π de A sur A^q telle que $\pi x = x^q$; π est un homomorphisme de A sur A^q; on a $\pi t = t^q$ pour tout point t de A. Nous appellerons π *l'homomorphisme de q-ième puissance de A.* Si A est de dimension d, on a $\nu(\pi) = \nu_i(\pi) = q^d$.

LEMME 2. *Soient A, B, C trois variétés abéliennes de même dimension, λ un homomorphisme de A sur B et μ un homomorphisme de A sur C. Supposons que $\mathcal{A}(A)$ soit isomorphe à Z et que l'on ait $\nu(\lambda) = \nu(\mu)$, $\nu_i(\lambda) = \nu_i(\mu) = 1$. Alors, pour que B soit isomorphe à C, il faut et il suffit qu'on ait $\mathfrak{g}(\lambda) = \mathfrak{g}(\mu)$.*

D'après le th. 17 de [11] n°34, si l'on a $\mathfrak{g}(\lambda) = \mathfrak{g}(\mu)$, B est isomorphe à C. Réciproquement supposons qu'il existe un isomorphisme η de B sur C. D'après le th. 27 de [11] n°52, il y a un homomorphisme α de C sur A. $\alpha\eta\lambda$ et $\alpha\mu$ sont contenus dans $\mathcal{A}(A)$. Comme $\mathcal{A}(A)$ est isomorphe à Z, il y a deux entiers n, n' autres que 0 tels que $n\alpha\eta\lambda = n'\alpha\mu$. On a alors $\nu(n\delta_A)\nu(\alpha)\nu(\eta)\nu(\lambda) = \nu(n'\delta_A)\nu(\alpha)\nu(\mu)$. Si l'on désigne par d la dimension de A, on a $\nu(n\delta_A) = n^{2d}$, $\nu(n'\delta_A) = n'^{2d}$. Comme $\nu(\eta) = 1$ et $\nu(\lambda) = \nu(\mu)$, on a $n = \pm n'$ et par conséquent $\eta\lambda = \pm\mu$; d'où résulte $\mathfrak{g}(\lambda) = \mathfrak{g}(\eta\lambda) = \mathfrak{g}(\mu)$.

LEMME 3. *Les notations A, B, C, λ, μ et les hypothèses étant celles du lemme 2, soient k un corps de définition pour A, B, C, λ, μ et σ un isomorphisme de k tel que $A^\sigma = A$, $B^\sigma = C$. On a alors $\lambda^\sigma = \pm\mu$.*

D'après le lemme 2, on a $\mathfrak{g}(\lambda^\sigma) = \mathfrak{g}(\mu)$. Il y a donc un automorphisme ε de C tel que $\lambda^\sigma = \varepsilon\mu$ en vertu du th. 17 de [11] n°34. Comme C est isogène à A, $\mathcal{A}(C)$ est isomorphe à Z; on en déduit que $\varepsilon = \pm 1$; ce qui prouve notre lemme.

2. Une variété abélienne de dimension 1 est une courbe algébrique de genre 1. Réciproquement, une courbe algébrique de genre 1, définie par rapport à un corps k, ayant un point rationnel par rapport à k a une structure de variété abélienne définie par rapport à k. Ci-après, par une *courbe elliptique définie par rapport à k*, nous entendrons une variété abélienne de dimension 1 définie par rapport à k.

Soit k un corps dont la caractéristique n'est ni 2 ni 3. Soient $r_2, r_3,$ deux éléments de k tels que $r_2{}^3 - 27\,r_3{}^2 \neq 0$ et E_1 la courbe définie par l'équation

(1) $$X_0 X_2{}^2 = 4X_1{}^3 - r_2 X_0{}^2 X_1 - r_3 X_0{}^3$$

dans le plan projectif. En prenant le point $(X_0, X_1, X_2)=(0, 0, 1)$ pour l'élément neutre de variété de groupe, nous pouvons considérer E_1 comme une variété abélienne définie par rapport à k. En posant $X=X_1/X_0$ et $Y=X_2/X_0$ dans l'équation (1), on a l'équation affine

$$(2) \qquad Y^2=4X^3-\gamma_2 X-\gamma_3 .$$

Nous conviendrons d'entendre par la courbe elliptique définie par l'équation (2) la courbe elliptique projective définie par l'équation (1), ayant le point $(0, 0, 1)$ comme l'élément neutre 0. La correspondance $(X, Y)\to(X, -Y)$ donne l'endomorphisme -1 de la courbe elliptique.

Soit E une courbe elliptique définie par rapport à k. Il est bien connu que E est isomorphe à une courbe elliptique définie par une équation $Y^2=4X^3-\gamma_2 X-\gamma_3$ où γ_2, γ_3 sont deux éléments de k. Nous appellerons $\gamma_2{}^3/(\gamma_2{}^3-27\gamma_3{}^2)$ l'invariant de E et le désignerons par j_E ou $j(E)$; $j(E)$ est contenu dans tout corps de définition pour E. Pour que deux courbes elliptiques soient isomorphes, il faut et il suffit qu'elles aient le même invariant. Si σ est un isomorphisme d'un corps de définition pour une courbe elliptique E, on a $j(E^\sigma)=j(E)^\sigma$.

Soient E la courbe elliptique définie par l'équation (2) et h la fonction sur E qui prend les coordonnées aux points de E (autrement dit, la fonction X_1/X_0 sur la courbe (1)). Nous appellerons h *la fonction canonique sur* E. On voit que

$$(3) \qquad h(u)=h(v) \Longleftrightarrow u=\pm v .$$

3. Soient E une courbe elliptique et $\mathfrak{g}$ un sous-groupe fini de E. D'après le th. 17 de [11] n°34, il y a une courbe elliptique E' et un homomorphisme λ de E sur E' tel qu'on ait $\mathfrak{g}=\mathfrak{g}(\lambda)$, $\nu_i(\lambda)=1$. D'après le même théorème, l'invariant $j(E')$ de E' ne dépend que de $E, \mathfrak{g}$ et non du choix de E' et de λ. Nous désignerons $j(E')$ par $j(E/\mathfrak{g})$.

Proposition 1. *Soient E, E' deux courbes elliptiques isogènes et k_0 le corps premier contenu dans un corps de définition pour E, E'. Alors $j(E')$ est algébrique sur $k_0(j_E)$.*

D'après la définition d'isogénéité, il existe un homomorphisme λ de E sur E'. Soient k un corps de définition pour E, E', λ et x un point générique de E par rapport à k; posons $q=\nu_i(\lambda)$. Alors il y a un sous-corps K de $k(x)$ tel que $k(x)$ soit une extension purement inséparable de K de degré q et K soit séparable sur $k(\lambda x)$. Comme $k(x)$ est de dimension 1 sur k, on a $K=k(x^q)$; par suite, en posant $\mu x^q=\lambda x$, on obtient un homomorphisme μ de E^q sur E' défini par rapport à k, pour lequel on a $\nu_i(\mu)=1$. Soit σ un isomorphisme du corps $k(j_E, j_{E'})$ fixant tous les éléments de $k_0(j_E)$; μ^σ est alors un homomorphisme de $(E^q)^\sigma$ sur E'^σ. On a $j(E'^\sigma)=j(E')^\sigma$ et $j(E^{q\sigma})$ $=(j(E)^q)^\sigma=j(E)^q$. Comme $E^{q\sigma}$ a l'invariant $j(E)^q=j(E^q)$, il y a un isomorphisme η de E^q sur $E^{q\sigma}$. $\mu^\sigma\eta$ est un homomorphisme de E^q sur E'^σ dont le noyau $\mathfrak{g}(\mu^\sigma\eta)$ est d'ordre $\nu(\mu)$. Comme on a $\nu_i(\mu^\sigma\eta)=1$, on a $j(E'^\sigma)=j(E^q/\mathfrak{g}(\mu^\sigma\eta))$. Il n'y a qu'un nombre fini de sous-groupes de E^q ayant l'ordre $\nu(\mu)$; par suite il n'y a qu' un nombre fini de $j(E')^\sigma$; ce qui montre que $j(E')$ est algébrique sur $k_0(j_E)$.

PROPOSITION 2. *Soient E, E' deux courbes elliptiques définies par rapport à un corps de caractéristique $p\neq 0$ et λ un homomorphisme de E sur E' tel qu'on ait $\nu_s(\lambda)=1$. Alors $\nu(\lambda)$ est une puissance q de p; il existe un isomorphisme ε de E^q sur E' tel qu'on ait $\lambda t=\varepsilon t^q$ pour tout $t\in E$.*

Soient k un corps de définition pour E, E', λ et x un point générique de E par rapport à k. Comme $\nu_s(\lambda)=1$, $k(x)$ est purement inséparable sur $k(\lambda x)$; par suite $\nu(\lambda)=[k(x):k(\lambda x)]$ est une puissance q de p. Comme $k(x)$ est de dimension 1 sur k, on a $k(\lambda x)=k(x^q)$. On obtient donc un isomorphisme ε de E^q sur E', défini par rapport à k, tel que $\varepsilon x^q=\lambda x$. On vérifie aisément $\varepsilon t^q=\lambda t$ pour tout $t\in E$.

PROPOSITION 3. *Soit E une courbe elliptique définie par rapport à un corps de caractéristique $p\neq 0$ telle qu'on ait $j(E)^{p^2}\neq j(E)$. On a alors $\nu_i(p\delta_E)=\nu_s(p\delta_E)=p$; $\mathfrak{g}(p, E)$ est d'ordre p.*

D'après le lemme 1, on a $\nu_i(p\delta_E)\geqq p$; $\nu_i(p\delta_E)$ est donc égal à p ou p^2. Supposons qu'on ait $\nu_i(p\delta_E)=p^2$. Alors on a $\nu_s(p\delta_E)=1$. D'après la prop. 2, E est isomorphe à E^{p^2}; par suite on a $j(E)=j(E^{p^2})=j(E)^{p^2}$; ce qui est en contradiction avec l'hypothèse de la proposition. On a donc $\nu_i(p\delta_E)=p$ et par suite $\nu_s(p\delta_E)=p$, de sorte que $\mathfrak{g}(p, E)$ est d'ordre p.

PROPOSITION 4. *Soit E une courbe elliptique telle que $j(E)$ soit transcendant sur le corps premier. $\mathcal{A}(E)$ est alors isomorphe à $\mathbf{Z}$.*

Cette proposition est bien connue au cas où la caractéristique du corps de base est 0. Supposons donc que la caractéristique soit autre que 0. $\mathcal{A}_0(E)$ est un corps (commutatif ou non-commutatif). D'après la prop. 3 et d'après le th. 14 de [11] n°31, il existe un homomorphisme φ de $\mathcal{A}_0(E)$ dans le corps des nombres p-adiques; φ est un isomorphisme puisque $\mathcal{A}_0(E)$ est un corps. $\mathcal{A}_0(E)$ est donc un corps commutatif. On a $[\mathcal{A}_0(E):\mathbf{Q}]\leqq 2$ puisque, d'après le cor. 2 du th. 36 de [11] n°69, tout élément de $\mathcal{A}_0(E)$ satisfait à une équation à coefficients rationnels de degré 2. Supposons que l'on ait $[\mathcal{A}_0(E):\mathbf{Q}]=2$. Soit l un nombre premier qui reste premier dans $\mathcal{A}_0(E)$. Soit $\mathfrak{g}$ un sous-groupe de E d'ordre l; Il y a une courbe elliptique E' et un homomorphisme λ de E sur E' tels que $\mathfrak{g}(\lambda)=\mathfrak{g}$. Pour un l approprié, on peut facilement voir que $\mathcal{A}(E')$ n'est pas isomorphe à $\mathcal{A}(E)^{2)}$. D'autre

---

2) C'est une conséquence des résultats dans [8].

part, d'après la prop. 1, si l'on désigne par k_0 le corps premier, $j(E)$ est algébrique sur $k_0(j(E'))$, de sorte que $j(E')$ est transcendant sur k_0; on peut en déduire que $\mathcal{A}(E')$ est isomorphe à $\mathcal{A}(E)$; ce qui est absurde. On a donc $[\mathcal{A}_0(E); \mathbf{Q}]=1$; par suite $\mathcal{A}(E)$ est isomorphe à $\mathbf{Z}$.

§ 2. Correspondances modulaires.

4. Soient γ une variable sur $\mathbf{Q}$ et E la courbe elliptique définie par l'équation

$$(4) \qquad Y^2=4X^3-\gamma X-\gamma .$$

Posons $j=j(E)$. On a alors $j=\gamma/(\gamma-27)$, $\gamma=27j/(j-1)$ et $\mathbf{Q}(\gamma)=\mathbf{Q}(j)$; j est donc une variable sur $\mathbf{Q}$. Nous désignons par F la clôture algébrique du corps $\mathbf{Q}(j)$. Soit h la fonction canonique sur E. Dans cette section nous employons ces notations E, γ, j, h, F toujours en ce sens.

Nous désignerons par $K_N{}^*(E)$ et par $K_N(E)$, ou simplement par $K_N{}^*$ et par K_N, respectivement les corps

$$\mathbf{Q}(j, t \mid t \in \mathfrak{g}(N, E))$$

$$\text{et } \mathbf{Q}(j, h(t) \mid t \in \mathfrak{g}(N, E))$$

pour chaque entier $N>0$. Ces corps sont galoisiens sur $\mathbf{Q}(j)$. Nous désignerons par $G_N(E)$ le groupe de Galois de $K_N(E)$ sur $\mathbf{Q}(j)$. Soit $\{t_1, t_2\}$ un système de générateurs de $\mathfrak{g}(N, E)$; on a alors

$$\mathfrak{g}(N, E)=\{\alpha t_1+\beta t_2 \mid \ 0\leqq\alpha<N, 0\leqq\beta<N\} .$$

Soit σ un élément de $G_N(E)$; σ est prolongé à un automorphisme de $K_N{}^*$ que nous désignons aussi par σ. $\{t_1{}^\sigma, t_2{}^\sigma\}$ est un système de générateurs de $\mathfrak{g}(N, E)$; il y a donc une matrice $\begin{pmatrix} a & b \\ c & d \end{pmatrix}$ à coefficients entiers telle qu'on ait

$$\begin{pmatrix} t_1{}^\sigma \\ t_2{}^\sigma \end{pmatrix}=\begin{pmatrix} a & b \\ c & d \end{pmatrix}\begin{pmatrix} t_1 \\ t_2 \end{pmatrix},$$

et par suite

$$(5) \qquad h(\alpha t_1+\beta t_2)^\sigma=h(\alpha' t_1+\beta' t_2), \qquad (\alpha' \ \beta')=(\alpha \ \beta)\begin{pmatrix} a & b \\ c & d \end{pmatrix}.$$

L'automorphisme σ est determiné par la matrice $\begin{pmatrix} a & b \\ c & d \end{pmatrix}$ et s'appellera *l'automorphisme de $K_N(E)$ correspondant à* $\begin{pmatrix} a & b \\ c & d \end{pmatrix}$ *par rapport à* $\{t_1, t_2\}$. Nous désignerons par $G_N{}^*$ le groupe des matrices de degré 2 à coefficients dans l'anneau $\mathbf{Z}/N\mathbf{Z}$ dont les déterminants sont inversibles et par $S_N{}^*$ le sous-groupe de $G_N{}^*$ consistant en les éléments unimodulaires (c.-à-d. de déterminant

$=1$). Nous désignerons respectivement par G_N et S_N les groupes $G_N{}^*/\{\pm I\}$ et $S_N{}^*/\{\pm I\}$. Comme ci-dessus nous pouvons faire correspondre à chaque $\sigma \in G_N(E)$, une matrice $\begin{pmatrix} a & b \\ c & d \end{pmatrix} \in G_N{}^*$. Tenant compte de (3), on vérifie facilement que l'application $\sigma \to \begin{pmatrix} a & b \\ c & d \end{pmatrix}$ donne un isomorphisme de $G_N(E)$ dans $G_N = G_N{}^*/\{\pm I\}$. Cet isomorphisme est surjectif:

Proposition 5. *En faisant correspondre une matrice $\begin{pmatrix} a & b \\ c & d \end{pmatrix}$ à $\sigma \in G_N(E)$ par la relation (5), on obtient un isomorphisme de $G_N(E)$ sur G_N.*

Soit $S_N(E)$ le sous-groupe de $G_N(E)$ correspondant au sous-groupe S_N de G_N par cet isomorphisme. $S_N(E)$ ne dépend pas du choix de $\{t_1, t_2\}$.

Proposition 6. *Le sous-corps de K_N correspondant au sous-groupe $S_N(E)$ de $G_N(E)$ est le corps $Q(j, \zeta_N)$ où ζ_N désigne une racine primitive N-ième d'unité. $Q(\zeta_N)$ est algébriquement fermé dans K_N.*

Dans §§ 2, 3, nous désignerons par ζ_N une racine primitive N-ième d'unité.

Proposition 7. *Soit $\begin{pmatrix} a & b \\ c & d \end{pmatrix}$ un élément de $G_N{}^*$ et σ l'automorphisme de K_N correspondant à $\begin{pmatrix} a & b \\ c & d \end{pmatrix}$. On a alors $\zeta_N{}^\sigma = \zeta_N{}^{ad-bc}$. En d'autres termes, si $ad-bc$ est congru à un nombre premier p modulo N, $\begin{pmatrix} a & b \\ c & d \end{pmatrix}$ donne la substitution de Frobenius $\left(\dfrac{Q(\zeta_N)/Q}{p} \right)$ dans $Q(\zeta_N)$.*

Nous démontrerons les prop. 5–7 dans le § 4.

5. Maintenant nous allons déterminer les sous-groupes de $G_N(E)$ correspondant à certains sous-corps de $K_N(E)$. Dans ce but nous fixons un système de générateurs $\{t_1, t_2\}$ de $\mathfrak{g}(N, E)$ et identifions $G_N(E)$ avec G_N par l'isomorphisme donné ci-dessus.

Proposition 8. *Soient $\mathfrak{g}, \mathfrak{g}'$ deux sous-groupes d'ordre N de E et σ un automorphisme de F sur $Q(j)$. Alors pour qu'on ait $j(E/\mathfrak{g})^\sigma = j(E/\mathfrak{g}')$, il faut et il suffit que $\mathfrak{g}^\sigma = \mathfrak{g}'$; $j(E/\mathfrak{g})$ est contenu dans $K_N(E)$.*

La première assertion est une conséquence immédiate de la prop. 4 et du lemme 2. Si σ est l'identité dans $K_N(E)$, on a $h(t^\sigma) = h(t)$ pour $t \in \mathfrak{g}(N, E)$; par suite on a $t^\sigma = \pm t$ pour $t \in \mathfrak{g}$, de sorte qu'on a $\mathfrak{g}^\sigma = \mathfrak{g}$ et $j(E/\mathfrak{g})^\sigma = j(E/\mathfrak{g})$. Il s'ensuit de là que $j(E/\mathfrak{g})$ est contenu dans $K_N(E)$.

Soient $\mathfrak{g}_{(1)}, \mathfrak{g}_{(2)}$ respectivement les sous-groupes de E engendrés par t_1 et par t_2; posons $j_{(1)} = j(E/\mathfrak{g}_{(1)})$, $j_{(2)} = j(E/\mathfrak{g}_{(2)})$. Nous avons le tableau suivant:

<table>
<tr><td align="center">Sous-corps de $K_N(E)$</td><td align="center">Sous-groupe de $G_N(E)$</td></tr>
<tr><td align="center">$Q(j, j_{(1)})$</td><td align="center">$\left\{\begin{pmatrix} a & 0 \\ c & d \end{pmatrix}\right\}\big/\{\pm I\}$</td></tr>
<tr><td align="center">$Q(j, j_{(2)})$</td><td align="center">$\left\{\begin{pmatrix} a & b \\ 0 & d \end{pmatrix}\right\}\big/\{\pm I\}$</td></tr>
<tr><td align="center">$Q(j, h(t_1))$</td><td align="center">$\left\{\begin{pmatrix} \pm 1 & 0 \\ c & d \end{pmatrix}\right\}\big/\{\pm I\}$</td></tr>
<tr><td align="center">$Q(j, h(t_2))$</td><td align="center">$\left\{\begin{pmatrix} a & b \\ 0 & \pm 1 \end{pmatrix}\right\}\big/\{\pm I\}$</td></tr>
<tr><td align="center">$Q(j, j_{(2)}, h(t_1))$</td><td align="center">$\left\{\begin{pmatrix} \pm 1 & 0 \\ 0 & d \end{pmatrix}\right\}\big/\{\pm I\}$</td></tr>
<tr><td align="center">$Q(j, j_{(1)}, h(t_2))$</td><td align="center">$\left\{\begin{pmatrix} a & 0 \\ 0 & \pm 1 \end{pmatrix}\right\}\big/\{\pm I\}$</td></tr>
</table>

où les groupes au côté droit désignent les sous-groupes de $G_N(E)$ correspondant aux sous-corps de $K_N(E)$ au côté gauche ; les lettres a, d désignent les éléments inversibles de Z/NZ et b, c désignent les éléments quelconques de Z/NZ. Soit en effet σ un élément de $G_N(E)$ donné par une matrice $\begin{pmatrix} a & b \\ c & d \end{pmatrix}$; on a alors $h(t_1^\sigma)=h(at_1+bt_2)$; d'où résulte $\pm t_1^\sigma=at_1+bt_2$. D'après la prop. 8, pour qu'on ait $j_{(1)}^\sigma=j_{(1)}$, il faut et il suffit qu'on ait $\mathfrak{g}_1^\sigma=\mathfrak{g}_1$, ou qu'on ait $b=0$. De plus, pour qu'on ait $h(t_1)^\sigma=h(t_1)$, il faut et il suffit qu'on ait $a=\pm 1$, $b=0$. Nous obtenons le tableau par ces relations.

6. Soient maintenant n un entier positif tel que $(n, N)=1$ et $\mathfrak{g}_\alpha$ $(1\leq\alpha\leq s)$ les sous-groupes cycliques de E d'ordre n ; posons $j_\alpha=j(E/\mathfrak{g}_\alpha)$. D'après la prop. 1, j est algébrique sur $Q(j_\alpha)$, de sorte que j_α est transcendant sur Q pour chaque α. Il existe donc un isomorphisme τ_α de $Q(j)$ sur $Q(j_\alpha)$ tel que $j^{\tau_\alpha}=j_\alpha$ pour chaque α. Posons $\gamma_\alpha=\gamma^{\tau_\alpha}$, $E_\alpha=E^{\tau_\alpha}$, $h_\alpha=h^{\tau_\alpha}(1\leq\alpha\leq s)$; E_α est alors défini par l'équation

$$(6) \qquad Y^2=4X^3-\gamma_\alpha X-\gamma_\alpha ;$$

h_α est la fonction canonique sur E_α. Comme on a $j(E_\alpha)=j_\alpha=j(E/\mathfrak{g}_\alpha)$, il existe un homomorphisme λ_α de E sur E_α tel que $\mathfrak{g}_\alpha=\mathfrak{g}(\lambda_\alpha)$ pour chaque α.

Proposition 9. *Il existe un isomorphisme σ_α de $K_N(E)$ sur $K_N(E_\alpha)$ tel que*

$$j^{\sigma_\alpha}=j_\alpha, \quad h(t)^{\sigma_\alpha}=h_\alpha(\lambda_\alpha t) \quad pour \ t\in\mathfrak{g}(N, E)$$

pour chaque α ; σ_α est déterminé par ces relations.

Prolongeons l'isomorphisme τ_α à un automorphisme de F que nous désignons aussi par τ_α. L'application $t\to t^{\tau_\alpha^{-1}}$ donne un isomorphisme de $\mathfrak{g}(N, E_\alpha)$ sur $\mathfrak{g}(N, E)$. D'autre part, comme $(n, N)=1$, l'application $t\to\lambda_\alpha t$

donne un isomorphisme de $\mathfrak{g}(N, E)$ sur $\mathfrak{g}(N, E_\alpha)$; par suite $t \to (\lambda_\alpha t)^{\tau_\alpha^{-1}}$ donne un automorphisme de $\mathfrak{g}(N, E)$; il existe donc une matrice $\begin{pmatrix} a & b \\ c & d \end{pmatrix}$ à coefficients entiers telle que

$$\begin{pmatrix} \lambda_\alpha t_1 \\ \lambda_\alpha t_2 \end{pmatrix}^{\tau_\alpha^{-1}} = \begin{pmatrix} a & b \\ c & d \end{pmatrix} \begin{pmatrix} t_1 \\ t_2 \end{pmatrix}.$$

D'après la prop. 5, il existe un automorphisme ρ_α de $K_N(E)$ donné par $\begin{pmatrix} a & b \\ c & d \end{pmatrix}$; on a alors $j^{\rho_\alpha}=j$, $h(t)^{\rho_\alpha}=h((\lambda_\alpha t)^{\tau_\alpha^{-1}})$. Posons $\sigma_\alpha=\rho_\alpha\tau_\alpha$; on a alors $j^{\sigma_\alpha}=j_\alpha$, $h(t)^{\sigma_\alpha}=h_\alpha(\lambda_\alpha t)$. Comme j, $h(t)$ engendrent $K_N(E)$ sur $\boldsymbol{Q}$, σ_α est déterminé par ces relations.

PROPOSITION 10. *Soit $u=(u_1,\cdots,u_m)$ un ensemble fini d'éléments de $K_N(E)$ tel que j soit contenu dans $\boldsymbol{Q}(u)$. Soient $\sigma_1,\cdots,\sigma_s$ les isomorphismes de K_N déterminés dans la prop. 9. Alors $\boldsymbol{Q}(u, j^{\sigma_\alpha})$ contient $\boldsymbol{Q}(u^{\sigma_\alpha})$; $(u^{\sigma_1},\cdots,u^{\sigma_s})$ est l'ensemble complet des conjugués de u^{σ_1} sur $\boldsymbol{Q}(u)$.*

Soit τ un automorphisme de la clôture algébrique F de $\boldsymbol{Q}(j)$ qui fixe u, j_α. Comme j est contenu dans $\boldsymbol{Q}(u)$, τ laisse fixe j; on a donc $E^\tau=E$, $E_\alpha{}^\tau=E_\alpha$. D'après le lemme 3, on a $\lambda_\alpha{}^\tau=\pm\lambda_\alpha$. Il en résulte que

$$h(t)^{\sigma_\alpha\tau}=h_\alpha(\lambda_\alpha t)^\tau=h_\alpha(\pm\lambda_\alpha t^\tau)=h_\alpha(\lambda_\alpha t^\tau)=h(t^\tau)^{\sigma_\alpha}=h(t)^{\tau\sigma_\alpha}$$

pour chaque $t\in\mathfrak{g}(N, E)$. On vérifie aisément $j^{\sigma_\alpha\tau}=j_\alpha=j^{\tau\sigma_\alpha}$. Comme j et $h(t)$ engendrent $K_N(E)$, on a $\sigma_\alpha\tau=\tau\sigma_\alpha$; par suite on a $(u^{\sigma_\alpha})^\tau=(u^\tau)^{\sigma_\alpha}=u^{\sigma_\alpha}$; ce qui montre que $\boldsymbol{Q}(u^{\sigma_\alpha})$ est contenu dans $\boldsymbol{Q}(u, j_\alpha)$. On déduit de là

$$[\boldsymbol{Q}(u, u^{\sigma_\alpha}) : \boldsymbol{Q}(u)]=[\boldsymbol{Q}(u, j_\alpha) : \boldsymbol{Q}(u)]\leqq[\boldsymbol{Q}(j, j_\alpha) : \boldsymbol{Q}(j)].$$

Soit σ un automorphisme de F sur $\boldsymbol{Q}(j)$; $\mathfrak{g}_1{}^\sigma$ est alors un des $\mathfrak{g}_\alpha$; par suite, d'après la prop. 8, les conjugués de j_1 sur $\boldsymbol{Q}(j)$ sont contenus dans $\{j_1,\cdots,j_s\}$; ce qui montre $[\boldsymbol{Q}(j, j_\alpha): \boldsymbol{Q}(j)]\leqq s$. Comme on a $\boldsymbol{Q}(u^{\sigma_\alpha})\supset\boldsymbol{Q}(j^{\sigma_\alpha})$ et $j^{\sigma_\alpha}\neq j^{\sigma_\beta}$ pour $\alpha\neq\beta$, $u^{\sigma_1},\cdots,u^{\sigma_s}$ sont différents l'un de l'autre. Notre proposition sera donc démontrée si nous faisons voir que u^{σ_α} est un conjugué de u^{σ_1} sur $\boldsymbol{Q}(u)$ pour chaque α. Soit $\{t_1', t_2'\}$ un système de générateurs de $\mathfrak{g}(nN, E)$; $\{Nt_1'\ Nt_2'\}$ est alors un système de générateurs de $\mathfrak{g}(n, E)$. D'après la prop. 5, il existe un automorphisme ρ de $K_n{}^*(E)$ sur $\boldsymbol{Q}(j)$ tel que $\mathfrak{g}_1{}^\rho=\mathfrak{g}_\alpha$. Soit $\begin{pmatrix} \alpha & \beta \\ \gamma & \delta \end{pmatrix}$ une matrice de $G_n{}^*$ à laquelle ρ correspond par rapport à $\{Nt_1', Nt_2'\}$. Comme on a $(n, N)=1$, il existe une matrice $\begin{pmatrix} a & b \\ c & d \end{pmatrix}$ de G_{nN}^* telle que

$$\begin{pmatrix} a & b \\ c & d \end{pmatrix}\equiv\begin{pmatrix} \alpha & \beta \\ \gamma & \delta \end{pmatrix} \quad \text{mod. } n,$$

$$\begin{pmatrix} a & b \\ c & d \end{pmatrix}\equiv\begin{pmatrix} 1 & 0 \\ 0 & 1 \end{pmatrix} \quad \text{mod. } N.$$

Désignons par σ l'automorphisme de $K_{nN}(E)$ correspondant à $\begin{pmatrix} a & b \\ c & d \end{pmatrix}$ par rapport à $\{t_1', t_2'\}$. On a alors $z^\sigma = z$ pour $z \in K_N(E)$ et $w^\sigma = w^\rho$ pour $w \in K_n(E)$; on a donc $E^\sigma = E$, $\mathfrak{g}_1{}^\sigma = \mathfrak{g}_\alpha$, $E_1{}^\sigma = E_\alpha$, $j_1{}^\sigma = j_\alpha$, $h_1{}^\sigma = h_\alpha$, $\lambda_1{}^\sigma = \pm \lambda_\alpha$ en vertu du lemme 3 et $t^\sigma = \pm t$ pour $t \in \mathfrak{g}(N, E)$; il s'ensuit de là que, pour $t \in \mathfrak{g}(N, E)$, on a

$$h(t)^{\sigma_1 \sigma} = h_1(\lambda_1 t)^\sigma = h_\alpha(\pm \lambda_\alpha(\pm t)) = h_\alpha(\lambda_\alpha t) = h(t)^{\sigma_\alpha}$$

et $j^{\sigma_1 \sigma} = j^{\sigma_\alpha}$. Comme j et $h(t)$ pour $t \in \mathfrak{g}(N, E)$ engendrent $K_N(E)$, on obtient $\sigma_1 \sigma = \sigma_\alpha$ sur $K_N(E)$; on a donc $(u^{\sigma_1})^\sigma = u^{\sigma_\alpha}$. Comme σ est l'identité sur $K_N(E)$, u^{σ_α} est un conjugué de u^{σ_1} sur $K_N(E)$; ceci achève la démonstration.

Proposition 11. *Les notations étant celles de la prop.* 10, *on a*

$$[Q(u, u^{\sigma_1}) : Q(u)] = [Q(u, u^{\sigma_1}) : Q(u^{\sigma_1})].$$

D'après le th. 27 de [11] n°52, il existe un homomorphisme μ de E_1 sur E tel que $\mu \lambda_1 = n \delta_E$. En appliquant la prop. 9 à E_1, μ, E, on peut montrer qu'il y a un isomorphisme τ de $K_N(E_1)$ sur $K_N(E)$ tel que

$$j_1{}^\tau = j, \quad h_1(t_1)^\tau = h(\mu t_1) \text{ pour } t_1 \in \mathfrak{g}(N, E_1).$$

On a alors $j^{\sigma_1 \tau} = j$, $h(t)^{\sigma_1 \tau} = h(nt)$ pour $t \in \mathfrak{g}(N, E)$; autrement dit, $\sigma_1 \tau$ est l'élément de $G_N(E)$ correspondant à $\begin{pmatrix} n & 0 \\ 0 & n \end{pmatrix}$. $\sigma_1 \tau$ est donc contenu dans le centre du groupe $G_N(E)$; il en résulte qu'on a $Q(u^{\sigma_1 \tau}) = Q(u)$. D'après la prop. 10, on a $[Q(u, u^{\sigma_1}) : Q(u)] = s =$ le nombre des sous-groupes cycliques de E d'ordre n. En appliquant ce résultat à u^{σ_1}, τ, on obtient $[Q(u^{\sigma_1}, u^{\sigma_1 \tau}) : Q(u^{\sigma_1})] = s$; ce qui prouve la proposition.

7. Soit L un sous-corps de $K_N(E)$ tel que $L \supset Q(j)$, $L \cap Q(\zeta_N) = Q$. Comme $Q(\zeta_N)$ est algébriquement fermé dans K_N, Q est algébriquement fermé dans L. Il y a donc une courbe complète Γ définie par rapport à Q, sans point multiple, telle qu'on ait $L = Q(u)$ pour un point générique u de Γ par rapport à Q. Le système $\{\Gamma, u\}$ s'appellera un *modèle de* L. Nous fixons pour le moment L et $\{\Gamma, u\}$. Nous allons définir maintenant certaines correspondances algébriques sur la courbe Γ.

Soient σ_1 l'isomorphisme de K_N déterminé à la prop. 9 et X_n le diviseur premier rationnel par rapport à Q sur $\Gamma \times \Gamma$ ayant $u \times u^{\sigma_1}$ comme point générique sur Q. D'après la prop. 10 et d'après le th. 12 de [9] chap. VIII, on a

$$(7) \qquad X_n \cdot (u \times \Gamma) = u \times \sum_{\alpha=1}^{s} u^{\sigma_\alpha}.$$

X_n s'appellera *la correspondance modulaire de degré n sur* Γ. D'après la prop. 11, on a

$$(8) \qquad\qquad d(X_n)=d'(X_n)=s \;^{3)} \,.$$

Soit ρ_n l'élément de $G_N(E)$ correspondant à la matrice $\begin{pmatrix} n & 0 \\ 0 & n \end{pmatrix}$. Il est clair que cet automorphisme de K_N ne dépend pas du choix d'un système de générateurs pour $\mathfrak{g}(N, E)$. Soit Y_n le lieu de $u \times u^{\rho n}$ sur $\Gamma \times \Gamma$ par rapport à $\boldsymbol{Q}$. Comme $\begin{pmatrix} n & 0 \\ 0 & n \end{pmatrix}$ est contenu dans le centre du groupe G_N^*, ρ_n donne un automorphisme de L; il en résulte que l'on a $\boldsymbol{Q}(u)=\boldsymbol{Q}(u^{\rho n})$, de sorte que Y_n donne une correspondance birationnelle de Γ. On a donc

$$(9) \qquad\qquad Y_n(u)=u^{\rho n} \,.$$

Si $n \equiv n'$ mod. N, on a $Y_n = Y_{n'}$.

8. En outre de X_n, Y_n nous avons besoin d'une autre correspondance pour exprimer la formule de congruence (II); pour la définir nous nous bornons à considérer le cas particulier où le corps L est explicitement donné ainsi qu'il suit.

Fixons un système de générateurs $\{t_1, t_2\}$ de $\mathfrak{g}(N, E)$ et identifions $G_N(E)$ avec G_N au moyen de l'isomorphisme défini par $\{t_1, t_2\}$. Nous désignerons par H_N le sous-groupe $\left\{ \begin{pmatrix} a & 0 \\ 0 & \pm 1 \end{pmatrix} \middle| (a, N)=1 \right\} \Big/ \{\pm I\}$ de G et par L_N le sous-corps de K_N correspondant à H_N. On voit que

$$H_N S_N = G_N \,, \qquad H_N \cap S_N = \{e\};$$

on a donc

$$L_N \cap \boldsymbol{Q}(\zeta_N)=\boldsymbol{Q} \,, \qquad L_N(\zeta_N)=K_N \,.$$

D'après le tableau dans **5,** on a

$$L_N = \boldsymbol{Q}(j, j_{(1)}, h(t_2)) \,.$$

Nous fixons un modèle de L_N et le désignons par $\{\Gamma_N, u\}$. Soit ψ l'automorphisme de K_N correspondant à $\begin{pmatrix} 0 & 1 \\ -1 & 0 \end{pmatrix}$ par rapport à $\{t_1, t_2\}$. Il est clair que $\boldsymbol{Q}(\zeta_N, u)=\boldsymbol{Q}(\zeta_N, u^\psi)=K_N$. Soit A le lieu de $u \times u^\psi$ sur $\Gamma_N \times \Gamma_N$ par rapport à $\boldsymbol{Q}(\zeta_N)$; A donne une correspondance birationnelle de Γ_N; on a donc

$$(10) \qquad\qquad A(u)=u^\psi \,.$$

PROPOSITION 12. *Soit φ_n l'automorphisme de $\boldsymbol{Q}(\zeta_N)$ tel que $\zeta_N^{\varphi n}=\zeta_N^n$. On a alors*

$$A^{\varphi n} \circ Y_n = A \,.$$

Soient τ_n et τ_n' les automorphismes de K_N qui correspondent respective-

---

3) Pour les définitions des notations $d(X)$, $d'(X)$, voir [10] p. 31.

ment à $\begin{pmatrix} 1 & 0 \\ 0 & n \end{pmatrix}$ et $\begin{pmatrix} n & 0 \\ 0 & 1 \end{pmatrix}$; d'après la prop. 7, on a $\tau_n = \tau_n' = \varphi_n$ sur $Q(\zeta_N)$. On vérifie aisément $\psi \tau_n = \tau_n' \psi$ et $\tau_n' \tau_n = \rho_n$. Comme τ_n' est contenu dans H_N, on a

$$(u \times u^\psi)^{\tau_n} = u^{\tau_n} \times u^{\psi \tau_n} = u^{\tau_n' \tau_n} \times u^{\tau_n' \psi} = u^{\rho_n} \times u^\psi;$$

$u^{\rho_n} \times u^\psi$ est donc un point générique de A^{φ_n} sur $Q(\zeta_N)$; d'où résultent les relations

$$A^{\varphi_n}(u^{\rho_n}) = u^\psi, \quad (A^{\varphi_n})'(u^\psi) = u^{\rho_n}.$$

Par suite on a

$$A^{\varphi_n}[Y_n(u)] = u^\psi = A(u), \quad Y_n'[(A^{\varphi_n})'(u^\psi)] = u = A'(u^\psi).$$

A et Y_n sont rationnels par rapport à $Q(\zeta_N)$ et u, u^ψ sont génériques sur Γ par rapport à $Q(\zeta_N)$. Il s'ensuit donc de la définition de $A^{\varphi_n} \circ Y_n$ ([10] II, § 1, n°5) que $A^{\tau_n} \circ Y_n = A$.

Soit $\mathfrak{h}$ un sous-groupe du groupe d'éléments inversibles de Z/NZ; nous supposons que $\mathfrak{h}$ contienne -1. Nous désignerons par $H_{N,\mathfrak{h}}$ le sous-groupe $\left\{ \begin{pmatrix} a & b \\ 0 & d \end{pmatrix} \middle| (a, N) = 1, d \in \mathfrak{h} \right\} \middle/ \{\pm I\}$ de G_N et par $M_{N,\mathfrak{h}}$ le sous-corps de K_N correspondant à $H_{N,\mathfrak{h}}$. Nous désignerons aussi par $H_{N,\mathfrak{h}}'$ le sous-groupe $\left\{ \begin{pmatrix} a & b \\ 0 & d \end{pmatrix} \middle| (d, N) = 1, a \in \mathfrak{h} \right\} \middle/ \{\pm I\}$ de G_N et par $M_{N,\mathfrak{h}}'$ le sous-corps de K_N correspondant à $H_{N,\mathfrak{h}}'$. On voit que

$$H_{N,\mathfrak{h}} S_N = H_{N,\mathfrak{h}}' S_N = G_N, \quad H_{N,\mathfrak{h}} \cap S_N = H_{N,\mathfrak{h}}' \cap S_N;$$

on a donc

(11) $$M_{N,\mathfrak{h}} \cap Q(\zeta_N) = M_{N,\mathfrak{h}}' \cap Q(\zeta_N) = Q, \quad M_{N,\mathfrak{h}}(\zeta_N) = M_{N,\mathfrak{h}}'(\zeta_N).$$

Selon que $\mathfrak{h}$ ne contient que ± 1 ou que $\mathfrak{h}$ contient tous les éléments inversibles de Z/NZ nous désignerons $H_{N,\mathfrak{h}}$, $H_{N,\mathfrak{h}}'$, $M_{N,\mathfrak{h}}$, $M_{N,\mathfrak{h}}'$ respectivement par $H_{N,1}$, $H_{N,1}'$, $M_{N,1}$, $M_{N,1}'$ ou par $H_{N,0}$, $H_{N,0}'$, $M_{N,0}$, $M_{N,0}'$. On a alors

$$H_{N,0} \supset H_{N,\mathfrak{h}} \supset H_{N,1}, \qquad H_{N,0}' \supset H_{N,\mathfrak{h}}' \supset H_{N,1}',$$

$$M_{N,0} \subset M_{N,\mathfrak{h}} \subset M_{N,1}, \qquad M_{N,0}' \subset M_{N,\mathfrak{h}}' \subset M_{N,1}'$$

pour chaque $\mathfrak{h}$; on voit aussi que $H_{N,0} = H_{N,0}'$ et $M_{N,0} = M_{N,0}'$.

Nous allons démontrer qu'il existe un isomorphisme de $M_{N,\mathfrak{h}}$ sur $M_{N,\mathfrak{h}}'$. $\{t_1, t_2\}$ étant le système de générateurs pour $\mathfrak{g}(N, E)$ que nous avons fixé, nous désignons par $\mathfrak{g}$ le groupe engendré par t_2. Posons $j' = j(E/\mathfrak{g})$. D'après la prop. 1, j' est transcendant sur Q et F est la clôture algébrique de $Q(j')$. Soit σ un automorphisme de F tel que $j^\sigma = j'$; posons $E' = E^\sigma$. Il existe alors un homomorphisme λ de E sur E' tel que $\mathfrak{g} = \mathfrak{g}(\lambda)$. D'après le th. 27 de [11] n°52, il existe un homomorphisme μ de E' sur E tel que $\mu\lambda = N\delta_E$, $\lambda\mu = N\delta_{E'}$.

Posons $t_1'=\lambda t_1$. On voit que $\mu t_1'=0$ et que t_1' est d'ordre N. Comme on a $\nu(\mu)=N$, t_1' engendre $\mathfrak{g}(\mu)$. $t_2{}^\sigma$ est un point de E' d'ordre N; par suite, d'après la prop. 5, il existe un automorphisme τ de F sur $Q(j')$ tel que $(t_2{}^\sigma)^\tau=\pm t_1'$; posons $\rho=\sigma\tau$; alors ρ est un automorphisme de F. On a $j^\rho=j'$, $E^\rho=E'$, $t_2{}^\rho=\pm t_1'$, de sorte qu'on a $\mathfrak{g}(\lambda^\rho)=\mathfrak{g}(\lambda)^\rho=\mathfrak{g}(\mu)$; on en conclut que E'^ρ est isomorphe à E; on a donc $j'^\rho=j$. En rappelant que E est la courbe définie par l'équation (4), on voit que $E'^\rho=E^{\sigma\rho}$ est définie par l'équation

$$Y^2=4X^3-\gamma^{\sigma\rho}X-\gamma^{\sigma\rho}.$$

On a $\gamma^{\sigma\rho}=27j^{\sigma\rho}/(j^{\sigma\rho}-1)=27j/(j-1)=\gamma$; ce qui montre que $E'^\rho=E$. Soient α et α' les automorphismes de F qui correspondent respectivement à $\begin{pmatrix} a & b \\ 0 & d \end{pmatrix}$ et $\begin{pmatrix} d & b \\ 0 & a \end{pmatrix}$ par rapport à $\{t_1, t_2\}$; on a alors $t_1{}^\alpha=\pm(at_1+bt_2)$ et $t_2{}^{\alpha'}=\pm at_2$. Nous allons maintenant montrer que $\rho\alpha=\alpha'\rho$ sur $M_{N,1}$. D'après le tableau dans **5**, on a $M_{N,1}=Q(j, h(t_2))$, $M_{N,0}=M_{N,0}'=Q(j, j')$. Comme α et α' fixent les éléments de $M_{N,0}$, on a $E^\alpha=E^{\alpha'}=E$, $E'^\alpha=E'^{\alpha'}=E'$. D'après le lemme 3, on a $\lambda^\alpha=\pm\lambda$, $\lambda^{\alpha'}=\pm\lambda$. On voit que $j^{\rho\alpha}=j'=j^{\alpha'\rho}$, $h^{\rho\alpha}=h^\rho=h^{\alpha'\rho}$ et

$$h(t_2)^{\rho\alpha}=[h^\rho(\pm t_1')]^\alpha=[h^\rho(\lambda t_1)]^\alpha=h^\rho(\lambda(at_1+bt_2))$$

$$=h^\rho(a\lambda t_1)=h^\rho(\pm at_1')=h(at_2)^\rho=h(t_2)^{\alpha'\rho}.$$

On obtient ainsi $j^{\rho\alpha}=j^{\alpha'\rho}$ et $h(t_2)^{\rho\alpha}=h(t_2)^{\alpha'\rho}$; il s'ensuit de là qu'on a $\rho\alpha=\alpha'\rho$ sur $M_{N,1}$. Si $a\in\mathfrak{h}$, α' fixe les éléments de $M_{N,\mathfrak{h}}$; on a alors $\rho\alpha=\rho$ sur $M_{N,\mathfrak{h}}$; autrement dit, si $a\in\mathfrak{h}$, α fixe les éléments de $M_{N,\mathfrak{h}}{}^\rho$; d'où résulte $M_{N,\mathfrak{h}}{}^\rho\subset M_{N,\mathfrak{h}}'$. D'autre part, comme on a bien $M_{N,0}{}^\rho=M_{N,0}=M_{N,0}'$ et comme $[M_{N,\mathfrak{h}}: M_{N,0}]$, $[M_{N,\mathfrak{h}}': M_{N,0}']$ sont égaux à l'indice du groupe $\mathfrak{h}$, on a

$$[M_{N,\mathfrak{h}}{}^\rho: M_{N,0}]=[M_{N,\mathfrak{h}}: M_{N,0}]=[M_{N,\mathfrak{h}}': M_{N,0}];$$

ce qui montre $M_{N,\mathfrak{h}}{}^\rho=M_{N,\mathfrak{h}}'$.

Soient maintenant $\{\Gamma_{N,\mathfrak{h}}, u\}$ un modèle du corps $M_{N,\mathfrak{h}}$ et B le lieu de $u\times u^\rho$ sur $\Gamma_{N,\mathfrak{h}}\times\Gamma_{N,\mathfrak{h}}$ par rapport à $Q(\zeta_N)$, D'après la relation (11) et $M_{N,\mathfrak{h}}{}^\rho=M_{N,\mathfrak{h}}'$, B donne une correspondance birationnelle de $\Gamma_{N,\mathfrak{h}}$. En vertu de la relation $\rho\alpha=\alpha'\rho$ que l'on a vue ci-dessus, on peut démontrer la proposition suivante de la même manière que la prop. 12.

PROPOSITION 13. *Soit φ_n l'automorphisme de $Q(\zeta_N)$ tel que $\zeta_N{}^{\varphi_n}=\zeta_N{}^n$. On a alors*

$$B^{\varphi_n}\circ Y_n=B.$$

§ 3. Formules de congruence.

9. Dans cette section nous démontrerons quelques relations de congruence pour les correspondances définies dans § 2. Pour cela, nous profiterons de la notion de réduction des variétés algébriques modulo p, et nous servirons des termes et des résultats dans [5–8].

Soient k un corps, $\mathfrak{p}$ un diviseur premier de k et $\tilde{k}$ le corps des restes modulo $\mathfrak{p}$. Si $\mathfrak{X}$ représente un objet algébro-géométrique défini par rapport à k, nous indiquerons par $\mathfrak{X}(\mathfrak{p})$ ou $\tilde{\mathfrak{X}}$ la réduction de $\mathfrak{X}$ modulo $\mathfrak{p}$.

Soit Γ une courbe complète sans point multiple, définie par rapport à k. Nous dirons que Γ *n'a pas de défaut pour* $\mathfrak{p}$, si Γ est $\mathfrak{p}$-simple et $\mathfrak{p}$-complet et la réduction $\tilde{\Gamma}$ n'a pas de point multiple; s'il en est ainsi, Γ et $\tilde{\Gamma}$ ont le même genre et Γ n'a pas de défaut comme variété abélienne lorsque Γ est une courbe elliptique.

Maintenant supposons que $\mathfrak{p}$ ne divise ni 2, ni 3. Soit j un élément de k. Pour qu'il existe une courbe elliptique E', définie par rapport à une extension k' de k, sans défaut pour chaque prolongement de $\mathfrak{p}$ dans k', telle que $j(E')=j$, il faut et il suffit que j soit $\mathfrak{p}$-entier; s'il en est ainsi, on a $j(E'(\mathfrak{p})) =j(E')(\mathfrak{p}')$ où $\mathfrak{p}'$ désigne un prolongement de $\mathfrak{p}$ dans k' (cf. [2] § 4).

Soient Γ, Γ' deux courbes complètes sans point multiple définies par rapport à k et X une correspondance entre Γ et Γ' rationnelle par rapport à k. Supposons que Γ, Γ' n'aient pas de défaut pour $\mathfrak{p}$. On obtient alors une correspondance $\tilde{X}=X(\mathfrak{p})$ entre $\tilde{\Gamma}$ et $\tilde{\Gamma}'$; on vérifie aisément $X'(\mathfrak{p})=X(\mathfrak{p})'$. Soient w et $\tilde{w}$ respectivement un point générique de Γ par rapport à k et un point générique de $\tilde{\Gamma}$ par rapport à $\tilde{k}$. Les produits $X\cdot(w\times\Gamma')$ et $\tilde{X}\cdot(\tilde{w}\times\tilde{\Gamma}')$ sont alors définis. La spécialisation $w\to\tilde{w}[\mathfrak{p}]$ définit un prolongement $\mathfrak{p}_1$ dans $k(w)$; d'après le th. 17 de [5], $\tilde{X}\cdot(\tilde{w}\times\tilde{\Gamma}')$ est la réduction de $X\cdot(w\times\Gamma')$ modulo $\mathfrak{p}_1$. Par suite on a $d(X)=d(\tilde{X})$ et semblablement $d'(X)=d'(\tilde{X})$. Soient J, J' respectivement une jacobienne de Γ et une jacobienne de Γ'; soient φ une application canonique de Γ dans J et φ' une application canonique de Γ' dans J'. Nous supposons que J, J', φ, φ' soient définis par rapport à k et que J, J' n'aient pas de défaut pour $\mathfrak{p}$. Alors, on peut démontrer que J est une jacobienne de $\tilde{\Gamma}$, que $\tilde{\varphi}$ est une application canonique de $\tilde{\Gamma}$ dans J et de même pour $\tilde{\Gamma}', J', \tilde{\varphi}'$. Soit ξ l'homomorphisme de J dans J' correspondant à X. Alors on vérifie aisément que $\tilde{\xi}$ est l'homomorphisme de J dans J' correspondant à $\tilde{X}$.

10. Soit maintenant E la courbe elliptique avec l'invariant transcendant j sur $\mathbf{Q}$, que nous avons considérée dans § 2. Soient N un entier positif et p un nombre premier qui ne divise pas $6N$. En regardant ce nombre premier p comme le nombre n dans **6,** nous obtenons les sous-groupes $\mathfrak{g}_\alpha$ de E d'ordre p, les courbes elliptiques E_α et les homomorphismes λ_α. Le nombre s des $\mathfrak{g}_\alpha$ est ici égal à $p+1$. Pour chaque α, soit μ_α un homomorphisme de E_α sur E tel que $\mu_\alpha\lambda_\alpha=p\delta_E$ et $\lambda_\alpha\mu_\alpha=p\delta_{E_\alpha}$. Nous nous servirons des mêmes notations que dans **6–8.**

Soit k_1 une extension de degré fini de $K_{pN}{}^*(E)$ telle que tous les λ_ν, μ_ν soient définis par rapport à k_1. Soit $\mathfrak{p}_1$ un diviseur premier de k_1 prolongeant le diviseur premier p de Q tel que $j(\mathfrak{p}_1)$ soit transcendant sur le corps premier. D'après un résultat classique[4], tous les j_α sont $\mathfrak{p}_1$-entiers. Il existe donc pour chaque α une courbe elliptique E_α' isomorphe à E_α qui n'a pas de défaut pour tout prolongement de $\mathfrak{p}_1$. Soient λ_α' un homomorphisme de E sur E_α' tel que $\mathfrak{g}(\lambda_\alpha')=\mathfrak{g}_\alpha$ et μ_α' un homomorphisme de E_α' sur E tel que $\mu_\alpha'\lambda_\alpha'=p\delta_E$, $\lambda_\alpha'\mu_\alpha'=p\delta_{E_\alpha'}$. Soient k une extension finie de k_1 par rapport à laquelle tous les E_α' et tous les λ_α', μ_α' sont définis et $\mathfrak{p}$ un prolongement de $\mathfrak{p}_1$ dans k. En réduisant modulo $\mathfrak{p}$, on a $\tilde{\mu}_\alpha'\tilde{\lambda}_\alpha'=p\delta_{\tilde{E}}$. D'après la prop. 3, on a $\nu_s(\tilde{\mu}_\alpha')\nu_s(\tilde{\lambda}_\alpha')=p$; il en résulte que l'on a $\nu_s(\tilde{\mu}_\alpha')=1$ ou $\nu_s(\tilde{\lambda}_\alpha')=1$. D'après la prop. 2, on a $\tilde{j}_\alpha=j^p$ ou $\tilde{j}_\alpha{}^p=j$; par suite tous les $\tilde{j}_\alpha$ sont transcendants sur le corps premier. Comme E_α est défini par l'équation (6) et $r_\alpha=27j_\alpha/(j_\alpha-1)$, E_α n'a pas de défaut pour $\mathfrak{p}$; $\tilde{E}_\alpha$ est alors défini par l'équation

$$(12) \qquad Y^2=4X^3-\tilde{r}_\alpha X-\tilde{r}_\alpha .$$

La réduction modulo $\mathfrak{p}$ donne un homomorphisme de $\mathfrak{g}(p, E)$ *sur* $\mathfrak{g}(p, \tilde{E})$. Comme le groupe $\mathfrak{g}(p, E)$ est d'ordre p^2 et comme, d'après la prop. 3, le groupe $\mathfrak{g}(p, \tilde{E})$ est d'ordre p, le noyau de cet homomorphisme est d'ordre p; donc il coïncide avec un des $\mathfrak{g}_\alpha$, mettons $\mathfrak{g}_1$. On a alors $\mathfrak{g}(\tilde{\lambda}_1)=\tilde{\mathfrak{g}}_1=\{0\}$. Si $\alpha>1$, on a $\mathfrak{g}(p, E)=\mathfrak{g}_1+\mathfrak{g}_\alpha$, de sorte qu'on a $\mathfrak{g}(\tilde{\lambda}_\alpha)=\tilde{\mathfrak{g}}_\alpha=\mathfrak{g}(p, \tilde{E})$. On a donc $\nu_s(\tilde{\lambda}_1)=1$ et $\nu_s(\tilde{\lambda}_\alpha)=p$ pour $\alpha>1$. Comme on a $\tilde{\mu}_\alpha\tilde{\lambda}_\alpha=p\delta_{\tilde{E}}$, d'après la prop. 3 on a $\nu_s(\tilde{\mu}_1)=p$ et $\nu_s(\tilde{\mu}_\alpha)=1$ pour $\alpha>1$. Il en résulte d'après la prop. 2, qu'il y a un isomorphisme ε de $\tilde{E}^p$ sur $\tilde{E}_1$ et un isomorphisme ε_α de $\tilde{E}_\alpha{}^p$ sur $\tilde{E}$ pour chaque $\alpha>1$ tels que

$$(13) \qquad \begin{aligned} \tilde{\lambda}_1\tilde{x}&=\varepsilon\tilde{x}^p &&\text{pour } \tilde{x}\in\tilde{E}, \\ \tilde{\mu}_\alpha\tilde{x}_\alpha&=\varepsilon_\alpha\tilde{x}_\alpha{}^p &&\text{pour } \tilde{x}_\alpha\in\tilde{E}_\alpha \qquad (\alpha>1). \end{aligned}$$

On a donc

$$(14) \qquad \begin{aligned} \tilde{j}_1&=j^p, & \tilde{r}_1&=\tilde{r}^p, \\ j&=\tilde{j}_\alpha{}^p, & \tilde{r}&=\tilde{r}_\alpha{}^p \qquad (\alpha>1). \end{aligned}$$

On en déduit que $\tilde{E}_1=\tilde{E}^p$, $\tilde{E}=\tilde{E}_\alpha{}^p$ $(\alpha>1)$ comme les équations de E et de E_α ont les formes (4) et (6); les isomorphismes ε, ε_α sont donc éguax à ± 1. Soient h et h_α respectivement les fonctions canoniques sur E et sur E_α comme dans **6~8**; on a alors $\tilde{h}_1=\tilde{h}^p$ et $\tilde{h}=\tilde{h}_\alpha{}^p$ pour $\alpha>1$. Il s'ensuit donc de la relation (13) que

$$(15) \qquad \begin{aligned} \tilde{h}_1(\tilde{\lambda}_1\tilde{t})&=\tilde{h}^p(\pm\tilde{t}^p)=\tilde{h}(\tilde{t})^p &&\text{pour } \tilde{t}\in\mathfrak{g}(N, \tilde{E}), \\ \tilde{h}(\tilde{\mu}_\alpha\tilde{t}_\alpha)&=\tilde{h}_\alpha{}^p(\pm\tilde{t}_\alpha{}^p)=\tilde{h}_\alpha(\tilde{t}_\alpha)^p &&\text{pour } \tilde{t}_\alpha\in\mathfrak{g}(N, \tilde{E}_\alpha), \qquad (\alpha>1). \end{aligned}$$

---

4)   Une démonstration moderne de ce résultat est donnée dans [2] §6.

En substituant $\tilde{\lambda}_\alpha \tilde{i}$ à $\tilde{i}_\alpha$ on obtient

$$(15') \qquad \tilde{h}_\alpha(\tilde{\lambda}_\alpha \tilde{i})^p = \tilde{h}(p\tilde{i}) \qquad\qquad \text{pour} \quad \tilde{i} \in \mathfrak{g}(N, \tilde{E}).$$

Soient σ_α pour $1 \leq \alpha \leq p+1$ et ρ_n les isomorphismes du corps $K_N(E)$ définis dans **6**. Les relations (14), (15), (15') montrent que K_N a un système de générateurs $(x) = (x_1, \cdots, x_c)$ sur $\boldsymbol{Q}$ tel que tous les $x_i, x_i^{\sigma_\alpha}, x_i^{\rho_p}$ soient $\mathfrak{p}$-entiers et que l'on ait

$$x_i^{\sigma_1} \equiv x_i^p \qquad \text{mod. } \mathfrak{p},$$

(16)

$$(x_i^{\sigma_\alpha})^p \equiv x_i^{\rho_p} \ (1 < \alpha \leq p+1) \qquad \text{mod. } \mathfrak{p}.$$

11. Nous allons maintenant nous occuper des correspondances définies dans **7**. Soient L un sous-corps de K_N tel que $L \supset \boldsymbol{Q}(j)$, $L \cap \boldsymbol{Q}(\zeta_N) = \boldsymbol{Q}$ et $\{\Gamma, u\}$ un modèle de L. Soient X_p, Y_p les correspondances sur Γ définies dans **7**. Γ n'a pas de défaut pour presque tous les diviseurs premiers de $\boldsymbol{Q}$. Pour chacun de ces diviseurs premiers de $\boldsymbol{Q}$, nous prenons un de ses prolongements dans F et le fixons; nous désignons le diviseur premier de F prolongeant (p) aussi par $\mathfrak{p}$. On vérifie facilement que le point u reste générique par rapport au corps premier modulo presque tous les $\mathfrak{p}$.

Soient $u_1, \cdots, u_z$ les coordonnées du point u; chaque u_i s'exprime sous la forme $u_i = f_i(x)/g_i(x)$ où f_i, g_i sont deux polynomes à coefficients entiers. Tous les $g_i(x)$, $g_i(x)^{\rho_n}$ ($1 \leq i \leq z$, $1 \leq n \leq N$, $(n, N) = 1$) sont des $\mathfrak{p}$-unités pour presque tous les diviseurs premiers $\mathfrak{p}$. La formule (16) montre que les $g_i(x)^{\sigma_\alpha}$ sont des $\mathfrak{p}$-unités pour ces $\mathfrak{p}$. Il s'ensuit donc de (16) que l'on a pour presque tous les $\mathfrak{p}$,

$$(17) \qquad \tilde{u}^{\sigma_1} = \tilde{u}^p, \quad (\tilde{u}^{\sigma_\alpha})^p = \tilde{u}^{\rho_p} \qquad\qquad (\alpha > 1).$$

Si $\tilde{u}$ est générique sur $\tilde{\Gamma}$ par rapport au corps premier, les produits d'intersection $\tilde{X}_p \cdot (\tilde{u} \times \tilde{\Gamma})$ et $\tilde{Y}_p \cdot (\tilde{u} \times \tilde{\Gamma})$ sont définis. Il résulte alors de (7), (9), (17) et du th. 17 de [5] que l'on a

$$\tilde{X}_p(\tilde{u}) = \tilde{u}^p + p\,\tilde{Y}_p(\tilde{u})^{1/p}.$$

$\tilde{Y}_n$ donne, pour chaque n, une correspondance birationnelle de $\tilde{\Gamma}$ pour presque tous les $\mathfrak{p}$. Soit Π le lieu de $\tilde{u} \times \tilde{u}^p$ sur $\tilde{\Gamma} \times \tilde{\Gamma}$ par rapport au corps premier. Alors on a

$$\tilde{X}_p(\tilde{u}) = (\Pi + \Pi' \circ \tilde{Y}_p)(\tilde{u});$$

on en déduit que $\tilde{X}_p - (\Pi + \Pi' \circ \tilde{Y}_p)$ est un cycle $\mathfrak{a} \times \tilde{\Gamma}$ où $\mathfrak{a}$ est un diviseur sur $\tilde{\Gamma}$ en vertu du th. 2 de [10]. Comme $\Pi + \Pi' \circ \tilde{Y}_p$ n'a pas de composant de la forme $\mathfrak{a} \times \tilde{\Gamma}$, on voit que $\mathfrak{a} > 0$. D'autre part on a $d(\tilde{X}_p) = d'(\tilde{X}_p) = p+1$ en vertu de (8) et $d'(\Pi + \Pi' \circ \tilde{Y}_p) = p+1$; d'où résulte $\mathfrak{a} = 0$. Nous obtenons ainsi *la première formule de congruence pour la correspondance modulaire*

(I)
$$\tilde{X}_p = \Pi + \Pi' \circ \tilde{Y}_p$$

pour presque tous les nombres premiers p.

12. Nous prenons maintenant pour L le corps L_N ou le corps $M_{N,\mathfrak{z}}$ définis dans **8**. Soient les notations A, B, φ_n comme dans **8**. D'après la prop. 7, φ_p donne un automorphisme de Frobenius $\left(\dfrac{Q(\zeta_N)/Q}{p} \right)$. Il s'ensuit de là que l'on a $\tilde{A}^{\varphi p} = \tilde{A}^p$, $\tilde{B}^{\varphi p} = \tilde{B}^p$. D'après la prop. 12 et la prop. 13, on a

(18)
$$\tilde{A}^p \circ \tilde{Y}_p = \tilde{A} \quad \text{sur} \quad \tilde{\Gamma}_N,$$

$$\tilde{B}^p \circ \tilde{Y}_p = \tilde{B} \quad \text{sur} \quad \tilde{\Gamma}_{N,\mathfrak{z}}.$$

Comme A est une correspondance birationnelle, on a $A' \circ A = \varDelta$ et par conséquent

(19)
$$\tilde{A}' \circ \tilde{A} = \tilde{\varDelta}.$$

D'autre part, on vérifie aisément

(20)
$$\Pi \circ \mathfrak{X} = \mathfrak{X}^p \circ \Pi \quad \text{et} \quad \mathfrak{X} \circ \Pi' = \Pi' \circ \mathfrak{X}^p$$

pour toute correspondance $\mathfrak{X}$. Il s'ensuit alors des formules (18), (19), (20) que l'on a

$$\tilde{A}' \circ \Pi' \circ \tilde{A} = \tilde{A}' \circ \Pi' \circ \tilde{A}^p \circ \tilde{Y}_p = \tilde{A}' \circ \tilde{A} \circ \Pi' \circ \tilde{Y}_p = \Pi' \circ \tilde{Y}_p$$

sur $\tilde{\Gamma}_N$. Il en est de même pour $\tilde{B}, \tilde{\Gamma}_{N,\mathfrak{z}}$. On obtient ainsi *la deuxième formule de congruence*:

(II)
$$\Pi' \circ \tilde{Y}_p = \tilde{A}' \circ \Pi' \circ \tilde{A} \quad \text{sur} \quad \tilde{\Gamma}_N,$$

$$\Pi' \circ \tilde{Y}_p = \tilde{B}' \circ \Pi' \circ \tilde{B} \quad \text{sur} \quad \tilde{\Gamma}_{N,\mathfrak{z}}$$

pour presque tous les nombres premiers p.

13. Nous allons récrire les formules (I), (II) en termes d'endomorphismes de la jacobienne.

Soient L, Γ comme dans **11**, J une jacobienne de Γ et φ une application canonique de Γ dans J. Comme Γ est défini par rapport à Q, on peut supposer d'après [1] ou d'après [12] que J soit défini par rapport à Q; de plus si Γ a un point rationnel par rapport à Q, on peut aussi supposer que φ soit défini par rapport à Q. Cela posé, nous désignerons par ξ_n et η_n les endomorphismes de J correspondant à X_n et à Y_n. Nous désignerons la jacobienne J de Γ par J_N ou $J_{N,\mathfrak{z}}$ selon que L est égal à L_N ou à $M_{N,\mathfrak{z}}$; nous désignerons aussi par α et β les endomorphismes de J_N et de $J_{N,\mathfrak{z}}$ correspondant à A et à B, respectivement. Nous démontrerons, dans **17**, que $\Gamma_N, \Gamma_{N,\mathfrak{z}}$ ont des points rationnels par rapport à Q.

Pour presque tous les nombres premiers p, J n'a pas de défaut, $\tilde{J}$ est une jacobienne de $\tilde{\Gamma}$ et $\tilde{\varphi}$ est une application canonique dans $\tilde{\Gamma}$. D'après (I), (II), nous obtenons donc les relations suivantes pour presque tous les nombres premiers p:

$$(\mathrm{I}') \qquad\qquad \tilde{\xi}_p = \pi + \pi'\tilde{\eta}_p \qquad \text{sur } J,$$

$$(\mathrm{II}') \qquad\qquad \pi'\tilde{\eta}_p = \tilde{\alpha}^{-1}\pi'\tilde{\alpha} \qquad \text{sur } J_N,$$

$$\pi'\tilde{\eta}_p = \tilde{\beta}^{-1}\pi'\tilde{\beta} \qquad \text{sur } J_{N,\mathfrak{h}}$$

où π désigne l'homomorphisme de p-ième puissance de J.

§ 4.   Fonctions modulaires elliptiques.

14.   Soient ω_1, ω_2 deux nombres complexes tels que $\Im(\omega_1/\omega_2) > 0$ et D le réseau dans le plan complexe $\boldsymbol{C}$ engendré par $\{\omega_1, \omega_2\}$.   Posons

$$g_2(D) = g_2(\omega_1, \omega_2) = 60\sum{}' \omega^{-4},$$

$$g_3(D) = g_3(\omega_1, \omega_2) = 140\sum{}' \omega^{-6},$$

$$\wp(z; D) = \wp(z; \omega_1, \omega_2) = z^{-2} + \sum{}' [(z-\omega)^{-2} - \omega^{-2}]$$

où $\sum'$ désigne la sommation étendue à tous les nombres $\omega \neq 0$ de D.   La fonction $\wp$ et sa dérivée $\wp'$ satisfont à l'équation

$$\wp'^2 = 4\wp^3 - g_2\wp - g_3.$$

$g_2(\omega_1, \omega_2)$ et $g_3(\omega_1, \omega_2)$ sont les formes modulaires d'espèce («Stufe») 1 respectivement de degré -4 et -6.   Elles ont les expressions

$$(21) \qquad g_2(\omega_1, \omega_2) = \left(\frac{\pi}{\omega_2}\right)^4 \left(\frac{1}{12} + 20\sum_{n=1}^{\infty} \frac{n^3 q^n}{1-q^n}\right),$$

$$g_3(\omega_1, \omega_2) = \left(\frac{\pi}{\omega_2}\right)^6 \left(\frac{1}{216} - \frac{7}{3}\sum_{n=1}^{\infty} \frac{n^5 q^n}{1-q^n}\right).$$

où $q = e^{2\pi i\tau}$, $\tau = \omega_1/\omega_2$.   Nous nous servirons des notations q, τ toujours en ce sens et posons aussi $q_N = e^{2\pi i\tau/N}$, $\zeta_N = e^{2\pi i/N}$.

La fonction $j(\tau) = g_2^3/(g_2^3 - 27g_3^2)$ est une fonction modulaire d'espèce 1; elle a l'expression

$$(22) \qquad\qquad j(\tau) = 12^{-3}(q^{-1} + 744 + \cdots)$$

dont les coefficients sont rationnels.   Soient N un entier > 1 et α, β deux entiers tels que $(\alpha, \beta) \not\equiv (0, 0)$ mod. N.   La fonction

$$\wp\left(\frac{\alpha\omega_1 + \beta\omega_2}{N}; \omega_1, \omega_2\right)$$

est une forme modulaire d'espèce N et de degré -2; elle a l'expression

$$(23) \qquad \wp\left(\frac{\alpha\omega_1+\beta\omega_2}{N};\ \omega_1,\ \omega_2\right)=\left(\frac{2\pi}{\omega_2}\right)^2\left\{-\frac{1}{12}+2\sum_{n=1}^{\infty}\frac{nq^n}{1-q^n}-\frac{\zeta_N{}^{\beta}q_N{}^{\alpha}}{(1-\zeta_N{}^{\beta}q_N{}^{\alpha})^2}\right.$$

$$\left.-\sum_{n=1}^{\infty}\frac{nq^n}{1-q^n}(\zeta_N{}^{n\beta}q_N{}^{n\alpha}+\zeta_N{}^{-n\beta}q_N{}^{-n\alpha})\right\},$$

$$(0\leq\alpha<N,\ (\alpha,\beta)\not\equiv(0,0)\ \mathrm{mod.}\ N).$$

Posons maintenant

$$f_{\alpha\beta}(\tau)=f(\alpha,\beta\ ;\ \tau)=\frac{g_2(\omega_1,\omega_2)}{g_3(\omega_1,\omega_2)}\ \wp\left(\frac{\alpha\omega_1+\beta\omega_2}{N}\ ;\ \omega_1,\ \omega_2\right).$$

Les fonctions $f_{\alpha\beta}(\tau)$ sont des fonctions modulaires d'espèce N. Pour qu'on ait $f_{\alpha\beta}=f_{\alpha'\beta'}$, il faut et il suffit que $(\alpha,\beta)\equiv\pm(\alpha',\beta')$ mod. N. Soient a,b,c,d quatre entiers rationnels tels que $ad-bc=1$; on vérifie alors facilement que

$$(24) \qquad f\left(\alpha,\beta\ ;\ \frac{a\tau+b}{c\tau+d}\right)=f(\alpha',\beta'\ ;\ \tau)\ \text{pour}\ (\alpha'\ \beta')=(\alpha\ \beta)\begin{pmatrix}a&b\\c&d\end{pmatrix}.$$

D'après (21), (23), on voit que tous les $f_{\alpha\beta}(\tau)$ ont des développements en q_N à coefficients dans $Q(\zeta_N)$ et qu'en particulier, $f_{10}(\tau)$ a un développement en q_N à coefficients rationnels; si $N\geq3$, $f_{1\beta}(\tau)$ a un développement en q_N

$$(25) \qquad f_{1\beta}(\tau)=r_{\beta}+s_{\beta}\zeta_N{}^{\beta}q_N+(\text{termes d'ordres plus hauts})$$

où r_{β},s_{β} sont deux nombres rationnels et $s_{\beta}\neq0$.

15. Soit maintenant τ_0 un nombre complexe tel que $j(\tau_0)$ soit transcendant sur Q. Posons $j_0=j(\tau_0)$ et $r_0=27j_0/(j_0-1)$; soient E la courbe elliptique définie par l'equation $Y^2=4X^3-r_0X-r_0$ et h la fonction canonique sur E. Soient $\omega_{01},\ \omega_{02}$ deux nombres complexes tels que $\omega_{01}/\omega_{02}=\tau_0$ et D le réseau engendré par $\{\omega_{01},\omega_{02}\}$. Il existe alors un isomorphisme analytique $u(z)$ de C/D sur E. Nous allons démontrer que l'on a

$$(26) \qquad h(u(z))=\frac{g_2(\omega_{01},\omega_{02})}{g_3(\omega_{01},\omega_{02})}\ \wp(z\ ;\ \omega_{01},\omega_{02})$$

pour tout $z\in C$. Posons $\mu=[g_3(D)/g_2(D)]^{1/2}$; on a alors $g_2(\mu D)=g_3(\mu D)$. Puisqu'on a $j_0=j(\tau_0)=g_2(\mu D)/[g_2(\mu D)-27g_3(\mu D)]$, on voit que $r_0=g_2(\mu D)=g_3(\mu D)$. Par suite, pour tout nombre complexe z, $u'(z)=(\wp(\mu z\ ;\ \mu D),\ \wp'(\mu z\ ;\ \mu D))$ est un point sur E; de plus $z\to u'(z)$ donne un isomorphisme analytique de C/D sur E. On a $h(u'(z))=\wp(\mu z\ ;\ \mu D)=\mu^{-2}\wp(z\ ;\ D)=\frac{g_2(D)}{g_3(D)}\ \wp(z\ ;\ D)$. Comme $\mathcal{A}(E)$ est isomorphe à Z, on a $u(z)=\pm u'(z)$; ce qui montre (26). Posons $t_1=u(\omega_{01}/N),\ t_2=u(\omega_{02}/N)$; $\{t_1,t_2\}$ est alors un système de générateurs pour $\mathfrak{g}(N,E)$. D'après (26), on vérifie aisément

$$(27) \qquad \begin{aligned}&j(E)=j(\tau_0)\,,\\&h(\alpha t_1+\beta t_2)=f_{\alpha\beta}(\tau_0)\,;\end{aligned}$$

on a donc

$$K_N(E) = Q(j(\tau_0),\ f_{\alpha\beta}(\tau_0)\,|\,0 \leq \alpha < N,\ 0 \leq \beta < N,\ (\alpha,\beta) \not\equiv (0,0) \bmod. N).$$

Soit $\mathfrak{g}_{(i)}$ le sous-groupe de E engendré par t_i pour $i=1,2$. Soient $D_{(1)}$, $D_{(2)}$ les réseaux respectivement engendrés par $\{\omega_{01}/N,\ \omega_{02}\}$ et $\{\omega_{01},\ \omega_{02}/N\}$. Alors l'application identique $z \to z$ sur le plan complexe C donne un homomorphisme de C/D sur $C/D_{(i)}$ dont le noyau est $D/D_{(i)}$ pour $i=1,2$. On peut en déduire que

$$(28) \qquad\qquad j(E/\mathfrak{g}_{(1)}) = j(\tau_0/N),\quad j(E/\mathfrak{g}_{(2)}) = j(N\tau_0).$$

16. Nous désignerons par $\mathfrak{M}_N$ le corps complet des fonctions modulaires elliptiques d'espèce N. $\mathfrak{M}_N$ est un corps de fonctions algébriques d'une variable sur C. $\mathfrak{M}_N$ est galoisien sur $\mathfrak{M}_1 = C(j(\tau))$; le groupe de Galois de $\mathfrak{M}_N/\mathfrak{M}_1$ est isomorphe au groupe S_N de telle façon que pour chaque $\begin{pmatrix} a & b \\ c & d \end{pmatrix} \in S_N{}^*$, l'application

$$f(\tau) \to f\left(\frac{a\tau+b}{c\tau+d}\right)$$

donne un automorphisme de $\mathfrak{M}_N/\mathfrak{M}_1$.

Nous désignerons par $\mathfrak{R}_N$ le corps $Q(j(\tau),\ f_{\alpha\beta}(\tau))$ engendré sur Q par $j(\tau)$ et tous les $f_{\alpha\beta}(\tau)$ $(0 \leq \alpha \leq N-1, 0 \leq \beta \leq N-1, (\alpha,\beta) \not\equiv (0,0) \bmod. N)$. D'après la relation (24), a, b, c, d, ayant les mêmes significations que dans cette relation-là, pour qu'on ait $f\left(\dfrac{a\tau+b}{c\tau+d}\right) = f(\tau)$ pour tous les $f \in \mathfrak{R}_N$, il faut et il suffit qu'on ait $\begin{pmatrix} a & b \\ c & d \end{pmatrix} \equiv \pm I \bmod. N$. Cela démontre que $\mathfrak{R}_N$ engendre $\mathfrak{M}_N$ sur C, c'est-à-dire

$$\mathfrak{M}_N = C(j(\tau),\ f_{\alpha\beta}(\tau)).$$

E étant la courbe elliptique avec l'invariant j transcendant sur Q, définie dans **4**, nous allons démontrer que $K_N(E)$ est isomorphe au corps $\mathfrak{R}_N$. Soit k le corps de constantes du corps $\mathfrak{R}_N$; $\mathfrak{R}_N$ est alors un corps de fonctions algébrique d'une variable sur k. Comme $\mathfrak{R}_N$ est engendré par l'adjonction à Q de quantités en nombre fini, il en est de même pour k; on peut donc prendre un nombre complexe τ_0 tel que $j(\tau_0)$ soit transcendant sur k. On voit que $(j(\tau_0),\ f_{\alpha\beta}(\tau_0))$ est une spécialisation de $(j(\tau),\ f_{\alpha\beta}(\tau))$ par rapport à C, et donc naturellement par rapport à k. D'après la définition de k, $(j(\tau), f_{\alpha\beta}(\tau))$ est de dimension 1 sur k. Comme $j(\tau_0)$ est transcendant sur k, $(j(\tau_0), f_{\alpha\beta}(\tau_0))$ est une spécialisation générique de $(j(\tau),\ f_{\alpha\beta}(\tau))$ par rapport à k et donc par rapport à Q, de sorte que $\mathfrak{R}_N = Q(j(\tau),\ f_{\alpha\beta}(\tau))$ est isomorphe à $Q(j(\tau_0),\ f_{\alpha\beta}(\tau_0))$. Nous avons vu dans **15**, que l'on a $Q(j(\tau_0),\ f_{\alpha\beta}(\tau_0)) = K_N(E_0)$ pour une courbe elliptique E_0 définie par l'équation

$$Y^2 = 4X^3 - r_0 X - r_0$$

où $\gamma_0 = 27j(\tau_0)/[j(\tau_0)-1]$. $\Re_N$ est donc isomorphe à $K_N(E_0)$; de plus on voit que $(j(\tau), f_{\alpha\beta}(\tau))$ est de dimension 1 sur $\boldsymbol{Q}$. L'argument ci-dessus est valable pour E_0 si seulement $j(\tau_0)$ est transcendant sur $\boldsymbol{Q}$. Par suite on peut exprimer les propositions sur $\Re_N$ en termes de courbe elliptique, au moyen de l'isomorphisme donné ainsi par une spécialisation $\tau \rightarrow \tau_0$ où τ_0 est un nombre complexe tel que $j(\tau_0)$ soit transcendant sur $\boldsymbol{Q}$. Nous appellerons cette méthode *le principe de spécialisation*.

17. Nous allons maintenant démontrer les propositions 5—7. Pour cela nous démontrons d'abord le lemme suivant:

LEMME 4. *Soient k_0 un corps, k_1 une extension séparable de k_0, x une variable sur k_1 et M une extension de $k_0(x)$, de degré fini, telle que $M \subset k_1(x)$. Alors il existe une extension k de k_0, de degré fini, telle que $M = k(x)$.*

Soit $\overline{k}_0$ la clôture algébrique de k_0. Comme $[M: k_0(x)]$ est fini, il existe un ensemble fini $(y) = (y_1, \cdots, y_s)$ d'éléments de k_1 tel que $M \subset k_0(x, y)$. Comme x est une variable sur $\overline{k}_0(y)$ et comme $\overline{k}_0(y)$ est une extension régulière de $\overline{k}_0$, $\overline{k}_0(x, y)$ est une extension régulière de $\overline{k}_0(x)$; par suite $\overline{k}_0(x)$ est algébriquement fermé dans $\overline{k}_0(x, y)$; il en résulte que $\overline{k}_0(x) = \overline{k}_0 M$; on a donc $M \subset \overline{k}_0(x)$. Il existe donc une extension k' de k_0, de degré fini, telle que $M \subset k'(x)$. Posons $k = M \cap k'$; on a alors $k(x) \subset M$ et $\overline{k} \cap M = \overline{k} \cap k'(x) \cap M = k' \cap M = k$. D'autre part, puisque k_1 est séparable sur k_0, on voit que M est une extension séparable de k; on en conclut que M est une extension régulière de k. On a donc

$$[k'(x): M] = [k'M: M] = [k': k] = [k'(x): k(x)];$$

ce qui montre $M = k(x)$.

$K_N(E)$ est identifié avec $\Re_N$ par l'isomorphisme donné dans **16**; nous fixerons cette identification. On a alors

$$j_{(1)} = j(\tau/N), \qquad j_{(2)} = j(N\tau),$$
$$h(t_1) = f_{10}(\tau), \qquad h(t_2) = f_{01}(\tau)$$

en vertu de (27), (28). La formule (5) prend maintenant la forme

$$f_{\alpha\beta}^{\sigma} = f_{\alpha'\beta'}, \qquad (\alpha'\ \beta') = (\alpha\ \beta)\begin{pmatrix} a & b \\ c & d \end{pmatrix};$$

cette formule signifie que pour chaque automorphisme σ de $\Re_N$ sur $\boldsymbol{Q}(j(\tau))$ il existe une telle matrice $\begin{pmatrix} a & b \\ c & d \end{pmatrix} \in G_N^*$. L'application $\sigma \rightarrow \begin{pmatrix} a & b \\ c & d \end{pmatrix}$ est un isomorphisme dans $G_N = G_N^*/\{\pm I\}$. Soit G_N' l'image de cet isomorphisme. Il est bien connu que chaque élément de S_N^* est représenté par une matrice $\begin{pmatrix} a & b \\ c & d \end{pmatrix}$ à coefficients entiers telle que $ad - bc = 1$ (non seulement $ad - bc \equiv 1$

mod. N). L'application $\tau \to \dfrac{a\tau+b}{c\tau+d}$ donne un automorphisme de $\mathfrak{M}_N$ sur $Q(j(\tau))$; par suite on voit que $G_N' \supset S_N$ en vertu de (24). Puisque le groupe quotient G_N/S_N est isomorphe au groupe $(Z/NZ)^\times$ d'éléments inversibles de Z/NZ, il existe un sous-groupe $\mathfrak{h}$ de $(Z/NZ)^\times$ tel qu'on ait

$$G_N' = \left\{ \begin{pmatrix} a & b \\ c & d \end{pmatrix} \,\middle|\, ad-bc \in \mathfrak{h} \right\} \Big/ \{\pm I\}.$$

Soit H' le sous-groupe

$$\left\{ \begin{pmatrix} \pm 1 & 0 \\ 0 & a \end{pmatrix} \,\middle|\, (a,N)=1 \right\} \Big/ \{\pm I\}$$

de G_N. On a alors

$$(29) \qquad G_N' \cap H' = \left\{ \begin{pmatrix} \pm 1 & 0 \\ 0 & a \end{pmatrix} \,\middle|\, \pm a \in \mathfrak{h} \right\},$$

$$(30) \qquad (G_N' \cap H')S_N = G_N', \quad (G_N' \cap H') \cap S_N = \{e\}.$$

Soit $\mathfrak{L}$ le sous-corps de $\mathfrak{K}_N$ correspondant à S_N; alors on a $Q(j) \subset \mathfrak{L} \subset C(j)$. Comme $\mathfrak{L}$ est une extension de $Q(j)$ de degré fini, il existe, d'après le lemme 4, une extension k de Q de degré fini telle que $\mathfrak{L}=k(j)$. Tous les éléments de $\mathfrak{K}_N \cap C$ sont fixés par S_N; on a donc $\mathfrak{K}_N \cap C \subset k(j)$; d'où résulte $\mathfrak{K}_N \cap C = k$. Dans 5, nous avons montré, en supposant que $G_N=G_N'$, que $Q(j, j_{(2)}, h(t_1))$ correspond à H'. Quand on ne suppose pas $G_N=G_N'$, on voit que $Q(j, j_{(2)}, h(t_1))$ correspond à $H' \cap G_N'$. On a $Q(j, j_2, h(t_1))=Q(j(\tau), j(N\tau), f_{10}(\tau))$ en vertu de notre identification. D'après (30), on a

$$Q(j(\tau),\, j(N\tau),\, f_{10}(\tau)) \cap k(j(\tau)) = Q(j),$$

$$\mathfrak{K}_N = k(j(\tau),\, j(N\tau),\, f_{10}(\tau)).$$

On peut regarder $\mathfrak{K}_N$ comme un sous-corps du corps $Q(\zeta_N)((q_N))$ des séries formelles en q_N à coefficients dans $Q(\zeta_N)$. L'application $q_N \to 0$ donne un diviseur premier $\mathfrak{P}$ du corps $\mathfrak{K}_N$. D'après la formule (25), on a

$$\frac{f_{1\beta} - r_\beta}{f_{1\beta'} - r_{\beta'}}\, (\mathfrak{P}) = \frac{s_\beta}{s_{\beta'}}\, \zeta_N^{\beta-\beta'}\, ;$$

ce qui montre que le corps des restes modulo $\mathfrak{P}$ est $Q(\zeta_N)$. $\mathfrak{P}$ induit un diviseur premier $\mathfrak{p}$ dans $Q(j(\tau), j(N\tau), f_{10}(\tau))$. Comme $j(\tau)$, $j(N\tau)$, $f_{10}(\tau)$ ont des développements en q_N à coefficients rationnels, le corps des restes modulo $\mathfrak{p}$ est Q. On a donc

$$[H' \cap G_N' : \{e\}] = [\mathfrak{K}_N : Q(j(\tau), j(N\tau), f_{10}(\tau))] \geq [Q(\zeta_N) : Q] = [H' : \{e\}];$$

d'où résulte $H' \cap G_N' = H'$; par suite on a $G_N = H'S_N = G_N'$; ce qui démontre la prop. 5.

Puisqu'on a $\mathfrak{K}_N \cap C = k$, on a $k \subset Q(\zeta_N)$. D'autre part, on a

$$[k:Q]\geq[k(j, j_{(2)}, h(t_1)):Q(j, j_{(2)}, h(t_1))]$$
$$=[\mathfrak{R}_N:Q(j(\tau), j(N\tau), f_{10}(\tau))]=[Q(\zeta_N):Q];$$

il s'ensuit de là que $k=Q(\zeta_N)$; ce qui prouve la prop. 6. Dans le corps $Q(\zeta_N)((q_N))$, l'application $\zeta_N\to\zeta_N{}^{a}, q_N\to q_N$ donne un automorphisme de $\mathfrak{R}_N$ sur $Q(j)$ pour chaque α premier avec N. D'après la formule (23), cet automorphisme coïncide avec l'automorphisme correspondant à $\begin{pmatrix}1 & 0\\0 & \alpha\end{pmatrix}$. Soit $\begin{pmatrix}a & b\\c & d\end{pmatrix}$ un élément de $G_N{}^*$ tel que $ad-bc=\alpha$; alors, puisque chaque élément de S_N fixe tous les éléments de $Q(\zeta_N)$, $\begin{pmatrix}a & b\\c & d\end{pmatrix}, \begin{pmatrix}1 & 0\\0 & \alpha\end{pmatrix}$ donnent le même automorphisme de $Q(\zeta_N)$; ce qui démontre la prop. 7.

Nous avons vu que $Q(j, j_{(2)}, h(t_1))$ a un diviseur premier $\mathfrak{p}$ tel que le corps des restes modulo $\mathfrak{p}$ est Q. Cela démontre que Γ_N *a un point rationnel par rapport à* Q comme $L_N=Q(j, j_{(1)}, h(t_2))$ est isomorphe à $Q(j, j_{(2)}, h(t_1))$. Puisque $M_{N,\mathfrak{l}}$ est un sous-corps de L_N, on voit de même que $\Gamma_{N,\mathfrak{l}}$ a un point rationnel par rapport à Q.

18. Nous disons que deux matrices $\begin{pmatrix}a & b\\c & d\end{pmatrix}, \begin{pmatrix}a' & b'\\c' & d'\end{pmatrix}$ à coefficients entiers sont équivalentes s'il existe une matrice $\begin{pmatrix}\alpha & \beta\\\gamma & \delta\end{pmatrix}$ à coefficients entiers telle que $\alpha\delta-\beta\gamma=1$ et $\begin{pmatrix}a' & b'\\c' & d'\end{pmatrix}=\begin{pmatrix}\alpha & \beta\\\gamma & \delta\end{pmatrix}\begin{pmatrix}a & b\\c & d\end{pmatrix}$. Si p est un nombre premier, il y a $p+1$ classes d'équivalence des matrices $\begin{pmatrix}a & b\\c & d\end{pmatrix}$ telles que $ad-bc=p$. Soient $\begin{pmatrix}a_\nu & b_\nu\\c_\nu & d_\nu\end{pmatrix}$ $(1\leq\nu\leq p+1)$ les représentants de ces classes; de plus, en supposant que p ne divise pas N, on peut les choisir de telle façon que

$$\begin{pmatrix}a_\nu & b_\nu\\c_\nu & d_\nu\end{pmatrix}\equiv\begin{pmatrix}1 & 0\\0 & p\end{pmatrix}\quad\text{mod. } N\quad(1\leq\nu\leq p+1).$$

Cela posé, soient $\begin{pmatrix}a_\nu' & b_\nu'\\c_\nu' & d_\nu'\end{pmatrix}$ pour $1\leq\nu\leq p+1$ les matrices telles que

$$\begin{pmatrix}a_\nu' & b_\nu'\\c_\nu' & d_\nu'\end{pmatrix}\begin{pmatrix}a_\nu & b_\nu\\c_\nu & d_\nu\end{pmatrix}=\begin{pmatrix}p & 0\\0 & p\end{pmatrix};$$

on a alors

$$(31)\qquad\begin{pmatrix}a_\nu' & b_\nu'\\c_\nu' & d_\nu'\end{pmatrix}\equiv\begin{pmatrix}p & 0\\0 & 1\end{pmatrix}\quad\text{mod. } N.$$

Soient $\tau_0, E, D, \{\omega_{01}, \omega_{02}\}$ et $u(z)$ comme dans **15**; posons

$$(32)\qquad\begin{pmatrix}\omega_{\nu 1}\\\omega_{\nu 2}\end{pmatrix}=p^{-1}\begin{pmatrix}a_\nu & b_\nu\\c_\nu & d_\nu\end{pmatrix}\begin{pmatrix}\omega_{01}\\\omega_{02}\end{pmatrix}\quad(1\leq\nu\leq p+1);$$

on a alors

$$(33) \qquad \begin{pmatrix} \omega_{01} \\ \omega_{02} \end{pmatrix} = \begin{pmatrix} a_\nu' & b_\nu' \\ c_\nu' & d_\nu' \end{pmatrix} \begin{pmatrix} \omega_{\nu 1} \\ \omega_{\nu 2} \end{pmatrix} \qquad (1 \leq \nu \leq p+1).$$

Soit D_ν le réseau engendré par $\{\omega_{\nu 1}, \omega_{\nu 2}\}$ pour chaque ν; posons $\mathfrak{g}_\nu = u(D_\nu/D)$. $\mathfrak{g}_\nu$ est un sous-groupe de E d'ordre p; on a $\mathfrak{g}_\nu \neq \mathfrak{g}_\mu$ pour $\nu \neq \mu$. Définissons E_ν, λ_ν comme dans **6** pour ces E, $\mathfrak{g}_\nu$; et posons $u_\nu(z) = \lambda_\nu(u(z))$; alors $u_\nu(z)$ donne un isomorphisme analytique de C/D_ν sur E_ν. Soit h_ν la fonction canonique sur E_ν. En appliquant la formule (26) à E_ν, on a

$$h_\nu(u_\nu(z)) = \frac{g_2(\omega_{\nu 1}, \omega_{\nu 2})}{g_3(\omega_{\nu 1}, \omega_{\nu 2})} \cdot \wp(z; \omega_{\nu 1}, \omega_{\nu 2}).$$

Posons $t_1 = u(\omega_{01}/N)$, $t_2 = u(\omega_{02}/N)$; on a alors $\alpha t_1 + \beta t_2 = u\left(\frac{\alpha \omega_{01} + \beta \omega_{02}}{N} \right)$ et par suite

$$h_\nu(\lambda_\nu(\alpha t_1 + \beta t_2)) = \frac{g_2(\omega_{\nu 1}, \omega_{\nu 2})}{g_3(\omega_{\nu 1}, \omega_{\nu 2})} \; \wp\left(\frac{\alpha \omega_{01} + \beta \omega_{02}}{N} ; \omega_{\nu 1}, \omega_{\nu 2} \right).$$

D'après les formules (31), (33), on a

$$\frac{\alpha \omega_{01} + \beta \omega_{02}}{N} \equiv \frac{p \alpha \omega_{\nu 1} + \beta \omega_{\nu 2}}{N} \qquad \text{mod. } D_\nu .$$

D'après (32) on a $\omega_{\nu 1}/\omega_{\nu 2} = \frac{a_\nu \tau_0 + b_\nu}{c_\nu \tau_0 + d_\nu}$. Nous obtenons donc

$$j(E_\nu) = j\left(\frac{a_\nu \tau_0 + b_\nu}{c_\nu \tau_0 + d_\nu} \right),$$

$$(34)$$

$$h_\nu(\lambda_\nu(\alpha t_1 + \beta t_2)) = f\left(p\alpha, \beta ; \frac{a_\nu \tau_0 + b_\nu}{c_\nu \tau_0 + d_\nu} \right).$$

Soient τ_p' l'automorphisme de $\mathfrak{R}_N$ correspondant à $\begin{pmatrix} p & 0 \\ 0 & 1 \end{pmatrix}$ et $\bar{\sigma}_\nu$ l'isomorphisme de $\mathfrak{R}_N$ défini par la relation

$$g^{\bar{\sigma}_\nu}(\tau) = g^{\tau p'}\left(\frac{a_\nu \tau + b_\nu}{c_\nu \tau + d_\nu} \right)$$

pour $g \in \mathfrak{R}_N$. On a alors

$$j^{\bar{\sigma}_\nu}(\tau_0) = j(E_\nu), \quad f_{\alpha\beta}{}^{\bar{\sigma}_\nu}(\tau_0) = h_\nu(\lambda_\nu(\alpha t_1 + \beta t_2)).$$

Par le principe de spécialisation, on peut donc identifier $(K_N, \sigma_\nu, K_N{}^{\bar{\sigma}_\nu})$ défini dans **6** avec $(\mathfrak{R}_N, \bar{\sigma}_\nu, \mathfrak{R}_N{}^{\bar{\sigma}_\nu})$.

Soient $\mathfrak{L}$ un sous-corps de $\mathfrak{R}_N$ tel que $\mathfrak{L} \cap Q(\zeta_N) = Q$, $\mathfrak{L} \supset Q(j(\tau))$ et $\{\Gamma, u\}$ un modèle du sous-corps L de K_N correspondant à $\mathfrak{L}$. Si l'on prend C comme le domaine universel, on peut regarder $C \cdot \mathfrak{L}$ comme le corps des fonctions sur Γ, en définissant $f(u) = f(\tau_0)$ pour $f \in \mathfrak{L}$.

Soit $\mathfrak{L}_N$ le sous-corps de $\mathfrak{R}_N$ correspondant au sous-groupe H_N de G_N; on a alors

$$\mathfrak{L}_N = Q(j(\tau),\ j(\tau/N),\ f_{01}(\tau)), \qquad \mathfrak{M}_N = C\mathfrak{L}_N;$$

$\mathfrak{L}_N$ peut être regardé comme le corps des fonctions sur Γ_N définies par rapport à Q. Puisque $\tau_p{}'$ est contenu dans H_N, on a

$$(35) \qquad g^{\bar\sigma_\nu}(\tau) = g\left(\frac{a_\nu\tau + b_\nu}{c_\nu\tau + d_\nu}\right) \qquad\qquad (1 \leq \nu \leq p+1)$$

pour $g \in \mathfrak{L}_N$; on a donc

$$(36) \qquad g^{\bar\sigma_\nu\prime}(\tau) = g'\left(\frac{a_\nu\tau + b_\nu}{c_\nu\tau + d_\nu}\right)\frac{p}{(c_\nu\tau + d_\nu)^2} \qquad\qquad (1 \leq \nu \leq p+1)$$

pour $g \in \mathfrak{L}_N$, où $f'(\tau)$ désigne la dérivée de $f(\tau)$.

Soit maintenant $f\,dg$ une forme différentielle sur Γ_N, de première espèce, où f, g sont deux éléments de $\mathfrak{M}_N$; $f(\tau)g'(\tau)$ est alors une forme parabolique de degré 2, d'espèce N. Et réciproquement, chacune de telles formes est obtenue de cette manière. Soit T_p l'opérateur de Hecke ([4]); on a alors

$$(37) \qquad (fg' \mid T_p)(\tau) = \sum_{\nu=1}^{p+1} f\left(\frac{a_\nu\tau + b_\nu}{c_\nu\tau + d_\nu}\right)g'\left(\frac{a_\nu\tau + b_\nu}{c_\nu\tau + d_\nu}\right)\frac{p}{(c_\nu\tau + d_\nu)^2}\ .$$

Il s'ensuit des formules (35), (36), en vertu d'une proposition de [8], que $fg' \mid T_p$ correspond à la forme différentielle $\delta X_p(f\,dg)$, si f, g sont dans $\mathfrak{L}_N$, où δX_p désigne la différentielle de la correspondance X_p. En d'autres termes, si l'on désigne par M^d une représentation de $\mathcal{A}(J_N)$ par les formes différentielles de première espèce, $M^d(\xi_p)$ peut être regardé comme une représentation de T_p pour les formes paraboliques de degré 2, d'espèce N.

Soient $\begin{pmatrix} \alpha & \beta \\ \gamma & \delta \end{pmatrix}$ une matrice à coefficients entiers telle que

$$\alpha\delta - \beta\gamma = 1, \quad \begin{pmatrix} \alpha & \beta \\ \gamma & \delta \end{pmatrix} \equiv \begin{pmatrix} p^{-1} & 0 \\ 0 & p \end{pmatrix} \qquad \mathrm{mod}\ N.$$

et ρ_p l'automorphisme de $\mathfrak{R}_N$ correspondant à $\begin{pmatrix} p & 0 \\ 0 & p \end{pmatrix}$. Alors, comme on a

$$\begin{pmatrix} p & 0 \\ 0 & p \end{pmatrix} \equiv \begin{pmatrix} p^2 & 0 \\ 0 & 1 \end{pmatrix}\begin{pmatrix} \alpha & \beta \\ \gamma & \delta \end{pmatrix} \qquad \mathrm{mod.}\ N,$$

on voit que

$$g^{\rho_p}(\tau) = g\left(\frac{\alpha\tau + \beta}{\gamma\tau + \delta}\right) \quad \text{pour } g \in \mathfrak{L}_N.$$

D'après la définition de Y_p, on peut donc regarder $M^d(\eta_p)$ comme une représentation de l'opérateur R_p dans [4].

§ 5. Résultats principaux.

19. Les notations étant celles de **13**, soient g le genre de Γ et M_l une représentation l-adique de $\mathcal{A}(J)$. Si $l \neq p$, on peut choisir une représentation l-adique $M_{l,p}$ de $\mathcal{A}(J(p))$ de telle façon que, pour chaque $\mu \in \mathcal{A}(J)$, on ait $M_l(\mu) = M_{l,p}(\mu(p))$. Nous désignerons $M_{l,p}$ aussi par M_l. On a alors d'après (I')

$$M_l(\xi_p) = M_l(\pi) + M_l(\pi' \eta_p).$$

Comme on a $\pi \pi' = p \delta_J$, U étant une indéterminée, on a

$$I_{2g} - M_l(\xi_p)U + M_l(\eta_p)pU^2 = [I_{2g} - M_l(\pi)U][I_{2g} - M_l(\pi' \eta_p)U]$$

où I_m désigne la matrice unité de m lignes. Selon que J est égal à J_N ou à $J_{N,\natural}$, on a en vertu de (II')

$$I_{2g} - M_l(\pi' \eta_p)U = M_l(\alpha)^{-1}[I_{2g} - M_l(\pi')U]M_l(\alpha)$$

ou

$$I_{2g} - M_l(\pi' \eta_p)U = M_l(\beta)^{-1}[I_{2g} - M_l(\pi')U]M_l(\beta).$$

D'après la formule $M_l(\pi') = E_l(\Theta)^{-1}\, {}^tM_l(\pi)E_l(\Theta)$ dans [11] n°76, on a

$$\det[I_{2g} - M_l(\xi_p)U + M_l(\eta_p)pU^2] = \det[I_{2g} - M_l(\pi)U]^2$$

pour Γ_N ou $\Gamma_{N,\natural}$. Soit M^d une représentation de $\mathcal{A}(J)$ par les formes différentielles de première espèce; alors M_l est équivalent à $M^d \oplus \overline{M}^d$ où $\overline{M}^d$ désigne la représentation imaginaire conjuguée de M^d. Comme $J_N, J_{N,\natural}, \xi_p, \eta_p$ sont définis par rapport à $\mathbf{Q}$, on peut choisir la représentation M^d de telle façon que $M^d(\xi_p) = \overline{M}^d(\xi_p)$ et $M^d(\eta_p) = \overline{M}^d(\eta_p)$; on a donc

$$\det[I_{2g} - M_l(\xi_p)U + M_l(\eta_p)pU^2] = \det[I_g - M^d(\xi_p)U + M^d(\eta_p)pU^2]^2$$

et par suite

$$\det[I_{2g} - M_l(\pi)U] = \det[I_g - M^d(\xi_p)U + M^d(\eta_p)pU^2].$$

Désignons par $\zeta(s, \Gamma, p)$ la fonction ζ de $\Gamma(p)$; alors on a

$$\zeta(s, \Gamma, p) = (1 - p^{-s})^{-1}(1 - p^{1-s})^{-1}\det[I_g - M^d(\xi_p)p^{-s} + M^d(\eta_p)p^{1-2s}].$$

La fonction ζ de la courbe Γ est définie par

$$\zeta(s, \Gamma) = \prod_p \zeta(s, \Gamma, p),$$

où le produit est étendu à tous les nombres premiers pour lesquels Γ n'a pas de défaut. Dans **18**, on a vu que $M^d(\xi_p)$, $M^d(\eta_p)$ ne sont autres que les représentations des opérateurs T_p, R_p pour les formes paraboliques de degré 2 (et d'espèce N).

Nous sommes ainsi parvenus à notre résultat principal:

La fonction ζ de la courbe Γ_N s'exprime sous la forme

$$\zeta(s,\, \Gamma_N)=f(s)\zeta(s)\zeta(s-1)\Phi(s)^{-1},$$

$$\Phi(s)=\prod_{t,\,\varepsilon}\det\Phi_{t,\varepsilon}(s)$$

où $\zeta(s)$ est la fonction ζ de Riemann, $f(s)$ est une fonction rationnelle de p^{-s} et $\Phi_{t,\varepsilon}(s)$ désigne le produit d'Euler introduit par E. Hecke [4], attaché aux formes paraboliques de degré 2, de diviseur t et de caractère $\varepsilon(n)$.

On obtient un résultat analogue aussi pour la fonction ζ de $\Gamma_{N,\mathfrak{h}}$; dans ce cas, il faut seulement limiter (t,ε) à certaines valeurs selon $\mathfrak{h}$ dans le produit pour $\Phi(s)$. Le cas où $\mathfrak{h}$ a l'index 1 ou 2 a été traité par M. Eichler [3].

20. Nous allons maintenant étudier les racines caractéristiques de T_p. Elles sont les racines de l'équation caractéristique de $M_l(\xi_p)=M_l(\tilde{\xi}_p)$ et par suite les racines caractéristiques d'une représentation régulière de $\tilde{\xi}_p$ dans l'algèbre $\mathcal{A}_0(\tilde{J}_N)$. Soit $\mathcal{B}$ l'algèbre sur C déduite de $\mathcal{A}_0(\tilde{J}_N)$ par extension à C du corps de base. Soit σ le prolongement dans $\mathcal{B}$ de la trace de $\mathcal{A}_0(\tilde{J}_N)$ (cf. [10] p. 58); $\sigma(\mu'\mu)$ est une forme d'Hermite positive non-dégénérée dans $\mathcal{B}$; posons $\|\mu\|=\sigma(\mu'\mu)^{1/2}$ pour $\mu\in\mathcal{B}$. D'après les relations $\pi\pi'=\pi'\pi=p\delta$ et $\tilde{\alpha}'=\tilde{\alpha}^{-1}$, on a

$$\|\pi\mu\|=\sqrt{p}\,\|\mu\|,\quad \|(\tilde{\alpha}^{-1}\pi'\tilde{\alpha})\mu\|=\sqrt{p}\,\|\mu\|;$$

par suite, d'après (I'), (II') on a

$$\|\tilde{\xi}_p\mu\|\leqq\|\pi\mu\|+\|(\tilde{\alpha}^{-1}\pi'\tilde{\alpha})\mu\|=2\sqrt{p}\,\|\mu\|.$$

Il en résulte que les valeurs absolues des racines d'une représentation régulière de $\tilde{\xi}_p$ dans $\mathcal{A}_0(\tilde{J}_N)$ ne dépassent pas $2\sqrt{p}$. Nous obtenons ainsi le résultat suivant:

Les valeurs absolues des racines caractéristiques de l'opérateur T_p de Hecke pour les formes paraboliques de degré 2 ne dépassent pas $2\sqrt{p}$ pour presque tous les nombres premiers p.

Université de Tokyo.

Bibliographie

[1] W. L. Chow, The jacobian variety of an algebraic curve, Amer. J. Math., 76 (1954), pp. 453–476.

[2] M. Deuring, Die Typen der Multiplikatorenringe elliptischer Funktionenkörper, Abh. Math. Sem. Univ. Hamburg, 14 (1941), pp. 197–272.

[3] M. Eichler, Quaternäre quadratische Formen und die Riemannsche Vermutung für die Kongruenzzetafunktion, Arch. Math., 5 (1954), pp. 355–366.

[4] E. Hecke, Über die Modulfunktionen und die Dirichletschen Reihen mit Eulerschen Produktentwicklung I, II, Math. Ann., 114 (1937), pp. 1–28, 316–351.

[5] G. Shimura, Reduction of algebraic varieties with respect to a discrete valuation of the basic field, Amer. J. Math., 77 (1955), pp. 134-176.

[6] G. Shimura, On complex multiplications, Proc. International Symposium on algebraic number theory, Tokyo-Nikko, (1955), pp. 23-30.

[7] Taniyama, Jacobian varieties and number fields, ibid. pp. 31-45.

[8] G. Shimura et Y. Taniyama, Complex multiplication of abelian varieties and its applications to number theory, à paraître dans J. Math. Soc. Japan.

[9] A. Weil, Foundations of algebraic geometry, New York, (1946).

[10] A. Weil, Sur les courbes algébriques et les variétés qui s'en déduisent, Paris, (1948).

[11] A. Weil, Variétés abéliennes et courbes algébriques, Paris, (1948).

[12] A. Weil, On algebraic groups and homogeneous spaces, Amer. J. Math., 77 (1955), pp. 493-512.

Modules des variétés abéliennes polarisées et fonctions modulaires

Séminaire Henri Cartan, École Normale Supérieure, 1957/58,
Fonctions Automorphes, Exposé 18-20 (1958)*

I. Exposé du 5 mai 1958

On pourrai regarder la théorie des fonctions modulaires de Siegel comme la théorie des modules des variétés abéliennes, ou plus précisément des variétés abéliennes polarisées de la famille principale. En effet, on peut faire correspondre à tout point z de l'espace de Siegel $\mathfrak{S}_n$ une variété abélienne projective $A(z)$ ayant $(z \ 1_n)$ comme matrice des périodes; et l'application $z \mapsto A(z)$ donne une correspondance biunivoque entre l'espace quotient $\mathfrak{S}_n/Sp(n, \mathbf{Z})$ et l'ensemble des structures de variété abélienne polarisée de la famille principale, comme Weil l'a signalé à la fin de l'exposé 2. L'objet de cet exposé est de représenter d'une manière concrète cette relation en démontrant que les "modules" des variétés $A(z)$ polarisées par les sections hyperplanes, regardés comme fonctions de z, engendrent les fonctions modulaires de Siegel. Dans ce but, on va d'abord s'occuper des notions de "modules" d'une variété abélienne polarisée et de variété de Kummer.

1. Variétés polarisées. Soit V une variété algébrique complète non-singulière en co-dimension 1, définie par rapport à un corps k, et X un diviseur sur V rationnel par rapport à k. Soit $L(X, k)$ l'ensemble des fonctions f sur V, définies par rapport à k, telles que l'on ait $(f) \succ -X$, où (f) désigne le diviseur de la fonction f. On désignera par $\ell(X)$ la dimension de l'espace vectoriel $L(X, k)$ sur k. Soit $\{f_0, \ldots, f_n\}$ une base de $L(X, k)$ sur k, et x un point générique de V par rapport à k; soit $V(X, f)$ le lieu de $\big(f_0(x), \ldots, f_n(x)\big)$ par rapport à k dans l'espace projectif de dimension n. On obtient alors une application rationnelle F de V sur $V(X, f)$, ayant k comme corps de définition, définie par $F(x) = \big(f_0(x), \ldots, f_n(x)\big)$. Nous dirons que la diviseur X est *ample* si cette application F est birationnelle et birégulière; cette définition ne dépend que de X et non du choix de k, f_i, x.

Avec les mêmes notations V, X, on désignera par $\mathscr{C}(X)$ l'ensemble de tous les diviseurs X' sur V pour lesquels il existe deux entiers strictement positifs m, m' tels que mX soit algébriquement équivalent à $m'X'$. L'ensemble $\mathscr{C}(X)$ s'appellera une *polarisation* de V si $\mathscr{C}(X)$ contient un diviseur ample. Nous entendrons par *variété polarisée* une variété algébrique V sur laquelle est donnée une polarisation $\mathscr{C}$ et la désignerons par $(V, \mathscr{C})$. Nous dirons qu'une variété polarisée $(V, \mathscr{C})$ est *définie par rapport* à un corps k si V est définie par rapport à k, et si $\mathscr{C}$ contient un diviseur rationnel par rapport à k. Soit σ un isomorphism de k sur un corps k'; soient $(V, \mathscr{C})$ une variété polarisée définie par rapport à

k, et X un diviseur dans $\mathscr{C}$ rationnel par rapport à k. On obtient alors une polarisation $\mathscr{C}(X^\sigma)$ de V^σ, que l'on désignera par $\mathscr{C}^\sigma$.

2. Variétés abéliennes polarisées. Rappelons d'abord quelques notions et propriétés des diviseurs sur une variété abélienne. Soient A une variété abélienne, X un diviseur sur A, et a un point de A. On désignera par X_a la transformée de X par la translation $x \mapsto x + a$ sur A. Le diviseur X est dit *non-dégénéré* s'il n'y a qu'un nombre fini d'éléments u de A tels que X_u soit linéairement équivalent à X.

Lemme 1 (Weil [6]). *Soit X un diviseur sur une variété abélienne A. S'il existe un entier $n > 0$ tel que nX soit ample, X est non-dégénéré. Réciproquement, si X est positif non-dégénéré, il existe un entier $n_0 > 0$ tel que nX soit ample pour $n > n_0$.*

Lemme 2. *Soient X un diviseur non-dégénéré sur une variété abélienne A et Y un diviseur sur A, algébriquement équivalent à X. Alors, il existe un point u de A tel que Y soit linéairement équivalent à X_u.*

En effet, en faisant correspondre u à la classe de $X_u - X$, on obtient un homomorphisme de A dans la variété de Picard de A; si X est non-dégénéré, cet homomorphisme est surjectif; il existe donc un point u de A tel que $Y - X$ soit linéairement équivalent à $X_u - X$; d'où résulte le lemme 2.

Lemme 3 (Nishi [2]). *Soient A une variété abélienne de dimension n et X un diviseur positif non-dégénéré sur A. Alors on a $\ell(X) = \deg(X^n)/n!$, où $\deg(X^n)$ désigne $\deg(X_{u_1} \cdots X_{u_n})$ pour n points $u_1, \ldots, u_n$ de A tels que le produit d'intersection soit défini.*

Lemme 4. *Soient A et A' deux variétés abéliennes de même dimension, λ un homomorphisme de A sur A', et X' un diviseur positif non-dégénéré sur A'. On a alors $\ell\big(\lambda^{-1}(X')\big) = \nu(\lambda)\ell(X')$, où $\nu(\lambda)$ désigne le degré de l'homomorphisme λ.*

C'est une conséquence immédiate du lemme 3.

Soient $(A, \mathscr{C})$ et $(A', \mathscr{C}')$ deux variétés abéliennes polarisées de même dimension, et λ un homomorphisme (isomorphisme) de A sur A'. Nous appellerons λ un homomorphisme (isomorphisme) de $(A, \mathscr{C})$ sur $(A', \mathscr{C}')$ s'il existe un diviseur X' dans $\mathscr{C}'$ tel que $\lambda^{-1}(X')$ soit contenu dans $\mathscr{C}$; s'il en est ainsi, $\lambda^{-1}(Y)$ est contenu dans $\mathscr{C}$ pour tout $Y \in \mathscr{C}'$. D'après notre définition et d'après Weil [6], pour toute variété abélienne polarisée $(A, \mathscr{C})$, il y a toujours un isomorphisme de $(A, \mathscr{C})$ sur une variété abélienne polarisée $(A', \mathscr{C}')$ telle que A' soit une variété projective et $\mathscr{C}'$ contienne les sections hyperplanes de A'.

Théorème 1. *Soient A et A' deux variétés abéliennes dans un espace projectif; soient $\mathscr{C}$ et $\mathscr{C}'$ respectivement les polarisations de A et de A' définies par les sections hyperplanes. Supposons que A et A' ne soient contenus dans aucun hyperplan et que les systèmes linéaires des sections hyperplanes de A et de $A'* *soient complets. Alors, pour que $(A', \mathscr{C}')$ soit isomorphe à $(A, \mathscr{C})$, il faut et il suffit que A' soit une transformée de A par une transformation projective.*

Soit H un hyperplan de l'espace ambiant pour A et A'; soient $Y = H \cdot A$, $Y' = H \cdot A'$. Supposons que A' soit une transformée de A par une transformation projective φ; désignons par F l'application de A sur A' induit par φ. Il existe alors un isomorphisme λ de A sur A' et un point $b \in A'$ tel que l'on ait $F(x) = \lambda(x) + b$ pour $x \in A$. Soit a un point de A tel que l'on ait $b = -\lambda(a)$. On vérifie aisément que $\lambda^{-1}(Y') = F^{-1}(Y')_a$; ce dernier est algébriquement équivalent à $F^{-1}(Y')$. Comme Y' est une section hyperplane de A et comme F est induit par une transformation projective, on voit que $F^{-1}(Y')$ est une section hyperplane de A, de sorte que $F^{-1}(Y')$ est algébriquement équivalent à Y; ceci démontre que $(A, \mathscr{C})$ est isomorphe à $(A', \mathscr{C}')$. Réciproquement, supposons qu'il existe un isomorphisme ψ de $(A, \mathscr{C})$ sur $(A', \mathscr{C}')$. D'après notre définition, il existe deux entiers positifs r, s tels que $r\psi^{-1}(Y')$ soit algébriquement équivalent à sY. Par suite, d'après le lemme 2 et le lemme 3, on a $r^n \ell(\psi^{-1}(Y')) = s^n \ell(Y)$, où n est la dimension de A. D'autre part, on voit, en vertu de notre hypothèse, que $\ell(\psi^{-1}(Y'))$ et $\ell(Y)$ sont égaux à $N + 1$, N étant la dimension de l'espace ambiant pour A. On a donc $r = s$; d'où résulte que $\psi^{-1}(Y')$ est algébriquement équivalent à Y. D'après le lemme 2, il existe un point u de A tel que $\psi^{-1}(Y')$ soit linéairement équivalent à Y_u; il existe donc une fonction g sur A telle qu'on ait $(g) = \psi^{-1}(Y')_{-u} - Y$. Soit $(X_0, \ldots, X_N)$ le système des coordonnées de l'espace ambiant pour A et soient f_i, f_i' les fonctions sur A et sur A' induites par $X_i / (\sum_j \gamma_j X_j)$, respectivement, où $\sum_j \gamma_j X_j = 0$ est l'équation du hyperplan H. Soit k un corps de définition pour A, A', ψ et g par rapport auquel u et les γ_j sont rationnels. D'après notre hypothèse, les f_i forment une base de $L(Y, k)$. Soit h_i la fonction sur A définie par $h_i(x) = f_i'(\psi(x + u))g(x)$ par rapport à k, x étant un point générique de A par rapport à k. On vérifie facilement que $h_i \in L(Y, k)$, de sorte que les h_i s'expriment sous les formes $h_i = \sum_j \xi_{ij} f_j$ $(0 \le i \le N)$, où les ξ_{ij} sont des éléments de k. Comme les h_i sont linéairement indépendants sur k, la matrice (ξ_{ij}) est inversible. Soit x un point générique de A par rapport à k; alors les $f_i(x)$ sont les coordonnées de x; et de plus, on voit que les $h_i(x)$ sont les coordonnées de $\psi(x + u)$. Il en résulte que A' est la transformée de A par la transformation projective donnée par la matrice (ξ_{ij}); ce qui complète la démonstration du théorème 1.

Soit V une variété algébrique dans l'espace projectif P^N de dimension N. Soient k un corps de définition pour V et (t_{ij}) une matrice de degré $N + 1$, dont les coordonnées sont $(N + 1)^2$ variables indépendantes sur k. Soient V' la transformée de V par la transformation projective $(a_i) \mapsto (\sum_{j=0}^N t_{ij} a_j)$, où $(a_i) \in P^N$, et z le point de Chow de V'. Comme le corps $k(z)$ est contenu dans $k(t)$, $k(z)$ est une extension régulière de k; donc on obtient une variété $\mathscr{F}$ comme le lieu de z par rapport à k. On peut facilement vérifier que $\mathscr{F}$ ne dépend que de V et non du choix de k et de (t). La variété $\mathscr{F}$ s'appellera la *famille projective* de V. Soient V et V' deux variétés algébriques dans un espace projectif; soient $\mathscr{F}$ et $\mathscr{F}'$ respectivelment les familles projectives de V et de V'.

alors, pour que l'on ait $\mathscr{F} = \mathscr{F}'$, il faut et il suffit que V' soit une transformée de V par une transformation projective.

Théorème 2. *Soient A une variété abélienne dans un espace projectif, $\mathscr{C}$ la polarisation de A définie par les sections hyperplanes et $\mathscr{F}$ la famille projective de A. Supposons que A ne soit contenu dans aucun hyperplane et que le système linéaire des sections hyperplanes de A soit complet. Soient k un corps de définition pour A et σ un isomorphisme de k sur un corps k'. Alors, pour que l'on ait $\mathscr{F}^\sigma = \mathscr{F}$, il faut et il suffit que $(A^\sigma, \mathscr{C}^\sigma)$ soit isomorphe à $(A, \mathscr{C})$.*

On vérifie facilement que $\mathscr{F}^\sigma$ est la famille projective de A^σ. Alors la théorème 2 est une conséquence du théorème 1.

3. Le corps du module d'une variété abélienne polarisée. Soit $(A, \mathscr{C})$ une variété abélienne polarisée, et soit $\mathscr{C}^*$ l'ensemble de tous les diviseurs amples dans $\mathscr{C}$. Chaque élément X de $\mathscr{C}^*$ définit une immersion birationnelle birégulière de A dans un espace projectif. Désignons par $\mathscr{F}(A, X)$ la famille projective de l'image de cette immersion. D'appès le lemme 2, si X' est algébriquement équivalent à X, on a $\mathscr{F}(A, X') = \mathscr{F}(A, X)$.

Proposition 1. *Les notations étant comme ci-dessus, soient K et K' respectivement les plus petits corps de définition pour $\mathscr{F}(A, X)$ et pour $\mathscr{F}(A, mX)$, où m est un entier positif. Alors, K est une extension purement inséparable de K'. De plus, si la caractéristique p du corps de base ne divise pas m, on a $K = K'$.*

Soit z un point générique de $\mathscr{F}(A, X)$ par rapport à K. D'après notre définition de la famille projective, z est le point de Chow d'une variété abélienne A_0 isomorphe à A. Soit X_0 une section hyperplane de A_0. On voit facilement $\mathscr{F}(A, X) = \mathscr{F}(A_0, X_0)$ et $\mathscr{F}(A, mX) = \mathscr{F}(A_0, mX_0)$. Comme A_0 est défini par rapport à $K(z)$ et comme A_0 n'est contenu dans aucun hyperplan, on peut prendre X_0 de telle façon qu'il soit rationnel par rapport à $K(z)$. Par suite $\mathscr{F}(A_0, mX_0)$ est défini par rapport à $K(z)$; d'où résulte qu'on a $K(z) \supset K'$. Soit σ un isomorphisme de $K(z)$ sur un corps, qui est l'identité sur K. On a alors $\mathscr{F}(A, X)^\sigma = \mathscr{F}(A, X)$; d'après le théorème 2, $(A, \mathscr{C})$ est isomorphe à $(A^\tau, \mathscr{C}^\tau)$, où τ est un prolongement de σ. D'après le même théorème on a $\mathscr{F}(A, mX)^\sigma = \mathscr{F}(A, mX)$; donc σ est l'identité sur K'. On en déduit que le corps composé KK' est une extension purement inséparable de K. D'après la même raison, KK' est purement inséparable sur K'. Comme $K(z)$ est régulier sur K, on doit avoir $KK' = K$; ce qui démontre la première assertion de la proposition. Supposons maintenant que la caractéristique p ne divise pas m. Soit z' un point générique de $\mathscr{F}(A, mX)$ par rapport à K'; il existe une variété abélienne A' dont le point de Chow est z'. Soit Y une section hyperplane de A' rationnelle par rapport à $K'(z')$. D'après notre construction, il existe un diviseur X' sur A' tel que mX' soit algébriquement équivalent à Y. On peut facilement démontrer qu'il existe un corps k séparablement engendré sur $K'(z')$ est une puissance p^r de p tels que $p^r X'$ soit rationnel par rapport à k. Comme m et

p^r sont premiers entre eux, il existe deux entiers a, b tels qu'on ait $am+bp^r = 1$; alors $aY + b(p^r X')$ est algébriquement équivalent à X'; et l'on voit facilement

$$\mathscr{F}(A', aY + b(p^r X')) = \mathscr{F}(A', X') = \mathscr{F}(A, X).$$

Comme $aY + b(p^r X')$ est rationnel par rapport à k, $\mathscr{F}(A, X)$ est défini par rapport à k; il en résulte qu'on a $k \supset K$. D'autre part, k est séparablement engendré sur K'; on peut en conclure $K = K'$, car on a vu plus haut que K est purement inséparable sur K'.

Soient les notations $(A, \mathscr{C})$, $\mathscr{C}^*$, $\mathscr{F}(A, X)$ comme ci-dessus. On peut démontrer qu'il existe un diviseur Y sur A tel que tout X dans $\mathscr{C}$ soit algébriquement équivalent à un multiple mY, où m est un entier positif. Soit K_m le corps minimum de définition pour $\mathscr{F}(A, mY)$ pour chaque entier positif m tel que mY soit ample. Soient m et m' deux entiers positifs tels que mY et $m'Y$ soient ample. Il existe un entier positif s tel que $smm'Y$ soit ample; où l'on peut supposer que p ne divise pas s. D'après la proposition 1, les corps K_m et $K_{m'}$ sont des extensions purement inséparables de $K_{smm'}$; d'après la même proposition, si p ne divise pas m, on a $K_{m'} = K_{smm'}$, de sorte qu'alors K_m est une extension purement inséparable de $K_{m'}$. Par suite, si p ne divise ni m, ni m', on a $K_m = K_{m'}$. Nous appellerons ce même corps le *corps du module* de la variété abélienne polarisée $(A, \mathscr{C})$; le corps du module K est une extension purement inséparable de K_m pour tout m; et l'on a $K = K_m$ si la caractéristique p ne divise pas m. Soient k un corps de définition pour $(A, \mathscr{C})$ contenant K et σ un isomorphisme de k sur un corps k'; d'après le théorème 2, pour que $(A, \mathscr{C})$ soit isomorphe à $(A^\sigma, \mathscr{C}^\sigma)$, il faut et il suffit que σ soit l'identité sur K. Dans le cas où la caractéristique est 0, le corps du module K est déterminé par cette propriété; on a $K = K_m$ pour chaque m; donc, K est engendré par le point de Chow de $\mathscr{F}(A, mY)$ sur le corps $\mathbf{Q}$ des nombres rationnels, pour tout m. De toute façon, on verrait, dans le théorème 1, que les points de Chow des $\mathscr{F}(A, mY)$ peuvent être regardés, sans restriction pour la caractéristique, comme des "modules" de $(A, \mathscr{C})$.

4. Variété de Kummer. Soient (A, C) une variété abélienne polarisée et G le groupe des automorphismes de $(A, \mathscr{C})$; on peut démontrer que G est d'ordre fini. Nous entendrons par une *variété de Kummer* de A une variété quotient de A par rapport à G, c'est-à-dire un couple (W, h) formé d'une variété W et d'une application rationnelle h de A sur W jouissant des propriétés suivantes:

1° *h est défini partout sur A;*

2° *$h = h \circ \gamma$ pour tout $\gamma \in G$;*

3° *si h' est une application rationnelle de A sur une variété W' satisfaisant à $h' = h' \circ \gamma$ pour tout $\gamma \in G$, il existe une application rationnelle g de W sur W' telle que l'on ait $h' = g \circ h$ et que g soit défini en tout point $h(a)$ pour lequel h' est défini en a.*

It existe toujours un tel (W, h) (Serre [3]); on peut construire W comme une variété projective. De plus, on peut démontrer, en vertu des résultats dans

Weil [4], qu'il existe une variété de Kummer (W, h) satisfaisant aux conditions suivantes (Matsusaka [1]):

(W1) *W est défini sur le corps K du module de $(A, \mathscr{C})$.*

(W2) *h est défini sur tout corps de définition pour $(A, \mathscr{C})$ contenant K.*

(W3) *Si σ est un isomorphisme d'un corps de définition pour $(A, \mathscr{C})$ contenant K, et si λ est un isomorphisme de $(A, \mathscr{C})$ sur $(A^\sigma, \mathscr{C}^\sigma)$, on a $h^\sigma \circ \lambda$.*

BIBLIOGRAPHIE

Les notions d'une variété polarisée, d'un corps du module et d'une variété de Kummer sont dues à Weil [5] et Matsusaka [1]; ce dernier a traité le cas général de variétés quelconques. Dans le cas de variété abélienne Matsusaka a considéré une structure d'une variété sur laquelle est fixé un point; mais on peut voir que cela n'engendre aucune différence essentielle entre [1] et cet exposé.

[1] MATSUSAKA (Teruhisa). Polarized varieties, fields of moduli and generalized Kummer varieties of polarized abelian variety, Amer. J. of Math., t. 80, 1958, p. 45–82.

[2] NISHI (Mieo). Some results on abelian varieties, Natural Science Report, Ochanomizu University, t. 9, 1958, p. 1–12.

[3] SERRE (Jean-Pierre). Sur la topologie des variétés algébriques en caractéristique p, Symposium de Topologie algébrique [Mexico. 1956] (à paraître).

[4] WEIL (André). The field of definition of a variety, Amer. J. of Math., t. 78, 1956, p. 509–524.

[5] WEIL (André). On the theory of complex multiplication, Proceedings of the International Symposium on algebraic number theory [Tokyo-Nikko. 1955], Tokyo, Science Council of Japan, 1956, p. 9–22.

[6] WEIL (André). On the projective embedding of abelian varieties, Algebraic geometry and topology, A Symposium in honor of S. Lefschets, Princeton, Princeton University Press, 1957, p. 177–181.

II. Exposé du 12 mai 1958

Dans cet exposé on s'occupera d'un système analytique de variétés algébriques et on démontrera que les coordonnées de Chow des membres du système se représentent par des fonctions analytiques.

5. Quelques lemmes. Il est commode d'introduire la notion d'un "point générique" pour les fonctions analytiques. Pour cela, on va d'abord démontrer le lemme suivant.

Lemme 5. *Soient U un ouvert connexe d'un espace numérique complexe $\mathbf{C}^n$, $\{f_i(z)\}$ un ensemble dénombrable de fonctions holomorphes sur U et k un sous-corps dénombrable du corps $\mathbf{C}$ des nombres complexes. Il existe alors une réunion dénombrable $H = \bigcup_{\nu=1}^{\infty} H_\nu$ de sous-ensembles analytiques H_ν de U, de dimension $n - 1$, telle que $\big(f_1(z_0),\, f_2(z_0),\, \dots \big)$ soit une spécialisation générique de $(f_1, f_2, \dots)$ par rapport à k pour tout $z_0 \in U - H$; cela veut dire qu'il existe un isomorphisme σ du corps $k(f_1, f_2, \dots)$ sur le corps $k\big(f_1(z_0), f_2(z_0), \dots \big)$ tel que $f_i^\sigma = f_i(z_0)$ et $a^\sigma = a$ pour $a \in k$.*

Nous choisissons et fixons, pour chaque entier positif N, un système $\{f_{i_1}, \dots, f_{i_p}\}$ de telle façon que $f_{i_1}, \dots, f_{i_p}$ soient algébriquement indépendants sur k et que $f_1, \dots, f_N$ soient algébriques sur le corps $k(f_{i_1}, \dots, f_{i_p})$. Soit F un polynôme en $X_1, \dots, X_p$, autre que 0, à coefficients dans k; soit $H(N, F)$ l'ensemble des points z de U tels qu'on ait $F\big(f_{i_1}(z), \dots, f_{i_p}(z)\big) = 0$. Comme $f_{i_1}, \dots, f_{i_p}$ sont algébriquement indépendants sur k, $H(N, F)$ est un sous-ensemble analytique de U, de dimension $n - 1$. Soit H_N la réunion de $H(N, F)$ pour tous les $F \neq 0$. Si $z_0 \in U - H_N$, on a $F\big(f_{i_1}(z_0), \dots, f_{i_p}(z_0)\big) \neq 0$ pour tout $F \neq 0$, de sorte qu'alors $f_{i_1}(z_0), \dots, f_{i_p}(z_0)$ sont algébriquement indépendants sur k. D'autre part, on voit que $\big(f_1(z_0), \dots, f_N(z_0)\big)$ est une spécialisation de $(f_1, \dots, f_N)$ par rapport à k. Si $z_0 \in U - H_N$, on a

$$\dim_k(f_1, \dots, f_N) = p = \dim_k \big(f_1(z_0), \dots, f_N(z_0)\big);$$

par suite $\big(f_1(z_0), \dots, f_N(z_0)\big)$ est une spécialisation générique de $(f_1, \dots, f_N)$ par rapport à k; on vérifie alors facilement que l'ensemble $H = \bigcup_{N=1}^{\infty} H_N$ jouit la propriété du lemme.

Les notations U, f_i et k étant les mêmes que ci-dessus, nous dirons qu'un point z_0 de U est un *point générique pour les f_i par rapport à k*, si $\big(f_1(z_0), f_2(z_0), \dots \big)$ est une spécialisation générique de $(f_1, f_2, \dots)$ par rapport à k. D'après le lemme 5, on voit que "presque tous" les points de U sont génériques pour les f_i par rapport à k. Si z_0 est générique pour les f_i par rapport à k, $\big(f_1(z'), f_2(z'), \dots \big)$ est une spécialisation de $\big(f_1(z_0), f_2(z_0), \dots \big)$ par rapport à k, pour tout point z' de U.

Lemme 6. *Soit $(u_1, \dots, u_n)$ un système de n indéterminées et soit $\mathbf{C}\{u\}$ l'anneau des séries formelles en $(u_1, \dots, u_n)$ à coefficients complexes. Soit $\big\{M_0(u) = 1, M_1(u), M_2(u), \dots \big\}$ la suite de tous les monômes $u_1^{e_1} \cdots u_n^{e_n}$, et soient $\sum_{\nu=0}^{\infty} c_{\nu\lambda} M_\nu(u)$, pour $1 \leq \lambda \leq r$, r éléments de $\mathbf{C}\{u\}$. Supposons* *qu'il existe $r + 1$ polynômes $G_0 = 1 + \sum_{\nu=1}^{p} a_\nu M_\nu(u)$, $G_\lambda = \sum_{\nu=0}^{q_\lambda} b_{\lambda\nu} M_\nu(u)$ $(1 \leq \lambda \leq r)$ à coefficients complexes, premiers entre eux, pour lesquels on a $G_0 \sum_{\nu=0}^{\infty} c_{\lambda\nu} M_\nu(u) = G_\lambda$ pour $1 \leq \lambda \leq r$. On a alors*

$$\mathbf{Q}(a_1, \ldots, a_p, b_{10}, \ldots, b_{\lambda\nu}, \ldots) = \mathbf{Q}(c_{10}, \ldots, c_{\lambda\nu}, \ldots),$$

$\mathbf{Q}$ *désignant le corps des nombres rationnels.*

Comme on a $\left(1 + \sum_{\nu=1}^{p} a_\nu M_\nu(u)\right)^{-1} = \sum_{m=0}^{\infty} \left(-\sum_{\nu=1}^{p} a_\nu M_\nu(u)\right)^m$, on voit facilement que le corps $\mathbf{Q}(a_\nu, b_{\lambda\nu})$ contient $Q(c_{\lambda\nu})$. Soit σ un automorphisme de $\mathbf{C}$ qui laisse fixes les éléments de $\mathbf{Q}(c_{\lambda\nu})$; σ peut être prolongé en un automorphisme du corps des fractions de $\mathbf{C}\{u\}$, que l'on désignera encore par σ, au moyen de la formule $\left(\sum_{\nu=0}^{\infty} d_\nu M_\nu(u)\right)^\sigma = \sum_{\nu=0}^{\infty} d_\nu^\sigma M_\nu(u)$ pour $\sum_{\nu=0}^{\infty} d_\nu M_\nu(u) \in \mathbf{C}\{u\}$. On a alors $G_\lambda/G_0 = G_\lambda^\sigma/G_0^\sigma$ pour tout λ. Comme $G_0, \ldots, G_r$ sont premiers entre eux, il existe r nombres complexes h_λ tels que G_0 et $\sum_{\lambda=1}^{r} h_\lambda G_\lambda$ soient premiers entre eux. D'après la relation $G_0 \sum_{\lambda=1}^{r} h_\lambda G_\lambda^\sigma = G_0^\sigma \sum_{\lambda=1}^{r} h_\lambda G_\lambda$ on voit que G_0^σ est un multiple de G_0. Comme G_0 et G_0^σ s'expriment sous les formes $G_0 = 1 + \sum_{\nu=1}^{p} a_\nu M_\nu(u)$, $G_0^\sigma = 1 + \sum_{\nu=1}^{p} a_\nu^\sigma M_\nu(u)$, on a $G_0 = G_0^\sigma$; d'où résute qu'on a $G_\lambda = G_\lambda^\sigma$ pour tout λ; en d'autres termes, σ laisse fixes les a_ν et les $b_{\lambda\nu}$. On en déduit que les a_ν et les $b_{\lambda\nu}$ appartiennent à $\mathbf{Q}(c_{\lambda\nu})$; ce qui complète la démonstration.

Lemme 7. *Soient U et V deux ouverts connexes respectivement de $\mathbf{C}^n$ et de $\mathbf{C}^m$; soient $F_\lambda(u, v)$, pour $1 \leq \lambda \leq r$, r fonctions holomorphes sur $U \times V$, où u et v désignent respectivement des points de U et de V. Supposons que, pour chaque $v_1 \in V$, il existe $r + 1$ polynômes $G_0(u; v_1), \ldots, G_r(u; v_1)$ en u satisfaisant aux conditions suivantes:*

- (G1) $F_\lambda(u, v_1) G_0(u; v_1) = G_\lambda(u; v_1)$ *pour $1 \leq \lambda \leq r$;*
- (G2) *les polynômes $G_0(u; v_1), \ldots, G_r(u; v_1)$ sont premiers entre eux;*
- (G3) *le polynôme $G_0(u; v_1)$ est de degré d, d étant un entier qui ne dépend pas de $v_1 \in V$.*

Alors, il existe des fonctions holomorphes $H_{\lambda\nu}(v)$ sur V telles que l'on ait, pour chaque $v_1 \in V$,

$$G_\lambda(u; v_1) = \gamma(v_1) \sum_{\nu=0}^{t} H_{\lambda\nu}(v_1) M_\nu(u) \qquad (0 \leq \lambda \leq r),$$

où $\gamma(v_1)$ est un nombre autre que 0 et les $M_\nu(u)$ sont des monômes en u à coefficient 1.

On va d'abord démontrer que $G_0(u; v_1)$ est partout $\neq 0$ sur U. D'après (G2), il existe r nombres complexes c_λ tels que $G_0(u; v_1)$ et $\sum_{\lambda=1}^{r} c_\lambda G_\lambda(u; v_1)$ soient premiers entre eux. On a, en vertu de (G1),

$$G_0(u; v_1) \sum_{\lambda=1}^{r} c_\lambda F_\lambda(u; v_1) = \sum_{\lambda=1}^{r} c_\lambda G_\lambda(u; v_1);$$

la fonction $\sum_{\lambda=1}^{r} c_\lambda F_\lambda(u; v_1)$ est holomorphe sur U; ce qui démontre que $G_0(u; v_1)$ est partout $\neq 0$ sur U.

On peut se borner à considérer le cas où U contient l'origine. Chaque fonction F_λ admet alors un dévelopement de la forme $F_\lambda(u; v) = \sum_{\nu=0}^{\infty} f_{\lambda\nu}(v) M_\nu(u)$ qui converge dans un voisinage de l'origine pour chaque $v \in V$, où

$$\{ M_0(u) = 1, \, M_1(u), \, M_2(u), \, \dots \}$$

est la suite de tous les monômes $u_1^{e_1} \cdots u_n^{e_n}$; on sait que les doefficients $f_{\lambda\nu}(v)$ sont des fonctions holomorphes sur V. Soit x un point générique de V pour les $f_{\lambda\nu}$ par rapport à $\mathbf{Q}$. Il existe $r+1$ polynômes $G_0(u; x), \dots, G_r(u; x)$ jouissant des propriétés (G1-3). Comme $G_0(u; x)$ n'est pas nul à l'origine, les $G_\lambda(u; x)$ s'expriment sous la forme

$$\xi G_0(u; x) = 1 + \sum_{\nu=1}^{p} \alpha_\nu M_\nu(u), \quad \xi G_\lambda(u; x) = \sum_{\nu=0}^{q_\lambda} \beta_{\lambda\nu} M_\nu(u) \qquad (1 \le \lambda \le r),$$

où $\xi \, (\neq 0)$, les α_ν et les $\beta_{\lambda\nu}$ sont des nombres complexes. On a, en vertu de (G1),

(i) $\quad \left(1 + \sum_{\nu=1}^{p} \alpha_\nu M_\nu(u)\right) \sum_{\nu=0}^{\infty} f_{\lambda\nu}(x) M_\nu(u) = \sum_{\nu=0}^{q_\lambda} \beta_{\lambda\nu} M_\nu(u) \quad (1 \le \lambda \le r)$

pour chaque λ; d'après le lemme 6, on a $\mathbf{Q}(f_{\lambda\nu}(x)) = \mathbf{Q}(\alpha_\nu, \beta_{\lambda\nu})$. Il existe donc des fonctions rationnelles R_ν, $S_{\lambda\nu}$ à coefficients dans $\mathbf{Q}$ tels qu'on ait $\alpha_\nu = R_\nu(f_{ij}(x))$, $\beta_{\lambda\nu} = S_{\lambda\nu}(f_{ij}(x))$. Posons, pour tout $v \in V$,

$$a_\nu(v) = R_\nu(f_{ij}(v)), \quad b_{\lambda\nu}(v) = S_{\lambda\nu}(f_{ij}(v));$$

les $a_\nu(v)$ et les $b_{\lambda\nu}(v)$ sont des fonctions méromorphes sur V. Comme x est générique pour les $f_{\lambda\nu}$, on a

$$\left(1 + \sum_{\nu=1}^{p} a_\nu(v) M_\nu(u)\right) \sum_{\nu=0}^{\infty} f_{\lambda\nu}(v) M_\nu(u) = \sum_{\nu=0}^{q_\lambda} b_{\lambda\nu}(v) M_\nu(u)$$

sur $U \times V$, pour chaque λ. Nous allons maintenant démontrer que les a_ν et les $b_{\lambda\nu}(v)$ sont holomorphes sur V. On peut supposer que $\{ M_0(u) = 1, M_1(u), \dots, M_p(u) \}$ est l'ensemble de tous les monômes $u_1^{e_1} \cdots u_n^{e_n}$ de degré $\le d$. Soit y un point de V; supposons que b_{ij} ne soit pas holomorphe en y. Alors il existe une suite $\{y_h\}_{h=1}^{\infty}$ de points de V qui converge vers y, et telle que l'on ait $\lim_{h\to\infty} b_{ij}(y_h)^{-1} = 0$. Soit B le sous-anneau des éléments $g(v)$ de $\mathbf{Q}(f_{\lambda\nu}(v))$ tels que $\lim_{h\to\infty} g(y_h)$ existe et soit fini. Posons $\varphi(g) = \lim_{h\to\infty} g(y_h)$ pour $g \in B$; alors φ est un homomorphisme de B dans $\mathbf{C}$; on a $\varphi(f_{\lambda\nu}) = f_{\lambda\nu}(y)$ puisque les $f_{\lambda\nu}$ sont holomorphes en y. On voit ainsi que $(f_{\lambda\nu}(y), 0)$ est une spécialisation de $(f_{\lambda\nu}, b_{ij}^{-1})$ par rapport à $\mathbf{Q}$. Comme x est générique pour les $f_{\lambda\nu}$, $(f_{\lambda\nu}(y), 0)$ est une spécialisation de $(f_{\lambda\nu}(x), \beta_{ij}^{-1})$ par rapport à $\mathbf{Q}$. On prolonge cette spécialisation en une spécialisation

$$\left(f_{\lambda\nu}(x), \beta_{ij}^{-1}, (1, \alpha_\nu, \beta_{\lambda\nu})\right) \longrightarrow \left(f_{\lambda\nu}(y), 0, (\alpha_0', \alpha_\nu', \beta_{\lambda\nu}')\right)$$

par rapport à $\mathbf{Q}$, où l'on considère $(1, \alpha_\nu, \beta_{\lambda\nu})$ et $(\alpha_0', \alpha_\nu', \beta_{\lambda\nu}')$ comme deux points de l'espace projectif de dimension $p + r + \sum_{\lambda=1}^{r} q_\lambda$. On vérifie facilement $\alpha_0' = 0$, car la spécialisation $(1, \alpha_\nu, \beta_{\lambda\nu}) \to (\alpha_0', \alpha_\nu', \beta_{\lambda\nu}')$ est compatible avec la spécialisation $\beta_{ij}^{-1} \to 0$. On a donc, d'après la relation (i),

$$\sum_{\nu=1}^{p} \alpha_\nu' M_\nu(u) \cdot \sum_{\nu=0}^{\infty} f_{\lambda\nu}(y) M_\nu(u) = \sum_{\nu=0}^{q_\lambda} \beta_{\lambda\nu}' M_\nu(u) \qquad (1 \le \lambda \le r).$$

Comme $(\alpha_0', \alpha_\nu', \beta_{\lambda\nu}')$ est un point de l'espace projectif, le polynôme $\sum_{\nu=1}^{p} \alpha_\nu' \cdot M_\nu(u)$ n'est pas 0. On a, en vertu de (G1),

(ii) $\quad G_0(u; y) \sum_{\nu=0}^{q_\lambda} \beta_{\lambda\nu}' M_\nu(u) = G_\lambda(u; y) \sum_{\nu=1}^{p} \alpha_\nu' M_\nu(u) \qquad (1 \le \lambda \le r).$

Comme $G_0(u; y), \dots, G_r(u; y)$ sont premiers entre eux, on voit que $\sum_{\nu=1}^{p} \alpha_\nu' \cdot M_\nu(u)$ est un multiple de $G_0(u; y)$. Le polynôme $G_0(u; y)$ est de degré d, et

$\sum_{\nu=1}^{p} \alpha'_\nu M_\nu(u)$ est un polynôme non nul de degré $\leq d$; il en résulte qu'on a $\sum_{\nu=1}^{p} \alpha'_\nu M_\nu(u) = \eta G_0(u; y)$, où η est un nombre complexe non nul; ce qui est absurde puisque $G_0(u; y)$ n'est pas nul à l'origine. Les fonctions $b_{\lambda\nu}$ sont donc holomorphes sur V. On peut démontrer, de la même manière, que les a_ν sont holomorphes sur V. Posons

$$H_0(u, v) = 1 + \sum_{\nu=1}^{p} a_\nu(v)M_\nu(u), \quad H_\lambda(u, v) = \sum_{\nu=0}^{q_\lambda} b_{\lambda\nu}(v)M_\nu(u)$$

pour $1 \leq \lambda \leq r$. On a alors $H_0(u, v)F_\lambda(u, v) = H_\lambda(u, v)$ pour $1 \leq \lambda \leq r$. Il en résulte qu'on a pour chaque $v_1 \in V$

(iii) $$G_0(u, v_1)H_\lambda(u, v_1) = G_\lambda(u, v_1)H_0(u, v_1) \qquad (1 \leq \lambda \leq r).$$

Comme $G_0(u, v_1), \ldots, G_\lambda(u, v_1)$ sont premiers entre eux, on voit que $H_0(u, v_1)$ est un multiple de $G_0(u, v_1)$. Le polynôme $G_0(u, v_1)$ est de degré d, et $H_0(u, v_1)$ est un polynôme autre que 0, de degré $\leq d$; on en déduit qu'il existe un nombre $\gamma(v_1) \neq 0$ tel que $G_0(u, v_1) = \gamma(v_1)H_0(u, v_1)$. On a alors $G_\lambda(u, v_1) = \gamma(v_1)H_\lambda(u, v_1)$ pour tout λ; ce qui complète la démonstration du lemme 7.

6. Systèmes analytiques de variétés algébriques. Soient U et S deux ouverts connexes respectivement de $\mathbf{C}^n$ et de $\mathbf{C}^m$; soient $f_i(u, s)$, pour $0 \leq i \leq N$, $N+1$ fonctions holomorphes sur $U \times S$, où u et s désignent respectivement les points de U et de S. Nous considérons les propriétés suivantes (S1-3) pour les fonctions f_i :

$$\text{(S1)} \qquad \text{rang} \begin{bmatrix} f_0(u, s) & f_1(u, s) & \cdots & f_N(u, s) \\ \dfrac{\partial f_0}{\partial u_1}(u, s) & \dfrac{\partial f_1}{\partial u_1}(u, s) & \cdots & \dfrac{\partial f_N}{\partial u_1}(u, s) \\ \cdots & \cdots & \cdots & \cdots \\ \dfrac{\partial f_0}{\partial u_n}(u, s) & \dfrac{\partial f_1}{\partial u_n}(u, s) & \cdots & \dfrac{\partial f_N}{\partial u_n}(u, s) \end{bmatrix} = n + 1$$

partout sur $U \times S$.

S'il en est ainsi, il n'y a aucun point (u, s) de $U \times S$ tel qu'on ait $f_i(u, s) = 0$ pour $0 \leq i \leq N$. On peut alors considérer $\big(f_0(u, s), \ldots, f_N(u, s)\big)$ comme un point de l'espace projectif P^N de dimension N; on le désignera par $f(u, s)$.

(S2) *Pour chaque point s de S, il existe une variété algébrique $V(s)$ dans P^N tel que l'ensemble $B(s) = \big\{ f(u, s) \,\big|\, u \in U \big\}$ soit un ouvert de $V(s)$ au sens de Zariski.*

D'après la condition (S1), la dimension de $V(s)$ est n.

(S3) *Le degré des variétés projectives $V(s)$ ne dépend pas du point $s \in S$.*

On désignera ce degré par d.

Théorème 3. *Soient les notations comme ci-dessus; supposons que les f_i satisfassent aux conditions (S1-3). Alors, pour tout point s_0 de S, il existe un voisinage W de s_0 et des fonctions holomorphes $\Phi_0(s), \ldots, \Phi_\kappa(s)$ sur W tels que $(\Phi_0(s), \ldots, \Phi_\kappa(s))$ soit le point de Chow de $V(s)$ pour tout $s \in W$.*

Nous allons rappeler, avant de démontrer ce théorème, la définition du point de Chow d'une variété algébrique.

Soit V une variété algébrique de dimension n, de degré d, dans l'espace projectif P^N de dimension N, et soit k un corps de définition pour V. Soient $t_j^{(i)}$, pour $1 \leq i \leq n$, $0 \leq j \leq N$, $n(N+1)$ variables indépendantes sur k; soit L la variété linéaire de dimension $N - n$ définie par les équations $\sum_{j=0}^{N} t_j^{(i)} X_j = 0$ $(1 \leq i \leq n)$, où les X_j désignent les coordonnées de P^N. On obtient alors d points $\left(x_0^{(i)}, \ldots, x_N^{(i)}\right)$ $(1 \leq i \leq d)$ comme les points de l'intersection $V \cap L$; on peut supposer que les $\left(x_0^{(i)}, \ldots, x_N^{(i)}\right)$ soient les conjugués algébriques de $\left(x_0^{(1)}, \ldots, x_N^{(1)}\right)$ sur $k(t)$ non seulement comme points de P^N, mais comme points de l'espace affine de dimension $N+1$. Soient $t_j^{(0)}$, pour $0 \leq j \leq N$, $N+1$ variables indépendantes sur $k(t)$; et $M_p(t^{(0)})$, pour $1 \leq p \leq \alpha$, les monômes en $(t_j^{(0)})$ à coefficients 1, de degré d. On voit alors facilement qu'il existe un élément ξ de $k\left(t^{(1)}, \ldots, t^{(n)}\right)$ tel que l'on ait

$$\xi \prod_{i=1}^{d} \left(\sum_{j=0}^{N} t_j^{(0)} x_j^{(i)} \right) = \sum_{p=1}^{\alpha} F_p\left(t^{(1)}, \ldots, t^{(n)}\right) M_p(t^{(0)}),$$

où les F_p sont des polynômes à coefficients dans k, premiers entre eux. On peut démontrer que le polynôme $\sum_{p=1}^{\alpha} F_p\left(t^{(1)}, \ldots, t^{(n)}\right) M_p(t^{(0)})$ est un polynôme homogène en $(t_0^{(i)}, \ldots, t_N^{(i)})$ de degré d, pour chaque $i \geq 0$. Soient $M_q(t)$ pour $0 \leq q \leq \beta$ les monômes à coefficient 1 en $(t_0^{(0)}, \ldots, t_N^{(n)})$, de degré d en $(t_0^{(i)}, \ldots, t_N^{(i)})$ pour chaque i. On a alors

$$\sum_{p=1}^{\alpha} F_p\left(t^{(1)}, \ldots, t^{(n)}\right) M_p(t^{(0)}) = \sum_{q=0}^{\beta} a_q M_q(t),$$

où les a_q sont des éléments de k; le point $(a_0, \ldots, a_\beta)$ de l'espace projectif de dimension β ne dépend que de la variété V et non du choix de k et de (t). On appellera ce point le *point de Chow* de V et on le désignera par $c(V)$; le polynôme $\sum_{q=0}^{\beta} a_q M_q(t)$ s'appellera la *forme de Chow* de V en les variables $(t_j^{(i)})$.

Nous allons maintenant démontrer le théorème 3. Nous désignerons par $L(t)$ la variété linéaire de dimension $N - n$ définie par $\sum_{j=0}^{N} t_{ij} X_j = 0$ $(1 \leq i \leq n)$ où les t_{ij} sont des nombres complexes. Fixons un point s_0 de S. Il existe un (t') tel que l'intersection $L(t') \cap V(s_0)$ consiste en exactement d points $p_1, \ldots, p_d$ de multiplicité 1 et que les p_α soient contenus dans $B(s_0)$; on peut supposer, sans perdre la généralité, que la première coordonnée de p_α soit $\neq 0$ pour tout α. On fixe un tel (t'). D'après la définition de $B(s)$, il existe un point $u^1 \in U$ tel qu'on ait $p_1 = f(u^1, s_0)$. Posons

$$\varphi_i(t, u, s) = \sum_{j=0}^{N} t_{ij} f_j(u, s) \qquad (1 \leq i \leq n);$$

on a alors $\varphi_i(t', u^1, s_0) = 0$ pour $1 \leq i \leq n$. Comme p_1 est un point de l'intersection $L(t') \cap V(s_0)$ de multiplicité 1, $L(t')$ est transversal à la variété linéaire tangente en p_1 à $V(s_0)$; on peut en déduire, en vertu de la condition (S1), qu'on a

(iv) $$\det\left[\left(\partial \varphi_i / \partial u_j\right)(t', u^1, s_0)\right]_{i,j=1}^{n} \neq 0.$$

On obtient donc, d'après le théorème des fonctions implicites, n fonctions $g_i(t, s)$ holomorphes en (t', s_0) telles que l'on ait

$$u_i^1 = g_i(t', s_0), \quad \sum_{j=0}^N t_{ij} f_j\big(g(t, s), s\big) = 0 \qquad (1 \le i \le n)$$

dans un voisinage de (t', s_0). Posons $h_j(t, s) = f_j\big(g(t, s), s\big)$ $(0 \le j \le N)$. On obtient, de la même manière, pour chaque point p_α, un système de $N + 1$ fonctions $h_j^{(\alpha)}(t, s)$ holomorphes en (t', s_0) telles que l'on ait

$$p_\alpha = \big(h_0^{(\alpha)}(t', s_0), \ldots, h_N^{(\alpha)}(t', s_0)\big), \quad \sum_{j=0}^N t_{ij} h_j^{(\alpha)}(t, s) = 0 \qquad (1 \le i \le n)$$

dans un voisinage de (t', s_0). Désignons par $p_\alpha(t, s)$, pour chaque α, le point de P^N dont les coordonnées sont $h_0^{(\alpha)}(t, s), \ldots, h_N^{(\alpha)}(t, s)$; alors $p_\alpha(t, s)$ est un point d'intersection $V(s) \cap L(t)$. Comme les p_α sont distincts, il existe un voisinage Y de (t', s_0) tel que les $p_\alpha(t, s)$ soient tous distincts pour chaque (t', s) de Y. De plus, comme on a $h_0^{(\alpha)}(t', s_0) \ne 0$ pour chaque α, on peut prendre le voisinage Y de telle façon que $h_0^{(\alpha)}(t', s) \ne 0$ $(1 \le \alpha \le d)$ pour tout $(t, s) \in Y$. Soient maintenant z_j, pour $0 \le j \le N$, $N + 1$ indéterminées et $M_\lambda(z)$, pour $0 \le \lambda \le r$, les monômes $z_0^{e_0} \cdots z_N^{e_N}$ de degré d; on posera $M_0(z) = z_0^d$. On a alors

$$\text{(v)} \qquad \prod_{\alpha=1}^d \Big(\sum_{j=0}^N z_j \cdot h_j^{(\alpha)}(t, s) / h_0^{(\alpha)}(t, s) \Big) = \sum_{\lambda=0}^r F_\lambda(t, s) M_\lambda(z),$$

où les $F_\lambda(t, s)$ sont des fonctions holomorphes sur Y; on a $F_0 = 1$. Soient T et W deux voisinages ouverts connexes respectivement de t' et de s_0 tels que l'on ait $T \times W \subset Y$. Soient w un point de W et k un corps minimum de définition pour $V(w)$. Soit $x \in T$ un point tel que la spécialisation

$$\big(F_\lambda(t, w), t_{ij}\big) \longrightarrow \big(F_\lambda(x, w), x_{ij}\big)$$

soit générique par rapport à k. En substituant (x, w) à (t, s) dans (v), on a

$$\prod_{\alpha=1}^d \Big(\sum_{j=0}^N z_j \cdot h_j^{(\alpha)}(x, w) / h_0^{(\alpha)}(x, w) \Big) = \sum_{\lambda=0}^r F_\lambda(x, w) M_\lambda(z).$$

On peut considérer (x_{ij}) comme un système de variables indépendantes par rapport à k. Soit $\sum_{\lambda=0}^r G_\lambda(x; w) M_\lambda(z)$ la forme de Chow de $V(w)$ en les variables (z_j, x_{ij}). Comme les $h_j^{(\alpha)}(x, w)$ sont les coordonées des points $p_\alpha(x, w)$ de l'intersection $L(x) \cap V(w)$, on vérifie facilement

$$G_0(x; w) \sum_{\lambda=0}^r F_\lambda(x, w) M_\lambda(z) = \sum_{\lambda=0}^r G_\lambda(x; w) M_\lambda(z);$$

d'où résulte qu'on a $G_0(x; w) F_\lambda(x, w) = G_\lambda(x; w)$. Comme les $G_\lambda(x; w)$ sont des polynômes à coefficients dans k, et comme x est générique pour les $F_\lambda(t, w)$ et les t_{ij} par rapport à k, on a $G_0(t; w) F_\lambda(t, w) = G_\lambda(t; w)$ $(1 \le \lambda \le r)$ dans T. D'après la propriété de la forme de Chow, les polynômes $G_0(t; w), \ldots, G_r(t; w)$ sont premiers entre eux, et $G_0(t; w)$ est de degré d^n. On peut donc appliquer le lemme 7 à ces fonctions $F_\lambda(t, s)$; il existe alors des fonctions holomorphes $H_{\lambda\nu}(s)$ sur W et un nombre $\gamma(w) \ne 0$ pour chaque $w \in W$, tels que l'on ait

$$G_\lambda(t; w) = \gamma(w) \sum_{\nu=0}^p H_{\lambda\nu}(w) M_\nu(t) \qquad (0 \le \lambda \le r),$$

où les $M_\nu(t)$ sont des monômes en t á coefficient 1. Alors, le polynôme $\sum_{\lambda,\nu} H_{\lambda\nu}(w) M_\nu(t) M_\lambda(z)$ donne la forme de Chow de $V(w)$ pour chaque $w \in W$; par suite les $H_{\lambda\nu}(w)$ sont les coordonées du point de Chow de $V(w)$ pour chaque $w \in W$; ce qui démontre le théorème 3.

Corollaire 1. *Les notations et les hypothèses étant comme ci-dessus, il y a des fonctions méromorphes $\Psi_1(s), \ldots, \Psi_\kappa(s)$ sur S telles que l'on ait*

$$c(V(s)) = (1, \Psi_1(s), \ldots, \Psi_\kappa(s)),$$

(en changeant l'ordre des indices, s'il y a lieu), chaque fois que les $\Psi_i(s)$ sont définis en s.

Prenons un recouvrement ouvert $\{W_\lambda\}$ de S tel que, pour chaque W_λ, il existe des fonctions holomorphes $\Phi_0(s; \lambda), \ldots, \Phi_\kappa(s; \lambda)$ sur W_λ pour lesquelles on a

$$c(V(s)) = (\Phi_0(s; \lambda), \ldots, \Phi_\kappa(s; \lambda)),$$

pour tout $s \in W$. On vérifie facilement que si $\Phi_i(s; \lambda)$ est identiquement nul dans W_λ pour un λ, $\Phi_i(s; \lambda)$ est identiquement nul pour tous les λ, car S est connexe. On voit que la valeur $\Phi_i(s; \lambda)/\Phi_j(s; \lambda)$ ne dépend que de s et non de λ si $\Phi_j(s; \lambda) \neq 0$. Supposons que $\Phi_0(s; \lambda)$ ne soit pas identiquement nul. On peut alors définir une fonction méromorphe $\Psi_i(s)$ sur S par $\Psi_i(s) = \Phi_i(s; \lambda)/\Phi_0(s; \lambda)$ pour chaque i. Soit s_0 un point de S tel que les $\Psi_i(s)$ soient définis en s_0. Il existe un ouvert W_λ du recouvrement contenant s_0. Comme les $\Phi_i(s; \lambda)$ donnent les coordonées du point $c(V(s))$, il existe un indice j tel que $\Phi_j(s_0; \lambda) \neq 0$. D'aprés la relation $\Phi_j(s_0; \lambda) = \Phi_0(s_0; \lambda)\Psi_j(s_0)$ on a $\Phi_0(s_0; \lambda) \neq 0$; on en déduit que $(1, \Psi_1(s_0), \ldots, \Psi_\kappa(s_0)) = c(V(s_0))$.

Corollaire 2. *Les notations et les hypothèses étant comme ci-dessus, soit k un sous-corps dénombrable de $\mathbf{C}$. Soit s_0 un point générique de S pour les $\Psi_i(s)$ par rapport à k. Alors, pour tout point s de S, $V(s)$ est une spécialisation de $V(s_0)$ par rapport à k.*

On se sert des mêmes notations que pour la démonstration du corollaire 1, et on suppose que $\Phi_0(s_0; \lambda)$ ne soit pas identiquement nul. Soit s' un point de S, et soit W un ouvert contenant s'. Soit s'' un point générique de W pour les $\Phi_i(s; \lambda)$ par rapport à k; alors, $c(V(s'))$ est une spécialisation de $c(V(s''))$ par rapport à k. Comme s'' est générique pour $\Phi_0(s; \lambda)$, $\Phi_0(s''; \lambda)$ n'est pas nul; donc les $\Psi_i(s)$ sont définis en s''. Par suite, si s_0 est générique pour les Ψ_i par rapport à k, $c(V(s''))$ est une spécialisation de $c(V(s_0))$ par rapport à k, de sorte que $c(V(s'))$ est une spécialisation de $c(V(s_0))$ par rapport à k; ce qui démontre le corollaire.

III. Exposé du 2 juin 1958

On se propose maintenant de démontrer les théorèmes principaux pour lesquels les exposés I, II sont des préliminaires. On aura besoin de fonctions thêta; pour commencer, on va rappeler quelques résultats concernant les formes de Riemann et les fonctions thêta.

7. Formes de Riemann et fonctions thêta. Soient D un lattice de $\mathbf{C}^n$ et $\{v_1, \ldots, v_{2n}\}$ une base de D par rapport à $\mathbf{Z}$. Pour que le torre complexe $\mathbf{C}^n/D$ ait une structure de variété abélienne, il faut et il suffit qu'il existe une forme $\mathbf{R}$-bilinéaire $\mathscr{E}(x, y)$ de $\mathbf{C}^n$ à valeurs dans $\mathbf{R}$ satisfaisant aux conditions suivantes:

(F1) *la valeur $\mathscr{E}(x, y)$ est entière pour tout $x \in D$, $y \in D$;*
(F2) $\mathscr{E}(x, y) = -\mathscr{E}(y, x)$;
(F3) *la forme $\mathscr{E}(x, \sqrt{-1}y)$ est une forme symétrique positive non-dégénérée.*

On appllera une telle forme $\mathscr{E}(x, y)$ une *forme de Riemann* sur $\mathbf{C}^n/D$. Soit ω la matrice dont les colonnes sont $v_1, \ldots, v_{2n}$. Il existe alors une matrice E de degré $2n$ telle qu'on ait $\mathscr{E}(\omega a, \omega b) = {}^t a E b$ pour deux vecteurs a, b de $\mathbf{R}^{2n}$. Les conditions (F1-F3) sont équivalentes aux conditions suivantes:

(E1) *les coefficients de E sont entiers;*
(E2) ${}^t E = -E$;
(E3) $\omega \cdot {}^t E^{-1} \cdot {}^t \omega = 0$;
(E4) $\sqrt{-1}\omega \cdot {}^t E^{-1} \cdot {}^t \overline{\omega}$ *est une matrice hermitienne positive non-dégénérée.*

Une forme de Riemann sur un tore $\mathbf{C}^n/D$ s'obtient à partir d'un diviseur analytique de $\mathbf{C}^n/D$ ainsi qu'il suit. Soit X un diviseur analytique positif sur $\mathbf{C}^n/D$. On sait qu'il existe une fonction f holomorphe sur $\mathbf{C}^n$ telle qu'on ait $(f) = X$, où (f) désigne le diviseur de la fonction f, et qu'on ait

$$f(u + d) = f(u) \exp\left[\ell_d(u) + c_d\right]$$

pour chaque $d \in D$, où $\ell_d(u)$ est une forme linéaire sur $\mathbf{C}^n$ et c_d est une constante; on appellera une telle fonction f une *fonction thêta* attachée à X. On peut démontrer qu'il existe deux matrices H, H_0 de degré $2n$ à coefficients dans $\mathbf{C}$ et un vecteur b de $\mathbf{C}^{2n}$ tels que

$$(\Theta) \qquad f(\omega x + \omega a) = f(\omega x) \exp\left[2\pi\sqrt{-1}({}^t a H x + (1/2){}^t a H_0 a + {}^t b a)\right],$$

pour tout $x \in \mathbf{R}^{2n}$ et tout $a \in \mathbf{Z}^{2n}$; on a de plus

$$H \equiv H_0 \pmod{\mathbf{Z}}, \quad {}^t H_0 = H_0.$$

Poson

$$E = H - {}^t H;$$

la matrice E ne dépend que de X et non du choix de f; on la désignera par $E(X)$. Si $\det(E) \neq 0$, la matrice donne une forme de Riemann sur $\mathbf{C}^n/D$. Toute forme de Riemann est ainsi obtenue à partir d'un diviseur analytique positif sur le tore.

Soit $\mathscr{L}$ l'ensemble des fonctions holomorphes f sur $\mathbf{C}^n$ satisfaisant à l'équation (Θ) où l'on fixe H, H_0 et b. L'ensemble $\mathscr{L}$ est un espace vectoriel de dimension $\det(E)^{1/2}$ sur $\mathbf{C}$. On peut choisir une base $\{v_1, \dots, v_{2n}\}$ de telle façon que la matrice E s'exprime sous la forme

$$E = \mu \begin{bmatrix} 0 & -\delta \\ \delta & 0 \end{bmatrix}, \quad \delta = \begin{bmatrix} \delta_1 & & & 0 \\ & \delta_2 & & \\ & & \ddots & \\ 0 & & & \delta_n \end{bmatrix}, \quad \delta_1 = 1,\ \delta_i | \delta_{i+1},$$

où $\mu, \delta_1, \dots, \delta_n$ sont des entiers positifs; μ et les δ_i ne dépendent pas du choix de $\{v_i\}$. Le théorème de plongement projectif de $\mathbf{C}^n/D$ par des fonctions thêta s'énonce:

Lemme 8. *Avec les même notations que ci-dessus, soit $\{\theta_0, \dots, \theta_N\}$ une base de $\mathscr{L}$. Si $\mu \geq 3$, on a*

$$\operatorname{rang} \begin{bmatrix} \theta_0(u) & \theta_1(u) & \cdots & \theta_N(u) \\ \dfrac{\partial \theta_0}{\partial u_1}(u) & \dfrac{\partial \theta_1}{\partial u_1}(u) & \cdots & \dfrac{\partial \theta_N}{\partial u_1}(u) \\ \cdots & \cdots & \cdots & \cdots \\ \dfrac{\partial \theta_0}{\partial u_n}(u) & \dfrac{\partial \theta_1}{\partial u_n}(u) & \cdots & \dfrac{\partial \theta_N}{\partial u_n}(u) \end{bmatrix} = n+1$$

partout sur $\mathbf{C}^n$. Soit Θ l'application de $\mathbf{C}^n$ dans l'espace projectif P^N de dimension N, définie par $\Theta(u) = \big(\theta_0(u), \dots, \theta_N(u)\big)$ pour $u \in \mathbf{C}^n$. Si $\mu \geq 3$, Θ induit un isomorphisme analytique de $\mathbf{C}^n/D$ sur une variété abélienne A dans P^N; de plus $\Theta(X)$ est une section hyperplane de A.

Soient A une variété abélienne de dimension n, définie par rapport à $\mathbf{C}$, et X un diviseur positif sur A. On sait qu'il existe un isomorphisme analytique η de A sur un tore complexe $\mathbf{C}^n/D$; $\eta(X)$ est alors un diviseur analytique positif sur $\mathbf{C}^n/D$. Posons $E_0(X) = E(\eta(X))$. Pour que l'on ait $\det E_0(X) \neq 0$, il faut et il suffit que X soit non-dégénéré. X et Y étant deux diviseurs positifs sur A, on a $E_0(X) = E_0(Y)$ si et seulement si X et Y sont algébriquement équivalents. Soient A' une autre variété abélienne de dimension n, X' un diviseur positif sur A' et η' un isomorphisme analytique de A' sur un tore $\mathbf{C}^n/D'$. Soient $\mathscr{E}$ et $\mathscr{E}'$ les formes de Riemann correspondant à $\eta(X)$ et $\eta'(X')$, respectivement, où l'on suppose que X, X' sont non-dégénérés. Soit λ un homomorphisme de A sur A'; il existe alors une matrice inversible M de degré n telle que l'on ait $M(D) \subset D'$, $M\eta(x) \equiv \eta'(\lambda x) \pmod{D'}$ pour $x \in A$. On voit que $\mathscr{E}^*(u, v) = \mathscr{E}'(Mu, Mv)$ est la forme de Riemann correspondant à $\eta\big(\lambda^{-1}(X')\big)$. Soient ω et ω' dont les colonnes forment des bases de D et de D', respectivement; comme on a $M(D) \subset D'$, il existe une matrice T de degré $2n$ à coefficients entiers telle que $M\omega = \omega'T$. Soient E et E' respectivement les matrices de $\mathscr{E}$ et de $\mathscr{E}'$ par rapport à ω et à ω'. La matrice ${}^tTE'T$ est alors la matrice de $\mathscr{E}^*$ par rapport à ω. Si λ est un homomorphisme de la variété abélienne polarisée $\big(A, \mathscr{C}(X)\big)$ sur la variété abélienne polarisée $\big(A', \mathscr{C}(X')\big)$, il existe, d'après la définition, deux entiers positifs r, s tels que $r\lambda^{-1}(X')$ soit algébriquement équivalent à sX; on a alors $r \cdot {}^tTE'T = sE$. Si λ est un isomorphisme, et si $\ell(X) =$

$\ell(X')$, on a ${}^t T E' T = E$, et T^{-1} est à coefficients entiers. Réciproquement, s'il existe une matrice complexe M de degré n et une matrice T de degré $2n$ à coefficients entiers telles que l'on ait $M\omega = \omega' T$, ${}^t T E' T = E$, $\det T = 1$, les variétés abéliennes polarisées $\big(A, \mathscr{C}(X)\big)$ et $\big(A', \mathscr{C}(X')\big)$ sont isomorphes.

8. Les groupes paramodulaires. On va rappeler quelques notions que l'on a déjà étudiées dans l'exposé 3. Soient $\delta_1, \ldots, \delta_n$ n entiers positifs tels que $\delta_1 = 1$, $\delta_i | \delta_{i+1}$ $(1 \le i \le n-1)$; et posons

$$\delta = \begin{bmatrix} \delta_1 & & & 0 \\ & \delta_2 & & \\ & & \ddots & \\ 0 & & & \delta_n \end{bmatrix}, \qquad E = \begin{bmatrix} 0 & -\delta \\ \delta & 0 \end{bmatrix}.$$

Soient u, v deux matrices de degré n à coefficents dans $\mathbf{C}$. Pour que la matrice $\omega = [u \ \ v]$ satisfasse aux conditions (E3-E4), il faut et il suffit que l'on ait

$$v\delta^{-1}\cdot{}^t u = u\delta^{-1}\cdot{}^t v, \quad \sqrt{-1}(v\delta^{-1}\cdot{}^t\overline{u} - u\delta^{-1}\cdot{}^t\overline{v}) \gg 0 \quad \text{(positive non-dégénérée);}$$

s'il en est ainsi, les deux matrices u, v sont inversibles. Posons $z = \delta v^{-1} u$; on voit alors que ω satisfait à (E3-E4) si et seulement si la matrice z est symétrique et à partie imaginaire positive non-dégénérée, c'est-à-dire si z est un point de l'espace de Siegel $\mathfrak{S}_n$. On notera désormais pour $z \in \mathfrak{S}_n$

$$\omega_\delta(z) = [z \ \ \delta];$$

alors $\omega_\delta(z)$ satisfait à (E3-E4). Soit $\Gamma'(\delta)$ le groupe des matrices T de degré $2n$ à coefficients entiers telles que l'on ait ${}^t T E T = E$. On vérifie facilemant que le groupe

$$\Gamma(\delta) = \left\{ \begin{bmatrix} 1 & 0 \\ 0 & \delta \end{bmatrix}^{-1} \cdot {}^t T^{-1} \begin{bmatrix} 1 & 0 \\ 0 & \delta \end{bmatrix} \ \middle| \ T \in \Gamma'(\delta) \right\}$$

est un sous-groupe de $Sp(n, \mathbf{R})$; on a évidemment $\Gamma(1_n) = Sp(n, \mathbf{Z})$. Soient z, z' deux points de $\mathfrak{S}_n$; supposons qu'il existe une matrice T de $\Gamma'(\delta)$ et une matrice M de degré n telles que $M\omega_\delta(z) = \omega_\delta(z')T$. On a alors

$$\begin{bmatrix} 1 & 0 \\ 0 & \delta \end{bmatrix}^{-1} \cdot {}^t T^{-1} \begin{bmatrix} 1 & 0 \\ 0 & \delta \end{bmatrix} \begin{bmatrix} z \\ 1 \end{bmatrix} = \begin{bmatrix} z' \\ 1 \end{bmatrix} \cdot {}^t M^{-1}.$$

En posant

$$U = \begin{bmatrix} 1 & 0 \\ 0 & \delta \end{bmatrix}^{-1} \cdot {}^t T^{-1} \begin{bmatrix} 1 & 0 \\ 0 & \delta \end{bmatrix} = \begin{bmatrix} a & b \\ c & d \end{bmatrix},$$

on a $U \in \Gamma(\delta)$ et $U(z) = (az+b)(cz+d)^{-1} = z'$. Réciproquement, si l'on a $z' = U(z)$ pour un élément U de $\Gamma(\delta)$, il existe un élément T de $\Gamma'(\delta)$ et une matrice M tels que $M\omega_\delta(z) = \omega_\delta(z')T$.

9. Les systémes analytiques de variétés abéliennes. Avec les mêmes notations que ci-dessus, soit $D_\delta(z)$ un lattice de $\mathbf{C}^n$ engendré par les colonnes de la matrice $\omega_\delta(z)$ sur $\mathbf{Z}$. Le fait que z est un point de $\mathfrak{S}_n$ entraîne que $\mathbf{C}^n/D_\delta(z)$ a une structure de variété abélienne; la matrice alternée E définit une forme de Riemann sur $\mathbf{C}^n/D_\delta(z)$. On va donner une base des fonctions thêta correspondant à un multiple de cette forme.

Soit μ un entier positif, et soit g un vecteur de $\mathbf{R}^n$ tel que les coordonnées de $\mu\delta g$ soient entières. Posons, pour $u \in \mathbf{C}^n$, $z \in \mathfrak{S}_n$,

$$\theta_{\mu,g}(u,\, z) = \sum \exp\left[2\mu\pi\sqrt{-1}\bigl(2^{-1}\cdot {}^t d z d + {}^t d u\bigr)\right],$$

où la somme est étendue à tous les vecteurs $d = a + g$, $a \in \mathbf{Z}^n$. On vérifie facilement que, pour $x \in \mathbf{R}^{2n}$, $w \in \mathbf{Z}^{2n}$, on a

$$\theta_{\mu,g}\bigl(\omega_\delta(z)(x+w),\, z\bigr) = \theta_{\mu,g}\bigl(\omega_\delta(z)x,\, z\bigr) \exp\left[2\pi\sqrt{-1}\bigl({}^t w H x + 2^{-1}\cdot {}^t w H_0 w\bigr)\right],$$

où

$$H = -\mu \begin{bmatrix} z & \delta \\ 0 & 0 \end{bmatrix}, \qquad H_0 = -\mu \begin{bmatrix} z & 0 \\ 0 & 0 \end{bmatrix}.$$

On a donc $H - {}^t H = \mu E$. Les $\theta_{\mu,g}$ sont donc des fonctions thêta correspondant à la matrice μE; de plus, les $\theta_{\mu,g}$ pour $\mu\delta g = [c_1 \ \cdots \ c_n]$, $0 \le c_i < \mu\delta_i$, forment une base de l'espace vectoriel des fonctions holomorphes sur $\mathbf{C}^n$ satisfaisant à l'équation (Θ) où $H = -\mu \begin{bmatrix} z & \delta \\ 0 & 0 \end{bmatrix}$, $H_0 = -\mu \begin{bmatrix} z & 0 \\ 0 & 0 \end{bmatrix}$, $b = 0$. Désignons par $\theta_0(u,\, z),\, \ldots,\, \theta_N(u,\, z)$ ces $\theta_{\mu,g}(u,\, z)$. Fixons un entier $\mu \ge 3$. D'après le lemme 8, l'application $u \mapsto \theta_z(u)$ définie par

$$\theta_z(u) = \theta(u,\, z) = \bigl(\theta_0(u,\, z),\, \ldots,\, \theta_N(u,\, z)\bigr)$$

induit un plongement de $\mathbf{C}^n/D_\delta(z)$ dans un espace projectif P^N. Soit $A_\delta(z)$ l'image de cette application; $A_\delta(z)$ est une variété abélienne. D'après un théorème de Poincaré où d'après le lemme 3, le degré de la variété projective $A_\delta(z)$ est $n!\mu^n \det(\delta)$. On voit ainsi que les fonctions $\theta_0(u,\, z),\, \ldots,\, \theta_N(u,\, z)$ sur $\mathbf{C}^n \times \mathfrak{S}_n$ satisfont aux conditions (S1-S3). On obtient donc, en appliquant le corollaire 1 du théorème 3, des fonctions méromorphes $\Psi_1(z),\, \ldots,\, \Psi_\kappa(z)$ sur $\mathfrak{S}_n$ telles que l'on ait

$$c\bigl(A_\delta(z)\bigr) = \bigl(1,\, \Psi_1(z),\, \ldots,\, \Psi_\kappa(z)\bigr)$$

chaque fois que les $\Psi_i(z)$ sont définis en z.

Proposition 2. *Soit $\mathscr{F}_\delta(z)$ la famille projective de $A_\delta(z)$ pour chaque z. Alors, les variétés $\mathscr{F}_\delta(z)$, pour $z \in \mathfrak{S}_n$, sont toutes de même dimension.*

Soit z un point de $\mathfrak{S}_n$, et soit f une transformation projective qui envoie $A_\delta(z)$ sur lui-même. Il existe alors un automorphisme ε de $A_\delta(z)$ et un point a de $A_\delta(z)$ tels que $f(x) = \varepsilon(x) + a$ pour $x \in A_\delta(z)$. Soit X une section hyperplane de $A_\delta(z)$; alors X est linéairement équivalent à $\varepsilon(X)_a$. L'automorphisme ε conserve la polarisation $\mathscr{C}(X)$; il n'y a qu'un nombre fini de tels automorphismes ε. De plus, comme $\varepsilon(X)$ est non-dégénéré, il n'y a qu'un nombre fini d'éléments a de $A_\delta(z)$ tels que $\varepsilon(X)_a$ soit linéairement équivalent à X; on en conclut qu'il n'y a qu'un nombre fini de transformations projectives qui laissent fixe $A_\delta(z)$. Il s'ensuit de là que $\mathscr{F}_\delta(z)$ est de dimension $(N+1)^2 - 1$, où N est la dimension de l'espace ambiant pour $A_\delta(z)$; ce qui démontre la proposition.

Théorème 4. *Il existe un sous-ensemble analytique Y de $\mathfrak{S}_n$ de codimension 1 et des fonctions méromorphes $\chi_1(z),\, \ldots,\, \chi_\lambda(z)$ sur $\mathfrak{S}_n$ jouissant des propriétés suivantes:*

18

(m1) *les $\mathscr{F}_\delta(z)$ pour $z \in \mathfrak{S}_n - Y$ sont de même degré;*
(m2) *on a $c(\mathscr{F}_\delta(z)) = (1, \chi_1(z), \ldots, \chi_\lambda(z))$ pour chaque point $z \in \mathfrak{S}_n - Y$.*

Les fonctions Ψ_i étant comme ci-dessus, soit W un sous-ensemble analytique de $\mathfrak{S}_n$ de codimension 1 tel que les Ψ_i sont holomorphes sur $\mathfrak{S}_n - W$, et soit z_0 un point générique de $\mathfrak{S}_n - W$ pour les Ψ_i par rapport à $\mathbf{Q}$. Comme on a

$$c(A_\delta(z_0)) = (1, \Psi_1(z_0), \ldots, \Psi_\kappa(z_0)),$$

la variété $A_\delta(z_0)$ est définie par rapport au corps $\mathbf{Q}(\Psi_i(z_0))$, de sorte que $\mathscr{F}_\delta(z_0)$ est défini par rapport au corps $\mathbf{Q}(\Psi_i(z_0))$. Il existe donc des éléments $\xi_1, \ldots, \xi_\lambda$ de $\mathbf{Q}(\Psi_i(z_0))$ tels que l'on ait $c(\mathscr{F}_\delta(z_0)) = (1, \xi_1, \ldots, \xi_\lambda)$. On obtient, au moyen de la spécialisation générique $(\Psi_i) \to (\Psi_i(z_0))$, des éléments $\chi_1, \ldots, \chi_\lambda$ du corps de fonctions $\mathbf{Q}(\Psi_i)$ tels que l'on ait $\chi_j(z_0) = \xi_j$ $(1 \le j \le \lambda)$; les fonctions χ_j sont méromorphes sur $\mathfrak{S}_n$; et l'on observera que la spécialisation $(\chi_j) \to (\chi_j(z_0))$ est générique par rapport à $\mathbf{Q}$. On sait qu'il existe un nombre fini d'éléments $a_1, \ldots, a_t$ de $\mathbf{Q}(\chi_j(z_0))$ jouissant de la propriété suivante: si $(\mathscr{F}', a_1', \ldots, a_t')$ est une spécialisation de $(\mathscr{F}_\delta(z_0), a_1, \ldots, a_t)$ par rapport à $\mathbf{Q}$ et si $a_k' \ne 0$ pour tout k, $\mathscr{F}'$ est une variété irréductible. Soit φ_k, pour chaque k, la fonction de $\mathbf{Q}(\chi_j)$ correspondant à a_k par la spécialisation générique $(\chi_j) \to (\chi_j(z_0))$; les φ_k sont méromorphes sur $\mathfrak{S}_n$. On voit facilement qu'il existe un sous-ensemble analytique Y de $\mathfrak{S}_n$ de codimension 1 tel que les Ψ_i, les χ_j, et les φ_k soient holomorphes sur $\mathfrak{S}_n - Y$ et qu'on ait $\varphi_k \ne 0$ $(1 \le k \le t)$ pour tout $z \in \mathfrak{S}_n - Y$. Soit z' un point de $\mathfrak{S}_n - Y$. Comme z_0 est générique pour les Ψ_i et comme les χ_j, les φ_k sont contenus dans $\mathbf{Q}(\Psi_i)$, $(\Psi_i(z_0), \chi_j(z_0), \varphi_k(z_0))$ est une spécialisation générique de $(\Psi_i, \chi_j, \varphi_k)$ par rapport à $\mathbf{Q}$. Il en résulte que $(\Psi_i(z'), \chi_j(z'), \varphi_k(z'))$ est une spécialisation de $(\Psi_i(z_0), \chi_j(z_0), \varphi_k(z_0))$ par rapport à $\mathbf{Q}$. Soit $(A', \mathscr{F}')$ une spécialisation de $(A_\delta(z_0), \mathscr{F}_\delta(z_0))$ sur cette spécialisation. Comme on a

$$c(A_\delta(z')) = (1, \Psi_1(z'), \ldots, \Psi_\kappa(z')),$$

on voit que $A' = A_\delta(z')$. On peut en déduire que $\mathscr{F}'$ contient la famille projective $\mathscr{F}_\delta(z')$ de $A_\delta(z')$. Comme $\mathscr{F}'$ est une spécialisation de $\mathscr{F}_\delta(z_0)$, $\mathscr{F}'$ a la même dimension et le même degré que $\mathscr{F}_\delta(z_0)$; par suite, $\mathscr{F}'$ a la même dimension que $\mathscr{F}_\delta(z')$. D'autre part, comme on a $\varphi_k(z') \ne 0$ pour tout k, la spécialisation $\mathscr{F}'$ de $\mathscr{F}_\delta(z_0)$ sur $(a_k) \to (\varphi_k(z'))$ est irréductible; on en conclut $\mathscr{F}' = \mathscr{F}_\delta(z')$, de sorte que $\mathscr{F}_\delta(z')$ a le même degré que $\mathscr{F}_\delta(z_0)$. Comme $(\chi_j(z'), \mathscr{F}_\delta(z'))$ est une spécialisation de $(\chi_j(z_0), \mathscr{F}_\delta(z_0))$, et comme on a $c(\mathscr{F}_\delta(z_0)) = (1, \chi_1(z_0), \ldots, \chi_\lambda(z_0))$, on voit que $c(\mathscr{F}_\delta(z')) = (1, \chi_1(z'), \ldots, \chi_\lambda(z'))$; ce qui complète la démonstration du théorème 4.

Proposition 3. *Les notations étant comme ci-dessus, pour qu'on ait $\mathscr{F}_\delta(z) = \mathscr{F}_\delta(z')$, il faut et il suffit qu'il existe un élément U de $\Gamma(\delta)$ tel que $z' = U(z)$.*

Soient z un point de $\mathfrak{S}_n$ et U un élément de $\Gamma(\delta)$. On a vu au n° 8 qu'il existe un élément T de $\Gamma'(\delta)$ et une matrice M tels que $M\omega_\delta(z) = \omega_\delta(U(z))T$. D'après ce qu'on a signalé au n° 7, les variétés abéliennes $A_\delta(z)$ et $A_\delta(U(z))$, polarisées par les sections hyperplanes, sont isomorphes; il suit de là et du théorème 1, que

$A_\delta(U(z))$ est une transformée de $A_\delta(z)$ par une transformation projective; on a donc $\mathscr{F}_\delta(z) = \mathscr{F}_\delta(U(z))$. Réciproquement, si l'on a $\mathscr{F}_\delta(z) = \mathscr{F}_\delta(z')$, on voit, en suivant le processus inverse, qu'il existe un élément U de $\Gamma(\delta)$ tel qu'on ait $z' = U(z)$.

Soit e le degré des variétés projectives $\mathscr{F}_\delta(z)$ pour $z \in \mathfrak{S}_n - Y$; et soit Y_0 l'ensemble des points z de $\mathfrak{S}_n$ tels que le degré de $\mathscr{F}_\delta(z)$ soit $< e$; Y_0 est alors un sous-ensemble de Y. Soit x un point de $\mathfrak{S}_n - Y_0$, et soit $c(\mathscr{F}_\delta(x)) = (b_0, \ldots, b_\lambda)$. On va démontrer que si $b_p \neq 0$, les fonctions $\chi_j(z)/\chi_p(z)$ sont holomorphes en x pour $0 \leq j \leq \lambda$, où l'on pose $\chi_0(z) = 1$. Supposons que χ_q/χ_p ne soit pas holomorphe en x. D'après le théorème 3, il existe un voisinage W de x et des fonctions holomorphes $\Phi_i(z)$ sur W tels que l'on ait $c(A_\delta(z)) = (\Phi_0(z), \ldots, \Psi_\kappa(z))$ pour tout $z \in W$. Les χ_j sont des fonctions rationnelles des fonctions $\Phi_i(z)$. Soit z_0 un point générique de W pour les Φ_i par rapport à $\mathbf{Q}$; on peut supposer $z_o \notin Y$. D'après le même argument que dans la démonstration du lemme 7, on peut démontrer que $(\Phi_i(x), \infty)$ est une spécialisation de $(\Phi_i(z_0), \chi_q(z_0)/\chi_p(z_0))$ par rapport à $\mathbf{Q}$. On prolonge cette spécialisation en une spécialisation

$$\left(\Phi_i(z_0), \chi_q(z_0)/\chi_p(z_0), A_\delta(z_0), \mathscr{F}_\delta(z_0)\right) \longrightarrow \left(\Phi_i(x), \infty, A', \mathscr{F}'\right).$$

On voit que $A' = A_\delta(x)$, car on a $c(A_\delta(x)) = (\Phi_i(x))$. Par suite, $\mathscr{F}'$ contient la famille projective $\mathscr{F}_\delta(x)$ de $A_\delta(x)$. Comme $\mathscr{F}_\delta(x)$ a la même dimension et le même degré que $\mathscr{F}_\delta(z_0)$, on a $\mathscr{F}' = \mathscr{F}_\delta(x)$; en d'autres termes, la spécialisation

$$c(\mathscr{F}_\delta(z_0)) = (\chi_j(z_0)) \longrightarrow c(\mathscr{F}_\delta(x)) = (b_j)$$

est compatible avec la spécialisation $\chi_q(z_0)/\chi_p(z_0) \to \infty$; ce qui est absurde, car on a $b_p \neq 0$. On a pu ainsi démontrer que les χ_j/χ_p sont holomorphes en x.

Soit maintenant S' l'ensemble des points z de $\mathfrak{S}_n$ tels que le degré de $\mathscr{F}_\delta(z)$ soit e et que la 0-ième coordonnée de $c(\mathscr{F}_\delta(z))$ ne soit pas 0; et posons $Y_1 = \mathfrak{S}_n - S'$. On a alors $Y \supset Y_1 \supset Y_0$. D'après ce qu'on vient de démontrer, les χ_i sont holomorphes sur $\mathfrak{S}_n - Y_1$ et l'on a

$$c(\mathscr{F}_\delta(z)) = (1, \chi_1(z), \ldots, \chi_\lambda(z))$$

pour tout $z \in \mathfrak{S}_n - Y_1$. D'après la proposition 3, on voit que l'ensemble Y_1 est invariant par le groupe $\Gamma(\delta)$.

Théorème 5. *Les fonctions χ_i du théorème 4 sont invariantes par le groupe paramodulaire $\Gamma(\delta)$. De plus, z et z' étant deux points de $\mathfrak{S}_n - Y_1$, si l'on a $\chi_i(z) = \chi_i(z')$ pour tout i, il existe un élément U de $\Gamma(\delta)$ tel que $z' = U(z)$.*

C'est une conséquence immédiate de la proposition 3. Nous avons ainsi construit des fonctions automorphes sur $\mathfrak{S}_n$ par rapport au groupe paramodulaire $\Gamma(\delta)$ au moyen des modules des variétés abéliennes polarisées.

Le système $\{A_\delta(z) \,|\, z \in \mathfrak{S}_n\}$ dépend du choix d'un entier $\mu \geq 3$. Désignons par $\chi_j^{(\delta,\mu)}$ les fonctions χ_i du théorème 4 obtenues à partir du système correspondant à μ. On va démontrer

$$(**) \qquad \mathbf{Q}(\chi_j^{(\delta,\mu)}) = \mathbf{Q}(\chi_j^{(\delta,\mu')})$$

pour tout μ, μ'. Soit z_0 un point générique de $\mathfrak{S}_n$ pour les $\chi_j^{(\delta,\mu)}$ et les $\chi_j^{(\delta,\mu')}$ par rapport à $\mathbf{Q}$. Soient $A_{\delta,\mu}(z_0)$ et $A_{\delta,\mu'}(z_0)$ respectivement les plongements projectifs de $\mathbf{C}^n/D_\delta(z_0)$ donnés par les fonctions thêta correspondant à μE et à $\mu' E$. Alors on voit facilement que les variétés abéliennes $A_{\delta,\mu}(z_0)$ et $A_{\delta,\mu'}(z_0)$, polarisées par les sections hyperplanes, sont isomorphes. Par suite, leurs corps de modules $\mathbf{Q}(\chi_j^{(\delta,\mu)}(z_0))$ et $\mathbf{Q}(\chi_j^{(\delta,\mu')}(z_0))$ coïncident. Comme z_0 est générique, on a la relation (**). On désignera par $K(\delta)$ ce même corps de (**).

C. L. Siegel a défini une fonction modulaire comme quotient de deux formes automorphes de même poids ayant des développements de Fourier, et démontré que toutes les fonctions modulaires sur $\mathfrak{S}_n$ forment un corps de fonctions de dimension $n(n+1)/2$ sur $\mathbf{C}$. Bailey a démontré, au moyen de la compactification de Satake, que si $n > 1$, toute fonction méromorphe sur $\mathfrak{S}_n$ invariante par $\Gamma(1_n)$ se représente comme quotient de deux formes automorphes de même poids ayant des développements de Fourier (cf. l'article signalé dans l'exposé 11). D'après Satake et Cartan, ce résultat est maintenant généralisé au cas de tout groupe commensurable au groupe modulaire $\Gamma(1_n)$ (cf. les exposés 11 et 17; voir en particulier le théorème fondamental de l'exposé 17). Nous supposons désormais $n > 1$ et désignons par $\mathfrak{M}(\delta)$ le corps des fonctions méromorphes sur $\mathfrak{S}_n$ invariantes par $\Gamma(\delta)$. D'après les résultats qu'on vient de signaler, $\mathfrak{M}(\delta)$ est un corps de fonctions de dimension $n(n+1)/2$ sur $\mathbf{C}$; et l'on a $\mathfrak{M}(\delta) \supset K(\delta)$. Le corps $\mathfrak{M}(1_n)$ est le corps de fonctions modulaires de Siegel.

Théorème 6. *Les nottions étant comme ci-dessus, on a $\mathfrak{M}(\delta) = \mathbf{C} \cdot K(\delta)$. De plus, les corps $K(\delta)$ et $\mathbf{C}$ sont linéairement disjoints sur $\mathbf{Q}$.*

D'après le théorème fondamental de l'exposé 17, il existe des formes automorphes $f_0(z), \ldots, f_M(z)$ sur $\mathfrak{S}_n$ par rapport à $\Gamma(\delta)$ telles que l'application $z \mapsto f(z) = (f_0(z), \ldots, f_M(z))$ induise un isomorphisme de l'espace qotient $\mathfrak{S}_n/\Gamma(\delta)$ sur un ouvert de Zariski V_0 d'une variété algébrique V dans l'espace projectif P^M. Soient χ_i les fonctions du théorème 4. Alors il existe des polynômes H_0, H_i à coefficients dans $\mathbf{C}$, de méme degré, tels que l'on ait $\chi_i(z) = H_i(f_j(z))/H_0(f_j(z))$. Soit k un sous-corps dénombrable de $\mathbf{C}$ contenant les coefficients de H_0, H_i, par rapport auquel V et $V-V_0$ soient rationnels. On va démontrer que le corps $k(\chi_i)$ contient les fonctions $f_j/f_{j'}$. Pour cela, supposons que $k(\chi_i)$ ne contienne pas f_p/f_q. Soit z_0 un point générique de $\mathfrak{S}_n - Y_1$ pour les f_j par rapport à k. Alors $k(\chi_i(z_0))$ ne contient pas $f_p(z_0)/f_q(z_0)$; par suite il existe un isomorphisme σ du corps $k(f_j(z_0))$ dans $\mathbf{C}$ tel que l'on ait $\chi_i(z_0)^\sigma = \chi_i(z_0)$ pour tout i et $f_p(z_0)^\sigma/f_q(z_0)^\sigma \neq f_p(z_0)/f_q(z_0)$. Comme V et $V - V_0$ sont rationnels par rapport à k, il existe un point z' de $\mathfrak{S}_n$ tel que $f_j(z') = f_j(z_0)^\sigma$ $(0 \leq j \leq M)$. On a alors $\chi_i(z') = \chi_i(z_0)^\sigma = \chi_i(z_0)$ pour tout i. Soit z_1 un point générique de $\mathfrak{S}_n - Y_1$ pour les Φ_i et les χ_i par rapport à $\mathbf{Q}$, où les Φ_i sont des fonctions sur un voisinage de z' comme dans le théorème 3; soit $(A', \mathscr{F}')$ une spécialisation de $(A_\delta(z_1), \mathscr{F}_\delta(z_1))$ sur la spécialisation $(\Phi_i(z_1), \chi_i(z_1)) \to (\Phi_i(z'), \chi_i(z'))$. On vérifie alors $A' = A_\delta(z')$ et $\mathscr{F}' \supset \mathscr{F}_\delta(z')$. D'autre part, on a $c(\mathscr{F}') = (\chi_i(z')) = (\chi_i(z_0)) = c(\mathscr{F}_\delta(z_0))$; d'où résulte $\mathscr{F}' = \mathscr{F}_\delta(z_0)$; ce qui entraîne que $\mathscr{F}'$ est irréductible. Comme $\mathscr{F}'$ et $\mathscr{F}_\delta(z')$ sont de même dimension, on a $\mathscr{F}' = \mathscr{F}_\delta(z')$ en vertu de la relation $\mathscr{F}' \supset \mathscr{F}_\delta(z')$. On a donc $\mathscr{F}_\delta(z') = \mathscr{F}_\delta(z_0)$. Il suit de là et de la proposition 3, qu'il existe un élément U de $\Gamma(\delta)$ tel que $z' = U(z_0)$; ce qui est absurde puisqu'on a $f_p(z')/f_q(z') \neq f_p(z_0)/f_q(z_0)$. On a ainsi démontré que $k(\chi_i)$ contient les $f_j/f_{j'}$; ce qui démontre $\mathfrak{M}(\delta) = \mathbf{C} \cdot K(\delta)$. La dernière assertion du théorème sera démontrée dans la section 11.

10. Fonction automorphes par rapport aux groupes de congruence.

Soit S_0 l'ensemble des points génériques de $\mathfrak{S}_n$ pour les Ψ_i et les $\theta_j(0, z)$, par rapport à $\mathbf{Q}$, où les Ψ_i sont des fonctions méromorhes sur $\mathfrak{S}_n$ telles que l'on ait $c(A_\delta(z)) = (1, \Psi_1(z), \ldots, \Psi_\kappa(z))$ chaque fois que les $\Psi_i(z)$ sont définis en z. On fixe un point z_0 de S_0 et on considère une variété de Kummer (W, h) de $A_\delta(z_0)$ jouissant des propriétés (W1-W3). La variété W est définie par rapport à $\mathbf{Q}(\chi_j(z_0))$, et la variété abélienne $A_\delta(z_0)$ et l'application h sont définies par rapport à $\mathbf{Q}(\Psi_i(z_0), \theta_j(0, z_0))$. On supposera que W soit une variété dans un espace projectif. Pour chaque point z' de S_0, on obtient une spécialisation générique

$$\big(\Psi_i(z_0), \theta_j(0, z_0)\big) \longrightarrow \big(\Psi_i(z'), \theta_j(0, z')\big)$$

par rapport à $\mathbf{Q}$. Soit (W', h') la spécialisation générique de (W, h) sur cette spécialisation. Alors (W', h') est une variété de Kummer de $A_\delta(z')$ jouissant des propriétés (W1-W3). On notera $W' = W(z')$ et $h'(t) = h(t, z')$ pour $t \in A_\delta(z')$. $W(z')$ est défini par rapport à $\mathbf{Q}(\chi_i(z'))$, et $h(*, z')$ est défini par rapport à $\mathbf{Q}(\Psi_i(z'), \theta_j(0, z'))$.

Soit b un vecteur de $\mathbf{R}^{2n}$; on le considérera comme une matrice à $2n$ lignes et une colonne. Soit $S(b)$ l'ensemble des points génériques de $\mathfrak{S}_n$ pour les Ψ_i, les $\theta_j(0, z)$ et les $\theta_j\big(\omega_\delta(z)b, z\big)$. On fixe un point z_0 de $S(b)$ et on considère le point $h\big(\theta_{z_0}(\omega_\delta(z_0)b), z_0\big)$ sur $W(z_0)$; ce point est rationnel par rapport au corps

$$\mathbf{Q}\bigg(\Psi_i(z_0), \theta_j\big(\omega_\delta(z_0)b, z_0\big), \theta_j(0, z_0) \bigg);$$

il existe donc des fonctions rationnelles φ_α à coefficients dans $\mathbf{Q}$ telles que les coordonnées du point $h\big(\theta_{z_0}(\omega_\delta(z_0)b), z_0\big)$ soient les

$$\varphi_\alpha\bigg(\Psi_i(z_0), \theta_j(\omega_\delta(z_0)b, z_0), \theta_j(0, z_0) \bigg) \qquad (0 \leq \alpha \leq \nu).$$

Posons

$$\xi_\alpha(z, b) = \varphi_\alpha\bigg(\Psi_i(z), \theta_j(\omega_\delta(z)b, z), \theta_j(0, z) \bigg)$$

pour chaque α; les $\xi_\alpha(z, b)$ sont des fonctions méromorphes sur $\mathfrak{S}_n$. On vérifie facilement que si z' est un point de $S(b)$, on a

$$h\big(\theta_{z'}(\omega_\delta(z')b), z'\big) = (\xi_0(z', b), \ldots, \xi_\nu(z', b)).$$

Théorème 7. *Les notations étant comme ci-dessus, soit U un élément de $\Gamma(\delta)$ et soit $T = \begin{bmatrix} 1 & 0 \\ 0 & \delta \end{bmatrix}^{-1} \cdot {}^t U^{-1} \begin{bmatrix} 1 & 0 \\ 0 & \delta \end{bmatrix}$. On a alors*

$$\xi_\alpha\big(U(z), Tb\big)/\xi_\beta\big(U(z), Tb\big) = \xi_\alpha(z, b)/\xi_\beta(z, b).$$

Soit z_0 un point générique de $\mathfrak{S}_n$ pour les fonctions $\Psi_i(z)$, $\Psi_i\big(U(z)\big)$, $\theta_j(0, z)$, $\theta_j\big(0, U(z)\big)$, $\xi_\alpha(z, b)$, $\xi_\alpha\big(U(z), Tb\big)$ par rapport à $\mathbf{Q}$; on peut prendre z_0 de telle façon que $z_0 \in S(b)$, $U(z_0) \in S(Tb)$. On définit un isomorphisme σ du corps $\mathbf{Q}\big(\Psi_i(z_0), \theta_j(0, z_0)\big)$ par $\Psi_i(z_0)^\sigma = \Psi_i\big(U(z_0)\big)$, $\theta_j(0, z_0)^\sigma = \theta_j\big(0, U(z_0)\big)$. On a alors $A_\delta\big(U(z_0)\big) = A_\delta(z_0)^\sigma$, $W\big(U(z_0)\big) = W(z_0)^\sigma = W(z_0)$ et $h^\sigma(t) = h\big(t, U(z_0)\big)$ pour $t \in A_\delta\big(U(z_0)\big)$, où h désigne l'application $x \mapsto h(x, z_0)$. Il existe une matrice M de degré n telle que l'on ait

$$(1) \qquad\qquad M\omega_\delta(z_0) = \omega_\delta\big(U(z_0)\big)T.$$

M donne un isomorphism de $\mathbf{C}^n/D_\delta(z_0)$ sur $\mathbf{C}^n/D_\delta\big(U(z_0)\big)$; il existe alors un isomorphisme λ de $A_\delta(z_0)$ sur $A_\delta\big(U(z_0)\big)$ tel que l'on ait

$$(2) \qquad\qquad \lambda\big(\theta(u, z_0)\big) = \theta\big(Mu, U(z_0)\big)$$

pour tout $u \in \mathbf{C}^n$. D'après la propriété (W3) de la variété de Kummer $\big(W(z_0),$ $h\big)$, on a $h(t, z_0) = h\big(\lambda t, U(z_0)\big)$ pour $t \in A_\delta(z_0)$. En vertu des relations (1), (2), on a

$$h\big(\theta\big(\omega_\delta(z_0)b, z_0\big), z_0\big) = h\big(\theta\big(\omega_\delta(U(z_0))Tb, U(z_0)\big), U(z_0)\big).$$

Comme z_0 est un point de $S(b)$ et comme $U(z_0)$ est un point de $S(Tb)$, on a

$$\xi_\alpha(z_0, b)/\xi_\beta(z_0, b) = \xi_\alpha\big(U(z_0), Tb\big)/\xi_\beta\big(U(z_0), Tb\big);$$

ce qui démontre le théorème, puisque z_0 est un point générique pour les fonction $\xi_\alpha(z, b)$, $\xi_\alpha\big(U(z), Tb\big)$.

Soit q un entier positif; on désignera par $\Gamma'(\delta, q)$ le sous-groupe de $\Gamma'(\delta)$ formé des éléments T tels que $T \equiv \pm 1_{2n} \pmod q$, et par $\Gamma(\delta, q)$ le sous-groupe $\left\{ \begin{bmatrix} 1 & 0 \\ 0 & \delta \end{bmatrix}^{-1} \cdot {}^t T^{-1} \begin{bmatrix} 1 & 0 \\ 0 & \delta \end{bmatrix} \,\middle|\, T \in \Gamma'(\delta, q) \right\}$ de $\Gamma(\delta)$. Soit $\mathfrak{M}(\delta, q)$ le corps des fonctions méromorphes sur $\mathfrak{S}_n$ invariantes par $\Gamma(\delta, q)$. Alors $\mathfrak{M}(\delta, q)$ est une extension galoisienne du corps $\mathfrak{M}(\delta)$ dont le groupe de Galois est isomorphe à $\Gamma(\delta)/\Gamma(\delta, q)$. Nous allons maintenant démontrer que le corps $\mathfrak{M}(\delta, q)$ est engendré par $\xi_\alpha(z, b)$ pour certaines valeurs b.

Proposition 4. *Soient b, b' deux vecteurs de $\mathbf{R}^{2n}$. Alors, pour qu'on ait*

$$\xi_\alpha(z, b)/\xi_\beta(z, b) = \xi_\alpha(z, b')/\xi_\beta(z, b')$$

pour tous les α, β, il faut et il suffit qu'on ait $b \equiv \pm b' \pmod{\mathbf{Z}^{2n}}$.

Soit z_0 un point de $S(b) \cap S(b')$, générique pour les $\xi_\alpha(z, b)$ et les $\xi_\alpha(z, b')$ par rapport à $\mathbf{Q}$. Il n'y a pas d'automorphisme de $A_\delta(z_0)$ autre que $x \mapsto \pm x$, puisque $A_\delta(z_0)$ est générique. Par suite, on a $h(t, z_0) = h(t', z_0)$ si et seulement si $t = \pm t'$. On en déduit que l'on a

$$h\big(\theta\big(\omega_\delta(z_0)b, z_0\big), z_0\big) = h\big(\theta\big(\omega_\delta(z_0)b', z_0\big), z_0\big)$$

si et seulement si $\omega_\delta(z_0)b \equiv \pm\omega_\delta(z_0)b' \pmod{D_\delta(z_0)}$. En d'autre termes, pour qu'on ait $\xi_\alpha(z_0, b)/\xi_\beta(z_0, b) = \xi_\alpha(z_0, b')/\xi_\beta(z_0, b')$, il faut et il suffit qu'on ait $b \equiv \pm b' \pmod{\mathbf{Z}^{2n}}$; ce qui démontre la proposition, car z_0 est générique pour les $\xi_\alpha(z, b)$ et les $\xi_\alpha(z, b')$.

Soient a_i $(1 \leq i \leq q^{2n})$ les vecteurs de $\mathbf{R}^{2n}$ tels que $a_i = q^{-1}(\gamma_1, \ldots, \gamma_{2n})$, $0 \leq \gamma_j < q$. D'après le théorème 7 et la proposition 3, U étant un élément de $\Gamma(\delta)$, pour qu'on ait

$$\xi_\alpha\big(U(z),\, a_i\big)/\xi_\beta\big(U(z),\, a_i\big) = \xi_\alpha(z,\, a_i)/\xi_\beta(z,\, a_i)$$

pour tous les α, β et tous les a_i, il faut et il suffit que U soit contenu dans $\Gamma(\delta, q)$. On peut en déduire facilement le théorème suivant:

Théorème 8. *Soit $K(\delta, q)$ le corps engendré sur $\mathbf{Q}$ par les fonctions χ_i et les $\xi_\alpha(z, a_i)/\xi_\beta(z, a_i)$ pour tous les α, β et tous les a_i. On a alors $\mathfrak{M}(\delta, q) = \mathbf{C} \cdot K(\delta, q)$.*

11. Corps de constantes des corps de fonctions automorphes. Les systèmes $\big\{A_\delta(z) \,\big|\, z \in \mathfrak{S}_n\big\}$ sont les plus grands systèmes de variétés abéliennes polarisées. En effet, toute variété abélienne polarisée définie par rapport à $\mathbf{C}$, est isomorphe à un membre de ces systèmes. En outre de ces systèmes, il y a divers systèmes de variétés abéliennes polarisées, qui engendrent aussi des fonctions automorphes attachées à certains groupes de transformations. Mais dans cet exposé, on va seulement démontrer un théorème concernant le corps de constantes.

Soit S un ouvert connexe de $\mathbf{C}^m$, et soit f une application holomorphe de S dans $\mathfrak{S}_n$. On fixe un δ et on pose $A(s) = A_\delta\big(f(s)\big)$. On obtient alors un système $\big\{A(s) \,\big|\, s \in S\big\}$ de variétés abéliennes polarisées. Soit $\mathscr{F}(s)$ la famille projective de $A(s)$ pour chaque $s \in S$. D'après le même procédé que pour $A_\delta(z)$, on obtient des fonctions méromorphes $\varphi_1(s), \ldots, \varphi_\kappa(s)$ telles que

$$c\big(A(s)\big) = \big(1,\, \varphi_1(s),\, \ldots,\, \varphi_\kappa(s)\big)$$

chaque fois que les $\varphi_i(s)$ sont définis en s; de plus, il existe un sous-ensemble Y' de S de codimension 1 et des fonctions méromorphes $\eta_1(s), \ldots, \eta_\nu(s)$ sur S telles qu'on ait

$$c\big(\mathscr{F}(s)\big) = \big(1,\, \eta_1(s),\, \ldots,\, \eta_\nu(s)\big)$$

pour $s \in S - Y'$.

Soit k un sous-corps dénombrable de $\mathbf{C}$. Le système $\big\{A(s) \,\big|\, s \in S\big\}$ est dit *complet* par rapport à k s'il satisfait à la condition suivante:

s_1 étant un point générique de S pour les φ_i par rapport à k, si B est une spécialisation générique de $A(s_1)$ par rapport à k, il existe un point s' tel que les variétés abéliennes $A(s')$ et B polarisées par les sections hyperplanes, soient isomorphes.

Théorème 9. *Supposons que le système $\big\{A(s) \,\big|\, s \in S\big\}$ soit complet par rapport à un corps dénombrable k. Alors, le corps $k(\eta_1, \ldots, \eta_\nu)$ est une extension régulière de k et l'on a $\dim_k k(\eta_1, \ldots, \eta_\nu) = \dim_{\mathbf{C}} \mathbf{C}(\eta_1, \ldots, \eta_\nu)$.*

On choisit $\eta_{i_1}, \ldots, \eta_{i_r}$ de telle façon que les η_{i_j} soient algébriquement indépendants sur $\mathbf{C}$ et les η_i algébriques sur $\mathbf{C}(\eta_{i_1}, \ldots, \eta_{i_r})$. On voit facilement qu'il existe un sous-corps k_0 de $\mathbf{C}$ engendré sur $\mathbf{Q}$ par un nombre fini d'éléments, tel que les η_i soient algébriques sur $k_0(\eta_{i_1}, \ldots, \eta_{i_r})$; on a slors

$$\dim_{k_0} k_0(\eta_1, \ldots, \eta_\nu) = r = \dim_{\mathbf{C}} \mathbf{C}(\eta_1, \ldots, \eta_\nu).$$

Soit s_0 un point générique de S pour les η_i par rapport à $\overline{k}_0$, où $\overline{k}_0$ désigne la clôture algébrique de k_0. Soit V le lieu de $c(\mathscr{F}(s_0))$ par rapport à $\overline{k}_0$ et soit k_1 le plus petit corps de définition pour V. Le corps k_1 est un sous-corps de $\overline{k}_0$ engendré sur $\mathbf{Q}$ par un nombre fini d'éléments. On voit que $(1, \eta_1, \ldots, \eta_\nu)$ est une spécialisation générique de $c(\mathscr{F}(s_0))$ par rapport à $\overline{k}_0$, et donc par rapport à k_1; on en déduit

$$\dim_{k_1} k_1(\eta_1, \ldots, \eta_\nu) = r = \dim_{\mathbf{C}} \mathbf{C}(\eta_1, \ldots, \eta_\nu).$$

En vertu du fait que k_1 est un corps de définition pour la variété V, on voit que le corps $k_1(\eta_1, \ldots, \eta_\nu)$ est une extension régulière de k_1. Notre théorème sera donc démontré si nous faisons voir que k contient k_1. Dans ce but, supposons que k ne contienne pas k_1; alors, K étant le corps composé $k \cdot k_1$, il existe un isomorphisme σ du corps K dans $\mathbf{C}$ tel que σ soit l'identité sur k et qu'on ait $V \neq V^\sigma$. Soit s_1 un point générique de S pour les φ_i par rapport à K; on vérifie aisément que $c(\mathscr{F}(s_1))$ est un point générique de V par rapport à K. Soit w un point générique de V^σ par rapport au corps composé $K \cdot K^\sigma$. Il existe alors un isomorphisme τ de $K(c(\mathscr{F}(s_1)))$ sur $K^\sigma(w)$ tel que l'on ait $c(\mathscr{F}(s_1))^\tau = w$ et $\tau = \sigma$ sur K. Prolongeons τ à un isomorphisme de $K(c(A(s_1)))$ que l'on désignera encore par τ. Soient $B = A(s_1)^\tau$ et $\mathscr{F}'$ la famille projective de B; on a alors $\mathscr{F}' = \mathscr{F}(s_1)^\tau$, $c(\mathscr{F}') = c(\mathscr{F}(s_1))^\tau = w$. Comme τ est l'identité sur k, B est une spécialisation générique de $A(s_1)$ par rapport à k. D'après notre hypothèse, il existe un point s' de S tel que les variétés abéliennes B et $A(s')$, polarisées par les sections hyperplanes, soient isomorphes. Il en résulte qu'on a $\mathscr{F}' = \mathscr{F}(s')$. On voit, en vertu du corollaire 2 du théorème 3, que $A(s')$ est une spécialisation de $A(s_1)$ par rapport à K. Soit $(A(s'), \mathscr{F}'')$ une spécialisation de $(A(s_1), \mathscr{F}(s_1))$ par rapport à K; on voit que $\mathscr{F}''$ contient la famille projective $\mathscr{F}(s')$ de $A(s')$. Comme on a $\mathscr{F}(s') = \mathscr{F}' = \mathscr{F}(s_1)^\tau$, $\mathscr{F}(s')$ a la même dimension et le même degré que $\mathscr{F}(s_1)$, et par conséquent que $\mathscr{F}''$; d'où résulte qu'on a $\mathscr{F}'' = \mathscr{F}'$. On a ainsi démontré que $\mathscr{F}'$ est une spécialisation de $\mathscr{F}(s_1)$ par rapport à K; en d'autres termes, $w = c(\mathscr{F}')$ est une spécialisation de $c(\mathscr{F}(s_1))$ par rapport à K; donc w est un point de V; ce qui est absurde, car w est un point générique de V^σ par rapport à $K \cdot K^\sigma$. Par suite k doit contenir k_1; ceci achève la démonstation.

Le système $\{A_\delta(z) \,|\, z \in \mathfrak{S}_n\}$ est complet par rapport à $\mathbf{Q}$, pour tout δ. En effet, soient z_1 un point de $\mathfrak{S}_n$ et B une spécialisation générique de $A_\delta(z_1)$ par rapport à $\mathbf{Q}$. Soient X et X' des sections hyperplanes de $A_\delta(z_1)$ et de B, respectivement; et soient $E(X)$ et $E(X')$ les matrices alternées correspondant aux diviseurs X et X', respectivement, par rapport à certains systèmes de matrices coordonnées. On voit alors facilement que les diviseurs élémentaires des matrices $E(X)$ et $E(X')$ sont les mêmes, en considérant les représentations ℓ-adiques. Il existe donc un membre $A_\delta(z')$ du système tel que les variétés abéliennes $A_\delta(z')$ et B polarisées par les sections hyperplanes, soient isomorphes, comme on l'a signalé au n° 7. On peut donc appliquer le théorème 9 aux corps $\mathbf{Q}$ et $K(\delta)$. Le corps $K(\delta)$ est alors une extension régulière de $\mathbf{Q}$, de dimension $n(n+1)/2$; ce qui démontre la dernière assertion du théorème 6.

Fonctions automorphes et correspondances modulaires

Proceedings of the International Congress of Mathematicians,
Edinburgh, 1958 (1959), 330-338

Il est bien connu que le groupe modulaire de Siegel est le groupe de transformations pour les périodes des fonctions abéliennes, mais on sait peu de chose des relations entre les fonctions modulaires et les abéliennes, sauf au cas de dimension 1. L'objet de cette conférence est d'énoncer quelques idées et résultats à ce sujet. Nous démontrerons d'abord que les 'modules' des variétés abéliennes polarisées, regardés comme fonctions des périodes, engendrent les fonctions modulaires de Siegel, et que les fonctions modulaires par rapport aux groupes de congruence s'obtiennent à partir des points t sur les variétés abéliennes tels que $qt = 0$ pour un entier q. On peut appliquer le même procédé à d'autres types de fonction automorphe, par exemple, aux fonctions de Hilbert. Ce sont non seulement la généralisation de la fonction $j(\tau)$ ou des '$\wp$-Teilwerte', mais les outils dont on se sert pour attaquer les problèmes arithmétiques des fonctions automorphes. On voit en effet que les opérateurs T_n, introduits par Hecke pour les formes modulaires elliptiques et généralisés par ses disciples pour divers formes automorphes, ne sont autres que les représentations de certaines correspondances algébriques, appelées correspondances modulaires, définies au moyen d'isogénies de variétées abéliennes. En se basant sur cette idée, on acquiert des formules de congruence pour les correspondances modulaires dans le cas d'une certaine classe de fonctions automorphes de dimension 1, que l'on rencontrera à la fin de cet exposé.

Pour commencer, on va rappeler les notions de variété abélienne polarisée et de ses modules (Weil[7], Matsusaka[3], Shimura[6]). Soient A une variété abélienne et X un diviseur sur A. On désignera par $\mathscr{C}(X)$ l'ensemble de tous les diviseurs X' sur A pour lesquels il existe deux entiers positifs m, m' tels que mX soit algébriquement équivalent à $m'X'$. $\mathscr{C}(X)$ s'appellera une *polarisation* de A si $\mathscr{C}(X)$ contient un diviseur ample. On entendra par une *variété abélienne polarisée* une variété abélienne A sur laquelle est donnée une polarisation $\mathscr{C}$, désignée par $(A, \mathscr{C})$. $(A, \mathscr{C})$ est dit isomorphe à une autre variété abélienne polarisée $(A', \mathscr{C}')$ s'il existe un isomorphisme de A sur A' qui envoie $\mathscr{C}$ sur $\mathscr{C}'$. On dira que $(A, \mathscr{C})$ est défini sur un corps k si A est défini sur k et si $\mathscr{C}$ contient un diviseur X rationnel sur k. Ainsi, soit σ un isomorphisme de k sur un corps k'; on désignera par $\mathscr{C}^\sigma$ la polarisation $\mathscr{C}(X^\sigma)$ de A^σ. On peut démontrer qu'il existe un sous-corps K de k jouissant de la propriété suivante: pour qu'un isomorphisme σ de k soit l'identité sur K, il faut et il suffit que $(A, \mathscr{C})$ soit isomorphe à $(A^\sigma, \mathscr{C}^\sigma)$. Si la caractéristique est 0, K est déterminé par cette condition, et s'appellera le *corps de modules* de $(A, \mathscr{C})$. On l'obtient au moyen de points de Chow ainsi qu'il suit. Soient X un diviseur ample dans $\mathscr{C}$ et A' l'image d'un plongement projectif de A donné par X; soient A^* la transformée de A' par une transformation projective générique sur k et z le point de Chow de A^*. Le lieu $\mathscr{F}$ de z sur k ne dépend que de A et de X; on appellera $\mathscr{F}$ la *famille projective* de (A, X). $(A, \mathscr{C})$ est isomorphe à $(A', \mathscr{C}(X'))$ si et seulement si les familles projectives de (A, X) et de (A', X') coïncident, où l'on suppose que les dimensions des systèmes linéaires définis par X et X' soient les mêmes. Il en résulte que le corps de modules de (A, X) est engendré sur $\mathbf{Q}$† par le point de Chow de $\mathscr{F}$, qui peut être donc considéré comme un 'module' de $(A, \mathscr{C})$. On peut définir de même les 'modules' au cas de caractéristique $\neq 0$; mais on n'en s'occupera pas dans cet exposé.

On va maintenant étudier les fonctions modulaires et para-modulaires. Soient $\delta_1, \ldots, \delta_n$ n entiers positifs tels que

$$\delta_1 = 1, \quad \delta_i \,|\, \delta_{i+1} \quad (1 \leqslant i \leqslant n-1);$$

soient δ la matrice diagonale ayant pour éléments les δ_i, et

$$E = \begin{pmatrix} 0 & -\delta \\ \delta & 0 \end{pmatrix}, \quad F = \begin{pmatrix} 1_n & 0 \\ 0 & \delta \end{pmatrix},$$

où 1_n est la matrice unité de degré n. On désignera par $\Gamma'(\delta)$ le groupe composé des matrices T de degré $2n$ à coefficients entiers telles qu'on ait ${}^t T E T = E$. On voit que le groupe

$$\Gamma(\delta) = \{F^{-1}\,{}^t T^{-1} F \mid T \in \Gamma'(\delta)\}$$

est un sous-groupe du groupe symplectique. $\Gamma(1_n) = \Gamma'(1_n)$ est le groupe modulaire de Siegel. Les éléments de $\Gamma(\delta)$ opèrent d'une manière ordinaire sur l'espace de Siegel S_n. Soit $D_\delta(z)$ un lattice de l'espace numérique complexe $\mathbf{C}^n$ engendré par les colonnes de la matrice $(z\ \delta)$ sur $\mathbf{Z}$. Si z est un point de S_n, le tore complexe $\mathbf{C}^n/D_\delta(z)$ a une structure de variété abélienne; la matrice alternée E donne une forme de Riemann sur $\mathbf{C}^n/D_\delta(z)$, pour ainsi dire, la matrice $(z\ \delta)$ détermine une structure de

---

† On désignera par $\mathbf{Z}$, $\mathbf{Q}$, $\mathbf{R}$ et $\mathbf{C}$ l'anneau des entiers rationnels, les corps des nombres rationnels, réels et complexes.

variété abélienne polarisée, qui est réalisée comme une variété projective $A_\delta(z)$ par les fonctions thêta correspondant à la matrice mE pour un entier $m \geqslant 3$. Fixons désormais un tel entier m, et désignons par $\mathscr{F}_\delta(z)$ la famille projective de $A_\delta(z)$ pour chaque z. Tandis que $A_\delta(z)$ dépend du choix d'une base des fonctions thêta, $\mathscr{F}_\delta(z)$ ne dépend que de z (et de m). On vérifie facilement que les $\mathscr{F}_\delta(z)$ pour $z \in S_n$ sont toutes de même dimension. Il existe, de plus, un sous-ensemble analytique Y de S_n de co-dimension 1 et des fonctions méromorphes $\phi_1(z), ..., \phi_\lambda(z)$ sur S_n jouissant des propriétés suivantes:

(m 1) les $\mathscr{F}_\delta(z)$ pour $z \in S_n - Y$ sont de même degré;

(m 2) pour chaque $z \in S_n - Y$, $(1, \phi_1(z), ..., \phi_\lambda(z))$ donne le point de Chow de la variété $\mathscr{F}_\delta(z)$.

Soient T un élément du groupe $\Gamma'(\delta)$ et $U = F^{-1}\,{}^t T^{-1} F$. La relation ${}^t TET = E$ entraîne que deux variétés $A_\delta(z)$ et $A_\delta(U(z))$, polarisées par les sections hyperplanes, sont isomorphes; d'où résulte $\mathscr{F}_\delta(z) = \mathscr{F}_\delta(U(z))$. Réciproquement, on peut démontrer que si l'on a $\mathscr{F}_\delta(z) = \mathscr{F}_\delta(z')$, il existe un élément U de $\Gamma(\delta)$ tel que $z' = U(z)$. On déduit de ceci, en vertu du théorème de plongement projectif de l'espace quotient $S_n/\Gamma(\delta)$ (Baily[1], Satake et Cartan[4]), le théorème suivant:

Théorème 1. *Si $n > 1$, le corps* $\mathbf{C}(\phi_i)$ *est le corps des fonctions méromorphes sur S_n invariantes par* $\Gamma(\delta)$.

En d'autres termes, les fonctions automorphes par rapport à $\Gamma(\delta)$ sont engendrées par les modules des variétés abéliennes polarisées 'de la famille δ'; en effet, d'après ce qu'on a vu plus haut, pour chaque point z' de S_n, le corps $\mathbf{Q}(\phi_i(z'))$ est le corps de modules de la variété $A_\delta(z')$.

Désignons par $L(\delta)$ le corps $\mathbf{C}(\phi_i)$; le corps $L(1_n)$ est le corps des fonctions modulaires de Siegel, même si $n = 1$.

Considérons maintenant les points t sur $A_\delta(z)$ tels qu'on ait $qt = 0$ pour un entier q; ils engendreront les fonctions automorphes par rapport aux groupes de congruence. Pour avoir ce résultat il faut rappeler la notion de variété de Kummer introduite par Weil[7]. Soit G le groupe des automorphismes d'une variété abélienne polarisée $(A, \mathscr{C})$; on sait que G est d'ordre fini. On entendra par une variété de Kummer de $(A, \mathscr{C})$ une variété quotient W de A par rapport à G. Bien entendu, W n'est pas uniquement déterminé; mais on peut construire, en vertu des résultats de Weil[8], une variété de Kummer W dans un espace projectif ainsi qu'une application naturelle h de A sur W satisfaisant aux conditions suivantes:

(W 1) W est défini sur le corps K de modules de $(A, \mathscr{C})$.

(W 2) h est défini sur tout corps de définition pour $(A, \mathscr{C})$ contenant K.

(W 3) Si σ est un isomorphisme d'un corps de définition pour $(A, \mathscr{C})$ contenant K, et si λ est un isomorphisme de $(A, \mathscr{C})$ sur $(A^\sigma, \mathscr{C}^\sigma)$, on a $h = h^\sigma \circ \lambda$.

On construit pour $A_\delta(z)$ un couple (W, h) jouissant des propriétés (W 1–W 3). Soit $\theta(u)$ l'isomorphisme de $\mathbf{C}^n/D_\delta(z)$ sur $A_\delta(z)$ où $u \in \mathbf{C}^n$. Soit b un vecteur de $\mathbf{R}^{2n}$ (une matrice à $2n$ lignes et une colonne). On peut considérer, en gros, le point $h(\theta(\omega_\delta(z)\,b))$ comme une fonction de z. Il est difficile d'éclaircir la situation pour tout point de S_n, puisqu'on ignore comment fabriquer (W, h) comme fonction de z. De toute façon, nous pouvons obtenir des fonctions méromorphes $\xi_\alpha(z, b)$ sur S_n dont les valeurs en z donnent les coordonnées du point $h(\theta(\omega_\delta(z)\,b))$ pour 'presque tout' point z de S_n. Ces fonctions satisfont à l'équation

$$(\xi) \qquad \xi_\alpha(U(z), Tb)/\xi_\beta(U(z), Tb) = \xi_\alpha(z, b)/\xi_\beta(z, b),$$

où U est un élément de $\Gamma(\delta)$ et $T = F^{-1}\,{}^t U^{-1} F$. C'est une conséquence de la propriété (W 3). Soit q un entier positif; on désignera par $\Gamma'(\delta, q)$ le sous-groupe formé des éléments T de $\Gamma'(\delta)$ tels que $T \equiv \pm 1_{2n} \pmod q$, et par $\Gamma(\delta, q)$ le sous-groupe $\{F^{-1}T^{-1}F \mid T \in \Gamma'(\delta, q)\}$ de $\Gamma(\delta)$. Soient a_i ($1 \leqslant i \leqslant q^{2n}$) les vecteurs de $\mathbf{R}^{2n}$ tels que les coordonnées de qa_i soient des entiers non-négatifs $< q$. D'après la relation (ξ), on voit qu'un élément U de $\Gamma(\delta)$ laisse invariantes les fonctions $\xi_\alpha(z, a_i)/\xi_\beta(z, a_i)$ si et seulement si U est contenu dans $\Gamma(\delta, q)$. Il s'ensuit de là le théorème suivant.

Théorème 2. Si $n > 1$, le corps engendré sur $\mathbf{C}$ par les ϕ_i et les

$$\xi_\alpha(z, a_i)/\xi_\beta(z, a_i)$$

est le corps des fonctions méromorphes sur S_n invariantes par $\Gamma(\delta, q)$.

Dans le cas $n = 1$, les fonctions qu'on vient de construire engendrent les fonctions modulaires elliptiques de 'Stufe' q.

Les systèmes $\{A_\delta(z) \mid z \in S_n\}$ sont les plus grands systèmes de variétés abéliennes polarisées; et chaque membre générique de ces systèmes n'a pas de multiplication complexe, c.-à-d. son anneau des endomorphismes est isomorphe à $\mathbf{Z}$. En considérant les variétés abéliennes dont les anneaux d'endomorphismes contiennent un certain anneau donné, on obtient un système de variétés abéliennes polarisées, auquel notre méthode est applicable également. Il vaudrait mieux, dans ce cas, généraliser quelque peu la notion de corps de modules ainsi qu'il suit. On se bornera au cas de caractéristique 0. Soit $\mathfrak{r}$ un anneau; on entendra par une variété abélienne polarisée de type $\mathfrak{r}$, une variété abélienne polarisée $(A, \mathscr{C})$ pour laquelle est donné un isomorphisme η de $\mathfrak{r}$ dans l'anneau des endomorphismes de A; on la désignera par $(A, \mathscr{C}, \eta)$. Un isomorphisme λ de $(A, \mathscr{C})$ sur $(A'\mathscr{C}')$ s'appellera un isomorphisme de $(A, \mathscr{C}, \eta)$ sur $(A', \mathscr{C}', \eta')$ si l'on a $\lambda\eta(r) = \eta'(r)\lambda$ pour tout $r \in \mathfrak{r}$. Soit k un corps de définition pour $(A, \mathscr{C})$ par rapport auquel tout élément de $\eta(\mathfrak{r})$ est défini; et soit σ un isomorphisme de k sur un corps k^σ. On obtient alors une variété abélienne polarisée $(A^\sigma, \mathscr{C}^\sigma, \eta^\sigma)$ de type $\mathfrak{r}$, en posant $\eta^\sigma(r) = \eta(r)^\sigma$. On peut démontrer qu'il existe un sous-corps K' de k pour lequel σ est l'identité sur K' si et seulement si $(A, \mathscr{C}, \eta)$ est isomorphe à $(A^\sigma, \mathscr{C}^\sigma, \eta^\sigma)$. On appellera K' le *corps de modules* de $(A, \mathscr{C}, \eta)$; le corps de modules de $(A, \mathscr{C})$ est un sous-corps de K'; ils coïncident si $\mathfrak{r} = \mathbf{Z}$; la réciproque n'est pas nécessairement vrai.

Considérons par exemple le cas de fonctions de Hilbert. Soit $\mathfrak{f}$ un corps totalement réel de degré $n > 1$ sur $\mathbf{Q}$, et soit $\mathfrak{r}$ l'anneau des entiers de $\mathfrak{f}$. On désignera par $a^{(1)}, \ldots, a^{(n)}$ les conjugués de $a \in \mathfrak{f}$. Soit $\mathfrak{a}$ un idéal de $\mathfrak{r}$; désignons par $\Gamma(\mathfrak{a})$ le groupe des transformations $\tau \to (a\tau + b)/(c\tau + d)$ où a, b, c, d sont quatre éléments tels que $a \in \mathfrak{r}$, $b \in \mathfrak{a}$, $c \in \mathfrak{a}^{-1}$, $d \in \mathfrak{r}$ et que $ad - bc = 1$. $\Gamma(\mathfrak{a})$ opère sur l'espace produit H_n de n demi-plans complexes $\mathrm{Im}\,(\tau) > 0$. Soit $(\tau) = (\tau_1, \ldots, \tau_n)$ un point de H_n; soit $D(\tau, \mathfrak{a})$ le lattice de $\mathbf{C}^n$ composé des vecteurs $(a^{(1)}\tau_1 + b^{(1)}, \ldots, a^{(n)}\tau_n + b^{(n)})$ où $a \in \mathfrak{r}$, $b \in \mathfrak{a}$. Le tore $\mathbf{C}^n/D(\tau, \mathfrak{a})$ a une structure de variété abélienne, sur laquelle toute forme de Riemann correspond à un nombre y de $\mathfrak{f}$ tel que $y^{(i)} < 0$ pour tout i; chaque élément a de $\mathfrak{r}$ définit un endomorphisme de $\mathbf{C}^n/D(\tau, \mathfrak{a})$ donné par la matrice diagonale ayant pour éléments les $a^{(i)}$. On obtient ainsi les systèmes de variétés abéliennes polarisées $A(\tau, \mathfrak{a}, y)$ de type $\mathfrak{r}$. Nous pouvons démontrer qu'il existe des fonctions méromorphes $\psi_\nu(\tau)$ sur H_n telles que $\mathbf{Q}(\psi_\nu(\tau'))$ soit le corps de modules de $A(\tau', \mathfrak{a}, y)$ pour presque tout point τ' de H_n. De plus, ces fonctions engendrent sur $\mathbf{C}$ toutes les fonctions méromorphes sur H_n invariantes par $\Gamma(\mathfrak{a})$. Si l'on ne tient pas compte des endomorphismes, on se procure un certain sous-corps de ce corps de fonctions comme 'corps de modules absolus'; pour qu'ils soient les mêmes, il faut et il suffit qu'on ait $(y\mathfrak{a})^\sigma \neq y\mathfrak{a}$ pour tout automorphisme $\sigma \neq 1$ de $\mathfrak{f}$.

On se propose maintenant d'étudier les corps de définition pour les corps de fonctions automorphes. Supposons qu'on ait défini un système $\{A(s)\}$ de variétés abéliennes polarisées de type $\mathfrak{o}$, où $\mathfrak{o}$ est un anneau, dont les membres dépendent d'une manière convenable de points s sur un sous-ensemble S ouvert connexe de $\mathbf{C}^m$; on obtient alors des fonctions méromorphes χ_ν sur S telles que $\mathbf{Q}(\chi_\nu(s'))$ soit le corps de modules de $A(s')$ pour presque tout $s' \in S$. Soit k un sous-corps dénombrable de $\mathbf{C}$. Le système est dit *complet* par rapport à k s'il satisfait à la condition suivante:

Soit s_0 un point de S tel que tout $A(s)$ soit une spécialisation de $A(s_0)$ sur k. Si B est une spécialisation générique de $A(s_0)$ sur k, il existe un point s_1 tel que deux variétés abéliennes polarisées $A(s_1)$ et B de type $\mathfrak{o}$, soient isomorphes.

Notre critérium de corps de définition s'énonce:

Théorème 3. Si le système $\{A(s)\}$ est complet par rapport à un corps dénombrable k, le corps $k(\chi_\nu)$ est une extension régulière de k et l'on a $\dim_k k(\chi_\nu) = \dim_{\mathbf{C}} \mathbf{C}(\chi_\nu).$

On vérifie facilement que les systèmes $\{A_\delta(z)\}$ et $\{A(\tau, \mathfrak{a}, y)\}$ sont tous complets par rapport à $\mathbf{Q}$.

Ce théorème est applicable même au cas de domaine fondamentale compact, où l'on ne peut pas se servir de série de Fourier. On sait à titre d'exemple une classe de fonctions automorphes de dimension 1 définie pour la première fois par Poincaré, qui sont en rapport avec une forme quadratique ternaire indéfinie, et qu'on trouve dans le livre de Fricke et Klein. Les recherches sont 'peu développées', comme Eichler a dit, dans l'arithmétique de ces fonctions. On va maintenant s'occuper de cette classe. Soit $\mathfrak{A}$ une algèbre de quaternions sur $\mathbf{Q}$ dont la norme est une forme quadratique indéfinie, et soit $\mathfrak{o}$ un ordre maximal de $\mathfrak{A}$. Comme $\mathfrak{A}$ contient un corps quadratique réel $\mathfrak{K}$, $\mathfrak{A}$ a une représentation de degré 2 à coefficients dans $\mathfrak{K}$, que l'on désignera par M. Soit γ un unité de $\mathfrak{o}$ tel que $\det M(\gamma) = 1$; γ donne une transformation $\tau \to \dfrac{a\tau + b}{c\tau + d}$ du demi-plan complexe H, où $M(\gamma) = \begin{pmatrix} a & b \\ c & d \end{pmatrix}$. Désignons par $\Gamma(\mathfrak{o})$ le groupe des transformations ainsi obtenues; si $\mathfrak{A}$ n'a pas de diviseur de zéro, le domaine fondamental est compact. Soient τ un point de H et $D(\tau)$ le lattice de $\mathbf{C}^2$ composé des vecteurs $M(\alpha)\begin{pmatrix} \tau \\ 1 \end{pmatrix}$ pour $\alpha \in \mathfrak{o}$. On obtient sur le tore $\mathbf{C}^2/D(\tau)$ une forme de Riemann correspondant à un élément w de $\mathfrak{A}$ tel que w^2 soit un nombre négatif de $\mathbf{Q}$. La matrice $M(\alpha)$ pour $\alpha \in \mathfrak{o}$ donne un endomorphisme de $\mathbf{C}^2/D(\tau)$. On peut ainsi définir un système $\{A(\tau)\}$ de variétés abéliennes polarisées de type $\mathfrak{o}$. Le corps de modules de $A(\tau)$ est donné par les valeurs de certaines fonctions méromorphes $g_i(\tau)$ sur H, qui engendrent toutes les fonctions automorphes par rapport à $\Gamma(\mathfrak{o})$. On vérifie que $\{A(\tau)\}$ est complet par rapport à $\mathbf{Q}$, de sorte que le corps $\mathbf{Q}(g_i)$ est une extension régulière de $\mathbf{Q}$ de dimension 1. Par suite il existe une courbe algébrique définie sur le corps rationnel donnant un modèle du corps des fonctions automorphes par rapport à $\Gamma(\mathfrak{o})$.

Commençons la théorie des correspondances modulaires par l'étude du groupe $\Gamma(\mathfrak{o}, q)$ composé des transformations obtenues à partir des unités $\gamma \in \mathfrak{o}$ tels que $\gamma \equiv \pm 1 \bmod q$, où q est un entier positif. Dans ce but, on modifie, eu égard aux endomorphismes, la définition de variété de Kummer et les propriétés (W 1–W 3). Soient W une variété de Kummer de $A(\tau)$ et h une application de $A(\tau)$ sur W ayant les propriétés modifiées. Les coordonnées du point $h(t)$, où t est un point sur $A(\tau)$ tel que $qt = 0$, regardées comme fonctions de τ, donnent des fonctions méromorphes f_j sur H, qui engendrent, avec les g_i, toutes les fonctions automorphes par rapport à $\Gamma(\mathfrak{o}, q)$. Prenons un point τ_0 sur H tel que tout $A(\tau)$ soit une spécialisation de $A(\tau_0)$ sur $\mathbf{Q}$, et posons $K_q = \mathbf{Q}(g_i(\tau_0), f_j(\tau_0))$. K_q est une extension galoisienne de K_1 dont le groupe de Galois est isomorphe au groupe G des éléments réguliers de l'anneau $\mathfrak{o}/q\mathfrak{o}$. ζ_q étant une racine primitive q-ième d'unité, $\mathbf{Q}(\zeta_q)$ est algébriquement fermé dans K_q; $K_1(\zeta_q)$ correspond à un sous-groupe des éléments α tels que $\det M(\alpha) \equiv 1 \bmod q$. Un élément α donne un automorphisme $\zeta_q \to \zeta_q^m$ sur $\mathbf{Q}(\zeta_q)$, où $m = \det M(\alpha)$. Si q est premier avec le discriminant de l'algèbre $\mathfrak{A}$ (ce que l'on suppose dans ce qui suit), K_q contient un sous-corps K_q' tel que $K_q = K_q'(\zeta_q)$ et $K_q' \cap \mathbf{Q}(\zeta_q) = \mathbf{Q}$; il existe donc une courbe algébrique C_q définie sur $\mathbf{Q}$, dont le corps des fonctions est le corps des fonctions automorphes par rapport à $\Gamma(\mathfrak{o}, q)$. Soit p un nombre premier. Les points u sur $A(\tau_0)$ tels que $pu = 0$ forment un groupe $\mathfrak{g}$ d'ordre p^4 invariant par $\mathfrak{o}$. Il existe exactement $p + 1$ sous-groupes de $\mathfrak{g}$ d'ordre p^2, invariants par $\mathfrak{o}$, qu'on notera par $\mathfrak{g}_1, \ldots, \mathfrak{g}_{p+1}$. Chaque $\mathfrak{g}_\nu$ correspond à un idéal $\mathfrak{o}\alpha_\nu$ de norme p de telle façon qu'il existe un homomorphisme λ_ν de $A(\tau_0)$ sur $A(\tau_\nu)$ dont le noyau est $\mathfrak{g}_\nu$, où

$$\tau_\nu = \frac{a_\nu \tau_0 + b_\nu}{c_\nu \tau_0 + d_\nu}, \quad M(\tau_\nu) = \begin{pmatrix} a_\nu & b_\nu \\ c_\nu & d_\nu \end{pmatrix}.$$

On définit un isomorphisme σ_ν de K_q par $g_i(\tau_0)^{\sigma_\nu} = g_i(\tau_\nu)$ et $h(t)^{\sigma_\nu} = h_\nu(\lambda_\nu t)$ pour $qt = 0$, où h_ν est l'application naturelle de $A(\tau_\nu)$ sur sa variété de Kummer. Soit x un point générique de C_q et soit X_p le lieu de $x \times x^{\sigma_1}$ par rapport à $\mathbf{Q}$; X_p s'appellera la *correspondance modulaire de degré p* sur C_q. Soit P un diviseur premier de p dans une clôture algébrique de K_q; on indiquera par la barre la réduction modulo P. La réduction modulo P donne un homomorphisme de $\mathfrak{g}$ sur le groupe $\bar{\mathfrak{g}}$ des éléments $\bar{u}$ sur $\bar{A}(\tau_0)$ tels que $p\bar{u} = 0$. Comme $\bar{\mathfrak{g}}$ est d'ordre p^2, le noyau de cet homomorphisme est un des $\mathfrak{g}_\nu$, mettons $\mathfrak{g}_1$. On voit alors que le noyau de $\bar{\lambda}_\nu$ est $\bar{\mathfrak{g}}$ ou $\{0\}$ selon que $\nu > 1$ ou $\nu = 1$. Désignons par μ_ν l'homomorphisme de $A(\tau_\nu)$ sur $A(\tau_0)$ tel que $\mu_\nu \lambda_\nu = p$; le noyau des $\bar{\mu}_\nu$ est d'ordre 1 ou p^2 selon que $\nu > 1$ ou $\nu = 1$. On en déduit que $\bar{A}(\tau_1)$ est isomorphe à $\bar{A}(\tau_0)^p$ et que

$\overline{A(\tau_0)}$ est isomorphe à $\overline{A(\tau_\nu)}^p$ pour $\nu > 1$; les homomorphismes $\overline{\lambda}_1$ et $\overline{\mu}_\nu$ pour $\nu > 1$ sont équivalents aux homomorphismes de p-ième puissance. Nous pouvons démontrer, d'après ces relations, deux formules de congruence pour la correspondance modulaire

$$\overline{X}_p = \Pi + \Pi' \circ \overline{Y}_p, \quad \Pi' \circ \overline{Y}_p = \overline{Z}' \circ \Pi' \circ \overline{Z}$$

sur $\overline{C}_q$ pour presque tous p, où Π est la correspondance $\overline{x} \to \overline{x}^p$ sur $\overline{C}_q$, Y_p est la correspondance birationnelle de C_q donné par $h(t) \to h(pt)$, Z est une certaine correspondance birationnelle de C_q et $'$ désigne l'anti-automorphisme de Rosati. Ces formules sont des généralisations de celles qui ont été **obtenues** pour les fonctions modulaires elliptiques (Eichler[2], Shimura[5]), puisque nos fonctions automorphes coïncident avec les fonctions modulaires elliptiques de 'Stufe' q, si $\mathfrak{o}$ est l'ensemble des matrices de degré 2 à coefficients entiers. La représentation de X_p par les formes différentielles de première espèce, n'est autre que l'opérateur T_p de Hecke, pour les formes paraboliques de poids 1. Par suite la fonction ζ de la courbe C_q s'exprime sous la forme

$$\zeta(s, C_q) = r(s)\, \zeta(s)\, \zeta(s-1)\, \Phi(s)^{-1}$$

où $\zeta(s)$ est la fonction ζ de Riemann, $r(s)$ est une fonction rationnelle de p^{-s} et $\Phi(s)$ désigne un produit d'Euler du type introduit par Hecke. D'après un résultat de Weil, on constate que les valeurs absolues des racines caractéristiques de l'opérateur T_p pour les formes paraboliques de poids 1 ne dépassent pas $2\sqrt{p}$ pour presque tous les nombres premiers p.

On signale que les formules de congruence sont démontrées pour les correspondances elles-mêmes non seulement pour les classes de correspondances. Ce fait nous semble bien significatif pour les formes automorphes de poids > 1.

On peut définir, de la même manière que ci-dessus, les correspondances modulaires pour les fonctions automorphes de plusieurs variables au moyen des isogénies de variétés abéliennes; on obtiendra alors les formules de congruence pour ces correspondances. Et il est à souhaiter en déduire quelque chose d'intéressant; le conférencier regrette qu'il n'a rien à dire sur ce que signifient ces formules.

BIBLIOGRAPHIE

[1] Baily, W. L. Satake's compactification of V_n. *Amer. J. Math.* 80, 348–364 (1958).

[2] Eichler, M. Quaternäre quadratische Formen und Riemannsche Vermutung für die Kongruenzzetafunktion. *Arch. Math.* 5, 355–366 (1954).

[3] Matsusaka, T. Polarized varieties, fields of moduli and generalized Kummer varieties of polarized abelian varieties. *Amer. J. Math.* 80, 45–82 (1958).

[4] Satake, I. et Cartan, H. Exposés 11–17 du *Séminaire H. Cartan*, 10 (1957–58).

[5] Shimura, G. Correspondances modulaires et les fonctions ζ de courbes algébriques. *J. Math. Soc. Japan*, 10, 1–28 (1958).

[6] Shimura, G. Modules des variétés abéliennes polarisées et fonctions modulaires. *Séminaire H. Cartan*, 10 (1957–58).

[7] Weil, A. On the theory of complex multiplication. *Proc. Int. Symp. Alg. number theory*. Tokyo-Nikko, 1955. Tokyo, Science Council of Japan, 9–22 (1956).

[8] Weil, A. The field of definition of a variety. *Amer. J. Math.* 78, 509–524 (1956).

On the theory of automorphic functions

Annals of Mathematics, 70 (1959), 101-144

(Received December 29, 1958)

We know that the elliptic modular function $j(\tau)$ gives the birational invariant of the elliptic curve with the analytic modulus τ, and that the elliptic modular functions belonging to the congruence-subgroups are obtained from the values of elliptic functions at the points of finite order on elliptic curves. This is not only the origin of those functions, but one of the most essential points to which we may ascribe the significance of elliptic modular functions in number-theory. It is important to generalize these facts, namely, to investigate in a more general case the relation between automorphic functions with one or more variables and systems of algebraic varieties, especially, systems of abelian varieties. The object of this paper is to give some results on this subject.

We shall deal with certain systems of polarized abelian varieties parametrized by holomorphic functions and show that there exist meromorphic functions whose values are considered as "moduli" of the members of the systems (Theorem 1). The theory developed here will be chiefly concerned with systems of abelian varieties with non-trivial endomorphisms, since I have already given elsewhere a theory for Siegel's modular functions [11]. I am particularly interested, in the present paper, in the determination of the fields of definition for fields of automorphic functions. This is the first problem which confronts us when we proceed beyond a formal treatment in the arithmetic theory of automorphic functions. We obtain a criterion (Theorem 2) applicable even in the case of compact fundamental domain where one can not employ Fourier expansions. The last part of the paper is devoted to the theory of a certain type of automorphic functions of one variable known in the literature as functions belonging to indefinite ternary quadratic forms [8], [5]; they occur as moduli of abelian varieties of dimension 2 whose endomorphism-rings are isomorphic to an order of an indefinite quaternion algebra. We get in this case the field of rational numbers as a field of definition for the function-field. We can also prove the congruence formulae for the modular correspondences, which give a generalization of what is obtained in [3], [10]. We shall treat these formulae in a subsequent paper together with the theory of the functions belonging to congruence-subgroups. Our method can be applied to other types of automorphic functions, for example, to Hilbert's modular functions, by considering a system of abelian varieties whose endomorphism-rings are isomorphic to the ring of integers in a totally real number-field. The theory in the last section will serve as a pattern when one studies these automorphic functions from our view-point.

NOTATIONS. We denote by Z, Q, R, and C, respectively the ring of rational integers, the fields of rational numbers, real numbers and complex numbers. R^n and C^n denote the vector spaces composed of all matrices with n rows and 1 column, with real coefficients and complex coefficients, respectively. P^n denotes the projective space of dimension n whose ground field will be given in each case. We denote by $\bar{u}$ the complex conjugate of u, where u may be a complex number, a matrix with complex coefficients, a function taking complex values, etc. 1_n denotes the unit matrix of degree n. It is well known that there exists a biregular embedding φ of $P^{n_1} \times \cdots \times P^{n_s}$ into P^N defined as follows: N is given by

$$N + 1 = (n_1 + 1) \cdots (n_s + 1) ;$$

for every $(\alpha_{i_1}) \times \cdots \times (\alpha_{i_s}) \in P^{n_1} \times \cdots \times P^{n_s}$,

$$\varphi((\alpha_{i_1}) \times \cdots \times (\alpha_{i_s}))_{j_1 j_2 \cdots j_s} = \alpha_{j_1} \alpha_{j_2} \cdots \alpha_{j_s} .$$

We call φ *the canonical embedding* of $P^{n_1} \times \cdots \times P^{n_s}$ in P^N. For every positive cycle V in $P^{n_1} \times \cdots \times P^{n_s}$ we understand by *the Chow point of* V the Chow point of $\varphi(V)$ in the ordinary sense, and denote it by $c(V)$; we understand also by the *degree* of V the degree of $\varphi(V)$ and denote it by $\deg(V)$. A being an abelian variety, we denote by $\mathcal{A}(A)$ and $\mathcal{A}_0(A)$ the ring of endomorphisms of A and the algebra $\mathcal{A}(A) \otimes Q$ defined in [12].

1. Polarized abelian varieties of type $\mathfrak{r}$

1. We first recall the notion of polarized variety [14], [7]. Let V be a complete variety, non-singular in codimension 1, defined over a field k, and X a divisor on V which is rational over k. We denote by $L(X, k)$ the set of all functions f on V, defined over k, such that $(f) > - X$, where (f) denotes the divisor of the function f. The dimension of the vector space $L(X, k)$ over k is independent of the choice of k; we denote this dimension by $l(X)$. Let $\{f_0, \cdots, f_n\}$ be a basis of $L(X, k)$ over k and x a generic point of V over k; let $V(X, f)$ the locus of $(f_0(x), \cdots, f_n(x))$ over k in P^n. Then we obtain a rational mapping F of V onto $V(X, f)$ defined by $F(x) = (f_0(x), \cdots, f_n(x))$ with respect to k. We say that the divisor X is *ample* if this mapping F is birational and biregular; this definition does not depend upon the choice of k, f_i. We call F a *biregular embedding* of

V into P^n defined by X.

V, X being as above, we denote by $C(X)$ the set of all divisors X' on V for which there exist two positive integers m, m' such that mX is algebraically equivalent to $m'X'$. The set $C(X)$ is called a *polarization* of V if it contains an ample divisor. We understand by a *polarized variety* a couple (V, C) formed by a variety V and a polarization C of V. We say that (V, C) is *defined over* a field k if V is defined over k and if C contains a divisor which is rational over k. Let σ be an isomorphism of k into a field k'; (V, C) being defined over k, we denote by C^σ the polarization $C(X^\sigma)$ of V^σ, where X is a divisor in C which is rational over k.

2. Before dealing with a polarized abelian variety, we give some of the known results about divisors on abelian varieties. X and t being respectively a divisor and a point on an abelian variety A, we denote by X_t the transform of X by the translation $x \to x + t$ on A. We call X *non-degenerate* if there are only finitely many points t on A such that X_t is linearly equivalent to X.

LEMMA 1 (Weil [13]). *Let X be a divisor on an abelian variety A. If there exists an integer $n > 0$ such that nX is ample, X is non-degenerate. Conversely, if X is a non-degenerate positive divisor, there exists an integer $n_0 > 0$ such that nX is ample for $n > n_0$.*

LEMMA 2. *Let X be a non-degenerate divisor on an abelian variety A and Y a divisor on A which is algebraically equivalent to X. Then there exists a point t on A such that Y is linearly equivalent to X_t.*

LEMMA 3. *Let A be an abelian variety of dimension n and X a non-degenerate positive divisor on A, then we have*

$$l(X) = \frac{\deg(X^n)}{n!} ,$$

where $\deg(X^n)$ denotes $\deg(X_{t_1} \cdots \cdots X_{t_n})$ for n points $t_1, \cdots, t_n$ of A such that the intersection-product is defined.

LEMMA 4. *Let A, A' be two abelian varieties of the same dimension, λ a homomorphism of A onto A' and X' a non-degenerate positive divisor on A'. We have then*

$$l(\lambda^{-1}(X')) = \nu(\lambda)l(X') ,$$

where $\nu(\lambda)$ denotes the degree of the homomorphism λ.

This is an immediate consequence of Lemma 3.

LEMMA 5. *Let C be a polarization of an abelian variety A and X, Y two divisors in C. Then $l(X) = l(Y)$ if and only if there exists a point t of A such that X_t is linearly equivalent to Y.*

If there exists a point t of A such that X_t is linearly equivalent to Y, we have $l(Y) = l(X_t) = l(X)$. Conversely, suppose that we have $l(X) = l(Y)$. By our definition of polarization, there exist two positive integers r, s such that rX is algebraically equivalent to sY. By Lemma 1, every divisor in C is non-degenerate. Hence, by Lemma 2, there exists a point u on A such that sY is linearly equivalent to rX_u. We have then by Lemma 3,

$$s^n l(Y) = l(sY) = l(rX_u) = r^n l(X) \ ,$$

n being the dimension of A, so that we have $r = s$; this shows that X is algebraically equivalent to Y. Again by Lemma 2, there exists a point v on A such that Y is linearly equivalent to X_v. This completes our proof.

3. When we consider a polarized abelian variety, we take into account endomorphisms of the variety. Let $\mathfrak{r}$ be a ring having a finite basis over $\mathbf{Z}$. We understand by a *polarized abelian variety of type* $\mathfrak{r}$ a triplet (A, C, θ) formed by a polarized abelian variety (A, C) and an isomorphism θ of $\mathfrak{r}$ into $\mathcal{A}(A)$. Let (A, C, θ) and (A', C', θ') be two polarized abelian varieties of type $\mathfrak{r}$, of the same dimension; a homomorphism (resp. an isomorphism) λ of A onto A' is called a *homomorphism* (resp. an *isomorphism*) of (A, C, θ) onto (A', C', θ') if there exists a divisor X' in C' such that $\lambda^{-1}(X')$ is contained in C and if $\lambda\theta(r) = \theta'(r)\lambda$ holds for all $r \in \mathfrak{r}$; if the first condition is satisfied, $\lambda^{-1}(Y)$ is contained in C for every $Y \in C'$. We say that (A, C, θ) is *defined over* a field k if (A, C) and all elements of $\theta(\mathfrak{r})$ are defined over k. This being so, let σ be an isomorphism of k into a field k'; we obtain then a polarized abelian variety $(A^\sigma, C^\sigma, \theta^\sigma)$ of type $\mathfrak{r}$ by putting $\theta^\sigma(r) = \theta(r)^\sigma$. If σ is the identity on a subfield k_1 of k, we call $(A^\sigma, C^\sigma, \theta^\sigma)$ a *generic specialization* of (A, C, θ) over k_1.

Fix now a basis $\{r_1, \cdots, r_d\}$ of $\mathfrak{r}$ over $\mathbf{Z}$. Let B be an abelian variety in P^N and θ an isomorphism of $\mathfrak{r}$ into $\mathcal{A}(B)$; and let k be a field of definition for B and for the $\theta(r_\nu)$. Take a matrix (t_{ij}) of degree $N + 1$ whose coordinates are $(N + 1)^2$ independent variables over k and d independent generic points $u_1, \cdots, u_d$ on B over $k(t_{ij})$. Let B_0 be the transform of B by the projective transformation $\varphi: (\xi_i) \rightarrow (\Sigma_j t_{ij}\xi_j)$ of P^N; let b be the Chow point of B_0. As $k(b)$ is contained in $k(t)$, $k(b)$ is regular over k. Denote by $\mathcal{F}(B)$ the locus of b over k; $\mathcal{F}(B)$ does not depend on the choice of k and (t_{ij}). Let W_ν be the graph of the rational mapping

$$x \longrightarrow \varphi[\theta(r_\nu)\varphi^{-1}(x) + u_\nu]$$

of B_0 onto itself and z_ν the Chow point of W_ν. We see that the field $k(b, z_1, \cdots, z_d)$ is contained in $k(u_1, \cdots, u_d, t_{ij})$, so that $k(b, z_1, \cdots, z_d)$ is a regular extension of k. Denote by $\mathcal{F}(B, \theta)$ the locus of $b \times z_1 \times \cdots \times z_d$ over k. We can easily verify that the variety $\mathcal{F}(B, \theta)$ does not depend upon the choice of k, (t_{ij}) and u_ν; but it may depend on the choice of $\{r_\nu\}$.

PROPOSITION 1. *Let A, A' be two abelian varieties in P^N and C, C' the polarization of A, A' defined by the hyperplane sections, respectively; let θ, θ' be isomorphisms of $\mathfrak{r}$ into $\mathcal{A}(A)$ and into $\mathcal{A}(A')$, respectively. Suppose that neither of A, A' is contained in any hyperplane and that the linear systems on A, A' defined by the hyperplane sections are complete. Then, (A, C, θ) is isomorphic to (A', C', θ') if and only if $\mathcal{F}(A, \theta) = \mathcal{F}(A', \theta')$.*

Let H be a hyperplane of P^N; put $Y = H \cdot A$, $Y' = H \cdot A'$. Suppose that there exists an isomorphism η of (A', C', θ') onto (A, C, θ). By our definition, $\eta^{-1}(Y)$ is contained in C'. From our assumption on A, A' follows that both $l(\eta^{-1}(Y))$ and $l(Y')$ are equal to $N + 1$. Hence, by Lemma 5, there exist a point t on A' and a function g on A' such that

$$(g) = \eta^{-1}(Y)_t - Y' \; .$$

Let $(X_0, \cdots, X_N)$ be the system of coordinates of P^N; let f_i, f_i' for $0 \leqq i \leqq N$ be the functions on A, A' induced by $X_i / \sum_j \gamma_j X_j$, respectively, where $\sum_j \gamma_j X_j = 0$ is the equation of the hyperplane H. Let k be a field of definition for (A, C, θ), (A', C', θ'), η and g such that t and the γ_j are rational over k. By our assumption, the f_i form a basis of $L(Y, k)$ and the f_i' form a basis of $L(Y', k)$. Let h_i be the function on A' defined by

$$h_i(x') = f_i(\eta(x' - t))g(x')$$

with respect to k, where x' is a generic point of A' over k. Then we see that the h_i are in $L(Y', k)$, so that there exist elements ξ_{ij} in k such that $h_i = \sum_j \xi_{ij} f_j'$ $(0 \leqq i \leqq N)$. Since the h_i are linearly independent over k, the matrix (ξ_{ij}) is non-singular. If x' is a generic point of A' over k, the $f_i'(x')$ give the coordinates of x', and the $h_i(x')$ the coordinates of $\eta(x' - t)$. Hence, denoting by ψ the projective transformation given by the matrix (ξ_{ij}), we see that $\psi(x') = \eta(x' - t)$. Put $a = -\eta(t)$; since η is an isomorphism of (A', C', θ') onto (A, C, θ), we have $\eta\theta'(r_\nu) = \theta(r_\nu)\eta$; by means of this relation, we have

$$(*) \qquad \psi[\theta'(r_\nu)x' + v] = \theta(r_\nu)\psi(x') + \eta(v) + a - \theta(r_\nu)a$$

for $x' \in A'$, $v \in A'$. Let $u_1, \cdots, u_d$ be d independent generic points of A

over k and φ a generic projective transformation over $k(u_1, \cdots, u_d)$; denote by A_0 the image $\varphi(A)$ and by W_ν the graph of the rational mapping

$$x \longrightarrow \varphi[\theta(r_\nu)\varphi^{-1}(x) + u_\nu]$$

of A_0 onto itself. Let b be the Chow point of A_0 and z_ν the Chow point of W_ν for each ν. Put for each ν,

$$v_\nu = \eta^{-1}[u_\nu + \theta(r_\nu)a - a];$$

it is easy to see that the v_ν are d independent generic points of A' over k and $\varphi \circ \psi$ is a generic projective transformation over $k(v_1, \cdots, v_d)$. We have $\varphi \circ \psi(A') = A_0$ and, by the relation (*),

$$\begin{aligned}
\varphi \circ \psi[\theta'(r_\nu)(\varphi \circ \psi)^{-1}(y) + v_\nu] &= \varphi[\theta(r_\nu)\varphi^{-1}(y) + \eta(v_\nu) + a - \theta(r_\nu)a] \\
&= \varphi[\theta(r_\nu)\varphi^{-1}(y) + u_\nu] \,,
\end{aligned}$$

so that W_ν is the graph of the mapping

$$y \longrightarrow \varphi \circ \psi[\theta'(r_\nu)(\varphi \circ \psi)^{-1}(y) + v_\nu]$$

of A_0 onto itself. From this and from our definition follows that $b \times z_1 \times \cdots \times z_d$ is a generic point of $\mathcal{F}(A', \theta')$ over k; we have therefore $\mathcal{F}(A, \theta) = \mathcal{F}(A', \theta')$. This proves "only if" part of our proposition. Conversely, suppose that we have $\mathcal{F}(A, \theta) = \mathcal{F}(A', \theta')$. Let k be a field of definition for (A, C, θ) and (A', C', θ'); we construct similarly as above, A_0, W_ν, b, z_ν from d independent generic points u_ν on A over k and a generic projective transformation φ over $k(u_1, \cdots, u_d)$, and A_0', W_ν', b', z_ν' for (A', C', θ') from d independent generic points u_ν' on A' over k and a generic projective transformation φ' over $k(u_1', \cdots, u_d')$. Then, two points $b \times z_1 \times \cdots \times z_d$ and $b' \times z_1' \times \cdots \times z_d'$ are both generic on the variety $\mathcal{F}(A, \theta) = \mathcal{F}(A', \theta')$ over k, so that there exists an isomorphism σ of the field $k(b, z_1, \cdots, z_d)$ over k such that

$$(b \times z_1 \times \cdots \times z_d)^\sigma = b' \times z_1' \times \cdots \times z_d' \,.$$

We have then $A_0' = A_0^\sigma$ and $W_\nu' = W_\nu^\sigma$. Extend σ to an isomorphism τ of a field of definition for φ, containing $k(u_1, \cdots, u_d)$, and put $\psi = (\varphi^\tau)^{-1} \circ \varphi'$. Then ψ gives a biregular mapping of A' onto A; hence there exists an isomorphism η of A' onto A and a point c on A such that $\psi(x') = \eta(x') + c$ for $x' \in A'$. We have, by the relation $W_\nu' = W_\nu^\tau$,

$$\varphi^\tau[\theta(r_\nu)(\varphi^\tau)^{-1}(y) + u_\nu^\tau] = \varphi'[\theta'(r_\nu)(\varphi')^{-1}(y) + u_\nu']$$

for $y \in A_0^\tau = A_0'$. Substituting $\varphi'(x')$ for y, where x' is in A', we have

$$\theta(r_\nu)(\eta(x') + c) + u_\nu^\tau = \eta[\theta'(r_\nu)x'] + \eta(u_\nu') + c \,.$$

This implies that the homomorphism $\theta(r_\nu)\eta - \eta\theta'(r_\nu)$ is a constant mapping, so that we have $\theta(r_\nu)\eta = \eta\theta'(r_\nu)$. Put $w = \eta^{-1}(c)$. By the relation $\psi = \eta + c$, we have $\eta^{-1}(Y) = \psi^{-1}(Y)_w$; hence $\eta^{-1}(Y)$ is algebraically equivalent to $\psi^{-1}(Y)$. As Y is a hyperplane section of A and as ψ is a projective transformation, we see that $\psi^{-1}(Y)$ is a hyperplane section of A', so that $\eta^{-1}(Y)$ is algebraically equivalent to Y'. Hence η is an isomorphism of (A', C', θ') onto (A, C, θ). This completes our proof.

If we take no account of endomorphisms, we have easily, from the above proof,

PROPOSITION 2. *The notations (A, C), (A', C') and the assumptions being the same as in Proposition 1, (A, C) is isomorphic to (A', C') if and only if $\mathcal{F}(A) = \mathcal{F}(A')$.*

2. Fields of moduli

4. Let (A, C, θ) be a polarized abelian variety of type $\mathfrak{r}$ and C^* the set of all ample divisors in C. Each divisor X in C^* defines a biregular embedding f_X of A in a projective space. Denote by A_X the image of A by f_X. We define a structure of abelian variety on A_X in such a way that f_X is an isomorphism of A onto A_X, and put $\theta_X(r) = f_X\theta(r)f_X^{-1}$ for $r \in \mathfrak{r}$. While A_X depends on the choice of a basis of the linear system determined by X, the varieties $\mathcal{F}(A_X)$ and $\mathcal{F}(A_X, \theta_X)$ depend only upon A, X, θ and a basis of $\mathfrak{r}$. We denote $\mathcal{F}(A_X)$ and $\mathcal{F}(A_X, \theta_X)$ by $\mathcal{F}(A, X)$ and $\mathcal{F}(A, X, \theta)$. If A is defined over a field k and X is rational over k, $\mathcal{F}(A, X)$ is defined over k; if, moreover, every element of $\theta(\mathfrak{r})$ is defined over k, then $\mathcal{F}(A, X, \theta)$ is defined over k.

PROPOSITION 3. *Let (A, C, θ) and (A', C', θ') be two polarized abelian varieties and X, X' respectively ample divisors in C, C'. If there exists an isomorphism η of A onto A' such that $\eta(X)$ is algebraically equivalent to X', then we have $\mathcal{F}(A, X) = \mathcal{F}(A', X')$. If moreover $\eta\theta(r) = \theta'(r)\eta$ holds for every $r \in \mathfrak{r}$, we have $\mathcal{F}(A, X, \theta) = \mathcal{F}(A', X', \theta')$.*

If $\eta(X)$ is algebraically equivalent to X', then we have $l(X) = l(X')$. Then our proposition is an immediate consequence of Propositions 1, 2.

PROPOSITION 4. *Let (A, C, θ) be a polarized abelian variety of type $\mathfrak{r}$ and X an ample divisor in C; let M be a field of definition for $\mathcal{F}(A, X)$ and K a field of definition for $\mathcal{F}(A, X, \theta)$. Then there exist an isomorphism η of A onto an abelian variety A_0 in P^N, a hyperplane section X_0 of A_0, a regular extension M_1 of M and a regular extension K_1 of K satisfying the following conditions:*

(i) $\eta(X)$ *is algebraically equivalent to* X_0;

(ii) A_0 *is defined over* M_1 *and* X_0 *is rational over* M_1;

(iii) A_0 *and* $\eta\theta(r)\eta^{-1}$ *for* $r \in \mathfrak{r}$ *are all defined over* K_1 *and* X_0 *is rational over* K_1.

Moreover, if k *is a field of definition for* (A, C, θ) *containing* M, K *and if* X *is rational over* k, *then we can take* M_1, K_1 *in such a way that* M_1, k *are independent over* M *and* K_1, k *are independent over* K.

Take f_X, A_X as above; we may assume that they are defined over k. Let $u_1, \cdots, u_d$ be d independent generic points of A over k and φ a generic projective transformation over $k(u_1, \cdots, u_d)$ of the ambient space for A_X. Denote by A_0 the image of A_X by φ; put $f = \varphi \circ f_X$ and

$$\lambda_\nu(y) = f[\theta(r_\nu)f^{-1}(y) + u_\nu]$$

for $y \in A_0$; denote by b the Chow point of A_0 and by z_ν the Chow point of the graph of λ_ν. By our definition, b is a generic point of $\mathfrak{F}(A, X)$ over k, and hence over M; $b \times z_1 \times \cdots \times z_d$ is a generic point of $\mathfrak{F}(A, X, \theta)$ over k, and hence over K. $M(b)$ and $K(b)$ are fields of definition for the variety A_0 since b is its Chow point. Let c be a generic point of A_0 over $k(b, z_1, \cdots, z_d)$. We can put into A_0 a structure of abelian variety by taking c as the origin. A_0 is then, as an abelian variety, defined over $M(b, c)$ and over $K(b, c)$. Put

$$M_1 = M(b, c), \quad K_1 = K(b, c, z_1, \cdots, z_d) \ .$$

Then, M_1 is a regular extension of M and K_1 is regular over K. As A and A_0 are abelian varieties, the mapping f is written in the form

$$f(x) = \eta(x) + a' \ ,$$

where η is an isomorphism of A onto A_0 and a' is a point on A_0. We see that the mapping λ_ν is written in the form

$$\lambda_\nu(y) = \eta\theta(r_\nu)\eta^{-1}(y) + a_\nu$$

for $y \in A_0$, where a_ν is a point on A_0, for each ν. As z_ν is the Chow point of the graph of λ_ν, the λ_ν are defined over K_1, so that $\eta\theta(r_\nu)\eta^{-1}$ are defined over K_1. Now let X_0 be a hyperplane section of A_0; as there is no hyperplane containing A_0, we can take X_0 in such a way that X_0 is rational over any field of definition for A_0. Then it is easy to see that η, A_0, X_0, M_1, K_1 satisfy the conditions in our proposition.

PROPOSITION 5. *(A, C, θ) and X being as in Proposition 4, let K be the smallest field of definition for $\mathfrak{F}(A, X, \theta)$ and k a field of definition for (A, C, θ) containing K. Then, K has the following property:*

(I) *σ being an isomorphism of k into a field k', (A, C, θ) is isomorphic* *to $(A^\sigma, C^\sigma, \theta^\sigma)$ if and only if σ is the identity on K.*

Extend σ to an isomorphism τ of an extension of k over which X is rational. If there exists an isomorphism η of (A, C, θ) onto $(A^\sigma, C^\sigma, \theta^\sigma)$, then $\eta(X)$ is contained in C^σ. We have $l(\eta(X)) = l(X) = l(X^\tau)$; so that by Proposition 3, we have

$$\mathcal{F}(A, X, \theta) = \mathcal{F}(A^\tau, X^\tau, \theta^\tau) = \mathcal{F}(A, X, \theta)^\sigma \; ;$$

from this follows that σ is the identity on K. Conversely, suppose that σ is the identity on K. We have then

$$\mathcal{F}(A, X, \theta) = \mathcal{F}(A, X, \theta)^\sigma = \mathcal{F}(A^\tau, X^\tau, \theta^\tau) \; .$$

By Proposition 1, (A, C, θ) is isomorphic to $(A^\sigma, C^\sigma, \theta^\sigma)$.

PROPOSITION 6. *(A, C, θ) and X being as in Proposition 4, let m be a positive integer such that mX is ample. Let K and K' be the smallest fields of definition for $\mathcal{F}(A, X, \theta)$ and $\mathcal{F}(A, mX, \theta)$, respectively. Then K is a purely inseparable extension of K'. Moreover, if the characteristic p of the basic field does not divide m, we have $K = K'$.*

Take A_0, X_0, K_1 and η for (A, C, θ) as in Proposition 4. Put $\theta_0(r) = \eta\theta(r)\eta^{-1}$ for $r \in \mathfrak{r}$. By Proposition 3, we have

$$\mathcal{F}(A, X, \theta) = \mathcal{F}(A_0, X_0, \theta_0), \quad \mathcal{F}(A, mX, \theta) = \mathcal{F}(A_0, mX_0, \theta_0) \; .$$

Since mX_0 is rational over K_1, $\mathcal{F}(A_0, mX_0, \theta_0)$ is defined over K_1. Hence we have $K' \subset K_1$.

Let k be a field of definition for (A, C, θ) and η containing K_1. By Proposition 5, for any isomorphism σ of k into a field k_1, σ is the identity on K if and only if σ is the identity on K'; from this follows that KK' is a purely inseparable extension of K, and of K'. Since K' is contained in K_1 and K_1 is a regular extension of K, we must have $KK' = K$. This proves the first assertion of the proposition. Now assume that the characteristic p does not divide m. By Proposition 4, there exist an isomorphism λ of A onto an abelian variety A' in P^N, a hyperplane section Y of A' and a regular extension K_1' such that $\lambda(mX)$ is algebraically equivalent to Y, A' and $\lambda\theta(r)\lambda^{-1}$ for $r \in \mathfrak{r}$ are all defined over K_1' and Y is rational over K_1'. Put $\theta'(r) = \lambda\theta(r)\lambda^{-1}$ for $r \in \mathfrak{r}$. We can easily verify that there exists a separably generated extension k of K_1' and a power p^a of p such that $p^a\lambda(X)$ is rational over k'. Since m and p are prime to each other, there exist two integers r, s such that $rm + sp^a = 1$; then $rY + sp^a\lambda(X)$ is algebraically equivalent to $\lambda(X)$. By Proposition 3, we have

$$\mathcal{F}(A, X, \theta) = \mathcal{F}(A', \lambda(X), \theta') = \mathcal{F}(A', rY + sp^a\lambda(X), \theta') \; .$$

Since $rX + sp^a\lambda(X)$ is rational over k', $\mathcal{F}(A, X, \theta)$ is defined over k'; this proves $k' \supset K$. On the other hand, k' is separably generated over K' and KK' is purely inseparable over K', so that we have $K = K'$; this completes our proof.

5. Let (A, C, θ) be a polarized abelian variety of type $\mathfrak{r}$. We know that there exists a divisor Y on A such that every X in C is algebraically equivalent to mY for a suitable positive integer m. Let K_m be the smallest field of definition for $\mathcal{F}(A, mY, \theta)$ for each positive integer m such that mY is ample. Let m and m' be two positive integers such that mY and $m'Y$ are ample. There exists, by Lemma 1, a positive integer s, prime to the characteristic p, such that $smm'Y$ is ample. By Proposition 6, the fields K_m and $K_{m'}$ are purely inseparable extension of $K_{smm'}$; if p does not divide m, we have $K_{m'} = K_{smm'}$ by the same proposition; hence K_m is a purely inseparable extension of $K_{m'}$ if m is prime to p. Therefore, if p does not divide m, nor m', we have $K_m = K_{m'}$. We call this same field *the field moduli of* (A, C, θ). It is necessary to show that this field does not depend upon the choice of a basis of $\mathfrak{r}$; we shall prove this in **6**. The field of moduli K is a purely inseparable extension of K_m for every m; we have $K = K_m$ if m is prime to p. Each field K_m has the property (I) of Proposition 5. Suppose now that the characteristic is 0. Then we have $K = K_m$ for every m; and the field of moduli K is determined by the property (I). Furthermore, we see that K is generated over the field Q of rational numbers by the Chow point of $\mathcal{F}(A, mY, \theta)$ for each m. If (A', C', θ') is isomorphic to (A, C, θ), the field of moduli of (A', C', θ') is the same as that of (A, C, θ).

6. Let now $\mathfrak{r}'$ be a subring of $\mathfrak{r}$ and θ' the restriction of θ on $\mathfrak{r}'$. Take a basis $\{s_1, \cdots, s_e\}$ of $\mathfrak{r}'$ over Z and construct for each ample divisor X in C the variety $\mathcal{F}'(A, X, \theta')$ with respect to this basis in the same manner as above. Let K and K' be respectively the smallest field of definition for $\mathcal{F}(A, X, \theta)$ and $\mathcal{F}'(A, X, \theta')$. We shall show that K contains K'. Take A_0, X_0, η and K_1 for A, X, θ, K as in Proposition 4. Put $\theta_0(r) = \eta\theta(r)\eta^{-1}$ for $r \in \mathfrak{r}$ and denote by θ_0' the restriction of θ_0 in $\mathfrak{r}'$. Then, by Proposition 3, we have $\mathcal{F}'(A, X, \theta') = \mathcal{F}'(A_0, X_0, \theta_0')$. By our choice of K_1, $\mathcal{F}'(A_0, X_0, \theta_0')$ is defined over K_1; this proves $K' \subset K_1$. By Proposition 5, for every isomorphism σ of K_1, if σ fixes every element of K, σ fixes every element of K'; hence KK' must be purely inseparable over K. Since K_1 is a regular extension of K, we have $KK' = K$, namely $K \supset K'$. If $\mathfrak{r}' = \mathfrak{r}$, we have also $K' \supset K$ and hence $K = K'$. This proves *the field of moduli of* (A, C, θ) *does not depend on the choice of a basis of* $\mathfrak{r}$.

Now assuming that $\mathfrak{r}'$ does not necessarily coincide with $\mathfrak{r}$, we shall show that K is separably algebraic over K'. By our definition, K' is a field of definition for $\mathcal{F}(A, X)$. Hence it is sufficient to prove that K is separably algebraic over the smallest field of definition M for $\mathcal{F}(A, X)$. Let k be a field of definition for (A, C, θ) containing K, over which X is rational. By Proposition 4, we can find A_0, X_0, η and M_1 with the properties as in that proposition, for A, X, M and k. Put $\theta_0(r) = \eta\theta(r)\eta^{-1}$ for $r \in \mathfrak{r}$. We have then

$$\mathcal{F}(A, X, \theta) = \mathcal{F}(A_0, X_0, \theta_0) .$$

We know that all endomorphisms of A_0 are defined over a finite separably algebraic extension M_2 of M_1; so $\mathcal{F}(A_0, X_0, \theta_0)$ is defined over M_2. Let w be a set of quantities generating M_2 over M and w' a generic specialization of w over k, such that w and w' are independent over k. Then there exists an isomorphism σ of $M(w)$ onto $M(w')$ such that $w^\sigma = w'$ and σ is the identity on k. We see that $\mathcal{F}(A, X, \theta) = \mathcal{F}(A_0^\sigma, X_0^\sigma, \theta_0^\sigma)$ and that $\mathcal{F}(A_0^\sigma, X_0^\sigma, \theta_0^\sigma)$ is defined over $M(w)^\sigma = M(w')$. Hence we have $K \subset M(w) \cap M(w')$. The fields $M(w)$, $M(w')$ are separably generated extension of M and they are independent over M since they are so over k and since k and $M(w, w')$ are independent over M. Therefore, from the relation $K \subset M(w) \cap M(w')$ follows that K is separably algebraic over M; this implies also that K is separably algebraic over K'. We have thus proved

PROPOSITION 7. *Let (A, C, θ) be a polarized abelian variety of type $\mathfrak{r}$. Let $\mathfrak{r}'$ be a subring of $\mathfrak{r}$ and θ' the restriction of θ on $\mathfrak{r}'$. Then the field of moduli of (A, C, θ) is a finite separably algebraic extension of the field of moduli of (A, C, θ').*

Now we consider the case where $\mathfrak{r}$ is the ring of integers Z and $\theta(1)$ is the identity of $\mathcal{A}(A)$. In this case, to define the field of moduli, it is not necessary to take into account the operation of $\theta(Z)$. In fact, notations being as above, let σ be an isomorphism of k which is the identity on M. Then we have $\mathcal{F}(A, X) = \mathcal{F}(A^\sigma, X^\sigma)$. By Proposition 2, there exists an isomorphism ξ of (A, C) onto (A^σ, C^σ). By our assumption on $\mathfrak{r}$ and θ, it is clear that ξ is an isomorphism of (A, C, θ) onto $(A^\sigma, C^\sigma, \theta^\sigma)$. By Proposition 5, σ is the identity on K. As K is separable over M, this proves $K = M$. Thus we have proved that $\mathcal{F}(A, X)$ and $\mathcal{F}(A, X, \theta)$ have the same field as their smallest fields of definition. Hence we can define the field of moduli of (A, C, θ) as the smallest field of definition for $\mathcal{F}(A, X)$ for a certain X in C; the choice of X is the same as in our previous definition. We call this field *the field of moduli of (A, C)*.

PROPOSITION 8. *Let (A, C, θ) be a polarized abelian variety of type $\mathfrak{r}$.*

Then the field of moduli of (A, C, θ) is a separably algebraic extension of the field of moduli of (A, C).

This is what we have proved above in Proposition 7.

PROPOSITION 9. *Let (A, C) and (A', C') be two polarized abelian varieties of the same dimension, and let K, K' be respectively the fields of moduli of (A, C), (A', C'). If there exists a separable homomorphism of A onto A', then KK' is algebraic over K.*

Let X be an ample divisor in C, X' an ample divisor in C', K_0 the smallest field of definition for $\mathcal{F}(A, X)$; let k be a field of definition for (A, C), (A', C') containing K_0 over which X, X' are rational. Then, we can find A_0, X_0, η, K_1 with the properties in Proposition 4, for A, X, K_0 and k. Let λ be a separable homomorphism of A onto A' and $\mathfrak{g}$ the kernel of $\lambda \circ \eta^{-1}$. Then there exist an abelian variety B defined over an algebraic extension K_2 of K_1, and a separable homomorphism μ of A_0 onto B whose kernel is $\mathfrak{g}$. It is easy to see that there exists an isomorphism ξ of B onto A' such that $\xi \circ \mu = \lambda \circ \eta^{-1}$. We can easily find a divisor Y on B which is algebraically equivalent to $\xi^{-1}(X')$ and is rational over an algebraic extension K_3 of K_2. We have $\mathcal{F}(A', X') = \mathcal{F}(B, Y)$ by Proposition 3; $\mathcal{F}(B, Y)$ is defined over K_3. Let w be a set of quantities generating K_3 over K_0 and w' a generic specialization of w over k such that w, w' are independent over k. Let σ be an isomorphism of $k(w)$ onto $k(w')$ such that $w^\sigma = w'$ and σ is the identity on k. We have then $\mathcal{F}(A', X') = \mathcal{F}(B^\sigma, Y^\sigma)$; so $\mathcal{F}(A', X')$ is defined over $K_0(w) = K_0(w')$. Hence the smallest field of definition K_0' for $\mathcal{F}(A', X')$ is contained in $K_0(w) \cap K_0(w')$. By our choice of w, w', two fields $K_0(w)$, $K_0(w')$ are independent over K_0, so that $K_0 K_0'$ is algebraic over K_0; this proves our proposition.

7. We shall now consider the case of dimension 1. Let A be an abelian variety of dimension 1, defined over k. Since any two points on A are algebraically equivalent to each other, there exists only one polarization C on A. Therefore, we understand by the field of moduli of A the field of moduli of the uniquely determined polarized abelian variety (A, C). Suppose that the characteristic of the basic field is different from 2, 3. A is then isomorphic to a curve in the projective plane, defined by an equation

$$X_0 X_2^2 = 4X_1^3 - \gamma_2 X_0^2 X_1 - \gamma_3 X_0^3$$

where γ_2, γ_3 are elements in K. Put $j = \gamma_2^3/(\gamma_2^3 - 27\gamma_3^2)$. Then it is well known that j is uniquely determined by A; so we write $j = j(A)$. We have $j(A) = j(A')$ if and only if A and A' are isomorphic. We know moreover that, for any j, there exists an abelian variety A_0 in the projective plane defined over $k_0(j)$ for which we have $j = j(A)$, where k_0 denotes the prime field. This shows that $\mathcal{F}(A, \mathfrak{z}(0))$ is defined over $k_0(j(A))$. Hence the field of moduli K of A is contained in $k_0(j(A))$. If σ is an isomorphism of $k_0(j(A))$ which fixes every element of K, A^σ is isomorphic to A by Proposition 5, so that we have $j(A) = j(A^\sigma) = j(A)^\sigma$. This shows that $k_0(j(A))$ is purely inseparable over K. On the other hand, by Proposition 4, there exists an abelian variety A', isomorphic to A, defined over a regular extension K_1 of K. We see that $j(A) = j(A')$ is contained in K_1. As $j(A)$ is purely inseparable over K, $j(A)$ is contained in K. Hence we have $k_0(j(A)) = K$.

3. Analytic systems of polarized abelian varieties

8. Let $\mathfrak{U}$ be a connected open subset of C^n and $\{f_i(z) \, (i = 1, 2, \cdots)\}$ a countable set of holomorphic functions on $\mathfrak{U}$; let k be a subfield of C composed of countably infinite elements. We say that a point $z_0 \in \mathfrak{U}$ is a *generic point of $\mathfrak{U}$ for $\{f_i(z)\}$ over k* if $(f_1(z_0), f_2(z_0), \cdots)$ is a generic specialization of $(f_1, f_2, \cdots)$ over k, namely, if there exists an isomorphism σ of the field $k(f_1, f_2, \cdots)$ onto the field $k(f_1(z_0), f_2(z_0), \cdots)$ such that $f_i^\sigma = f_i(z_0)$ for every i and $a^\sigma = a$ for every $a \in k$.

LEMMA 6. *The notations $\mathfrak{U}$, f_i, k being as above, there exists a join $\mathfrak{F}$ of countably infinite analytic subsets of $\mathfrak{U}$ of codimension 1, such that every point in $\mathfrak{U} - \mathfrak{F}$ is generic for the f_i over k.*

This is an easy consequence of our definition. For a proof, see [11, Exposé 19]. We shall also use "generic points" for meromorphic functions by considering only points where the functions are holomorphic.

Let $\{F_i(z)\}$ be a countable set of analytic mappings of $\mathfrak{U}$ into projective spaces; we denote by P^{N_i} the space on which F_i takes the values, for each i. We know that the quotients of the coordinates of each F_i define meromorphic functions on $\mathfrak{U}$. We denote by $k(F_1, F_2, \cdots)$ the field of functions generated over k by these functions. For every point $y \in \mathfrak{U}$, $k(F_1(y), F_2(y), \cdots)$ denotes, as a usual notation in algebaric geometry, the field of numbers generated over k by the quotients of the coordinates of the points $F_i(y)$ in P^{N_i}. We say that a point $z \in \mathfrak{U}$ *is generic for the F_i over k*, if for every $y \in \mathfrak{U}$, $(F_1(y), F_2(y), \cdots)$ is a specialization of $(F_1(z), F_2(z), \cdots)$ over k.

LEMMA 7. *The notations $\mathfrak{U}$, F_i, k being as above, there exists a join $\mathfrak{F}$ of countably infinite analytic subsets of $\mathfrak{U}$ of codimension 1, such that every point in $\mathfrak{U} - \mathfrak{F}$ is generic for the F_i over k.*

Let $F_{\alpha\beta}^{(i)}$ for $0 \leq \alpha \leq N_i$, $0 \leq \beta \leq N_i$ be meromorphic functions defined as the quotient (α^{th} coordinate of F_i)/(β^{th} coordinate of F_i). There may occur some pair of α, β for which $F_{\alpha\beta}^{(i)}$ has no meaning, since β^{th} coordinate of F_i may be identically equal to 0. We can assume, after reordering of the coordinates of P^{N_i}, that the $F_{a0}^{(i)}$ are really meromorphic functions on $\mathfrak{U}$. Now fix a positive integer n; let $\mathfrak{U}_n$ be the set of points of $\mathfrak{U}$ where the $F_{a0}^{(i)}$ for $1 \leq i \leq n$, $0 \leq a \leq N_i$ are all holomorphic and $\mathfrak{U}'_n$ the set of generic points of $\mathfrak{U}_n$ for these functions over k. Let z be a point of $\mathfrak{U}'_n$ and y a point of $\mathfrak{U}$. For each i, we can find $\beta(i)$ such that $\beta(i)^{\text{th}}$ coordinate of $F_i(y)$ is not 0. Then, $F_{a\beta(i)}^{(i)}$ has a meaning and is equal to $F_{a0}^{(i)}/F_{\beta(i)0}^{(i)}$. As z is generic for the $F_{\gamma0}^{(i)}$, we have $F_{\gamma0}^{(i)}(z) \neq 0$ for every γ. We see that

$$(\cdots, F_{a\beta(i)}^{(i)}(y), \quad \cdots, \quad F_{N_i\beta(i)}^{(i)}(y), \cdots), \qquad 1 \leq i \leq n,$$

is a specialization of

$$(\cdots, F_{a0}^{(i)}/F_{\beta(i)0}^{(i)}, \cdots, F_{aN_i}^{(i)}/F_{\beta(i)0}^{(i)}, \cdots), \qquad 1 \leq i \leq n,$$

over k. The latter is a generic specialization of

$$(\cdots, F_{a0}^{(i)}(z)/F_{\beta(i)0}^{(i)}(z), \cdots, F_{aN_i}^{(i)}(z)/E_{\beta(i)0}^{(i)}(z), \cdots), \quad 1 \leq i \leq n,$$

over k. Hence, we see that $(F_1(y), \cdots, F_n(y))$ is a specialization of $(F_1(z), \cdots, F_n(z))$ over k. By Lemma 1, $\bigcup_{n=1}^{\infty}(\mathfrak{U} - \mathfrak{U}'_n)$ is contained in a join $\mathfrak{F}$ of countably infinite analytic subsets of $\mathfrak{U}$ of codimension 1. We can then easily verify that this $\mathfrak{F}$ has the property of our lemma.

When we consider a finite number of analytic mappings F_i $(0 \leq i \leq s)$ of $\mathfrak{U}$ into affine spaces or projective spaces, we can substitute for them one analytic mapping of $\mathfrak{U}$ into a projective space, as follows. If F_i is an analytic mapping into an affine space S^n of dimension n, then F_i can be considered as an analytic mapping of $\mathfrak{U}$ into P^n by identifying a point $(\xi_1, \cdots, \xi_n)$ of S^n with the point $(1, \xi_1, \cdots, \xi_n)$ of P^n. F_i being thus regarded as a mapping into P^{n_i} for each i, we take the canonical embedding φ of $P^{n_1} \times \cdots \times P^{n_s}$ into P^N. Put

$$G(z) = \varphi(F_1(z) \times \cdots \times F_s(z)) \, .$$

Then $G(z)$ is an analytic mapping of $\mathfrak{U}$ onto P^N. We call $G(z)$ *the gathering* of the $F_i(z)$. For every $z \in \mathfrak{U}$ and for every subfield k of C, we have

$$k(G(z)) = k(F_1(z), \cdots, F_s(z)) \, .$$

For $y \in \mathfrak{U}$, $z \in \mathfrak{U}$, $G(y)$ is a specialization of $G(z)$ over k if and only if $(F_1(y), \cdots, F_s(y))$ is a specialization of $(F_1(z), \cdots, F_s(z))$ over k. Hence z is generic for G over k, if and only if z is generic for the F_i over k.

We see that, if z is generic for $G(z)$ over k, then, for every function g in $k(G)$, g is defined at z, and the mapping $g \to g(z)$ gives an isomorphism of $k(G)$ onto $k(G(z))$. We call this isomorphism *the canonical isomorphism* of $k(G)$ onto $k(G(z))$.

9. We recall a theorem of [11, Exposé 19], which is fundamental for our whole theory in this section. Let $\mathfrak{U}$ and $\mathfrak{Z}$ be respectively connected open subsets of C^n and C^m; let $f_i(u, z)$, for $0 \leq i \leq N$, be $N + 1$ holomorphic functions on $\mathfrak{U} \times \mathfrak{Z}$, where u and z denote respectively the points of $\mathfrak{U}$ and $\mathfrak{Z}$. We consider the following properties for the f_i:

(S1) *We have, everywhere on* $\mathfrak{U} \times \mathfrak{Z}$,

$$
\mathrm{rank}
\begin{bmatrix}
f_0(u, z) & f_1(u, z) & \cdots & f_N(u, z) \\
\dfrac{\partial f_0}{\partial u_1}(u, z) & \dfrac{\partial f_1}{\partial u_1}(u, z) & \cdots & \dfrac{\partial f_N}{\partial u_1}(u, z) \\
\cdots & \cdots & \cdots & \cdots \\
\dfrac{\partial f_0}{\partial u_n}(u, z) & \dfrac{\partial f_1}{\partial u_n}(u, z) & \cdots & \dfrac{\partial f_N}{\partial u_n}(u, z)
\end{bmatrix}
= n + 1 .
$$

If this is so, there is no point (u, z) of $\mathfrak{U} \times \mathfrak{Z}$ such that $f_i(u, z) = 0$ for all i. Hence we can regard

$$
(f_0(u, z), \cdots, f_N(u, z))
$$

as a point of P^N; we denote this point by $f(u, z)$.

(S2) *For every* $z \in \mathfrak{Z}$, *there exists an algebraic variety* $V(z)$ *in* P^N *such that the set*

$$
B(z) = \{ f(u, z) \mid u \in \mathfrak{U} \}
$$

is an open subset of $V(z)$ *in the sense of Zariski's topology.*

(S3) *The degree of the projective varieties* $V(z)$ *does not depend on* $z \in \mathfrak{Z}$.

We have then

THEOREM S (Theorem 3 of [11, Exposé 19]). *If the* $f_i(u, z)$ *satisfy the conditions* (S1–3), *the mapping* $z \to c(V(z))$ *is everywhere holomorphic on* $\mathfrak{Z}$.

10. Let D be a discrete subgroup of C^n of rank 2n. Let $\mathfrak{E}(x, y)$ be an R-bilinear form defined for $x \in C^n$, $y \in C^n$, with values in R. We call $\mathfrak{E}(x, y)$ a *Riemann form* on the complex torus C^n/D if it satisfies the following conditions:

(F1) *the values* $\mathfrak{E}(x, y)$ *for* $x \in D$, $y \in D$, *are integers;*

(F2) $\mathfrak{E}(x, y) = - \mathfrak{E}(y, x)$;

(F3) *the form* $\mathfrak{E}(x, \sqrt{-1}\,y)$ *is a positive definite symmetric form.*

C^n/D has a structure of abelian variety if and only if there exists a Riemann form on it. Every Riemann form on C^n/D is obtained from a non-degenerate analytic divisor of C^n/D, and every non-degenerate analytic divisor of C^n/D defines a Riemann form on C^n/D; hence, if A is an abelian variety isomorphic to C^n/D, every Riemann form on C^n/D determines a polarization of A. A modern approach for these facts can be found in [15]. Let $\{u_1, \cdots, u_{2n}\}$ be a basis of D over Z and ω the matrix whose columns are $u_1, \cdots, u_{2n}$. Then, for every R-bilinear form $\mathfrak{E}(x, y)$ on C^n, there exists a matrix E of degree $2n$ such that

$$\mathfrak{E}(\omega a, \omega b) = {}^t a E b$$

for $a \in R^{2n}$, $b \in R^{2n}$. The conditions (F1–3) are equivalent to the following conditions:

(G1) *the coefficients of E are integers;*

(G2) ${}^t E = - E$;

(G3) $\omega\, {}^t E^{-1}\, {}^t \omega = 0$;

(G4) $\sqrt{-1}\, \omega^t E^{-1}\, {}^t \overline{\omega}$ *is a positive definite hermitian matrix.*

Now let $\delta_1, \cdots, \delta_n$ be n positive integers such that

$$\delta_1 = 1, \qquad \delta_i \mid \delta_{i+1} \qquad (1 \leqq i \leqq n - 1);$$

let δ be the diagonal matrix having the δ_i as the diagonal elements; and put

$$E = \begin{pmatrix} 0 & -\delta \\ \delta & 0 \end{pmatrix}.$$

Denote by $\mathfrak{S}_n$ the set of all symmetric matrices s with complex coefficients, of degree n, such that $\mathrm{Im}(s)$ is positive definite. For every $s \in \mathfrak{S}_n$, denote by $D_\delta(s)$ the discrete subgroup of C^n generated over Z by the columns of the matrix $(s\ \delta)$. We see easily that the matrix E_δ defines a Riemann form on $C^n/D_\delta(s)$; the complex torus $C^n/D_\delta(s)$, attached with this form, determines a structure of polarized abelian variety. In [11, Exposé 20] it was shown that, for each δ, there exists a system of abelian varieties $A_\delta(s)$ parametrized by holomorphic functions $\alpha_0(u, s), \cdots, \alpha_N(u, s)$ on $C^n \times \mathfrak{S}_n$ satisfying the conditions (S1–3). We denote by $\Lambda_\delta(u, s)$ the point

$$(\alpha_0(u, s), \cdots, \alpha_N(u, s))$$

in P^N. This system has the following properties:

(M1) $A_\delta(s)$ *is an abelian variety in P^N of degree g, whose hyperplane sections define a complete linear system, where g is independent of s. There is no hyerplane containing $A_\delta(s)$.*

(M2) *For every $s \in \mathfrak{S}_n$, the mapping $u \to \Lambda_\delta(u, s)$ gives an analytic isomorphism of $C^n/D_\delta(s)$ onto $A_\delta(s)$.*

(M3) *The polarization of $A_\delta(s)$ determined by the hyperplane sections* *corresponds to the Riemann form given by $E_{\cdot}$.*

Applying Theorem S to this case, we obtain the following theorem.

THEOREM T. *The mapping $s \to c(A_\delta(s))$ is everywhere holomorphic on* $\mathfrak{S}_n$.

11. Let $\mathfrak{Z}$ be a connected open subset of C^m. We consider a system $\{E, w(z)\}$ having the following properties.

(R1) *E is a skew-symmetric matrix of degree $2n$ with integral coefficients.*

(R2) *The mapping $z \to w(z)$ is an analytic mapping of $\mathfrak{Z}$ into the space of complex matrices with n rows and $2n$ columns.*

(R3) *The group $D(z)$ generated by the columns of $w(z)$ over Z is a discrete subgroup of C^n of rank $2n$.*

(R4) *The matrix E defines a Riemann form on the complex torus $C^n/D(z)$ for every $z \in \mathfrak{Z}$; namely, E and $w(z)$ satisfy the conditions (G3–4).*

Then, we can find a matrix U of degree $2n$ with integral coefficients such that $\det U = \pm 1$ and

$$
{}^{t}UEU = \begin{pmatrix} 0 & -\delta \\ \delta & 0 \end{pmatrix}
$$

where δ is a diagonal matrix of degree n as given in **10**. Denote by $u(z)$ and $v(z)$ the matrices of degree n whose columns are respectively the first n columns and the last n columns of $w(z)U$; so we can write

$$
w(z)U = (u(z)\ v(z)) .
$$

As E and $w(z)$ satisfy (G3–4), we have

$$
v(z)\delta^{-1}\ {}^{t}u(z) = u(z)\delta^{-1}\ {}^{t}v(z)
$$

and

$$
\sqrt{-1}[v(z)\delta^{-1}\ \overline{{}^{t}u(z)} - u(z)\delta^{-1}\ \overline{{}^{t}v(z)}]
$$

is a positive definite hermitian matrix. This implies in particular that $v(z)$ has the inverse; put

$$
s(z) = \delta v(z)^{-1}u(z)
$$

for $z \in \mathfrak{Z}$. Then we see that $z \to s(z)$ is an analytic mapping of $\mathfrak{Z}$ into $\mathfrak{S}_n$; and we have

$$
(s(z)\ \delta)U^{-1} = \delta v(z)^{-1}w(z) .
$$

Hence, the linear mapping $x \to v(z)^{-1}x$ of C^n onto itself gives an isomorphism of $C^n/D(z)$ onto $C^n/D_\delta(s(z))$ for each $z \in \mathfrak{Z}$. Put

$$A(z) = A_\delta(s(z)), \ \Lambda(x, z) = \Lambda(\delta v(z)^{-1}x, s(z))$$

for $x \in C^n$, $z \in \mathfrak{Z}$. Then, $\Lambda(x, z)$ is an analytic mapping of $C^n \times \mathfrak{Z}$ into P^N; and for each $z \in \mathfrak{Z}$, the mapping $x \to \Lambda(x, z)$ gives an analytic isomorphism of $C^n/D(z)$ onto $A(z)$; the polarization of $A(z)$ determined by the hyperplane sections corresponds to the Riemann form on $C^n/D(z)$ given by the matrix E. By Theorem T the mapping $z \to c(A(z))$ is an analytic mapping of $\mathfrak{Z}$ into a projective space.

We shall now consider an analytic system of polarized abelian varieties of type $\mathfrak{r}$. Let $\mathfrak{r}$ be a ring having a finite basis over Z and $\chi(r)$ a representation of $\mathfrak{r}$. We assume that χ satisfies the following conditions.

(R5) $\chi(r)$ *is a faithful representation of* $\mathfrak{r}$ *by complex matrices of degree* n.

(R6) *For every* $r \in \mathfrak{r}$, *the linear mapping* $x \to \chi(r)x$ *of* C^n *into itself gives an endomorphism of* $C^n/D(z)$.

Then, there exists an endomorphism $\theta_z(r)$ of $A(z)$ such that

$$\theta_z(r)\Lambda(x, z) = \Lambda(\chi(r)x, z) \ .$$

We see easily that $r \to \theta_z(r)$ defines an isomorphism of $\mathfrak{r}$ into $\mathcal{A}(A(z))$ for every $z \in \mathfrak{Z}$. Thus we have obtained a system of polarized abelian varieties $(A(z), C_s, \theta_s)$ of type $\mathfrak{r}$, where C_s denotes the polarization determined by the hyperplane sections of $A(z)$.

PROPOSITION 10. *$(A(z), C_s, \theta_s)$ being as above, let $T(z)$ be the graph of the law of composition of $A(z)$ and $V(r, z)$ the graph of $\theta_z(r)$ for each $z \in \mathfrak{Z}$, $r \in \mathfrak{r}$. Then the mappings $z \to c(A(z))$, $z \to c(T(z))$, $z \to c(V(r, z))$ are everywhere holomorphic on $\mathfrak{Z}$.*

We see that $T(z)$ is the locus of $\Lambda(u, z) \times \Lambda(v, z) \times \Lambda(u + v, z)$ for $u \in C^n$, $v \in C^n$. Let φ be the canonical embedding of $P^N \times P^N \times P^N$ into P^M; put $U(z) = \varphi(T(z))$ and consider the gathering of the mappings $\Lambda(u, z)$, $\Lambda(v, z)$ and $\Lambda(u + v, z)$; we obtain then a system of holomorphic functions $\{f_i(w, z); 0 \leq i \leq M\}$ on $C^n \times C^n \times \mathfrak{Z}$ parametrizing $U(z)$, namely

$$U(z) = \{f(w, z) \mid w \in C^n \times C^n\}$$

where $f(w, z)$ denotes the point $(f_0(w, z), \cdots, f_M(w, z))$ in P^M. We can easily verify that the $f_i(w, z)$ satisfy the condition (S1). We shall show that $\deg(U(z))$ is independent of s. It is sufficient to prove that, for any $z^0 \in \mathfrak{Z}$, there exists a neighborhood of z^0 in which we have $\deg U(z) = \deg U(z^0)$. To prove this, fix a point z^0 of $\mathfrak{Z}$ and put $d = \deg U(z^0)$. Let L be a linear variety of dimension $M - 2n$ such that the intersection

$L \cdot U(z^0)$ consists of exactly d distinct points $p^1, \cdots, p^d$ of multiplicity 1. Let

$$\sum_{j=0}^{M} t_{ij} X_j = 0 , \qquad (1 \le i \le 2n)$$

be a system of equations for L; put

$$\xi_i(w, z) = \sum_{j=0}^{M} t_{ij} f_j(w, z) , \qquad (1 \le i \le 2n).$$

There exists a point $w^1 \in C^n \times C^n$ such that

$$p^1 = f(w^1, z^0) .$$

We have then

$$\xi_i(w^1, z^0) = 0 , \qquad (1 \le i \le 2n).$$

Since p^1 is a point of $L \cdot U(z^0)$ of multiplicity 1, L is transversal to the tangent linear variety to $U(z^0)$ at p^1. Hence, by virtue of the fact that the $f_i(w, z)$ satisfy (S1), we have

$$\det\!\left(\frac{\partial \xi_i}{\partial w_j}(w^1, z^0)\right) \neq 0 .$$

Therefore, by the theorem of implicit functions, we obtain $2n$ functions $g_i(z)$ holomorphic at z^0 such that

$$w_i^1 = g_i(z^0) , \qquad (1 \le i \le 2n),$$

$$\sum_{j=0}^{M} t_{ij} f_j(g(z), z) = 0 , \qquad (1 \le i \le 2n)$$

in a neighborhood of z^0. We get, by the same argument, for each α, a system of $2n$ functions $g_i^\alpha(z)$ holomorphic at z^0 such that

$$p^\alpha = f(g^\alpha(z^0), z^0) ,$$

$$\sum_{j=0}^{M} t_{ij} f_j(g^\alpha(z), z) = 0 , \qquad (1 \le i \le 2n)$$

in a neighborhood of z^0. Put, for each α,

$$p^\alpha(z) = f(g^\alpha(z), z) .$$

Then the $p^\alpha(z)$ are contained in $L \cap U(z)$ for every z in a certain neighborhood of z^0. Since the p^α are distinct, there exists a neighborhood Y of z^0 such that the $p^\alpha(z)$ are all distinct for each $z \in Y$. We can take Y in such a way that we have

$$\det\left(\frac{\partial \xi_i}{\partial w_j}(g^\alpha(z), z)\right) \neq 0$$

for every $z \in Y$ and every α, since this holds at z^0. Then $p^\alpha(z)$ is a point of the intersection $L \cdot U(z)$ of multiplicity 1. Hence, for $z \in Y$, we have deg $U(z) = d$ if there is no point of $L \cap U(z)$ other than the $p^{\alpha}(z)$. Suppose that we can not have deg $U(z) = d$ in any neighborhood of z^0. Then we can find a sequence $\{z^{\nu}\}$ in Y converging to z^0 such that deg $U(z^{\nu}) > d$ for every ν; $L \cap U(z^{\nu})$ contains a point q^{ν} different from the $p^{\alpha}(z^{\nu})$. Take a point $a^{\nu} \in C^n \times C^n$ such that

$$q^{\nu} = f(a^{\nu}, z^{\nu})$$

for each ν. We can take all the a^{ν} in a compact set of $C^n \times C^n$, since we have

$$f(w, z) = f(w', z)$$

if $w - w' \in D(z) \times D(z)$. Then, by choosing a suitable subsequence of $\{a^{\nu}\}$, we may assume that $\{a^{\nu}\}$ converges to a point $a^0 \in C^n \times C^n$. Since q^{ν} is contained in L, we have

$$\sum_{j=0}^{M} t_{ij} f_j(a^{\nu}, z^{\nu}) = 0 , \qquad (1 \leqq i \leqq 2n),$$

so that we have

$$\sum_{j=0}^{M} t_{ij} f_j(a^0, z^0) = 0 , \qquad (1 \leqq i \leqq 2n).$$

We see that the point $f(a^0, z^0)$ is contained in $L \cap U(z^0)$; this point must coincide with one of the p^{α}, say p^1. Then, $a^0 - w^1$ is contained in $D(z^0) \times D(z^0)$. By our construction, there exists a set of generators $\{x_1(z), \cdots, x_{4n}(z)\}$ of $D(z) \times D(z)$ such that the $x_i(z)$ are holomorphic functions of z. We have

$$a^0 - w^1 = \sum_{i=1}^{4n} \gamma_i x_i(z^0)$$

for suitable integers γ_i. Put

$$b^{\nu} = a^{\nu} - \sum_{i=1}^{4n} \gamma_i x_i(z^{\nu}) .$$

Then b^{ν} converges to w^1, and we have

$$f(a^{\nu}, z^{\nu}) = f(b^{\nu}, z^{\nu}) .$$

Put $c^{\nu} = g^1(z^{\nu})$; then $\{c^{\nu}\}$ converges to w^1, and we have

$$\sum_{j=0}^{M} t_{ij}[f_j(b^{\nu}, z^{\nu}) - f_j(c^{\nu}, z^{\nu})] = 0 , \qquad (1 \leqq i \leqq 2n).$$

By a well known theorem of differential calculus, we can find points $e(j, k, \nu)$ such that

$$\sum_{j,k} t_{ij} \frac{\partial f_j}{\partial w_k}(e(j, k, \nu), z^{\nu})(b_k^{\nu} - c_k^{\nu}) = 0 , \qquad (1 \leqq i \leqq 2n),$$

$$\lim_{\nu \to \infty} e(j, k, \nu) = w^1 , \qquad (0 \leqq j \leqq M, 1 \leqq k \leqq 2n).$$

As we have

$$\det\left(\sum_{j=0}^{M} t_{ij}\frac{\partial f_j}{\partial w_k}(w^1,\, z^0)\right) \neq 0\ ,$$

there exists a natural number N_0 such that

$$b_k^\nu - c_k^\nu = 0 \qquad\qquad (1 \leq k \leq 2n)$$

for every $\nu > N_0$. Then we must have

$$f(a^\nu,\, z^\nu) = f(b^\nu,\, z^\nu) = f(c^\nu,\, z^\nu)\ ,$$

namely $q^\nu = p^1(z^\nu)$ for every $\nu > N_0$. This is a contradiction since q^ν is different from the $p^\alpha(z^\nu)$. Hence there exists a neighborhood of z^0 in which $\deg U(z) = d$ holds. Thus we have proved our system $\{U(z);\ f(w,\, z)\}$ satisfies the condition (S3). By Theorem S, the mapping $z \to c(T(z)) = c(U(z))$ is everywhere holomorphic on $\mathfrak{Z}$. We can prove, by the same argument, that the mapping $z \to c(V(r,\, z))$ is everywhere holomorphic on $\mathfrak{Z}$. As to $c(\dot{A}(z))$, our assertion is an immediate consequence of Theorem T; hence our proposition is proved.

12. Notations being the same as in Proposition 10, take a basis $\{r_1,\, \cdots,\, r_d\}$ of $\mathfrak{r}$; let $\Psi(z)$ the gathering of the mappings $c(A(z))$, $\Lambda(0,\, z)$, $c(T(z))$ and $c(V(r_\nu,\, z))$ for $1 \leq \nu \leq d$. As $\Lambda(0,\, z)$ is the origin of $A(z)$, the abelian variety $A(z)$ is defined over the field $Q(c(A(z)),\, \Lambda(0,\, z))$ and the $\theta_z(r_\nu)$ are defined over

$$Q(c(A(z)),\, \Lambda(0,\, z),\, c(V(r_\nu,\, z)))\ .$$

As $\{r_\nu\}$ is a basis of $\mathfrak{r}$, every element of $\theta(\mathfrak{r})$ is defined over $Q(\Psi(z))$. Hence the polarized abelian variety $(A(z),\, C_z,\, \theta_z)$ of type $\mathfrak{r}$ is defined over $Q(\Psi(z))$ for every $z \in \mathfrak{Z}$.

Now we construct the variety $\mathcal{F}(A(z),\, \theta_z)$ as in **3**, with respect to the basis $\{r_\nu\}$, for each $z \in \mathfrak{Z}$ and put

$$\mathcal{F}(z) = \mathcal{F}(A(z),\, \theta_z)\ .$$

We see that the variety $\mathcal{F}(z)$ is defined over $Q(\Psi(z))$ for every $z \in \mathfrak{Z}$.

LEMMA 8. *n and N being respectively the dimensions of $A(z)$ and the ambient space for $A(z)$, the variety $\mathcal{F}(z)$ is of dimension $(N+1)^2 - 1 + dn$.*

Let k be a field of definition for $(A(z),\, C_z,\, \theta_z)$ and (t_{ij}) a matrix of degree $N + 1$ whose coordinates are $(N + 1)^2$ independent variables over k; denote by φ the projective transformation of P^N given by (t_{ij}); let $u_1,\, \cdots,\, u_d$ be d independent generic points of $A(z)$ over $k(t_{ij})$. Put $B = \varphi(A(z))$ and denote by W_ν the graph of the mapping

$$y \to \varphi[\theta_x(r_\nu)\varphi^{-1}(y) + u_\nu]$$

of B onto itself. Then, by our definition, $c(B) \times c(W_1) \times \cdots \times c(W_d)$ is a generic point of $A(z)$ over k. We first prove that $k(c(B))$ is of dimension $(N+1)^2 - 1$ over k. We see that $k(c(B)) \subset k(t_{ij}/t_{00})$ and $k(t_{ij}/t_{00})$ is the smallest field of definition for φ containing k. Let σ be an isomorphism of $k(t_{ij}/t_{0j})$ which is the identity on $k(c(B))$; put $\psi = \varphi^{-1} \circ \varphi^\sigma$; we have then $\psi(A(z)) = A(z)$ since we have $B^\sigma = B$ and $A(z)^\sigma = A(z)$. There exist an automorphism λ of $A(z)$ and an element a of $A(z)$ such that $\psi(x) = \lambda x + a$ for $x \in A(z)$. X being a hyperplane section of $A(z)$, we see that X is linearly equivalent to $\lambda(X)_a$, as ψ is a projective transformation; hence λ is an automorphism of $(A(z), C_z)$. By Corollary 1 of Theorem 6 in [7], there exist only a finite number of automorphisms of a polarized abelian variety. Moreover, as $\lambda(X)$ is non-degenerate, there are only a finite number of points a of $A(z)$ such that $\lambda(X)_a$ is linearly equivalent to X. On the other hand, if two projective transformations of P^N coincide on $A(z)$, then they coincide on the whole space P^N, as the hyperplane sections of $A(z)$ determine a complete linear system of $A(z)$ of dimension $N+1$. Hence there are only a finite number of isomorphisms σ of $k(t_{ij}/t_{00})$ which is the identity on $k(c(B))$. We have therefore

$$\dim_k k(c(B)) = \dim_k k(t_{ij}/t_{00}) = (N+1)^2 - 1 .$$

Now let τ be an isomorphism of $k(t_{ij}/t_{00}, u_1, \cdots, u_d)$ which is the identity on $k(t_{ij}/t_{00}, c(W_1), \cdots, c(W_d))$. We have then $W_\nu^\tau = W_\nu$, $\varphi^\tau = \varphi$ and $\theta_x(r_\nu)^\tau = \theta_x(r_\nu)$, so that we have

$$\varphi[\theta_x(r_\nu)\varphi^{-1}(y) + u_\nu] = \varphi[\theta_x(r_\nu)\varphi^{-1}(y) + u_\nu^\tau]$$

for every $y \in B$; this shows $u_\nu^\tau = u_\nu$ for every ν. Hence τ is the identity on $k(t_{ij}/t_{00}, u_1, \cdots, u_d)$. From this follows

$$\dim_k k(c(B), c(W_1), \cdots, c(W_d)) = \dim_k k(t_{ij}/t_{00}, c(W_1), \cdots, c(W_d))$$
$$= \dim_k k(t_{ij}/t_{00}, u_1, \cdots, u_d) = (N+1)^2 - 1 + dn .$$

This proves our lemma.

THEOREM 1. *Notations being as above, there exist an analytic subset $\mathfrak{S}$ of $\mathfrak{Z}$ of codimension 1 and a set of meromorphic functions $\{f_1, \cdots, f_a\}$ on $\mathfrak{Z}$ such that*

(i) *the varieties $\mathfrak{F}(z)$ for $z \in \mathfrak{Z} - \mathfrak{S}$ are of the same dimension and of the same degree;*

(ii) *the f_i are holomorphic on $\mathfrak{Z} - \mathfrak{S}$, and we have*

$$c(\mathscr{F}(z)) = (1, f_1(z), \cdots, f_\alpha(z))$$

for every $z \in \mathscr{B} - \mathscr{G}$.

To prove this theorem, we need some elementary facts of the theory of specialization over a place, on which we have given in Appendix necessary definitions and lemmas. $T(z)$, $V(r, z)$, $\Psi(z)$ being as above, take and fix a generic point z_0 of $\mathscr{B}$ for Ψ; z_0 is then generic for $c(A(z))$, $\Lambda(0, z)$, $c(T(z))$ and the $c(V(r_\nu, z))$. By Lemma 6 of Appendix, there exist a finite number of elements α_i in $Q(\Psi(z_0))$ such that, for every place $\mathfrak{q}$ of $Q(\Psi(z_0))$, the cycle $\mathfrak{q}(\mathscr{F}(z_0))$ is a variety if $\mathfrak{q}(\alpha_i) \neq 0$ for every i. Take the functions a_i in $Q(\Psi)$ which correspond to the α_i by the canonical isomorphism of $Q(\Psi)$ onto $Q(\Psi(z_0))$; we have then $\alpha_i = a_i(z_0)$. Now let $\mathscr{B}'$ be the set of points y where the a_i are holomorphic and $a_i(y) \neq 0$ for every i. Let y be a point of $\mathscr{B}$ and $\mathfrak{o}$ the set of all the functions $F(z)$ in $Q(\Psi)$ which are holomorphic at y; the mapping $F \to F(y)$ is a homomorphism of $\mathfrak{o}$ into C. We extend this homomorphism to a place $\mathfrak{p}_0$ of $Q(\Psi)$ taking values in C. Denote by $\mathfrak{p}$ the place of $Q(\Psi(z_0))$ corresponding to $\mathfrak{p}_0$ by the canonical isomorphism of $Q(\Psi)$ onto $Q(\Psi(z_0))$. Then we have $\mathfrak{p}(\Psi(z_0)) = \Psi(y)$, so that we have

$$\mathfrak{p}(A(z_0)) = A(y) , \qquad \mathfrak{p}(\Lambda(0, z_0)) = \Lambda(0, y) ,$$
$$\mathfrak{p}(T(z_0)) = T(y) , \qquad \mathfrak{p}(V(r_\nu, z_0)) = V(r_\nu, y) , \qquad (1 \leqq \nu \leqq d).$$

Suppose that y is contained in $\mathscr{B}'$. Then we have $a_i(y) \neq 0$ for every i; hence we have $\mathfrak{p}(\alpha_i) \neq 0$ for every i, so that $\mathfrak{p}(\mathscr{F}(z_0))$ is a variety.

Let ξ be the projective transformation of P^N defined by a generic matrix (t_{ij}) over $Q(\Psi(z_0))$ and ξ' the projective transformation of P^N defined by a generic matrix (t'_{ij}) over $\mathfrak{p}(Q(\Psi(z_0)))$. It is easy to see that there exists an extension $\mathfrak{p}_1$ of $\mathfrak{p}$ on $Q(\Psi(z_0), t_{ij})$ such that $\mathfrak{p}_1(t_{ij}) = t'_{ij}$. Let $u_1, \cdots, u_d$ be d independent generic points of $A(z_0)$ over $Q(\Psi(z_0), t_{ij})$ and $u'_1, \cdots, u'_d$ d independent generic points of $A(y)$ over $\mathfrak{p}_1(Q(\Psi(z_0), t_{ij}))$. As we have

$$\mathfrak{p}_1(A(z_0) \times \cdots \times A(z_0)) = A(y) \times \cdots \times A(y) ,$$

there exists, by Lemma 3 of Appendix, an extension $\mathfrak{P}$ of $\mathfrak{p}_1$ such that $\mathfrak{P}(u_\nu) = u'_\nu$ for every ν. Put now $B = \xi(A(z_0))$, $B' = \xi'(A(y))$. Then it is easy to see $\mathfrak{p}_1(B) = B'$. Let λ_ν be the rational mapping of B into itself defined by

$$\lambda_\nu(x) = \xi[\theta_{z_0}(r_\nu)\xi^{-1}(x) + u_\nu]$$

and λ'_ν the mapping of B' into itself defined by

$$\lambda'_\nu(x') = \xi'[\theta_y(r_\nu)\xi'^{-1}(x') + u'_\nu] .$$

Applying Lemma 4 of Appendix to the law of composition, $\theta_{z_0}(r_\nu)$, ξ and

ξ^{-1}, we see that if (x', x_1') is a specialization of $(x, \lambda_\nu(x))$ over $\mathfrak{P}$, we have $x_1' = \lambda_\nu'(x')$. Let W_ν and W_ν' be respectively the graphs of λ_ν and λ_ν'. Then, by Lemma 5 of Appendix, we have $\mathfrak{P}(W_\nu) = W_\nu'$. Thus we have proved

$$\mathfrak{P}[c(B) \times c(W_1) \times \cdots \times c(W_a)] = c(B') \times c(W_1') \times \cdots \times c(W_a') \, .$$

We see that the point on the right hand is a generic point of $\mathcal{F}(y)$ over $\mathfrak{p}(Q(\Psi(z_0)))$. As $\mathfrak{p}(\mathcal{F}(z_0))$ is a variety defined over $\mathfrak{p}(Q(\Psi(z_0)))$, we have therefore $\mathfrak{p}(\mathcal{F}(z_0)) \supset \mathcal{F}(y)$. On the other hand, by Lemma 8, $\mathcal{F}(z)$ is of dimension $(N+1)^2 - 1 + dn$ for every $z \in \mathfrak{Z}$, so that $\mathfrak{p}(\mathcal{F}(z_0))$ and $\mathcal{F}(y)$ are of the same dimension. Hence we have $\mathfrak{p}(\mathcal{F}(z_0)) = \mathcal{F}(y)$. Now put

$$c(\mathcal{F}(z_0)) = (1, \beta_1, \cdots, \beta_m) \, ,$$

changing the order of coordinates, if necessary; the β_i are elements in $Q(\Psi(z_0))$. Take the functions b_i corresponding to the β_i by the canonical isomorphism of $Q(\Psi)$ onto $Q(\Psi(z_0))$. Then we have $b_i(z_0) = \beta_i$ for each i. Now let $\mathfrak{Z}''$ be the set of all points of $\mathfrak{Z}'$ where the b_i are holomorphic. If $y \in \mathfrak{Z}''$, we have $\mathfrak{p}(\beta_i) = b_i(y)$ for the above place $\mathfrak{p}$, so that we have

$$c(\mathcal{F}(y)) = \mathfrak{p}(c(\mathcal{F}(z_0))) = (1, b_1(y), \cdots, b_m(y)) \, .$$

By our construction, $\mathfrak{Z} - \mathfrak{Z}''$ is an analytic subset of $\mathfrak{Z}$. This proves our theorem.

COROLLARY. *There exist an analytic subset $\mathfrak{S}$ of $\mathfrak{Z}$ of codimension 1 and a set of meromorphic functions $\{f_1, \cdots, f_a\}$ on $\mathfrak{Z}$ such that $Q(f_1(z), \cdots, f_a(z))$ is the field of moduli of $(A(z), C_z, \theta_z)$ for every $z \in \mathfrak{Z} - \mathfrak{S}$.*

This is an immediate consequence of Theorem 1, since $Q(c(\mathcal{F}(z)))$ is the field of moduli of $(A(z), C_z, \theta_z)$.

4. Fields of definition

13. Let k be a subfield of C composed of countably infinite elements. We say that the system $\{(A(z), C_z, \theta_z) \,|\, z \in \mathfrak{Z}\}$ constructed in **11** is *complete with respect to k* if there exists a join $\mathfrak{F}$ of countably infinite analytic subsets of $\mathfrak{Z}$ of codimension 1, satisfying the following condition:

(D) *For any point z in $\mathfrak{Z} - \mathfrak{F}$, if (A', C', θ') is a generic specialization of $(A(z), C_z, \theta_z)$ over k, there exists a point y in $\mathfrak{Z}$ such that $(A(y), C_y, \theta_y)$ is isomorphic to (A', C', θ').*

THEOREM 2. *Let k be a subfield of C composed of countably infinite elements. Let $(f_1, \cdots, f_a)$ be a set of meromorphic functions in Theorem 1. If the system $(A(z), C_z, \theta_z)$ is complete with respect to k, then the field $k(f_1, \cdots, f_a)$ is a regular extension of k; and we have*

$$\dim_k k(f_1, \cdots, f_\alpha) = \dim_C C(f_1, \cdots, f_\alpha) \, .$$

In other words, the fields C and $k(f_1, \cdots, f_\alpha)$ are linearly disjoint over k.

Choose $f_{i_1}, \cdots, f_{i_\beta}$ in such a way that the f_{i_ν} are algebraically independent over C and the f_i are algebraic over $C(f_{i_\nu})$. We see easily that there exists a subfield k_0 of C, finitely generated over Q, such that the f_i are algebraic over $k_0(f_{i_\nu})$. We have then

$$\dim_{k_0} k_0(f_1, \cdots, f_\alpha) = \beta = \dim_C C(f_1, \cdots, f_\alpha) \, .$$

Let z_0 be a generic point of $\mathfrak{Z} - \mathfrak{G}$ for the f_j over k_0^*, where k_0^* denotes the algebraic closure of k_0. Let $\mathfrak{M}$ be the locus of $c(\mathcal{F}(z_0))$ over k_0^* and k_1 the smallest field of definition for $\mathfrak{M}$. The field k_1 is a subfield of k_0^*, finitely generated over Q. We see that $(1, f_1, \cdots, f_\alpha)$ is a generic specialization of $c(\mathcal{F}(z_0))$ over k_0^*, and hence over k_1; from this follows

$$\dim_{k_1} k_1(f_1, \cdots, f_\alpha) = \beta = \dim_C C(f_1, \cdots, f_\alpha) \, .$$

The fact that k_1 is the field of definition for $\mathfrak{M}$ implies that $k_1(f_1, \cdots, f_\alpha)$ is a regular extension of k_1. Therefore, our theorem is proved if we show that k contains k_1. For this purpose, suppose that k_1 is not contained in k; and put $K = kk_1$; then there exists an isomorphism σ of K into C such that σ is the identity on k and we have $\mathfrak{M}^\sigma \neq \mathfrak{M}$. The notations $\Psi(z)$, $T(z)$, $V(r, z)$ being the same as in the proof of Theorem 1, take a generic point z of $\mathfrak{Z}$ for the f_i, Ψ over K. We may assume that z is not contained in $\mathfrak{G} \cup \mathfrak{F}$, where $\mathfrak{F}$ is a subset of $\mathfrak{Z}$ as in the above condition (D). We see that $c(\mathcal{F}(z))$ is a generic point of $\mathfrak{M}$ over K. Let q be a generic point of $\mathfrak{M}^\sigma$ over KK^σ. Then there exists an isomorphism τ of $K(c(\mathcal{F}(z)))$ onto $K^\sigma(q)$ such that $c(\mathcal{F}(z))^\tau = q$ and $\tau = \sigma$ on K. We extend τ to an isomorphism of $K(\Psi(z))$, which we denote again by τ. We have then

$$\mathcal{F}(z)^\tau = \mathcal{F}(A(z)^\tau, \theta_z^\tau) \, .$$

Since τ is the identity on k, $(A(z)^\tau, C_z^\tau, \theta_z^\tau)$ is a generic specialization of $(A(z), C_z, \theta_z)$ over k. By our assumption of completeness, there exists a point y in $\mathfrak{Z}$ such that $(A(z)^\tau, C_z^\tau, \theta_z^\tau)$ is isomorphic to $(A(y), C_y, \theta_y)$. By Proposition 1, we have

$$\mathcal{F}(z)^\tau = \mathcal{F}(y) \, .$$

We can find, by the same procedure as in the proof of Theorem 1, a place $\mathfrak{p}$ of $K(\Psi(z))$ such that $\mathfrak{p}(g(z)) = g(y)$ for every function g in $K(\Psi)$ which is holomorphic at y; we have then $\mathfrak{p}(\Psi(z)) = \Psi(y)$. By a similar argument as in the proof of Theorem 1, we can find a point x in $\mathcal{F}(z)$, a generic point x' of $\mathcal{F}(y)$ over $\mathfrak{p}(K(\Psi(y)))$ and an extension $\mathfrak{q}$ of $\mathfrak{p}$ in $K(\Psi(z), x)$ such that $\mathfrak{q}(x) = x'$. We see that $\mathfrak{p}(\mathcal{F}(z))$ contains $\mathfrak{q}(x) = x'$. This implies

$$\mathfrak{p}(\mathcal{F}(z)) \supset \mathcal{F}(y) \, ,$$

since $\mathfrak{q}(\mathcal{F}(z))$ is rational over $\mathfrak{p}(K(\Psi(z)))$ and x' is generic on $\mathcal{F}(y)$ over $\mathfrak{p}(K(\Psi(z)))$. On the other hand, $\mathcal{F}(y) = \mathcal{F}(z)^\tau$ has the same dimension and the same degree as $\mathcal{F}(z)$, so that $\mathcal{F}(y)$ and $\mathfrak{p}(\mathcal{F}(z))$ are of the same dimension and of the same degree. Hence we have $\mathfrak{p}(\mathcal{F}(z)) = \mathcal{F}(y)$; as $\mathfrak{p}$ is the identity on K, $\mathcal{F}(y) = \mathcal{F}(z)^\tau$ is a specialization of $\mathcal{F}(z)$ over K; in other words, $q = c(\mathcal{F}(z))^\tau$ is a specialization of $c(\mathcal{F}(z))$ over K. As $c(\mathcal{F}(z))$ is a generic point of $\mathfrak{M}$ over K, q must be contained in $\mathfrak{M}$. This is a contradiction, because q is a generic point of $\mathfrak{M}^\sigma$ over KK^σ and we have supposed $\mathfrak{M}^\sigma \neq \mathfrak{M}$. Therefore k must contain k_1; this completes our proof.

5. Automorphic functions belonging to a quaternion algebra

14. We shall apply our theory to a system of abelian varieties whose endomorphism-rings are isomorphic to an order of a quaternion algebra. We begin with some elementary facts on quaternion algebras.

Let Φ be a *quaternion algebra* over a field k, by which we understand a central simple algebra over k of degree 2. It is well known that there exists an involution $\alpha \to \alpha'$ on Φ such that $(X - \alpha)(X - \alpha')$ is the principal polynomial of α; the involution $\alpha \to \alpha'$ is uniquely determined by this property; so we call it *the canonical involution of* Φ. Put

$$\mathrm{tr}(\alpha) = \alpha + \alpha', \ \mathrm{nr}(\alpha) = \alpha\alpha' \, .$$

$\mathrm{tr}(\alpha)$ and $\mathrm{nr}(\alpha)$ are contained in k and called the *trace* and the *norm* of α. Suppose now that Φ has a subfield $\mathfrak{k}$ which is of degree 2 over k. The canonical involution gives a non-trivial automorphism of $\mathfrak{k}$. Since Φ is central simple over k, there exists an element λ of Φ such that $\lambda^{-1}\alpha\lambda = \alpha'$ for every α in $\mathfrak{k}$. If λ has this property, λ^2 is contained in k; put $\lambda^2 = q$. Every element ξ of Φ is written in the form $\xi = \alpha + \beta\lambda$ where α, β are elements in $\mathfrak{k}$. We have then

$$\binom{1}{\lambda}(\alpha + \beta\lambda) = \begin{pmatrix} \alpha & \beta \\ q\beta' & \alpha' \end{pmatrix}\binom{1}{\lambda} \, .$$

Put

$$\chi(\xi) = \begin{pmatrix} \alpha & \beta \\ q\beta' & \alpha' \end{pmatrix} ;$$

then $\xi \to \chi(\xi)$ is a faithful representation of Φ. We see easily

$$\xi' = \alpha' - \beta\lambda, \ \mathrm{tr}(\xi) = \mathrm{tr}\,\chi(\xi), \ \mathrm{nr}(\xi) = \det\chi(\xi) \, .$$

PROPOSITION 11. *Every inner automorphism of* Φ *commutes with the canonical involution of* Φ; *namely, for any regular element* ρ *of* Φ,

$$(\rho^{-1}\alpha\rho)' = \rho^{-1}\alpha'\rho$$

holds for every $\alpha \in \Phi$.

This is proved by the fact that $\rho\rho'$ is contained in k and hence commutes with all elements of Φ.

PROPOSITION 12. *Let* ρ *be a regular element of* Φ; *put*

$$\alpha^* = \rho^{-1}\alpha'\rho$$

for every $\alpha \in \Phi$. *Then* $\alpha \to \alpha^*$ *is an involution of* Φ *if and only if* $\rho^2 \in k$. *Every involution of* Φ *is obtained in this way.*

It is clear that $\alpha \to \alpha^* = \rho^{-1}\alpha'\rho$ gives an anti-automorphism of Φ. By Proposition 11, we have $(\alpha^*)^* = \rho^{-2}\alpha\rho^2$. Hence $\alpha \to \alpha^*$ is an involution of Φ if and only if ρ^2 is contained in k. Now let $\alpha \to \alpha^*$ be an involution of Φ. Since the mapping $\alpha \to (\alpha^*)'$ is an automorphism, there exists a regular element ρ of Φ such that $(\alpha^*)' = \rho^{-1}\alpha\rho$; we have then by Proposition 11, $\alpha^* = (\rho^{-1}\alpha\rho)' = \rho^{-1}\alpha'\rho$. This proves our proposition.

We say that an involution $*$ of Φ is *defined by* an element ρ of Φ if $\alpha^* = \rho^{-1}\alpha'\rho$ for every $\alpha \in \Phi$, and call ρ a *defining element* of the involution $*$.

15. Now consider the case where the ground field k is the field Q of rational numbers. We call a quaternion algebra Φ over Q *definite* if $\mathrm{nr}(\alpha) > 0$ for every $\alpha \neq 0$ in Φ, otherwise we call Φ *indefinite*. In the following treatment, only indefinite quaternion algebras will be our object. A quaternion algebra Φ over Q is indefinite if and only if any one of the following conditions is satisfied.

(Q1) Φ has a real quadratic subfield.

(Q2) Φ has a faithful representation by matrices of degree 2 with real coefficients.

PROPOSITION 13. *Let* Φ *be an indefinite quaternion algebra over* Q *and* $*$ *an involution of* Φ *defined by an element* ρ *of* Φ. *Then* $\mathrm{tr}(\alpha\alpha^*) > 0$ *for every* $\alpha \neq 0$, *if and only if* ρ^2 *is a negative number of* Q.

If ρ is contained in Q, we have $\alpha^* = \alpha'$ for every $\alpha \in \Phi$. Hence we have $\mathrm{tr}(\alpha\alpha^*) = \mathrm{tr}(\alpha\alpha') = 2\mathrm{nr}(\alpha)$. As Φ is indefinite, if $\mathrm{tr}(\alpha\alpha^*) > 0$ for $\alpha \neq 0$, ρ is not contained in Q. By Proposition 2, ρ^2 is contained in Q, so that $\rho' = -\rho$. We see then $\mathrm{tr}(\rho\rho^*) = -2\rho^2$. This shows that ρ^2 is negative if $\mathrm{tr}(\alpha\alpha^*)$ is a positive form. Conversely suppose that ρ^2 is a negative number of Q. Then $Q(\rho)$ is an imaginary quadratic subfield of Φ, and there exists an element τ in Φ such that $\tau^{-1}\rho\tau = -\rho$. We have

$$\Phi = Q(\rho) + Q(\rho)\tau \,, \qquad\qquad \tau^2 \in Q.$$

Let a, b be two elements of $Q(\rho)$; then we have

$$(a + b\tau)(a + b\tau)' = aa' - bb'\tau^2 \,;$$

from this follows that τ^2 is positive since Φ is indefinite. We see then

$$\mathrm{tr}((a + b\tau)\rho^{-1}(a + b\tau)'\rho) = 2(aa' + bb'\tau^2) > 0$$

for $a + b\tau \neq 0$. This completes our proof.

An involution $*$ of an indefinite quaternion algebra Φ over Q is called a *positive involution* if $\mathrm{tr}(\alpha\alpha^*) > 0$ for every $\alpha \neq 0$ in Φ.

16. Let Φ be a quaternion algebra over Q. We understand by an *order* of Φ, as usual, a subring of Φ, containing Z, which is a free Z-module of rank 4; an order $\mathfrak{o}$ is called *maximal* if there is no order other than itself, which contains $\mathfrak{o}$. The following theorem is due to Eichler [2], [1, Satz 9].

THEOREM E. *Let $\mathfrak{o}$ be a maximal order of an indefinite quaternion algebra Φ over Q. Then,*

(i) *for any other maximal order $\mathfrak{o}'$, there exists an element ξ of Φ such that $\mathfrak{o}' = \xi^{-1}\mathfrak{o}\xi$;*

(ii) *every $\mathfrak{o}$-ideal is principal, namely, if $\mathfrak{a}$ is a left (resp. right) $\mathfrak{o}$-ideal of Φ, there exists an element α of Φ such that $\mathfrak{a} = \mathfrak{o}\alpha$ (resp. $\mathfrak{a} = \alpha\mathfrak{o}$);*

(iii) *there exists an element γ in $\mathfrak{o}$ such that $\mathrm{nr}(\gamma) = -1$.*

We say that an order $\mathfrak{o}$ of an indefinite quaternion algebra Φ is *of type E* if $\mathfrak{o}$ satisfies the following conditions.

(E1) *If $\mathfrak{m}$ is a free Z-submodule of Φ of rank 4 such that*

$$\alpha\,\mathfrak{m} \subset \mathfrak{m} \Longleftrightarrow \alpha \in \mathfrak{o} \,,$$

then, there exists an element μ in Φ for which $\mathfrak{m} = \mathfrak{o}\mu$.

(E2) *$\mathfrak{o}$ contains an element of norm -1.*

By Theorem E, every maximal order is of type E.

17. Φ being as before an indefinite quaternion algebra over Q, let A be an abelian variety of dimension 2 defined over C such that there exists an isomorphism θ of Φ into $\mathcal{A}_0(A)$. We shall determine the structure of such an abelian variety A.

Take and fix an analytic isomorphism of A onto a complex torus C^2/D of dimension 2, where D denotes a discrete subgroup of C^2 of rank 4. Every element of $\mathcal{A}(A)$ corresponds to a matrix of degree 2 with coefficients in C. We denote by $\chi(\alpha)$ the matrix corresponding to $\theta(\alpha)$ for every $\alpha \in \Phi$. Since any two representations of Φ of degree 2 are equivalent, we can choose a coordinate system of C^2 in such a way that the representation $\chi(\alpha)$ determined by this coordinate system is equal to a representation of Φ given previously. So we assume hereafter that $\chi(\alpha)$ *are real matrices for all* $\alpha \in \Phi$.

Let $\mathfrak{o}$ be the subset of Φ consisting of all the elements α of Φ such that $\theta(\alpha)$ is an endomorphism of A, namely,

$$\mathfrak{o} = \theta^{-1}[\mathcal{A}(A) \cap \theta(\Phi)] \; .$$

Then $\mathfrak{o}$ is an order of Φ. We see that $\chi(\alpha)D \subset D$ for every $\alpha \in \mathfrak{a}$; so D is considered as an $\mathfrak{o}$-module. Fix an element $\mathfrak{x}$ of D and consider an additive mapping $\mathfrak{o} \ni \alpha \to \chi(\alpha)\mathfrak{x} \in D$; this gives clearly an $\mathfrak{o}$-homomorphism of $\mathfrak{o}$ into D. Let D' be the image of this homomorphism; then, for a suitable $\mathfrak{x}$, D' is a discrete subgroup of C^2 of rank 4. This is proved by the following proposition.

PROPOSITION 14. *Let* $\mathfrak{m}$ *be a free submodule of* Φ *of rank 4 over* Z *and* $\mathfrak{x} = \begin{pmatrix} x_1 \\ x_2 \end{pmatrix}$ *an element of* C^2; *put* $D_1 = \{\chi(\alpha)\mathfrak{x} \mid \alpha \in \mathfrak{m}\}$. *Then,* D_1 *is a discrete subgroup of* C^2 *of rank 4 if and only if* $x_2 \neq 0$, $\mathrm{Im}(x_1/x_2) \neq 0$.

First we remark the following fact: there is no real matrix $M \neq 0$ of degree 2 such that $M\mathfrak{x} = 0$ if and only if $x_2 \neq 0$, $\mathrm{Im}(x_1/x_2) \neq 0$. Let now $\{\alpha_i (1 \leq i \leq 4)\}$ be a basis of $\mathfrak{m}$ over Z. As χ is irreducible, the $\chi(\alpha_i)$ are linearly independent over R (even over C). D is generated over Z by the $\chi(\alpha_i)$; it is a discrete subgroup of C^2 of rank 4 if and only if the four vectors $\chi(\alpha_i)\mathfrak{x}$ are linearly independent over R. Suppose that there exist real numbers r_i such that $\sum_{i=1}^4 r_i\chi(\alpha_i)\mathfrak{x} = 0$. If $x_2 \neq 0$ and $\mathrm{Im}(x_1/x_2) \neq 0$, then we have by the above remark $\sum_{i=1}^4 r_i\chi(\alpha_i) = 0$; as the $\chi(\alpha_i)$ are linearly independent over R, we have $r_i = 0$ for all i; this shows that the $\chi(\alpha_i)\mathfrak{x}$ are linearly independent over R. If $x_2 = 0$ or if $\mathrm{Im}(x_1/x_2) = 0$, there exists a real matrix $M \neq 0$ such that $M\mathfrak{x} = 0$. We can find real numbers s_i such that $M = \sum_{i=1}^4 s_i\chi(\alpha_i)$, as the $\chi(\alpha_i)$ give a basis of the algebra of all real matrices of degree 2. Then we have $\sum_{i=1}^4 s_i\chi(\alpha_i)\mathfrak{x} = 0$; we remark that there exists an i for which $s_i \neq 0$, as we have $M \neq 0$. Hence the $\chi(\alpha_i)\mathfrak{x}$ are not linearly independent over R; this completes our proof.

Now take an element $\mathfrak{x}$ of D as in Proposition 14; this is possible since D is a discrete subgroup of C^2 of rank 4. Then, $D' = \{\chi(\alpha)\mathfrak{x} \mid \alpha \in \mathfrak{o}\}$ is a discrete subgroup of C^2 of rank 4, so that D' has a finite index in D; hence there exists a positive integer m such that $m\mathfrak{x}' \in D'$ for every $\mathfrak{x}' \in D$. Fixing such an integer m, we obtain a mapping $D \ni \mathfrak{x}' \to \alpha \in \mathfrak{o}$ by the relation $m\mathfrak{x}' = \chi(\alpha)\mathfrak{x}$. This gives an $\mathfrak{o}$-isomorphism of D into $\mathfrak{o}$; let

$\mathfrak{m}$ be the image of this mapping; then $\mathfrak{m}$ is a left ideal of $\mathfrak{o}$. Putting $\mathfrak{x}_0 = m^{-1}\mathfrak{x}$, we have

$$D = \{\chi(\alpha)\mathfrak{x}_0 \mid \alpha \in \mathfrak{m}\} \ .$$

If we have $\alpha\mathfrak{m} \subset \mathfrak{m}$ for an element α of Φ, then we have $\chi(\alpha)D \subset D$; $\chi(\alpha)$ gives an endomorphism of C^2/D and hence corresponds to an element of $\mathcal{A}(A)$. By our definition of $\mathfrak{o}$, we have $\alpha \in \mathfrak{o}$. We have therefore

$$\mathfrak{o} = \{\alpha \mid \alpha \in \Phi, \alpha\mathfrak{m} \subset \mathfrak{m}\} \ .$$

18. We shall now determine the polarizations of A. Before doing this, we recall the relation between a polarization of an abelian variety A and an involution of $\mathcal{A}_0(A)$. Let A be an abelian variety and B its Picard variety. Every divisor Y on A which is algebraically equivalent to 0 defines a point of B which we denote by $\mathrm{Cl}(Y)$. For every element λ of $\mathcal{A}(A)$, there exists an element μ of $\mathcal{A}(B)$ defined by

$$\mu(\mathrm{Cl}(Y)) = \mathrm{Cl}(\lambda^{-1}(Y)) \ .$$

We call μ the transpose of λ and denote it by ${}^t\lambda$. The mapping $\lambda \to {}^t\lambda$ is an anti-isomorphism of $\mathcal{A}(A)$ into $\mathcal{A}(B)$; we extend this anti-isomorphism to an anti-isomorphism of $\mathcal{A}_0(A)$ into $\mathcal{A}_0(B)$; we denote also by ${}^t\alpha$ the image of $\alpha \in \mathcal{A}_0(A)$ by this anti-isomorphism. Let X be a divisor on A; define the mapping φ_X of A into B by the relation

$$\varphi_X(u) = \mathrm{Cl}(X_u - X)$$

for $u \in A$; then φ_X gives a homomorphism of A into B; X is non-degenerate if and only if $\varphi_X(A) = B$. Assuming that X is non-degenerate, for every $\alpha \in \mathcal{A}_0(A)$, there exists an element α^* of $\mathcal{A}_0(A)$ such that, following the notations of [12, no. 54],

$$\varphi_X^{-1} \, {}^t\alpha\varphi_X = \alpha^* \ .$$

The mapping $\alpha \to \alpha^*$ of $\mathcal{A}_0(A)$ onto itself is an involution of $\mathcal{A}_0(A)$ and we have

$$\mathrm{tr}(\alpha\alpha^*) > 0$$

for every $\alpha \neq 0$ of $\mathcal{A}_0(A)$, where $\mathrm{tr}(\beta)$ denotes the trace of l-adic representations of β. For proofs of these results we refer to [6]. We see that two non-degenerate divisors give the same involution if they are contained in the same polarization of A; hence, we can say that every polarization of A defines an involution of $\mathcal{A}_0(A)$. Now suppose that the characteristic of the ground field is 0. Let A be an abelian variety of dimension n defined over C. A is isomorphic to a complex torus C^n/D, where D is a discrete subgroup of C^n of rank $2n$. Fixing an isomorphism of A onto C^n/D, every divisor X on A corresponds to a Riemann form $E(\mathfrak{x}, \mathfrak{y})$ on C^n/D; and every element $\alpha \in \mathcal{A}_0(A)$ corresponds to a matrix $M(\alpha)$ of degree n. If X is a non-degenerate divisor of A and we put, as above,

$$\alpha^* = \varphi_{\bar{x}}^{-1}\,{}^t\alpha\varphi_x \, ,$$

then we have

$$E(M(\alpha)\mathfrak{x}, \mathfrak{y}) = E(\mathfrak{x}, M(\alpha^*)\mathfrak{y})$$

for the Riemann form E corresponding to the divisor X.

Now coming back to our abelian variety A of dimension 2 and the complex torus C^2/D, consider a polarization C of A; take a divisor X in C; X corresponds to a Riemann form $E(\mathfrak{x}, \mathfrak{y})$ on C^2/D. The polarization C determines an involution of $\mathcal{A}_0(A)$. We shall assume in the following treatment that *this involution fixes as set the subring* $\theta(\Phi)$; this is trivially the case whenever we have $\mathcal{A}_0(A) = \theta(\Phi)$. We denote by $*$ the involution of Φ corresponding to that of $\theta(\Phi)$ given by C; we have then

$$E(\chi(\alpha)\mathfrak{x}, \mathfrak{y}) = E(\mathfrak{x}, \chi(\alpha^*)\mathfrak{y})$$

for every $\alpha \in \Phi$ and

$$\mathrm{tr}(\alpha\alpha^*) > 0$$

for every $\alpha \neq 0$ in Φ. Hence, by Proposition 13, there exists an element ρ of Φ such that

$$\alpha^* = \rho^{-1}\alpha'\rho$$

and ρ^2 is a negative number of Q. Put

$$E_0(\alpha, \beta) = E(\chi(\alpha)\mathfrak{x}_0, \chi(\beta)\mathfrak{x}_0) \, ,$$

for $\alpha \in \Phi$, $\beta \in \Phi$, where $\mathfrak{x}_0$ is the vector defined in **17**. We see that the value $E_0(\alpha, \beta)$ is an integer for $\alpha \in \mathfrak{m}$, $\beta \in \mathfrak{m}$, and that $\xi \to E_0(\xi, 1)$ is a Q-linear mapping of Φ into Q; so there exists an element $\tau \in \Phi$ such that $E_0(\xi, 1) = \mathrm{tr}(\tau\xi)$. We have then

$$E_0(\alpha, \beta) = \mathrm{tr}(\tau\beta^*\alpha) \quad = \mathrm{tr}(\tau\rho^{-1}\beta'\rho\alpha) \, .$$

As $E(\mathfrak{x}, \mathfrak{y})$ is a skew-symmetric form, we have

$$-\mathrm{tr}(\tau\alpha) = -\, E_0(\alpha, 1) = E_0(1, \alpha) = \mathrm{tr}(\tau\rho^{-1}\alpha'\rho)$$
$$= \mathrm{tr}(\rho^{-1}\alpha\rho\tau') = \mathrm{tr}(\rho\tau'\rho^{-1}\alpha)$$

for every $\alpha \in \Phi$; from this follows $-\tau = \rho\tau'\rho^{-1}$. We have then $\rho^{-1}\tau = -\tau'\rho^{-1} = (\rho^{-1}\tau)'$; this shows that $\rho^{-1}\tau$ is contained in Q. Put $\rho^{-1}\tau = c_1$. Then we have $\tau = c_1\rho$ and

$$E(\chi(\alpha)\mathfrak{x}_0, \chi(\beta)\mathfrak{x}_0) = c_1 \mathrm{tr}(\rho\alpha\beta')$$

for $\alpha \in \Phi$, $\beta \in \Phi$.

We have thus shown that a polarized abelian variety A is determined by a left ideal $\mathfrak{m}$ of $\mathfrak{o}$, a vector $\mathfrak{x}_0$ and an involution $*$ of Φ. We shall now construct a polarized abelian variety A from these data. Fix a real representation $\chi(\alpha)$ of Φ of degree 2; take a free submodule $\mathfrak{m}$ of Φ of rank 4 over $\boldsymbol{Z}$ and a vector $\mathfrak{x}_0 = \begin{pmatrix} x_1 \\ x_2 \end{pmatrix}$ such that $x_2 \neq 0$, $\mathrm{Im}(x_1/x_2) \neq 0$. Put

$$D = \{\chi(\alpha)\mathfrak{x}_0 \mid \alpha \in \mathfrak{m}\} \ .$$

Then by Proposition 14, D is a discrete subgroup of $\boldsymbol{C}^2$ of rank 4, so that $\boldsymbol{C}^2/D$ is a complex torus. Let $*$ be a positive involution of Φ defined by an element ρ. Denote by Φ_R the scalar extension over $\boldsymbol{R}$ of the algebra Φ; the trace, the norm, the involutions ', $*$ and the representation χ are all extended on Φ_R in a natural manner. We denote them in Φ_R by the same notations as in Φ. We have also $\mathrm{tr}(\alpha\alpha^*) > 0$ for $\alpha \neq 0$ in Φ_R, since Φ is dense in Φ_R. We see that $\alpha \to \chi(\alpha)\mathfrak{x}_0$ gives an $\boldsymbol{R}$-isomorphism of Φ_R onto $\boldsymbol{C}^2$. Hence there exists an element $\mu \in \Phi_R$ such that

$$\sqrt{-1}\,\mathfrak{x}_0 = \chi(\mu)\mathfrak{x}_0 \ .$$

We have $\mu^2 = -1$ and $\mu' = -\mu$. Put now

$$E(\chi(\alpha)\mathfrak{x}_0, \chi(\beta)\mathfrak{x}_0) = c \, \mathrm{tr}(\rho\alpha\beta')$$

for $\alpha \in \Phi_R$, $\beta \in \Phi_R$, where c is an integer $\neq 0$ such that $c\,\mathrm{tr}(\rho\alpha\beta')$ is an integer whenever α and β are contained in $\mathfrak{m}$; such a constant c really exists since $\mathfrak{m}$ is a free submodule of finite rank. Then, by the relation $\rho' = -\rho$, $E(\mathfrak{x}, \mathfrak{y})$ is an alternating $\boldsymbol{R}$-bilinear form on $\boldsymbol{C}^2$; and the values on D are integers. By our definition, we have

$$E(\chi(\alpha)\mathfrak{x}_0, \sqrt{-1}\,\chi(\beta)\mathfrak{x}_0) = c\,\mathrm{tr}(\rho\alpha\mu'\beta') \ .$$

Using the relations $\mathrm{tr}(\xi) = \mathrm{tr}(\xi')$, $\mathrm{tr}(\xi\eta) = \mathrm{tr}(\eta\xi)$, $\rho' = -\rho$ and $\mu' = -\mu$, we have

$$\mathrm{tr}(\rho\alpha\mu'\beta') = \mathrm{tr}(\beta\mu\alpha'\rho') = \mathrm{tr}(\rho'\beta\mu\alpha') = \mathrm{tr}(\rho\beta\mu'\alpha') \ ;$$

this shows that $E(\mathfrak{x}, \sqrt{-1}\,\mathfrak{y})$ is symmetric. We shall now prove that $E(\mathfrak{x}, \sqrt{-1}\,\mathfrak{x})$ is a definite form. Put $\rho^2 = -s$; then s is a positive rational number and we have $(\sqrt{s}\,\rho^{-1})^2 = -1$. Since Φ_R is central simple over $\boldsymbol{R}$, there exists an element $\gamma \in \Phi_R$ such that

$$\mu = \gamma(\sqrt{s}\,\rho^{-1})\gamma^{-1} \ .$$

We have then

$$-\text{tr}(\rho\alpha\mu'\alpha') = \text{nr}(\gamma)^{-1}\sqrt{s}\ \text{tr}((\alpha\gamma)(\alpha\gamma)^*)\ .$$

If we take c in such a way that $c\,\text{nr}(\gamma)^{-1} < 0$, then the form $E(\mathfrak{x}, \sqrt{-1}\,\mathfrak{x})$ is positive definite. We obtain thus a Riemann form E on C^2/D; hence there exists an analytic isomorphism of C^2/D onto an abelian variety A defined over C; E determines a polarization C on A. Now let $\mathfrak{o}$ be the set of all elements α of Φ such that $\alpha\mathfrak{m} \subset \mathfrak{m}$; then $\mathfrak{o}$ is an order of Φ and $\mathfrak{m}$ is a left $\mathfrak{o}$-ideal (which is not necessarily contained in $\mathfrak{o}$). For every α of Φ, we have $\chi(\alpha)D \subset D$ if and only if $\alpha \in \mathfrak{o}$; hence, for evey $\alpha \in \mathfrak{o}$, $\chi(\alpha)$ gives an endomorphism of C^2/D; denote by $\theta(\alpha)$ the corresponding endomorphism of A; θ is extended to an isomorphism of Φ into $\mathcal{A}_0(A)$ which we denote again by θ. We see easily the relation

$$\text{tr}(\rho\alpha(\xi\beta)') = \text{tr}(\rho(\rho^{-1}\xi'\rho)\alpha\beta')\ ,$$

from which follows

$$E(\mathfrak{x}, \chi(\xi)\mathfrak{y}) = E(\chi(\xi^*)\mathfrak{x}, \mathfrak{y})\ ;$$

this shows that the involution $*$ corresponds to the involution of $\mathcal{A}_0(A)$ determined by the polarization C. Thus we have proved the following two theorems.

THEOREM 3. *Let Φ be an indefinite quaternion algebra over Q and χ a representation of Φ by real matrices of degree 2; let $\mathfrak{m}$ be a free submodule of Φ of rank 4 over Z and $\mathfrak{x}_0 = \begin{pmatrix} x_1 \\ x_2 \end{pmatrix}$ a vector of C^2 such that $x_2 \neq 0$, $\text{Im}(x_1/x_2) \neq 0$; put*

$$\mathfrak{o} = \{\alpha \mid \alpha\mathfrak{m} \subset \mathfrak{m}\},\ D = \{\chi(\alpha)\mathfrak{x}_0 \mid \alpha \in \mathfrak{m}\}\ .$$

Then there exists an analytic isomorphism of the complex torus C^2/D onto an abelian variety A defined over C; $\alpha \to \chi(\alpha)$ gives an isomorphism θ of Φ into $\mathcal{A}_0(A)$ and we have

$$\theta(\mathfrak{o}) = \mathcal{A}(A) \cap \theta(\Phi)\ ;$$

$\mathfrak{o}$ is an order of Φ and $\mathfrak{m}$ is a left $\mathfrak{o}$-ideal. Conversely, A being an abelian variety of dimension 2 defined over C, if there exists an isomorphism of Φ into $\mathcal{A}_0(A)$, then A is obtained from a module $\mathfrak{m}$ and a vector $\mathfrak{x}_0$ in this manner.

THEOREM 4. *Notations being as in the above theorem, let $*$ be a positive involution of Φ and ρ a defining element of $*$. Let $E(\mathfrak{x}, \mathfrak{y})$ be the R-bilinear form on C^2 for which we have*

$$E(\chi(\alpha)\mathfrak{x}_0, \chi(\beta)\mathfrak{x}_0) = \text{tr}(\rho\alpha\beta')$$

for $\alpha \in \Phi$, $\beta \in \Phi$. Then there exists an integer $c \neq 0$ such that cE gives a Riemann form on C^2/D; the involution $*$ corresponds to the involution of $\mathcal{A}_0(A)$ determined by this Riemann form. Every Riemann form of C^2/D giving the involution $*$ on Φ is obtained in this manner.

Suppose now that the order $\mathfrak{o}$ in Theorem 3 is of type E. Then, there exists an element μ such that $\mathfrak{m} = \mathfrak{o}\mu$. Put $\mathfrak{x}_1 = \chi(\mu)\mathfrak{x}_0$; we have then

$$D = \{\chi(\alpha)\mathfrak{x}_1 \mid \alpha \in \mathfrak{o}\} .$$

Put now $\mathfrak{x}_1 = \begin{pmatrix} y_1 \\ y_2 \end{pmatrix}$; we have of course $y_2 \neq 0$, $\mathrm{Im}(y_1/y_2) \neq 0$. We can take $\mathfrak{x}_1$ in such a way that $\mathrm{Im}(y_1/y_2) > 0$. In fact, if $\mathrm{Im}(y_1/y_2) < 0$, then take an element γ in $\mathfrak{o}$ such that $\mathrm{nr}(\gamma) = -1$; this is possible by the property (E2). Put $\mathfrak{x}_2 = \chi(\gamma)\mathfrak{x}_1$; we have then

$$D = \{\chi(\alpha)\mathfrak{x}_2 \mid \alpha \in \mathfrak{o}\}$$

with $\mathfrak{x}_2 = \begin{pmatrix} z_1 \\ z_2 \end{pmatrix}$, $z_2 \neq 0$, $\mathrm{Im}(z_1/z_2) > 0$.

Put $\tau = z_1/z_2$ and $D' = \left\{\chi(\alpha)\begin{pmatrix} \tau \\ 1 \end{pmatrix} \mid \alpha \in \mathfrak{o}\right\}$. Then the linear mapping $\mathfrak{x} \to z_2^{-1}\mathfrak{x}$ of C^2 onto itself gives an isomorphism of C^2/D onto C^2/D' which commutes with the operations $\chi(\alpha)$ for $\alpha \in \mathfrak{o}$. Thus, if $\mathfrak{o}$ is of type E, the discrete group D is always reduced to the form

$$D = \left\{\chi(\alpha)\begin{pmatrix} \tau \\ 1 \end{pmatrix} \mid \alpha \in \mathfrak{o}\right\}$$

with $\mathrm{Im}(\tau) > 0$.

19. Let Φ be an indefinite quaternion algebra over Q. We fix a positive involution $*$ of Φ, a defining element ρ of $*$ and an order $\mathfrak{o}$ of Φ. We shall define an analytic system of polarized abelian varieties of type $\mathfrak{o}$, with the polarization determined by $*$.

Let now $\mathfrak{H}$ denote the upper halfplane, namely,

$$\mathfrak{H} = \{\tau \mid \tau \in C, \mathrm{Im}(\tau) > 0\} .$$

Put for each $\tau \in \mathfrak{H}$,

$$D(\tau) = \left\{\chi(\alpha)\begin{pmatrix} \tau \\ 1 \end{pmatrix} \mid \alpha \in \mathfrak{o}\right\} .$$

Then, $D(\tau)$ is a discrete subgroup of C^2 of rank 4 for every $\tau \in \mathfrak{H}$. Take a basis $\{\alpha_1, \alpha_2, \alpha_3, \alpha_4\}$ of $\mathfrak{o}$ over Z and denote by $w(\tau)$ the matrix whose columns are the $\chi(\alpha_i)\begin{pmatrix} \tau \\ 1 \end{pmatrix}$ for $1 \leq i \leq 4$. Let $E = (e_{ij})$ be the matrix of degree 4 given by

$$e_{ij} = \mathrm{ctr}(\rho\alpha_i\alpha_j')$$

where c is an integer $\neq 0$ such that $\mathrm{ctr}(\rho\alpha\beta')$ is an integer for every $\alpha \in \mathfrak{o}, \beta \in \mathfrak{o}$. By Theorem 3 and Theorem 4, E defines a Riemann form on $\boldsymbol{C}^2/D(\tau)$ for every τ, so that the system $\{E, w(\tau)\}$ satisfies the conditions (R1-4) in **11**. We obtain therefore a system of abelian varieties $A(\tau)$ parametrized by an analytic mapping $\Lambda(x, \tau)$ of $\boldsymbol{C}^2 \times \mathfrak{H}$ into a projective space; for every τ, the mapping $x \to \Lambda(x, \tau)$ is an analytic isomorphism of $\boldsymbol{C}^2/D(\tau)$ onto $A(\tau)$; and the form $\mathrm{ctr}(\rho\alpha\beta')$ (or the matrix E) corresponds to the polarization of $A(\tau)$ determined by the hyperplane sections, which we denote by C_τ. Moreover, we have $\chi(\alpha)D(\tau) \subset D(\tau)$ for every $\alpha \in \mathfrak{o}$; namely, χ satisfies the conditions (R5-6) in **11**. Hence, each element α of $\mathfrak{o}$ gives an endomorphism $\theta_\tau(\alpha)$ of $A(\tau)$ such that

$$\theta_\tau(\alpha)\Lambda(x, \tau) = \Lambda(\chi(\alpha)x, \tau) \ .$$

Thus we obtain a system $\{(A(\tau), C_\tau, \theta_\tau)\}$ of polarized abelian varieties of type $\mathfrak{o}$. Extend the isomorphism θ_τ of $\mathfrak{o}$ into $\mathcal{A}(A)$ to an isomorphism of Φ into $\mathcal{A}_0(A)$, which we denote again by θ_τ. Then, by Theorem 3, we have

$$\theta_\tau(\mathfrak{o}) = \theta_\tau(\Phi) \cap \mathcal{A}(A) \ .$$

THEOREM 5. *Notations being as above, suppose that $\mathfrak{o}$ is of type E. Then, the system $\{(A(\tau), C_\tau, \theta_\tau) \mid \tau \in \mathfrak{H}\}$ is complete with respect to $\boldsymbol{Q}$.*

Let τ be a point of $\mathfrak{H}$ and (A', C', θ') be a generic specialization of $(A(\tau), C_\tau, \theta_\tau)$ over $\boldsymbol{Q}$. By our definition of a generic specialization, there exists an isomorphism σ of a field k of definition for $(A(\tau), C_\tau, \theta_\tau)$ such that

$$(A', C', \theta') = (A(\tau)^\sigma, C_\tau^\sigma, \theta_\tau^\sigma) \ .$$

Denoting by the same notation θ' the extension of θ' in Φ, we see easily

$$\theta'(\mathfrak{o}) = \theta'(\Phi) \cap \mathcal{A}(A') \ .$$

Let X be a hyperplane section of $A(\tau)$ which is rational over k. Let B be the Picard variety of $A(\tau)$ and φ the homomorphism of $A(\tau)$ onto B defined by $\varphi(t) = \mathrm{Cl}(X_t - X)$ for $t \in A(\tau)$. Then we have

$$\theta_\tau(\alpha^*) = \varphi^{-1}\,{}^t\theta_\tau(\alpha)\varphi \ .$$

We see that B^σ is the Picard variety of A'; and we have

$$\theta'(\alpha^*) = (\varphi^\sigma)^{-1}\,{}^t\theta'(\alpha)\varphi^\sigma$$

and

$$\varphi^\sigma(u) = \mathrm{Cl}(X_u^\sigma - X^\sigma)$$

for $u \in A'$. Hence we see that the involution $*$ corresponds to the polarization $C' = C_\tau^\sigma$ of $A' = A(\tau)^\sigma$ which contains X^σ. By Theorem 3, Theorem 4 and the remark below Theorem 4, there exists a point τ' of $\mathfrak{H}$ such that A' is isomorphic to $C^2/D(\tau')$, $\theta'(\alpha)$ corresponds to $\chi(\alpha)$, and the polarization C' corresponds to the form $\mathrm{tr}(\rho\alpha\beta')$. Hence (A', C', θ') is isomorphic to a member $(A(\tau'), C_{\tau'}, \theta_{\tau'})$ of our system; this proves our theorem.

20. Let τ_0. τ_1 be two points of $\mathfrak{H}$ and λ be a homomorphism of $A(\tau_0)$ into $A(\tau_1)$; λ corresponds to a linear mapping M of C^2 into itself such that $MD(\tau_0) \subset D(\tau_1)$. Assume that λ commutes with the operations of elements in $\mathfrak{o}$, namely,

$$M\chi(\alpha) = \chi(\alpha)M$$

for every $\alpha \in \Phi$. Then, as the representation χ is absolutely irreducible, M is a scalar matrix; put $M = gI_2$, where g is a complex number. Put $\mathfrak{x}_0 = \begin{pmatrix} \tau_0 \\ 1 \end{pmatrix}$, $\mathfrak{x}_1 = \begin{pmatrix} \tau_1 \\ 1 \end{pmatrix}$. Since $M\mathfrak{x}_0$ is contained in $D(\tau_1)$, there exists an element $\gamma \in \mathfrak{o}$ such that

$$M\mathfrak{x}_0 = \chi(\gamma)\mathfrak{x}_1 \; ;$$

we have then

$$MD(\tau_0) = \{\chi(\alpha)\mathfrak{x}_1 \mid \alpha \in \mathfrak{o}\gamma\} \, ,$$

and

$$\chi(\gamma)\begin{pmatrix} \tau_1 \\ 1 \end{pmatrix} = g\begin{pmatrix} \tau_0 \\ 1 \end{pmatrix} \, .$$

As we have $\mathrm{Im}(\tau_0) > 0$, $\mathrm{Im}(\tau_1) > 0$, we see

(P) $\mathrm{nr}(\gamma) = \det \chi(\gamma) > 0.$

Let E_0, E_1 be the forms defined by

$$E_0(\chi(\alpha)\mathfrak{x}_0, \chi(\beta)\mathfrak{x}_0) = \mathrm{tr}(\rho\alpha\beta') \, ,$$

$$E_1(\chi(\alpha)\mathfrak{x}_1, \chi(\beta)\mathfrak{x}_1) = \mathrm{tr}(\rho\alpha\beta') \, ,$$

for $\alpha \in \Phi$, $\beta \in \Phi$. We have then

$$E_1(M\chi(\alpha)\mathfrak{x}_0, M\chi(\beta)\mathfrak{x}_0) = E_1(\chi(\alpha\gamma)\mathfrak{x}_1, \chi(\beta\gamma)\mathfrak{x}_1) = \mathrm{nr}(\gamma)\,\mathrm{tr}(\rho\alpha\beta') \, .$$

This shows

$$E_1(M\mathfrak{y}, M\mathfrak{z}) = \mathrm{nr}(\gamma)E_0(\mathfrak{y}, \mathfrak{z}) \, ,$$

namely, λ^{-1} sends the polarization C_{τ_1} into C_{τ_0} if $\lambda \neq 0$. In particular, every isomorphism of $A(\tau_0)$ onto $A(\tau_1)$ gives an isomorphism of $(A(\tau_0), C_{\tau_0}, \theta_{\tau_0})$ onto $(A(\tau_1), C_{\tau_1}, \theta_{\tau_1})$ if and only if it commutes with the operation of $\mathfrak{o}$. Suppose now that the above λ is an isomorphism. Then we have

$$MD(\tau_0) = D(\tau_1) \; ;$$

from this follows $\mathfrak{o}\gamma = \mathfrak{o}$; hence γ is a unit of $\mathfrak{o}$; so we have, taking account of (P),

$$\mathrm{nr}(\gamma) = 1 \; .$$

21. Let ξ be an element of Φ such that $\mathrm{nr}(\xi) > 0$; put

$$\chi(\xi) = \begin{pmatrix} a & b \\ c & d \end{pmatrix} \; .$$

Then, a, b, c, d are four real numbers and $ad - bc > 0$. We define the operation of ξ on $\mathfrak{H}$ by

$$\xi[\tau] = \frac{a\tau + b}{c\tau + d} \; ;$$

$\tau \to \xi[\tau]$ gives an analytic automorphism of $\mathfrak{H}$. Now denote by $\Gamma(\mathfrak{o})$ or simply by Γ, the group composed of all units γ of $\mathfrak{o}$ such that $\mathrm{nr}(\gamma) = 1$. Then $\Gamma/\{\pm 1\}$ is considered as a group of transformations on $\mathfrak{H}$.

PROPOSITION 15. *$(A(\tau_0), C_{\tau_0}, \theta_{\tau_0})$ and $(A(\tau_1), C_{\tau_1}, \theta_{\tau_1})$ are isomorphic to each other if and only if there exists an element $\gamma \in \Gamma(\mathfrak{o})$ such that $\tau_0 = \gamma[\tau_1]$.*

We have seen that if there exists an isomorphism of $(A(\tau_0), C_{\tau_0}, \theta_{\tau_0})$ onto $(A(\tau_1), C_{\tau_1}, \theta_{\tau_1})$, then there exist an element $\gamma \in \Gamma(\mathfrak{o})$ and a complex number g such that

$$\chi(\gamma)\begin{pmatrix} \tau_1 \\ 1 \end{pmatrix} = g\begin{pmatrix} \tau_0 \\ 1 \end{pmatrix} \; .$$

This shows that $\tau_0 = \gamma[\tau_1]$. Conversely, suppose that there exists an element $\gamma \in \Gamma$ such that $\tau_0 = \gamma[\tau_1]$. Put

$$\chi(\gamma) = \begin{pmatrix} a & b \\ c & d \end{pmatrix}, \quad g = c\tau_1 + d;$$

we see then easily

$$\chi(\gamma)\begin{pmatrix} \tau_1 \\ 1 \end{pmatrix} = g\begin{pmatrix} \tau_0 \\ 1 \end{pmatrix} \; .$$

As γ is a unit of $\mathfrak{o}$, we have $gD(\tau_0) = D(\tau_1)$. Hence $\mathfrak{x} \to g\mathfrak{x}$ gives an isomorphism of $C^2/D(\tau_0)$ onto $C^2/D(\tau_1)$; it is clear that this isomorphism commutes with the operation of $\mathfrak{o}$. By the above remark, this gives an isomorphism of $(A(\tau_0), C_{\tau_0}, \theta_{\tau_0})$ onto $(A(\tau_1), C_{\tau_1}, \theta_{\tau_1})$.

22. Now we construct the variety $\mathcal{F}(A(\tau),\, \theta_\tau)$ with respect to the basis $\{\alpha_i\}$ of $\mathfrak{o}$, as in **3**, and put

$$\mathcal{F}(\tau) = \mathcal{F}(A(\tau),\, \theta_\tau)\ .$$

By Theorem 1, there exists a set of meromorphic functions $\{f_1,\, \cdots,\, f_m\}$ on $\mathfrak{H}$ such that

$$c(\mathcal{F}(\tau)) = (1,\, f_1(\tau),\, \cdots,\, f_m(\tau))$$

for every $\tau \in \mathfrak{H} - \mathfrak{G}$, where $\mathfrak{G}$ is a discrete subset of $\mathfrak{H}$. Let γ be an element of $\Gamma(\mathfrak{o})$. Take a generic point τ_0 of $\mathfrak{H}$ for the $f_i(\tau)$ and the $f_i(\gamma[\tau])$ over $\boldsymbol{Q}$. We can assume that τ_0 is not contained in $\mathfrak{G} \cup \gamma^{-1}[\mathfrak{G}]$. Then we have

$$c(\mathcal{F}(\tau_0)) = (1,\, f_1(\tau_0),\, \cdots,\, f_m(\tau_0))$$

$$c(\mathcal{F}(\gamma[\tau_0])) = (1,\, f_1(\gamma[\tau_0]),\, \cdots,\, f_m(\gamma[\tau_0]))\ .$$

By Proposition 15, we have $c(\mathcal{F}(\tau_0)) = c(\mathcal{F}(\gamma[\tau_0]))$, hence $f_i(\tau_0) = f_i(\gamma[\tau_0])$, for every i. As τ_0 is generic for the f_i and the $f_i(\gamma[\tau])$, we have

$$f_i(\tau) = f_i(\gamma[\tau]) \qquad\qquad (1 \leqq i \leqq m),$$

namely, the f_i are invariant under the group $\Gamma(\mathfrak{o})$.

It is known that the group $\Gamma(\mathfrak{o})$ is a properly discontinuous group of transformations on $\mathfrak{H}$ and that the quotient space $\mathfrak{H}/\Gamma$ is of finite measure; $\mathfrak{H}/\Gamma$ is compact if and only if Φ is a division algebra ([4], [5], [8]). Hence we can speak of automorphic functions on $\mathfrak{H}$ with respect to the group $\Gamma(\mathfrak{o})$; we understand by an automorphic function on $\mathfrak{H}$ with respect to $\Gamma(\mathfrak{o})$ a meromorphic function on $\mathfrak{H}$ which is invariant under $\Gamma(\mathfrak{o})$ and meromorphic at every parabolic cusp as function of a local parameter given in a usual manner. Denote by $\mathfrak{K}(\mathfrak{o})$ the field of all the automorphic functions on $\mathfrak{H}$ with respect to $\Gamma(\mathfrak{o})$. We shall now prove

THEOREM 6. *The functions f_i being as above, we have*

$$\mathfrak{K}(\mathfrak{o}) = \boldsymbol{C}(f_1,\, \cdots,\, f_m)\ .$$

If $\mathfrak{o}$ is of type E, $\boldsymbol{Q}(f_1,\, \cdots,\, f_m)$ is a regular extension of $\boldsymbol{Q}$ of dimension 1.

First we consider the case where Φ is a division algebra. Then, as there is no cusp, every meromorphic function on $\mathfrak{H}$ invariant under $\Gamma(\mathfrak{o})$ is contained in $\mathfrak{K}(\mathfrak{o})$. Hence we have $\boldsymbol{C}(f_1,\, \cdots,\, f_m) \subset \mathfrak{K}(\mathfrak{o})$. $\mathfrak{G}$ being as above, let τ_0, τ_1 be two points of $\mathfrak{H} - \mathfrak{G}$. By Proposition 15, if there is no element γ of $\Gamma(\mathfrak{o})$ such that $\gamma[\tau_0] = \tau_1$, then we have $\mathcal{F}(\tau_0) \neq \mathcal{F}(\tau_1)$, namely there exists an i such that $f_i(\tau_0) \neq f_i(\tau_1)$. From this fact easily follows $\mathfrak{K}(\mathfrak{o}) = \boldsymbol{C}(f_1,\, \cdots,\, f_m)$. Suppose now that Φ is not a division algebra; then Φ is isomorphic to the total matrix algebra of degree 2 over Q; every maximal order is conjugate to the order composed of all matrices with integral coefficients. Hence, there exists a real matrix T of degree 2 such that $T\chi(\alpha)T^{-1}$ has integral coefficients for every $\alpha \in \mathfrak{o}$; we may assume $\det T > 0$. Fix a point τ_0 of $\mathfrak{H}$ and put

$$T\begin{pmatrix} \tau_0 \\ 1 \end{pmatrix} = \begin{pmatrix} x_1 \\ x_2 \end{pmatrix},$$

$$D' = \left\{ \begin{pmatrix} ax_1 + bx_2 \\ cx_1 + dy_2 \end{pmatrix} \middle| a, b, c, d \in \mathbf{Z} \right\}.$$

Then we see that $TD \subset D'$, and C^2/D' is isomorphic to $C/D_0 + C/D_0$ where D_0 denotes the discrete subgroup of C defined by

$$D_0 = \{ax_1 + bx_2 \mid a, b \in \mathbf{Z}\} \ .$$

Denote by $j(\tau)$ the classical modular function which gives the moduli of elliptic curves as in **7**, and put $j_0 = j(x_1/x_2)$. Then C/D_0 is isomorphic to an elliptic curve B, with $j(B) = j_0$, defined over $Q(j_0)$. The linear mapping $\mathfrak{x} \to T\mathfrak{x}$ gives a homomorphism of C^2/D onto C^2/D'; hence T corresponds to a homomorphism of $A(\tau_0)$ onto $B \times B$. Take a polarization C' of $B \times B$ defined over $Q(j_0)$. Then, the field of moduli of $(B \times B, C')$ is contained in $Q(j_0)$. By Proposition 8 and Proposition 9, the $f_i(\tau_0)$ are algebraic over $Q(j_0)$. Expressing $T = \begin{pmatrix} a & b \\ c & d \end{pmatrix}$, we define

$$T[\tau] = \frac{a\tau + b}{c\tau + d} \ ;$$

then we have $j_0 = j(T[\tau_0])$. Applying this to a generic point of $\mathfrak{H}$ for the f_i and $j(T[\tau])$ over Q, we see that the f_i are algebraic over $Q(j(T[\tau]))$. Then, by the theory of elliptic modular functions, the f_i are automorphic functions with respect to $\Gamma(\mathfrak{o})$; hence we have $C(f_1, \cdots, f_m) \subset \mathfrak{R}(\mathfrak{o})$. Then, by the same reason as in the case where Φ is a division algebra, we have $C(f_1, \cdots, f_m) = \mathfrak{R}(\mathfrak{o})$. The last assertion of our theorem follows from Theorem 2 and Theorem 5.

23. The functions f_i in the above theorem are determined by $\mathfrak{o}$, $*$ and a basis of $\mathfrak{o}$, while the function field $\mathfrak{R}(\mathfrak{o})$ depends only upon $\mathfrak{o}$. By the result of **6**, the field $Q(f_1, \cdots, f_m)$ does not depend upon the choice of a basis of $\mathfrak{o}$. We denote by $K(\mathfrak{o}, \rho)$ the field $Q(f_1, \cdots, f_m)$. By the above theorem, for any ρ, ρ_1, the fields $K(\mathfrak{o}, \rho)$ and $K(\mathfrak{o}, \rho_1)$ are birationally equivalent over the field C of complex numbers. We shall now prove

$$K(\mathfrak{o}, \rho) = K(\mathfrak{o}, \rho_1) \ .$$

Let (A, C, θ) be a polarized abelian variety of dimension 2 of type $\mathfrak{o}$ such that C corresponds to the involution defined by ρ; let C_1 be the polarization of A corresponding to the involution defined by ρ_1. Denote by K and K_1 the field of moduli of (A, C, θ) and (A, C_1, θ), respectively. The equality $K(\mathfrak{o}, \rho) = K(\mathfrak{o}, \rho_1)$ is proved if we show $K = K_1$. Let k be a field of definition for (A, C, θ) and (A, C_1, θ) containing K and K_1; let σ be an isomorphism of k into a field, which fixes every element of K. Then, by Proposition 5, there exists an isomorphism η of (A, C, θ) onto $(A^\sigma, C^\sigma, \theta^\sigma)$. Let B be the Picard variety of A. Take a divisor $X \in C_1$; then we obtain a homomorphism φ of A onto B such that

$$\varphi(u) = \mathrm{Cl}(X_u - X)$$

for $u \in A$. As C_1 corresponds to ρ_1, we have

$$\varphi^{-1}\,{}^t\theta(\alpha)\varphi = \theta(\rho_1^{-1}\alpha'\rho_1)\ .$$

We see that B^σ is the Picard variety of A^σ and

$$\varphi^\sigma(v) = \mathrm{Cl}(X_v^\sigma - X^\sigma)$$

for $v \in A^\sigma$, where we denote again by σ a certain extension of σ. We see moreover that the relation

$$(\varphi^\sigma)^{-1}\,{}^t\theta^\sigma(\alpha)\varphi^\sigma = \theta^\sigma(\rho_1^{-1}\alpha'\rho_1)$$

holds; hence ρ_1 defines the involution determined by the polarization C_1^σ, so that (A, C_1, θ) and $(A^\sigma, C_1^\sigma, \theta^\sigma)$ are isomorphic to two members of the same system $\{(A_\tau, C_\tau, \theta_\tau)\}$ obtained from $\mathfrak{o}$ and ρ_1. On the other hand η is an isomorphism of A onto A^σ such that

$$\eta\theta(\alpha) = \theta^\sigma(\alpha)\eta$$

for every $\alpha \in \mathfrak{o}$. By the remark of 20, η sends the polarization C_1 onto C_1^σ, so that (A, C_1, θ) and $(A^\sigma, C_1^\sigma, \theta^\sigma)$ are isomorphic. By Proposition 5, σ fixes every element of K_1. This proves $K \supset K_1$. By the same argument, we have $K_1 \supset K$, so that $K = K_1$.

APPENDIX

Specialization theory with respect to a place

We understand by a *place* $\mathfrak{p}$ of a field k, as usual, a homomorphism of k into a field, which maps the identity on the identity, where we admit that $\mathfrak{p}$ takes the value ∞. We call the image $\mathfrak{p}(k)$ *the residue field of* $\mathfrak{p}$. The following lemma is well known.

LEMMA 1. *Every homomorphism of a subring of a field k into a field k', which maps the identity on the identity, is extended to a place of k*

whose residue field is contained in the algebraic closure of k'.

We fix two "universal domains" K, K' with the same or different characteristics and now deal with only places defined on a subfield of K taking values in K'. We denote by $P^n(K)$ and $P^n(K')$ respectively the projective space of dimension n defined with respect to K and K'. We say that a place $\mathfrak{p}_1$ of a field k_1 is an *extension* of a place $\mathfrak{p}$ of a field k if $k_1 \supset k$ and $\mathfrak{p}_1(a) = \mathfrak{p}(a)$ for every $a \in k$. Let $(a_1, \cdots, a_n)$ be a set of elements in K and $(a'_1, \cdots, a'_n)$ a set of elements in K'; some of the a_i, a'_i may be ∞. We say that $(a'_1, \cdots, a'_n)$ is a specialization of $(a_1, \cdots, a_n)$ over a place $\mathfrak{p}$ of a field k, if there exists an extension $\mathfrak{p}_1$ of $\mathfrak{p}$ such that $a'_i = \mathfrak{p}_1(a_i)$ for $1 \leq i \leq n$. If the a_i are all contained in k, $(a_1, \cdots, a_n)$ has a unique specialization $(\mathfrak{p}(a_1), \cdots, \mathfrak{p}(a_n))$ over $\mathfrak{p}$. Let $(b_0, \cdots, b_n)$ be a point in $P^n(K)$ and $(b'_0, \cdots, b'_n)$ a point in $P^n(K')$. We say that (b') is a specialization of (b) over $\mathfrak{p}$, if there exists a suffix i such that $b_i \neq 0$, $b'_i \neq 0$ and $(b'_0/b'_i, \cdots, b'_n/b'_i)$ is a specialization of $(b_0/b_i, \cdots, b_n/b_i)$ over $\mathfrak{p}$ in the above sense. If (b) is rational over k, (b) has a unique specialization over $\mathfrak{p}$, which we denote by $\mathfrak{p}(b)$. Let $V_1, \cdots, V_s$ be positive cycles in product-varieties of projective spaces $P^{n_\nu}(K)$ and $V'_1, \cdots, V'_s$ positive cycles in product-varieties of $P^{n_\nu}(K')$. We say that $(V'_1, \cdots, V'_s)$ is a specialization of $(V_1, \cdots, V_s)$ over $\mathfrak{p}$ if there exists an extension $\mathfrak{p}_1$ of $\mathfrak{p}$ such that $c(V'_i) = \mathfrak{p}_1(c(V_i))$ for $1 \leq i \leq s$. If V is rational over k, V has a unique specialization over $\mathfrak{p}$, which we denote by $\mathfrak{p}(V)$. We can also define specialization of cycles in affine varieties or more generally in abstract varieties; but that discussion is omitted.

I have given in [9] the theory of specialization over $\mathfrak{p}$ only in the case where $\mathfrak{p}$ is defined by a discrete valuation of rank 1; we call such a $\mathfrak{p}$ a *discrete place*. The general case is easily reduced to this case, thus satisfying our need in the present paper, by means of the following lemma.

LEMMA 2. *Let $\mathfrak{p}$ be a place of k and $(V'_1, \cdots, V'_s)$ a specialization of a set of positive cycles $(V_1, \cdots, V_s)$ over $\mathfrak{p}$. Then there exist a field k_1 and a discrete place $\mathfrak{p}_1$ of k_1 such that the V_i are rational over k_1 and $V'_i = \mathfrak{p}_1(V_i)$ for $1 \leq i \leq s$.*

Let (x') be a specialization of a set of elements (x) in K over $\mathfrak{p}$. Call k_0 the prime field in k and $\mathfrak{p}_0$ the restriction of $\mathfrak{p}$ on k_0. Then (x') is a specialization of (x) over $\mathfrak{p}_0$. We see that $\mathfrak{p}_0$ is discrete. Hence, by Proposition 26 of [9], there exists a discrete place $\mathfrak{p}_1$ of $k(x)$ such that $\mathfrak{p}_1(x) = (x')$. Applying this fact to the coordinates of the $c(V_i)$ we get our lemma.

Now we can prove that the specialization over $\mathfrak{p}$ preserves inclusion, sum, intersection-product, direct product, projection, with suitable conditions as in [9], for non-discrete $\mathfrak{p}$. For example, let us consider the case of intersection-product. Let V be a projective variety defined over k and X, Y positive rational cycles on V. Suppose that $\mathfrak{p}(V)$ has only one component and its multiplicity is equal to 1. (We have called in [9] such a V $\mathfrak{p}$-simple.) Then $\mathfrak{p}(V)$ is a projective variety defined over $\mathfrak{p}(k)$. Let (X', Y') be a specialization of (X, Y) over $\mathfrak{p}$. By Lemma 2, there exists a discrete place $\mathfrak{p}_1$ for which we have $\mathfrak{p}(V) = \mathfrak{p}_1(V)$, $X' = \mathfrak{p}_1(X)$, $Y' = \mathfrak{p}_1(Y)$. By the theory in [9], we have $X' \subset \mathfrak{p}(V)$, $Y' \subset \mathfrak{p}(V)$. We assume, for the sake of simplicity, that V, $\mathfrak{p}(V)$ are non-singular. Then X', Y' are positive cycles on $\mathfrak{p}(V)$. Suppose that the intersection-products $X \cdot Y$ on V and $X' \cdot Y'$ on $\mathfrak{p}(V)$ are both defined. Let Z' be a specialization of $X \cdot Y$ over the specialization $(X, Y, V) \to (X', Y', \mathfrak{p}(V))$ with respect to $\mathfrak{p}$. We can take, by Lemma 2, $\mathfrak{p}_1$ in such a way that we have $Z' = \mathfrak{p}_1(X \cdot Y)$. Then, by Theorem 17 of [9], we have $\mathfrak{p}_1(X \cdot Y) = \mathfrak{p}_1(X) \cdot \mathfrak{p}_1(Y)$, namely $Z' = X' \cdot Y'$. This is, as we have said, that the specialization over $\mathfrak{p}$ preserves intersection-product.

Propositions 6, 7, 8 in [9] hold also for non-discrete case. This is easily verified by the fact that, if an ideal $\mathfrak{a}$ of a (discrete or non-discrete) valuation ring is finitely generated, $\mathfrak{a}$ is a principal ideal.

LEMMA 3. *Let $\mathfrak{p}$ be a place of a field k. Let B be a variety in a product-variety of projective spaces, defined over k and x a generic point of V over k. Then, as a point set, $\mathfrak{p}(V)$ is equal to the set of all specializations of x over $\mathfrak{p}$.*

First we consider the case where V is a variety in $P^n(K)$. If x' is a specialization of x over $\mathfrak{p}$, then by the above remark, x' is contained in $\mathfrak{p}(V)$; hence we have only to prove that every point of $\mathfrak{p}(V)$ is a specialization of x over $\mathfrak{p}$. Let r be the dimension of V; let t_{ij}, for $1 \leqq i \leqq r$, $0 \leqq j \leqq n$ be $r(n+1)$ independent variables over k and L the linear variety in $P^n(K)$ defined by the equations

$$\sum_j t_{ij} X_j = 0 , \qquad\qquad (1 \leqq i \leqq r);$$

similarly, let t'_{ij} be $r(n+1)$ independent variables over $\mathfrak{p}(k)$ and L' the linear variety defined by the equations

$$\sum_j t'_{ij} X_j = 0 , \qquad\qquad (1 \leqq i \leqq r).$$

It is easy to see that there exists an extension $\mathfrak{P}$ of $\mathfrak{p}$ such that $\mathfrak{P}(t_{ij}) = t'_{ij}$ for every i, j. We can easily verify $\mathfrak{P}(L) = L'$. Let $\{x_1, \cdots, x_g\}$ be the components of $V \cdot L$ and $\{x'_1, \cdots, x'_h\}$ the components of $\mathfrak{p}(V) \cdot L'$. Then by the above remark, we see, after reordering the x_i, that $g = h$ and $(x'_1, \cdots, x'_g)$ is a specialization of $(x_1, \cdots, x_g)$ over $\mathfrak{p}$. Let x' be a point in $\mathfrak{p}(V)$. As L' is generic over $\mathfrak{p}(k)$ and $\mathfrak{p}(V)$ is rational over $\mathfrak{p}(k)$, x' is a specialization over $\mathfrak{p}(k)$ of one of the x'_i, say x'_1; x'_1 is a specialization of x_1

over $\mathfrak{p}$ and x_1 is a specialization of x over k. From this follows that x' is a specialization of x over $\mathfrak{p}$. Thus we have proved our lemma in the case where V is a variety in $P^n(K)$. In the general case, take the canonical embedding φ of the ambient space of V into $P^n(K)$. It is easy to see that x' is a specialization of x over $\mathfrak{p}$ if and only if $\varphi'(x')$ is a specialization of $\varphi(x)$ over $\mathfrak{p}$, where φ' denotes the canonical embedding of the ambient space of $\mathfrak{p}(V)$. Applying Lemma 2 to V, $\varphi(V)$, we have $\mathfrak{p}(\varphi(V)) = \varphi'(\mathfrak{p}(V))$. Therefore, our lemma in the general case follows from the above special case.

In the following lemmas, varieties are assumed to be defined in product-spaces of projective spaces.

LEMMA 4. *Let U, V be two varieties and f a rational mapping of U into V; denote by T the graph of f. Let k be a field of definition for U, V, T, and $\mathfrak{p}$ a place of k. Suppose that $\mathfrak{p}(U)$, $\mathfrak{p}(V)$ are varieties and $\mathfrak{p}(T)$ is the graph of a rational mapping f' of $\mathfrak{p}(U)$ into $\mathfrak{p}(V)$. Let u, u' be points on U, $\mathfrak{p}(U)$, respectively, such that f is defined at u and f' is defined at u'. If (u', v') is a specialization of $(u, f(u))$ over $\mathfrak{p}$, then we have $v' = f'(u')$.*

As $u \times f(u)$ is contained in T, $u' \times v'$ is contained in $\mathfrak{p}(T)$. Then, from our assumption that f' is defined at u' follows $v' = f'(u')$.

LEMMA 5. *Let the notations U, V, T, f, k, $\mathfrak{p}$ be the same as in Lemma 4. Suppose that $\mathfrak{p}(U)$, $\mathfrak{p}(V)$ are varieties and that $\mathfrak{p}(V)$ has no multiple point. Let f' be a rational mapping of $\mathfrak{p}(U)$ into $\mathfrak{p}(V)$ and T' the graph of f'. Suppose that f, f' are everywhere defined and that, for any point u of U, if (u', v') is a specialization of $(u, f(u))$, we have $v' = f'(u')$. Then we have $T' = \mathfrak{p}(T)$.*

Let u be a generic point of U over k; then $u \times f(u)$ is a generic point of T over k. By Lemma 3, every point of $\mathfrak{p}(T)$ is a specialization of $u \times f(u)$ over $\mathfrak{p}$. By our assumption of lemma, such a specialization has a form $u' \times f'(u')$, so that $\mathfrak{p}(T)$, as a point set, is contained in T'. As $\mathfrak{p}(T)$ and T' have the same dimension, $\mathfrak{p}(T)$ coincides with T' as point set. Hence we have $\mathfrak{p}(T) = mT'$, with a positive multiplicity m. By virtue of Lemma 2, we can apply Theorem 19 of [9] to our case. We have then $\mathfrak{p}(U) = \mathfrak{p}(\mathrm{pr}_U(T)) = \mathrm{pr}_{\mathfrak{p}(U)}(\mathfrak{p}(T)) = m\mathfrak{p}(U)$, so that we have $m = 1$. This proves our lemma.

By virtue of Lemma 2, Theorem 26 of [9] is now written in the following form.

LEMMA 6. *Let V be a variety defined over a field k. Then there exist a finite number of elements a_i of k such that, for every place $\mathfrak{p}$ of k,*

$\mathfrak{p}(V)$ *is a variety if* $\mathfrak{p}(a_\iota) \neq 0$ *for every* $i.$

UNIVERSITY OF TOKYO

REFERENCES

1. M. EICHLER, *Über die Einheiten der Divisionsalgebren*, Math. Ann., 114 (1937), 635-654.
2. ————, *Über die Idealklassenzahl hyperkomplexer Systeme*, Math. Zeit., 43 (1938), 481-494.
3. ————, *Quaternäre quadratische Formen und die Riemannsche Vermutung für Kongruenzzetafunktion*, Arch. Math., 5 (1954), 355-366.
4. ————, On Modular Correspondences, Tata Institute, 1957.
5. R. FRICKE und F. KLEIN, Vorlesungen über die Theorie der Automorphen Funktionen, Leipzig, 1897.
6. S. LANG, Abelian Varieties, Interscience Tracts, New York, 1959.
7. T. MATSUSAKA, *Polarized varieties, fields of moduli and generalized Kummer varieties of polarized abelian varieties*, Amer. J. Math., 80 (1958), 45-82.
8. H. POINCARÉ, Les fonctions fuchsiennes et l'arithmétique (1887), Oeuvres, t. II, 463-511.
9. G. SHIMURA, *Reduction of algebraic varieties with respect to a discrete valuation of the basic field*, Amer. J. Math., 77 (1955), 134-176.
10. ————, *Correspondances modulaires et les fonctions ζ de courbes algébriques*, J. Math. Soc. Japan, 10 (1958), 1-28.
11. ————, *Modules des variétés abéliennes polarisées et fonctions modulaires*, in Séminaire H. Cartan, E. N. S. 1957/58, Exposés 18-20.
12. A. Weil, Variétés Abéliennes et Courbes Algébriques, Hermann et $C^{\iota e}$, Paris, 1948.
13. ————, *On the projective embedding of abelian varieties*, in Algebraic Geometry and Topology, A symposium in honor of S. Lefschetz, Princeton, 1957, 177-181.
14. ————, *On the theory of complex multiplication*, in Proceedings of the International Symposium on Algebraic Number Theory (Tokyo-Nikko, 1955), Tokyo, Science Council of Japan, 1956, 9-22.
15. ————, Variétés Kählériennes, Hermann, Paris, 1958.

Sur les intégrales attachées aux formes automorphes

Journal of the Mathematical Society of Japan, 11 (1959), 291-311

(Reçu le 24 avril, 1959)

Dans un travail récent [1] M. Eichler a donné une théorie des intégrales attachées aux formes automorphes d'une variable, et en l'appliquant aux correspondances modulaires, il a obtenu la formule de la trace. Si le degré des formes est 2, les formes paraboliques ne sont autres que les formes différentielles de première espèce; et l'on peut définir la variété jacobienne au moyen des périodes des intégrales. Il est à souhaiter d'obtenir l'analogue pour l'intégrale d'Eichler de degré > 2. L'objet du présent mémoire est de démontrer que, dans le « cas arithmétique », on peut construire, pour chaque entier pair positif n, une variété abélienne à partir des formes paraboliques de degré n.

Nous considérerons l'intégrale d'une certaine forme différentielle vectorielle attachée à une forme parabolique, qu'Eichler a déjà signalée dans [1]; les périodes d'une telle intégrale définissent un cocycle par rapport à une représentation tensorielle M du groupe fuchsien. Si la représentation M est à coefficients entiers, on peut distinguer les formes paraboliques dont les périodes ont des parties réelles entières; celle-ci forment un lattice D dans l'espace S des formes paraboliques. Nous verrons que la partie imaginaire de la métrique de Petersson prend des valeurs rationnelles sur D, ce qui démontre que le tore complexe S/D a une structure de variété abélienne. Les opérateurs de Hecke peuvent alors être regardés comme des endomorphismes de la variété abélienne. On devra observer que si les coefficients de M ne sont pas entiers, il n'y a aucune méthode canonique pour définir un tel lattice D pour lequel S/D devienne une variété abélienne, tandis que la formule de la trace pour les opérateurs de Hecke est obtenue dans le cas général. Du moins, cette variété abélienne S/D est sans doute une des choses dont on aura besoin lorsqu'on cherchera à approfondir la théorie arithmétique des formes automorphes.

Le lecteur s'apercevra de la connexion de notre étude avec les théories de A. Weil [6] et de E. Hecke [3]; on pourrait les regarder comme des cas particuliers des intégrales des formes différentielles à valeurs dans un espace fibré vectoriel auquel est fixée une forme hermitienne invariante par le groupe de fibre, dont on rencontre des exemples aussi dans le cas des formes automorphes de plusieurs variables; mais on ne s'en occupera pas dans ce mémoire.

NOTATIONS. On désignera par Z, Q, R et C respectivement l'anneau des entiers rationnels, les corps des nombres rationnels, réels et complexes; R^n, C^n désignent alors les espaces vectoriels de dimension n sur R, C, respectivement, dont on considère les élements comme des matrices à n lignes et une colonne; $GL(2, R)$ désigne le groupe formé des matrices inversibles de degré 2 à coefficients réels, $GL_+(2, R)$ le groupe des éléments de $GL(2, R)$ à determinant positif, $SL(2, R)$ le groupe des éléments de $GL(2, R)$ à determinant 1, et $SL(2, Z)$ le groupe des éléments de $SL(2, R)$ à coefficients entiers. On désignera par ${}^t X$ la transposée d'une matrice X, et par I_n la matrice unité de degré n. Si z est un nombre complexe ou une fonction à valeurs complexes ou une forme différentielle à valeurs complexes, etc., on indiquera par $\bar{z}, \mathrm{Re}(z)$ et $\mathrm{Im}(z)$ le conjugué complexe, la partie réelle et la partie imaginaire de z; l'unité imaginaire $\sqrt{-1}$ sera notée par i.

§ 1. Les représentations tensorielles de $GL(2, R)$.

Soit $\begin{pmatrix} u \\ v \end{pmatrix}$ un vecteur de R^2; on désignera par $\begin{pmatrix} u \\ v \end{pmatrix}^n$, pour chaque entier positif n, le vecteur de R^{n+1} dont les composantes sont $u^n, u^{n-1}v, \cdots, uv^{n-1}, v^n$. Nous conviendrons d'entendre par $\begin{pmatrix} u \\ v \end{pmatrix}^0$ le nombre 1. Soit σ un élément de $GL(2, R)$; posons

$$\begin{pmatrix} u_1 \\ v_1 \end{pmatrix} = \sigma \begin{pmatrix} u \\ v \end{pmatrix}.$$

On définit, pour chaque entier non-négatif n, une matrice réelle $M_n(\sigma)$ de degré $n+1$ par

$$\begin{pmatrix} u_1 \\ v_1 \end{pmatrix}^n = M_n(\sigma)\begin{pmatrix} u \\ v \end{pmatrix}^n.$$

On voit que M_n donne une représentation de $GL(2, R)$, pour chaque n. $\begin{pmatrix} w \\ z \end{pmatrix}$ étant un autre vecteur de R^2, posons

$$\begin{pmatrix} w_1 \\ z_1 \end{pmatrix} = \sigma \begin{pmatrix} w \\ z \end{pmatrix}.$$

On a evidemment

$$(u_1 z_1 - v_1 w_1)^n = (\det \sigma)^n (uz - vw)^n.$$

Définissons une matrice P_n de degré $n+1$ par

$$(uz - vw)^n = {}^t\begin{pmatrix} u \\ v \end{pmatrix}^n P_n \begin{pmatrix} w \\ z \end{pmatrix}^n.$$

On a alors, pour tout n et pour tout $\sigma \in GL(2, R)$:

$$(1) \qquad {}^{t}M_n(\sigma)P_n M_n(\sigma) = (\det \sigma)^n P_n ,$$

$$(2) \qquad {}^{t}P_n = (-1)^n P_n$$

et

$$M_n(-I_2) = (-1)^n I_{n+1} .$$

Par suite, si n est pair, M_n donne une représentation du groupe quotient $\mathrm{GL}(2, \boldsymbol{R})/\{\pm I_2\}$.

§2. La métrique de Petersson.

On désigne par $\mathcal{H}$ le demi-plan complexe $\mathrm{Im}(z) > 0$, où z est une variable complexe. Pour chaque élément $\sigma = \begin{pmatrix} a & b \\ c & d \end{pmatrix}$ de $\mathrm{GL}_+(2, \boldsymbol{R})$, on définit une transformation $z \to \sigma(z)$ de $\mathcal{H}$ par

$$\sigma(z) = \frac{az+b}{cz+d} .$$

Alors, le groupe des transformations analytiques de $\mathcal{H}$ est isomorphe au groupe $\mathrm{SL}(2, \boldsymbol{R})/\{\pm I_2\}$. On posera

$$J(\sigma, z) = (cz+d)^{-1} .$$

On voit facilement qu'on a

$$J(\sigma\tau, z) = J(\sigma, \tau(z))J(\tau, z)$$

pour tout $\sigma, \tau \in \mathrm{GL}_+(2, \boldsymbol{R})$. Avec les notations du §1, on a

$$(3) \qquad \begin{pmatrix} \sigma(z) \\ 1 \end{pmatrix}^n = J(\sigma, z)^n M_n(\sigma) \begin{pmatrix} z \\ 1 \end{pmatrix}^n .$$

Soit G un sous-groupe de $\mathrm{SL}(2, \boldsymbol{R})$. On supposera dans ce qui suit, que G, comme groupe de transformations de $\mathcal{H}$, est un groupe fuchsien, c'est à dire que G est un sous-groupe discret de $\mathrm{SL}(2, \boldsymbol{R})$ pour lequel $\mathcal{H}/G$ ait une mesure finie. Soit m un entier positif. On entendra par une *forme parabolique de degré m* par rapport à G, comme d'ordinaire, une fonction f sur $\mathcal{H}$ jouissant des propriétés (P1-3) suivantes:

(P1) *f est holomorphe sur $\mathcal{H}$;*

(P2) *$f(\sigma(z))J(\sigma, z)^m = f(z)$ pour tout $\sigma \in G$.*

Pour exprimer la condition (P3), considérons un point parabolique s de G; s est un nombre réel ou le point à l'infini ∞. Posons

$$\rho = \begin{pmatrix} -s & 1 \\ -1 & 0 \end{pmatrix} \quad \text{ou} \quad \rho = I_2$$

selon que s est un nombre réel ou ∞. Alors chaque transformation de G ayant s comme point fixe est une puissance τ^k d'un élément

$$\tau = \rho \begin{pmatrix} 1 & h \\ 0 & 1 \end{pmatrix} \rho^{-1}$$

de G, où h est un nombre positif. Posons

$$q = \exp(2\pi i h^{-1} z).$$

Alors, la condition (P3) s'énonce :

(P3) *Pour tout point parabolique s de G, la fonction $f(\rho(z))J(\rho, z)^m$ est holomorphe comme fonction de q et nulle en $q = 0$.*

On désignera par $S_m(G)$ l'espace linéaire des formes paraboliques de degré m par rapport au groupe G. Dans ce mémoire, nous ne nous occuperons que des formes paraboliques de degré *pair*.

D'après Petersson [4], on peut introduire comme suit une métrique dans l'espace $S_m(G)$. Posons d'abord

$$dv = \frac{i}{2y^2} \, dz d\bar{z} = \frac{1}{y^2} \, dx dy,$$

où $z = x + iy$; dv est alors une forme différentielle invariante par toute transformation de $SL(2, \boldsymbol{R})$. On définit le produit scalaire (f, g) de deux éléments f, g de $S_m(G)$ par

$$(f, g) = \int_{\mathcal{F}} f(z)\overline{g(z)} y^m dv$$

où $\mathcal{F}$ est un domaine fondamental quelconque de G; on sait que cette intégrale converge pour tout $f, g \in S_m(G)$ et (f, g) est une forme hermitienne positive non-dégénérée.

Considérons maintenant la forme différentielle vectorielle $d\mathfrak{z}_n$ sur $\mathcal{H}$ définie par

$$d\mathfrak{z}_n = \begin{pmatrix} z \\ 1 \end{pmatrix}^n dz = \begin{pmatrix} z^n dz \\ z^{n-1} dz \\ \vdots \\ z dz \\ dz \end{pmatrix}.$$

On a alors, en vertu de (3), pour tout $\sigma \in GL_+(2, \boldsymbol{R})$,

$$(4) \qquad d\mathfrak{z}_n \circ \sigma = (\det \sigma)J(\sigma, z)^{n+2} M_n(\sigma) d\mathfrak{z}_n,$$

$\omega \circ \sigma$ désignant la transformée de la forme différentielle ω par la transformation σ. Si $f(z)$ est un élément de $S_{n+2}(G)$, on a

$$(5) \qquad (f d\mathfrak{z}_n) \circ \sigma = M_n(\sigma) f d\mathfrak{z}_n$$

pour chaque $\sigma \in G$. Soit g un autre élément de $S_{n+2}(G)$. On voit, compte tenu de la relation (1), que la forme différentielle

$${}^t(f d\mathfrak{z}_n) P_n(\overline{g d\mathfrak{z}_n})$$

est invariante par le groupe G. On vérifie aisément

$$^t(fd\mathfrak{z}_n)P_n(\overline{gd\mathfrak{z}_n}) = -(2i)^{n+1}f\cdot\bar{g}\cdot y^{n+2}dv\ ;$$

il en résulte qu'on a

$$(6) \qquad -(2i)^{n+1}(f,g) = \int_{\mathscr{F}}{}^t(fd\mathfrak{z}_n)P_n(\overline{gd\mathfrak{z}_n})\ .$$

Posons, pour deux éléments f, g de $S_{n+2}(G)$,

$$\Lambda(f,g) = 2^{n-1}i[(f,g)-(g,f)]\ .$$

On vérifie facilement que $\Lambda(f,g)$ est une forme R-bilinéaire alternée et $\Lambda(f, ig)$ est une forme symétrique positive non-dégénérée.

Nous aurons à nous servir par la suite de la partie réelle $\mathrm{Re}(fd\mathfrak{z}_n)$ de la forme différentielle $fd\mathfrak{z}_n$. Comme $M_n(\sigma)$ est une matrice réelle, on a

$$(7) \qquad \mathrm{Re}(fd\mathfrak{z}_n)\circ\sigma = M_n(\sigma)\mathrm{Re}(fd\mathfrak{z}_n)$$

pour tout $\sigma\in G$; par suite, la forme ${}^t\mathrm{Re}(fd\mathfrak{z}_n)P_n\mathrm{Re}(gd\mathfrak{z}_n)$ est invariante par G. Supposons que n soit un entier pair non-négatif; la matrice P_n est alors symétrique; on obtient donc

$$4\,{}^t\mathrm{Re}(fd\mathfrak{z}_n)P_n\mathrm{Re}(gd\mathfrak{z}_n) = {}^t(fd\mathfrak{z}_n)P_n(\overline{gd\mathfrak{z}_n})-{}^t(gd\mathfrak{z}_n)P_n(\overline{fd\mathfrak{z}_n})\ .$$

D'après la formule (6), $\Lambda(f,g)$ s'exprime sous la forme

$$(8) \qquad \Lambda(f,g) = (-1)^{\frac{n}{2}+1}\int_{\mathscr{F}}{}^t\mathrm{Re}(fd\mathfrak{z}_n)P_n\mathrm{Re}(gd\mathfrak{z}_n)\ .$$

§ 3. Groupes de cohomologie.

G étant comme ci-dessus, désignons par Y l'ensemble de toutes les transformations paraboliques de G. Soit M une représentation de G par des matrices de degré k à coefficients réels. On va définir un certain groupe de cohomologie attaché à M. Soit $\mathfrak{x}$ une application de G dans R^k; on appelle $\mathfrak{x}$ un *cocycle parabolique* par rapport à M s'il satisfait aux conditions suivantes:

(C1) $\mathfrak{x}(\sigma\tau) = \mathfrak{x}(\sigma)+M(\sigma)\mathfrak{x}(\tau)$ *pour tout* $\sigma,\tau\in G$;

(C2) *pour chaque* $\tau\in Y$, *il existe un vecteur* $\mathfrak{a}\in R^k$ *tel que l'on ait* $\mathfrak{x}(\tau) = \mathfrak{a}-M(\tau)\mathfrak{a}$; *le vecteur* $\mathfrak{a}$ *peut dépendre de* τ.

On désignera par $Z(M, R)$ l'espace linéaire des cocycles paraboliques par rapport à G. On voit facilement que si $\mathfrak{x}$ satisfait à la condition (C1), on a

$$\mathfrak{x}(1) = 0\ ,$$
$$(9) \qquad \mathfrak{x}(\sigma^{-1}) = -M(\sigma)^{-1}\mathfrak{x}(\sigma)\ .$$

Soit $\mathfrak{a}$ un vecteur de R^k; posons

$$\mathfrak{y}(\sigma) = \mathfrak{a}-M(\sigma)\mathfrak{a}$$

pour tout $\sigma \in G$. Il est facile de voir que η satisfait aux conditions (C1-2). Un tel cocycle η s'appellera un *cobord* par rapport à M. On désigne par $B(M, \boldsymbol{R})$ l'espace linéaire des cobords par rapport à M; et l'on pose

$$H(M, \boldsymbol{R}) = Z(M, \boldsymbol{R})/B(M, \boldsymbol{R}).$$

On appelle ce groupe quotient le groupe de cohomologie de G par rapport à M.

Supposons maintenant que les coefficients de $M(\sigma)$ soient entiers pour tout $\sigma \in G$. Un cocycle parabolique $\mathfrak{x}$ par rapport à M sera dit *entier* si les composantes du vecteur $\mathfrak{x}(\sigma)$ sont entières pour tout $\sigma \in G$. On désigne par $Z(M, \boldsymbol{Z})$ (resp. $B(M, \boldsymbol{Z})$) le module des cocycles paraboliques (resp. cobords) entiers par rapport à M; et on pose

$$H(M, \boldsymbol{Z}) = Z(M, \boldsymbol{Z})/B(M, \boldsymbol{Z}).$$

On voit que $H(M, \boldsymbol{Z})$ est canoniquement isomorphe à un sous-module de $H(M, \boldsymbol{R})$; on considère donc désormais $H(M, \boldsymbol{Z})$ comme un sous-module de $H(M, \boldsymbol{R})$.

PROPOSITION 1. *Toute base de $H(M, \boldsymbol{Z})$ sur $\boldsymbol{Z}$ donne une base de $H(M, \boldsymbol{R})$ sur $\boldsymbol{R}$.*

On sait que G a un système de générateurs en nombre fini $\{\sigma_1, \cdots, \sigma_d\}$. Si $\mathfrak{x}$ est un cocycle parabolique, on voit, en vertu de la condition (C1), que les vecteurs $\mathfrak{x}(\sigma)$ pour $\sigma \in G$ sont déterminés par les d vecteurs $\mathfrak{x}(\sigma_\nu)$. On peut donc considérer $\mathfrak{x}$ comme un élément de $\boldsymbol{R}^{dk} = \boldsymbol{R}^k \times \cdots \times \boldsymbol{R}^k$ en identifiant $\mathfrak{x}$ avec le vecteur $(\mathfrak{x}(\sigma_1), \cdots, \mathfrak{x}(\sigma_d)) \in \boldsymbol{R}^k \times \cdots \times \boldsymbol{R}^k$; $\mathfrak{x}$ appartient à $Z(M, \boldsymbol{Z})$ pourvu que les composantes du vecteur $(\mathfrak{x}(\sigma_1), \cdots, \mathfrak{x}(\sigma_d))$ soient entières. On voit facilement que les conditions (C1-2) s'expriment sous la forme

$$\sum_{\nu=1}^{d} L_{\mu\nu}\mathfrak{x}(\sigma_\nu) = 0 \qquad (\mu = 1, 2, \cdots)$$

où les $L_{\mu\nu}$ sont des matrices de degré k à coefficients entiers. Par conséquent, $Z(M, \boldsymbol{R})$ correspond à un sous-espace V linéaire de $\boldsymbol{R}^{dk}$ défini sur le corps $\boldsymbol{Q}$ des nombres rationnels; et les éléments de $Z(M, \boldsymbol{Z})$ correspondent aux vecteurs de V dont les composantes sont entières; il en est de même pour $B(M, \boldsymbol{R})$ et $B(M, \boldsymbol{Z})$. On peut en déduire l'assertion de la proposition.

Supposons que Y ne soit pas vide et fixons un élément τ_0 de Y. On désigne par $Z(M, \boldsymbol{R}, \tau_0)$ (resp. $B(M, \boldsymbol{R}, \tau_0), Z(M, \boldsymbol{Z}, \tau_0), B(M, \boldsymbol{Z}, \tau_0)$) l'ensemble des éléments $\mathfrak{x}$ de $Z(M, \boldsymbol{R})$ (resp. $B(M, \boldsymbol{R}), Z(M, \boldsymbol{Z}), B(M, \boldsymbol{Z})$) tels que $\mathfrak{x}(\tau_0) = 0$; et l'on pose

$$H(M, \boldsymbol{R}, \tau_0) = Z(M, \boldsymbol{R}, \tau_0)/B(M, \boldsymbol{R}, \tau_0),$$

$$H(M, \boldsymbol{Z}, \tau_0) = Z(M, \boldsymbol{Z}, \tau_0)/B(M, \boldsymbol{Z}, \tau_0).$$

$H(M, \boldsymbol{Z}, \tau_0)$ s'identifie canoniquement avec un sous-module de $H(M, \boldsymbol{R}, \tau_0)$. On peut alors démontrer, de même que plus haut, que toute base de $H(M, \boldsymbol{Z}, \tau_0)$

sur Z forme une base de $H(M, \boldsymbol{R}, \tau_0)$ sur $\boldsymbol{R}$. Soit $\mathfrak{x}$ un élément de $Z(M, \boldsymbol{R})$. D'après la condition (C2), il existe un vecteur $\mathfrak{a} \in \boldsymbol{R}^k$ tel que

$$\mathfrak{x}(\tau_0) = \mathfrak{a} - M(\tau_0)\mathfrak{a}.$$

Posons $\mathfrak{y}(\sigma) = \mathfrak{a} - M(\sigma)\mathfrak{a}$ pour tout $\sigma \in G$; on voit alors que $\mathfrak{x} - \mathfrak{y}$ est contenu dans $Z(M, \boldsymbol{R}, \tau_0)$; on en conclut

$$Z(M, \boldsymbol{R}) = B(M, \boldsymbol{R}) + Z(M, \boldsymbol{R}, \tau_0).$$

Il s'ensuit de là que $H(M, \boldsymbol{R}, \tau_0)$ est canoniquement isomorphe à $H(M, \boldsymbol{R})$. On constate de même que $H(M, \boldsymbol{Z}, \tau_0)$ est canoniquement isomorphe à un sous-module H' de $H(M, \boldsymbol{Z})$ d'indice fini.

§ 4. Périodes des intégrales.

G et M étant comme dans le § 3, supposons qu'il existe une matrice réelle P de degré k telle que l'on ait

$$(10) \qquad {}^t M(\sigma) P M(\sigma) = P$$

pour tout $\sigma \in G$. Désignons par $F(M)$ l'espace linéaire des formes différentielles vectorielles

$$\omega = \begin{pmatrix} \omega_1 \\ \vdots \\ \omega_k \end{pmatrix}$$

sur $\mathcal{H}$ jouissant des propriétés (D1–3) suivantes:

(D1) *Chaque composante ω_ν de ω est une forme différentielle fermée de degré 1, analytique par rapport à x, y, et à valeurs réelles.*

(D2) *On a $\omega \circ \sigma = M(\sigma)\omega$ pour tout $\sigma \in G$.*

La condition (D1) entraîne que l'intégrale

$$\int_z^{z'} \omega$$

ne dépend que de z, z' et non du choix du chemin d'intégration.

(D3) *Pour chaque point parabolique s de G, la valeur limite*

$$\lim_{z \to s} \int_{z_0}^{z} \omega$$

lorsque z tend vers s, en restant dans un domaine fondamental de G, est déterminée et finie, z_0 étant un point quelconque de $\mathcal{H}$.

Soit ω un élément de $F(M)$; fixons un point z_0 de $\mathcal{H}$ et posons

$$(11) \qquad \mathfrak{f}(z) = \int_{z_0}^{z} \omega + \mathfrak{a},$$

où $\mathfrak{a}$ est une «constante d'intégration» qui est un vecteur quelconque de $\boldsymbol{R}^k$. Alors $\mathfrak{f}(z)$ est une fonction sur $\mathcal{H}$ à valeurs dans $\boldsymbol{R}^k$. En vertu de la propriété

(D2), on obtient, pour chaque $\sigma \in G$,

$$(12) \qquad \mathfrak{f}(\sigma(z)) = M(\sigma)\mathfrak{f}(z) + \mathfrak{x}(\sigma),$$

où $\mathfrak{x}(\sigma)$ est le vecteur de $\boldsymbol{R}^k$ donné par

$$(13) \qquad \mathfrak{x}(\sigma) = \int_{z_0}^{\sigma(z_0)} \omega + \mathfrak{a} - M(\sigma)\mathfrak{a}.$$

On vérifie facilement que $\mathfrak{x}$ satisfait à la condition (C1) de §3. Soit τ une transformation parabolique de G, et soit s le point fixe de τ. D'après la propriété (D3), $\lim_{z \to s} \mathfrak{f}(z)$ est déterminé et fini; nous poserons

$$\mathfrak{f}(s) = \lim_{z \to s} \mathfrak{f}(z).$$

On a alors $\mathfrak{f}(s) = \mathfrak{f}(\tau(s)) = M(\tau)\mathfrak{f}(s) + \mathfrak{x}(\tau)$, de sorte qu'on a

$$(14) \qquad \mathfrak{x}(\tau) = [1 - M(\tau)]\mathfrak{f}(s).$$

Il en résulte que $\mathfrak{x}$ est un élément de $Z(M, \boldsymbol{R})$. Le cocycle parabolique $\mathfrak{x}$ s'appellera la période de l'intégrale (11). On voit facilement que la classe de cohomologie de $\mathfrak{x}$ ne dépend que de ω et non du choix de z_0 et $\mathfrak{a}$.

Soit η un autre élément de $F(M)$; $\mathfrak{b}$ étant une constante d'intégration, posons

$$(15) \qquad \mathfrak{g}(z) = \int_{z_0}^{z} \eta + \mathfrak{b}.$$

On obtient alors

$$(16) \qquad \mathfrak{g}(\sigma(z)) = M(\sigma)\mathfrak{g}(z) + \mathfrak{y}(\sigma),$$

où $\mathfrak{y}$ est un cocycle parabolique donné par une formule analogue à (13). Considérons maintenant la forme différentielle ${}^t\omega P\eta$; on voit que ${}^t\omega P\eta$ est invariant par G. Posons

$$(\omega, \eta) = \int_{\mathscr{F}} {}^t\omega P\eta \,,$$

où $\mathscr{F}$ est un domaine fondamental quelconque de G. On va maintenant exprimer la valeur (ω, η) par les périodes $\mathfrak{x}(\sigma), \mathfrak{y}(\sigma)$, On voit d'abord qu'on a ${}^t\omega P\eta = d({}^t\mathfrak{f}P d\mathfrak{g})$, de sorte qu'on a

$$(\omega, \eta) = \int_{\partial\mathscr{F}} {}^t\mathfrak{f}P d\mathfrak{g} \,,$$

où $\partial\mathscr{F}$ désigne le bord de $\mathscr{F}$. On peut supposer que $\mathscr{F}$ est un polygone dont les côtés se correspondent deux à deux, en sens inverse, par des éléments de G; pour ainsi dire, on a

$$\partial\mathscr{F} = \sum_{\mu} [E_\mu - \sigma_\mu(E_\mu)] \,,$$

où les E_μ sont des côtés qui forment la moitié du contour de $\mathscr{F}$ et les σ_μ

sont des éléments de G. On a alors

$$(\omega, \eta) = \sum_{\mu} \left(\int_{E_\mu} {}^t\mathfrak{f}Pd\mathfrak{g} - \int_{\sigma_\mu(E_\mu)} {}^t\mathfrak{f}Pd\mathfrak{g} \right).$$

Au moyen de (10), (12), (16), on obtient

$$\int_{\sigma_\mu(E_\mu)} {}^t\mathfrak{f}Pd\mathfrak{g} = \int_{E_\mu} {}^t(\mathfrak{f} \circ \sigma_\mu)Pd(\mathfrak{g} \circ \sigma_\mu)$$

$$= \int_{E_\mu} {}^t\mathfrak{f}Pd\mathfrak{g} + \int_{E_\mu} {}^t\mathfrak{x}(\sigma_\mu)PM(\sigma_\mu)d\mathfrak{g}.$$

En vertu des relations (9) et (10), on a enfin

$$(17) \qquad (\omega, \eta) = \sum_{\mu} {}^t\mathfrak{x}(\sigma_\mu^{-1})P\int_{E_\mu} d\mathfrak{g}.$$

Pour calculer plus explicitement, on adopte un domaine fondamental obtenu à partir d'une « dissection canonique » de la surface de Riemann comme suit. Soit $\mathcal{R}$ la surface de Riemann du corps des fonctions automorphes par rapport à G, et soit g le genre de $\mathcal{R}$. Soient $p_1, \cdots, p_j$ les points de $\mathcal{R}$ correspondant aux points elliptiques de G; et soient $q_1, \cdots, q_h$ les points de $\mathcal{R}$ correspondant aux points paraboliques de G. On prend un point r_0 de $\mathcal{R}$ autre que p_μ, q_ν; on coupe $\mathcal{R}$ par $2g$ courbes fermées passant r_0 qui forment un système de générateurs canoniques du groupe fondamental de $\mathcal{R}$, où l'on suppose que ces courbes ne traversent aucun des points p_μ, q_ν. On obtient alors un polygone à $4g$ côtés contenant les p_μ, q_ν dans son intérieur. On coupe en outre ce polygone le long de $j+h$ segments joignant un sommet au p_μ, q_ν; on obtient ainsi un polygone $\mathcal{G}$ à $4g+2j+2h$ côtés. Prenons pour domaine fondamental de G un polygone $\mathcal{F}$ dans $\mathcal{H}$ correspondant au polygone $\mathcal{G}$, et désignons par t_μ, s_ν les sommets de $\mathcal{F}$ correspondant aux p_μ, q_ν. D'après notre construction, le bord $\partial\mathcal{F}$ de $\mathcal{F}$ est la somme de $4g+2j+2h$ côtés

$$A_1, B_1, -\alpha_1(A_1), -\beta_1^{-1}(B_1), \cdots, A_g, B_g, -\alpha_g(A_g), -\beta_g^{-1}(B_g),$$

$$C_1, -\gamma_1(C_1), \cdots, C_j, -\gamma_j(C_j),$$

$$D_1, -\delta_1(D_1), \cdots, D_h, -\delta_h(D_h),$$

où les $\alpha_\lambda, \beta_\lambda, \gamma_\mu, \delta_\nu$ sont des éléments de G tels que

$$\delta_h \cdots \delta_1 \gamma_j \cdots \gamma_1 \beta_g^{-1} \alpha_g^{-1} \beta_g \alpha_g \cdots \beta_1^{-1} \alpha_1^{-1} \beta_1 \alpha_1 = 1,$$

$$\gamma_\mu^{e_\mu} = 1 \qquad\qquad (\mu = 1, \cdots, j);$$

γ_μ est une transformation elliptique qui laisse fixe t_μ; e_μ est l'ordre de ramification en t_μ; et δ_μ est une transformation parabolique qui laisse fixe le point parabolique s_μ. Posons

$$\sigma_\lambda = \beta_\lambda^{-1}\alpha_\lambda^{-1}\beta_\lambda\alpha_\lambda, \quad \tau_\lambda = \sigma_\lambda \cdots \sigma_1 \qquad (\lambda = 1, \cdots, g).$$

Soit u_0 le point de départ de A_1; et posons

$$u_\lambda = \tau_\lambda(u_0), \qquad\qquad (\lambda = 1, \cdots, g),$$
$$v_0 = \gamma_f \cdots \gamma_1(u_g).$$

Alors, $A_\lambda, B_\lambda, C_\mu, D_\nu$ joignent respectivement $u_{\lambda-1}$ à $\alpha_\lambda^{-1}\beta_\lambda\alpha_\lambda(u_{\lambda-1})$, $\alpha_\lambda^{-1}\beta_\lambda\alpha_\lambda(u_{\lambda-1})$ à $\beta_\lambda\alpha_\lambda(u_{\lambda-1})$, $\gamma_{\mu-1}\cdots\gamma_1(u_g)$ à t_μ, $\delta_{\nu-1}\cdots\delta_1(v_0)$ à s_ν. On a donc, au moyen de (17),

$$(\omega, \eta) = \sum_{\lambda=1}^{g}{}' \,{}^t\mathfrak{x}(\alpha_\lambda^{-1})P[\mathfrak{g}(\alpha_\lambda^{-1}\beta_\lambda\alpha_\lambda(u_{\lambda-1}))-\mathfrak{g}(u_{\lambda-1})]$$

$$+ \sum_{\lambda=1}^{g}{}' \,{}^t\mathfrak{x}(\beta_\lambda)P[\mathfrak{g}(\beta_\lambda\alpha_\lambda(u_{\lambda-1}))-\mathfrak{g}(\alpha_\lambda^{-1}\beta_\lambda\alpha_\lambda(u_{\lambda-1}))]$$

$$+ \sum_{\mu=1}^{f}{}' \,{}^t\mathfrak{x}(\gamma_\mu^{-1})P[\mathfrak{g}(t_\mu)-\mathfrak{g}(\gamma_{\mu-1}\cdots\gamma_1(u_g))]$$

$$+ \sum_{\nu=1}^{h}{}' \,{}^t\mathfrak{x}(\delta_\nu^{-1})P[\mathfrak{g}(s_\nu)-\mathfrak{g}(\delta_{\nu-1}\cdots\delta_1(v_0))] .$$

Comme γ_μ laisse fixe le point t_μ, on a

$$\mathfrak{g}(t_\mu) = \mathfrak{g}(\gamma_\mu{}^m(t_\mu)) = M(\gamma_\mu{}^m)\mathfrak{g}(t_\mu)+\mathfrak{y}(\gamma_\mu{}^m)$$

pour $m = 0, 1, \cdots, e_\mu-1$; on en déduit

$$e_\mu\mathfrak{g}(t_\mu) = \sum_{m=1}^{e_\mu-1} M(\gamma_\mu{}^m)\mathfrak{g}(t_\mu)+\sum_{m=1}^{e_\mu-1}\mathfrak{y}(\gamma_\mu{}^m) .$$

D'autre part, on a

$$0 = \mathfrak{x}(1) = \mathfrak{x}(\gamma_\mu^{-e_\mu}) = \sum_{m=1}^{e_\mu-1} M(\gamma_\mu^{-m})\mathfrak{x}(\gamma_\mu^{-1}) ;$$

de là et de (9), on obtient

$$(18) \qquad {}^t\mathfrak{x}(\gamma_\mu^{-1})P\mathfrak{g}(t_\mu) = {}^t\mathfrak{x}(\gamma_\mu^{-1})P[e_\mu^{-1}\sum_{m=1}^{e_\mu-1}\mathfrak{y}(\gamma_\mu{}^m)] .$$

Posons $\mathfrak{g}(u_0) = \mathfrak{r}$; on a alors $\mathfrak{g}(\rho(u_0)) = M(\rho)\mathfrak{r}+\mathfrak{y}(\rho)$ pour tout $\rho\in G$. Au moyen de cette relation et (18), on a

$$(19) \qquad (\omega, \eta) = \sum_{\lambda=1}^{g}{}' \,{}^t\mathfrak{x}(\alpha_\lambda^{-1})P[\mathfrak{y}(\alpha_\lambda^{-1}\beta_\lambda\alpha_\lambda\tau_{\lambda-1})-\mathfrak{y}(\tau_{\lambda-1})]$$

$$+ \sum_{\lambda=1}^{g}{}' \,{}^t\mathfrak{x}(\beta_\lambda)P[\mathfrak{y}(\beta_\lambda\alpha_\lambda\tau_{\lambda-1})-\mathfrak{y}(\alpha_\lambda^{-1}\beta_\lambda\alpha_\lambda\tau_{\lambda-1})]$$

$$+ \sum_{\mu=1}^{f}{}' \,{}^t\mathfrak{x}(\gamma_\mu^{-1})P[e_\mu^{-1}\sum_{m=1}^{e_\mu-1}\mathfrak{y}(\gamma_\mu{}^m)-\mathfrak{y}(\gamma_{\mu-1}\cdots\gamma_1\tau_g)]$$

$$- \sum_{\nu=1}^{h}{}' \,{}^t\mathfrak{x}(\delta_\nu^{-1})P\mathfrak{y}(\delta_{\nu-1}\cdots\delta_1\gamma_f\cdots\gamma_1\tau_g)+\sum_{\nu=1}^{h}{}' \,{}^t\mathfrak{x}(\delta_\nu^{-1})P\mathfrak{g}(s_\nu)+\psi(\mathfrak{r}),$$

où $\psi(\mathfrak{r})$ est le nombre donné par

$$\psi(\mathfrak{r}) = \sum_{\lambda=1}^{g} {}^{t}\mathfrak{x}(\alpha_{\lambda}^{-1})P[M(\alpha_{\lambda}^{-1}\beta_{\lambda}\alpha_{\lambda}\tau_{\lambda-1})-M(\tau_{\lambda-1})]\mathfrak{r}$$

$$+ \sum_{\lambda=1}^{g} {}^{t}\mathfrak{x}(\beta_{\lambda})P[M(\beta_{\lambda}\alpha_{\lambda}\tau_{\lambda-1})-M(\alpha_{\lambda}^{-1}\beta_{\lambda}\alpha_{\lambda}\tau_{\lambda-1})]\mathfrak{r}$$

$$- \sum_{\mu=1}^{j} {}^{t}\mathfrak{x}(\gamma_{\mu}^{-1})PM(\gamma_{\mu-1}\cdots\gamma_{1}\tau_{g})\mathfrak{r}$$

$$- \sum_{\nu=1}^{h} {}^{t}\mathfrak{x}(\delta_{\nu}^{-1})PM(\delta_{\nu-1}\cdots\delta_{1}\gamma_{j}\cdots\gamma_{1}\tau_{g})\mathfrak{r} \,.$$

On va maintenant démontrer $\psi(\mathfrak{r}) = 0$. D'après la relation (10), on a

$$\psi(\mathfrak{r}) = {}^{t}\psi(\mathfrak{r}) = {}^{t}\mathfrak{r}^{t}P\big(\sum_{\lambda=1}^{g} \mathfrak{c}_{\lambda} - \sum_{\mu=1}^{j} \mathfrak{b}_{\mu} - \sum_{\nu=1}^{h} \mathfrak{e}_{\nu} \big) \,,$$

où $\mathfrak{c}_{\lambda}, \mathfrak{b}_{\mu}, \mathfrak{e}_{\nu}$ sont les vecteurs donnés par

$$\mathfrak{c}_{\lambda} = M(\tau_{\lambda-1}^{-1})[M(\alpha_{\lambda}^{-1}\beta_{\lambda}^{-1}\alpha_{\lambda})\mathfrak{x}(\alpha_{\lambda}^{-1})-\mathfrak{x}(\alpha_{\lambda}^{-1})$$

$$+M(\alpha_{\lambda}^{-1}\beta_{\lambda}^{-1})\mathfrak{x}(\beta_{\lambda})-M(\alpha_{\lambda}^{-1}\beta_{\lambda}^{-1}\alpha_{\lambda})\mathfrak{x}(\beta_{\lambda})] \,,$$

$$\mathfrak{b}_{\mu} = M(\tau_{g}^{-1}\gamma_{1}^{-1}\cdots\gamma_{\mu-1}^{-1})\mathfrak{x}(\gamma_{\mu}^{-1}) \,,$$

$$\mathfrak{e}_{\nu} = M(\tau_{g}^{-1}\gamma_{1}^{-1}\cdots\gamma_{j}^{-1}\delta_{1}^{-1}\cdots\delta_{\nu-1}^{-1})\mathfrak{x}(\delta_{\nu}^{-1}) \,.$$

On voit facilement

$$\mathfrak{c}_{\lambda} = -M(\tau_{\lambda-1}^{-1})\mathfrak{x}(\alpha_{\lambda}^{-1}\beta_{\lambda}^{-1}\alpha_{\lambda}\beta_{\lambda}) = \mathfrak{x}(\tau_{\lambda-1}^{-1})-\mathfrak{x}(\tau_{\lambda}^{-1}) \,,$$

$$-\mathfrak{b}_{\mu} = \mathfrak{x}(\tau_{g}^{-1}\gamma_{1}^{-1}\cdots\gamma_{\mu-1}^{-1})-\mathfrak{x}(\tau_{g}^{-1}\gamma_{1}^{-1}\cdots\gamma_{\mu}^{-1}) \,,$$

$$-\mathfrak{e}_{\nu} = \mathfrak{x}(\tau_{g}^{-1}\gamma_{1}^{-1}\cdots\gamma_{j}^{-1}\delta_{1}^{-1}\cdots\delta_{\nu-1}^{-1})-\mathfrak{x}(\tau_{g}^{-1}\gamma_{1}^{-1}\cdots\gamma_{j}^{-1}\delta_{1}^{-1}\cdots\delta_{\nu}^{-1}) \,.$$

D'après la relation $\tau_{g}^{-1}\gamma_{1}^{-1}\cdots\gamma_{j}^{-1}\delta_{1}^{-1}\cdots\delta_{h}^{-1} = 1$, on obtient

$$\psi(\mathfrak{r}) = 0 \,.$$

§5. L'isomorphisme de $S_m(G)$ sur le groupe de cohomologie.

Soit n un entier pair non-négatif; soit U une matrice inversible de degré $n+1$ à coefficients réels; posons

$$P = {}^{t}U^{-1}P_n U^{-1}, \quad M(\sigma) = UM_n(\sigma)U^{-1}$$

pour $\sigma \in \mathrm{GL}(2, \boldsymbol{R})$, M_n étant la représentation définie au §1; M est une représentation réelle de $\mathrm{GL}(2, \boldsymbol{R})$, et on a

$${}^{t}M(\sigma)PM(\sigma) = P$$

pour tout $\sigma \in \mathrm{SL}(2, \boldsymbol{R})$. Soit f un élément de $S_{n+2}(G)$; considérons la forme différentielle vectorielle $\mathrm{Re}(Ufd\mathfrak{z}_n)$ sur $\mathcal{H}$, où $d\mathfrak{z}_n$ est la forme dèfinie au §2.

Comme $fd_{\partial n}$ est holomorphe sur $\mathcal{H}$, on a $d(\mathrm{Re}(Ufd_{\partial n}))=0$; en vertu de (5), on a

$$\mathrm{Re}(Ufd_{\partial n})\circ\sigma = M(\sigma)\mathrm{Re}(Ufd_{\partial n})$$

pour tout $\sigma\in G$; la propriété (P3) de la forme parabolique f entraine la condition (D3) pour $\mathrm{Re}(Ufd_{\partial n})$. Il en résulte que $\mathrm{Re}(Ufd_{\partial n})$ est un élément de $F(M)$. On a vu au §4 que la période de l'intégrale d'un élément de $F(M)$ définit un élément de $H(M, \boldsymbol{R})$. Désignons par $\varphi(f)$ l'élément de $H(M, \boldsymbol{R})$ correspondant à la forme $\mathrm{Re}(Ufd_{\partial n})$; φ est alors une application $\boldsymbol{R}$-linéaire de $S_{n+2}(G)$ dans $H(M, \boldsymbol{R})$.

Théorème 1. *Les notations étant comme ci-dessus, φ est un isomorphisme de $S_{n+2}(G)$ sur $H(M, \boldsymbol{R})$.*

D'après un résultat d'Eichler [1], les espaces $S_{n+2}(G)$ et $H(M, \boldsymbol{R})$ sont de la même dimension sur $\boldsymbol{R}$. Il nous faut donc seulement démontrer que le noyau de φ est $\{0\}$. Soient f, g deux éléments de $S_{n+2}(G)$; posons $\omega = \mathrm{Re}(Ufd_{\partial n})$, $\eta = \mathrm{Re}(Ugd_{\partial n})$. On a alors, d'après la formule (8) de §2,

$$(20) \qquad \Lambda(f, g) = (-1)^{\frac{n}{2}+1}(\omega, \eta).$$

La valeur $\Lambda(f, g)$ est donc donnée par (19). Supposons qu'on ait $\varphi(f)=0$. Alors on peut choisir la constante $\mathfrak{a}$ de l'intégrale (11) de telle manière que le cocycle parabolique $\mathfrak{x}$ déterminé par (13) soit 0. D'après (19) et (20), on a $\Lambda(f, g)=0$ pour tout $g\in S_{n+2}(G)$; en particulier, on a $\Lambda(f, if)=0$, de sorte qu'on a $f=0$; ce qui complète la démonstration.

§6. Les variétés abéliennes attachées aux formes paraboliques.

Considérons maintenant le cas où la représentation M_n de G satisfait à la condition suivante.

(A) *Il existe une matrice inversible U de degré $n+1$ à coefficients réels telle que ${}^t U^{-1}P_n U^{-1}$ et $UM_n(\sigma)U^{-1}$ pour tout $\sigma\in G$ soient des matrices à coefficients entiers.*

Posons

$$P = {}^t U^{-1}P_n U^{-1}, \quad M(\sigma) = UM_n(\sigma)U^{-1}$$

pour tout $\sigma\in \mathrm{GL}(2, \boldsymbol{R})$. Soit φ l'isomorphisme de $S_{n+2}(G)$ sur $H(M, \boldsymbol{R})$ du théorème 1; posons

$$D_{n+2}(G) = \varphi^{-1}[H(M, Z)].$$

On voit alors, d'après la proposition 1, que $D_{n+2}(G)$ est un lattice dans l'espace vectoriel $S_{n+2}(G)$, de sorte que l'espace quotient $S_{n+2}(G)/D_{n+2}(G)$ est un tore complexe. Nous allons maintenant démontrer notre théorème principal.

Théorème 2. *Les notations et les hypothèses étant comme ci-dessus, il existe un entier positif c tel que la forme c(f, g) donne une forme de Riemann sur* $S_{n+2}(G)/D_{n+2}(G)$.

Il en résulte que $S_{n+2}(G)/D_{n+2}(G)$ *a une structure de variété abélienne.*

On entend ici par une forme de Riemann sur le tore complexe $S_m(G)/D_m(G)$ une forme hermitienne positive sur $S_m(G)$ dont la partie imaginaire est à valeurs entières sur $D_m(G) \times D_m(G)$; on s'en réfère à Weil [7].

Le théorème sera démontré si nous faisons voir que la valeur $\Lambda(f, g)$ est rationnelle pour tout $f, g \in D_{n+2}(G)$. Pour cela, on se sert de la formule (19). Soient f, g deux éléments de $D_{n+2}(G)$. Posons $\omega = \mathrm{Re}(Ufd_{3n})$, $\eta = \mathrm{Re}(Ugd_{3n})$, et définissons pour ω, η les fonctions $\mathfrak{f}, \mathfrak{g}$ et les cocycles paraboliques $\mathfrak{x}, \mathfrak{y}$ comme dans le §4. D'après la définition de $D_{n+2}(G)$, on peut prendre les vecteurs $\mathfrak{a}, \mathfrak{b}$ dans (11), (15) de telle façon que $\mathfrak{x}(\sigma), \mathfrak{y}(\sigma)$ soient des vecteurs à composantes entières pour tout $\sigma \in G$. $\mathfrak{x}, \mathfrak{y}$ étant définis ainsi, la valeur $\Lambda(f, g)$ est donnée par (19) et (20). Il est clair que les quatre premières sommes de (19) sont des nombres rationnelles. Comme les δ_ν sont des transformations paraboliques, il existe, pour chaque ν, un vecteur $\mathfrak{v}_\nu$ tel que l'on ait

$$\mathfrak{x}(\delta_\nu{}^{-1}) = [1 - M(\delta_\nu{}^{-1})]\mathfrak{v}_\nu.$$

Comme les composantes du vecteur $\mathfrak{x}(\delta_\nu{}^{-1})$ sont entières, on peut supposer que les composantes de $\mathfrak{v}_\nu$ sont des nombres rationnels. On voit alors que la cinquième somme de (19) est égale à

$$\sum_{\nu=1}^{h} {}^t\mathfrak{v}_\nu{}^t(1 - M(\delta_\nu{}^{-1}))P\mathfrak{g}(s_\nu) = \sum_{\nu=1}^{h} {}^t\mathfrak{v}_\nu P(1 - M(\delta_\nu))\mathfrak{g}(s_\nu)$$

$$= \sum_{\nu=1}^{h} {}^t\mathfrak{v}_\nu P\mathfrak{y}(\delta_\nu);$$

ce nombre est évidemment rationnel. Quant au dernier terme $\psi(\mathfrak{r})$ de (19), on a déjà vu qu'il est égal à 0. Il en résulte que la valeur $\Lambda(f, g)$ est rationnelle pour $f, g \in D_{n+2}(G)$; ceci achève la démonstration.

On voit que, si $n = 0$, la variété abélienne $S_2(G)/D_2(G)$ n'est autre que la variété jacobienne du corps des fonctions automorphes par rapport à G.

Soit G' un sous-groupe de G d'indice fini; on observe que $S_m(G)$ est contenu dans $S_m(G')$ et $D_m(G)$ est contenu dans $D_m(G')$. Par suite il existe une isogénie de la variété abélienne $S_m(G)/D_m(G)$ sur une sous-variété abélienne de $S_m(G')/D_m(G')$.

§7. L'algèbre des transformations.

Soit $\mathfrak{G}$ un groupe quelconque; deux sous-groupes G, G' de $\mathfrak{G}$ sont dits commensurables si l'intersection $G \cap G'$ est d'indice fini dans G et dans G'. Fixons un sous-groupe G de $\mathfrak{G}$; soit $\tilde{G}$ l'ensemble de tous les éléments ρ de $\mathfrak{G}$ tels que $\rho^{-1}G\rho$ soit commensurable avec G. Il est facile de voir que $\tilde{G}$

est un sous-groupe de $\mathfrak{G}$ contenant G. Nous nous proposons maintenant de construire, d'après une idée de A. Weil, une algèbre $\mathfrak{A}$ à partir des éléments de $\tilde{G}$. Soit $\mathfrak{M}$ l'ensemble des classes doubles $G\rho G$ pour $\rho \in \tilde{G}$, et soit $\mathfrak{A}$ le $\mathbf{Z}$-module libre engendré par les éléments de $\mathfrak{M}$. On va définir une loi de multiplication sur $\mathfrak{A}$. On observe d'abord que, pour tout $\rho \in G$, le nombre des classes à droite $G\alpha$ contenues dans $G\rho G$ est fini et égal à l'indice $[G : G \cap \rho^{-1}G\rho]$; en effet, comme $\rho^{-1}G\rho$ est commensurable avec G, il existe un nombre fini d'éléments $\{\gamma_\nu\}$ tel que

$$G = \bigcup_\nu (G \cap \rho^{-1}G\rho)\gamma_\nu .$$

On a alors

$$G\rho G = \bigcup_\nu G\rho\gamma_\nu .$$

Il est facile de voir que les $G\rho\gamma_\nu$ sont distincts si les $(G \cap \rho^{-1}G\rho)\gamma_\nu$ sont distincts.

Lemme 1. *Soient $A, B = G\psi G, C$ trois éléments de $\mathfrak{M}$ et $G\omega$ une classe à droite suivant G contenue dans C. Soit $\{\psi_1 G, \cdots, \psi_s G\}$ le système complet des classes à gauche suivant G contenues dans $G\psi^{-1}G$. Alors, le nombre de ν tel que l'on ait $G\omega\psi_\nu G = A$ ne dépend que de A, B, C et non du choix de $\omega, \{\psi_\nu\}$.*

Il est clair que le nombre en question ne dépend pas du choix de $\{\psi_\nu\}$. Considérons donc une autre classe à droite $G\omega'$ contenue dans C. Alors, il existe un élément $\gamma \in G$ tel que $G\omega = G\omega'\gamma$. On voit que l'ensemble $\{\gamma\psi_1 G, \cdots, \gamma\psi_s G\}$ coïncide, pris dans sa totalité, avec $\{\psi_1 G, \cdots, \psi_s G\}$, et on a $(G\omega')(\gamma\psi_\nu G) = G\omega\psi_\nu G$, d'où résulte l'assertion du lemme.

On désignera par $(A \cdot B, C)$ le nombre du lemme 1.

Lemme 2. *Soient $A = G\rho G, B = G\psi G$ deux éléments de $\mathfrak{M}$; soient $\{G\rho_1, \cdots, G\rho_r\}$ et $\{G\psi_1, \cdots, G\psi_s\}$ respectivement les systèmes complets des classes à droite suivant G contenues dans A et B. Alors, le nombre de (μ, ν) tel que $G\rho_\mu\psi_\nu = G\rho\psi$ est égal à $(A \cdot B, G\rho\psi G)$.*

Supposons qu'on ait $G\rho_\mu\psi_\nu = G\rho\psi$; on a alors $G\rho\psi\psi_\nu^{-1} = G\rho_\mu$ et donc $G\rho\psi\psi_\nu^{-1}G = G\rho_\mu G = G\rho G$. Réciproquement, si l'on a $G\rho\psi\psi_\nu^{-1}G = G\rho G$, la classe à droite $G\rho\psi\psi_\nu^{-1}$ coïncide avec $G\rho_\mu$ pour un et un seul μ, de sorte qu'alors on a $G\rho\psi = G\rho_\mu\psi_\nu$. Par suite, le nombre de (μ, ν) tel que $G\rho_\mu\psi_\nu = G\rho\psi$ est égal au nombre de ν tel que

$$(G\rho\psi)\psi_\nu^{-1}G = G\rho G ,$$

c'est à dire $(A \cdot B, G\rho\psi G)$, compte tenu du fait que $\{\psi_1^{-1}G, \cdots, \psi_s^{-1}G\}$ donne le système complet des classes à gauche suivant G contenues dans $G\psi^{-1}G$.

Les notations étant comme dans le lemme 2, on a

$$G\rho G\psi G = \bigcup_{\mu, \nu} G\rho_\mu\psi_\nu .$$

D'après le lemme 2, la classe $G\rho_\alpha\psi_\beta$ figure justement $(A\cdot B, G\rho_\alpha\psi_\beta G)$ fois parmi $\{G\rho_\mu\psi_\nu\}$. Or on définit le produit $A\cdot B$ par

$$A\cdot B = \sum (A\cdot B, C)C,$$

où la somme est étendue à toutes les classes $C = G\tau G$ contenues dans $G\rho G\psi G$. En prolongeant par linéarité cette opération sur $\mathfrak{A}\times\mathfrak{A}$, on obtient une structure d'anneau sur $\mathfrak{A}$. Pour démontrer l'associativité, considérons un espace $\mathfrak{E}$ sur lequel $\tilde{G}$ opère à gauche. Désignons par $\mathfrak{D} = \mathfrak{D}(\mathfrak{E}/G)$ le $\boldsymbol{Z}$-module libre engendré par les éléments de $\mathfrak{E}/G$. Chaque élément $A = G\rho G$ de $\mathfrak{M}$ donne alors un endomorphisme de $\mathfrak{D}$ comme suit. Notons π l'application canonique de $\mathfrak{E}$ sur $\mathfrak{E}/G$. Soit p un point de $\mathfrak{E}/G$ et soit ξ un point de $\mathfrak{E}$ tel que $\pi(\xi) = p$. $\{\rho_\mu\}$ étant comme plus haut, posons

$$A(p) = \sum_\mu \pi(\rho_\mu\xi).$$

On voit aisément que cette définition ne dépend pas du choix de $\{\rho_\mu\}$. On prolonge, par linéarité, l'opération sur $\mathfrak{A}\times\mathfrak{D}$. D'après notre définition et d'après le lemme 2, on a $A\cdot B(p) = A(B(p))$ pour tout $A, B\in\mathfrak{M}$ et tout $p\in\mathfrak{E}/G$. On obtient ainsi une représentation de l'anneau $\mathfrak{A}$ par des endomorphismes du module $\mathfrak{D}$. Prenons en particulier $\tilde{G}$ pour $\mathfrak{E}$. On vérifie alors facilement que la représentation est fidèle, ce qui démontre l'associativité de l'anneau $\mathfrak{A}$. On appelle $\mathfrak{A}$ *l'anneau des transformations de G par rapport à* $\mathfrak{G}$.

Considérons le cas où G est un groupe fuchsien sur $\mathcal{H}$ et $\mathfrak{G} = \mathrm{SL}(2, \boldsymbol{R})/\{\pm I_2\}$. Soit $\mathcal{H}^*$ la réunion de $\mathcal{H}$ et de l'ensemble des points paraboliques pour G. On observe alors que $\tilde{G}$ opère sur $\mathcal{H}^*$. Soit $\mathfrak{E}$ une courbe algébrique dont le corps des fonctions est le corps des fonctions automorphes par rapport à G. Alors, l'espace quotient $\mathcal{H}^*/G$ s'identifie avec $\mathfrak{E}$ et le module $\mathfrak{D}(\mathcal{H}^*/G)$ avec le module des diviseurs de la courbe $\mathfrak{E}$. Soit π l'application canonique de $\mathcal{H}^*$ sur $\mathfrak{E}$. $\mathfrak{M}, \mathfrak{A}$ étant définis pour G comme plus haut, soit $A = G\rho G$ un élément de $\mathfrak{M}$. On vérifie facilement que l'ensemble des points de $\mathfrak{E}\times\mathfrak{E}$ de la forme $\pi(z)\times\pi(\rho(z))$ pour $z\in\mathcal{H}^*$ est une sous-variété X de $\mathfrak{E}\times\mathfrak{E}$, de dimension 1; X ne dépend que de A et non du choix de ρ. On peut démontrer sans peine que l'application $A\to X$ donne un isomorphisme de l'anneau $\mathfrak{A}$ dans l'anneau des correspondances algébriques de la courbe $\mathfrak{E}$. L'opération de A sur $\mathfrak{D}(\mathcal{H}^*/G)$ donnée plus haut coïncide avec l'opération de X définie dans Weil [8]. On voit de plus que la transposée X' de X correspond à la classe $G\rho^{-1}G$.

§8. Les opérateurs de Hecke.

Soit G un sous-groupe discret de $\mathrm{SL}(2, \boldsymbol{R})$ tel que $\mathcal{H}/G$ ait une mesure finie. Nous allons maintenant définir des représentations de l'anneau des transformations de G dans $S_m(G)$.

Soit ρ un élément de $\mathrm{GL}_+(2, \boldsymbol{R})$ tel que $\rho^{-1}G\rho$ soit commensurable avec G. Comme on a vu au §7, il existe un système $\{\rho_\mu\}$ des éléments en nombre fini tel que l'on ait

$$G\rho G = \bigcup_\mu G\rho_\mu$$

et que les $G\rho_\mu$ soient distincts. Soit f un élément de $S_m(G)$; on définit $g = f \cdot \mathfrak{X}$ par

$$g(z) = (\det \rho)^{m-1} \sum_\mu f(\rho_\mu(z)) J(\rho_\mu, z)^m .$$

On constate que g est un élément de $S_m(G)$ et que $\mathfrak{X}$ ne dépend que de la classe $G\rho G$. On posera $\mathfrak{X} = \mathfrak{X}(\rho) = \mathfrak{X}(G\rho G)$. On a, pour tout $\rho \in \mathrm{GL}_+(2, \boldsymbol{R})$ et pour tout nombre positif a,

$$\mathfrak{X}(a\rho) = a^{m-2}\mathfrak{X}(\rho) .$$

Soit $\mathfrak{A}$ l'anneau des transformations de G par rapport à $\mathrm{SL}(2, \boldsymbol{R})$. Alors, on déduit de notre définition de $\mathfrak{A}$ et de $\mathfrak{X}$, que l'application $G\rho G \to \mathfrak{X}(G\rho G)$ donne une représentation de $\mathfrak{A}$ dans $S_m(G)$.

Proposition 2. *Soit ρ un élément de* $\mathrm{GL}_+(2, \boldsymbol{R})$; *posons*

$$\rho^* = (\det \rho)\rho^{-1} .$$

Alors, on a, pour tout $f, g \in S_m(G)$,

$$(f \cdot \mathfrak{X}(\rho), g) = (f, g \cdot \mathfrak{X}(\rho^*)) .$$

Posons $m = n+2$, $\mathfrak{X} = \mathfrak{X}(\rho)$, $\mathfrak{X}^* = \mathfrak{X}(\rho^*)$ et $q = \det \rho$. On a alors

$$(21) \qquad (f \cdot \mathfrak{X}) d_{\mathfrak{z}n} = q^n \sum_\mu M_n(\rho_\mu^{-1})(f d_{\mathfrak{z}n}) \circ \rho_\mu ,$$

$\{\rho_\mu\}$ étant comme plus haut. On peut supposer qu'on a $\rho^{-1}\rho_\mu \in G$ pour tout μ. Soit $\mathscr{F}$ un domaine fondamental de G. D'après la formule (6), on a

$$-(2i)^{n+1}(f \cdot \mathfrak{X}, g) = q^n \sum_\mu \int_{\mathscr{F}} {}^t[(f d_{\mathfrak{z}n}) \circ \rho_\mu]\, {}^t M_n(\rho_\mu^{-1}) P_n \overline{g d_{\mathfrak{z}n}}$$

$$= \sum_\mu \int_{\rho_\mu(\mathscr{F})} {}^t(f d_{\mathfrak{z}n}) P_n M_n(\rho_\mu)(\overline{g d_{\mathfrak{z}n}}) \circ \rho_\mu^{-1}$$

$$= \sum_\mu \int_{\rho_\mu(\mathscr{F})} {}^t(f d_{\mathfrak{z}n}) P_n M_n(\rho)(\overline{g d_{\mathfrak{z}n}}) \circ \rho^{-1} .$$

Soit $\mathscr{F}_1$ un domaine fondamental de $G \cap \rho G\rho^{-1}$; la dernière somme est alors égale à $\int_{\mathscr{F}_1} {}^t(f d_{\mathfrak{z}n}) P_n M_n(\rho)(\overline{g d_{\mathfrak{z}n}}) \circ \rho^{-1}$. On obtient de même, en désignant par $\mathscr{F}_2$ un domaine fondamental de $G \cap \rho^* G\rho^{*-1}$,

$$-(2i)^{n+1}(f,\ g\cdot\mathfrak{X}^*)=\int_{\mathfrak{F}_2}{}^t[(fd\mathfrak{z}_n)\circ\rho^{*-1}]^t M_n(\rho^*)P_n\overline{g d\mathfrak{z}_n}\,.$$

Comme on a $G\cap\rho^*G\rho^{*-1}=\rho^{-1}(G\cap\rho G\rho^{-1})\rho$, $\rho(\mathfrak{F}_2)$ est un domaine fondamental de $G\cap\rho G\rho^{-1}$. On en déduit l'assertion de la proposition.

On va maintenant étudier l'effet de l'opérateur $\mathfrak{X}=\mathfrak{X}(\rho)$ sur les périodes. Soit f un élément de $S_m(G)$ et $g=f\cdot\mathfrak{X}$. Les notations U,P,M étant comme dans le §5, posons $m=n+2$ et

$$\omega=\mathrm{Re}(Ufd\mathfrak{z}_n)\,,\quad \eta=\mathrm{Re}(Ugd\mathfrak{z}_n)\,.$$

$\{\rho_\mu\}$ étant comme précédemment, on a

$$\eta=q^n\sum_\mu M(\rho_\mu^{-1})\omega\circ\rho_\mu\,,$$

ou $q=\det\rho$. Définissons les fonctions $\mathfrak{f},\mathfrak{g}$ et les cocycles paraboliques $\mathfrak{x},\mathfrak{y}$ comme dans le §4, avec les constantes d'intégration $\mathfrak{a},\mathfrak{b}$ et le point de départ z_0. On a alors, pour chaque $\sigma\in G$,

$$\begin{aligned}
\mathfrak{y}(\sigma)-[1-M(\sigma)]\mathfrak{b}&=\int_{z_\bullet}^{\sigma(z_\bullet)}\eta\\
&=q^n\sum_\mu M(\rho_\mu^{-1})\int_{\rho_\mu(z_\bullet)}^{\rho_\mu\sigma(z_\bullet)}\omega\\
&=q^n\sum_\mu M(\rho_\mu^{-1})[\mathfrak{f}(\rho_\mu\sigma(z_0))-\mathfrak{f}(\rho_\mu(z_0))]\,.
\end{aligned}$$

D'après notre définition de ρ_μ, il existe un indice $\kappa(\mu)$ et un élément τ_μ de G tels que $\rho_\mu\sigma=\tau_\mu\rho_{\kappa(\mu)}$; et $\mu\to\kappa(\mu)$ donne une permutation de $\{1,\cdots,s\}$, où s est le nombre des éléments ρ_μ. On a donc

$$\begin{aligned}
\mathfrak{f}(\rho_\mu\sigma(z_0))&=\mathfrak{f}(\tau_\mu\rho_{\kappa(\mu)}(z_0))\\
&=M(\tau_\mu)\mathfrak{f}(\rho_{\kappa(\mu)}(z_0))+\mathfrak{x}(\tau_\mu)\,.
\end{aligned}$$

Par suite on a

$$\mathfrak{y}(\sigma)=q^n\sum_\mu M(\rho_\mu^{-1})\mathfrak{x}(\tau_\mu)+\mathfrak{r}(\sigma)\,,$$

où $\mathfrak{r}(\sigma)$ est donné par

$$\mathfrak{r}(\sigma)=[1-M(\sigma)]\mathfrak{b}+q^n\sum_\mu M(\rho_\mu^{-1})[M(\tau_\mu)\mathfrak{f}(\rho_{\kappa(\mu)}(z_0))-\mathfrak{f}(\rho_\mu(z_0))]\,.$$

Posons

$$\mathfrak{c}=\mathfrak{b}-q^n\sum_\mu M(\rho_\mu^{-1})\mathfrak{f}(\rho_\mu(z_0))\,.$$

On a alors, au moyen de la relation $\rho_\mu^{-1}\tau_\mu=\sigma\rho_{\kappa(\mu)}^{-1}$,

$$\mathfrak{r}(\sigma)=[1-M(\sigma)]\mathfrak{c}\,;$$

autrement dit, $\mathfrak{r}(\sigma)$ est un cobord. Par conséquent, si l'on pose

$$\mathfrak{i}(\sigma)=q^n\sum_\mu M(\rho_\mu^{-1})\mathfrak{x}(\tau_\mu)\,,$$

$\mathfrak{k}(\sigma)$ est un cocycle parabolique et les classes de cohomologie de $\mathfrak{h}$ et de $\mathfrak{k}$ sont les mêmes. Considérons maintenant le cas où M_n satisfait à la condition (A) du § 6. On obtient alors

Théorème 3. *Les notations étant comme ci-dessus, supposons que les matrices P et $M(\sigma)$ pour $\sigma \in G$ soient des matrices à coefficients entiers. Si les coefficients de $M((\det \rho)\rho^{-1})$ sont entiers, $\mathfrak{X}(\rho)$ donne un endomorphisme de la variété abélienne $S_m(G)/D_m(G)$.*

En effet, on a

$$q^n \sum_\mu M(\rho_\mu^{-1})\mathfrak{x}(\tau_\mu) = \sum_\mu M(q\rho_\mu^{-1})\mathfrak{x}(\tau_\mu) \, .$$

On voit que, si $M(q\rho^{-1})$ et les $M(\sigma)$ pour $\sigma \in G$ sont à coefficients entiers, il l'est aussi pour les $M(\rho_\mu)$. Par suite, on observe que, si $\mathfrak{x}(\sigma)$ est un élément de $Z(M, \mathbf{Z})$, $\mathfrak{k}(\sigma)$ est contenu dans $Z(M, \mathbf{Z})$; on en déduit que $\mathfrak{X}(\rho)$ envoie $D_m(G)$ dans lui-même; ceci démontre le théorème.

Corollaire. *Les notations et les hypothèses étant comme dans le théorème 3, les racines caractéristiques de $\mathfrak{X}(\rho)$ sont des entiers algébriques de degré $2h(m)$, où $h(m)$ désigne la dimension de $S_m(G)$ sur C.*

C'est une conséquence immédiate du théorème 3. On voit d'ailleurs, que, si $M(\rho)$ et $M(\rho^*)$ sont tous les deux à coefficients entiers, $\mathfrak{X}(\rho)$ et $\mathfrak{X}(\rho^*)$ sont des endomorphismes de la variété $S_m(G)/D_m(G)$, et d'après la proposition 2, ils sont les conjugués l'un de l'autre par rapport à la forme de Riemann $c(f, g)$. En particulier, si l'on a $\mathfrak{X}(\rho) = \mathfrak{X}(\rho^*)$, (ce qui est le cas pour les opérateurs de Hecke ordinaires de « Stufe » 1), les racines caractéristiques de $\mathfrak{X}(\rho)$ sont des entiers algébriques totalement réels de degré $h(m)$.

§ 9. Exemples.

On peut appliquer le théorème 2 à tout sous-groupe de $SL(2, \mathbf{Z})$ d'indice fini, puisque $M_n(\sigma)$ est à coefficients entiers pour tout $\sigma \in SL(2, \mathbf{Z})$ et tout $n \geq 0$. Il se trouve donc qu'on obtient une variété abélienne $S_m(G)/D_m(G)$ pour tout sous-groupe G du groupe $SL(2, \mathbf{Z})$ d'indice fini et pour chaque entier pair positif m. Si ρ est un élément de $GL_+(2, \mathbf{R})$ à coefficients entiers, $M_n(\rho)$ et $M_n((\det \rho)\rho^{-1})$ sont des matrices à coefficients entiers. Par suite, d'après le théorème 3, les opérateurs $\mathfrak{X}(\rho), \mathfrak{X}(\rho^*)$ donnent des endomorphismes de la variété abélienne $S_m(G)/D_m(G)$.

Supposons qu'on ait $\tau^{-1}G\tau = G$ pour un élément $\tau \in SL(2, \mathbf{Z})$. Alors, on obtient un automorphisme $\mathfrak{U}$ de $S_m(G)$ en posant

$$f \cdot \mathfrak{U} = f(\tau(z))J(\tau, z)^m$$

pour $f \in S_m(G)$, et on vérifie de même que pour $\mathfrak{X}(\rho)$ que $\mathfrak{U}$ envoie $D_m(G)$ sur $D_m(G)$, de sorte que $\mathfrak{U}$ est un automorphisme de $S_m(G)/D_m(G)$. En particulier, si G est un sous-groupe invariant de SL(2, Z) d'indice fini, tout élément de SL(2, Z) engendre ainsi un automorphisme de $S_m(G)/D_m(G)$. Considérons par exemple le cas où G est le groupe $G(3)$ formé des éléments σ de SL(2, Z) tels que $\sigma \equiv I_2$ (mod. 3). On sait que $S_1(G(3))$ est de dimension 1, et est engendré par $\varDelta(z)^{1/3}$, où $\varDelta(z)$ désigne le discriminant de fonctions elliptiques. On voit que $\begin{pmatrix} 1 & 1 \\ 0 & 1 \end{pmatrix}$ donne un automorphisme d'ordre 3 sur $S_4(G(3))/D_4(G(3))$. On en déduit que la variété abélienne $S_4(G(3))/D_4(G(3))$ est une courbe elliptique à multiplications complexes. Il semble que ce résultat se rattache au fait que la série de Dirichlet correspondant à $\varDelta(z)^{1/3}$ est le produit de deux fonctions L attachées à des « Grössencharakteren » du corps $Q(e^{2\pi i/3})$. Plus généralement, quel que soit le sous-groupe G de SL(2, Z), et même s'il n'y a pas de multiplications complexes, on a le droit de penser qu'il existe une relation profonde entre l'arithmétique des séries de Dirichlet attachées aux formes automorphes et nos variétés abéliennes $S_m(G)/D_m(G)$.

Considérons maintenant le cas où $G = \mathrm{SL}(2, Z)$; posons $G(1) = \mathrm{SL}(2, Z)$; $G(1)$ est engendré par deux éléments

$$\sigma = \begin{pmatrix} 0 & 1 \\ -1 & 0 \end{pmatrix}, \qquad \tau = \begin{pmatrix} 1 & 1 \\ 0 & 1 \end{pmatrix}.$$

Comme on a vu au §3, on peut prendre $H(M_n, R, \tau)$ au lieu de $H(M_n, R)$; et $H(M_n, Z, \tau)$ s'identifie avec un sous-module de $H(M_n, Z)$ d'indice fini. Désignons par $D_m'(G(1))$ le sous-module de $D_m(G(1))$ correspondant à $H(M_{m-2}, Z, \tau)$. Alors, $S_m(G(1))/D_m'(G(1))$ est une variété abélienne isogène à $S_m(G(1))/D_m(G(1))$. Tout élément $\mathfrak{x}$ de $Z(M_n, R, \tau)$ est déterminé par un seul vecteur $\mathfrak{x}(\sigma)$. Posons

$$V_n = \{\mathfrak{x}(\sigma) \mid \mathfrak{x} \in Z(M_n, R, \tau)\},$$
$$W_n = \{\mathfrak{x}(\sigma) \mid \mathfrak{x} \in B(M_n, R, \tau)\}.$$

On voit facilement que V_n est le sous-espace de R^{n+1} formé des vecteurs $\mathfrak{x}$ tels que l'on ait

(22) $$[1 + M_n(\sigma)]\mathfrak{x} = 0,$$

(22') $$[1 + M_n(\sigma\tau) + M_n(\sigma\tau)^2]\mathfrak{x} = 0,$$

et W_n est formé des vecteurs $[1 - M_n(\sigma)]\mathfrak{y}$, $\mathfrak{y}$ étant un vecteur tel que

$$[1 - M_n(\tau)]\mathfrak{y} = 0.$$

D'après notre définition, V_n/W_n est canoniquement isomorphe à $S_{n+2}(G(1))$. On vérifie, par un calcul simple, que la forme alternée $\varLambda(f, g)$ sur $S_{n+2}(G(1))$ correspond à la forme bilinéaire $A(\mathfrak{a}, \mathfrak{b})$ sur V_n/W_n donné par

$$A(\mathfrak{a}, \mathfrak{b}) = c \, {}^t\mathfrak{a}[P_n M_n(\sigma\tau) - M_n(\sigma\tau)P_n]\mathfrak{b},$$

où c est un nombre rationnel.

On va maintenant s'occuper de $S_{12}(G(1))$; $S_{12}(G(1))$ est de dimension 1 et engendré par la fonction Δ :

$$\Delta(z) = q \prod_{n=1}^{\infty} (1-q^n)^{24} : \quad q = e^{2\pi i z} .$$

Posons

$$\mathfrak{h}(z) = \int_{\infty}^{z} \Delta(z) d\mathfrak{z}_{10} .$$

On a alors, pour chaque $\gamma \in G(1)$,

$$\mathfrak{h}(\gamma(z)) = M_{10}(\gamma)\mathfrak{h}(z) + \mathfrak{w}(\gamma) ,$$

où $\mathfrak{w}(\gamma)$ est un cocycle parabolique à *coefficients complexes*. On a evidemment $\mathfrak{w}(\tau) = 0$. On obtient, compte tenu de $\sigma(0) = \infty$,

$$\mathfrak{w}(\sigma) = -\int_{0}^{\infty} \Delta(z) d\mathfrak{z}_{10} ,$$

de sorte que les composantes de $\mathfrak{w}(\sigma)$ sont données par

$$(23) \qquad\qquad i^{1-\lambda}\int_{0}^{\infty} \Delta(iy)y^{10-\lambda}dy \qquad\qquad (0 \leq \lambda \leq 10) .$$

Posons $\Delta(z) = \sum_{n=1}^{\infty} a_n q^n$, $\Phi(s) = \sum_{n=1}^{\infty} a_n n^{-s}$ et $R(s) = (2\pi)^{-s}\Gamma(s)\Phi(s)$; on a alors

$$R(s) = \int_{0}^{\infty} \Delta(iy)y^{s-1}dy .$$

On sait que $R(s)$ est une fonction entière satisfaisant à l'équation fonctionnelle $R(s) = R(12-s)$. D'après (23), les six premières composantes de $\mathfrak{w}(\sigma)$ sont égales à

$$iR(11), \ R(10), \ -iR(9), \ -R(8), \ iR(7), \ R(6) .$$

Comme $\mathfrak{w}(\sigma)$ satisfait à (22) et (22′) pour $n = 10$, on a

$$R(10) = \frac{12}{5} R(6), \ R(8) = \frac{5}{4} R(6), \ R(9) = \frac{14}{9} R(7) .$$

Cette relation est analogue à une relation bien connue pour les valeurs de la fonction ζ de Riemann en les nombres entiers pairs positifs.

Or, on voit que $20R(6)^{-1}\Delta$ et $9R(7)^{-1}i\Delta$ engendrent $D_{12}{}'(G(1))$. Par suite, la courbe elliptique $S_{12}(G(1))/D_{12}{}'(G(1))$ a le module (analytique)

$$\frac{9}{20} \frac{R(6)}{R(7)} i .$$

On peut aborder de même les cas des formes de degré > 12 ou de « Stufe » > 1; cependant un calcul pareil ne nous offre rien sur le caractère algébro-géométrique de la variété $S_m(G)/D_m(G)$.

Nous terminons cette étude en signalant qu'il existe une plus large classe de groupes qui satisfont à la condition (A); c'est la classe des groupes fuchsiens attachés aux formes quadratiques ternaires à coefficients entiers (Poincaré [5], Fricke et Klein [2, p. 502]), parmi lesquels le groupe modulaire SL(2, Z) se trouve comme cas particulier.

Université de Tokyo.

Bibliographie

[1]　M. Eichler, Eine Verallgemeinerung der Abelschen Integrale, Math. Z., 67 (1957), 267-298.

[2]　R. Fricke et F. Klein, Vorlesungen über die Theorie der Automorphen Funktionen I, Leipzig, 1897.

[3]　E. Hecke, Grundlagen einer Theorie der Integralgruppen und der Integralperioden bei den Normalteilern der Modulgruppen, Math. Ann., 116 (1938), 469-510.

[4]　H. Petersson, Konstruktion der sämtlichen Lösungen einer Riemannschen Funktionalgleichung durch Dirichletreihen mit Eulerscher Produktentwicklung I, Math. Ann., 116 (1938), 401-412.

[5]　H. Poincaré, Les fonctions fuchsiennes et l'arithmétique, œuvres, t. II (1887), 463-511.

[6]　A. Weil, Généralisation des fonctions abéliennes, J. Math. pures appl., IX. sér. 17 (1938), 47-87.

[7]　A. Weil, Variétés kählériennes, Hermann, Paris, 1958.

[8]　A. Weil, Sur les courbes algébriques et les variétés qui s'en déduisent, Hermann, Paris, 1948.

On specializations of abelian varieties

Jointly with Shoji Koizumi

Scientific Papers of the College of General Education, University of Tokyo, 9 (1959), 187-211

(Received October 3, 1959)

In the study of specializations of Albanese or Picard varieties as well as in the arithmetic of automorphic functions, there arises a problem which is stated as follows: *Let A and B be abelian varieties defined over a field k with a prime divisor $\mathfrak{p}$. Suppose that there exists a homomorphism of A onto B, defined over k. If A is without defect for $\mathfrak{p}$, then is there an abelian variety which is isomorphic to B over k and is without defect for $\mathfrak{p}$?* Here we say that an abelian variety A is without defect for $\mathfrak{p}$, if the specialization of A with respect to $\mathfrak{p}$ is an abelian variety $\tilde{A}$ and the specialization of the graph of composition-law on A gives that on $\tilde{A}$. The main purpose of the present paper is to solve this problem; an affirmative answer is stated in Theorem 4 (§ 6). Besides this, we shall give, in the first part of the paper, some results which belong to foundations of specialization-theory. We now give a summary of the contents.

Among the fundamental results on abelian varieties, it is known that a rational mapping f of a variety V into an abelian variety A is defined at any simple point on V. In § 1, we shall give a generalization of this result as Theorem 1, which asserts that, if A is without defect for $\mathfrak{p}$ and if the specialization of V with respect to $\mathfrak{p}$ has only one component $\tilde{V}$ of multiplicity 1, then f is defined at any simple point $\bar{a}$ of $\tilde{V}$. Though this result is not needed for the rest of the paper, we have included it in view of future applications. In § 2, we study the expansion of a function f on a variety V, which is defined and finite at a simple point $\bar{a}$ of the specialization $\tilde{V}$ of V, by power-series in local parameters at $\bar{a}$; every coefficient of the power-series is $\mathfrak{p}$-integral and the expansion of the specialization $\tilde{f}$ of f is obtained by the specialization of coefficients of the series for f. This result is used for the specialization-theory of a function-module on V, which is the object of § 3, where we prove that a function-module on V and its specialization are of the same dimension. § 4 is devoted to the study of a problem concerning projective embeddings of an abelian variety without defect for $\mathfrak{p}$ and its field of definition. Problems of this kind are considered for any $\mathfrak{p}$-variety; and one can have a solution generalizing Weil's theory [13], with a certain condition of unramifiedness of $\mathfrak{p}$; in the present paper, however, we have restricted ourselves within the case of abelian varieties. The concept of abstract varieties enables us to construct a group-variety from a pre-group (Weil [10, 11]). In § 5, we shall give a construction of a group $\mathfrak{p}$-variety without defect from a pre-group $\mathfrak{p}$-variety without defect, following the idea of [10]; here is one of the reasons why we have preferred in our treatment abstract and $\mathfrak{p}$-varieties in the sense of [7] to projective varieties. § 6 contains the main theorem, which we have already explained above. In Appendix, it is proved that the specialization ring in the field of functions on V at a simple point $\bar{a}$ of $\hat{V}$ is a regular local ring.

Throughout the paper, we shall freely use the terminologies and results of [7].

§ 1. Specialization of rational mappings

Let k be a field with a discrete valuation of rank 1 $\{\mathfrak{o}, \mathfrak{p}, \tilde{k}\}$ where $\mathfrak{o}, \mathfrak{p}$ and $\tilde{k}$ denote respectively the valuation-ring, the maximal ideal of $\mathfrak{o}$ and the residue-field $\mathfrak{o}/\mathfrak{p}$. We shall consider two kinds of algebraic geometry: the one is the geometry under a universal domain K containing k, and the other is the geometry under another universal domain $\bar{K}$ containing $\tilde{k}$. Throughout the paper, by letters with bars such as $\bar{V}, \bar{x}, \bar{\varphi}, \cdots$, we mean geometric objects in the geometry under $\bar{K}$, and by $V, x, \varphi, \cdots$, those under K. On the other hand, we shall always denote by $\hat{V}, \hat{x}, \hat{\varphi}, \cdots$, specializations of $V, x, \varphi, \cdots$, with respect to $\mathfrak{o}$, where $V, x, \varphi, \cdots$ may be or may not be defined over k; and we shall write $V \to \hat{V}$ ref. $\mathfrak{o}$, etc. If $F(X)$ is a polynomial in $\mathfrak{o}[X]$, we denote by $\hat{F}(X)$ the polynomial in $\tilde{k}[X]$ obtained from F taking the coefficients of F modulo $\mathfrak{p}$. Let $\mathfrak{a}$ be an ideal of $k[X]$; we shall write

$$\mathfrak{a}_0 = \mathfrak{a} \cap \mathfrak{o}[X], \quad \tilde{\mathfrak{a}}_0 = \{\hat{P}(X) \mid P(X) \in \mathfrak{a}_0\}\ .$$

Let V be a $\mathfrak{p}$-variety[1], x a point of V and $\bar{x}$ a specialization of x with respect to $\mathfrak{o}$; take affine representatives $x_\alpha, \bar{x}_\alpha$ of $x, \bar{x}$ and consider the set of elements $F(x_\alpha)/G(x_\alpha)$ such that $\hat{G}(\bar{x}_\alpha) \neq 0$, where $F(X)$ and $G(X)$ are polynomials in $\mathfrak{o}[X]$. This set forms a local ring and is independent of the choice of representatives $x_\alpha, \bar{x}_\alpha$; we denote this local ring by $[x \to \bar{x}; \mathfrak{o}]$.

---

1) Any affine or projective variety defined over k can be regarded as a $\mathfrak{p}$-variety in a natural way; so we shall identify an affine or a projective variety with the $\mathfrak{p}$-varity in this sense.

Now let us consider specialization of rational mappings. Let V and W be $\mathfrak{p}$-varieties and f a rational mapping of V into W defined over k. Let x be a generic point of V over k and $y = f(x)$. We say that f is *defined at* a point $\bar{a}$ on $\tilde{V}$ if there exists a point $\bar{b}$ on $\tilde{W}$ such that

$$[y \to \bar{b}\,;\,\mathfrak{o}] \subset [x \to \bar{a}\,;\,\mathfrak{o}].$$

We can easily verify that $\bar{b}$ is determined only by f and $\bar{a}$; so we write $f(\bar{a}) = \bar{b}$. Suppose that V and W are $\mathfrak{p}$-simple. If $\bar{A}$ is a subvariety in $\tilde{V}$, and if f is defined at a point in $\bar{A}$, we say that f is *defined along* $\bar{A}$. If $\tilde{x}$ is a generic point of $\tilde{V}$ over $\tilde{k}, [x \to \tilde{x}\,;\,\mathfrak{o}]$ is a discrete valuation ring of rank 1 (Prop. 5 of [7]). Hence f is defined along $\tilde{V}$ whenever W is $\mathfrak{p}$-complete. Assuming f to be defined along $\tilde{V}$, put $f(\tilde{x}) = \tilde{y}$ for a generic point $\tilde{x}$ of $\tilde{V}$ over $\tilde{k}$. Then we obtain a rational mapping $\tilde{f}$ of $\tilde{V}$ into $\tilde{W}$ defined by $\tilde{f}(\tilde{x}) = \tilde{y}$ with respect to $\tilde{k}$. We call $\tilde{f}$ the *specialization of f with respect to* $\mathfrak{o}$. We see easily that if f is defined at a point $\bar{a}$ of $\tilde{V}$, $\tilde{f}$ is also defined at $\bar{a}$ and $f(\bar{a}) = \tilde{f}(\bar{a})$.

Now we begin with the study of behaviour of a rational mapping f of V at a simple point $\bar{a}$ on $\tilde{V}$.

PROPOSITION 1. *Let V be a $\mathfrak{p}$-simple $\mathfrak{p}$-variety and f a rational mapping, defined over k, of V into an affine space S, such that f is defined along $\tilde{V}$. Put*

$$S = \{a \mid a \in V, f \text{ is not defined at } a\}\,, \quad \bar{S} = \{\bar{a} \mid \bar{a} \in \tilde{V}, f \text{ is not defined at } \bar{a}\}\,.$$

Then the following assertions hold.

i) *S (resp. $\bar{S}$) is a k-normal (resp. $\tilde{k}$-normal) bunch of subvarieties on V (resp. $\tilde{V}$), and $\tilde{S} \subseteq \bar{S}$.*

ii) *A simple point $\bar{a}$ of $\tilde{V}$ is contained in $\bar{S}$ if and only if $\bar{a}$ is contained in $\tilde{S}$. In particular, if $\tilde{V}$ is non-singular, we have $\bar{S} = \tilde{S}$.*

iii) *There exists a k-normal bunch $\mathfrak{F}$ of subvarieties on V such that $\mathfrak{F} \supset S$, $\tilde{\mathfrak{F}} \supset \bar{S}$ and $\mathfrak{F} \neq V$.*

Proof. It is not difficult to reduce our proposition to the case where V is an affine variety and f is a numerical function, i.e., a rational mapping of V into the affine 1-space; so we shall deal only with such a case. Let x be a generic point of V over k. Define two ideals in $\mathfrak{o}[X]$ or in $k[X]$ by

$$\mathfrak{a}_1 = \{P(X) \mid P(X) \in \mathfrak{o}[X],\, P(x)f(x) \in \mathfrak{o}[x]\}\,,$$

$$\mathfrak{a} = \{P(X) \mid P(X) \in k[X],\, P(x)f(x) \in k[x]\}\,.$$

We have then $\mathfrak{a}_0 \supset \mathfrak{a}_1$ and $\tilde{\mathfrak{a}}_0 \supset \tilde{\mathfrak{a}}_1$. We see easily that

$$S = \text{the set of zero points of } \mathfrak{a},$$

$$\widetilde{S} = \text{the set of zero points of } \widetilde{\mathfrak{a}_0},$$

$$\overline{S} = \text{the set of zero points of } \widetilde{\mathfrak{a}_1}.$$

This proves the assertion i). Let $\bar{x}$ be a generic point of $\widetilde{V}$ over $\tilde{k}$. Since $\overline{S}$ does not contain $\bar{x}$, there exists a polynomial $P(X) \in \mathfrak{a}_1$ such that $\widetilde{P}(\bar{x}) \neq 0$. If we denote by $\mathcal{F}$ the set of points on V where $P(X)$ vanishes, $\mathcal{F}$ satisfies our requirements in iii). The numerical function f defines naturally a rational mapping f_1 of V into the projective 1-space P^1. Let T be the graph of f_1; let S^n and $\bar{S}^n$ be the ambient spaces for V and $\widetilde{V}$. As f is defined along $\widetilde{V}$, any component of $\widetilde{T}$ is not contained in $\bar{S}^n \times \overline{\infty}$, so that the intersection-product $\widetilde{T} \cdot (\bar{S}^n \times \overline{\infty})$ is defined; and $\widetilde{T} \cap (\bar{S}^n \times \overline{\infty})$ is a specialization of $T \cap (S^n \times \infty)$ with respect to $\mathfrak{o}$. Let $\bar{a}$ be a simple point on $\widetilde{V}$. If $\bar{a} \times \overline{\infty}$ is not contained in $\widetilde{T} \cap (\bar{S}^n \times \overline{\infty})$, every specialization of $f(x)$ over $x \to \bar{a}$ ref. $\mathfrak{o}$ is finite; since $[x \to \bar{a}; \mathfrak{o}]$ is integrally closed by Theorem of Appendix, $f(x)$ is contained in $[x \to \bar{a}; \mathfrak{o}]$, so that f is defined at $\bar{a}$. Therefore, if f is not defined at $\bar{a}$, then $\bar{a} \times \overline{\infty}$ must be contained in $\widetilde{T} \cap (\bar{S}^n \times \overline{\infty})$; and we can find a point a in $T \cap (S \times \infty)$ such that $a \to \bar{a}$ ref. $\mathfrak{o}$. We see that f is not defined at a since $a \times \infty$ is a specialization of $x \times f(x)$ over k. Hence we have $a \in S$, so that $\bar{a} \in \widetilde{S}$. This proves ii).

Remark. 1) By the above discussion, we see that every simple component of $S, \bar{S}, \widetilde{S}$ is of codimension 1.

2) The assumption that the image of f is embedded in an affine variety is not necessary for i) and iii); but the assertion ii) requires the assumption.

3) In the proof of ii), we have only needed that $[x \to \bar{a}; \mathfrak{o}]$ is integrally closed. Since $[x \to \bar{a}; \mathfrak{o}]$ is integrally closed if $\bar{a}$ is $\tilde{k}$-normal, (Hironaka [2]), we can replace, in the assertion ii), the simplicity of $\bar{a}$ by the $\tilde{k}$-normality of $\bar{a}$.

4) By ii) and 3), we know that, $\widetilde{V}$ being a $\tilde{k}$-normal affine variety, we have $\widetilde{f}(\bar{x}) \in \tilde{k}[\bar{x}]$ if $f(x) \in k[x]$ is finite on V. This means that the defining ideal of $\widetilde{V}$ is the specialization of that of V with respect to $\mathfrak{o}$. When V and $\widetilde{V}$ are projective varieties, the defining ideal of $\widetilde{V}$ is the intersection of the specialization of the defining ideal of V and an irrelevant ideal in $\tilde{k}[X]$. This is a natural way to the equality of arithmetic genera of V and $\widetilde{V}$ (Igusa [3]).

Let G be a group (resp. an abelian) variety defined over k, having a structure of a $\mathfrak{p}$-variety. Then the notion of *group* (resp. *abelian*) $\mathfrak{p}$-*variety*, denoted by the same letter G, is defined by the combination of the structure of group (resp. abelian) variety and the structure of $\mathfrak{p}$-variety.

Definition 1. Let G be a group $\mathfrak{p}$-variety; let $\varphi : G \times G \to G$ and $\psi : G \to G$ be respectively the group-composition function and the rational mapping which corresponds a point of G to its inverse. We say that G is a *group $\mathfrak{p}$-variety without defect* if the following conditions are satisfied:

1) G is $\mathfrak{p}$-*simple, i.e., $\tilde{G}$ is a variety;*
2) φ *is everywhere defined on $\tilde{G} \times \tilde{G}$;*
3) ψ *is everywhere defined on $\tilde{G}$.*

Moreover, *if G is $\mathfrak{p}$-complete,* we say that G is an *abelian $\mathfrak{p}$-variety without defect.*

From the definition we can easily see that if G is a group (resp. an abelian) $\mathfrak{p}$-variety without defect, $\tilde{G}$ is considered in a natural way to be a group (resp. an abelian) variety defined over $\tilde{k}$ and the specialization of φ and ψ give the corresponding mappings of $\tilde{G}$.

Let G and G' be group $\mathfrak{p}$-varieties, both without defect, and λ a homomorphism of G into G' defined over k. If λ is defined along $\tilde{G}$, λ is everywhere defined on $\tilde{G}$ and $\tilde{\lambda}$ induces a homomorphism $\tilde{G} \to \tilde{G}'$. We shall say that G and G' are isomorphic to each other (with respect to the structure of group $\mathfrak{p}$-varieties) if there exists a surjective isomorphism λ of G onto G' such that λ is defined along $\tilde{G}$ and that $\tilde{\lambda}$ is also an isomorphism of $\tilde{G}$ onto $\tilde{G}'$. We note that if both G and G' are abelian $\mathfrak{p}$-varieties without defect, any group-isomorphism between G and G' in a usual sense is always an isomorphism between abelian $\mathfrak{p}$-varieties G and G'.

PROPOSITION 2. *Let f be a rational mapping, defined over k, of a $\mathfrak{p}$-simple $\mathfrak{p}$-variety V into a group $\mathfrak{p}$-variety G without defect, and let F be a rational mapping of $V \times V$ into G defined by $F(x, y) = f(x)f(y)^{-1}$. Let $\bar{a}$ be a simple point of $\tilde{V}$. Then f is defined at $\bar{a}$ if and only if F is defined at $(\bar{a}, \bar{a})$ and $F(\bar{a}, \bar{a}) = \bar{e}$, where $\bar{e}$ is the identity element of $\tilde{G}$.*

We omit the proof because it is easy and is quite similar to the discussion in $n°$ 15 of Weil [10]; we shall make use of the idea given there in the following treatment.

Notations being as in Prop. 2, let G_α be an affine representative of G such that the corresponding representative $\tilde{G}_\alpha$ of $\tilde{G}$ has the representative of $\bar{e}$. Let $\bar{a}$ be a simple point of $\tilde{V}$. If we denote by F_α the rational mapping $V \times V \to G_\alpha$ induced by F, F is defined at $(\bar{a}, \bar{a})$ and $F(\bar{a}, \bar{a}) = \bar{e}$ if and only if every coordinate-function of F_α is defined at $(\bar{a}, \bar{a})$. Suppose that F_α is not defined at $(\bar{a}, \bar{a})$. By ii) of Prop. 1, there exists a point (a_1, a_2) on $V \times V$ such that F_α is not defined at (a_1, a_2) and $(a_1, a_2) \to (\bar{a}, \bar{a})$ ref. $\mathfrak{o}$. By Remark 1) below Prop. 1, there exists a simple subvariety X of $V \times V$, of codimension 1, containing (a_1, a_2), where F_α is not defined. Let $\tilde{X}$ be a specialization of X over $(a_1, a_2) \to (\bar{a}, \bar{a})$ ref. $\mathfrak{o}$. Let Δ and $\bar{\Delta}$ denote respectively the diagonals on $V \times V$ and on $\tilde{V} \times \tilde{V}$. Then the support of $\tilde{X}$ does not contain $\bar{\Delta}$ because F_α is defined along $\bar{\Delta}$. Hence, both the intersection-products $X \cdot \Delta$ and $\tilde{X} \cdot \bar{\Delta}$ are defined and

$(\bar{a}, \bar{a})$ is contained in a component of $\tilde{X} \cdot \bar{\mathit{\Delta}}$; so there exists a point (a, a) in $X \cap \mathit{\Delta}$ such that $(a, a) \to (\bar{a}, \bar{a})$ ref. $\mathfrak{o}$. As $(\bar{a}, \bar{a})$ is simple on $V \times V$, (a, a) is simple on $V \times V$. F_α is not defined at (a, a) since (a, a) is contained in X. Thus we have shown that if f is not defined at a simple point $\bar{a}$ of $\bar{V}$, there exists a simple point a of V where f is not defined. In view of the results in $n°$ 15 of Weil [10], we have:

THEOREM 1. *Let A be an abelian $\mathfrak{p}$-variety and f a rational mapping, defined over k, of a $\mathfrak{p}$-simple $\mathfrak{p}$-variety V into A. Suppose that A is without defect as group $\mathfrak{p}$-variety and that f is defined along $\tilde{V}$. Then f is defined at any simple point on $\tilde{V}$.*

REMARK. 1) A is not necessarily without defect as abelian $\mathfrak{p}$-variety even if it is without defect as group $\mathfrak{p}$-variety.

2) f is defined along $\tilde{V}$ whenever A is without defect as abelian $\mathfrak{p}$-variety.

§ 2. Specialization-ring at a simple point

Let V be a $\mathfrak{p}$-variety, x a generic point of V over k and $\bar{a}$ a point of $\tilde{V}$ which is simple on V. We can prove that $[x \to \bar{a}; \mathfrak{o}]$ is a regular local ring (Theorem in Appendix); it is not so easy, however, to prove this result. If $\bar{a}$ is $\tilde{k}$-rational, the situation becomes easier and we can consider expansion by power-series in local parameters; so in this section we shall concern ourselves only with such a case. First we give a definition of local parameters at $\bar{a}$, generalizing a definition in Koizumi [4]. Let n be the dimension of V. We say that a set of n rational functions $\{\tau_1, \cdots, \tau_n\}$ on V is a *set of local parameters* on V at $\bar{a}$, if the following conditions are satisfied:

i) *the τ_i are defined and finite at $\bar{a}$.*

ii) *Let $V_\alpha, x_\alpha, \bar{a}_\alpha$ be respectively representatives of $V, x, \bar{a}$ and S^N the ambient space for V_α. Then there exists a set of N polynomials $F_i(X_1, \cdots, X_N, T_1, \cdots, T_n)$ in $\mathfrak{o}[X, T]$ such that $F_i(x, \tau(x)) = 0$ for $1 \le i \le N$ and*

$$\det \left(\frac{\partial \tilde{F}_i}{\partial X_j} (\bar{a}, \tau(\bar{a})) \right) \ne 0.$$

We say that $\{\tau_1, \cdots, \tau_n\}$ is defined over k if the τ_i are all defined over k. We can verify that the condition ii) is independent of the choice of representatives $V_\alpha, x_\alpha, \bar{a}_\alpha$. The existence of a set of local parameters on V at $\bar{a}$ is a direct consequence of the definition of simple point[2].

PROPOSITION 3. *With the same notations and assumptions as above, if $\bar{a}$ is $\tilde{k}$-rational, the specialization-ring $\mathfrak{R} = [x \to \bar{a}; \mathfrak{o}]$ is a regular local ring of dimen-*

---

2) For a more detailed treatment, see [8].

sion $n+1$. *More precisely, if $\{\tau_1, \cdots, \tau_n\}$ is a set of local parameters on V at $\bar{a}$, defined over k, such that $\tau_i(\bar{a})=0$ for $i=1, \cdots, n$, and π is a generator of the maximal ideal $\mathfrak{p}$ of $\mathfrak{o}$, then the maximal ideal $\mathfrak{M}$ of $\mathfrak{R}$ is generated by $\{\pi, \tau_1(x), \cdots, \tau_n(x)\}$.*

Proof. It is almost obvious that $[x \to \bar{a}; \mathfrak{o}]$ is a local ring of dimension $\geqq n+1$ and that there exists a set of local parameters $\{\tau_1, \cdots, \tau_n\}$ at $\bar{a}$, defined over k, such that $\tau_i(\bar{a})=0$ for $i=1, \cdots, n$. Hence our proposition is proved if we show that $\mathfrak{M}$ is generated by $\{\pi, \tau_1(x), \cdots, \tau_n(x)\}$. We may assume that V is an affine variety in the affine space S^N and that $\bar{a}$ is the origin $\overline{0}$. Then we have clearly $\mathfrak{M} = \mathfrak{R}x_1 + \cdots + \mathfrak{R}x_N + \mathfrak{R}\pi$, where the x_i are the coordinates of x; so we have only to prove that every x_i is contained in $\mathfrak{R}t_1 + \cdots + \mathfrak{R}t_n + \mathfrak{R}\pi$, where $t_i = \tau_i(x)$. Let the $F_i(X, T)$ be polynomials in the condition ii). If we express $F_i(X, T)$ in the form

$$F_i(X, T) = F_{i1}(X, T)X_1 + \cdots + F_{iN}(X, T)X_N + F_{i0}(T),$$

where $F_{ij}(X, T) \in \mathfrak{o}[X, T]$ for $1 \leqq i \leqq N$, $0 \leqq j \leqq N$, then we have

$$\frac{\partial \tilde{F}_i}{\partial X_j}(\overline{0}, \overline{0}) = \tilde{F}_{ij}(\overline{0}, \overline{0}),$$

and hence $\det(\tilde{F}_{ij}(\overline{0}, \overline{0})) \neq 0$. As we have $\tilde{F}_i(\overline{0}, \overline{0}) = 0$, $F_{i0}(T)$ is expressed in the form

$$F_{i0}(T) = \sum_{\nu=1}^{n} T_\nu F_{i0}^{(\nu)}(T) + \pi \alpha_i,$$

where $F_{i0}^{(\nu)}(T) \in \mathfrak{o}[T]$ and $\alpha_i \in \mathfrak{o}$. We have then

$$F_{i1}(x, t)x_1 + \cdots + F_{iN}(x, t)x_N = -\sum_{\nu=1}^{n} F_{i0}^{(\nu)}(t)t_\nu - \pi \alpha_i \qquad (1 \leqq i \leqq N).$$

Solving these equations with respect to $x_1, \cdots, x_N$, we see that the x_i are contained in $\mathfrak{R}t_1 + \mathfrak{R}t_2 + \cdots + \mathfrak{R}t_n + \mathfrak{R}\pi$; this completes our proof.

In addition to the assumptions in Prop. 3, assume that *there exists a k-rational point a on V such that $a \to \bar{a}$ ref. $\mathfrak{o}$.* We can choose a set of local parameters $\tau_1, \cdots, \tau_n$ at $\bar{a}$ in such a way that $\tau_i(a)=0$ for $1 \leqq i \leqq n$. Put $\mathfrak{R}_1 = [x \to a; k]$ and call $\mathfrak{M}_1$ the maximal ideal of $\mathfrak{R}_1$. Then, since $\tau_1, \cdots, \tau_n$ is a set of local parameters on V at a, we have $\mathfrak{M}_1 = \mathfrak{R}_1 t_1 + \cdots + \mathfrak{R}_1 t_n$. Put $\mathfrak{l} = \mathfrak{R}t_1 + \cdots + \mathfrak{R}t_n$. Then, by the same argument as the above proof, we see that $x_1 - a_1, \cdots, x_N - a_N$ are contained in $\mathfrak{l}$; and the ring of quotients of $\mathfrak{R}$ with respect to the prime ideal $\mathfrak{l}$, coincides with $\mathfrak{R}_1$. Hence the residue class ring $\mathfrak{R}/\mathfrak{l}$ is canonically isomorphic to a subring of $\mathfrak{R}_1/\mathfrak{M}_1$; the ring $\mathfrak{R}_1/\mathfrak{M}_1$ admits k as a complete set of representatives, and $\mathfrak{o}$ is contained in a complete set of representatives for $\mathfrak{R}/\mathfrak{l}$. Since $\mathfrak{R}/\mathfrak{l}$ is not a field and $\mathfrak{o}$ is a maximal ring in k, we conclude that $\mathfrak{o}$ *is a complete set of representatives of $\mathfrak{R}/\mathfrak{l}$.*

Proposition 4. *Notations and assumptions being as above, let z be an element of $\mathfrak{R}$. Then there exist homogeneous forms $H_i(T_1, \cdots, T_n)$ of degree $i = 0, 1, 2, \cdots$ in $\mathfrak{o}[T]$ such that*

$$z \equiv H_0 + H_1(t) + \cdots + H_\rho(t) \qquad (\mathrm{mod.}\ \mathfrak{t}^{\rho+1}).$$

Such forms $H_i(T)$ are uniquely determined by z.

Proof. The existence of $H_i(T)$ is due to that $\mathfrak{o}$ is a complete set of representatives of $\mathfrak{R}/\mathfrak{t}$ and the uniqueness follows from the fact that we have in $\mathfrak{R}_1$,

$$z \equiv H_0 + H_1(t) + \cdots + H_\rho(t) \qquad (\mathrm{mod.}\ \mathfrak{M}_1{}^{\rho+1}).$$

§ 3. Specialization of function-modules[3]

Let V be a variety defined over k. We call a subset L of the field of rational functions on V a *function-module* on V if L is a vector space of finite dimension over the field of constant functions on V[4]. L is said to be *defined over k* if L has a base over K consisting of the functions defined over k. If L is a function-module defined over k, we denote by L_k the subset of L composed of the elements in L defined over k. Let $\{f_1, \cdots, f_m\}$ be a base of L such that the f_i are defined over k. Then we see that $L_k = kf_1 + \cdots + kf_m$. Now assume that V has a structure of $\mathfrak{p}$-simple $\mathfrak{p}$-variety. L being a function-module on V defined over k, we denote by $\widetilde{L}_k$ the set of rational functions $\bar{f}$ on $\widetilde{V}$ such that there exists an element f of L_k having $\bar{f}$ as a specialization with respect to $\mathfrak{o}$. Then $\widetilde{L}_k$ is clearly a vector space over $\tilde{k}$, and $\dim_{\tilde{k}}\widetilde{L}_k$ is not greater than $\dim_k L_k$; in fact, the specializations $\tilde{f}_1, \cdots, \tilde{f}_r$ of functions $f_1, \cdots, f_r$ can not be linearly independent over $\tilde{k}$ unless the f_i are so over k. We obtain, more precisely,

Proposition 5. $\dim_{\tilde{k}}\widetilde{L}_k = \dim_k L_k$.

Proof. Denote by $\mathcal{B}$ the totality of bases of L_k over k, consisting of functions defined and finite along $\widetilde{V}$. Let $a \in V$ and $\bar{a} \in \widetilde{V}$ be points, both simple on V, such that $a \to \bar{a}$ ref. $\mathfrak{o}$ and that every function in L_k is defined and finite at $\bar{a}$. Let $\{k', \mathfrak{o}', \mathfrak{p}'\}$ be a prolongation of $\{k, \mathfrak{o}, \mathfrak{p}\}$ such that $a \to \bar{a}$ ref. $\mathfrak{o}'$ and that a is rational over k'. Take a set of local parameters $\{\tau_1, \cdots, \tau_n\}$ on V at $\bar{a}$, defined over k', such that $\tau_i(a) = 0$ for $1 \leq i \leq n$. For every $\{f\} = \{f_1, \cdots, f_m\}$ in $\mathcal{B}$, by virtue of Prop. 4, we get an expansion of f_ν by power-series:

$$f_\nu = \sum_{(i)} f_{(i)}{}^{(\nu)} \tau_1{}^{i_1} \cdots \tau_n{}^{i_n}, \qquad\qquad (1 \leq \nu \leq m),$$

---

3) Another approach to the same subject will be found in [8].

4) The field of constant functions on V may be identified with the universal domain K.

where the $f_{(i)}^{(\nu)}$ are elements of $\mathfrak{o}'$. Since $f_1, \cdots, f_m$ are linearly independent over k', there exists a set of m indices among $\{(i_1, \cdots, i_n)\}$, briefly denoted by $1, 2, \cdots, m$, such that

$$
\begin{vmatrix}
f_1^{(1)} & f_2^{(1)} & \cdots & f_m^{(1)} \\
f_1^{(2)} & f_2^{(2)} & \cdots & f_m^{(2)} \\
& \cdots & \cdots & \\
f_1^{(m)} & f_2^{(m)} & \cdots & f_m^{(m)}
\end{vmatrix} \neq 0 .
$$

If $\{g\} = \{g_1, \cdots, g_m\}$ is another element of $\mathscr{B}$, there exists a non-singular matrix $M_{(g, f)}$ of degree m with coefficients in k such that

$$
\begin{pmatrix} g_1 \\ \vdots \\ g_m \end{pmatrix} = M_{(g, f)} \begin{pmatrix} f_1 \\ \vdots \\ f_m \end{pmatrix} .
$$

We observe that

$$
\det(g_j^{(\nu)}) = \det M_{(g, f)}(f_j^{(\nu)}) \neq 0 ,
$$

where $g_j^{(\nu)}$ is the corresponding coefficient of the expansion for g_j. Let v' be the normalized exponential valuation of k' defined by $\mathfrak{p}'$. Put $\mu(f_1, \cdots, f_m) = v'(\det(f_j^{(\nu)}))$. Then we have $\mu(f_1, \cdots, f_m) \geq 0$ because every $f_j^{(\nu)}$ is contained in $\mathfrak{o}'$. Now put

$$
\mu_0 = \operatorname*{Min}_{\{f\} \in \mathscr{B}} \mu(f_1, \cdots, f_m) ;
$$

and let $\{h_1, \cdots, h_m\}$ be an element of $\mathscr{B}$ such that

$$
\mu_0 = \mu(h_1, \cdots, h_m) .
$$

Then, $\tilde{h}_1, \cdots, \tilde{h}_m$ are linearly independent over $\tilde{k}$. In fact, if there can be found m elements $c_1, \cdots, c_m$ in $\mathfrak{o}$, not all non-units of $\mathfrak{o}$, such that $\tilde{c}_1\tilde{h}_1 + \cdots + \tilde{c}_m\tilde{h}_m = 0$, then $g = \pi^{-1}(c_1 h_1 + \cdots + c_m h_m)$ is defined and finite along $\tilde{V}$, where π is a prime element of $\mathfrak{o}$. If $\tilde{c}_1 \neq 0$, $\{g, h_2, \cdots, h_m\}$ is contained in $\mathscr{B}$ and we have

$$
\mu(h_1, \cdots, h_m) > \mu(g, h_2, \cdots, h_m) .
$$

This contradicts the definition of $(h_1, \cdots, h_m)$; so our proposition is proved.

We shall say that a rational function f on V, defined over k, is $\mathfrak{p}$-*finite* if f is defined and finite along $\tilde{V}$; $\mathfrak{p}$-finite functions $f_1, \cdots, f_r$ are said to be *linearly $\mathfrak{p}$-independent* if $\tilde{f}_1, \cdots, \tilde{f}_r$ are linearly independent over $\tilde{k}$. L and L_k being as above, let $\{f_1, \cdots, f_m\}$ be a base of L_k, consisting of $\mathfrak{p}$-finite functions. Then, $f_1, \cdots, f_m$ are linearly $\mathfrak{p}$-independent if and only if for every $\mathfrak{p}$-finite function $h = c_1 f_1 + \cdots + c_m f_m \in L_k$ with $c_i \in k$, we have $c_i \in \mathfrak{o}$. On the other hand, when $f_1, \cdots, f_m$ are linearly $\mathfrak{p}$-independent, a function $h = c_1 f_1 + \cdots + c_m f_m$ with $c_i \in k$ is $\mathfrak{p}$-finite if and only if $c_i \in \mathfrak{o}$ for every i. Now we shall define the specialization of a function-module. Let L be as before a function-module on a $\mathfrak{p}$-simple $\mathfrak{p}$-variety V, defined over k. We denote by $\widetilde{L}$ the set of functions $\bar{f}$ on $\widetilde{V}$ such that there exists an element f of L having $\bar{f}$ as a specialization with respect to $\mathfrak{o}$; we call $\widetilde{L}$ the *specialization of L with respect to* $\mathfrak{o}$. We see that $\widetilde{L}$ is a function-module on $\widetilde{V}$, defined over $\tilde{k}$, whose dimension is equal to that of L.

We shall apply our result to linear systems on a variety. Let V be a $\mathfrak{p}$-simple $\mathfrak{p}$-complete $\mathfrak{p}$-variety and X a k-rational divisor on V. If both V and $\widetilde{V}$ are non-singular in codimension 1, we can define two function-modules $L(X)$ on V and $\bar{L}(\widetilde{X})$ on $\widetilde{V}$, defined over k and $\tilde{k}$, respectively, such that

$$L(X) = \{f \mid f \text{ is a function on } V, (f) > -X\},$$

$$\bar{L}(\widetilde{X}) = \{\bar{f} \mid \bar{f} \text{ is a function on } \widetilde{V}, (\bar{f}) > -\widetilde{X}\}.$$

We denote by $\widetilde{L}(X)$ the specialization of $L(X)$ with respect to $\mathfrak{o}$. In view of Theorem 20 of [7], we see that $\widetilde{L}(X)$ is contained in $\bar{L}(\widetilde{X})$. If we denote by $l(X)$ (resp. $\bar{l}(\widetilde{X})$) the dimension of $L(X)$ (resp. $\bar{L}(\widetilde{X})$), the above discussion leads us to the inequality $l(X) \leq \bar{l}(\widetilde{X})$.

§4. Projective embedding of an abelian $\mathfrak{p}$-variety without defect

Let A be an abelian $\mathfrak{p}$-variety without defect. If X_1 is a positive k-rational and non-degenerate divisor on A, the specialization $\widetilde{X}_1$ of X_1 with respect to $\mathfrak{o}$ is also non-degenerate on $\widetilde{A}$; and by a result of Nishi [5], we have $l(X_1) = \bar{l}(\widetilde{X}_1)$. From Weil [12], we know that for a sufficiently large integer s, the divisors $sX_1 = X$ and $s\widetilde{X}_1 = \widetilde{X}$ are ample on A and $\widetilde{A}$ respectively; and we have $l(X) = \bar{l}(\widetilde{X})$. Hence, if $\{f_0, f_1, \cdots, f_m\}$ is a linearly $\mathfrak{p}$-independent base of $L(X)$, then $\{\bar{f}_0, \bar{f}_1, \cdots \bar{f}_m\}$ is a base of $\widetilde{L}(X) = \bar{L}(\widetilde{X})$. Take a generic point x of A over k and consider the locus A_1 of $(f_0(x), \cdots, f_m(x))$ over k in the projective space P^m; then we obtain a birational mapping τ of A onto A_1 defined by $\tau(x) = (f_0(x), \cdots, f_m(x))$ with respect to k. We can put into A_1 a structure of abelian variety so that τ is an isomorphism of A onto A_1. As A_1 is a projective variety, A_1 is naturally endowed with a structure of $\mathfrak{p}$-variety. We shall now prove that A_1 is an abelian $\mathfrak{p}$-variety without defect. Since $\widetilde{X}$ is ample on $\widetilde{A}$, and since $\{\bar{f}_0, \cdots, \bar{f}_m\}$ is a base of $\bar{L}(\widetilde{X})$, for any point $\bar{a}$ of $\widetilde{A}$, there is a function among $\bar{f}_0, \cdots, \bar{f}_m$, say $\bar{f}_0$, such that $\bar{a}$ is not contained in the support of $(\bar{f}_0) + \widetilde{X}$. Then we see easily that the functions f_i/f_0 are all defined and finite at $\bar{a}$; so τ is defined at $\bar{a}$. Call T the graph of τ; then, as τ is everywhere defined on $\widetilde{A}$, the specialization of T with respect to $\mathfrak{o}$ is a variety and coincides with the locus of $\bar{\tau}$. Let $\bar{A}_1$ be the image of $\widetilde{A}$ by $\bar{\tau}$. Recalling that $\widetilde{X}$ is ample on $\widetilde{A}$ and $\{\bar{f}_0, \cdots, \bar{f}_m\}$ is a base of $\bar{L}(\widetilde{X})$, we see that

$\tilde{\tau}$ gives a birational biregular mapping of $\tilde{A}$ onto $\overline{A}_1$. It follows from this that A_1 is an abelian $\mathfrak{p}$-variety without defect and $\tilde{A}_1 = \overline{A}_1$. This solves the problem about a projective embedding of an abelian $\mathfrak{p}$-variety without defect. We shall now consider the problem concerning the field of definition for A_1.

THEOREM 2. *Let A be an abelian variety defined over k; suppose the following conditions are satisfied.*

D1) *There are a prolongation $\{k', \mathfrak{o}', \mathfrak{p}'\}$ of $\{k, \mathfrak{o}, \mathfrak{p}\}$, an abelian $\mathfrak{p}'$-variety A^* without defect, defined over k', and an isomorphism θ of A onto A^*, defined over k'.*

D2) *There are a $\mathfrak{p}$-simple $\mathfrak{p}$-variety V and a surjective rational mapping φ of V to A, both defined over k, such that the specialization $\widetilde{\theta \circ \varphi}$ of $\theta \circ \varphi$ with respect to $\mathfrak{o}'$ induces a surjective rational mapping of $\tilde{V}$ to $\tilde{A}^*$.*

Then there exists a projective abelian variety A_1, defined over k, without defect with respect to a natural structure of $\mathfrak{p}$-variety, which is k-isomorphic to A.

Proof. Let X be a positive k-rational divisor on A such that the divisor $X^* = \theta(X)$ on A^* and its specialization $\tilde{X}^*$ with respect to $\mathfrak{o}'$ on $\tilde{A}^*$ are both ample on A^* and $\tilde{A}^*$, respectively. Then $L(X^*)$ is a function-module on A^*, defined over k', whose specialization is $\overline{L}(\tilde{X})$. Now consider a function-module L^0 on V composed of all functions g of the form $g = f^* \circ \theta \circ \varphi$ for some $f^* \in L(X^*)$; then, L^0 coincides with the set of all functions $f \circ \varphi$ for $f \in L(X)$, because θ is an isomorphism; this implies that L^0 is defined over k. Let $\{g_0, \cdots, g_m\}$ be a linearly $\mathfrak{p}$-independent base of L_k^0; for every i, there exists a function f_i^* in $L_{k'}(X^*)$ such that $g_i = f_i^* \circ \theta \circ \varphi$. Since $\widetilde{\theta \circ \varphi}$ is surjective, we have $\tilde{g}_i = \tilde{f}_i^* \circ (\widetilde{\theta \circ \varphi})$. As $\tilde{g}_0, \cdots, \tilde{g}_m$ are linearly independent over the constant field, we see that $\tilde{f}_0^*, \cdots, \tilde{f}_m^*$ are also linearly independent over the constant field. It follows that $\{f_0^*, \cdots, f_m^*\}$ is a linearly $\mathfrak{p}'$-independent base of $L(X^*)$. Using this base $\{f_0^*, \cdots, f_m^*\}$, we obtain a projective abelian variety A_1 defined over k', without defect as an abelian $\mathfrak{p}'$-variety. On the other hand, we see that $\{f_0^* \circ \theta, \cdots, f_m^* \circ \theta\}$ is a base of $L_k(X)$. Hence A_1 is a projective embedding of A defined over k. This concludes that A_1 satisfies our requirements.

REMARK. The assumption D2) in Theorem 2 is indispensable. For instance: for a prime number $p > 2$, the elliptic curve $y^2 = 4x^3 - px - p$ defined over the field Q of rational numbers does not admit any model, without defect for p, defined over Q, while it is birationally equivalent to the elliptic curve $y^2 = 4x^3 - p^{1/3}x - 1$, which is without defect for the prime divisor $(p^{1/3})$.

§ 5. Construction of group $\mathfrak{p}$-varieties without defect

In this section we shall translate a part of $n°$ 32-33 of Weil [10] to the case of $\mathfrak{p}$-varieties. We first restate the definition of pre-group given in Weil [11]. A *pre-group* is a pair (V, f) of two geometric objects such that:

P 1) *V is a variety;*

P 2) *f is a rational mapping of $V \times V$ into V which defines a normal law of composition on V, namely, if x, y, z are independent generic points of V over a common field k of definition for V and f, we have* i) $k(x, y) = k(x, f(x, y)) = k(y, f(x, y))$ *and* ii) $f(x, f(y, z)) = f(f(x, y), z)$.

In these circumstances we say that k is a field of definition for a pre-group (V, f) or that (V, f) is defined over k. Two pre-groups (V, f) and (V', f') are said to be isomorphic to each other if there exists a birational correspondence between V and V' by which f corresponds to f'.

Let (V, f) be a pre-group. If V has a structure of $\mathfrak{v}$-variety, the pair (V, f) of the $\mathfrak{p}$-variety V and f is called a *pre-group $\mathfrak{v}$-variety*. In particular, when V is an affine variety with the natural structure of $\mathfrak{p}$-variety, (V, f) is called an *affine pre-group $\mathfrak{v}$-variety*.

Definition 2. *A pre-group $\mathfrak{p}$-variety (V, f) is said to be without defect if the following two conditions are satisfied.*

PD 1) *V is $\mathfrak{p}$-simple;*

PD 2) *f is defined along $\tilde{V} \times \tilde{V}$ and $\tilde{f}$ induces a normal law of composition on $\tilde{V}$.*

Hence, if (V, f) is without defect, $(\tilde{V}, \tilde{f})$ is a pre-group defined over $\tilde{k}$.

Two pre-group $\mathfrak{p}$-varieties (V, f) and (V_1, f_1), both without defect, are said to be $(k\text{-})isomorphic$ to each other if there exists a birational mapping σ, defined over k, of V onto V_1 such that:

PDI 1) *f_1 is the translation of f by σ;*

PDI 2) *σ is defined along $\tilde{V}$ and $\tilde{\sigma}$ induces a birational mapping of $\tilde{V}$ onto $\tilde{V}_1$.*

From the conditions PDI 1-2), we can see that $\tilde{f}_1$ is the translation of $\tilde{f}$ by $\tilde{\sigma}$, so that $(\tilde{V}, \tilde{f})$ is isomorphic to $(\tilde{V}_1, \tilde{f}_1)$. It is not difficult to show that, for any pre-group $\mathfrak{p}$-variety (V, f) without defect, there is an affine pre-group $\mathfrak{p}$-variety without defect, isomorphic to (V, f). The purpose in this section is to prove that for any pre-group $\mathfrak{p}$-variety (V, f) without defect, we can construct a group $\mathfrak{p}$-variety without defect, which is isomorphic to (V, f). Since the method is quite similar to Weil's construction of group variety in [10], we shall only state two preliminary propositions and the main theorem without proof. Let (V, f) be a pre-group defined over k and x, y two independent generic points of V over k. Put $z = f(x, y)$. We obtain then two rational mappings φ and $\psi \colon V \times V \to V$, defined by

$$y = \varphi(x, z), \quad x = \psi(z, y)$$

with respect to k.

PROPOSITION 6. *Let (V, f) be a pre-group $\mathfrak{p}$-variety without defect. Then there exist a prolongation $\{k', \mathfrak{o}', \mathfrak{p}'\}$ of $\{k, \mathfrak{o}, \mathfrak{p}\}$ and a frontier $\mathscr{F}$ on V, normally algebraic over k', having the following properties:*

i) *for any point a of $V - \mathscr{F}$ and any generic point x of V over $k'(a)$, $f(x, a)$ and $\psi(x, a)$ are defined; $\varphi(x, f(x, a))$ is defined and equal to a; and $\psi(f(x, a), a)$ and $f(\psi(x, a), a)$ are defined and equal to x;*

ii) *for any point $\bar{a}$ of $\tilde{V} - \mathscr{F}$ and any generic point $\bar{x}$ of $\tilde{V}$ over $\tilde{k}'(\bar{a})$, all assertions, replaced x, a by $\bar{x}, \bar{a}$, in* i) *are true.*

PROPOSITION 7. *Let (V, f) be a pre-group $\mathfrak{p}$-variety without defect; let $\{k', \mathfrak{o}', \mathfrak{p}'\}$ and $\mathscr{F}$ be a prolongation of $\{k, \mathfrak{o}, \mathfrak{p}\}$ and a frontier on V, having the properties in Prop. 6. Let x be a generic point of V over k' and $\bar{x}$ a generic point of $\tilde{V}$ over $\tilde{k}$; put $\mathfrak{o}_1 = [x \to \bar{x}; \mathfrak{o}']$. Let y be a generic point of V over $k'(x)$ and T_x the locus of $y \times f(x, y)$ on $V \times V$ over $k'(x)$; and let $\tilde{T}_x$ be the specialization of T_x with respect to $\mathfrak{o}_1$. Then, T_x is a birational correspondence of V onto itself, and:*

i) *if (a, b) is a point of T_x, such that both a and b are points in $V - \mathscr{F}$, the points a and b are regularly corresponding points of V by T_x;*

ii) *if $(\bar{a}, \bar{b})$ is a point of the support of $\tilde{T}_x$, such that both $\bar{a}$ and $\bar{b}$ are points in $\tilde{V} - \tilde{\mathscr{F}}$, $\bar{a}$ and $\bar{b}$ are regularly corresponding points of $\tilde{V}$ by T_x.*

The above two propositions are translations of the Lemmas 6, 7 in [10, p. 52-53]. After these, we are now in a position of stating the construction theorem.

THEOREM 3. *Let (V, f) be a pre-group $\mathfrak{p}$-variety without defect. Then there exist a prolongation $\{K, \mathfrak{O}, \mathfrak{P}\}$ of $\{k, \mathfrak{o}, \mathfrak{p}\}$ and a group $\mathfrak{P}$-variety G, defined over K, without defect, which is K-isomorphic to (V, f). G is uniquely determined by (V, f) up to an isomorphism.*

We shall only give an outline of the proof. At first we may assume that (V, f) is an affine pre-group $\mathfrak{p}$-variety because we can always find an affine model isomorphic to (V, f). $\{k', \mathfrak{o}', \mathfrak{p}'\}$ and $\mathscr{F}$ having the same meanings as in Prop. 6, let $x, t_1, \cdots, t_N$ be independent generic points of V over k' and $\bar{t}_1, \cdots, \bar{t}_N$ be independent generic points of $\tilde{V}$ over $\tilde{k}$, for a sufficiently large N. Put, for $1 \leq \alpha \leq N$, $G_\alpha = V$, $\mathscr{F}_\alpha = \mathscr{F}$ and $x_\alpha = f(t_\alpha, x)$. We define $\{K, \mathfrak{O}, \mathfrak{P}\}$ by

$$K = k'(t_1, \cdots, t_N), \qquad \mathfrak{O} = [(t_1, \cdots, t_N) \to (\bar{t}_1, \cdots, \bar{t}_N); \mathfrak{o}'].$$

Then, $\mathfrak{O}$ is a discrete valuation ring of rank 1 and $\{K, \mathfrak{O}, \mathfrak{P}\}$ is a prolongation of $\{k, \mathfrak{o}, \mathfrak{p}\}$. If we denote by $T_{\beta\alpha}$ the locus of (x_α, x_β) on $G_\alpha \times G_\beta$ with respect to K, we know that $[G_\alpha, \mathscr{F}_\alpha, \tilde{\mathscr{F}}_\alpha, T_{\beta\alpha}]$ defines a $\mathfrak{P}$-variety, and that this is just the one which we want to construct. The uniqueness is a consequence of the fact that any two group $\mathfrak{p}$-varieties, both without defect, which are isomorphic to each other with respect to the structure of pre-group $\mathfrak{p}$-variety, are also isomorphic with respect to the structure of group $\mathfrak{p}$-variety.

Corollary. *(V, f) and G being as in Theorem 3, suppose that G is an abelian $\mathfrak{P}$-variety without defect. Then there exists a projective abelian $\mathfrak{p}$-variety A, defined over k, without defect, k-isomorphic to (V, f).*

Proof. By Weil [11], we know that there exists an abelian variety, defined over k, which is derived from (V, f) by a birational transformation defined over k. Applying then Theorem 2 to the present case we obtain our result.

§ 6.　Specialization of homomorphic images of an abelian $\mathfrak{p}$-variety without defect

In this section we want to prove:

Theorem 4. *Let A be an abelian $\mathfrak{p}$-variety without defect, and let λ be a surjective homomorphism of A onto another abelian variety B, where both λ and B are defined over k. Then there exists an abelian $\mathfrak{p}$-variety B_1 without defect, which is birationally equivalent to B over k.*

Let A_1 and A_2 be abelian varieties and λ a homomorphism of A_1 onto A_2. We shall call λ a *solid homomorphism*, if, for a common field k_1 of definition for A_1, A_2 and λ, and for a generic point x of A_1 over $k_1, k_1(x)$ is a regular extension of $k_1(\lambda(x))$. Now any surjective homomorphism λ of A onto B is decomposed into two surjective homomorphisms, $\lambda_1 : A \to B^*$, $\lambda_2 : B^* \to B$ and $\lambda = \lambda_2 \circ \lambda_1$, where B^* is an abelian variety, λ_1 is an isogeny and λ_2 is a solid homomorphism. Hence, if the problem concerning the common field of definition for B_1 and the birational correspondence between B and B_1, is left out of consideration, it is sufficient to prove Theorem 4 in two special cases where 1) λ is an isogeny or 2) λ is a solid homomorphism. If this is done, the theorem would be a consequence of Theorem 2.

At first we shall make some preliminary considerations. X and u being a cycle on an abelian variety A and a point on A, we denote by X_u the transform of X by the translation $x \to x+u$ on A. Now let A be an abelian variety embedded in a projective space and C an abelian subvariety of A, both defined over k. Then $\{C_u \mid u \in A\}$ forms an algebraic system of positive cycles on A. Let B and Γ be the Chow variety associated with the algebraic system $\{C_u\}$ and the corresponding graph of $\{C_u\}$ to B. Then B is a model of the factor group variety of A by C, Γ is the graph of the natural homomorphism of A onto B; and both B and Γ are defined over k (Chow [1]). Moreover, if A is an abelian $\mathfrak{p}$-variety without defect, we can easily see that the specialization $\tilde{C}$ of C with respect to $\mathfrak{p}$ is a multiple of an abelian subvariety $\bar{C}$ of $\tilde{A}$, i.e., $\tilde{C} = s\bar{C}$ for a positive integer s. Let $\bar{B}$ and Γ (resp. $\bar{B}'$ and Γ') be the Chow variety associated with the algebraic system $\{\tilde{C}_{\bar{u}} | \bar{u} \in A\}$ (resp. $\{\bar{C}_{\bar{u}} | \bar{u} \in A\}$) and the corresponding graph of $\{\tilde{C}_{\bar{u}}\}$ to $\bar{B}$ (resp. $\{\bar{C}_{\bar{u}}\}$ to $\bar{B}'$). Denote by p the characteristic of $\tilde{k}$. In the following discussion, the case $p \neq 0$ is essential; in fact, Theorem 4 itself is rather trivial if p is equal to 0. The following proposition is concerned with the case $p \neq 0$; it is also true, however, in the case $p = 0$, if we put 1 in place of any exponent of p.

PROPOSITION 8. *With the above notations, if p is positive, we have*

i) $\tilde{\Gamma} = \Gamma$;

ii) *there exists an integer e such that $s = p^e$; i. e., $\tilde{C} = p^e \bar{C}$;*

iii) $\bar{B}$ *is an abelian variety and is the image of $\bar{B}'$ by the rational mapping $\mathbf{p}^e$, where $\mathbf{p}$ is a rational mapping defined by*

$$\mathbf{p}(x_0, \cdots, x_n) = (x_0{}^p, \cdots, x_n{}^p)$$

for every point $(x_0, \cdots, x_n)$ in the projective space;

iv) $\tilde{B} = p^{(m-1)e} \bar{B}$, *where m is the dimension of B, and the specialization of the graph of the composition-law in B is a multiple of that of $\bar{B}$.*

Proof. The assertion i) and the fact that the support of $\tilde{B}$ coincides with $\bar{B}$, are directly derived from our definition. Put $\tilde{B} = r\bar{B}$, $s = s'p^e$, where s' is prime to p, $\deg \bar{B} = \bar{b}$, $\deg \bar{C} = \bar{c}$, $\deg B = \deg \tilde{B} = b$ and $\deg C = \deg \tilde{C} = c$. Then we have $b = r\bar{b}$ and $c = s\bar{c}$. Let $P, P', \bar{P}, \bar{P}'$ be respectively the ambient projective spaces for $A, B, \tilde{A}, \tilde{B}$; and let $L, M, \bar{L}, \bar{M}$ be respectively independent generic linear varieties of dual dimension to $C, B, \tilde{C}, \tilde{B}$ in $P, P', \bar{P}, \bar{P}'$. Then $(\tilde{\Gamma}, \bar{L}, \bar{M})$ is a specialization of (Γ, L, M) with respect to $\mathfrak{o}$; and the intersections $\Gamma \cdot (L \times M)$ in $P \times P'$ and $\tilde{\Gamma} \cdot (\bar{L} \times \bar{M})$ in $\bar{P} \times \bar{P}'$ are defined. Comparing the degrees of both 0-cycles $\Gamma \cdot (L \times M)$ and $\tilde{\Gamma} \cdot (\bar{L} \times \bar{M})$, we have the equality $bc = sr\bar{b}\bar{c} = s'rp^e\bar{b}\bar{c} = p^{me}\bar{b}\bar{c}$, and hence $s' = 1$, $r = p^{(m-1)e}$. The remaining part of the proposition can be shown immediately.

In order to apply Theorem 3 to the proof of Theorem 4, we shall first prove the existence of a pre-group $\mathfrak{p}$-variety without defect, k-isomorphic to B.

PROPOSITION 9. *Besides the assumptions in Theorem 4, suppose that λ is an isogeny or a solid homomorphism. Then there exists a pre-group $\mathfrak{p}$-variety without defect, which is k-isomorphic to B.*

Proof. On account of Theorem 2, we may assume that A is a projective variety. In the following, under the titles I) or S), λ will be considered an isogeny or a solid homomorphism. We first fix a generic point x of A over k and a generic point $\bar{x}$ of $\tilde{A}$ over $\tilde{k}$.

i) I) There is a positive integer n and an isogeny $\lambda': B \to A$ such that $n\delta_A = \lambda' \circ \lambda$, where δ_A is the identity mapping of A onto itself. Put $\lambda(x) = y$, $nx = z$ and $n\bar{x} = \bar{z}$.

S) Denote by C the kernel of λ and by B^* the canonical model of the factor group variety of A by C, defined above Prop. 8. B^* is an abelian variety, defined over k, embedded in a projective space; and there is a k-isomorphism $\kappa: B \to B^*$. Applying Prop. 8 to our case, we have

$\tilde{C} = p^e \bar{C}$, where $\bar{C}$ is an abelian subvariety of A,

$\tilde{B}^* = p^{(m-1)e} \bar{B}^*$, where $\bar{B}^*$ is an abelian variety and $m = \dim B$.

We can easily see that the specialization $\widetilde{\kappa \circ \lambda}$ of $\kappa \circ \lambda$ is a homomorphism $\tilde{A} \to \bar{B}^*$. Put $\lambda(x) = y, \kappa(y) = z$ and $\widetilde{\kappa \circ \lambda}(\bar{x}) = \bar{z}$.

ii) If we denote by $\mathfrak{o}^*$ and $\mathfrak{p}^*$ the specialization ring $[x \to \bar{x}; \mathfrak{o}]$ and the maximal ideal of $\mathfrak{o}^*$, $\mathfrak{o}^*$ is a discrete valuation ring of rank 1 in $k(x)$ and $\{k(x), \mathfrak{o}^*, \mathfrak{p}^*\}$ is a prolongation of $\{k, \mathfrak{o}, \mathfrak{p}\}$. Put furthermore $\mathfrak{o}_1 = \mathfrak{o}^* \cap k(y)$; and denote by $\mathfrak{p}_1$ the maximal ideal of $\mathfrak{o}_1$. Then, $\mathfrak{o}_1$ is a discrete valuation ring of rank 1 in $k(y)$; and $\{k(x), \mathfrak{o}^*, \mathfrak{p}^*\}$ is a prolongation of $\{k(y), \mathfrak{o}_1, \mathfrak{p}_1\}$.

Assertion (I). $\mathfrak{o}_1$ *is the integral closure of the specialization ring* $[z \to \bar{z}; \mathfrak{o}]$ *in* $k(y)$.

Proof of (I). Since $\mathfrak{o}_1$ is integrally closed in $k(y)$ and contains $[z \to \bar{z}; \mathfrak{o}]$, $\mathfrak{o}_1$ contains the integral closure of $[z \to \bar{z}; \mathfrak{o}]$ in $k(y)$. Now we shall prove that every element t in $\mathfrak{o}_1$ is integral with respect to $[z \to \bar{z}; \mathfrak{o}]$, namely, that $+\infty$ is not a specialization of t over $z \to \bar{z}$ with respect to $\mathfrak{o}$. Let $\bar{t}$ be a specialization of t over $z \to \bar{z}$ with respect to $\mathfrak{o}$; let $\bar{x}'$ be an isolated specialization of x over the specialization $(z, t) \to (\bar{z}, \bar{t})$ ref. $\mathfrak{o}$. Then, by Th. 6 of [7], we have

$$\dim_{\tilde{k}(\bar{z})}(\bar{x}') \geqq \dim_{\tilde{k}(\bar{z}, \bar{t})}(\bar{x}') \geqq \dim_{k(z, t)}(x) = \dim_{k(z)}(x).$$

On the other hand, we have $k(x) \supset k(z)$, $\tilde{k}(\bar{x}') \supset \tilde{k}(\bar{z})$ and $\dim_k(z) = \dim_{\tilde{k}}(\bar{z})$. Therefore we must have $\dim_{\tilde{k}}(\bar{x}') \geqq \dim_k(x)$. This shows that $\bar{x}'$ is a generic point of $\tilde{A}$ over $\tilde{k}$. Hence $[x \to \bar{x}'; \mathfrak{o}]$ coincides with $\mathfrak{o}^*$ which contains t; so $\bar{t}$ is not ∞; this proves the assertion.

iii) I) Let A_0 be an affine representative of A and z^0 the representative of z on A_0.

S) Let B_0^* be an affine representative of B^* and z^0 the representative of z on B_0^*.

I, S) Denote by K and $\tilde{K}$ the field $k(y)$ and the residue field $\mathfrak{o}_1/\mathfrak{p}_1$.

Assertion (II). *There exists a set* $(t) = (t_1, \cdots, t_s)$ *of quantities in* K *such that every* t_i *is integral over* $\mathfrak{o}[z^0]$ *and that* $k(z^0, t) = K$, $\tilde{k}(\bar{z}^0, \bar{t}) = \tilde{K}$, *where tilde means the specialization with respect to* $\mathfrak{o}_1$.

Proof of (II). From the assertion (I) we know that there is a set $(u) = (u_1, \cdots, u_s)$ of quantities in K such that every u_i is integral with respect to $[z^0 \to \tilde{z}^0\,; \mathfrak{o}]$ and that $\tilde{k}(\tilde{z}^0, \tilde{u}) = \tilde{K}, k(z^0, u) = K$. For each u_i, we can find a polynomial $f_i(U)$ in $k(z^0)[U]$ such that

$$f_i(u_i) = 0, \qquad f_i(U) = U^M + \frac{P_1(z^0)}{Q(z^0)} U^{M-1} + \cdots + \frac{P_M(z^0)}{Q(z^0)},$$

where the P_j and Q are polynomials with coefficients in $\mathfrak{o}$ and $\tilde{Q}(\tilde{z}^0) \neq 0$. If we put $t_i = Q(z^0)u_i$, (t) is a set of quantities which we wanted to find.

iv) Let V be the locus of (t, z^0) over k in an affine space. Since V is birationally equivalent to B, V itself can be considered to be a pre-group (V, f) in a natural way. We shall now prove that (V, f) is a pre-group $\mathfrak{p}$-variety without defect.

ASSERTION (III) *The support of $\tilde{V}$ is a variety.*

Proof of (III). Let $(t) \to (\bar{a})$ be a specialization over any finite specialization $(z^0) \to (\bar{b})$ ref. $\mathfrak{o}$. Since the quantities t_i are integral over $\mathfrak{o}[z^0]$, (a) is finite and we have $\dim_{\tilde{k}}(\bar{a}, \bar{b}) = \dim_{\tilde{k}}(\bar{b})$. In particular this implies that the support of $\tilde{V}$ is not empty. Let $(\bar{t}, \bar{z}^0)$ and $(\bar{t}_1, \bar{z}^0{}_1)$ be generic points of any two (same or different) components of $\tilde{V}$ over the algebraic closure of $\tilde{k}$. In order to conclude our assertion it is sufficient to show that $(\bar{t}_1, \bar{z}^0{}_1)$ is a generic specialization of $(\bar{t}, \bar{z}^0)$ over $\tilde{k}$ and that $\tilde{k}(\bar{t}, \bar{z}^0)$ is a regular extension of $\tilde{k}$. From the fact that $\dim_{\tilde{k}}(\bar{t}, \bar{z}^0) = \dim_{\tilde{k}}(\bar{z}^0)$, $\dim_{\tilde{k}}(\bar{t}_1, \bar{z}^0{}_1) = \dim_{\tilde{k}}(\bar{z}^0{}_1)$, we know that both $\dim_{\tilde{k}}(\bar{z}^0)$ and $\dim_{\tilde{k}}(\bar{z}^0{}_1)$ are equal to $m = \dim B$. Consider isolated specializations $\bar{x}$ and $\bar{x}_1$ of x respectively over $(t, z^0) \to (\bar{t}, \bar{z}^0)$ ref. $\mathfrak{o}$ and $(t, z^0) \to (\bar{t}_1, \bar{z}^0{}_1)$ ref. $\mathfrak{o}$. We see then, as in the proof of (I), that both $\bar{x}$ and $\bar{x}_1$ are generic on $\tilde{A}$ over $\tilde{k}$. From this it follows that every coordinate of the point (t, z^0) is contained in the specialization-ring $[x \to \bar{x}\,; \mathfrak{o}] = [x \to \bar{x}_1\,; \mathfrak{o}]$; so we have $\tilde{k}(\bar{x}) \supset \tilde{k}(\bar{t}, \bar{z}^0)$; this shows that $\tilde{k}(\bar{t}, \bar{z}^0)$ is regular over $\tilde{k}$. At the same time, we observe that $(\bar{t}_1, \bar{z}^0{}_1)$ is a generic specialization of $(\bar{t}, \bar{z}^0)$ over $\tilde{k}$; so the assertion is proved.

ASSERTION (IV) *V is $\mathfrak{p}$-simple.*

Proof of (IV). Using the notations (t, z^0), $(\bar{t}, \bar{z}^0)$ in the above proof, it is sufficient, on account of Theorem 12 of [7], to show the equalities:

I) $[\tilde{k}(\bar{t}, \bar{z}^0) : \tilde{k}(\bar{z}^0)] = [k(t, z^0) : k(z^0)]$,

S) $[\tilde{k}(\bar{t}, \bar{z}^0) : \tilde{k}(\bar{z}^0)] = p^{(m-1)e}$.

I) By Assertion (II) and by a property of specialization, we have

$$[k(x) : k(t, z^0)] \geqq [\tilde{k}(\bar{x}) : \tilde{k}(\bar{t}, \bar{z}^0)], \qquad [k(t, z^0) : k(z^0)] \geqq [\tilde{k}(\bar{t}, \bar{z}^0) : \tilde{k}(\bar{z}^0)].$$

On the other hand, we have $[k(x) : k(z^0)] = \nu(n\delta_A) = \nu(n\delta_{\tilde{A}}) = [\tilde{k}(\bar{x}) : \tilde{k}(\bar{z}^0)]$. Hence we have the above equality.

S) From the fact that the specialization $\tilde{B}^*$ of B^* is equal to $p^{(m-1)e}\bar{B}^*$ follows the inequality $[\tilde{k}(\bar{t}, \bar{z}^0) : \tilde{k}(\bar{z}^0)] \leqq p^{(m-1)e}$. On the other hand since the specialization $(\tilde{C})_{\bar{x}} = p^e(\bar{C})_{\dot{x}}$ of C_x is rational over $\tilde{k}(\bar{t}, \bar{z}^0)$, the opposite inequality must hold.

ASSERTION (V) (V, f) *is a pre-group $\mathfrak{p}$-variety without defect.*

Proof of (V). We shall only prove the case S) as the other case will be obtained by substituting B^* for A in the following proof.

Let $(t_1, z^0{}_1) \times (t_2, z^0{}_2) \times (t_3, z^0{}_3)$ be a generic point of the graph of the composition-law in V over k. Then, $z^0{}_1 \times z^0{}_2 \times z^0{}_3$ is a generic point of the graph of the composition-law in B_0^* over k. We know that $\bar{B}_0^*$ is a pre-group defined over the algebraic closure $\tilde{k}_c$ of $\tilde{k}$, and $\tilde{B}_0^* = p^{(m-1)e}\bar{B}_0^*$. Let $\bar{z}^0{}_1 \times \bar{z}^0{}_2 \times \bar{z}^0{}_3$ be a generic point of the graph of the composition-law in $\bar{B}_0^*$ over $\tilde{k}_c$; then $(\bar{z}^0{}_1, \bar{z}^0{}_2, \bar{z}^0{}_3)$ is a specialization of $(z^0{}_1, z^0{}_2, z^0{}_3)$ with respect to $\mathfrak{o}$. If we extend this to a specialization

$$((t_1, z^0{}_1), (t_2, z^0{}_2), (t_3, z^0{}_3)) \rightarrow ((\bar{t}_1, \bar{z}^0{}_1), (\bar{t}_2, \bar{z}^0{}_2), (\bar{t}_3, \bar{z}^0{}_3)) \text{ ref. } \mathfrak{o},$$

then, for every i, $\bar{t}_i$ is finite and algebraic over $\tilde{k}(\bar{z}^0{}_i)$. It is not difficult to see that the locus of $((\bar{t}_1, \bar{z}^0{}_1), (\bar{t}_2, \bar{z}^0{}_2), (\bar{t}_3, \bar{z}^0{}_3))$ over $\tilde{k}_c$ is a simple component of the specialization of the composition-law on V, and that it defines a normal law of composition on $\tilde{V}$; so the assertion is proved.

Proof of Theorem 4. If λ is an isogeny or a solid homomorphism, we can obtain, by Prop. 9, a pre-group $\mathfrak{p}$-variety (V, f) without defect, which is k-isomorphic to B. Then, by Theorem 3, there exist a prolongation $\{K, \mathfrak{O}, \mathfrak{P}\}$ of $\{k, \mathfrak{o}, \mathfrak{p}\}$ and a group $\mathfrak{P}$-variety B_1, without defect, which is K-isomorphic to (V, f). By the uniqueness of group variety isomorphic to a given pre-group, B_1 is isomorphic to B; so there exists a homomorphism λ_1 of A onto B_1. Since both A and B_1 are without defect as group $\mathfrak{P}$-variety, λ_1 is everywhere defined on $\tilde{A}$. It follows from this and the fact that A is $\mathfrak{p}$-complete, that B_1 is $\mathfrak{P}$-complete; so B_1 is an abelian $\mathfrak{P}$-variety without defect. In the general case, we decompose λ into two homomorphisms, one of which is an isogeny and the other is a solid homomorphism. Applying our result in special cases to these two homomorphisms, we can find a prolongation $\{k', \mathfrak{o}', \mathfrak{p}'\}$ of $\{k, \mathfrak{o}, \mathfrak{p}\}$ and an abelian $\mathfrak{p}$-variety B', without defect, isomorphic to B. Now apply Theorem 2 to the present case, considering $\{B, B', A\}$ to be $\{A, A^*, V\}$ in that theorem; then we obtain a projective abelian $\mathfrak{p}$-variety, which is without defect and birationally equivalent to B over k. Thus Theorem 4 is completely proved.

REMARK. Let B be an abelian subvariety of an abelian $\mathfrak{p}$-variety A; and suppose that A is without defect. Then the specialization $\tilde{B}$ of B is a multiple of an abelian subvariety $\bar{B}$ of $\tilde{A}$: we have $\tilde{B} = p^e\bar{B}$, where p is the characteristic of $\tilde{k}$ (we put 1 in place of p^e if $p = 0$). The multipilicity p^e is independent of the choice of models of A, which are of course assumed to be without defect. On the other hand, since there is a homomorphism of A onto B, by Theorem 4, we know that B is birationally equivalent to an abelian $\mathfrak{p}$-variety without defect. Thus a question arises whether B itself is always an abelian variety without defect. The following example will show that this is not so, namely, B is not necessarily without defect.

Example. Let E be an elliptic curve, embedded in a projective space, over a field of characteristic 0, such that the specialization $\tilde{E}$ of E with respect to $\mathfrak{p}$ is an elliptic curve over $\tilde{k}$, having no point of order p, where p is the characteristic of $\tilde{k}$. Assume that every point t on E of order p is rational over k. Let T_1 and T_2 be two distinct subgroups of E, of order p; and let E_i be the canonical model (by the Chow variety) of the factor group variety of E by T_i; and denote by λ_i the natural homomorphism of E onto E_i. Then, E_1 and E_2 are abelian $\mathfrak{p}$-varieties without defect; and they have the same specialization $\tilde{E}^{(p)}$. Take a point x generic on E over k, and put $x_1 = \lambda_1(x), x_2 = \lambda_2(x)$. Call E_0 the locus of $x_1 \times x_2$ over k on $E_1 \times E_2$. We see then that the specialization of E_0 is $p\tilde{\Delta}^{(p)}$, where $\tilde{\Delta}^{(p)}$ is the diagonal on $\tilde{E}^{(p)} \times \tilde{E}^{(p)}$, so that E_0 is not without defect, while $E_1 \times E_2$ is without defect.

APPENDIX

First we recall some terminologies and elementary results on local rings. We shall call a commutative ring $\mathfrak{R}$ with an identity element a *local ring* if $\mathfrak{R}$ is Noetherian and has a unique maximal ideal. A local ring having no zero-divisor is called a *local domain*. Let $\mathfrak{R}$ be a local ring and $\mathfrak{m}$ the maximal ideal of $\mathfrak{R}$. We call a set of generators $\{u_1, \cdots, u_r\}$ of $\mathfrak{m}$ a *minimal base* of $\mathfrak{m}$ if no proper subset of $\{u_1, \cdots, u_r\}$ generates $\mathfrak{m}$. A set of elements $\{u_1, \cdots, u_r\}$ in $\mathfrak{R}$ is a minimal base of $\mathfrak{m}$ if and only if $\{u_1, \cdots, u_r\}$ gives a base of the module $\mathfrak{m}/\mathfrak{m}^2$ over $\mathfrak{R}/\mathfrak{m}$. Hence the number of elements in any minimal base of $\mathfrak{m}$ is determined by $\mathfrak{R}$. A local ring $\mathfrak{R}$ is said to be *regular* if this number is equal to the dimension of $\mathfrak{R}$. If $\mathfrak{R}$ is regular, every minimal base $\{u_1, \cdots, u_r\}$ of $\mathfrak{m}$ satisfies the following condition:

(R) *If $F(X_1, \cdots, X_r)$ is a homogeneous polynomial in $(X_1, \cdots, X_r)$ of degree ν with coefficients in $\mathfrak{R}$ and if*

$$F(u_1, \cdots, u_r) \in \mathfrak{m}^{\nu+1},$$

then every coefficient of F is contained in $\mathfrak{m}$.

Conversely, if a set of generators $\{u_1, \cdots, u_r\}$ of the maximal ideal $\mathfrak{m}$ of a local ring $\mathfrak{R}$ satisfies this condition, then $\mathfrak{R}$ is regular and $\{u_1, \cdots, u_r\}$ is a minimal base of $\mathfrak{m}$. Every regular local ring has no zero-divisor and is inte-

207

grally closed. If $\mathfrak{R}$ is a local ring and $\mathfrak{m}$ is the maximal ideal of $\mathfrak{R}$, then the powers of $\mathfrak{m}$ define a topology on $\mathfrak{R}$. With respect to this topology, $\mathfrak{R}$ has a completion $\mathfrak{R}^*$, which is a local ring containing $\mathfrak{R}$ as subring and subspace, and in which $\mathfrak{R}$ is dense. $\mathfrak{R}$ is regular if and only if its completion $\mathfrak{R}^*$ is regular; and every minimal base of $\mathfrak{m}$ gives a minimal base of the maximal ideal $\mathfrak{m}^*$ of $\mathfrak{R}^*$.

Our main purpose is to prove the following theorem.

THEOREM. *Notations being as in § 1, let V be a $\mathfrak{v}$-variety and x a generic point of V over k. If a point $\bar{a}$ of $\tilde{V}$ is simple on V, then the local ring $[x \to \bar{a}\,;\,\mathfrak{o}]$ is regular.*

To prove this we need several lemmas. First we generalize the concept of specialization (cf. Northcott [6], Shimura [7]). Let $\mathfrak{R}$ be a local domain and $\mathfrak{m}$ the maximal ideal; let K be the quotient field of $\mathfrak{R}$ and $\tilde{K}$ the residue-field $\mathfrak{R}/\mathfrak{m}$. Let $(x_1, \cdots, x_n)$ be a set of n elements in an extension field of K and $(\xi_1, \cdots, \xi_n)$ a set of n elements in an extension field of $\tilde{K}$. We say that (ξ) is a *specialization of* (x) *over* $\mathfrak{R}$, if the natural homomorphism of $\mathfrak{R}$ onto $\tilde{K} = \mathfrak{R}/\mathfrak{m}$ can be extended to a homomorphism of $\mathfrak{R}[x]$ onto $\tilde{K}[\xi]$ which maps (x) on (ξ). For any polynomial $F(X)$ with coefficients in $\mathfrak{R}$, we denote by $\tilde{F}(X)$ the polynomial with coefficients in $\tilde{K}$ obtained from F considering the coefficients of F modulo $\mathfrak{m}$. (ξ) being a specialization of (x) over $\mathfrak{R}$, we denote by

$$[(x) \to (\xi)\,;\,\mathfrak{R}]$$

the set of elements $F(x)/G(x)$ such that $\tilde{G}(\xi) \neq 0$, where $F(X)$ and $G(X)$ are polynomials in $\mathfrak{R}[X]$. This set is also a local domain.

LEMMA 1. *Let $\mathfrak{R}$ be a regular local ring and K the quotient field of $\mathfrak{R}$. Let a be an element in an algebraic extension of K and α a specialization of a over $\mathfrak{R}$. If there exists a polynomial $F(X)$ in $\mathfrak{R}[X]$ such that $F(a) = 0$ and $\tilde{F}'(\alpha) \neq 0$, then $[a \to \alpha\,;\,\mathfrak{R}]$ is a regular local ring, where F' denotes the derivative of F.*

Proof. Put $\mathfrak{S} = [a \to \alpha\,;\,\mathfrak{R}]$. Let $\mathfrak{m}$ and $\mathfrak{M}$ denote respectively the maximal ideals of $\mathfrak{R}$ and $\mathfrak{S}$; and let $\tilde{K}$ be the residue field $\mathfrak{R}/\mathfrak{m}$. Let $\bar{F}_0(X) = 0$ be an irreducible equation for α over $\tilde{K}$. As we have $\tilde{F}(\alpha) = 0$, $\tilde{F}(X)$ is divisible by $\bar{F}_0(X)$; so there exists a polynomial $F_1(X)$ in $\mathfrak{R}[X]$ such that $\tilde{F}(X) = \bar{F}_0(X)\tilde{F}_1(X)$. By the assumption $\tilde{F}'(\alpha) \neq 0$, we must have $\tilde{F}_1(\alpha) \neq 0$. Let b be an element of $\mathfrak{M}$; then we can find two polynomials $P(X)$ and $Q(X)$ in $\mathfrak{R}[X]$, such that $b = P(a)/Q(a)$, $\tilde{P}(\alpha) = 0$, $\tilde{Q}(\alpha) \neq 0$. There exists a polynomial $G(X)$ in $\mathfrak{R}[X]$ such that $\tilde{P}(X) = \bar{F}_0(X)\tilde{G}(X)$. Let $\{u_1, \cdots, u_r\}$ be a minimal base of $\mathfrak{m}$. Since we have $\tilde{P}(X)\tilde{F}_1(X) = \tilde{F}(X)\tilde{G}(X)$, there exists r polynomials $H_1(X), \cdots, H_r(X)$ in $\mathfrak{R}[X]$ such that

$$P(X)F_1(X) = F(X)G(X) + \sum_{i=1}^{r} u_i H_i(X) .$$

We have then $b = [\sum_i u_i H_i(a)]/[Q(a)F_1(a)]$. We observe $\tilde{Q}(\alpha)\tilde{F}_1(\alpha) \neq 0$, so that r elements $H_i(a)/[Q(a)F_1(a)]$ are all contained in $\mathfrak{S}$. Hence b is contained in $\mathfrak{S}u_1 + \cdots + \mathfrak{S}u_r$. This shows that $\mathfrak{M}$ is generated by $\{u_1, \cdots, u_r\}$. Let $\mathfrak{R}^*$ be the completion of $\mathfrak{R}$ and K^* the quotient field of $\mathfrak{R}^*$. Then, by Theorem 1 of Northcott [6], there exists an isomorphism of $K(a)$ into the algebraic closure of K^*, such that, if a' is the image of a, α is a specialization of a' over $\mathfrak{R}^*$. For our purpose, we may put $a = a'$, so that α is a specialization of a over $\mathfrak{R}^*$. We can easily verify that α is a proper specialization of a over $\mathfrak{R}^*$, in the sense of [6], [7]. Then, by Theorem 3 of [6], a is integral over $\mathfrak{R}^*$. Hence we can find an irreducible polynomial $M(X)$ in $\mathfrak{R}^*[X]$ with the leading coefficient 1 such that $M(a) = 0$. Now we shall show that $\{u_1, \cdots, u_r\}$ has the property (R) for $\mathfrak{M}$. Let $\sum c_{(i)} X_1^{i_1} \cdots X_r^{i_r}$ be a homogeneous polynomial of degree ν with $c_{(i)}$ in $\mathfrak{S}$ such that

$$\sum_{(i)} c_{(i)} u_1^{i_1} \cdots u_r^{i_r} \in \mathfrak{M}^{\nu+1} .$$

Then there exists a homogeneous polynomial $\sum_{(j)} d_{(j)} X_1^{j_1} \cdots X_r^{j_r}$ of degree $\nu+1$ with $d_{(j)}$ in $\mathfrak{S}$ such that

$$\sum_{(i)} c_{(i)} u_1^{i_1} \cdots u_r^{i_r} = \sum_{(j)} d_{(j)} u_1^{j_1} \cdots u_r^{j_r} .$$

We can find a polynomial $\Phi(X)$ in $\mathfrak{R}[X]$ such that $\tilde{\Phi}(\alpha) \neq 0$ and the elements $\Phi(a)c_{(i)}$, $\Phi(a)d_{(j)}$ are contained in $\mathfrak{R}[a]$. Since a satisfies the equation $M(X) = 0$ with the leading coefficient 1, there exist elements $c_{(i)\mu}$, $d_{(j)\mu}$ in $\mathfrak{R}^*$ such that

$$\Phi(a)c_{(i)} = \sum_{\mu=0}^{s-1} c_{(i)\mu} a^\mu, \quad \Phi(a)d_{(j)} = \sum_{\mu=0}^{s-1} d_{(j)\mu} a^\mu ,$$

where s is the degree of $M(X)$. We have then

$$\sum_{\mu=0}^{s-1} \left[\sum_{(i)} c_{(i)\mu} u_1^{i_1} \cdots u_r^{i_r} - \sum_{(j)} d_{(j)\mu} u_1^{j_1} \cdots u_r^{j_r}\right] a^\mu = 0 ,$$

so that we get, for every μ,

$$\sum_{(i)} c_{(i)\mu} u_1^{i_1} \cdots u_r^{i_r} = \sum_{(j)} d_{(j)\mu} u_1^{j_1} \cdots u_r^{j_r} \in (\mathfrak{m}^*)^{\nu+1} .$$

As $\{u_1, \cdots, u_r\}$ satisfies the condition (R) for $\mathfrak{m}^*$, we have $c_{(i)\mu} \in \mathfrak{m}^*$, so that the $c_{(i)}$ are contained in the maximal ideal of $[a \rightarrow \alpha; \mathfrak{R}^*]$. Since the $c_{(i)}$ are elements of $\mathfrak{S}$, we have $c_{(i)} \in \mathfrak{M}$ for every (i). Thus we have shown that $\{u_1, \cdots, u_r\}$ satisfies the condition (R) for $\mathfrak{M}$. This proves our lemma.

LEMMA 2. *Let $\mathfrak{R}$ be a local domain and $\mathfrak{m}$ the maximal ideal of $\mathfrak{R}$. Let K be the quotient field of $\mathfrak{R}$ and $\tilde{K}$ the residue-field $\mathfrak{R}/\mathfrak{m}$. Let t be a variable over*

K and τ a variable over $\tilde{K}$. If the local domain $[t \to \tau; \mathfrak{R}]$ is regular, so is $\mathfrak{R}$.

Proof. Put $\mathfrak{S} = [t \to \tau; \mathfrak{R}]$; let $\mathfrak{M}$ be the maximal ideal of $\mathfrak{S}$. We can easily verify that $\mathfrak{M} = \mathfrak{S}\mathfrak{m}$ and $\mathfrak{m} = \mathfrak{M} \cap \mathfrak{R}$. Let $\{u_1, \cdots, u_r\}$ be a minimal base of $\mathfrak{m}$. We shall prove that $\{u_1, \cdots, u_r\}$ gives a base of $\mathfrak{M}/\mathfrak{M}^2$ over $\mathfrak{S}/\mathfrak{M}$. Suppose that $\sum a_i u_i \in \mathfrak{M}^2$ for r elements a_i of $\mathfrak{S}$. Then we can find a polynomial $f(t)$ in $\mathfrak{R}[t]$ such that $\tilde{f}(\tau) \neq 0$, $f(t)a_i \in \mathfrak{R}[t]$ and

$$f(t) \sum_i a_i u_i = \sum_{i,j} g_{ij}(t) u_i u_j,$$

where the $g_{ij}(t)$ are elements of $\mathfrak{R}[t]$. Put

$$f(t)a_i = \sum_\nu a_{i\nu} t^\nu, \quad g_{ij}(t) = \sum_\nu b_{ij\nu} t^\nu,$$

where the $a_{i\nu}$ and the $b_{ij\nu}$ are elements of $\mathfrak{R}$. Then we have $\sum_i a_{i\nu} u_i = \sum_{i,j} b_{ij\nu} u_i u_j$ for every ν; since $\{u_1, \cdots, u_r\}$ is a base of $\mathfrak{m}/\mathfrak{m}^2$ over $\mathfrak{R}/\mathfrak{m}$, we have $a_{i\nu} \in \mathfrak{m}$, so that the a_i are contained in $\mathfrak{M}$. Thus we have proved that $\{u_1, \cdots, u_r\}$ is a minimal base of $\mathfrak{M}$, so that $\{u_1, \cdots, u_r\}$ satisfies the condition (R) for the ideal $\mathfrak{M}$. Then it is obvious that $\{u_1, \cdots, u_r\}$ satisfies (R) for the ideal $\mathfrak{m}$. This proves our lemma.

LEMMA 3. *Notations being as in Lemma 2, let α be an element which is algebraic over $\tilde{K}$. If $\mathfrak{R}$ is regular, so is $[t \to \alpha; \mathfrak{R}]$.*

Proof. Put $\mathfrak{S} = [t \to \alpha; \mathfrak{R}]$; let $\mathfrak{M}$ be the maximal ideal of $\mathfrak{S}$. We can find a polynomial $F(X)$ in $\mathfrak{R}[X]$, such that $\tilde{F}(X) = 0$ is an irreducible equation for α over $\tilde{K}$; we may assume that F and $\tilde{F}$ have the same degree d. Let $\{u_1, \cdots, u_r\}$ be a minimal base of $\mathfrak{m}$; put $u_0 = F(t)$. We shall prove that $\{u_0, u_1, \cdots, u_r\}$ satisfies the condition (R) for $\mathfrak{M}$. Let x be an element of $\mathfrak{M}$; then we can find two polynomials $P(t)$, $Q(t)$ in $\mathfrak{R}[t]$ such that $x = P(t)/Q(t)$, $\tilde{P}(\alpha) = 0$ and $\tilde{Q}(\alpha) \neq 0$. There exists a polynomial $G(t)$ in $\mathfrak{R}[t]$ such that $\tilde{P}(X) = \tilde{F}(X)\tilde{G}(X)$. We see that $P(t) - F(t)G(t)$ is contained in $\mathfrak{m}[t]$. This shows that $u_0, u_1, \cdots, u_r$ generate $\mathfrak{M}$. Let $\sum a_{(i)} X_0^{i_0} \cdots X_r^{i_r}$ be a homogeneous polynomial of degree ν with a_i in $\mathfrak{S}$ such that

$$\sum_{(i)} a_{(i)} u_0^{i_0} u_1^{i_1} \cdots u_r^{i_r} \in \mathfrak{M}^{\nu+1}.$$

Then there exists a homogeneous polynomial $\sum b_{(j)} X_0^{j_0} X_1^{j_1} \cdots X_r^{j_r}$ of degree $\nu+1$ with $b_{(j)}$ in $\mathfrak{S}$ such that

$$\sum_{(i)} a_{(i)} u_0^{i_0} u_1^{i_1} \cdots u_r^{i_r} = \sum_{(j)} b_{(j)} u_0^{j_0} u_1^{j_1} \cdots u_r^{j_r}.$$

We can find a polynomial $f(t)$ in $\mathfrak{R}[t]$ such that $\tilde{f}(\alpha) \neq 0$ and the elements $f(t)a_{(i)}$, $f(t)b_{(j)}$ are contained in $\mathfrak{R}[t]$. Since $u_0 = F(t)$ is a polynomial in t of degree d, there exist polynomials $a_{(i)\mu}(t)$, $b_{(i)\mu}(t)$ in $\mathfrak{R}[t]$, all of degree less than d, such that

$$f(t)a_{(i)} = \sum_{\mu=0}^{m} a_{(i)\mu}(t)u_0{}^{\mu}, \quad f(t)b_{(j)} = \sum_{\mu=0}^{m} b_{(j)\mu}(t)u_0{}^{\mu}.$$

We have then

$$\sum_{(i)} a_{(i)0}u_0{}^{i_0}\cdots u_r{}^{i_r} = \sum_{\mu=0}^{m}\sum_{(j)} b_{(j)\mu}u_0{}^{j_0+\mu}u_1{}^{j_1}\cdots u_r{}^{j_r} - \sum_{\mu=1}^{m}\sum_{(i)} a_{(i)\mu}u_0{}^{i_0+\mu}u_1{}^{i_1}\cdots u_r{}^{i_r}.$$

We can rewrite this equation in the form

$$\sum_{\lambda=0}^{\nu} u_0{}^{\lambda}\Phi_{\lambda}(t, u_1, \cdots, u_r) = \sum_{\lambda=0}^{n} u_0{}^{\lambda}\Psi_{\lambda}(t, u_1, \cdots, u_r),$$

where Φ_{λ} is a polynomial with coefficients in $\mathfrak{R}$ of degree $<d$ in t and homogeneous in $(u_1, \cdots, u_r)$ of degree $\nu-\lambda$ and Ψ_{λ} is a polynomial with coefficients in $\mathfrak{R}$ of degree $<d$ in t and of degree $\geq \nu-\lambda+1$ in $(u_1, \cdots, u_r)$. We have then, for every λ,

$$\Phi_{\lambda}(t, u_1, \cdots, u_r) = \Psi_{\lambda}(t, u_1, \cdots, u_r).$$

We see that the right hand side is contained in $\mathfrak{m}^{\nu-\lambda+1}[t]$. Since $\{u_1, \cdots, u_r\}$ satisfies (R) for $\mathfrak{m}$, the coefficients of the polynomial $\Phi_{\lambda}(T, U_1, \cdots, U_r)$ are contained in $\mathfrak{m}$. It follows from this that the $a_{(i)}$ are contained in $\mathfrak{M}$. Hence $\{u_0, u_1, \cdots, u_r\}$ satisfies (R) for $\mathfrak{M}$.

LEMMA 4. *The notations* $\mathfrak{o}, \mathfrak{p}, k, \tilde{k}$ *being as in* §1, *let* $t_1, \cdots, t_n$ *be* n *independent variables over* k *and* $\bar{a}_1, \cdots, \bar{a}_n$ *be* n *elements in an extension of* $\tilde{k}$. *Then,* $[(t) \to (\bar{a}); \mathfrak{o}]$ *is a regular local ring.*

Proof. Let s be the dimension of $(\bar{a})$ over $\tilde{k}$; if s is not 0, we may, after reordering the $\bar{a}_i$ if necessary, assume that $\bar{a}_1, \cdots, \bar{a}_s$ are independent variables over $\tilde{k}$ and $(\bar{a})$ is algebraic over $\tilde{k}(\bar{a}_1, \cdots, \bar{a}_s)$. Put $\mathfrak{o}' = [(t_1, \cdots, t_s) \to (\bar{a}_1, \cdots, \bar{a}_s); \mathfrak{o}]$. Then $\mathfrak{o}'$ is a discrete valuation ring of rank 1; and we have

$$[(t) \to (\bar{a}); \mathfrak{o}] = [(t_{s+1}, \cdots, t_n) \to (\bar{a}_{s+1}, \cdots, \bar{a}_n); \mathfrak{o}'].$$

Hence by lemma 3, $[(t) \to (\bar{a}); \mathfrak{o}]$ is a regular local ring.

We shall now prove Theorem 1. We may assume that V is an affine variety. Let r and n be respectively the dimensions of V and the ambient space for V. Let t_{ij}, for $0 \leq i \leq r$, $1 \leq j \leq n$, be $(r+1)n$ independent variables over $k(x)$; and let $\bar{t}_{ij}$, for $0 \leq i \leq r$, $1 \leq j \leq n$, be $(r+1)n$ independent variables over $\tilde{k}(\bar{a})$. Put $y_i = \sum_j t_{ij}x_j$, $\bar{b}_i = \sum_j \bar{t}_{ij}\bar{a}_i$ and

$$\mathfrak{o}' = [(t_{ij}) \to (\bar{t}_{ij}); \mathfrak{o}],$$

$$\mathfrak{R} = [(y_1, \cdots, y_r) \to (\bar{b}_1, \cdots, \bar{b}_r); \mathfrak{o}'],$$

$$\mathfrak{S} = [y_0 \to \bar{b}_0; \mathfrak{R}].$$

Then, $\mathfrak{o}'$ is a discrete valuation ring. Since $y_1, \cdots, y_r$ are independent variables over $k(t_{ij})$, $\mathfrak{R}$ is a regular local ring by virtue of Lemma 4. By the proof of

Theorem 15 of [7], $\bar{b}_0$ is a proper specialization of y_0 over $\mathfrak{R}$ of multiplicity 1. Let

$$F(Y_0, Y_1, \cdots, Y_r) = 0$$

be an irreducible equation for $(y_0, y_1, \cdots, y_r)$ over $k(t_{ij})$. We may assume that all coefficients of F are contained in $\mathfrak{o}'$ and at least one of them is equal to 1. Then, we have $\tilde{F}(Y_0, \bar{b}_1, \cdots, \bar{b}_r) \neq 0$; for otherwise, $\bar{c}$ being a variable over $\tilde{k}(\bar{t}_{ij}, \bar{b}_i)$, $\bar{c}$ would be a specialization of y_0 over $\mathfrak{R}$; this contradicts the fact that $\bar{b}_0$ is a proper specialization of y_0 over $\mathfrak{R}$. Since $\bar{b}_0$ is of multiplicity 1, $\bar{b}_0$ is a simple root of the equation

$$\tilde{F}(Y_0, \bar{b}_1, \cdots, \bar{b}_r) = 0 ;$$

so we have $\partial\tilde{F}/\partial Y_0(\bar{b}_0, \bar{b}_1, \cdots, \bar{b}_r) \neq 0$. By Lemma 1, this proves that $\mathfrak{S}$ is a regular local ring; in particular, $\mathfrak{S}$ is integrally closed in its quotient field $k(t, y)$. By Proposition 16 of [7], we have $k(t, x) = k(t, y)$ and x is finite over $\mathfrak{S}$. Since $\mathfrak{S}$ is integrally closed, the coordinates of x must be contained in $\mathfrak{S}$. This shows

$$\mathfrak{S} = [x \rightarrow a; \mathfrak{o}'] .$$

Put $\mathfrak{T} = [x \rightarrow \bar{a}; \mathfrak{o}]$; then it is easy to see $\mathfrak{S} = [(t_{ij}) \rightarrow (\bar{t}_{ij}); \mathfrak{T}]$. Hence, by Lemma 2, $\mathfrak{T}$ is a regular local ring; so our theorem is proved.

We profit by this opportunity to revise some points in [7].

(1) The proof of Proposition 17 is omitted by reason of that it is a translation of Proposition 19 of [9] Chap. V. The first part which asserts $\partial\bar{F}/\partial Z(\eta, \zeta) \neq 0$ is proved in fact in the same way as in Weil's book. It is hardly possible, however, to prove the remaining part by the same argument as in [9]. The above Theorem 1, or Lemma 1, with their proofs, will supply this gap.

(2) p. 150, the lowest line. "Obviously, (η) is" should be read "Obviously, (ξ) is".

(3) p. 151, the first line. (τ_{ij}, τ_i) should be read $(\delta_{ij}, \varepsilon_i)$.

(4) p. 155. Corollary of Theorem 10 should be as follows:

Corollary. *Let V be a variety defined over k and $\mathfrak{B}$ a component of $\bar{V}$. If $\mathfrak{B}$ is simple on V, then we have $\mu(V, \mathfrak{B}) = 1$ and $[\mathfrak{B} : \kappa]_i = 1$.*

References

[1]	Chow, W. L., On the quotient variety of an abelian variety, *Proc. Nat. Acad. Sci. U. S. A.*, **38**, 1039-1044 (1952).

[2]	Hironaka, H., A note on algebraic geometry over ground rings. The invariance of Hilbert characteristic functions under the specialization process, *Illinois J. Math.*, **2**, 355-366 (1958).

[3] Igusa, J., Arithmetic genera of normal varieties in an algebraic family, *Proc. Nat. Acad. Sci. U.S.A.*, **41**, 34-37 (1955).

[4] Koizumi, S., On the differential forms of the first kind on algebraic varieties, *J. Math. Soc. Japan*, **1**, 273-280 (1949).

[5] Nishi, M., Some results on abelian varieties, *Nat. Sci. Rep., Ochanomizu Univ.*, **9**, 1-12 (1958).

[6] Northcott, D. G., Specializations over a local domain, *Proc. London Math. Soc.* (III), **1**, 129-137 (1951).

[7] Shimura, G., Reduction of algebraic varieties with respect to a discrete valuation of the basic field, *Amer. J. Math.*, **77**, 134-176 (1955).

[8] Shimura, G. and Y. Taniyama, Complex multiplication of abelian varieties and its applications to number theory, *to appear*.

[9] Weil, A., Foundations of algebraic geometry, *Amer. Math. Soc. Colloquium Publications*, (New York, 1946).

[10] ————, *Variétés abéliennes et courbes algébriques*, (Hermann, Paris, 1948).

[11] ————, On algebraic groups of transformations, *Amer. J, Math.*, **77**, 355-391 (1955).

[12] ————, On the projective embedding of abelian varieties, Algebraic geometry and topology, *A symposium in honor of S. Lefschetz*, (Princeton, 1957).

[13] ————, The field of definition of a variety, *Amer. J. Math.*, **78**, 509-524 (1956).

60a

On vector differential forms attached to automorphic forms

Jointly with Michio Kuga

Journal of the Mathematical Society of Japan, 12 (1960), 258-270

(Received Sept. 18, 1959)

In recent works [2], [3], it was found that the integral of certain vector differential forms, attached to automorphic forms with respect to a Fuchsian group G, is important in the arithmetic theory of modular correspondences. Those vector differential forms ω are defined on the upper half plane and satisfy the transformation formula

$$(1) \qquad \omega \circ \sigma = M(\sigma)\omega$$

for every element σ of the group G, where $M(\sigma)$ is a tensor representation of G. The object of the present paper is to determine all holomorphic forms satisfying this relation (1). M being of degree $2m-1$, we can attach to every cusp form of degree $\leqq 2m$ a holomorphic form ω with the representation M (Theorem 1). Conversely, any holomorphic form satisfying (1) is expressed as a sum of the forms thus obtained from cusp forms of degree $\leqq 2m$; and this expression gives a direct decomposition of the vector space $\mathfrak{F}$ of such holomorphic forms (Theorem 2). Hence the dimension of the vector space $\mathfrak{F}$ is easily obtained if we know the dimension of the linear space of cusp forms for each degree. We note that the integral of the form attached to a cusp form of degree $<2m$ has a period cohomologous to 0, in the sense described in [3]. This fact distinguishes among such forms the forms attached to cusp forms of degree $2m$, which were the object of the investigation in [3].

§1. Cusp forms with respect to a Fuchsian group.

Let $\mathscr{H}$ denote the upper half plane, the set of all complex numbers with positive imaginary parts. Every element $\sigma = \begin{pmatrix} a & b \\ c & d \end{pmatrix}$ of $SL(2, \boldsymbol{R})$ operates on $\mathscr{H}$, as usual:

$$\sigma(z) = \frac{az+b}{cz+d};$$

we put

$$J(\sigma, z) = (cz+d)^{-1}.$$

For every differential form ω on $\mathcal{H}$ we shall denote by $\omega \circ \sigma$ the transform of ω by σ; so if ω is expressed in the form $\omega = f(z)dz$ for a function $f(z)$ on $\mathcal{H}$, we have $\omega \circ \sigma = f(\sigma(z))J(\sigma,z)^2 dz$.

Let G be a discrete subgroup of $SL(2, \boldsymbol{R})$ such that $SL(2, \boldsymbol{R})/G$ has a finite total volume, measured by an invariant volume element. Then, G, as group of transformations on $\mathcal{H}$, is a Fuchsian group; namely, G operates discontinuously on $\mathcal{H}$ and $\mathcal{H}/G$ has a fundamental domain $\mathcal{D}$ with a finite Poincaré area. If we denote by $\mathcal{H}^*$ the join of $\mathcal{H}$ and the "cusps" of G, the quotient space $\mathcal{H}^*/G$, with a suitable analytic structure, can be regarded as a compact Riemann surface.

A *cusp* of G is the fixed point of a parabolic element of G, which is a real number or the point at infinity ∞. Let s be a cusp of G. Put

$$\rho = \begin{pmatrix} -s & 1 \\ -1 & 0 \end{pmatrix} \quad \text{or} \quad \rho = \begin{pmatrix} 1 & 0 \\ 0 & 1 \end{pmatrix}$$

according as s is a real number or ∞. Then the set of all elements of G having s as fixed point is the free cyclic group generated by an element τ of G, which is of the form

$$\tau = \rho \begin{pmatrix} 1 & h \\ 0 & 1 \end{pmatrix} \rho^{-1},$$

where h is a positive real number.

Let ν be an integer. We shall understand, by an *automorphic form of degree ν with respect to G*, a function $f(z)$ on $\mathcal{H}$ satisfying the following conditions (A 1-3).

(A 1) *$f(z)$ is meromorphic on $\mathcal{H}$.*

(A 2) *For every $\sigma \in G$, we have $f(\sigma(z))J(\sigma, z)^\nu = f(z)$.*

Consider a cusp s of G; the transformation ρ and the positive number h being defined for s as above, we see that, if f satisfies (A 1-2), the function $f(\rho(z))J(\rho, z)^\nu$ is invariant under the translation $z \to z+h$. Hence, if we put

$$q = \exp(2\pi i h^{-1} z),$$

there exists a function $g(q)$ meromorphic in the domain $0 < |q| < 1$ such that

$$f(\rho(z))J(\rho, z)^\nu = g(q).$$

The condition (A 3) is now stated as follows.

(A 3) *For every cusp s of G, the function $g(q)$, defined as above, is meromorphic at $q = 0$.*

An automorphic form $f(z)$ with respect to G is called a *cusp form with respect to G*, if the following conditions are satisfied.

(A 1′) *$f(z)$ is holomorphic on $\mathcal{H}$.*

(A 3′) *For every cusp s of G, the function $g(q)$, defined as above, is holomorphic and takes the value 0 at $q = 0$.*

We denote by $S_\nu(G)$ the set of all cusp forms of degree ν with respect to G. In this paper we shall only deal with the forms of *even* degree.

§ 2. M_n-forms and M_n-vectors.

Let

$$GL(2, C) \ni \sigma \to M_n(\sigma) \in GL(n+1, C)$$

be the representation of $GL(2, C)$ by symmetric contravariant tensors of order n, so that the equality

$$\sigma\begin{pmatrix} u \\ v \end{pmatrix} = \begin{pmatrix} z \\ w \end{pmatrix}$$

is led to

$$M_n(\sigma)\begin{pmatrix} u^n \\ u^{n-1}v \\ \vdots \\ uv^{n-1} \\ v^n \end{pmatrix} = \begin{pmatrix} z^n \\ z^{n-1}w \\ \vdots \\ zw^{n-1} \\ w^n \end{pmatrix}.$$

For instance, we have

$$(2) \qquad M_n\left(\begin{pmatrix} 1 & z \\ 0 & 1 \end{pmatrix}\right) = \begin{pmatrix} 1 & nz & \dfrac{n(n-1)}{2}z^2 & \cdots & z^n \\ 0 & 1 & (n-1)z & \cdots & z^{n-1} \\ & & \cdots & & \cdots \\ 0 & \cdots & & \cdots & 1 \end{pmatrix}.$$

As this matrix $M_n\left(\begin{pmatrix} 1 & z \\ 0 & 1 \end{pmatrix}\right)$ will be often used in our investigation, we denote it briefly by $L_n(z)$:

$$L_n(z) = M_n\left(\begin{pmatrix} 1 & z \\ 0 & 1 \end{pmatrix}\right).$$

We have then, for every $\tau \in SL(2, R)$,

$$(3) \qquad L_n(\tau(z))^{-1}M_n(\tau)L_n(z) = M_n\left(\begin{pmatrix} J & 0 \\ c & J^{-1} \end{pmatrix}\right),$$

where $J = J(\tau, z) = (cz+d)^{-1}$. In particular, if r is a real number and $\tau = \begin{pmatrix} 1 & r \\ 0 & 1 \end{pmatrix}$, we have

$$(4) \qquad L_n(\tau(z)) = M_n(\tau)L_n(z).$$

Let f be an automorphic form of degree $n+2$ with respect to G. In [3] we have studied the vector differential form

$$(5) \qquad \omega = L_n(z)\begin{pmatrix} 0 \\ \vdots \\ 0 \\ f \end{pmatrix} dz = \begin{pmatrix} fz^n dz \\ fz^{n-1}dz \\ \vdots \\ fdz \end{pmatrix}$$

which satisfies, for every $\sigma \in G$,

$$\omega \circ \sigma = M_n(\sigma)\omega \,.$$

This is an example of M_n-form, whose definition is given as follows. A column vector of dimension $n+1$

$$\omega = \begin{pmatrix} \omega_0 \\ \omega_1 \\ \vdots \\ \omega_n \end{pmatrix}$$

is called an M_n-*form with respect to G,* if it satisfies the following conditions (M 1-3).

(M 1) *Each component ω_k is a meromorphic differential form on $\mathcal{H}$.*

(M 2) *For every $\sigma \in G$, we have $\omega \circ \sigma = M_n(\sigma)\omega$.*

Let s be a cusp of G; and let ρ and h be as in §1. Then, if ω satisfies (M 1-2), we can easily verify, using (4), that the form

$$L_n(z)^{-1} M_n(\rho)^{-1} \omega \circ \rho$$

is invariant under the translation $z \to z+h$. Therefore, if we put $q = \exp(2\pi i h^{-1} z)$, there exist $n+1$ functions $f_0(q), \cdots, f_n(q)$, meromorphic in $0 < |q| < 1$, such that

$$\tag{6} L_n(z)^{-1} M_n(\rho)^{-1} \omega \circ \rho = \begin{pmatrix} f_0(q)dq \\ \vdots \\ f_n(q)dq \end{pmatrix}.$$

Now the condition (M 3) is stated as follows.

(M 3) *For every cusp s of G, the functions $f_k(q)$ defined by (6) are meromorphic at $q = 0$.*

An M_n-form ω with respect to G is called a *cusp M_n-form with respect to G* if the following conditions (M 1') and (M 3') are satisfied.

(M 1') *Every component of ω is holomorphic on $\mathcal{H}$.*

(M 3') *For every cusp s of G, the functions $f_k(q)$ defined by (6) are holomorphic at $q = 0$.*

We can prove that the form ω defined by (5) is an M_n-form with respect to G; it is a cusp M_n-form if and only if $f(z)$ is a cusp form. This fact is a special case of the following Theorem 1. We shall denote by $\mathfrak{F}_n(G)$ the set of all cusp M_n-forms with respect to G.

Considering functions in place of differential forms, we get the following definition. A column vector of dimension $n+1$

$$\mathfrak{g} = \begin{pmatrix} g_0 \\ \vdots \\ g_n \end{pmatrix}$$

is called an M_n-*vector with respect to G,* if it satisfies the following conditions

(V 1-3).

    (V 1) *Every component g_k is a meromorphic function on $\mathcal{H}$.*

    (V 2) *For every $\sigma \in G$, we have $\mathfrak{g} \circ \sigma = M_n(\sigma)\mathfrak{g}$.*

The notations s, ρ, h being as above, if $\mathfrak{g}$ satisfies (V 1-2), there exist $n+1$ functions $F_0(q), \cdots, F_n(q)$, meromorphic in $0 < |q| < 1$, such that

$$
(7) \qquad L_n(z)^{-1} M_n(\rho)^{-1} \mathfrak{g} \circ \rho = \begin{pmatrix} F_0(q) \\ \vdots \\ F_n(q) \end{pmatrix}.
$$

    (V 3) *For every cusp s of G, the functions $F_k(q)$ defined by (7) are meromorphic at $q = 0$.*

    An M_n-vector $\mathfrak{g}$ with respect to G is called a *cusp M_n-vector with respect to G* if the following conditions (V 1′) and (V 3′) are satisfied.

    (V 1′) *Every component of $\mathfrak{g}$ is holomorphic on $\mathcal{H}$.*

    (V 3′) *For every cusp s of G, the functions $F_k(q)$ defined by (7) are holomorphic and take the value 0 at $q = 0$.*

    We shall denote by $\mathfrak{B}_n(G)$ the set of all cusp M_n-vectors with respect to G.

§ 3. Main results.

    We shall now state our results in the following theorems, for which the proofs will be given in § 4. We first introduce some notations. For every integer k and a non-negative integer j, we shall write

$$
\binom{k}{j} = \begin{cases} 1 & \text{for } j = 0, \\ \dfrac{k(k-1)\cdots(k-j+1)}{j!} & \text{for } j > 0. \end{cases}
$$

Consider a triplet (n, ν, k) of integers such that

    i) n is even and non-negative;

    ii) ν is even and $-(n-2) \leq \nu \leq n+2$;

    iii) $0 \leq k \leq n - \dfrac{\nu + n - 2}{2}$.

For such a triplet (n, ν, k), we put

$$
\alpha_{n, \nu, k} = \begin{cases} 0 & \text{for } \nu + k - 1 < 0, \\ \dfrac{\left(k + \dfrac{\nu + n - 2}{2}\right)!}{k!\,(\nu + k - 1)!} & \text{for } \nu + k - 1 \geq 0 \end{cases}
$$

and

$$
\gamma_{n, \nu, k} = \begin{cases} 0 & \text{for } \nu + k - 1 < 0, \\ \dfrac{\left(k + \dfrac{\nu + n}{2}\right)!}{k!\,(\nu + k - 1)!} & \text{for } \nu + k - 1 \geq 0. \end{cases}
$$

For fixed n and ν, we denote $\alpha_{n,\nu,k}$ and $\gamma_{n,\nu,k}$ simply by α_k and γ_k.

LEMMA 1. *Let t be an integer such that $0 \leq t \leq n$ and $f_t, f_{t+1}, \cdots, f_n$ be $n-t+1$ meromorphic functions on $\mathcal{H}$. If*

$$\omega = L_n(z) \begin{pmatrix} 0 \\ \vdots \\ 0 \\ f_t \\ \vdots \\ f_n \end{pmatrix} dz$$

is an M_n-form with respect to G, then f_t is an automorphic form of degree $2t+2-n$ with respect to G. Moreover, if ω is a cusp M_n-form, f_t is a cusp form.

THEOREM 1. *Let n and ν be two even integers such that $n > 0$ and $-(n-2) \leq \nu \leq n+2$; put $\mu = \dfrac{n+2-\nu}{2}$. Then, for every automorphic form f of degree ν with respect to G, the vector differential form*

$$(8) \qquad \omega = L_n(z) \begin{pmatrix} 0 \\ \vdots \\ 0 \\ \alpha_0 f \\ \alpha_1 f' \\ \vdots \\ \alpha_\mu f^{(\mu)} \end{pmatrix} dz$$

is an M_n-form with respect to G, where $\alpha_0 = \alpha_{n,\nu,0}, \cdots, \alpha_k = \alpha_{n,\nu,k}$; $f', \cdots, f^{(\mu)}$ denote the derivatives $df/dz, \cdots, d^\mu f/dz^\mu$; and the number of 0 in the column is $n-\mu$. Moreover, in order that ω is a cusp M_n-form, it is necessary and sufficient that f is a cusp form.

Remark that, if $\nu \leq 0$, we have $\alpha_0 = \alpha_1 = \cdots = \alpha_{-\nu} = 0$. We denote by $\mathfrak{S}^n_\nu(G)$ the set of all M_n-forms ω of the form (8), where f is a cusp form of degree ν. If $\nu \leq 0$, the set $\mathfrak{S}^n_\nu(G)$ consists only of the zero element. If $\nu > 0$, we have $\alpha_0 \neq 0$, so that the vector space $\mathfrak{S}^n_\nu(G)$ is canonically isomorphic to the vector space $S_\nu(G)$ by the mapping $f \to \omega$.

THEOREM 2. *The vector space $\mathfrak{F}_n(G)$ of all cusp M_n-forms is the direct sum of the vector spaces $\mathfrak{S}^n_\nu(G)$ for even ν such that $2 \leq \nu \leq n+2$:*

$$\mathfrak{F}_n(G) = \mathfrak{S}^n_2(G) + \cdots + \mathfrak{S}^n_n(G) + \mathfrak{S}^n_{n+2}(G).$$

Hence, if we denote by $d_\nu(G)$ the dimension of the vector space $S_\nu(G)$, the dimension of the vector space $\mathfrak{F}_n(G)$ is equal to

$$d_2(G) + \cdots + d_n(G) + d_{n+2}(G).$$

The number $d_\nu(G)$ is easily obtained by means of Riemann-Roch Theorem.

We note that from Lemma 1 and Theorem 1 follows a result of Bol [1], which asserts the $(n-1)$-th derivative of an automorphic form of degree $-(n-2)$ to be an automorphic form of degree n. In fact, consider the case $\nu = -(n-2)$ in Theorem 1; we have then

$$\alpha_0 = \alpha_1 = \cdots \alpha_{n-2} = 0, \; \alpha_{n-1} \neq 0 ;$$

so the vector

$$L_n(z) \begin{pmatrix} 0 \\ \vdots \\ 0 \\ \alpha_{n-1} f^{(n-1)} \\ \alpha_n f^{(n)} \end{pmatrix} dz$$

is an M_n-form for every automorphic form f of degree $-(n-2)$. Hence, by Lemma 1, $f^{(n-1)}$ is an automorphic form of degree n.

THEOREM 3. *Let the integers n, ν, μ be the same as in Theorem 1. Then, for every automorphic form f of degree ν with respect to G, the vector function*

$$(9) \qquad \mathfrak{f} = L_n(z) \begin{pmatrix} 0 \\ \vdots \\ 0 \\ \gamma_0 f \\ \gamma_1 f' \\ \vdots \\ \gamma_{\mu-1} f^{(\mu-1)} \end{pmatrix}$$

is an M_n-vector with respect to G, where $\gamma_k = \gamma_{n, \nu, k}$; and the number of 0 in the column is $n - \mu + 1$. Moreover, in order that $\mathfrak{f}$ is a cusp M_n-vector, it is necessary and sufficient that f is a cusp form.

Denote by $\mathfrak{T}^n{}_\nu(G)$ the set of all M_n-vectors $\mathfrak{f}$ of the form (9), where f is a cusp form of degree ν. We see easily $\mathfrak{T}^n{}_\nu(G) = \{0\}$ for $\nu \leq 0$ and $\nu = n+2$. If $0 < \nu \leq n$, we have $\gamma_0 \neq 0$, so that the vector space $\mathfrak{T}^n{}_\nu(G)$ is canonically isomorphic to the vector space $S_\nu(G)$ by the mapping $f \to \mathfrak{f}$.

THEOREM 4. *The vector space $\mathfrak{B}^n(G)$ of all cusp M_n-vectors is the direct sum of the vector spaces $\mathfrak{T}^n{}_\nu(G)$ for even ν such that $2 \leq \nu \leq n$:*

$$\mathfrak{B}^n(G) = \mathfrak{T}^n{}_2(G) + \cdots + \mathfrak{T}^n{}_n(G) .$$

Now we consider the differential $d\mathfrak{f}$ of an M_n-vector $\mathfrak{f}$. If $\mathfrak{f}$ is an M_n-vector with respect to G, then we can easily prove that $d\mathfrak{f}$ is an M_n-form with respect to G; if $\mathfrak{f}$ is a cusp M_n-vector, then $d\mathfrak{f}$ is a cusp M_n-form. More precisely, we have

THEOREM 5. *The integers n, ν, μ being as in Theorem 1, let f be an auto-*

morphic form of degree ν with respect to G. Define an M_n-form ω and an M_n-vector $\mathfrak{f}$ by (8) and (9). Then we have

$$d\mathfrak{f} = \mu(n - \mu + 1)\omega .$$

Remark that $\mu(n-\mu+1) \neq 0$ if $\mu \geq 1$. Hence, if $0 < \nu \leq n$, the mapping $\mathfrak{f} \to d\mathfrak{f}$ gives an isomorphism of $\mathfrak{T}^n_\nu(G)$ onto $\mathfrak{S}^n_\nu(G)$.

From the last theorem, we can conclude that, if $0 < \nu < n+2$ and if $\omega \in \mathfrak{S}^n_\nu(G)$, the period of the integral $\int^z \omega$ is cohomologous to 0 in the sense of [3]. On the other hand, Theorem 1 of [3] claims that the period of $\int^z \omega$ is not cohomologous to 0 for every element $\omega \neq 0$ of $\mathfrak{S}^n_{n+2}(G)$. Therefore, we obtain the following result.

THEOREM 6. *Let $\mathfrak{N}_n(G)$ denote the set of all cusp M_n-forms with respect to G, whose integrals have the periods cohomologous to 0. Then, the factor space $\mathfrak{F}_n(G)/\mathfrak{N}_n(G)$ is canonically isomorphic to $S_{n+2}(G)$.*

Put, similarly as in [3], for $\omega, \eta \in \mathfrak{F}_n(G)$,

$$(\omega, \eta) = i\int_{\mathscr{D}} {}^t\omega P_n \bar{\eta} ,$$

where P_n is the symmetric matrix introduced in §1 of [3] and $\mathscr{D}$ is a fundamental domain of G. Then (ω, η) is a Hermitian form on $\mathfrak{F}_n(G)$. By the above considerations, we see that two subspaces $\mathfrak{S}^n_{n+2}(G)$ and $\mathfrak{S}^n_2(G) + \cdots + \mathfrak{S}^n_n(G)$ of $\mathfrak{F}_n(G)$ are transversal to each other with respect to this form (ω, η), and (ω, η) is a zero form on the latter space, while it is a definite form on the former space (§2 of [3]).

§4. Proofs of Theorems.

LEMMA 2. *Let $f_0, \cdots, f_n$ be $n+1$ meromorphic functions on $\mathscr{H}$; put*

$$\mathfrak{f} = \begin{pmatrix} f_0 \\ \vdots \\ f_n \end{pmatrix}, \quad \omega = L_n(z)\mathfrak{f}dz .$$

Then, ω satisfies the condition (M 2) *if and only if*

$$(\mathfrak{f} \circ \sigma)J^2 = M_n\left(\begin{pmatrix} J & 0 \\ c & J^{-1} \end{pmatrix}\right)\mathfrak{f}$$

holds for every $\sigma = \begin{pmatrix} a & b \\ c & d \end{pmatrix} \in G$, where $J = J(\sigma, z) = (cz+d)^{-1}$.

This follows from the relation (3) of §2.

Let $\tau = \begin{pmatrix} a & b \\ c & d \end{pmatrix}$ be an element of $SL(2, \boldsymbol{R})$ and $J = (cz+d)^{-1}$; we have then

$$(10) \qquad M_n\left(\begin{pmatrix} J & 0 \\ c & J^{-1} \end{pmatrix}\right) = J^n \begin{pmatrix} 1 & & & \\ cJ^{-1} & J^{-2} & & \\ \cdots & \cdots & \cdots & \\ c^n J^{-n} & nc^{n-1}J^{-n-1} & \cdots & J^{-2n} \end{pmatrix}.$$

In the matrix (10), the elements above the diagonal are all 0; the $(r+1)$-th diagonal element is J^{n-2r}; and the $(r+1)$-th row is

$$(10') \qquad \left(c^r J^{n-r}, \binom{r}{1} c^{r-1} J^{n-r-1}, \binom{r}{2} c^{r-2} J^{n-r-2}, \cdots, J^{n-2r}, 0, \cdots, 0\right).$$

We shall now prove Lemma 1. Suppose that $f_0 = \cdots = f_{t-1} = 0$ in Lemma 2 and $\omega = L_n(z)\mathfrak{f}dz$ is an M_n-form with respect to G. Then, by Lemma 2 and by (10), we have, for every $\sigma \in G$,

$$(f_t \circ \sigma) J(\sigma, z)^{2t+2-n} = f_t;$$

so f_t satisfies the condition (A 2) for $\nu = 2t+2-n$. Let s be a cusp of G; ρ, h and q being defined for s as in § 2, there exist $n+1$ meromorphic functions $g_0(q), \cdots, g_n(q)$ in $|q| < 1$, such that

$$L_n(z)^{-1} M_n(\rho)^{-1} L_n(\rho(z))(\mathfrak{f} \circ \rho) J(\rho, z)^2 dz = \begin{pmatrix} g_0(q)dq \\ \vdots \\ g_n(q)dq \end{pmatrix}.$$

By the relation (3), putting $J = J(\rho, z) = (cz+d)^{-1}$, we have

$$M_n\left(\begin{pmatrix} J & 0 \\ c & J^{-1} \end{pmatrix}\right)^{-1} (\mathfrak{f} \circ \rho) J^2 = 2\pi i h^{-1} q \begin{pmatrix} g_0(q) \\ \vdots \\ g_n(q) \end{pmatrix},$$

so that by (10),

$$(11) \qquad (f_t \circ \rho) J^{2t+2-n} = 2\pi i h^{-1} q g_t(q).$$

This shows that f_t satisfies (A 3). Hence f_t is an automorphic form of degree $2t+2-n$ with respect to G. Furthermore, if ω is a cusp M_n-form, f_t must be holomorphic on $\mathcal{H}$, since f_t is the $(t+1)$-th component of $L_n(-z)\omega/dz$; and as $g_t(q)$ is holomorphic at $q = 0$ by virtue of (M 3'), the relation (11) shows that f_t satisfies (A 3'). This completes the proof of Lemma 1.

Lemma 3. *If f is an automorphic form of degree ν with respect to G, we have, for every $\sigma = \begin{pmatrix} a & b \\ c & d \end{pmatrix} \in G$,*

$$(f^{(k)} \circ \sigma) J^2 = \sum_{j=0}^{k} \binom{k}{j} \binom{\nu+k-1}{j} j! \, c^j J^{j+2-2k-\nu} f^{(k-j)},$$

where $J = J(\sigma, z) = (cz+d)^{-1}$.

This is easily obtained by the induction on k.

Now we shall prove Theorem 1. Notations being as in that theorem, by

Lemma 2, ω satisfies the condition (M 2) if we have, for every $\sigma \in G$,

$$
(12) \qquad J^2 \begin{pmatrix} 0 \\ \vdots \\ 0 \\ \alpha_0 f \circ \sigma \\ \alpha_1 f' \circ \sigma \\ \vdots \\ \alpha_\mu f^{(\mu)} \circ \sigma \end{pmatrix} = M_n\left(\begin{pmatrix} J & 0 \\ c & J^{-1} \end{pmatrix}\right) \begin{pmatrix} 0 \\ \vdots \\ 0 \\ \alpha_0 f \\ \alpha_1 f' \\ \vdots \\ \alpha_\mu f^{(\mu)} \end{pmatrix} ,
$$

where $J = J(\sigma, z) = (cz+d)^{-1}$. Put $t = n - \mu$. By (10), we see that the first t components of the vectors in both sides are equal to 0; and by (10′), if $r \geq t$, the $(r+1)$-th component of the vector on the right hand side of (12) is equal to

$$
(13) \qquad \sum_{u=t}^{r} \binom{r}{u} c^{r-u} J^{n-r-u} \alpha_{u-t} f^{(u-t)} ;
$$

hence the equality (12) is proved if we show that (13) is equal to $J^2 \alpha_{r-t} f^{(r-t)} \circ \sigma$. By Lemma 3, we have

$$
J^2 \alpha_{r-t} f^{(r-t)} \circ \sigma = \alpha_{r-t} \sum_{j=0}^{r-t} \binom{r-t}{j} \binom{\nu+r-t-1}{j} j! \, c^j J^{j+2-2(r-t)-\nu} f^{(r-t-j)}
$$

$$
= \alpha_{r-t} \sum_{u=t}^{r} \binom{r-t}{r-u} \binom{\nu+r-t-1}{r-u} (r-u)! \, c^{r-u} J^{e(u)} f^{(u-t)} ,
$$

where $e(u) = r - u + 2 - 2(r-t) - \nu$. Since $\nu = 2t - (n-2)$, we have $e(u) = n - r - u$. On the other hand, we can easily verify

$$
\alpha_{r-t} \binom{r-t}{r-u} \binom{\nu+r-t-1}{r-u} (r-u)! = \alpha_{u-t} \binom{r}{u} .
$$

This proves the equality (10). Hence ω satisfies (M 2). The condition (M 1) is of course satisfied. Now consider a cusp s of G. Since ω satisfies (M 1-2), ρ and q being as in §1, there exist $n+1$ meromorphic functions $f_0(q), \cdots, f_n(q)$ in $0 < |q| < 1$ such that

$$
L_n(z)^{-1} M_n(\rho)^{-1} \omega \circ \rho = \begin{pmatrix} f_0(q) dq \\ \vdots \\ f_n(q) dq \end{pmatrix} .
$$

By (A 3), there exists a meromorphic function $g(q)$ in $|q| < 1$ such that

$$
(14) \qquad f(\rho(z)) = g(q) J(\rho, z)^{-\nu} .
$$

Differentiating this successively, we get, for every k,

$$
(15) \qquad f^{(k)}(\rho(z)) = J^{a(k)} \sum_{u} F_{ku}(q) z^u ,
$$

where $a(k)$ is an integer and the $F_{ku}(q)$ are meromorphic functions in $|q| < 1$. Comparing both sides of the equality

$$(16) \qquad \begin{pmatrix} f_0(q)dq \\ \vdots \\ f_n(q)dq \end{pmatrix} = L_n(z)^{-1} M_n(\rho)^{-1} L_n(\rho(z)) \begin{pmatrix} 0 \\ \vdots \\ 0 \\ \alpha_0 f \circ \rho \\ \vdots \\ \alpha_\mu f^{(\mu)} \circ \rho \end{pmatrix} J^2 \frac{h}{2\pi i} \frac{dq}{q},$$

we observe that $f_k(q)$ is written in the form

$$(17) \qquad f_k(q) = J^{b(k)} \sum_u H_{ku}(q) z^u .$$

where $b(k)$ is an integer and the $H_{ku}(q)$ are meromorphic functions in $|q| < 1$. Hence there exists an integer m such that

$$\lim_{q \to 0} q^m f_k(q) = 0$$

for every k. This shows that the $f_k(q)$ are meromorphic at $q = 0$. Thus we have proved that ω is an M_n-form. Furthermore, suppose that f is a cusp form. Then the function $g(q)$ of (14) takes the value 0 at $q = 0$; so, in the expression (15), we may assume that the $F_{ku}(q)$ take the value 0 at $q = 0$. Comparing again both sides of (16), we see that the functions $H_{ku}(q)$ in the expression (17) are holomorphic at $q = 0$, so that we have

$$\lim_{q \to 0} q f_k(q) = 0$$

for every k. This shows that the $f_k(q)$ are holomorphic at $q = 0$. Hence ω is a cusp M_n-form. We can similarly show that if ω is a cusp M_n-form, f satisfies (A 3'). Theorem 1 is then completely proved.

We can prove Theorem 3 in a quite similar way. We shall now prove Theorem 5. Differentiating both sides of

$$L_n(z + w) = L_n(z)L_n(w)$$

with respect to w, and then putting $w = 0$, we obtain

$$(18) \qquad L_n'(z) = L_n(z)L_n'(0) .$$

From (2) we see that

$$(19) \qquad L_n'(0) = \begin{pmatrix} 0 & n & 0 & \cdots & 0 & 0 \\ 0 & 0 & n-1 & \cdots & 0 & 0 \\ & & \cdots & & \cdots & \cdots \\ 0 & \cdots & & \cdots & 2 & 0 \\ 0 & \cdots & & \cdots & 0 & 1 \\ 0 & \cdots & & \cdots & 0 & 0 \end{pmatrix} .$$

Notations being as in Theorem 5, we have, using (18) and (19),

$$
d\mathfrak{f} = d\left[L_n(z) \begin{pmatrix} 0 \\ \vdots \\ 0 \\ \gamma_0 f \\ \vdots \\ \gamma_{\mu-1} f^{(\mu-1)} \end{pmatrix} \right] = \left[L_n{}'(z) \begin{pmatrix} 0 \\ \vdots \\ 0 \\ \gamma_0 f \\ \vdots \\ \gamma_{\mu-1} f^{(\mu-1)} \end{pmatrix} + L_n(z) \begin{pmatrix} 0 \\ \vdots \\ 0 \\ \gamma_0 f' \\ \vdots \\ \gamma_{\mu-1} f^{(\mu)} \end{pmatrix} \right] dz
$$

$$
= L_n(z) \begin{pmatrix} 0 \\ \vdots \\ 0 \\ \alpha_0{}' f \\ \alpha_1{}' f' \\ \vdots \\ \alpha_\mu{}' f^{(\mu)} \end{pmatrix} dz ,
$$

where $\alpha_0{}' = \mu\gamma_0$, $\alpha_1{}' = (\mu-1)\gamma_1 + \gamma_0$, $\cdots$, $\alpha_{\mu-1}{}' = \gamma_{\mu-1} + \gamma_{\mu-2}$, $\alpha_\mu{}' = \gamma_{\mu-1}$. We can easily verify $\alpha_k{}' = \mu(n-\mu+1)\alpha_k$ for $0 \le k \le \mu$. This proves Theorem 5.

It remains to prove Theorem 2 and Theorem 4. We need for that purpose

LEMMA 4. *Suppose that the Fuchsian group G has no cusp. Let n be a positive even integer and $r = \dfrac{n}{2}$. Then there is no cusp M_n-form ω with respect to G of the type*

$$
\omega = L_n(z) \begin{pmatrix} 0 \\ \vdots \\ 0 \\ 1 \\ f_r \\ \vdots \\ f_n \end{pmatrix} dz ,
$$

where $f_r, \cdots, f_n$ are meromorphic functions on $\mathcal{H}$.

PROOF. First we remark that f_r must be everywhere holomorphic on $\mathcal{H}$. By lemma 2 and by (10'), we have, for every $\sigma = \begin{pmatrix} a & b \\ c & d \end{pmatrix} \in G$,

$$
f_r(\sigma(z)) J(\sigma, z)^2 = f_r(z) + rcJ(\sigma, z) .
$$

Put $\eta = f_r(z)dz$. Then η is a holomorphic differential form on $\mathcal{H}$ satisfying

$$
(20) \qquad\qquad\qquad \eta \circ \sigma = \eta - rd(\log J(\sigma, z))
$$

for every $\sigma \in G$. Consider the integral of η along the boundary $\mathscr{B}$ of a fundamental domain of G; then we find, taking account of the relation (20),

$$(21) \qquad \frac{2}{r} \int_{\mathcal{B}} \eta = 2g - 2 + \sum_{\lambda} \left(1 - \frac{1}{m_\lambda} \right),$$

where g is the genus of the Riemann surface $\mathcal{H}/G$ and the m_λ denote the orders of ramification at the elliptic points of G. It is well known that the number on the right hand side of (21) is positive. On the other hand, as η is holomorphic, we must have $\int_{\mathcal{B}} \eta = 0$; thus we are led to contradiction if we assume the existence of a cusp M_n-form of the type described in our lemma.

Now we are ready to prove Theorem 2. First we remark that $S_\nu(G) = \{0\}$ for $\nu < 0$ and $S_0(G) = \{0\}$ or $= C$ according as G has a cusp or not. Let ω be a cusp M_n-form with respect to G; put

$$L_n(z)^{-1} \omega = \begin{pmatrix} f_0(z) \\ \vdots \\ f_n(z) \end{pmatrix} dz .$$

Let t be the first integer such that $f_t \neq 0$. Then, by Lemma 1, f_t is a cusp form with respect to G of degree $2t - n + 2$. By the above remark, we must have $t \geq \frac{n-2}{2}$. If $t = \frac{n-2}{2}$, f_t is a cusp form of degree 0; then, G has no cusp and f_t is a constant. This is impossible, however, in view of Lemma 4. Hence we have $2t - n + 2 > 0$. Put $\nu = 2t - n + 2$. Then we have $\alpha_{n,\nu,0} \neq 0$; put $f = \alpha_{n,\nu,0}^{-1} f_t$. Let η_ν be the cusp M_n-form defined for the cusp form f by (8). Then we see that the first $t+1$ components of $L_n(z)^{-1}(\omega - \eta_\nu)$ are all 0. Applying the same argument to the form $\omega - \eta_\nu$, we can find an element $\eta_{\nu+2}$ of $\mathfrak{S}^n_{\nu+2}(G)$ such that the first $t+2$ components of $L_n(z)^{-1}(\omega - \eta_\nu - \eta_{\nu+2})$ are all 0. Repeating this procedure, we get the expression

$$\omega = \sum_{\lambda=\nu}^{n+2} \eta_\lambda$$

where η_λ is an element of $\mathfrak{S}^n_\lambda(G)$ for every λ. It is easy to see that this expression gives a decomposition of the vector space $\mathfrak{F}_n(G)$ as the direct sum of the vector spaces $\mathfrak{S}^n_\lambda(G)$ for $2 \leq \lambda \leq n+2$. Thus we have proved Theorem 2. Theorem 4 can be proved in a quite similar way.

University of Tokyo.

References

[1] Bol, G., Invarianten linearer Differentialgleichungen, Abh, Math. Seminar Hamburger Univ., 16 (1948), 1–28.

[2] Eichler, M., Eine Verallgemeinerung der Abelschen Integrale, Math. Z., 67 (1957), 267–298.

[3] Shimura, G., Sur les intégrales attachées aux formes automorphes, J. Math. Soc. Japan, 11 (1959), 291–311.

61b

On the zeta functions of the algebraic curves uniformized by certain automorphic functions

Journal of the Mathematical Society of Japan, 13 (1961), 275-331

(Received March 31, 1961)

Introduction. After Hasse and Weil, we can attach a zeta-function to every algebraic variety defined over an algebraic number field. In contrast with its importance, our knowledge of the zeta-function of this kind is little. At present, as far as I know, the zeta-function is determined only in the following two cases.

I) Abelian varieties with sufficiently many complex multiplications [**30, 3, 27**].

II) Algebraic curves uniformized by modular functions belonging to congruence-subgroups [**6, 22**].

Here we note that the determination of the zeta-function of a curve is essentially the same as the determination of the zeta-function of its jacobian. Now, in all these cases, the zeta-functions are meromorphic on the whole complex plane and satisfy functional equations, as conjectured by Hasse.

The purpose of the present paper is to supply a new class of algebraic curves, for which Hasse's conjecture is true, and of which the curves of II) are particular cases. Our principal result is as follows. Let $\mathcal{O}$ be an indefinite quaternion algebra over the rational number field Q, and $\mathfrak{o}$ a maximal order in $\mathcal{O}$. Take a positive integer N which is prime to the discriminant of $\mathcal{O}$ and denote by Γ_N the group of units γ of $\mathfrak{o}$, with positive reduced norm, such that $\gamma \equiv 1 \bmod. N\mathfrak{o}$. As $\mathcal{O}$ has a faithful representation by real matrices of degree 2, Γ_N is considered as a Fuchsian group on the upper half plane $\mathfrak{H}$. If $\mathcal{O}$ has no zero-divisor, $\Gamma_N \backslash \mathfrak{H}$ is compact, while if $\mathcal{O}$ is the total matric algebra of degree 2 over Q, Γ_N is nothing but the principal congruence-subgroup of $SL(2, Z)$ of level N. Now, according to Eichler [**7**], we can develop the theory of Hecke's operators for cusp-forms with respect to Γ_N. We obtain then Dirichlet-series $D(s)$, meromorphic on the whole plane, having Euler-products, and satisfying functional equations. Let $\mathfrak{R}_N$ be the field of automorphic functions with respect to Γ_N. We can find an algebraic curve $\mathfrak{C}_N$, defined over Q, whose function-field is identified with $\mathfrak{R}_N$. Our main theorem asserts that the zeta-function of $\mathfrak{C}_N$ is determined by the Dirichlet-series $D(s)$ for cusp-forms of degree 2.

We shall now explain our method by giving a summary of the contents.[1] §§ 1.1~1.4 introduce the ring of modular correspondences for the group Γ_N; in § 1.5, we consider the representations of modular correspondences in the vector spaces of cusp-forms. Each representation yields a Dirichlet series $D(s)$ with an Euler-product. We can express $D(s)$ as a certain integral on the idèle-group of $\mathcal{O}$; then the Poisson summation formula on the adèle-space lead to the functional equation for $D(s)$ (Theorem 1 of § 1.6). Now we consider the one-parameter system $\{A_z \,|\, z \in \mathfrak{H}\}$ of polarized abelian varieties of dimension 2, whose endomorphism-rings are isomorphic to $\mathfrak{o}$; such a system has been constructed in a previous paper [23] (quoted hereafter as [AF]). We have shown in [AF] that the moduli $f_i(z)$ of A_z, considered as functions of z, generate the field of automorphic functions $\mathfrak{R}_1$. We construct a quotient variety V_z of A_z by the automorphisms ± 1, which is called the Kummer variety of A_z, as well as a natural mapping h_z of A_z onto V_z with a suitable property. Then, t_z being a point on A_z of order N, the coordinates of $h_z(t_z)$, regarded as functions of z, give automorphic functions $g_j(z)$ with respect to Γ_N; the functions f_i and g_j generate over C the field $\mathfrak{R}_N$. These facts are proved in §§ 2.1~3.4. The field $Q(f_i, g_j)$ is a Galois extension of $Q(f_i)$; we determine in § 4.1 the Galois group. §§ 4.2~4.3 concern the relation of modular correspondences and isogenies of A_z. Taking a generic member A_z of our system, we consider the isogenies λ_ν of A_z onto other members A_{z_ν}, whose kernels are isomorphic to $\mathfrak{o}/\mathfrak{q}$ for a given left $\mathfrak{o}$-ideal $\mathfrak{q}$. Then the correspondence $(f_i(z), g_j(z)) \to (f_i(z_\nu), g_j(z_\nu))$ determines an algebraic correspondence $X_\mathfrak{q}$ of the curve $\mathfrak{C}_N$. If $\mathfrak{o}/\mathfrak{q}$ is of order p^2 for a prime number p, and if p does not divide the discriminant of $\mathcal{O}$, then, by the reduction modulo p, we obtain from the λ_ν one purely inseparable isogeny and p separable isogenies. This fact is the key to the congruence-relations for $X_\mathfrak{q}$, which are fundamental in our whole theory, and whose proof is the object of §§ 5.1~5.5. Our principal result is then easily derived from those relations. The idea is almost the same as [22], where the author treated the one-parameter system of elliptic curves. The present situation is, however, more complicated than [22], since the abelian varieties A_z are not necessarily defined over the field of moduli. We can overcome this difficulty by the use of "normalized Kummer variety".

The present investigation may be thus regarded as a continuation of [22] and [AF]. It is also considered as an example of a more general theory, which is definitely non-abelian in character, and which one may expect to be constructed in future; but as for this, I have only mentioned some related problems at

---

1) Our results were partly announced in the memoir "Fonctions automorphes et correspondances modulaires", Proc. Int. Cong. Math. 1958, 330–338. In the last half of this article, the reader will also find a brief and easy account of the theory.

the end of the paper.

Notation. We shall use the same notation as in [AF]. In particular, we denote by $c(V)$ the Chow-point of an algebraic variety V. The notation concerning abelian varieties will be the same as Weil [29]; so, if λ is an isogeny of an abelian variety A, all being defined over a field k, we denote by $\nu_i(\lambda)$ and $\nu_s(\lambda)$ the inseparable and separable factors of the degree of $k(x)$ over $k(\lambda x)$, respectively, where x is a generic point of A over k.

§ 1. Analytic theory of modular correspondences.

1.1. Ring of transformations. We first recall the definition of ring of transformations introduced in [24, §7]. Let $\mathfrak{G}$ be a group; two subgroups G and G' of $\mathfrak{G}$ are called *commensurable* if the intersection $G \cap G'$ is of finite index in G and in G'. Fix a subgroup G of $\mathfrak{G}$; let $\tilde{G}$ be the set of all elements α of $\mathfrak{G}$ such that $\alpha^{-1}G\alpha$ is commensurable with G. It can be easily verified that $\tilde{G}$ is a subgroup of $\mathfrak{G}$ containing G. For every element α of G, we see easily that

$$(1) \qquad [G:G \cap \alpha^{-1}G\alpha] = [\alpha G\alpha^{-1} : \alpha G\alpha^{-1} \cap G]$$

= the number of right cosets $G\beta$ contained in $G\alpha G$;

and a similar equality holds for the left cosets in $G\alpha G$.

LEMMA 1.1. *If the number of right cosets in $G\alpha G$ is equal to the number of left cosets in $G\alpha G$, then there exists a common system of representatives for right and left cosets in $G\alpha G$.*

PROOF. Let $G\beta$ and γG be a right coset and a left coset contained in $G\alpha G$. As we have $G\beta G = G\alpha G = G\gamma G$, the intersection $G\beta \cap \gamma G$ is not empty. Taking an element δ in $G\beta \cap \gamma G$, we get $G\beta = G\delta$ and $\gamma G = \delta G$; our lemma is a consequence of this fact.

Now fix a sub-semi-group S of $\tilde{G}$ containing G; we can take for example $\tilde{G}$ itself as S. Let $\mathfrak{R}$ denote the free $\mathbf{Z}$-module generated by the $G\alpha G$ for $\alpha \in S$. We shall now define a law of multiplication on the module $\mathfrak{R}$. For any two elements α and β of S, let $\{G\alpha_i\}$ and $\{G\beta_k\}$ be the complete systems of right cosets contained in $G\alpha G$ and in $G\beta G$, respectively. γ being an element of S, we can easily verify that the number of (i, k) such that $G\alpha_i\beta_k = G\gamma$ depends only on $G\alpha G, G\beta G, G\gamma G$, and is independent of the choice of $\{\alpha_i\}, \{\beta_k\}$, and γ. Putting $\sigma = G\alpha G, \tau = G\beta G, \rho = G\gamma G$, we denote this number by $\mu(\sigma \cdot \tau ; \rho)$. Define the product $\sigma \cdot \tau$ by

$$\sigma \cdot \tau = \sum \mu(\sigma \cdot \tau ; \rho)\rho ,$$

where the sum is extended over all the $\rho = G\gamma G$ contained in $G\alpha G\beta G$. We extend this by linearity to a law of multiplication on $\mathfrak{R}$; the module $\mathfrak{R}$

then becomes an associative ring; the identity element is the coset $G = G1G$. We denote this ring by $\mathcal{R}(G, S)$ and call it the *ring of transformations of G with respect to S*. For every $\sigma = G\alpha G$, we denote by $\deg(\sigma)$ the number given by (1) and put for $\sigma_\nu = G\alpha_\nu G$ and for $c_\nu \in \mathbf{Z}$,

$$\deg\left(\sum_\nu c_\nu \sigma_\nu\right) = \sum_\nu c_\nu \deg(\sigma_\nu).$$

By our definition, we see easily, for every $\xi, \eta \in \mathcal{R}$,

$$\deg(\xi + \eta) = \deg(\xi) + \deg(\eta),$$

$$\deg(\xi \cdot \eta) = \deg(\xi) \cdot \deg(\eta);$$

and, for $\sigma = G\alpha G = \bigcup G\alpha_i$, $\tau = G\beta G = \bigcup G\beta_k$, $\rho = G\gamma G$,

(2) $\deg(\rho)\mu(\sigma \cdot \tau; \rho) = $ the number of (i, k) such that $G\alpha_i\beta_k G = G\gamma G$.

PROPOSITION 1.2. *If there exists an anti-automorphism $\alpha \to \alpha^*$ of the semigroup S which maps GαG onto GαG itself for every $\alpha \in S$, then the ring $\mathcal{R}(G, S)$ is commutative.*

PROOF. Considering the anti-automorphism on $G\alpha G$, we see that the number of left cosets in $G\alpha G$ and the number of right cosets in $G\alpha G$ are the same. Hence, by Lemma 1.1, for every $\alpha, \beta \in S$, we can find sets of elements $\{\alpha_i\}$ and $\{\beta_k\}$ such that $G\alpha G = \bigcup G\alpha_i = \bigcup \alpha_i G$, $G\beta G = \bigcup G\beta_k = \bigcup \beta_k G$ are disjoint sums. We have then $G\alpha G = \bigcup G\alpha_i^*$ and $G\beta G = \bigcup G\beta_k^*$. Put $\sigma = G\alpha G, \tau = G\beta G$. By the relation (2), we have, for every $\rho = G\gamma G$ contained in $G\alpha G\beta G$,

$\deg(\rho)\mu(\sigma \cdot \tau; \rho) = $ the number of (i, k) such that $G\alpha_i\beta_k G = G\gamma G$,

$\deg(\rho)\mu(\tau \cdot \sigma; \rho) = $ the number of (k, i) such that $G\beta_k^*\alpha_i^* G = G\gamma G$.

Applying the anti-automorphism $\alpha \to \alpha^*$ to each double coset, we observe that these two numbers coincide; so we have $\mu(\sigma \cdot \tau; \rho) = \mu(\tau \cdot \sigma; \rho)$. This proves our proposition.

1.2. Arithmetic of indefinite quaternion algebras. Let $\varPhi$ be an indefinite quaternion algebra over $\mathbf{Q}$ (cf. [AF, §5, no. 14]). We denote by $\alpha \to \alpha'$ the canonical involution of $\varPhi$ and put $N(\alpha) = \alpha\alpha', \mathrm{tr}(\alpha) = \alpha + \alpha'$. Let $\mathfrak{o}$ be a maximal order in $\varPhi$; put, for any base $\{u_i\}$ of $\mathfrak{o}$ over $\mathbf{Z}$,

$$d(\varPhi) = |\det(\mathrm{tr}(u_i u_j))|^{1/2}.$$

This number is independent of the choice of $\mathfrak{o}$ and $\{u_i\}$; it is a square-free positive integer. Throughout the present paper, we shall use these notations always in this sense.

Now we fix once for all a maximal order $\mathfrak{o}$. For every integral right, left, two-sided $\mathfrak{o}$-ideal $\mathfrak{a}$, we denote by $N_1(\mathfrak{a})$ the number of elements in $\mathfrak{o}/\mathfrak{a}$ and put

$N(\mathfrak{a}) = N_1(\mathfrak{a})^{1/2}$; we can define in a natural manner $N(\mathfrak{a})$ and $N_1(\mathfrak{a})$ for any $\mathfrak{o}$-ideal which is not necessarily integral. $N(\mathfrak{a})$ is a positive integer for any integral $\mathfrak{o}$-ideal $\mathfrak{a}$; and we have

$$N(\alpha\mathfrak{o}) = N(\mathfrak{o}\alpha) = |N(\alpha)|.$$

The two-sided $\mathfrak{o}$-ideals form a commutative group, which is a direct product of the infinite cyclic groups generated by the prime ideals. Every prime ideal $\mathfrak{p}$ divides one and only one rational prime p; and we have

$$\mathfrak{p} = p\mathfrak{o} \quad \text{if} \quad p \nmid d(\Phi),$$

$$\mathfrak{p}^2 = p\mathfrak{o} \quad \text{if} \quad p \mid d(\Phi).$$

Therefore, every integral two-sided $\mathfrak{o}$-ideal $\mathfrak{a}$ is written in the form

$$(3) \qquad\qquad \mathfrak{a} = a_0 \mathfrak{p}_1 \cdots \mathfrak{p}_s,$$

where a_0 is a rational integer and the $\mathfrak{p}_i$ are distinct prime ideals dividing $d(\Phi)$. By Eichler [5], every one-sided $\mathfrak{o}$-ideal is principal. For our later use, we state here a lemma which is a particular case of Eichler [4, Satz 5].

LEMMA 1.3. *Let $\mathfrak{a}$ be an integral two-sided $\mathfrak{o}$-ideal; let β be an element of $\mathfrak{o}$ and b an element of $\mathbf{Z}$ such that $b \equiv N(\beta)\,\mathrm{mod.}\,\mathfrak{a}$. Then there exists an element β_0 of $\mathfrak{o}$ such that*

$$\beta_0 \equiv \beta\,\mathrm{mod.}\,\mathfrak{a}, \quad N(\beta_0) = b.$$

Taking $\mathfrak{a}$ to be $\mathfrak{o}$, we obtain

LEMMA 1.4. *For every rational integer b, there exists an element β of $\mathfrak{o}$ such that $N(\beta) = b$.*

1.3. Ring of modular correspondences. We denote by Γ the group of all units γ of $\mathfrak{o}$ such that $N(\gamma) = 1$. By Lemma 1.4, $\mathfrak{o}$ contains an element ε such that $N(\varepsilon) = -1$; for any such element ε, $\Gamma \cup \Gamma\varepsilon$ is the group of all units in $\mathfrak{o}$. Let $\mathfrak{a} = \alpha\mathfrak{o}$ be an integral two-sided $\mathfrak{o}$-ideal; we denote by $\Gamma_\mathfrak{a} = \Gamma_\alpha$ the subgroup of Γ consisting of the elements γ such that $\gamma \equiv 1\,\mathrm{mod.}\,\mathfrak{a}$.

PROPOSITION 1.5. *Let α be an element of $\mathfrak{o}$ such that $N(\alpha) = m \neq 0$. Then $\alpha^{-1}\Gamma\alpha$ contains Γ_m.*

PROOF. If γ is an element of Γ_m, we have

$$\alpha\gamma\alpha^{-1} = \alpha\gamma\alpha' m^{-1} \equiv 1\,\mathrm{mod.}\,\alpha\mathfrak{o}\alpha'.$$

This shows that $\alpha\gamma\alpha^{-1}$ is contained in $\mathfrak{o}$. As we have $N(\alpha\gamma\alpha^{-1}) = 1$, $\alpha\gamma\alpha^{-1}$ is contained in Γ, so that $\gamma \in \alpha^{-1}\Gamma\alpha$, Q. E. D.

It follows from Proposition 1.5 that, for every regular element α of Φ, Γ and $\alpha^{-1}\Gamma\alpha$ are commensurable. Let $\varDelta$ (resp. $\varDelta_0$) be the set of all the elements α of Φ (resp. $\mathfrak{o}$) such that $N(\alpha) > 0$. Now we shall consider the ring $\mathscr{R}(\Gamma, \varDelta)$.

Our first task is to characterize the double cosets $\Gamma\alpha\Gamma$ by their "elementary divisors ". For every rational prime number p, let Q_p and Z_p denote respectively the field of p-adic numbers and the ring of p-adic integers; and put

$$(4) \qquad \varPhi_p = \varPhi \otimes_Q Q_p, \quad \mathfrak{o}_p = \mathfrak{o} \otimes_Z Z_p$$

If $p \nmid d(\varPhi)$, $\varPhi_p$ is isomorphic to the total matric algebra $M_2(Q_p)$ over Q_p of degree 2; and for a suitable choice of isomorphism, $\mathfrak{o}_p$ corresponds to the ring $M_2(Z_p)$ of matrices with entries in Z_p. If $p \mid d(\varPhi)$, $\varPhi_p$ is a division algebra over Q_p. For each prime factor p of $d(\varPhi)$, we fix a prime element π_p in $\mathfrak{o}_p$, which satisfies $N(\pi_p) = p$.

Let α be an element of $\varDelta_0$. We now define the elementary divisors of α. First consider a prime p which does not divide $d(\varPhi)$. Then, regarding α as an element of $M_2(Z_p)$, we can find two units ε_1 and ε_2 of $M_2(Z_p)$ so that $\varepsilon_1 \alpha \varepsilon_2$ is of the following form:

$$\varepsilon_1 \alpha \varepsilon_2 = \begin{pmatrix} p^{c_1} & 0 \\ 0 & p^{c_2} \end{pmatrix},$$

where c_1 and c_2 are non-negative integers such that $c_1 \leq c_2$. When p divides $d(\varPhi)$, $\mathfrak{o}_p \alpha$ is a power $(\mathfrak{o}_p \pi_p)^c$. We call then

$$\{ \cdots, (p^{c_1}, p^{c_2}), \cdots, \pi_p^c, \cdots \}$$

the *elementary divisors* of α.

PROPOSITION 1.6. *Let α and β be two elements of $\varDelta_0$. Then the following four conditions are equivalent to each other.*

   i) *$\varGamma \alpha \varGamma = \varGamma \beta \varGamma$.*

   ii) *α and β have the same elementary divisors.*

   iii) *$\mathfrak{o}/\mathfrak{o}\alpha$ and $\mathfrak{o}/\mathfrak{o}\beta$ are isomorphic as $\mathfrak{o}$-modules.*

   iv) *There exists an element γ of $\varGamma$ such that $\mathfrak{o}\alpha\gamma = \mathfrak{o}\beta$.*

PROOF. The equivalences i) $\Leftrightarrow$ iv), ii) $\Leftrightarrow$ iii) and the implication i) $\Rightarrow$ ii) are obvious. Therefore our proposition is proved if we show ii) $\Rightarrow$ iv). Suppose the condition ii) holds. Then, for each prime p, we can find two units $\varepsilon_1^{(p)}, \varepsilon_2^{(p)}$ of $\mathfrak{o}_p$ such that $\varepsilon_1^{(p)} \alpha \varepsilon_2^{(p)} = \beta$. We may assume, without loss of generality, that $N(\varepsilon_1^{(p)}) = N(\varepsilon_2^{(p)}) = 1$; put $q = N(\alpha)$; we have clearly $N(\beta) = q$. We can find two elements a_1 and a_2 of $\mathfrak{o}$ such that, for every prime factor p of q,

$$a_1 \equiv \varepsilon_1^{(p)}, \quad a_2 \equiv \varepsilon_2^{(p)} \mod. q\mathfrak{o}_p.$$

This relation holds for any p, since if p does not divide q, we have $q\mathfrak{o}_p = \mathfrak{o}_p$. Hence we have

$$N(a_1) \equiv N(a_2) \equiv 1 \mod. (q).$$

Now by Lemma 1.3, there exist two elements γ_1 and γ_2 of $\mathfrak{o}$ such that

$$N(\gamma_1) = N(\gamma_2) = 1,$$

$$\gamma_1 \equiv a_1, \quad \gamma_2 \equiv a_2 \quad \text{mod. } q\mathfrak{o}.$$

We have then $\gamma_1 \alpha \gamma_2 \equiv \beta$ mod. $q\mathfrak{o}$. It follows that $\mathfrak{o}\alpha\gamma_2 = \mathfrak{o}\beta$ since both $\mathfrak{o}\alpha\gamma$ and $\mathfrak{o}\beta$ contain $q\mathfrak{o}$. We have thus proved ii)$\Rightarrow$iv).

We call the elementary divisors of α also the *elementary divisors* of $\Gamma\alpha\Gamma$ or of the integral left $\mathfrak{o}$-ideal $\mathfrak{o}\alpha$. It is easy to see that α and α' have the same elementary divisors; so by Proposition 1.6, we have $\Gamma\alpha\Gamma = \Gamma\alpha'\Gamma$; this holds not only for $\alpha \in \Delta_0$ but also for every $\alpha \in \Delta$. Applying Proposition 1.2 to the present case, we obtain

PROPOSITION 1.7. *The ring $\mathfrak{R}(\Gamma, \Delta)$ is commutative.*

We shall now determime the structure of $\mathfrak{R}(\Gamma, \Delta_0)$. It is easy to see that $\Gamma\alpha \to \mathfrak{o}\alpha$ gives a one-to-one correspondence between the right cosets contained in Δ_0 and the integral left $\mathfrak{o}$-ideals. By Proposition 1.6, $\Gamma\alpha$ and $\Gamma\beta$ belongs to the same double coset $\Gamma\alpha\Gamma$ if and only if $\mathfrak{o}/\mathfrak{o}\alpha$ and $\mathfrak{o}/\mathfrak{o}\beta$ are isomorphic. Thus we observe that $\deg(\Gamma\alpha\Gamma)$ is equal to the number of integral left $\mathfrak{o}$-ideals $\mathfrak{b}$ such that $\mathfrak{o}/\mathfrak{b}$ is isomorphic to $\mathfrak{o}/\mathfrak{o}\alpha$. In particular, when $N(\alpha) = p$ is a prime number, we have

$$\deg(\Gamma\alpha\Gamma) = \begin{cases} p+1 & \text{if} \quad p \mid d(\Phi), \\ 1 & \text{if} \quad p \nmid d(\Phi). \end{cases}$$

PROPOSITION 1.8. *Put $\sigma = \Gamma\alpha\Gamma, \tau = \Gamma\beta\Gamma, \rho = \Gamma\gamma\Gamma$. Then $\mu(\sigma \cdot \tau; \rho)$ is equal to the number of integral left $\mathfrak{o}$-ideals $\mathfrak{b}$ such that: i) $\mathfrak{b} \supset \mathfrak{o}\gamma$; ii) $\mathfrak{o}/\mathfrak{b}$ is isomorphic to $\mathfrak{o}/\mathfrak{o}\beta$; iii) $\mathfrak{b}/\mathfrak{o}\gamma$ is isomorphic to $\mathfrak{o}/\mathfrak{o}\alpha$.*

PROOF. Let $\Gamma\alpha\Gamma = \bigcup\Gamma\alpha_i$ and $\Gamma\beta\Gamma = \bigcup\Gamma\beta_k$ be disjoint sums. Then $\mu(\sigma \cdot \tau; \rho)$ is the number of (i, k) such that $\Gamma\alpha_i\beta_k = \Gamma\gamma$. We note that, for each k, there exists only one or no i such that $\Gamma\alpha_i\beta_k = \Gamma\gamma$. Now if $\Gamma\alpha_i\beta_k = \Gamma\gamma$ holds, we have $\mathfrak{o} \supset \mathfrak{o}\beta_k \supset \mathfrak{o}\alpha_i\beta_k = \mathfrak{o}\gamma$; and $\mathfrak{o}/\mathfrak{o}\beta_k$ is isomorphic to $\mathfrak{o}/\mathfrak{o}\beta$, and $\mathfrak{o}\beta_k/\mathfrak{o}\gamma$ is isomorphic to $\mathfrak{o}/\mathfrak{o}\alpha$; so the integral left $\mathfrak{o}$-ideal $\mathfrak{o}\beta_k$ satisfies the conditions i, ii, iii). Conversely, suppose that an integral left $\mathfrak{o}$-ideal $\mathfrak{b}$ satisfies i, ii, iii). By ii), we have $\mathfrak{b} = \mathfrak{o}\beta_k$ for some k. Put $\gamma\beta_k^{-1} = \alpha_0$; we have then $\alpha_0 \in \Delta_0$, and, by virtue of iii), $\mathfrak{o}/\mathfrak{o}\alpha_0$ is isomorphic to $\mathfrak{o}/\mathfrak{o}\alpha$. We have therefore $\Gamma\alpha_0 = \Gamma\alpha_i$ for some i. It follows that $\Gamma\alpha_i\beta_k = \Gamma\gamma$. This proves our proposition.

PROPOSITION 1.9. *If $N(\alpha)$ and $N(\beta)$ are relatively prime,*

$$(\Gamma\alpha\Gamma)(\Gamma\beta\Gamma) = \Gamma\alpha\beta\Gamma.$$

PROOF. Using the same notation as in the preceding proposition, assume that $N(\alpha)$ and $N(\beta)$ are relatively prime. Let $\mathfrak{b}_1$ and $\mathfrak{b}_2$ be integral left $\mathfrak{o}$-ideals satisfying the conditions i, ii, iii). By ii), $\mathfrak{b}_1 + \mathfrak{b}_2/\mathfrak{b}_2$ is isomorphic to a submodule of $\mathfrak{o}/\mathfrak{o}\beta$. On the other hand, $\mathfrak{b}_1/\mathfrak{b}_1 \cap \mathfrak{b}_2$ is isomorphic to a submodule of $\mathfrak{o}/\mathfrak{o}\alpha$. Hence we must have $\mathfrak{b}_1 + \mathfrak{b}_2 = \mathfrak{b}_2, \mathfrak{b}_1 = \mathfrak{b}_1 \cap \mathfrak{b}_2$, namely, $\mathfrak{b}_1 = \mathfrak{b}_2$. This proves $\mu(\sigma \cdot \tau; \rho) = 1$ by virtue of Proposition 1.8. It is easy to see that, for every

$\alpha_1 \in \Gamma\alpha\Gamma$ and $\beta_1 \in \Gamma\beta\Gamma$, the element $\alpha_1\beta_1$ have the same elementary divisors as $\alpha\beta$. Hence $(\Gamma\alpha\Gamma)(\Gamma\beta\Gamma)$ has the only component $\Gamma\alpha\beta\Gamma$; this proves our proposition.

Now fix our attention to one prime number p. We observe that the $\Gamma\alpha\Gamma$, for which $N(\alpha)$ is a power of p, generate a subring of $\mathfrak{R}(\Gamma, \varDelta_0)$, which we denote by $\mathfrak{R}_p$. Let $T(p^m)$ be the sum of $\Gamma\alpha\Gamma$ such that $N(\alpha) = p^m$. If p is a factor of $d(\varPhi)$, we have $T(p^m) = T(p)^m$, so that the ring $\mathfrak{R}_p$ is the polynomial ring $\mathbf{Z}[T(p)]$. Now suppose that p does not divide $d(\varPhi)$. Let $T(p^\lambda, p^\mu)$ denote the element $\Gamma\alpha\Gamma$ of $\mathfrak{R}_p$ whose elementary divisors are (p^λ, p^μ).

PROPOSITION 1.10. *If $p \nmid d(\varPhi)$, the following relations hold.*

(5) $$T(p,p)T(p^\lambda, p^\mu) = T(p^{\lambda+1}, p^{\mu+1}).$$

(6) $$T(p)T(p^m) = T(1, p^{m+1}) + (p+1)T(p,p)T(p^{m-1}) \quad for \quad m \geq 1.$$

PROOF. The first equality is obvious. Let $c_{\lambda\mu}$ be the multiplicity of $T(p^\lambda, p^\mu)$ in the product $T(1, p)T(p^m)$. Fix an element $\alpha_{\lambda\mu}$ of $\mathfrak{o}$ whose elementary divisors are (p^λ, p^μ). Then, by Proposition 1.8, $c_{\lambda\mu}$ is the number of integral left $\mathfrak{o}$-ideals $\mathfrak{b}$ such that: i) $\mathfrak{b} \supset \mathfrak{o}\alpha_{\lambda\mu}$; ii) $\mathfrak{o}/\mathfrak{b}$ is isomorphic to $\mathfrak{o}/\mathfrak{o}\alpha_{01}$. If $1 \leq \lambda \leq \mu$, $\mathfrak{o}\alpha_{\lambda\mu}$ is contained in $p\mathfrak{o}$; so in this case, the condition i) is a consequence of ii), so that we have $c_{\lambda\mu} = p+1$. If $\lambda = 0$, we have $\mu = m+1$; in this case, $\mathfrak{o}\alpha_{0\mu}$ is not contained in $p\mathfrak{o}$; hence we must have $\mathfrak{b} = \mathfrak{o}\alpha_{0\mu} + p\mathfrak{o}$. This implies $c_{0\mu} = 1$. The relation (6) follows from these facts.

We can also verify that

$$T(1, p)T(1, p^m) = T(1, p^{m+1}) + \begin{cases} pT(p, p^m) & (m > 1), \\ (p+1)T(p,p) & (m=1), \end{cases}$$

$$\deg(T(1, p^m)) = p^{m-1}(p+1) \quad (m \geq 1);$$

and the ring $\mathfrak{R}_p$ is the polynomial ring $\mathbf{Z}[T(1, p), T(p, p)]$. It follows that the ring $\mathfrak{R}(\Gamma, \varDelta)$ is an integral domain.

PROPOSITON 1.11. *Let $T(n)$ be the sum of $\Gamma\alpha\Gamma$ for $\alpha \in \varDelta_0, N(\alpha) = n$. Then, the formal Dirichlet-series $\sum\limits_{n=1}^{\infty} T(n)n^{-s}$ is decomposed into an Euler-product:*

$$\sum_{n=1}^{\infty} T(n)n^{-s} = \prod_{p | d(\varPhi)} [1 - T(p)p^{-s}]^{-1} \prod_{p \nmid d(\varPhi)} [1 - T(p)p^{-s} + T(p,p)p^{1-2s}]^{-1}.$$

PROOF. By Proposition 1.9, we have $\sum\limits_{n=1}^{\infty} T(n)n^{-s} = \prod\limits_{p}(\sum\limits_{m=0}^{\infty} T(p^m)p^{-ms})$. If p is a factor of $d(\varPhi)$, we have

$$\sum_{m=0}^{\infty} T(p^m)p^{-ms} = \sum_{m=0}^{\infty} (T(p)p^{-s})^m = [1 - T(p)p^{-s}]^{-1}.$$

Now suppose that p does not divide $d(\varPhi)$; putting $X = p^{-s}$, we observe

$$\sum_{m=0}^{\infty} T(p^m)X^m = 1 + \sum_{m=1}^{\infty} T(1, p^m)X^m + T(p, p)X^2 \sum_{\nu=0}^{\infty} T(p^\nu)X^\nu .$$

Then, by Proposition 1.10, we can easily verify

$$[1 - T(1, p)X + pT(p, p)X^2] \sum_{m=0}^{\infty} T(p^m)X^m = 1 .$$

Our proposition is thereby proved.

By Proposition 1.11, we see easily

$$(7) \qquad\qquad\qquad \deg T(n) = \sum{}' d ,$$

where the sum is extended over all positive divisors d of n which are prime to $d(\mathcal{O})$.

1.4. Congruence-subgroups of Γ. We begin with

PROPOSITION 1.12. *Let $\mathfrak{a}$ and $\mathfrak{b}$ be integral two-sided $\mathfrak{o}$-ideals which are relatively prime. Then we have $\Gamma = \Gamma_\mathfrak{a}\Gamma_\mathfrak{b}$.*

PROOF. Let α be an element of Γ. We can find an element β of $\mathfrak{o}$ such that $\beta \equiv 1 \bmod. \mathfrak{b}$ and $\beta \equiv \alpha \bmod. \mathfrak{a}$. We have then $N(\beta) \equiv 1 \bmod. \mathfrak{ab}$. By Lemma 1.3, there exists an element γ of Γ such that $N(\gamma) = 1$ and $\gamma \equiv \beta \bmod. \mathfrak{ab}$. As $\gamma \equiv \beta \equiv 1 \bmod. \mathfrak{b}, \gamma$ is contained in $\Gamma_\mathfrak{b}$; and as $\gamma \equiv \alpha \bmod. \mathfrak{a}, \alpha\gamma^{-1}$ is contained in $\Gamma_\mathfrak{a}$. We have therefore $\alpha = \alpha\gamma^{-1} \cdot \gamma \in \Gamma_\mathfrak{a}\Gamma_\mathfrak{b}$; this proves our proposition.

Now we fix an integral two-sided $\mathfrak{o}$-ideal $\mathfrak{a}$.

PROPOSITION 1.13. *Let α and β be two elements of Δ_0 whose norms are prime to $\mathfrak{a}$. Then we have $\Gamma_\mathfrak{a}\alpha = \Gamma_\mathfrak{a}\beta$ if and only if $\Gamma\alpha = \Gamma\beta$ and $\alpha \equiv \beta \bmod. \mathfrak{a}$.*

This is an easy consequence of the definition of $\Gamma_\mathfrak{a}$.

PROPOSITION 1.14. *Let α be an element of Δ_0 such that $N(\alpha)$ is prime to $\mathfrak{a}$. Then the following assertions hold.*

i) $\Gamma\alpha\Gamma = \Gamma\alpha\Gamma_\mathfrak{a} = \Gamma_\mathfrak{a}\alpha\Gamma$.

ii) $\Gamma_\mathfrak{a}\alpha\Gamma_\mathfrak{a} = \{\beta \mid \beta \in \Gamma\alpha\Gamma, \beta \equiv \alpha \bmod. \mathfrak{a}\}$.

iii) *If $\Gamma_\mathfrak{a}\alpha\Gamma_\mathfrak{a} = \bigcup_\nu \Gamma_\mathfrak{a}\alpha_\nu$ is a disjoint sum, then $\Gamma\alpha\Gamma = \bigcup_\nu \Gamma \alpha_\nu$ is a disjoint sum.*

PROOF. By Propositions 1.5 and 1.12 we have $\Gamma = (\Gamma \cap \alpha^{-1}\Gamma\alpha)\Gamma_\mathfrak{a}$. Multiplying by $\alpha^{-1}\Gamma\alpha$, we obtain $\alpha^{-1}\Gamma\alpha\Gamma = \alpha^{-1}\Gamma\alpha\Gamma_\mathfrak{a}$, so that $\Gamma\alpha\Gamma = \Gamma\alpha\Gamma_\mathfrak{a}$; the relation $\Gamma\alpha\Gamma = \Gamma_\mathfrak{a}\alpha\Gamma$ is similarly proved. The assertions ii, iii) follow from this and Proposition 1.13.

Now let $\Delta_\mathfrak{a}$ be the subset of Δ_0 consisting of the elements whose norms are prime to $\mathfrak{a}$. Assume that $\mathfrak{a}$ *is prime to* $d(\mathcal{O})$. Then we have $\mathfrak{a} = a\mathfrak{o}$ for a positive integer a; and $\mathfrak{o}/\mathfrak{a}$ is isomorphic to the total matric ring of degree 2 over $\mathbf{Z}/a\mathbf{Z}$. Identifying $\mathfrak{o}/\mathfrak{a}$ with the matric ring, let $\Delta_\mathfrak{a}^*$ be the set of elements α in $\Delta_\mathfrak{a}$ such that

$$\text{(8)} \qquad \alpha \equiv \begin{pmatrix} 1 & 0 \\ 0 & c \end{pmatrix} \qquad \text{mod. } \mathfrak{a}.$$

If the relation (8) holds, we have clearly $N(\alpha) \equiv c \bmod. \mathfrak{a}$.

PROPOSITION 1.15. *Notations and assumptions being as above, the correspondence $\Gamma_\mathfrak{a} \alpha \Gamma_\mathfrak{a} \to \Gamma \alpha \Gamma$ gives a surjective isomorphism of $\mathcal{R}(\Gamma_\mathfrak{a}, \Delta_\mathfrak{a}^*)$ onto $\mathcal{R}(\Gamma, \Delta_\mathfrak{a})$*

PROOF. Denote by φ the linear mapping of $\mathcal{R}(\Gamma_\mathfrak{a}, \Delta_\mathfrak{a}^*)$ into $\mathcal{R}(\Gamma, \Delta)$ given by the correspondence. First we prove that φ is surjective. Let α be an element of $\Delta_\mathfrak{a}$; put $N(\alpha) = c$. As c is prime to $\mathfrak{a}$, there exists an integer b such that $bc \equiv 1 \bmod. \mathfrak{a}$. Let β be an element of $\mathfrak{o}$ such that $\beta \equiv \begin{pmatrix} 1 & 0 \\ 0 & b \end{pmatrix} \bmod. \mathfrak{a}$. Then we have $N(\alpha\beta) \equiv 1 \bmod. \mathfrak{a}$. By Lemma 1.3, there exists an elemet γ of Γ such that $\gamma \equiv \alpha\beta \bmod. \mathfrak{a}$. We have then $\gamma^{-1}\alpha \equiv \begin{pmatrix} 1 & 0 \\ 0 & c \end{pmatrix} \bmod. \mathfrak{a}$, so that $\gamma^{-1}\alpha$ is contained in $\Delta_\mathfrak{a}^*$. This proves that φ is surjective. Suppose that $\Gamma\alpha\Gamma = \Gamma\beta\Gamma$ for two elements α, β of $\Delta_\mathfrak{a}^*$. Since $N(\alpha)$ is equal to $N(\beta)$, we get $\alpha \equiv \beta \bmod. \mathfrak{a}$, so that by ii) of Proposition 1.14, we obtain $\Gamma_\mathfrak{a}\alpha\Gamma_\mathfrak{a} = \Gamma_\mathfrak{a}\beta\Gamma_\mathfrak{a}$; this proves that φ is one-to-one. Now let $\Gamma_\mathfrak{a}\alpha\Gamma_\mathfrak{a} = \bigcup \Gamma_\mathfrak{a}\alpha_i$ and $\Gamma_\mathfrak{a}\beta\Gamma_\mathfrak{a} = \bigcup_k \Gamma_\mathfrak{a}\beta_k$ be disjoint sums for $\alpha, \beta \in \Delta_\mathfrak{a}^*$. We have then $\Gamma\alpha\Gamma = \bigcup \Gamma\alpha_i$ and $\Gamma\beta\Gamma = \bigcup \Gamma\beta_k$; and these are disjoint sums by virtue of iii) of Proposition 1.14. The product $(\Gamma\alpha\Gamma)(\Gamma\beta\Gamma)$ in the ring $\mathcal{R}(\Gamma, \Delta_\mathfrak{a})$ is a linear combination of the $\Gamma\alpha_i\beta_k\Gamma$. Fix a pair (λ, ν) and put $\gamma = \alpha_\lambda\beta_\nu$. If we have $\Gamma_\mathfrak{a}\alpha_i\beta_k = \Gamma_\mathfrak{a}\gamma$, then obviously $\Gamma\alpha_i\beta_k = \Gamma\gamma$. Conversely, suppose that $\Gamma\alpha_i\beta_k = \Gamma\gamma$. As $\alpha_i \equiv \alpha_\lambda$ and $\beta_k \equiv \beta_\nu \bmod. \mathfrak{a}$, we have $\alpha_i\beta_k \equiv \gamma \bmod. \mathfrak{a}$. Hence by Proposition 1.13, we obtain $\Gamma_\mathfrak{a}\alpha_i\beta_k \equiv \Gamma_\mathfrak{a}\gamma$. This proves that φ is an isomorphism.

The ring $\mathcal{R}(\Gamma, \Delta)$ was first introduced by Hecke [12] in the case where $\mathcal{O}$ is the total matric ring $M_2(\mathbf{Q})$; this is generalized by Eichler [7] to the case of quaternion algebra. Recently, Tamagawa [26] has given a theory for arbitrary division algebras over $\mathbf{Q}$. The result of §§ 1.3–4 is essentially contained in these works. Hecke considered, in the case $\mathcal{O} = M_2(\mathbf{Q})$, the representation of $T(n)$ by modular forms and constructed Dirichlet-series whose coefficients are those representations of $T(n)$; the series have Euler-products and satisfy functional equations; this work was completed by Petersson [17]. Eichler [7] and Selberg (unpublished?) considered a similar problem in the case of quaternion algebras. Tamagawa treated the case of general division algebras; his work is, however, concerned with automorphic "functions" but not with "forms". On the other hand, Godement [10] gave a fairly general theory of zeta-functions attached to division algebras, which is applicable to both the cases "functions" and "forms"; but in this work, there remain unexplained some essential aspects in the case of automorphic forms. Therefore, we shall now give a treatment in the case of automorphic forms attached to quaternion algebras.

REMARK. Prof. Eichler kindly communicated to the author that $\mathrm{tr}\,(T(n))$ in the case of division quaternion algebra is a linear combination of similar traces in the classical case; this would be another way to Theorem 1 of §1.6.

1.5. Cusp-forms. Let $\mathfrak{H}$ denote the upper half-plane. For every matrix $x = \begin{pmatrix} a & b \\ c & d \end{pmatrix}$ with real entries and for $z \in C$, we shall put

$$j(x, z) = cz + d .$$

And when $\det x \neq 0$, we put

$$x[z] = \frac{az+b}{cz+d} ;$$

we have then, if $\det x \neq 0$ and $\det y \neq 0$,

$$j(xy, z) = j(x, y[z])j(y, z) .$$

If $\det x > 0, z \to x[z]$ gives an analytic automorphism of $\mathfrak{H}$.

Fix once for all a faithful representation χ of $\varPhi$ by real matrices of degree 2. We identify every element ξ of $\varPhi$ with the matrix $\chi(\xi)$; then the notation $\xi[z]$ does not contradict the one introduced in [AF, no. 21].

$\varGamma_a$ being as in §1.2, we call as usual a function $f(z)$ on $\mathfrak{H}$ a *cusp-form of degree κ with respect to $\varGamma_a$*, where κ is a positive integer, if:

 i) $f(z)$ *is holomorphic on* $\mathfrak{H}$;

 ii) $f(\sigma[z])j(\sigma, z)^{-\kappa} = f(z)$ *for every* $\sigma \in \varGamma_a$;

 iii) $f(z)$ *vanishes at every cusp of* $\varGamma_a$.

We denote by $S_\kappa(\varGamma_a)$ the set of such $f(z)$. Let $\varGamma_a \alpha \varGamma_a = \bigcup \varGamma_a \alpha_\nu$ be a disjoint expression of an element of $\mathfrak{R}(\varGamma_a, \varDelta)$. For every $f \in S_\kappa(\varGamma_a)$, we define a function g by

$$g(z) = N(\alpha)^{\kappa-1} \sum_\nu f(\alpha_\nu[z])j(\alpha_\nu, z)^{-\kappa} .$$

It can be easily verified that g is an element of $S_\kappa(\varGamma_a)$ and does not depend on the choice of $\{\alpha_\nu\}$. We denote by $(\varGamma_a \alpha \varGamma_a)_\kappa$ the linear mapping $f \to g$ of $S_\kappa(\varGamma_a)$ into itself thus obtained, and write $g = f \mid (\varGamma_a \alpha \varGamma_a)_\kappa$. By our definition of $\mathfrak{R}(\varGamma_a, \varDelta)$ we can conclude that $\varGamma_a \alpha \varGamma_a \to (\varGamma_a \alpha \varGamma_a)_\kappa$ is a representation of the ring $\mathfrak{R}(\varGamma_a, \varDelta)$ in the vector space $S_\kappa(\varGamma_a)$. We shall give another expression for $(\varGamma_a \alpha \varGamma_a)_\kappa$. Put $\alpha'_\nu = \beta_\nu$ for each ν. We have then $\varGamma_a \alpha' \varGamma_a = \bigcup_\nu \beta_\nu \varGamma_a$; and as $\alpha_\nu = N(\alpha)\beta_\nu^{-1}$, we get $f(\alpha_\nu[z]) = f(\beta_\nu^{-1}[z])$ and $j(\alpha_\nu, z) = N(\alpha)j(\beta_\nu^{-1}, z)$. Hence,

$$(9) \qquad f \mid (\varGamma_a \alpha \varGamma_a)_\kappa = N(\alpha)^{-1} \sum_\nu f(\beta_\nu^{-1}[z])j(\beta_\nu^{-1}, z)^{-\kappa} .$$

If b is a positive integer, we see easily

$$(10) \qquad f \mid (\varGamma_a b \varGamma_a)_\kappa = b^{\kappa-2}f .$$

Let $\{f_1, \cdots, f_m\}$ be a base of $S_\kappa(\Gamma_0)$ over C and f be the column-vector whose components are $f_1, \cdots, f_m$. Then, for every element $\Gamma_0\alpha\Gamma_0$ of $\mathfrak{R}(\Gamma_0, \varDelta)$, we obtain a matrix $\mathfrak{X}_\kappa(\Gamma_0\alpha\Gamma_0)$ with entries in C such that

$$(11) \qquad f\,|\,(\Gamma_0\alpha\Gamma_0)_\kappa = \mathfrak{X}_\kappa(\Gamma_0\alpha\Gamma_0)f\,.$$

Restricting α to Γ, we observe that $\alpha \to \mathfrak{X}_\kappa(\Gamma_0\alpha\Gamma_0)$ gives a representation of Γ, whose kernel contains Γ_0; we denote $\mathfrak{X}_\kappa(\Gamma_0\alpha\Gamma_0)$ simly by $L(\alpha)$ for every $\alpha \in \Gamma$.

Now suppose that $\mathfrak{a}$ is prime to $d(\mathcal{O})$. For every integer b which is prime to $\mathfrak{a}$, we can find, by Lemma 1.3, an element γ of Γ such that $\gamma \equiv \begin{pmatrix} b^{-1} & 0 \\ 0 & b \end{pmatrix} \bmod.\,\mathfrak{a}$; and $L(\gamma)$ is determined only by b; so we put

$$(12) \qquad R_\kappa(b\,;\mathfrak{a}) = L(\gamma)\,.$$

$T(n)$ and $T(\mathfrak{p}, \mathfrak{p})$ being as in § 1.3, let $T(n\,;\mathfrak{a})$ and $T(\mathfrak{p}, \mathfrak{p}\,;\mathfrak{a})$ be the elements of $\mathfrak{R}(\Gamma_0, \varDelta_0^*)$ corresponding to $T(n)$ and $T(\mathfrak{p}, \mathfrak{p})$ by the isomorphism of Proposition 1.15. Denote by $\mathfrak{X}_\kappa(n\,;\mathfrak{a})$ and $\mathfrak{X}_\kappa(\mathfrak{p}, \mathfrak{p}\,;\mathfrak{a})$ the matrices determined for $T(n\,;\mathfrak{a})$ and $T(\mathfrak{p}, \mathfrak{p}\,;\mathfrak{a})$ as in (11). If γ is an element of Γ such that $\gamma \equiv \begin{pmatrix} p^{-1} & 0 \\ 0 & p \end{pmatrix} \bmod.\,\mathfrak{a}$, we have $p\gamma \equiv \begin{pmatrix} 1 & 0 \\ 0 & p^2 \end{pmatrix} \bmod.\,\mathfrak{a}$, so that $T(\mathfrak{p}, \mathfrak{p}\,;\mathfrak{a}) = \Gamma_0 p\gamma\Gamma_0$. Using the relations (10) and (12), we obtain $\mathfrak{X}_\kappa(\mathfrak{p}, \mathfrak{p}\,;\mathfrak{a}) = p^{\kappa-2}R_\kappa(\mathfrak{p}\,;\mathfrak{a})$. Therefore, by Propositions 1.11 and 1.15, we get (formally for the moment)

$$(13) \quad \sum{}' \mathfrak{X}_\kappa(\Gamma_0\alpha\Gamma_0)N(\alpha)^{-s} = \sum_{(n,\mathfrak{a})=1} \mathfrak{X}_\kappa(n\,;\mathfrak{a})n^{-s}$$
$$= \prod_{\mathfrak{p}|d}[1-\mathfrak{X}_\kappa(\mathfrak{p}\,;\mathfrak{a})p^{-s}]^{-1} \prod_{\mathfrak{p}\nmid d\mathfrak{a}} [1-\mathfrak{X}_\kappa(\mathfrak{p}\,;\mathfrak{a})p^{-s}+R_\kappa(\mathfrak{p}\,;\mathfrak{a})p^{\kappa-1-2s}]^{-1}\,,$$

where $d = d(\mathcal{O})$, and the first sum is extended over all $\Gamma_0\alpha\Gamma_0$ with $\alpha \in \varDelta_0^*$. In order to examine the convergence, define, for every $f \in S_\kappa(\Gamma_0)$, a function f^* on $G_0 = \mathrm{SL}\,(2, \boldsymbol{R})$ by $f^*(u) = f(u[i])j(u, i)^{-\kappa}$, where $i = \sqrt{-1}$. Then, we can easily verify $f^*(\gamma u) = f^*(u)$ for every $\gamma \in \Gamma_0$; and if $g = f\,|\,(\Gamma_0\alpha\Gamma_0)_\kappa$, we have

$$(14) \qquad g^*(u) = N(\alpha)^{\kappa-1}\sum_\nu f^*(\alpha_\nu u)\,,$$

the α_ν being as above. We may consider f^* as function on $\Gamma_0\backslash G_0$. Since $\Gamma_0\backslash G_0$ is compact, f^* attains its maximum at some point of G_0. Hence, using the relation (14), we observe that the absolute values of the characteristic roots of $\mathfrak{X}_\kappa(\Gamma_0\alpha\Gamma_0)$ do not exceed $N(\alpha)^{\kappa-1}\deg\,(\Gamma_0\alpha\Gamma_0)$. As we shall see a little later, for a suitable base of $S_\kappa(\Gamma_0)$, the $\mathfrak{X}_\kappa(\Gamma_0\alpha\Gamma_0)$ become diagonal matrices. Therefore, on account of (7), the Dirichlet-series (13) converges absolutely for $\mathrm{Re}\,(s) > \kappa+1$.

For our later use, it is necessary to consider an operator defined by an element with negative norm. Let ε be a unit of $\mathfrak{o}$ such that $N(\varepsilon) = -1$. For every $f \in S_\kappa(\Gamma_0)$, define $g = f\,|\,T(\varepsilon)_\kappa$ by

$$(15) \qquad g(z) = \overline{f(\overline{\varepsilon[\bar{z}]})}j(\varepsilon, z)^{-\kappa}\,,$$

where bars indicate the complex conjugate. It can be easily verified that $T(\varepsilon)_\kappa$ is an R-linear mapping of $S_\kappa(\Gamma_\mathfrak{a})$ onto itself satisfying $(af)\,|\,T(\varepsilon)_\kappa = \bar{a}(f\,|\,T(\varepsilon)_\kappa)$ for $a \in C$. When $\mathfrak{a}$ is prime to $d(\varPhi)$, we can take ε so that $\varepsilon \equiv \begin{pmatrix} -1 & 0 \\ 0 & 1 \end{pmatrix}$ mod. $\mathfrak{a}$. Then we have $\varepsilon^2 \in \Gamma_\mathfrak{a}, \varepsilon^{-1}\alpha\varepsilon \equiv \alpha$ mod. $\mathfrak{a}$ for every $\alpha \in \varDelta_\mathfrak{a}^*$. Therefore, $T(\varepsilon)_\kappa^2 = 1$ and $T(\varepsilon)_\kappa$ commutes with $(\Gamma_\mathfrak{a}\alpha\Gamma_\mathfrak{a})_\kappa$ for every $\alpha \in \varDelta_\mathfrak{a}^*$.

After Petersson [17], we define the inner product of the elements f and g of $S_\kappa(\Gamma_\mathfrak{a})$ by

$$(f,g) = \int_D f(z)\overline{g(z)}\,\mathrm{Im}\,(z)^{\kappa-2}\,|\,dz d\bar{z}\,|\,,$$

where D is a fundamental domain for $\Gamma_\mathfrak{a}$. Then, by Proposition 2 of [24], $T^* = (\Gamma_\mathfrak{a}\alpha'\Gamma_\mathfrak{a})_\kappa$ is the adjoint of $T = (\Gamma_\mathfrak{a}\alpha\Gamma_\mathfrak{a})_\kappa$; namely, we have $(f\,|\,T, g) = (f, g\,|\,T^*)$. Let α be an element of $\varDelta_\mathfrak{a}^*$; put $N(\alpha) = b$. Let γ be an element of Γ such that $\gamma \equiv \begin{pmatrix} b^{-1} & 0 \\ 0 & b \end{pmatrix}$ mod. $\mathfrak{a}$. We have then $\alpha'\gamma \equiv \gamma\alpha' \equiv \alpha$ mod. $\mathfrak{a}$, so that

$$(16) \qquad \mathfrak{T}_\kappa(\Gamma_\mathfrak{a}\alpha\Gamma_\mathfrak{a}) = R_\kappa(N(\alpha)\,;\,\mathfrak{a})\mathfrak{T}_\kappa(\Gamma_\mathfrak{a}\alpha'\Gamma_\mathfrak{a}) = \mathfrak{T}_\kappa(\Gamma_\mathfrak{a}\alpha'\Gamma_\mathfrak{a})R_\kappa(N(\alpha)\,;\,\mathfrak{a})\,.$$

If follows that $(\Gamma_\mathfrak{a}\alpha\Gamma_\mathfrak{a})_\kappa$ commutes with its adjoint. Therefore, the operators $(\Gamma_\mathfrak{a}\alpha\Gamma_\mathfrak{a})_\kappa$ for $\alpha \in \varDelta_\mathfrak{a}^*$ form a commutative ring of normal operators. Therefore, we can find a base of $S_\kappa(\Gamma_\mathfrak{a})$ with respect to which $(\Gamma_\mathfrak{a}\alpha\Gamma_\mathfrak{a})_\kappa$ is represented by a diagonal matrix for every $\alpha \in \varDelta_\mathfrak{a}^*$. By Theorem 3 of [24], the characteristic roots of $(\Gamma_\mathfrak{a}\alpha\Gamma_\mathfrak{a})_\kappa$ are algebraic integers for every even κ and every $\alpha \in \varDelta_0$; they are totally real if $\mathfrak{a} = \mathfrak{o}$. Furthermore, we can find a base $\{f_1, \cdots, f_m\}$ of $S_\kappa(\Gamma_\mathfrak{a})$ whose members are invariant under $T(\varepsilon)_\kappa$. With respect to this base, *the $(\Gamma_\mathfrak{a}\alpha\Gamma_\mathfrak{a})_\kappa$ are represented by real matrices for all $\alpha \in \varDelta_\mathfrak{a}^*$.*

1.6. Functional equations for Dirichlet-series. Our method is the one due to Iwasawa-Tate; besides we shall use the ideas of Fujisaki [9], Godement [10] and Tamagawa [26].

We assume, until the end of this §, that $\varPhi$ is a division algebra. Let $\mathfrak{A}$ and $\mathfrak{J}$ denote respectively the adèle-ring and the idèle-group of the quaternion algebra $\varPhi$; we identify, in the usual manner, $\varPhi$ with a subring of $\mathfrak{A}$ and the multiplicative group $\varPhi^*$ of $\varPhi$ with a subgroup of $\mathfrak{J}$. For every $x \in \mathfrak{A}$, we denote by x_p the p-component of x; in particular, x_∞ will denote the component at the infinite prime. Let $\mathfrak{b}_p$ be the different of $\mathfrak{o}_p$ with respect to Z_p. Define a Haar measure dm_p of the additive group $\varPhi_p$ by the condition $m_p(\mathfrak{o}_p) = N_1(\mathfrak{b}_p)^{-\frac{1}{2}}$ for each finite prime p, where $N_1(\mathfrak{b}_p)$ denotes the number of elements in $\mathfrak{o}_p/\mathfrak{b}_p$; and define a Haar measure dm_∞ of $\varPhi_\infty = M_2(R)$ by the usual Euclidean volume element. Then the product $dm(x) = \prod_p dm_p(x)$ gives a Haar measure on $\mathfrak{A}$ satisfying $m(\mathfrak{A}/\varPhi) = 1$. For every element a of $\mathfrak{J}$, we define a positive number $|a|$ by

$$dm(ax) = |a|^2 dm(x)\,.$$

$|a|$ is also given by $|a|=\prod_p |N(a_p)|_p$, where N denotes as before the reduced norm. Let U_p be the group of units of $\mathfrak{o}_p$; define a Haar measure dm_p^* of the multiplicative group $\mathit{\Phi}_p^*$ by $m_p^*(U_p)=1$. Take and fix any Haar measure dm_∞^* on $\mathit{\Phi}_\infty^*=\mathrm{GL}(2,\boldsymbol{R})$; then the product $dm^*(x)=\prod_p dm_p^*(x_p)$ gives a Haar measure on $\mathfrak{J}$ and we have $dm^*(x)=c\,|x|^{-2}dm(x)$ for a suitable constant c. We shall write

$$G=\mathrm{GL}(2,\boldsymbol{R}),$$
$$G_+=\{x\mid x\in\mathrm{GL}(2,\boldsymbol{R}),\det x>0\},$$
$$K=\mathrm{SO}(2,\boldsymbol{R}),$$
$$U=U_0\times G;\quad U_0=\prod_{p:\,\mathrm{finite}}U_p.$$

We want to express the Dirichlet-series (13) by an integral on the idèle-group $\mathfrak{J}$. Let $\mathfrak{a}$ be an integral two-sided $\mathfrak{o}$-ideal which is prime to $d(\mathit{\Phi})$. We denote by $\mathfrak{G}_\mathfrak{a}$ the group of regular elements of the ring $\mathfrak{o}/\mathfrak{a}$ and by $\mathfrak{S}_\mathfrak{a}$ the subgroup of $\mathfrak{G}_\mathfrak{a}$ consisting of the residue-classes of the elements α such that $N(\alpha)\equiv 1\bmod.\mathfrak{a}$. Let $U_\mathfrak{a}$ be the subgroup of U consisting of the elements u such that $u_p\equiv 1\bmod.\mathfrak{a}_p$, where $\mathfrak{a}_p=\mathfrak{o}_p\mathfrak{a}$. Then, $U/U_\mathfrak{a}$ is canonically isomorphic to $\mathfrak{G}_\mathfrak{a}$; and as $\mathfrak{a}$ is prime to $d(\mathit{\Phi})$, $\mathfrak{G}_\mathfrak{a}$ is isomorphic to the group of matrices with entries in $\boldsymbol{Z}/(\boldsymbol{Z}\cap\mathfrak{a})$. Fixing such an isomorphism, let $\mathfrak{R}$ denote the subgroup of $\mathfrak{G}_\mathfrak{a}$ consisting of the elements of the form $\begin{pmatrix}c & 0\\ 0 & 1\end{pmatrix}$. Then every element x of $\mathfrak{G}_\mathfrak{a}$ is written uniquely in the following form:

$$(17)\qquad x=x_1x_2,\quad x_1\in\mathfrak{S}_\mathfrak{a},\quad x_2\in\mathfrak{R}.$$

Fix a unit ε of $\mathfrak{o}$ such that $N(\varepsilon)=-1$ and $\varepsilon\equiv\begin{pmatrix}-1 & 0\\ 0 & 1\end{pmatrix}\bmod.\mathfrak{a}$. Let $\{f_1,\cdots,f_m\}$ be a base of $S_\kappa(\Gamma_\mathfrak{a})$ over C whose members are invariant under $T(\varepsilon)_\kappa$; and denote by f the column-vector with the components $f_1,\cdots,f_m$. Then, as is seen in the preceding section, we obtain a representation $L(\gamma)$ of Γ by $f(\gamma[z])j(\gamma,z)^{-\kappa}=L(\gamma)f$; and $L(\gamma)=1$ for $\gamma\in\Gamma_\mathfrak{a}$. Now by Lemma 1.3, $\Gamma/\Gamma_\mathfrak{a}$ is canonically isomorphic to $\mathfrak{S}_\mathfrak{a}$. Hence we can consider L as a representation of $\mathfrak{S}_\mathfrak{a}$; so, for every element x of $\mathfrak{G}_\mathfrak{a}$ of the form (17), we define $L^*(x)$ by

$$L^*(x)=L(x_1)^{-1}.$$

Furthermore, using the isomorphism $u\to x$ of $U/U_\mathfrak{a}$ onto $\mathfrak{G}_\mathfrak{a}$, we define

$$L^*(u)=L^*(x).$$

$L^*(u)$ is not necessarily a representation of U; only we have

$$(18)\qquad L^*(\gamma u)=L^*(u)L(\gamma)^{-1}$$

for every $\gamma\in\Gamma$. Define a column-vector function f_0 on U by

$$f_0(u) = \begin{cases} L^*(u)f(u_\infty[i])j(u_\infty, i)^{-\varepsilon} & \text{if } \det(u_\infty) > 0, \\[2mm] L^*(\varepsilon u)f(\varepsilon u_\infty[i])j(\varepsilon u_\infty, i)^{-\varepsilon} & \text{if } \det(u_\infty) < 0. \end{cases}$$

Then, on account of (18), we have, for every unit γ of $\mathfrak{o}$,

$$(19) \qquad\qquad f_0(\gamma u) = f_0(u).$$

As every right $\mathfrak{o}$-ideal is principal, we have

$$(20) \qquad\qquad \mathfrak{J} = \Phi^* U.$$

Therefore, on account of (19), we can define a function $F(x)$ on $\mathfrak{J}$ by

$$F(\alpha u) = f_0(u) \text{ for } \alpha \in \Phi^*, \quad u \in U.$$

We have then, for every $\alpha \in \Phi^*$,

$$(21) \qquad\qquad F(\alpha x) = F(x).$$

Define a function $\varphi(x) = \Pi_p \varphi_p(x)$ on $\mathfrak{A}$ as follows.

i) For every finite prime p,

$$\varphi_p(x_p) = \begin{cases} 1, & \text{if } x_p \in \mathfrak{o}_p \text{ and } x_p \equiv \begin{pmatrix} c & 0 \\ 0 & 1 \end{pmatrix} \bmod. \ \mathfrak{a}_p, (c, \mathfrak{a}_p) = 1, \\[2mm] 0, & \text{otherwise.} \end{cases}$$

ii) $\varphi_\infty(w) = j(w, i)^\varepsilon \exp\{-\pi \operatorname{tr}(w^t w)\}$, where ${}^t w$ denotes the transpose of w.

We need furthermore a function $\psi_k(x)$ on $\mathfrak{J}$, for any integer k, defined by $\psi_k(x) = (x_\infty[i])^k$.

Now consider the integral

$$\zeta(s, f, \varphi, k, y) = \int_\mathfrak{J} F(yx)\psi_k(x)\varphi(x) |x|^s dm^*(x),$$

where y is an element of $\mathfrak{J}$ such that $y_p = 1$ for all finite p and $\det(y_\infty) > 0$. As $F(x) = O(|x|^{-\varepsilon/2})$, we observe that this integral converges absolutely for large $\operatorname{Re}(s)$; first we transform it as follows.

$$\zeta(s, f, \varphi, k, y) = \int_\mathfrak{J} F(x)\psi_k(y^{-1}x)\varphi(y^{-1}x) |y^{-1}x|^s dm^*(x)$$

$$= \sum_{\{\alpha\}} \int_{\alpha U} \ ,$$

where the sum is extended over the representatives α for $\Phi^*/(\Gamma \cup \Gamma\varepsilon)$; we may take α so that $N(\alpha) > 0$. By the relation (21),

$$\int_{\alpha U} = \int_U F(x)\psi_k(y^{-1}\alpha x)\varphi(y^{-1}\alpha x) |y^{-1}x|^s dm^*(x)$$

$$= B(\alpha)\int_{G_+} + B'(\alpha)\int_{\varepsilon^{-1}G_+} \ ,$$

where

$$B(\alpha) = \Pi_p \int_{U_p} \varphi_p(\alpha x_p)L^*(x_p)dm^*(x_p), \quad B'(\alpha) = \Pi_p \int_{U_p} \varphi_p(\alpha x_p)L^*(\varepsilon x_p)dm^*(x_p).$$

By our definition of φ_p, $B(\alpha)$ does not vanish only when $\alpha \in \mathfrak{o}$ and α is prime to $\mathfrak{a}$. If that is so, we can find an element γ of Γ so that $\alpha\gamma \equiv \begin{pmatrix} c & 0 \\ 0 & 1 \end{pmatrix}$ mod. $\mathfrak{a}$. Therefore, taking $\alpha\gamma$ in place of α, we have only to consider $B(\alpha)$ and $B'(\alpha)$ for the elements α satisfying $\alpha \equiv \begin{pmatrix} c & 0 \\ 0 & 1 \end{pmatrix}$ mod. $\mathfrak{a}$. Then, recalling the definition of L^* and our choice of ε, we get

$$B(\alpha) = B'(\alpha) = e(\mathfrak{a})1_m,$$

where $e(\mathfrak{a})$ is a positive number depending only on $\mathfrak{a}$.

Now the integral on G_+ is equal to

$$N(\alpha)^{-s} \int_{G_+} f(w[i]) j(y^{-1}\alpha, w[i])^s (y^{-1}\alpha w[i])^k \times$$
$$\exp\{-\pi \operatorname{tr}({}^t(y^{-1}\alpha)(y^{-1}\alpha)w^t w)\} \cdot \det(y^{-1}\alpha w)^s dm^*(w).$$

We observe that the integrand is invariant under the right multiplication of the elements of K; so it is considered as a function on G_+/K. By the correspondence $w \to w^t w$, G_+/K is identified with the space P of positive symmetric matrices of degree 2. Let $g(Y)$ be a function on P defined by

$$g(w^t w) = f(w[i]) j(y^{-1}\alpha, w[i])^s (y^{-1}\alpha w[i])^k \quad \text{for} \quad w \in G_+.$$

Then the above integral is equal to

$$(22) \qquad N(\alpha)^{-s} \int_P g(Y) \exp\{-\pi \operatorname{tr}(A^{-1}Y)\} \det(A^{-1}Y)^{s/2} dY,$$

where $A = (\alpha^{-1}y)^t(\alpha^{-1}y)$. We have $Dg = 0$ for any invariant differential operator D on $P = G/K$, so that by virtue of the result of Selberg [18, pp. 58-59], (20) is equal to

$$c_1 N(\alpha)^{-s} \pi^{-s} \Gamma\left(\frac{s}{2}\right) \Gamma\left(\frac{s-1}{2}\right) g(A)$$
$$= c_1 \pi^{-s} \Gamma\left(\frac{s}{2}\right) \Gamma\left(\frac{s-1}{2}\right) N(\alpha)^{-s} i^k f(\alpha^{-1}y[i]) j(\alpha^{-1}, y[i])^{-k} j(y, i)^{-k},$$

where c_1 is a constant depending only upon our choice of invariant measure of $P = G/K$. The integral on $\varepsilon^{-1}G_+$ is transformed by $w \to \varepsilon^{-1}w$ to the integral

$$(23) \qquad \int_{G_+} f(w[i]) j(y^{-1}\alpha\varepsilon^{-1}, w[i])^s (y^{-1}\alpha\varepsilon^{-1}w[i])^k \times$$
$$\exp\{-\pi \operatorname{tr}({}^t(y^{-1}\alpha\varepsilon^{-1})(y^{-1}\alpha\varepsilon^{-1})w^t w)\} \det(y^{-1}w)^s dm^*(w).$$

We note that $\det(y^{-1}\alpha\varepsilon^{-1}) < 0$; so take an element u such that $u^t u = 1$, $\det u = -1$. Let $g'(Y)$ be a function on P defined by

$$g'(w^t w) = f(w[i]) j(y^{-1}\alpha\varepsilon^{-1}, w[i])^s (y^{-1}\alpha\varepsilon^{-1}w[i])^k \quad \text{for} \quad w \in G_+$$

Then the integral (23) is equal to

$$N(\alpha)^{-s}\int_P g'(Y)\exp\{-\pi\,\mathrm{tr}\,(B^{-1}Y)\}\det(B^{-1}Y)^{s/2}dY,$$

where $B=(\varepsilon\alpha^{-1}yu)^t(\varepsilon\alpha^{-1}yu)$. Then, by the same reason as above, this is equal to

$$c_1\pi^{-s}\Gamma\Big(\frac{s}{2}\Big)\Gamma\Big(\frac{s-1}{2}\Big)N(\alpha)^{-s}g'(B)$$

$$=c_1\pi^{-s}\Gamma\Big(\frac{s}{2}\Big)\Gamma\Big(\frac{s-1}{2}\Big)N(\alpha)^{-s}(-i)^k f(\varepsilon\alpha^{-1}y[-i])j(\varepsilon\alpha^{-1},y[-i])^{-k}j(y,-i)^{-k}.$$

Put now

$$\xi(s)=(2\pi)^{-s}\Gamma(s),\quad c_2=2\sqrt{\pi}\,c_1\varepsilon(\mathfrak{a}).$$

For any point z on the upper half plane $\mathfrak{H}$, we can find an element $y\in G_+$ such that $y[i]=z$ and $j(y,i)=1$. For such y, the above calculation shows that
$$\zeta(s,f,\varphi,k,y)$$

$$=c_2\xi(s-1)\sum_{\{\mathfrak{a}\}}N(\alpha)^{-s}\{i^k f(\alpha^{-1}[z])j(\alpha^{-1},z)^{-k}+(-i)^k f(\varepsilon\alpha^{-1}[\bar z])j(\varepsilon\alpha^{-1},\bar z)^{-k}\}.$$

Define the matrix $\mathfrak{T}_k(\Gamma_\mathfrak{a}\alpha\Gamma_\mathfrak{a})$ as in (11) of §1.5. By our choice of f, the $\mathfrak{T}_k(\Gamma_\mathfrak{a}\alpha\Gamma_\mathfrak{a})$ are real matrices for every $\alpha\in\Delta_\mathfrak{a}^*$. Hence, using the relation (9) of §1.5, we obtain

(24) $$\zeta(s,f,\varphi,k,y)=c_2\xi(s-1)\sum_{\mathfrak{a}\in\Delta_\mathfrak{a}^*}\mathfrak{T}_k(\Gamma_\mathfrak{a}\alpha\Gamma_\mathfrak{a})N(\alpha)^{1-s}\,\mathrm{Re}\,(i^k f(z)).$$

We shall now consider the functional equation. Let $\hat\varphi(x)=\prod_p\hat\varphi_p(x)$ be the Fourier-transform of φ. We can easily verify that: i) the support of $\hat\varphi_p$ is contained in $(\mathfrak{b}_p\mathfrak{a}_p)^{-1}$; ii) $\hat\varphi_\infty(w)=i^\kappa\varphi_\infty({}^t w)$. Hence, if $\det(w)\neq 0$, we get

(25) $$\hat\varphi_\infty(w')=(w[i])^\kappa\varphi_\infty(w),$$

where w' denotes as before the transform of w by the canonical involution. Now by the Poisson summation formula, we get, for every x and y of $\mathfrak{X}$,

$$\sum_{\mathfrak{a}\in\mathcal{O}}\varphi(y^{-1}\alpha x)=|yx^{-1}|^2\sum_{\mathfrak{a}\in\mathcal{O}}\hat\varphi(x^{-1}\alpha y).$$

As we have $\varphi(0)=\hat\varphi(0)=0$ and as $\mathcal{O}$ is a division algebra,

(26) $$\sum_{\mathfrak{a}\in\mathcal{O}^*}\varphi(y^{-1}\alpha x)=|yx^{-1}|^2\sum_{\mathfrak{a}\in\mathcal{O}^*}\hat\varphi(x^{-1}\alpha y).$$

By virtue of (21), we have

$$\zeta(s,f,\varphi,0,y)=\int_{\mathcal{O}^*\backslash\mathfrak{X}}F(x)|y^{-1}x|^s\sum_{\mathfrak{a}\in\mathcal{O}^*}\varphi(y^{-1}\alpha x)dm^*(x).$$

Then, the usual technique decomposing this into two parts for $|x|\geq 1$ and $|x|\leq 1$, together with the formula (26), shows that the function $\zeta(s,f,\varphi,0,y)$ can be holomorphically prolongated on the whole s-plane and is equal to

$$\int_{\mathfrak{X}}F(x^{-1})|yx|^{2-s}\hat\varphi(xy)dm^*(x).$$

Now y being such that $y_p = 1$ and $\det(y_\infty) > 0$, transform this by the canonical involution $x \to x'$. By our definition of F, we have

$$F(x'^{-1}) = |x|^\kappa F(x),$$

and by (25),

$$\hat{\varphi}_\infty(x'y) = (y'x[i])^\kappa \varphi_\infty(y'x),$$

so that

$$\zeta(s, f, \varphi, 0, y) = \int_{\mathfrak{I}} F(x)|x|^{2+\kappa-s}|y|^{2-\kappa}\psi_\kappa(y'x)\varphi_1(y'x)dm^*(x),$$

where $\varphi_1(x) = \prod_p \varphi_{1p}(x_p)$ is defined by: i) $\varphi_{1p} = \hat{\varphi}_p$ for every finite p; ii) $\varphi_{1\infty} = \varphi_\infty$. Transform x into $y'^{-1}x$ and observe that

$$F(y'^{-1}x) = |y|^\kappa F(yx).$$

We obtain then

(27) $$\zeta(s, f, \varphi, 0, y) = \zeta(2+\kappa-s, f, \varphi_1, \kappa, y),$$

which is the functional equation for the Dirichlet-series (13).

In order to find a more explicit form, we restrict ourselves to the case $\mathfrak{a} = \mathfrak{o}$. Since the group Γ contains -1, the vector space $S_\kappa(\Gamma)$ reduces to $\{0\}$ if κ is odd; so we assume henceforth κ is *even*. Put $D(s) = \sum_{n=1}^{\infty} \mathfrak{T}_\kappa(n; \mathfrak{a})n^{-s}$. By (24),

$$\zeta(s, f, \varphi, 0, y) = c_2 \xi(s-1)D(s-1)\operatorname{Re}(f(z)).$$

As φ_p is a characteristic function of $\mathfrak{o}_p$, we have

$$\varphi_{1p}(x_p) = \hat{\varphi}_p(x_p) = \begin{cases} N_1(\mathfrak{b}_p)^{-1/2} & \text{if } x_p \in \mathfrak{b}_p^{-1}, \\ 0 & \text{if } x_p \notin \mathfrak{b}_p^{-1}. \end{cases}$$

Hence we obtain

(28) $$\zeta(s, f, \varphi_1, \kappa, y) = c_2(-1)^{\kappa/2}N_1(\mathfrak{b})^{-1/2}\xi(s-1)\sum \mathfrak{T}_\kappa(\Gamma\alpha\Gamma)N(\alpha)^{1-s}\operatorname{Re}(f(z)),$$

where the sum is extended over all the $\Gamma\alpha\Gamma$ such that $\alpha \in \mathfrak{b}^{-1}, N(\alpha) > 0$. Let δ be an element of $\mathfrak{o}$ such that $\mathfrak{o}\delta = \mathfrak{b}$ and $N(\delta) > 0$. Then,

$$N(\delta)^2 = N_1(\mathfrak{b}) = d(\Phi)^2 = \prod_{p|d(\Phi)} p^2,$$

$$\mathfrak{T}_\kappa(\Gamma\delta\Gamma) = \prod_{p|d(\Phi)} \mathfrak{T}_\kappa(p; \mathfrak{o}).$$

If p is a prime factor of $d(\Phi)$, we have

$$\mathfrak{T}_\kappa(p; \mathfrak{o})^2 = \mathfrak{T}_\kappa(\Gamma p\Gamma) = p^{\kappa-2}1_m.$$

Therefore $\mathfrak{T}_\kappa(\Gamma\delta\Gamma)$ is invertible; and (28) is equal to

$$c_2(-1)^{\kappa/2}N_1(\mathfrak{b})^{-1/2}\xi(s-1)N(\delta)^{s-1}\mathfrak{T}_\kappa(\Gamma\delta\Gamma)^{-1}D(s-1)\operatorname{Re}(f(z)).$$

Putting

$$\Lambda_\kappa = d(\Phi)^{\kappa/2-1}\mathfrak{T}_\kappa(\Gamma\delta\Gamma)^{-1},$$

the functional equation (27) is now written in the following form.

$$d(\Phi)^{s/2}\xi(s)D(s)=(-1)^{\kappa/2}d(\Phi)^{(\kappa-s)/2}\xi(\kappa-s)\Lambda_\kappa D(\kappa-s).$$

We state the result as

THEOREM 1. *Let $\mathfrak{a}$ be an integral two-sided $\mathfrak{o}$-ideal which is prime to $d(\Phi)$. Let $\mathfrak{T}_\kappa(n;\mathfrak{a})$ be the representation of the operators $T(n;\mathfrak{a})$ in the vector space $S_\kappa(\Gamma_\mathfrak{a})$ of cusp-forms of degree κ with respect to $\Gamma_\mathfrak{a}$ (cf. §1.5). Then the Dirichlet series $\sum\limits_{(n,\mathfrak{a})=1} \mathfrak{T}_\kappa(n;\mathfrak{a})n^{-s}$ converges absolutely for $\mathrm{Re}(s)>\kappa+1$ and has an Euler-product:*

$$\sum_{(n,\mathfrak{a})=1} \mathfrak{T}_\kappa(n;\mathfrak{a})n^{-s} = \prod_{p\mid d} [1-\mathfrak{T}_\kappa(p;\mathfrak{a})p^{-s}]^{-1} \prod_{p\nmid d\mathfrak{a}} [1-\mathfrak{T}_\kappa(p;\mathfrak{a})p^{-s}+R_\kappa(p;\mathfrak{a})p^{\kappa-1-2s}]^{-1},$$

where $d=d(\Phi)$. Put

$$H_\kappa(s;\mathfrak{a}) = d(\Phi)^{s/2}(2\pi)^{-s}\Gamma(s) \sum_{(n,\mathfrak{a})=1} \mathfrak{T}_\kappa(n;\mathfrak{a})n^{-s}.$$

Then, $H_\kappa(s;\mathfrak{a})$ is a holomorphic function on the whole s-plane, and satisfies the functional equation (27). When $\mathfrak{a}=\mathfrak{o}$, the functional equation is written in the form

$$H_\kappa(s;\mathfrak{o}) = \Lambda H_\kappa(\kappa-s;\mathfrak{o}),$$

where

$$\Lambda = (-1)^{\kappa/2} \prod_{p\mid d(\Phi)} p^{\kappa/2-1}\mathfrak{T}_\kappa(p;\mathfrak{o}).$$

We note that $\Lambda^2=1$. If we transform the $\mathfrak{T}_\kappa(n;\mathfrak{o})$ into diagonal matrices, then the diagonal elements of $\sum\limits_{n=1}^{\infty} \mathfrak{T}_\kappa(n;\mathfrak{o})n^{-s}$ are Dirichlet series which belong to the type $\{\lambda,k,\gamma\}$ of Hecke [11] for $\lambda=d(\Phi)^{1/2}, k=\kappa$.

As in the classical case, we may conjecture that the absolute values of the characteristic roots of $\mathfrak{T}_\kappa(p;\mathfrak{a})$ do not exceed $2p^{(\kappa-1)/2}$. We shall show in §6.2 that, if $\kappa=2$, this is true for almost all p.

§2. Kummer varieties.

2.1. Quotient of an abelian variety. Let A be an abelian variety and G a finite group of automorphisms of A; let k be a field of definition for A and for the elements of G. Then, we can construct a couple (V,h) formed by a projective variety V and a rational mapping h of A onto V, both defined over k, satisfying the following conditions.

(Q1) h is everywhere defined on A.

(Q2) $h(u)=h(v)$ if and only if there exists an element $\gamma \in G$ such that $\gamma(u)=v$.

(Q3) If h' is a rational mapping of A into a variety V' satisfying $h'\circ\gamma = h'$ for every $\gamma \in G$, then there exists a rational mapping g of V into V' such that $h'=g\circ h$ and g is defined at a point $h(x)$ whenever h' is defined at a point

$x \in A$.

A proof following Serre's idea is given in [25] (cf. also Matsusaka [15], Serre [20], Weil [31]). (V, h) is uniquely determined by these conditions up to biregular birational mappings; we call (V, h) a *quotient of A by G*, defined over k. We note that in the condition (Q3), if k' is a field of definition for V' and h' containing k, then g is defined over k'.

2.2. Normalized Kummer varieties.

Let $\mathfrak{r}$ be a ring having a finite basis over Z and $\mathscr{P} = (A, C, \theta)$ a polarized abelian variety of type $\mathfrak{r}$ (cf. [AF, no. 3]). For the sake of simplicity, we assume always that the ring $\mathfrak{r}$ has an identity element 1 and $\theta(1)$ is the identity element of $\mathscr{A}(A)$. Let $\mathcal{Q}$ denote the group of automorphisms of $\mathscr{P}$; then $\mathcal{Q}$ is a subgroup of the group of automorphisms of (A, C). Hence by [15] and [31], $\mathcal{Q}$ is a finite group. We can therefore construct a quotient of A by $\mathcal{Q}$. Let K be the field of moduli of $\mathscr{P}$. We call a quotient (V_0, h_0) of A by $\mathcal{Q}$ a *normalized Kummer variety* of $\mathscr{P}$ if it satisfies the following conditions.

(K1) V_0 *is defined over K.*

(K2) *If k is a field of definition for $\mathscr{P}$ containing K, h_0 is defined over k.*

(K3) k *being as in (K2), if σ is an isomorphism of k into the universal domain leaving invariant the elements of K, then we have*

$$h_0^\sigma \circ \xi = h_0$$

for every isomorphism ξ of $\mathscr{P}$ onto $\mathscr{P}^\sigma$.

Remark that if σ is the identity on K, there always exists an isomorphism of $\mathscr{P}$ onto $\mathscr{P}^\sigma$ (cf. Proposition 5 of [AF]). We shall now show the existence of normalized Kummer variety. For this purpose we need a result due to W. L. Chow [1], which we state as

LEMMA 2.1. *Let A and B be two abelian varieties and λ a homomorphism of A into B; let k be a field of definition for A and B. Then λ is defined over a separably algebraic extension of k.*

PROPOSITION 2.2. *Let $\mathscr{P} = (A, C, \theta)$ and $\mathscr{P}_1 = (A_1, C_1, \theta_1)$ be two polarized abelian varieties of type $\mathfrak{r}$, and η an isomorphism of $\mathscr{P}$ onto $\mathscr{P}_1$. Let K be the field of moduli of $\mathscr{P}_1$, and (V, h) a quotient of A_1 by the group of automorphisms of $\mathscr{P}_1$ such that V is defined over K. Suppose that there exists a separably generated extension M of K satisfying the following conditions:*

i) A_1 *and h are defined over M;*

ii) *if σ is an isomorphism of M into the universal domain leaving invariant the elements of K, then we have $h^\sigma \circ \xi = h$ for every isomorphism ξ of $\mathscr{P}_1$ onto $\mathscr{P}_1^\sigma$. Then $(V, h \circ \eta)$ is a normalized Kummer variety of $\mathscr{P}$; in particular, (V, h) is a normalized Kummer variety of $\mathscr{P}_1$.*

PROOF. Since $\mathscr{P}$ is isomorphic to $\mathscr{P}_1$, K is also the field of moduli of $\mathscr{P}$.

It is easy to verify that $(V, h \circ \eta)$ is a quotient of A by the group of automorphisms of $\mathcal{P}$. Let k be a field of definition for $\mathcal{P}$ containing K, and τ an isomorphism of k into the universal domain leaving invariant the elements of K. By Lemma 2.1, there exists a separably algebraic extension k_1 of kM over which η is defined. We take an extension k_2 of k_1 over which $\mathcal{P}_1$ is defined, and extend τ to an isomorphism of k_2 which we denote by σ. Then, for every isomorphism ξ of $\mathcal{P}$ onto $\mathcal{P}^\sigma$, $\eta^\sigma \circ \xi \circ \eta^{-1}$ is an isomorphism of $\mathcal{P}_1$ onto $\mathcal{P}_1^\sigma$. By our assumption ii), we have $h^\sigma \circ \eta^\sigma \circ \xi \circ \eta^{-1} = h$, so that $(h \circ \eta)^\sigma \circ \xi = h \circ \eta$. If we assume that σ is the identity on k, we have $\mathcal{P} = \mathcal{P}^\sigma$, so that we can take ξ to be the identity mapping of $\mathcal{P}$ onto itself. We have then $(h \circ \eta)^\sigma = h \circ \eta$. This shows that $h \circ \eta$ is defined over k, since $h \circ \eta$ is defined over a separably generated extension k_1 of k. Therefore $(V, h \circ \eta)$ satisfies the conditions (K1, 2, 3).

Let $\mathcal{P} = (A, C, \theta)$ be a polarized abelian variety of type $\mathfrak{r}$ and K the field of moduli of $\mathcal{P}$. By the definition of field of moduli, there exists an ample divisor X in C such that the field of moduli K is the smallest field of definition for the variety $\mathcal{T}(A, X, \theta)$ (cf. no. 5 of [AF]). By Proposition 4 of [AF], there exist a regular extension M of K, an abelian variety A_1 in a projective space P^n, a hyperplane section X_1 of A_1 and an isomorphism η of A onto A_1 satisfying the following conditions:

(A'1) $\eta(X)$ is algebraically equivalent to X_1;

(A'2) A_1 and $\eta\theta(r)\eta^{-1}$ for $r \in \mathfrak{r}$ are all defined over M, and X_1 is rational over M.

Put $\theta_1(r) = \eta\theta(r)\eta^{-1}$ for $r \in \mathfrak{r}$, $C_1 = C(X_1)$, and $\mathcal{P}_1 = (A_1, C_1, \theta_1)$. Then η is an isomorphism of $\mathcal{P}$ onto $\mathcal{P}_1$. Denote by $\mathcal{Q}_1$ the group of automorphisms of $\mathcal{P}_1$. By virtue of Proposition 2.2, if we construct a quotient (V, h) of A_1 by $\mathcal{Q}_1$, satisfying the conditions i) and ii) of the proposition for the present $\mathcal{P}_1$ and M, then $(V, h \circ \eta)$ gives a normalized Kummer variety of $\mathcal{P}$. Therefore we shall now proceed in the construction of such a quotient. By Lemma 2.1, we can find a finite Galois extension M_1 of M such that every element of $\mathcal{Q}_1$ is defined over M_1; let $\mathcal{G}$ denote the Galois group of M_1 over M. Take any quotient (V, h) of A_1 by $\mathcal{Q}_1$ defined over M_1. We see easily that for every $\gamma \in \mathcal{Q}_1$ and for every $\sigma \in \mathcal{G}$, γ^σ is an automorphism of $\mathcal{P}_1$, so that $\gamma^\sigma \in \mathcal{Q}_1$. Hence, for every $\sigma \in \mathcal{G}$, (V^σ, h^σ) is a quotient of A_1 by $\mathcal{Q}_1$, defined over M_1. On account of (Q3), there exist rational mappings g_σ and g'_σ, of V into V^σ and of V^σ into V, such that $g_\sigma \circ h = h^\sigma$, $g'_\sigma \circ h^\sigma = h$, and, g_σ is everywhere defined on V and g'_σ is everywhere defined on V^σ. We see then that g_σ is a birational mapping of V onto V^σ which is everywhere biregular on V and $g'_\sigma = (g_\sigma)^{-1}$. By the remark of § 2.1, g_σ is defined over M_1. Put $f_{\tau,\sigma} = g_\tau \circ (g_\sigma)^{-1}$ for $\sigma, \tau \in \mathcal{G}$. Then we have

$$f_{\tau,\sigma}\circ f_{\sigma,\rho}=f_{\tau,\rho}\,.$$

Morever, if $\omega\in\mathcal{G}$, we have

$$g_\sigma^\omega\circ g_\omega\circ h=g_\sigma^\omega\circ h^\omega=(g_\sigma\circ h)^\omega=h^{\sigma\omega}=g_{\sigma\omega}\circ h\,,$$

so that $g_\sigma^\omega\circ g_\omega=g_{\sigma\omega}$; hence we have

$$f_{\tau\omega,\sigma\omega}=g_{\tau\omega}\circ(g_{\sigma\omega})^{-1}=g_\tau^\omega\circ g_\omega\circ(g_\sigma^\omega\circ g_\omega)^{-1}=g_\tau^\omega\circ(g_\sigma^\omega)^{-1}=(f_{\tau,\sigma})^\omega\,.$$

Thus the mappings $f_{\tau,\sigma}$ satisfy the conditions of Theorem 1 of Weil [**32**]. Therefore, by the result of [**32**], there exist a projective variety V_1, defined over M, and a birational mapping f of V onto V_1, defined over M_1, such that f is everywhere biregular and $f_{\tau,\sigma}=(f^\tau)^{-1}\circ f^\sigma$ for $\tau,\sigma\in\mathcal{G}$. Put $h_1=f\circ h$; then we see that h_1 is defined over M_1. Moreover, for every $\rho\in\mathcal{G}$, we have

$$h_1^\rho=f^\rho\circ h^\rho=f\circ f_{1,\rho}\circ h^\rho=f\circ(g_\rho)^{-1}\circ h^\rho=f\circ h=h_1\,;$$

this shows that h_1 is defined over M. We can easily verify that (V_1,h_1) satisfies (Q1, 2, 3) for A_1 and $\mathcal{Q}_1$, so that (V_1,h_1) is a quotient of A_1 by $\mathcal{Q}_1$, defined over M.

Now let t be a point in an affine space such that $M=K(t)$. As M is a regular extension of K, t has a locus over K, which we denote by T. For every generic point u of T over K, we consider the isomorphism of $K(t)$ onto $K(u)$ over K which maps t onto u, and denote by $\mathcal{P}_u, A_u, \mathcal{Q}_u, V_u, h_u$ the transform of $\mathcal{P}_1, A_1, \mathcal{Q}_1, V_1, h_1$ by the isomorphism. It is clear that $\mathcal{Q}_u$ is the group of automorphisms of $\mathcal{P}_u$ and (V_u, h_u) is a quotient of A_u by $\mathcal{Q}_u$, defined over $K(u)$. Let u and v be two generic points of T over K. As K is the field of moduli of $\mathcal{P}_1$, we see that, by Proposition 5 of [AF], both $\mathcal{P}_u$ and $\mathcal{P}_v$ are isomorphic to $\mathcal{P}_1$. Hence there exists an isomorphism λ of $\mathcal{P}_u$ onto $\mathcal{P}_v$; and so we can find a birational mapping φ of V_u onto V_v, everywhere biregular on V_u, such that

$$\varphi\circ h_u=h_v\circ\lambda\,.$$

If λ' is another isomorphism of $\mathcal{P}_u$ onto $\mathcal{P}_v$, $\lambda'\circ\lambda^{-1}$ is contained in $\mathcal{Q}_v$; we have therefore $h_v=h_v\circ(\lambda'\circ\lambda^{-1})$ and hence

$$\varphi\circ h_u=h_v\circ\lambda=h_v\circ(\lambda'\circ\lambda^{-1})\circ\lambda=h_v\circ\lambda'\,.$$

This shows that the mapping φ does not depend on the choice of λ; so we write $\varphi=\varphi_{v,u}$. We shall now prove that $\varphi_{v,u}$ is defined over $K(u,v)$. As A_u and A_v are defined over $K(u,v)$, we can find, on account of Lemma 2.1, a finite separably algebraic extension L of $K(u,v)$ over which λ is defined. Then, by the remark in §2.1, $\varphi_{v,u}$ is defined over L. Let σ be an isomorphism of L over $K(u,v)$ into the algebraic closure of $K(u,v)$. Then λ^σ is an isomorphism of $\mathcal{P}_u$ onto $\mathcal{P}_v$, so that we have

$$(\varphi_{v,u})^\sigma\circ h_u=(\varphi_{v,u}\circ h_u)^\sigma=(h_v\circ\lambda)^\sigma=h_v\circ\lambda^\sigma=\varphi_{v,u}\circ h_u\,.$$

Hence we have $\varphi_{v,u}{}^{\sigma}=\varphi_{v,u}$; this shows that $\varphi_{v,u}$ is defined over $K(u,v)$. Now, u,v,w being three generic points of T over K, we have

(1) $$\varphi_{w,v}\circ\varphi_{v,u}=\varphi_{w,u}\,.$$

In fact, if λ is an isomorphism of $\mathcal{P}_u$ onto $\mathcal{P}_v$ and μ is an isomorphism of $\mathcal{P}_v$ onto $\mathcal{P}_w$, $\mu\circ\lambda$ is an isomorphism of $\mathcal{P}_u$ onto $\mathcal{P}_w$, so that $\varphi_{w,u}\circ h_u=h_w\circ\mu\circ\lambda=\varphi_{w,v}\circ h_v\circ\lambda=\varphi_{w,v}\circ\varphi_{v,u}\circ h_u$; this proves the relation (1). Applying the result of [32] to V_u and $\varphi_{v,u}$, we obtain a projective variety V_0, defined over K and a birational mapping φ_t of V_0 onto V_t, which is biregular on V_0 and defined over $K(t)$, such that

(2) $$\varphi_{v,u}=\varphi_v\circ(\varphi_u)^{-1}\,,$$

where φ_w denotes, for any generic point w of T over K, the transform of φ_t by the isomorphism of $K(t)$ onto $K(w)$ over K which maps t onto w. In [32], only independent generic points are considered; but once we obtain the relation (2) for independent generic points, we have the same formula for any two generic points by virtue of the relation (1), since (1) holds for any three generic points. Put now

$$h_0=(\varphi_t)^{-1}\circ h_t\,.$$

We note that $\mathcal{P}_1=\mathcal{P}_t$, $V_1=V_t$, $h_1=h_t$. It is easy to see that (V_0,h_0) is a quotient of A_1 by $\mathcal{Q}_1$, and h_0 is defined over M. Let σ be an isomorphism of M into the universal domain leaving invariant the elements of K, and ξ an isomorphism of $\mathcal{P}_1$ onto $\mathcal{P}_1^{\sigma}$. Putting $t^{\sigma}=u$, we have $\mathcal{P}_1^{\sigma}=\mathcal{P}_u$, $h_t^{\sigma}=h_u$, $\varphi_t^{\sigma}=\varphi_u$; and by the property of $\varphi_{u,v}$, we have $\varphi_{u,t}\circ h_t=h_u\circ\xi$. Hence,

$$h_0^{\sigma}\circ\xi=(\varphi_t^{-1}\circ h_t)^{\sigma}\circ\xi=\varphi_u^{-1}\circ h_u\circ\xi=\varphi_u^{-1}\circ\varphi_{u,t}\circ h_t=\varphi_t^{-1}\circ h_t=h_0\,.$$

Thus we have proved that $\{\mathcal{P}_1,M,(V_0,h_0)\}$ satisfies the conditions i), ii) of Proposition 2.2. Therefore, by that proposition, $(V_0,h_0\circ\eta)$ is a normalized Kummer variety of $\mathcal{P}$.

Now we consider about the uniqueness of normalized Kummer variety.

PROPOSITION 2.3. *Let $\mathcal{P}=(A,C,\theta)$ be a polarized abelian variety of type $\mathfrak{r}$, (V,h) and (V_1,h_1) two normalized Kummer variety of $\mathcal{P}$, and K the field of moduli of $\mathcal{P}$. Then, there exists a biregular birational mapping α of V onto V_1 such that $h_1=\alpha\circ h$; the mapping α is defined over a purely inseparable extension of K. Moreover, if A is defined over a separably generated extension of K, α is defined over K.*

PROOF. The existence of a biregular birational mapping α such that $h_1=\alpha\circ h$ follows directly from the property (Q3) of quotient. Let k be a field of definition for $\mathcal{P}$ containing K. Then, h and h_1 are defined over k, so that α is defined over k. Let σ be an isomorphism of k into the universal domain leaving invariant the elements of K. Then, for any isomorphism ξ of $\mathcal{P}$ onto

$\mathcal{P}^\sigma$, we have $h = h^\sigma \circ \xi$, $h_1 = h_1^\sigma \circ \xi$, and hence $\alpha \circ h = h_1 = h_1^\sigma \circ \xi = \alpha^\sigma \circ h^\sigma \circ \xi = \alpha^\sigma \circ h$; so we have $\alpha^\sigma = \alpha$. This proves the first assertion of our proposition. If k is separably generated over K, the relation $\alpha^\sigma = \alpha$ shows that α is defined over K; this proves the last assertion.

2.3. Homomorphisms of polarized abelian varieties.

Let $\mathcal{P} = (A, C, \theta)$ and $\mathcal{P}' = (A', C', \theta')$ be two polarized abelian varieties of type $\mathfrak{r}$, of the same dimension, and λ a homomorphism of A onto A'. We call λ a *homomorphism* of $\mathcal{P}$ onto $\mathcal{P}'$ if we have $\lambda^{-1}(X') \in C$ for every $X' \in C'$ and $\lambda\theta(r) = \theta'(r)\lambda$ for every $r \in \mathfrak{r}$. The following proposition is an easy consequence of this definition.

PROPOSITION 2.4. *Let* $\mathcal{P} = (A, C, \theta)$, $\mathcal{P}_1 = (A_1, C_1, \theta_1)$, $\mathcal{P}_2 = (A_2, C_2, \theta_2)$ *be three polarized abelian varieties of type* $\mathfrak{r}$, *of the same dimension; let* $\lambda_1, \lambda_2, \mu$ *be respectively homomorphisms of* A *onto* A_1, *of* A *onto* A_2, *of* A_1 *onto* A_2 *such that* $\mu \circ \lambda_1 = \lambda_2$. *If any two of* $\lambda_1, \lambda_2, \mu$ *are homomorphisms of polarized abelian varieties of type* $\mathfrak{r}$, *then so is the remaining one.*

PROPOSITION 2.5. *Let* $\mathcal{P} = (A, C, \theta)$, $\mathcal{P}_1 = (A_1, C_1, \theta_1)$, $\mathcal{P}_2 = (A_2, C_2, \theta_2)$ *be three polarized abelian varieties of type* $\mathfrak{r}$, *of the same dimension; let* λ_i, *for* $i = 1, 2$, *be a separable homomorphism of* $\mathcal{P}$ *onto* $\mathcal{P}_i$ *and* $\mathfrak{g}_i$ *the kernel of* λ_i. *Then the following two assertions hold.*

i) *If* $\mathfrak{g}_1 = \mathfrak{g}_2$, *then there exists an isomorphism* η *of* $\mathcal{P}_1$ *onto* $\mathcal{P}_2$ *such that* $\eta \circ \lambda_1 = \lambda_2$.

ii) *Suppose that* $\nu(\lambda_1) = \nu(\lambda_2)$ *and there exists an element* a *in* $\mathfrak{r}$ *for which we have, for each* i,

$$\beta \in \mathcal{A}(A), \quad \beta(\mathfrak{g}_i) = \{0\} \Leftrightarrow \beta \in \theta(a\mathfrak{r}).$$

Under these assumptions, if $\mathcal{P}_1$ *is isomorphic to* $\mathcal{P}_2$, *then we have* $\mathfrak{g}_1 = \mathfrak{g}_2$.

PROOF. The assertion i) is an immediate consequence of Proposition 2.4; so we shall prove ii). Assumptions being as in ii), let ϵ be an isomorphism of $\mathcal{P}_2$ onto $\mathcal{P}_1$. Since $\theta(a)(\mathfrak{g}_1) = \{0\}$, there exists a homomorphism μ of A_1 into A such that $\mu \circ \lambda_1 = \theta(a)$. As $\nu(\lambda_1)1_A$ is contained in $\theta(a\mathfrak{r})$, we have $\nu(\theta(a)) \neq 0$, so that μ is an isogeny. We see that $\mu \circ \epsilon \circ \lambda_2 \in \mathcal{A}(A)$ and $\mu \circ \epsilon \circ \lambda_2(\mathfrak{g}_2) = \{0\}$. Hence there exists an element $r \in \mathfrak{r}$ such that $\mu \circ \epsilon \circ \lambda_2 = \theta(ar)$; we obtain then $\mu \circ \epsilon \circ \lambda_2 = \mu \circ \lambda_1 \circ \theta(r)$, so that $\epsilon \circ \lambda_2 = \lambda_1 \circ \theta(r)$. From this and the assumption $\nu(\lambda_1) = \nu(\lambda_2)$, it follows that $\nu(\theta(r)) = 1$; this shows that $\theta(r)$ is an automorphism of A. As λ_1 commutes with the operation of $\mathfrak{r}$, we have $\epsilon \circ \lambda_2 = \theta_1(r) \circ \lambda_1$; this implies $\mathfrak{g}_1 = \mathfrak{g}_2$, since ϵ and $\theta_1(r)$ are isomorphisms.

2.4. Fields obtained from the points of finite order.

PROPOSITION 2.6. *Let* $\mathcal{P} = (A, C, \theta)$ *be an abelian variety of type* $\mathfrak{r}$ *and* (V, h) *a normalized Kummer variety of* $\mathcal{P}$. *Let* $\mathfrak{a}$ *be a subset of* $\mathfrak{r}$ *such that* $\theta(\mathfrak{a})$ *contains a regular element of* $\mathcal{A}_0(A)$; *denote by* $\mathfrak{g}(\mathfrak{a}, A)$ *the set of points* t *on* A *such* *that $\theta(\mathfrak{a})t = 0$ for every $a \in \mathfrak{a}$. Let K be the field of moduli of $\mathscr{P}$ and $K_\mathfrak{a}$ the field generated over K by the points $h(t)$ for $t \in \mathfrak{g}(\mathfrak{a}, A)$. Then, $K_\mathfrak{a}$ is a normal algebraic extension of K.*

PROOF. As $\theta(\mathfrak{a})$ contains a regular element of $\mathcal{A}_0(A)$, the field $K_\mathfrak{a}$ is algebraic over K. Let σ be an isomorphism of the universal domain into itself, which leaves invariant the elements of K. Then, there exists an isomorphism ε of $\mathscr{P}$ onto $\mathscr{P}^\sigma$, for which we have $h^\sigma \circ \varepsilon = h$. Denote by $\mathfrak{g}(\mathfrak{a}, A^\sigma)$ the set of points u on A^σ such that $\theta^\sigma(a)u = 0$ for every $a \in \mathfrak{a}$. It is easy to see that $t \to t^\sigma$ gives an isomorphism of $\mathfrak{g}(\mathfrak{a}, A)$ onto $\mathfrak{g}(\mathfrak{a}, A^\sigma)$ and $u \to \varepsilon^{-1}u$ gives an isomorphism of $\mathfrak{g}(\mathfrak{a}, A^\sigma)$ onto $\mathfrak{g}(\mathfrak{a}, A)$; hence $t \to \varepsilon^{-1}t^\sigma$ gives an automorphism of $\mathfrak{g}(\mathfrak{a}, A)$. By the relation $h^\sigma \circ \varepsilon = h$, we have $h(t)^\sigma = h(\varepsilon^{-1}t^\sigma)$. Therefore, $h(t) \to h(t)^\sigma$ is a permutation of the points $\{h(t) \mid t \in \mathfrak{g}(\mathfrak{a}, A)\}$. This proves our proposition.

PROPOSITION 2.7. *Let $\mathscr{P} = (A, C, \theta)$ and $\mathscr{P}' = (A', C', \theta')$ be two polarized abelian varieties of type $\mathfrak{r}$, of the same dimension, defined over a field of characteristic 0; let (V, h) be a normalized Kummer variety of $\mathscr{P}$. Let λ be a homomorphism of $\mathscr{P}$ onto $\mathscr{P}'$ and $\mathfrak{g}$ the kernel of λ. Suppose that every automorphism of $\mathscr{P}$ leaves invariant $\mathfrak{g}$ as a whole. Let K and K' be respectively the fields of moduli of $\mathscr{P}$ and $\mathscr{P}'$. Then K' is contained in the field $K(h(t) \mid t \in \mathfrak{g})$.*

PROOF. Let σ be an isomorphism of the universal domain into itself, which is the identity on $K(h(t) \mid t \in \mathfrak{g})$. There exists an isomorphism ε of $\mathscr{P}$ onto $\mathscr{P}^\sigma$, for which we have $h^\sigma \circ \varepsilon = h$. For every $t \in \mathfrak{g}$, we have $h(t) = h(t)^\sigma = h(\varepsilon^{-1}t^\sigma)$, so that by the property (Q2) of quotient, there exists an automorphism η of $\mathscr{P}$ such that $\varepsilon^{-1}t^\sigma = \eta t$. By our assumption, this shows that $\varepsilon^{-1}t^\sigma$ is contained in $\mathfrak{g}$. Hence $\mathfrak{g}$ is the kernel of $\lambda^\sigma \circ \varepsilon$. By i) of Proposition 2.5, $\mathscr{P}'^\sigma$ is isomorphic to $\mathscr{P}'$; so σ must be the identity on K'. This proves that K' is contained in $K(h(t) \mid t \in \mathfrak{g})$.

PROPOSITION 2.8. *Notations and assumptions being as in Proposition 2.7, let (V', h') be a normalized Kummer variety of $\mathscr{P}'$; and put $K_1 = K(h(t) \mid t \in \mathfrak{g})$. Then, for every point u on A, $K_1(h(u))$ contains $K'(h'(\lambda(u)))$.*

PROOF. Let σ be an isomorphism of the universal domain into itself leaving invariant the elements of $K_1(h(u))$. ε being as in the proof of Proposition 2.7, we have $h(u) = h(u)^\sigma = h(\varepsilon^{-1}u^\sigma)$, so that there exists an automorphism ξ of $\mathscr{P}$ such that $\varepsilon^{-1}u^\sigma = \xi u$. Put $\varepsilon_0 = \varepsilon\xi$. Then ε_0 is also an isomorphism of $\mathscr{P}$ onto $\mathscr{P}'$. Applying the argument of the proof of Proposition 2.7 to ε_0, we observe that $\lambda^\sigma \circ \varepsilon_0$ has the same kernel as λ. Hence, by i) of Proposition 2.5, there exists an isomorphism α of $\mathscr{P}'$ onto $\mathscr{P}'^\sigma$ such that $\lambda^\sigma \circ \varepsilon_0 = \alpha \circ \lambda$. We have then

$$h'(\lambda(u))^\sigma = h'^\sigma(\lambda^\sigma(u^\sigma)) = h'^\sigma(\lambda^\sigma(\varepsilon_0(u))) = h'(\lambda(u)).$$

On account of Proposition 2.7, it follows that σ is the identity on $K'(h'(\lambda(u)))$;

this proves our proposition.

§3. Automorphic functions attached to an indefinite quaternion algebra.

We shall now consider the functions obtained from the points of finite order on the abelian varieties belonging to an analytic system. We shall only deal with the system attached to an indefinite quaternion algebra ([AF, §5]), though our method is applicable to a more general case.

3.1. The analytic system $S = \{\mathcal{P}(z) \mid z \in \mathfrak{H}\}$. First we recall the results of [AF, §5] with a few changes of notations. Let $\mathcal{O}$ be an indefinite quaternion algebra over $\boldsymbol{Q}$. We fix once for all a faithful representation χ of $\mathcal{O}$ by real matrices of degree 2. Let $\mathfrak{H}$ denote the upper half complex plane defined by $\mathrm{Im}(z) > 0$. For every $z \in \mathfrak{H}$, we denote by $e(z)$ the column-vector $\begin{pmatrix} z \\ 1 \end{pmatrix}$. Let $\mathfrak{o}$ be an order in $\mathcal{O}$; for every $z \in \mathfrak{H}$, put

$$D(z) = \chi(\mathfrak{o})e(z) = \{\chi(\alpha)e(z) \mid \alpha \in \mathfrak{o}\}.$$

Then, $D(z)$ is a lattice in $\boldsymbol{C}^2$ and $\boldsymbol{C}^2/D(z)$ has a structure of abelian variety. This was shown by constructing a Riemann form on $\boldsymbol{C}^2/D(z)$ as follows (cf. [AF, no. 18]): Let ρ be an element of $\mathcal{O}$ such that ρ^2 is a negative rational number. Put, for every $\alpha \in \mathcal{O}$,

$$\alpha^* = \rho^{-1}\alpha'\rho.$$

Then $\alpha \to \alpha^*$ gives an involution of $\mathcal{O}$; and we have $\mathrm{tr}(\alpha\alpha^*) > 0$ for every $\alpha \neq 0$ of $\mathcal{O}$. Define an $\boldsymbol{R}$-bilinear form $E(x, y)$ on $\boldsymbol{C}^2$ in such a way that

$$E(\chi(\alpha)e(z), \quad \chi(\beta)e(z)) = \mathrm{tr}(\rho\alpha\beta')$$

holds for every $\alpha \in \mathcal{O}$ and $\beta \in \mathcal{O}$. Then, for a suitable integer $c \neq 0$, cE defines a non-degenerate Riemann form on the complex torus $\boldsymbol{C}^2/D(z)$.

Now we can construct (cf. [AF, no. 19]) a system of polarized abelian varieties $\{\mathcal{P}(z) \mid z \in \mathfrak{H}\}$ of type $\mathfrak{o}$ parametrized by an analytic mapping $\Lambda(x, z)$ of $\boldsymbol{C}^2 \times \mathfrak{H}$ into a projective space P^N. The polarized abelian variety $\mathcal{P}(z) = (A(z), C_z, \theta_z)$ of type $\mathfrak{o}$ is defined as follows.

   i) For every $z \in \mathfrak{H}$, $x \to \Lambda(x, z)$ is an analytic isomorphism of $\boldsymbol{C}^2/D(z)$ onto the abelian variety $A(z)$.

   ii) C_z is the polarization of $A(z)$ determined by the hyperplane sections; and it corresponds to the Riemann form $cE(x, y)$.

   iii) For every $\alpha \in \mathfrak{o}$, $\theta_z(\alpha)$ is the endomorphism of $A(z)$ corresponding to $\chi(\alpha)$; namely, we have

$$\theta_z(\alpha)\Lambda(x, z) = \Lambda(\chi(\alpha)x, z).$$

We shall denote the system $\{\mathcal{P}(z) \mid z \in \mathfrak{H}\}$ by $S(\mathfrak{o}, *)$ or simply by S. For every $z \in \mathfrak{H}$, we put

$$\mathcal{F}(z) = \mathcal{F}(A(z), \theta_z).$$

The definition of $\mathcal{F}(A, \theta)$ is given in [AF, no. 4] or §3.4 of the present paper. If we denote by $c(\mathcal{F}(z))$ the Chow-point of the variety $\mathcal{F}(z)$, the field $Q(c(\mathcal{F}(z)))$ is the field of moduli of $\mathcal{P}(z)$. Now, there exist a discrete subset $\mathfrak{W}$ of $\mathfrak{H}$ and a set of meromorphic functions $\{f_1, \cdots, f_m\}$ on $\mathfrak{H}$ such that

$$(3) \qquad c(\mathcal{F}(z)) = (1, f_1(z), \cdots, f_m(z))$$

for every $z \in \mathfrak{H} - \mathfrak{W}$. Denote by $\Gamma = \Gamma(\mathfrak{o})$ the group composed of all units γ of $\mathfrak{o}$ such that $N(\gamma) = 1$. Then, Γ is a Fuchsian group on $\mathfrak{H}$. Let $\mathfrak{R}(\mathfrak{o})$ denote the field of automorphic functions on $\mathfrak{H}$ with respect to Γ. Then, Theorem 6 of [AF] asserts that the meromorphic functions f_i determined by (3) generate the function-field $\mathfrak{R}(\mathfrak{o})$; namely, we have

$$\mathfrak{R}(\mathfrak{o}) = C(f_1, \cdots, f_m).$$

Furthermore, (3) implies that, for every $z \in \mathfrak{H} - \mathfrak{W}, Q(f_1(z), \cdots, f_m(z))$ is the field of moduli of $\mathcal{P}(z)$.

PROPOSITION 3.1. *If $Q(c(\mathcal{F}(z)))$ is not algebraic over Q, we have $\mathcal{A}_0(A(z)) = \theta_z(\Phi)$, $\mathcal{A}(A(z)) = \theta_z(\mathfrak{o})$; and the automorphisms of $\mathcal{P}(z)$ are $\pm 1_z$, where 1_z denotes the identity element of $\mathcal{A}(A(z))$.*

PROOF. Suppose first that $A(z)$ is simple; then $\mathcal{A}_0(A(z))$ is a division algebra. Since $\mathcal{A}_0(A(z))$ has a rational representation of degree 4, we must have $[\mathcal{A}_0(A(z)) : Q] \leqq 4$; this shows $\mathcal{A}_0(A(z)) = \theta_z(\Phi)$. Now consider the case where $A(z)$ is not simple; $A(z)$ is then isogenous to a product $E_1 \times E_2$ of two elliptic curves E_1 and E_2. If E_1 is not isogenous to E_2, $\mathcal{A}_0(A(z))$ is isomorphic to the direct sum of $\mathcal{A}_0(E_1)$ and $\mathcal{A}_0(E_2)$; this is impossible since $\mathcal{A}_0(A(z))$ contains a central simple algebra $\theta_z(\Phi)$ of degree 2 over Q. Hence $A(z)$ must be isogenous to the product $E_1 \times E_1$; and so $\mathcal{A}_0(A(z))$ is isomorphic to the total matrix ring of degree 2 over $\mathcal{A}_0(E_1)$. By the same argument as in the proof of Theorem 6 of [AF], we can show that $Q(c(\mathcal{F}(z)))$ is algebraic over $Q(j(E_1))$, where $j(E)$ denotes the birational invariant of an elliptic curve E. Since $Q(c(\mathcal{F}(z)))$ is not algebraic over Q, $j(E_1)$ can not be algebraic over Q, so that $\mathcal{A}_0(E_1)$ is isomorphic to Q. If follows from this that $\mathcal{A}_0(A(z)) = \theta_z(\Phi)$. Recall that $A(z)$ is isomorphic to $C^2/D(z)$ and $D(z) = \chi(\mathfrak{o})e(z)$. Then the equality $\mathcal{A}_0(A(z)) = \theta_z(\Phi)$ implies $\mathcal{A}(A(z)) = \theta_z(\mathfrak{o})$. Now let α be an automorphism of $\mathcal{P}(z)$; since α commutes with every element of $\theta_z(\Phi), \alpha$ is contained in the center Q of $\theta_z(\Phi)$. Hence α must be equal to 1_z or -1_z. Our proposition is thereby proved.

3.2. Functions obtained from the points of finite order. We shall now make use of the mapping $\Psi(z)$ attached to the analytic system S, whose definition is given in [AF, no. 12]. This mapping has the following properties.

(Ψ1) $\Psi(z)$ is an analytic mapping of $\mathfrak{H}$ into a projective space.

(Ψ2) For every $z \in \mathfrak{H}$, $\mathscr{P}(z)$ is defined over $Q(\Psi(z))$.

(Ψ3) If z_1 and z_2 are two points of $\mathfrak{H}$ and if $\mathfrak{p}$ is a place of $Q(\Psi(z_1))$ taking values in C such that $\mathfrak{p}(\Psi(z_1)) = \Psi(z_2)$, then we have

$$\mathfrak{p}(A(z_1)) = A(z_2), \quad \mathfrak{p}(T(z_1)) = T(z_2), \quad \mathfrak{p}(U(\alpha, z_1)) = U(\alpha, z_2),$$

where $T(z)$ denotes the graph of the law of composition on $A(z)$, and $U(\alpha, z)$ denotes the graph of $\theta_z(\alpha)$. (For the definition of places and the notation $\mathfrak{p}(V)$, see Appendix of [AF] and [25 Chap. III].)

Let $\mathfrak{H}_1$ be the set of all generic points of $\mathfrak{H} - \mathfrak{W}$ for Ψ over Q. Take and fix a point z_0 of $\mathfrak{H}_1$. Let (V_0, h_0) be a normalized Kummer variety of $\mathscr{P}(z_0)$. If z is a point of $\mathfrak{H}_1$, there exists an isomorphism σ of $Q(\Psi(z_0))$ onto $Q(\Psi(z))$ such that $\Psi(z_0)^\sigma = \Psi(z)$; we have then $\mathscr{P}(z_0)^\sigma = \mathscr{P}(z)$. Put

$$V(z) = V_0^\sigma, \quad h_z = h_0^\sigma .$$

(V_z, h_z) is obviously a normalized Kummer variety of $\mathscr{P}(z)$. Now let $z \to u(z)$ be an analytic mapping of $\mathfrak{H}$ into C^2. Put

$$\Theta_u(z) = \Lambda(u(z), z),$$

and consider, for every $z \in \mathfrak{H}_1$, the point $h_z(\Theta_u(z))$ lying on the variety $V(z)$. As h_z is defined over $Q(\Psi(z))$, the quotients of the coordinates of this point are contained in $Q(\Psi(z), \Theta_u(z))$. Let $\mathfrak{H}_1(u)$ be the set of all generic points of $\mathfrak{H} - \mathfrak{W}$ for Ψ and Θ_u over Q. Take and fix a point z_1 of $\mathfrak{H}_1(u)$. Then there exist elements $y_1, \cdots, y_M$ of $Q(\Psi(z_1), \Theta_u(z_1))$ such that

$$h_{z_1}(\Theta_u(z_1)) = (1, y_1, \cdots, y_M).$$

Let $g_1, \cdots, g_M$ be the elements of $Q(\Psi, \Theta_u)$ corresponding to $y_1, \cdots, y_M$ by the canonical isomorphism of $Q(\Psi, \Theta_u)$ onto $Q(\Psi(z_1), \Theta_u(z_1))$. Then $g_1, \cdots, g_M$ are meromorphic functions on $\mathfrak{H}$; and for every $z \in \mathfrak{H}_1(u)$, we have

$$h_z(\Lambda(u(z), z)) = (1, g_1(z), \cdots, g_M(z)).$$

Now consider the case where $u(z)$ is defined by

$$u(z) = \chi(\xi)e(z),$$

where ξ is an element of $\mathcal{O}$. In this case we denote by $g_\nu(\xi, z)$ the meromorphic function $g_\nu(z)$ and put $\mathfrak{H}_1(\xi) = \mathfrak{H}_1(u)$. Then, for every $z \in \mathfrak{H}_1(\xi)$, we have

$$(4) \qquad h_z(\Lambda(\chi(\xi)e(z), z)) = (1, g_1(\xi, z), \cdots, g_M(\xi, z)).$$

PROPOSITION 3.2. *Notations being as above, for every* $\xi \in \mathcal{O}$ *and* $\gamma \in \Gamma(\mathfrak{o})$, *we have*

$$g_\nu(\xi\gamma, z) = g_\nu(\xi, \gamma[z]) \qquad (1 \leq \nu \leq M).$$

PROOF. Let z be a point of $\mathfrak{H}_1 \cap \gamma^{-1}(\mathfrak{H}_1)$. Putting

$$z' = \gamma[z], \quad \chi(\gamma) = \begin{pmatrix} a & b \\ c & d \end{pmatrix}, \quad s = cz + d,$$

we get $s \cdot e(z') = \chi(\gamma)e(z)$. We can find an isomorphism η of $\mathcal{P}(z)$ onto $\mathcal{P}(z')$ such that $\eta \Lambda(sx, z) = \Lambda(x, z')$ (cf. the proof of Proposition 15 of [AF]). As z and z' are generic for Ψ over Q, there exists an isomorphism σ of $Q(\Psi(z))$ onto $Q(\Psi(z'))$ such that $\Psi(z)^\sigma = \Psi(z')$; we have then $\mathcal{P}(z)^\sigma = \mathcal{P}(z')$. Since $\mathcal{P}(z)$ is isomorphic to $\mathcal{P}(z')$, σ must leave invariant the elements of the field of moduli of $\mathcal{P}(z)$. Therefore, by the property (K3) of normalized Kummer variety, the equality $h_z^\sigma \circ \eta = h_z$ holds. By our construction of h_z, we have $h_z^\sigma = h_{z'}$, so that

$$h_z(\Lambda(sx, z)) = h_{z'}(\eta \Lambda(sx, z)) = h_{z'}(\Lambda(x, z')).$$

Substituting $\chi(\xi)e(z')$ for x, we obtain

$$h_z(\Lambda(\chi(\xi\gamma)e(z), z)) = h_{z'}(\Lambda(\chi(\xi)e(z'), z')).$$

If we take z sufficiently generic, this proves the relation of our proposition.

PROPOSITION 3.3. *Let ξ and ξ_1 be two elements of Φ. Then*

$$g_\nu(\xi, z) = g_\nu(\xi_1, z)$$

holds for every ν if and only if

$$\xi \equiv \pm \xi_1 \quad \mathrm{mod.\,o}.$$

PROOF. If $\xi \equiv \pm \xi_1 \,\mathrm{mod.\,o}$, we have

$$\chi(\xi)e(z) \equiv \pm \chi(\xi_1)e(z) \quad \mathrm{mod.}\ D(z).$$

This implies $\Lambda(\chi(\xi)e(z), z) = \pm \Lambda(\chi(\xi_1)e(z), z)$. Since $h_z(\pm x) = h_z(x)$, we obtain

(5) $$h_z(\Lambda(\chi(\xi)e(z), z)) = h_z(\Lambda(\chi(\xi_1)e(z), z)).$$

This proves the "if" part of our proposition. Conversely, suppose that $g_\nu(\xi, z) = g_\nu(\xi_1, z)$ holds for every ν. Then, for every $z \in \mathfrak{H}_1(\xi) \cap \mathfrak{H}_1(\xi_1)$, the equality (5) holds. By Proposition 3.1, we have $\Lambda(\chi(\xi)e(z), z) = \pm \Lambda(\chi(\xi_1)e(z), z)$. Hence, we must have $\chi(\xi)e(z) \equiv \pm \chi(\xi_1)e(z) \,\mathrm{mod.}\ D(z)$, and so $\xi \equiv \pm \xi_1 \,\mathrm{mod.\,o}$. This completes our proof.

3.3. Automorphic functions belonging to congruence-subgroups. Let $\mathfrak{a}$ be an integral right $\mathfrak{o}$-ideal. We denote by $\Gamma_\mathfrak{a}^*$ the subgroup of $\Gamma(\mathfrak{o})$ composed of the elements γ such that $\gamma \equiv \pm 1 \,\mathrm{mod.\,a}$. Then $\Gamma_\mathfrak{a}^*$ is of finite index in $\Gamma(\mathfrak{o})$. We denote by $\mathfrak{K}(\mathfrak{a})$ the field of automorphic functions on $\mathfrak{H}$ with respect to $\Gamma_\mathfrak{a}^*$.

PROPOSITION 3.4. *Let $\mathfrak{a} = \alpha \mathfrak{o}$ be an integral right $\mathfrak{o}$-ideal. Then we have*

$$\mathfrak{K}(\mathfrak{o}\,;\mathfrak{a}) = C(f_i(z), g_j(\alpha^{-1}, z) \mid 1 \leq i \leq m, 1 \leq j \leq M),$$

where the f_i and the g_j are the meromorphic functions defined by (3) and (4) of § 3.2.

PROOF. Let γ be an element of $\Gamma(\mathfrak{o})$. By Propositions 3.2 and 3.3,

$g_j(\alpha^{-1}, \gamma[z]) = g_j(\alpha^{-1}, z)$ holds for every j if and only if $\gamma \equiv \pm 1 \bmod. \mathfrak{a}$, On the other hand, we have obviously

$$\theta_i(\alpha)\Lambda(\chi(\alpha^{-1})e(z), z) = 0;$$

hence, by Proposition 2.6, for every $z \in \mathfrak{H}_1(\alpha^{-1})$, the coordinates $g_j(\alpha^{-1}, z)$ of the point $h_i(\Lambda(\chi(\alpha^{-1})e(z), z))$ are algebraic over the field of moduli $\mathbf{Q}(f_i(z))$ of $\mathcal{P}(z)$. If follows that the functions $g_j(\alpha^{-1}, z)$ are algebraic over the function-field $\mathbf{Q}(f_i)$. This proves our proposition, since the equality $\Re(\mathfrak{o}) = C(f_i)$ holds.

REMARK. Suppose that $\mathfrak{a} = \alpha\mathfrak{o}$ is an integral two-sided $\mathfrak{o}$-ideal. Then, we see that, for every $\beta \in \mathfrak{o}, g_j(\beta\alpha^{-1}, z)$ is invariant under $\Gamma(\mathfrak{o}; \mathfrak{a})$ and algebraic over $\mathbf{Q}(f_i)$. Hence we can write also

$$\Re(\mathfrak{o}; \mathfrak{a}) = C(f_i(z), g_j(\beta\alpha^{-1}, z) \mid 1 \leq i \leq m, 1 \leq j \leq M, \beta \in \mathfrak{o}).$$

3.4. Generic members of the system S. In order to make our later discussion easy, we recall here the definition of the variety $\mathcal{F}(A, \theta)$ introducing some new notations. Let $\mathcal{P} = (A, C, \theta)$ be a polarized abelian variety of type $\mathfrak{r}$. We assume that A is a variety in a projective space P^N of dimension N and C is the polarization determined by the hyperplane sections. Fix a basis $\{r_1, \cdots, r_d\}$ of $\mathfrak{r}$ over $\mathbf{Z}$. Let φ be a non-degenerate projective transformation in P^N and $v_1, \cdots, v_d$ be d points on A. Let W_ν be the graph of the rational mapping

$$x \to \varphi[\theta(r_\nu)\varphi^{-1}(x) + v_\nu]$$

of A into itself. Put

$$T(\varphi, v_1, \cdots, v_d) = c(\varphi(A)) \times c(W_1) \times \cdots \times c(W_d)$$

and $M = (N+1)^2 - 1$. We can regard φ as a point in the projective space P^M; then T defines a rational mapping of $P^M \times A \times \cdots \times A$ into a certain product of projective spaces. The image $T(P^M \times A \times \cdots \times A)$ is nothing but the variety $\mathcal{F}(A, \theta)$; we call it the projective family of (A, θ) with respect to $\{r_1, \cdots, r_d\}$. Let U be the set of the elements in P^M which are non-degenerate as projective transformations of P^N. Then U is an open subset of P^M in the sense of Zariski-topology. We observe that T is defined at every point on $U \times A \times \cdots \times A$. Denote by $\mathcal{F}^*(A, \theta)$ the set-theoretical image $T(U \times A \times \cdots \times A)$. By Lemma 8 of [AF], we have $\dim \mathcal{F}(A, \theta) = \dim(P_M \times A \times \cdots \times A)$, so that $\mathcal{F}^*(A, \theta)$ contains an open subset of $\mathcal{F}(A, \theta)$ in the sense of Zariski-topology.

PROPOSITION 3.5. *Let k be a subfield of C composed of countably infinite elements. $\mathcal{P}(z), \mathcal{F}(z), \Psi(z)$ being as in §3.1 and §3.2, let z_0 be a generic point for Ψ over k and y an arbitrary point on $\mathfrak{H}$. Then we have $\dim_k c(\mathcal{F}(z_0) \geq \dim_k c(\mathcal{F}(y))$. Suppose that*

(6) $$\dim_k c(\mathcal{F}(z_0)) = \dim_k c(\mathcal{F}(y)).$$

Then, there exists an isomorphism σ of the universal domain over k such that

$\mathcal{P}(z_0)^\sigma$ *is isomorphic to* $\mathcal{P}(y)$.

PROOF. Take a place $\mathfrak{p}$ of the field $k(\Psi(z_0))$ such that, for every function f in $k(\Psi(z))$ holomorphic at y, $\mathfrak{p}(f(z_0))=f(y)$. The existence of such a place $\mathfrak{p}$ is shown in the proof of Theorem 1 of [AF]. We have then $\mathfrak{p}(\mathcal{F}(z_0))\supset\mathcal{F}(y)$. Put

$$K_0 = k(c(\mathcal{F}(z_0))), \quad K = k(c(\mathcal{F}(y))).$$

Then, the varieties $\mathcal{F}(z_0)$ and $\mathcal{F}(y)$ are respectively defined over K_0 and K. We can find a point a on $\mathcal{F}(z_0)$ and a point b on $\mathcal{F}(y)$ such that $a \to b$ ref. $\mathfrak{p}$, and, a is algebraic over K_0 and b is algebraic over K. Moreover, we can take a and b so that $a \in \mathcal{F}^*(A_{z_0}, \theta_{z_0}), b \in \mathcal{F}^*(A_y, \theta_y)$. Then, by the definition of $\mathcal{F}^*$, there exist non-degenerate projective transformations φ, ψ and points $u_1, \cdots, u_d$ on $A(z_0), v_1, \cdots, v_d$ on $A(y)$ such that

$$T(\varphi, u_1, \cdots, u_d)=a, \quad T(\psi, v_1, \cdots, v_d)=b.$$

Put $B_1 = \varphi(A(z_0)), B_2 = \psi(A(y))$. As B_1 is defined over $k(a)$, we can put into B_1 a structure of abelian variety defind over an algebraic extension L_1 of $k(a)$; we can then define easily a polarized abelian variety $\mathcal{P}_1 = (B_1, C_1, \theta_1)$ of type $\mathfrak{o}$, defined over L_1, such that φ defines, up to a constant, an isomorphism of $\mathcal{P}(z_0)$ onto $\mathcal{P}_1$. We can find similarly an algebraic extension L_2 of $k(b)$ and a polarized abelian variety $\mathcal{P}_2 = (B_2, C_2, \theta_2)$, defined over L_2, such that ψ defines, up to a constant, an isomorphism of $\mathcal{P}(y)$ onto $\mathcal{P}_2$. We have clearly

$$\mathcal{F}(A_{z_0}, \theta_{z_0}) = \mathcal{F}(B_1, \theta_1), \quad \mathcal{F}(A_y, \theta_y) = \mathcal{F}(B_2, \theta_2).$$

It follows that $K_0 \subset L_1, K \subset L_2$. We have therefore $\dim_k a = \dim_k K_0, \dim_k b = \dim_k K$; as b is a specialization of a over k, we have $\dim_k K_0 \geqq \dim_k K$; this proves our first assertion. Now the assumption (6) implies

$$\dim_k a = \dim_k b.$$

Hence b must be a generic specialization of a over k; so there exists an isomorphism σ of $k(a)$ onto $k(b)$ such that $a^\sigma = b$. Extend this isomorphism to an isomorphism of the universal domain and denote it again by σ. On account of the definition of $T, a^\sigma = b$ implies $B_1^\sigma = B_2$. We may assume, without any loss of generality, that σ maps the origin of B_1 onto the origin of B_2. Then, again by the definition of T and by the equality $a^\sigma = b$, we see that $\mathcal{P}_1^\sigma = (B_1^\sigma, C_1^\sigma, \theta_1^\sigma)$ coincides with $\mathcal{P}_2 = (B_2, C_2, \theta_2)$. This proves our proposition.

REMARK. Proposition 3.5 holds for a more general system of polarized abelian varieties constructed in [AF, no. 11]; in the above proof, we have only to substitute $\mathfrak{o}$ and $\mathfrak{H}$ for $\mathfrak{r}$ and $\mathfrak{Z}$. We also note that in the present case of quaternion algebra, we have

$$\dim_k c(\mathcal{F}(z_0))=1,$$

on account of Theorem 6 of [AF].

§4. Algebro-geometric theory of modular correspondences.

4.1. Determination of Galois groups. $\mathfrak{o}$ being as before an order in Φ, we assume henceforth that $\mathfrak{o}$ is *maximal*. Let $\mathfrak{a}$ be an integral two-sided $\mathfrak{o}$-ideal; in this § we denote by $G_\mathfrak{a} = G(\mathfrak{a})$ the multiplicative group of regular elements of the ring $\mathfrak{o}/\mathfrak{a}$ and by $S_\mathfrak{a} = S(\mathfrak{a})$ the subgroup of $G_\mathfrak{a}$ consisting of the residue-classes of the elements α such that $N(\alpha) \equiv 1 \bmod. \mathfrak{a} \cap Z$. It is clear that $S_\mathfrak{a}$ is a normal subgroup of $G_\mathfrak{a}$. By Lemma 1.4, the mapping $\beta \to N(\beta)$ is an isomorphism of $G_\mathfrak{a}/S_\mathfrak{a}$ onto the multiplicative group of regular elements of $Z/(Z \cap \mathfrak{a})$. We note that if $\mathfrak{a}$ is of the form (3) of §1.2, we have $Z \cap \mathfrak{a} = (a_0 N(\mathfrak{p}_1 \cdots \mathfrak{p}_s))$. As before let Γ denote the group of units γ in $\mathfrak{o}$ such that $N(\gamma) = 1$, and $\Gamma_\mathfrak{a}^*$ the subgroup of Γ consisting of the elements γ such that $\gamma \equiv \pm 1 \bmod. \mathfrak{a}$. Applying Lemma 1.3 to the case $b = 1$, we observe that every element of $S_\mathfrak{a}$ has a representative in Γ. It follows that $\Gamma/\Gamma_\mathfrak{a}^*$ is canonically isomorphic to $S_\mathfrak{a}/\{\pm 1\}$.

Let $\mathscr{P}(z)$ be a member of our system S and $K_1 = K_{1,z}$ the field of moduli of $\mathscr{P}(z)$. By Proposition 3.5, we have $\dim_Q K_1 = 0$ or 1. Let (V, h) be a normalized Kummer variety of $\mathscr{P}(z)$. For every integral two-sided $\mathfrak{o}$-ideal $\mathfrak{a}$, we put

$$\mathfrak{g}(\mathfrak{a}, A_z) = \{ t \in A_z \mid \theta_z(\alpha)t = 0 \ \text{ for every } \ \alpha \in \mathfrak{a} \} ,$$

and denote by $K_\mathfrak{a} = K_{\mathfrak{a},z}$ the field generated over K_1 by the points $h(t)$ for $t \in \mathfrak{g}(\mathfrak{a}, A_z)$. By Proposition 2.3, $K_\mathfrak{a}$ does not depend on the choice of (V, h); and by Proposition 2.6, $K_\mathfrak{a}$ is a Galois extension of K_1. Our purpose in this § is to determine the Galois group of $K_\mathfrak{a}$ over K_1 in the case $\dim_Q K_1 = 1$.

PROPOSITION 4.1. *Notations being as above, there exists an element t of* $\mathfrak{g}(\mathfrak{a}, A_z)$ *satisfying the following conditions:*

   i) $\mathfrak{g}(\mathfrak{a}, A_z) = \theta_z(\mathfrak{o})t$;

   ii) $\theta_z(\alpha)t = 0 \Rightarrow \alpha \in \mathfrak{a}$.

PROOF. As every $\mathfrak{o}$-ideal is a principal ideal, there exists an element α_0 such that $\mathfrak{a} = \mathfrak{o}\alpha_0$. We have then also $\mathfrak{a} = \alpha_0 \mathfrak{o}$, since $\mathfrak{a}$ is a two-sided $\mathfrak{o}$-ideal and $\mathfrak{o}$ is maximal. Put $t = \Lambda(\chi(\alpha_0^{-1})e(z), z)$. It is easy to verify this point satisfies the conditions.

If t_0 is a point of $\mathfrak{g}(\mathfrak{a}, A_z)$ satisfying the conditions i) and ii), we observe that the mapping $\alpha \to \theta_z(\alpha)t_0$ gives an isomorphism of $\mathfrak{o}/\mathfrak{a}$ onto $\mathfrak{g}(\mathfrak{a}, A_z)$. It follows from this fact that the conditions i) and ii) of the above proposition are equivalent to each other. We call an element t of $\mathfrak{g}(\mathfrak{a}, A_z)$ satisfying these conditions a *primitive element* of $\mathfrak{g}(\mathfrak{a}, A_z)$.

Now let $\mathcal{G}_\mathfrak{a}$ denote the Galois group of $K_\mathfrak{a}$ over K_1.

PROPOSITION 4.2. *Let t_0 be a primitive element of* $\mathfrak{g}(\mathfrak{a}, A_z)$. *Then, for every element σ of $\mathcal{G}_\mathfrak{a}$, there exists an element $\alpha_\sigma \in \mathfrak{o}$ such that*

(1) $$h(\theta_z(\beta)t_0)^\sigma = h(\theta_z(\beta\alpha_\sigma)t_0)$$

for every $\beta \in \mathfrak{o}$. If K_1 is not algebraic over $\boldsymbol{Q}$, such an element α_σ is uniquely determined modulo $\mathfrak{a}$ up to the factors ± 1.

PROOF. Let σ be an element of $\mathcal{G}_\mathfrak{a}$; extend σ to an isomorphism of the universal domain into itself and denote it again by σ. Since σ leaves invariant the elements of K_1, there exists an isomorphism η of $\mathcal{P}$ onto $\mathcal{P}^\sigma$; by the property (K3) of normalized Kummer variety, we have $h^\sigma \circ \eta = h$. We see easily that $\eta^{-1}t_0^\sigma$ is contained in $\mathfrak{g}(\mathfrak{a}, A_z)$. Hence, by the property i) of Proposition 4.1, there exists an element $\alpha_\sigma \in \mathfrak{o}$ such that $\eta^{-1}t_0^\sigma = \theta_z(\alpha_\sigma)t_0$. We have then, for every $\beta \in \mathfrak{o}$,

$$h(\theta_z(\beta)t_0)^\sigma = h^\sigma(\theta_z^\sigma(\beta)t_0^\sigma) = h(\eta^{-1}\theta_z^\sigma(\beta)t_0^\sigma) = h(\theta_z(\beta)\eta^{-1}t_0^\sigma) = h(\theta_z(\beta\alpha_\sigma)t_0).$$

This proves the first assertion. Suppose that K_1 is not algebraic over $\boldsymbol{Q}$ and we have $h(t_0)^\sigma = h(\theta_z(\gamma)t_0)$ for an element $\gamma \in \mathfrak{o}$; we have then, by Proposition 3.1, $\theta_z(\alpha_\sigma)t_0 = \pm\theta_z(\gamma)t_0$. As t_0 is a primitive element of $\mathfrak{g}(\mathfrak{a}, A_z)$, we have $\alpha_\sigma \equiv \pm\gamma \bmod. \mathfrak{a}$; this completes the proof.

If K_1 is not algebraic over $\boldsymbol{Q}$, we observe that the mapping $\sigma \to \alpha_\sigma$ gives an isomorphism of $\mathcal{G}_\mathfrak{a}$ into $G_\mathfrak{a}/\{\pm 1\}$; this isomorphism depends on the choice of a primitive element t_0 of $\mathfrak{g}(\mathfrak{a}, A_z)$. If we choose another primitive element of $\mathfrak{g}(\mathfrak{a}, A_z)$, the isomorphism is transformed by an inner automorphism of $G_\mathfrak{a}/\{\pm 1\}$.

THEOREM 2. *Suppose that the field of moduli K_1 of $\mathcal{P}(z)$ is not algebraic over $\boldsymbol{Q}$. Then the following assertions hold.*

i) *The Galois group $\mathcal{G}_\mathfrak{a}$ of $K_\mathfrak{a}$ over K_1 is isomorphic to $G_\mathfrak{a}/\{\pm 1\}$ by the correspondence $\sigma \to \alpha_\sigma$ defined by the relation (1) of Proposition 4.2.*

ii) *Let a be the smallest positive integer divisible by $\mathfrak{a}$ and ζ_a a primitive a-th root of unity. Then, $K_1(\zeta_a)$ is the subfield of $K_\mathfrak{a}$ corresponding to the subgroup $S_\mathfrak{a}/\{\pm 1\}$ of $G_\mathfrak{a}/\{\pm 1\}$.*

iii) *$\boldsymbol{Q}(\zeta_a)$ is algebraically closed in $K_\mathfrak{a}$.*

iv) *If α_σ is a representative of the element of $G_\mathfrak{a}/\{\pm 1\}$ corresponding to an element σ of $\mathcal{G}_\mathfrak{a}$, we have $\zeta_a^\sigma = \zeta_a^{N(\alpha_\sigma)}$.*

PROOF. By virtue of Proposition 3.5, our theorem is established if we prove the assertions i–iv) for any one of the points z of $\mathfrak{H}$ such that $\dim_{\boldsymbol{Q}} c(\mathcal{F}(z)) = 1$. Therefore, in the course of our proof, we may assume, as occasion demands, the point z to be sufficiently generic. For convenience' sake, we use the letter y instead of z for a sufficiently generic point of $\mathfrak{H}$, reserving the letter z for the variable. Now the f_i being as in (3) of §3.1, we have

$$K_1 = \boldsymbol{Q}(f_1(y), \cdots, f_m(y)).$$

Let α be an element of $\mathfrak{o}$ such that $\mathfrak{a} = \alpha\mathfrak{o}$; put $t_0 = \Lambda(\chi(\alpha^{-1})e(y), y)$. Then t_0 is a primitive element of $\mathfrak{g}(\mathfrak{a}, A_y)$; and we have $h_y(\theta_y(\beta)t_0) = h_y(\Lambda(\chi(\beta\alpha^{-1})e(y), y))$, so that

$$K_a = K_1(g_j(\beta\alpha^{-1}, y) \mid 1 \leq j \leq M, \beta \in \mathfrak{o}),$$

where the g_j are the functions determined by (4) of §3.2. Let $\bar{\gamma}$ be an element of S_a. By Lemma 1.3, there exists an element γ of $\mathfrak{o}$ such that $N(\gamma) = 1$ and $\bar{\gamma}$ is the class of γ modulo a. If y is sufficiently generic, K_a is isomorphic to the function-field

$$(2) \qquad Q(f_i(z), g_j(\beta\alpha^{-1}, z) \mid 1 \leq i \leq m, 1 \leq j \leq M, \beta \in \mathfrak{o}),$$

and K_1 corresponds to $Q(f_i(z))$. As a is a two-sided $\mathfrak{o}$-ideal and $\mathfrak{o}$ is maximal, we have $\alpha\mathfrak{o}\alpha^{-1} = \mathfrak{o}$, so that $\alpha\gamma\alpha^{-1}$ is a unit of $\mathfrak{o}$. Put $\gamma_1 = \alpha\gamma\alpha^{-1}$. By proposition 3.2, we have

$$g_j(\beta\alpha^{-1}, \gamma_1[z]) = g_j(\beta\gamma\alpha^{-1}, z).$$

Therefore, the mapping $F(z) \to F(\gamma_1[z])$ gives an automorphism of the field (2) over $Q(f_i(z))$. If we transform this onto K_a/K_1, we observe that $h_y(\theta_y(\beta)t_0) \to h_y(\theta_y(\beta\gamma)t_0)$ gives an element of $\mathcal{G}_a$. In other words, $S_a/\{\pm 1\}$ is contained in the image of the isomorphism $\sigma \to \alpha_\sigma$.

Let Y be a divisor contained in the polarization C_y. Then Y corresponds to a Riemann form E_1 on $C^2/D(y)$ defined by

$$E_1(\chi(\xi)e(y), \chi(\eta)e(y)) = \mathrm{tr}\,(\rho_1\xi\eta'),$$

where ρ_1 is an element of Φ. As $E_1(u, v)$ is an integer for every u, v in $D(y) = \chi(\mathfrak{o})e(y)$, $\mathrm{tr}\,(\rho_1\mathfrak{o})$ is an ideal of Z. Let q be a positive integer such that $\mathrm{tr}\,(\rho_1\mathfrak{o}) = qZ$. Then $q^{-1}E_1$ is also a Riemann form. Let X be a divisor on A_y corresponding to $q^{-1}E_1$. Then qX is algebraically equivalent to Y, so that $X \in C_y$. Put $q^{-1}\rho_1 = \rho, q^{-1}E_1 = E$. Then there exists an element $\xi_0 \in \mathfrak{o}$ such that $\mathrm{tr}\,(\rho\xi_0) = 1$. Now, a being as in ii) of our theorem, consider the symbol $e_{X,a}$ defined in Weil [29, no. 75]. By the formula (7) of [25, p. 25], we have

$$e_{X,a}(t_2, t_1) = \exp\,(2\pi i a E(x_1, x_2)),$$

where t_1 and t_2 are points on A_y such that $at_1 = at_2 = 0$ and x_1, x_2 are vectors in C^2 corresponding to t_1 and t_2. Hence, for every β_1 and β_2 of $\mathfrak{o}$ we have

$$
\begin{aligned}
(3) \qquad e_{X,a}(\theta_y(\beta_2)t_0, \theta_y(\beta_1)t_0) &= \exp\,(2\pi i a E(\chi(\beta_1\alpha^{-1})e(y), \chi(\beta_2\alpha^{-1})e(y))) \\
&= \exp\,(2\pi i a N(a)^{-1}\,\mathrm{tr}\,(\rho\beta_1\beta_2')).
\end{aligned}
$$

Let σ be an element of $\mathcal{G}_a$; extend σ to an isomorphism of the universal domain into itself and denote it again by σ. η and α_σ being as in the proof of Proposition 4.2, $\eta^{-1}(X^\sigma)$ is algebraically equivalent to X, so that $e_{X^\sigma,a}(s_2, s_1) = e_{X,a}(\eta^{-1}s_2, \eta^{-1}s_1)$; we have therefore

$$(4) \quad e_{X,a}(\theta_y(\beta_2)t_0, \theta_y(\beta_1)t_0)^\sigma = e_{X^\sigma,a}(\theta_y^\sigma(\beta_2)t_0^\sigma, \theta_y^\sigma(\beta_1)t_0^\sigma) = e_{X,a}(\theta_y(\beta_2\alpha_\sigma)t_0, \theta_y(\beta_1\alpha_\sigma)t_0).$$

Now assume that $a = a\mathfrak{o}$ and put $\zeta_a = e_{X,a}(t_0, \theta_y(\xi_0)t_0)$. By (3), we have $\zeta_a = \exp\,(2\pi i a^{-1})$, so that ζ_a is a primitive a-th root of unity. Substituting ξ_0 and

1 for β_1 and β_2 in (4), we obtain, on account of (3), $\zeta_a^\sigma = \zeta_a^{N(\alpha_\sigma)}$; this proves that ζ_a is contained in K_a and the assertion iv) in the case $a = a\mathfrak{o}$.

Coming back to the general case, put $a = \alpha\alpha_1$ and

$$u_0 = \Lambda(\chi(a^{-1})e(y), y).$$

Then α_1 is an element of $\mathfrak{o}$ and we have $t_0 = \theta_y(\alpha_1)u_0$. Obviously K_a contains K_a. Taking $a\mathfrak{o}$ in place of a, we obtain an isomorphism $\sigma \to \beta_\sigma$ of the Galois group $\mathcal{G}_a$ of K_a over K_1 into $G_a/\{\pm 1\}$ by means of the relation $h_y(\theta_y(\beta)u_0)^\sigma = h_y(\theta_y(\beta\beta_\sigma)u_0)$. By what we have just proved, ζ_a is contained in K_a and $\zeta_a^\sigma = \zeta^{N(\beta_\sigma)}$. As $\alpha_1 = a\alpha^{-1}$, we have $\alpha_1\mathfrak{o}\alpha_1^{-1} = \mathfrak{o}$. Put $\alpha_1\beta_\sigma\alpha_1^{-1} = \gamma_\sigma$. We have then, $h_y(\theta_y(\beta)t_0)^\sigma = h_y(\theta_y(\beta\gamma_\sigma)t_0)$, so that $\gamma_\sigma \equiv \pm\alpha_\sigma \bmod. \mathfrak{a}$. Hence an element σ of $\mathcal{G}_a$ leaves invariant the elements of K_a if and only if $\alpha_1\beta_\sigma\alpha_1^{-1} \equiv \pm 1 \bmod. \mathfrak{a}$, namely, $\beta_\sigma \equiv \pm 1 \bmod. \mathfrak{a}$. Now if $\beta_\sigma \equiv \pm 1 \bmod. \mathfrak{a}$, we have $N(\beta_\sigma) \equiv 1 \bmod. a\mathbf{Z}$, so that $\zeta_a^\sigma = \zeta_a$. This shows that if an element σ of $\mathcal{G}_a$ leaves invariant the elements of K_a, we have $\zeta_a^\sigma = \zeta_a$. It follows that ζ_a is contained in K_a; and we have, for every element σ of $\mathcal{G}_a$, $\zeta_a^\sigma = \zeta_a^{N(\beta_\sigma)} = \zeta_a^{N(\alpha_\sigma)}$. This proves the assertion iv) in the general case. In particular, σ is the identity on $K_1(\zeta_a)$ if and only if the class of $\alpha_\sigma \bmod. \mathfrak{a}$ is contained in S_a. Let G_a' denote the image of the isomorphism $\sigma \to \alpha_\sigma$. We have proved above $G_a' \supset S_a/\{\pm 1\}$. As $K_1(\zeta_a)$ corresponds to $S_a/\{\pm 1\}$, we have

$$[G_a' : S_a/\{\pm 1\}] = [K_1(\zeta_a) : K_1].$$

If we denote by $\varphi(a)$ the order of the multiplicative group of $\mathbf{Z}/a\mathbf{Z}$, we get $[G_a/\{\pm 1\} : S_a/\{\pm 1\}] = [G_a : S_a] = \varphi(a)$. On the other hand, by Theorem 6 of [AF], $\mathbf{Q}$ is algebraically closed in K_1, so that

$$[K_1(\zeta_a) : K_1] = [\mathbf{Q}(\zeta_a) : \mathbf{Q}] = \varphi(a).$$

It follows that $G_a' = G_a/\{\pm 1\}$. Thus we have proved the assertions i) and ii). Now let k_1 be the set of elements of K_a which is algebraic over $\mathbf{Q}$. Then k_1 contains $\mathbf{Q}(\zeta_a)$. We have seen above that every element σ of $\mathcal{G}_a$ corresponding to S_a is obtained from an isomorphism $F(z) \to F(\gamma_1[z])$ of the function-field (2). Obviously, this isomorphism leaves invariant the elements of k_1. Therefore k_1 must be contained in the subfield of K_a corresponding to S_a; so we have $k_1 \subset K_1(\zeta_a)$, and hence $k_1 = \mathbf{Q}(\zeta_a)$. Our theorem is thereby completely proved.

In the above proof, we have used an isomorphism between K_a and the function-field (2). Therefore, we may consider Theorem 2 as a statement concerning the Galois-group of the function-field (2) over $\mathbf{Q}(f_i)$. If we extend the constant field to the complex number field, the function-fields (2) and $\mathbf{Q}(f_i)$ yield $\mathfrak{K}(\mathfrak{o}; \mathfrak{a})$ and $\mathfrak{K}(\mathfrak{o})$. The relation between the fields and the groups is illustrated by the following table.

$$
\begin{array}{ccccc}
 & & & & 1 \\
 & & & & | \\
 & & & CK \cong \Re(\mathfrak{o};\mathfrak{a})\cdots\cdots\cdots\Gamma_{\mathfrak{a}}{}^* \\
1\cdots\cdots\cdots\cdots K_{\mathfrak{a}} & & | & \\
| & & CK_1 \cong \Re(\mathfrak{o})\cdots\cdots\cdots\Gamma \\
S_{\mathfrak{a}}/\{\pm 1\}\cdots\cdots K_1(\zeta_{\mathfrak{a}}) & & | & \\
G_{\mathfrak{a}}/\{\pm 1\}\cdots\cdots K_1 & C & \\
| & Q(\zeta_{\mathfrak{a}}) & \\
Q & &
\end{array}
$$

4.2. Transformations of $\mathscr{P}(z)$. Let $\mathfrak{q}=\mathfrak{o}\alpha$ be an integral left $\mathfrak{o}$-ideal; suppose that $N(\alpha)>0$; and let $\{\mathfrak{q}_1,\cdots,\mathfrak{q}_m\}$ be the set of integral left $\mathfrak{o}$-ideals having the same elementary divisors as $\mathfrak{q}$. $\{\mathfrak{q}_1,\cdots,\mathfrak{q}_m\}$ corresponds to a double coset $\Gamma\alpha\Gamma$; and if we take elements α_i so that $\mathfrak{q}_i=\mathfrak{o}\alpha_i$ and $N(\alpha_i)>0$ for every i, $\Gamma\alpha\Gamma=\bigcup_{i=1}^{m}\Gamma\alpha_i$ is a disjoint sum. Put $q=N(\mathfrak{q})$. We have then $\mathfrak{o}\supset\mathfrak{q}\supset q\mathfrak{o}$, and $\mathfrak{o}/\mathfrak{q}$ is $\mathfrak{o}$-isomorphic to $\mathfrak{o}\alpha'/q\mathfrak{o}$. As we have $\Gamma\alpha\Gamma=\Gamma\alpha'\Gamma$, $\mathfrak{o}\alpha'$ coincides with one of the $\mathfrak{q}_i$. Therefore, $\mathfrak{o}/q\mathfrak{o}$ contains exactly m $\mathfrak{o}$-submodules isomorphic to $\mathfrak{o}/\mathfrak{q}$.

Now let $\mathscr{P}(y)=(A_y,C_y,\theta_y)$ be a member of the system S. By Proposition 4.1, $\mathfrak{g}(\mathfrak{q},A_y)$ is isomorphic to $\mathfrak{o}/q\mathfrak{o}$ as $\mathfrak{o}$-module. Hence $\mathfrak{g}(\mathfrak{q},A_y)$ has exactly m subgroups $\mathfrak{g}_i$ which are $\mathfrak{o}$-isomophic to $\mathfrak{o}/\mathfrak{q}$.

PROPOSITION 4.3. *Notations being as above, there exist m members $\mathscr{P}(y_i)$ $(1\leq i\leq m)$ of S and a homomorphism λ_i of $\mathscr{P}(y)$ onto $\mathscr{P}(y_i)$ for each i such that the kernel of λ_i is $\mathfrak{g}_i$.*

PROOF. The elements α_i being as above, put $\alpha_i[y]=y_i$; then there exists a complex number a_i such that $\chi(\alpha_i)e(y)=a_ie(y_i)$. We have obviously

$$D(y_i)=\chi(\mathfrak{o})e(y_i)=a_i^{-1}\chi(\mathfrak{q}_i)e(y).$$

Hence the linear mapping $x\to qa_i^{-1}x$ gives a homomorphism of $C^2/D(y)$ onto $C^2/D(y_i)$; denote by λ_i the homomorphism of $A(y)$ onto $A(y_i)$ corresponding to this linear mapping. Since λ_i commutes with the operation of $\mathfrak{o}$, λ_i is a homomorphism of $\mathscr{P}(y)$ onto $\mathscr{P}(y_i)$ (cf. [AF, no. 20]). Put $t_0=A(q^{-1}e(y),y)$; then t_0 is a primitive element of $\mathfrak{g}(\mathfrak{q},A_y)$; and $\theta_y(\mathfrak{q}_i)t_0$ for $1\leq i\leq m$ give the submodules of $\mathfrak{g}(\mathfrak{q},A_y)$ which are $\mathfrak{o}$-isomorphic to $\mathfrak{o}/\mathfrak{q}$. We see easily that the kernel of λ_i is $\theta_y(\mathfrak{q}_i)t_0$. This proves our proposition.

Let $\mathfrak{a}$ be an integral two-sided $\mathfrak{o}$-ideal. We shall now consider the fields $K_{1,z}$ and $K_{\mathfrak{a},z}$ for the points y_i determined in Proposition 4.3. The α_i being as above, suppose that y is generic for $\Psi(z),\Psi(\alpha_i[z])$ over Q; then there exists an isomorphism ρ_i of $Q(\Psi(y))$ onto $Q(\Psi(y_i))$ such that $\Psi(y)^{\rho_i}=\Psi(y_i)$. We have then

$$(6)\qquad\qquad \mathscr{P}(y_i)^{\rho_i}=\mathscr{P}(y_i),\quad V(y)^{\rho_i}=V(y_i),\quad h_y^{\rho_i}=h_{y_i}.$$

It is easy to see that ρ_i induces an isomorphisms of $K_{\mathfrak{a},y}$ onto $K_{\mathfrak{a},y_i}$ and of $K_{1,y}$ onto K_{1,y_i}. Now suppose that q is prime to $\mathfrak{a}$. Then, the mapping $t\to$

$(\lambda_i t)^{\rho_i - 1}$ gives an $\mathfrak{o}$-automorphism of $\mathfrak{g}(\mathfrak{a}, A_y)$. Hence, if t_0 is a primitive element of $\mathfrak{g}(\mathfrak{a}, A_y)$, there exists an element γ_i of $\mathfrak{o}$, prime to $\mathfrak{a}$, such that

$$(\lambda_i \theta_y(\beta) t_0)^{\rho_i - 1} = \theta_y(\beta \gamma_i) t_0$$

for every $\beta \in \mathfrak{o}$. By Theorem 2, there exists an automorphism τ_i of $K_{\mathfrak{a}, y}$ over $K_{1, y}$ such that

$$h_y(\theta_y(\beta) t_0)^{\tau_i} = h_y(\theta_y(\beta \gamma_i) t_0)$$

for every $\beta \in \mathfrak{o}$. Put $\sigma_i = \tau_i \rho_i$. We have then

$$h_y(\theta_y(\beta) t_0)^{\sigma_i} = h_{y_i}(\lambda_i \theta_y(\beta) t_0) .$$

We have thus proved the following proposition.

PROPOSITION 4.4. *Let $\mathfrak{a}$ be an integral two-sided $\mathfrak{o}$-ideal and y a point of $\mathfrak{H}$; define $\mathscr{P}(y_i)$ and λ_i as in Proposition 4.3. Suppose that q and $\mathfrak{a}$ are relatively prime. If y is sufficiently generic, there exists, for each i, an isomorphism σ_i of $K_{\mathfrak{a}, y}$ onto $K_{\mathfrak{a}, y_i}$ such that*

(7) $$c(\mathscr{F}(y))^{\sigma_i} = c(\mathscr{F}(y_i)) ,$$

(8) $$h_y(t)^{\sigma_i} = h_{y_i}(\lambda_i t)$$

for every $t \in \mathfrak{g}(\mathfrak{a}, A_y)$.

Since $K_{\mathfrak{a}, y}$ is generated over $\boldsymbol{Q}$ by the points $c(\mathscr{F}(y))$ and $h_y(t)$ for $t \in \mathfrak{g}(\mathfrak{a}, A_y)$, the isomorphism σ_i is uniquely determined by (7) and (8).

From now on, we assume y to be so generic that we may apply Proposition 4.4 to $\mathscr{P}(y)$ for any pair of q and $\mathfrak{a}$. By Propositions 2.7 and 2.8, $K_{\mathfrak{a}, y_i}$ is contained in $K_{q\mathfrak{a}, y}$; this implies in particular that $K_{\mathfrak{a}, y_i}$ is algebraic over $K_{1, y}$.

PROPOSITION 4.5. *Notations being as in Proposition 4.4, every conjugate of $c(\mathscr{F}(y_1))$ over $K_{1, y}$ is of the form $c(\mathscr{F}(y_i))$ for $1 \leq i \leq m$.*

PROOF. Let τ be an isomorphism of the universal domain into itself leaving invariant the elements of $K_{1, y}$. Then, there exists an isomorphism η of $\mathscr{P}(y)$ onto $\mathscr{P}(y)^{\tau}$. We see that $\lambda_1^{\tau} \circ \eta$ is a homomorphism of $\mathscr{P}(y)$ onto $\mathscr{P}(y_1)^{\tau}$, and, the kernel of $\lambda_1^{\tau} \circ \eta$ is contained in $\mathfrak{g}(q, A_y)$ and is $\mathfrak{o}$-isomorphic to $\mathfrak{o}/q$. Therefore, the kernel of $\lambda_1^{\tau} \circ \eta$ coincides with one of the $\mathfrak{g}_i$. Then, by i) of Proposition 2.5, $\mathscr{P}(y_1)^{\tau}$ is isomorphic to one of the $\mathscr{P}(y_i)$. It follows that $\mathscr{F}(y_1)^{\tau}$ coincides with one of the $\mathscr{F}(y_i)$ on account of Proposition 1 of [AF]; this proves our proposition.

Now let $\mathfrak{b}$ be the set of elements β such that $\beta \mathfrak{o} \subset q$. Then $\mathfrak{b}$ is an integral two-sided $\mathfrak{o}$-ideal; and we have $q \supset \mathfrak{b} \supset q\mathfrak{o}$. As $\mathfrak{g}_i$ is $\mathfrak{o}$-isomorphic to $\mathfrak{o}/q$, we have

(9) $$\theta_y(\beta) \mathfrak{z}_i = \{0\} \Leftrightarrow \beta \in \mathfrak{b} .$$

We shall use this relation in the proof of the following proposition.

PROPOSITION 4.6. *Let $u = (u_1, \cdots, u_l)$ be a set of quantities such that $K_{\mathfrak{a}, y} \supset \boldsymbol{Q}(u) \supset K_{1, y}$. Let $\sigma_1, \cdots, \sigma_m$ be the isomorphisms of $K_{\mathfrak{a}, y}$ onto the $K_{\mathfrak{a}, y_i}$ deter-*

mined by (7) *and* (8) *of Proposition* 4.4. *Then*:

  i) $Q(u, c(\mathcal{F}(y_i)))$ *contains* $Q(u^{\sigma_i})$ *for each* i;

  ii) $(u^{\sigma_1}, \cdots, u^{\sigma_m})$ *is the complete set of conjugates of* u^{σ_1} *over* $Q(u)$.

PROOF. Let τ be an isomorphism of the universal domain into itself leaving invariant the elements of $Q(c(\mathcal{F}(y)), c(\mathcal{F}(y_1)))$. Then there exist an isomorphism ϵ of $\mathcal{P}(y)$ onto $\mathcal{P}(y)^\tau$ and an isomorphism ϵ_1 of $\mathcal{P}(y_1)$ onto $\mathcal{P}(y_1)^\tau$; by the property (K3) of normalized Kummer variety, we have

$$(10) \qquad h_y^\tau \circ \epsilon = h_{y'}, \quad h_{y_1}^\tau \circ \epsilon_1 = h_{y_1}.$$

We observe that the kernel of $\lambda_1^\tau \circ \epsilon$ is $\mathfrak{o}$-isomorphic to $\mathfrak{o}/\mathfrak{q}$. We can apply, on account of (9), Proposition 2.5 to the homomorphisms $\epsilon_1 \circ \lambda_1$ and $\lambda_1^\tau \circ \epsilon$ of $\mathcal{P}(y)$ onto $\mathcal{P}(y_1)^\tau$; then by ii) of that proposition, $\epsilon_1 \circ \lambda_1$ and $\lambda_1^\tau \circ \epsilon$ have the same kernel; hence by i) of the same proposition, there exists an automorphism η of $\mathcal{P}(y_1)^\tau$ such that $\eta \circ \epsilon_1 \circ \lambda_1 = \lambda_1^\tau \circ \epsilon$. By Proposition 3.1, η must be ± 1, so that

$$(11) \qquad \pm \epsilon_1 \circ \lambda_1 = \lambda_1^\tau \circ \epsilon.$$

Let t_0 be a primitive element of $\mathfrak{g}(\mathfrak{a}, A_y)$ and α_τ be an element of $\mathfrak{o}$ such that

$$(12) \qquad h_y(\theta_y(\beta)t_0)^\tau = h_y(\theta_y(\beta\alpha_\tau)t_0)$$

for every $\beta \in \mathfrak{o}$. Then, by the relation (10), we have

$$(13) \qquad \epsilon^{-1}(\theta_y(\beta)t_0)^\tau = \pm \theta_y(\beta\alpha_\tau)t_0.$$

The relations (8), (10), (11), (12), (13) yield

$$(14) \quad h_y(\theta_y(\beta)t_0)^{\sigma_1\tau} = h_{y_1}(\lambda_1\theta_y(\beta)t_0)^\tau = h_{y_1}^\tau(\lambda_1^\tau \circ \epsilon \circ \epsilon^{-1}(\theta_y(\beta)t_0)^\tau) = h_{y_1}^\tau(\epsilon_1\lambda_1\theta_y(\beta\alpha_\tau)t_0)$$

$$= h_{y_1}(\lambda_1\theta_y(\beta\alpha_\tau)t_0) = h_y(\theta_y(\beta\alpha_\tau)t_0)^{\sigma_1} = h_y(\theta_y(\beta)t_0)^{\tau\sigma_1}.$$

On the other hand, we have $c(\mathcal{F}(y))^{\sigma_1\tau} = c(\mathcal{F}(y_1)) = c(\mathcal{F}(y))^{\tau\sigma_1}$. This combined with (14) shows $\sigma_1\tau = \tau\sigma_1$, since $K_{\mathfrak{a},y}$ is generated by $c(\mathcal{F}(y))$ and $h_y(\theta_y(\beta)t_0)$. Therefore, if τ leaves invariant the elements of $Q(u, c(\mathcal{F}(y_1)))$, we have $u^{\sigma_1\tau} = u^{\tau\sigma_1} = u^{\sigma_1}$; this proves the assertion i). From i) we obtain

$$[Q(u, u^{\sigma_1}) : Q(u)] = [Q(u, c(\mathcal{F}(y_1)) : Q(u)]$$

$$\leqq [Q(c(\mathcal{F}(y)), c(\mathcal{F}(y_1))) : Q(c(\mathcal{F}(y)))].$$

By Proposition 4.5, the right hand side of this inequality is not greater than m. If $i \neq j$, the kernel of λ_i and λ_j are different from each other; hence, by ii) of Proposition 2.5 and (9), $\mathcal{P}(y_i)$ and $\mathcal{P}(y_j)$ are not isomorphic. It follows that m points $c(\mathcal{F}(y_i))$ are different from each other. Therefore, our proposition is completely proved if we show that u^{σ_i} is a conjugate of u^{σ_1} over $K_{\mathfrak{a},y}$ for every i. Let s be a primitive element of $\mathfrak{g}(\mathfrak{q}\mathfrak{a}, A_y)$ and α an element of $\mathfrak{o}$ such that $\mathfrak{a} = \alpha\mathfrak{o}$. Put $r = \theta_y(\alpha)s$. Then r is a primitive element of $\mathfrak{g}(\mathfrak{q}, A_y)$. The ideals $\mathfrak{q}_i$ being as in the first part of this §, we may, after reordering if necessary, assume that $\theta_v(\mathfrak{q}_i)r$ is the kernel of λ_i for each i. Now by iv) of Proposition 1.6, there exists a unit r_1 of $\mathfrak{o}$ such that

$$(15) \qquad \mathfrak{q}_1 r_1 = \mathfrak{q}_i .$$

As q is prime to $\mathfrak{a}$, and as we have $\alpha^{-1}\mathfrak{o}\alpha = \mathfrak{o}$, there exists an element r of $\mathfrak{o}$ such that

$$(16) \qquad r \equiv \alpha^{-1} r_1 \alpha \ \mathrm{mod}.\, q\mathfrak{o}, \quad r \equiv 1 \ \mathrm{mod}.\, \mathfrak{a} .$$

By Theorem 2, there exists an automorphism ρ of $K_{\mathfrak{q}\mathfrak{a},v}$ over $K_{1,v}$ such that

$$(17) \qquad h_v(\theta_v(\beta)s)^\rho = h_v(\theta_v(\beta r)s)$$

for every $\beta \in \mathfrak{o}$. Obviously qs is a primitive element of $\mathfrak{g}(\mathfrak{a}, A_v)$; and by (16) we have $\beta r \equiv \beta \ \mathrm{mod}.\,\mathfrak{a}$, so that $\theta_v(\beta r)qs = \theta_v(\beta)qs$. Hence we obtain

$$h_v(\theta_v(\beta)qs)^\rho = h_v(\theta_v(\beta r)qs) = h_v(\theta_v(\beta)qs) .$$

This shows that ρ leaves invariant the elements of $K_{\mathfrak{a},v}$. Extend ρ to an isomorphism of the universal domain into itself and denote it again by ρ. As ρ is the identity on $K_{1,v}$, there exists an isomorphism ξ of $\mathscr{P}(y)$ onto $\mathscr{P}(y)^\rho$: and we have $h_v^\rho \circ \xi = h_v$. By (16) and (17), we find

$$h_v(\xi^{-1}(\theta_v(\beta)r)^\rho) = h_v(\theta_v(\beta)r)^\rho = h_v(\theta_v(\beta\alpha r \alpha^{-1})r) = h_v(\theta_v(\beta r_1)r) ,$$

and hence $\xi^{-1}(\theta_v(\beta)r)^\rho = \pm\theta_v(\beta r_1)r$. We have therefore by (15)

$$\xi^{-1}(\theta_v(\mathfrak{q}_1)r)^\rho = \theta_v(\mathfrak{q}_1 r_1)r = \theta_v(\mathfrak{q}_i)r .$$

This show that the kernel of $\lambda_1^\rho \circ \xi$ coincides with the kernel of λ_i. By i) of Proposition 2.5, there exists an isomorphism ξ_1 of $\mathscr{P}(y_i)$ onto $\mathscr{P}(y_1)^\rho$ such that

$$(18) \qquad \xi_1 \circ \lambda_i = \lambda_1^\rho \circ \xi .$$

It follows that $c(\mathscr{F}(y_1))^\rho = c(\mathscr{F}(y_i))$ by virtue of Proposition 1 of [AF], and hence

$$(19) \qquad c(\mathscr{F}(y))^{\sigma_1 \rho} = c(\mathscr{F}(y_i)) = c(\mathscr{F}(y))^{\sigma_i} .$$

Consider the isomorphisms ρ_i satisfying the relation (6). We have obviously $\mathscr{P}(y_1)^\rho = \mathscr{P}(y_i)^{\rho_i^{-1}\rho_1\rho}$. Applying the property (K3) of normalized Kummer variety to $\mathscr{P}(y_i)$, we have $(h_{v_i})^{\rho_i^{-1}\rho_1\rho} \circ \xi_1 = h_{v_i}$, namely,

$$(20) \qquad h_{v_1}^\rho \circ \xi_1 = h_{v_i} .$$

As ρ leaves invariant the elements of $K_{\mathfrak{a},v}$, we have, for every $t \in \mathfrak{g}(\mathfrak{a}, A_v)$, $h_v(t) = h_v^\rho(t^\rho) = h_v(\xi^{-1}t^\rho)$, so that

$$(21) \qquad \xi^{-1}t^\rho = \pm t .$$

By (8), (18), (20), (21), we obtain, for every $t \in \mathfrak{g}(\mathfrak{a}, A_v)$,

$$h_v(t)^{\sigma_1 \rho} = h_{v_1}(\lambda_1 t)^\rho = h_{v_1}^\rho(\lambda_1^\rho \xi \xi^{-1} t^\rho) = h_{v_1}^\rho(\xi_1 \lambda_i t) = h_{v_i}(\lambda_i t) = h_v(t)^{\sigma_i} .$$

This together with (19) shows $\sigma_1 \rho = \sigma_i$. Hence we have $u^{\sigma_i} = (u^{\sigma_1})^\rho$. It follows that u^{σ_i} is a conjugate of u^{σ_1} over $K_{a,y}$. This completes the proof.

Proposition 4.7. *Notations being as in Proposition 4.4, let τ_q be the automorphism of $K_{a,y}$ corresponding to the element q of G_a. Then there exists an isomorphism τ of the universal domain into itself such that $\tau = \sigma_1$ on $K_{a,y}$ and $\tau = \sigma_1^{-1}\tau_q$ on K_{a,y_1}.*

Proof. As the kernel of λ_1 is contained in $\mathfrak{g}(q, A_y)$, there exists a homomorphism μ of $A(y_1)$ onto A_y such that $\mu \circ \lambda = q1_y$, where 1_y denotes the identity of $\mathcal{A}(A_y)$. By Proposition 2.4, μ is a homomorphism of $\mathcal{P}(y_1)$ onto $\mathcal{P}(y)$. We observe that the kernel of μ is $\mathfrak{o}$-isomorphic to $\mathfrak{o}/\mathfrak{q}$. Therefore, if y is sufficiently generic, we can apply Proposition 4.4 to $\mathcal{P}(y_1)$; we obtain then an isomorphism σ of K_{a,y_1} onto $K_{a,y}$ such that

$$(22) \qquad c(\mathcal{F}(y_1))^{\sigma} = c(\mathcal{F}(y)),$$

$$(23) \qquad h_{y_1}(t)^{\sigma} = h_y(\mu t),$$

for every $t \in \mathfrak{g}(a, A(y))$. Extend σ to an isomorphism of the universal domain and denote it again by σ. By (22), there exists an isomorphism ε of $\mathcal{P}(y)$ onto $\mathcal{P}(y_1)^{\sigma}$. The isomorphisms ρ_i being as in (6), we have $\mathcal{F}(y_1)^{\sigma} = \mathcal{P}(y)^{\rho_1\sigma}$, so that by the property (K3) of normalized Kummer variety, we obtain $h_{y_1}^{\sigma} \circ \varepsilon = h_y^{\rho_1\sigma} \circ \varepsilon = h_y$. Hence, for every $t \in \mathfrak{g}(a, A_y)$, we get, by (23),

$$h_y(t) = h_{y_1}^{\sigma}(\varepsilon t) = h_{y_1}((\varepsilon t)^{\sigma^{-1}})^{\sigma} = h_y(\mu(\varepsilon t)^{\sigma^{-1}}).$$

It follows that $t = \pm\mu(\varepsilon t)^{\sigma^{-1}}$, namely

$$(24) \qquad t^{\sigma} = \pm\mu^{\sigma}\varepsilon t.$$

Now we observe that $\mu^{\sigma} \circ \varepsilon$ is a homomorphism of $\mathcal{P}(y)$ onto $\mathcal{P}(y)^{\sigma}$ whose kernel is $\mathfrak{o}$-isomorphic to $\mathfrak{o}/\mathfrak{q}$. By i) of Proposition 2.5, there exists, for some i, an isomorphism η of $\mathcal{P}(y_i)$ onto $\mathcal{P}(y)^{\sigma}$ such that

$$(25) \qquad \eta \circ \lambda_i = \mu^{\sigma} \circ \varepsilon.$$

We have then

$$(26) \qquad c(\mathcal{F}(y))^{\sigma} = c(\mathcal{F}(y_i)).$$

As we have $\mathcal{P}(y)^{\sigma} = \mathcal{P}(y_i)^{\rho_i^{-1}\sigma}$, we obtain, by the property (K3),

$$(27) \qquad h_y^{\sigma} \circ \eta = h_{y_i}^{\rho_i^{-1}\sigma} \circ \eta = h_{y_i}.$$

In the proof of Proposition 4.6, we have constructed an isomorphism ρ of the universal domain such that ρ leaves invariant the elements of $K_{a,y}$ and $\sigma_1\rho = \sigma_i$. Put now $\tau = \sigma\rho^{-1}$. Then, by (26), we have

$$c(\mathcal{F}(y))^{\sigma\rho^{-1}} = c(\mathcal{F}(y_i))^{\rho^{-1}} = c(\mathcal{F}(y))^{\sigma_i\rho^{-1}} = c(\mathcal{F}(y))^{\sigma_1},$$

and by (24), (25), (27), (8), for every $t \in \mathfrak{g}(a, A_y)$,

$$h_y(t)^{\sigma\rho^{-1}} = h_y^{\sigma}(t^{\sigma})^{\rho^{-1}} = h_y^{\sigma}(\mu^{\sigma}\varepsilon t)^{\rho^{-1}} = h_y^{\sigma}(\eta\lambda_i t)^{\rho^{-1}} = h_{y_i}(\lambda_i t)^{\rho^{-1}} = h_y(t)^{\sigma_i\rho^{-1}} = h_y(t)^{\sigma_1}.$$

Hence we have $\tau = \sigma_1$ on $K_{a,y}$. Now, by (22),

$$c(\mathcal{F}(y_1))^{\sigma\rho^{-1}} = c(\mathcal{F}(y))^{\rho^{-1}} = c(\mathcal{F}(y)) = c(\mathcal{F}(y))^{\tau_q} = c(\mathcal{F}(y_1))^{\sigma_1^{-1}\tau_q},$$

and by (23), (8), for every $t \in \mathfrak{g}(\mathfrak{a}, A(y_1))$,

$$h_{y_1}(qt)^{\sigma\rho^{-1}} = h_y(q\mu t)^{\rho^{-1}} = h_y(q\mu t) = h_y(\mu t)^{\tau_q} = h_{y_1}(\lambda_1 \mu t)^{\sigma_1^{-1}\tau_q} = h_{y_1}(qt)^{\sigma_1^{-1}\tau_q}.$$

This shows that $\tau = \sigma_1^{-1}\tau_q$ on $K_{\mathfrak{a}, y_1}$; our proposition is thereby proved.

u being as in Proposition 4.6, we have, by Proposition 4.7, $u^\tau = u^{\sigma_1}$, $u^{\sigma_1 \tau} = u^{\tau_q}$. As τ_q is contained in the center of the group $\mathcal{G}_\mathfrak{a}$, we have $Q(u) = Q(u^{\tau_q})$; hence τ gives an automorphism of $Q(u, u^{\sigma_1})$ and maps $Q(u)$ onto $Q(u^{\sigma_1})$. We have therefore, by Proposition 4.6,

$$(28) \qquad [Q(u^{\sigma_1}, u) : Q(u)] = [Q(u, u^{\sigma_1}) : Q(u^{\sigma_1})] = m.$$

4.3. Modular correspondences.

Let L be a subfield of $K_{\mathfrak{a}, y}$ such that

$$(29) \qquad L \supset K_{1, y}, \quad L \cap Q(\zeta_a) = Q,$$

where a is the smallest positive integer divisible by $\mathfrak{a}$ and ζ_a is a primitive a-th root of unity. By Theorem 2, Q is algebraically closed in L. Hence we can find a complete non-singular curve $\mathfrak{C} = \mathfrak{C}_L$ defined over Q such that we have $L = Q(u)$ for a generic point u of $\mathfrak{C}$ over Q. We call $\{\mathfrak{C}, u\}$ a model of L. We shall now define certain algebraic correspondences on the curve $\mathfrak{C}$.

Fix an integral left $\mathfrak{o}$-deal $\mathfrak{q}$ and put $N(\mathfrak{q}) = q$; suppose that q is prime to $\mathfrak{a}$. Define the isomorphisms $\{\sigma_1, \cdots, \sigma_m\}$ as in Proposition 4.4. Let $X_\mathfrak{q}$ be the locus of $u \times u^{\sigma_1}$ on $\mathfrak{C} \times \mathfrak{C}$ over Q. We have then, using the notation of Weil [28],

$$(30) \qquad X_\mathfrak{q}(u) = u^{\sigma_1} + \cdots + u^{\sigma_m}.$$

We call $X_\mathfrak{q}$ the *modular correspondence on $\mathfrak{C}$ associated with* $\mathfrak{q}$.

Let n be an integer prime to $\mathfrak{a}$ and τ_n be the automorphism of $K_{\mathfrak{a}, y}$ corresponding to the element n of $G_\mathfrak{a}$. As τ_n is contained in the center of $G_\mathfrak{a}$, τ_n induces an automorphism on $L = Q(u)$. Let Y_n be the locus of $u \times u^{\tau_n}$ on $\mathfrak{C} \times \mathfrak{C}$ over Q. Obviously, Y_n gives a birational correspondence on $\mathfrak{C}$. By Proposition 4.7, we obtain

$$X_\mathfrak{q}' \circ Y_q = X_\mathfrak{q},$$

$$d(X_\mathfrak{q}) = d'(X_\mathfrak{q}) = m,$$

the notations being as in [28].

Now assume that $\mathfrak{a}$ is prime to $d(\mathcal{O})$. Then we have $\mathfrak{a} = \mathfrak{o}N$ for a positive integer N: and $\mathfrak{o}/\mathfrak{a}$ is isomorphic to the total matric ring $M_2(Z/NZ)$. Fix an isomorphism of $\mathfrak{o}/\mathfrak{a}$ onto $M_2(Z/NZ)$; then $G_\mathfrak{a}$ is identified with the group of regular elements of $M_2(Z/NZ)$. Let H be the subgroup of $G_\mathfrak{a}$ consisting of the matrices $\begin{pmatrix} a & 0 \\ 0 & \pm 1 \end{pmatrix}$ for $(a, N) = 1$. Let L_N be the subfield of $K_{\mathfrak{a}, y}$ corresponding to H. Then we see easily

$$L_N(\zeta_N) = K_{\mathfrak{a},\mathfrak{y}}, \quad L_N \cap Q(\zeta_N) = Q.$$

Let $\{\mathfrak{C}_N, u\}$ be a model of L_N. Put $\psi = \begin{pmatrix} 0 & 1 \\ -1 & 0 \end{pmatrix}$. It is clear that

$$K_{\mathfrak{a},\mathfrak{y}} = Q(u, \zeta_N) = Q(u^\psi, \zeta_N).$$

Let Z be the locus of $u \times u^\psi$ over $Q(\zeta_N)$; then Z is a birational correspondence on $\mathfrak{C}_N$.

PROPOSITION 4.8 *Let φ_n be an automorphism of $Q(\zeta_N)$ such that $\zeta_N^{\varphi_n} = \zeta_N^n$. Then we have*

$$Z^{\varphi_n} \circ Y_n = Z.$$

As the proof is quite similar to that of Proposition 12 of [22], we omit it.

§5. Congruence-relations for modular correspondences.

5.1. In the following treament, we shall make use of the theory of reduction modulo p of algebraic varieties (cf. [21, 25]). We shall use mainly the terminology of [25]; and a place (or valuation) will mean a *discrete* one. We recall here only one definition: let U be a variety defined over a field with a place p; U is then called *p-simple* if the reduction of U modulo p has only one component and its multiplicity is 1.

Let k be a field with a place p; we denote by $p(U)$ or $\tilde{U}$ the reduction of any object U modulo p. By [21, Proposition 5, Theorem 15] (see also [14, Appendix]), we obtain

LEMMA R. *Let U be a variety defined over k, which is p-simple. Let x be a generic point of U over k and ξ a generic point of $\tilde{U}$ over $\tilde{k}$. Then, the specialization-ring $[x \rightarrow \xi; p]$ is a discrete valuation ring.*

Hence there exists one and only one extension p_1 of p in $k(x)$ such that $p_1(x) = \xi$; we call p_1 the place determined by the specialization $x \rightarrow \xi$ ref. p.

PROPOSITION 5.1. *Let A be a projective abelian variety, defined over k, satisfying the following conditions:*

i) *there is no hyperplane containing A;*

ii) *the linear system on A defined by hyperplane sections is complete.*

Suppose that A is without defect for p in the sense of [25, §11]. Then, $\tilde{A}$ satisfies the conditions i, ii).

PROOF. We can find a prolongation $\{k_1, p_1\}$ of $\{k, p\}$ and a hyperplane section X of A, rational over k_1, so that $\tilde{X} = p(X)$ is a hyperplane section of $\tilde{A}$. Let $L(X; k_1)$ and $L(\tilde{X}; \tilde{k}_1)$ be respectively the set of functions f on A defined over k_1 such that $(f) > -X$ and the set of functions g on $\tilde{A}$ defined over $\tilde{k}_1$ such that $(g) > -\tilde{X}$; denote by $l(X)$ and $l(\tilde{X})$ the dimensions of $L(X; k_1)$ over k_1 and of $L(\tilde{X}; \tilde{k}_1)$ over $\tilde{k}_1$. By our assumption, if we denote by n the dimension of the ambient space for A, we have $l(X) = n+1$. By the result of [25, pp. 86–87], $L(X; k_1)$ has a base $\{f_0, \cdots, f_n\}$ over k_1 such that $p(f_0), \cdots, p(f_n)$ are linearly independent over $\tilde{k}_1$. Hence, no hyperplane contains $\tilde{A}$. Now by Nishi [16], we have $l(X) = l(\tilde{X})$. It follows that $\tilde{A}$ satisfies ii).

Let $\mathcal{P} = (A, C, \theta)$ be a polarized abelian variety of type $\mathfrak{r}$, defined over k. Suppose that A is without defect for p. Take a divisor X in C which is rational over k. Then, by [25, §11, Proposition 14], $\tilde{X}$ is non-degenerate divisor on $\tilde{A}$; so $\tilde{X}$ determines a polarization on $\tilde{A}$, which we denote by $\tilde{C}$. Let $\tilde{\theta}(\alpha)$ be the reduction of $\theta(\alpha)$ modulo p for every $\alpha \in \mathfrak{r}$. In this way, we obtain a polarized abelian variety $\tilde{\mathcal{P}} = (\tilde{A}, \tilde{C}, \tilde{\theta})$ of type $\mathfrak{r}$, defined over k.

PROPOSITION 5.2. *Let $\mathcal{P} = (A, C, \theta)$ be a polarized abelian variety of type $\mathfrak{r}$, defined over k. Suppose that A is a projective variety satisfying i, ii) of Proposition 5.1 and that A is without defect for p. Then we have*

$$p[\mathcal{H}(A, \theta)] \supset \mathcal{H}(\tilde{A}, \tilde{\theta}).$$

Moreover, if $\mathcal{H}(A, \theta)$ is p-simple, we have

$$p[\mathcal{H}(A, \theta)] = \mathcal{H}(\tilde{A}, \tilde{\theta}).$$

PROOF. The first assertion is proved in a straightforward way; the argument is the same as in the proof of [AF, Theorem 1]. Now we note that [AF, Lemma 8] is valid for any polarized abelian variety of type $\mathfrak{r}$ whenever the variety satisfies the conditions i, ii) of Proposition 5.1. Therefore, by the proposition, $\mathcal{H}(A, \theta)$ and $\mathcal{H}(\tilde{A}, \tilde{\theta})$ are of the same dimension; so $\mathcal{H}(\tilde{A}, \tilde{\theta})$ is a component of the cycle $p[\mathcal{H}(A, \theta)]$. Hence, if $\mathcal{H}(A, \theta)$ is p-simple, we must have the equality of our proposition.

5.2. Fix a sufficiently generic point y on $\mathfrak{H}$ and denote $\mathcal{P}_y = (A_y, C_y, \theta_y)$, (V_y, h_y) simply by $\mathcal{P} = (A, C, \theta)$, (V, h), respectively. Let U be the locus of $c(\mathcal{H}(y))$ over Q. Let k_0 be a field of definition for $\mathcal{P}$, which is finitely generated over Q. Fix a set of independent variables $(t_1, \cdots, t_s)$ in k_0 such that k_0 is algebraic over $Q(t_1, \cdots, t_s)$. For each prime number p, we obtain a place of Q: $a \to a \bmod p$. We extend this to a place p_0 of k_0 as follows: first define a place p_1 of $Q(t_1, \cdots, t_s)$ so that $p_1 = p$ on Q and $p_1(t_1), \cdots, p_1(t_s)$ are independent variables over Z/pZ; then extend p_1 to a place p_0 of k_0. Such a place p_0 is not necessarily unique. We choose and fix once for all a place p_0 for each p. By the result of [25, §12], the following assertions hold for almost all p.

P1) A is without defect for p_0.

P2) $\mathcal{F} = \mathcal{H}(A, \theta)$ is p_0-simple.

P3) U is p_0-simple.

P4) $p_0(c(\mathcal{F}))$ is not algebraic over Z/pZ.

P5) V is p_0-simple.

P6) h is everywhere defined on $p_0(A)$.

Here and henceforth, by the terms "for almost all", we understand "for all except a finite number of ".

PROPOSITION 5.3. *If the conditions P1, 2, 4) are satisfied and if p is prime to $d(\Phi)$, we have $\nu_i(p\delta) = \nu_s(p\delta) = p^2$, where δ is the identity element of $\mathcal{A}(\tilde{A})$.*

PROOF. As p is prime to $d(\Phi)$, we can find two integral left $\mathfrak{o}$-ideals $\mathfrak{b}_1$ and $\mathfrak{b}_2$ such that $\mathfrak{o} = \mathfrak{b}_1 + \mathfrak{b}_2$, $p\mathfrak{o} = \mathfrak{b}_1 \cap \mathfrak{b}_2$, $N(\mathfrak{b}_1) = N(\mathfrak{b}_2) = p$. Take a generic point x of $\tilde{A}$ over $\tilde{k}_0$. Let K_i, for $i = 1, 2$, be the composite of the fields $\tilde{k}_0(\tilde{\theta}(\beta)x)$ for $\beta \in \mathfrak{b}_i$. We have obviously

$$(1) \qquad \tilde{k}_0(x) = K_1 K_2, \quad \tilde{k}_0(x) \supset K_i \supset \tilde{k}_0(px).$$

By [25, § 7.2, Proposition 10], we have $[\tilde{k}_0(x) : K_i] = p^2$; and by a well-known theorem, we have $p^4 \geq \nu_i(p\delta) \geq p^2$, so that $1 \leq \nu_s(p\delta) \leq p^2$. On the other hand, we observe that the points of order p on $\tilde{A}$ form an $\mathfrak{o}$-module. Since there is no $\mathfrak{o}$-module of order p, we must have $\nu_s(p\delta) = 1$ or p^2, so that $\nu_i(p\delta) = p^4$ or p^2. Suppose that $\nu_i(p\delta) = p^4$. Then, $\tilde{k}_0(x)$ is purely inseparable over K_i and hence $K_i \supset \tilde{k}_0(x^{p^\iota})$. Putting $M = K_1 \cap K_2$, we see that $M \supset \tilde{k}_0(x^{p^\iota})$. If $M \neq \tilde{k}_0(x^{p^\iota})$, we must have $[K_1 : M] = [K_2 : M] \leq p$, so that $[K_1 K_2 : K_1] \leq p$; but this is a contradiction in view of (1); so we must have $M = \tilde{k}_0(x^{p^\iota})$. As $K_i \supset \tilde{k}_0(px)$, we have $M \supset \tilde{k}_0(px)$; and considering the degrees, we find $\tilde{k}_0(x^{p^\iota}) = \tilde{k}_0(px)$. Hence there exists an isomorphism ε of $\tilde{A}^{p^\iota}$ onto $\tilde{A}$ such that $\varepsilon\pi = p$, where π denotes the p^2-th power homomorphism of $\tilde{A}$ onto $\tilde{A}^{p^\iota}$. Put $\tilde{\theta}^{p^\iota}(\alpha) = \tilde{\theta}(\alpha)^{p^\iota}$ for every $\alpha \in \mathfrak{o}$. We have then $\mathcal{F}(\tilde{A}, \tilde{\theta})^{p^\iota} = \mathcal{F}(\tilde{A}^{p^\iota}, \tilde{\theta}^{p^\iota})$. We see easily that π is a homomorphism of $\tilde{\mathcal{P}}$ onto $\tilde{\mathcal{P}}^{p^\iota}$. Then, by Proposition 2.4, ε is an isomorphism of $\tilde{\mathcal{P}}^{p^\iota}$ onto $\tilde{\mathcal{P}}$; so we have, by [AF, Proposition 1], $\mathcal{F}(\tilde{A}, \tilde{\theta}) = \mathcal{F}(\tilde{A}, \tilde{\theta})^{p^\iota}$. On account of Proposition 5.2, this shows that $p_0(c(\mathcal{F}))$ is algebraic over Z/pZ; this contradicts the condition P4). Therefore we must have $\nu_i(p\delta) = p^2$.

5.3. Let p be a prime number which does not divide $d(\Phi)$. There are exactly $p+1$ integral left $\mathfrak{o}$-ideals $\mathfrak{q}$ such that $N(\mathfrak{q}) = p^2$; we denote them by $\mathfrak{q}_0, \cdots, \mathfrak{q}_p$. For these $\mathfrak{q}_i$, we define $\mathcal{P}(y_i)$, $V(y_i)$, h_{y_i} and λ_i for $0 \leq i \leq p$ as in § 4.2; we denote $\mathcal{P}(y_i)$, $V(y_i)$, h_{y_i} simply by $\mathcal{P}_i = (A_i, C_i, \theta_i)$, V_i, h_i. Let $\mathfrak{g}_i$ denote the kernel of λ_i. As $\mathfrak{g}_i$ is contained in $\mathfrak{g}(p, A)$, there exists a homomorphism μ_i of A_i onto A such that $\mu_i \circ \lambda_i = p$; then μ_i is a homomorphism of $\mathcal{P}_i$ onto $\mathcal{P}$.

THEOREM 3. *Let p be a prime number which satisfies the conditions P1~6) and is prime to $d(\Phi)$. Let k be an extension of k_0 such that the $\mathcal{P}_i$ and the λ_i are defined over k; and let $\mathbf{p}$ be an extension of $\mathbf{p}_0$ in k. Then, reordering the $\mathcal{P}_i$ suitably, the following relations hold.*

$$\mathbf{p}(c(\mathcal{F}(y_0))) = \mathbf{p}(c(\mathcal{F}(y)))^p,$$

$$\mathbf{p}(c(\mathcal{F}(y_i))) = \mathbf{p}(c(\mathcal{F}(y)))^{1/p} \qquad (i > 0),$$

$$\mathbf{p}(h_0(\lambda_0 t)) = \mathbf{p}(h(t))^p,$$

$$\mathbf{p}(h_i(\lambda_i t)) = \mathbf{p}(h(pt))^{1/p} \qquad (i > 0),$$

for every point t of A rational over k.

We prove this theorem in this and the following sections.

PROPOSITION 5.4. *There exist an extension $\{k_1, \mathbf{p}_1\}$ of $\{k, \mathbf{p}\}$ and polarized abelian varieties $\mathcal{P}_i{}^* = (A_i{}^*, C_i{}^*, \theta_i{}^*)$ of type* $\mathfrak{o}$, *defined over k_1, having the following properties.*

    i) *For each i, there exists an isomorphism η_i of $\mathcal{P}_i$ onto $\mathcal{P}_i{}^*$.*

    ii) $\eta_i = $ *(a projective transformation)$+$const.*

    iii) $A_i{}^*$ *is without defect for $\mathbf{p}_1$.*

    iv) *Reordering suitably,*

$$\tilde{A}_0{}^* = \tilde{A}^p, \quad \tilde{\theta}_0{}^* = \tilde{\theta}^p, \quad \tilde{A}_i{}^{*p} = \tilde{A}, \quad \tilde{\theta}_i{}^{*p} = \tilde{\theta} \qquad (i > 0).$$

    v) *Let π be the p-th power homomorphism of $\tilde{A}$ onto $\tilde{A}^p$ and π' the p-th power homomorphism of $\tilde{A}^{1/p}$ onto $\tilde{A}$; then*

$$\mathbf{p}_1(\eta_0 \circ \lambda_0) = \pi, \quad \mathbf{p}_1(\mu_i \circ \eta_i^{-1}) = \pi' \qquad (i > 0).$$

PROOF. Without loss of generality, we may assume that the points in $\mathfrak{g}(p, A)$ are rational over k. As A_i is isogenous to A, we can find, by [14, Theorem 4], an abelian variety B_i and an isomorphism ξ_i of A_i onto B_i, both defined over k, such that B_i is without defect for $\mathbf{p}$. Put $\alpha_i = \xi_i \circ \lambda_i$, $\beta_i = \mu_i \circ \xi_i^{-1}$; we have then $\beta_i \circ \alpha_i = p$. Now the reduction modulo p defines a homomorphism of $\mathfrak{g}(p, A)$ onto $\mathfrak{g}(p, \tilde{A})$, By Proposition 5.3, the kernel $\mathfrak{g}^*$ of this homomorphism is of order p^2; and we observe that $\mathfrak{g}^*$ is an $\mathfrak{o}$-module. Hence $\mathfrak{g}^*$ must coincide with one of the $\mathfrak{g}_i$, say $\mathfrak{g}_0$. Then we have $\mathfrak{g}_0 = \{0\}$. As $\mathfrak{g}(p, A) = \mathfrak{g}_0 + \mathfrak{g}_i$ for $i > 0$, $\mathfrak{g}_i$ is of order p^2 for $i > 0$. By [25, § 11, Proposition 3], $\tilde{\mathfrak{g}}_i$ is the kernel of $\tilde{\alpha}_i$ for every i. It follows that $\tilde{\alpha}_0$ is purely inseparable and $\tilde{\alpha}_i$ is separable for $i > 0$. As we have $\tilde{\beta}_i \circ \tilde{\alpha}_i = p$, we see, on account of Proposition 5.3, that $\tilde{\beta}_0$ is separable and $\tilde{\beta}_i$ is purely inseparable for $i > 0$. Let x be a generic point of $\tilde{A}$ over $\tilde{k}$. We have then $\tilde{k}(x) \supset \tilde{k}(x^p) \supset \tilde{k}(px)$. By Proposition 5.3, $\tilde{k}(x^p)$ is the maximal separable extension of $\tilde{k}(px)$ in $\tilde{k}(x)$. We have therefore

$$(2) \qquad\qquad \tilde{k}(x^p) = \tilde{k}(\tilde{\alpha}_0 x);$$

a similar consideration shows, for $i > 0$,

$$(3) \qquad\qquad \tilde{k}(x_i{}^p) = \tilde{k}(\tilde{\beta}_i x_i)$$

for a generic point x_i of $\tilde{B}_i$ over $\tilde{k}$. Putting $x_i = \tilde{\alpha}_i x$, we obtain

$$(4) \qquad\qquad \tilde{k}((\tilde{\alpha}_i x)^p) = \tilde{k}(px).$$

Now take a hyperplane section X of A and a hyperplane section X_i of A_i for each i, all defined over k. By our construction of $A(z)$, we see that $\lambda_i^{-1}(X_i) \equiv pX$, $\mu_i^{-1}(X) \equiv pX_i$, where $\equiv$ denotes algebraic equivalence. Put $Y_i = \xi_i(X_i)$; then we have $\alpha_i^{-1}(Y_i) \equiv pX$, $\beta_i^{-1}(X) \equiv pY_i$. By virtue of [25, § 11, Proposition 14], we see that

(5) $$\tilde{\alpha}_i^{-1}(\tilde{Y}_i) \equiv p\tilde{X}, \quad \tilde{\beta}_i^{-1}(\tilde{X}) \equiv p\tilde{Y}_i.$$

Let π be the p-th power homomorphism of $\tilde{A}$ onto $\tilde{A}^p$; by (2), there exists an isomorphism ε of $\tilde{A}^p$ onto $\tilde{B}_0$ such that $\tilde{\alpha}_0 = \varepsilon\pi$. We have then, $\tilde{\alpha}_0^{-1}(\varepsilon(\tilde{X}^p)) = \pi^{-1}(\tilde{X}^p) = p\tilde{X}$; hence on account of (5), $\varepsilon(\tilde{X}^p) \equiv \tilde{Y}_0$. By Proposition 5.1, $\tilde{X}$ is ample; therefore, $\tilde{X}^p$ is ample, and hence $\tilde{Y}_0$ is ample. By the result of [14, §4], we can find a projective embedding C_0 of B_0 by Y_0, whose reduction modulo p is a projective embedding of $\tilde{B}_0$ by $\tilde{Y}_0$; as $Y_0 = \xi_0(X_0)$, C_0 is a projective transform of A_0. We can take C_0 as B_0; namely, we may assume that B_0 is a projective transform of A_0, and ξ_0 differs from a projective transformation only by a translation. Define a polarized abelian variety $\mathcal{P}_0' = (B_0, C_0', \theta_0')$ of type $\mathfrak{o}$ so that ξ_0 is an isomorphism of $\mathcal{P}_0$ onto $\mathcal{P}_0'$; then C_0' is determined by the hyperplane sections. As $\alpha_0 = \xi_0 \circ \lambda_0$, α_0 is a homomorphism of $\mathcal{P}$ onto $\mathcal{P}_0'$, so that $\tilde{\alpha}_0$ is a homomorphism of $\tilde{\mathcal{P}}$ onto $\tilde{\mathcal{P}}_0'$. Since π is a homomorphism of $\tilde{\mathcal{P}}$ onto $\tilde{\mathcal{P}}^p$ and $\tilde{\alpha}_0 = \varepsilon \circ \pi$, we see that, by Proposition 2.4, ε is an isomorphism of $\tilde{\mathcal{P}}^p$ onto $\tilde{\mathcal{P}}_0'$. Now by Proposition 5.1 and by the proof of [AF, Proposition 1], there exists a projective transformation $\bar{\psi}$ and a point $\bar{a}$ on $\tilde{B}_0$ such that $\bar{\psi}(u) = \varepsilon(u) + \bar{a}$ for $u \in \tilde{A}^p$. We can find a projective transformation ψ, rational over k, and a point a of B_0 so that $(\psi, a) \to (\bar{\psi}, \bar{a})$ ref. p. Put $k_1 = k(a)$ and extend this specialization to a place p_1 of k_1. Put $A_0^* = \psi^{-1}(B_0)$, $\zeta_0 = \psi - a$; and define a polarized abelian variety $\mathcal{P}_0^* = (A_0^*, C_0^*, \theta_0^*)$ of type $\mathfrak{o}$, so that ζ_0 is an isomorphism of $\mathcal{P}_0^*$ onto $\mathcal{P}_0'$. Then A_0^* is without defect for p_1 and $\tilde{A}_0^* = \bar{\psi}^{-1}(\tilde{B}_0) = \varepsilon^{-1}(\tilde{B}_0) = \tilde{A}^p$, $\tilde{\zeta}_0 = \varepsilon$. Hence $(\tilde{A}_0^*, \tilde{\theta}_0^*)$ coincides with $(\tilde{A}^p, \tilde{\theta}^p)$. Moreover, putting $\eta_0 = \zeta_0^{-1} \circ \xi_0$, we have $p_1(\eta_0 \circ \lambda_0) = \tilde{\zeta}_0^{-1} \circ \tilde{\alpha}_0 = \varepsilon^{-1} \circ \tilde{\alpha}_0 = \pi$. Thus $\mathcal{P}_0^* = (A_0^*, C_0^*, \theta_0^*)$ satisfies i-v) of our proposition. Consider now B_i for $i > 0$. Let π_i' be the p-th power homomorphism of $\tilde{B}_i$ onto $\tilde{B}_i^p$. Then, by (3), there exists an isomorphism ε_i of $\tilde{B}_i^p$ onto $\tilde{A}$ such that $\tilde{\beta}_i = \varepsilon_i \circ \pi_i'$. By the same argument as above, we get $\tilde{X} \equiv \varepsilon_i(\tilde{Y}_i^p)$; it follows that $\tilde{Y}_i$ is ample. Therefore, by the same reasoning as above, we may assume that B_i is a projective transform of A_i and $\xi_i = $ (a projective transformation)+ const. Define a polarized abelian variety $\mathcal{P}_i' = (B_i, C_i', \theta_i')$ so that ξ_i is an isomorphism of $\mathcal{P}_i$ onto $\mathcal{P}_i'$. Then, ε_i is an isomorphism of $\tilde{\mathcal{P}}_i'^p$ onto $\tilde{\mathcal{P}}$, so that $\varepsilon_i^{1/p}$ is an isomorphism of $\tilde{\mathcal{P}}_i'$ onto $\tilde{\mathcal{P}}^{1/p}$. In the same way as above, we can find, taking a suitable extension of $\{k_1, p_1\}$, if necessary, a polarized abelian variety $\mathcal{P}_i^* = (A_i^*, C_i^*, \theta_i^*)$ and an isomorphism ζ_i of $\mathcal{P}_i'$ onto $\mathcal{P}_i^*$ such that: i) $\zeta_i = $ (a projective transformation)+const.; ii) A_i^* is without defect for p_1 and $\tilde{A}_i^* = \tilde{A}^{1/p}$, $\tilde{\zeta}_i = \varepsilon_i^{1/p}$. Putting $\eta_i = \zeta_i \circ \xi_i$, we obtain $\mathcal{P}_i^*$ and η_i having the properties i-v); our proposition is thereby proved.

Now $\mathcal{P}_i^*$ being as in the above proposition, by [AF, Proposition 1], we have

$$(6) \qquad \mathfrak{F}(y_i) = \mathfrak{F}(A_i, \theta_i) = \mathfrak{F}(A_i^*, \theta_i^*).$$

By Proposition 5.2, we get

$$(7) \qquad p_1(\mathfrak{F}(A, \theta)) = \mathfrak{F}(\tilde{A}, \tilde{\theta}),$$

$$(8) \qquad p_1(\mathfrak{F}(A_i^*, \theta_i^*)) \supset \mathfrak{F}(\tilde{A}_i^*, \tilde{\theta}_i^*).$$

By iv) of Proposition 5.4 and by (6),

$$(9) \qquad \mathfrak{F}(\tilde{A}_0^*, \tilde{\theta}_0^*) = \mathfrak{F}(\tilde{A}^p, \tilde{\theta}^p) = \mathfrak{F}(\tilde{A}, \tilde{\theta})^p = p(\mathfrak{F}(y))^p,$$

$$(10) \qquad \mathfrak{F}(\tilde{A}_i^*, \tilde{\theta}_i^*)^p = \mathfrak{F}(\tilde{A}_i^{*p}, \tilde{\theta}_i^{*p}) = \mathfrak{F}(\tilde{A}, \tilde{\theta}) = p(\mathfrak{F}(y)).$$

The relations (6), (8), (9) lead to

$$p(\mathfrak{F}(y_0)) \supset p(\mathfrak{F}(y))^p.$$

As y is sufficiently generic, $\mathfrak{F}(y)$ and $\mathfrak{F}(y_0)$ have the same dimension and the same degree. Therefore we must have

$$(11) \qquad p(\mathfrak{F}(y_0)) = p(\mathfrak{F}(y))^p.$$

By (6), (8), (10) and a similar consideration, we obtain

$$(11') \qquad p(\mathfrak{F}(y_i))^p = p(\mathfrak{F}(y)).$$

The relations (11) and (11') prove the first two equalities of Theorem 3.

5.4. Let the notations be the same as in Proposition 5.4. For the sake of simplicity, we denote k_1 and p_1 again by k and p. The ambient space for A is denoted by P^n. By our construction of $\mathfrak{F}(y_i)$, there exists an isomorphism ρ_i of a field of definition for $\mathcal{P}$ such that $\mathcal{P}^{\rho_i} = \mathcal{P}_i$, $V^{\rho_i} = V_i$, $h^{\rho_i} = h_i$. Extend ρ_0 to an isomorphism of k and denote it by ρ; put $M = kk^\rho$. We denote by K and K_i the fields of moduli of $\mathcal{P}$ and $\mathcal{P}_i$, respectively.

Now fix a basis $\{r_\nu\}$ of $\mathfrak{o}$ over Z, and consider the mapping T defined in §3.4; we use the same notation T for varieties in P^n and in $\tilde{P}^n$. Let φ be a projective transformation of P^n, generic over M, and $v_1, \cdots, v_d$ be independent generic points of A over $M(\varphi)$. Put

$$B = \varphi(A), \quad z = T(\varphi, v_1, \cdots, v_d).$$

Then, $M(z) \subset M(\varphi, v_1, \cdots, v_d)$; and $\mathfrak{F}(A, \theta)$ is the locus of z over M, and hence over K. Let w be a generic point of B over $M(\varphi, v_1, \cdots, v_d)$. Put into B a structure of abelian variety by taking w as its origin; then, B is defined over $K(z, w)$ as abelian variety. We can find an isomorphism ξ of A onto B and a point a on B such that $\xi(x) = \varphi(x) + a$ for $x \in A$. Define a polarized abelian variety $\mathcal{P}' = (B, \theta', C')$ of type $\mathfrak{o}$ so that ξ is an isomorphism of $\mathcal{P}$ onto $\mathcal{P}'$; then $\mathcal{P}'$ is defined over $K(z, w)$ (cf. [AF, Proof of Proposition 1]). Now extend ρ to an isomorphism of $k(\varphi, v_1, \cdots, v_d, w)$, which we denote again by ρ; we may assume that $(\varphi^\rho, v_1^\rho, \cdots, v_d^\rho, w^\rho)$ and $(\varphi, v_1, \cdots, v_d, w)$ are independent over M, and

$$\dim_M(\varphi^\rho, v_1{}^\rho, \cdots, v_d{}^\rho, w^\rho) = \dim_{k^\rho}(\varphi^\rho, v_1{}^\rho, \cdots, v_d{}^\rho, w^\rho).$$

Let $\mathfrak{p}$ be an extension of p in M. Let $\bar\varphi$ be a projective transformation of $\tilde P^n$, generic over $\tilde M$, and $\bar v_1, \cdots, \bar v_d$ be independent generic points of $\tilde A$ over $\tilde M(\bar\varphi)$; and let $\bar w$ be a generic point of $\bar\varphi(\tilde A)$ over $\tilde M(\bar\varphi, \bar v_1, \cdots, \bar v_d)$. Then, we obtain a specialization

$$(12) \qquad\qquad (\varphi, v_1, \cdots, v_d, w) \to (\bar\varphi, \bar v_1, \cdots, \bar v_d, \bar w) \text{ ref. } \mathfrak{p}.$$

Now consider A_i^* of Proposition 5.4. By ii) of the proposition, there exists a projective transformation ψ, defined over k, and a point b on A_0^* such that $\eta_0 = \psi + b$; we have then $A_0^* = \psi(A_0)$. Put $\chi = \varphi^\rho \circ \psi^{-1}$, $u_\nu = \eta_0(v_\nu{}^\rho) - b + \theta_0^*(r_\nu)b$. Then, we have $B^\rho = \chi(A_0^*)$; χ is generic over M, and $u_1, \cdots, u_d$ are independent generic on A_0^* over $M(\chi)$. Furthermore, we see easily

$$M(\varphi^\rho, v_1{}^\rho, \cdots, v_d{}^\rho) = M(\chi, u_1, \cdots, u_d);$$

by the definition of T and by our choice of u_ν, we obtain

$$(13) \qquad\qquad T(\chi, u_1, \cdots, u_d) = T(\varphi^\rho, v_1{}^\rho, \cdots, v_d{}^\rho) = z^\rho.$$

Note that: $\bar\varphi^p$ is generic over $\tilde M$; $\bar v_1{}^p, \cdots, \bar v_d{}^p$ are independent generic on $\tilde A_0^* = \tilde A^p$ over $\tilde M(\bar\varphi^p)$; $\bar w^p$ is generic on $\bar\varphi(\tilde A)^p$ over $\tilde M(\bar\varphi^p, \bar v_1{}^p, \cdots, \bar v_d{}^p)$. We see easily that the following specialization holds.

$$(14) \qquad\qquad (\chi, u_1, \cdots, u_d, w^\rho) \to (\bar\varphi^p, \bar v_1{}^p, \cdots, \bar v_d{}^p, \bar w^p) \text{ ref. } \mathfrak{p}.$$

As $M(\varphi, v_1, \cdots, v_d, w)$ and $M(\chi, u_1, \cdots, u_d, w^\rho)$ are linearly disjoint over M, the specialization (12) and (14) are compatible:

$$(16) \quad (\varphi, v_1, \cdots, v_d, w, \chi, u_1, \cdots, u_d, w^\rho) \to (\bar\varphi, \bar v_1, \cdots, \bar v_d, \bar w, \bar\varphi^p, \bar v_1{}^p, \cdots, \bar v_d{}^p, \bar w^p) \text{ ref. } \mathfrak{p}.$$

Extend this to a place $\mathfrak{P}$ of $M(\varphi, v_1, \cdots, v_d, w, \chi, u_1, \cdots, u_d, w^\rho)$. We can easily verify $\xi^\rho \circ \lambda_0 \circ \xi^{-1} = \chi \circ \eta_0 \circ \lambda_0 \circ \varphi^{-1} + \text{const.}$, so that by (16) and v) of Proposition 5.4,

$$(17) \quad \mathfrak{P}(\xi^\rho \circ \lambda_0 \circ \xi^{-1}) = \bar\varsigma^p \circ \pi \circ \bar\varsigma^{-1} = \text{the } p\text{-th power homomorphism of } \mathfrak{P}(B).$$

Put $\mathfrak{P}(z) = \bar z$. We have then

$$(18) \qquad\qquad T(\bar\varsigma, \bar v_1, \cdots, \bar v_d) = \bar z.$$

By the definition of T, we have $\bar z = c(\mathfrak{P}(B)) \times \cdots$, so that $\mathfrak{P}(B)$ is defined over $\tilde Q(\bar z)$ if we leave the structure of abelian variety out of consideration. Now by (13), (16), (18),

$$\mathfrak{P}(z^\rho) = \mathfrak{P}(T(\chi, u_1, \cdots, u_d)) = T(\bar\varsigma^p, \bar v_1{}^p, \cdots, \bar v_d{}^p) = \bar z^p.$$

Hence

$$(19) \qquad\qquad \mathfrak{P}(z, w, z^\rho, w^\rho) = (\bar z, \bar w, \bar z^p, \bar w^p).$$

Put $f = c(\mathcal{F}(A, \theta))$, $\tilde f = p(f)$. We have then $f^\rho = c(\mathcal{F}(A_0, \theta_0))$, $p(f^\rho) = \tilde f^p$ by (11). By our assumption P3) and P4), $\tilde f$ is a generic point of $\hat U$ over $\tilde Q = Z/pZ$. By

(18), z is generic on $\mathfrak{F}(\tilde{A}, \tilde{\vartheta})$ over $\tilde{M}$, and hence over $\tilde{Q}(\tilde{f})$; and w is generic on $\bar{\varphi}(\tilde{A}) = \mathfrak{P}(B)$ over $\tilde{M}(z)$, and hence over $\tilde{Q}(\tilde{f}, \tilde{z})$. Therefore, by Lemma R, the specialization

$$(f, z, w) \rightarrow (\tilde{f}, \tilde{z}, \tilde{w}) \text{ ref. } p$$

determines a place $\mathfrak{P}_1$ of $K(z, w)$. We see then easily that the specialization

$$(f^\rho, z^\rho, w^\rho) \rightarrow (\tilde{f}^p, \tilde{z}^p, \tilde{w}^p) \text{ ref. } p$$

determines a place $\mathfrak{P}_2$ of $K^\rho(z^\rho, w^\rho)$, satisfying $\mathfrak{P}_2(a^\rho) = \mathfrak{P}_1(a)^p$ for *every* $a \in K(z, w)$. Now, by (19), $\mathfrak{P} = \mathfrak{P}_1$ on $K(z, w)$ and $\mathfrak{P} = \mathfrak{P}_2$ on $K^\rho(z^\rho, w^\rho)$; hence, we have, for *every* $a \in K(z, w)$,

(20) $$\mathfrak{P}(a^\rho) = \mathfrak{P}(a)^p .$$

Since V is defined over K, we have, by (20),

$$\mathfrak{P}(V_0) = \mathfrak{P}(V^\rho) = \mathfrak{P}(V)^p .$$

Put $g = h \circ \xi^{-1}$. By Proposition 2.2, (V, g) is a normalized Kummer variety of $\mathscr{P}'$; and by the property (K2) of normalized Kummer variety, g is defined over $K(z, w)$ since $\mathscr{P}'$ is defined over $K(z, w)$. By the assumption P6), g is everywhere defined on $\mathfrak{P}(B)$; and by (20), we have

(21) $$\mathfrak{P}(g^\rho) = \mathfrak{P}(g)^p .$$

Therefore, if t is a point on A, rational over k, we have, by (17) and (21),

$$p(h_0(\lambda_0 t)) = \mathfrak{P}(h^\rho(\lambda_0 t)) = \mathfrak{P}(g^\rho(\xi^\rho \lambda_0 \xi^{-1} \xi t)) = \tilde{g}^p((\tilde{\xi} \tilde{t})^p) = \tilde{g}(\tilde{\xi} \tilde{t})^p = p(h(t))^p .$$

This proves the third equality of Theorem 3. Consider now $\mathscr{P}_i$ for $i > 0$. Fix an $i > 0$ and put $\sigma = \rho_i$. By the same argument as above, we extend σ to $L = k(\varphi, v_1, \cdots, v_d, w)$ suitably and find an extension $\mathfrak{Q}$ of p in LL^σ such that: i) B and B^σ are without defect for $\mathfrak{Q}$; ii) $\mathfrak{Q}(\xi \circ \mu_i \circ (\xi^\sigma)^{-1}) = $ the p-th power homomorphism of $\mathfrak{Q}(B^\sigma)$; iii) $\mathfrak{Q}(g^\sigma) = \mathfrak{Q}(g)^{1/p}$. Then, for every point t on A, rational over k, putting $s = \xi^\sigma \lambda_i t$, we obtain

$$h(pt) = h(\mu_i \lambda_i t) = g(\xi \mu_i (\xi^\sigma)^{-1} s) ,$$

so that

$$p(h(pt)) = \mathfrak{Q}(g(\xi \mu_i (\xi^\sigma)^{-1} s)) = \mathfrak{Q}(g^\sigma(s))^p = \mathfrak{Q}(h^\sigma((\xi^\sigma)^{-1} s))^p = p(h_i(\lambda_i t))^p .$$

This proves the fourth equality of Theorem 3 and completes the proof.

5.5. Congruence-relations. Fix an integral two-sided $\mathfrak{o}$-ideal $\mathfrak{a}$. We take a field k_0 of § 5.2 so that the points in $\mathfrak{g}(\mathfrak{a}, A)$ are rational over k_0. Put

$$b = c(\mathscr{F}(y)) \times h(t_1) \times \cdots \times h(t_m) ,$$

where $t_1, \cdots, t_m$ are the points in $\mathfrak{g}(\mathfrak{a}, A)$. Let F be the algebraic closure of Q in $K_{\mathfrak{a}, y}$; and let $\mathfrak{B}$ be the locus of b over F. Then, for almost all p, the following assertions hold.

P7) $\mathfrak{B}$ is p_0-simple.

P8) $p_0(b)$ is generic on $p_0(\mathfrak{B})$ over $p_0(F)$.

Now let σ_i, for $0 \leq i \leq p$, be the isomorphisms of $K_{\mathfrak{a},\mathfrak{y}}$ onto $K_{\mathfrak{a},\mathfrak{y}_i}$, determined by Proposition 4.4 for the ideals $\mathfrak{q}_i$ of §5.3. Let τ_p be the automorphism of $K_{\mathfrak{a},\mathfrak{y}}$ over $K_{1,\mathfrak{y}}$ corresponding to the element p of $G_{\mathfrak{a}}$. By Theorem 3, if p satisfies P1~6) and is prime to $d(\mathcal{O})\mathfrak{a}$, we have

$$(22) \qquad p(b^{\sigma_0}) = p(b)^p, \quad p(b^{\sigma_i}) = p(b^{\tau_p})^{1/p} \qquad (i > 0).$$

Let p_1 be the restriction of p on F. If p satisfies P7, 8), then, by Lemma R, the specialization

$$b \to p(b) \text{ ref. } p_1$$

determines a place of $F(b) = K_{\mathfrak{a},\mathfrak{y}}$; and the specializations

$$b^{\sigma_0} \to p(b)^p, \quad b^{\sigma_i} \to p(b^{\tau_p})^{1/p} \text{ ref. } p_1$$

determine respectively places on $K_{\mathfrak{a},\mathfrak{y}_0}$ and on $K_{\mathfrak{a},\mathfrak{y}_i}$. These places are of course restrictions of the place p. Therefore, we observe that, if p satisfies P1~8) and is prime to $d(\mathcal{O})\mathfrak{a}$,

$$(23) \qquad p(a^{\sigma_0}) = p(a)^p, \quad p(a^{\sigma_i}) = p(a^{\tau_p})^{1/p} \qquad (i > 0)$$

hold for *every* $a \in K_{\mathfrak{a},\mathfrak{y}}$.

Let L be a subfield of $K_{\mathfrak{a}}$ satisfying (29) of §4.3 and $\{\mathfrak{C}, u\}$ a model of L. For almost all p, the following assertions hold.

P9) $\mathfrak{C}$ is p-simple and $p(\mathfrak{C})$ has no multiple point.

P10) $p_0(u)$ is a generic point of $p(\mathfrak{C})$ over $\mathbf{Z}/p\mathbf{Z}$.

Let p be a prime number which satisfies P1~10) and is prime to $d(\mathcal{O})\mathfrak{a}$. Let $\mathfrak{q}$ be an integral left $\mathfrak{o}$-ideal such that $N(\mathfrak{q}) = p$ and $X_{\mathfrak{q}}$ the modular correspondence on $\mathfrak{C}$ associated with $\mathfrak{q}$ (cf. §4.3). Now we want to consider the reduction of $X_{\mathfrak{q}}$ modulo p. As $\mathfrak{C}$ is defined over $\mathbf{Q}$, $\widetilde{\mathfrak{C}}$ is defined over $\mathbf{Z}/p\mathbf{Z}$. Let Π and Π' be respectively the loci of $\tilde{u} \times \tilde{u}^p$ and of $\tilde{u}^p \times \tilde{u}$ on $\widetilde{\mathfrak{C}} \times \widetilde{\mathfrak{C}}$ over $\mathbf{Z}/p\mathbf{Z}$. The relation (23) shows

$$p(u^{\sigma_0}) = p(u)^p, \quad p(u^{\sigma_i}) = p(u^{\tau_p})^{1/p} \qquad (i > 0),$$

so that, by (30) of §4.3 and by [21, Theorem 19], we have

$$\widetilde{X}_{\mathfrak{q}}(\tilde{u}) = \tilde{u}^p + p\widetilde{Y}_p(\tilde{u})^{1/p} = \Pi(\tilde{u}) + \Pi' \circ \widetilde{Y}_p(\tilde{u}).$$

It follows that $\widetilde{X}_{\mathfrak{q}} - (\Pi + \Pi' \circ \widetilde{Y}_p)$ is of the form $\mathfrak{e} \times \widetilde{\mathfrak{C}}$, where $\mathfrak{e}$ is a divisor on $\widetilde{\mathfrak{C}}$. Since $\Pi + \Pi' \circ \widetilde{Y}^p$ has no component of the form $e \times \widetilde{\mathfrak{C}}$, we conclude that $\mathfrak{e} > 0$. On the other hand, we have

$$d'(\widetilde{X}_{\mathfrak{q}}) = d'(X_{\mathfrak{q}}) = p + 1 = d'(\Pi + \Pi' \circ Y_p);$$

so we must have $\mathfrak{e} = 0$. Thus we have proved:

THEOREM 4. *Notations being as above, let p be a prime number which satisfies P1~10) and is prime to $d(\mathcal{O})\mathfrak{a}$. Then we have*

$$\tilde{X}_q = \varPi + \varPi' \circ \tilde{Y}_p$$

on the reduction $\tilde{\mathbb{C}}$ of the curve $\mathbb{C}$ modulo p.

Now suppose that $\mathfrak{a}$ is prime to $d(\mathcal{O})$; let L_N be the subfield of $K_{\mathfrak{a},\mathfrak{y}}$ given in §4.3, and $\{\mathbb{C}_N, u\}$ a model of L_N. Notations being as in Proposition 4.8, consider the case $n = p$. We observe then $\tilde{Z}^p \circ \tilde{Y}_p = \tilde{Z}$, so that

$$\tilde{Z}' \circ \varPi' \circ \tilde{Z} = \tilde{Z}' \circ \varPi' \circ \tilde{Z}^p \circ \tilde{Y}_p = \tilde{Z}' \circ \tilde{Z} \circ \varPi' \circ \tilde{Y}_p = \varPi' \circ \tilde{Y}_p .$$

We obtain thus:

THEOREM 5. *Notations being as above, we have*

$$\varPi' \circ \tilde{Y}_p = \tilde{Z}' \circ \varPi' \circ \tilde{Z}$$

on $\tilde{\mathbb{C}}_N$.

5.6. Let J_N be a jacobian variety of $\mathbb{C}_N$, and φ a canonical mapping of $\mathbb{C}_N$ onto J_N. As $\mathbb{C}_N$ is defined over Q, we may assume that J_N is defined over Q; φ may not be defined over Q; but we may assume that $\varphi^\sigma = \varphi + \text{const.}$ for any isomorphism σ over Q. Every correspondence X on $\mathbb{C}_N$ determines an endomorphism ξ of J_N by the relations

$$X(x) = \sum_\nu x_\nu , \quad \xi(\varphi(x)) = \sum_\nu \varphi(x_\nu) + \text{const.}$$

(cf. [**29**, no. 43]). We see easily that ξ is defined over any field of definition for X. Let ξ_p, η_p, ζ be the endomorphisms of J_N determined by X_q, Y_p, Z, respectively. Now J_N is without defect for almost all p, and $\tilde{J}_N$ is a jacobian variety of $\tilde{\mathbb{C}}_N$; more precisely, as remarked by Igusa, Chow's construction of jacobian is compatible with the specialization; so we may assume that J_N is without defect and $\tilde{J}_N$ is a jacobian variety of $\tilde{\mathbb{C}}_N$ for every prime p satisfying P9). Let π be the p-th power endomorphism of $\tilde{J}_N$ and $\pi' = p\pi^{-1}$. Then, Theorems 4 and 5 yield the relations

$$(24) \qquad\qquad\qquad \tilde{\xi}_p = \pi + \pi' \circ \tilde{\eta}_p ,$$

$$(25) \qquad\qquad\qquad \pi' \circ \tilde{\eta}_p = \tilde{\zeta}^{-1} \circ \pi' \circ \tilde{\zeta} .$$

§ 6. The zeta-functions of algebraic curves.

6.1. Transference to the upper half plane. Let $\mathfrak{a} = N\mathfrak{o}$ be an integral two-sided $\mathfrak{o}$-ideal which is prime to $d(\mathcal{O})$. $\varGamma_N$ being as in §1.3, let $\Re_N$ denote the field of automorphic functions with respect to $\varGamma_N$. Put

$$\mathfrak{F}_N = Q(f_i(z), \; g_j(N^{-1}\beta, z) \mid 1 \le i \le m, \; 1 \le j \le M, \; \beta \in \mathfrak{o}),$$

where the f_i and the g_j are the functions determined by (3) of §3.1 and (4) of §3.2. We have seen that $C\mathfrak{F}_N = \Re_N$; and if y is sufficiently generic, the mapping $f(z) \to f(y)$ gives an isomorphism of $\mathfrak{F}_N$ onto $K_{\mathfrak{a},\mathfrak{y}}$. Let $\mathfrak{L}_N$ be the subfield of $\mathfrak{F}_N$ corresponding to the subfield L_N of $K_{\mathfrak{a},\mathfrak{y}}$. We have then

$$\mathfrak{L}_N(\zeta_N)=\mathfrak{F}_N, \quad C\mathfrak{L}_N=\mathfrak{R}_N,$$

where ζ_N is a primitive N-th root of unity. For every $f\in\mathfrak{F}_N$, define a function f_1 on $\mathfrak{C}_N$ by $f_1(u)=f(y)$. Identifying f_1 with f, $\mathfrak{L}_N$ is regarded as the field of functions on the curve $\mathfrak{C}_N$ defined over Q; and then $\mathfrak{R}_N$ is identified with the field of functions on $\mathfrak{C}_N$, the universal domain being C.

Let α be an element of $\mathfrak{o}$ such that $N(\alpha)>0$; suppose that α is prime to N. Let the notations be as in Propositions 4.3 and 4.4. Consider the coordinates of the points $c(\mathscr{F}(y))$ and $h_y(\theta(\beta)t)$ for $t=\Lambda(N^{-1}e(y),y)$. We see easily that

$$f_j(y)^{\sigma_\nu}=f_j(\alpha_\nu[y]), \quad g_j(N^{-1}\beta,y)^{\sigma_\nu}=g_j(N^{-1}\beta\alpha_\nu{}',\alpha_\nu[y]),$$

where the α_ν are representatives for $\Gamma_N\backslash\Gamma_N\alpha\Gamma_N$; namely we have

$$\Gamma_N\alpha\Gamma_N=\bigcup_{\nu=1}^{m}\Gamma_N\alpha_\nu.$$

As $\alpha_\nu'\equiv\alpha'$ mod. $N\mathfrak{o}$, we have $g_j(N^{-1}\beta\alpha_\nu',\alpha_\nu[y])=g_j(N^{-1}\beta\alpha',\alpha_\nu[y])$. By Theorem 2 of §4.1, there exists an automorphism ρ of $\mathfrak{F}_N$ over $\mathfrak{F}_1$ defined by $g_j(N^{-1}\beta,z)^\rho=g_j(N^{-1}\beta\alpha',z)$. Then, for every $f\in\mathfrak{F}_N$,

$$(1)\qquad\qquad f(y)^{\sigma_\nu}=f^\rho(\alpha_\nu[y]).$$

Now define an isomorphism σ_ν of $\mathfrak{F}_N$ by

$$f^{\sigma_\nu}(z)=f^\rho(\alpha_\nu[z]).$$

Then, (1) shows, for every $f\in\mathfrak{F}_N$,

$$(2)\qquad\qquad f^{\sigma_\nu}(y)=f(y)^{\sigma_\nu}.$$

Now fix an isomorphism of $\mathfrak{o}/N\mathfrak{o}$ onto $M_2(Z/NZ)$ and define with respect to this isomorphism the set $\Delta_a{}^*$ of §1.4 and the field L_N of §4.3. Then, if $\alpha\in\Delta_a{}^*$, the automorphism ρ is the identity on $\mathfrak{L}_N$; hence, for every $f\in\mathfrak{L}_N$,

$$f^{\sigma_\nu}(z)=f(\alpha_\nu[z]).$$

If we denote by the prime the derivation with respect to z, we obtain

$$(f^{\sigma_\nu})'(z)=f'(\alpha_\nu[z])j(\alpha_\nu,z)^{-2}N(\alpha).$$

New let gdf be a differential form on $\mathfrak{C}_N$ of the first kind, f and g being elements of $\mathfrak{R}_N$; $g(z)f'(z)$ is then a cusp-form of degree 2 with respect to Γ_N, namely, $f(z)g'(z)\in S_2(\Gamma_N)$. Conversely, every element of $S_2(\Gamma_N)$ is obtained in this manner. If f and g are contained in $\mathfrak{L}_N$, we have

$$(3)\qquad gf'\mid(\Gamma_N\alpha\Gamma_N)_2=N(\alpha)\sum_\nu g(\alpha_\nu[z])f'(\alpha_\nu[z])j(\alpha_\nu,z)^{-2}=\sum_\nu g^{\sigma_\nu}(f^{\sigma_\nu})'.$$

Let $\mathscr{D}_0(\mathfrak{C}_N)$ and $\mathscr{D}_0(J_N)$ denote the sets of differential forms of the first kind, of degree 1, on $\mathfrak{C}_N$ and J_N, respectively. Then, φ being a canonical mapping of $\mathfrak{C}_N$ onto J_N, $\omega\to\omega\circ\varphi$ gives an isomorphism of $\mathscr{D}_0(J_N)$ onto $\mathscr{D}_0(\mathfrak{C}_N)$. Put $\omega\circ\varphi=gdf$. Then, by (2), (3) and [25, §2.9, Proposition 9], we observe that

$fg' \mid (\Gamma_N \alpha \Gamma_N)_2$ corresponds to $\omega \circ \xi \circ \varphi$, where ξ denotes the endomorphism of J_N determined by X_q for $q = \mathfrak{o}\alpha$. Therefore, if we denote by $M^d(\xi)$ the representation of $\xi \in \mathcal{A}(J_N)$ in $\mathcal{D}_0(J_N)$, we have, for a suitable choice of bases,

$$(4) \qquad M^d(\xi) = \mathfrak{T}_2(\Gamma_N \alpha \Gamma_N).$$

Let γ be an element of Γ such that $\gamma \equiv \begin{pmatrix} p^{-1} & 0 \\ 0 & p \end{pmatrix}$ mod. $\mathfrak{a}$. Let τ_p be the automorphism of $\mathfrak{F}_N$ over $\mathfrak{F}_1$ defined by

$$g_j(N^{-1}\beta, z)^{\tau_p} = g_j(N^{-1}\beta p, z).$$

By Proposition 3.2 and by the definition of L_N, we see easily

$$f^{\tau_p}(z) = f(\gamma[z])$$

for every $f \in \mathfrak{L}_N$. It follows that $\Gamma_N \gamma \Gamma_N$ corresponds to Y_p defined in §4.3; and hence

$$(5) \qquad M^d(\eta_p) = \mathfrak{T}_2(\Gamma_N \gamma \Gamma_N) = R_2(p ; \mathfrak{a}),$$

notations being as in §1.5.

6.2. The zeta-function of $\mathfrak{C}_N$. Let p be a prime number satisfying P9) for $\mathfrak{C} = \mathfrak{C}_N$. Denote by $\zeta(s, \mathfrak{C}_N, p)$ the zeta-function of $p(\mathfrak{C}_N)$ over Z/pZ; we have

$$\zeta(s, \mathfrak{C}_N, p) = \frac{\det[1 - M_l(\pi_p)p^{-s}]}{(1 - p^{-s})(1 - p^{1-s})},$$

where π_p is the p-th power endomorphism of $p(J_N)$ and M_l is an l-adic representation of $\mathcal{A}_0(\tilde{J}_N)$. Now the zeta-function of the algebraic curve $\mathfrak{C}_N$ over $\boldsymbol{Q}$ is defined by

$$\zeta(s, \mathfrak{C}_N) = \prod' \zeta(s, \mathfrak{C}_N, p),$$

where the product is extended over all the prime numbers p satisfying P9) for $\mathfrak{C} = \mathfrak{C}_N$. U being an indeterminate, the relations (24) and (25) of §5.6 imply

$$(6) \qquad 1 - M_l(\tilde{\xi}_p)U + M_l(\tilde{\eta}_p)pU^2 = [1 - M_l(\pi_p)U][1 - M_l(\tilde{\zeta}^{-1}\pi_p'\tilde{\zeta})U].$$

By the same argument as in [22, §5], we obtain

$$\det[1 - M_l(\pi_p)p^{-s}] = \det[1 - M^d(\xi_p) + M^d(\eta_p)p^{1-2s}].$$

By (4) and (5), the right hand side is equal to

$$\det[1 - \mathfrak{T}_2(p ; \mathfrak{a})p^{-s} + R_2(p ; \mathfrak{a})p^{1-2s}].$$

Therefore we obtain the following result.

MAIN THEOREM. *Let Φ be an indefinite quaternion algebra over $\boldsymbol{Q}$, and $\mathfrak{o}$ a maximal order in Φ. Let N be a positive integer which is prime to the discriminant of Φ, and Γ_N be the group of units γ of $\mathfrak{o}$, with positive reduced norm, satisfying $\gamma \equiv 1$ mod. $N\mathfrak{o}$. Regarding Γ_N as a Fuchsian group on the upper half plane, we can find an algebraic curve $\mathfrak{C}_N$ defined over $\boldsymbol{Q}$, such that: the field of* *functions on $\mathfrak{S}_N$ is the field of automorphic functions with respect to Γ_N; and the zeta-function of $\mathfrak{S}_N$ over $\mathbf{Q}$ is written in the form*

$$\zeta(s, \mathfrak{S}_N) = f(s)\zeta(s)\zeta(s-1)D(s)^{-1},$$

$$D(s) = \det\left[\sum_{(n,N)=1} \mathfrak{X}_2(n\,;\,N_0)n^{-s}\right],$$

where $f(s)$ is a product of rational functions of p^{-s} for a finite number of primes p, $\zeta(s)$ is Riemann's zeta-function, and $\mathfrak{X}_2(n\,;\,N_0)$ is a representation of a certain algebraic correspondence by cusp-forms of degree 2 with respect to Γ_N, given in § 1.5.

By Theorem 1, and by Hecke's theory in the case $\varPhi = M_2(\mathbf{Q})$, we can conclude:

COROLLARY. *The zeta-function of $\mathfrak{S}_N$ over $\mathbf{Q}$ is meromorphic on the whole s-plane and satisfies a functional equation.*

Our theorem is a generalization of the previous results of [6, 22], obtained in the case where $\varPhi = M_2(\mathbf{Q})$.

The relation (24) together with the argument of [22, no. 20] gives also in a general case the following result.[2]

THEOREM 6. *Notations and assumptions being as in Main Theorem, the absolute values of the characteristic roots of $\mathfrak{X}_2(p\,;\,N_0)$ do not exceed $2\sqrt{p}$ for almost all prime numbers p.*

In the case $\varPhi = M_2(\mathbf{Q})$, a more precise result is obtained by Igusa [13].

6.3. Concluding remarks. I) We begin with an interpretation of the congruence-relations. Let l be a prime number and $\mathfrak{g}_l$ the set of points on J_N whose orders are powers of l; and let k_l be the extension of $\mathbf{Q}$ generated by the coordinates of all $t \in \mathfrak{g}_l$. We denote by G_l the Galois group of k_l over $\mathbf{Q}$. Then, as every element of G_l induces an automorphism of $\mathfrak{g}_l$, we obtain a representation $\mathfrak{M}_l$ of G_l by matrices whose coefficients are l-adic integers. Let p be a prime number different from l and $\mathfrak{p}$ its extension in k_l; let $\sigma_\mathfrak{p}$ be a Frobenius substitution for $\mathfrak{p}/p$. Then, as is shown in [25, 27], if J_N is without defect for p, then p is unramified in k_l; and we obtain, for a suitable choice of l-adic coordinates,

$$\mathfrak{M}_l(\sigma_\mathfrak{p}) = M_l(\pi_p),$$

where π_p is the p-th power endomorphism of $p(J_N)$. Hence $D(s) = \prod \det[1 - M_l(\pi_p)p^{-s}]^{-1}$ gives an analogue of Artin's L-function for the infinite extension k_l of $\mathbf{Q}$. By (6) of § 6.2, $\operatorname{tr} M_l(\pi_p)^n$ is easily obtained from the trace of certain modular correspondences. Therefore, if we know the trace of $\mathfrak{X}_2(\Gamma_N \alpha \Gamma_N)$, this determines the characteristic polynomial of $\mathfrak{M}_l(\sigma_\mathfrak{p})$; and the former is obtained by the trace-formula of Eichler and Selberg. Thus the congruence-relations,

---

2) In [22, no. 20], the relation (25) was needed. But this is not necessary; only the relation (24) proves the inequality, in view of the relation $\eta_p'\eta_p = 1$.

or the above main theorem, may be regarded as a reciprocity-law for the extensions k_l over Q, which are not necessarily abelian, even may be non-solvable.

II) There are many systems of polarized abelian varieties whose moduli are given by the automorphic functions with respect to some discontinuous groups. Our method is certainly applicable to those systems. Even in the case of dimension one, we have more Fuchsian groups, defined arithmetically, than treated in the present paper. In fact, take an algebraic number field k whose conjugates are all real, and take a quaternion algebra $\mathfrak{A}$ over k which is un-ramified at exactly one infinite prime spot of k. The unit-groups obtained from $\mathfrak{A}$ in the same way as for $\mathcal{O}$, yield also Fuchsian groups; and we can at-tach to them certain analytic systems of abelian varieties. Some new difficulties may arise in treating them; it is sure, however, that we can investigate in detail the arithmetic of the curves uniformized by the automorphic functions with respect to those groups, by using modular correspondences.

The theory of modular correspondences, with their congruence-relations, is the only tool, which we know at present, to calculate the zeta-function of algebraic curves in a certain degree of generality. This connection does not seem accidental, though one may find another approach to it. Therefore, it is important to determine the algebraic curves which are uniformized by automorphic functions "defined arithmetically" This would be a difficult problem; but a recent work of Selberg [19] and Weil [33] suggest that one may anticipate something in this direction.

III) The zeta-function of the curve $\mathfrak{C}_N$ is concerned only with the cusp-forms of degree 2. Now, in [24], it was shown that, for each even degree κ, we can define an abelian variety by means of the "periods" of certain integrals attached to cusp-forms of degree κ. This abelian variety admit doubtlessly an algebro-geometric interpretation; and what arithmetic does it dominate? We can expect from this not only a solution of Ramanujan's conjecture but also something more interesting; and needless to say, a similar problem in the case of automorphic forms with more than one variables is no less important.

University of Tokyo.

References

[1] W. L. Chow, Abelian varieties over function-fields, Trans. Amer. Math. Soc., 78 (1955) 253-275.

[2] M. Deuring, Algebren, Berlin, 1935.

[3] M. Deuring, Die Zetafunktion einer algebraischen Kurve vom Geschlechte Eins, I, II, III, IV, Nachr. Akad. Wiss. Göttingen, (1953), 85-94; (1955), 13-42; (1956),

37-76; (1957), 55-80.

[4] M. Eichler, Allgemeine Kongruenzklasseneinteilungen der Ideale einfacher Algebren über algebraischen Zahlkörpern und ihre L-Reihen, J. Reine Angew. Math., 179 (1938), 227-251.

[5] M. Eichler, Über die Idealklassenzahl hyperkomplexer Systeme, Math. Z., 43 (1938), 481-494.

[6] M. Eichler, Quaternäre quadratische Formen und die Riemannsche Vermutung für die Kongruenzzetafunktion, Arch. Math., 5 (1954), 355-366.

[7] M. Eichler, Modular correspondences and their representations. J. Indian Math. Soc., 20 (1956), 163-206.

[8] M. Eichler, Eine Verallgemeinerung der Abelschen Integrale, Math. Z., 67 (1957), 267-298.

[9] G. Fujisaki, On the zeta-function of the simple algebra over the field of rational numbers, J. Fac. Sci. Univ. Tokyo. Sec. I, vol. VII, Part 5 (1958), 567-604.

[10] R. Godement, Les fonctions ζ des algèbres simples II, Séminaire Bourbaki (février 1959, 176), 1-20.

[11] E. Hecke, Über die Bestimmung Dirichletscher Reihen durch ihre Funktionalgleichung, Math. Ann., 112 (1936), 664-699.

[12] E. Hecke, Über Modulfunktionen und die Dirichletschen Reihen mit Eulerscher Produktentwicklung, I, II, Math. Ann., 114 (1937), 1-28, 316-351.

[13] J. Igusa, Kroneckerian model of fields of elliptic modular functions, Amer. J. Math., 81 (1959), 561-577.

[14] S. Koizumi and G. Shimura, On specializations of abelian varieties, Sci. Papers. Coll. Gen. Ed. Univ. Tokyo, 9 (1959), 187-211.

[15] T. Matsusaka, Polarized varieties, fields of moduli and generalized Kummer varieties of polarized abelian varieties, Amer. J. Math., 80 (1958), 45-82.

[16] M. Nishi, Some results on abelian varieties, Nat. Sci. Rep., Ochanomizu Univ., 9 (1958), 1-12.

[17] H. Petersson, Konstruktion der sämtilichen Lösungen einer Riemannschen Funktionalgleichung durch Dirichlet-Reihen mit Eulerscher Produktentwickelung, I, II, III, Math. Ann., 116 (1939), 401-412; 117 (1940), 39-64, 277-300.

[18] A. Selberg, Harmonic analysis and discontinuous groups in weakly symmetric Riemannian spaces with applications to Dirichlet series, J. Indian Math. Soc., 20 (1956), 47-87.

[19] A. Selberg, On discontinuous groups in higher-dimensional symmetric spaces, Contributions to Function Theory, Tata Institute of Fundamental Research, Bombay, 1960, 147-164.

[20] J.-P. Serre, Groupes algébriques et corps de classes, Hermann, Paris, 1959.

[21] G. Shimura, Reduction of algebraic varieties with respect to a discrete valuation of the basic field, Amer. J. Math., 77 (1955), 134-176.

[22] G. Shimura, Correspondances modulaires et les fonctions ζ de courbes algébriques, J. Math. Soc. Japan, 10 (1958), 1-28.

[23]=[AF] G. Shimura, On the theory of automorphic functions, Ann. Math., 70 (1959), 101-144.

[24] G. Shimura, Sur les intégrales attachées aux formes automorphes, J. Math. Soc. Japan, 11 (1959), 291-311.

[25] G. Shimura and Y. Taniyama, Complex multiplication of abelian varieties and its applications to number theory, Publ. Math. Soc. Japan, No. 6, 1961.

[26] T. Tamagawa, On ζ-functions of a division algebra, to appear in Amer. J. Math.

[27] Y. Taniyama, L-functions of number fields and zeta functions of abelian varieties, J. Math. Soc. Japan, 9 (1957), 330-366.

[28] A. Weil, Sur les courbes algébriques et les variétés qui s'en déduisent Hermann, Paris 1948.

[29] A. Weil, Variétés abéliennes et courbes algébriques, Hermann, Paris, 1948.

[30] A. Weil, Jacobi sums as "Grössencharaktere", Trans. Amer. Math. Soc., 73 (1952), 487-495.

[31] A. Weil, On the theory of complex multiplication, Proc. Int. Symp. Alg. Nb. Th., Tokyo-Nikko, 1955, 9-22.

[32] A. Weil, The field of definition of a variety, Amer. J. Math., 78 (1956), 509-524.

[33] A. Weil, On discrete subgroups of Lie groups, Ann. Math. 72 (1960), 369-384,

[34] A. Weil, Adèles and algebraic groups, Lecture note, The Institute for Advanced Study, Princeton, 1961.

Added in proof. Theorem 1 of §1 lacks the explicit form of functional equation in the case $\mathfrak{a} \neq \mathfrak{o}$; it is given only by (27), which includes terms of the form $\mathrm{Re}\,(i^k f(z))$. It is not difficult to make it into the form containing only holomorphic functions; then a more explicit form can be obtained. Furthermore, it is better to deal rather with the representations of $\mathfrak{G}_a$ than with those of $\mathfrak{S}_a$. Thus, in this respect, the view-point of Godement [10] will be a more appropriate one. The author would like to give a treatment for this in a more general case on some occasion. Recently, in the case $\Phi = M_2(Q)$, the relation between Hecke's Euler-product and Artin's L-function for the extension $\mathfrak{F}_N/\mathfrak{F}_1(\zeta_N)$ is investigated in the paper: Rangachari, Modulare Korrespondenzen und L-Reihen, J. Reine Angew. Math., 205 (1961), 119-155. A similar consideration seems also meaningful in the case of division quaternion algebras.

On Dirichlet series and abelian varieties attached to automorphic forms

Annals of Mathematics, 76 (1962), 237-294

(Received November 20, 1961)

Introduction

Since Hecke [13] had given a general theory of constructing Dirichlet series with Euler-product and functional equation out of elliptic modular forms of any level, several authors considered its generalization for other types of automorphic forms. In the case of the Hilbert modular group of level one, Herrmann [14] succeeded in this problem; he has shown the necessity of considering not only the product of the upper half-planes but also the domain $\mathfrak{F}_r$ consisting of the points $(z_1, \cdots, z_r)$ of the r-dimensional complex vector space C^r such that $\mathrm{Im}(z_1) \neq 0, \cdots, \mathrm{Im}(z_r) \neq 0$, and h distinct discontinuous groups commensurable to each other, h being the class number of the totally real number field in the problem. On the other hand, the unit-group of an order in an indefinite quaternion algebra over the rational number field Q yields a fuchsian group. In this case, Eichler [6] defined Hecke's operators as representations of algebraic correspondences, called modular correspondences, and proved a formula for the trace of the operators. The trace-formula was proved also by Selberg [22] in a more general formulation. Recently, Godement [9] has given a theory of zeta-functions attached to division algebras; namely, he has shown the possibility of applying the adele-idele method of Iwasawa-Tate [15, 28] to automorphic functions and forms with respect to the unit-group of a division algebra over Q. The case of non-holomorphic automorphic functions of this type has been investigated by Tamagawa [27]. In [25] I have treated cusp-forms with respect to the unit-group of an indefinite quaternion algebra over Q.

Now, the purpose of Part I of the present paper is to develop an analogous theory for automorphic forms on the domain $\mathfrak{F}_r$ mentioned above with respect to the unit-groups of an indefinite quaternion algebra. Let F be a totally real algebraic number field of degree t, and D be a quaternion algebra over F. Denote by r the number of infinite prime spots of F unramified in D, and suppose that $r > 0$. Let $D_{\infty 1}$ and $D_{\infty 2}$ denote the product of r copies of the total matric algebra $M_2(R)$ over the real number field R and the product of $t - r$ copies of the division ring K of real quaternions, respectively. Then, $D \otimes_Q R$ is isomorphic to $D_{\infty 1} \times D_{\infty 2}$. Let $\mathfrak{o}$ be a maximal order in D and Γ be the group of units in $\mathfrak{o}$. Let γ_1 and γ_2

be the projections of an element γ of Γ on $D_{\infty 1}$ and $D_{\infty 2}$, respectively. Put

$$\gamma_1 = \left(\begin{pmatrix} a_1 & b_1 \\ c_1 & d_1 \end{pmatrix}, \cdots, \begin{pmatrix} a_r & b_r \\ c_r & d_r \end{pmatrix} \right)$$

and define the operation of γ on the domain $\mathfrak{F}_r$ by

$$\gamma(z_1, \cdots, z_r) = \left(\frac{a_1 z_1 + b_1}{c_1 z_1 + d_1}, \cdots, \frac{a_r z_r + b_r}{c_r z_r + d_r} \right).$$

Then Γ gives a discontinuous group of transformations on $\mathfrak{F}_r$; and $\Gamma \backslash \mathfrak{F}_r$ is of finite measure; $\Gamma \backslash \mathfrak{F}_r$ is compact if and only if D is a division algebra. If D is the total matric algebra $M_2(F)$ over F, then Γ is commensurable with the Hilbert modular group over F. We take natural numbers $n_1, \cdots,$ n_r and adopt $\prod_{i=1}^{r}(c_i z_i + d_i)^{n_i}$ as an automorphic factor. Besides this, we consider two representations ρ and Ψ of Γ by complex matrices; ρ is a representation of the multiplicative group of regular elements in $\mathfrak{o}/\mathfrak{a}$, where $\mathfrak{a}$ is an integral two-sided $\mathfrak{o}$-ideal, and Ψ is a tensor representation of the multiplicative group of regular elements in $D_{\infty 2}$. The automorphic forms which we shall consider are holomorphic vector-functions $f(z)$ defined on $\mathfrak{F}_r$ satisfying

$$(0) \qquad f(\gamma(z)) = L(\gamma)\rho(\gamma) \otimes \Psi(\gamma_2) \prod_{i=1}^{r}(c_i z_i + d_i)^{n_i} f(z)$$

for every $\gamma \in \Gamma$, where $L(\gamma)$ is a suitable scalar factor (cf. § 3.2). As in the case of Herrmann, we have to consider the non-connected domain $\mathfrak{F}_r$ with 2^r connected components, and h distinct unit groups $\Gamma_1, \cdots, \Gamma_h$ in D, h being the class-number of D. Denoting by S_λ the set of functions satisfying (0) for $\gamma \in \Gamma_\lambda$, we define the Hecke operators in this case as linear transformations in $S = S_1 \times \cdots \times S_h$. Then we get a Dirichlet series whose coefficients are linear transformations in S. Our principal aim in Part I is to prove that this Dirichlet series has an Euler-product and satisfies a functional equation (Theorem 1 of § 4.10). The functional equation is proved by virtue of the Poisson summation formula on the adele space of D, as in [8, 9, 15, 25, 27, 28, 31]. We have considered only holomorphic forms; but our method is certainly applicable to non-holomorphic automorphic forms or functions of any level, which are eigen-functions of invariant differential operators. The only difference will occur in the so-called Γ-factor of the functional equation. It would be worthwhile mentioning that the case of a matric algebra, namely, the case of the Hilbert modular group, can be treated by the same method, though one has to restrict oneself to cusp-forms. A factor in the functional equation is related to certain non-abelian gaussian sums (§4.12), further generalizations of which seem very interesting. It is also desirable to clarify the meaning of the formulas (25), (26), (28') in a wider context.

Let us come to the subject of Part II, which is concerned only with automorphic forms of one variable. In a previous paper [24], for every "arithmetical" fuchsian group Δ, and for each even integer $m > 0$, an abelian variety $A_m(\Delta)$ has been constructed from the periods of certain integrals attached to cusp-forms of degree m with respect to Δ. Here a fuchsian group Δ is called arithmetical if the representation of SL(2, R) by symmetric tensors of degree $m - 2$ induces on Δ a representation with integral coefficients. This is not satisfied, however, by the fuchsian groups Γ obtained from D as above (taking $r = 1$), if $F \neq Q$. The purpose of Part II is to show that we can also construct abelian varieties analogous to $A_m(\Delta)$ for the groups Γ of this kind. For every f satisfying (0), we consider the linear differential form ω given by

$$\omega = f(z) \otimes \begin{pmatrix} z^n \\ z^{n-1} \\ \vdots \\ \vdots \\ 1 \end{pmatrix} dz \ ,$$

where $n = n_1 - 2$. Then, for a suitable representation χ of Γ, we have

$$\omega \circ \gamma = \chi(\gamma)\omega$$

for $\gamma \in \Gamma$. The coefficients of χ belong to a certain algebraic number field K. Considering the periods of the integral of $\mathrm{Re}(\omega)$, we obtain an isomorphism of the vector-space of cusp-forms onto the cohomology group of Γ with respect to χ (Theorem 2 of § 7.2). If we take all the conjugates of χ over Q into account, we can define cohomology groups with integral coefficients. Then, by the same argument as in [24], we obtain an abelian variety, on which Hecke's operators act as endomorphisms (Theorem 3 of § 8).

Thus the theories of Part I and Part II seem to point in rather different directions. However, we have shown, in [25], that the Dirichlet series in the case $F = Q$ and $n_1 = 2$ give L-functions of infinite algebraic extensions of Q generated by the points of finite order on the abelian varieties obtained from cusp-forms in this case. Though the case $n_1 > 2$ still presents difficulties, it may be expected that there is a deep relation between the Dirichlet series of Part I and the abelian varieties of Part II.

NOTATION. We denote by Z, Q, R, C, and K, respectively, the ring of rational integers, the rational number field, the real number field, the complex number field, and the division ring of real quaternions. For a prime number p, Q_p and Z_p denote respectively the field of p-adic numbers and the ring of p-adic integers. When A is a ring with an identity element, we denote by $M_n(A)$ the ring of matrices of degree n with entries in A, and by A^* the multiplicative group of regular elements in A. Then we put, as usual, $\mathrm{GL}(n, A) = M_n(A)^*$; if A is commutative, $\mathrm{SL}(n, A)$ denotes the subgroup of $\mathrm{GL}(n, A)$ consisting of the elements of determinant 1. For every $x \in R^*$, we put $\mathrm{sgn}(x) = x/|x|$. $U \otimes V$ means the tensor-product of U and V, where U and V may be vectors, vector spaces, or matrices.

PART I. DIRICHLET SERIES ATTACHED TO AUTOMORPHIC FORMS

1. Arithmetic of indefinite quaternion algebras

1.1. *Indefinite quaternion algebras.* Let F be a totally real algebraic number field of finite degree. Put $[F:Q] = t$. We treat exclusively a quaternion algebra D over F; i.e. a central simple algebra over F such that $[D:F] = 4$. Let $\mathfrak{g}$ be the ring of algebraic integers in F. For every prime ideal $\mathfrak{p}$ of F, we denote by $\mathfrak{g}_\mathfrak{p}$ and $F_\mathfrak{p}$ the $\mathfrak{p}$-completions of $\mathfrak{g}$ and F, respectively, and put $D_\mathfrak{p} = D \otimes_F F_\mathfrak{p}$. As F is totally real, F has t infinite prime spots $\mathfrak{p}_{\infty,1}, \cdots, \mathfrak{p}_{\infty,t}$. We denote by $D^{(\nu)}$ the completion of D by $\mathfrak{p}_{\infty,\nu}$; we obtain then an isomorphism:

$$(1) \qquad D \otimes_Q R \cong D^{(1)} \times \cdots \times D^{(t)} .$$

As $D^{(\nu)}$ is a central simple algebra over R such that $[D^{(\nu)}:R] = 4$, we may assume, reordering suitably, $D^{(\nu)} = M_2(R)$ for $1 \leq \nu \leq r$ and $D^{(\nu)} = K$ for $\nu > r$. In the present investigation, we always assume D to be indefinite, namely, $r > 0$; the case $t = r$ is admitted; so the matric algebra $M_2(F)$ is included as a special case. Identifying both sides of (1), we denote by $a^{(\nu)}$ the projection of an element a of D onto the νth factor; so, $a^{(\nu)} \in M_2(R)$ for $1 \leq \nu \leq r$, and $a^{(\nu)} \in K$ for $\nu > r$. Restricting a to F, the mappings $a \to a^{(\nu)}$, for $1 \leq \nu \leq t$, give all the isomorphisms of F into R.

The algebra D (resp. $D_\mathfrak{p}$, $D^{(\nu)}$) has an involution $a \to a'$, which is uniquely determined by the property that $(X - a)(X - a')$ is the principal polynomial of a over F (resp. $F_\mathfrak{p}$, R). For every $a \in D$(resp. $D_\mathfrak{p}$, $D^{(\nu)}$), we put

$$N(a) = aa', \qquad \mathrm{tr}(a) = a + a' .$$

When $1 \leq \nu \leq r$ and $a \in D^{(\nu)} = M_2(R)$, we have $N(a) = \det(a)$; and $\mathrm{tr}(a)$ coincides with the usual trace of the matrix. For every $a \in D$, we put

$$N_{D/Q}(a) = N_{F/Q}\big(N(a)\big) , \qquad \mathrm{tr}_{D/Q}(a) = \mathrm{tr}_{F/Q}\big(\mathrm{tr}(a)\big) .$$

1.2. *Ideal-theory in D.*[1] We understand by a $\mathfrak{g}$-*lattice* in D a finitely generated $\mathfrak{g}$-module $\mathfrak{m}$ in D such that $F\mathfrak{m} = D$. We call a subring $\mathfrak{o}$ of D

---

[1] For the ideal-theory in an algebra, we refer to Deuring [3], Jacobson [16].

287

an order in D if it is a g-lattice in D containing g. Let $\mathfrak{a}$ be a g-lattice in D; let $\mathfrak{o}_r$ (resp. $\mathfrak{o}_l$) be the set of all elements α of D such that $\mathfrak{a}\alpha \subset \mathfrak{a}$ (resp. $\alpha\mathfrak{a} \subset \mathfrak{a}$). Then $\mathfrak{o}_r$ and $\mathfrak{o}_l$ are orders in D. We call $\mathfrak{o}_r$ (resp. $\mathfrak{o}_l$) the right (resp. left) order of $\mathfrak{a}$. An order $\mathfrak{o}$ in D is called *maximal* if there is no order containing $\mathfrak{o}$ other than itself. Every order in D is contained in a maximal one. Let $\mathfrak{a}$ be a g-lattice in D and $\mathfrak{o}$ an order in D. We call $\mathfrak{a}$ a *right* (resp. *left*) $\mathfrak{o}$-*ideal* if we have $\mathfrak{ao} \subset \mathfrak{a}$ (resp. $\mathfrak{oa} \subset \mathfrak{a}$); $\mathfrak{a}$ is called a *two-sided $\mathfrak{o}$-ideal* if $\mathfrak{a}$ is both a right $\mathfrak{o}$-ideal and a left $\mathfrak{o}$-ideal. $\mathfrak{o}_r$ and $\mathfrak{o}_l$ being the right and left orders of a g-lattice $\mathfrak{a}$, if we have $\mathfrak{a} \subset \mathfrak{o}_r$, then $\mathfrak{a} \subset \mathfrak{o}_l$, and conversely; we call such $\mathfrak{a}$ *integral*. A g-lattice $\mathfrak{a}$ is called *normal* if both its right and left orders are maximal. It is known that if one of the left and right orders of $\mathfrak{a}$ is maximal, then $\mathfrak{a}$ is normal. Therefore, if $\mathfrak{o}$ is a maximal order, every right, left or two-sided $\mathfrak{o}$-ideal is normal. $\mathfrak{a}$ and $\mathfrak{b}$ being normal g-lattices in D, the product $\mathfrak{ab}$ is called *proper* if the right order of $\mathfrak{a}$ coincides with the left order of $\mathfrak{b}$. The normal g-lattices in D form a groupoid with respect to the operation of proper product. An integral two-sided $\mathfrak{o}$-ideal $\mathfrak{p}$ is called *prime* if the residue-class-ring $\mathfrak{o}/\mathfrak{p}$ is simple. If $\mathfrak{o}$ is a maximal order in D, the two-sided $\mathfrak{o}$-ideals form a commutative group, which is a direct product of the infinite cyclic groups generated by the prime ideals.

We can define a mapping $\mathfrak{a} \to N(\mathfrak{a})$ which assigns to every normal g-lattice $\mathfrak{a}$ in D an ideal $N(\mathfrak{a})$ in F and has the following three properties.

(i) $N(\mathfrak{ab}) = N(\mathfrak{a})N(\mathfrak{b})$ if $\mathfrak{ab}$ is a proper product.

(ii) If $\mathfrak{a}$ is integral and $\mathfrak{o}$ is the right or left order of $\mathfrak{a}$, and if $\mathfrak{o}/\mathfrak{a}$, as g-module, is isomorphic to $\mathfrak{g}/\mathfrak{b}_1 \oplus \cdots \oplus \mathfrak{g}/\mathfrak{b}_s$ for some ideals $\mathfrak{b}_i$ of g, then $N(\mathfrak{a})^2 = \mathfrak{b}_1 \cdots \mathfrak{b}_s$.

(iii) For every regular element α of D and for any maximal order $\mathfrak{o}$, we have $N(\alpha\mathfrak{o}) = N(\mathfrak{o}\alpha) = (N(\alpha))$.

We put, for every g-lattice $\mathfrak{a}$ in D,

$$N_{D/Q}(\mathfrak{a}) = N_{F/Q}(N(\mathfrak{a})) .$$

If $\mathfrak{a}$ is integral and $\mathfrak{o}$ is the right or left order of $\mathfrak{a}$, then $N_{D/Q}(\mathfrak{a})^2$ is the number of elements in $\mathfrak{o}/\mathfrak{a}$.

Let $\mathfrak{o}$ be a maximal order in D. We put

$$\mathfrak{d}(\mathfrak{o}/\mathfrak{g}) = \{\alpha \in D \,|\, \mathrm{tr}(\alpha\mathfrak{o}) \subset \mathfrak{g}\}^{-1} ,$$

$$\mathfrak{d}(\mathfrak{o}/Z) = \{\alpha \in D \,|\, \mathrm{tr}_{D/Q}(\alpha\mathfrak{o}) \subset Z\}^{-1} ,$$

$$\mathfrak{d}(F/Q) = \mathfrak{d}(\mathfrak{g}/Z) = \{\xi \in F \,|\, \mathrm{tr}_{F/Q}(\xi\mathfrak{g}) \subset Z\}^{-1} .$$

Then $\mathfrak{d}(\mathfrak{o}/\mathfrak{g})$ and $\mathfrak{d}(\mathfrak{o}/Z)$ are integral two-sided $\mathfrak{o}$-ideals, and

$$\mathfrak{d}(\mathfrak{o}/Z) = \mathfrak{d}(\mathfrak{o}/\mathfrak{g})\mathfrak{d}(\mathfrak{g}/Z) .$$

We put

$$N\big(\mathfrak{d}(\mathfrak{o}/\mathfrak{g})\big) = d(D/F), \; N\big(\mathfrak{d}(\mathfrak{o}/Z)\big) = d(D/Q) .$$

These are integral ideals in F and independent of the choice of $\mathfrak{o}$. Every prime ideal $\mathfrak{P}$ of $\mathfrak{o}$ divides exactly one prime ideal $\mathfrak{p}$ of F; and we have

$$\mathfrak{p}\mathfrak{o} = \mathfrak{P} \quad \text{if } \mathfrak{p}\nmid d(D/F) ,$$
$$\mathfrak{p}\mathfrak{o} = \mathfrak{P}^2 \quad \text{if } \mathfrak{p}\mid d(D/F) .$$

We note that $d(D/F)$ is a square-free ideal of F. For a given maximal order $\mathfrak{o}$ in D, two right (resp. left) $\mathfrak{o}$-ideals $\mathfrak{a}$ and $\mathfrak{b}$ are called *equivalent* if there exists a regular element δ of D such that $\delta\mathfrak{a} = \mathfrak{b}$ (resp. $\mathfrak{a}\delta = \mathfrak{b}$). We understand, by an *equivalence-class of right* (resp. *left*) *$\mathfrak{o}$-ideals*, a maximal set of mutually equivalent right (resp. left) $\mathfrak{o}$-ideals. The number of equivalence-classes of right $\mathfrak{o}$-ideals is finite and equal to the number of equivalence-classes of left $\mathfrak{o}$-ideals; moreover, this number is independent of the choice of $\mathfrak{o}$; so we call it the *class-number of D*.

Here we quote some results due to Eichler [4, 5] which we specialize in the following form.

LEMMA 1.1. *Let $\mathfrak{o}$ be a maximal order in D. Then the mapping $\mathfrak{a} \to N(\mathfrak{a})$ gives a one-to-one correspondence between the equivalence-classes of right* (resp. *left*) *$\mathfrak{o}$-ideals and the ideal-classes modulo $\prod_{\nu > r} \mathfrak{p}_{\infty,\nu}$ in F.*

LEMMA 1.2. *Let $\mathfrak{o}$ be a maximal order in D and $\mathfrak{a}$ an integral two-sided $\mathfrak{o}$-ideal. Let b be an element of $\mathfrak{g}$ such that $b^{(\nu)} > 0$ for $\nu > r$ and let α be an element of $\mathfrak{o}$ such that $N(\alpha) \equiv b \bmod^* \mathfrak{a} \cap \mathfrak{g}$. Then there exists an element β of $\mathfrak{o}$ such that $\beta \equiv \alpha \bmod \mathfrak{a}$ and $N(\beta) = b$.*

Here mod* means the multipicative congruence. Taking $\mathfrak{a}$ to be $\mathfrak{o}$, we obtain

LEMMA 1.3. *Let $\mathfrak{o}$ be a maximal order in D and b an element of $\mathfrak{g}$ such that $b^{(\nu)} > 0$ for $\nu > r$. Then there exists an element β of $\mathfrak{o}$ such that $N(\beta) = b$.*

1.3. *The invariants and similarity of normal $\mathfrak{g}$-lattices.* Let $\mathfrak{o}$ be a maximal order in D and $\mathfrak{M}$ a right $\mathfrak{o}$-module.[2] We say that $\mathfrak{M}$ is *bounded* if $\mathfrak{M}$ is finitely generated over $\mathfrak{o}$ and $\mathfrak{M}\alpha = \{0\}$ for some regular element α of $\mathfrak{o}$. Every bounded $\mathfrak{o}$-module $\mathfrak{M}$ is expressed as a direct sum of indecomposable $\mathfrak{o}$-modules $\mathfrak{M}_1, \cdots, \mathfrak{M}_s$. Let $\mathfrak{a}_i$ be the set of elements x of $\mathfrak{o}$ such that $\mathfrak{M}_i x = \{0\}$. It is known that $\mathfrak{a}_i$ is a power $\mathfrak{P}_i^{e_i}$ of a prime ideal $\mathfrak{P}_i$ of $\mathfrak{o}$. Let $\mathfrak{p}_i$ be the prime ideal of F divisible by $\mathfrak{P}_i$. We call then

---

[2] We consider only those modules $\mathfrak{M}$ such that the identity element of $\mathfrak{o}$ gives on $\mathfrak{M}$ the identity mapping. For the theory of bounded $\mathfrak{o}$-modules, we refer to Jacobson [16, Chap. 6].

$\{\mathfrak{p}_1^{a_1}, \cdots, \mathfrak{p}_s^{a_s}\}$ *the set of invariants of* $\mathfrak{M}$. Two bounded right $\mathfrak{o}$-modules are isomorphic if and only if they have the same set of invariants.

Let $\mathfrak{a}$ be a normal $\mathfrak{g}$-lattice in D. We define *the set of invariants of* $\mathfrak{a}$ by the following two properties.

(i) If $\mathfrak{a}$ is integral and $\mathfrak{o}$ is the right order of $\mathfrak{a}$, then the set of invariants of $\mathfrak{a}$ is the set of invariants of the right $\mathfrak{o}$-module $\mathfrak{o}/\mathfrak{a}$.

(ii) Let ξ be a regular element of F. If $\{\mathfrak{p}_1^{a_1}, \cdots, \mathfrak{p}_s^{a_s}\}$ is the set of invariants of $\mathfrak{a}$, and if $(\xi)_i$ denotes the $\mathfrak{p}_i^{\text{th}}$ component of the ideal $\xi\mathfrak{g}$, then $\{(\xi)_1^{a_1}\mathfrak{p}_1^{a_1}, \cdots, (\xi)_s^{a_s}\mathfrak{p}_s^{a_s}\}$, removing the trivial components which are equal to (1), is contained in the set of invariants of $\xi\mathfrak{a}$, where $a_i = 1$ or 2 according as $\mathfrak{p}_i$ divides $d(D/F)$ or not.

We say that two normal $\mathfrak{g}$-lattices in D are *similar* if they have the same set of invariants. Let $\mathfrak{o}$ be a maximal order in D and $\mathfrak{a}_1, \mathfrak{a}_2$ be integral right $\mathfrak{o}$-ideals. Then $\mathfrak{a}_1$ and $\mathfrak{a}_2$ are similar if and only if the right $\mathfrak{o}$-modules $\mathfrak{o}/\mathfrak{a}_1$ and $\mathfrak{o}/\mathfrak{a}_2$ are isomorphic. We can also define the set of invariants of left modules. But for any normal $\mathfrak{g}$-lattice $\mathfrak{a}$ in D, the set of invariants of $\mathfrak{a}$ as right module coincides with the set of invariants of $\mathfrak{a}$ as left module.

Let $\mathfrak{p}$ be a prime ideal of F. For every $\mathfrak{g}$-lattice $\mathfrak{a}$ in D, we put $\mathfrak{a}_\mathfrak{p} = \mathfrak{g}_\mathfrak{p}\mathfrak{a}$; then $\mathfrak{a}_\mathfrak{p}$ is a $\mathfrak{g}_\mathfrak{p}$-lattice in $D_\mathfrak{p}$. If $\mathfrak{o}$ is a maximal order in D, then $\mathfrak{o}_\mathfrak{p}$ is a maximal order in $D_\mathfrak{p}$. For any $\mathfrak{p}$, the right or left $\mathfrak{o}_\mathfrak{p}$-ideals are all principal ideals. When $\mathfrak{p}$ does not divide $d(D/F)$, the algebra $D_\mathfrak{p}$ is isomorphic to $M_2(F_\mathfrak{p})$; and $\mathfrak{o}_\mathfrak{p}$ corresponds to $M_2(\mathfrak{g}_\mathfrak{p})$ for a suitable choice of isomorphism. If $\mathfrak{p}$ divides $d(D/F)$, $D_\mathfrak{p}$ is a division algebra and every $\mathfrak{o}_\mathfrak{p}$-ideal is a power of a maximal ideal.

Now using the $\mathfrak{p}$-completion, the set of invariants of a right $\mathfrak{o}$-ideal $\mathfrak{a}$ is also determined as follows. If $\mathfrak{p}$ divides $d(D/F)$, $\mathfrak{a}_\mathfrak{p}$ is a power $\mathfrak{P}^{c(\mathfrak{p})}$ of a maximal ideal $\mathfrak{P}$ of $\mathfrak{o}_\mathfrak{p}$. When $\mathfrak{p}$ does not divide $d(D/F)$, take an element $\alpha_\mathfrak{p}$ such that $\mathfrak{a}_\mathfrak{p} = \alpha_\mathfrak{p}\mathfrak{o}_\mathfrak{p}$. We can find two units ε_1 and ε_2 of $\mathfrak{o}_\mathfrak{p}$ so that

$$(2) \qquad \varepsilon_1\alpha_\mathfrak{p}\varepsilon_2 = \begin{pmatrix} \pi^{c_1(\mathfrak{p})} & 0 \\ 0 & \pi^{c_2(\mathfrak{p})} \end{pmatrix},$$

where π is a prime element of $\mathfrak{g}_\mathfrak{p}$ and $c_1(\mathfrak{p})$, $c_2(\mathfrak{p})$ are integers such that $c_1(\mathfrak{p}) \leq c_2(\mathfrak{p})$. Then the set of invariants of $\mathfrak{a}$ is

$$\{\cdots, \mathfrak{p}^{c_1(\mathfrak{p})}, \mathfrak{p}^{c_2(\mathfrak{p})}, \cdots, \mathfrak{p}^{c(\mathfrak{p})}, \cdots\}.$$

This consideration shows in particular that two integral right $\mathfrak{o}$-ideals $\mathfrak{a}$ and $\mathfrak{b}$ are similar if and only if $\mathfrak{o}/\mathfrak{a}$ and $\mathfrak{o}/\mathfrak{b}$ are isomorphic as $\mathfrak{g}$-modules. In the expression (2), we may assume that $N(\varepsilon_1) = 1$. In fact, if this is not so, put

$$\eta = \begin{pmatrix} N(\varepsilon_1) & 0 \\ 0 & 1 \end{pmatrix}$$

and take $\eta^{-1}\varepsilon_1$ and $\varepsilon_2\eta$ in place of ε_1 and ε_2.

PROPOSITION 1.4. *Let $\mathfrak{o}$ be a maximal order in D. Two right $\mathfrak{o}$-ideals $\mathfrak{a}$ and $\mathfrak{b}$ are similar if and only if there exists a unit γ of $\mathfrak{o}$ such that $N(\gamma) = 1$ and $\gamma\mathfrak{a} = \mathfrak{b}$.*

PROOF. It is sufficient to prove our proposition in the case where both $\mathfrak{a}$ and $\mathfrak{b}$ are integral, since the general case is easily reduced to this one. If there exists a unit γ of $\mathfrak{o}$ such that $\gamma\mathfrak{a} = \mathfrak{b}$, then the mapping $x \to \gamma x$ gives an isomorphism of $\mathfrak{o}/\mathfrak{a}$ onto $\mathfrak{o}/\mathfrak{b}$, so that $\mathfrak{a}$ and $\mathfrak{b}$ are similar; this proves the "if" part. Now suppose that $\mathfrak{a}$ and $\mathfrak{b}$ have the same set of invariants. By the above consideration, there exists, for each $\mathfrak{p}$, a unit $\gamma_\mathfrak{p}$ of $\mathfrak{o}_\mathfrak{p}$ such that $\gamma_\mathfrak{p}\mathfrak{a}_\mathfrak{p} = \mathfrak{b}_\mathfrak{p}$. Moreover, as remarked just now, we may assume that $N(\gamma_\mathfrak{p}) = 1$. Let $\mathfrak{c}$ be a two-sided $\mathfrak{o}$-ideal contained in $\mathfrak{a} \cap \mathfrak{b}$. Then we can find an element δ of $\mathfrak{o}$ such that $\delta \equiv \gamma_\mathfrak{p} \bmod \mathfrak{c}_\mathfrak{p}$ for every $\mathfrak{p}$. We have then $N(\delta) \equiv 1 \bmod \mathfrak{c}$. By Lemma 1.2, there exists an element γ of $\mathfrak{o}$ such that $\gamma \equiv \delta \bmod \mathfrak{c}$ and $N(\gamma) = 1$. As we have $\gamma \equiv \gamma_\mathfrak{p} \bmod \mathfrak{c}_\mathfrak{p}$ for every $\mathfrak{p}$, we get $\gamma\mathfrak{a}_\mathfrak{p} = \mathfrak{b}_\mathfrak{p}$ for every $\mathfrak{p}$, so that $\gamma\mathfrak{a} = \mathfrak{b}$. This proves the "only if" part and completes the proof.

PROPOSITION 1.5. *Let $\mathfrak{c}$ be a normal $\mathfrak{g}$-lattice; let $\mathfrak{o}_1$ and $\mathfrak{o}_2$ be respectively the right order and the left order of $\mathfrak{c}$. Denote by Γ_1 and Γ_2 the groups of units in $\mathfrak{o}_1$ and $\mathfrak{o}_2$, respectively. Let α be a regular element of D. Then the correspondence $\beta\Gamma_2 \to \beta\mathfrak{c}$ gives a one-to-one correspondence between the cosets $\beta\Gamma_2$ contained in $\Gamma_1\alpha\Gamma_2$ and the right $\mathfrak{o}_1$-ideals which are similar to $\alpha\mathfrak{c}$.*

PROOF. If β is an element of $\Gamma_1\alpha\Gamma_2$, there exist elements $\gamma_1 \in \Gamma_1$ and $\gamma_2 \in \Gamma_2$ such that $\beta = \gamma_1\alpha\gamma_2$; so we have $\beta\mathfrak{c} = \gamma_1\alpha\mathfrak{c}$. By Proposition 1.4, $\beta\mathfrak{c}$ is similar to $\alpha\mathfrak{c}$. By the same proposition, every right $\mathfrak{o}_1$-ideal similar to $\alpha\mathfrak{c}$ is obtained in this manner. Now if we have $\beta_1\mathfrak{c} = \beta_2\mathfrak{c}$, then $\beta_1^{-1}\beta_2$ must be a unit of $\mathfrak{o}_2$, so that $\beta_1\Gamma_2 = \beta_2\Gamma_2$. Our proposition is thereby proved.

1.4. *Adeles and ideles of D.* We denote by $\mathfrak{A}$ and $\mathfrak{J}$ the ring of adeles and the group of ideles of the algebra D, respectively; we identify, in the usual manner, D with a subring of $\mathfrak{A}$, and the multiplicative group D^* of regular elements of D with a subgroup of $\mathfrak{J}$. For every $x \in \mathfrak{A}$, we denote by $x_\mathfrak{p}$ the $\mathfrak{p}$-component of x if $\mathfrak{p}$ is a finite prime and by $x^{(\nu)}$ the $\mathfrak{p}_{\infty,\nu}$-component of x. D is discrete in $\mathfrak{A}$ and $\mathfrak{A}/D$ is compact; D^* is discrete in $\mathfrak{J}$. For every $x \in \mathfrak{J}$, we define a positive real number $|x|$ by

$$(3) \qquad\qquad |x| = \prod_\mathfrak{p} |N(x_\mathfrak{p})|_\mathfrak{p} \cdot \prod_{\nu=1}^{t} |N(x^{(\nu)})| \,,$$

where $|\ |_\mathfrak{p}$ denotes the valuation of $F_\mathfrak{p}$ such that $|\pi_\mathfrak{p}| = N(\mathfrak{p})^{-1}$ for every prime element $\pi_\mathfrak{p}$. Denote by $\mathfrak{J}_0$ the subgroup of $\mathfrak{J}$ consisting of the elements x such that $|x| = 1$; we observe that D^* is a subgroup of $\mathfrak{J}_0$. The following lemma is well-known (cf. [8, 31]).

LEMMA 1.6. *The factor space $D^*\backslash \mathfrak{J}_0$ is of finite measure with respect to a Haar measure of $\mathfrak{J}_0$. Moreover, if D is a division ring, $D^*\backslash \mathfrak{J}_0$ is compact.*

For every $x \in \mathfrak{A}$, we denote by x' the element of $\mathfrak{A}$ such that $(x')_\mathfrak{p} = (x_\mathfrak{p})'$ and $x'^{(\nu)} = x^{(\nu)\prime}$ for every $\mathfrak{p}$ and ν; obviously, $x \to x'$ is an involution of $\mathfrak{A}$. We put also $N(x) = xx'$ for $x \in \mathfrak{A}$.

Let $\mathfrak{o}$ be a maximal order in D. We denote by $U(\mathfrak{o})$ the subgroup of $\mathfrak{J}$ consisting of the ideles x such that $x_\mathfrak{p}$ is a unit of $\mathfrak{o}_\mathfrak{p}$ for every prime ideal $\mathfrak{p}$ of F. For every $x \in \mathfrak{J}$, there exists a uniquely determined right $\mathfrak{o}$-ideal $\mathfrak{a}$ such that $\mathfrak{a}_\mathfrak{p} = x_\mathfrak{p}\mathfrak{o}_\mathfrak{p}$. If we denote by $\bigcap x_\mathfrak{p}\mathfrak{o}_\mathfrak{p}$ this right $\mathfrak{o}$-ideal $\mathfrak{a}$, then $x \to \bigcap x_\mathfrak{p}\mathfrak{o}_\mathfrak{p}$ gives a one-to-one correspondence between $\mathfrak{J}/U(\mathfrak{o})$ and the set of right $\mathfrak{o}$-ideals. Furthermore, $x \to \bigcap x_\mathfrak{p}\mathfrak{o}_\mathfrak{p}$ gives a one-to-one correspondence between $D^*\backslash \mathfrak{J}/U(\mathfrak{o})$ and the equivalence-classes of right $\mathfrak{o}$-ideals.

Let $\mathfrak{K}$ be the group of ideal-classes modulo $\prod_{\nu > r} \mathfrak{p}_{\infty,\nu}$ in F and h the order of $\mathfrak{K}$. By Lemma 1.1, every element of $\mathfrak{K}$ corresponds to an equivalence-class of right $\mathfrak{o}$-ideals; so h is the class-number of D. Let $\mathfrak{k}_1 = 1, \cdots, \mathfrak{k}_h$ be the elements of $\mathfrak{k}$. Take and fix, for each λ, a right $\mathfrak{o}$-ideal $\mathfrak{x}_\lambda$ such that $N(\mathfrak{x}_\lambda) \in \mathfrak{k}_\lambda$ and an element x_λ of $\mathfrak{J}$ such that $\mathfrak{x}_\lambda = \bigcap x_{\lambda\mathfrak{p}}\mathfrak{o}_\mathfrak{p}$. Put $U = U(\mathfrak{o})$ and

$$\mathfrak{o}_\lambda = \mathfrak{x}_\lambda\mathfrak{x}_\lambda^{-1}, \qquad \mathfrak{x}_{\lambda\mu} = \mathfrak{x}_\lambda\mathfrak{x}_\mu^{-1}, \qquad x_{\lambda\mu} = x_\lambda x_\mu^{-1}, \qquad U_\lambda = U(\mathfrak{o}_\lambda) .$$

Obviously, $\mathfrak{o}_\lambda$ is the left order of $\mathfrak{x}_\lambda$. Let Γ_λ be the group of units in $\mathfrak{o}_\lambda$. We can easily verify that

$$(4) \qquad\qquad\qquad \Gamma_\lambda = D^* \cap U_\lambda ,$$

$$(5) \qquad\qquad\qquad U_\lambda = x_\lambda U x_\lambda^{-1} ,$$

$$(6) \qquad\qquad\qquad \mathfrak{J} = \bigcup_{\lambda=1}^{h} D^* x_\lambda U ,$$

and the last expression is a disjoint sum.

2. The Hecke-ring of D

2.1. First we generalize a little the notion of "ring of transformations" introduced in [24]. Let G be a group. Suppose that a system of subgroups $\{\Gamma_\lambda | \lambda \in \Lambda\}$ of G and a subset Δ of G are given and satisfy the following two conditions.

(R1) Δ is a semi-group containing every Γ_λ.

(R2) For every $\alpha \in \Delta$, and for every λ and μ, $\alpha\Gamma_\lambda\alpha^{-1}$ is commensurable with Γ_μ.

Let $\mathfrak{R}_{\mu\lambda}$ be the free Z-module generated by the cosets $\Gamma_\mu\alpha\Gamma_\lambda$ for $\alpha \in \Delta$; so, every element of $\mathfrak{R}_{\mu\lambda}$ is a formal finite sum $\sum_k c_k(\Gamma_\mu\alpha_k\Gamma_\lambda)$ with $c_k \in Z$,

$\alpha_k \in \Delta$. We will now define a bilinear mapping of $\mathfrak{R}_{\nu\mu} \times \mathfrak{R}_{\mu\lambda}$ into $\mathfrak{R}_{\nu\lambda}$. Let $\Gamma_\mu \alpha \Gamma_\lambda = \bigcup_i \alpha_i \Gamma_\lambda$ and $\Gamma_\nu \beta \Gamma_\mu = \bigcup_j \beta_j \Gamma_\mu$ be disjoint sums. For any $\xi \in \Gamma_\nu \beta \Gamma_\mu \alpha \Gamma_\lambda$, the number of (i, j) such that $\beta_j \alpha_i \Gamma_\lambda = \xi \Gamma_\lambda$ is determined only by $\sigma = \Gamma_\mu \alpha \Gamma_\lambda$, $\tau = \Gamma_\nu \beta \Gamma_\mu$, $\rho = \Gamma_\nu \xi \Gamma_\lambda$; it does not depend on the choice of $\{\alpha_i\}$, $\{\beta_j\}$, ξ. We denote this number by $\mu(\tau \cdot \sigma; \rho)$ and put

$$\tau \cdot \sigma = \sum \mu(\tau \cdot \sigma; \rho)\rho ,$$

where the sum is taken over all $\rho = \Gamma_\nu \xi \Gamma_\lambda$ contained in $\Gamma_\nu \beta \Gamma_\mu \alpha \Gamma_\lambda$. Extend this by linearity to the bilinear mapping of $\mathfrak{R}_{\nu\mu} \times \mathfrak{R}_{\mu\lambda}$ into $\mathfrak{R}_{\nu\lambda}$. We can easily verify that this operation is associative; namely, for every $\rho \in \mathfrak{R}_{\lambda\kappa}$, $\sigma \in \mathfrak{R}_{\mu\lambda}$, $\tau \in \mathfrak{R}_{\nu\mu}$, we have $\tau \cdot (\sigma \cdot \rho) = (\tau \cdot \sigma) \cdot \rho$. In particular, $\mathfrak{R}_{\lambda\lambda}$ is an associative ring, which we denote by $\mathfrak{R}(\Gamma_\lambda; \Delta)$. In [24, 25], right cosets were considered; here we prefer left cosets in view of later applications.

2.2. Fix a maximal order $\mathfrak{o}$ in D and define $\mathfrak{o}_\lambda$, $\mathfrak{x}_\lambda$, $\mathfrak{x}_{\lambda\mu}$, Γ_λ as in §1.4. Then it is easily seen that $\{\Gamma_1, \cdots, \Gamma_h\}$ and D^* satisfies the conditions (R1, 2); so we can define $\mathfrak{R}_{\mu\lambda}$ for this system. In order to know the structure of $\{\mathfrak{R}_{\mu\lambda}\}$, let us classify the normal $\mathfrak{g}$-lattices in D by similarity. We call each class a *similarity-class of normal* $\mathfrak{g}$-*lattices*. Let $\mathfrak{E}$ denote the set of similarity-classes of normal $\mathfrak{g}$-lattices. For every element e of $\mathfrak{E}$ we can speak of the set of invariants of e and its norm $N(e)$ by which we understand the set of invariants and the norm of the $\mathfrak{g}$-lattices belonging to e; and we put $N_{D/Q}(e) = N_{F/Q}(N(e))$. We call an element e of $\mathfrak{E}$ *integral* if the $\mathfrak{g}$-lattices belonging to e are integral. Let $\mathfrak{o}_0$ be a maximal order in D and $\mathfrak{a}$ a two-sided $\mathfrak{o}_0$-ideal. For an element e of $\mathfrak{E}$, we write $e \subset \mathfrak{a}$ if the right $\mathfrak{o}_0$-ideals belonging to e are contained in $\mathfrak{a}$. We say that two elements e_1 and e_2 of $\mathfrak{E}$ are *relatively prime* if there is no prime ideal $\mathfrak{p}$ such that $\mathfrak{b}_{1\mathfrak{p}} \neq \mathfrak{o}_{0\mathfrak{p}}$, $\mathfrak{b}_{2\mathfrak{p}} \neq \mathfrak{o}_{0\mathfrak{p}}$, for the right $\mathfrak{o}_0$-ideals $\mathfrak{b}_1$ belonging to e_1 and the right $\mathfrak{o}_0$-ideals $\mathfrak{b}_2$ belonging to e_2; this definition does not depend on the choice of $\mathfrak{o}_0$. We say also that an element e of $\mathfrak{E}$ is *prime to* a two-sided $\mathfrak{o}_0$-ideal $\mathfrak{a}$ if e and the similarity-class of $\mathfrak{a}$ are relatively prime.

Let α be an element of D^*. When $\alpha \mathfrak{x}_{\lambda\mu}$ belongs to an element e of $\mathfrak{E}$, we write $\Gamma_\mu \alpha \Gamma_\lambda = T_{\mu\lambda}(e)$; by Proposition 1.5, e is uniquely determined by $\Gamma_\mu \alpha \Gamma_\lambda$ and independent of the choice of α. (It depends, of course, on the choice of $\mathfrak{x}_\lambda$ which we have fixed.) By our definition of $\mathfrak{x}_{\lambda\mu}$,

$$(7) \qquad\qquad\qquad N(e) \in \mathfrak{t}_\lambda \mathfrak{t}_\mu^{-1} .$$

Conversely, for every λ and for every $e \in \mathfrak{E}$, determine μ by the relation (7); we can then find an element α of D^* such that $\alpha \mathfrak{x}_{\lambda\mu}$ belongs to e. We have then $\Gamma_\mu \alpha \Gamma_\lambda = T_{\mu\lambda}(e)$. Similarly, for every μ and every e, there exists an element α of D^* such that $\Gamma_\mu \alpha \Gamma_\lambda = T_{\mu\lambda}(e)$.

Let a be an element of F^*, and e an element of $\mathfrak{E}$. Take a normal $\mathfrak{g}$-lattice $\mathfrak{a}$ belonging to e. Then the similarity-class of $a\mathfrak{a}$ is determined only by a and e; it does not depend on the choice of $\mathfrak{a}$. We denote this similarity-class by ae. The maximal orders in D form an element of $\mathfrak{E}$, which we denote by 1; and put $a1 = a$ for every $a \in F^*$. We see easily that

$$T_{\mu\lambda}(ae) = T_{\mu\mu}(a)\,T_{\mu\lambda}(e) = T_{\mu\lambda}(e)\,T_{\lambda\lambda}(a)$$

for every $a \in F^*$ and every $e \in \mathfrak{E}$.

PROPOSITION 2.1. *Let e, f, g be elements of $\mathfrak{E}$. If we fix a right $\mathfrak{o}_\nu$-ideal $\mathfrak{c}$ belonging to g, then the multiplicity of $T_{\nu\lambda}(g)$ in $T_{\nu\mu}(f) \cdot T_{\mu\lambda}(e)$ is equal to the number of right $\mathfrak{o}_\nu$-ideals $\mathfrak{b}$ belonging to f such that $\mathfrak{c}\mathfrak{b}^{-1}$ belongs to e.*

PROOF. Let α, β, ξ be elements of D^* such that $\Gamma_\mu\alpha\Gamma_\lambda = T_{\mu\lambda}(e)$, $\Gamma_\nu\beta\Gamma_\mu = T_{\nu\mu}(f)$, $\Gamma_\nu\xi\Gamma_\lambda = T_{\nu\lambda}(g)$; and let $\Gamma_\mu\alpha\Gamma_\lambda = \bigcup_i \alpha_i\Gamma_\lambda$, $\Gamma_\nu\beta\Gamma_\mu = \bigcup_j \beta_j\Gamma_\mu$ be disjoint sums. By our definition, the multiplicity c of $T_{\nu\lambda}(g)$ in the product $T_{\nu\mu}(f) \cdot T_{\mu\lambda}(e)$ is equal to the number of (i, j) such that $\beta_j\alpha_i\Gamma_\lambda = \xi\Gamma_\lambda$. If we have $\beta_j\alpha_i\Gamma_\lambda = \xi\Gamma_\lambda$, then $\alpha_i\Gamma_\lambda = \beta_j^{-1}\xi\Gamma_\lambda$, so that i is uniquely determined by j. It is easy to see that c is the number of (i, j) such that $\beta_j\alpha_i\mathfrak{x}_{\lambda\nu} = \xi\mathfrak{x}_{\lambda\nu}$. Let $\mathfrak{c}$ be a right $\mathfrak{o}_\nu$-ideal belonging to g. We may assume that $\xi\mathfrak{x}_{\lambda\nu} = \mathfrak{c}$; in fact, if this is not so, we take a suitable element of $\Gamma_\nu\xi\Gamma_\lambda$ in place of ξ. Now put $\beta_j\mathfrak{x}_{\mu\nu} = \mathfrak{b}_j$; then $\mathfrak{b}_j$ belongs to f. If we have $\beta_j\alpha_i\mathfrak{x}_{\lambda\nu} = \xi\mathfrak{x}_{\lambda\nu}$, we get

$$\mathfrak{c}\mathfrak{b}_j^{-1} = \beta_j\alpha_i\mathfrak{x}_{\lambda\nu}\mathfrak{x}_{\nu\mu}^{-1}\mathfrak{x}_{\nu\mu}\mathfrak{x}_{\mu\nu}^{-1}\beta_j^{-1} = \beta_j(\alpha_i\mathfrak{x}_{\lambda\mu})\beta_j^{-1} \, .$$

As $\alpha_i\mathfrak{x}_{\lambda\mu}$ belongs to e, $\mathfrak{c}\mathfrak{b}_j^{-1}$ belongs to e. Conversely, let $\mathfrak{b}$ be a right $\mathfrak{o}_\nu$-ideal belonging to f such that $\mathfrak{c}\mathfrak{b}^{-1}$ belongs to e. By Proposition 1.5, we have $\mathfrak{b} = \mathfrak{b}_j$ for some j, so that $\beta_j^{-1}\mathfrak{c}\mathfrak{x}_{\nu\mu} = \beta_j^{-1}(\mathfrak{c}\mathfrak{b}^{-1})\beta_j$. As $\mathfrak{c}\mathfrak{b}^{-1}$ belongs to e, $\beta_j^{-1}\mathfrak{c}\mathfrak{x}_{\nu\mu}$ belongs to e; hence by Proposition 1.5, $\beta_j^{-1}\mathfrak{c}\mathfrak{x}_{\nu\mu} = \alpha_i\mathfrak{x}_{\lambda\mu}$ for some i. We have then $\mathfrak{c} = \beta_j\alpha_i\mathfrak{x}_{\lambda\nu}$. Our proposition is a consequence of this consideration.

2.3. *Hecke ring* $\mathfrak{R}$. As in §1.4, put $U = U(\mathfrak{o})$. It can easily be seen that, for every $x \in \mathfrak{F}$, the groups xUx^{-1} and U are commensurable; so we can define the ring $\mathfrak{R}(U; \mathfrak{F})$.

PROPOSITION 2.2. *The ring $\mathfrak{R}(U; \mathfrak{F})$ is commutative.*

PROOF. Consider the involution $x \to x'$ of the group $\mathfrak{F}$. For every $x \in \mathfrak{F}$ and every $\mathfrak{p}$, we can find units $\varepsilon_\mathfrak{p}$ and $\eta_\mathfrak{p}$ of $\mathfrak{o}_\mathfrak{p}$ such that $\varepsilon_\mathfrak{p}x_\mathfrak{p}\eta_\mathfrak{p} = x'_\mathfrak{p}$. It follows that $(UxU)' = Ux'U = UxU$. By [25, Proposition 1.2], the ring $\mathfrak{R}(U; \mathfrak{F})$ is commutative.

For every UxU, let e be the similarity-class of $\bigcap x_\mathfrak{p}\mathfrak{o}_\mathfrak{p}$; we observe that

e is uniquely determined by UxU and independent of the choice of x; and $UxU \to e$ gives a one-to-one correspondence between $U\backslash\mathfrak{I}/U$ and $\mathfrak{E}$. When e corresponds to UxU, we write $T(e) = UxU$. Now we want to clarify the relation between $T(e)$ and $T_{\mu\lambda}(e)$.

PROPOSITION 2.3. *Let α be an element of D^* and $\Gamma_\mu\alpha\Gamma_\lambda = \bigcup_i \alpha_i\Gamma_\lambda$ be a disjoint sum. Then, we have $U_\mu\alpha U_\lambda = \Gamma_\mu\alpha U_\lambda$; and $Ux_\mu^{-1}\alpha x_\lambda U = \bigcup_i x_\mu^{-1}\alpha_i x_\lambda U$ is a disjoint sum.*

PROOF. Let u be an element of U_μ; put $\mathfrak{a} = \bigcap u_\mathfrak{p}\alpha x_{\lambda\mathfrak{p}}x_{\mu\mathfrak{p}}^{-1}\mathfrak{o}_{\mu\mathfrak{p}}$. Then, as $\mathfrak{a}_\mathfrak{p} = u_\mathfrak{p}(\alpha x_{\lambda\mu})_\mathfrak{p}$, the right $\mathfrak{o}_\mu$-ideal $\mathfrak{a}$ is similar to $\alpha x_{\lambda\mu}$. Therefore, by Proposition 1.4, there exists an element γ of Γ_μ such that $\mathfrak{a} = \gamma\alpha x_{\lambda\mu}$. We have then $u_\mathfrak{p}\alpha x_{\lambda\mathfrak{p}} = \gamma\alpha x_{\lambda\mathfrak{p}}$, so that $u_\mathfrak{p}\alpha x_{\lambda\mathfrak{p}}\mathfrak{o}_\mathfrak{p} = \gamma\alpha x_{\lambda\mathfrak{p}}\mathfrak{o}_\mathfrak{p}$. Hence, there exists a unit $v_\mathfrak{p}$ of $\mathfrak{o}_\mathfrak{p}$ such that $u_\mathfrak{p}\alpha x_{\lambda\mathfrak{p}} = \gamma\alpha x_{\lambda\mathfrak{p}}v_\mathfrak{p}$; so we have $u_\mathfrak{p}\alpha = \gamma\alpha x_{\lambda\mathfrak{p}}v_\mathfrak{p}x_{\lambda\mathfrak{p}}^{-1}$. This shows $u_\mathfrak{p}\alpha \in \Gamma_\mu\alpha U_\lambda$; it follows that $U_\mu\alpha U_\lambda = \Gamma_\mu\alpha U_\lambda$. As Γ_λ is contained in U_λ, we have $\Gamma_\mu\alpha U_\lambda = \bigcup_i \alpha_i U_\lambda$ and hence

$$Ux_\mu^{-1}\alpha x_\lambda U = x_\mu^{-1}(U_\mu\alpha U_\lambda)x_\lambda = x_\mu^{-1}(\Gamma_\mu\alpha U_\lambda)x_\lambda = \bigcup_i x_\mu^{-1}\alpha_i x_\lambda U \ .$$

If $x_\mu^{-1}\alpha_i x_\lambda U = x_\mu^{-1}\alpha_j x_\lambda U$, we have $\alpha_i U_\lambda = \alpha_j U_\lambda$. By (4) of §1.4, we must have $\alpha_i\Gamma_\lambda = \alpha_j\Gamma_\lambda$, so that $i = j$. This completes our proof.

PROPOSITION 2.4. *Let e, f, g_k denote elements of $\mathfrak{E}$. If $T_{\nu\mu}(f) \cdot T_{\mu\lambda}(e) = \sum_k c_k T_{\nu\lambda}(g_k)$, then $T(f) \cdot T(e) = \sum_k c_k T(g_k)$.*

PROOF. Consider the correspondence $\Gamma_\mu\alpha\Gamma_\lambda \to Ux_\mu^{-1}\alpha x_\lambda U$. As we have $x_{\mu\mathfrak{p}}^{-1}\alpha x_{\lambda\mathfrak{p}}\mathfrak{o}_\mathfrak{p} = x_{\mu\mathfrak{p}}^{-1}(\alpha x_{\lambda\mu\mathfrak{p}})x_{\mu\mathfrak{p}}$, the right $\mathfrak{o}$-ideal $\bigcap x_{\mu\mathfrak{p}}^{-1}\alpha x_{\lambda\mathfrak{p}}\mathfrak{o}_\mathfrak{p}$ is similar to $\alpha x_{\lambda\mu}$. Therefore, if $\Gamma_\mu\alpha\Gamma_\lambda = T_{\mu\lambda}(e)$ for an element e of $\mathfrak{E}$, we have $Ux_\mu^{-1}\alpha x_\lambda U = T(e)$. Let $T_{\mu\lambda}(e) = \Gamma_\mu\alpha\Gamma_\lambda = \bigcup_i \alpha_i\Gamma_\lambda$, $T_{\nu\mu}(f) = \Gamma_\nu\beta\Gamma_\mu = \bigcup_j \beta_j\Gamma_\mu$ be disjoint sums. By Proposition 2.3, $Ux_\mu^{-1}\alpha x_\lambda U = \bigcup_i x_\mu^{-1}\alpha_i x_\lambda U$ and $Ux_\nu^{-1}\beta x_\mu U = \bigcup_j x_\nu^{-1}\beta_j x_\mu U$ are disjoint sums. We have then $Ux_\nu^{-1}\beta x_\mu U x_\mu^{-1}\alpha x_\lambda U = \bigcup_{i,j} x_\nu^{-1}\beta_j\alpha_i x_\lambda U$, so that $T(f) \cdot T(e)$ is a linear combination of $Ux_\nu^{-1}\beta_j\alpha_i x_\lambda U$. Let ξ be an element of $\Gamma_\nu\beta\Gamma_\mu\alpha\Gamma_\lambda$. If we have $\beta_j\alpha_i\Gamma_\lambda = \xi\Gamma_\lambda$, then $\beta_j\alpha_i U_\lambda = \xi U_\lambda$, so that $(x_\nu^{-1}\beta_j x_\mu)(x_\mu^{-1}\alpha_i x_\lambda)U = x_\nu^{-1}\xi x_\lambda U$. Conversely, if the last equality holds, we have $\beta_j\alpha_i U_\lambda = \xi U_\lambda$. As $\Gamma_\lambda = D^* \cap U_\lambda$, we must have $\beta_j\alpha_i\Gamma_\lambda = \xi\Gamma_\lambda$. Hence the multiplicity of $\Gamma_\nu\xi\Gamma_\lambda$ in $T_{\nu\mu}(f) \cdot T_{\mu\lambda}(e)$ is equal to that of $Ux_\nu^{-1}\xi x_\lambda U$ in $T(f) \cdot T(e)$. This proves our proposition.

We call, after Tamagawa, the ring $\mathfrak{R}(U; \mathfrak{I})$ the *Hecke-ring* of D. We denote it simply by $\mathfrak{R}$. Let $U_\mathfrak{p}$ be the group of units in $\mathfrak{o}_\mathfrak{p}$. By the structure of U and $\mathfrak{I}$, we can easily verify that $\mathfrak{R}$ is the tensor-product of the rings $\mathfrak{R}(U_\mathfrak{p}; D_\mathfrak{p}^*)$. In particular, we obtain

PROPOSITION 2.5. *Let e and f be elements of $\mathfrak{E}$; suppose that e and f are relatively prime. Then, there exists an element g of $\mathfrak{E}$ such that*

*the set of invariants of g is the join of the sets of invariants of e and f;
and we have $T(f) \cdot T(e) = T(g)$.*

This can be also proved by virtue of Propositions 2.1 and 2.4, in the same way as in [25, Proposition 1.9].

Let $\mathfrak{n}$ be an integral ideal of F. We put $T(\mathfrak{n}) = \sum T(e)$, where the sum is extended over all integral elements e of $\mathfrak{E}$ such that $N(e) = \mathfrak{n}$. Let $\mathfrak{p}$ be a prime ideal of F. If $\mathfrak{p}$ divides $d(D/F)$, we have $T(\mathfrak{p}^a) = T(e)$ for the element e of $\mathfrak{E}$, of which the set of invariants is $\{\mathfrak{p}^a\}$. Now suppose that $\mathfrak{p}$ does not divide $d(D/F)$. We put $T(\mathfrak{p}^a, \mathfrak{p}^b) = T(e)$ for the element e, of which the set of invariants is $\{\mathfrak{p}^a, \mathfrak{p}^b\}$.

PROPOSITION 2.6. *Notation being as above, the following relations hold.*
(i) *If $\mathfrak{p}$ divides $d(D/F)$, $T(\mathfrak{p}^a)T(\mathfrak{p}^b) = T(\mathfrak{p}^{a+b})$.*
(ii) *If $\mathfrak{p}$ does not divide $d(D/F)$,*

$$T(\mathfrak{p}, \mathfrak{p})T(\mathfrak{p}^a, \mathfrak{p}^b) = T(\mathfrak{p}^{a+1}, \mathfrak{p}^{b+1}) \,,$$
$$T(\mathfrak{p})T(\mathfrak{p}^m) = T(1, \mathfrak{p}^{m+1}) + (N(\mathfrak{p}) + 1)T(\mathfrak{p}, \mathfrak{p})T(\mathfrak{p}^{m-1}) \,.$$

These formulas are shown by the same argument as in [25, Proposition 1.10], on account of Propositions 2.1 and 2.4. Since we may regard $T(\mathfrak{p}^a)$ and $T(\mathfrak{p}^a, \mathfrak{p}^b)$ as elements of $\mathfrak{R}(U_\mathfrak{p}; D_\mathfrak{p}^*)$, Proposition 2.6 can be also proved by "local consideration".

2.4. *Formal Dirichlet series.* Let $\mathfrak{a}$ be an integral two-sided $\mathfrak{o}$-ideal. Denote by $I(\mathfrak{a})$ the group of ideals of F which are prime to $N(\mathfrak{a})$. Let ξ_0 be a homomorphism of $I(\mathfrak{a})$ into C^*. Consider now a formal Dirichlet series

$$\mathscr{D}(s) = \sum_{\mathfrak{n}}' \xi_0(\mathfrak{n})N(\mathfrak{n})^{-s}T(\mathfrak{n}) = \sum_{e}' \xi_0(N(e))N_{D/Q}(e)T(e) \,,$$

where the first sum is extended over all the integral ideals $\mathfrak{n}$ of F which are prime to $N(\mathfrak{a})$, and the second sum is extended over all the integral elements e of $\mathfrak{E}$ which are prime to $\mathfrak{a}$. By Propositions 2.5 and 2.6, the series $\mathscr{D}(s)$ is expressed in the form of an Euler-product

$$\mathscr{D}(s) = \prod{}' [1 - \xi_0(\mathfrak{p})N(\mathfrak{p})^{-s}T(\mathfrak{p})]^{-1}$$
$$\times \prod{}'' [1 - \xi_0(\mathfrak{p})N(\mathfrak{p})^{-s}T(\mathfrak{p}) + \xi_0(\mathfrak{p})^2 N(\mathfrak{p})^{1-2s}T(\mathfrak{p}, \mathfrak{p})]^{-1} \,,$$

where the first product is taken over all the prime ideals $\mathfrak{p}$ dividing $d(D/F)$ but not $N(\mathfrak{a})$, and the second product is taken over all the prime ideals $\mathfrak{p}$ which are prime to $N(\mathfrak{a})d(D/F)$.

3. Automorphic forms

3.1. *Tensor representation of* GL(2, *C*). For every vector $\binom{u}{v}$ of C^2, and for every integer $n \geq 0$, we denote by $\binom{u}{v}^n$ the column-vector of C^{n+1} with the components u^n, $u^{n-1}v$, $\cdots$, uv^{n-1}, v^n; in particular, $\binom{u}{v}^0$ denotes the one-dimensional vector with the component 1. Define, for every $\sigma \in M_2(C)$, a matrix $\Phi_n(\sigma)$ by

$$\left(\sigma\binom{u}{v}\right)^* = \Phi_n(\sigma)\binom{u}{v}^* .$$

Then, for every n, $\sigma \to \Phi_n(\sigma)$ gives a representation of GL(2, C). Define a matrix P_n of degree $n + 1$ by

$$(8) \qquad \det\begin{pmatrix} u & w \\ v & z \end{pmatrix}^* = {}^t\binom{u}{v}^* P_n \binom{w}{z}^* .$$

We see easily that

$$^tP_n = (-1)^n P_n ,$$
$$^t\Phi_n(\sigma)P_n\Phi_n(\sigma) = \det(\sigma)^n P_n .$$

We note that $P_0 = 1$ and $\Phi_0(\sigma) = 1$ for every $\sigma \in M_2(C)$.

Now fix an injection of C into K and consider C as a subfield of K. Take an element j of K such that $j^2 = -1$, $j\alpha j^{-1} = \bar{\alpha}$ for $\alpha \in C$. Every element of K is written in the form $\alpha + \beta j$ with α, β in C. We put

$$\iota(\alpha + \beta j) = \begin{pmatrix} \alpha & \beta \\ -\bar{\beta} & \bar{\alpha} \end{pmatrix} ,$$

and, for every $\xi \in K$,

$$\Phi_n(\xi) = \Phi_n(\iota(\xi)) .$$

3.2. *Representations of D^*.*[3] Now the $D^{(\nu)}$ being as in §1.1, consider the sets

$$\mathfrak{W} = \{\xi \in D \,|\, \mathrm{tr}(\xi) = 0\} ,$$
$$\mathfrak{W}^{(\nu)} = \{\xi \in D^{(\nu)} \,|\, \mathrm{tr}(\xi) = 0\} .$$

Obviously, $\mathfrak{W}$(resp. $\mathfrak{W}^{(\nu)}$) is a vector space of dimension 3 over F(resp. R); and we have

$$\mathfrak{W} \otimes_Q R = \mathfrak{W}^{(1)} \times \cdots \times \mathfrak{W}^{(t)} .$$

If $\{w_i\} = \{w_1, w_2, w_3\}$ is a basis of $\mathfrak{W}$ over F, $\{w_i^{(\nu)}\}$ is a basis of $\mathfrak{W}^{(\nu)}$ over R. We observe that every element α of D gives a mapping $\xi \to \alpha\xi\alpha'$ of $\mathfrak{W}$ into itself. Let $\Psi_2(\alpha)$ be the matrix representing this mapping with respect to $\{w_i\}$. Similarly, we get, for every $\beta \in D^{(\nu)}$, a matrix $\Psi_2^{(\nu)}(\beta)$ representing $\xi \to \beta\xi\beta'$ with respect to $\{w_i^{(\nu)}\}$. Then we see easily that, for every $\alpha \in D$, $\Psi_2^{(\nu)}(\alpha^{(\nu)})$ is the νth conjugate of $\Psi_2(\alpha)$ over Q.

Let Θ_2 be the symmetric matrix which represents the quadratic form

---

[3] The results of this section will be used only in §3.5 and Part II; so the reader who is interested only in the main theorem of Part I may omit it.

$N(\xi) = \xi\xi'$ on $\mathfrak{W}$ with respect to the basis $\{w_i\}$. Then we have,

$$'\Psi_2(\alpha)\Theta_2\Psi_2(\alpha) = N(\alpha)^2\Theta_2 \qquad (\alpha \in D) .$$

It is easy to see that the νth conjugate $\Theta_2^{(\nu)}$ of Θ_2 represents the quadratic form $N(\xi)$ on $\mathfrak{W}^{(\nu)}$ with respect to $\{w_i^{(\nu)}\}$. We have therefore

$$'\Psi_2^{(\nu)}(\beta)\Theta_2^{(\nu)}\Psi_2^{(\nu)}(\beta) = N(\beta)^2\Theta_2^{(\nu)} \qquad (\beta \in D^{(\nu)}) .$$

The symmetric matrix $\Theta_2^{(\nu)}$ is clearly indefinite for $1 \leq \nu \leq r$, and positive definite for $\nu > r$. Now, as $D^{(\nu)}$ is $M_2(R)$ or K, there exists, for each ν, a complex matrix $U_2^{(\nu)}$ such that

$$\Psi_2^{(\nu)}(\beta) = U_2^{(\nu)}\Phi_2(\beta)U_2^{(\nu)-1} \qquad (\beta \in D^{(\nu)}) .$$

We may take $U_2^{(\nu)}$ so as to be real matrices for $1 \leq \nu \leq r$. Since Φ_2 is an irreducible representation of $SL(2, R)$, there exist real numbers $b_2^{(\nu)}$ such that

$$\Theta_2^{(\nu)} = b_2^{(\nu)}{}^tU_2^{(\nu)-1}P_2U_2^{(\nu)-1} \qquad (1 \leq \nu \leq r) .$$

Now by the representation-theory of orthogonal groups, we know that, for each even integer n, Φ_n is obtained as an irreducible component of a certain Kronecker product $\Phi_2 \otimes \cdots \otimes \Phi_2$. The decomposition into irreducible components can be done within the operation in Q. Therefore, consider a tensor product $\mathfrak{W} \otimes \cdots \otimes \mathfrak{W}$ as a representation space of D^* and decompose it into irreducible components $\mathfrak{B}_1, \cdots, \mathfrak{B}_s$. Then we have in a natural manner

$$\mathfrak{B}_j \otimes_Q R = \mathfrak{B}_j^{(1)} \times \cdots \times \mathfrak{B}_j^{(t)} \qquad (1 \leq j \leq s) ,$$

$$\mathfrak{W}^{(\nu)} \otimes \cdots \otimes \mathfrak{W}^{(\nu)} = \mathfrak{B}_1^{(\nu)} \times \cdots \times \mathfrak{B}_s^{(\nu)} \qquad (1 \leq \nu \leq t) ;$$

and the $\mathfrak{B}_j^{(\nu)}$ are the irreducible components of $W^{(\nu)} \otimes \cdots \otimes W^{(\nu)}$. The representation Φ_n of $D^{(\nu)*}$ can be obtained as such an irreducible component. For each even $n \geq 0$, fix an irreducible component $\mathfrak{B}_j^{(1)}$ from which the representation Φ_n of $GL(2, R)$ is obtained; we take a basis $\{v_i\}$ of $\mathfrak{B}_j$ over F and denote by $\Psi_n(\alpha)$ (resp. $\Psi_n^{(\nu)}(\alpha)$) the matrix representing the operation of α on $\mathfrak{B}_j$ (resp. $\mathfrak{B}_j^{(\nu)}$) with respect to $\{v_i\}$ (resp. $\{v_i^{(\nu)}\}$). We note that the matrices $\Psi_n(\alpha)$ (resp. $\Psi_n^{(\nu)}(\alpha)$) can be defined not only for regular elements of D (resp. $D^{(\nu)}$) but also for any element α of D (resp. $D^{(\nu)}$). The matrices $\Psi_n^{(\nu)}(\alpha)$ thus obtained, have the following properties.

(Ψ1) For every $\alpha \in D$, $\Psi_n(\alpha)$ is a matrix of degree $n + 1$ with coefficients in F.

(Ψ2) For every $\beta \in D^{(\nu)}$, $\Psi_n^{(\nu)}(\alpha)$ is a real matrix of degree $n + 1$.

(Ψ3) For each ν, there exists a matrix $U_n^{(\nu)}$, which is of real coefficients if $1 \leq \nu \leq r$, such that

$$\Psi_n^{(\nu)}(\beta) = U_n^{(\nu)}\Phi_n(\beta)U_n^{(\nu)-1} \qquad (\beta \in D^{(\nu)})$$

($\Psi4$) For every $\alpha \in D$, $\Psi_n^{(\nu)}(\alpha^{(\nu)})$ is the νth conjugate of $\Psi_n(\alpha)$.

From the quadratic form $N(\xi)$ on $\mathfrak{W}$, we obtain naturally a quadratic form q on $\mathfrak{W} \otimes \cdots \otimes \mathfrak{W}$ by $q(\xi_1 \otimes \cdots \otimes \xi_u) = \prod_{i=1}^{u} N(\xi_i)$. Restricting this to the subspaces $\mathfrak{W}_j$, we get, for each even $n \geq 0$, a matrix Θ_n with the following properties.

($\Theta1$) Θ_n is a symmetric matrix of degree $n + 1$ with coefficients in F;

($\Theta2$) Let $\Theta_n^{(\nu)}$ denote the νth conjugate of Θ_n; then $\Theta_n^{(\nu)}$ is indefinite for $1 \leq \nu \leq r$ and positive definite for $\nu > r$.

($\Theta3$) For every $\alpha \in D$, ${}^t\Psi_n(\alpha)\Theta_n\Psi_n(\alpha) = N(\alpha)^n\Theta_n$.

($\Theta4$) For every $\beta \in D^{(\nu)}$, ${}^t\Psi_n^{(\nu)}(\beta)\Theta_n^{(\nu)}\Psi_n^{(\nu)}(\beta) = N(\beta)^n\Theta_n^{(\nu)}$.

By the same reasoning as in the case $n = 2$, there exist real numbers $b_n^{(\nu)}$ such that

($\Theta5$) $$\Theta_n^{(\nu)} = b_n^{(\nu)t}U_n^{(\nu)-1}P_nU^{(\nu)-1} \qquad\qquad (1 \leq \nu \leq r) .$$

3.3. *Automorphic forms.* For every $w = \begin{pmatrix} a & b \\ c & d \end{pmatrix} \in M_2(R)$ and for every $z \in C$, we put

$$j(w, z) = cz + d ,$$

and, when $\det(w) \neq 0$, we define the operation of w on C by

$$w(z) = \frac{az + b}{cz + d} .$$

Let $\mathfrak{F}_r$ denote the domain in C^r defined by

$$\mathfrak{F}_r = \{(z_1, \cdots, z_r) \in C^r \,|\, \mathrm{Im}(z_1) \neq 0, \cdots, \mathrm{Im}(z_r) \neq 0\} .$$

Obviously, $\mathfrak{F}_r$ has 2^r connected components. Define the operation of an element x of the idele-group $\mathfrak{J}$ on $\mathfrak{F}_r$ by

$$x(z) = \left(x^{(1)}(z_1), \cdots, x^{(r)}(z_r)\right) \qquad\qquad \text{for } z = (z_1, \cdots, z_r) .$$

We define an invariant measure $dm(z)$ on $\mathfrak{F}_r$ by

$$dm(z) = \left(\frac{i}{2}\right)^r \mathrm{Im}(z_1)^{-2} \cdots \mathrm{Im}(z_r)^{-2}dz_1 \wedge d\bar{z}_1 \wedge \cdots \wedge dz_r \wedge d\bar{z}_r .$$

Fix a maximal order $\mathfrak{o}$ in D; and let Γ be the group of units in $\mathfrak{o}$. Then, Γ gives a properly discontinuous transformation group on $\mathfrak{F}_r$; moreover, Lemma 1.6 shows that $m(\Gamma\backslash\mathfrak{F}_r)$ is finite; and $\Gamma\backslash\mathfrak{F}_r$ is compact if D is a division algebra. If D is not a division algebra, namely, if $D = M_2(F)$, then Γ is commensurable with the so-called Hilbert modular group, and $\Gamma\backslash\mathfrak{F}_r$ is not compact.

Let $\mathfrak{a}$ be an integral two-sided $\mathfrak{o}$-ideal; and let ρ be a representation of the multiplicative group $(\mathfrak{o}/\mathfrak{a})^*$ by complex matrices. Denote by $V_\mathfrak{a}$ the set of ideles x such that $x_\mathfrak{p}$ is a unit of $\mathfrak{o}_\mathfrak{p}$ for every $\mathfrak{p}$ dividing $N(\mathfrak{a})$, and by $W_\mathfrak{a}$ the set of ideles $x \in V_\mathfrak{a}$ such that $x_\mathfrak{p} \equiv 1 \bmod \mathfrak{a}_\mathfrak{p}$ for every $\mathfrak{p}$ dividing $N(\mathfrak{a})$. Then $V_\mathfrak{a}/W_\mathfrak{a}$ is canonically isomorphic to $(\mathfrak{o}/\mathfrak{a})^*$; so we consider ρ also as a representation of $V_\mathfrak{a}$ which is trivial on $W_\mathfrak{a}$.

Take a set of non-negative integers $\{n_1, \cdots, n_t\}$ such that $n_\nu > 0$ for $1 \leqq \nu \leqq r$. Put, for every element x of the adele-ring $\mathfrak{A}$,

$$J(x, z) = \prod_{\nu=1}^{r} j(x^{(\nu)}, z_\nu)^{n_\nu} \qquad (z = (z_1, \cdots, z_r) \in \mathfrak{F}_r) ,$$

$$\Phi(x) = \Phi_{n_{r+1}}(x^{(r+1)}) \otimes \cdots \otimes \Phi_{n_t}(x^{(t)}) ,^4$$

$$L(x) = \prod_{\nu=1}^{t} |N(x^{(\nu)})|^{n_\nu/2} ,$$

and for every $x \in V_\mathfrak{a}$,

$$\chi(x) = L(x)^{-1}\rho(x) \otimes \Phi(x) .$$

Now we understand by an *integral automorphic form* of type $(\Gamma; \rho; \{n_\nu\})$ a vector function $f(z)$ on $\mathfrak{F}_r$ taking values in the representation-space of $\rho \otimes \Phi$ (considered as complex vector space) and satisfying the following conditions (A1, 2), and besides, (A3) when $D = M_2(F)$.

(A1) *The components of $f(z)$ are holomorphic functions on $\mathfrak{F}_r$.*

(A2) *For every $\gamma \in \Gamma$, $f(\gamma(z)) = \chi(\gamma)J(\gamma, z)f(z)$.*

Now suppose that $D = M_2(F)$; then we have $r = t$, so that $\chi(x) = L(x)^{-1}\rho(x)$. Let $\Gamma_\mathfrak{a}$ be the subgroup of Γ consisting of elements γ such that $\gamma \equiv 1 \bmod \mathfrak{a}$. The condition (A3) is then as follows.

(A3) *$f(z)$ is regular at every parabolic point of $\Gamma_\mathfrak{a}$.*

In other words, the components of $f(z)$ have Fourier expansions without terms of negative exponent at every parabolic point of $\Gamma_\mathfrak{a}$ (cf. Maass [19, 20], Gundlach [10], Shimizu [23]). It has been shown in [10, 23] that if $r > 1$, the condition (A3) is a consequence of (A1) and (A2). A similar result holds also in the case of Siegel modular forms of degree > 1 [Koecher 17].

An integral automorphic form $f(z)$ is called a *cusp form* if either D is a division algebra or $D = M_2(F)$ and the Fourier expansions of $f(z)$ at parabolic points have vanishing constant terms. We denote by $S(\Gamma; \rho; \{n_\nu\})$ the set of cusp forms of type $(\Gamma; \rho; \{n_\nu\})$. The vector space $S(\Gamma; \rho; \{n_\nu\})$ is of finite demension; an explicit formula for the dimension in the case

---

[4] For a set of non-negative integers $\{m_1, \cdots, m_t\}$, put

$$\Psi(x) = \Phi_{m_1}(x^{(1)}) \otimes \cdots \otimes \Phi_{m_t}(x^{(t)}) .$$

By a theorem of Borel [1], the restriction of Ψ to Γ is irreducible if $m_{r+1} = \cdots = m_t = 0$. One may conjecture that this is so for any set $\{m_\nu\}$. Of course a similar conjecture may be made about arithmetically defined discontinuous subgroups of any semi-simple Lie groups with compact components.

where $n_{r+1} = \cdots = n_t = 0$ is given in Shimizu [23].

We note that $f(z)\prod_{\nu=1}^{r}|\mathrm{Im}(z_\nu)|^{n_\nu/2}$ is bounded on $\mathfrak{F}_r$ if f is a cusp-form. This is easily shown by the same argument as in Hecke [13, I, Satz 5].

Now we observe that, if ρ is the sum of irreducible representations $\rho_1, \cdots, \rho_s$, then the vector space $S(\Gamma; \rho; \{n_\nu\})$ is the sum of the spaces $S(\Gamma; \rho_i; \{n_\nu\})$ for $1 \leq i \leq s$. Assume that ρ itself is irreducible and $S(\Gamma; \rho; \{n_\nu\})$ contains a function f other than 0. Then, for every unit ε of F, the condition (A2) implies $f(z) = \prod_{\nu=1}^{t}\mathrm{sgn}(\varepsilon^{(\nu)})^{n_\nu}\rho(\varepsilon)f(z)$. As ε is contained in the center of V_a, and ρ is irreducible, the matrix $\rho(\varepsilon)$ must be a scalar matrix; so we have

$$(10) \qquad \rho(\varepsilon) = \prod_{\nu=1}^{t}\mathrm{sgn}(\varepsilon^{(\nu)})^{n_\nu} \qquad \text{for every unit } \varepsilon \text{ of } F.$$

Therefore, we assume henceforth that the representation ρ of V_a satisfies (10). By making this assumption, we lose nothing of importance.

Let $\mathfrak{x}_\lambda, \mathfrak{o}_\lambda, x_\lambda, \Gamma_\lambda$ be as in §1.4; we can take the $\mathfrak{x}_\lambda$ so that $\mathfrak{x}_{\lambda\mathfrak{p}} = \mathfrak{o}_\mathfrak{p}$ for every prime factor $\mathfrak{p}$ of $N(\mathfrak{a})$. Hereafter we always assume this, and choose x_λ so that $x_{\lambda\mathfrak{p}} = 1$ for every prime factor $\mathfrak{p}$ of $N(\mathfrak{a})$ and $x_\lambda^{(\nu)} = 1$ for every ν; and we put $x_{\lambda\mu} = x_\lambda x_\mu^{-1}$. Then, the groups Γ_λ are contained in V_a; and hence we can define the set $S(\Gamma_\lambda; \rho; \{n_\nu\})$. We put

$$S(\rho; \{n_\nu\}) = S(\Gamma_1; \rho; \{n_\nu\}) \times \cdots \times S(\Gamma_h; \rho; \{n_\nu\}) .$$

For the sake of simplicity, we fix ρ and $\{n_\nu\}$, and write simply

$$S_\lambda = S(\Gamma_\lambda; \rho; \{n_\nu\}) \qquad\qquad (1 \leq \lambda \leq h) ,$$
$$S = S(\rho; \{n_\nu\}) = S_1 \times \cdots \times S_h .$$

When we speak of an element $f = (f_1, \cdots, f_h)$ of S, f_λ means the projection of f on the λth factor S_λ.

REMARK 1. Let Γ_a^0 be the group of units γ of $\mathfrak{o}$ such that $N(\gamma) = 1$ and $\gamma \equiv 1 \bmod \mathfrak{a}$. Then Γ_a^0 is a properly discontinuous transformation group on $\mathfrak{F}_r$ and $\Gamma_a^0 \backslash \mathfrak{F}_r$ is of finite measure. Therefore, it would be meaningful to consider the automorphic forms on $\mathfrak{F}_r$ with respect to Γ_a^0, i.e., the holomorphic vector functions $f(z)$ on $\mathfrak{F}_r$ satisfying $f(\gamma(z)) = \Phi(\gamma)J(\gamma, z)f(z)$ for every $\gamma \in \Gamma_a^0$. However, the functions of this type are essentially included in the above formulation. In fact, let W be the vector space of such functions f. For every $f \in W$ and for every $\gamma \in \Gamma$, define f^γ by $f^\gamma(z) = L(\gamma)\Phi(\gamma)^{-1}J(\gamma, z)^{-1}f(\gamma(z))$. Then, $f \to f^\gamma$ gives a C-linear mapping of W into itself. We get in this way a representation ρ_1 of Γ in the vector space W. Let E be the group of totally positive units of F. Then it is clear that ρ_1 is trivial on $E\Gamma_a^0$. Now, for every integral two-sided $\mathfrak{o}$-ideal $\mathfrak{b}$, denote by $\Gamma_\mathfrak{b}$ the group of units γ of $\mathfrak{o}$ such that $\gamma \equiv 1 \bmod \mathfrak{b}$, and put $E_\mathfrak{b} = \Gamma_\mathfrak{b} \cap E$. By a theorem of Chevalley [2, Théorème 1], there exists a positive rational integer b with the following property: every unit α of F such that $\alpha \equiv 1 \bmod b\mathfrak{g}$ is totally positive, and every element ε of E_α such that $\varepsilon \equiv 1 \bmod b\mathfrak{g}$ is the square of an element of E_α. Taking such an integer b, let γ be an element of $\Gamma_{b\mathfrak{a}}$; then there exists an element ε of E_α such that $\varepsilon^2 = N(\gamma)$; obviously $\varepsilon^{-1}\gamma$ is contained in $\Gamma_\mathfrak{a}^0$, so that $\gamma = \varepsilon \cdot \varepsilon^{-1}\gamma \in E\Gamma_\mathfrak{a}^0$. This shows $\Gamma_{b\mathfrak{a}} \subset \Gamma_\mathfrak{a}^0$. It follows that ρ_1 is trivial on $\Gamma_{b\mathfrak{a}}$. The group $\Gamma/\Gamma_{b\mathfrak{a}}$ may be regarded as a subgroup of $(\mathfrak{o}/b\mathfrak{a})^*$. Therefore, if we take a suitable representation ρ of $(\mathfrak{o}/b\mathfrak{a})^*$, then ρ_1 can be found as a component of the restriction of ρ to $\Gamma/\Gamma_{b\mathfrak{a}}$. Thus the investigation of W is reduced to that of $S(\Gamma; \rho; \{n_\nu\})$.

REMARK 2. Let Γ_+ be the subgroup of Γ consisting of the elements γ such that $N(\gamma^{(\nu)}) > 0$ for every ν. Let the $\mathfrak{H}_i$, for $1 \leq i \leq 2^r$, be the connected components of the domain $\mathfrak{F}_r$. Then Γ_+ operates on each $\mathfrak{H}_i$; so we can consider automorphic forms with respect to Γ_+ defined on each $\mathfrak{H}_i$. Let $S(\Gamma_+; \mathfrak{H}_i)$ be the set of holomorphic vector functions f on $\mathfrak{H}_i$ satisfying $f(\gamma(z)) = \chi(\gamma)J(\gamma, z)f(z)$ for every $\gamma \in \Gamma_+$. If, for some i and j, there exists an element δ of Γ such that $\delta(\mathfrak{H}_i) = \mathfrak{H}_j$, then the mapping $f \to \chi(\delta)^{-1}J(\delta, z)^{-1}f(\delta(z))$ gives an isomorphism of $S(\Gamma_+; \mathfrak{H}_j)$ onto $S(\Gamma_+; \mathfrak{H}_i)$. When there is not such a δ, we say that $\mathfrak{H}_i$ and $\mathfrak{H}_j$ are independent. Let E_1 (resp. E_2) be the group of units ε of F such that $\varepsilon^{(\nu)} > 0$ for $\nu > r$ (resp. for every ν). Put $[E_1 : E_2] = 2^s$; then, by Lemma 1.3, $[\Gamma : \Gamma_+] = 2^s$. We observe that there exist exactly 2^{r-s} independent $\mathfrak{H}_i$; and $S(\Gamma; \rho; \{n_\nu\})$ is canonically isomorphic to the direct sum of the 2^{r-s} vector spaces $S(\Gamma_+; \mathfrak{H}_i)$ for such $\mathfrak{H}_i$. If $F = \mathbf{Q}$, we have $2^{r-s} = 1$. Therefore, in this case, we may treat Γ_+ and the half plane $\mathfrak{H}_1$ instead of Γ and $\mathfrak{F}_1$.

3.4. *The representation of the Hecke-ring in* S. Let $\mathfrak{E}_\mathfrak{a}$ be the set of elements e of $\mathfrak{E}$ which are prime to $\mathfrak{a}$, and $\mathfrak{R}_\mathfrak{a}$ the submodule of $\mathfrak{R}$ generated by the $T(e)$ for $e \in \mathfrak{E}_\mathfrak{a}$. It is easy to see that $\mathfrak{R}_\mathfrak{a} = \mathfrak{R}(U; V_\mathfrak{a})$, so that it is a subring of $\mathfrak{R}$. Now we want to define a representation of $\mathfrak{R}_\mathfrak{a}$ in the vector space S. Let e be an element of $\mathfrak{E}_\mathfrak{a}$ and $T_{\mu\lambda}(e) = \Gamma_\mu \alpha \Gamma_\lambda = \bigcup_i \alpha_i \Gamma_\lambda$. Then, the α_i are contained in $V_\mathfrak{a}$. For every $f_\lambda \in S_\lambda$, define $g_\mu = \mathfrak{T}_{\mu\lambda}(e)f_\lambda$ by

$$g_\mu(z) = \sum_i \chi(\alpha_i)J(\alpha_i^{-1}, z)^{-1}f_\lambda(\alpha_i^{-1}(z)) \, .$$

We see easily that g_μ is contained in S_μ; so $\mathfrak{T}_{\mu\lambda}(e)$ gives a C-linear mapping of S_λ into S_μ. For every $f = (f_1, \cdots, f_h)$ of S, define an element $g = \mathfrak{T}(e)f$ of S by

$$g = (g_1, \cdots, g_h) : g_\mu = \mathfrak{T}_{\mu\lambda}(e)f_\lambda \, .$$

This definition is in fact possible, because, for every μ, there exists one and only one λ such that $T_{\mu\lambda}(e)$ has a meaning (cf. §2.2). By our definition of multiplication in the ring $\mathfrak{R}$, $T(e) \to \mathfrak{T}(e)$ gives a representation of $\mathfrak{R}_\mathfrak{a}$

in the vector space S. We denote by $\mathfrak{X}(\mathfrak{n})$ and $\mathfrak{X}(\mathfrak{p}^a, \mathfrak{p}^b)$ the linear mappings corresponding to $T(\mathfrak{n})$ and $T(\mathfrak{p}^a, \mathfrak{p}^b)$, respectively.

3.5. *Inner product.* Now assume that n_ν is even for $\nu > r$. Notation being as in § 3.2, put

$$U = U^{(r+1)}_{n_{r+1}} \otimes \cdots \otimes U^{(t)}_{n_t} \,,$$

$$\Psi(x) = U\Phi(x)U^{-1} \,,$$

$$\Theta = \Theta^{(r+1)}_{n_{r+1}} \otimes \cdots \otimes \Theta^{(t)}_{n_t} \,.$$

We observe that it makes essentially no difference if we consider Ψ in place of Φ in the definition of $S(\Gamma; \rho; \{n_\nu\})$ and $\mathfrak{X}_{\mu\lambda}(e)$. Therefore, in this section, we consider S_λ as the set of cusp-forms f satisfying $f(\gamma(z)) = L(\gamma)^{-1}\rho(\gamma) \otimes \Psi(\gamma)J(\gamma, z)f(z)$ for $\gamma \in \Gamma_\lambda$ instead of (A2); and we modify the definition of $\mathfrak{X}_{\mu\lambda}(e)$ correspondingly.

As ρ is a representation of a finite group $(\mathfrak{o}/\mathfrak{a})^*$, there exists a positive definite hermitian matrix θ such that

$${}^t\rho(v)\theta\rho(v) = \theta \qquad\qquad\qquad\qquad (v \in V_a) \,.$$

Fix such a matrix θ once for all, and put, for every f_λ and g_λ of S_λ,

$$<f_\lambda, g_\lambda> = \int_{B_\lambda} {}^t f_\lambda(z)(\theta \otimes \Theta)\overline{g_\lambda(z)}\,|\operatorname{Im}(z_1)^{n_1} \cdots \operatorname{Im}(z_r)^{n_r}|\, dm(z) \,,$$

where B_λ is a fundamental domain for Γ_λ, and $dm(z)$ is the volume element of $\mathfrak{F}_r$ given in § 3.3. We see easily that this integral converges. Then, for two elements $f = (f_1, \cdots, f_h)$ and $g = (g_1, \cdots, g_h)$ of S, we put

$$<f, g> = \sum_{\lambda=1}^{h} <f_\lambda, g_\lambda> \,.$$

This is a hermitian form on the vector space S, which is obviously positive non-degenerate.

Let us consider the mapping $UxU \to Ux^{-1}U$. By the same argument as in the proof of [25, Proposition 1.2], and by the commutativity of $\mathfrak{R}$, we can easily prove that this determines an automorphism of the ring $\mathfrak{R} = \mathfrak{R}(U; \mathfrak{J})$. This is also given as follows. Let e be an element of $\mathfrak{E}$ and $\{\mathfrak{p}_1^{c_1}, \cdots, \mathfrak{p}_s^{c_s}\}$ be the set of invariants of e. We can find an element of $\mathfrak{E}$, which we denote by e^{-1}, having $\{\mathfrak{p}_1^{-c_1}, \cdots, \mathfrak{p}_s^{-c_s}\}$ as its set of invariants. Then the automorphism of $\mathfrak{R}$ is given by $T(e) \to T(e^{-1})$. If we have $T_{\mu\lambda}(e) = \Gamma_\mu \alpha \Gamma_\lambda$, then $T_{\lambda\mu}(e^{-1}) = \Gamma_\lambda \alpha^{-1} \Gamma_\mu$.

PROPOSITION 3.1. *For every element e of $\mathfrak{E}_a$ and for every f, g of S,*

$$<\mathfrak{X}(e)f, g> = <f, \mathfrak{X}(e^{-1})g> \,.$$

PROOF. It is sufficient to prove $<\mathfrak{X}_{\mu\lambda}(e)f_\lambda, g_\mu> = <f_\lambda, \mathfrak{X}_{\lambda\mu}(e^{-1})g_\mu>$. Let $T_{\mu\lambda}(e) = \Gamma_\mu \alpha \Gamma_\lambda$, and let $\Gamma_\mu = \bigcup_i \gamma_i(\alpha_\lambda \Gamma \alpha^{-1} \cap \Gamma_\mu)$ be a disjoint sum; put

$\alpha_i = \gamma_i\alpha$. Then $\Gamma_\mu\alpha\Gamma_\lambda = \bigcup_i \alpha_i\Gamma_\lambda$ is a disjoint sum. Let B_λ and B_μ be respectively fundamental domains for Γ_λ and Γ_μ. By our definition, we have

$$<\mathfrak{T}_{\mu\lambda}(e)f_\lambda,\, g_\mu>$$

$$= \sum_i \int_{B_\mu} {}^t f_\lambda(\alpha_i^{-1}(z))^t \chi(\alpha_i)(\theta \otimes \Theta)\overline{g_\mu(z)}J(\alpha_i^{-1},\, z)^{-1}\,|\,\mathrm{Im}(z_1)^{n_1} \cdots \mathrm{Im}(z_r)^{n_r}\,|\,dm(z)$$

$$= \sum_i \int_{\alpha_i^{-1}(B_\mu)} {}^t f_\lambda(z)^t \chi(\alpha_i)(\theta \otimes \Theta)\overline{g_\mu(\alpha_i(z))}J(\alpha_i^{-1},\, \alpha_i(z))^{-1}$$

$$\cdot\,|\,\mathrm{Im}(\alpha_i^{(1)}(z_1))^{n_1} \cdots \mathrm{Im}(\alpha_i^{(r)}(z_r))^{n_r}\,|\,dm(z)$$

$$= \sum_i \int_{\alpha_i^{-1}(B_\mu)} {}^t f_\lambda(z)(\theta \otimes \Theta)\chi(\alpha^{-1})\overline{g_\mu(\alpha(z))}J(\alpha,\, z)^{-1}\,|\,\mathrm{Im}(z_1)^{n_1}\cdots \mathrm{Im}(z_r)^{n_r}\,|\,dm(z).$$

Now $\bigcup_i \alpha_i^{-1}(B_\mu)$ is a fundamental domain for $\Gamma_\lambda \cap \alpha^{-1}\Gamma_\mu\alpha$. Let $\Gamma_\lambda = \bigcup_j \delta_j(\Gamma_\lambda \cap \alpha^{-1}\Gamma_\mu\alpha)$ be a disjoint sum; and put $\delta_j\alpha^{-1} = \beta_j$; then $\Gamma_\lambda\alpha^{-1}\Gamma_\mu = \bigcup_j \beta_j\Gamma_\mu$ is a disjoint sum. And we observe that the integral on $\bigcup_i \alpha_i^{-1}(B_\mu)$ is equal to the integral on $\bigcup_j \delta_j^{-1}(B_\lambda)$. Therefore, the above integral is equal to

$$\sum_j \int_{B_\lambda} {}^t f_\lambda(\delta_j^{-1}(z))(\theta \otimes \Theta)\overline{\chi(\alpha^{-1})g_\mu(\beta_j^{-1}(z))}\overline{J(\alpha,\, \delta_j^{-1}(z))}^{-1}$$

$$\cdot\,|\,\mathrm{Im}(\delta_j^{(1)-1}(z_1))^{n_1} \cdots \mathrm{Im}(\delta_j^{(r)-1}(z_r))^{n_r}\,|\,dm(z)$$

$$= \sum_j \int_{B_\lambda} {}^t f_\lambda(z)(\theta \otimes \Theta)\overline{\chi(\beta_j)g_\mu(\beta_j^{-1}(z))}J(\beta_j^{-1},\, z)^{-1}\,|\,\mathrm{Im}(z_1)^{n_1} \cdots \mathrm{Im}(z_r)^{n_r}\,|\,dm(z)$$

$$= <f_\lambda,\, \mathfrak{T}_{\lambda\mu}(e^{-1})g_\mu> .$$

Our proposition is thereby proved.

By the commutativity of $\mathfrak{R}$ and by the above proposition, the ring $\mathfrak{R}_a$ gives a commutative ring of normal operators on S. We obtain therefore

PROPOSITION 3.2. *If n_ν is even for $\nu > r$, the elements of $\mathfrak{R}_a$ are represented by diagonal matrices with respect to a suitable basis of S.*

3.6. *The operator $\mathfrak{Y}$.* Let the $\mathfrak{k}_\lambda$ be as in § 1.4. For every $\lambda \in \{1, \cdots, h\}$, define $\bar\lambda$ by the relation $\mathfrak{k}_\lambda^{-1} = \mathfrak{k}_{\bar\lambda}$. Then $\lambda \to \bar\lambda$ gives a permutation of $\{1, \cdots, h\}$ of order 2. As $N(\mathfrak{x}_\lambda) = N(\mathfrak{x}_\lambda')$, the lattices $(\mathfrak{x}_\lambda')^{-1}$ and $\mathfrak{x}_{\bar\lambda}$ belong to the same equivalence-class; so there exists an element δ_λ of D^* such that $\delta_\lambda(\mathfrak{x}_\lambda')^{-1} = \mathfrak{x}_{\bar\lambda}$. We see that such an element δ_λ belongs to V_a; so we can consider $\rho(\delta_\lambda)$. It is easy to verify $\mathfrak{o}_{\bar\lambda} = \delta_\lambda\mathfrak{o}_\lambda\delta_\lambda^{-1}$ and $\Gamma_{\bar\lambda} = \delta_\lambda\Gamma_\lambda\delta_\lambda^{-1}$. Now, for every element f_λ of S_λ, define $g_{\bar\lambda} = \mathfrak{Y}_\lambda f_\lambda$ by

$$g_{\bar\lambda}(z) = \chi(\delta_\lambda)J(\delta_\lambda^{-1},\, z)^{-1}f_\lambda(\delta_\lambda^{-1}(z)) .$$

Then we can easily verify that $g_{\bar\lambda}$ is contained in $S_{\bar\lambda}$; and hence $\mathfrak{Y}_\lambda$ gives a C-linear mapping of S_λ into $S_{\bar\lambda}$. Futhermore we observe that $\mathfrak{Y}_\lambda$ does not depend on the choice of δ_λ, and $\mathfrak{Y}_{\bar\lambda} = \mathfrak{Y}_\lambda^{-1}$. We define the operator $\mathfrak{Y}$

on S by

$$\mathfrak{D}(f_1, \cdots, f_h) = (g_1, \cdots, g_h), \qquad g_{\bar{\lambda}} = \mathfrak{D}_\lambda f_\lambda .$$

Obviously, we have $\mathfrak{D}^2 = 1$.

PROPOSITION 3.3. *Let α be an element of D^*; let e_1 and e_2 be respectively the similarity-classes of $\alpha \mathfrak{x}_{\lambda\mu}$ and $\alpha \mathfrak{x}'_{\mu\lambda}$. Then*

$$\mathfrak{D}_\mu \mathfrak{X}_{\mu\lambda}(e_1) = \mathfrak{X}_{\overline{\mu\lambda}}(e_2)\mathfrak{D}_\lambda .$$

PROOF. Take δ_λ, for each λ, as above. Then we have $\mathfrak{x}'_{\mu\lambda} = \delta_\lambda^{-1}\mathfrak{x}_{\overline{\lambda\mu}}\delta_\mu$, so that $\alpha\mathfrak{x}'_{\mu\lambda} = \delta_\mu^{-1}(\delta_\mu\alpha\delta_\lambda^{-1})\mathfrak{x}_{\overline{\lambda\mu}}\delta_\mu$. Therefore, $\delta_\mu\alpha\delta_\lambda^{-1}\mathfrak{x}_{\overline{\lambda\mu}}$ belongs to e_2. Now we have $\Gamma_{\bar{\mu}}\delta_\mu\alpha\delta_\lambda^{-1}\Gamma_{\bar{\lambda}} = \delta_\mu\Gamma_\mu\alpha\Gamma_\lambda\delta_\lambda^{-1}$. Let $\Gamma_\mu\alpha\Gamma_\lambda = \bigcup_i \alpha_i\Gamma_\lambda$ be a disjoint sum; put $\delta_\mu\alpha_i\delta_\lambda^{-1} = \beta_i$. Then $\Gamma_{\bar{\mu}}(\delta_\mu\alpha\delta_\lambda^{-1})\Gamma_{\bar{\lambda}} = \bigcup_i \beta_i\Gamma_{\bar{\lambda}}$ is a disjoint sum, and $\delta_\mu\alpha_i\Gamma_\lambda = \beta_i\delta_\lambda\Gamma_\lambda$ for every i. Hence, by our definition of the operators $\mathfrak{X}_{\mu\lambda}(e)$ and $\mathfrak{D}_\lambda$ on S_λ, we obtain the formula of the proposition.

REMARK. Notation being as above, we have $\Gamma_{\bar{\lambda}}\delta_\lambda\Gamma_\lambda = \delta_\lambda\Gamma_\lambda$ and $\delta_\lambda\mathfrak{x}_{\lambda\bar{\lambda}} = \delta_\lambda N(\mathfrak{x}_\lambda)\delta_\lambda^{-1}$. Therefore, if e_λ denotes the similarity-class of $N(\mathfrak{x}_\lambda)\mathfrak{o}$, we have $\mathfrak{D}_\lambda = \mathfrak{X}_{\bar{\lambda}\lambda}(e_\lambda)$. By the same argument as in the proof of Proposition 3.1, we can easily show that $\mathfrak{D}$ is self-adjoint with respect to $<f, g>$.

3.7. *Transference to the idele-group.* To every element $f = (f_1, \cdots, f_h)$ of S, we attach a function $\mathfrak{f}$ on the idele-group $\mathfrak{F}$ as follows. Notation being as in § 1.4, we define a function $\mathfrak{f}_\lambda$ on U_λ by

$$\mathfrak{f}_\lambda(v) = \chi(v)^{-1} J(v, i)^{-1} f_\lambda(v(i)) ,$$

where i denotes the point $(i, \cdots, i)$ of $\mathfrak{F}_r$. We see easily $\mathfrak{f}_\lambda(\gamma v) = \mathfrak{f}_\lambda(v)$ for every γ of Γ_λ. Hence, we can extend the function $\mathfrak{f}_\lambda$ to $D^* U_\lambda$ by putting $\mathfrak{f}_\lambda(\alpha v) = \mathfrak{f}_\lambda(v)$ for $\alpha \in D^*$ and $v \in U_\lambda$. Then define a function $\mathfrak{f}$ on $\mathfrak{F}$ by $\mathfrak{f}(x) = \mathfrak{f}_\lambda(x x_\lambda^{-1})$ for $x \in D^* x_\lambda U$. Since $\mathfrak{F}$ is a disjoint sum of the $D^* x_\lambda U$ for $1 \leq \lambda \leq h$, this determines the function $\mathfrak{f}$ on $\mathfrak{F}$. We see easily that $\mathfrak{f}$ is continuous on $\mathfrak{F}$ and satisfies

$$(11) \qquad\qquad \mathfrak{f}(\alpha x) = \mathfrak{f}(x) \qquad\qquad \text{for } \alpha \in D^* .$$

We note that the functions $\mathfrak{f}_\lambda$ have the property

$$(12) \quad \mathfrak{f}_\lambda(x'^{-1}) = \left\{\prod_{\nu=1}^r \operatorname{sgn} N(x^{(\nu)})^{n_\nu}\right\} L(x)\rho(x') \otimes \Phi(x)^{-1} J(x, i)^{-1} f_\lambda(x(i))$$

$$(x \in U_\lambda) .$$

PROPOSITION 3.4. *The function $\mathfrak{f}$ is bounded.*

PROOF. Let w be an element of $\mathfrak{F}$ such that $w_\mathfrak{p} = 1$ for every prime ideal $\mathfrak{p}$ of F and $w^{(\nu)}$ is a positive real number for every ν. Then we see easily that $\mathfrak{f}(wx) = \mathfrak{f}(x)$. By (11), $\mathfrak{f}$ is considered as continuous function on $D^*\backslash\mathfrak{F}$. By Lemma 1.6, $D^*\backslash\mathfrak{F}_0$ is compact if D is a division algebra, so that $\mathfrak{f}$ must be bounded on $\mathfrak{F}$. If $D = M_2(F)$, our assertion follows from the fact that $\prod_{\nu=1}^{r}|\operatorname{Im}(z_\nu)|^{n_\nu/2}f(z)$ is bounded on $\mathfrak{F}_r$.

4. Functional equations

4.1. *Haar measures on $\mathfrak{A}$ and $\mathfrak{F}$.* Put $\mathfrak{b} = \mathfrak{d}(\mathfrak{o}/Z)$; define a Haar measure $dm_\mathfrak{p}$ of the additive group $D_\mathfrak{p}$ by the condition $m_\mathfrak{p}(\mathfrak{o}_\mathfrak{p}) = N_1(\mathfrak{b}_\mathfrak{p})^{-1/2}$, where $N_1(\mathfrak{b}_\mathfrak{p})$ denotes the number of elements in $\mathfrak{o}_\mathfrak{p}/\mathfrak{b}_\mathfrak{p}$ for any integral $\mathfrak{o}_\mathfrak{p}$-ideal $\mathfrak{b}_\mathfrak{p}$. For every ν, define a Haar measure $dm_\nu(x)$ of $D^{(\nu)}$ by

$$dm_\nu\begin{pmatrix}\alpha & \beta \\ \gamma & \delta\end{pmatrix} = d\alpha d\beta d\gamma d\delta \qquad (1 \leq \nu \leq r),$$

$$dm_\nu(\alpha + \beta i + \gamma j + \delta ij) = 4 \cdot d\alpha d\beta d\gamma d\delta \qquad (\nu > r),$$

where $i = \sqrt{-1}$ and j is the element of K given in §3.1. Then the product $dm(x) = \prod_\mathfrak{p} dm_\mathfrak{p}(x_\mathfrak{p})\prod_\nu dm_\nu(x^{(\nu)})$ gives a Haar measure of the additive group $\mathfrak{A}$ satisfying $m(\mathfrak{A}/D) = 1$. We see easily $dm(ax) = |a|^2 dm(x)$ for every $a \in \mathfrak{F}$, where $|a|$ is the positive real number defined in §1.4. Let $U_\mathfrak{p}$ be the group of units in $\mathfrak{o}_\mathfrak{p}$; define a Haar measure $dm_\mathfrak{p}^*$ of the multiplicative group $D_\mathfrak{p}^*$ by the condition $m_\mathfrak{p}^*(U_\mathfrak{p}) = 1$. Take and fix any Haar measure dm_ν^* of the group $D^{(\nu)*}$. Then the product $dm^*(x) = \prod_\mathfrak{p} dm_\mathfrak{p}^*(x_\mathfrak{p}) \cdot \prod_\nu dm_\nu^*(x)$ gives a Haar measure of $\mathfrak{F}$; and we have $dm^*(x) = c|x|^{-2}dm(x)$ for a suitable constant c.

Let τ_p, for every prime number p, be the complex-valued function on Q_p defined by $\tau_p(\sum_{n \geq n_0} c_p p^n) = \exp(2\pi i \sum_{n<0} c_n p^n)$, where $0 \leq c_n < p$. For every prime ideal $\mathfrak{p}$ of F, define a function $\tau_\mathfrak{p}$ on $D_\mathfrak{p}$ by $\tau_\mathfrak{p}(x) = \tau_p[\operatorname{tr}_{D_\mathfrak{p}/Q_p}(x)]$; and define a function $\tau^{(\nu)}$ on $D^{(\nu)}$ by $\tau^{(\nu)}(x) = \exp[-2\pi i \operatorname{tr}(x)]$. We put then, for $x \in \mathfrak{A}$, $\tau(x) = \prod_\mathfrak{p} \tau_\mathfrak{p}(x_\mathfrak{p})\prod_\nu \tau^{(\nu)}(x^{(\nu)})$.

4.2. *The matrices $\sigma_\mathfrak{m}(x; \omega)$.* Let U_0 be the subgroup of U consisting of the elements x of U such that $x^{(\nu)} = 1$ for every ν. Let ω be a continuous representation of U_0 by complex matrices. Then, ω is written in the form $\omega(x) = \prod_\mathfrak{p}\omega_\mathfrak{p}(x_\mathfrak{p})$, where $\omega_\mathfrak{p}$ is a representation of $U_\mathfrak{p}$; and there exists an integral two-sided $\mathfrak{o}$-ideal $\mathfrak{m}$ such that $\omega_\mathfrak{p}(x_\mathfrak{p}) = 1$ for $x_\mathfrak{p} \equiv 1 \bmod \mathfrak{m}_\mathfrak{p}$. We say then that ω is defined modulo $\mathfrak{m}$ and call $\mathfrak{m}$ a defining-ideal of ω. Among the defining-ideals of ω, there exists one which divides all the others; we call it *the conductor of ω.* Now consider the integral

$$\sigma(x_\mathfrak{p}; \omega) = \int_{\sigma_\mathfrak{p}} \tau_\mathfrak{p}(x_\mathfrak{p} y_\mathfrak{p})\omega_\mathfrak{p}(y_\mathfrak{p})^{-1}dm_\mathfrak{p}^*(y_\mathfrak{p}) \qquad (x_\mathfrak{p} \in D_\mathfrak{p}).$$

It can be easily seen that $\sigma_\mathfrak{p}(x_\mathfrak{p}; \omega)$ and $\sigma_\mathfrak{q}(x_\mathfrak{q}; \omega)$ commute with each other for any $\mathfrak{p}$ and $\mathfrak{q}$. Therefore, for every defining-ideal $\mathfrak{m}$ of ω, we can define a matrix-valued function $\sigma_\mathfrak{m}(x; \omega)$ on $\mathfrak{A}$ by

$$\sigma_\mathfrak{m}(x; \omega) = \prod_{\mathfrak{p}|N(\mathfrak{m})}\sigma_\mathfrak{p}(x_\mathfrak{p}; \omega).$$

PROPOSITION 4.1. *Let $\mathfrak{m}$ be a defining-ideal of ω. Let v be an element* *of $\mathfrak{J}$ such that $v_\mathfrak{p} \in U_\mathfrak{p}$ for every prime factor $\mathfrak{p}$ of $N(\mathfrak{m})$. Put $\omega_\mathfrak{m}(v) = \prod_{\mathfrak{p}|N(\mathfrak{m})} \omega_\mathfrak{p}(v_\mathfrak{p})$. Then*

$$\sigma_\mathfrak{m}(xv; \omega) = \sigma_\mathfrak{m}(x; \omega)\omega_\mathfrak{m}(v) ,$$

$$\sigma_\mathfrak{m}(vx; \omega) = \omega_\mathfrak{m}(v)\sigma_\mathfrak{m}(x; \omega) .$$

This follows easily from our definition of $\sigma_\mathfrak{m}(x; \omega)$ and the property $\tau_\mathfrak{p}(xy) = \tau_\mathfrak{p}(yx)$.

PROPOSITION 4.2. *Let $\mathfrak{a}$ be the conductor of ω. Suppose that there exists, for each prime factor $\mathfrak{p}$ of $N(\mathfrak{a})$, a right or left $\mathfrak{o}_\mathfrak{p}$-ideal $\mathfrak{b}_\mathfrak{p}$ with the following properties:*

(i) *$\mathfrak{b}_\mathfrak{p} \supset \mathfrak{a}_\mathfrak{p}$ and $\mathfrak{p}N(\mathfrak{b}_\mathfrak{p}) = N(\mathfrak{a}_\mathfrak{p})$;*

(ii) *put $Y_\mathfrak{p} = \{y_\mathfrak{p} | y_\mathfrak{p} \in U_\mathfrak{p}, y_\mathfrak{p} \equiv 1 \bmod \mathfrak{b}_\mathfrak{p}\}$; then the restriction of $\omega_\mathfrak{p}$ to $Y_\mathfrak{p}$ contains no unit representation.*

Let x be an element of $\mathfrak{J}$ such that $x_\mathfrak{p} \in \mathfrak{b}_\mathfrak{p}^{-1}\mathfrak{a}_\mathfrak{p}^{-1}$ for every prime factor $\mathfrak{p}$ of $N(\mathfrak{a})$. Then, $\sigma_\mathfrak{a}(x: \omega) \neq 0$ only when $\mathfrak{o}_\mathfrak{p}x_\mathfrak{p} = \mathfrak{b}_\mathfrak{p}^{-1}\mathfrak{a}_\mathfrak{p}^{-1}$ for every prime factor $\mathfrak{p}$ of $N(\mathfrak{a})$.

PROOF. Let $\mathfrak{b}_\mathfrak{p}$ be a right $\mathfrak{o}_\mathfrak{p}$-ideal with the properties (i), (ii). Assume that $\mathfrak{o}_\mathfrak{p}x_\mathfrak{p} \neq \mathfrak{b}_\mathfrak{p}^{-1}\mathfrak{a}_\mathfrak{p}^{-1}$. Then we have $x_\mathfrak{p}^{-1}\mathfrak{o}_\mathfrak{p} \supsetneqq \mathfrak{a}_\mathfrak{p}\mathfrak{b}_\mathfrak{p}$, so that there exists a unit ε of $\mathfrak{o}_\mathfrak{p}$ such that $\varepsilon x_\mathfrak{p}^{-1}\mathfrak{b}_\mathfrak{p}^{-1} \supset \mathfrak{b}_\mathfrak{p}$. Let $U_\mathfrak{p} = \bigcup_i Y_\mathfrak{p}w_i$ be a disjoint expression. We have then, for every $y_\mathfrak{p} \in Y_\mathfrak{p}$, $x_\mathfrak{p}\varepsilon^{-1}y_\mathfrak{p}w_i \equiv x_\mathfrak{p}\varepsilon^{-1}w_i \bmod \mathfrak{b}_\mathfrak{p}^{-1}$, so that $\tau_\mathfrak{p}(x_\mathfrak{p}\varepsilon^{-1}y_\mathfrak{p}w_i) = \tau_\mathfrak{p}(x_\mathfrak{p}\varepsilon^{-1}w_i)$. Therefore, we obtain

$$\sigma_\mathfrak{p}(x_\mathfrak{p}; \omega) = \int_{\mathfrak{o}_\mathfrak{p}} \tau_\mathfrak{p}(x_\mathfrak{p}\varepsilon^{-1}u_\mathfrak{p})\omega_\mathfrak{p}(\varepsilon^{-1}u_\mathfrak{p})^{-1}dm_\mathfrak{p}^*(u_\mathfrak{p})$$

$$= \sum_i \int_{Y_\mathfrak{p}} \tau_\mathfrak{p}(x_\mathfrak{p}\varepsilon^{-1}y_\mathfrak{p}w_i)^{-1}\omega_\mathfrak{p}(\varepsilon^{-1}y_\mathfrak{p}w_i)^{-1}dm_\mathfrak{p}^*(y_\mathfrak{p})$$

$$= \sum_i \omega_\mathfrak{p}(w_i)^{-1}\tau_\mathfrak{p}(x_\mathfrak{p}\varepsilon^{-1}w_i)\int_{Y_\mathfrak{p}} \omega_\mathfrak{p}(y_\mathfrak{p})^{-1}dm_\mathfrak{p}^*(y_\mathfrak{p})\omega_\mathfrak{p}(\varepsilon) .$$

As the restriction of $\omega_\mathfrak{p}$ to $Y_\mathfrak{p}$ does not contain the unit representation, we have $\int_{Y_\mathfrak{p}} \omega_\mathfrak{p}(y_\mathfrak{p})^{-1}dm_\mathfrak{p}^*(y_\mathfrak{p}) = 0$, so that $\sigma_\mathfrak{p}(x_\mathfrak{p}; \omega) = 0$. By the same argument, we can show this also in case where $\mathfrak{b}_\mathfrak{p}$ is a left $\mathfrak{o}_\mathfrak{p}$-ideal. Our proposition is thereby proved.

4.3. *Grössen-character of F.* Let $\mathfrak{J}_F$ be the idele-group of F. $\mathfrak{J}_F$ can be regarded as a subgroup of $\mathfrak{J}$ in a natural way. By a *Grössen-character* of F, we understand a continuous homomorphism ξ of $\mathfrak{J}_F$ into the multiplicative group of complex numbers with absolute value 1 such that

$$(13) \qquad\qquad \xi(\alpha) = 1 \qquad\qquad \text{for } \alpha \in F^* .$$

Every Grössen-character ξ of F can be written in the form $\xi(x) = \prod_{\mathfrak{p}}\xi_{\mathfrak{p}}(x_{\mathfrak{p}})\prod_{\nu}\xi^{(\nu)}(x^{(\nu)})$, where $\xi_{\mathfrak{p}}$ is a character of $F_{\mathfrak{p}}^*$ and $\xi^{(\nu)}$ is a character of R^*. By the continuity of ξ, there exists an integral ideal $\mathfrak{c}$ of F such that: if $\mathfrak{p}$ divides $\mathfrak{c}$, $\xi_{\mathfrak{p}}(x_{\mathfrak{p}}) = 1$ for $x_{\mathfrak{p}} \equiv 1 \bmod \mathfrak{c}_{\mathfrak{p}}$, and if $\mathfrak{p}$ does not divide $\mathfrak{c}$, $\xi_{\mathfrak{p}}(x_{\mathfrak{p}}) = 1$ for every unit $x_{\mathfrak{p}}$ of $\mathfrak{g}_{\mathfrak{p}}$. Among such ideals $\mathfrak{c}$, there exists an ideal $\mathfrak{c}_0$ which divides any other one; we call $\mathfrak{c}_0$ *the conductor of ξ*. As $\xi^{(\nu)}$ is a charactor of R^*, there exist, for each ν, a real number a_ν and an integer $b_\nu = 0$ or 1 such that

$$(14) \qquad \xi^{(\nu)}(y) = |y|^{ia_\nu}\operatorname{sgn}(y)^{b_\nu} .$$

The numbers a_ν and b_ν can not be arbitrary, in view of the relation (13). Let $\mathfrak{c}_0$ be the conductor of a Grössen-character ξ. Denote by $I(\mathfrak{c}_0)$ the group of ideals of F which are prime to $\mathfrak{c}_0$. For every ideal $\mathfrak{n}$ contained in $I(\mathfrak{c}_0)$, we can find an element x of $\mathfrak{J}_F$ such that $\mathfrak{g}_{\mathfrak{p}}x_{\mathfrak{p}} = \mathfrak{g}_{\mathfrak{p}}\mathfrak{n}$ for every $\mathfrak{p}$. Put $\xi_0(\mathfrak{n}) = \prod'\xi_{\mathfrak{p}}(x_{\mathfrak{p}})$, where the product is taken over all the $\mathfrak{p}$ which are prime to $\mathfrak{c}_0$. It is easy to see that $\xi_0(\mathfrak{n})$ does not depend on the choice of x. We get in this way a character of $I(\mathfrak{c}_0)$, which is just a Grössen-character in the sense of Hecke [11]. We call ξ_0 *the ideal-character attached to ξ*.

Now, $x \to N(x)$ gives a continuous homomorphism of $\mathfrak{J}$ into $\mathfrak{J}_F$. Put

$$(15) \qquad \begin{aligned} \xi_1(x) &= \xi\big(N(x)\big) & &\text{for } x \in \mathfrak{J} , \\ \xi_{1\mathfrak{p}}(x_{\mathfrak{p}}) &= \xi_{\mathfrak{p}}\big(N(x_{\mathfrak{p}})\big) & &\text{for } x_{\mathfrak{p}} \in D_{\mathfrak{p}}^* , \\ \xi_1^{(\nu)}(x^{(\nu)}) &= \xi^{(\nu)}\big(N(x^{(\nu)})\big) & &\text{for } x^{(\nu)} \in D^{(\nu)*} . \end{aligned}$$

Then ξ_1 is a character of $\mathfrak{J}$; and $\xi_1(x) = \prod_{\mathfrak{p}}\xi_{1\mathfrak{p}}(x_{\mathfrak{p}})\cdot\prod_{\nu}\xi_1^{(\nu)}(x^{(\nu)})$,

$$(16) \qquad \xi_1(\alpha) = 1 \qquad\qquad \text{for } \alpha \in D^* .$$

Let U_F^0 be the subgroup of $\mathfrak{J}_F \cap U$ consisting of the elements x such that $x^{(\nu)} > 0$ for $\nu > r$. Put $N(x_\lambda) = w_\lambda$. Assume that the representation ρ of V_a is irreducible. Then, as every element x of U_F^0 is contained in the center of U, $\rho(x)$ is a scalar matrix. Regarding $\rho(x)$ for $x \in U_F^0$ as complex number, we put, for $x \in U_F^0$,

$$(16') \qquad \rho^F(x) = \rho(x)\cdot\prod_{\nu=1}^r \operatorname{sgn}(x^{(\nu)})^{n_\nu} .$$

By (10), we have $\rho^F(\varepsilon) = 1$ for every unit ε of F such that $\varepsilon^{(\nu)} > 0$ for $\nu > r$. Now note that $\mathfrak{J}_F = \bigcup_{\lambda=1}^h F^*w_\lambda U_F^0$ is a disjoint expression, and define $\rho^F(x)$ for $x \in \mathfrak{J}_F$ by

$$\rho^F(\alpha w_\lambda u) = \rho^F(u) \qquad\qquad (\alpha \in F^*, u \in U_F^0) .$$

This definition is certainly possible. We get thus a Grössen-character ρ^F of F. As $x' = N(x)x^{-1}$, we have, for an element x of $\mathfrak{J}$ such that $N(x) \in U_F^0$,

$$(16'')\qquad \rho(x') = \prod_{\nu=1}^{r}\operatorname{sgn}(N(x^{(\nu)}))^{n_\nu}\rho^{r}(N(x))\rho(x)^{-1}\ .$$

If we denote by ρ_0^{r} the ideal-character attached to ρ^{r}, then, for every λ and μ, we have $\rho_0^{r}(N(\mathfrak{x}_{\lambda\mu})) = 1$. If α is an element of F^* which is prime to $\mathfrak{a}$ and such that $\alpha^{(\nu)} > 0$ for $\nu > r$, then we have

$$(16''')\qquad \rho_0^{r}((\alpha^{-1})) = \rho(\alpha)\prod_{\nu=1}^{r}\operatorname{sgn}(\alpha^{(\nu)})^{n_\nu}\ .$$

4.4. The operators $\mathfrak{X}(e;\xi)$. Let ξ be a Grössen-character of F and ξ_0 be the ideal-character attached to ξ; define a character ξ_1 of $\mathfrak{J}$ by (15). For the sake of simplicity, we assume hereafter that $\mathfrak{a}$ *is a defining ideal of ξ_1*. This is not an essential assumption; in fact, if this does not hold, then we consider a suitable multiple of $\mathfrak{a}$ in place of $\mathfrak{a}$. Put $\eta(x) = \prod_{\mathfrak{p}}\xi_{1\mathfrak{p}}(x_{\mathfrak{p}})$ for $x \in \mathfrak{J}$. Now using the symbol $\sigma_{\mathfrak{a}}(x;\eta^{-1}\rho)$, we define an operator $\mathfrak{X}(e;\xi)$ on S, for every $e \in \mathfrak{E}$, as follows. Let e be an element of $\mathfrak{E}$ and $T_{\mu\lambda}(e) = \Gamma_\mu\alpha\Gamma_\lambda = \bigcup_i\alpha_i\Gamma_\lambda$. For every $f_\lambda \in S_\lambda$, define $g_\mu = \mathfrak{X}_{\mu\lambda}(e;\xi)f_\lambda$ by

$$g_\mu(z) = \xi_0(N(\mathfrak{x}_{\lambda\mu}))\sum_i\eta(\alpha_i)L(\alpha_i)^{-1}\sigma_{\mathfrak{a}}(\alpha_i;\eta^{-1}\rho)\otimes\Phi(\alpha_i)J(\alpha_i^{-1},z)^{-1}f_\lambda(\alpha_i^{-1}(z))\ .$$

On account of Proposition 4.1, we see that g_μ is contained in S_μ, so that $\mathfrak{X}_{\mu\lambda}(e;\xi)$ gives a C-linear mapping of S_λ into S_μ. Then, for every $f = (f_1,\cdots,f_h)$ of S, define an element $g = \mathfrak{X}(e;\xi)f$ by

$$g = (g_1,\cdots,g_h)\ ,\qquad g_\mu = \mathfrak{X}_{\mu\lambda}(e;\xi)f_\lambda\ .$$

PROPOSITION 4.3. *Let e_0 and e_1 be elements of $\mathfrak{E}$. Suppose that e_0 is prime to $\mathfrak{a}$. If we have $T(e_0)T(e_1) = \sum_e c_e T(e)$, then*

$$\xi_0(N(e_0))\mathfrak{X}(e_0)\mathfrak{X}(e_1;\xi) = \mathfrak{X}(e_1;\xi)\xi_0(N(e_0))\mathfrak{X}(e_0) = \sum_e c_e\mathfrak{X}(e;\xi)\ .$$

This follows from the definition of multiplication in $\mathfrak{R}$, the commutativity of the ring $\mathfrak{R}$, and Proposition 4.1.

4.5. Dirichlet series as integrals on the idele-group. The notation ξ, ξ_0, ξ_1 being as in § 4.4, we now consider Dirichlet series

$$\mathscr{D}(s;\xi) = \sum_e\xi_0(N(e))N_{D/Q}(e)^{-s}\mathfrak{X}(e) = \sum_{\mathfrak{n}}\xi_0(\mathfrak{n})N(\mathfrak{n})^{-s}\mathfrak{X}(\mathfrak{n})\ ,$$

where e ranges over all integral elements of $\mathfrak{E}$ which are prime to $\mathfrak{a}$ and $\mathfrak{n}$ ranges over all integral ideals of F which are prime to $N(\mathfrak{a})$. The principal aim in Part I is to establish a functional equation for this Dirichlet series. We first transform it into certain integrals on the idele-group $\mathfrak{J}$.

Let m_ν and k_ν, for $1 \leq \nu \leq r$, be integers such that $0 \leq m_\nu$ and $0 \leq k_\nu \leq n_\nu$. For such k_ν, define a matrix-valued function φ on the adele-ring $\mathfrak{A}$ by

$$\varphi(x) = \{\textstyle\prod_{\mathfrak{p}}\varphi_{\mathfrak{p}}(x_{\mathfrak{p}})\}\rho(1)\otimes\varphi_\infty(x) \ ,$$

$$\varphi_r(x_r) = \begin{cases} \tau_{\mathfrak{p}}(x_{\mathfrak{p}}) & \text{for}\ \ x_{\mathfrak{p}} \in \mathfrak{d}_{\mathfrak{p}}^{-1}\mathfrak{a}_{\mathfrak{p}}^{-1}\ , \\ 0 & \text{for}\ \ x_{\mathfrak{p}} \notin \mathfrak{d}_{\mathfrak{p}}^{-1}\mathfrak{a}_{\mathfrak{p}}^{-1}\ , \end{cases}$$

$$\varphi_\infty(x) = \{\textstyle\prod_{\nu=1}^{r}\varphi^{(\nu)}(x^{(\nu)})\}\varphi^{(r+1)}(x^{(r+1)})\otimes \cdots \otimes\varphi^{(t)}(x^{(t)}) \ ,$$

$$\varphi^{(\nu)}(x) = \exp\{-\pi\mathrm{tr}(x^t x)\}\det(x)^{m_\nu}(ai + b)^{k_\nu}(ci + d)^{n_\nu - k_\nu} \quad \text{for}\ \ x = \begin{pmatrix} a & b \\ c & d \end{pmatrix},$$

$$(1 \le \nu \le r) \ ,$$

$$\varphi^{(\nu)}(x) = \exp\{-2\pi N(x)\}\Phi_{n_\nu}(x) \qquad\qquad (r < \nu \le t) \ .$$

Let y be an element of $\mathfrak{F}$ such that $y_{\mathfrak{p}} = 1$ for every $\mathfrak{p}$ and $y^{(\nu)} = 1$ for $\nu > r$. For every element f of $S(\rho; \{n_\nu\})$, define the function $\mathfrak{f}$ on $\mathfrak{A}$ as in § 3.7, and put, for $1 \le \mu \le h$,

$$\zeta_\mu = \zeta_\mu(s, \xi, f, \{k_\nu, m_\nu\}, y) = \int_{\mathfrak{F}}|x|^s\xi_1(x)\varphi(x_\mu^{-1}xx_\mu)\mathfrak{f}(yxx_\mu)dm^*(x) \ .$$

As $\mathfrak{f}$ is bounded (cf. Proposition 3.4), we observe that this integral converges for suitably large $\mathrm{Re}(s)$. If we substitute a constant for $\mathfrak{f}$, we find that this is so for $\mathrm{Re}(s) > 2$. We transform it in the following way.

$$\zeta_\mu = \int_{\mathfrak{F}}|y^{-1}xx_\mu^{-1}|^s\xi_1(y^{-1}xx_\mu^{-1})\varphi(x_\mu^{-1}y^{-1}x)\mathfrak{f}(x)dm^*(x)$$

$$= \sum_{\lambda=1}^{h}\int_{D^*x_\lambda U} = \sum_{\lambda=1}^{h}\sum_{(\alpha)}\int_{\alpha x_\lambda U} \ ,$$

where the integrands of the last integrals are the same as the first one, and the sum $\sum_{(\alpha)}$ is extended over the representatives α for D^*/Γ_λ. By (11) and (16), we have

$$\int_{\alpha x_\lambda U} = \int_{U_\lambda}|y^{-1}xx_{\lambda\mu}|^s\xi_1(y^{-1}xx_{\lambda\mu})\varphi(x_\mu^{-1}y^{-1}\alpha xx_\lambda)\mathfrak{f}_\lambda(x)dm^*(x)$$

$$= |x_{\lambda\mu}|^s\xi_1(x_{\lambda\mu})\Big\{(\textstyle\prod_{\mathfrak{p}}B_{\lambda\mathfrak{p}})\otimes\int_{G_2}\Big\}\Big|_{G_1} \ ,$$

where

$$B_{\lambda\mathfrak{p}} = \int_{U_{\lambda\mathfrak{p}}}\xi_{1\mathfrak{p}}(x_{\mathfrak{p}})\varphi_{\mathfrak{p}}(x_{\mu\mathfrak{p}}^{-1}\alpha x_{\mathfrak{p}}x_{\lambda\mathfrak{p}})\rho_{\mathfrak{p}}(x_{\mathfrak{p}}^{-1})dm^*(x_{\mathfrak{p}}) \ ,$$

$$G_1 = D^{(1)*} \times \cdots \times D^{(r)*} = \mathrm{GL}(2, \boldsymbol{R}) \times \cdots \times \mathrm{GL}(2, \boldsymbol{R}) \quad (r\ \text{copies}) \ ,$$

$$G_2 = D^{(r+1)*} \times \cdots \times D^{(t)*} = \boldsymbol{K}^* \times \cdots \times \boldsymbol{K}^* \qquad (t - r\ \text{copies}) \ ,$$

$$\int_{G_2} = \Phi(\alpha)\textstyle\prod_{\nu=r+1}^{t}\int_{K^*}N(u)^{s+(n_\nu/2)+ia_\nu}\exp\{-2\pi N(\alpha^{(\nu)}u)\}dm^*(u) \ ,$$

$$\int_{G_1} = \int_{G_1} \prod_{\nu=1}^{r} |N(y^{-1}x)^{(\nu)}|^{s+ia_\nu} |N(x^{(\nu)})|^{n_\nu/2} \{\operatorname{sgn} N(y^{-1}x^{(\nu)})\}^{b_\nu}$$
$$\cdot \exp\{-\pi\operatorname{tr}((y^{-1}\alpha x)^{(\nu)t}(y^{-1}\alpha x)^{(\nu)})\}\{N(y^{-1}\alpha x)^{(\nu)}\}^{m_\nu}\{(y^{-1}\alpha x)^{(\nu)}(i)\}^{k_\nu}$$
$$\cdot j((y^{-1}\alpha)^{(\nu)}, x^{(\nu)}(i))^{n_\nu} f_\lambda(x^{(1)}(i), \cdots, x^{(r)}(i)) dm^*(x^{(1)}, \cdots, x^{(r)}) .$$

Here the a_ν and b_ν are the numbers determined by (14).

By our definition of $\varphi_\mathfrak{p}$, the integral on $\alpha x_\lambda U$ does not vanish only when $\alpha\mathfrak{x}_{\lambda\mu} \subset \mathfrak{x}_\mu \mathfrak{d}^{-1}\mathfrak{a}^{-1}\mathfrak{x}_\mu^{-1}$; so we assume this holds. Then we have, if $\mathfrak{p}$ is prime to $N(\mathfrak{a})$, $B_{\lambda\mathfrak{p}} = 1$; and if $\mathfrak{p}$ divides $N(\mathfrak{a})$,

$$B_{\lambda\mathfrak{p}} = \int_{U_\mathfrak{p}} \tau_\mathfrak{p}(\alpha u_\mathfrak{p})\xi_{1\mathfrak{v}}(u_\mathfrak{p})\rho_\mathfrak{p}(u_\mathfrak{p})^{-1} dm^*(u_\mathfrak{p}) = \sigma_\mathfrak{p}(\alpha; \eta^{-1}\rho) .$$

Therefore, we obtain

$$\prod_\mathfrak{p} B_{\lambda\mathfrak{p}} = \begin{cases} 0 & \text{if } \alpha\mathfrak{x}_{\lambda\mu} \not\subset \mathfrak{x}_\mu\mathfrak{d}^{-1}\mathfrak{a}^{-1}\mathfrak{x}_\mu^{-1}, \\ \sigma_\mathfrak{a}(\alpha; \eta^{-1}\rho) & \text{if } \alpha\mathfrak{x}_{\lambda\mu} \subset \mathfrak{x}_\mu\mathfrak{d}^{-1}\mathfrak{a}^{-1}\mathfrak{x}_\mu^{-1} . \end{cases}$$

Now the integral on G_2 is equal to

$$c_2 \prod_{\nu=r+1}^{t} (2\pi N(\alpha^{(\nu)}))^{-[s+(n_\nu/2)+ia_\nu]}\Gamma\left(s + \frac{n_\nu}{2} + ia_\nu\right)\Phi(\alpha) ,$$

where c_2 is a positive constant depending on the choice of Haar measure on K^*. In order to calculate the integral on G_1, consider the connected component of the identity element of G_1. We can find $2^r \cdot r$ orthogonal matrices $u_\beta^{(\nu)}(1 \leq \nu \leq r, 1 \leq \beta \leq 2^r)$ such that

$$G_1 = \bigcup_{\beta=1}^{2^r} G_0 u_\beta : \quad u_\beta = (u_\beta^{(1)}, \cdots, u_\beta^{(r)}) .$$

The integral on G_1 is then decomposed into the sum of the integrals on $G_0 u_\beta$ for $1 \leq \beta \leq 2^r$, and

$$\int_{G_0 u_\beta} =$$

$$(17) \quad \int_{G_0} \prod_{\nu=1}^{r} |N(y^{-1}x)|^{s+ia_\nu} |N(x^{(\nu)})|^{n_\nu/2}\{\operatorname{sgn} N(yu_\beta)^{(\nu)}\}^{b_\nu}\{N(y^{-1}\alpha x u_\beta)^{(\nu)}\}^{m_\nu}$$
$$\cdot \exp\{-\pi\operatorname{tr}((y^{-1}\alpha x)^{(\nu)t}(y^{-1}\alpha x)^{(\nu)})\}\{(y^{-1}\alpha x u_\beta)^{(\nu)}(i)\}^{k_\nu}$$
$$\cdot j((y^{-1}\alpha)^{(\nu)}, x^{(\nu)}u_\beta^{(\nu)}(i))^{n_\nu} f_\lambda(x^{(1)}u_\beta^{(1)}(i), \cdots, x^{(r)}u_\beta^{(r)}(i)) dm^*(x^{(1)}, \cdots, x^{(r)}) .$$

As the $u_\beta^{(\nu)}$ are orthogonal, we see that $u_\beta^{(\nu)}(i) = \pm i$; therefore, the integrand of the right hand side of (17) is invariant under the right translation $x^{(\nu)} \to x^{(\nu)}v^{(\nu)}$ for $v^{(\nu)} \in \mathrm{SO}(2, R)$; so it may be considered as a function on G_0/K, where K is the product of r copies of $\mathrm{SO}(2, R)$. By the mapping

$$(x^{(1)}, \cdots, x^{(r)}) \longrightarrow (x^{(1)t}x^{(1)}, \cdots, x^{(r)t}x^{(r)}) ,$$

G_0/K is identified with the product P of r copies of the space of positive symmetric matrices of degree 2. Let $g_{\lambda\beta}(X)$ be the function on P defined by

$$(18) \quad g_{\lambda\beta}(x^{(1)t}x^{(1)}, \cdots, x^{(r)t}x^{(r)})$$

$$= \prod_{\nu=1}^{r} \{(y^{-1}\alpha x u_\beta)^{(\nu)}(i)\}^{k_\nu} j((y^{-1}\alpha)^{(\nu)}, x^{(\nu)}u_\beta^{(\nu)}(i))^{n_\nu} f_\lambda(x^{(1)}u_\beta^{(1)}(i), \cdots, x^{(r)}u_\beta^{(r)}(i))$$

$$((x^{(1)}, \cdots, x^{(r)}) \in G_0) .$$

Then the above integral (17) is equal to

$$\prod_{\nu=1}^{r} |N(\alpha^{(\nu)})|^{-[s+(n_\nu/2)+ia_\nu]} |N(y^{(\nu)})|^{n_\nu/2} \{\operatorname{sgn} N(yu_\beta)^{(\nu)}\}^{b_\nu+m_\nu} \operatorname{sgn} N(\alpha^{(\nu)})^{m_\nu}$$

$$(19) \quad \times \int_P \prod_{\nu=1}^{r} \det(A_\nu^{-1}X_\nu)^{(1/2)[s+(n_\nu/2)+m_\nu+ia_\nu]} \exp\{-\pi \operatorname{tr}(A_\nu^{-1}X_\nu)\}$$

$$\times g_{\lambda\beta}(X_1, \cdots, X_r) dm^*(X_1, \cdots, X_r) ,$$

where $A_\nu = (\alpha^{-1}y)^{(\nu)\cdot t}(\alpha^{-1}y)^{(\nu)}$ and $dm^*(X_1, \cdots, X_r)$ is an invariant measure of $P = G_0/K$. The function $g_{\lambda\beta}$ is obviously holomorphic as a function of $(z_1, \cdots, z_r) = (x^{(1)}u_\beta^{(1)}(i), \cdots, x^{(r)}u_\beta^{(r)}(i))$. It follows that $\Omega g_{\lambda\beta} = 0$ for any invariant differential operator Ω on $P = G_0/K$. In general, if a function $H(X_1, \cdots, X_r)$ on P is a simultaneous eigenfunction of all the invariant differential operators Ω_κ, then we have

$$(20) \quad \int_P \prod_{\nu=1}^{r} \det(A_\nu^{-1}X_\nu)^{s_\nu} \exp\{-\pi\operatorname{tr}(A_\nu^{-1}X_\nu)\} H(X_1, \cdots, X_r) dm^*(X_1, \cdots X_r)$$

$$= \lambda H(A_1, \cdots, A_r) .$$

If $\Omega_\kappa H = \mu_\kappa H$, the eigen-value λ is determined by the μ_κ (cf. Selberg [22]). If $\mu_\kappa = 0$ for any Ω_κ, and if we take the measure

$$dm^*(X_1, \cdots, X_r) = \prod_{\nu=1}^{r} \{\det(X_\nu)^{-3/2} \prod_{i \leq j} dx_{ij}^\nu\} \qquad \text{for } X^\nu = (x_{ij}^\nu) ,$$

then the eigen-value λ is equal to $2^r \prod_{\nu=1}^{r} (2\pi)^{1-2s_\nu}\Gamma(2s_\nu - 1)$. This is shown by taking a suitable testing function H. Applying this to our case, we see that (19) is equal to

$$c_1 \prod_{\nu=1}^{r} |N(\alpha^{(\nu)})|^{-[s+(n_\nu/2)+ia_\nu]} |N(y^{(\nu)})|^{n_\nu/2} \{\operatorname{sgn} N(yu_\beta)^{(\nu)}\}^{b_\nu+m_\nu} \operatorname{sgn} N(\alpha^{(\nu)})^{m_\nu}$$

$$\times (2\pi)^{-[s+(n_\nu/2)+ia_\nu+m_\nu-1]}\Gamma\left(s + \frac{n_\nu}{2} + ia_\nu + m_\nu - 1\right) g_{\lambda\beta}(A_1, \cdots, A_r) ,$$

where c_1 is a positive constant depending on the choice of Haar measure. Let $v = (v^{(1)}, \cdots, v^{(r)})$ be an element of G_1 such that the $v^{(\nu)}$ are orthogonal and $\det(\alpha^{-1}yv)^{(\nu)} > 0$ for $1 \leq \nu \leq r$. Put $v_\beta^{(\nu)} = v^{(\nu)}u_\beta^{(\nu)}$. Then, by (18),

$$g_{\lambda\beta}(A_1, \cdots, A_r)$$

$$= \prod_{\nu=1}^{r} v_\beta^{(\nu)}(i)^{k_\nu} j((y^{-1}\alpha)^{(\nu)},(\alpha^{-1}yv_\beta)^{(\nu)}(i))^{n_\nu} f_\lambda((\alpha^{-1}yv_\beta)^{(1)}(i), \cdots, (\alpha^{-1}yv_\beta)^{(r)}(i)) .$$

Thus we have finished the integral on G_1. Put now $c_0 = c_1 c_2$ and

$$(21) \quad \Gamma(s; \{n_\nu\}; \xi; \{m_\nu\}) = (2\pi)^{-ts+r-\sum_{\nu=1}^{t}[(n_\nu/2)+ia_\nu]-\sum_{\nu=1}^{r} m_\nu}$$

$$\times \prod_{\nu=1}^{r} \Gamma\left(s + \frac{n_\nu}{2} + ia_\nu + m_\nu - 1\right) \prod_{\nu=r+1}^{t} \Gamma\left(s + \frac{n_\nu}{2} + ia_\nu\right) .$$

Consider a transformation ε of $\mathfrak{F}_r$ onto itself such that

$$\varepsilon(z_1, \cdots, z_r) = \big(\varepsilon_1(z_1), \cdots, \varepsilon_r(z_r)\big)$$

and, for each ν, $\varepsilon_\nu(z_\nu)$ is identically equal to z_ν or $\bar{z}_\nu$. There exist 2^r such transformations, which we denote by the ε_β for $1 \leq \beta \leq 2^r$; we put $\varepsilon_\beta(z_1, \cdots, z_r) = (\varepsilon_{\beta1}(z_1), \cdots, \varepsilon_{\beta r}(z_r))$ and $\mathrm{sgn}(\varepsilon_{\beta\nu}) = 1$ or -1 according as $\varepsilon_{\beta\nu}(z_\nu) = z_\nu$ or $\bar{z}_\nu$.

Now, for every point $z = (z_1, \cdots, z_r)$ on $\mathfrak{F}_r$, there exists an element y of $\mathfrak{F}$ such that

$$(22) \qquad \begin{cases} y_\mathfrak{p} = 1 & \text{for every } \mathfrak{p}, \\ y^{(\nu)} = 1 & \text{for every } \nu > r, \\ y^{(\nu)}(i) = z_\nu & \text{for } 1 \leq \nu \leq r. \end{cases}$$

For such an element y, put

$$M_\beta(y) = \prod_{\nu=1}^{r} |\, N(y^{(\nu)})\,|^{n_\nu/2} j(y^{(\nu)}, \varepsilon_{\beta\nu}(i))^{-n_\nu} .$$

Then the above calculation shows

$$\zeta_\mu = \zeta_\mu(s, \xi, f, \{k_\nu, m_\nu\}, y)$$
$$= c_0 \Gamma(s; \{n_\nu\}; \xi; \{m_\nu\}) \sum_{\lambda=1}^{h} \sum_{\{\alpha\}} N_{D/}\, (\alpha \mathfrak{x}_{\lambda\mu})^{-s} \xi_0\big(N(\mathfrak{x}_{\lambda\mu})\big)\eta(\alpha)L(\alpha)^{-1}\sigma_\alpha(\alpha; \eta^{-1}\rho)$$
$$\otimes \Phi(\alpha) \sum_{\beta=1}^{2^r} \Big\{\prod_{\nu=1}^{r} i^{k_\nu} \mathrm{sgn}(\varepsilon_{\beta\nu})^{b_\nu + k_\nu + m_\nu}\Big\} M_\beta(y)J(\alpha^{-1}, \varepsilon_\beta(z))^{-1}f_\lambda\big(\alpha^{-1}(\varepsilon_\beta(z))\big) ,$$

where $\sum_{\{\alpha\}}$ is the sum extended over all representatives α of D^*/Γ_λ satisfying $\alpha\mathfrak{x}_{\lambda\mu} \subset \mathfrak{x}_\mu \mathfrak{d}^{-1}\mathfrak{a}^{-1}\mathfrak{x}_\mu^{-1}$. Put

$$\mathscr{D}_1(s; \xi) = \sum N_{D/Q}(e)^{-s}\mathfrak{T}(e; \xi) ,$$

where e ranges over all the elements e of $\mathfrak{E}$ such that $e \subset \mathfrak{d}^{-1}\mathfrak{a}^{-1}$. Then, by our definition of $\mathfrak{T}(e; \xi)$, we obtain

$$\zeta = \zeta(s, \xi, f, \{k_\nu, m_\nu\}, y) = (\zeta_1, \cdots, \zeta_h)$$
$$(23) \qquad = c_0\Gamma(s; \{n_\nu\}; \xi; \{m_\nu\}) \sum_{\beta=1}^{2^r} \Big\{\prod_{\nu=1}^{r} i^{k_\nu} \mathrm{sgn}(\varepsilon_{\beta\nu})^{b_\nu + k_\nu + m_\nu}\Big\}$$
$$\times M_\beta(y)\big(\mathscr{D}_1(s; \xi)f\big)(\varepsilon_\beta(z)) .$$

4.6. *Fourier tranforms in $M_2(R)$.* For every function $g(x)$ on $M_2(R)$ satisfying a suitable condition of integrability, we define its Fourier transform $\hat{g}$ by

$$\hat{g}(x) = \int_{M_2(R)} g(y) \exp\{2\pi i \mathrm{tr}(xy)\}dm(y) ,$$

where $dm(y)$ is the volume element given in § 4.1 for $D^{(\nu)} = M_2(R)$. Put, for

$$x = \begin{pmatrix} x_{11} & x_{12} \\ x_{21} & x_{22} \end{pmatrix} \in M_2(R)$$

and for non-negative integers $k, l, m,$

$$\varphi_{kl}^m(x) = \det(x)^m (x_{11}i + x_{12})^k (x_{21}i + x_{22})^l \exp\{-\pi \mathrm{tr}(x^t x)\} ,$$
$$\psi_{kl}^m(x) = \mathrm{tr}(x^t x)\varphi_{kl}^m(x) .$$

Using a well-known formula

$$\int_{R^2} (u + vi)^n \exp\{-\pi(u^2 + v^2)\} \exp\{2\pi i(ux + vy)\}dudv$$
$$= i^n(x + yi)^n \exp\{-\pi(x^2 + y^2)\} ,$$

we obtain

$$(25) \qquad \hat{\varphi}_{kl}^0(x) = \int \varphi_{kl}^0(y) \exp\{2\pi i\,\mathrm{tr}(xy)\}dm(y) = (-1)^k \varphi_{lk}^0(x') .$$

If we restrict ourselves to the case where D is a division algebra, this
formula is sufficient for our purpose. In order to treat the case $D = M_2(F)$,
we use a small trick. Applying $\partial^2/\partial x_{11}\partial x_{22} - \partial^2/\partial x_{12}\partial x_{21}$ to (25) successively,
we find easily the formulas for $\hat{\varphi}_{kl}^m$. From those formulas we can deduce
the following equality:

$$(26) \quad \begin{aligned}
& 4\pi^2 \hat{\varphi}_{0n}^3(x') + 6n\pi i \hat{\varphi}_{1,n-1}^2(x') + 2(n-1)\hat{\varphi}_{0n}^1(x') - 2n(n-1)\hat{\varphi}_{2,n-2}^1(x') \\
& = -4\pi^2 \varphi_{n0}^3(x) + 6\pi \psi_{n0}^1(x) - 2(n+5)\varphi_{n0}^1(x) - n(n-1)\varphi_{n-2,2}^1(x) .
\end{aligned}$$

Here we understand that $\varphi_{kl}^m = 0$ if either k or l is negative.

4.7. *Fourier transforms in* $\mathfrak{A}$. Notation being as in § 4.5, put

$$\hat{\varphi}(x) = \int_{\mathfrak{A}} \overline{\tau(xy)}\varphi(y)dm(y) ,$$
$$\hat{\varphi}_{\mathfrak{p}}(x_{\mathfrak{p}}) = \int_{D_{\mathfrak{p}}} \overline{\tau_{\mathfrak{p}}(x_{\mathfrak{p}}y_{\mathfrak{p}})}\varphi_{\mathfrak{p}}(y_{\mathfrak{p}})dm_{\mathfrak{p}}(y_{\mathfrak{p}}) ,$$
$$\hat{\varphi}^{(\nu)}(x^{(\nu)}) = \int_{D^{(\nu)}} \overline{\tau^{(\nu)}(x^{(\nu)}y^{(\nu)})}\varphi^{(\nu)}(y^{(\nu)})dm_{\nu}(y^{(\nu)}) .$$

We have then

$$\hat{\varphi}(x) = \{\textstyle\prod_{\mathfrak{p}} \hat{\varphi}_{\mathfrak{p}}(x_{\mathfrak{p}})\}\rho(1) \otimes \{\textstyle\prod_{\nu=1}^r \hat{\varphi}^{(\nu)}(x^{(\nu)})\}\hat{\varphi}^{(r+1)}(x^{(r+1)}) \otimes \cdots \otimes \hat{\varphi}^{(t)}(x^{(t)}) .$$

By our definition of $\varphi_{\mathfrak{p}}$, we have

$$(27) \qquad \hat{\varphi}_{\mathfrak{p}}(x_{\mathfrak{p}}) = \begin{cases} N_1(\mathfrak{d}_{\mathfrak{p}})^{1/2}N_1(\mathfrak{a}_{\mathfrak{p}}) & \text{if } x_{\mathfrak{p}} - 1 \in \mathfrak{a}_{\mathfrak{p}} , \\ 0 & \text{if } x_{\mathfrak{p}} - 1 \notin \mathfrak{a}_{\mathfrak{p}} . \end{cases}$$

We have given $\hat{\varphi}^{(\nu)}(x^{(\nu)})$ for $1 \leq \nu \leq r$ in § 4.6. Now let us prove, for
$\nu > r$,

$$(28) \qquad \hat{\varphi}^{(\nu)}(w') = i^{n_\nu}\varphi^{(\nu)}(w) ,$$

namely

$$(28')\qquad \int_K \exp\left[2\pi i\,\mathrm{tr}(\alpha\beta')\right]\exp\left[-2\pi N(\alpha)\right]\Phi_n(\alpha)dm(\alpha)$$
$$= i^*\exp\left[-2\pi N(\beta)\right]\Phi_n(\beta)\ .$$

Recalling the definition of Φ_n, we see that this is equivalent to

$$\int_{C^2}\exp\left[2\pi i(u\bar w+\bar u w+v\bar z+\bar v z)\right]\exp\left[-2\pi(u\bar u+v\bar v)\right]$$
$$\times\,(uX+vY)^k(-\bar v X+\bar u Y)^l dm(u,v)$$
$$= i^{k+l}\exp\left[-2\pi(w\bar w+z\bar z)\right](wX+zY)^k(-\bar z X+\bar w Y)^l$$
$$(k,l=0,1,2,\cdots)\ ,$$

where u,v,w,z are complex variables, X and Y are indeterminates, and $dm(u,v)=-du\wedge d\bar u\wedge dv\wedge d\bar v$. This formula is clearly true for $k=l=0$. Then applying the operators $X\partial/\partial\bar w+Y\partial/\partial\bar z$ and $-X\partial/\partial z+Y\partial/\partial w$ successively, we get the equality in the general case.

4.8. *Functional equation in the case of a division algebra.* In this section we assume that D is a division algebra. Taking the m_ν to be 0, consider the function $\varphi(x)$ defined in § 4.5 and its Fourier transform $\hat\varphi(x)$. Then, by the Poisson summation formula (cf. [8, 9, 15, 28, 31]), we have, for every x and y of $\Im$,

$$\sum_{\alpha\in D}\varphi(y^{-1}\alpha x)=|yx^{-1}|^2\sum_{\alpha\in D}\hat\varphi(x^{-1}\alpha y)\ .$$

As $\varphi(0)=\hat\varphi(0)=0$, and as D is a division algebra, we get

$$(29)\qquad \sum_{\alpha\in D^*}\varphi(y^{-1}\alpha x)=|yx^{-1}|^2\sum_{\alpha\in D^*}\hat\varphi(x^{-1}\alpha y)\ .$$

Now, by (11) and (16),

$$\zeta_\mu=\zeta_\mu(s,\xi,f,\{k_\nu,0\},y)$$
$$=\int_{D^*\backslash\Im}|y^{-1}xx_\mu^{-1}|^s\,\xi_1(y^{-1}xx_\mu^{-1})\{\textstyle\sum_{\alpha\in D}\varphi(x_\mu^{-1}y^{-1}\alpha x)\}\mathfrak{f}(x)dm^*(x)\ .$$

Decompose this integral into the part for $|x|\geqq 1$ and the part for $|x|\leqq 1$ and consider the transformation $x\to x^{-1}$. Then, on account of (29), we observe that ζ_μ can be continued holomorphically on the whole s-plane and is equal to

$$\int_{\Im}|yxx_\mu|^{2-s}\,\xi_1(yxx_\mu)^{-1}\hat\varphi(xyx_\mu)\mathfrak{f}(x^{-1})dm^*(x)\ .$$

Transform this by the canonical involution $x\to x'$. Then, by the relation $\Im=\bigcup_{\lambda=1}^h D^*x'_\lambda U^{-1}$, we obtain

$$\zeta_\mu=\sum_{\lambda=1}^h\int_{D'x'_\lambda{}^{-1}U}|yxx_\mu|^{2-s}\xi_1(yxx_\mu)^{-1}\hat\varphi(x'yx_\mu)\mathfrak{f}(x'^{-1})dm^*(x)\ ,$$

and, on account of $x_\lambda'^{-1}Ux_\lambda' = x_\lambda Ux_\lambda^{-1} = U_\lambda$,

$$\zeta_\mu = \sum_{\lambda=1}^h \int_{D^*U_\lambda} |\, yxx_\lambda^{-1}x_\mu\,|^{2-s}\xi_1(yxx_\lambda^{-1}x_\mu)^{-1}\hat\varphi(x_\lambda^{-1}x'yx_\mu)\mathfrak{f}(x'^{-1}x_\lambda)dm^*(x)\ .$$

We have $\mathfrak{f}(x'^{-1}x_\lambda) = \mathfrak{f}_\lambda(x'^{-1})$ for $x \in D^*U_\lambda$, so that

$$\zeta_\mu = \sum_{\lambda=1}^h |\, x_{\mu\lambda}\,|^{2-s}\,\xi_1(x_{\mu\lambda})^{-1}\int_{D^*U_\lambda} |\, yx\,|^{2-s}\,\xi_1(yx)^{-1}\hat\varphi(x_\lambda^{-1}x'yx_\mu)\mathfrak{f}_\lambda(x'^{-1})dm^*(x)$$

$$= \sum_{\lambda=1}^h |\, x_{\mu\lambda}\,|^{2-s}\,\xi_1(x_{\mu\lambda})^{-1}\sum_{\{a\}}\int_{U_\lambda} |\, yx\,|^{2-s}\xi_1(yx)^{-1}$$
$$\times\ \hat\varphi(x_\lambda^{-1}x'\alpha'yx_\mu)\mathfrak{f}_\lambda(x'^{-1})dm^*(x)\ .$$

In view of (12), this can be calculated in the same way as in § 4.5; so the integral on U_λ is expressed in the form

$$\int_{U_\lambda} = \left[\left(\prod_\mathfrak{p} B'_{\lambda\mathfrak{p}}\right)\otimes\int_{\sigma_2}\right]\cdot\int_{\sigma_1}\ .$$

Now assume that the representation ρ is irreducible and define ρ^r and ρ_0^r as in § 4.3. Then, by (16″) of § 4.3, we have

$$B'_\lambda = \int_{U_\mathfrak{p}}\xi_{1\mathfrak{p}}(x_\mathfrak{p})^{-1}\hat\varphi_\mathfrak{p}(x_{\lambda\mathfrak{p}}^{-1}x'\alpha'_\mathfrak{p}x_{\mu\mathfrak{p}})\rho_\mathfrak{p}^r(N(x_\mathfrak{p}))\rho_\mathfrak{p}(x_\mathfrak{p})^{-1}dm_\mathfrak{p}^*(x_\mathfrak{p})\ .$$

This vanishes if $(\alpha\mathfrak{x}'_{\mu\lambda})_\mathfrak{p} \not\subset \mathfrak{o}_{\mu\mathfrak{p}}$; so assume $\alpha\mathfrak{x}'_{\mu\lambda} \subset \mathfrak{o}_\mu$. Then, if $\mathfrak{p}$ is prime to $N(\mathfrak{a})$, we have $B'_{\lambda\mathfrak{p}} = N_1(\mathfrak{d}_\mathfrak{p})^{1/2}$, and if $\mathfrak{p}$ divides $N(\mathfrak{a})$,

$$B'_{\lambda\mathfrak{p}} = \int_{U_\mathfrak{p}}\xi_{1\mathfrak{p}}(x_\mathfrak{p})^{-1}\hat\varphi_\mathfrak{p}(x'_\mathfrak{p}\alpha')\rho_\mathfrak{p}^r(N(x_\mathfrak{p}))\rho_\mathfrak{p}(x_\mathfrak{p})^{-1}dm_\mathfrak{p}^*(x_\mathfrak{p})\ .$$

By (27), this does not vanish only when $\alpha \in U_\mathfrak{p}$. If $\alpha \in U_\mathfrak{p}$, we have

$$B'_{\lambda\mathfrak{p}} = N_1(\mathfrak{d}_\mathfrak{p})^{1/2}N_1(\mathfrak{a}_\mathfrak{p})q(\mathfrak{a}_\mathfrak{p})^{-1}\xi_{1\mathfrak{p}}(\alpha)\rho_\mathfrak{p}^r(N(\alpha))^{-1}\rho_\mathfrak{p}(\alpha)\ ,$$

where $q(\mathfrak{a}_\mathfrak{p})$ is the number of elements in $(\mathfrak{o}_\mathfrak{p}/\mathfrak{a}_\mathfrak{p})^*$. Therefore, we get

$$(30)\qquad \prod_\mathfrak{p} B'_{\lambda\mathfrak{p}} = N_{D/\mathbf{Q}}(\mathfrak{d}\mathfrak{a}^2)q(\mathfrak{a})^{-1}\{\prod_{\mathfrak{p}|N(\mathfrak{a})}\xi_{1\mathfrak{p}}(\alpha)\rho_\mathfrak{p}^r(N(\alpha))^{-1}\}\rho(\alpha)\ ,$$

where $q(\mathfrak{a})$ is the number of elements in $(\mathfrak{o}/\mathfrak{a})^*$, if $\alpha\mathfrak{x}'_{\mu\lambda} \subset \mathfrak{o}_\mu$ and $\alpha \in V_\mathfrak{a}$; and otherwise we have $\prod_\mathfrak{p} B'_{\lambda\mathfrak{p}} = 0$.

Suppose that y satisfies (22). Then we have $(y'^{-1})_\mathfrak{p} = 1$ for every $\mathfrak{p}$, $(y'^{-1})^{(\nu)} = 1$ for $\nu > r$, and $(y'^{-1})^{(\nu)}(i) = z_\nu$ for $1 \leq \nu \leq r$. In view of (12), (16″), (25), (28), (30), we obtain, in the same manner as in § 4.5,

$$\zeta_\mu(s, \xi, f, \{k_\nu, 0\}, y)$$
$$= c_0d_0\Gamma(2-s;\{n_\nu\};\xi^{-1};\{0\})\sum_{\lambda=1}^h\sum_{\{a\}}N_{D/Q}(\alpha\mathfrak{x}'_{\mu\lambda})^{2-s}\xi_0(N(\alpha\mathfrak{x}'_{\mu\lambda}))^{-1}\rho_0^r(N(\alpha\mathfrak{x}'_{\mu\lambda}))$$
$$\times\ L(\alpha)^{-1}\rho(\alpha)\otimes\Phi(\alpha)\sum_{\beta=1}^{2^r}\{\prod_{\nu=1}^r i^{k_\nu}\operatorname{sgn}(\varepsilon_{\beta\nu})^{b_\nu+k_\nu}\}$$
$$\times\ M'_\beta(y)J(\alpha^{-1}, \varepsilon_\beta(z))^{-1}f_\lambda(\alpha^{-1}(\varepsilon_\beta(z)))\ ,$$

where

$$d_0 = i^{\sum_{\nu=1}^{t} n_\nu} N_{D/}\,(\mathfrak{b}\mathfrak{a}^2)q(\mathfrak{a})^{-1}\,,$$

$$M_\beta'(y) = M_\beta(y'^{-1}) \prod_{\nu=1}^{r} \operatorname{sgn} N(y^{(\nu)})^{n_\nu}\,,$$

and $\sum_{\{\alpha\}}$ is the sum extended over all representatives α of D^*/Γ_λ satisfying $\alpha\mathfrak{x}_{\mu\lambda}' \subset \mathfrak{o}_\mu$ and $\alpha \in V_\alpha$. We see easily $M_\beta'(y) = M_\beta(y)$. Put now

$$\xi^* = \rho^{\mathcal{F}} \cdot \xi^{-1}\,,$$

$$\mathscr{D}(s; \xi^*) = \sum N_{D/Q}(e)^{-s}\xi_0^*\big(N(e)\big)\mathfrak{X}(e)\,,$$

where e ranges over all integral elements of $\mathfrak{E}$ which are prime to $\mathfrak{a}$. Then, by Proposition 3.3,

$$(31)\quad \begin{aligned}\zeta &= (\zeta_1, \cdots, \zeta_h) = c_0 d_0 \Gamma(2 - s; \{n_\nu\}; \xi^*; \{0\})\\ &\times \sum_{\beta=1}^{2^r} \Big\{\prod_{\nu=1}^{r} i^{k_\nu}\operatorname{sgn}(\varepsilon_{\beta\nu})^{b_\nu + k_\nu}\Big\}M_\beta(y)\big(\mathscr{Y}\mathscr{D}(2 - s; \xi^*)\mathscr{Y}f\big)(\varepsilon_\beta(z))\,.\end{aligned}$$

Put simply $\Gamma(s; \{n_\nu\}; \xi) = \Gamma(s; \{n_\nu\}; \xi; \{0\})$. (31) combined with (23) shows

$$(32)\quad \begin{aligned}\Gamma(s; \{n_\nu\}; \xi) &\sum_{\beta=1}^{2^r} \Big\{\prod_{\nu=1}^{r} \operatorname{sgn}(\varepsilon_{\beta\nu})^{b_\nu + k_\nu}\Big\}M_\beta(y)\big(\mathscr{D}_1(s; \xi)f\big)(\varepsilon_\beta(z))\\ &= d_0\Gamma(2 - s; \{n_\nu\}; \xi^*) \sum_{\beta=1}^{2^r} \Big\{\prod_{\nu=1}^{r} \operatorname{sgn}(\varepsilon_{\beta\nu})^{b_\nu + k_\nu}\Big\}\\ &\quad \cdot M_\beta(y)\big(\mathscr{Y}\mathscr{D}(2 - s; \xi^*)\mathscr{Y}f\big)(\varepsilon_\beta(z))\,.\end{aligned}$$

We will deduce from this the following equality:

$$(33)\quad \Gamma(s; \{n_\nu\}; \xi)\mathscr{D}_1(s; \xi) = d_0\Gamma(2 - s; \{n_\nu\}; \xi^*)\mathscr{Y}\mathscr{D}(2 - s; \xi^*)\mathscr{Y}\,.$$

We give two proofs, which are essentially the same in their nature.

FIRST PROOF. Take y so that $j(y^{(\nu)}, i) = 1$ for $1 \leqq \nu \leqq r$; this is possible even with the condition (22). Then $M_\beta(y)$ is independent of β. Consider the equality (32) for any system $\{k_\nu\}$ such that $k_\nu = 0$ or 1. There exist exactly 2^r such systems. Sum up the equalities (32) for those systems $\{k_\nu\}$. Then we find (33).

SECOND PROOF. Fix any system $\{k_\nu\}$. Let w be an element of $\mathfrak{F}$ such that $w_\mathfrak{p} = 1$ for every $\mathfrak{p}$, $w^{(\nu)} = 1$ for every $\nu > r$, and $w^{(\nu)} \in \mathrm{SO}(2, R)$ for $1 \leqq \nu \leqq r$. In (32), substitute yw for y. Then z remains invariant; and $M_\beta(y)$ is multiplied by $J(w, \varepsilon_\beta(i))^{-1}$. Call $(32)_w$ the equality thus obtained. Put $w_\nu = \begin{pmatrix} p_\nu & -q_\nu \\ q_\nu & p_\nu \end{pmatrix}$ and $\omega_\nu = (q_\nu i + p_\nu)^{n_\nu}$. Then $J(w, \varepsilon_\beta(i))^{-1} = \prod_{\nu=1}^{r}\omega_\nu^{-\operatorname{sgn}(\varepsilon_{\beta\nu})}$. Multiplying $(32)_w$ by $\omega_1 \cdots \omega_r$, consider its two sides as polynomials of $\omega_1, \cdots, \omega_r$. As the ω_ν are arbitrary complex numbers with absolute value 1, the constant terms must coincide. Then we get (33).

In both proofs, we observe that both sides of (33) are holomorphic on the whole s-plane.

4.9. *Functional equation in the case of* $M_2(F)$. Now we take D to be $M_2(F)$ and want to prove (33) in this case. Here we cannot use (29) with the same φ as in the case of a division algebra. We use (26) instead.

Namely, we substitute the function

$$\theta(x) = 4\pi^2 \varphi^3_{0n}(x) + 6n\pi i \varphi^2_{1,n-1}(x) + 2(n-1)\varphi^1_{0n}(x) - 2n(n-1)\varphi^1_{2,n-2}(x)$$

for $\varphi^{(1)}$, where $n = n_1$. The Fourier transform $\hat\theta$ of θ is given by (26); we note that $\theta(x) = \hat\theta(x) = 0$ if $\det(x) = 0$. Now denote by φ_* the function obtained from φ by this exchange. Then the Fourier transform $\hat\varphi_*$ of φ_* is obtained from $\hat\varphi$ by exchanging $\hat\varphi^{(1)}$ for $\hat\theta$. We have $\varphi^*(x) = \hat\varphi_*(x) = 0$ if $\det(x^{(1)}) = 0$, so that the formula (29) is true for φ_* and $\hat\varphi_*$; namely, we have, for $x, y \in \mathfrak{S}$,

$$(34) \qquad \sum_{\alpha \in D^*} \varphi_*(y^{-1}\alpha x) = |yx^{-1}|^2 \sum_{\alpha \in D^*} \hat\varphi_*(x^{-1}\alpha y) \ .$$

Therefore, with these φ_* and $\hat\varphi_*$, we can proceed in the same way as in § 4.5 and in § 4.8. Take $k_2 = \cdots = k_r = 0$, (and of course $m_2 = \cdots = m_r = 0$,) and put

$$\zeta^* = (\zeta^*_1, \cdots, \zeta^*_h) \ ,$$

$$\zeta^*_\mu = \int_{\mathfrak{S}} |x|^s \xi_1(x) \varphi_*(x^{-1}_\mu x x_\mu) \mathfrak{f}(y x x_\mu) dm^*(x) \ .$$

Then we obtain

$$(35) \qquad \zeta^* = c_0 \Gamma_*(s) \sum_{\beta=1}^{2^r} \{\textstyle\prod_{\nu=1}^r \operatorname{sgn}(\varepsilon_{\beta\nu})^{l_\nu}\} M_\beta(y)(\mathscr{D}_1(s; \xi)f)(\varepsilon_\beta(z)) \ ,$$

where the l_ν are integers and $\Gamma_*(s)$ is the function obtained from $\Gamma(s; \{n_\nu\}; \xi)$ by replacing the factor $\Gamma(s + (n/2) + ia_1 - 1)$ with

$$(2\pi)^{-1}\Big\{\Gamma\Big(s + \frac{n}{2} + ia_1 + 2\Big) - 3n\Gamma\Big(s + \frac{n}{2} + ia_1 + 1\Big)$$
$$+ 2(n^2 - 1)\Gamma\Big(s + \frac{n}{2} + ia_1\Big)\Big\} \ .$$

By virtue of the relation $\Gamma(s + 1) = s\Gamma(s)$, this is equal to $\Gamma(s + (n/2) + ia_1 - 1)$ multiplied by

$$(36) \qquad (2\pi)^{-1}\Big(s - \frac{3n}{2} + ia_1 + 2\Big)\Big(s - \frac{n}{2} + ia_1 - 1\Big)\Big(s + \frac{n}{2} + ia_1 - 1\Big) \ .$$

Now applying the argument of § 4.8 to the present case, we get

$$(37) \qquad \begin{aligned} \zeta^* &= c_0 d_0 \Gamma_{**}(2 - s) \\ &\cdot \sum_{\beta=1}^r \{\textstyle\prod_{\nu=1}^r \operatorname{sgn}(\varepsilon_{\beta\nu})^{l'_\nu}\} M_\beta(y)(\mathscr{Y}\mathscr{D}(2 - s; \xi^*)\mathscr{Y}f)(\varepsilon_\beta(z)) \ , \end{aligned}$$

where the l'_ν are integers, and $\Gamma_{**}(s)$ differs from $\Gamma(s; \{n_\nu\}; \xi^*)$ by a simple factor, which we now want to give. In order to get this factor, we have to consider the integral of the form

$$(38) \qquad \int_P \textstyle\prod_{\nu=1}^r \det(A_\nu^{-1} X_\nu)^{s_\nu} \exp\{-\pi\operatorname{tr}(A_\nu \,{}^1 X_\nu)\}$$
$$\cdot \operatorname{tr}(A_1^{-1} X_1) H(X_1, \cdots, X_r) dm^*(X_1, \cdots, X_r)$$

besides (20), since $\hat{\theta}(x')$ includes the term $6\pi\psi^1_{*0}(x)$. If $\Omega_\kappa H = 0$ for every invariant differential operator Ω_κ, (38) is equal to

$$2^r \cdot 2\pi^{-1}s_1 \prod_{\nu=1}^r (2\pi)^{1-2s_\nu}\Gamma(2s_\nu - 1) \cdot H \, .$$

This is shown by choosing a suitable testing function H. Then, using this fact, we get, in the same way as above,

$$\Gamma_{**}(s)\Gamma(s; \{n_\nu\}; \xi^*)^{-1}$$
$$= -(2\pi)^{-1}\left(s - \frac{n}{2} - ia_1 - 1\right)\left(s + \frac{3n}{2} - ia_1 - 4\right)\left(s + \frac{n}{2} - ia_1 - 1\right) .$$

If we substitute $2 - s$ for s, this becomes (36). Therefore, dividing (35) and (37) by this common factor, we find

$$\Gamma(s; \{n_\nu\}; \xi) \sum_{\beta=1}^{2r} \left\{\prod_{\nu=1}^r \operatorname{sgn}(\varepsilon_{\beta\nu})^{1_\nu}\right\} M_\beta(y)(\mathscr{D}_1(s; \xi)f)(\varepsilon_\beta(z))$$
$$= d_0\Gamma(2 - s; \{n_\nu\}; \xi^*)$$
$$\cdot \sum_{\beta=1}^{2r} \left\{\prod_{\nu=1}^r \operatorname{sgn}(\varepsilon_{\beta\nu})^{1'_\nu}\right\} M_\beta(y)(\mathscr{Y}\mathscr{D}(2 - s; \xi^*)\mathscr{Y}f)(\varepsilon_\beta(z)) \, .$$

We then obtain (33) in the case $D = M_2(F)$. As we have fixed the form of φ_*, only the argument of the second proof of § 4.8 can be applicable to the present case.

This method can only prove that both sides of (33) in the case $D = M_2(F)$ are meromorphic on the whole plane, since we have used the division by a polynomial of s. It is plausible that the functions are holomorphic. This is in fact true in some special cases [Herrmann, 14].

4.10. *Statement of the principal result.* Taking D to be arbitrary, put

$$\mathscr{W}(s; \xi) = \sum_e N_{D/Q}(e)^{-s}\mathfrak{T}(e; \xi) \, ,$$
$$\mathscr{D}_0(s; \xi) = \sum_{e_0} N_{D/Q}(e_0)^{-s}\xi_0(N(e_0))\mathfrak{T}(e_0) \, ,$$

where e ranges over all the elements e of $\mathfrak{E}$ such that $e \subset \mathfrak{d}^{-1}\mathfrak{a}^{-1}$ and the set of invariants of e contains only powers of prime ideals dividing $N(\mathfrak{a})$, and e_0 ranges over all the elements e_0 of $\mathfrak{E}$ which are prime to $\mathfrak{a}$ and satisfy $e_0 \subset \mathfrak{d}^{-1}$. Then, by Propositions 2.5 and 4.3,

$$(39) \qquad\qquad \mathscr{D}_1(s; \xi) = \mathscr{W}(s; \xi)\mathscr{D}_0(s; \xi) \, .$$

Now we need

PROPOSITION 4.4. *Let g be the exponent of the group $\mathfrak{K}$ of ideal-classes modulo $\prod_{\nu>r} \mathfrak{p}_{\infty,\nu}$ in F. Let $\mathfrak{p}$ be a prime ideal of F which is prime to $N(\mathfrak{a})$. If $\mathfrak{p}$ does not divide $d(D/F)$, we have $\mathfrak{T}(\mathfrak{p}, \mathfrak{p})^g = \rho_0^F(\mathfrak{p}^{-g})$; and if $\mathfrak{p}$ divides $d(D/F)$, then $\mathfrak{T}(\mathfrak{p})^{2g} = \rho_0^F(\mathfrak{p}^{-g})$.*

PROOF. Suppose that $\mathfrak{p}$ does not divide $d(D/F)$. Then there exists only one integral right $\mathfrak{o}_\lambda$-ideal whose set of invariants is $\{\mathfrak{p}^\sigma, \mathfrak{p}^\sigma\}$; it is just $\mathfrak{p}^\sigma\mathfrak{o}_\lambda$. We can find an element α of $\mathfrak{g}$ such that $\mathfrak{p}^\sigma = \mathfrak{g}\alpha$ and $\alpha^{(\nu)} > 0$ for $\nu > r$. We have then $T_{\lambda\lambda}(\mathfrak{p}^\sigma, \mathfrak{p}^\sigma) = \Gamma_\lambda\alpha\Gamma_\lambda = \alpha\Gamma_\lambda$. Therefore, by our definition of $\mathfrak{X}_{\mu\lambda}$, we have $\mathfrak{X}_{\lambda\lambda}(\mathfrak{p}, \mathfrak{p})^\sigma f_\lambda = \prod_{\nu=1}^{r} \operatorname{sgn}(\alpha^{(\nu)})^{n_\nu}\rho(\alpha) \otimes \Psi(1)f_\lambda$ for every $f_\lambda \in S$. By (16''') of § 4.3, we get $\mathfrak{X}(\mathfrak{p}, \mathfrak{p})^\sigma = \rho_0^F(\mathfrak{p}^{-\sigma})$. The last assertion can be similarly proved.

If $\mathfrak{P}$ denotes the maximal two-sided ideal of $\mathfrak{o}_\mathfrak{p}$, then $\mathfrak{d}_\mathfrak{p}$ is a power $\mathfrak{P}^\kappa$. Put

$$(40) \qquad \mathfrak{X}(\mathfrak{d}_\mathfrak{p}) = \begin{cases} \mathfrak{X}(\mathfrak{p}^\kappa) & \text{if } \mathfrak{p} \mid d(D|F) , \\ \mathfrak{X}(\mathfrak{p}^\kappa, \mathfrak{p}^\kappa) & \text{if } \mathfrak{p} \nmid d(D|F) , \end{cases}$$

$$\mathfrak{X}(\mathfrak{d}_1) = \prod_{\mathfrak{p}\nmid N(\mathfrak{a})} \mathfrak{X}(\mathfrak{d}_\mathfrak{p}) , \qquad \mathfrak{d}_1 = \prod_{\mathfrak{p}\nmid N(\mathfrak{a})} \mathfrak{d}_\mathfrak{p} .$$

By Proposition 4.4, $\mathfrak{X}(\mathfrak{d}_1)$ is invertible; and it can be easily verified that

$$(41) \qquad N_{D/Q}(\mathfrak{d}_1)^{-s}\xi_0\big(N(\mathfrak{d}_1)\big)\mathfrak{X}(\mathfrak{d}_1)\mathscr{D}_0(s; \xi) = \mathscr{D}(s; \xi) .$$

By, (33), (39), (41), we obtain thus

THEOREM 1. *Let D be an indefinite quaternion algebra over a totally real algebraic number field F (cf. § 1.1). Let $\mathfrak{o}$ be a maximal order in D and $\mathfrak{a}$ be an integral two-sided $\mathfrak{o}$-ideal. Let ρ be an irreducible representation of $(\mathfrak{o}/\mathfrak{a})^*$ by complex matrices. Let the n_ν for $1 \leq \nu \leq t$ be integers such that $n_\nu > 0$ for $1 \leq \nu \leq r$ and $n_\nu \geq 0$ for $\nu > r$. Let $S(\rho; \{n_\nu\})$ be the product of h vector spaces of cusp-forms with respect to mutually commensurable unit groups in D (cf. § 3.3), h being the class-number of D. Let ξ be a Grössen-character of F in the sense of § 4.3, and ξ_0 be the ideal-character attached to ξ. Define ρ^F as in § 4.3. Suppose that $\mathfrak{a}$ is a defining ideal for the character $\xi(N(x))$ of the idele-group of D. Define the operator $\mathfrak{X}(e)$, $\mathfrak{X}(\mathfrak{n})$, $\mathfrak{X}(e; \xi)$ and $\mathfrak{Y}$ in $S(\rho; \{n_\nu\})$ as in §§ 3.4, 3.6, 4.4. Put*

$$\mathscr{D}(s; \xi) = \sum_{e_0} N_{D/Q}(e_0)^{-s}\xi_0\big(N(e_0)\big)\mathfrak{X}(e_0) = \sum_{\mathfrak{n}} N(\mathfrak{n})^{-s}\xi_0(\mathfrak{n})\mathfrak{X}(\mathfrak{n}) ,$$

$$\mathscr{W}(s; \xi) = \sum_{e} N_{D/Q}(e)^{-s}\mathfrak{X}(e; \xi) ,$$

$$\mathscr{E}(s; \xi) = N_{D/Q}(\mathfrak{d}_1)^{s/2}\Gamma(s; \{n_\nu\}; \xi)\mathscr{D}(s; \xi) ,$$

$$\Lambda_\xi = i^m N_{N/Q}(\mathfrak{d}_2\mathfrak{a}^2)\xi_0\big(N(\mathfrak{d}_1)\big)q(\mathfrak{a})^{-1}\mathfrak{X}(\mathfrak{d}_1) .$$

Here the notations are as follows: $\mathfrak{d} = \mathfrak{d}(\mathfrak{o}/Z) = \mathfrak{d}_1\mathfrak{d}_2 = \prod_\mathfrak{p} \mathfrak{d}_\mathfrak{p}$ is the different of $\mathfrak{o}$ with respect to Z; $\mathfrak{d}_1 = \prod_{\mathfrak{p}\nmid N(\mathfrak{a})}\mathfrak{d}_\mathfrak{p}$, $\mathfrak{d}_2 = \prod_{\mathfrak{p}\mid N(\mathfrak{a})}\mathfrak{d}_\mathfrak{p}$; $\sum_{e_0}$ is the sum extended over all the integral elements e_0 of $\mathfrak{E}$ which are prime to $\mathfrak{a}$; $\sum_{\mathfrak{n}}$ is the sum extended over all the integral ideals $\mathfrak{n}$ of F which are prime to $N(\mathfrak{a})$; $\sum_e$ is the sum extended over all the elements e of $\mathfrak{E}$ such that $e \subset \mathfrak{d}^{-1}\mathfrak{a}^{-1}$ and the set of invariants of e contains only powers of prime factors of $N(\mathfrak{a})$; $\Gamma(s; \{n_\nu\}; \xi)$ is the function defined by (21) of § 4.5 for

$m_\nu = 0$; $m = \sum_{\nu=1}^{t} n_\nu$; $q(\mathfrak{a})$ *is the number of regular elements in* $\mathfrak{o}/\mathfrak{a}$; $\mathfrak{T}(\mathfrak{d}_1)$ *is the operator in* $S(\rho; \{n_\nu\})$ *defined by* (40).

Then: $\mathscr{D}(s;\xi)$ *and* $\mathscr{W}(s;\xi)$ *converges for* $\mathrm{Re}(s) > 2$; $\mathscr{E}(s;\xi)$ *and* $\mathscr{W}(s;\xi)\mathscr{E}(s;\xi)$ *can be continued holomorphically (resp. meromorphically) to the whole s-plane if D is a division algebra (resp.* $D = M_2(F)$)); $\mathscr{D}(s;\xi)$ *can be expressed in the form of an Euler-product*

$$\mathscr{D}(s; \xi) = \prod_{\substack{\mathfrak{p} \mid d \\ \mathfrak{p} \nmid N(\mathfrak{a})}} [1 - \xi_0(\mathfrak{p})N(\mathfrak{p})^{-s}\mathfrak{T}(\mathfrak{p})]^{-1}$$

$$\times \prod_{\mathfrak{p} \nmid dN(\mathfrak{a})} [1 - \xi_0(\mathfrak{p})N(\mathfrak{p})^{-s}\mathfrak{T}(\mathfrak{p}) + \xi_0(\mathfrak{p})^2 N(\mathfrak{p})^{1-2s}\mathfrak{T}(\mathfrak{p}, \mathfrak{p})]^{-1} ,$$

where $d = d(D/F)$ *is the discriminant-ideal of D over F; and the following functional equation holds.*

$$\mathscr{W}(s; \xi)\mathscr{E}(s; \xi) = \Lambda_\xi \mathfrak{Y} \mathscr{E}(2 - s; \rho^r \xi^{-1})\mathfrak{Y} .$$

Now as $\Gamma(s; \{n_\nu\}; \xi)^{-1}$ is holomorphic on the whole s-plane, we observe that, if D is a division algebra, then $\mathscr{D}(s; \xi)$ and $\mathscr{W}(s; \xi)\mathscr{D}(s; \xi)$ are holomorphic on the whole plane. It may be conjectured that this is also true for $D = M_2(F)$ though our argument in § 4.9 is too weak to prove it. The consideration of § 4.9 shows at most (or at least) that

$$\left(s - \frac{3n_1}{2} + ia_1 + 2\right)\left(s - \frac{n_1}{2} + ia_1 - 1\right)\left(s + \frac{n_1}{2} + ia_1 - 1\right)$$
$$\cdot \Gamma(s; \{n_\nu\}; \xi)\mathscr{W}(s; \xi)\mathscr{D}(s; \xi) ,$$

is holomorphic on the whole plane. If we modify φ_* by replaing $\varphi_\mathfrak{p}$ with $\hat{\varphi}_\mathfrak{p}$, then we find that

$$\left(s - \frac{3n_1}{2} + ia_1 + 2\right)\left(s - \frac{n_1}{2} + ia_1 - 1\right)\left(s + \frac{n_1}{2} + ia_1 - 1\right)$$
$$\cdot \Gamma(s; \{n_\nu\}; \xi)\mathscr{D}(s; \xi)$$

is holomorphic on the whole plane. Note that the inverse of $s\Gamma(s)$ is holomorphic on the whole plane. Therefore, $\mathscr{W}(s; \xi)\mathscr{D}(s; \xi)$ and $\mathscr{D}(s; \xi)$ can have their poles only at $3n_1/2 - ia_1 - 2, n_1/2 - ia_1 + 1$. On the other hand, $\mathscr{W}(s; \xi)\mathscr{D}(s; \xi)$ and $\mathscr{D}(s; \xi)$ are holomorphic for $\mathrm{Re}(s) > 2$. It follows that, if $n_1 \geq 3$, $\mathscr{W}(s; \xi)\mathscr{D}(s; \xi)$ and $\mathscr{D}(s; \xi)$ have no pole. We can take any one of the n_ν in place of n_1. Hence we can say that, if $n_\nu \geq 3$ holds for at least one of the n_ν, the functions $\mathscr{W}(s; \xi)\mathscr{D}(s; \xi)$ and $\mathscr{D}(s; \xi)$ are holomorphic on the whole plane.

4.11. *Further analysis of* $\mathscr{W}(s;\xi)$. First consider the case $\mathfrak{a} = \mathfrak{o}$ in Theorem 1. Then we have $\mathfrak{d}_1 = \mathfrak{d}$, $\mathfrak{d}_2 = \mathfrak{o}$, $\Lambda_\xi = i^m \xi_0(N(\mathfrak{d}))\mathfrak{T}(\mathfrak{d})$, $\rho^r = 1$ and $\mathscr{W}(s; \xi) = 1$. Therefore the functional equation can be written in the form

$$\mathcal{E}(s;\xi) = \Lambda_\xi \mathfrak{Y} \mathcal{E}(2-s;\xi^{-1})\mathfrak{Y} .$$

Substituting $2-s$ and ξ^{-1} for s and ξ, we get $(\mathfrak{T}(\mathfrak{d}_1)\mathfrak{Y})^2 = 1$.

Now consider the general case where $\mathfrak{a}$ is not assumed to be equal to $\mathfrak{o}$. For every prime factor $\mathfrak{p}$ of $N(\mathfrak{a})$, put

$$\sigma'_\mathfrak{p}(x; \eta^{-1}\rho) = \sigma_\mathfrak{p}(x_\mathfrak{p}; \eta^{-1}\rho) \prod_{\substack{\mathfrak{q}|N(\mathfrak{a})\\ \mathfrak{q}\neq\mathfrak{p}}} \xi_{1\mathfrak{q}}(x_\mathfrak{q})^{-1}\rho_\mathfrak{q}(x_\mathfrak{q}) \qquad (x\in\mathfrak{J}) .$$

Let e be an element of $\mathfrak{E}$ whose set of invariants contains only powers of one prime ideal $\mathfrak{p}$, and $T_{\mu\lambda}(e) = \Gamma_\mu \alpha \Gamma_\lambda = \bigcup_i \alpha_i \Gamma_\lambda$. Define the operator $\mathfrak{T}(e;\xi;\mathfrak{p})$ on S by

$$\mathfrak{T}(e;\xi;\mathfrak{p})(f_1,\cdots,f_h) = (g_1,\cdots,g_h) ,$$
$$g_\mu(z) = \mathfrak{T}_{\mu\lambda}(e;\xi;\mathfrak{p})f_\lambda$$
$$= \xi_0(N(\mathfrak{x}_{\lambda\mu})) \sum_i \eta(\alpha_i)L(\alpha_i)^{-1}\sigma'_\mathfrak{p}(\alpha_i; \eta^{-1}\rho)$$
$$\otimes \Psi(\alpha_i) J(\alpha_i^{-1}, z)^{-1} f_\lambda(\alpha_i^{-1}(z)) ,$$

and put

$$\mathcal{W}_\mathfrak{p}(s;\xi) = \sum N_{D/Q}(e)^{-s}\mathfrak{T}(e;\xi;\mathfrak{p}) ,$$

where the sum is extended over all the elements e of $\mathfrak{E}$ such that $e \subset \mathfrak{b}^{-1}\mathfrak{a}^{-1}$ and the set of invariants of e contains only powers of $\mathfrak{p}$. Then we can easily verify that the following relation holds:

$$\mathcal{W}(s;\xi) = \prod_{\mathfrak{p}|N(\mathfrak{a})} \mathcal{W}_\mathfrak{p}(s;\xi) .$$

By the functional equations for $\mathcal{D}(s;\xi)$ and $\mathcal{D}(s;\xi^*)$, we obtain

$$(42) \qquad\qquad \mathcal{W}(2-s;\xi^*) = \Lambda_{\xi^*}\mathfrak{Y}\mathcal{W}(s;\xi)^{-1}\Lambda_\xi\mathfrak{Y} .$$

Obviously, $\mathcal{W}_\mathfrak{p}(s;\xi)$ is a power-series of $N(\mathfrak{p})^{-s}$. If we represent $\mathcal{W}_\mathfrak{p}(s;\xi)$ by a matrix with respect to a basis of S, then it is plausible that the entries of that matrix are *rational functions of* $N(\mathfrak{p})^{-s}$. We now show that this is in fact true in the following two cases.

(I) $\mathfrak{a}$ *has only one prime factor.*

(II) $\mathfrak{a}$ *is the conductor of* $\xi_1^{-1}\rho$; *and* $\omega = \xi_1^{-1}\rho$ *satisfies the condition of Proposition* 4.2.

Case (I). In this case $N(\mathfrak{a})$ has only one prime factor $\mathfrak{p}$. $\mathcal{W}(s;\xi)$ and $\mathcal{W}(s;\xi^*)$ are expressed in the form

$$\mathcal{W}(s;\xi) = \sum_{n=n_0}^\infty N(\mathfrak{p})^{-ns}A_n , \qquad \mathcal{W}(s;\xi^*) = \sum_{n=n_0'}^\infty N(\mathfrak{p})^{-ns}B_n ,$$

where n_0 and n_0' are integers, and A_n, B_n are linear mappings of S into itself. Put, for $z \in C$,

$$A(z) = \sum_{n=n_0}^\infty z^n A_n , \qquad B(z) = \sum_{n=n_0'}^\infty z^n B_n .$$

Then these power series converge in a neighborhood of $z = 0$ except the origin, so that they define meromorphic functions at $z = 0$. As $\mathscr{W}(s; \xi)$ and $\mathscr{W}(s; \xi^*)$ are meromorphic on the whole s-plane, we observe that $A(z)$ and $B(z)$ can be continued meromorphically to the whole z-plane, and $\mathscr{W}(s; \xi) = A(z)$, $\mathscr{W}(s; \xi^*) = B(z)$ hold for $z = N(\mathfrak{p})^{-s}$. By the relation (42), we get $B(N(\mathfrak{p})^{-2}z^{-1}) = \Lambda_{\xi^{-1}}\mathfrak{Y}A(z)^{-1}\Lambda_\xi\mathfrak{Y}$. The left-hand side of this equality is meromorphic at $z = \infty$. Therefore, the entries of $A(z)$ (considered as a matrix) are rational functions of z. Thus we have obtained the desired result.

Case (II). Let e be an element of $\mathfrak{E}$ such that $e \subset \mathfrak{b}^{-1}\mathfrak{a}^{-1}$ and the set of invariants of e consists only of powers of prime factors of $N(\mathfrak{a})$. Put $T_{\mu\lambda}(e) = \Gamma_\mu\alpha\Gamma_\lambda$, and suppose that $\sigma_\mathfrak{a}(\alpha; \eta^{-1}\rho) \neq 0$. By Proposition 4.2, we have $\mathfrak{o}_\lambda\alpha = \mathfrak{b}_\mathfrak{p}^{-1}\mathfrak{a}_\mathfrak{p}^{-1}$ for $\mathfrak{p} \mid N(\mathfrak{a})$. As $(\alpha\mathfrak{x}_{\lambda\mu})_\mathfrak{q} = \mathfrak{o}_{\mu\mathfrak{q}}$ for every $\mathfrak{q} \nmid N(\mathfrak{a})$, we have $\alpha\mathfrak{x}_{\lambda\mu} = \mathfrak{x}_\mu^{-1}\mathfrak{b}_2^{-1}\mathfrak{a}^{-1}\mathfrak{x}_\mu$. Therefore, $\mathfrak{X}(e; \xi) \neq 0$ only for the similarity-class e_0 of $\mathfrak{b}_2^{-1}\mathfrak{a}^{-1}$; and $T_{\mu\lambda}(e_0) = \Gamma_\mu\alpha\Gamma_\lambda = \alpha\Gamma_\lambda$. Hence $\mathscr{W}(s; \xi)$ contains only one term:

$$\mathscr{W}(s; \xi) = \mathfrak{X}(e_0; \xi)N_{D/Q}(\mathfrak{b}_2\mathfrak{a})^s .$$

Then the relation (42) shows

$$(43) \qquad \mathfrak{X}(e_0; \xi^*) = (-1)^m N_{D/Q}(\mathfrak{a})^2 q(\mathfrak{a})^{-2}\rho_0^F\big(N(\mathfrak{b}_1)\big)\mathfrak{X}(\mathfrak{b}_1)\mathfrak{Y}\mathfrak{X}(e_0; \xi)^{-1}\mathfrak{X}(\mathfrak{b}_1)\mathfrak{Y} .$$

Putting

$$\mathcal{E}_0(s; \xi) = N_{D/Q}(\mathfrak{b}\mathfrak{a})^{s/2}\Gamma\big(s; \{n_\nu\}; \xi\big)\mathcal{D}(s; \xi) ,$$
$$\Lambda_\xi' = i^m N_{D/Q}(\mathfrak{a})q(\mathfrak{a})^{-1}\xi_0\big(N(\mathfrak{b}_1)\big)\mathfrak{X}(e_0; \xi)^{-1}\mathfrak{X}(\mathfrak{b}_1) ,$$

the functional equation for $\mathcal{D}(s; \xi)$ is now written in the form

$$\mathcal{E}_0(s; \xi) = \Lambda_\xi'\mathfrak{Y}\mathcal{E}_0(2 - s; \xi^*)\mathfrak{Y} .$$

4.12. *Relation between functional equation and Gaussian sums.* Denote by F_q the finite field with q elements, where q is a power of a prime number p; we identify F_p with $Z/(p)$. Put $\zeta = e^{2\pi i/p}$ and, for $\alpha \in F_q$,

$$e[\alpha] = \zeta^{\text{tr}_{F_q/F_p}^{(\alpha)}} .$$

Let ω be an irreducible representation of $GL(2, F_q)$ by complex matrices. Put

$$(44) \qquad\qquad B(\omega) = \sum e[\text{tr}(X)]\omega(X) ,$$

where X runs over all elements of $GL(2, F_q)$. We see easily that $B(\omega)\omega(Y) = \omega(Y)B(\omega)$ for every $Y \in GL(2, F_q)$. As ω is irreducible, $B(\omega)$ must be a scalar matrix; so we can put

$$B(\omega) = b(\omega)\omega(1)$$

with a complex number $b(\omega)$. If we put $\chi(X) = \text{tr}\omega(X)$, we get

$$\chi(1)b(\omega) = \sum e[\mathrm{tr}(X)]\chi(X) ,$$

the sum being as above. Therefore, $b(\omega)$ depends only on the equivalence-class of irreducible representations. The number $b(\omega)$ may be considered as a generalized gaussian sum. Lamprecht [*Struktur und Relationen allgemeiner Gaussscher Summen in endlichen Ringen* I, II, J. reine angew. Math., 197 (1957), 1–48] (quoted hereafter as [L]) investigated such sums in a more general case. Let H be a subgroup of GL(2, F_q) consisting of the elements of the form $\begin{pmatrix} a & 0 \\ b & 1 \end{pmatrix}$. Then, if the restriction of ω to H contains no unit representation, we have, by [L, Satz 4],

$$(45) \qquad |\,b(\omega)\,| = q^2 , \qquad b(\omega)b(\bar{\omega}) = \omega(-1)q^4 ,$$

where $\bar{\omega}$ means the complex conjugate representation of ω. This is easily shown by an analogy of the calculation of absolute values of the usual gaussian sums together with the argument of the proof of Proposition 4.2. Now we show that $b(\omega)$ is related with our functional equation. For the sake of simplicity, we consider only the case where $F = Q$, $\mathfrak{a} = \mathfrak{p}\mathfrak{o}$, and $p \nmid d(D/Q)$ in Theorem 1. We have then $h = 1$, $\mathfrak{Y} = 1$, and by Proposition 4.4, $\mathfrak{X}(\mathfrak{b}_1)^2 = \rho_0^F(N(\mathfrak{b}_1))^{-1}$. As $\mathfrak{o}/\mathfrak{a}$ is isomorphic to $M_2(Z/(p))$, we consider ρ_p as a representation of GL(2, F_p). As ρ is irreducible, $\xi_{1p}^{-1}\rho_p$ is irreducible. Put $\omega(x_p) = \xi_p(\det(x_p))^{-1}\rho_p(x_p)$ for $x_p \in U_p$, and consider ω also as a representation of GL(2, F_p). Transforming by the canonical involution, we find

$$B(\omega) = q((p))\int_{U_p} \tau_p(p^{-1}u_p)\xi_{1p}(u_p)^{-1}\rho_{1p}^F(u_p)\rho_p(u_p)^{-1}dm_p^*(u_p) ,$$

so that $\mathfrak{X}(e_0; \xi^*) = q((p))^{-1}b(\omega)$, on account of our definition of $\mathfrak{X}(e; \xi)$. Similarly we get $\mathfrak{X}(e_0; \xi) = q((p))^{-1}b(\bar{\omega})$. Then (45) is a consequence of (43). Of course, in this proof, we must assume the existence of certain automorphic forms.

Thus, it is interesting to investigate $b(\omega)$ in connection with the functional equation. The determination of the factor $\mathcal{W}(s; \xi)$ is certainly related to the sums of this kind. For the group $G = \mathrm{GL}(n, F_q)$, [L, Satz 7] shows

$$\sum_{X \in G} \mathrm{tr}(X)\psi(\det(X)) = q^{n(n-1)/2}\Big(\sum_{\alpha \in F_q^*} e[\alpha]\psi(\alpha)\Big)^n$$

for every character ψ of F_q^*. The group GL(2, F_q) has exactly $q^2 - 1$ inequivalent irreducible representations; there are exactly q representations among them, whose restriction to H contains the unit representation. It can be verified that $|\,b(\omega)\,| = q$ if ω is the unit representation or the representation of degree q, and $|\,b(\omega)\,| = q^{3/2}$ for the other $q - 2$ representations. It seems also meaningful to investigate similar sums for the groups of similarity-transformations of quadratic forms, hermitian forms and skew-symmetric forms over finite fields.

Finally we note that Theorem 1 gives us some Dirichlet series, which seem to be new, even in the case $D = M_2(Q)$. Consider for example

$$\Delta(z) = x \prod_{n=1}^{\infty} (1 - x^n)^{24} = \sum_{n=1}^{\infty} c_n x^n \; ; \qquad\qquad x = e^{2\pi i z} \; .$$

This is a cusp-form of degree 12; so the function $R(s) = (2\pi)^{-s}\Gamma(s)\sum_{n=1}^{\infty} c_n n^{-s}$ is holomorphic on the whole plane and satisfies $R(12 - s) = R(s)$. Now let ξ_0 be an arbitrary character of $(Z/(a))^*$, where a is a positive integer. We can find a Grössen-character ξ of Q with the conductor (a) such that ξ_0 is the ideal-character attached to ξ. Put

$$R_0(s; \xi) = \sum_{(n,a)=1} \xi_0(n) c_n n^{-s} \; .$$

Then Theorem 1, together with the consideration in the last part of § 4.10, shows that $R_0(s; \xi)$ is holomorphic on the whole plane and satisfies a functional equation. As the present case is included in the *case* II of § 4.11, we find, putting $R(s; \xi) = a^s (2\pi)^{-s}\Gamma(s)R_0(s; \xi)$, that there exists a complex number $\lambda(\xi)$ such that

$$R(s; \xi) = \lambda(\xi)R(12 - s; \xi^{-1}) \; .$$

If a is a prime number p, we get, by the above consideration,

$$\lambda(\xi) = p\left(\sum_{a=1}^{p-1} \zeta^a \bar{\xi}_0(a)\right)^{-2} \; .$$

If $\xi_0(x) = \left(\dfrac{x}{p}\right)$, we have $\lambda(\xi) = 1$ or -1 according as $p \equiv 1$ or $-1 \bmod (4)$.

Therefore, in this case, the functions $R_0(s; \xi)$ belong to the class of Dirichlet series investigated by Hecke [12].

Part II. Abelian Varieties Attached to Automorphic Forms

§ 5. Automorphic forms with respect to a fuchsian group

5.1. In the remainder of this paper, we consider only the automorphic functions with one variable; so we write simply $\mathfrak{F}_1 = \mathfrak{F}$, namely

$$\mathfrak{F} = \{z \in C \mid \mathrm{Im}(z) \neq 0\} \; .$$

We define two connected components $\mathfrak{H}^+$ and $\mathfrak{H}^-$ of $\mathfrak{F}$ by

$$\mathfrak{H}^+ = \{z \in C \mid \mathrm{Im}(z) > 0\} \; , \qquad \mathfrak{H}^- = \{z \in C \mid \mathrm{Im}(z) < 0\} \; ,$$

and mean by the notation $\mathfrak{H}$ any one of the domains $\mathfrak{H}^+$ and $\mathfrak{H}^-$.

Though our interest lies in the investigation of arithmetically defined fuchsian groups, we begin with general fuchsian groups. By a *fuchsian group*, we understand a discrete subgroup Γ of SL(2, R) such that $\mathfrak{H}^+/\Gamma$

is of finite measures.[6] We always assume $\Gamma \ni -1$ and put $\bar{\Gamma} = \Gamma/\{\pm 1\}$. If Γ is a fuchsian group, $\mathfrak{H}^-/\Gamma$ is also of finite measure. In general, the Riemann surfaces $\mathfrak{H}^+/\Gamma$ and $\mathfrak{H}^-/\Gamma$ are not isomorphic. Moreover, the theory of Part I proves the necessity of considering automorphic forms on $\mathfrak{F}$. This is the reason why we introduce not only $\mathfrak{H}^+$ but also $\mathfrak{H}^-$.

D being as in § 1.1, suppose that $r = 1$. Let $\mathfrak{o}$ be a maximal order in D and Γ be the group of units γ in $\mathfrak{o}$ such that $N(\gamma) = 1$. Then Γ, considered as group of transformations on $\mathfrak{H}$, is a fuchsian group. If D is a division algebra, $\mathfrak{H}/\Gamma$ is compact; if not, we have $D = M_2(Q)$ on account of $r = 1$; and Γ coincides with SL(2, Z) by a suitable choice of basis. The principal aim of Part II is to construct abelian varieties from cusp forms with respect to Γ. This was already done in the case $\Gamma = $ SL(2, Z) in a previous paper [24]. Therefore, in the present investigation, *we consider only the fuchsian groups Γ such that $\mathfrak{H}/\Gamma$ is compact.*

5.2. *Automorphic forms.* Let Γ be a fuchsian group. Let χ be a representation of Γ by complex matrices such that $\chi(-1) = 1$, and m a non-negative even integer. By a cusp-form of type $(\Gamma; \mathfrak{H}; \chi; m)$, we understand a column-vector function $f(z)$ on $\mathfrak{H}$ taking values in the representation space of χ and satisfying the following conditions:

(S_1) *the components of $f(z)$ are holomorphic functions on $\mathfrak{H}$;*

(S_2) *for every $\gamma \in \Gamma$, $f(\gamma(z)) = \chi(\gamma)j(\gamma, z)^m f(z)$.*

We denote by $S(\Gamma; \mathfrak{H}; \chi; m)$ the vector space of such functions f. Besides this, we consider a column-vector differential form ω on $\mathfrak{H}$ satisfying the following conditions:

(W_1) *the components of ω are holomorphic differential forms of degree one on $\mathfrak{H}$;*

(W_2) *for every $\gamma \in \Gamma$, $\omega \circ \gamma = \chi(\gamma)\omega$.*[7]

We denote by $W(\Gamma; \mathfrak{H}; \chi)$ the set of such forms ω. Fixing Γ and $\mathfrak{H}$, we denote $S(\Gamma; \mathfrak{H}; \chi; m)$ and $W(\Gamma; \mathfrak{H}; \chi)$ simply by $S(\chi; m)$ and $W(\chi)$, respectively.

Now suppose that there exists a hermitian matrix Y, not necessarily positive, such that, for every $\gamma \in \Gamma$,

$$(51) \qquad {}^t\chi(\gamma)\, Y\overline{\chi(\gamma)} = Y .$$

Then, we define a hermitian form $Y(f, g)$ on $S(\chi; m)$ and a hermitian form $Y(\omega, \eta)$ on $W(\chi)$ by

---

[6] It would be proper to write $\Gamma\backslash\mathfrak{H}^+$ since we write the elements of Γ on the left. However, as we do not use the operation on the right, there will be no fear of confusion if we write $\mathfrak{H}^+/\Gamma$.

[7] We mean by $\omega \circ \gamma$ the transform of ω by γ, so, if $\omega = f(z)dz$, then $\omega \circ \gamma = f(\gamma(z))j(\gamma, z)^{-2}dz$.

$$Y(f, g) = \int_B {}^t f Y \bar{g} \operatorname{Im}(z)^{m-2} dx dy , \qquad (z = x + iy) ,$$

$$Y(\omega, \eta) = i \int_B {}^t \omega \wedge Y \bar{\eta} ,$$

where $B = \mathfrak{H}/\Gamma$. It is clear that $Y(f, g)$ and $Y(\omega, \eta)$ are well-defined and determine hermitian forms on $S(\chi; m)$ and $W(\chi)$, respectively, which are positive non-degenerate if Y is positive non-degenerate.

If χ has real coefficients and Y is a real symmetric matrix, we have, for every ω and η of $W(\chi)$, $4 \,{}^t\operatorname{Re}(\omega) \wedge Y\operatorname{Re}(\eta) = {}^t\omega \wedge Y\bar{\eta} - {}^t\eta \wedge Y\bar{\omega}$, so that

$$(52) \qquad Y(\omega, \eta) - Y(\eta, \omega) = 4i \int_B {}^t\operatorname{Re}(\omega) \wedge Y\operatorname{Re}(\eta) .$$

5.3. *The injection ι.* Now recall the notation $\binom{u}{v}^n$, $\Phi_n(\sigma)$ introduced in § 3.1. For every $\sigma \in \mathrm{GL}(2, R)$, we have

$$(53) \qquad \left[\binom{z}{1}^n dz \right] \circ \sigma = \det(\sigma) j(\sigma, z)^{n+2} \Phi_n(\sigma) \binom{z}{1}^n dz .$$

Suppose $m \geq 2$ and put $n = m - 2$. For every $f \in S(\chi; m)$, define $\iota(f)$ by

$$\iota(f) = f \otimes \binom{z}{1}^n dz .$$

Using (53), we can easily verify that $\iota(f)$ is an element of $W(\chi \otimes \Phi_n)$. From now on, we always mean by m and n two even integers such that $n \geq 0$ and $m = n + 2$. The matrix P_n being as in § 3.1, if Y satisfies (51), then we have, for every $\gamma \in \Gamma$,

$${}^t(\chi(\gamma) \otimes \Phi_n(\gamma))(Y \otimes P_n)(\overline{\chi(\gamma) \otimes \Phi_n(\gamma)}) = Y \otimes P_n ,$$

and for every f and g of $S(\chi; m)$, on account of (8) of § 3.1,

$$(54) \qquad Y \otimes P_n(\iota(f), \iota(g)) = 2^{n+1} i^n Y(f, g) .$$

6. Cohomology groups of fuchsian groups

χ being as before a representation of a fuchsian group Γ, assume now that χ has real coefficients. Denote by V the representation-space for χ, which we identify with the column-vector space R^k. We can define homology groups $H_p(\Gamma, V)$ and cohomology groups $H^p(\Gamma, V)$ for every integer $p \geq 0$. In the present investigation, we denote $H^p(\Gamma, V)$ by $H^p(\Gamma, \chi, R)$. In particular, $H^1(\Gamma, \chi, R)$ is defined as follows. Let $\mathfrak{Z}(\chi, \Gamma, R)$ be the set of the mappings $\mathfrak{x}$ of Γ into V such that $\mathfrak{x}(\sigma\tau) = \mathfrak{x}(\sigma) + \chi(\sigma)\mathfrak{x}(\tau)$ for every σ and τ of Γ. For every $\mathfrak{a} \in V$, the mapping $\mathfrak{h}$ defined by $\mathfrak{h}(\gamma) = [1 - \chi(\gamma)]\mathfrak{a}$ for $\gamma \in \Gamma$, is an element of $\mathfrak{Z}(\Gamma, \chi, R)$; denote by $\mathfrak{B}(\Gamma, \chi, R)$ the set of such $\mathfrak{y}$. Then, $H^1(\Gamma, \chi, R) = \mathfrak{Z}(\Gamma, \chi, R)/\mathfrak{B}(\Gamma, \chi, R)$. Every element of $\mathfrak{Z}(\Gamma, \chi, R)$ is called a cocycle. We note that $H^0(\Gamma, \chi, R)$ is isomorphic to the vector space $\{\mathfrak{v} \in V \mid \chi(\gamma)\mathfrak{v} = \mathfrak{v} \text{ for every } \gamma \in \Gamma\}$.

PROPOSITION 6.1. *Suppose that $\bar{\Gamma}$ has no element of finite order other than the identity element. Let g be the genus of the Riemann surface $\mathfrak{H}/\Gamma$. Let χ be a real representation of Γ of degree k and Y be a real non-singular symmetric matrix satisfying (51). Then we have*

$$\dim H^1(\Gamma, \chi, R) = 2k(g - 1) + 2 \dim H^0(\Gamma, \chi, R) \,.$$

PROOF. Using the representation χ of Γ, we can construct a fibre bundle $\mathscr{B}$ with the base space $\mathscr{R} = \mathfrak{H}/\Gamma$ and the typical fibre V. Consider the cohomology groups $H^p(\mathscr{R}, \mathscr{B})$ in the sense of Steenrod [26, p. 156]. If we consider the sheaf $\mathscr{B}'$ of germs of locally constant sections of $\mathscr{B}$, then $H^p(\mathscr{R}, \mathscr{B})$ can be identified with the cohomology groups $H^p(\mathscr{R}, \mathscr{B}')$ in sheaf-theory. Take a simplicial division of $\mathscr{R}$ and denote by a_p the number of p-simplices. Then we find easily

$$\sum_{p=0}^{2} (-1)^p \dim H^p(\mathscr{R}, \mathscr{B}) = k \sum_{p=0}^{2} (-1)^p a_p \,.$$

The right hand side is of course $k(2 - 2g)$. Now, by a theorem of Eilenberg-McLane, $H^p(\mathscr{R}, \mathscr{B})$ is isomorphic to $H^p(\Gamma, \chi, R)$. The existence of Y implies the equivalence of χ and ${}^t\chi^{-1}$. It follows that $\dim H^0(\mathscr{R}, \mathscr{B}) = \dim H^2(\mathscr{R}, \mathscr{B})$. We obtain therefore the equality in the proposition.

PROPOSITION 6.2. *Let χ be a real representation of Γ and Y be a real symmetric matrix satisfying (51). If Y is positive non-degenerate, $\dim H^0(\Gamma, \chi \otimes \Phi_n, R) = 0$ for every $n > 0$.*

PROOF. We can find an element γ_0 of Γ whose characteristic roots are a and a^{-1} for a real number $a > 1$. Then the characteristic roots of $\Phi_n(\gamma_0)$ are $a^n, a^{n-2}, \cdots, a^2, 1, a^{-2}, \cdots, a^{-n}$. Let u be a non-zero vector such that $\Phi_n(\gamma_0)u = u$. On the other hand, the characteristic roots of $\chi(\gamma_0)$ are complex numbers of absolute value 1. Therefore, if w is a non-zero vector invariant under $\chi \otimes \Phi_n(\gamma_0)$, then w must be contained in $V \otimes u$, where V is the representation space of χ; put $w = v \otimes u$. If, furthermore, $\chi \otimes \Phi_n(\gamma)w = w$ for every $\gamma \in \Gamma$, we must have $\Phi_n(\gamma)u \in Ru$ for every $\gamma \in \Gamma$. This means that the one-dimensional space Ru is invariant under $\Phi_n(\Gamma)$ and contradicts the irreducibility of Φ_n on Γ [Borel, 1]. Hence there is no $\chi \otimes \Phi_n(\Gamma)$-invariant vector other than 0, such that $\dim H^0(\Gamma, \chi \otimes \Phi_n, R) = 0$.

7. A theorem of duality

7.1. *Periods of integrals.* Let χ be a real representation of Γ of degree k. We consider, for every element ω of $W(\chi)$, a function $\mathfrak{f}(z)$ defined by

$$\mathfrak{f}(z) = \int_{z_0}^{z} \operatorname{Re}(\omega) + \mathfrak{a} ,$$

where z_0 is an arbitrarily fixed point on $\mathfrak{H}$ and $\mathfrak{a}$ is a vector of $\boldsymbol{R}^k$. As ω is holomorphic, the integral does not depend on the choice of the path of integration. By the property $\omega \circ \gamma = \chi(\gamma)\omega$, we can easily verify the relation

$$\mathfrak{f}(\gamma(z)) = \chi(\gamma)\mathfrak{f}(z) + \mathfrak{x}(\gamma) \qquad\qquad (\gamma \in \Gamma) ,$$

where $\mathfrak{x}$ is an element of $\mathfrak{Z}(\Gamma, \chi, \boldsymbol{R})$. Furthermore, the cohomology-class of $\mathfrak{x}$ is uniquely determined by ω; it is independent of the choice of z_0 and $\mathfrak{a}$. We denote the cohomology-class of $\mathfrak{x}$ by $\varphi(\omega)$. Then φ is an $\boldsymbol{R}$-linear mapping of $W(\chi)$ into $H^1(\Gamma, \chi, \boldsymbol{R})$.

Let Y be a real symmetric matrix such that (51) holds. Then there exist rational numbers a_λ and elements $\alpha_\lambda, \beta_\lambda$ of Γ with the following properties: let ω and η be elements of $W(\chi)$, and $\mathfrak{x}, \mathfrak{y}$ be respectively cocycles belonging to $\varphi(\omega)$ and $\varphi(\eta)$; then we have

$$(55) \qquad \int_{\mathfrak{H}/\Gamma} {}^t\operatorname{Re}(\omega) \wedge Y\operatorname{Re}(\eta) = \sum_{\lambda=1}^{r} a_\lambda {}^t\mathfrak{x}(\alpha_\lambda) Y\mathfrak{y}(\beta_\lambda) .$$

This result is proved in [24, § 5].

7.2. Let us consider three mappings ι, d, φ:

$$\iota: S(\chi; m) \longrightarrow W(\chi \otimes \Phi_n) ,$$
$$d: S(\chi \otimes \Phi_n; 0) \longrightarrow W(\chi \otimes \Phi_n) ,$$
$$\varphi: W(\chi \otimes \Phi_n) \longrightarrow H^1(\Gamma, \chi \otimes \Phi_n, \boldsymbol{R}) .$$

The definitions of ι and φ are given in §§ 5.3 and 7.1; the second mapping d is the usual differentiation.

THEOREM 2. *Let χ be a real representation of Γ and Y be a real symmetric matrix such that ${}^t\chi(\gamma)Y\chi(\gamma) = Y$ for every $\gamma \in \Gamma$. Suppose that Y is positive non-degenerate. Then the following assertions hold.*
  (i) *$W(\chi \otimes \Phi_n)$ is the direct sum of $\iota[S(\chi; m)]$ and $d[S(\chi \otimes \Phi_n; 0)]$.*
  (ii) *φ is an $\boldsymbol{R}$-isomorphism of $\iota[S(\chi; m)]$ onto $H^1(\Gamma, \chi \otimes \Phi_n, \boldsymbol{R})$.*
  (iii) *φ is 0 on $d[S(\chi \otimes \Phi_n; 0)]$.*

PROOF. Let $\mathfrak{g}$ be an element of $S(\chi \otimes \Phi_n; 0)$. Obviously, we have $\operatorname{Reg}(z) = \int_{z_0}^{z} \operatorname{Re}(d\mathfrak{g}) + \mathfrak{g}(z_0)$, and $\operatorname{Reg}(\gamma(z)) = \chi \otimes \Phi_n(\gamma)\operatorname{Reg}(z)$, so that $\varphi(d\mathfrak{g}) = 0$. This proves the assertion (iii). Now we show that φ is injective on $\iota[S(\chi; m)]$. Let f and g be elements of $S(\chi; m)$. Put $\omega = \iota(f)$, $\eta = \iota(g)$. Then, by (52) and (54),

$$(56) \qquad \int_{\mathfrak{H}/\Gamma} {}^t\operatorname{Re}(\omega) \wedge (Y \otimes P_n)\operatorname{Re}(\eta) = (2i)^n \operatorname{Im} Y(f, g) .$$

Putting $g = if$, we get

$$\int_{\mathfrak{H}/\Gamma} {}^t\mathrm{Re}(\omega) \wedge (Y \otimes P_n)\mathrm{Re}(i\omega) = -(2i)^n Y(f, f) .$$

If $\varphi(\omega) = 0$, the left hand side is 0 in view of (55). By our assumption on Y, the form $Y(f, g)$ is positive non-degenerate. Therefore, we must have $f = 0$ and hence $\omega = 0$. This proves the injectivity of φ on $\iota[S(\chi, m)]$. Consequently, $\iota[S(\chi; m)]$ and $d[S(\chi \otimes \Phi_n; 0)]$ contains only 0 in common. We prove the remaining part of the theorem in two steps.

(I) We first assume that $\bar{\Gamma}$ has no element of finite order other than the identity. Let g be the genus of the Riemann surface $\mathfrak{H}/\Gamma$, and k be the degree of χ. We now prove

$$(57) \qquad \dim_C S(\chi; m) \geqq k(m - 1)(g - 1) + \dim H^0(\Gamma, \chi \otimes \Phi_n, R) .$$

To prove this, take a square matrix Ξ whose entries are meromorphic functions on $\mathfrak{H}$, such that $\Xi(\gamma(z)) = \chi(\gamma)\Xi(z)$ for every $\gamma \in \Gamma$ and $\det \Xi \neq 0$. Such a matrix-function Ξ is easily constructed by means of "fonction zêta-fuchsiennes." Let u be a meromorphic automorphic function on $\mathfrak{H}$ with respect to Γ, other than the constants; put $v = (du/dz)^{m/2}$. For every $f \in S(\chi; m)$, consider $g = \Xi^{-1}fv^{-1}$. Then g is a column-vector whose components can be regarded as meromorphic functions on the Riemann surface $\mathfrak{H}/\Gamma$. As f is holomorphic, Ξgv is everywhere finite. Conversely, if there exists a meromorphic vector g on $\mathfrak{H}/\Gamma$ such that Ξgv is everywhere finite, then Ξgv is an element of $S(\chi; m)$. Now, the dimension of the vector space of such functions g is given by Weil's generalization of the Riemann-Roch theorem [29]. Using the notation of [29], we see that $I(\Xi) = 0$ and $I(v) = m(g - 1)$. Therefore, we obtain

$$(58) \qquad \dim_C S(\chi; m) = k(m - 1)(g - 1) + q ,$$

where q is the dimension of the vector space of column-vectors Ω, whose components are meromorphic differentials on $\mathfrak{H}/\Gamma$, such that ${}^t\Xi^{-1}\Omega v^{-1}$ is everywhere finite. As $q \geqq 0$, we get, on account of Proposition 6.2, the inequality (57) in the case $m > 2$. Suppose that $m = 2$, and put $s = \dim H^0(\Gamma, \chi, R)$. As χ is an orthogonal representation, χ is decomposed into the sum of two representations χ_1 and χ_2, where χ_1 is a unit representation of degree s and χ_2 is a representation containing no unit representation. Then $S(\chi; 2)$ is isomorphic to the direct sum of $S(\chi_1; 2)$ and $S(\chi_2; 2)$. Applying the equality (58) to $S(\chi_2; 2)$, we get

$$(59) \qquad \dim_C S(\chi_2; 2) \geqq (k - s)(g - 1) .$$

On the other hand, $S(\chi_1; 2)$ is isomorphic to the direct product of s copies of the space of differential forms of the first kind on $\mathfrak{H}/\Gamma$. Hence we have $\dim_C S(\chi_1; 2) = sg$, which proves, together with (59), the inequality (57) in the case $m = 2$. Therefore, by Proposition 6.1, we obtain

$$2 \dim_C S(\chi; m) \geqq \dim H^1(\Gamma, \chi \otimes \Phi_n, R) .$$

As the injectivity of φ on $\iota[S(\chi; m)]$ is already proved, this inequality must be an equality; and hence the assertion (ii) is proved. The assertion (i) will be proved if we show

$$(60) \qquad \dim W(\chi \otimes \Phi_n) = \dim S(\chi; m) + \dim d[S(\chi \otimes \Phi_n; 0)] .$$

To prove this, take a square matrix Δ, whose entries are meromorphic functions on $\mathfrak{H}$, such that $\Delta(\gamma(z)) = \chi \otimes \Phi_n(\gamma)\Delta$ for every $\gamma \in \Gamma$ and $\det \Delta \neq 0$. For every element ω of $W(\chi \otimes \Phi_n)$, $\eta = \Delta^{-1}\omega$ is a column-vector, whose components are differential forms on $\mathfrak{H}/\Gamma$, such that $\Delta\eta$ is everywhere finite. Conversely, every element ω of $W(\chi \otimes \Phi_n)$ is obtained from such η by $\omega = \Delta\eta$. Therefore, applying again Weil's generalization of the Riemann-Roch theorem, we find

$$(61) \qquad \dim W(\chi \otimes \Phi_n) = k(m - 1)(g - 1) + p ,$$

where p is the dimension of the vector space $\mathfrak{B}$ of column-vectors $\mathfrak{g}$, whose components are meromorphic functions on $\mathfrak{H}/\Gamma$, such that $^t\Delta^{-1}\mathfrak{g}$ is everywhere finite. We observe that $\mathfrak{g} \to (Y \otimes P_n)^{-1} \cdot {}^t\Delta^{-1}\mathfrak{g}$ gives an isomorphism of $\mathfrak{B}$ onto $S(\chi \otimes \Phi_n; 0)$; so we have $p = \dim S(\chi \otimes \Phi_n; 0)$. Now consider the homomorphism $f \to df$ of $S(\chi \otimes \Phi_n; 0)$ onto $d[S(\chi \otimes \Phi_n; 0)]$. If $df = 0$, then f must be a constant (complex) vector satisfying $\chi \otimes \Phi_n(\gamma)f = f$ for every $\gamma \in \Gamma$; conversely, such a constant vector is contained in the kernel of the homomorphism. It follows that

$$(62) \qquad p = \dim_C d[S(\chi \otimes \Phi_n; 0)] + \dim H^0(\Gamma, \chi \otimes \Phi_n, R) .$$

This combined with the equality (57) and (61) proves (60).

(II) We now remove the assumption that $\bar{\Gamma}$ has no element of finite order other than the identity. For any fuchsian group Γ, we can find a subgroup Γ_0 of Γ of finite index, satisfying the assumption. By the result of (I), our theorem is true for Γ_0. For the sake of simplicity put $\psi = \chi \otimes \Phi_n$. Let ω be an element of $W(\Gamma; \mathfrak{H}; \psi)$. As ω is clearly contained in $W(\Gamma_0; \mathfrak{H}; \psi)$, ω is written in the form $\omega = \iota(f) + dg$, where $f \in S(\Gamma_0; \mathfrak{H}; \chi; m)$ and $g \in S(\Gamma_0; \mathfrak{H}; \psi; 0)$. Now let $\Gamma = \bigcup_{\nu=1}^e \gamma_\nu \Gamma_0$ be a disjoint expression. Put

$$h = \sum_{\nu=1}^e \chi(\gamma_\nu)j(\gamma_\nu^{-1}, z)^{-m}f \circ \gamma_\nu^{-1} , \qquad k = \sum_{\nu=1}^e \psi(\gamma_\nu)g \circ \gamma_\nu^{-1} .$$

It is easy to verify that $h \in S(\Gamma; \mathfrak{H}; \chi; m)$, $k \in S(\Gamma; \mathfrak{H}; \psi; 0)$; and

$$e\omega = \sum_{\nu=1}^{e} \psi(\gamma_\nu)\omega \circ \gamma_\nu^{-1}$$
$$= \sum_{\nu=1}^{e} \psi(\gamma_\nu)\mathfrak{c}(f) \circ \gamma_\nu^{-1} + \sum_{\nu=1}^{e} \psi(\gamma_\nu)dg \circ \gamma_\nu^{-1} = \mathfrak{c}(h) + dk \ .$$

This proves (i) of our theorem for the group Γ. Our theorem will be completely proved if we show the surjectivity of φ for the group Γ. Let η be an element of $W(\Gamma_0; \mathfrak{H}; \psi)$; put $\xi = \sum_{\nu=1}^{e} \psi(\gamma_\nu)\eta \circ \gamma_\nu^{-1}$. Then $\xi \in W(\Gamma; \mathfrak{H}; \psi)$. z_0 being a point of $\mathfrak{H}$, put $\mathfrak{f}(z) = \int_{z_0}^{z} \mathrm{Re}(\xi)$, $\mathfrak{g}(z) = \int_{z_0}^{z} \mathrm{Re}(\eta)$, $\mathfrak{x}(\gamma) = \int_{z_0}^{\gamma(z_0)} \mathrm{Re}(\xi)$ for $\gamma \in \Gamma$, $\mathfrak{y}(\gamma_0) = \int_{z_0}^{\gamma_0(z_0)} \mathrm{Re}(\eta)$ for $\gamma_0 \in \Gamma_0$. Then $\mathfrak{x}$ and $\mathfrak{y}$ are cocycles belonging to $\varphi(\xi)$ and $\varphi(\eta)$, respectively. Fix an element γ of Γ. There exist elements δ_ν of Γ_0 such that $\gamma^{-1}\gamma_\nu = \gamma_{\mu(\nu)}\delta_\nu$, where $\nu \to \mu(\nu)$ is a permutation of $\{1, \cdots, e\}$. We have

$$\mathfrak{x}(\gamma) = \sum_{\nu=1}^{e} \psi(\gamma_\nu)[\mathfrak{g}(\gamma_\nu^{-1}\gamma(z_0)) - \mathfrak{g}(\gamma_\nu^{-1}(z_0))]$$
$$= \sum_{\nu=1}^{e} \psi(\gamma_\nu)[\mathfrak{g}(\delta_\nu\gamma_{\mu(\nu)}^{-1}(z_0)) - \mathfrak{g}(\gamma_\nu^{-1}(z_0))]$$
$$= \sum_{\nu=1}^{e} \psi(\gamma_\nu)[\psi(\delta_\nu)\mathfrak{g}(\gamma_{\mu(\nu)}^{-1}(z_0)) + \mathfrak{y}(\delta_\nu) - \mathfrak{g}(\gamma_\nu^{-1}(z_0))] \ .$$

Put $\mathfrak{c} = \sum_{\nu=1}^{e} \psi(\gamma_\nu)\mathfrak{g}(\gamma_\nu^{-1}(z_0))$. Then we see easily

$$\mathfrak{x}(\gamma) = \sum_{\nu=1}^{e} \psi(\gamma_\nu)\mathfrak{y}(\delta_\nu) + [\psi(\gamma) - 1]\mathfrak{c} \ .$$

In particular, if $\mathfrak{y} \in \mathfrak{Z}(\Gamma, \psi, R)$, we have

$$\sum_{\nu=1}^{e} \psi(\gamma_\nu)\mathfrak{y}(\delta_\nu) = \sum_{\nu=1}^{e} \psi(\gamma_\nu)\mathfrak{y}(\gamma_\nu^{-1}\gamma\gamma_{\mu(\nu)})$$
$$= e\mathfrak{y}(\gamma) + [\psi(\gamma) - 1]\sum_{\nu=1}^{e} \mathfrak{y}(\gamma_\nu) \ .$$

Now as φ is a surjective mapping of $W(\Gamma_0; \mathfrak{H}; \psi)$ onto $H^1(\Gamma_0, \psi, R)$ and as $\mathfrak{Z}(\Gamma, \psi, R) \subset \mathfrak{Z}(\Gamma_0, \psi, R)$, there exists, for every $\mathfrak{y} \in \mathfrak{Z}(\Gamma, \psi, R)$, an element η of $W(\Gamma_0; \mathfrak{H}; \psi)$ such that $\mathfrak{y} \in \varphi(\eta)$. Then the above calculation shows $e\mathfrak{y} \in \varphi(\xi)$ for an element $\xi \in W(\Gamma; \mathfrak{H}; \psi)$. This proves the surjectivity of φ for Γ, and completes our proof.

REMARK 1. In the particular case $\chi = 1$, Theorem 2 has been proved in [18]; that result includes the case of fuchsian groups with cusps. By combining this and the consideration of (II) of the above proof, one would get the same result as Theorem 2 for fuchsian groups Γ with cusps in the case where the kernel of χ is of finite index in Γ.

REMARK 2. Let $\gamma_1, \cdots, \gamma_q$ be representatives of the conjugate classes of elements of finite order in $\bar{\Gamma}$. Let e_ν be the order of γ_ν. Let $e^{2\pi i a_{\mu\nu}/e_\nu}$, for $1 \leq \mu \leq k(n + 1)$, be the characteristic roots of $\chi \otimes \Phi_n(\gamma_\nu)$. For any real number x, denote by $[x]$ the largest integer which is not greater than x and put $\langle x \rangle = x - [x]$. Applying the argument of (I) of the above proof to $W(\Gamma; \mathfrak{H}; \chi \otimes \Phi_n)$, we get

$$\dim W(\chi \otimes \Phi_n) = k(m-1)(g-1)$$
$$+ \sum_{\mu,\nu} \left\langle \frac{d_{\mu\nu}}{e_\nu} \right\rangle + \dim S(\chi \otimes \Phi_n; 0) .$$

Then, by (62) (which is valid also in this case) and (i) of Theorem 2, we find

$$\dim S(\chi; m) = k(m-1)(g-1)$$
$$+ \sum_{\mu,\nu} \left\langle \frac{d_{\mu\nu}}{e_\nu} \right\rangle + \dim H^0(\Gamma, \chi \otimes \Phi_n, R) .$$

Therefore, we have $\dim S(\chi; m) > 0$, whenever $g \geqq 2$.

Let D be an indefinite quaternion algebra over Q and $\mathfrak{o}$ a maximal order in D. Let Γ_D be the group of units γ of $\mathfrak{o}$ such that $N(\gamma) = 1$. Then the genus g of $\mathfrak{H}^+/\Gamma_D$ is given in Eichler [6]. For example, we obtain $g = 3$ for $d(D/Q) = (3 \cdot 13)$ and $g = 4$ for $d(D/Q) = (2 \cdot 37)$. (In these cases Γ_D has no element of finite order other than the identity.) Hence, for any prime number p, we can find an indefinite quaternion algebra D such that $p \nmid d(D/Q)$ and the genus of $\mathfrak{H}^+/\Gamma_D$ is greater than 1.

8. Construction of abelian varieties

8.1. Now consider the couple (χ, Y) formed by a real representation χ of Γ and a real symmetric matrix Y such that ${}^t\chi(\gamma) Y \chi(\gamma) = Y$ for every $\gamma \in \Gamma$. Two such couples (χ_1, Y_1) and (χ_2, Y_2) are called *equivalent* if there exists a real matrix U such that $\chi_2(\gamma) = U^{-1}\chi_1(\gamma) U$ and $Y_2 = {}^tUY_1U$. When q such couples (χ_ν, Y_ν) for $1 \leqq \nu \leqq q$ are given, we define their direct sum (χ, Y) by

$$\chi(\gamma) = \begin{pmatrix} \chi_1(\gamma) & & \\ & \ddots & \\ & & \chi_q(\gamma) \end{pmatrix}, \qquad Y = \begin{pmatrix} Y_1 & & \\ & \ddots & \\ & & Y_q \end{pmatrix}.$$

Now suppose that, for a couple (χ, Y), the coefficients of $\chi(\gamma)$ are rational integers for every $\gamma \in \Gamma$. Let $\mathfrak{Z}(\Gamma, \chi, Z)$ denote the set of cocycles $\mathfrak{x} \in \mathfrak{Z}(\Gamma, \chi, R)$ such that the components of $\mathfrak{x}(\gamma)$ are rational integers for every $\gamma \in \Gamma$; and put

$$\mathfrak{B}(\Gamma, \chi, Z) = \mathfrak{Z}(\Gamma, \chi, Z) \cap \mathfrak{B}(\Gamma, \chi, R) ,$$
$$H^1(\Gamma, \chi, Z) = \mathfrak{Z}(\Gamma, \chi, Z)/\mathfrak{B}(\Gamma, \chi, Z) .$$

Then $H^1(\Gamma, \chi, Z)$ can be considered as a lattice in the vector space $H^1(\Gamma, \chi, R)$ in a natural manner.

THEOREM 3. *Let χ be a real representation of Γ and Y a real symmetric matrix such that ${}^t\chi(\gamma) Y \chi(\gamma) = Y$ for every $\gamma \in \Gamma$. Suppose that the* *following conditions are satisfied.*

(i) *There exist even integers $n_\nu \geq 0$ and couples (χ_ν, Y_ν), for $1 \leq \nu \leq q$, such that Y_ν is positive non-degenerate and (χ, Y) is equivalent to the direct sum of $(\chi_\nu \otimes \Phi_{n_\nu}, (-1)^{n_\nu/2} Y_\nu \otimes P_{n_\nu})$ for $1 \leq \nu \leq q$.*

(ii) *Y and the $\chi(\gamma)$ for all $\gamma \in \Gamma$ have integral coefficients.*
Then the mapping φ of $W(\chi)$ into $H^1(\Gamma, \chi, R)$ is surjective; and the kernel N of φ is a complex vector subspace of $W(\chi)$. After transporting the complex structure of $W(\chi)/N$ to $H^1(\Gamma, \chi, R)$ by the mapping φ, the torus $H^1(\Gamma, \chi, R)/H^1(\Gamma, \chi, Z)$ has a structure of abelian variety.

PROOF. We observe that $W(\chi)$ can be identified with the direct sum of the $W(\chi_\nu \otimes \Phi_{n_\nu})$, and $H^1(\Gamma, \chi, R)$ with the direct sum of the $H^1(\Gamma, \chi_\nu \otimes \Phi_{n_\nu}, R)$. Therefore the surjectivity of φ is a consequence of Theorem 2; and by that theorem, the kernel N of φ is the direct sum of the $d[S(\chi_\nu \otimes \Phi_{n_\nu}; 0)]$. Hence N is a complex vector subspace of $W(\chi)$; so $W(\chi)/N$ is a complex vector space. As φ gives an R-isomorphism of $W(\chi)/N$ onto $H^1(\Gamma, \chi, R)$, we can put a complex structure on $H^1(\Gamma, \chi, R)$ so that φ is a C-isomorphism of $W(\chi)/N$ onto $H^1(\Gamma, \chi, R)$. Then $H^1(\Gamma, \chi, R)/H^1(\Gamma, \chi, Z)$ becomes a complex torus. Now we want to show that this complex torus has a structure of abelian variety. To prove this, consider the hermitian form $Y(\omega, \eta)$ on $W(\chi)$. If $\mathfrak{x} \in \varphi(\omega)$ and $\mathfrak{y} \in \varphi(\eta)$, we have, by (52) and (55),

$$(63) \qquad \operatorname{Im} Y(\omega, \eta) = 2 \sum_{\lambda=1}^{s} a_\lambda \, {}^t\mathfrak{x}(\alpha_\lambda) Y \mathfrak{y}(\beta_\lambda) \ .$$

If $\varphi(\omega) = 0$, we have $\operatorname{Im} Y(\omega, \eta) = 0$ for every $\eta \in W(\chi)$. Putting $i\eta$ in place of η, we obtain $\operatorname{Re} Y(\omega, \eta) = 0$, so that $Y(\omega, \eta) = 0$ for every η. Therefore $Y(\omega, \eta)$ can be regarded as a hermitian form on $W(\chi)/N$. We now show that this is positive non-degenerate. As $W(\chi)$ is the direct sum of N and the $c[S(\chi_\nu; n_\nu + 2)]$ for $1 \leq \nu \leq q$, it is sufficient to show that $(-1)^{n_\nu/2} Y_\nu \otimes P_{n_\nu}(\omega, \eta)$ is positive non-degenerate on $c[S(\chi_\nu; n_\nu + 2)]$ for each ν. But this is obvious in view of (54). Consider $Y(\omega, \eta)$ as a hermitian form on $H^1(\Gamma, \chi, R)$ by identifying $W(\chi)/N$ with $H^1(\Gamma, \chi, R)$. Then (63) shows that the values of $\operatorname{Im} Y(\omega, \eta)$ on the lattice $H^1(\Gamma, \chi, Z)$ are rational numbers. Therefore a suitable multiple of $Y(\omega, \eta)$ gives a positive non-degenerate hermitian form on $H^1(\Gamma, \chi, R)$ whose imaginary part takes integral values on $H^1(\Gamma, \chi, Z)$, that is, a Riemann form. This completes our proof.

We denote by $A(\Gamma, \mathfrak{H}, \chi)$ the abelian variety $H^1(\Gamma, \chi, R)/H^1(\Gamma, \chi, Z)$ thus obtained. The varieties $A(\Gamma, \mathfrak{H}^+, \chi)$ and $A(\Gamma, \mathfrak{H}^-, \chi)$ are not necessarily isomorphic. Let ω be the mapping of $\mathfrak{H}^-$ onto $\mathfrak{H}^+$ defined by $w(z) = \bar{z}$. For every $\omega \in W(\Gamma, \mathfrak{H}^+, \chi)$, put $\omega^* = \bar{\omega} \circ w$. Then $\omega \to \omega^*$ is an R-isomorphism of $W(\Gamma, \mathfrak{H}^+, \chi)$ onto $W(\Gamma, \mathfrak{H}^-, \chi)$; and $(a\omega)^* = \bar{a}\omega^*$ for $a \in C$. We see easily $\int_z^{\gamma(z)} \mathrm{Re}(\omega) = \int_{\bar z}^{\gamma(\bar z)} \mathrm{Re}(\omega^*)$ for every $\gamma \in \Gamma$, $z \in \mathfrak{H}^+$, so that $\varphi(\omega) = \varphi(\omega^*)$. It follows that the complex structures of $A(\Gamma, \mathfrak{H}^-, \chi)$ and $A(\Gamma, \mathfrak{H}^+, \chi)$ are complex conjugate to each other.

8.2. *Arithmetically defined fuchsian groups.* Let D be an indefinite quaternion algebra over a totally real algebraic number field F, as in § 1.1. Suppose that $r = 1$. Define $\mathfrak{o}$, $\mathfrak{o}_\lambda$ and Γ_λ as in § 1.4. Let $\mathfrak{a}$ be an integral two-sided $\mathfrak{o}$-ideal; define $V_\mathfrak{a}$, $W_\mathfrak{a}$ as in § 3.3. Let ρ be a representation of $(\mathfrak{o}/\mathfrak{a})^*$; we consider ρ also as a representation of $V_\mathfrak{a}$. Take a set of non-negative even integers $\{n_1, \cdots, n_t\}$. Notation being as in §§ 3.2, 3.3, put, for every $x \in V_\mathfrak{a}$,

$$L(x) = \prod_{\nu=1}^t N(x^{(\nu)})^{n_\nu/2} \,,$$

$$\psi(x) = L(x)^{-1}\rho(x) \otimes \Psi_{n_1}^{(1)}(x^{(1)}) \otimes \cdots \otimes \Psi_{n_t}^{(t)}(x^{(t)}) \,.$$

Now assume that the coefficients of ρ belong to a totally real algebraic number field F_1. Let K be the composite of F_1 and the $F^{(\nu)}$ such that $n_\nu > 0$; then K is a totally real algebraic number field. Let K^k denote the space of k-dimensional column-vectors with coefficients in K. Now k being the degree of ρ, we consider K^k the representation space of ρ. Let $\mathfrak{g}_K$ be the ring of integers in K. Take any $\mathfrak{g}_K$-lattice[9] $\mathfrak{N}_1$ in K^k and put $\mathfrak{N} = \sum \rho(\alpha)\mathfrak{N}_1$, where α runs over all representatives of $(\mathfrak{o}/\mathfrak{a})^*$. Then $\mathfrak{N}$ is a $\mathfrak{g}_K$-lattice in K^k; and $\rho(x)\mathfrak{N} \subset \mathfrak{N}$ for every $x \in V_\mathfrak{a}$. Notation being as in § 3.2, let $\mathfrak{L}_{n\lambda}$ be the projection of $(\mathfrak{W} \cap \mathfrak{x}_\lambda) \otimes \cdots \otimes (\mathfrak{W} \cap \mathfrak{x}_\lambda)$ to $\mathfrak{B}_j$. Then $\mathfrak{L}_{n\lambda}$ is a $\mathfrak{g}$-lattice in $\mathfrak{B}_j$; and, of course $\mathfrak{L}_{n\lambda} \otimes \mathfrak{g}_K$ is a $\mathfrak{g}_K$-lattice in $\mathfrak{B}_j \otimes K$. Note that the coefficients of $\psi(\alpha)$ for $\alpha \in D^* \cap V_\mathfrak{a}$ belong to K. Put $s = k + (n_1 + 1) + \cdots + (n_t + 1)$ and consider K^s as the representation space of $\psi \,|\, D^* \cap V_\mathfrak{a}$. Put

$$\mathfrak{M}_\lambda = \mathfrak{N} \otimes (\mathfrak{L}_{n_1\lambda}^{(1)} \otimes \mathfrak{g}_K) \otimes \cdots \otimes (\mathfrak{L}_{n_t\lambda}^{(t)} \otimes \mathfrak{g}_K) \,,$$

and consider it a $\mathfrak{g}_K$-lattice in K^s. By our construction, if $\alpha \in \mathfrak{o}_\lambda \cap D^*$, we have $L(\alpha)\psi(\alpha)\mathfrak{M}_\lambda \subset \mathfrak{M}_\lambda$. Let $a \to a^{(\beta)}$, for $1 \leq \beta \leq q$, denote the isomorphisms of K into R. Then, we can define the conjugates $u^{(\beta)}$, $L^{(\beta)}$, $\psi^{(\beta)}$ of an element u of K^s, L, and of ψ in a natural manner. Consider a submodule

$$\mathfrak{M}_{\lambda 0} = \{(u^{(1)}, \cdots, u^{(q)}) \,|\, u \in \mathfrak{M}\}$$

of $K^{(1)s} \times K^{(2)s} \times \cdots \times K^{(q)s} \subset R^s \times \cdots \times R^s = R^{sq}$. Then we observe that $\mathfrak{M}_{\lambda 0}$ is a lattice in R^{sq}; namely, there exists a real matrix T_λ of degree sq

---

[9] By a $\mathfrak{g}$-lattice (resp. $\mathfrak{g}_K$-lattice) in a vector space U over F (resp. K), we understand a finitely generated $\mathfrak{g}$-submodule (resp. $\mathfrak{g}_K$-submodule) $\mathfrak{N}$ of U such that $F\mathfrak{N} = U$ (resp. $K\mathfrak{N} = U$).

such that $T_\lambda \mathfrak{M}_{\lambda 0} = Z^{\cdot q}$. Therefore, if we put

$$(65) \qquad \chi_\lambda(\alpha) = T_\lambda \begin{bmatrix} \psi^{(1)}(\alpha) \\ & \ddots \\ & & \psi^{(q)}(\alpha) \end{bmatrix} T_\lambda^{-1} \qquad (\alpha \in D^* \cap V_a) ,$$

$$L_\lambda(\alpha) = T_\lambda \begin{bmatrix} L^{(1)}(\alpha) \\ & \ddots \\ & & L^{(q)}(\alpha) \end{bmatrix} T_\lambda^{-1} \qquad (\alpha \in D^*) ,$$

then the coefficients of $\chi_\lambda(\alpha)$, $L_\lambda(\alpha)$ are rational numbers; and if $\alpha \in \mathfrak{o}_\lambda$, the coefficients of $L_\lambda(\alpha)\chi_\lambda(\alpha)$ are integers. In particular, if $\gamma \in \Gamma_\lambda$, the coefficients of $L_\lambda(\gamma)$ and $\chi_\lambda(\gamma)$ are integers. Now put

$$(66) \qquad \begin{aligned} \theta &= \sum{}' {}^t\rho(\alpha)\rho(\alpha) , \\ X &= c \cdot \theta \otimes \Theta_{n_1}^{(1)} \otimes \cdots \otimes \Theta_{n_t}^{(t)} , \end{aligned}$$

where $\sum'$ means the sum extended over all elements of $(\mathfrak{o}/\mathfrak{a})^*$, the matrices $\Theta_{n_\nu}^{(\nu)}$ are as in § 3.2, and c is an element of K which we determine a little later. We have then

$${}^t\psi(\alpha)X\psi(\alpha) = X \qquad (\alpha \in D^* \cap V_a) .$$

Put

$$Y_\lambda = T_\lambda \begin{bmatrix} X^{(1)} \\ & \ddots \\ & & X^{(q)} \end{bmatrix} T_\lambda^{-1} .$$

Then, Y_λ is a rational matrix and

$${}^t\chi_\lambda(\alpha) Y_\lambda \chi_\lambda(\alpha) = Y \qquad (\alpha \in D^* \cap V_a) .$$

Put

$$\begin{aligned} \Gamma_\lambda^+ &= \{\gamma \in \Gamma_\lambda \mid N(\gamma^{(1)}) > 0\} , \\ \Gamma_\lambda^0 &= \{N(\gamma^{(1)})^{-1/2}\gamma^{(1)} \mid \gamma \in \Gamma_\lambda^+\} . \end{aligned}$$

Then Γ_λ^0 is a fuchsian group in the sense of § 5.1. Now assume, as in § 3.3, (10), that $\rho(\varepsilon) = 1$ for every unit ε of F. Put

$$\psi_0^{(\beta)}(\delta) = \psi^{(\beta)}(\gamma) , \quad \chi_\lambda^0(\delta) = \chi_\lambda(\gamma) \qquad \text{for} \quad \delta = N(\gamma^{(1)})^{-1/2}\gamma^{(1)}, \, \gamma \in \Gamma_\lambda^+ .$$

Then $\psi_0^{(\beta)}$, χ_λ^0 are representations of Γ_λ^0 with integral coefficients, and

$${}^t\chi_\lambda^0(\delta) Y_\lambda \chi_\lambda^0(\delta) = Y_\lambda \qquad (\delta \in \Gamma_\lambda^0) .$$

Now we show that the couple $(\chi_\lambda^0, Y_\lambda)$ for the fuchsian group Γ_λ^0, satisfies the conditions (i), (ii) of Theorem 3, for a suitable choice of c. The condition (ii) is obviously satisfied if we take, if necessary, a suitable integral multiple of c in place of c. For every β, $\psi_0^{(\beta)}$ and $X^{(\beta)}$ have the form, for $\delta = N(\gamma^{(1)})^{-1/2}\gamma^{(1)}$, $\gamma \in \Gamma_\lambda^+$,

$$\psi_0^{(\beta)}(\delta) = \psi^{(\beta)}(\gamma) = \rho^{(\beta)}(\gamma) \otimes N(\gamma^{(1)})^{-l_1/2}\Psi_{l_1}^{(1)}(\gamma^{(1)}) \otimes \cdots \otimes N(\gamma^{(t)})^{-l_t/2}\Psi_{l_t}^{(t)}(\gamma^{(t)}) ,$$

$$X^{(\beta)} = c^{(\beta)}\theta^{(\beta)} \otimes \Theta_{l_1}^{(1)} \otimes \cdots \otimes \Theta_{l_t}^{(t)} ,$$

where $(l_1, \cdots, l_t)$ is a suitable permutation of $(n_1, \cdots, n_t)$. By (66) and (Θ2) of § 3.2, $\theta^{(\beta)} \otimes \Theta_{l_2}^{(2)} \otimes \cdots \cdot \otimes \Theta_{l_t}^{(t)}$ is positive definite. By (Ψ3) and (Θ5) of § 3.2, $(\Psi_{l_1}^{(1)}, \Theta_{l_1}^{(1)})$ is equivalent to $(\Phi_{l_1}, b_{l_1}^{(1)}P_{l_1})$. Therefore, if we choose c so that $(-1)^{l_1/2}c^{(\beta)}b_{l_1}^{(1)} > 0$ for every β, then the condition (i) of Theorem 3 is satisfied. Applying the theorem to the present case, we obtain abelian varieties $A(\Gamma_\lambda^0, \mathfrak{H}^+, \chi_\lambda^0)$ and $A(\Gamma_\lambda^0, \mathfrak{H}^-, \chi_\lambda^0)$.

9. The operators $\mathfrak{T}_{\mu\lambda}$ as endomorphisms of abelian varieties

9.1. Let E_1 (resp. E_2) be the group of units u of F such that $u^{(\nu)} > 0$ for $\nu > 1$ (resp. for every ν). Put $[E_1 : E_2] = \varepsilon(D)$. Then, $\varepsilon(D) = 1$ or 2. Notation being as in § 8.2, we have $[\Gamma_\lambda : \Gamma_\lambda^+] = \varepsilon(D)$, on account of Lemma 1.3. By our construction, the abelian varieties $A(\Gamma_\lambda^0, \mathfrak{H}, \chi_\lambda^0)$ are essentially obtained from several $S(\Gamma_\lambda; \rho; \{m_\nu\})$, where $\{m_1 - 2, m_2, \cdots, m_t\}$ is a permutation of $\{n_1, \cdots, n_t\}$. In § 3.4, we have defined the operators $\mathfrak{T}_{\mu\lambda}(e)$ on $S(\Gamma_\lambda; \rho; \{m_\nu\})$. Our purpose in this section is to consider them as homomorphisms of $A(\Gamma_\lambda^0, \mathfrak{H}, \chi_\lambda^0)$. First we note that if $\omega \in W(\Gamma_\lambda^0, \mathfrak{H}, \chi_\lambda^0)$, then $\omega \circ \gamma^{(1)} = \chi_\lambda(\gamma)\omega$ for every $\gamma \in \Gamma_\lambda^+$. Let e be an element of $\mathfrak{E}_a$ and $T_{\mu\lambda}(e) = \Gamma_\mu\alpha\Gamma_\lambda$. When $\varepsilon(D) = 2$, we take α so that $N(\alpha^{(1)}) > 0$; this is obviously possible. Let $\Gamma_\mu^+\alpha\Gamma_\lambda^+ = \bigcup_i \alpha_i\Gamma_\lambda^+$ be a disjoint expression; then, on account of Propositions 1.4 and 1.5, $\Gamma_\mu\alpha\Gamma_\lambda = \bigcup_i \alpha_i\Gamma_\lambda$ is a disjoint expression. By our choice of α, we have $N(\alpha_i^{(1)}) > 0$ when $\varepsilon(D) = 2$; if $\varepsilon(D) = 1$, the $N(\alpha_i^{(1)})$ have the same sign as $N(\alpha^{(1)})$. For every $\omega \in W(\Gamma_\lambda^0, \mathfrak{H}, \chi_\lambda^0)$, we define $\eta = \mathfrak{T}_{\mu\lambda}^0(e)\omega$ by

$$(68) \qquad \eta = T_\mu T_\lambda^{-1} \sum_i \chi_\lambda(\alpha_i)\omega \circ \alpha_i^{(1)-1} ,$$

where T_μ and T_λ are as in (65). It is easy to verify $\eta \in W(\Gamma_\mu^0, \mathfrak{H}', \chi_\mu^0)$, where

$$\mathfrak{H}' = \begin{cases} \mathfrak{F} - \mathfrak{H} , & \text{if } \varepsilon(D) = 1 \text{ and } N(\alpha^{(1)}) < 0 , \\ \mathfrak{H} , & \text{otherwise} . \end{cases}$$

Consider now the relation between $\varphi(\omega)$ and $\varphi(\eta)$. z_0 being a point of $\mathfrak{H}$, and $\mathfrak{a}$ being a real vector, put $\eta_1 = L_\lambda(\alpha)\eta$, and

$$\mathfrak{j}(z) = \int_{z_0}^z \mathrm{Re}(\omega) + \mathfrak{a} ,$$

$$\mathfrak{j}(\delta) = \int_{z_0}^{\delta(z_0)} \mathrm{Re}(\omega) + [1 - \chi_\lambda^0(\delta)]\mathfrak{a} \qquad\qquad (\delta \in \Gamma_\lambda^0) ,$$

$$\mathfrak{y}(\varepsilon) = \int_{z_0}^{\varepsilon(z_0)} \mathrm{Re}(\eta_1) \qquad\qquad (\varepsilon \in \Gamma_\mu^0) \; .$$

Then we have $\mathfrak{f}(\delta(z)) = \chi_\lambda^0(\delta)\mathfrak{f}(z) + \mathfrak{z}(\delta)$ for every $\delta \in \Gamma_\lambda^0$. Let γ be an element of Γ_μ^+. There exists a permutation $i \to j(i)$ and elements γ_i of Γ_λ^+ such that $\gamma^{-1}\alpha_i = \alpha_{j(i)}\gamma_i^{-1}$. Put $\delta_i = N(\gamma_i^{(1)})^{-1/2}\gamma_i^{(1)}$, $\varepsilon = N(\gamma^{(1)})^{1/2}\gamma^{(1)}$; we have then

$$\mathfrak{y}(\varepsilon) = L_\mu(\alpha)T_\mu T_\lambda^{-1} \sum_i \chi_\lambda(\alpha_i)[\mathfrak{f}(\alpha_i^{-1}\gamma(z_0)) - \mathfrak{f}(\alpha_i^{-1}(z_0))]$$
$$= L_\mu(\alpha)T_\mu T_\lambda^{-1} \sum_i \chi_\lambda(\alpha_i)[\chi_\lambda(\gamma_i)\mathfrak{f}(\alpha_{j(i)}^{-1}(z_0)) + \mathfrak{z}(\delta_i) - \mathfrak{f}(\alpha_i^{-1}(z_0))] \; .$$

Putting $\mathfrak{c} = T_\mu T_\lambda^{-1} \sum_i \chi_\lambda(\alpha_i)\mathfrak{f}(\alpha_i^{-1}(z_0))$, we obtain

$$\mathfrak{y}(\varepsilon) = L_\mu(\alpha)T_\mu T_\lambda^{-1} \sum_i \chi_\lambda(\alpha_i)\mathfrak{z}(\delta_i) + [\chi_\mu(\gamma) - 1]\mathfrak{c} \; .$$

Therefore, if $\varphi(\omega) = 0$, then $\varphi(\eta_1) = 0$. Hence $L_\mu(\alpha)\mathfrak{T}_{\mu\lambda}^0(e)$ gives a C-linear mapping of $H^1(\Gamma_\lambda^0, \chi_\lambda^0, R)$ into $H^1(\Gamma_\mu^0, \chi_\mu^0, R)$. Furthermore, note that if $\alpha\mathfrak{x}_{\lambda\mu} \subset \mathfrak{o}_\mu$, then $L(\alpha)\psi(\alpha)\mathfrak{M}_\lambda \subset \mathfrak{M}_\mu$, and $T_\lambda\mathfrak{M}_{\lambda 0} = Z^{sq}$. Hence we observe that if $\varphi(\omega) \in H^1(\Gamma_\lambda^0, \chi_\lambda^0, Z)$, then $\varphi(\eta_1) \in H^1(\Gamma_\mu^0, \chi_\mu^0, Z)$. It follows that $L_\mu(\alpha)\mathfrak{T}_{\mu\lambda}^0(e)$ induces a homomorphism of $A(\Gamma_\lambda^0, \mathfrak{H}, \chi_\lambda^0)$ into $A(\Gamma_\mu^0, \mathfrak{H}', \chi_\mu^0)$, if e is an integral element of $\mathfrak{E}_a$. More generally, for every element e of $\mathfrak{E}_a$, we see that a suitable integral multiple of $\mathfrak{T}_{\mu\lambda}^0(e)$ induces a homomorphism of $A(\Gamma_\lambda^0, \mathfrak{H}, \chi_\lambda^0)$ into $A(\Gamma_\mu^0, \mathfrak{H}', \chi_\mu^0)$.

Now put

$$A_\lambda = \begin{cases} A(\Gamma_\lambda^0, \mathfrak{H}^+, \chi_\lambda^0) & \text{if } \varepsilon(D) = 2 \; , \\ A(\Gamma_\lambda^0, \mathfrak{H}^+, \chi_\lambda^0) \times A(\Gamma_\lambda^0, \mathfrak{H}^-, \chi_\lambda^0) & \text{if } \varepsilon(D) = 1 \; . \end{cases}$$
$$A = A(\rho; \{n_\nu\}) = A_1 \times \cdots \times A_h \; .$$

For any two abelian varieties B_1 and B_2, denote by $\mathcal{H}(B_1, B_2)$ the set of homomorphisms of B_1 into B_2; and put $\mathcal{H}_0(B_1, B_2) = \mathcal{H}(B_1, B_2) \otimes Q$, $\mathcal{A}_0(B_1) = \mathcal{H}_0(B_1, B_1)$ (cf. Weil's *Variétés abéliennes*). Then, for every $e \in \mathfrak{E}_a$, we can regard $\mathfrak{T}_{\mu\lambda}^0(e)$ as an element of $\mathcal{H}_0(A_\lambda, A_\mu)$. As e and λ determine μ by the condition that $\mathfrak{T}_{\mu\lambda}^0(e)$ has a meaning, we can define an element $\mathfrak{T}^0(e)$ of $\mathcal{A}_0(A)$ in a natural manner from $\mathfrak{T}_{\mu\lambda}^0(e)$. As a consequence, we obtain

PROPOSITION 9.1. *Notation being as in §§ 3.3, 3.4, suppose that the n_ν are all even. Then the characteristic roots of $\mathfrak{T}(e)$ are algebraic numbers for every $e \in \mathfrak{E}_a$.*

By Proposition 3.2, the $\mathfrak{T}(e)$ are represented by diagonal matrices for a suitable choice of basis. If we make the $\mathfrak{T}(e)$ so, the above proposition shows that the diagonal elements of $\mathcal{D}(s; \xi)$ are Dirichlet series with algebraic coefficients for algebraic valued Grössen-characters ξ.

9.2. We note that $\mathcal{A}_0(A(\Gamma_\lambda^0, \mathfrak{H}, \chi_\lambda^0))$ contains K in a natural manner. In fact, for any $a \in K$, put

$$\tilde{a} = T_\lambda \begin{bmatrix} a^{(1)}1_s & & \\ & \ddots & \\ & & a^{(q)}1_s \end{bmatrix} T_\lambda^{-1} .$$

It is clear that $\tilde{a}$ is a rational matrix and $\chi_\lambda(\alpha)\tilde{a} = \tilde{a}\chi_\lambda(\alpha)$ for every $\alpha \in D^* \cap V_a$. Therefore, $\omega \to \tilde{a}\omega$ gives an automorphism or 0-mapping of $W(\Gamma_\lambda^0, \mathfrak{H}, \chi_\lambda^0)$ according as $a \neq 0$ or $a = 0$. Moreover, we see easily that a suitable multiple of a gives an endomorphism of $H^1(\Gamma_\lambda^0, \chi_\lambda^0, R)/H^1(\Gamma_\lambda^0, \chi_\lambda^0, Z)$. We get in this way an isomorphism $a \to \tilde{a}$ of K into $\mathcal{A}_0(A(\Gamma_\lambda^0, \mathfrak{H}, \chi_\lambda^0))$.

9.3. *Equivalence-classes of fibre-bundles.* Finally we give an interpretation of the abelian variety $A(\Gamma, \mathfrak{H}, \chi)$ obtained in Theorem 3. For the sake of simplicity, assume that $\bar{\Gamma}$ has no element of finite order other than the identity. Let k be the degree of χ and $(R/Z)^k$ be the product of k copies of the additive group R/Z; $M_k(Z)$ gives a ring of endomorphisms of $(R/Z)^k$ in a natural manner. Let $\mathfrak{G}$ be the subset of $M_k(Z) \times (R/Z)^k$ given by

$$\mathfrak{G} = \{(B, u) \mid B \in M_k(Z), \det B = \pm 1, u \in (R/Z)^k\} .$$

Define the operation of the elements of $\mathfrak{G}$ on $(R/Z)^k$ by

$$(B, u)v = Bv + u \qquad\qquad \text{for } v \in (R/Z)^k .$$

Then $\mathfrak{G}$ may be considered as a group of homeomorphisms of $(R/Z)^k$. Now fixing a representation χ of $\bar{\Gamma}$ with integral coefficients, suppose that $\Gamma \ni \gamma \to (\chi(\gamma), u(\gamma)) \in \mathfrak{G}$ is a homomorphism of Γ into $\mathfrak{G}$. We see easily

$$(70) \qquad\qquad u(\gamma_1\gamma_2) = u(\gamma_1) + \chi(\gamma_1)u(\gamma_2) .$$

Conversely, if there exists a mapping $u(\gamma)$ of $\bar{\Gamma}$ into $(R/Z)^k$ satisfying (70), then $\gamma \to (\chi(\gamma), u(\gamma))$ is a homomorphism of $\bar{\Gamma}$ into $\mathfrak{G}$. As $\bar{\Gamma}$ may be considered as the fundamental group of the Riemann surface $\mathcal{R} = \mathfrak{H}/\bar{\Gamma}$, we obtain, from such a homomorphism of $\bar{\Gamma}$, a fibre bundle having $\mathcal{R}$ as base and $(R/Z)^k$ as typical fibre. Now, roughly speaking, *the abelian variety $A(\Gamma, \mathfrak{H}, \chi)$ is a connected component of the set of equivalence-classes of such fibre bundles.* To show this we first prove

LEMMA 9.2. *Let (a_{ij}) be an $m \times n$ matrix with integral coefficients. Then there exists a positive integer N with the following properties: for any real numbers $x_1, \cdots, x_n$ such that $\sum_{i=1}^n a_{ij}x_j \equiv 0 \bmod Z(1 \le i \le m)$, we can find real numbers $y_1, \cdots, y_n$ such that $Nx_j \equiv y_j \bmod Z(1 \le j \le n)$ and $\sum_{j=1}^n a_{ij}y_j = 0\ (1 \le i \le m)$.*

PROOF. Put $A = (a_{ij})$. We can find unimodular matrices U and V of degree m and n, respectively, such that the components of UAV are 0 except the diagonal elements $c_1, \cdots, c_r$, which are positive integers such that c_i is a multiple of c_{i-1}. Put $N = c_r$. Denote by x the column vector having $x_1, \cdots, x_n$ as components; put $z = V^{-1}x$; and let $z_1, \cdots, z_n$ be the components of z. As $Ax \equiv 0 \bmod Z$, we have $c_i z_i \equiv 0 \bmod Z$ for $1 \leqq i \leqq n$. Now put $w_1 = \cdots = w_r = 0$, $w_{r+h} = N z_{r+h}$ for $1 \leqq h \leqq n - r$, and denote by w the vector with the components w_i. Then we see easily $Nz \equiv w \bmod Z$ and $UAVw = 0$. Put $y = Vw$. We have then $y \equiv V(Nz) = Nx$ and $Ay = 0$. Therefore, the components of y satisfy the condition of our lemma.

Now let $\mathfrak{Z}$ be the set of u satisfying (70) and $\mathfrak{B}$ the subset of $\mathfrak{Z}$ consisting of the mappings $u(\gamma)$ given by $u(\gamma) = [1 - \chi(\gamma)]v$ for an element v of $(R/Z)^k$. As Γ has a finite number of generators, we can put on $\mathfrak{Z}$ and $\mathfrak{B}$ structures of topological groups in a natural manner. Let $\mathfrak{Z}_0$ and $\mathfrak{B}_0$ denote the connected components of the identity in $\mathfrak{Z}$ and $\mathfrak{B}$, respectively. By the above lemma (or by its proof), $\mathfrak{Z}/\mathfrak{Z}_0$ and $\mathfrak{B}/\mathfrak{B}_0$ are finite groups. Put $H_0 = \mathfrak{Z}_0/\mathfrak{B}_0$. Then H_0 can be regarded as a connected component of the set of equivalence-classes of the fibre bundles of the above type. Now, for every element $\mathfrak{x}$ of $\mathfrak{Z}(\Gamma, \chi, R)$ (resp. $\mathfrak{B}(\Gamma, \chi, R)$), we obtain an element of $\mathfrak{Z}$ (resp. $\mathfrak{B}$) by taking residue-classes modulo Z. Here note that $\mathfrak{x}(-1) = 0$ since we have assumed $\chi(-1) = 1$. By Lemma 9.2, the images of $\mathfrak{Z}(\Gamma, \chi, R)$ and $\mathfrak{B}(\Gamma, \chi, R)$ by this homomorphism are $\mathfrak{Z}_0$ and $\mathfrak{B}_0$. Obviously, $\mathfrak{Z}(\Gamma, \chi, Z)$ is mapped onto 0, and $\mathfrak{Z}(\Gamma, \chi, R)/\mathfrak{Z}(\Gamma, \chi, Z)$ is isomorphic to $\mathfrak{Z}_0$. On the other hand $H^1(\Gamma, \chi, R)/H^1(\Gamma, \chi, Z)$ is isomorphic to

$$\mathfrak{Z}(\Gamma, \chi, R)/\{\mathfrak{B}(\Gamma, \chi, R) + \mathfrak{Z}(\Gamma, \chi, Z)\}\ .$$

As $\mathfrak{B}(\Gamma, \chi, R) + \mathfrak{Z}(\Gamma, \chi, Z)$ is mapped onto $\mathfrak{B}_0$, we can conclude that $H^1(\Gamma, \chi, R)/H^1(\Gamma, \chi, Z)$ is isomorphic to $H_0 = \mathfrak{Z}_0/\mathfrak{B}_0$. This establishes the above interpretation of $A(\Gamma, \mathfrak{H}, \chi)$.

OSAKA UNIVERSITY

REFERENCES

1. A. BOREL, *Density properties for certain subgroups of semi-simple groups without compact components*, Ann. of Math., 72 (1960), 179–188.
2. C. CHEVALLEY, *Deux théorèmes d'arithmétique*, J. Math. Soc. Japan, 3 (1951), 36–44.
3. M. DEURING, Algebren, Ergebnisse der Math., no. 4, Berlin, 1935.
4. M. EICHLER, *Allgemeine Kongruenzklasseneinteilungen der Ideale einfacher Abgebren über algebraischen Zahlkörpern und ihre L-Reihen*, J. Reine Angew. Math., 179 (1938) 227–251.
5. ———, *Über die Idealklassenzahl hyperkomplexer Systeme*, Math. Z., 43 (1938), 481–494.
6. ———, *Modular correspondences and their representations*, J. Indian Math. Soc., 20 (1956), 163–206.

7. ———, *Eine Verallgemeinerung der Abelschen Integrale*, Math. Z., 67 (1957), 267–298.

8. G. FUJISAKI, *On the zeta-functions of the simple algebra over the field of rational numbers*, J. Fac. Sci. Univ. Tokyo, Sec. I, vol. VII, Part 5 (1958), 567–604.

9. R. GODEMENT, Les fonctions ζ des algèbres simples, II, Séminaire Bourbaki (février 1959, 176), 1–20.

10. K. B. GUNDLACH, *Über die Darstellung der ganzen Spitzenformen zu den Idealstufen der Hilbertschen Modulgruppe und die Abschätzung ihrer Fourierkoeffizienten*, Acta Math, 92 (1954), 309–345.

11. E. HECKE, *Eine neue Art von Zetafunktionen und ihre Beziehungen zur Verteilung der Primzahlen*, I, II, Math. Z., 1 (1918), 357–376, 6 (1920), 11–51.

12. ———, *Über die Bestimmung Dirichletscher Reihen durch ihre Funktionalgleichung*, Math. Ann., 112 (1936), 664–699.

13. ———, *Über Modulfunktionen und die Dirichletschen Reihen mit Eulerscher Produktentwicklung*, I, II, Math. Ann., 114 (1937), 1–28, 316–351.

14. O. HERRMANN, *Über Hilbertsche Modulfunktionen und die Dirichletschen Reihen mit Eulerscher Produktentwicklung*, Math. Ann., 127 (1954) 357–400.

15. K. IWASAWA, *A note on L-functions*. Proc. Int. Congress. Math. Cambr. Mass. U.S.A. (1950), 322.

16. N. JACOBSON, The theory of rings, Math. Surveys no. 2, New York, 1943.

17. M. KOECHER, *Zur Theorie der Modulformen n-ten Grades*, I, II, Math. Z., 59 (1954), 399–416, 61 (1955) 455–466.

18. M. KUGA and G. SHIMURA, *On vector differential forms attached to automorphic forms*. J. Math. Soc. Japan, 12 (1960), 258–270.

19. H. MAASS, *Über Gruppen von hyperabelschen Transformationen*, Sitzungsber. Heidelberg Akad. Wiss. Math.-nat. Klasse, 2. Abh. (1940).

20. ———, *Zur Theorie der automorphen Funktionen von n Veränderlichen*, Math. Ann., 117 (1940) 538–578.

21. H. PETERSSON, *Konstruktion der sämtlichen Lösungen einer Riemannschen Funktionalgleichung durch Dirichlet-Reihen mit Eulerscher Produktentwicklung*, I, II, III, Math. Ann., 116 (1939), 401–412, 117 (1940), 39–64, 277–300.

22. A. SELBERG, *Harmonic analysis and discontinuous groups in weakly symmetric riemannian spaces with applications to Dirichlet series*, J. Indian Math. Soc., 20 (1956), 47–87.

23. H. SHIMIZU, *On discontinuous groups operating on the product of the upper half-planes*, to appear.

24. G. SHIMURA, *Sur les intégrales attachées aux formes automorphes*, J. Math. Soc. Japan, 11 (1959), 291–311.

25. ———, *On the zeta-functions of the algebraic curves uniformized by certain automorphic funtions*, J. Math. Soc. Japan, 13 (1961), 275–331.

26. N. E. STEENROD, The topology of fibre bundles, Princeton Math. series, no. 14, 1951.

27. T. TAMAGAWA, *On ζ-functions of a division algebra*, to appear.

28. J. TATE, Fourier analysis in number fields and Hecke's zeta-functions, Thesis, Princeton Univ. (1950).

29. A. WEIL, *Généralisation des fonctions abéliennes*, J. Math. pures appl., IX. sér. 17 (1938), 47–87.

30. ———, Discontinuous subgroups of classical groups, Lecture note, Univ. of Chicago, 1958.

31. ———, Adeles and algebraic groups, Lecture note, Institute for advanced study, Princeton, 1961.

On the class-fields obtained by complex multiplication of abelian varieties

Osaka Mathematical Journal, 14 (1962), 33-44

By complex multiplication of abelian varieties, we get certain class-fields over a totally imaginary quadratic extension F of a totally real algebraic number field F_0. The corresponding ideal-groups are explicitly given in Main Theorems of [3]. On this subject, one may ask how large class-fields over F can be constructed by such a means. An answer to the question is given in [4, 5], to a certain degree, in terms of local characters attached to Grössen-characters. However, this does not give any information, for example, about unramified class-fields over F so obtained. The purpose of the present paper is to give some results concerning this problem, which are almost directly derived from the defining-relation for the ideal-groups mentioned above.

In general the ideal-class group $\mathfrak{R}$ of F is approximately decomposed into the ideal-class group $\mathfrak{R}_0$ of F_0 and its complementary part $\mathfrak{R}_1$. Adjoining the absolute class-field over F_0 to F, we get the unramified class-field over F corresponding to $\mathfrak{R}/\mathfrak{R}_1$. Now, roughly speaking, the unramified class-field over F corresponding to $\mathfrak{R}/\mathfrak{R}_0$ is generated by the fields of moduli of certain polarized abelian varieties. The ramified class-fields over F are found in a similar situation, if we consider the points of finite order on the varieties. In §2, we show these facts under a condition on F, which is satisfied whenever F is normal over the rational number field. We shall prove that the class-fields over F_0 and complex multiplication yield at least a subfield B of the maximal abelian extension A of F such that $A \subset B(\sqrt{x} \mid x \in B)$ (Theorem 1); B contains the absolute class-field over F (Theorem 2). If F is an imaginary cyclotomic field, the results are stated in a little preciser and simpler form, as we shall see in §3. The object of the final §4 is the investigation of a special kind of CM-types, by which we can prove, without any condition on F, similar results for the class-fields over F obtained from complex multiplication of an abelian variety whose endomorphism-algebra contains a quadratic extension of F (Theorem 4). In all these cases, if the class-number of F is odd, the absolute class-field over F is contained in the composite of the absolute class-field over F_0 and the fields of moduli of certain polarized abelian varieties which we can specify in each case.

The author wishes to express his sincere thanks to Prof. M. Eichler whose communications gave him a chance to consider the problem, and also to Dr. Y. Akagawa, T. Honda, N. Nobusawa for their valuable discussions; in particular, the paper [2] is noticed by Nobusawa.

Notation and Convention. Q and C denote respectively the field of rational numbers and the field of complex numbers. For every $x \in C$, we denote by x^ρ the complex conjugate of x. Any algebraic number field will be considered as a subfield of C. If K is an algebraic number field of finite degree and $\mathfrak{b}$ is an integral ideal of K, $I_\mathfrak{b}(K)$ denotes the group of all ideals prime to $\mathfrak{b}$, and $P_\mathfrak{b}(K)$ the subgroup of $I_\mathfrak{b}(K)$ consisting of all principal ideals (a) such that $a \in K$, $a \equiv 1 \bmod \mathfrak{b}$. For every positive integer b, the ideal (b) generated by b (in some algebraic number field) will be often denoted simply by b. Further we denote by $C_\mathfrak{b}(K)$ the class-field over K corresponding to the ideal-group $P_\mathfrak{b}(K)$, namely, the ray-class-field modulo $\mathfrak{b}$ over K. In particular, $C_1(K)$ is the absolute class-field (Hilbert's class-field) over K.

§ 1. Preliminaries. Let F_0 be a totally real algebraic number field of finite degree, and F a totally imaginary quadratic extension of F_0. Define, for every positive integer b, a subgroup $I_b(F/F_0)$ of $I_b(F)$ by

$$(1) \quad I_b(F/F_0) = \{\mathfrak{a} \in I_b(F) \,|\, \mathfrak{a}/\mathfrak{a}^\rho = (a) \text{ for some } a \in F$$
$$\text{such that } aa^\rho = 1, \ a \equiv 1 \bmod (b)\}.$$

We see easily that

$$(2) \quad I_b(F/F_0) \supset P_b(F) \cdot \{\mathfrak{a} \in I_b(F) \,|\, \mathfrak{a}^\rho = \mathfrak{a}\} \supset P_b(F) I_b(F_0).$$

Consider the case $b=1$. If $\mathfrak{a} \in I_1(F/F_0)$, we have $\mathfrak{a}/\mathfrak{a}^\rho = (a)$ for some $a \in F$ such that $aa^\rho = 1$. By Hilbert's lemma, there exists an element w of F such that $a = w^\rho/w$. Then $(w\mathfrak{a})^\rho = w\mathfrak{a}$. It follows that

$$(3) \quad I_1(F/F_0) = P_1(F) \cdot \{\mathfrak{a} \in I_1(F) \,|\, \mathfrak{a}^\rho = \mathfrak{a}\}.$$

Let $(F; \{\sigma_1, \cdots, \sigma_n\})$ be a CM–type and $(F^*; \{\tau_j\})$ be its dual (cf. [3, §§ 5.2, 8.3]). Let $\mathfrak{b}$ be an integral ideal of F^*, and b the smallest positive integer divisible by $\mathfrak{b}$. We denote by $I_\mathfrak{b}(F; \{\sigma_i\})$ the subgroup of $I_b(F)$ consisting of all ideals $\mathfrak{a}$ such that there exists an element u of F^* for which we have

$$(4) \qquad \prod_{i=1}^{n} \mathfrak{a}^{\sigma_i} = (u), \quad N(\mathfrak{a}) = uu^{\rho}, \quad u \equiv 1 \bmod \mathfrak{b}.$$

Further we denote by $C_b(F/F_0)$ and $C_{\mathfrak{b}}(F; \{\sigma_i\})$ the class-fields over F corresponding to the ideal-groups $I_b(F/F_0)$ and $I_{\mathfrak{b}}(F; \{\sigma_i\})$, respectively. If $\mathfrak{a} \in I_b(F; \{\sigma_i\})$, we have $N(\mathfrak{a}) \equiv 1 \bmod (b)$. It follows that $C_b(F; \{\sigma_i\})$ *contains the cyclotomic field* $\mathbf{Q}(\zeta)$ *for a primitive b-th root of unity* ζ.

Now Main Theorems 1 and 2 of [3] assert that if $(K^*; \{\psi_\alpha\})$ is a primitive CM-type, we get the class-fields $C_{\mathfrak{b}}(K^*; \{\psi_\alpha\})$ by means of complex multiplication of an abelian variety belonging to the dual of $(K^*; \{\psi_\alpha\})$. This result holds in a little more general form:

Proposition 1. *The assertions of Main Theorems 1 and 2 of* [3] *are true even in case where* $(K^*; \{\psi_\alpha\})$ *is not primitive.*

Proof. Let $(K^*; \{\psi_\alpha\})$ be a CM-type which is not necessarily primitive. Let $(K; \{\varphi_\lambda\})$ be the dual of $(K^*; \{\psi_\alpha\})$, and $(K_1^*; \{\chi_\nu\})$ be the dual of $(K; \{\varphi_\lambda\})$. Then $(K; \{\varphi_\lambda\})$ and $(K_1^*; \{\chi_\nu\})$ are primitive; and $(K; \{\varphi_\lambda\})$ is the dual of $(K_1^*; \{\chi_\nu\})$ (cf. [3, §8.3]). Let L be a Galois extension of $\mathbf{Q}$ containing K^*. Then K and K_1^* are subfields of L. Let G be the Galois group of L over $\mathbf{Q}$, and H^*, H_1^* be respectively the subgroups of G corresponding to K^*, K_1^* by Galois theory. We have $K^* \supset K_1^*$, $H^* \subset H_1^*$, in view of the result of [3, §8.3]. Extend ψ_α and χ_ν to, elements of G and denote them again by the same letters. We have then

$$(5) \qquad \bigcup_{\alpha} H^* \psi_\alpha = \bigcup_{\nu} H_1^* \chi_\nu.$$

Let $\mathfrak{b}$ be an integral ideal of K and b the smallest positive integer divisible by $\mathfrak{b}$. Considering an abelian variety belonging to $(K; \{\varphi_\lambda\})$, we get the class-field $C_{\mathfrak{b}}(K_1^*; \{\chi_\nu\})$ over K_1^*. The composite of K^* and $C_{\mathfrak{b}}(K_1^*; \{\chi_\nu\})$ is a class-field over K^*; and by the "theorem of translation" of class-field theory, the corresponding ideal-group is the group of ideals $\mathfrak{a} \in I_b(K^*)$ such that $N_{K^*/K_1^*}(\mathfrak{a}) \in I_{\mathfrak{b}}(K_1^*; \{\chi_\nu\})$. By the relation (5), this ideal-group is just $I_{\mathfrak{b}}(K^*; \{\psi_\alpha\})$; so we get our proposition.

For convenience, we state here a part of [3, §8.3, Prop. 28] as

Proposition 2. *Let* $(F; \{\sigma_i\})$ *be a CM-type and* $(F^*; \{\tau_j\})$ *its dual.* *Then* F^* *is generated over* $\mathbf{Q}$ *by the elements* $\sum_{i=1}^{n} x^{\sigma_i}$ *for* $x \in F$.

§2. **Class-fields obtained from two CM-types.** F and F_0 being as in §1, let $(F; \{\sigma_1, \cdots, \sigma_n\})$ be a CM-type such that σ_1 is the identity mapping of F. Consider the condition:

(A) *If* $(F^*; \{\tau_j\})$ *is the dual of* $(F; \{\sigma_i\})$, *then* $F \supset F^*$.

This is satisfied whenever F is normal over Q. Now we observe that $(F ; \{\rho, \sigma_2, \cdots, \sigma_n\})$ is a CM type. Let $(F_1^* ; \{\wp_\lambda\})$ be the dual of this CM type. If $(F ; \{\sigma_i\})$ satisfies the condition (A), we have $F_1^* \subset F$. In fact, by Proposition 2, for every $x \in F$, we see that $\sum_{i=1}^n x^{\tau_i} \in F^* \subset F$, so that $x^\rho + \sum_{i=2}^n x^{\sigma_i} = x^\rho - x + \sum_{i=1}^n x^{\sigma_i} \in F$; this implies, again by Proposition 2, $F_1^* \subset F$.

Proposition 3. *Notation being as above, suppose that the condition (A) is satisfied. Then, for every positive integer b, we have*

$$I_b(F ; \{\sigma_i\}) \cap I_b(F ; \{\rho, \sigma_2, \cdots, \sigma_n\}) \subset I_b(F/F_0).$$

Proof. If $\mathfrak{a} \in I_b(F ; \{\sigma_i\}) \cap I_b(F ; \{\rho, \sigma_2, \cdots, \sigma_n\})$, we have $\mathfrak{a}\mathfrak{a}^{\sigma_2} \cdots \mathfrak{a}^{\sigma_n} = (u)$, $\mathfrak{a}^\rho \mathfrak{a}^{\sigma_1} \cdots \mathfrak{a}^{\sigma_n} = (v)$, $N(\mathfrak{a}) = uu^\rho = vv^\rho$ for an element u of F^* and an element v of F_1^* such that $u \equiv 1 \bmod (b)$, $v \equiv 1 \bmod (b)$. Put $a = u/v$. By our assumption and by the above consideration, a is an element of F; and we have $a/a^\rho = (a)$, $aa^\rho = 1$, $a \equiv 1 \bmod (b)$. This proves our proposition.

Theorem 1. *Let F_0 be a totally real algebraic number field of degree $n > 1$, and F a totally imaginary quadratic extension of F_0. Then, the composite D_b of $C_b(F/F_0)$ and $C_b(F_0)$ contains the class-field over F corresponding to the ideal-group $\{\mathfrak{a} \in I_b(F) \mid \mathfrak{a}^2 \in P_b(F)\}$. Let further $(F ; \{\sigma_1, \cdots, \sigma_n\})$ be a CM-type such that σ_1 is the identity mapping of F. Suppose that the condition (A) is satisfied. Then, for every positive integer b, the composite of $C_b(F ; \{\sigma_i\})$ and $C_b(F ; \{\rho, \sigma_2, \cdots, \sigma_n\})$ contains $C_b(F/F_0)$.*

In other words, *if there exists a CM-type satisfying the condition (A), then, adjoining the ray-class-field modulo (b) over F_0, we get, by complex multiplication of abelian varieties, at least a subfield D_b of the ray-class-field $C_b(F)$ modulo (b) over F such that the Galois group of $C_b(F)/D_b$ is of exponent 1 or 2.*

Proof. The composite of $C_b(F_0)$ and F is the class-field over F corresponding to the ideal-group $\{\mathfrak{a} \in I_b(F) \mid \mathfrak{a}\mathfrak{a}^\rho \in P_b(F_0)\}$. If $\mathfrak{a}\mathfrak{a}^\rho \in P_b(F_0)$ and $\mathfrak{a}/\mathfrak{a}^\rho \in P_b(F)$, we have $\mathfrak{a}^2 \in P_b(F)$. This proves the first assertion. The second assertion is an immediate consequence of Proposition 3.

REMARK 1. If F is a non-abelian imaginary extension of Q of degree 4, the condition (A) is never satisfied by any CM-type $(F ; \{\sigma_i\})$. In §4, we shall give an example of a primitive CM-type $(F ; \{\sigma_i\})$ satisfying (A) with an F which is not normal over Q.

The author is ignorant of the difference between the maximal abelian

extension $A = \bigcup\limits_{b=1}^{\infty} C_b(F)$ and $B = \bigcup\limits_{b=1}^{\infty} C_b(F_0)C_b(F; \{\sigma_i\})C_b(F; \{\rho, \sigma_2, \cdots, \sigma_n\})$[1].
If we put $D = \bigcup\limits_{b=1}^{\infty} D_b$, we have $A \supset B \supset D$, and $A \subset D(\sqrt{x} \mid x \in D)$. We can at least prove:

Theorem 2. *F, F_0 and D_b being as in Theorem 1, the absolute class-field over F is contained in D_b for a suitable b.*

Proof. Let $E_1, \cdots, E_r$ be cyclic unramified extentions of F such that the composite of them is the maximal one among the unramified abelian extensions of F whose degrees are powers of 2. By [2, Satz 1b], we can find, for each i, a cyclic extension E_i' of F containing E_i such that $[E_i' : E_i] = 2$. Let b be a positive integer such that the ideal-groups corresponding to the E_i' are all defined modulo (b). Now let E_0 be the maximal one among the unramified abelian extensions of F of odd degree. Let $\mathfrak{H}, \mathfrak{R}, \mathfrak{L}$ denote respectively the subgroups of $I_b(F)$ corresponding to $E_0, E_0E_1 \cdots E_r, E_0E_1' \cdots E_r'$. We have clearly $I_b(F) \supset \mathfrak{H} \supset \mathfrak{R} \supset \mathfrak{L} \supset P_b(F)$. If $\mathfrak{a} \in I_b(F)$ and $\mathfrak{a}^2 \in P_b(F)$, then $\mathfrak{a}^2 \in \mathfrak{L}$. As $\mathfrak{H}/\mathfrak{L}$ is the 2-Sylow subgroup of $I_b(F)/\mathfrak{L}$, we obtain $\mathfrak{a} \in \mathfrak{H}$. By our construction of the E_i', we must have $\mathfrak{a} \in \mathfrak{R}$. This shows that $\mathfrak{R}$ contains the ideal-group $\{\mathfrak{a} \in I_b(F) \mid \mathfrak{a}^2 \in P_b(F)\}$. It follows that D_b contains the field $E_0E_1 \cdots E_r$, the absolute class-field over F.

If either one or both of the groups

$$\{\mathfrak{a} \in I_b(F) \mid \mathfrak{a}\mathfrak{a}^\rho \in P_b(F_0)\}/P_b(F), \qquad I_b(F/F_0)/P_b(F)$$

have odd orders, then $D_b = C_b(F)$.

Lemma 1. *F and F_0 being as in Theorem 1, let h and h_0 be respectively the class-numbers of F and F_0. Then h is a multiple of h_0, and h/h_0 is the order of the group $\{\mathfrak{a} \in I_1(F) \mid \mathfrak{a}\mathfrak{a}^\rho \in P_1(F_0)\}/P_1(F)$.*

Proof. Let K be the absolute class-field over F_0. As the infinite prime spots of F_0 ramify in F, F is not contained in K, so that $[FK : F] = [K : F_0] = h_0$. Our lemma follows easily from this and class-field theory.

We call h/h_0 the *relative class-number of F*. Then we can conclude that, *if the relative class-number of F is odd, D_1 is the absolute class-field over F.* Further we obtain

Proposition 4. *F and F_0 being as in Theorem 1, let h and h_0 be respectively the class-numbers of F and F_0. Suppose that every prime ideal of F ramified in F/F_0 is a principal ideal. Then we have*

---

1) It would be meaningful to take account of the infinite prime spots of F_0, though we have not used them in the present investigation.

$$I_1(F/F_0) = P_1(F)I_1(F_0), \quad [C_1(F/F_0):F] \geq h/h_0.$$

Moreover, if h_0 is odd, the composite D_1 of $C_1(F/F_0)$ and $C_1(F_0)$ is the absolute class-field over F.

Proof. The equality $I_1(F/F_0)=P_1(F)I_1(F_0)$ follows easily from our assumption and the relation (3) of §1. Now the injection of $I_1(F_0)$ into $I_1(F)$ gives a homomorphism of $I_1(F_0)/P_1(F_0)$ onto $I_1(F/F_0)/P_1(F)$; so we have $[I_1(F/F_0):P_1(F)]\leq h_0$, and hence $[I_1(F):I_1(F/F_0)]\geq h/h_0$, which implies $[C_1(F/F_0):F]\geq h/h_0$. If h_0 is odd, the order of the group $I_1(F/F_0)/P_1(F)$ must be odd; as remarked above, this implies $D_1=C_1(F)$.

§3. Class-fields over cyclotomic fields. Let F be an imaginary cyclotomic field and F_0 the maximal real subfield of F. As F is normal over $\mathbf{Q}$, we can apply to F the result of §2. In particular, we get the following assertion. *If the relative class-number of an imaginary cyclotomic field F is odd, then the absolute class-field over F is generated by the absolute class-field over the maximal real subfield of F and the unramified class-fields over F obtained from the fields of moduli of certain two polarized abelian varieties having subfields of F as endomorphism algebras.* Several criteria for the oddness of relative class-number of imaginary cyclotomic fields are given in [1, Satz 38, 42, 46].

F being still an imaginary cyclotomic field, if $(F; \{\sigma_i\})$ is primitive, the dual of $(F; \{\sigma_i\})$ is $(F; \{\sigma_i^{-1}\})$ in virtue of [3, §8.4, (1)]. By (1) and (4) of §1, we see easily

$$(6) \qquad I_b(F/F_0) \cap I_b(F; \{\sigma_i\}) = I_b(F/F_0) \cap I_b(F; \{\tau_i\})$$

for any two primitive CM-types $(F; \{\sigma_i\})$ and $(F; \{\tau_i\})$. For every automorphism γ of F and for every $(F; \{\sigma_i\})$, we have

$$(7) \qquad I_b(F; \{\sigma_i\}) = I_b(F; \{\gamma\sigma_i\}).$$

Theorem 3. *Let F be an imaginary cyclic extension of $\mathbf{Q}$ of degree $2n$ and F_0 the maximal real subfield of F; let σ be a generator of the Galois group of F over $\mathbf{Q}$. Then we have*

$$C_b(F; \{1, \sigma, \cdots, \sigma^{n-1}\}) \supset C_b(F/F_0),$$
$$C_b(F; \{1, \sigma, \cdots, \sigma^{n-1}\}) \supset C_b(F; \{\tau_i\})$$

for every positive integer b and for every primitive CM-type $(F; \{\tau_i\})$. Moreover, if every prime ideal of F ramified in F/F_0 is a principal ideal, then, we have, for every CM-type $(F; \{\tau_i\})$,

$$C_1(F; \{1, \sigma, \cdots, \sigma^{n-1}\}) = C_1(F/F_0) \supset C_1(F; \{\tau_i\}).$$

Proof. It is easy to see that $(F; \{1, \sigma, \cdots, \sigma^{n-1}\})$ is a primitive CM-type. By (7), we have $I_b(F; \{1, \sigma, \cdots, \sigma^{n-1}\}) = I_b(F; \{\sigma^n, \sigma, \sigma^2, \cdots, \sigma^{n-1}\})$. Then by Proposition 3, we have $I_b(F; \{1, \sigma, \cdots, \sigma^{n-1}\}) \subset I_b(F/F_0)$. This proves the first inclusion. The second inclusion follows from this and (6). Now assume that every prime ideal of F ramified in F/F_0 is a principal ideal. By Proposition 4, $I_1(F/F_0) = P_1(F)I_1(F_0)$. We can easily verify that $I_1(F_0) \subset I_1(F; \{\tau_i\})$ for every CM-type $(F; \{\tau_i\})$, so that $I_1(F/F_0) \subset I_1(F; \{\tau_i\})$, which implies $C_1(F/F_0) \supset C_1(F; \{\tau_i\})$. Apply this to the case $\{\tau_i\} = \{1, \sigma, \cdots, \sigma^{n-1}\}$. As we have already seen the inverse inclusion, we must have $C_1(F; \{1, \sigma, \cdots, \sigma^{n-1}\}) = C_1(F/F_0)$.

In general, for every positive integer b, we see that

$$I_b(F; \{\sigma_i\}) \supset P_b(F) \cdot \{\mathfrak{a} \in I_b(F_0) \mid N(\mathfrak{a}) \equiv 1 \bmod (b)\} \ .$$

If $(F; \{\sigma_i\}) = (F; \{1, \sigma, \cdots, \sigma^{n-1}\})$, *the factor group*

$$I_b(F; \{\sigma_i\}) / [P_b(F) \cdot \{\mathfrak{a} \in I_b(F_0) \mid N(\mathfrak{a}) \equiv 1 \bmod (b)\}]$$

is of exponent 1 *or* 2. In fact, in this case, if $\mathfrak{a} \in I_b(F; \{\sigma_i\})$, we have $\mathfrak{a} \in I_b(F/F_0)$ by Theorem 3, so that $\mathfrak{a}/\mathfrak{a}^\rho \in P_b(F)$; on the other hand, it is clear that $N_{F_0/\mathbf{Q}}(\mathfrak{a}\mathfrak{a}^\rho) \equiv 1 \bmod (b)$; therefore, we have

$$\mathfrak{a}^2 = (\mathfrak{a}/\mathfrak{a}^\rho)(\mathfrak{a}\mathfrak{a}^\rho) \in P_b(F) \cdot \{\mathfrak{a} \in I_b(F_0) \mid N(\mathfrak{a}) \equiv 1 \bmod (b)\} \ .$$

Let l^ν be a power of an odd prime number l and ζ a primitive l^ν-th root of unity. Put $F = \mathbf{Q}(\zeta)$, $F_0 = \mathbf{Q}(\zeta + \zeta^{-1})$. Then F is cyclic over $\mathbf{Q}$ and every prime ideal of F ramified in F/F_0 is a principal ideal. Therefore, we can apply Proposition 4 and Theorem 3 to the present case. In particular, *if the class-number of F_0 is odd, then, the field of moduli of a certain polarized abelian variety having F as endomorphism-algebra, together with the absolute class-field over F_0, generates the absolute class-field over F.* By a theorem of Kummer, the class-number of $F = \mathbf{Q}(\zeta)$ is odd if and only if the relative class-number of F is odd (cf. [1, Satz 45]). Hence, the class-number of $F_0 = \mathbf{Q}(\zeta + \zeta^{-1})$ is odd whenever the relative class-number of F is odd; the table of [1] shows that the relative class-number of $\mathbf{Q}(\zeta)$ is odd for $l^\nu < 100$, $l^\nu \neq 29$.

Remark 2. In Theorem 3, it may happen that $C_1(F/F_0) \neq C_1(F; \{\tau_i\})$ for some $\{\tau_i\}$. In fact, let l be a prime number ≥ 5 and ζ a primitive l-th root of unity. Choose as τ_i the automorphism of F defined by $\zeta^{\tau_i} = \zeta^i$ for $1 \leq i \leq n = (l-1)/2$. As observed in [3, §8.4, (1)], $(F; \{\tau_i\})$ is primitive; further by [3, §15.4, Example 2)], we have $C_1(F; \{\tau_i\}) = F$, so that $I_1(F; \{\tau_i\}) = I_1(F)$. Therefore, $C_1(F/F_0) \neq C_1(F; \{\tau_i\})$ if the relative class-number of F is greater than 1; the latter is of course the case for many l.

Now if we put $\{\sigma_i\} = \{\tau_1\sigma, \tau_2, \cdots, \tau_n\}$, we must have $I_1(F\,;\,\{\sigma_i\}) \subset I_1(F/F_0)$ in view of Proposition 3. We can prove that this CM-type $(F\,;\,\{\sigma_i\})$ is primitive. In fact, if $l \neq 17$, the trick of [3, §8.4, (1)] is applicable; and if $l = 17$, this is shown by means of [3, §8.2, Prop. 26]. Then, by Theorem 3 and by what we have just proved, we get $I_1(F\,;\,\{\sigma_i\}) = I_1(F/F_0)$, which implies $C_1(F\,;\,\{\sigma_i\}) = C_1(F/F_0)$. In general, it is not necessarily true that there exists an automorphism γ of F such that $\{\gamma\sigma_i\} = \{1, \sigma, \cdots, \sigma^{n-1}\}$.

§4. A CM-type obtained from two CM-types. The argument of §2 is powerless when F has no CM-type satisfying (A). In order to treat such a case, we consider a special kind of CM-type. We begin with an easy

Lemma 2. *Let F be a totally imaginary quadratic extension of a totally real algebraic number field F_0. Let L be the smallest normal extension of $\mathbf{Q}$ containing F, and G the Galois group of L over $\mathbf{Q}$. Then ρ, considered as an element of G, belongs to the center of G; and L is a totally imaginary quadratic extension of a totally real subfield.*

Proof. We can find an element z of F such that $F = F_0(z)$ and z^2 is a totally negative element of F_0. For every $\gamma \in G$, $(z^\gamma)^2$ is a totally negative element of F_0, so that $z^{\gamma\rho} = -z^\gamma = (-z)^\gamma = z^{\rho\gamma}$. Further, for every $x \in F_0$, we have $x^{\gamma\rho} = x^\gamma = x^{\rho\gamma}$. Therefore, for every γ, $\delta \in G$ and for every $x \in F_0$, we have $(x^\delta)^{\gamma\rho} = (x^\delta)^{\rho\gamma}$, $(z^\delta)^{\gamma\rho} = (z^\rho)^{\delta\gamma} = (z^\delta)^{\rho\gamma}$. These relations imply $y^{\gamma\rho} = y^{\rho\gamma}$ for every $y \in L$, since L is generated by F_0^δ and z^δ; this proves the first assertion. If we denote by L_0 the set of elements y of L such that $y^\rho = y$, we have $(y^\gamma)^\rho = y^{\rho\gamma} = y^\gamma$ for every $y \in L_0$. It follows that L_0 is totally real; this proves the last assertion.

Let F_0 be a totally real algebraic number field of degree $n > 1$. Let F and M be totally imaginary quadratic extensions of F_0. We assume $F \neq M$. Let K be the composite of F and M. Obviously, K contains a totally real algebraic number field K_0 such that $[K_0 : F_0] = 2$. Let $(F\,;\,\{\sigma_i\})$ and $(M\,;\,\{\tau_i\})$ be CM-types. We assume $\sigma_i = \tau_i$ on F_0. This is not an essential restriction, since for any $\{\sigma_i\}$ and $\{\tau_i\}$, we can reorder them so that $\sigma_i = \tau_i$ on F_0.

Now fix an integer r such that $1 \leq r \leq n$, and define $2n$ isomorphisms $\alpha_1, \beta_1, \cdots, \alpha_n, \beta_n$ of K into C by

$$(8) \quad \begin{cases} \alpha_i = \sigma_i \text{ on } F, \ \alpha_i = \tau_i \text{ on } M \text{ for } 1 \leq i \leq n, \\ \beta_j = \sigma_j \text{ on } F, \ \beta_j = \tau_j\rho \text{ on } M \text{ for } 1 \leq j \leq r, \\ \beta_k = \sigma_k\rho \text{ on } F, \ \beta_k = \tau_k \text{ on } M \text{ for } r < k \leq n. \end{cases}$$

It can be easily seen that $(K\,;\,\{\alpha_1, \beta_1, \cdots, \alpha_n, \beta_n\})$ is a CM-type. We assume henceforth that σ_1 is the identity mapping of F and τ_1 is the identity mapping of M, and consider only the case $r=1$.

Let $(M^*; \{\chi_\mu\})$ be the dual of $(M; \{\tau_i\})$; let M^{**} be the field generated over Q by the elements $\sum_{\nu=2}^{n} x^{\tau_\nu}$ for $x\in M$. By Proposition 2, M^* is generated over Q by the elements $\sum_{i=1}^{n} x^{\tau_i}$ for $x\in M$. It follows that

$$(9) \qquad\qquad M^*M = M^{**}M .$$

Proposition 5. *Let $(K^*; \{\varphi_\lambda\})$ be the dual of $(K; \{\alpha_i, \beta_i\})$. Then we have $K^* = FM^{**}$.*

Proof. Put $g(y)=\sum_{i=1}^{n}(y^{\alpha_i}+y^{\beta_i})$ for $y\in K$. By Proposition 2, K^* is generated over Q by the elements $g(y)$ for $y\in K$. For any $y\in K$, we see easily $y^{\alpha_1}+y^{\beta_1}=\mathrm{Tr}_{K/F}(y)$, $y^{\alpha_\nu}+y^{\beta_\nu}=\mathrm{Tr}_{K/M}(y)^{\tau_\nu}$ for $\nu>1$, so that

$$g(y) = \mathrm{Tr}_{K/F}(y)+\sum_{\nu=2}^{n} \mathrm{Tr}_{K/M}(y)^{\tau_\nu} .$$

This implies $K^*\subset FM^{**}$. Now take elements z and w so that $F=F_0(z)$, $M=F_0(w)$, $z^2\in F_0$, $w^2\in F_0$. If $x\in F_0$, we have

$$g(x) = 2\mathrm{Tr}_{F_0/Q}(x), \ g(xz) = 2xz, \ g(xw) = 2\sum_{\nu=2}^{n}(xw)^{\tau_\nu} .$$

These relations show that K^* contains F and M^{**}; this completes the proof.

Proposition 6. *M^{**} is a totally imaginary quadratic extension of a totally real algebraic number field containing F_0. Moreover, for every $x\in M$, we have $x^{\tau_2}\cdots x^{\tau_n}\in M^{**}$; and for every ideal $\mathfrak{c}$ of M, $\mathfrak{c}^{\tau_2}\cdots \mathfrak{c}^{\tau_n}$ is an ideal of M^{**}.*

Proof. If $x\in F_0$, we have $x=\mathrm{Tr}_{F_0/Q}(x)-\sum_{\nu=2}^{n} x^{\tau_\nu}\in M^{**}$, so that $F_0\subset M^{**}$. Now let L be the smallest normal extension of Q containing K and G the Galois group of L over Q. Denote by H the set of elements $\gamma\in G$ such that $\{\tau_2\gamma, \cdots, \tau_n\gamma\}$ coincides with $\{\tau_2, \cdots, \tau_n\}$ on M as a whole. Then, by the same argument as in the proof of [3, §8.3, Prop. 28], we can prove that $H=\{\gamma\in G\,|\,x^\gamma=x \text{ for every } x\in M^{**}\}$. Using this fact, the second and last assertions are proved in the same manner as in the proof of [3, §8.3, Prop. 28]. Now, by the definition of CM-type, $\tau_2\rho$ does not coincide with any τ_ν on M; so ρ is not contained in H. If we put $H_1=H\cup H\rho$, then H_1 is a subgroup of G on account of Lemma 2. Call M_1 the subfield of L corresponding to H_1 by Galois theory. Then we see easily that M_1 is totally real and M^{**} is a totally imaginary quadratic extension of M_1.

Consider a particular case where F_0 is normal over $\boldsymbol{Q}$. Take an element w of M so that $M = F_0(w)$. Choose $n-1$ elements $x_2, \cdots, x_n$ of F_0 in such a way that $\det(x_\mu^{\tau_\nu})_{\mu,\nu=2,\cdots,n} \neq 0$. We have $\sum_{\nu=2}^{n} x_\mu^{\tau_\nu} w^{\tau_\nu} \in M^{**}$ for every μ, so that $w^{\tau_2}, \cdots, w^{\tau_n}$ are contained in M^{**} since $F_0 \subset M^{**}$. This shows that M^{**} is the composite of $M^{\tau_2}, \cdots, M^{\tau_n}$. We can similarly prove that $F_0 M^*$ is the composite of $M^{\tau_1}, \cdots, M^{\tau_n}$. Now assume that the composite of $M^{\tau_1}, M^{\tau_2}, \cdots, M^{\tau_n}$ is of degree 2^n over F_0. This is the case for example, if there exists a prime ideal $\mathfrak{p}$ of F_0 of absolute degree 1 such that $\mathfrak{p}$ is inertial in M while the conjugates of $\mathfrak{p}$, other than $\mathfrak{p}$ itself, decompose in M. Then, we have $[F_0 M^* : \boldsymbol{Q}] = 2^n \cdot n$, and hence $[M^* : \boldsymbol{Q}] \geq 2^n$.[2] This gives an example of CM-type $(M : \{\tau_i\})$ such that $[M^* : \boldsymbol{Q}] > [M : \boldsymbol{Q}]$ for the dual $(M^* ; \{X_\mu\})$ of $(M ; \{\tau_i\})$. This shows also that the case $[K^* : F] > 2$ may happen.

Coming back to the general case, we get

Proposition 7. *Three CM-types* $(F ; \{\sigma_i\})$, $(M ; \{\tau_i\})$, $(K ; \{\alpha_i, \beta_i\})$ *being as above, let* $(K^* ; \{\varphi_\lambda\})$ *be the dual of* $(K ; \{\alpha_i, \beta_i\})$. *Then, for every positive integer* b, *the composite of* $C_b(M)$ *and* $C_b(K ; \{\alpha_i, \beta_i\})$ *contains the class-field* H_b *over* F *corresponding to the ideal group* $I_b(F) \cap P_b(K^*)$.

Proof. Let $\mathfrak{a}$ be an ideal of K. In the same way as in the proof of Proposition 5, we see that

$$(10) \qquad \mathfrak{a}^{\alpha_1}\mathfrak{a}^{\beta_1} \cdots \mathfrak{a}^{\alpha_n}\mathfrak{a}^{\beta_n} = N_{K/F}(\mathfrak{a}) \prod_{\nu=2}^{n} N_{K/M}(\mathfrak{a})^{\tau_\nu} .$$

The composite of $C_b(M)$ and $C_b(K ; \{\alpha_i, \beta_i\})$ is a class-field over K; denote by $\mathfrak{H}$ the corresponding ideal-group. If $\mathfrak{a} \in \mathfrak{H}$, we have $\mathfrak{a}^{\alpha_1}\mathfrak{a}^{\beta_1} \cdots \mathfrak{a}^{\alpha_n}\mathfrak{a}^{\beta_n} \in P_b(K^*)$ and $N_{K/M}(\mathfrak{a}) \in P_b(M)$; so we see that $\prod_{\nu=2}^{n} N_{K/M}(\mathfrak{a})^{\tau_\nu} \in P_b(M^{**})$ in view of Proposition 6. By (10) and by Proposition 5, we have $N_{K/F}(\mathfrak{a}) \in P_b(K^*)$. This shows that $\mathfrak{H}$ is contained in the ideal-group corresponding to the composite of K and H_b; our proposition is thereby proved.

If we put $m = [K^* : F]$, we see easily that

$$P_b(F) \subset I_b(F) \cap P_b(K^*) \subset \{\mathfrak{a} \in I_b(F) \mid \mathfrak{a}^m \in P_b(F)\} .$$

Therefore, the exponent of the Galois group of $C_b(F)/H_b$ is a divisor of $m = [K^* : F]$.

---

2) In reality we can show that $[M^* : \boldsymbol{Q}] = 2^n$.

If we fix M (and hence F_0) and consider M and $C_b(M)$ auxiliary, Proposition 7 may be regarded as a statement concerning the class-fields over the variant field F, which can be obtained by complex multiplication of abelian varieties having a certain overfield K^* of F as endomorphism-algebra.[3] In order to get a more transparent result, we consider a restrictive case.

Proposition 8. $(M : \{\tau_i\})$ *satisfies the condition* (A) *if and only if* $M \supset M^{**}$; *and if this is satisfied, we have* $M = M^{**}$, $K = K^*$.

Proof. The first assertion is a direct consequence of (9). If $M \supset M^{**}$, we must have $M = M^{**}$ on account of Proposition 6, so that $K^* = FM = K$ by Proposition 5.

In particular, if M is normal over Q, then $(K; \{\alpha_i, \beta_i\})$ satisfies (A); in this case, K is normal over Q if and only if F is normal over Q; and if we take as F a non-abelian extension of Q of degree 4, we see easily that $(K; \{\alpha_i, \beta_i\})$ is primitive.

For any totally real algebraic number field F_0, we can find a CM-type $(M; \{\tau_i\})$ such that $[M : F_0] = 2$ and $M \supset M^*$. In fact, for any positive integer s, put $M = F_0(\sqrt{-s})$ and define $\tau_1, \cdots, \tau_n$ so that $(\sqrt{-s})^{\tau_\nu} = \sqrt{-s}$. Then it is easy to see $M^* = Q(\sqrt{-s})$.

Theorem 4. *Let* F_0 *be a totally real algebraic number field of degree* $n > 1$. *Let* F *and* M *be distinct totally imaginary quadratic extensions of* F_0, *and* K *the composite of* F *and* M. *Let* $(F; \{\sigma_i\})$ *and* $(M; \{\tau_i\})$ *be CM-types such that* σ_1 *is the identity on* F, τ_1 *is the identity on* M, *and* $\sigma_i = \tau_i$ *on* F_0. *Define a CM-type* $(K; \{\alpha_i, \beta_i\})$ *by the relation* (8) *with* $r = 1$. *Suppose that* $(M; \{\tau_i\})$ *satisfies the condition* (A) *of* §2. *Then, for every positive integer* b, $C_b(K; \{\alpha_i, \beta_i\})$ *contains the class-field* $C_b(F/F_0)$ *over* F.

Proof. For every ideal $\mathfrak{a}$ of K, the equality (10) is also written in the form

$$(11) \qquad \mathfrak{a}^{\alpha_1 \beta_1} \cdots \mathfrak{a}^{\alpha_n \beta_n} = N_{K/F}(\mathfrak{a}) N_{K/M}(\mathfrak{a})^{-1} \prod_{i=1}^{n} N_{K/M}(\mathfrak{a})^{\tau_i} .$$

By our assumption and Proposition 8, we have $K = K^*$. Hence, if $\mathfrak{a} \in I_b(K; \{\alpha_i, \beta_i\})$, there exists an element u of K such that

$$\mathfrak{a}^{\alpha_1 \beta_1} \cdots \mathfrak{a}^{\alpha_n \beta_n} = (u), \quad N(\mathfrak{a}) = uu^\rho, \quad u \equiv 1 \bmod (b) .$$

---

3) In fact, the abelian varieties belonging to $(K; \{\alpha_i, \beta_i\})$ are special members of an analytic family of polarized abelian varieties whose moduli are given by certain automorphic functions of one variable.

Put $v = N_{K/F}(u)$, $\mathfrak{c} = N_{K/F}(\mathfrak{a})$. Now take $N_{K/F}$ of the both sides of (11). We note that for any ideal $\mathfrak{e}$ of M, $N_{K/F}(\mathfrak{e}) = N_{M/F_0}(\mathfrak{e})$; especially,

$$N_{K/F}(N_{K/M}(\mathfrak{a})) = N_{M/F_0}(N_{K/M}(\mathfrak{a})) = N_{K/F_0}(\mathfrak{a}) = N_{F/F_0}(\mathfrak{c}) = \mathfrak{c}\mathfrak{c}^{\rho},$$

and

$$N_{K/F}\left(\prod_{i=1}^{n} N_{K/M}(\mathfrak{a})^{\tau_i}\right) = N_{M/F_0}\left(\prod_{i=1}^{n} N_{K/M}(\mathfrak{a})^{\tau_i}\right) = \prod_{i=1}^{n} N_{K/M}(\mathfrak{a})^{\tau_i} N_{K/M}(\mathfrak{a})^{\tau_i \rho}$$
$$= N_{M/Q}(N_{K/M}(\mathfrak{a})) = N_{K/Q}(\mathfrak{a}) = (uu^{\rho}).$$

Therefore, we obtain from (11), $(v) = \mathfrak{c}^2(\mathfrak{c}\mathfrak{c}^{\rho})^{-1}(uu^{\rho})$. Put $w = v(uu^{\rho})^{-1}$. Then w is an element of F, and $\mathfrak{c}/\mathfrak{c}^{\rho} = (w)$, $ww^{\rho} = 1$, $w \equiv 1 \bmod (b)$. Thus we have shown $N_{K/F}[I_b(K; \{\alpha_i, \beta_i\})] \subset I_b(F/F_0)$. This proves our theorem.

By means of Theorem 4, we obtain several assertions concerning the class-fields over F similar to those given in §2. In particular, the absolute class-field $C_1(F)$ is contained in the composite $C_b(K; \{\alpha_i, \beta_i\})$ and $C_b(F_0)$ for a suitable positive integer b. As another example of specializations (or degenerations) of Theorem 4, we get the following conclusion: *F and F_0 being as in Theorem 4, let s be a positive integer such that $\sqrt{-s} \notin F$. Then the field of moduli of a certain polarized abelian variety having $F(\sqrt{-s})$ as endomorphism-algebra, together with the absolute class-field over F_0, generates a class-field over $F(\sqrt{-s})$ containing the absolute class-field over F, if the class-number of F is odd.*

Osaka University

(Received March 19, 1962)

References.

[1] H. Hasse: Über die Klassenzahl Abelscher Zahlkörper, Akademie-Verlag, Berlin, 1952.

[2] H. Richter: *Über die Lösbarkeit einiger Nicht-Abelscher Einbettungs-probleme*, Math. Ann. **112** (1936), 69–84.

[3] G. Shimura and Y. Taniyama: Complex multiplication of abelian varieties and its applications to number theory, Publications of Math. Soc. Japan, No. **6**, 1961.

[4] Y. Taniyama: *L-functions of number fields and zeta functions of abelian varieties*, J. Math. Soc. Japan 9 (1957), 330–366.

[5] A. Weil: *On a certain type of characters of the idèle-class group of an algebraic number-field*, Proceedings of the International Symposium on Algebraic Number Theory, Tokyo-Nikko, 1955 (1956), 1–7.

Arithmetic of alternating forms and quaternion hermitian forms

Journal of the Mathematical Society of Japan, 15 (1963), 33-65

(Received June 21, 1962)

Hecke's Dirichlet series obtained from modular forms can be regarded as zeta-functions attached to the general linear group $GL(2, \boldsymbol{Q})$ over the rational number field $\boldsymbol{Q}$. In general, we may expect to obtain zeta-functions of this kind for a fairly wide class of algebraic groups defined over $\boldsymbol{Q}$. In order to realize this, it is necessary to develop, in the first place, the theory of elementary divisors for any algebraic group G in question. This is actually done in the case where G is the multiplicative group of a semi-simple algebra. Further, the case of the orthogonal group is investigated in detail by M. Eichler [3]. In both cases, there are fundamental theorems, due to Eichler [4, 5] and M. Kneser [6], which may be called the approximation theorem in the group G, from which one can easily derive an important conclusion about the class-number for G. This approximation theorem plays an essential role also in the theory of Hecke-rings attached to quaternion algebras [8, 9]. In fact, by means of the theorem, we can prove the isomorphism between the Hecke-ring defined by the idele-group of a quaternion algebra D and the Hecke-ring defined by the unit-groups of maximal orders in D (cf. [9, § 2]).

The purpose of the present paper is to give an extension of the theory of elementary divisors for the group of similitudes of a hermitian form over a quaternion algebra, and to prove an approximation theorem for this group. Let F be the quotient field of a Dedekind domain $\mathfrak{g}$ and A a quaternion (not necessarily division) algebra over F. Let V be a left A-module which is isomorphic to the product of n copies of A. We consider an A-valued non-degenerate hermitian form $f(x, y)$ on V with respect to the canonical involution of A (cf. § 2.2). Let G be the group consisting of all A-automorphisms σ of V such that $f(x\sigma, y\sigma) = N(\sigma)f(x, y)$ for $x \in V$, $y \in V$ with $N(\sigma) \in F$. Take a maximal oder $\mathfrak{o}$ in A. Let L be a $\mathfrak{g}$-lattice in V such that $\mathfrak{o}L \subset L$. We denote by $N(L)$ the two-sided $\mathfrak{o}$-ideal generated by $f(x, y)$ for $x \in L$, $y \in L$, and call it the norm of L. We say that L is maximal if L is a maximal one among the lattices with the same norm. As in [3], our theory is mostly concerned with maximal lattices in V. If A is the total matric algebra of degree 2 over F, then G is isomorphic to the group of similitudes of an alternating form over F with $2n$ variables. We treat this case in §1 and §2.5, and prove fundamental propositions concerning the existence of canonical bases for maximal lattices and their elementary divisors. We give in §3 similar propositions in case where A is a division quaternion algebra over a $\mathfrak{p}$-adic field. These propositions correspond to the results of the same kind obtained in the case of orthogonal groups [Eicher, **3**] and of quaternion anti-hermitian forms [Tsukamoto, **10**]. In §4, we consider the global theory, namely the case where F is an algebraic number field. Our principal aim is to prove approximation-theorems for G (Theorems 1 and 2 of §4.6) in case where A is indefinite. As an application of the theorems, we can show that the classes of maximal lattices in each genus are in one-to-one correspondence with the ideal-classes modulo $\mathfrak{l}$ in F for a suitable product $\mathfrak{l}$ of infinite prime spots of F (Theorem 3). If we denote by G^0 the unitary group of f, i.e. the subgroup of G composed of the elements σ such that $N(\sigma)=1$, then each genus with respect to G^0 consists of only one class (§4.9). Finally we give a result on global set of elementary divisors of maximal lattices (Theorem 4). As explained in the beginning, our theory can be considered as preliminaries for the theory of the Hecke-ring of G. In fact, by means of our propositions and theorems, we can develop such a theory, which is a generalization of the theory in [**9**, §2]. As for this, we have only given Proposition 4.11. A further investigation of the Hecke-ring of G will be made in a subsequent paper.

NOTATION. We denote by $\boldsymbol{Z}$, $\boldsymbol{Q}$, $\boldsymbol{R}$, $\boldsymbol{C}$ and $\boldsymbol{K}$, respectively, the ring of rational integers, the rational number field, the real number field, the complex number field, and the division ring of real quaternions. For a ring S with an identity element, $M_m(S)$ denotes the ring of matrices of degree m with entries in S; the identity matrix of degree m is denoted by 1_m; and the transpose of a matrix X is denoted by ${}^tX$. We mean by δ_{ij} the usual Kronecker's delta, namely $\delta_{ij}=0$ or 1 according as $i \neq j$ or $i=j$.

§1. Arithmetic of alternating forms.

1.1. Alternating form and symplectic group. Let F be an arbitrary field and W a vector space over F of finite dimension. We denote by $E(W)$ the ring of all F-linear mappings of W into itself, and by $GL(W)$ the group of regular elements of $E(W)$. We write the operation of an element of $E(W)$ on the right; so we have $(ax)\sigma = a(x\sigma)$ for $a \in F$, $x \in W$, $\sigma \in E(W)$. Let $g(x,y)$ be a non-degenerate alternating form on W. We denote by $G(W,g)$ the subgroup of $GL(W)$ consisting of the elements σ of $GL(W)$ for which there exists a number $N(\sigma)$ of F such that $g(x\sigma, y\sigma) = N(\sigma)g(x,y)$ for every $x,y \in W$, and denote by $G^0(W,g)$ the symplectic group associated to g, namely, the subgroup of $G(W,g)$ consisting of the elements σ such that $N(\sigma)=1$.

1.2. Lattices in a vector space. Let $\mathfrak{g}$ be a Dedekind domain and F the quotient field of $\mathfrak{g}$. Let W be a vector space over F. By a $\mathfrak{g}$-*lattice* in W, we understand a finitely generated $\mathfrak{g}$-submodule L of W such that $FL=W$. Let $\mathfrak{p}$ be a prime ideal of $\mathfrak{g}$. We denote by $F_{\mathfrak{p}}$ and $\mathfrak{g}_{\mathfrak{p}}$ the $\mathfrak{p}$-completions of F and $\mathfrak{g}$, respectively. Put $W_{\mathfrak{p}}=W\underset{F}{\otimes}F_{\mathfrak{p}}$. For every $\mathfrak{g}$-lattice L in W, put $L_{\mathfrak{p}}=\mathfrak{g}_{\mathfrak{p}}L$; then $L_{\mathfrak{p}}$ is a $\mathfrak{g}_{\mathfrak{p}}$-lattice in $W_{\mathfrak{p}}$. The following lemma is well-known.

LEMMA 1.1. *Let L be a $\mathfrak{g}$-lattice in W. Take, for each prime ideal $\mathfrak{p}$ of $\mathfrak{g}$, a $\mathfrak{g}_{\mathfrak{p}}$-lattice $M^{\mathfrak{p}}$ in $W_{\mathfrak{p}}$. Then there exists a $\mathfrak{g}$-lattice M in W such that $M_{\mathfrak{p}}=M^{\mathfrak{p}}$ for every $\mathfrak{p}$, if and only if $M^{\mathfrak{p}}=L_{\mathfrak{p}}$ for all except a finite number of $\mathfrak{p}$. If such a lattice M exists, we have $M=\bigcap_{\mathfrak{p}}(M_{\mathfrak{p}}\cap W)$.*

LEMMA 1.2. *Let $L_{\mathfrak{p}}$ be a $\mathfrak{g}_{\mathfrak{p}}$-lattice in $W_{\mathfrak{p}}$; let σ and τ be elements of $GL(W_{\mathfrak{p}})$. Suppose that $L_{\mathfrak{p}}(\sigma-\tau)\subset\mathfrak{p}L_{\mathfrak{p}}\sigma$. Then we have $L_{\mathfrak{p}}\sigma=L_{\mathfrak{p}}\tau$.*

PROOF. Let $x_1,\cdots,x_m$ be generators of $L_{\mathfrak{p}}$ over $\mathfrak{g}_{\mathfrak{p}}$. Put $M=L_{\mathfrak{p}}\sigma$, $K=L_{\mathfrak{p}}\tau$. Then we have $x_i\sigma-x_i\tau\in\mathfrak{p}M\subset M$ for every i. As M and K are respectively generated by the $x_i\sigma$ and the $x_i\tau$, we get $K\subset M$. Further we have $M\subset K+\mathfrak{p}M$. From this we obtain inductively $M\subset K+\mathfrak{p}^eM$ for every positive integer e. This implies $M\subset K$, so that $M=K$.

1.3. Canonical base of a lattice with respect to an alternating form. We first prove a generalization of a well-known theorem of Frobenius.

PROPOSITION 1.3. *Let $\mathfrak{g}$ be a Dedekind domain and F the quotient field of $\mathfrak{g}$. Let W be a vector space of dimension $2n$ over F and $g(x,y)$ a non-degenerate alternating form on W. Let M be a $\mathfrak{g}$-lattice in W. Then there exist a base $\{y_1,\cdots,y_n,z_1,\cdots,z_n\}$ of W over F and (fractional) $\mathfrak{g}$-ideals $\mathfrak{a}_1,\cdots,\mathfrak{a}_n$ such that*

$$g(y_i,y_j)=g(z_i,z_j)=0, \qquad g(y_i,z_j)=\delta_{ij},$$

$$M=\mathfrak{g}y_1+\mathfrak{g}y_2+\cdots+\mathfrak{g}y_n+\mathfrak{a}_1z_1+\mathfrak{a}_2z_2+\cdots+\mathfrak{a}_nz_n,$$

$$\mathfrak{a}_1\supset\mathfrak{a}_2\supset\cdots\supset\mathfrak{a}_n.$$

The ideals $\mathfrak{a}_i$ are uniquely determined by M and g.

PROOF. We prove this by induction on n. For every $x\in M$, put $\mathfrak{a}_x=g(x,M)$. Obviously, $\mathfrak{a}_x$ is a $\mathfrak{g}$-ideal. As M is a $\mathfrak{g}$-lattice, there exists a maximal one among the $\mathfrak{a}_x$, say $\mathfrak{a}_1$; and take an element y_1 of M so that $\mathfrak{a}_1=g(y_1,M)$. As we have $\mathfrak{g}=g(y_1,\mathfrak{a}_1^{-1}M)$, there exists an element z_1 of $\mathfrak{a}_1^{-1}M$ such that $g(y_1,z_1)=1$. Put $\mathfrak{b}=g(M,z_1)$. As $\mathfrak{b}\ni g(y_1,z_1)=1$, we have $\mathfrak{b}\supset\mathfrak{g}$, so that $\mathfrak{a}_1\mathfrak{b}=\mathfrak{a}_1g(M,z_1)\supset\mathfrak{a}_1$. Assume that $\mathfrak{b}\neq\mathfrak{g}$. Then $\mathfrak{a}_1g(M,z_1)\neq\mathfrak{a}_1$, and hence there exist an element u of M and an element α of $\mathfrak{a}_1$ such that $g(u,\alpha z_1)\notin\mathfrak{a}_1$. Put $\beta=-g(u,\alpha z_1)$, $\gamma=g(y_1,u)$. We have then $g(y_1+\alpha z_1,u-\gamma z_1)=\beta$. Since $\gamma\in\mathfrak{a}_1$ and $\mathfrak{a}_1z_1\subset M$, the element $u-\gamma z_1$ is contained in M. We note that $g(y_1+\alpha z_1,\mathfrak{a}_1z_1)=\mathfrak{a}_1$. Therefore, we have

$$g(y_1+\alpha z_1,M)\supset\mathfrak{a}_1+\mathfrak{g}\beta\supset\mathfrak{a}_1, \qquad \mathfrak{a}_1+\mathfrak{g}\beta\neq\mathfrak{a}_1.$$

This contradicts the maximality of $\mathfrak{a}_1$. Hence we must have $g(M, z_1) = \mathfrak{g}$. Now define a submodule M' of M by $M' = \{v \in M \mid g(y_1, v) = g(z_1, v) = 0\}$. For every $w \in M$, put $\xi = g(y_1, w)$, $\eta = g(z_1, w)$, $w_0 = w + \eta y_1 - \xi z_1$. Then $\xi \in \mathfrak{a}_1$, $\eta \in \mathfrak{g}$, and we have $g(y_1, w_0) = g(z_1, w_0) = 0$, so that $w_0 \in M'$. This shows that $M = \mathfrak{g} y_1 + \mathfrak{a}_1 z_1 + M'$. Applying our induction to M', we get an expression $M' = \mathfrak{g} y_2 + \cdots + \mathfrak{g} y_n + \mathfrak{a}_2 z_2 + \cdots + \mathfrak{a}_n z_n$ with the properties $\mathfrak{a}_2 \supset \cdots \supset \mathfrak{a}_n$, $g(y_i, y_j) = g(z_i, z_j) = 0$, $g(y_i, z_j) = \delta_{ij}$ for $2 \leq i \leq n$, $2 \leq j \leq n$. Therefore, the first assertion is proved if we show $\mathfrak{a}_1 \supset \mathfrak{a}_2$. Let u and v be elements of M'. We have $g(y_1 + u, M) \supset g(y_1 + u, \mathfrak{a}_1 z_1 + \mathfrak{g} v) \supset \mathfrak{a}_1 + \mathfrak{g} g(u, v) \supset \mathfrak{a}_1$. By the maximality of $\mathfrak{a}_1$, we must have $g(u, v) \in \mathfrak{a}_1$, namely $g(M', M') \subset \mathfrak{a}_1$. This implies $\mathfrak{a}_1 \supset \mathfrak{a}_2$ and completes the proof of the first assertion. The invariance of the ideals $\mathfrak{a}_i$ is easily shown by "localization". Namely, for every prime ideal $\mathfrak{p}$ of $\mathfrak{g}$, consider $W_{\mathfrak{p}} = W \underset{F}{\otimes} F_{\mathfrak{p}}$ and a $\mathfrak{g}_{\mathfrak{p}}$-lattice $M_{\mathfrak{p}} = \mathfrak{g}_{\mathfrak{p}} M$ in $W_{\mathfrak{p}}$. Then the invariance is an immediate consequence of the theory of elementary divisors over a principal ideal domain (cf. [2, §5.1, Theorem 1]). We can also prove the invariance more directly with no use of localization.

We call the ideals $\mathfrak{a}_i$ of Proposition 1.3 *the invariant factors of M* (with respect to g), and call $\{y_1, \cdots, y_n, z_1, \cdots, z_n\}$ a *canonical base of M* (with respect to g).

1.4. Maximal lattices. Let F, $\mathfrak{g}$, W, g be the same as in Proposition 1.3. For every $\mathfrak{g}$-lattice M in W, we see that the first member $\mathfrak{a}_1$ of the invariant factors of M is the $\mathfrak{g}$-ideal generated by $g(x, y)$ for $x, y \in M$. We put $N_g(M) = \mathfrak{a}_1$ and call $N_g(M)$ the *norm* of M with respect to g. For simplicity, we fix g and write $N(M) = N_g(M)$. We say that M is *maximal* (with respect to g) if M is a maximal one among the $\mathfrak{g}$-lattices in W with the same norm (with respect to g). It is clear that $N(M\sigma) = N(M)N(\sigma)$ for every $\sigma \in G(W, g)$. If M is maximal, $M\sigma$ is maximal for every $\sigma \in G(W, g)$. By Proposition 1.3, we see easily that M is maximal if and only if the invariant factors of M are all equal to $N(M)$. Furthermore, if M is a $\mathfrak{g}$-lattice in W and $\mathfrak{a}$ is a $\mathfrak{g}$-ideal such that $\mathfrak{a} \supset N(M)$, we can find a maximal lattice L in W such that $L \supset M$, $N(L) = \mathfrak{a}$.

PROPOSITION 1.4. *Let M_1 and M_2 be maximal lattices in W. Then we have $M_1 \sigma = M_2$ for an element σ of $G(W, g)$, if and only if $N(M_1)^{-1} N(M_2)$ is a principal ideal.*

PROOF. If $M_1 \sigma = M_2$ for an element $\sigma \in G(W, g)$, we have $N(M_2) = N(M_1\sigma) = N(M_1)N(\sigma)$; this proves the 'only if' part. Conversely, put $\mathfrak{a}_i = N(M_i)$ and $\mathfrak{a}_1^{-1} \mathfrak{a}_2 = \mathfrak{g}\alpha$ with $\alpha \in F$. Let $\{y_1, \cdots, y_n, z_1, \cdots, z_n\}$ and $\{u_1, \cdots, u_n, v_1, \cdots, v_n\}$ be respectively canonical bases of M_1 and M_2. Define an element σ of $E(W)$ by $y_i \sigma = u_i$, $z_i \sigma = \alpha v_i$ for $1 \leq i \leq n$. Then we see easily $\sigma \in G(W, g)$, $N(\sigma) = \alpha$, $M_1 \sigma = M_2$. This proves the 'if' part.

We say that maximal lattices M_1 and M_2 in W are *equivalent* if $M_1 = M_2\sigma$ for an element σ of $G(W, g)$, and call a maximal set of mutually equivalent maximal lattices a *class* of maximal lattices. By Proposition 1.4, we observe that *the mapping $M \to N(M)$ gives a one-to-one correspondence between the classes of maximal lattices in W and the ideal-classes of F.*

1.5. Invariant factors of elements of $G(W, g)$. Notation being as in §§ 1.3 –4, suppose that g is a principal ideal domain.

PROPOSITION 1.5. *Let L and M be maximal lattices in W. Let α be an element of F such that $N(M) = \alpha N(L)$. Put $N(L) = \mathfrak{a}$. Then there exist a canonical base $\{y_1, \cdots, y_n, z_1, \cdots, z_n\}$ of L and elements $a_1, \cdots, a_n, b_1, \cdots, b_n$ of F such that*

$$L = g y_1 + \cdots + g y_n + \mathfrak{a} z_1 + \cdots + \mathfrak{a} z_n ,$$

$$M = g a_1 y_1 + \cdots + g a_n y_n + \mathfrak{a} b_1 z_1 + \cdots + \mathfrak{a} b_n z_n ,$$

$$\alpha = a_1 b_1 = \cdots = a_n b_n ,$$

$$g a_1 \supset \cdots \supset g a_n \supset g b_n \supset \cdots \supset g b_1 .$$

PROOF. We proceed by induction on n. Put $\mathfrak{c} = \{c \in F \mid cM \subset L\}$. It is easy to see that $\mathfrak{c}$ is a g-ideal. As g is a principal ideal domain, we have $\mathfrak{c} = g c_0$ for an element c_0 of F. Put $M' = c_0 M$. Then M' is a maximal lattice, and $N(M') = c_0^2 \alpha \mathfrak{a}$, $g = \{c \in F \mid cM' \subset L\}$. If we prove our proposition for M' and $c_0^2 \alpha$, we get easily the assertion for M and α. Therefore, we may assume that $M = M'$, namely, $g = \{c \in F \mid cM \subset L\}$. The last relation implies that $L \supset M$ and M contains an element $y_1 \neq 0$ such that $L/g y_1$ is a free g-module. Put $M_1 = M + \alpha L$. Then M_1 is a g-lattice in W. As $L \supset M$ and $N(M) = \alpha N(L)$, we must have $\alpha \in g$; hence we see easily $N(M_1) = \alpha N(L) = N(M)$. As M is maximal, we must have $M = M_1$, so that $M \supset \alpha L$. Now taking a canonical base of L, and expressing y_1 in a linear form of the base, we find that $g(y_1, L) = \mathfrak{a}$. By the proof of Proposition 1.3, we can find an element z_1 of $\mathfrak{a}^{-1} L$ such that $g(y_1, z_1) = 1$; and if we put $U = \{x \in W \mid g(y_1, x) = g(z_1, x) = 0\}$, $L_0 = L \cap U$, we get $L = g y_1 + \mathfrak{a} z_1 + L_0$. We see easily that L_0 is a maximal lattice in U and $N(L_0) = \mathfrak{a}$. As $\alpha L \subset M$, we have $\alpha \mathfrak{a} z_1 \subset M$. For every $x \in M$, we have $g(y_1, x) \in N(M) = \alpha \mathfrak{a}$, $g(z_1, x) \in \mathfrak{a}^{-1} N(L) = g$. Hence if we put $g(y_1, x) = \xi \alpha$, $g(z_1, x) = \eta$, then $\xi \in \mathfrak{a}$, $\eta \in g$. Put $x_0 = x + \eta y_1 - \xi \alpha z_1$. We have then $x_0 \in M$ and $g(y_1, x_0) = g(z_1, x_0) = 0$, so that $x_0 \in U \cap M$. This proves that $M = g y_1 + \mathfrak{a} \alpha z_1 + M_0$, if we put $M_0 = U \cap M$. As M is maximal, M_0 must be a maximal lattice in U such that $N(M_0) = \mathfrak{a} \alpha$. Applying our induction assumption to L_0 and M_0, we find a canonical base $\{y_2, \cdots, y_n, z_2, \cdots, z_n\}$ of L_0 and elements $a_2, \cdots, a_n, b_2, \cdots, b_n$ of F such that

$$L_0 = \mathfrak{g} y_2 + \cdots + \mathfrak{g} y_n + \mathfrak{a} z_2 + \cdots + \mathfrak{a} z_n ,$$

$$M_0 = \mathfrak{g} a_2 y_2 + \cdots + \mathfrak{g} a_n y_n + \mathfrak{a} b_2 z_2 + \cdots + \mathfrak{a} b_n z_n ,$$

$$\alpha = a_2 b_2 = \cdots = a_n b_n ,$$

$$\mathfrak{g} a_2 \supset \cdots \supset \mathfrak{g} a_n \supset \mathfrak{g} b_n \supset \cdots \supset \mathfrak{g} b_2 .$$

As $L_0 \supset M_0$, we have $\mathfrak{g} \supset \mathfrak{g} a_2$, so that $\mathfrak{g} b_2 \supset \mathfrak{g} \alpha$. Putting $a_1 = 1$ and $b_1 = \alpha$, we obtain our assertion for L and M.

PROPOSITION 1.6. *Let L be a maximal lattice in W. Let $\{u_1, \cdots, u_n, v_1, \cdots, v_n\}$ be a canonical base of L. Denote by Γ^0 the subgroup of $G^0(W, g)$ consisting of elements γ of $G^0(W, g)$ such that $L\gamma = L$, and by Δ the set of elements σ of $G(W, g)$ such that $u_i \sigma = a_i u_i$, $v_i \sigma = b_i v_i$ for $1 \leq i \leq n$ with elements a_i, b_i of F and $\mathfrak{g} a_1 \supset \cdots \supset \mathfrak{g} a_n \supset \mathfrak{g} b_n \supset \cdots \supset \mathfrak{g} b_1$. Then we have $G(W, g) = \Gamma^0 \cdot \Delta \cdot \Gamma^0$.*

PROOF. Let σ be an element of $G(W, g)$. Put $M = L\sigma$, $\alpha = N(\sigma)$, and apply Proposition 1.5 to this $\{L, M, \alpha\}$. Then we get a canonical base $\{y_i, z_i\}$ of L and elements a_i, b_i of F with the properties of that proposition. Define two elements γ and τ of $E(W)$ by $u_i \gamma = y_i$, $v_i \gamma = z_i$, $u_i \tau = a_i u_i$, $v_i \tau = b_i v_i$. We see easily that $\gamma \in \Gamma^0$ and $\tau \in \Delta$, $N(\tau) = \alpha$. Further we have $L\tau\gamma = L\sigma$. Hence if we put $\varepsilon \tau \gamma = \sigma$, we have $L\varepsilon = L$, $\varepsilon \in G(W, g)$, $N(\varepsilon) = 1$, so that $\varepsilon \in \Gamma^0$. It follows that $\sigma = \varepsilon \tau \gamma \in \Gamma^0 \cdot \Delta \cdot \Gamma^0$. Our proposition is thereby proved.

§2. Hermitian forms over a quaternion algebra.

2.1. Quaternion algebras. By a *quaternion algebra* over a field F, we understand a central simple algebra A over F such that $[A : F] = 4$. Every quaternion algebra A over F has an involution $a \to a'$, which is uniquely determined by the property that $(X - a)(X - a')$ is the principal polynomial of a over F. We call it the *canonical involution* of A and always denote it by $a \to a'$. For every $a \in A$, we put

$$N(a) = aa' , \qquad \mathrm{Tr}(a) = a + a' .$$

If A is not a division algebra, A is isomorphic to $M_2(F)$; and for every $a \in M_2(F)$, $N(a)$ is just the determinant of a and $\mathrm{Tr}(a)$ is the trace of a. Hereafter we assume that the characteristic of F is different from 2. Then, for an element a of A, we have $a = a'$ if and only if $a \in F$.

If F is the quotient field of a Dedekind domain $\mathfrak{g}$, we can develop ideal-theory in A. Here we recall only the definition of different and norm of ideals. Let $\mathfrak{o}$ be a maximal order in A. The different $\mathfrak{D} = \mathfrak{D}(\mathfrak{o}/\mathfrak{g})$ of $\mathfrak{o}$ with respect to $\mathfrak{g}$ is the integral two-sided $\mathfrak{o}$-ideal defined by

$$\mathfrak{D}^{-1} = \{x \in A \mid \mathrm{Tr}(x\mathfrak{o}) \subset \mathfrak{g}\} .$$

Let $\mathfrak{a}$ be a right (resp. left) $\mathfrak{o}$-ideal. We denote by $N(\mathfrak{a})$ the $\mathfrak{g}$-ideal generated by the elements $N(a)$ for $a \in \mathfrak{a}$. If we put $\mathfrak{a}' = \{x' \mid x \in \mathfrak{a}\}$, then $\mathfrak{a}'\mathfrak{a} = N(\mathfrak{a})\mathfrak{o}$ (resp. $\mathfrak{a}\mathfrak{a}' = N(\mathfrak{a})\mathfrak{o}$).

2.2. Q-hermitian forms. Let A be a quaternion algebra over a field F. By an *A-space of dimension n*, we understand a left A-module V isomorphic to the product of n copies of A; and we put $n = \dim_A V$. We call a set of elements $\{x_1, \cdots, x_n\}$ of V a *base of V over A* if $V = Ax_1 + \cdots + Ax_n$.

Let V be an A-space of dimension n. We understand by a *Q-hermitian form* on V an F-bilinear mapping f of $V \times V$ into A satisfying

$$f(ax, y) = af(x, y), \qquad f(x, y)' = f(y, x)$$

for $a \in A$, $x \in V$, $y \in V$. We call f *non-degenerate* if $f(x, V) = \{0\}$ implies $x = 0$.

PROPOSITION 2.1. *Let A be a quaternion algebra over a field F and V be an A-space of dimension n. For every Q-hermitian form $f(x, y)$ on V, there exists a base $\{x_1, \cdots, x_n\}$ of V over A such that $f(x_i, x_j) = \alpha_i \delta_{ij}$ for $1 \leq i \leq n$, $1 \leq j \leq n$ with $\alpha_i \in F$. Moreover, suppose that f is non-degenerate and A satisfies the following condition:*

(D) *For every $\alpha \in F$, there exists an element a of A such that $N(a) = \alpha$. Then there exists a base $\{y_1, \cdots, y_n\}$ of V over A such that $f(y_i, y_j) = \delta_{ij}$.*

This is well-known and in fact easily proved. If $A = M_2(F)$, the condition (D) is clearly satisfied.

Let V be an A-space. We denote by $E(V, A)$ the ring of all F-linear mappings σ of V into itself satisfying $(ax)\sigma = a(x\sigma)$ for every $a \in A$, $x \in V$, and by $GL(V, A)$ the group of regular elements of $E(V, A)$. Let f be a non-degenerate Q-hermitian form on V. We denote by $G(V, f)$ the subgroup of $GL(V, A)$ consisting of the elements σ for which there exists a number $N(\sigma)$ of F such that $f(x\sigma, y\sigma) = N(\sigma)f(x, y)$ for every $x \in V$, $y \in V$; and put $G^0(V, f) = \{\sigma \in G(V, f) \mid N(\sigma) = 1\}$. $G^0(V, f)$ is clearly a normal subgroup of $G(V, f)$. If ξ is a non-zero element of F, we have $f(\xi x, \xi y) = \xi^2 f(x, y)$; so we often consider ξ as an element of $G(V, f)$. If $\dim_A V = 1$, $E(V, A)$ is isomorphic to A, and $G(V, f)$ is isomorphic to the group of regular elements of A; for every $\sigma \in G(V, f)$, $N(\sigma)$ coincides with $N(\sigma)$ of σ considered as an element of A.

Fix a base $\{x_1, \cdots, x_n\}$ of V over A. Every element σ of $E(V, A)$ is represented by a matrix (s_{ij}) of $M_n(A)$ with respect to $\{x_i\}$:

$$(1) \qquad x_i\sigma = \sum_{j=1}^{n} s_{ij}x_j \qquad (1 \leq i \leq n).$$

For every element $S = (s_{ij})$ of $M_n(A)$, we put $S' = (t_{ij})$ with $t_{ij} = s_{ji}'$. Then $S \to S'$ is an involution of $M_n(A)$. Let $f(x, y)$ be a Q-hermitian form on V. Define an element $H = (h_{ij})$ of $M_n(A)$ by $h_{ij} = f(x_i, x_j)$. Then we have $H' = H$. An element σ of $GL(V, A)$ belongs to $G(V, f)$ if and only if we have $SHS' = \alpha H$ with $\alpha \in F$ for the matrix S corresponding to σ; and then we have $N(\sigma) = \alpha$.

2.3. Elementary theory of maximal lattices. Let $\mathfrak{g}$ be a Dedekind domain and F the quotient field of $\mathfrak{g}$. Let A be a quaternion algebra over F and V an A-space of dimension n. Take a non-degenerate Q-hermitian form f on V. Let L be a $\mathfrak{g}$-lattice in V. Put $\mathfrak{o} = \{a \in A \mid aL \subset L\}$. Then $\mathfrak{o}$ is an order in A. We call $\mathfrak{o}$ *the order of* L and say that L is *normal* if $\mathfrak{o}$ is a maximal order in A. Assume that L is normal. We denote by $N_f(L)$ the two-sided $\mathfrak{o}$-ideal generated by the elements $f(x, y)$ for $x \in L$, $y \in L$, and call $N_f(L)$ the norm of L with respect to f. We denote $N_f(L)$ simply by $N(L)$ when we fix f and there is no fear of confusion.

Now, for every prime ideal $\mathfrak{p}$ of $\mathfrak{g}$, consider the $\mathfrak{p}$-completion $F_\mathfrak{p}$ and $\mathfrak{g}_\mathfrak{p}$ of F and $\mathfrak{g}$. Put $A_\mathfrak{p} = A \underset{F}{\otimes} F_\mathfrak{p}$, $V_\mathfrak{p} = V \underset{F}{\otimes} F_\mathfrak{p}$. Then $V_\mathfrak{p}$ can be considered as an $A_\mathfrak{p}$-space of dimension n in a natural manner. Further f is uniquely extended to a non-degenerate Q-hermitian form on $V_\mathfrak{p}$, which we denote again by f. The following proposition is an easy consequence of our definition.

PROPOSITION 2.2. *Let L be a $\mathfrak{g}$-lattice in V. If $\mathfrak{o}$ is the order of L, then $\mathfrak{o}_\mathfrak{p}$ $(= \mathfrak{g}_\mathfrak{p}\mathfrak{o})$ is the order of $L_\mathfrak{p}$ $(= \mathfrak{g}_\mathfrak{p}L)$. L is normal if and only if $L_\mathfrak{p}$ is normal for every prime ideal $\mathfrak{p}$ of $\mathfrak{g}$. If L is normal, we have $N(L)_\mathfrak{p} = N(L_\mathfrak{p})$.*

Let L be a normal lattice in V and $\mathfrak{o}$ the order of L. We call L *maximal* (with respect to f) if L is a maximal one among the normal lattices with the same order $\mathfrak{o}$ and the same norm $N(L)$.

PROPOSITION 2.3. *Let L be a $\mathfrak{g}$-lattice in V and σ an element of $G(V, f)$. Then $L\sigma$ is a $\mathfrak{g}$-lattice in V with the same order as L. If L is normal, so is $L\sigma$; and we have $N(L\sigma) = N(L)N(\sigma)$. Moreover, if L is maximal, so is $L\sigma$.*

This is also an easy consequence of definition. Further, by Lemma 1.1 and Proposition 2.2, we obtain

PROPOSITION 2.4. *A normal $\mathfrak{g}$-lattice in V is maximal if and only if $L_\mathfrak{p}$ is maximal for every prime ideal $\mathfrak{p}$ of $\mathfrak{g}$.*

Hereafter, we call a normal maximal $\mathfrak{g}$-lattice in V simply a *maximal lattice* in V.

PROPOSITION 2.5. *Let L be a normal $\mathfrak{g}$-lattice in V with the order $\mathfrak{o}$. Let $\mathfrak{a}$ be a right $\mathfrak{o}$-ideal and $\mathfrak{o}_1$ the left order of $\mathfrak{a}$. Then $\mathfrak{a}L$ is a normal $\mathfrak{g}$-lattice in V with the order $\mathfrak{o}_1$, and $N(\mathfrak{a}L) = \mathfrak{a}N(L)\mathfrak{a}^{-1} \cdot N(\mathfrak{a})$. Moreover, if L is maximal, so is $\mathfrak{a}L$.*

PROOF. The first assertion is clear. Let $x = \sum_i a_i x_i$ and $y = \sum_j b_j y_j$ be elements of $\mathfrak{a}L$ where $a_i, b_j \in \mathfrak{a}$ and $x_i, y_j \in L$. Then we have $f(x, y) = \sum_{i,j} a_i f(x_i, y_j) b_j' \in \mathfrak{a}N(L)\mathfrak{a}'$. As $\mathfrak{a}\mathfrak{a}' = \mathfrak{o}_1 N(\mathfrak{a})$, we get $\mathfrak{a}N(L)\mathfrak{a}' = \mathfrak{a}N(L)\mathfrak{a}^{-1} \cdot N(\mathfrak{a})$. Therefore we obtain $N(\mathfrak{a}L) \subset \mathfrak{a}N(L)\mathfrak{a}^{-1} \cdot N(\mathfrak{a})$. Substituting $\mathfrak{a}^{-1}$ and $\mathfrak{a}L$ for $\mathfrak{a}$ and L, we get the inverse inclusion, so that the equality $N(\mathfrak{a}L) = \mathfrak{a}N(L)\mathfrak{a}^{-1}N(\mathfrak{a})$ holds. The last assertion follows easily from this relation.

PROPOSITION 2.6. *Let $\{x_1, \cdots, x_n\}$ be a base of V over A such that $f(x_i, x_j)$*

$=\alpha_i\delta_{ij}$ with $\alpha_i\in F$. *Let $\mathfrak{o}$ be a maximal order in A and $\mathfrak{b}_1,\cdots,\mathfrak{b}_n$ be left $\mathfrak{o}$-ideals such that $\alpha_1 N(\mathfrak{b}_1)=\cdots=\alpha_n N(\mathfrak{b}_n)$. Then $L=\mathfrak{b}_1 x_1+\cdots+\mathfrak{b}_n x_n$ is a maximal lattice with the order $\mathfrak{o}$ and $N(L)=\alpha_1 N(\mathfrak{b}_1)\mathfrak{o}$.*

PROOF. It is clear that L is a $\mathfrak{g}$-lattice in V with the order $\mathfrak{o}$, and $N(L)$ $=\alpha_i N(\mathfrak{b}_i)\mathfrak{o}$. Let M be a $\mathfrak{g}$-lattice with the order $\mathfrak{o}$ such that $M\supset L$ and $N(M)$ $=N(L)$. Let $y=\sum_{i=1}^{n} b_i x_i$ be an element of M with $b_i\in A$. We have $b_i\alpha_i b_i'$ $=f(y,\mathfrak{b}_i x_i)\subset N(M)=N(L)=\alpha_i\mathfrak{b}_i b_i'$, so that $b_i\in\mathfrak{b}_i$. This implies $y\in L$ and hence $M=L$. Therefore L is maximal.

PROPOSITION 2.7. *Let L and M be maximal lattices in V with the same order. Let $\mathfrak{a}$ be a $\mathfrak{g}$-ideal. If $L\supset M$ and $N(M)\supset\mathfrak{a}N(L)$, then $M\supset\mathfrak{a}L$.*

PROOF. As $\mathfrak{a}N(L)\subset N(M)\subset N(L)$, $\mathfrak{a}$ is an integral ideal. Put $K=M+\mathfrak{a}L$. Then K is a $\mathfrak{g}$-lattice in V with the same order as L. We have $f(K,K)$ $\subset f(M,M)+\mathfrak{a}f(M,L)+\mathfrak{a}f(L,M)+\mathfrak{a}^2 f(L,L)\subset N(M)$, so that $N(K)=N(M)$. By the maximality of M, we must have $K=M$, and hence $\mathfrak{a}L\subset M$.

PROPOSITION 2.8. *Let M be a normal $\mathfrak{g}$-lattice in V. Then there exists a maximal lattice L, with the same order as M, such that $N(L)=N(M)$ and $L\supset M$.*

PROOF. Let $\mathfrak{o}$ be the order of M. Take a base $\{x_1,\cdots,x_n\}$ of V over A such that $x_i\in M$ for every i. Put $K=\{y\,|\,f(y,x_i)\in N(M)$ for every $i\}$. It is easy to see that K is a $\mathfrak{g}$-lattice in V with order $\mathfrak{o}$. Now, let L be a $\mathfrak{g}$-lattice in V with order $\mathfrak{o}$ such that $L\supset M$ and $N(L)=N(M)$. If $y\in L$, we have $f(y,x_i)$ $\in N(L)=N(M)$, so that $y\in K$. Hence L is contained in K. As K is a $\mathfrak{g}$-lattice, the ascending chain condition holds for the $\mathfrak{g}$-lattices contained in K. Therefore, we can find a maximal one among the lattices containing M, with order $\mathfrak{o}$ and with norm $N(M)$. This proves our proposition.

2.4. The relation between alternating form and Q-hermitian form. We now consider the case $A=M_2(F)$ for an arbitrary field F. Let e_{ij} $(i=1,2;$ $j=1,2)$ be the matrix units of A. We note that $e_{11}'=e_{22}$, $e_{12}'=-e_{12}$, $e_{21}'=-e_{21}$. Let V be an A-space of dimension n and f a Q-hermitian form on V. Put $W_i=e_{ii}V$ for $i=1,2$. Then W_i is a vector space over F of dimension n for $i=1,2$; and V is the direct sum of W_1 and W_2. If $x,y\in W_1$, we have $f(x,y)$ $=f(e_{11}x,e_{11}y)=e_{11}f(x,y)e_{22}\in Fe_{12}$. Hence we can define an F-bilinear mapping g of $W_1\times W_1$ into F by

$$(2) \qquad\qquad f(x,y)=g(x,y)e_{12}\,.$$

As $e_{12}'=-e_{12}$, we have $g(y,x)e_{12}=f(y,x)=f(x,y)'=-g(x,y)e_{12}$, so that g is an alternating form on W_1. If $\sigma\in E(V,A)$, we have $W_1\sigma\subset W_1$; so the restriction of σ to W_1 gives rise to an element of $E(W_1)$, which we denote by σ_1.

PROPOSITION 2.9. *Notation being as above, the mapping $\sigma\to\sigma_1$ gives an isomorphism of $E(V,A)$ and $GL(V,A)$ onto $E(W_1)$ and $GL(W_1)$, respectively. Moreover, suppose that f is non-degenerate. Then g is non-degenerate; and $\sigma\to\sigma_1$*

gives an isomorphism of $G(V,f)$ and $G^0(V,f)$ onto $G(W_1,g)$ and $G^0(W_1,g)$, respectively; and further we have $N(\sigma)=N(\sigma_1)$ for $\sigma \in G(V,f)$.

This can be proved in an almost straightforward way. By Proposition 2.1, there exists a base $\{x_1, \cdots, x_n\}$ of V over A such that $f(x_i, x_j)=\delta_{ij}$. Using matricial representation with respect to this base, the mapping $\sigma \to \sigma_1$ is given explicitly as follows. First note that $\{e_{11}x_i, e_{12}x_i \ (1 \le i \le n)\}$ is a base of W_1 over F. Let σ be an element of $E(V, A)$ and $S=(s_{ij})$ the element of $M_n(A)$ determined by (1) of §2.2. Put $e_{11}x_i=y_i$, $e_{12}x_i=z_i$ and $s_{ij}=\begin{pmatrix} a_{ij} & b_{ij} \\ c_{ij} & d_{ij} \end{pmatrix}$ with

$a_{ij}, b_{ij}, c_{ij}, d_{ij}$ in F. Then we have $y_i\sigma=\sum\limits_{j=1}^{n}a_{ij}y_j+\sum\limits_{j=1}^{n}b_{ij}z_j$, $z_i\sigma=\sum\limits_{j=1}^{n}c_{ij}y_j+\sum\limits_{j=1}^{n}d_{ij}z_j$.

Now define an isomorphism ι of $M_n(A)$ onto $M_{2n}(F)$ by $\iota(s_{ij})=\begin{pmatrix} (a_{ij}) & (b_{ij}) \\ (c_{ij}) & (d_{ij}) \end{pmatrix}$. Then σ_1 is represented by $\iota(S)$ with respect to the base $\{y_i, z_i\}$. As $f(x_i, x_j)=\delta_{ij}$, we get $g(y_i, y_j)=g(z_i, z_j)=0$, $g(y_i, z_j)=-\delta_{ij}$. Put $J=\begin{pmatrix} 0 & 1_n \\ -1_n & 0 \end{pmatrix}$. Then we have $\iota(S')=\begin{pmatrix} {}^t(d_{ij}) & -{}^t(b_{ij}) \\ -{}^t(c_{ij}) & {}^t(a_{ij}) \end{pmatrix}=J\cdot{}^t\iota(S)J^{-1}$. Therefore, if $SS'=\alpha 1_n$ with $\alpha \in F$, we have $\iota(S)J\cdot{}^t\iota(S)=\alpha J$. This will give a 'non-intrinsic' proof of Proposition 2.8.

2.5. Paraphrase of the result of §1. The notation $A=M_2(F)$, V, f, W_i, g being as in §2.4, suppose that F is the quotient field of a Dedekind domain $\mathfrak{g}$. For every $\mathfrak{g}$-ideal $\mathfrak{c}$, put $\mathfrak{o}(\mathfrak{c})=\mathfrak{g}e_{11}+\mathfrak{c}e_{12}+\mathfrak{c}^{-1}e_{21}+\mathfrak{g}e_{22}=\begin{pmatrix} \mathfrak{g} & \mathfrak{c} \\ \mathfrak{c}^{-1} & \mathfrak{g} \end{pmatrix}$. Then, $\mathfrak{o}(\mathfrak{c})$ is a maximal order in A; and for every maximal order $\mathfrak{o}$ in A, there exist an element a of A and a $\mathfrak{g}$-ideal $\mathfrak{c}$ such that $a\mathfrak{o}a^{-1}=\mathfrak{o}(\mathfrak{c})$. Fix a $\mathfrak{g}$-ideal $\mathfrak{c}$ and put $\mathfrak{o}=\mathfrak{o}(\mathfrak{c})$. Let L be a $\mathfrak{g}$-lattice in V with the order $\mathfrak{o}$. Put $e_{ii}L=M_i$. Then we have $M_i=L\cap W_i$, $L=M_1+M_2$; and M_i is a $\mathfrak{g}$-lattice in W_i. Further we have $M_1=\mathfrak{c}e_{12}M_2$, $M_2=\mathfrak{c}^{-1}e_{21}M_1$. Now, by Proposition 1.3, there exist $\mathfrak{g}$-ideals $\mathfrak{a}_1, \cdots, \mathfrak{a}_n$ and a base $\{y_i, z_i\}$ of W_1 over F with the properties of that proposition for $M=M_1$. Then we have

$$M_2=\mathfrak{c}^{-1}e_{21}y_1+\cdots+\mathfrak{c}^{-1}e_{21}y_n+\mathfrak{c}^{-1}\mathfrak{a}_1e_{21}z_1+\cdots+\mathfrak{c}^{-1}\mathfrak{a}_ne_{21}z_n.$$

Put $x_i=y_i-e_{21}z_i$, $\mathfrak{b}_i=\mathfrak{g}e_{11}+\mathfrak{c}\mathfrak{a}_ie_{12}+\mathfrak{c}^{-1}e_{21}+\mathfrak{a}_ie_{22}=\begin{pmatrix} \mathfrak{g} & \mathfrak{a}_i \\ \mathfrak{c}^{-1} & \mathfrak{c}^{-1}\mathfrak{a}_i \end{pmatrix}$ for $1 \le i \le n$. Then the $\mathfrak{b}_i$ are left $\mathfrak{o}$-ideals; and we see easily $L=\mathfrak{b}_1x_1+\cdots+\mathfrak{b}_nx_n$, $\mathfrak{b}_1 \supset \cdots \supset \mathfrak{b}_n$, $f(x_i, x_j)=\delta_{ij}$. Further we have $N(L)=N(\mathfrak{b}_1)\mathfrak{o}$, $N(\mathfrak{b}_1)=\mathfrak{c}^{-1}\mathfrak{a}_1=\mathfrak{c}^{-1}N_g(M_1)$. Therefore, if L is maximal, we must have $\mathfrak{b}_1=\cdots=\mathfrak{b}_n$, and hence $\mathfrak{a}_1=\cdots=\mathfrak{a}_n$, so that M_1 is maximal with respect to g. Thus we obtain

PROPOSITION 2.10. *Let F be the quotient field of a Dedekind domain $\mathfrak{g}$, and $A=M_2(F)$. Let V be an A-space of dimension n and f be a non-degenerate Q-hermitian form on V. Let L be a normal lattice in V and $\mathfrak{o}$ the order of L. Then there exist left $\mathfrak{o}$-ideals $\mathfrak{b}_1, \cdots, \mathfrak{b}_n$ and a base $\{x_1, \cdots, x_n\}$ of V over A such* *that*

$$L = \mathfrak{b}_1 x_1 + \cdots + \mathfrak{b}_n x_n , \qquad \mathfrak{b}_1 \supset \cdots \supset \mathfrak{b}_n ,$$

$$f(x_i, x_j) = \delta_{ij} ;$$

and we have $N(L) = N(\mathfrak{b}_1)\mathfrak{o}$. *Moreover, if L is maximal,* $\mathfrak{b}_1 = \cdots = \mathfrak{b}_n$.

PROPOSITION 2.11. *Notation being as in Proposition 2.10, let L_1 and L_2 be maximal lattices in V with the same order $\mathfrak{o}$. If $L_1\sigma = L_2$ for an element σ of $G(V, f)$, then $\alpha N(L_1) = N(L_2)$ for an element α of F. Conversely, if $\alpha N(L_1) = N(L_2)$ with $\alpha \in F$, we can find an element σ of $G(V, f)$ such that $L_1\sigma = L_2$, $N(\sigma) = \alpha$.*

PROOF. The first assertion is obvious. Now suppose that $N(L_2) = \alpha N(L_1)$ with $\alpha \in F$. We may assume that $\mathfrak{o} = \mathfrak{o}(\mathfrak{c})$ for a $\mathfrak{g}$-ideal $\mathfrak{c}$. Put $M_i^1 = e_{ii}L_1$, $M_i^2 = e_{ii}L_2$ for $i = 1, 2$. By the above consideration, M_1^1 and M_1^2 are maximal lattices in W_1, and $N_g(M_1^1)\alpha = N_g(M_1^2)$. By Proposition 1.4 and its proof, there exists an element σ_1 of $G(W_1, g)$ such that $M_1^1\sigma_1 = M_1^2$, $N(\sigma_1) = \alpha$. Let σ be an element of $G(V, f)$ corresponding to σ_1 by the mapping of Proposition 2.9. We get then $N(\sigma) = \alpha$, and $L_1\sigma = L_2$, since $L_1 = \mathfrak{o}M_1^1$, $L_2 = \mathfrak{o}M_1^2$. This completes our proof. We can also derive our proposition more directly from Proposition 2.10.

PROPOSITION 2.12. *Notation being as in Proposition 2.10, suppose that $\mathfrak{g}$ is a principal ideal domain. Put $\mathfrak{o} = M_2(\mathfrak{g})$. Let L and M be maximal lattices in V with the order $\mathfrak{o}$. Let η and α be elements of F such that $N(L) = \eta\mathfrak{o}$ and $N(M) = \alpha N(L)$. Then there exist a base $\{x_1, \cdots, x_n\}$ of V over A and elements $a_1, \cdots, a_n, b_1, \cdots, b_n$ of F such that*

$$f(x_i, x_j) = \eta\delta_{ij} ,$$

$$L = \mathfrak{o}x_1 + \cdots + \mathfrak{o}x_n ,$$

$$M = \mathfrak{o}e_1 x_1 + \cdots + \mathfrak{o}e_n x_n , \qquad e_i = \begin{pmatrix} a_i & 0 \\ 0 & b_i \end{pmatrix} \qquad (0 \leq i \leq n) ,$$

$$\mathfrak{g}a_1 \supset \cdots \supset \mathfrak{g}a_n \supset \mathfrak{g}b_n \supset \cdots \supset \mathfrak{g}b_1 ,$$

$$\alpha = a_1 b_1 = \cdots = a_n b_n .$$

PROOF. Put $L_1 = e_{11}L$, $M_1 = e_{11}M$. Then L_1 and M_1 are maximal lattices in W_1 with respect to g, and $N(L_1) = \mathfrak{g}\eta$, $N(M_1) = \mathfrak{g}\alpha\eta$. Applying Proposition 1.5 to this $\{L_1, M_1, \alpha\}$, we obtain $\{y_i, z_i\}$ and $\{a_i, b_i\}$ with the properties of that proposition for L_1 and M_1. Put $x_i = y_i - e_{21}\eta z_i$. Then we can easily verify that $f(x_i, x_j) = \eta\delta_{ij}$, $L = \mathfrak{o}x_1 + \cdots + \mathfrak{o}x_n$, $M = \mathfrak{o}e_1 x_1 + \cdots + \mathfrak{o}e_n x_n$ with $e_i = \begin{pmatrix} a_i & 0 \\ 0 & b_i \end{pmatrix}$. This proves our proposition.

Notation being as in Proposition 2.12, we call $\{\mathfrak{g}a_1, \cdots, \mathfrak{g}a_n, \mathfrak{g}b_1, \cdots, \mathfrak{g}b_n\}$ *the set of elementary divisors of M relative to L* and denote it by $\{L : M\}$. We get an assertion for $\{V, f\}$ which is a paraphrase of Proposition 1.6. Instead of stating it, we give the following proposition.

PROPOSITION 2.13. *Notation and assumption being as in Proposition 2.12,* *let L, M, K be maximal lattices in V with the order $\mathfrak{o}$. Then there exists an element σ of $G^0(V,f)$ such that $L\sigma = L$ and $M\sigma = K$, if and only if $\{L:M\} = \{L:K\}$.*

PROOF. The 'only if' part is clear. Put $N(L) = \mathfrak{o}\eta$, $N(M) = \mathfrak{o}\alpha\eta$, $N(K) = \mathfrak{o}\beta\eta$ with $\alpha, \beta, \eta \in F$. By Proposition 2.12, we get a base $\{x_i\}$ of V over A and a set of elements $\{a_i, b_i\}$ of F for M with the properties of that proposition, and a base $\{u_i\}$ of V over A and a set of elements $\{c_i, d_i\}$ of F with the corresponding properties for K. If $\{L:M\} = \{L:K\}$, we have $\mathfrak{g}a_i = \mathfrak{g}c_i$, $\mathfrak{g}b_i = \mathfrak{g}d_i$, so that $\mathfrak{g}\alpha = \mathfrak{g}\beta$. Hence we may put $\alpha = \beta$. We have then $a_i b_i = c_i d_i$. Let ε_i, for each i, be a unit of $\mathfrak{g}$ such that $\varepsilon_i a_i = c_i$; then we have $\varepsilon_i^{-1} b_i = d_i$. Define an element σ of $E(V, A)$ by $x_i\sigma = \left(\begin{smallmatrix} \varepsilon_i & 0 \\ 0 & \varepsilon_i^{-1} \end{smallmatrix} \right) u_i$. Then we see easily $\sigma \in G^0(V,f)$, $L\sigma = L$, $M\sigma = K$. This proves the 'if' part.

REMARK. Notation being as in Proposition 2.13, suppose that $L \supset M$, $L \supset K$. Then the following three conditions are equivalent to each other.

   i) $\{L:M\} = \{L:K\}$.

   ii) L/M and L/K are isomorphic as $\mathfrak{g}$-modules.

   iii) L/M and L/K are isomorphic as $\mathfrak{o}$-modules.

§3. Local theory of Q-hermitian forms.

3.1. Quaternion algebras over local fields.

By a $\mathfrak{p}$-*adic number field*, we understand a finite extension of the p-adic number field, for any prime number p. In this §3, $F_\mathfrak{p}$, $\mathfrak{g}_\mathfrak{p}$, $\mathfrak{p}$ denote respectively a $\mathfrak{p}$-adic number field, the ring of $\mathfrak{p}$-integers in $F_\mathfrak{p}$, the maximal ideal of $\mathfrak{g}_\mathfrak{p}$. It is well-known that there exist, up to isomorphism, only two quaternion algebras over $F_\mathfrak{p}$, the matric algebra $M_2(F_\mathfrak{p})$ and a division algebra. The latter is written as $(B, S, \pi) = B + Bu$ in the usual notation of cyclic algebra, where B is the unique unramified quadratic extension of $F_\mathfrak{p}$, S is the Frobenius automorphism of B over $F_\mathfrak{p}$, π is a prime element of $F_\mathfrak{p}$, and $u\beta u^{-1} = \beta^S$ for $\beta \in B$, $u^2 = \pi$. We denote this division quaternion algebra over $F_\mathfrak{p}$ by $D_\mathfrak{p}$.

For every maximal order $\mathfrak{o}_\mathfrak{p}$ in $M_2(F_\mathfrak{p})$, there exists an element w such that $w\mathfrak{o}_\mathfrak{p}w^{-1} = M_2(\mathfrak{g}_\mathfrak{p})$. Every one-sided $\mathfrak{o}_\mathfrak{p}$-ideal is principal. Every two-sided $\mathfrak{o}_\mathfrak{p}$-ideal $\mathfrak{a}_\mathfrak{p}$ is written in the form $\mathfrak{a}_\mathfrak{p} = \mathfrak{p}^\nu \mathfrak{o}_\mathfrak{p}$ with $\nu \in \mathbf{Z}$, and conversely. Further $\mathfrak{D}(\mathfrak{o}_\mathfrak{p}/\mathfrak{g}_\mathfrak{p}) = \mathfrak{o}_\mathfrak{p}$. As for $D_\mathfrak{p}$, it has only one maximal order $\mathfrak{o}_\mathfrak{p} = \{x \in D_\mathfrak{p} \mid N(x) \in \mathfrak{g}_\mathfrak{p}\}$; and every one-sided $\mathfrak{o}_\mathfrak{p}$-ideal is principal and equal to a power of the maximal ideal $\mathfrak{P}$, so that it is a two-sided $\mathfrak{o}_\mathfrak{p}$-ideal. We have $\mathfrak{P}^2 = \mathfrak{p}\mathfrak{o}_\mathfrak{p}$, $\mathfrak{P} = \mathfrak{D}(\mathfrak{o}_\mathfrak{p}/\mathfrak{g}_\mathfrak{p})$.

PROPOSITION 3.1. *Let $A_\mathfrak{p}$ be a quaternion algebra over $F_\mathfrak{p}$ and $\mathfrak{o}_\mathfrak{p}$ a maximal order in $A_\mathfrak{p}$. For every two-sided $\mathfrak{o}_\mathfrak{p}$-ideal $\mathfrak{a}_\mathfrak{p}$, we have $\mathrm{Tr}(\mathfrak{a}_\mathfrak{p}) = \mathfrak{a}_\mathfrak{p} \cap F_\mathfrak{p}$.*

PROOF. This is clear if $A_\mathfrak{p} = M_2(F_\mathfrak{p})$. Therefore suppose that $A_\mathfrak{p} = D_\mathfrak{p}$. Then every $\mathfrak{o}_\mathfrak{p}$-ideal $\mathfrak{a}_\mathfrak{p}$ is written in the form $\mathfrak{a}_\mathfrak{p} = \mathfrak{P}^e \cdot (\mathfrak{a}_\mathfrak{p} \cap F_\mathfrak{p})$ with $e = 0$ or

-1. We have $\mathrm{Tr}(\mathfrak{a}_\mathfrak{p}) = \mathrm{Tr}(\mathfrak{P}^e) \cdot (\mathfrak{a}_\mathfrak{p} \cap F_\mathfrak{p})$ and $\mathrm{Tr}(\mathfrak{o}_\mathfrak{p}) \subset \mathrm{Tr}(\mathfrak{P}^e) \subset \mathrm{Tr}(\mathfrak{P}^{-1}) \subset \mathfrak{g}_\mathfrak{p}$. Therefore, it is sufficient to prove $\mathrm{Tr}(\mathfrak{o}_\mathfrak{p}) = \mathfrak{g}_\mathfrak{p}$. As $\mathrm{Tr}(\mathfrak{o}_\mathfrak{p})$ is an integral $\mathfrak{g}_\mathfrak{p}$-ideal, there exists an integer $\nu \geq 0$ such that $\mathrm{Tr}(\mathfrak{o}_\mathfrak{p}) = \mathfrak{p}^\nu \mathfrak{g}_\mathfrak{p}$. We get then $\mathrm{Tr}(\mathfrak{p}^{-\nu}\mathfrak{o}_\mathfrak{p}) \subset \mathfrak{g}_\mathfrak{p}$, so that $\mathfrak{p}^{-\nu}\mathfrak{o}_\mathfrak{p} \subset \mathfrak{D}(\mathfrak{o}_\mathfrak{p}/\mathfrak{g}_\mathfrak{p})^{-1} = \mathfrak{P}^{-1}$, which implies $\nu = 0$, since $\mathfrak{P}^2 = \mathfrak{p}\mathfrak{o}_\mathfrak{p}$. This completes our proof.

PROPOSITION 3.2. *$A_\mathfrak{p}$ and $\mathfrak{o}_\mathfrak{p}$ being as in Proposition 3.1, let β be a non-zero element of $\mathfrak{g}_\mathfrak{p}$ and a an element of $\mathfrak{o}_\mathfrak{p}$ such that $\beta^{-1}N(a) \equiv 1 \bmod \mathfrak{p}^\lambda$ where λ is a positive integer. Then there exists an element b of $\mathfrak{o}_\mathfrak{p}$ such that $N(b) = \beta$, $b \equiv a \bmod \mathfrak{p}^\lambda \mathfrak{o}_\mathfrak{p}$.*

PROOF. We first consider the case $A_\mathfrak{p} = M_2(F_\mathfrak{p})$. We may then put $\mathfrak{o}_\mathfrak{p} = M_2(\mathfrak{g}_\mathfrak{p})$. Let π be a prime element of F. We can find elements x, y of $\mathfrak{o}_\mathfrak{p}$ such that $N(x) = N(y) = 1$, $xay = \begin{pmatrix} \pi^\mu \varepsilon & 0 \\ 0 & \pi^\nu \eta \end{pmatrix}$ where $0 \leq \mu \leq \nu$, and ε, η are units of $\mathfrak{g}_\mathfrak{p}$. As $\beta^{-1}N(a) \equiv 1 \bmod \mathfrak{p}^\lambda$, we have $\beta = \pi^{\mu+\nu}\delta$ with a unit δ of $\mathfrak{g}_\mathfrak{p}$, and $\delta \equiv \varepsilon\eta \bmod \mathfrak{p}^\lambda$. Put $b = x^{-1}\begin{pmatrix} \pi^\mu \varepsilon & 0 \\ 0 & \pi^\nu \varepsilon^{-1}\delta \end{pmatrix}y^{-1}$. As $\varepsilon^{-1}\delta \equiv \eta \bmod \mathfrak{p}^\lambda$, and as x^{-1}, $y^{-1} \in \mathfrak{o}_\mathfrak{p}$, we have $b \equiv a \bmod \mathfrak{p}^\lambda$, and clearly $N(b) = \pi^{\mu+\nu}\delta = \beta$. This proves our assertion for $A_\mathfrak{p} = M_2(F_\mathfrak{p})$. Now put $A_\mathfrak{p} = D_\mathfrak{p}$. Let Π be a prime element in $\mathfrak{o}_\mathfrak{p}$; put $N(\Pi) = \pi$; then π is a prime element in $\mathfrak{g}_\mathfrak{p}$. Put $a = \Pi^\nu e$ with a unit e of $\mathfrak{o}_\mathfrak{p}$. As $\beta^{-1}N(a) \equiv 1 \bmod \mathfrak{p}^\lambda$, we have $\beta = \pi^\nu \varepsilon$ with a unit ε of $\mathfrak{g}_\mathfrak{p}$, and $N(e) \equiv \varepsilon \bmod \mathfrak{p}^\lambda$. Now we construct inductively a sequence $\{e_0, e_1, \cdots, e_n, \cdots\}$ of units of $\mathfrak{o}_\mathfrak{p}$ such that $e_0 = e$, $N(e_n) \equiv \varepsilon \bmod \mathfrak{p}^{\lambda+n}$, $e_{n+1} \equiv e_n \bmod \mathfrak{p}^{\lambda+n}\mathfrak{o}_\mathfrak{p}$. Assume that e_n is already defined. Put $\varepsilon - N(e_n) = \pi^{\lambda+n} \cdot \gamma$ with $\gamma \in \mathfrak{g}_\mathfrak{p}$. By Proposition 3.1, we have $\mathrm{Tr}(e_n'\mathfrak{o}_\mathfrak{p}) = \mathrm{Tr}(\mathfrak{o}_\mathfrak{p}) = \mathfrak{g}_\mathfrak{p}$, so that there exists an element d of $\mathfrak{o}_\mathfrak{p}$ such that $\mathrm{Tr}(e_n'd) = \gamma$. Put $e_{n+1} = e_n + \pi^{\lambda+n}d$. Then we have $N(e_{n+1}) = N(e_n) + \pi^{\lambda+n}\mathrm{Tr}(e_n'd) + \pi^{2(\lambda+n)}N(d) \equiv \varepsilon \bmod \mathfrak{p}^{\lambda+n+1}$. We get thus a sequence $\{e_n\}$ with the required property. As $e_{n+1} \equiv e_n \bmod \mathfrak{p}^{\lambda+n}\mathfrak{o}_\mathfrak{p}$, this converges to a unit h of $\mathfrak{o}_\mathfrak{p}$, for which we have $N(h) = \varepsilon$, $h \equiv e \bmod \mathfrak{p}^\lambda \mathfrak{o}_\mathfrak{p}$. Put $b = \Pi^\nu h$. Then we have $N(b) = \beta$, $b \equiv a \bmod \mathfrak{p}^\lambda \mathfrak{o}_\mathfrak{p}$. This completes our proof.

PROPOSITION 3.3. *$A_\mathfrak{p}$ and $\mathfrak{o}_\mathfrak{p}$ being as in Proposition 3.1, let ξ be an element of $\mathfrak{g}_\mathfrak{p}$. Then there exists an element x of $\mathfrak{o}_\mathfrak{p}$ such that $N(x) = \xi$. In particular, $A_\mathfrak{p}$ satisfies the condition (D) of Proposition 2.1.*

PROOF. If $A_\mathfrak{p} = M_2(F_\mathfrak{p})$, we may put $\mathfrak{o}_\mathfrak{p} = M_2(\mathfrak{g}_\mathfrak{p})$, so that our assertion is obvious. If $A_\mathfrak{p} = D_\mathfrak{p}$, it is well-known that any quadratic extension of $F_\mathfrak{p}$ is isomorphic to a subfield of $D_\mathfrak{p}$, over $F_\mathfrak{p}$. Hence, for every $\xi \in F_\mathfrak{p}$, there exists an element x of $D_\mathfrak{p}$ such that $N(x) = \xi$. If $\xi \in \mathfrak{g}_\mathfrak{p}$, we have $x \in \mathfrak{o}_\mathfrak{p}$ automatically. This completes our proof.

3.2. Canonical bases of maximal lattices. Let $A_\mathfrak{p}$ be a quaternion algebra over $F_\mathfrak{p}$, and $V_\mathfrak{p}$ be an $A_\mathfrak{p}$-space of dimension n. Take a non-degenerate Q-hermitian form f on $V_\mathfrak{p}$. By Proposition 3.3 and Proposition 2.1, we see that, for every regular element $H = (h_{ij})$ of $M_n(A_\mathfrak{p})$ such that $H' = H$, there exists a base $\{x_1, \cdots, x_n\}$ of $V_\mathfrak{p}$ over $A_\mathfrak{p}$ for which $f(x_i, x_j) = h_{ij}$. In particular we get

PROPOSITION 3.4. $V_\mathfrak{p}$ *has a base* $\{x_1, \cdots, x_m, y_1, \cdots, y_m, z\}$ *over* $A_\mathfrak{p}$ *such that*

$$f(x_i, x_j) = f(y_i, y_j) = f(x_i, z) = f(y_i, z) = 0,$$

$$f(x_i, y_j) = a\delta_{ij}, \qquad f(z, z) = \beta \qquad (1 \leq i \leq m,\ 1 \leq j \leq m),$$

where a *is a regular element of* $A_\mathfrak{p}$, β *is a non-zero element of* $F_\mathfrak{p}$, *the last member* z *(and hence* β*) occurring only in the case where* n *is odd.*

We call a base $\{x_1, \cdots, x_m, y_1, \cdots, y_m, z\}$ of $V_\mathfrak{p}$ over $A_\mathfrak{p}$ with the property of the above proposition a *canonical base* of $V_\mathfrak{p}$. Proposition 3.4 implies that, if $n > 1$, $V_\mathfrak{p}$ contains an element x such that $f(x, x) = 0$, $A_\mathfrak{p} x \cong A_\mathfrak{p}$.

Now we want to study the arithmetic of maximal lattices in $V_\mathfrak{p}$. If $A_\mathfrak{p} = M_2(F_\mathfrak{p})$, we can apply the theory of § 2.5 and § 1 to the present case, since $\mathfrak{g}_\mathfrak{p}$ is a principal ideal domain; and this is sufficient for our later use. Therefore, we have only to consider the case $A_\mathfrak{p} = D_\mathfrak{p}$. From now on, until the end of this § 3.2, $V_\mathfrak{p}$ is a $D_\mathfrak{p}$-space of dimension n, and $\mathfrak{o}_\mathfrak{p}$ denotes the unique maximal order in $D_\mathfrak{p}$.

PROPOSITION 3.5. *Let* L *be a maximal lattice in* $V_\mathfrak{p}$. *Let* a *be an element of* $D_\mathfrak{p}$ *such that* $N(L) = \mathfrak{o}_\mathfrak{p} a$, *and* β *be an element of* $F_\mathfrak{p}$ *such that* $N(L) \cap F_\mathfrak{p} = \mathfrak{g}_\mathfrak{p} \beta$. *Then there exists a canonical base* $\{x_1, \cdots, x_m, y_1, \cdots, y_m, z\}$ *of* $V_\mathfrak{p}$ *such that*

(3) $$L = \mathfrak{o}_\mathfrak{p} x_1 + \mathfrak{o}_\mathfrak{p} y_1 + \cdots + \mathfrak{o}_\mathfrak{p} x_m + \mathfrak{o}_\mathfrak{p} y_m + \mathfrak{o}_\mathfrak{p} z,$$

(4) $$f(x_i, y_j) = a\delta_{ij}, \qquad f(z, z) = \beta,$$

where the term $\mathfrak{o}_\mathfrak{p} z$ *and* β *occur only when* n *is odd. Conversely, let* a *be a regular element of* $D_\mathfrak{p}$ *and* β *an element of* $F_\mathfrak{p}$ *such that* $(\mathfrak{o}_\mathfrak{p} a) \cap F_\mathfrak{p} = \mathfrak{g}_\mathfrak{p} \beta$. *Let* $\{x_i, y_i, z\}$ *be a canonical base of* $V_\mathfrak{p}$ *satisfying* (4). *Then the lattice* L *defined by* (3) *is normal, maximal and* $N(L) = \mathfrak{o}_\mathfrak{p} a$ *or* $\mathfrak{o}_\mathfrak{p} \beta$ *according as* $n > 1$ *or* $n = 1$.

PROPOSITION 3.6. *Let* L *be a* $\mathfrak{g}_\mathfrak{p}$-*lattice in* $V_\mathfrak{p}$. *Let* $\mathfrak{b}$ *be an* $\mathfrak{o}_\mathfrak{p}$-*ideal such that* $N(L) \subset \mathfrak{b}$. *Then there exists a maximal lattice* M *such that* $M \supset L$, $N(M) = \mathfrak{b}$ *or* $N(M) = \mathfrak{o}_\mathfrak{p} \cdot (\mathfrak{b} \cap F_\mathfrak{p})$ *according as* $n > 1$ *or* $n = 1$.

We first show that, if Proposition 3.5 is true for n, then Proposition 3.6 is true for n. Let the notation be as in Proposition 3.6. By Proposition 2.8, we may assume that L is maximal. Put $N(L) = \mathfrak{o}_\mathfrak{p} a$, $N(L) \cap F_\mathfrak{p} = \mathfrak{g}_\mathfrak{p} \beta$ with $a \in D_\mathfrak{p}$, $\beta \in F_\mathfrak{p}$. If $n = 1$, we put $a = \beta$. By Proposition 3.5, there exists a canonical base $\{x_i, y_i, z\}$ of $V_\mathfrak{p}$ satisfying (3) and (4). Put $\mathfrak{b} = \mathfrak{o}_\mathfrak{p} b$, $\mathfrak{b} \cap F_\mathfrak{p} = \mathfrak{g}_\mathfrak{p} \gamma$, $a = cb$, $\beta = \varepsilon \gamma$. Then $c \in \mathfrak{o}_\mathfrak{p}$, $\varepsilon \in \mathfrak{g}_\mathfrak{p}$. By Proposition 3.3, we can find an element e of $\mathfrak{o}_\mathfrak{p}$ such that $N(e) = \varepsilon$. Put $M = \sum_{i=1}^{m} \mathfrak{o}_\mathfrak{p} c^{-1} x_i + \sum_{i=1}^{m} \mathfrak{o}_\mathfrak{p} y_i + \mathfrak{o}_\mathfrak{p} e^{-1} z$. As $f(c^{-1} x_i, y_j) = b\delta_{ij}$ and $f(e^{-1} z, e^{-1} z) = \gamma$, we see, from Proposition 3.5, that M is a maximal lattice and $N(M) = \mathfrak{b}$ or $\mathfrak{o}_\mathfrak{p} \cdot (\mathfrak{b} \cap F_\mathfrak{p})$ according as $n > 1$ or $n = 1$. By our construction of M, we have $M \supset L$. This proves Proposition 3.6.

Now we want to prove the converse part of Proposition 3.5. Define L as in Proposition 3.5. It is clear that L has $\mathfrak{o}_\mathfrak{p}$ as its order and $N(L)=\mathfrak{o}_\mathfrak{p}a$ or $\mathfrak{o}_\mathfrak{p}\beta$ according as $n>1$ or $n=1$. Let M be a lattice with order $\mathfrak{o}_\mathfrak{p}$ such that $M\supset L$, $N(M)=N(L)$. Let $u=\sum_{i=1}^{m}(c_ix_i+d_iy_i)+ez$, with $c_i, d_i, e\in D_\mathfrak{p}$, be an element of M, the term ez occurring only when n is odd. As the x_i and y_i are contained in M, we have

$$d_ia' = f(u, x_i)\in N(M)=\mathfrak{o}_\mathfrak{p}a\,, \qquad ac_i=f(u, y_i)\in N(M)=\mathfrak{o}_\mathfrak{p}a\,.$$

This implies $d_i\in\mathfrak{o}_\mathfrak{p}$, $c_i\in\mathfrak{o}_\mathfrak{p}$. It follows that $ez=u-\sum_{i=1}^{m}(c_ix_i+d_iy_i)\in M$. Hence we have $N(e)\beta=f(ez, ez)\in N(M)\cap F_\mathfrak{p}=\mathfrak{g}_\mathfrak{p}\beta$, so that $N(e)\in\mathfrak{g}_\mathfrak{p}$, and hence $e\in\mathfrak{o}_\mathfrak{p}$. Therefore u must be contained in L; so we have $M=L$; this proves the maximality of L.

Let us prove the direct part of Proposition 3.5 by induction on n. If $n=1$, take a base x of $V_\mathfrak{p}$ over $D_\mathfrak{p}$. Then L is written in the form $L=\mathfrak{a}x$ with an $\mathfrak{o}_\mathfrak{p}$-ideal $\mathfrak{a}$. We have $N(L)=\mathfrak{o}_\mathfrak{p}N(\mathfrak{a})f(x, x)$, so that $N(\mathfrak{a})f(x, x)=\mathfrak{g}_\mathfrak{p}\beta$. By Proposition 3.3, there exists an element b of $D_\mathfrak{p}$ such that $N(b)=\beta f(x, x)^{-1}$. Put $z=bx$. Then $f(z, z)=\beta$, $N(b)\mathfrak{z}_\mathfrak{p}=N(\mathfrak{a})$ and hence $\mathfrak{o}_\mathfrak{p}b=\mathfrak{a}$. We have therefore $L=\mathfrak{o}_\mathfrak{p}z$. This proves the case $n=1$. Now suppose that $n>1$. By the remark after Proposition 3.4, $V_\mathfrak{p}$ contains an element $x\neq 0$ such that $f(x, x)=0$. Put $\mathfrak{c}=\{c\in D_\mathfrak{p}\,|\,cx\in L\}$. Obviously, $\mathfrak{c}$ is an $\mathfrak{o}_\mathfrak{p}$-ideal, so it is written in the form $\mathfrak{c}=\mathfrak{o}_\mathfrak{p}c_0$. Put $c_0x=x_1$. Then we see that

$$(5) \qquad\qquad\qquad \mathfrak{o}_\mathfrak{p}=\{c\in D_\mathfrak{p}\,|\,cx_1\in L\}$$

and $f(x_1, x_1)=0$. Put $\mathfrak{b}=f(x_1, L)$. It is clear that $\mathfrak{b}$ is an $\mathfrak{o}_\mathfrak{p}$-ideal. If $b\in N(L)\mathfrak{b}^{-1}$ and $u\in L$, we have $f(bx_1, u)=bf(x_1, u)\in\mathfrak{b}N(L)\mathfrak{b}^{-1}=N(L)$. Therefore, if $b, c\in N(L)\mathfrak{b}^{-1}$ and $u, v\in L$, we have

$$(6) \qquad\qquad f(bx_1+u, cx_1+v)=f(bx_1, v)+f(u, cx_1)+f(u, v)\in N(L)\,.$$

Put $M=N(L)\mathfrak{b}^{-1}x_1+L$. The relation (6) shows that $N(M)=N(L)$. As L is maximal, we must have $L=M$, so that $N(L)\mathfrak{b}^{-1}x_1\subset L$. By (5), we have $N(L)\mathfrak{b}^{-1}\subset\mathfrak{o}_\mathfrak{p}$, so that $N(L)\subset\mathfrak{b}$. As $\mathfrak{b}=f(x_1, L)\subset N(L)$, we must have $\mathfrak{b}=N(L)=\mathfrak{o}_\mathfrak{p}a$. Hence there exists an element y of L such that $f(x_1, y)=a$. By Proposition 3.1, we have $\mathrm{Tr}(\mathfrak{o}_\mathfrak{p}a)=N(L)\cap F_\mathfrak{p}\ni -f(y, y)$. Therefore we can find an element t of $\mathfrak{o}_\mathfrak{p}$ such that $\mathrm{Tr}(ta)=-f(y, y)$. Put $y_1=tx_1+y$. Then $y_1\in L$, and we have $f(x_1, y)=a$, $f(y_1, y_1)=0$. Put

$$U=\{u\in V_\mathfrak{p}\,|\,f(x_1, u)=f(y_1, u)=0\}\,,$$

$$K=U\cap L\,.$$

For every $w\in V$, if we put $a^{-1}f(x_1, w)=d$ and $f(w, y_1)a^{-1}=c$, we see easily $w-cx_1-d'y_1\in U$. This implies $V_\mathfrak{p}=D_\mathfrak{p}x_1+D_\mathfrak{p}y_1+U$. If $w\in L$, then $f(x_1, w)$ and $f(w, y_1)$ are contained in $N(L)=\mathfrak{o}_\mathfrak{p}a$, so that c and d are contained in $\mathfrak{o}_\mathfrak{p}$.

We have therefore, $L = \mathfrak{o}_\mathfrak{p} x_1 + \mathfrak{o}_\mathfrak{p} y_1 + K$. Obviously, K is a lattice in U with the order $\mathfrak{o}_\mathfrak{p}$, and $N(K) \subset N(L)$. As L is maximal, K must be maximal; so we can apply our induction to K. By the assumption of induction, Proposition 3.6 is true for $n-2$. Therefore, if $n > 3$, we must have $N(K) = N(L) = \mathfrak{o}_\mathfrak{p} a$, while if $n = 3$, $N(K) = \mathfrak{o}_\mathfrak{p} \beta$. This completes our proof.

We call the base $\{x_i, y_i, z\}$ in the above proposition a *canonical base* of L.

PROPOSITION 3.7. *Let L and M be maximal lattices in $V_\mathfrak{p}$. If $L\sigma = M$ for an element σ of $G(V_\mathfrak{p}, f)$, then $N(L)^{-1}N(M)$ is an even power of the maximal ideal of $\mathfrak{o}_\mathfrak{p}$, namely $N(L)^{-1}N(M) = \mathfrak{o}_\mathfrak{p}\alpha$ for an element α of $F_\mathfrak{p}$. Conversely, if $\alpha N(L) = N(M)$ with $\alpha \in F$, there exists an element σ of $G(V_\mathfrak{p}, f)$ such that $L\sigma = M$, $N(\sigma) = \alpha$.*

PROOF. If $L\sigma = M$ for some $\sigma \in G(V_\mathfrak{p}, f)$, we have $N(L)N(\sigma) = N(M)$, so that $N(L)^{-1}N(M) = N(\sigma)\mathfrak{o}_\mathfrak{p}$. This proves the first assertion. Conversely, suppose that $\alpha N(L) = N(M)$ with $\alpha \in F_\mathfrak{p}$. Put $N(L) = \mathfrak{o}_\mathfrak{p} a$, $N(L) \cap F_\mathfrak{p} = \mathfrak{g}_\mathfrak{p}\beta$ with $a \in D_\mathfrak{p}$, $\beta \in F_\mathfrak{p}$. Then we have $N(M) = \mathfrak{o}_\mathfrak{p}\alpha a$, $N(M) \cap F_\mathfrak{p} = \mathfrak{g}_\mathfrak{p}\alpha\beta$. By Proposition 3.5, there exists a canonical base $\{x_i, y_i, z\}$ of L such that $f(x_i, y_j) = a\delta_{ij}$, $f(z, z) = \beta$, and a canonical base $\{u_i, v_i, w\}$ of M such that $f(u_i, v_j) = \alpha a\delta_{ij}$, $f(w, w) = \alpha\beta$. Define an element σ of $E(V_\mathfrak{p}, D_\mathfrak{p})$ by $x_i\sigma = u_i$, $y_i\sigma = v_i$, $z\sigma = w$. Then we see easily that $\sigma \in G(V_\mathfrak{p}, f)$, $L\sigma = M$ and $N(\sigma) = \alpha$. This completes our proof.

PROPOSITION 3.8. *Let L be a maximal lattice in $V_\mathfrak{p}$. If $n > 1$, there exists a base $\{u_1, \cdots, u_n\}$ of $V_\mathfrak{p}$ over $D_\mathfrak{p}$ such that $L = \mathfrak{o}_\mathfrak{p} u_1 + \cdots + \mathfrak{o}_\mathfrak{p} u_n$, $f(u_i, u_i) = 0$ for $1 \leq i \leq n$.*

PROOF. Take a canonical base $\{x_i, y_i, z\}$ of L. If n is even, our assertion is a consequence of the relation $f(x_i, x_i) = f(y_i, y_i) = 0$. Suppose that n is odd. The elements a and β being as in Proposition 3.5, we get, by Proposition 3.1, $\beta \in F_\mathfrak{p} \cap \mathfrak{o}_\mathfrak{p} a = \mathrm{Tr}(\mathfrak{o}_\mathfrak{p} a)$. Hence there exists an element b of $\mathfrak{o}_\mathfrak{p}$ such that $\mathrm{Tr}(ba) = \beta$. Put $w = z + bx_1 - y_1$. Then we have $f(w, w) = 0$ and $L = \mathfrak{o}_\mathfrak{p} x_1 + \mathfrak{o}_\mathfrak{p} y_1 + \cdots + \mathfrak{o}_\mathfrak{p} x_m + \mathfrak{o}_\mathfrak{p} y_m + \mathfrak{o}_\mathfrak{p} w$. This proves our proposition.

PROPOSITION 3.9. *Let L and M be maximal lattices in $V_\mathfrak{p}$. Put $N(L) = h\mathfrak{o}_\mathfrak{p}$, $N(L) \cap F_\mathfrak{p} = \eta\mathfrak{g}_\mathfrak{p}$ with $h \in D_\mathfrak{p}$, $\eta \in F_\mathfrak{p}$, and suppose that $N(M) = \alpha N(L)$ for an element α of $F_\mathfrak{p}$. Then there exist a canonical base $\{x_i, y_i, z\}$ of $V_\mathfrak{p}$ and elements a_i, b_i, c of $D_\mathfrak{p}$ such that*

$$L = \mathfrak{o}_\mathfrak{p} x_1 + \mathfrak{o}_\mathfrak{p} y_1 + \cdots + \mathfrak{o}_\mathfrak{p} x_m + \mathfrak{o}_\mathfrak{p} y_m + \mathfrak{o}_\mathfrak{p} z \,,$$

$$M = \mathfrak{o}_\mathfrak{p} a_1 x_1 + \mathfrak{o}_\mathfrak{p} b_1 y_1 + \cdots + \mathfrak{o}_\mathfrak{p} a_m x_m + \mathfrak{o}_\mathfrak{p} b_m y_m + \mathfrak{o}_\mathfrak{p} c z \,,$$

$$f(x_i, y_j) = h\delta_{ij}\,, \qquad f(z, z) = \eta \,,$$

$$a_1 h b_1' = \cdots = a_m h b_m' = \alpha h\,, \qquad cc' = \alpha \,,$$

$$\mathfrak{o}_\mathfrak{p} a_1 \supset \cdots \supset \mathfrak{o}_\mathfrak{p} a_m \supset \mathfrak{o}_\mathfrak{p} c \supset \mathfrak{o}_\mathfrak{p} b_m \supset \cdots \supset \mathfrak{o}_\mathfrak{p} b_1 \,,$$

where z and c occur only when n is odd.

PROOF. We proceed by induction on n. If $n=1$, this is obvious. Suppose that $n>1$. Put $\mathfrak{e}=\{e\in D_\mathfrak{p}\mid eM\subset L\}$. As $\mathfrak{e}$ is an $\mathfrak{o}_\mathfrak{p}$-ideal, we have $\mathfrak{e}=\mathfrak{o}_\mathfrak{p}e_0$ for an element $e_0\in\mathfrak{o}_\mathfrak{p}$. Put $e_0M=M_1$; then $N(M_1)=N(e_0)N(L)$. If we prove our proposition for M_1, we get easily the assertion for M. In fact, suppose that we get a canonical base $\{x_i,y_i,z\}$ of $V_\mathfrak{p}$ and elements r_i, s_i, t of $D_\mathfrak{p}$ such that

$$L=\sum_{i=1}^{m}(\mathfrak{o}_\mathfrak{p}x_i+\mathfrak{o}_\mathfrak{p}y_i)+\mathfrak{o}_\mathfrak{p}z,\quad M_1=\sum_{i=1}^{m}(\mathfrak{o}_\mathfrak{p}r_ix_i+\mathfrak{o}_\mathfrak{p}s_iy_i)+\mathfrak{o}_\mathfrak{p}tz,\quad f(x_i,y_i)=h,\quad f(z,z)=\eta,\quad r_ihs'_i$$
$$=N(e_0)\alpha h,\quad tt'=N(e_0)\alpha,\quad \mathfrak{o}_\mathfrak{p}r_1\supset\cdots\supset\mathfrak{o}_\mathfrak{p}r_m\supset\mathfrak{o}_\mathfrak{p}t\supset\mathfrak{o}_\mathfrak{p}s_m\supset\cdots\supset\mathfrak{o}_\mathfrak{p}s_1.$$

Put $a_i=e_0^{-1}r_i$, $b_i=h^{-1}e_0^{-1}hs_i$, $c=e_0^{-1}t$. Then we can easily verify that $\{x_i,y_i,z\}$ and $\{a_i,b_i,c\}$ have the properties of our proposition for M and L. Therefore we may assume that $M=M_1$, namely $\mathfrak{o}_\mathfrak{p}=\{e\in D_\mathfrak{p}\mid eM\subset L\}$. Let Π be a prime element of $\mathfrak{o}_\mathfrak{p}$. By Proposition 3.8, M contains an element x_1 such that $f(x_1,x_1)=0$, $\Pi^{-1}x_1\notin L$. Namely, the relation (5) holds for this $\{x_1,L\}$. Hence, applying the proof of Proposition 3.5 to the present case, we get an element y_1 of L such that $f(x_1,y_1)=h$, $f(y_1,y_1)=0$. By Proposition 2.7, we have $\alpha L\subset M$, so that $\alpha y_1\in M$. Put $U=\{u\in V_\mathfrak{p}\mid f(x_1,u)=f(y_1,u)=0\}$, $L_0=U\cap L$, $M_0=U\cap M$. Then, as in the proof of Proposition 3.5, we obtain

$$V_\mathfrak{p}=D_\mathfrak{p}x_1+D_\mathfrak{p}y_1+U,\qquad L=\mathfrak{o}_\mathfrak{p}x_1+\mathfrak{o}_\mathfrak{p}y_1+L_0;$$

and L_0 is a maximal lattice in U such that $N(L_0)=h\mathfrak{o}_\mathfrak{p}$ or $\eta\mathfrak{o}_\mathfrak{p}$ according as $n>1$ or $n=1$. Let $w=dx_1+ey_1+w_0$, with $d\in\mathfrak{o}_\mathfrak{p}$, $e\in\mathfrak{o}_\mathfrak{p}$, $w_0\in L_0$, be an element of M. Then we have $e=f(w,x_1)\in N(M)=\mathfrak{o}_\mathfrak{p}\alpha h$, so that $e\in\mathfrak{o}_\mathfrak{p}\alpha$, $ey_1\in\mathfrak{o}_\mathfrak{p}\alpha y_1\subset M$, and hence $w_0=w-dx_1-ey_1\in M\cap U=M_0$. This implies $M=\mathfrak{o}_\mathfrak{p}x_1+\mathfrak{o}_\mathfrak{p}\alpha y_1+M_0$. We observe that M_0 is a maximal lattice in U such that $N(M_0)=\alpha N(L_0)$. Therefore we can apply our induction to L_0 and M_0. Then we obtain a canonical base $\{x_2,\cdots,x_m,y_2,\cdots,y_m,z\}$ of U and elements $a_2,\cdots,a_m,b_2,\cdots,b_m,c$ of $D_\mathfrak{p}$ such that $L_0=\sum_{i=2}^{m}(\mathfrak{o}_\mathfrak{p}x_i+\mathfrak{o}_\mathfrak{p}y_i)+\mathfrak{o}_\mathfrak{p}z$, $M_0=\sum_{i=2}^{m}(\mathfrak{o}_\mathfrak{p}a_ix_i+\mathfrak{o}_\mathfrak{p}b_iy_i)+\mathfrak{o}_\mathfrak{p}cz$, $f(x_i,y_i)=h$ for $2\leqq i\leqq m$, $f(z,z)=\eta$, $a_2hb'_2=\cdots=a_mhb'_m=\alpha h$, $cc'=\alpha$, $\mathfrak{o}_\mathfrak{p}a_2\supset\cdots\supset\mathfrak{o}_\mathfrak{p}a_m\supset\mathfrak{o}_\mathfrak{p}c\supset\mathfrak{o}_\mathfrak{p}b_m\supset\cdots\supset\mathfrak{o}_\mathfrak{p}b_2$. As $L_0\supset M_0$, we have $\mathfrak{o}_\mathfrak{p}\ni a_2$, so that $\mathfrak{o}_\mathfrak{p}b_2\supset\mathfrak{o}_\mathfrak{p}\alpha$. Putting $a_1=1$ and $b_1=\alpha$, we obtain our assertion for L and M. This completes the proof.

PROPOSITION 3.10. *Let L be a maximal lattice in $V_\mathfrak{p}$. Put $N(L)=h\mathfrak{o}_\mathfrak{p}$, $N(L)\cap F_\mathfrak{p}=\eta\mathfrak{g}_\mathfrak{p}$ with $h\in D_\mathfrak{p}$, $\eta\in F_\mathfrak{p}$. Let $\{u_i,v_i,w\}$ be a canonical base of L such that $f(u_i,v_j)=h\delta_{ij}$, $f(w,w)=\eta$. Denote by Γ^0 the subgroup of $G^0(V_\mathfrak{p},f)$ consisting of the elements $\gamma\in G^0(V_\mathfrak{p},f)$ such that $L\gamma=L$, and by Δ the set of elements σ of $G(V_\mathfrak{p},f)$ such that $u_i\sigma=a_iu_i$, $v_i\sigma=b_iv_i$, $w\sigma=cw$ with elements a_i, b_i, c of $D_\mathfrak{p}$ satisfying the relation*

$$\mathfrak{o}_\mathfrak{p}a_1\supset\cdots\supset\mathfrak{o}_\mathfrak{p}a_m\supset\mathfrak{o}_\mathfrak{p}c\supset\mathfrak{o}_\mathfrak{p}b_m\supset\cdots\supset\mathfrak{o}_\mathfrak{p}b_1.$$

Then we have $G(V_\mathfrak{p},f)=\Gamma^0\cdot\Delta\cdot\Gamma^0$.

PROOF. Let τ be an element of $G(V_\mathfrak{p},f)$. Put $M=L\tau$, $\alpha=N(\tau)$, and apply Proposition 3.9 to this $\{L,M,\alpha\}$. Then we get a canonical base $\{x_i,y_i,z\}$ of

L and elements a_i, b_i, c of $D_\mathfrak{p}$ with the properties of that proposition. Define two elements γ and σ of $E(V_\mathfrak{p}, A_\mathfrak{p})$ by $u_i\gamma = x_i$, $v_i\gamma = y_i$, $w\gamma = z$, $u_i\sigma = a_iu_i$, $v_i\sigma = b_iv_i$, $w\sigma = cw$. We see easily that $\gamma \in \Gamma^0$ and $\sigma \in \Delta$, $N(\sigma) = \alpha$. Further we have $L\sigma\gamma = L\tau$. Hence if we put $\varepsilon\sigma\gamma = \tau$, we have $L\varepsilon = L$, $\varepsilon \in G(V_\mathfrak{p}, f)$, $N(\varepsilon) = 1$, so that $\varepsilon \in \Gamma^0$. It follows that $\tau = \varepsilon\sigma\gamma \in \Gamma^0 \cdot \Delta \cdot \Gamma^0$. Our proposition is thereby proved.

Notation being as in Proposition 3.9, we call $\{\mathfrak{o}_\mathfrak{p}a_1, \cdots, \mathfrak{o}_\mathfrak{p}a_m, \mathfrak{o}_\mathfrak{p}c, \mathfrak{o}_\mathfrak{p}b_1, \cdots, \mathfrak{o}_\mathfrak{p}b_m\}$ *the set of elementary divisors of M relative to L* and denote it by $\{L : M\}$.

PROPOSITION 3.11. *Let L, M, K be maximal lattices in $V_\mathfrak{p}$ such that $N(M) = \alpha N(L)$, $N(K) = \beta N(L)$ with α, $\beta \in F_\mathfrak{p}$. Then, there exists an element σ of $G^0(V_\mathfrak{p}, f)$ such that $L\sigma = L$ and $M\sigma = K$, if and only if $\{L : M\} = \{L : K\}$.*

By virtue of Proposition 3.9, this can be proved by the same argument as in the proof of Proposition 2.13. When $L \supset M$ and $L \supset K$, the equality $\{L : M\} = \{L : K\}$ holds if and only if L/M and L/K are isomorphic as $\mathfrak{o}_\mathfrak{p}$-modules.

3.3. Local approximation theorem. Let $A_\mathfrak{p}$ be a quaternion algebra over $F_\mathfrak{p}$, which may be or may not be a division algebra. Let $\mathfrak{o}_\mathfrak{p}$ be a maximal order in $A_\mathfrak{p}$.

PROPOSITION 3.12. *Let $V_\mathfrak{p}$ be an $A_\mathfrak{p}$-space of dimension n and f be a non-degenerate Q-hermitian form on $V_\mathfrak{p}$. Let L be a maximal lattice in $V_\mathfrak{p}$ such that $N(L) = \mathfrak{o}_\mathfrak{p}$. Then there exists a base $\{x_1, \cdots, x_n\}$ of $V_\mathfrak{p}$ over $A_\mathfrak{p}$ such that $f(x_i, x_j) = \delta_{ij}$ and $L = \mathfrak{o}_\mathfrak{p}x_1 + \cdots + \mathfrak{o}_\mathfrak{p}x_n$.*

PROOF. By Proposition 2.1 and Proposition 3.3, $V_\mathfrak{p}$ has a base $\{y_1, \cdots, y_n\}$ over $A_\mathfrak{p}$ such that $f(y_i, y_j) = \delta_{ij}$. Put $M = \mathfrak{o}_\mathfrak{p}y_1 + \cdots + \mathfrak{o}_\mathfrak{p}y_n$. By Proposition 2.6, M is a maximal lattice in $V_\mathfrak{p}$ and $N(M) = \mathfrak{o}_\mathfrak{p}$. By Proposition 2.11 (if $A_\mathfrak{p} = M_2(F_\mathfrak{p})$) and by Proposition 3.7 (if $A_\mathfrak{p} = D_\mathfrak{p}$), there exists an element σ of $G^0(V_\mathfrak{p}, f)$ such that $L = M\sigma$. Putting $x_i = y_i\sigma$ for $1 \leq i \leq n$, we get the desired result.

PROPOSITION 3.13. *Let $V_\mathfrak{p}$ and $U_\mathfrak{p}$ be $A_\mathfrak{p}$-spaces of the same dimension; let f and h be non-degenerate Q-hermitian forms on $V_\mathfrak{p}$ and on $U_\mathfrak{p}$, respectively. Let L and M be maximal lattices in $V_\mathfrak{p}$ and in $U_\mathfrak{p}$, respectively, such that $N_f(L) = N_h(M) = \mathfrak{o}_\mathfrak{p}$. Let τ be an $A_\mathfrak{p}$-linear mapping of $V_\mathfrak{p}$ into $U_\mathfrak{p}$ such that $L\tau \subset M$, $f(x, y) \equiv h(x\tau, y\tau) \bmod \mathfrak{p}^\lambda\mathfrak{o}_\mathfrak{p}$ for every x, $y \in L$, where λ is an integer ≥ 0. Then there exists an $A_\mathfrak{p}$-isomorphism σ of $V_\mathfrak{p}$ onto $U_\mathfrak{p}$ such that $L\sigma = M$, $f(x, y) = h(x\sigma, y\sigma)$ for every x, $y \in V_\mathfrak{p}$ and $L(\sigma - \tau) \subset \mathfrak{p}^\lambda M$.*

PROOF. Our proposition is clear if $\lambda = 0$; so we assume $\lambda \geq 1$. Let n be the common dimension of $V_\mathfrak{p}$ and $U_\mathfrak{p}$. We proceed by induction on n. By Proposition 3.12, there exists a base $\{u_1, \cdots, u_n\}$ of $U_\mathfrak{p}$ over $A_\mathfrak{p}$ such that $M = \mathfrak{o}_\mathfrak{p}u_1 + \cdots + \mathfrak{o}_\mathfrak{p}u_n$, $h(u_i, u_j) = \delta_{ij}$; and L contains an element v such that $f(v, v) = 1$. Put $v\tau = \sum_{i=1}^n a_iu_i$ with $a_i \in \mathfrak{o}_\mathfrak{p}$. Then $1 = f(v, v) \equiv h(v\tau, v\tau) = \sum_{i=1}^n N(a_i) \bmod \mathfrak{p}^\lambda\mathfrak{o}_\mathfrak{p}$.

Therefore $N(a_i)$ is a unit of $\mathfrak{g}_\mathfrak{p}$ for some i, say 1. Put $\beta = 1 - \sum_{i=2}^{n} N(a_i)$. Then $N(a_1) \equiv \beta \bmod \mathfrak{p}^\lambda \mathfrak{o}_\mathfrak{p}$, and hence β is a unit of $\mathfrak{g}_\mathfrak{p}$. By Proposition 3.2, there exists an element b of $\mathfrak{o}_\mathfrak{p}$ such that $b \equiv a_1 \bmod \mathfrak{p}^\lambda \mathfrak{o}_\mathfrak{p}$ and $N(b) = \beta$. Put $w = bu_1 + \sum_{i=2}^{n} a_i u_i$. Then $h(w, w) = 1$, $w \equiv v\tau \bmod \mathfrak{p}^\lambda M$, and $w \in M$. Put

$$V^0 = \{x \in V_\mathfrak{p} \mid f(x, v) = 0\}, \qquad U^0 = \{x \in U_\mathfrak{p} \mid h(x, w) = 0\},$$

$$L^0 = L \cap V^0, \qquad M^0 = M \cap U^0.$$

As $f(v, v) = h(w, w) = 1$, we obtain

$$V_\mathfrak{p} = A_\mathfrak{p} v + V^0, \qquad U_\mathfrak{p} = A_\mathfrak{p} w + V^0,$$

$$L = \mathfrak{o}_\mathfrak{p} v + L^0, \qquad M = \mathfrak{o}_\mathfrak{p} w + M^0.$$

It can be easily seen that L and M are respectively maximal lattices in V^0 and U^0; and $N_f(L^0) = N_h(M^0) = \mathfrak{o}_\mathfrak{p}$. Now define an $A_\mathfrak{p}$-linear mapping ρ of V^0 into U^0 by $x\tau = tw + x\rho$ for $x \in V^0$, where $t \in A_\mathfrak{p}$. We see easily $L^0\rho \subset M^0$. If $x \in L^0$ and $x\tau = tw + x\rho$, we have $0 = f(v, x) \equiv h(v\tau, x\tau) \equiv h(w, tw + x\rho) = t \bmod \mathfrak{p}^\lambda \mathfrak{o}_\mathfrak{p}$. This shows $x\tau \equiv x\rho \bmod \mathfrak{p}^\lambda M$ for $x \in L^0$. If further $y \in L^0$, we have $f(x, y) \equiv h(x\tau, y\tau) \equiv h(x\rho, y\rho) \bmod \mathfrak{p}^\lambda \mathfrak{o}_\mathfrak{p}$. Therefore we can apply induction to L^0, M^0, ρ. Namely there exists an $A_\mathfrak{p}$-isomorphism σ^0 of V^0 onto U^0 such that $L^0\sigma^0 = M^0$, $f(x, y) = h(x\sigma^0, y\sigma^0)$ for every $x, y \in V^0$, and $L^0(\sigma^0 - \rho) \subset \mathfrak{p}^\lambda M^0$. Now define an $A_\mathfrak{p}$-isomorphism σ of $V_\mathfrak{p}$ onto $U_\mathfrak{p}$ by $v\sigma = w$ and $x\sigma = x\sigma^0$ for every $x \in V^0$. Then we have clearly $L\sigma = M$, $f(x, y) = h(x\sigma, y\sigma)$ for every $x, y \in V_\mathfrak{p}$. Furthermore, $v\sigma = w \equiv v\tau \bmod \mathfrak{p}^\lambda M$; and if $x \in L^0$, $x\sigma = x\sigma^0 \equiv x\rho \equiv x\tau \bmod \mathfrak{p}^\lambda M$. Therefore $L(\sigma - \tau) \subset \mathfrak{p}^\lambda M$. This completes our proof.

§4. Global theory of Q-hermitian forms.

In this section, we always mean by F an algebraic number field of finite degree, and by $\mathfrak{g}$ the ring of integers in F. For every prime ideal $\mathfrak{p}$ of F, $F_\mathfrak{p}$ and $\mathfrak{g}_\mathfrak{p}$ denote respectively the $\mathfrak{p}$-completions of F and $\mathfrak{g}$. We denote by $\mathfrak{p}_{\infty_\kappa}$ for $1 \leqq \kappa \leqq v$ the infinite prime spots of F and by F_κ the completion of F with respect to $\mathfrak{p}_{\infty_\kappa}$.

4.1. Quaternion algebras over an algebraic number field. Let A be a quaternion algebra over F. For each prime ideal $\mathfrak{p}$ of F, and for each infinite prime spot $\mathfrak{p}_{\infty_\kappa}$ of F, we put

$$A_\mathfrak{p} = A \underset{F}{\otimes} F_\mathfrak{p}, \qquad A_\kappa = A \underset{F}{\otimes} F_\kappa.$$

A finite or infinite prime spot of F is called *ramified* in A/F if the corresponding completion $A_\mathfrak{p}$ or A_κ is a division algebra. Let $\mathfrak{o}$ be a maximal order in A. Let $\mathfrak{D} = \mathfrak{D}(\mathfrak{o}/\mathfrak{g})$ be the different of $\mathfrak{o}$ with respect to $\mathfrak{g}$. Then we have

$\mathfrak{D} = \prod_{i=1}^{s} \mathfrak{O}_i$, $\mathfrak{O}_i^2 = \mathfrak{q}_i$, where the $\mathfrak{q}_i$ are all the prime ideals of F which are ramified in A/F, and $\mathfrak{O}_i$ is a prime $\mathfrak{o}$-ideal. Every two-sided $\mathfrak{o}$-ideal $\mathfrak{a}$ is written in the form $\mathfrak{a} = \prod \mathfrak{O}_i^{e_i} \cdot \mathfrak{a}_0$, where $e_i = 0$ or 1, and $\mathfrak{a}_0$ is a $\mathfrak{g}$-ideal.

PROPOSITION 4.1. *Let* $\mathfrak{p}_{\infty 1}, \cdots, \mathfrak{p}_{\infty u}$ *be the infinite prime spots of* F *ramified in* A/F, *and* ξ *be a non-zero element of* F. *Then there exists an element* x *of* A *such that* $N(x) = \xi$, *if and only if* $\xi \equiv 1 \bmod \mathfrak{p}_{\infty 1} \cdots \mathfrak{p}_{\infty u}$.

PROOF. Consider $N(x) = xx'$ as a quadratic form on A over F. By Hasse's theorem, the equation $xx' = \xi$ has a solution if and only if it is solvable in every local fields. Our proposition is therefore an immediate consequence of Proposition 3.3.

We call A *definite* (or *totally definite*) if all the infinite prime spots of F are ramified in A/F, and call A *indefinite* otherwise. If A is definite, then F must be totally real and $A_\kappa = K$ for every infinite prime spot $\mathfrak{p}_{\infty \kappa}$ of F. Now the following two fundamental lemmas are due to Eichler; they are originally given in a more general case (cf. [5, Satz 5]).

LEMMA 4.2. *Suppose that* A *is indefinite. Let* $\mathfrak{o}$ *be a maximal order in* A *and let* $\mathfrak{p}_{\infty 1}, \cdots, \mathfrak{p}_{\infty u}$ *be the infinite prime spots ramified in* A/F. *Let* $\mathfrak{b}$ *and* $\mathfrak{c}$ *be left* $\mathfrak{o}$*-ideals. Then there exists an element* x *of* A *such that* $\mathfrak{b} = \mathfrak{c}x$, *if and only if* $N(\mathfrak{b})$ *and* $N(\mathfrak{c})$ *belong to the same ideal-class modulo* $\mathfrak{p}_{\infty 1} \cdots \mathfrak{p}_{\infty u}$ *of* F.

LEMMA 4.3. *Notation and assumption being as in Lemma 4.2, let* $\mathfrak{a}$ *be an integral two-sided* $\mathfrak{o}$*-ideal. Let* β *be an element of* $\mathfrak{g}$ *and* b *an element of* $\mathfrak{o}$ *such that* $\beta \equiv 1 \bmod \mathfrak{p}_{\infty 1} \cdots \mathfrak{p}_{\infty u}$, $N(b) \equiv \beta \bmod^* (\mathfrak{a} \cap F)$. *Then there exists an element* b_0 *of* $\mathfrak{o}$ *such that* $b \equiv b_0 \bmod \mathfrak{a}$, $N(b_0) = \beta$.

Here mod* means the multiplicative congruence. Lemma 4.2 is easily derived from Lemma 4.3 (cf. [5, p. 239]). Our later discussion will prove this fact as a particular case.

4.2. Hasse principle for Q-hermitian forms.

In view of Proposition 3.3, there exists, among the quaternion algebras over local fields $F_\mathfrak{p}$ and F_κ, only one which does not satisfy the condition (D) of Proposition 2.1; it is the division ring K of real quaternions. Let V be a K-space of dimension n and f a non-degenerate Q-hermitian form on V. Then there exists a base $\{x_1, \cdots, x_n\}$ of V over K such that $f(x_i, x_j) = \varepsilon_i \delta_{ij}$ for $1 \leq i \leq n$, $1 \leq j \leq n$ and $\varepsilon_i = 1$ for $1 \leq i \leq \nu$, $\varepsilon_i = -1$ for $\nu < i \leq n$. The integer ν is uniquely determined by f. We put $\nu = \nu(f)$.

Let A be a quaternion algebra over F and let $\mathfrak{p}_{\infty 1}, \cdots, \mathfrak{p}_{\infty u}$ be all the infinite prime spots of F ramified in A/F. Consider an A-space V and a non-degenerate Q-hermitian form f on V. Put $V_\kappa = V \otimes_F F_\kappa$ for $1 \leq \kappa \leq u$. Then V_κ can be considered as an A_κ-space in a natural manner; and f is uniquely extended to a non-degenerate Q-hermitian form f_κ on V_κ. As A_κ is isomorphic to K, we can define $\nu(f_\kappa)$. We put $\nu_\kappa(f) = \nu(f_\kappa)$. Now, by Ramanathan [7], the structure of $\{V, f\}$ is completely determined by the $\nu_\kappa(f)$. We state this result in the following form.

LEMMA 4.4. *Let f and g be non-degenerate Q-hermitian forms on an A-space V. There exists an element σ of $GL(V, A)$ such that $f(x\sigma, y\sigma) = g(x, y)$ for every x, y of V, if and only if $\nu_\kappa(f) = \nu_\kappa(g)$ for every infinite prime spot $\mathfrak{p}_{\infty\kappa}$ of F ramified in A/F.*

This can be proved easily by means of Proposition 4.1 and the approximation theorem in the number field F.

4.3. Adele-group of $G(V, f)$. Let A be a quaternion algebra over F and V an A-space of dimension n. For each prime ideal $\mathfrak{p}$ of F and for each infinite prime spot $\mathfrak{p}_{\infty\kappa}$ of F, we put

$$V_\mathfrak{p} = V \underset{F}{\otimes} F_\mathfrak{p}, \qquad V_\kappa = V \underset{F}{\otimes} F_\kappa.$$

Then $V_\mathfrak{p}$ (resp. V_κ) can be considered as an $A_\mathfrak{p}$-space (resp. A_κ-space) in a natural manner. Let f be a non-degenerate Q-hermitian form on V. We extend f to non-degenerate Q-hermitian forms on $V_\mathfrak{p}$ and on V_κ, and denote them again by f. Put now $G = G(V, f)$, $G_\mathfrak{p} = G(V_\mathfrak{p}, f)$, $G_\kappa = G(V_\kappa, f)$. Then $G_\mathfrak{p}$, G_κ are locally compact topological groups with usual topology. Let L be a $\mathfrak{g}$-lattice in V. For each $\mathfrak{p}$, denote by $\mathfrak{U}_\mathfrak{p}$ the set of elements τ of $G_\mathfrak{p}$ such that $L_\mathfrak{p}\tau = L_\mathfrak{p}$. Then $\mathfrak{U}_\mathfrak{p}$ is a compact subgroup of $G_\mathfrak{p}$. Put

$$\mathfrak{U}_L = \prod_\mathfrak{p} \mathfrak{U}_\mathfrak{p} \times \prod_\kappa G_\kappa.$$

By the product topology, $\mathfrak{U}_L$ is a locally compact group. Now we define the *adele-group* $\mathfrak{G}$ of $G(V, f)$ as the set of elements $(\sigma_\mathfrak{p}, \sigma_\kappa)$ of $\prod_\mathfrak{p} G_\mathfrak{p} \times \prod_\kappa G_\kappa$ such that $\sigma_\mathfrak{p} \in \mathfrak{U}_\mathfrak{p}$ for all except a finite number of $\mathfrak{p}$. Define a topology of $\mathfrak{G}$ so that $\mathfrak{U}_L$ is an open subgroup of $\mathfrak{G}$. Then $\mathfrak{G}$ becomes a locally compact group. The topological group $\mathfrak{G}$ is determined independently of the choice of L. By the injection $\sigma \to \cdots, \sigma, \sigma, \cdots)$, G can be considered as a discrete subgroup of $\mathfrak{G}$. By a general theorem of Borel [1], $\mathfrak{G}$ is the union of a finite number of double cosets $\mathfrak{U}_L \xi G$ with $\xi \in \mathfrak{G}$ (cf. also Weil [11, 12]).

4.4. Classes and Genera of maximal lattices. Notation being as in §4.3, let $\mathfrak{o}$ be a maximal order in A, and $\mathfrak{o}_\mathfrak{p} = \mathfrak{g}_\mathfrak{p}\mathfrak{o}$. We denote by $\mathfrak{L}(\mathfrak{o})$ the set of all maximal lattices in V with the order $\mathfrak{o}$. Let L and M be two members of $\mathfrak{L}(\mathfrak{o})$. We say that L and M belong to the same *genus*, if there exists, for each prime ideal $\mathfrak{p}$ of F, an element $\sigma_\mathfrak{p}$ of $G(V_\mathfrak{p}, f)$ such that $L_\mathfrak{p}\sigma_\mathfrak{p} = M_\mathfrak{p}$. Further we say that L and M belong to the same *class*, if there exists an element σ of $G(V, f)$ such that $L\sigma = M$.

PROPOSITION 4.5. *If $n > 1$, for every two-sided $\mathfrak{o}$-ideal $\mathfrak{a}$, there exists a member L of $\mathfrak{L}(\mathfrak{o})$ such that $N(L) = \mathfrak{a}$.*

PROOF. Take an arbitrary maximal lattice M in V with the order $\mathfrak{o}$. There exist only a finite number of $\mathfrak{p}$ such that $N(M_\mathfrak{p}) \neq \mathfrak{a}_\mathfrak{p}$. For each one of such $\mathfrak{p}$, take a maximal lattice $L^\mathfrak{p}$ in $V_\mathfrak{p}$ with the order $\mathfrak{o}_\mathfrak{p}$, such that $N(L^\mathfrak{p}) = \mathfrak{a}_\mathfrak{p}$. This is possible by Propositions 2.6 and 3.5. Put $L^\mathfrak{p} = M_\mathfrak{p}$ for every $\mathfrak{p}$ such that $N(M_\mathfrak{p}) = \mathfrak{a}_\mathfrak{p}$. Then, by Lemma 1, there exists a $\mathfrak{g}$-lattice L in V such that $L_\mathfrak{p} = L^\mathfrak{p}$ for any $\mathfrak{p}$. It is clear that L is a member of $\mathfrak{L}(\mathfrak{o})$ and $N(L) = \mathfrak{a}$.

PROPOSITION 4.6. *Let $\mathfrak{o}$ be a maximal order in A. If $n = 1$, $\mathfrak{L}(\mathfrak{o})$ consists of only one genus, $\mathfrak{L}(\mathfrak{o})$ itself. If $n > 1$, there are exactly 2^s genera in $\mathfrak{L}(\mathfrak{o})$, where s is the number of prime ideals ramified in A/F.*

PROOF. The $\mathfrak{Q}_i$ being as in §4.1, we have $N(L) = \mathfrak{Q}_1^{e_1} \cdots \mathfrak{Q}_s^{e_s} \cdot \mathfrak{a}$ for every $L \in \mathfrak{L}(\mathfrak{o})$, where $e_i = 0$ or 1, and $\mathfrak{a}$ is a $\mathfrak{g}$-ideal. By Proposition 2.11 and Proposition 3.7, the genus of L is determined only by $\{e_1, \cdots, e_s\}$. This together with Proposition 4.5 proves our proposition.

We denote, for any set of integers $\{e_1, \cdots, e_s\}$ such that $e_i = 0$ or 1, by $\mathfrak{L}(\mathfrak{o}; \{e_i\})$ the genus of L such that $N(L) = \prod_{i=1}^{s} \mathfrak{Q}_i^{e_i} \cdot \mathfrak{a}$ with an ideal $\mathfrak{a}$ of F. We call especially $\mathfrak{L}(\mathfrak{o}; \{0, \cdots, 0\})$ *the principal genus with the order $\mathfrak{o}$* and denote it by $\mathfrak{L}_0(\mathfrak{o})$.

Fix a member L of $\mathfrak{L}(\mathfrak{o})$ and define $\mathfrak{U}_L$ as in §4.3. For every element $\xi = (\xi_\mathfrak{p}, \xi_\kappa)$ of the adele-group $\mathfrak{G}$, put $L\xi = \bigcap_\mathfrak{p} (L_\mathfrak{p}\xi_\mathfrak{p} \cap V)$. By Lemma 1.1, $L\xi$ is a $\mathfrak{g}$-lattice in V; and $(L\xi)_\mathfrak{p} = L_\mathfrak{p}\xi_\mathfrak{p}$. By Propositions 2.2, 2.3, 2.4, we see that $L\xi$ is a member of $\mathfrak{L}(\mathfrak{o})$. Further, by our definition, $L\xi$ belongs to the same genus as L. Conversely, if M is a maximal lattice belonging to the same genus as L, we can find an element ξ of $\mathfrak{G}$ such that $L\xi = M$. If $\xi \in G$, the notation $L\xi$ is just the same as the transform of L by ξ; so there is no fear of confusion. We have $L\xi = L\eta$ if and only if $\mathfrak{U}_L\xi = \mathfrak{U}_L\eta$. Therefore, the mapping $\xi \to L\xi$ gives a one-to-one mapping of $\mathfrak{U}_L \backslash \mathfrak{G}$ onto the genus of L. Moreover, we note that this gives a one-to-one correspondence between $\mathfrak{U}_L \backslash \mathfrak{G} / G$ and the classes in the genus. By the fact remarked at the end of §4.3, this implies that each genus consists of a finite number of classes. Further, by Proposition 2.5, we observe that the number of classes in $\mathfrak{L}(\mathfrak{o}; \{e_i\})$ depends only on $\{e_i\}$ and is independent of the choice of $\mathfrak{o}$.

4.5. An existence theorem in the case $n = 2$. Let A be a quaternion algebra over F. We denote by $\mathfrak{q}_1, \cdots, \mathfrak{q}_s$ all the prime ideals of F which are ramified in A/F, and by $\mathfrak{p}_{\infty 1}, \cdots, \mathfrak{p}_{\infty u}$ all the infinite prime spots of F which are ramified in A. We put

$$\mathfrak{d} = \prod_{h=1}^{s} \mathfrak{q}_h, \qquad \mathfrak{u} = \prod_{\kappa=1}^{u} \mathfrak{p}_{\infty\kappa}.$$

Let V be an A-space of dimension n and f a non-degenerate Q-hermitian form on V. Fix a maximal order $\mathfrak{o}$ in A; and for each $\mathfrak{q}_h$, let $\mathfrak{Q}_h$ be the prime o-ideal such that $\mathfrak{O}_h^2 = \mathfrak{q}_h$. We put $\mathfrak{g}_\mathfrak{p}\mathfrak{p} = \mathfrak{p}$ for every prime ideal $\mathfrak{p}$ of F, and $\mathfrak{o}_{\mathfrak{q}_h}\mathfrak{O}_h = \mathfrak{O}_h$ for every h, when there is no fear of confusion.

LEMMA 4.7. *Let $\mathfrak{p}$ be a prime ideal of F which is unramified in A/F. Let $\mathfrak{o}_\mathfrak{p}$ be a maximal order in $A_\mathfrak{p}$. Let a be a regular element of $A_\mathfrak{p}$ and δ be an element of $\mathfrak{g}_\mathfrak{p}$. Then there exists an element d of $\mathfrak{o}_\mathfrak{p} \cap a\mathfrak{o}_\mathfrak{p}a^{-1}$ such that $N(d) = \delta$.*

PROOF. We may assume that $A_\mathfrak{p} = M_2(F_\mathfrak{p})$ and $\mathfrak{o}_\mathfrak{p} = M_2(\mathfrak{g}_\mathfrak{p})$. Then we can find units ε, η of $\mathfrak{o}_\mathfrak{p}$ such that $\varepsilon a \eta = \begin{pmatrix} \alpha & 0 \\ 0 & \beta \end{pmatrix}$ with $\alpha, \beta \in F_\mathfrak{p}$. Put $d = \varepsilon^{-1}\begin{pmatrix} 1 & 0 \\ 0 & \delta \end{pmatrix}\varepsilon$. Then we have $N(d) = \delta$ and $d \in \mathfrak{o}_\mathfrak{p}$. Further we get

$$a^{-1}da = a^{-1}\varepsilon^{-1}\begin{pmatrix} 1 & 0 \\ 0 & \delta \end{pmatrix}\varepsilon a = \eta\begin{pmatrix} \alpha & 0 \\ 0 & \beta \end{pmatrix}^{-1}\begin{pmatrix} 1 & 0 \\ 0 & \delta \end{pmatrix}\begin{pmatrix} \alpha & 0 \\ 0 & \beta \end{pmatrix}\eta^{-1} = \eta\begin{pmatrix} 1 & 0 \\ 0 & \delta \end{pmatrix}\eta^{-1} \in \mathfrak{o}_\mathfrak{p},$$

so that $d \in a\mathfrak{o}_\mathfrak{p}a^{-1}$, which completes the proof.

PROPOSITION 4.8. *Suppose that A is indefinite and $n = 2$. Let L be a $\mathfrak{g}$-lattice in V written in the form $L = \mathfrak{o}x + \mathfrak{c}^{-1}y$, where $\mathfrak{c}$ is an integral right $\mathfrak{o}$-ideal, $f(x, x) = 1$, $f(x, y) = 0$, $f(y, y) = r$, $N(\mathfrak{c}) = r\mathfrak{g}$ with an element r of $\mathfrak{g}$. Suppose that r is prime to $\mathfrak{b}$. Let $\mathfrak{p}_1, \cdots, \mathfrak{p}_r$ be distinct prime ideals which are prime to $\mathfrak{b}$, and let α be a non-zero element of $\mathfrak{g}$ such that $\alpha \equiv 1 \bmod \mathfrak{p}_{\times_\kappa}$ whenever $r \equiv 1 \bmod \mathfrak{p}_{\infty_\kappa}$ for $1 \leq \kappa \leq u$. Put $\mathfrak{g}_{\mathfrak{p}_i}\alpha = \mathfrak{p}_i^{\mu_i}$ for $1 \leq i \leq r$ and $\mathfrak{g}_{\mathfrak{q}_h}\alpha = \mathfrak{q}_h^{\lambda_h}$ for $1 \leq h \leq s$. Let ξ_i, η_i, for $1 \leq i \leq r$, and ν_h, for $1 \leq h \leq s$, be integers such that*

$$0 \leq \xi_i \leq \eta_i \leq \mu_i - \eta_i \leq \mu_i - \xi_i \leq \mu_i, \qquad 0 \leq \nu_h \leq \lambda_h.$$

Then there exists an element σ of $G(V, f)$ such that $L\sigma \subset L$, $N(\sigma) = \alpha$,

$$\{L_{\mathfrak{p}_i} : L_{\mathfrak{p}_i}\sigma\} = \{\mathfrak{p}_i^{\xi_i}, \mathfrak{p}_i^{\eta_i}, \mathfrak{p}_i^{\mu_i - \xi_i}, \mathfrak{p}_i^{\mu_i - \eta_i}\} \qquad \text{for} \quad 1 \leq i \leq r,$$

$$\{L_{\mathfrak{q}_h} : L_{\mathfrak{q}_h}\sigma\} = \{\mathfrak{O}_h^{\nu_h}, \mathfrak{O}_h^{2\lambda_h - \nu_h}\} \qquad \text{for} \quad 1 \leq h \leq s.$$

PROOF. For simplicity, we denote the indices $\mathfrak{p}_i$ and $\mathfrak{q}_h$ respectively by i and h; for example, L_i means $L_{\mathfrak{p}_i}$ and $\mathfrak{g}_h$ means $\mathfrak{g}_{\mathfrak{q}_h}$. By Proposition 2.6, L is maximal and $N(L) = \mathfrak{o}$. Now, for each $\mathfrak{p}_i$, we identify $\mathfrak{o}_i$ with $M_2(\mathfrak{g}_i)$, and fix an element π_i of $\mathfrak{g}$ such that $\mathfrak{p}_i = \mathfrak{g}_i\pi_i$. Put $\mathfrak{g}_i r = \mathfrak{p}_i^{e_i}$. Without any loss of generality, we may assume that $\mathfrak{c}_i^{-1}$ is written in the form $\mathfrak{c}_i^{-1} = \begin{pmatrix} \mathfrak{p}_i^{-c_i} & \mathfrak{p}_i^{-d_i} \\ \mathfrak{p}_i^{-c_i} & \mathfrak{p}_i^{-d_i} \end{pmatrix}$, where c_i and d_i are integers such that $c_i \geq d_i \geq 0$ and $c_i + d_i = e_i$. Consider the ideal-class modulo $\mathfrak{u}$ containing the inverse of the ideal

$$\prod_{i=1}^r \mathfrak{p}_i^{2\mu_i + 2 + c_i - d_i + \eta_i - \xi_i} \prod_{h=1}^s \mathfrak{q}_h^{\nu_h}.$$

We can find an integral ideal $\mathfrak{a}$ in that class which is prime to $r\alpha\prod_{i=1}^r \mathfrak{p}_i \cdot \mathfrak{b}$. We get then

$$\mathfrak{a} \cdot \prod_{i=1}^r \mathfrak{p}_i^{2\mu_i + 2 + c_i - d_i + \eta_i - \xi_i} \prod_{h=1}^s \mathfrak{q}_h^{\nu_h} = (\beta), \qquad \beta \equiv 1 \bmod \mathfrak{u}$$

for an element β of $\mathfrak{g}$. By Lemma 4.3, there exists an element a_1 of $\mathfrak{o}$ such that $N(a_1) = \beta$ and

$$(11) \qquad a_1 \equiv \begin{pmatrix} \pi_i^{\mu_i+1} & 0 \\ 0 & \beta\pi_i^{-\mu_i-1} \end{pmatrix} \mod \beta\mathfrak{p}_i^{\mu_i+e_i+1}\mathfrak{o}_i \qquad (1 \leq i \leq r).$$

Let ε be a unit of F such that $\alpha(1-\varepsilon^2) \equiv 1 \mod \mathfrak{u}$. Such an ε really exists, because $\mathfrak{u}$ is not the product of all the infinite prime spots of F. Then, by our assumption on α, we have

$$\alpha - \varepsilon^{2m}N(a_1)\gamma \equiv 1 \mod \mathfrak{u}$$

for a suitably large integer m. Fix such an m. By Lemma 4.3, there exists an element b_1 of $\mathfrak{o}$ such that $N(b_1) = \alpha - \varepsilon^{2m}N(a_1)\gamma$. By our choice of β, we observe that $\mathfrak{g}_iN(b_1) = \mathfrak{p}_i^{\mu_i}$ for $1 \leq i \leq r$. Put $a = \varepsilon^m a_1$. For every prime ideal $\mathfrak{r}$ of F, let $\bar{\mathfrak{o}}_\mathfrak{r}$ denote the right order of $(\mathfrak{c}^{-1}a^{-1})_\mathfrak{r}$. Let $\{\mathfrak{r}\}$ be the set of prime ideals $\mathfrak{r}$ such that $(\mathfrak{r}, \mathfrak{b} \cdot \prod_{i=1}^{r}\mathfrak{p}_i) = 1$, $\bar{\mathfrak{o}}_\mathfrak{r} \neq \mathfrak{o}_\mathfrak{r}$. Obviously $\{\mathfrak{r}\}$ is a finite set. For each $\mathfrak{r}$, take an element $b_\mathfrak{r}$ of $\bar{\mathfrak{o}}_\mathfrak{r} \cap \mathfrak{o}_\mathfrak{r}$ such that $N(b_\mathfrak{r}) = N(b_1)$. This is possible by virtue of Lemma 4.7. Now by Lemma 4.3, we can find an element b of $\mathfrak{o}$ such that $N(b) = N(b_1)$, $b \equiv b_\mathfrak{r} \mod (\bar{\mathfrak{o}}_\mathfrak{r} \cap \mathfrak{o}_\mathfrak{r})$ for $\mathfrak{r} \in \{\mathfrak{r}\}$, and

$$(12) \qquad b \equiv \begin{pmatrix} 0 & \pi_i^{\xi_i} \\ -N(b_1)\pi_i^{-\xi_i} & 0 \end{pmatrix} \mod \beta\mathfrak{p}_i^{\mu_i+e_i+1}\mathfrak{o}_i \qquad (1 \leq i \leq r).$$

Then we have $N(b)+\gamma N(a) = \alpha$, and by (11) and (12),

$$a^{-1}ba = a_1^{-1}ba_1 = \beta^{-1}a_1'ba_1 \equiv \begin{pmatrix} 0 & \beta\pi_i^{\xi_i-2\mu_i-2} \\ -N(b)\beta^{-1}\pi_i^{-\xi_i+2\mu_i+2} & 0 \end{pmatrix} \mod \mathfrak{p}_i^{\mu_i+e_i+1}\mathfrak{o}_i,$$

so that

$$(13) \qquad a^{-1}b'a \equiv \begin{pmatrix} 0 & \theta\pi_i^{\eta_i+c_i-d_i} \\ \psi\pi_i^{\mu_i-\eta_i-c_i+d_i} & 0 \end{pmatrix} \mod \mathfrak{p}_i^{\mu_i+e_i+1}\mathfrak{o}_i$$

with units θ and ψ of $\mathfrak{g}_i$. It follows that

$$(14) \qquad \mathfrak{c}_i^{-1}(a^{-1}b'a) = \begin{pmatrix} \mathfrak{p}_i^{\mu_i-\eta_i-c_i} & \mathfrak{p}_i^{\eta_i-d_i} \\ \mathfrak{p}_i^{\mu_i-\eta_i-c_i} & \mathfrak{p}_i^{\eta_i-d_i} \end{pmatrix}.$$

Hence we have $\mathfrak{c}_i^{-1}(a^{-1}b'a) \subset \mathfrak{c}_i^{-1}$. By our choice of $b_\mathfrak{r}$, we have $b' \in \bar{\mathfrak{o}}_\mathfrak{r}$ for every $\mathfrak{r}$ such that $(\mathfrak{r}, \mathfrak{b} \cdot \prod_{i=1}^{r}\mathfrak{p}_i) = 1$, and hence $\mathfrak{c}_\mathfrak{r}^{-1}(a^{-1}b'a) \subset \mathfrak{c}_\mathfrak{r}^{-1}$ for any such $\mathfrak{r}$. Further it is obvious that $\mathfrak{c}_h^{-1}(a^{-1}b'a) \subset \mathfrak{c}_h^{-1}$. Therefore, we have

$$(15) \qquad \mathfrak{c}^{-1}(a^{-1}b'a) \subset \mathfrak{c}^{-1}.$$

As $a \in \mathfrak{o} \subset \mathfrak{c}^{-1}$ and $N(\mathfrak{c}) = \gamma\mathfrak{g}$, we have

$$(15') \qquad \mathfrak{c}^{-1}\gamma a' \subset \mathfrak{o}.$$

Moreover, by (11) we have

$$(16) \qquad \mathfrak{c}_i^{-1}\gamma a' \subset \mathfrak{p}_i^{\mu_i+1}\mathfrak{o}_i.$$

Now define an element σ of $E(V, A)$ by

$$(17) \qquad x\sigma = bx+ay, \qquad y\sigma = -\gamma a'x+a^{-1}b'ay.$$

By the relation $N(b)+\gamma N(a) = \alpha$, we can easily verify that $\sigma \in G(V, f)$ and

$N(\sigma)=\alpha$. Further by (15), (15′), we have $L\sigma\subset L$. By Proposition 2.7, we have $\mathfrak{p}_i^{\mu_i}L_i=\alpha L_i\subset L_i\sigma$, so that

$$(18) \qquad\qquad \mathfrak{p}_i^{\mu_i+1}L_i\subset\mathfrak{p}_iL_i\sigma\,.$$

Put $M_i=\mathfrak{o}_ibx+\mathfrak{c}_i^{-1}a^{-1}b'ay$. Then we have, by (11), (16), (17), (18) and by Lemma 1.2, $M_i=L_i\sigma$, so that $L_i/L_i\sigma=L_i/M_i\cong\mathfrak{o}_i/\mathfrak{o}_ib+\mathfrak{c}_i^{-1}/\mathfrak{c}_i^{-1}a^{-1}b'a$ (as $\mathfrak{o}_i$-modules). By (12) and (14), $L_i/L_i\sigma$ has the desired elementary divisors. Let us now consider $\mathfrak{q}_h$ for $1\leq h\leq s$. As $N(\sigma)=\alpha$ and $\mathfrak{o}_h\alpha=\mathfrak{O}_h^{2\lambda_h}$, we have, by Proposition 2.7, $L_h\sigma\supset\mathfrak{O}_h^{2\lambda_h}L_h$. As $N(a_1)=\beta$ and $\mathfrak{g}_h\beta=\mathfrak{q}_h^{\nu_h}$, we have $\mathfrak{o}_ha=\mathfrak{O}_h^{\nu_h}$. If $\lambda_h=0$, we have $L_h\sigma=L_h$. Suppose that $\lambda_h>0$. If $\nu_h=\lambda_h$, we have $\mathfrak{o}_ha=\mathfrak{O}_h^{\lambda_h}$ and hence $N(b)=\alpha-\gamma N(a)\in\mathfrak{O}_h^{2\lambda_h}$. It follows that a, b, $-\gamma a'$, $a^{-1}b'a$ are contained in $\mathfrak{O}_h^{\lambda_h}$. Hence we have $x\sigma$, $y\sigma\in\mathfrak{O}_h^{\lambda_h}L_h$, so that $L_h\sigma\subset\mathfrak{O}_h^{\lambda_h}L_h$. As $L_h\sigma$ is maximal and $N(L_h\sigma)=\mathfrak{O}_h^{2\lambda_h}=N(\mathfrak{O}_h^{\lambda_h}L_h)$, we must have $L_h\sigma=\mathfrak{O}_h^{\lambda_h}L_h$. Then $L_h/L_h\sigma\cong\mathfrak{o}_h/\mathfrak{O}_h^{\lambda_h}+\mathfrak{o}_h/\mathfrak{O}_h^{\lambda_h}$. It remains to consider the case $\lambda_h>\nu_h\geq0$. As $N(b)+\gamma N(a)=\alpha$, $\mathfrak{o}_ha=\mathfrak{O}_h^{\nu_h}$, $\mathfrak{o}_h\alpha=\mathfrak{O}_h^{2\lambda_h}$, we must have $\mathfrak{o}_hb=\mathfrak{O}_h^{\nu_h}$. It follows that $a'b^{-1}$ is a unit of $\mathfrak{o}_h$. We note that

$$y\sigma+\gamma(a'b^{-1})x\sigma=(a^{-1}b'a+\gamma a'b^{-1}a)y=a^{-1}b^{-1}(bb'+\gamma baa'b^{-1})ay$$
$$=a^{-1}b^{-1}(N(b)+\gamma N(a))ay=\alpha(a^{-1}b^{-1}a)y\,.$$

Therefore we have $L_h\sigma=\mathfrak{o}_hx\sigma+\mathfrak{o}_hy\sigma=\mathfrak{o}_hx\sigma+\mathfrak{o}_h\alpha(a^{-1}b^{-1}a)y$. On the other hand, as $b^{-1}a$ is a unit of $\mathfrak{o}_h$ and as $x=b^{-1}x\sigma-b^{-1}ay$, we have $L_h=\mathfrak{o}_hx+\mathfrak{o}_hy=\mathfrak{o}_hb^{-1}x\sigma+\mathfrak{o}_hy$. Hence $L_h/L_h\sigma\cong\mathfrak{o}_hb^{-1}/\mathfrak{o}_h+\mathfrak{o}_h/\mathfrak{o}_h(a^{-1}b^{-1}a)\alpha=\mathfrak{o}_h/\mathfrak{O}_h^{\nu_h}+\mathfrak{o}_h/\mathfrak{O}_h^{2\lambda_h-\nu_h}$ (as $\mathfrak{o}_h$-modules). This completes our proof.

4.6. Global approximation theorem. As in §4.2, we define $\nu_\kappa(f)$ for $1\leq\kappa\leq u$, and reorder the $\mathfrak{p}_{\nu_\kappa}$ so that $\nu_\kappa(f)\neq n/2$ for $1\leq\kappa\leq t$ and $\nu_\kappa(f)=n/2$ for $t<\kappa\leq u$. Put

$$\mathfrak{t}=\mathfrak{t}_f=\prod_{\kappa=1}^{t}\mathfrak{p}_{\infty\kappa}\,.$$

For every $\sigma\in G(V,f)$, we have $N(\sigma)\equiv1\bmod\mathfrak{t}$. If n is odd, we have $t=u$ and $\mathfrak{t}=\mathfrak{u}$.

Now we are ready to state and prove our main theorems.

THEOREM 1. *Suppose that A is indefinite. Let L be a maximal lattice belonging to the principal genus $\mathfrak{L}_0(\mathfrak{o})$. Let $\mathfrak{p}_1,\cdots,\mathfrak{p}_r$ be prime ideals of F; and let σ_i, for each i, be an element of $G(V_{\mathfrak{p}_i},f)$ such that $L_{\mathfrak{p}_i}\sigma_i\subset L_{\mathfrak{p}_i}$. Let α be an element of $\mathfrak{g}$. Suppose that*

$$\alpha^{-1}N(\sigma_i)\equiv1\quad\bmod\mathfrak{p}_i^{\lambda_i},\qquad(1\leq i\leq r)\,,$$

$$\alpha\equiv1\quad\bmod\mathfrak{t}\,,$$

where the λ_i are positive integers. Then there exists an element σ of $G(V,f)$ such that $L\sigma\subset L$, $N(\sigma)=\alpha$, $L_{\mathfrak{p}_i}(\sigma-\sigma_i)\subset\mathfrak{p}_i^{\lambda_i}L_{\mathfrak{p}_i}$ $(1\leq i\leq r)$.

We prove Theorem 1 in several steps. For simplicity, we denote $L_{\mathfrak{p}_i}$, $\mathfrak{g}_{\mathfrak{p}_i}$, etc. by L_i, $\mathfrak{g}_i$, etc.

ASSERTION 1. If Theorem 1 is proved when $N(L) = \mathfrak{o}$ for every maximal order $\mathfrak{o}$, then it is true for any L belonging to the principal genus.

In fact, if L belongs to $\mathfrak{L}_0(\mathfrak{o})$, we can find a left $\mathfrak{o}$-ideal $\mathfrak{x}$ such that $N(L) = N(\mathfrak{x})\mathfrak{o}$. Then by Proposition 2.5, $\mathfrak{x}^{-1}L$ is a maximal lattice, and $N(\mathfrak{x}^{-1}L) = \mathfrak{o}_1$, where $\mathfrak{o}_1$ is the right order of $\mathfrak{x}$. We see easily that if Theorem 1 is true for $\mathfrak{x}^{-1}L$, then it is true for L.

ASSERTION 2. If Theorem 1 is true for L, then, for every $\tau \in G(V,f)$, Theorem 1 is true for $L\tau$.

This is clear.

ASSERTION 3. If Theorem 1 is proved for a certain $L^0 \in \mathfrak{L}_0(\mathfrak{o})$ such that $N(L^0) = \mathfrak{o}$, then for any $L \in \mathfrak{L}_0(\mathfrak{o})$ such that $N(L) = \mathfrak{o}$, we have $L = L^0\tau$ for an element $\tau \in G^0(V,f)$.

Let β be a non-zero element of $\mathfrak{g}$ such that $\beta L \subset L^0$. Take an integral $\mathfrak{g}$-ideal $\mathfrak{a}$ such that $\beta L \supset \mathfrak{a}L^0$. Let $\mathfrak{p}_1, \cdots, \mathfrak{p}_r$ be the prime factors of $\mathfrak{a}$. By Proposition 2.11 and by Proposition 3.7, there exists, for each $\mathfrak{p}_i$, an element τ_i of $G^0(V_i, f)$ such that $L_i^0\tau_i = L_i$. Put $\sigma_i = \beta\tau_i$. Then we have $N(\sigma_i) = \beta^2$. Applying Theorem 1 to L^0 and these σ_i, we get an element σ of $G(V,f)$ such that $L^0\sigma \subset L^0$, $N(\sigma) = \beta^2$, $L_i^0(\sigma - \sigma_i) \subset \mathfrak{p}_i\mathfrak{a}L_i^0$ for $1 \leq i \leq r$. As we have $\mathfrak{p}_i\mathfrak{a}L_i^0 \subset \mathfrak{p}_i\beta L_i = \mathfrak{p}_iL_i^0\sigma_i$, we see, from Lemma 1.2, $L_i^0\sigma = L_i^0\sigma_i = \beta L_i$. If $\mathfrak{p}$ is a prime ideal which does not divide $\mathfrak{a}$, we have $L_{\mathfrak{p}}^0 = \beta L_{\mathfrak{p}}$, since $L^0 \supset \beta L \supset \mathfrak{a}L^0$. It follows that $N(L_{\mathfrak{p}}^0) = \beta^2 N(L_{\mathfrak{p}})$, and hence β is a $\mathfrak{p}$-unit. As $L^0\sigma \subset L^0$ and $N(\sigma) = \beta^2$, we have $L_{\mathfrak{p}}^0\sigma = L_{\mathfrak{p}}^0 = \beta L_{\mathfrak{p}}$. Therefore, we have $L_{\mathfrak{p}}^0\sigma = \beta L_{\mathfrak{p}}$ for every prime ideal $\mathfrak{p}$ of F, so that $L^0\sigma = \beta L$. Putting $\tau = \beta^{-1}\sigma$, we get $L^0\tau = L$.

ASSERTION 4. In order to prove Theorem 1, we may exchange the Q-hermitian form f for θf for any non-zero element θ of F.

In fact, put $g(x,y) = \theta f(x,y)$ for $(x,y) \in V \times V$. Then g is a non-degenerate Q-hermitian form. We see easily $G(V,f) = G(V,g)$, $G^0(V,f) = G^0(V,g)$, and, for every $\sigma \in G(V,f)$, $N(\sigma)$ is common for f and g. Further, for every $\mathfrak{g}$-lattice L in V, we have $N_g(L) = \theta N_f(L)$. When L is normal, L is maximal with respect to f if and only if L is maximal with respect to g. The genera and the classes of normal maximal lattices do not change by exchanging f for g. Finally we note that $\mathfrak{t}_f = \mathfrak{t}_g$. Therefore we get Assertion 4.

Now we proceed by induction on n. If $n = 1$, Theorem 1 is just a restatement of Lemma 4.3. Assume that $n > 1$ and Theorem 1 is true for $\dim_A V < n$.

ASSERTION 5. If $N(L) = \mathfrak{o}$ and L contains an element x such that $f(x,x) = 1$, then Theorem 1 is true for this L and for $\alpha = 1$.

As $N(\sigma_i) \equiv 1 \bmod \mathfrak{p}_i^{\lambda_i}$ and $N(L) = \mathfrak{o}$, we have $f(u\sigma_i, v\sigma_i) \equiv f(u,v) \bmod \mathfrak{p}_i^{\lambda_i}\mathfrak{o}_i$ for

$u, v \in L$. By Proposition 3.13, there exists an element τ_i of $G(V_i, f)$ such that $N(\tau_i)=1$, $L_i\tau_i = L_i$, $L_i(\sigma_i - \tau_i) \subset \mathfrak{p}_i^{\lambda_i} L_i$. Put

$$W = \{v \in V \mid f(x, v) = 0\}, \qquad M = W \cap L.$$

Then we have $V = Ax + W$, $L = \mathfrak{o}x + M$; and M is a maximal lattice in W such that $N(M) = \mathfrak{o}$. Put $x\tau_i = a_i x + y_i$ with $a_i \in \mathfrak{o}_i$ and $y_i \in M_i$. As $f(x\tau_i, x\tau_i) = f(x, x)$ $= 1$, we have $N(a_i) + f(y_i, y_i) = 1$. Now reorder the $\mathfrak{p}_i$ so that $N(a_i) \neq 0$ for $1 \leq i \leq h$, $N(a_i) = 0$ for $h < i \leq r$, where h is an integer such that $0 \leq h \leq r$. Put $\mathfrak{g}_i N(a_i) = \mathfrak{p}_i^{\mu_i}$ for $1 \leq i \leq h$. We can find a regular element a of A such that $a \in \mathfrak{o}$, and

$$a \equiv a_i \pmod{\mathfrak{o}_i \mathfrak{p}_i^{\lambda_i + \mu_i}} \qquad \text{for } 1 \leq i \leq h,$$

$$a \equiv a_i \pmod{\mathfrak{o}_i \mathfrak{p}_i^{\lambda_i}} \qquad \text{for } h < i \leq r.$$

We have then

$$N(a) \equiv N(a_i) \pmod{\mathfrak{p}_i^{\lambda_i + \mu_i}} \qquad \text{for } 1 \leq i \leq h,$$

$$N(a) \equiv 0 \pmod{\mathfrak{p}_i^{\lambda_i}} \qquad \text{for } h < i \leq r,$$

and hence $\mathfrak{g}_i N(a) = \mathfrak{p}_i^{\mu_i}$ for $1 \leq i \leq h$. Put $\mathfrak{g}_i N(a) = \mathfrak{p}_i^{\mu_i}$ for $i > h$. We have then $\mu_i \geq \lambda_i \geq 1$ for $i > h$. Now, as $1 - N(a) \equiv 1 \bmod \mathfrak{p}_i^{\mu_i}$, we can find, by Proposition 3.2, for each $i > h$, an element ε_i of $\mathfrak{o}_i$ such that $\varepsilon_i \equiv 1 \bmod \mathfrak{o}_i \mathfrak{p}_i^{\mu_i}$, $N(\varepsilon_i) = 1 - N(a)$. Take an element y of M so that

$$y \equiv y_i \pmod{\mathfrak{p}_i^{\lambda_i + \mu_i} M_i} \qquad \text{for } 1 \leq i \leq h,$$

$$y \equiv \varepsilon_i y_i \pmod{\mathfrak{p}_i^{\lambda_i + \mu_i} M_i} \qquad \text{for } h < i \leq r.$$

Then we can easily verify that $f(y, y) \equiv 1 - N(a) \bmod \mathfrak{p}_i^{\lambda_i + \mu_i}$ for every i. Since $\mathfrak{u}$ is not the product of all infinite prime spots of F, the projection of the set $\{\beta \mid \beta = 1 + \xi, \xi \in \prod_{i=1}^{r} \mathfrak{p}_i^{\mu_i + \lambda_i}\}$ on $F_1 \times \cdots \times F_u$ is dense. Hence there exists an element β of $\mathfrak{g}$ such that $1 - \beta^2 \equiv 1 \bmod \mathfrak{u}$, $\beta \equiv 1 \bmod \prod_{i=1}^{r} \mathfrak{p}_i^{\mu_i + \lambda_i}$. For a suitably large integer k, $1 - \beta^{2k} f(y, y) \equiv 1 \bmod \mathfrak{u}$. Put $w = \beta^k y$ for such an integer k. Then we have

$$w \equiv y \pmod{\prod_{i=1}^{r} \mathfrak{p}_i^{\lambda_i + \mu_i} M},$$

$$1 - f(w, w) \equiv 1 - f(y, y) \equiv N(a) \pmod{\prod_{i=1}^{r} \mathfrak{p}_i^{\lambda_i + \mu_i}},$$

$$1 - f(w, w) \equiv 1 \pmod{\mathfrak{u}}.$$

As $N(a)\mathfrak{g}_i = \mathfrak{p}_i^{\mu_i}$, we have $N(a)^{-1}(1 - f(w, w)) \equiv 1 \bmod \prod_{i=1}^{r} \mathfrak{p}_i^{\lambda_i}$. Therefore, by Lemma 4.3, there exists an element b of $\mathfrak{o}$ such that

$$N(b) = 1 - f(w, w), \qquad b \equiv a \pmod{\prod_{i=1}^{r} \mathfrak{p}_i^{\lambda_i} \mathfrak{o}}.$$

Put $u = bx + w$. Then $f(u, u) = N(b) + f(w, w) = 1$, and

$$u \equiv ax+y \equiv a_ix+y_i = x\tau_i \quad \mathrm{mod}\ \mathfrak{p}_i^{\lambda_i}L_i \qquad (1 \leqq i \leqq r).$$

Hence, putting

$$U = \{z \in V \mid f(u,z)=0\}, \qquad K = U \cap L,$$

we have $V = Au + U$, $L = \mathfrak{o}u + K$; and K is a maximal lattice in U such that $N(K) = \mathfrak{o}$. By Lemma 4.4, (U,f) and (W,f) are isomorphic. Therefore we can find an element ρ of $G^0(V,f)$ such that $x\rho = u$, $W\rho = U$. We see easily that $M\rho$ is a maximal lattice in U and $N(M\rho) = \mathfrak{o}$. By our induction assumption and Assertion 3, there exists an element φ of $G^0(U,f)$ such that $M\rho\varphi = K$. Exchanging ρ for $\rho\varphi$ on W, we may assume that $M\rho = K$ for ρ itself. Then we have $L\rho = L$, and $x\tau_i\rho^{-1} \equiv x \bmod \mathfrak{p}_i^{\lambda_i}L_i$. For every $z \in W_i$, denote by $z\psi_i$ the projection of $z\tau_i\rho^{-1}$ onto W_i defined by the decomposition $V_i = A_ix + W_i$. Then ψ_i can be considered as an element of $E(W_i, A_i)$. If $z \in M_i$, we have $f(z\tau_i\rho^{-1}, x) \equiv f(z\tau_i\rho^{-1}, x\tau_i\rho^{-1}) = f(z,x) = 0 \bmod \mathfrak{p}_i^{\lambda_i}\mathfrak{o}_i$. It follows that $M_i(\tau_i\rho^{-1}-\psi_i) \subset \mathfrak{p}_i^{\lambda_i}\mathfrak{o}_ix$, and hence $f(z_1\psi_i, z_2\psi_i) \equiv f(z_1,z_2) \bmod \mathfrak{p}_i^{\lambda_i}\mathfrak{o}_i$ for $z_1 \in M_i$, $z_2 \in M_i$. By Proposition 3.13, there exists an element θ_i of $G^0(W_i,f)$ such that $M_i\theta_i \subset M_i$ and $M_i(\psi_i-\theta_i) \subset \mathfrak{p}_i^{\lambda_i}M_i$. Applying our induction assumption to M and the θ_i, we find an element θ of $G^0(W,f)$ such that $M\theta \subset M$, $M_i(\theta-\theta_i) \subset \mathfrak{p}_i^{\lambda_i}M_i$ for $1 \leqq i \leqq r$. Define an element σ of $E(V,A)$ by $x\sigma = u$, $z\sigma = z\theta\rho$ for $z \in W$. Then we have $\sigma \in G(V,f)$ and $N(\sigma) = 1$, $L\sigma \subset L$. Further, we have

$$x\sigma = u \equiv x\tau_i \equiv x\sigma_i \quad \mathrm{mod}\ \mathfrak{p}_i^{\lambda_i}L_i \qquad (1 \leqq i \leqq r),$$

and if $z \in M_i$,

$$z\sigma = z\theta\rho \equiv z\theta_i\rho \equiv z\psi_i\rho \equiv z\tau_i \equiv z\sigma_i \quad \mathrm{mod}\ \mathfrak{p}_i^{\lambda_i}L_i \qquad (1 \leqq i \leqq r).$$

Therefore $L_i(\sigma-\sigma_i) \subset \mathfrak{p}_i^{\lambda_i}L_i$. This completes the proof of Assertion 5.

Proposition 4.9. *Let L be a maximal lattice belonging to $\mathfrak{L}_0(\mathfrak{o})$. Let α be an element of $\mathfrak{g}$ such that $\alpha \equiv 1 \bmod \mathfrak{t}$, and let $\mathfrak{p}_1, \cdots, \mathfrak{p}_r$ be prime ideals of F. Let σ_i, for $1 \leqq i \leqq r$, be an element of $G(V_{\mathfrak{p}_i}, f)$ such that $N(\sigma_i)\mathfrak{g}_{\mathfrak{p}_i} = \alpha\mathfrak{g}_{\mathfrak{p}_i}$, $L_{\mathfrak{p}_i}\sigma_i \subset L_{\mathfrak{p}_i}$. Then there exists an element σ of $G(V,f)$ such that $N(\sigma) = \alpha$, $L\sigma \subset L$, and $L_{\mathfrak{p}_i}/L_{\mathfrak{p}_i}\sigma$ is isomorphic to $L_{\mathfrak{p}_i}/L_{\mathfrak{p}_i}\sigma_i$ as $\mathfrak{o}_{\mathfrak{p}_i}$-modules for $1 \leqq i \leqq r$.*

Assertion 6. If Theorem 1 is true, then Proposition 4.9 is true.

In fact, as $N(\sigma_i)^{-1}\alpha$ is a $\mathfrak{p}_i$-unit for each i, there exists, by Proposition 3.3 and Proposition 3.12, an element τ_i of $G(V_i,f)$ such that $L_i\tau_i = L_i$, $N(\tau_i) = N(\sigma_i)^{-1}\alpha$. Then we have $L_i\tau_i\sigma_i \subset L_i$, $N(\tau_i\sigma_i) = \alpha$. By Theorem 1, there exists an element σ of $G(V,f)$ such that $L\sigma \subset L$, $N(\sigma) = \alpha$, $L_i(\sigma-\tau_i\sigma_i) \subset \mathfrak{p}_i^{\lambda_i+1}L_i = \mathfrak{p}_i\alpha L_i$ for every i. By Proposition 2.7, we have $\alpha L_i \subset L_i\tau_i\sigma_i$. Therefore, by Lemma 1.2, we have $L_i\sigma = L_i\tau_i\sigma_i = L_i\sigma_i$. This proves our assertion.

Now exchanging f for θf with a suitable θ of F, if necessary, we may assume that $\nu_\kappa(f) > n/2$ for $1 \leqq \kappa \leqq t$, $\nu_\kappa(f) = n/2$ for $\kappa < t \leqq u$. By Assertion 4, this does not influence the validity of our proof of Theorem 1.

Assertion 7. There exists a member L of $\mathfrak{L}_0(\mathfrak{o})$ satisfying the following conditions: i) $N(L)=\mathfrak{o}$; ii) L contains an element x such that $f(x,x)=1$; iii) Proposition 4.9 is true for L.

To prove this, let $U=Ax+Ay$ be an A-space of dimension 2. We can find an element γ of $\mathfrak{g}$ such that $(\gamma,\mathfrak{b})=1$, $\gamma\equiv 1 \bmod \mathfrak{p}_{\infty_\kappa}$ for $1\leq\kappa\leq t$, $\gamma\equiv -1 \bmod \mathfrak{p}_{\infty_\kappa}$ for $t<\kappa\leq u$. Define a Q-hermitian form f_0 on U by $f_0(x,x)=1$, $f_0(x,y)=0$, $f_0(y,y)=\gamma$. Now let W be an A-space of dimension $n-2$ and f_1 be a non-degenerate Q-hermitian form on W such that $\nu_\kappa(f_1)=\nu_\kappa(f)-2$ for $1\leq\kappa\leq t$ and $\nu_\kappa(f_1)=\nu_\kappa(f)-1$ for $t<\kappa\leq u$. Then, by Lemma 4.4, (V,f) is isomorphic to the direct sum of (U,f_0) and (W,f_1). Therefore, we may assume that $V=U+W$ $=Ax+Ay+W$, $W=\{z\in V\,|\,f(x,z)=f(y,z)=0\}$, $f=f_0$ on $U\times U$ and $f=f_1$ on $W\times W$. Further we see easily that $\mathfrak{i}(f_1)$ is a factor of $\mathfrak{i}(f)$. Let $\mathfrak{c}$ be an integral right $\mathfrak{o}$-ideal such that $N(\mathfrak{c})=\mathfrak{g}\gamma$, and let M be a maximal lattice in W such that $N(M)=\mathfrak{o}$. Put

$$K=\mathfrak{o}x+\mathfrak{c}^{-1}y, \qquad L=K+M.$$

Then, K is a maximal lattice in U, L is a maximal lattice in V; and $N(K)$ $=N(L)=\mathfrak{o}$. Now let the notation be as in Proposition 4.9. The structure of the $\mathfrak{o}_i$-module $L_i/L_i\sigma_i$ is determined by Proposition 2.12 and Proposition 3.9. In view of those propositions, we can find an element τ_i of $G(U_i,f_0)$ and an element ρ_i of $G(W_i,f_1)$ such that $N(\tau_i)=\alpha$, $N(\rho_i)=\alpha$, and

$$L_i/L_i\sigma_i\cong K_i/K_i\tau_i\oplus M_i/M_i\rho_i \qquad (1\leq i\leq r),$$

where $\cong$ means $\mathfrak{o}_i$-isomorphism. As $\alpha\equiv 1 \bmod \mathfrak{i}(f)$, we have $\alpha\equiv 1 \bmod \mathfrak{i}(f_1)$. By Assertion 6 and by our assumption of induction, there exists an element ρ of $G(W,f_1)$ such that $N(\rho)=\alpha$, $M\rho\subset M$,

$$M_i/M_i\rho\cong M_i/M_i\rho_i \qquad (1\leq i\leq r).$$

By Proposition 4.8, there exists an element τ of $G(U,f_0)$ such that $N(\tau)=\alpha$, $K\tau\subset K$, $K_i/K_i\tau\cong K_i/K_i\tau_i$ $(1\leq i\leq r)$. Define an element σ of $E(V,A)$ by $z\sigma=z\tau$ for $z\in U$ and $w\sigma=w\rho$ for $w\in W$. Then it is clear that this σ has the required properties of Proposition 4.9. Our assertion is thereby proved.

Assertion 8. For every maximal order $\mathfrak{o}$ in A, there exists a member L of $\mathfrak{L}_0(\mathfrak{o})$ for which Theorem 1 is true and $N(L)=\mathfrak{o}$.

We take as L the one which satisfies the conditions i-iii) of Assertion 7. By Assertion 5, Theorem 1 is true for $\alpha=1$, for this L. Now let the notation be as in Theorem 1. Then $\mathfrak{g}_iN(\sigma_i)=\mathfrak{g}_i\alpha$. By Assertion 7, there exists an element τ of $G(V,f)$ such that $N(\tau)=\alpha$, $L\tau\subset L$ and $L_i/L_i\tau$ is isomorphic to $L_i/L_i\sigma_i$ as $\mathfrak{o}_i$-module for $1\leq i\leq r$. By Proposition 2.13 and Proposition 3.11, there exists, for each i, an element ε_i of $G^0(V_i,f)$ such that $L_i\varepsilon_i=L_i$, $L_i\tau\varepsilon_i$ $=L_i\sigma_i$. Then we have $L_i\sigma_i\varepsilon_i^{-1}\tau^{-1}=L_i$ and $N(\sigma_i\varepsilon_i^{-1}\tau^{-1})\equiv 1 \bmod \mathfrak{p}_i^{\lambda_i}$. By Assertion 5, there exist elements ρ and η of $G^0(V,f)$ such that $L\rho=L$, $L\eta=L$, $L_i(\rho-\sigma_i\varepsilon_i^{-1}\tau^{-1})$ $\subset\mathfrak{p}_i^{\lambda_i}L_i$, $L_i(\eta-\varepsilon_i)\subset\mathfrak{p}_i^{\lambda_i}L_i$. Put $\sigma=\rho\tau\eta$. We have then $N(\sigma)=\alpha$, $L\sigma\subset L$. If

$z \in L_i$, we have

$$z\sigma \equiv z\rho\tau\eta \equiv z\rho\tau\varepsilon_i \equiv z(\sigma_i\varepsilon_i^{-1}\tau^{-1})\tau\varepsilon_i \equiv z\sigma_i \quad \mathrm{mod}\ \mathfrak{p}_i^{\lambda_i}L_i,$$

so that $L_i(\sigma-\sigma_i)\subset\mathfrak{p}_i^{\lambda_i}L_i$ for every i. This proves our assertion.

By Assertions 1, 2, 3 and 8, our Theorem 1 is completely proved.

We get a little weaker result than Theorem 1 for general maximal lattices, namely:

THEOREM 2. *Suppose that A is indefinite. Let M be a maximal lattice in V not necessarily belonging to the principal genus. Let $\mathfrak{p}_1, \cdots, \mathfrak{p}_r$ be distinct prime ideals of F and let α be an element of $\mathfrak{g}$. Let σ_i, for $1 \leq i \leq r$, be an element of $G(V_{\mathfrak{p}_i}, f)$. Suppose that $\alpha \equiv 1 \bmod \mathfrak{t}$, $M_{\mathfrak{p}_i}\sigma_i \subset M_{\mathfrak{p}_i}$, $N(\sigma_i) = \alpha$ for $1 \leq i \leq r$. Then, for any set of positive integers $\{\lambda_1, \cdots, \lambda_r\}$, there exists an element σ of $G(V, f)$ such that $M\sigma \subset M$, $N(\sigma) = \alpha$, $M_{\mathfrak{p}_i}(\sigma-\sigma_i) \subset \mathfrak{p}_i^{\lambda_i}M_{\mathfrak{p}_i}$ for $1 \leq i \leq r$.*

PROOF. Let $\mathfrak{o}$ be the order of M. Take a member L of $\mathfrak{L}_0(\mathfrak{o})$. Let $\{\mathfrak{p}_{r+1}, \cdots, \mathfrak{p}_w\}$ be the set of prime ideals $\mathfrak{p}$ of F such that $M_{\mathfrak{p}} \neq L_{\mathfrak{p}}$ and $\mathfrak{p} \notin \{\mathfrak{p}_1, \cdots, \mathfrak{p}_r\}$. For each $\mathfrak{p}_{r+i}$, we can find, in view of Proposition 2.10 and Proposition 3.5, an element σ_{r+i} of $G(V_{r+i}, f)$ such that $M_{r+i}\sigma_{r+i} \subset M_{r+i}$, $N(\sigma_{r+i}) = \alpha$. Take an integral ideal $\mathfrak{a}$ of F such that $\mathfrak{a}L \subset M$, $\mathfrak{a}M \subset L$, $\mathfrak{a}L_k\sigma_k \subset L_k$ for $1 \leq k \leq w$. We may assume that the prime factors of $\mathfrak{a}$ belong to $\{\mathfrak{p}_1, \cdots, \mathfrak{p}_r, \mathfrak{p}_{r+1}, \cdots, \mathfrak{p}_w\}$. For a suitably large positive integer h, $\mathfrak{a}^h$ is a principal ideal $\mathfrak{g}\beta$; we have then $\beta L \subset M$, $\beta M \subset L$, $\beta L_k\sigma_k \subset L_k$ for $1 \leq k \leq w$. By Theorem 1, there exists an element τ of $G(V, f)$ such that $L\tau \subset L$, $N(\tau) = \beta^2\alpha$, $L_k(\tau-\beta\sigma_k) \subset \beta^3\mathfrak{p}_k^{\lambda_k}L_k$ for $1 \leq k \leq w$. Put $\sigma = \beta^{-1}\tau$. Then $N(\sigma) = \alpha$ and $M_k(\sigma-\sigma_k) = \beta^{-1}M_k(\tau-\beta\sigma_k) \subset \beta^{-2}L_k(\tau-\beta\sigma_k) \subset \beta\mathfrak{p}_k^{\lambda_k}L_k \subset \mathfrak{p}_k^{\lambda_k}M_k$ for $1 \leq k \leq w$. As $M_k\sigma_k \subset M_k$, this implies $M_k\sigma \subset M_k$ for $1 \leq k \leq w$. If $\mathfrak{p} \notin \{\mathfrak{p}_1, \cdots, \mathfrak{p}_w\}$, we have $L_{\mathfrak{p}} = M_{\mathfrak{p}}$, and β is a $\mathfrak{p}$-unit. We have therefore $M_{\mathfrak{p}}\sigma = L_{\mathfrak{p}}\beta^{-1}\sigma = L_{\mathfrak{p}}\tau \subset L_{\mathfrak{p}} = M_{\mathfrak{p}}$. Hence $M_{\mathfrak{p}}\sigma \subset M_{\mathfrak{p}}$ for any prime ideal $\mathfrak{p}$ of F. It follows that $M\sigma \subset M$. This completes the proof.

4.7. Class-number theorem. For every maximal lattice L in V, put $N^0(L) = N(L) \cap F$. Then $N^0(L)$ is a $\mathfrak{g}$-ideal. If L is a member of $\mathfrak{L}(\mathfrak{o}; \{e_i\})$, we have

$$N(L) = N^0(L) \cdot \prod_{i=1}^{s} \mathfrak{Q}_i^{-e_i}.$$

THEOREM 3. *Suppose that A is indefinite. Then, for every maximal order $\mathfrak{o}$ in A and for every genus $\mathfrak{L}(\mathfrak{o}; \{e_i\})$ of maximal lattices in V, the mapping $L \to N^0(L)$ gives a one-to-one correspondence between the classes of maximal lattices in $\mathfrak{L}(\mathfrak{o}; \{e_i\})$ and the ideal-classes modulo $\mathfrak{t}$ in F.*

Therefore, the number of classes in the genus $\mathfrak{L}(\mathfrak{o}; \{e_i\})$ is equal to the number of ideal-classes modulo $\mathfrak{t}$ in F.

PROOF. Let L and M be members of $\mathfrak{L}(\mathfrak{o}; \{e_i\})$. If we have $L\rho = M$ for an element $\rho \in G(V, f)$ we have $N(L)N(\rho) = N(M)$, so that $N^0(L)N(\rho) = N^0(M)$. As $N(\rho) \equiv 1 \bmod \mathfrak{t}$, the ideals $N^0(L)$ and $N^0(M)$ belong to the same ideal-class modulo $\mathfrak{t}$ in F. Conversely, suppose that $\alpha N^0(L) = N^0(M)$ for an element $\alpha \in F$

and $\alpha \equiv 1 \bmod \mathfrak{l}$. Let β be an element of $\mathfrak{g}$ such that $\beta M \subset L$. Let $\mathfrak{a}$ be an integral ideal of F such that $\beta M \subset \mathfrak{a}L$. Let $\mathfrak{p}_1, \cdots, \mathfrak{p}_r$ be the prime factors of $\mathfrak{a}$. As we have $N(\beta M) = \beta^2 \mathfrak{a} N(L)$, we find, for each $\mathfrak{p}_i$, by Proposition 2.11 and Proposition 3.7, an element σ_i of $G(V_i, f)$ such that $L_i \sigma_i = \beta M_i$, $N(\sigma_i) = \beta^2 \mathfrak{a}$. Since $\beta^2 \mathfrak{a} \equiv 1 \bmod \mathfrak{l}$, we can apply Theorem 2 to $\{L, \sigma_i, \beta^2 \mathfrak{a}\}$. Then we get an element σ of $G(V, f)$ such that $L\sigma \subset L$, $N(\sigma) = \beta^2 \mathfrak{a}$, $L_i(\sigma - \sigma_i) \subset \mathfrak{p}_i \mathfrak{a} L_i$. As $\mathfrak{p}_i \mathfrak{a} L_i \subset \mathfrak{p}_i \beta M_i = \mathfrak{p}_i L_i \sigma_i$, we have $L_i \sigma = L_i \sigma_i = \beta M_i$ by Lemma 1.2. If $\mathfrak{p}$ is a prime ideal which does not divide $\mathfrak{a}$, we have $L_\mathfrak{p} = \beta M_\mathfrak{p}$, because $L \supset \beta M \supset \mathfrak{a}L$. It follows that $N(L_\mathfrak{p}) = \beta^2 \mathfrak{a} N(L_\mathfrak{p})$, and hence $\beta^2 \mathfrak{a}$ is a $\mathfrak{p}$-unit. As $L\sigma \subset L$ and $N(\sigma) = \beta^2 \mathfrak{a}$, we must have $L_\mathfrak{p}\sigma = L_\mathfrak{p} = \beta M_\mathfrak{p}$. Therefore we have $L_\mathfrak{p}\sigma = \beta M_\mathfrak{p}$ for any prime ideal $\mathfrak{p}$ of F, so that $L\sigma = \beta M$. Putting $\tau = \beta^{-1}\sigma$, we get $L\tau = M$. In view of Proposition 4.5, this proves our theorem.

4.8. Classes with respect to $G^0(V, f)$. If we take $G^0(V, f)$ instead of $G(V, f)$, we find that the class-number of each genus is equal to one. In fact, by Theorem 3 and its proof, we obtain easily

PROPOSITION 4.10. *Suppose that A is indefinite. Let L and M be maximal lattices in V with the same order. Then, there exists an element σ of $G^0(V, f)$ such that $L\sigma = M$, if and only if $N(L) = N(M)$.*

Notation being as in § 4.3, let $\mathfrak{G}^0$ be the subgroup of $\mathfrak{G}$ consisting of elements $(\sigma_\mathfrak{p}, \sigma_\kappa)$ such that $\sigma_\mathfrak{p} \in G^0(V_\mathfrak{p}, f)$ for every $\mathfrak{p}$ and $\sigma_\kappa \in G^0(V_\kappa, f)$ for every κ. Then $\mathfrak{G}^0$ can be regarded as the adele-group of $G^0(V, f)$. Put $G^0 = G^0(V, f)$ and $\mathfrak{U}_L^0 = \mathfrak{U}_L \cap \mathfrak{G}^0$. Then Proposition 4.10 implies the equality

$$\mathfrak{G}^0 = \mathfrak{U}_L^0 \cdot G^0$$

for every maximal lattice L in V.

4.9. Elementary divisors of lattices. Let L and M be members of the same genus $\mathfrak{L}(\mathfrak{o}; \{e_i\})$. For every prime ideal $\mathfrak{p}$ of F, we can define, as in § 2.5 and § 3.2, the set of elementary divisors $\{L_\mathfrak{p} : M_\mathfrak{p}\}$. We put $\{L : M\}_\mathfrak{p} = \{L_\mathfrak{p} : M_\mathfrak{p}\}$ and call it the $\mathfrak{p}$-part of the set of elementary divisors of M relative to L. *The (global) set of elementary divisors of M relative to L is defined as the join of $\{L : M\}_\mathfrak{p}$ for all prime ideals $\mathfrak{p}$ of F and denoted by $\{L : M\}$.*

THEOREM 4. *Suppose that A is indefinite. Let L, M, K be maximal lattices in V belonging to the same genus. Then, we have $\{L : M\} = \{L : K\}$ if and only if there exists an element τ of $G^0(V, f)$ such that $L\tau = L$ and $M\tau = K$.*

PROOF. The 'if' part is clear. Suppose that $\{L : M\} = \{L : K\}$. Let Ψ be the set of prime ideals $\mathfrak{p}$ of F for which $L_\mathfrak{p} = M_\mathfrak{p} = K_\mathfrak{p}$ does not hold. By Lemma 1.1, Ψ is a finite set. By Proposition 2.13 and Proposition 3.11, there exists, for each $\mathfrak{p} \in \Psi$, an element $\tau_\mathfrak{p}$ of $G^0(V_\mathfrak{p}, f)$ such that $L_\mathfrak{p}\tau_\mathfrak{p} = L_\mathfrak{p}$ and

$M_\mathfrak{p}\gamma_\mathfrak{p}=K_\mathfrak{p}$. Take a positive integer c such that $\mathfrak{p}^c M_\mathfrak{p}\subset L_\mathfrak{p}$ and $\mathfrak{p}^c L_\mathfrak{p}\subset K_\mathfrak{p}$ for every $\mathfrak{p}\in\Psi$. By Theorem 2, there exists an element γ of $G(V,f)$ such that $N(\gamma)=1$, $L\gamma\subset L$, $L_\mathfrak{p}(\gamma-\gamma_\mathfrak{p})\subset\mathfrak{p}^{2c+1}L_\mathfrak{p}$ for every $\mathfrak{p}\in\Psi$. Then obviously $L\gamma=L$, and $M_\mathfrak{p}(\gamma-\gamma_\mathfrak{p})\subset\mathfrak{p}^{-c}L_\mathfrak{p}(\gamma-\gamma_\mathfrak{p})\subset\mathfrak{p}\cdot\mathfrak{p}^c L_\mathfrak{p}\subset\mathfrak{p}K_\mathfrak{p}=\mathfrak{p}M_\mathfrak{p}\gamma_\mathfrak{p}$. By Lemma 1.2, we have $M_\mathfrak{p}\gamma=M_\mathfrak{p}\gamma_\mathfrak{p}=K_\mathfrak{p}$ for every $\mathfrak{p}\in\Psi$. If $\mathfrak{p}\notin\Psi$, we have $M_\mathfrak{p}\gamma=L_\mathfrak{p}\gamma=L_\mathfrak{p}=K_\mathfrak{p}$. Hence $M_\mathfrak{p}\gamma=K_\mathfrak{p}$ holds for any prime ideal $\mathfrak{p}$ of F, so that $M\gamma=K$. This proves the 'only if' part.

PROPOSITION 4.11. *Suppose that A is indefinite. Let L and M be maximal lattices in V belonging to the same genus. Define the subgroups $\mathfrak{U}_L$ and $\mathfrak{U}_M$ of the adele-group $\mathfrak{G}$ as in* §4.3. *Put $\Gamma_L=\mathfrak{U}_L\cap G$, $\Gamma_M=\mathfrak{U}_M\cap G$. Then we have, for every $\xi\in\mathfrak{G}$,*

$$\mathfrak{U}_L\xi\mathfrak{U}_M=\mathfrak{U}_L\xi\Gamma_M=\Gamma_L\xi\mathfrak{U}_M.$$

PROOF. It is clear that $\mathfrak{U}_L\xi\mathfrak{U}_M\supset\mathfrak{U}_L\xi\Gamma_M$. Let u be an element of $\mathfrak{U}_M$. As $Mu=M$, we see easily, on account of the definition of $\mathfrak{U}_M$, that $\{M:L\xi u\}=\{M:L\xi\}$. By Theorem 4, there exists an element γ of Γ_M such that $L\xi u=L\xi\gamma$. It follows that $\mathfrak{U}_L\xi u=\mathfrak{U}_L\xi\gamma\subset\mathfrak{U}_L\xi\Gamma_M$. This shows $\mathfrak{U}_L\xi\mathfrak{U}_M\subset\mathfrak{U}_L\xi\Gamma_M$, and hence $\mathfrak{U}_L\xi\mathfrak{U}_M=\mathfrak{U}_L\xi\Gamma_M$. Similarly we get $\mathfrak{U}_M\xi^{-1}\mathfrak{U}_L=\mathfrak{U}_M\xi^{-1}\Gamma_L$, so that $\mathfrak{U}_L\xi\mathfrak{U}_M=\Gamma_L\xi\mathfrak{U}_M$. This completes the proof.

The above theorem and proposition are generalization of [9, Proposition 1.4, Proposition 2.3]. These are necessary for our future investigation of the Hecke-ring of G.

Osaka University

References

[1] A. Borel, Some properties of adele groups attached to algebraic groups, Bull. Amer. Math. Soc., 67 (1961), 583-585.

[2] N. Bourbaki, Algèbre, Chap. 9, Formes sesquilinéaires et formes quadratiques, Hermann, Paris, 1959.

[3] M. Eichler, Quadratische Formen und orthogonale Gruppen, Berlin-Göttingen-Heidelberg (Springer), 1952.

[4] M. Eichler, Die Ähnlichkeitsklassen indefiniter Gitter, Math. Z., 55 (1952), 216-252.

[5] M. Eichler, Allgemeine Kongruenzklasseneinteilungen der Ideale einfacher Algebren über algebraischen Zahlkörpern und ihre L-Reihen, J. Reine Angew. Math., 179 (1938), 227-251.

[6] M. Kneser, Klassenzahlen indefiniter quadratischer Formen in drei oder mehr Veränderlichen, Arch. Math., 7 (1956), 323-332.

[7] K. G. Ramanathan, Quadratic forms over involutorial division algebras, J. Indian Math. Soc., 20 (1956), 227-257.

[8] G. Shimura, On the zeta-functions of the algebraic curves uniformized by certain automorphic functions, J. Math. Soc. Japan, 13 (1961), 275-331.

[9] G. Shimura, On Dirichlet series and abelian varieties attached to automorphic

forms, Ann. Math., **76** (1962), 237-294.

[10] T. Tsukamoto, On the local theory of quaternionic anti-hermitian forms, J. Math. Soc. Japan, **13** (1961), 387-400.

[11] A. Weil, Discontinuous subgroups of classical groups, lecture note, Univ. of Chicago, 1958.

[12] A. Weil, Adeics and algebraic groups, lecture note, Institute for Advanced Study, Princeton, 1961.

On analytic families of polarized abelian varieties and automorphic functions

Annals of Mathematics, 78 (1963), 149-192

(Received June 22, 1962)

There are several types of automorphic functions whose values occur as moduli of abelian varieties. In the arithmetical theory of those functions, it is important to investigate them in this connection with abelian varieties. In a previous paper [14] (quoted hereafter as [AF]), we have shown that, for any analytic family of polarized abelian varieties, there exist meromorphic functions whose values are considered as moduli of the members of the family. The purpose of the present paper is to continue this investigation. We now explain our problems by giving a summary of the contents.

A polarization C of an abelian variety A determines a positive involution on the endomorphism-algebra $\mathcal{A}_0(A)$ of A. If the universal domain is the complex number field C, the analytic coordinate-system of A determines a representation of $\mathcal{A}_0(A)$ by complex matrices whose degree is the dimension of A. Thus our first problem is as follows. Take a division algebra L over the rational number field Q with a positive involution ρ and a representation Φ of L by complex matrices; determine all the analytic families of polarized abelian varieties $(A,\ C)$ such that $\mathcal{A}_0(A)$ contains L, and C gives the involution ρ on L, and Φ is equivalent with the representation by an analytic coordinate-system. Now the classification of division algebras with positive involution was done by Albert, who also proved the existence of A such that $\mathcal{A}_0(A) = L$ for a given L [1, 2, 3]. Each analytic family Σ of polarized abelian varieties, for any $\{L, \Phi, \rho\}$, is shown to be parametrized by points of a symmetric domain $\mathcal{H}$, and two members of Σ are isomorphic if and only if the corresponding points on $\mathcal{H}$ are transformed onto each other by an element of a discontinuous group Γ of transformations on $\mathcal{H}$. The classification of $\mathcal{H}$ and Γ obtained from a given $\{L, \Phi, \rho\}$ has been treated by Pyatetskii-Shapiro [9] from a somewhat different view-point than ours; in this work, however, the interpretation of the points of $\Gamma\backslash\mathcal{H}$ was left vague. In §2 of the present paper, we determine, using a notion introduced in [AF], all Σ as analytic families of polarized abelian varieties with given $\{L, \Phi, \rho\}$, and establish a one-to-one correspondence between Σ and $\Gamma\backslash\mathcal{H}$ in an explicit form (Theorems 1, 2). Then applying the theory of [AF],

149

we get some automorphic functions $f_1, \cdots, f_\kappa$ on $\mathcal{H}$ with respect to Γ, whose values generate the field of moduli of each (at least generic) member of Σ (Theorems 3, 4); and we prove that the field $C(f_1, \cdots, f_\kappa)$ can be defined over an algebraic number field (Theorem 4, (iii)). It is conjecturable that $C(f_1, \cdots, f_\kappa)$ is the field of *all* automorphic functions on $\mathcal{H}$ with respect to Γ.[1] If L is a totally real algebraic number field, this is true by virtue of Baily's work [4, 5]. For general L, we have only proved that the transcendency-degree of $C(f_1, \cdots, f_\kappa)$ is equal to the dimension of $\mathcal{H}$, and any two *generic* points of $\Gamma \backslash \mathcal{H}$ can be separated by the functions $f_1, \cdots, f_\kappa$ (Theorem 4, (i), (ii)). Considering the points of finite order on the members of Σ, we get automorphic functions with respect to congruence-subgroups of Γ (§3.5).

For a given $\{L, \Phi, \rho\}$, and for a generic member A of Σ, one might expect $\mathcal{A}_0(A) = L$; but this is not necessarily true as was already noted by Albert. The object of §4 is to determine $\mathcal{A}_0(A)$ for a generic member A of Σ. There are five exceptional cases where $\mathcal{A}_0(A)$ does not coincide with L (Theorem 5, Propositions 14, 15, 17, 18, 19).

The group Γ is defined as a group of linear transformations leaving invariant a ρ-skew-symmetric form T over L. There is a little larger discontinuous group Γ_0 containing Γ whose elements are similitudes of T. In order to treat the automorphic functions with respect to Γ_0, we introduce, in the final §5, a notion of weak polarization. Then the values of the automorphic functions with respect to Γ_0 generate the fields of moduli of weakly polarized abelian varieties (Theorem 7).

Notation. We denote by $\boldsymbol{Z}, \boldsymbol{Q}, \boldsymbol{R}, \boldsymbol{C}$ and $\boldsymbol{K}$, respectively, the ring of rational integers, the rational number field, the real number field, the complex number field, and the division ring of real quaternions. For a ring S with an identity element, $M_m(S)$ denotes the ring of matrices of degree m with entries in S; the identity matrix of degree m is denoted by 1_m; and the transpose of a matrix X is denoted by $^t X$. When A is an abelian variety, we denote by $\mathcal{A}(A)$ the ring of endomorphisms of A, and by $\mathcal{A}_0(A)$ the algebra obtained from $\mathcal{A}(A)$ by extending the coefficients to $\boldsymbol{Q}$. The Chow-point of an algebraic variety V is denoted by $c(V)$ (cf. [AF, *Notation*]).

1. Algebras with involutions

1.1. *Positive involutions.* Let L be a semi-simple algebra over $\boldsymbol{Q}$ (resp. $\boldsymbol{R}$). By an *involution* of L, we understand, as usual, an anti-auto-

---

[1] When $\mathcal{H}$ is of dimension one and $\Gamma \backslash \mathcal{H}$ is not compact, an *automorphic function* must be defined as meromorphic at every cusp of Γ with respect to a suitable local parameter.

morphism ρ of L over Q (resp. R) such that $(x^\rho)^\rho = x$ for every $x \in L$. An element s of L is called ρ-*symmetric* (resp. ρ-*skew-symmetric*) if $s^\rho = s$ (resp. $s^\rho = -s$). Let $\mathrm{tr}(x)$ denote the reduced trace of an element x of L with respect to Q (resp. R). We call an involution ρ of L *positive* if $\mathrm{tr}(xx^\rho) > 0$ for every non-zero element x of L. When ρ is positive, a ρ-symmetric element s is called ρ-*positive* if $\mathrm{tr}(xsx^\rho) > 0$ for every non-zero element x of L.

Every involution ρ of L is extended to an involution of $M_r(L)$, which we denote again by ρ, by putting $(x_{ij})^\rho = {}^t(x_{ij}^\rho)$ for $(x_{ij}) \in M_r(L)$ with x_{ij} in L. If ρ is positive on L, then it is also positive on $M_r(L)$.

Let L_1 be a semi-simple algebra over R with a positive involution ρ. Then, obviously, ρ maps every simple component of L_1 onto itself. Now, any simple algebra over R is of the form $M_r(R)$, $M_r(C)$, or $M_r(K)$. For these matric algebras, we define positive involutions $*$ as follows:

 (i) $M_r(R)$: $X^* = {}^t X$;

 (ii) $M_r(C)$: $X^* = {}^t \bar{X}$, where the bar denotes the complex conjugate;

 (iii) $M_r(K)$: $X^* = {}^t \bar{X}$, where the bar denotes the quaternionic conjugate.

If L denotes any one of these simple algebras, for any $*$-symmetric and $*$-positive element T of L, we obtain a positive involution ρ of L by $X^\rho = TX^*T^{-1}$; and any positive involution of L is obtained in this manner. As T is $*$-symmetric and $*$-positive, we can find a $*$-symmetric and $*$-positive element U such that $U^2 = T$. We have then $U^{-1}X^\rho U = (U^{-1}XU)^*$. In other words,

LEMMA 1. *Let L be a simple algebra over R with a positive involution ρ. Then there exists an isomorphism ψ of L onto a matric algebra belonging to the above three types such that $\psi(x^\rho) = \psi(x)^*$ for every $x \in L$.*

Now we consider a representation of $M_r(K)$ in $M_{2r}(C)$. First we identify C with a subfield of K and take an element j of K such that $jxj^{-1} = \bar{x}$ for every $x \in C$ and $j^2 = -1$. Then we have $K = C + Cj$; and every element X of $M_r(K)$ is written in the form $X = A + Bj$ with A, B in $M_r(C)$. Put

$$(1) \qquad \iota(X) = \begin{pmatrix} A & B \\ -\bar{B} & \bar{A} \end{pmatrix}.$$

Then ι is an irreducible representation of $M_r(K)$ in $M_{2r}(C)$. Henceforth we fix j and an identification of C with a subfield of K, and always denote by ι this representation. It is easy to verify

$$(2) \qquad \iota(X^*) = \iota(X)^*$$

for every $X \in M_r(K)$. Therefore, X is $*$-symmetric (resp. $*$-skew-symmetric) if and only if $\iota(X)$ is $*$-symmetric (resp. $*$-skew-symmetric). Moreover, when X is $*$-symmetric, X is $*$-positive if and only if $\iota(X)$ is $*$-positive.

1.2. *Classification of division algebras with positive involutions.*[2] Let L be a division algebra over Q with a positive involution ρ. Denote by Z the center of L. Then ρ induces a positive involution on Z. Let F be the set of elements z of Z such that $z^\rho = z$. As we have $\mathrm{tr}(z^2) > 0$ for every non-zero element z of F, the field F must be a totally real algebraic number field; and the following two cases may occur.

(i) $Z = F$;

(ii) $[Z:F] = 2$, and Z is totally imaginary.

We say that ρ is of *the first kind* or *the second kind* according as $Z = F$ or $Z \neq F$. When ρ is of the second kind, we denote Z by F_0.

Now suppose that ρ is of the first kind. Then ρ is an anti-automorphism of L over F. By Hasse-Brauer-Noether's theorem, $L = F$ or L is a quaternion division algebra over F, i.e., a central division algebra over F such that $[L:F] = 4$. Assuming L to be a quaternion algebra over F, L has an involution σ over F such that $x + x^\sigma = \mathrm{tr}_{L/F}(x)$ for every $x \in L$, where $\mathrm{tr}_{L/F}$ denotes the reduced trace with respect to F. We call σ the *canonical involution* of L; σ may be or may not be positive. As both ρ and σ are anti-automorphisms of L over F, there is an element a of L such that $ax^\sigma a^{-1} = x^\rho$ for every $x \in L$; and as ρ is an involution, we have $a^\sigma = \varepsilon a$, $\varepsilon = \pm 1$, so that $a^2 \in F$. Let $[F:Q] = g$, and let $L_1, \cdots, L_g$ be the simple components of $L_R = L \otimes_Q R$. The L_i are isomorphic to $M_2(R)$ or K, since F is totally real. The involutions ρ and σ are extended to the involutions of L_R in a natural manner, which we denote again by the same notation; and ρ is positive on L_R. By Lemma 1, we can identify each L_i with $M_2(R)$ or K in such a way that $x_i^\rho = x_i^*$ for every $x_i \in L_i$. As σ is the canonical involution, we observe that σ induces on each L_i also the "canonical involution"; namely, $x_i^\sigma + x_i = \mathrm{tr}(x_i)$ for $x_i \in L_i$. Let a_i be the component of a on L_i. We have then $a_i^\sigma = \varepsilon a_i$, $a_i x_i^\sigma a_i^{-1} = x_i^*$ for $x_i \in L_i$. If $L_i = K$ for some i, we get $x_i^\sigma = x_i^*$ so that a_i must be contained in R. This implies $\varepsilon = 1$. Then we have $a^\sigma = a$, so that a is contained in F, and hence $\sigma = \rho$. As the canonical involution of $M_2(R)$ is not positive, we must have $L_k = K$ for every k. Thus we have proved that either $L_i = K$ for every i, or $L_i = M_2(R)$ for every i; and in the former case, the canonical involution is the only positive involution. Now suppose that $L_i = M_2(R)$ for every i. Then ε must be -1, since σ

---

[2] The result of this section is essentially due to Albert [2, 3].

is not positive; so we have $a_i^\sigma = -a_i$. Substituting a_i for x_i in the formula $a_i x_i^\sigma a_i^{-1} = x_i^*$, we get $a_i^* = -a_i$. Hence we have $a_i = \begin{pmatrix} 0 & \alpha_i \\ -\alpha_i & 0 \end{pmatrix}$ with α_i in $\boldsymbol{R}$, so that $a_i^2 = -\alpha_i^2$. This shows that $-a^2$ is a totally positive element of F. Conversely, let a be any element of L such that $-a^2$ is a totally positive element of F. Then we have $a^\sigma = -a$, $\det (a_i) = a_i a_i^\sigma = -a_i^2 > 0$. Put $b = \begin{pmatrix} 0 & 1 \\ -1 & 0 \end{pmatrix}$. Then $x^* = b^{-1} x^\sigma b$ for every $x \in M_2(\boldsymbol{R})$. Define an involution ρ of L (and of L_R) by $y^\rho = a y^\sigma a^{-1}$. We have then $x_i^\rho = a_i x_i^\sigma a_i^{-1} = (a_i b) x_i^* (a_i b)^{-1}$ for $x_i \in L_i$. It is easy to see that $a_i b$ is $*$-symmetric. As we have $\det (a_i b) = \det (a_i) > 0$, $a_i b$ or $-a_i b$ is $*$-positive. Hence ρ is a positive involution. Summing up whole results, we get the following three propositions.

PROPOSITION 1. *Every division algebra over $\boldsymbol{Q}$ with a poitive involution belongs to the following four types of algebras.*

(Type I) Totally real algebraic number field F.

(Type II) Central simple algebra L over F such that the simple components of $L_R = L \otimes_Q \boldsymbol{R}$ are all isomorphic to $M_2(\boldsymbol{R})$.

(Type III) Central simple algebra L over F such that the simple components of $L_R = L \otimes_Q \boldsymbol{R}$ are all isomorphic to $\boldsymbol{K}$.

(Type IV) Central simple algebra over a totally imaginary quadratic extension F_0 of F.

We call the algebras of (*Type* II) and (*Type* III) respectively a *totally indefinite quaternion algebra over F* and a *totally definite quaternion algebra over F*. Obviously, the rational number field $\boldsymbol{Q}$ belongs to (*Type* I), and any totally imaginary quadratic extension of a totally real algebraic number field belongs to (*Type* IV).

PROPOSITION 2. *Let L be a totally indefinite quaternion (not necessarily division) algebra over F, and σ the canonical involution of L. Then, for every element a of L such that a^2 is a totally negative element of F, the involution ρ of L defined by $x^\rho = a x^\sigma a^{-1}$ is positive. Conversely, every positive involution of L is obtained from such an element a of L in this manner.*

PROPOSITION 3. *If L is a totally definite quaternion algebra over F, the canonical involution of L is the only positive involution of L.*

In order that an algebra L of (*Type* IV) have a positive involution, L must satisfy a certain restrictive condition; a complete investigation of this, is given in [2, 3].[3]

---

[3] If an algebra L of (*Type* IV) has an involution of the second kind over F, then we can easily verify that L has a *positive* involution.

1.3. *The endomorphism-algebra of an abelian variety.* Let A be an abelian variety defined over a field with any characteristic, and B its Picard variety. Every divisor Y on A algebraically equivalent to 0, defines a point of B, which we denote by $Cl(Y)$. For every element λ of $\mathcal{A}(A)$, there exists an element ${}^t\lambda$ of $\mathcal{A}(B)$ defined by ${}^t\lambda\, Cl(Y) = Cl(\lambda^{-1}(Y))$. Then $\lambda \to {}^t\lambda$ is extended in a natural manner to an anti-isomorphism of $\mathcal{A}_0(A)$ onto $\mathcal{A}_0(B)$. Let X be a non-degenerate divisor on A; we obtain an isogeny φ_X of A onto B by $\varphi_X(u) = Cl(X_u - X)$, where X_u denotes the transform of X by the translation $x \to x + u$ on A. Assuming X to be positive non-degenerate, put for every $\alpha \in \mathcal{A}_0(A)$,

$$(3) \qquad\qquad \alpha^\rho = \varphi_X^{-1} \cdot {}^t\alpha \varphi_X \; .$$

Then ρ is a positive involution of $\mathcal{A}_0(A)$. Now let C be a polarization of A. Then we observe that any two divisors in C determine the same involution of $\mathcal{A}_0(A)$. Therefore, any polarized abelian variety (A, C) determines a semi-simple algebra over Q with a positive involution. If A is simple, $\mathcal{A}_0(A)$ is a division algebra; so $\mathcal{A}_0(A)$ belongs to one of the four types given in Proposition 1.

Suppose that A is an abelian variety of dimension n defined over C. A is then isomorphic to a complex torus C^n/D, where D is a discrete subgroup of C^n of rank $2n$. Fixing an isomorphism of A onto C^n/D, every positive non-degenerate divisor X on A corresponds to a non-degenerate Riemann form $E(\mathfrak{x}, \mathfrak{y})$ on C^n/D; and every element a of $\mathcal{A}_0(A)$ is represented by a matrix $\psi(a)$ of degree n with entries in C. If X corresponds to $E(\mathfrak{x}, \mathfrak{y})$ and a^ρ is determined by (3), we have

$$(4) \qquad\qquad E\big(\psi(a)\mathfrak{x}, \mathfrak{y}\big) = E\big(\mathfrak{x}, \psi(a^\rho)\mathfrak{y}\big) \; .$$

We call $a \to \psi(a)$ the *representation of $\mathcal{A}_0(A)$ by an analytic coordinate-system of A.*

1.4. *Formulation of the first problem.* Let L be a semi-simple algebra over Q with a positive involution ρ, and Φ a representation of L by complex matrices of degree n. We say that a triplet $\mathcal{P} = (A, C, \theta)$ is a *polarized abelian variety of type $\{L, \Phi, \rho\}$,* or more briefly, $\mathcal{P}$ *belongs to* $\{L, \Phi, \rho\}$, if the following conditions (i-iii) are satisfied.

(i) *A is an abelian variety of dimension n, defined over C.*

(ii) *θ is an isomorphism of L into $\mathcal{A}_0(A)$; and the representation of $\theta(x)$ for $x \in L$ by an analytic coordinate-system of A is equivalent to Φ.*

(iii) *C is a polarization of A; and the involution of $\mathcal{A}_0(A)$ determined by C coincides on $\theta(L)$ with the involution $\theta(x) \to \theta(x^\rho)$.*

Now our first problem is to determine analytic families of polarized abelian varieties belonging to a given system $\{L, \Phi, \rho\}$. We will then show that the moduli of members of each family give the automorphic functions belonging to a certain discontinuous group. In the present investigation, we restrict ourselves to the case where L is simple and belongs to the four types of Proposition 1. Of course one can consider a more general case where L is a total matric algebra $M_r(L_1)$ over an algebra L_1 belonging to those four types. However, by a simple calculation, we can easily verify that this generalization does not produce any new discontinuous group. Therefore, by our restriction, we would not lose anything of importance.

2. Determination of analytic families of polarized abelian varieties

2.1. *A necessary condition for* Φ. Let L be an algebra belonging to any one of the four types of Proposition 1. In view of our later need, we do not assume that L is a division algebra. F and F_0 being as in Proposition 1, we put

$$[F \colon \boldsymbol{Q}] = g \,,$$
$$[L \colon F_0] = q^2 \qquad\qquad \text{for } (Type\ IV)\,.$$

Then $L_R = L \otimes_{\boldsymbol{Q}} \boldsymbol{R}$ is isomorphic to

$$
\begin{aligned}
&(Type\ \mathrm{I}) && \boldsymbol{R} \times \cdots \times \boldsymbol{R}\,,\\
&(Type\ \mathrm{II}) && M_2(\boldsymbol{R}) \times \cdots \times M_2(\boldsymbol{R})\,,\\
&(Type\ \mathrm{III}) && \boldsymbol{K} \times \cdots \times \boldsymbol{K}\,,\\
&(Type\ \mathrm{IV}) && M_q(\boldsymbol{C}) \times \cdots \times M_q(\boldsymbol{C})\,,
\end{aligned}
$$

where the number of factors is g in each case. We identify L_R with these products of matric algebras, and denote by π_ν the projection of L_R onto the νth factor. Let ρ be a positive involution of L. Extend ρ to a positive involution of L_R and denote it again by ρ. By Lemma 1, we can take the coordinate-system in such a way that we have

$$(5) \qquad\qquad \pi_\nu(x^\rho) = \pi_\nu(x)^* $$

for every $x \in L_R$ and for every ν. Put for each ν,

$$
\begin{aligned}
&(Type\ \mathrm{I, II, IV}) && \chi_\nu = \pi_\nu\,,\\
&(Type\ \mathrm{III}) && \chi_\nu = \iota \circ \pi_\nu\,,
\end{aligned}
$$

where ι is the injection of $\boldsymbol{K}$ into $M_2(\boldsymbol{C})$ determined by (1) of §1.1. By (5) and (2) of §1.1, we have, for every $x \in L_R$,

(6)
$$\chi_\nu(x^\rho) = \chi_\nu(x)^* \ .$$

All the inequivalent absolutely irreducible representations of L are given by

$$(Type\ \mathrm{I, II, III}) \quad \chi_1, \cdots, \chi_g\ ,$$
$$(Type\ \mathrm{IV}) \qquad \chi_1, \cdots, \chi_g, \bar\chi_1, \cdots, \bar\chi_g\ ,$$

where $\bar\chi_\nu$ denotes the complex conjugate of χ_ν.

Let Φ be a representation of L by complex matrices of degree n containing no 0-representation. Φ can not be arbitrary in order to insure that there exists a polarized abelian variety of type $\{L, \Phi, \rho\}$. In fact, suppose that $\mathscr{P} = (A, C, \theta)$ belongs to $\{L, \Phi, \rho\}$. Then L has a rational representation of degree $2n$, which is equivalent to the sum of Φ and its complex conjugate $\bar\Phi$. Therefore, $\Phi + \bar\Phi$ contains every absolutely irreducible representations of L with the same multiplicities. Furthermore, if L is a division algebra, $[L: \boldsymbol{Q}]$ must divide $2n$. When L is not a division algebra, this is not necessarily true. For the sake of simplicity, we assume that $2n$ is a multiple of $[L: \boldsymbol{Q}]$, and put

(7)
$$2n = [L: \boldsymbol{Q}]m \ .$$

As we have said, this is not any restriction when L is a division algebra. Then our representation Φ must be as follows.

(*Type* I, II, III) Φ *is equivalent to a certain multiple of the sum of all the* χ_ν; *the multiplicity is equal to* $m/2$ *for* (*Type* I), *and it is* m *for* (*Type* II, III).

(*Type* IV) *Let* r_ν *and* s_ν *be respectively the multiplicities of* χ_ν *and* $\bar\chi_\nu$ *in* Φ. *Then, for every* ν, *we have*

(8)
$$r_\nu + s_\nu = mq \ .$$

2.2. Assume that Φ satisfies the above condition. In order to fix our idea, we take an explicit form of Φ as follows. For any matrices $X = (\xi_{ij}) \in M_u(\boldsymbol{C})$ and $Y \in M_v(\boldsymbol{C})$, we put

$$X \otimes Y = \begin{bmatrix} \xi_{11}Y & \cdots & \xi_{1u}Y \\ \cdots & \cdots & \cdots \\ \xi_{u1}Y & \cdots & \xi_{uu}Y \end{bmatrix} \ .$$

Define Φ_ν for each ν by

$$(Type\ \mathrm{I}) \qquad \Phi_\nu(x) = \chi_\nu(x) \otimes 1_{m/2}\ ,$$
$$(Type\ \mathrm{II, III}) \quad \Phi_\nu(x) = \chi_\nu(x) \otimes 1_m\ ,$$
$$(Type\ \mathrm{IV}) \qquad \Phi_\nu(x) = \begin{bmatrix} \chi_\nu(x) \otimes 1_{r_\nu} & 0 \\ 0 & \bar\chi_\nu(x) \otimes 1_{s_\nu} \end{bmatrix}\ ,$$

and put

$$\Phi(x) = \begin{bmatrix} \Phi_1(x) & & \\ & \cdot \cdot \cdot & \\ & & \Phi_g(x) \end{bmatrix} .$$

Now we will determine the polarized abelian varieties of type $\{L, \Phi, \rho\}$. Let $\mathscr{P} = (A, C, \theta)$ belong to $\{L, \Phi, \rho\}$. Take a complex torus C^n/D isomorphic to A. We may choose the coordinate system so that $\theta(a)$ is represented by the matrix $\Phi(a)$ for every $a \in L$. Then QD is considered as a left L-module. In view of (7), we have

$$(9) \qquad QD = \sum_{i=1}^{m} \Phi(L)\mathfrak{x}_i$$

for a suitable set of vectors $\mathfrak{x}_i$; and there exists a free Z-submodule $\mathfrak{M}$ of rank $2n$ in $L^m = L \times \cdots \times L$ such that

$$(10) \qquad D = \{\sum_{i=1}^{m} \Phi(a_i)\mathfrak{x}_i \mid (a_1, \cdots, a_m) \in \mathfrak{M}\} .$$

Let $E(\mathfrak{x}, \mathfrak{y})$ be a non-degenerate Riemann form on C^n/D. Then, E must satisfy the conditions:

($E1$) E is an R-valued R-bilinear skew-symmetric form on $C^n \times C^n$.

($E2$) E is Z-valued on $D \times D$.

($E3$) $E(\mathfrak{x}, \sqrt{-1}\,\mathfrak{y})$ is symmetric and positive non-degenerate.

Furthermore, if E corresponds to a divisor in a polarization C, then, by (4) of §1.3 and by (iii) of §1.4, E must satisfy

($E4$) $E(\mathfrak{x}, \Phi(a)\mathfrak{y}) = E(\Phi(a^\rho)\mathfrak{x}, \mathfrak{y})$ for every $a \in L$.

For each i and j, the mapping $a \to E(\Phi(a)\mathfrak{x}_i, \mathfrak{x}_j)$ is a Q-linear mapping of L into Q. Hence there exists an element t_{ij} of L such that $E(\Phi(a)\mathfrak{x}_i, \mathfrak{x}_j) = \operatorname{tr}(at_{ij})$, where tr denotes the reduced trace of L to Q. By ($E4$), we obtain

$$(11) \qquad E\big(\sum_{i=1}^{m} \Phi(a_i)\mathfrak{x}_i, \sum_{j=1}^{m} \Phi(b_j)\mathfrak{x}_j\big) = \operatorname{tr}\big(\sum_{i,j=1}^{m} a_i t_{ij} b_j^\rho\big) .$$

This relation holds also for $a_i, b_j \in L_R$. Put $T = (t_{ij})$; T is an element of $M_m(L)$; and by ($E1$), we have

$$(12) \qquad T^\rho = -T .$$

By ($E2$), we must have

$$(12') \qquad \operatorname{tr}(\mathfrak{M}T\mathfrak{M}^\rho) \subset Z ,$$

where we consider the elements of $\mathfrak{M}$ as row-vectors with components in L. Thus from $\mathscr{P} = (A, C, \theta)$ we obtain m vectors $\mathfrak{x}_1, \cdots, \mathfrak{x}_m$ of C^n, a lattice $\mathfrak{M}$ in L^m, and a ρ-skew-symmetric element T of $M_m(L)$. As D is a lattice in C^n, we must have

$$(13) \qquad C^n = \sum_{i=1}^{m} \Phi(L_R)\mathfrak{x}_i ,$$

158 GORO SHIMURA

and the mapping $(a_1, \cdots, a_m) \to \sum_{i=1}^{m} \Phi(a_i)\mathfrak{x}_i$ gives an isomorphism of $L_R^m = L_R \times \cdots \times L_R$ onto C^n. Hence there exists one and only one matrix $H = (h_{ij}) \in M_m(L_R)$ such that

$$(14) \qquad \sqrt{-1}\, \mathfrak{x}_i = \sum_{j=1}^{m} \Phi(h_{ij})\mathfrak{x}_j \qquad\qquad (1 \leqq i \leqq m) .$$

Then, we have, for $a_i, b_j \in L_R$,

$$E\big(\textstyle\sum_{i=1}^{m} \Phi(a_i)\mathfrak{x}_i,\ \sqrt{-1}\sum_{j=1}^{m}\Phi(b_j)\mathfrak{x}_j\big) = \mathrm{tr}\big(\textstyle\sum_{i\cdot j\cdot k=1}^{m} a_i t_{ik} h_{jk}^{\rho} b_j^{\rho}\big) .$$

By ($E3$), we observe that

$$(15) \qquad\qquad TH^{\rho} \text{ is } \rho\text{-symmetric and } \rho\text{-positive.}$$

By (12), we get

$$(16) \qquad\qquad TH^{\rho} = (TH^{\rho})^{\rho} = -HT .$$

Conversely suppose that m vectors $\mathfrak{x}_1, \cdots, \mathfrak{x}_m$ of C^n, a ρ-skew-symmetric element T of $M_m(L)$ and a lattice $\mathfrak{M}$ in L^m are given. If the $\mathfrak{x}_i$ satisfy (13), we can define a lattice D in C^n by (10); so we get a complex torus C^n/D; and (14) determines a matrix $H = (h_{ij})$. Define $E(\mathfrak{x}, \mathfrak{y})$ by (11). If (12') and (15) are satisfied, we can easily verify that E satisfies ($E1$, 2, 3, 4). We get in this way a polarized abelian variety of type $\{L, \Phi, \rho\}$. Therefore, $\mathcal{P} = (A, C, \theta)$ is completely determined by $\{\mathfrak{x}_1, \cdots, \mathfrak{x}_m; \mathfrak{M}; T\}$; and the data $\{\mathfrak{x}_1, \cdots, \mathfrak{x}_m; \mathfrak{M}; T\}$ determine a polarized abelian variety of type $\{L, \Phi, \rho\}$ if and only if the conditions (12'), (13) and (15) are satisfied. Thus we are led to the task of describing the conditions (13) and (15) in a more explicit form.

2.3. First we write each vector $\mathfrak{x}_i$ in the form

$$^t\mathfrak{x}_i = (^t\mathfrak{x}_i^1 \cdots {}^t\mathfrak{x}_i^{\nu} \cdots {}^t\mathfrak{x}_i^g)$$

with $\mathfrak{x}_i^{\nu} \in C^{n/g}$. Define a matrix X_{ν} for each ν as follows.

(*Type* I) Put

$$(17) \qquad\qquad X_{\nu} = \begin{bmatrix} \mathfrak{x}_1^{\nu} \cdots \mathfrak{x}_m^{\nu} \\ \overline{\mathfrak{x}}_1^{\nu} \cdots \overline{\mathfrak{x}}_m^{\nu} \end{bmatrix} .$$

(*Type* II) Put $\mathfrak{x}_i^{\nu} = \begin{bmatrix} u_i^{\nu} \\ v_i^{\nu} \end{bmatrix}$ with $u_i^{\nu}, v_i^{\nu} \in C^m$, and

$$(18) \qquad\qquad X_{\nu} = \begin{bmatrix} u_1^{\nu} \cdots u_m^{\nu} & v_1^{\nu} \cdots v_m^{\nu} \\ \overline{u}_1^{\nu} \cdots \overline{u}_m^{\nu} & \overline{v}_1^{\nu} \cdots \overline{v}_m^{\nu} \end{bmatrix} .$$

(*Type* III) Put $\mathfrak{x}_i^{\nu} = \begin{bmatrix} u_i^{\nu} \\ v_i^{\nu} \end{bmatrix}$ with $u_i^{\nu}, v_i^{\nu} \in C^m$, and

$$\mathfrak{X}_{\nu} = (u_1^{\nu} \cdots u_m^{\nu}) , \qquad \mathfrak{Y}_{\nu} = (v_1^{\nu} \cdots v_m^{\nu}) ,$$

$$(19) \qquad X_\nu = \begin{bmatrix} \mathfrak{X}_\nu & \mathfrak{Y}_\nu \\ \overline{\mathfrak{Y}}_\nu & -\overline{\mathfrak{X}}_\nu \end{bmatrix}.$$

(*Type* IV) Put ${}^t\mathfrak{x}_i^\nu = ({}^tu_{i1}^\nu \cdots {}^tu_{iq}^\nu \; {}^tv_{i1}^\nu \cdots {}^tv_{iq}^\nu)$ with $u_{ik}^\nu \in C^{r_\nu}$, $v_{ik}^\nu \in C^{s_\nu}$, and

$$(20) \qquad X_\nu = \begin{bmatrix} u_{11}^\nu \cdots u_{m1}^\nu & u_{12}^\nu \cdots u_{m2}^\nu \cdots & u_{1q}^\nu \cdots u_{mq}^\nu \\ \overline{v}_{11}^\nu \cdots \overline{v}_{m1}^\nu & \overline{v}_{12}^\nu \cdots \overline{v}_{m2}^\nu \cdots & \overline{v}_{1q}^\nu \cdots \overline{v}_{mq}^\nu \end{bmatrix}.$$

Then, in all cases, the matrices X_ν for $1 \leq \nu \leq g$ determine the vectors $\mathfrak{x}_1, \cdots, \mathfrak{x}_m$, and conversely. It is easy to see that the condition (13) is equivalent to

$$(21) \qquad \det(X_\nu) \neq 0 \qquad (1 \leq \nu \leq g).$$

2.4. To express (15) in a more explicit form, we define representations ω_ν of $M_m(L_R)$ as follows. For every element $B = (b_{ij})$ of $M_m(L_R)$ with $b_{ij} \in L_R$, write

$$\chi_\nu(b_{ij}) = (\beta_{ijhk}^\nu)_{h, k=1, \ldots, w},$$

where $\beta_{ijhk}^\nu \in C$, and

$$w = \begin{cases} 1 & \text{for } (Type\ \text{I}), \\ 2 & \text{for } (Type\ \text{II, III}), \\ q & \text{for } (Type\ \text{IV}). \end{cases}$$

Put then $B_{hk}^\nu = (\beta_{ijhk}^\nu)_{i,j=1, \ldots, m}$, and

$$\omega_\nu(B) = \begin{bmatrix} B_{11}^\nu & \cdots & B_{1w}^\nu \\ \cdots & \cdots & \cdots \\ B_{w1}^\nu & \cdots & B_{ww}^\nu \end{bmatrix}.$$

By (6) of §2.1, we see easily $\omega_\nu(B^\rho) = \omega_\nu(B)^*$. Obviously, an element B of $M_m(L_R)$ is ρ-symmetric and ρ-positive if and only if $\omega_\nu(B)$ is $*$-symmetric and $*$-positive for every ν.

Now put $\omega_\nu(T) = T_\nu$, $\omega_\nu(H) = H_\nu$. We have then $T_\nu^* = -T_\nu$; and on account of (16), the condition (15) is equivalent to

(22) $\quad -H_\nu T_\nu$ is $*$-symmetric and $*$-positive.

By our definition of X_ν and H_ν, the relation (14) is now written in the form

$$(23) \qquad \sqrt{-1}\, J_\nu X_\nu = X_\nu \cdot {}^tH_\nu,$$

where

$$(Type\ \text{I}) \qquad J_\nu = \begin{bmatrix} 1_{m/2} & 0 \\ 0 & -1_{m/2} \end{bmatrix},$$

$$(Type\ II,\ III)\quad J_\nu = \begin{bmatrix} 1_m & 0 \\ 0 & -1_m \end{bmatrix},$$

$$(Type\ IV)\qquad J_\nu = \begin{bmatrix} 1_{r_\nu} & 0 \\ 0 & -1_{s_\nu} \end{bmatrix}.$$

Assuming (21) to be true, we get ${}^tH_\nu = \sqrt{-1}\,X_\nu^{-1}J_\nu X_\nu$, so that the condition (22) holds if and only if $-\sqrt{-1}\,{}^tX_\nu J_\nu \cdot {}^t X_\nu^{-1} T_\nu$ is $*$-symmetric and $*$-positive; in other words, $\bar{X}_\nu T_\nu^{-1}\cdot {}^t X_\nu$ commutes with J_ν and $\sqrt{-1}\,J_\nu \bar{X}_\nu T_\nu^{-1}\cdot {}^t X_\nu$ is $*$-positive. Thus (22) is equivalent to the following condition:

$$(24)\qquad \sqrt{-1}\,\bar{X}_\nu T_\nu^{-1}\cdot {}^t X_\nu = \begin{bmatrix} \mathfrak{A}_\nu & 0 \\ 0 & -\mathfrak{B}_\nu \end{bmatrix},$$

where $\mathfrak{A}_\nu$ and $\mathfrak{B}_\nu$ are positive hermitian matrices such that

$$(Type\ I)\qquad \mathfrak{A}_\nu,\ \mathfrak{B}_\nu \in M_{m/2}(C),$$
$$(Type\ II,\ III)\quad \mathfrak{A}_\nu,\ \mathfrak{B}_\nu \in M_m(C),$$
$$(Type\ IV)\qquad \mathfrak{A}_\nu \in M_{r_\nu}(C),\qquad \mathfrak{B}_\nu \in M_{s_\nu}(C).$$

Therefore the ρ-skew-symmetric element T of $M_m(L)$ must satisfy the condition

(25) $\sqrt{-1}\,\omega_\nu(T)^{-1}$ *has the same signature as* J_ν *for every* ν.

This is always satisfied for $(Type\ I, II, III)$ whenever T is ρ-skew-symmetric and non-degenerate; for $(Type\ IV)$, however, (25) gives an essential restriction of T.

2.5. We now want to transform the "period-matrices" X_ν, satisfying (24) into a normalized form.

$(Type\ I, II)$ As T_ν^{-1} is a non-degenerate real skew-symmetric matrix, there exists, for each ν, a real matrix W_ν such that

$$W_\nu T_\nu^{-1}\cdot {}^t W_\nu = \begin{bmatrix} 0 & 1_l \\ -1_l & 0 \end{bmatrix},\qquad l = \begin{cases} m/2 & \text{for } (Type\ I), \\ m & \text{for } (Type\ II), \end{cases}$$

Fixing such a matrix W_ν, we can put $X_\nu W_\nu^{-1} = \begin{pmatrix} U_\nu & V_\nu \\ \bar{U}_\nu & \bar{V}_\nu \end{pmatrix}$, since X_ν is of the form (17), (18), and W_ν is of real coefficients. Then the condition (24) is equivalent to

(26) $U_\nu\,{}^tV_\nu - V_\nu\,{}^tU_\nu = 0;$ *and* $\sqrt{-1}(\bar{U}_\nu\,{}^tV_\nu - \bar{V}_\nu\,{}^tU_\nu)$ *is positive hermitian.*

This implies, in particular, that U_ν and V_ν are invertible. Put $z_\nu = V_\nu^{-1}U_\nu$. Then (26) is equivalent to

(27) ${}^tz_\nu = z_\nu;$ *and* $\mathrm{Im}(z_\nu)$ *is positive symmetric.*

(*Type* III) We can find an element W of $M_m(L_R)$ such that

$$\pi_\nu(WT^{-1}W^\rho) = \sqrt{-1}\,1_m \qquad\qquad (1 \leqq \nu \leqq g)\,,$$

where π_ν denotes the projection of $M_m(L_R) = M_m(K) \times \cdots \times M_m(K)$ onto the νth factor. Put $W_\nu = \omega_\nu(W)$. Then

$$\sqrt{-1}\,W_\nu T_\nu^{-1}\cdot{}^t\bar{W}_\nu = \begin{bmatrix} -1_m & 0 \\ 0 & 1_m \end{bmatrix}.$$

We observe that $X_\nu \bar{W}_\nu^{-1}$ is written in the form

$$X_\nu \bar{W}_\nu^{-1} = \begin{bmatrix} U_\nu & V_\nu \\ \bar{V}_\nu & -\bar{U}_\nu \end{bmatrix}.$$

Then (24) is equivalent to

(28) $\quad U_\nu{}^t V_\nu + V_\nu{}^t U_\nu = 0;$ *and* $-U_\nu{}^t\bar{U}_\nu + V_\nu{}^t\bar{V}_\nu$ *is positive hermitian.*

This implies in particular, that V_ν is invertible. If we put $z_\nu = -V_\nu^{-1}U_\nu$, then (28) is equivalent to

(29) $\quad {}^t z_\nu = -z_\nu;$ *and* $1 - z_\nu{}^t\bar{z}_\nu$ *is positive hermitian.*

(*Type* IV) Assuming (25) to be true, take, for each ν, a matrix W_ν in $M_{mq}(C)$ such that

$$W_\nu(\sqrt{-1}\,T_\nu^{-1})\,{}^t\bar{W}_\nu = \begin{bmatrix} 1_{r_\nu} & 0 \\ 0 & -1_{s_\nu} \end{bmatrix}.$$

Put

$$X_\nu \bar{W}_\nu^{-1} = \begin{bmatrix} \mathfrak{U}_\nu & \mathfrak{V}_\nu \\ \mathfrak{W}_\nu & \mathfrak{Y}_\nu \end{bmatrix}.$$

Then (24) is equivalent to

(30) $\quad \bar{\mathfrak{U}}_\nu{}^t\mathfrak{W}_\nu - \bar{\mathfrak{V}}_\nu{}^t\mathfrak{Y}_\nu = 0,\ \bar{\mathfrak{W}}_\nu{}^t\mathfrak{U}_\nu - \bar{\mathfrak{Y}}_\nu{}^t\mathfrak{V}_\nu = 0;$ *and*
$\qquad\ \ \bar{\mathfrak{U}}_\nu{}^t\mathfrak{U}_\nu - \bar{\mathfrak{V}}_\nu{}^t\mathfrak{V}_\nu,\ \bar{\mathfrak{Y}}_\nu{}^t\mathfrak{Y}_\nu - \bar{\mathfrak{W}}_\nu{}^t\mathfrak{W}_\nu$ *are positive hermitian.*

It follows that $\mathfrak{U}_\nu$ and $\mathfrak{Y}_\nu$ are invertible, and $\mathfrak{U}_\nu^{-1}\mathfrak{V}_\nu = \overline{{}^t(\mathfrak{Y}_\nu^{-1}\mathfrak{W}_\nu)}$. Therefore, putting $z_\nu = \overline{\mathfrak{U}_\nu^{-1}\mathfrak{V}_\nu} = {}^t(\mathfrak{Y}_\nu^{-1}\mathfrak{W}_\nu)$, we observe that (30) is equivalent to

(31) $\qquad\quad 1 - z_\nu{}^t\bar{z}_\nu$ *is positive hermitian.*

Now any matrix Λ commuting with the elements of $\Phi(L_R)$ is of the form

(32) $$\Lambda = \begin{bmatrix} \Lambda_1' & & \\ & \cdot\ \cdot & \\ & & \cdot\ \Lambda_g' \end{bmatrix},$$

where

$$(33)\quad \begin{cases} (Type\ \mathrm{I}) & \Lambda'_\nu \in M_{m/2}(C)\ . \\[2mm] (Type\ \mathrm{II,\ III}) & \Lambda'_\nu = \begin{bmatrix} \Sigma_\nu & 0 \\ 0 & \Sigma_\nu \end{bmatrix}, & \Sigma_\nu \in M_m(C)\ , \\[4mm] (Type\ \mathrm{IV}) & \Lambda'_\nu = \begin{bmatrix} 1_q \otimes \Sigma_\nu & 0 \\ 0 & 1_q \otimes \Delta_\nu \end{bmatrix}, & \Sigma_\nu \in M_{r_\nu}(C),\ \Delta_\nu \in M_{s_\nu}(C)\ . \end{cases}$$

Define, for each ν, a matrix Λ_ν by

$$(34)\quad \begin{cases} (Type\ \mathrm{I}) & \Lambda_\nu = \begin{bmatrix} \Lambda'_\nu & 0 \\ 0 & \bar{\Lambda}'_\nu \end{bmatrix}, \\[4mm] (Type\ \mathrm{II,\ III}) & \Lambda_\nu = \begin{bmatrix} \Sigma_\nu & 0 \\ 0 & \bar{\Sigma}_\nu \end{bmatrix}, \\[4mm] (Type\ \mathrm{IV}) & \Lambda_\nu = \begin{bmatrix} \Sigma_\nu & 0 \\ 0 & \bar{\Delta}_\nu \end{bmatrix}. \end{cases}$$

If we change χ_i for $\Lambda\chi_i$, X_ν is transformed into $\Lambda_\nu X_\nu$. This transformation does not give any effect upon T, $\mathfrak{M}$ and the structure of a given polarized abelian variety. Therefore, transforming by a suitable Λ, we may assume that $Y_\nu = X_\nu \bar{W}_\nu^{-1}$ is of the following "normalized form".

$$(35)\quad \begin{cases} (Type\ \mathrm{I}) & Y_\nu = \begin{bmatrix} z_\nu & 1_{m/2} \\ \bar{z}_\nu & 1_{m/2} \end{bmatrix}, \\[4mm] (Type\ \mathrm{II}) & Y_\nu = \begin{bmatrix} z_\nu & 1_m \\ \bar{z}_\nu & 1_m \end{bmatrix}, \\[4mm] (Type\ \mathrm{III}) & Y_\nu = \begin{bmatrix} -z_\nu & 1_m \\ 1_m & \bar{z}_\nu \end{bmatrix}, \\[4mm] (Type\ \mathrm{IV}) & Y_\nu = \begin{bmatrix} 1_{r_\nu} & z_\nu \\ {}^t\bar{z}_\nu & 1_{s_\nu} \end{bmatrix}. \end{cases}$$

2.6. *Analytic families*. Let $\mathcal{H}^1_r$, $\mathcal{H}^2_r$, $\mathcal{H}^3_{r,s}$ denote respectively the space of complex matrices z defined by

$\mathcal{H}^1_r = \{z \in M_r(C) \mid {}^tz = z;\ \mathrm{Im}(z)$ is positive symmetric$\}$,

$\mathcal{H}^2_r = \{z \in M_r(C) \mid {}^tz = -z;\ 1 - z{}^t\bar{z}$ is positive hermitian$\}$,

$\mathcal{H}^3_{r,s} = \{z \mid$ complex matrix with r rows and s columns; $1 - z{}^t\bar{z}$ is positive hermitian$\}$.

Put then

$$(36)\quad \begin{aligned} (Type\ \mathrm{I}) &\quad \mathcal{H}(\Phi) = \mathcal{H}^1_{m/2} \times \cdots \times \mathcal{H}^1_{m/2}\ , \\[2mm] (Type\ \mathrm{II}) &\quad \mathcal{H}(\Phi) = \mathcal{H}^1_m \times \cdots \times \mathcal{H}^1_m\ , \\[2mm] (Type\ \mathrm{III}) &\quad \mathcal{H}(\Phi) = \mathcal{H}^2_m \times \cdots \times \mathcal{H}^2_m\ , \\[2mm] (Type\ \mathrm{IV}) &\quad \mathcal{H}(\Phi) = \mathcal{H}^3_{r_1,s_1} \times \cdots \times \mathcal{H}^3_{r_g,s_g}\ , \end{aligned}$$

where the number of factors is g in each case.

Now, for every point $z = (z_1, \cdots, z_g)$ of $\mathcal{H}(\Phi)$, define $Y_\nu = Y_\nu(z)$ by (35) and put $X_\nu = X_\nu(z, T) = Y_\nu \bar{W}_\nu$. Then, the matrices X_ν determine m vectors $\mathfrak{x}_1 = \mathfrak{x}_1(z, T), \cdots, \mathfrak{x}_m = \mathfrak{x}_m(z, T)$, from which we obtain a lattice $D = D(z, T, \mathfrak{M})$ by (10) of §2.2. The fact that z is a point of $\mathcal{H}(\Phi)$ implies that the complex torus $C^n/D(z, T, \mathfrak{M})$ has a structure of abelian variety; and if (12′) is satisfied, (11) gives a non-degenerate Riemann form on it. In this way, we get a polarized abelian variety of type $\{L, \Phi, \rho\}$. Conversely, our consideration shows that any polarized abelian variety of type $\{L, \Phi, \rho\}$ is obtained in this manner. We have thus proved

THEOREM 1. *Let T be a non-degenerate ρ-skew-symmetric element of $M_m(L)$ and let $\mathfrak{M}$ be a free $\mathbf{Z}$-submodule of $L^m = L \times \cdots \times L$ of rank $2n$ $(=m[L: \mathbf{Q}])$. Suppose that T satisfies (25) when L is of (Type IV). Then, for every point z of $\mathcal{H}(\Phi)$, we obtain a polarized abelian variety of type $\{L, \Phi, \rho\}$ in the above described manner; and every polarized abelian variety of type $\{L, \Phi, \rho\}$ is obtained from suitable $z, T, \mathfrak{M}$.*

Thus, if we fix T and $\mathfrak{M}$, we get an analytic family of polarized abelian varieties of type $\{L, \Phi, \rho\}$ parametrized by the points of $\mathcal{H}(\Phi)$. We denote by $\Sigma(T, \mathfrak{M})$ this family and by $\mathscr{P}(z, T, \mathfrak{M})$ the member of $\Sigma(T, \mathfrak{M})$ determined by $z, T, \mathfrak{M}$.

REMARK 1. If L is of (Type III) and $m = 1$, $\mathcal{H}(\Phi)$ consists of only one point. Hence, in this case each family $\Sigma(T, \mathfrak{M})$ consists of only one member.

REMARK 2. If $r = 0$ or $s = 0$, the space $\mathcal{H}_{r,s}^3$ has no meaning. But in this case we understand by $\mathcal{H}_{r,s}^3$ a space which consists of only one point; and when L is of (Type IV), we put $Y_\nu(z) = 1_{mq}$ if $r_\nu s_\nu = 0$. Then, the above consideration is meaningful even in the case $r_\nu s_\nu = 0$ for some ν. In particular, if $r_\nu s_\nu = 0$ for every ν, then $\Sigma(T, \mathfrak{M})$ consists of only one member.

REMARK 3. The condition (12′) is not so important, since if this is not satisfied, a suitable multiple of T satisfies the condition.

2.7. *Discontinuous groups.* Let $G_r^1, G_r^2, G_{r,s}^3$ be respectively the groups defined by

$$G_r^1 = \left\{ V \in M_{2r}(\mathbf{R}) \,\middle|\, {}^t V \begin{bmatrix} 0 & 1_r \\ -1_r & 0 \end{bmatrix} V = \begin{bmatrix} 0 & 1_r \\ -1_r & 0 \end{bmatrix} \right\},$$

$$G_r^2 = \left\{ V = \iota(U) \,\middle|\, U \in M_r(K), {}^t \bar{V} \begin{bmatrix} 1_r & 0 \\ 0 & -1_r \end{bmatrix} V = \begin{bmatrix} 1_r & 0 \\ 0 & -1_r \end{bmatrix} \right\},$$

$$G_{r,s}^3 = \left\{ V \in M_{r+s}(\mathbf{C}) \,\middle|\, {}^t \bar{V} \begin{bmatrix} 1_r & 0 \\ 0 & -1_s \end{bmatrix} V = \begin{bmatrix} 1_r & 0 \\ 0 & -1_s \end{bmatrix} \right\}.$$

Define $G(\Phi)$ by

$$
\begin{aligned}
(Type\ \mathrm{I}) \quad & G(\Phi) = G^1_{m/2} \times \cdots \times G^1_{m/2}\ , \\
(Type\ \mathrm{II}) \quad & G(\Phi) = G^1_m \times \cdots \times G^1_m\ , \\
(Type\ \mathrm{III}) \quad & G(\Phi) = G^2_m \times \cdots \times G^2_m\ , \\
(Type\ \mathrm{IV}) \quad & G(\Phi) = G^3_{r_1,s_1} \times \cdots \times G^3_{r_g,s_g}\ ,
\end{aligned}
\tag{37}
$$

where the number of factors is g in each case. The group $G(\Phi)$ operates on the space $\mathcal{H}(\Phi)$ as follows. Let $V = (V_1, \cdots, V_g)$ be an element of $G(\Phi)$; put each V_ν in the form

$$
V_\nu = \begin{bmatrix} a_\nu & b_\nu \\ c_\nu & d_\nu \end{bmatrix},
$$

where $a_\nu, b_\nu, c_\nu, d_\nu$ are square matrices for $(Type\ \mathrm{I, II, III})$ and $a_\nu \in M_{r_\nu}(C)$, $d_\nu \in M_{s_\nu}(C)$ for $(Type\ \mathrm{IV})$. Now, for $z = (z_1, \cdots, z_g) \in \mathcal{H}(\Phi)$, define $V(z)$ by

$$
\begin{aligned}
V(z) &= (z'_1, \cdots, z'_g)\ , \\
z'_\nu &= (a_\nu z_\nu + b_\nu)(c_\nu z_\nu + d_\nu)^{-1} \qquad && \text{for } (Type\ \mathrm{I, II, III})\ , \\
z'_\nu &= (\bar{a}_\nu z_\nu + \bar{b}_\nu)(\bar{c}_\nu z_\nu + \bar{d}_\nu)^{-1} \qquad && \text{for } (Type\ \mathrm{IV})\ .
\end{aligned}
$$

When $r_\nu s_\nu = 0$ for some ν, we put $z'_\nu = z_\nu$. Then $G(\Phi)$ operates on $\mathcal{H}(\Phi)$ transitively; and $\mathcal{H}(\Phi)$ is isomorphic to a homogeneous space $G(\Phi)/H$ with a compact subgroup H of $G(\Phi)$.

Fix a ρ-skew-symmetric element T of $M_m(L)$ and a lattice $\mathfrak{M}$ in L^m. We consider the elements of L^m as row-vectors with components in L; then L^m can be regarded as a left L-module and a right $M_m(L)$-module in a natural manner. Put

$$
\begin{aligned}
G(T) &= \{U \in M_m(L_R) \mid UTU^\rho = T\}\ , \\
\Gamma(T, \mathfrak{M}) &= \{U \in M_m(L) \mid UTU^\rho = T,\ \mathfrak{M}U = \mathfrak{M}\}\ .
\end{aligned}
\tag{38}
$$

For every $U \in G(T)$, put

$$
\psi(U) = \{{}^t\bar{W}_1^{-1}\omega_1(U) \cdot {}^t\bar{W}_1, \cdots, {}^t\bar{W}_g^{-1}\omega_g(U) \cdot {}^t\bar{W}_g\}\ ;
$$

then $U \to \psi(U)$ gives an isomorphism of $G(T)$ onto $G(\Phi)$. For every $U \in G(T)$ and for $z \in \mathcal{H}(\Phi)$, we write $U(z) = \psi(U)(z)$. $\Gamma(T, \mathfrak{M})$ is a discrete subgroup of a semi-simple Lie group $G(T)$, and $\Gamma(T, \mathfrak{M})\backslash G(T)$ is of finite measure.

2.8. *Isomorphism-classes of polarized abelian varieties.* Let $\mathscr{P} = (A, C, \theta)$ and $\mathscr{P}' = (A', C', \theta')$ be polarized abelian varieties of type

$\{L, \Phi, \rho\}$. An isomorphism (resp. a homomorphism) λ of A onto A' is called an isomorphism (resp. a homomorphism) of $\mathscr{P}$ onto $\mathscr{P}'$ if $\lambda\theta(a) = \theta'(a)\lambda$ for every $a \in L$ and λ^{-1} sends C' into C.

Suppose that there exists a homomorphism λ of $\mathscr{P}(z, T, \mathfrak{M})$ onto $\mathscr{P}(z', T', \mathfrak{M}')$. Denote by Λ the complex matrix which represents λ with respect to the coordinate system of C^n. We have then $\Lambda\Phi(a) = \Phi(a)\Lambda$ for every $a \in L$, so that Λ is of the form (32), (33). As Λ maps $D(z, T, \mathfrak{M})$ into $D(z', T', \mathfrak{M}')$, there exists an element $U = (u_{ij})$ of $M_m(L)$ such that

$$\Lambda\mathfrak{x}_i(z, T) = \sum_{j=1}^m \Phi(u_{ij})\mathfrak{x}_j(z', T') \qquad (1 \leq i \leq m) ,$$
$$\mathfrak{M}U \subset \mathfrak{M}' .$$

Defining Λ_ν as in (34) of §2.5, we obtain

$$(39) \qquad\qquad \Lambda_\nu X_\nu(z, T) = X_\nu(z', T') \cdot {}^t\omega_\nu(U) .$$

Let E and E' be respectively the Riemann forms on $C^n/D(z, T, \mathfrak{M})$ and $C^n/D(z', T', \mathfrak{M}')$ defined by

$$E\left(\sum_{i=1}^m a_i\mathfrak{x}_i, \sum_{j=1}^m b_j\mathfrak{x}_j\right) = \operatorname{tr}_{L/Q}\left(\sum_{i,j=1}^m a_i t_{ij} b_j^\rho\right) ,$$
$$E'\left(\sum_{i=1}^m a_i\mathfrak{x}_i', \sum_{j=1}^m b_j\mathfrak{x}_j'\right) = \operatorname{tr}_{L/Q}\left(\sum_{i,j=1}^m a_i t_{ij}' b_j^\rho\right) ,$$

where $\mathfrak{x}_i = \mathfrak{x}_i(z, T)$, $\mathfrak{x}_i' = \mathfrak{x}_i(z', T')$. As λ^{-1} sends C' into C, we have $E'(\Lambda\mathfrak{u}, \Lambda\mathfrak{v}) = \delta E(\mathfrak{u}, \mathfrak{v})$ with a positive rational number δ. We have therefore

$$(40) \qquad\qquad UT'U^\rho = \delta T .$$

If λ is an isomorphism, we get $\Lambda D(z, T, \mathfrak{M}) = D(z' T', \mathfrak{M}')$, so that

$$(41) \qquad\qquad \mathfrak{M}U = \mathfrak{M}' .$$

Now suppose that $T = T'$ and $\mathfrak{M} = \mathfrak{M}'$. Then δ must be equal to 1, so that $U \in \Gamma(T, \mathfrak{M})$; and (39) implies $z = U(z')$.

Conversely, suppose that there exists an element U of $\Gamma(T, \mathfrak{M})$ such that $U(z') = z$. Define Λ_ν by (39). Then it can be easily verified that Λ_ν is of the form (34) of §2.5. Defining Λ by (33) and (32), we see that Λ gives an isomorphism of $\mathscr{P}(z, T, \mathfrak{M})$ onto $\mathscr{P}(z', T, \mathfrak{M})$. We get thus:

THEOREM 2. *Two members $\mathscr{P}(z, T, \mathfrak{M})$ and $\mathscr{P}(z', T, \mathfrak{M})$ of the same analytic family $\Sigma(T, \mathfrak{M})$ are isomorphic if and only if $z = U(z')$ for an element U of $\Gamma(T, \mathfrak{M})$.*

Therefore, the isomorphism-classes of the members of $\Sigma(T, \mathfrak{M})$ are in one-to-one correspondence with the points of the quotient space $\Gamma(T, \mathfrak{M})\backslash\mathscr{H}(\Phi)$.

2.9. Let T_1 and T_2 be non-degenerate ρ-skew-symmetric elements of $M_m(L)$; let $\mathfrak{M}_1$ and $\mathfrak{M}_2$ be lattices in L_m. We say that couples $\{T_1, \mathfrak{M}_1\}$ and $\{T_2, \mathfrak{M}_2\}$ are *equivalent* if there exist an element U of $M_m(L)$ and a positive rational number δ such that $UT_2U^\rho = \delta T_1$, $\mathfrak{M}_1 U = \mathfrak{M}_2$. We call $\{T, \mathfrak{M}\}$ *primitive* if $\mathrm{tr}_{L/Q}(\mathfrak{M}T\mathfrak{M}^\rho) = Z$. Obviously, every equivalence-class of $\{T, \mathfrak{M}\}$ contains a primitive couple. The following proposition is an easy consequence of our definition.

PROPOSITION 4. $\mathscr{P}(z_1, T_1, \mathfrak{M}_1)$ and $\mathscr{P}(z_2, T_2, \mathfrak{M}_2)$ *are isomorphic only if* $\{T_1, \mathfrak{M}_1\}$ *and* $\{T_2, \mathfrak{M}_2\}$ *are equivalent. Conversely, if* $\{T_1, \mathfrak{M}_1\}$ *and* $\{T_2, \mathfrak{M}_2\}$ *are equivalent, every member of* $\Sigma(T_1, \mathfrak{M}_1)$ *is isomorphic to a member of* $\Sigma(T_2, \mathfrak{M}_2)$.

3. Automorphic functions with respect to $\Gamma(T, \mathfrak{M})$

3.1. First we supply the theory of [AF] with several propositions. We mean by $\mathfrak{r}$ a ring having a finite basis over Z.

PROPOSITION 5. *Let* $\mathscr{P} = (A, C, \theta)$ *be a polarized abelian variety of type* $\mathfrak{r}$ *in the sense of* [AF, no. 3], *and* K *the field of moduli of* $\mathscr{P}$. *Then there exists a polarized abelian variety* $\mathscr{P}'$, *isomorphic to* $\mathscr{P}$, *defined over a finite algebraic extension of* K.

This result is essentially given in the proof of [15, Prop. 3.5]. In fact, putting $\mathscr{P}$ in place of $\mathscr{P}(z_0)$ in that proof, $\mathscr{P}_1 = (B_1, C_1, \theta_1)$ constructed there has the required property of $\mathscr{P}'$.

PROPOSITION 6. *Let* $\mathscr{P} = (A, C, \theta)$ *and* $\mathscr{P}' = (A, C', \theta)$ *be two polarized abelian varieties of type* $\mathfrak{r}$ *with the same* A *and* θ. *Let* K *and* K' *be respectively the fields of moduli of* $\mathscr{P}$ *and* $\mathscr{P}'$. *Suppose that* $Q\,\theta(\mathfrak{r}) = \mathcal{A}_0(A)$. *Then* KK' *is purely inseparable over* K *and over* K'. *If the universal domain is of characteristic* 0, *we have* $K = K'$.

PROOF. Let σ be an isomorphism of the universal domain into itself which leaves invariant the elements of K. Then there exists an isomorphism ε of $\mathscr{P}$ onto $\mathscr{P}^\sigma$. Let X be a divisor in C and Y a divisor in C'. Let B be the Picard variety of A. Consider the isogenies φ_X, φ_Y of A to B given in §1.3. By our assumption $Q\,\theta(\mathfrak{r}) = \mathcal{A}_0(A)$, there exists an element $\alpha \in \mathfrak{r}$ and a positive integer r such that $r\varphi_Y = \varphi_X\theta(\alpha)$. B^σ is considered as the Picard variety of A^σ and $(\varphi_X)^\sigma = \varphi_{X^\sigma}$, $(\varphi_Y)^\sigma = \varphi_{Y^\sigma}$; so we have $r\varphi_{Y^\sigma} = \varphi_{X^\sigma}\theta^\sigma(\alpha)$. As ε^{-1} sends C^σ into C, we have ${}^t\varepsilon\varphi_{X^\sigma}\varepsilon = \varphi_X$. We have therefore $r{}^t\varepsilon\varphi_{Y^\sigma}\varepsilon = {}^t\varepsilon\varphi_{X^\sigma}\theta^\sigma(\alpha)\varepsilon = {}^t\varepsilon\varphi_{X^\sigma}\varepsilon\theta(\alpha) = \varphi_X\theta(\alpha) = r\varphi_Y$. It follows that $\varepsilon^{-1}(Y^\sigma)$ is algebraically equivalent to Y. Hence ε is an isomorphism of $\mathscr{P}'$ onto $\mathscr{P}'^\sigma$; so σ must be the identity on K'. This proves that KK' is purely inseparable over K. If the characteristic of the universal domain is 0, we have $K \supset K'$. As the inclusion $K \subset K'$ is proved similarly, we get $K = K'$.

PROPOSITION 7. *Let* $\mathscr{P} = (A, C)$ *and* $\mathscr{P}' = (A, C')$ *be two polarized abelian varieties with the same* A; *let* K *and* K' *be respectively the fields of moduli of* $\mathscr{P}$ *and* $\mathscr{P}'$. *Then* KK' *is algebraic over* K *and over* K'.

This is only a particular case of [AF, Prop. 9].

PROPOSITION 8. *Let* A *be an abelian variety in the projective space* P^N. *Suppose that*:

(i) *there is no hyperplane containing* A;

(ii) *the linear system on* A *defined by hyperplane sections is complete.* *Let* C *be the polarization of* A *given by hyperplane sections, and* K *the field of moduli of* (A, C). *Let* α *be a projective transformation of* P^N, *generic over a field of definition for* A; *put* $M = (N + 1)^2 - 1$. *Then we have*

$$\dim_{k_0} c(\alpha(A)) = M + \dim_{k_0} K \,,$$

where k_0 *denotes the prime field and* $c(V)$ *denotes the Chow-point of an algebraic variety* V.

PROOF. Let k be a field of definition for A, and α a projective transformation of P^N, generic over k. By Proposition 5, there exists a projective transformation β of P^N such that $\beta(A)$ is defined over an algebraic extension K_1 of K. Let γ be a projective transformation of P^N, generic over kK_1. Let K_2 be the smallest field of definition for the variety $\mathscr{F}(A)$ defined in [AF, no. 3]. As $\mathscr{F}(A) = \mathscr{F}(\gamma\beta(A))$, $\mathscr{F}(A)$ is defined over $k_0(c(\gamma\beta(A)))$, so that $k_0(c(\gamma\beta(A))) \supset K_2$. By [AF, no. 5], K is purely inseparable over K_2. Now we note that [AF, no. 12, Lemma 8] is valid for any polarized abelian variety of type $\mathfrak{r}$ whenever the variety satisfies the assumptions (i), (ii) of our proposition. Hence $\mathscr{F}(A)$ is of dimension M. By our definition of $\mathscr{F}(A)$, the variety $\mathscr{F}(A) = \mathscr{F}(\beta(A))$ is a locus of $c(\gamma\beta(A))$ over K_1. We have therefore $\dim_{K_1} c(\gamma\beta(A)) = M$, so that

$$\begin{aligned}
\dim_{k_0} k_0(c(\gamma\beta(A))) &= \dim_{k_0} K_2(c(\gamma\beta(A))) \\
&= \dim_{k_0} K_2 + \dim_{K_2} c(\gamma\beta(A)) \\
&= \dim_{k_0} K + M \,.
\end{aligned}$$

This proves our proposition, since $\gamma\beta(A)$ is a generic specialization of $\alpha(A)$ over k, and hence over k_0.

3.2. We now assume (until the end of § 3.5) that $\dim \mathscr{H}(\Phi) \geq 1$. We note that the vectors $\mathfrak{x}_i(z, T)$ given in §2.6 are holomorphic functions of the coordinates of z. Therefore we can apply the theory of [AF, no.

11–12] to the present case. Namely, there exist a holomorphic mapping $\Xi(u, z)$ of $C^n \times \mathcal{H}(\Phi)$ into a projective space P^N and a family of abelian varieties $A(z, T, \mathfrak{M})$ in P^N with the following properties.

(Ξ1) For every $z \in \mathcal{H}(\Phi)$, the mapping $u \to \Xi(u, z)$ gives an isomorphism of $C^n/D(z, T, \mathfrak{M})$ onto $A(z, T, \mathfrak{M})$.

(Ξ2) The hyperplane sections of $A(z, T, \mathfrak{M})$ defines a complete linear system and a polarization of $A(z, T, \mathfrak{M})$ corresponding to the Riemann form (11) on $C^n/D(z, T, \mathfrak{M})$.

There are many ways of choosing projective embedding Ξ of $C^n/D(z, T, \mathfrak{M})$. For our later use, we choose one among them as follows. Take a rational number δ so that $\{\delta T, \mathfrak{M}\}$ is primitive in the sense of §2.9. Let X be the positive divisor whose Riemann form E is given by

$$E\left(\sum_{i=1}^{m} a_i \mathfrak{x}_i(z, T), \sum_{j=1}^{m} b_j \mathfrak{x}_j(z, T)\right) = \delta \, \mathrm{tr}_{L/Q}\left(\sum_{i,j=1}^{m} a_i t_{ij} b_j^\sigma\right) .$$

Then the complete linear system determined by $3X$ is ample. We assume hereafter that the hyperplane-sections of $A(z, T, \mathfrak{M})$ is just that linear system. This is really possible, in view of our construction of $\Xi(u, z)$ in [AF, no. 10]. By our choice of X, for every prime number l, the l-adic representation $E_l(X)$ of the divisor X is not reduced to 0 modulo (l).

Denote by C_z the polarization of $A_z = A(z, T, \mathfrak{M})$ determined by the hyperplane sections, and by θ_z the isomorphism of L into $\mathcal{A}_0(A_z)$ defined by

$$\theta_z(\alpha)\Xi(u, z) = \Xi(\Phi(\alpha)u, z) .$$

Then we may write $\mathcal{P}(z, T, \mathfrak{M}) = (A_z, C_z, \theta_z)$. Obviously we obtain

PROPOSITION 9. *Let $\mathfrak{o}$ be the set of elements α of L such that $\alpha\,\mathfrak{M} \subset \mathfrak{M}$. Then $\theta_z(\mathfrak{o}) = \mathcal{A}(A_z) \cap \theta_z(L)$.*

$\mathfrak{o}$ being determined as in Proposition 9, $\mathcal{P}(z) = \mathcal{P}(z, T, \mathfrak{M})$ is a polarized abelian variety of type $\mathfrak{o}$ in the sense of [AF, no. 3]. By the result of [AF, no. 12], there exists a holomorphic mapping $\Psi(z) = \Psi(z, T, \mathfrak{M})$ of $\mathcal{H}(\Phi)$ into a complex projective space with the following properties:

(Ψ1) For every $z \in \mathcal{H}(\Phi)$, $\mathcal{P}(z)$ is defined over $Q(\Psi(z))$.

(Ψ2) If z_1 and z_2 are two points of $\mathcal{H}(\Phi)$, and if $\mathfrak{p}$ is a place of $Q(\Psi(z_1))$ taking values in C such that $\mathfrak{p}(\Psi(z_1)) = \Psi(z_2)$, then $(A_{z_2}, C_{z_2}, \theta_{z_2})$ is the reduction of $(A_{z_1}, C_{z_1}, \theta_{z_1})$ modulo $\mathfrak{p}$ (cf. [15, §5.1]).

Now fix a base (α_i) of $\mathfrak{o}$ over Z. With respect to this base, define the projective family $\mathcal{F}(A_z, \theta_z)$ for every member (A_z, C_z, θ_z) of $\Sigma(T, \mathfrak{M})$ (cf. [AF, no. 3], [15, §3.4]). We write

$$\mathcal{F}(z) = \mathcal{F}(z, T, \mathfrak{M}) = \mathcal{F}(A_z, \theta_z) .$$

Denote by $c(V)$ the Chow-point of an algebraic variety (or a cycle) V.

By [AF, Lemma 8], the $\mathcal{F}(z, T, \mathfrak{M})$ are of the same dimension for all $z \in \mathcal{H}(\Phi)$. Applying [AF, Theorem 1] to our case, we get

THEOREM 3. *There exist an analytic subset $\mathcal{W}$ of $\mathcal{H}(\Phi)$ of codimension one and a set of meromorphic functions $f_1, \cdots, f_\kappa$ on $\mathcal{H}(\Phi)$ with the following properties:*

(i) *the varieties $\mathcal{F}(z, T, \mathfrak{M})$ for $z \in \mathcal{H}(\Phi) - \mathcal{W}$ are of the same degree;*

(ii) *the f_i are holomorphic on $\mathcal{H}(\Phi) - \mathcal{W}$, and we have*

$$c\big(\mathcal{F}(z, T, \mathfrak{M})\big) = (1, f_1(z), \cdots, f_\kappa(z)) \text{ for } z \in \mathcal{H}(\Phi) - \mathcal{W}$$

The property (ii) implies, in particular that $\boldsymbol{Q}(f_1(z), \cdots, f_\kappa(z))$ *is the field of moduli of* $\mathcal{P}(z, T, \mathfrak{M})$.

By Theorem 2 and by [AF, Proposition 1],

(45) $\mathcal{F}(z, T, \mathfrak{M}) = \mathcal{F}(z', T, \mathfrak{M}) \Longleftrightarrow z = U(z')$ for some $U \in \Gamma(T, \mathfrak{M})$.

Therefore, the f_i are invariant under $\Gamma(T, \mathfrak{M})$. Moreover we get

THEOREM 4. *Let the set $\mathcal{W}$ and the functions $f_1, \cdots, f_\kappa$ be as in Theorem 3. Then the following assertions hold.*

(i) *There exists a $\Gamma(T, \mathfrak{M})$-invariant subset $\mathcal{Y}$ of $\mathcal{W}$ with the property that: the f_i are holomorphic on $\mathcal{H}(\Phi) - \mathcal{Y}$; and for any two points z, z' of $\mathcal{H}(\Phi) - \mathcal{Y}$, we have $f_i(z) = f_i(z')$ $(1 \leq i \leq \kappa)$ if and only if $z = U(z')$ for some $U \in \Gamma(T, \mathfrak{M})$.*

(ii) *The transcendency-degree of the field $C(f_1, \cdots, f_\kappa)$ over C is equal to the dimension of $\mathcal{H}(\Phi)$.*

(iii) *There exists an algebraic number field k of finite degree such that the fields $k(f_1, \cdots, f_\kappa)$ and C are linearly disjoint over k.*

We will prove Theorem 4 in the following sections.

3.3. We begin with an elementary

LEMMA 2. *Let U be an open subset of C^r, and $g_0, g_1, \cdots, g_\lambda$ be meromorphic functions on U. Suppose that $g_1, \cdots, g_\lambda$ are holomorphic at a point y of U and g_0 is not holomorphic at y. Then, $(0, g_1(y), \cdots, g_\lambda(y))$ is a specialization of $(g_0^{-1}, g_1, \cdots, g_\lambda)$ over C.*

PROOF. As g_0 is not holomorphic at y, we can find a sequence $y_1, \cdots,$ $y_h, \cdots$ of points of U such that $\lim_{h \to \infty} y_h = y$ and $\lim_{h \to \infty} g_0(y_h)^{-1} = 0$. Let B be the set of functions ψ of $C(g_0, g_1, \cdots, g_\lambda)$ such that $\lim_{h \to \infty} \psi(y_h)$ exists and is finite. Then $\psi \to \lim_{h \to \infty} \psi(y_h)$ gives a homomorphism of B into C which maps $(g_0^{-1}, g_1, \cdots, g_\lambda)$ onto $(0, g_1(y), \cdots, g_\lambda(y))$. This proves our lemma.

The functions f_i being as in Theorem 3, put $f_0 = 1$ and denote by $\mathcal{Z}_p$ for $0 \leq p \leq \kappa$ the set of points z of $\mathcal{H}(\Phi)$ where $f_0/f_p, f_1/f_p, \cdots, f_\kappa/f_p$ are holomorphic. Put $\mathcal{W}_0 = \bigcap_{p=0}^{\kappa}(\mathcal{H}(\Phi) - \mathcal{Z}_p)$. Then $\mathcal{W}_0$ is an analytic subset of $\mathcal{H}(\Phi)$ of codimension 1. For every point y of $\mathcal{H}(\Phi) - \mathcal{W}_0$, we can find an index p such that f_i/f_p are holomorphic at y for $0 \leq i \leq \kappa$. It is easy to see that the point $(f_0/f_p(y), \cdots, f_\kappa/f_p(y))$ in the projective space P^κ is determined only by y and independent of the choice of p. We denote this point by $f(y) = f(y, T, \mathfrak{M})$. Then $z \to f(z)$ gives a holomorphic mapping of $\mathcal{H}(\Phi) - \mathcal{W}_0$ into P^κ. Theorem 3 asserts that $\mathcal{W} \supset \mathcal{W}_0$ and $f(z) = c(\mathcal{F}(z))$ for $z \in \mathcal{H}(\Phi) - \mathcal{W}$. Let e be the degree of the varieties $\mathcal{F}(z)$ for $z \in \mathcal{H}(\Phi) - \mathcal{W}$. We now prove a more precise result than (ii) of Theorem 3.

PROPOSITION 10. *Let y be a point of $\mathcal{H}(\Phi)$ such that $\deg \mathcal{F}(y) = e$. Then $y \in \mathcal{H}(\Phi) - \mathcal{W}_0$ and $f(y) = c(\mathcal{F}(y))$.*

PROOF. Put $c(\mathcal{F}(y)) = (b_0, b_1, \cdots, b_\kappa)$. We want to show that if $b_p \neq 0$, the functions f_j/f_p for $0 \leq j \leq \kappa$ are holomorphic at y. Assume that $b_p \neq 0$ and f_i/f_p is not holomorphic for some i. Then, by Lemma 2, $(0, \Psi(y))$ is a specialization of $(f_p/f_i, \Psi)$ over C. Let z_0 be a generic point of $\mathcal{H}(\Phi) - \mathcal{W}$ for $\Psi, f_1, \cdots, f_\kappa$ over Q in the sense of [AF, no. 8]. Extend the specialization $(f_p/f_i(z_0), \Psi(z_0)) \to (0, \Psi(y))$ ref. Q to a specialization

$$(45') \qquad \left(f_p/f_i(z_0), \Psi(z_0), A(z_0), \mathcal{F}(z_0)\right) \to \left(0, \Psi(y), A', \mathcal{F}'\right) \qquad \text{ref. } Q.$$

By the property $(\Psi2)$, we have $A' = A(y)$. Hence by [15, Prop. 5.2], we get $\mathcal{F}' \supset \mathcal{F}(y)$. As $\mathcal{F}'$ is a specialization of $\mathcal{F}(z_0)$, $\mathcal{F}'$ is of the same dimension and of the same degree as $\mathcal{F}(z_0)$. On the other hand, by our assumption on y, $\mathcal{F}(y)$ is of the same dimension and of the same degree as $\mathcal{F}(z_0)$. Therefore we must have $\mathcal{F}' = \mathcal{F}(y)$. It follows that the specialization $c(\mathcal{F}(z_0)) = (f_0(z_0), \cdots, f_\kappa(z_0)) \to c(\mathcal{F}(y)) = (b_0, \cdots, b_\kappa)$ is compatible with $(45')$; so we must have $b_p = 0$, which is a contradiction. Thus we have proved that f_j/f_p are holomorphic at y for $0 \leq j \leq \kappa$ if $b_p \neq 0$. Then, considering again the specialization of $A(z_0)$ and $\mathcal{F}(z_0)$ over

$$(f_j/f_p(z_0), \Psi(z_0)) \to (f_j/f_p(y), \Psi(y)) \qquad \text{ref. } Q \, ,$$

we find $c(\mathcal{F}(y)) = (f_0/f_p(y), \cdots, f_\kappa/f_p(y))$. This completes our proof.

Let $\mathcal{Y}$ be the set of points y of $\mathcal{H}(\Phi)$ such that $\deg \mathcal{F}(y) \neq e$. As $\mathcal{F}(y) = \mathcal{F}(U(y))$ for $U \in \Gamma(T, \mathfrak{M})$, the set $\mathcal{Y}$ is $\Gamma(T, \mathfrak{M})$-invariant. By Theorem 3 and Proposition 10, we have $\mathcal{W} \supset \mathcal{Y} \supset \mathcal{W}_0$; and if $y \in \mathcal{H}(\Phi) - \mathcal{Y}$, we have $f(y) = c(\mathcal{F}(y))$. Therefore the assertion (i) of Theorem 4 follows from (45).

3.4. Now let $\bar{Q}$ denote the algebraic closure of Q. Let z_0 be a generic point of $\mathcal{H}(\Phi) - \mathcal{W}$ for Ψ and the f_i with respect to $\bar{Q}$. Let Δ be the set of elements U of $\Gamma(T, \mathfrak{M})$ such that $U(z) = z$ for every $z \in \mathscr{H}(\Phi)$; and put $\bar{\Gamma}(T, \mathfrak{M}) = \Gamma(T, \mathfrak{M})/\Delta$. We can choose z_0 in such a way that there is no element of $\bar{\Gamma}(T, \mathfrak{M})$, other than the identity, leaving z_0 invariant. Then there exists an open neighborhood $\mathscr{V}$ of z_0 such that $\mathscr{V} \subset \mathscr{H}(\Phi)$ $- \mathscr{W}$ and $\gamma(\mathscr{V}) \cap \mathscr{V} = \varnothing$ for every $\gamma \neq 1$ of $\bar{\Gamma}(T, \mathfrak{M})$. Let V be the locus of the point $c(\mathscr{F}(z_0)) = f(z_0)$ over $\bar{Q}$. The mapping $z \to f(z)$ gives a holomorphic mapping of $\mathscr{H}(\Phi) - \mathscr{W}_0$ into the algebraic variety V. As we have $f(z) = c(\mathscr{F}(z))$ for $z \in \mathscr{V}$, this mapping is one-to-one on $\mathscr{V}$. This means that the variety V contains a one-to-one holomorphic image of an open subset of $\mathscr{H}(\Phi)$. Hence we have

$$(46) \qquad\qquad \dim V \geqq \dim \mathscr{H}(\Phi) .$$

Let the $\Xi_i(u, z)$ for $0 \leq i \leq N$ be the coordinates of $\Xi(u, z)$. Consider a non-zero matrix of degree $N + 1$ as a point of a projective space P^M of dimension $M = (N + 1)^2 - 1$. Let R be the subset of P^M consisting of all non-degenerate matrices; R is then an open subset of P^M in the sense of Zariski-topology. Put, for $u \in C^n$, $z \in \mathscr{H}(\Phi)$, $\beta = (b_{ij}) \in R$,

$$\xi_i(u, z, \beta) = \sum_{j=0}^{N} b_{ij} \Xi_j(u, z) ,$$
$$\xi(u, z, \beta) = (\xi_0(u, z, \beta), \cdots, \xi_N(u, z, \beta)) .$$

Then $(u, z, \beta) \to \xi(u, z, \beta)$ is a holomorphic mapping of $C^n \times \mathscr{H}(\Phi) \times R$ into P^N; and for every $(z, \beta) \in \mathscr{H}(\Phi) \times R$, $u \to \xi(u, z, \beta)$ gives an isomorphism of $C^n/D(z, T, \mathfrak{M})$ onto the projective transform $\beta(A(z, T, \mathfrak{M}))$ of $A(z, T, \mathfrak{M})$ by β. By virtue of [AF, no. 9, Theorem S], the mapping $(z, \beta) \to c(\beta(A(z, T, \mathfrak{M})))$ is everywhere holomorphic. We denote this mapping by $p_{T,M}(z, \beta)$.

z_0 being as above, let α be a generic point of R over a field of definition for $A(z_0)$. Put $A' = \alpha(A(z_0))$. Let S be the locus of $c(A')$ over $\bar{Q}$. We can find non-zero elements $a_1, \cdots, a_r$ of $Q(c(A'))$ such that for every place $\mathfrak{q}$ of $Q(c(A'))$, A' is without defect for $\mathfrak{q}$ in the sense of [16, §11.1] if $\mathfrak{q}(a_i) \neq 0$ for every i. Now let S_0 be a non-empty Zariski open subset of S with the property that, for every $x \in S_0$ and for every i, $(0, x)$ is not a specialization of $(a_i, c(A'))$ over $\bar{Q}$; of course such S_0 exists. Now let x be a point of S_0; x is the Chow-point of a cycle B, since x is a specialization of the Chow-point $c(A')$ of a variety A'. We can find a place $\mathfrak{q}$ of $\bar{Q}(c(A'))$ such that $\mathfrak{q}(c(A')) = x = c(B)$ and $\mathfrak{q}(v) = v$ for $v \in \bar{Q}$. Then, by our choice of S_0, A' is without defect for $\mathfrak{q}$, and B is the reduction modulo $\mathfrak{q}$ of A'. Let C be a polarization of A' given by hyperplane sections. We can find an isomorphism θ of L into $\mathscr{A}_0(A')$ so that $\mathscr{P} = (A', C, \theta)$ is isomorphic to $\mathscr{P}(z_0)$, since A' is a projective transform of $A(z_0)$. Moreover, we may assume that $\mathscr{P}$ is defined over a finite algebraic extension K of

$Q(c(A'))$. Let X be a divisor on $A(z_0)$ such that $3X$ is algebraically equivalent to a hyperplane section of $A(z_0)$ (cf. §3.2). Then there exists a finite algebraic extension K_1 of K and a divisor Y on A', rational over K_1, which is algebraically equivalent to $\alpha(X)$. Extend q to a place of K_1, which we denote again by q. Let (B, C', θ') be the reduction modulo q of $\mathscr{P}$ (cf. [15, §5.1]). Then q(Y) is contained in C' and 3 q(Y) is algebraically equivalent to a hyperplane section of B. Moreover, for a suitable choice of l-adic coordinate systems, we get, by [16, §11.1, Prop. 14], $E_l(\mathrm{q}(Y)) = E_l(Y) = E_l(X)$. Therefore, according to what was said in §3.2, $E_l(\mathrm{q}(Y))$ is not reduced to 0 modulo (l), for every l. As q$(v) = v$ for every $v \in \bar{\boldsymbol{Q}}$, we see easily that (B, C', θ') belongs to $\{L, \Phi, \rho\}$. Hence there exist a couple $\{T_1, \mathfrak{M}_1\}$ and an isomorphism η of (B, C', θ') onto a member $\mathscr{P}(z_1, T_1, \mathfrak{M}_1)$ of $\Sigma(T_1, \mathfrak{M}_1)$. By the property of $E_l(\mathrm{q}(Y))$ just mentioned, and by our construction of $A(z_1, T_1, \mathfrak{M}_1)$ of §3.2, we observe that B must be a projective transform $\beta(A(z_1, T_1, \mathfrak{M}_1))$ for an element $\beta \in R$. Therefore we have $c(B) = p_{T_1, \mathfrak{M}_1}(z_1, \beta)$. This shows that S_0 is contained in the join of $p_{T_1, \mathfrak{M}_1}(\mathscr{H}(\Phi) \times R)$ for several $\{T_1, \mathfrak{M}_1\}$. But there are only countably infinite couples $\{T_1, \mathfrak{M}_1\}$. It follows that $\dim S \leqq \dim (\mathscr{H}(\Phi) \times R) = \dim \mathscr{H}(\Phi) + M$. On the other hand, by Proposition 8, $\dim S = \dim_{\boldsymbol{Q}} c(A')$ $= \dim_{\boldsymbol{Q}} c(\mathscr{F}(z_0)) + M$. Hence we must have $\dim V = \dim_{\boldsymbol{Q}} c(\mathscr{F}(z_0)) \leqq \dim \mathscr{H}(\Phi)$. This together with (46) proves

$$\dim V = \dim_{\boldsymbol{Q}} c(\mathscr{F}(z_0)) = \dim \mathscr{H}(\Phi) \ .$$

Therefore, $z \to f(z) = c(\mathscr{F}(z))$ gives an analytic isomorphism of an open subset $\mathscr{V}$ of $\mathscr{H}(\Phi)$ onto an open subset of V. From this we can conclude

$$\dim_C \boldsymbol{C}(f_1, \cdots, f_\kappa) = \dim_{\boldsymbol{Q}} \boldsymbol{Q}(f_1, \cdots, f_\kappa) = \dim \mathscr{H}(\Phi) \ .$$

This proves the assertions (ii), (iii) of Theorem 4.

By an *automorphic function with respect to* $\Gamma(T, \mathfrak{M})$, we understand a meromorphic function f on $\mathscr{H}(\Phi)$ invariant under the operation of the elements of $\Gamma(T, \mathfrak{M})$ except when $\dim \mathscr{H}(\Phi) = 1$ and $\Gamma(T, \mathfrak{M})\backslash\mathscr{H}(\Phi)$ is not compact; in the exceptional case, we impose on f a condition that f is meromorphic with respect to a local parameter at every cusp of $\Gamma(T, \mathfrak{M})$. Now, if the factor space $\Gamma(T, \mathfrak{M})\backslash\mathscr{H}(\Phi)$ is compact, it follows from (i), (ii) of Theorem 4 that $\boldsymbol{C}(f_1, \cdots, f_\kappa)$ *is the field of all automorphic functions with respect to* $\Gamma(T, \mathfrak{M})$. Even in the case where $\Gamma(T, \mathfrak{M})\backslash\mathscr{H}(\Phi)$ is not compact, if we get a suitable analytically normal compactification of $\Gamma(T, \mathfrak{M})\backslash\mathscr{H}(\Phi)$, we can prove that $\boldsymbol{C}(f_1, \cdots, f_\kappa)$ is the field of all automorphic functions with respect to $\Gamma(T, \mathfrak{M})$. Such a compactification is first considered by Satake [11, 12] and recently by Pyatetski-Shapiro [10] in a fairly wide class of discontinuous groups. In the case of $\Gamma(T, \mathfrak{M})$

of ($Type$ I), Baily [4, 5] proved that $\Gamma(T, \mathfrak{M})\backslash\mathscr{H}(\Phi)$ is compactified to a normal analytic space, isomorphic to a projective variety. Therefore, if L is of ($Type$ I), $C(f_1, \cdots, f_\kappa)$ is the field of all automorphic functions with respect to $\Gamma(T, \mathfrak{M})$. (cf. [13, exposé 20; 7]). It is conjecturable that the same conclusion is true for ($Type$ II, III, IV). As for the field of definition for $C(f_1, \cdots, f_\kappa)$, we have proved that we can take $\boldsymbol{Q}$ as the field k of (iii) of Theorem 4 in several cases [13 exposé 20; 14]. Recently, Katayama [7] has shown that $\boldsymbol{Q}$ can be taken as k in general for ($Type$ I).

3.5. *Automorphic functions with respect to congruence-subgroups.* Let $\mathscr{D}$ be the set of all generic points of $\mathscr{H}(\Phi) - \mathscr{W}$ for Ψ over $\boldsymbol{Q}$. Fix a point y of $\mathscr{D}$. Let (V_0, h_0) be a normalized Kummer variety of $\mathscr{P}(y)$ in the sense of [15, §2.2]. If z is a point of $\mathscr{D}$, there exists an isomorphism σ of $\boldsymbol{Q}(\Psi(y))$ onto $\boldsymbol{Q}(\Psi(z))$ such that $\Psi(y)^\sigma = \Psi(z)$; we have then $\mathscr{P}(y)^\sigma = \Psi(z)$. Put

$$V_z = V(z) = V_0^\sigma, \quad h_z = h_0^\sigma .$$

(V_z, h_z) is obviously a normalized Kummer variety of $\mathscr{P}(z)$. Now let $b = (b_1, \cdots, b_m)$ be an element of $L^m = L \times \cdots \times L$. Put

$$x_b(z) = \Xi\left(\sum_{i=1}^m \Phi(b_i)\mathfrak{x}_i(z, T), z\right) .$$

The quotients of the coordinates of the point $h_z(x_b(z))$ on V_z are contained in the field $\boldsymbol{Q}(\Psi(z), x_b(z))$ for every $z \in \mathscr{D}$, since h_z is defined over $\boldsymbol{Q}(\Psi(z))$. Let $\mathscr{D}(b)$ be the set of all generic points of $\mathscr{H}(\Phi) - \mathscr{W}$ for Ψ and x_b over $\boldsymbol{Q}$. Fix a point z_1 of $\mathscr{D}(b)$. Then there exist elements $v_1, \cdots, v_\lambda$ of $\boldsymbol{Q}(\Psi(z_1), x_b(z_1))$ such that

$$h_{z_1}\big(x_b(z_1)\big) = (1, v_1, \cdots, v_\lambda) .$$

Let $g_1(b, z), \cdots, g_\lambda(b, z)$ be the elements of $\boldsymbol{Q}(\Psi, x_b)$ corresponding to $v_1, \cdots, v_\lambda$ by the canonical isomorphism of $\boldsymbol{Q}(\Psi, x_b)$ onto $\boldsymbol{Q}(\Psi(z_1), x_b(z_1))$. Then $g_1, \cdots, g_\lambda$ are meromorphic functions on $\mathscr{H}(\Phi)$; and for every $z \in \mathscr{D}(b)$, we have

$$h_z\big(\Xi(\sum_{i=1}^m \Phi(b_i)\mathfrak{x}_i(z, T), z)\big) = (1, g_1(b, z), \cdots, g_\lambda(b, z)) .$$

PROPOSITION 11. *For every* $b \in L^m$ *and* $U \in \Gamma(T, \mathfrak{M})$, *we have*

$$g_\nu(bU, z) = g_\nu(b, U(z)) \qquad\qquad (1 \leqq \nu \leqq \lambda) .$$

PROOF. Let z be a point of $\mathscr{D} \cap U(\mathscr{D})$. Put $z = U(z')$. Define Λ_ν by (39) of §2.8. As we have seen in §2.8, there exists an isomorphism η of $\mathscr{P}(z)$ onto $\mathscr{P}(z')$ such that $\eta\Xi(u, z) = \Xi(\Lambda u, z')$, where Λ is the linear transformation determined by (33) and (32). As z and z' are generic for Ψ over $\boldsymbol{Q}$, there exists an isomorphism σ of $\boldsymbol{Q}(\Psi(z))$ onto $\boldsymbol{Q}(\Psi(z'))$ such that $\Psi(z)^\sigma = \Psi(z')$; we have then $\mathscr{P}(z)^\sigma = \mathscr{P}(z')$. As $\mathscr{P}(z)$ is isomorphic to $\mathscr{P}(z')$, σ must leave invariant the elements of the field of moduli of $\mathscr{P}(z)$. Therefore, by the property of normalized Kummer variety [15, §2.2, (K3)], the equality $h_z^\sigma \circ \eta = h_z$ holds. By our construction of h_z, we have $h_z^\sigma = h_{z'}$, so that

$$h_{z'}\big(\Xi(\Lambda u, z')\big) = h_z^\sigma\big(\eta \Xi(u, z)\big) = h_z\big(\Xi(u, z)\big) \ .$$

Substituting $\sum_{i=1}^m \Phi(b_i)\mathfrak{x}_i(z, T)$ for u, and putting $bU = b' = (b_1', \cdots, b_m')$, we obtain

$$h_{z'}\big(\Xi(\textstyle\sum_{i=1}^m \Phi(b_i')\mathfrak{x}_i(z', T), z')\big) = h_z\big(\Xi(\textstyle\sum_{i=1}^m \Phi(b_i)\mathfrak{x}_i(z, T), z)\big) \ .$$

If we take z' sufficiently generic, this proves the relation of our proposition.

By the consideration of §2.8, any automorphism of $\mathscr{P}(z)$ is obtained as follows. Let U be an element of $\Gamma(T, \mathfrak{M})$ such that $U(z) = z$. Define Λ_ν by $\Lambda_\nu X_\nu(z, T) = X_\nu(z, T) \cdot {}^t\omega_\nu(U)$, and Λ by (32) and (33). Then there exists an automorphism η of $\mathscr{P}(z)$ such that $\Xi(\Lambda u, z) = \eta\Xi(u, z)$. Let Δ be the set of elements U of $\Gamma(T, \mathfrak{M})$ such that $U(z) = z$ for all $z \in \mathscr{H}(\Phi)$. For every $z \in \mathscr{H}(\Phi)$, the mapping $U \to \eta$ gives an isomorphism of Δ into the group of automorphisms of $\mathscr{P}(z)$. If z is so generic that $U(z) = z$ implies $U \in \Delta$, this isomorphism is surjective. Fix a sufficiently generic point z_0 of $\mathscr{H}(\Phi)$. Denote by η_U the automorphism of $\mathscr{P}(z_0)$ corresponding to $U \in \Delta$. Let a and b be two elements of L^m. If we have $h_{z_0}(x_a(z_0)) = h_{z_0}(x_b(z_0))$, then $\eta_U(x_a(z_0)) = x_b(z_0)$ for some $U \in \Delta$, since (V_z, h_z) is a quotient variety of A_z with respect to the group of automorphisms of $\mathscr{P}(z)$. Putting $aU = a' = (a_1', \cdots, a_m')$, we get $x_{a'}(z_0) = x_b(z_0)$, so that $aU = a' \equiv b \bmod \mathfrak{M}$. Conversely, if $a' = aU \equiv b \bmod \mathfrak{M}$, we have $\eta_U(x_a(z_0)) = x_b(z_0)$, and hence $h_{z_0}(x_a(z_0)) = h_{z_0}(x_b(z_0))$. Therefore, we get

PROPOSITION 12. *Let Δ be the set of elements U of $\Gamma(T, \mathfrak{M})$ such that $U(z) = z$ for every $z \in \mathscr{H}(\Phi)$. Let a and b be two elements of L^m. Then $g_\nu(a, z) = g_\nu(b, z)$ holds for every ν if and only if $aU \equiv b \bmod \mathfrak{M}$ for some $U \in \Delta$.*

The group Δ will be determined in §4.9.

Let $\mathfrak{N}$ be a $\mathbf{Z}$-lattice in L^m containing $\mathfrak{M}$. Denote by $\Gamma(T, \mathfrak{N}/\mathfrak{M})$ the set of elements U of $\Gamma(T, \mathfrak{M})$ such that $aU \equiv a \bmod \mathfrak{M}$ for every $a \in \mathfrak{N}$. It is clear that $\Gamma(T, \mathfrak{N}/\mathfrak{M})$ is a subgroup of $\Gamma(T, \mathfrak{M})$ of finite index. If $\mathfrak{N}V = \mathfrak{N}$ for every $V \in \Gamma(T, \mathfrak{M})$, then $\Gamma(T, \mathfrak{N}/\mathfrak{M})$ is a normal subgroup of $\Gamma(T, \mathfrak{M})$; in particular, for every positive integer s, $\Gamma(T, s^{-1}\mathfrak{M}/\mathfrak{M})$ is a normal subgroup of $\Gamma(T, \mathfrak{M})$, which may be called the principal congruence subgroup of $\Gamma(T, \mathfrak{M})$ of level s.

PROPOSITION 13. *Let $\mathfrak{N}$ be a $\boldsymbol{Z}$-lattice in L^m containing $\mathfrak{M}$, and $b^1, \cdots, b^\mu$ be the representatives for $\mathfrak{N}$ modulo $\mathfrak{M}$. Let $f_1, \cdots, f_\kappa$ be as in Theorem* 3. *Put*

$$\mathfrak{k}(T, \mathfrak{M}) = C(f_1, \cdots, f_\kappa) ,$$

$$\mathfrak{k}(T, \mathfrak{N}/\mathfrak{M}) = C\big(f_1, \cdots, f_\kappa, g_1(b^\nu, z), \cdots, g_\lambda(b^\nu, z) \mid 1 \leqq \nu \leqq \mu\big) .$$

Then $\mathfrak{k}(T, \mathfrak{N}/\mathfrak{M})$ is a field of automorphic functions with respect to $\Gamma(T, \mathfrak{N}/\mathfrak{M})$, which is a finite algebraic extension of $\mathfrak{k}(T, \mathfrak{M})$. If an element U of $\Gamma(T, \mathfrak{M})$ leaves invariant all the functions belonging to $\mathfrak{k}(T, \mathfrak{N}/\mathfrak{M})$, then $U \in \Gamma(T, \mathfrak{N}/\mathfrak{M})\Delta$.

This follows immediately from Propositions 11 and 12.

By Proposition 13, if $\mathfrak{k}(T, \mathfrak{M})$ is the field of all automorphic functions with respect to $\Gamma(T, \mathfrak{M})$, then $\mathfrak{k}(T, \mathfrak{N}/\mathfrak{M})$ is the field of all automorphic functions with respect to $\Gamma(T, \mathfrak{N}/\mathfrak{M})$. Therefore, we can at least conclude that, if $\Gamma(T, \mathfrak{M})\backslash\mathcal{H}(\Phi)$ is compact or if L is of (*Type* I), then $\mathfrak{k}(T, \mathfrak{N}/\mathfrak{M})$ is the field of all automorphic functions with respect to $\Gamma(T, \mathfrak{N}/\mathfrak{M})$.

4. The endomorphism-algebras of generic members of analytic families $\Sigma(T, \mathfrak{M})$.

Hereafter we assume that L is a division algebra.

4.1. First we recall some of our notation.

$$g = [F\colon \boldsymbol{Q}] ,$$

$$n = \dim A(z, T, \mathfrak{M}) ,$$

$$2n = m[L\colon \boldsymbol{Q}] ,$$

$$q^2 = [L\colon F_0], \ r_\nu + s_\nu = mq \qquad \text{for (} Type \text{ IV) .}$$

Denote by d the dimension of $\mathcal{H}(\Phi)$. Then we have

$$(Type \text{ I}) \qquad n = \frac{m}{2} g , \qquad d = \frac{1}{2} \cdot \frac{m}{2}\left(\frac{m}{2} + 1\right)g ,$$

$$(Type \text{ II}) \qquad n = 2mg , \qquad d = \frac{1}{2} m(m + 1)g ,$$

$$(Type \text{ III}) \qquad n = 2mg , \qquad d = \frac{1}{2} m(m - 1)g ,$$

$$(Type \text{ IV}) \qquad n = q^2 mg , \qquad d = \sum_{\nu=1}^{g} r_\nu s_\nu .$$

By Theorem 4, and by [15, Prop. 3.5 and p. 305, Remark], we have, for every point z of $\mathcal{H}(\Phi)$,

$$(46') \qquad\qquad \dim_Q c(\mathcal{F}(z, T, \mathfrak{M})) \leqq d .$$

We call $\mathscr{P}(z, T, \mathfrak{M})$ a generic member of $\Sigma(T, \mathfrak{M})$ if the equality $\dim_Q c(\mathscr{F}(z, T, \mathfrak{M})) = d$ holds. This is so whenever z is a generic point of $\mathscr{H}(\Phi) - \mathscr{W}_0$ for the f_i over Q. Now we want to prove

THEOREM 5. *If $\mathscr{P}(z, T, \mathfrak{M}) = (A, C, \theta)$ is a generic member of $\Sigma(T, \mathfrak{M})$, then $\theta(L) = \mathscr{A}_0(A)$ holds except in the following five cases:*

(a) *L is of (Type III), $m = 1$.*

(b) *L is of (Type III), $m = 2$, and there exists a totally positive element α such that $N(T) = \alpha^2$, where N denotes the reduced norm of $M_2(L)$ to F.*

(c) *L is of (Type IV), $\sum_{\nu=1}^{g} r_\nu s_\nu = 0$.*

(d) *L is of (Type IV), $m = 2$, $q = 1$, $r_\nu = s_\nu = 1 \ (1 \leqq \nu \leqq g)$.*

(e) *L is of (Type IV), $m = 1$, $q = 2$, $r_\nu = s_\nu = 1 \ (1 \leqq \nu \leqq g)$.*

The result for these exceptional cases will be given in Propositions 14, 15, 17, 18, 19.

4.2. *Case* (c). Exchanging χ_ν for $\bar{\chi}_\nu$, if necessary, we may assume that $r_1 = \cdots = r_g = mq$ and $s_1 = \cdots = s_g = 0$. Put $k = mq^2$. Then, for suitable g isomorphisms $\sigma_1, \cdots, \sigma_g$ of F_0 into C, we have $\chi_\nu(\alpha) = \alpha^{\sigma_\nu} 1_k$ for $\alpha \in F_0$. As $\sigma_1, \cdots, \sigma_g$ together with their complex conjugates form the set of all isomorphisms of F into C, the system $(F_0; \{\sigma_\nu\})$ is a CM-type in the sense of [16, §5.2].

PROPOSITION 14. *Suppose that L is of (Type IV) and $\sum_{\nu=1}^{g} r_\nu s_\nu = 0$. If (A, C, θ) belongs to $\{L, \Phi, \rho\}$, then A is isogenous to the product of $k \ (= mq^2)$ copies of an abelian variety of dimension g belonging to the CM-type $(F_0; \{\sigma_\nu\})$ defined as above.*

PROOF. Let C^n/D be a complex torus isomorphic to A. We can find k vectors $\mathfrak{y}_1, \cdots, \mathfrak{y}_k$ of C^n such that $QD = \sum_{i=1}^{k} \Phi(F_0)\mathfrak{y}_i$. Let $\mathfrak{g}$ be the ring of integers in F_0. Put $D' = \sum_{i=1}^{k} \Phi(\mathfrak{g})\mathfrak{y}_i$. Then C^n/D is isogenous to C^n/D'. Put $\mathfrak{U}_i = \Phi(F_0 \otimes_Q R)\mathfrak{y}_i$, $D_i = \Phi(\mathfrak{g})\mathfrak{y}_i$ for $1 \leqq i \leqq k$. Then, by our assumption $s_1 = \cdots = s_g = 0$, the $\mathfrak{U}_i$ are complex vector subspaces of C^n. Obviously, we have $C^n = \mathfrak{U}_1 + \cdots + \mathfrak{U}_k$, $D' = D_1 + \cdots + D_k$, and C^n/D is isomorphic to $\mathfrak{U}_1/D_1 \times \cdots \times \mathfrak{U}_k/D_k$. By [16, §6.2, Theorem 3], each $\mathfrak{U}_i/D_i$ is an abelian variety belonging to the CM-type $(F_0; \{\sigma_\nu\})$. This completes the proof.

REMARK 4. An abelian variety of dimension g belonging to $(F_0; \{\sigma_\nu\})$ may be or may not be simple. A criterion for this is given in [16, §8.2].

4.3. *Case* (a). Let (A, C, θ) belong to $\{L, \Phi, \rho\}$ and let C^n/D be a torus isomorphic to A. As $m = 1$, we get $D = \Phi(\mathfrak{M})\mathfrak{x}_0$ for a vector $\mathfrak{x}_0$ and a lattice $\mathfrak{M}$ in L. The Riemann form (11) of §2.2 is now written in the form

$$E\big(\Phi(a)\mathfrak{x}_0,\ \Phi(b)\mathfrak{x}_0\big) = \mathrm{tr}_{L/Q}(atb^\rho) \qquad\qquad (a, b \in L_R)$$

with a ρ-skew-symmetric element t of L. As L is a totally definite quaternion algebra over F, ρ must be the canonical involution of L, by virtue of Proposition 3 of §1.2. As $C^n = \Phi(L_R)\mathfrak{x}_0$, there exists an element of L_R such that $\sqrt{-1}\,\mathfrak{x}_0 = \Phi(h)\mathfrak{x}_0$. We see easily $h^2 = -1$. By (15) of §2.2, th^ρ is ρ-symmetric. It follows that $h^{-1}t$ is contained in the center of L_R. Put $a = h^{-1}t$, $t^2 = -tt^\rho = s$; obviously, s is an element of F, which must be totally negative, because L is a totally definite quaternion algebra. Now consider the linear mapping of C^n into itself given by $\mathfrak{x} \to \sqrt{-1}\,\Phi(a)\mathfrak{x}$. As a is contained in the center of L_R, this linear mapping sends $QD = \Phi(L)\mathfrak{x}_0$ into itself. Therefore, this gives rise to an element λ of $\mathcal{A}_0(A)$. As we have $-\Phi(a^2)\mathfrak{x}_0 = (\sqrt{-1}\,\Phi(a))^2\mathfrak{x}_0 = \Phi(t)^2\mathfrak{x}_0$ and as $\xi \to \Phi(\xi)\mathfrak{x}_0$ is an isomorphism of L_R onto $C^n = \Phi(L_R)\mathfrak{x}_0$, we must have $-\Phi(a^2) = \Phi(t^2)$, so that $-a^2 = t^2 = s$. Now identify F with a subfield of $\mathcal{A}_0(A)$ by θ. Then we have $\lambda^2 = s$. Therefore $F(\lambda)$ is isomorphic to the quadratic subfield $F(t)$ of L. As $\sqrt{-1}\,\chi_\nu(a)$ is a scalar matrix for every ν, A belongs to just the *Case* (c) with $r_\nu = 2$, $s_\nu = 0$, taking F_0 to be $F(\lambda)$. By Proposition 14, we obtain

PROPOSITION 15. *Suppose that L is of (Type III) and $m = 1$. If (A, C, θ) belongs to $\{L, \Phi, \rho\}$, then A is isogenous to the product of two copies of an abelian variety of dimension g belonging to a CM-type $(F_0; \{\sigma_\nu\})$, where F_0 is isomorphic to a quadratic subfield of L.*

4.4. Now let us discuss the general case, leaving *Cases* (b) (d) (e) until later. First we prove

PROPOSITION 16. *If $\mathcal{P}(z, T, \mathfrak{M}) = (A, C, \theta)$ is a generic member of $\Sigma(T, \mathfrak{M})$, then the commutor of $\theta(L)$ in $\mathcal{A}_0(A)$ is contained in $\theta(L)$, except Cases (a) (b) (c) of Theorem 5.*

PROOF. In view of [15, Prop. 3.5], it is sufficient to prove the conclusion for at least one generic member of $\Sigma(T, \mathfrak{M})$. Let λ be an element of $\mathcal{A}_0(A)$ commuting with every element of $\theta(L)$. Let Λ be the linear mapping of C^n into itself representing λ. As Λ maps $D(z, T, \mathfrak{M})$ into itself, there exists an element $U = (u_{ij})$ of $M_m(L)$ such that

$$(47) \qquad \Lambda\mathfrak{x}_i(z, T) = \sum_{j=1}^{m} \Phi(u_{ij})\mathfrak{x}_j(z, T) \qquad\qquad (1 \leq i \leq m)\ .$$

Λ must be of the form (32), (33), since Λ commutes with $\Phi(\alpha)$ for every $\alpha \in L$. Define Λ_ν by (34). We have then

$$(48) \qquad \Lambda_\nu X_\nu(z) = X_\nu(z)\cdot {}^t\omega_\nu(U)\ .$$

Putting ${}^t\bar{W}_\nu^{-1}\omega_\nu(U)\cdot{}^t\bar{W}_\nu = \begin{pmatrix} a_\nu & b_\nu \\ c_\nu & d_\nu \end{pmatrix}$, $z = (z_1, \cdots, z_g)$, we get

$$(49) \qquad \begin{aligned} a_\nu z_\nu + b_\nu &= z_\nu(c_\nu z_\nu + d_\nu) && \text{for } (Type \text{ I, II, III}) . \\ \bar{a}_\nu z_\nu + \bar{b}_\nu &= z_\nu(\bar{c}_\nu z_\nu + \bar{d}_\nu) && \text{for } (Type \text{ IV}), \ r_\nu s_\nu > 0 . \end{aligned}$$

For a given U of $M_m(L)$, these equations define an analytic set $\mathfrak{S}(U)$ in $\mathcal{H}(\Phi)$. Suppose that $\mathfrak{S}(U) = \mathcal{H}(\Phi)$. Now we treat four $Types$ I, II, III, IV separately.

($Type$ I, II). We see easily that if (49) holds everywhere on $\mathcal{H}(\Phi)$, then $b_\nu = c_\nu = 0$ and $a_\nu z_\nu = z_\nu d_\nu$ holds for every $z = (z_1, \cdots, z_q)$ of $\mathcal{H}(\Phi)$. It is easy to see that $a_\nu z_\nu = z_\nu d_\nu$ holds for every $z_\nu \in \mathcal{H}_r^1$, if and only if $a_\nu = d_\nu$ and a_ν is a scalar matrix. Hence U must be contained in the center of $M_m(L)$. Namely we have $U = \eta 1_m$ with an element $\eta \in F$. Therefore, $\mathfrak{S}(U)$ is of codimension >0 in $\mathcal{H}(\Phi)$ if $U \notin F$. As $M_m(L)$ contains only countably infinite elements, we can take a generic member $\mathscr{P}(z_0, T, \mathfrak{M})$ so that $z_0 \notin \mathfrak{S}(U)$ for every $U \notin F$. For this member, we must have $\Lambda \mathfrak{x}_i(z_0, T) = \Phi(\eta) \mathfrak{x}_i(z_0, T)$ with an element η of F, so that $\Lambda(\sum_{i=1}^m \Phi(\alpha_i) \mathfrak{x}_i(z_0, T)) = \Phi(\eta)(\sum_{i=1}^m \Phi(\alpha_i) \mathfrak{x}_i(z_0, T))$ for every $\alpha_i \in L_R$. Therefore Λ must coincide with $\Phi(\eta)$. This shows that the commutor of $\theta(L)$ in $\mathcal{A}_0(A(z_0))$ is just $\theta(F)$. Our proposition is thereby proved for ($Type$ I, II).

($Type$ IV). Suppose that $r_\nu s_\nu > 0$ for some ν, say 1. If (49) holds everywhere on $\mathcal{H}(\Phi)$, then we have $b_1 = c_1 = 0$ and $a_1 = \beta 1_{r_1}$, $d_1 = \beta 1_{s_1}$ with $\beta \in C$, so that $\omega_1(U) = \beta 1_{mq}$. It follows that U is contained in the center of $M_m(L)$. By the same argument as for ($Type$ I, II), we get the assertion of our proposition for ($Type$ IV).

($Type$ III). If $m > 2$, the argument and the conclusion are quite similar to the above; so suppose that $m = 2$. Then (48) is written in the form

$$(50) \qquad \begin{bmatrix} a_\nu & b_\nu \\ c_\nu & d_\nu \end{bmatrix} \begin{bmatrix} z_\nu & 1_2 \\ 1_2 & -\bar{z}_\nu \end{bmatrix} = \begin{bmatrix} z_\nu & 1_2 \\ 1_2 & -\bar{z}_\nu \end{bmatrix} \begin{bmatrix} {}^t\Sigma_\nu & 0 \\ 0 & {}^t\bar{\Sigma}_\nu \end{bmatrix} ,$$

where Σ_ν is a matrix determined by (34). If (49) holds everywhere on $\mathcal{H}(\Phi)$, we have $b_\nu = c_\nu = 0$ and $a_\nu z_\nu = z_\nu d_\nu$ for every $z_\nu \in \mathcal{H}_2^2$; then (50) shows $\bar{a}_\nu = d_\nu = {}^t\Sigma_\nu$. As z_ν is a skew-symmetric matrix, we have

$$a_\nu = \bar{d}_\nu = {}^t\bar{\Sigma}_\nu = \begin{pmatrix} \alpha_\nu & \beta_\nu \\ -\bar{\beta}_\nu & \bar{\alpha}_\nu \end{pmatrix} , \qquad \text{with } \alpha_\nu, \beta_\nu \in C .$$

Recall that $W_\nu = \omega_\nu(W)$ for an element W of $M_2(L_R)$ such that $\pi_\nu(WT^{-1}W^\rho) = \sqrt{-1} \, 1_2$, where π_ν is the projection of $M_2(L_R) = M_2(K) \times \cdots \times M_2(K)$ onto the νth factor. Then we have

$$\omega_\nu((W^\rho)^{-1} U W^\rho) = \begin{bmatrix} a_\nu & 0 \\ 0 & \bar{a}_\nu \end{bmatrix} \quad \text{and} \quad \pi_\nu((W^\rho)^{-1} U W^\rho) = \begin{pmatrix} \alpha_\nu & \beta_\nu \\ -\bar{\beta}_\nu & \bar{\alpha}_\nu \end{pmatrix} .$$

Now exchanging $\{T, \mathfrak{M}\}$ for some equivalent couple, we may assume that T is diagonal. Then we can take W so as to be a diagonal matrix of $M_2(L_R)$. Put

$$U = \begin{pmatrix} x & y \\ u & v \end{pmatrix}, \qquad T = \begin{pmatrix} t & 0 \\ 0 & s \end{pmatrix}, \qquad (W^\rho)^{-1} = \begin{pmatrix} e & 0 \\ 0 & f \end{pmatrix},$$

Recalling that we have fixed an identification of C with a subfield of K, denote by $\mathfrak{C}$ the subalgebra $C \times \cdots \times C$ of $L_R = K \times \cdots \times K$. Let α, β, ζ denote the elements of $\mathfrak{C}$ such that $\pi_\nu(\alpha) = \alpha_\nu$, $\pi_\nu(\beta) = \beta_\nu$, $\pi_\nu(\zeta) = -\sqrt{-1}$. Then $ete^\rho = fsf^\rho = \zeta$, $x = e^{-1}\alpha e$, $y = e^{-1}\beta f$, $u = -f^{-1}\beta^\rho e$, $v = f^{-1}\alpha^\rho f$. Put $N(\xi) = \xi\xi^\rho$ for $\xi \in L_R$. Then $uy = -N(\beta)$. As u and y are elements of L, this shows that $-uy$ is a totally positive element of F; put $-uy = w$. Suppose that $\beta \neq 0$. Then $uy \neq 0$ and as $N(y) = N(e^{-1}f)N(\beta)$, $N(u) = N(e^{-1}f)^{-1}N(\beta)$, we obtain $N(e^{-1}f)^2 = N(yu^{-1}) = N(y)^2 w^{-2}$. As $t = e^{-1}fs(e^{-1}f)^\rho$, we get $N(ts) = N(e^{-1}f)^2 N(s)^2 = (N(ys)w^{-1})^2$. This is a contradiction, since we have excluded the *Case* (b). Hence we must have $\beta = 0$, so that $u = y = 0$; namely, U is of the form $U = \begin{pmatrix} x & 0 \\ 0 & v \end{pmatrix}$. The relation $x = e^{-1}\alpha e$ shows $x \in e^{-1}\mathfrak{C}e \cap L$. On the other hand, we have $t = e^{-1}\zeta(e^\rho)^{-1} = e^{-1}(N(e)^{-1}\zeta)e$, so that $t \in e^{-1}\mathfrak{C}e \cap L$. It follows that x commutes with t. As $t^\rho = -t$, $F(t)$ is a quadratic subfield of the quaternion algebra L. Hence $x \in F(t)$. As our purpose is to prove that $\lambda \in \theta(L)$, we can put $\lambda - \theta(\eta)$ in place of λ for any $\eta \in F$. By this exchange, $\Lambda - \Phi(\eta)$ takes the place of Λ, and $U - \eta 1_2$ takes the place of U. For a suitable choice of η, we may assume that $x = \xi t$ with $\xi \in F$. Then we have $v^\rho = f^{-1}\alpha f = f^{-1} exe^{-1}f = f^{-1}e(\xi t)e^{-1}f$. Suppose that $\alpha \neq 0$; we have then $xv \neq 0$, $\xi \neq 0$, and $s = f^{-1}et(f^{-1}e)^\rho = (f^{-1}e)(N(f^{-1}e)t)$ $(f^{-1}e)^{-1} = N(f^{-1}e)\xi^{-1}v^\rho$, so that $N(e^{-1}f) = \xi^{-1}v^\rho s^{-1}$. This shows that $N(e^{-1}f)$ is contained in L. As $N(e^{-1}f)$ belongs to the center of L_R, $N(e^{-1}f)$ must be an element of F, which is obviously totally positive. This is a contradiction, since $N(ts) = N(e^{-1}f)^2 N(s)^2$. Therefore we must have $\alpha = 0$, and hence $U = 0$. Namely, we have shown that if z is sufficiently generic, the commutor of $\theta(L)$ in $\mathcal{A}_0(A)$ is contained in $\theta(F)$. Thus Proposition 16 is completely proved.

4.5. *Case* (b). We use the same notation as in the proof of Proposition 16. In *Case* (b), there exists a totally positive element ξ of F such that $N(ts) = \xi^2$. As $e^{-1}fs(e^{-1}f)^\rho = t$, we have $N(e^{-1}f)^2 = N(ts^{-1}) = (\xi N(s)^{-1})^2$. As ξ is totally positive, we obtain $N(e^{-1}f) = \xi N(s)^{-1}$. Put

$$U = \begin{bmatrix} t & 0 \\ 0 & -\xi N(s)^{-1}s \end{bmatrix}, \qquad \alpha = N(e^{-1})\zeta, \qquad \pi_\nu(\alpha) = \alpha_\nu.$$

180 GORO SHIMURA

Then

$$(W^\rho)^{-1}UW^\rho = \begin{bmatrix} \alpha & 0 \\ 0 & \alpha^\rho \end{bmatrix}.$$

Put

$$\Sigma_\nu = \begin{bmatrix} \bar\alpha_\nu & 0 \\ 0 & \alpha_\nu \end{bmatrix},$$

and define Λ_ν, Λ by (32), (33), (34). Then the equalities (48) and (47) hold for every $z \in \mathcal{H}(\Phi)$. As Λ commutes with every element of $\Phi(L)$, Λ maps $QD(z, T, \mathfrak{M})$ into itself. Therefore Λ defines an element λ of $\mathcal{A}_0(A(z))$. We note that $\Lambda^2 = -\Phi(N(t)) = \Phi(t^2)$; so $F(\lambda)$ is isomorphic to $F(t)$. If we put $F_0 = F(\lambda)$, then $\mathcal{A}_0(A(z))$ contains an algebra isomorphic to $L \otimes_F F_0$, since the elements of F_0 commute with the elements of $\theta(L)$. As F_0 is isomorphic to $F(t)$ and as $F(t)$ is a quadratic subfield of L over F, $L \otimes_F F_0$ is isomorphic to a total matrix algebra $M_2(F_0)$. It follows that $A(z)$ is isogenous to the product of two copies of an abelian variety B whose endomorphism algebra $\mathcal{A}_0(B)$ contains a field isomorphic to F_0. Thus we have proved

PROPOSITION 17. *In Case* (b) *of Theorem 5, every member $A(z, T, \mathfrak{M})$ of $\Sigma(T, \mathfrak{M})$ is isogenous to the product of two copies of an abelian variety B such that $\mathcal{A}_0(B)$ contains a field isomorphic to a quadratic subfield of L over F.*

REMARK 5. $\mathcal{A}_0(B)$ is not isomorphic to F_0, but larger than F_0. In fact, it is not difficult to see that B belongs to *Case* (d) of Theorem 5. Then our later result (Proposition 18) will show that $\mathcal{A}_0(B)$ contains a totally indefinite quaternion algebra over F.

REMARK 6. *For every algebra L of (Type III), there exists a non-degenerate ρ-skew-symmetric element T of $M_2(L)$ for which there is no totally positive element ξ of F such that $N(T) = \xi^2$.* In fact let $\mathfrak{p}_1, \cdots, \mathfrak{p}_r$ be the prime ideals of F at which L ramifies. Let $\mathfrak{p}_0$ be a prime ideal of F other than $\mathfrak{p}_1, \cdots, \mathfrak{p}_r$. We can find an ideal $\mathfrak{a}$ prime to $\mathfrak{p}_0\mathfrak{p}_1 \cdots \mathfrak{p}_r$ and a totally positive element α of F such that $\mathfrak{p}_0\mathfrak{p}_1 \cdots \mathfrak{p}_r\mathfrak{a} = (\alpha)$. Similarly we find an ideal $\mathfrak{f}$ prime to $\mathfrak{p}_0\mathfrak{p}_1 \cdots \mathfrak{p}_r$ and a totally positive element β of F such that $\mathfrak{p}_1 \cdots \mathfrak{p}_r\mathfrak{f} = (\beta)$. Then $F(\sqrt{-\alpha})$ and $F(\sqrt{-\beta})$ are totally imaginary quadratic extensions of F; and the prime ideals $\mathfrak{p}_1, \cdots, \mathfrak{p}_r$ ramify in these fields. Hence L has two subfields isomorphic to $F(\sqrt{-\alpha})$, $F(\sqrt{-\beta})$. Namely, L contains two elements t and u such that $t^2 = -\alpha$ and $u^2 = -\beta$. Put $T = \begin{pmatrix} t & 0 \\ 0 & u \end{pmatrix}$. Then T is ρ-skew-symmetric and $N(T)$

$= \alpha\beta$. As we have $\mathfrak{p}_0\mathfrak{p}_1^2 \cdots \mathfrak{p}_r^2\mathfrak{a}\mathfrak{f} = (\alpha\beta)$, there is no element ξ of F such that $\alpha\beta = \xi^2$. This proves our assertion.

4.6. **PROOF OF THEOREM 5.** Exclude *Cases* (a) (b) (c). Let $\mathscr{P}(z, T, \mathfrak{M})$ $= (A, C, \theta)$ be a generic member of $\Sigma(T, \mathfrak{M})$. By Proposition 16, the center of $\mathcal{A}_0(A)$ is written as $\theta(K)$ for a subfield K of the center of L. This implies that $\mathcal{A}_0(A)$ is a simple algebra. Therefore, A is isogenous to the product of several copies of a simple abelian variety B. Put $\dim A = h \cdot \dim B$, $\mathcal{A}_0(B) = S$, and identify L as a subalgebra of $M_h(S)$. Now we need a result of ring-theory (cf. Jacobson [6, p. 106, Th. 19]).

LEMMA 3. *Let $\mathfrak{A}$ be a central simple algebra over a field $\mathfrak{F}$, and $\mathfrak{B}$ a simple subalgebra of $\mathfrak{A}$ containing the identity element of $\mathfrak{A}$. If $\mathfrak{C}$ is the commutor of $\mathfrak{B}$ in $\mathfrak{A}$, then the following assertions hold.*

(i) *The commutor of $\mathfrak{C}$ in $\mathfrak{A}$ is $\mathfrak{B}$.*

(ii) $[\mathfrak{B}\colon \mathfrak{F}][\mathfrak{C}\colon \mathfrak{F}] = [\mathfrak{A}\colon \mathfrak{F}]$.

(iii) *If $\mathfrak{B}^{-1} \otimes_{\mathfrak{F}} \mathfrak{A}$ is isomorphic to a matric algebra $M_u(\mathfrak{D})$ over a division algebra $\mathfrak{D}$, then $\mathfrak{C}$ is isomorphic to a matric algebra $M_s(\mathfrak{D})$, where $\mathfrak{B}^{-1}$ denotes the algebra anti-isomorphic to $\mathfrak{B}$.*

(*Type* I, II, III). As F is the center of L, we have $K \subset F$, so that K is totally real; hence S is a division algebra of (*Type* I, II, III). By Proposition 16, the commutor of L in $M_h(S)$ is F. Taking $M_h(S)$, F, L, K to be $\mathfrak{A}$, $\mathfrak{B}$, $\mathfrak{C}$, $\mathfrak{F}$ of Lemma 3, we find

$$(51) \qquad\qquad [L\colon K][F\colon K] = h^2[S\colon K] \;;$$

and if $S \otimes_K F$ is isomorphic to a total matric algebra $M_u(\mathfrak{D})$ over a division algebra $\mathfrak{D}$, then L is isomorphic to $M_s(\mathfrak{D})$. As L is a division algebra, L must be isomorphic to $\mathfrak{D}$, and hence $S \otimes_K F$ is isomorphic to $M_u(L)$. Therefore, the following four cases may occur.

	L	S	$[K\colon Q]$	$2 \dim B/[S\colon Q]$
(1°)	*Type* I	*Type* I	g/h	m
(2°)	*Type* I	*Type* II	$g/2h$	$m/2$
(3°)	*Type* II	*Type* II	g/h	m
(4°)	*Type* III	*Type* III	g/h	m

Let C' be any polarization of B. Let $\mathfrak{K}$ and $\mathfrak{K}'$ be respectively the field of moduli of (A, C, θ) and (B, C'). Then by [AF, Propositions 8, 9], $\dim_Q\mathfrak{K} = \dim_Q\mathfrak{K}'$. By (46') of §4.1, $\dim_Q \mathfrak{K}'$ is not greater than

$$\frac{1}{2}\frac{m}{2}\left(\frac{m}{2} + 1\right)\frac{g}{h},\; \frac{1}{2} \cdot \frac{m}{2}\left(\frac{m}{2} + 1\right)\frac{g}{2h},\; \frac{1}{2}m(m + 1)\frac{g}{h},\; \frac{1}{2}m(m - 1)\frac{g}{h},$$

according as L and S belong to the above cases $(1°)$, $(2°)$, $(3°)$, $(4°)$. On the other hand, as (A, C, θ) is a generic member of $\Sigma(T, \mathfrak{M})$, $\dim_Q \mathfrak{K}$ is

$$\frac{1}{2} \cdot \frac{m}{2}\left(\frac{m}{2} + 1\right)g, \; \frac{1}{2}\frac{m}{2}\left(\frac{m}{2} + 1\right)g, \; \frac{1}{2}m(m + 1)g, \; \frac{1}{2}m(m - 1)g \;,$$

according to these four cases. Therefore we must have $L = S$. This proves Theorem 5 when L is of ($Type$ I, II, III).

($Type$ IV). In this case F_0 is the center of L; so $K \subset F_0$. Applying Lemma 3 to the present case, we have

$$[L\colon K][F_0\colon K] = h^2[S\colon K] \;;$$

and $S \otimes_K F_0$ is isomorphic to a matric algebra $M_u(L)$. We obtain then $[F_0\colon K] = hu$, $[S\colon K] = q^2 u^2$. Therefore, the following cases may occur.

	L	S	u	q	$[K\colon Q]$	$2 \dim B/[S\colon Q]$
$(5°)$	$Type$ IV	$Type$ I	1	1	$2g/h$	m
$(6°)$	$Type$ IV	$Type$ II	1	2	$2g/h$	m
$(7°)$	$Type$ IV	$Type$ II	2	1	g/h	$m/2$
$(8°)$	$Type$ IV	$Type$ III	1	2	$2g/h$	m
$(9°)$	$Type$ IV	$Type$ III	2	1	g/h	$m/2$
$(10°)$	$Type$ IV	$Type$ IV	u	q	$2g/hu$	m/u

If K is totally real, we must have $r_\nu = s_\nu$ for every ν. Hence mq is even, and we get $\sum_{\nu=1}^{g} r_\nu s_\nu = m^2 q^2 g/4$. Moreover, as K is contained in F, hu must be even. If we exclude $Cases$ (d) (e), mq is not less than 4 in the cases $(5°, 6°, 7°, 8°, 9°)$. Then comparing the dimension of the fields of moduli of (A, C, θ) and (B, C'), we can conclude that these five cases do not occur. Suppose that S is of ($Type$ IV), namely, K is imaginary. Put $f = g/hu$; and let $\sigma_1, \cdots, \sigma_f, \bar{\sigma}_1, \cdots, \bar{\sigma}_f$ be the inequivalent absolutely irreducible representations of S; let r'_λ and s'_λ be respectively the multiplicities of σ_λ and $\bar{\sigma}_\lambda$ in the representation of S by an analytic coordinate-system of B. We have then $r'_\lambda + s'_\lambda = (m/u)\cdot qu = mq$ for every λ. Now reordering χ_ν suitably, we may assume that

$$\chi_{\lambda+kf}(x) \otimes 1_u = \sigma_\lambda(x) \qquad \text{for } x \in K, 1 \leq \lambda \leq f, 0 \leq k \leq hu - 1 \;.$$

Restricting χ_ν and σ_λ to K, we find $r'_\lambda = r_{\lambda+kf}$, $s'_\lambda = s_{\lambda+kf}$ for $1 \leq \lambda \leq f$, $0 \leq k \leq hu - 1$. Therefore we have $\sum_{\lambda=1}^{f} r'_\lambda s'_\lambda = (1/hu)\sum_{\nu=1}^{g} r_\nu s_\nu$ Comparing the dimension of fields of moduli, we must have $hu = 1$, so that $L = S$. This completes the proof of Theorem 5.

4.7. $Case$ (d). Exchanging $(T, \mathfrak{M})$ for an equivalent couple, we may assume that T is diagonal, so put $T = \begin{pmatrix} u & 0 \\ 0 & v \end{pmatrix}$, $-u^{-1}v = \xi$. As $u^\rho = -u$, $v^\rho = -v$, we have $\xi^\rho = \xi$, so that $\xi \in F$. As $r_1 = s_1 = \cdots = r_g = s_g = 1$, ξ must be totally positive in view of (25) of §2.4. Let M be a quaternion algebra over F having F_0 as a quadratic subfield and such that $M = F_0 + F_0 w$, $w^2 = \xi$, $wxw^{-1} = x^\rho$ for every $x \in F_0$. Of course such an algebra exists and is totally indefinite, since $F(w)$ is a totally real algebraic number field or the direct sum of two copies of F. Let $y \to y'$ be the canonical involution of M; and put $y^\tau = u^{-1}y'u$ for $y \in M$. As $u^2 = -uu^\rho$ is a totally negative element of F, we see, by Proposition 2, that τ is a positive involution of M. Let ψ be a reduced representation of M with respect to $\boldsymbol{Q}$. Now consider an analytic family $\Sigma(u_0, \mathfrak{M}_0)$ of polarized abelian varieties belonging to $\{M, \psi, \tau\}$. We take a lattice $\mathfrak{M}_0$ in M and a τ-skew-symmetric element u_0 so that

$$
\begin{aligned}
u_0 &= u\,, \\
\mathfrak{M}_0 &= \{a + bw \mid (a, b) \in \mathfrak{M}\}\,.
\end{aligned}
\tag{52}
$$

Then we have

$$
\begin{aligned}
\mathrm{tr}_{M/Q}\big((a + bw)u(c + dw)^\tau\big) &= \mathrm{tr}_{F_0/Q}\big(u(ac^\rho - bd^\rho\xi)\big) \\
&= \mathrm{tr}_{F_0/Q}\Big((ab)T\begin{pmatrix} c^\rho \\ d^\rho \end{pmatrix}\Big)\,.
\end{aligned}
\tag{53}
$$

Let $\mathscr{P} = \mathscr{P}(z_0, u_0, \mathfrak{M}_0) = (A, C, \theta)$ be a generic member of $\Sigma(u_0, \mathfrak{M}_0)$. As F_0 is a subfield of M and $\tau = \rho$ on F_0, if we denote by θ' the restriction of θ to F_0 and put $\mathscr{P}' = (A, C, \theta')$, then $\mathscr{P}'$ is obviously of type $\{F_0, \Phi, \rho\}$. The relations (52) and (53) show that $\mathscr{P}'$ is isomorphic to a member of $\Sigma(T, \mathfrak{M})$. As $\mathscr{P}$ is a generic member, the field of moduli of $\mathscr{P}$ is of dimension g over $\boldsymbol{Q}$. By [AF, Proposition 8] and by Proposition 7 of §3.1, the field of moduli of $\mathscr{P}'$ is of dimension g over $\boldsymbol{Q}$, so that $\mathscr{P}'$ is a generic member of $\Sigma(T, \mathfrak{M})$. As $\theta(M) \subset \mathcal{A}_0(A)$, we have $\mathcal{A}_0(A) \neq \theta(F_0)$. Apply the argument of §4.6 to the present case. Then, comparing the dimension of fields of moduli, we observe that only the *Cases* (5°), (7°) may occur. Therefore we obtain the following result.

PROPOSITION 18. *Assumptions being as in Case* (d) *of Theorem 5, if* (A, C, θ) *is a generic member of* $\Sigma(T, \mathfrak{M})$, *then* $\mathcal{A}_0(A)$ *is a totally indefinite quaternion algebra over* $\theta(F)$ *having* $\theta(F_0)$ *as a quadratic subfield.*

REMARK 7. The quaternion algebra $\mathcal{A}_0(A)$ may be or may not be a division algebra. If this is not a division algebra, A is isogenous to the product of two copies of a simple abelian variety B such that $\mathcal{A}_0(B)$ is isomorphic to F.

4.8. *Case* (e). Assumptions being as in *Case* (e) of Theorem 5, let

(A, C, θ) be a polarized abelian variety of type $\{L, \Phi, \rho\}$. Put $Y = \{y \in L \mid y^\rho = y\}$. Then $Y \otimes_Q R$ is identified with a subset $\{(y_1, \cdots, y_g) \mid y_\nu^* = y_\nu \in M_2(C)\}$ of $L \otimes_Q R = M_2(C) \times \cdots \times M_2(C)$; and Y is dense in $Y \otimes_Q R$. Hence we can find an element $y^\rho = y$ of L not belonging to F_0. Fixing such an element y, put $K_0 = F_0(y)$, $K = F(y)$. We observe that $x \to x^\rho$ gives a positive involution of K_0, and K is the set of elements of K_0 invariant under this involution. Hence K is a totally real algebraic number field of degree $2g$, and K_0 is a totally imaginary quadratic extension of K. Let $\sigma_1, \cdots, \sigma_{2g}$ and their complex conjugates $\bar\sigma_1, \cdots, \bar\sigma_{2g}$ be all the isomorphisms of K_0 into C. By the assumption $r_1 = s_1 = \cdots = r_g = s_g = 1$, the restriction Φ_0 of Φ to K_0 is equivalent to the direct sum of the σ_i and $\bar\sigma_i$. Therefore, denoting by θ_0 the restriction of θ to K_0, we see that (A, C, θ_0) belongs to $\{K_0, \Phi_0, \rho\}$. Thus we can reduce the problem to *Case* (d), taking K and K_0 in place of F and F_0. Hence by Proposition 18, $\mathcal{A}_0(A)$ contains a totally indefinite quaternion algebra N over K. If $[\mathcal{A}_0(A): Q] = 8g$, then we must have $\theta(L) = \mathcal{A}_0(A) = N$. But the center of N is totally real, while the center of L is totally imaginary. We have therefore $[\mathcal{A}_0(A): Q] > 8g$, and $\mathcal{A}_0(A) \neq \theta(L)$. Now apply the argument of §4.6 to the present case. If (A, C, θ) is a generic member, we see easily, taking into account the dimension of fields of moduli, that only the case (6°) may occur, with $h = 2$. We obtain thus

PROPOSITION 19. *In Case (e) of Theorem 5, if (A, C, θ) is a generic member of $\Sigma(T, \mathfrak{M})$, then A is isogenous to the product of two copies of a simple abelian variety B such that $\mathcal{A}_0(B)$ is a totally indefinite quaternion algebra over F.*

4.9. *Determination of Δ.*

PROPOSITION 20. *Let $\mathcal{P}(z)$ be a generic member of $\Sigma(T, \mathfrak{M})$. Then the automorphisms of $\mathcal{P}(z)$ are given by ± 1 if L belongs to (Type I, II, III), and by $\theta(\zeta)$ for the roots of unity ζ in F_0 such that $\zeta\mathfrak{M} = \mathfrak{M}$ if L is of (Type IV), except Cases (a), (b), (c) of Theorem 5.*

Proof. By Proposition 16, every element η of $\mathcal{A}(A)$ commuting with the elements of $\theta(L)$ belongs to $\theta(L)$, hence to the center of $\theta(L)$. If η is an automorphism of $\mathcal{P}(z)$, it must be of finite order. Therefore, $\eta = \pm 1$ if the center of L is totally real. Suppose that L is of (Type IV). Then η must be of the form $\theta(\zeta)$ for a root of unity ζ contained in F_0. By the consideration of §2.8, we must have $\zeta\mathfrak{M} = \mathfrak{M}$. Conversely if ζ is a root of unity in F_0 satisfying $\zeta\mathfrak{M} = \mathfrak{M}$, it is easily seen that $\theta(\zeta)$ is an automorphism of $\mathcal{P}(z)$. This proves our proposition.

Let Δ be the subgroup of $\Gamma(T, \mathfrak{M})$ defined in Proposition 12. By Proposition 20, we get, except *Cases* (a) (b) (c) of Theorem 5,

(*Type* I, II, III) $\Delta = \{\pm 1_m\}$

(*Type* IV) $\Delta = \{\zeta 1_m \mid \zeta$ is a root of unity in $F_0, \zeta \mathfrak{M} = \mathfrak{M}\}$.

5. Weak polarization

5.1. Let F and L be as before a totally real algebraic number field and a division algebra belonging to (*Type* I, II, III, IV). Suppose an abelian variety A and an isomorphism θ of L into $\mathcal{A}_0(A)$ to be given. Put

$$(56) \qquad \mathfrak{o} = \theta^{-1}[\mathcal{A}(A) \cap \theta(L)] .$$

Then $\mathfrak{o}$ is an order in L. Let B be a Picard variety of A. For any divisor X on A, we denote by φ_X the homomorphism of A into B defined by $\varphi_X(u) = \mathrm{Cl}(X_u - X)$ (cf. §1.3). For two divisors X, Y of A and for an element b of F, we write $Y \equiv X \cdot b$ if we have $\varphi_Y = \varphi_X \theta(b)$.

PROPOSITION 21. *Let X be a positive non-degenerate divisor on A such that the involution of $\mathcal{A}_0(A)$ determined by X leaves the elements of $\theta(F)$ invariant. Let b be a totally positive element of F. Then there exists a positive non-degenerate divisor Y on A and a positive integer m such that $Y \equiv mX \cdot b$.*

Proof. By our assumption on X, we have $\varphi_X^{-1} \cdot {}^t\theta(a)\varphi_X = \theta(a)$ for every $a \in F$, so that ${}^t\theta(a)\varphi_X \theta(a) = \varphi_X \theta(a^2)$. By a theorem of Siegel [17, Satz 1], we can find a positive integer m and four elements c_1, c_2, c_3, c_4 of $\mathfrak{o} \cap F$ such that $mb = \sum_{i=1}^4 c_i^2$. Put $Y = \sum_{i=1}^4 \theta(c_i)^{-1}(X)$. Then $\varphi_Y = \sum_{i=1}^4 {}^t\theta(c_i)\varphi_X \theta(c_i) = \sum_{i=1}^4 \varphi_X \theta(c_i^2) = m\varphi_X \theta(b)$. This proves our proposition.

(A, θ) being as above, let C be a polarization of A such that the involution determined by C leaves invariant the elements of $\theta(F)$. Take a divisor X belonging to C. For every totally positive element b of F, we can find, by Proposition 21, a positive integer m and a positive non-degenerate divisor Y on A such that $Y \equiv mX \cdot b$. Let C' be a polarization of A containing Y. It is easily seen that C' does not depend on the choice of X; we denote C' by $C \cdot b$. Let E_+ denote the set of all totally positive units of F. Now, by a *weak polarization* (or briefly, a *w-polarization*) of A, we understand the join

$$\mathcal{E} = \bigcup_{b \in E_+} C \cdot b$$

obtained from a polarization C of A, and call the structure $(A, \mathcal{E}, \theta)$ a *w-polarized abelian variety of type* $\mathfrak{o}$. We see easily that the divisors of $\mathcal{E}$ give one and the same involution of $\mathcal{A}_0(A)$, which we call the involution determined by $\mathcal{E}$. We say that $\mathcal{P}_0 = (A, \mathcal{E}, \theta)$ is *defined over* a field k if A and all elements of $\theta(\mathfrak{o})$ are defined over k, and $\mathcal{E}$ contains a divisor X rational over k. This being so, let σ be an isomorphism of k into a field k'. Put $\theta^\sigma(x) = \theta(x)^\sigma$ for every $x \in \mathfrak{o}$ and define a w-polarization $\mathcal{E}^\sigma$ on A^σ so as to contain X^σ. Then we obtain a w-polarized abelian variety $(A^\sigma, \mathcal{E}^\sigma, \theta^\sigma)$, which we denote by $\mathcal{P}_0^\sigma$.

Let $\mathcal{P}_0 = (A, \mathcal{E}, \theta)$ and $\mathcal{P}_1 = (A', \mathcal{E}', \theta')$ be two w-polarized abelian varieties of type $\mathfrak{o}$, of the same dimension. An isogeny (resp. isomorphism) λ of A onto A' is called an *isogeny* (resp. *isomorphism*) of $\mathcal{P}_0$ onto $\mathcal{P}_1$ if there exists a divisor X' in $\mathcal{E}'$ such that $\lambda^{-1}(X')$ is contained in $\mathcal{E}$ and if $\lambda\theta(a) = \theta'(a)\lambda$ for all $a \in \mathfrak{o}$; if these conditions are satisfied, $\lambda^{-1}(Y)$ is contained in $\mathcal{E}$ for every $Y \in \mathcal{E}'$.

5.2. *The field of moduli of a w-polarized abelian variety.* We assume hereafter that the universal domain is of characteristic 0. $L, F, \mathfrak{o}$ being as in §5.1, let $(A, \mathcal{E}, \theta)$ be a w-polarized abelian variety of type $\mathfrak{o}$. Put

$$(57) \qquad E_1 = \begin{cases} \{e^2 \mid e \text{ is a unit of } \mathfrak{o} \cap F\} & \text{for } (\textit{Type I, II, III}) , \\ \{N_{F_0/F}(e) \mid e \text{ is a unit of } \mathfrak{o} \cap F_0\} & \text{for } (\textit{Type IV}) . \end{cases}$$

Obviously we have $E_+ \supset E_1 \supset \{a^2 \mid a \in E_+\}$, so that E_+/E_1 is a finite abelian group.

THEOREM 6. *Notation being as above, suppose that $\theta(L) = \mathcal{A}_0(A)$. Let C be a polarization of A contained in $\mathcal{E}$, and let K be the field of moduli of the polarized abelian variety (A, C, θ) of type $\mathfrak{o}$. (By Proposition 6, K is determined by $\mathcal{E}$ and is independent of the choice of C.). Then there exists a subfield K_0 of K with the following property: for every field of definition k of $\mathcal{P}_0 = (A, \mathcal{E}, \theta)$ and for every isomorphism σ of k into the universal domain, σ is the identity on K_0 if and only if $\mathcal{P}_0^\sigma$ is isomorphic to $\mathcal{P}_0$. Moreover, K is a Galois extension of K_0, and the Galois group of K over K_0 is isomorphic to a subgroup of E_+/E_1.*

PROOF. Let k be a field of definition for (A, C, θ) and σ an isomorphism of k into the universal domain. Suppose that there exists an isomorphism μ of $\mathcal{P}_0$ onto $\mathcal{P}_0^\sigma$. Let X be a divisor in C, rational over k. Then there exist two positive integers m, m' and an element e of E_+ such that $m\mu^{-1}(X^\sigma) \equiv m'X \cdot e$; it follows that μ is an isomorphism of $(A, C \cdot e, \theta)$ onto $(A^\sigma, C^\sigma, \theta^\sigma)$. By Proposition 6, we must have $K^\sigma = K$, so that σ induces an automorphism of K. Let G be the set of automorphisms of K obtained in this manner. We can easily verify that G forms a group. Considering the degree of φ_X and φ_X^σ, we find $m = m'$, so that $\mu^{-1}(X^\sigma) \equiv X \cdot e$. If we fix σ and μ, the element e does not depend on the choice of X. If λ is another isomorphism of $\mathcal{P}_0$ onto $\mathcal{P}_0^\sigma$, then $\mu^{-1}\lambda$ is an automorphism of A commuting with every element of $\theta(L)$; so there exists a unit d of $\mathfrak{o} \cap F$ ($\mathfrak{o} \cap F_0$ for ($Type$ IV)) such that $\lambda = \mu \cdot \theta(d)$; and conversely, for every such a unit d, $\mu \cdot \theta(d)$ is an isomorphism of $\mathscr{P}_0$ onto $\mathscr{P}_0^\sigma$. In fact, if $\lambda = \mu \cdot \theta(d)$, we have

$$'\lambda \varphi_X^\sigma \lambda = {}^t\theta(d) \cdot {}^t\mu \varphi_X \mu \theta(d) = {}^t\theta(d)\varphi_X \theta(e)\theta(d) = \varphi_X \theta(d'de) \ ,$$

where $\theta(a) \to \theta(a')$ is the involution of $\mathscr{A}_0(A)$ determined by $\mathcal{E}$. We have hence $\lambda^{-1}(X^\sigma) \equiv X \cdot (d'de)$; and we observe that e and $d'de$ belong to the same class of elements of E_+ modulo E_1. On account of this consideration, we can easily verify that $\mu \to e$ gives an isomorphism of G into E_+/E_1. It follows that G is a finite group. Let K_0 be the subfield of K consisting of the elements of K fixed by every element of G. Then K_0 has the property of our theorem. In fact, let τ be an isomorphism of k into the universal domain. If $\mathscr{P}_0^\tau$ is isomorphic to $\mathscr{P}_0$, then the restriction of τ to K belongs to G; hence τ is the identity on K_0. Conversely, if τ is the identity on K_0, we must have $K^\tau = K$, since K is a Galois extension of K_0. By our definition of G, there exists an isomorphism σ of k such that $\mathscr{P}_0^\sigma$ is isomorphic to $\mathscr{P}_0$ and $\sigma = \tau$ on K. As K is the field of moduli of (A, C, θ), the equality $\sigma = \tau$ on K implies that $(A^\sigma, C^\sigma, \theta^\sigma)$ is isomorphic to $(A^\tau, C^\tau, \theta^\tau)$. It follows that $\mathscr{P}_0^\tau$ is isomorphic to $\mathscr{P}_0^\sigma$, and hence to $\mathscr{P}_0$. This completes the proof.

We call the field K_0 of Theorem 6 *the field of moduli* of the w-polarized abelian variety $\mathscr{P}_0 = (A, \mathcal{E}, \theta)$ of type $\mathfrak{o}$.

5.3. *Groups of similitudes of T.* Define groups $\mathfrak{G}_r^1$, $\mathfrak{G}_r^2$, $\mathfrak{G}_{r,s}^3$ by

$$\mathfrak{G}_r^1 = \left\{ V \in M_{2r}(\boldsymbol{R}) \,\middle|\, {}^t V \begin{pmatrix} 0 & 1_r \\ -1_r & 0 \end{pmatrix} V = c \begin{pmatrix} 0 & 1_r \\ -1_r & 0 \end{pmatrix} \right.$$
$$\left. \text{for a real number } c \neq 0 \right\} \ ,$$

$$\mathfrak{G}_r^2 = \left\{ V = \iota(U) \,\middle|\, U \in M_r(\boldsymbol{K}), \ {}^t\bar{V} \begin{pmatrix} 1_r & 0 \\ 0 & -1_r \end{pmatrix} V = c \begin{pmatrix} 1_r & 0 \\ 0 & -1_r \end{pmatrix} \right.$$
$$\left. \text{for a real number } c \neq 0 \right\} \ ,$$

$$\mathfrak{G}_{r,s}^3 = \left\{ V \in M_{r+s}(\boldsymbol{C}) \,\middle|\, {}^t\bar{V} \begin{pmatrix} 1_r & 0 \\ 0 & -1_s \end{pmatrix} V = c \begin{pmatrix} 1_r & 0 \\ 0 & -1_s \end{pmatrix} \right.$$
$$\left. \text{for a real number } c \neq 0 \right\} \ .$$

For the group $\mathfrak{G}_{r,s}^3$, if $r \neq s$, the multiplicator c must be positive, while if $r = s$, it may be negative.

Put now

$$\mathscr{D}_r^{1+} = \left\{ \begin{pmatrix} z & 1_r \\ \bar{z} & 1_r \end{pmatrix} \middle| z \in \mathscr{H}_r^1 \right\},$$

$$\mathscr{D}_r^{2+} = \left\{ \begin{pmatrix} -z & 1_r \\ 1_r & \bar{z} \end{pmatrix} \middle| z \in \mathscr{H}_r^2 \right\},$$

$$\mathscr{D}_{r,s}^{3+} = \left\{ \begin{cases} \left\{ \begin{pmatrix} 1_r & z \\ {}^t\bar{z} & 1_s \end{pmatrix} \middle| z \in \mathscr{H}_{r,s}^3 \right\} & (rs > 0), \\ \{1_{r+s}\} & (rs = 0), \end{cases} \right.$$

$$\mathscr{D}_r^{1-} = \left\{ \begin{pmatrix} z & 1_r \\ \bar{z} & 1_r \end{pmatrix} \middle| -z \in \mathscr{H}_r^1 \right\},$$

$$\mathscr{D}_r^{2-} = \left\{ \begin{pmatrix} 1_r & z \\ \bar{z} & -1_r \end{pmatrix} \middle| z \in \mathscr{H}_r^2 \right\},$$

$$\mathscr{D}_{r,r}^{3-} = \left\{ \begin{pmatrix} z & 1_r \\ 1_r & {}^t\bar{z} \end{pmatrix} \middle| z \in \mathscr{H}_{r,r}^3 \right\},$$

$$\mathscr{D}_r^1 = \mathscr{D}_r^{1+} \cup \mathscr{D}_r^{1-},$$

$$\mathscr{D}_r^2 = \mathscr{D}_r^{2+} \cup \mathscr{D}_r^{2-},$$

$$\mathscr{D}_{r,s}^3 = \begin{cases} \mathscr{D}_{r,s}^{3+} & (r \neq s), \\ \mathscr{D}_{r,r}^{3+} \cup \mathscr{D}_{r,r}^{3-} & (r = s). \end{cases}$$

We can put a complex structure into each domain, taking the coordinates of z as holomorphic parameters. We define the operation of the elements of the groups $\mathfrak{G}_r^1$, $\mathfrak{G}_r^2$, $\mathfrak{G}_{r,s}^3$, on $\mathscr{D}_r^1$, $\mathscr{D}_r^2$, $\mathscr{D}_{r,s}^3$, respectively, as follows. Let U be an element of $\mathfrak{G}(= \mathfrak{G}_r^1, \mathfrak{G}_r^2, \mathfrak{G}_{r,s}^3)$ and Y be a point of $\mathscr{D}(= \mathscr{D}_r^1, \mathscr{D}_r^2, \mathscr{D}_{r,s}^3)$. Then there exist a point Y' of $\mathscr{D}$ and invertible matrices u, v such that $Y \cdot {}^t U = \begin{pmatrix} u & 0 \\ 0 & v \end{pmatrix} Y'$ where $u, v \in M_r(C)$ if $\mathfrak{G} = \mathfrak{G}_r^1$ or $\mathfrak{G}_r^2$, and, $u \in M_r(C)$, $v \in M_s(C)$ if $\mathfrak{G} = \mathfrak{G}_{r,s}^3$. Y' is uniquely determined by U and Y; put $Y' = U(Y)$. Every element U of $\mathfrak{G}$ gives in this way an analytic automorphism of $\mathscr{D}$.

Define a group $\mathfrak{G}(\Phi)$ and a domain $\mathscr{D}(\Phi)$ by

$$(Type\ \mathrm{I}) \qquad \mathfrak{G}(\Phi) = \mathfrak{G}_{m/2}^1 \times \cdots \times \mathfrak{G}_{m/2}^1,$$

$$\mathscr{D}(\Phi) = \mathscr{D}_{m/2}^1 \times \cdots \times \mathscr{D}_{m/2}^1,$$

$$(Type\ \mathrm{II}) \qquad \mathfrak{G}(\Phi) = \mathfrak{G}_m^1 \times \cdots \times \mathfrak{G}_m^1,$$

$$\mathscr{D}(\Phi) = \mathscr{D}_m^1 \times \cdots \times \mathscr{D}_m^1,$$

$$(Type\ \mathrm{III}) \qquad \mathfrak{G}(\Phi) = \mathfrak{G}_m^2 \times \cdots \times \mathfrak{G}_m^2,$$

$$\mathscr{D}(\Phi) = \mathscr{D}_m^2 \times \cdots \times \mathscr{D}_m^2,$$

$$(Type\ IV)\quad \mathfrak{G}(\Phi) = \mathfrak{G}^3_{r_1,s_1} \times \cdots \times \mathfrak{G}^3_{r_g,s_g}\ ,$$
$$\mathscr{D}(\Phi) = \mathscr{D}^3_{r_1,s_1} \times \cdots \times \mathscr{D}^3_{r_g,s_g}\ ,$$

where the number of factors is g in each case. For $Y = (Y_1, \cdots, Y_g) \in \mathscr{D}(\Phi)$ and $U = (U_1, \cdots, U_g) \in \mathfrak{G}(\Phi)$, we put $U(Y) = (U_1(Y_1), \cdots, U_g(Y_g))$. Take a ρ-skew-symmetric element T of $M_m(L)$ and a lattice $\mathfrak{M}$ in L^m. Define groups $G_0(T)$, $\Gamma_0(T, \mathfrak{M})$, $\Gamma_+(T, \mathfrak{M})$ by

$$G_0(T) = \{U \in M_m(L_R) \mid UTU^\rho = cT \text{ for a regular element } c \in F_R\}\ ,$$
$$\Gamma_0(T, \mathfrak{M}) = \{U \in M_m(L) \mid \mathfrak{M}U = \mathfrak{M},\ UTU^\rho = cT \text{ for a non-zero } c \in F\}\ ,$$
$$\Gamma_+(T, \mathfrak{M}) = \{U \in \Gamma_0(T, \mathfrak{M}) \mid UTU^\rho = cT \text{ for a totally positive } c \in F\}\ .$$

We note that, if $U \in \Gamma_0(T, \mathfrak{M})$, the multiplicator c must be a unit. For every $U \in G_0(T)$, put

$$\psi(U) = ({}^t\bar{W}_1^{-1}\omega_1(U){}^t\bar{W}_1, \cdots, {}^t\bar{W}_g^{-1}\omega_g(U){}^t\bar{W}_g)\ .$$

Then $U \to \psi(U)$ gives an isomorphism of $G_0(T)$ onto $\mathfrak{G}(\Phi)$. For every $U \in G_0(T)$ and for $Y \in \mathscr{D}(\Phi)$, we write $U(Y) = \psi(U)(Y)$.

5.4. *Analytic families* $\Sigma_0(T, \mathfrak{M})$. Let $Y = (Y_1, \cdots, Y_g)$ be a point of $\mathscr{D}(\Phi)$. Put

$$\varepsilon_\nu(Y) = \begin{cases} 1 & \text{if } Y_\nu \in \mathscr{D}^{1+}_r,\ \mathscr{D}^{2+}_r,\ \mathscr{D}^{3+}_{r,s}\ , \\ -1 & \text{if } Y_\nu \in \mathscr{D}^{1-}_r,\ \mathscr{D}^{2-}_r,\ \mathscr{D}^{3-}_{r,s}\ , \end{cases}$$
$$\varepsilon(Y) = (\varepsilon_1(Y), \cdots, \varepsilon_g(Y))\ .$$

Let $\mathfrak{E}(\Phi)$ be the set of $\varepsilon = (\varepsilon_1, \cdots, \varepsilon_g)$ with $\varepsilon_\nu = \pm 1$ such that $\varepsilon = \varepsilon(Y)$ for some $Y \in \mathscr{D}(\Phi)$. The number of elements in $\mathfrak{E}(\Phi)$ is 2^g for (*Type* I, II, III) and 2^h for (*Type* IV) if we denote by h the number of ν such that $r_\nu = s_\nu$. Put, for each $\varepsilon \in \mathfrak{E}(\Phi)$,

$$\mathscr{D}_\varepsilon(\Phi) = \{Y \in \mathscr{D}(\Phi) \mid \varepsilon(Y) = \varepsilon\}\ .$$

Then the $\mathscr{D}_\varepsilon(\Phi)$ for $\varepsilon \in \mathfrak{E}(\Phi)$ are the connected components of $\mathscr{D}(\Phi)$. Now fix a couple $(T, \mathfrak{M})$, and put $\mathfrak{o} = \{a \in L \mid a\mathfrak{M} \subset \mathfrak{M}\}$. For every ε of $\mathfrak{E}(\Phi)$, we can find an element α_ε of F such that $\alpha_\varepsilon \mathfrak{M} \subset \mathfrak{M}$, and the ν-th conjugate of α_ε has sign ε_ν for each ν. We fix, once and for all, such an element α_ε for each ε of $\mathfrak{E}(\Phi)$.

Let $Y = (Y_1, \cdots, Y_g)$ be a point of $\mathscr{D}(\Phi)$ and $\varepsilon = \varepsilon(Y)$. Put $X_\nu = Y_\nu\bar{W}_\nu$. Then the matrices X_ν determine m vectors $\mathfrak{x}_1, \cdots, \mathfrak{x}_m$ by (17), (18), (19), (20) of §2.3, from which we obtain a lattice D by (10) of §2.2. Define an R-bilinear form E_ε on C^n by

$$E_\varepsilon(\textstyle\sum_{i=1}^m \Phi(a_i)\mathfrak{x}_i, \sum_{j=1}^m \Phi(b_j)\mathfrak{x}_j) = \mathrm{tr}\,(\alpha_\varepsilon \sum_{i,j=1}^m a_i t_{ij} b_j^\sigma)$$

for a_i, $b_j \in L_R$. Then we observe that E_ε is a Riemann form on C^n/D; and the polarization C containing the divisor which corresponds to E_ε, determines the involution $\theta(a) \to \theta(a^\rho)$ on $\theta(L)$, if we denote by $\theta(a)$ the endomorphism of C^n/D corresponding to $a \in L$. Therefore we can consider the w-polarization containing C. We get in this way a w-polarized abelian variety of type $\mathfrak{o}$, which we denote by $\mathscr{P}_0(Y, T, \mathfrak{M})$. Further we denote by $\Sigma_0(T, \mathfrak{M})$ this family of w-polarized abelian varieties $\mathscr{P}_0(Y, T, \mathfrak{M})$ parametrized by the point Y of $\mathscr{D}(\Phi)$.

PROPOSITION 22. *Two members* $\mathscr{P}_0(Y, T, \mathfrak{M})$ *and* $\mathscr{P}_0(Y', T, \mathfrak{M})$ *of the same family* $\Sigma_0(T, \mathfrak{M})$ *are isomorphic if and only if* $Y = U(Y')$ *for an element* U *of* $\Gamma_0(T, \mathfrak{M})$.

As the proof is quite similar to that of Theorem 2, we omit it. We can also obtain a statement for $\mathscr{P}_0(Y, T, \mathfrak{M})$ analogous to Proposition 4, modifying the notion of equivalence for T, $\mathfrak{M}$ according to our definition of w-polarization.

Let Δ_0 be the set of elements U of $\Gamma_0(T, \mathfrak{M})$ such that $U(z) = z$ for all $z \in \mathscr{H}(\Phi)$. By the same argument as in §3.5 and §4.9, we get, except *Cases* (a) (b) (c) of Theorem 5,

(*Type* I, II, III)　$\Delta_0 = \{\delta 1_m \mid \delta$ is a unit of $\mathfrak{o} \cap F\}$,

(*Type* IV)　　　$\Delta_0 = \{\delta 1_m \mid \delta$ is a unit of $\mathfrak{o} \cap F_0\}$.

5.5. *Automorphic functions with repect to* $\Gamma_+(T, \mathfrak{M})$. It is easy to see that every element of $\Gamma_+(T, \mathfrak{M})$ operates on any component $\mathscr{D}_\varepsilon(\Phi)$ of $\mathscr{D}(\Phi)$; and $\Gamma_0(T, \mathfrak{M})\backslash\mathscr{D}(\Phi)$ can be identified with the disjoint sum of $\Gamma_+(T, \mathfrak{M})\backslash\mathscr{D}_\varepsilon(\Phi)$ for several ε. Now we want to investigate the relation between the automorphic functions on $\mathscr{D}_\varepsilon(\Phi)$ with respect to $\Gamma_+(T, \mathfrak{M})$ and the field of moduli of w-polarized abelian variety.

For every $U \in \Gamma_0(T, \mathfrak{M})$, define $\mu = \mu(U)$ by $UTU^\rho = \mu T$. Put $E_0 = \mu(\Gamma_+(T, \mathfrak{M}))$. Then E_0 is a subgroup of the group E_+ of all totally positive units of F; and E_0 contains the group E_1 defined in §5.2, since $\mathfrak{o} = \{a \in L \mid a\mathfrak{M} \subset \mathfrak{M}\}$.

For the sake of simplicity, we consider $\Gamma_0(T, \mathfrak{M})\backslash\mathscr{D}_\varepsilon(\Phi)$ only in the case $\varepsilon_\nu = 1$ for every ν; other cases can be treated similarly. If $\varepsilon_\nu = 1$ for every ν, we can identify $\mathscr{D}_\varepsilon(\Phi)$ with $\mathscr{H}(\Phi)$ by the correspondence $z = (z_\nu) \to Y = (Y_\nu(z))$, where $Y_\nu = Y_\nu(z)$ are matrices defined by (35). Therefore we use $\mathscr{H}(\Phi)$ instead of $\mathscr{D}_\varepsilon(\Phi)$. Let $f_1, \cdots, f_\kappa$ be the functions defined in Theorem 3.

PROPOSITION 23. *Let* z *be a point of* $\mathscr{H}(\Phi)$ *generic for* $f_1, \cdots, f_\kappa$ *over* $\mathbf{Q}$, *and let* $(A, C, \theta) = \mathscr{P}(z, T, \mathfrak{M})$. *Then, the notations being as in Theorem 6, E_0/E_1 is contained in the isomorphic image of the Galois group of*

K over K_0 in E_+/E_1.

PROOF. In view of [15, Prop. 3.5], it is sufficient to prove our proposition for a *sufficiently generic* point z of $\mathcal{H}(\Phi)$. Let $U \in \Gamma_+(T, \mathfrak{M})$ and $\mu = \mu(U)$. Put $w = U(z)$. Then, by Proposition 22, $\mathcal{P}_0(w, T, \mathfrak{M})$ is isomorphic to $\mathcal{P}_0(z, T, \mathfrak{M})$. If we take z so as to be generic for $\Psi(U(\ ))$ for $U \in \Gamma_+(T, \mathfrak{M})$, there exists an isomorphism σ of $Q(\Psi(z))$ onto $Q(\Psi(w))$ such that $\Psi(z)^\sigma = \Psi(w)$. Then $\mathcal{P}(z)^\sigma = \mathcal{P}(w)$ and $f_\nu(z)^\sigma = f_\nu(w)$. By our definition of field of moduli of w-polarized abelian variety, σ must be the identity on K_0. Let X_1 and X_2 be respectively the divisors on $A(z)$ and $A(w)$ determined by E_ε, and let λ be the isomorphism of $A(z)$ onto $A(w)$ given by U; then we get $\lambda^{-1}(X_2) \equiv X_1 \cdot \mu$. As $\mathcal{P}(z)^\sigma = \mathcal{P}(w)$, the divisor X_1^σ must be algebraically equivalent to X_2; so we get $\lambda^{-1}(X_1^\sigma) \equiv X_1 \cdot \mu$. This proves that μ is contained in the image of the Galois group.

For every function f on $\mathcal{H}(\Phi)$ and for every $U \in \Gamma_+(T, \mathfrak{M})$, define f^U by $f^U(z) = f(U(z))$.

PROPOSITION 24. *Put* $\mathfrak{k} = Q(f_1, \cdots, f_\kappa)$. *Then every element* U *of* $\Gamma_+(T, \mathfrak{M})$ *gives an automorphism of* $\mathfrak{k}$ *by the operation* $f \rightarrow f^U$. *Put further* $\mathfrak{k}_0 = \{f \in \mathfrak{k} \mid f^U = f \text{ for every } U \in \Gamma_+(T, \mathfrak{M})\}$. *Then* $\mathfrak{k}$ *is a Galois extension of* $\mathfrak{k}_0$ *whose Galois group is isomorphic to* E_0/E_1.

PROOF. If z is sufficiently generic, $Q(f_1(z), \cdots, f_\kappa(z))$ is the field of moduli of $\mathcal{P}(z, T, \mathfrak{M})$, and $Q(f_1^U(z), \cdots, f_\kappa^U(z))$ is the field of moduli of $\mathcal{P}(U(z), T, \mathfrak{M})$. If $U \in \Gamma_+(T, \mathfrak{M})$, $\mathcal{P}_0(z, T, \mathfrak{M})$ and $\mathcal{P}_0(U(z), T, \mathfrak{M})$ are isomorphic. Hence, by Proposition 6, we have $Q(f_1(z), \cdots, f_\kappa(z)) = Q(f_1^U(z), \cdots, f_\kappa^U(z))$. This proves the first assertion. Now, if $\mu(U) \in E_1$, there exists a unit e of $\mathfrak{o} \cap F$ (resp. $\mathfrak{o} \cap F_0$ for $(Type \text{ IV})$) such that $\mu(U) = e^2$ (resp. $\mu(U) = N_{F_0/F}(e)$). Put $V = e^{-1} U$. Then $V \in \Gamma(T, \mathfrak{M})$; and $f^U = f^V = f$ for every $f \in \mathfrak{k}$. Conversely, suppose that $f^U = f$ for every $f \in \mathfrak{k}$. Let z be a point on $\mathcal{H}(\Phi)$ generic for the functions f^W for $f \in \mathfrak{k}$, $W \in \Gamma_-(T, \mathfrak{M})$ over Q. As we have $f_\nu^U(z) = f_\nu(z)$ for every ν, there exists an element V of $\Gamma(T, \mathfrak{M})$ such that $U(z) = V(z)$. Hence $V^{-1}U = b1_m$ for a unit b of $\mathfrak{o} \cap F$ (resp. $\mathfrak{o} \cap F_0$); so we have $\mu(U) \in E_1$. This proves our proposition.

THEOREM 7. *Exclude Cases* (a) (b) (c) *of Theorem 5, and suppose that* $E_+ = E_0$. *Then, for every point* z *on* $\mathcal{H}(\Phi)$ *generic for* $f_1, \cdots, f_\kappa$ *over* Q, *the field of moduli of the* w-*polarized abelian variety* $\mathcal{P}_0(z, T, \mathfrak{M})$ *is* $Q(f(z) \mid f \in \mathfrak{k}_0)$, *where* $\mathfrak{k}_0$ *is the field defined in Proposition* 24.

PROOF. In view of [15, Proposition 3.5], it is sufficient to prove our assertion for a sufficiently generic point z of $\mathcal{H}(\Phi)$. Let U be an element of $\Gamma_+(T, \mathfrak{M})$. Put $w = U(z)$ and define an isomorphism σ of $Q(\Psi(z))$ onto $Q(\Psi(w))$ as in Proof of Proposition 23. Then we have $f_\nu(z)^\sigma = f_\nu(w) =$

192 GORO SHIMURA

$f_v^v(z)$. Therefore we have $g(z)^\sigma = g^v(z)$ for every $g \in \mathfrak{k}$. K and K_0 being as in Theorem 6, we note that $g \to g(z)$ gives an isomorphism of $\mathfrak{k}$ onto K. If $g(z)$ is contained in K_0, we have $g(z)^\sigma = g(z)$. Take z so as to be generic over Q for the functions f^W for $f \in \mathfrak{k}$, $W \in \Gamma_+(T, \mathfrak{M})$. Then the equality $g^v(z) = g(z)$ implies $g^v = g$. As this holds for every $U \in \Gamma_+(T, \mathfrak{M})$, g must belong to $\mathfrak{k}_0$. Thus we have proved that, if $g(z) \in K_0$, then $g \in \mathfrak{k}_0$. It follows that $[K: K_0] \geq [\mathfrak{k}: \mathfrak{k}_0]$. But by Propositions 23 and 24, we have $[K: K_0] = [E_+: E_1] = [\mathfrak{k}: \mathfrak{k}_0]$. Therefore, we must have $K_0 = Q(f(z) \mid f \in \mathfrak{k}_0)$.

For general T and $\mathfrak{M}$, the condition $E_+ = E_0$ is not necessarily always satisfied. We can verify, however, that this condition is satisfied in many cases where $\mathfrak{o}$ is a maximal order in L and $\mathfrak{M}$ is suitably chosen.

Osaka University

References

1. A. A. Albert, *On the construction of Riemann matrices*, I, II, Ann. of Math., 35 (1934), 1-28; 36 (1935), 376-394.
2. ———, *A solution of the principal problem in the theory of Riemann matrices*, Ann. of Math., 35 (1934), 500-515.
3. ———, *Involutorial simple algebras and real Riemann matrices*, Ann. of Math., 36 (1935) 886-964.
4. W. L. Baily, *On Satake's compactification of V_n*, Amer. J. Math., 80 (1938), 348-364.
5. ———, *On the Hilbert-Siegel modular space*, Amer. J. Math., 81 (1959) 846-874.
6. N. Jacobson, The Theory of Rings, New York 1943.
7. K. Katayama, *On the Hilbert-Siegel modular group and abelian varieties*, J. Fac. Sci., Univ. of Tokyo, Sec. I, 9 (1962), 261-291.
8. S. Lang, Abelian Varieties, Interscience Tracts, New York, 1959.
9. I. I. Pyatetskii-Shapiro, *The classification of modular groups* (in Russian), Dokl. Akad. Nauk SSSR, 110 (1956), 19-22.
10. ———, The Geometry of Classical Domains and the Theory of Automorphic Functions (in Russian) 1961.
11. I. Satake, *On the compactification of the Siegel space*, J. Ind. Math., Soc., 20 (1956), 259-281.
12. ———, *On compactifications of the quotient spaces for arithmetically defined discontinuous groups*, Ann. of Math., 72 (1960), 555-580.
13. Séminaire H. Cartan, E. N. S., 1957/1958, Fonctions automorphes.
14.=[AF]. G. Shimura, *On the theory of automorphic functions*, Ann. of Math., 70 (1959), 101-144.
15. ———, *On the zeta-functions of the algebraic curves uniformized by certain automorphic functions*, J. Math. Soc. Japan, 13 (1961), 275-331.
16. ———, and Y. Taniyama, Complex multiplication of abelian varieties and its applications to number theory, Publ. Math. Soc. Japan, No. 6, 1961.
17. C. L. Siegel, *Additive Theorie der Zahlkörper* II, Math. Ann., 88 (1923), 184-210.
18. A. Weil, Variétés abéliennes et courbes algébriques, Hermann, Paris, 1948.
19. ———, Discontinuous subgroups of classical groups, lecture note, University of Chicago, 1958.
20. ———, Adeles and algebraic groups, lecture note, Institute for Advanced Study, Princeton, 1961.

On the cohomology groups attached to certain vector valued differential forms on the product of the upper half planes

Jointly with Yozo Matsushima

Annals of Mathematics, 78 (1963), 417-449

(Received August 29, 1962)
(Revised January 14, 1963)

Introduction

This is a continuation of the paper [8] by S. Murakami and one of the authors of the present paper. We shall study in this paper the cohomology group $H(\Gamma, X, \rho)$, which is the subject of the theory developed in [8], in the case where X is the product of N copies of the upper half plane. Let G be a connected Lie group such that $G = G_1 \times \cdots \times G_N \times G_{N+1}$, where $G_1, \cdots, G_N$ are N copies of the special linear group SL(2, $\boldsymbol{R}$), and G_{N+1} is a connected compact Lie group, possibly equal to the identity element. The group G operates naturally on X (cf. § 1). Let Γ be a discrete subgroup of G with compact quotient space G/Γ, and ρ a representation of G in a complex vector space F. Let $A^r(\Gamma, X, \rho)$ be the complex vector space of all F-valued r-forms ω defined on X such that

$$\omega \circ L_\gamma = \rho(\gamma)\omega$$

for all $\gamma \in \Gamma$, where L_γ denotes the transformation of X defined by γ, and $\omega \circ L_\gamma$ denotes the r-form obtained by transforming ω by L_γ. If we consider $\rho(\gamma)$ as an operator on the space of all F-valued forms on X, it commutes with the operator d of the exterior differentiation, and therefore d defines a coboundary operator of the graded module $A(\Gamma, X, \rho) = \sum_r A^r(\Gamma, X, \rho)$. The cohomology groups of the complex $A(\Gamma, X, \rho)$ will be denoted by $H^r(\Gamma, X, \rho)$.

In [8], the laplacian is defined for the forms in $A(\Gamma, X, \rho)$, and it is shown that each cohomology class in $H^r(\Gamma, X, \rho)$ is represented by a unique "harmonic" form, and that each harmonic form decomposes into the sum of harmonic forms of type (p, q). Therefore, if we denote by $b^r(\Gamma, X, \rho)$ and $h^{p,q}(\Gamma, X, \rho)$ respectively the dimensions of the complex vector spaces of all harmonic r-forms and of all harmonic forms of type (p, q), we have

$$\dim_C H^r(\Gamma, X, \rho) = b^r(\Gamma, X, \rho) = \sum_{p+q=r} h^{p,q}(\Gamma, X, \rho) \ .$$

If the representation ρ of G decomposes into the sum of the representations $\rho_1, \cdots, \rho_m$, the cohomology group $H^r(\Gamma, X, \rho)$ decomposes also into the direct sum of the cohomology groups $H^r(\Gamma, X, \rho_i)$. Therefore it is sufficient to treat the case where ρ is an absolutely irreducible representation of G. We know that the representation ρ is then of the type $\rho_{m_1} \otimes \cdots \otimes \rho_{m_N} \otimes \sigma$, $(m_i \geqq 0)$. Here $\rho_{m_i}(i = 1, \cdots, N; m_i \geqq 1)$ denotes the representation of SL $(2, \mathbf{R})$ in the complex vector space of all symmetric tensors of order m_i constructed over $\mathbf{C}^2$, and ρ_0 denotes the trivial representation of SL $(2, \mathbf{R})$ and σ is an absolutely irreducible representation of the compact factor G_{N+1}.

The main result of this paper is then stated as follows: Let Γ be a discrete subgroup of G satisfying certain conditions which will be stated in §1, and let $\rho = \rho_{m_1} \otimes \cdots \otimes \rho_{m_N} \otimes \sigma$. We denote by d the dimension of the vector space consisting of all elements u in the representation space S of σ such that $\sigma(\gamma_{N+1})u = u$ for all $\gamma \in \Gamma$, where γ_{N+1} denotes the G_{N+1}-component of γ. Then we shall obtain the following results:

$$h^{p,q}(\Gamma, X, \rho) = 0 \qquad \text{if } p \neq q \text{ and } p + q \neq N,$$

$$h^{p,p}(\Gamma, X, \rho) = \begin{cases} 0 & \text{if } 2p \neq N \text{ and } (m_1, \cdots, m_N) \neq (0, \cdots, 0), \\ \binom{N}{p}d & \text{if } 2p \neq N \text{ and } (m_1, \cdots, m_N) = (0, \cdots, 0), \end{cases}$$

$$h^{N-q,q}(\Gamma, X, \rho) = \begin{cases} \binom{N}{q}h^{N,0}(\Gamma, X, \rho) & \text{if } (m_1, \cdots, m_N) \neq (0, \cdots, 0), \\ \binom{N}{q}\{d\delta_{N-q,q} + h^{N,0}(\Gamma, X, \rho)\} & \\ & \text{if } (m_1, \cdots, m_N) = (0, \cdots, 0). \end{cases}$$

Moreover, $h^{N,0}(\Gamma, X, \rho)$ is equal to the dimension of the complex vector space of all S-valued holomorphic automorphic forms on X with respect to Γ for the automorphic factor

$$J_{m_1+2, \cdots, m_N+2, \sigma}(s, z) = \prod_{j=1}^{N} (c_j z_j + d_j)^{m_j+2} \sigma(s_{N+1}) .$$

Here $z = (z_1, \cdots, z_N)$ denotes a point of X and $s = (s_1, \cdots, s_N, s_{N+1})$ is an element of G with $s_j = \begin{pmatrix} a_j\, b_j \\ c_j\, d_j \end{pmatrix}(1 \leq j \leq N)$.

Further, we shall prove the following formula:

$$h^{N,0}(\Gamma, X, \rho) = \begin{cases} (-1)^N n \prod_{j=1}^{N} (m_j + 1)(1 + (-1)^N h^{N,0}(\Gamma\backslash X)) & \\ & \text{if } (m_1, \cdots, m_N) \neq (0, \cdots, 0), \\ (-1)^N n(1 + (-1)^N h^{N,0}(\Gamma\backslash X)) - (-1)^N d & \\ & \text{if } (m_1, \cdots, m_N) = (0, \cdots, 0), \end{cases}$$

where n is the degree of the representation σ, and $h^{N,0}(\Gamma\backslash X)$ denotes the dimension of the complex vector space consisting of all holomorphic N-forms on X which are invariant by Γ.

By the above formulas the dimension $b^r(\Gamma, X, \rho)$ of the cohomology group $H^r(\Gamma, X, \rho)$ may be computed. In particular, if ρ is the trivial representation of G, $b^r(\Gamma, X, \rho)$ is equal to the rth Betti number of the compact complex manifold $\Gamma\backslash X$; and we obtain the following result on the Betti numbers: Let X_u be the product of N copies of the Riemann sphere. Then

$$b^r(\Gamma\backslash X) = b^r(X_u) \qquad\qquad \text{for } r \neq N,$$
$$b^N(\Gamma\backslash X) = b^N(X_u) + E(X_u)h^{N,0}(\Gamma\backslash X),$$

where $E(X_u) = 2^N$ is the Euler characteristic of X_u.

The outline of the paper is as follows. In §1, after explaining the assumptions on the discrete subgroup Γ, we shall prove a decomposition theorem for harmonic forms which is a special case of the Hodge-Chern decomposition theorem for harmonic forms of the type described above (cf. [2], [8]); we became aware of this result by a communication of A. Weil. This decomposition theorem shows that, if

$$\omega = \sum_{k_1<\cdots<k_p, l_1<\cdots<l_q} f_{k_1\cdots k_p \bar{l}_1\cdots\bar{l}_q} dz_{k_1} \wedge \cdots \wedge dz_{k_p} \wedge d\bar{z}_{l_1} \wedge \cdots \wedge d\bar{z}_{l_q}$$

is harmonic, then each monomial $f_{k_1\cdots k_p \bar{l}_1\cdots\bar{l}_q} dz_{k_1} \wedge \cdots \wedge dz_{k_p} \wedge d\bar{z}_{l_1} \wedge \cdots \wedge d\bar{z}_{l_q}$ is also harmonic. We denote by $\mathfrak{h}(k_1, \cdots, k_p; l_1, \cdots, l_q; \rho)$ the complex vector space of all harmonic forms of the type $f_{k_1\cdots k_p \bar{l}_1\cdots\bar{l}_q} dz_{k_1} \wedge \cdots \wedge dz_{k_p} \wedge d\bar{z}_{l_1} \wedge \cdots \wedge d\bar{z}_{l_q}$ $(k_1 < \cdots < k_p, l_1 < \cdots < l_q)$. In §3 we prove a vanishing theorem for automorphic forms which will be used to prove the vanishing of harmonic forms of some types. In §4 we shall compute the dimension of the space $\mathfrak{h}(k_1, \cdots, k_p; l_1, \cdots, l_p; \rho)$ for the representation $\rho = \rho_{m_1} \otimes \cdots \otimes \rho_{m_N} \otimes \sigma$ of G. The space $\mathfrak{h}(k_1, \cdots, k_p; l_1, \cdots, l_q; \rho)$ reduces to (0) except in the following two cases:

(1) $p = q$ and $(k_1, \cdots, k_p) = (l_1, \cdots, l_q)$;

(2) $p + q = N$ and $(k_1, \cdots, k_p, l_1, \cdots, l_q)$ is a permutation of $(1, \cdots, N)$. The dimension of the space $\mathfrak{h}(k_1, \cdots, k_p; k_1, \cdots, k_p; \rho)$ will be given in Proposition 4.1. We shall show that, if $(k_1, \cdots, k_p, l_1, \cdots, l_q)$ is a permutation of $(1, \cdots, N)$, then the space $\mathfrak{h}(k_1, \cdots, k_p; l_1, \cdots, l_q; \rho)$ is canonically isomorphic to the complex vector space of all holomorphic automorphic forms of certain type. The §§ 5 and 6 are devoted to a proof of the equality of the dimensions of the spaces $\mathfrak{h}(1, \cdots, N; \rho)$ and $\mathfrak{h}(k_1, \cdots, k_p; l_1, \cdots, l_q, \rho)$, where $(k_1, \cdots, k_p, l_1, \cdots, l_q)$ is a permutation of $(1, \cdots, N)$. For this purpose we shall prove in § 5 a vanishing theorem for the cohomology groups attached to the complex vector bundle $E(J_{n_1\cdots n_N\sigma})$ over $\Gamma\backslash X$, which is defined by the automorphic factor $J_{n_1\cdots n_N\sigma}$. We shall determine in § 6 the Chern classes of the complex vector bundle $E(J_{n_1\cdots n_N\sigma})$ and, using the Riemann-Roch-Hirzebruch theorem, we shall

conclude that the dimensions of $\mathfrak{h}(1, \cdots, N; \rho)$ and $\mathfrak{h}(k_1, \cdots, k_p; l_1, \cdots, l_q; \rho)$ are equal. The main theorem and some of its applications are given in § 7, under the condition that Γ operates freely on X. In the final § 8, we shall discuss the case where Γ may have non-trivial elements of finite order.

In concluding the introduction, the authors wish to express their gratitude to Professor A. Weil for his valuable communications.

1. Preliminaries and a decomposition theorem for harmonic forms

A. Let G_i $(i = 1, \cdots, m)$ be a connected non-compact simple Lie group, and let $G_0 = G_1 \times \cdots \times G_m$. We call a discrete subgroup Γ_0 of G_0 *irreducible*, if the projection of Γ_0 into any partial factor of G_0 different from G_0 itself is not discrete in the partial factor. It is known that every discrete subgroup Δ of G_0 with compact quotient space G_0/Δ is commensurable with a discrete subgroup Γ' of G_0 of the type $\Gamma' = \Gamma^{(1)} \times \cdots \times \Gamma^{(s)}$, where each $\Gamma^{(i)}$ is an irreducible discrete subgroup of a partial factor $G^{(i)}$ of G_0 with $G_0 = G^{(1)} \times \cdots \times G^{(s)}$ (cf. [13]).

PROPOSITION 1.1. *Let Γ_0 be an irreducible discrete subgroup of G_0 with compact quotient space G_0/Γ_0. Then the projection of Γ_0 into each partial factor of G_0 different from G_0 is dense in the partial factor.*

PROOF. Let $G_{i_1} \times \cdots \times G_{i_p} (i_1 < \cdots < i_p \leqq m, p \neq m)$ be a partial factor of G_0. By assumption, the projection $\Gamma_{i_1 \cdots i_p}$ of Γ_0 into $G_{i_1} \times \cdots \times G_{i_p}$ is not discrete. Let $\bar{\Gamma}_{i_1 \cdots i_p}$ be the closure of $\Gamma_{i_1 \cdots i_p}$ in $G_{i_1} \times \cdots \times G_{i_p}$ and let Δ be the connected component of $\bar{\Gamma}_{i_1 \cdots i_p}$ containing the identity element. Since $\Gamma_{i_1 \cdots i_p}$ is not discrete, we have $\Delta \neq \{e\}$, and, since $\gamma \Gamma_{i_1 \cdots i_p} \gamma^{-1} = \Gamma_{i_1 \cdots i_p}$ for every $\gamma \in \Gamma_0$, we have also $\gamma \Delta \gamma^{-1} = \Delta$. It follows from a theorem of Borel [1] that Δ is a normal subgroup of G_0. Therefore Δ is a partial factor of $G_{i_1} \times \cdots \times G_{i_p}$. For simplicity let $\Delta = G_{i_1} \times \cdots \times G_{i_t}$ with $t \leqq p$. Suppose that $t < p$. Then $\bar{\Gamma}_{i_1 \cdots i_p}$ is of the form $\bar{\Gamma}_{i_1 \cdots i_p} = \Delta \times \Gamma'$, where Γ' is a discrete subgroup of $G_{i_{t+1}} \times \cdots \times G_{i_p}$. It is clear that Γ' contains the projection of Γ_0 into $G_{i_{t+1}} \times \cdots \times G_{i_p}$, and hence the projection is a discrete subgroup of $G_{i_{t+1}} \times \cdots \times G_{i_p}$. This contradicts the assumption, and hence we must have $t = p$. Thus $\bar{\Gamma}_{i_1 \cdots i_p} = G_{i_1} \times \cdots \times G_{i_p}$, and this shows that $\Gamma_{i_1 \cdots i_p}$ is dense in $G_{i_1} \times \cdots \times G_{i_p}$.

B. We denote by $H_i (i = 1, \cdots, N)$ the upper half plane consisting of all complex numbers z with $\mathrm{Im}\,(z) > 0$ and put $X = H_1 \times \cdots \times H_N$. Let $G = G_1 \times \cdots \times G_N \times G_{N+1}$, where each G_i for $1 \leqq i \leqq N$ is a copy of the special linear group SL $(2, \boldsymbol{R})$, and G_{N+1} is a connected compact Lie group, possibly equal to the identity element. Then $K = K_1 \times \cdots \times K_N \times G_{N+1}$

is a maximal compact subgroup of G, where each K_i is the special orthogonal group SO(2). The group G operates transitively on X in the following manner: Let $z = (z_1, \cdots, z_N) \in X$ and $s = (s_1, \cdots, s_N, s_{N+1}) \in G$, where $s_i \in G_i$. Then

$$s \cdot z = (s_1 \cdot z_1, \cdots, s_N \cdot z_N) ,$$

$$s_i \cdot z_i = (a_i z_i + b_i)(c_i z_i + d_i)^{-1} , \qquad s_i = \begin{pmatrix} a_i & b_i \\ c_i & d_i \end{pmatrix} .$$

Throughout this paper, z_0 will denote the point of X, all of whose coordinates are equal to $i(i^2 = -1)$. The isotropy subgroup of G at the point z_0 is equal to the maximal compact group K, and X can be identified with the quotient space G/K.

Let Γ be a discrete subgroup of G. Let $G_0 = G_1 \times \cdots \times G_N$ and let Γ_0 be the projection of Γ into the partial factor G_0 of G. Since $G = G_0 \times G_{N+1}$ and G_{N+1} is compact, it is easily seen that Γ_0 is a discrete subgroup of G_0. Let Z be the center of G_0. Clearly, Z consists of the 2^N elements whose components are $\pm \begin{pmatrix} 1 & 0 \\ 0 & 1 \end{pmatrix}$. We consider now the following three conditions on the discrete subgroup Γ of G:

(C_1) *The projection of Γ onto Γ_0 is bijective, and the quotient space G/Γ is compact.*

(C_2) *Γ_0 is an irreducible discrete subgroup of G_0.*

(C_3) *$\Gamma_0/(\Gamma_0 \cap Z)$ has no element of finite order other than the identity element.*

If Γ satisfies (C_1) and (C_2), then by Proposition 1.1, the projection of Γ_0 into any partial factor of G_0 different from G_0 is dense in the partial factor. Further, (C_1) and (C_3) imply that the quotient space $\Gamma \backslash X$ is a compact complex manifold without singularities.

EXAMPLE. Let Γ_0 be an irreducible discrete subgroup of G_0 with compact quotient space G_0/Γ_0. Let σ be a unitary representation of Γ_0 of degree m; and put $G_{N+1} = \mathrm{U}(m)$. Then $\Gamma = \{(\gamma, \sigma(\gamma)) \,|\, \gamma \in \Gamma_0\}$ is a discrete subgroup of $G = G_0 \times G_{N+1}$ satisfying the conditions (C_1) and (C_2). The groups investigated in [10, 12] are examples of Γ which are arithmetically defined and satisfies (C_1) and (C_2).

In the remainder of § 1 and in § 2, Γ will denote a discrete subgroup of G satisfying the condition (C_1), and in §§ 3-4, we shall assume that Γ satisfies the conditions (C_1) and (C_2). Further in §§ 5-7, we shall assume that Γ satisfies (C_1), (C_2) and (C_3).

Now denote by $\mathfrak{g}$ the Lie algebra of all left invariant vector fields on G. The Lie algebra $\mathfrak{g}$ can be identified with the Lie algebra consisting of the elements $Y = (Y_1, \cdots, Y_N, Y_{N+1})$, where $Y_1, \cdots, Y_N$ are real 2×2 matrices of trace 0, and Y_{N+1} is an element of the Lie algebra $\mathfrak{g}_{N+1}$ of G_{N+1}. Put

$$(1.1) \quad \begin{cases} W = \dfrac{1}{2} \begin{pmatrix} 0 & -i \\ i & 0 \end{pmatrix}, \\[2mm] X = \dfrac{1}{4} \begin{pmatrix} 1 & i \\ i & -1 \end{pmatrix}, \\[2mm] \bar{X} = \dfrac{1}{4} \begin{pmatrix} 1 & -i \\ -i & -1 \end{pmatrix}, \end{cases}$$

and

$$W_i = (0, \cdots, 0, \overset{i}{W}, 0, \cdots, 0),$$

$$X_i = (0, \cdots, 0, \overset{i}{X}, 0, \cdots, 0),$$

$$\bar{X}_i = (0, \cdots, 0, \overset{i}{\bar{X}}, 0, \cdots, 0),$$

for $i = 1, \cdots, N$. Further let $Y_1', \cdots, Y_l'$ be a basis of the Lie algebra $\mathfrak{g}_{N+1}$ such that $\varphi_{N+1}(Y_i', Y_j') = -\delta_{ij}$, where φ_{N+1} denotes the Killing form of $\mathfrak{g}_{N+1}$. Put

$$Y_a = (0, \cdots, 0, Y_a') \qquad\qquad (1 \leqq a \leqq l).$$

Then the elements $W_i, X_i, \bar{X}_i, Y_a$ form a basis of the complexification $\mathfrak{g}^c$ of $\mathfrak{g}$, and we have

$$(1.2) \quad \begin{cases} [W_i, X_j] = \delta_{ij} X_j, \\ [W_i, \bar{X}_j] = -\delta_{ij} \bar{X}_j, \\ [X_i, \bar{X}_j] = \dfrac{\delta_{ij}}{2}, \\ [X_i, X_j] = [\bar{X}_i, \bar{X}_j] = 0, \\ [W_i, Y_a] = [X_i, Y_a] = [\bar{X}_i, Y_a] = 0, \end{cases}$$

where $i, j = 1, \cdots, N$ and $a = 1, \cdots, l$. Let $\mathfrak{n}^+$ (resp. $\mathfrak{n}^-$) be the subspace of $\mathfrak{g}^c$ spanned by $\{X_1, \cdots, X_N\}$ (resp. $\{\bar{X}_1, \cdots, \bar{X}_N\}$), and $\mathfrak{k}^c$ the subalgebra of $\mathfrak{g}^c$ spanned by $\{W_1, \cdots, W_N, Y_1, \cdots, Y_l\}$. Then

$$\mathfrak{g}^c = \mathfrak{n}^+ + \mathfrak{n}^- + \mathfrak{k}^c,$$

$$[\mathfrak{k}^c, \mathfrak{n}^+] = \mathfrak{n}^+, [\mathfrak{k}^c, \mathfrak{n}^-] = \mathfrak{n}^-,$$

$$[\mathfrak{n}^+, \mathfrak{n}^-] \subset \mathfrak{k}^c, [\mathfrak{n}^+, \mathfrak{n}^+] = [\mathfrak{n}^-, \mathfrak{n}^-] = (0).$$

In the following treatment, we consider the elements of $\mathfrak{g}^c$ as (complex) left invariant vector fields on G.

C. Retaining the notation used in the introduction, we denote by $A^{p,q}(\Gamma, X, \rho)$ the subspace of all $\omega \in A^r(\Gamma, X, \rho)$ of type (p, q). Then we have $A^r(\Gamma, X, \rho) = \sum_{p+q=r} A^{p,q}(\Gamma, X, \rho)$ and $A(\Gamma, X, \rho) = \sum_{p,q} A^{p,q}(\Gamma, X, \rho)$. We denote by d' (resp. d'') as usual the exterior differentiation with respect to the local complex coordinates (resp. the conjugates of the local complex coordinates). The operator d' (resp. d'') is a coboundary operator of degree $(1, 0)$ (resp. $(0, 1)$) of the bigraded complex $A(\Gamma, X, \rho)$. In view of our assumption on Γ, taking a suitable positive definite hermitian inner product on the vector space $A(\Gamma, X, \rho)$, we can introduce the operators $\delta, \delta', \delta''$ on $A(\Gamma, X, \rho)$ of degree $-1, (-1, 0), (0, -1)$ respectively, which are adjoint to d, d', d'' (cf. [8]). The laplacian Δ is defined by $\Delta = d\delta + \delta d$. A form $\omega \in A^r(\Gamma, X, \rho)$ is called *harmonic*, if $\Delta\omega = 0$. Every cohomology class of $H^r(\Gamma, X, \rho)$ is represented by a unique harmonic form. Moreover, we have $\Delta = \Delta' + \Delta''$, where $\Delta' = d'\delta' + \delta'd'$ and $\Delta'' = d''\delta'' + \delta''d''$. A form ω is harmonic if and only if $\Delta'\omega = \Delta''\omega = 0$; and this condition is equivalent to the condition that $d'\omega = d''\omega = \delta'\omega = \delta''\omega = 0$. It follows from this that every harmonic r-form decomposes into the sum of harmonic forms of type (p, q) with $p + q = r$.

Let π be the projection of G onto $X = G/K$. Then $\pi(s) = sz_0$ for all $s \in G$. To each $\omega \in A^r(\Gamma, X, \rho)$ we associate an F-valued r-form ω^0 on G by putting

$$\omega_s^0 = \rho(s^{-1})(\omega \circ \pi)_s$$

for all $s \in G$, where F denotes the representation space of ρ. The r-forms ω^0 are characterized, among the F-valued forms on G, by the following conditions (cf. [8]).

$$(1.3) \qquad \omega^0 \circ L_\gamma = \omega^0 \qquad\qquad \text{for all } \gamma \in \Gamma ,$$

$$(1.4) \qquad i(X)\omega^0 = 0 \qquad\qquad \text{for all } X \in \mathfrak{k} ,$$

$$(1.5) \qquad \theta(X)\omega^0 = -\rho(X)\omega^0 \qquad\qquad \text{for all } X \in \mathfrak{k} .$$

Here $i(X)$ and $\theta(X)$ denote respectively the operators of interior product and Lie derivation with respect to the vector field $X \in \mathfrak{k}$, $\mathfrak{k}$ being the subalgebra of $\mathfrak{g}$ corresponding to K, and $X \to \rho(X)$ denotes the representation of the Lie algebra $\mathfrak{g}$ of G defined by the representation ρ of G.

Now the vector fields $\{X_1, \cdots, X_N, \bar{X}_1, \cdots, \bar{X}_N, W_1, \cdots, W_N, Y_1, \cdots, Y_l\}$ form, at each point $s \in G$, a basis of the complexification of the tangent vector space of G. Since $\{W_1, \cdots, W_N, Y_1, \cdots, Y_l\} = \mathfrak{k}^c$ and $i(X)\omega^0 = 0$ for all $X \in \mathfrak{k}^c$, we see that the form ω^0, and hence the form ω, is determined uniquely by the system of F-valued functions $\{\omega_{i_1 \cdots i_p \bar{j}_1 \cdots \bar{j}_q}\}$, putting

$$\omega_{i_1\cdots i_p \bar{j}_1 \cdots \bar{j}_q} = \omega^0(X_{i_1}, \cdots, X_{i_p}, \bar{X}_{j_1}, \cdots, \bar{X}_{j_q}) .$$

The form ω is of type (p, q) if and only if $\omega_{i_1 \cdots i_a \bar{j}_1 \cdots \bar{j}_b} = 0$ for $(a, b) \neq (p, q)$.

We see easily from (1.3) and (1.5) that these functions are characterized by the following conditions:

$$(1.6) \quad \omega_{i_1 \cdots i_p \bar{j}_1 \cdots \bar{j}_q}(\gamma \cdot s) = \omega_{i_1 \cdots i_p \bar{j}_1 \cdots \bar{j}_q}(s) \qquad \text{for all } \gamma \in \Gamma \text{ and } s \in G ,$$

$$(1.7) \quad W_k \omega_{i_1 \cdots i_p \bar{j}_1 \cdots \bar{j}_q} = -\rho(W_k)\omega_{i_1 \cdots i_p \bar{j}_1 \cdots \bar{j}_q} + \Big(\sum_{u=1}^p \delta_{k i_u} - \sum_{v=1}^q \delta_{k j_v}\Big)\omega_{i_1 \cdots i_p \bar{j}_1 \cdots \bar{j}_q}$$
$$(k = 1, \cdots, N) ,$$

$$(1.8) \quad X_a \omega_{i_1 \cdots i_p \bar{j}_1 \cdots \bar{j}_q} = -\rho(Y_a)\omega_{i_1 \cdots i_p \bar{j}_1 \cdots \bar{j}_q} \qquad (a = 1, \cdots, l) .$$

Let $\omega \in A^{p,q}(\Gamma, X, \rho)$. The systems of F-valued functions on G corresponding to $d'\omega$, $d''\omega$, $\delta'\omega$, $\delta''\omega$, $\Delta'\omega$ and $\Delta''\omega$ are given as follows (cf. [8]):

$$(1.9) \quad (d'\omega)_{i_1 \cdots i_{p+1} \bar{j}_1 \cdots \bar{j}_q} = \sum_{u=1}^{p+1} (-1)^{u+1}(X_{i_u} + \rho(X_{i_u}))\omega_{i_1 \cdots \hat{i}_u \cdots i_{p+1} \bar{j}_1 \cdots \bar{j}_q} ,$$

$$(1.10) \quad (d''\omega)_{i_1 \cdots i_p \bar{j}_1 \cdots \bar{j}_{q+1}} = \sum_{v=1}^{q+1} (-1)^{p+v+1}(\bar{X}_{j_v} + \rho(\bar{X}_{j_v}))\omega_{i_1 \cdots i_p \bar{j}_1 \cdots \hat{\bar{j}}_v \cdots \bar{j}_{q+1}} ,$$

$$(1.11) \quad (\delta'\omega)_{i_1 \cdots i_{p-1} \bar{j}_1 \cdots \bar{j}_q} = \sum_{k=1}^N (-\bar{X}_k + \rho(\bar{X}_k))\omega_{k i_1 \cdots i_{p-1} \bar{j}_1 \cdots \bar{j}_q} ,$$

$$(1.12) \quad (\delta''\omega)_{i_1 \cdots i_p \bar{j}_1 \cdots \bar{j}_{q-1}} = (-1)^p \sum_{k=1}^N (-X_k + \rho(X_k))\omega_{i_1 \cdots i_p \bar{k} \bar{j}_1 \cdots \bar{j}_{q-1}} .$$

$$(1.13) \quad (\Delta'\omega)_{i_1 \cdots i_p \bar{j}_1 \cdots \bar{j}_q} = \sum_{k=1}^N (-\bar{X}_k + \rho(\bar{X}_k))(X_k + \rho(X_k))\omega_{i_1 \cdots i_p \bar{j}_1 \cdots \bar{j}_q}$$
$$+ \sum_{k=1}^N \sum_{u=1}^p (-1)^u([X_{i_u}, \bar{X}_k] - \rho([X_{i_u}, \bar{X}_k]))\omega_{k i_1 \cdots \hat{i}_u \cdots i_p \bar{j} \cdots \bar{j}_q} ,$$

$$(1.14) \quad (\Delta''\omega)_{i_1 \cdots i_p \bar{j}_1 \cdots \bar{j}_q} = \sum_{k=1}^N (-X_k + \rho(X_k))(\bar{X}_k + \rho(\bar{X}_k))\omega_{i_1 \cdots i_p \bar{j}_1 \cdots \bar{j}_q}$$
$$+ \sum_{k=1}^N \sum_{v=1}^q (-1)^v([\bar{X}_{j_v}, X_k] - \rho([\bar{X}_{j_v}, X_k]))\omega_{i_1 \cdots i_p \bar{k} \bar{j}_1 \cdots \hat{\bar{j}}_v \cdots \bar{j}_q} .$$

Since $[X_i, \bar{X}_k] = (\delta_{ik}/2) W_i$, we can write (1.13) and (1.14) in the form

$$(1.13') \quad (\Delta'\omega)_{i_1 \cdots i_p \bar{j}_1 \cdots \bar{j}_q} = \sum_{k=1}^N (-\bar{X}_k + \rho(\bar{X}_k))(X_k + \rho(X_k))\omega_{i_1 \cdots i_q \bar{j}_1 \cdots \bar{j}_q}$$
$$- \frac{1}{2} \sum_{u=1}^p (W_{i_u} - \rho(W_{i_u}))\omega_{i_1 \cdots i_p \bar{j}_1 \cdots \bar{j}_q} ,$$

$$(1.14') \quad (\Delta''\omega)_{i_1 \cdots i_p \bar{j}_1 \cdots \bar{j}_q} = \sum_{k=1}^N (-X_k + \rho(X_k))(\bar{X}_k + \rho(\bar{X}_k))\omega_{i_1 \cdots i_p \bar{j}_1 \cdots \bar{j}_q}$$
$$+ \frac{1}{2} \sum_{v=1}^q (W_{j_v} - \rho(W_{j_v}))\omega_{i_1 \cdots i_p \bar{j}_1 \cdots \bar{j}_q} .$$

Now let $\omega \in A^{p,q}(\Gamma, X, \rho)$ and let $\{\omega_{i_1 \cdots i_p \bar{j}_1 \cdots \bar{j}_q}\}$ be the corresponding system of F-valued functions on G. Further let

$$\omega = \sum_{\substack{i_1 < \cdots < i_p \\ j_1 < \cdots < j_q}} f_{i_1 \cdots i_p \bar{j}_1 \cdots \bar{j}_q} dz_{i_1} \wedge \cdots \wedge dz_{i_p} \wedge d\bar{z}_{j_1} \wedge \cdots \wedge d\bar{z}_{j_q} .$$

For any system of indices $(k_1, \cdots, k_p; l_1, \cdots, l_q)$ such that $k_1 < \cdots < k_p$ and $l_1 < \cdots < l_q$, put

$$\omega(k_1, \cdots, k_p; l_1, \cdots, l_q) = f_{k_1 \cdots k_p \bar{l}_1 \cdots \bar{l}_q} dz_{k_1} \wedge \cdots \wedge dz_{k_p} \wedge d\bar{z}_{l_1} \wedge \cdots \wedge d\bar{z}_{l_q} .$$

It is easily verified that each $\omega(k_1, \cdots, k_p; l_1, \cdots, l_q)$ belongs to $A^{p,q}(\Gamma, X, \rho)$. If $\{\eta_{i_1 \cdots i_p \bar{j}_1 \cdots \bar{j}_q}\}$ denotes the system of F-valued functions corresponding to $\omega(k_1, \cdots, k_p; l_1, \cdots, l_q)$, then $\eta_{k_1 \cdots k_p \bar{l}_1 \cdots \bar{l}_q} = \omega_{k_1 \cdots k_p \bar{l}_1 \cdots \bar{l}_q}$ and $\eta_{i_1 \cdots i_p \bar{j}_1 \cdots \bar{j}_q} = 0$ for the indices $(i_1, \cdots, i_p; j_1, \cdots, j_q) \neq (k_1, \cdots, k_p; l_1, \cdots, l_q)$, where $i_1 < \cdots < i_p, j_1 < \cdots < j_q$.

We see from (1.13′) and (1.14′) that ω is harmonic if and only if each $\omega(k_1, \cdots, k_p; l_1, \cdots, l_q)$ is harmonic. We say that a form $\omega \in A^{p,q}(\Gamma, X, \rho)$ is of type $(k_1, \cdots, k_p; l_1, \cdots, l_q)$ if $\omega = \omega(k_1, \cdots, k_p; l_1, \cdots, l_q)$. We denote by $\mathfrak{h}^{p,q}(\rho)$ (resp. $\mathfrak{h}(k_1, \cdots, k_p; l_1, \cdots, l_q; \rho)$) the vector space of all harmonic forms of type (p, q) (resp. of type $(k_1, \cdots, k_p; l_1, \cdots, l_q)$). Then by the above consideration, we get

PROPOSITION 1.2. *Let Γ be a discrete subgroup of G satisfying the condition* (C_1). *Then the vector space $\mathfrak{h}^{p,q}(\rho)$ is the direct sum of the subspaces $\mathfrak{h}(k_1, \cdots, k_p; l_1, \cdots, l_q; \rho)$.*

2. The representation $\rho_{m_1 \cdots m_N \sigma}$ and harmonic forms of type $(k_1, \cdots, k_p; l_1, \cdots, l_q)$

The vectors

$$e = \begin{pmatrix} 1 \\ i \end{pmatrix}, \qquad \bar{e} = \begin{pmatrix} 1 \\ -i \end{pmatrix}$$

form a basis of the vector space C^2 and we have

$$We = \frac{1}{2} e, \qquad W\bar{e} = -\frac{1}{2} \bar{e},$$

$$Xe = 0, \qquad X\bar{e} = \frac{1}{2} e,$$

$$\bar{X}e = \frac{1}{2} \bar{e}, \qquad \bar{X}\bar{e} = 0,$$

where W, X and $\bar{X}$ denote the basis of the Lie algebra $\mathfrak{sl}(2, C)$ defined by (1.1).

Let ρ_m be the representation of $GL(2, C)$ in the vector space of all symmetric tensors S_m. Put

$$e_a = \underbrace{e \cdots \cdots e}_{a} \cdot \underbrace{\bar{e} \cdots \cdots \bar{e}}_{m - a},$$

where $\cdot$ denotes the symmetric product of the vectors. Then, $\{e_0, e_1, \cdots, e_m\}$ form a basis of S_m, and

$$\rho_m(W)e_a = \left(a - \frac{m}{2} \right) e_a,$$

$$\rho_m(X)e_a = \frac{m - a}{2} e_{a+1},$$

$$\rho_m(\bar{X})e_a = \frac{a}{2} e_{a-1},$$

where we put $e_{m+1} = 0$ and $e_{-1} = 0$.

Let σ be a representation of the compact group G_{N+1} in a complex vector space S and $\rho_{m_1\cdots m_N\sigma}$ be the representation of $G = G_1 \times \cdots \times G_N \times G_{N+1}$ $(G_1 = \cdots = G_N = \mathrm{SL}\,(2,\,\boldsymbol{R}))$ in the vector space $F = S_{m_1} \otimes \cdots \otimes S_{m_N} \otimes S$ defined by

$$\rho_{m_1\cdots m_N\sigma} = \rho_{m_1} \otimes \cdots \otimes \rho_{m_N} \otimes \sigma \,.$$

We note that every absolutely irreducible representation of G is of the type $\rho_{m_1\cdots m_N\sigma}$. Let n be the dimension of S and $\{u_1, \cdots, u_n\}$ be a basis of S. Put

$$e_{a_1\cdots a_N t} = e_{a_1} \otimes \cdots \otimes e_{a_N} \otimes u_t \,.$$

Then $\{e_{a_1\cdots a_N t}; 0 \leq a_i \leq m_i, 1 \leq t \leq n\}$ is a basis of F, and

$$(2.1) \quad \begin{cases} \rho_{m_1\cdots m_N\sigma}(W_i)e_{a_1\cdots a_N t} = \left(a_i - \dfrac{m_i}{2}\right)e_{a_1\cdots a_N t}\,, \\[2mm] \rho_{m_1\cdots m_N\sigma}(X_i)e_{a_1\cdots a_N t} = \dfrac{m_i - a_i}{2}\,e_{a_1\cdots a_i+1\cdots a_N t}\,, \\[2mm] \rho_{m_1\cdots m_N\sigma}(\bar{X}_i)e_{a_1\cdots a_N t} = \dfrac{a_i}{2}\,e_{a_1\cdots a_i-1\cdots a_N t}\,, \\[2mm] \rho_{m_1\cdots m_N\sigma}(Y_b)e_{a_1\cdots a_N t} = \sum_{d=1}^n \sigma_t^d(Y_b)e_{a_1\cdots a_N d}\,, \end{cases}$$

where $\sigma(Y_b)u_t = \sum_{d=1}^n \sigma_t^d(Y_b)u_d$ and we put $e_{a_1\cdots a_N t} = 0$ if one of the indices a_i is equal to -1 or to $m_i + 1$.

Now let ω be a harmonic form of type $(k_1, \cdots, k_p;\, l_1, \cdots, l_q)$ for the representation $\rho = \rho_{m_1\cdots m_N\sigma}$. Then ω is determined completely by the F-valued function $\omega_{k_1\cdots k_p \bar{l}_1\cdots \bar{l}_q}$ on G satisfying the conditions (1.6), (1.7) and (1.8). Denote $\omega_{k_1\cdots k_p \bar{l}_1\cdots \bar{l}_q}$ simply by ψ and put

$$\psi = \sum_{a_1,\cdots,a_N,t} \psi(a_1, \cdots, a_N, t)e_{a_1\cdots a_N t}\,,$$
$$\psi(a_1, \cdots, a_N) = \sum_{t=1}^n \psi(a_1, \cdots, a_N, t)u_t \,.$$

Since ω is harmonic, we have $d'\omega = d''\omega = \delta'\omega = \delta''\omega = 0$, and we obtain, by $(1.9) - (1.12)$, the following equations:

$$\begin{aligned} X_i\psi &= -\rho_{m_1\cdots m_N\sigma}(X_i)\psi & (i \neq k_u,\, u = 1, \cdots, p)\,, \\ \bar{X}_j\psi &= -\rho_{m_1\cdots m_N\sigma}(\bar{X}_j)\psi & (j \neq l_v,\, v = 1, \cdots, q)\,, \\ \bar{X}_{k_u}\psi &= \rho_{m_1\cdots m_N\sigma}(\bar{X}_{k_u})\psi & (u = 1, \cdots, p)\,, \\ X_{l_v}\psi &= \rho_{m_1\cdots m_N\sigma}(X_{l_v})\psi & (v = 1, \cdots, q)\,. \end{aligned}$$

We see from these equations and (2.1) that the system $\{\psi(a_1, \cdots, a_N)\}$ of the S-valued functions on G satisfies the following equations:

$$(2.2) \quad X_i \psi(a_1, \cdots, a_N) = -\frac{1}{2}(m_i - a_i + 1)\psi(a_1, \cdots, a_i - 1, \cdots, a_N)$$

$$(i \neq k_u),$$

$$(2.3) \quad \bar{X}_j \psi(a_1, \cdots, a_N) = -\frac{1}{2}(a_j + 1)\psi(a_1, \cdots, a_j + 1, \cdots, a_N) \quad (i \neq l_v),$$

$$(2.4) \quad \bar{X}_{k_u} \psi(a_1, \cdots, a_N) = \frac{1}{2}(a_{k_u} + 1)\psi(a_1, \cdots, a_{k_u} + 1, \cdots, a_N)$$

$$(u = 1, \cdots, p),$$

$$(2.5) \quad X_{l_v} \psi(a_1, \cdots, a_N) = \frac{1}{2}(m_{l_v} - a_{l_v} + 1)\psi(a_1, \cdots, a_{l_v} - 1, \cdots, a_N)$$

$$(v = 1, \cdots, q),$$

$$(2.6) \quad \begin{cases} W_i \psi(a_1, \cdots, a_N) = \frac{1}{2}\{m_i - 2a_i + 2(\sum_{u=1}^{p} \delta_{ik_u} - \sum_{v=1}^{q} \delta_{il_v})\}\psi(a_1, \cdots, a_N) \\ \qquad\qquad\qquad\qquad\qquad\qquad\qquad\qquad\qquad\qquad (i = 1, \cdots, N), \\ Y_b \psi(a_1, \cdots, a_N) = -\sigma(Y_b)\psi(a_1, \cdots, a_N) \qquad\qquad (b = 1, \cdots, l). \\ \psi(a_1, \cdots, a_N)(\gamma s) = \psi(a_1, \cdots, a_N)(s) \qquad\qquad \text{for all } \gamma \in \Gamma,\, s \in G, \end{cases}$$

where we put $\psi(a_1, \cdots, a_N) = 0$ if $a_i = -1$ or $a_i = m_i + 1$ for some a_i.

Conversely, given a system $\{\psi(a_1, \cdots, a_N)\}$ of S-valued functions on G satisfying the above conditions, there exists a harmonic form of type $(k_1, \cdots, k_p; l_1, \cdots, l_q)$ such that the corresponding system of the S-valued functions on G is equal to the given $\{\psi(a_1, \cdots, a_N)\}$.

Now for any element $w = \sum_{t=1}^{n} \alpha_t u_t (\alpha_t \in C)$ of S, put $\bar{w} = \sum_{t=1}^{n} \bar{\alpha}_t u_t$. For the system $\{\psi(a_1, \cdots, a_N)\}$ corresponding to a harmonic form of type $(k_1, \cdots, k_p; l_1, \cdots, l_q)$, put

$$(2.7) \qquad\qquad \varphi(a_1, \cdots, a_N) = \overline{\psi(m_1 - a_1, \cdots, m_N - a_N)}.$$

Then taking the conjugate complex of the above equations, we see that the system $\{\varphi(a_1, \cdots, a_N)\}$ defines a harmonic form belonging to $A^{p,q}(\Gamma, X, \rho_{m_1 \cdots m_N \sigma^*})$ of type $(l_1, \cdots, l_q; k_1, \cdots, k_p)$, where σ^* denotes the representation of G_{N+1} contragredient to σ. Therefore we get the isomorphism:

$$(2.8) \quad \mathfrak{h}(k_1, \cdots, k_p; l_1, \cdots, l_q; \rho_{m_1 \cdots m_N \sigma}) \cong \mathfrak{h}(l_1, \cdots, l_q; k_1, \cdots, k_p; \rho_{m_1 \cdots m_N \sigma^*}).$$

3. A vanishing theorem for automorphic forms

Throughout the following sections of the paper, we assume that Γ satisfies the conditions (C_1) and (C_2).

As before, let σ be a (not necessarily irreducible) representation of the compact factor G_{N+1} in a complex vector space S. Let n_i for $i = 1, \cdots,$ N be integers. For every $s \in G$ and $z \in X$, put

$$J_{n_1 \cdots n_N \sigma}(s, z) = \prod_{i=1}^{N} (c_i z_i + d_i)^{n_i} \sigma(s_{N+1}) \, ,$$

where $s = (s_1, \cdots, s_N, s_{N+1})$, $s_i = \begin{pmatrix} a_i & b_i \\ c_i & d_i \end{pmatrix}$, $z = (z_1, \cdots, z_N)$. The function $J_{n_1 \cdots n_N \sigma}$ on $G \times X$ is a $\mathrm{GL}(n, C)$-valued automorphic factor in the sense of [8].

An S-valued C^∞-function f on X is called an *automorphic form* (with respect to Γ) for the automorphic factor $J_{n_1 \cdots n_N \sigma}$, if

$$f(\gamma z) = J_{n_1 \cdots n_N \sigma}(\gamma, z) f(z) \qquad\qquad \text{for all } \gamma \in \Gamma \, .$$

Note that we do not assume here f is holomorphic. We denote by $\mathcal{F}(\Gamma, n_1, \cdots, n_N, \sigma)$ the set of all automorphic forms for the automorphic factor $J_{n_1 \cdots n_N \sigma}$.

THEOREM 3.1. *Let $f \in \mathcal{F}(\Gamma, n_1, \cdots, n_N, \sigma)$. Suppose that f is holomorphic in the variable z_1. If $n_1 \leq 0$ and if there exists an index k such that $n_k \neq n_1$, then f vanishes identically.*

PROOF. We first suppose that $n_1 = 0$ and put $g(z) = \prod_{i=1}^{N} \mathrm{Im}\,(z_i)^{n_i/2} f(z)$. Since $n_1 = 0$, the function g is also holomorphic in z_1. For any $\gamma \in \Gamma$, we have

$$(3.1) \qquad g(\gamma z) = \prod_{i=1}^{N} (c_i z_i + d_i)^{n_i} \, | \, c_i z_i + d_i \, |^{-n_i} \sigma(\gamma_{N+1}) g(z) \, ,$$

where $\gamma = (\gamma_1, \cdots, \gamma_N, \gamma_{N+1})$ and $\gamma_i = \begin{pmatrix} a_i & b_i \\ c_i & d_i \end{pmatrix} (i = 1, \cdots, N)$. Take a basis $\{u_1, \cdots, u_n\}$ of S and put $\sigma(s_{N+1}) u_t = \sum_{r=1}^{N} \sigma_{rt}(s_{N+1}) u_r$ with $\sigma_{rt}(s_{N+1}) \in C$. We may assume that the matrix $(\sigma_{rt}(s_{N+1}))$ is unitary for every $s_{N+1} \in G_{N+1}$. For an element $w = \sum_{t=1}^{n} \alpha_t u_t (\alpha_t \in C)$ of S, put $\| w \|^2 = \sum_{t=1}^{n} | \alpha_t |^2$. We see from (3.1) that $\| g(\gamma z) \| = \| g(z) \|$ for all $\gamma \in \Gamma$ and $z \in X$. Since the quotient space $\Gamma \backslash X$ is compact, the function $\| g(z) \|$ attains its maximum at a point $z^0 = (z_1^0, \cdots, z_N^0)$. Put

$$\gamma z^0 = (z_1^\gamma, \cdots, z_N^\gamma), \quad g^\gamma(z_1) = g(z_1, z_2^\gamma, \cdots, z_N^\gamma) \, .$$

Then $g^\gamma(z_1)$ is an S-valued holomorphic function defined on the upper half-plane and $\| g^\gamma(z_1) \| \leq \| g^\gamma(z_1^\gamma) \|$. Therefore, by the maximum principle, g^γ is a constant, and this implies that

$$(3.2) \qquad\qquad \frac{\partial g}{\partial z_1}(z_1, z_2^\gamma, \cdots, z_N^\gamma) = 0 \, .$$

Since the discrete subgroup Γ satisfies the conditions (C_1) and (C_2), the set of points $\{(z_2^\gamma, \cdots, z_n^\gamma) \mid \gamma \in \Gamma\}$ is a dense subset of the product of $N - 1$ upper half-planes $H_2 \times \cdots \times H_N$ (cf. Proposition 1.1). Therefore (3.2) implies that $\dfrac{\partial g}{\partial z_1}(z) = 0$ for all $z \in X$. Since g is holomorphic in z_1, this shows that g does not contain the variable z_1. It follows from the definition of g that the function f also does not contain the variable z_1. Remarking that $J_{n_1 \cdots n_N \sigma}(s, z)$ does not contain the variables s_1 and z_1, we have

$$(3.3) \quad f(\gamma_2 z_2, \cdots, \gamma_N z_N) = J_{n_1 \cdots n_N \sigma}(\gamma_2, \cdots, \gamma_{N+1}, z_2, \cdots, z_N) f(z_2, \cdots, z_N)$$

for all $\gamma = (\gamma_1, \cdots, \gamma_N, \gamma_{N+1}) \in \Gamma$. Let $(s_2, \cdots, s_N)$ be an element of $G_2 \times \cdots \times G_N$. By (C_1), (C_2) and Proposition 1.1, there exists a sequence $\{\gamma^{(j)}\}$ of elements of Γ such that the sequence $\{(\gamma_2^{(j)}, \cdots, \gamma_N^{(j)})\}$ converges to $(s_2, \cdots, s_N)$. Since G_{N+1} is compact, we may assume that the sequence $\{\gamma_{N+1}^{(j)}\}$ converges to an element $s_{N+1} \in G_{N+1}$. Then $\{(\gamma_2^{(j)}, \cdots, \gamma_N^{(j)}, \gamma_{N+1}^{(j)})\}$ converges to $(s_2, \cdots, s_N, s_{N+1})$, and we obtain from (3.3)

$$(3.4) \qquad f(sz_0) = \prod_{k=2}^{N} (c_j i + d_j)^{n_j} \sigma(s_{N+1}) f(z_0) \;,$$

where $s_j = \begin{pmatrix} a_j & b_j \\ c_j & d_j \end{pmatrix} (j = 1, \cdots, N)$. We now show that $\sigma(s_{N+1}) f(z_0) = f(z_0)$. For this purpose, let Γ_{N+1} be the projection of Γ into G_{N+1} and $\bar{\Gamma}_{N+1}$ the closure of Γ_{N+1}. Denote by L_{N+1} the subset of G_{N+1} consisting of all the elements δ satisfying the following condition: there exists a sequence $\{\delta^{(j)}\}$ of elements of Γ such that the sequence $\{(\delta_2^{(j)}, \cdots, \delta_N^{(j)}, \delta_{N+1}^{(j)})\}$ converges to the element $(1, \cdots, 1, \delta)$. It can easily be seen that L_{N+1} is a subgroup of $\bar{\Gamma}_{N+1}$ and $\lambda L_{N+1} \lambda^{-1} = L_{N+1}$ for every $\lambda \in \Gamma_{N+1}$. Therefore the closure $\bar{L}_{N+1}$ of L_{N+1} is a normal subgroup of $\bar{\Gamma}_{N+1}$. Moreover, it follows from the definition of L_{N+1} and from (3.3) that

$$(3.5) \qquad f(z_0) = \sigma(\delta) f(z_0)$$

holds for all $\delta \in L_{N+1}$ and hence for all $\delta \in \bar{L}_{N+1}$. Let $(s_2, \cdots, s_N)$ be an element of $G_2 \times \cdots \times G_N$ and let, as above, $\{\gamma^{(j)}\}$ be a sequence of elements of Γ such that $(\gamma_2^{(j)}, \cdots, \gamma_N^{(j)}, \gamma_{N+1}^{(j)}) \to (s_2, \cdots, s_N, s_{N+1})$. Then s_{N+1} belongs to $\bar{\Gamma}_{N+1}$, and it is easily seen that s_{N+1} is uniquely determined modulo $\bar{L}_{N+1}$ by $(s_2, \cdots, s_N)$. Denoting by $\bar{\delta}(s_2, \cdots, s_N)$ the class of s_{N+1} modulo $\bar{\Gamma}_{N+1}$, we see easily that $\bar{\delta}$ is a homomorphism of $G_2 \times \cdots \times G_N$ into $\bar{\Gamma}_{N+1}/\bar{L}_{N+1}$. Now let T be the linear subspace of S spanned by the elements of the form $\sigma(\lambda) f(z_0)$ for $\lambda \in \Gamma_{N+1}$. Then, for every $\nu \in \bar{\Gamma}_{N+1}$, $\sigma(\nu) f(z_0)$ belongs to T, and hence T is spanned by the elements of the form $\sigma(\nu) f(z_0)$ for $\nu \in \bar{\Gamma}_{N+1}$. Therefore the linear transformations $\sigma(\nu)$ for $\nu \in \bar{\Gamma}_{N+1}$ leave invariant the subspace T of S, and hence σ induces a representation σ_T of $\bar{\Gamma}_{N+1}$ in the vector space T. Let $\delta \in \bar{L}_{N+1}$. For every $\lambda \in \Gamma_{N+1}$, we have

$$\sigma(\delta) \sigma(\lambda) f(z_0) = \sigma(\lambda) \sigma(\lambda^{-1} \delta \lambda) f(z_0) \;.$$

Since $\bar{L}_{N+1}$ is a normal subgroup of $\bar{\Gamma}_{N+1}$, we have $\lambda^{-1}\delta\lambda \in \bar{L}_{N+1}$, so that $\sigma(\lambda^{-1}\delta\lambda)f(z_0) = f(z_0)$ by (3.5). Hence we have $\sigma(\delta)\sigma(\lambda)f(z_0) = \sigma(\lambda)f(z_0)$. This implies that $\sigma_T(\delta) = 1$ for every $\delta \in \bar{L}_{N+1}$. Therefore σ_T defines a representation $\bar{\sigma}_T$ of the factor group $\bar{\Gamma}_{N+1}/\bar{L}_{N+1}$. If we put $\tau(s_2, \cdots, s_N) = \bar{\sigma}_T(\bar{\delta}(s_2, \cdots, s_N))$, then τ is a representation of $G_2 \times \cdots \times G_N$, and we get, by (3.4),

$$f(sz_0) = \prod_{j=2}^{N} (c_j i + d_j)^{n_j} \tau(s_2, \cdots, s_N) f(z_0) \,.$$

Now we show that τ is a differentiable representation of $G_2 \times \cdots \times G_N$. For this purpose, let $\gamma_{N+1}^{(1)}, \cdots, \gamma_{N+1}^{(m)}$ be elements of Γ_{N+1} such that $\{\sigma(\gamma_{N+1}^{(1)})f(z_0), \cdots, \sigma(\gamma_{N+1}^{(m)})f(z_0)\}$ is a basis of T. Let $\gamma^{(1)}, \cdots, \gamma^{(m)}$ be elements of Γ whose G_{N+1}-components are $\gamma_{N+1}^{(1)}, \cdots, \gamma_{N+1}^{(m)}$, respectively. Then we get by (3.4)

$$f(s\gamma^{(p)}z_0) = \prod_{j=2}^{N} (c_j^{(p)} i + d_j^{(p)})^{n_j} \tau(s_2, \cdots, s_N)\sigma(\gamma_{N+1}^{(p)})f(z_0) \,,$$

where

$$s_j\gamma_j^{(p)} = \begin{pmatrix} a_j^{(p)} & b_j^{(p)} \\ c_j^{(p)} & d_j^{(p)} \end{pmatrix} \qquad\qquad (j = 1, \cdots, N) \,.$$

Therefore

$$\tau(s_2, \cdots, s_N)\sigma(\gamma_{N+1}^{(p)})f(z_0) = \prod_{j=2} (c_j^{(p)}i + d_j^{(p)})^{-n_j} f(s\gamma^{(p)}z_0) \,,$$

and the right hand side of this equality is an S-valued differentiable function of $(s_2, \cdots, s_N)$. Since $\{\sigma(\gamma_{N+1}^{(1)})f(z_0), \cdots, \sigma(\gamma_{N+1}^{(m)})f(z_0)\}$ is a basis of T, it follows from this that the components of the matrix which represents the linear transformation $\tau(s_2, \cdots, s_N)$ of T are differentiable; this means that the representation τ is differentiable. Now, since G_{N+1} is compact, every matrix $\tau(s_2, \cdots, s_N)$ is bounded and therefore τ is a differentiable and bounded representation of $G_2 \times \cdots \times G_N$. Since the factors $G_2, \cdots, G_N$ are non-compact and simple, τ must be the trivial representation. Therefore, $\sigma(s_{N+1})f(z_0) = f(z_0)$ as we have said, and (3.4) is now written as

$$f(sz_0) = \prod_{j=2}^{N} (c_j i + d_j)^{n_j} f(z_0)$$

for every $s = (s_1, \cdots, s_N)$. Let

$$s_k = \begin{pmatrix} \cos\theta_k & -\sin\theta_k \\ \sin\theta_k & \cos\theta_k \end{pmatrix}$$

and $s_j = 1$ for $j \neq k$. We get then $f(z_0) = \exp(in_k\theta_k)f(z_0)$ for any θ_k. By the assumption $n_k \neq 0$, we must have $f(z_0) = 0$, which implies $f(sz_0) = 0$. For any $z \in X$, there exists an $s = (s_1, \cdots, s_N, 1)$ such that $z = sz_0$; so we have $f(z) = 0$ for any z. This proves our theorem in the case $n_1 = 0$.

Suppose now that $n_1 < 0$, and put $f^r(z) = f(z) \otimes \cdots \otimes f(z)$. Then f^r

is an $S \otimes \cdots \otimes S$-valued automorphic form for the factor $J_{rn_1 \cdots rn_N \xi}$, where $\xi = \sigma \otimes \cdots \otimes \sigma$. To prove that f vanishes identically, it suffices to prove that f^r vanishes identically. If r is a sufficiently large even integer, there exists a non-zero *holomorphic* automorphic form $h(z_1, \cdots, z_N)$ with respect to Γ for the automorphic factor

$$J_{-rn_1, \cdots, -rn_1}(s, z) = \prod_{j=1}^{N} (c_j z_j + d_j)^{-rn_1}$$

(cf. [5, 10]). Then $h \cdot f^r$ is an automorphic form for the factor $J_{n_1' \cdots n_N' \xi}$, where $n_j' = r(n_j - n_1)(j = 1, \cdots, N)$. Then $n_1' = 0$ and $n_k' \neq 0$; and moreover, $h \cdot f^r$ is holomorphic in z_1. Then as we have already shown, $h \cdot f^r$ vanishes identically, so that $f^r \equiv 0$, since $h \not\equiv 0$. Theorem 3.1 is thereby completely proved.

For every $f \in \mathcal{F}(\Gamma, n_1, \cdots, n_N, \sigma)$, put

$$f^0(s) = J_{n_1 \cdots n_N \sigma}(s, z_0)^{-1} f(s z_0) .$$

Then the S-valued function f^0 on G satisfies the following two conditions:

$$(3.6) \qquad f^0(\gamma s) = f^0(s) \qquad\qquad\qquad \text{for all } \gamma \in \Gamma, s \in G ,$$

$$(3.7) \qquad f^0(sk) = \prod_{j=1}^{N} \exp\left(-in_j\theta_j\right)\sigma(k_{N+1}^{-1})f^0(s) \qquad \text{for all } k \in K, s \in G ,$$

where $k = (k_1, \cdots, k_N, k_{N+1})$ and

$$k_j = \begin{pmatrix} \cos\theta_j & -\sin\theta_j \\ \sin\theta_j & \cos\theta_j \end{pmatrix} \qquad\qquad (j = 1, \cdots, N) .$$

The condition (3.7) is equivalent to the set of the following two conditions:

$$(3.7') \qquad\qquad W_j f^0 = \frac{n_j}{2} f^0 \qquad\qquad\qquad (j = 1, \cdots, N) ,$$

$$(3.7'') \qquad\qquad Y_b f^0 = -\sigma(Y_b) f^0 \qquad\qquad\qquad (b = 1, \cdots, l) .$$

Conversely, given an S-valued function f^0 on G satisfying the conditions (3.6) and (3.7), there exists a function f of $\mathcal{F}(\Gamma, n_1, \cdots, n_N, \sigma)$ such that the S-valued function on G corresponding to f is equal to the given f^0. We denote by $\mathcal{F}^0(\Gamma, n_1, \cdots, n_N, \sigma)$ the set of all S-valued C^∞-functions on G satisfying the conditions (3.6), (3.7') and (3.7''). It can be proved easily that an element f of $\mathcal{F}^0(\Gamma, n_1, \cdots, n_N, \sigma)$ is holomorphic in the variable z_j if and only if $\bar{X}_j f^0 = 0$.

For later use, it is convenient to modify Theorem 3.1 in the following form.

THEOREM 3.2. *Let* $f^0 \in \mathcal{F}^0(\Gamma, n_1, \cdots, n_N, \sigma)$. *Suppose that* $\bar{X}_j f^0 = 0$, $n_j \leq 0$ *for an index* j, *and that there exists an index* k *such that* $n_j \neq n_k$. *Then* f^0 *vanishes identically.*

4. The dimension of the space $\mathfrak{h}(k_1, \cdots, k_p; l_1, \cdots, l_q; \rho_{m_1 \cdots m_N \sigma})$

Now we come back to the study of harmonic forms ω of type $(k_1, \cdots, k_p;$ $l_1, \cdots, l_q)$, and put $\psi = \omega_{k_1 \cdots k_p \bar{l}_1 \cdots \bar{l}_q}$. We retain the notation introduced in the preceding sections.

PROPOSITION 4.1. *Let d be the dimension of the vector space consisting of all $u \in S$ such that $\sigma(\gamma_{N+1})u = u$ for every $\gamma = (\gamma_1, \cdots, \gamma_N, \gamma_{N+1})$ of Γ. Then*

$$\dim_C \mathfrak{h}(k_1, \cdots, k_p; k_1, \cdots, k_p; \rho_{m_1 \cdots m_N \sigma}) = \begin{cases} 0 & \text{if } (m_1, \cdots, m_N) \neq (0, \cdots, 0), \\ d & \text{if } (m_1, \cdots, m_N) = (0, \cdots, 0). \end{cases}$$

PROOF. Let $\omega \in \mathfrak{h}(k_1, \cdots, k_p; k_1, \cdots, k_p; \rho_{m_1 \cdots m_N \sigma})$. To simplify the notation, we assume that $(k_1, \cdots, k_p) = (1, \cdots, p)$. It follows from (2.6) that $\psi(0, \cdots, 0, m_{p+1}, \cdots, m_N) \in \mathcal{F}^0(\Gamma, n_1, \cdots, n_N, \sigma)$, where $n_i = m_i (i = 1, \cdots, p)$ and $n_j = -m_j (j = p+1, \cdots, N)$. Suppose that $p < N$. Then by (2.3), $\bar{X}_j \psi(0, \cdots, 0, m_{p+1}, \cdots, m_N) = 0$ for each $j > p$. Suppose that $(m_1, \cdots, m_N) \neq (0, \cdots, 0)$. Then, by Theorem 3.2, we get $\psi(0, \cdots, 0, m_{p+1}, \cdots, m_N) = 0$. By (2.2), we obtain inductively $\psi(0, \cdots, 0, a_{p+1}, \cdots, a_N) = 0$ for any $(a_{p+1}, \cdots, a_N)$. Then by (2.4), we get $\psi(a_1, \cdots, a_N) = 0$ for any $(a_1, \cdots, a_N)$, so that $\psi = 0$, and hence $\omega = 0$. Now let $m_1 = \cdots = m_N = 0$. Then $\psi = \psi(0, \cdots, 0)$, and by (2.2)–(2.6) we have $X_j \psi = \bar{X}_j \psi = W_j \psi = 0$ for $j = 1, \cdots, N$. This shows that $\psi(s) = \psi(s_1, \cdots, s_N, s_{N+1})$ is independent of the variables $s_1, \cdots, s_N$. Therefore, we have $\psi(s_1, \cdots, s_N, s_{N+1}) = \psi(1, \cdots, 1, s_{N+1})$, and by (2.6), $\psi(1, \cdots, 1, s_{N+1}) = \sigma(s_{N+1}^{-1})\psi(1, \cdots, 1, 1)$. Thus we get $\psi(s) = \sigma(s_{N+1}^{-1})\psi(1)$ for all $s = (s_1, \cdots, s_{N+1}) \in G$. On the other hand, $\psi(\gamma \cdot 1) = \psi(1)$ by (2.6) and hence $\sigma(\gamma_{N+1}^{-1})\psi(1) = \psi(1)$ for all $\gamma = (\gamma_1, \cdots, \gamma_{N+1})$ of Γ. Conversely, let u be an element of S such that $\sigma(\gamma_{N+1})u = u$ for every $\gamma \in \Gamma$. Put $\psi(s) = \sigma(s_{N+1}^{-1})u$ for all $s = (s_1, \cdots, s_{N+1}) \in G$. Then it is easily verified that the S-valued function ψ defined on G satisfies the equations (2.2)–(2.6) with $m_1 = \cdots = m_N = 0$. We get in this way an isomorphism of $\mathfrak{h}(k_1, \cdots, k_p; k_1, \cdots, k_p; \rho_{0 \cdots 0 \sigma})$ onto the vector space consisting of all $u \in S$ such that $\sigma(\gamma_{N+1})u = u$ for every $\gamma \in \Gamma$.

Consider now the case $p = N$. In this case, we have

$$\dim \mathfrak{h}(1, \cdots, N; 1, \cdots, N; \rho_{m_1 \cdots m_N \sigma}) = \dim H^{2N}(\Gamma, X, \rho_{m_1 \cdots m_N \sigma}).$$

Further if we denote by σ^* the representation of G_{N+1} contragredient to σ, we have $\dim H^{2N}(\Gamma, X, \rho_{m_1 \cdots m_N \sigma}) = \dim H^0(\Gamma, X, \rho_{m_1 \cdots m_N \sigma^*})$. By a result of [8, Part I, § 6], $\dim H^0(\Gamma, X, \rho_{m_1 \cdots m_N \sigma^*}) = 0$ if $(m_1, \cdots, m_N) \neq (0, \cdots, 0)$; and if $(m_1, \cdots, m_N) = (0, \cdots, 0)$, $\dim H^0(\Gamma, X, \rho_{m_1 \cdots m_N \sigma^*})$ is equal to the maximal number d^* of linearly independent elements $u^* \in S^*$ such that

$\sigma^*(\gamma_{N+1})u^* = u^*$, where S^* denotes the dual space of S. It is clear that $d^* = d$. Thus Proposition 4.1 is completely proved.

Hereafter, we assume that $\{k_1, \cdots, k_p\} \neq \{l_1, \cdots, l_q\}$.

Now divide the set of indices $\{1, \cdots, N\}$ into the sum of the disjoint subsets I_1, I_2, I_3, I_4 defined by

$$I_1 = \{k_1, \cdots, k_p\} \cap \{l_1, \cdots, l_q\}\,,$$
$$I_2 = \{k_1, \cdots, k_p\} - I_1\,,$$
$$I_3 = \{l_1, \cdots, l_q\} - I_1\,,$$
$$I_4 = \{1, \cdots, N\} - I_1 \cup I_2 \cup I_3\,.$$

By the assumption $\{k_1, \cdots, k_p\} \neq \{l_1, \cdots, l_q\}$, we have either $I_2 \neq \varnothing$ or $I_3 \neq \varnothing$.

PROPOSITION 4.2. *If* $\{k_1, \cdots, k_p\} \neq \{l_1, \cdots, l_q\}$ *and* $I_4 \neq \varnothing$, *then* $\mathfrak{h}(k_1, \cdots, k_p; l_1, \cdots, l_q; \rho_{m_1 \cdots m_N \sigma}) = (0)$.

PROOF. Suppose first that $I_2 \neq \varnothing$. Take the $\psi(b_1, \cdots, b_N)$ with the index $(b_1, \cdots, b_N)$ such that $b_j = m_j$ for $j \in I_3 \cup I_4$ and $b_i = 0$ for $i \in I_1 \cup I_2$. Then by (2.6), $\psi(b_1, \cdots, b_N) \in \mathcal{F}^0(\Gamma, n_1, \cdots, n_N, \sigma)$, where $n_i = m_i$ for $i \in I_1$, $n_i = m_i + 2$ for $i \in I_2$, $n_i = -m_i - 2$ for $i \in I_3$, $n_i = -m_i$ for $i \in I_4$. Now, by (2.3) we have $\bar{X}_j\psi(b_1, \cdots, b_N) = 0$ for every $j \in I_4$. Since $n_j \leq 0$ for $j \in I_4$ and $n_k > 0$ for $k \in I_2$, and $I_4 \neq \varnothing$, $I_2 \neq \varnothing$, we get $\psi(b_1, \cdots, b_N) = 0$ by Theorem 3.2. By (2.2) we obtain inductively $\psi(a_1, \cdots, a_N) = 0$ for every $(a_1, \cdots, a_N)$ such that $a_i = 0$ for $i \in I_1 \cup I_2$. Then by (2.4) we get inductively $\psi(a_1, \cdots, a_N) = 0$ for any $(a_1, \cdots, a_N)$, so that $\psi = 0$. Now let $I_2 = \varnothing$; then $I_3 \neq \varnothing$. The system of S-valued functions $\{\varphi(a_1, \cdots, a_N)\}$ defined by (2.7) corresponds to a harmonic form of type $(l_1, \cdots, l_q; k_1, \cdots, k_p)$; and, for the index $(l_1, \cdots, l_q; k_1, \cdots, k_p)$ we have $I_2 \neq \varnothing$. Therefore, as we have already proved, $\varphi = 0$ and hence $\psi = 0$. This completes our proof.

PROPOSITION 4.3. *Suppose that* $\{k_1, \cdots, k_p\} \neq \{l_1, \cdots, l_q\}$, $I_1 \neq \varnothing$ *and* $I_4 = \varnothing$. *Then* $\mathfrak{h}(k_1, \cdots, k_p; l_1, \cdots, l_q; \rho_{m_1 \cdots m_N \sigma}) = (0)$.

PROOF. As in the proof of Proposition 4.2, we may assume $I_2 \neq \varnothing$. Take the $\psi(b_1, \cdots, b_N)$ with the index $(b_1, \cdots, b_N)$ such that $b_i = 0$ for every $i \in I_2$ and $b_i = m_i$ for every $i \in I_1 \cup I_3$. Here note that $I_4 = \varnothing$. Then $\psi(b_1, \cdots, b_N) \in \mathcal{F}^0(\Gamma, n_1, \cdots, n_N, \sigma)$, with $n_i = -m_i$ for $i \in I_1$, $n_i = m_i + 2$ for $i \in I_2$, $n_i = -m_i - 2$ for $i \in I_3$. By (2.4), we have $\bar{X}_j\psi(b_1, \cdots, b_N) = 0$ for every $j \in I_1$. Since $n_j \leq 0$ for every $j \in I_1$ and $n_k > 0$ for every $k \in I_2$, and since $I_1 \neq \varnothing$ and $I_2 \neq \varnothing$, we get $\psi(b_1, \cdots, b_N) = 0$ by Theorem 3.2. Then by (2.5), we obtain inductively $\psi(a_1, \cdots, a_N) = 0$ for every $(a_1, \cdots, a_N)$ such that $a_i = 0$ for $i \in I_2$. Then by (2.4) we get inductively

$\psi(a_1, \cdots, a_N) = 0$ for any $(a_1, \cdots, a_N)$. This proves the proposition.

It remains only to treat the case where $(k_1, \cdots, k_p, l_1, \cdots, l_q)$ $(k_1 < \cdots < k_p, l_1 < \cdots < l_q)$ is a permutation of $(1, \cdots, N)$. We consider first the case $p = N$, that is the case $(k_1, \cdots, k_p) = (1, \cdots, N)$. We have clearly $\mathfrak{h}(1, \cdots, N; \rho_{m_1 \cdots m_N \sigma}) = \mathfrak{h}^{N,0}(\rho_{m_1 \cdots m_N \sigma})$.

PROPOSITION 4.4.[1]   *The vector space $\mathfrak{h}^{N,0}(\rho_{m_1 \cdots m_N \sigma})$ is canonically isomorphic to the complex vector space of all holomorphic automorphic forms on X with respect to Γ for the automorphic factor $J_{m_1+2, \cdots, m_N+2, \sigma}$.*

PROOF. It follows from (2.6) that $\psi(0, \cdots, 0) \in \mathcal{F}^0(\Gamma, m_1 + 2, \cdots, m_N + 2, \sigma)$. By (2.3) and (2.4), we get $\bar{X}_j \psi(0, \cdots, 0) = 0$ for every j, so that $\psi(0, \cdots, 0)$ corresponds to a holomorphic automorphic form for the automorphic factor $J_{m_1+2, \cdots, m_N+2, \sigma}$. Now we show that $\psi(a_1, \cdots, a_N) = 0$ if $(a_1, \cdots, a_N) \neq (0, \cdots, 0)$. In fact, by (2.3) and (2.4), we get $\bar{X}_j \psi(a_1, \cdots, a_N) = -(1/2)(a_j + 1)\psi(a_1, \cdots, a_j + 1, \cdots, a_N)$ and $\bar{X}_j \psi(a_1, \cdots, a_N) = (1/2)(a_j + 1)\psi(a_1, \cdots, a_j + 1, \cdots, a_N)$, and hence $\psi(a_1, \cdots, a_j + 1, \cdots, a_N) = 0$; this implies that $\psi(b_1, \cdots, b_N) = 0$ if one of the b_j is $\neq 0$. Conversely, let f be an S-valued holomorphic automorphic form for $J_{m_1+2, \cdots, m_N+2, \sigma}$ and let f^0 be the corresponding S-valued function on G. Put $\psi(0, \cdots, 0) = f^0$ and $\psi(a_1, \cdots, a_N) = 0$ for $(a_1, \cdots, a_N) \neq (0, \cdots, 0)$. Then it is easily verified that the system of functions $\{\psi(a_1, \cdots, a_N)\}$ thus defined corresponds to a harmonic form of type $(1, \cdots, N)$. Proposition 4.4 is thus proved.

Let $p < N$ and let $(k_1, \cdots, k_p, l_1, \cdots, l_q)$ be a permutation of $(1, \cdots, N)$. We denote by $\bar{H}_j$ the complex manifold with the same underlying differentiable manifold as the upper half plane H_j and with the conjugate complex structure. Put

$$X_{k_1 \cdots k_p \bar{l}_1 \cdots \bar{l}_q} = H_{k_1} \times \cdots \times H_{k_p} \times \bar{H}_{l_1} \times \cdots \times \bar{H}_{l_q} .$$

Let ω be a harmonic form of type $(k_1, \cdots, k_p; l_1, \cdots, l_q)$. Then we may consider ω as an element of $A^{N,0}(\Gamma, X_{k_1 \cdots k_q \bar{l}_1 \cdots \bar{l}_q}, \rho_{m_1 \cdots m_N \sigma})$. Now the notion of the harmonic form depends only on the riemannian manifold structure of the underlying differentiable manifold and of the representation ρ (cf. [8]). Since X and $X_{k_1 \cdots k_p \bar{l}_1 \cdots \bar{l}_q}$ are the same as riemannian manifolds, it follows that ω is a harmonic form of type $(1, \cdots, N)$ belonging to $A^{N,0}(\Gamma, X_{k_1 \cdots k_p \bar{l}_1 \cdots \bar{l}_q}, \rho_{m_1 \cdots m_N \sigma})$. By Proposition 4.4, ω is written in the form

$$\omega = (e_{0 \cdots 0} \otimes f)dz_{k_1} \wedge \cdots \wedge dz_{k_p} \wedge d\bar{z}_{l_1} \wedge \cdots \wedge d\bar{z}_{l_q} ,$$

where $f = f(z_{k_1}, \cdots, z_{k_p}, \bar{z}_{l_1}, \cdots, \bar{z}_{l_q})$ is holomorphic in the variables $z_{k_1}, \cdots, z_{k_p}, \bar{z}_{l_1}, \cdots, \bar{z}_{l_q}$, i.e., $\partial f/\partial \bar{z}_{k_u} = \partial f/\partial z_{l_v} = 0$ $(u = 1, \cdots, p, v = 1, \cdots, q)$,

---

[1] cf. [3], [8 Part II, §6], [11], [12].

and

$$(4.1) \quad f(\gamma z, \overline{\gamma z}) = \prod_{u=1}^{p} (c_{k_u} z_{k_u} + d_{k_u})^{m_{k_u}+2} \prod_{v=1}^{q} (c_{l_v} \overline{z}_{l_v} + d_{l_v})^{m_{l_v}+2} \sigma(\gamma_{N+1}) f(z, \overline{z})$$

for every $\gamma \in \Gamma$, where $\gamma = (\gamma_1, \cdots, \gamma_N, \gamma_{N+1})$ and

$$\gamma_i = \begin{pmatrix} a_i & b_i \\ c_i & d_i \end{pmatrix} \qquad (i = 1, \cdots, N) .$$

Thus we get the following

PROPOSITION 4.5. *Let* $(k_1, \cdots, k_p, l_1, \cdots, l_q)$ *be a permutation of* $(1, \cdots, N)$. *Then the vector space* $\mathfrak{h}(k_1, \cdots, k_p; l_1, \cdots, l_q; \rho_{m_1 \cdots m_N \sigma})$ *is canonically isomorphic to the vector space of all S-valued functions f on X satisfying (4.1) and* $(\partial f/\partial \overline{z}_{k_u}) = (\partial f/\partial z_{l_v}) = 0$ *for* $u = 1, \cdots, p, v = 1, \cdots, q$.

We denote by $h^{p,q}(\Gamma, X, \rho_{m_1 \cdots m_N \sigma})$ the dimension of the vector space of all harmonic forms of type (p, q) (with the representation $\rho_{m_1 \cdots m_N \sigma}$). Summing up the above considerations, we get the following result: Let d be as in Proposition 4.1. Then

$$h^{p,q}(\Gamma, X, \rho_{m_1 \cdots m_N \sigma}) = 0 \qquad \text{if } p \neq q \text{ and } p + q \neq N ,$$

$$h^{p,p}(\Gamma, X, \rho_{m_1 \cdots m_N \sigma}) = \begin{cases} 0 & \text{if } 2p \neq N \text{ and } (m_1, \cdots, m_N) \neq (0, \cdots, 0) , \\ \binom{N}{p} d & \text{if } 2p \neq N \text{ and } (m_1, \cdots, m_N) = (0, \cdots, 0) , \end{cases}$$

$$h^{N-q,q}(\Gamma, X, \rho_{m_1 \cdots m_N \sigma})$$
$$= \begin{cases} \sum_{(k_1 \cdots k_{N-q} l_1 \cdots l_q)} h^{N,0}(\Gamma, X_{k_1 \cdots k_{N-q} \overline{l}_1 \cdots \overline{l}_q}, \rho_{m_1 \cdots m_N \sigma}) \\ \qquad\qquad\qquad\qquad \text{if } (m_1, \cdots, m_N) \neq (0, \cdots, 0) , \\ \binom{N}{q} \delta_{N-q,q} d + \sum_{(k_1 \cdots k_{N-q} l_1 \cdots l_q)} h^{N,0}(\Gamma, X_{k_1 \cdots k_{N-q} \overline{l}_1 \cdots \overline{l}_q}, \rho_{m_1 \cdots m_N \sigma}) \\ \qquad\qquad\qquad\qquad \text{if } (m_1, \cdots, m_N) = (0, \cdots, 0) , \end{cases}$$

where the summations are taken over all indices $(k_1, \cdots, k_{N-q}, l_1, \cdots, l_q)$ such that $k_1 < \cdots < k_{N-q}, l_1 < \cdots < l_q$ and $(k_1, \cdots, k_{N-q}, l_1, \cdots, l_q)$ is a permutation of $(1, \cdots, N)$, q being a fixed integer $(0 \leq q \leq N)$. We shall prove in § 6 that $h^{N,0}(\Gamma, X_{k_1 \cdots k_{N-q} \overline{l}_1 \cdots \overline{l}_q}, \rho_{m_1 \cdots m_N \sigma}) = h^{N,0}(\Gamma, X, \rho_{m_1 \cdots m_N \sigma})$ for all such $(k_1, \cdots, k_{N-q}, l_1, \cdots, l_q)$.

5. Complex vector bundle $E(J_{n_1 \cdots n_N \sigma})$ and a vanishing theorem

Let G_0, Z, Γ_0 be as in § 1.B. In §§ 5–7, we assume that Γ satisfies (C_1), (C_2) and (C_3). Therefore, $\Gamma \backslash X$ is a compact complex manifold without singularity. Further we consider the following conditions on $\rho_{m_1 \cdots m_N \sigma}$ and

$J_{n_1\cdots n_N\sigma}$:

(D$_1$) $\rho_{m_1\cdots m_N\sigma}(\gamma) = 1$ *for all* $\gamma \in \Gamma_0 \cap Z$;

(D$_2$) $J_{n_1\cdots n_N\sigma}(\gamma, z) = 1$ *for all* $\gamma \in \Gamma_0 \cap Z$ *and all* $z \in X$.

It is clear that, if $n_i = m_i + 2$ for every i, (D$_1$) and (D$_2$) are equivalent. For instance, if all the m_i are even and σ is trivial, (D$_1$) is satisfied. Hereafter, we treat only the representations $\rho_{m_1\cdots m_N\sigma}$ satisfying (D$_1$) and the automorphic factors $J_{n_1\cdots n_N\sigma}$ satisfying (D$_2$).

Now let us recall the definition of the vector bundle $E(J_{n_1\cdots n_N\sigma})$ given in [8]. This is a complex vector bundle over $\Gamma\backslash X$ with typical fibre S, and defined as follows. Suppose that $J_{n_1\cdots n_N\sigma}$ satisfies (D$_2$). To simplify the notation, we write J in place of $J_{n_1\cdots n_N\sigma}$. We let Γ operate on $X \times S$ by the rule

$$\gamma(z, u) = \big(\gamma z, \, J(\gamma, z)u\big)$$

for every $\gamma \in \Gamma$, $(z, u) \in X \times S$, and denote by $E(J)$ the quotient of $X \times S$ by the operation of Γ. Let $[z, u]$ be the image in $E(J)$ of the element $(z, u) \in X \times S$, and ϖ_0 be the canonical projection of X onto $\Gamma\backslash X$. We can define the projection of $E(J)$ onto $\Gamma\backslash X$ by $[z, u] \to \varpi_0(z)$. Then $E(J)$ has a structure of the complex vector bundle over the complex manifold $\Gamma\backslash X$ with typical fibre S. We denote by $E(J)$ the sheaf of germs of holomorphic sections of $E(J)$.

The aim of this section is to prove

THEOREM 5.1.[2] *Suppose that $n_i \leq 0$ for $i = 1, \cdots, N$. Then the cohomology group $H^q(\Gamma\backslash X, E(J_{n_1\cdots n_N\sigma}))$ vanishes for $0 < q < N$. Moreover, let d be the dimension of the complex vector space of all the elements u of S such that $\sigma(\gamma_{N+1})u = u$ for all $\gamma = (\gamma_1, \cdots, \gamma_{N+1}) \in \Gamma$. Then*

$$\dim H^0(\Gamma\backslash X, E(J_{n_1\cdots n_N\sigma})) = \begin{cases} 0 & \text{if } (n_1, \cdots, n_N) \neq (0, \cdots, 0), \\ d & \text{if } (n_1, \cdots, n_N) = (0, \cdots, 0). \end{cases}$$

Let $C^q(\Gamma, X, J)$ be the complex vector space of all S-valued forms η of type $(0, q)$ defined on X such that

$$\eta \circ L_\gamma = J(\gamma, z)\eta$$

for every $\gamma \in \Gamma$ and $z \in X$. Since $J(\gamma, z)$ is holomorphic in z, we see that, if $\eta \in C^q(\Gamma, X, J)$, then $d''\eta \in C^{q+1}(\Gamma, X, J)$. Therefore the operator d'' is a coboundary operator of the graded module $C = \sum_{q=0}^{N} C^q(\Gamma, X, J)$, and, by a theorem of Dolbeault (cf. [4]) the cohomology group $H^q(\Gamma\backslash X, E(J))$ is isomorphic to the cohomology group $H^q(C)$ of the complex C. In particular, the holomorphic automorphic forms for the factor J can be

---

[2] If $n_i < 0$ for $i = 1, \cdots, N$, and if σ is the trivial representation, this assertion is a direct consequence of the vanishing theorem of Kodaira [6].

identified with the elements of $H^0(\Gamma\backslash X, E(J))$. As in the case of auto-morphic forms, we associate to each form η of $C^q(\Gamma, X, J)$ an S-valued q-form η^0 on G defined by

$$\eta_s^0 = J(s, z_0)^{-1}(\eta\circ\pi)_s$$

for all $s \in G$, π being the projection of G onto X. The S-valued forms η^0 on G thus defined are characterized among the S-valued forms on G by the following conditions (cf. § 3 and [8]).

$$(5.1) \qquad \eta^0\circ L_\gamma = \eta^0 \qquad\qquad \text{for every } \gamma \in \Gamma ,$$

$$(5.2) \qquad \theta(W_j)\eta^0 = \frac{1}{2}n_j\eta^0 \qquad\qquad (j = 1, \cdots, N) ,$$

$$(5.3) \qquad \theta(Y_b)\eta^0 = -\sigma(Y_b)\eta^0 \qquad\qquad (b = 1, \cdots, l) ,$$

$$(5.4) \qquad i(W)\eta^0 = 0 \qquad\qquad \text{for every } W \in \mathfrak{k} ,$$

$$(5.5) \qquad i(X_i)\eta^0 = 0 \qquad\qquad (i = 1, \cdots, N) .$$

From (5.4) and (5.5), we see that η^0 is completely determined by the system $\{\eta_{\bar{j}_1\cdots\bar{j}_q}\}$ of the S-valued functions on G defined by

$$\eta_{\bar{j}_1\cdots\bar{j}_q} = \eta^0(\bar{X}_{j_1}, \cdots, \bar{X}_{j_q}) \qquad\qquad (j_1, \cdots, j_q = 1, \cdots, N) .$$

In terms of $\{\eta_{\bar{j}_1\cdots\bar{j}_q}\}$ the conditions (5.1)–(5.3) are written as follows:

$$(5.1') \qquad \eta_{\bar{j}_1\cdots\bar{j}_q}(\gamma s) = \eta_{\bar{j}_1\cdots\bar{j}_q}(s) \qquad\qquad \text{for every } \gamma \in \Gamma \text{ and } s \in G ,$$

$$(5.2') \qquad W_k\eta_{\bar{j}_1\cdots\bar{j}_q} = \frac{1}{2}\Big(n_k - 2\sum_{t=1}^q \delta_{kj_t}\Big)\eta_{\bar{j}_1\cdots\bar{j}_q} ,$$

$$(5.3') \qquad Y_b\eta_{\bar{j}_1\cdots\bar{j}_q} = -\sigma(Y_b)\eta_{\bar{j}_1\cdots\bar{j}_q} .$$

The system of functions $(d''\eta)_{\bar{j}_1\cdots\bar{j}_q}$ corresponding to $d''\eta$ is given by the following formula [8]:

$$(5.6) \qquad (d''\eta)_{\bar{j}_1\cdots\bar{j}_{q+1}} = \sum_{t=1}^{q+1}(-1)^{t-1}X_{j_t}\eta_{\bar{j}_1\cdots\hat{\bar{j}}_t\cdots\bar{j}_{q+1}} .$$

Now, choosing a suitable positive definite hermitian inner product on $C^q(\Gamma, X, J)$ for $q = 0, 1, \cdots, N$, we can define the adjoint operator ∂'' of d'' and the laplacian $\square'' = d''\partial'' + \partial''d''$ (cf. [8]; for the general theory, see [4]). We say that η is harmonic if $\square''\eta = 0$. We know that each cohomology class of $H^q(C)$ is represented by a unique harmonic form. In terms of $\{\eta_{\bar{j}_1\cdots\bar{j}_q}\}$, the forms $\partial''\eta$ and $\square''\eta$ are given by the following formulas:

$$(5.7) \qquad (\partial''\eta)_{\bar{j}_1\cdots\bar{j}_{q-1}} = -\sum_{k=1}^N X_k\eta_{\bar{k}\bar{j}_1\cdots\bar{j}_{q-1}} ,$$

$$(5.8) \qquad (\square''\eta)_{\bar{j}_1\cdots\bar{j}_q} = -\sum_{k=1}^N X_k\bar{X}_k\eta_{\bar{j}_1\cdots\bar{j}_q}$$
$$+ \sum_{k=1}^N \sum_{t=1}^q (-1)^t[\bar{X}_{j_t}, X_k]\eta_{\bar{k}\bar{j}_1\cdots\hat{\bar{j}}_t\cdots\bar{j}_q} .$$

Since $[\bar{X}_{j_t}, X_k] = -(1/2)\delta_{j_t k} W_{j_t}$, we have

$$(5.8') \qquad (\square''\eta)_{\bar{j}_1\cdots\bar{j}_q} = \Big(-\sum_{k=1}^{N} X_k \bar{X}_k + \frac{1}{2}\sum_{t=1}^{q} W_j\Big)\eta_{\bar{j}_1\cdots\bar{j}_q} .$$

We say that a form η is of type $(l_1, \cdots, l_q)(l_1 < \cdots < l_q)$ if $\eta_{\bar{j}_1\cdots\bar{j}_q} = 0$ for every $(j_1, \cdots, j_q) \neq (l_1, \cdots, l_q)(j_1 < \cdots < j_q)$. It follows from $(5.8')$ that each harmonic q-form η decomposes uniquely into the sum of harmonic forms of type $(l_1, \cdots, l_q)$. We denote by $\mathfrak{h}^q$ (resp. $\mathfrak{h}(l_1, \cdots, l_q)$) the vector space of all harmonic q-forms (resp. harmonic forms of type $(l_1, \cdots, l_q)$). Then $\mathfrak{h}^q = \sum_{l_1<\cdots<l_q} \mathfrak{h}(l_1, \cdots, l_q)$.

Now let η be a q-form of type $(l_1, \cdots, l_q)$. Since η is harmonic if and only if $d''\eta = \partial''\eta = 0$, it follows from (5.6) and (5.7) that the following two conditions are necessary and sufficient for η be harmonic:

$$(5.9) \qquad \bar{X}_j \eta_{\bar{l}_1\cdots\bar{l}_q} = 0 \qquad\qquad (j \neq l_t, t = 1, \cdots, q) ,$$

$$(5.10) \qquad X_{l_t} \eta_{\bar{l}_1\cdots\bar{l}_q} = 0 \qquad\qquad (t = 1, \cdots, q) .$$

For each $(j_1, \cdots, j_q) \neq (l_1, \cdots, l_q)$, we have $(\square''\eta)_{\bar{j}_1\cdots\bar{j}_q} = 0$, and from $(5.2')$ and $(5.8')$ we get

$$(\square''\eta)_{\bar{l}_1\cdots\bar{l}_q} = -\sum_{k=1}^{N} X_k \bar{X}_k \eta_{\bar{l}_1\cdots\bar{l}_q} + \frac{1}{4}\Big(\sum_{t=1}^{q} n_{l_t} - 2q\Big)\eta_{\bar{l}_1\cdots\bar{l}_q} .$$

Since $-X_k \bar{X}_k = -\bar{X}_k X_k - [X_k, \bar{X}_k] = -\bar{X}_k X_k - (1/2) W_k$, we obtain from $(5.2')$

$$-\sum_{k=1}^{N} X_k \bar{X}_k \eta_{\bar{l}_1\cdots\bar{l}_q} = -\sum_{k=1}^{N} \bar{X}_k X_k \eta_{\bar{l}_1\cdots\bar{l}_q} - \frac{1}{4}\Big(\sum_{k=1}^{N} n_k - 2q\Big)\eta_{\bar{l}_1\cdots\bar{l}_q} .$$

Therefore we have

$$(\square''\eta)_{\bar{l}_1\cdots\bar{l}_q} = -\sum_{k=1}^{N} \bar{X}_k X_k \eta_{\bar{l}_1\cdots\bar{l}_q} - \frac{1}{4}\Big(\sum_{k \neq l_t} n_k\Big)\eta_{\bar{l}_1\cdots\bar{l}_q} .$$

Let $\langle u, v\rangle$ $(u, v \in S)$ denote the positive definite hermitian inner product of the vector space S invariant by the representation σ of G_{N+1}; and put $\|u\|^2 = \langle u, u\rangle$. The inner product $(\square''\eta, \eta)$ is given by

$$(5.11) \qquad \begin{aligned} (\square''\eta, \eta) = \frac{1}{q!}\Big\{&\sum_{k=1}^{N} \int_{\Gamma\backslash G} \| X_k \eta_{\bar{l}_1\cdots\bar{l}_q} \|^2 \, dv \\ &- \frac{1}{4}\sum_{k \neq l_t} n_k \int_{\Gamma\backslash G} \| \eta_{\bar{l}_1\cdots\bar{l}_q} \|^2 \, dv\Big\} , \end{aligned}$$

where dv denotes a certain volume element on $\Gamma\backslash G$ [8]. Note that, the functions $X_k \eta_{\bar{l}_1\cdots\bar{l}_q}$ and $\eta_{\bar{l}_1\cdots\bar{l}_q}$ being left invariant by Γ, we may consider them as functions on $\Gamma\backslash G$. Since the first term of the right hand side of (5.11) is non-negative, it follows that, if $\Box''\eta = 0$, we must have $\sum_{k \neq l_t} n_k \geq 0$. Hence, if $\sum_{k \neq l_t} n_k$ is negative, η must vanish. Now suppose that $n_i \leq 0$ for $i = 1, \cdots, N$. Then, if $q < N$ and $n_k < 0$ for an index k different from $l_1, \cdots, l_q$, we have $\mathfrak{h}(l_1, \cdots, l_q) = 0$. Suppose now that $q < N$ and $n_k = 0$ for all k different from $l_1, \cdots, l_q$. The formulas (5.1), (5.2') and (5.3') show that, in the notation of §3, the function $\eta_{i_1 \cdots i_q}$ belongs to $\mathcal{F}^0(\Gamma, u_1, \cdots, u_N, \sigma)$, where $u_k = 0$ for $k \neq l_t$, $t = 1, \cdots$, q, and $u_{l_t} = n_{l_t} - 2$ for $t = 1, \cdots, q$. By (5.9), we have $\bar{X}_k \eta_{i_1 \cdots i_q} = 0$ for $k \neq l_t$. Therefore, if $q > 0$, we get, by Theorem 3.2, $\eta_{i_1 \cdots i_q} = 0$ and hence $\eta = 0$. From what we have shown, it follows that $\mathfrak{h}^q = 0$ for $0 < q < N$ if $n_j \leq 0$ for every j. Now suppose that $q = 0$. We have already proved that $\eta = 0$ if $n_k < 0$ for an index k. Therefore we assume that $n_1 = \cdots = n_N = 0$. Then it follows from (5.11) that

$$\sum_{k=1}^{N} \int_{\Gamma \backslash G} \| X_k \eta^0 \|^2 \, dv = 0 \, ,$$

which implies that $X_k \eta^0 = 0$ for $k = 1, \cdots, N$. On the other hand, by (5.2') and (5.9), $W_k \eta^0 = \bar{X}_k \eta^0 = 0$ for $k = 1, \cdots, N$. These relations show that the function $\eta^0(s_1, \cdots, s_N, s_{N+1})$ on G is independent of the variables $s_1, \cdots, s_N$, and hence $\eta^0(s) = \eta^0(1, \cdots, 1, s_{N+1})$. By (5.3') we have $\eta^0(1, \cdots, 1, s_{N+1}) = \sigma(s_{N+1}^{-1})\eta^0(1, \cdots, 1)$. Further, by (5.1'), $\eta^0(\gamma_1, \cdots, \gamma_N, \gamma_{N+1}) = \eta^0(1, \cdots, 1)$ for every $\gamma = (\gamma_1, \cdots, \gamma_{N+1}) \in \Gamma$. Therefore we get $\sigma(\gamma_{N+1})\eta^0(1) = \eta^0(1)$ for every γ. Conversely, let u be an element of S such that $\sigma(\gamma_{N+1})u = u$ for every $\gamma \in \Gamma$. If we put $\eta^0(s_1, \cdots, s_{N+1}) = \sigma(s_{N+1}^{-1})u$, it is easily verified that the function η^0 satisfies the equations (5.1'), (5.2'), (5.3') and (5.9), hence η^0 corresponds to a harmonic 0-form η. From the above considerations, we see that dim $H^0(\Gamma \backslash X, E(J_{0 \cdots 0 \sigma})) = d$. Thus Theorem 5.1 is completely proved.

Now let $E^*(J_{n_1 \cdots n_N \sigma})$ be the dual vector bundle of $E(J_{n_1 \cdots n_N \sigma})$ and K the canonical line bundle of the complex manifold $\Gamma \backslash X$. Then we see that $E^*(J_{n_1 \cdots n_N \sigma}) = E(J_{-n_1, \cdots, -n_N, \sigma^*})$ and $K = E(J_{2 \cdots 2\varepsilon})$, where σ^* (resp. ε) denotes the representation of G_{N+1} contragredient to σ (resp. the trivial representation of G_{N+1}). By the duality theorem of Serre [4] we get the isomorphism:

$$H^q(\Gamma \backslash X, E(J_{n_1 \cdots n_N \sigma})) \cong H^{N-q}(\Gamma \backslash X, E(J_{-n_1+2, \cdots, -n_N+2, \sigma^*})) \, .$$

From this and Theorem 5.1, we obtain

THEOREM 5.2. *Suppose that $n_i \geq 2$ for $i = 1, \cdots, N$. Then the cohomology group $H^q(\Gamma \backslash \bar{X}, E(J_{n_1 \cdots n_N \sigma}))$ vanishes for $0 < q < N$. Moreover, let d be the dimension of the complex vector space of all the elements u of S such that $\sigma(\gamma_{N+1})u = u$ for all $\gamma = (\gamma_1, \cdots, \gamma_{N+1}) \in \Gamma$. Then*

$$\dim H^N\big(\Gamma\backslash X,\, \boldsymbol{E}(J_{n_1\cdots n_N\sigma})\big) = \begin{cases} 0 & \text{if } (n_1,\, \cdots,\, n_N) \neq (2,\, \cdots,\, 2)\,, \\ d & \text{if } (n_1,\, \cdots,\, n_N) = (2,\, \cdots,\, 2)\,. \end{cases}$$

6. Chern classes of the complex vector bundle $E(J)$

In this section we shall prove

$$(6.1) \qquad h^{N,0}(\Gamma,\, X,\, \rho_{m_1\cdots m_N\sigma}) = h^{N,0}(\Gamma,\, X_{k_1\cdots k_p \bar{l}_1\cdots \bar{l}_q},\, \rho_{m_1\cdots m_N\sigma})\,.$$

To simplify the notation, we assume that $k_j = j$ for $j = 1,\, \cdots,\, p$ and $l_j = p + j$ for $j = 1,\, \cdots,\, q(q = N - p)$, and put $X' = X_{1\cdots p\,\overline{p+1}\cdots \overline{N}}$. We denote by Y the common underlying differentiable manifold of X and X'. Then the complex manifolds $\Gamma\backslash X$ and $\Gamma\backslash X'$ have the common underlying differentiable manifold $\Gamma\backslash Y$. We denote by $u,\, v,\, \cdots$ the points of Y; in particular, u_0 will denote the point of Y whose coordinates in X are $(i,\, \cdots,\, i)$. Further, put

$$J(s,\, u) = \prod_{j=1}^{N} \big(c_j z_j(u) + d_j\big)^{m_j+2}\sigma(s_{N+1})\,,$$

$$J'(s,\, u) = \prod_{j=1}^{p} \big(c_j z_j(u) + d_j\big)^{m_j+2} \prod_{k=p+1}^{N} \big(c_k \bar{z}_k(u) + d_k\big)^{m_k+2}\sigma(s_{N+1})\,,$$

where $s = (s_1,\, \cdots,\, s_N,\, s_{N+1}) \in G$ and

$$s_j = \begin{pmatrix} a_j & b_j \\ c_j & d_j \end{pmatrix} \qquad\qquad\qquad \text{for } j = 1,\, \cdots,\, N\,.$$

Let $E(J)$ (resp. $E(J')$) be the complex vector bundle over $\Gamma\backslash X$ (resp. $\Gamma\backslash X'$) defined by the automorphic factor J (resp. J') as in § 5. Then, by Propositions 4.4 and 4.5, we have

$$h^{N,0}(\Gamma,\, X,\, \rho_{m_1\cdots m_N\sigma}) = \dim H^0\big(\Gamma\backslash X,\, \boldsymbol{E}(J)\big)\,,$$

$$h^{N,0}(\Gamma,\, X',\, \rho_{m_1\cdots m_N\sigma}) = \dim H^0\big(\Gamma\backslash X',\, \boldsymbol{E}(J')\big)\,.$$

Hence by Theorem 5.2, we get, if $(m_1,\, \cdots,\, m_N) \neq (0,\, \cdots,\, 0)$,

$$h^{N,0}(\Gamma,\, X,\, \rho_{m_1\cdots m_N\sigma}) = \chi\big(E(J)\big) = \sum_{q=0}^{N} (-1)^q \dim H^q(\Gamma\backslash X,\, \boldsymbol{E}(J))\,,$$

$$h^{N,0}(\Gamma,\, X',\, \rho_{m_1\cdots m_N\sigma}) = \chi\big(E(J')\big) = \sum_{q=0}^{N} (-1)^q \dim H^q\big(\Gamma\backslash X',\, \boldsymbol{E}(J')\big)\,;$$

and if $(m_1,\, \cdots,\, m_N) = (0,\, \cdots,\, 0)$,

$$h^{N,0}(\Gamma,\, X,\, \rho_{m_1\cdots m_N\sigma}) = \chi\big(E(J)\big) - (-1)^N d\,,$$

$$h^{N,0}(\Gamma,\, X',\, \rho_{m_1\cdots m_N\sigma}) = \chi\big(E(J')\big) - (-1)^N d\,.$$

Therefore, to prove (6.1), it is sufficient to show that $\chi(E(J)) = \chi(E(J'))$. Now, we know that $\Gamma\backslash X$ and $\Gamma\backslash X'$ are projective algebraic manifolds [7]. Then by the generalized Riemann-Roch theorem of Hirzebruch [4], $\chi(E(J)) = \chi(E(J'))$ will be proved if we show that the complex vector bundles $E(J)$ and $E(J')$ have the same Chern classes, and the complex manifolds $\Gamma\backslash X$ and $\Gamma\backslash X'$ have the same Chern classes.

Now the principal $GL(n, C)$-bundle over $\Gamma\backslash Y$ ($n = \dim_C S$) associated to $E(J)$ is defined as follows. Define the action of $\Gamma \times K$ on $G \times GL(n, C)$ by

$$(\gamma, k)(s, a) = (\gamma s k^{-1}, J(\gamma, su_0)a) ,$$

for $\gamma \in \Gamma$, $k \in K$, $s \in G$, $a \in GL(n, C)$. Let P be the quotient of $G \times GL(n, C)$ by this operation of $\Gamma \times K$; and denote by $[s, a]$ the image in P of the element $(s, a) \in G \times GL(n, C)$. Then P has a principal $GL(n, C)$-bundle structure associated to the vector bundle $E(J)$. The projection of P onto $\Gamma\backslash Y$ is defined by $[s, a] \to \Gamma sK$, and the action of $GL(n, C)$ on P is given by $[s, a]b = [s, ab]$ for every $b \in GL(n, C)$.

In the above definition, if we use J' instead of J, we obtain the principal $GL(n, C)$-bundle P' associated to the vector bundle $E(J')$.

Let τ be the representation of K defined by

$$\tau(k) = \prod_{j=1}^{N} \exp\left(i(m_j + 2)\theta_j\right)\sigma(k_{N+1}) ,$$

where $k = (k_1, \cdots, k_N, k_{N+1})$ and

$$k_j = \begin{pmatrix} \cos\theta_j & -\sin\theta_j \\ \sin\theta_j & \cos\theta_j \end{pmatrix} .$$

We have then $\tau(k) = J(k, u_0)$ for all $k \in K$. We define the action of $\Gamma \times K$ on $G \times U(n)$ by

$$(\gamma, k)(s, m) = (\gamma s k^{-1}, \tau(k)m) ,$$

where $(\gamma, k) \in \Gamma \times K$ and $(s, m) \in G \times U(n)$. Let P_0 be the quotient of $G \times U(n)$ by $\Gamma \times K$. Denote by $\{s, m\}$ the image of (s, m) in P_0. Then P_0 has a principal $U(n)$-bundle structure over $\Gamma\backslash Y$. From the definition of P_0 we see that P_0 is associated to the principal K-bundle $\Gamma\backslash G$ over $\Gamma\backslash Y$ by the representation τ of K.

We show that the $U(n)$-bundle P_0 is a reduced bundle of the $GL(n, C)$-bundle P. Let φ be the mapping of $G \times U(n)$ into $G \times GL(n, C)$ defined by $\varphi(s, m) = (s, J(s, u_0)m)$. It can easily be seen that $\varphi((\gamma, k)(s, m)) = (\gamma, k)\varphi(s, m)$ for every $(\gamma, k) \in \Gamma \times K$ and $(s, m) \in G \times U(n)$. Therefore we can define a one-to-one mapping $\bar{\varphi}$ of P_0 into P by $\bar{\varphi}(\{s, m\}) = [s, J(s, u_0)m]$. It is then easily verified that $\bar{\varphi}(xa) = \bar{\varphi}(x)a$ for every $x \in P_0$ and $a \in U(n)$ and that the diagram

$$\begin{array}{ccc} P_0 & \xrightarrow{\bar{\varphi}} & P \\ \downarrow & & \downarrow \\ \Gamma\backslash Y & \xrightarrow{\text{id}} & \Gamma\backslash Y \end{array}$$

is commutative. These show that P_0 is a reduced bundle of P.

Now the Chern classes of the GL(n, C)-bundle P are the Chern classes of the U(n)-bundle P_0. We are going to show that the Chern class $c_k(P_0)$ of P_0 is the cohomology class of $H^{2k}(\Gamma \backslash Y, R)$ represented by a 2-form ξ_k on $\Gamma \backslash Y$ such that

$$(6.2) \qquad \xi_k \circ \alpha = \binom{n}{k} \left(\frac{1}{8\pi i} \right)^k \left(\sum_{j=1} (m_j + 2)\omega_j \wedge \bar{\omega}_j \right)^k \qquad (k = 1, \cdots, n) .$$

Here α is the canonical projection of G onto $\Gamma \backslash Y$; the kth power of the 2-form is the power in the sense of exterior product; ω_j denotes the complex valued left invariant 1-form on G such that $\omega_j(X_k) = \delta_{jk}$, $\omega_j(\bar{X}_k) = 0$ and $i(X)\omega_j = 0$ for all $X \in \mathfrak{k}^c$; and $\bar{\omega}_j$ is the complex conjugate of ω_j. Note that the complex structure of X defines uniquely an endomorphism I of the Lie algebra $\mathfrak{g}$, such that $I\mathfrak{k} = (0)$, $I\mathfrak{m} = \mathfrak{m}$ and $I^2 X = -X$ for all $X \in \mathfrak{m}$, where $\mathfrak{m}$ denotes the subspace of $\mathfrak{g}$ consisting of all the elements of $\mathfrak{g}$ orthogonal to $\mathfrak{k}$ with respect to the Killing form φ of $\mathfrak{g}$. Extending I to an endomorphism of $\mathfrak{g}^c$, the elements X_j and $\bar{X}_j (j = 1, \cdots, N)$ are characterized by the properties that $IX_j = iX_j$, $I\bar{X}_j = -i\bar{X}_j$ and $\varphi(X_k, \bar{X}_j) = \delta_{kj}$.

In order to study the Chern classes of the U(n)-bundle P_0, we define a connection of the principal K-bundle G over $Y = G/K$ as follows. Let $\mathfrak{m}$ be the subspace of $\mathfrak{g}$ defined as above. Then we have $\mathfrak{g} = \mathfrak{m} + \mathfrak{k}$, $[\mathfrak{m}, \mathfrak{m}] \subset \mathfrak{k}$, $[\mathfrak{k}, \mathfrak{m}] = \mathfrak{m}$, $\mathfrak{m}^c = \mathfrak{n}^+ + \mathfrak{n}^-$. We can define a connection of the principal K-bundle G by the condition that each vector field belonging to $\mathfrak{m}$ is horizontal (cf. [9]). The curvature form Ω of this connection is a $\mathfrak{k}$-valued 2-form on G such that $\Omega(X, Y) = -[X, Y]$ for $X, Y \in \mathfrak{m}$. Moreover Ω is a left invariant form on G. Hence the form Ω may be considered as a 2-form $\bar{\Omega}$ on $\Gamma \backslash G$.

Now each vector field belonging to $\mathfrak{m}$ is projectable onto $\Gamma \backslash G$, since it is left invariant. Therefore we can define a connection C of the principal K-bundle $\Gamma \backslash G$ over $\Gamma \backslash Y$ by the condition that the projections onto $\Gamma \backslash G$ of the vector fields belonging to $\mathfrak{m}$ are horizontal. The curvature form of the connection C is the 2-form $\bar{\Omega}$. Since the U(n)-bundle P_0 is associated to the principal K-bundle $\Gamma \backslash G$ by the representation of K, the connection C of $\Gamma \backslash G$ induces a connection C_0 of P_0 (cf. [9]). Let Ω_0 be the curvature form of the connection C_0 and let θ be the mapping of $\Gamma \backslash G$ into P_0 defined by $\theta(\varpi(s)) = \{s, 1_n\}$, where ϖ denotes the projection of G onto $\Gamma \backslash G$ and 1_n denotes the unit matrix of degree n. If we denote by ω and ω_0 the connection forms of the connections C and C_0 respectively, we have $(\omega_0 \circ \theta)(X) = \tau(\omega(X))$ for every vector field X on $\Gamma \backslash G$. We see then that $\Omega_0 \circ \theta = \tau \bar{\Omega}$. Hence, putting $\bar{\theta} = \theta \circ \varpi$, we have

$$(\Omega_0 \circ \bar{\theta})(X, Y) = -\tau(\Omega(X, Y))$$

for all $X, Y \in \mathfrak{g}$. From the definition of the representation τ, we see

$$\tau(W_j) = -\frac{1}{2}(m_j + 2) \cdot 1_n \qquad\qquad (j = 1, \cdots, N).$$

Since $[X_i, \bar{X}_j] = (1/2)\delta_{ij} W_j$, we get

$$(\Omega_0 \circ \bar{\theta})(X_i, \bar{X}_j) = -\frac{1}{4}\delta_{ij}(m_j + 2) \cdot 1_n \,;$$

and since $[X_i, X_j] = [\bar{X}_i, \bar{X}_j] = 0$, we obtain

$$(\Omega_0 \circ \bar{\theta})(X_i, X_j) = (\Omega_0 \circ \bar{\theta})(\bar{X}_i, \bar{X}_j) = 0 \,.$$

Moreover, $i(X)(\Omega_0 \circ \bar{\theta}) = 0$ for all $X \in \mathfrak{k}^c$. Therefore we get

$$(6.3)\qquad \Omega_0 \circ \bar{\theta} = -\frac{1}{4}\Big(\sum_{j=1}^{N}(m_j + 2)\omega_j \wedge \bar{\omega}_j\Big)1_n \,.$$

Since Ω_0 is $\mathrm{U}(n)$-valued, we write it in matrix form

$$\Omega_0 = (\Omega_{ij})_{i,j=1\cdots n}$$

with complex valued 2-forms Ω_{ij}. It follows from (6.2) that $\Omega_{ij} = 0$ for $i \neq j$ and $\Omega_{11} = \cdots = \Omega_{nn}$, and that

$$(6.4)\qquad \Omega_{kk} \circ \bar{\theta} = -\frac{1}{4}\sum_{j=1}^{N}(m_j + 2)\omega_j \wedge \bar{\omega}_j \qquad\qquad (k = 1, \cdots, n).$$

Now let ξ be the form on P_0 defined by

$$\xi = \det\Big(1_n - \frac{1}{2\pi i}\Omega_0\Big) \,.$$

Then ξ is invariant by the action of $\mathrm{U}(n)$ on P_0, and we may consider ξ as a form on the base $\Gamma \backslash Y$. If we write ξ in the form

$$\xi = 1 + \xi_1 + \cdots + \xi_n \,,$$

where the form ξ_k is a form of degree $2k$, then ξ_k represents the kth Chern class $c_k(P_0)$. Since $\Omega_{ij} = 0$ for $i \neq j$ and $\Omega_{11} = \cdots = \Omega_{nn}$, we get by (6.4)

$$\xi \circ \alpha = 1 + \xi_1 \circ \alpha + \cdots + \xi_n \circ \alpha = \Big(1 + \frac{1}{8\pi i}\sum_{j=1}^{N}(m_j + 2)\omega_j \wedge \bar{\omega}_j\Big)^n \,,$$

from which follows (6.2).

Now we consider the Chern classes of the bundle P'. For this purpose, let τ' be the representation of K defined by

$$\tau'(k) = \prod_{j=1}^{p} \exp\big(i(m_j + 2)\theta_j\big) \prod_{l=p+1}^{N} \exp\big(-i(m_l + 2)\theta_l\big)\sigma(k_{N+1}) \,,$$

where $k = (k_1, \cdots, k_N, k_{N+1})$ and

$$k_j = \begin{pmatrix} \cos\theta_j & -\sin\theta_j \\ \sin\theta_j & \cos\theta_j \end{pmatrix} \qquad (j = 1, \cdots, N) .$$

We have then $\tau'(k) = J'(k, u_0)$ for $k \in K$. Using τ' instead of τ, we can define the principal $U(n)$-bundle P_0' associated to the principal K-bundle $\Gamma \backslash G$, and we can show that the $U(n)$-bundle P_0' is a reduced bundle of P'. By the same argument as above, we can see that the Chern class $c_k(P_0')$ of P_0' is represented by a $2k$-form ξ_k' such that

$$\xi_k' \circ \alpha = \binom{n}{k} \left(\frac{1}{8\pi i} \right)^k \left\{ \sum_{j=1}^{p} (m_j + 2)\eta_j \wedge \bar{\eta}_j - \sum_{l=p+1}^{N} (m_l + 2)\eta_l \wedge \bar{\eta}_l \right\}^k .$$

Here the 1-forms η_j and $\bar{\eta}_j$ on G are given as follows. Let I' be the endomorphism of the Lie algebra $\mathfrak{g}$ defined by the complex structure of X' and let Y_j and $\bar{Y}_j$ be the elements of $\mathfrak{g}^c$ such that $I'Y_j = iY_j$, $I'\bar{Y}_j = -i\bar{Y}_j$ and $\varphi(Y_k, \bar{X}_j) \delta_{kj}$. The η_j are the complex valued left invariant 1-forms on G such that $\eta_j(Y_i) = \delta_{ji}$, $\eta_j(\bar{Y}_i) = 0$ and $i(X)\eta_j = 0$ for all $X \in \mathfrak{k}^c$; and $\bar{\eta}_j$ is the complex conjugate of η_j. By the definition of the complex structure of X', we have $X_j = Y_j$ and $\bar{X}_j = \bar{Y}_j$ for $j = 1, \cdots, p$, and $\bar{X}_j = Y_j$, $X_j = \bar{Y}_j$ for $j = p + 1, \cdots, N$. It follows that $\eta_j = \omega_j$ and $\bar{\eta}_j = \bar{\omega}_j$ for $j = 1, \cdots, p$, and $\eta_j = \bar{\omega}_j$, $\bar{\eta}_j = \omega_j$ for $j = p + 1, \cdots, N$. We have therefore

$$\xi_k' \circ \alpha = \binom{n}{k} \left\{ \frac{1}{8\pi i} \sum_{j=1}^{N} (m_j + 2)\omega_j \wedge \bar{\omega}_j \right\}^k ,$$

and hence $\xi_k' \circ \alpha = \xi_k \circ \alpha$, which implies $\xi_k' = \xi_k$. Thus the Chern classes of the $U(n)$-bundles P_0 and P_0' are the same.

In the next place, we show that the complex manifolds $\Gamma \backslash X$ and $\Gamma \backslash X'$ have the same Chern classes. The Chern classes of $\Gamma \backslash X$ (resp. $\Gamma \backslash X'$) are the Chern classes of the complex tangent bundle T of $\Gamma \backslash X$ (resp. T' of $\Gamma \backslash X'$). We see that T and T' are the complex vector bundles over $\Gamma \backslash X$ and $\Gamma \backslash X'$ defined respectively by the $GL(n, \mathbf{C})$-valued automorphic factors

$$(6.5) \qquad J_0(s, u) = \mathrm{diag}\, [(c_1 z_1(u) + d_1)^{-2}, \cdots, (c_N z_N(u) + d_N)^{-2}] ,$$

$$(6.5') \qquad J_0'(s, u) = \mathrm{diag}\, [(c_1 z_1(u) + d_1)^{-2}, \cdots, (c_p z_p(u) + d_p)^{-2},$$
$$(c_{p+1}\bar{z}_{p+1}(u) + d_{p+1})^{-2}, \cdots, (c_N\bar{z}_N(u) + d_N)^{-2}] ,$$

where $\mathrm{diag}\, [a_1, \cdots, a_N]$ denotes the diagonal matrix of degree N whose diagonal elements are $a_1, \cdots, a_N$, and $s = (s_1, \cdots, s_N, s_{N+1})$,

$$s_i = \begin{pmatrix} a_i & b_i \\ c_i & d_i \end{pmatrix} \qquad (i = 1, \cdots, N) .$$

Let $\eta = 1 + \eta_1 + \cdots + \eta_N$ and $\eta' = 1 + \eta'_1 + \cdots + \eta'_N$ be the forms on $\Gamma \backslash Y$ representing the Chern classes of $\Gamma \backslash X$ and $\Gamma \backslash X'$, respectively. Then, in the same way as above, we see that

$$\eta \circ \alpha = \det \left(\operatorname{diag} \left[1 + \frac{1}{2\pi i} \omega_1 \wedge \bar{\omega}_1, \cdots, 1 + \frac{1}{2\pi i} \omega_N \wedge \bar{\omega}_N \right] \right),$$

$$\eta' \circ \alpha = \det \left(\operatorname{diag} \left[1 + \frac{1}{2\pi i} \omega_1 \wedge \bar{\omega}_1, \cdots, 1 + \frac{1}{2\pi i} \omega_p \wedge \bar{\omega}_p, \right.\right.$$

$$\left.\left. 1 - \frac{1}{2\pi i} \bar{\omega}_{p+1} \wedge \omega_{p+1}, \cdots, 1 - \frac{1}{2\pi i} \bar{\omega}_N \wedge \omega_N \right] \right),$$

and hence $\eta = \eta'$, so that the Chern classes of $\Gamma \backslash X$ and $\Gamma \backslash X'$ are the same. This, together with the above result proves the equality (6.1).

7. Main theorems

From the formulas given in the last part of § 4 and the result of § 6, we obtain the following main theorem.

THEOREM 7.1. *Let Γ be a discrete subgroup of G satisfying the conditions* (C_1), (C_2) *and* (C_3) *of § 1. Let $\rho_{m_1 \cdots m_N \sigma}$ be a representation of G defined in § 2, satisfying the condition* (D_1) *of § 5. Denote by $h^{p,q}(\Gamma, X, \rho_{m_1 \cdots m_N \sigma})$ the dimension of the complex vector space of all harmonic forms of type (p, q) for $\rho_{m_1 \cdots m_N \sigma}$. Further, let $b^r(\Gamma, X, \rho_{m_1 \cdots m_N \sigma})$ be the dimension of the complex vector space $H^r(\Gamma, X, \rho_{m_1 \cdots m_N \sigma})$, and let d be the dimension of the complex vector space of all the elements u of S such that $\sigma(\gamma_{N+1})u = u$ for every $\gamma = (\gamma_1, \cdots, \gamma_{N+1}) \in \Gamma$, S being the representation space of σ. Then we have*

$$b^r(\Gamma, X, \rho_{m_1 \cdots m_N \sigma}) = \sum_{p+q=r} h^{p,q}(\Gamma, X, \rho_{m_1 \cdots m_N \sigma}),$$

$$h^{p,q}(\Gamma, X, \rho_{m_1 \cdots m_N \sigma}) = 0 \qquad \text{if } p \neq q \text{ and } p + q \neq N,$$

$$h^{p,p}(\Gamma, X, \rho_{m_1 \cdots m_N \sigma}) = \begin{cases} 0 & \text{if } 2p \neq N \text{ and } (m_1, \cdots, m_N) \neq (0, \cdots, 0), \\ \binom{N}{p} d & \text{if } 2p \neq N \text{ and } (m_1, \cdots, m_N) = (0, \cdots, 0), \end{cases}$$

$$h^{N-q,q}(\Gamma, X, \rho_{m_1 \cdots m_N \sigma}) = \begin{cases} \binom{N}{q} h^{N,0}(\Gamma, X, \rho_{m_1 \cdots m_N \sigma}) \\ \qquad \text{if } (m_1, \cdots, m_N) \neq (0, \cdots, 0), \\ \binom{N}{q} \{ \delta_{N-q,q} d + h^{N,0}(\Gamma, X, \rho_{m_1 \cdots m_N \sigma}) \} \\ \qquad \text{if } (m_1, \cdots, m_N) = (0, \cdots, 0). \end{cases}$$

Moreover, $h^{N,0}(\Gamma, X, \rho_{m_1 \cdots m_N \sigma})$ is equal to the dimension of the complex vector space of all S-valued holomorphic automorphic forms with respect to Γ for the automorphic factor $J_{m_1+2, \cdots, m_N+2, \sigma}$ (cf. Proposition 4.4).

From Theorem 7.1 we get immediately the following results: When $(m_1, \cdots, m_N) \neq (0, \cdots, 0)$, we have

$$(7.1) \qquad \begin{cases} b^r(\Gamma, X, \rho_{m_1\cdots m_N\sigma}) = 0 & \text{if } r \neq N, \\ b^N(\Gamma, X, \rho_{m_1\cdots m_N\sigma}) = 2^N \cdot h^{N,0}(\Gamma, X, \rho_{m_1\cdots m_N\sigma}), \end{cases}$$

When $(m_1, \cdots, m_N) = (0, \cdots, 0)$, we have

$$(7.2) \qquad \begin{cases} b^{2p+1}(\Gamma, X, \rho_{0\cdots 0\sigma}) = 0, \\[4pt] b^{2p}(\Gamma, X, \rho_{0\cdots 0\sigma}) = \binom{N}{p}d & \text{if } 2p \neq N, \\[4pt] b^N(\Gamma, X, \rho_{0\cdots 0\sigma}) = \begin{cases} 2^N h^{N,0}(\Gamma, X, \rho_{0\cdots 0\sigma}) & \text{if } N \text{ is odd}. \\[4pt] \binom{N}{\frac{N}{2}}d + 2^N h^{N,0}(\Gamma, X, \rho_{0\cdots 0\sigma}) & \text{if } N \text{ is even}. \end{cases} \end{cases}$$

Further, if σ is the trivial representation of G_{N+1}, $b^r(\Gamma, X, \rho_{0\cdots 0\sigma})$ and $h^{N,0}(\Gamma, X, \rho_{0\cdots 0\sigma})$ are respectively equal to the rth Betti number $b^r(\Gamma \backslash X)$, and the number $h^{N,0}(\Gamma \backslash X)$ of the linearly independent holomorphic N-forms of the complex manifold $\Gamma \backslash X$. Therefore we get

THEOREM 7.2. *Let X_u be the product of N copies of the Riemann sphere. Then we have*

$$b^r(\Gamma \backslash X) = b^r(X_u) \qquad\qquad (r \neq N),$$
$$b^N(\Gamma \backslash X) = b^N(X_u) + E(X_u)h^{N,0}(\Gamma \backslash X),$$

where $E(X_u) = 2^N$ denotes the Euler characteristic of X_u.

As for the number $h^{N,0}(\Gamma, X, \rho_{m_1\cdots m_N\sigma})$, the following theorem holds.

THEOREM 7.3. *The notation being as in Theorems 7.1 and 7.2, let n be the degree of the representation σ of the compact factor G_{N+1}. Then we have*

$$(7.3) \qquad \begin{cases} h^{N,0}(\Gamma, X, \rho_{m_1\cdots m_N\sigma}) = (-1)^N \cdot n \cdot \prod_{k=1}^N (m_k + 1)(1 + (-1)^N h^{N,0}(\Gamma \backslash X)) \\ \qquad\qquad\qquad\qquad\qquad\qquad\qquad \text{if } (m_1, \cdots, m_N) \neq (0, \cdots, 0), \\ h^{N,0}(\Gamma, X, \rho_{0\cdots 0\sigma}) = (-1)^N n(1 + (-1)^N h^{N,0}(\Gamma \backslash X)) - (-1)^N d. \end{cases}$$

Moreover, $1 + (-1)^N h^{N,0}(\Gamma \backslash X)$ is equal to the arithmetic genus of $\Gamma \backslash X$.

From this theorem, it follows easily that, if N is odd, $h^{N,0}(\Gamma \backslash X) > 1$.

To prove Theorem 7.3, we need

LEMMA. *Let F be a locally constant sheaf of complex vector space of dimension m over a compact manifold M. Then we have*

$$\sum_r (-1)^r \dim_C H^r(M, F) = m \sum_r (-1)^r \dim_C H^r(M, C).$$

This is an easy consequence of the definition of the cohomology groups $H^r(M, F)$ with coefficients in the sheaf F; so we omit the proof.

PROOF OF THEOREM 7.3. Let F be the representation space of $\rho_{m_1\cdots m_N\sigma}$, and $E(\rho_{m_1\cdots m_N\sigma})$ be the vector bundle over $\Gamma\backslash X$ with typical fibre F associated with the principal Γ-bundle X over $\Gamma\backslash X$ by the representation $\rho_{m_1\cdots m_N\sigma}$ of Γ. Let F be the sheaf of germs of locally constant sections of the vector bundle $E(\rho_{m_1\cdots m_N\sigma})$. Then F is a locally constant sheaf over $\Gamma\backslash X$ whose stalks are complex vector spaces isomorphic to F. The cohomology group $H^r(\Gamma, X, \rho_{m_1\cdots m_N\sigma})$ may be identified with the cohomology group $H^r(\Gamma\backslash X, F)$ (cf. [8]). By the above lemma, we get

$$(7.4) \qquad \sum_{r=0}^{2N} (-1)^r b^r(\Gamma, X, \rho_{m_1\cdots m_N\sigma}) = \dim_C F \cdot \sum_{r=0}^{2N} (-1)^r b^r(\Gamma\backslash X) .$$

Since $\dim_C F = n \prod_{k=1}^{N} (m_k + 1)$, we get, from (7.1), (7.2) and (7.4)

$$(-1)^N 2^N h^{N,0}(\Gamma, X, \rho_{m_1\cdots m_N\sigma}) = n \prod_{k=1}^{N} (m_k + 1) 2^N (1 + (-1)^N h^{N,0}(\Gamma\backslash X)) ,$$
$$\text{if } (m_1, \cdots, m_N) \neq (0, \cdots, 0) ,$$

$$2^N(d + (-1)^N h^{N,0}(\Gamma, X, \rho_{0\cdots 0\sigma})) = n 2^N (1 + (-1)^N h^{N,0}(\Gamma\backslash X)) ,$$

which implies (7.3). Further, by Theorem 7.1, we have

$$\sum_{p=0}^{N} (-1)^p h^{p,0}(\Gamma\backslash X) = 1 + (-1)^N h^{N,0}(\Gamma\backslash X) ,$$

so that $1 + (-1)^N h^{N,0}(\Gamma\backslash X)$ is equal to the arithmetic genus of $\Gamma\backslash X$; this completes the proof.

As an application of Theorem 5.1 we prove the following

THEOREM 7.4. *The assumptions and the notation being as in Theorem 7.1, let Θ be the sheaf of germs of holomorphic vector fields on the complex manifold $\Gamma\backslash X$. Then the cohomology group $H^q(\Gamma\backslash X, \Theta)$ vanishes for $q \neq N$. Moreover, we have*

$$\dim_C H^N(\Gamma\backslash X, \Theta) = (-1)^N 3N(1 + (-1)^N h^{N,0}(\Gamma\backslash X)) .$$

PROOF. Since the complex tangent bundle T of $\Gamma\backslash X$ is equal to the vector bundle over $\Gamma\backslash X$ defined by the automorphic factor (6.5), we see that T is the Whitney sum of the bundles $E(J_k)$ for $k = 1, \cdots, N$, where $E(J_k)$ denotes the complex line bundle over $\Gamma\backslash X$ defined by the automorphic factor $J_k(s, u) = (c_k z_k(u) \dotplus d_k)^{-2}$. Therefore we have $H^q(\Gamma\backslash X, \Theta) = \sum_{k=1}^{N} H^q(\Gamma\backslash X, E(J_k))$, and by Theorem 5.1 $H^q(\Gamma\backslash X, E(J_k)) = (0)$ for $q \neq N$, which implies that $H^q(\Gamma\backslash X, \Theta) = (0)$ for $q \neq N$. Now let J_k' denote the automorphic factor $J_{m_1\cdots m_N}$ with $m_k = 4$ and $m_j = 2$ for $j \neq k$, and let K be the canonical line bundle of $\Gamma\backslash X$. Then the line bundle $E(J_k)$ is isomorphic to $E^*(J_k') \otimes K$, where $E^*(J_k')$ is the dual bundle of $E(J_k')$. Therefore, by the duality theorem of Serre, $H^N(\Gamma\backslash X, E(J_k))$ is isomorphic to $H^0(\Gamma\backslash X, E(J_k'))$. The vector space $H^0(\Gamma\backslash X, E(J_k'))$ may be identified with the vector space of all holomorphic automorphic forms for the automorphic factor J'_k, and hence, by Proposition 4.4 and Theorem 7.3, $\dim_C H^0(\Gamma\backslash X, E(J'_k)) = (-1)^N 3(1 + (-1)^N h^{N,0}(\Gamma\backslash X))$. We have therefore

$$\dim_C H^N(\Gamma\backslash X, \Theta) = \sum_{k=1}^N \dim_C H^0(\Gamma\backslash X, E(J'_k))$$
$$= (-1)^N 3N(1 + (-1)^N h^{N,0}(\Gamma\backslash X)) .$$

This completes the proof.

8. Discrete groups with elements of finite order

Let G_0, Z, Γ_0 be as in §1. Some of the results in §7 can be generalized to the case where $\Gamma_0/(\Gamma_0 \cap Z)$ has elements of finite order other than the identity element. As we have already seen in §4, the assertions concerning $h^{p,q}(\Gamma, X, \rho)$ of Theorem 7.1 are true without the condition (C_3), and with only (C_1) and (C_2), if $p + q \neq N$. The validity of the assertion for $h^{N-q,q}$ depends on the equality (6.1). By Proposition 4.4, $h^{N,0}(\Gamma, X, \rho_{m_1\cdots m_N\sigma})$ is equal to the dimension of the vector space of holomorphic automorphic forms for the automorphic factor $J_{m_1+2,\cdots,m_N+2,\sigma}$. Now the latter is given by [10, Theorem 11] with a certain condition on the m_i and σ. Let $\Gamma(z)$, for every $z \in X$, be the subgroup of Γ consisting of the elements γ such that $\gamma z = z$, and let $\Gamma_0(z)$ be the projection of $\Gamma(z)$ into G_0. Let $\{w_1, \cdots, w_t\}$ be a set of representatives, in the sense of Γ-equivalence, of the points w on X such that $\Gamma_0(w)$ is not contained in Z. Let ν_i be the order of $\Gamma_0(w_i)/(\Gamma_0(w_i) \cap Z)$. Let F be the representation space for $\rho_{m_1\cdots m_N\sigma}$ and let $F(w_i)$ be the subspace of F consisting of the vectors v such that $\rho_{m_1\cdots m_N\sigma}(\gamma)v = v$ for every $\gamma \in \Gamma(w_i)$. Then applying [10, Theorem 11] to 2^N spaces $X_{k_1\cdots k_p \bar{l}_1\cdots \bar{l}_q}$, we find, on account of Proposition 4.4, that, if σ is trivial, and all the m_i are even and ≥ 2, then

$$(8.1)\qquad \begin{aligned} b^N(\Gamma, X, \rho_{m_1\cdots m_N\sigma}) &= \sum h^{N,0}(\Gamma, X_{k_1\cdots k_p\bar{l}_1\cdots\bar{l}_q}, \rho_{m_1\cdots m_N\sigma}) \\ &= c\cdot\dim F + (-1)^N \sum_{i=1}^t \dim F(w_i) . \end{aligned}$$

Here the first summation is extended over 2^N spaces $X_{k_1\cdots k_p\bar{l}_1\cdots\bar{l}_q}$ which are obtained from $H_1 \times \cdots \times H_N$ by exchanging the complex structure of $H_{l_1}, \cdots, H_{l_q}$ for their conjugate complex structures; and c is given by

$$(8.2)\qquad \begin{aligned} c &= (2\pi)^{-N} v(\Gamma\backslash X) - (-1)^N \sum_{i=1}^t \nu_i^{-1} , \\ v(\Gamma\backslash X) &= \int_{\Gamma\backslash X} \prod_{\lambda=1}^N \frac{dx_\lambda dy_\lambda}{y_\lambda^2} \qquad (z_\lambda = x_\lambda + \sqrt{-1}\, y_\lambda) . \end{aligned}$$

According to a communication of Shimizu, (8.1) is also true for any σ if all the m_i are positive, and if the condition (D_1) of §5 is satisfied.

In general, one may conjecture that, if (C_1), (C_2) and (D_1) are satisfied, then

$$(8.3) \quad b^N(\Gamma, X, \rho_{m_1 \cdots m_N \sigma}) = c \cdot \dim F + (-1)^N \sum_{i=1}^{t} \dim F(w_i) + \mu ,$$

$$\mu = \begin{cases} 0 & \text{if } (m_1, \cdots, m_N) \neq (0, \cdots, 0) , \\ 2^N d & \text{if } (m_1, \cdots, m_N) = (0, \cdots, 0) \text{ and } N \text{ is odd} , \\ \left\{ \binom{N}{N/2} - 2^N \right\} d & \text{if } (m_1, \cdots, m_N) = (0, \cdots, 0) \text{ and } N \text{ is even} , \end{cases}$$

where d is as in Theorem 7.1. If this were true, we would obtain

$$(8.4) \quad b^N(\Gamma, X, \varepsilon) = c + (-1)^N t + \begin{cases} 2^N & (N: \text{ odd}) , \\ \binom{N}{N/2} - 2^N & (N: \text{ even}) , \end{cases}$$

where ε is the trivial representation of G, so that

$$(8.5) \quad \begin{aligned} \sum_{r=0}^{2N}(-1)^r b^r(\Gamma, X, \varepsilon) &= (-1)^N c + t \\ &= (-1)^N (2\pi)^{-N} v(\Gamma \backslash X) + \sum_{i=1}^{t} (1 - \nu_i^{-1}) , \end{aligned}$$

which is in fact true and well-known in the case $N = 1$.

As for the equality (6.1), we can not deduce it from [10, Theorem 11] if $\Gamma_0/(\Gamma_0 \cap Z)$ has elements of finite order other than the identity element.

In the general case, it is not known whether (6.1) is true or not.

OSAKA UNIVERSITY

REFERENCES

1. A. BOREL, *Density properties for certain subgroups of semi-simple groups without compact components*, Ann. of Math., 72 (1960), 179-188.

2. S. S. CHERN, On a generalization of Kähler geometry, Algebraic Geometry and Topology, A symposium in honor of S. Lefschetz, Princeton, 1957.

3. M. EICHLER, *Eine Verallgemeinerung der Abelschen Integrale*, Math. Z., 67 (1957), 267-298.

4. F. HIRZEBRUCH, Neue Topologische Methoden in der algebraischen Geometrie, Springer-Verlag, Berlin-Göttingen-Heidelberg, 1956.

5. ———, Automorphe Formen und der Satz von Riemann-Roch, Symposium de topologia algebraica, Univ. de Mexico, 1958, 129-143.

6. K. KODAIRA, *On a differential-geometric method in the theory of analytic stacks*, Proc. Nat. Acad. Sci. U. S. A., 39 (1953), 1268-1273.

7. ———, *On Kähler varieties of restricted type*, Ann. of Math. 60 (1954), 28-48.

8. Y. MATSUSHIMA and S. MURAKAMI, *On vector bundle valued harmonic forms and automorphic forms on symmetric riemannian manifolds*, Ann. of Math., 78 (1963), 365-416.

9. K. NOMIZU, Lie groups and differential geometry, Publications of Math. Soc. Japan, No. 2, 1956.

10. H. SHIMIZU, *On discontinuous groups operating on the product of the upper half-planes*, Ann. of Math. 77 (1963), 33-71.

11. G. SHIMURA, *Sur les intégrales attachées aux formes automorphes*, J. Math. Soc. Japan, 11 (1959), 291-311.

12. ———, On Dirichlet series and abelian varieties attached to automorphic forms, Ann. of Math., 76 (1962), 237-294.

13. A. WEIL, *On discrete subgroups of Lie groups*: II, Ann. of Math., 75 (1962), 578-602.

On modular correspondences for $Sp(n, \mathbf{Z})$ and their congruence relations

Proceedings of the National Academy of Sciences, 49 (1963), 824-828

ON MODULAR CORRESPONDENCES FOR Sp(n, Z) AND THEIR CONGRUENCE RELATIONS

By Goro Shimura

DEPARTMENT OF MATHEMATICS, OSAKA UNIVERSITY, AND PRINCETON UNIVERSITY

Communicated by D. C. Spencer, April 2, 1963

M. Eichler[1] and the author[2,3] gave the congruence relation $T_p \equiv \Pi_p + \Pi_p' \mathrm{o} R_i$ mod (p) for the modular correspondences of the field of elliptic modular functions and of the fields of certain automorphic functions of one variable. It is natural to seek an analogy for the automorphic functions with more than one variables. In this paper we investigate the modular correspondences for the field of Siegel's modular functions and give some congruence relations for them.

1. We denote as usual by $\mathbf{Z}$ the ring of rational integers. Let

$$G = Sp(n, \mathbf{Z}), \quad J = \begin{pmatrix} 0 & 1_n \\ -1_n & 0 \end{pmatrix},$$

where 1_n is the unit matrix of size n. Let S be the set of all matrices B of size $2n$ with entries in $\mathbf{Z}$ such that ${}^tBJB = rJ$ with $r \in \mathbf{Z}$, $r > 0$. We write then $r(B) = r$. Let L be the free $\mathbf{Z}$-module generated by the double cosets GBG for all $B \in S$. We define a law of multiplication in L as follows.[3,4] First we note that each GBG with B in S contains only a finite number of right and left cosets. Let $B \in S$, $C \in S$, and let $GBG = \cup_i GB_i$ and $GCG = \cup_j GC_j$ be disjoint expressions. For every $D \in S$, the number of (i, j) such that $GB_i C_j = GD$ is determined only by GBG, GCG, GDG. Denote this number by $\mu(GBG \cdot GCG; \ GDG)$ and define the product $GBG \cdot GCG$ by

$$GBG \cdot GCG = \sum \mu(GBG \cdot GCG; \ GDG) GDG,$$

where the summation is taken over all the GDG contained in $GBGCG$. Extending this by linearity to a law of multiplication of L, we obtain an associative ring. We can prove that L *is a commutative integral domain.*

2. The representatives for $G \backslash S / G$ are given by the diagonal matrices B with diagonal elements $d_1, \ldots, d_n, e_1, \ldots, e_n$ such that $d_i | d_{i+1}, d_n | e_n, e_{i+1} | e_i, d_i e_i = r(B)$.[5] For this B, denote GBG (as an element of L) by $T(d_1, \ldots, d_n, e_n, \ldots, e_1)$. It can be easily seen that

$$T(a_1, \ldots, a_{2n}) T(b_1, \ldots, b_{2n}) = T(a_1 b_1, \ldots, a_{2n} b_{2n}) \text{ if } (a_{2n}, b_{2n}) = 1. \tag{1}$$

Our first problem is to give an Euler product for the formal Dirichlet series $D(s)$, with coefficients in the ring L, defined by

$$D(s) = \sum (GBG) r(B)^{-s},$$

where the summation is taken over all the GBG with $B \in S$. If we denote by $D_p(s)$ the sum of all terms $(GBG) r(B)^{-s}$ such that $r(B)$ is a power of a prime number p, then (1) implies

$$D(s) = \Pi_p D_p(s).$$

Therefore, let us fix our attention to one prime number p, and denote by L_p the subring of L generated by the GBG for which $r(B)$ is a power of p. Then we obtain

THEOREM 1. L_p *is a polynomial ring over* $\mathbf{Z}$ *generated by the following* $n + 1$ *elements:*

$$T(1, \ldots, 1, p, \ldots, p), \ T(\underbrace{1, \ldots, 1}_{n-i}, \underbrace{p, \ldots, p}_{2i}, \underbrace{p^2, \ldots, p^2}_{n-i}) \quad (1 \leqq i \leqq n).$$

These $n + 1$ *elements are algebraically independent.*

Putting $X = p^{-s}$, we have

$$D_p(s) = \sum T(p^{c_1}, \ldots, p^{c_{2n}}) X^{c_1 + c_{2n}} \quad (c_1 \leqq \ldots \leqq c_{2n}).$$

If $n = 1$, we get the relation $D_p(s) = [1 - T(1,p)X + pT(p,p)X^2]^{-1}$, which is essentially due to E. Hecke.[6]

THEOREM 2. *If* $n = 2$,

$$\begin{aligned}
D_p(s) = &[1 - p^2 T(p,p,p,p) X^2] \times \\
&[1 - T(1,1,p,p)X + \{pT(1,p,p,p^2) + p(p^2 + 1)T(p,p,p,p)\}X^2 \\
&\quad - p^3 T(p,p,p^2,p^2)X^3 + p^6 T(p^2,p^2,p^2,p^2)X^4]^{-1}.
\end{aligned}$$

In general, it is plausible that $D_p(s) = E(X)/F(X)$ with polynomials $E(X)$ and $F(X)$ in X with coefficients in $\mathbf{Z}$ of degree $2^n - 2$ and 2^n, respectively.

3. Let H_n be the space of all complex matrices of size n whose imaginary parts are positive definite. For every $U = \begin{pmatrix} a & b \\ c & d \end{pmatrix}$ in S with square matrices a, b, c, d of size n, and for $z \in H_n$, we put $U(z) = (az + b)(cz + d)^{-1}$. We can construct an analytic family of polarized abelian varieties $\mathcal{P}_z = (A_z, \mathcal{C}_z)$ parametrized by the points of H_n. Here A_z is an abelian variety of dimension n having $(z \ 1_n)$ as a period matrix and $\mathcal{C}_z$ is a polarization of A_z corresponding to the principal matrix J.[7, 8] $\mathcal{P}_z$ is isomorphic to $\mathcal{P}_w$ if and only if $z = U(w)$ for some $U \in G$. Let K_z be the field of moduli[9] of $\mathcal{P}_z$. Then there exist meromorphic functions $f_1(z), \ldots, f_m(z)$ on H_n and an analytic subset Y of H_n of codimension 1 such that $K_z = \mathbf{Q}(f_1(z), \ldots,$

$f_m(z))$ for $z \in H_n - Y$, and $\mathbf{C}(f_1, \ldots, f_m)$ is the field of all Siegel's modular functions on H_n.[7–9] Here $\mathbf{Q}$ and $\mathbf{C}$ denote, respectively, the rational number field and the complex number field. Let (W_z, h_z) be a normalized Kummer variety[3] of $\wp_z$. For a positive integer N, denote by $g(N, A_z)$ the set of all points t on A_z such that $Nt = 0$, and by $K_{N,z}$ the field generated over K_z by the coordinates of $h_z(t)$ for all $t \in g(N, A_z)$. Let $G_N = \{B \in G | B \equiv 1 \bmod (N)\}$. Then there exist meromorphic functions $\varphi_1(z), \ldots, \varphi_\lambda(z)$ and a join Y_N of countably infinite analytic subsets of H_n of codimension 1 with the following properties:

$$K_{N,z} = \mathbf{Q}(f_1(z), \ldots, f_m(z), \varphi_1(z), \ldots, \varphi_\lambda(z)) \text{ for } z \in H_n - Y_N;$$

and $\mathbf{C}(f_1, \ldots, f_m, \varphi_1, \ldots, \varphi_\lambda)$ is the set of all meromorphic automorphic functions on H_n with respect to G_N.[7, 8]

Let $G_N{}^*$ be the group of all invertible matrices U of size $2n$ with entries in $\mathbf{Z}/(N)$ such that ${}^t U J U = rJ$ with $r \in \mathbf{Z}/(N)$. Then we obtain

THEOREM 3. *Let* $K = \mathbf{Q}(f_1, \ldots, f_m)$, $K_N = \mathbf{Q}(f_1, \ldots, f_m, \varphi_1, \ldots, \varphi_\lambda)$ *and let* ζ *be a primitive N-th root of unity. Then, K_N is a galois extension of K, whose galois group $\mathcal{G}$ is isomorphic to* $G_N{}^*/\{\pm 1\}$, *and* $\mathbf{Q}(\zeta)$ *is the algebraic closure of* $\mathbf{Q}$ *in* K_N. *Moreover, if an element σ of $\mathcal{G}$ corresponds to an element U of $G_N{}^*$, and if ${}^t U J U = rJ$, then* $\zeta^\sigma = \zeta^r$.

By this theorem and by the same argument as in references 2 and 3, we can conclude the following

THEOREM 4. *Let* $\mathcal{K}_N$ *be the field of all meromorphic automorphic functions on H_n with respect to G_N. Then there exists a subfield $\mathcal{K}'$ of $\mathcal{K}_N$ such that $\mathbf{C}$ and $\mathcal{K}'$ are linearly disjoint over $\mathbf{Q}$. In other words, the function field $\mathcal{K}_N$ has a model defined over* $\mathbf{Q}$.

4. Now we realize the elements of the ring L defined in section 1 as algebraic correspondences. Let $B \in S$ and let $GBG = GB_1 \cup \ldots \cup GB_s$ be a disjoint expression. Roughly speaking, $f(z) \to (f(B_1(z)), \ldots, f(B_s(z)))$ for $f \in \mathcal{K}_1$ gives the algebraic correspondence attached to GBG which is to be defined. Fix a point z on H_n which is generic for the f_μ over $\mathbf{Q}$ in the sense of references 7 and 9. Put $r = r(B)$, $B_i(z) = z_i$ for $1 \leq i \leq s$. The principal matrix J induces naturally an alternating form $e_r(x, y)$ on $g(r, A_z)$ with values in $\mathbf{Z}/(r)$.

LEMMA 1. *Let* $GBG = T(d_1, \ldots, d_{2n})$ *and let* $\{g_1, \ldots, g_{s'}\}$ *be the set of all subgroups g of $g(r, A_z)$ satisfying the following conditions: i) g is isomorphic to the product of the cyclic groups $\mathbf{Z}/(d_i)$ for $1 \leq i \leq 2n$; ii) $e_r(x, y) = 0$ for $(x, y) \in g \times g$. Then, $s = s'$, and after reordering the g_i suitably, we can find an isogeny λ_i of A_z to A_{z_i} whose kernel is* g_i.

LEMMA 2. *Suppose that r is prime to N. Then, for each i, there exists an isomorphism σ_i of $K_{N,z}$ onto K_{N,z_i} such that $f_\mu(z)^{\sigma_i} = f_\mu(z_i)$ for $1 \leq \mu \leq m$ and $h_z(t)^{\sigma_i} = h_{z_i}(\lambda_i t)$ for $t \in g(N, A_z)$, where the λ_i are the isogenies of Lemma 1.*

LEMMA 3. *Let the σ_i be as in Lemma 2. Let $u = (u_1, \ldots, u_l)$ be a set of elements in $K_{N,z}$ such that $\mathbf{Q}(u) \supset K_z$. Then $K_{z_i}(u)$ contains $\mathbf{Q}(u^{\sigma_i})$ for each i, and $(u^{\sigma_1}, \ldots, u^{\sigma_s})$ is the complete set of conjugates of u^{σ_1} over $\mathbf{Q}(u)$.*

Now let V be a projective variety with a generic point u over $\mathbf{Q}$ such that $K_z = \mathbf{Q}(u)$. Then, we may consider the field of Siegel's modular functions of degree n as the function field of the variety V. For $GBG = T(d_1, \ldots, d_{2n})$, take the isomorphisms σ_i of Lemma 2, and denote by the same notation $T(d_1, \ldots, d_{2n})$ the locus of $u \times u^{\sigma 1}$ over $\mathbf{Q}$ on $V \times V$. We call the algebraic correspondences on V thus obtained from the elements of L *the modular correspondences* (for Siegel's modular functions).

5. Let V_p be the reduction of V modulo p. We consider only those p such that V_p is an irreducible variety with multiplicity 1. We denote by $T(d_1, \ldots, d_{2n})_p$ the reduction of $T(d_1, \ldots, d_{2n})$ modulo p. Let Y and Z be cycles on $V_p \times V_p$ of the same dimension as V_p. We denote by $Y \circ Z$ the projection of $(Y \times V_p) \cdot (V_p \times Z)$ onto the product of the first and the second factors of $V_p \times V_p \times V$. We write $Y \equiv Z$ if $Y \cdot (x \times V_p) = Z \cdot (x \times V_p)$ and $Y \cdot (V_p \times x) = Z \cdot (V_p \times x)$ for a sufficiently generic point x of V_p.

Let $GBG = T(p^{a_1}, \ldots, p^{a_{2n}})$ and $r = r(B) = p^c$. Consider the alternating form $e_r(x, y)$ on $g(r, A_z)$ introduced in section 4. We can find a subgroup g^* of $g(r, A_z)$ with the following properties: (i) g^* is isomorphic to the product of n copies of $\mathbf{Z}/(r)$; (ii) $e_r(x, y) = 0$ for $(x, y) \in g^* \times g^*$. Let $g_1, \ldots, g_s$ be as in Lemma 1. Write $g_i \sim g_j$ if $g_i \cap g^*$ and $g_j \cap g^*$ are isomorphic as abelian groups. Divide $\{g_1, \ldots, g_s\}$ into classes by this relation $\sim$, and denote the classes by $F_1, \ldots, F_q$.

THEOREM 5. *Let the notation be as above. Then, for almost all prime numbers p, there exist q positive cycles $U_\nu (1 \leq \nu \leq q)$ on $V_p \times V_p$ of the same dimension as V_p such that*

$$T(p^{a_1}, \ldots, p^{a_{2n}})_p \equiv U_1 + \ldots + U_q.$$

To define the cycles U_ν, consider the B_i. z_i, λ_i, σ_i defined in section 4. Let u be a generic point of V over $\mathbf{Q}$ such that $K_z = \mathbf{Q}(u)$. Let k be a field of definition for A_z, A_{z_i}, λ_i, over which the points of $g(r, A_z)$ and u, u^{σ_i} are rational. Let (w_μ) be a set of independent variables over $\mathbf{Q}$ such that k is an algebraic extension of $\mathbf{Q}(w_\mu)$. For every prime number p, we extend the natural homomorphism of $\mathbf{Z}$ onto $\mathbf{Z}/(p)$ to a place of $\mathbf{Q}(w_\mu)$ which maps the w_μ onto independent variables over $\mathbf{Z}/(p)$, and extend it further to a place P of k. For each p we fix such a place P and denote by $P(Y)$ the reduction modulo P of an object Y. If A_z and A_{z_i} have nondegenerate reduction modulo P (this is the case for almost all p), we obtain from λ_i an isogeny $P(\lambda_i)$ of $P(A_z)$ to $P(A_{z_i})$, whose kernel is $P(g_i)$. Now the reduction modulo P defines a *surjective* homomorphism of $g(r, A_z)$ to $g(r, P(A_z))$. The kernel of this homomorphism has the properties (i), (ii) of the above g^*. We take the kernel as g^*. Since g_i is the kernel of λ_i, we can divide the isogenies (A_{z_i}, λ_i) into q classes corresponding to $F_1, \ldots, F_q$. Then, for each F_ν, we can define a positive cycle U_ν on $V_p \times V_p$ satisfying

$$U_\nu \cdot (P(u) \times V_p) = \sum P(u) \times P(u^{\sigma_i}),$$

where the summation is taken over all i such that $g_i \in F_\nu$.

Let Δ_p, Π_p and Π_p' denote, respectively, the loci of $x \times x$, $x \times x^p$ and $x^p \times x$ on $V_p \times V_p$, where x is a generic point of V_p over the prime field. As a special case of Theorem 5, we can find $n + 1$ positive cycles $\Phi_{\nu p}$ of $V_p \times V_p$ such that

$$T(1, \ldots, 1, p, \ldots, p)_p \equiv \sum_{\nu=0}^{n} \Phi_{\nu p}, \quad \Phi_{0p} = \Pi_p, \quad \Phi_{np} = \Pi_p'.$$

If $n = 1$, we obtain $T(1, p)_p = \Pi_p + \Pi_p'$, which gives a factorization[1, 2]

$$\Delta_p - T(1, p)_p X + p T(p, p)_p X^2 = [\Delta_p - \Pi_p X] \circ [\Delta_p - \Pi_p' X].$$

If $n = 2$, we get the following

THEOREM 6. *Suppose that $n = 2$. Then, for almost all p, there exist three positive cycles Φ_p, Ψ_p, Ω_p on $V_p \times V_p$, of the same dimension as V_p, satisfying the following relations:*

$$T(1,1,p,p)_p \equiv \Pi_p + \Phi_p + \Pi_p',$$

$$T(1,p,p,p^2)_p \equiv (p^2 - 1)\Delta_p + \Psi_p + \Omega_p,$$

$$p\Psi_p = \Pi_p \circ \Phi_p = \Phi_p \circ \Pi_p, \quad p\Omega_p = \Pi_p' \circ \Phi_p = \Phi_p \circ \Pi_p'.$$

These relations give a factorization of the denominator of $D_p(s)$ in Theorem 2 into the product

$$(\Delta_p - \Pi_p X) \circ (\Delta_p - \Phi_p X + p^3 X^2) \circ (\Delta_p - \Pi_p' X).$$

One may conjecture that a similar factorization of the denominator of $D_p(s)$ holds also for $n > 2$; the degree of the factors would be $n!/(j!(n - j)!)$ for $0 \leqq j \leqq n$.

[1] Eichler, M., "Quaternäre quadratische Formen und die Riemannsche Vermutung für die Kongruenzzetafunktion," *Arch. Math.*, **5**, 355–366 (1954).

[2] Shimura, G., "Correspondances modulaires et les fonctions ƶ de courbes algébriques," *J. Math. Soc. Japan*, **10**, 1–28 (1958).

[3] Shimura, G., "On the zeta-functions of the algebraic curves uniformized by certain automorphic functions," *J. Math. Soc. Japan*, **13**, 275–331 (1961).

[4] Shimura, G., "Sur les intégrales attachées aux formes automorphes," *J. Math. Soc. Japan*, **11**, 291–311 (1959).

[5] Shimura, G., "Arithmetic of alternating forms and quaternion hermitian forms," to appear in *J. Math. Soc. Japan*.

[6] Hecke, E., "Uber Modulfunktionen und die Dirichletschen Reihen mit Eulerscher Produktentwicklung," I, II, *Math. Ann.*, **114**, 1–28, 316–351 (1937).

[7] Cartan, Séminaire H., E.N.S. (1957–58) exposés 18–20.

[8] Shimura, G., "On analytic families of polarized abelian varieties and automorphic functions," to appear in *Ann. Math.*

[9] Shimura, G., "On the theory of automorphic functions," *Ann. Math.*, **70**, 101–144 (1959).

On the fields of definition for fields of automorphic functions

Proceedings of the 1963 Number Theory Conference,
Boulder Colorado, August 5-24, 1963, 26-32*

1. Let us first recall the classical theory of complex multiplication. Let K be an imaginary quadratic field and $\mathfrak{a}$ a fractional ideal in K. Regarding K as a subfield of the complex number field $\mathbf{C}$, we get a complex torus $\mathbf{C}/\mathfrak{a}$, which is analytically isomorphic to an elliptic curve $E(\mathfrak{a}) : y^2 = 4x^3 - g_2 x - g_3$, where g_2 and g_3 are complex numbers. The number $j = g_2^3/(g_2^3 - 27g_3^2)$ is called the invariat of $E(\mathfrak{a})$, and g_2, g_3 can be taken from the field $K(j)$, so that $K(j)$ is a field of definition for the field of elliptic functions with periods in $\mathfrak{a}$. Now the first main theorem of complex multiplication is stated as follows:

$K(j)$ is the absolute class-field over K. The reciprocity-law for $K(j)/K$ is described by the relation

$$(1) \qquad E(\mathfrak{a})^\sigma \cong E(\mathfrak{a}\mathfrak{b}^{-1}) \quad if \quad \sigma = \left(\frac{K(j)/K}{\mathfrak{b}} \right).$$

Here $E(\mathfrak{a})^\sigma$ is the curve $y^2 = 4x^3 - g_2^\sigma x - g_3^\sigma$. A similar relation can be found for the ray-class-field over K, if we take the points of finite order on $E(\mathfrak{a})$ into account.

2. The above result has a generalization for abelian varieties (= algebraic varieties analytically isomorphic to complex tori) of higher dimension. Let F be a totally real algebraic number field of degree g, and K a totally imaginary quadratic extension of F. Let $\tau_1, \ldots, \tau_g, \tau_1\rho, \ldots, \tau_g\rho$ be all the isomorphisms of K tino $\mathbf{C}$, where ρ is the complex conjugation. Let $\mathbf{C}^g$ denote the complex vector space of dimension g. For every fractional ideal $\mathfrak{a}$ in K, let $D(\mathfrak{a})$ be the submodule of $\mathbf{C}^g$ defined by

$$D(\mathfrak{a}) = \left\{ (x^{\tau_1}, \ldots, x^{\tau_g}) \,\middle|\, x \in \mathfrak{a} \right\}.$$

Then $\mathbf{C}^g/D(\mathfrak{a})$ is a complex torus. Moreover we can find an abelian varity $A(\mathfrak{a})$, isomorphic to $\mathbf{C}^g/D(\mathfrak{a})$, whose equations have coefficients in an algebraic number field. Let K^* be the field generated over the rational number field $\mathbf{Q}$ by the elements $\sum_{\nu=1}^g x^{\tau_\nu}$ for all x in K. Then K^* is a totally imaginary quadratic extension of a totally real algebraic number field. Let $2h$ be the degree of K^* over $\mathbf{Q}$. There exist h isomorphisms $\varphi_1, \ldots, \varphi_h$ of K^* into $\mathbf{C}$ with the following properties:

(i) *$\varphi_1, \ldots, \varphi_h, \varphi_1\rho, \ldots, \varphi_h\rho$ are all the isomorphisms of K^* into $\mathbf{C}$.*

(ii) *For every ideal $\mathfrak{b}$ in K^*, $\prod_{\nu=1}^h \mathfrak{b}^{\varphi_\nu}$ is an ideal in K.*

Now let H be the ideal-group in K^* defined by

$$H = \left\{ \mathfrak{b} \,\middle|\, \prod_{\nu=1}^h \mathfrak{b}^{\varphi_\nu} = (\beta), \ N(\mathfrak{b}) = \beta\beta^\rho \text{ for some } \beta \text{ in } K \right\}.$$

Let k be the unramified class-field over K^* corresponding to H. Then, k *is the composite of K^* and the field of moduli of a "polarized abelian variety" defined on $A(\mathfrak{a})$, and*

$$(2) \qquad A(\mathfrak{a})^\sigma \cong A\left(\mathfrak{a} \cdot \left(\textstyle\prod_{\nu=1}^{h} \mathfrak{b}^{\varphi_\nu}\right)^{-1}\right) \quad \text{if} \quad \sigma = \left(\frac{k/K^*}{\mathfrak{b}}\right).$$

The meaning of $A(\mathfrak{a})^\sigma$ is as follows: Write the equations $P_\lambda = 0$ for the variety $A(\mathfrak{a})$ and extend σ to an isomorphism of a suitably large field containing the coefficients of the P_λ; then $A(\mathfrak{a})^\sigma$ is the variety defined by $P_\lambda^\sigma = 0$. We note that (1) is a particular case of (2). The details of the theory were given in [G. Shimura and Y. Taniyama, Complex multiplication of abelian varieties and its applications to number theory, Publ. Math. Soc. Japan, No.6, 1961].

3. We now take a symmetric domain W in place of $\mathbf{C}^g$ and a discontinuous group Γ in place of $D(\mathfrak{a})$. Our purpose is to give an analogue of (1) and (2) for the variety W/Γ.

Let F, K, $\tau_1, \ldots, \tau_g$, ρ be as in **2**. Let m be a positive integer > 1. Let $T = (t_{ij})$ be an invertible matrix of size m with entries in K such that $t_{ij}^\rho = t_{ji}$. Let (r_ν, s_ν) be the signature of the complex hermitian matrix T^{τ_ν} for $1 \leq \nu \leq g$. We assume that $\sum_{\nu=1}^{g} r_\nu s_\nu > 0$. Let $\mathfrak{r}$ be the ring of integers in K, and K^m the vector space of m-dimensional row vectors with components in K. By an $\mathfrak{r}$-*lattice in* K^m we understand a finitely generated $\mathfrak{r}$-submodule of K^m which generates K^m over K. For an $\mathfrak{r}$-lattice L in K^m, put

$$\Gamma(L) = \left\{ B \in GL_m(K) \,\middle|\, BT \cdot {}^t B^\rho = T, \; LB = L \right\}.$$

Then $\Gamma(L)$ gives a properly discontinuous group of transformation on a symmetric domain $W = W_1 \times \cdots \times W_g$, where W_ν is the space of complex matrices z with r_ν rows and s_ν columns such that $1 - {}^t\bar{z}z$ is positive definite; if $r_\nu s_\nu = 0$ for some ν, we mean by W_ν the set consisting of only one point. For the sake of simplicity, assume that $W/\Gamma(L)$ is compact, though we can proceed without this assumption. Let $\mathfrak{K}(L)$ be the field of all meromorphic functions on W invariant under $\Gamma(L)$. Let K' be the field generated over $\mathbf{Q}$ by the elements $\sum_{\nu=1}^{g} \left(r_\nu x^{\tau_\nu} + s_\nu x^{\tau_\nu\rho}\right)$ for all $x \in K$.

Theorem 1. *Suppose that m is even and $r_\nu = s_\nu$ for every ν. Then we have $K' = \mathbf{Q}$. The function-field $\mathfrak{K}(L)$ has a subfield $\mathfrak{K}_0$ which generates $\mathfrak{K}(L)$ over $\mathbf{C}$ and such that $\mathfrak{K}_0$ and $\mathbf{C}$ are linearly disjoint over $\mathbf{Q}$. Roughly speaking, $\mathfrak{K}(L)$ can be defined over $\mathbf{Q}$.*

Assume that $r_\nu \neq s_\nu$ for some ν. Then K' is a totally imaginary quadratic extension of a totally real algebraic number field. Let $2h$ be the degree of K' over $\mathbf{Q}$. Let $\varphi_1, \ldots, \varphi_h, \varphi_1\rho, \ldots, \varphi_h\rho$ be all the isomorphisms of K' into $\mathbf{C}$. Then there exist h integers $u_1, \ldots, u_h$, depending on K, r_ν, s_ν, with which we can state our main result as follows:

Theorem 2. *Let H be the ideal-group in K' defined by*

$$H = \left\{ \mathfrak{r} \,\middle|\, \textstyle\prod_{\lambda=1}^{h}(\mathfrak{r}^{\varphi_\lambda})^{u_\lambda}(\mathfrak{r}^{\varphi_\lambda\rho})^{-u_\lambda} = (a) \text{ for some } a \in K \text{ such that } aa^\rho = 1 \right\}$$

$$\textit{if } m \textit{ is even,}$$

$$H = \left\{ \mathfrak{x} \ \middle| \ \prod_{\lambda=1}^{h} (\mathfrak{x}^{\varphi_\lambda})^{u_\lambda+1}(\mathfrak{x}^{\varphi_\lambda \rho})^{-u_\lambda} = (a) \ \text{for some } a \in K \ \text{such that } N(\mathfrak{x}) = aa^\rho \right\}$$

$$\textit{if } m \textit{ is odd.}$$

Let k_0 be the unramified class-field over K' corresponding to the ideal-group H. Then, for a suitable choice of L, there exists a projective variety $V(L)$ defined over k_0, whose function-field over $\mathbf{C}$ may be identified with $\mathfrak{K}(L)$. Moreover, if m is even and $\sigma = \left(\dfrac{k_0/K'}{\mathfrak{a}} \right)$ for an ideal $\mathfrak{a}$ in K', we have

(3) $\ V(L)^\sigma = V(M)$ for a lattice M such that $[M/L] = \prod_{\lambda=1}^{h} (\mathfrak{a}^{\varphi_\lambda})^{u_\lambda}(\mathfrak{a}^{\varphi_\lambda \rho})^{-u_\lambda}$, where $[M/L]$ is the ideal in K generated by the $\det(S)$ for all $S \in GL_m(K)$ such that $MS \subset L$.

Thus we get an analogue of (1) and (2) for the variety $W/\Gamma(L)$ as (3). A somewhat more complicated relation holds for odd m.

4. The abelian varieties (or elliptic curves) $A(\mathfrak{a})$ and $A(\mathfrak{b})$ are isomorphic if and only if $\mathfrak{a}$ and $\mathfrak{b}$ belong to the same ideal class. In the case of $W/\Gamma(L)$, if the lattices L and M are tranformed to each other by a similitude of T, then $W/\Gamma(L)$ and $W/\Gamma(M)$ are isomorphic. But the converse is not rue. It may happen that the conjugate $V(L)^\sigma$ of the variety $V(L)$ over K' are all birationally equivalent over $\mathbf{C}$ even if $[k_0 : K'] > 1$. For example, this is the case if m is even and prime to the class number of K. Hence there is a difference between (2) and (3) in the sense that (3) concerns *the field of definition* while (2) concerns *the field of moduli*.

Now let us investigate the field of moduli of $W/\Gamma(L)$ in the case where W is one-dimensional, so that $V(L)$ is an algebraic curve. In this case $\Gamma(L)$ is commensurable with the group of units in an order in a quaternion algebra over F. To state our result, we define the field of moduli $\mathfrak{F}(Y)$ of an algebraic curve Y defined over a subfield of $\mathbf{C}$ by the following property:

If k is a field of definition for Y, then $k \supset \mathfrak{F}(Y)$. If further σ is an isomorphism of k into $\mathbf{C}$, then σ is the identity on $\mathfrak{F}(Y)$ if and only if Y and Y^σ are birationally equivalent.

We can show that such a field $\mathfrak{F}(Y)$ exists and is uniquely determined by the birational equivalence class of Y.

Let F be a real cyclic extension of $\mathbf{Q}$ of degree g, and D be a quaternion algebra over F. Suppose that exactly $g - 1$ infinite prime spots of F, say $\mathfrak{p}_{\infty 2}, \ldots, \mathfrak{p}_{\infty g}$, ramify in D. Let $\mathfrak{o}$ be a maximal order in D, and Γ the group of units in $\mathfrak{o}$ with reduced norm 1. Then Γ gives a properly discontinuous group on the upper half plane W_0, and W_0/Γ is compact. Therefore we can find an algebraic curve analytically isomorphic to W_0/Γ. Then we obtain

Theorem 3. *Suppose that $[F : \mathbf{Q}] = 2$ or $[F : \mathbf{Q}]$ is odd. Let H be the group of ideals in F generated by the following ideals: (i) the prime ideals in F ramified in D; (ii) the principal ideals (a) such that a is totally positive; (iii) the squares of ideals in F. Then the field of moduli $\mathfrak{F}(W_0/\Gamma)$ of the curve*

W_0/Γ is contained in the class-field over F corresponds to the ideal group H. Moreover, if the multiplicative group of regular elements of D has no element of finite order other than ± 1, then the composite of F and $\mathfrak{F}(W_0/\Gamma)$ coincides with that class-field.

We note that the index of H in the whole ideal group is the (modified) type number of D. It is probable that the above theorem is true without any condition on $[F : \mathbf{Q}]$.

Arithmetic of unitary groups

Annals of Mathematics, 79 (1964), 369-409

(Received January 23, 1963)

The purpose of this paper is to develop the theory of elementary divisors, to prove the approximation theorem, and to determine the class number for the following two types of algebraic groups:

(i) the unitary group of a hermitian form over an algebraic number field;

(ii) the unitary group of a hermitian form over a quaternion algebra, having an algebraic number field as center, with respect to the canonical involution.

The latter includes the symplectic group as a special case, since we admit the total matric algebra of degree two as a quaternion algebra. Of course, the approximation theorem is obtained only in the case where the hermitian form is indefinite. A problem of the same kind was investigated for the group of regular elements of a simple algebra by Eichler [5], and for the orthogonal group by Eichler [6, 7] and Kneser [9]. In [11], we treated the groups of type (ii) with the same intent and obtained a somewhat weaker result than in the present paper.

We now explain our result by giving a summary of each section. Let F be an algebraic number field of finite degree and K a quadratic extension of F. Let V be a vector space over K of dimension n and $\Phi(x, y)$ a non-degenerate hermitian form on V with respect to the non-trivial automorphism of K over F. We denote by $U(V, \Phi)$ and $G(V, \Phi)$ respectively the unitary group of Φ and the group of similitudes of Φ (cf. 1.1 below). Further, we denote by $SU(V, \Phi)$ the subgroup of $U(V, \Phi)$ consisting of the elements with determinant 1. If n is even, we can consider the group of direct similitudes $H(V, \Phi)$ (cf. 2.2). In the first two sections, we give preliminaries for these four groups, and study elementary properties of maximal lattices, whose definition is as follows. Let $\mathfrak{g}$ and $\mathfrak{r}$ be the ring of integers in F and in K, respectively. Let L be an $\mathfrak{r}$-lattice in V. We denote by $\mu(L)$ the ideal in F generated by the elements $\Phi(x, x)$ for all $x \in L$, and call it the norm of L. We say that L is maximal if L is a maximal one among the $\mathfrak{r}$-lattices with the same norm. Let $\mathfrak{p}$ be a prime ideal of F, and $F_\mathfrak{p}$ the completion of F with respect to $\mathfrak{p}$; and let $K_\mathfrak{p} = K \otimes_F F_\mathfrak{p}$ and $V_\mathfrak{p} = V \otimes_F F_\mathfrak{p}$. In §§ 3 and 4, we treat the local theory of elementary divisors for the lattices in $V_\mathfrak{p}$. If $\mathfrak{p}$ decomposes in F, $K_\mathfrak{p}$ is isomorphic to $F_\mathfrak{p} \times F_\mathfrak{p}$, and therefore things are much easier. Section 3 is concerned with this case. In § 4, we discuss the case where $K_\mathfrak{p}$ is a field.

The existence of a canonical base for two maximal $\mathfrak{r}_\mathfrak{p}$-lattices is proved
(4.11). The same problem was treated by Bruhat [1] and Tamagawa
(unpublished). The former assumed the characteristic of the residue field
to be different from 2. It is of course indispensable to include the case
of prime factors of 2 in order to get the global theorem. There is another
important point, which seems to have been hitherto overlooked, in connection with the global problem: since the approximation theorem can be
proved only for $SU(V, \Phi)$ and not for $U(V, \Phi)$, the theory of elementary
divisors must be formulated in terms of $SU(V, \Phi)$, even in the local
case. This is why we have given the local theorems of elementary divisors (4.19, 4.21) for $SU(V, \Phi)$. To obtain these results, we need an
analysis of the determinants of the elements of $U(V, \Phi)$ leaving invariant
an $\mathfrak{r}_\mathfrak{p}$-lattice in $V_\mathfrak{p}$ (4.18). As an application of this proposition, we can
determine the local classes of maximal lattices with respect to $SU(V, \Phi)$
(4.23). Now in § 5, we get the global result. First the approximation
theorem (5.12) for $SU(V, \Phi)$ is proved in the case where $n > 1$ and Φ is
indefinite. The idea is to reduce the problem to the case of dimension
2, where the theorem is already given by Eichler. As an easy consequence,
it turns out that, with respect to $SU(V, \Phi)$, every genus of g-lattice in
V consists of only one class (5.19).[1] Then we determine the class number
of every genus of $\mathfrak{r}$-lattices in V with respect to $U(V, \Phi)$, $G(V, \Phi)$, and
$H(V, \Phi)$ under the same assumption on Φ (5.24). For the last two groups,
we restrict ourselves to the case of maximal lattices. The local theory of
elementary divisors, together with the approximation theorem, gives us
the global theorem of elementary divisors (5.30). In the final § 6, we
investigate the quaternion hermitian forms. Let A be a quaternion algebra
over F. For a hermitian form Φ over A with respect to the canonical
involution, we define the groups $G(V, \Phi)$ and $U(V, \Phi)$ in the same way
as above. Then we can prove the approximation theorem (6.7), the class
number theorems (6.9, 6.14), and the global theorem of elementary divisors
(6.20) for these groups when Φ is indefinite. This part is a sharpening
of the result of [11] and a complement to it.

Notation. R and C are as usual the real number field and the complex
number field, respectively. K is the division ring of real quaternions. If
S is an associative ring with an identity element, we denote by $M_m(S)$
and GL(m, S) the ring of $m \times m$ matrices with coefficients in S and the
group of invertible elements of $M_m(S)$, respectively. The identity element

---

[1] The part 5.0–5.19 is independent of the theory of maximal lattices. Therefore the
reader who is interested only in the approximation theorem may omit the discussions
concerning maximal lattices.

of $M_m(S)$ is denoted by 1_m. The δ_{ij} mean the usual Kronecker's symbol, namely, $\delta_{ij} = 1$ or 0 according as $i = j$ or $i \neq j$. Further notation will be given in the beginning of each section.

1. Preliminaries

1.0. In this section, we mean by F a field, by S a semi-simple algebra over F of finite dimension, by V a free left S-module of finite rank, and by n the rank of V over S. We denote by $E(V, S)$ the ring of all S-endomorphisms of V; every element β of $E(V, S)$ will operate on the right, so that $a(x\beta) = (ax)\beta$ for $a \in S$, $x \in V$, $\beta \in E(V, S)$. We denote by $GL(V, S)$ the group of all regular elements in $E(V, S)$. By an *involution* of S over F, we understand as usual an anti-automorphism ρ of S over F such that $(a^\rho)^\rho = a$ for all $a \in S$.

1.1. Let ρ be an involution of S over F. An F-bilinear mapping Φ of $V \times V$ into S is called a *ρ-hermitian form* if $\Phi(x, y)^\rho = \Phi(y, x)$, $\Phi(ax, by) = a\Phi(x, y)b^\rho$ for $x \in V$, $y \in V$, $a \in S$, $b \in S$. We say that Φ is *non-degenerate* if $\Phi(x, V) = \{0\}$ only for $x = 0$. For a non-degenerate ρ-hermitian form Φ on V, we denote by $G(V, \Phi)$ the group consisting of all the elements α of $GL(V, S)$ satisfying $\Phi(x\alpha, y\alpha) = \mu(\alpha)\Phi(x, y)$ with $\mu(\alpha) \in F$. It is clear that $\alpha \to \mu(\alpha)$ gives a homomorphism of $G(V, \Phi)$ into the multiplicative group of F. We denote by $U(V, \Phi)$ its kernel, namely

$$U(V, \Phi) = \{\alpha \in GL(V, S) \mid \Phi(x\alpha, y\alpha) = \Phi(x, y)\} \ .$$

$U(V, \Phi)$ is called the *unitary group* of Φ.

1.2. Suppose that ρ does not leave invariant any simple component of S. Then we can decompose S into the direct sum $S = R + R^\rho$ for a suitable subalgebra R. Let $\{y_i\}$ be a basis of V over S and let $\Phi(y_i, y_j) = a_{ij} + b_{ij}$ with $a_{ij} \in R$, $b_{ij} \in R^\rho$. Then $a_{ij}^\rho = b_{ji}$. As Φ is non-degenerate, the matrix (a_{ij}) in $M_n(R)$ is invertible. Put $(a_{ij})^{-1} = (c_{ij})$ with $c_{ij} \in R$. Let e be the identity element of R. Put $x_i = \sum_{j=1}^{n} (c_{ij} + \delta_{ij}e^\rho)y_j$ for $1 \leq i \leq n$. It is easy to verify that $\{x_i\}$ is a basis of V over S and $\Phi(x_i, x_j) = \delta_{ij}$.

1.3. The notation and the assumption being as in 1.2, put $W = eV$, $W' = e^\rho V$. If $W \ni x$ and $W \ni y$, then $\Phi(x, y) = \Phi(ex, ey) = e\Phi(x, y)e^\rho = 0$. On the other hand, if $W \ni x$ and $W' \ni y$, we have $\Phi(x, y) = \Phi(x, y)e \in R$. Let β and β' be the restrictions of an element α of $G(V, \Phi)$ to W and W', respectively. We see easily that $\beta \in GL(W, R)$, $\beta' \in GL(W', R^\rho)$, and $\Phi(x\beta, y\beta') = \mu(\alpha)\Phi(x, y)$ for $(x, y) \in W \times W'$. Conversely, if an element (β, β') of $GL(W, R) \times GL(W', R^\rho)$ satisfies $\Phi(x\beta, y\beta') = c\Phi(x, y)$ for $(x, y) \in W \times W'$ with an element c of F, this gives rise to an element α

of V such that $\mu(\alpha) = c$. By 1.2, we can find a basis $\{x_i\}$ of V over S such that $\Phi(x_i, x_j) = \delta_{ij}$. Put $x_i = u_i + v_i$ with $u_i \in W$, $v_i \in W'$. Then $\Phi(u_i, v_j) = \delta_{ij} e$. Using these bases $\{u_i\}$ of W and $\{v_i\}$ of W', we can verify easily that the group $G(V, \Phi)$ is isomorphic to the group

$$\{(X, Y) \in \mathrm{GL}(n, R) \times \mathrm{GL}(n, R^\rho) \mid XY^\rho = c1_n \text{ with } c \in F\},$$

where we put $Y^\rho = (\eta_{ji}^\rho)$ for $Y = (\eta_{ij}) \in M_n(R)$.

1.4. By a *quaternion algebra* over a field k, we understand a central simple algebra A over k such that $[A:k] = 4$. Every quaternion algebra A over a field k has an involution $a \to a^\iota$ such that $a + a^\iota$ is the reduced trace of a to k; we call this involution the *canonical involution* of A.

Let K be the center of our algebra S with an involution ρ. Suppose that $[K:F] = 2$, and the restriction of ρ to K gives the non-trivial automorphism of K over F. Further suppose that $[S:F] = 8$. If S is simple, K is a field and S is a quaternion algebra over K; denote by ι the canonical involution of S. If S is not simple, S is the sum of two quaternion algebras R and R^ρ over F. Denote by ι the involution of S whose restrictions to R and to R^ρ are their canonical involutions. In either case, we call ι the *canonical involution* of S.

1.5. PROPOSITION. *Let S and ι be as above, and let u be a regular element of S such that $u^\rho = u$. Put $B = \{a \in S \mid a^\iota = ua^\rho u^{-1}\}$. Then B is a quaternion algebra over F, and $S = B \otimes_F K$.*

Let F' be an algebraic closure of F. Put $K' = K \otimes_F F'$, $S' = S \otimes_F F'$, $B' = B \otimes_F F'$. Then S' is a semi-simple algebra over F', K' is the center of S', and $[K': F'] = 2$. We can extend ρ to an involution of S' over F', and ι to the canonical involution of S'; we denote these involutions of S' by the same letters ρ and ι. It is clear that $B' = \{a \in S' \mid a^\iota = ua^\rho u^{-1}\}$. Since F' is algebraically closed, S' is isomorphic to $M_2(F') \times M_2(F')$. Therefore, by a straightforward calculation, we find that B' is isomorphic to $M_2(F')$ and $S' = B' \otimes_{F'} K'$. Hence we obtain our assertions.

2. Hermitian forms over a commutative semi-simple algebra of rank 2

2.0. In this section, F denotes a field, and K denotes either a separable quadratic extension of F or the sum of two copies of F. We denote by ρ the non-trivial automorphism of K over F. For every $a \in K$, we put, as usual, $N_{K/F}(a) = aa^\rho$, $\mathrm{Tr}_{K/F}(a) = a + a^\rho$. We denote by F^* and K^* the multiplicative groups of regular elements in F and in K, respectively. Further we mean by V a free left K-module of rank n, and by Φ a non-degenerate ρ-hermitian form on V.

2.1. Let $\{x_i\}$ be a basis of V over K. Then $\det(\Phi(x_i, x_j)) \in F^*$. Let $d(\Phi)$ be the image of $\det(\Phi(x_i, x_j))$ by the natural homomorphism of F^* into $F^*/N_{K/F}(K^*)$. It is easy to see that $d(\Phi)$ is determined only by Φ and independent of the choice of $\{x_i\}$.

2.2. The ring $E(V, K)$ is isomorphic to $M_n(K)$. Since K is commutative, we can consider the determinant of the elements of $M_n(K)$, and hence of the elements of $E(V, K)$. For every $\alpha \in E(V, K)$, we denote by $\det(\alpha)$ the determinant of α. If $\alpha \in G(V, \Phi)$, we have

$$(2.2.1) \qquad\qquad N_{K/F}(\det(\alpha)) = \mu(\alpha)^n .$$

We define a subgroup $SU(V, \Phi)$ of $U(V, \Phi)$ by

$$SU(V, \Phi) = \{\alpha \in U(V, \Phi) \,|\, \det(\alpha) = 1\} .$$

Furthermore, if n is even, we define a subgroup $H(V, \Phi)$ of $G(V, \Phi)$ by

$$H(V, \Phi) = \{\alpha \in G(V, \Phi) \,|\, \det(\alpha) = \mu(\alpha)^{n/2}\} .$$

2.3. PROPOSITION. *The mapping* $\alpha \to \det(\alpha)$ *gives a surjective isomorphism of* $U(V, \Phi)/SU(V, \Phi)$ *to the group* $\{a \in K^* \,|\, N_{K/F}(a) = 1\}$.

2.4. PROPOSITION. *Suppose that n is odd. Let c be an element of F. Then, there exists an element α of $G(V, \Phi)$ such that $\mu(\alpha) = c$ if and only if $c \in N_{K/F}(K^*)$.*

2.5. PROPOSITION. *Suppose that n is even. Then $G(V, \Phi) = H(V, \Phi) \cdot U(V, \Phi)$ and $SU(V, \Phi) = H(V, \Phi) \cap U(V, \Phi)$.*

These propositions are well-known in the case where K is a field. The usual proof applies also to the case $K = F \times F$, on account of the considerations of 1.2, 1.3.

2.6. PROPOSITION. *Suppose that $n = 2$. Let ι be the canonical involution of $E(V, K)$ (cf. 1.4). Put $\mathscr{B} = \{\alpha \in E(V, K) \,|\, \Phi(x\alpha^\iota, y) = \Phi(x, y\alpha)\}$. Then $\mathscr{B}$ is a quaternion algebra over F, and $E(V, K) = \mathscr{B} \otimes_F K$. Let $\mathscr{B}^*$ be the group of regular elements of $\mathscr{B}$. Then $G(V, \Phi) = K^* \cdot \mathscr{B}^*$, $H(V, \Phi) = \mathscr{B}^*$, $SU(V, \Phi) = \{\alpha \in \mathscr{B} \,|\, \alpha\alpha^\iota = 1\}$.*

Consider the representation of $E(V, K)$ and Φ by matrices with coefficients in K with respect to a basis of V over K. Then the first two assertions follow immediately from 1.5. To prove the remaining part, we note that $\det(\alpha) = \alpha\alpha^\iota$ for every $\alpha \in E(V, K)$. If $\alpha \in \mathscr{B}^*$, we have $\Phi(x\alpha, y\alpha) = \Phi(x\alpha\alpha^\iota, y) = \det(\alpha)\Phi(x, y)$. Since Φ is ρ-hermitian, $\det(\alpha) \in F$, so that $\alpha \in G(V, \Phi)$ and $\mu(\alpha) = \det(\alpha) = \alpha\alpha^\iota$. Hence $\mathscr{B}^* \subset H(V, \Phi)$. For every $\alpha \in G(V, \Phi)$, we have

$$\det(\alpha)^{-1}\Phi(x\alpha^{\iota}, y) = \Phi(x\alpha^{-1}, y\alpha\alpha^{-1}) = \mu(\alpha)^{-1}\Phi(x, y\alpha) .$$

Hence if $\alpha \in H(V, \Phi)$, we have $\det(\alpha) = \mu(\alpha)$, so that $\alpha \in \mathscr{B}^*$. This proves $H(V, \Phi) = \mathscr{B}^*$, from which follows easily the last equality of the proposition. Let $\alpha \in G(V, \Phi)$. By (2.2.1), $N_{K/F}(\mu(\alpha)^{-1}\det(\alpha)) = 1$. Hence there exists an element b of K such that $\mu(\alpha)^{-1}\det(\alpha) = b^{-1}b^\rho$. Put $\beta = b\alpha$. Then $\det(\beta) = b^2\det(\alpha) = bb^\rho\mu(\alpha) = \mu(\beta)$; so $\beta \in H(V, \Phi) = \mathscr{B}^*$. This shows $G(V, \Phi) = K^* \cdot \mathscr{B}^*$.

We call $\mathscr{B}$ *the quaternion algebra attached to* Φ.

2.7. PROPOSITION. *Let the assumption and the notation be as in* 2.6. *Then there exist an F-linear bijection ξ of V to $\mathscr{B}$ and a non-zero element c of F such that:*

(i) $\xi(x\alpha) = \xi(x)\alpha$ *for every* $x \in V$ *and* $\alpha \in \mathscr{B}$;

(ii) $\Phi(x, x) = c \cdot \det(\xi(x))$ *for every* $x \in V$.

Since $n = 2$, V is a vector space over F of dimension 4. Consider V as a right $\mathscr{B}$-module. Since $\mathscr{B}$ is a quaternion algebra, any two right $\mathscr{B}$-modules (on which the identity element of $\mathscr{B}$ operates as the identity mapping) of the same dimension over F are isomorphic. Therefore, V is isomorphic to $\mathscr{B}$ as right $\mathscr{B}$-module. Let ξ be a $\mathscr{B}$-isomorphism of V onto $\mathscr{B}$. Then we have $\xi(x\alpha) = \xi(x)\alpha$ for $x \in V$, $\alpha \in \mathscr{B}$. Put $u = \xi^{-1}(1)$. If $\xi(x) = \alpha$, then $x = u\alpha$, and $\Phi(x, x) = \Phi(u\alpha, u\alpha) = \det(\alpha)\Phi(u, u) = \det(\xi(x))\Phi(u, u)$. Putting $c = \Phi(u, u)$, we get (ii). Assume that $c = 0$. Then $\Phi(x, x) = 0$ for every $x \in V$. It follows easily from this that $\Phi(x, y) = 0$ for every $(x, y) \in V \times V$, which is a contradiction; hence $c \neq 0$.

2.8. PROPOSITION. *The algebra $\mathscr{B}$ defined in* 2.6 *is a division algebra if and only if there is no element x of V such that $x \neq 0$ and $\Phi(x, x) = 0$.* This follows immediately from (ii) of 2.7.

2.9. Now we assume that F is the quotient field of a Dedekind domain $\mathfrak{g}$, and denote by $\mathfrak{r}$ the ring of all $\mathfrak{g}$-integral elements in K. If K is a field, $\mathfrak{r}$ is a Dedekind domain; and if $K = F \times F$, then $\mathfrak{r} = \mathfrak{g} \times \mathfrak{g}$. By a $\mathfrak{g}$-*lattice* in a finite dimensional vector space W over F, we mean a finitely generated $\mathfrak{g}$-module L in W such that $FL = W$. Every fractional ideal in F with respect to $\mathfrak{g}$ is a $\mathfrak{g}$-lattice in F and nothing else; we call it a $\mathfrak{g}$-*ideal*. Further, we call a $\mathfrak{g}$-lattice $\mathfrak{a}$ in K an $\mathfrak{r}$-ideal if $\mathfrak{r}\mathfrak{a} \subset \mathfrak{a}$. Then all the $\mathfrak{r}$-ideals form a commutative group with respect to the multiplication. We define an $\mathfrak{r}$-ideal $\mathfrak{b}$ by

$$\mathfrak{b}^{-1} = \{a \in K \,|\, \mathrm{Tr}_{K/F}(a\mathfrak{r}) \subset \mathfrak{g}\} .$$

If $K = F \times F$, we have $\mathfrak{b} = \mathfrak{r}$; and if K is a field, $\mathfrak{b}$ is the "different" of $\mathfrak{r}$ with respect to $\mathfrak{g}$. For every $\mathfrak{r}$-ideal $\mathfrak{a}$, we put $N_{K/F}(\mathfrak{a}) = \mathfrak{a}\mathfrak{a}^\rho \cap F$.

2.10. We call a $\mathfrak{g}$-lattice L in V an $\mathfrak{r}$-*lattice* if $\mathfrak{r}L \subset L$. For every $\mathfrak{r}$-lattice L in V, we denote by $\mu(L)$ the $\mathfrak{g}$-ideal generated by the elements $\Phi(x, x)$ for $x \in L$, and call $\mu(L)$ the *norm* of L (with respect to Φ). We say that L is *maximal* (with respect to Φ) if L is a maximal one among the $\mathfrak{r}$-lattices with the same norm. Further we denote by $\mu_0(L)$ the $\mathfrak{r}$-ideal generated by the elements $\Phi(x, y)$ for $x \in L$, $y \in L$.

2.11. PROPOSITION. *Let L be an $\mathfrak{r}$-lattice in V. Then*

$$\mu(L)\mathfrak{r} \subset \mu_0(L) \subset \mu(L)\mathfrak{d}^{-1}$$

and

$$\mathrm{Tr}_{K/F}\big(\mu_0(L)\big) \subset \mu(L) \ .$$

If $x \in L$, $y \in L$, $a \in \mathfrak{r}$, then

$$\mathrm{Tr}_{K/F}\big(a\Phi(x, y)\big) = \Phi(ax + y, ax + y) - \Phi(ax, ax) - \Phi(y, y) \in \mu(L) \ .$$

Therefore $\Phi(x, y)\mu(L)^{-1} \subset \mathfrak{d}^{-1}$, so that $\Phi(x, y) \subset \mathfrak{d}^{-1}\mu(L)$.

2.12. PROPOSITION. *Let L be an $\mathfrak{r}$-lattice in V and α be an element of $G(V, \Phi)$. Then $L\alpha$ is an $\mathfrak{r}$-lattice in V, and $\mu(L\alpha) = \mu(L)\mu(\alpha)$. If L is maximal, so is $L\alpha$.*

This is an easy consequence of our definition of $\mu(\alpha)$ and $\mu(L)$.

2.13. PROPOSITION. *Let L and M be $\mathfrak{r}$-lattices in V. Suppose that M is maximal. Let c be an element of $\mathfrak{g}$. If $L \supset M$ and $\mu(M) \supset c\mu(L)$, then $M \supset cL$.*

Put $H = M + cL$. Then H is an $\mathfrak{r}$-lattice in V. Let $u = x + cy$ be an element of H with $x \in M$ and $y \in L$. Then

$$\Phi(u, u) = \Phi(x, x) + c \cdot \mathrm{Tr}_{K/F}\big(\Phi(x, y)\big) + c^2\Phi(y, y) \ .$$

By our assumption and 2.11, we see that $\Phi(u, u) \in \mu(M)$. It follows that $\mu(H) = \mu(M)$. Since M is maximal, we must have $H = M$, so that $M \supset cL$.

2.14. PROPOSITION. *Let M be an $\mathfrak{r}$-lattice in V. Then there exists a maximal $\mathfrak{r}$-lattice L in V such that $\mu(L) = \mu(M)$ and $L \supset M$.*

Take a basis $\{x_i\}$ of V over K contained in M. Let H be the set of vectors y of V such that $\Phi(y, x_i) \in \mu(M)\mathfrak{d}^{-1}$ for every i. H is clearly an $\mathfrak{r}$-lattice in V. Let L be an $\mathfrak{r}$-lattice such that $L \supset M$ and $\mu(L) = \mu(M)$. If $y \in L$, we have $\Phi(y, x_i) \in \mu(M)\mathfrak{d}^{-1}$ by 2.11, so that $y \in H$. Hence $L \subset H$. Since H is a $\mathfrak{g}$-lattice, the ascending chain condition holds for the $\mathfrak{g}$-lattices contained in H. Therefore, we can find a maximal one among the $\mathfrak{r}$-lattices containing M and with the norm $\mu(M)$. This proves our proposition.

2.15. For two $\mathfrak{r}$-lattices L and M in V, we denote by $[L/M]$ the

$\mathfrak{r}$-ideal generated by the $\det(\alpha)$ for all $\alpha \in E(V, K)$ such that $L\alpha \subset M$. $[L/M]$ has the following properties:

(i) $[L/M][M/N] = [L/N]$;

(ii) If $L \supset M$ and the factor module L/M is $\mathfrak{r}$-isomorphic to $\mathfrak{r}/\mathfrak{b}$ for an $\mathfrak{r}$-ideal $\mathfrak{b}$, then $[L/M] = \mathfrak{b}$.

(iii) For every $\alpha \in \mathrm{GL}(V, K)$, we have

$$(2.15.1) \qquad\qquad [L/L\alpha] = \det(\alpha)\mathfrak{r} \ .$$

3. Maximal lattices in the case $K = F \times F$

3.0. In this section we assume that $K = F \times F$, and F is the quotient field of a principal ideal domain $\mathfrak{g}$. We denote by e and e^ρ the idempotents of K such that $e + e^\rho$ is the identity element of K.

3.1. PROPOSITION. *Let a be an element of K^* and b an element of F^* satisfying $N_{K/F}(a) = b^n$. Then there exists an element α of $G(V, \Phi)$ such that $\det(\alpha) = a$, $\mu(\alpha) = b$.*

As observed in 1.3, $G(V, \Phi)$ is isomorphic to the group

$$\{(X, Y) \in \mathrm{GL}(n, F) \times \mathrm{GL}(n, F) \,|\, X \cdot {}^t Y = c1_n \text{ with } c \in F\} \ .$$

If α corresponds to (X, Y), then $\det(\alpha) = \det(X)e + \det(Y)e^\rho$, $\mu(\alpha)1_n = X \cdot {}^t Y$. Our assertion follows quite easily from this fact.

3.2. PROPOSITION. *For every $\mathfrak{r}$-lattice L in V, there exist a basis $\{z_i\}$ of V over K and $\mathfrak{g}$-ideals $\mathfrak{a}_1, \cdots, \mathfrak{a}_n$ such that $L = \sum_{i=1}^n (\mathfrak{g}e + \mathfrak{a}_i e^\rho)z_i$, $\Phi(z_i, z_j) = \delta_{ij}$, $\mathfrak{a}_1 \supset \cdots \supset \mathfrak{a}_n$; and we have $\mu(L) = \mathfrak{a}_1$. L is maximal if and only if $\mathfrak{a}_1 = \cdots = \mathfrak{a}_n$.*

Put $W = eV$, $W' = e^\rho V$, $M = eL$, $M' = e^\rho L$. Then $V = W + W'$, $L = M + M'$; and M (resp. M') is a $\mathfrak{g}$-lattice in W (resp. W'). As observed in 1.3, $\Phi(x, y) \in Fe$ for $x \in W$ and $y \in W'$. Hence we can define an F-bilinear mapping Ψ of $W \times W'$ into F by $\Phi(x, y) = \Psi(x, y)e$ for $(x, y) \in W \times W'$. Put $N' = \{y \in W' \,|\, \Psi(M, y) \subset \mathfrak{g}\}$. Then N' is a $\mathfrak{g}$-lattice in W'. Applying the fundamental theorem of elementary divisors to the $\mathfrak{g}$-lattices N' and M', we can find a basis $\{y_i\}$ of W' over F, and $\mathfrak{g}$-ideals $\mathfrak{a}_1, \cdots, \mathfrak{a}_n$ such that $N' = \sum_{i=1}^n \mathfrak{g}y_i$, $M' = \sum_{i=1}^n \mathfrak{a}_i y_i$, $\mathfrak{a}_1 \supset \cdots \supset \mathfrak{a}_n$. On the other hand, taking a basis of M over $\mathfrak{g}$, we observe that $M = \{x \in W \,|\, \Psi(x, N') \subset \mathfrak{g}\}$. Therefore, if we select a basis $\{x_i\}$ of W so that $\Psi(x_i, y_j) = \delta_{ij}$, we must have $M = \sum_{i=1}^n \mathfrak{g}x_i$. Put now $z_i = x_i + y_i$. Then we have $L = \sum_{i=1}^n (\mathfrak{g}e + \mathfrak{a}_i e^\rho)z_i$ and $\Phi(z_i, z_j) = \delta_{ij}$. It is clear that $\mu(L) = \mathfrak{a}_1$. Now put $L' = \sum_{i=1}^n (\mathfrak{g}e + \mathfrak{a}_1 e^\rho)z_i$. Then L' is an $\mathfrak{r}$-lattice in V, $L \subset L'$, and $\mu(L') = \mathfrak{a}_1$. Hence if L is maximal, we must have $L = L'$, so that $\mathfrak{a}_1 = \cdots = \mathfrak{a}_n$. It remains to prove that, if $L = \sum_{i=1}^n (\mathfrak{g}e + \mathfrak{a}e^\rho)z_i$ with a $\mathfrak{g}$-ideal $\mathfrak{a}$, and if

$\Phi(z_i, z_j) = \delta_{ij}$, then L is maximal. To show this, take any $\mathfrak{r}$-lattice M in V such that $L \subset M$ and $\mu(M) = \mu(L) = \mathfrak{a}$. Let u be an element of M, and let $u = \sum_{i=1}^{n} a_i z_i$ with $a_i \in K$. By 2.11, we have $a_j(\mathfrak{g}e + \mathfrak{g}e^\rho) = \Phi(u, (\mathfrak{g}e + \mathfrak{a}e^\rho)z_j) \subset \mu(M)\mathfrak{r} = \mathfrak{a}\mathfrak{r}$, so that $a_j \in \mathfrak{g}e + \mathfrak{a}e^\rho$, and hence $u \in L$. Therefore we get $L = M$. This proves the maximality of L and completes our proof.

3.3. PROPOSITION. *Let L and M be maximal $\mathfrak{r}$-lattices in V. Then there exists an element α of $G(V, \Phi)$ such that $L\alpha = M$. If $\mu(L) = \mu(M)$, we can take α from $U(V, \Phi)$. Moreover, if $[L/M] = \mathfrak{r}$, then $\mu(L) = \mu(M)$ and we can take α from $SU(V, \Phi)$.*

Put $\mu(L) = \mathfrak{a}$, $\mu(M) = \mathfrak{b}$ and $\mathfrak{a}^{-1}\mathfrak{b} = \mathfrak{g}c$ with $c \in F$. By 3.2, there exist bases $\{z_i\}$ and $\{w_i\}$ of V over K such that $\Phi(z_i, z_j) = \Phi(w_i, w_j) = \delta_{ij}$, $L = \sum_{i=1}^{n} (\mathfrak{g}e + \mathfrak{a}e^\rho)z_i$, $M = \sum_{i=1}^{n} (\mathfrak{g}e + \mathfrak{b}e^\rho)w_i$. Define an element α of $GL(V, K)$ by $z_i\alpha = (e + ce^\rho)w_i$ for $1 \leq i \leq n$. Then we see that $\alpha \in G(V, \Phi)$, $\mu(\alpha) = c$, $L\alpha = M$. If $\mathfrak{a} = \mathfrak{b}$, we can take $c = 1$; then $\alpha \in U(V, \Phi)$. If $[L/M] = \mathfrak{r}$, $\det(\alpha)$ must be a unit of $\mathfrak{r}$, because of (2.15.1). Hence, by (2.2.1), $\mu(\alpha)$ is a unit of $\mathfrak{g}$. It follows that $\mu(L) = \mu(M)$. As just proved, there exists an element β of $U(V, \Phi)$ such that $L\beta = M$. By (2.15.1) and (2.2.1), we see that $\det(\beta)$ is a unit of $\mathfrak{r}$ and $N_{K/F}(\det(\beta)) = 1$. Define an element γ of $GL(V, K)$ by $z_1\gamma = \det(\beta)^{-1}z_1$, $z_i\gamma = z_i$ for $i > 1$. Then $\gamma \in U(V, \Phi)$, $\det(\gamma) = \det(\beta)^{-1}$, $L\gamma = L$. Hence $L\gamma\beta = M$, $\det(\gamma\beta) = 1$, $\mu(\gamma\beta) = 1$. This completes our proof.

3.4. PROPOSITION. *Let L and M be maximal $\mathfrak{r}$-lattices in V. Put $\mu(L) = \mathfrak{a}$, $\mu(M) = \mathfrak{a}c$ with a $\mathfrak{g}$-ideal $\mathfrak{a}$ and an element c of F. Then there exist a basis $\{z_i\}$ of V over K and elements $a_1, \cdots, a_n, b_1, \cdots, b_n$ of F such that $\Phi(z_i, z_j) = \delta_{ij}$, $L = \sum_{i=1}^{n} (\mathfrak{g}e + \mathfrak{a}e^\rho)z_i$, $M = \sum_{i=1}^{n} (\mathfrak{g}a_ie + \mathfrak{a}b_ie^\rho)z_i$, $\mathfrak{g}a_1 \supset \cdots \supset \mathfrak{g}a_n$, $\mathfrak{g}b_n \supset \cdots \supset \mathfrak{g}b_1$, $a_1b_1 = \cdots = a_nb_n = c$.*

Let W, W', Ψ be as in the proof of 3.2. Since L and M are maximal, the proof of 3.2 shows that

$$e^\rho L = \{y \in W' \mid \Psi(eL, y) \subset \mathfrak{a}\}, \qquad e^\rho M = \{y \in W' \mid \Psi(eM, y) \subset \mathfrak{a}c\}.$$

Applying the theory of elementary divisors to eL and eM, we can find a basis $\{x_i\}$ of W over F and elements a_i of F such that $eL = \sum_{i=1}^{n} \mathfrak{g}x_i$, $eM = \sum_{i=1}^{n} \mathfrak{g}a_ix_i$, $\mathfrak{g}a_1 \supset \cdots \supset \mathfrak{g}a_n$. Let $\{y_i\}$ be the basis of W' over F determined by $\Psi(x_i, y_j) = \delta_{ij}$. Then $e^\rho L = \sum_{i=1}^{n} \mathfrak{a}y_i$, $e^\rho M = \sum_{i=1}^{n} \mathfrak{a}a_i^{-1}cy_i$. Put $z_i = x_i + y_i$, $b_i = a_i^{-1}c$. Then we get the desired result.

3.5. PROPOSITION. *Let L be a maximal $\mathfrak{r}$-lattice in V, and $\{u_i\}$ a basis of V over K such that $\Phi(u_i, u_j) = \delta_{ij}$ and $L = \sum_{i=1}^{n} (\mathfrak{g}e + \mathfrak{a}e^\rho)u_i$ where $\mathfrak{a} = \mu(L)$. Denote by Γ the subgroup of $SU(V, \Phi)$ consisting of the elements*

γ *such that* $L\gamma = L$, *and by* Δ *the set of elements* α *of* $G(V, \Phi)$ *such that* $u_i\alpha = (a_ie + b_ie^\rho)u_i$ *for* $1 \leq i \leq n$ *with elements* a_i, b_i *of* F *and* $\mathfrak{g}a_1 \supset \cdots \supset \mathfrak{g}a_n$. *Then* $G(V, \Phi) = \Gamma \cdot \Delta \cdot \Gamma$.

Let $\alpha \in G(V, \Phi)$, $M = L\alpha$, $c = \mu(\alpha)$. Applying 3.4 to this $\{L, M, c\}$, we get a basis $\{z_i\}$ of V over K and elements a_i, b_i of F with the properties given in that proposition. Define two elements γ and β of $GL(V, K)$ by $u_i\gamma = z_i$ and $u_i\beta = (a_ie + b_ie^\rho)u_i$. We see easily that $\gamma \in U(V, \Phi)$, $L\gamma = L$, $\beta \in \Delta$, $\mu(\beta) = c$. Further we have $L\beta\gamma = M = L\alpha$. Therefore if we put $\varepsilon = \alpha(\beta\gamma)^{-1}$, then $\alpha = \varepsilon\beta\gamma$, $L\varepsilon = L$, $\varepsilon \in G(V, \Phi)$, and $\mu(\varepsilon) = 1$, so that $\varepsilon \in U(V, \Phi)$. Now define elements σ and τ of $GL(V, K)$ by $u_1\sigma = \det(\gamma)^{-1}u_1$, $u_1\tau = \det(\varepsilon)^{-1}u_1$, $u_i\sigma = u_i\tau = u_i$ for $i > 1$. Since $N_{K/F}(\det(\gamma)) = N_{K/F}(\det(\varepsilon)) = 1$, σ and τ are contained in $U(V, \Phi)$; and since $L\gamma = L\varepsilon = L$, $\det(\gamma)$ and $\det(\varepsilon)$ are units of $\mathfrak{r}$, so that $L\sigma = L\tau = L$. Further we have $\det(\sigma) = \det(\gamma)^{-1}$, $\det(\tau) = \det(\varepsilon)^{-1}$, $\alpha = (\varepsilon\sigma)\sigma^{-1}\beta\tau^{-1}(\tau\gamma)$. We see easily that $\varepsilon\sigma \in \Gamma$, $\tau\gamma \in \Gamma$, $\sigma^{-1}\beta\tau^{-1} \in \Delta$. This proves our proposition.

3.6. Notation being as in 3.4, we call $\{(a_ie + b_ie^\rho)\mathfrak{r}\}$ *the set of elementary divisors of* M *relative to* L, and denote it by $\{L : M\}$. If $\{L : M\} = \{c_1, \cdots, c_n\}$, we have $[L/M] = c_1 \cdots c_n$. It follows that $N_{K/F}([L/M]) = (\mu(L)^{-1}\mu(M))^n$.

3.7. PROPOSITION. *Let* L, M, N *be maximal* $\mathfrak{r}$-*lattices in* V. *If there exists an element* α *of* $G(V, \Phi)$ *such that* $L\alpha = L$ *and* $M\alpha = N$, *then* $\{L : M\} = \{L : N\}$. *Conversely if* $\{L : M\} = \{L : N\}$, *there exists an element* α *of* $SU(V, \Phi)$ *such that* $L\alpha = L$ *and* $M\alpha = N$.

The first assertion is obvious. Suppose now that $\{L : M\} = \{L : N\}$; then $\mu(M) = \mu(N)$. Put $\mu(L) = a$, $\mu(M) = ac$ with $c \in F$. By 3.4, we get a basis $\{z_i\}$ of V over K and elements a_i, b_i of F with the properties in 3.4. Further we get a basis $\{u_i\}$ and elements c_i, d_i of F with the corresponding properties for L and N, with the same c. Our assumption implies $(a_ie + b_ie^\rho)\mathfrak{r} = (c_ie + d_ie^\rho)\mathfrak{r}$. Define an element β of $GL(V, K)$ by $z_i\beta = u_i$ for $1 \leq i \leq n$. Then $\beta \in U(V, \Phi)$, $L\beta = L$, $M\beta = N$. As in the proof of 3.5, define an element γ of $GL(V, K)$ by $z_1\gamma = \det(\beta)^{-1}z_1$, $z_i\gamma = z_i$ for $i > 1$. Then $\det(\gamma) = \det(\beta)^{-1}$, $L\gamma = L$, $M\gamma = M$, $\gamma \in U(V, \Phi)$. Put $\alpha = \gamma\beta$. Then $\alpha \in U(V, \Phi)$, $\det(\alpha) = 1$, $L\alpha = L$, $M\alpha = N$. This completes our proof.

3.8. PROPOSITION. *Suppose that* $n = 2$. *Let* $\mathscr{B}$ *be the quaternion algebra attached to* Φ (cf. 2.6), *and let* L *be an* $\mathfrak{r}$-*lattice in* V. *Put* $\mathfrak{o} = \{\alpha \in \mathscr{B} \mid L\alpha \subset L\}$. *Then* L *is maximal if and only if* $\mathfrak{o}$ *is a maximal order in* $\mathscr{B}$.

By 3.2, there exist a basis $\{z_1, z_2\}$ of V over K and $\mathfrak{g}$-ideals a_1, a_2 such that $\Phi(z_i, z_j) = \delta_{ij}$, $L = \sum_{i=1}^{2} (\mathfrak{g}e + a_ie^\rho)z_i$, $a_1 \supset a_2$. To every $\alpha \in E(V, K)$, we let correspond an element $((a_{ij}), (b_{ij}))$ of $M_2(F) \times M_2(F)$ by the relation

$z_i\alpha = \sum_{j=1}^{2}(a_{ij}e + b_{ij}e^\rho)z_j$. By our definition of $\mathscr{B}$, we see easily that $\mathscr{B}$ corresponds to

$$\left\{\left(\begin{pmatrix} a & b \\ c & d \end{pmatrix}, \begin{pmatrix} d & -c \\ -b & a \end{pmatrix}\right) \,\middle|\, a, b, c, d \in F\right\},$$

and hence $\mathfrak{o}$ corresponds to

$$\left\{\left(\begin{pmatrix} a & b \\ c & d \end{pmatrix}, \begin{pmatrix} d & -c \\ -b & a \end{pmatrix}\right) \,\middle|\, a \in \mathfrak{g},\, d \in \mathfrak{g},\, b \in \mathfrak{g},\, c \in \mathfrak{a}_1^{-1}\mathfrak{a}_2\right\}.$$

If L is maximal, we have $\mathfrak{a}_1 = \mathfrak{a}_2$, so that $\mathfrak{o}$ is isomorphic to $M_2(\mathfrak{g})$, and hence $\mathfrak{o}$ is a maximal order in $\mathscr{B}$. If $\mathfrak{a}_1 \neq \mathfrak{a}_2$, then we get a strictly larger order than $\mathfrak{o}$ by exchanging the condition $b \in \mathfrak{g}$ for the condition $b \in \mathfrak{a}_2^{-1}\mathfrak{a}_1$. Therefore, if $\mathfrak{o}$ is maximal, we must have $\mathfrak{a}_1 = \mathfrak{a}_2$, so that L is maximal.

4. Hermitian forms and maximal lattices over $\mathfrak{p}$-adic fields

4.0. In this section, F denotes a field which is locally compact with respect to a discrete valuation; $\mathfrak{g}$ denotes the valuation ring, and $\mathfrak{p}$ the maximal ideal of $\mathfrak{g}$. We assume that the characteristic of F is different from 2; we admit the case where F is of characteristic 0 and $\mathfrak{p}$ divides 2. Further K, $\mathfrak{r}$, and $\mathfrak{P}$ denote respectively a separable quadratic extension of F, the ring of integers in K, and the maximal ideal of $\mathfrak{r}$. Then the $\mathfrak{r}$-ideal $\mathfrak{d}$ defined in 2.2 is a power of $\mathfrak{P}$; more precisely, $\mathfrak{d} = \mathfrak{r}$ if $\mathfrak{pr} = \mathfrak{P}$, and $\mathfrak{d} = \mathfrak{P}^e$ with a positive integer e if $\mathfrak{pr} = \mathfrak{P}^2$; and in the latter case, $e = 1$ if $2 \notin \mathfrak{p}$. As before, V denotes a vector space over K of dimension n, and Φ denotes a non-degenerate ρ-hermitian form on V. We call Φ *isotropic* if $\Phi(x, x) = 0$ for some non-zero element x of V; otherwise we call Φ *anisotropic*.

4.1. PROPOSITION. *If $n > 2$, every Φ is isotropic. If $n = 1$, every Φ is anisotropic. If $n = 2$, Φ is isotropic or anisotropic according as $-d(\Phi)$ is the identity element of $F^*/N_{K/F}(K^*)$ or not.*

4.2. PROPOSITION. *Let Φ and Φ' be non-degenerate ρ-hermitian forms on V. Then there exists an element α of $\mathrm{GL}(V, K)$ such that $\Phi(x\alpha, y\alpha) = \Phi'(x, y)$, if and only if $d(\Phi) = d(\Phi')$ (cf. 2.1).*

These propositions are well-known.

4.3. By 4.2 and by the usual technique, we can find a direct decomposition $V = \sum_{i=1}^{m}(Kx_i + Ky_i) + W$, $n = 2m + \dim W$, where $\Phi(x_i, y_j) = \delta_{ij}$, $\Phi(x_i, x_j) = \Phi(y_i, y_j) = \Phi(x_i, W) = \Phi(y_i, W) = 0$, W is a subspace of V of dimension 0, 1 or 2, and the restriction of Φ to W is anisotropic if $\dim W$ is 1 or 2. We call such a decomposition a *Witt decomposition* of V, and call W a *kernel subspace* of V with respect to Φ. When $\dim W = 2$, we often need the quaternion algebra attached to the restriction of Φ to W, defined in 2.6. We denote the algebra by $\mathscr{B}(W)$, or simply by $\mathscr{B}$, and the canonical involution of $\mathscr{B}$ by ι.

4.4. PROPOSITION. *Let c be an element of F^*. There exists an element α of $G(V, \Phi)$ such that $\mu(\alpha) = c$ if and only if, either n is even, or n is odd and $c \in N_{K/F}(K^*)$.*

This is easily obtained by applying 4.2 to Φ and $c\Phi$.

4.5. PROPOSITION. *Suppose that Φ is anisotropic (and hence $n = 1$ or 2). Let $\mathfrak{a}$ be a $\mathfrak{g}$-ideal, and let $L = \{x \in V \mid \Phi(x, x) \in \mathfrak{a}\}$. Then L is a maximal $\mathfrak{r}$-lattice; and if $n = 2$, or if $n = 1$ and $\Phi(y, y)\mathfrak{a}\mathfrak{r}$ is an even power of $\mathfrak{P}$ for some y, then $\mu(L) = \mathfrak{a}$. If $n = 1$ and $\Phi(y, y)\mathfrak{a}\mathfrak{r}$ is an odd power of $\mathfrak{P}$, then $\mu(L) = \mathfrak{p}\mathfrak{a}$. Conversely, every maximal $\mathfrak{r}$-lattice L is written in this form with $\mathfrak{a} = \mu(L)$.*

Suppose that $n = 1$. Then we have $V = Ky$ for a non-zero element y of V. Every $\mathfrak{r}$-lattice L in V is written in the form $L = \mathfrak{b}y$ for an $\mathfrak{r}$-ideal $\mathfrak{b}$. Then $\mu(L) = \Phi(y, y) \cdot N_{K/F}(\mathfrak{b})$. It follows that every $\mathfrak{r}$-lattice L is maximal, and L is determined by $\mu(L)$. Let $\mathfrak{a}$ be a $\mathfrak{g}$-ideal and let ν be the smallest integer such that $\mathfrak{P}^{2\nu} \subset \Phi(y, y)^{-1}\mathfrak{a}\mathfrak{r}$. Then we see easily that $\mathfrak{P}^\nu = \{a \in K \mid \Phi(ay, ay) \in \mathfrak{a}\}$, so that $\mathfrak{P}^\nu y = \{x \in V \mid \Phi(x, x) \in \mathfrak{a}\}$. Our assertion for $n = 1$ follows easily from these considerations.

Now consider the case $n = 2$. Let $\mathscr{B}$ be the quaternion algebra attached to Φ. By 2.8, $\mathscr{B}$ is a division algebra. For every $\alpha \in \mathscr{B}$, define $\nu(\alpha)$ by $\nu(\alpha) = \nu$ if $\alpha \neq 0$ and $\det(\alpha)\mathfrak{g} = \mathfrak{p}^\nu$, and $\nu(0) = \infty$. Then $\nu(\alpha\beta) = \nu(\alpha) + \nu(\beta)$, $\nu(a + \beta) \geq \mathrm{Min}(\nu(\alpha), \nu(\beta))$, and there exists an element π in $\mathscr{B}$ such that $\nu(\pi) = 1$. Put $\mathfrak{o} = \{\alpha \in \mathscr{B} \mid \nu(\alpha) \geq 0\}$, $\mathfrak{Q} = \{\alpha \in \mathscr{B} \mid \nu(\alpha) \geq 1\}$. Then $\mathfrak{o}$ is the unique maximal order in $\mathscr{B}$, and $\mathfrak{Q}$ is the maximal ideal in $\mathfrak{o}$. Further, every right or left $\mathfrak{o}$-ideal $\mathfrak{b}$ (i.e., a $\mathfrak{g}$-lattice in $\mathscr{B}$ such that $\mathfrak{b}\mathfrak{o} \subset \mathfrak{b}$ or $\mathfrak{o}\mathfrak{b} \subset \mathfrak{b}$) is a power $\mathfrak{Q}^\nu$ for an integer ν, and then $\mathfrak{b} = \mathfrak{Q}^\nu = \{\alpha \in \mathscr{B} \mid \nu(\alpha) \geq \nu\}$. By 2.7, there exists a $\mathscr{B}$-isomorphism ξ of V onto $\mathscr{B}$ satisfying $\Phi(x, x) = c \cdot \det(\xi(x))$ for every $x \in V$, where $c = \Phi(\xi^{-1}(1), \xi^{-1}(1))$. Now let $\mathfrak{a}$ be a $\mathfrak{g}$-ideal and let $L = \{x \in V \mid \Phi(x, x) \in \mathfrak{a}\}$. Let ν_0 be an integer such that $c^{-1}\mathfrak{a} = \mathfrak{p}^{\nu_0}$. Then $L = \{x \in V \mid \nu(\xi(x)) \geq \nu_0\}$. It follows that L is a $\mathfrak{g}$-lattice, and $\xi(L) = \mathfrak{Q}^{\nu_0}$. By the definition of L, we have $\mathfrak{r}x \subset L$ for every $x \in L$, so that L is an $\mathfrak{r}$-lattice. As $\xi^{-1}(\pi^{\nu_0}) \in L$, we get $\mu(L) = \mathfrak{a}$. By our definition of L, it is clear that L is maximal, and every maximal $\mathfrak{r}$-lattice whose norm is $\mathfrak{a}$ must coincide with L. This completes our proof.

4.6. PROPOSITION. *Suppose that $n = 2$ and Φ is anisotropic. Define $\mathscr{B}$ as in 2.6. Let $\mathfrak{o}$ be the maximal order in $\mathscr{B}$ and let ξ be a $\mathscr{B}$-isomorphism* *of V onto $\mathcal{B}$ as given in 2.7. Then every $\mathfrak{g}$-lattice L in V is a maximal $\mathfrak{r}$-lattice if and only if $\xi(L)$ is an $\mathfrak{o}$-ideal.*

This is included in the proof of 4.5.

4.7. PROPOSITION. *Let L be a maximal $\mathfrak{r}$-lattice in V. Put $\mathfrak{b} = \mu(L)\mathfrak{d}^{-1}$. Then there exists a Witt decomposition $V = \sum_{i=1}^{m}(Kx_i + Ky_i) + W$ such that $L = \sum_{i=1}^{m}(\mathfrak{r}x_i + \mathfrak{b}y_i) + M$, where M is a maximal $\mathfrak{r}$-lattice in W defined by $M = \{z \in W \mid \Phi(z, z) \in \mu(L)\}$. Conversely, let $\mathfrak{a}$ be a $\mathfrak{g}$-ideal, $V = \sum_{i=1}^{m}(Kx_i + Ky_i) + W$ be a Witt decomposition. Let*

$$M = \{z \in W \mid \Phi(z, z) \in \mathfrak{a}\}, \quad L = \sum_{i=1}^{m}(\mathfrak{r}x_i + \mathfrak{a}\mathfrak{b}^{-1}y_i) + M .$$

Then L is a maximal $\mathfrak{r}$-lattice in V. If $n > 1$, we have $\mu(L) = \mathfrak{a}$.

4.8. PROPOSITION. *Suppose that $n > 1$. Let L be an $\mathfrak{r}$-lattice in V and $\mathfrak{a}$ be a $\mathfrak{g}$-ideal such that $\mu(L) \subset \mathfrak{a}$. Then there exists a maximal $\mathfrak{r}$-lattice L' such that $L' \supset L$, $\mu(L') = \mathfrak{a}$.*

If $n = 1$, 4.7 is included in 4.5. Therefore we assume $n > 1$. Before proving 4.7, we show that, if 4.7 is true for n, then 4.8 is true for n. Let the notation be as in 4.8. By 2.14, we may assume that L is maximal. Put $\mu(L) = \mathfrak{c}$. Applying 4.7 to L, we obtain a Witt decomposition $V = \sum_{i=1}^{m}(Kx_i + Ky_i) + W$ such that $L = \sum_{i=1}^{m}(\mathfrak{r}x_i + \mathfrak{c}\mathfrak{b}^{-1}y_i) + M$, where $M = \{z \in W \mid \Phi(z, z) \in \mathfrak{c}\}$. By 4.5, if we put $M' = \{z \in W \mid \Phi(z, z) \in \mathfrak{a}\}$, then M' is a maximal $\mathfrak{r}$-lattice in W, and clearly $M' \supset M$. Put $L' = \sum_{i=1}^{m}(\mathfrak{r}x_i + \mathfrak{a}\mathfrak{b}^{-1}y_i) + M'$. By the converse part of 4.7, L' is maximal, $\mu(L') = \mathfrak{a}$, and clearly $L' \supset L$. This proves 4.8.

Next we prove the converse part of 4.7. Define L as in 4.7. L is clearly an $\mathfrak{r}$-lattice and $\mu(L) = \mathfrak{a}$. Let L' be an $\mathfrak{r}$-lattice in V such that $L' \supset L$, and $\mu(L') = \mathfrak{a}$. Let $u = \sum_{i=1}^{m}(c_i x_i + d_i y_i) + z$ (with $c_i \in K, d_i \in K, z \in W$) be an element of L'. Then by 2.11, we have $\mathfrak{a}\mathfrak{b}^{-1}c_i = \Phi(u, \mathfrak{a}\mathfrak{b}^{-1}y_i) \in \mu(L')\mathfrak{b}^{-1}$, so that $c_i \in \mathfrak{r}$. Similarly, $d_i = \Phi(u, x_i) \in \mu(L')\mathfrak{b}^{-1} = \mathfrak{a}\mathfrak{b}^{-1}$, and hence $\sum_{i=1}^{m}(c_i x_i + d_i y_i) \in L \subset L'$. It follows that $z \in L' \cap W$. As $\mu(L' \cap W) \subset \mathfrak{a}$ and $M \subset L' \cap W$, we must have $L' \cap W = M$. Therefore we get $z \in M$, so that $u \in L$. This shows $L = L'$, which proves the maximality of L.

Now let us prove the direct part of 4.7 by induction on n. If Φ is anisotropic, our assertion is included in 4.5. Suppose that Φ is isotropic. Then there exists a non-zero vector x_1 in V such that $\Phi(x_1, x_1) = 0$. By taking a suitable multiple of x_1, we may assume that $\mathfrak{r} = \{a \in K \mid ax_1 \in L\}$. Put $\mathfrak{c} = \Phi(x_1, L)$. Then $\mathfrak{c}$ is an $\mathfrak{r}$-ideal and by 2.11, $\mathfrak{c} \subset \mu(L)\mathfrak{b}^{-1} = \mathfrak{b}$, hence $\mathfrak{b}\mathfrak{c}^{-1} \supset \mathfrak{r}$. If $a \in \mathfrak{b}\mathfrak{c}^{-1}$ and $u \in L$, we have

$$\Phi(ax_1 + u, ax_1 + u) = \mathrm{Tr}_{K/F}\big(a\Phi(x_1, u)\big) + \Phi(u, u) \in \mu(L) .$$

This shows that $\mu(\mathfrak{b}\mathfrak{c}^{-1}x_1 + L) = \mu(L)$. By the maximality of L, we have

$bc^{-1}x_1 \subset L$, so that $bc^{-1} \subset r$. As we have already seen the inverse inclusion $bc^{-1} \supset r$, we get $bc^{-1} = r$, so that $\Phi(x_1, L) = b$. Let b be an element of K such that $b = rb$. We can find an element y of L such that $\Phi(x_1, y) = b$. As $\Phi(y, y) \in \mu(L) = \mathrm{Tr}_{K/F}(\mu(L)b^{-1}) = \mathrm{Tr}_{K/F}(br)$, there exists an element c of r such that $\Phi(y, y) = \mathrm{Tr}_{K/F}(bc)$. Put $y_1 = (b^\rho)^{-1}(y - cx_1)$. Then $by_1 \in L$, and $\Phi(x_1, y_1) = 1$, $\Phi(y_1, y_1) = 0$. Put $V' = \{z \in V \mid \Phi(x_1, z) = \Phi(y_1, z) = 0\}$, $L' = L \cap V'$. We can easily verify that $V = Kx_1 + Ky_1 + V'$ and L' is an r-lattice in V'. If $L \ni u = fx_1 + gy_1 + u'$ with $f \in K$, $g \in K$, $u' \in V'$, then

$$g^\rho = \Phi(x_1, u) \in \Phi(x_1, L) = b \ ,$$
$$f = \Phi(u, y_1) = \Phi(u, by_1)(b^\rho)^{-1} \in \mu(L)b^{-1}(b^\rho)^{-1} = r \ .$$

Hence $fx_1 + gy_1 \in L$, so that $u' \in L \cap V' = L'$. It follows that $L = rx_1 + by_1 + L'$. Obviously, $\mu(L') \subset \mu(L)$. As L is maximal, L' must be maximal among the r-lattices L_1 in V' satisfying $\mu(L_1) \subset \mu(L)$. Therefore if $n = 2$ or 3, we get our assertion by virtue of 4.5. If $n > 3$, by our induction-assumption, 4.8 is true for $n - 2$. Therefore we can find a maximal r-lattice L'' in V' such that $L'' \supset L'$ and $\mu(L'') = \mu(L)$. Put $L_1 = rx_1 + by_1 + L''$. Then $\mu(L_1) = \mu(L)$. By the maximality of L, we have $L = L_1$, so that $L' = L''$. Hence L' is maximal and $\mu(L') = \mu(L)$. Applying our induction to L', we get our assertion for n.

4.9. Proposition. *Let L be a maximal r-lattice in V. If Φ is isotropic, there exists a basis $\{u_i\}$ of V over K such that $L = \sum_{i=1}^{n} ru_i$ and $\Phi(u_i, u_i) = 0$ for every i.*

Let the decomposition $L = \sum_{i=1}^{m} (rx_i + by_i) + M$ be as in 4.7. For every $z \in M$, we have $\Phi(z, z) \in \mu(L) = \mathrm{Tr}_{K/F}(b)$. Hence there exists an element c of b such that $\Phi(z, z) = \mathrm{Tr}_{K/F}(c)$. Put $w = x_1 - cy_1 + z$. Then $w \in L$ and $\Phi(w, w) = 0$. Our assertion follows easily from this consideration.

4.10. Proposition. *Let L and M be maximal r-lattices in V. Put $b = \mu(L)b^{-1}$, $gc = \mu(L)^{-1}\mu(M)$ with $c \in F$. Then there exist a Witt decomposition $V = \sum_{i=1}^{m} (Kx_i + Ky_i) + W$, and elements a_i, b_i of K such that $L = \sum_{i=1}^{m} (rx_i + by_i) + L^*$, $M = \sum_{i=1}^{m} (ra_ix_i + bb_iy_i) + M^*$, $a_1b_1^\rho = \cdots = a_mb_m^\rho = c$, $ra_1 \supset \cdots \supset ra_m \supset rb_m \supset \cdots \supset rb_1$, where $L^* = \{z \in W \mid \Phi(z, z) \in \mu(L)\}$, $M^* = \{z \in W \mid \Phi(z, z) \in \mu(M)\}$. (For the uniqueness of $\{ra_i, rb_i\}$, see 4.20 below.)*

Proof proceeds by induction on n. If Φ is anisotropic, this is trivial in view of 4.5. Suppose that Φ is isotropic. Let g be an element of K. Then $gM = \{gx \mid x \in M\}$ is a maximal r-lattice in V, and $\mu(gM) = N_{K/F}(g)\mu(M)$. If we could prove our assertion for gM, then it is easy to verify that the assertion is true for M. Therefore, replacing M with gM for a suitable g, we may assume that $\mathfrak{r} = \{a \in K \mid aM \subset L\}$; this implies in particular that $M \subset L$. Applying 4.9 to M, we can find an element x_1 of M such that $\Phi(x_1, x_1) = 0$, $\mathfrak{r} = \{a \in K \mid ax_1 \in L\}$. Now we can apply the argument of the proof of 4.7 to this x_1. Then we get an element y_1 of $\mathfrak{b}^{-1}L$ such that $\Phi(y_1, y_1) = 0$, $\Phi(x_1, y_1) = 1$. Put $V' = \{z \in V \mid \Phi(x_1, z) = \Phi(y_1, z) = 0\}$, $L' = L \cap V'$, $M' = M \cap V'$. As in the proof of 4.7, we have $V = Kx_1 + Ky_1 + V'$, $L = \mathfrak{r}x_1 + \mathfrak{b}y_1 + L'$, and L' is a maximal one among the $\mathfrak{r}$-lattices L_1' in V' such that $\mu(L_1') \subset \mu(L)$. By 2.13, $M \supset cL$, so that $c\mathfrak{b}y_1 \subset M$. Let $w = dx_1 + ey_1 + z$ (with $d \in K$, $e \in K$, $z \in V'$) be an element of M. Then $db = \Phi(w, by_1) \subset \mu(L)\mathfrak{b}^{-1} = \mathfrak{b}$, so that $d \in \mathfrak{r}$; and $e = \Phi(w, x_1) \subset \mu(M)\mathfrak{b}^{-1} = c\mathfrak{b}$. Therefore $dx_1 + ey_1 \in \mathfrak{r}x_1 + \mathfrak{b}cy_1 \subset M$, so that $z \in M \cap V' = M'$. It follows that $M = \mathfrak{r}x_1 + \mathfrak{b}cy_1 + M'$. By the maximality of M, M' is a maximal one among the $\mathfrak{r}$-lattices M_1' in V' such that $\mu(M_1') \subset \mu(M)$. Applying our induction to L' and M', we get our assertion.

4.11. Proposition. *Let L and M be maximal $\mathfrak{r}$-lattices in V. Then there exists an element α of $G(V, \Phi)$ such that $L\alpha = M$, if and only if either n is even, or, n is odd and $\mu(L)^{-1}\mu(M) = N_{K/F}(\mathfrak{f})$ for some $\mathfrak{r}$-ideal $\mathfrak{f}$. Moreover, let c be an element of F such that $\mu(L)^{-1}\mu(M) = \mathfrak{g}c$. Then we can take α so that $\mu(\alpha) = c$, if n is even, or, if n is odd and $c \in N_{K/F}(K^*)$.*

If $L\alpha = M$ for some $\alpha \in G(V, \Phi)$, we have $\mu(L)^{-1}\mu(M) = \mu(\alpha)\mathfrak{g}$, and $\mu(\alpha)^n = N_{K/F}(\det(\alpha))$ (2.2.1). Hence, if $n = 2m + 1$,

$$\mu(\alpha)\mathfrak{g} = N_{K/F}(\det(\alpha)\mu(\alpha)^{-m}\mathfrak{r}) .$$

This proves the "only if" part of the first assertion. To prove the "if" part, take an element c of F such that $\mu(M) = c\mu(L)$, and apply 4.10 to L, M and c. Then we get a Witt decomposition $V = \sum_{i=1}^{m} (Kx_i + Ky_i) + W$, maximal $\mathfrak{r}$-lattices L^*, M^* in W, and elements a_i, b_i of K, with the properties given in 4.10. If n is even, W is of dimension 0 or 2. When $\dim W = 2$, we have $\mu(L^*) = \mu(L)$, $\mu(M^*) = \mu(M)$ by 4.5. Let $\mathscr{B} = \mathscr{B}(W)$ and ι be as in 4.3. Since $\mathscr{B}$ is a quaternion algebra over a $\mathfrak{p}$-adic field, we can find an element β of $\mathscr{B}$ such that $\beta\beta^\iota = c$. Then $\Phi(z\beta, z\beta) = c\Phi(z, z)$ for $z \in W$. It follows that $L^*\beta = M^*$. Now, in the case of even n, define an element α of $GL(V, K)$ by $x_i\alpha = a_i x_i$, $y_i\alpha = b_i y_i$, and $z\alpha = z\beta$ for $z \in W$ if $\dim W = 2$. Then $\alpha \in G(V, \Phi)$, $\mu(\alpha) = c$, $L\alpha = M$. This proves the "if" part of the first assertion and the second assertion for even n. Assume that $n = 2m + 1$ and $c\mathfrak{g} = N_{K/F}(\mathfrak{f})$ for some $\mathfrak{r}$-ideal $\mathfrak{f}$. Let f be an element of K such that $\mathfrak{f} = \mathfrak{r}f$. It is clear that $fL^* = M^*$. Further we can find a unit h of $\mathfrak{g}$ such that $ff^\rho = ch$. Define an element α of $GL(V, K)$ by $x_i\alpha = a_i h x_i$, $y_i\alpha = b_i y_i$, $z\alpha = fz$ for $z \in W$. Then $\alpha \in G(V, \Phi)$, $\mu(\alpha) = ch$,

$L\alpha = M$. If $c \in N_{K/F}(K^*)$, we can take f so that $ff^\rho = c$. Then we have $\mu(\alpha) = c$. This completes the proof.

4.12. Remark. In the above proposition, we have considered the following two conditions for an element c of F^*:

(i) $c\mathfrak{g} = N_{K/F}(\mathfrak{f})$ for some $\mathfrak{r}$-ideal $\mathfrak{f}$;

(ii) $c \in N_{K/F}(K^*)$.

Obviously, (ii) implies (i). If K is unramified over F, (i) implies (ii), so that (i) and (ii) are equivalent. If K is ramified over F, (i) is always satisfied, but (ii) is an essential restriction on c.

4.13. Proposition. *Let L and M be maximal $\mathfrak{r}$-lattices in V. Then $\mu(L) = \mu(M)$ if and only if there exists an element α of $U(V, \Phi)$ such that $L\alpha = M$.*

This is only a special case of 4.11.

4.14. Proposition. *Let L and M be maximal $\mathfrak{r}$-lattices in V. Then, $\mu(L) = \mu(M)$ if and only if $[L/M] = \mathfrak{r}$.*

If $\mu(L) = \mu(M)$, we have, by 4.13, $M = L\alpha$ for some $\alpha \in U(V, \Phi)$. Since $N_{K/F}(\det(\alpha)) = \mu(\alpha)^n = 1$, $\det(\alpha)$ is a unit of $\mathfrak{r}$. Hence $[L/M] = \det(\alpha)\mathfrak{r} = \mathfrak{r}$. Conversely, suppose that $[L/M] = \mathfrak{r}$. Put $\mu(L)^{-1}\mu(M) = \mathfrak{g}c$ with an element $c \in F$. Assume that $c \in \mathfrak{g}$, and apply 4.10 to L, M and c. Then using the notation of 4.10, we find $\mathfrak{r} = [L/M] = \mathfrak{r}c^m[L^*/M^*]$. As $c \in \mathfrak{g}$, we have $\mu(L) \supset \mu(M)$, so that $L^* \supset M^*$ and hence $[L^*/M^*] \subset \mathfrak{r}$. If $m > 0$, we must have $\mathfrak{g}c = \mathfrak{g}$. If $m = 0$, we have $[L^*/M^*] = \mathfrak{r}$, so that $L = L^* = M^* = M$. In either case, we have $\mu(L) = \mu(M)$. If $c \notin \mathfrak{g}$, we have $c^{-1} \in \mathfrak{g}$. Exchanging L and M, by the same argument, we find again $\mu(L) = \mu(M)$. This completes the proof.

4.15. Proposition. *Let L be an $\mathfrak{r}$-lattice in V. Then there exist direct decompositions $V = \sum_{i=1}^t W_i$, $L = \sum_{i=1}^t L_i$ with the following properties:*

(i) *the W_i are subspaces of V of dimension 1 or 2 over K;*

(ii) $L_i = W_i \cap L$;

(iii) $\Phi(x, y) = 0$ *for $x \in W_i$, $y \in W_j$ if $i \neq j$;*

(iv) $\mu_0(L_1) \supset \mu_0(L_2) \supset \cdots \supset \mu_0(L_t)$ *(cf 2.10).*

Moreover, if K is unramified over F, we can take all the W_i to be one-dimensional. If K is ramified over F and $\mathfrak{p} \not\ni 2$, then, we can take the decompositions so that $L_i = \mathfrak{r}x_i + \mathfrak{r}y_i$ with $\Phi(x_i, x_i) = \Phi(y_i, y_i) = 0$ whenever $\dim W_i = 2$.

This is a result due to Jacobowitz [8, Prop. 4.3, Prop. 8.1]. For reader's convenience, we sketch a proof, which proceeds by induction on n. By 2.11, $\mu(L)\mathfrak{r} \subset \mu_0(L) \subset \mu(L)\mathfrak{d}^{-1}$. Suppose that $\mu(L)\mathfrak{r} = \mu_0(L)$. This is the case if K is unramified over F. Let x be an element of L such that $\mu(L) = \Phi(x, x)\mathfrak{g}$. Put $W = \{w \in V \mid \Phi(x, w) = 0\}$. Then we see that $V = Kx + W$, $L = \mathfrak{r}x + (W \cap L)$. Applying the induction, we get the desired result. Next suppose that $\mu(L)\mathfrak{r} \neq \mu_0(L)$. Then $\mu(L)\mathfrak{r} = \mu_0(L)\mathfrak{P}^f$ for a positive integer f. Let x, y be elements of L such that $\mu_0(L) = \Phi(x, y)\mathfrak{r}$. Put $W = \{w \in V \mid \Phi(x, w) = \Phi(y, w) = 0\}$. Put

$$\Phi(x, x) = a , \qquad \Phi(y, y) = b , \qquad \Phi(x, y) = c ,$$

and observe that $c^{-1}\begin{pmatrix} a & c \\ c^\rho & b \end{pmatrix}$ is an invertible element of $M_2(\mathfrak{r})$. Then we find $V = Kx + Ky + W$, $L = \mathfrak{r}x + \mathfrak{r}y + (W \cap L)$ and therefore we can apply our induction. Suppose that K is ramified over F and $\mathfrak{p} \ni 2$. Then we have $\mathfrak{d} = \mathfrak{P}$, so that $\mu(L)\mathfrak{r} = \mu_0(L)\mathfrak{P}$ in the case $\mu(L)\mathfrak{r} \neq \mu_0(L)$. Since $-\det\begin{pmatrix} a & c \\ c^\rho & b \end{pmatrix} \in N_{K/F}(K^*)$, the restriction of Φ to $Kx + Ky$ is isotropic (4.1). Hence we can find an element $u = gx + hy$ in $\mathfrak{r}x + \mathfrak{r}y$ such that $\Phi(u, u) = 0$, and either g or h is a unit of $\mathfrak{r}$. Then we find $\Phi(u, L) = \mu_0(L)$. Applying the argument of the proof of 4.7, we get an element v such that $\mathfrak{r}x + \mathfrak{r}y = \mathfrak{r}u + \mathfrak{r}v$ and $\Phi(v, v) = 0$. This completes the proof.

4.16. LEMMA. *Let E be the group of units of $\mathfrak{r}$ and let*

$$E_0 = \{e \in E \mid N_{K/F}(e) = 1\} , \qquad E_1 = \{e^{-1}e^\rho \mid e \in E\} .$$

Then $[E_0 : E_1] = 2$ or 1 according as K is ramified or not.

Consider the surjective mapping $K^* \ni a \to a^{-1}a^\rho \in E_0$. The inverse image of E_1 is clearly $F^* \cdot E$. Therefore we get $[E_0 : E_1] = [K^* : F^* \cdot E]$, which proves the assertion.

4.17. LEMMA. *There exists an element $d \in K$ such that $d^\rho = -d$, $\mathfrak{d} = d\mathfrak{r}$.*

In fact, we can find an element u of K such that $\mathfrak{r} = \mathfrak{g} + \mathfrak{g}u$. Let $f(X) = X^2 + aX + b = 0$ be a minimal equation for u over F. Then, it is well-known that $\mathfrak{d} = f'(u)\mathfrak{r}$, where $f'(X) = 2X + a$. Putting $d = f'(u)$, we get our assertion.

4.18. PROPOSITION. *Let L be an $\mathfrak{r}$-lattice in V. Let E_0 and E_1 be as in 4.16, and $E_0^2 = \{e^2 \mid e \in E_0\}$. Denote by E_L the set of $\det(\alpha)$ for all the $\alpha \in U(V, \Phi)$ such that $L\alpha = L$. Then $E_0 \supset E_L \supset E_0^2$. If K is unramified over F or if n is odd, we have $E_L = E_0$. If K is ramified and $\mathfrak{p} \ni 2$, then $E_L \supset E_1$. Moreover, suppose that L is maximal. Then*

$$E_L = \begin{cases} E_1 & \text{if K is ramified over F, and $\dim W = 0$,} \\ E_0 & \text{otherwise,} \end{cases}$$

where W is a kernel subspace of V with respect to Φ.

It is clear that $E_0 \supset E_L$. Let $e \in E_0$. Let $V = \sum_{i=1}^t W_i$, $L = \sum_{i=1}^t L_i$

be as in 4.15. Suppose that some W_i is one-dimensional. Define an element α of $GL(V, K)$ by $x\alpha = ex$ for $x \in W_i$ and $y\alpha = y$ for $y \in W_j$ with $j \neq i$. Then $\alpha \in U(V, \Phi)$, $\det(\alpha) = e$ and $L\alpha = L$. This shows $E_0 = E_L$. Hence if K is unramified over F or if n is odd, we have $E_L = E_0$. If all the W_i are two-dimensional, define an element β of $GL(V, K)$ by $x\beta = ex$ for $x \in W_1$ and $y\beta = y$ for $y \in W_i$ with $i > 1$. Then $\beta \in U(V, \Phi)$, $\det(\beta) = e^2$, $L\beta = L$. This shows $E_L \supset E_0^2$. Suppose that K is ramified over F, $\mathfrak{p} \not\ni 2$, and all the W_i are two-dimensional. By 4.15, we may assume that $L_1 = \mathfrak{r}x_1 + \mathfrak{r}y_1$ and $\Phi(x_1, x_1) = \Phi(y_1, y_1) = 0$. Let f be a unit of $\mathfrak{r}$. Define an element γ of $GL(V, K)$ by $x_1\gamma = f^{-1}x_1$, $y_1\gamma = f^{\rho}y_1$, $z\gamma = z$ for $z \in \sum_{i=2}^{t} W_i$. Then

$$\gamma \in U(V, \Phi), \qquad \det(\gamma) = f^{-1}f^{\rho}, \qquad L\gamma = L.$$

This shows that $E_1 \subset E_L$. Now suppose that L is maximal. By 4.7, there exists a Witt decomposition $V = \sum_{i=1}^{m} (Kx_i + Ky_i) + W$ such that $L = \sum_{i=1}^{m} (\mathfrak{r}x_i + \mathfrak{b}y_i) + L^*$, where $\mathfrak{b} = \mu(L)\mathfrak{b}^{-1}$ and $L^* = \{z \in W \mid \Phi(z, z) \in \mu(L)\}$. We have only to consider the case of even n. Suppose that $\dim W = 2$. Let $\mathscr{B} = \mathscr{B}(W)$ and ι be as in 4.3. For every $e \in E_0$, we can find an element g of K such that $e = g^{-1}g^{\rho}$, and an element σ of $\mathscr{B}$ such that $\sigma\sigma^{\iota} = gg^{\rho}$, since F is a $\mathfrak{p}$-adic field. Define an element τ of $GL(V, K)$ by $u\tau = u$ for $u \in \sum_{i=1}^{m} (Kx_i + Ky_i)$, $z\tau = g^{-1}z\sigma$ for $z \in W$. Then we see easily that $\tau \in U(V, \Phi)$, $\det(\tau) = g^{-2}\sigma\sigma^{\iota} = e$, $L^*\tau = g^{-1}L^*\sigma$. Since

$$\mu(L^* \tau) = (gg^{\rho})^{-1}\mu(L^*)\sigma\sigma^{\iota} = \mu(L^*),$$

we must have $L^*\tau = L^*$ on account of 4.5. Therefore $L\tau = L$, and hence $e \in E_L$; so we get $E_0 = E_L$. It remains to consider the case where $\dim W = 0$. In this case, $L = \sum_{i=1}^{m} (\mathfrak{r}x_i + \mathfrak{b}y_i)$ with $\mathfrak{b} = \mu(L)\mathfrak{b}^{-1}$. By the same argument as above, we can show that $E_L \supset E_1$. By 4.17, we can find an element h of K such that $\mathfrak{b} = \mathfrak{r}h$, $h^{\rho} = -h$. For every element $\alpha \in E(V, K)$, put

$$\begin{aligned}
x_i\alpha &= \sum_{j=1}^{m} a_{ij}x_j + \sum_{j=1}^{m} b_{ij}hy_j, \\
hy_i\alpha &= \sum_{j=1}^{m} c_{ij}x_j + \sum_{j=1}^{m} d_{ij}hy_j
\end{aligned} \qquad (1 \leq i \leq m),$$

$$A = (a_{ij}), \quad B = (b_{ij}), \quad C = (c_{ij}), \quad D = (d_{ij}).$$

Then $\alpha \to \begin{pmatrix} A & B \\ C & D \end{pmatrix}$ is an isomorphism of $GL(V, K)$ to $GL(2m, K)$; and $\alpha \in U(V, \Phi)$ if and only if $A \cdot {}^t B^{\rho} - B \cdot {}^t A^{\rho} = 0$, $A \cdot {}^t D^{\rho} - B \cdot {}^t C^{\rho} = 1_m$, $C \cdot {}^t D^{\rho} - D \cdot {}^t C^{\rho} = 0$; moreover, $L\alpha \subset L$ if and only if A, B, C, D are contained in $M_m(\mathfrak{r})$. For convenience, we identify α with the corresponding matrix $\begin{pmatrix} A & B \\ C & D \end{pmatrix}$. Let Γ be the subgroup of $U(V, \Phi)$ consisting of the elements γ such that $L\gamma = L$. We see easily that the matrices

$$\xi = \begin{pmatrix} A & 0 \\ 0 & D \end{pmatrix} \qquad (A \cdot {}^t D^\rho = 1_m,\ A \in M_m(\mathfrak{r}),\ D \in M_m(\mathfrak{r})),$$

$$\eta = \begin{pmatrix} 1_m & B \\ 0 & 1_m \end{pmatrix} \qquad (B \in M_m(\mathfrak{r}),\ B = {}^t B^\rho),$$

$$\zeta = \begin{pmatrix} 0 & 1_m \\ -1_m & 0 \end{pmatrix}$$

are contained in Γ. By the standard technique (which is analogous to and much simpler than that of Witt [12, Satz C]), it can be easily shown that Γ is generated by the elements ξ, η, ζ. We have clearly $\det(\xi) = \det(A) \cdot \det(A^\rho)^{-1}$, $\det(\eta) = \det(\zeta) = 1$. This shows that $E_L \subset E_1$, so that $E_L = E_1$. In view of 4.16, this completes the proof.

4.19. PROPOSITION. *Let L be a maximal $\mathfrak{r}$-lattice in V. Let $V = \sum_{i=1}^m (Ku_i + Kv_i) + Z$ be a Witt decomposition such that $L = \sum_{i=1}^m (\mathfrak{r}u_i + \mathfrak{b}v_i) + L'$ where $\mathfrak{b} = \mu(L)\mathfrak{d}^{-1}$ and $L' = \{z \in Z \mid \Phi(z, z) \in \mu(L)\}$. Denote by Γ_0 the subgroup of $SU(V, \Phi)$ consisting of the elements γ such that $L\gamma = L$, and by Δ the set of elements α of $G(V, \Phi)$ such that $u_i\alpha = a_iu_i$, $v_i\alpha = b_iv_i$ for $1 \le i \le m$, and $Z\alpha \subset Z$, with elements a_i, b_i of K satisfying $\mathfrak{r}a_1 \supset \cdots \supset \mathfrak{r}a_m \supset \mathfrak{r}b_m \supset \cdots \supset \mathfrak{r}b_1$. Then we have $G(V, \Phi) = \Gamma_0 \cdot \Delta \cdot \Gamma_0$.*

Let $\alpha \in G(V, \Phi)$, $\mu(\alpha) = c$, $M = L\alpha$. Applying 4.10 to the present case, we get a Witt decomposition $V = \sum_{i=1}^m (Kx_i + Ky_i) + W$ and elements a_i, b_i of K with the properties given in that proposition. By 4.2 (or by Witt's theorem), there exists a K-isomorphism λ of Z onto W such that $\Phi(x\lambda, y\lambda) = \Phi(x, y)$ for $(x, y) \in Z \times Z$. By 4.4, there exists an element θ of $G(Z, \Phi)$ such that $\mu(\theta) = c$. Define two elements β and γ of $\mathrm{GL}(V, K)$ as follows: $u_i\beta = a_iu_i$, $v_i\beta = b_iv_i$, $u_i\gamma = x_i$, $v_i\gamma = y_i$ for $1 \le i \le m$, and $z\beta = z\theta$, $z\gamma = z\lambda$ for $z \in Z$. Then $\beta \in \Delta$, $\mu(\beta) = c$, $\gamma \in U(V, \Phi)$. By 4.5, we have $L'\beta\gamma = M^*$, $L'\gamma = L^*$. Therefore, $L\gamma = L$ and $L\beta\gamma = M = L\alpha$. Put $\varepsilon = \alpha(\beta\gamma)^{-1}$. Then $\varepsilon \in U(V, \Phi)$, $L\varepsilon = L$ and $\alpha = \varepsilon\beta\gamma$. Now let E_L be as in 4.18. We have clearly $\det(\gamma) \in E_L$, $\det(\varepsilon) \in E_L$. By the proof of 4.18, we can find elements σ and τ of $U(V, \Phi) \cap \Delta$ such that $L\sigma = L\tau = L$, and $\det(\sigma) = \det(\gamma)^{-1}$, $\det(\tau) = \det(\varepsilon)^{-1}$. Then $\sigma\gamma \in \Gamma_0$, $\varepsilon\tau \in \Gamma_0$, $\tau^{-1}\beta\sigma^{-1} \in \Delta$, $\alpha = \varepsilon\tau(\tau^{-1}\beta\sigma^{-1})\sigma\gamma$. This proves $G(V, \Phi) = \Gamma_0 \cdot \Delta \cdot \Gamma_0$.

4.20. Let Y be a vector space over K of dimension r. Let C and D be $\mathfrak{r}$-lattices in Y. Then there exist a basis $\{y_i\}$ of Y over K and $\mathfrak{r}$-ideals $\mathfrak{a}_i$ such that $C = \sum_{i=1}^r \mathfrak{r}y_i$, $D = \sum_{i=1}^r \mathfrak{a}_iy_i$, $\mathfrak{a}_1 \supset \cdots \supset \mathfrak{a}_r$. The $\mathfrak{a}_i$ are uniquely determined by C and D; they do not depend on the choice of $\{y_i\}$. We call $\{\mathfrak{a}_1, \cdots, \mathfrak{a}_r\}$ *the set of elementary divisors of D relative to C*, and denote it simply by $\{C : D\}$.

Let the notation be as in 4.10. We now determine $\{L : M\}$. Put $\mu(L)^{-1}\mu(M) = \mathfrak{p}^{\nu}$.

Case a: $\dim W = 0$. In this case there is no question; we have

$$\{L : M\} = \{\mathfrak{r}a_1, \cdots, \mathfrak{r}a_m, \mathfrak{r}b_m, \cdots, \mathfrak{r}b_1\} .$$

Case b: $\dim W = 2$, K is ramified over F. We see easily that $\mathfrak{P}^{\nu}L^* = M^*$, so that $\{L^* : M^*\} = \{\mathfrak{P}^{\nu}, \mathfrak{P}^{\nu}\}$. Since $\mathfrak{r}a_m \supset \mathfrak{r}b_m$ and $\mathfrak{r}a_m b_m = \mathfrak{r}c = \mathfrak{P}^{2\nu}$, we must have $\mathfrak{r}a_m \supset \mathfrak{P}^{\nu} \supset \mathfrak{r}b_m$. Therefore,

$$\{L : M\} = \{\mathfrak{r}a_1, \cdots, \mathfrak{r}a_m, \mathfrak{P}^{\nu}, \mathfrak{P}^{\nu}, \mathfrak{r}b_m, \cdots, \mathfrak{r}b_1\} .$$

Case c: $\dim W = 2$, K is unramified over F. If ν is even, we have $\mathfrak{P}^{\nu/2}L^* = M^*$, and hence

$$\{L : M\} = \{\mathfrak{r}a_1, \cdots, \mathfrak{r}a_m, \mathfrak{P}^{\nu/2}, \mathfrak{P}^{\nu/2}, \mathfrak{r}b_m, \cdots, \mathfrak{r}b_1\} .$$

If ν is odd, put $\nu = 2k + 1$. Then $\mathfrak{P}^k L^* \supsetneqq M^* \supsetneqq \mathfrak{P}^{k+1}L^*$, and hence $\{L^* : M^*\} = \{\mathfrak{P}^k, \mathfrak{P}^{k+1}\}$. Therefore

$$\{L : M\} = \{\mathfrak{r}a_1, \cdots, \mathfrak{r}a_m, \mathfrak{P}^k, \mathfrak{P}^{k+1}, \mathfrak{r}b_m, \cdots, \mathfrak{r}b_1\} .$$

Case d: $\dim W = 1$. If K is ramified over F, we get $\mathfrak{P}^{\nu}L^* = M^*$. Suppose that K is unramified over F; if ν is even we get $\mathfrak{P}^{\nu/2}L^* = M^*$. If ν is odd, put $\nu = 2k + 1$; then $\mathfrak{P}^k L^* \supset M^* \supset \mathfrak{P}^{k+1}L^*$. In any case we have $\mathfrak{r}a_m \supset [L^*/M^*] \supset \mathfrak{r}b_m$. Hence

$$\{L : M\} = \{\mathfrak{r}a_1, \cdots, \mathfrak{r}a_m, [L^*/M^*], \mathfrak{r}b_m, \cdots, \mathfrak{r}b_1\} .$$

Concluding all the *Cases* a, b, c, d, we observe that $\{L : M\}$ is determined by $\{\mathfrak{r}a_i, \mathfrak{r}b_i, [L^*/M^*]\}$. Conversely, the latter is determined by $\{L : M\}$. This implies in particular that $\{\mathfrak{r}a_i, \mathfrak{r}b_i\}$ are uniquely determined by L and M, and independent of the choice of the Witt decomposition.

4.21. PROPOSITION. *Let L, M, N be maximal $\mathfrak{r}$-lattices in V. If there exists an element α of $G(V, \Phi)$ such that $L\alpha = L$ and $M\alpha = N$, then $\{L : M\} = \{L : N\}$. Conversely, if $\{L : M\} = \{L : N\}$, there exists an element α of $SU(V, \Phi)$ such that $L\alpha = L$ and $M\alpha = N$.*

The first assertion is obvious. To prove the converse part, suppose that $\{L : M\} = \{L : N\}$; then $\mu(M) = \mu(N)$. Put $\mathfrak{b} = \mu(L)\mathfrak{d}^{-1}$. By 4.10, there exists a Witt decomposition $V = \sum_{i=1}^{m}(Kx_i + Ky_i) + Z$ and elements a_i, b_i of K such that

$$L = \sum_{i=1}^{m}(\mathfrak{r}x_i + \mathfrak{b}y_i) + L_1^*$$

and

$$M = \sum_{i=1}^{m}(\mathfrak{r}a_i x_i + \mathfrak{b}b_i y_i) + M^* ;$$

and the a_i, b_i, L_1^*, M^* have the properties described in that proposition. Similarly, we get the decomposition $V = \sum_{i=1}^{m} (Ku_i + Kv_i) + W$, $L = \sum_{i=1}^{m} (\mathfrak{r}u_i + \mathfrak{b}v_i) + L_2^*$ and $N = \sum_{i=1}^{m} (\mathfrak{r}d_iu_i + \mathfrak{b}e_iv_i) + N^*$ with the corresponding properties. As $\{L : M\} = \{L : N\}$, we have $\mathfrak{r}a_i = \mathfrak{r}d_i$, $\mathfrak{r}b_i = \mathfrak{r}e_i$ for every i. By 4.2, there exists a K-isomorphism λ of Z onto W such that $\Phi(x\lambda, y\lambda) = \Phi(x, y)$ for every $(x, y) \in Z \times Z$. Define an element β of $GL(V, K)$ by $x_i\beta = u_i$, $y_i\beta = v_i$, $z\beta = z\lambda$ for $z \in Z$. Then $\beta \in U(V, \Phi)$, $L\beta = L$, $M\beta = N$. Let E_L be as in 4.18. Then $\det(\beta) \in E_L$. By the proof of 4.18, we can find an element γ of $U(V, \Phi)$ such that, $L\gamma = L$, $\det(\gamma) = \det(\beta)^{-1}$, and moreover $\mathfrak{r}x_i\gamma = \mathfrak{r}x_i$, $\mathfrak{r}y_i\gamma = \mathfrak{r}y_i$, $L_1^*\gamma = L_1^*$. The last equality implies that the restriction of γ to Z is contained in $U(Z, \Phi)$. Hence $M\gamma = M$. It follows that $L\gamma\beta = L$, $M\gamma\beta = N$, and $\gamma\beta \in SU(V, \Phi)$. This completes the proof.

4.22. Let E_0 and E_L be as in 4.18. Decompose all $\mathfrak{r}$-lattices in V into classes with respect to the operation of the elements of $U(V, \Phi)$. Fix a class C. We see that the E_L are the same for all $L \in C$; so we put $E_C = E_L$ for $L \in C$. In view of 2.3 and our definition of E_L, we observe that C is divided into $[E_0 : E_C]$ classes with respect to $SU(V, \Phi)$. If we restrict ourselves to maximal $\mathfrak{r}$-lattices, then by 4.13, C consists of the maximal $\mathfrak{r}$-lattices with the same norm. Hence, by 4.18, we obtain

4.23. PROPOSITION. *Let L and M be maximal $\mathfrak{r}$-lattices in V. If $\mu(L) = \mu(M)$, there exists an element α of $SU(V, \Phi)$ such that $L\alpha = M$, except when K is ramified over F and V has the trivial kernel subspace. In this exceptional case, all the maximal $\mathfrak{r}$-lattices with the same norm are divided into two classes with respect to $SU(V, \Phi)$.*

4.24. PROPOSITION. *Suppose that n is even. Let L be a maximal $\mathfrak{r}$-lattice in V. Then, for every unit c of $\mathfrak{g}$, there exists an element α of $H(V, \Phi)$ such that $L\alpha = L$ and $\mu(\alpha) = c$.*

Let $V = \sum_{i=1}^{m} (Kx_i + Ky_i) + W$ and $L = \sum_{i=1}^{m} (\mathfrak{r}x_i + \mathfrak{b}y_i) + M$ be as in 4.7. If $\dim W = 0$, define an element α of $GL(V, K)$ by $x_i\alpha = cx_i$, $y_i\alpha = y_i$ for $1 \leq i \leq m$. Then $\alpha \in H(V, \Phi)$, $L\alpha = L$, $\mu(\alpha) = c$. If $\dim W = 2$, consider $\mathscr{B} = \mathscr{B}(W)$ and ι defined in 4.3. Since F is a $\mathfrak{p}$-adic field, $\mathscr{B}$ contains an element β such that $\beta\beta^\iota = c$. By 4.5, we have $M\beta = M$. Define α by $x_i\alpha = cx_i$, $y_i\alpha = y_i$ for $1 \leq i \leq m$ and $z\alpha = z\beta$ for $z \in W$. Then α has the required property.

4.25. Suppose that n is even, K is ramified over F, and V has the trivial kernel subspace with respect to Φ. Put $n = 2m$. Let E_0 and E_1 be as in 4.16. Let L and M be maximal $\mathfrak{r}$-lattices in V. By 4.11, there exists an element α of $G(V, \Phi)$ such that $L\alpha = M$. We denote by $e(L/M)$ the element of the factor group E_0/E_1 represented by $\det(\alpha)\mu(\alpha)^{-m}$. This does not depend on the choice of α. In fact, if $L\beta = M$ for another element β of $G(V, \Phi)$, we have $L\alpha\beta^{-1} = L$, so that $\mu(\alpha\beta^{-1})$ is a unit of $\mathfrak{g}$. By 4.24, there exists an element γ of $H(V, \Phi)$ such that $\mu(\alpha\beta^{-1}) = \mu(\gamma)$ and $L\gamma = L$. Then we have $L\gamma\beta\alpha^{-1} = L$ and $\mu(\gamma\beta\alpha^{-1}) = 1$. By 4.18, $\det(\gamma\beta\alpha^{-1}) \in E_1$. Hence

$$(\det(\alpha)\mu(\alpha)^{-m})^{-1}\det(\beta)\mu(\beta)^{-m} = \det(\gamma\beta\alpha^{-1})\mu(\gamma\beta\alpha^{-1})^{-m} = \det(\gamma\beta\alpha^{-1}) \in E_1 \,.$$

Thus $e(L/M)$ is determined only by L and M. It is clear that $e(L/M) = e(M/L)$ and $e(L/M)e(M/N) = e(L/N)$. Moreover, $e(L/M) = 1$ if and only if there exists an element α of $H(V, \Phi)$ such that $L\alpha = M$. When $\mu(L) = \mu(M)$, $e(L/M) = 1$ if and only if $L\beta = M$ for some $\beta \in SU(V, \Phi)$. We obtain thus

4.26. PROPOSITION. *Suppose that n is even. If either K is unramified over F or V has a non-trivial kernel subspace, then $H(V, \Phi)$ operates transitively on the set of all maximal $\mathfrak{r}$-lattices in V. If K is ramified over F and V has the trivial kernel subspace, then, with respect to $H(V, \Phi)$, all the maximal $\mathfrak{r}$-lattices are divided into two classes.*

4.27. PROPOSITION. *Suppose that $n = 2$ and Φ is isotropic. Define $\mathscr{B}$ as in 2.6. Let L be an $\mathfrak{r}$-lattice in V. Put $\mathfrak{o} = \{\alpha \in \mathscr{B} \mid L\alpha \subset L\}$. Then L is maximal if and only if $\mathfrak{o}$ is a maximal order in the algebra $\mathscr{B}$.*

If L is maximal, then, by 4.7, there exists a Witt decomposition $V = Kx + Ky$ such that $L = \mathfrak{r}x + \mathfrak{b}y$ with $\mathfrak{b} = \mu(L)\mathfrak{d}^{-1}$. Consider the isomorphism

$$E(V, K) \ni \alpha \longrightarrow \begin{pmatrix} A & B \\ C & D \end{pmatrix} \in M_2(K)$$

defined in the proof of 4.18. Then $\alpha \in \mathscr{B}$ if and only if

$$\begin{pmatrix} D & -B \\ -C & A \end{pmatrix}\begin{pmatrix} 0 & -h \\ h & 0 \end{pmatrix} = \begin{pmatrix} 0 & -h \\ h & 0 \end{pmatrix}\begin{pmatrix} A^\rho & C^\rho \\ B^\rho & D^\rho \end{pmatrix}.$$

Obviously, this equality holds if and only if A, B, C, D are contained in F. Therefore $\mathscr{B}$ corresponds to $M_2(F)$, and hence $\mathfrak{o}$ corresponds to $M_2(\mathfrak{g})$. This implies that $\mathfrak{o}$ is a maximal order in $\mathscr{B}$. Conversely, suppose that $\mathfrak{o}$ is a maximal order in $\mathscr{B}$. By 4.8, there exists a maximal $\mathfrak{r}$-lattice M such that $\mu(M) = \mu(L)$ and $M \supset L$. Put $\mathfrak{o}' = \{\alpha \in \mathscr{B} \mid M\alpha \subset M\}$. As proved above, $\mathfrak{o}'$ is a maximal order in $\mathscr{B}$. Let ξ and c be as in 2.7. Then $\xi(L)$ is a right $\mathfrak{o}$-ideal and $\xi(M)$ is a right $\mathfrak{o}'$-ideal. We have $N(\xi(M)) = c^{-1}\mu(M) = c^{-1}\mu(L) = N(\xi(L))$, and $\xi(L) \subset \xi(M)$. As $\mathscr{B} = M_2(F)$, there exist elements α, β, γ such that $\xi(L) = \alpha\mathfrak{o}$, $\xi(M) = \beta\mathfrak{o}'$, $\gamma\mathfrak{o}\gamma^{-1} = \mathfrak{o}'$. Then $\alpha\mathfrak{o} \supset \beta\gamma\mathfrak{o}\gamma^{-1}$, so that $\mathfrak{o}\gamma \supset \alpha^{-1}\beta\gamma\mathfrak{o}$. But $N(\mathfrak{o}\gamma) = \det(\gamma)\mathfrak{g} = N(\alpha^{-1}\beta\gamma\mathfrak{o})$. Hence we must have $\mathfrak{o}\gamma = \alpha^{-1}\beta\gamma\mathfrak{o}$, so that $L = M$. It follows that L is maximal.

4.28. REMARK. Let the assumption and the notation be as in 4.27 and its proof. Let L and M be maximal $\mathfrak{r}$-lattices in V with the same norm. Then $L\alpha = M$ for some $\alpha \in SU(V, \Phi)$ if and only if $\xi(L)$ and $\xi(M)$ have the same *left* order. Further, by 4.23 and 4.27, the number of maximal orders in $M_2(F)$ containing $\mathfrak{r}$ is 1 or 2 according as K is unramified over F or not. This is a classical result.

5. Global theory of hermitian forms and lattices

5.0. In this section, we mean by F an algebraic number field of finite degree, and by K a quadratic extension of F. As before, ρ is the non-trivial automorphism of K over F. The rings of integers in F and in K are denoted by $\mathfrak{g}$ and $\mathfrak{r}$ respectively. For every valuation v of F, we denote by F_v the completion of F with respect to v, and put $K_v = K \otimes_F F_v$. In particular, $F_{\mathfrak{p}}$ and F_λ for $1 \leq \lambda \leq r$ denote respectively the completions of F with respect to a prime ideal $\mathfrak{p}$ and the infinite places $\mathfrak{p}_{\infty\lambda}$; then $K_{\mathfrak{p}} = K \otimes_F F_{\mathfrak{p}}$, $K_\lambda = K \otimes_F F_\lambda$. We denote by $\mathfrak{g}_{\mathfrak{p}}$ the ring of $\mathfrak{p}$-integers in $F_{\mathfrak{p}}$, and put $\mathfrak{r}_{\mathfrak{p}} = \mathfrak{g}_{\mathfrak{p}}\mathfrak{r}$. Further we denote by the same letter ρ the non-trivial automorphism of K_v over F_v for every v.

5.1. Let W be a vector space over F of finite dimension, and L a $\mathfrak{g}$-lattice in W. We always put $W_{\mathfrak{p}} = W \otimes_F F_{\mathfrak{p}}$, $L_{\mathfrak{p}} = \mathfrak{g}_{\mathfrak{p}}L$. Then $L_{\mathfrak{p}}$ is a $\mathfrak{g}_{\mathfrak{p}}$-lattice in $W_{\mathfrak{p}}$. In particular, for every $\mathfrak{g}$-ideal (resp. $\mathfrak{r}$-ideal) $\mathfrak{a}$, we put $\mathfrak{a}_{\mathfrak{p}} = \mathfrak{g}_{\mathfrak{p}}\mathfrak{a}$; obviously, $\mathfrak{a}_{\mathfrak{p}}$ is a $\mathfrak{g}_{\mathfrak{p}}$-ideal (resp. $\mathfrak{r}_{\mathfrak{p}}$-ideal). The following two lemmas are well-known.

5.2. LEMMA. *Let W be a vector space over F of finite dimension, and L a $\mathfrak{g}$-lattice in W. Let $M^{\mathfrak{p}}$, for every prime ideal $\mathfrak{p}$ of F, be a $\mathfrak{g}_{\mathfrak{p}}$-lattice in $W_{\mathfrak{p}}$. Then, there exists a $\mathfrak{g}$-lattice M in W such that $M_{\mathfrak{p}} = M^{\mathfrak{p}}$ for every $\mathfrak{p}$, if and only if $M^{\mathfrak{p}} = L_{\mathfrak{p}}$ for all except a finite number of $\mathfrak{p}$. If such a lattice M exists, we have $M = \bigcap_{\mathfrak{p}} (M^{\mathfrak{p}} \cap W)$.*

5.3. LEMMA. *Let $W_{\mathfrak{p}}$ be a vector space over $F_{\mathfrak{p}}$ of finite dimension, and $L_{\mathfrak{p}}$ a $\mathfrak{g}_{\mathfrak{p}}$-lattice in $W_{\mathfrak{p}}$. Let α and β be elements of $\mathrm{GL}(W_{\mathfrak{p}}, F_{\mathfrak{p}})$. If $L_{\mathfrak{p}}(\alpha - \beta) \subset \mathfrak{p}L_{\mathfrak{p}}\alpha$, then $L_{\mathfrak{p}}\alpha = L_{\mathfrak{p}}\beta$.*

5.4. Let V be a vector space over K of dimension n, and Φ a non-degenerate ρ-hermitian form on V. For every valuation v of K, put $V_v = V \otimes_F F_v$. Then V_v has a natural structure of left K_v-module; and Φ is uniquely extended to a non-degenerate ρ-hermitian form on V_v, which we denote again by Φ. We often consider $\mathrm{GL}(V, K)$ as a subgroup of

$\mathrm{GL}(V_v, K_v)$ in an obvious manner. If v corresponds to the infinite place $\mathfrak{p}_{\infty\lambda}$, we write V_λ for V_v.

5.5. PROPOSITION. *Let L be an $\mathfrak{r}$-lattice in V. Then, for every prime ideal $\mathfrak{p}$ of F, $\mu(L_\mathfrak{p}) = \mu(L)_\mathfrak{p}$.*

From the definition of $\mu(L)$, it follows immediately that $\mu(L)_\mathfrak{p} \subset \mu(L_\mathfrak{p})$. Now let $x_1, \cdots, x_m$ be elements of L and $a_1, \cdots, a_m$ be elements of $\mathfrak{g}_\mathfrak{r}$. Then

$$\Phi\left(\sum_{i=1}^m a_i x_i, \sum_{i=1}^m a_i x_i\right) = \sum_{i=1}^m a_i a_i^\rho \Phi(x_i, x_i) + \sum_{i<j} \mathrm{Tr}_{K_\mathfrak{p}/F_\mathfrak{p}}\left(a_i a_j^\rho \Phi(x_i, x_j)\right) .$$

By 2.11, this is contained in $\mu(L)_\mathfrak{p}$. Hence we get $\mu(L_\mathfrak{r}) \subset \mu(L)_\mathfrak{p}$, so that $\mu(L)_\mathfrak{p} = \mu(L_\mathfrak{p})$.

5.6. PROPOSITION. *An $\mathfrak{r}$-lattice L in V is maximal if and only if $L_\mathfrak{p}$ is maximal for every prime ideal $\mathfrak{p}$ of F.*

This is an easy consequence of 5.2 and 5.5.

5.7. We re-order the infinite places of F so that $F_\lambda = R$ and $K_\lambda = C$ for $1 \leq \lambda \leq t$, and K_λ is not a field for $t < \lambda \leq r$; it may happen that $t = 0$ or $t = r$. Then, for $1 \leq \lambda \leq t$, V_λ is a complex vector space, and the form Φ on V_λ is a hermitian form in the complex sense. With respect to a suitable basis of V_λ, Φ is represented by a real diagonal matrix. Let $J_\lambda(V, \Phi)$ be the number of negative diagonal elements of that matrix. Then the $J_\lambda(V, \Phi)$ are invariants of the structure (V, Φ).

5.8. PROPOSITION (Landherr [10]). *Let V and W be a vector space over K of dimension n and m, respectively. Let Φ and Ψ be non-degenerate ρ-hermitian forms on V and W, respectively. Then there exists a K-linear injection α of W into V such that $\Phi(x\alpha, y\alpha) = \Psi(x, y)$ for every $(x, y) \in W \times W$, if and only if the following condition is satisfied.*
Case I: $n > m$. $0 \leq J_\lambda(V, \Phi) - J_\lambda(W, \Psi) \leq n - m$ for $1 \leq \lambda \leq t$.
Case II: $n = m$. $J_\lambda(V, \Phi) = J_\lambda(W, \Psi)$ for $1 \leq \lambda \leq t$, and $d(\Phi)^{-1} d(\Psi)$ is the identity element of $F_\mathfrak{p}^* / N_{K_\mathfrak{p}/F_\mathfrak{p}}(K_\mathfrak{p}^*)$ for every prime ideal $\mathfrak{p}$ of F (cf. 2.1).

5.9. PROPOSITION. *Let V be a vector space over K of odd dimension; let Φ and Φ' be non-degenerate ρ-hermitian forms on V. If $J_\lambda(V, \Phi) = J_\lambda(V, \Phi')$ for $1 \leq \lambda \leq t$, then there exists an element c of F and a K-automorphism α of V such that $\Phi(x\alpha, y\alpha) = c\Phi'(x, y)$ for every $(x, y) \in V \times V$.*

Let $\{v\}$ be the set of all places v of F such that K_v is a field and $d(\Phi)d(\Phi')$ is not the identity element of $F_v^* / N_{K_v/F_v}(K_v^*)$. By the theory of norm residue (or by the theory of cyclic algebra), the number of places in $\{v\}$ is finite and even. Hence there exists an element c of F such that $\{v\} = \{v \mid c \notin N_{K_v/F_v}(K_v^*)\}$. As $J_\lambda(V, \Phi) = J_\lambda(V, \Phi')$ for $1 \leq \lambda \leq t$, we see that $\mathfrak{p}_{\infty\lambda} \notin \{v\}$ for $1 \leq \lambda \leq t$. Therefore $c \equiv 1 \mod \prod_{\lambda=1}^t \mathfrak{p}_{\infty\lambda}$, and hence

$J_\lambda(V, c\Phi') = J_\lambda(V, \Phi)$ for $1 \leq \lambda \leq t$. Let $2m + 1$ be the dimension of V over K. Then $d(\Phi)d(c\Phi') = c^{2m+1}d(\Phi)d(\Phi') \in N_{K_v/F_v}(K_v^*)$ for all v. Applying 5.8 to Φ and $c\Phi'$, we find an element α of $GL(V, K)$ such that $\Phi(x\alpha, y\alpha) = c\Phi'(x, y)$.

5.10. PROPOSITION (Dieudonné [4]). *Let V be a vector space over K of dimension n, and Φ a non-degenerate ρ-hermitian form on V. Let $\mathfrak{u}$ be the product of the infinite places $\mathfrak{p}_{\infty\lambda}$ of F such that $J_\lambda(V, \Phi) \neq n/2$. Let c be a non-zero element of F. Then, $c = \mu(\alpha)$ for some $\alpha \in G(V, \Phi)$, if and only if, $c \in N_{K/F}(F^*)$ when n is odd, and $c \equiv 1 \bmod \mathfrak{u}$ when n is even.*

If n is odd, this is a special case of 2.4. If n is even, apply 5.8 to Φ and $c\Phi$. Then we get the desired result.

5.11. V and Φ being as above, we call Φ *definite* if $K_v = C$ for every archimedean valuation v (so that $r = t$) and $J_\lambda(V, \Phi) = 0$ or n for every λ; otherwise we call Φ *indefinite*.

5.12. THEOREM. *Let V be a vector space over K of dimension n, and Φ a non-degenerate ρ-hermitian form on V. Suppose that $n > 1$ and Φ is indefinite. Let P be a finite set of prime ideals of F and let $\sigma_{\mathfrak{p}}$ be an element of $SU(V_{\mathfrak{p}}, \Phi)$ for each $\mathfrak{p} \in P$. Let L be a $\mathfrak{g}$-lattice in V and e be a positive integer. Then there exists an element σ of $SU(V, \Phi)$ such that $L_{\mathfrak{p}}(\sigma - \sigma_{\mathfrak{p}}) \subset \mathfrak{p}^e L_{\mathfrak{p}}$ for every $\mathfrak{p} \in P$, and $L_{\mathfrak{q}}\sigma = L_{\mathfrak{q}}$ for every prime ideal $\mathfrak{q}$ of F not belonging to P.*

We prove this theorem in 5.13–17.

5.13. If $n = 2$, 5.12 is an immediate consequence of Eichler [5, Satz 5]. In fact, define quaternion algebras $\mathcal{B}$ and $\mathcal{B}_{\mathfrak{p}}$ by

$$\mathcal{B} = \{\alpha \in E(V, K) \mid \Phi(x\alpha^\iota, y) = \Phi(x, y\alpha)\},$$

$$\mathcal{B}_{\mathfrak{p}} = \{\alpha \in E(V_{\mathfrak{p}}, K_{\mathfrak{p}}) \mid \Phi(x\alpha^\iota, y) = \Phi(x, y\alpha)\},$$

where ι denotes the canonical involution of $E(V, K)$ and $E(V_{\mathfrak{p}}, K_{\mathfrak{p}})$. Then, by 2.6, $SU(V, \Phi) = \{\alpha \in \mathcal{B} \mid \alpha\alpha^\iota = 1\}$, $SU(V_{\mathfrak{p}}, \Phi) = \{\alpha \in \mathcal{B}_{\mathfrak{p}} \mid \alpha\alpha^\iota = 1\}$, and $\mathcal{B}_{\mathfrak{p}} = \mathcal{B} \otimes_F F_{\mathfrak{p}}$. Since Φ is indefinite, $\mathcal{B}$ is an indefinite quaternion algebra over F. Let ξ be a B-isomorphism of V into $\mathcal{B}$ given in 2.7. Put $\xi(L) = \mathfrak{a}$. Take a maximal order $\mathfrak{o}$ in $\mathcal{B}$. We can find an element $f \neq 0$ of $\mathfrak{g}$ such that $f\sigma_{\mathfrak{p}} \in \mathfrak{o}_{\mathfrak{p}}$ for every $\mathfrak{p} \in P$. Let P_1 be the set of prime ideals $\mathfrak{p}$ such that $\mathfrak{a}_{\mathfrak{p}} \neq \mathfrak{o}_{\mathfrak{p}}$, and P_2 the set of prime factors of f. Put $Q = P \cup P_1 \cup P_2$. We can find a positive integer d such that $d > e$, $\mathfrak{p}^d \mathfrak{a}_{\mathfrak{p}} \subset \mathfrak{o}_{\mathfrak{p}} \subset \mathfrak{p}^{-d} \mathfrak{a}_{\mathfrak{p}}$ for $\mathfrak{p} \in Q$, and an element β of $\mathfrak{o}$ such that $\beta \equiv f\sigma_{\mathfrak{p}} \bmod f^3 \mathfrak{p}^{3d} \mathfrak{o}_{\mathfrak{p}}$ for $\mathfrak{p} \in P$, $\beta \equiv f$ $\bmod f^3 \mathfrak{p}^{3d} \mathfrak{o}_{\mathfrak{p}}$ for $\mathfrak{p} \in Q - P$. Then we have $\beta\beta^\iota \equiv f^2 \bmod f^3 \prod_{\mathfrak{p} \in Q} \mathfrak{p}^{3d}$. By [5, Satz 5], there exists an element α of $\mathfrak{o}$ such that $\alpha \equiv \beta \bmod f \prod_{\mathfrak{p} \in Q} \mathfrak{p}^{3d}\mathfrak{o}$, $\alpha\alpha^\iota = f^2$. Put $\sigma = f^{-1}\alpha$. Then $\sigma\sigma^\iota = 1$, so that $\sigma \in SU(V, \Phi)$, and for every $\mathfrak{p} \in P$, $\mathfrak{a}_\mathfrak{p}(\sigma - \sigma_\mathfrak{p}) \subset \mathfrak{p}^{-d}\mathfrak{o}_\mathfrak{p}(f^{-1}\alpha - f^{-1}\beta + f^{-1}\beta - \sigma_\mathfrak{p}) \subset \mathfrak{p}^{2d}\mathfrak{o}_\mathfrak{p} \subset \mathfrak{p}^d\mathfrak{a}_\mathfrak{p}$. For $\mathfrak{p} \in Q - P$, we have $\mathfrak{a}_\mathfrak{p}(\sigma - 1) \subset f^{-1}\mathfrak{p}^{-d}\mathfrak{o}_\mathfrak{p}(\alpha - \beta + \beta - f) \subset \mathfrak{p}^d\mathfrak{a}_\mathfrak{p}$. By 5.3, we have $\mathfrak{a}_\mathfrak{p}\sigma = \mathfrak{a}_\mathfrak{p}$ for $\mathfrak{p} \in Q - P$. If $\mathfrak{p} \notin Q$, then $\mathfrak{p}$ does not divide f, so that f and α are units of $\mathfrak{o}_\mathfrak{p}$. By our definition of P_1, we have $\mathfrak{a}_\mathfrak{p} = \mathfrak{o}_\mathfrak{p}$. Hence we have $\mathfrak{a}_\mathfrak{p}\sigma = \mathfrak{a}_\mathfrak{p}$. As $\xi(L_\mathfrak{r}) = \mathfrak{a}_\mathfrak{p}$, we get our theorem for $n = 2$.

5.14. Suppose that $n > 2$. Let U be a subspace of V of dimension 2 on which Φ is non-degenerate and indefinite. Put $U' = \{z \in V \mid \Phi(z, U) = 0\}$, and for every prime ideal $\mathfrak{p}$ of F, put

$$H(\mathfrak{p}, U) = \{\alpha \in SU(V_\mathfrak{p}, \Phi) \mid U_\mathfrak{p}\alpha = U_\mathfrak{p}, z\alpha = z \text{ for every } z \in U'\} \,.$$

5.15. PROPOSITION. *Let $U^\mathfrak{p}$ be a free $K_\mathfrak{p}$-submodule of rank 2 of $V_\mathfrak{p}$, on which Φ is non-degenerate. Then there exist a vector subspace U of V of dimension 2 and an element α of $U(V_\mathfrak{p}, \Phi)$ such that $U_\mathfrak{p}\alpha = U^\mathfrak{p}$ and Φ is non-degenerate and indefinite on U.*

This is an easy consequence of 5.8.

5.16. PROPOSITION. *$SU(V_\mathfrak{p}, \Phi)$ is the closure of the group generated by the subgroups $H(\mathfrak{p}, U)$ for all subspaces U of V of dimension 2 on which Φ is non-degenerate and indefinite.*

Let H be the closure of the subgroup of $SU(V_\mathfrak{p}, \Phi)$ generated by the $H(\mathfrak{p}, U)$ for all U with the above property. H is a normal subgroup of $SU(V_\mathfrak{p}, \Phi)$, on account of 5.15. Assume that $K_\mathfrak{p}$ is a field. As $n > 2$, Φ is isotropic in $V_\mathfrak{p}$. Then our assertion follows immediately from 5.15 and Dieudonné [2, p. 66, Prop. 19]. Next we consider the case where $K_\mathfrak{p} = F_\mathfrak{p} \times F_\mathfrak{p}$. Then $SU(V_\mathfrak{p}, \Phi)$ is isomorphic to $\mathrm{SL}(n, F_\mathfrak{p})$ (1.3). Let Z be the center of $\mathrm{SL}(n, F_\mathfrak{p})$. It is well-known that $\mathrm{SL}(n, F_\mathfrak{p})/Z$ is simple, and Z is the set of all scalar matrices $\zeta 1_n$ with n^{th} roots of unity ζ in $F_\mathfrak{p}$. Let α_k for $1 \leq k \leq n - 1$ be the diagonal matrix whose i^{th} diagonal element a_{ki} is given as follows: $a_{kk} = \zeta^k$, $a_{k,k+1} = \zeta^{-k}$, $a_{ki} = 1$ for $i \neq k$, $i \neq k + 1$. Then we have $\alpha_1 \cdots \alpha_{n-1} = \zeta 1_n$. This shows that Z is contained in the image of H in $\mathrm{SL}(n, F_\mathfrak{p})$. As H is not commutative, and as $\mathrm{SL}(n, F_\mathfrak{p})/Z$ is simple, we must have $H = SU(V_\mathfrak{r}, \Phi)$. This completes the proof.

5.17. PROOF OF 5.12 (*in the case $n > 2$*). Let the notation be as in 5.12. Suppose that we can find, for every positive integer d, and for each $\mathfrak{p} \in P$, an element $\tau^\mathfrak{p}$ of $SU(V, \Phi)$ such that $L_\mathfrak{p}(\tau^\mathfrak{p} - \sigma_\mathfrak{p}) \subset \mathfrak{p}^d L_\mathfrak{p}$, $L_\mathfrak{q}(\tau^\mathfrak{p} - 1) \subset \mathfrak{q}^d L_\mathfrak{q}$ for $\mathfrak{q} \in P - \{\mathfrak{p}\}$, $L_\mathfrak{q}\tau^\mathfrak{p} = L_\mathfrak{q}$ for $\mathfrak{q} \notin P$. Then, for a suitably large d, the product of the $\tau^\mathfrak{p}$ for $\mathfrak{p} \in P$ (regardless of the order) has the required property of σ of 5.12. Hence it suffices to prove the existence of such $\tau^\mathfrak{p}$. Fix a prime ideal $\mathfrak{p}$ belonging to P. Further, in view of 5.16, it is sufficient to prove the assertion in the case where $\sigma_\mathfrak{p}$ belongs to $H(\mathfrak{p}, U)$ for some

U. Put $U' = \{z \in V \mid \Phi(z, U) = 0\}$. Then $V = U + U'$. Let M and M' be $\mathfrak{g}$-lattices in U and U', respectively. Let P_1 be the set of prime ideals $\mathfrak{q}$ of F such that $M_{\mathfrak{c}} + M'_{\mathfrak{q}} \neq L_{\mathfrak{q}}$. Put $Q = P \cup P_1$. We can find a positive integer f such that $f > d$ and $\mathfrak{q}^f L_{\mathfrak{q}} \subset M_{\mathfrak{q}} + M'_{\mathfrak{q}} \subset \mathfrak{q}^{-f} L_{\mathfrak{q}}$ for $\mathfrak{q} \in Q$. Since our theorem is true for $n = 2$ (5.13), there exists an element α of $SU(U, \Phi)$ such that $M_{\mathfrak{p}}(\alpha - \sigma_{\mathfrak{r}}) \subset \mathfrak{p}^{3f} M_{\mathfrak{p}}$, $M_{\mathfrak{q}}(\alpha - 1) \subset \mathfrak{q}^{3f} M_{\mathfrak{q}}$ for $\mathfrak{q} \in Q - \{\mathfrak{p}\}$ and $M_{\mathfrak{q}}\alpha = M_{\mathfrak{q}}$ for $\mathfrak{q} \notin Q$. Define an element β of $SU(V, \Phi)$ by $\beta = \alpha$ on U and $\beta = 1$ on U'. Then we have

$$L_{\mathfrak{p}}(\beta - \sigma_{\mathfrak{p}}) \subset \mathfrak{p}^{-f}(M_{\mathfrak{p}} + M'_{\mathfrak{p}})(\beta - \sigma_{\mathfrak{p}}) \subset \mathfrak{p}^{2f} M_{\mathfrak{p}} \subset \mathfrak{p}^f L_{\mathfrak{p}} ,$$

and similarly $L_{\mathfrak{q}}(\beta - 1) \subset \mathfrak{q}^f L_{\mathfrak{q}}$ for $\mathfrak{q} \in Q - \{\mathfrak{p}\}$. By 5.3, we have $L_{\mathfrak{q}}\beta = L_{\mathfrak{q}}$ for $\mathfrak{q} \in Q - P$. If $\mathfrak{q} \notin Q$, then $L_{\mathfrak{q}} = M_{\mathfrak{q}} + M'_{\mathfrak{q}}$, so that $L_{\mathfrak{q}}\beta = L_{\mathfrak{q}}$ for $\mathfrak{q} \notin Q$. This completes the proof.

5.18. By a *class* of $\mathfrak{g}$-lattices in V with respect to $G(V, \Phi)$ (resp. $U(V, \Phi)$, $SU(V, \Phi)$, $H(V, \Phi)$), we understand a maximal set of $\mathfrak{g}$-lattices in V whose members are transformed to each other by elements of $G(V, \Phi)$ (resp. $U(V, \Phi)$, $SU(V, \Phi)$, $H(V, \Phi)$). By a *genus* of $\mathfrak{g}$-lattices in V with respect to $G(V, \Phi)$ (resp. $U(V, \Phi)$, $SU(V, \Phi)$, $H(V, \Phi)$), we understand a maximal set Λ of $\mathfrak{g}$-lattices in V with the following property: if L and M are members of Λ, then, for every $\mathfrak{p}$, there exists an element $\alpha_{\mathfrak{p}}$ of $G(V, \Phi)$ (resp. $U(V_{\mathfrak{r}}, \Phi)$, $SU(V_{\mathfrak{r}}, \Phi)$, $H(V_{\mathfrak{p}}, \Phi)$) such that $L_{\mathfrak{p}}\alpha_{\mathfrak{p}} = M_{\mathfrak{p}}$.

5.19. THEOREM. *Suppose that $n > 1$ and Φ is indefinite. Then, with respect to the group $SU(V, \Phi)$, every genus of $\mathfrak{g}$-lattices in V consists of only one class.*

Let L and M be $\mathfrak{g}$-lattices in V belonging to the same genus with respect to $SU(V, \Phi)$. Let P be the set of prime ideals $\mathfrak{p}$ of F such that $L_{\mathfrak{p}} \neq M_{\mathfrak{p}}$. For each $\mathfrak{p} \in P$, there exists an element $\alpha_{\mathfrak{p}}$ of $SU(V_{\mathfrak{p}}, \Phi)$ such that $L_{\mathfrak{p}}\alpha_{\mathfrak{p}} = M_{\mathfrak{p}}$. By 5.12, there exists an element α of $SU(V, \Phi)$ such that $L_{\mathfrak{p}}(\alpha - \alpha_{\mathfrak{p}}) \subset \mathfrak{p}L_{\mathfrak{p}}\alpha_{\mathfrak{p}}$ for $\mathfrak{p} \in P$ and $L_{\mathfrak{q}}\alpha = L_{\mathfrak{q}}$ for $\mathfrak{q} \notin P$. Then, by 5.3, $L_{\mathfrak{p}}\alpha = L_{\mathfrak{p}}\alpha_{\mathfrak{p}} = M_{\mathfrak{p}}$ for $\mathfrak{p} \in P$, and $L_{\mathfrak{q}}\alpha = L_{\mathfrak{q}} = M_{\mathfrak{q}}$ for $\mathfrak{q} \notin P$. Therefore $(L\alpha)_{\mathfrak{p}} = M_{\mathfrak{p}}$ for every prime ideal $\mathfrak{p}$ of F, hence $L\alpha = M$. This proves our theorem.

5.20. Let L and M be $\mathfrak{g}$-lattices in V belonging to the same genus with respect to $G(V, \Phi)$. If L is an $\mathfrak{r}$-lattice, M is an $\mathfrak{r}$-lattice. Moreover, by 2.12 and 5.6, if L is maximal, so is M. Now let us consider the genera and classes consisting of $\mathfrak{r}$-lattices and then maximal $\mathfrak{r}$-lattices.

5.21. PROPOSITION. *Let L be an $\mathfrak{r}$-lattice in V. Let $\Lambda_1, \Lambda_2, \Lambda_3$ be the genera of L with respect to $U(V, \Phi)$, $G(V, \Phi)$, $H(V, \Phi)$, respectively. (Λ_3 can be defined only if n is even.) Then:*

(i) *If $M \in \Lambda_1$, $N_{K/F}([L/M]) = \mathfrak{g}$. Conversely, for every $\mathfrak{r}$-ideal $\mathfrak{a}$ satis-*

fying $N_{K/F}(\mathfrak{a}) = \mathfrak{g}$, *there exists an* $M \in \Lambda_1$ *such that* $[L/M] = \mathfrak{a}$.

(ii) *If* $M \in \Lambda_2$, $N_{K/F}([L/M]) = (\mu(L)^{-1}\mu(M))^n$. *Conversely, if* $\mathfrak{a}$ *is an* $\mathfrak{r}$-*ideal and* $\mathfrak{b}$ *is a* $\mathfrak{g}$-*ideal satisfying* $N_{K/F}(\mathfrak{a}) = \mathfrak{b}^n$, *there exists an* $M \in \Lambda_2$ *such that* $[L/M] = \mathfrak{a}$.

(iii) *If* $M \in \Lambda_3$, $[L/M] = \mathfrak{b}^{n/2}\mathfrak{r}$ *for some* $\mathfrak{g}$-*ideal* $\mathfrak{b}$. *Conversely, for every* $\mathfrak{g}$-*ideal* $\mathfrak{b}$, *there exists an* $M \in \Lambda_3$ *such that* $[L/M] = \mathfrak{b}^{n/2}\mathfrak{r}$.

Let $\mathfrak{p}$ be a prime ideal of F and $\alpha_{\mathfrak{p}}$ an element of $G(V_{\mathfrak{p}}, \Phi)$. Then $[L_{\mathfrak{p}}/L_{\mathfrak{p}}\alpha_{\mathfrak{p}}] = \det(\alpha_{\mathfrak{p}})\mathfrak{r}_{\mathfrak{p}}$, $\mu(L_{\mathfrak{p}})^{-1}\mu(M_{\mathfrak{p}}) = \mu(\alpha_{\mathfrak{p}})$, and $N_{K_{\mathfrak{p}}/F_{\mathfrak{p}}}(\det(\alpha_{\mathfrak{p}})) = \mu(\alpha_{\mathfrak{p}})^n$. From this we obtain quite easily the first assertions of (i), (ii), (iii). Let $\mathfrak{a}$ be an $\mathfrak{r}$-ideal such that $N_{K/F}(\mathfrak{a}) = \mathfrak{g}$. We can find an $\mathfrak{r}$-ideal $\mathfrak{c}$ such that $\mathfrak{a} = \mathfrak{c}^{-1}\mathfrak{c}^{\rho}$. Let P be the set of prime ideals $\mathfrak{p}$ of F such that $\mathfrak{a}_{\mathfrak{p}} \neq \mathfrak{r}_{\mathfrak{p}}$. For every $\mathfrak{p} \in P$, let $c_{\mathfrak{p}}$ be an element of $K_{\mathfrak{p}}$ such that $\mathfrak{c}_{\mathfrak{p}} = \mathfrak{r}_{\mathfrak{p}}c_{\mathfrak{p}}$. By 2.3, there exists an element $\alpha_{\mathfrak{p}}$ of $U(V_{\mathfrak{p}}, \Phi)$ such that $\det(\alpha_{\mathfrak{p}}) = c_{\mathfrak{p}}^{-1}c_{\mathfrak{p}}^{\rho}$. Then $[L_{\mathfrak{p}}/L_{\mathfrak{p}}\alpha_{\mathfrak{p}}] = \mathfrak{a}_{\mathfrak{p}}$. By 5.2, there exists a $\mathfrak{g}$-lattice M such that $M_{\mathfrak{p}} = L_{\mathfrak{p}}\alpha_{\mathfrak{p}}$ for every $\mathfrak{p} \in P$, and $M_{\mathfrak{q}} = L_{\mathfrak{q}}$ for $\mathfrak{q} \notin P$. M belongs to Λ_1 and $[L/M] = \mathfrak{a}$. This proves the second assertion of (i). Next, let $\mathfrak{a}$ and $\mathfrak{b}$ be as in (ii). Let P be the set of prime ideals $\mathfrak{p}$ of F such that $\mathfrak{a}_{\mathfrak{p}} \neq \mathfrak{r}_{\mathfrak{p}}$. Let $b_{\mathfrak{p}}$, for each $\mathfrak{p} \in P$, be an element of $F_{\mathfrak{p}}$ such that $\mathfrak{g}_{\mathfrak{p}}b_{\mathfrak{p}} = \mathfrak{b}_{\mathfrak{p}}$. If n is odd, we take this $b_{\mathfrak{p}}$ as follows: let $n = 2m + 1$; take an element $d_{\mathfrak{p}}$ of $K_{\mathfrak{p}}$ such that $\mathfrak{r}_{\mathfrak{p}}d_{\mathfrak{p}} = \mathfrak{b}_{\mathfrak{p}}^{-m}\mathfrak{a}$ and put $b_{\mathfrak{p}} = N_{K_{\mathfrak{p}}/F_{\mathfrak{p}}}(d_{\mathfrak{p}})$. By 3.1 and 4.4, there exists an element $\alpha_{\mathfrak{p}}$ of $G(V_{\mathfrak{p}}, \Phi)$ such that $\mu(\alpha_{\mathfrak{p}}) = b_{\mathfrak{p}}$, for each $\mathfrak{p} \in P$. By 5.2, we can find a $\mathfrak{g}$-lattice M' in V such that $M'_{\mathfrak{p}} = L_{\mathfrak{p}}\alpha_{\mathfrak{p}}$ for every $\mathfrak{p} \in P$, and $M'_{\mathfrak{q}} = L_{\mathfrak{q}}$ for $\mathfrak{q} \notin P$. Then $M' \in \Lambda_2$, and $\mu(L)^{-1}\mu(M') = \mathfrak{b}$. Put $[L/M'] = \mathfrak{a}'$. Then we have $N_{K/F}(\mathfrak{a}') = \mathfrak{b}^n$, so that $N_{K/F}(\mathfrak{a}'\mathfrak{a}^{-1}) = \mathfrak{g}$. By the second assertion of (i) there exists an $\mathfrak{r}$-lattice M belonging to the same genus as M' with respect to $U(V, \Phi)$ such that $[M'/M] = (\mathfrak{a}')^{-1}\mathfrak{a}$. Then $M \in \Lambda_2$ and $[L/M] = \mathfrak{a}$. This proves the second assertion of (ii). Let $\mathfrak{b}$ be a $\mathfrak{g}$-ideal, and P the set of prime ideals $\mathfrak{p}$ of F such that $\mathfrak{b}_{\mathfrak{p}} \neq \mathfrak{g}_{\mathfrak{p}}$. To prove (iii), put $n = 2m$. For each $\mathfrak{p} \in P$, take an element $b_{\mathfrak{p}}$ of $F_{\mathfrak{p}}$ such that $\mathfrak{b}_{\mathfrak{p}} = \mathfrak{g}_{\mathfrak{p}}b_{\mathfrak{p}}$. Then we can find an element $\alpha_{\mathfrak{p}}$ of $H(V_{\mathfrak{p}}, \Phi)$ such that $\mu(\alpha_{\mathfrak{p}}) = b_{\mathfrak{p}}$ (and hence $\det(\alpha_{\mathfrak{p}}) = b_{\mathfrak{p}}^m$). In fact, if $K_{\mathfrak{p}}$ is not a field, this follows from 3.1; and if $K_{\mathfrak{p}}$ is a field, this follows from 4.4 and 2.5. By 5.2, there exists a $\mathfrak{g}$-lattice M in V such that $M_{\mathfrak{p}} = L_{\mathfrak{p}}\alpha_{\mathfrak{p}}$ for every $\mathfrak{p} \in P$ and $M_{\mathfrak{q}} = L_{\mathfrak{q}}$ for $\mathfrak{q} \notin P$. Then $M \in \Lambda_3$ and $[L/M] = \mathfrak{b}^m\mathfrak{r}$. This completes our proof.

5.22. Let Λ be a genus of $\mathfrak{r}$-lattices in V with respect to $U(V, \Phi)$, and L a member of Λ. For every prime ideal $\mathfrak{p}$ of F, put $E_{0\mathfrak{p}} = \{e \in \mathfrak{r}_{\mathfrak{p}} \mid ee^{\rho} = 1\}$ and

$$E_{\mathfrak{p}}(\Lambda) = \{\det(\alpha) \mid \alpha \in U(V_{\mathfrak{p}}, \Phi), \; L_{\mathfrak{p}}\alpha = L_{\mathfrak{p}}\} \, .$$

It is clear that $E_{\mathfrak{p}}(\Lambda)$ depends only upon Λ and is independent of the choice of L. In view of 3.2 and 4.18, $E_{0\mathfrak{p}} \neq E_\mathfrak{p}(\Lambda)$ only when n is even and $\mathfrak{p}$ is ramified in K. Suppose that n is even. We call $\mathfrak{p}$ *irregular for* Λ if $E_{0\mathfrak{p}} \neq E_\mathfrak{p}(\Lambda)$, and denote by $\mathfrak{E}(\Lambda)$ the product of the factor groups $E_{0\mathfrak{p}}/E_\mathfrak{p}(\Lambda)$ for all irregular prime ideals for Λ. By 4.18, $\mathfrak{E}(\Lambda)$ is a product of cyclic groups of order 2. Let L and M be members of Λ. For every irregular $\mathfrak{p}$, take an element $\alpha_\mathfrak{p}$ of $U(V_\mathfrak{p}, \Phi)$ such that $L_\mathfrak{p}\alpha_\mathfrak{p} = M_\mathfrak{p}$. Then we denote by $e_\Lambda(L/M)$ the element of $\mathfrak{E}(\Lambda)$ whose components are represented by the $\det(\alpha_\mathfrak{r})$. Further let c be an element of K such that $cc^\rho = 1$. We denote by $f_\Lambda(c)$ the element of $\mathfrak{E}(\Lambda)$ whose components are the cosets $cE_\mathfrak{r}(\Lambda)$.

5.23. The notation being as in 5.22, suppose that the members of Λ are *maximal*. Then, by 4.18, a prime ideal $\mathfrak{p}$ of F is irregular for Λ if and only if $\mathfrak{p}$ is ramified in K, and $V_\mathfrak{p}$ has the trivial kernel subspace with respect to Φ. We call such a $\mathfrak{p}$ *irregular for* Φ. Put $E_{1\mathfrak{p}} = \{e^{-1}e^\rho \,|\, e \in K_\mathfrak{p}, e\mathfrak{r}_\mathfrak{p} = \mathfrak{r}_\mathfrak{p}\}$. By 4.18, $E_\mathfrak{p}(\Lambda) = E_{1\mathfrak{p}}$, if $\mathfrak{p}$ is irregular for Φ. When Λ consists of maximal $\mathfrak{r}$-lattices, we denote $\mathfrak{E}(\Lambda)$ and $f_\Lambda(c)$ simply by $\mathfrak{E}$ and $f(c)$, since they depend only upon the structure (V, Φ). If s is the number of irregular prime ideals for Φ, $\mathfrak{E}$ is of order 2^s. Let L and M be maximal $\mathfrak{r}$-lattices in V. In 4.25, we have defined an element $e(L_\mathfrak{p}/M_\mathfrak{p})$ of $E_{0\mathfrak{p}}/E_{1\mathfrak{p}}$. We denote by $e(L/M)$ the element of $\mathfrak{E}$ whose components are $e(L_\mathfrak{p}/M_\mathfrak{p})$. If L and M belong to the same genus Λ with respect to $U(V, \Phi)$, we have $e(L/M) = e_\Lambda(L/M)$. If α is an element of $G(V, \Phi)$, then $e(L/L\alpha) = f(\det(\alpha)\mu(\alpha)^{-m})$, where $n = 2m$.

5.24. THEOREM. *Let V be a vector space over K of dimension n, and Φ a non-degenerate ρ-hermitian form on V. Suppose that $n > 1$ and Φ is indefinite. Let $\mathfrak{C}$ be the group of ideal-classes in K and let $\mathfrak{C}_0$ be the subgroup of $\mathfrak{C}$ which consists of the ideal-classes containing ideals $\mathfrak{a}$ such that $\mathfrak{a}^\rho = \mathfrak{a}$. Let $\mathfrak{u}$ be the product of the infinite places $\mathfrak{p}_{\infty\lambda}$ of F such that $J_\lambda(V, \Phi) \neq n/2$ (cf. 5.7). Let $h(\mathfrak{u})$ be the number of ideal-classes modulo $\mathfrak{u}$ in F. Let $\mathfrak{E}$ be as in 5.23 and $\mathfrak{E}(\Lambda)$ be as in 5.22. Let E_0 be the group of units u of $\mathfrak{r}$ such that $N_{K/F}(u) = 1$. Then:*

    (i) *With respect to $U(V, \Phi)$, every genus Λ of $\mathfrak{r}$-lattices (not necessarily maximal) contains exactly $[\mathfrak{C} : \mathfrak{C}_0]$ or $[\mathfrak{C} : \mathfrak{C}_0][\mathfrak{E}(\Lambda) : f_\Lambda(E_0)]$ classes according as n is odd or even.*

    (ii) *With respect to $G(V, \Phi)$, every genus of maximal $\mathfrak{r}$-lattices contains exactly $[\mathfrak{C} : 1]$ or $h(\mathfrak{u})[\mathfrak{C} : \mathfrak{C}_0][\mathfrak{E} : f(E_0)]$ classes according as n is odd or even.*

    (iii) *With respect to $H(V, \Phi)$, every genus of maximal $\mathfrak{r}$-lattices contains exactly $h(\mathfrak{u})$ classes.*

    We shall prove this in 5.28.

5.25. Proposition. *Let L, Λ_1, Λ_2, Λ_3 be the same as in 5.21, and Λ_0 be the genus of L with respect to $SU(V, \Phi)$. Suppose that L is maximal. Then:*

(i) *If n is odd, Λ_0 consists of all the maximal $\mathfrak{r}$-lattices M such that $[L/M] = \mathfrak{r}$.*

(ii) *If n is even, Λ_0 consists of all the maximal $\mathfrak{r}$-lattices M such that $[L/M] = \mathfrak{r}$ and $e(L/M) = 1$.*

(iii) *Λ_1 consists of all the maximal $\mathfrak{r}$-lattices M such that $\mu(L) = \mu(M)$; for every maximal $\mathfrak{r}$-lattice M, $\mu(L) = \mu(M)$ if and only if $N_{K/F}([L/M]) = \mathfrak{g}$.*

(iv) *If n is odd, Λ_2 consists of all the maximal $\mathfrak{r}$-lattices M such that $\mu(L)^{-1}\mu(M) = N_{K/F}(\mathfrak{a})$ for some $\mathfrak{r}$-ideal $\mathfrak{a}$.*

(v) *If n is even, Λ_2 consists of all the maximal $\mathfrak{r}$-lattices in V.*

(vi) *Λ_3 consists of all the maximal $\mathfrak{r}$-lattices M such that $e(L/M) = 1$, and $[L/M] = \mathfrak{b}^{n/2}\mathfrak{r}$ for some $\mathfrak{g}$-ideal $\mathfrak{b}$.*

The assertions (i) and (ii) follow from 3.3, 4.13, 4.14, 4.23, 4.25. Let M be a maximal $\mathfrak{r}$-lattice in V. Let $\mathfrak{p}$ be a prime ideal of F. If $K_\mathfrak{p} = F_\mathfrak{p} \times F_\mathfrak{p}$, then $N_{K/F}([L/M])_\mathfrak{p} = (\mu(L_\mathfrak{p})^{-1}\mu(M_\mathfrak{p}))^n$ (3.6). Hence $\mu(L_\mathfrak{p}) = \mu(M_\mathfrak{p})$ if and only if $N_{K/F}([L/M])_\mathfrak{p} = \mathfrak{g}_\mathfrak{p}$. This assertion is true also in the case where $K_\mathfrak{p}$ is a field, on account of 4.14. Therefore $\mu(L) = \mu(M)$ if and only if $N_{K/F}([L/M]) = \mathfrak{g}$. If $\mu(L) = \mu(M)$, $M \in \Lambda_1$ by 3.3 and 4.13. On account of 5.21, we obtain (iii). The assertions (iv) and (v) follow from 3.3 and 4.11. Suppose that n is even and put $n = 2m$. Let M be a maximal $\mathfrak{r}$-lattice in V such that $[L/M] = \mathfrak{b}^m\mathfrak{r}$ for some $\mathfrak{g}$-ideal $\mathfrak{b}$. By (iii) of 5.21, there exists an $M' \in \Lambda_3$ such that $[L/M'] = \mathfrak{b}^m\mathfrak{r}$. Then $[M/M'] = \mathfrak{r}$. As $M' \in \Lambda_3$, we must have $e(L/M') = 1$. Therefore, if $e(L/M) = 1$, we get $e(M'/M) = 1$, so that M and M' belong to the same genus with respect to $SU(V, \Phi)$. Hence $M \in \Lambda_3$. This proves (vi).

5.26. Remark. Suppose that n is even. Let s be the number of irregular prime ideals for Φ. Then (vi) of 5.25 shows that all the maximal $\mathfrak{r}$-lattices in V are divided into 2^s genera with respect to $H(V, \Phi)$. Fix a maximal $\mathfrak{r}$-lattice L in V. Then, all the maximal $\mathfrak{r}$-lattices M in V such that $[L/M] = \mathfrak{r}$ are divided into 2^s genera with respect to $SU(V, \Phi)$.

5.27. Proposition. *Suppose that $n > 1$ and Φ is indefinite. Let $\mathfrak{u}$ be as in 5.24. If n is odd, the following assertions (i), (ii) hold.*

(i) *With respect to $U(V, \Phi)$, two members L and M of the same genus belong to the same class if and only if there exists an element a of K such that $[L/M] = \mathfrak{r}a$, $aa^\rho = 1$.*

(ii) *Two maximal $\mathfrak{r}$-lattices L and M belong to the same class with respect to $G(V, \Phi)$ if and only if there exist an element a of K and an* *element b of F such that $[L/M] = \mathfrak{r}a$, $aa^\rho = b^n$.*

If n is even and $n = 2m$, the following assertions (iii), (iv), (v) hold.

(iii)   *With respect to $U(V, \Phi)$, two members L and M of the same genus Λ belong to the same class if and only if there exists an element a of K such that $[L/M] = \mathfrak{r}a$, $aa^\rho = 1$ and $f_\Lambda(a) = e_\Lambda(L/M)$.*

(iv)   *Two maximal $\mathfrak{r}$-lattices L and M belong to the same class with respect to $G(V, \Phi)$ if and only if there exist an element a of K and an element b of F such that $[L/M] = \mathfrak{r}a$, $aa^\rho = b^n$, $f(b^{-m}a) = e(L/M)$, $b \equiv 1 \bmod \mathfrak{u}$.*

(v)   *With respect to $H(V, \Phi)$, two maximal $\mathfrak{r}$-lattices L and M in the same genus belong to the same class if and only if $\mu(L)$ and $\mu(M)$ belong to the same ideal-class modulo $\mathfrak{u}$ in F.*

The "only if" parts can be proved in a straightforward way; so we have only to prove the "if" parts. Let L and M be members of a genus Λ with respect to $U(V, \Phi)$. Suppose that $[L/M] = \mathfrak{r}a$, $aa^\rho = 1$ for some $a \in K$. By 2.3, we can find an element α of $U(V, \Phi)$ such that $\det(\alpha) = a$. Then $[L\alpha/M] = \mathfrak{r}$. If n is even and $f_\Lambda(a) = e_\Lambda(L/M)$, we get $e_\Lambda(L\alpha/M) = 1$. For every prime ideal $\mathfrak{p}$ of F, there exists an element $\beta_\mathfrak{p}$ of $U(V_\mathfrak{p}, \Phi)$ such that $M_\mathfrak{p}\beta_\mathfrak{p} = L_\mathfrak{p}\alpha$. Since $[L\alpha/M] = \mathfrak{r}$, $\det(\beta_\mathfrak{p})$ is a unit of $\mathfrak{r}_\mathfrak{p}$. Further, since $e_\Lambda(L\alpha/M) = 1$, $\det(\beta_\mathfrak{p}) \in E_\mathfrak{p}(\Lambda)$ for every $\mathfrak{p}$. Hence there exists an element $\gamma_\mathfrak{p}$ of $U(V_\mathfrak{p}, \Phi)$ such that $M_\mathfrak{p}\gamma_\mathfrak{p} = M_\mathfrak{p}$ and $\det(\gamma_\mathfrak{p}) = \det(\beta_\mathfrak{p})^{-1}$. Then $\gamma_\mathfrak{p}\beta_\mathfrak{p} \in SU(V_\mathfrak{p}, \Phi)$ and $M_\mathfrak{p}\gamma_\mathfrak{p}\beta_\mathfrak{p} = L_\mathfrak{p}\alpha$. This shows that $L\alpha$ and M belong to the same genus with respect to $SU(V, \Phi)$. By 5.19, we have $M = L\alpha\sigma$ for some $\sigma \in SU(V, \Phi)$. This proves (i) and (iii). Now let L and M be maximal $\mathfrak{r}$-lattices in V. Suppose that $[L/M] = \mathfrak{r}a$, $aa^\rho = b^n$ for some $a \in K$ and some $b \in F$; if n is even, suppose moreover that $f(b^{-m}a) = e(L/M)$, $b \equiv 1 \bmod \mathfrak{u}$. By 5.10, there exists an element α of $G(V, \Phi)$ such that $\mu(\alpha) = b$. Put $c = \det(\alpha)$. Then $[L\alpha/M] = \mathfrak{r}c^{-1}a$, $N_{K/F}(c^{-1}a) = 1$, and, if n is even, $e(L\alpha/M) = e(L\alpha/L)e(L/M) = f(b^m c^{-1})f(b^{-m}a) = f(c^{-1}a)$. By (iii) of 5.25, and by (i) and (iii) which we have just proved, there exists an element β of $U(V, \Phi)$ such that $L\alpha\beta = M$. This proves (ii) and (iv). Finally suppose that n is even, L and M have the same genus with respect to $H(V, \Phi)$, and $\mu(L)b = \mu(M)$, $b \equiv 1 \bmod \mathfrak{u}$ for some $b \in F$. By 5.21, $[L/M] = b^m\mathfrak{r}$. By 5.10 and 2.5, there exists an element α of $H(V, \Phi)$ such that $\mu(\alpha) = b$. Then $[L\alpha/M] = \mathfrak{r}$ and $e(L\alpha/M) = 1$. By (ii) of 5.25 and 5.19, there exists an element β of $SU(V, \Phi)$ such that $L\alpha\beta = M$. This proves (v).

5.28.   PROOF OF 5.24.   Let L, Λ_1, Λ_2, Λ_3 be the same as in 5.21. Let I_K and I_F be the groups of all ideals in K and in F, respectively. Let J be the group of $\mathfrak{r}$-ideals $\mathfrak{a}$ such that $N_{K/F}(\mathfrak{a}) = \mathfrak{g}$, and J_0 the group of principal ideals $\mathfrak{r}a$ such that $aa^\rho = 1$. By (i) of 5.21 and (i) of 5.27, if n is odd,

$M \to [L/M]$ gives a ono-to-one correspondence between the classes in Λ_1 and the factor group J/J_0. Further the mapping $I_K \ni \mathfrak{c} \to \mathfrak{c}^{-1}\mathfrak{c}^\rho \in J$ gives an isomorphism of $\mathfrak{C}/\mathfrak{C}_0$ onto J/J_0. Hence we obtain (i) for odd n. If n is even, we put $\Lambda = \Lambda_1$ and consider the mapping

$$\Lambda \ni M \longrightarrow \big([L/M],\, e_\Lambda(L/M)\big) \in J \times \mathfrak{E}(\Lambda) \ .$$

We can easily verify that this is surjective. Let $\mathfrak{H}$ be the subgroup of $J \times \mathfrak{E}(\Lambda)$ consisting of the elements $(\mathfrak{r}a, f_\Lambda(a))$ for all $a \in K$ such that $aa^\rho = 1$. By (iii) of 5.27, the classes in Λ are in one-to-one correspondence with the elements of $(J \times \mathfrak{E}(\Lambda))/\mathfrak{H}$. Let $\{\mathfrak{r}_i\}$ be a complete set of representatives for J/J_0, and $\{\omega_j\}$ a complete set of representatives for $\mathfrak{E}(\Lambda)/f_\Lambda(E_0)$. Then $\{(\mathfrak{r}_i,\, \omega_j)\}$ is a complete set of representatives for $(J \times \mathfrak{E}(\Lambda))/\mathfrak{H}$ (cf. the proof of (ii) for even n below). Hence we get (i) for even n. Now suppose that L is maximal. First we consider the case of odd n and put $n = 2m + 1$. Let $\mathfrak{a} \in I_K$ and $\mathfrak{b} = N_{K/F}(\mathfrak{a})$. Since $N_{K/F}(\mathfrak{b}^m\mathfrak{a}) = \mathfrak{b}^n$, there exists, by (ii) of 5.21, an $M \in \Lambda_2$ such that $[L/M] = \mathfrak{b}^m$. Then $\mu(L)^{-1}\mu(M) = \mathfrak{b}$. This implies that $M \to (\mu(L)^{-1}\mu(M))^{-m}[L/M]$ gives a surjective mapping of Λ_2 to I_K. Let $M \in \Lambda_2$, $N \in \Lambda_2$. If M and N belong to the same class, we have $[M/N] = \mathfrak{r}a$, $N_{K/F}(a) = b^n$ for some $a \in K$ and some $b \in F$ ((ii) of 5.27). Then $\mu(M)^{-1}\mu(N) = \mathfrak{g}b$, so that $(\mu(M)^{-1}\mu(N))^{-m}[M/N]$ is a principal ideal. Conversely, suppose that $(\mu(M)^{-1}\mu(N))^{-m}[M/N] = \mathfrak{r}c$ for some $c \in K$. Put $cc^\rho = b$. Applying $N_{K/F}$, we find $\mu(M)^{-1}\mu(N) = \mathfrak{g}b$, so that $[M/N] = \mathfrak{r}b^mc$, and $N_{K/F}(b^mc) = b^n$. By (ii) of 5.27, M and N belong to the same class. This proves (ii) of 5.24 for odd n. Suppose that n is even and put $n = 2m$. Let J_1 be the group of $\mathfrak{r}$-ideals $\mathfrak{a}$ such that $N_{K/F}(\mathfrak{a}) = \mathfrak{b}^n$ for some $\mathfrak{g}$-ideal $\mathfrak{b}$, and let J_2 be the group of principal ideals $\mathfrak{r}a$ such that $N_{K/F}(a) = b^n$ for some $b \in F$ satisfying $b \equiv 1 \bmod \mathfrak{u}$. Now consider the mapping $\Lambda_2 \ni M \to ([L/M],\, e(L/M)) \in J_1 \times \mathfrak{E}$. By (ii) of 5.21, and by the definition of $e(L/M)$, we see that this is surjective. Let $\mathfrak{K}$ be the subgroup of $J_1 \times \mathfrak{E}$ consisting of the elements $(\mathfrak{r}a, f(b^{-m}a))$ for the elements $a \in K$, $b \in F$ such that $aa^\rho = b^n$, $b \equiv 1 \bmod \mathfrak{u}$. By (iv) of 5.27, the number of classes in Λ_2 is equal to the index $[J_1 \times \mathfrak{E} : \mathfrak{K}]$. Let $\{\mathfrak{y}_i\}$ be a complete set of representatives for J_1/J_2 and $\{\theta_j\}$ be a complete set of representatives for $\mathfrak{E}/f(E_0)$. Then $\{(\mathfrak{y}_i,\, \theta_j)\}$ gives a complete set of representatives for $(J_1 \times \mathfrak{E})/\mathfrak{K}$. In fact, let $(\mathfrak{y}, \theta) \in J_1 \times \mathfrak{E}$. Then $\mathfrak{y} = \mathfrak{y}_i a$, $aa^\rho = b^n$, $b \equiv 1 \bmod \mathfrak{u}$ for some i, some $a \in K$ and $b \in F$. Since $f(b^{-m}a) \in \mathfrak{E}$, we have $f(b^{-m}a)^{-1}\theta = \theta_j f(u)$ for some j and some $u \in E_0$. Put $c = au$. Then $\mathfrak{r}a = \mathfrak{r}c$, $cc^\rho = b^n$, $b^{-m}c = b^{-m}au$. Hence $(\mathfrak{y}, \theta) = (\mathfrak{y}_i, \theta_j) \cdot (\mathfrak{r}c, f(b^{-m}c))$. This shows $J_1 \times \mathfrak{E} = \bigcup_{i,j} (\mathfrak{y}_i, \theta_j)\mathfrak{K}$. Suppose that $(\mathfrak{y}_i, \theta_j) = (\mathfrak{y}_h, \theta_k)(\mathfrak{r}a, f(b^{-m}a))$ with the property $aa^\rho = b^n$, $b \equiv 1 \bmod \mathfrak{u}$. Then obviously $i = h$, and hence

a is a unit of $\mathfrak{r}$, so that b is a unit of $\mathfrak{g}$. Since $N_{K/F}(b^{-m}a) = 1$, we have $b^{-m}a \in E_0$. Hence $j = k$. Thus we get $[J_1 \times \mathfrak{C} : \mathfrak{R}] = [J_1 : J_2][\mathfrak{C} : f(E_0)]$. Consider the mapping $I_K \times I_F \ni (\mathfrak{c}, \mathfrak{b}) \to \mathfrak{b}^m \mathfrak{c}^{-1}\mathfrak{c}^\rho \in J_1$. This is surjective; in fact, if $N_{K/F}(\mathfrak{a}) = \mathfrak{b}^n$, we have $N_{K/F}(\mathfrak{b}^{-m}\mathfrak{a}) = \mathfrak{g}$, so that there exists a $\mathfrak{c} \in I_K$ such that $\mathfrak{b}^{-m}\mathfrak{a} = \mathfrak{c}^{-1}\mathfrak{c}^\rho$, hence $\mathfrak{a} = \mathfrak{b}^m \mathfrak{c}^{-1}\mathfrak{c}^\rho$. Suppose that $\mathfrak{b}^m \mathfrak{c}^{-1}\mathfrak{c}^\rho = \mathfrak{r}a$, $aa^\rho = b^n$, $b \equiv 1 \bmod \mathfrak{u}$ for some $a \in K$ and some $b \in F$. As $N_{K/F}(\mathfrak{b}^{-m}\mathfrak{a}) = 1$, there exists an element c of K such that $\mathfrak{b}^{-m}\mathfrak{a} = c^{-1}c^\rho$. Then we have $(b^{-1}\mathfrak{b})^m (c^{-1}\mathfrak{c})^{-1}(c^{-1}\mathfrak{c})^\rho = \mathfrak{r}$. Applying $N_{K/F}$, we find $(b^{-1}\mathfrak{b})^n = \mathfrak{g}$, so that $\mathfrak{b} = \mathfrak{g}b$, hence $(c^{-1}\mathfrak{c})^\rho = c^{-1}\mathfrak{c}$. This shows that the class of $\mathfrak{c}$ belongs to $\mathfrak{C}_0$ and $\mathfrak{b}$ belongs to the identity class modulo $\mathfrak{u}$ in F. Conversely, if $(c^{-1}\mathfrak{c})^\rho = c^{-1}\mathfrak{c}$ for some $c \in K$, and if $\mathfrak{b} = \mathfrak{g}b$ for some $b \in F$ such that $b \equiv 1 \bmod \mathfrak{u}$, then $\mathfrak{b}^m \mathfrak{c}^{-1}\mathfrak{c}^\rho = (b^m c^{-1}c^\rho)\mathfrak{r}$, and $N_{K/F}(b^m c^{-1}c^\rho) = b^n$, so that $\mathfrak{b}^m \mathfrak{c}^{-1}\mathfrak{c}^\rho \in J_1$. Thus we get $[J_1 : J_2] = h(\mathfrak{u})[\mathfrak{C} : \mathfrak{C}_0]$. This proves (ii) for even n. It remains to prove (iii). By (iii) of 5.21, for every $\mathfrak{b} \in I_F$, there exists an $M \in \Lambda_3$ such that $[L/M] = \mathfrak{r}\mathfrak{b}^m$. By (ii) of 5.21, we have $\mathfrak{b}^n = N_{K/F}([L/M]) = (\mu(L)^{-1}\mu(M))^n$, so that $\mu(M) = \mathfrak{b}\mu(L)$. This shows that $M \to \mu(M)$ is a surjective mapping of Λ_3 to I_F. We note that $e(M/N) = 1$ for any two members M and N of Λ_3. Hence, from (v) of 5.27, we obtain (iii) of our theorem.

5.29. Let L and M be maximal $\mathfrak{r}$-lattices in V. Let $\mathfrak{p}$ be a prime ideal of F. In 3.6 and 4.20, we have defined the set of elementary divisors of $M_\mathfrak{p}$ relative to $L_\mathfrak{p}$, which has been denoted by $\{L_\mathfrak{p} : M_\mathfrak{p}\}$. Now we define the (global) *set of elementary divisors of M relative to L* as the join of $\{L_\mathfrak{p} : M_\mathfrak{p}\}$ for all prime ideals $\mathfrak{p}$, and denote it by $\{L : M\}$. If $L \supset M$, the set $\{L : M\}$ is completely determined by the structure of the $\mathfrak{r}$-module L/M.

5.30. Theorem. *Let V and Φ be as in 5.24. Suppose that $n > 1$ and Φ is indefinite. Let L, M, N be maximal $\mathfrak{r}$-lattices in V. If there exists an element α of $G(V, \Phi)$ such that $L\alpha = L$ and $M\alpha = N$, then $\{L : M\} = \{L : N\}$. Conversely, if $\{L : M\} = \{L : N\}$, then there exists an element β of $SU(V, \Phi)$ such that $L\beta = L$, $M\beta = N$.*

The first assertion is clear. Suppose that $\{L : M\} = \{L : N\}$. Let P be the set of prime ideals $\mathfrak{p}$ of F such that $L_\mathfrak{p} \neq M_\mathfrak{p}$ or $L_\mathfrak{p} \neq N_\mathfrak{p}$. For every $\mathfrak{p} \in P$, apply 3.7 or 4.21 to $L_\mathfrak{p}$, $M_\mathfrak{p}$, $N_\mathfrak{p}$. Then we find an element $\alpha_\mathfrak{p}$ of $SU(V_\mathfrak{p}, \Phi)$ such that $L_\mathfrak{p}\alpha_\mathfrak{p} = L_\mathfrak{p}$, $M_\mathfrak{p}\alpha_\mathfrak{p} = N_\mathfrak{p}$. Let k be a positive integer such that $\mathfrak{p}^k M_\mathfrak{p} \subset L_\mathfrak{p}$ and $\mathfrak{p}^k L_\mathfrak{p} \subset N_\mathfrak{p}$ for every $\mathfrak{p} \in P$. By 5.12, there exists an element β of $SU(V, \Phi)$ such that $L_\mathfrak{p}(\beta - \alpha_\mathfrak{p}) \subset \mathfrak{p}^{3k} L_\mathfrak{p}$ for every $\mathfrak{p} \in P$, and $L_\mathfrak{q}\beta = L_\mathfrak{q}$ for every $\mathfrak{q} \notin P$. Then, by 5.3, $L_\mathfrak{p}\beta = L_\mathfrak{p}$ for every $\mathfrak{p} \in P$, hence $L\beta = L$. Further, $M_\mathfrak{p}(\beta - \alpha_\mathfrak{p}) \subset \mathfrak{p}^{-k} L_\mathfrak{p}(\beta - \alpha_\mathfrak{p}) \subset \mathfrak{p}^{2k} L_\mathfrak{p} \subset \mathfrak{p}^k N_\mathfrak{p} = \mathfrak{p}^k M_\mathfrak{p}\alpha_\mathfrak{p}$ for every $\mathfrak{p} \in P$. Again by 5.3, we have $M_\mathfrak{p}\beta = M_\mathfrak{p}\alpha_\mathfrak{p} = N_\mathfrak{p}$. If $\mathfrak{q} \notin P$, we get $M_\mathfrak{q}\beta = L_\mathfrak{q}\beta = L_\mathfrak{q} = N_\mathfrak{q}$. Therefore $M\beta = N$. This proves our theorem.

5.31. Proposition. *Suppose that $n = 2$. Let $\mathscr{B}$ be the quaternion algebra attached to Φ (cf. 2.6). Let L be an $\mathfrak{r}$-lattice in V. Put $\mathfrak{o} = \{\alpha \in \mathscr{B} \mid L\alpha \subset L\}$. Then L is maximal if and only if $\mathfrak{o}$ is a maximal order in $\mathscr{B}$.*

This is an easy consequence of 3.8, 4.6, 4.27 and 5.6.

6. Hermitian forms over a quaternion algebra

6.0. Let F be an algebraic number field of finite degree, and $\mathfrak{g}$ the ring of integers in F. Let A be a quaternion algebra over F; A may be a division algebra or isomorphic to a total matric algebra $M_2(F)$. We denote by x^ι the image of an element x of A by the canonical involution of A. Let V be a free left A-module of rank n, and Φ a non-degenerate ι-hermitian form on V in the sense of 1.1, which we have called a Q-hermitian form in [11]. As in 1.1, we can define the groups $G(V, \Phi)$ and $U(V, \Phi)$ for this Φ. Our purpose in this section is to prove an approximation theorem for $U(V, \Phi)$ in a little more general case than in [11] and thereby to complement the theory of [11].

6.1. Lemma. *Let G be an affine group variety (in the sense of Weil) defined over a field k. Let Y be a non-empty Zariski open set in G defined over k. Suppose that the function-field of G over k is contained in a purely transcendental extension of k. Suppose further that k has infinitely many elements. Then, every element of G rational over k is the product of two points on Y rational over k.*

We can find a polynomial f with coefficients in k such that

$$Y \supset \{y \in G \mid f(y) \neq 0\} \neq \varnothing \ .$$

Let $x = (x_1, \cdots, x_m)$ be a generic point of G over k. By our assumption, there exists a set of independent variables $\{t_1, \cdots, t_s\}$ over k such that $k(x) \subset k(t_1, \cdots, t_s)$. Let u be an element of G rational over k. Let x_i' for $1 \leq i \leq m$ be the coordinates of ux^{-1}. We can find polynomials $a(t)$, $b_i(t)$, $c_i(t)$ with coefficients in k such that $x_i = b_i(t)/a(t)$, $x_i' = c_i(t)/a(t)$. For a suitably large positive integer e, the $a(t)^e f(x)$ and $a(t)^e f(ux^{-1})$ are contained in the polynomial ring $k[t]$. Put $g(t) = a(t)^e f(x)$, $h(t) = a(t)^e f(ux^{-1})$. As k has infinitely many elements, there exists a set of elements $(v_1, \cdots, v_s)$ in k such that $a(v) \neq 0$, $g(v) \neq 0$, $h(v) \neq 0$. Put $y_i = b_i(v)/a(v)$, $y = (y_1, \cdots, y_m)$. Then $y \in G$, and $f(y) \neq 0$, $f(uy^{-1}) \neq 0$. Therefore $y \in Y$, $uy^{-1} \in Y$. As $u = (uy^{-1})y$, this proves our lemma.

6.2. As in § 5, we consider the completions $F_\mathfrak{p}$ and F_λ and put

$$A_\mathfrak{p} = A \otimes_F F_\mathfrak{p}, \quad V_\mathfrak{p} = V \otimes_F F_\mathfrak{p}, \quad A_\lambda = A \otimes_F F_\lambda, \quad V_\lambda = V \otimes_F F_\lambda$$

$$(1 \leq \lambda \leq r) \ ,$$

Then $A_{\mathfrak{p}}$ and A_{λ} are quaternion algebras, and ι is extended to their canonical involutions. We extend Φ naturally to ι-hermitian forms on the $A_{\mathfrak{p}}$-module $V_{\mathfrak{p}}$ and the A_{λ}-module V_{λ}; we denote the forms again by the same letter Φ.

6.3. Proposition. *Let L be a $\mathfrak{g}$-lattice in V, and let P be a finite set of prime ideals of F. Let $\alpha_{\mathfrak{p}}$, for each $\mathfrak{p} \in P$, be an element of $U(V_{\mathfrak{p}}, \Phi)$. Then, for every positive integer m, there exists an element α of $U(V, \Phi)$ such that $L_{\mathfrak{p}}(\alpha - \alpha_{\mathfrak{p}}) \subset \mathfrak{p}^m L_{\mathfrak{p}}$ for every $\mathfrak{p} \in P$.*

Take a basis $\{x_i\}$ of V over A and represent every A-endomorphism σ of V by a matrix $(s_{ij}) \in M_n(A)$ determined by $x_i \sigma = \sum_{j=1}^{n} s_{ij} x_j$ $(1 \leq i \leq n)$. Define an involution $T = (t_{ij}) \to T' = (u_{ij})$ of $M_n(A)$ by $u_{ij} = t_{ji}^{\iota}$. If we put $h_{ij} = \Phi(x_i, x_j)$ and $H = (h_{ij})$, then, by the correspondence $\sigma \to (s_{ij})$, $U(V, \Phi)$ can be identified with the subgroup of $GL(n, A)$ consisting of the elements S such that $SHS' = H$. Over the universal domain Ω, $U(V, \Phi)$ becomes the symplectic group $Sp(n, \Omega)$, so that $U(V, \Phi)$ is an absolutely irreducible affine group variety. Put now

$$\mathfrak{Y} = \{Y \in M_n(A) \mid 1 + Y \text{ is invertible}, \ YH = -HY'\},$$
$$\mathfrak{S} = \{S \in U(V, \Phi) \mid 1 + S \text{ is invertible}\}.$$

We see easily that $Y \to S = (1 - Y)(1 + Y)^{-1}$ is a bijection of $\mathfrak{Y}$ to $\mathfrak{S}$ and its inverse is given by $S \to Y = (1 - S)(1 + S)^{-1}$. As $\mathfrak{Y}$ is a Zariski open set in a vector space over F, this shows that the function field of $U(V, \Phi)$ over F is a purely transcendental extension of F (of dimension $n(2n + 1)$). Put

$$\mathfrak{Y}_{\mathfrak{p}} = \{Y \in M_n(A_{\mathfrak{p}}) \mid 1 + Y \text{ is invertible}, \ YH = -HY'\},$$
$$\mathfrak{S}_{\mathfrak{p}} = \{S \in U(V_{\mathfrak{p}}, \Phi) \mid 1 + S \text{ is invertible}\}.$$

Applying 6.1 to $U(V_{\mathfrak{p}}, \Phi)$ and $\mathfrak{S}_{\mathfrak{p}}$, we can find elements $T_{\mathfrak{p}}$ and $U_{\mathfrak{p}}$ of $\mathfrak{S}_{\mathfrak{p}}$ such that $\alpha_{\mathfrak{p}} = T_{\mathfrak{p}} U_{\mathfrak{p}}$. Put $Y_{\mathfrak{p}} = (1 - T_{\mathfrak{p}})(1 + T_{\mathfrak{p}})^{-1}$, $Z_{\mathfrak{p}} = (1 - U_{\mathfrak{p}})(1 + U_{\mathfrak{p}})^{-1}$. Since $\{Y \in M_n(A) \mid YH = -HY'\}$ is a vector space over F, we can find elements Y and Z of $\mathfrak{Y}$ which are sufficiently near to $Y_{\mathfrak{p}}$ and $Z_{\mathfrak{p}}$ with respect to $\mathfrak{p}$-topology for every $\mathfrak{p} \in P$. Putting $T = (1 - Y)(1 + Y)^{-1}$, $U = (1 - Z)(1 + Z)^{-1}$ and $\alpha = TU$, we get an element α of $U(V, \Phi)$ with the required property.

6.4. Now we fix a basis $\{u_1, \cdots, u_n\}$ of V over A such that $\Phi(u_i, u_j) = h_i \delta_{ij}$ with $h_i \in F$. Further we take and fix a quadratic extension K of F such that $F \subset K \subset A$. For every prime ideal $\mathfrak{p}$ of F, we put $K_{\mathfrak{p}} = F_{\mathfrak{p}} \cdot K$. Then $K_{\mathfrak{p}}$ is a commutative semi-simple subalgebra of $A_{\mathfrak{p}}$ of dimension 2 over $F_{\mathfrak{p}}$. Let $\{x_1, \cdots, x_n\}$ be a basis of $V_{\mathfrak{p}}$ over $A_{\mathfrak{p}}$ such that $\Phi(x_i, x_j) = h_i \delta_{ij}$. Denote by $H_{\mathfrak{p}}(\{x_i\})$ the subgroup of $U(V_{\mathfrak{p}}, \Phi)$ consisting of the elements α such that $x_i\alpha = \sum_{j=1}^{m} a_{ij}x_j$ $(1 \leq i \leq n)$ with $a_{ij} \in K_\mathfrak{p}$.

6.5. PROPOSITION. *If $n > 1$, $U(V_\mathfrak{p}, \Phi)$ is generated by the subgroups $H_\mathfrak{p}(\{x_i\})$ for all choices of $\{x_i\}$ satisfying $\Phi(x_i, x_j) = h_i\delta_{ij}$.*

Let H be the subgroup of $U(V_\mathfrak{p}, \Phi)$ generated by all the $H_\mathfrak{p}(\{x_i\})$. Let α be an element of $U(V_\mathfrak{p}, \Phi)$. We see easily that $\alpha^{-1}H_\mathfrak{p}(\{x_i\})\alpha = H_\mathfrak{p}(\{x_i\alpha\})$. It follows that H is a normal subgroup of $U(V_\mathfrak{p}, \Phi)$. Now by Dieudonné [3], $U(V_\mathfrak{p}, \Phi)/\{\pm1\}$ is a simple group. It is clear that $H_\mathfrak{p} \supset \{\pm1\}$ and $H_\mathfrak{p} \neq \{\pm1\}$. Therefore we get $U(V_\mathfrak{p}, \Phi) = H$.

6.6. Let K denote the real division quaternion algebra. We re-order the infinite places $\mathfrak{p}_{\infty\lambda}$ of F so that $A_\lambda = K$ if and only if $1 \leq \lambda \leq s$; s may be 0. For every such λ, we denote by $J_\lambda(V, \Phi)$ the number of the elements h_i such that $h_i \equiv -1 \bmod \mathfrak{p}_{\infty\lambda}$. Of course $J_\lambda(V, \Phi)$ does not depend on the choice of the basis $\{u_i\}$. We call Φ *definite* if $A_\lambda = K$ for all infinite places $\mathfrak{p}_{\infty\lambda}$ of F and $J_\lambda(V, \Phi) = 0$ or n for every λ; otherwise we call Φ *indefinite*.

Put $V^* = \sum_{i=1}^{n} Ku_i$ and denote by Φ^* the restriction of Φ to V^*. Further put $a^\rho = a^\iota$ for $a \in K$. Then Φ^* is a non-degenerate ρ-hermitian form on the vector space V^* over K. As $A_\lambda = K$ for $1 \leq \lambda \leq s$, K_λ is isomorphic to C for $1 \leq \lambda \leq s$. Now we can select K so that K_λ is not isomorphic to C for $\lambda > s$. If $A = M_2(F)$, this is obvious. Suppose that A is a division algebra. Put $B = \{b \in A \mid b^\iota = -b\}$, $B_\lambda = \{b \in A_\lambda \mid b^\iota = -b\}$. We see easily that $B_\lambda = B \otimes_F F_\lambda$ and B is dense in $B_1 \times \cdots \times B_r$ by the injection $b \to (b, \cdots, b)$. If $\lambda > s$, A_λ can be identified with $M_2(R)$ or $M_2(C)$. Then we can find an element b of B which is sufficiently near to $\begin{pmatrix} 0 & 1 \\ 1 & 0 \end{pmatrix}$ in B_λ for every $\lambda > s$. Then $b^2 = -bb^\iota \equiv 1 \bmod \mathfrak{p}_{\infty\lambda}$ for every $\lambda > s$. Put $K = F(b)$. As $b^\iota = -b$, $b \notin F$. Hence K is a quadratic extension of F; by our choice of b, K_λ is not isomorphic to C for $\lambda > s$.

Hereafter we assume that K_λ is isomorphic to C if and only if $1 \leq \lambda \leq s$. We have clearly $J_\lambda(V, \Phi) = J_\lambda(V^*, \Phi^*)$ for $1 \leq \lambda \leq s$. Therefore, if Φ is indefinite, Φ^* is indefinite.

6.7. THEOREM. *Let V and Φ be as in 6.0. Suppose that Φ is indefinite. Let P be a finite set of prime ideals of F. Let $\sigma_\mathfrak{p}$, for each $\mathfrak{p} \in P$, be an element of $U(V_\mathfrak{p}, \Phi)$. Let L be a $\mathfrak{g}$-lattice in V and m a positive integer. Then there exists an element σ of $U(V, \Phi)$ such that $L_\mathfrak{p}(\sigma - \sigma_\mathfrak{p}) \subset \mathfrak{p}^m L_\mathfrak{p}$ for every $\mathfrak{p} \in P$, and $L_\mathfrak{q}\sigma = L_\mathfrak{q}$ for every prime ideal $\mathfrak{q}$ of F which is not contained in P.*

If $n=1$, the indefiniteness of Φ implies that A is an indefinite quaternion algebra, namely, A_λ is not isomorphic to K for at least one λ. Then our assertion follows immediately from Eichler [5, Satz 5], by an argument similar to 5.13.

Suppose that our assertion is true for one particular lattice L^0. Then it is true for any other $\mathfrak{g}$-lattice L in V. In fact, notation being as in our theorem, put $P' = P \cup \{\mathfrak{p} \mid L_\mathfrak{p} \neq L_\mathfrak{p}^0\}$. There exists a positive integer k such that $\mathfrak{p}^k L_\mathfrak{p} \subset L_\mathfrak{p}^0 \subset \mathfrak{p}^{-k} L_\mathfrak{p}$ for every $\mathfrak{p} \in P'$. Applying our theorem to L^0, we find an element σ of $U(V, \Phi)$ such that $L_\mathfrak{p}^0(\sigma - \sigma_\mathfrak{p}) \subset \mathfrak{p}^{m+2k} L_\mathfrak{p}^0$ for every $\mathfrak{p} \in P$, $L_\mathfrak{p}^0(\sigma - 1) \subset \mathfrak{p}^{1+2k} L_\mathfrak{p}^0$ for every $\mathfrak{p} \in P' - P$, and $L_\mathfrak{p}^0 \sigma = L_\mathfrak{p}^0$ for $\mathfrak{p} \notin P'$. Then we can easily verify that σ has the required property for L.

Now suppose that $n > 1$. Let $\mathfrak{r}$ be the ring of integers in K, and $\mathfrak{o}$ an order in A containing $\mathfrak{r}$. Put $L^* = \sum_{i=1}^n \mathfrak{r} u_i$ and $L = \mathfrak{o} L^* = \sum_{i=1}^n \mathfrak{o} u_i$. By the above consideration, it is sufficient to prove our theorem for this $\mathfrak{g}$-lattice L. Moreover, by the same reasoning as in the proof of 5.12 (cf. 5.17), it is sufficient to prove our assertion in the case where at most one of the $\sigma_\mathfrak{p}$ is not the identity. Namely, we single out any $\mathfrak{p}$ of P and put $P_1 = P - \{\mathfrak{p}\}$; our purpose is to show the existence of an element σ of $U(V, \Phi)$ such that $L_\mathfrak{p}(\sigma - \sigma_\mathfrak{p}) \subset \mathfrak{p}^m L$, $L_\mathfrak{q}(\sigma - 1) \subset \mathfrak{q}^m L_\mathfrak{q}$ for every $\mathfrak{q} \in P_1$, and $L_\mathfrak{q} \sigma = L_\mathfrak{q}$ for every $\mathfrak{q} \notin P$. Further, by 6.5, it is sufficient to prove this in the case where $\sigma_\mathfrak{p} \in H_\mathfrak{p}(\{x_i\})$ for some basis $\{x_i\}$ of $V_\mathfrak{p}$ over $A_\mathfrak{p}$ such that $\Phi(x_i, x_j) = h_i \delta_{ij}$. Now define an $A_\mathfrak{p}$-automorphism ξ of $V_\mathfrak{p}$ by $u_i = x_i \xi$ for $1 \leq i \leq n$. Then $\xi \in U(V_\mathfrak{p}, \Phi)$ and $\xi^{-1} \sigma_\mathfrak{p} \xi \in H_\mathfrak{p}(\{u_i\})$. Let h be a positive integer such that

$$(1) \quad \mathfrak{p}^h L_\mathfrak{p} \subset L_\mathfrak{p} \xi \subset \mathfrak{p}^{-h} L_\mathfrak{p}, \quad \mathfrak{p}^h L_\mathfrak{p} \subset L_\mathfrak{p} \sigma_\mathfrak{p} \subset \mathfrak{p}^{-h} L_\mathfrak{p}, \quad \mathfrak{p}^h L_\mathfrak{p} \subset L_\mathfrak{p} \xi^{-1} \sigma_\mathfrak{p} \xi \subset \mathfrak{p}^{-h} L_\mathfrak{p}.$$

By 6.3, we can find an element η of $U(V, \Phi)$ such that

$$(2) \qquad L_\mathfrak{p}(\eta - \xi) \subset \mathfrak{p}^{4h+m} L_\mathfrak{p}, \quad L_\mathfrak{q}(\eta - 1) \subset \mathfrak{q} L_\mathfrak{q} \qquad \text{for } \mathfrak{q} \in P_1.$$

Then $L_\mathfrak{p}(\eta - \xi) \subset \mathfrak{p} L_\mathfrak{p} \xi$, so that $L_\mathfrak{p} \eta = L_\mathfrak{p} \xi$ by 5.3, and similarly

$$(3) \qquad L_\mathfrak{q} \eta = L_\mathfrak{q} \qquad \text{for } \mathfrak{q} \in P_1.$$

By (1), we have

$$(4) \qquad \mathfrak{p}^h L_\mathfrak{p} \subset L_\mathfrak{p} \eta \subset \mathfrak{p}^{-h} L_\mathfrak{p},$$

and by (1), (2), (4),

$$(5) \qquad L_\mathfrak{p}(\eta^{-1} - \xi^{-1}) = L_\mathfrak{p} \xi^{-1}(\xi - \eta)\eta^{-1} \subset \mathfrak{p}^{2h+m} L.$$

Put $Q = P \cup \{\mathfrak{q} \mid L_\mathfrak{q} \eta \neq L_\mathfrak{q}\}$. Let k be a positive integer such that

$$(6) \qquad \mathfrak{q}^k L_\mathfrak{q} \subset L_\mathfrak{q} \eta \subset \mathfrak{q}^{-k} L_\mathfrak{q} \qquad \text{for } \mathfrak{q} \in Q - P.$$

Now the restriction of $\xi^{-1} \sigma_\mathfrak{p} \xi$ to $V_\mathfrak{p}^*$ belongs to $SU(V_\mathfrak{p}^*, \Phi^*)$. By 5.12, there exists an element ζ^* of $SU(V^*, \Phi^*)$ such that

$$(7) \qquad L_\mathfrak{p}^*(\zeta^* - \xi^{-1} \sigma_\mathfrak{p} \xi) \subset \mathfrak{p}^{2h+m} L_\mathfrak{p}^*,$$

$$(8) \qquad L_{\mathfrak{q}}^*(\zeta^* - 1) \subset \mathfrak{q}^m L_{\mathfrak{q}}^* \qquad\qquad \text{for } \mathfrak{q} \in P_1 ,$$

$$(9) \qquad L_{\mathfrak{q}}^*(\zeta^* - 1) \subset \mathfrak{q}^{2k+1} L_{\mathfrak{q}}^* \qquad\qquad \text{for } \mathfrak{q} \in Q - P ,$$

$$(10) \qquad L_{\mathfrak{q}}^* \zeta^* = L_{\mathfrak{q}}^* \qquad\qquad \text{for } \mathfrak{q} \notin Q .$$

By our definition of V^*, Φ^*, we can find an element ζ of $U(V, \Phi)$ whose restriction to V^* is ζ^*. Since $L = \mathfrak{o}L^*$, the relations (7), (8), (9), (10) are true even after substituting L and ζ for L^* and ζ^*. Put $\sigma = \eta\zeta\eta^{-1}$. This σ has the required property. In fact, by (1), (2), (4), (5), (7),

$$L_{\mathfrak{p}}(\sigma - \sigma_{\mathfrak{p}}) = L_{\mathfrak{p}}[\eta(\zeta - \xi^{-1}\sigma_{\mathfrak{p}}\xi)\eta^{-1} + \eta\xi^{-1}\sigma_{\mathfrak{p}}\xi(\eta^{-1} - \xi^{-1}) + (\eta - \xi)\xi^{-1}\sigma\,]$$
$$\subset \mathfrak{p}^m L_{\mathfrak{p}} ,$$

and for $\mathfrak{q} \in P_1$, by (3), (8),

$$L_{\mathfrak{q}}(\sigma - 1) = L_{\mathfrak{q}}\eta(\zeta - 1)\eta^{-1} = L_{\mathfrak{q}}(\zeta - 1)\eta^{-1} \subset \mathfrak{q}^m L_{\mathfrak{q}} .$$

Further for $\mathfrak{q} \in Q - P$, by (6) and (9)

$$L_{\mathfrak{q}}(\sigma - 1) = L_{\mathfrak{q}}\eta(\zeta - 1)\eta^{-1} \subset \mathfrak{q} L_{\mathfrak{q}} ,$$

so that, by 5.3, $L_{\mathfrak{q}}\sigma = L_{\mathfrak{q}}$ for $\mathfrak{q} \in Q - P$. Finally if $\mathfrak{q} \notin Q$, $L_{\mathfrak{q}}\eta = L_{\mathfrak{q}}\zeta = L_{\mathfrak{q}}$, hence $L_{\mathfrak{q}}\sigma = L_{\mathfrak{q}}$. Our theorem is thereby proved.

6.8. We can define genera and classes of $\mathfrak{g}$-lattices in V with respect to the present $G(V, \Phi)$ and $U(V, \Phi)$ in the same way as in 5.18. Then, we obtain the following theorem from 6.7 by the same argument as in the proof of 5.19.

6.9. THEOREM. *Let V and Φ be as in 6.0. Suppose that Φ is indefinite. Then, with respect to $U(V, \Phi)$, every genus of $\mathfrak{g}$-lattices in V consists of only one class.*

6.10. Let $\mathfrak{o}$ be a maximal order in A. We call a $\mathfrak{g}$-lattice L in V an $\mathfrak{o}$-lattice if $\mathfrak{o}L \subset L$. If L is an $\mathfrak{o}$-lattice and M belongs to the same genus as L, then M is an $\mathfrak{o}$-lattice. Denote by $\mu(L)$ the $\mathfrak{g}$-ideal generated by the elements $\Phi(x, x)$ for $x \in L$. We call L μ-*maximal* if L is a maximal one among the $\mathfrak{o}$-lattices L with the same $\mu(L)$. Let $\mu_0(L)$ be the two-sided $\mathfrak{o}$-ideal generated by the elements $\Phi(x, y)$ for $x \in L$, $y \in L$. We call L μ_0-*maximal* if L is a maximal one among the $\mathfrak{o}$-lattices L with the same $\mu_0(L)$. Let L and M be $\mathfrak{g}$-lattices in V belonging to the same genus with respect to $G(V, \Phi)$. If L is an $\mathfrak{o}$-lattice, then M is an $\mathfrak{o}$-lattice; and if L is μ-maximal (resp. μ_0-maximal), so is M.

6.11. PROPOSITION. *Let L be an $\mathfrak{o}$-lattice in V. Then, for every prime ideal $\mathfrak{p}$ of F, there exist direct decompositions $V_{\mathfrak{p}} = \sum_{i=1}^{t} W_i$ and $L_{\mathfrak{p}} = \sum_{i=1}^{t} L_i$ with the following properties:*

(i) *the W_i are subspaces of $V_\mathfrak{p}$ of dimension 1 or 2 over $A_\mathfrak{p}$;*

(ii) $L_i = W_i \cap L$;

(iii) $\Phi(x, y) = 0$ *for* $x \in W_i$, $y \in W_j$ *if* $i \neq j$;

(iv) $\mu_0(L_1) \supset \cdots \supset \mu_0(L_t)$.

Moreover, if $A_\mathfrak{p}$ is a total matric algebra, we can take all the W_i to be one-dimensional. If $A_\mathfrak{p}$ is a division algebra, we can take the decomposition so that $L_i = \mathfrak{o}_\mathfrak{p}x_i + \mathfrak{o}_\mathfrak{p}y_i$ with $\Phi(x_i, x_i) = \Phi(y_i, y_i) = 0$ whenever $\dim W_i = 2$.

If $A_\mathfrak{p} = M_2(F_\mathfrak{p})$, this is a special case of [11, Prop. 2.10]. If $A_\mathfrak{p}$ is a division algebra, this is included in Jacobowitz [8, Prop. 4.3, Prop. 6.1], and can be proved in the same way as in 4.15.

6.12. PROPOSITION. *Let L be an $\mathfrak{o}$-lattice in V. Then, for every prime ideal $\mathfrak{p}$ of F, and for every unit c of $\mathfrak{g}_\mathfrak{p}$, there exists an element α of $G(V_\mathfrak{p}, \Phi)$ such that $L_\mathfrak{p}\alpha = L_\mathfrak{p}$ and $\mu(\alpha) = c$.*

Let $V_\mathfrak{p} = \sum_{i=1}^{t} W_i$ and $L = \sum_{i=1}^{t} L_i$ be as in 6.11. We can find a unit b of $\mathfrak{o}_\mathfrak{p}$ such that $bb^\iota = c$. If W_i is one-dimensional, we define the operation of α on W_i by $z\alpha = bz$ for $z \in W_i$. If W_i is two-dimensional, and $L_i = \mathfrak{o}_\mathfrak{p}x_i + \mathfrak{o}_\mathfrak{p}y_i$ with $\Phi(x_i, x_i) = \Phi(y_i, y_i) = 0$, then we define α by $x_i\alpha = cx_i$, $y_i\alpha = y_i$. Then α has the required property.

6.13. PROPOSITION. *Let $\mathfrak{u}$ be the product of infinite places $\mathfrak{p}_{\infty\lambda}$ of F such that $J_\lambda(V, \Phi) \neq n/2$ (cf. 6.6). Let c be an element of F. Then, there exists an element α of $G(V, \Phi)$ such that $\mu(\alpha) = c$ if and only if $c \equiv 1 \bmod \mathfrak{u}$.*

This is obtained by applying Ramanathan's theorem (which is stated as Lemma 4.4 in [11]) to $\Phi(x, y)$ and $c\Phi(x, y)$.

6.14. THEOREM. *Let V and Φ be as in 6.0, and $\mathfrak{u}$ be as in 6.13. Suppose that Φ is indefinite. Let $\mathfrak{o}$ be a maximal order in A and Λ a genus of $\mathfrak{o}$-lattices in V with respect to $G(V, \Phi)$. Then $\Lambda \ni L \to \mu(L)$ gives a one-to-one correspondence between the classes (with respect to $G(V, \Phi)$) in Λ and the ideal-classes modulo $\mathfrak{u}$ in F.*

First we note that, for every $\mathfrak{g}$-ideal $\mathfrak{a}$, there exists an $L \in \Lambda$ such that $\mu(L) = \mathfrak{a}$. This can be proved in the same way as in 5.21. Now let L and M be members of Λ. Suppose that $L\alpha = M$ for some $\alpha \in G(V, \Phi)$. Then $\mu(L)\mu(\alpha) = \mu(M)$. Since $\mu(\alpha) \equiv 1 \bmod \mathfrak{u}$ (6.13), the ideals $\mu(L)$ and $\mu(M)$ belong to the same ideal-class modulo $\mathfrak{u}$. Conversely, suppose that $\mu(L)c = \mu(M)$ with an element c of F such that $c \equiv 1 \bmod \mathfrak{u}$. By 6.13, there exists an element α of $G(V, \Phi)$ such that $\mu(\alpha) = c$. Then $\mu(L\alpha) = \mu(M)$. Since $L\alpha$ and M belong to Λ, there exists, for every prime ideal $\mathfrak{p}$ of F, an element $\beta_\mathfrak{p}$ of $G(V_\mathfrak{p}, \Phi)$ such that $L_\mathfrak{p}\alpha = M_\mathfrak{p}\beta_\mathfrak{p}$. $\mu(\beta_\mathfrak{p})$ must be a unit of $\mathfrak{g}_\mathfrak{p}$, on account of $\mu(L\alpha) = \mu(M)$. By 6.12, there exists an element $\gamma_{\mathfrak{p}}$ of $G(V_{\mathfrak{p}}, \Phi)$ such that $M_{\mathfrak{p}}\gamma_{\mathfrak{p}} = M_{\mathfrak{p}}$ and $\mu(\gamma_{\mathfrak{p}}) = \mu(\beta_{\mathfrak{p}})^{-1}$. Then $\gamma_{\mathfrak{p}}\beta_{\mathfrak{p}} \in U(V_{\mathfrak{p}}, \Phi)$ and $L_{\mathfrak{p}}\alpha = M_{\mathfrak{p}}\gamma_{\mathfrak{p}}\beta_{\mathfrak{p}}$ for every $\mathfrak{p}$. This shows that $L\alpha$ and M belong to the same genus with respect to $U(V, \Phi)$. By 6.9, we have $L\alpha\varepsilon = M$ for some $\varepsilon \in U(V, \Phi)$. This completes the proof.

6.15. In [11], we have investigated the structure of μ_0-maximal lattices, with the notation $N(L)$ instead of the present $\mu_0(L)$. Now let us study the relation between the μ-maximality and the μ_0-maximality. If $n = 1$, it is easy to see that they are the same. Suppose that $n > 1$. Let $\mathfrak{o}$ be a maximal order in A. Let $\mathfrak{q}_1, \cdots, \mathfrak{q}_g$ be all the prime ideals of F where A is ramified. Let $\mathfrak{Q}_i$ be the two sided $\mathfrak{o}$-ideal such that $\mathfrak{Q}_i^2 = \mathfrak{q}_i\mathfrak{o}$. Then all the μ_0-maximal $\mathfrak{o}$-lattices are divided into 2^g genera $\mathfrak{L}(\mathfrak{o}; \{\varepsilon_i\})$ with respect to $G(V, \Phi)$ defined by

$$\mathfrak{L}\big(\mathfrak{o}; \{\varepsilon_1, \cdots, \varepsilon_g\}\big) = \left\{ L \mid \mu_0(L) = (\mu_0(L) \cap F) \prod_{i=1}^{g} \mathfrak{Q}_i^{-\varepsilon_i} \right\},$$

where $\varepsilon_i = 0$ or 1 [11, § 4.4].

6.16. PROPOSITION. *Let L be an $\mathfrak{o}$-lattice. Then $\mu_0(L) \subset \mu(L) \prod_{i=1}^{g} \mathfrak{Q}_i^{-1}$. If L is μ_0-maximal, then $\mu(L) = \mu_0(L) \cap F$.*

The first assertion can be proved in the same way as in 2.11. If $L \in \mathfrak{L}(\mathfrak{o}; \{\varepsilon_i\})$, we have $(\mu_0(L) \cap F) \prod_{i=1}^{g} \mathfrak{Q}_i^{-\varepsilon_i} = \mu_0(L) \subset \mu(L) \prod_{i=1}^{g} \mathfrak{Q}_i^{-1}$. Since $\mu(L)$ and $\mu_0(L) \cap F$ are $\mathfrak{g}$-ideals, we must have $\mu_0(L) \cap F \subset \mu(L)$. On the other hand, from our definition, it follows easily that $\mu(L) \subset \mu_0(L) \cap F$. Hence we get $\mu(L) = \mu_0(L) \cap F$.

6.17. PROPOSITION. *Suppose that $n > 1$. Then an $\mathfrak{o}$-lattice in V is μ-maximal if and only if $L \in \mathfrak{L}(\mathfrak{o}; \{1, \cdots, 1\})$.*

Let $L \in \mathfrak{L}(\mathfrak{o}; \{1, \cdots, 1\})$ and let M be an $\mathfrak{o}$-lattice such that $M \supset L$ and $\mu(M) = \mu(L)$. Then $\mu_0(L) \subset \mu_0(M) \subset \mu(M) \prod_{i=1}^{g} \mathfrak{Q}_i^{-1} = \mu(L) \prod_{i=1}^{g} \mathfrak{Q}_i^{-1} = \mu_0(L)$, by 6.16, so that $\mu_0(L) = \mu_0(M)$. Since L is μ_0-maximal, we get $L = M$. Hence L is μ-maximal. Conversely, let L be a μ-maximal $\mathfrak{o}$-lattice. Since $\mu_0(L) \subset \mu(L) \prod_{i=1}^{g} \mathfrak{Q}_i^{-1}$, we can find, by virtue of [11, § 1.4 and Prop. 3.6], a μ_0-maximal $\mathfrak{o}$-lattice M such that $M \supset L$ and $\mu_0(M) = \mu(L) \prod_{i=1}^{g} \mathfrak{Q}_i^{-1}$. Then we have $\mu(L) = \mu_0(M) \cap F$, and hence $M \in \mathfrak{L}(\mathfrak{o}; \{1, \cdots, 1\})$. By 6.16, we have $\mu(M) = \mu_0(M) \cap F = \mu(L)$. Since L is μ-maximal, we have $L = M$. This completes the proof.

6.18. PROPOSITION. *Let $\mathfrak{o}$ be a maximal order in A. Let L and M be μ-maximal (resp. μ_0-maximal) $\mathfrak{o}$-lattices in V. Then L and M belong to the same genus with respect to $U(V, \Phi)$ if and only if $\mu(L) = \mu(M)$ (resp. $\mu_0(L) = \mu_0(M)$).*

This is an easy consequence of [11, Prop. 2.11, Prop. 3.7], 6.15, 6.16.

6.19. Let L and M be μ_0-maximal $\mathfrak{o}$-lattices belonging to the same genus with respect to $G(V, \Phi)$. In [11, § 4.9], we have defined the set of elementary divisors $\{L : M\}$ in the same way as in 5.29. With this notion, we now get

6.20. THEOREM. *Suppose that Φ is indefinite. Let L, M, N be μ_0-maximal $\mathfrak{o}$-lattices in V belonging to the same genus with respect to $G(V, \Phi)$. Then, $\{L : M\} = \{L : N\}$ if and only if there exists an element α of $U(V, \Phi)$ such that $L\alpha = L$ and $M\alpha = N$.*

This follows easily from 6.7 and [11, Prop. 2.13, Prop. 3.11] by the same argument as in the proof of 5.30 (cf. also [11, Theorem 4]).

Osaka University and Princeton University

REFERENCES

1. F. BRUHAT, *Sur les représentations des groupes classiques p-adiques*: I, II, Amer. J. Math., 83 (1961), 321-338, 343-368.
2. J. DIEUDONNÉ, Sur les groupes classiques, Actualités Sci. Ind., $n°$ 1040, Hermann, Paris, 1948.
3. ————, *On the structure of unitary groups*, Trans. Amer. Math. Soc., 72 (1952), 367-385.
4. ————, *Sur les multiplicateurs des similitudes*, Rend. Circ. Mat. Palermo, Ser. II, 3 (1954), 398-408.
5. M. EICHLER, *Allgemeine Kongruenzklasseneinteilungen der Ideale einfacher Algebren über algebraischen Zahlkörpern und ihre L-Reihen*, J. Reine Angew. Math., 179 (1938), 227-251.
6. ————, Quadratische Formen und orthogonale Gruppen, Springer, Berlin-Göttingen-Heidelberg, 1952.
7. ————, *Die Ähnlichkeitsklassen indefiniter Gitter*, Math. Z., 55 (1952), 216-252.
8. R. JACOBOWITZ, *Hermitian forms over local fields*, Amer. J. Math., 84 (1962), 441-465.
9. M. KNESER, *Klassenzahlen indefiniter quadratischer Formen in drei oder mehr Veränderlichen*, Archiv der Math., 7 (1956), 323-332.
10. W. LANDHERR, *Äquivalenz Hermitscher Formen über einem beliebigen algebraischen Zahlkörper*, Abh. Math. Sem. Hamb., 11 (1936), 245-248.
11. G. SHIMURA, *Arithmetic of alternating forms and quaternion hermitian forms*, J. Math. Soc. Japan, 15 (1963), 33-65.
12. E. WITT, *Eine Identität zwischen Modulformen zweiten Grades*, Abh. Math. Sem. Hamb., 14 (1941), 323-337.

On the field of definition for a field of automorphic functions

Annals of Mathematics, 80 (1964), 160-189

In a previous paper [5], we treated discontinuous groups and automorphic functions in connection with analytic families of polarized abelian varieties with a given algebra of endomorphisms. We showed that the fields of automorphic functions in question are defined over some algebraic number fields; but the precise determination of those number fields was left untouched. The discontinuous group investigated in [5] is the group $\Gamma(T, \mathfrak{M})$ of automorphisms of a structure $(V, T, \mathfrak{M})$ formed by a vector space V over a division algebra L over the rationals with a positive involution, an anti-hermitian form T on V, and a lattice $\mathfrak{M}$ in V. The purpose of the present paper is to determine the field of definition for the field of automorphic functions with respect to $\Gamma(T, \mathfrak{M})$ in the case where L belongs to the following three types of algebras:

(i) a totally real algebraic number field F,

(ii) a totally indefinite quaternion algebra over F,

(iii) a totally imaginary quadratic extension of F.

To obtain a transparent result, one has to select a suitable lattice $\mathfrak{M}$ in V; we note that $\Gamma(T, \mathfrak{M})$ and $\Gamma(T, \mathfrak{N})$ are commensurable for any two lattices $\mathfrak{M}$ and $\mathfrak{N}$ in the same space V. After such a selection, we shall show that the rational number field Q can be taken as a field of definition in the cases (i), (ii) (Theorem 3.3). In the case (iii), the matter is not so simple; certain class-fields come out. It will be worthwhile to state the result in more detail, since this may be regarded as a generalization of the theory of complex multiplication of abelian varieties [7].

Let F be a totally real algebraic number field of degree g, and K a totally imaginary quadratic extension of F. Let $\tau_1, \cdots, \tau_g, \tau_1\rho, \cdots, \tau_g\rho$ be all the isomorphisms of K into the complex number field C, where ρ is the complex conjugation. Let $T = (t_{ij})$ be an invertible matrix of size m with entries in K, such that $t_{ij}^\rho = -t_{ji}$. Let (r_ν, s_ν) be the signature of $-\sqrt{-1}\,T^{\tau_\nu}$ for $1 \leq \nu \leq g$; then r_ν and s_ν are non-negative integers such that $r_\nu + s_\nu = m$. We assume that $\sum_{\nu=1}^{g} r_\nu s_\nu > 0$. For a lattice $\mathfrak{M}$ in $K^m = K \times \cdots \times K$, put

$$\Gamma(T, \mathfrak{M}) = \{B \in \mathrm{GL}_m(K) \mid BTB^\rho = T,\ \mathfrak{M}B = \mathfrak{M}\} \,,$$

where $B^\rho = {}^t(b_{ij}^\rho)$ for $B = (b_{ij})$. Then $\Gamma(T, \mathfrak{M})$ gives a properly discontinuous group of transformations on a symmetric domain

$$\mathfrak{K} = \mathfrak{K}_1 \times \cdots \times \mathfrak{K}_g \, ,$$

where $\mathfrak{K}_\nu$ is the space of complex matrices z with r_ν rows and s_ν columns such that $1 - {}^t\bar{z}z$ is positive hermitian; if $r_\nu s_\nu = 0$, we mean by $\mathfrak{K}_\nu$ the set consisting of only one point. For the sake of simplicity, assume that $\mathfrak{K}/\Gamma(T, \mathfrak{M})$ is compact, though we can proceed without this assumption. Let $\mathfrak{K}(T, \mathfrak{M})$ be the field of all meromorphic functions on $\mathfrak{K}$ invariant under $\Gamma(T, \mathfrak{M})$. Let K' be the field generated over Q by the elements $\sum_{\nu=1}^{g} (r_\nu x^{\tau_\nu} + s_\nu x^{\tau_\nu \rho})$ for all $x \in K$. If m is even and $r_\nu = s_\nu$ for every ν, we get $K' = Q$; and $\mathfrak{K}(T, \mathfrak{M})$ has a subfield $\mathfrak{K}_0$ such that $\mathfrak{K}(T, \mathfrak{M}) = C\mathfrak{K}_0$, and $\mathfrak{K}_0$, C are linearly disjoint over Q. Roughly speaking, $\mathfrak{K}(T, \mathfrak{M})$ can be defined over Q.

Exclude this case; namely, assume that $r_\nu \neq s_\nu$ for some ν. Then K' is a totally imaginary quadratic extension of a totally real algebraic number field. Let $\sigma_1, \cdots, \sigma_{g'}, \sigma_1 \rho, \cdots, \sigma_{g'} \rho$ be all the isomorphisms of K' into C. Then there exist integers $v_1, \cdots, v_{g'}$ with which we can state our main result as follows (Theorem 5.10). Define an ideal-group H in K' by

$$H = \{\mathfrak{x} \mid \textstyle\prod_{\lambda=1}^{g'} (\mathfrak{x}^{\sigma_\lambda}/\mathfrak{x}^{\sigma_\lambda \rho})^{v_\lambda} = (a) \text{ for some } a \in K \text{ such that } a a^\rho = 1\} \text{ if } m \text{ is}$$

even,

$$H = \{\mathfrak{x} \mid \textstyle\prod_{\lambda=1}^{g'} \mathfrak{x}^{\sigma_\lambda} \cdot (\mathfrak{x}^{\sigma_\lambda}/\mathfrak{x}^{\sigma_\lambda \rho})^{v_\lambda} = (a) \text{ for some } a \in K \text{ such that } N(\mathfrak{x}) = a a^\rho\} \text{ if}$$

m is odd.

Let k_0 be the unramified class-field over K' corresponding to the ideal group H. Then, for a suitable choice of $\mathfrak{M}$, there exists a projective variety $\mathcal{V}(T, \mathfrak{M})$ defined over k_0, whose function-field over C may be identified with $\mathfrak{K}(T, \mathfrak{M})$; roughly speaking, $\mathfrak{K}(T, \mathfrak{M})$ can be defined over k_0. Moreover, if m is even and $\sigma = \left(\dfrac{k_0/K'}{\mathfrak{a}}\right)$ for an ideal $\mathfrak{a}$ in K', we have $\mathcal{V}(T, \mathfrak{M})^\sigma = \mathcal{V}(T, \mathfrak{N})$ for a lattice $\mathfrak{N}$ such that

$$[\mathfrak{N}/\mathfrak{M}] = \textstyle\prod_{\lambda=1}^{g'} \left(\dfrac{\mathfrak{a}^{\sigma_\lambda}}{\mathfrak{a}^{\sigma_\lambda \rho}}\right)^{v_\lambda} \, ,$$

where $[\mathfrak{N}/\mathfrak{M}]$ is the ideal in K generated by the $\det(S)$ for all $S \in \mathrm{GL}_m(K)$ such that $\mathfrak{N}S \subset \mathfrak{M}$. Thus the reciprocity law for the abelian extension k_0/K' can be given in terms of the varieties $\mathcal{V}(T, \mathfrak{M})$. A similar but slightly more complicated relation holds for odd m.

To obtain this result, we need a fairly deep theorem in the arithmetic of the unitary group

$$U(T) = \{B \in \mathrm{GL}_m(K) \mid BTB^\rho = T\} \, .$$

In [6] we gave a one-to-one correspondence between the classes in a genus of lattices in K^m with respect to $U(T)$ and a certain group of ideal-classes in K. This plays an essential role in the proof of the above reciprocity law. In addition, we make use of a fundamental relation [7, p. 127, (1)] in the theory of complex multiplication of abelian varieties. As for the cases (i) and (ii), our assertion is a consequence of the criterion [1, Th. 2] combined with the fact that, with respect to the unitary group over an algebra L of type (i) and (ii), each genus of lattices in V consists of only one class [4, 6].

The investigation of [5] includes also the cases where the basic algebra L is a totally definite quaternion algebra over F or a central division algebra over K with a positive involution. We would be able to get a similar result in those cases, if we knew a theorem concerning the classes in a genus of lattices with respect to the unitary group of an anti-hermitian form over the algebra L. And such a theorem would be a consequence of the approximation-theorem in the group, which is expected to be true.

Notation. As usual Z, Q, R, C denote respectively the ring of rational integers, the rational number field, the real number field, the complex number field. $\bar{Q}$ denotes the field of all algebraic numbers. If p is a prime number, Z_p and Q_p denote the ring of p-adic integers and the field of p-adic numbers, respectively. For an associative ring S with an identity element, we denote by $M_m(S)$ and $\mathrm{GL}_m(S)$ the ring of $m \times m$ matrices with entries in S and the group of invertible elements of $M_m(S)$, respectively. If X is a positive cycle in a projective space, $c(X)$ is the Chow point of X. For an abelian variety A, we denote by $\mathrm{End}\,(A)$ the ring of endomorphisms of A and put $\mathrm{End}_Q\,(A) = \mathrm{End}\,(A) \otimes Q$.

1. The anti-hermitian structure determined by a polarized abelian variety

1.1. Let k be a field and L a semi-simple algebra over k. Let ρ be an involution of L over k, and V a left L-module. By a *ρ-hermitian* (resp. *ρ-anti-hermitian*) *form* on V, we understand a k-bilinear mapping h of $V \times V$ into L satisfying $h(ax, by) = ah(x, y)b^\rho$, $h(x, y)^\rho = h(y, x)$ (resp. $h(x, y)^\rho = -h(y, x)$) for $a \in L$, $b \in L$, $(x, y) \in V \times V$. Further, by a *ρ-alternating form* on V, we understand a k-bilinear mapping f of $V \times V$ into k satisfying $f(x, y) = -f(y, x)$, $f(ax, y) = f(x, a^\rho y)$ for $(x, y) \in V \times V$, $a \in L$.

1.2. Lemma. *Let L, k, ρ, V be as above. Suppose that the simple components of the center of L are separable over k. Denote by $\mathrm{Tr}_{L/k}$ the reduced trace from L to k. Let h be a ρ-anti-hermitian form on V; put $f(x, y) = \mathrm{Tr}_{L/k}\,(h(x, y))$. Then f is a ρ-alternating form on V. Conversely, for every ρ-alternating form on V, there exists one and only one ρ-anti-hermitian form h on V such that $f(x, y) = \mathrm{Tr}_{L/k}\,(h(x, y))$.*

First we note that $\mathrm{Tr}_{L/k}\,(a) = \mathrm{Tr}_{L/k}\,(a^\rho)$ for every $a \in L$. Then the direct part of our lemma can be proved in a straightforward way. To prove the converse part, suppose that f is ρ-alternating, and consider the k-linear mapping: $L \ni a \to f(ax, y) \in k$ for fixed x and y. Since $\mathrm{Tr}_{L/k}(ab)$ is a non-degenerate symmetric form on $L \times L$, there exists an element h of L such that $f(ax, y) = \mathrm{Tr}_{L/k}(ah)$ for every $a \in L$. Such an element h is uniquely determined by x and y. Write $h = h(x, y)$. Then we can easily verify that $h(x, y)$ is ρ-anti-hermitian.

1.3. Let L, k, ρ, V be as above. Suppose that $k = \boldsymbol{Q}$. Let F be an algebraic number field contained in the center of L, with the same identity element as L. We assume that $a^\rho = a$ for every $a \in F$. For every prime ideal $\mathfrak{p}$ of F, we denote by $F_\mathfrak{p}$ the $\mathfrak{p}$-completion of F, and put $L_\mathfrak{p} = L \otimes_F F_\mathfrak{p}$, $V_\mathfrak{p} = V \otimes_F F_\mathfrak{p}$. Further, for every rational prime p, we denote by $\boldsymbol{Q}_p$ the p-adic number field, and put $F_p = F \otimes_{\boldsymbol{Q}} \boldsymbol{Q}_p$, $L_p = L \otimes_{\boldsymbol{Q}} \boldsymbol{Q}_p$, $V_p = V \otimes_{\boldsymbol{Q}} \boldsymbol{Q}_p$. If $\mathfrak{p}_1, \cdots, \mathfrak{p}_s$ are all the prime factors of p in F, we can identify F_p with the product $F_{\mathfrak{p}_1} \times \cdots \times F_{\mathfrak{p}_s}$, and therefore L_p with $L_{\mathfrak{p}_1} \times \cdots \times L_{\mathfrak{p}_s}$, and V_p with $V_{\mathfrak{p}_1} \times \cdots \times V_{\mathfrak{p}_s}$. We extend ρ to the involutions of L_p over F_p and of $L_\mathfrak{p}$ over $F_\mathfrak{p}$, which we denote again by ρ. Let $h(x, y)$ be a ρ-anti-hermitian form on V, and let $f(x, y) = \mathrm{Tr}_{L/\boldsymbol{Q}}(h(x, y))$. We extend $h(x, y)$ to an F_p-bilinear mapping $h_p(x, y)$ of $V_p \times V_p$ into L_p, and $f(x, y)$ to a $\boldsymbol{Q}_p$-bilinear mapping $f_p(x, y)$ of $V_p \times V_p$ into $\boldsymbol{Q}_p$. It is clear that h_p is ρ-anti-hermitian and f_p is ρ-alternating. Let $f_i(x, y)$ and $h_i(x, y)$ be the restrictions of $f_p(x, y)$ and $h_p(x, y)$ to $V_{\mathfrak{p}_i} \times V_{\mathfrak{p}_i}$, respectively.

1.4. LEMMA. $h_i(x, y)$ is the $F_{\mathfrak{p}_i}$-linear extension of $h(x, y)$ to $V_{\mathfrak{p}_i} \times V_{\mathfrak{p}_i}$, and $f_i(x, y) = \mathrm{Tr}_i(h_i(x, y))$, where Tr_i denotes the reduced trace from $L_{\mathfrak{p}_i}$ to $\boldsymbol{Q}_p$.

The first assertion is obvious. Let $a = (a_1, \cdots, a_s)$ be an element of $L_p = L_{\mathfrak{p}_1} \times \cdots \times L_{\mathfrak{p}_s}$ with $a_i \in L_{\mathfrak{p}_i}$. Then $\mathrm{Tr}_{L_p/\boldsymbol{Q}_p}(a) = \sum_{i=1}^{s} \mathrm{Tr}_i(a_i)$. Therefore, if $(x, y) \in V_{\mathfrak{p}_i} \times V_{\mathfrak{p}_i}$, then $h_p(x, y) = h_i(x, y) \in L_{\mathfrak{p}_i}$, so that $f_i(x, y) = \mathrm{Tr}_{L_p/\boldsymbol{Q}_p}(h_p(x, y)) = \mathrm{Tr}_i(h_i(x, y))$, which proves the second assertion.

1.5. Let F, L, ρ be as in 1.3. Suppose that F is totally real and ρ is positive, namely $\mathrm{Tr}_{L/\boldsymbol{Q}}(aa^\rho) > 0$ for every element a of L other than 0. Let Φ be a representation of L by complex matrices of size n, which maps the identity element of L to the identity matrix. Let $\mathscr{P} = (A, \mathcal{C}, \theta)$ be a polarized abelian variety of type $\{L, \Phi, \rho\}$ in the sense of [5, 1.4]. This means that:

(i) A is an abelian variety of dimension n, defined over $\boldsymbol{C}$;

(ii) θ is an isomorphism of L into $\mathrm{End}_{\boldsymbol{Q}}(A)$, and the representation of $\theta(a)$ for $a \in L$ by an analytic coordinate-system of A is equivalent to Φ;

(iii) $\mathcal{C}$ is a polarization of A, and the involution of $\mathrm{End}_{\boldsymbol{Q}}(A)$ determined by $\mathcal{C}$ coincides with $\theta(a) \to \theta(a^\rho)$ on $\theta(L)$.

For every isomorphism σ of $\boldsymbol{C}$ into itself, we get naturally a triple $\mathscr{P}^\sigma =$

$(A^\sigma, C^\sigma, \theta^\sigma)$, where $\theta^\sigma(a) = \theta(a)^\sigma$ for $a \in L$. If we put $\Phi^\sigma(a) = \Phi(a)^\sigma$ for $a \in L$, then $\mathcal{P}^\sigma$ is a polarized abelian variety of type $\{L, \Phi^\sigma, \rho\}$. In fact, we first note that C^σ determines the involution $\theta^\sigma(a) \rightarrow \theta^\sigma(a^\rho)$ on $\theta^\sigma(L)$. This can be verified in a straightforward way. Let $\mathfrak{D}(A)$ and $\mathfrak{D}(A^\sigma)$ be respectively the space of invariant linear differential forms on A and on A^σ. Let the $\varphi_{ij}(a)$ be the entries of the matrix $\Phi(a)$. By our condition (ii) for θ, we can find a basis $\{\omega_1, \cdots, \omega_n\}$ of $\mathfrak{D}(A)$ such that $\omega_i \circ \theta(a) = \sum_{j=1}^{n} \varphi_{ij}(a)\omega_j$. Here we denote by $\omega \circ \lambda$ the transform of the differential form ω by an endomorphism λ. Applying σ to this relation, we get $\omega_i^\sigma \circ \theta^\sigma(a) = \sum_{j=1}^{n} \varphi_{ij}(a)^\sigma \omega_j^\sigma$. Since $\{\omega_i^\sigma\}$ is a basis of $\mathfrak{D}(A^\sigma)$, this implies that $\mathcal{P}^\sigma$ is of type $\{L, \Phi^\sigma, \rho\}$. From this consideration, we obtain easily

1.6. PROPOSITION. *Let K' be the algebraic number field generated over Q by the elements* $\mathrm{tr}\,(\Phi(a))$ *for all elements a of the center of L. Let $\mathcal{P}$ be a polarized abelian variety of type $\{L, \Phi, \rho\}$, and let σ be an isomorphism of C into itself. Then $\mathcal{P}^\sigma$ is of type $\{L, \Phi, \rho\}$ if and only if σ leaves invariant every element of K'.*

1.7. PROPOSITION. *Let K' and $\mathcal{P}$ be as in 1.6. Let k be the field of moduli of $\mathcal{P}$* (cf. [1, p. 110]). *Then k contains K'.*

If an isomorphism σ of C into itself leaves invariant every element of k, then $\mathcal{P}^\sigma$ is isomorphic to $\mathcal{P}$. By 1.6, σ must leave invariant every element of K'. This proves $k \supset K'$.

1.8. Let $\mathcal{P} = (A, C, \theta)$ be as in 1.5. Take any rational prime p. Let $g_p(A)$ be the group of all points x on A such that $p^\nu x = 0$ for a positive integer ν. Then there exists an isomorphism of $g_p(A)$ onto the product of $2n$ copies of Q_p/Z_p. If we fix such an isomorphism $\mathfrak{w}$, we get a representation $R_p: \mathrm{End}_Q(A) \rightarrow M_{2n}(Q_p)$ [8, No. 31]. Let X be a basic polar divisor in C. This means that every divisor Y in C is algebraically equivalent to a positive multiple of X (cf. [7, 4.2, Prop. 15]). Let U_p be the group consisting of all $p^{\nu\mathrm{th}}$ roots of unity in C for all positive integers ν. Choose an isomorphism t of U_p onto Q_p/Z_p. With respect to $\mathfrak{w}$ and t, X determines an alternating matrix E_p of $M_{2n}(Z_p)$ [8, No. 76]. By our assumption (iii) on C, we have $E_p R_p(\theta(a)) = {}^t R_p(\theta(a^\rho)) E_p$. If we change $\mathfrak{w}$ and t, then $R_p(\lambda)$ and E_p are transformed to $R_p'(\lambda) = T^{-1} R_p(\lambda) T$ and $E_p' = c_p \cdot {}^t T E_p T$ for an invertible element T of $M_{2n}(Z_p)$ and a unit c_p of Z_p. Now let V^p be the vector space of column vectors of dimension $2n$ over Q_p, and D^p be the set of all elements of V^p with coordinates in Z_p. By means of the operation of $R_p(\theta(a))$, we may consider V^p as a left L_p-module, and further, putting $f^p(x, y) = -{}^t x E_p y$ for $(x, y) \in V^p \times V^p$, we get a Q_p-bilinear mapping of $V^p \times V^p$ into Q_p. We see easily that f^p is ρ-alternating. In this way, for every prime number p, we get a structure (V^p, f^p, D^p) formed by a left L_p-module V^p, a ρ-alternating form f^p, and a $\mathbf{Z}_p$-lattice D^p in V^p. Such a structure is determined by $\mathcal{P}$ uniquely up to isomorphisms and the multiplication of f^p by a unit of $\mathbf{Z}_p$.

1.9. Let $\mathbf{C}^n/D$ be a complex torus analytically isomorphic to A, where D is a lattice in $\mathbf{C}^n$. Put $V = \mathbf{Q} \cdot D$. By the operation of $\Phi(a)$, V can be considered as a left L-module. Let $f(x, y)$ be the Riemann form of the divisor X. Then f is a ρ-alternating form on V. By 1.2, there exists a ρ-anti-hermitian form h on V such that $f(x, y) = \mathrm{Tr}_{L/Q}(h(x, y))$. Thus we get a structure (V, h, D) formed by a left L-module V, a ρ-anti-hermitian form h on V, and a $\mathbf{Z}$-lattice D in V. $\mathcal{P}$ determines this structure uniquely up to isomorphisms. We call (V, h, D) *the anti-hermitian structure determined by* $\mathcal{P}$, and denote it simply by (h, D) if there is no fear of confusion.

1.10. Let $V_p = V \otimes_Q \mathbf{Q}_p$, and let f_p be the $\mathbf{Q}_p$-linear extension of f to $V_p \times V_p$; put $D_p = \mathbf{Z}_p \cdot D$. Then the structure (V_p, f_p, D_p) is isomorphic to (V^p, f^p, D^p) for a suitable choice of $\mathfrak{w}$ and $\mathfrak{t}$. To see this, let $\{u_1, \cdots, u_{2n}\}$ be a basis of D over $\mathbf{Z}$, and let H_p be the set of all elements x of $\mathbf{Q}$ such that $p^\nu x \in \mathbf{Z}$ for a positive integer ν. Then $\mathbf{Q}_p/\mathbf{Z}_p$ is canonically isomorphic to $H_p/\mathbf{Z}$, and $g_p(A)$ corresponds to $(\sum_{i=1}^{2n} H_p u_i)/D$. Therefore $\{u_i\}$ determines an isomorphism $\mathfrak{w}$ of $g_p(A)$ to $(\mathbf{Q}_p/\mathbf{Z}_p)^{2n}$. If $a \in L$ and $\Phi(a)u_j = \sum_{i=1}^{2n} r_{ij}(a)u_i$ with $r_{ij}(a)$ in $\mathbf{Q}$, then, with respect to $\mathfrak{w}$, we get the matrix $(r_{ij}(a))$ as the above $R_p(\theta(a))$. Further, we see that the mapping $H_p \ni x \to \exp(2\pi\sqrt{-1}\,x) \in U_p$ gives an isomorphism of $\mathbf{Q}_p/\mathbf{Z}_p$ to U_p. Denote by $\mathfrak{t}$ the inverse of this isomorphism. Then with respect to the present $\mathfrak{w}$ and $\mathfrak{t}$, we get the matrix $^t(f(u_i, u_j))$ as E_p, on account of the formula [7, p. 25, (7)]. Hence (V_p, f_p, D_p) is isomorphic to (V^p, f^p, D^p).

Since $a^\rho = a$ for $a \in F$, $h(x, y)$ is F-bilinear. Let $\mathfrak{p}$ be a prime ideal of F. Put $V_{\mathfrak{p}} = V \otimes_F F_{\mathfrak{p}}$, $L_{\mathfrak{p}} = L \otimes_F F_{\mathfrak{p}}$. Let $h_{\mathfrak{p}}$ be the $F_{\mathfrak{p}}$-linear extension of h to $V_{\mathfrak{p}} \times V_{\mathfrak{p}}$. Let $\mathfrak{g}$ be the ring of integers in F, and $\mathfrak{g}_{\mathfrak{p}}$ be the ring of $\mathfrak{p}$-integers in $F_{\mathfrak{p}}$. Now assume that $\theta(\mathfrak{g}) \subset \mathrm{End}(A)$. This is the case if and only if $\Phi(\mathfrak{g})D \subset D$, in other words, D is a $\mathfrak{g}$-lattice in V. Put $D_{\mathfrak{p}} = \mathfrak{g}_{\mathfrak{p}} \cdot D$. We get in this way a structure $(V_{\mathfrak{p}}, h_{\mathfrak{p}}, D_{\mathfrak{p}})$. If $\mathfrak{p}_1, \cdots, \mathfrak{p}_s$ are all the prime factors of p in F, $\mathfrak{g} \otimes_Z \mathbf{Z}_p$ may be identified with $\mathfrak{g}_{\mathfrak{p}_1} \times \cdots \times \mathfrak{g}_{\mathfrak{p}_s}$. Hence, by the identification of V_p with $V_{\mathfrak{p}_1} \times \cdots \times V_{\mathfrak{p}_s}$, D_p corresponds to $D_{\mathfrak{p}_1} \times \cdots \times D_{\mathfrak{p}_s}$. Therefore, by 1.4, the structures $(V_{\mathfrak{p}_i}, h_{\mathfrak{p}_i}, D_{\mathfrak{p}_i})$ for $1 \leq i \leq s$ are completely determined by (V_p, f_p, D_p), and hence by (V^p, f^p, D^p), up to isomorphisms and the multiplication of $h_{\mathfrak{p}_i}$ by a unit of $\mathbf{Z}_p$.

1.11. PROPOSITION. *Let* F, L, ρ, Φ *be as in* 1.5 *and let* $\mathfrak{g}$ *be the ring of integers in* F. *Let* $\mathcal{P} = (A, \mathcal{C}, \theta)$ *be a polarized abelian variety of type* $\{L, \Phi, \rho\}$. *Suppose that* $\theta(\mathfrak{g}) \subset \mathrm{End}(A)$. *Let* K' *be the field defined in* 1.6. *Let* σ *be an isomorphism of* $\mathbf{C}$ *into itself which leaves invariant every element of* K'. *Let*

(V, h, D) and (V^*, h^*, D^*) be respectively the anti-hermitian structures determined by $\mathscr{P}$ and $\mathscr{P}^\sigma$. Then, for every prime ideal $\mathfrak{p}$ of F, there exist an $L_\mathfrak{p}$-isomorphism $\lambda_\mathfrak{p}$ of $V_\mathfrak{p}$ onto $V_\mathfrak{p}^*$ and a unit c_p of Z_p, such that $D_\mathfrak{p}\lambda_\mathfrak{p} = D_\mathfrak{p}^*$ and $h_\mathfrak{p}^*(x\lambda_\mathfrak{p}, y\lambda_\mathfrak{p}) = c_p h_\mathfrak{p}(x, y)$ for all $(x, y) \in V_\mathfrak{p} \times V_\mathfrak{p}$, where p is the rational prime divisible by $\mathfrak{p}$.

Choose an isomorphism $\mathfrak{w}$ of $g_p(A)$ onto $(Q_p/Z_p)^{2n}$ and an isomorphism $\mathfrak{t}$ of U_p onto Q_p/Z_p. Let X be a basic polar divisor in $\mathcal{C}$, and let $R_p(\lambda)$ and E_p be the representations of an element λ of $\mathrm{End}_Q(A)$ and X with respect to $\mathfrak{w}$ and $\mathfrak{t}$. Then X^σ is a basic polar divisor in $\mathcal{C}^\sigma$. It is clear that $\mathfrak{w} \circ \sigma^{-1}$ is an isomorphism of $g_p(A^\sigma)$ to $(Q_p/Z_p)^{2n}$, and $\mathfrak{t} \circ \sigma^{-1}$ is an isomorphism of U_p onto Q_p/Z_p. Let $R_p^*(\mu)$ and E_p^* be the representations of an element μ of $\mathrm{End}_Q(A^\sigma)$ and X^σ with respect to $\mathfrak{w} \circ \sigma^{-1}$ and $\mathfrak{t} \circ \sigma^{-1}$. Then we see easily that $R_p^*(\lambda^\sigma) = R_p(\lambda)$ for $\lambda \in \mathrm{End}_Q(A)$ and $E_p^* = E_p$. Therefore $\mathscr{P}$ and $\mathscr{P}^\sigma$ give the same structure (V^p, f^p, D^p). This combined with the last statement of 1.10 proves our proposition.

2. Discontinuous groups and analytic families of polarized abelian varieties

2.1. Let L be a simple algebra over Q, with a positive involution ρ, belonging to the four types of [5, Prop. 1], and Φ be a representation of L by complex matrices of size n given in [5, 2.2]. Put $2n = [L : Q]m$. We extend ρ to an involution of $L_R = L \otimes_Q R$ and denote it again by ρ. For every matrix $B = (b_{ij})$ with b_{ij} in L_R, we put $B^\rho = {}^t(b_{ij}^\rho)$. Let V be the vector space of all m-dimensional row-vectors with components in L; we consider V as a left L-module. Let T be an invertible element of $M_m(L)$ such that $T^\rho = -T$. We get a ρ-anti-hermitian form $T(x, y)$ on V by putting $T(x, y) = xTy^\rho$ for $(x, y) \in V \times V$. Put

$$U_R(T) = \{S \in M_m(L_R) \mid STS^\rho = T\} ,$$
$$U(T) = \{S \in M_m(L) \mid STS^\rho = T\} .$$

Then we get a symmetric domain $\mathcal{H}$ (with complex structure) as the quotient of $U_R(T)$ by a maximal compact subgroup; an explicit form for $\mathcal{H}$ is given in [5, 2.6, 2.7]. Let $\mathfrak{M}$ be a Z-lattice in V. We assume that $(T, \mathfrak{M})$ is *primitive* in the sense of [5, 2.9], i.e.,

$$(2.1.1) \qquad \{\mathrm{Tr}_{L/Q}(xTy^\rho) \mid (x, y) \in \mathfrak{M} \times \mathfrak{M}\} = Z .$$

In [5], we have constructed, for given $\{L, \Phi, \rho\}$ and $(V, T, \mathfrak{M})$, an analytic family $\Sigma = \{\mathscr{P}_z \mid z \in \mathcal{H}\}$ of polarized abelian varieties parametrized by the points of $\mathcal{H}$ with the following properties:

(2.1.2) Every member $\mathcal{P}_z$ of Σ is of type $\{L, \Phi, \rho\}$ and determines the anti-hermitian structure $(V, T, \mathfrak{M})$ in the sense of 1.9.

(2.1.3) If a polarized abelian variety $\mathcal{P}$ is of type $\{L, \Phi, \rho\}$ and determines the anti-hermitian structure $(V, T, \mathfrak{M})$, then $\mathcal{P}$ is isomorphic to a member of Σ.

In order that Σ be not empty, T should satisfy [5, 2.4, (25)] when L is an algebra of (Type IV) of [5, Prop. 1]. We denote Σ by $\Sigma(T, \mathfrak{M})$ or $\Sigma(L, \Phi, \rho; T, \mathfrak{M})$, and $\mathcal{P}_z$ by $\mathcal{P}(z, T, \mathfrak{M})$, when such a specification is necessary. We note that

(2.1.4) Every polarized abelian variety of type $\{L, \Phi, \rho\}$ is isomorphic to a member of $\Sigma(L, \Phi, \rho; T, \mathfrak{M})$ for some T and $\mathfrak{M}$.

We let every element of $M_m(L)$ operate on V on the right, and define a subgroup $\Gamma(T, \mathfrak{M})$ of $U(T)$ by

$$\Gamma(T, \mathfrak{M}) = \{S \in U(T) \,|\, \mathfrak{M}S = \mathfrak{M}\} \,.$$

Since $\mathcal{H}$ is a quotient space of $U_R(T)$, every element of $U_R(T)$ operates naturally on $\mathcal{H}$. By this operation, $\Gamma(T, \mathfrak{M})$ gives a properly discontinuous group of transformations of $\mathcal{H}$.

(2.1.5) Two members $\mathcal{P}_z$ and $\mathcal{P}_w$ of the same family $\Sigma(T, \mathfrak{M})$ are isomorphic if and only if $z = S(w)$ for some $S \in \Gamma(T, \mathfrak{M})$ [5, Theorem 2].

2.2. Let $\mathcal{P} = (A, \mathcal{C}, \theta)$ be of type $\{L, \Phi, \rho; T, \mathfrak{M}\}$. Put

$$(2.2.1) \qquad\qquad \mathfrak{o} = \{a \in L \,|\, a\mathfrak{M} \subset \mathfrak{M}\} \,.$$

Then $\mathfrak{o}$ is an order in L, and $\theta(\mathfrak{o}) = \mathrm{End}\,(A) \cap \theta(L)$. We fix a basis $\{a_i\}$ of $\mathfrak{o}$ over Z. Let X be a basic polar divisor in $\mathcal{C}$. We denote by $\mathcal{F}(\mathcal{P})$ the variety $\mathcal{F}(A, 3X, \theta)$ constructed in [1, p. 107] with respect to the basis $\{a_i\}$ (cf. also [2, 3.4]). $\mathcal{F}(\mathcal{P})$ is determined by the structure $\mathcal{P}$, and independent of the choice of X. If $\mathcal{P}_z = (A_z, \mathcal{C}_z, \theta_z)$ is a member of $\Sigma(T, \mathfrak{M})$, we put

$$(2.2.2) \qquad\qquad \mathcal{F}(\mathcal{P}_z) = \mathcal{F}_z = \mathcal{F}(z, T, \mathfrak{M}) \,.$$

As in [5, 3.2], we may assume that A_z is a projective embedding of itself given by the completeli near system of $3X$ for a basic polar divisor X in $\mathcal{C}_z$. Let $\nu = \nu(T, \mathfrak{M})$ be the dimension of the ambient space for the A_z. Then $\mathcal{F}(\mathcal{P}_z)$ is a subvariety of a product R of several projective spaces; and R is determined by $(T, \mathfrak{M})$. We embed R biregularly into a projective space P^N in a standard way, and identify $\mathcal{F}_z$ with its image by this embedding. Write N as $N(T, \mathfrak{M})$. For a variety or a positive cycle Y in a projective space, denote by $c(Y)$ the Chow point of Y. By [1, Lemma 8], $\mathcal{F}_z$ is of dimension $(\nu + 1)^2 - 1 + n[L:Q]$. Hence $c(\mathcal{F}(z, T, \mathfrak{M}))$ is a point in a projective space $P^{l(\nu, N, e)}$, where the dimension $l(\nu, N, e)$ depends only upon $\nu = \nu(T, \mathfrak{M})$, $N = N(T, \mathfrak{M})$, and $e = \deg(\mathcal{F}(z, T, \mathfrak{M}))$.

2.3. Proposition. *Let d be the complex dimension of $\mathcal{H}$. Then, for every $z \in \mathcal{H}$, we have $\dim_Q c(\mathcal{F}(z, T, \mathfrak{M})) \leqq d$. If $\dim_Q c(\mathcal{F}(z, T, \mathfrak{M})) = \dim_Q c(\mathcal{F}(w, T, \mathfrak{M})) = d$ for some z and w of $\mathcal{H}$, then there exists an isomorphism σ of $\overline{C}$ into itself over Q such that $\mathcal{P}(z, T, \mathfrak{M})^\sigma$ is isomorphic to $\mathcal{P}(w, T, \mathfrak{M})$, where $\overline{Q}$ is the algebraic closure of Q.*

This follows easily from [5, Th. 4] and [2, Prop. 3.5 and Remark of p. 305]. We call $\mathcal{P}(z, T, \mathfrak{M})$ a *generic member* of $\Sigma(T, \mathfrak{M})$ if $\dim_Q c(\mathcal{F}(z, T, \mathfrak{M})) = \dim \mathcal{H}$.

2.4. Proposition. *Let $\mathcal{P}_u$ be a generic member of $\Sigma(T, \mathfrak{M})$ and $\mathcal{P}_z$ be any member of $\Sigma(T, \mathfrak{M})$. Then there exists a specialization $\mathcal{F}'$ of $\mathcal{F}_u$ over $\overline{Q}$ such that $\mathcal{F}' \supset \mathcal{F}_z$.*

The proof of [1, Th. 1] contains (essentially) the proof of this proposition (cf. also the proof of 2.6 below).

2.5. Let $\mathcal{P}_u$ be a generic member of $\Sigma(T, \mathfrak{M})$. Denote by $e(T, \mathfrak{M})$ the degree of $\mathcal{P}_u$. This does not depend on the choice of u, on account of 2.3. By 2.4, $\deg(\mathcal{F}(z, T, \mathfrak{M})) \leqq e(T, \mathfrak{M})$ for every $z \in \mathcal{H}$. We denote by $\mathcal{V}(T, \mathfrak{M})$ the locus of $c(\mathcal{F}_u)$ over $\overline{Q}$, and by $\mathcal{I}(T, \mathfrak{M})$ the set of all points z on $\mathcal{H}$ such that $\deg(\mathcal{F}(z, T, \mathfrak{M})) = e(T, \mathfrak{M})$.

2.6. Proposition. *$\mathcal{I}(T, \mathfrak{M})$ is an open set in $\mathcal{H}$.*

The variety $\mathcal{V}(T, \mathfrak{M})$ is the set of $c(W)$ for all specializations W of $\mathcal{F}_u$ over $\overline{Q}$. Let Z be the subset of $\mathcal{V}(T, \mathfrak{M})$ consisting of all $c(W)$ such that W is not irreducible. Then Z is Zariski closed over $\overline{Q}$ in $\mathcal{V}(T, \mathfrak{M})$. Let $y \in \mathcal{I}(T, \mathfrak{M})$, and let $f_1, \cdots, f_\kappa$ be as in [5, Th. 3]. Put $f_0 = 1$. By [5, Prop. 10], there exists an index $p \, (0 \leqq p \leqq \kappa)$ such that $f_0/f_p, \cdots, f_\kappa/f_p$ are holomorphic at y and $c(\mathcal{F}_y) = (f_0/f_p(y), \cdots, f_\kappa/f_p(y))$. Since $c(\mathcal{F}_y) \in \mathcal{V}(T, \mathfrak{M}) - Z$, there exists a homogeneous polynomial G in the coordinates of the ambient space for $\mathcal{V}(T, \mathfrak{M})$ such that $G(c(\mathcal{F}_y)) \neq 0$ and $G(x) = 0$ for every $x \in Z$. Put $g(z) = G(f_0/f_p(z), \cdots, f_\kappa/f_p(z))$. There exists a neighborhood U of y such that $f_0/f_p, \cdots, f_\kappa/f_p$ are holomorphic in U and $g(z) \neq 0$ for $z \in U$. Now let us show that $U \subset \mathcal{I}(T, \mathfrak{M})$. Let $w \in U$. We make use of the function Ψ of [5, 3.2] (cf. also [1, No. 12]). Let v be a generic point on $\mathcal{H}$ for $\Psi, f_0/f_p, \cdots, f_\kappa/f_p$ and g over $\overline{Q}$ in the sense of [1, No. 8]. By the same argument as in the proof of [1, Th. 1], we have a specialization $\mathcal{F}'$ of $\mathcal{F}_v$ which is compatible with the specialization

$$\big(\Psi(v), f_0/f_p(v), \cdots, f_\kappa/f_p(v), g(v)\big) \to \big(\Psi(w), f_0/f_p(w), \cdots, f_\kappa/f_p(w), g(w)\big)$$

over $\overline{Q}$. Then we have $\mathcal{F}' \supset \mathcal{F}_w$ by [2, Prop. 5.2]. Since $c(\mathcal{F}_v) = (f_0/f_p(v), \cdots, f_\kappa/f_p(v))$, we must have $c(\mathcal{F}') = (f_0/f_p(w), \cdots, f_\kappa/f_p(w))$, so that $G(c(\mathcal{F}')) = g(w) \neq 0$. By our choice of G, $\mathcal{F}'$ is irreducible. Since $\mathcal{F}' \supset \mathcal{F}_w$, we get $\mathcal{F}' =$

$\mathcal{F}_w$, and hence $\deg(\mathcal{F}_w) = \deg(\mathcal{F}') = \deg(\mathcal{F}_v) = e(T, \mathfrak{M})$. Therefore $w \in \mathcal{I}(T, \mathfrak{M})$. This shows the inclusion $U \subset \mathcal{I}(T, \mathfrak{M})$ and completes our proof.

2.7. By [5, Prop. 10], the mapping $z \to c(\mathcal{F}(z, T, \mathfrak{M}))$ is a holomorphic mapping of $\mathcal{I}(T, \mathfrak{M})$ into $\mathcal{V}(T, \mathfrak{M})$, and by the discussion of [5, 3.4], the image contains a non-empty open subset of $\mathcal{V}(T, \mathfrak{M})$ in the sense of usual topology. Let $\mathcal{V}^*(T, \mathfrak{M})$ be the set of all generic points on $\mathcal{V}(T, \mathfrak{M})$ over $\bar{Q}$, and $\mathcal{V}_0(T, \mathfrak{M})$ be the set of all simple points on $\mathcal{V}(T, \mathfrak{M})$, the universal domain being C. Put

$$\mathfrak{X}(T, \mathfrak{M}) = \mathcal{V}_0(T, \mathfrak{M}) \cap \{c(\mathcal{F}(z, T, \mathfrak{M})) \mid z \in \mathcal{I}(T, \mathfrak{M})\}.$$

2.8. PROPOSITION. $\mathfrak{X}(T, \mathfrak{M})$ *is an open subset of* $\mathcal{V}_0(T, \mathfrak{M})$ *containing* $\mathcal{V}^*(T, \mathfrak{M})$. *Moreover, let* $(T', \mathfrak{M}')$ *be another couple such that* $\mathfrak{o} = \{a \in L \mid a\mathfrak{M}' \subset \mathfrak{M}'\}$, $\nu(T, \mathfrak{M}) = \nu(T', \mathfrak{M}')$, $N(T, \mathfrak{M}) = N(T', \mathfrak{M}')$, $e(T, \mathfrak{M}) = e(T', \mathfrak{M}')$. *Then we have* $\mathcal{V}(T, \mathfrak{M}) = \mathcal{V}(T', \mathfrak{M}')$ *if and only if the structures* $(T, \mathfrak{M})$ *and* $(T', \mathfrak{M}')$ *are isomorphic.*

Let $w \in \mathcal{I}(T, \mathfrak{M})$, and $c(\mathcal{F}(w, T, \mathfrak{M})) \in \mathcal{V}_0(T, \mathfrak{M})$. Since $\Gamma(T, \mathfrak{M})$ is properly discontinuous on $\mathcal{H}$, we can find a neighborhood U of w in $\mathcal{I}(T, \mathfrak{M})$ so that $\gamma(U) \cap U \neq \varnothing$ with $\gamma \in \Gamma(T, \mathfrak{M})$ holds only for $\gamma(w) = w$. If $c(\mathcal{F}_z) = c(\mathcal{F}_w)$ for some $z \in U$, then, by [5, 3.2, (45)], there exists an element γ of $\Gamma(T, \mathfrak{M})$ such that $\gamma(z) = w$. By our choice of U, we must have $z = w$. Therefore, the holomorphic mapping $z \to c(\mathcal{F}_z)$ of U into $\mathcal{V}(T, \mathfrak{M})$ takes the value $c(\mathcal{F}_w)$ only at $z = w$. By a classical theorem of Poincaré-Osgood [Osgood, Lehrbuch der Funktionentheorie II_1, 2 Auflage, 1929, p. 139, Satz 2], we see that the image of U contains an open neighborhood of $c(\mathcal{F}_w)$ in $\mathcal{V}_0(T, \mathfrak{M})$. Hence $\mathfrak{X}(T, \mathfrak{M})$ is an open set in $\mathcal{V}_0(T, \mathfrak{M})$. Let $c \in \mathcal{V}^*(T, \mathfrak{M})$. Then there exists an isomorphism σ of $\bar{Q}(c(\mathcal{F}_u))$ to $\bar{Q}(x)$ over $\bar{Q}$ such that $c(\mathcal{F}_u)^\sigma = x$. Extend this isomorphism to an isomorphism of C into itself, and denote it again by σ. By 1.6 and (2.1.4), $\mathcal{P}_u^\sigma$ is isomorphic to a member $\mathcal{P}(v, T', \mathfrak{M}')$ of $\Sigma(T', \mathfrak{M}')$ for some $(T', \mathfrak{M}')$. If $\mathcal{P}(v, T', \mathfrak{M}') = (A_v, C_v, \theta_v)$, we have $\theta_v(\mathfrak{o}) = \mathrm{End}\,(A_v) \cap \theta_v(L)$, so that $\mathfrak{o} = \{a \in L \mid a\mathfrak{M}' \subset \mathfrak{M}'\}$. We have clearly $x = c(\mathcal{F}_u)^\sigma = c(\mathcal{F}(v, T', \mathfrak{M}'))$, $\nu(T', \mathfrak{M}') = \nu(T, \mathfrak{M})$, $N(T', \mathfrak{M}') = N(T, \mathfrak{M})$. Since $\dim_Q Q(x) = \dim \mathcal{H}$, $\mathcal{P}(v, T', \mathfrak{M}')$ is a generic member of $\Sigma(T', \mathfrak{M}')$, so that $e(T', \mathfrak{M}') = e(T, \mathfrak{M})$. Since x is a generic point of $\mathcal{V}(T', \mathfrak{M}')$ over $\bar{Q}$ and $x \in \mathcal{V}(T, \mathfrak{M})$, we have $\mathcal{V}(T', \mathfrak{M}') = \mathcal{V}(T, \mathfrak{M})$. Let Δ be the set of all $(T', \mathfrak{M}')$ such that $\mathfrak{o} = \{a \in L \mid a\mathfrak{M}' \subset \mathfrak{M}'\}$, $\nu(T', \mathfrak{M}') = \nu(T, \mathfrak{M})$, $N(T', \mathfrak{M}') = N(T, \mathfrak{M})$, $e(T', \mathfrak{M}') = e(T, \mathfrak{M})$, and $\mathcal{V}(T', \mathfrak{M}') = \mathcal{V}(T, \mathfrak{M})$. Then the above consideration shows that $\mathcal{V}^*(T, \mathfrak{M})$ is contained in the join of the $\mathfrak{X}(T', \mathfrak{M}')$ for all $(T', \mathfrak{M}') \in \Delta$. It is clear that we have $\mathcal{V}(T, \mathfrak{M}) = \mathcal{V}(T', \mathfrak{M}')$ if the structures $(T, \mathfrak{M})$ and $(T', \mathfrak{M}')$ are isomorphic. Now assume that $\mathfrak{X}(T, \mathfrak{M})$ and $\mathfrak{X}(T', \mathfrak{M}')$ have a common point y for some $(T', \mathfrak{M}') \in \Delta$. Then $y = c(\mathcal{F}(z, T, \mathfrak{M})) = c(\mathcal{F}(z', T', \mathfrak{M}'))$ for some points z and z' of $\mathcal{H}$. Since $\nu(T, \mathfrak{M}) =$

$\nu(T', \mathfrak{M}')$, we see, by virtue of [1, Prop. 1], that $\mathcal{P}(z, T, \mathfrak{M})$ and $\mathcal{P}(z', T', \mathfrak{M}')$ are isomorphic. By our discussion of 1.9, the structures $(T, \mathfrak{M})$ and $(T', \mathfrak{M}')$ must be isomorphic. Therefore, our proof will be completed if we prove the following lemma.

2.9. Lemma. *Let k be a subfield of C with countably many elements, and W an affine or a projective variety defined over k. Let Y be the set of all generic points of U over k (with coordinates in C). Then Y is connected in the sense of the usual topology.*

It is sufficient to prove only the affine case; so we assume that W is an affine variety and denote by C^n the ambient space for W. Let W_0 be the set of all simple points on W. Let S be the set of all polynomial functions on C^n, with coefficients in k, not identically vanishing on W. For every $f \in S$, put $Z_f = \{x \in W_0 \mid f(x) = 0\}$. Then $Y = W_0 - \bigcup_{f \in S} Z_f$. If W is of dimension t, then W_0 is a connected manifold of topological dimension $2t$, and Z_f is a *closed* subset of W_0 whose topological dimension is $\leq 2t - 2$, for each $f \in S$. Since S is a countable set, the topological dimension of $\bigcup_{f \in S} Z_f$ is $\leq 2t - 2$, on account of the *sum theorem* for dimension [Hurewicz and Wallman, Dimension theory, Princeton Math. Ser. No. 4, 1941, p. 30, Th. III 2]. By Cor. 1 to Th. IV (4) of the same book, we know that in every m-dimensional manifold, a subset of dimension $\leq m - 2$ has a connected complement. Therefore Y is connected.

3. Discontinuous groups over an algebra of (Type I, II)

3.1. Suppose that L belongs to (Type I, II, III) of [5, Prop. 1]. Then the center of L is a totally real algebraic number field F, and the restriction of Φ to F is equivalent to a multiple of a regular representation of F over $\mathbf{Q}$. Therefore $\mathrm{tr}\,(\Phi(a)) \in \mathbf{Q}$ for every $a \in F$. By 1.6, this implies that, if $\mathcal{P}$ is of type $\{L, \Phi, \rho\}$ and σ is an isomorphism of C into itself, then $\mathcal{P}^\sigma$ is of the same type $\{L, \Phi, \rho\}$.

3.2. **Proposition.** *Let $\mathcal{P}$ be a member of $\Sigma = \Sigma(L, \Phi, \rho; T, \mathfrak{M})$ and let $\mathfrak{o} = \{a \in L \mid a\mathfrak{M} \subset \mathfrak{M}\}$. Suppose that L is of (Type I) or (Type II) and $\mathfrak{o}$ is a maximal order in L. Then for every isomorphism σ of C into itself, $\mathcal{P}^\sigma$ is isomorphic to a member of the same family Σ.*

By 3.1 and (2.1.4), $\mathcal{P}^\sigma$ is isomorphic to a member of $\Sigma(T', \mathfrak{M}')$ for some T' and $\mathfrak{M}'$. Putting $\mathcal{P} = (A, C, \theta)$, we have $\theta(\mathfrak{o}) = \mathrm{End}\,(A) \cap \theta(L)$, so that $\theta^\sigma(\mathfrak{o}) = \mathrm{End}\,(A^\sigma) \cap \theta^\sigma(L)$. This implies $\mathfrak{o} = \{a \in L \mid a\mathfrak{M}' \subset \mathfrak{M}'\}$. Now suppose that L is of (Type I). This means that L coincides with the center F and the involution ρ is the identity mapping. Then a ρ-anti-hermitian form on V is an F-valued alternating form on V, and $\mathfrak{o}$ is the ring of all integers in F. Put $k = m/2$.

By [4, Prop. 1.3], we may assume, after exchanging the coordinate system, that $T = T'$, $\mathfrak{M} = \sum_{i=1}^{k} \mathfrak{o} x_i + \sum_{i=1}^{k} \mathfrak{a}_i y_i$, $\mathfrak{M}' = \sum_{i=1}^{k} \mathfrak{o} x_i + \sum_{i=1}^{k} \mathfrak{b}_i y_i$, where $\{x_i, y_i\}$ is a basis of V over F such that $T(x_i, y_j) = \delta_{ij}$, $T(x_i, x_j) = T(y_i, y_j) = 0$, and the $\mathfrak{a}_i$, $\mathfrak{b}_i$ are ideals in F such that $\mathfrak{a}_1 \supset \cdots \supset \mathfrak{a}_k$, $\mathfrak{b}_1 \supset \cdots \supset \mathfrak{b}_k$. By 1.11, for every prime ideal $\mathfrak{p}$ of F, there exist an element $R_\mathfrak{p}$ of $M_m(F_\mathfrak{p})$ and a unit $c_\mathfrak{p}$ of $\mathfrak{o}_\mathfrak{p}$ such that $R_\mathfrak{p} T \cdot {}^t R_\mathfrak{p} = c_\mathfrak{p} T$ and $\mathfrak{M}_\mathfrak{p} R_\mathfrak{p} = \mathfrak{M}'_\mathfrak{p}$. Put $S_\mathfrak{p} = \begin{bmatrix} 1_k & 0 \\ 0 & c_\mathfrak{p}^{-1} \cdot 1_k \end{bmatrix} \cdot R_\mathfrak{p}$. Then $S_\mathfrak{p} T \cdot {}^t S_\mathfrak{p} = T$ and $\mathfrak{M}_\mathfrak{p} S_\mathfrak{p} = \mathfrak{M}'_\mathfrak{p}$. By the uniqueness of the $\mathfrak{a}_i$ and $\mathfrak{b}_i$ [4, Prop. 1.3], we must have $\mathfrak{a}_i = \mathfrak{b}_i$ for every i, so that $\mathfrak{M} = \mathfrak{M}'$. This proves our proposition for (Type I).

Next suppose that L is of (Type II). Then L is a totally indefinite quaternion algebra over a totally real algebraic number field F. Let ι be the canonical involution of L. By [5, Prop. 2], there exists an element a of L such that $a^\iota = -a$ and $u^\rho = a u^\iota a^{-1}$ for every $u \in L$. Put $f(x, y) = T(x, y)a$ for $(x, y) \in V \times V$. Then we see easily that $f(x, y)$ is an ι-hermitian form on $V \times V$. Since L is totally indefinite, any two non-degenerate ι-hermitian forms on V are transformed to each other by an L-automorphism of V. This is a special case of Ramanathan's theorem, and is easily derived from the fact that every element c of F can be written in the form $c = bb^\iota$ by an element b of L (cf. [4, Prop. 4.1, Lemma 4.4]). Therefore we may put $T = T'$. By 1.11, for every prime ideal $\mathfrak{p}$ of F, there exist an $L_\mathfrak{p}$-automorphism $\lambda_\mathfrak{p}$ of $V_\mathfrak{p}$ and a $\mathfrak{p}$-adic unit $c_\mathfrak{p}$ in $F_\mathfrak{p}$ such that $T(x\lambda_\mathfrak{p}, y\lambda_\mathfrak{p}) = c_\mathfrak{p} T(x, y)$ and $\mathfrak{M}_\mathfrak{p}\lambda_\mathfrak{p} = \mathfrak{M}'_\mathfrak{p}$. By [6, 6.12], there exists an $L_\mathfrak{p}$-automorphism $\mu_\mathfrak{p}$ of $V_\mathfrak{p}$ such that $\mathfrak{M}_\mathfrak{p}\mu_\mathfrak{p} = \mathfrak{M}_\mathfrak{p}$ and $f(x\mu_\mathfrak{p}, y\mu_\mathfrak{p}) = c_\mathfrak{p}^{-1} f(x, y)$. Then we have $f(x\mu_\mathfrak{p}\lambda_\mathfrak{p}, y\mu_\mathfrak{p}\lambda_\mathfrak{p}) = f(x, y)$ and $\mathfrak{M}_\mathfrak{p}\mu_\mathfrak{p}\lambda_\mathfrak{p} = \mathfrak{M}'_\mathfrak{p}$. By [6, 6.9], there exists an L-automorphism λ of V such that $f(x\lambda, y\lambda) = f(x, y)$ and $\mathfrak{M}\lambda = \mathfrak{M}'$. It follows that $(T, \mathfrak{M})$ and $(T, \mathfrak{M}')$ are isomorphic. Hence $\mathscr{P}^\sigma$ is isomorphic to a member of $\Sigma(T, \mathfrak{M})$. This completes the proof.

3.3. **Theorem.** *Let $f_1, \cdots, f_\kappa$ be as in [5, Th. 3]. Put $\mathfrak{o} = \{a \in L \mid a\mathfrak{M} \subset \mathfrak{M}\}$. Suppose that $\mathfrak{o}$ is a maximal order in L, and L is a totally real algebraic number field F or a totally indefinite quaternion algebra over F. Then, $\boldsymbol{Q}$ is algebraically closed in $\boldsymbol{Q}(f_1, \cdots, f_\kappa)$, and*

$$\dim_{\boldsymbol{Q}} \boldsymbol{Q}(f_1, \cdots, f_\kappa) = \dim_C C(f_1, \cdots, f_\kappa) = \dim \mathscr{H} \ .$$

This is an immediate consequence of 3.2 and [1, Th. 2]. If $\mathscr{P}_z$ is a generic member of Σ, then the field of moduli of $\mathscr{P}_z$ is isomorphic to $\boldsymbol{Q}(f_1, \cdots, f_\kappa)$. Therefore we get

3.4. **Corollary.** *The assumption being as in 3.3, let $\mathscr{P}_z$ be a generic member of $\Sigma(T, \mathfrak{M})$. Then the field of moduli of $\mathscr{P}_z$ is a regular extension of $\boldsymbol{Q}$ of dimension d ($=\dim \mathscr{H}$).*

3.5. REMARK. As was mentioned in the last part of [5, 3.4], if L is of (Type I) or if $\Gamma(T, \mathfrak{M})\backslash\mathcal{H}$ is compact, then $C(f_1, \cdots, f_\kappa)$ is the field of *all* automorphic functions with respect to $\Gamma(T, \mathfrak{M})$. Hence, the above theorem implies that, *in these cases, under the assumption of* 3.3, *the field of all automorphic functions with respect to* $\Gamma(T, \mathfrak{M})$ *has* Q *as a field of definition.* When L is a division algebra of (Type II), then $\Gamma(T, \mathfrak{M})\backslash\mathcal{H}$ is compact if and only if $m = 1$. In this case $\Gamma(T, \mathfrak{M})$ is nothing but the group of all units in $\mathfrak{o}$ with the reduced norm (to F) 1, and $\mathcal{H}$ is analytically isomorphic to the product of $[F:Q]$ copies of the upper half plane.

4. Arithmetical preliminaries on unitary groups

4.1. Let F be a totally real algebraic number field of degree g, and K be a totally imaginary quadratic extension of F. We consider any algebraic number field as a subfield of C, and denote by ρ the complex conjugation. Then the restriction of ρ to K is a positive involution. Let $\{\tau_1, \cdots, \tau_g\}$ be a set of g isomorphisms of K into C, whose restrictions to F are different from each other. Then $\{\tau_1, \cdots, \tau_g, \tau_1\rho, \cdots, \tau_g\rho\}$ is the set of all isomorphisms of K into C. In other words, $(K; \{\tau_\nu\})$ is a CM-type in the sense of [7]. Let m be a positive integer, and Φ be a representation of K by complex matrices of size mg. Let r_ν and s_ν be the multiplicities of τ_ν and $\tau_\nu\rho$ in Φ, respectively. We assume

$$(4.1.1) \qquad\qquad r_\nu + s_\nu = m \qquad\qquad \text{for every } \nu .$$

4.2. Let S be a Galois extension of Q containing K, and G be the Galois group of S over Q. Assume that S is a totally imaginary quadratic extension of a totally real field. Then the restriction of ρ to S is contained in the center of G. By [3, § 4, Lemma 2], we can take as S, for example, the *smallest* Galois extension of Q containing K. Denote by G_Q the group ring of G over Q. For every element $\xi = \sum_{\gamma \in G} a_\gamma \gamma$ (with $a_\gamma \in Q$) of G_Q, we put $\xi^* = \sum_\gamma a_\gamma \gamma^{-1}$. Let H_K be the subgroup of G corresponding to K. We choose g elements of G whose restrictions to K are $\tau_1, \cdots, \tau_g$, and denote them by the same letters $\tau_1, \cdots, \tau_g$. The integers r_ν and s_ν being as above, we define two elements η_K and ω of G_Q by

$$(4.2.1) \qquad \begin{cases} \eta_K = \sum_{\gamma \in H_K} \gamma \, , \\ \omega = \sum_{\nu=1}^g (r_\nu \eta_K \tau_\nu + s_\nu \eta_K \tau_\nu \rho) \, . \end{cases}$$

Further put

$$(4.2.2) \qquad\qquad H_K' = \{\gamma \in G \mid \omega\gamma = \omega\} \, .$$

4.3. PROPOSITION. *Let K' be the field generated over Q by the elements $\sum_{\nu=1}^g (r_\nu x^{\tau_\nu} + s_\nu x^{\tau_\nu\rho})$ for all $x \in K$. Then K' is the subfield of S corresponding* *to the subgroup H'_K of G. If $r_\nu = s_\nu$ for every ν, then $K' = \mathbf{Q}$. Otherwise, K' is a totally imaginary quadratic extension of a totally real algebraic number field.*

The first assertion is an easy consequence of the representation theory of algebras. The second assertion is clear. If $r_\nu \neq s_\nu$ for some ν, then ρ is not contained in H'_K. It is easily seen that $H'_K \cup H'_K\rho$ is a subgroup of G. Let F' be the subfield of S corresponding to this subgroup. Then we see easily that F' is totally real and K' is a totally imaginary quadratic extension of F' (cf. [7, 8.1, Lemma 3]).

4.4 PROPOSITION. *K' being as in 4.3, suppose that $r_\nu \neq s_\nu$ for some ν. Put $2g' = [K':\mathbf{Q}]$. Let $\sigma_1, \cdots, \sigma_{g'}$ be elements of G such that $G = \bigcup_{\lambda=1}^{g'}(H'_K\sigma_\lambda \cup H'_K\sigma_\lambda\rho)$. Put $\eta'_K = \sum_{\gamma \in H'_K} \gamma$. Then there exist non-negative integers t_λ, u_λ such that*

$$(4.4.1) \qquad \omega^* = \sum_{\lambda=1}^{g'}(t_\lambda \eta'_K \sigma_\lambda + u_\lambda \eta'_K \sigma_\lambda \rho),$$

$$(4.4.2) \qquad t_\lambda + u_\lambda = m \qquad\qquad (1 \leq \lambda \leq m).$$

Moreover, let Φ' be a representation of K' by complex matrices of size mg' such that the multiplicities of σ_λ and $\sigma_\lambda\rho$ in Φ' are t_λ and u_λ respectively. Then Φ' is uniquely determined, up to equivalence, by only K and Φ, and does not depend on the choice of S.

The existence of t_λ and u_λ satisfying (4.4.1) is clear from the definition of H'_K and η'_K. If we express ω in the form $\omega = \sum_{\gamma \in G} a_\gamma \gamma$ with $a_\gamma \in \mathbf{Z}$, then $a_\gamma + a_{\gamma\rho} = m$. Hence we obtain (4.4.2). The uniqueness of Φ' can be proved in a straightforward way.

We call $(K'; \Phi')$ the *dual* of $(K; \Phi)$. If $m = 1$, this is just the dual of a CM-type.

4.5. PROPOSITION. *Let the notation be as in 4.4. For every element x of K', $\prod_{\lambda=1}^{g'}(x^{\sigma_\lambda})^{t_\lambda}(x^{\sigma_\lambda\rho})^{u_\lambda}$ belongs to K, and for every ideal $\mathfrak{x}$ of K', $\prod_{\lambda=1}^{g'}(\mathfrak{x}^{\sigma_\lambda})^{t_\lambda}(\mathfrak{x}^{\sigma_\lambda\rho})^{u_\lambda}$ is an ideal of K.*

This can be proved in the same way as in [7, 8.3, Prop. 29].

4.6. Let N be a totally real algebraic extension of F of degree m, and let M be the composite of K and N. Then M is a totally imaginary quadratic extension of N. Let S be the smallest Galois extension of $\mathbf{Q}$ containing M, and G be the Galois group of S over $\mathbf{Q}$. Then S is a totally imaginary quadratic extension of a totally real field, and ρ (considered as an element of G) belongs to the center of G [3, Lemma 2]. For these S and G, define $\tau_1, \cdots, \tau_g, H_K, G_Q,$ H'_K, η_K, ω as in 4.2, and define $\sigma_1, \cdots, \sigma_{g'}, \eta'_K, t_\lambda, u_\lambda$ as in 4.4 when $K' \neq \mathbf{Q}$. Let H_M be the subgroup of G corresponding to M. We can find mg elements $\alpha_1, \cdots, \alpha_{mg}$ of G satisfying the following two conditions:

(4.6.1) $\;\; G = \bigcup_{i=1}^{mg} (H_M \alpha_i \cup H_M \alpha_i \rho)$; in other words, $(M; \{\alpha_1, \cdots, \alpha_{mg}\})$ is a CM-type in the sense of [7].

(4.6.2) The restriction of $\{\alpha_1, \cdots, \alpha_{mg}\}$ to K gives r_ν times τ_ν and s_ν times $\tau_\nu \rho$.

To see the existence of $\{\alpha_i\}$, let $H_K = H_M \xi_1 \cup \cdots \cup H_M \xi_m$ be a disjoint expression. Put

$$\alpha_{\nu i} = \begin{cases} \xi_i \tau_\nu & \text{for } 1 \leq i \leq r_\nu \,, \; 1 \leq \nu \leq g \,, \\ \xi_i \tau_\nu \rho & \text{for } r_\nu < i \leq m, \; 1 \leq \nu \leq g \,. \end{cases}$$

Write $\{\alpha_{\nu i} \,|\, 1 \leq \nu \leq g, 1 \leq i \leq m\}$ as $\{\alpha_1, \cdots, \alpha_{mg}\}$. Then we see easily that this satisfies (4.6.1) and (4.6.2). Now define two elements η_M and ψ by

$$\eta_M = \sum_{\gamma \in H_M} \gamma \,, \qquad \psi = \sum_{i=1}^{mg} \eta_M \alpha_i \,,$$

and put

$$H'_M = \{\gamma \in G \,|\, \psi \gamma = \psi\} \,,$$
$$\eta'_M = \sum_{\gamma \in H'_M} \gamma \,.$$

Let M' be the subfield of S corresponding to the subgroup H'_M of G. By our definition of H'_M, we can find elements β_j of G such that $\psi^* = \sum_j \eta'_M \beta_j$. Then $(M'; \{\beta_j\})$ is the dual of $(M; \{\alpha_i\})$ in the sense of [7, 8.3]. We note that M' is generated by the elements $\sum_{i=1}^{mg} x^{\alpha_i}$ for all $x \in M$, over $\boldsymbol{Q}$ [7, 8.3, Prop. 28]. By the property (4.6.2), we have $K' \subset M'$.

4.7. PROPOSITION. *The notation being as in 4.6, put*

$$v_\lambda = \begin{cases} t_\lambda - (m/2) & \text{if } m \text{ is even and } K' \neq \boldsymbol{Q} \,, \\ t_\lambda - (m+1)/2 & \text{if } m \text{ is odd.} \end{cases}$$

Let $\mathfrak{x}$ be an ideal of M', and let $\mathfrak{y} = N_{M'/K'}(\mathfrak{x})$, $\mathfrak{z} = \prod_{\lambda=1}^{g'} (\mathfrak{y}^{\sigma_\lambda})^{v_\lambda}$. Then $\prod_j \mathfrak{x}^{\beta_j}$ is an ideal of M, and

$$N_{M/K}\left(\prod_j \mathfrak{x}^{\beta_j}\right) = \begin{cases} N_{K'/\boldsymbol{Q}}(\mathfrak{y})^{m/2} & \text{if } m \text{ is even and } K' = \boldsymbol{Q} \,, \\ N_{K'/\boldsymbol{Q}}(\mathfrak{y})^{m/2} \cdot \mathfrak{z} \cdot (\mathfrak{z}^\rho)^{-1} & \text{if } m \text{ is even and } K' \neq \boldsymbol{Q} \,, \\ N_{K'/\boldsymbol{Q}}(\mathfrak{y})^{(m-1)/2} \cdot \prod_{\lambda=1}^{g'} \mathfrak{y}^{\sigma_\lambda} \cdot \mathfrak{z} \cdot (\mathfrak{z}^\rho)^{-1} & \text{if } m \text{ is odd.} \end{cases}$$

The first assertion is included in [7, 8.3, Prop. 29]. Let $H_K = H_M \xi_1 \cup \cdots \cup H_M \xi_m$ be a disjoint expression. Then we have

$$\sum_j \eta'_M \beta_j \left(\sum_{i=1}^{m} \xi_i\right) = \psi^*\left(\sum_{i=1}^{m} \xi_i\right) = \sum_{j=1}^{mg} \alpha_j^{-1} \eta_M \left(\sum_{i=1}^{m} \xi_i\right)$$
$$= \sum_{j=1}^{mg} \alpha_j^{-1} \eta_K = \omega^*$$
$$= \begin{cases} \sum_{\lambda=1}^{g'} (t_\lambda \eta'_K \sigma_\lambda + u_\lambda \eta'_K \sigma_\lambda \rho) & \text{if } K' \neq \boldsymbol{Q} \,, \\ (m/2) \sum_{\gamma \in G} \gamma & \text{if } K' = \boldsymbol{Q} \,. \end{cases}$$

From this we obtain

$$N_{M/K}(\Pi, \mathfrak{x}^{\beta j}) = \begin{cases} \prod_{\lambda=1}^{g'} (\mathfrak{y}^{\sigma_\lambda})^{t_\lambda}(\mathfrak{y}^{\sigma_\lambda \rho})^{u_\lambda} & \text{if } K' \neq Q , \\ N_{K'/Q}(\mathfrak{y})^{m/2} & \text{if } K' = Q . \end{cases}$$

Since $t_\lambda + u_\lambda = m$ for every m, we get the relation of our proposition.

4.8. REMARK. When $K' \neq Q$, we may assume, after exchanging σ_λ for $\sigma_\lambda \rho$ if necessary, that $t_\lambda \geqq u_\lambda$ for every λ. Then the integers v_λ of 4.7 are non-negative for every λ.

4.9. Take an element e of M such that $e^\rho = -e$, and put

$$h_e(x, y) = \mathrm{Tr}_{M/K} (exy^\rho) \qquad\qquad \text{for } (x, y) \in M \times M .$$

If we consider M as a vector space of dimension m over K, then h_e is a ρ-anti-hermitian form on M.

Let V be the space of all m-dimensional row-vectors with components in K, and T an invertible element of $M_m(K)$ such that $T^\rho = -T$. As in 2.1, we put $T(x, y) = xTy^\rho$ for $(x, y) \in V \times V$. Let us now treat the problem to find out M and e such that (M, h_e) is isomorphic to a given (V, T).

4.10. PROPOSITION. *For a given (V, T) and for any finite algebraic extension P of K', there exist an extension M of K of degree m, an element e of M, and mg isomorphisms $\alpha_1, \cdots, \alpha_{mg}$ of M into C with the following properties:*

(i) M is the composite of K and a totally real algebraic extension N of F of degree m.

(ii) The structure (M, h_e) is isomorphic to (V, T) over K.

(iii) $\{\alpha_1, \cdots, \alpha_{mg}\}$ satisfies (4.6.1) and (4.6.2).

(iv) Let M' be the field generated over Q by the elements $\sum_{i=1}^{mg} x^{\alpha_i}$ for all $x \in M$. Then M' and P are linearly disjoint over K'.

To prove this we need several lemmas.

4.11. LEMMA. *Let Ω be a field of characteristic 0, and U an invertible symmetric matrix of size m with entries in Ω. Let x_{ij} for $1 \leqq i \leqq m, 1 \leqq j \leqq m$ and y be $m^2 + 1$ independent variables over Ω. Put $X = (x_{ij})$ and*

$$\varphi(y) = \varphi(y, X; U) = \det (y \cdot 1_m - XU \cdot {}^tX) .$$

Then $\varphi(y, X; U)$ is irreducible in $\Omega[y, x_{ij}]$, and the Galois group of $\varphi(y)$ over $\Omega(x_{ij})$ is isomorphic to the symmetric group of m letters.

It is sufficient to prove our lemma in the case where Ω is algebraically closed, since the general case can be easily reduced to this case. If Ω is algebraically closed, we can find an element W of $M_m(\Omega)$ such that $WU \cdot {}^tW = 1_m$. Then $\varphi(y, XW; U) = \varphi(y, X; 1_m)$ Therefore it is sufficient to treat only the case $U = 1_m$. Put $\varphi_m(y, X) = \varphi(y, X; 1_m)$. We now prove by induction on m that φ_m is irreducible in $\Omega[y, x_{ij}]$ and its Galois group over $\Omega(x_{ij})$ is the symmetric group of m letters. This is clear if $m = 1$. Suppose that $m > 1$ and $\varphi_m(y, X) = p(y, X)q(y, X)$ with two elements p and q of $\Omega[y, x_{ij}]$ other than constants. Since φ_m has the leading coefficient 1 as a polynomial in y, we may assume that p and q contain y non-trivially and have the leading coefficient 1 as polynomials in y. Let X' be the matrix of size $m - 1$ obtained by removing the last row and the last column from X. Put $X_0 = \begin{pmatrix} X' & 0 \\ 0 & x_{mm} \end{pmatrix}$. Then $\varphi_m(y, X_0) = \varphi_{m-1}(y, X')(y - x_{mm}^2) = p(y, X_0)q(y, X_0)$. By our induction, $\varphi_{m-1}(y, X')$ is irreducible. Hence we may assume, exchanging p and q if necessary, that

$$\varphi_{m-1}(y, X') = p(y, X_0), \qquad y - x_{mm}^2 = q(y, X_0) .$$

Let $p(X)$ and $q(X)$ be respectively the constant terms of $p(y, X)$ and $q(y, X)$ viewed as polynomials in y. Then we have

$$p(X)q(X) = (-1)^m \det(X)^2 , \qquad p(X_0) = (-1)^{m-1} \det(X')^2 ,$$
$$q(X_0) = -x_{mm}^2 .$$

By the last two relations, $p(X)$ and $q(X)$ are not constants. Since $\det(X)$ is irreducible in $\Omega[x_{ij}]$, the first relation implies that $p(X)$ and $q(X)$ differ from $\det(X)$ only by constant factors. But this is a contradiction in view of the relation $q(X_0) = -x_{mm}^2$. Therefore φ_m must be irreducible in $\Omega[y, x_{ij}]$, so that its Galois group G is a transitive permutation group of m letters. By the assumption of induction, the Galois group of $\varphi_m(y, X_0)$ over $\Omega(X_0)$ is a symmetric group of $m - 1$ letters. Since $\varphi_m(y, X_0)$ is a specialization of $\varphi_m(y, X)$, G must contain a symmetric group of $m - 1$ letters as a subgroup (cf. for example, van der Waerden, Algebra I, 5 Auflage, p. 198). Since G is transitive, G must coincide with the whole symmetric group of m letters. This completes our proof.

4.12. Lemma. *Let R be an algebraic number field of finite degree and P be a finite algebraic extension of R. Let $x_1, \cdots, x_n, z$ be $n + 1$ independent variables and $\psi(z, x_1, \cdots, x_n)$ an irreducible polynomial in $P[z, x_1, \cdots, x_n]$. Let $R(x_1, \cdots, x_n, y_1, \cdots, y_s)$ be an algebraic extension of $R(x_1, \cdots, x_n)$. Suppose that R is algebraically closed in $R(x_1, \cdots, x_n, y_1, \cdots, y_s)$. Then there exist rational numbers $a_1, \cdots, a_n$ and algebraic numbers $b_1, \cdots, b_s$ such that:*

(i) *$(a_1, \cdots, a_n, b_1, \cdots, b_s)$ is a specialization of $(x_1, \cdots, x_n, y_1, \cdots, y_s)$ over P;*

(ii) *$\psi(z, a_1, \cdots, a_n)$ is irreducible over P;*

(iii) *$R(b_1, \cdots, b_s)$ and P are linearly disjoint over R.*

We can find rational numbers $d_1, \cdots, d_s$ such that $R(x_1, \cdots, x_n, y_1, \cdots, y_s) =$

$R(x_1, \cdots, x_n, \sum_{i=1}^{s} d_i y_i)$. Put $u = \sum_{i=1}^{s} d_i y_i$. There exist polynomials $f(x_1, \cdots, x_n)$ and $g_i(x_1, \cdots, x_n, z)$ with coefficients in R such that $y_i = f(x_1, \cdots, x_n)^{-1} g_i(x_1, \cdots, x_n, u)(1 \leq i \leq s)$. Let $h(u, x_1, \cdots, x_n) = 0$ be an irreducible equation for u over $R(x_1, \cdots, x_n)$. Since R is algebraically closed in $R(x_1, \cdots, x_n, u)$, $h(z, x_1, \cdots, x_n)$ is irreducible over $P(x_1, \cdots, x_n)$. By Hilbert's irreducibility theorem, (cf. Hilbert, Über die irreduzibilität ganzer rationaler Funktionen mit ganzzahligen Koeffizienten, Werke 2, 264–286), there exist rational numbers $a_1, \cdots, a_n$ such that $\psi(z, a_1, \cdots, a_n)$ and $h(z, a_1, \cdots, a_n)$ are irreducible over P, $\deg_z h(z, a_1, \cdots, a_n) = \deg_z h(z, x_1, \cdots, x_n)$, and $f(a_1, \cdots, a_n) \neq 0$. Let c be a root of $h(z, a_1, \cdots, a_n) = 0$. Put $b_i = f(a_1, \cdots, a_n)^{-1} g_i(c, a_1, \cdots, a_n)$. Then $(a_1, \cdots, a_n, b_1, \cdots, b_s, c)$ is a specialization of $(x_1, \cdots, x_n, y_1, \cdots, y_s, u)$ over P, and $c = \sum_{i=1}^{s} d_i b_i$. Therefore $R(c) = R(b_1, \cdots, b_s)$ and $P(c) = P(b_1, \cdots, b_s)$. By our choice of $a_1, \cdots, a_n$, we have $[P(c) : P] = \deg_z h(z, x_1, \cdots, x_n) = [R(c) : R]$. Therefore $R(c)$ and P are linearly disjoint over R. This proves our lemma.

4.13. Let $\{a_1, \cdots, a_g\}$ be a basis of F over $\mathbf{Q}$. Take gm^2 independent variables $x_{ij}^p(i, j = 1, \cdots, m; p = 1, \cdots, g)$ and define matrices X_p and Y_ν of size m by

$$X_p = (x_{ij}^p) \qquad\qquad (p = 1, \cdots, g) ,$$
$$Y_\nu = \sum_{p=1}^{g} a_p^{\tau_\nu} X_p \qquad\qquad (\nu = 1, \cdots, g) .$$

Take an element ζ of K such that $\zeta^\rho = -\zeta$. Then $(\zeta T)^\rho = \zeta T$. We can find a matrix W of $M_m(K)$ such that $W(\zeta T)W^\rho$ is a diagonal matrix with diagonal elements in F. Put $U = W(\zeta T)W^\rho$ and

$$\psi_\nu(z) = \psi_\nu(z, X_1, \cdots, X_g) = \det(z1_m - Y_\nu U^{\tau_\nu} \cdot {}^t Y_\nu) \qquad (\nu = 1, \cdots, g) .$$

Let K_0 be the smallest Galois extension of $\mathbf{Q}$ containing K. Let S_0 be the splitting field of $\psi_1(z) \cdots \psi_g(z)$ over $K_0(x_{ij}^p)$. Let $u_{\nu 1}, \cdots, u_{\nu m}$ be the roots of $\psi_\nu(z)$ in S_0. Put

$$v_{pkl} = \sum_{\nu=1}^{g} \left[\sum_{i=1}^{r_\nu} (a_p \zeta^k)^{\tau_\nu} u_{\nu i}^l + \sum_{i=r_\nu+1}^{m} (a_p \zeta^k)^{\tau_\nu \rho} u_{\nu i}^l \right]$$
$$(p = 1, \cdots, g; k = 0, 1; l = 0, 1, \cdots, m - 1) .$$

By the definition of K' (cf. 4.3), we see that K' is generated by the v_{pk0} over $\mathbf{Q}$.

4.14. LEMMA. *K' is algebraically closed in $\mathbf{Q}(x_{ij}^p, v_{pkl})$.*

Since $\psi_1 \cdots \psi_g$ is a polynomial in $\mathbf{Q}[x_{ij}^p, z]$, S_0 is a Galois extension of $\mathbf{Q}(x_{ij}^p)$. Let γ be an automorphism of K_0 over K'. Extend γ to an automorphism of $K_0(x_{ij}^p)$ over $K'(x_{ij}^p)$ and denote it again by γ. Since γ is the identity mapping on K', we have, on account of (4.2.2),

$$(4.14.1) \qquad \begin{cases} r_\mu = r_\nu \;\; \text{and} \;\; s_\mu = s_\nu & \text{if } \tau_\nu \gamma = \tau_\mu \text{ on } K , \\ r_\mu = s_\nu \;\; \text{and} \;\; s_\mu = r_\nu & \text{if } \tau_\nu \gamma = \tau_\mu \rho \text{ on } K . \end{cases}$$

If $\tau_\nu\gamma = \tau_\mu$ or $\tau_\mu\rho$ on K, we have $\psi_\nu^\gamma = \psi_\mu$. By 4.11, we see easily $[S_0:K_0(x_{ij}^p)]=(m!)^g$. Hence we can extend γ to an automorphism of S_0, which we denote again by γ, so that

$$u_{\nu i}^\gamma = u_{\mu i} \qquad\qquad \text{if } \tau_\nu\gamma = \tau_\mu \text{ on } K,$$
$$u_{\nu i}^\gamma = u_{\mu, m+1-i} \qquad\qquad \text{if } \tau_\nu\gamma = \tau_\mu\rho \text{ on } K.$$

Then, on account of (4.14.1), we have $v_{pkl}^\gamma = v_{pkl}$ for every p, k, l. Now by 4.11, we have $[\bar{Q}S_0 : \bar{Q}(x_{ij}^p)] = (m!)^g$, so that S_0 and $\bar{Q}$ are linearly disjoint over K_0. Therefore K_0 is algebraically closed in S_0. Hence if K^* is the algebraic closure of K' in $Q(x_{ij}^p, v_{pkl})$, then K^* must be contained in K_0. In the above we have proved that every automorphism γ of K_0 over K' can be extended to an automorphism of S_0 over $Q(x_{ij}^p, v_{pkl})$. Therefore we have $K^* = K'$.

4.15. *Proof of* 4.10. Let P be any finite algebraic extension of K'. To prove 4.10, we may assume that P contains K_0. Further, we may assume that τ_1 is the identity mapping on K. By 4.12 and 4.14, there exist g matrices $B_p = (b_{ij}^p)$ $(p = 1, \cdots, g)$ with rational entries, and algebraic numbers c_{pkl} such that:

(i) (b_{ij}^p, c_{pkl}) is a specialization of (x_{ij}^p, v_{pkl}) over P;

(ii) $\psi_1(z, B_1, \cdots, B_g)$ is irreducible over P;

(iii) $K'(c_{pkl})$ and P are linearly disjoint over K'.

Now let $(y_{\nu i})$ be a specialization of $(u_{\nu i})$ over $(x_{ij}^p, v_{pkl}) \to (b_{ij}^p, c_{pkl})$ ref. P. Put $D_\nu = \sum_{p=1}^g a_p^{\tau_\nu}B_p$. Then $y_{\nu 1}, \cdots, y_{\nu m}$ are the roots of the equation $\det[z1_m - D_\nu U^{\tau_\nu}\cdot{}^tD_\nu] = 0$. Put $N = F(y_1)$, $M = K(y_1)$. By (ii), we have $[M:K] = [N:F] = m$. We can define isomorphisms $\alpha_{\nu i}$ of M into C, for $1 \leqq \nu \leqq g, 1 \leqq i \leqq m$, by

$$y_1^{\alpha_{\nu i}} = y_{\nu i}$$
$$\alpha_{\nu i} = \begin{cases} \tau_\nu \\ \tau_\nu\rho \end{cases} \qquad \text{on } K \qquad \begin{array}{l} \text{for } 1 \leqq i \leqq r_\nu, \\ \text{for } r_\nu < i \leqq s_\nu. \end{array}$$

We have then

$$\sum_{\nu=1}^g \sum_{i=1}^m (a_p\zeta^k y_1^l)^{\alpha_{\nu i}}$$
$$= \sum_{\nu=1}^g \left[\sum_{i=1}^{r_\nu} (a_p\zeta^k)^{\tau_\nu}y_{\nu i}^l + \sum_{i=r_\nu+1}^m (a_p\zeta^k)^{\tau_\nu\rho}y_{\nu i}^l\right] = c_{pkl}.$$

Therefore $Q(c_{pkl})$ is just the field M' defined for M and $\{\alpha_{\nu i}\}$ as in 4.6.

Put $E = D_1 U \cdot {}^tD_1$. Since F is totally real and ${}^tE = E$, we see easily that $N = F(y_1)$ is totally real. Further we observe that $E \to y_1$ gives an isomorphism of $K[E]$ to $K(y_1) = M$ over K, since $\det[z1_m - E]$ is irreducible over K. Identify M with $K[E]$ by this isomorphism. Recall that $U = W(\zeta T)W^\rho$. We see that the involution $Y \to Y^\rho$ of $M_m(K)$ induces a non-trivial automorphism of M over N. Let f be a non-zero vector in V. Put $h(A) = \zeta^{-1} f A E f^\rho$ for $A \in M = K[E]$. Since h is a K-linear mapping of M into K, there exists an element e of M such that $h(A) = \mathrm{Tr}_{M/K}(eA)$. Every element of M commutes with E, so that

$$\mathrm{Tr}_{M/K}(eA_1 A_2^\rho) = \zeta^{-1} f A_1 E A_2^\rho f^\rho = (f A_1 D_1 W) T (f A_2 D_1 W)^\rho$$

for every $(A_1, A_2) \in M \times M$. This shows that the mapping $A \to f A D_1 W$ gives an isomorphism of the structure (M, h_e) to (V, T). We have thus proved that $M, \{\alpha_{\nu_i}\}, e$ satisfy the conditions (i-iv) of 4.10.

4.16. Let F, K be as in 4.1; and V, T be as in 4.9. For every prime ideal $\mathfrak{p}$ of F, we denote by $F_\mathfrak{p}$ the $\mathfrak{p}$-completion of F, and put $K_\mathfrak{p} = K \otimes_F F_\mathfrak{p}$, $V_\mathfrak{p} = V \otimes_F F_\mathfrak{p}$. Extend ρ to an automorphism of $K_\mathfrak{p}$ over $F_\mathfrak{p}$ and denote it again by ρ. We define groups $G(T)$ and $G_\mathfrak{p}(T)$ by

$$G(T) = \{ S \in \mathrm{GL}_m(K) \mid STS^\rho = \mu T \text{ with } \mu \in F \},$$
$$G_\mathfrak{p}(T) = \{ S \in \mathrm{GL}_m(K_\mathfrak{p}) \mid STS^\rho = \mu T \text{ with } \mu \in F_\mathfrak{p} \},$$

and put $\mu = \mu(S)$ if $STS^\rho = \mu T$. Further we define $U(T)$ and $U_\mathfrak{p}(T)$ by

$$U(T) = \{ S \in G(T) \mid \mu(T) = 1 \},$$
$$U_\mathfrak{p}(T) = \{ S \in G_\mathfrak{p}(T) \mid \mu(T) = 1 \}.$$

As in 2.1, we put $T(x, y) = x T y^\rho$ for $(x, y) \in V_\mathfrak{p} \times V_\mathfrak{p}$.

Let $\mathfrak{g}$ and $\mathfrak{r}$ be the rings of integers in F and in K, respectively. Let $\mathfrak{M}$ be a $\mathbf{Z}$-lattice in V. We say that $\mathfrak{M}$ is an $\mathfrak{r}$-lattice if $\mathfrak{r}\mathfrak{M} \subset \mathfrak{M}$. For every prime ideal $\mathfrak{p}$ of F, we denote by $\mathfrak{g}_\mathfrak{p}$ the ring of integers in $F_\mathfrak{p}$, and put $\mathfrak{r}_\mathfrak{p} = \mathfrak{g}_\mathfrak{p}\mathfrak{r}$, $\mathfrak{M}_\mathfrak{p} = \mathfrak{g}_\mathfrak{p}\mathfrak{M}$.

4.17. We fix, once and for all, an element ζ of K such that $\zeta^\rho = -\zeta$. Since $(\zeta T)^\rho = \zeta T$, ζT defines a non-degenerate ρ-hermitian form on V. Then the groups $G(T)$ and $U(T)$ are just the groups $G(V, \zeta T)$ and $U(V, \zeta T)$ defined in [6, 1.1]. Henceforward, we often use the terminology and the result of [6]. For an $\mathfrak{r}$-lattice $\mathfrak{M}$ in V, we denote by $\mu_T(\mathfrak{M})$ the $\mathfrak{r}$-ideal generated by the elements $\zeta T(x, x)$ for all x in $\mathfrak{M}$. $\mathfrak{M}$ is called *maximal* (with respect to T) if there is no $\mathfrak{r}$-lattice $\mathfrak{N}$ other than $\mathfrak{M}$ such that $\mathfrak{N} \supset \mathfrak{M}$ and $\mu_T(\mathfrak{N}) = \mu_T(\mathfrak{M})$ [6, 2.10]. If $\mathfrak{M}$ and $\mathfrak{N}$ are two $\mathfrak{r}$-lattices, $[\mathfrak{M}/\mathfrak{N}]$ denotes the $\mathfrak{r}$-ideal generated by the $\det(S)$ for all $S \in M_m(K)$ such that $\mathfrak{M}S \subset \mathfrak{N}$ [6, 2.15]. By a *class* of $\mathfrak{r}$-lattices in V with respect to $G(T)$ (resp. $U(T)$), we understand a maximal set of $\mathfrak{r}$-lattices in V whose members are transformed to each other by elements of $G(T)$ (resp. $U(T)$). By a *genus* of $\mathfrak{r}$-lattices in V with respect to $G(T)$ (resp. $U(T)$), we understand a maximal set Λ of $\mathfrak{r}$-lattices in V with the following property: if $\mathfrak{M}$ and $\mathfrak{N}$ are members of Λ, then, for every $\mathfrak{p}$, there exists an element $R_\mathfrak{p}$ of $G_\mathfrak{p}(T)$ (resp. $U_\mathfrak{p}(T)$) such that $\mathfrak{M}_\mathfrak{p}R_\mathfrak{p} = \mathfrak{N}_\mathfrak{p}$.

4.18. We say that an r-lattice $\mathfrak{M}$ in V is of *type* (U) (with respect to T) if the following conditions $(U_{\mathfrak{p}1})$ and $(U_{\mathfrak{p}2})$ are satisfied for every prime ideal $\mathfrak{p}$ of F.

$(U_{\mathfrak{p}1})$ For every unit b of $\mathfrak{r}_{\mathfrak{p}}$ such that $bb^{\rho} = 1$, there exists an element R of $U_{\mathfrak{p}}(T)$ such that $\mathfrak{M}_{\mathfrak{p}}R = \mathfrak{M}_{\mathfrak{p}}$ and $\det(R) = b$.

$(U_{\mathfrak{p}2})$ Let p be the prime number divisible by $\mathfrak{p}$. For every unit c of $\mathbf{Z}_p$, there exists an element S of $G_{\mathfrak{p}}(T)$ such that $\mathfrak{M}_{\mathfrak{p}}S = \mathfrak{M}_{\mathfrak{p}}$ and $\mu(S) = c$.

4.19. PROPOSITION. *If m is even, there exists an r-lattice in V of type* (U).

Let P be the set of all the prime ideals $\mathfrak{p}$ in F such that:

(i) $\mathfrak{p}$ is ramified in K, and

(ii) $V_{\mathfrak{p}}$ has the trivial kernel subspace with respect to ζT in the sense of [6, 4.3].

By [6, 3.2], $(U_{\mathfrak{p}1})$ and $(U_{\mathfrak{p}2})$ are satisfied by any $\mathfrak{r}_{\mathfrak{p}}$-lattice in $V_{\mathfrak{p}}$, if $\mathfrak{p}$ decomposes in K. If $\mathfrak{p}$ does not decompose in K and $\mathfrak{p} \notin P$, every maximal $\mathfrak{r}_{\mathfrak{p}}$-lattice in $V_{\mathfrak{p}}$ satisfies $(U_{\mathfrak{p}1})$ and $(U_{\mathfrak{p}2})$ on account of [6, 4.18 and 4.24]. Now for each $\mathfrak{p} \in P$, we can find (cf. [6, 4.2 and 4.3]) a basis $\{x_1, \cdots, x_m\}$ of $V_{\mathfrak{p}}$ over $K_{\mathfrak{p}}$ such that

$$\zeta T(x_i, x_j) = \begin{cases} 0 & \text{for } i \neq j\,, \\ 1 & \text{for } i = j \leq m/2\,, \\ -1 & \text{for } i = j > m/2\,. \end{cases}$$

Put $\mathfrak{L}_{\mathfrak{p}} = \sum_{i=1}^{m} \mathfrak{r}_{\mathfrak{p}}x_i$. Then $\mathfrak{L}_{\mathfrak{p}}$ is an $\mathfrak{r}_{\mathfrak{p}}$-lattice in $V_{\mathfrak{p}}$. Let b be a unit of $\mathfrak{r}_{\mathfrak{p}}$ such that $bb^{\rho} = 1$. Let R be a diagonal matrix of degree m whose diagonal elements are $b, 1, \cdots, 1$. Then $R \in U_{\mathfrak{p}}(T)$, $\det(R) = b$, and $\mathfrak{L}_{\mathfrak{p}}R = \mathfrak{L}_{\mathfrak{p}}$. Hence $\mathfrak{L}_{\mathfrak{p}}$ satisfies $(U_{\mathfrak{p}1})$. Let c be a unit of $\mathbf{Z}_p$. Put $e = (c + 1)/2$, $f = (c - 1)/2$. Then e and f are elements of $\mathbf{Z}_p$. In fact, if $p \neq 2$, this is clear. If $p = 2$, we have $c \equiv 1 \equiv -1 \bmod (2\mathbf{Z}_2)$, since c is a unit of $\mathbf{Z}_2$; hence e and f are contained in $\mathbf{Z}_2$. Put $k = m/2$ and

$$S = \begin{pmatrix} e1_k & f1_k \\ f1_k & e1_k \end{pmatrix}.$$

Then $S \in G_{\mathfrak{p}}(T)$ and $\mu(S) = e^2 - f^2 = c$. Since e and f are contained in $\mathfrak{r}_{\mathfrak{p}}$, and $\mu(S)$ is a unit of $\mathfrak{r}_{\mathfrak{p}}$, we have $\mathfrak{L}_{\mathfrak{p}}S = \mathfrak{L}_{\mathfrak{p}}$, so that $\mathfrak{L}_{\mathfrak{p}}$ satisfies $(U_{\mathfrak{p}2})$. Now take a maximal r-lattice $\mathfrak{N}$ in V. Then there exists an r-lattice $\mathfrak{M}$ in V such that

$$\mathfrak{M}_{\mathfrak{p}} = \begin{cases} \mathfrak{N}_{\mathfrak{p}} & \text{for } \mathfrak{p} \notin P\,, \\ \mathfrak{L}_{\mathfrak{p}} & \text{for } \mathfrak{p} \in P\,. \end{cases}$$

By our definition of P and our choice of $\mathfrak{L}_{\mathfrak{p}}$, we see that $\mathfrak{M}$ is of type (U). This proves the proposition.

4.20. PROPOSITION. *Suppose that m is odd, $m > 1$, and T is indefinite in the sense of [6, 5.11]. Put $m = 2l + 1$. Let $\mathfrak{M}$ and $\mathfrak{N}$ be maximal r-lattices,* *and c be a totally positive element of F. Then, there exists an element R of $G(T)$ such that $\mathfrak{M}R = \mathfrak{N}$ and $\mu(R) = c$, if and only if there exists an element b of K such that $c^{-1}[\mathfrak{M}/\mathfrak{N}] = \mathfrak{r}b$ and $bb^\rho = c$.*

For every $R \in G(T)$, we have $N_{K/F}(\det(R)) = \mu(R)^m$. Assume that $\mathfrak{M}R = \mathfrak{N}$ and $\mu(R) = c$. Put $b = c^{-l}\det(R)$. Then $c^{-1}[\mathfrak{M}/\mathfrak{N}] = \mathfrak{r}b$ and $bb^\rho = c$. This proves the 'only if' part. Conversely, assume that $c^{-1}[\mathfrak{M}/\mathfrak{N}] = \mathfrak{r}b$ and $bb^\rho = c$ for some $b \in K$. By [6, 2.4], there exists an element S of $G(T)$ such that $\mu(S) = c$. Then $[\mathfrak{M}S/\mathfrak{N}] = [\mathfrak{M}/\mathfrak{N}] \cdot \det(S)^{-1} = \mathfrak{r}c^l b \cdot \det(S)^{-1}$ and $N_{K/F}(c^l b \cdot \det(S)^{-1}) = 1$. By [6, 5.25 (iii) and 5.27 (i)], there exists an element R of $U(T)$ such that $\mathfrak{M}SR = \mathfrak{N}$. Since $\mu(SR) = \mu(S) = c$, this proves the 'if' part.

4.21. Suppose that m is odd. By an *oriented* $\mathfrak{r}$-*lattice* in V, we mean a couple $(\mathfrak{M}, a)$ formed by a maximal $\mathfrak{r}$-lattice $\mathfrak{M}$ in V and a totally positive element a of F. We say that two oriented $\mathfrak{r}$-lattices $(\mathfrak{M}, a)$ and $(\mathfrak{N}, b)$ are *semi-equivalent* if $a\mu_\mathfrak{r}(\mathfrak{M}) = b\mu_\mathfrak{r}(\mathfrak{N})$, and $\mathfrak{M}$, $\mathfrak{N}$ belong to the same genus with respect to $G(T)$; they are called *equivalent* if there exists an element R of $G(T)$ such that $\mathfrak{M}R = \mathfrak{N}$ and $b\mu(R) = a$, namely the structures $(\mathfrak{M}, aT)$ and $(\mathfrak{N}, bT)$ are isomorphic. By a *class* of oriented $\mathfrak{r}$-lattices, we mean a maximal set of mutually equivalent oriented $\mathfrak{r}$-lattices.

Let $\mathfrak{B}$ be the set of all couples $(\mathfrak{a}, c)$ formed by an ideal $\mathfrak{a}$ in K and a totally positive element c of F such that $N_{K/F}(\mathfrak{a}) = (c)$. Define the product of two elements $(\mathfrak{a}, c)$ and $(\mathfrak{a}', c')$ of $\mathfrak{B}$ by

$$(\mathfrak{a}, c)(\mathfrak{a}', c') = (\mathfrak{a}\mathfrak{a}', cc') \,.$$

Then $\mathfrak{B}$ becomes a group. Let $\mathfrak{B}_0$ be the subgroup of $\mathfrak{B}$ consisting of the couples $(\mathfrak{r}b, bb^\rho)$ for all non-zero elements b of K. Then $\mathfrak{B}/\mathfrak{B}_0$ is nothing but the group $\mathfrak{C}(K)$ of the classes of hermitian forms introduced in [7, 14.5]. It can be easily verified that $[\mathfrak{B} : \mathfrak{B}_0]$ is finite.

Let $(\mathfrak{M}, a)$ be an oriented $\mathfrak{r}$-lattice, and Λ be the set of all oriented $\mathfrak{r}$-lattices semi-equivalent to $(\mathfrak{M}, a)$. If $(\mathfrak{N}, b) \in \Lambda$, then by [6, 5.21 (ii)], $N_{K/F}([\mathfrak{M}/\mathfrak{N}]) = (\mu_\mathfrak{r}(\mathfrak{M})^{-1}\mu_\mathfrak{r}(\mathfrak{N}))^m = (ab^{-1})^m\mathfrak{g}$. Put $m = 2l + 1$. Then $N_{K/F}((a^{-1}b)^l[\mathfrak{M}/\mathfrak{N}]) = (ab^{-1})\mathfrak{g}$. Therefore $((a^{-1}b)^l[\mathfrak{M}/\mathfrak{N}], ab^{-1})$ is an element of $\mathfrak{B}$.

4.22. PROPOSITION. *Suppose that m is odd, $m > 1$, and T is indefinite. Put $m = 2l + 1$. Let $(\mathfrak{M}, a)$ and Λ be as above. Then the mapping*

$$\Lambda \ni (\mathfrak{N}, b) \to \left((a^{-1}b)^l[\mathfrak{M}/\mathfrak{N}], ab^{-1}\right) \in \mathfrak{B}$$

gives a one-to-one correspondence between the classes in Λ and $\mathfrak{B}/\mathfrak{B}_0$.

If $(\mathfrak{a}, c) \in \mathfrak{B}$, $N_{K/F}(c^l\mathfrak{a}) = c^m$. By [6, 5.21 (ii)], there exists a maximal $\mathfrak{r}$-lattice $\mathfrak{N}$, belonging to the same genus as $\mathfrak{M}$ with respect to $G(T)$, such that $[\mathfrak{M}/\mathfrak{N}] =$

$c'\mathfrak{a}$. Put $ac^{-1} = b$. Then $(a^{-1}b)'[\mathfrak{M}/\mathfrak{N}] = \mathfrak{a}$ and $ab^{-1} = c$. This proves the surjectivity of our mapping. Then our proposition follows easily from 4.20.

4.23. PROPOSITION. *Suppose that m is even and T is indefinite. Let $\mathfrak{M}$ be an $\mathfrak{r}$-lattice of type* (U), *and Λ the genus of $\mathfrak{M}$. Let J be the group of all the ideals $\mathfrak{a}$ in K such that $N_{K/F}(\mathfrak{a}) = \mathfrak{g}$ and J_0 be the group of all principal ideals $\mathfrak{x}\mathfrak{a}$ such that $\mathfrak{a}\mathfrak{a}^\rho = 1$. Then the mapping*

$$\Lambda \ni \mathfrak{N} \to [\mathfrak{M}/\mathfrak{N}] \in J$$

gives a one-to-one correspondence between the classes in Λ and J/J_0.

This is a special case of [6, 5.24, 5.28]. We note that there is no irregular prime ideal for Λ on account of $(U_{\mathfrak{F}1})$, so that $[\mathfrak{E}_\Lambda : f_\Lambda(E_0)] = 1$, the notation and terminology being as in [6, 5.22].

5. Discontinuous groups over a field of (Type IV)

5.1. Let F, K, $\{\tau_\nu\}$, Φ, ρ be as in 4.1 and V, T be as in 4.9. We now take K as the algebra L of §2, and treat the families $\Sigma(K, \Phi, \rho; T, \mathfrak{M})$ and the discontinuous groups $\Gamma(T, \mathfrak{M})$, where $\mathfrak{M}$ is a Z-lattice in V satisfying (2.1.1).

Let r_ν and s_ν be as in 4.1. By [5, 2.1 and 2.4 (25)], the family $\Sigma(K, \Phi, \rho; T, \mathfrak{M})$ is not empty if and only if (4.1.1) and the following condition are satisfied.

(5.1.1) *The complex hermitian matrix $\sqrt{-1}\, T^{\tau_\nu}$ has the same signature as*

$$\begin{bmatrix} -1_{r_\nu} & 0 \\ 0 & 1_{s_\nu} \end{bmatrix} \qquad\qquad \textit{for every } \nu\,.$$

Hereafter we assume (4.1.1) and (5.1.1). As for the symmetric domain $\mathcal{H}$ parametrizing $\Sigma(T, \mathfrak{M})$, we have $\dim \mathcal{H} = \sum_{\nu=1}^{g} r_\nu s_\nu$.

5.2. PROPOSITION. *Let K' be as in 4.3, and σ an isomorphism of C into itself which leaves every element of K' invariant. Then, for every member $\mathscr{P}$ of $\Sigma(T, \mathfrak{M})$, $\mathscr{P}^\sigma$ is isomorphic to a member of $\Sigma(cT, \mathfrak{M}')$ for a totally positive element c of F and a Z-lattice $\mathfrak{M}'$. Moreover, if m is even, we can take $c=1$.*

By 1.6 and (2.1.4), $\mathscr{P}^\sigma$ is isomorphic to a member of $\Sigma(T', \mathfrak{M}')$ for some $(T', \mathfrak{M}')$. Since $\Sigma(T', \mathfrak{M}')$ is not empty, $\sqrt{-1}\, T'^{\tau_\nu}$ has the same signature as $\sqrt{-1}\, T^{\tau_\nu}$ for every ν. Let ζ be as in 4.17. Applying [6, 5.9] to ζT and $\zeta T'$, we can conclude, if m is odd, that there exist an element P of $GL_m(K)$ and an element c of F such that $PT'P^\rho = cT$. On account of the signature, c must be totally positive. This proves the case of odd m. Now by 1.11, for every prime ideal $\mathfrak{p}$ of F, there exist an element $R_\mathfrak{p}$ of $GL_m(K_\mathfrak{p})$ and an element $c_\mathfrak{p}$ of $F_\mathfrak{p}$ such that $R_\mathfrak{p}T'R_\mathfrak{p}^\rho = c_\mathfrak{p}T$. Suppose that m is even. By a well-known fact on hermitian forms over $\mathfrak{p}$-adic fields (cf. [6, 3.1 and 4.4]), there exists an element $S_\mathfrak{p}$ of $\mathrm{GL}_m(K_\mathfrak{p})$ such that $S_\mathfrak{p} T S_\mathfrak{p}^\rho = c_\mathfrak{p}^{-1} T$. Therefore T and T' are equivalent over $K_\mathfrak{p}$ for every $\mathfrak{p}$. As observed above, T and T' are equivalent over the completion of K with respect to any infinite prime spot. Applying Hasse's principle on hermitian forms (due to Landherr) to T and T' (or to ζT and $\zeta T'$), we find an element U of $\mathrm{GL}_m(K)$ such that $U T' U^\rho = T$. This completes the proof.

5.3. PROPOSITION. *Let $\mathscr{P}$ be a member of $\Sigma(T, \mathfrak{M})$ and let $\mathscr{P}^\sigma$ be isomorphic to a member of $\Sigma(cT, \mathfrak{M}')$, σ being as in 5.2. Suppose that $\mathfrak{M}$ is an $\mathfrak{r}$-lattice. Then $\mathfrak{M}'$ is an $\mathfrak{r}$-lattice belonging to the same genus as $\mathfrak{M}$ with respect to $G(T)$, and $\mu_T(\mathfrak{M}) = c\mu_T(\mathfrak{M}')$. Moreover, if m is even, $c = 1$, and $\mathfrak{M}$ satisfies $(\mathrm{U}_{\mathfrak{p}2})$, then $\mathfrak{M}'$ belongs to the same genus as $\mathfrak{M}$ with respect to $U(T)$.*

If $\mathscr{P} = (A, \mathcal{C}, \theta)$, we have $\theta(\mathfrak{r}) = \theta(K) \cap \mathrm{End}(A)$, so that $\theta^\sigma(\mathfrak{r}) = \theta^\sigma(K) \cap \mathrm{End}(A^\sigma)$. Hence $\mathfrak{M}'$ is an $\mathfrak{r}$-lattice. By 1.11, for every $\mathfrak{p}$, there exist an element R of $\mathrm{GL}_m(K_\mathfrak{p})$ and a unit e of $Z_\mathfrak{p}$ such that $\mathfrak{M}_\mathfrak{p} R = \mathfrak{M}'_\mathfrak{p}$ and $R(cT)R^\rho = eT$. Then $R \in G_\mathfrak{p}(T)$ and $\mu(R) = c^{-1}e$. Therefore $\mathfrak{M}$ and $\mathfrak{M}'$ belong to the same genus with respect to $G(T)$. By [6, 2.12 and 5.5], we have $c\mu_T(\mathfrak{M}') = \mu_T(\mathfrak{M})$. Now suppose that m is even, $c = 1$, and $\mathfrak{M}$ satisfies $(\mathrm{U}_{\mathfrak{p}2})$. Then there exists an element S of $G_\mathfrak{p}(T)$ such that $\mathfrak{M}_\mathfrak{p} S = \mathfrak{M}_\mathfrak{p}$ and $\mu(S) = e$. Hence $\mathfrak{M}_\mathfrak{p} S R = \mathfrak{M}'_\mathfrak{p}$ and $S R \in U_\mathfrak{p}(T)$. Therefore $\mathfrak{M}$ and $\mathfrak{M}'$ belong to the same genus with respect to $U(T)$.

5.4. LEMMA. *Let U be a variety in a projective space and k be an algebraically closed field. Let e be a positive integer smaller than $\deg(U)$. Let W be the set of the $c(X)$ for all irreducible varieties X of degree e such that, for some positive cycle Y, $X + Y$ is a specialization of U over k. Then W is contained in a Zariski closed set with respect to k whose dimension is less than $\dim_k c(U)$.*

Here we recall that $c(X)$ denotes the Chow point of X. We prove this by induction on $\dim_k c(U)$. If $\dim_k c(U) = 0$, W is empty. Assume that $\dim_k c(U) > 0$. Let Z be the locus of $c(U)$ over k. Then Z is the set of the $c(V)$ for all specializations V of U over k. Let Z' be the set of $c(V)$ for all reducible V. It is well known that Z' is a join of a finite number of varieties $Z_1, \cdots, Z_r$ defined over k. Let $c(V_i)$ be a generic point of Z_i over k. Let $V'_1, \cdots, V'_s$ be the irreducible components of the V_i, of degree $> e$, and $V''_1, \cdots, V''_t$ be the irreducible components of the V_i, of degree e. Let W'_i be the set of the $c(X)$ for all irreducible varieties X of degree e such that, for some positive cycle Y, $X + Y$ is a specialization of V'_i over k. Further, let W''_i be the locus of $c(V''_i)$ over k. It can be easily seen that W is contained in $W'_1 \cup \cdots \cup W'_s \cup W''_1 \cup \cdots \cup W''_t$. Applying our induction to the V'_i, we obtain easily our lemma.

5.5. PROPOSITION. *Let $\mathcal{P}_u$ be a generic member of $\Sigma(T, \mathfrak{M})$, and $\mathfrak{K}_u$ the field of moduli of $\mathcal{P}_u$. Let k_0 be the algebraic closure of $\mathbf{Q}$ in $\mathfrak{K}_u$. Then k_0 is a finite algebraic extension of K', and it is the smallest field of definition for the algebraic variety $\mathcal{V}(T, \mathfrak{M})$ defined in 2.5.*

By 1.7, $\mathfrak{K}_u$ contains K', so that k_0 contains K'. Since $\mathcal{V}(T, \mathfrak{M})$ is the locus of $c(\mathcal{F}_u)$ over $\bar{\mathbf{Q}}$, and since $\mathfrak{K}_u = \mathbf{Q}(c(\mathcal{F}_u))$, we see that k_0 is the smallest field of definition for $\mathcal{V}(T, \mathfrak{M})$.

Now our purpose is to determine k_0.

5.6. PROPOSITION. *Let k_0 be as in 5.5, and σ an isomorphism of k_0 into C which leaves every element of K' invariant. Then the following assertions hold.*

(i) *If m is odd and $\mathfrak{M}$ is a maximal $\mathfrak{r}$-lattice, we have $\mathcal{V}(T, \mathfrak{M})^\sigma = \mathcal{V}(cT, \mathfrak{N})$ for an oriented $\mathfrak{r}$-lattice $(\mathfrak{N}, c)$ which is semi-equivalent to $(\mathfrak{M}, 1)$. The class of $(\mathfrak{N}, c)$ is uniquely determined by σ.*

(ii) *If m is even and $\mathfrak{M}$ is an $\mathfrak{r}$-lattice satisfying $(\mathrm{U}_{\mathfrak{f}_2})$, then $\mathcal{V}(T, \mathfrak{M})^\sigma = \mathcal{V}(T, \mathfrak{N})$ for an $\mathfrak{r}$-lattice $\mathfrak{N}$ belonging to the same genus as $\mathfrak{M}$ with respect to $U(T)$. The class of $\mathfrak{N}$ with respect to $U(T)$ is uniquely determined by σ.*

Let $\mathcal{P}_u$ be as in 5.5. Extend σ to an isomorphism of C into itself and denote it again by σ. By 5.2, $\mathcal{P}_u^\sigma$ is isomorphic to a member $\mathcal{P}(v, cT, \mathfrak{N})$ of $\Sigma(cT, \mathfrak{N})$ for some $(c, \mathfrak{N})$. By 5.3, if $\mathfrak{M}$ is an $\mathfrak{r}$-lattice, $\mathfrak{N}$ is also an $\mathfrak{r}$-lattice. By the same argument as in the proof of 2.8, we obtain $\mathcal{V}(T, \mathfrak{M})^\sigma = \mathcal{V}(cT, \mathfrak{N})$. The assertions of our proposition, except the uniqueness of the class of $(\mathfrak{N}, c)$ or $\mathfrak{N}$, follow easily from this, 5.2, and 5.3. The uniqueness follows from 2.8.

5.7. By 4.10, we can find M, N, $\{\alpha_i\}$ and e satisfying the conditions (i–iii) of that proposition. We need not consider the condition (iv) for the moment. Let $\mathfrak{M}$ be an $\mathfrak{r}$-lattice in V. Fix an isomorphism of (V, T) to (M, h_e). Let $\mathfrak{L}$ be the $\mathfrak{r}$-lattice in M corresponding to $\mathfrak{M}$ by this isomorphism. For each $x \in M$, let $\Phi_0(x)$ be the diagonal matrix of size mg with diagonal elements $x^{\alpha_1}, \cdots, x^{\alpha_{mg}}$. Let w be the mg-dimensional row-vector whose components are all 1. Put $D(\mathfrak{L}) = \{w\Phi_0(x) \mid x \in \mathfrak{L}\}$. By [7, 6.2, Th. 3, Th. 4], the complex torus $C^{mg}/D(\mathfrak{L})$ has a structure of abelian variety, and we can define a Riemann form E on it by

$$E\big(w\Phi_0(a),\, w\Phi_0(b)\big) = \mathrm{Tr}_{M/Q}(eab^\rho) = \mathrm{Tr}_{K/Q}\big(h_e(a, b)\big)$$

for $(a, b) \in M \times M$; and $x \to \Phi_0(x)$ gives an isomorphism of M into $\mathrm{End}_Q(C^n/D(\mathfrak{L}))$. In this way we get a polarized abelian variety $(A, \mathcal{C}, \theta)$ of type $\{M, \Phi_0, \rho; e, \mathfrak{L}\}$, where ρ denotes the non-trivial automorphism of M over N. Since $\{\alpha_i\}$ satisfies (4.6.2), the restriction of Φ_0 to K is equivalent to Φ. Therefore if we denote by θ_1 the restriction of θ to K, $(A, \mathcal{C}, \theta_1)$ is of type $\{K, \Phi, \rho; T, \mathfrak{M}\}$. Hence

(A, C, θ_1) is isomorphic to a member $\mathscr{P}(y)$ of $\Sigma(T, \mathfrak{M})$. We identify $\mathscr{P}(y)$ with (A, C, θ_1).

Let $R \in U(T)$, $y' = R(y)$. By the consideration of [5, 2.7], there exists an isogeny λ of $\mathscr{P}(y, T, \mathfrak{M})$ to $\mathscr{P}(y', T, \mathfrak{M}) = (A', C', \theta_1')$. We can define an isomorphism θ' of M into $\mathrm{End}_Q(A')$ by $\theta'(x)\lambda = \lambda\theta(x)$ for $x \in M$. It can be easily seen that (A', C', θ') belongs to $\{M, \Phi_0, \rho\}$. We note that the set $\{R(y) \mid R \in U(T)\}$ is dense in $\mathscr{H}$; this is easily shown by using the Cayley parametrization for the unitary group and the approximation theorem in algebraic number fields.

5.8. PROPOSITION. *Let $\mathscr{P} = (A, C, \theta)$ and θ_1 be as above, and $(M'; \{\beta_j\})$ be the dual of $(M; \{\alpha_i\})$. Then there exists a Galois extension k of M' and a positive integer b with the following property:*

Let $\mathfrak{p}$ be a prime ideal of M' relatively prime to b, and $\mathfrak{P}$ be a prime ideal of k dividing $\mathfrak{p}$. Let σ be an isomorphism of C into itself which induces on k a Frobenius automorphism for $\mathfrak{P}/\mathfrak{p}$. Let $(A^\sigma, C^\sigma, \theta_1^\sigma)$ be isomorphic to a member of $\Sigma(cT, \mathfrak{N})$. Then there exists an element d of K such that

$$[\mathfrak{M}/\mathfrak{N}] = d \cdot N_{M/K}\left(\prod_j \mathfrak{p}^{\beta_j}\right)^{-1}, \qquad dd^\rho = c^{-m} \cdot N(\mathfrak{p})^m .$$

Let $\mathfrak{o}$ be the ring of all integers in M. By [7, 7.1, Prop. 7, 17.1], there exists a polarized abelian variety $\mathscr{P}' = (A', C', \theta')$ of type $\{M, \Phi_0, \rho\}$, isogenous to $\mathscr{P}$, such that $\theta'(\mathfrak{o}) = \theta'(M) \cap \mathrm{End}(A')$. Let μ be an isogeny of $\mathscr{P}'$ to $\mathscr{P}$. We may assume that $\mathscr{P}'$ is defined over a Galois extension k of M' [7, 12.4, Prop. 26]. There exists a positive integer f such that $\mu(t) = 0$ with $t \in A'$ only if $ft = 0$. By [7, 7.5, Prop. 21], we can find a point t_0 on A such that

$$a \in f\mathfrak{o} \Longleftrightarrow \theta'(a)t_0 = 0 .$$

We take k so large that t_0 is rational over k. By [7, 15.2–3, 16.3], there exists a positive integer f' with the following property: Let $\mathfrak{p}$ be a prime ideal of M' not dividing ff', and $\mathfrak{P}$ be a prime factor of $\mathfrak{p}$ in k; put $\mathfrak{q} = \prod_j \mathfrak{p}^{\beta_j}$; let σ be a Frobenius automorphism of k for $\mathfrak{P}/\mathfrak{p}$; then there exists an isogeny λ of $\mathscr{P}'$ to $\mathscr{P}'^\sigma$ such that

$$\lambda t_0 = t_0^\sigma , \qquad \lambda^{-1}(Y^\sigma) \equiv N(\mathfrak{p}) \cdot Y , \qquad \mathrm{Ker}(\lambda) \cong \mathfrak{o}/\mathfrak{q} ,$$

where Y is a basic polar divisor in C', and $\equiv$ means algebraic equivalence. In [7, Ch. IV], the abelian variety in question is assumed to be simple. However, the proof given in [7, 15.2–3] is applicable also to a non-simple abelian variety, without any modification. Let $\mathfrak{p}$, $\mathfrak{P}$, $\mathfrak{q}$, σ, λ be as above, and let $\mathfrak{g}(f, A') = \{t \in A' \mid ft = 0\}$. Then $\mathrm{Ker}(\mu) \subset \mathfrak{g}(f, A')$. By [7, 7.5, Prop. 20], we have $\mathfrak{g}(f, A') = \theta'(\mathfrak{o})t_0$, so that there exists a submodule $\mathfrak{m}$ of $\mathfrak{o}$ such that $\mathfrak{o} \supset \mathfrak{m} \supset f\mathfrak{o}$ and $\mathrm{Ker}(\mu) = \theta'(\mathfrak{m})t_0$. Then we have $\mathrm{Ker}(\mu^\sigma) = \theta'^\sigma(\mathfrak{m})t_0^\sigma = \theta'^\sigma(\mathfrak{m})\lambda t_0 = \lambda\theta'(\mathfrak{m})t_0 = \lambda(\mathrm{Ker}(\mu))$, so that, if $\mu t = 0$, then $\mu^\sigma(\lambda t) = 0$. Hence there exists an isogeny λ_0 of A to

A^σ such that $\mu^\sigma\lambda = \lambda_0\mu$. It can be easily seen that λ_0 is an isogeny of $\mathscr{P}$ to $\mathscr{P}^\sigma$ [2, 2.3, Prop. 2.4]; and if X is a basic polar divisor in $\mathcal{C}$, we have

$$\lambda_0^{-1}(X^\sigma) \equiv N(\mathfrak{p})\cdot X .$$

Since $\mathrm{Ker}\,(\mu)$ and $\mathrm{Ker}\,(\mu^\sigma)$ have the same order, and since $\lambda_0\mu = \mu^\sigma\lambda$, we see that $\mathrm{Ker}\,(\lambda_0)$ and $\mathrm{Ker}\,(\lambda)$ have the same order. If $\lambda t = 0$, then $\lambda_0(\mu t) = \mu^\sigma\lambda t = 0$, so that μ gives an $\mathfrak{r}$-homomorphism of $\mathrm{Ker}\,(\lambda)$ into $\mathrm{Ker}\,(\lambda_0)$. If $\lambda t = 0$ and $\mu t = 0$, we have $N_{M/.}(\mathfrak{q})t = 0$ and $ft = 0$, so that $t = 0$, since $\mathfrak{p}$ is prime to f. Therefore μ gives an $\mathfrak{r}$-isomorphism of $\mathrm{Ker}\,(\lambda)$ onto $\mathrm{Ker}\,(\lambda_0)$. It follows that $\mathrm{Ker}\,(\lambda_0)$ is isomorphic to $\mathfrak{o}/\mathfrak{q}$ as $\mathfrak{r}$-module. Now $(A, \mathcal{C}, \theta_1)$ determines the structure $(V, T, \mathfrak{M})$. Let $(A^\sigma, \mathcal{C}^\sigma, \theta_1^\sigma)$ determine $(V, cT, \mathfrak{N})$ as in our proposition. Let R be the linear transformation of V corresponding to λ_0. Then $R(cT)R^\rho = N(\mathfrak{p})\cdot T$, and $\mathrm{Ker}\,(\lambda_0)$ is isomorphic to $\mathfrak{N}/\mathfrak{M}R$ as $\mathfrak{r}$-module. We have therefore $[\mathfrak{N}/\mathfrak{M}R] = N_{M/K}(\mathfrak{q})$. Put $d = \det(R)$. Then we get $[\mathfrak{M}/\mathfrak{N}] = d\cdot N_{M/K}(\mathfrak{q})^{-1}$ and $c^m dd^\rho = N(\mathfrak{p})^m$. Our proposition is thereby proved.

5.9. Let $\{\sigma_\lambda\}$ be as in 4.4, and v_λ be as in 4.7. Suppose that m is even. Let H be the group of ideals in K' defined by

$$(5.9.1)\qquad H = \left\{ \mathfrak{x} \;\middle|\; \prod_{\lambda=1}^{g'}\left(\frac{\mathfrak{x}^{\sigma_\lambda}}{\mathfrak{x}^{\sigma_\lambda\rho}}\right)^{v_\lambda} = (a)\ \text{ for some } a \in K \text{ such that } aa^\rho = 1\right\} .$$

Let x be an element of K'. Put $y = \prod_{\lambda=1}^{g'}(x^{\sigma_\lambda}/x^{\sigma_\lambda\rho})^{v_\lambda}$. By 4.5, y is an element of K, and $yy^\rho = 1$. This shows that H contains all the principal ideals of K'. Therefore H is an ideal group in K' defined modulo (1).

Next suppose that m is odd. Define a group H of ideals in K' by

$(5.9.2)$

$$H = \left\{ \mathfrak{x} \;\middle|\; \prod_{\lambda=1}^{g'}\mathfrak{x}^{\sigma_\lambda}\cdot\left(\frac{\mathfrak{x}^{\sigma_\lambda}}{\mathfrak{x}^{\sigma_\lambda\rho}}\right)^{v_\lambda} = (a)\ \text{ for some } a \in K \text{ such that } N(\mathfrak{x}) = aa^\rho\right\} .$$

By the same argument as above, we see that H is an ideal group in K' defined modulo (1).

5.10. THEOREM. *Let F, K, g, $\{\tau_\nu\}$, r_ν, s_ν, m, Φ, ρ be as in 4.1, and V, T be as in 4.9. Suppose that $m > 1$ and $\sum_{\nu=1}^{g} r_\nu s_\nu > 0$. Let K' be as in 4.3. When $K' \neq Q$, define isomorphisms $\sigma_1, \cdots, \sigma_{g'}$ of K' into C as in 4.4, and define integers $v_1, \cdots, v_{g'}$ as in 4.7. Let $\mathfrak{M}$ be an $\mathfrak{r}$-lattice in V, and $\mathfrak{V}(T, \mathfrak{M})$ the algebraic variety defined in 2.5. Let k_0 be the smallest field of definition for $\mathfrak{V}(T, \mathfrak{M})$ (cf. 5.5). Then k_0 is determined as follows.*

(Case I: m even; $r_\nu = s_\nu$ for every ν). Suppose that $\mathfrak{M}$ is of type (U) (cf. 4.18). Then $k_0 = K' = Q$.

(Case II: m even; $r_\nu \neq s_\nu$ for some ν). Suppose that $\mathfrak{M}$ is of type (U). Then k_0 is an unramified class-field over K' corresponding to the ideal group

H defined by (5.9.1). Moreover, if $\sigma = \left(\dfrac{k_0/K'}{\mathfrak{a}}\right)$, we have $\mathcal{V}(T, \mathfrak{M})^\sigma = \mathcal{V}(T, \mathfrak{N})$ for an $\mathfrak{r}$-lattice $\mathfrak{N}$, belonging to the same genus as $\mathfrak{M}$ with respect to $U(T)$, such that $[\mathfrak{N}/\mathfrak{M}]$ and $\prod_{\lambda=1}^{g'}(\mathfrak{a}^{\sigma_\lambda}/\mathfrak{a}^{\sigma_\lambda\rho})^{v_\lambda}$ respresent the same element of J/J_0, where J and J_0 are the groups defined in 4.23.

(Case III: m odd). *Suppose that $\mathfrak{M}$ is maximal (cf. 4.17). Then k_0 is an unramified class-field over K' corresponding to the ideal group H defined by (5.9.2). Moreover, if $\sigma = \left(\dfrac{k_0/K'}{\mathfrak{a}}\right)$, we have $\mathcal{V}(T, \mathfrak{M})^\sigma = \mathcal{V}(cT, \mathfrak{N})$ for an oriented $\mathfrak{r}$-lattice $(\mathfrak{N}, c)$, semi-equivalent to $(\mathfrak{M}, 1)$ (cf. 4.21), such that $(c^{-l}[\mathfrak{N}/\mathfrak{M}], c)$ and $(\prod_{\lambda=1}^{g'}\mathfrak{a}^{\sigma_\lambda}\cdot(\mathfrak{a}^{\sigma_\lambda}/\mathfrak{a}^{\sigma_\lambda\rho})^{v_\lambda}, N(\mathfrak{a}))$ represent the same element of $\mathfrak{B}/\mathfrak{B}_0$, where $l = (m-1)/2$ and $\mathfrak{B}, \mathfrak{B}_0$ are the groups defined in 4.21.*

We first consider even m. Let Λ be the genus of $\mathfrak{M}$ with respect to $U(T)$. Let $\nu(T, \mathfrak{M})$ and $N(T, \mathfrak{M})$ be as in 2.2, and $e(T, \mathfrak{M})$ be as in 2.5. Let $\mathfrak{N}$ be an $\mathfrak{r}$-lattice in V such that

$$(5.10.1) \qquad \mathfrak{N} \in \Lambda, \ \nu(T, \mathfrak{N}) = \nu(T, \mathfrak{M}), \ N(T, \mathfrak{N}) = N(T, \mathfrak{M}),$$

$$(5.10.2) \qquad e(T, \mathfrak{N}) > e(T, \mathfrak{M}).$$

Let $W(\mathfrak{N})$ be the set of the $c(X)$ for all irreducible varieties X of degree $e(T, \mathfrak{M})$ such that $X + Y$ is a specialization of $\mathcal{F}(v, T, \mathfrak{N})$ over $\bar{\mathbf{Q}}$ for some positive cycle Y, where $\mathcal{P}(v, T, \mathfrak{N})$ is a generic member of $\Sigma(T, \mathfrak{N})$. By 5.4, the join of $W(\mathfrak{N})$ for all $\mathfrak{N}$ satisfying (5.10.1,2) is contained in a Zariski closed set W_1 over $\bar{\mathbf{Q}}$ whose dimension is less than $\dim \mathcal{K} (=\sum_{v=1}^{g} r_v s_v)$. W_1 has the same ambient space as $\mathcal{V}(T, \mathfrak{M})$. Put

$$W_2 = \bigcup \left(\mathcal{V}(T, \mathfrak{M}) \cap \mathcal{V}(T, \mathfrak{N})\right),$$

where $\mathfrak{N}$ runs through all $\mathfrak{r}$-lattices $\mathfrak{N}$ satisfying (5.10.1) and

$$(5.10.3) \qquad e(T, \mathfrak{N}) = e(T, \mathfrak{M}),$$

$$(5.10.4) \qquad \mathcal{V}(T, \mathfrak{N}) \neq \mathcal{V}(T, \mathfrak{M}).$$

The condition (5.10.4) is satisfied if and only if $\mathfrak{N}$ and $\mathfrak{M}$ belong to distinct classes with respect to $U(T)$, on account of 2.8. Let W be the join of all the conjugates of $W_1 \cup W_2$ over $\mathbf{Q}$. By 2.8, there exists a point y in $\mathcal{J}(T, \mathfrak{M})$ such that

$$(5.10.5) \qquad c\big(\mathcal{F}(y, T, \mathfrak{M})\big) \in \mathcal{V}_0(T, \mathfrak{M}) - W,$$

notation being as in 2.5–8. Let k_1 be $\mathbf{Q}$ or the class-field over K' corresponding to the ideal group H defined by (5.9.1), according as $K' = \mathbf{Q}$ or $K' \neq \mathbf{Q}$. By 4.10, we can find M, N, $\{\alpha_i\}$ and e satisfying the conditions (i–iv) of that proposition, taking as P the composite of k_0 and k_1. By the consideration of 5.7, we may take y as follows: there exists a polarized abelian variety $(A, \mathcal{C}, \theta)$ of type $\{M, \Phi_0, \rho\}$ such that $\mathcal{P}(y, T, \mathfrak{M}) = (A, \mathcal{C}, \theta_1)$, where Φ_0 is as in 5.7 and

θ_1 is the restriction of θ to K. Let $\mathfrak{a}$ be an ideal in K', and let $\tau = \left(\dfrac{k_0/K'}{\mathfrak{a}}\right)$. Since M' and k_1 are linearly disjoint over K', we can extend τ to an isomorphism σ of C into itself which leaves every element of M' invariant. By 5.3, $\mathcal{P}(y, T, \mathfrak{M})^\sigma$ is isomorphic to a member $\mathcal{P}(z, T, \mathfrak{N})$ of $\Sigma(T, \mathfrak{N})$ for some $\mathfrak{N} \in \Lambda$. We have then $\nu(T, \mathfrak{N}) = \nu(T, \mathfrak{M})$, $N(T, \mathfrak{N}) = N(T, \mathfrak{M})$, $\mathcal{F}(y, T, \mathfrak{M})^\sigma = \mathcal{F}(z, T, \mathfrak{N})$, and

$$e(T, \mathfrak{M}) = \deg \mathcal{F}(y, T, \mathfrak{M}) = \deg \mathcal{F}(z, T, \mathfrak{N}) \leq e(T, \mathfrak{N}) \, .$$

Assume that $\deg \mathcal{F}(z, T, \mathfrak{N}) < e(T, \mathfrak{N})$. Then we have $c(\mathcal{F}(z, T, \mathfrak{N})) \in W(\mathfrak{N})$ on account of 2.4, so that $c(\mathcal{F}(y, T, \mathfrak{M})) \in W(\mathfrak{N})^{\sigma^{-1}} \subset W$, which contradicts (5.10.5). Therefore we must have $e(T, \mathfrak{M}) = e(T, \mathfrak{N})$, so that $c(\mathcal{F}(z, T, \mathfrak{N})) \in \mathcal{V}(T, \mathfrak{N})$. Hence $c(\mathcal{F}(y, T, \mathfrak{M})) \in \mathcal{V}(T, \mathfrak{N})^{\sigma^{-1}}$. By 5.6, $\mathcal{V}(T, \mathfrak{N})^{\sigma^{-1}} = \mathcal{V}(T, \mathfrak{L})$ for some $\mathfrak{L} \in \Lambda$. Then $c(\mathcal{F}(y, T, \mathfrak{M}))$ is contained in $\mathcal{V}(T, \mathfrak{L}) \cap \mathcal{V}(T, \mathfrak{M})$. By (5.10.5), we must have $\mathcal{V}(T, \mathfrak{L}) = \mathcal{V}(T, \mathfrak{M})$, so that $\mathcal{V}(T, \mathfrak{M})^\sigma = \mathcal{V}(T, \mathfrak{N})$.

For (A, C, θ), take k and b with the property of 5.8. Let Ω be a Galois extension of K' containing k, k_0 and k_1. We can find a prime ideal $\mathfrak{p}$ of M' not dividing b, and a prime factor $\mathfrak{P}$ of $\mathfrak{p}$ in Ω such that σ induces on Ω a Frobenius automorphism for $\mathfrak{P}/\mathfrak{p}$. Put $N_{M'/K'}(\mathfrak{p}) = \mathfrak{q}$. By 4.7 and 5.8, there exists an element d of K such that

$$(5.10.6) \qquad [\mathfrak{M}/\mathfrak{N}] = \begin{cases} (d \cdot N(\mathfrak{p})^{-m/2}) & \text{if } K' = \mathbf{Q}\,, \\[2mm] d \cdot N(\mathfrak{p})^{-m/2} \prod_{\lambda=1}^{g'} \left(\dfrac{\mathfrak{q}^{\sigma_\lambda \rho}}{\mathfrak{q}^{\sigma_\lambda}}\right)^{v_\lambda} & \text{if } K' \neq \mathbf{Q}\,, \end{cases}$$

$$(5.10.7) \qquad\qquad dd^\rho = N(\mathfrak{p})^m \, .$$

Put $a = d \cdot N(\mathfrak{p})^{-m/2}$. Then $aa^\rho = 1$.

Suppose that $k' = \mathbf{Q}$. By 4.23, (5.10.6) shows that $\mathfrak{M}$ and $\mathfrak{N}$ belong to the same class with respect to $U(T)$. Hence $\mathcal{V}(T, \mathfrak{M})^\sigma = \mathcal{V}(T, \mathfrak{M})$ for every isomorphism σ which is the identity on M'. Since M' and k_0 are linearly disjoint over $\mathbf{Q}$, this shows $k_0 = \mathbf{Q}$.

Next suppose that $K' \neq \mathbf{Q}$. By our choice of $\mathfrak{p}$, we have $\left(\dfrac{k_1/K'}{\mathfrak{a}}\right) = \left(\dfrac{k_1/K'}{\mathfrak{q}}\right)$, so that $\mathfrak{a}^{-1}\mathfrak{q} \in H$, hence $\prod_{\lambda=1}^{g'} (\mathfrak{a}^{\sigma_\lambda}/\mathfrak{a}^{\sigma_\lambda \rho})^{v_\lambda}$ and $\prod_{\lambda=1}^{g'} (\mathfrak{q}^{\sigma_\lambda}/\mathfrak{q}^{\sigma_\lambda \rho})^{v_\lambda}$ represent the same element of J/J_0. Therefore, by (5.10.6), $[\mathfrak{N}/\mathfrak{M}]$ and $\prod_{\lambda=1}^{g'} (\mathfrak{a}^{\sigma_\lambda}/\mathfrak{a}^{\sigma_\lambda \rho})^{v_\lambda}$ represent the same element of J/J_0. Hence we observe that

$$\sigma = 1 \text{ on } k_1 M' \Longleftrightarrow \left(\dfrac{k_1/K'}{\mathfrak{a}}\right) = 1 \Longleftrightarrow \mathfrak{a} \in H$$

$$\Longleftrightarrow \prod_{\lambda=1}^{g'} \left(\dfrac{\mathfrak{a}^{\sigma_\lambda}}{\mathfrak{a}^{\sigma_\lambda \rho}}\right)^{v_\lambda} \in J_0 \qquad\qquad (5.9.1)$$

$$\Longleftrightarrow [\mathfrak{N}/\mathfrak{M}] \in J_0$$

$$\Longleftrightarrow \mathcal{V}(T, \mathfrak{M}) = \mathcal{V}(T, \mathfrak{N}) \qquad\qquad (2.8,\ 4.23)$$

$$\Longleftrightarrow \sigma = 1 \text{ on } k_0 M' \, .$$

It follows that $k_0 M' = k_1 M'$. Since M' and $k_0 k_1$ are linearly disjoint over K', we obtain $k_0 = k_1$. This proves the assertions in (Case II).

It remains to consider (Case III). In this case, instead of the genus of $\mathfrak{M}$, we have to consider the set Λ' of all oriented r-lattices $(\mathfrak{N}, c)$ semi-equivalent to $(\mathfrak{M}, 1)$, and take into account the $\mathcal{O}(cT, \mathfrak{N})$ for all $(\mathfrak{N}, c) \in \Lambda'$. Then, in view of the results 4.7, 5.8, 4.22, we can prove the assertions of (Case III) by the same argument as in (Case II).

5.11. COROLLARY. *Let T and $\mathfrak{M}$ be as in 5.10. Let $f_1, \cdots, f_\kappa$ be as in* [5, Th. 3]. *Then the field k_0 determined in 5.10 is algebraically closed in* $Q(f_1, \cdots, f_\kappa)$, *and*

$$\dim_{k_0} k_0(f_1, \cdots, f_\kappa) = \dim_C C(f_1, \cdots, f_\kappa) = \dim \mathcal{K} .$$

This follows immediately from 5.5, since $Q(f_1, \cdots, f_\kappa)$ is isomorphic to the field of moduli of a generic member of $\Sigma(T, \mathfrak{M})$.

5.12. REMARK. We can make the same remark as in 3.5. Namely, if $\Gamma(T, \mathfrak{M})\backslash\mathcal{K}$ is compact, the field of all automorphic functions with respect to $\Gamma(T, \mathfrak{M})$ has k_0 as a field of definition. If $m > 2$, we know that $\Gamma(T, \mathfrak{M})\backslash\mathcal{K}$ is compact if and only if $r_\nu s_\nu = 0$ for some ν. If $m = 2$, $\Gamma(T, \mathfrak{M})$ is commensurable with a group of units in an order in a quaternion algebra L over F. Then $\Gamma(T, \mathfrak{M})\backslash\mathcal{K}$ is compact if and only if L is a division algebra. If L is totally indefinite, this is commensurable to the group mentioned in the last part of 3.5.

OSAKA UNIVERSITY AND PRINCETON UNIVERSITY

REFERENCES

1. G. SHIMURA, *On the theory of automorphic functions*, Ann. of Math. 70 (1959), 101-144.
2. ———, *On the zeta-functions of the algebraic curves uniformized by certain automorphic functions*, J. Math. Soc. Japan 13 (1961), 275-331.
3. ———, *On the class-fields obtained by complex multiplication of abelian varieties*, Osaka Math. J. 14 (1962), 33-44.
4. ———, *Arithmetic of alternating forms and quaternion hermitian forms*, J. Math. Soc. Japan 15 (1963), 33-65.
5. ———, *On analytic families of polarized abelian varieties and automorphic functions*, Ann. of Math. 78 (1963), 194-192.
6. ———, *Arithmetic of unitary groups*, Ann. of Math. 79 (1964), 369-409.
7. ——— and Y. TANIYAMA, Complex multiplication of abelian varieties and its applications to number theory, Publ. Math. Soc. Japan, No. 6, 1961.
8. A. WEIL, Variétés abéliennes et courbes algébriques, Hermann, Paris, 1948.

(Received June 24, 1963)

Class-fields and automorphic functions

Annals of Mathematics, 80 (1964), 444-463

Introduction

Let F be a totally real algebraic number field of degree g, and D a quaternion algebra over F. Then we have an isomorphism

$$(1) \qquad D \otimes_Q R \cong M_2(R) \times \cdots \times M_2(R) \times K \times \cdots \times K ,$$

where $M_2(R)$ is the total matric algebra of degree 2 over R, and K is the division ring of real quaternions. Let t be the number of copies of $M_2(R)$ in (1); then the number of copies of K is $g - t$. We assume that $t > 0$. For every maximal order $\mathfrak{o}$ in D, put

$$(2) \qquad \Gamma(\mathfrak{o}) = \{a \in \mathfrak{o} \mid aa^{\iota} = 1\} ,$$

where ι is the canonical involution of D. The projection of $\Gamma(\mathfrak{o})$ to the factor $M_2(R) \times \cdots \times M_2(R)$ (t copies) of (1) is contained in $\mathrm{SL}_2(R) \times \cdots \times \mathrm{SL}_2(R)$. In this way we may consider $\Gamma(\mathfrak{o})$ as a discontinuous group of transformations operating on the product $\mathfrak{H}^t$ of t copies of the upper half plane $\mathfrak{H} = \{z \in C \mid \mathrm{Im}\,(z) > 0\}$. Let $Y(\mathfrak{o})$ be an algebraic variety whose function field over C is isomorphic to the field of automorphic functions on $\mathfrak{H}^t$ with respect to $\Gamma(\mathfrak{o})$. If $t = g$, we can take $Y(\mathfrak{o})$ so as to be defined over Q [4, 8]. Now assume that $1 \le t < g$. The purpose of the present paper is to show the following fact: a certain class-field occurs as *the bottom field*[1] for $Y(\mathfrak{o})$, which bounds the fields of definition for the varieties birationally equivalent to $Y(\mathfrak{o})$; and the reciprocity-law for the class-field is described in terms of the conjugate class of $\Gamma(\mathfrak{o})$. If $t = 1$, $Y(\mathfrak{o})$ is an algebraic curve, and the bottom field for $Y(\mathfrak{o})$ is *the field of moduli*[1] of $Y(\mathfrak{o})$. Our result will be stated as Main Theorems I and II (4.3, 4.9) in the text. Since we need many symbols in order to explain this in the general case, we give result here only in the case $t = 1$.

THEOREM 1. *Let $I(D/F)$ be the ideal-group in F generated by the following three kinds of ideals:*

(i) *all the principal ideals (a) such that a is totally positive;*

(ii) *the squares of all the ideals in F;*

(iii) *all the prime ideals in F which are ramified in D.*[2]

Let $k(D)$ be the class-field over F corresponding to $I(D/F)$. Then the field of

---

[1] For a precise definition, see 2.1.

[2] A prime ideal $\mathfrak{p}$ in F is said to be ramified in D if $D \otimes_F F_{\mathfrak{p}}$ is a division algebra, where F is the completion of F with respect to $\mathfrak{p}$.

moduli of the algebraic curve $Y(\mathfrak{o})$ is contained in $k(D)$. Moreover, if the multiplicative group of regular elements in D has no element of finite order other than ± 1, then $k(D)$ is exactly the composite of F and the field of moduli of $Y(\mathfrak{o})$.

To get an explicit form of the reciprocity-law for $k(D)/F$, we consider the conjugate class of $\Gamma(\mathfrak{o})$ with respect to the inner automorphisms of $SL_2(R)$. For every right $\mathfrak{c}$-ideal $\mathfrak{x}$ in D, we denote by $N(\mathfrak{x})$ the ideal in F generated by the elements xx^ι for all $x \in \mathfrak{x}$. Let $\mathfrak{o}_1$ and $\mathfrak{o}_2$ be two maximal orders in D. We can find a right $\mathfrak{o}_1$-ideal $\mathfrak{x}$ which is a left $\mathfrak{o}_2$-ideal at the same time. Denote by $\{\mathfrak{o}_1 : \mathfrak{o}_2\}$ the ideal-class of $N(\mathfrak{x})$ modulo $I(D/F)$. The symbol $\{\mathfrak{o}_1 : \mathfrak{o}_2\}$ has the following properties:

(i) $\{\mathfrak{o}_1 : \mathfrak{o}_2\}\{\mathfrak{o}_2 : \mathfrak{o}_3\} = \{\mathfrak{o}_1 : \mathfrak{o}_3\}$;

(ii) $\{\mathfrak{o}_1 : \mathfrak{o}_2\} = 1$ if and only if there exists an element α of $SL_2(R)$ such that $\alpha\Gamma(\mathfrak{o}_1)\alpha^{-1} = \Gamma(\mathfrak{o}_2)$.

Now the reciprocity-law is described as follows.

THEOREM 2. *Let $\mathfrak{a}$ be an ideal in F, and σ an isomorphism of a field of definition for the curve $Y(\mathfrak{o})$ into C such that $\sigma = \left(\dfrac{k(D)/F}{\mathfrak{a}}\right)$ on $k(D)$. Then $Y(\mathfrak{o})^\sigma$ is birationally equivalent to $Y(\mathfrak{o}')$ with a maximal order $\mathfrak{o}'$ in D such that $\{\mathfrak{o} : \mathfrak{o}'\}$ is the class of $\mathfrak{a}$ modulo $I(D/F)$.*

One may regard the above theorems as an analogue of the classical theory of complex multiplication, where the absolute class-field over an imaginary quadratic field occurs as the field of moduli of an elliptic curve, and the reciprocity-law is given in terms of isogenies of elliptic curves.

If $t > 1$, our result has a feature quite similar to the theory of complex multiplication of abelian varieties of higher dimension. In fact, we get a class-field over a certain totally real field F^* which is not necessarily equal to F.

In a previous paper [8], we investigated certain algebraic varieties $\mathcal{V}(T,\mathfrak{M})$ uniformized by automorphic functions. The variety $Y(\mathfrak{o})$ is a special case of those varieties. In the main theorem of [8], we obtained a class-field over a totally imaginary quadratic extension K' of a totally real algebraic number field as a field of definition for $\mathcal{V}(T, \mathfrak{M})$. If $\mathcal{V}(T, \mathfrak{M}) = Y(\mathfrak{o})$, taking the bottom field instead of the field of definition, we can reduce the class-field over K' to a class-field over F^*. In the present investigation, we have included also some results concerning the bottom field for the variety $\mathcal{V}(T, \mathfrak{M})$ which is not necessarily of the type $Y(\mathfrak{o})$ (2.5, 2.7).

Throughout the paper, we shall use the same notation and terminology as in [7, 8]. In particular, $M_n(S)$ means the total matrix ring of degree n over an associative ring S. In §1, the lattices in a vector space are denoted by roman letters L, M, etc., in accordance with [7]. But in §§ 2–4, they are denoted by German letters $\mathfrak{M}$, $\mathfrak{N}$, etc., in accordance with [8].

1. Supplement to [7]

1.1. Let F be an algebraic number field of finite degree, and K a quadratic extension of F. We denote by ρ the non-trivial automorphism of K over F. Let V be a vector space over K of dimension m, and $\Phi(x, y)$ a non-degenerate hermitian form on V. As in [7], we define the group $\mathrm{G}(V, \Phi)$ of similitudes of Φ and its subgroups $\mathrm{U}(V, \Phi)$, $\mathrm{SU}(V, \Phi)$, $\mathrm{H}(V, \Phi)$ [7, 1.1, 2.2]. For simplicity, we put $\mathrm{G}(\Phi) = \mathrm{G}(V, \Phi)$, $\mathrm{U}(\Phi) = \mathrm{U}(V, \Phi)$, $\mathrm{SU}(\Phi) = \mathrm{SU}(V, \Phi)$, $\mathrm{H}(\Phi) = \mathrm{H}(V, \Phi)$, and $\mathrm{G}_{\mathfrak{p}}(\Phi) = \mathrm{G}(V_{\mathfrak{p}}, \Phi)$, $\mathrm{U}_{\mathfrak{p}}(\Phi) = \mathrm{U}(V_{\mathfrak{p}}, \Phi)$, $\mathrm{SU}_{\mathfrak{p}}(\Phi) = \mathrm{SU}(V_{\mathfrak{p}}, \Phi)$, $\mathrm{H}_{\mathfrak{p}}(\Phi) = \mathrm{H}(V_{\mathfrak{p}}, \Phi)$ for every prime ideal $\mathfrak{p}$ in F [7, 5.4]. We denote by $\mathfrak{g}$ and $\mathfrak{r}$ the rings of integers in F and in K, respectively.

1.2. For every prime ideal $\mathfrak{p}$ in F, let $E_{\mathfrak{p}}$ denote the group of units in $\mathfrak{r}_{\mathfrak{p}}$ and let

$$E_{0\mathfrak{p}} = \{e \in E_{\mathfrak{p}} \mid ee^{\rho} = 1\}, \quad E_1 = \{e^{-1}e^{\rho} \mid e \in E_{\mathfrak{p}}\} \, .$$

Suppose that m is even. Let P be the set of all the prime ideals $\mathfrak{p}$ in F ramified in K and such that $V_{\mathfrak{p}}$ has the trivial kernel subspace (cf. [7, 4.3]. Such a $\mathfrak{p}$ is called irregular for Φ in [7, 5.23].) We denote by $\mathfrak{E}$ the product of the factor groups $E_{0\mathfrak{p}}/E_{1\mathfrak{p}}$ for all $\mathfrak{p} \in P$. For an element x of K such that $xx^{\rho} = 1$, we denote by $f(x)$ the element of $\mathfrak{E}$ whose components are represented by x. Further we denote by E_0 the group of units u in $\mathfrak{r}$ such that $uu^{\rho} = 1$.

1.3. PROPOSITION. *Suppose that m is even. Let the notation be as above. If either P is empty, or P consists of only one prime ideal $\mathfrak{p}$ and $\mathfrak{p}$ does not divide 2, then the equality $f(E_0) = \mathfrak{E}$ holds.*

If P is empty, there is no question. Suppose that P has only one element $\mathfrak{p}$, and $\mathfrak{p}$ does not divide 2. Then $[E_{0\mathfrak{p}} : E_{1\mathfrak{p}}] = 2$ [7, 4.16], and $\mathfrak{E} = E_{0\mathfrak{p}}/E_{1\mathfrak{p}}$. If $-1 \in E_{1\mathfrak{p}}$, we have $e^{\rho} = -e$ for some unit e in $\mathfrak{r}_{\mathfrak{p}}$. This is a contradiction, since $\mathfrak{p}$ is ramified in K and $\mathfrak{p}$ does not divide 2. Therefore $-1 \notin E_{1\mathfrak{p}}$. It follows that $f(E_0) = \mathfrak{E}$, since $-1 \in E_0$.

1.4. PROPOSITION. *Let L be a maximal $\mathfrak{r}$-lattice in V, and Λ the genus of L with respect to $\mathrm{U}(\Phi)$. Let J be the group of all ideals $\mathfrak{a}$ in K such that $N_{K/F}(\mathfrak{a}) = \mathfrak{g}$ and J_0 the group of all principal ideals $\mathfrak{r}\alpha$ such that $\alpha\alpha^{\rho} = 1$. Suppose that Φ is indefinite, m is even, and $\mathfrak{E} = f(E_0)$. Then the mapping $\Lambda \ni M \to [L/M] \in J$ gives a one-to-one correspondence between the classes (with respect to $\mathrm{U}(\Phi)$) in Λ and the factor group J/J_0.*

Suppose that $[L/M] = \mathfrak{r}\alpha$ with $\alpha\alpha^{\rho} = 1$. Since $\mathfrak{E} = f(E_0)$, there exists a unit u of $\mathfrak{r}$ such that $uu^\rho = 1$, $f(u) = f(a)e(L/M)$. Then $[L/M] = \mathfrak{r}(au)$, $(au)(au)^\rho = 1$, $f(au) = e(L/M)$. By [7, 5.27, (iii)], L and M belong to the same class with respect to $U(\Phi)$. Our assertion follows from this and the consideration of [7, 5.28].

1.5. PROPOSITION. *Let the notation and the assumption be the same as in 1.4. Let Λ' be the genus of L with respect to $G(\Phi)$. Let J^* be the group of all ideals $\mathfrak{a}$ in K such that $N_{K/F}(\mathfrak{a}) = \mathfrak{b}^m$ for some ideal $\mathfrak{b}$ in F, and J_0^* the group of all principal ideals $\mathfrak{r}a$ such that $aa^\rho = b^m$ for some $b \in F$ satisfying $b \equiv 1 \bmod \mathfrak{u}$, where $\mathfrak{u}$ is the product of certain infinite places of F defined in [7, 5.24]. Then the mapping $\Lambda' \ni M \to [L/M] \in J^*$ gives a one-to-one correspondence between the classes (with respect to $G(\Phi)$) in Λ' and the factor group J^*/J_0^*.*

This follows from [7, 5.28] by the same argument as in 1.4.

1.6. PROPOSITION. *Suppose that m is even and Φ is indefinite. Let $\mathfrak{u}$ be as in 1.5. Let L be a maximal $\mathfrak{r}$-lattice in V, and v a unit in $\mathfrak{g}$ such that $v \equiv 1 \bmod \mathfrak{u}$. Then there exists an element α of $H(\Phi)$ such that $L\alpha = L$ and $\mu(\alpha) = v$.*

By [7, 5.10], there exists an element σ of $G(\Phi)$ such that $\mu(\sigma) = v$. Put $c = \det(\sigma)$. Then $cc^\rho = v^m$. By [7, 2.3], there exists an element τ of $U(\Phi)$ such that $\det(\tau) = c^{-1}v^{m/2}$. Then $\mu(\sigma\tau) = v$, $\det(\sigma\tau) = v^{m/2}$, so that $\sigma\tau \in H(\Phi)$. Further we have $[L/L\sigma\tau] = \mathfrak{r}$, $e(L/L\sigma\tau) = 1$. By (ii) of [7, 5.25] and [7, 5.19], there exists an element β of $SU(\Phi)$ such that $L\sigma\tau\beta = L$. Since $\sigma\tau\beta \in H(\Phi)$, $\mu(\sigma\tau\beta) = v$, this proves our proposition.

1.7. PROPOSITION. *Suppose that m is odd, $m > 1$, and Φ is indefinite. Let L be a maximal $\mathfrak{r}$-lattice in V, and u a unit in $\mathfrak{r}$. Then there exists an element α of $G(\Phi)$ such that $L\alpha = L$ and $\mu(\alpha) = uu^\rho$.*

By [7, 2.4], there exists an element σ of $G(\Phi)$ such that $\mu(\sigma) = uu^\rho$. Put $\det(\sigma) = c$. Then $cc^\rho = (uu^\rho)^m$, so that $N_{K/F}(cu^{-m}) = 1$. Now $[L/L\sigma] = \mathfrak{r}c = \mathfrak{r}(cu^{-m})$, $\mu(L) = \mu(L\sigma)$. By (iii) of [7, 5.25] and (i) of [7, 5.27], there exists an element τ of $U(\Phi)$ such that $L\sigma\tau = L$. Since $\mu(\sigma\tau) = uu^\rho$, this completes the proof.

1.8. For every $\mathfrak{g}$-lattice L in V, we put

$$\Gamma(L) = \{\gamma \in U(\Phi) \mid L\gamma = L\},$$
$$\Gamma_0(L) = \{\gamma \in \Gamma(L) \mid \det(\gamma) = 1\}.$$

1.9. PROPOSITION. *Suppose that Φ is indefinite. Let L and M be maximal $\mathfrak{r}$-lattices in V belonging to the same genus with respect to $G(\Phi)$. If*

$m > 2$, *the following three conditions are equivalent to each other.*

(i) *There exists an ideal* $\mathfrak{a}$ *in* K *such that* $\mathfrak{a}L = M$.

(ii) $\Gamma(L) = \Gamma(M)$.

(iii) $\Gamma_0(L) = \Gamma_0(M)$.

If $m = 2$, *these conditions are equivalent provided that* $\mu(L_\mathfrak{p})^{-1}\mu(M_\mathfrak{p})$ *is an even power of the maximal ideal of* $\mathfrak{r}_\mathfrak{p}$ *whenever* $\mathfrak{p}$ *remains prime in* K *and* Φ *is anisotropic in* $V_\mathfrak{p}$.

The implication (i) $\Rightarrow$ (ii) $\Rightarrow$ (iii) is clear. Suppose that $\Gamma_0(L) = \Gamma_0(M)$. For every prime ideal $\mathfrak{p}$ in F, put

$$\Gamma_0(L_\mathfrak{p}) = \{\gamma \in \mathrm{SU}_\mathfrak{p}(\Phi) \mid L_\mathfrak{p}\gamma = L_\mathfrak{p}\} \ .$$

By the *approximation theorem* [7, 5.12], we see easily $\Gamma_0(L_\mathfrak{p}) = \Gamma_0(M_\mathfrak{p})$ for every $\mathfrak{p}$. We shall derive from this that there exists an element $e_\mathfrak{p}$ of $K_\mathfrak{p}$ such that $e_\mathfrak{p}L_\mathfrak{p} = M_\mathfrak{p}$. To prove this, first suppose that $\mathfrak{p}$ does not decompose in K. By our assumption, there exists an element α of $\mathrm{G}_\mathfrak{p}(\Phi)$ such that $L_\mathfrak{p}\alpha = M_\mathfrak{p}$. Let $\mathfrak{d}$ be the different of K with respect to F. Put $\mathfrak{h} = \mu(L_\mathfrak{p})\mathfrak{d}^{-1}$, $c = \mu(\alpha)$. By [7, 4.10], there exist a Witt decomposition $V_\mathfrak{p} = \sum_{i=1}^{t}(K_\mathfrak{p}x_i + K_\mathfrak{p}y_i) + W$ and elements a_i, b_i of $K_\mathfrak{p}$ such that $L_\mathfrak{p} = \sum_{i=1}^{t}(\mathfrak{r}_\mathfrak{p}x_i + \mathfrak{h}y_i) + L^*$, $M_\mathfrak{p} = \sum_{i=1}^{t}(\mathfrak{r}_\mathfrak{p}a_ix_i + \mathfrak{h}b_iy_i) + M^*, a_1b_1^\rho = \cdots = a_tb_t^\rho = c, \mathfrak{r}_\mathfrak{p}a_1 \supset \cdots \supset \mathfrak{r}_\mathfrak{p}a_t \supset \mathfrak{r}_\mathfrak{p}b_t \supset \cdots \supset \mathfrak{r}_\mathfrak{p}b_1,$ where $L^* = \{w \in W \mid \Phi(w, w) \in \mu(L_\mathfrak{p})\}$, $M^* = \{w \in W \mid \Phi(w, w) \in \mu(M_\mathfrak{p})\}$. Suppose that $V_\mathfrak{p} \neq W$, so that $t > 0$. By [7, 4.17], there exists an element h of $K_\mathfrak{p}$ such that $\mathfrak{h} = \mathfrak{r}_\mathfrak{p}h$, $h^\rho = -h$. Define a $K_\mathfrak{p}$-linear automorphism η of $V_\mathfrak{p}$ by $x_1\eta = hy_1$, $y_1\eta = -h^{-1}x_1$, $z\eta = z$ for $z \in \sum_{i=2}^{t}(K_\mathfrak{p}x_i + K_\mathfrak{p}y_i) + W$. Then $\eta \in \Gamma_0(L_\mathfrak{p}) = \Gamma_0(M_\mathfrak{p})$, so that $M_\mathfrak{p}\eta = M_\mathfrak{p}$. It follows that $\mathfrak{r}_\mathfrak{p}a_1h = \mathfrak{h}b_1$, and hence $\mathfrak{r}_\mathfrak{p}a_1 = \mathfrak{r}_\mathfrak{p}b_1$. Therefore $\mathfrak{r}_\mathfrak{p}a_1 = \cdots = \mathfrak{r}_\mathfrak{p}a_t = \mathfrak{r}_\mathfrak{p}b_t = \cdots = \mathfrak{r}_\mathfrak{p}b_1$. Hence $M_\mathfrak{p} = a_1L_\mathfrak{p}$. If $t = 0$, we have $L_\mathfrak{p} = L^*$, $M_\mathfrak{p} = M^*$. If $\mu(L_\mathfrak{p})^{-1}\mu(M_\mathfrak{p})$ is an even power of the maximal ideal of $\mathfrak{r}_\mathfrak{p}$, we find an element a of $K_\mathfrak{p}$ such that $aa^\rho\mathfrak{r}_\mathfrak{p} = c\mathfrak{r}$. Then we get $\mu(M_\mathfrak{p}) = \mu(aL_\mathfrak{p})$, so that $aL_\mathfrak{p} = M_\mathfrak{p}$.

Next assume that $\mathfrak{p}$ decomposes in K. By [7, 3.4], there exist an $\mathfrak{r}_\mathfrak{p}$-ideal $\mathfrak{c}$, a basis $\{z_i\}$ of $V_\mathfrak{p}$ over $K_\mathfrak{p}$, and elements e_i of $K_\mathfrak{p}$ such that $\Phi(z_i, z_j) = \delta_{ij}$, $L_\mathfrak{p} = \sum_{i=1}^{m}\mathfrak{c}z_i$, $M_\mathfrak{p} = \sum_{i=1}^{m}\mathfrak{c}e_iz_i$. Let ξ be a $K_\mathfrak{p}$-linear automorphism of $V_\mathfrak{p}$ such that $z_1\xi = z_2$, $z_2\xi = z_1$, $z_i\xi = z_i$ for $i > 2$. Then $\xi \in \Gamma_0(L_\mathfrak{p}) = \Gamma_0(M_\mathfrak{p})$, so that $M_\mathfrak{p}\xi = M_\mathfrak{p}$, and hence $\mathfrak{r}_\mathfrak{p}e_1 = \mathfrak{r}_\mathfrak{p}e_2$. Similarly we get $\mathfrak{r}_\mathfrak{p}e_1 = \mathfrak{r}_\mathfrak{p}e_i$ for $i > 2$. Therefore $M_\mathfrak{p} = e_1L_\mathfrak{p}$. We have thus proved that, for every $\mathfrak{p}$, there exists an element $e_\mathfrak{p}$ of $K_\mathfrak{p}$ such that $e_\mathfrak{p}L_\mathfrak{p} = M_\mathfrak{p}$. This implies (i) and completes the proof.

1.10. REMARK. If $m = 2$ and Φ is anisotropic in $V_\mathfrak{p}$, then $\Gamma_0(L_\mathfrak{p}) = \Gamma_0(M_\mathfrak{p})$ for any two maximal $\mathfrak{r}_\mathfrak{p}$-lattices $L_\mathfrak{p}$ and $M_\mathfrak{p}$ in $V_\mathfrak{p}$. In fact, let B be the quaternion algebra attached to Φ, which is defined in [7, 2.6], and ξ a B-isomorphism of

V into B with the property of [7, 2.7]. Let $\mathfrak{o}_\mathfrak{p}$ be the unique maximal order in $B_\mathfrak{p}$. Then, by [7, 2.6, 4.6], $\Gamma_0(L_\mathfrak{p}) = \{\gamma \in \mathfrak{o}_\mathfrak{p} \mid \gamma\gamma^\iota = 1\}$, where ι is the canonical involution of $B_\mathfrak{p}$.

1.11. PROPOSITION. *Suppose that Φ is indefinite and $m > 1$. When $m = 2$, suppose further that for every prime ideal $\mathfrak{p}$ in F which remains prime in K, Φ is isotropic in $V_\mathfrak{p}$. Let L and M be maximal $\mathfrak{r}$-lattices in V of the same genus with respect to $\mathrm{G}(\Phi)$, and α an element of $\mathrm{G}(\Phi)$. Then the following conditions are equivalent to each other.*

(i) *There exists an ideal $\mathfrak{a}$ in K such that $\mathfrak{a}L\alpha = M$.*

(ii) *$\alpha^{-1}\Gamma(L)\alpha = \Gamma(M)$.*

(iii) *$\alpha^{-1}\Gamma_0(L)\alpha = \Gamma_0(M)$.*

This is an immediate consequence of 1.10.

1.12. PROPOSITION. *Suppose that m is odd, $m > 1$, and Φ is indefinite; let $m = 2l + 1$. Let L and M be maximal $\mathfrak{r}$-lattices of the same genus with respect to $\mathrm{G}(\Phi)$. Let $\mathfrak{C}$ be the ideal-class group of K and $\mathfrak{C}_m$ the subgroup of $\mathfrak{C}$ consisting of all the classes represented by the ideals of the form $\mathfrak{a}^{l+1} \cdot (\mathfrak{a}^\rho)^{-l}$, where $\mathfrak{a}$ is an ideal in K. Then, there exists an element α of $\mathrm{G}(\Phi)$ such that $\alpha^{-1}\Gamma(L)\alpha = \Gamma(M)$ if and only if the ideal-class of $[L/M]$ belongs to $\mathfrak{C}_m$.*

We note that an ideal-class belongs to $\mathfrak{C}_m$ if it contains an ideal in F. Assume that $\alpha^{-1}\Gamma(L)\alpha = \Gamma(M)$ for some $\alpha \in \mathrm{G}(\Phi)$. By 1.11, we have $\mathfrak{a}L\alpha = M$ for some ideal $\mathfrak{a}$ in K. Put $b = \det(\alpha)$. Then we have $[L/M] = b \cdot \mathfrak{a}^m = b(\mathfrak{a}\mathfrak{a}^\rho)^l\mathfrak{a}^{l+1}(\mathfrak{a}^\rho)^{-l}$. This proves the 'only if' part. Conversely, assume that $[L/M] = d \cdot \mathfrak{a}^{l+1}(\mathfrak{a}^\rho)^{-l}$ for some $d \in K$ and some ideal $\mathfrak{a}$ in K. Put $\mu(L)^{-1}\mu(M) = \mathfrak{b}$, $dd^\rho = e$, $f = e^l d$. By [7, 5.21, (ii)], we have $\mathfrak{b}^m = N_{K/F}([L/M]) = dd^\rho\mathfrak{a}\mathfrak{a}^\rho$, and hence $[\mathfrak{a}\mathfrak{b}^{-l}L/M] = \mathfrak{r}f$, $ff^\rho = e^m$. By [7, 5.27, (ii)] there exists an element β of $\mathrm{G}(\Phi)$ such that $\mathfrak{a}\mathfrak{b}^{-l}L\beta = M$. This proves the 'if' part, in view of 1.11.

1.13. PROPOSITION. *Suppose that m is even, Φ is indefinite, and $f(E_0) = \mathfrak{E}$, the notation being as in 1.2. Let J, J_0 be as in 1.4, and $\mathfrak{u}$ be as in 1.5. Let P_∞ be a product of infinite places of F including all the factors of $\mathfrak{u}$. Let $J_m(P_\infty)$ be the subgroup of J generated by J_0 and all the ideals of the form $\mathfrak{b}^{m/2}\mathfrak{a}^m$, where $\mathfrak{a}$ is an ideal in K and $\mathfrak{b}$ is an element of F such that $N_{K/F}(\mathfrak{a}) = (\mathfrak{b}^{-1})$, $\mathfrak{b} \equiv 1 \bmod P_\infty$. Let L and M be maximal $\mathfrak{r}$-lattices of the same genus with respect to $\mathrm{U}(\Phi)$. Then $[L/M] \in J_m(P_\infty)$ if and only if there exists an element α of $\mathrm{G}(\Phi)$ and an ideal $\mathfrak{a}$ in K such that $\mathfrak{a}L\alpha = M$ and $\mu(\alpha) \equiv 1 \bmod P_\infty$.*

Assume that $[L/M] = de^{m/2}\mathfrak{a}^m$, $dd^\rho = 1$, $N_{K/F}(\mathfrak{a}) = (e^{-1})$, $e \equiv 1 \bmod P_\infty$ for some $d \in K$, $e \in F$ and some ideal $\mathfrak{a}$ in K. Then $[\mathfrak{a}L/M] = \mathfrak{r}de^{m/2}$, $N_{K/F}(de^{m/2}) = e^m$.

By 1.5, we have $\mathfrak{a}L\beta = M$ for some $\beta \in G(\Phi)$. Then $e\mathfrak{g} = N_{K/F}(\mathfrak{a})^{-1} = \mu(\mathfrak{a}L)^{-1}\mu(M) = \mu(\beta)\mathfrak{g}$. Hence $e\mu(\beta)^{-1}$ is a unit in $\mathfrak{g}$. By 1.6, There exists an element α of $G(\Phi)$ such that $L\alpha = L$, $\mu(\alpha) = e\mu(\beta)^{-1}$. Then we have $\mathfrak{a}L\alpha\beta = M$, $\mu(\alpha\beta) = e$. Conversely, assume that $\mathfrak{a}L\alpha = M$, $\mu(\alpha) \equiv 1 \bmod P_\infty$ for some $\alpha \in G(\Phi)$ and some ideal $\mathfrak{a}$ in K. Put $\det(\alpha) = b$, $\mu(\alpha) = c$. Then $N_{K/F}(bc^{-m/2}) = 1$, $[L/M] = (bc^{-m/2})(c^{m/2}\mathfrak{a}^m)$. Since L and M have the same genus with respect to $U(\Phi)$, we get $N_{K/F}([L/M]) = \mathfrak{g}$, so that $N_{K/F}(\mathfrak{a}) = (c^{-1})$. This shows $[L/M] \in J_m(P_\infty)$ and completes the proof.

1.14. PROPOSITION. *Let $J_m(P_\infty)$ be as in 1.13. Suppose that $m = 2$, and all the factors of P_∞ are ramified in K. Let $I(K/F; P_\infty)$ be the group of ideals in F generated by the following three kinds of ideals:*

(i) *all the principal ideals (x) such that $x \equiv 1 \bmod P_\infty$;*

(ii) *the squares of all ideals in F;*

(iii) *the prime ideals of F ramified in K.*

Then, for an ideal $\mathfrak{a}$ in K, $N_{K/F}(\mathfrak{a}) \in I(K/F; P_\infty)$ if and only if $\mathfrak{a}^{-1}\mathfrak{a}^\rho \in J_2(P_\infty)$.

If $\mathfrak{a}^{-1}\mathfrak{a}^\rho \in J_2(P_\infty)$, we have $\mathfrak{a}^{-1}\mathfrak{a}^\rho = x^{-1}x^\rho y^{-1}\mathfrak{b}^2$ for some $x \in K$, $y \in F$ and some ideal $\mathfrak{b}$ in K such that $N_{K/F}(\mathfrak{b}) = (y)$, $y \equiv 1 \bmod P_\infty$. Since $y^{-1}\mathfrak{b}^2 = \mathfrak{b} \cdot (\mathfrak{b}^\rho)^{-1}$, we have $(x^{-1}\mathfrak{a}\mathfrak{b})^\rho = x^{-1}\mathfrak{a}\mathfrak{b}$, so that $N_{K/F}(x^{-1}\mathfrak{a}\mathfrak{b}) \in I(K/F; P_\infty)$. Therefore $N_{K/F}(\mathfrak{a}) = N_{K/F}(x^{-1}\mathfrak{a}\mathfrak{b})N_{K/F}(x)y^{-1} \in I(K/F; P_\infty)$. Conversely, suppose that $N_{K/F}(\mathfrak{a}) \in I(K/F; P_\infty)$. By our definition of $I(K/F; P_\infty)$, we can find an element x of F and an ideal $\mathfrak{b}$ in K such that $x \equiv 1 \bmod P_\infty$, $\mathfrak{b}^\rho = \mathfrak{b}$, $\mathfrak{a}\mathfrak{a}^\rho = x\mathfrak{b}^2$. Then $\mathfrak{a}^{-1}\mathfrak{a}^\rho = \mathfrak{a}^{-2}(\mathfrak{a}\mathfrak{a}^\rho) = x(\mathfrak{a}^{-1}\mathfrak{b})^2$, so that $N_{K/F}(\mathfrak{a}^{-1}\mathfrak{b}) = (x^{-1})$. This shows $\mathfrak{a}^{-1}\mathfrak{a}^\rho \in J_2(P_\infty)$ and completes the proof.

1.15. PROPOSITION. *Suppose that F is totally real and K is totally imaginary. Let W be the group of all the roots of unity contained in K. Then, for every $\mathfrak{g}$-lattice L in V, the factor group $\Gamma(L)/\Gamma_0(L)$ is isomorphic to a subgroup of W.*

Let $\gamma \in \Gamma(L)$ and $c = \det(\gamma)$. Then $cc^\rho = 1$. Since $L\gamma = L$, c is a unit of $\mathfrak{r}$. For every isomorphism σ of K into C, we have $|c^\sigma| = 1$. Hence c is a root of unity. We get therefore our assertion.

1.16. PROPOSITION. *Let F, K and W be as in 1.15. Suppose that $W = \{\pm 1\}$ and m is even. Suppose further that there exists a prime ideal $\mathfrak{q}$ of F not dividing 2, ramified in K, and such that $V_\mathfrak{q}$ has the trivial kernel subspace. Let L be a maximal $\mathfrak{r}$-lattice in V. Then $\Gamma(L) = \Gamma_0(L)$.*

Let $\gamma \in \Gamma(L)$. By 1.15, we have $\det(\gamma) = \pm 1$. Suppose that $\det(\gamma) = -1$. By [7, 4.18], there exists a unit e of $\mathfrak{r}_\mathfrak{q}$ such that $-1 = e^{-1}e^\rho$. Then $K_\mathfrak{q} = F_\mathfrak{q}(e)$ and $e^2 \in F_\mathfrak{q}$. Since $\mathfrak{q}$ does not divide 2, $K_\mathfrak{q}$ is unramified over $F_\mathfrak{q}$; this is a contradiction. Hence we get our proposition.

2. The bottom field for the variety $\mho(T, \mathfrak{M})$

2.1. Let V be an algebraic variety defined over a subfield of C. Suppose that there exists a subfield $\mathfrak{f}$ of C with the following property:

(B) *If k is a field of definition for V, then $k \supset \mathfrak{f}$. Moreover, if σ is an isomorphism of k into C, then V is birationally equivalent to V^σ if and only if σ is the identity mapping on $\mathfrak{f}$.*

Such a field $\mathfrak{f}$ is uniquely determined by V, if it exists; it is uniquely determined even by the birational equivalence class of V. We call $\mathfrak{f}$ *the bottom field for V*. If V is defined over an algebraic number field, then it is easily seen that the bottom field for V exists.

Taking the *biregular* equivalence in place of the birational equivalence, we can define another bottom field for V. However, if V is non-singular and is a minimal model, this makes no difference. For example, let $\mathcal{H}$ be a bounded symmetric domain and Γ a properly discontinuous group of transformations on $\mathcal{H}$ with compact quotient and without fixed point. Then $\mathcal{H}/\Gamma$ is analytically isomorphic to a non-singular projective variety V. Following the argument of Igusa [3, Th. 6], we observe that V is a minimal model. Hence, for such a variety, we can define the bottom field in either way. By virtue of the results of Calabi, Vesentini, Selberg, and Weil, we know that $\mathcal{H}/\Gamma$ has no continuous deformation in almost all cases, except the one-dimensional case. Therefore, in general, there are no moduli for $\mathcal{H}/\Gamma$ in the usual sense. This is why we do not call $\mathfrak{f}$ the field of moduli of V.

Every algebraic curve defined over a subfield of C has the bottom field. In fact, let J be the jacobian variety of an algebraic curve Y, and $\mathcal{C}$ the canonical polarization of J. Then, by Torelli's theorem, we see that the field of moduli of the polarized abelian variety $(J, \mathcal{C})$ has the property (B) for the curve Y. In this case we call it *the field of moduli of Y*.

2.2. **Lemma.** *Let $\mathfrak{o}$ be a ring with a finite basis over $\mathbf{Z}$. Let $\mathcal{P} = (A, \mathcal{C}, \theta)$ and $\mathcal{P}' = (A', \mathcal{C}', \theta')$ be polarized abelian varieties of type $\mathfrak{o}$ in the sense of [4, no. 3], defined over a field of characteristic 0. Let λ be an isogeny of A onto A' such that $\lambda\theta(x) = \theta'(x)\lambda$ for every $x \in \mathfrak{o}$. Let $\mathfrak{a}$ be an ideal of $\mathfrak{o}$. Let k and k' be the fields of moduli of $\mathcal{P}$ and $\mathcal{P}'$, respectively. Suppose that $\mathbf{Q} \cdot \theta(\mathfrak{o}) = \mathrm{End}_{\mathbf{Q}}(A)$ and $\mathrm{Ker}(\lambda) = \{t \in A \mid \theta(\mathfrak{a})t = 0\}$. Then $k \supset k'$.*

We can find a polarization $\mathcal{C}''$ of A' such that λ^{-1} sends $\mathcal{C}''$ to $\mathcal{C}$. Put $\mathcal{P}'' = (A', \mathcal{C}'', \theta')$. By [6, p. 166, Prop. 6], k' is the field of moduli of $\mathcal{P}''$. Let σ be an isomorphism of the universal domain into itself, leaving the elements of k

invariant. Then there exists an isomorphism μ of $\mathscr{P}$ onto $\mathscr{P}^\sigma$. We see easily that

$$\mathrm{Ker}\,(\lambda^\sigma) = \{u \in A^\sigma \mid \theta^\sigma(\mathfrak{a})u = 0\}\,,$$

and hence $\mathrm{Ker}\,(\lambda^\sigma \cdot \mu) = \mathrm{Ker}\,(\lambda)$. By [5, Prop. 2.5], $\mathscr{P}''$ and $\mathscr{P}''^\sigma$ are isomorphic. Therefore σ is the identity on k'. This proves $k \supset k'$.

2.3. From now on, F and K denote respectively a totally real algebraic number field of degree g, and a totally imaginary quadratic extension of F. Let V be a vector space over K of dimension m and $T(x, y)$ a non-degenerate anti-hermitian form on V. Take an element ζ of K such that $\zeta^\rho = -\zeta$. Then ζT is a hermitian form. We put $\mathrm{G}(T) = \mathrm{G}(\zeta T)$, $\mathrm{U}(T) = \mathrm{U}(\zeta T)$.

Let $\mathfrak{M}$ be an $\mathfrak{r}$-lattice in V. In [8], under a certain condition on T and $\mathfrak{M}$, we defined a variety $\mathcal{V}(T, \mathfrak{M})$, and investigated the field of definition for it. Let us now consider the bottom field for $\mathcal{V}(T, \mathfrak{M})$. First we supply

2.4. PROPOSITION. *The assertions of Cases* I *and* II *of* [8, 5.10] *are true if $\mathfrak{M}$ is a maximal $\mathfrak{r}$-lattice and $\mathfrak{E} = f(E_0)$ (cf. 1.2).*

If m is even and $\mathfrak{M}$ is maximal, then $\mathfrak{M}$ satisfies $(\mathrm{U}_{\mathfrak{p}2})$ of [8, 4.18] for every prime ideal $\mathfrak{p}$ of F, on account of [7, 3.2, 4.24]. In the proof of *Cases* I and II of [8, 5.10], we needed [8, 4.23], so that $\mathfrak{M}$ was assumed to be of type (U). Now, if $\mathfrak{M}$ is maximal and $\mathfrak{E} = f(E_0)$, we obtain 1.4, which takes the place of [8, 4.23]. Then the proof of [8, 5.10] is valid without any modification.

2.5.0. Let the notation be as in [8, 5.10]. We exclude the case $K' = Q$. Let P_∞ be the product of *all* infinite places of F. Define a group H^* of ideals in K' by

$$H^* = \left\{\mathfrak{x} \mid \prod\nolimits_{\lambda=1}^{g'} \left(\frac{\mathfrak{x}^{\sigma_\lambda}}{\mathfrak{x}^{\sigma_\lambda \rho}}\right)^{v_\lambda} \in J_m(P_\infty)\right\} \qquad \text{if } m \text{ is even,}$$

where $J_m(P_\infty)$ is as in 1.13,

$$H^* = \left\{\mathfrak{x} \mid \text{the ideal class of } \prod\nolimits_{\lambda=1}^{g'} \mathfrak{x}^{\sigma_\lambda} \cdot \left(\frac{\mathfrak{x}^{\sigma_\lambda}}{\mathfrak{x}^{\sigma_\lambda \rho}}\right)^{v_\lambda} \text{ in } K \text{ belongs to } \mathfrak{C}_m\right\}$$

$$\text{if } m \text{ is odd,}$$

where $\mathfrak{C}_m$ is as in 1.12.

2.5. THEOREM. *Let $\mathcal{V}(T, \mathfrak{M})$ be as in* [8, 5.10]. *Let k^* be the class-field over K' corresponding to the ideal group H^* defined as above. Suppose that $\mathfrak{M}$ is maximal. When m is even, suppose further that $f(E_0) = \mathfrak{E}$. Then the bottom field for $\mathcal{V}(T, \mathfrak{M})$ is contained in k^*.*

Let k_0 be as in [8, 5.10]. Then $k_0 \supset k^*$. Let $\mathfrak{a}$ be an ideal in K' and

$\sigma = \left(\dfrac{k_0/K'}{\mathfrak{a}} \right)$. Suppose that m is even. By [8, 5.10] and 2.4, we have $\mathcal{V}(T, \mathfrak{M})^\sigma = \mathcal{V}(T, \mathfrak{N})$ for a maximal $\mathfrak{r}$-lattice $\mathfrak{N}$ such that

$$[\mathfrak{M}/\mathfrak{N}] \cdot \prod_{\lambda=1}^{g'} \left(\frac{\mathfrak{a}^{\sigma_\lambda}}{\mathfrak{a}^{\sigma_{\lambda^\rho}}} \right)^{v_\lambda} \in J_0 \ .$$

Now assume that σ is the identity on k^*. Then $\mathfrak{a} \in H^*$, hence $[\mathfrak{M}/\mathfrak{N}] \in J_m(P_\infty)$. By 1.13, there exist an element α of $G(T)$ and an ideal $\mathfrak{b}$ in K such that $\mathfrak{b}\mathfrak{M}\alpha = \mathfrak{N}, \mu(\alpha) \equiv 1 \bmod P_\infty$. Let $\mathcal{P}(z, T, \mathfrak{M}) = (A, C, \theta)$ be a generic member of $\Sigma(T, \mathfrak{M})$ (cf. [8, 2.3]). In [6, 2.7], we defined the operation of the elements of $U(T)$ on the domain $\mathcal{H}$. If $\alpha \in G(T)$ and $\mu(\alpha)$ is totally positive, then we can let α operate on $\mathcal{H}$ in a natural manner. In fact

$$\left((\mu(\alpha)^{\sigma_1})^{-1/2}\alpha^{\sigma_1}, \ \cdots, \ (\mu(\alpha)^{\sigma_g})^{-1/2}\alpha^{\sigma_g} \right)$$

belong to the unitary group of T over R. Then it operates on $\mathcal{H}$. Let $z' = \alpha^{-1}(z)$, $\mathcal{P}(z', T, \mathfrak{N}) = (A', C', \theta')$. Let h be a positive integer such that $h\mathfrak{b}^{-1} \subset \mathfrak{r}$. Then $\mathfrak{M}(h\alpha) = h\mathfrak{b}^{-1}\mathfrak{N} \subset \mathfrak{N}$. By using the argument of [6, 2.8], from $h\alpha$ we obtain an isogeny λ of A to A' such that $\lambda\theta(x) = \theta'(x)\lambda$ for every $x \in K$, and

$$\mathrm{Ker}\,(\lambda) = \{ t \in A \mid \theta(h\mathfrak{b}^{-1})t = 0 \} \ .$$

By [6, p. 176, Th.5], we have $\theta(\mathfrak{r}) = \mathrm{End}\,(A)$. Therefore we may apply 2.2 to the present case; denoting by $\mathfrak{K}$ and $\mathfrak{K}'$ the fields of moduli of $\mathcal{P}(z, T, \mathfrak{M})$ and $\mathcal{P}(z', T, \mathfrak{N})$, we get $\mathfrak{K} \supset \mathfrak{K}'$. Applying the same argument to α^{-1}, we can prove $\mathfrak{K}' \supset \mathfrak{K}$, so that $\mathfrak{K} = \mathfrak{K}'$. This shows that $\mathcal{V}(T, \mathfrak{M})$ and $\mathcal{V}(T, \mathfrak{N})$ are birationally equivalent. Hence σ must be the identity on the bottom field for $\mathcal{V}(T, \mathfrak{M})$ by means of 1.12. This proves our theorem for even m. The case of odd m can be proved similarly.

2.6. Remark. If $\mathcal{H}/\Gamma(\mathfrak{M})$ is compact, then we can simplify the above proof. In this case, the function field of $\mathcal{V}(T, \mathfrak{M})$ may be identified with the field of all automorphic functions on $\mathcal{H}$ with respect to $\Gamma(\mathfrak{M})$ [6, pp. 172–173]. Therefore, if we have $\mathfrak{b}\mathfrak{M}\alpha = \mathfrak{N}$, then $\alpha^{-1}\Gamma(\mathfrak{M})\alpha = \Gamma(\mathfrak{N})$, so that $\mathcal{H}/\Gamma(\mathfrak{M})$ is analytically isomorphic to $\mathcal{H}/\Gamma(\mathfrak{N})$. It follows immediately that $\mathcal{V}(T, \mathfrak{M})$ and $\mathcal{V}(T, \mathfrak{N})$ are birationally equivalent.

2.7. Corollary. *Let the notation and the assumption be the same as in 2.5. Let $\mathfrak{C}$ be the ideal-class group of K and $\mathfrak{C}_0$ the subgroup of $\mathfrak{C}$ which consists of the ideal-classes containing ideals $\mathfrak{a}$ such that $\mathfrak{a}^\rho = \mathfrak{a}$. Then the bottom field for $\mathcal{V}(T, \mathfrak{M})$ is contained in K' if m is prime to $[\mathfrak{C} : \mathfrak{C}_0]$.*

Suppose that m is even. Let $\mathfrak{x} \in J$. Then we can find an ideal $\mathfrak{a}$ such that

$\mathfrak{a}^{-1}\mathfrak{a}^\rho = \mathfrak{x}$. If m is prime to $[\mathfrak{C}:\mathfrak{C}_0]$, then there exist ideals $\mathfrak{b}$, $\mathfrak{c}$, and an element d of K such that $\mathfrak{a} = \mathfrak{b}^m\mathfrak{c}d$, $\mathfrak{c}^\rho = \mathfrak{c}$. We have therefore $\mathfrak{x} = (\mathfrak{b}^{-1}\mathfrak{b}^\rho)^m(d^{-1}d^\rho) \in J_m(P_\infty)$. This shows $J = J_m(P_\infty)$. Hence $k^* = K'$. Next suppose that m is odd, and put $\mathfrak{C}^* = \{c^m \mid c \in \mathfrak{C}\}$. It can be easily verified that $\mathfrak{C}_m = \mathfrak{C}_0 \cdot \mathfrak{C}^*$. Therefore, if m is prime to $[\mathfrak{C}:\mathfrak{C}_0]$ we get

$$\mathfrak{C} = \mathfrak{C}_m \text{ and hence } k^* = K'.$$

2.8. Suppose that $\mathfrak{H}/\Gamma(\mathfrak{M})$ and $\mathfrak{H}/\Gamma(\mathfrak{N})$ are compact, and $\Gamma(\mathfrak{M})$ and $\Gamma(\mathfrak{N})$ have no fixed points. Then, as remarked in 2.1, $\mathfrak{H}/\Gamma(\mathfrak{M})$ is a minimal model. Therefore, if $\mathcal{V}(T, \mathfrak{M})$ and $\mathcal{V}(T, \mathfrak{N})$ are birationally equivalent, then $\mathfrak{H}/\Gamma(\mathfrak{M})$ and $\mathfrak{H}/\Gamma(\mathfrak{N})$ are analytically isomorphic. Hence there exists an analytic automorphism ξ of $\mathfrak{H}$ such that $\xi\Gamma(\mathfrak{M})\xi^{-1} = \Gamma(\mathfrak{N})$. Under a suitable condition, we can take such a ξ from $G(T)$.[8] Then applying the converse part of 1.13 or 1.12, we know that $[\mathfrak{M}/\mathfrak{N}] \in J_m(P_\infty)$ or the ideal class of $[\mathfrak{M}/\mathfrak{N}]$ belongs to $\mathfrak{C}_m$, according as m is even or odd. Therefore, in such a case, the bottom field, together with K', generates exactly the classfield k^*. But here we do not discuss this problem in detail. We shall treat only the case $m = 2$ in §4.

3. Unit groups in a quaternion algebra

3.1. Let F and $\mathfrak{g}$ be as in 1.1. Let D be a quaternion algebra over F. We denote by ι the canonical involution of D and put $N(x) = xx^\iota$ for $x \in D$. For every $\mathfrak{g}$-lattice $\mathfrak{a}$ in D, we denote by $N(\mathfrak{a})$ the ideal in F generated by the elements $N(x)$ for all $x \in \mathfrak{a}$. For a maximal order $\mathfrak{o}$ in D, we put

$$\Gamma(\mathfrak{o}) = \{a \in \mathfrak{o} \mid N(a) = 1\} \, .$$

3.2. PROPOSITION. *Suppose that D is indefinite (i.e., $D \otimes_Q R$ has at least one simple component isomorphic to $M_2(R)$ or $M_2(C)$). Let $\mathfrak{o}$ and $\mathfrak{o}'$ be maximal orders in D. Then $\Gamma(\mathfrak{o}) = \Gamma(\mathfrak{o}')$ if and only if $\mathfrak{o} = \mathfrak{o}'$.*

Suppose that $\Gamma(\mathfrak{o}) = \Gamma(\mathfrak{o}')$. For every prime ideal $\mathfrak{p}$ in F, denote by $F_\mathfrak{p}$ and $\mathfrak{g}_\mathfrak{p}$ the $\mathfrak{p}$-completions of F and $\mathfrak{g}$, and put $D_\mathfrak{p} = D \otimes_F F_\mathfrak{p}$ and $\mathfrak{a}_\mathfrak{p} = \mathfrak{g}_\mathfrak{p}\mathfrak{a}$ for every $\mathfrak{g}$-lattice $\mathfrak{a}$ in D. It is sufficient to prove $\mathfrak{o}_\mathfrak{p} = \mathfrak{o}'_\mathfrak{p}$ for every $\mathfrak{p}$. If $D_\mathfrak{p}$ is a division algebra, this is clear from the uniqueness of maximal order. Hence we may assume that $D_\mathfrak{p} = M_2(F_\mathfrak{p})$, $\mathfrak{o}_\mathfrak{p} = M_2(\mathfrak{g}_\mathfrak{p})$. Let $\Gamma(\mathfrak{o}_\mathfrak{p})$ (resp. $\Gamma(\mathfrak{o}'_\mathfrak{p})$) denote the set of all the elements a in $\mathfrak{o}_\mathfrak{p}$(resp. $\mathfrak{o}'_\mathfrak{p}$) such that $N(a) = 1$. By Eichler's approximation-theorem [2, satz 5], we get $\Gamma(\mathfrak{o}_\mathfrak{p}) = \Gamma(\mathfrak{o}'_\mathfrak{p})$. We can find an element b in D such that $\mathfrak{o}'_\mathfrak{p} = b\mathfrak{o}_\mathfrak{p}b^{-1}$. Then $\Gamma(\mathfrak{o}_\mathfrak{p}) = \Gamma(\mathfrak{o}'_\mathfrak{p}) = b\Gamma(\mathfrak{o}_\mathfrak{p})b^{-1}$. We note that $\Gamma(\mathfrak{o}_\mathfrak{p})=$

---

[8] For this problem, a result of Borel [1, p. 24] on "the group of transformations" is useful. Besides, we have to consider the permutation of the irreducible factors of $\mathfrak{H}$ (cf. the proof of 4.9 below).

$SL_2(\mathfrak{g}_\mathfrak{p})$. Taking the elementary divisors of b, we can easily show that b is a scalar multiple of a unit of $\mathfrak{o}_\mathfrak{p}$. It follows that $\mathfrak{o}_\mathfrak{p} = \mathfrak{o}'_\mathfrak{p}$.

3.3. Let the notation and the assumption be as in 3.2. Denote by $I(F)$ the group of all ideals in F, and by $I(D/F)$ the subgroup of $I(F)$ generated by the following three kinds of ideals:

(i) the principal ideals (a) such that a is totally positive;

(ii) the squares of all ideals in F;

(iii) the prime ideals of F ramified in D.

Then we see easily that the subgroup of $I(F)$ generated by the ideals belonging to (ii) and (iii) coincides with the set of $N(\mathfrak{h})$ for all two-sided $\mathfrak{o}$-ideals $\mathfrak{h}$, where $\mathfrak{o}$ is any fixed maximal order in D. Let $\mathfrak{o}$ and $\mathfrak{o}'$ be maximal orders in D, and $\mathfrak{x}$ a $\mathfrak{g}$-lattice in D whose right order is $\mathfrak{o}$ and left order is $\mathfrak{o}'$. Then the class of $N(\mathfrak{x})$ modulo $I(D/F)$ is uniquely determined by $\mathfrak{o}$ and $\mathfrak{o}'$, and independent of the choice of $\mathfrak{x}$. We denote by $\{\mathfrak{o} : \mathfrak{o}'\}$ the element of $I(F)/I(D/F)$ represented by $N(\mathfrak{x})$.

3.4. PROPOSITION. *Let the notation and the assumption be as in 3.2 and 3.3. Then the symbol $\{\mathfrak{o} : \mathfrak{o}'\}$ has the following properties:*

(i) $\{\mathfrak{o}_1 : \mathfrak{o}_2\}\{\mathfrak{o}_2 : \mathfrak{o}_3\} = \{\mathfrak{o}_1 : \mathfrak{o}_3\}$.

(ii) $\{\mathfrak{o}_1 : \mathfrak{o}_2\} = 1$ *if and only if there exists an element b of D such that $N(b)$ is totally positive and $b\Gamma(\mathfrak{o}_1)b^{-1} = \Gamma(\mathfrak{o}_2)$.*

The first property is clear from our definition. Now let $\mathfrak{x}$ be a right $\mathfrak{o}_1$-ideal whose left order is $\mathfrak{o}_2$. Suppose that $b\Gamma(\mathfrak{o}_1)b^{-1} = \Gamma(\mathfrak{o}_2)$ and $N(b)$ is totally positive for some b in D. By 3.2, we have $b\mathfrak{o}_1b^{-1} = \mathfrak{o}_2$. Then $b^{-1}\mathfrak{x}$ is a two-sided $\mathfrak{o}_1$-ideal. Therefore $N(b^{-1}\mathfrak{x}) \in I(D/F)$, and hence $N(\mathfrak{x}) \in I(D/F)$. Conversely, if $N(\mathfrak{x}) \in I(D/F)$, we can find a totally positive element c of F and a two-sided $\mathfrak{o}_1$-ideal $\mathfrak{h}$ such that $N(\mathfrak{x})=cN(\mathfrak{h})$. By Eichler's theorem, there exists an element u of D such that $\mathfrak{x} = u\mathfrak{h}$. Then $cN(u)^{-1}$ is a unit of $\mathfrak{g}$. By Eichler's approximation theorem [2, Satz 5], there exists a unit v of $\mathfrak{o}_1$ such that $N(v) = cN(u)^{-1}$. Then $\mathfrak{x} = uv\mathfrak{h}$, $N(uv) = c$. Since $\mathfrak{o}_2$ is the left order of $\mathfrak{x}$, we have $\mathfrak{o}_2=(uv)\mathfrak{o}_1(uv)^{-1}$. This completes our proof.

3.5. PROPOSITION. *Let P_∞ be the product of all real infinite places of F. Then there exist infinitely many totally imaginary quadratic extensions K of F with the following properties:*

(i) *D splits over K.*

(ii) *$I(K/F; P_\infty) \subset I(D/F)$ (cf. 1.14 and 3.3).*

(iii) *There exists one and only one prime ideal $\mathfrak{q}$ of F which is ramified in K and unramified in D; $\mathfrak{q}$ is not a divisor of 2.*

(iv) *K contains no root of unity other than ± 1.*

Let P be the set of prime ideals in F which are either ramified in D or divisors of 2. For each $\mathfrak{p} \in P$, we can find an element $a_\mathfrak{p}$ of F such that $F_\mathfrak{p}(\sqrt{a_\mathfrak{p}})$ is unramified over $F_\mathfrak{p}$; we may assume that $a_\mathfrak{p}$ is a $\mathfrak{p}$-unit. There exists a positive integer e such that for any $\mathfrak{p} \in P$, $F_\mathfrak{p}(\sqrt{a_\mathfrak{p}}) = F_\mathfrak{p}(\sqrt{b_\mathfrak{p}})$ whenever $a_\mathfrak{p} \equiv b_\mathfrak{p} \bmod \mathfrak{p}^e$. Now we can find a totally positive element c of F such that $c \equiv -a_\mathfrak{p} \bmod \mathfrak{p}^e$ for every $\mathfrak{p} \in P$. By virtue of generalized Dirichlet's theorem, we may take the ideal (c) so as to be a prime ideal in F. Let $K = F(\sqrt{-c})$. Then K is totally imaginary and every $\mathfrak{p}$ in P remains prime in K. Hence K satisfies (i). The condition (iii) holds with $\mathfrak{q} = (c)$; (ii) is clear from our construction. Since there are infinitely many choice of $\mathfrak{q}$, (iv) is satisfied for a suitable c. This completes the proof.

3.6. PROPOSITION. *Let K be a quadratic extension of F such that $D \otimes_F K = M_2(K)$. Denote by ρ the non-trivial automorphism of K over F. Then there exists an invertible element w in $M_2(K)$ such that ${}^t w^\rho = w$ and $D = \{x \in M_2(K) \mid x^\iota = w \cdot {}^t x^\rho w^{-1}\}$, where ι is the canonical involution of $M_2(K)$.*

First we note that the canonical involution ι of $M_2(K)$ induces on D the canonical involution of D. Let us define an involution σ of $D \otimes_F K$ by putting $x^\sigma = x^\rho$ for $x \in K$ and $x^\sigma = x^\iota$ for $x \in D$. Then σ is an involution of $M_2(K)$ of the second kind. Hence we can find an element w of $M_2(K)$ such that ${}^t w^\rho = w$ and $x^\sigma = w \cdot {}^t x^\rho w^{-1}$ for every $x \in M_2(K)$. We see that $D \subset \{x \in M_2(K) \mid x^\iota = w \cdot {}^t x^\rho w^{-1}\}$. The right hand side is a quaternion algebra over F [7, 1.5]. Therefore we get the equality of the proposition.

3.7. Let K and w be as in 3.6. Let V be the vector space of two-dimensional row vectors with components in K. For $x \in V$ and $y \in V$, put $\Phi(x, y) = xw \cdot {}^t y^\rho$. Then Φ is a non-degenerate ρ-hermitian form on V, and D is the quaternion algebra attached to Φ in the sense of [7, 2.6, p. 374]. Therefore, by [7, 2.7], there exist an F-linear bijection ξ of V to D and a non-zero element c of F such that

(3.7.1) $\xi(x\alpha) = \xi(x)\alpha$ for every $x \in V$ and $\alpha \in D$.

(3.7.2) $\Phi(x, x) = c \cdot N(\xi(x))$ for every $x \in V$.

3.8. PROPOSITION. *Let the notation be as in 3.7. Let $\mathfrak{r}$ be the ring of integers in K. Let $\mathfrak{M}$ and $\mathfrak{N}$ be maximal $\mathfrak{r}$-lattices in V belonging to the same genus with respect to $U(\Phi)$. Let*

$$\mathfrak{o} = \{\alpha \in D \mid \mathfrak{M}\alpha \subset \mathfrak{M}\}, \qquad \mathfrak{o}' = \{\alpha \in D \mid \mathfrak{N}\alpha \subset \mathfrak{N}\}.$$

Then there exists an ideal $\mathfrak{a}$ in K such that

$$[\mathfrak{M}/\mathfrak{N}] = \mathfrak{a}(\mathfrak{a}^\rho)^{-1},$$

$$\{\mathfrak{o} : \mathfrak{o}'\} = \textit{the class of } N_{K/F}(\mathfrak{a}) \textit{ modulo } I(D/F).$$

By [7, 5.31], $\mathfrak{o}$ and $\mathfrak{o}'$ are maximal orders in D. Let P be the set of all prime ideals $\mathfrak{p}$ in F such that $\mathfrak{M}_\mathfrak{p} \neq \mathfrak{N}_\mathfrak{p}$. For each $\mathfrak{p} \in P$, there exists an element $a_\mathfrak{p}^{\cdot}$ of $U_\mathfrak{p}(\Phi)$ such that $\mathfrak{M}_\mathfrak{p} a_\mathfrak{p} = \mathfrak{N}_\mathfrak{p}$. By [7, 2.6], there exists an element $b_\mathfrak{p}$ of $K_\mathfrak{p}$ and an element $c_\mathfrak{p}$ of $D_\mathfrak{p}$ such that $a_\mathfrak{p} = b_\mathfrak{p} c_\mathfrak{p}$. Since $\mu(a_\mathfrak{p}) = 1$, we get $b_\mathfrak{p} b_\mathfrak{p}^\rho N(c_\mathfrak{p}) = 1$. Hence $[\mathfrak{M}/\mathfrak{N}]_\mathfrak{p} = b_\mathfrak{p}^2 N(c_\mathfrak{p}) \mathfrak{r}_\mathfrak{p} = b_\mathfrak{p}(b_\mathfrak{p}^\rho)^{-1} \mathfrak{r}_\mathfrak{p}$. On the other hand, we have $\mathfrak{o}_\mathfrak{p}' = c_\mathfrak{p}^{-1} \mathfrak{o}_\mathfrak{p} c_\mathfrak{p}$, so that $\mathfrak{o}_\mathfrak{p}' c_\mathfrak{p}^{-1} = c_\mathfrak{p}^{-1} \mathfrak{o}_\mathfrak{p}$. We can find a $\mathfrak{g}$-lattice $\mathfrak{x}$ in D such that $\mathfrak{x}_\mathfrak{p} = c_\mathfrak{p}^{-1} \mathfrak{o}_\mathfrak{p}$ for every $\mathfrak{p} \in P$ and $\mathfrak{x}_\mathfrak{p} = \mathfrak{o}_\mathfrak{p}$ for every $\mathfrak{p} \notin P$. Such an $\mathfrak{x}$ has $\mathfrak{o}$ and $\mathfrak{o}'$ as its right order and left order. Similarly, there exists an ideal $\mathfrak{a}$ in K such that $\mathfrak{a}_\mathfrak{p} = b_\mathfrak{p} \mathfrak{r}_\mathfrak{p}$ for every $\mathfrak{p} \in P$ and $\mathfrak{a}_\mathfrak{p} = \mathfrak{r}_\mathfrak{p}$ for every $\mathfrak{p} \notin P$. Then $[\mathfrak{M}/\mathfrak{N}] = \mathfrak{a}(\mathfrak{a}^\rho)^{-1}$, $N(\mathfrak{x}) = N_{K/F}(\mathfrak{a})$. This proves our proposition.

4. The class-field over a totally real field occurring as the bottom field for a variety

4.1. Let F be a totally real algebraic number field of degree g. We denote by $p_{\infty 1}, \cdots, p_{\infty g}$ the infinite places of F. Let D be a quaternion algebra over F which is unramified at $p_{\infty 1}, \cdots, p_{\infty t}$, and ramified at $p_{\infty t+1}, \cdots, p_{\infty g}$. We assume that D is indefinite, namely $t \geq 1$. Let $D_R = D \otimes_Q R$. Then we may identify D_R with the product of t copies of $M_2(R)$ and $g - t$ copies of the division ring K of real quaternions:

$$(4.1.1) \qquad D_R = M_2(R) \times \cdots \times M_2(R) \times K \times \cdots \times K.$$

For every maximal order $\mathfrak{o}$ in D, we define $\Gamma(\mathfrak{o})$ as in 3.1. Let $h(x)$ be the projection of an element x of D_R on the partial product $M_2(R) \times \cdots \times M_2(R)$ (t copies). Then $h(\Gamma(\mathfrak{o}))$ is contained in $\mathrm{SL}_2(R) \times \cdots \times \mathrm{SL}_2(R)$ (t copies). Now let $\mathfrak{H}$ denote the usual upper half plane $\{z \in C \mid \mathrm{Im}\,(z) > 0\}$, and $\mathfrak{H}^t$ the product of t copies of $\mathfrak{H}$. For every element

$$\alpha = (\alpha_1, \cdots, \alpha_t), \quad \alpha_i = \begin{pmatrix} a_i & b_i \\ c_i & d_i \end{pmatrix}$$

of $\mathrm{SL}_2(R) \times \cdots \times \mathrm{SL}_2(R)$, we define the operation of α on $\mathfrak{H}^t$ by

$$\alpha(z_1, \cdots, z_t) = (\alpha_1(z_1), \cdots, \alpha_t(z_t)), \; \alpha_i(z_i) = \frac{a_i z_i + b_i}{c_i z_i + d_i}.$$

Identifying $\Gamma(\mathfrak{o})$ with $h(\Gamma(\mathfrak{o}))$, we may regard $\Gamma(\mathfrak{o})$ as a discontinuous group operating on $\mathfrak{H}^t$. The quotient space $\mathfrak{H}^t/\Gamma(\mathfrak{o})$ is compact if and only if D is a division algebra. If $t = g$, then the field of automorphic functions on $\mathfrak{H}^t$ with respect to $\Gamma(\mathfrak{o})$ can be defined over Q [8, 3.3 and 3.5]. Therefore we assume hereafter that $t < g$. We shall mean by $Y(\mathfrak{o})$ a projective variety whose function-field over C is isomorphic (over C) to the field of all automorphic functions on $\mathfrak{H}^t$ with respect to $\Gamma(\mathfrak{o})$.

4.2. Let L be a Galois extension of Q containing F, and G the Galois group of L over Q. Let $\tau_1, \cdots, \tau_g$ be elements of G whose restrictions to F correspond to $p_{\infty 1}, \cdots, p_{\infty g}$. Let H_F be the subgroup of G corresponding to F. Then $G = \bigcup_{\nu=1}^{g} H_F \tau_\nu$. Let F^* be the field generated over Q by the elements $\sum_{\nu=1}^{t} a^{\tau_\nu}$ for all $a \in F$, and H_F^* the subgroup of G corresponding to F^*. We see easily that

$$H_F^* = \{\gamma \in G \mid (\bigcup_{\nu=1}^{t} H_F \tau_\nu)\gamma = \bigcup_{\nu=1}^{t} H_F \tau_\nu\} \ .$$

Put $[F^* : Q] = h$. Let $\theta_1, \cdots, \theta_h$ be elements of G such that $G = \bigcup_{\mu=1}^{h} H_F^* \theta_\mu$. After re-ordering $\theta_1, \cdots, \theta_h$ suitably, we can find an expression

$$(4.2.1) \qquad\qquad \bigcup_{\nu=1}^{t} \tau_\nu^{-1} H_F = \bigcup_{\mu=1}^{v} H_F^* \theta_\mu$$

for an integer v such that $1 \leq v < h$. It can be easily seen that the restrictions of $\theta_{v+1}, \cdots, \theta_h$ to F^* depend only on F and $p_{\infty 1}, \cdots, p_{\infty t}$, and they are independent of the choice of G. For every ideal $\mathfrak{a}$ in F^*, we see that $\prod_{\mu=1}^{v} \mathfrak{a}^{\theta_\mu}$ is an ideal in F (cf. [9, p. 71, Prop. 29]). Now let $I(D/F)$ be the ideal-group in F defined in 3.3, and $J(F^*)$ the ideal-group in F^* defined by

$$(4.2.2) \qquad\qquad J(F^*) = \{\mathfrak{a} \mid \prod_{\mu=1}^{v} \mathfrak{a}^{\theta_\mu} \in I(D/F)\} \ .$$

4.3. MAIN THEOREM I. *Let the notation and the assumption be as in 4.1 and 4.2. Let $k(D)$ be the class-field over F^* corresponding to the ideal-group $J(F^*)$. Let $\mathfrak{o}$ be a maximal order in D. Then the bottom field (cf. 2.1) for the variety $Y(\mathfrak{o})$ is contained in $k(D)$. Moreover, let $\mathfrak{a}$ be an ideal in F^*, and let σ be an isomorphism of a field of definition for $Y(\mathfrak{o})$ into C such that $\sigma = \left(\dfrac{k(D)/F^*}{\mathfrak{a}}\right)$ on $k(D)$. Then $Y(\mathfrak{o})^\sigma$ is birationally equivalent to $Y(\mathfrak{o}')$ with a maximal order $\mathfrak{o}'$ in D such that $\{\mathfrak{o} : \mathfrak{o}'\}$ is the class of $\prod_{\mu=1}^{v} \mathfrak{a}^{\theta_\mu}$ modulo $I(D/F)$ (cf. 3.3).*

We shall prove this in the following 4.4–4.7.

4.4. Let K be a totally imaginary quadratic extension of F satisfying the conditions of 3.5. Define V and Φ as in 3.7. Let L be a Galois extension of Q containing K, and G the Galois group of L over Q. Let $\tau_1, \cdots, \tau_g$ be as in 4.2. Then Φ^{τ_ν} is indefinite for $1 \leq \nu \leq t$ and definite for $\nu > t$. We can find an element ζ of K such that $\zeta^\rho = -\zeta$ and $\sqrt{-1}(\zeta\Phi)^{\tau_\nu}$ is negative definite for $\nu > t$. Put $T = \zeta\Phi$. We are going to apply [8, 5.10] to the present case. With the notation of [8, 5.10], we have $m = 2$, $r_\nu = s_\nu = 1$ for $1 \leq \nu \leq t$, $r_\nu = 2$ for $\nu > t$. As in [8, 4.2 and 4.3], we consider the group ring G_Q, and define H_K, η_K, ω, ω^*, H_K' and K'. Then we verify easily that K' is generated by the elements $\sum_{\nu>t} a^{\tau_\nu}$ for all $a \in K$ over $\boldsymbol{Q}$, and

$$H'_K = \{\gamma \in G \mid (\sum_{\nu>t} \eta_K \tau_\nu)\gamma = \sum_{\nu>t} \eta_K \tau_\nu\} \ .$$

Put $[K' : \boldsymbol{Q}] = 2g'$ and define $\sigma_1, \cdots, \sigma_{g'}, \eta_{K'}$ as in [8, 4.4]. Then, exchanging σ_λ for $\sigma_\lambda\rho$ if necessary, we have

$$\omega^* = \sum_{\lambda=1}^{u} (\eta'_K \sigma_\lambda + \eta'_K \sigma_\lambda\rho) + 2\sum_{\lambda=u+1}^{g'} \eta'_K \sigma_\lambda$$

for an integer u such that $1 \leq u < g'$, so that

$$(4.4.1) \qquad \sum_{\nu=t+1}^{g} \eta_K \tau_\nu = \sum_{\lambda=u+1}^{g'} \sigma_\lambda^{-1} \eta'_K \ .$$

Therefore, if we define v_λ as in [8, 4.7], then

$$(4.4.2) \qquad v_1 = \cdots = v_u = 0, \ v_{u+1} = \cdots = v_{g'} = 1 \ .$$

Now define $H_F, H_F^*, \theta_1, \cdots, \theta_h$ as in 4.2. We see that $F^* \subset K'$, $H_F = H_K \cup H_K\rho$, $H_F^* \supset H'_K \cup H'_K\rho$, so that

$$(4.4.3) \qquad \mathbf{U}_{\lambda>u}(H'_K \cup H'_K\rho)\sigma_\lambda = \mathbf{U}_{\nu>t} \tau_\nu^{-1} H_F = \mathbf{U}_{\mu>v} H_F^*\theta_\mu \ .$$

For every ideal $\mathfrak{y}$ in K', put

$$\psi(\mathfrak{y}) = \prod_{\lambda=u+1}^{g'} \mathfrak{y}^{\sigma_\lambda} \ .$$

By (4.4.1), we see that $\psi(\mathfrak{y})$ is an ideal in K, and by (4.4.3),

$$(4.4.4) \qquad N_{K/F}(\psi(\mathfrak{y})) = \prod_{\lambda=u+1}^{g'} \mathfrak{y}^{\sigma_\lambda}\mathfrak{y}^{\sigma_\lambda\rho} = \prod_{\mu=v+1}^{h} N_{K'/F^*}(\mathfrak{y})^{\theta_\mu}$$
$$= N_{K'/\boldsymbol{Q}}(\mathfrak{y})[\prod_{\mu=1}^{v} N_{K'/F^*}(\mathfrak{y})^{\theta_\mu}]^{-1} \ .$$

Let $J(K')$ be the ideal-group in K' defined by

$$(4.4.5) \qquad J(K') = \{\mathfrak{y} \mid N_{K/F}(\psi(\mathfrak{y})) \in I(D/F)\} \ .$$

By (4.4.4) and (4.2.2), we have

$$J(K') = \{\mathfrak{y} \mid N_{K'/F^*}(\mathfrak{y}) \in J(F^*)\} \ ,$$

so that $K'k(D)$ is the class-field over K' corresponding to $J(K')$.

Let J_0 be as in 1.4, and let H be the group of ideals in K' defined by

$$H = \{\mathfrak{y} \mid \psi(\mathfrak{y})^{-1}\psi(\mathfrak{y})^\rho \in J_0\} \ .$$

In view of (4.4.2), this is just the ideal-group defined by [8, (5.9.1)]. Let k_0 be the class-field over K' corresponding to H. By 1.14 and (ii) of 3.5, we have $H \subset J(K')$, so that $k_0 \supset K'k(D)$.

4.5. Let ξ be as in 3.7. Fix a maximal order $\mathfrak{o}$ in D. Put $\mathfrak{M} = \xi^{-1}(\mathfrak{o})$. By [7, 5.31], $\mathfrak{M}$ is a maximal $\mathfrak{r}$-lattice in V. Further by [7, 2.6] and 1.16, we have

$$(4.5.1) \qquad \Gamma(\mathfrak{M}) = \Gamma_0(\mathfrak{M}) = \Gamma(\mathfrak{o}) \ .$$

By 2.4, 1.3 and (iii) of 3.5, we can apply [8, 5.10] to $\mathfrak{M}$, and then obtain, for

$$\sigma = \left(\frac{k_0/K'}{\mathfrak{a}}\right),$$

(4.5.2) $\mathcal{V}(T, \mathfrak{M})^\sigma = \mathcal{V}(T, \mathfrak{N})$

with a maximal $\mathfrak{r}$-lattice $\mathfrak{N}$, belonging to the same genus as $\mathfrak{M}$ with respect to $U(T)$, and such that

(4.5.3) $[\mathfrak{M}/\mathfrak{N}]\psi(\mathfrak{a})\psi(\mathfrak{a}^\rho)^{-1} \in J_0$.

Let $\mathfrak{o}' = \{\alpha \in D \mid \mathfrak{N}\alpha \subset \mathfrak{N}\}$. By [7, 2.6] and 1.16, we have

(4.5.4) $\Gamma(\mathfrak{N}) = \Gamma_0(\mathfrak{N}) = \Gamma(\mathfrak{o}')$.

By 3.8, there exists an ideal $\mathfrak{b}$ in K such that $[\mathfrak{M}/\mathfrak{N}] = \mathfrak{b}(\mathfrak{b}^\rho)^{-1}$ and $\{\mathfrak{o} : \mathfrak{o}'\}$ is the class of $N_{K/F}(\mathfrak{b})$ modulo $I(D/F)$. By (4.5.3), there exists an element c of K such that

$$\mathfrak{b}(\mathfrak{b}^\rho)^{-1}\psi(\mathfrak{a})\psi(\mathfrak{a}^\rho)^{-1} = c^{-1}c^\rho\mathfrak{r} \ .$$

Then $(c\mathfrak{b}\psi(\mathfrak{a}))^\rho = c\mathfrak{b}\psi(\mathfrak{a})$. By (ii) of 3.5, $N_{K/F}(c\mathfrak{b}\psi(\mathfrak{a})) \in I(D/F)$. From this and 4.4.4, we obtain

(4.5.5) $\{\mathfrak{o} : \mathfrak{o}'\}$ is the class of $\prod_{\mu=1}^{v} N_{K'/F^*}(\mathfrak{a})^{\theta_\mu}$ modulo $I(D/F)$.

In [6] and [8], we considered $\Gamma(\mathfrak{M})$ as a group of transformations on the product $\mathfrak{D}^t$ of t copies of the unit disc $\mathfrak{D} = \{z \in C \mid |z| < 1\}$. Then $\mathcal{V}(T, \mathfrak{M})$ is a projective model (not necessarily non-singular) of the field of all automorphic functions on $\mathfrak{D}^t$ with respect to $\Gamma(\mathfrak{M})$. On the other hand, in 4.1, we have defined the operation of $\Gamma(\mathfrak{o})$ on $\mathfrak{H}^t$. By choosing a suitable analytic isomorphism of $\mathfrak{D}^t$ onto $\mathfrak{H}^t$, we see easily that both operations are compatible. Therefore we may regard $\mathcal{V}(T, \mathfrak{M})$ as a projective model of the field of automorphic functions on $\mathfrak{H}^t$ with respect to $\Gamma(\mathfrak{o})$.

Now, using $Y(\mathfrak{o})$ instead of $\mathcal{V}(T, \mathfrak{M})$ (cf. 4.1), we have thus proved that

(4.5.6) *if* $\sigma = \left(\dfrac{k_0/K'}{\mathfrak{a}}\right)$, $Y(\mathfrak{o})^\sigma$ *is birationally equivalent to* $Y(\mathfrak{o}')$ *with a maximal order* $\mathfrak{o}'$ *in* D *satisfying* (4.5.5).

If σ is the identity mapping on $K'k(D)$, then $\mathfrak{a} \in J(K')$, hence $N_{K'/F^*}(\mathfrak{a}) \in J(F^*)$. By 4.5.5 and (ii) of 3.4, $Y(\mathfrak{o})^\sigma$ is birationally equivalent to $Y(\mathfrak{o})$. This shows that $K'k(D)$ *contains the bottom field for* $Y(\mathfrak{o})$.

4.6. PROPOSITION. *For every finite algebraic extension* M *of* F^*, *there exists a totally imaginary quadratic extension* K *of* F, *satisfying the conditions of* 3.5, *and such that* K' *and* M *are linearly disjoint over* F^*.

To prove our proposition, we may assume that M is a Galois extension of Q containing F. In the construction of K in 3.5, we can take c so that $N_{F/Q}(c)$ is a prime number which is not ramified in M/Q. Let L be the smallest Galois extension of Q containing K and M. Let G_L (resp. G_M) be the Galois group of L (resp. M) over Q. Take $\tau_1, \cdots, \tau_g$ as in 4.4. Put $c_i = c^{\tau_i}$, $d = \sqrt{-c}$, $d_i = d^{\tau_i}$. Then $L = M(d_1, \cdots, d_g)$. Since the principal ideals (c_i) and (c_j) in M are relatively prime, and since $N_{F/Q}(c)$ is not ramified in M, we see that $[L:M]=2^g$. For every $\alpha \in G_L$, consider its restriction β to M; β determines a permutation of $(c_1, \cdots, c_g)$. If $c_i^\beta = c_j$, then $d_i^\alpha = \varepsilon_i d_j$ with $\varepsilon_i = \pm 1$. Then α is completely determined by β and $\varepsilon_1, \cdots, \varepsilon_g$. Since $[L:M] = 2^g$, we can conclude that, for every $\beta \in G_M$, and for every arrangement $(\varepsilon_1, \cdots, \varepsilon_g)$ of ± 1, there exists an element α of G_L such that $\alpha = \beta$ on M and $d_i^\alpha = \varepsilon_i d_j$ for $c_i^\beta = c_j$. We apply this to the case where β is the identity mapping on F^*. Then, by our definition of F^*, $\{\tau_{t+1}\beta, \cdots, \tau_g\beta\}$ coincides with $\{\tau_{t+1}, \cdots, \tau_g\}$ on F, as a whole. We can find an element α of G_L so that $\alpha = \beta$ on M and $d_i^\alpha = d_j$ for $c_i^\beta = c_j$. For every $x \in F$ and $y \in F$, we have

$$(4.6.1) \qquad \left(\sum_{\nu > t}(x + yd)^{\tau_\nu}\right)^\alpha = \sum_{\nu > t} x^{\tau_\nu \beta} + \sum_{\nu > t} y^{\tau_\nu \beta} d_\nu^\alpha.$$

If $\tau_\nu\beta = \tau_\mu$ on F, then $c_\nu^\beta = c_\mu$, so that $d_\nu^\alpha = d_\mu = d^{\tau\mu}$. Hence (4.6.1) is equal to $\sum_{\mu > t} x^{\tau\mu} + \sum_{\mu > t} (yd)^{\tau\mu} = \sum_{\mu > t} (x + yd)^{\tau\mu}$. This shows that α is the identity mapping on K'. Therefore, every automorphism of M over F^* can be extended to an element of G_L which induces an identity mapping on K'. It follows that M and K' are linearly disjoint over F^*.

4.7. Coming back to the stage of 4.5, by virtue of 4.6, we can take K so that K' and $k(D)$ are linearly disjoint over F^*. Take another K_1 with the property of 3.5 so that K_1' and $K'k(D)$ are linearly disjoint over F^*. By the assertion at the end of 4.5, the bottom field for $Y(\mathfrak{o})$ is contained in $K'k(D) \cap K_1'k(D)$. By our choice of K_1, we have $K'k(D) \cap K_1'k(D) = k(D)$. This proves the first assertion of 4.3. The remaining part of 4.3 follows from our choice of K and (4.5.6).

4.8. Let $\mathfrak{p}_1, \cdots, \mathfrak{p}_s$ be all the prime ideals in F which are ramified in D. Then $g - t + s$ is even. Conversely, if $g - t + s$ is even, for every s prime ideals $\mathfrak{p}_1, \cdots, \mathfrak{p}_s$ in F, there exists one and only one (up to isomorphism) quaternion algebra D over F which is ramified exactly at $\mathfrak{p}_1, \cdots, \mathfrak{p}_s, p_{\infty t+1}, \cdots, p_{\infty g}$. Now let σ be an automorphism of F. If σ leaves invariant the set $\{\mathfrak{p}_1, \cdots, \mathfrak{p}_s; p_{\infty 1}, \cdots, p_{\infty t}\}$ as a whole, then we can extend σ to an automorphism of D. Conversely, if D has an automorphism ω over Q, then the restriction of ω to F must leave invariant $\{\mathfrak{p}_1, \cdots, \mathfrak{p}_s; p_{\infty 1}, \cdots, p_{\infty t}\}$ as a whole.

4.9. MAIN THEOREM II. *Let the notation be as in 4.3. Let $\mathfrak{p}_1, \cdots, \mathfrak{p}_s$ be all the prime ideals in F which are ramified in D. Suppose that:*

(i) There is no automorphism of F, other than the identity mapping, which leaves invariant $\{\mathfrak{p}_1, \cdots, \mathfrak{p}_s; p_{\infty 1}, \cdots, p_{\infty t}\}$ as a whole.

(ii) The multiplicative group of regular elements in D has no element of finite order other than ± 1.

Then $k(D)$ is exactly the composite of F^* and the bottom field for $Y(\mathfrak{o})$.

In view of 4.8, we see that (i) is equivalent to

(iii) *D has no automorphism over $\boldsymbol{Q}$ other than inner automorphisms.*

Let k be a field of definition for $Y(\mathfrak{o})$ containing $k(D)$, and σ an isomorphism of k into $\boldsymbol{C}$. By 4.3, if $\sigma = \left(\dfrac{k(D)/F^*}{\mathfrak{a}} \right)$ on $k(D)$, then $Y(\mathfrak{o})^\sigma$ is birationally equivalent to $Y(\mathfrak{o}')$ with a maximal order $\mathfrak{o}'$ such that $\{\mathfrak{o} : \mathfrak{o}'\}$ is the class of $\prod_{\mu=1}^{v} \mathfrak{a}^{\theta_\mu}$ modulo $I(D/F)$. Now suppose that σ is the identity mapping on the bottom field for $Y(\mathfrak{o})$. Then $Y(\mathfrak{o})$ and $Y(\mathfrak{o}')$ are birationally equivalent. As remarked in 2.1, $\mathfrak{H}^t/\Gamma(\mathfrak{o})$ and $\mathfrak{H}^t/\Gamma(\mathfrak{o}')$ are minimal models. It follows that $\mathfrak{H}^t/\Gamma(\mathfrak{o})$ and $\mathfrak{H}^t/\Gamma(\mathfrak{o}')$ are analytically isomorphic. Therefore we can find an analytic automorphism α of $\mathfrak{H}^t$ such that $\alpha^{-1}\Gamma(\mathfrak{o})\alpha = \Gamma(\mathfrak{o}')$. To describe this relation more precisely, denote by $x \to h_i(x)$ the projection of D_R onto the ith factor of the right hand side of (4.1.1). Then there exist a permutation $i \to j(i)$ of $\{1, 2, \cdots, t\}$ and t elements $a_1, \cdots, a_t$ of $\mathrm{SL}_2(\boldsymbol{R})$ such that

$$\alpha(z_1, \cdots, z_t) = (w_1, \cdots, w_t), \quad w_{j(i)} = a_i(z_i)$$

for $(z_1, \cdots, z_t) \in \mathfrak{H}^t$. Then the precise meaning of $\alpha^{-1}\Gamma(\mathfrak{o})\alpha = \Gamma(\mathfrak{o}')$ is as follows: there exists a one-to-one correspondence $\Gamma(\mathfrak{o}) \ni \gamma \leftrightarrow \gamma' \in \Gamma(\mathfrak{o}')$ such that

$$a_i^{-1} h_j(\gamma) a_i = h_i(\gamma') \qquad\qquad \text{if } j = j(i) \ .$$

Now we observe that $\Gamma(\mathfrak{o})$ generates D over $\boldsymbol{Q}$. Hence $\gamma \to \gamma'$ can be extended to an automorphism ω of D over $\boldsymbol{Q}$. Namely, for every $x \in D$, we have $a_i^{-1} h_j(x) a_i = h_i(x^\omega)$ if $j = j(i)$. By our assumption, and by 4.8, ω must be the identity mapping on F. Hence $j(i) = i$ for every i, and we can find an element b of D such that $x^\omega = b^{-1}xb$ for every $x \in D$. It follows that $b^{-1}\Gamma(\mathfrak{o})b = \Gamma(\mathfrak{o}')$, and $a_i^{-1} h_i(x) a_i = h_i(b)^{-1} h_i(x) h_i(b)$ for every $x \in D$. Since $h_i(D)$ generates $M_2(\boldsymbol{R})$ over $\boldsymbol{R}$, we have $a_i^{-1} y a_i = h_i(b)^{-1} y h_i(b)$ for every $y \in M_2(\boldsymbol{R})$. Therefore $a_i^{-1} h_i(b)$ is a scalar matrix. It follows that $N(b)^{\tau_i} = \det(h_i(b)) > 0$ for $1 \leq i \leq t$. On the other hand, we have $N(x)^{\tau_i} > 0$ for $i > t$ with any element x of D other than 0. Hence $N(b)$ is totally positive. By 3.4, we get $\{\mathfrak{o} : \mathfrak{o}'\} = 1$, so that $\mathfrak{a} \in J(F^*)$, hence σ is the identity mapping on $k(D)$. This proves our theorem.

4.10. Suppose that $t = 1$ and τ_1 is the identity mapping on F. Then we have obviously $F = F^*$, $J(F^*) = I(D/F)$. Furthermore, the assumption (i) of 4.9 is satisfied. Therefore we get the theorems in the Introduction as a special case of *main theorems* I and II.

Next suppose that $t = g - 1$ and τ_g is the identity mapping. Then we have $F = F^*$ and $\prod_{\mu=1}^{v} \mathfrak{a}^{\theta_\mu} = N_{F/Q}(\mathfrak{a})\mathfrak{a}^{-1}$, so that $J(F^*) = I(D/F)$. It can be easily seen that (i) of 4.9 is satisfied. Therefore, in this case, we can simplify the assertions of *main theorems* accordingly.

4.11. In view of 4.8, we can take $\mathfrak{p}_1, \cdots, \mathfrak{p}_s$ so as to be principal ideals generated by totally positive numbers. Then the ideal-group $I(D/F)$ is generated by only the ideals belonging to (i) and (ii) of 3.3. If F is a real quadratic field, this ideal-group is the one called "principal genus", and the corresponding class-field is contained in a certain cyclotomic field.

4.12. For a given F, there exist only a finite number of positive integers n such that $[F(e^{2\pi i/n}) : F] = 2$. Let $\mathfrak{p}$ be a prime ideal in F which decomposes in $F(e^{2\pi i/n})$ for any such n. As remarked in 4.8, we can take D so as to be ramified at $\mathfrak{p}$. Then the multiplicative group of regular elements of D has no element of finite order other than ± 1.

OSAKA UNIVERSITY AND PRINCETON UNIVERSITY

REFERENCE

1. A. BOREL, "Ensembles fondamentaux pour les groupes arithmétiques," in Colloque sur la théorie des groupes algébriques, Bruxelles, 1962, 23-40.
2. M. EICHLER, *Allgemeine Kongruenzklasseneinteilungen der Ideale einfacher Algebren über algebraischen Zahlkörpern und ihre L-Reihen*, J. Reine Angew. Math., 179 (1938). 227-251.
3. J. IGUSA, *On the structure of a certain class of kaehler varieties*, Amer. J. Math., 76 (1954), 669-678.
4. G. SHIMURA, *On the theory of automorphic functions*, Ann. of Math., 70 (1959), 101-144.
5. ———, *On the zeta-functions of the algebraic curves uniformized by certain automorphic functions*, J. Math. Soc. Japan, 13 (1961), 275-331.
6. ———, *On analytic families of polarized abelian varieties and automorphic functions*, Ann. of Math., 78 (1963), 149-192.
7. ———, *Arithmetic of unitary groups*, Ann. of Math., 79 (1964), 369-409.
8. ———, *On the field of definition for a field of automorphic functions*, Ann. of Math., 80 (1964), 160-189.
9. ——— and Y. TANIYAMA, Complex multiplication of abelian varieties and its applications to number theory, Publ. Math. Soc. Japan, No. 6, 1961.

(Received October 23, 1963)

On purely transcendental fields of automorphic functions of several variables

Osaka Journal of Mathematics, 1 (1964), 1-14

(Received December 2, 1963)

The purpose of this paper is to give some examples of arithmetically defined discontinuous groups Γ operating on a complex ball

$$H^r = \left\{ (z_1, \cdots, z_r) \in C^r \,\middle|\, |z_1|^2 + \cdots + |z_r|^2 < 1 \right\}$$

such that the field of all automorphic functions[1] on H^r with respect to Γ is a purely transcendental extension of C of dimension r. To get such a Γ, we consider the field $K = Q(\zeta)$ generated by a primitive m-th root of unity ζ, and take a hermitian matrix S of size $r+1$ with entries in K such that S itself has exactly r positive and one negative characteristic roots while all the other conjugates of S over Q are definite. Let $U_0(S)$ be the group of all complex matrices X such that ${}^t\overline{X}SX = S$. Let Γ be the subgroup of $U_0(S)$ consisting of the matrices whose entries are algebraic integers in K. Since H^r is isomorphic to the quotient space of $U_0(S)$ with respect to a maximal compact subgroup, Γ operates naturally on H^r. In our examples, the automorphic functions with respect to Γ give moduli of algebraic curves $Y: y^m = p(x)$, where $p(x)$ is a polynomial in $C[x]$. Then the following table describes our examples.

	K	r	S	H^r/Γ	Y
(1)	$Q(1^{1/3})$	2	diag $[1, 1, -1]$	non-compact	$y^3 = p_4(x)$
(2)	$Q(1^{1/3})$	3	diag $[1, 1, 1, -1]$	non-compact	$y^3 = p_6(x)$
(3)	$Q(1^{1/4})$	2	diag $[1, 1, -1]$	non-compact	$y^4 = p_2(x)p_3(x)^2$
(4)	$Q(1^{1/5})$	1	diag $[1, (1-\sqrt{5})/2]$	compact	$y^5 = p_3(x)$
(5)	$Q(1^{1/5})$	2	diag $[1, 1, (1-\sqrt{5})/2]$	compact	$y^5 = p_5(x)$
(6)	$Q(1^{1/7})$	1	diag $\left[1, -\dfrac{\sin(3\pi/7)}{\sin(2\pi/7)}\right]$	compact	$y^7 = p_3(x)$

---

1) By an automorphic function, we always mean a *meromorphic* function which is invariant under the operation of the group in question.

Here $1^{1/m}$ denotes a primitive m-th root of unity, $\mathrm{diag}\,[a_1,\cdots,a_s]$ the diagonal matrix with diagonal elements $a_1,\cdots,a_s$, and $p_n(x)$ a polynomial of degree n and without multiple root.

Theorem. *In these six cases, the field of all automorphic functions on H^r with respect to Γ is a purely transcendental extension of C of dimension r.*

It would be worth while mentioning the following point. There was previously no known example of a discontinuous group Γ operating on a bounded symmetric domain D of dimension >1 such that D/Γ is compact and the field of all automorphic functions on D with respect to Γ is purely transcendental over C. The case (5) gives actually such a discontinuous group.

Picard [3] investigated the curve $y^3 = p_4(x)$ and observed that moduli of such curves give automorphic functions on H^2. But it seems that he did not determine the whole field of automorphic functions.

To prove our theorem, we consider the canonically polarized jacobian variety J of the algebraic curve Y. It turns out that J belongs to an analytic family Σ treated in our previous paper [6]. In the above cases, if Y is a generic curve of the given type, J is a generic member of Σ. Now the moduli of Y are, roughly speaking, the same as the moduli of J. Then the main theorem of [6] shows that the moduli of Y are given by the automorphic functions with respect to a certain discontinuous group Γ. In order to determine the explicit form of Γ, we need some analysis of lattices in a vector space over K with a hermitian form, which was one of the subjects investigated in [7]. In the Appendix, we give a supplement to it.

In the present paper, we treated the moduli of Y only at *generic* points. It would be interesting to study the moduli of Y in more detail, for example, from the view-point of Igusa [2], who investigated the moduli of algebraic curves of genus two.

1. First we recall some results of [6]. Let F be a totally real algebraic number field of degree g, and K a totally imaginary quadratic extension of F. We denote by ρ the complex conjugation. Let Φ be a representation of K by complex matrices of size h. We say that a triplet $\mathcal{P} = (A, C, \theta)$ is a polarized abelian variety of type $\{K, \Phi, \rho\}$ if the following conditions are satisfied.

(i) A is an abelian variety of dimension h, defined over C.

(ii) θ is an isomorphism of K into $\mathrm{End}_Q(A)$; and the representation of $\theta(x)$ for $x \in K$ by an analytic coordinate system of A is equivalent to Φ.

(iii) $\mathcal{C}$ is a polarization of A; and the involution of $\mathrm{End}_Q(A)$ determined by $\mathcal{C}$ coincides with $\theta(x)\to\theta(x^\rho)$ on $\theta(K)$.

Let $\sigma_1,\cdots,\sigma_g,\ \sigma_1\rho,\cdots,\sigma_g\rho$ be all the isomorphisms of K into C, and let r_ν (resp. s_ν) be the multiplicity of σ_ν (resp. $\sigma_\nu\rho$) in Φ. In order to insure the existence of $\mathcal{P}$ of type $\{K,\Phi,\rho\}$, the following relation should be satisfied [6, 2.1]:

$$(1.1) \qquad h = g(r_\nu + s_\nu) \qquad (1\leq\nu\leq g).$$

Hereafter we assume (1.1) and put $u=h/g$.

Let $\mathcal{P}=(A,\mathcal{C},\theta)$ be of type $\{K,\Phi,\rho\}$. Take a complex torus C^h/D isomorphic to A, where D is a lattice in C^h. We may choose the coordinate system of C^h so that $\theta(a)$ is represented by the matrix $\Phi(a)$ on C^h for every $a\in K$. Let K^u be the vector space of all u-dimensional row vectors with components in K. Then we find u vectors $x_1,\cdots,x_u$ in C^h such that $QD=\sum_{i=1}^u \Phi(K)x_i$. For every $a=(a_1,\cdots,a_u)$ in K^u, put $x(a)=\sum_{i=1}^u \Phi(a_i)x_i$. Then the mapping $a\to x(a)$ is an isomorphism of K^u onto QD. Let L be the inverse image of D by this mapping.

Let $E(x,y)$ be a Riemann form on C^h/D corresponding to a basic polar divisor in $\mathcal{C}$. Then there exists an anti-hermitian form $T(a,b)$ on K^u such that

$$(1.2) \qquad E(x(a),x(b)) = Tr_{K/Q}(T(a,b)) \qquad ((a,b)\in K^u\times K^u).$$

The structure $\{K^u, T, L\}$ is uniquely determined by $\mathcal{P}$ up to isomorphism. We say that $\mathcal{P}$ is of type $\{K,\Phi,\rho;T,L\}$. We note that T can not be arbitrary; it must satisfy the following condition [6, p. 160, (25)]:

(1.3) *The hermitian matrix* $\sqrt{-1}\,T^{\sigma_\nu}$ *has the same signature as* $\begin{bmatrix} -1_{r_\nu} & 0 \\ 0 & 1_{s_\nu} \end{bmatrix}$
for every ν, where 1_r denotes the identity matrix of degree r.

Let H_ν be the space of all complex matrices z with r_ν rows and s_ν columns such that $1-{}^t\bar{z}z$ is positive hermitian, and let

$$H = H_1 \times \cdots \times H_g.$$

Then we get an analytic family $\Sigma(K,\Phi,\rho;T,L)=\{\mathcal{P}_z|z\in H\}$ of polarized abelian varieties $\mathcal{P}_z$ of type $\{K,\Phi,\rho;T,L\}$ parametrized by the point z on H. Every $\mathcal{P}$ of type $\{K,\Phi,\rho;T,L\}$ is isomorphic to a member of $\Sigma(K,\Phi,\rho;T,L)$.

Now we let every element of $M_u(K)$ operate on K^u on the right, and define a group $\Gamma(T,L)$ by

$$\Gamma(T,L) = \left\{ X\in GL_u(K) \,\middle|\, T(aX,bX) = T(a,b),\ LX = L \right\}.$$

Then $\Gamma(T, L)$ gives a properly discontinuous group of transformations on H [6, 2.7]. In [6, Th. 3], we get meromorphic functions $f_1, \cdots, f_\kappa$ on H and an analytic subset W of H of codimension one, such that $Q(f_1(z), \cdots, f_\kappa(z))$ is the field of moduli of $\mathcal{P}_z$ for every $z \in H - W$. As remarked in [6, p. 172], if $H/\Gamma(T, L)$ is compact, $C(f_1, \cdots, f_\kappa)$ is the field of all automorphic functions on H with respect to $\Gamma(T, L)$. Even if $H/\Gamma(T, L)$ is not compact, the last statement is true in view of [6, Th. 4] and a recent result of Baily and Borel on the compactification of $H/\Gamma(T, L)$.

Proposition 1. *Let $\mathcal{P}$ be of type $\{K, \Phi, \rho ; T, L\}$ and k_0 the field of moduli of $\mathcal{P}$. If $\dim_Q k_0 = \sum_{\nu=1}^g r_\nu s_\nu$, then $Q(f_1, \cdots, f_\kappa)$ is isomorphic to k_0.*

This follows from [6, Theorem 4, (iii)] and [5, Prop. 3.5 and p. 305, Remark].

2. Let m and n be positive integers. Let Y be an algebraic curve defined by $y^m = p(x)$, where $p(x)$ is a polynomial in $C[x]$, of degree n and without multiple root. If d is the greatest common divisor of m and n, the genus h of Y is given by

$$h = \frac{1}{2}\Big[(m-1)(n-1)-(d-1)\Big].$$

The vector space of differential forms of the first kind on Y is spanned by the $x^a dx/y^b$ with integers a and b satisfying $0 \leq a < n$, $0 < b < m$, $bn - am - m - d \geq 0$.

If m divides $n+1$, take a complex number c so that $p(c) \neq 0$, and put $v = 1/(x-c)$, $n+1 = me$. Then we get $(v^e y)^m = v \cdot v^n p(v^{-1} + c)$. This shows that Y is birationally equivalent to the curve $y^m = q(x)$ with a polynomial $q(x)$ in $C[x]$ of degree $n+1$ and without multiple root.

Hereafter we assume that m does not divide $n+1$, $h > 1$, and m is an odd prime number. Let J be the jacobian variety of Y, and φ a canonical mapping of Y into J. We fix a primitive m-th root of unity ζ. Let ζ_0 be the birational correspondence of Y with itself given by $(x, y) \rightarrow (x, \zeta y)$. Denote by $\theta(\zeta)$ the automorphism of J corresponding to ζ_0. We see that $\zeta \rightarrow \theta(\zeta)$ can be extended naturally to an isomorphim θ of $Q(\zeta)$ into $\mathrm{End}_Q(J)$. Let C be the canonical polarization of J, and ρ the automorphism of $Q(\zeta)$ such that $\zeta^\rho = \zeta^{-1}$. The involution of $\mathrm{End}_Q(J)$ determined by C gives the automorphism $\theta(a) \rightarrow \theta(a^\rho)$ on $\theta(Q(\zeta))$. In this way we get a polarized abelian variety of type $\{Q(\zeta), \Phi, \rho\}$ in the sense of §1, for a certain representation Φ of degree h. In view of the explicit form of differential forms of the first kind given above, we see that, for every integer b such that $0 < b < m$, the matrix $\Phi(\zeta)$ has ζ^{-b} as a characteristic root with multiplicity $[(bn-d)/m]$, where $[\alpha]$ denotes the largest non-negative integer $\leq \alpha$.

3. Let Y^* be another curve defined by $y^m = p^*(x)$ for a polynomial $p^*(x)$ in $C[x]$ of degree n and without multiple root. From Y^*, we obtain a polarized abelian variety $\mathcal{P}^* = (J^*, C^*, \theta^*)$ of type $\{Q(\zeta), \Phi, \rho\}$ in the same way as above; we note that the representation Φ is the same for fixed m and n. Let ζ_0^* be the birational correspondence of Y^* with itself given by $(x, y) \rightarrow (x, \zeta y)$.

Proposition 2. *$\mathcal{P}$ is isomorphic to $\mathcal{P}^*$ if and only if there exists a birational mapping λ of Y to Y^* such that $\lambda \zeta_0 = \zeta_0^* \lambda$.*

The 'if' part is obvious. Let φ^* be a canonical mapping of Y^* to J^*. Suppose that there exists an isomorphism μ of $\mathcal{P}$ to $\mathcal{P}^*$. By Torelli's theorem, there exists a birational mapping λ of Y to Y^* such that $\varphi^* \lambda = \pm \mu \varphi + a$, where a is a point of J^*. Since $\mu \theta(\zeta) = \theta^*(\zeta)\mu$, we see easily that $\lambda^{-1} \zeta_0^* \lambda$ and ζ_0 correspond to the same automorphism $\theta(\zeta)$ of J. By our assumption $h > 1$, we must have $\lambda^{-1} \zeta_0^* \lambda = \zeta_0$. Our proposition is thereby proved.

Proposition 3. *Let k_0 be the composite of $Q(\zeta)$ and the field of moduli of $\mathcal{P}$. Then k_0 is the subfield of C which is uniquely determined by the following properties.*

(M_1) *If k is a field of definition for Y and ζ_0, then $k \supset k_0$. If furthermore σ is an isomorphism of k into C, over $Q(\zeta)$, then σ is the identity mapping on k_0 if and only if there exists a birational mapping λ of Y to Y^σ such that $\lambda \zeta_0 = \zeta_0^\sigma \lambda$.*

(M_2) *$k_0 \supset Q(\zeta)$.*

This follows immediately from Prop. 2 and the definition of the field of moduli of $\mathcal{P}$ [4, p. 110].

Proposition 4. *Let λ be a birational mapping of Y to Y^* such that $\lambda \zeta_0 = \zeta_0^* \lambda$, and let $\lambda(x, y) = (u, v)$. Then u, v are rational expressions of x, y of the following form.*

(I) *If m divides n, $u = (ax+b)/(cx+d)$, $v = ey/(cx+d)^{n/m}$.*

(II) *If m does not divide n, $u = ax+b$, $v = ey$.*

Here a, b, c, d, e are complex numbers.

Let $u = \sum_{i=0}^{m-1} r_i(x) y^i$, $v = \sum_{i=0}^{m-1} s_i(x) y^i$ with $r_i(x)$ and $s_i(x)$ in $C(x)$. Since $\lambda \zeta_0 = \zeta_0^* \lambda$, we have $\sum_{i=0}^{m-1} r_i(x) \zeta^i y^i = \sum_{i=0}^{m-1} r_i(x) y^i$, $\sum_{i=0}^{m-1} s_i(x) \zeta^i y^i$

$=\zeta \sum_{i=0}^{m-1} s_i(x)y^i$, so that $u=r_0(x)$, $v=s_1(x)y$. Since λ is one-to-one, r_0 must be linear fractional: $r_0(x)=(ax+b)/(cx+d)$. Write $s_1(x)$ as $s_1(x)$ $=s(x)/t(x)$ with polynomials $s(x)$ and $t(x)$ which are relatively prime. Then we get

$$s(x)^m(cx+d)^n p(x) = t(x)^m(cx+d)^n p^*((ax+b)/(cx+d)).$$

We see that $(cx+d)^n p^*((ax+b)/(cx+d))$ is a polynomial in x of degree n or $n-1$, without multiple root. It follows that $s(x)$ is a constant. Recall that we excluded the case $m\mid n+1$. Then we get easily our assertions.

Suppose that m divides n. We see easily that the transformation of (I) of Prop. 4 always gives a birational mapping of Y to another curve $v^m=p^*(u)$ with a polynomial $p^*(u)$ of degree n or $n-1$, without multiple root. If a/c is not a root of $p(x)$, $p^*(u)$ is of degree n.

If m does not divide n, it is clear that the transformation of (II) of Prop. 4 gives a birational mapping of Y to a curve of the same type.

4. Let q be a polynomial in $C[x]$ of degree $\leq n$, other than 0, and let $q(x)= \sum_{i=0}^{n} q_i x^i$. Let P^n be the projective space of dimension n. Denote by $[q]$ the point $(q_0,\cdots,q_n)$ in P^n. Let $\alpha=\begin{pmatrix} a & b \\ c & d \end{pmatrix}$ be a generic point of $GL_2(C)$ over $Q(q_0,\cdots,q_n)$ and let q^α be the polynomial determined by

$$q^\alpha(x) = (cx+d)^n q((ax+b)/(cx+d)).$$

We denote by $W(q)$ the locus of $[q^\alpha]$ over $Q(q_0,\cdots,q_n)$. It can be easily seen that $W(q)$ is a variety determined only by q, and independent of the choice of α. By Prop. 4 and by a standard argument, we get

Proposition 5. *Suppose that m divides n. Let Y and Y^* be as in §§ 2, 3. Then $W(p)=W(p^*)$ if and only if there exists a birational mapping λ of Y to Y^* such that $\lambda\zeta_0=\zeta_0^*\lambda$.*

From this and Prop. 3, we obtain

Proposition 6. *Suppose that m divides n. Let c be the Chow point of $W(p)$. Then $Q(\zeta, c)$ is the field k_0 of Prop. 3.*

Let $p(x)= \sum_{i=0}^{n} p_i x^i$. If $p_0,\cdots, p_n$ are algebraically independent over Q, we see easily that $Q(c)$ is the field of all quotients of homogeneous invariants, in the classical sense, of the binary form $\sum_{i=0}^{n} p_i x^i y^{n-i}$. In particular, if $n=5$, it is known that every invariant is a polynomial of certain invariants A, B, C, R of degree $4, 8, 12, 18$; and R^2 is a polynomial of A, B, C [1]. Then it is clear that $Q(c)=Q(B/A^2, C/A^3)$. If $n=6$, by the same argument, the classical result [1] shows that $Q(c)$ is a purely transcendental extension of Q of dimension 3 (cf. also [2]).

In the next place, suppose that m does not divide n. Choosing a suitable transformation of the type (II) of Prop. 4, we can transform Y to the curve $Y': y^m = x^n + x^{n-2} + \sum_{i=0}^{n-3} p_i x^{n-3-i}$. Suppose that the p_i are algebraically independent over Q. Then, by Prop. 4, we see that $Q(\zeta, p_0^2, p_1, p_2^2, p_3, \cdots)$ is the field k_0 of Prop. 3.

5. Let us now consider the curve $Y: y^m = p(x)$ in the special case $m=n=5$. Then $h=6$, and

$$y^{-2}dx, \ y^{-3}dx, \ xy^{-3}dx, \ y^{-4}dx, \ xy^{-4}dx, \ x^2y^{-4}dx$$

form a basis of the vector space of differential forms of the first kind. Let (J, C, θ) be as in §2. Define an isomorphism θ' of $Q(\zeta)$ into $\mathrm{End}_{Q}(J)$ so that $\theta'(\zeta)=\theta(\zeta^3)$. Hereafter we consider $\mathcal{P}'=(J, C, \theta')$ instead of $\mathcal{P}$. $\mathcal{P}'$ is of type $\{Q(\zeta), \Phi', \rho\}$, for a representation Φ' such that $\Phi'(\zeta)$ is the diagonal matrix with diagonal elements $\zeta, \zeta, \zeta^{-1}, \zeta^3, \zeta^3, \zeta^3$. It is easy to verify that $\mathcal{P}$ and $\mathcal{P}'$ have the same field of moduli. Let $K=Q(\zeta)$, $\zeta=e^{2\pi i/5}$, and let σ_ν, for $\nu=1, 2$, be the automorphism of K such that $\zeta^{\sigma_\nu}=\zeta^\nu$. With the notation r_ν and s_ν of §1, we have $r_1=2$, $s_1=1$, $r_2=0$, $s_2=3$. Define an anti-hermitian form T and a lattice L in K^3 as in §1, for the present $\mathcal{P}'$. The family $\Sigma(K, \Phi', \rho; T, L)$ is parametrized by the point in a domain

$$(5.1) \qquad\qquad H = \left\{(z, w) \in C^2 \Big| |z|^2 + |w|^2 < 1\right\}.$$

In view of (1.3), $\sqrt{-1}\,T^{\sigma_2}$ is positive definite. Hence $H/\Gamma(T, L)$ is compact.

Now take $p(x) = \sum_{i=0}^{5} p_i x^i$ so that the p_i are algebraically independent over Q. Let k_0 be the field of moduli of $\mathcal{P}'$. By [8, 1.7], k_0 contains $K=Q(\zeta)$. The consideration of §4 shows that k_0 is a purely transcendental extension of $Q(\zeta)$ of dimension 2. By Prop. 1, $Q(f_1, \cdots, f_s)$ is a purely transcendental extension of $Q(\zeta)$ of dimension 2.

6. Our next task is to determine T and L explicitly. Let C^h/D and E be as in §1. In our case of $\mathcal{P}'=(J, C, \theta')$, it is essential that J is a jacobian variety. Since every jacobian variety is self-dual, we have

$$D = \left\{x \in C^h \Big| E(x, D) \subset Z\right\},$$

so that by (1.2),

$$(6.1) \qquad L = \left\{ a \in K^3 \,\middle|\, Tr_{K/Q}(T(a, L)) \subset Z \right\}.$$

Put $\eta = \zeta^2 - \zeta^{-2}$, $S = \eta^{-1}T$, $\mathfrak{r} = Z[\zeta]$. We see that $\theta(\mathfrak{r}) \subset \mathrm{End}(J)$, and hence L is an $\mathfrak{r}$-lattice in K^3. Since $\eta^3\mathfrak{r}$ is the different of K with respect to Q, and since $\eta^4\mathfrak{r} = 5\mathfrak{r}$, we have

$$L = \left\{ a \in K^3 \,\middle|\, S(a, L) \subset 5^{-1}\mathfrak{r} \right\}.$$

From this relation we can derive the structure of S and L as follows. Let $\{e_1,\ e_2,\ e_3\}$ be a basis of K^3, and S_0 a hermitian form on K^3 represented by the diagonal matrix with diagonal elements $1, 1,\ (1-\sqrt{5})/2$ with respect to $\{e_i\}$. Then S_0 and S have the same signature at every infinite place of $Q(\sqrt{5})$. Let $\mathfrak{a} = 5^{-1/2}\mathfrak{r}$, $L_0 = \mathfrak{a}e_1 + \mathfrak{a}e_2 + \mathfrak{a}e_3$. Then L_0 is an $\mathfrak{r}$-lattice in K^3, and we have

$$L_0 = \left\{ a \in K^3 \,\middle|\, S_0(a, L_0) \subset 5^{-1}\mathfrak{r} \right\}.$$

By Prop. 8 of Appendix, there exists a K-linear automorphism τ of K^3 such that $S_0(x\tau, y\tau) = S(x, y)$. Therefore we may put $S = S_0$ without loss of generality. By Prop. 6 of Appendix, L and L_0 are μ_0-maximal $\mathfrak{r}$-lattices and $\mu_0(L) = \mu_0(L_0) = 5^{-1}\mathfrak{r}$. By Prop. 5 of Appendix, L and L_0 belong to the same genus with respect to $U(S_0)$. Now $Q(\zeta)$ has the class number 1. Hence by [7, 5.24, (i)], there exists an element α of $U(S_0)$ such that $L_0\alpha = L$. Therefore taking a suitable coordinate system, we may identify $\Gamma(T, L)$ with the group

$$\left\{ \tau \in GL(K^3) \,\middle|\, S_0(x\tau, y\tau) = S_0(x, y),\ L_0\tau = L_0 \right\}.$$

Combining this and the result of §5, we get the assertion of our main theorem in the case (5).

7. We can treat the remaining cases by the same procedure, except the case (3). Let Y be the curve defined by $y^4 = p(x)q(x)^2$, where p and q are polynomials without multiple root, and $\deg(p) = 2$, $\deg(q) = 3$; we assume that p and q have no common root. The genus of Y is 3, and $y^{-1}dx$, $q(x)y^{-3}dx$, $xq(x)y^{-3}dx$ form a basis of differential forms of the first kind. As in §2, from this Y we get $\mathscr{P} = (J, C, \theta)$ of type $\{Q(i),\ \Phi,\ \rho\}$, where $\Phi(i)$ is the diagonal natrix with diagonal elements $i, i, -i$. Define T and L as in §1. Then $\Sigma(Q(i),\ \Phi,\ \rho;\ T, L)$ is again parametrized by H of (5.1). Let $p(x) = \sum_{\lambda=0}^{2} p_\lambda x^\lambda$, $q(x) = \sum_{\mu=0}^{3} q_\mu x^\mu$, and let k_0 be the field of moduli of $\mathscr{P}$. Suppose that the p_λ and q_μ are algebraically independent

over Q. Then we see easily that $Q(i) \subset k_0 \subset Q(i, p_\lambda, q_\mu)$, $\dim_Q k_0 = 2$. By virtue of Castelnuovo's theorem (cf. [9]), this, together with Prop. 1, shows that the field of automorphic functions on H with respect to $\Gamma(T, L)$ is purely transcendental over C.

To determine T anp L, we employ the same argument as in §6. In this case, 2 is the only prime ramified in $Q(i)$. Therefore, the present situation is somewhat different from §6. But the consideration in the last part of Appendix is sufficient to determine T and L explicitly from the relation similar to (6.1). Thus we get the whole result of our theorem.

Appendix

Let F be an algebraic number field of finite degree, and K a quadratic extension of F. We denote by $\mathfrak{g}$ and $\mathfrak{r}$ the ring of integers in F and in K respectively, and by ρ the non-trivial automorphism of K over F. Let V be a vector space over K of dimension n, and $S(x, y)$ a non-degenerate hermitian form: $V \times V \to K$, with respect to ρ. For every $\mathfrak{r}$-lattice L in V, we denote by $\mu'_0(L)$ (resp. $\mu_0(L)$) the ideal in F (resp. K) generated by the elements $S(x, x)$ (resp. $S(x, y)$) for all $x \in L$ (resp. $x \in L$, $y \in L$). L is called *maximal* (resp. μ_0-*maximal*) if there is no $\mathfrak{r}$-lattice M in V, other than L, such that $L \subset M$ and $\mu'_0(L) = \mu'_0(M)$) (resp. $\mu_0(L) = \mu_0(M)$). For every prime ideal $\mathfrak{p}$ of F, let $\mathfrak{g}_\mathfrak{p}$ and $F_\mathfrak{p}$ denote the completions of $\mathfrak{g}$ and F with respect to $\mathfrak{p}$. Then we put $K_\mathfrak{p} = K \otimes_F F_\mathfrak{p}$, $\mathfrak{r}_\mathfrak{p} = \mathfrak{r}\mathfrak{g}_\mathfrak{p}$, $V_\mathfrak{p} = V \otimes_F F_\mathfrak{p}$; ρ and S can be extended naturally to $K_\mathfrak{p}$ and $V_\mathfrak{p}$. We can define similarly μ, μ_0, the maximality, and the μ_0-maximality for $\mathfrak{r}_\mathfrak{p}$-lattices in $V_\mathfrak{p}$. In [7] we investigated maximal lattices. Here we supply some results on μ_0-maximal lattices which are necessary for the proof of our theorem.

Let $\mathfrak{b}$ be the different of K with respect to F. By [7, 2.11], for every $\mathfrak{r}$-lattice L in V, we have

$$(A.1) \qquad \mu(L)\mathfrak{r} \subset \mu_0(L) \subset \mu'(L)\mathfrak{b}^{-1},$$

$$(A.2) \qquad Tr_{K/F}(\mu_0(L)) \subset \mu(L).$$

Therefore, if $\mathfrak{p}$ is unramified in K, we have $\mu_0(L)_\mathfrak{p} = \mu'(L)\mathfrak{r}_\mathfrak{p}$, and hence there is no distinction between maximality and μ_0-maximality for the $\mathfrak{r}_\mathfrak{p}$-lattices in $V_\mathfrak{p}$. If V is one-dimensional, it is clear that every $\mathfrak{r}$-lattice L is maximal and μ_0-maximal, and $\mu_0(L) = \mu'(L)\mathfrak{r}$.

Proposition 1. *Let L be a μ_0-maximal $\mathfrak{r}_\mathfrak{p}$-lattice in $V_\mathfrak{p}$ such that $\mu_0(L) = \mu(L)\mathfrak{b}_\mathfrak{p}^{-1}$. Then L is maximal.*

Let M be an $\mathfrak{r}_\mathfrak{p}$-lattice such that $L \subset M$ and $\mu(M) = \mu_0'(L)$. Then $\mu_0(L) \subset \mu_0(M) \subset \mu(M) \mathfrak{d}_\mathfrak{p}^{-1} = \mu(L) \mathfrak{d}_\mathfrak{p}^{-1} = \mu_0(L)$, so that $\mu_0(L) = \mu_0(M)$. Since L is μ_0-maximal, we get $L = M$; this shows that L is maximal.

Proposition 2. *If $\mathfrak{p}$ does not divide 2, every maximal $\mathfrak{r}_\mathfrak{p}$-lattice in $V_\mathfrak{p}$ is μ_0-maximal.*

By our assumption, for every ideal $\mathfrak{a}_\mathfrak{p}$ in $K_\mathfrak{p}$, we have

$$(A.3) \qquad Tr_{K_\mathfrak{p}/F_\mathfrak{p}}(\mathfrak{a}_\mathfrak{p}) = \mathfrak{a}_\mathfrak{p} \cap F_\mathfrak{p}.$$

Hence, from (A. 1) and (A. 2), we obtain, for every $\mathfrak{r}_\mathfrak{p}$-lattice L in $V_\mathfrak{p}$,

$$(A.4) \qquad Tr_{K_\mathfrak{p}/F_\mathfrak{p}}(\mu_0(L)) = \mu_0(L) \cap F_\mathfrak{p} = \mu_0'(L).$$

Then our assertion is obvious.

Proposition 3. *Suppose that $n = 2$, $\mathfrak{p}$ does not divide 2, and S is anisotropic in $V_\mathfrak{p}$. Then every μ_0-maximal $\mathfrak{r}_\mathfrak{p}$-lattice in $V_\mathfrak{p}$ is maximal.*

If $\mathfrak{p}$ is unramified in K, there is no problem; so we assume that $\mathfrak{p}$ is ramified in K. Let L be a μ_0-maximal $\mathfrak{r}_\mathfrak{p}$-lattice in $V_\mathfrak{p}$. Since $\mathfrak{p}$ does not divide 2, the relation (A. 1) implies that $\mu_0(L) = \mu_0'(L)\mathfrak{r}_\mathfrak{p}$ or $\mu_0(L) = \mu(L)\mathfrak{d}_\mathfrak{p}^{-1}$. If $\mu_0(L) = \mu_0'(L)\mathfrak{d}_\mathfrak{p}^{-1}$, L is maximal by virtue of Prop. 1. Assume that $\mu_0(L) = \mu_0'(L)\mathfrak{r}_\mathfrak{p}$. Then there exists an element x of L such that $\mu_0(L) = (S(x, x))$. Put $L' = \{y \in L \mid S(x, y) = 0\}$. We can easily verify that $L = \mathfrak{r}_\mathfrak{p} x + L'$. Since V is two-dimensional, we have $L' = \mathfrak{r}_\mathfrak{p} y$ for some y. Since L is μ_0-maximal and $\mathfrak{p}$ is ramified in K, we must have $(S(y, y)) = \mu_0(L)$. Now put $M = \{u \in V \mid S(u, u) \in \mu_0'(L)\}$. By [7, 4. 5], M is a maximal $\mathfrak{r}_\mathfrak{p}$-lattice in $V_\mathfrak{p}$. We have clearly $L \subset M$. Let $u = ax + by \in M$ with a, b in $K_\mathfrak{p}$. Then

$$aa^\rho S(x, x) + bb^\rho S(y, y) \in \mu_0'(L) = (S(x, x)).$$

Put $c = S(x, x)^{-1} S(y, y)$. Then c is a unit in $\mathfrak{g}_\mathfrak{p}$, and $aa^\rho + bb^\rho c \in \mathfrak{g}_\mathfrak{p}$. Let π be a prime element of $\mathfrak{r}_\mathfrak{p}$. Assume that $u \notin L$. Then $\pi^e a$ and $\pi^e b$ are units in $\mathfrak{r}_\mathfrak{p}$ with a positive integer e, and

$$(\pi^e a)(\pi^e a)^\rho + (\pi^e b)(\pi^e b)^\rho c \equiv 0 \qquad \mathrm{mod}\,(\pi\pi^\rho)^e \mathfrak{g}_\mathfrak{p}.$$

It follows that $-c$ is the norm of an element in $K_\mathfrak{p}$. But this is a contradiction, since S is anisotropic in $V_\mathfrak{p}$. Therefore $u \in L$, and hence $M = L$. This proves our proposition.

Proposition 4. *Suppose that $\mathfrak{p}$ does not decompose in K. When $\mathfrak{p}$ divides 2, suppose further that $\mathfrak{p}$ is unramified in K. Let L be a μ_0-*

maximal $\mathfrak{r}_\mathfrak{p}$-lattice in $V_\mathfrak{p}$. Put $\mathfrak{b}=\mu_0(L)$. Then there exists a Witt decomposition $V_\mathfrak{p}=\sum_{i=1}^m (K_\mathfrak{p}x_i+K_\mathfrak{p}y_i)+W$ (cf. [7, 4. 3]) such that $L=\sum_{i=1}^m (\mathfrak{r}_\mathfrak{p}x_i+\mathfrak{b}y_i)+M$, where M is a maximal $\mathfrak{r}_\mathfrak{p}$-lattice in W given by $M=\{z\in W\,|\,S(z, z)\in\mu(L)\}$. Conversely, let $\mathfrak{b}$ be an ideal in $K_\mathfrak{p}$, and $V_\mathfrak{p}=\sum_{i=1}^m (K_\mathfrak{p}x_i+K_\mathfrak{p}y_i)+W$ be a Witt decomposition. Let

$$M = \left\{z\in W\,\middle|\,S(z, z)\in\mathfrak{b}\cap F_\mathfrak{p}\right\}, \quad L=\sum_{i=1}^m (\mathfrak{r}_\mathfrak{p}x_i+\mathfrak{b}\,y_i)+M\,.$$

Then L is a μ_0-maximal $\mathfrak{r}_\mathfrak{p}$-lattice in $V_\mathfrak{p}$.

The converse part can be proved in a straightforward way. The proof of the direct part is similar to the proof of [7, 4. 7]; so here we only sketch a proof. Assume that S is isotropic in $V_\mathfrak{p}$. Then we can find an element x in $V_\mathfrak{p}$ such that $S(x, x)=0$ and $\mathfrak{r}_\mathfrak{p}=\{a\in K_\mathfrak{p}\,|\,ax\in L\}$. Put $\mathfrak{a}=S(x, L)$. We get easily $\mu_0(\mathfrak{a}^{-1}\mathfrak{b}x+L)=\mu_0(L)$, so that $\mathfrak{a}^{-1}\mathfrak{b}x+L=L$ by virtue of the μ_0-maximality of L. We have therefore $L\supset\mathfrak{a}^{-1}\mathfrak{b}x$, and hence $\mathfrak{a}^{-1}\mathfrak{b}\subset\mathfrak{r}_\mathfrak{p}$. It follows that $S(x, L)=\mathfrak{b}$. Therefore we find an element $u\in L$ such that $\mathfrak{b}=(S(x, u))$. Our assumption implies $\mu(L)=Tr_{K_\mathfrak{p}/F_\mathfrak{p}}(\mathfrak{b})$ $=Tr_{K_\mathfrak{p}/F_\mathfrak{p}}(S(x, u)\mathfrak{r}_\mathfrak{p})$. Using this fact, we can find an element λ of $\mathfrak{r}_\mathfrak{p}$ such that $S(x+\lambda u, x+\lambda u)=0$. Put $y=x+\lambda u$, $L'=\{z\in L\,|\,S(x, z)=S(y, z)=0\}$. Then we have $L=\mathfrak{r}_\mathfrak{p}x+\mathfrak{b}y+L'$. Applying induction to L', we get our assertion, in view of Prop. 3.

Let $U(S)$ be the group of all K-linear automorphisms σ of V such that $S(x\sigma, y\sigma)=S(x, y)$. As in [7, 5. 18] we define genera of $\mathfrak{r}$-lattices in V.

Proposition 5. *Suppose that every prime factor of 2 in F is unramified in K. Let L be a μ_0-maximal $\mathfrak{r}$-lattice in V. Then the genus of L with respect to $U(S)$ consists of all μ_0-maximal $\mathfrak{r}$-lattices M such that $\mu_0(M)=\mu_0(L)$.*

This follows directly from [7, 3. 3] and Prop. 4 by the same argument as in the proof of [7, 5. 25].

Proposition 6. *Suppose that every prime factor of 2 in F is unramified in K. Let $\mathfrak{a}$ be an ideal in F, and L an $\mathfrak{r}$-lattice in V. Suppose that $L=\{x\in V\,|\,S(x, L)\subset\mathfrak{a}\mathfrak{r}\}$. Then L is μ_0-maximal, and $\mu_0(L)=\mathfrak{a}\mathfrak{r}$.*

Our assertion is clear if $n=1$. Suppose that $n>1$. For every $\mathfrak{r}$-lattice M in V, define M^* by $M^*=\{x\in V\,|\,S(x, M)\subset\mathfrak{a}\mathfrak{r}\}$. We see that $M\subset M^*$ if and only if $\mu_0(M)\subset\mathfrak{a}\mathfrak{r}$. Since $L=L^*$, we have $\mu_0(L)\subset\mathfrak{a}\mathfrak{r}$. If $M_1\subset M_2$, then $M_1^*\supset M_2^*$. Now let $L\subset M$, $\mu_0(M)\subset\mathfrak{a}\mathfrak{r}$. Then we have $M^*\subset L^*=L\subset M\subset M^*$, so that $L=M$. This shows that L is μ_0-maximal.

By [7, 3.2] and by Prop. 4, we can easily find a μ_0-maximal $\mathfrak{r}$-lattice L' such that $L \subset L'$ and $\mu_0(L') = \mathfrak{a}\mathfrak{r}$. Then the above argument shows again $L = L'$. This proves our proposition.

Proposition 7. *Let $\mathfrak{p}$ be a prime ideal in F which remains prime in K. Suppose that there exist an ideal $\mathfrak{a}$ in $F_\mathfrak{p}$ and an $\mathfrak{r}_\mathfrak{p}$-lattice L in $V_\mathfrak{p}$ such that $L = \{x \in V_\mathfrak{p}^* | S(x, L) \subset \mathfrak{a}\mathfrak{r}_\mathfrak{p}\}$. Then the structure $(V_\mathfrak{p}, S)$ is uniquely determined by $\mathfrak{a}$. More precisely, if n is odd and $\mathfrak{a} = a\mathfrak{g}_\mathfrak{p}$, $d(S)$ is the class of $(-1)^{(n-1)/2}a$ modulo $N_{K_\mathfrak{p}/F_\mathfrak{p}}(K_\mathfrak{p}^*)$ (cf. [7, 2.1 and 4.2]). If n is even, S is maximally isotropic in $V_\mathfrak{p}$, namely, $V_\mathfrak{p}$ has the trivial kernel subspace with respect to S (cf. [7, 4.3]).*

By Prop. 4, we find a Witt decomposition $V_\mathfrak{p} = \sum_{i=1}^{m}(K_\mathfrak{p}x_i + K_\mathfrak{p}y_i) + W$ such that $L = \sum_{i=1}^{m}(\mathfrak{r}_\mathfrak{p}x_i + \mathfrak{b}y_i) + M$, $M = \{u \in W | S(u, u) \in \mathfrak{a}\}$. By our assumption on L, we have

$$(A.5) \qquad M = \left\{ u \in W \Big| S(u, M) \subset \mathfrak{a}\mathfrak{r}_\mathfrak{p} \right\}.$$

If n is odd, we have $W = K_\mathfrak{p}z$, $W = \mathfrak{r}_\mathfrak{p}z$ for some z. Hence (A.5) implies that $\mathfrak{a} = S(z, z)\mathfrak{g}_\mathfrak{p}$. Since $\mathfrak{p}$ is unramified in K, every unit in $\mathfrak{g}_\mathfrak{p}$ is the norm of an element of $K_\mathfrak{p}$. Therefore we get our assertion for odd n. Next assume that n is even and W is two-dimensional. Since $\mu_0(M) = \mu(M)\mathfrak{r}_\mathfrak{p} = \mathfrak{a}\mathfrak{r}_\mathfrak{p}$, we find, using the argument of the proof of Prop. 3, an expression $M = \mathfrak{r}_\mathfrak{p}u + \mathfrak{r}_\mathfrak{p}v$ with $S(u, v) = 0$. On account of (A.5), we see that $\mathfrak{a} = S(u, u)\mathfrak{g}_\mathfrak{p} = S(v, v)\mathfrak{g}_\mathfrak{p}$; hence $S(u, u)^{-1}S(v, v)$ is a unit in $\mathfrak{g}_\mathfrak{p}$. Therefore we find an element c in $K_\mathfrak{p}$ such that $cc^\rho = -S(u, u)^{-1}S(v, v)$. Then we get $S(cu + v, cu + v) = 0$, which is a contradiction. Hence S must be maximally isotropic in $V_\mathfrak{p}$.

Proposition 8. *Suppose that there is no or only one prime ideal in F which is ramified in K. Suppose that there exist an ideal $\mathfrak{a}$ in F and an $\mathfrak{r}$-lattice L such that $L = \{x \in V | S(x, L) \subset \mathfrak{a}\mathfrak{r}\}$. Then the structure (V, S) is uniquely determined, up to isomorphism, by $\mathfrak{a}$ and the signature of S at infinite prime spots of F.*

Let $\mathfrak{q}$ be a possible prime ideal in F which is ramified in K. By Prop. 7, the structure $(V_\mathfrak{p}, S)$ is uniquely determined by $\mathfrak{a}$ if $\mathfrak{p} \neq \mathfrak{q}$. If we assign a fixed signature to each infinite prime spot of F, then the structure $(V_\mathfrak{q}, S)$ is automatically determined by virtue of the product formula of norm residue symbol. This proves our proposition.

If a prime factor of 2 in F is ramified in K, we can not apply Prop. 5. However, under a suitable condition, we may treat such a case.

For example, let us consider the case where $F=\mathbf{Q}$, $K=\mathbf{Q}(i)$, $i^2=-1$, $n=3$. Let $\mathfrak{p}=(2)$, $\mathfrak{P}=(1+i)$, and let L be an $\mathfrak{r}_{\mathfrak{p}}$-lattice in $V_{\mathfrak{p}}$ such that

$$(A.6) \qquad L = \left\{ x \,\middle|\, S(x, L)\subset(2^{-1}) \right\}.$$

Assume that S is represented in $V_{\mathfrak{p}}$ by the diagonal matrix with diagonal elements $1, 1, -1$. Now by [7, 4. 15], the following two cases may occur.

(I) $\quad L = \mathfrak{r}_{\mathfrak{p}}x+\mathfrak{r}_{\mathfrak{p}}y+\mathfrak{r}_{\mathfrak{p}}z$, $S(x, y) = S(y, z)=S(z, x) = 0$.

(II) $\quad L = \mathfrak{r}_{\mathfrak{p}}x+\mathfrak{r}_{\mathfrak{p}}y+\mathfrak{r}_{\mathfrak{p}}z$, $S(x, z) = S(y, z) = 0$,

$\qquad S(x, x)\in \mathfrak{P}S(x, y)$, $S(y, y)\in \mathfrak{P}S(x, y)$.

Put $S(x, x)=a$, $S(y, y)=b$, $S(z, z)=c$, $S(x, y)=d$. In the case (I), by (A.6), we have $(a)=(b)=(c)=(2^{-1})$. Put $2a=a'$, $2b=b'$, $2c=c'$. By our assumption on S, $-a'b'c'$ must be the norm of an element of $K_{\mathfrak{p}}$ (cf. [7, 4. 2]). Since -1 is not a norm residue, we may assume, exchanging the order of x, y, z if necessary, that $a'=b'=c'=-1$ or $a'=b'=1$, $c'=-1$. The former case can be reduced to the latter case by the transformation $u=e^{-1}(x-(1+i)y)$, $v=e^{-1}((1-i)x+y)$, $w=z$, where e is an element of $\mathfrak{r}_{\mathfrak{p}}$ such that $ee^{\rho}=-3$.

In the case (II), by (A.6), we have $(c)=(d)=(2^{-1})$, so that $a\in \mathfrak{g}_{\mathfrak{p}}$, $b\in \mathfrak{a}_{\mathfrak{p}}$. Hence $dd^{\rho}-ab$ is the norm of an elemet of $K_{\mathfrak{p}}$. Therefore, by [7, 4. 1], S is isotropic in $K_{\mathfrak{p}}x+K_{\mathfrak{p}}y$. It follows that c is the norm of an element of $K_{\mathfrak{p}}$, on account of our assumption on S. Hence we may assume $\mathfrak{r}_{\mathfrak{p}}z=\mathfrak{P}^{-1}w$ with $S(w, w)=1$. Put $M=\mathfrak{r}_{\mathfrak{p}}x+\mathfrak{r}_{\mathfrak{p}}y$. Then $\mu_0'(M)\subset \mathfrak{g}_{\mathfrak{p}}$ $=2\mu_0(M)\subset\mu_0'(M)$, so that $2\mu_0(M)=\mu_0'(M)$. Applying the argument of the proof of Prop. 6 to M, we see that M is μ_0-maximal, so that by Prop. 1, M is maximal. By [7, 4. 7], we have $M=\mathfrak{P}^{-1}u+\mathfrak{P}^{-1}v$ with $S(u, u)=S(v, v)$ $=0$, $S(u, v)=1$. Put $r=u+w$, $s=v-w$, $t=u-v+w$. Then $S(r, r)=S(s, s)$ $=1$, $S(t, t)=-1$, $S(r, s)=S(s, t)=S(t, r)=0$, $L=\mathfrak{P}^{-1}r+\mathfrak{P}^{-1}s+\mathfrak{P}^{-1}t$. Therefore L is reduced to the case (I).

This result, combined with Prop. 4 and a localization of Prop. 6, shows that every $\mathfrak{r}$-lattice L in V satisfying (A.6) belongs to one and the same genus with respect to $U(S)$.

Osaka University and Princeton University

References

[1] A. Clebsch: Theorie der binären algebraischen Formen, Leipzig, 1872.

[2] J. Igusa: *Arithmetic variety of moduli for genus two*, Ann. of Math. 72 (1960), 612-649.

[3] E. Picard: *Sur des fonctions de deux variables indépendantes analogues aux fonctions modulaires*,　Acta Mathematica **2** (1883), 114–135.

[4] G. Shimura: *On the theory of automorphic functions*,　Ann. of Math. **70** (1959), 101–144.

[5] ————: *On the zeta-functions of the algebraic curves uniformized by certain automorphic functions*,　J. Math. Soc. Japan **13** (1961), 275–331.

[6] ————: *On analytic families of polarized abelian varieties and automorphic functions*,　Ann. of Math. **78** (1963), 149–192.

[7] ————: *Arithmetic of unitary groups*, Ann. of Math. **79** (1964), 369–409.

[8] ————: *On the field of definition for a field of automorphic functions*,　Ann. of Math. **80** (1964), 160–189.

[9] O. Zariski: *On Castelnuovo's criterion of rationality $p_a = P_2 = 0$ of an algebraic surface*,　Illinois J. Math. **2** (1958), 303–315.

The zeta function of an algebraic variety and automorphic functions

Summer Research Institute on Algebraic Geometry,
Woods Hole, Massachusetts, July 6-31, 1964 (mimeographed notes of lectures)*

One of our colleagues asked me not to release the "latest" pictures in this conference, but to rerun some classical ones. Following his suggestion, at least in the first half of this lecture, I will tell the old story of what happened to the zeta-function of an algebraic curve uniformized by modular functions. Then I'd like to talk about its application to the law of reciprocity in non-solvable extensions, and indicate briefly some generalizations.

1. Introduction. Let V be an algebraic variety defined over an algebraic number field k. For every prime ideal $\mathfrak{p}$ of k, let $V(\mathfrak{p})$ denote the reduction of V modulo $\mathfrak{p}$ and $k(\mathfrak{p})$ the residue field of k modulo $\mathfrak{p}$. For each $\mathfrak{p}$ we can define the zeta-function $Z\big(u; V(\mathfrak{p})/k(\mathfrak{p})\big)$ by

$$Z\big(0; V(\mathfrak{p})/k(\mathfrak{p})\big) = 1, \qquad d\big[\log Z\big(u; V(\mathfrak{p})/k(\mathfrak{p})\big)\big]/du = \sum_{m=1}^{\infty} N_m u^{m-1},$$

where N_m is the number of points on $V(\mathfrak{p})$ rational over the extension of $k(\mathfrak{p})$ of degree m. The zeta-function of V over k, denoted by $\zeta(s; V/k)$, is then defined by

$$\zeta(s; V/k) = \prod_{\mathfrak{p}} Z\big(N(\mathfrak{p})^{-s}; V(\mathfrak{p})/k(\mathfrak{p})\big),$$

the product being taken over all the prime ideals $\mathfrak{p}$ of k. If we assume Weil's conjecture to be true for $V(\mathfrak{p})$, then, for all except a finite number of $\mathfrak{p}$, we have

$$Z\big(u; V(\mathfrak{p})/k(\mathfrak{p})\big) = \frac{H_{\mathfrak{p}}^{(1)}(u) \cdots H_{\mathfrak{p}}^{(2n-1)}(u)}{H_{\mathfrak{p}}^{(0)}(u) \cdots H_{\mathfrak{p}}^{(2n)}(u)}.$$

Here $n = \dim(V)$ and $H_{\mathfrak{p}}^{(i)}(u)$ is a polynomial, with the constant term 1, whose roots are algebraic integers of absolute value $N(\mathfrak{p})^{-i/2}$. Moreover, we may expect that the degree of $H_{\mathfrak{p}}^{(i)}$ is independent of $\mathfrak{p}$ for each i. Then it will be meaningful to consider a function

$$\zeta^{(i)}(s; V/k) = \prod_{\mathfrak{p}}' H_{\mathfrak{p}}^{(i)}\big(N(\mathfrak{p})^{-s}\big)^{-1},$$

where the product is taken over all good $\mathfrak{p}$'s. Now the conjecture of Hasse-Weil (in the generalized sense) may be stated as follows: *For every i, $\zeta^{(i)}(s; V/k)$ is meromorphically continued to the whole s-plane and satisfies a functional equation.* For example, if V is an abelian variety (resp. a curve), we have

$$\zeta^{(1)}(s; V/k) = \prod_{\mathfrak{p}}' \det\big[1 - M_\ell(\pi_{\mathfrak{p}})N(\mathfrak{p})^{-s}\big]^{-1},$$

where $\pi_{\mathfrak{p}}$ is the $N(\mathfrak{p})$-th power endomorphism of $V(\mathfrak{p})$ (resp. the jacobian of $V(\mathfrak{p})$), and $M_\ell(\pi_{\mathfrak{p}})$ is its ℓ-adic representation. Therefore, the determination of $\zeta^{(1)}(s; V/k)$ is, roughly speaking, the determination of $M_\ell(\pi_{\mathfrak{p}})$ as a function of $\mathfrak{p}$.

At present, there are two known classes of varieties V for which the Hasse-Weil conjecture is true:

(I) abelian varieties with sufficiently many complex multiplications;

(II) algebraic curves uniformized by certain automorphic functions of one variable.

In Case (I), V is an abelian variety of dimension n such that $\mathrm{End}_{\mathbf{Q}}(V)$ is isomorphic to an algebraic number field F of degree $2n$. (We may consider a somewhat more general case where $\mathrm{End}_{\mathbf{Q}}(V)$ is not necessarily a field. For simplicity, we assume here $\mathrm{End}_{\mathbf{Q}}(V)$ to be a field.) It can be shown that there exists an element $\mu_{\mathfrak{p}}$ in $\mathrm{End}_{\mathbf{Q}}(V)$ whose reduction modulo $\mathfrak{p}$ is $\pi_{\mathfrak{p}}$. Moreover, we can determine the prime ideal decomposition of $(\mu_{\mathfrak{p}})$ in F. These facts, together with a simple class-field theoretical consideration, show that $\mathfrak{p} \mapsto \mu_{\mathfrak{p}}$ is essentially a Grössen-character of k. From this it follows that $\zeta^{(1)}(s; V/k)$ is the product of $2n$ Hecke L-functions with Grössen-characters. Detailed accounts of the theory can be found in Taniyama [17]; a comparatively easy and systematic exposition is given in [14, Ch. IV, §18]; for partial or related results see Deuring [1] and Weil [18].

Among many factors which make the calculation of $\zeta^{(1)}$ possible in Case (I), it is most important that $\pi_{\mathfrak{p}}$ can be lifted up to an element of $\mathrm{End}_{\mathbf{Q}}(V)$ for every $\mathfrak{p}$. Of course we cannot expect this in general. However, in Case (II), we can show, roughly speaking, that $\pi_{\mathfrak{p}} + \pi_{\mathfrak{p}}^{*}$ belongs to the original $\mathrm{End}_{\mathbf{Q}}(\text{Jacobian of } V)$ for a certain involution $*$. This fact makes it possible to prove the Hasse-Weil conjecture for the curves of (II). To exlain these in detail, we need some preliminaries on automorphic forms and Hecke operators.

2. Discontinuous groups and automorphic forms on the upper half plane. Let H be the complex upper half plane, that is,

$$H = \left\{ z \in \mathbf{C} \,\middle|\, \mathrm{Im}(z) > 0 \right\}.$$

Every $\alpha = \begin{bmatrix} a & b \\ c & d \end{bmatrix} \in GL_2(\mathbf{R})$ with $\det(\alpha) > 0$ acts on H by $\alpha(z) = (az + b)/(cz + d)$. There is a natural measure on H invariant under this action. A discrete subgroup Γ of $SL_2(\mathbf{R})$ is called a *Fuchsian group of the first kind,* if $\Gamma \backslash H$ is of finite measure. Hereafter we fix such a Γ. An element $\alpha = \begin{bmatrix} a & b \\ c & d \end{bmatrix}$ of Γ, other than ± 1, is called *parabolic* if $z \mapsto (az + b)/(cz + d)$ has only one fixed point on the whole z-sphere. If that is so, the fixed point should be a real number or the point at infinity, and is called a *cusp* of Γ. Let s be a cusp of Γ. All the elements of Γ that leave s invariant are ± 1 or parabolic. Together with ± 1, they form a group which is the product of $\{\pm 1\}$ and an infinite cyclic group generated by an element τ of the form $\tau = \rho \begin{bmatrix} 1 & 1 \\ 0 & 1 \end{bmatrix} \rho^{-1}$ with an element ρ of $SL_2(\mathbf{R})$ such that $\rho(\infty) = s$. Let H^* be the union of H and all the cusps of Γ. We can introduce a complex structure on $\Gamma \backslash H^*$ so that $\Gamma \backslash H^*$ is a compact Riemann surface. To be more precise, a base of neighborhoods of s in H^* is given by $\{s\} \cup \rho(\{z \in \mathbf{C} \,|\, \mathrm{Im}(z) > r\})$ for $r > 0$, and $\exp\left[2\pi i \rho^{-1}(z)\right]$ is a local analytic coordinate around s modulo Γ. Therefore $\Gamma \backslash H$ is compact if and only if Γ has no parabolic elements.

Now, for every $\alpha = \begin{bmatrix} a & b \\ c & d \end{bmatrix} \in GL_2(\mathbf{R})$ with $\det(\alpha) > 0$, set

$$j(\alpha, z) = \det(\alpha)^{-1/2}(cz + d).$$

We can easily verify that $j(\alpha, z)^{-2} = (d/dz)\alpha(z)$, and

$$j(\alpha\beta, z) = j(\alpha, \beta(z))j(\beta, z).$$

Let m be an integer. An *automorphic form of weight m with respect to Γ* is a meromorphic function f on H satisfying the following conditions (A1, 2).

(A1) $$f(\alpha(z))j(\alpha, z)^{-m} = f(z) \quad \text{for every} \quad \alpha \in \Gamma.$$

To describe condition (A2), take a cusp s of Γ and elements τ, ρ of $SL_2(\mathbf{R})$ as above. If f satisfies (A1), then the function $f(\rho(z))j(\rho, z)^{-m}$ is invariant under the translation $z \mapsto z + 1$. Hence there exists a function $F_s(q)$, meromorphic in $0 < |q| < 1$, such that $f(\rho(z))j(\rho, z)^{-m} = F_s(e^{2\pi i z})$. Then (A2) is stated as follows:

(A2) *For every cusp s of Γ, $F_s(q)$ is meromorphic at $q = 0$.*

An automorphic form of weight 0 is called an *automorphic function*. If g is a meromorphic function on the compact Riemann surface $\Gamma\backslash H^*$ and φ is a natural projection of H^* to $\Gamma\backslash H^*$, then $g \circ \varphi$ is an automorphic function with respect to Γ; conversely every automorphic function with respect to Γ can be obtained in this way.

An automorphic form of weight $m > 0$ is called a *cusp form* if f is holomorphic on the whole H, and the following condition is satisfied:

(A3) *For every cusp s of Γ, $F_s(q)$ is holomorphic at $q = 0$.*

We denote by $S_m(\Gamma)$ the set of all cusp forms of weight m with respect to Γ. The dimension of $S_m(\Gamma)$ can be easily determined by means of the Riemann-Roch theorem for $\Gamma\backslash H^*$ In particular, there is a canonical isomorphism $f \mapsto \omega$ of $S_2(\Gamma)$ onto the vector space of all the differential forms of the first kind on $\Gamma\backslash H^*$, defined by $\omega \circ \varphi = f(z)dz$. Therefore the dimension of $S_2(\Gamma)$ over $\mathbf{C}$ is exactly the genus of $\Gamma\backslash H^*$.

3. Hecke operators. Let Δ be a subset of $GL_2(\mathbf{R})$, closed under multiplication and containing Γ. Suppose that $\det(\alpha) > 0$ for every $\alpha \in \Delta$, and the following condition is satisfied:

(3.1) *For every $\alpha \in \Delta$, the double coset $\Gamma\alpha\Gamma$ contains only a finite number of right and left cosets with respect to Γ.*

Let $R(\Gamma, \Delta)$ be the module consisting of all the formal finite sums $\sum_\lambda c_\lambda \Gamma\alpha_\lambda\Gamma$ with $\alpha_\lambda \in \Delta$ and $c_\lambda \in \mathbf{C}$. We can introduce a law of multiplication in $R(\Gamma, \Delta)$ as follows. Let $\alpha, \beta \in \Delta$, and let $\Gamma\alpha\Gamma = \bigcup_i \Gamma\alpha_i$ and $\Gamma\beta\Gamma = \bigcup_j \Gamma\beta_j$ be disjoint expressions. Then for every $\Gamma\xi\Gamma$ with $\xi \in \Delta$, the number of (i, j) such that $\Gamma\alpha_i\beta_j = \Gamma\xi$ is uniquely determined by the double cosets $\Gamma\alpha\Gamma$, $\Gamma\beta\Gamma$, $\Gamma\xi\Gamma$; it is independent of the choice of representatives $\{\alpha_i\}$, $\{\beta_j\}$, and ξ. Call this number $\mu(\Gamma\alpha\Gamma \cdot \Gamma\beta\Gamma; \Gamma\xi\Gamma)$ and set

$$(3.2) \qquad \Gamma\alpha\Gamma \cdot \Gamma\beta\Gamma = \sum_{\Gamma\xi\Gamma} \mu(\Gamma\alpha\Gamma \cdot \Gamma\beta\Gamma;\ \Gamma\xi\Gamma)\Gamma\xi\Gamma.$$

Extending this to the whole $R(\Gamma,\ \Delta)$ by linearity, we obtain an associative ring.

Every element of $R(\Gamma,\ \Delta)$ operates on $S_m(\Gamma)$ in the following way: Let $\Gamma\alpha\Gamma = \bigcup_{i=1}^{d} \Gamma\alpha_i$ be a disjoint expression. For every $f \in S_m(\Gamma)$, define $g = f|T_m(\Gamma\alpha\Gamma)$ by

$$g(z) = \det(\alpha)^{m/2-1} \sum_{i=1}^{d} f(\alpha_i(z)) j(\alpha_i,\ z)^{-m}.$$

It can easily be shwon that $g \in S_m(\Gamma)$. From (3.2) we see that $\Gamma\alpha\Gamma \mapsto T_m(\Gamma\alpha\Gamma)$ defines a representation of $R(\Gamma,\ \Delta)$ by linear transformations on $S_m(\Gamma)$.

Let us now consider a special case where Γ is $SL_2(\mathbf{Z})$, and Δ is the set of all integral matrices of size 2 with positive determinant. It can be shown that $R(\Gamma,\ \Delta)$ is a commutative integral domain. The representatives for $\Gamma\backslash\Delta/\Gamma$ are given by the matrices of the form $\alpha = \begin{bmatrix} a & 0 \\ 0 & d \end{bmatrix}$, where $a > 0$ and a divides d. For this α, let $T(a, d)$ denote $\Gamma\alpha\Gamma$ as an element of $R(\Gamma,\ \Delta)$. Then we have

$$(3.3) \qquad T(a, d)T(a', d') = T(aa', dd') \quad \text{if} \quad (d, d') = 1.$$

Therefore $R(\Gamma,\ \Delta)$ is generated by $T(p^\lambda,\ p^\mu)$ with $\mu \geq \lambda \geq 0$ for all the prime numbers p. Let $T(p^n)$ be the sum of all $T(p^\lambda,\ p^\mu)$ such that $\lambda + \mu = n$ and $\mu \geq \lambda \geq 0$. Then we can prove that $R(\Gamma,\ \Delta)$ is commutative and

$$T(p)T(p^n) = T(p^{n+1}) + pT(p, p)T(p^{n-1}) \qquad (n > 0).$$

It follows from this that, for an indeterminate u,

$$(3.4) \qquad \sum_{n=1}^{\infty} T(p^n)u^n = \left[1 - T(p)u + pT(p, p)u^2\right]^{-1}.$$

Now we consider a formal Dirichlet series $D(s;\ \Gamma)$ with coefficients in $R(\Gamma,\ \Delta)$:

$$D(s;\ \Gamma) = \sum_{\alpha \in \Gamma\backslash\Delta/\Gamma} \Gamma\alpha\Gamma \cdot \det(\alpha)^{-s} = \sum_{a|d} T(a, d)(ad)^{-s},$$

where the sum is extended over all the double cosets $\Gamma\alpha\Gamma$ with $\alpha \in \Delta$. By virtue of (3.3) and (3.4), we get an Euler product:

$$D(s;\ \Gamma) = \prod_p \left[1 - T(p)p^{-s} + T(p, p)p^{1-2s}\right]^{-1}.$$

Let us define *the principal congruence subgroup Γ_N of level N* by

$$(3.5) \qquad \Gamma_N = \left\{ \alpha \in SL_2(\mathbf{Z}) \,\middle|\, \alpha \equiv 1 \ (\mathrm{mod}\ N) \right\},$$

for every positive integer N. A holomorphic automorphic form (resp. an automorphic function) with respect to Γ_N is usually called a *modular form* (resp. *modular function*) *of level N*. Let Δ_N be the set of all matrices α in Δ such that

$$\alpha \equiv \begin{bmatrix} 1 & 0 \\ 0 & d \end{bmatrix} \ (\mathrm{mod}\ N), \quad (d, N) = 1,$$

and let Δ_N^* be the set of all α in Δ such that $\left(\det(\alpha),\ N\right) = 1$. Then we obtain an isomorphism of $R(\Gamma_N,\ \Delta_N)$ onto $R(\Gamma,\ \Delta_N^*)$ by

$$(3.6) \qquad \Gamma_N\alpha\Gamma_N \mapsto \Gamma\alpha\Gamma \qquad (\alpha \in \Delta_N).$$

Therefore, if we set $D^N(s) = \sum_{\alpha \in \Gamma_N\backslash\Delta_N/\Gamma_N} \Gamma_N\alpha\Gamma_N \cdot \det(\alpha)^{-s}$, then

$$D^N(s) = \prod_{p\nmid N} \left[1 - T^N(p)p^{-s} + T^N(p, p)p^{1-2s}\right]^{-1},$$

where $T^N(p)$ and $T^N(p, p)$ are the elements of $R(\Gamma_N, \Delta_N)$ corresponding to $T(p)$ and $T(p, p)$ by the isomorphism of (3.6), respectively.

Taking the representation of $R(\Gamma_N, \Delta_N)$ on $S_m(\Gamma_N)$, i. e., the map $\Gamma_N \alpha \Gamma_N \mapsto T_m(\Gamma_N \alpha \Gamma_N)$, we obtain a Dirichlet series with matrix coefficients

$$(3.7) \qquad D_m^N(s) = \sum_{\alpha \in \Gamma_N \backslash \Delta_N / \Gamma_N} T_m(\Gamma_N \alpha \Gamma_N) \det(\alpha)^{-s}$$

$$= \prod_{p \nmid N} \left[1 - T_m^N(p)p^{-s} + T_m^N(p, p)p^{1-2s} \right]^{-1},$$

which converges absolutely for sufficiently large $\mathrm{Re}(s)$. Moreover, $D_m^N(s)$ can be continued holomorphically to the whole s-plane, and satisfies a functional equation. If $N = 1$, the functional equation has the following form:

$$D_m^*(s) = D_m^*(m - s) \quad \text{with} \quad D_m^*(s) = \Gamma(s)(2\pi)^{-s} D_m^1(s).$$

With respect to a suitable basis $\{f_1, \ldots, f_h\}$ of $S_m(\Gamma_1)$, the $T_m(\Gamma_1 \alpha \Gamma_1)$ for all $\alpha \in \Delta$ can be represented by diagonal matrices simultaneously. Put

$$f_j(z) = \sum_{n=1}^\infty a_n^{(j)} e^{2\pi i n z}.$$

Then one can prove that the diagonal elements of $D_m^1(s)$ are

$$\sum_{n=1}^\infty a_n^{(j)} n^{-s} = \prod_p \left[1 - a_p^{(j)} p^{-s} + p^{m-1-2s} \right]^{-1} \qquad (1 \leq j \leq h).$$

In particular, $S_{12}(\Gamma_1)$ is one-dimensional and generated by

$$(3.8) \qquad \Delta(z) = q \prod_{n=1}^\infty (1 - q^n)^{24} = \sum_{n=1}^\infty a_n q^n, \quad q = e^{2\pi i z}.$$

In 1916, Ramanujan conjectured the existence of an Euler product for the series $\sum_{n=1}^\infty a_n n^{-s}$ with the coefficients a_n of (3.8) and the inequality $|a_p| < 2p^{11/2}$ for every prime number p. The Euler product was established by Mordell. In [3, 4], Hecke completed a general theory of constructing Dirichlet series with Euler product and functional equation out of modular forms. Thus the operators $T_m(\Gamma \alpha \Gamma)$ are called *Hecke operators*. This work was continued by Petersson, who generalized Ramanujan's conjecture to the following form: *For every prime number p not dividing N, the characteristic roots of $T_m^N(p)$ have absolute values $\leq 2p^{(m-1)/2}$.*

In the above I gave a survey of (a part of) Hecke's theory; as for $R(\Gamma, \Delta)$ and its generalizations, I refer the reader to [11], [12], [13] and Tamagawa [15], [16].

4. Modular correspondences and their congruence relations. Let Γ be a Fuchsian group of the first kind, and V a projective nonsingular curve analytically isomorphic to $\Gamma \backslash H^*$. Let Δ be as in the beginning of §3. Denote by φ the natural projection map of H^* to V. Given $\alpha \in \Delta$, let $X = \{\varphi(z) \times \varphi(\alpha(z)) \mid z \in H^*\}$. It can be shown that X is a curve on $V \times V$, and if $\Gamma \alpha \Gamma = \bigcup_{i=1}^d \Gamma \alpha_i$ is a disjoint union, then

$$(4.1) \qquad X \cdot \left(\varphi(z) \times V \right) = \varphi(z) \times \sum_{i=1}^d \varphi(\alpha_i(z)).$$

In view of our definition (3.2) of the law of multiplication in $R(\Gamma, \Delta)$, we see that $\Gamma \alpha \Gamma \mapsto X$ defines a homomorphism of $R(\Gamma, \Delta)$ into the ring of algebraic correspondences of V. We call X a *modular correspondence* of V.

Now let us take the group Γ_N defined by (3.5) as our Γ. Let V_N be a projective non-singular model for $\Gamma_N\backslash\mathfrak{H}^*$, and X_p^N, Y_p^N be the modular correspondences of V_N obtained from the elements $T^N(p)$, $T^N(p, p)$ of $R(\Gamma_N, \Delta_N)$, respectively.

Theorem ([10]). *A model V_N for $\Gamma_N\backslash\mathfrak{H}^*$ can be taken so that V_N, X_p^N, Y_p^N are defined over $\mathbf{Q}$, and, for all except a finite number of p, we have*

$$(4.2) \qquad \left(X_p^N\right)_p = \Pi_p + \Pi'_p \circ \left(Y_p^N\right)_p, \qquad \Pi'_p \circ \left(Y_p^N\right)_p = \left(Z\right)_p^{-1} \circ \Pi'_p \circ \left(Z\right)_p$$

on $\left(V_N \times V_N\right)_p$. Here $(\ )_p$ means reduction modulo p, Π_p is the locus of $x \times x^p$ on $\left(V_N \times V_N\right)_p$, that is, the Frobenius correspondence, Π'_p is the transpose of Π_p, and Z is a certain birational automorphism of V_N independent of p.

We shall sketch the proof in the following section.

Let J_N be the jacobian variety of V_N, and let ξ_p, η_p, ζ be the elements of $\mathrm{End}(J_N)$ corresponding to X_p, Y_p, Z. From our theorem it follows easily that

$$(4.3) \qquad\qquad\qquad (\xi_p)_p = \pi_p + \pi'_p \circ (\eta_p)_p,$$

$$(4.4) \qquad\qquad\qquad \pi'_p \circ (\eta_p)_p = (\zeta)_p^{-1} \circ \pi'_p \circ (\zeta)_p.$$

We can define ℓ-adic coordinate systems on J_N and $(J_N)_p$ so that $M_\ell(\lambda) = M_\ell((\lambda)_p)$ for every λ of $\mathrm{End}(J_N)$ defined over $\mathbf{Q}$. Then, u being an indeterminate, from (4.3) and (4.4) we obtain

$$(4.5) \qquad \det\left[1 - M_\ell(\xi_p)u + M_\ell(\eta_p)pu^2\right] = \det\left[1 - M_\ell(\pi_p)u\right]^2.$$

Let $M^d(\lambda)$ be a representation of $\lambda \in \mathrm{End}(J_N)$ in $H^{1,0}(J_N)$. It is well-known that M_ℓ is equivalent to the direct sum of M^d and its complex conjugate. Since ξ_p and η_p are defined over $\mathbf{Q}$, we may assume that $M^d(\xi_p)$ and $M^d(\eta_p)$ have rational coefficients, so that

$$\det\left[1 - M^d(\xi_p)u + M^d(\eta_p)pu^2\right] = \det\left[1 - M_\ell(\pi_p)u\right].$$

Now ξ_p (resp. η_p) corresponds to X_p (resp. Y_p), and X_p (resp. Y_p) is obtained from $T^N(p)$ (resp. $T^N(p, p)$). As remarked at the end of §2, $S_2(\Gamma_N)$ is canonically isomorphic to $H^{1,0}(V_N)$. Therefore $M^d(\xi_p)$ and $M^d(\eta_p)$ are essentially the same as $T_2^N(p)$ and $T_2^N(p, p)$. Therefore we obtain

$$\det\left[1 - M_\ell(\pi_p)p^{-s}\right] = \det\left[1 - T_2^N(p)p^{-s} + T_2^N(p, p)p^{1-2s}\right].$$

The right-hand side is exactly the determinant of the inverse of the p-factor of the Euler product (3.7) for $m = 2$. We have thus proved that $\zeta^{(1)}(s, V_N/\mathbf{Q})$ is equal to $\det\left[D_2^N(s)\right]$ up to a finite number of p-factors. Therefore the Hasse-Weil conjecture is assured for the curve V_N.

Combining (4.3) with Weil's result which asserts that the characteristic roots of $M_\ell(\pi_p)$ have the absolute value $p^{1/2}$, we know that the characteristic roots of $T_2^N(p)$ have absolute values not greater than $2p^{1/2}$ for almost all p.

5. Proof of the congruence relations. Let $g_2(\omega_1, \omega_2)$, $g_3(\omega_1, \omega_2)$, and $\wp(u; \omega_1, \omega_2)$ be the functions of complex variables ω_1, ω_2, and u under the condition $\mathrm{Im}(\omega_1/\omega_2) > 0$, defined by

$$g_2(\omega_1, \omega_2) = 60 \sum{}' \omega^{-4}, \qquad g_3(\omega_1, \omega_2) = 140 \sum{}' \omega^{-6},$$
$$\wp(u; \omega_1, \omega_2) = u^{-2} + \sum{}' \left[(u - \omega)^{-2} - \omega^{-2} \right],$$

where $\sum'$ means the sum extended over all $\omega, \neq 0$, in $\mathbf{Z}\omega_1 + \mathbf{Z}\omega_2$. Define functions $j(z)$ and $f_{ab}^N(z)$ on H by

$$j(z) = g_2(\omega_1, \omega_2)^3 / \left[g_2(\omega_1, \omega_2)^3 - 27 g_3(\omega_1, \omega_2)^2 \right],$$
$$f_{ab}^N(z) = g_2(\omega_1, \omega_2) g_3(\omega_1, \omega_2)^{-1} \wp\left((a\omega_1 + b\omega_2)/N; \omega_1, \omega_2 \right),$$
$$\left(z = \omega_1/\omega_2; \ a, b \in \mathbf{Z}; \ (a, b) \not\equiv (0, 0) \pmod{N} \right).$$

It is well-known that $\mathbf{C}(j)$ is the field of all modular functions of level 1. By a simple calculation, we observe that, for every $\alpha \in \Gamma_1$,

$$f_{ab}^N(\alpha(z)) = f_{ab}^N(z) \text{ for all } (a, b) \quad \Longleftrightarrow \quad \alpha \in \{\pm 1\}\Gamma_N.$$

From this it follows that all the f_{ab}^N and j generate the field of all modular functions of level N.

Roughly speaking, the modular functions of level N are obtained from the invariant of a generic elliptic curve and points of finite order on the curve. To be more precise, for $z \in H$, determine γ by $j(z) = \gamma/(\gamma - 27)$ and call $E(z)$ the elliptic curve $y^2 = 4x^3 - \gamma x - \gamma$. Then $E(z)$ has the invariant $j(z)$. Let h_z be the function on $E(z)$ such that $h_z(x, y) = x$ for $(x, y) \in E(z)$. Let us fix a point z_0 such that $j(z_0)$ is transcendental over $\mathbf{Q}$, and set $j_0 = j(z_0)$, $E_0 = E(z_0)$, $h_0 = h_{z_0}$, and

$$K_N = \mathbf{Q}\left(j_0, h_0(t) \,\middle|\, t \in E_0, Nt = 0 \right).$$

Then the field K_N is isomorphic to $\mathbf{Q}(j, f_{ab}^N)$. Moreover, K_N is a Galois extension of $\mathbf{Q}(j_0)$, and the Galois group is isomorphic to $GL_2(\mathbf{Z}/N\mathbf{Z})/\{\pm 1\}$. Let L_N be the subfield of K_N corresponding to the subgroup $\left\{ \pm \begin{bmatrix} a & 0 \\ 0 & 1 \end{bmatrix} \,\middle|\, (a, N) = 1 \right\}/\{\pm 1\}$. Then $L_N(e^{2\pi i/N}) = K_N$, and L_N has $\mathbf{Q}$ as its constant field. Therefore, if we take a curve V_N whose function-field over $\mathbf{Q}$ is L_N, then V_N is a model for $\Gamma_N \backslash H^*$ and defined over $\mathbf{Q}$.

Now let us consider a disjoint expression

$$T^N(p) = \Gamma_N \alpha \Gamma_N = \bigcup_{i=1}^{p+1} \Gamma_N \alpha_i \text{ with } \alpha = \begin{bmatrix} 1 & 0 \\ 0 & p \end{bmatrix}$$

for a prime number p not dividing N. Set $z_i = \alpha_i(z)$, $j_i = j(z_i)$, $E_i = E(z_i)$, and $h_i = h_{z_i}$ for $1 \leq i \leq p + 1$. Since $\det(\alpha_i) = p$, we can find an isogeny λ_i of E_0 to E_i whose kernel is of order p. Then the $\mathrm{Ker}(\lambda_i)$ for $1 \leq i \leq p+1$ are exactly all the subgroups of order p of E_0. In view of (4.1), the modular correspondence X_p^N can be described by the mapping

$$\left(j(z_0), f_{ab}^N(z_0) \right) \mapsto \left\{ \left(j(z_i), f_{ab}^N(z_i) \right) \,\middle|\, 1 \leq i \leq p+1 \right\},$$

or rather

(5.1) $$\left(j_0, h_0(t) \right) \mapsto \left\{ \left(j_i, h_i(\lambda_i t) \right) \,\middle|\, 1 \leq i \leq p+1 \right\},$$

where $t \in E_0$, $Nt = 0$. In this way X_p^N can be connected to the isogenies of elliptic curves.

In the next place, we extend the prime (p) to a prime divisor P of a suitably large field containing j_0, j_i, etc., so that $P(j_0)$ is transcendental over $\mathbf{Z}/p\mathbf{Z}$. Let $(\ )_P$ mean reduction modulo P. Since $(E_0)_P$ has exactly p points of order p, we have $\big(\mathrm{Ker}(\lambda_i)\big)_P = \{0\}$ for exactly one i, say $i = 1$. Then $\big(\mathrm{Ker}(\lambda_i)\big)_P$ is of order p for $i > 1$. Hence $(\lambda_1)_P$ is a purely inseparable isogeny of degree p. It follows easily that $(E_1)_P = (E_0)_P^p$, and hence

$$(5.2) \qquad j \equiv j_0^p, \;\; h_1(\lambda_1 t) \equiv h_0(t)^p \pmod{P} \qquad (t \in E_0,\, Nt = 0).$$

Let μ_i be an isogeny of E_i to E_0 such that $\mu_i \lambda_i = p$. Then we see that μ_i is purely inseparable for $i > 1$, so that $(E_0)_P = (E_i)_P^p$ and

$$(5.3) \qquad j_0 \equiv j_i^p, \;\; h_0(\mu_i s) \equiv h_i(s)^p \pmod{P} \qquad (s \in E_i,\, Ns = 0,\, i > 1).$$

Substituting $\lambda_i t$ for s in (5.3), we obtain

$$(5.4) \qquad j_i \equiv j_0^{1/p}, \;\; h_i(\lambda_i t) \equiv h_0(pt)^{1/p} \pmod{P} \qquad (t \in E_0,\, Nt = 0,\, i > 1).$$

By (5.2) and (5.4), reduction modulo P of (5.1) is

$$(5.5) \quad \big(j_0,\, h_0(t)\big)_P \mapsto \big(j_0^p,\, h_0(t)^p\big)_P + p \cdot \big(j_0^{1/p},\, h_0(pt)^{1/p}\big)_P \quad (t \in E_0,\, Nt = 0).$$

It can be shown that the correspondence $\big(j_0,\, h_0(t)\big) \mapsto \big(j_0,\, h_0(pt)\big)$ gives exactly Y_p^N on V_N. Therefore from (5.5) we obtain $\big(X_p^N\big)_p = \Pi_p + \Pi_p' \circ \big(Y_p^N\big)_p$.

The first relation in our theorem was proved by Eichler [2] for the field of modular functions with respect to the group

$$(5.6) \qquad \Gamma_0(N) = \left\{ \begin{bmatrix} a & b \\ c & d \end{bmatrix} \in SL_2(\mathbf{Z}) \;\middle|\; c \in N\mathbf{Z} \right\}.$$

The result was generalized by [10], whose method I explained in the above. The result for $\Gamma_0(N)$ or any congruence subgroup of Γ_1 can be derived essentially from that for Γ_N. However, as we shall see later, it is convenient to state the result for a special congruence subgroup such as $\Gamma_0(N)$. For details I refer the reader to [2] and [10]. Igusa [6] showed that the reduction process works well for *all* primes p not dividing N. This fact is useful in our later discussion.

6. The unit group of a quaternion algebra. Let Φ be an indefinite quaternion algebra over $\mathbf{Q}$, that is, an algebra such that $\Phi \otimes_{\mathbf{Q}} \mathbf{R}$ is isomorphic to the matrix algebra $M_2(\mathbf{R})$. Let $\mathfrak{o}$ be a maximal order in Φ, that is, a maximal one among the subrings of Φ which are finitely generated $\mathbf{Z}$-modules. We consider Φ as a subring of $M_2(\mathbf{R})$, and set

$$\Gamma(\mathfrak{o}) = \big\{ \alpha \in \mathfrak{o} \;\big|\; \det(\alpha) = 1 \big\}, \quad \Gamma_N(\mathfrak{o}) = \big\{ \alpha \in \Gamma(\mathfrak{o}) \,\big|\, \alpha - 1 \in N\mathfrak{o} \big\},$$

where N is a positive integer. Then the groups $\Gamma(\mathfrak{o})$ and $\Gamma_N(\mathfrak{o})$, viewed as subgroups of $SL_2(\mathbf{R})$, are Fuchsian groups of the first kind. If $\Phi = M_2(\mathbf{Q})$, we can take $\mathfrak{o} = M_2(\mathbf{Z})$; then these are nothing but Γ_1 and Γ_N of §3. If Φ is a division algebra, then they have compact quotient spaces.

Suppose that N is prime to the discriminant of Φ. Then the ring $\mathfrak{o}/N\mathfrak{o}$ is isomorphic to the matrix ring $M_2(\mathbf{Z}/N\mathbf{Z})$. Let $\Delta_N(\mathfrak{o})$ be the set of elements

α in $\mathfrak{o}$ such that $\det(\alpha) > 0$ and α modulo $N\mathfrak{o}$ corresponds to $\begin{bmatrix} 1 & 0 \\ 0 & d \end{bmatrix}$ with $d \in (\mathbf{Z}/N\mathbf{Z})^\times$. Then we can show that $R\big(\Gamma_N(\mathfrak{o}), \Delta_N(\mathfrak{o})\big)$ has the same structure as $R(\Gamma_N, \Delta_N)$ of §3. Taking the representation of $R\big(\Gamma_N(\mathfrak{o}), \Delta_N(\mathfrak{o})\big)$ on $S_m\big(\Gamma_N(\mathfrak{o})\big)$, we can construct a Dirichlet series $D_m^N(s;\, \Phi)$ with a functional equation and an Euler product analogous to (3.7). Furthermore, $\Gamma_N(\mathfrak{o})\backslash H$ has a model $V_N(\mathfrak{o})$ defined over $\mathbf{Q}$, provided that N is prime to the discriminant of Φ; also, $\zeta^{(1)}\big(s; V_N(\mathfrak{o})/\mathbf{Q}\big)$ differs from $\det\big[D_2^N(s;\, \Phi)\big]$ only by a finite number of p-factors.

This result can be proved as follows. First we define, for every $z \in H$, a complex torus A_z of dimension 2 by

$$(6.1) \qquad A_z = \mathbf{C}^2/L_z, \quad L_z = \left\{ \alpha \begin{bmatrix} z \\ 1 \end{bmatrix} \,\middle|\, \alpha \in \mathfrak{o} \right\},$$

where an element of $\mathfrak{o}$ is considered an element of $M_2(\mathbf{R})$. We can prove that A_z has a structure of abelian variety. Furthermore, every element of $\mathfrak{o}$ defines an endomorphism of A_z in a natural manner. The A_z form an analytic family $\left\{ A_z \,\middle|\, z \in H \right\}$ of abelian varieties endowed with the structure consisting of endomorphisms and polarization, parametrized by the points of H. The moduli of an abelian variety A_z with such a structure are given by the values of automorphic functions with respect to $\Gamma(\mathfrak{o})$ at z. The automorphic functions with respect to the congruence subgroup $\Gamma_N(\mathfrak{o})$ can be obtained from the coordinates of the points of order N on A_z. Here again we can connect Hecke operators (or modular correspondences on $\Gamma_N(\mathfrak{o})\backslash H$) with the isogenies of A_z. Using the same idea as in §5, we obtain congruence relations for modular correspondences on $\Gamma_N(\mathfrak{o})\backslash H$, though the present case involves more technical difficulties than the case of elliptic modular functions. A full detail of the theory is given in [12].

7. A law of reciprocity in non-solvable extensions. Let us consider the group $\Gamma_0(N)$ of (5.6) for a particular case $N = 11$. It is known [5] that $S_2\big(\Gamma_0(11)\big)$ is one-dimensional and generated by

$$\big[\Delta(z)\Delta(11z)\big]^{1/12} = q \prod_{n=1}^\infty (1 - q^n)^2 (1 - q^{11n})^2 \qquad (q = e^{2\pi i z}).$$

Write this as $\sum_{n=1}^\infty c_n q^n$ and set $D(s) = \sum_{n=1}^\infty c_n n^{-s}$ By Hecke's theory we have

$$(7.1) \qquad D(s) = (1 - 11^{-s})^{-1} \prod_{p \neq 11} \big(1 - c_p p^{-s} + p^{1-2s}\big)^{-1},$$

$$D^*(s) = D^*(2 - s) \quad \text{with} \quad D^*(s) = \Gamma(s)(2\pi)^{-s} 11^{s/2} D(s).$$

The field of automorphic functions with respect to $\Gamma_0(11)$ is generated by $j(z)$ and $j(11z)$. The field $\mathbf{Q}\big(j(z),\, j(11z)\big)$ has a model

$$(7.2) \qquad E : y^2 = 4x^3 - (4 \cdot 31/3)x - (2501/27).$$

Therefore, by virtue of the congruence relation, we know that if π_p is the p-th power endomorphism of $(E)_p$, then, with an indeterminate X, we have

$$(7.3) \qquad \det\big[X - M_\ell(\pi_p)\big] = X^2 - c_p X + p,$$

where c_p is determined by

$$(7.4) \qquad \sum_{n=1}^{\infty} c_n q^n = q \prod_{n=1}^{\infty} (1 - q^n)^2 (1 - q^{11n})^2.$$

In view of the result of Igusa mentioned at the end of §5, relation (7.3) is true for all $p \neq 11$.

Given a prime ℓ, let K_ℓ denote the field generated over $\mathbf{Q}$ by all the coordinates of the points of order ℓ on the elliptic curve E of (7.2). Then K_ℓ is a Galois extension of $\mathbf{Q}$, and every element of the Galois group $G(K_\ell/\mathbf{Q})$ gives an automorphism of the group of points of order ℓ on E, and hence we obtain an isomorphism S_ℓ of $G(K_\ell/\mathbf{Q})$ onto a subgroup of $GL_2(\mathbf{Z}/\ell\mathbf{Z})$. Let p be a prime number other than 11 and ℓ. Let P be a prime ideal in K_ℓ dividing p, and σ_P a Frobenius automorphism of K_ℓ over $\mathbf{Q}$ for P. By taking suitable ℓ-adic coordinate systems on E and $(E)_p$, we find that $S_\ell(\sigma_P) \equiv M_\ell(\pi_p) \pmod{\ell}$, so that

$$(7.5) \qquad \det\left[X - S_\ell(\sigma_P)\right] \equiv X^2 - c_p X + p \pmod{\ell}.$$

It follows, in particular, that $S_\ell\big(G(K_\ell/\mathbf{Q})\big)$ contains an element whose characteristic polynomial is $X^2 - c_p X + p$. Using this fact, I found that $S_\ell\big(G(K_\ell/\mathbf{Q})\big) = GL_2(\mathbf{Z}/\ell\mathbf{Z})$ at least for $7 \leq \ell \leq 97$.

This is noteworthy, for there was previously no known example of non-solvable extension for which the law of reciprocity is given explicitly (in any sense); here are such examples. Indeed, we have obtained a Galois extension K_ℓ of $\mathbf{Q}$ whose Galois group is isomorphic to $GL_2(\mathbf{Z}/\ell\mathbf{Z})$, and of which the law of reciprocity is given by (7.5), where the c_p are coefficients of the Dirichlet series (7.1) with Euler product and functional equation; they are easily obtained from (7.4) as many as we need. Moreover, we can determine Artin's L-functions of the extension $K_\ell/\mathbf{Q}$ for a fairly large number of non-abelian characters. A more detailed account of the result will be published elsewhere.

We can expect a result of the same kind for other congruence subgroups and also for the Fuchsian group discussed in §6; see [12, pp. 328-329]. But here my emphasis is laid on the explicitness and comprehensibility, and not on generality. It will be an important task to reorganize and generalize the result from a new viewpoint.

8. Change of model and extension of basic field. In the case of an abelian variety A with sufficiently many complex multiplications, we can determine $\zeta^{(1)}(s; A/k)$ for almost any field of definition k of A. In contrast to this, in the case of the curve $\Gamma_N \backslash H^*$ (or $\Gamma_N(\mathfrak{o}) \backslash H$) we have determined $\zeta^{(1)}(s; V_N/\mathbf{Q})$ only for a particular model V_N over $\mathbf{Q}$. It is a highly nontrivial problem to prove the Hasse-Weil conjecture for an arbitrary model V' birationally equivalent to V_N, with an arbitrary algebraic number field k as the basic field. However, if k is abelian over $\mathbf{Q}$, a part of this problem can be solved in the following way. For every abelian character χ of $\mathbf{Q}$, we define a Dirichelt series

$$D_m^N(s; \chi; \varPhi) = \sum_{n=1}^{\infty} \chi(n) B_n n^{-s}$$

when $D_m^N(s; \Phi) = \sum_{n=1}^{\infty} B_n n^{-s}$. Then $D_m^N(s; \chi; \Phi)$ can be continued holomorphically to the whole s-plane and satisfies a functional equation [13, Theorem 1]. It is easy to see that $\zeta^{(1)}(s; V_N(\mathfrak{o})/k)$ is a product of $\det\left[D_2^N(s; \chi; \Phi)\right]$ for several χ's, up to a finite number of p-factors. A discussion from somewhat different viewpoints can be found in Rangachari [9] and Konno [7].

As an explicit example, the c_n being as in (7.1) and (7.4), set

$$D(s; \chi) = \sum_{n=1}^{\infty} \chi(n)c_n n^{-s}, \qquad D^*(s; \chi) = \Gamma(s)(2\pi)^{-s}(11d^2)^{s/2}D(s, \chi)$$

for a primitive Dirichlet character χ modulo $|d|$. Assuming that d is prime to 11, we can prove

$$D^*(s; \chi) = \chi(11)W(\chi)^2 D^*(2 - s; \overline{\chi}),$$

where $W(\chi) = |d|^{-1/2} \sum_{a=1}^{d} \chi(a)e^{2\pi i a/|d|}$. In particular, if $\chi(n) = \left(\dfrac{d}{n}\right)$ with $0 \lessgtr d \in \mathbf{Z}$, then $W(\chi)^2 = \chi(-1)$, so that

$$D^*(s; \chi) = \chi(-11)D^*(2 - s; \chi).$$

Now let E be defined by (7.2), and E' an elliptic curve over $\mathbf{Q}$, isomorphic to E over $\mathbf{C}$ but not over $\mathbf{Q}$. Since ± 1 are the only automorphisms of E, every isomorphism λ of E to E' is defined over a quadratic extension $\mathbf{Q}(\sqrt{d})$ for some $d \in \mathbf{Q}$. Here we take d to be the discriminant of $\mathbf{Q}(\sqrt{d})$. Then it can easily be verified that $\zeta^{(1)}(s; E'/\mathbf{Q})$ is, up to a finite number of p-factors, equal to $D(s; \chi)$ with $\chi(n) = \left(\dfrac{d}{n}\right)$. Therefore the Hasse-Weil conjecture for E' over $\mathbf{Q}$ is assured.

9. The zeta-function of a fibre variety. So far only automorphic forms of weight 2 have been related to the zeta-function of a curve. Now we can construct a certain fibre variety $W_N^{(h)}$ whose zeta-function is expressed by the Dirichlet series $D_m^N(s; \Phi)$ considered in §6 for $m \geq 2$. To construct such a fibre variety, take Φ and $\mathfrak{o}$ as in §6; assume that Φ is a division algebra. Let $GL_2^+(\mathbf{R}) = \left\{ \alpha \in GL_2(\mathbf{R}) \,\middle|\, \det(\alpha) > 0 \right\}$. For a positive integer h, let F_h be the product of h copies of $M_2(\mathbf{R})$, viewed as a right and left $M_2(\mathbf{R})$-module in a natural manner. The product $GL_2^+(\mathbf{R}) \times F_h$ forms a group with respect to the law of multiplication:

$$(\xi, u)(\eta, v) = (\xi\eta, v\xi' + u) \qquad (\xi, \eta \in GL_2^+(\mathbf{R}); \; u, v \in F_h),$$

where ξ' denotes the image of ξ under the canonical involution of $M_2(\mathbf{R})$. We let $GL_2^+(\mathbf{R}) \times F_h$ act on $H \times F_h$ by the rule:

$$(\alpha, u)(z, v) = (\alpha(z), v\alpha' + u) \qquad (\alpha \in GL_2^+(\mathbf{R}); \; z \in H; \; u, v \in F_h).$$

Define a bijection $x_h : H \times F_h \to H \times \mathbf{C}^{2h}$ by

$$x_h(z; u_1, \ldots, u_h) = \left(z; u_1\begin{bmatrix} z \\ 1 \end{bmatrix}, \ldots, u_h\begin{bmatrix} z \\ 1 \end{bmatrix}\right) \qquad (z \in H; \; u_i \in M_2(\mathbf{R})).$$

We introduce a complex structure on $H \times F_h$ so that x_h is a complex analytic isomorphism. Then every element of $GL_2^+(\mathbf{R}) \times F_h$ acts on $H \times F_h$ as a complex analytic automorphism. Let $\mathfrak{o}^h$ denote the product of h copies of $\mathfrak{o}$. The product $\Gamma_N(\mathfrak{o}) \times \mathfrak{o}^h$ is a subgroup of $GL_2^+(\mathbf{R}) \times F_h$, and gives a properly discontinuous group of transformations on $H \times F_h$ with a compact quotient. Set

$$W_N^{(h)} = \left(\Gamma_N(\mathfrak{o}) \times \mathfrak{o}^h\right) \backslash (H \times F_h).$$

Assume hereafter that $\Gamma_N(\mathfrak{o})$ has no element of finite order other than the identity element. Then $W_N^{(h)}$ is a compact complex manifold. Furthermore, we can easily verify that $W_N^{(h)}$ is a fibre variety of which base is $V_N = \Gamma_N(\mathfrak{o}) \backslash H$ and each fibre is the product of h copies of the abelian variety A_z of (6.1).

Kuga [8] determined completely the cohomology groups of a certain class of fibre varieties, which includes $W_N^{(h)}$ as a special case. In the case of $W_N^{(h)}$, it turns out that every cohomology group is canonically isomorphic to a direct sum of $S_m(\Gamma_N(\mathfrak{o}))$ for some m's. He proved also that $W_N^{(h)}$ can be embedded in a projective space.

Let $s \mapsto w(s)$ denote the natural projection map of $H \times F_h$ to $W_N^{(h)}$. Assume that N is prime to the discriminant of Φ. Let $\Delta_N(\mathfrak{o})$ be as in §6. For simplicity, set $\Gamma = \Gamma_N(\mathfrak{o})$ and $\Delta = \Delta_N(\mathfrak{o})$. For every $\alpha \in \Delta$, set

$$X(\Gamma\alpha\Gamma) = \left\{ w(s) \times w\big((\alpha, 0)s\big) \,\big|\, s \in H \times F_h \right\}.$$

We can easily verify that $X(\Gamma\alpha\Gamma)$ is a subvariety of $W_N^{(h)} \times W_N^{(h)}$, and also that $\Gamma\alpha\Gamma \mapsto X(\Gamma\alpha\Gamma)$ defines a homomorphism of $R(\Gamma, \Delta)$ into the ring of correspondences on $W_N^{(h)}$, the latter being defined suitably.

We can find two elements α_p and β_p of Δ, for each prime number p, such that $\det(\alpha_p) = p$ and $p^{-1}\beta_p \in \Gamma(\mathfrak{o})$. Set

$$X_p^{(h)} = X(\Gamma\alpha_p\Gamma), \qquad Y_p^{(h)} = X(\Gamma\beta_p\Gamma).$$

Now we can find a projective model of $W_N^{(h)}$ so that $W_N^{(h)}$, $X_p^{(h)}$, and $Y_p^{(h)}$ are all defined over $\mathbf{Q}$, and moreover that

$$\left(X_p^{(h)}\right)_p = \Pi_p + \Pi_p^* \quad \text{and} \quad p \cdot \left(Y_p^{(h)}\right)_p = \Pi_p \circ \Pi_p^* \quad \text{on} \quad \left(W_N^{(h)} \times W_N^{(h)}\right)_p$$

for all but a finite number of p. Here $(\ )_p$ means reduction modulo p, Π_p is the locus of $x \times x^p$ on $\left(W_N^{(h)} \times W_N^{(h)}\right)_p$, and Π_p^* is a certain correspondence such that $\left(\Pi_p^*\right)^n$ has the same number of fixed points as Π_p^n for $0 < n \in \mathbf{Z}$. Combining these congruence relations with the result of Kuga concerning the cohomology groups of $W_N^{(h)}$ mentioned above, we find that

$$\zeta\big(s; W_N^{(h)}/\mathbf{Q}\big) \doteq \prod_{b=0}^{4h} \left\{ \zeta\big(s - (b/2)\big)\zeta\big(s - (b+2)/2\big) \right\}^{e(h,b,0)}$$

$$\prod_{b=0}^{4h}\prod_{i=0}^{b} \det\left[D_{i+2}^N\big(s - (b-i)/2; \Phi\big) \right]^{(-1)^{b+1}e(h,b,i)}$$

Here = means the equality up to a finite number of p-factors, and the $e(h, b, i)$ are nonnegative integers, depending only on h, b, and i. A full detail of this result will appear as a paper in collaboration with Kuga.

A formula of this kind was first given empirically by M. Sato for a certain fibre variety whose base is $\Gamma_N \backslash H^*$ and fibres are the product of copies of elliptic curves modulo ± 1. Such a fibre variety had been proposed by Kuga as the one which would describe Ramanujan's function in terms of Hasse's zeta function.

REFERENCES

1. M. Deuring, Die Zetafunktion einer algebraischen Kurve vom Geschlechte Eins I, II, III, IV, Nachr. Akad. Wiss. Göttingen, (1953) 85–94, (1955) 13–42, (1956) 37–56, (1957) 55–80.
2. M. Eichler, Quaternäre quadratische Formen und die Riemannsche Vermutung für die Kongruenzzetafunktion, Arch. Math. 5 (1954), 355–366.
3. E. Hecke, Über die Bestimmung Dirichletscher Reihen durch ihre Funktionalgleichung, Math. Ann. 112 (1936), 664–699.
4. ———, Über die Modulfunktionen und die Dirichletscher Reihen mit Eulerscher Produktentwicklung I, II, Math. Ann. 114 (1937), 1–28, 316–351.
5. ———, Analytische Arithmetik der positiven quadratischen Formen, Dansk. Vidensk. Selsk. Math.-fys. Meddel. XVII, 12 (1940).
6. J. Igusa, Kroneckerian model of fields of elliptic modular functions, Amer. J. Math. 81 (1959), 561–577.
7. S. Konno, On Artin's L-functions of the algebraic curves uniformaized by certain automorphic functions, J. Math. Soc. Japan, 15 (1963), 89–100.
8. M. Kuga, Automorphic forms and fibre varieties, Lecture notes, Chicago University, 1964, to appear.
9. S. S. Rangachari, Modulare Korrespondenzen und L-Reihen, J. Reine Angew. Math. 205 (1961), 119–155.
10. G. Shimura, Correspondances modulaires et les fonctions zeta de courbes algébriques, J. Math. Soc. Japan, 10 (1958), 1–28.
11. ———, Sur les intégrales attachées aux formes automorphes, J. Math. Soc. Japan, 11 (1959), 291–311.
12. ———, On the zeta functions of the algebraic curves uniformized by certain automorphic functions, J. Math. Soc. Japan, 13 (1961), 275–331.
13. ———, On Dirichlet series and abelian varieties attached to automorphic forms, Ann. of Math. 76 (1962), 237–294.
14. G. Shimura and Y. Taniyama, Complex multiplication of abelian varieties and its applications to number theory, Publ. Math. Soc. Japan, No. 6, 1961.
15. T. Tamagawa, On Selberg's trace formula, J. Fac. Sci. Univ. Tokyo, Sect. I, vol. III, Part 2, 363–386.
16. ———, On the ζ-functions of a division algebra, Ann. of Math. 77 (1963), 387–405.
17. Y. Taniyama, L-functions of number fields and zeta functions of abelian varieties, J. Math. Soc. Japan, 9 (1957), 330–366.
18. A. Weil, Jacobi sums as "Grössencharaktere", Trans. Amer. Math. Soc. 75 (1952), 487–495.

On the field of definition for a field of automorphic functions: II

Annals of Mathematics, 81 (1965), 124-165

Let L be a division algebra over the rational number field Q with a positive involution ρ, and V a vector space over L of dimension m. For a non-degenerate anti-hermitian form T in V with respect to ρ, we can define the unitary group $U(T)$ in the usual manner. Extending the basic field to the real number field R, we get a Lie group $U_R(T)$, whose quotient by a maximal compact subgroup yields a bounded symmetric domain $\mathfrak{H}$. Let $\mathfrak{M}$ be a lattice in V, and let $\Gamma(T, \mathfrak{M})$ be the subgroup of $U(T)$ consisting of the elements which leave $\mathfrak{M}$ invariant. As was shown in [9], $\mathfrak{H}/\Gamma(T, \mathfrak{M})$ is the space of moduli for a certain family of abelian varieties with some additional structure. In the previous paper [11] with the same title, referred hereafter as [FI], we investigated the field of definition for the field of automorphic functions on $\mathfrak{H}$ with respect to $\Gamma(T, \mathfrak{M})$. The problem of determining a lower bound for the fields of definition has been treated in [12]. In both cases, certain class-fields came out, and the reciprocity-law for those class-fields was described in terms of $\Gamma(T, \mathfrak{M})$. The purpose of this second part is to discuss the same problem for the congruence subgroups Γ of $\Gamma(T, \mathfrak{M})$, and to determine the field of definition for the fibre variety $\mathfrak{V}$ of which the base is $\mathfrak{H}/\Gamma$ (assumed to be compact) and the fibres are abelian varieties of the family. It has been proved by Kuga [3] that such a fibre variety $\mathfrak{V}$ is a projective variety.

Here we give a brief account of our results and method of proof.[1] First we shall consider a structure $\mathfrak{A} = (A, C, \theta; t_1, \cdots, t_s)$ formed by an abelian variety A, a polarization C of A, an isomorphism θ of L into $\mathrm{End}_Q(A)$, and the points $t_1, \cdots, t_s$ of order N on A, the universal domain being the complex number field C. *The field of moduli of $\mathfrak{A}$* can be defined as the subfield $\mathfrak{f}$ of C with the following property: For an isomorphism σ of C into itself, $\mathfrak{A}^\sigma$ is isomorphic to $\mathfrak{A}$ if and only if σ is the identity mapping on $\mathfrak{f}$. If Γ_N is the principal congruence subgroup of $\Gamma(T, \mathfrak{M})$ of level N, we get a family $\{\mathfrak{A}_z \mid z \in \mathfrak{H}\}$ of such structures, for which the space of moduli is given by $\mathfrak{H}/\Gamma_N$. Then there exist meromorphic functions $h_1, \cdots, h_\lambda$ on $\mathfrak{H}$ such that:

(i) $C(h_1, \cdots, h_\lambda)$ is the field of all automorphic functions on $\mathfrak{H}$ with re-

---

[1] For simplicity, we state a rather vague and weaker result than given in the text, adopting the notation which will not necessarily be retained in the text.

spect to Γ_N;

(ii) $Q(h_1(z), \cdots, h_\lambda(z))$ is the field of moduli of $\mathfrak{Q}_z$ for all z on $\mathfrak{H}$ not belonging to a subset of codimension ≥ 1;

(iii) if k_N is the algebraic closure of Q in $Q(h_1, \cdots, h_\lambda)$, then the fields $k_N(h_1, \cdots, h_\lambda)$ and C are linearly disjoint over k_N (Theorems 2.3, 4.5).

In the case $N = 1$, this was given in [9]; and in the general case, [9, Prop. 2.3] is essentially equivalent with the present one, if we take account of a recent result of Baily and Borel [16] on compactification. But we feel it convenient, in view of our later application, to reformulate and restate the result. In § 5, we shall prove that the fibre variety $\mathfrak{W}$ mentioned above has a model defined over the algebraic number field k_N (Theorem 5.11). This fact will play an essential role in a forthcoming joint work with Kuga, concerning a number theoretical investigation of $\mathfrak{W}$ and automorphic forms with respect to Γ_N. In §§ 6–7, we show that $k_N = Q(e^{2\pi i/N})$, if L is a totally real algebraic number field F or a totally indefinite quaternion algebra over such a field F. In general, the field of moduli of $\mathfrak{Q} = (A, \mathcal{C}, \theta; t_1, \cdots, t_s)$ is a Galois extension of the field of moduli of $(A, \mathcal{C}, \theta)$. The determination of k_N can be done by describing explicitly the Galois group (Theorems 6.8, 7.2). The field k_N is not necessarily the *smallest* field of definition for the field of automorphic functions. If $L = F$, or if L is an indefinite quaternion algebra and N is prime to the discriminant of L, then Q may be taken as a field of definition (Theorems 6.11, 7.3).

More interesting will be the case where L is a totally imaginary quadratic extension of F. As in the text, let us use K instead of L. With respect to an analytic coordinate system of the abelian variety A, we get a representation of K by complex matrices. Let K^* be the field generated by the values of the trace of this representation. We shall prove that k_N is a class-field over K^* (Theorem 9.4). If $N = 1$, we have seen in [FI] that k_N is unramified over K^* for a suitable choice of $\mathfrak{M}$. In the case $N > 1$, we obtain naturally a ray-class field corresponding to the ideal-group defined modulo a multiple of N. In the final § 10, the result of [12] will be generalized. Namely, an unramified class-field over a totally real algebraic number field occurs as the bottom field of the variety $\mathfrak{H}/\Gamma_N$, if $\mathfrak{H}$ is the product of copies of the upper half plane of dimension one (Theorems 10.6, 10.7). In this case, we get, even if $N > 1$, the same unramified class-field as in [12].

Besides the notation introduced in [FI], we shall use the following one: for an ideal $\mathfrak{c}$ in an algebraic number field, $\mathrm{mod}^\times \mathfrak{c}$ will mean the usual multiplicative congruence modulo $\mathfrak{c}$.

1. Structure of an abelian variety with polarization, endomorphisms and points of finite order

1.1. Let $\mathfrak{r}$ be an associative ring with an identity element and a finite basis over Z. Let $\mathcal{P} = (A, \mathcal{C}, \theta)$ be a polarized abelian variety of type $\mathfrak{r}$ in the sense of [5, No. 3], by which we mean a structure formed by an abelian variety A, a polarization $\mathcal{C}$ of A, and an isomorphism θ of $\mathfrak{r}$ into End (A); we assume that θ maps the identity element of $\mathfrak{r}$ to the identity element of End (A). For a finite set $\{t_1, \cdots, t_s\}$ of points on A, we consider a structure

$$\mathfrak{Q} = \mathcal{P}(t_1, \cdots, t_s) = (A, \mathcal{C}, \theta; t_1, \cdots, t_s) \ .$$

We take account of the order of $t_1, \cdots, t_s$; for example, if $t \neq t'$, $\mathcal{P}(t, t')$ and $\mathcal{P}(t', t)$ are distinct. Let $\mathfrak{Q}' = (A', \mathcal{C}', \theta'; t_1', \cdots, t_s') = \mathcal{P}'(t_1', \cdots, t_s')$ be another structure of the same kind. An isomorphism (resp. isogeny) λ of A to A' is called an *isomorphism* (resp. *isogeny*) of $\mathfrak{Q}$ to $\mathfrak{Q}'$, if λ^{-1} sends $\mathcal{C}'$ into $\mathcal{C}$, $\lambda\theta(a) = \theta'(a)\lambda$ for all $a \in \mathfrak{r}$, and $\lambda t_i = t_i'$ for every i. We say that $\mathfrak{Q}$ is defined over a field k, if A is defined over k as an abelian variety, $\mathcal{C}$ contains a divisor rational over k, the $\theta(a)$ for all $a \in \mathfrak{r}$ are defined over k, and the t_i are rational over k. This being so, let σ be an isomorphism of k into a field k'. Then we get naturally a structure

$$\mathfrak{Q}^\sigma = \mathcal{P}^\sigma(t_1^\sigma, \cdots, t_s^\sigma) = (A^\sigma, \mathcal{C}^\sigma, \theta^\sigma; t_1^\sigma, \cdots, t_s^\sigma)$$

defined over k'.

1.2. Fix a basis $\{r_1, \cdots, r_d\}$ of $\mathfrak{r}$ over Z. Suppose that A is embedded in a projective space P^N. Let $\mathfrak{Q}$ be defined over k. Take a generic matrix (g_{ij}) of degree $N + 1$ over k, and $d + 1$ independent generic points $u_1, \cdots, u_d, v$ on A over $k(g_{ij})$. Denote by g the projective transformation in P^N defined by $(x_i) \rightarrow (\sum_{j=0}^{N} g_{ij}x_j)$. Put $B = g(A)$. For each ν, let W_ν be the graph of the rational mapping

$$x \rightarrow g[\theta(r_\nu)g^{-1}(x) + u_\nu]$$

of B into itself. Denote by $\mathcal{F}(A, \theta; \{t_i\})$ the locus of

(1.1.1) $\quad c(B) \times c(W_1) \times \cdots \times c(W_d) \times g(v) \times g(v + t_1) \times \cdots \times g(v + t_s)$

over k. (For the definition of $c(*)$, see [FI, *Notation*].) It can easily be verified that $\mathcal{F}(A, \theta, \{t_i\})$ does not depend on the choice of $k, g, u_1, \cdots, u_d$ and v; but it may of course depend on $\{r_\nu\}$.

1.3. PROPOSITION. *Let A and A' be abelian varieties in P^N. Let $\mathcal{C}$ (resp. $\mathcal{C}'$) be the polarization of A (resp. A') defined by the hyperplane sections, θ (resp. θ') an isomorphism of $\mathfrak{r}$ into End (A) (resp. End (A')), and $t_1, \cdots, t_s$ (resp. $t_1', \cdots, t_s'$) points on A (resp. A'). Suppose that A and A' satisfy the following two conditions.*

(i) *There is no hyperplane containing either A or A'.*

(ii) *The linear systems on A and A' defined by hyperplane sections are complete.*

Then $\mathfrak{Q} = (A, \mathcal{C}, \theta; t_1, \cdots, t_s)$ and $\mathfrak{Q}' = (A', \mathcal{C}', \theta'; t_1', \cdots, t_s')$ are isomorphic if and only if $\mathcal{F}(A, \theta; \{t_i\}) = \mathcal{F}(A', \theta'; \{t_i'\})$.

This can be proved by a slight modification of the proof of [5, Prop. 1]. In fact, suppose that there exists an isomorphism λ of $\mathfrak{Q}'$ to $\mathfrak{Q}$. By the proof of [5, Prop. 1], we can find a projective transformation f of P^N such that $f(x) = \lambda(x) + b$ for $x \in A'$ with a constant b on A. Let g be an arbitrary projective transformation in P^N. Put $h = g \circ f$. Let $w \in A'$ and $v = f(w)$. Then $h(w) = g(v)$, and $h(w + t_i') = g(v + t_i)$ for every i. This, combined with the proof of [5, Prop. 1], shows that $\mathcal{F}(A, \theta; \{t_i\})$ and $\mathcal{F}(A', \theta'; \{t_i'\})$ have the same generic point over a suitable field.

Conversely, suppose that $\mathcal{F}(A, \cdots) = \mathcal{F}(A', \cdots)$. Again applying the proof of [5, Prop. 1], we get an isomorphism η of $(A', \mathcal{C}', \theta')$ to $(A, \mathcal{C}, \theta)$ and a projective transformation f such that $f(x) = \eta(x) + a$ for $x \in A'$ with a constant a on A. Applying the argument of [5, p. 106] to the present case, we may take f so that $v = f(w)$ and $v + t_i = f(w + t_i')$ for some $v \in A$ and $w \in A'$. Then we get $\eta t_i' = t_i$. This completes the proof.

1.4. Let $\mathfrak{Q} = (A, \mathcal{C}, \theta; t_1, \cdots, t_s)$ be as in 1.1. Take an ample divisor X in $\mathcal{C}$. Let $\lambda: A \to A^*$ be a projective embedding of A into a projective space P^N by X. We can define a structure $\mathfrak{Q}^* = (A^*, \mathcal{C}^*, \theta^*; t_1^*, \cdots, t_s^*)$ so that λ gives an isomorphism of $\mathfrak{Q}$ to $\mathfrak{Q}^*$. Put $\mathcal{F} = \mathcal{F}(A^*, \theta^*; \{t_i\})$. Let k_0 be the field generated by $c(\mathcal{F})$ over the prime field. Then k_0 has the following property:

(M) *Let σ and τ be isomorphisms of the universal domain into itself. Then $\mathfrak{Q}^\sigma$ and $\mathfrak{Q}^\tau$ are isomorphic if and only if $\sigma = \tau$ on k_0.*

In fact, $\mathfrak{Q}^\sigma$ and $\mathfrak{Q}^\tau$ are respectively isomorphic to $\mathfrak{Q}^{*\sigma}$ and $\mathfrak{Q}^{*\tau}$. By 1.3, $\mathfrak{Q}^{*\sigma}$ and $\mathfrak{Q}^{*\tau}$ are isomorphic if and only if $\mathcal{F}^\sigma = \mathcal{F}^\tau$. Hence we get (M).

If the universal domain is of characteristic 0, the field k_0 is uniquely determined by the property (M); we call k_0 *the field of moduli of* $\mathfrak{Q}$. In the case of positive characteristic, the same procedure as in [5, No. 5] may be applied to the structure $\mathfrak{Q}$. However, for the sake of simplicity, we assume hereafter that *the universal domain is of characteristic* 0. We note that if $\sum_{i=1}^s \theta(\mathfrak{r})t_i = \sum_{j=1}^q \theta(\mathfrak{r})u_j$, then $\mathcal{P}(t_1, \cdots, t_s)$ and $\mathcal{P}(u_1, \cdots, u_q)$ have the same field of moduli.

1.5. PROPOSITION. *Let $\mathfrak{Q}$ be as in 1.1, and k_0 the field of moduli of $\mathfrak{Q}$. Suppose that there is no automorphism of $\mathfrak{Q}$ other than the identity mapping. Then there exists a structure $\mathfrak{Q}'$ which is isomorphic to $\mathfrak{Q}$ and defined over k_0. Such a $\mathfrak{Q}'$ is uniquely determined up to an isomorphism defined over k_0.*

Put $\mathcal{P} = (A, \mathcal{C}, \theta)$. We may assume that A is the projective embedding of A itself by an ample divisor in $\mathcal{C}$. Let k be a field of definition for $\mathfrak{Q}$, and let the notation be as in (1.1.1). We can find a projective transformation g' in P^N and points $u_1', \cdots, u_d', v'$ so that the point (1.1.1) has a specialization, algebraic over k_0, which is compatible with the specialization

$$(g, u_1, \cdots, u_d, v) \rightarrow (g', u_1', \cdots, u_d', v') \qquad \text{(over } k) ,$$

since $\mathcal{F}(A, \theta; \{t_i\})$ is defined over k_0. Then, taking $g'(A)$ as the underlying variety, we can find a structure which is isomorphic to $\mathfrak{Q}$ and defined over a finite algebraic extension of k_0. For the sake of simplicity, we may assume that $\mathfrak{Q}$ is already defined over a finite Galois extension k_1 of k_0. Let G be the Galois group of k_1 over k_0. For every $\sigma \in G$, we find, in view of (M), an isomorphism f_σ of $\mathfrak{Q}$ to $\mathfrak{Q}^\sigma$. Since there is no automorphism of $\mathfrak{Q}$ other than the identity mapping, f_σ is uniquely determined by σ. Therefore, we verify easily that $f_{\sigma\tau} = (f_\sigma)^\tau \circ f$ for $\sigma \in G$, $\tau \in G$. Put $f_{\tau,\sigma} = f_\tau \circ f_\sigma^{-1}$. Then the $f_{\tau,\sigma}$ satisfy the conditions (i), (ii) of Weil [15, Th. 1]. Hence, by [15, Th. 3], there exists a variety B, defined over k_0, and a biregular mapping f of B to A, defined over k_1, such that $f_\sigma = f^\sigma \circ f^{-1}$ for every $\sigma \in G$. Let e be the neutral element of A. Then e^σ is the neutral element of A^σ, so that $f_\sigma(e) = e^\sigma$. Put $y = f^{-1}(e)$. Then y is rational over k_1, and $y^\sigma = (f^\sigma)^{-1}(e^\sigma) = (f^\sigma)^{-1} \circ f_\sigma(e) = f^{-1}(e) = y$. Therefore y is rational over k_0. Taking y as the neutral element, we may put into B a structure of abelian variety defined over k_0. Then f is an isomorphism of B to A. Define a structure $\mathfrak{Q}' = (B, \mathcal{C}', \theta'; t_1', \cdots, t_s')$ so that f gives an isomorphism of $\mathfrak{Q}'$ to $\mathfrak{Q}$. Since $f(t_i') = t_i$ and $f_\sigma(t_i) = t_i^\sigma$, we see easily that the t_i' are rational over k_0. For every $r \in \mathfrak{r}$, we have $f_\sigma \circ \theta(r) = \theta^\sigma(r) \circ f_\sigma$ and $f \circ \theta'(r) = \theta(r) \circ f$. From this we can easily derive that $\theta'(r)$ is defined over k_0. Let X be a divisor in $\mathcal{C}$ rational over k_1. Put $Y = \sum_{\sigma \in G} f^{-1}(X)^\sigma$. Then Y is a divisor on B rational over k_0. Furthermore, $f(Y) = \sum_{\sigma \in G} f_\sigma^{-1}(X^\sigma)$, and $f(Y)$ is a divisor in $\mathcal{C}$, since f_σ^{-1} sends $\mathcal{C}^\sigma$ to $\mathcal{C}$. Hence $Y \in \mathcal{C}'$. Therefore $\mathfrak{Q}'$ has the required property. The uniqueness of $\mathfrak{Q}'$ is obvious.

1.6. PROPOSITION. *Let $\mathcal{P} = (A, \mathcal{C}, \theta)$ be as above, and m a positive integer. Let $\{t_1, \cdots, t_s\}$ be a set of generators for the group of all the points t on A such that $mt = 0$, and let $\mathfrak{Q} = \mathcal{P}(t_1, \cdots, t_s)$. Then the field of moduli of $\mathfrak{Q}$ contains a primitive m^{th} root of unity.*

Let k_0 be the field of moduli of $\mathfrak{Q}$, and σ an isomorphism of the universal domain into itself, leaving every element of k_0 invariant. Then there exists an isomorphism λ of $\mathfrak{Q}$ onto $\mathfrak{Q}^\sigma$. We see that $\lambda t = t^\sigma$ for every t on A such that $mt = 0$. Let X be a basic polar divisor in $\mathcal{C}$ in the sense of [13, 4.2, p. 30]. The notation being as in Weil [14, § XI], if $mt = mt' = 0$, we have

$$e_{X,m}(t, t')^\sigma = e_{X^\sigma,m}(t^\sigma, t'^\sigma) = e_{X,m}(t, t') \,,$$

since $\lambda^{-1}(X^\sigma)$ is algebraically equivalent to X, and $\lambda t = t^\sigma$, $\lambda t' = t'^\sigma$. We can take t and t' so that $e_{X,m}(t, t')$ is a primitive m^{th} root of unity; in fact this is easily seen from the formula [13, p. 25, (7)], since X is the basic polar divisor. Our proposition is thereby proved.

1.7. Let $\mathscr{P} = (A, \mathcal{C}, \theta)$ be as above, and let $\mathfrak{h}$ be a finite subgroup of A. We consider the following condition on $\mathfrak{h}$.

(1.7.1) Let α be an automorphism of $\mathfrak{h}$ such that, for each $t \in \mathfrak{h}$, there exists an automorphism ε_t of $\mathscr{P}$ such that $\alpha(t) = \varepsilon_t(t)$. Then there exists an automorphism ε of $\mathscr{P}$ such that $\alpha(t) = \varepsilon(t)$ for all $t \in \mathfrak{h}$.

1.8. PROPOSITION. *Let $\mathscr{P}$ be as above, and (V, h) a normalized Kummer variety of $\mathscr{P}$ in the sense of [6, 2.2]. Let k_1 be the field of moduli of $\mathscr{P}$. Let $\mathfrak{h} = \{t_1, \cdots, t_s\}$ be a finite subgroup of A, satisfying the condition (1.7.1), and let $\mathfrak{Q} = \mathscr{P}(t_1, \cdots, t_s)$. Then $k_1(h(t_1), \cdots, h(t_s))$ is the field of moduli of $\mathfrak{Q}$.*

Let σ be an isomorphism of the universal domain into itself. Suppose that there exists an isomorphism λ of $\mathfrak{Q}$ to $\mathfrak{Q}^\sigma$. Then σ leaves invariant every element of k_1, and $\lambda t_i = t_i^\sigma$. Hence $h(t_i)^\sigma = h^\sigma(t_i^\sigma) = h^\sigma(\lambda t_i) = h(t_i)$. Therefore σ is the identity mapping on $k_1(h(t_1), \cdots, h(t_s))$. Conversely, suppose that σ is the identity mapping on $k_1(h(t_1), \cdots, h(t_s))$. Then there exists an isomorphism λ of $\mathscr{P}$ to $\mathscr{P}^\sigma$. We have $h(t_i) = h^\sigma(t_i^\sigma) = h(\lambda^{-1}t_i^\sigma)$. Hence, for each i, there exists an automorphism ε_i of $\mathscr{P}$ such that $\lambda^{-1}(t_i^\sigma) = \varepsilon_i(t_i)$. It is clear that $t \to \lambda^{-1}(t^\sigma)$ is an automorphism of $\mathfrak{h}$. By our assumption (1.7.1), there exists an automorphism ε of $\mathscr{P}$ such that $\lambda^{-1}(t^\sigma) = \varepsilon(t)$ for all $t \in \mathfrak{h}$. Then $\lambda\varepsilon$ is an isomorphism of $\mathfrak{Q}$ to $\mathfrak{Q}^\sigma$. This proves our proposition.

1.9. REMARK. In the proof of 1.8, we observe that the mapping $x \to \lambda^{-1}(x^\sigma)$ commutes with the operation of $\theta(a)$ for all $a \in \mathfrak{r}$. Therefore, if $\theta(a)\mathfrak{h} \subset \mathfrak{h}$ for all $a \in \mathfrak{r}$, then, for the validity of 1.8, we may modify (1.7.1) by restricting α to those which commute with $\theta(a)$ for all $a \in \mathfrak{r}$.

1.10. REMARK. Suppose that $\mathscr{P}$ has only ± 1 as automorphisms. Then for every positive integer m, the group $\mathfrak{h} = \{t \in A \mid mt = 0\}$ satisfy the condition (1.7.1).

1.11. PROPOSITION. *Let $\mathscr{P}$ and $\mathfrak{r}$ be as in 1.1, and $\mathfrak{a}$ a right ideal of $\mathfrak{r}$ containing an element which is not a zero-divisor. Let $\mathfrak{h} = \{t \in A \mid \theta(a)t = 0$ for every $a \in \mathfrak{a}\}$ and let $\{t_1, \cdots, t_s\}$ be a set of generators of $\mathfrak{h}$ over $\theta(\mathfrak{r})$. Let k_1 and k_0 be the fields of moduli of $\mathscr{P}$ and $\mathscr{P}(t_1, \cdots, t_s)$, respectively. Then k_0 is a finite Galois extension of k_1.*

First we note that k_0 is a finitely generated extension over the prime field.

Now by our assumption on $\mathfrak{a}$, $\mathfrak{h}$ is a finite set. Let $\mathfrak{h} = \{u_1, \cdots, u_m\}$. For every permutation π of the u_i, put $\mathfrak{Q}_\pi = \mathscr{P}(\pi(u_1), \cdots, \pi(u_m))$. In view of (M), we see easily that the $\mathfrak{Q}_\pi$ have k_0 as the field of moduli. Let σ be an isomorphism of the universal domain into itself, which leaves every element of k_1 invariant. Then there exists an isomorphism λ of $\mathscr{P}$ to $\mathscr{P}^\sigma$. We see that $t \to \lambda^{-1} t^\sigma$ gives a permutation π of the u_i. Then λ is an isomorphism of $\mathfrak{Q}_\pi$ to $\mathscr{P}^\sigma(u_1^\sigma, \cdots, u_m^\sigma)$. Since $\mathscr{P}^\sigma(u_1^\sigma, \cdots, u_m^\sigma)$ have k_0^σ as the field of moduli, this shows $k_0^\sigma = k_0$. Our proposition is thereby proved.

1.12. Proposition. *Let* $\mathfrak{Q} = (A, \mathcal{C}, \theta; \{t_i\})$ *be as in 1.1, and* λ *an isogeny of* $\mathfrak{Q}$ *to another structure* $\mathfrak{Q}' = (A', \mathcal{C}', \theta'; \{t_i'\})$. *Let* $u_1, \cdots, u_q$ *be points on* A *such that* $\mathrm{Ker}\,(\lambda) = \sum_{j=1}^q \theta(\mathfrak{r}) u_j$. *Then the field of moduli of* $\mathfrak{Q}'$ *is contained in the field of moduli of* $(A, \mathcal{C}, \theta; \{t_i, u_j\})$.

Let σ be an isomorphism of the universal domain into itself which leaves invariant every element of the field of moduli of $(A, \mathcal{C}, \theta; \{t_i, u_j\})$. Then there exists an isomorphism μ of $\mathfrak{Q}$ to $\mathfrak{Q}^\sigma$ such that $\mu(t_i) = t_i^\sigma$ and $\mu(u_j) = u_j^\sigma$. We have $\mu(\mathrm{Ker}\,(\lambda)) = \sum_j \theta^\sigma(\mathfrak{r}) u_j^\sigma = \mathrm{Ker}\,(\lambda^\sigma)$. Hence there exists an isomorphism α of A to A' such that $\alpha\lambda = \lambda^\sigma \mu$. By [6, Prop. 2.4], α is an isogeny of $(A', \mathcal{C}', \theta')$ to $(A', \mathcal{C}', \theta')^\sigma$. Moreover, we have $\alpha t_i' = \alpha\lambda t_i = \lambda^\sigma \mu t_i = \lambda^\sigma t_i^\sigma = (\lambda t_i)^\sigma = t_i'^\sigma$. Hence α is an isogeny of $\mathfrak{Q}'$ to $\mathfrak{Q}'^\sigma$. Therefore σ is the identity mapping on the field of moduli of $\mathfrak{Q}'$. This proves our proposition.

2. Analytic families

2.1. Let us first recall some result of [5, Nos. 11, 12]. Let $\mathfrak{Z}$ be a connected open subset of a complex vector space, and $\mathfrak{r}$ a ring as in 1.1. We consider a system $\{E, w(z), \Phi(r)\}$ with the following properties.

(2.1.1) E is an alternating matrix of degree $2n$ with integral coefficients;

(2.1.2) the mapping $z \to w(z)$ is a holomorphic mapping of $\mathfrak{Z}$ into the space of complex $(n \times 2n)$-matrices;

(2.1.3) the group D_z generated by the columns of $w(z)$ over Z is a discrete subgroup of C^n of rank $2n$;

(2.1.4) E defines a Riemann form on the complex torus C^n/D_z;

(2.1.5) Φ is a faithful representation of $\mathfrak{r}$ by complex matrices of size n;

(2.1.6) for every $r \in \mathfrak{r}$, the linear transformation in C^n given by $\Phi(r)$ is an endomorphism of C^n/D_z.

Then we get an analytic family of polarized abelian varieties of type $\mathfrak{r}$:

$$\Sigma = \{\mathscr{P}_z = (A_z, \mathcal{C}_z, \theta_z) \mid z \in \mathfrak{Z}\} \,.$$

The $\mathscr{P}_z$ are parametrized as follows by a holomorphic mapping $\Lambda(x, z)$ of $C^n \times \mathfrak{Z}$ into a projectives space P^N.

(2.1.7) For every $z \in \mathfrak{Z}$, $x \to \Lambda(x, z)$ gives an analytic isomorphism of C^n/D_z to an abelian variety A_z in P^N.

(2.1.8) $\mathcal{C}_z$ is a polarization of A_z determined by the hyperplane sections, and it corresponds to the Riemann form given by E.

(2.1.9) For every $r \in \mathfrak{r}$, $\theta_z(r)\Lambda(x, z) = \Lambda(\Phi(r)x, z)$.

We may assume that the entries of E have no non-trivial common divisor; in fact, if this is not so, we exchange E for $c^{-1}E$ for a suitable positive integer c. Then, in view of the construction of Λ in [4, expose 20] and [5, No. 11], we may assume

(2.1.10) A_z is a projective embedding of itself by the linear system of a divisor corresponding to $3E$.

In [5, No. 12], we gave a holomorphic mapping Ψ of $\mathfrak{Z}$ into a projective space with the following properties:

(2.1.11) For every $z \in \mathfrak{Z}$, $\mathcal{P}_z$ is defined over $Q(\Psi(z))$;

(2.1.12) let y and w be points of $\mathfrak{Z}$, and $\mathfrak{p}$ a place of the field $Q(\Psi(y))$ taking values in C such that $\mathfrak{p}(\Psi(y)) = \Psi(w)$. Then

$$\mathfrak{p}(A_y) = A_w , \qquad \mathfrak{p}(T_y) = T_w , \qquad \mathfrak{p}(U_y(r)) = U_w(r) ,$$

where T_z denotes the graph of the law of composition on A_z, and $U_z(r)$ the graph of $\theta_z(r)$. (For the definition of places and $\mathfrak{p}(V)$, we refer to [5, Appendix] and [13, Chap. III].)

2.2. Let $z \to x_i(z)$ $(1 \leq i \leq s)$ be holomorphic mappings of $\mathfrak{Z}$ into C^n. Put $t_i(z) = \Lambda(x_i(z), z)$. Then we get a structure

$$\mathfrak{Q}_z = \mathcal{P}_z(t_1(z), \cdots, t_s(z)) .$$

We call Σ^* the family $\{\mathfrak{Q}_z \mid z \in \mathfrak{Z}\}$. Fix a basis of $\mathfrak{r}$ over Z, and define the variety $\mathcal{F}(A_z, \theta_z; \{t_i(z)\})$ as in 1.2. For simplicity, we put

$$\mathcal{F}_z = \mathcal{F}(A_z, \theta_z; \{t_i(z)\}) .$$

It is clear that the $\mathcal{F}_z$ for all $z \in \mathfrak{Z}$ have the common ambient space. By [5, Lemma 8], we see easily that the $\mathcal{F}_z$ are of dimension $(N + 1)^2 - 1 + dn + n$.

2.3. THEOREM. *There exist meromorphic functions $h_1, \cdots, h_\lambda$ on $\mathfrak{Z}$ and an analytic subset $\mathfrak{W}$ of $\mathfrak{Z}$ of codimension 1 such that*

(i) *For $z \in \mathfrak{Z} - \mathfrak{W}$, the varieties $\mathcal{F}_z$ have the same degree.*

(ii) $c(\mathcal{F}_z) = (1, h_1(z), \cdots, h_\lambda(z))$ *for $z \in \mathfrak{Z} - \mathfrak{W}$.*

The proof is quite similar to the proof of [5, Theorem 1]. Let $\Psi(z)$ be as in 2.1. Let u be a generic point of $\mathfrak{Z}$ for the functions Ψ, t_j over Q in the sense of [5, No. 8]. There exist a finite number of elements α_i, other than 0, in $Q(\Psi(u), t_j(u))$ such that, for every place $\mathfrak{q}$ of $Q(\Psi(u), t_j(u))$, $\mathfrak{q}(\mathcal{F}_z)$ is an irreducible variety if $q(\alpha_i) \neq 0$ for every i. Let a_i be the function in $Q(\Psi, t_j)$ corresponding to α_i by the canonical isomorphism of $Q(\Psi, t_j)$ to $Q(\Psi(u), t_j(u))$. Let $\mathfrak{Y}$ be the set of points y in $\mathfrak{Z}$ where the a_i are holomorphic and $a_i(y) \neq 0$ for every i. Let $y \in \mathfrak{Y}$. Then we can find a place $\mathfrak{p}$ of $Q(\Psi(u), t_j(u))$ such that $\mathfrak{p}(g(u)) = g(y)$ for every function g in $Q(\Psi, t_j)$ which is holomorphic at y. By means of the argument of [5, p. 123], we see that, for a suitable extension $\mathfrak{P}$ of $\mathfrak{p}$ and for a suitable generic point v of $\mathcal{F}_u$ over $Q(\Psi(u), t_j(u))$, $\mathfrak{P}(v)$ is a generic point of $\mathcal{F}_y$ over $\mathfrak{p}(Q(\Psi(u), t_j(u)))$. Since $\mathfrak{p}(\mathcal{F}_u)$ is a variety defined over $\mathfrak{p}(Q(\Psi(u), t_j(u)))$, we have $\mathfrak{p}(\mathcal{F}_u) \supset \mathcal{F}_y$. Now $\mathfrak{p}(\mathcal{F}_u)$ and $\mathcal{F}_y$ are irreducible varieties of the same dimension; so we have $\mathfrak{p}(\mathcal{F}_u) = \mathcal{F}_y$. By the same argument as in the last part of the proof of [5, Th. 1], we get the above assertion.

2.4. PROPOSITION. *Let k be a subfield of C with countably many elements. Let u be a generic point of $\mathfrak{Z}$ for the functions Ψ and t_j over k, and y an arbitrary point of $\mathfrak{Z}$. Then we have $\dim_k c(\mathcal{F}_u) \geqq \dim_k c(\mathcal{F}_y)$. Moreover, if $\dim_k c(\mathcal{F}_u) = \dim_k c(\mathcal{F}_y)$, there exists an isomorphism σ of C into itself over k such that $\mathfrak{Q}_u^\sigma$ is isomorphic to $\mathfrak{Q}_y$.*

This is a generalization of [6, Prop. 3.5] and can be proved similarly. The number $\dim_k c(\mathcal{F}_u)$ is uniquely determined by k. If $\dim_k c(\mathcal{F}_y) = \dim_k c(\mathcal{F}_u)$, then we call $\mathfrak{Q}_y$ a *generic member* of the family Σ^*, over k.

2.5. PROPOSITION. *Let k be as in 2.4, $\mathfrak{Q}_v$ a generic member of Σ^* over k, and $\mathfrak{Q}_y$ any member of Σ^*. Then there exists a specialization $\mathcal{F}'$ of $\mathcal{F}_v$ over k such that $\mathcal{F}' \supset \mathcal{F}_y$.*

Let u be as in 2.4. Apply the proof of 2.3 to the present case, taking k instead of Q. Then the assertion $\mathfrak{p}(\mathcal{F}_u) \supset \mathcal{F}_y$ holds without the condition $a_i(y) \neq 0$. Now by 2.4, $\mathcal{F}_v$ is a generic specialization of $\mathcal{F}_u$ over k. Therefore $\mathfrak{p}(\mathcal{F}_u)$ is a specialization of $\mathcal{F}_v$ over k. This proves our proposition.

2.6. Let the functions h_i be as in 2.3 and let $h_0 = 1$. Denote by $\mathfrak{Z}_p$ for $0 \leqq p \leqq \lambda$ the set of points z of $\mathfrak{Z}$ where $h_0/h_p, \cdots, h_\lambda/h_p$ are holomorphic. Put $\mathfrak{W}_0 = \bigcap_{p=0}^{\lambda} (\mathfrak{Z} - \mathfrak{Z}_p)$. Then $\mathfrak{W}_0$ is an analytic subset of $\mathfrak{Z}$ of codimension >1.[2] Then we find a holomorphic mapping h of $\mathfrak{Z} - \mathfrak{W}_0$ into P^λ which coincides with the mapping $y \to (h_0/h_p(y), \cdots, h_\lambda/h_p(y))$ on $\mathfrak{Z}_p$. From 2.3, we see that $\mathfrak{W}_0 \subset \mathfrak{W}$, and $h(z) = c(\mathcal{F}_z)$ for $z \in \mathfrak{Z} - \mathfrak{W}$. Let e be the degree of the varieties $\mathcal{F}_z$ for $z \in \mathfrak{Z} - \mathfrak{W}$. Then we get

2.7. PROPOSITION. *Let y be a point of $\mathfrak{Z}$ such that $\deg(\mathcal{F}_y) = e$. Then*

---

[2] In [9, p. 170], it is wrongly stated that $\mathfrak{W}_0$ is of codimension 1. Since the image space is the projective space, this should be corrected as in the present text. However, this correction has no effect upon our discussion.

$y \in \mathfrak{Z} - \mathfrak{W}_0$ and $h(y) = c(\mathfrak{F}_\nu)$.

This is a generalization of [9, Prop. 10]; the proof can be given by the same argument, if we take the points $t_j(z)$ into account.

3. Anti-hermitian structure and points of finite order

3.1. Let F be a totally real algebraic number field of finite degree, and L a semi-simple algebra over F, with a positive involution ρ. Let Φ be a representation of L by complex matrices of size n, which maps the identity element of L to the identity matrix. Let $\mathscr{P} = (A, C, \theta)$ be a polarized abelian variety of type $\{L, \Phi, \rho\}$ (cf. [9, 1.4] and [FI, 1.5]). We consider the anti-hermitian structure (V, h, D) determined by $\mathscr{P}$ [FI, 1.9]. Put $f(x, y) = \mathrm{Tr}_{L/Q}\big(h(x, y)\big)$. As in [FI, 1.10], for every prime number p, we obtain a structure (V_p, f_p, D_p). Let $g_p(A)$ be the group of all points x on A such that $p^\nu x = 0$ for some positive integer ν. Let U_p be the set of all $p^{\nu\text{-th}}$ roots of unity in C for all positive integers ν. Denote by Q_p and Z_p the field of p-adic numbers and the ring of p-adic integers, respectively. By our definition of D_p and V_p, we have a canonical homomorphism $\mathfrak{x}$ of V_p onto $g_p(A)$ with kernel D_p. Every basis $\{u_1, \cdots, u_n\}$ of D over Z determines an isomorphism $\mathfrak{y}$ of V_p/D_p onto $(Q_p/Z_p)^{2n}$. Then $\mathfrak{y} \circ \mathfrak{x}^{-1}$ coincides with the isomorphism $\mathfrak{w}$ given in [FI, 1.10]. Further let us define an isomorphism $\mathfrak{t}$ of U_p onto Q_p/Z_p as in [FI, 1.10].

3.2. Let K' be as in [FI, 1.6], and σ an isomorphism of C into itself leaving every element of K' invariant. Then $\mathscr{P}^\sigma = (A^\sigma, C^\sigma, \theta^\sigma)$ is of type $\{L, \Phi, \rho\}$ [FI, 1.6]. Let $\mathscr{P}^\sigma$ determine an anti-hermitian structure (V^*, h^*, D^*). Put $f^*(u, v) = \mathrm{Tr}_{L/Q}\big(h^*(u, v)\big)$. Let $\mathfrak{x}^*$ be a canonical homomorphism of V_p^* onto $g_p(A^\sigma)$ with kernel D_p^*. Since $\zeta \to \zeta^\sigma$ gives an automorphism of U_p, there exists a unit c_p of Z_p such that

$$(3.2.1) \qquad\qquad \mathfrak{t}(\zeta^\sigma) = c_p\mathfrak{t}(\zeta) \ . \qquad\qquad (\zeta \in U_p) \ .$$

We see that $x \to x^\sigma$ is an isomorphism of $g_p(A)$ onto $g_p(A^\sigma)$, and $\theta^\sigma(a)x^\sigma = (\theta(a)x)^\sigma$. Therefore we can find an L_p-linear isomorphism α of V_p onto V_p^* such that

$$(3.2.2) \qquad\qquad D_p\alpha = D_p^*$$

$$(3.2.3) \qquad\qquad \mathfrak{x}(u)^\sigma = \mathfrak{x}^*(u\alpha) \qquad\qquad (u \in V_p) \ .$$

Let X be a basic polar divisor in C. Using the notation $e_{X, N}(s, t)$ of Weil [14, § XI], f_p corresponds to an alternating matrix E_p with coefficients in Z_p, by the following relation:

$$\mathfrak{t}\big(e_{X, q}(s, t)\big) \equiv q \cdot {}^t\mathfrak{w}(s)E_p\mathfrak{w}(t) \mod Z_p \ ,$$

where q is a power of p, and s, t are points on A such that $qs = qt = 0$. Since we have $e_{x,q}(s, t)^\sigma = e_{x^\sigma,q}(s^\sigma, t^\sigma)$, if u and v are points in V_p such that $qu \in D_p$ and $qv \in D_p$, then

$$c_p q \cdot f_p(u, v) \equiv \mathfrak{t}[e_{x,q}(\mathfrak{x}(u), \mathfrak{x}(v))^\sigma] = \mathfrak{t}[e_{x^\sigma,q}(\mathfrak{x}^*(u\alpha), \mathfrak{x}^*(v\alpha))]$$
$$\equiv q \cdot f_p^*(u\alpha, v\alpha) \quad \mathrm{mod}\ Z_p .$$

Hence we obtain

$$(3.2.4) \qquad\qquad f_p^*(u\alpha, v\alpha) = c_p f_p(u, v) \qquad\qquad ((u, v) \in V_p \times V_p) .$$

3.3. Let $\mathfrak{g}$ be the ring of integers in F. Hereafter we assume

$$(3.3.1) \qquad\qquad \theta(\mathfrak{g}) \subset \mathrm{End}\,(A) .$$

For every prime ideal $\mathfrak{p}$ in F, we get a structure $(V_\mathfrak{p}, h_\mathfrak{p}, D_\mathfrak{p})$ and $(V_\mathfrak{p}^*, h_\mathfrak{p}^*, D_\mathfrak{p}^*)$ [FI, 1.10]. The notation being as in 3.2, we see that α induces naturally an $L_\mathfrak{p}$-linear isomorphism $\alpha_\mathfrak{p}$ of $V_\mathfrak{p}$ onto $V_\mathfrak{p}^*$ such that

$$(3.3.2) \qquad\qquad D_\mathfrak{p} \alpha_\mathfrak{p} = D_\mathfrak{p}^* .$$

Further, by [FI, 1.2 and 1.4], we have

$$(3.3.3) \qquad\qquad h_\mathfrak{p}^*(u\alpha_\mathfrak{p}, v\alpha_\mathfrak{p}) = c_p h_\mathfrak{p}(u, v) \qquad\qquad ((u, v) \in V_\mathfrak{p} \times V_\mathfrak{p}) ,$$

where p is a prime number divisible by $\mathfrak{p}$.

Denote by $g_\mathfrak{p}(A)$ the group of all points t on A such that $\theta(\mathfrak{p}^\nu)t = 0$ for some positive integer ν. Let t be a point of finite order on A, and x an element of V corresponding to t, which is uniquely determined modulo D. The element of $V_\mathfrak{p}/D_\mathfrak{p}$ represented by x is uniquely determined by t. We write the element as $v_\mathfrak{p}(t)$; for convenience, we shall mean also by $v_\mathfrak{p}(t)$ any element of $V_\mathfrak{p}$ whose residue-class modulo $D_\mathfrak{p}$ is $v_\mathfrak{p}(t)$. Then (3.2.3) implies

$$(3.3.4) \qquad\qquad v_\mathfrak{p}(t)\alpha_\mathfrak{p} \equiv v_\mathfrak{p}^*(t^\sigma) \quad \mathrm{mod}\ D_\mathfrak{p}^*$$

for every point t on A of finite order, where $v_\mathfrak{p}^*$ is the corresponding function for A^σ.

3.4. Let $t_1, \cdots, t_s$ be elements of finite order on A. They determine elements $x_1, \cdots, x_s$ of V, uniquely modulo D. Put $\mathfrak{Q} = \mathscr{P}(t_1, \cdots, t_s)$. Then we say that the structure $\mathfrak{Q}$ is of type $(V, h, D; x_1, \cdots, x_s)$. Let $\mathfrak{Q}^\sigma$ be of type $(V^*, h^*, D^*; x_1^*, \cdots, x_s^*)$. From (3.3.4) we obtain easily

$$(3.4.1) \qquad\qquad x_i \alpha_\mathfrak{p} \equiv x_i^* \quad \mathrm{mod}\ D_\mathfrak{p}^* \qquad\qquad (1 \leq i \leq s)$$

for every prime ideal $\mathfrak{p}$ in F.

4. Automorphic functions with respect to a congruence subgroup

4.1. First we give a list of symbols which are, for the most part, those used in [FI].

L: division algebra over Q with a positive involution.

ρ: positive involution of L.

Φ: representation of L by complex matrices of size n.

F: the set of all elements x in the center of L such that $x^\rho = x$.

V: the vector space of all m-dimensional row-vectors with components in L, considered as a left L-module; $2n = m[L:Q]$.

T: non-degenerate ρ-anti-hermitian form on V.

$U(T)$: the group of all L-linear automorphisms α of V such that $T(x\alpha, y\alpha) = T(x, y)$.

$U_R(T)$: the group of all L_R-linear automorphisms α of V_R such that $T(x\alpha, y\alpha) = T(x, y)$, where $L_R = L \otimes_Q R$, $V_R = V \otimes_Q R$.

$\mathcal{H}$: bounded symmetric domain, isomorphic to the quotient space of $U_R(T)$ by a maximal compact subgroup.

$\mathfrak{M}$: lattice in V.

$\Gamma(T, \mathfrak{M})$: the subgroup of $U(T)$ consisting of all the elements α such that $\mathfrak{M}\alpha = \mathfrak{M}$.

Now we shall apply the result of § 2 to the family

$$\Sigma = \Sigma(L, \Phi, \rho; T, \mathfrak{M}) = \{\mathscr{P}_z = (A_z, \mathcal{C}_z, \theta_z) \mid z \in \mathcal{H}\}$$

of [FI, 2.1]. In view of [9, Th. 1 and 3.2], our family is certainly a special case of the family investigated in 2.1.

Let $\mathfrak{x}_i(z, T)$ $(1 \leq i \leq m)$ be as in [9, 2.6, p. 163]; with the notation of 2.1, they are just the columns of $w(z)$. For every element $b = (b^1, \cdots, b^m) \in V_R$ (with $b^i \in L_R$), put

$$(4.1.1) \qquad \mathfrak{x}(z, b) = \sum_{i=1}^m \Phi(b^i)\mathfrak{x}_i(z, T) \ .$$

For every $z \in \mathcal{H}$, the mapping $b \to \mathfrak{x}(z, b)$ is an R-isomorphism of V_R onto C^n, from which we obtain an isomorphism of $V_R/\mathfrak{M}$ onto A_z. It is clear that $V/\mathfrak{M}$ corresponds to the set of all points on A_z of finite order. Since the $\mathfrak{x}_i(z, T)$ are holomorphic in z [9, 3.2], we see that the mapping $z \to \mathfrak{x}(z, b)$ is holomorphic for every b.

4.2. PROPOSITION. *Let α be an L-linear automorphism of V such that $T(x\alpha, y\alpha) = \delta T(x, y)$ with a totally positive number δ of F. Let $z \in \mathcal{H}$, and $\alpha^{-1}(z) = w$. Then there exists one and only one C-linear automorphism $\Lambda(z, \alpha)$ of C^n such that*

$$\Lambda(z, \alpha)\mathfrak{x}(z, b) = \mathfrak{x}(\alpha^{-1}(z), b\alpha) \qquad\qquad (b \in V_R) .$$

Moreover, if $\mathfrak{M}\alpha \subset \mathfrak{M}$, *then* $\Lambda(z, \alpha)$ *defines an isogeny of* $\mathcal{P}_z$ *to* $\mathcal{P}_w$. *Conversely, if* $\mathcal{P}_z$ *and* $\mathcal{P}_w$ *are members of* Σ, *every isogeny of* $\mathcal{P}_z$ *to* $\mathcal{P}_w$ *can be obtained in this manner. For a fixed* α, $\Lambda(z, \alpha)$ *is holomorphic in* z.

Here the operation of α on $\mathcal{H}$ should be understood as in [9, 5.3]. Our proposition follows immediately from the consideration of [9, 2.8], except the last assertion. (With the notation of [9, 2.8], Λ and U correspond to the present $\Lambda(z, \alpha)$ and α.) We can find n elements $b_1, \cdots, b_n$ of V_R such that $\mathfrak{x}(z, b_1), \cdots, \mathfrak{x}(z, b_n)$ are linearly independent over C. Since the $\mathfrak{x}(z, b_i)$ and $\mathfrak{x}(\alpha^{-1}(z), b_i\alpha)$ are holomorphic in z, from the relation

$$\Lambda(z, \alpha)\mathfrak{x}(z, b_i) = \mathfrak{x}(\alpha^{-1}(z), b_i\alpha) ,$$

it follows that $\Lambda(z, \alpha)$ is holomorphic in z.

4.3. Let $v_1, \cdots, v_s$ be elements of V. Put

$$\mathfrak{N} = \mathfrak{M} + \textstyle\sum_{i=1}^{s} \boldsymbol{Z}v_i .^{3}$$

Let $t(z, b)$ be the point on A_z corresponding to $\mathfrak{x}(z, b)$, and let $t_i(z) = t(z, v_i)$ for $1 \leq i \leq s$. Then we get a structure

$$\mathfrak{Q}_z = \mathcal{P}_z(t_1(z), \cdots, t_s(z)) .$$

We denote the family $\{\mathfrak{Q}_z \mid z \in \mathcal{H}\}$ by $\Sigma^*(L, \Phi, \rho; T, \mathfrak{M}; \{v_i\})$ or simply by Σ^*. Every member of Σ^* is of type $(V, T, \mathfrak{M}; \{v_i\})$ in the sense of 3.4. Conversely, given a structure $\mathfrak{Q} = \mathcal{P}(t_1, \cdots, t_m)$, if $\mathcal{P}$ is of type $\{L, \Phi, \rho\}$ and $\mathfrak{Q}$ is of type $(V, T, \mathfrak{M}; \{v_i\})$, then $\mathfrak{Q}$ is isomorphic to a member of Σ^*.

4.4. PROPOSITION. *The notation being as above, let*

$$\Gamma(T, \mathfrak{N}/\mathfrak{M}) = \{\alpha \in \Gamma(T, \mathfrak{M}) \mid x\alpha \equiv x \bmod \mathfrak{M} \textit{ for every } x \in \mathfrak{N}\} .$$

Then $\mathfrak{Q}_z$ *and* $\mathfrak{Q}_w$ *are isomorphic, if and only if* $z = \alpha(w)$ *for an element* α *of* $\Gamma(T, \mathfrak{N}/\mathfrak{M})$.

Let $\alpha \in \Gamma(T, \mathfrak{M})$ and $z = \alpha(w)$. Let λ be an isomorphism of $\mathcal{P}_z$ to $\mathcal{P}_w$ corresponding to $\Lambda(z, \alpha)$ (cf. 4.2). Then we have

$$(4.4.1) \qquad\qquad \lambda(t(z, b)) = t(w, b\alpha) . \qquad\qquad (b \in V) .$$

If $\alpha \in \Gamma(T, \mathfrak{N}/\mathfrak{M})$, we have $v_i\alpha \equiv v_i \bmod \mathfrak{M}$ for every i, so that $t(w, v_i\alpha) = t(w, v_i)$. By (4.4.1), we get $\lambda(t_i(z)) = t_i(w)$. It follows that λ is an isomorphism of $\mathfrak{Q}_z$ to $\mathfrak{Q}_w$. Conversely, suppose that there exists an isomorphism λ of $\mathfrak{Q}_z$ to $\mathfrak{Q}_w$. Then, by 4.2, we get an element α of $\Gamma(T, \mathfrak{M})$ such that $z = \alpha(w)$ and

---

[3] We may also put $\mathfrak{N} = \mathfrak{M} + \sum_{i=1}^{s} \mathfrak{o}v_i$, where $\mathfrak{o} = \{a \in L \mid a\mathfrak{M} \subset \mathfrak{M}\}$. Then $\mathfrak{N}$ is an $\mathfrak{o}$-lattice. The following Proposition 4.4 and Theorem 4.5 are true for this $\mathfrak{N}$, without modifying the proofs.

(4.4.1) holds. Since $\lambda\big(t_i(z)\big) = t_i(w)$, we have $v_i\alpha \equiv v_i \bmod \mathfrak{M}$ for every i, so that $\alpha \in \Gamma(T, \mathfrak{N}/\mathfrak{M})$. This completes the proof.

4.5. Theorem. *Define the functions h_j as in 2.3 for the present family Σ^*, taking the ring $\{a \in L \mid a\mathfrak{M} \subset \mathfrak{M}\}$ as $\mathfrak{r}$. Then the following assertions hold.*

(i) *$C(h_1, \cdots, h_\lambda)$ is the field of all automorphic functions on $\mathcal{H}$ with respect to $\Gamma(T, \mathfrak{N}/\mathfrak{M})$.*

(ii) *If k is the algebraic closure of Q in $Q(h_1, \cdots, h_\lambda)$, the fields $k(h_1, \cdots, h_\lambda)$ and C are linearly disjoint over k.*

By 1.3, 2.3 and 4.4, for an element α of $\Gamma(T, \mathfrak{M})$, we have $h_j = h_j \circ \alpha (1 \leq j \leq \lambda)$ if and only if $\alpha \in \Gamma(T, \mathfrak{N}/\mathfrak{M})$. Let the functions f_i be as in [9, Th. 3]. Take a generic point u of $\mathcal{H}$ for the f_i and h_j over Q so that the equalities in [9, Th. 3, (ii)] and (ii) of 2.3 are true with $u = z$. Then $Q\big(f_1(u), \cdots, f_\kappa(u)\big)$ and $Q\big(h_1(u), \cdots, h_\lambda(u)\big)$ are the fields of moduli of $\mathscr{P}_u$ and $\mathscr{Q}_u$, respectively. By 1.11, the latter is a finite algebraic extension of the former. Since u is generic, this shows that $Q(h_1, \cdots, h_\lambda)$ is a finite algebraic extension of $Q(f_1, \cdots, f_\kappa)$. From this and [9, Th. 4, (iii)], we obtain our assertion (ii). Now we can say that $C(f_1, \cdots, f_\kappa)$ is the field of all automorphic functions on $\mathcal{H}$ with respect to $\Gamma(T, \mathfrak{M})$. If $\mathcal{H}/\Gamma(T, \mathfrak{M})$ is compact, this follows from only [9, Th. 4]. If $\mathcal{H}/\Gamma(T, \mathfrak{M})$ is not compact, we have to combine [9, Th. 4] with a recent result of Baily and Borel [16] on the projective embedding of $\mathcal{H}/\Gamma(T, M)$ (cf. also [9, pp. 172–173]). Then the assertion (i) is clear from our discussion.

The above theorem implies that, roughly speaking, the field of all automorphic functions on $\mathcal{H}$ with respect to $\Gamma(T, \mathfrak{N}/\mathfrak{M})$ has a model defined over an algebraic number field k. In the following §§ 6 9, we shall give a precise description for k, when L belongs to (Type I, II, IV) of [9, Prop. 1]; in the case of (Type IV), we shall have to assume that L is commutative.

5. Fibre varieties

5.1. In this section, we shall define a fibre variety $\mathcal{W}$, of which the base space is $\mathcal{H}/\Gamma(T, \mathfrak{N}/\mathfrak{M})$ and each fibre is the product of several copies of the abelian variety A_z. Such a fibre variety $\mathcal{W}$ was first suggested by M. Kuga. From number-theoretical view-point, it is important to have a model for $\mathcal{W}$, defined over an algebraic number field. Our purpose is to define such a model, as an application of our results in the previous sections.

Let the notation be as in 4.1. Let q be a positive integer. We denote by V_R^q the product of q copies of V_R. Let α be an L_R-linear automorphism of V_R. We define the operation of α on V_R^q by

$$(u_1, \cdots, u_q)\alpha = (u_1\alpha, \cdots, u_q\alpha)$$

for $(u_1, \cdots, u_q) \in V_R^q$ with u_i in V_R. Further we define a law of multiplication on $U_R(T) \times V_R^q$ by

$$(\alpha, x)(\beta, y) = (\alpha\beta, x\beta + y) \qquad (\alpha, \beta \in U_R(T); x, y \in V_R^q) .$$

Then $U_R(T) \times V_R^q$ becomes a group with respect to this law. We let operate every element of $U_R(T) \times V_R^q$ on $\mathcal{K} \times V_R^q$ as follows:

$$(z, x)(\alpha, y) = (\alpha^{-1}(z), x\alpha + y) \qquad (z \in \mathcal{K}, \alpha \in U_R(T), x, y \in V_R^q) .$$

We see easily that $\mathcal{K} \times V_R^q$ is isomorphic to a quotient space of $U_R(T) \times V_R^q$ with respect to a compact subgroup.

Let Γ be a subgroup of $\Gamma(T, \mathfrak{M})$ of finite index. We assume

(5.1.1) Γ *operates freely on* $\mathcal{K}$, *and* $\mathcal{K}/\Gamma$ *is compact.*

This implies, in particular, that Γ has no element, other than the identity matrix, which gives the identity transformation on $\mathcal{K}$. Let $\mathfrak{M}^q$ be the product of q copies $\mathfrak{M}$. Then $\Gamma \times \mathfrak{M}^q$ is a discrete subgroup of $U_R(T) \times V_R^q$, which operates freely on $\mathcal{K} \times V_R^q$. We put

$$\mathcal{V} = \mathcal{K}/\Gamma , \qquad \mathcal{W}_q = (\mathcal{K} \times V_R^q)/(\Gamma \times \mathfrak{M}^q) .$$

Then $\mathcal{V}$ and $\mathcal{W}_q$ are compact manifolds. We obtain a commutative diagram:

(5.1.2)

$$\begin{array}{ccc} \mathcal{K} \times V_R^q & \xrightarrow{\ \psi\ } & \mathcal{W}_q \\ {\scriptstyle \pi_0}\downarrow & & \downarrow{\scriptstyle \pi} \\ \mathcal{K} & \xrightarrow[\ \varphi\ ]{} & \mathcal{V} \end{array}$$

Here ψ and φ are projections defined in an obvious way; $\pi_0(z, x) = z$; and π is determined by $\pi \circ \psi = \varphi \circ \pi_0$.

Let $\xi(z, b)$ be as in (4.1.1). For $u = (u_1, \cdots, u_q) \in V_R^q$, put

$$\xi^q(z, u) = (\xi(z, u_1), \cdots, \xi(z, u_q)) .$$

Then $z \to \xi^q(z, v)$ is a holomorphic mapping of $\mathcal{K}$ into C^{nq}. Define a mapping ω of $\mathcal{K} \times V_R^q$ to $\mathcal{K} \times C^{nq}$ by

$$\omega(z, u) = (z, \xi^q(z, u)) \qquad (z \in \mathcal{K}, u \in V_R^q) .$$

Then ω is a real analytic isomorphism of $\mathcal{K} \times V_R^q$ to $\mathcal{K} \times C^{nq}$. By this mapping ω, we transfer the complex structure of $\mathcal{K} \times C^{nq}$ to $\mathcal{K} \times V_R^q$.

5.2. PROPOSITION. *Let α be as in 4.2, and let $w \in V_R^q$. Then the mapping $(z, u) \to (\alpha^{-1}(z), u\alpha + w)$ is a complex analytic automorphism of $\mathcal{K} \times V_R^q$.*

Let u_i and w_i be components of u and w. By 4.2, we have

$$\xi(\alpha^{-1}(z), u_i\alpha + w_i) = \Lambda(z, \alpha)\xi(z, u_i) + \Lambda(z, \alpha)\xi(z, w_i\alpha^{-1}) .$$

Since $\Lambda(z, \alpha)$ and $\mathfrak{x}(z, w_i \alpha^{-1})$ are holomorphic in z, we get our assertion.

5.3. In view of 5.2, we may regard $\mathcal{W}_q$ as a complex manifold. Now a recent result of Kuga [3] shows that $\mathcal{W}_q$ has a structure of projective variety. On the other hand, it is well-known that $\mathcal{V}$ has a structure of projective variety. Therefore, we choose embeddings of $\mathcal{V}$ and $\mathcal{W}_q$ into projective spaces, and from now on, regard them as projective varieties. Then the projection π of $\mathcal{W}_q$ to $\mathcal{V}$ in the diagram (5.1.2) is a rational mapping of $\mathcal{W}_q$ to $\mathcal{V}$. For every $y \in \mathcal{V}$, put

$$B_y = \pi^{-1}(y) \ .$$

From our construction of $\mathcal{W}_q$, we see that B_y is isomorphic to the product A_z^q of q copies of an abelian variety A_z with a point z on $\mathcal{H}$ such that $\varphi(z) = y$. If we denote by Z the graph of π, we get

(5.3.1) $$Z \cdot (\mathcal{W}_q \times y) = B_y \times y \ ,$$

where $\cdot$ means the intersection product of algebraic cycles.

5.4. Let $\mathfrak{o} = \{a \in L \mid a\mathfrak{M} \subset \mathfrak{M}\}$. Then $\theta_z(\mathfrak{o}) = \theta_z(L) \cap \mathrm{End}\,(A_z)$. Let $a = (a_{ij})$ with $a_{ij} \in \mathfrak{o}$ be an element of the total matric ring $M_q(\mathfrak{o})$. For $u = (u_1, \cdots, u_q) \in V_R^q$, put $au = (v_1, \cdots, v_q)$ with $v_i = \sum_j a_{ij} u_j$. This operation corresponds to an endomorphism of A_z^q, which we denote by $\theta_z^q(a)$. Then θ_z^q is an isomorphism of $M_q(\mathfrak{o})$ into $\mathrm{End}\,(A_z^q)$. For every $a \in M_q(\mathfrak{o})$, put

(5.4.1) $$\Theta(a) = \{\psi(z, u) \times \psi(z, au) \mid z \in \mathcal{H}, u \in V_R^q\} \ .$$

Then $\Theta(a)$ is a subvariety of $\mathcal{W}_q \times \mathcal{W}_q$, and if $y = \varphi(z)$, $\Theta(a) \cdot (B_y \times B_y)$ corresponds to the graph of $\theta_z^q(a)$ on A_z^q. We denote this endomorphism of B_y by $\theta_y'(a)$. For every $b \in V$, put

(5.4.2) $$\mathcal{S}(b) = \{\psi(z, b) \mid z \in \mathcal{H}\} \ .$$

We see that $\mathcal{S}(b)$ is a subvariety of $\mathcal{W}_q$. The restriction of π to $\mathcal{S}(b)$ gives a rational mapping of $\mathcal{S}(b)$ onto $\mathcal{V}$. If $b\gamma \equiv b \bmod \mathfrak{M}^q$ for every $\gamma \in \Gamma$, this mapping is one-to-one, so that it is biregular. In particular, $\mathcal{S}(0)$ is the locus of the origin of B_y for $y \in \mathcal{V}$.

5.5. LEMMA. *Let X be a non-singular projective variety, defined over a subfield of C, which is a minimal model in the sense that every rational mapping of a variety U into X is defined at every simple point of U. Suppose that there exist a variety X', defined over a field k, and a birational mapping f of X to X'. Then there exists a projective variety X^* and a birational mapping f^* of X^* to X', both defined over k, such that $f^{-1} \circ f^*$ is a biregular isomorphism of X^* to X.*

If σ is an isomorphism of a field of definition for X and f, then X^σ is a non-singular minimal model. Hence $f^{-1} \circ f^\sigma$ is a biregular isomorphism of X^σ to X. Therefore, applying [15, Th. 5, Th. 7, Th. 2], we get our lemma.

5.6. Let X, X', f, X^*, f^* be as in 5.5, and let Y, Y', g, Y^*, g^* be varieties and mappings which are set in the the same situation as X, X', f, X^*, f^*, with the same basic field k. Suppose that there exist a rational mapping h of X to Y and a rational mapping h' of X' to Y' such that $h' \circ f = g \circ h$, and h' is defined over k. Then we find a rational mapping h^* of X^* to Y^* such that $h' \circ f^* = g^* \circ h^*$. Since f^*, h' and g^* are all defined over k, h^* should be defined over k. Moreover, we have clearly $h \circ (f^{-1} \circ f^*) = (g^{-1} \circ g^*) \circ h^*$.

5.7. We note that $\mathcal{V}$ and $\mathcal{W}_q$ are minimal models, since Igusa's device [2] is applicable to these varieties.

Let $\mathfrak{R}, v_1, \cdots, v_s, t_i(z), \mathfrak{Q}_z, \Sigma^*$ be as in 4.3. We take $\Gamma(T, \mathfrak{R}/\mathfrak{M})$ as Γ, assuming that (5.1.1) is satisfied. As remarked in 5.4, for every $b \in \mathfrak{R}^q$, the restriction of π to $\mathfrak{S}(b)$ is a biregular isomorphism to $\mathcal{V}$. Therefore, we can define a point $t_b(y)$ on B_y by

$$(5.7.1) \qquad\qquad t_b(y) = \mathfrak{S}(b) \cdot B_y \ .$$

Now we confine ourselves to the case $q = 1$. We denote $\mathcal{W}_1$ simply by $\mathcal{W}$, and put $t_i'(y) = \mathfrak{S}(v_i) \cdot B_y = t_{v_i}(y)$. Then the structure $(B_y, \theta_y', \{t_i'(y)\})$ is isomorphic to $(A_z, \theta_z, \{t_i(z)\})$, where $z = \wp(y)$.

Let $\mathfrak{Q}_u$ be a generic member of Σ^* over Q, and $\mathfrak{K}$ the field of moduli of $\mathfrak{Q}_u$. Let k be as in (ii) of 4.5. Then k is the algebraic closure of Q in $\mathfrak{K}$. Using the functions $h_1, \cdots, h_\lambda$ in 4.5, we have $\mathfrak{K} = k(h_1(u), \cdots, h_\lambda(u))$. Let $\mathcal{V}'$ be the locus of $(h_1(u), \cdots, h_\lambda(u))$ over k. By virtue of 4.5, we see that there exists a birational mapping g of $\mathcal{V}$ to $\mathcal{V}'$ such that

$$(5.7.2) \qquad\qquad g\bigl(\wp(z)\bigr) = \bigl(h_1(z), \cdots, h_\lambda(z)\bigr)$$

for every point z of $\mathfrak{H}$ where the h_i are holomorphic. Applying 5.5 to $\mathcal{V}$, $\mathcal{V}'$ and g, we get a projective variety $\mathcal{V}^*$ and a birational mapping g^* of $\mathcal{V}^*$ to $\mathcal{V}'$, both defined over k, such that $g^{-1} \circ g^*$ is everywhere biregular. Therefore, taking $\mathcal{V}^*$ in place of $\mathcal{V}$, we assume hereafter that $\mathcal{V}$ is defined over k, and the birational mapping g of $\mathcal{V}$ to $\mathcal{V}'$ is defined over k. Then we have

$$(5.7.3) \qquad Q\bigl(g(\wp(z))\bigr) = \text{the field of moduli of } \mathfrak{Q}_z \text{ for } z \in \mathfrak{H} - \mathfrak{W} \ ,$$

where $\mathfrak{W}$ is as in 2.3.

5.8. Under the assumption (5.1.1), $\mathfrak{Q}_{z_0}$ has the identity mapping as the only automorphism, for every $z_0 \in \mathfrak{H}$. In fact, by 4.2, every automorphism of $\mathfrak{Q}_{z_0}$ corresponds to an element α of $\Gamma(T, \mathfrak{M})$ such that $\alpha(z_0) = z_0$, $v_i \alpha \equiv v_i \bmod \mathfrak{M}$ for every i. By (5.1.1), α should be the identity matrix.

In particular, we may apply 1.5 to the generic member $\mathfrak{Q}_u$. Namely, there exists a structure $\mathfrak{Q} = (A,\, \mathcal{C},\, \theta;\, t_1,\, \cdots,\, t_s)$ which is isomorphic to $\mathfrak{Q}_u$ and defined over $\mathfrak{R}$.

Take a basis $\{a_1,\, \cdots,\, a_d\}$ of $\mathfrak{o}$ over $\mathbf{Z}$. Let k_1 be a finitely generated extension of k, over which $\mathfrak{V}$, π, $\mathfrak{S}(0)$, $\mathfrak{S}(v_1)$, $\cdots$, $\mathfrak{S}(v_s)$, $\Theta(a_1)$, $\cdots$, $\Theta(a_d)$ are defined. Let $y \in \mathcal{H}$. Since $\mathfrak{S}(0) \cdot B_y$ is the origin of B_y, and since B_y is defined by (5.3.1), we see that B_y is defined over $k_1(y)$ as an abelian variety. Furthermore, since $\Theta(a) \cdot (B_y \times B_y)$ is the graph of $\theta_y'(a)$, and $t_i'(y) = \mathfrak{S}(v_i) \cdot B_y$, we see that every element of $\theta_y'(\mathfrak{o})$ is defined over $k_1(y)$, and the points $t_i'(y)$ are rational over $k_1(y)$.

Now we take y so as to be generic on $\mathfrak{V}$ over k_1, and so that $y = \varphi(u)$ with a point u of $\mathcal{H}$, for which $\mathfrak{Q}_u$ is a generic member of Σ^* over k_1. Such a choice of y and u is of course possible.

Exclude the cases (a, b, c, d, e) in [9, Th. 5]. Then we have $\theta_u(\mathfrak{o}) = \mathrm{End}\,(A_u)$, so that $\theta_y'(\mathfrak{o}) = \mathrm{End}\,(B_y)$. From this we can conclude that every polarization of B_y is defined over $k_1(y)$. In fact, since B_y is defined over $k_1(y)$, we can find a polar divisor X on B_y, rational over $k_1(y)$. Let Y be another polar divisor on B_y. Denoting by B' the Picard variety of B_y, we can attach to X and Y isogenies φ_X and φ_Y of B_y to B' in the usual manner (cf. [9, 1.3]). Then $p\varphi_Y = \varphi_X \theta_y'(a)$ for some positive integer p and some $a \in \mathfrak{o}$. Let σ be an isomorphism of C into itself over $k_1(y)$. Then we have $p\varphi_{Y^\sigma} = \varphi_X \theta_y'(a) = p\varphi_Y$, so that Y^σ is algebraically equivalent to Y. This shows that the polarization of B_y determined by Y is rational over $k_1(y)$.

Let $\mathcal{C}'$ be the polarization of B_y such that $\big(B_y,\, \mathcal{C}',\, \theta_y';\, \{t_i'(y)\}\big)$ is isomorphic to $\mathfrak{Q}_u = \big(A_u,\, \mathcal{C}_u,\, \theta_u;\, \{t_i(u)\}\big)$. Substituting u for z in (5.7.3), we have $\mathfrak{R} = Q(y) = k(y) \subset k_1(y)$. Since $\mathfrak{Q}_u$ is isomorphic to $\mathfrak{Q} = (A,\, \mathcal{C},\, \theta;\, \{t_i\})$, there exists an isomorphism λ of $\mathfrak{Q}$ to $\big(B_y,\, \mathcal{C}',\, \theta_y';\, \{t_i'(y)\}\big)$. Such an isomorphism is unique, for $\mathfrak{Q}$ has only one automorphism. As remarked above, $\big(B_y,\, \mathcal{C}',\, \theta_y';\, \{t_i'(y)\}\big)$ is defined over $k_1(y)$, while $\mathfrak{Q}$ is defined over $k(y)$. In view of the uniqueness of λ, we have $\lambda^\sigma = \lambda$ for every isomorphism σ of C into itself over $k_1(y)$. This implies that λ is defined over $k_1(y)$.

Let x be a generic point of A over $k_1(y)$. Then $\lambda(x)$ is a generic point of B_y over $k_1(y)$. Since y is generic on $\mathfrak{V}$ over k_1, we observe that $\lambda(x)$ is a generic point of $\mathfrak{V}$ over k_1. Hence we have $k_1(\lambda(x)) \supset k_1(y)$, for $y = \pi(\lambda(x))$, so that $k_1(\lambda(x)) = k_1(x,\, y)$. If d is the dimension of $\mathfrak{V}$, $k(y)$ is a regular extension of k of dimension d. Since A is defined over $k(y)$, we see that $k(x,\, y)$ is a regular extension of k of dimension $d + n$, where $n = \dim A$. By our choice of y, the fields $k(y)$ and k_1 are linearly disjoint over k. Since x is generic on A over $k_1(y)$, the fields $k(x,\, y)$ and $k_1(y)$ are linearly disjoint over $k(y)$. Combining both linear disjointness, we conclude that k_1 and $k(x, y)$ are linearly disjoint over k. Let $\mathcal{W}'$ be the locus of (x, y) over k. Then there exists a birational mapping f, defined over k_1, of $\mathcal{W}$ to $\mathcal{W}'$ such that $f(\lambda(x)) = (x, y)$. Applying 5.5 to $\mathcal{W}$, $\mathcal{W}'$ and f, we obtain a projective variety $\mathcal{W}^*$ and a birational mapping f^* of $\mathcal{W}^*$ to $\mathcal{W}'$, both defined over k, such that $f^{-1} \circ f^*$ is everywhere biregular.

Let π' be the rational mapping of $\mathcal{W}'$ to $\mathcal{V}$ defined by $\pi'(x, y) = y$ over k. Put $\pi^* = \pi' \circ f^*$. Then π^* is a rational mapping of $\mathcal{W}^*$ to $\mathcal{V}$, defined over k, and $\pi^* = \pi \circ (f^{-1} \circ f^*)$. Taking $\mathcal{W}^*$, f^* and π^* in place of $\mathcal{W}$, f and π, we may assume hereafter that $\mathcal{W}$, f and π are all defined over k.

5.9. Still we use the same notation k_1, y, x, λ as in 5.8, assuming that $\mathcal{V}$, $\mathcal{W}$ and π are defined over k. Let e be a point on A, rational over $k(y)$. We see easily that f^{-1} is defined at (e, y) and $f^{-1}(e, y) = \lambda(e)$, and hence $k(\lambda(e)) \subset k(y)$. On the other hand, we have $\pi(\lambda(e)) = y$, so that $k(\lambda(e)) = k(y)$. Now let us take the origin of A as e. Then $\lambda(e)$ is the origin of B_y, so that $\lambda(e) = \mathcal{S}(0) \cdot B_y$. As remarked in 5.4, π gives a biregular mapping of $\mathcal{S}(0)$ to $\mathcal{V}$. Since $\mathcal{S}(0)$ is defined over k_1 and y is generic on $\mathcal{V}$ over k_1, the point $\lambda(e)$ is generic on $\mathcal{S}(0)$ over k_1. Since $k(\lambda(e)) = k(y)$, this implies that $\mathcal{S}(0)$ is defined over k. Similarly, we find that $\mathcal{S}(v_1)$, $\cdots$, $\mathcal{S}(v_s)$ are all defined over k.

Let $a \in \mathfrak{o}$. We see easily that $\lambda(x) \times \theta'_y(a)\lambda(x) \in \Theta(a)$. Since $\lambda(x)$ is generic on $\mathcal{W}$ over k_1, and since $\Theta(a)$ is defined over k_1 and of dimension $n + d$ over k_1, we observe that $\lambda(x) \times \theta'_y(a)\lambda(x)$ is generic on $\Theta(a)$ over k_1. On the other hand, we see that f^{-1} is defined at $(\theta(a)x, y)$, and $f^{-1}(\theta(a)x, y) = \theta'_y(a)\lambda(x)$. Since $\theta(a)$ is defined over $k(y)$, and f is defined over k, we have $k(\theta'_y(a)\lambda(x)) \subset k(x, y) = k(\lambda(x))$. Hence we have $\dim_k(\lambda(x) \times \theta'_y(a)\lambda(x)) = n + d$. It follows that $\Theta(a)$ is defined over k.

Therefore, in the consideration of 5.8, we may take k as k_1. Then we find that, for every $p \in \mathcal{V}$, the abelian variety B_p and the $\theta'_p(a)$ for $a \in \mathfrak{o}$ are defined over $k(p)$, and the points $t'_i(p)$ are rational over $k(p)$. For a generic point y of $\mathcal{V}$ over k, we can conclude that $(B_y, C', \theta'_y; \{t'_i(y)\})$ is defined over $k(y)$, and λ is defined over $k(y)$.

5.10. Let us now consider $\mathcal{W}_q$ for $q > 1$. Let $(A, C, \theta; \{t_i\})$ be as in 5.8. Put $A^q = A \times \cdots \times A$ (q copies). Let θ^q be an isomorphism of the total matrix algebra $M_q(L)$ into $\operatorname{End}_Q(A^q)$ defined as follows: for $a = (a_{ij}) \in M_q(\mathfrak{o})$ with $a_{ij} \in \mathfrak{o}$,

$$\theta^q(a)(u_1, \cdots, u_q) = (u'_1, \cdots, u'_q), \quad u'_i = \sum_j \theta(a_{ij})u_j .$$

Let X be a divisor in C, rational over $k(y)$. Let C^q be the polarization of A^q containing the divisor

$$\sum_{i=1}^{q} (A \times \cdots \times A \times \overset{i}{\check{X}} \times A \times \cdots \times A) \, .$$

Further, for $1 \leq i \leq s$, put $t_i^q = t_i \times \cdots \times t_i$. Then we can verify easily that $(A^q, \mathcal{C}^q, \theta^q; \{t_i^q\})$ is defined over $k(y)$ and has no automorphism other than the identity mapping. Taking $\mathcal{W}_q$ and $(A^q, \mathcal{C}^q, \theta^q; \{t_i^q\})$ in place of $\mathcal{W}$ and $(A, \mathcal{C}, \theta; \{t_i\})$, and repeating the argument of 5.8 and 5.9, we can conclude that $\mathcal{W}_q$, π, $\theta(a)$ for $a \in M_q(\mathfrak{o})$, $\mathfrak{S}(b)$ for $b \in \mathfrak{N}^q$ are defined over k. Here we should note that the points $t_1^q, \cdots, t_s^q$ generate, over $\theta^q(M_q(\mathfrak{o}))$, all the points on A^q corresponding to the elements of $\mathfrak{N}^q$.

We can reason also in the following way. Let $\mathcal{W}_1^q$ be the product of q copies of $\mathcal{W}_1$, and let

$$\mathcal{W}_q' = \{(x_1, \cdots, x_q) \in \mathcal{W}_1^q \mid \pi(x_1) = \cdots = \pi(x_q)\} \, ,$$
$$\iota\big(\psi(z, u)\big) = \psi(z, u_1) \times \cdots \times \psi(z, u_q)$$

$$\text{for } z \in \mathcal{K}, \ u = (u_1, \cdots, u_q) \in V_R^q \, .$$

It can be easily verified that $\mathcal{W}_q'$ is a non-singular subvariety of $\mathcal{W}_1^q$, and ι is an analytic isomorphism of $\mathcal{W}_q$ to $\mathcal{W}_q'$. Therefore we may take $\mathcal{W}_q'$ as a model for $\mathcal{W}_q$. Then our assertion concerning $\mathcal{W}_q$ and its fibres can be derived quite easily. Thus we have proved the following theorem except the uniqueness.

5.11. Theorem. *Suppose that $\{L, \Phi\}$ does not belong to the five exceptional cases* (a), (b), (c), (d), (e) *of* [9, p. 176, Th. 5]. *Let* $\mathfrak{o} = \{a \in L \mid a\mathfrak{M} \subset \mathfrak{M}\}$, *and let $\mathfrak{N}$ be such that $\Gamma = \Gamma(T, \mathfrak{N}/\mathfrak{M})$ satisfies* (5.1.1). *Define* $\mathcal{V}$, $\mathcal{W}_q$, π *as in* 5.1, *and* $\Theta(a)$, $\mathfrak{S}(b)$ *as in* 5.4. *Let k be as in* (ii) *of* 4.5. *Then we can find projective non-singular models for $\mathcal{V}$ and $\mathcal{W}_q$ so that k is a field of rationality for $\mathcal{V}$, $\mathcal{W}_q$, π, the $\Theta(a)$ for $a \in M_q(\mathfrak{o})$, and the $\mathfrak{S}(b)$ for $b \in \mathfrak{N}^q$. Such models for $\mathcal{V}$, $\mathcal{W}_q$, π, $\Theta(a)$, $\mathfrak{S}(b)$ are unique up to a biregular isomorphism defined over k.*

As a consequence of the first assertion, we notice that, for every $p \in \mathcal{V}$, the field $k(p)$ is a field of rationality for B_p, the $\theta_p'(a)$ for $a \in M_q(\mathfrak{o})$, and the $t_b(p)$ for $b \in \mathfrak{N}^q$ (cf. (5.7.1)).

To prove the uniqueness, let $\{\mathcal{V}', \mathcal{W}_q', \pi', \Theta'(a), \mathfrak{S}'(b)\}$ be another system of models rational over k. Then there exists a biregular isomorphism g of $\mathcal{V}$ to $\mathcal{V}'$ such that $g(v(z)) = v'(z)$ for $z \in \mathcal{K}$, where v' is the mapping of $\mathcal{K}$ to $\mathcal{V}'$ corresponding to v. Let k_1 be a field of definition for g, containing k, and y a generic point of $\mathcal{V}$ over k_1. Put $r = g(y)$. Then $k_1(y) = k_1(r)$. Let $z \in \mathcal{K}$ be such that $y = v(z)$, and let $B_y = \pi^{-1}(y)$, $B_r' = \pi'^{-1}(r)$. Then both B_y and B_r' are isomorphic to A_z. Define structures $\mathfrak{A}_y = (B_y, \mathcal{C}_y, \theta_y; \{t_i(y)\})$ and $\mathfrak{A}_r' = (B_r', \mathcal{C}_r', \theta_r'; \{t_i'(r)\})$ so as to be isomorphic to $(A_z, \mathcal{C}_z, \theta_z; \{t_i(z)\})$. Let μ be an isomorphism of $\mathfrak{A}_y$ to $\mathfrak{A}_r'$. As remarked above, we see that $(B_y, \theta_y, \{t_i(y)\})$ and $(B_r', \theta_r', \{t_i'(r)\})$ are defined over $k(y)$ and $k(r)$, respectively. By the reasoning given in 5.8, every polarization of B'_r is defined over $k(r)$. Therefore $k(r)$ contains the field of moduli of $\mathfrak{Q}'_r$, which is the same as the field of moduli of $\mathfrak{Q}_y$, i.e., $k(y)$. Hence we have $k(y) \subset k(r)$. Since k_1 and $k(r)$ are linearly disjoint over k, the equality $k_1(y) = k_1(r)$ implies that $k(y) = k(r)$. It follows that g is defined over k. Now $\mathfrak{Q}_y$ resp. $\mathfrak{Q}$ has the identity mapping as the only automorphism. From this we can conclude that μ is defined over $k(y)$. Let x be a generic point of B_y over $k(y)$. Then μx is a generic point of B'_r over $k(y)$, and $k(x) = k(x, y) = k(\mu x, y) = k(\mu x, r) = k(\mu x)$, since $\pi(x) = y$ and $\pi'(\mu x) = r$. We observe that x resp. μx is a generic point of $\mathfrak{W}_q$ resp. $\mathfrak{W}'_q$ over k. Hence there exists a birational mapping f of $\mathfrak{W}_q$ to $\mathfrak{W}'_q$, defined over k, such that $f(x) = \mu x$. Since $\mathfrak{W}_q$ and $\mathfrak{W}'_q$ are minimal models, f is a biregular isomorphism. It is clear that $\pi' \circ f = g \circ \pi$. Since μ sends $\mathfrak{Q}_y$ to $\mathfrak{Q}'_r$, and y is a generic point of $\mathfrak{V}$ over k, we see that f sends the $\Theta(a)$ and $\mathfrak{S}(b)$ to the $\Theta'(a)$ and $\mathfrak{S}'(b)$. This completes the proof of uniqueness.

5.12. PROPOSITION. *Suppose that $q = 1$. Let $\big(B_p, \theta'_p, \{t'_i(p)\}\big)$ for $p \in \mathfrak{V}$ be as in 5.4 and 5.7. Then, with the same assumption and the notation as in 5.11, there exists a non-empty subset $\mathfrak{U}$ of $\mathfrak{V}$, which is Zariski-open with respect to k, and has the following property: For $p = \varphi(z)$ with $z \in \mathfrak{U}$,*

　(i) *there exists a polarization $\mathcal{C}'_p$ of B_p, rational over $k(p)$, such that $\big(B_p, \mathcal{C}'_p, \theta'_p; \{t'_i(p)\}\big)$ is isomorphic to $\mathfrak{Q}_z$;*

　(ii) *$k(p)$ is the composite of k and the field of moduli of $\mathfrak{Q}_z$.*

Moreover, if $\mathfrak{V}$ is of dimension one, we may take the whole $\mathfrak{V}$ as $\mathfrak{U}$.

To prove this we need the holomorphic mapping $\Psi(z)$ satisfying (2.1.11) and (2.1.12) for the present family $\Sigma(L, \Phi, \rho; T, \mathfrak{M})$. Let u be a generic point on $\mathfrak{H}$ for Ψ, $t_i(z)$, φ over $\mathbf{Q}$, and let $y = \varphi(u)$. As remarked at the end of 5.9, there exists a polarization $\mathcal{C}'$ of B_y, rational over $k(y)$, such that $\big(B_y, \mathcal{C}', \theta'_y; \{t'_i(y)\}\big)$ is isomorphic to $\mathfrak{Q}_u$. Let X be a divisor in $\mathcal{C}'$, rational over $k(y)$. We can find a non-empty Zariski open set $\mathfrak{U}$ of $\mathfrak{V}$, with respect to k, such that for every $p \in \mathfrak{U}$, X has a *unique* specialization X', rational over $k(p)$, over the specialization $y \to p$ ref. k. It is clear that we may put $\mathfrak{V} = \mathfrak{U}$ if $\mathfrak{V}$ is of dimension one. Let $z \in \mathfrak{H}$ and $p = \varphi(z)$. We can find a place $\mathfrak{p}$ of the field $k(\Psi(u), \varphi(u), t_i(u))$ so that $\mathfrak{p}(\Psi(u)) = \Psi(z)$, $\mathfrak{p}(\varphi(u)) = \varphi(z)$, $\mathfrak{p}(t_i(u)) = t_i(z)$, and $\mathfrak{p}(a) = a$ for $a \in k$. By reduction modulo $\mathfrak{p}$, we get $\big(B_p, \theta'_p; \{t'_i(p)\}\big)$ from $\big(B_y, \theta'_y; \{t'_i(y)\}\big)$, and $\mathfrak{Q}_z$ from $\mathfrak{Q}_u$ in view of (2.1.12). Here we note that A_u and A_z are projective varieties, and $\mathcal{C}_u$, $\mathcal{C}_z$ are determined by hyperplane sections; hence $\mathcal{C}_u$ is sent to $\mathcal{C}_z$ by the reduction process. Let ξ be an isomorphism of $\mathfrak{Q}_u$ to $\big(B_y, \mathcal{C}', \theta'_y; \{t'_i(y)\}\big)$, and η the reduction of ξ modulo $\mathfrak{p}$. Then η is an isomorphism of $\big(A_z, \theta_z; \{t_i(z)\}\big)$ to $\big(B_p, \theta'_p; \{t'_i(p)\}\big)$. Put $Y = \mathfrak{p}(X)$. Then Y determines a polarization $\mathcal{C}'_p$ on B_p, and η gives an isomorphism of $\mathfrak{Q}_z$ to $\left(B_p, \mathcal{C}_p', \theta_p'; \{t_i'(p)\}\right)$. In particular, if $p \in \mathfrak{U}$, as remarked above, Y is rational over $k(p)$. This proves (i). Now let k_0 be the field of moduli of $\mathfrak{Q}_z$. Then we see that $k_0 \subset k(p)$, since $(B_p, \mathcal{C}_p', \cdots)$ is defined over $k(p)$. Let σ be an isomorphism of $k(p)$ into C, over $k_0 \cdot k$. Put $r = p^\sigma$. Then $r \in \mathfrak{U}$, $B_p^\sigma = B_r$, $\theta_p'^\sigma = \theta_r'$, $t_i'(p)^\sigma = t_i'(r)^\sigma$. We observe that Y^σ is a specialization of X over $y \to r$ ref. k. Since $r \in \mathfrak{U}$, such a specialization of X is unique. Therefore, if $r = \varphi(w)$ for a point w of $\mathcal{H}$, we see that $\mathfrak{Q}_w$ should be isomorphic to $\left(B_p, \mathcal{C}_p', \theta_p'^\sigma; \{t_i'(p)^\sigma\}\right)$. Since σ is the identity mapping on k_0, we can conclude that $\mathfrak{Q}_z$ is isomorphic to $\mathfrak{Q}_w$. Hence by 4.4, $p^\sigma = r = \varphi(w) = \varphi(z) = p$. This shows $k(p) \subset k_0 \cdot k$, so that $k(p) = k_0 \cdot k$, which is the assertion (ii).

6. Unitary groups over a quaternion algebra

6.1. In this section, F denotes an algebraic number field, and L a quaternion algebra over F. We denote by ι the canonical involution of L. As before, let V be the vector space of all m-dimensional row-vectors with components in L. Now we consider a non-degenerate ι-hermitian form $S(x, y)$ on V in the sense of [FI, 1.1], and define groups $G(S)$ and $U(S)$ by

$$G(S) = \{\alpha \in GL(V, L) \mid S(x\alpha, y\alpha) = \mu(\alpha)S(x, y) \text{ with } \mu(\alpha) \in F\},$$
$$U(S) = \{\alpha \in G(S) \mid \mu(\alpha) = 1\},$$

where $GL(V, L)$ is the group of all L-linear automorphisms of V. For each prime ideal $\mathfrak{p}$ in F, we denote by $F_\mathfrak{p}$ the $\mathfrak{p}$-completion of F, and put $L_\mathfrak{p} = L \otimes_F F_\mathfrak{p}$, $V_\mathfrak{p} = V \otimes_F F_\mathfrak{p}$. We can extend $S(x, y)$ naturally to a form on $V_\mathfrak{p}$. Then $G_\mathfrak{p}(S)$ and $U_\mathfrak{p}(S)$ can be defined similarly over $F_\mathfrak{p}$. We denote by $\mathfrak{u}$ the product of infinite places of F defined in [10, 6.13].

Let $\mathfrak{o}$ be a maximal order in L, and $\mathfrak{g}$ the ring of integers in F. A $\mathfrak{g}$-lattice $\mathfrak{M}$ in V is called an $\mathfrak{o}$-lattice if $\mathfrak{o}\mathfrak{M} \subset \mathfrak{M}$.

6.2. PROPOSITION. *Suppose that S is indefinite in the sense of* [10, 6.6]. *Let $\mathfrak{M}$ be an $\mathfrak{o}$-lattice in V, and c an element of $\mathfrak{g}$ such that $c \equiv 1 \bmod \mathfrak{u}$. Then there exists an element α of $G(S)$ such that $\mathfrak{M}\alpha \subset \mathfrak{M}$, and $\mu(\alpha) = c$.*

By [10, 6.13], there exists an element β of $G(S)$ such that $\mu(\beta) = c$. On the other hand, for every prime ideal $\mathfrak{p}$ in F, there exists an element $\alpha_\mathfrak{p}$ of $G_\mathfrak{p}(S)$ such that $\mathfrak{M}_\mathfrak{p}\alpha_\mathfrak{p} \subset \mathfrak{M}_\mathfrak{p}$ and $\mu(\alpha_\mathfrak{p}) = c$. This is easily shown by the same argument as in [10, 6.12]. Now let P be the set of all prime ideals $\mathfrak{p}$ in F such that $\mathfrak{M}_\mathfrak{p}\beta \neq \mathfrak{M}_\mathfrak{p}$. Then we can find an $\mathfrak{o}$-lattice $\mathfrak{N}$ in V such that $\mathfrak{N}_\mathfrak{p} = \mathfrak{M}_\mathfrak{p}\alpha_\mathfrak{p}$ for $\mathfrak{p} \in P$ and $\mathfrak{N}_\mathfrak{p} = \mathfrak{M}_\mathfrak{p}$ for $\mathfrak{p} \notin P$. We have $\mathfrak{N}_\mathfrak{p} = \mathfrak{M}_\mathfrak{p}\beta$ for $\mathfrak{p} \notin P$ and $\mathfrak{N}_\mathfrak{p} = (\mathfrak{M}_\mathfrak{p}\beta)\beta^{-1}\alpha_\mathfrak{p}$ for $\mathfrak{p} \in P$. Therefore $\mathfrak{N}$ and $\mathfrak{M}\beta$ belong to the same genus with respect to $U(S)$. By [10, 6.9], there exists an element γ of $U(S)$ such that

$\mathfrak{N} = \mathfrak{M}\beta\gamma$. Since $\mu(\beta\gamma) = c$ and $\mathfrak{N} \subset \mathfrak{M}$, the element $\beta\gamma$ has the required property.

6.3. PROPOSITION. *Let $\mathfrak{M}$ be an $\mathfrak{o}$-lattice in V, $\mathfrak{a}$ an integral two-sided $\mathfrak{o}$-ideal, and $\mathfrak{p}$ a prime ideal in F. For every unit c in $\mathfrak{g}_\mathfrak{p}$ such that $c \equiv 1 \bmod \mathfrak{a}_\mathfrak{p} \cap F_\mathfrak{p}$, there exists an element α of $G_\mathfrak{p}(S)$ such that $\mu(\alpha) = c$, $\mathfrak{M}_\mathfrak{p}\alpha = \mathfrak{M}_\mathfrak{p}$, $\mathfrak{M}_\mathfrak{p}(\alpha - 1) \subset \mathfrak{a}_\mathfrak{p}\mathfrak{M}_\mathfrak{p}$.*

By [7, Prop. 3.2], there exists an element b of $L_\mathfrak{p}$ such that $bb^\iota = c$, $b \equiv 1 \bmod \mathfrak{a}_\mathfrak{p}$.[4] Using this element b, define α as in the proof of [10, 6.12]. We see easily that α has the required property.

6.4. Let $\mathfrak{M}$ be an $\mathfrak{o}$-lattice in V, and $\mathfrak{a}$ an integral two-sided $\mathfrak{o}$-ideal. For every prime ideal $\mathfrak{p}$ in F, we define groups $\Gamma_\mathfrak{p}(\mathfrak{M})$ and $\Gamma_\mathfrak{p}(\mathfrak{M}; \mathfrak{a})$ by

$$(6.4.1) \qquad \Gamma_\mathfrak{p}(\mathfrak{M}) = \{\alpha \in G_\mathfrak{p}(S) \mid \mathfrak{M}_\mathfrak{p}\alpha = \mathfrak{M}_\mathfrak{p}\} \,,$$

$$(6.4.2) \qquad \Gamma_\mathfrak{p}(\mathfrak{M}; \mathfrak{a}) = \{\alpha \in \Gamma_\mathfrak{p}(\mathfrak{M}) \mid \mathfrak{M}_\mathfrak{p}(\alpha - 1) \subset \mathfrak{a}_\mathfrak{p}\mathfrak{M}_\mathfrak{p}\} \,.$$

Let $N(\mathfrak{a})$ denote the ideal in F generated by xx^ι for all $x \in \mathfrak{a}$. Then put

$$(6.4.3) \qquad \mathfrak{G}(\mathfrak{M}; \mathfrak{a}) = \textstyle\prod_{\mathfrak{p} \mid N(\mathfrak{a})} [\Gamma_\mathfrak{p}(\mathfrak{M})/\Gamma_\mathfrak{p}(\mathfrak{M}; \mathfrak{a})] \,.$$

For simplicity, we mean by $(\alpha_\mathfrak{p})$ the element of $\mathfrak{G}(\mathfrak{M}; \mathfrak{a})$ whose components are represented by the elements $\alpha_\mathfrak{p}$ in $\Gamma_\mathfrak{p}(\mathfrak{M})$. We denote by $\mathfrak{G}_0(\mathfrak{M}; \mathfrak{a})$ the subgroup of $\mathfrak{G}(\mathfrak{M}; \mathfrak{a})$ consisting of the elements $(\alpha_\mathfrak{p})$ satisfying the following condition:

(6.4.4) There exists a rational integer c such that $\mu(\alpha_\mathfrak{p}) \equiv c \bmod \mathfrak{a}_\mathfrak{p}$ for every $\mathfrak{p}$ dividing $N(\mathfrak{a})$.

Further we denote by $\mathfrak{S}(\mathfrak{M}; \mathfrak{a})$ the subgroup of $\mathfrak{G}_0(\mathfrak{M}; \mathfrak{a})$ consisting of the elements $(\alpha_\mathfrak{p})$ such that $\mu(\alpha_\mathfrak{p}) \equiv 1 \bmod \mathfrak{a}_\mathfrak{p}$.

6.5. PROPOSITION. *If $\alpha \in \Gamma_\mathfrak{p}(\mathfrak{M}; \mathfrak{a})$, $\mu(\alpha) \equiv 1 \bmod \mathfrak{a}_\mathfrak{p}$.*

Let μ_0 be as in [10, 6.10]. Then there exist elements x and y of $\mathfrak{a}^{-1}\mathfrak{M}_\mathfrak{p}$ such that $\mu_0(\mathfrak{a}^{-1}\mathfrak{M}_\mathfrak{p}) = S(x, y)\mathfrak{o}_\mathfrak{p}$. Let $\alpha \in \Gamma_\mathfrak{p}(\mathfrak{M}; \mathfrak{a})$. Then $\mu(\alpha)S(x, y) = S(x\alpha, y\alpha) \equiv S(x, y) \bmod \mu_0(\mathfrak{a}^{-1}\mathfrak{M}_\mathfrak{p})\mathfrak{a}_\mathfrak{p}$. Hence $\mu(\alpha) \equiv 1 \bmod \mathfrak{a}_\mathfrak{p}$.

6.6. PROPOSITION. *If S is indefinite, every element of $\mathfrak{G}_0(\mathfrak{M}; \mathfrak{a})$ (resp. $\mathfrak{S}(\mathfrak{M}; \mathfrak{a})$) is represented by an element α of $G(S)$ (resp. $U(S)$) such that $\mathfrak{M}\alpha \subset \mathfrak{M}$ and $\mu(\alpha)$ is a rational integer prime to $N(\mathfrak{a})$.*

Let $(\alpha_\mathfrak{p})$ be an element of $\mathfrak{G}_0(\mathfrak{M}; \mathfrak{a})$ and let c be as in (6.4.4). We may assume $c > 0$. Since $\mu(\alpha_\mathfrak{p})$ is a $\mathfrak{p}$-unit, c is prime to $N(\mathfrak{a})$. By 6.2, there exists an element α of $G(S)$ such that $\mathfrak{M}\alpha \subset \mathfrak{M}$ and $\mu(\alpha) = c$. Then for every $\mathfrak{p}$

---

[4] If $\mathfrak{a}_\mathfrak{p} = (\mathfrak{a}_\mathfrak{p} \cap F_\mathfrak{p})\mathfrak{o}_\mathfrak{p}$, we get the element b immediately from [7, Prop. 3.2], putting $\beta = c$, $a = 1$. Suppose that $\mathfrak{a}_\mathfrak{p} \cap F_\mathfrak{p} = \mathfrak{p}^\lambda$, $\mathfrak{a}_\mathfrak{p} \neq \mathfrak{p}^\lambda\mathfrak{o}_\mathfrak{p}$. Then $L_\mathfrak{p}$ is a division algebra, and $\mathfrak{a}_\mathfrak{p} = \mathfrak{P}^{2\lambda-1}$ for the maximal ideal $\mathfrak{P}$ in $\mathfrak{o}_\mathfrak{p}$. Applying [7, Prop. 3.2] to this case with the exponent λ, we get an element b such that $bb^\iota = c$ and $b \equiv 1 \bmod \mathfrak{p}^\lambda\mathfrak{o}_\mathfrak{p}$. Then $b \equiv 1 \bmod \mathfrak{a}_\mathfrak{p}$ automatically.

dividing $N(\mathfrak{a})$, $\mathfrak{M}_\mathfrak{p}\alpha = \mathfrak{M}_\mathfrak{p}$, and $\mu(\alpha_\mathfrak{p}^{-1}\alpha) \equiv 1 \bmod \mathfrak{a}_\mathfrak{p}$. By 6.3, there exists an element $\beta_\mathfrak{p}$ of $G_\mathfrak{p}(S)$ such that $\mu(\beta_\mathfrak{p}) = \mu(\alpha_\mathfrak{p}^{-1}\alpha)$, $\mathfrak{M}_\mathfrak{p}\beta_\mathfrak{p} = \mathfrak{M}_\mathfrak{p}$, $\mathfrak{M}_\mathfrak{p}(\beta_\mathfrak{p} - 1) \subset \mathfrak{a}_\mathfrak{p}\mathfrak{M}_\mathfrak{p}$. By [10, 6.7], there exists an element β of $U(S)$ such that $\mathfrak{M}_\mathfrak{p}(\beta - \beta_\mathfrak{p}\alpha_\mathfrak{p}\alpha^{-1}) \subset \mathfrak{p}\mathfrak{a}_\mathfrak{p}\mathfrak{M}_\mathfrak{p}$ if $\mathfrak{p} \mid N(\mathfrak{a})$, and $\mathfrak{M}_\mathfrak{r}\beta = \mathfrak{M}_\mathfrak{p}$ if $\mathfrak{p} \nmid N(\mathfrak{a})$. Then $\mathfrak{M}\beta = \mathfrak{M}$, $\mathfrak{M}\alpha\beta \subset \mathfrak{M}$, $\mu(\alpha\beta) = c$, $\mathfrak{M}_\mathfrak{p}(\alpha_\mathfrak{p}\alpha^{-1}\beta^{-1} - 1) \subset \mathfrak{a}_\mathfrak{p}\mathfrak{M}_\mathfrak{p}$ if $\mathfrak{p} \mid N(\mathfrak{a})$. Therefore $\alpha\beta$ represents the element $(\alpha_\mathfrak{p})$ in $\mathfrak{G}_0(\mathfrak{M}; \mathfrak{a})$. If $(\alpha_\mathfrak{p}) \in \mathfrak{S}(\mathfrak{M}; \mathfrak{a})$, we may put $c = 1$. Then $\alpha\beta \in U(S)$.

6.7. Now let us assume that F is totally real and L is totally indefinite (cf. [9, p. 153]). Let ρ be a positive involution of L. By [9, Prop. 2], there exists an element s of L such that $s^\iota = -s$ and $u^\rho = su^\iota s^{-1}$ for every $u \in L$. Let T be a ρ-anti-hermitian form in V. Put $S(x, y) = T(x, y)s$. Then S is an ι-hermitian form.

Let $\mathfrak{o}$ be a maximal order in L and $\mathfrak{M}$ an $\mathfrak{o}$-lattice in V. Let $\mathfrak{a}$ be an integral two-sided $\mathfrak{o}$-ideal. We can find m elements $x_1, \cdots, x_m$ in V so that

$$(6.7.1) \qquad\qquad \mathfrak{a}^{-1}\mathfrak{M} = \mathfrak{M} + \sum_{i=1}^{m} \mathfrak{o}x_i \,.$$

We now consider the family $\Sigma^* = \Sigma^*(L, \Phi, \rho; T, \mathfrak{M}; \{x_i\})$. Let $\mathfrak{Q} = (A, \mathcal{C}, \theta; t_1, \cdots, t_m)$ be a generic member of Σ^*. Put $\mathscr{P} = (A, \mathcal{C}, \theta)$,

$$(6.7.2) \qquad\qquad \mathfrak{h}(\mathfrak{a}; A) = \{t \in A \mid \theta(\mathfrak{a})t = 0\} \,.$$

Then $\mathfrak{h}(\mathfrak{a}; A) = \sum_{i=1}^{m} \theta(\mathfrak{o})t_i$. Let k_1 and $k_\mathfrak{a}$ be the fields of moduli of $\mathscr{P}$ and $\mathfrak{Q}$ respectively. By 1.11, $k_\mathfrak{a}$ is a finite Galois extension of k_1. Let $\mathfrak{G}$ be the Galois group. We are going to determine the structure of $\mathfrak{G}$.

Let σ be an isomorphism of C into itself, leaving every element of k_1 invariant. Then there exists an isomorphism λ of $\mathscr{P}$ to $\mathscr{P}^\sigma$. Of course $\mathscr{P}^\sigma$ is of type $\{L, \Phi, \rho; T, \mathfrak{M}\}$. By 3.3, there exists an element $\alpha_\mathfrak{p}$ of $G_\mathfrak{p}(S)$ such that

$$(6.7.3) \qquad \begin{aligned} \mathfrak{M}_\mathfrak{p}\alpha_\mathfrak{p} &= \mathfrak{M}_\mathfrak{p}, \qquad \mu(\alpha_\mathfrak{p}) = c_p \in \mathbf{Z}_p, \\ v_\mathfrak{p}(t)\alpha_\mathfrak{p} &\equiv v_\mathfrak{p}(\lambda^{-1}t^\sigma) \quad \bmod \mathfrak{M}_\mathfrak{p} \end{aligned} \qquad (t \in \mathfrak{h}(\mathfrak{a}; A)) \,,$$

where p is a prime number divisible by $\mathfrak{p}$, c_p is determined by (3.2.1), and $v_\mathfrak{p}$ is the mapping defined in 3.3. By [9, Prop. 20], the automorphism group of $\mathscr{P}$ is $\{\pm 1\}$. Therefore we see that σ determines an element $(\alpha_\mathfrak{p})$ of $\mathfrak{G}_0(\mathfrak{M}; \mathfrak{a})/\{\pm 1\}$. Let τ be another isomorphism of C over k_1, and μ an isomorphism of $\mathscr{P}$ to $\mathscr{P}^\tau$. If $\sigma = \tau$ on $k_\mathfrak{a}$, then $\mathfrak{Q}^\sigma$ is isomorphic to $\mathfrak{Q}^\tau$, in view of (M) of 1.4. We get therefore from τ the same element of $\mathfrak{G}_0(\mathfrak{M}; \mathfrak{a})/\{\pm 1\}$. In this way we obtain a mapping $\sigma \to (\alpha_\mathfrak{p})$ of $\mathfrak{G}$ into $\mathfrak{G}_0(\mathfrak{M}; \mathfrak{a})/\{\pm 1\}$. To prove that this is a homomorphism, take again τ and μ without assuming $\sigma = \tau$ on $k_\mathfrak{a}$. We find an element $(\beta_\mathfrak{p})$ of $\mathfrak{G}_0(\mathfrak{M}; \mathfrak{a})$ such that $v_\mathfrak{p}(t)\beta_\mathfrak{p} \equiv v_\mathfrak{p}(\mu^{-1}t^\tau) \bmod \mathfrak{M}_\mathfrak{p}$ for $t \in \mathfrak{h}(\mathfrak{a}; A)$. Then $\lambda^\tau\mu$ is an isomorphism of $\mathscr{P}$ to $\mathscr{P}^{\sigma\tau}$, and $v_\mathfrak{p}((\lambda^\tau\mu)^{-1}t^{\sigma\tau}) = v_\mathfrak{p}(\mu^{-1}(\lambda^{-1}t^\sigma)^\tau) \equiv v_\mathfrak{p}(\lambda^{-1}t^\sigma)\beta_\mathfrak{p} \equiv v_\mathfrak{p}(t)\alpha_\mathfrak{p}\beta_\mathfrak{p} \bmod \mathfrak{M}_\mathfrak{p}$ for $t \in \mathfrak{h}(\mathfrak{a}; A)$. Next assume that $\varepsilon\alpha_\mathfrak{p} \in \Gamma_\mathfrak{p}(\mathfrak{M}; \mathfrak{a})$ for every $\mathfrak{p}$ dividing $N(\mathfrak{a})$, where $\varepsilon = \pm 1$. Then we see that $v_\mathfrak{p}(\lambda^{-1}t^\sigma) = \varepsilon v_\mathfrak{p}(t)$, so that $\varepsilon\lambda t = t^\sigma$ for every $t \in \mathfrak{h}(\mathfrak{a}; A)$. Hence $\mathfrak{Q}^\sigma$ is isomorphic to $\mathfrak{Q}$. It follows that σ is the identity mapping on $k_\mathfrak{a}$. Thus we have obtained an isomorphism of $\mathfrak{G}$ into $\mathfrak{G}_0(\mathfrak{M}; \mathfrak{a})/\{\pm 1\}$. We now prove a more precise result.

6.8. **THEOREM.** *Let* $\mathfrak{o}$, $\mathfrak{a}$, $\mathfrak{M}$, x_i *be as in* 6.7, *and let* $\mathfrak{Q} = (A, C, \theta; t_1, \cdots, t_m)$ *be a generic member of* $\Sigma^*(L, \Phi, \rho; T, \mathfrak{M}; \{x_i\})$; *put* $\mathfrak{P} = (A, C, \theta)$. *Let* k_1 *and* $k_\mathfrak{a}$ *be the fields of moduli of* $\mathfrak{P}$ *and* $\mathfrak{Q}$, *respectively. Let* $\mathfrak{G}_0(\mathfrak{M}; \mathfrak{a})$ *and* $\mathfrak{S}(\mathfrak{M}; \mathfrak{a})$ *be as in* 6.4, *and* $\mathfrak{G}$ *the Galois group of* $k_\mathfrak{a}$ *over* k_1. *Then the following assertions hold.*

(i) $\mathfrak{G}$ *is isomorphic to* $\mathfrak{G}_0(\mathfrak{M}; \mathfrak{a})/\{\pm 1\}$ *by the mapping* $\sigma \to (\alpha_\mathfrak{p})$ *defined by the relation* (6.7.3).

(ii) *Let* a *be the smallest positive integer divisible by* $\mathfrak{a}$, *and* ζ *a primitive* a^{th} *root of unity. Then* $k_1(\zeta)$ *is a subfield of* $k_\mathfrak{a}$ *corresponding to* $\mathfrak{S}(\mathfrak{M}; \mathfrak{a})/\{\pm 1\}$.

(iii) $\boldsymbol{Q}(\zeta)$ *is algebraically closed in* $k_\mathfrak{a}$.

(iv) *If* $(\alpha_\mathfrak{p})$ *corresponds to an element* σ *of* $\mathfrak{G}$ *and* c *is an integer as in* (6.4.4), *then* $\zeta^\sigma = \zeta^c$.

Let U_p be as in 3.1. Then we get a decomposition $\zeta = \prod_p \zeta_p$ with $\zeta_p \in U_p$. The notation being as in (iv), we have $\zeta_p^\sigma = \zeta_p^c$, since $\mu(\alpha_\mathfrak{p})$ is given by the number c_p in (3.2.1). Hence $\zeta^\sigma = \zeta^c$. If $(\alpha_\mathfrak{p})$ belongs to $\mathfrak{S}(\mathfrak{M}; \mathfrak{a})$, $\zeta^\sigma = \zeta$. Conversely, if $\zeta^\sigma = \zeta$, $\mu(\alpha_\mathfrak{p}) \equiv 1 \bmod \mathfrak{a}_\mathfrak{p}$. Hence $(\alpha_\mathfrak{p}) \in \mathfrak{S}(\mathfrak{M}; \mathfrak{a})$. Therefore $k_1(\zeta)$ corresponds to $\mathfrak{S}(\mathfrak{M}; \mathfrak{a})$.

To prove the surjectivity of our isomorphism, first take an element $(\alpha_\mathfrak{p})$ of $\mathfrak{S}(\mathfrak{M}; \mathfrak{a})$. By 6.6, we may assume that $\alpha_\mathfrak{p} = \alpha$ for an element α of $U(S)$ such that $\mathfrak{M}\alpha \subset \mathfrak{M}$; then $\alpha \in \Gamma(T, \mathfrak{M})$. Let y be a generic point of $\mathfrak{K}$ for the functions $\Psi(z)$, $\Psi(\alpha(z))$, $t(z, b)$, $t(\alpha(z), b)$ for all $b \in V$ over $\bar{\boldsymbol{Q}}$ (cf. 2.1 and 4.3). In view of 2.4, we may assume $\mathfrak{Q} = \mathfrak{Q}_y$. Put $w = \alpha(y)$. Then w is generic for Ψ and $t(z, b)$ for all $b \in V$ over $\bar{\boldsymbol{Q}}$. Hence there exists an isomorphism σ of C into itself over $\bar{\boldsymbol{Q}}$ such that $\Psi(y)^\sigma = \Psi(w)$, $\mathfrak{P}_y^\sigma = \mathfrak{P}_w$, $t(y, b)^\sigma = t(w, b)$. By 4.2, we find an isomorphism λ of $\mathfrak{P}_y$ to $\mathfrak{P}_w$ such that $\lambda^{-1}(t(w, b)) = t(y, b\alpha)$ for every $b \in V$. Then $v_\mathfrak{p}(\lambda^{-1}t^\sigma) = v_\mathfrak{p}(t)\alpha$ for every $t \in \mathfrak{h}(\mathfrak{a}; A)$. This means that every element of $\mathfrak{S}(\mathfrak{M}; \mathfrak{a})$ corresponds to an element of $\mathfrak{G}$. Moreover, $x^\sigma = x$ for every $x \in \bar{\boldsymbol{Q}}$. Therefore, if an element τ of $\mathfrak{G}$ corresponds to an element of $\mathfrak{S}(\mathfrak{M}; \mathfrak{a})$, then τ leaves every element of $k_\mathfrak{a} \cap \bar{\boldsymbol{Q}}$ invariant. Hence $k_\mathfrak{a} \cap \bar{\boldsymbol{Q}} \subset k_1(\zeta)$. Since $\boldsymbol{Q}$ is algebraically closed in k_1 [FI, 3.4], $\boldsymbol{Q}(\zeta)$ is algebraically closed in $k_1(\zeta)$. It follows that $\boldsymbol{Q}(\zeta) = k_\mathfrak{a} \cap \bar{\boldsymbol{Q}}$. Finally take an element $(\alpha_\mathfrak{p})$ of $\mathfrak{G}_0(\mathfrak{M}; \mathfrak{a})$ and define c as in (6.4.4); then c is prime to a. Since $\boldsymbol{Q}$ is algebraically closed in k_1, there exists an isomorphism σ of C over k_1 such that $\zeta^\sigma = \zeta^c$. Let $(\beta_\mathfrak{p})$ be an element of $\mathfrak{G}(\mathfrak{M}; \mathfrak{a})$ corresponding to σ. Then $(\alpha_\mathfrak{p}\beta_\mathfrak{p}^{-1}) \in \mathfrak{S}(\mathfrak{M}; \mathfrak{a})$. This proves that our isomorphism is surjective, and the proof is completed.

6.9. Remark. Define the functions $h_1, \cdots, h_\lambda$ as in 2.3 for the present family Σ^*. Let $f_1, \cdots, f_\kappa$ be as in [9, Th. 3]. In the above theorem, if $\mathfrak{Q} = \mathfrak{Q}_y$, then $g \to g(y)$ gives an isomorphism of $\boldsymbol{Q}(h_j)$ onto $k_\mathfrak{a}$; and $\boldsymbol{Q}(f_i)$ corresponds to k_1. Therefore, we may regard 6.8 as a theorem concerning the Galois group of $\boldsymbol{Q}(h_j)$ over $\boldsymbol{Q}(f_i)$; it is also a generalization of [6, p. 307, Th. 2]. The relation between the fields and the groups is quite similar to the diagram of [6, p. 310].

6.10. Since L is totally indefinite, we can find, by [7, Lemma 4.4], a basis $\{u_1, \cdots, u_t, v_1, \cdots, v_t, w\}$ of V over L such that

$$(6.10.1) \qquad S(u_i, v_j) = \delta_{ij}, \qquad S(w, w) = 1,$$
$$S(u_i, u_j) = S(v_i, v_j) = S(u_i, w) = S(v_i, w) = 0,$$

where w occurs if and only if m is odd. Define on $\mathfrak{o}$-lattice $\mathfrak{M}$ by

$$(6.10.2) \qquad \mathfrak{M} = \sum_{i=1}^{t} (\mathfrak{o}u_i + \mathfrak{o}v_i) + \mathfrak{o}w.$$

First assume that m is even. Let $\mathfrak{H}(\mathfrak{M}; \mathfrak{a})$ be the subgroup of $\mathfrak{G}_0(\mathfrak{M}; \mathfrak{a})$ consisting of the elements $(\alpha_\mathfrak{p})$ such that $u_i\alpha_\mathfrak{p} = u_i$, $v_i\alpha_\mathfrak{p} = \mu(\alpha_\mathfrak{p})v_i$. Then we see easily that

$$(6.10.3) \qquad \mathfrak{H}(\mathfrak{M}; \mathfrak{a}) \cap \mathfrak{S}(\mathfrak{M}; \mathfrak{a}) = \{1\},$$
$$(6.10.4) \qquad \mathfrak{H}(\mathfrak{M}; \mathfrak{a})\mathfrak{S}(\mathfrak{M}; \mathfrak{a}) = \mathfrak{G}_0(\mathfrak{M}; \mathfrak{a}).$$

Next assume that m is odd, and $\mathfrak{a}$ is prime to the discriminant of L over F. Then $\mathfrak{a} = \mathfrak{o}\mathfrak{a}_0$ for some ideal $\mathfrak{a}_0$ in F, and $L_\mathfrak{p}$ may be identified with $M_2(F_\mathfrak{p})$ if $\mathfrak{p} \mid \mathfrak{a}_0$. Let $\mathfrak{H}(\mathfrak{M}; \mathfrak{a})$ be the subgroup of $\mathfrak{G}_0(\mathfrak{M}; \mathfrak{a})$ consisting of the elements $(\alpha_\mathfrak{p})$ such that $u_i\alpha_\mathfrak{p} = u_i$, $v_i\alpha_\mathfrak{p} = cv_i$, $w\alpha_\mathfrak{p} = \begin{pmatrix} 1 & 0 \\ 0 & c \end{pmatrix}w$ with $c = \mu(\alpha_\mathfrak{p})$. Then (6.10.3) and (6.10.4) are satisfied.

Let k' be the subfield of $k_\mathfrak{a}$ corresponding to $\mathfrak{H}(\mathfrak{M}; \mathfrak{a})\{\pm 1\}/\{\pm 1\}$. Then (6.10.3) and (6.10.4) implies that $k' \cap \boldsymbol{Q}(\zeta) = \boldsymbol{Q}$ and $k'(\zeta) = k_\mathfrak{a}$. From this and 4.5 we obtain the following theorem.

6.11. Theorem. *Let $\mathfrak{M}$ be an $\mathfrak{o}$-lattice defined by (6.10.2) for a basis $\{u_i, v_i, w\}$ of V satisfying (6.10.1). Let $\mathfrak{a}$ be an integral two-sided $\mathfrak{o}$-ideal. When m is odd, suppose that $\mathfrak{a}$ is prime to the discriminant of L over F. Define $h_1, \cdots, h_\lambda$ as in 2.3 for the family Σ^* of 6.7 with the present $\mathfrak{M}$ and $\mathfrak{a}$. Let ζ be as in 6.8. Then there exists a subfield $\mathfrak{K}$ of $\boldsymbol{Q}(h_1, \cdots, h_\lambda)$ such that*

    (i) $\mathfrak{K}(\zeta) = \boldsymbol{Q}(h_1, \cdots, h_\lambda)$;

    (ii) *The composite $\mathfrak{K} \cdot \boldsymbol{C}$ is the field of all automorphic functions on $\mathfrak{H}$ with respect to $\Gamma(T; \mathfrak{a}^{-1}\mathfrak{M}/\mathfrak{M})$;*

    (iii) *$\mathfrak{K}$ and $\boldsymbol{C}$ are linearly disjoint over $\boldsymbol{Q}$.*

In other words, the field of all automorphic functions with respect to

$\Gamma(T, \mathfrak{a}^{-1}\mathfrak{M}/\mathfrak{M})$ *has a model defined over* $\boldsymbol{Q}$.

7. Symplectic groups

7.1. In this section, we consider the case $L = F$. Let F be a totally real algebraic number field, and V a vector space of dimension m over F. Let T be a non-degenerate alternating form on V. We define groups $G(T)$ and $U(T)$ by

$$G(T) = \{\alpha \in GL(V, F) \mid T(x\alpha, y\alpha) = \mu(\alpha)T(x, y) \text{ with } \mu(\alpha) \in F\},$$
$$U(T) = \{\alpha \in G(T) \mid \mu(\alpha) = 1\}.$$

Then $U(T)$ is just the symplectic group of T. For every prime ideal $\mathfrak{p}$ in F, we define $V_\mathfrak{p}$, $G_\mathfrak{p}(T)$ and $U_\mathfrak{p}(T)$ in the same manner as in 6.1.

Let $\mathfrak{g}$ be the ring of integers in F, $\mathfrak{M}$ a $\mathfrak{g}$-lattice in V, and $\mathfrak{a}$ an integral ideal in F. For every prime ideal $\mathfrak{p}$ in F, we define groups $\Gamma_\mathfrak{p}(\mathfrak{M})$, $\Gamma_\mathfrak{p}(\mathfrak{M}; \mathfrak{a})$, $\mathfrak{G}(\mathfrak{M}; \mathfrak{a})$ and $\mathfrak{G}_0(\mathfrak{M}; \mathfrak{a})$ by (6.4.1), (6.4.2), (6.4.3) and (6.4.4), taking T and $\mathfrak{a}$ in place of S and $N(\mathfrak{a})$; further $\mathfrak{S}(\mathfrak{M}; \mathfrak{a})$ can be defined similarly. Then we can prove, for the groups $G(T)$ and $U(T)$, the assertions corresponding to 6.2, 6.3, 6.5 and 6.6; in fact, they follow easily from the existence of a canonical basis for $\mathfrak{M}$ [7, Prop. 1.3] and the approximation theorem [10, 6.7].

Take m elements $x_1, \cdots, x_m$ in V so that

$$\mathfrak{a}^{-1}\mathfrak{M} = \mathfrak{M} + \sum_{i=1}^{m} \mathfrak{g}x_i .$$

For the family $\Sigma^* = \Sigma^*(F, \Phi, \rho; T, \mathfrak{M}; \{x_i\})$, we now obtain the following theorem.

7.2. THEOREM. *Let* $\mathfrak{Q} = (A, C, \theta; t_1, \cdots, t_m)$ *be a generic member of* Σ^*, *and let* $\mathcal{P} = (A, C, \theta)$. *Let* k_1 *and* $k_\mathfrak{a}$ *be the fields of moduli of* $\mathcal{P}$ *and* $\mathfrak{Q}$, *respectively. Then the assertions corresponding to* (i), (ii), (iii), (iv) *of* 6.8 *hold for the present* k_1 *and* $k_\mathfrak{a}$.

Since the proof is quite similar to that of 6.8, we do not repeat it; we may also make a remark similar to 6.9.

7.3. THEOREM. *Define* $h_1, \cdots, h_\lambda$ *as in* 2.3 *for the family* Σ^*. *Let* a *be the smallest positive integer divisible by* $\mathfrak{a}$, *and* ζ *a primitive* a^{th} *root of unity. Then there exists a subfield* $\mathfrak{K}$ *of* $\boldsymbol{Q}(h_1, \cdots, h_\lambda)$ *with the properties* (i), (ii), (iii) *of* 6.11, *the symbols being understood as the ones in the present case.*

In fact, we can define a subgroup $\mathfrak{H}(\mathfrak{M}; \mathfrak{a})$ of $\mathfrak{G}_0(\mathfrak{M}; \mathfrak{a})$ satisfying (6.10.3) and (6.10.4) by means of a canonical basis of $\mathfrak{M}$ [7, Prop. 1.3]. Then our assertion follows from 4.5 and a consideration similar to 6.10.

In the special case $F = \boldsymbol{Q}$, our Theorems 7.2 and 7.3 has been announced as Theorems 3 and 4 in [8].[5]

8. Unitary groups over a number field

8.1. Let F be an algebraic number field of finite degree, and K a quadratic extension of F. We denote by ρ the non-trivial automorphism of K over F. The rings of integers in F and in K are denoted by $\mathfrak{g}$ and $\mathfrak{r}$, respectively. For every prime ideal $\mathfrak{p}$ in F, we denote by $F_\mathfrak{p}$ and $\mathfrak{g}_\mathfrak{p}$ the $\mathfrak{p}$-completions of F and $\mathfrak{g}$, and put $K_\mathfrak{p} = K \otimes_F F_\mathfrak{p}$ and $\mathfrak{r}_\mathfrak{p} = \mathfrak{g}_\mathfrak{p}\mathfrak{r}$.

8.2. LEMMA. *Let ν be a non-negative integer. Then there exists an integer $\lambda \geqq \nu$ with the following property.*

For every unit u in $\mathfrak{r}_\mathfrak{p}$ such that $uu^\rho = 1$ and $u \equiv 1 \bmod \mathfrak{p}^\lambda \mathfrak{r}_\mathfrak{p}$, there exists a unit v in $\mathfrak{r}_\mathfrak{p}$ such that $u = v^{-1}v^\rho$ and $v \equiv 1 \bmod \mathfrak{p}^\nu \mathfrak{r}_\mathfrak{p}$.

We can put $\lambda = \nu$ in either of the following cases:

(i) *$\mathfrak{p}$ remains prime in K;*

(ii) *$\mathfrak{p}$ does not divide 2, and $\nu > 0$.*

Define λ so that $\mathfrak{p}^\lambda = 2\mathfrak{p}^{\nu+1}$. Let u be as in our assertion. Then $v = (1 + u^\rho)/2$ has the required property. If $\mathfrak{p}$ does not divide 2 and $\nu > 0$, we may put $\lambda = \nu$. Now suppose that $\mathfrak{p}$ remains prime in K. If $uu^\rho = 1$ and $u \equiv 1 \bmod \mathfrak{p}^\nu \mathfrak{r}_\mathfrak{p}$, there exists a unit x in $\mathfrak{r}_\mathfrak{p}$ such that $u = x^{-1}x^\rho$ [10, 4.16]. Then $x^\rho \equiv x \bmod \mathfrak{p}^\nu \mathfrak{r}_\mathfrak{p}$. By induction on ν, we can easily prove that $x \equiv y \bmod \mathfrak{p}^\nu \mathfrak{r}_\mathfrak{p}$ for some $y \in \mathfrak{g}_\mathfrak{p}$. Then $v = y^{-1}x$ has the required property.

8.3. LEMMA. *Let ν be a non-negative integer. Then there exists an integer $\lambda \geqq \nu$ with the following property.*

For every unit b in $\mathfrak{g}_\mathfrak{p}$ such that $b \equiv 1 \bmod \mathfrak{p}^\lambda$, there exists a unit c in $\mathfrak{r}_\mathfrak{p}$ such that $c \equiv 1 \bmod \mathfrak{p}^\nu \mathfrak{r}_\mathfrak{p}$ and $cc^\rho = b$.

If $\mathfrak{p}$ is unramified in K, we may put $\lambda = \nu$.

The existence of λ is well-known. If $\mathfrak{p}$ is unramified in K, we can prove our assertion easily by a standard argument, constructing a convergent sequence by induction.

8.4. LEMMA. *Let ν be a non-negative integer and m a positive integer. Then there exists an integer $\kappa \geqq \nu$ with the following property.*

For every unit u in $\mathfrak{r}_\mathfrak{p}$ such that $uu^\rho = 1$ and $u \equiv 1 \bmod \mathfrak{p}^\kappa \mathfrak{r}_\mathfrak{p}$, there exists a unit w in $\mathfrak{r}_\mathfrak{p}$ such that $ww^\rho = 1$, $w \equiv 1 \bmod \mathfrak{p}^\nu \mathfrak{r}_\mathfrak{p}$ and $w^m = u$.

We obtain w as follows: take an element v so that $u = v^{-1}v^\rho$, find x so that $x^m = v$, and put $w = x^{-1}x^\rho$. Then our assertion is an easy consequence of

---

[5] The second sentence of [8, Th. 4] should read: Then there exists a subfield $\mathfrak{K}'$ of $\mathfrak{K}_N$ such that C and $\mathfrak{K}'$ are linearly disjoint over $\mathbf{Q}$, and $\mathfrak{K}_N = \mathfrak{K}' \cdot C$.

8.2 and a well-known fact concerning m^{th} roots in a $\mathfrak{p}$-adic field.

8.5. Let V be a vector space over K of dimension m, and $S(x, y)$ a non-degenerate ρ-hermitian form on V. We define groups $G(S)$, $U(S)$ and $SU(S)$ by

$$G(S) = \{\alpha \in GL(V, K) \mid S(x\alpha, y\alpha) = \mu(\alpha)S(x, y) \text{ with } \mu(\alpha) \in F\} \,,$$
$$U(S) = \{\alpha \in G(S) \mid \mu(\alpha) = 1\} \,,$$
$$SU(S) = \{\alpha \in U(S) \mid \det(\alpha) = 1\} \,,$$

where $GL(V, K)$ is the group of all K-linear automorphisms of V. For every prime ideal $\mathfrak{p}$ in F, we put $V_{\mathfrak{p}} = V \otimes_F F_{\mathfrak{p}}$, and define $G_{\mathfrak{p}}(S)$, $U_{\mathfrak{p}}(S)$, $SU_{\mathfrak{p}}(S)$ similarly.[6]

8.6. Suppose that m is even. Let us fix an integral ideal $\mathfrak{c}$ in F and a genus Λ of $\mathfrak{r}$-lattices in V with respect to $G(S)$ (cf. [10, 5.18]). For every prime ideal $\mathfrak{p}$ in F, let $D_{\mathfrak{p}}$ be the subgroup of $K_{\mathfrak{p}}^* \times F_{\mathfrak{p}}^*$ defined by[7]

$$D_{\mathfrak{p}} = \{(a, b) \in K_{\mathfrak{p}}^* \times F_{\mathfrak{p}}^* \mid aa^\rho = b^m\} \,.$$

Take a member $\mathfrak{L}$ of Λ, and define a subgroup $D_{\mathfrak{p}}(\Lambda; \mathfrak{c})$ of $D_{\mathfrak{p}}$ by

$$D_{\mathfrak{p}}(\Lambda; \mathfrak{c}) = \{(\det(\alpha), \mu(\alpha)) \mid \alpha \in G_{\mathfrak{p}}(S), \mathfrak{L}_{\mathfrak{p}}\alpha = \mathfrak{L}_{\mathfrak{p}}, \mathfrak{L}_{\mathfrak{p}}(\alpha - 1) \subset \mathfrak{c}_{\mathfrak{p}}\mathfrak{L}_{\mathfrak{p}}\} \,.$$

The group $D_{\mathfrak{p}}(\Lambda; \mathfrak{c})$ does not depend on the choice of $\mathfrak{L}$. Let $\mathfrak{d}$ be the relative discriminant of K with respect to F. We denote by $D(\Lambda; \mathfrak{c})$ the product of the factor groups $D_{\mathfrak{p}}/D_{\mathfrak{p}}(\Lambda; \mathfrak{c})$ for all $\mathfrak{p}$ dividing $\mathfrak{c}\mathfrak{d}$:

$$D(\Lambda; \mathfrak{c}) = \prod_{\mathfrak{p}\mid\mathfrak{c}\mathfrak{d}} \{D_{\mathfrak{p}}/D_{\mathfrak{p}}(\Lambda; \mathfrak{c})\} \,.$$

For every a in K such that $aa^\rho = 1$, we denote by $f(a)$ the element of $D(\Lambda; \mathfrak{c})$ whose components are the cosets containing $(a, 1)$.

8.7. Assuming still m to be even, by a *frame of level* $\mathfrak{c}$ in V, or simply, a $\mathfrak{c}$-*frame in* V, we understand a structure $(\mathfrak{L}; x_1, \cdots, x_m)$ formed by an $\mathfrak{r}$-lattice $\mathfrak{L}$ in V and m elements $x_1, \cdots, x_m$ in V such that $\mathfrak{c}^{-1}\mathfrak{L} = \mathfrak{L} + \mathfrak{r}x_1 + \cdots + \mathfrak{r}x_m$. We say that two $\mathfrak{c}$-frames $\mathfrak{L}^* = (\mathfrak{L}; x_1, \cdots, x_m)$ and $\mathfrak{M}^* = (\mathfrak{M}; y_1, \cdots, y_m)$ belong to the same *genus*, if, for every prime ideal $\mathfrak{p}$ in F, there exists an element $\alpha_{\mathfrak{p}}$ of $G_{\mathfrak{p}}(S)$ such that $\mathfrak{L}_{\mathfrak{p}}\alpha_{\mathfrak{p}} = \mathfrak{M}_{\mathfrak{p}}$ and $x_i\alpha_{\mathfrak{p}} \equiv y_i \bmod \mathfrak{M}_{\mathfrak{p}}$ for $1 \leq i \leq m$.

This being so, assume that $\mathfrak{L}$ and $\mathfrak{M}$ are members of Λ. We denote by $d(\mathfrak{L}^*/\mathfrak{M}^*)$ the element of $D(\Lambda; \mathfrak{c})$ whose components are represented by $(\det(\alpha_{\mathfrak{p}}), \mu(\alpha_{\mathfrak{p}}))$. It is easy to see that $d(\mathfrak{L}^*/\mathfrak{M}^*)$ does not depend on the choice of the $\alpha_{\mathfrak{p}}$.

8.8. PROPOSITION. *Suppose that S is indefinite. Let Λ, $\mathfrak{L}^*$ and $\mathfrak{M}^*$ be as*

---

[6] In [10], $G_{\mathfrak{p}}(S)$, $U_{\mathfrak{p}}(S)$, $SU_{\mathfrak{p}}(S)$ has been denoted by $G(V_{\mathfrak{p}}, S)$, $U(V_{\mathfrak{p}}, S)$, $SU(V_{\mathfrak{p}}, S)$, with a letter Φ instead of S.

[7] $K_{\mathfrak{p}}^*$ and $F_{\mathfrak{p}}^*$ are the groups of invertible elements in $K_{\mathfrak{p}}$ and $F_{\mathfrak{p}}$, respectively.

above. There exists an element α of $U(S)$ such that $\mathfrak{L}\alpha = \mathfrak{M}$ and $x_i\alpha \equiv y_i \bmod \mathfrak{M}$ for every i, if and only if there exists an element a of K such that $aa^\rho = 1$, $[\mathfrak{L}/\mathfrak{M}] = \mathfrak{r}a$ and $d(\mathfrak{L}^*/\mathfrak{M}^*) = f(a)$.[8]

The *only if* part is easily seen by putting $\det(\alpha) = a$. To prove the *if* part, suppose that $[\mathfrak{L}/\mathfrak{M}] = \mathfrak{r}a$, $d(\mathfrak{L}^*/\mathfrak{M}^*) = f(a)$, $aa^\rho = 1$ for some $a \in K$. By [10, 2.3], we can find an element β of $U(S)$ such that $\det(\beta) = a$. Put $\mathfrak{N} = \mathfrak{L}\beta$, $z_i = x_i\beta$, $\mathfrak{N}^* = (\mathfrak{N}; z_1, \cdots, z_m)$. Then $[\mathfrak{N}/\mathfrak{M}] = \mathfrak{r}$, $d(\mathfrak{N}^*/\mathfrak{M}^*) = 1$. For each prime divisor $\mathfrak{p}$ of $\mathfrak{cd}$, there exists an element $\alpha_\mathfrak{p}$ of $G_\mathfrak{p}(S)$ such that $\mathfrak{N}_\mathfrak{p}\alpha_\mathfrak{p} = \mathfrak{M}_\mathfrak{p}$, $z_i\alpha_\mathfrak{p} \equiv y_i \bmod \mathfrak{M}_\mathfrak{p}$ and $(\det(\alpha_\mathfrak{p}), \mu(\alpha_\mathfrak{p})) \in D_\mathfrak{p}(\Lambda; \mathfrak{c})$. In view of our definition of $D_\mathfrak{p}(\Lambda; \mathfrak{c})$, we may assume that $\alpha_\mathfrak{p} \in SU_\mathfrak{p}(S)$. Let $\mathfrak{q}$ be a prime ideal in F such that $\mathfrak{N}_\mathfrak{q} \neq \mathfrak{M}_\mathfrak{q}$ and $\mathfrak{q} \nmid \mathfrak{cd}$. Since $\mathfrak{N}$ and $\mathfrak{M}$ belong to the same genus with respect to $G(S)$, there exists an element $\gamma_\mathfrak{q}$ of $G_\mathfrak{q}(S)$ such that $\mathfrak{N}_\mathfrak{q}\gamma_\mathfrak{q} = \mathfrak{M}_\mathfrak{q}$. Since $[\mathfrak{N}/\mathfrak{M}] = \mathfrak{r}$, $\det(\gamma_\mathfrak{q})$ is a unit of $\mathfrak{r}_\mathfrak{q}$, and $\mu(\gamma_\mathfrak{q})$ is a unit of $\mathfrak{g}_\mathfrak{q}$. By 8.9 below, we find an element $\delta_\mathfrak{q}$ of $G_\mathfrak{q}(S)$ such that $\mathfrak{N}_\mathfrak{q}\delta_\mathfrak{q} = \mathfrak{N}_\mathfrak{q}$, $\det(\delta_\mathfrak{q}) = \det(\gamma_\mathfrak{q})^{-1}$, $\mu(\delta_\mathfrak{q}) = \mu(\gamma_\mathfrak{q})^{-1}$. Then $\mathfrak{N}_\mathfrak{q}\delta_\mathfrak{q}\gamma_\mathfrak{q} = \mathfrak{M}_\mathfrak{q}$ and $\delta_\mathfrak{q}\gamma_\mathfrak{q} \in SU_\mathfrak{q}(S)$. Now by [10, 5.12], there exists an element α of $SU(S)$ such that

$$\mathfrak{N}_\mathfrak{p}(\alpha - \alpha_\mathfrak{p}) \subset \mathfrak{pc}_\mathfrak{p}\mathfrak{M}_\mathfrak{p} \qquad \text{for } \mathfrak{p} \mid \mathfrak{cd},$$
$$\mathfrak{N}_\mathfrak{q}(\alpha - \delta_\mathfrak{q}\gamma_\mathfrak{q}) \subset \mathfrak{q}\mathfrak{M}_\mathfrak{q} \qquad \text{for } \mathfrak{q} \nmid \mathfrak{cd},\ \mathfrak{N}_\mathfrak{q} \neq \mathfrak{M}_\mathfrak{q},$$
$$\mathfrak{N}_\mathfrak{q}\alpha = \mathfrak{N}_\mathfrak{q} \qquad \text{for } \mathfrak{q} \nmid \mathfrak{cd},\ \mathfrak{N}_\mathfrak{q} = \mathfrak{M}_\mathfrak{q}.$$

By [10, 5.3] we have $\mathfrak{N}_\mathfrak{q}\alpha = \mathfrak{M}_\mathfrak{q}$ for every prime ideal $\mathfrak{q}$ in F, so that $\mathfrak{N}\alpha = \mathfrak{M}$. If $\mathfrak{p} \mid \mathfrak{c}$, we have $\mathfrak{c}_\mathfrak{p}z_i(\alpha - \alpha_\mathfrak{p}) \subset \mathfrak{N}_\mathfrak{p}(\alpha - \alpha_\mathfrak{p}) \subset \mathfrak{c}_\mathfrak{p}\mathfrak{M}_\mathfrak{p}$, hence $z_i\alpha \equiv z_i\alpha_\mathfrak{p} \equiv y_i \bmod \mathfrak{M}_\mathfrak{p}$. If $\mathfrak{p} \nmid \mathfrak{c}$, we have $z_i \in \mathfrak{N}_\mathfrak{p}$ and $y_i \in \mathfrak{M}_\mathfrak{p}$, so that $z_i\alpha - y_i \in \mathfrak{M}_\mathfrak{p}$. Therefore $z_i\alpha \equiv y_i \bmod \mathfrak{M}$ for every i. This completes the proof.

8.9. PROPOSITION. *For every prime ideal $\mathfrak{p}$ in F, let*

$$D_\mathfrak{p}(\mathfrak{c}) = \{(a, b) \in D_\mathfrak{p} \mid a \text{ is a unit in } \mathfrak{r}_\mathfrak{p},\ a \equiv b \equiv 1 \bmod \mathfrak{c}_\mathfrak{p}\mathfrak{r}_\mathfrak{p}\}.$$

Then we have $D_\mathfrak{p}(\Lambda; \mathfrak{c}) \subset D_\mathfrak{p}(\mathfrak{c})$. Moreover, if $\mathfrak{p}$ is unramified in K, $D_\mathfrak{p}(\Lambda; \mathfrak{c}) = D_\mathfrak{p}(\mathfrak{c})$.

Let $\mathfrak{L}$ be a member of Λ. Let $\alpha \in G_\mathfrak{p}(S)$, $\mathfrak{L}_\mathfrak{p}\alpha = \mathfrak{L}_\mathfrak{p}$, $\mathfrak{L}_\mathfrak{p}(\alpha - 1) \subset \mathfrak{c}_\mathfrak{p}\mathfrak{L}_\mathfrak{p}$. Represent α by a matrix with respect to a basis of $\mathfrak{L}_\mathfrak{p}$ over $\mathfrak{r}_\mathfrak{p}$. Then it is clear that $\det(\alpha)$ is a unit in $\mathfrak{r}_\mathfrak{p}$, and $\det(\alpha) \equiv 1 \bmod \mathfrak{c}_\mathfrak{p}\mathfrak{r}_\mathfrak{p}$. Define $\mu_0(\mathfrak{L})$ as in [10, 2.10]. There exist elements x and y of $\mathfrak{L}_\mathfrak{p}$ such that $\mu_0(\mathfrak{L}_\mathfrak{p}) = S(x, y)\mathfrak{r}_\mathfrak{p}$. Then $\mu(\alpha)S(x, y) = S(x\alpha, y\alpha) \equiv S(x, y) \bmod S(x, y)\mathfrak{c}_\mathfrak{p}\mathfrak{r}_\mathfrak{p}$. Hence $\mu(\alpha) \equiv 1 \bmod \mathfrak{c}_\mathfrak{p}$. This proves that $D_\mathfrak{p}(\Lambda; \mathfrak{c}) \subset D_\mathfrak{p}(\mathfrak{c})$. Now suppose that $\mathfrak{p}$ is unramified in K. Let $(a, b) \in D_\mathfrak{p}(\mathfrak{c})$. By 8.3, there exists a unit u in $\mathfrak{r}_\mathfrak{p}$ such that $u \equiv 1 \bmod \mathfrak{c}_\mathfrak{p}\mathfrak{r}_\mathfrak{p}$ and $uu^\rho = b$. Put $v = u^{-m}a$. Then $v \equiv 1 \bmod \mathfrak{c}_\mathfrak{p}\mathfrak{r}_\mathfrak{p}$ and $vv^\rho = 1$. By [10, 3.2 and

---

[8] For the definition of the notation $[\mathfrak{L}/\mathfrak{M}]$, see [10, 2.15].

4.15], there exists a basis $\{x_i\}$ of $\mathfrak{L}_\mathfrak{p}$ over $\mathfrak{r}_\mathfrak{p}$ such that $S(x_i, x_j) = c_i \delta_{ij}$ with $c_i \in F_\mathfrak{p}$. Define an element α of $G_\mathfrak{p}(S)$ by $x_1 \alpha = u v x_1$, $x_i \alpha = u x_i$ for $i > 1$. Then $\mathfrak{L}_\mathfrak{p} \alpha = \mathfrak{L}_\mathfrak{p}$, $\mathfrak{L}_\mathfrak{p}(\alpha - 1) \subset \mathfrak{c}_\mathfrak{p} \mathfrak{L}_\mathfrak{p}$, $\det(\alpha) = a$, $\mu(\alpha) = b$. This proves $D_\mathfrak{p}(\mathfrak{c}) = D_\mathfrak{p}(\Lambda; \mathfrak{c})$.

8.10. PROPOSITION. *If $\mathfrak{p}$ is ramified in K, we have $D_\mathfrak{p}(\mathfrak{c}\mathfrak{p}^e) \subset D_\mathfrak{p}(\Lambda; \mathfrak{c})$ for a suitably large positive integer e.*

Let $(a, b) \in D_\mathfrak{p}(\mathfrak{c}\mathfrak{p}^e)$. Take a unit u in $\mathfrak{r}_\mathfrak{p}$ so that $u u^\rho = b$. Put $v = u^{-m} a$. Then $v v^\rho = 1$. Take a unit w in $\mathfrak{r}_\mathfrak{p}$ so that $w w^\rho = 1$ and $w^m = v$. By virtue of 8.3 and 8.4, we can find u and w so that $u \equiv w \equiv 1 \bmod \mathfrak{c}_\mathfrak{p} \mathfrak{r}_\mathfrak{p}$, assuming that e is sufficiently large. Define an element α of $G_\mathfrak{p}(S)$ so that $x\alpha = u w x$ for all x in V. Then $\mathfrak{L}_\mathfrak{p} \alpha = \mathfrak{L}_\mathfrak{p}$, $\mathfrak{L}_\mathfrak{p}(1 - \alpha) \subset \mathfrak{c}_\mathfrak{r} \mathfrak{L}_\mathfrak{p}$, $\det(\alpha) = a$, $\mu(\alpha) = b$. This proves our proposition.

8.11. PROPOSITION. *Let $\mathfrak{L}$ be a member of Λ. Suppose that $\mathfrak{p}$ is ramified in K, $V_\mathfrak{p}$ has a non-trivial kernel subspace with respect to S (cf. [10, 4.3]), and $\mathfrak{L}_\mathfrak{p}$ is a maximal $\mathfrak{r}_\mathfrak{p}$-lattice in $V_\mathfrak{p}$ (cf. [10, 2.10]). If $\mathfrak{p} \nmid \mathfrak{c}$, or if $\mathfrak{p} \mid \mathfrak{c}$ and $\mathfrak{p}$ does not divide 2, then $D_\mathfrak{p}(\Lambda; \mathfrak{c}) = D_\mathfrak{p}(\mathfrak{c})$.*

If $\mathfrak{p} \nmid \mathfrak{c}$, our assertion is a consequence of [10, 4.18 and 4.24]. Now suppose that $\mathfrak{p} \mid \mathfrak{c}$ and $\mathfrak{p}$ does not divide 2. By [10, 4.7] there exists a Witt decomposition $V_\mathfrak{p} = \sum_{i=1}^t (K_\mathfrak{p} x_i + K_\mathfrak{p} y_i) + W$ such that $\mathfrak{L}_\mathfrak{p} = \sum_{i=1}^t (\mathfrak{r}_\mathfrak{p} x_i + \mathfrak{b} y_i) + \mathfrak{M}$, where $t = (m - 2)/2$, $\mathfrak{b} = \mu(\mathfrak{L}_\mathfrak{p})\mathfrak{d}^{-1}$, $\mathfrak{M} = \{z \in W \mid S(z, z) \in \mu(\mathfrak{L}_\mathfrak{p})\}$. Let $(a, b) \in D_\mathfrak{p}(\mathfrak{c})$. By 8.2, there exists a unit u in $\mathfrak{r}_\mathfrak{p}$ such that $u^{-1} u^\rho = b^{-m/2} a$ and $u \equiv 1 \bmod \mathfrak{c}_\mathfrak{p} \mathfrak{r}_\mathfrak{p}$. Let $\mathfrak{B} = \mathfrak{B}(W)$ be the quaternion algebra attached to the restriction of S to W (cf. [10, 2.6 and 4.3]), and ι the canonical involution of $\mathfrak{B}$. By [7, Prop. 3.2] and by [10, 4.6], there exists an element β of $\mathfrak{B}$ such that $\mathfrak{M}(\beta - 1) \subset \mathfrak{c}_\mathfrak{p} \mathfrak{M}$ and $\beta \beta^\iota = b u u^\rho$. Define an element α of $G_\mathfrak{p}(S)$ by $x_i \alpha = b x_i$, $y_i \alpha = y_i$ for $1 \leqq i \leqq t$ and $z \alpha = u^{-1} z \beta$ for every $z \in W$. Then $\mu(\alpha) = b$, $\det(\alpha) = a$, $\mathfrak{L}_\mathfrak{p} \alpha = \mathfrak{L}_\mathfrak{p}$ and $\mathfrak{L}_\mathfrak{p}(1 - \alpha) \subset \mathfrak{c}_\mathfrak{p} \mathfrak{L}_\mathfrak{p}$. This completes the proof.

8.12. PROPOSITION. *Let $\mathfrak{L}$ be a member of Λ. Suppose that $\mathfrak{p}$ is ramified in K and $\mathfrak{L}_\mathfrak{p}$ is a maximal $\mathfrak{r}_\mathfrak{p}$-lattice in $V_\mathfrak{p}$. Then for every unit a in $\mathfrak{g}_\mathfrak{p}$ such that $a \equiv 1 \bmod \mathfrak{c}_\mathfrak{p}$, there exists an element α of $G_\mathfrak{p}(S)$ such that $\mathfrak{L}_\mathfrak{p} \alpha = \mathfrak{L}_\mathfrak{p}$, $\mathfrak{L}_\mathfrak{p}(1 - \alpha) \subset \mathfrak{c}_\mathfrak{p} \mathfrak{L}_\mathfrak{p}$, $\det(\alpha) = a^{m/2}$, $\mu(\alpha) = a$.*

This can be proved by taking a Witt decomposition as in the proof of 8.11. In the present case, we may dispense with the restrictive conditions which are necessary for dealing with an element u such that $u u^\rho = 1$.

8.13. Now we consider the case where m is odd; put $m = 2l + 1$. For every $\alpha \in G(S)$ or $\alpha \in G_\mathfrak{p}(S)$, put

$$\kappa(\alpha) = \mu(\alpha)^{-l} \det(\alpha) \, .$$

Then $\kappa(\alpha)$ determines $\mu(\alpha)$ and $\det(\alpha)$ by the following relation:

$$\mu(\alpha) = \kappa(\alpha)\kappa(\alpha)^\rho , \qquad \det(\alpha) = \mu(\alpha)^l \kappa(\alpha) .$$

Therefore, we have

$$SU(S) = \{\alpha \in G(S) \mid \kappa(\alpha) = 1\} ,$$
$$SU_\mathfrak{p}(S) = \{\alpha \in G_\mathfrak{p}(S) \mid \kappa(\alpha) = 1\} .$$

8.14. PROPOSITION. *For every invertible element u in K (resp. $K_\mathfrak{p}$), there exists an element α of $G(S)$ (resp. $G_\mathfrak{p}(S)$) such that $\kappa(\alpha) = u$.*

By [10, 2.4], there exists an element β of $G(S)$ (resp. $G_\mathfrak{p}(S)$) such that $\mu(\beta) = uu^\rho$. Put $v = \kappa(\beta)^{-1}u$. Then $vv^\rho = 1$. By [10, 2.3], there exists an element γ of $U(S)$ (resp. $U_\mathfrak{p}(S)$) such that $\det(\gamma) = v$. Then $\kappa(\beta\gamma) = u$.

8.15. Let $\mathfrak{c}$ be an integral ideal in F. For every prime ideal $\mathfrak{p}$ in F, define a multiplicative group $I_\mathfrak{p}(\mathfrak{c})$ by

$$I_\mathfrak{p}(\mathfrak{c}) = \{u \in K_\mathfrak{p} \mid \text{unit in } \mathfrak{r}_\mathfrak{p},\ u \equiv 1 \bmod \mathfrak{c}_\mathfrak{p}\mathfrak{r}_\mathfrak{p}\} .$$

We denote by $I(\mathfrak{c})$ the product of the factor groups $K_\mathfrak{p}^*/I_\mathfrak{p}(\mathfrak{c})$ for all $\mathfrak{p}$ dividing $\mathfrak{c}$. For every element u of K, we denote by $h(u)$ the element of $I(\mathfrak{c})$ whose components are the cosets containing u.

8.16. PROPOSITION. *Let $\mathfrak{L}_\mathfrak{p}$ be an $\mathfrak{r}_\mathfrak{p}$-lattice in $V_\mathfrak{p}$, and $I_\mathfrak{p}(\mathfrak{c})$ be as above. Then*

$$I_\mathfrak{p}(\mathfrak{c}) = \{\kappa(\alpha) \mid \alpha \in G_\mathfrak{p}(S),\ \mathfrak{L}_\mathfrak{p}\alpha = \mathfrak{L}_\mathfrak{p},\ \mathfrak{L}_\mathfrak{p}(\alpha - 1) \subset \mathfrak{c}_\mathfrak{p}\mathfrak{L}_\mathfrak{p}\} .$$

By [10, 3.2 and 4.15] there exists a decomposition $\mathfrak{L}_\mathfrak{p} = \mathfrak{r}_\mathfrak{p}x + \mathfrak{M}$ such that $S(x, x) \neq 0$ and $S(x, y) = 0$ for $y \in \mathfrak{M}$. For $u \in I_\mathfrak{p}(\mathfrak{c})$, define an element α of $G_\mathfrak{p}(S)$ so that $x\alpha = u^{1-l}(u^\rho)^l x$, $y\alpha = uy$ for every $y \in \mathfrak{M}$. Then $\kappa(\alpha) = u$, $\mathfrak{L}_\mathfrak{p}\alpha = \mathfrak{L}_\mathfrak{p}$, $\mathfrak{L}_\mathfrak{p}(\alpha - 1) \subset \mathfrak{c}_\mathfrak{p}\mathfrak{L}_\mathfrak{p}$. Conversely, for such an element α, we can prove that $\kappa(\alpha) \in I_\mathfrak{p}(\mathfrak{c})$ by the same argument as in 8.9.

8.17. By an *oriented frame of level* $\mathfrak{c}$, or simply, an *oriented $\mathfrak{c}$-frame*, we understand a structure $(\mathfrak{L}; a; x_1, \cdots, x_m)$ formed by an $\mathfrak{r}$-lattice $\mathfrak{L}$ in V, a non-zero element a in K, and m elements $x_1, \cdots, x_m$ in V such that $\mathfrak{c}^{-1}\mathfrak{L} = \mathfrak{L} + \mathfrak{r}x_1 + \cdots + \mathfrak{r}x_m$. We say that two oriented $\mathfrak{c}$-frames $\mathfrak{L}^* = (\mathfrak{L}; a; x_1, \cdots, x_m)$ and $\mathfrak{M}^* = (\mathfrak{M}; b; y_1, \cdots, y_m)$ belong to the same *genus*, if, for every prime ideal $\mathfrak{p}$ in F, there exists an element $\alpha_\mathfrak{p}$ of $G_\mathfrak{p}(S)$ such that $\mathfrak{L}_\mathfrak{p}\alpha_\mathfrak{p} = \mathfrak{M}_\mathfrak{p}$ and $x_i\alpha_\mathfrak{p} \equiv y_i \bmod \mathfrak{M}_\mathfrak{p}$ for $1 \leq i \leq m$. When $\mathfrak{L}^*$, $\mathfrak{M}^*$ and $\alpha_\mathfrak{p}$ are in this situation, we denote by $i(\mathfrak{L}^*/\mathfrak{M}^*)$ the element of $I(\mathfrak{c})$ whose components are represented by $\kappa(\alpha_\mathfrak{p})$. By 8.16, this does not depend on the choice of $\alpha_\mathfrak{p}$. $\mathfrak{L}^*$ and $\mathfrak{M}^*$ are called *equivalent*, if there exists an element α of $G(S)$ such that

$$(8.17.1) \qquad \mathfrak{L}\alpha = \mathfrak{M} , \quad \mu(\alpha)b = a , \quad x_i\alpha \equiv y_i \bmod \mathfrak{M} \qquad \text{for } 1 \leq i \leq m .$$

8.18. PROPOSITION. *Suppose that S is indefinite and $m > 1$. Let $\mathfrak{L}^* =$*

$(\mathfrak{L}; a; x_1, \cdots, x_m)$ *and* $\mathfrak{M}^* = (\mathfrak{M}; b; y_1, \cdots, y_m)$ *be oriented frames of level* $\mathfrak{c}$, *belonging to the same genus. Then they are equivalent if and only if there exists an element* u *of* K *such that*

$$(a^{-1}b)^{\iota}[\mathfrak{L}/\mathfrak{M}] = \mathfrak{r}u , \qquad ab^{-1} = uu^{\rho} , \qquad i(\mathfrak{L}^*/\mathfrak{M}^*) = h(u) .$$

The *only if* part is easily shown by putting $u = \kappa(\alpha)$ for an element α of $G(S)$ satisfying (8.17.1). To prove the *if* part, let u be as in our assertion. By 8.14, there exists an element β of $G(S)$ such that $\kappa(\beta) = u$. Put $\mathfrak{N} = \mathfrak{L}\beta$, $c = \mu(\beta)^{-1}a$, $z_i = x_i\beta$, $\mathfrak{N}^* = (\mathfrak{N}; c; z_1, \cdots, z_m)$. Then $[\mathfrak{N}/\mathfrak{M}] = \mathfrak{r}$, $c = b$, $i(\mathfrak{N}^*/\mathfrak{M}^*) = 1$. Let $\mathfrak{p}$ be a prime factor of $\mathfrak{c}$. There exists an element $\alpha_{\mathfrak{p}}$ of $G_{\mathfrak{p}}(S)$ such that $\mathfrak{N}_{\mathfrak{p}}\alpha_{\mathfrak{p}} = \mathfrak{M}_{\mathfrak{p}}$, $z_i\alpha_{\mathfrak{p}} \equiv y_i \bmod \mathfrak{M}_{\mathfrak{p}}$. Then $\kappa(\alpha_{\mathfrak{p}}) \in I_{\mathfrak{p}}(\mathfrak{c})$. By 8.16, there exists an element $\beta_{\mathfrak{p}}$ of $G_{\mathfrak{p}}(S)$ such that $\mathfrak{M}_{\mathfrak{p}}\beta_{\mathfrak{p}} = \mathfrak{M}_{\mathfrak{p}}$, $\mathfrak{M}_{\mathfrak{p}}(1 - \beta_{\mathfrak{p}}) \subset \mathfrak{c}_{\mathfrak{p}}\mathfrak{M}_{\mathfrak{p}}$ and $\kappa(\beta_{\mathfrak{p}}) = \kappa(\alpha_{\mathfrak{p}})^{-1}$. Then $\mathfrak{N}_{\mathfrak{p}}\alpha_{\mathfrak{p}}\beta_{\mathfrak{p}} = \mathfrak{M}_{\mathfrak{p}}$, $z_i\alpha_{\mathfrak{p}}\beta_{\mathfrak{p}} \equiv y_i \bmod \mathfrak{M}_{\mathfrak{p}}$, and $\alpha_{\mathfrak{p}}\beta_{\mathfrak{p}} \in SU_{\mathfrak{p}}(S)$. Let $\mathfrak{q}$ be a prime ideal in F such that $\mathfrak{q} \nmid c$ and $\mathfrak{N}_{\mathfrak{q}} \neq \mathfrak{M}_{\mathfrak{q}}$. There exists an element $\alpha_{\mathfrak{q}}$ of $G_{\mathfrak{q}}(S)$ such that $\mathfrak{N}_{\mathfrak{q}}\alpha_{\mathfrak{q}} = \mathfrak{M}_{\mathfrak{q}}$. Since $[\mathfrak{N}/\mathfrak{M}] = \mathfrak{r}$, $\det(\alpha_{\mathfrak{q}})$ and $\mu(\alpha_{\mathfrak{q}})$ are units, so that $\kappa(\alpha_{\mathfrak{q}})$ is a unit in $\mathfrak{r}_{\mathfrak{q}}$. By 8.16, there exists an element $\beta_{\mathfrak{q}}$ of $G_{\mathfrak{q}}(S)$ such that $\mathfrak{M}_{\mathfrak{q}}\beta_{\mathfrak{q}} = \mathfrak{M}_{\mathfrak{q}}$, $\kappa(\beta_{\mathfrak{q}}) = \kappa(\alpha_{\mathfrak{q}})^{-1}$. Then $\mathfrak{N}_{\mathfrak{q}}\alpha_{\mathfrak{q}}\beta_{\mathfrak{q}} = \mathfrak{M}_{\mathfrak{q}}$, $\alpha_{\mathfrak{q}}\beta_{\mathfrak{q}} \in SU_{\mathfrak{q}}(S)$. By [10, 5.12], there exists an element γ of $SU(S)$ such that

$$\mathfrak{N}_{\mathfrak{p}}(\gamma - \alpha_{\mathfrak{p}}\beta_{\mathfrak{p}}) \subset \mathfrak{p}\mathfrak{c}_{\mathfrak{p}}\mathfrak{M}_{\mathfrak{p}} \qquad \text{for } \mathfrak{p} \mid \mathfrak{c} ,$$
$$\mathfrak{N}_{\mathfrak{q}}(\gamma - \alpha_{\mathfrak{q}}\beta_{\mathfrak{q}}) \subset \mathfrak{q}\mathfrak{M}_{\mathfrak{q}} \qquad \text{for } \mathfrak{q} \nmid \mathfrak{c},\ \mathfrak{N}_{\mathfrak{q}} \neq \mathfrak{M}_{\mathfrak{q}} ,$$
$$\mathfrak{N}_{\mathfrak{q}}\gamma = \mathfrak{M}_{\mathfrak{q}} \qquad \text{for } \mathfrak{q} \nmid \mathfrak{c},\ \mathfrak{N}_{\mathfrak{q}} = \mathfrak{M}_{\mathfrak{q}} .$$

By [10, 5.3], we have $\mathfrak{N}_{\mathfrak{q}}\gamma = \mathfrak{M}_{\mathfrak{q}}$ for all prime ideals $\mathfrak{q}$ in F, so that $\mathfrak{N}\gamma = \mathfrak{M}$. If $\mathfrak{p} \mid \mathfrak{c}$, we have $z_i\gamma \equiv y_i \bmod \mathfrak{M}_{\mathfrak{p}}$ for every i. Hence $\alpha = \beta\gamma$ satisfies (8.17.1). This completes the proof.

9. Class-field as field of definition

9.1. Let F, K, $\mathfrak{g}$, $\mathfrak{r}$ be as in 8.1. From now on we assume that F is totally real, and K is totally imaginary; we put $[F:Q] = g$. We denote by ρ the complex conjugation. Let $\{\tau_1, \cdots, \tau_g\}$ be a set of g isomorphisms of K into C, whose restrictions to F are different from each other. Let m be a positive integer, and Φ a representation of K by complex matrices of size mg. Let r_{ν} and s_{ν} be the multiplicities of τ_{ν} and $\tau_{\nu}\rho$ in Φ, respectively. We assume $m > 1$, $r_{\nu} + s_{\nu} = m$ for every ν, and $\sum_{\nu=1}^{g} r_{\nu}s_{\nu} > 0$.

Let V be a vector space over K of dimension m, and $T(x, y)$ a non-degenerate ρ-anti-hermitian form, satisfying the condition [FI, (5.1.1.)]. Let $\mathfrak{M}$ be an $\mathfrak{r}$-lattice in V, and $\mathfrak{c}$ an integral ideal in F. Take m elements $x_1, \cdots, x_m$ in V so that

$$\mathfrak{c}^{-1}\mathfrak{M} = \mathfrak{M} + \mathfrak{r}x_1 + \cdots + \mathfrak{r}x_m .$$

Then we consider the family $\Sigma^*(K, \Phi, \rho; T, \mathfrak{M}; \{x_i\}) = \{\mathfrak{Q}_z \mid z \in \mathcal{H}\}$ defined in

4.3. For simplicity, we denote this family by $\Sigma^*(T, \mathfrak{M}; \{x_i\})$. We put $\mathfrak{Q}_z = (A_z, \mathcal{C}_z, \theta_z; t_1(z), \cdots, t_m(z))$, and $\mathfrak{h}(\mathfrak{c}, A_z) = \{t \in A_z \mid \theta_z(\mathfrak{c})t = 0\}$. Then $\mathfrak{h}(\mathfrak{c}, A_z) = \sum_{i=1}^{m} \theta_z(\mathfrak{r})t_i(z)$.

For every member $\mathfrak{Q}_z$ of $\Sigma^*(T, \mathfrak{M}; \{x_i\})$, we define $\mathcal{F}_z$ as in 2.2. Let $\mathfrak{Q}_u$ be a generic member of Σ^* over $\boldsymbol{Q}$. We denote by $\mathcal{V}(T, \mathfrak{M}; \{x_i\})$ the locus of $c(\mathcal{F}_u)$ over $\boldsymbol{Q}$. By 2.3 and 4.5, the variety $\mathcal{V}(T, \mathfrak{M}; \{x_i\})$ may be regarded as a model of the field of all automorphic functions on $\mathcal{H}$ with respect to $\Gamma(T, \mathfrak{c}^{-1}\mathfrak{M}/\mathfrak{M})$. Let $\mathfrak{k}_u$ be the field of moduli of $\mathfrak{Q}_u$. Then $\mathfrak{k}_u = \boldsymbol{Q}(c(\mathcal{F}_u))$. Let k_0 be the algebraic closure of $\boldsymbol{Q}$ in $\mathfrak{k}_u$. Then k_0 *is the smallest field of definition for the variety* $\mathcal{V}(T, \mathfrak{M}; \{x_i\})$. Now our purpose is to determine the structure of k_0.

9.2. Let K' be the field generated over $\boldsymbol{Q}$ by the elements $\sum_{\nu=1}^{g} (r_\nu x^{\tau\nu} + s_\nu x^{\tau\nu\rho})$ for all $x \in K$. By [FI, 4.3], $K' = \boldsymbol{Q}$ or K' is a totally imaginary quadratic extension of a totally real field. In the latter case, put $[K' : \boldsymbol{Q}] = 2g'$. Then we define g' isomorphisms $\sigma_1, \cdots, \sigma_{g'}$ of K' into $\boldsymbol{C}$ and integers $t_\lambda, u_\lambda, v_\lambda$ as in [FI, 4.4 and 4.7].

We fix once and for all an element ζ of K such that $\zeta^\rho = -\zeta$. Then ζT is a ρ-hermitian form on V. We write $U(\zeta T)$ and $G(\zeta T)$ as $U(T)$ and $G(T)$. Let Λ be the genus of $\mathfrak{M}$ with respect to $G(T)$. Let $\mathfrak{b}$ be the relative discriminant of K with respect to F. We are going to define an ideal-group $H_{\mathfrak{c}}$ in K. For every ideal $\mathfrak{a}$ in K', put

$$(9.2.1) \qquad \psi(\mathfrak{a}) = \begin{cases} \mathfrak{r} & \text{if } m \text{ is even and } K' = \boldsymbol{Q}, \\[2ex] \prod_{\lambda=1}^{g'} \left(\dfrac{\mathfrak{a}^{\sigma_\lambda}}{\mathfrak{a}^{\sigma_\lambda\rho}}\right)^{v_\lambda} & \text{if } m \text{ is even and } K' \neq \boldsymbol{Q}, \\[2ex] \prod_{\lambda=1}^{g'} \mathfrak{a}^{\sigma_\lambda}\left(\dfrac{\mathfrak{a}^{\sigma_\lambda}}{\mathfrak{a}^{\sigma_\lambda\rho}}\right)^{v_\lambda} & \text{if } m \text{ is odd .} \end{cases}$$

First suppose that m is even. Define $D_{\mathfrak{p}}(\Lambda; \mathfrak{c})$ as in 8.6. Let $H_{\mathfrak{c}}$ be the set of all ideals $\mathfrak{a}$ in K', prime to $N(\mathfrak{c}\mathfrak{b})$, satisfying the following condition:

(9.2.2) *There exists an element a of K such that* $\psi(\mathfrak{a}) = (a)$, $aa^\rho = 1$, *and* $(aN(\mathfrak{a})^{m/2}, N(\mathfrak{a})) \in D_{\mathfrak{p}}(\Lambda; \mathfrak{c})$ *for every prime ideal* $\mathfrak{p}$ *in F dividing* $\mathfrak{c}\mathfrak{b}$.

By 8.9 and 8.10, there exists an integral ideal $\mathfrak{e}$ in F such that $\mathfrak{e} \subset \mathfrak{c}$ and $D_{\mathfrak{p}}(\mathfrak{e}) \subset D_{\mathfrak{p}}(\Lambda; \mathfrak{c})$ for every $\mathfrak{p}$. Let x be an element of K' such that $x \equiv 1 \bmod^\times (N(\mathfrak{e}))$. Put $a = 1$ or $a = \prod_{\lambda=1}^{g'} (x^{\sigma_\lambda}/x^{\sigma_\lambda\rho})^{v_\lambda}$ according as $K' = \boldsymbol{Q}$ or $K' \neq \boldsymbol{Q}$. By [FI, 4.5], a belongs to K. Further we have $aa^\rho = 1$, $a \equiv N(x) \equiv 1 \bmod^\times(N(\mathfrak{e}))$. Hence $(x) \in H_{\mathfrak{c}}$. This shows that $H_{\mathfrak{c}}$ is an ideal-group in K', whose conductor is a divisor of $N(\mathfrak{e})$.

Next suppose that m is odd. Let $H_{\mathfrak{c}}$ be the set of all ideals $\mathfrak{a}$ in K', prime to $N(\mathfrak{c})$, satisfying the following condition:

(9.2.3) *There exists an element a of K such that* $\psi(\mathfrak{a}) = (a)$, $aa^\rho = N(\mathfrak{a})$,

$a \equiv 1 \bmod^\times \mathfrak{c}.$

Then $H_\mathfrak{c}$ is an ideal-group in K', whose conductor is a divisor of $N(\mathfrak{c})$.

9.3. Remark. Suppose that m is even, and the following conditions are satisfied:

(9.3.1) *For every prime ideal $\mathfrak{p}$ in F dividing $\mathfrak{d}$, $V_\mathfrak{p}$ has a non-trivial kernel subspace with respect to ζT (cf. [10, 4.3]), and $\mathfrak{M}_\mathfrak{p}$ is a maximal $\mathfrak{r}_\mathfrak{p}$-lattice in $V_\mathfrak{p}$ (cf. [10, 2.10]).*

(9.3.2) $(\mathfrak{c}, \mathfrak{d}, 2) = (1).$

Then by 8.9 and 8.11, we have $D_\mathfrak{p}(\Lambda; \mathfrak{c}) = D_\mathfrak{p}(\mathfrak{c})$. Therefore the condition (9.2.2) can be replaced by the following one:

(9.3.3) *There exists an element a of K such that $\psi(\mathfrak{a}) = (a)$, $aa^\rho = 1$, $N(\mathfrak{a}) \equiv a \equiv 1 \bmod^\times \mathfrak{c}$.*

If $K' = \boldsymbol{Q}$, we can simplify still more. Suppose that $\mathfrak{M}_\mathfrak{p}$ is a maximal $\mathfrak{r}_\mathfrak{p}$-lattice in $V_\mathfrak{p}$ for all the prime ideals $\mathfrak{p}$ ramified in K. Then, by 8.9 and 8.12, we have (without assuming (9.3.1) and (9.3.2))

$$H_\mathfrak{c} = \{(a) \mid a \in \boldsymbol{Q}, \; |a| \equiv 1 \bmod^\times \mathfrak{c} \cap \boldsymbol{Q}\}.$$

Therefore, if c is the smallest positive integer divisible by $\mathfrak{c}$, the cyclotomic field $\boldsymbol{Q}(e^{2\pi i/c})$ is the class-field corresponding to $H_\mathfrak{c}$.

9.4. Theorem. *Let the notation be as in 9.1 and 9.2. Let $\mathfrak{Q}_u$ be a generic member of $\Sigma^*(K, \Phi, \rho; T, \mathfrak{M}; \{x_i\})$ over $\boldsymbol{Q}$, and k_0 the algebraic closure of $\boldsymbol{Q}$ in the field of moduli of $\mathfrak{Q}_u$. Then k_0 is the class-field over K' corresponding to the ideal-group $H_\mathfrak{c}$ in K'.*

As remarked at the end of 9.1, *the field of all automorphic functions on $\mathfrak{K}$ with respect to $\Gamma(T, \mathfrak{c}^{-1}\mathfrak{M}/\mathfrak{M})$ has a model defined over k_0.* We shall prove this theorem in the following 9.5–9.8. In the case $\mathfrak{c} = \mathfrak{r}$, the assertion of 9.4 is included in [FI, 5.10], under the assumption that $\mathfrak{M}$ is maximal or of type (U).

9.5. Let $\mathfrak{Q}_z$ be a member of $\Sigma^*(T, \mathfrak{M}; \{x_i\})$, and σ an isomorphism of C into itself which leaves every element of K' invariant. By [FI, 5.2], $\mathfrak{Q}_z^\sigma$ is isomorphic to a member of $\Sigma^*(cT, \mathfrak{N}; \{y_i\})$ for a totally positive element c of K, an $\mathfrak{r}$-lattice $\mathfrak{N}$ in V, and m elements $y_1, \cdots, y_m$ in V. Since $\mathfrak{h}(\mathfrak{c}, A_z) = \sum_{i=1}^m \theta_z(\mathfrak{r})t_i(z)$, we have $\mathfrak{h}(\mathfrak{c}, A_z^\sigma) = \sum_{i=1}^m \theta_z^\sigma(\mathfrak{r})t_i(z)^\sigma$, so that $\mathfrak{c}^{-1}\mathfrak{N} = \mathfrak{N} + \mathfrak{r}y_1 + \cdots + \mathfrak{r}y_m$.

If $\mathfrak{k}_z$ is the field of moduli of $\mathfrak{Q}_z$, it is clear that $\mathfrak{k}_z^\sigma$ is the field of moduli of $\mathfrak{Q}_z^\sigma$. Suppose now that $\mathfrak{Q}_z$ is a generic member of our family over $\boldsymbol{Q}$. Then we have $\dim_{\boldsymbol{Q}} \mathfrak{k}_z^\sigma = \dim_{\boldsymbol{Q}} \mathfrak{k}_z = \dim \mathfrak{K}$, so that $\mathfrak{Q}_z^\sigma$ is isomorphic to a generic member of $\Sigma^*(cT, \mathfrak{N}; \{y_i\})$. From this we can conclude that $\mathfrak{V}(T, \mathfrak{M}; \{x_i\})^\sigma = \mathfrak{V}(cT, \mathfrak{N}; \{y_i\})$, by 2.4 and by an argument similar to the proof of [FI, 2.8]. If m is even, we may put $c = 1$ in view of [FI, 5.2].

9.6. Let us take M, N, $\{\alpha_i\}$ and e as in [FI, 4.10], reserving the choice of P for later discussion. Then define a polarized abelian variety $(A, \mathcal{C}, \theta)$ belonging to the CM-type $(M; \{\alpha_i\})$ and a member $\mathscr{P}_y = (A, \mathcal{C}, \theta_1)$ of $\Sigma(T, \mathfrak{M})$ as in [FI, 5.7].

9.7. PROPOSITION. *There exists a positive integer b and a Galois extension k of M' with the following property:*

Let $\mathfrak{p}$ be a prime ideal in M' prime to b, and $\mathfrak{P}$ a prime ideal in k dividing $\mathfrak{p}$. Let σ be an isomorphism of C into itself which induces on k a Frobenius automorphism for $\mathfrak{P}/\mathfrak{p}$. Let $\mathcal{Q}_y^\sigma$ be isomorphic to a member of $\Sigma^*(cT, \mathfrak{N}; \{y_i\})$ (cf. 9.5 above). Then there exists an element γ of $G(T)$ such that $\mathfrak{N} \supset \mathfrak{M}c\gamma$, $[\mathfrak{N}/\mathfrak{M}c\gamma] = N_{M/K}(\prod_j \mathfrak{p}^{\beta_j})$, $\mu(\gamma)c = N(\mathfrak{p})$, $x_i\gamma \equiv y_i \bmod \mathfrak{N}$.

Let $\mathfrak{o}$, $\mathscr{P}' = (A', \mathcal{C}', \theta')$, μ and f be as in the proof of [FI, 5.8]. Taking $f \cdot N(c)$ in place of f, we repeat the argument of that proof. Then we find an isogeny λ of $\mathscr{P}'$ to $\mathscr{P}'^\sigma$ such that:

(9.7.1) $\lambda^{-1}(Y^\sigma)$ is algebraically equivalent to $N(\mathfrak{p})Y$ for a basic polar divisor Y in $\mathcal{C}'$;

(9.7.2) Ker (λ) is isomorphic to $\mathfrak{o}/(\prod_j \mathfrak{p}^{\beta_j})$ as $\mathfrak{r}$-module;

(9.7.3) $\lambda t_0 = t_0^\sigma$,

where t_0 is a point on A' such that $\theta'(\mathfrak{o})t_0 = \{t \in A' \mid fN(c)t = 0\}$. Further we obtain an isogeny λ_0 of $\mathscr{P}_y$ to $\mathscr{P}_y^\sigma$ such that $\mu^\sigma\lambda = \lambda_0\mu$. Let $t \in \mathfrak{h}(c, A)$. Take an element u on A' so that $\mu(u) = t$. Then we have $\mu(\theta'(c)u) = 0$, so $f\theta'(c)u = 0$. Hence there exists an element $a \in \mathfrak{o}$ such that $\theta'(a)t_0 = u$. We get therefore

$$\lambda_0 t = \lambda_0\mu\theta'(a)t_0 = \mu^\sigma\lambda\theta'(a)t_0 = \mu^\sigma\theta'(a)^\sigma\lambda t_0 = \mu^\sigma\theta'(a)^\sigma t_0^\sigma = t^\sigma\,.$$

Namely we have

(9.7.4) $\lambda_0 t = t^\sigma$ for every $t \in \mathfrak{h}(c, A)$.

Now the isogeny λ_0 corresponds to a K-linear automorphism γ of V. As in the proof of [FI, 5.8], we see that $\gamma \in G(T)$, $\mathfrak{M}c\gamma \subset \mathfrak{N}$, $[\mathfrak{N}/\mathfrak{M}c\gamma] = N_{M/K}(\prod_j \mathfrak{p}^{\beta_j})$, $\mu(\gamma)c = N(\mathfrak{p})$. From (9.7.4), we obtain $x_i\gamma \equiv y_i \bmod \mathfrak{N}$. This completes the proof.

9.8. Let $\mathcal{Q}_u$ be a generic member of $\Sigma^*(T, \mathfrak{M}; \{x_i\})$, and k_0 the algebraic closure of $\boldsymbol{Q}$ in the field of moduli of $\mathcal{Q}_u$. Let k_c be the class-field over K' corresponding to the ideal-group H_c. We are going to prove $k_0 = k_c$. We choose M, N, $\{\alpha_i\}$ and e in 9.6, by taking the composite $k_0 k_c$ as the field P in [FI, 4.10]. Let $\mathscr{P}_y$ be as in 9.6. We can find such a $\mathscr{P}_y$ so that the following assertion is true.

(9.8.1) Let σ be an isomorphism of C into itself over K'. If $\mathcal{Q}_y^\sigma$ is isomorphic to a member of $\Sigma^*(cT, \mathfrak{N}; \{y_i\})$, then $\mathcal{V}(T, \mathfrak{M}; \{x_i\})^\sigma = \mathcal{V}(cT, \mathfrak{N}; \{y_i\})$.

This can be proved by the same technique as in the proof of [FI, 5.10].[9]

Let $\mathfrak{x}$ be an ideal in K', prime to $N(c\mathfrak{b})$, and let

$$\tau = \left(\frac{k_c/K'}{\mathfrak{x}} \right) .$$

Since M' and k_c are linearly disjoint over K', we can extend τ to an isomorphism σ of C into itself over M'. Let $\mathfrak{Q}_y^\sigma$ be isomorphic to a member of $\Sigma^*(cT, \mathfrak{N}; \{y_i\})$. By (9.8.1), we have

$$(9.8.2) \qquad \mathcal{V}(T, \mathfrak{M}; \{x_i\})^\sigma = \mathcal{V}(cT, \mathfrak{N}; \{y_i\}) .$$

If m is even, we may put $c = 1$ (9.5). By the consideration of 3.3 and 3.4, we see that, if m is even, two c-frames $\mathfrak{M}^* = (\mathfrak{M}; x_1, \cdots, x_m)$ and $\mathfrak{N}^* = (\mathfrak{N}; y_1, \cdots, y_m)$ belong to the same genus in the sense of 8.7; if m is odd, two oriented c-frames $\mathfrak{M}^* = (\mathfrak{M}; 1; x_1, \cdots, x_m)$ and $\mathfrak{N}^* = (\mathfrak{N}; c; y_1, \cdots, y_m)$ belong to the same genus in the sense of 8.17.

Let k and b be as in 9.7, and R be a Galois extension of K containing k, k_0 and k_c. We can find a prime ideal q of M', not dividing $bN(c\mathfrak{b})$, and a prime factor $\mathfrak{Q}$ of q in R such that σ induces on R a Frobenius automorphism for $\mathfrak{Q}/q$. Put $\mathfrak{y} = N_{M'/K'}(q)$. By 9.7 and [FI, 4.7], there exists an element γ of $G(T)$ such that

$$(9.8.3) \qquad \mathfrak{N} \supset \mathfrak{M}\gamma ,$$

$$(9.8.4) \qquad [\mathfrak{N}/\mathfrak{M}\gamma] = \begin{cases} N(\mathfrak{y})^{m/2} & \text{if } m \text{ is even and } K' = \boldsymbol{Q} , \\[2ex] N(\mathfrak{y})^{m/2} \prod_{\lambda=1}^{g'} \left(\dfrac{\mathfrak{y}^{\sigma_\lambda}}{\mathfrak{y}^{\sigma_\lambda \rho}} \right)^{v_\lambda} & \text{if } m \text{ is even and } K' \neq \boldsymbol{Q} , \\[2ex] N(\mathfrak{y})^{(m-1)/2} \prod_{\lambda=1}^{g'} \mathfrak{y}^{\sigma_\lambda} \left(\dfrac{\mathfrak{y}^{\sigma_\lambda}}{\mathfrak{y}^{\sigma_\lambda \rho}} \right)^{v_\lambda} & \text{if } m \text{ is odd} , \end{cases}$$

$$(9.8.5) \qquad \mu(\gamma)c = N(\mathfrak{y}) ,$$

$$(9.8.6) \qquad x_i\gamma \equiv y_i \mod \mathfrak{N} .$$

If $\mathfrak{p}$ is a prime ideal in F, dividing $c\mathfrak{b}$, then $\mathfrak{p}$ is prime to $N(\mathfrak{y})$, so that (9.8.3) and (9.8.4) imply $\mathfrak{N}_\mathfrak{p} = \mathfrak{M}_\mathfrak{p}\gamma$. Therefore, if m is even, $d(\mathfrak{M}^*/\mathfrak{N}^*)$ is represented by $(\det(\gamma), N(\mathfrak{y}))$, and if m is odd, $i(\mathfrak{M}^*/\mathfrak{N}^*)$ is represented by $\kappa(\gamma)$ (cf. 8.7, 8.17).

---

[9] To prove this fact, we have to formulate and prove propositions analogous to [FI, 2.6 and 2.8]. This can be done by following step by step the procedure of the proofs of the original propositions. In particular, we know that $\mathcal{V}(T, \mathfrak{M}; \{x_i\}) = \mathcal{V}(T', \mathfrak{M}'; \{x_i'\})$ if and only if there exists an element α of $GL(V, K)$ such that $T'(x\alpha, y\alpha) = T(x, y)$, $\mathfrak{M}\alpha = \mathfrak{M}'$, $x_i\alpha \equiv x_i' \mod \mathfrak{M}'$, under the condition $\nu(T, \mathfrak{M}) = \nu(T', \mathfrak{M}')$, $N(T, \mathfrak{M}) = N(T', \mathfrak{M}')$, $e(T, \mathfrak{M}) = e(T', \mathfrak{M}')$ (cf. [FI, 2.8]).

By our choice of q, we have

$$\left(\frac{k_c/K'}{\mathfrak{x}}\right) = \left(\frac{k_c/K'}{\mathfrak{y}}\right),$$

so that $\mathfrak{x}^{-1}\mathfrak{y} \in H_c$. Hence we see that $\sigma = 1$ on $k_c M' \Leftrightarrow \mathfrak{x} \in H_c \Leftrightarrow \mathfrak{y} \in H_c$. Suppose that m is even. Let $\mathfrak{y} \in H_c$. Then there exists an element a of K such that $\psi(\mathfrak{y}) = (a)$, $aa^\rho = 1$, $(aN(\mathfrak{y})^{m/2}, N(\mathfrak{y})) \in D_\mathfrak{p}(\Lambda; c)$ for every $\mathfrak{p} \mid c\mathfrak{b}$, where $\psi(\mathfrak{y})$ is as in (9.2.1). Put $w = a \cdot \det(\gamma)^{-1}N(\mathfrak{y})^{m/2}$. Then $ww^\rho = 1$, $[\mathfrak{N}/\mathfrak{M}] = \mathfrak{x}w$, $d(\mathfrak{N}*/\mathfrak{M}*) = f(w)$. By 8.8 and footnote 9, we have $\mathcal{V}(T, \mathfrak{M}; \{x_i\}) = \mathcal{V}(T, \mathfrak{N}; \{y_i\})$, so that $\sigma = 1$ on $k_0 M'$, in view of (9.8.2), since k_0 is the smallest field of definition for $\mathcal{V}(T, \mathfrak{M}; \{x_i\})$. We see easily that our argument is reversible. Therefore, if $\sigma = 1$ on $k_0 M'$, then $\mathfrak{y} \in H_c$, and hence $\sigma = 1$ on $k_c M'$. Since M' and $k_0 k_c$ are linearly disjoint over K', we obtain $k_0 = k_c$.

Next suppose that m is odd. Put $l = (m - 1)/2$. Let $\mathfrak{y} \in H_c$. Then there exists an element a of K such that $\psi(\mathfrak{y}) = (a)$, $aa^\rho = N(\mathfrak{y})$, $a \equiv 1 \bmod^\times c$. Put $w = \kappa(\gamma)^{-1}a$. Then $c^{-l}[\mathfrak{N}/\mathfrak{M}] = \mathfrak{x}w$, $ww^\rho = c$, $i(\mathfrak{N}*/\mathfrak{M}*) = h(w)$. By 8.18, we have $\mathcal{V}(T, \mathfrak{M}; \{x_i\}) = \mathcal{V}(cT, \mathfrak{N}; \{y_i\})$, so that $\sigma = 1$ on $k_0 M'$. By following the reverse argument, we see that, if $\sigma = 1$ on $k_0 M'$, then $\sigma = 1$ on $k_c M'$. Hence we get $k_0 = k_c$. Our theorem is thereby proved.

9.9. As before, let $\mathfrak{Q}_u$ be a generic member of $\Sigma^*(T, \mathfrak{M}; \{x_i\})$ over $\boldsymbol{Q}$. Let $\mathfrak{R}$ (resp. $\mathfrak{R}_1$) be the field of moduli of $\mathfrak{Q}_u$ (resp. $\mathscr{P}_u$). By 1.11, $\mathfrak{R}_c$ is a Galois extension of $\mathfrak{R}_1$. Let $\mathscr{G}$ be the Galois group of $\mathfrak{R}_c$ over $\mathfrak{R}_1$. We are going to give an analysis of $\mathscr{G}$, though it is not so precise as 6.8 and 7.2. Put, for every prime ideal $\mathfrak{p}$ in F dividing c,

$$\begin{aligned}
\Gamma_\mathfrak{p}(\mathfrak{M}) &= \{\alpha \in G_\mathfrak{p}(T) \mid \mathfrak{M}_\mathfrak{p}\alpha = \mathfrak{M}_\mathfrak{p}\}, \\
\Gamma_\mathfrak{p}(\mathfrak{M}; c) &= \{\alpha \in \Gamma_\mathfrak{p}(\mathfrak{M}) \mid \mathfrak{M}_\mathfrak{p}(\alpha - 1) \subset c_\mathfrak{p}\mathfrak{M}_\mathfrak{p}\}, \\
\mathfrak{G}(\mathfrak{M}; c) &= \prod_{\mathfrak{p}\mid c} [\Gamma_\mathfrak{p}(\mathfrak{M})/\Gamma_\mathfrak{p}(\mathfrak{M}; c)]\,; \\
\Delta &= \{\zeta 1_m \mid \zeta \text{ is a root of unity in } K\}\,.
\end{aligned}$$

Let $\mathfrak{G}_0(\mathfrak{M}; c)$ be the subgroup of $\mathfrak{G}(\mathfrak{M}; c)$ consisting of the elements $(\alpha_\mathfrak{p})$ satisfying

(9.9.1) There exists a rational integer e such that $\mu(\alpha_\mathfrak{p}) \equiv e \bmod c_\mathfrak{p}$ for every prime factor $\mathfrak{p}$ of c.

Let Δ' be the image of Δ by an obvious homomorphism of Δ into $\mathfrak{G}_0(\mathfrak{M}; c)$. Then we get, by the same procedure as in 6.7, an isomorphism of $\mathscr{G}$ into $\mathfrak{G}_0(\mathfrak{M}; c)/\Delta'$. Let $\mathfrak{U}(\mathfrak{M}; c)$ be the subgroup of $\mathfrak{G}_0(\mathfrak{M}; c)$ consisting of the elements represented by the elements of $\Gamma(T, \mathfrak{M})$. Then the image of $\mathscr{G}$ in $\mathfrak{G}_0(\mathfrak{M}; c)/\Delta'$ contains $\mathfrak{U}(\mathfrak{M}; c)/\Delta'$; and an element σ of $\mathscr{G}$ corresponds to an element of $\mathfrak{U}(\mathfrak{M}; c)/\Delta'$ if and only if σ leaves invariant every element of $k_0\mathfrak{R}_1$, where k_0 is as in 9.4. Furthermore, if we put

$$\mathfrak{S}(\mathfrak{M}; c) = \begin{cases} \{(\alpha_\mathfrak{p}) \in \mathfrak{G}_0(\mathfrak{M}; c) \mid (\det(\alpha_\mathfrak{p}), \mu(\alpha_\mathfrak{p})) \in D_\mathfrak{p}(\Lambda; c)\} & (m:\ \text{even}), \\ \{(\alpha_\mathfrak{p}) \in \mathfrak{G}_0(\mathfrak{M}; c) \mid \kappa(\alpha_\mathfrak{p}) \equiv 1 \bmod c_\mathfrak{p} \mathfrak{r}_\mathfrak{p}\} & (m:\ \text{odd}), \end{cases}$$

then $\mathfrak{S}(\mathfrak{M}; c) \subset \mathfrak{U}(\mathfrak{M}; c)$ on account of [10, 5.12].

10. Class-field as bottom field

10.1. Let F and $\mathfrak{g}$ be as in 8.1. Let D be a quaternion algebra over F. We denote by ι the canonical involution of D, and put $N(x) = xx^\iota$ for $x \in D$. Let $\mathfrak{o}$ be a maximal order in D, and $\mathfrak{a}$ an integral two-sided $\mathfrak{o}$-ideal. We put

$$\Gamma(\mathfrak{o}; \mathfrak{a}) = \{x \in \mathfrak{o} \mid N(x) = 1,\ x \equiv 1 \bmod \mathfrak{a}\}\ .$$

We denote by $N(\mathfrak{a})$ the ideal in F generated by $N(x)$ for all $x \in \mathfrak{a}$.

10.2. PROPOSITION. *Suppose that D is indefinite. Let $\mathfrak{o}$ and $\mathfrak{o}'$ be maximal orders in D. Let $\mathfrak{a}$ (resp. $\mathfrak{a}'$) be an integral two-sided $\mathfrak{o}$-ideal (resp. $\mathfrak{o}'$-ideal). Suppose that $N(\mathfrak{a}) = N(\mathfrak{a}')$. Then $\Gamma(\mathfrak{o}; \mathfrak{a}) = \Gamma(\mathfrak{o}'; \mathfrak{a}')$ if and only if $\mathfrak{o} = \mathfrak{o}'$.*

If $\mathfrak{o} = \mathfrak{o}'$, we have $\mathfrak{a} = \mathfrak{a}'$, so that our assertion is trivial. Conversely, suppose that $\Gamma(\mathfrak{o}; \mathfrak{a}) = \Gamma(\mathfrak{o}'; \mathfrak{a}')$. Put $D_\mathfrak{p} = D \otimes_F F_\mathfrak{p}$ and $\mathfrak{b}_\mathfrak{p} = \mathfrak{g}_\mathfrak{p} \mathfrak{b}$ for every prime ideal $\mathfrak{p}$ in F and for every $\mathfrak{g}$-lattice $\mathfrak{b}$ in D. It is sufficient to prove $\mathfrak{o}_\mathfrak{p} = \mathfrak{o}'_\mathfrak{p}$ for every $\mathfrak{p}$. If $D_\mathfrak{p}$ is a division algebra, this follows from the uniqueness of maximal order. Therefore, we consider the case $D_\mathfrak{p} = M_2(F_\mathfrak{p})$. Then we may assume that $\mathfrak{o}_\mathfrak{p} = M_2(\mathfrak{g}_\mathfrak{p})$ and $\mathfrak{o}'_\mathfrak{p} = \mathfrak{g}_\mathfrak{p} e_{11} + \mathfrak{p}^\lambda e_{12} + \mathfrak{p}^{-\lambda} e_{21} + \mathfrak{g}_\mathfrak{p} e_{22}$ for an integer λ, where the e_{ij} are the usual matrix units. Now we have $\mathfrak{a}_\mathfrak{p} = \mathfrak{p}^\mu \mathfrak{o}_\mathfrak{p}$, $\mathfrak{a}'_\mathfrak{p} = \mathfrak{p}^\mu \mathfrak{o}'_\mathfrak{p}$ for an integer $\mu \geqq 0$. Put

$$\Gamma_\mathfrak{p} = \{x \in \mathfrak{o}_\mathfrak{p} \mid N(x) = 1,\ x \equiv 1 \bmod \mathfrak{a}_\mathfrak{p}\}\ ,$$
$$\Gamma'_\mathfrak{p} = \{x \in \mathfrak{o}'_\mathfrak{p} \mid N(x) = 1,\ x \equiv 1 \bmod \mathfrak{a}'_\mathfrak{p}\}\ .$$

Let $x \in \Gamma_\mathfrak{p}$. By Eichler's approximation theorem [1, Satz 5], for every positive integer n, there exists an element y_n of $\mathfrak{o}$ such that $N(y_n) = 1$, $y_n \equiv x \bmod \mathfrak{p}^n \mathfrak{a}_\mathfrak{p}$, and $y_n \equiv 1 \bmod \mathfrak{a}_\mathfrak{q}$ for all prime ideals $\mathfrak{q}$ in F, other than $\mathfrak{p}$. Then $y_n \in \Gamma(\mathfrak{o}; \mathfrak{a}) = \Gamma(\mathfrak{o}'; \mathfrak{a}') \subset \Gamma'_\mathfrak{p}$. Since $\Gamma'_\mathfrak{p}$ is closed by $\mathfrak{p}$-topology, we have $x = \lim y_n \in \Gamma'_\mathfrak{p}$, so that $\Gamma_\mathfrak{p} \subset \Gamma'_\mathfrak{p}$. If $a \in \mathfrak{p}^\mu$, we have $1 + a e_{12} \in \Gamma_\mathfrak{p}$, hence $1 + a e_{12} \in \Gamma'_\mathfrak{p}$. By our definition of $\Gamma'_\mathfrak{p}$, we have $1 + a e_{12} \equiv 1 \bmod \mathfrak{p}^\mu \mathfrak{o}'_\mathfrak{p}$, so that $a e_{12} \in \mathfrak{p}^\mu \mathfrak{o}'_\mathfrak{p}$. This shows that $\mathfrak{p}^\mu e_{12} \subset \mathfrak{p}^\mu \mathfrak{o}'_\mathfrak{p}$, hence $e_{12} \in \mathfrak{o}'_\mathfrak{p}$. Similarly $e_{21} \in \mathfrak{o}'_\mathfrak{p}$. Therefore λ should be 0. It follows that $\mathfrak{o}'_\mathfrak{p} = \mathfrak{o}_\mathfrak{p}$. This completes the proof.

10.3. PROPOSITION. *Let $\mathfrak{o}, \mathfrak{o}', \mathfrak{a}, \mathfrak{a}'$ be as in 9.2. Let the symbol $\{\mathfrak{o} : \mathfrak{o}'\}$ be as in [12, 3.3]. Suppose that D is indefinite and $N(\mathfrak{a}) = N(\mathfrak{a}')$. Then $\{\mathfrak{o} : \mathfrak{o}'\} = 1$ if and only if there exists an element b of D such that $N(b)$ is totally posi-*

tive and $b\Gamma(\mathfrak{o}; \mathfrak{a})b^{-1} = \Gamma(\mathfrak{o}'; \mathfrak{a}')$.

By [12, 3.2, 3.4], if $\{\mathfrak{o} : \mathfrak{o}'\} = 1$, there exists an element b of D such that $N(b)$ is totally positive and $b\mathfrak{o}b^{-1} = \mathfrak{o}'$. Then we have $b\mathfrak{a}b^{-1} = \mathfrak{a}'$, hence $b\Gamma(\mathfrak{o}; \mathfrak{a})b^{-1} = \Gamma(\mathfrak{o}'; \mathfrak{a}')$. Conversely, suppose that $b\Gamma(\mathfrak{o}; \mathfrak{a})b^{-1} = \Gamma(\mathfrak{o}'; \mathfrak{a}')$ and $N(b)$ is totally positive for some $b \in D$. Since $b\Gamma(\mathfrak{o}; \mathfrak{a})b^{-1} = \Gamma(b\mathfrak{o}b^{-1}; b\mathfrak{a}b^{-1})$ and $N(b\mathfrak{a}b^{-1}) = N(\mathfrak{a})$, we have $b\mathfrak{o}b^{-1} = \mathfrak{o}'$, in view of 10.2. By [12, 3.2, 3.4], we get $\{\mathfrak{o} : \mathfrak{o}'\} = 1$.

10.4. Now we assume that F is totally real. Let g be the degree of F over Q, and t the number of infinite places of F, where D is unramified. As in [12, 4.1], $\Gamma(\mathfrak{o}; \mathfrak{a})$ may be regarded as a discontinuous group operating on the product $\mathfrak{H}^t$ of t copies of the upper half plane. We shall mean by $Y(\mathfrak{o}; \mathfrak{a})$ a projective variety whose function-field over C is isomorphic (over C) to the field of all automorphic functions on $\mathfrak{H}^t$ with respect to $\Gamma(\mathfrak{o}; \mathfrak{a})$. Our purpose is to determine the bottom field for the variety $Y(\mathfrak{o}; \mathfrak{a})$ in the sense of [12, 2.1].

10.5. First suppose that $t = g$. Then D is totally indefinite, so that our group $\Gamma(\mathfrak{o}; \mathfrak{a})$ is nothing but $\Gamma(T; \mathfrak{o}/\mathfrak{a})$ investigated in § 5 (with $m = 1$, $L = D$, $V = D$). Therefore, by 6.8, we may assume that $Y(\mathfrak{o}; \mathfrak{a})$ is defined over $Q(\zeta)$ for a root of unity ζ. Moreover, by 6.8, if x is a generic point of $Y(\mathfrak{o}; \mathfrak{a})$ over $Q(\zeta)$, every automorphism of $Q(\zeta)$ can be extended to an automorphism of $Q(\zeta, x)$. It follows from this easily that Q *is the bottom field for* $Y(\mathfrak{o}; \mathfrak{a})$ *if* $t = g$.

10.6. THEOREM. *Let the notation be as in* 10.1, 10.4 *and* [12, 4.1–3]. *Let* $\mathfrak{o}$ *be a maximal order in* D *and* $\mathfrak{c}$ *an integral ideal in* F. *Suppose that* $0 < t < g$. *Then the bottom field for* $Y(\mathfrak{o}; \mathfrak{co})$ *is contained in* $k(D)$. *Moreover, let* $\mathfrak{x}$ *be an ideal in* F^*, *and let* σ *be an isomorphism of a field of definition for* $Y(\mathfrak{o}; \mathfrak{co})$ *into* C *such that*

$$\sigma = \left(\frac{k(D)/F^*}{\mathfrak{x}} \right) \qquad \text{on } k(D) \, .$$

Then $Y(\mathfrak{o}; \mathfrak{co})^\sigma$ *is birationally equivalent to* $Y(\mathfrak{o}'; \mathfrak{co}')$ *with a maximal order* $\mathfrak{o}'$ *in* D *such that* $\{\mathfrak{o} : \mathfrak{o}'\}$ *is the class of* $\prod_{\mu=1}^{v} \mathfrak{x}^{q_\mu}$ *modulo* $I(D/F)$.

Take a totally imaginary quadratic extension K of F satisfying the conditions of [12, 3.5]. We denote by $\mathfrak{r}$ the ring of integers in K. Let V, Φ and ξ be as in [12, 3.7]. We identify D with a subalgebra of the algebra of all K-linear endomorphisms of V. Let $\mathfrak{M} = \xi^{-1}(\mathfrak{o})$. By [10, 5.31], $\mathfrak{M}$ is a maximal $\mathfrak{r}$-lattice in V. By [12, 1.16] and [10, 2.6], we have

$$\Gamma(\mathfrak{o}; \mathfrak{co}) = \{\alpha \in G(\Phi) \,|\, \mathfrak{M}\alpha = \mathfrak{M}, \, \mathfrak{M}(1 - \alpha) \subset \mathfrak{cM}\} \, .$$

Therefore, if we take two elements x_1, x_2 of V so that $\mathfrak{c}^{-1}\mathfrak{M} = \mathfrak{M} + \mathfrak{r}x_1 + \mathfrak{r}x_2$, then we may adopt $\mathcal{O}(T, \mathfrak{M}, \{x_i\})$ as $Y(\mathfrak{o}; \mathfrak{co})$, where $T = \zeta\Phi$ for an element ζ of K such that $\zeta^\rho = -\zeta$ (cf. [12, 4.4]). Let K' and k_0 be as in [12, 4.4], and $\mathfrak{z}$ an ideal in K'. Let σ be an isomorphism of C into itself which induces $\left(\dfrac{k_0/K'}{\mathfrak{z}}\right)$ on k_0. By [FI, 5.10] and 9.5, we have $\mathcal{O}(T, \mathfrak{M})^\sigma = \mathcal{O}(T, \mathfrak{N})$ and $\mathcal{O}(T, \mathfrak{M}; \{x_i\})^\sigma = \mathcal{O}(T, \mathfrak{N}; \{y_i\})$ with a maximal $\mathfrak{r}$-lattice $\mathfrak{N}$ in V and elements y_1 and y_2 of V such that $\mathfrak{c}^{-1}\mathfrak{N} = \mathfrak{N} + \mathfrak{r}y_1 + \mathfrak{r}y_2$, and [12, 4.5.3] is satisfied with $\mathfrak{z}$ in place of $\mathfrak{a}$. Let $\mathfrak{o}' = \{\alpha \in D \mid \mathfrak{N}\alpha \subset \mathfrak{N}\}$. Then $\mathfrak{o}'$ satisfies [12, 4.5.5] with $\mathfrak{z}$ in place of $\mathfrak{a}$. We may consider $\mathcal{O}(T, \mathfrak{N}, \{y_i\})$ as $Y(\mathfrak{o}'; \mathfrak{co}')$. Therefore, by 10.3 and the same argument as in [12, 4.5], we see that the composite $K'k(D)$ contains the bottom field for $Y(\mathfrak{o}; \mathfrak{co})$. Then our theorem follows easily from the consideration in [12, 4.7].

10.7. Theorem. *Let the notation and the assumption be as in 10.6. Let $\mathfrak{p}_1, \cdots, \mathfrak{p}_s$ be all the prime ideals in F which are ramified in D, and $p_{\infty 1}, \cdots, p_{\infty t}$ be all the infinite places of F which are unramified in D. Let $\mathfrak{c}$ be an integral ideal in F. Suppose that:*

(i) *there is no automorphism of F, other than the identity mapping, which leaves invariant $\{\mathfrak{p}_1, \cdots, \mathfrak{p}_s; p_{\infty 1}, \cdots, p_{\infty t}\}$ as a whole;*

(ii) *for every maximal order $\mathfrak{o}$ in D, the group $\Gamma(\mathfrak{o}; \mathfrak{co})$ has no element of finite order other than ± 1.*

Then $k(D)$ is exactly the composite of F^* and the bottom field for $Y(\mathfrak{o}; \mathfrak{co})$, for every maximal order $\mathfrak{o}$ in D.

We note that D is spanned by the elements of $\Gamma(\mathfrak{o}; \mathfrak{co})$ over $\boldsymbol{Q}$. This follows easily from (for example) Eichler's approximation theorem [1, Satz 5]. Therefore, in view of 10.3, our theorem can be proved by the same argument as in the proof of [12, 4.9].

10.8. Remark. Let J be the set of positive integers n such that $[F(e^{2\pi i/n}) : F] = 2$, and D splits over $F(e^{2\pi i/n})$. It is clear that J is a finite set. If $\mathfrak{c}$ does not divide n for every $n \in J$, then the condition (ii) of 10.7 is satisfied. In particular, if $N(\mathfrak{c})$ is sufficiently large, or if $\mathfrak{c}$ is prime to n for every $n \in J$, the condition is satisfied.

Princeton University

References

1. M. Eichler, *Allgemeine Kongruenzklasseneinteilungen der Ideale einfacher Algebren über algebraischen Zahlkörpern und ihre L-Reihen*, J. Reine Angew. Math., 179 (1938), 227-251.
2. J. Igusa, *On the structure of a certain class of Kaehler varieties*, Amer. J. Math., 76 (1954), 669-678.

3. M. KUGA, Automorphic forms and fiber varieties, lecture note, Chicago University, 1964, to appear.

4. SÉMINAIRE H. CARTAN, E.N.S., 1957/1958, Fonctions automorphes.

5. G. SHIMURA, *On the theory of automorphic functions*, Ann. of Math., 70 (1959), 101-144.

6. ————, *On the zeta-functions of the algebraic curves uniformized by certain automorphic functions*, J. Math. Soc. Japan, 13 (1961), 275-331.

7. ————, *Arithmetic of alternating forms and quaternion hermitian forms*, J. Math. Soc. Japan, 15 (1963), 33-65.

8. ————, *On modular correspondences for* Sp (n, Z) *and their congruence relations*, Proc. Nat. Acad. Sci. U.S.A., 49 (1963), 824-828.

9. ————, *On analytic families of polarized abelian varieties and automorphic functions*, Ann. of Math., 78 (1963), 149-192.

10. ————, *Arithmetic of unitary groups*, Ann. of Math., 79 (1964). 369-409.

11. ————, *On the field of definition for a field of automorphic functions*, Ann. of Math., 80 (1964), 160-189, referred to as [FI].

12. ————, *Class-fields and automorphic functions*, Ann. of Math., 80 (1964), 444-463.

13. ———— and Y. TANIYAMA, Complex multiplication of abelian varieties and its applications to number theory, Publ. Math. Soc. Japan, No. 6, 1961.

14. A. WEIL, Variétés abéliennes et courbes algébriques, Hermann, Paris, 1948.

15. ————, *The field of definition of a variety*, Amer. J. Math., 78 (1956), 509-524.

16. W. L. BAILY and A. BOREL, *On the compactification of arithmetically defined quotients of bounded symmetric domains*, Bull. Amer. Math. Soc., 70 (1964), 588-593.

(Received May 4, 1964)

On the zeta funciton of a fibre variety
whose fibres are abelian varieties

Jointly with Michio Kuga

Annals of Mathematics, 82 (1965), 478-539

Introduction

Let Φ be an indefinite division quaternion algebra over the rational number field Q, which is, by definition, a division algebra over Q such that $\Phi \otimes R$ is isomorphic to the total matric algebra $M_2(R)$ of degree 2 over the real number field R. Let $\mathfrak{o}$ be a maximal order in Φ, i.e., a maximal one among the subrings of Φ which are free Z-modules of rank 4 over Z, where Z is the ring of rational integers. Regarding Φ as a subring of $M_2(R)$, let

$$\Gamma_b = \{\gamma \in \mathfrak{o} \mid \det(\gamma) = 1, \quad \gamma \equiv 1 \bmod b\mathfrak{o}\}$$

for every positive integer b.[1] It is known that Γ_b, as a discrete subgroup of $SL_2(R)$, operates naturally on the upper half plane $\mathfrak{H}$, and the quotient $\mathfrak{H}/\Gamma_b$ is compact. Moreover, $\mathfrak{H}/\Gamma_b$ may be regarded as a parameter space of a family of abelian varieties [12]. In fact, for every $z \in \mathfrak{H}$, define a lattice D_z in the two-dimensional complex vector space C^2 by $D_z = \mathfrak{o} \cdot \begin{pmatrix} z \\ 1 \end{pmatrix}$. Then the complex torus C^2/D_z has a structure of abelian variety. Let A_z be a projective embedding of C^2/D_z. Two points z and y of $\mathfrak{H}$ are equivalent with respect to Γ_b if and only if A_z and A_y are isomorphic as abelian varieties with certain additional structures of polarization, endomorphisms, and points of finite order. Assuming $b > 2$, we can construct a fiber space $W_{m,b}$, of which the base is $\mathfrak{H}/\Gamma_b$, and each fibre is the product of m copies of A_z, where m is a fixed non-negative integer. If $m = 0$, we may identify $W_{m,b}$ with $\mathfrak{H}/\Gamma_b$.[2] By [8] we know that $W_{m,b}$ has a structure of projective variety. Furthermore we can show that $W_{m,b}$ has a projective model defined over a cyclotomic field $Q(\zeta)$ with $\zeta = e^{2\pi i/b}$ [17, II]. This model can be characterized by certain properties, up to birational biregular isomorphisms over $Q(\zeta)$. For simplicity, we denote the model itself by $W_{m,b}$. Now the factor group Γ_1/Γ_b defines naturally a group of birational automorphisms

---

* Partial support by N.S.F. grant GP-3454 for the second named author.

[1] In the text we shall treat the problem more generally by taking a two-sided ideal $\mathfrak{b}$ of $\mathfrak{o}$ in place of $b\mathfrak{o}$.

[2] By virtue of a theorem of Eichler, any two maximal orders in Φ are transformed to each other by an inner automorphism $x \to axa^{-1}$ with an element a of Φ with $\det(a) > 0$. Therefore the variety $W_{m,b}$ depends essentially on Φ, m, b, and not on the choice of $\mathfrak{o}$.

of $W_{m,b}$ defined over $Q(\zeta)$. For every complex valued character ψ of Γ_1/Γ_b, we can define an L-function, $L(s; W_{m,b}, Q(\zeta), \psi)$ of $W_{m,b}$ with respect to the character ψ, in the sense of Hasse-Weil. Our main purpose of the present paper is to determine $L(s; W_{m,b}, Q(\zeta), \psi)$ in the case where ψ has real values. Taking ψ to be the character of a regular representation of Γ_1/Γ_b, we get the zeta function of $W_{m,b}$ over $Q(\zeta)$ as a special case.

To state our result more explicitly, we have to introduce the Hecke ring in Φ, i.e., a ring generated by the double cosets $\Gamma_1\xi\Gamma_1$ with elements ξ of $\mathfrak{o}$ such that $\det(\xi) > 0$. Let G_b be the group of invertible elements of $\mathfrak{o}/b\mathfrak{o}$, and ρ a representation of G_b in a complex vector space U of finite dimension. For every positive integer k, let $\mathfrak{S}_k(\Gamma_1, \rho)$ denote the vector space of all holomorphic mappings $\mathfrak{f}$ of $\mathfrak{H}$ into U satisfying $\mathfrak{f}(\gamma(z)) = \rho(\gamma)\mathfrak{f}(z)(cz + d)^k$ for every $\gamma = \begin{pmatrix} * & * \\ c & d \end{pmatrix}$ $\in \Gamma_1$. Let $\Delta_b = \{\xi \in \mathfrak{o} \mid \det(\xi) > 0, (b, \det(\xi)) = 1\}$. If $\xi \in \Delta_b$, one can define the action $(\Gamma_1\xi\Gamma_1)_{k,\rho}$ of $\Gamma_1\xi\Gamma_1$ on $\mathfrak{S}_k(\Gamma_1, \rho)$ in the following way. Let $\Gamma_1\xi\Gamma_1 = \bigcup_{i=1}^{r} \Gamma_1\xi_i$ be a disjoint union, and let $\xi_i = \begin{pmatrix} * & * \\ c_i & d_i \end{pmatrix}$. For every $\mathfrak{f} \in \mathfrak{S}_k(\Gamma_1, \rho)$, we put

$$\mathfrak{f} \mid (\Gamma_1\xi\Gamma_1)_{k,\rho} = \det(\xi)^{k-1} \sum_{i=1}^{r} \rho(\xi_i)^{-1}\mathfrak{f}(\xi_i(z))(c_iz + d_i)^{-k}.$$

In this way we get a representation of the Hecke ring by complex linear transformations. Define a Dirichlet series $D(s; b, k, \rho)$ by

$$D(s; b, k, \rho) = \sum (\Gamma_1\xi\Gamma_1)_{k,\rho} \det(\xi)^{-s},$$

where the summation is taken over all the distinct double cosets $\Gamma_1\xi\Gamma_1$ with $\xi \in \Delta_b$. Then $D(s; b, k, \rho)$ can be expressed as an Euler product. Moreover, it can be continued holomorphically on the whole s-plane, and satisfies a functional equation [14, 15]. Taking Γ_b in place of Γ_1, we can define similarly a Hecke ring generated by $\Gamma_b\xi\Gamma_b$, the vector space $\mathfrak{S}_k(\Gamma_b)$ of holomorphic automorphic forms of weight k with respect to Γ_b (without any representation), the action $(\Gamma_b\xi\Gamma_b)_k$ of $\Gamma_b\xi\Gamma_b$ on $\mathfrak{S}_k(\Gamma_b)$, and a Dirichlet series (cf. 7, 9)

$$D(s; b, k) = \sum (\Gamma_b\xi\Gamma_b)_k \det(\xi)^{-s}$$

with an Euler product and functional equation. Then we can state our results roughly as follows.

THEOREM 1. *Let ψ be a character of Γ_1/Γ_b with real values, and ρ a representation of G_b whose character is induced from ψ.[3] When $m = 0$, suppose that ψ is essentially a character of $\Gamma_1/\{\pm 1\} \cdot \Gamma_b$. Then, up to a finite number of p-factors, $L(s; W_{m,b}, Q(\zeta), \psi)$ is equal to a product of the following form:*

---

[3] One may regard naturally Γ_1/Γ_b as a subgroup of G_b.

$$\prod_{j=0}^{2m} [Z(s - j; \mathbf{Q}(\zeta))Z(s - j - 1; \mathbf{Q}(\zeta))]^{r \cdot a(m, 2j, 0)}$$
$$\times \prod_{i=0}^{4m} \prod_{\nu=0}^{i} \det \left[D\left(s - \frac{i - \nu}{2}; b, \nu + 2, \rho \right) \right]^{(-1)^{i+1} a(m, i, \nu)}.$$

Here $Z(s; \mathbf{Q}(\zeta))$ is Dedekind's zeta function of $\mathbf{Q}(\zeta)$; the $a(m, i, \nu)$ are non-negative integers depending only on m, i, ν (cf. 6.2); r is the multiplicity of the identity character in ψ.

THEOREM 2. *If b is prime to the discriminant of Φ (cf. 1.2), one can find the model $W_{m,b}$ so as to be defined over $\mathbf{Q}$, and the zeta function of $W_{m,b}$ over $\mathbf{Q}$ is, up to a finite number of p-factors, equal to a product of the following form:*

$$\prod_{j=0}^{2m} [\zeta(s - j)\zeta(s - j - 1)]^{a(m, 2j, 0)}$$
$$\times \prod_{i=0}^{4m} \prod_{\nu=0}^{i} \det \left[D\left(s - \frac{i - \nu}{2}; b, \nu + 2 \right) \right]^{(-1)^{i+1} a(m, i, \nu)}.$$

Here $\zeta(s)$ is Riemann's zeta function, and the $a(m, i, \nu)$ are the same as in Theorem 1.

For a fixed m, the largest ν with non-vanishing $a(m, i, \nu)$ is $2m$.

THEOREM 3. *Let p be a prime number and α an element of $\mathfrak{o}$ such that $\det(\alpha) = p$. For every representation ρ of $G_b/\{\pm 1\}$, the absolute values of the characteristic roots of $(\Gamma_1 \alpha \Gamma_1)_{2,\rho}$ and $(\Gamma_b \alpha \Gamma_b)_2$ do not exceed $2p^{1/2}$ except for a finite number of exceptional p's. Moreover, if Weil's conjecture [22] about the zeta functions of algebraic varieties over finite fields is true, then, for every representation ρ of G_b and for every $k > 2$, the absolute values of the characteristic roots of $(\Gamma_1 \alpha \Gamma_1)_{k,\rho}$ and $(\Gamma_b \alpha \Gamma_b)_k$ do not exceed $2p^{(k-1)/2}$ except for a finite number of exceptional p's.*

For more precise statements, see 6.8, 6.11, 6.12, 7.7, 7.8, 7.10 in the text. These are generalizations or analogues of previously known results concerning the zeta function and L-functions of an algebraic curve uniformized by automorphic functions [4, 7, 9, 11, 14]. In view of the results of Eichler [24] and Shimizu [25], there are certain linear relations between our Dirichlet series $D(s; b, k, \rho)$ (or $D(s; b, k)$) and Hecke's Dirichlet series attached to cusp forms with respect to congruence subgroups of $\mathrm{SL}_2(\mathbf{Z})$. Therefore, assuming Weil's conjecture to be true, one may derive an estimate for the Fourier coefficients of certain cusp forms.

We shall now explain our method of proof by giving a summary of contents. First (§ 1), we recall briefly the theory of Hecke rings and operators, and prove an equality 1.17 which connects each Euler factor of $D(s; b, k, \rho)$ with the $(\Gamma_b \xi \Gamma_b)_k$. In § 2, we introduce a notion of fibre system of abelian varieties $\{W, V, \varphi\}$, of which $\{W_{m,b}, \mathfrak{H}/\Gamma_b, \varphi_{m,b}\}$, with the above $W_{m,b}$, $\mathfrak{H}/\Gamma_b$ and the natural projection $\varphi_{m,b}$ of $W_{m,b}$ to $\mathfrak{H}/\Gamma_b$, is an example. We call a correspondence X on the fibre variety W a fibre correspondence if it acts as an isogeny on each fibre (2.4). Then we give a formula (2. 7. 6) which expresses the number of fixed points of X as the sum of the numbers of fixed points of those isogenies which are actually endomorphisms of the fibres. Then § 3 concerns the family of abelian varieties $\{A_z \mid z \in \mathfrak{H}\}$ and the construction of the fiber variety $W_{m,b}$. Choosing a model defined over $Q(\zeta)$, we discuss the action of G_b on a generic fibre. The considerations of § 1–3 are put together in § 4. Every double coset $\Gamma_b \xi \Gamma_b$ defines naturally a fibre correspondence $X_m(\Gamma_b \xi \Gamma_b)$ on $W_{m,b} \times W_{m,b}$. We shall prove that $X_m(\Gamma_b \xi \Gamma_b)$ is defined over $Q(\zeta)$ (4.7). Then we define certain birational correspondences $Y_m(\eta)$ between $W_{m,b}$ and its transform $W_{m,b}^{\sigma}$ under an automorphism σ of $Q(\zeta)$, and prove various formulas satisfied by $X_m(\Gamma_b \xi \Gamma_b)$ and $Y_m(\eta)$. The main result of § 5 is the congruence relation 5.14, which is crucial in the whole theory. Let α be an element of $\mathfrak{o}$ such that det (α) is prime number p and $\mathfrak{p}$ a prime ideal of $Q(\zeta)$ dividing p. Then the congruence relation gives the decomposition of the reduction of $X_m(\Gamma_b \alpha \Gamma_b)$ modulo $\mathfrak{p}$ into a sum of two correspondences Z_α and Z_α^*. A formula of the same kind has been known for certain algebraic curves uniformized by automorphic functions [4, 11, 14]. In the present case, Z_α differs from the Frobenius p^{th} power correspondence only by a biregular isomorphism, and Z_α^* is a fiber correspondence whose action on each fibre is, up to an automorphism, the *adjoint isogeny* of the pth power isogeny. If $m > 0$, Z_α^* is far from the transpose of the Frobenius correspondence, though the former behaves similarly to the latter in a certain respect. Therefore our relation 5.14 reduces to the known relation if $m = 0$ (and if b is prime to the discriminant); but for $m > 0$, it provides a generalization which seems somewhat different from what might have been expected. Our principal aims are achieved in §§ 6, 7. First we restate, as Proposition 6.2, a main result of [8] which gives $I_0(X_m(\Gamma_b \xi \Gamma_b))$ as a sum of traces of $(\Gamma_b \xi \Gamma_b)_k$, where $I_0(X)$ denotes the intersection number of a correspondence X with the diagonal. Applying the formula 2.7 mentioned above to Z_α and Z_α^*, we get $I_0(Z_{\alpha'}^n) = I_0(Z_\alpha^{*n})$ for $n > 0$ (6.4). The congruence relation 5.14, together with these facts (6.2, 6.4) and the formula 1.17, connects the L-function of the reduced variety $\mathfrak{p}(W_{m,b})$ with the Euler p-factor of $D(s; b, k, \rho)$ (6.8). Taking the infinite product for all good p's we get the global L-function $L(s; W_{m,b}, Q(\zeta), \psi)$

(Th. 6.11). The estimate for the characteristic roots of Hecke operators is obtained easily from this result and the rough estimate 1.20. In § 7, we treat the case where b is prime to the discriminant of Φ. In this case some of the $Y_m(\eta)$ satisfy the criterion of Weil [23], with respect to the Galois group of $Q(\zeta)$ over Q, so that we can find a model for $W_{m,b}$ defined over $Q(7.4)$, whose zeta function over Q can be expressed in terms of $D(s; b, k)$ (Th. 7.10).

We should mention at this point the material contribution of Mikio Sato to this subject. In fact, following a suggestion of the first named author, Sato considered a fibre variety W_m' whose base is the upper half plane modulo a congruence subgroup of $SL_2(Z)$, and each fibre is the product, of m copies of an elliptic curve, modulo ± 1. Then he compared the number of rational points of the reduction of W_m' modulo p over the prime field with the traces of Hecke operators of degree p (computed by Selberg and Eichler), and observed that there should exist a certain identity between them, which led him to propose a product of several Hecke's Dirichlet series (analogous to the one given in 7.8) as the zeta function of W_m'. As far as we know, he has not published anything about this except a short note.[4] We are grateful to Mikio Sato for this important observation which inspired us to a great extent.

Notation. As usual Z, Q, R and C denote respectively the ring of rational integers, the rational number field, the real number field, and the complex number field. For an associative ring S with an identity element, we denote by $M_r(S)$ and $GL_r(S)$ the ring of $r \times r$ matrices with entries in S and the group of invertible elements of $M_r(S)$, respectively.

---

[4] A note (in Japanese), taken by Hijikata, in a lecture at University of Tokyo, Sugaku no ayumi, vol. 10, no. 2 (1963) 2-11. In this note, a result concerning the zeta function of W_m' is stated somewhat vaguely with neither proofs nor comments on possible difficulties. It seems that one may regard it as a sort of conjecture, because it is hardly feasible to complete the task with only the ideas explained there. To get the zeta function of W_m', it is indispensable to consider the Hecke operators of degree p^n (the ones analogous to our $(\Gamma_b \xi \Gamma_b)_k$ with $\det(\xi) = p^n$) for *all* positive integers n. But Sato gives only an idea for getting the identity between the number of rational points over *the prime field* and the traces of operators of degree p. His argument does not work well for the operators of degree p^n with $n > 1$ and larger finite fields. Actually he does not define any correspondence on the fibre variety; only the congruence relations, which are essentially the same as [11, p. 15, (13), (14), (15)], are employed. There is another delicate point which has not been considered in the note. It is well known that all the fixed points of Frobenius correspondence and its transpose are of multiplicity one. For the fibre variety, one is confronted by the fixed points with multiplicity greater than one arising from the "adjoint correspondence" (Z_α^* in our Introduction) of the Frobenius coorespondence. (One can hardly imagine a proof which uses only the fixed points of Frobenius correspondence but not of their adjoints.) Moreover, the fibres of W_m' are not abelian varieties but Kummer varieties. Therefore singular points may produce another type of difficulty. We hope that our method of proof in the present investigation will be helpful in overcoming these difficulties.

1. Automorphic forms and Hecke operators with respect to unit groups in an indefinite quaternion algebra

1.1. *Hecke ring.* Let G be an arbitrary group and Γ a subgroup of G. Let Δ be a subset of G containing Γ, closed under multiplication, satisfying the following condition.

$$(1.1.1) \qquad [\Gamma : (\alpha^{-1}\Gamma\alpha) \cap \Gamma] < \infty \qquad \textit{for every } \alpha \in \Delta .$$

Then for every $\alpha \in \Delta$, one has $\Gamma\alpha\Gamma = \bigcup_i \Gamma\alpha_i$ for a finite number of elements α_i. Let $R(\Gamma, \Delta)$ be the free $\boldsymbol{Z}$-module consisting of all the formal finite sums $\sum_\lambda c_\lambda \Gamma\alpha_\lambda\Gamma$ with $c_\lambda \in \boldsymbol{Z}$, $\alpha_\lambda \in \Delta$. The module $R(\Gamma, \Delta)$ is called *the Hecke ring associated to Γ and Δ*, when furnished with the law of multiplication defined as follows: if $\Gamma\alpha\Gamma = \bigcup_i \Gamma\alpha_i$ and $\Gamma\beta\Gamma = \bigcup_j \Gamma\beta_j$, then $(\Gamma\alpha\Gamma)\cdot(\Gamma\beta\Gamma) = \sum_\xi c_\xi(\Gamma\xi\Gamma)$, $c_\xi =$ the number of (i, j) such that $\Gamma\alpha_i\beta_j = \Gamma\xi$, where the summation is taken over all the $\Gamma\xi\Gamma \subset \Gamma\alpha\Gamma\beta\Gamma$. This law is associative (cf. [13], [8, Ch. III]). If $\Gamma\xi\Gamma = \bigcup_h \Gamma\xi_h$, one has

$$(1.1.2) \qquad \sum_{i,j} \Gamma\alpha_i\beta_j = \sum_{\xi,h} c_\xi \Gamma\xi_h$$

in the free $\boldsymbol{Z}$-module generated by all the $\Gamma\eta$ with $\eta \in \Delta$.

1.2. *Quaternion algebra.* Let Φ be an indefinite division quaternion algebra over $\boldsymbol{Q}$, i.e., a division algebra over $\boldsymbol{Q}$ such that $\Phi \otimes_Q \boldsymbol{R}$ is isomorphic to $M_2(\boldsymbol{R})$. Throughout the paper, we shall fix such a Φ and identify $\Phi \otimes_Q \boldsymbol{R}$ with $M_2(\boldsymbol{R})$. Then every element of Φ will be regarded as a real matrix of size 2. For every $\alpha = \begin{pmatrix} a & b \\ c & d \end{pmatrix}$ in $M_2(\boldsymbol{R})$, put

$$(1.2.1) \qquad \alpha' = \begin{pmatrix} d & -b \\ -c & a \end{pmatrix} .$$

Then $\alpha + \alpha' = \mathrm{tr}(\alpha)$, $\alpha\alpha' = \det(\alpha)$; and $\alpha \to \alpha'$ induces the main involution of Φ.

For every prime number p, let $\boldsymbol{Q}_p$ denote the field of p-adic numbers. It is well-known that $\Phi \otimes_Q \boldsymbol{Q}_p$ is a division algebra for only a finite number of p. We denote by $d(\Phi)$ the product of all such p, and call $d(\Phi)$ *the discriminant of* Φ.

1.3. *Unit groups and Hecke rings in Φ.* We fix, once for all, a maximal order $\mathfrak{o}$ in Φ (cf. Introduction and footnote 2). Let $\mathfrak{b}$ be an integral two-sided $\mathfrak{o}$-ideal, and b_0 the positive integer determined by

$$(1.3.1) \qquad \mathfrak{b} \cap \boldsymbol{Z} = b_0\boldsymbol{Z} .$$

We put

$$\Gamma_1 = \{\gamma \in \mathfrak{o} \mid \det(\gamma) = 1\} ,$$
$$\Gamma_\mathfrak{b} = \{\gamma \in \Gamma_1 \mid \gamma \equiv 1 \bmod \mathfrak{b}\} ,$$

$$\Delta_1 = \{\alpha \in \mathfrak{o} \mid \det(\alpha) > 0\},$$
$$\Delta_\mathfrak{b} = \{\alpha \in \Delta_1 \mid (\det(\alpha), b_0) = 1\}.$$

It can be easily verified that Γ_1 and Δ_1 resp. $\Gamma_\mathfrak{b}$ and $\Delta_\mathfrak{b}$ satisfy the condition (1.1.1). Hence we may consider the Hecke rings $R(\Gamma_1, \Delta_1)$ and $R(\Gamma_\mathfrak{b}, \Delta_\mathfrak{b})$. The ring $R(\Gamma_1, \Delta_1)$ is commutative [14, Prop. 1.7].

1.4. PROPOSITION. *If $\alpha \in \Delta_1$, the correspondence $\Gamma_1\alpha\Gamma_1 \supset \Gamma_1\xi \to \mathfrak{o}\xi$ is one-to-one between the right cosets $\Gamma_1\xi$ contained in $\Gamma_1\alpha\Gamma_1$ and the integral left $\mathfrak{o}$-ideals $\mathfrak{o}\xi$ such that $\mathfrak{o}/\mathfrak{o}\xi$ is $\mathfrak{o}$-isomorphic to $\mathfrak{o}/\mathfrak{o}\alpha$.*

This follows easily from [14, Prop. 1.6]. One should also notice that every left $\mathfrak{o}$-ideal is principal by virtue of Eichler's theorem [3, Satz 5]. With the notation of the above proposition, suppose, in particular, that $\det(\alpha) = p$ with a prime number p which does not divide $d(\Phi)$. Then there are exactly $p + 1$ integral left $\mathfrak{o}$-ideals c such that $[\mathfrak{o} : c] = p^2$; and they are in one-to-one correspondence with the right cosets $\Gamma_1\xi$ contained in $\Gamma_1\alpha\Gamma_1$.

1.5. PROPOSITION. *Let $\alpha \in \Delta_\mathfrak{b}$. Then the following assertions hold.*

(1.5.1) $$\Gamma_1\alpha\Gamma_1 = \Gamma_1\alpha\Gamma_\mathfrak{b} = \Gamma_\mathfrak{b}\alpha\Gamma_1.$$

(1.5.2) $$\Gamma_\mathfrak{b}\alpha\Gamma_\mathfrak{b} = \{\xi \mid \xi \in \Gamma_1\alpha\Gamma_1, \quad \xi \equiv \alpha \bmod \mathfrak{b}\}.$$

(1.5.3) *If $\Gamma_\mathfrak{b}\alpha\Gamma_\mathfrak{b} = \bigcup_i \Gamma_\mathfrak{b}\alpha_i$ is a disjoint union, then $\Gamma_1\alpha\Gamma_1 = \bigcup_i \Gamma_1\alpha_i$ is a disjoint union.*

(1.5.4) *One has $\Gamma_\mathfrak{b}\alpha\Gamma_\mathfrak{b} = \Gamma_\mathfrak{b}\beta\Gamma_\mathfrak{b}$ if and only if $\Gamma_1\alpha\Gamma_1 = \Gamma_1\beta\Gamma_1$ and $\alpha \equiv \beta \bmod \mathfrak{b}$.*

The first three assertions are restatements of the assertions of [14, Prop. 1.14], and (1.5.4) follows directly from (1.5.2).

1.6. PROPOSITION. *The correspondence $\Gamma_\mathfrak{b}\alpha\Gamma_\mathfrak{b} \to \Gamma_1\alpha\Gamma_1$ gives a homomorphism of $R(\Gamma_\mathfrak{b}, \Delta_\mathfrak{b})$ into $R(\Gamma_1, \Delta_1)$.*

This can be proved in the same way as in the proof of [14, Prop. 1.15], by means of our definition of $R(\Gamma, \Delta)$, [14, Prop. 1.13] and 1.5.

1.7. We denote by $G_\mathfrak{b}$ the group of invertible elements in $\mathfrak{o}/\mathfrak{b}$, and by $S_\mathfrak{b}$ the subgroup of $G_\mathfrak{b}$ consisting of the residue classes of the elements α such that $\det(\alpha) \equiv 1 \bmod \mathfrak{b}$.

1.8. PROPOSITION. *The canonical homomorphism of $\mathfrak{o}$ to $\mathfrak{o}/\mathfrak{b}$ gives an isomorphism of $\Gamma_1/\Gamma_\mathfrak{b}$ onto $S_\mathfrak{b}$.*

The only non-trivial point is the surjectivity of the isomorphism of $\Gamma_1/\Gamma_\mathfrak{b}$ to $S_\mathfrak{b}$. This follows from Eichler's approximation theorem [3, Satz 5].

1.9. PROPOSITION. *If $\alpha \in \Delta_1$, $\varepsilon \in \Gamma_1$ and $c \in \mathbf{Z}$, $c \neq 0$, then*

$$\Gamma_\mathfrak{b}\alpha c\varepsilon\Gamma_\mathfrak{b} = (\Gamma_\mathfrak{b}\alpha\Gamma_\mathfrak{b})\cdot(\Gamma_\mathfrak{b}c\varepsilon\Gamma_\mathfrak{b}),$$
$$\Gamma_\mathfrak{b}c\varepsilon\alpha\Gamma_\mathfrak{b} = (\Gamma_\mathfrak{b}c\varepsilon\Gamma_\mathfrak{b})\cdot(\Gamma_\mathfrak{b}\alpha\Gamma_\mathfrak{b}).$$

This follows easily from the definition of the law of multiplication and the equality $\Gamma_\mathfrak{b}c\varepsilon = c\varepsilon\Gamma_\mathfrak{b}$.

1.10. PROPOSITION. *Let $N = [\Gamma_1 : \Gamma_\mathfrak{b}]$ and let $\{\gamma_1, \cdots, \gamma_N\}$ be a complete set of representatives for $\Gamma_1/\Gamma_\mathfrak{b}$. Then, for every $\alpha \in \Delta_\mathfrak{b}$, we have $\Gamma_1\alpha\Gamma_1 = \bigcup_{i=1}^{N}\Gamma_\mathfrak{b}\alpha\gamma_i\Gamma_\mathfrak{b} = \bigcup_{i=1}^{N}\Gamma_\mathfrak{b}\gamma_i\alpha\Gamma_\mathfrak{b}$, where both $\bigcup_{i=1}^{N}$ are disjoint unions.*

This is an easy consequence of (1.5.1) and (1.5.4).

1.11. PROPOSITION. *Let p be a prime number not dividing b_0, and α an element of $\mathfrak{o}$ such that $\det(\alpha) = p$. Then there exists an element δ of Γ_1 such that $\alpha'\delta \equiv \delta\alpha' \equiv \alpha$, $p\delta \equiv \alpha^2 \bmod \mathfrak{b}$. Moreover, for such an element δ, $\Gamma_\mathfrak{b}\alpha\Gamma_\mathfrak{b}$ commutes with $\Gamma_\mathfrak{b}p\delta\Gamma_\mathfrak{b}$.*

PROOF. Let ξ be an element of $\mathfrak{o}$ such that $\alpha'\xi \equiv \alpha \bmod \mathfrak{b}$. Then $\det(\xi) \equiv 1 \bmod \mathfrak{b}$, hence, by 1.8, there exists an element δ of Γ_1 such that $\delta \equiv \xi \bmod \mathfrak{b}$. We have then $\alpha'\delta \equiv \alpha$, $p\delta \equiv \alpha\alpha'\delta \equiv \alpha^2 \bmod \mathfrak{b}$, and $p\delta\alpha' \equiv \alpha^2\alpha' \equiv p\alpha \bmod \mathfrak{b}$, hence $\delta\alpha' \equiv \alpha \bmod \mathfrak{b}$. Now $p\delta\alpha \equiv \alpha^3 \equiv \alpha p\delta \bmod \mathfrak{b}$, and $\Gamma_1 p\delta\alpha\Gamma_1 = \Gamma_1 p\alpha\Gamma_1 = \Gamma_1\alpha p\delta\Gamma_1$. By (1.5.4) we have $\Gamma_\mathfrak{b}p\delta\alpha\Gamma_\mathfrak{b} = \Gamma_\mathfrak{b}\alpha p\delta\Gamma_\mathfrak{b}$, which, combined with 1.9, yields the commutativity of $\Gamma_\mathfrak{b}\alpha\Gamma_\mathfrak{b}$ with $\Gamma_\mathfrak{b}p\delta\Gamma_\mathfrak{b}$.

1.12. We denote by $\mathfrak{H}$ the complex upper half plane $\{z \in \mathbf{C} \mid \mathrm{Im}(z) > 0\}$, and by $\mathrm{GL}_2^+(\mathbf{R})$ the group of elements in $\mathrm{M}_2(\mathbf{R})$ with positive determinant. For $\alpha = \begin{pmatrix} a & b \\ c & d \end{pmatrix} \in \mathrm{GL}_2^+(\mathbf{R})$ and $z \in \mathfrak{H}$, we put

(1.12.1)
$$\alpha(z) = (az + b)/(cz + d),$$
$$j(\alpha, z) = cz + d.$$

Then $\Gamma_\mathfrak{b}$, regarded as a discrete subgroup of $\mathrm{SL}_2(\mathbf{R})$, gives a properly discontinuous group of transformations on $\mathfrak{H}$ with compact quotient $\Gamma_\mathfrak{b}\backslash\mathfrak{H}$.

1.13. *Automorphic forms and Hecke operators.* For a non-negative integer k, we denote by $\mathfrak{S}_k(\Gamma_\mathfrak{b})$ the vector space of all holomorphic functions $f(z)$ on $\mathfrak{H}$ such that $f(\gamma(z))j(\gamma, z)^{-k} = f(z)$ for every $\gamma \in \Gamma_\mathfrak{b}$. Let $\alpha \in \Delta_1$ and let $\Gamma_\mathfrak{b}\alpha\Gamma_\mathfrak{b} = \bigcup_{i=1}^{d}\Gamma_\mathfrak{b}\alpha_i$ be a disjoint expression. For every $f \in \mathfrak{S}_k(\Gamma_\mathfrak{b})$, we define a function $g = f \mid (\Gamma_\mathfrak{b}\alpha\Gamma_\mathfrak{b})_k$ by

(1.13.1)
$$g(z) = \det(\alpha)^{k-1}\sum_{i=1}^{d}f(\alpha_i(z))j(\alpha_i, z)^{-k}.$$

It can be easily verified that $g \in \mathfrak{S}_k(\Gamma_\mathfrak{b})$, and $\Gamma_\mathfrak{b}\alpha\Gamma_\mathfrak{b} \to (\Gamma_\mathfrak{b}\alpha\Gamma_\mathfrak{b})_k$ defines a representation of the ring $R(\Gamma_\mathfrak{b}, \Delta_1)$ in the vector space $\mathfrak{S}_k(\Gamma_\mathfrak{b})$.

1.14. *Automorphic forms and Hecke operators relative to a representation.* Let ψ be a representation of $\mathfrak{S}_{\mathfrak{b}}$ by complex matrices of degree d. We denote by $\mathfrak{S}_k(\Gamma_1, \psi)$ the vector space of all holomorphic mappings $\mathfrak{f}$ of $\mathfrak{H}$ to C^d such that $\mathfrak{f}(\gamma(z))j(\gamma, z)^{-k} = \psi(\gamma)\mathfrak{f}(z)$ for every $\gamma \in \Gamma_1$, where we understand that $\psi(\gamma) = \psi$(the class of $\gamma \bmod \mathfrak{b}$), and $\mathfrak{f}$ is a column vector function.

Let ρ be a representation of $G_{\mathfrak{b}}$ by complex matrices, and ρ' the restriction of ρ to $S_{\mathfrak{b}}$. For every $\alpha \in \Delta_{\mathfrak{b}}$, we put $\rho(\alpha) = \rho$(the class of $\alpha \bmod \mathfrak{b}$). Let $\alpha \in \Delta_{\mathfrak{b}}$, and let $\Gamma_1 \alpha \Gamma_1 = \bigcup_{i=1}^d \Gamma_1 \alpha_i$ be a disjoint union. For $\mathfrak{f} \in \mathfrak{S}_k(\Gamma_1, \rho')$, we define $\mathfrak{g} = \mathfrak{f} \mid (\Gamma_1 \alpha \Gamma_1)_{k,\rho}$ by

$$\mathfrak{g}(z) = \det(\alpha)^{k-1} \sum_{i=1}^d \rho(\alpha_i)^{-1} \mathfrak{f}(\alpha_i(z)) j(\alpha_i, z)^{-k}.$$

It can be easily verified that $\mathfrak{g} \in \mathfrak{S}_k(\Gamma_1 \, \rho')$, and $\Gamma_1 \alpha \Gamma_1 \to (\Gamma_1 \alpha \Gamma_1)_{k,\rho}$ defines a representation of $R(\Gamma_1, \Delta_{\mathfrak{b}})$. By (1.5.2) and (1.5.3), we have

$$(1.14.1) \qquad \mathfrak{f} \mid (\Gamma_1 \alpha \Gamma_1)_{k,\rho} = \rho(\alpha)^{-1} \mathfrak{f} \mid (\Gamma_{\mathfrak{b}} \alpha \Gamma_{\mathfrak{b}})_k.$$

If a representation ψ of $\mathfrak{S}_{\mathfrak{b}}$ is the direct sum of representations $\psi_1, \cdots, \psi_\lambda$, then $\mathfrak{S}_k(\Gamma_1, \psi)$ may be identified, in an obvious way, with the direct sum of the $\mathfrak{S}_k(\Gamma_1, \psi_i)$ for $i = 1, \cdots, \lambda$. Further, if a representation ρ of $G_{\mathfrak{b}}$ is the direct sum of representations $\rho_1, \cdots, \rho_\lambda$, then $(\Gamma_1 \alpha \Gamma_1)_{k,\rho}$ may be identified with the direct sum of the $(\Gamma_1 \alpha \Gamma_1)_{k,\rho_i}$ in an obvious sense.

1.15. PROPOSITION. *Let ψ be a representation of $\mathfrak{S}_{\mathfrak{b}}$ by complex matrices, and ρ a representation of $G_{\mathfrak{b}}$ induced from ψ. Put $\chi(\alpha) = \mathrm{tr}\,(\rho(\alpha))$ for $\alpha \in G_{\mathfrak{b}}$ or $\alpha \in \Delta_{\mathfrak{b}}$. Let $N = [\Gamma_1 : \Gamma_{\mathfrak{b}}]$ and let $\{\gamma_1, \cdots, \gamma_N\}$ be a complete set of representatives for $\Gamma_1/\Gamma_{\mathfrak{b}}$. Then, for every $\xi \in \Delta_{\mathfrak{b}}$, we have*

$$\mathrm{tr}\,[(\Gamma_1 \xi \Gamma_1)_{k,\rho}] = N^{-1} \sum_{\lambda=1}^N \chi(\xi' \gamma_\lambda) \cdot \mathrm{tr}\,[(\Gamma_{\mathfrak{b}} \gamma_\lambda^{-1} \Gamma_{\mathfrak{b}})_k (\Gamma_{\mathfrak{b}} \xi \Gamma_{\mathfrak{b}})_k].$$

PROOF. For simplicity, let us denote, for an element α of $\Delta_{\mathfrak{b}}$, the residue-class of α modulo $\mathfrak{b}$ by the same letter α. Let $r = [G_{\mathfrak{b}} : S_{\mathfrak{b}}]$, and let $\{\sigma_1, \cdots, \sigma_r\}$ be a set of representatives for $G_{\mathfrak{b}}/S_{\mathfrak{b}}$. Then we may assume that ρ has the following form:

$$\rho(\tau) = \begin{bmatrix} \psi(\sigma_1 \tau \sigma_1^{-1}) & \psi(\sigma_1 \tau \sigma_2^{-1}) & \cdots & \psi(\sigma_1 \tau \sigma_r^{-1}) \\ \psi(\sigma_2 \tau \sigma_1^{-1}) & \psi(\sigma_2 \tau \sigma_2^{-1}) & \cdots & \psi(\sigma_2 \tau \sigma_r^{-1}) \\ \cdots & \cdots & & \cdots \\ \psi(\sigma_r \tau \sigma_1^{-1}) & \psi(\sigma_r \tau \sigma_2^{-1}) & \cdots & \psi(\sigma_r \tau \sigma_r^{-1}) \end{bmatrix}.$$

Here we put $\psi(\tau) = 0$ if $\tau \notin S_{\mathfrak{b}}$. Since $S_{\mathfrak{b}}$ is a normal subgroup of $G_{\mathfrak{b}}$, we see that $\chi(\tau) = 0$ if $\tau \notin S_{\mathfrak{b}}$. Now we have to divide our proof into two cases according as $\xi \equiv 1$ or $\xi \not\equiv 1 \bmod \mathfrak{b}$.

Case I. $\xi \not\equiv 1 \bmod \mathfrak{b}$. In this case, $\chi(\xi' \gamma_\lambda) = 0$ for every λ. Hence it is sufficient to prove $\mathrm{tr}\,[(\Gamma_1 \xi \Gamma_1)_{k,\rho}] = 0$. Put

$$\psi_i(\tau) = \psi(\sigma_i \tau \sigma_i^{-1}) \qquad\qquad (\tau \in S_{\mathfrak{b}}).$$

Let $\mathfrak{f} \in \mathfrak{S}_k(\Gamma_1, \rho)$. Then $^t\mathfrak{f} = (^t\mathfrak{f}_1, \cdots, ^t\mathfrak{f}_r)$ with $\mathfrak{f}_i \in \mathfrak{S}_k(\Gamma_1, \psi_i)$. Put $^t(\mathfrak{f} \mid (\Gamma_1 \xi \Gamma_1)_{k,\rho}) = {}^t\mathfrak{g} = (^t\mathfrak{g}_1, \cdots, ^t\mathfrak{g}_r)$ with $\mathfrak{g}_i \in \mathfrak{S}_k(\Gamma_1, \psi_i)$. By (1.14.1) we have

$$\mathfrak{g}_i = \psi(\sigma_i \xi^{-1} \sigma_j^{-1}) \mathfrak{f}_j \mid (\Gamma_{\mathfrak{b}} \xi \Gamma_{\mathfrak{b}})_k ,$$

where i is determined by j according to the relation $S_{\mathfrak{b}} \sigma_i = S_{\mathfrak{b}} \sigma_j \xi$. Now $\mathfrak{S}_k(\Gamma_1, \rho)$ can be identified with $\mathfrak{S}_k(\Gamma_1, \psi_1) \oplus \cdots \oplus \mathfrak{S}_k(\Gamma_1, \psi_r)$ in a natural way. The above consideration shows that $(\Gamma_1 \xi \Gamma_1)_{k,\rho}$ maps $\mathfrak{S}_k(\Gamma_1, \psi_j)$ to $\mathfrak{S}_k(\Gamma_1, \psi_i)$. Since $\xi \notin S_{\mathfrak{b}}$, we always have $i \neq j$. Hence $\operatorname{tr}[(\Gamma_1 \xi \Gamma_1)_{k,\rho}] = 0$.

Case II. $\xi \equiv 1 \bmod \mathfrak{b}$. Let $\{\rho^1, \cdots, \rho^m\}$ be a maximal set of inequivalent irreducible representations of $S_{\mathfrak{b}}$. Put $\theta^a(\tau) = \operatorname{tr}(\rho^a(\tau))$. From the orthogonality relation of characters, one can easily derive the following relation:

$$(1.15.1) \qquad N^{-1} \sum_{\lambda=1}^{N} \theta^a(\xi^{-1}\gamma_\lambda)\rho_b(\gamma_\lambda^{-1}) = \begin{cases} 0 & (a \neq b), \\ \theta^a(1)^{-1}\rho^a(\xi^{-1}) & (a = b). \end{cases}$$

Now $\Gamma_1 \ni \gamma \to (\Gamma_{\mathfrak{b}} \gamma \Gamma_{\mathfrak{b}})_k$ gives a representation of $\Gamma_1/\Gamma_{\mathfrak{b}}$ in $\mathfrak{S}_k(\Gamma_{\mathfrak{b}})$. Regard this as a representation of $S_{\mathfrak{b}}$, and call t_a the multiplicity of ρ^a in it. It can be easily verified that $\dim \mathfrak{S}_k(\Gamma_1, \rho^a) = t_a$. Let $\{\mathfrak{f}_i^a \mid 1 \leq i \leq t_a\}$ be a basis of $\mathfrak{S}_k(\Gamma_1, \rho^a)$ over C. Put

$$\mathfrak{f} = \begin{bmatrix} \mathfrak{f}^1 \\ \vdots \\ \mathfrak{f}^m \end{bmatrix}, \qquad \mathfrak{f}^a = \begin{bmatrix} \mathfrak{f}_1^a \\ \vdots \\ \mathfrak{f}_{t_a}^a \end{bmatrix} \qquad\qquad (1 \leq a \leq m).$$

Then we see easily that the components of $\mathfrak{f}$ form a basic of $\mathfrak{S}_k(\Gamma_{\mathfrak{b}})$. Define a complex matrix $U(\Gamma_{\mathfrak{b}} \alpha \Gamma_{\mathfrak{b}})$ by

$$U(\Gamma_{\mathfrak{b}} \alpha \Gamma_{\mathfrak{b}})\mathfrak{f} = \mathfrak{f} \mid (\Gamma_{\mathfrak{b}} \alpha \Gamma_{\mathfrak{b}})_k \qquad\qquad (\alpha \in \Delta_{\mathfrak{b}}).$$

By our construction of $\mathfrak{f}$, we have, for every $\gamma \in \Gamma_1$,

$$(1.15.2) \qquad U(\Gamma_{\mathfrak{b}} \gamma \Gamma_{\mathfrak{b}}) = \begin{bmatrix} U^1(\gamma) & & \\ & \ddots & \\ & & U^m(\gamma) \end{bmatrix}, \qquad U^a(\gamma) = \begin{bmatrix} \rho^a(\gamma) & & \\ & \ddots & \\ & & \rho^a(\gamma) \end{bmatrix}$$

$$(t_a \text{ copies of } \rho^a(\gamma); 1 \leq a \leq m).$$

Since $\xi \equiv 1 \bmod \mathfrak{b}$, $\rho^a(\xi^{-1})$ is meaningful; and we see easily that the mapping $\mathfrak{h} \to \rho^a(\xi^{-1})\mathfrak{h} \mid (\Gamma_{\mathfrak{b}} \xi \Gamma_{\mathfrak{b}})_k$ maps $\mathfrak{S}_k(\Gamma_1, \rho^a)$ into itself. Therefore we may put

$$(1.15.3) \qquad \rho^a(\xi^{-1})\mathfrak{f}_i^a \mid (\Gamma_{\mathfrak{b}} \xi \Gamma_{\mathfrak{b}})_k = \sum_{j=1}^{t_a} c_{ij}^a \mathfrak{f}_j^a \qquad\qquad (1 \leq i \leq t_a)$$

with $c_{ij}^a \in C$. Let u_a be the multiplicity of ρ^a in the restriction of ρ to $S_{\mathfrak{b}}$. In view of (1.14.1) and the definition of $(\Gamma_1 \xi \Gamma_1)_{k,\rho}$, we have

$$(1.15.4) \qquad \operatorname{tr}[(\Gamma_1 \xi \Gamma_1)_{k,\rho}] = \sum_{a=1}^{m} \left(u_a \sum_{i=1}^{t_a} c_{ii}^a \right).$$

On the other hand, we have

$$(1.15.5) \quad \begin{aligned} N^{-1} \sum_\lambda \chi(\xi'\gamma_\lambda) \cdot \mathrm{tr}\,[(\Gamma_\mathfrak{b}\gamma_\lambda^{-1}\Gamma_\mathfrak{b})_k(\Gamma_\mathfrak{b}\xi\Gamma_\mathfrak{b})_k] \\ = \sum_{a=1}^{m} u_a \cdot \mathrm{tr}\,[N^{-1} \sum_\lambda \theta^a(\xi'\gamma_\lambda) U(\Gamma_\mathfrak{b}\gamma_\lambda\Gamma_\mathfrak{b})^{-1} U(\Gamma_\mathfrak{b}\xi\Gamma_\mathfrak{b})]. \end{aligned}$$

Let η be an element of Γ_1 such that $\eta \equiv \xi \bmod \mathfrak{b}$. By (1.15.1) and (1.15.2), we have

$$N^{-1} \sum_\lambda \theta^a(\xi'\gamma_\lambda) U(\Gamma_\mathfrak{b}\gamma_\lambda\Gamma_\mathfrak{b})^{-1} = \theta^a(1)^{-1} \cdot \begin{pmatrix} 0 & & \\ & U^a(\eta^{-1}) & \\ & & 0 \end{pmatrix} \begin{pmatrix} \text{non-zero part for only} \\ (a,\,a)\text{-block} \end{pmatrix}.$$

Put then, with the function $\mathfrak{f}$ defined above,

$$\begin{pmatrix} 0 & & \\ & U^a(\eta^{-1}) & \\ & & 0 \end{pmatrix} U(\Gamma_\mathfrak{b}\xi\Gamma_\mathfrak{b})\mathfrak{f} = \begin{pmatrix} 0 \\ \mathfrak{h}^a \\ 0 \end{pmatrix}, \qquad \mathfrak{h}^a = \begin{bmatrix} \mathfrak{h}^a_1 \\ \vdots \\ \mathfrak{h}^a_{t_a} \end{bmatrix}.$$

Then, by (1.15.2) and (1.15.3),

$$\mathfrak{h}^a_i = \rho^a(\eta^{-1})\mathfrak{f}^a_i \mid (\Gamma_\mathfrak{b}\xi\Gamma_\mathfrak{b})_k = \rho^a(\xi^{-1})\mathfrak{f}^a_i \mid (\Gamma_\mathfrak{b}\xi\Gamma_\mathfrak{b})_k = \sum_j c^a_{ij}\mathfrak{f}^a_j.$$

Hence we have

$$(1.15.6) \quad \begin{pmatrix} 0 & & \\ & U^a(\eta^{-1}) & \\ & & 0 \end{pmatrix} U(\Gamma_\mathfrak{b}\xi\Gamma_\mathfrak{b})\mathfrak{f} = \bar{C}^a\mathfrak{f}$$

with a complex matrix $\bar{C}^a$ of the form

$$\bar{C}^a = \begin{pmatrix} 0 & & \\ & C_a & \\ & & 0 \end{pmatrix}, \qquad C^a = \begin{pmatrix} c^a_{11}\rho^a(1) & \cdots & c^a_{1\mu}\rho^a(1) \\ \cdots & \cdots & \cdots \\ c^a_{\mu 1}\rho^a(1) & \cdots & c^a_{\mu\mu}\rho^a(1) \end{pmatrix} \quad (\mu = t_a).$$

Since the components of $\mathfrak{f}$ form a basis of $\mathfrak{S}_k(\Gamma_\mathfrak{b})$ (as remarked above), we get, from (1.15.6),

$$\begin{pmatrix} 0 & & \\ & U^a(\eta^{-1}) & \\ & & 0 \end{pmatrix} U(\Gamma_\mathfrak{b}\xi\Gamma_\mathfrak{b}) = \bar{C}^a,$$

so that (1.15.5) is equal to

$$\sum_{a=1}^{m} u_a \theta^a(1)^{-1} \mathrm{tr}\,(\bar{C}^a) = \sum_{a=1}^{m} u_a \theta^a(1)^{-1}\theta^a(1) \sum_{i=1}^{t_a} c^a_{ii}.$$

In view of (1.15.4), this completes the proof.

1.16. Let x, y, u be indeterminates. We consider the ring of formal power-series in u with coefficients in $C[x, y]$. Define polynomials $F_n(x, y)(n = 1, 2, \cdots)$ by means of the equality

$$(1.16.1) \qquad -(d/du) \log (1 - xu + yu^2) = \sum_{n=1}^{\infty} F_n(x, y) u^{n-1} .$$

The coefficients of the F_n are rational integers. If X and Y are commuting matrices, we can show easily

$$(1.16.2) \quad (d/du) \log [\det (1 - Xu + Yu^2)^{-1}] = \sum_{n=1}^{\infty} \mathrm{tr} \, (F_n(X, Y)) u^{n-1} .$$

1.17. PROPOSITION. *Let ψ, ρ, χ, N, $\{\gamma_1, \cdots, \gamma_N\}$ be as in 1.15, and p a prime number which does not divide $d(\Phi) b_0$. Let α be an element of Δ_1 such that $\det (\alpha) = p$, and δ an element of Γ_1 determined in 1.11 (modulo $\mathfrak{b}$). Then*

$$N^{-1} \sum_{n=1}^{\infty} u^{n-1} \sum_{\lambda=1}^{N} \chi(\alpha'^n \gamma_\lambda) \, \mathrm{tr} \, [(\Gamma_\mathfrak{b} \gamma_\lambda^{-1} \Gamma_\mathfrak{b})_k \cdot F_n((\Gamma_\mathfrak{b} \alpha \Gamma_\mathfrak{b})_k , \, p(\Gamma_\mathfrak{b} p \delta \Gamma_\mathfrak{b})_k)]$$
$$= (d/du) \log [\det (1 - (\Gamma_1 \alpha \Gamma_1)_{k,\rho} u + p(\Gamma_1 p \Gamma_1)_{k,\rho} u^2)^{-1}] .$$

PROOF. First we note that $F_n((\Gamma_\mathfrak{b} \alpha \Gamma_\mathfrak{b})_k, \, p(\Gamma_\mathfrak{b} p \delta \Gamma_\mathfrak{b})_k)$ is meaningful, since $(\Gamma_\mathfrak{b} \alpha \Gamma_\mathfrak{b})$ and $(\Gamma_\mathfrak{b} p \delta \Gamma_\mathfrak{b})$ are commutative (1.11). Similarly $F_n((\Gamma_1 \alpha \Gamma_1)_{k,\rho}, p(\Gamma_1 p \Gamma_1)_{k,\rho})$ is meaningful. By (1.16.2), the right hand side of the equality to be proved is equal to

$$(1.17.1) \qquad \sum_{n=1}^{\infty} u^{n-1} \, \mathrm{tr} \, [F_n((\Gamma_1 \alpha \Gamma_1)_{k,\rho}, \, p(\Gamma_1 p \Gamma_1)_{k,\rho})] .$$

Now we observe, from our definition of F_n, that $F_n(x, y) = \sum a_{ij}^n x^i y^j$ with non-vanishing a_{ij}^n only for $i + 2j = n$. For such i and j, we have $(\Gamma_\mathfrak{b} \alpha \Gamma_\mathfrak{b})^i (\Gamma_\mathfrak{b} p \delta \Gamma_\mathfrak{b})^j = \sum_h b_h (\Gamma_\mathfrak{b} \xi_h \Gamma_\mathfrak{b})$ with $b_h \in \mathbf{Z}$ and elements ξ_h such that $\xi_h \equiv \alpha^i (p\delta)^j \equiv \alpha^n$ mod $\mathfrak{b}$ and $\det (\xi_h) = p^n$. Therefore we have

$$(1.17.2) \qquad F_n(\Gamma_\mathfrak{b} \alpha \Gamma_\mathfrak{b}, \, p(\Gamma_\mathfrak{b} p \delta \Gamma_\mathfrak{b})) = \sum_\mu c_{n\mu} (\Gamma_\mathfrak{b} \xi_{n\mu} \Gamma_\mathfrak{b})$$

with $c_{n\mu} \in \mathbf{Z}$ and elements $\xi_{n\mu}$ of Δ_1 such that $\det (\xi_{n\mu}) = p^n$ and $\xi_{n\mu} \equiv \alpha^n$ mod $\mathfrak{b}$. By 1.6, we have

$$(1.17.3) \qquad F_n(\Gamma_1 \alpha \Gamma_1, \, p(\Gamma_1 p \Gamma_1)) = \sum_\mu c_{n\mu} (\Gamma_1 \xi_{n\mu} \Gamma_1) .$$

Our equality in question follows now easily from (1.17.1, 2, 3) and 1.15.

1.18. *The case $k = 0$.* We observe that $\mathfrak{S}_0(\Gamma_\mathfrak{b})$ is the set of all constant functions on $\mathfrak{H}$, hence may be identified with C. From (1.13.1) we find easily

$$(\Gamma_\mathfrak{b} \alpha \Gamma_\mathfrak{b})_0 = \det (\alpha)^{-1} \cdot d(\Gamma_\mathfrak{b} \alpha \Gamma_\mathfrak{b}) ,$$

where $d(\Gamma_\mathfrak{b} \alpha \Gamma_\mathfrak{b})$ is the number of right cosets $\Gamma_\mathfrak{b} \alpha_i$ in $\Gamma_\mathfrak{b} \alpha \Gamma_\mathfrak{b}$. In particular, the notation being as in 1.17, we have

$$d(\Gamma_\mathfrak{b} \alpha \Gamma_\mathfrak{b}) = p + 1, \qquad d(\Gamma_\mathfrak{b} p \delta \Gamma_\mathfrak{b}) = 1 .$$

Substituting $p + 1$ and p for x and y of (1.16.1), we get

$$(1.18.1) \qquad F_n(d(\Gamma_\mathfrak{b} \alpha \Gamma_\mathfrak{b}), \, p \cdot d(\Gamma_\mathfrak{b} p \delta \Gamma_\mathfrak{b})) = 1 + p^n \qquad (n = 1, 2, \cdots) .$$

Let r be the multiplicity of the identity representation in ψ. Then it is easy to verify that

$$(1.18.2) \qquad N^{-1}\sum_{\lambda=1}^{N}\chi(\alpha'^n\gamma_\lambda) = \begin{cases} r \cdot [G_{\mathfrak{b}} : S_{\mathfrak{b}}] & \text{if } p^n \equiv 1 \bmod (b_0), \\ 0 & \text{otherwise.} \end{cases}$$

Let $t = \varphi(b_0)$ with Euler's function φ, and $t = fg$ with the smallest positive integer f such that $p^f \equiv 1 \bmod (b_0)$. From (1.18.1) and (1.18.2) we get easily

$$\begin{aligned}
(1.18.3) \qquad & N^{-1}\sum_{n=1}^{\infty} u^{n-1}\sum_{\lambda=1}^{N}\chi(\alpha'^n\gamma_\lambda)F_n\big(d(\Gamma_{\mathfrak{b}}\alpha\Gamma_{\mathfrak{b}}),\ p\cdot d(\Gamma_{\mathfrak{b}}p\delta\Gamma_{\mathfrak{b}})\big) \\
& = rfg \cdot \sum_{m=1}^{\infty} u^{mf-1}(1 + p^{mf}) \\
& = (d/du)\log\left[(1 - u^f)(1 - p^f u^f)\right]^{-r_0}.
\end{aligned}$$

1.19. PROPOSITION. *The coefficients of the formal power series of 1.17 are real numbers.*

PROOF. If $\xi \in S_{\mathfrak{b}}$, then $\xi' \equiv \xi^{-1}\bmod \mathfrak{b}$, so that $\operatorname{tr}(\psi_i(\xi'))$ is the complex conjugate of $\operatorname{tr}(\psi_i(\xi))$, the ψ_i being as in the proof of 1.15. If $\xi \notin S_{\mathfrak{b}}$, we have $\chi(\xi) = \chi(\xi') = 0$. Hence, in any case,

$$(1.19.1) \qquad\qquad \chi(\xi') \text{ is the complex conjugate of } \chi(\xi) .$$

For $f,\ g \in S_k(\Gamma_{\mathfrak{b}})$, put

$$\langle f, g\rangle = \int f(z)\overline{g(z)}\cdot \operatorname{Im}(z)^{k-2}\,|\,dzd\bar{z}\,| .$$

The integral is taken on $\mathfrak{H}/\Gamma_{\mathfrak{b}}$. For every $\xi \in \Delta_1$, we can easily verify

$$\langle f\,|\,(\Gamma_{\mathfrak{b}}\xi\Gamma_{\mathfrak{b}})_k,\ g\rangle = \langle f,\ g\,|\,(\Gamma_{\mathfrak{b}}\xi'\Gamma_{\mathfrak{b}})_k\rangle .$$

Hence $\operatorname{tr}(\Gamma_{\mathfrak{b}}\xi'\Gamma_{\mathfrak{b}})_k$ is the complex conjugate of $\operatorname{tr}(\Gamma_{\mathfrak{b}}\xi\Gamma_{\mathfrak{b}})_k$. Now if $\Gamma_1\xi\Gamma_1 = \bigcup_i\Gamma_{\mathfrak{b}}\xi_i\Gamma_{\mathfrak{b}}$, then $\Gamma_1\xi\Gamma_1 = \Gamma_1\xi'\Gamma_1 = \bigcup_i\Gamma_{\mathfrak{b}}\xi_i'\Gamma_{\mathfrak{b}}$, so that

$$\sum_i\chi(\xi_i')\operatorname{tr}(\Gamma_{\mathfrak{b}}\xi_i\Gamma_{\mathfrak{b}})_k = \sum_i\chi(\xi_i)\operatorname{tr}(\Gamma_{\mathfrak{b}}\xi_i'\Gamma_{\mathfrak{b}})_k .$$

In view of (1.19.1), this number is a real number. Let the $\xi_{n\mu}$ be as in (1.17.2). By 1.10, we have $\Gamma_1\xi_{n\mu}\Gamma_1 = \bigcup_{i=1}^{N}\Gamma_{\mathfrak{b}}\gamma_i'\xi_{n\mu}\Gamma_{\mathfrak{b}}$. Since $\alpha^n \equiv \xi_{n\mu}\bmod \mathfrak{b}$, we have $\chi(\alpha'^n\gamma_i) = \chi(\xi_{n\mu}'\gamma_i)$, so that

$$\sum_{i=1}^{N}\chi(\alpha'^n\gamma_i)\operatorname{tr}\left[(\Gamma_{\mathfrak{b}}\gamma_i^{-1}\Gamma_{\mathfrak{b}})_k(\Gamma_{\mathfrak{b}}\xi_{n\mu}\Gamma_{\mathfrak{b}})_k\right] = \sum_{i=1}^{N}\chi(\xi_{n\mu}'\gamma_i)\operatorname{tr}(\Gamma_{\mathfrak{b}}\gamma_i'\xi_{n\mu}\Gamma_{\mathfrak{b}})_k .$$

By the above consideration, the right hand side is a real number. This proves our proposition.

Another simpler proof can be obtained by observing that for every $\xi \in \Delta_1$, $(\Gamma_1\xi\Gamma_1)_{k,\rho}$ is a self-adjoint operator with respect to a metric in $\mathfrak{S}_k(\Gamma_1, \rho)$.

1.20. PROPOSITION. *For every $\alpha \in \Delta_{\mathfrak{b}}$, the absolute value of any characteristic root of $(\Gamma_1\alpha\Gamma_1)_{k,\rho}$ (resp. $(\Gamma_{\mathfrak{b}}\alpha\Gamma_{\mathfrak{b}})_k$) does not exceed $\det(\alpha)^{(k/2)-1}d(\Gamma_1\alpha\Gamma_1)$ (resp. $\det(\alpha)^{(k/2)-1}d(\Gamma_{\mathfrak{b}}\alpha\Gamma_{\mathfrak{b}}))$.*

PROOF. For every $\mathfrak{f} \in \mathfrak{S}_k(\Gamma_1, \rho)$, define a function $\mathfrak{f}^*$ on $\mathrm{SL}_2(R)$ by $\mathfrak{f}^*(u) = \mathfrak{f}(u(\sqrt{-1}))j(u, \sqrt{-1})^{-k}$ for $u \in \mathrm{SL}_2(R)$. Let $\Gamma_1 \alpha \Gamma_1 = \bigcup_{\nu=1}^{d} \Gamma_1 \alpha_\nu$, $\mathfrak{g} = \mathfrak{f} \mid (\Gamma_1 \alpha \Gamma_1)_{k,\rho}$. Putting $\beta_\nu = \det(\alpha)^{-1/2}\alpha_\nu$, we get $\mathfrak{g}^*(u) = \det(\alpha)^{(k/2)-1} \sum_{\nu=1}^{d} \rho(\alpha_\nu)^{-1}\mathfrak{f}^*(\beta_\nu u)$. We may consider $\mathfrak{f}$ as a mapping of $\mathfrak{H}$ to the representation space V of ρ. We can find a hermitian form (x, y) on V invariant under $\rho(G_\mathfrak{b})$. Put $(x, x)^{1/2} = |x|$ for $x \in V$. It can be easily seen that $|\mathfrak{f}^*(\gamma u)| = |\mathfrak{f}^*(u)|$ for every $\gamma \in \Gamma_1$. Since $\Gamma_1 \backslash \mathrm{SL}_2(R)$ is compact, $\mathfrak{f}^*(u)$ attains its maximum at some point v in $\mathrm{SL}_2(R)$. If $\mathfrak{g} = \lambda \mathfrak{f}$ with a complex number λ, then we get $|\lambda \mathfrak{f}^*(v)| = |\mathfrak{g}^*(v)| \leqq \det(\alpha)^{(k/2)-1} \cdot d \cdot |\mathfrak{f}^*(v)|$, which proves our assertion about $(\Gamma_1 \alpha \Gamma_1)_{k,\rho}$. The same type of reasoning applies to $(\Gamma_\mathfrak{b} \alpha \Gamma_\mathfrak{b})_k$ (cf. also (1.14.1)).

The consideration of [14, p. 286] concerning the characteristic roots of $\mathfrak{T}_\kappa(\Gamma_a \alpha \Gamma_a)$ is erroneous; it should be corrected as in the above proposition. It turns out consequently that the Dirichlet series [14, p. 286, (13)] converges for $\mathrm{Re}(s) > (\kappa/2) + 1$.

2. Fibre correspondences on a fibre variety

2.1. *Proper correspondences.* Let R and S be complete non-singular varieties of the same dimension n. By a *proper correspondence*, or a *proper n-cycle*, on $R \times S$, we understand an n-cycle X on $R \times S$ such that $|X| \cap (a \times S)$ and $|X| \cap (R \times b)$ have no component of dimension > 0 for every $a \in R$ and $b \in S$, where $|X|$ is the support of X. Let k be a field of rationality for R, S and X. Then X is proper if and only if, for every point (a, b) in $|X|$, $k(a, b)$ is algebraic over $k(a)$ and over $k(b)$. For every 0-cycle c on R, we define a 0-cycle $X[c]$ by $X[c] = \mathrm{pr}_S[X \cdot (c \times S)]$. We denote by ${}^tX$ the proper n-cycle on $S \times R$ which is the transform of X by the birational transformation $(r, s) \to (s, r)$ of $R \times S$ to $S \times R$.

If X and Y are proper n-cycles on $R \times S$, and if $X[a] = Y[a]$ for every $a \in R$ (or, for a generic point a on R over a field of rationality for R, S, X, Y), then $X = Y$.

Let T be another complete non-singular variety of dimension n. Let X be a proper n-cycle on $R \times S$, and Y a proper n-cycle on $S \times T$. By a standard argument, we can verify easily that $X \times T$ and $R \times Y$ intersect properly on $R \times S \times T$, and if we put

$$(2.1.1) \qquad Z = \mathrm{pr}_{R \times T}[(X \times T) \cdot (R \times Y)],$$

then Z is a proper n-cycle on $R \times T$, and $Z[a] = Y[X[a]]$ for every $a \in R$ (cf. [19, VIII$_4$, Prop. 10]). We write $Z = Y \circ X$. This law of composition is obviously associative. Therefore all the proper n-cycles on $R \times R$ form an associative ring with respect to this law, which we call *the ring of proper corre-*

spondences on R.

2.2. We say that $\{V, W, \varphi, f\}$ is a *fibre system of abelian varieties* defined over a field k, if the following conditions are satisfied.

(2.2.1) V and W are non-singular projective varieties, φ is a rational mapping of W to V, f is a rational mapping of V to W, all defined over k.

(2.2.2) φ is everywhere defined on W, f is everywhere defined on V, and $\varphi \circ f =$ the identity mapping of V.

(2.2.3)   If Z is the graph of φ, then Z and $W \times a$ intersect properly on $W \times V$ for every $a \in V$, and $Z \cdot (W \times a) = \varphi^{-1}(a) \times a$ with a variety $\varphi^{-1}(a)$.

(2.2.4) $\varphi^{-1}(a)$ is an abelian variety with the origin $f(a)$, for every $a \in V$. We see that $\varphi^{-1}(a)$ is defined over $k(a)$ as abelian variety.

In the above definition we can include the case where the fibres $\varphi^{-1}(a)$ are 0-dimensional. Then φ is a biregular isomorphism of W to V and $f = \varphi^{-1}$.

2.3. Let $\{V, W, \varphi, f\}$ be a fibre system of abelian varieties defined over k. For an integer $m \geq 0$, define W_m, φ_m, f_m as follows: If $m > 0$, let W^m be the product of m copies of W, and

$$W_m = \{(x_1, \cdots, x_m) \in W^m \mid \varphi(x_1) = \cdots = \varphi(x_m)\},$$
$$\varphi_m(x_1, \cdots, x_m) = \varphi(x_1) \qquad\qquad ((x_1, \cdots, x_m) \in W_m),$$
$$f_m(u) = \big(f(u), \cdots, f(u)\big) \in W_m \qquad\qquad (u \in V).$$

If $m = 0$, put $W_0 = V$, $\varphi_0 = f_0 =$ the identity mapping of V. Let us prove that $\{V, W_m, \varphi_m, f_m\}$ is a fibre system of abelian varieties. Take a generic point y of V over k and m independent generic points $x_1, \cdots, x_m$ of $\varphi^{-1}(y)$ over $k(y)$. Let W'_m resp. Z_m the locus of $x_1 \times \cdots \times x_m$ resp. $x_1 \times \cdots \times x_m \times y$ over k. We see easily that $\varphi^{-1}(a)^m \times a = Z_m \cap (W^m \times a)$ for every $a \in V$. By Weil [19, VIII$_3$, Th. 6], we have $\varphi^{-1}(y)^m \times y = Z_m \cdot (W^m \times y)$ on $W^m \times V$. Specialize y to a point a of V. By [10, Th. 23], we get

(2.3.1) $$\varphi^{-1}(a)^m \times a = Z_m \cdot (W^m \times a).$$

We see that $W_m = \bigcup_{a \in V} \varphi^{-1}(a)^m = W'_m$, and hence Z_m is the graph of φ_m. By (2.3.1) and [19, VI$_3$, Th. 6], every point of $\varphi^{-1}(a)^m \times a$ is simple on Z_m. Since Z_m and W_m $(= W'_m)$ are biregularly isomorphic, this implies that W_m is non-singular. Thus $\{V, W_m, \varphi_m, f_m\}$ is a fibre system of abelian varieties, and $\varphi_m^{-1}(a)$ is the product of m copies of $\varphi^{-1}(a)$ for every $a \in V$.

2.4. *Fibre correspondences.* Let $\{V, W, \varphi, f\}$ and $\{V', W', \varphi', f'\}$ be fibre systems of abelian varieties. Suppose that $\dim(V) = \dim(V')$ and $\dim(W) = \dim(W')$. Let Y resp. X be a subvariety of $V \times V'$ resp. a positive cycle on

$W \times W'$ of the same dimension as V resp. W. Suppose that Y is proper. We call X a *fibre correspondence on $W \times W'$ associated with Y over* a field k, if the following condition is satisfied:

(2.4.1) *k is a field of rationality for $\{V, W, \varphi, f\}$, $\{V'\ W', \varphi', f'\}$, X and Y. There exist a generic point $u \times v$ of Y over k, a generic point x of $\varphi^{-1}(u)$ over $k(u)$, and an isogeny λ of $\varphi^{-1}(u)$ to $\varphi'^{-1}(v)$, such that X is the prime rational cycle with the generic point $x \times \lambda x$ over k (in the sense of* [19, VIII$_1$]).

We shall prove later that X is proper (cf. 5.4 below). The notation being as above, put $X[x] = \sum_i c_i z_i$ with $c_i \in Z$ and $y_i \in W'$. Put $\varphi'(z_i) = v_i$. We may assume $z_1 = \lambda x$, $v_1 = v$. Then there exists an isomorphism σ_i of $k(x, z_1)$ to $k(x, z_i)$ over $k(x)$ such that $z_1^{\sigma_i} = z_i$. We have then $u^{\sigma_i} = u$, $v_1^{\sigma_i} = v_i$. Extend σ_i to an isomorphism of an extension of $k(x, z_1)$ over which λ is defined, and denote it again by σ_i. Put $\lambda^{\sigma_i} = \lambda_i$. Then λ_i is an isogeny of $\varphi^{-1}(u)$ to $\varphi'^{-1}(v_i)$, and $z_i = (\lambda x)^{\sigma_i} = \lambda_i x$. Since Y is proper and (u, v_i) is a specialization of (u, v) over k, we see that v_i is algebraic over $k(u)$. Hence we can find a finite algebraic extension K of $k(u)$ over which the λ_i are defined. Let y be any generic point of $\varphi^{-1}(u)$ over $k(u)$. Since K is algebraic over $k(u)$, we see that x and y are generic on $\varphi^{-1}(u)$ over K, so that there exists an isomorphism τ of $K(x)$ to $K(y)$ over K such that $x^\tau = y$. Then $X[y] = X[x^\tau] = X[x]^\tau = \sum_i c_i z_i^\tau = \sum_i c_i(\lambda_i x)^\tau = \sum_i c_i \lambda_i y$. In particular take ax as y for a non-zero integer a. Then we have $\lambda_i(ax) = a z_i$, so that $X[ax] = \sum_i c_i \cdot (a z_i)$. Here we have to distinguish two kinds of multiplication by an integer: $a z_i$ should be understood with respect to the additive operation on the abelian variety $\varphi'^{-1}(v_i)$; and $c_i \cdot (a z_i)$ means the 0-cycle with the coefficient c_i and the component $a z_i$.

2.5. *Anti-Frobenius fibre correspondence.* Let $\{V, W, \varphi, f\}$ be defined over a field k of characteristic $p > 0$. Let q be a power of p. We shall denote by U^q the transform of an algebro-geometric object U by the qth power automorphism of the universal domain. Let Π_0 resp. Π be the Frobenius correspondence of degree q on $V \times V^q$ resp. $W \times W^q$, i.e., the locus of $u \times u^q$ resp. $x \times x^q$ over k, where u resp. x is a generic point of V resp. W over k. It is obvious that Π is a fibre correspondence associated with Π_0.

Let x be a generic point of W over k. Put $u = \varphi(x)$, $y = x^q$. There exists an isogeny π' of $\varphi^{-1}(u)^q$ to $\varphi^{-1}(u)$, defined over $k(u)$, such that $\pi' y = qx$. Then

(2.5.1) $k(y, \pi' y)$ is a regular extension of k.

In fact, since π' is defined over $k(u)$, and $\varphi(\pi' y) = \varphi(qx) = u$, we have $k(y, \pi' y) = k(u, y) \subset k(u, x) = k(x)$, which implies (2.5.1). Let Π^* be the locus of $y \times \pi' y$ on $W^q \times W$ over k. Π^* is a fibre correspondence associated with ${}^t\Pi_0$, so that Π^* is proper. we call Π^* *the anti-Frobenius fibre correspondence on* $W_q \times W$ *of degree* q. Notice that Π^* is different from ${}'\Pi$ unless the fibres are 0-dimensional. Let us now prove

(2.5.2) $k(y, \pi'y)$ is a purely inseparable extension of $k(y)$ of degree $q^{\dim(V)}$.

(2.5.3)   Let r be the dimension of $\varphi^{-1}(u)$, and p^a the number of points t on $\varphi^{-1}(u)$ such that $pt = 0$. Then

$$[k(y, \pi'y) : k(\pi'y)]_s = q^a ,$$
$$[k(y, \pi'y) : k(\pi'y)]_i = q^{r-a} .$$

Since y is generic on $\varphi^{-1}(u)^q$ over $k(u^q)$, the fields $k(y)$ and $k(u)$ are linearly disjoint over $k(u^q)$. The assertion (2.5.2) follows from this and the equality $k(y, \pi'y) = k(u, y)$. Since we have $k(\pi'y) = k(u, \pi'y)$, the assertion (2.5.3) is a consequence of a well-known property of π'.

Let $\{V', W', \varphi', f'\}$ be another fibre system of abelian varieties defined over k, and X a fibre correspondence on $W \times W'$ associated with a correspondence Y on $V \times V'$. Let Π' resp. Π'^* be the Frobenius, resp. anti-Frobenius, fibre correspondence of degree q on $W' \times W'^q$ resp. $W'^q \times W'$. Then

(2.5.4) $\Pi' \circ X = X^q \circ \Pi ,$

(2.5.5) $X \circ \Pi^* = \Pi'^* \circ X^q .$

The relation (2.5.4) is true for any proper correspondence X (which is not necessarily a fibre correspondence) on $W \times W'$, and can be proved easily. To prove (2.5.5), take a field k_1 containing k, over which X and Y are rational, and take a generic point x of W over k_1. Put $X[x] = \sum_i c_i z_i$ with $c_i \in \mathbf{Z}$ and $z_i \in W'$. As proved in 2.4, we have $X[qx] = \sum_i c_i \cdot (qz_i)$. Put $n = \dim(V)$. By (2.5.2), we have $X \circ \Pi^*[x^q] = q^n \cdot X[qx] = \sum_i q^n c_i \cdot (qz_i) = \Pi'^*[\sum_i c_i \cdot z_i^q] = \Pi'^* \circ X^q[x^q]$, from which we obtain (2.5.5).

Let r be another power of p. Let Π_1 and Π_1^* be the Frobenius, resp. anti-Frobenius, fibre correspondence of degree r on $W^q \times W^{qr}$ resp. $W^{qr} \times W^q$. Then one can easily verify that $\Pi_1 \circ \Pi$ resp. $\Pi^* \circ \Pi_1^*$ is the Frobenius resp. anti-Frobenius fibre correspondence of degree qr on $W \times W^{qr}$ resp. $W^{qr} \times W$. As for $\Pi^* \circ \Pi_1^*$, one should notice (2.5.2).

2.6. Our next task is to compute the intersection of a fibre correspondence with the diagonal. First we have to introduce some notations. Let U be a variety with a generic point y over a field k. Then we denote by D_U the locus of $y \times y$ on $U \times U$ over k, and call it *the diagonal on* $U \times U$. After [19], we denote by $\{A \cdot B\}_U$ the intersection-product of two cycles A and B on U. If U is an abelian variety and μ is an endomorphism of U defined over k, we put, following [21, No. 27],

$$\nu(\mu) = \begin{cases} [k(y) : k(\mu y)] & \text{if } \mu \text{ is an isogeny,} \\ 0 & \text{otherwise.} \end{cases}$$

Furthermore, we denote by $G(\mu)$ the graph of μ on $U \times U$.

2.7. PROPOSITION. *Let* $\{V, W, \varphi, f\}$, X, Y, k, u, v, x, λ *be as in* (2.4.1), *with* $\{V', W', \varphi', f'\} = \{V, W, \varphi, f\}$. *Put* $A_t = \varphi^{-1}(t)$ *for every* $t \in V$. *Suppose that the following conditions are satisfied.*

(2.7.1) λ *is defined over* $k(u, v)$.

(2.7.2) D_V *and* Y *intersect properly on* $V \times V$.

(2.7.3) *Every point of* $D_V \cap Y$ *is simple on* Y.

Then X *is a variety defined over* k; *and for each point* (b, b) *of* $D_V \cap Y$, *there is a unique isogeny* μ_b *of* A_b *to itself such that* $G(\mu_b)$ *is the specialization of* $G(\lambda)$ *over the specialization* $(u, v) \to (b, b)$ *ref.* k. *Suppose moreover that:*

(2.7.4) *For every* $(b, b) \in D_V \cap Y$, $\delta_b - \mu_b$ *is an isogeny of* A_b *to itself, where* δ_b *is the identity mapping of* A_b.

Put

$$\{D_V \cdot Y\}_{V \times V} = \sum_b c_b \cdot (b \times b) \qquad ((b, b) \in D_V \cap Y)$$

with positive integers c_b. *Then* X *and* D_W *intersect properly on* $W \times W$, *and the following formulas hold.*

(2.7.5) $\quad \{D_W \cdot X\}_{W \times W} = \sum_b c_b \cdot \{G(\delta_b) \cdot G(\mu_b)\}_{A_b \times A_b} \qquad ((b, b) \in D_V \cap Y),$

(2.7.6) $\quad \deg \{D_W \cdot X\}_{W \times W} = \sum_b c_b \cdot \nu(\delta_b - \mu_b) \qquad ((b, b) \in D_V \cap Y).$

PROOF. From (2.7.1) we see that $k(x, \lambda x) = k(x, \lambda x, u, v) = k(x, v)$. Since v is algebraic over $k(u)$ and x is generic on A_u over $k(u)$, we get

(2.7.7) $\qquad [k(x, \lambda x) : k(x)] = [k(u, v) : k(u)].$

Since $k(x)$ is regular over $k(u)$, $k(u, v, x)$ is regular over $k(u, v)$. On the other hand, $k(u, v)$ is regular over k, so that $k(x, \lambda x)$ $(= k(x, u, v))$ is regular over k. Therefore X is a variety defined over k.

Let $(b, b) \in D_V \cap Y$. Extend the specialization $(u, v) \to (b, b)$ ref. k to a place $\mathfrak{p}$ of $k(u, v)$. Then $\mathfrak{p}(A_u) = \mathfrak{p}(A_v) = A_b$. By Koizumi [5, Th. 3], the abelian varieties A_u and A_v are without defect for $\mathfrak{p}$ in the sense of [18, § 11.1], hence one obtains a well-defined isogeny $\mathfrak{p}(\lambda)$ of $\mathfrak{p}(A_u)$ to $\mathfrak{p}(A_v)$ [18, § 11.1, Prop. 12]. Write $\mathfrak{p}(\lambda)$ as μ_b. Then μ_b is an isogeny of A_b to itself. Let X_0 be the locus of $(x, \lambda x, u, v)$ over k on $W \times W \times V \times V$. By Weil [19, VIII$_3$, Th. 6] and (2.7.1), we have

$$(2.7.8) \qquad \{X_0 \cdot (W \times W \times u \times v)\}_{W \times W \times Y} = G(\lambda) \times u \times v \, .$$

Since X is proper, we see easily that X_0 and $W \times W \times b \times b$ intersect properly on $W \times W \times Y$. By [18, § 11, Prop. 12] we have $\mathfrak{p}(G(\lambda)) = G(\mu_b)$. Hence applying $\mathfrak{p}$ to (2.7.8), we get, by [10, Th. 17] and (2.7.3),

$$(2.7.9) \qquad \{X_0 \cdot (W \times W \times b \times b)\}_{W \times W \times Y} = G(\mu_b) \times b \times b \, .$$

This implies that μ_b is uniquely determined by b and independent of the choice of $\mathfrak{p}$.

Next assume (2.7.4). Then $G(\delta_b)$ and $G(\mu_b)$ intersect properly on $A_b \times A_b$. From (2.7.9) we obtain

$$X \cap D_W = \bigcup_b \left(G(\delta_b) \cap G(\mu_b) \right) .$$

Therefore X and D_W intersect properly on $W \times W$.

The points x and u $(= \varphi(x))$ being as above, take a generic point x_1 of W over $k(x)$ and put $u_1 = \varphi(x_1)$. Let Z be the locus of (x, x_1, u, u_1) over k, and D the locus of (x, x, u, u) over k, each being considered on $W \times W \times V \times V$. We observe that the projection of $W \times W \times V \times V$ to $W \times W$ defines a biregular isomorphism of Z to $W \times W$ which sends D and X_0 to D_W and X. Therefore, (2.7.5) is equivalent to

$$(2.7.10) \qquad \{D \cdot X_0\}_Z = \sum_b c_b\left(\{G(\delta_b) \cdot G(\mu_b)\}_{A_b \times A_b} \times b \times b\right).$$

Let x' be a generic point of A_u over $k(x)$, and U the locus of (x, x', u, u) over k on $W \times W \times V \times V$. We see that $D \subset U \subset W \times W \times D_V$. The projection of $W \times W \times V \times V$ to $W \times W \times V$ defines a biregular isomorphism of $W \times W \times D_V$ to $W \times W \times V$, which sends U to the variety Z_2 defined in 2.3. Transforming the relation (2.3.1) with $m = 2$ and $a = b$ by this isomorphism, we get

$$(2.7.11) \qquad \{U \cdot (W \times W \times b \times b)\}_{W \times W \times D_V} = A_b \times A_b \times b \times b \, .$$

Since U is isomorphic to Z_2, U is non-singular. Taking D and D_V in place of X_0 and Y in (2.7.9), we get

$$(2.7.12) \qquad \{D \cdot (W \times W \times b \times b)\}_{W \times W \times D_V} = G(\delta_b) \times b \times b \, .$$

By [19, VIII$_4$, Th. 10], (2.7.11) and (2.7.12), we have

$$(2.7.13) \qquad \begin{aligned} \{D \cdot (A_b \times A_b \times b \times b)\}_U &= \{D \cdot (W \times W \times b \times b)\}_{W \times W \times D_V} \\ &= G(\delta_b) \times b \times b \, . \end{aligned}$$

By [19, VIII$_4$, Th. 10], we have

$$(2.7.14) \qquad \{D \cdot X_0\}_Z = \{D \cdot \{X_0 \cdot U\}_Z\}_U \, .$$

We see that each component of $X_0 \cap U$ has the form $G(\mu_b) \times b \times b$ with

$(b, b) \in D_V \cap Y$, so that

$$(2.7.15) \qquad \{X_0 \cdot U\}_Z = \sum_b d_b(G(\mu_b) \times b \times b)$$

with positive integers d_b. We have now

$$(2.7.16) \quad \begin{aligned} \{D \cdot (G(\mu_b) &\times b \times b)\}_U \\ &= \{(G(\mu_b) \times b \times b) \cdot \{D \cdot (A_b \times A_b \times b \times b)\}_U\}_{A_b \times A_b \times b \times b} \quad [19, \text{VIII}_4, \text{Th.10}] \\ &= \{(G(\mu_b) \times b \times b) \cdot (G(\delta_b) \times b \times b)\}_{A_b \times A_b \times b \times b} \qquad\qquad (2.7.13) \\ &= (\{G(\mu_b) \cdot G(\delta_b)\}_{A_b \times A_b}) \times b \times b \, . \end{aligned}$$

From this, (2.7.14) and (2.7.15) we obtain the formula (2.7.10) with d_b in place of c_b. Therefore, our assertion will be proved if we can show $d_b = c_b$.

For this purpose we consider the locus S of $(f(u), x, u, u)$ over k. By [19, VIII_3, Th. 6] we have

$$\{S \cdot (W \times W \times u \times u)\}_{W \times W \times D_V} = f(u) \times A_u \times u \times u \, .$$

Specializing u to b, we get

$$\{S \cdot (W \times W \times b \times b)\}_{W \times W \times D_V} = f(b) \times A_b \times b \times b \, .$$

Therefore, taking S in place of D in (2.7.13) and (2.7.16), we get

$$\{S \cdot (A_b \times A_b \times b \times b)\}_U = f(b) \times A_b \times b \times b$$

and

$$\begin{aligned} \{S \cdot (G(\mu_b) \times b \times b)\}_U &= (\{G(\mu_b) \cdot (f(b) \times A_b)\}_{A_b \times A_b}) \times b \times b \\ &= f(b) \times f(b) \times b \times b \, . \end{aligned}$$

Hence taking S in place of D in (2.7.14), we get

$$\{S \cdot X_0\}_Z = \{S \cdot \{X_0 \cdot U\}_Z\}_U = \sum_b d_b(f(b) \times f(b) \times b \times b) \, .$$

Therefore our proof will be completed if we can show that

$$(2.7.17) \qquad \text{pr}_{V \times V}\{S \cdot X_0\}_Z = \{D_V \cdot Y\}_{V \times V} \, .$$

To prove this, denote by Y' the locus of $(f(u), f(v), u, v)$ over k. We see easily that

$$(2.7.18) \qquad \{X_0 \cdot (f(V) \times W \times V \times V)\}_{W \times W \times V \times V} = eY'$$

with a positive integer e. Take the projection to the first factor W. By [19, VIII_3, Th. 8], we have

$$[k(x, \lambda x) : k(x)]\{W \cdot f(V)\}_W = e[k(u, v) : k(u)]f(V) \, .$$

By (2.7.7), we have $e = 1$. Let T be the locus of $(f(u), x_1, u, u_1)$ over k. We see easily that

$$(2.7.19) \qquad \{Z \cdot (f(V) \times W \times V \times V)\}_{W \times W \times V \times V} = e'T$$

with a positive integer e'. Take the projection to $W \times W$. By [19, VIII_3, Th. 8], we find $\{(W \times W) \cdot (f(V) \times W)\}_{W \times W} = e'(f(V) \times W)$, so that $e' = 1$. By [19, VIII_4, Th. 10] and (2.6.19) (with $e' = 1$), we get

$$(2.7.20) \quad \{Z \cdot (f(V) \times W \times D_V)\}_{W \times W \times V \times V} = \{T \cdot (f(V) \times W \times D_V)\}_{f(V) \times W \times V \times V}.$$

We see easily that the right hand side equals $e''S$ with a positive integer e''. Take the projection to the factor $W \times V \times V$ (exclusion of the first factor $f(V)$). By [19, VIII_3, Th. 8], we have

$$(2.7.21) \qquad \{\mathrm{pr}_{W \times V \times V}[T] \cdot (W \times D_V)\}_{W \times V \times V} = e'' \cdot \mathrm{pr}_{W, V \times V}[S].$$

Let h be the mapping of $W \times V \times V$ to itself such that

$$h(r,\, s,\, t) = (r,\, t,\, s) \qquad\qquad \text{for} \quad (r,\, t,\, s) \in W \times V \times V.$$

Then we see that $\mathrm{pr}_{W \times V \times V}[T] = h(W_0 \times V)$, where W_0 is the locus of $(x,\, u)$ over k on $W \times V$. Therefore, taking the projection of (2.7.21) to the product of the first and last factors (exclusion of the middle factor V), we get, by [19, VIII_3, Th. 8], $\{W_0 \cdot (W \times V)\}_{W \times V} = e''W_0$, so that $e'' = 1$. Hence, by (2.7.20),

$$\{Z \cdot (f(V) \times W \times D_V)\}_{W \times W \times V \times V} = S.$$

From this and [19, VIII_4, Th. 10], we obtain

$$\{X_0 \cdot S\}_\mathbf{z} = \{X_0 \cdot (f(V) \times W \times D_V)\}_{W \times W \times V \times V}.$$

Again by [19, VIII_4 Th. 10] and (2.7.18) (with $e = 1$) we have

$$\{X_0 \cdot (f(V) \times W \times D_V)\}_{W \times W \times V \times V} = \{Y' \cdot (f(V) \times W \times D_V)\}_{f(V) \times W \times V \times V}.$$

Therefore, by [19, VIII_3, Th. 8], we get

$$\begin{aligned}
\mathrm{pr}_{V \times V}[\{S \cdot X_0\}_\mathbf{z}] &= \mathrm{pr}_{V \times V}[\{Y' \cdot (f(V) \times W \times D_V)\}_{f(V) \times W \times V \times V}] \\
&= \{\mathrm{pr}_{V \times V}(Y') \cdot D_V\}_{V \times V} \\
&= \{Y \cdot D_V\}_{V \times V},
\end{aligned}$$

which is (2.7.17) and completes our proof of (2.7.5). Now applying [21, p. 24, Cor. 2] to $G(\delta_b)$ and $G(\mu_b)$, we get

$$\nu(\delta_b - \mu_b) = \deg \{G(\delta_b) \cdot G(\mu_b)\}_{A_b \times A_b}.$$

Hence from (2.7.5), we obtain (2.7.6).

2.8. Let the notation and assumption be the same as in 2.7. Let X^* be a fibre correspondence on $W \times W$ associated with $'Y$ over k. Then there exists an isogeny λ^* of A_v to A_u such that X^* is the locus of $\lambda x \times \lambda^* \lambda x$. Here notice that λx is a generic point of A_v over $k(v)$.

Let E be a divisor on A_u rational over $k(u)$. For every generic point t of V over k, we denote by E_t the divisor on A_t which is the transform of E by the isomorphism of $k(u)$ to $k(t)$ sending u to t.

2.9. PROPOSITION. *The notation and assumption being as in 2.7 and 2.8, suppose that there exist a positive non-degenerate divisor E on A_u, rational over $k(u)$, and a positive integer r such that $\lambda^*\lambda x = rx$ and $\lambda^{-1}(E_v)$ is algebraically equivalent to rE_u.[5] Suppose further that V is of dimension one. Then X^* and D_W intersect properly on $W \times W$, and*

$$\deg\{D_W \cdot X\}_{W \times W} = \deg\{D_W \cdot X^*\}_{W \times W}.$$

PROOF. First we observe that λ^* and ${}^tY$ satisfy the conditions (2.7.1–3), and $\{D_V \cdot {}^tY\}_{V \times V} = \{D_V \cdot Y\}_{V \times V}$. Let $(b, b) \in D_V \cap Y$, and let μ_b be as in 2.7. Let $\mathfrak{p}$ be a place of $k(u, v)$ considered in the proof of 2.7. Since $\dim(V) = 1$, both specialization rings $[u \to b; k]$ and $[v \to b; k]$ are discrete valuation rings. Hence if σ is an isomorphism of $k(u)$ to $k(v)$ over k such that $u^\sigma = v$, then $\mathfrak{p}(a^\sigma) = \mathfrak{p}(a)$ for every $a \in k(u)$. Hence $\mathfrak{p}(E_u) = \mathfrak{p}(E_v)$. Put $\mathfrak{p}(\lambda^*) = \mu_b^*$, $\mathfrak{p}(E_u) = F$. Then $\mu_b\mu_b^* = r\delta_b$ and $\mu_b^{-1}(F) = \mathfrak{p}(\lambda^{-1}(E_v))$. Hence $\mu_b^{-1}(F)$ is algebraically equivalent to rF on A_b. This can be shown, for example, by taking suitable l-adic coordinate systems on A_u and A_b, and applying [18, § 11.1, Prop. 14, (ii)] to E_u, $\lambda^{-1}(E_v)$ and F. By Lemma 2.10 below, we have $\nu(\delta_b - \mu_b) = \nu(\delta_b - \mu_b^*)$. This implies in particular that X^* satisfies (2.7.4). Hence, by 2.7, X^* and D_W intersect properly on $W \times W$, and $\deg\{D_W \cdot X^*\}_{W \times W}$ is given by (2.7.6) with μ_b^* in place of μ_b. Since $\nu(\delta_b - \mu_b) = \nu(\delta_b - \mu_b^*)$, this proves our proposition.

2.10. LEMMA. *Let A be an abelian variety and E a positive non-degenerate divisor on A. Let μ and μ^* be endomorphisms of A such that $\mu^*\mu = r\delta$, and $\mu^{-1}(E)$ is algebraically equivalent to rE with a positive integer r, where δ is the identity mapping of A. Then $\nu(\delta - \mu) = \nu(\delta - \mu^*)$.*

PROOF. Let φ be the isogeny of A to its Picard variety, attached to the divisor E. Then one gets an involution $\alpha \to \varphi^{-1} \cdot {}^t\alpha\varphi$ of $\mathrm{End}_Q(A)$. By our assumption on $\mu^{-1}(E)$, we find ${}^t\mu\varphi\mu = r\varphi$, so that $r\delta = \varphi^{-1} \cdot {}^t\mu\varphi\mu$, hence $\mu^* = \varphi^{-1} \cdot {}^t\mu\varphi$. This proves our lemma, since the involution does not change the value of ν.

2.11. PROPOSITION. *Let $\{V, W, \varphi, f\}$ be a fibre system of abelian varieties defined over a finite field with q elements. Let Π be the Frobenius correspondence on $W \times W$ of degree q, and Π^* the anti-Frobenius fibre correspondence on $W \times W$ of degree q. Then D_W intersects properly with both Π and Π^* on $W \times W$, and*

$$\deg\{D_W \cdot \Pi\}_{W \times W} = \deg\{D_W \cdot \Pi^*\}_{W \times W}.$$

(In the present proposition, $\dim(V)$ can be arbitrary.)

---

[5] Under this condition, X^* may be called the *adjoint* of X.

PROOF. Apply the argument of the proof 2.9 to Π and Π^*, with $q = r$. The conditions (2.7.1–3) are satisfied by Π in view of Lemma 2.12 below. Let μ_b and μ_b^* be as in the proof of 2.9. Then we see easily that μ_b is the Frobenius isogeny of A_b to itself of degree q, and $\mu_b^* \mu_b = q\delta_b$. We can find a positive non-degenerate divisor E on A_b, rational over k. Then $\mu_b^{-1}(E) = qE$. By 2.10, we get $\nu(\delta_b - \mu_b) = \nu(\delta_b - \mu_b^*)$ which, together with 2.7, proves our assertion.

2.12. LEMMA. *Let U be an affine variety defined over a field k of characteristic $p > 0$. Let x be a generic point of U, and f a rational mapping of U to U such that $k(f(x)) = k(x^q)$ with a power q of p. Let Z be the graph of f, and D the locus of $x \times x$ on $U \times U$. Let b be a simple point of U such that $(b, b) \in Z$ and f is defined at b. Then (b, b) is simple on Z, and the varieties Z and D are transversal at (b, b) on $U \times U$.*

PROOF. The first assertion is obvious, since f is defined at b. Let S be the ambient affine space for U, and let $n = \dim(S)$, $r = \dim(U)$. Put $f(x) = (y_1, \cdots, y_n)$. Since f is defined at b, and b is simple on U, we see that $y_\mu \in [x \to b; k] \cap k(x^q) = [x^q \to b^q; k]^6$ for every μ. Hence there exist polynomials $G_\mu(X)$ and $H_\mu(X)$ in $k[X]$ such that $y_\mu = G_\mu(x^q)/H_\mu(x^q)$, $H_\mu(b^q) \neq 0$ $(1 \leq \mu \leq n)$. Since b is simple on U, we can find $n - r$ polynomials $F_\lambda(X)$ in $k[X]$ for $1 \leq \lambda \leq n - r$ such that $F_\lambda(x) = 0$ and rank $(\partial F_\lambda / \partial X_i(b)) = n - r$. Put $Q_\mu(X, Y) = H_\mu(X^q)Y_\mu - G_\mu(X^q)$. Then

$$\sum_{i=1}^n \frac{\partial Q_\mu}{\partial X_i}(b, b)(X_i - b_i) - \sum_{j=1}^n \frac{\partial Q_\mu}{\partial Y_j}(b, b)(Y_j - b_j) = H_\mu(b^q)(Y_\mu - b_\mu)$$

where $(b_1, \cdots, b_n) = b$. Therefore the tangent linear variety to Z at (b, b) is defined by the linear equations

$$\sum_{i=1}^n \partial F_\lambda / \partial X_i(b)(X_i - b_i) = 0 \qquad (1 \leq \lambda \leq n - r),$$
$$Y_\mu - b_\mu = 0 \qquad (1 \leq \mu \leq n).$$

Obviously, this coincides with the tangent linear variety to $U \times b$ at (b, b). Now D and $U \times b$ are transversal on $U \times U$ at (b, b). It follows that D and Z are transversal at (b, b) on $U \times U$.

3. The fiber variety W_m and its fibres

3.1. *Families of abelian varieties.* Let Φ, $\mathfrak{o}$, $\mathfrak{b}$, Γ_1, $\Gamma_\mathfrak{b}$ be the same as in 1.2 and 1.3, and $\mathfrak{H}$ the upper half plane (1.12). For every $z \in \mathfrak{H}$, put

---

[6] We denote by $[x \to b; k]$ the set of all quotients $F(x)/G(x)$ formed by two polynomials F and G with coefficients in k such that $G(b) \neq 0$. The equality in question follows from the fact that $[x^q \to b^q; k]$ is integrally closed.

$$(3.1.1) \qquad D_z = \left\{ \alpha \begin{pmatrix} z \\ 1 \end{pmatrix} \,\middle|\, \alpha \in \mathfrak{o} \right\}.$$

Then D_z is a lattice in C^2, so that C^2/D_z is a complex torus. For a suitable element τ of Φ such that τ^2 is a negative rational number, we can define a Riemann form E on C^2/D_z by

$$E\left(\xi \begin{pmatrix} z \\ 1 \end{pmatrix}, \, \eta \begin{pmatrix} z \\ 1 \end{pmatrix} \right) = \operatorname{tr}(\tau \xi \eta') \qquad\qquad (\xi, \, \eta \in M_2(\boldsymbol{R})).$$

Let A_z be an abelian variety isomorphic to C^2/D_z, and $\mathcal{C}_z$ be the polarization of A_z determined by E. For every $\alpha \in \Phi$, the linear transformation of C^2 given by the matrix α defines an element of $\operatorname{End}_Q(A_z)$; we denote this element by $\theta_z(\alpha)$. Then $\theta_z(\mathfrak{o}) = \theta_z(\Phi) \cap \operatorname{End}(A_z)$. Put

$$\mathcal{P}_z = (A_z, \, \mathcal{C}_z, \, \theta_z).$$

In this way we get an analytic family $\Sigma = \{\mathcal{P}_z \mid z \in \mathfrak{H}\}$ of polarized abelian varieties of type $\mathfrak{o}$. For the detail of construction of Σ, we refer to [12, § 5], [16], or [8, Ch. IV]. Two members $\mathcal{P}_z$ and $\mathcal{P}_w$ of Σ are isomorphic if and only if $z = \gamma(w)$ for some $\gamma \in \Gamma_1$.

By a result of Eichler, every $\mathfrak{o}$-ideal is principal. Hence we can find an element β_0 such that

$$(3.1.2) \qquad \mathfrak{b} = \mathfrak{o}\beta_0 = \beta_0\mathfrak{o}.$$

Let t_z be the point on A_z corresponding to $\beta_0^{-1}\begin{pmatrix} z \\ 1 \end{pmatrix}$. We have then

$$\theta_z(\mathfrak{o})t_z = \{t \in A_z \mid \theta_z(\mathfrak{b})t = 0\}.$$

For every $z \in \mathfrak{H}$, we consider a structure

$$\mathcal{Q}_z = \mathcal{P}_z(t_z) = (A_z, \, \mathcal{C}_z, \, \theta_z; \, t_z)$$

introduced in [17, II, § 1]. The $\mathcal{Q}_z$ form a family Σ^*, of which two members $\mathcal{Q}_z$ and $\mathcal{Q}_w$ are isomorphic if and only if $z = \gamma(w)$ for some $\gamma \in \Gamma_\mathfrak{b}$ [17, II, Prop. 4.4].

3.2. *Fibre varieties.* We are going to define a fibre variety of which the base is $\Gamma_\mathfrak{b}\backslash\mathfrak{H}$, and each fibre is the product of m copies of A_z. This is a special case of fibre varieties investigated in [8].

From now on, we always assume:

(3.2.1) $\Gamma_\mathfrak{b}$ *has no element of finite order other than the identity element.*

This is the case, for example, if $2 \notin \mathfrak{b}$ and $3 \notin \mathfrak{b}$, since every element of finite order in Γ_1 is of order 1, 2, 3, 4 or 6. The condition (3.2.1) implies in particular that $-1 \notin \Gamma_\mathfrak{b}$. Therefore we have

$$(3.2.2) \qquad 2 \notin \mathfrak{b} .$$

For an integer $m \geq 0$, let Ψ_m denote the product of m copies of $M_2(R)$, viewed as a right and left $M_2(R)$-module in a natural manner; we understand that Ψ_0 is the module $\{0\}$. The product $GL_2^+(R) \times \Psi_m$ forms a group with respect to the law of multiplication:

$$(3.2.3) \qquad (\xi, u)(\eta, v) = (\xi\eta, v\eta' + u) \qquad (\xi, \eta \in GL_2^+(R); u, v \in \Psi_m),$$

where η' is as in (1.2.1). We let $GL_2^+(R) \times \Psi_m$ act on $\mathfrak{H} \times \Psi_m$ by the rule:

$$(3.2.4) \qquad (\alpha, u)(z, v) = (\alpha(z), v\alpha' + u) \qquad (\alpha \in GL_2^+(R); u, v \in \Psi_m; z \in \mathfrak{H}).$$

Define a mapping $\mathfrak{x}_m$ of $\mathfrak{H} \times \Psi_m$ onto $\mathfrak{H} \times C^{2m}$ by

$$(3.2.5) \qquad \mathfrak{x}_m(z, u_1, \cdots, u_m) = \left(z, u_1\begin{pmatrix} z \\ 1 \end{pmatrix}, \cdots, u_m\begin{pmatrix} z \\ 1 \end{pmatrix}\right) \qquad (z \in \mathfrak{H}; u_i \in M_2(R)),$$

and introduce a complex structure in $\mathfrak{H} \times \Psi_m$ so that $\mathfrak{x}_m$ is a complex analytic isomorphism. Then every element of $GL_2^+(R) \times \Psi_m$ acts on $\mathfrak{H} \times \Psi_m$ as a complex analytic automorphism. In particular, the action of $(\alpha, 0)$ (with α in $GL_2^+(R)$) on $\mathfrak{H} \times C^{2m}$ is expressed by

$$(3.2.6) \qquad (z, x) \rightarrow (\alpha(z), j(\alpha, z)^{-1} \det(\alpha)x) \qquad (z \in \mathfrak{H}; x \in C^{2m}) .$$

Let $\mathfrak{o}^m$ denote the product of m copies of $\mathfrak{o}$. As a subgroup of $GL_2^+(R) \times \Psi_m$, $\Gamma_\mathfrak{b} \times \mathfrak{o}^m$ gives a properly discontinuous group of transformations on $\mathfrak{H} \times \Psi_m$ with a compact quotient. We put

$$(3.2.7) \qquad V = V_\mathfrak{b} = \Gamma_\mathfrak{b}\backslash\mathfrak{H} ,$$

$$(3.2.8) \qquad W_m = W_{m,\mathfrak{b}} = (\Gamma_\mathfrak{b} \times \mathfrak{o}^m)\backslash(\mathfrak{H} \times \Psi_m) .$$

In view of (3.2.1), V is a compact Riemann surface, and W_m is a compact complex manifold of complex dimension $2m + 1$. Now by [8, Th. II–6–8], W_m may be embedded in a projective space. (For the construction of W_m, cf. also [8, Ch. IV].) Obviously W_0 may be identified with V. Hereafter we regard V and W_m as projective non-singular varieties. From our construction, we obtain a commutative diagram

$$(3.2.9) \qquad \begin{array}{ccc} \mathfrak{H} \times \Psi_m & \xrightarrow{\;w_m\;} & W_m \\ \varphi_m^* \downarrow & & \downarrow \varphi_m \\ \mathfrak{H} & \xrightarrow{\;v\;} & V \end{array}$$

where v, w_m, φ_m^* are the projections defined in an obvious way, and φ_m is determined by the commutativity.

3.3. *Fibres of W_m.* Define a mapping f_m of V to W_m by

$$(3.3.1) \qquad f_m(v(z)) = w_m(z, 0) .$$

We see easily that $\{V, W_m, \varphi_m, f_m\}$ is a fibre system of abelian varieties in the sense of 2.2. For $y = v(z)$ with $z \in \mathfrak{H}$, the fibre $\varphi_m^{-1}(y)$ is isomorphic to the product of m copies of the abelian variety A_z defined in 3.1. Moreover, $\{V, W_m, \varphi_m, f_m\}$ is isomorphic to the one obtained from $\{V, W_1, \varphi_1, f_1\}$ by the procedure of 2.3.

For every $a = (a_{ij}) \in M_m(\mathfrak{o})$ with $a_{ij} \in \mathfrak{o}$ and $u = (u_1, \cdots, u_m) \in \Psi_m$, put $au = (u'_1, \cdots, u'_m)$ with $u'_i = \sum_{j=1}^m a_{ij} u_j$. Then we can define an endomorphism $\theta_{m,y}(a)$ of $A_{m,y}$ by

$$(3.3.2) \qquad \theta_{m,y}(a) w_m(z, u) = w_m(z, au) \qquad (u \in \Psi_m) .$$

Furthermore, put

$$(3.3.3) \qquad \Theta_m(a) = \{w_m(z, u) \times w_m(z, au) \mid z \in \mathfrak{H}, \, u \in \Psi_m\} .$$

Then $\Theta_m(a)$ is a subvariety of $W_m \times W_m$, and $\Theta_m(a) \cdot (A_{m,y} \times A_{m,y})$ is the graph of $\theta_{m,y}(a)$, the intersection product being taken on

$$(W_m)_2 = \{(x, x') \in W_m \times W_m \mid \varphi_m(x) = \varphi_m(x')\} .$$

Let $(\mathfrak{b}^{-1})^m$ denote the product of m copies of $\mathfrak{b}^{-1}$ viewed as a submodule of Ψ_m. For every $\xi \in (\mathfrak{b}^{-1})^m$, put

$$(3.3.4) \qquad T(\xi) = \{w_m(z, \xi) \mid z \in \mathfrak{H}\} .$$

We see easily that $T(\xi)$ is a subvariety of W_m, and the restriction of φ_m to $T(\xi)$ gives a biregular isomorphism of $T(\xi)$ to V. We put

$$(3.3.5) \qquad t_m(y, \xi) = w_m(z, \xi) = T(\xi) \cdot A_{m,y} \qquad (z \in \mathfrak{H}; \, y = v(z))$$

When $m = 1$, we write simply A_y, θ_y, $t(y, \xi)$ for $A_{1,y}$, $\theta_{1,y}$, $t_1(y, \xi)$, and put

$$(3.3.6) \qquad t_y = t(y, \beta_0^{-1}) ,$$

where β_0 is as in (3.1.2). Let $\mathcal{C}_y$ be the polarization of A_y corresponding to the polarization $\mathcal{C}_z$ of A_z. Put

$$(3.3.7) \qquad \mathscr{P}_y = (A_y, \mathcal{C}_y, \theta_y) ,$$

$$(3.3.8) \qquad \mathcal{Q}_y = \mathscr{P}_y(t_y) = (A_y, \mathcal{C}_y, \theta_y; t_y) .$$

Then, for $y = v(z)$, $\mathscr{P}_y$ resp. $\mathcal{Q}_y$ is isomorphic to $\mathscr{P}_z$ resp. $\mathcal{Q}_z$.

3.4. *Good model for* W_m. Let b_0 be the positive integer such that

$$(3.4.1) \qquad \mathfrak{b} \cap Z = b_0 Z ,$$

and let

$$(3.4.2) \qquad \zeta = e^{2\pi i/b_0} .$$

In view of (3.2.2), we have

$$(3.4.3) \qquad b_0 > 2 .$$

By [17, II, Prop. 1.6], the field of moduli of $\mathfrak{Q}_y$ contains ζ for every $y \in V$. Moreover, if $\mathfrak{Q}_y$ is a generic member, $Q(\zeta)$ is the algebraic closure of Q in the field of moduli of $\mathfrak{Q}_y$ [17, II, Th. 6.8]. Therefore, by [17, II, Th. 5.11, Prop. 5.12], we may take W_m and V so that the following conditions are satisfied:

(3.4.4) $W_m,\ V,\ \varphi_m,\ \Theta(a)\ \textit{for }a \in \mathrm{M}_m(\mathfrak{o}),\ T(\xi)\ \textit{for }\xi \in (\mathfrak{b}^{-1})^m\ \textit{are all defined over }Q(\zeta)$.

(3.4.5)   *For every $y \in V$, the abelian variety $A_{m,y}$, the endomorphisms $\theta_{m,y}(a)$ for $a \in \mathrm{M}_m(\mathfrak{o})$, and the points $t_m(y, \xi)$ for $\xi \in (\mathfrak{b}^{-1})^m$ are all rational over $Q(\zeta, y)$.* (This is a consequence of (3.4.4)).

(3.4.6)   *$\mathfrak{Q}_y$ is defined over $Q(\zeta, y)$, and $Q(\zeta, y)$ is the field of moduli of Q_y for every $y \in V$.*

Such a system of models $\{V,\ W_m,\ \varphi_m,\ \Theta(a),\ T(\xi)\}$ is unique up to a biregular isomorphism defined over $Q(\zeta)$. We notice also that the assertions concerning $W_m,\ \varphi_m$ etc. for $m > 1$ can be easily derived from the assertions for $W_1,\ \varphi_1$ etc., by means of the construction of 2.3. Hereafter, we always consider the models $W_1,\ \varphi_1,\ V$, etc. with the above properties, and adopt as the models for $W_m,\ \varphi_m$, etc. the ones obtained from $W_1,\ \varphi_1$, etc. by means of the procedure of 2.3. This agreement has no bad effect on generality, in view of the uniqueness of the models. We identify W_0 with V, and w_0 with v; φ_0 is then the identity mapping of V.

3.5. *Generic fibres.* Let y be a generic point of V over $Q(\zeta)$. Then we know that $\mathrm{End}(A_y) = \theta_y(\mathfrak{o})$, $\mathrm{End}_Q(A_y) = \theta_y(\Phi)$, and the automorphisms of $\mathscr{P}_y$ are ± 1 (cf. [14, Prop. 3.1] or [16, Th. 5, Prop. 20]). Hence $\mathrm{End}(A_{m,y}) = \mathrm{M}_m(\mathfrak{o})$ and $\mathrm{End}_Q(A_{m,y}) = \mathrm{M}_m(\Phi)$. In view of (3.2.1), $\mathfrak{Q}_x$ has the identity mapping as the only automorphism, for every $x \in V$ [17, II, 4.2, 5.8].

Denote by K_y resp. $K_{y,\mathfrak{b}}$ the field of moduli of $\mathscr{P}_y$ resp. $\mathfrak{Q}_y$. Then $K_{y,\mathfrak{b}}$ is a Galois extension of K_y. Let $\mathfrak{G}_\mathfrak{b}$ be the Galois group, and let $G_\mathfrak{b}$ and $S_\mathfrak{b}$ be as in 1.7. By [17, II, Th. 6.8] or by [14, p. 307, Th. 2], there is an isomorphism of $\mathfrak{G}_\mathfrak{b}$ onto $G_\mathfrak{b}/\{\pm 1\}$ with the following property: *if α is an element of $\mathfrak{o}$ corresponding to an element σ of $\mathfrak{G}_\mathfrak{b}$ by the isomorphism, then there is an isomorphism ε of $\mathscr{P}_y$ to $(\mathscr{P}_y)^\sigma$ such that*

(3.5.1) $t(y, \xi)^\sigma = \varepsilon\big(t(y, \xi\alpha)\big)$ $(\xi \in \mathfrak{b}^{-1})$,

(3.5.2) $\zeta^\sigma = \zeta^{\det(\alpha)}$.

Moreover, the following assertions hold:

(3.5.3)   *$Q(\zeta)$ is the algebraic closure of Q in $K_{y,\mathfrak{b}}$,*

(3.5.4)   *The subfield $K_y(\zeta)$ of $K_{y,\mathfrak{b}}$ corresponds to the subgroup $S_\mathfrak{b}$ of $G_\mathfrak{b}$.*

Since $\mathfrak{A}_y$ has the only automorphism, σ and ε are uniquely determined by the residue class of α modulo $\mathfrak{b}$ through (3.5.1), while the couples (ε, α) and $(-\varepsilon, -\alpha)$ correspond to the same σ. If ε_m is an isomorphism of $A_{m,y}$ to $(A_{m,y})^\sigma$ obtained by arranging m copies of ε diagonally, then

$$(3.5.5) \qquad t_m(y, \xi)^\sigma = \varepsilon_m\big(t_m(y, \xi\alpha)\big) \qquad\qquad \big(\xi \in (\mathfrak{b}^{-1})^m\big) .$$

Since $y \to t_m(y, \xi)$ is a birational injection of V into W_m, defined over $Q(\zeta)$, we get, for every $\xi \in (\mathfrak{b}^{-1})^m$,

$$(3.5.6) \quad t_m(y, \xi)^\sigma = t_m(y^\sigma, \xi) \; \textit{if } \sigma \textit{ is the identity mapping on } Q(\zeta) .$$

If $\alpha \in \Gamma_1$, the element σ of $\mathfrak{G}_\mathfrak{b}$ corresponding to α can be obtained as follows. Take $z \in \mathfrak{H}$ so that $v(z) = y$. Put $v(\alpha(z)) = x$. Now the action of $(\alpha, 0)$ on $\mathfrak{H} \times \Psi_m$ defines a biregular automorphism of W_m. On the fibre $A_{m,y}$, this gives an isomorphism ε_m to $A_{m,x}$. From (3.2.4) and (3.2.6), it can be easily seen that ε_1 is an isomorphism of $\mathcal{P}_y$ to $\mathcal{P}_x$. Then x should be a generic point of V over $Q(\zeta)$, since $Q(\zeta, y)$ is algebraic over $Q(\zeta, x)$ [12, Prop. 9]. Hence there exists an isomorphism σ of $Q(\zeta, y)$ to $Q(\zeta, x)$ over $Q(\zeta)$ such that $y^\sigma = x$. We have then $\mathcal{P}_x = (\mathcal{P}_y)^\sigma$ and $\mathfrak{A}_x = (\mathfrak{A}_y)^\sigma$, so that σ is the identity mapping on K_y, in view of the definition of the field of moduli of $\mathcal{P}_y$ [17, II, 1.4, (M)] Again from (3.2.4), we get $\varepsilon_m(t_m(y, \xi\alpha)) = t_m(y^\sigma, \xi)$ for $\xi \in (\mathfrak{b}^{-1})^m$, since $\alpha\alpha' = 1$. This combined with (3.5.6) yields (3.5.5). In other words, *if σ corresponds to an element α of Γ_1 through (3.5.1) and (3.5.2), then*

$$(3.5.7) \qquad y^\sigma = v\big(\alpha(z)\big)$$

for any $z \in \mathfrak{H}$ such that $v(z) = y$; and ε_m is obtained from the action of $(\alpha, 0)$ on $\mathfrak{H} \times \Psi_m$.

4. Modular correspondences on the fibre variety W_m

4.1. Let $\Gamma_\mathfrak{b}$ and Δ_1 be the same as in 1.3. For the subgroup $\Gamma_\mathfrak{b} \times \mathfrak{o}^m$ of $GL_2^+(R) \times \Psi_m$, we can define $R(\Gamma_\mathfrak{b} \times \mathfrak{o}^m, \Delta_1 \times \mathfrak{o}^m)$ (cf. 1.1). For simplicity, let us put $\Gamma = \Gamma_\mathfrak{b}$, $\Delta = \Delta_1$, $\Gamma^* = \Gamma_\mathfrak{b} \times \mathfrak{o}^m$, $\Delta^* = \Delta \times \mathfrak{o}^m$, and write an element $(\alpha, 0)$ of $GL_2^+(R) \times \Psi_m$ as α^*. It can be easily verified that $R(\Gamma^*, \Delta^*)$ is spanned by the elements of the form $\Gamma^*\alpha^*\Gamma^*$ with α in Δ.

4.2. PROPOSITION. *Let $\alpha, \beta \in \Delta$. Then the following assertions hold.*

$(4.2.1) \quad \Gamma^*\alpha^* = (\Gamma\alpha) \times \mathfrak{o}^m, \; \Gamma^*\alpha^*\Gamma^* = (\Gamma\alpha\Gamma) \times \mathfrak{o}^m .$

$(4.2.2) \quad$ *If $\Gamma\alpha\Gamma = \bigcup_i \Gamma\alpha_i$ is a disjoint union, then, $\Gamma^*\alpha^*\Gamma^* = \bigcup_i \Gamma^*\alpha_i^*$ is a disjoint union.*

$(4.2.3) \quad$ *If $(\Gamma\alpha\Gamma)\cdot(\Gamma\beta\Gamma) = \sum_\lambda c_\lambda(\Gamma\xi_\lambda\Gamma)$, then*

$$(\Gamma^*\alpha^*\Gamma^*)\cdot(\Gamma^*\beta^*\Gamma^*) = \sum_\lambda c_\lambda(\Gamma^*\xi_\lambda^*\Gamma^*) .$$

This can be shown in a straightforward way. From this proposition we see immediately that $\Gamma\alpha\Gamma \to \Gamma^*\alpha^*\Gamma^*$ gives an isomorphism of $R(\Gamma, \Delta)$ onto $R(\Gamma^*, \Delta^*)$.

4.3. *Remark.* Since $\mathfrak{H}$ has an invariant measure, we see that the number of right cosets in $\Gamma\alpha\Gamma$ is equal to the number of left cosets in $\Gamma\alpha\Gamma$. This is no longer true for $\Gamma^*\alpha^*\Gamma^*$. In reality, one has: the number of cosets $\sigma\Gamma^*$ in $\Gamma^*\alpha^*\Gamma^*$ $= \det(\alpha)^{2m} \cdot$(the number of cosets $\Gamma^*\tau$ in $\Gamma^*\alpha^*\Gamma^*$).

4.4. *Correspondences* $X_m(\Gamma\alpha\Gamma)$. For an element α of Δ, put

$$(4.4.1) \qquad X_m = \{w_m(s) \times w_m(\alpha^*s) \mid s \in \mathfrak{H} \times \Psi_m\} \qquad (m \geq 0) .$$

It can be easily verified that X_m is a subvariety of $W_m \times W_m$. X_m depends only on $\Gamma\alpha\Gamma$, and is independent of the choice of the representative α for the double coset $\Gamma\alpha\Gamma$. We write therefore

$$(4.4.2) \qquad X_m = X_m(\Gamma\alpha\Gamma) \qquad (m \geq 0) .$$

It is clear that X_m is a proper cycle on $W_m \times W_m$ in the sense of 2.1. If $\Gamma\alpha\Gamma = \bigcup_{i=1}^{d} \Gamma\alpha_i$ is a disjoint union, one has

$$X_m[w_m(s)] = \sum_{i=1}^{d} w_m(\alpha_i^*s) \qquad (m \geq 0;\ s \in \mathfrak{H} \times \Psi_m) .$$

Furthermore, if $(\Gamma\alpha\Gamma)\cdot(\Gamma\beta\Gamma) = \sum_\lambda c_\lambda \Gamma\xi_\lambda\Gamma$, then, in view of (1.1.2) and Prop. 4.2, we get $X_m(\Gamma\alpha\Gamma)[X_m(\Gamma\beta\Gamma)[a]] = \sum_\lambda c_\lambda X_m(\Gamma\xi_\lambda\Gamma)[a]$ for every $a \in W_m$, so that $X_m(\Gamma\alpha\Gamma)\circ X_m(\Gamma\beta\Gamma) = \sum_\lambda c_\lambda X_m(\Gamma\xi_\lambda\Gamma)$ for $m \geq 0$. In other words, $\Gamma\alpha\Gamma \to X_m(\Gamma\alpha\Gamma)$ gives a homomorphism of $R(\Gamma, \Delta)$ into the ring of proper correspondences on W_m. It is clear that this homomorphism is injective if $m > 0$.

4.5. X_m *as a fibre correspondence.* To explain more visibly the action of X_m, take a point z on $\mathfrak{H}$ and put $v(z) = u$, $v(\alpha_i(z)) = u_i$ $(1 \leq i \leq d)$. Then $X_0[u] = u_1 + \cdots + u_d$. By the correspondence X_m, the fibre $A_{m,u}$ is mapped onto d fibres $A_{m,u_1}, \cdots, A_{m,u_d}$. Now we can define an isogeny λ_{mi} of $A_{m,u}$ to A_{m,u_i} by

$$(4.5.1) \qquad \lambda_{mi}(w_m(z, x)) = w_m(\alpha_i(z), x\alpha_i') \qquad (x \in \Psi_m) .$$

Then, if $t \in A_{m,u}$,

$$(4.5.2) \qquad X_m[t] = \lambda_{m1}t + \cdots + \lambda_{md}t .$$

Write λ_{1i} simply as λ_i. Then we notice that

$$(4.5.3) \qquad \mathrm{Ker}(\lambda_i) = \{w_1(z, x) \mid x \in \mathfrak{o}\cdot(\alpha_i')^{-1}\} ;$$

so if we put $a\mathfrak{o} = \{\xi \in \mathfrak{o} \mid \xi\mathfrak{o} \subset \mathfrak{o}\alpha'\}$ with an element a of $\mathfrak{o}$, then for every i,

$$(4.5.4) \qquad \theta_u(\eta)(\mathrm{Ker}(\lambda_i)) = 0 \Longleftrightarrow \eta \in a\mathfrak{o} \qquad (\eta \in \mathfrak{o}) .$$

In view of 1.4 and 1.5.3, we find that

(4.5.5) $\mathrm{Ker}(\lambda_1), \cdots, \mathrm{Ker}(\lambda_d)$ *are exactly all the distinct subgroups of* A_u *which are* $\mathfrak{o}$-*isomorphic to* $\mathrm{Ker}(\lambda_1)$. *In particular, if* $\det(\alpha) = p$ *for a prime number* p *which does not divide* $d(\Phi)$, *then* $d = p + 1$, *and the* $\mathrm{Ker}(\lambda_i)$ *are all the distinct subgroups of* A_u *which are* $\mathfrak{o}$-*invariant and of order* p^2 *(cf.* [14, § 4.2]).

4.6. PROPOSITION. *Let the notation be as in 4.4. and 4.5. Suppose that* $\det(\alpha)$ *is prime to* $\mathfrak{b}$, *and* u *is generic on* V *over* $\boldsymbol{Q}(\zeta)$. *Then the isogeny* λ_{mi} *is defined over* $\boldsymbol{Q}(\zeta, u, u_i)$ *for every* i.

PROOF. First we notice that λ_{mi} can be obtained by arranging m copies of λ_i diagonally. Therefore, it is sufficient to prove our assertion in the case $m = 1$. By [12, Prop. 9], $\boldsymbol{Q}(\zeta, u)$ is algebraic over $\boldsymbol{Q}(\zeta, u_i)$, so that the u_i are generic on V over $\boldsymbol{Q}(\zeta)$. From (4.5.1), we see easily that λ_i is an isogeny of $\mathscr{P}_u$ to $\mathscr{P}_{u_i}$. Since $\alpha_i' \equiv \alpha' \bmod \mathfrak{b}$, we have, by (4.5.1),

$$(4.6.1) \qquad \lambda_i\big(t(u, \xi)\big) = t(u_i, \xi\alpha') \qquad\qquad (\xi \in \mathfrak{b}^{-1}) .$$

Let σ be an isomorphism of C into itself over $\boldsymbol{Q}(\zeta, u, u_i)$. Then λ_i and λ_i^σ are isogenies of $\mathscr{P}_u$ to $\mathscr{P}_{u_i}$. By (4.5.4) and [14, Prop. 2.5, (ii)], there exists an automorphism η of $\mathscr{P}_{u_i}$ such that $\lambda_i^\sigma = \eta\lambda_i$. Since u_i is generic on V, we have $\eta = \pm 1$. Since $t(u, \xi)$ and $t(u_i, \xi\alpha')$ are rational over $\boldsymbol{Q}(\zeta, u, u_i)$, we have, by (4.6.1),

$$\eta\lambda_i\big(t(u, \xi)\big) = \lambda_i^\sigma\big(t(u, \xi)^\sigma\big) = t(u_i, \xi\alpha') = \lambda_i\big(t(u, \xi)\big) .$$

In view of (3.2.2) and our assumption that $\det(\alpha)$ is prime to $\mathfrak{b}$, we have $2\lambda_i(t(u, \xi)) \neq 0$ for a suitable element ξ of $\mathfrak{b}^{-1}$. Hence we get $\eta = 1$, so that $\lambda_i^\sigma = \lambda_i$. This proves that λ_i is defined over $\boldsymbol{Q}(\zeta, u, u_i)$.

4.7. PROPOSITION. *Let the notation and assumption be as in 4.6. Then* $X_0(\Gamma\alpha\Gamma)$ *is the locus of* $u \times u_1$ *over* $\boldsymbol{Q}(\zeta)$. *If* x *is a generic point of* $A_{m,u}$ *over* $\boldsymbol{Q}(\zeta, u)$, $X_m(\Gamma\alpha\Gamma)$ *is the locus of* $x \times \lambda_{m1}x$ *over* $\boldsymbol{Q}(\zeta)$.

PROOF. Since u and u_i are generic on V over $\boldsymbol{Q}(\zeta)$, there exists an isomorphism τ_i of $\boldsymbol{Q}(\zeta, u)$ to $\boldsymbol{Q}(\zeta, u_i)$ over $\boldsymbol{Q}(\zeta)$ such that $u^{\tau_i} = u_i$. Then $(\mathscr{P}_u)^{\tau_i} = \mathscr{P}_{u_i}$, and for $\xi \in (\mathfrak{b}^{-1})^m$, we have

$$(4.7.1) \qquad t_m(u, \xi)^{\tau_i} = t_m(u_i, \xi) ,$$

so that $(\mathscr{Q}_u)^{\tau_i} = \mathscr{Q}_{u_i}$, and by (4.6.1),

$$(4.7.2) \qquad \lambda_{mi}\big(t_m(u, \xi)\big) = t_m(u, \xi\alpha')^{\tau_i} .$$

Let σ be an isomorphism of C into itself over $\boldsymbol{Q}(\zeta, u)$. We see that σ gives an automorphism of an $\mathfrak{o}$-module

$$(4.7.3) \qquad \{t \in A_u \mid \theta_u(\det(\alpha))t = 0\} .$$

By (4.5.5), $\mathrm{Ker}\,(\lambda_1)^\sigma$ should coincide with one of the $\mathrm{Ker}(\lambda_i)$. If $\mathrm{Ker}(\lambda_1)^\sigma = \mathrm{Ker}(\lambda_i)$, then there exists an isomorphism ε of $(\mathscr{P}_{u_1})^\sigma$ to $\mathscr{P}_{u_i}$ such that $\lambda_i = \varepsilon\lambda_1^\sigma$ [14, Prop. 2.5, (i)]. We obtain, by (4.7.2),

$$(4.7.4) \qquad \begin{aligned} t(u,\, \xi\alpha')^{\tau_i} &= \lambda_i\big(t(u,\, \xi)\big) = \varepsilon\lambda_1^\sigma\big(t(u,\, \xi)\big) = \varepsilon\big(\lambda_1(t(u,\, \xi))^\sigma\big) \\ &= \varepsilon\big(t(u,\, \xi\alpha')^{\tau_1\sigma}\big)\,. \end{aligned}$$

Since $\det(\alpha)$ is prime to $\mathfrak{b}$, we can take ξ in such a way that $\xi \in \mathfrak{b}^{-1}$ and $\xi\alpha' \equiv \beta_0^{-1}$ mod $\mathfrak{o}$, where β_0 is as in (3.1.2). Then (4.7.4) shows that ε is an isomorphism of $(\mathscr{Q}_u)^{\tau_i}$ to $(\mathscr{Q}_u)^{\tau_1\sigma}$. Hence $\tau_i = \tau_1\sigma$ on the field of moduli of $\mathscr{Q}_u$, i.e., $Q(\zeta,\, u)$ (cf. [17, II, 1.4, (M)]). It follows that $u_1^\sigma = u^{\tau_1\sigma} = u^{\tau_i} = u_i$.

Now let k be a field of rationality for $X_0(\Gamma\alpha\Gamma)$ and $X_m(\Gamma\alpha\Gamma)$. To prove our proposition, we may assume that u is generic on V over k. Since $X_0(\Gamma\alpha\Gamma)[u] = u_1 + \cdots + u_d$, $X_0(\Gamma\alpha\Gamma)$ is the locus of $u \times u_1$ over k. Let X' be the locus of $u \times u_1$ over $Q(\zeta)$. For every isomorphism σ of C into itself over $Q(\zeta,\, u)$, we have proved that $u_1^\sigma = u_i$ for some i. Hence u_1 is algebraic over $Q(\zeta,\, u)$, and $X'[u]$ is a partial sum of $u_1 + \cdots + u_d$. From this it follows easily that $X' = X_0(\Gamma\alpha\Gamma)$. Next take a generic point x of $A_{m,u}$ over $k(u)$. Then $X_m(\Gamma\alpha\Gamma)[x] = \lambda_{m1}x + \cdots + \lambda_{md}x$, so that $X_m(\Gamma\alpha\Gamma)$ is the locus of $x \times \lambda_{m1}x$ over k. Let X'_m be the locus of $x \times \lambda_{m1}x$ over $Q(\zeta)$. Obviously $X'_m \supset X_m(\Gamma\alpha\Gamma)$. Combining 4.6 with the relation $\varphi_m(x) = u$ and $\varphi_m(\lambda_{m1}x) = u_1$, we obtain $Q(\zeta,\, x,\, \lambda_{m1}x) = Q(\zeta,\, x,\, u_1)$, so that

$$[Q(\zeta,\, x,\, \lambda_{m1}x) : Q(\zeta,\, x)] \leqq [Q(\zeta,\, u,\, u_1) : Q(\zeta,\, u)] = d\,.$$

Therefore we get $X'_m = X_m(\Gamma\alpha\Gamma)$, which proves our proposition.

4.8. *Remark.* The idea of 4.7 is a modification of the discussion in [14, 4.2–4.3]. The isogenies λ_i of [14, 4.2] are the same as the present λ_i, at most up to the factors ± 1. However, the isomorphisms τ_i defined in the proof of 4.7 are slightly different from the σ_i defined in [14, Prop. 4.4].

4.9. PROPOSITION. *Let K_v and $K_{v,\mathfrak{b}}$ be as in 3.4. Let u and v be generic points of V over $Q(\zeta)$, and τ an isomorphism of $Q(\zeta,\, u)$ $(=K_{u,\mathfrak{b}})$ to $Q(\zeta,\, v)$ $(= K_{v,\mathfrak{b}})$ over $Q(\zeta)$ such that $u^\tau = v$. If an automorphism σ of $K_{u,\mathfrak{b}}$ over K_u corresponds to an element α of $\mathfrak{o}$ through (3.5.1) and (3.5.2) with $y = u$, then the automorphism $\tau^{-1}\sigma\tau$ of $K_{v,\mathfrak{b}}$ over K_v corresponds to α by (3.5.1) and (3.5.2) with $y = v$.*

In fact, assume that (3.5.1) and (3.5.2) are true with $y = u$ and an isomorphism ε of $\mathscr{P}_u$ to $(\mathscr{P}_u)^\sigma$. Then ε^τ is an isomorphism of $\mathscr{P}_v$ to $(\mathscr{P}_v)^{\tau^{-1}\sigma\tau}$. Moreover, we have $\zeta^{\tau^{-1}\sigma\tau} = \zeta^\sigma = \zeta^{\det(\alpha)}$, and for $\xi \in \mathfrak{b}^{-1}$,

$$\varepsilon^\tau\big(t(u^\tau,\, \xi\alpha)\big) = t(u,\, \xi)^{\sigma\tau} = t(u^\tau,\, \xi)^{\tau^{-1}\sigma\tau}\,.$$

Since $v = u^\tau$, this proves our proposition.

4.10. *The birational correspondences* $Y_m(\alpha)$. For every rational integer r, prime to $\mathfrak{b}$, we denote by ρ_r the automorphism of $\mathbf{Q}(\zeta)$ such that $\zeta^{\rho_r} = \zeta^r$. Then V^{ρ_r}, $W_m^{\rho_r}$, $\varphi_m^{\rho_r}$ are meaningful. For every $\xi \in (\mathfrak{b}^{-1})^m$, $y \to t_m(y, \xi)$ gives a rational mapping of V into W_m. We denote by $t_m^{\rho_r}(\ , \xi)$ the transform of this mapping by ρ_r; so $x \to t_m^{\rho_r}(x, \xi)$ is a rational mapping of V^{ρ_r} into $W_m^{\rho_r}$. If $m = 1$, $t_1^{\rho_r}$ will be denoted by t^{ρ_r}.

Let u be a generic point of V over $\mathbf{Q}(\zeta)$, and α an element of $\mathfrak{o}$ such that $\det(\alpha)$ is prime to $\mathfrak{b}$. Put $r = \det(\alpha)$. As remarked in 3.5, from α we obtain an automorphism σ of $\mathbf{Q}(\zeta, u)$ and an isomorphism ε of $\mathscr{P}_u$ to $(\mathscr{P}_u)^\sigma$ such that $\sigma = \rho_r$ on $\mathbf{Q}(\zeta)$, and

$$(4.10.1) \qquad t(u, \xi)^\sigma = \varepsilon\big(t(u, \xi\alpha)\big) \qquad (\xi \in \mathfrak{b}^{-1}) \, .$$

We notice that σ and ε are uniquely determined by the residue class of α modulo $\mathfrak{b}$. Now u^σ is a point of V^{ρ_r}. Denote by $Y_0(\alpha)$ the locus of $u \times u^\sigma$ on $V \times V^{\rho_r}$. Since $\mathbf{Q}(\zeta, u) = \mathbf{Q}(\zeta, u^\sigma)$, we see that $Y_0(\alpha)$ is a birational correspondence of V to V^{ρ_r}. In view of 4.9, we see that $Y_0(\alpha)$ is determined by the residue-class of α modulo $\mathfrak{b}$, and independent of the choice of u. The relation (4.10.1) can be written as

$$(4.10.2) \qquad t^{\rho_r}\big(Y_0(\alpha)[u], \xi\big) = \varepsilon\big(t(u, \xi\alpha)\big) \qquad (\xi \in \mathfrak{b}^{-1}) \, .$$

Now let us prove that ε *is defined over* $\mathbf{Q}(\zeta, u)$. For this purpose, take an isomorphism τ of C into itself, over $\mathbf{Q}(\zeta, u)$. Then ε^τ is an isomorphism of $\mathscr{P}_u$ to $(\mathscr{P}_u)^\sigma$, and $\varepsilon^\tau(t(u, \xi\alpha)) = t(u, \xi)^\sigma = \varepsilon(t(u, \xi\alpha))$ for $\xi \in \mathfrak{b}^{-1}$. Since $\mathfrak{A}_u$ has the only automorphism, we get $\varepsilon^\tau = \varepsilon$, which proves that ε is defined over $\mathbf{Q}(\zeta, u)$.

Let x be a generic point of $A_{m,u}$ over $\mathbf{Q}(\zeta, u)$, and let ε_m be an isomorphism of $A_{m,u}$ to $(A_{m,u})^\sigma$ which is obtained by arranging m copies of ε diagonally. Let $Y_m(\alpha)$ denote the locus of $x \times \varepsilon_m x$ on $W_m \times W_m^{\rho_r}$ over $\mathbf{Q}(\zeta)$. Since $\mathbf{Q}(\zeta, u) = \mathbf{Q}(\zeta, u^\sigma)$ and $\varphi_m(x) = u$, $\varphi_m^{\rho_r}(\varepsilon_m x) = u^\sigma$, we have $\mathbf{Q}(\zeta, x) = \mathbf{Q}(\zeta, u, x) = \mathbf{Q}(\zeta, u^\sigma, \varepsilon_m x) = \mathbf{Q}(\zeta, \varepsilon_m x)$. Therefore $Y_m(\alpha)$ is a birational correspondence of W_m to $W_m^{\rho_r}$. Since W_m and $W_m^{\rho_r}$ are minimal models, $Y_m(\alpha)$ is everywhere biregular. By (3.5.5), we have

$$(4.10.3) \qquad t_m^{\rho_r}\big(Y_0(\alpha)[u], \xi\big) = Y_m(\alpha)[t_m(u, \xi\alpha)] \qquad (\xi \in (\mathfrak{b}^{-1})^m) \, .$$

In view of 4.9, we see that $Y_m(\alpha)$ is determined by the residue class of α modulo $\mathfrak{b}$, and independent of the choice of u and x. From our definition, we see easily

$$(4.10.4) \qquad \varphi_m^{\rho_r} \circ Y_m(\alpha) = Y_0(\alpha) \circ \varphi_m \, . \qquad (m \geqq 0) \, .$$

Let β be another element of $\mathfrak{o}$ such that $\det(\beta)$ is prime to $\mathfrak{b}$. By a straightforward argument, we can easily show (with $r = \det(\alpha)$)

$$(4.10.5) \qquad Y_m(\beta)^{\rho_r} \circ Y_m(\alpha) = Y_m(\beta\alpha) \qquad (m \geq 0) .$$

If $\alpha \in \Gamma_1$ then

$$(4.10.6) \qquad X_m(\Gamma\alpha\Gamma) = Y_m(\alpha) \qquad (m \geq 0) .$$

This follows easily from our definition and the consideration at the end of 3.5.

4.11. PROPOSITION. *Let ψ, χ and α be elements of Δ satisfying $\alpha\psi \equiv \chi\alpha$ mod $\mathfrak{b}$. Suppose that $\det(\psi)$ and $\det(\alpha)$ are prime to $\mathfrak{b}$, and put $r = \det(\psi)$. Then*

$$X_m(\Gamma\alpha\Gamma)^{\rho_r} \circ Y_m(\psi) = Y_m(\chi) \circ X_m(\Gamma\alpha\Gamma) \qquad (m \geq 0) .$$

PROOF. Let $u, u_i, \lambda_{mi}, \lambda_i$ be as in 4.5, and let σ and ε be as in (4.10.1) with ψ in place of α. Extend σ to an automorphism of the algebraic closure of $\boldsymbol{Q}(u)$, and denote it again by σ. Then $\lambda_i^\sigma \circ \varepsilon$ is an isogeny of $\mathscr{P}_u$ to $(\mathscr{P}_{u_i})^\sigma$, and $\mathrm{Ker}(\lambda_i^\sigma \circ \varepsilon) = \varepsilon^{-1}(\mathrm{Ker}(\lambda_i^\sigma))$. In view of (4.5.5), we observe that the $\mathrm{Ker}(\lambda_i^\sigma \circ \varepsilon)$ coincide with the $\mathrm{Ker}(\lambda_i)$ as a whole. Consequently, we have $\mathrm{Ker}(\lambda_1^\sigma \circ \varepsilon) = \mathrm{Ker}(\lambda_i)$ for some i. By [14, Prop. 2.5], there exists an isomorphism η of $\mathscr{P}_{u_i}$ to $(\mathscr{P}_{u_1})^\sigma$ such that $\eta\lambda_i = \lambda_1^\sigma\varepsilon$. We have then, for $\xi \in \mathfrak{b}^{-1}$,

$$
\begin{aligned}
\eta\big(t(u_i,\ \xi\psi\alpha')\big) &= \eta\lambda_i\big(t(u,\ \xi\psi)\big) && (4.6.1)\\
&= \lambda_1^\sigma\varepsilon\big(t(u,\ \xi\psi)\big) \\
&= \lambda_1^\sigma\big(t(u,\ \xi)^\sigma\big) && (\text{4.10.1 with } \psi \text{ in place of } \alpha)\\
&= t(u_1,\ \xi\alpha')^\sigma && (4.6.1) .
\end{aligned}
$$

Substituting ξ for $\xi\alpha'$, we get, since $\psi\alpha' \equiv \alpha'\chi$ mod $\mathfrak{b}$,

$$(4.11.1) \qquad \eta\big(t(u_i,\ \xi\chi)\big) = t(u_1,\ \xi)^\sigma \qquad (\xi \in \mathfrak{b}^{-1}) .$$

Now applying the definition of $Y_m(\alpha)$ in 4.10 to $Y_m(\chi)$, we find an automorphism τ of $\boldsymbol{Q}(\zeta,\ u_i)$ and an isomorphism δ of $\mathscr{P}_{u_i}$ to $(\mathscr{P}_{u_i})^\tau$ such that $Y_0(\chi)[u_i] = u_i^\tau$ and $\delta(t(u_i,\ \xi\chi)) = t(u_i,\ \xi)^\tau$ for $\xi \in \mathfrak{b}^{-1}$. From the last relation and (4.11.1), we get

$$\eta\delta^{-1}\big(t(u_i,\ \xi)^\tau\big) = t(u_1,\ \xi)^\sigma \qquad (\xi \in \mathfrak{b}^{-1}) ,$$

so that $\eta\delta^{-1}$ is an isomorphism of $(\mathscr{Q}_{u_i})^\tau$ to $(\mathscr{Q}_{u_1})^\sigma$. Let τ_i be as in the proof of 4.7. Then $(\mathscr{Q}_{u_i})^\tau = (\mathscr{Q}_u)^{\tau_i\tau}$ and $(\mathscr{Q}_{u_1})^\sigma = (\mathscr{Q}_u)^{\tau_1\sigma}$. By the property [17, II, 1.4, (M)] of the field of moduli, we have $\tau_i\tau = \tau_1\sigma$ on $\boldsymbol{Q}(\zeta,\ u)$, so that $(\mathscr{Q}_{u_i})^\tau = (\mathscr{Q}_{u_1})^\sigma$, $u_i^\tau = u_1^\sigma$. Therefore $\eta\delta^{-1}$ is an automorphism of $(\mathscr{Q}_{u_1})^\sigma$, hence $\eta = \delta$. Since $X_0(\Gamma\alpha\Gamma)^{\rho_r}$ is the locus of $u^\sigma \times u_1^\sigma$ over $\boldsymbol{Q}(\zeta)$ and $Y_0(\psi)[u] = u^\sigma$, we see that $X_0(\Gamma\alpha\Gamma)^{\rho_r} \circ Y_0(\psi) = $ the locus of $u \times u_1^\sigma$ over $\boldsymbol{Q}(\zeta)$. On the other hand, since $Y_0(\chi)[u_i] = u_i^\tau$, and $X_0(\Gamma\alpha\Gamma)$ is the locus of $u \times u_i$ over $\boldsymbol{Q}(\zeta)$, we see that $Y_0(\chi) \circ X_0(\Gamma\alpha\Gamma) = $ the locus of $u \times u_i^\tau$ over $\boldsymbol{Q}(\zeta)$. Therefore, from the equality $u_i^\tau = u_1^\sigma$, we obtain the relation of our proposition for $m = 0$. To prove the case

$m > 0$, take a generic point x of $A_{m,u}$ over $Q(\zeta, u)$. Then we can extend σ to an isomorphism which maps x to $\varepsilon_m x$, where ε_m in an isomorphism of $A_{m,u}$ to $(A_{m,u})^\sigma$ obtained by arranging m copies of ε diagonally. Hence we see that $X_m(\Gamma \alpha \Gamma)^{\rho_r}$ is the locus of $\varepsilon_m x \times \lambda_{m1}{}^\sigma \varepsilon_m x$. Since $Y_m(\psi)[x] = \varepsilon_m x$, we have $X_m(\Gamma \alpha \Gamma)^{\rho_r} \circ Y_m(\psi) =$ the locus of $x \times \lambda_{m1}{}^\sigma \varepsilon_m x$ over $Q(\zeta)$. Define an isomorphism δ_m of A_{m,u_t} to $(A_{m,u_t})^\tau$ so that $Y_m(\chi)[z] = \delta_m z$ for $z \in A_{m,u_t}$ (cf. 4.10). Then $Y_m(\chi) \circ X_m(\Gamma \alpha \Gamma)$ $=$ the locus of $x \times \delta_m \lambda_{m1} x$ over $Q(\zeta)$. Since $\eta = \delta$ and $\eta \lambda_i = \lambda_1{}^\tau \varepsilon$, we get $\delta_m \lambda_{mi}$ $= \lambda_{m1}{}^\sigma \varepsilon_m$. This proves our proposition for $m > 0$.

4.12. PROPOSITION. *Let α and β be elements of Δ such that $\Gamma_1 \alpha \Gamma_1 = \Gamma_1 \beta \Gamma_1$. Suppose that $\det(\alpha)$ is prime to $\mathfrak{b}$. Then*

$$Y_m(\alpha') \circ X_m(\Gamma \alpha \Gamma) = Y_m(\beta') \circ X_m(\Gamma \beta \Gamma) \qquad (m \geqq 0) .$$

PROOF. By (1.5.1), we have $\alpha = \gamma \beta \delta$ for an element $\gamma \in \Gamma_1$ and an element $\delta \in \Gamma$. Since $(\Gamma \gamma' \Gamma)(\Gamma \alpha \Gamma) = \Gamma \gamma' \alpha \Gamma = \Gamma \beta \Gamma$ (cf. 1.9), we have

$$\begin{aligned}
Y_m(\beta') \circ X_m(\Gamma \beta \Gamma) &= Y_m(\beta') \circ X_m(\Gamma \gamma' \Gamma) \circ X_m(\Gamma \alpha \Gamma) \\
&= Y_m(\beta') \circ Y_m(\gamma') \circ X_m(\Gamma \alpha \Gamma) && (4.10.6) \\
&= Y_m(\beta' \gamma') \circ X_m(\Gamma \alpha \Gamma) && (4.10.5) \\
&= Y_m(\alpha') \circ X_m(\Gamma \alpha \Gamma) .
\end{aligned}$$

For the last equality, we used the fact $\alpha' \equiv \beta' \gamma' \bmod \mathfrak{b}$.

4.13. PROPOSITION. *For every $\alpha \in \Delta$, ${}^t X_0(\Gamma \alpha \Gamma) = X_0(\Gamma \alpha' \Gamma)$.*

PROOF. This follows easily from our definition (4.4.1) of $X_0(\Gamma \alpha \Gamma)$ and the relation $\alpha'(\alpha(z)) = z$ $(z \in \mathfrak{H})$.

5. Congruence relations

5.1. Let k be a field, and Λ a set of places of k. We assume that every place in Λ corresponds to a discrete valuation of rank one. For each $\mathfrak{p} \in \Lambda$, we shall consider reduction modulo $\mathfrak{p}$ of algebro-geometric objects defined with respect to k. For the general theory of reduction modulo $\mathfrak{p}$, we refer the reader to [10] and [18, Ch. III]. The latter will serve as a rapid survey of the theory. For an object X rational over k, we denote by $\mathfrak{p}(X)$ the reduction of X modulo $\mathfrak{p}$; in case $\mathfrak{p}$ is fixed, $\mathfrak{p}(X)$ will be also denoted by $\tilde{X}$. We say that a proposition $P(\mathfrak{p})$ concerning the places $\mathfrak{p}$ in Λ holds for *almost all* $\mathfrak{p}$ in Λ, if there exist a finite number of non-zero elements $a_1, \cdots, a_r$ of k such that $P(\mathfrak{p})$ holds for all the $\mathfrak{p}$ in Λ with $\mathfrak{p}(a_1) \neq 0, \cdots, \mathfrak{p}(a_r) \neq 0$. For every $\mathfrak{p}$ in Λ, we shall denote by $\mathfrak{p}(k)$ the residue field of k modulo $\mathfrak{p}$.

5.2. LEMMA. *Let V be an affine or a projective variety defined over k, and x a generic point of V over k. Let t be an element of $k(x)$, other than 0. Put*

$$Z = \{a \in V \mid (x,\, t) \to (a,\, 0) \text{ ref. } k\},$$
$$Z_{\mathfrak{p}} = \{\alpha \in \mathfrak{p}(V) \mid (x,\, t) \to (\alpha,\, 0) \text{ ref. } \mathfrak{p}\}.$$

(*For the notation, cf.* [18, § 9.2].) *Then* $Z_{\mathfrak{p}} = \mathfrak{p}(Z)$ *for almost all* $\mathfrak{p}$ *in* Λ.

PROOF. Let D be the projective line, and T the locus of $(x,\, t)$ in $V \times D$ over k. Denote by φ resp. $\varphi_{\mathfrak{p}}$ the projection map of $V \times D$ resp. $\mathfrak{p}(V) \times \mathfrak{p}(D)$ to V resp. $\mathfrak{p}(V)$. It is clear that

$$Z = \varphi[T \cap (V \times 0)], \qquad Z_{\mathfrak{p}} = \varphi_{\mathfrak{p}}[\mathfrak{p}(T) \cap (\mathfrak{p}(V) \times \mathfrak{p}(0))].$$

We note that $\mathfrak{p}(\varphi(X)) = \varphi_{\mathfrak{p}}(\mathfrak{p}(X))$ for every Zariski k-closed subset X of $V \times D$. Therefore our assertion is an easy consequence of [18, § 12.3, Prop. 19].

5.3. PROPOSITION. *Let* $\{V,\, W,\, \varphi,\, f\}$ *be a fibre system of abelian varieties defined over* k *(cf. 2.2). Then, for almost all* $\mathfrak{p}$ *in* Λ*, the following assertions hold.*

(5.3.1) φ *is everywhere defined on* $\mathfrak{p}(W)$*, and* f *is everywhere defined on* $\mathfrak{p}(V)$ *(for the terminology, cf.* [18, § 10.1]).

(5.3.2) $\{\mathfrak{p}(V),\, \mathfrak{p}(W),\, \mathfrak{p}(\varphi),\, \mathfrak{p}(f)\}$ *is a fibre system of abelian varieties defined over* $\mathfrak{p}(k)$.

(5.3.3) *If* S *is the graph of* φ*,* $\mathfrak{p}(S)$ *is the graph of* $\mathfrak{p}(\varphi)$.

We say that $\{V,\, W,\, \varphi,\, f\}$ *behaves well for* $\mathfrak{p}$ if these three conditions are satisfied.

PROOF. By [18, § 12.3, Prop. 23, Prop. 24], [10, § 6, Th. 26] and [18, §10.1, Prop. 2], for almost all $\mathfrak{p}$ in Λ, $\mathfrak{p}(V)$ and $\mathfrak{p}(W)$ are non-singular projective varieties defined over $\mathfrak{p}(k)$, and the assertions (5.3.1) and (5.3.3) hold. For every finite algebraic extension K of k, let $\Lambda(K)$ be the set of extensions in K of all the $\mathfrak{p}$ in Λ with these properties. Let r be an integer such that $0 \leq r \leq \dim(V)$. We are going to prove the following assertion (A_r) by induction on r:

(A_r) *Let* U *be a subvariety of* V *of dimension* r*, defined over a finite algebraic extension* K *of* k*. Then, for almost all* $\mathfrak{p}$ *in* $\Lambda(K)$*, the following assertion holds:* $\mathfrak{p}(S)$ *and* $\mathfrak{p}(W) \times \alpha$ *intersect properly on* $\mathfrak{p}(W) \times \mathfrak{p}(V)$*, and* $\mathfrak{p}(S) \cdot (\mathfrak{p}(W) \times \alpha)$ *is an abelian variety with multiplicity one and with the origin* $f(\alpha) \times \alpha$*, for all* α *in* $\mathfrak{p}(U)$.

Let x be a generic point of such a variety U over K. Then $S \cdot (W \times x)$ is an abelian variety defined over $K(x)$. By [18, § 12.3, Prop. 19, Prop. 25], there exist a finite number of non-zero elements $t_1,\, \cdots,\, t_n$ of $K(x)$ such that: if $\mathfrak{q}$ is a place of $K(x)$ and $\mathfrak{q}(t_1) \neq 0,\, \cdots,\, \mathfrak{q}(t_n) \neq 0$, then

$$\mathfrak{q}(S) \cap (\mathfrak{q}(W) \times \mathfrak{q}(x)) = \mathfrak{q}(S \cap (W \times x)),$$

and $q(S \cdot (W \times x))$ is an abelian variety with multiplicity one and with the origin $q(f(x) \times x)$. Put

$$Z_i = \{a \in U \mid (x, t_i) \rightarrow (a, 0) \text{ ref. } K\} \qquad (1 \leq i \leq n) .$$

By 5.2, we have, for almost all $\mathfrak{p}$ in $\Lambda(K)$,

$$\mathfrak{p}(Z_i) = \{\alpha \in \mathfrak{p}(U) \mid (x, t_i) \rightarrow (\alpha, 0) \text{ ref. } \mathfrak{p}\} \qquad (1 \leq i \leq n) .$$

For such a $\mathfrak{p}$, let $\alpha \in \mathfrak{p}(U) - \bigcup_{i=1}^{n}\mathfrak{p}(Z_i)$. Then we can find a specialization $(x, t_1, \cdots, t_n) \rightarrow (\alpha, \beta_1, \cdots, \beta_n)$ ref. $\mathfrak{p}$ with $\beta_1 \neq 0, \cdots, \beta_n \neq 0$. By [10, § 5, Prop. 26], there exists a discrete place q of $K(x)$ such that $q((x, t_1, \cdots, t_n)) = (\alpha, \beta_1, \cdots, \beta_n)$ and $q = \mathfrak{p}$ on K. By our choice of the t_i, we have

$$\mathfrak{p}(S) \cap \big(\mathfrak{p}(W) \times \alpha\big) = q(S) \cap \big(q(W) \times q(x)\big) = q\big(S \cap (W \times x)\big) .$$

Since the reduction process does not augment the dimension of varieties [10, § 3, Prop. 19], this implies that $\mathfrak{p}(S)$ and $\mathfrak{p}(W) \times \alpha$ intersect properly on $\mathfrak{p}(W) \times \mathfrak{p}(V)$. Then, by [10, § 4, Th. 17], we get

$$\mathfrak{p}(S) \cdot \big(\mathfrak{p}(W) \times \alpha\big) = q\big(S \cdot (W \times x)\big) .$$

By the property of the t_i, $\mathfrak{p}(S) \cdot (\mathfrak{p}(W) \times \alpha)$ is an abelian variety with multiplicity one and with the origin $f(\alpha) \times \alpha$. Now let the U_j be the components of the Z_i, and let L be a finite algebraic extension of K over which the U_j are defined. Since the U_j are of dimension less than $\dim(U)$, we can conclude (A_r), by combining the above result for the points in $\mathfrak{p}(U) - \bigcup_{i=1}^{n}\mathfrak{p}(Z_i)$ with the assertions (A_s) for $s < r$. This completes our proof, since (A_r) for $r = \dim(V)$ implies (5.3.2).

5.4. PROPOSITION. *Let $\{V, W, \varphi, f\}$ and $\{V', W', \varphi', f'\}$ be fibre systems of abelian varieties defined over k. Let Y be a subvariety of $V \times V'$, defined over k, which is proper in the sense of 2.1, and X a fibre correspondence on $W \times W'$ associated with Y over k. Then X is proper. Moreover, suppose that both $\{V, W, \varphi, f\}$ and $\{V', W', \varphi', f'\}$ behave well for a place $\mathfrak{p}$ in Λ, and $\mathfrak{p}(Y)$ is proper. Then $\mathfrak{p}(X)$ is proper.*

PROOF. Let (a, b) be a specialization of $(x, \lambda x)$ over k. Extend this specialization to a discrete place P of $k(x, \lambda x)$. Put $P(u) = y$, $P(v) = z$. Then $P(\varphi^{-1}(u)) = \varphi^{-1}(y)$, $P(\varphi^{-1}(v)) = \varphi^{-1}(z)$. By Koizumi [5, Th. 3], $\varphi^{-1}(u)$ and $\varphi^{-1}(v)$ have no defect for P in the sense of [18, § 11.1]. Therefore, by [18, § 11.1, Prop. 12], $P(\lambda)$ is an isogeny of $\varphi^{-1}(y)$ to $\varphi^{-1}(z)$, and $b = P(\lambda x) = P(\lambda) a$. Since Y is proper, z is algebraic over $k(y)$. Since $P(\lambda)$ is defined over an algebraic extension of $k(y, z)$, and since $y = \varphi(a)$, we see that b is algebraic over $k(a)$. On the other hand, we can find an isogeny μ of $\varphi^{-1}(z)$ to $\varphi^{-1}(y)$ such that $\mu P(\lambda) = g \cdot (\text{identity map of } \varphi^{-1}(y))$ for a positive integer g. Then we have $\mu b = g a$. Now μ is defined over an algebraic extension of $k(y, z)$. Since Y is proper, y is algebraic over $k(z)$. Therefore ga is algebraic over $k(b)$, and hence a is algebraic over $k(b)$. This proves that X is proper. Let $\mathfrak{p} \in \Lambda$, and let (α, β) be a specialization of $(x, \lambda x)$ over $\mathfrak{p}$. Extend this specialization to a discrete place of $k(x, \lambda x)$ and apply the same argument as above, employing again [5, Th. 3]. Then we find that $\mathfrak{p}(X)$ is proper.

5.5. PROPOSITION. *Let U, V, W be projective non-singular varieties of the same dimension n, defined over k, and let X resp. Y be positive n-cycles on $U \times V$ resp. $V \times W$. Suppose that $\mathfrak{p}(U)$, $\mathfrak{p}(V)$, $\mathfrak{p}(W)$ are non-singular, and $\mathfrak{p}(X)$, $\mathfrak{p}(Y)$ are proper, for a place $\mathfrak{p}$ of Λ. Then $\mathfrak{p}(Y \circ X) = \mathfrak{p}(Y) \circ \mathfrak{p}(X)$.*

This follows easily from (2.1.1) and [10, § 4, Th. 17, 18, 19].

5.6. Let $\mathfrak{o}$ and $\mathfrak{b}$ be as in 1.2, and let V, W_m, φ_m, f_m, $t_m(y, \xi)$ be as in 3.2 and 3.3. Furthermore let b_0 and ζ be as in 3.4. We are going to consider reduction modulo $\mathfrak{p}$ for a prime divisor $\mathfrak{p}$ of $Q(\zeta)$. For every prime divisor $\mathfrak{p}$ of $Q(\zeta)$, we shall denote by $\kappa_{\mathfrak{p}}$ the residue field modulo $\mathfrak{p}$.

Let $\Lambda'(\mathfrak{b})$ be the set of all the prime divisors $\mathfrak{p}$ of $Q(\zeta)$ such that the following conditions are satisfied:

(5.6.1)   *$\mathfrak{p}$ is prime to $b_0 d(\Phi)$, where $d(\Phi)$ is the discriminant of Φ.*

(5.6.2)   *For every automorphism σ of $Q(\zeta)$, the fibre system $\{V^\sigma, W_1^\sigma, \varphi_1^\sigma, f_1^\sigma\}$ behaves well for $\mathfrak{p}$.*

(5.6.3)   *For every element α of $\mathfrak{o}$ such that $\det(\alpha)$ is prime to b_0, the birational mapping of V to V^τ (resp. W_1 to W_1^τ) with the graph $Y_0(\alpha)$ (resp. $Y_1(\alpha)$) is everywhere defined on $\mathfrak{p}(V)$ (resp. $\mathfrak{p}(W_1)$), where τ is an automorphism of $Q(\zeta)$ such that $\zeta^\tau = \zeta^{\det(\alpha)}$.*

By 5.3 and [18, § 12.7, Prop. 24], $\Lambda'(\mathfrak{b})$ contains all except a finite number of prime divisors of $Q(\zeta)$. If (5.6.2) is satisified, then for every positive integer m, we see that the fibre system $\{V^\sigma, W_m^\sigma, \varphi_m^\sigma, f_m^\sigma\}$ behaves well for $\mathfrak{p}$, in view of our agreement at the end of 3.4. If (5.6.3) is satisfied, then, for every integer $m \geq 0$, $\mathfrak{p}(Y_m(\alpha))$ is the graph of a biregular isomorphism of $\mathfrak{p}(W_m)$ to $\mathfrak{p}(W_m^\tau)$.

5.7. Let us fix a generic point x of W_1 and put $\varphi_1(x) = u$. Let K be the algebraic closure of $Q(\zeta, x)$. For each $\mathfrak{p} \in \Lambda'(\mathfrak{b})$, we can find a place $\mathfrak{p}_0$ of K which is an extension of $\mathfrak{p}$, with the following properties:

(5.7.1)   *$\mathfrak{p}_0(x)$ is a generic point of $\mathfrak{p}(W_1)$ over $\kappa_{\mathfrak{p}}$;*

(5.7.2)   *$\mathfrak{p}_0(u)$ is a generic point of $\mathfrak{p}(V)$ over $\kappa_{\mathfrak{p}}$.*

Such a $\mathfrak{p}_0$ is not unique. For each $\mathfrak{p} \in \Lambda'(\mathfrak{b})$ we take and fix, once and for all, such a $\mathfrak{p}_0$. The restriction of $\mathfrak{p}_0$ to a finite algebraic extension of $Q(\zeta, x)$ is discrete, while $\mathfrak{p}_0$ itself is not discrete.

Let us choose and fix, once for all, a basic polar divisor Z in C_u (in the sense of [18, p. 30]), rational over K. Let A be a projective embedding of A_u by means of the linear system defined by $3Z$, and let $\mathfrak{Q} = (A, C, \theta, t)$ be a structure, with this A, isomorphic to $\mathfrak{Q}_u = (A_u, C_u, \theta_u, t_u)$. Put $\mathscr{P} = (A, C, \theta)$. We may assume that $\mathfrak{Q}$ is defined over K. In the subsequent consideration, we shall fix such a $\mathfrak{Q}$.

5.8. Let $\alpha_1, \cdots, a_r$ be representatives of all the invertible residue classes of $\mathfrak{o}$ modulo $\mathfrak{b}$. For each α_i, we can find an automorphism ω_i of K such that $u^{\omega_i} = Y_0(\alpha_i)[u]$, $\zeta^{\omega_i} = \zeta^{\det(\alpha_i)}$(cf. 3.5 and 4.10). For each α_i, we fix once for all such an ω_i. Notice that $Q(\zeta, u) = Q(\zeta, u^{\omega_i})$.

Let $\Lambda''(\mathfrak{b})$ be the subset of $\Lambda'(\mathfrak{b})$ consisting of all the $\mathfrak{p}$ such that the following conditions are satisfied:

(5.8.1) A^{ω_i} *is without defect for* $\mathfrak{p}_0$, *for every* i.

(5.8.2) $\mathfrak{p}_0\big(\mathscr{F}(A^{\omega_i}, \theta^{\omega_i}, t^{\omega_i})\big) = \mathscr{F}\big(\mathfrak{p}_0(A^{\omega_i}), \mathfrak{p}_0(\theta^{\omega_i}), \mathfrak{p}_0(t^{\omega_i})\big)$ *for every* i.

Here $\mathscr{F}$ is the variety defined in [17, II, § 1.2] with respect to a fixed basis of $\mathfrak{o}$. In view of [18, § 12.1, Lemma 5], and modifying the consideration of [14, § 5.1] suitably, we can verify easily that $\Lambda''(\mathfrak{b})$ contains all except a finite number of prime divisors of $Q(\zeta)$.

5.9. Let c be the Chow point of $\mathscr{F}(A, \theta, t)$, and U the locus of c over $Q(\zeta)$. Then $Q(c)$ is the field of moduli of $\mathfrak{Q}$ [17, II, § 1.4], and hence of $\mathfrak{Q}_u$, so that $Q(c) = Q(\zeta, u)$. Therefore, U is birationally equivalent with V over $Q(\zeta)$. Let $\Lambda(\mathfrak{b})$ be the subset of $\Lambda''(\mathfrak{b})$ consisting of all the $\mathfrak{p}$ satisfying the following conditions:

(5.9.1) U^{ω_i} *is* $\mathfrak{p}$-*simple for every* i;

(5.9.2) $\mathfrak{p}_0(c^{\omega_i})$ *is a generic point of* $\mathfrak{p}(U^{\omega_i})$ *over* $\kappa_\mathfrak{p}$, *for every* i;

(5.9.3) $\kappa_\mathfrak{p}(\mathfrak{p}_0(c^{\omega_i})) = \kappa_\mathfrak{p}(\mathfrak{p}_0(u^{\omega_i})) = \kappa_\mathfrak{p}(\mathfrak{p}_0(u))$ *for every* i.

Then $\Lambda(\mathfrak{b})$ contains all except a finite number of prime divisors of $Q(\zeta)$.

5.10. L**EMMA.** *Let* p *be an odd prime number which does not divide* $d(\Phi)$. *Then there are infinitely many imaginary quadratic extensions* F *of* Q *such that* Φ *splits over* F *and* p *decomposes in* F.

P**ROOF.** Let $d = d(\Phi)$, and let e be a square-free integer prime to pd. One can find a positive integer g such that

$$g \equiv de \bmod (de)^2 , \qquad g \equiv -1 \bmod (p) .$$

Then $F = Q(\sqrt{-g})$ has the required property. One obtains infinitely many such F, by choosing e suitably.

5.11. Proposition. *The notation being as in 5.7, let $\mathfrak{p} \in \Lambda'(\mathfrak{b})$, and p the prime number divisible by $\mathfrak{p}$. Then:*

(i) $\operatorname{End}_Q(\mathfrak{p}_0(A_u)) = \tilde{\theta}_u(\Phi)$;

(ii) $\mathfrak{p}_0(\mathcal{Q}_u)$ *has no automorphism other than the identity mapping;*

(iii) $\mathfrak{p}_0(A_u)$ *has exactly p^2 points t such that $pt = 0$.*

Here $\tilde{\theta}_u$ is an isomorphism of Φ into $\operatorname{End}_Q(\mathfrak{p}_0(A_u))$ obtained by $\tilde{\theta}_u(a) = \mathfrak{p}_0(\theta_u(a))$. (cf. also [14, Prop. 5.3].)

Proof. Take F as in 5.10 for the prime number p. There exists an elliptic curve E defined over an algebraic number field such that $\operatorname{End}_Q(E)$ is isomorphic to F. Put $B = E \times E$. Since Φ splits over F, there exists an isomorphism of Φ into $\operatorname{End}_Q(B)$. By [12, p. 133, Th. 3], B is isomorphic to a complex torus C^2/D with $D = \left\{ \alpha \begin{pmatrix} x_1 \\ x_2 \end{pmatrix} \middle| \alpha \in \mathfrak{m} \right\}$, where x_1 and x_2 are complex numbers such that $x_2 \neq 0$, $\operatorname{Im}(x_1/x_2) \neq 0$, and $\mathfrak{m}$ is a Z-lattice in Φ. Since $\mathfrak{o}$ contains an element with determinant -1 (a result due to Eichler), we see easily that C^2/D is isogenous to a torus C^2/D_z with D_z defined by (3.1.1) for a point z of $\mathfrak{H}$ (cf. [12, p. 134]). Put $y = v(z)$. Then A_y is isogenous to B, and y is algebraic over Q, on account of (3.4.6). Since p decomposes in F, $\operatorname{End}_Q(\mathfrak{p}_0(E))$ is isomorphic to F, and $\mathfrak{p}_0(E)$ has exactly p points t such that $pt = 0$. This fact is due to Deuring, cf. also [18, p. 114, Th. 2, Proof, pp. 62–63]. Therefore $\operatorname{End}_Q(\mathfrak{p}_0(A_y))$ is isomorphic to $M_2(F)$, and $\mathfrak{p}_0(A_y)$ has exactly p^2 points of order 1 or p. Since $\mathfrak{p}_0(A_y)$ is a specialization of $\mathfrak{p}_0(A_u)$, and there are infinitely many choices of F, we get the first and third assertions. It follows also that $\mathfrak{p}_0(\mathscr{P}_u)$ has ± 1 as only automorphisms. Since p is prime to b_0, $\mathfrak{p}_0(t_u)$ has the same order as t_u [18, § 11.1, Prop. 13]. From this we obtain the second assertion.

5.12. Proposition. *Let $\mathfrak{p}$ and p be as in 5.11. Let s be a generic point of $\mathfrak{p}(V)$ over $\kappa_{\mathfrak{p}}$, and κ an extension of $\kappa_{\mathfrak{p}}(s)$. Let r be a generic point of $\tilde{\varphi}_1^{-1}(s)$ over κ, and L a field such that $\kappa(pr) \subset L \subset \kappa(r)$ and $\kappa(r)$ is a purely inseparable extension of L of degree p^2. Then $L = \kappa(r^p)$.*

Proof. It is well known that $\kappa(pr) \subset \kappa(r^p)$, and $[\kappa(r) : \kappa(pr)]_s$ is the number of points t on $\tilde{\varphi}_1^{-1}(s)$ such that $pt = 0$. By (iii) of 5.11, the latter is p^2. It follows that $\kappa(r^p)$ is the maximal separable extension of $\kappa(pr)$ in $\kappa(r)$. On the other hand, we observe from our assumption that L is a separable extension of $\kappa(pr)$ of degree p^2. Hence $L = \kappa(r^p)$.

5.13. Let $\mathfrak{p} \in \Lambda(\mathfrak{b})$ and let p be the prime number divisible by $\mathfrak{p}$. Denote by ρ the automorphism of $Q(\zeta)$ such that $\zeta^\rho = \zeta^p$. For every power q of p, and for every algebro-geometric object X defined in the universal domain of characteristic p, we shall write X^q the transform of X by the qth power automorphism of the universal domain. For example, we have $\mathfrak{p}(W_m^{\mathfrak{p}}) = \mathfrak{p}(W_m)^p$ with this convention.

For every integer $m \geq 0$, we denote by Π_{mp} and Π_{mp}^* the Frobenius correspondence of degree p on $\mathfrak{p}(W_m) \times \mathfrak{p}(W_m)^p$ and the anti-Frobenius fibre correspondence on $\mathfrak{p}(W_m)^p \times \mathfrak{p}(W_m)$ of degree p, respectively (cf. 2.5). Recall that $\Pi_{0p}^* = {}'\Pi_{0p}$, and Π_{mp}^* is the locus of $r' \times \pi'r'$ on $\mathfrak{p}(W_m)^p \times \mathfrak{p}(W_m)$ over κ_p, where r' is a generic point of $\mathfrak{p}_0(A_{m,u})^p$ $(= \tilde{\varphi}_m^{-1}(u)^p)$ over $\kappa_p(\tilde{u})$, and π' is an isogeny of $\mathfrak{p}_0(A_{m,u})^p$ to $\mathfrak{p}_0(A_{m,u})$ such that $\pi'(t^p) = pt$ for $t \in \mathfrak{p}_0(A_{m,u})$. By (2.5.2), we have

$$(5.13.1) \qquad \Pi_{mp}^*[r'] = p \cdot (\pi'r') .$$

5.14. THEOREM. *Let* $\mathfrak{p} \in \Lambda(\mathfrak{b})$, *and let* p *be the prime number divisible by* $\mathfrak{p}$. *Let* α *be an element of* $\mathfrak{o}$ *such that* $\det(\alpha) = p$. *Then the following equality holds:*

$$\mathfrak{p}(X_m(\Gamma \alpha \Gamma)) = \mathfrak{p}({}^tY_m(\alpha')) \circ \Pi_{mp} + \Pi_{mp}^* \circ \mathfrak{p}(Y_m(\alpha)) \qquad (m \geq 0) .$$

Proof will be given in four steps. We remind the reader that a generic point x of W_1 over $\mathbf{Q}(\zeta)$ has been fixed and $u = \varphi_1(x)$; we have also chosen a place $\mathfrak{p}_0$ (cf. 5.7). Throughout the proof, $\mathfrak{p}_0(U)$ will be often denoted by $\tilde{U}$, as remarked in 5.1, for an algebro-geometric object U.

Step 1. Put $X_m = X_m(\Gamma \alpha \Gamma)$ for $m \geq 0$ and $X_0[u] = u_1 + \cdots + u_d$ as in 4.5. We note that $d = p + 1$, since $\det(\alpha) = p$. Define an isogeny λ_{mi} of $A_{m,u}$ to A_{m,u_i} by (4.5.1), and put $\lambda_i = \lambda_{1i}$. For any abelian variety B, let us denote by $g(p, B)$ the set of points t on B such that $pt = 0$. By [18, § 11.1, Prop. 13], the place $\mathfrak{p}_0$ defines a surjective homomorphism of $g(p, A_u)$ to $g(p, \mathfrak{p}_0(A_u))$. Now $\mathrm{Ker}\,(\lambda_1), \cdots, \mathrm{Ker}\,(\lambda_{p+1})$ are exactly all the distinct $\mathfrak{o}$-invariant subgroups of $g(p, A_u)$ of order p^2 (4.5.5). Since $g(p, \mathfrak{p}_0(A_u))$ is of order p^2 (5.11), we have, after re-ordering the λ_i,

$$(5.14.1) \qquad \mathfrak{p}_0(\mathrm{Ker}\,(\lambda_1)) = \{0\} ,$$

$$(5.14.2) \qquad \mathfrak{p}_0(\mathrm{Ker}\,(\lambda_i)) = g(p, \mathfrak{p}_0(A_u)) \qquad (2 \leq i \leq p + 1) .$$

By [18, § 11.1, Prop. 13], we have $\mathfrak{p}_0(\mathrm{Ker}\,(\lambda_j)) = \mathrm{Ker}\,(\tilde{\lambda}_j)$ for every j. Now we can find an isogeny λ_i' of $\mathscr{P}_{u_i}$ to $\mathscr{P}_u$ such that

$$(5.14.3) \qquad \lambda_i' \lambda_i = p \cdot (\text{the identity mapping of } A_u).$$

By (iii) of 5.11, we have

$$(5.14.4) \qquad \mathfrak{p}_0(\mathrm{Ker}\,(\lambda_1')) = g(p, \mathfrak{p}_0(A_{u_1})) ,$$

$$(5.14.5) \qquad \mathfrak{p}_0(\mathrm{Ker}\,(\lambda_i')) = \{0\} \qquad (2 \leq i \leq p + 1) .$$

Step 2. Put $v_1 = Y_0(\alpha')[u_1]$. Then there exist an automorphism σ of $Q(\zeta, u)$ and an isomorphism ε of $\mathcal{P}_{u_1}$ to $(\mathcal{P}_{u_1})^\sigma$ such that $\sigma = \rho$ on $Q(\zeta)$, $u_1^\sigma = v_1$, and

$$(5.14.6) \qquad t^p(v_1, \xi) = t(u_1, \xi)^\sigma = \varepsilon\big(t(u_1, \xi\alpha')\big) \qquad (\xi \in \mathfrak{b}^{-1})$$

(cf. 3.5, 4.10). Put $\mu = \varepsilon\lambda_1$, $\mu' = \lambda_1'\varepsilon^{-1}$. By (4.6.1) and (5.14.6), we have

$$(5.14.7) \qquad \mu\big(t(u, \xi)\big) = t^p(v_1, \xi) \qquad (\xi \in \mathfrak{b}^{-1}) .$$

Now μ and μ' are defined over a finite algebraic extension k of $Q(\zeta, u)$. Hence $\tilde{\mu}$ and $\tilde{\mu}'$ are defined over $\tilde{k} = \mathfrak{p}_0(k)$. In view of (5.7.2), $\tilde{k}$ is algebraic over $\kappa_\mathfrak{p}(\tilde{u})$. It follows from this and (5.7.1) that $\tilde{x}$ is generic on $\mathfrak{p}_0(A_u)$ over $\tilde{k}$. Now we have $p\tilde{x} = \tilde{\mu}'\tilde{\mu}\tilde{x}$, so that $\tilde{k}(p\tilde{x}) \subset \tilde{k}(\tilde{\mu}\tilde{x}) \subset \tilde{k}(\tilde{x})$. By (5.14.1), $\tilde{k}(\tilde{x})$ is a purely inseparable extension of $\tilde{k}(\tilde{\mu}\tilde{x})$ of degree p^2. By 5.12, we get $\tilde{k}(\tilde{\mu}\tilde{x}) = \tilde{k}(\tilde{x}^p)$. Hence there exists an isomorphism χ of $\mathfrak{p}_0(A_{u_1}^\sigma)$ to $\mathfrak{p}_0(A_u)^p$ such that $\chi\tilde{\mu} = \pi$, where π is the pth power isogeny of $\mathfrak{p}_0(A_u)$ to $\mathfrak{p}_0(A_u)^p$. By [14, Prop. 2.4], we see that χ is an isomorphism of $\mathfrak{p}_0((\mathcal{P}_{u_1})^\sigma)$ to $\mathfrak{p}_0(\mathcal{P}_u)^p$. Furthermore we have, by (5.14.7),

$$\chi\big(\mathfrak{p}_0\big(t^p(v_1, \xi)\big)\big) = \chi\tilde{\mu}\big(\mathfrak{p}_0\big(t(u, \xi)\big)\big) = \pi\big(\mathfrak{p}_0\big(t(u, \xi)\big)\big) = \mathfrak{p}_0\big(t(u, \xi)\big)^p .$$

Therefore χ is an isomorphism of $\mathfrak{p}_0((\mathfrak{Q}_{u_1})^\sigma)$ to $\mathfrak{p}(\mathfrak{Q}_u)^p$. Let τ_i be the isomorphism of $Q(\zeta, u)$ to $Q(\zeta, u_i)$ over $Q(\zeta)$ such that $u^{\tau_i} = u_i$ (cf. Proof of 4.7). Put $\sigma_1 = \tau_1\sigma$. Then $u^{\sigma_1} = v_1$, $(\mathfrak{Q}_u)^{\sigma_1} = (\mathfrak{Q}_{u_1})^\sigma$, $\sigma_1 = \rho$ on $Q(\zeta)$. By (5.6.3), we have $\tilde{v}_1 = \mathfrak{p}(Y_0(\alpha'))[\tilde{u}_1]$.

Let $\mathfrak{Q}$ be as in 5.7. By (5.8.1), $\mathfrak{p}_0(\mathfrak{Q})$ is meaningful, and isomorphic to $\mathfrak{p}_0(\mathfrak{Q}_u)$. Therefore we see that $\mathfrak{p}_0(\mathfrak{Q})^p \cong \mathfrak{p}_0(\mathfrak{Q}_u)^p \cong \mathfrak{p}_0((\mathfrak{Q}_{u_1})^\sigma)$. By (5.6.2), $\mathfrak{p}_0((A_{u_1})^\sigma)$ is defined over $\kappa_\mathfrak{p}(\tilde{v}_1)$. Hence $\mathfrak{p}_0((\mathfrak{Q}_{u_1})^\sigma)$ is defined over a finite algebraic extension κ of $\kappa_\mathfrak{p}(\tilde{v}_1)$. In view of (5.8.2) and [17, II, § 1.4], we see that $\tilde{c}^p$ is purely inseparable over κ, and hence algebraic over $\kappa_\mathfrak{p}(\tilde{v}_1)$. It follows from this, (5.9.3) and (5.7.2) that $\tilde{v}_1$ is a generic point of $\tilde{V}^p$ $(= \mathfrak{p}_0(V^p))$ over $\kappa_\mathfrak{p}$.

Since $\tilde{u}$ is generic on $\tilde{V}$ over $\kappa_\mathfrak{p}$, we can find an isomorphism τ of $\kappa_\mathfrak{p}(\tilde{u})$ to $\kappa_\mathfrak{p}(\tilde{v}_1)$ such that $\tilde{u}^\tau = \tilde{v}_1$ and $a^\tau = a^p$ for $a \in \kappa_\mathfrak{p}$. Then $\mathfrak{p}_0(u^{\sigma_1}) = \tilde{v}_1 = \tilde{u}^\tau$ and $\mathfrak{p}_0(z^{\sigma_1}) = \mathfrak{p}_0(z^p) = \tilde{z}^p = \tilde{z}^\tau$ for $z \in Q(\zeta)$, hence

$$(5.14.8) \qquad \mathfrak{p}_0(w^{\sigma_1}) = \mathfrak{p}_0(w)^\tau$$

for every $w \in Q(\zeta)$. In fact, if $w = H(u)$ with a polynomial H whose coefficients are $\mathfrak{p}$-integers in $Q(\zeta)$, this is obvious. Then one can easily verify (5.14.8) for every w in the specialization ring $[u \to \tilde{u}; \mathfrak{p}]$ (in the sense of [18, p. 78, p. 83]). By [10, Prop. 5] and [6, Appendix, Theorem], this specialization ring is a discrete valuation ring. Hence (5.14.8) holds for every $w \in Q(\zeta, u)$.

From (5.14.8) we obtain $\mathfrak{p}_0((\mathfrak{Q}_u)^{\sigma_1}) = \mathfrak{p}_0(\mathfrak{Q}_u)^\tau$. Extend τ to an isomorphism of $\mathfrak{p}_0(K)$ and denote it by the same letter τ. Since $\mathfrak{Q} \cong \mathfrak{Q}_u$, we have

$$\mathfrak{p}_0(\mathfrak{Q})^\tau \cong \mathfrak{p}_0(\mathfrak{Q}_u)^\tau = \mathfrak{p}_0\big((\mathfrak{Q}_u)^{\sigma_1}\big) = \mathfrak{p}_0\big((\mathfrak{Q}_{u_1})^\sigma\big) \cong \mathfrak{p}_0(\mathfrak{Q}_u)^p \cong \mathfrak{p}_0(\mathfrak{Q})^p .$$

By [14, Prop. 5.1], [17, II, Prop. 1.3] and (5.8.2), we have $\mathfrak{p}_0(c)^\tau = \mathfrak{p}_0(c)^p$. By (5.9.3) we have $a^\tau = a^p$ for every $a \in \kappa_p(\tilde{u})$. Therefore we get

(5.14.9) $$\tilde{v}_1 = \tilde{u}^\tau = \tilde{u}^p ,$$

(5.14.10) $$\mathfrak{p}_0(w^{\sigma_1}) = \mathfrak{p}_0(w)^p \qquad\qquad \text{for every } w \in \boldsymbol{Q}(\zeta, u) .$$

If follows that $\mathfrak{p}_0\big((\mathfrak{Q}_u)^{\sigma_1}\big) = \mathfrak{p}_0(\mathfrak{Q}_u)^p$, and hence χ is an automorphism of $\mathfrak{p}_0(\mathfrak{Q}_u)^p$. By (ii) of 5.11, χ is the identity mapping, so that

(5.14.11) $$\tilde{\mu} = \pi .$$

Let ε_m be the isogeny of A_{m,u_1} obtained by arranging m copies of ε diagonally, and π_m the pth power isogeny of $\mathfrak{p}_0(A_{m,u})$ to $\mathfrak{p}_0(A_{m,u})^p$. Then, from (5.14.11) we obtain easily

(5.14.12) $$\mathfrak{p}_0(\varepsilon_m \lambda_{m1}) = \pi_m \qquad\qquad (m > 0) .$$

Step 3. Let us consider the u_i and λ_i for $i > 1$. Put $Y_0(\alpha)[u] = v$. There exist an automorphism ω of $\boldsymbol{Q}(\zeta, u)$ and an isomorphism η of $\mathscr{P}_u$ to $(\mathscr{P}_u)^\omega$ such that $u^\omega = v$, $\omega = \rho$ on $\boldsymbol{Q}(\zeta)$, and

(5.14.13) $$t^p(v, \xi) = t(u, \xi)^\omega = \eta\big(t(u, \xi\alpha)\big) \qquad\qquad (\xi \in \mathfrak{b}^{-1})$$

(cf. 3.5, 4.10). Put $\nu_i = \eta\lambda_i'$, $\nu_i' = \lambda_i\eta^{-1}$, where λ_i' is as in (5.14.3). By (4.6.1), (5.14.13) and (5.14.3), we have, for $\xi_1 \in \mathfrak{b}^{-1}$,

$$\nu_i\big(t(u_i, p\xi_1)\big) = \nu_i\lambda_i\big(t(u, \xi_1\alpha)\big) = \nu_i\lambda_i\eta^{-1}\big(t^p(v, \xi_1)\big) = t^p(v, p\xi_1) .$$

Since p is prime to $\mathfrak{b}$, we get, by substituting ξ for $p\xi_1$,

(5.14.14) $$\nu_i\big(t(u_i, \xi)\big) = t^p(v, \xi) \qquad\qquad (\xi \in \mathfrak{b}^{-1}) .$$

Let τ_i be an isomorphism of $\boldsymbol{Q}(\zeta, u)$ to $\boldsymbol{Q}(\zeta, u_i)$ over $\boldsymbol{Q}(\zeta)$ such that $u^{\tau_i} = u_i$ (cf. Proof of 4.7). Put $\sigma_i = \tau_i^{-1}\omega$. Then $u_i^{\sigma_i} = v$, $(\mathfrak{Q}_{u_i})^{\sigma_i} = (\mathfrak{Q}_u)^\omega$, $\sigma_i = \rho$ on $\boldsymbol{Q}(\zeta)$.

Now ν_i and ν_i' are defined over a finite algebraic extension k of $\boldsymbol{Q}(\zeta, u)$. Hence $\tilde{\nu}_i$ and $\tilde{\nu}_i'$ are defined over $\tilde{k} = \mathfrak{p}_0(k)$. In view of (5.7.2), $\tilde{k}$ is algebraic over $\kappa_p(\tilde{u})$. It follows from this and (5.7.1) that $\tilde{x}$ is generic on $\mathfrak{p}_0(A_u)$ over $\tilde{k}$. Put $y_i = \lambda_i x$. Then $p\tilde{x} = \tilde{\lambda}_i'\tilde{y}_i$, so that $\tilde{y}_i$ is generic on $\mathfrak{p}_0(A_{u_i})$ over $\tilde{k}$. Since $p\tilde{y}_i = \tilde{\nu}_i'\tilde{\nu}_i\tilde{y}_i$, we have $\tilde{k}(p\tilde{y}_i) \subset \tilde{k}(\tilde{\nu}_i\tilde{y}_i) \subset \tilde{k}(\tilde{y}_i)$. By (5.14.5), we have $\text{Ker}\,(\tilde{\nu}_i) = \{0\}$, so that $\tilde{k}(\tilde{y}_i)$ is a purely inseparable extension of $\tilde{k}(\tilde{\nu}_i\tilde{y}_i)$ of degree p^2. By 5.12, we get $\tilde{k}(\tilde{\nu}_i\tilde{y}_i) = \tilde{k}(\tilde{y}_i^p)$. Hence there exists an isomorphism ψ of $\mathfrak{p}_0(A_u^\omega)$ to $\mathfrak{p}_0(A_{u_i})^p$ such that $\psi\tilde{\nu}_i = \varpi$ where ϖ is the pth power isogeny of $\mathfrak{p}_0(A_{u_i})$ to $\mathfrak{p}_0(A_{u_i})^p$. By [14, Prop. 2.4], ψ is an isomorphism of $\mathfrak{p}_0((\mathscr{P}_u)^\omega)$ to $\mathfrak{p}_0(\mathscr{P}_{u_i})^p$.

Furthermore, by (5.14.14), we have

$$\psi\big(\mathfrak{p}_0(t^p(v,\xi))\big) = \psi\tilde{\nu}_i\big(\mathfrak{p}_0(t(u_i,\xi))\big) = \varpi\big(\mathfrak{p}_0(t(u_i,\xi))\big) = \mathfrak{p}_0\big(t(u_i,\xi)\big)^p$$

for every $\xi \in \mathfrak{b}^{-1}$, so that ψ is an isomorphism of $\mathfrak{p}_0((\mathcal{Q}_u)^\omega)$ to $\mathfrak{p}_0(\mathcal{Q}_{u_i})^p$.

Therefore we have $\mathfrak{p}_0(\mathcal{Q}^\omega) \cong \mathfrak{p}_0((\mathcal{Q}_u)^\omega) \cong \mathfrak{p}_0(\mathcal{Q}_{u_i})^p$, $\mathcal{Q}$ being as in 5.7. By (5.6.2), $\mathfrak{p}_0(A_{u_i})$ is defined over $\kappa_\mathfrak{p}(\tilde{u}_i)$. Hence $\mathfrak{p}_0(\mathcal{Q}_{u_i})^p$ is defined over an algebraic extension of $\kappa_\mathfrak{p}(\tilde{u}_i)$. In view of (5.8.2) and [17, II, § 1.4], we see that $\mathfrak{p}_0(c^\omega)$ is algebraic over $\kappa_\mathfrak{p}(\tilde{u}_i)$. It follows from this, (5.9.3) and (5.7.2), that $\tilde{u}_i$ is generic on $\tilde{V}$ over $\kappa_\mathfrak{p}$.

By (5.6.3), we have $\tilde{v} = \mathfrak{p}_0(Y_0(\alpha))[\tilde{u}]$, so that by (5.7.2), $\tilde{v}$ is a generic point of $\tilde{V}^p$ over $\kappa_\mathfrak{p}$. Therefore one can find an isomorphism δ_i of $\kappa_\mathfrak{p}(\tilde{u}_i)$ to $\kappa_\mathfrak{p}(\tilde{v})$ such that $\tilde{u}_i^{\delta_i} = \tilde{v}$, $a^{\delta_i} = a^p$ for every $a \in \kappa_\mathfrak{p}$. Then we have $\mathfrak{p}_0(u_i^{\sigma_i}) = \mathfrak{p}_0(v) = \mathfrak{p}_0(u_i)^{\delta_i}$ and $\mathfrak{p}_0(z^{\sigma_i}) = z^p = \mathfrak{p}_0(z)$ for every $z \in Q(\zeta)$, hence (by the same reasoning as in (5.14.8)),

$$(5.14.15) \qquad\qquad \mathfrak{p}_0(w^{\sigma_i}) = \mathfrak{p}_0(w)^{\delta_i} \qquad\qquad \text{for every } w \in Q(\zeta, u_i).$$

It follows that $\mathfrak{p}_0((\mathcal{Q}_u)^\omega) = \mathfrak{p}_0((\mathcal{Q}_{u_i})^{\sigma_i}) = \mathfrak{p}_0(\mathcal{Q}_{u_i})^{\delta_i}$. Extend δ_i to an isomorphism of $\mathfrak{p}_0(K)$, and denote it again by δ_i. On the other hand, ω coincides, on $Q(\zeta, u)$, with exactly one of the ω_j defined in 5.8, say ω_1. Denote ω_1 simply by ω. Then we see that

$$\mathfrak{p}_0(\mathcal{Q}^\omega)^{\delta_i^{-1}} \cong \mathfrak{p}_0((\mathcal{Q}_u)^\omega)^{\delta_i^{-1}} = \mathfrak{p}_0(\mathcal{Q}_{u_i}) \cong \mathfrak{p}_0((\mathcal{Q}_u)^\omega)^{1/p} \cong \mathfrak{p}_0(\mathcal{Q}^\omega)^{1/p} \;.$$

By [14, Prop. 5.1], [17, II, Prop. 1.3] and (5.8.2), we have $\mathfrak{p}_0(c^\omega)^{\delta_i^{-1}} = \mathfrak{p}_0(c^\omega)^{1/p}$. By (5.9.3), we see that $a^{\delta_i^{-1}} = a^{1/p}$ for every $a \in \kappa_\mathfrak{p}(\mathfrak{p}_0(u^\omega)) = \kappa_\mathfrak{p}(\tilde{v})$. Therefore we get

$$(5.14.16) \qquad \tilde{u}_i = \tilde{v}^{\delta_i^{-1}} = \tilde{v}^{1/p}, \qquad \tilde{u}_i^{\delta_i} = \tilde{v} = \tilde{u}_i^p \qquad (2 \le i \le p + 1) \;.$$

Hence $a^{\delta_i} = a^p$ for every $a \in \kappa_\mathfrak{p}(\tilde{u}_i)$, and (5.14.15) implies

$$(5.14.17) \qquad\qquad \mathfrak{p}_0(w^{\sigma_i}) = \mathfrak{p}_0(w)^p \qquad\qquad \text{for every } w \in Q(\zeta, u_i).$$

It follows that $\mathfrak{p}_0((\mathcal{Q}_u)^\omega) = \mathfrak{p}_0((\mathcal{Q}_{u_i})^{\sigma_i}) = \mathfrak{p}_0(\mathcal{Q}_{u_i})^p$, and hence ψ is an automorphism of $\mathfrak{p}_0(\mathcal{Q}_{u_i})^p$. Since $\tilde{u}_i$ is generic on $\tilde{V}$ over $\kappa_\mathfrak{p}$, ψ is the identity mapping, on account of (ii) of 5.11. Therefore we have

$$(5.14.18) \qquad\qquad \tilde{\nu}_i = \varpi, \qquad \tilde{\nu}'_i = \varpi', $$

where ϖ' is the isogeny of $\mathfrak{p}_0(A_u^\omega)$ to $\mathfrak{p}_0(A_u^\omega)^{1/p}$ such that $\varpi'\varpi = p \cdot$(the identity map of $\mathfrak{p}_0(A_u^\omega)$). Let η_m resp. ϖ'_m be the isogeny obtained by arranging m copies of η resp. ϖ diagonally. From (5.14.18) we get

$$(5.14.19) \qquad\qquad \mathfrak{p}_0(\lambda_{mi}\eta_m^{-1}) = \varpi'_m \qquad\qquad (2 \le i \le p + 1) \;.$$

Step 4. For $m \ge 0$, put

$$T_m = \mathfrak{p}({}^t Y_m(\alpha')) \circ \Pi_{m\mathfrak{p}} + \Pi_{m\mathfrak{p}}^* \circ \mathfrak{p}(Y_m(\alpha)) \, .$$

By (5.6.3), (5.14.9), (5.14.16) and [10, Th. 17, 19], we have

$$T_0[\tilde{u}] = \mathfrak{p}({}^t Y_0(\alpha'))[\tilde{u}^p] + p \cdot \mathfrak{p}(Y_0(\alpha))[\tilde{u}]^{1/p}$$
$$= \tilde{u}_1 + (\tilde{u}_2 + \cdots + \tilde{u}_{p+1}) = \mathfrak{p}_0(X_0[u]) = \mathfrak{p}(X_0)[\tilde{u}] \, .$$

Hence there exists a 0-cycle $\mathfrak{c}$ on $\tilde{V}$ such that

$$\mathfrak{p}(X_0) = T_0 + \mathfrak{c} \times \tilde{V} \qquad\qquad \text{(cf. [20, p. 32, Th. 2]) } .$$

Since $\mathfrak{p}(X_0)$ is a positive cycle and T_0 has no component of the form $r \times \tilde{V}$ with a point r on $\tilde{V}$, we see that $\mathfrak{c}$ is a positive cycle. If w is sufficiently generic on $\tilde{V}$, we have, in view of 4.13,

$$\deg \left({}^t\mathfrak{p}(X_0)[w]\right) = \deg \left({}^t X_0[u]\right) = p + 1 = \deg \left({}^t T_0[w]\right) ,$$

so that $\deg (\mathfrak{c}) = 0$, hence $\mathfrak{c} = 0$. This proves the equality of our theorem for $m = 0$.

This fact implies especially that $\mathfrak{p}(X_0)$ is proper. Therefore, by 5.4, $\mathfrak{p}(X_m)$ is proper for $m > 0$. Let y be a generic point of $A_{m,u}$ over K, and s a generic point of $\mathfrak{p}_0(A_{m,u})$ over $\mathfrak{p}_0(K)$. We can find a place $\mathfrak{q}$ of $K(y)$ such that $\mathfrak{q} = \mathfrak{p}_0$ on K, and $\mathfrak{q}(y) = s$. By (4.5.2) and [10, Th. 17, 19], we have

$$\mathfrak{p}(X_m)[s] = \mathfrak{q}(X_m[y]) = \mathfrak{q}(\lambda_{m1}y + \cdots + \lambda_{m,p+1}y) \, .$$

On the other hand, by (5.14.12), we get

$$\mathfrak{q}(\lambda_{m1}y) = \tilde{\varepsilon}_m^{-1}\pi_m(s) = \mathfrak{p}({}^t Y_m(\alpha')) \circ \Pi_{m\mathfrak{p}}[s] \, .$$

Furthermore, by (5.14.19) and (5.13.1), we have

$$\mathfrak{q}(\lambda_{m2}y + \cdots + \lambda_{m,p+1}y) = p \cdot \left(\varpi_m' \tilde{\eta}_m(s)\right) = \Pi_{m\mathfrak{p}}^* \circ \mathfrak{p}(Y_m(\alpha))[s] \, .$$

Therefore $\mathfrak{p}(X_m)[s] = T_m[s]$. Since $\mathfrak{p}(X_m)$ and T_m are proper cycles defined over $\kappa_{\mathfrak{p}}$, and s is generic on $\tilde{W}_m$ over $\kappa_{\mathfrak{p}}$, we get $\mathfrak{p}(X_m) = T_m$. Our proof of Th. 5.14 is thus completed.

5.15. Let us fix one $\mathfrak{p}$ in $\Lambda(\mathfrak{b})$ and put, for simplicity,

$$(5.15.1) \qquad \tilde{X}_m(\Gamma\psi\Gamma) = \mathfrak{p}(X_m(\Gamma\psi\Gamma)) \, , \qquad\quad \tilde{Y}_m(\psi) = \mathfrak{p}(Y_m(\psi))$$

$$(5.15.2) \qquad Z_{m\mathfrak{p}}(\alpha) = {}^t \tilde{Y}_m(\alpha') \circ \Pi_{m\mathfrak{p}} \, , \qquad\quad Z_{m\mathfrak{p}}^*(\alpha) = \Pi_{m\mathfrak{p}}^* \circ \tilde{Y}_m(\alpha) \, ,$$

where α is as in 5.14. Then we obtain

$$(5.15.3) \qquad\qquad\qquad \Pi_{m\mathfrak{p}}^* \circ \Pi_{m\mathfrak{p}} = p \cdot \tilde{X}_m(\Gamma p \Gamma) \, ,$$

$$(5.15.4) \qquad\qquad\qquad \Pi_{m\mathfrak{p}} \circ \Pi_{m\mathfrak{p}}^* = p \cdot \tilde{X}_m(\Gamma p \Gamma)^p \, .$$

These relations are obvious if $m = 0$, since $X_0(\Gamma p \Gamma)$ is the identity correspondence and $\Pi_{0\mathfrak{p}}^* = {}^t\Pi_{0\mathfrak{p}}$. If $m > 0$, they follow easily from (5.13.1) and the definition of $X_m(\Gamma p \Gamma)$, $\Pi_{m\mathfrak{p}}$, $\Pi_{m\mathfrak{p}}^*$.

5.16. PROPOSITION. *Let the notation be as in 5.14 and 5.15. Let δ be an element of Γ_1 such that $\alpha'\delta \equiv \delta\alpha' \equiv \alpha$, $p\delta \equiv \alpha^2 \bmod \mathfrak{b}$. (cf. 1.11). Then*

$$Z_{m\mathfrak{p}}(\alpha) \circ Z_{m\mathfrak{p}}^*(\alpha) = Z_{m\mathfrak{p}}^*(\alpha) \circ Z_{m\mathfrak{p}}(\alpha) = p \cdot \tilde{X}_m(\Gamma p \delta \Gamma) \,.$$

PROOF. By (4.10.5), we get (ρ being as in 5.13),

$$(5.16.1) \qquad Y_m(\alpha') \circ Y_m(\delta) = Y_m(\alpha) = Y_m(\delta)^\rho \circ Y_m(\alpha') \,.$$

We have hence

$$
\begin{aligned}
p \cdot \tilde{X}_m(\Gamma p \delta \Gamma) &= \tilde{Y}_m(\delta) \circ \big(p \cdot \tilde{X}_m(\Gamma p \Gamma)\big) && \text{(1.9, 4.10.6)} \\
&= {}^t\tilde{Y}_m(\alpha') \circ \tilde{Y}_m(\alpha) \circ \big(p \cdot \tilde{X}_m(\Gamma p \Gamma)\big) && \text{(5.16.1)} \\
&= {}^t\tilde{Y}_m(\alpha') \circ \big(p \cdot \tilde{X}_m(\Gamma p \Gamma)\big)^\rho \circ \tilde{Y}_m(\alpha) && \text{(4.11)} \\
&= Z_{m\mathfrak{p}}(\alpha) \circ Z_{m\mathfrak{p}}^*(\alpha) && \text{(5.15.4)} \,,
\end{aligned}
$$

$$
\begin{aligned}
Z_{m\mathfrak{p}}^*(\alpha) \circ Z_{m\mathfrak{p}}(\alpha) &= \Pi_{m\mathfrak{p}}^* \circ \tilde{Y}_m(\alpha) \circ {}^t\tilde{Y}_m(\alpha') \circ \Pi_{m\mathfrak{p}} \\
&= \Pi_{m\mathfrak{p}}^* \circ \tilde{Y}_m(\delta)^\rho \circ \Pi_{m\mathfrak{p}} && \text{(5.16.1)} \\
&= \Pi_{m\mathfrak{p}}^* \circ \Pi_{m\mathfrak{p}} \circ \tilde{Y}_m(\delta) \\
&= p \cdot \tilde{X}_m(\Gamma p \Gamma) \circ \tilde{X}_m(\Gamma \delta \Gamma) && \text{(5.15.3, 4.10.6)} \\
&= p \cdot \tilde{X}_m(\Gamma p \delta \Gamma) \,, && \text{q.e.d.}
\end{aligned}
$$

5.17. PROPOSITION. *For every positive integer n, let $\Pi_{m\mathfrak{p}}^{(n)}$ and $\Pi_{m\mathfrak{p}}^{*(n)}$ denote the Frobenius correspondence of degree p^n on $\tilde{W}_m \times \tilde{W}_m^{p^n}$ and the anti-Frobenius fibre correspondence of degree p^n on $\tilde{W}_m^{p^n} \times \tilde{W}_m$. Then, the notation being as in 5.14 and 5.15, one has*

$$Z_{m\mathfrak{p}}(\alpha)^n = {}^t\tilde{Y}_m(\alpha'^n) \circ \Pi_{m\mathfrak{p}}^{(n)} \,, \qquad Z_{m\mathfrak{p}}^*(\alpha)^n = \Pi_{m\mathfrak{p}}^{*(n)} \circ \tilde{Y}_m(\alpha^n) \,,$$

$$\mathfrak{p}\big(F_n(X_m(\Gamma \alpha \Gamma),\, pX_m(\Gamma p \delta \Gamma))\big) = Z_{m\mathfrak{p}}(\alpha)^n + Z_{m\mathfrak{p}}^*(\alpha)^n \qquad (n = 1, 2, \cdots) \,.$$

Here the $F_n(x, y)$ are the polynomials defined in 1.16.

PROOF. The first two equalities follow easily from (4.10.5), (2.5.4), (2.5.5), and the remark at the end of 2.5. The last one follows from 5.5, 5.14 and 5.16.

6. The zeta-function and *L*-functions of W_m

6.1. *Intersection number.* Let U be a non-singular projective variety defined over a field k (with any characteristic). By virtue of Chow [2], one knows that the rational equivalence classes of cycles on U form a ring. Using this result, to any cycles X and Y on U with complementary dimensions (i.e., such that $\dim(X) + \dim(Y) = \dim(U)$), one can assign a well-defined intersection number $I_U(X, Y)$. We shall write it $I(X, Y)$ when the variety U is obvious from context. If the universal domain is C, $I(X, Y)$ is the same as the topological intersection number of X and Y on U (cf. Weil [19], Borel and Haefliger [1]).

Let X be a cycle on $U \times U$ with the same dimension as U, and let D_U be the diagonal on $U \times U$ (cf. 2.6). We shall put

$$I_0(X) = I_{U \times U}(X, D_U) .$$

6.2. PROPOSITION [8, Th. III–4–6, Cor. IV–2–4]. *Let $(\Gamma_\mathfrak{b} \alpha \Gamma_\mathfrak{b})_k$ be as in 1.13. Let W_m and $X_m(\Gamma_\mathfrak{b} \alpha \Gamma_\mathfrak{b})$ be as in 3.2 and 4.4. Let $d(\Gamma_\mathfrak{b} \alpha \Gamma_\mathfrak{b})$ be the number of right cosets $\Gamma_\mathfrak{b} \xi$ in $\Gamma_\mathfrak{b} \alpha \Gamma_\mathfrak{b}$. Define integers $a(m, i, \nu)$ as follows:*

$$a(0, 0, 0) = 1,$$

$$a(m, i, \nu) = \begin{cases} \left[\dbinom{2m}{\frac{i+\nu}{2}} \dbinom{2m}{\frac{i-\nu}{2}} - \dbinom{2m}{\frac{i+\nu}{2}+1} \dbinom{2m}{\frac{i-\nu}{2}-1} \right] & \text{if} \quad i \equiv \nu \bmod (2) , \\ 0 & \text{if} \quad i \not\equiv \nu \bmod (2) . \end{cases}$$

$$(m = 1, 2, \cdots; \; 0 \le i \le 4m; \; 0 \le \nu \le i) .$$

Then, for every $m \ge 0$, one has

$$I_0\big(X_m(\Gamma_\mathfrak{b} \alpha \Gamma_\mathfrak{b})\big) = 2 \sum_{j=0}^{2m} a(m, 2j, 0) \det (\alpha)^j d(\Gamma_\mathfrak{b} \alpha \Gamma_\mathfrak{b})$$
$$+ 2 \sum_{i=0}^{4m} \sum_{\nu=0}^{i} (-1)^{i+1} a(m, i, \nu) \det (\alpha)^{(i-\nu)/2} \, \mathrm{Re} \left[\mathrm{tr} \, (\Gamma_\mathfrak{b} \alpha \Gamma_\mathfrak{b})_{\nu+2} \right] .$$

Here $\dbinom{n}{r}$ denotes the binomial coefficient; we understand $\dbinom{n}{r} = 0$ if $r > n$ or $r < 0$. Therefore $a(m, i, \nu) = 0$ if $i + \nu > 4m$; hence, for a fixed m, the largest ν with non-vanishing $a(m, i, \nu)$ is $2m$.

The above formula has been proved by interpreting each cohomology group of W_m as a direct sum of $\mathfrak{S}_k(\Gamma_\mathfrak{b})$, and applying the Lefschetz fixed point theorem to the correspondence $X_m(\Gamma_\mathfrak{b} \alpha \Gamma_\mathfrak{b})$; for detail see [8]. One can get another proof by comparing the right hand side of (2.7.6) with the trace formula of Eichler-Selberg for $(\Gamma_\mathfrak{b} \alpha \Gamma_\mathfrak{b})_\nu$.[7]

6.3. PROPOSITION. *Let U be a projective non-singular variety defined over a field k, and $\mathfrak{p}$ a place of k. Suppose that $\mathfrak{p}(U)$ is a non-singular variety with multiplicity one. Let X and Y be cycles on U, rational over k, such that $\dim (X) + \dim (Y) = \dim (U)$. Then*

$$I_U(X, Y) = I_{\mathfrak{p}(U)}\big(\mathfrak{p}(X), \mathfrak{p}(Y)\big) .$$

PROOF. By Chow [2], there exist a purely transcendental extension $k(t)$ of k and a cycle X_t defined over $k(t)$ such that X is a specialization of X_t over k, and $X_t \cdot Z$ can be defined for every cycle Z on U rational over k. The construction of X_t can be done as follows [2, pp. 458–460]. Let S be the ambient projective space for V, and let $N = \dim (S)$, $n = \dim (U)$. Put $X = \sum_i c_i B_i$ with integers c_i and varieties B_i. Take a generic linear subspace L_v over k in S of dimension $N - n - 1$, defined over a purely transcendental extension $k(v)$

---

[7] One has to generalize 2.7 suitably, since the condition (2. 7. 3) is not necessarily satisfied by $X_m(\Gamma_\mathfrak{b} \alpha \Gamma_\mathfrak{b})$.

over k. Denote by B_i^* the variety which is the union of all the linear subspaces of dimension $N - n$ obtained by joining L_v with every point of B_i, and put $X_v^* = \sum_i c_i B_i^*$. It can be easily shown that X_v^* is rational over $k(v)$. Let T_u be a generic projective transformation in S over $k(v)$, defined over a purely transcendental extension $k(u, v)$ over $k(v)$. Put

$$R_{u,v}(X) = T_u(X_v^*) \cdot V - X_v^* \cdot V + X \,.$$

Now taking $R_{u,v}(X)$ and $k(u, v)$ in place of X and k, we can define $R_{u_1,v_1}(R_{u,v}(X))$ similarly. After a finite number of times of operations, we can adopt $R_{u_r,v_r}(\cdots(R_{u_1,v_1}(R_{u,v}(X)))\cdots)$ as the cycle X_t with $t = (u, v, u_1, v_1, \cdots, u_r, v_r)$. The number of times depends only on $\dim(X)$ and $\dim(V)$ (cf. [2, p. 458]).

We can apply the same kind of operations to $\mathfrak{p}(X)$ on $\mathfrak{p}(U)$ which are compatible with reduction modulo $\mathfrak{p}$. In fact, take a generic linear variety L_η in $\mathfrak{p}(S)$ over $\mathfrak{p}(k)$ and a generic projective transformation T_ξ of $\mathfrak{p}(S)$ over $\mathfrak{p}(k)(\eta)$. Then the specialization $(u, v) \rightarrow (\xi, \eta)$ ref. $\mathfrak{p}$ defines a place $\mathfrak{q}$ of $k(u, v)$. If we can prove $\mathfrak{q}(X_v^*) = \mathfrak{p}(X)_\eta^*$, then we will get $\mathfrak{q}(R_{u,v}(X)) = R_{\xi,\eta}(\mathfrak{p}(X))$. To show this, by taking a suitable finite algebraic extension k' of k and a prolongation $\mathfrak{p}'$ of $\mathfrak{p}$ in k', we may reduce the situation to the special case where X is a variety defined over k. Then it is clear that $\mathfrak{q}(X_v^*) = c \cdot \mathfrak{p}(X)_\eta^*$ with a positive integer c. As was remarked in Chow [2, p. 459], X (resp. $\mathfrak{p}(X)$) is a component of $X_v^* \cdot V$ (resp. $\mathfrak{p}(X)_\eta^* \cdot \mathfrak{p}(X)$) with multiplicity one. Since $\mathfrak{q}(X_v^* \cdot V) = \mathfrak{q}(X_v^*) \cdot \mathfrak{p}(V)$, we get $c = 1$. Thus we have $\mathfrak{q}(R_{u,v}(X)) = R_{\xi,\eta}(\mathfrak{p}(X))$. Repeating this procedure, we find a prolongation $\mathfrak{P}$ of $\mathfrak{p}$ in $k(t)$ such that $\mathfrak{P}(X_t)$ has the same property as X on $\mathfrak{p}(U)$. Then we have $\mathfrak{P}(X_t \cdot Y) = \mathfrak{P}(X_t) \cdot \mathfrak{p}(Y)$, so that $I(X, Y) = \deg(X_t \cdot Y) = \deg(\mathfrak{P}(X_t) \cdot \mathfrak{p}(Y)) = I(\mathfrak{p}(X), \mathfrak{p}(Y))$.

6.4. Proposition. *Let the notation be as in 5.14, 5.15, and 5.17, and let β be an element of $\mathfrak{o}$ such that $\det(\beta) = p^n$. Then*

$$I_0\big({}^t\widetilde{Y}_m(\beta) \circ \Pi_{m\mathfrak{p}}^{(n)}\big) = I_0\big(\Pi_{m\mathfrak{p}}^{*(n)} \circ \widetilde{Y}_m(\beta)\big) \,.$$

Proof. Let r be a generic point of W_m over $\kappa_\mathfrak{p}$. We can extend $\mathfrak{p}$ to a place $\mathfrak{q}$ of $Q(\zeta, r)$ so that $\mathfrak{q}(r) = \tilde{r}$ is generic on $\widetilde{W}_m$ over $\kappa_\mathfrak{p}$. In the course of proof, tilde will mean reduction modulo $\mathfrak{q}$. Put $\varphi_m(r) = s$, $Y_0(\beta)[s] = t$. Then there exist an automorphism σ of $Q(\zeta, s)$ and an isomorphism ε_m of $A_{m,s}$ to $(A_{m,s})^\sigma$ such that $\zeta^\sigma = \zeta^q$, $s^\sigma = t$ and $Y_m(\beta)$ is the locus of $r \times \varepsilon_m r$ over $Q(\zeta)$. Since ε_1 is an isomorphism of $\mathscr{P}_s$ to $(\mathscr{P}_s)^\sigma$, and ε_m is obtained by arranging m copies of ε_1 diagonally, we can find a positive non-degenerate divisor E_s on $A_{m,s}$ such that $\varepsilon_m^{-1}(E_s^\sigma)$ is algebraically equivalent to E_s. Let π_m be the qth power isogeny of $\tilde{\varphi}_m^{-1}(\tilde{t}^{1/q})$ to $(\tilde{\varphi}_m^q)^{-1}(\tilde{t})$, and π_m^* the isogeny of $(\tilde{\varphi}_m^q)^{-1}(\tilde{t})$ to $\tilde{\varphi}_m^{-1}(\tilde{t}^{1/q})$ such that $\pi_m^* \pi_m(w) = qw$ for $w \in \tilde{\varphi}_m^{-1}(\tilde{t}^{1/q})$. Here notice that $\tilde{\varphi}_m^q = \mathfrak{q}(\varphi_m^\sigma)$. Then ${}^t\widetilde{Y}_m(\beta) \circ \Pi_{m\mathfrak{p}}^{(n)}$

is the locus of $w \times \tilde{\varepsilon}_m^{-1} \pi_m w$ over $\kappa_{\mathfrak{p}}$ with a generic point w of $\tilde{\varphi}_m^{-1}(\tilde{t}^{1/q})$ over $\kappa_{\mathfrak{p}}(\tilde{t}^{1/q})$, and $\Pi_{m\mathfrak{p}}^{*(n)} \circ \tilde{Y}_m(\beta)$ is the locus of $v \times \pi_m^* \varepsilon_m v$ over $\kappa_{\mathfrak{p}}$ with a generic point v of $\tilde{\varphi}_m^{-1}(s)$. We observe that ${}^t({}^t\tilde{Y}_0(\beta) \circ \Pi_{0\mathfrak{p}}^{(n)}) = \Pi_{0\mathfrak{p}}^{*(n)} \circ \tilde{Y}_0(\beta)$ and $\tilde{\varepsilon}_m^{-1} \pi_m \pi_m^* \tilde{\varepsilon}_m(w) = qw$.

Now we have $(s, t) \rightarrow (\tilde{s}, \tilde{t})$ ref. $\mathfrak{p}$. Since $\tilde{s}$ is generic on $\tilde{W}_m$ and $\tilde{t}$ is generic on $\tilde{W}_m^q$, we see that both $[s \rightarrow \tilde{s}; \mathfrak{p}]$ and $[t \rightarrow \tilde{t}; \mathfrak{p}]$ are valuation rings. Moreover, in view of (5.6.3), we have $[s \rightarrow \tilde{s}; \mathfrak{p}] = [t \rightarrow \tilde{t}; \mathfrak{p}]$ and $\kappa_{\mathfrak{p}}(\tilde{s}) = \kappa_{\mathfrak{p}}(\tilde{t})$. Let τ be an isomorphism of $\kappa_{\mathfrak{p}}(\tilde{s})$ to $\kappa_{\mathfrak{p}}(\tilde{t})$ such that $\tilde{s}^\tau = \tilde{t}$ and $a^\tau = a^q$ for every $a \in \kappa_{\mathfrak{p}}$. Then we can easily verify that $\mathfrak{q}(b^\sigma) = \mathfrak{q}(b)^\tau$ for every $b \in Q(\zeta, s)$. It follows that $\mathfrak{q}(E_s^\sigma) = \mathfrak{q}(E_s)^\tau$. Put $F = \mathfrak{q}(E_s)$, $F' = (F^\tau)^{1/q}$. Then F resp. F' is a positive non-degenerate divisor on $\tilde{\varphi}_m^{-1}(\tilde{s})$ resp. $\tilde{\varphi}_m^{-1}(\tilde{t}^{1/q})$, and F' is the transform of F by an isomorphism of $\kappa_{\mathfrak{p}}(\tilde{s})$ to $\kappa_{\mathfrak{p}}(\tilde{t}^{1/q})$ over $\kappa_{\mathfrak{p}}$ which sends $\tilde{s}$ to $\tilde{t}^{1/q}$. We see furthermore that $(\tilde{\varepsilon}_m^{-1} \pi_m)^{-1}(F)$ is algebraically equivalent to $\pi_m^{-1}(\mathfrak{q}(E_s^\sigma)) = \pi_m^{-1}(F^\tau) = q \cdot (F^\tau)^{1/q} = qF'$.

We are going to apply 2.9 to the present case, by taking ${}^t\tilde{Y}_m(\beta) \circ \Pi_{m\mathfrak{p}}^{(n)}$ and $\Pi_{m\mathfrak{p}}^{*(n)} \circ \tilde{Y}_m(\beta)$ as X and X^*. Since ε_m is defined over $Q(\zeta, s) = Q(\zeta, t)$, and $[s \rightarrow \tilde{s}; \mathfrak{p}]$ is a valuation ring, we see that ε_m is defined over $\kappa_{\mathfrak{p}}(\tilde{s})$. Therefore, (2.7.1) is satisfied for $\tilde{\varepsilon}_m^{-1} \pi_m$. In view of 2.12, the conditions (2.7.2) and (2.7.3) are satisfied for ${}^t\tilde{Y}_0(\beta) \circ \Pi_{0\mathfrak{p}}^{(n)}$. Let $(b, b) \in D_{\tilde{v}} \cap ({}^t\tilde{Y}_0(\beta) \circ \Pi_{0\mathfrak{p}}^{(n)})$. Since $(\tilde{s}, \tilde{t}^{1/q}) \rightarrow (b, b)$ ref. $\kappa_{\mathfrak{p}}$, there exists a place $\mathfrak{r}$ of $\kappa_{\mathfrak{p}}(\tilde{s}, \tilde{t}^{1/q})$ such that $\mathfrak{r}(\tilde{s}) = \mathfrak{r}(\tilde{t}^{1/q}) = b$, and $\mathfrak{r}(z) = z$ for $z \in \kappa_{\mathfrak{p}}$. Put $\mathfrak{r}(\tilde{\varepsilon}_m) = \eta_b$, $\mathfrak{r}(\pi_m) = \pi_b$. Obviously π_b is the qth power homomorphism of $\tilde{\varphi}_m^{-1}(b)$ to $\tilde{\varphi}_m^{-1}(b)^q$, and η_b^{-1} is an isomorphism of $\tilde{\varphi}_m^{-1}(b)^q$ to $\tilde{\varphi}_m^{-1}(b)$. Therefore, by 2.12, the graph of $\eta_b^{-1}\pi_b$ intersects properly with the graph of the identity mapping δ_b of $\tilde{\varphi}_m^{-1}(b)$. In other words, $\delta_b - \eta_b^{-1}\pi_b$ is an isogeny of $\tilde{\varphi}_m^{-1}(b)$ to itself, so that (2.7.4) is satisfied. Hence we may apply 2.9 to the present case, and we get our assertion.

6.5. *Definition of zeta-function and L-functions.* Let k be a finite field with q elements, and V a projective non-singular variety defined over k. Let G be a finite group, and let $\sigma \rightarrow Y(\sigma)$ be an isomorphism of G to a group of birational automorphisms of V. We assume that every $Y(\sigma)$ is defined over k. Let ψ be a complex valued character of G. We define an L-function $L(u; V, k, \psi)$, with a variable u, by

$$(6.5.1) \qquad\qquad L(0; V, k, \psi) = 1 ,$$

$$(6.5.2) \qquad \begin{aligned} &(d/du) \log L(u; V, k, \psi) \\ &\qquad = [G : 1]^{-1} \sum_{n=1}^{\infty} u^{n-1} \sum_{\sigma \in G} \psi(\sigma) \cdot I_0\big(Y(\sigma^{-1}) \circ \Pi^n\big) , \end{aligned}$$

where Π is the qth power correspondence of V to itself, and I_0 is as in 6.1. If ψ_0 is the character of a regular representation of G, then the corresponding L-function is the zeta-function $Z(u; V, k)$ of V over k:

(6.5.3) $Z(u; V, k) = L(u; V, k, \psi_0)$.

6.6. *Remark.* In the above we have assumed that $\sigma \to Y(\sigma)$ is an isomorphism. Suppose now that this is a homomorphism which may not be one-to-one. Let N be the kernel, and let $G' = G/N$. Let ψ' be a character of G', and ψ the character of G defined by $\psi(\sigma) = \psi'(\sigma N)$. Then we see easily that the function $L(u; V, k, \psi)$ defined by (6.5.1–2) with this G and ψ, is exactly equal to $L(u; V, k, \psi')$.

6.7. *Preliminaries for the statement of main theorem.* Let Φ be an indefinite quaternion algebra over Q (cf. 1.2), $\mathfrak{o}$ a maximal order in Φ, and $\mathfrak{b}$ an integral two-sided $\mathfrak{o}$-ideal. Define $\Gamma_\mathfrak{b}$ as in 1.3 and the variety $W_m = W_{m,\mathfrak{b}}$ as in 3.2 under the assumption (3.2.1). Take the projective model for W_m described in 3.4. W_m is defined over a cyclotomic field $Q(\zeta)$, where ζ is a primitive $b_0{}^{\text{th}}$ root of unity for a positive integer b_0 such that $\mathfrak{b} \cap Z = b_0 Z$.

For an element α of $\mathfrak{o}$ such that $(\det(\alpha), b_0) = 1$, let $Y_m(\alpha)$ be the birational biregular mapping of W_m to W_m^σ defined in 4.10, where σ is an automorphism of $Q(\zeta)$ such that $\zeta^\sigma = \zeta^{\det(\alpha)}$. Let $G_\mathfrak{b}$ be the group of all invertible elements in $\mathfrak{o}/\mathfrak{b}$, and $S_\mathfrak{b}$ the subgroup of $G_\mathfrak{b}$ consisting of all the residue-classes of the elements α such that $\det(\alpha) \equiv 1 \bmod \mathfrak{b}$. Since $Y_m(\alpha)$ depends only on the residue-class of α modulo $\mathfrak{b}$, we may write $Y_m(\alpha)$ as $Y_m(\sigma)$ for an element σ of $G_\mathfrak{b}$ represented by α. There is a natural homomorphism of $G_\mathfrak{b}$ onto the Galois group of $Q(\zeta)$ over Q, whose kernel is $S_\mathfrak{b}$; namely the residue-class of α sends ζ to $\zeta^{\det(\alpha)}$. For simplicity, denote the automorphism of $Q(\zeta)$ corresponding to an element σ of $G_\mathfrak{b}$ by the same letter σ. Then (4.10.5) is now written as

(6.7.1) $Y_m(\sigma\tau) = Y_m(\sigma)^\tau \circ Y_m(\tau)$ $(\sigma, \tau \in G_\mathfrak{b})$.

In particular, $\sigma \to Y_m(\sigma)$ gives a homomorphism of $S_\mathfrak{b}$ to a group of birational automorphisms of W_m. If $m = 0$, the kernel is $\{\pm 1\}$; otherwise this homomorphism is injective.

Let ψ be a complex valued character of $S_\mathfrak{b}$, and χ the character of $G_\mathfrak{b}$ induced from ψ. Suppose that ψ is actually a character of $S_\mathfrak{b}/\{\pm 1\}$ if $m = 0$. If an element σ of $G_\mathfrak{b}$ (resp. $S_\mathfrak{b}$) is the residue class of an element ξ of $\mathfrak{o}$, we shall put $\chi(\xi) = \chi(\sigma)$ (resp. $\psi(\xi) = \psi(\sigma)$). In view of (1.19.1), the following two conditions are equivalent:

(6.7.2) $\chi(\xi)$ *is a real number for every* ξ.

(6.7.3) $\chi(\xi') = \chi(\xi)$ *for every* ξ.

6.8. MAIN THEOREM I. *Let the notation and the assumption be as in* 6.7. *Suppose that the conditions* (6.7.2, 3) *are satisfied. Let r be the multi-*

plicity of the identity character in ψ. Let the integers $a(m, i, \nu)$ be as in 6.2. Let p be a prime number such that every prime ideal in $\mathbf{Q}(\zeta)$ dividing p belongs to the set $\Lambda(\mathfrak{b})$ defined in 5.9. Put $N(\mathfrak{p}) = p^f$ with such a prime ideal $\mathfrak{p}$. Let α be an element of $\mathfrak{o}$ such that $\det(\alpha) = p$. Then, with an indeterminate u, and for $m \geqq 0$, one has

$$\prod_{\mathfrak{p}|p} L\big(u^f;\, \mathfrak{p}(W_m),\, \kappa_{\mathfrak{p}},\, \psi\big)$$
$$= \prod_{j=0}^{2m} \Big[\prod_{\mathfrak{p}|p}(1 - N(\mathfrak{p})^j u^f)(1 - N(\mathfrak{p})^{j+1}u^f)\Big]^{-r \cdot a(m, 2j, 0)}$$
$$\times \prod_{i=0}^{4m}\prod_{\nu=0} \det\big[1 - (\Gamma_1 \alpha \Gamma_1)_{\nu+2,\rho}\, p^{(i-\nu)/2}u + (\Gamma_1 p \Gamma_1)_{\nu+2,\rho}\, p^{i-\nu+1}u^2\big]^{(-1)^i a(m, i, \nu)} .$$

Here $\mathfrak{p}(W_m)$ is the reduction of W_m modulo $\mathfrak{p}$; $\kappa_{\mathfrak{p}}$ is the residue field of $\mathfrak{p}$; $(\Gamma_1 \xi \Gamma_1)_{k,\rho}$ is defined as in 1.14 with respect to a representation ρ of $G_{\mathfrak{b}}$ with χ as its character. $L(v, \mathfrak{p}(W_m), \kappa_{\mathfrak{p}}, \psi)$ is defined with respect to a homomorphism $\sigma \to \mathfrak{p}(Y(\sigma))$ of $S_{\mathfrak{b}}$ to a group of birational automorphisms of $\mathfrak{p}(W_m)$.

PROOF. For simplicity, let us denote W_m, $X_m(\Gamma\xi\Gamma)$, $Y_m(\sigma)$, $\Pi_{m\mathfrak{p}}^{(n)}$, and $\Pi_{m\mathfrak{p}}^{*(n)}$ by W, $X(\Gamma\xi\Gamma)$, $Y(\sigma)$, $\Pi_{\mathfrak{p}}^n$, and $\Pi_{\mathfrak{p}}^{*n}$. Fix a prime ideal $\mathfrak{p}$ in $\mathbf{Q}(\zeta)$ dividing p. Put $N = [S_{\mathfrak{b}} : 1]$. First we prove

$$(6.8.1) \qquad \begin{aligned} &(d/du) \log\Big[\prod_{\mathfrak{q}|p} L(u^f;\, \mathfrak{q}(W),\, \kappa_{\mathfrak{q}},\, \psi)\Big] \\ &\qquad = N^{-1}\sum_{n=1}^{\infty} u^{n-1}\sum_{\sigma}' \chi(\sigma) I_0\big(\mathfrak{p}({}^t Y(\sigma)) \circ \Pi_{\mathfrak{p}}^n\big) . \end{aligned}$$

Here $\sum_{\sigma}'$ is the summation taken over all the σ of $G_{\mathfrak{b}}$ such that $\mathfrak{p}^{\sigma} = \mathfrak{p}$ and $a^{\sigma} = a^{p^n} \bmod \mathfrak{p}$. Let $\tau \in G_{\mathfrak{b}}$ and $\sigma \in S_{\mathfrak{b}}$. From (6.7.1), we get

$$(6.8.2) \qquad {}^t Y(\tau^{-1}\sigma\tau) = {}^t Y(\tau) \circ Y(\sigma^{-1})^{\tau} \circ Y(\tau) .$$

If $\mathfrak{p}^{\tau} = \mathfrak{q}$, there exists an isomorphism τ' of $\kappa_{\mathfrak{p}}$ to $\kappa_{\mathfrak{q}}$ such that $(a \bmod \mathfrak{p})^{\tau'} = a^{\tau} \bmod \mathfrak{q}$. Then $\mathfrak{p}(X)^{\tau'} = \mathfrak{q}(X^{\tau})$ for every algebro-geometric object X defined over $\mathbf{Q}(\zeta)$ (for which reduction modulo $\mathfrak{p}$ is meaningful). We have therefore

$$(6.8.3)\ \ \mathfrak{q}({}^t Y(\tau)) \circ [\mathfrak{p}({}^t Y(\sigma)) \circ \Pi_{\mathfrak{p}}^{nf}]^{\tau'} \circ \mathfrak{q}(Y(\tau)) = \mathfrak{q}({}^t Y(\tau)) \circ \mathfrak{q}({}^t Y(\sigma)^{\tau}) \circ (\Pi_{\mathfrak{p}}^{nf})^{\tau'} \circ \mathfrak{q}(Y(\tau)) .$$

It can be easily seen that $(\Pi_{\mathfrak{p}}^{nf})^{\tau'}$ is the $p^{nf\,\text{th}}$ power correspondence of $\mathfrak{q}(W^{\tau})$ to itself. Hence we have

$$(6.8.4) \qquad (\Pi_{\mathfrak{p}}^{nf})^{\tau'} \circ \mathfrak{q}(Y(\tau)) = \mathfrak{q}(Y(\tau)) \circ \Pi_{\mathfrak{q}}^{nf} .$$

By (6.8. 2, 3, 4), we obtain

$$\mathfrak{q}({}^t Y(\tau^{-1}\sigma\tau)) \circ \Pi_{\mathfrak{q}}^{nf} = \mathfrak{q}({}^t Y(\tau)) \circ [\mathfrak{p}({}^t Y(\sigma)) \circ \Pi_{\mathfrak{p}}^{nf}]^{\tau'} \circ \mathfrak{q}(Y(\tau)) .$$

From this it follows easily that

$$(6.8.5) \qquad I_0\big(\mathfrak{q}({}^t Y(\tau^{-1}\sigma\tau)) \circ \Pi_{\mathfrak{q}}^{nf}\big) = I_0\big(\mathfrak{p}({}^t Y(\sigma)) \circ \Pi_{\mathfrak{p}}^{nf}\big) .$$

Let $T = \{\tau\}$ be a set of representatives for $G_{\mathfrak{h}}$ modulo $S_{\mathfrak{h}}$. We have then

$$(6.8.6) \qquad \chi(\sigma) = \begin{cases} \sum_{\tau \in T} \psi(\tau^{-1}\sigma\tau) & \text{if } \sigma \in S_{\mathfrak{h}}, \\ 0 & \text{if } \sigma \notin S_{\mathfrak{h}}. \end{cases}$$

Therefore we have

$$\begin{aligned}
& N^{-1} \sum_{n=1}^{\infty} u^{n-1} \sum_{\sigma}' \chi(\sigma) I_0\big(\mathfrak{p}(^t Y(\sigma)) \circ \Pi_{\mathfrak{p}}^n\big) \\
&= N^{-1} \sum_{n=1}^{\infty} u^{nf-1} \sum_{\tau \in T} \sum_{\sigma \in S_{\mathfrak{h}}} \psi(\tau^{-1}\sigma\tau) I_0\big(\mathfrak{p}^{\tau}(^t Y(\tau^{-1}\sigma\tau)) \circ \Pi_{\mathfrak{p}}^{nf}\big) \quad (6.8.5, 6.8.6) \\
&= N^{-1} \sum_{\tau \in T} u^{f-1} \sum_{n=1}^{\infty} u^{f(n-1)} \sum_{\sigma \in S_{\mathfrak{h}}} \psi(\sigma) I_0\big(\mathfrak{p}^{\tau}(^t Y(\sigma)) \circ \Pi_{\mathfrak{p}}^{nf}\big) \\
&= \sum_{\tau \in T} f^{-1}(d/du) \log L(u^f; \mathfrak{p}^{\tau}(W), \kappa_{\mathfrak{p}\tau}, \psi).
\end{aligned}$$

Now, when τ runs through T, $\mathfrak{p}^{\tau}$ runs through exactly f times the set of all prime ideals in $\mathbf{Q}(\zeta)$ dividing p. Hence we obtain (6.8.1).

Let $\{\gamma_1, \cdots, \gamma_N\}$ be a set of representatives for Γ_1 modulo $\Gamma_{\mathfrak{h}}$ (cf. 1.8). Let α be an element of $\mathfrak{o}$ such that $\det(\alpha) = p$. Then the right hand side of (6.8.1) is equal to each of the following two:

$$(6.8.7) \qquad N^{-1} \sum_{n=1}^{\infty} u^{n-1} \sum_{j=1}^{N} \chi(\alpha'^n \gamma_j) I_0(^t \tilde{Y}(\alpha'^n \gamma_j) \circ \Pi_{\mathfrak{p}}^n),$$

$$(6.8.8) \qquad N^{-1} \sum_{n=1}^{\infty} u^{n-1} \sum_{j=1}^{N} \chi(\gamma_j' \alpha^n) I_0(^t \tilde{Y}(\gamma_j' \alpha^n) \circ \Pi_{\mathfrak{p}}^n).$$

Here and henceforth we denote reduction modulo $\mathfrak{p}$ by tilde. By 6.4, we see that (6.8.8) equals

$$(6.8.9) \qquad N^{-1} \sum_{n=1}^{\infty} u^{n-1} \sum_{j=1}^{N} \chi(\gamma_j' \alpha^n) I_0(\Pi_{\mathfrak{p}}^{*n} \circ \tilde{Y}(\gamma_j' \alpha^n)).$$

By (2.5.5) and (4.10.5), we get

$$\Pi_{\mathfrak{p}}^{*n} \circ \tilde{Y}(\gamma_j' \alpha^n) = {}^t \tilde{Y}(\gamma_j) \circ \Pi_{\mathfrak{p}}^{*n} \circ \tilde{Y}(\alpha^n).$$

Therefore, by (6.8.1) and the assumption (6.7.3), we have

$$\begin{aligned}
& (d/du) \log \big[\Pi_{\mathfrak{q}|p} L(u^f; \mathfrak{q}(W), \kappa_{\mathfrak{q}}, \psi)\big] \\
&= (1/2)[(6.8.7) + (6.8.9)] \\
&= (2N)^{-1} \sum_{n=1}^{\infty} u^{n-1} \sum_{j=1}^{N} \chi(\alpha'^n \gamma_j) I_0(^t \tilde{Y}(\gamma_j) \circ U_n)
\end{aligned}$$

where $U_n = {}^t \tilde{Y}(\alpha'^n) \circ \Pi_{\mathfrak{p}}^n + \Pi_{\mathfrak{p}}^{*n} \circ \tilde{Y}(\alpha^n)$. By 5.17 and (4.10.6), we have

$$^t \tilde{Y}(\gamma_j) \circ U_n = \tilde{X}(\Gamma\gamma_j'\Gamma) \circ F_n\big(\tilde{X}(\Gamma\alpha\Gamma), p \cdot \tilde{X}(\Gamma p\delta\Gamma)\big).$$

Applying 6.2 and 6.3 to this correspondence, we find

$$\begin{aligned}
& (d/du) \log \big[\textstyle\prod_{\mathfrak{q}|p} L(u^f; \mathfrak{q}(W), \kappa_{\mathfrak{q}}, \psi)\big] \\
&= \sum_{i=0}^{2m} a(m, 2i, 0) P_i + \sum_{i=0}^{4m} \sum_{\nu=0} (-1)^{i+1} a(m, i, \nu) Q_{i\nu}
\end{aligned}$$

with

$$P_i = N^{-1} \sum_{n=1}^{\infty} u^{n-1} \sum_{j=1}^{N} \chi(\alpha'^n \gamma_j) p^{ni} F_n\big(d(\Gamma\alpha\Gamma), p \cdot d(\Gamma p\delta\Gamma)\big),$$

$$\begin{aligned}
Q_{i\nu} = N^{-1} \sum_{n=1}^{\infty} u^{n-1} \sum_{j=1}^{\infty} \chi(\alpha'^n \gamma_j) p^{n(i-\nu)/2} \cdot \\
\operatorname{Re}\big\{ \operatorname{tr}\big[(\Gamma\gamma_j\Gamma)_{\nu+2} \cdot F_n\big((\Gamma\alpha\Gamma)_{\nu+2}, p \cdot (\Gamma p\delta\Gamma)_{\nu+2}\big)\big]\big\}.
\end{aligned}$$

In view of (6.7.2) and 1.19, $Q_{i,v}$ is given by 1.17, and P_i is given by (1.18.3). Hence we get the formula of our theorem.

6.9. *Remark.* The condition (6.7.2) is satisfied if $\mathfrak{b}$ is a square-free two-sided $\mathfrak{o}$-ideal prime to the discriminant $d(\Phi)$. In fact, in such a case, we have $\mathfrak{b} = \mathfrak{o}b_0$ and $G_{\mathfrak{b}}$ (resp. $S_{\mathfrak{b}}$) is isomorphic to $\mathrm{GL}_2(Z/b_0Z)$ (resp. $\mathrm{SL}_2(Z/b_0Z)$). Then it is easy to see that ξ and ξ' belong to the same conjugate class of $G_{\mathfrak{b}}$.

If ψ is the character of a regular representation of $S_{\mathfrak{b}}$ (resp. $S_{\mathfrak{b}}/\{\pm 1\}$), then χ is the character of a regular representation of $G_{\mathfrak{b}}$ (resp. $G_{\mathfrak{b}}/\{\pm 1\}$), so that (6.7.2) is satisfied for these characters with no assumption on $\mathfrak{b}$.

Therefore, if $m > 0$, $\prod_{\mathfrak{p}|p} Z(u^f; \mathfrak{p}(W_m), \kappa_\mathfrak{p})$ is given by the right hand side of the formula of 6.8 with $r = 1$ and a regular representation of $G_{\mathfrak{b}}$ as ρ. Similarly, for a regular representation ρ of $G_{\mathfrak{b}}/\{\pm 1\}$, we get $\prod_{\mathfrak{p}|p} Z(u^f; \mathfrak{p}(W_m'), \kappa_\mathfrak{p})$, where W_m' is the quotient variety of W_m by $\{Y_m(1), Y_m(-1)\}$. In particular, if $m = 0$, we have $W_0' = W_0 = V$, hence we get $\prod_{\mathfrak{p}|p} Z(u^f; \mathfrak{p}(V), \kappa_\mathfrak{p})$.

6.10. Let us now define the global zeta function $Z(s; W_m, Q(\zeta))$ and L-functions $L(s; W_m, Q(\zeta), \psi)$ of W_m (in the sense of Hasse-Weil) with a complex variable s by

$$Z(s; W_m, Q(\zeta)) = \prod_{\mathfrak{p} \in \Lambda(\mathfrak{b})} Z(N(\mathfrak{p})^{-s}; \mathfrak{p}(W_m), \kappa_\mathfrak{p}) \ ,$$
$$L(s; W_m, Q(\zeta), \psi) = \prod_{\mathfrak{r} \in \Lambda(\mathfrak{b})} L(N(\mathfrak{p})^{-s}; \mathfrak{p}(W_m), \kappa_\mathfrak{p}, \psi) \ .$$

On the other hand, the notation being as in 1.14, define a Dirichlet series $D(s; \mathfrak{b}, k, \rho)$ by

$$D(s; \mathfrak{b}, k, \rho) = \sum (\Gamma_1 \xi \Gamma_1)_{k,\rho} \det(\xi)^{-s} \ ,$$

where the summation is taken over all the distinct double cosets $\Gamma_1 \xi \Gamma_1$ with $\xi \in \Delta_{\mathfrak{b}}$. Then we know [15, Th. 1] that if $k > 0$, $D(s; \mathfrak{b}, k, \rho)$ can be continued holomorphically to the whole s-plane and satisfies a functional equation. Moreover, it can be expressed as an Euler-product:

$$D(s; \mathfrak{b}, k, \rho) = \prod_{p|d(\Phi),\, p \nmid b_0} [1 - (\Gamma_1 \alpha_p \Gamma_1)_{k,\rho} p^{-s}]^{-1}$$
$$\times \prod_{p \nmid b_0 d(\Phi)} [1 - (\Gamma_1 \alpha_p \Gamma_1)_{k,\rho} p^{-s} + (\Gamma_1 p \Gamma_1)_{k,\rho} p^{1-2s}]^{-1} \ ,$$

where, for each prime number p, α_p is an element of $\mathfrak{o}$ such that $\det(\alpha_p) = p$.

6.11. MAIN THEOREM II. *Let the notation and the assumption be as in 6.7, 6.8, and 6.10. Let $Z(s; Q(\zeta))$ be Dedekind's zeta function of $Q(\zeta)$. Then* $L(s; W_m, Q(\zeta), \psi)$

$$= \prod_{p \in \Lambda^*} f(p^{-s}) \cdot \prod_{j=0}^{2m} [Z(s-j; Q(\zeta)) Z(s-j-1; Q(\zeta))]^{r \cdot a(m, 2j, 0)}$$
$$\times \prod_{i=0}^{4m} \prod_{\nu=0}^{t} \det \left[D\left(s - \frac{i-\nu}{2}; \mathfrak{b}, \nu+2, \rho\right) \right]^{(-1)^{t+1} a(m, i, \nu)} \ .$$

Here Λ^* is a finite set of prime numbers, and $f(p^{-s})$ is a rational fnnction of p^{-s}, both depending on Φ, m, $\mathfrak{b}$, ψ.

This follows immediately from 6.8. In particular, for a regular representation ρ of $G_\mathfrak{b}$ (or $G_\mathfrak{b}/\{\pm 1\}$ if m = 0), we get $Z(s;\, W_m,\, Q(\zeta))$.

6.12. COROLLARY. *For every representation ρ of $G_\mathfrak{b}/\{\pm 1\}$, the absolute values of the characteristic roots of $(\Gamma_1\alpha_p\Gamma_1)_{2,\rho}$ do not exceed $2p^{1/2}$ except for a finite number of exceptional p's. Moreover, if Weil's conjecture [22, p. 507] is true, then, for every representation ρ of $G_\mathfrak{b}$ and for every integer $k > 2$ the absolute values of the characteristic roots of $(\Gamma_1\alpha_p\Gamma_1)_{k,\rho}$ do not exceed $2p^{(k-1)/2}$ except for a finite number of exceptional p's. The exceptional p's are contained in a finite set which depends only on Φ and $\mathfrak{b}$.*

PROOF. Let μ be an irreducible representation of $S_\mathfrak{b}$. We can regard $\mathfrak{S}_k(\Gamma_\mathfrak{b})$ as a representation space of $S_\mathfrak{b} = \Gamma_1/\Gamma_\mathfrak{b}$. Then we see easily that $\dim \mathfrak{S}_k(\Gamma_1,\, \mu)$ is equal to the multiplicity of μ in this representation. Hence if ρ_1 is a regular representation of $S_\mathfrak{b}$, we have

$$\dim \mathfrak{S}_k(\Gamma_1,\, \rho_1) = \sum_\varphi \mathrm{tr}\,\big(\mu(1)\big)\cdot \dim \mathfrak{S}_k(\Gamma_1,\, \mu) = \dim \mathfrak{S}_k(\Gamma_\mathfrak{b})\ ,$$

so that $\dim \mathfrak{S}_k(\Gamma_1,\, \rho) = [G_\mathfrak{b} : S_\mathfrak{b}]\cdot \dim \mathfrak{S}_k(\Gamma_\mathfrak{b})$ if ρ is a regular representation of $G_\mathfrak{b}$. Let $B_j(W_m)$ be the jth Betti number of the variety W_m (regarded as a complex manifold). By [8, Ch. IV, § 2, [B]],

$$B_j(W_m) = a(m,\, j,\, 0) + a(m,\, j - 2,\, 0)$$
$$+ \sum_{\nu=0}^{j-1} 2a(m,\, j - 1,\, \nu)\,\dim \mathfrak{S}_{\nu+2}(\Gamma_\mathfrak{b})\ . \qquad (0 \leq j \leq 4m + 2)$$

Therefore, assuming Weil's conjecture to be true, and comparing the degrees with respect to u of the numerators and denominators of both sides of the equality in (6.8), we find that there are no cancellations on the right hand side of the equality, with a regular representation of $G_\mathfrak{b}$ as ρ. Let θ be an irreducible representation of $G_\mathfrak{b}$ such that $\mathfrak{S}_k(\Gamma_1,\, \theta)$ is not trivial. Then $\theta(p) = \theta_0(p)\cdot\theta(1)$ with a root of unity $\theta_0(p)$. Hence we have $(\Gamma_1 p\Gamma_1)_{k,\theta} = p^{k-2}\theta_0(p)$. Let λ be a characteristic root of $(\Gamma_1\alpha_p\Gamma_1)_{k,\theta}$. Since θ is an irreducible constituent of a regular representation of $G_\mathfrak{b}$, in view of the above consideration, we see that, for $k = \nu + 2$ and $i = \nu$,

$$(6.12.1) \qquad [1 - \lambda u + \theta_0(p)p^{i+1}u^2]^{(-1)^i a(m,i,i)}$$

is a factor of $\prod_{\mathfrak{p}|p} Z(u^f;\, \mathfrak{p}(W_m),\, \kappa_\mathfrak{p})$. (Notice that $a(m,\, i,\, i) > 0$ for $0 \leq i \leq 2m$). Let u_1 and u_2 be the roots of (6.12.1). If Weil's conjecture is true, then both $|u_1|^{-1}$ and $|u_2|^{-1}$ have the values

$$p^{1/2},\, p^{3/2},\, \cdots,\, p^{(4m+1)/2} \quad \text{or} \quad 1,\, p,\, p^2,\cdots,\, p^{2m+1}\ ,$$

according as i is even or odd. On the other hand, from (6.12.1), we see that $|u_1|^{-1}|u_2|^{-1} = p^{i+1}$. If $|u_1| = |u_2|$, we have $|\lambda| \leq |u_1|^{-1} + |u_2|^{-1} \leq 2p^{(i+1)/2} = 2p^{(k-1)/2}$, which is our required result. Assume that $|u_1| < |u_2|$. Then we have $|u_1|^{-1} \geq p^{(k+1)/2}$, $|u_2|^{-1} \leq p^{(k-3)/2}$, so that $|\lambda| \geq |u_1|^{-1} - |u_2|^{-1} \geq p^{(k+1)/2} - p^{(k-3)/2} > p^{(k-2)/2}(p+1)$, if $p > 2$. This contradicts 1.20. For $m = 0$, our variety W_0 is an algebraic curve, and Weil's conjecture is actually true. Therefore we get our assertions.

6.13. *Remark.* In view of the results of Eichler [24] and Shimizu [25], we know that there are some overlaps, or more precisely, linear relations between the Dirichlet series $D(s; \mathfrak{b}, k, \rho)$ and Hecke's Dirichlet series attached to cusp forms with respect to congruence subgroups of $SL_2(\mathbf{Z})$. Then, from 6.12, one may derive an estimate for the Fourier coefficients of certain cusp forms, assuming Weil's conjecture to be true.

6.14. *Remark.* It can be easily verified that $\mathfrak{S}_k(\Gamma_\mathfrak{b})$ is generated by the components of the eigenfunctions of $(\Gamma_1 \xi \Gamma_1)_{k,\rho}$ for any fixed ξ and for a regular representation ρ of $G_\mathfrak{b}$. From this fact and from (1.14.1), one sees that the characteristic roots of $(\Gamma_\mathfrak{b} \xi \Gamma_\mathfrak{b})_k$ differ from the characteristic roots of $(\Gamma_1 \xi \Gamma_1)_{k,\rho}$ only by multiplication of roots of unity. Therefore, 6.12 may be regarded as assertions concerning the characteristic roots of $(\Gamma_\mathfrak{b} \xi \Gamma_\mathfrak{b})_k$. Now, as was observed in [8, 13, 15], $\mathfrak{S}_k(\Gamma_\mathfrak{b})$ is isomorphic to a certain cohomology group H of $\Gamma_\mathfrak{b}$ defined algebraically, and H has a $\mathbf{Z}$-lattice which is stable under the operation corresponding to $(\Gamma_\mathfrak{b} \xi \Gamma_\mathfrak{b})_k$. Therefore *the characteristic roots of* $(\Gamma_\mathfrak{b} \xi \Gamma_\mathfrak{b})_k$ *(and hence of* $(\Gamma_1 \xi \Gamma_1)_{k,\rho})$ *are algebraic integers.* This assures that *the inverse roots and poles of our local L-functions are algebraic integers.*

6.15. *Congruence subgroups of general type.* We call a subgroup Γ' of Γ_1 a *congruence subgroup*, if $\Gamma_\mathfrak{a} \subset \Gamma'$ for an integral two-sided $\mathfrak{o}$-ideal $\mathfrak{a}$. For such a Γ', there exists an integral two-sided $\mathfrak{o}$-ideal $\mathfrak{b}$ such that $\Gamma_\mathfrak{b} \subset \Gamma'$ and the condition (3.2.1) is satisfied. Let $V_\mathfrak{b}$ be an algebraic curve defined in 3.2 and 3.4 with this $\mathfrak{b}$. Then we can find an algebraic curve V', which is a model of $\Gamma' \backslash \mathfrak{H}$, and for which the following diagram is commutative:

$$\begin{array}{ccc} \Gamma_\mathfrak{b} \backslash \mathfrak{H} & \longrightarrow & V_\mathfrak{b} \\ \downarrow & & \downarrow \\ \Gamma' \backslash \mathfrak{H} & \longrightarrow & V' \; . \end{array}$$

Since $V_\mathfrak{b}$ and the $Y_0(\alpha)$ are defined over $\mathbf{Q}(\zeta)$, we may take V' so that both V' and the projection map of $V_\mathfrak{b}$ to V' are defined over $\mathbf{Q}(\zeta)$. Then $V_\mathfrak{b}$ is a Galois covering of V' and the Galois group is isomorphic to $\Gamma' \cdot \{\pm 1\} / \Gamma_\mathfrak{b} \cdot \{\pm 1\}$. Let

ρ be a representation of $G_{\mathfrak{b}}$ whose character is induced from the identity character of $\Gamma'\{\pm 1\}/\Gamma_{\mathfrak{b}}$. The formula of 6.8, with this ρ, and with $r = 1$ obviously, gives $\prod_{\mathfrak{p}|p} Z(u^{\mathfrak{f}}; \mathfrak{p}(V'), \kappa_{\mathfrak{p}})$. Hence the global zeta function of V' over $\boldsymbol{Q}(\zeta)$ is given by

$$
(6.15.1) \quad
\begin{aligned}
&Z\big(s;\ V',\ \boldsymbol{Q}(\zeta)\big) \\
&= \prod_{p \in \Lambda_*} f(p^{-s}) \cdot Z\big(s;\ \boldsymbol{Q}(\zeta)\big) Z\big(s - 1;\ \boldsymbol{Q}(\zeta)\big) \cdot \det\big[D(s;\ \mathfrak{b},\ 2,\ \rho)\big]^{-1}.
\end{aligned}
$$

For a given Γ', the choice of $\mathfrak{b}$ is not unique. One may take any multiple of $\mathfrak{b}$ in place of $\mathfrak{b}$. Therefore (6.15.1) gives the zeta functions of V' over various cyclotomic fields $\boldsymbol{Q}(\zeta)$.

The same type of discussion can be made for the model W' of $(\Gamma' \times \mathfrak{o}^m)\backslash(\mathfrak{H} \times \Psi_m)$, which is a quotient of $W_{m,\mathfrak{b}}$ by a finite group of automorphisms. If $m > 0$ and Γ' has elements of finite order other than the identity element, W' has singularities.

7. The level prime to the discriminant

7.1. Let Φ, $\mathfrak{o}$, $\mathfrak{b}$ and b_0 be as in 1.3. Suppose that $\mathfrak{b}$ is prime to the discriminant $d(\Phi)$ (cf. § 1.2). Then it is known that $\mathfrak{b} = \mathfrak{o}b_0$, and $\mathfrak{o}/\mathfrak{b}$ is isomorphic to $\mathrm{M}_2(\boldsymbol{Z}/b_0\boldsymbol{Z})$. We fix once for all a homomorphism $h_{\mathfrak{b}}$ of $\mathfrak{o}$ to $\mathrm{M}_2(\boldsymbol{Z}/b_0\boldsymbol{Z})$ with the kernel $\mathfrak{b}$, and put

$$
\Delta_{\mathfrak{b}}^* = \left\{ \xi \in \Delta_{\mathfrak{b}} \,\Big|\, h_{\mathfrak{b}}(\xi) = \begin{pmatrix} 1 & 0 \\ 0 & d \end{pmatrix}, \quad d \in \boldsymbol{Z}/b_0\boldsymbol{Z} \right\}.
$$

Then we obtain a surjective isomorphism of $R(\Gamma_{\mathfrak{b}}, \Delta_{\mathfrak{b}}^*)$ to $R(\Gamma_1, \Delta_{\mathfrak{b}})$ by means of the correspondence $\Gamma_{\mathfrak{b}}\xi\Gamma_{\mathfrak{b}} \rightarrow \Gamma_1\xi\Gamma_1$ [14, Prop. 1.15]. This implies in particular that for every positive integer d prime to b_0, there exists an element ξ of $\Delta_{\mathfrak{b}}^*$ such that $\det(\xi) = d$.

7.2. PROPOSITION. *Suppose that $\mathfrak{b}$ is prime to $d(\Phi)$. For every integer $k \geqq 0$, there exists a basis of $\mathfrak{S}_k(\Gamma_{\mathfrak{b}})$ over $\boldsymbol{C}$, with respect to which $(\Gamma_{\mathfrak{b}}\xi\Gamma_{\mathfrak{b}})_k$ can be represented by a real matrix if $\xi \in \Delta_{\mathfrak{b}}^*$ and $h_{\mathfrak{b}}(\xi)$ is of the form $h_{\mathfrak{b}}(\xi) = \begin{pmatrix} a & 0 \\ 0 & d \end{pmatrix}$ with $a, d \in \boldsymbol{Z}/b_0\boldsymbol{Z}$.*

Proof is essentially given in [14, pp. 286–287]. For reader's convenience, we reproduce it here. By Eichler [3, Satz 5], there exists a unit η of $\mathfrak{o}$ such that $\det(\eta) = -1$, $h_{\mathfrak{b}}(\eta) = \begin{pmatrix} -1 & 0 \\ 0 & 1 \end{pmatrix}$. For $f \in \mathfrak{S}_k(\Gamma_{\mathfrak{b}})$, define $g = f \,|\, U(\eta)$ by

$$
g(z) = \overline{f(\eta(\bar{z}))} j(\eta,\ z)^{-k},
$$

where the bars indicate the complex conjugate. Then $U(\eta)$ is a semi-linear mapping of $\mathfrak{S}_k(\Gamma_{\mathfrak{b}})$ to itself, satisfying $(af) \,|\, U(\eta) = \bar{a}(f \,|\, U(\eta))$ for $a \in \boldsymbol{C}$. Since $\eta^2 \in \Gamma_{\mathfrak{b}}$, we have $U(\eta)^2 = 1$. Put

$$\mathfrak{S}' = \{ f \in \mathfrak{S}_k(\Gamma_\mathfrak{b}) \mid f = f \mid U(\eta) \} \, .$$

Then we see easily that $\mathfrak{S}'$ is an R-linear subspace of $\mathfrak{S}_k(\Gamma_\mathfrak{b})$, and $\mathfrak{S}_k(\Gamma_\mathfrak{b}) = \mathfrak{S}' + i\mathfrak{S}'$. Let $\xi \in \Delta_\mathfrak{b}$ be such that $h_\mathfrak{b}(\xi) = \begin{pmatrix} a & 0 \\ 0 & d \end{pmatrix}$. Then we have $\eta \xi \eta^{-1} \equiv \xi \bmod \mathfrak{b}$, so that $\Gamma_\mathfrak{b} \eta \xi \eta^{-1} \Gamma_\mathfrak{b} = \Gamma_\mathfrak{b} \xi \Gamma_\mathfrak{b}$ by (1.5.2). Using this relation, we can easily verify that $U(\eta)$ commutes with $(\Gamma_\mathfrak{b} \xi \Gamma_\mathfrak{b})_k$. Hence $(\Gamma_\mathfrak{b} \xi \Gamma_\mathfrak{b})_k$ maps $\mathfrak{S}'$ into itself. Therefore a basis of $\mathfrak{S}'$ over R has the required property,

7.3. For every integer r prime to b_0, there exists, by 1.8, an element η_r of Γ_1 such that $h_\mathfrak{b}(\eta_r) = \begin{pmatrix} r^{-1} & 0 \\ 0 & r \end{pmatrix}$. By (1.5.4), $\Gamma_\mathfrak{b} \eta_r \Gamma_\mathfrak{b}$ depends only on $\mathfrak{b}$ and r; it is independent of the choice of η_r. Let σ be a complex valued character of $GL_1(Z/b_0 Z)$. We denote by $\mathfrak{S}_k^*(\Gamma_\mathfrak{b}, \sigma)$ the subspace of $\mathfrak{S}_k(\Gamma_\mathfrak{b})$ consisting of the elements f such that

$$(7.3.1) \qquad f \mid (\Gamma_\mathfrak{b} \eta_r \Gamma_\mathfrak{b})_k = \sigma(r) f \, .$$

Then $\mathfrak{S}_k(\Gamma_\mathfrak{b})$ is the direct sum of the $\mathfrak{S}_k^*(\Gamma_\mathfrak{b}, \sigma)$ for all the characters σ of $GL_1(Z/b_0 Z)$. By (1.5.4) and 1.9, we see that $\Gamma_\mathfrak{b} \eta_r \Gamma_\mathfrak{b}$ and $\Gamma_\mathfrak{b} \xi \Gamma_\mathfrak{b}$ are commutative if $\xi \in \Delta_\mathfrak{b}$ and $h_\mathfrak{b}(\xi) = \begin{pmatrix} u & 0 \\ 0 & v \end{pmatrix}$ with $u, v \in Z/b_0 Z$. Hence, for such an element ξ, $(\Gamma_\mathfrak{b} \xi \Gamma_\mathfrak{b})_k$ maps $\mathfrak{S}_k^*(\Gamma_\mathfrak{b}, \sigma)$ into itself. We denote the restriction of $(\Gamma_\mathfrak{b} \xi \Gamma_\mathfrak{b})_k$ to $\mathfrak{S}_k^*(\Gamma_\mathfrak{b}, \sigma)$ by $(\Gamma_\mathfrak{b} \xi \Gamma_\mathfrak{b})_{k,\sigma}^*$. Then we have

$$\operatorname{tr}\left[(\Gamma_\mathfrak{b} \eta_r \Gamma_\mathfrak{b})_k^{-1} (\Gamma_\mathfrak{b} \xi \Gamma_\mathfrak{b})_k\right] = \sum_\tau \tau(r)^{-1} \cdot \operatorname{tr}(\Gamma_\mathfrak{b} \xi \Gamma_\mathfrak{b})_{k,\tau}^* \, ,$$

where τ runs over all the characters of $GL_1(Z/b_0 Z)$. Hence

$$(7.3.2) \qquad \sum_{(r, b_0)=1} \sigma(r) \cdot \operatorname{tr}\left[(\Gamma_\mathfrak{b} \eta_r \Gamma_\mathfrak{b})_k^{-1} (\Gamma_\mathfrak{b} \xi \Gamma_\mathfrak{b})_k\right] = t \cdot \operatorname{tr}(\Gamma_\mathfrak{b} \xi \Gamma_\mathfrak{b})_{k,\sigma}^* \, .$$

where $t = \varphi(b_0)$ with Euler's function φ.

7.4. PROPOSITION. *Let $\{V, W_m, \varphi_m, f_m\}$ be the fibre system of abelian varieties, defined over $Q(\zeta)$, which is constructed in 3.2, 3.3 and 3.4. If $\mathfrak{b}$ is prime to the discriminant of Φ, there exists a fibre system $\{V', W_m', \varphi_m', f_m'\}$ of abelian varieties and a biregular isomorphism z_0 (resp. z_m) of V' (resp. W_m') to V (resp. W_m) with the following properties.*

(7.4.1) $\{V', W_m', \varphi_m', f_m'\}$ is defined over Q.

(7.4.2) z_0 and z_m are defined over $Q(\zeta)$.

(7.4.3) $z_0 \circ \varphi_m' = \varphi_m \circ z_m$, $z_m \circ f_m' = f_m \circ z_0$.

(7.4.4) If $\xi \in \Delta_\mathfrak{b}$, $h_\mathfrak{b}(\xi) = \begin{pmatrix} r & 0 \\ 0 & 1 \end{pmatrix}$, and α is an automorphism of $Q(\zeta)$ such that $\zeta^\alpha = \zeta^r$, then $Y_0(\xi) = z_0^\alpha \circ z_0^{-1}$, $Y_m(\xi) = z_m^\alpha \circ z_m^{-1}$.

Such $\{V', W_m', \varphi_m', f_m'\}$, z_0, and z_m are unique up to a biregular isomorphism over Q.

PROOF. Let $G(\boldsymbol{Q}(\zeta)/\boldsymbol{Q})$ be the Galois group of $\boldsymbol{Q}(\zeta)$ over $\boldsymbol{Q}$. For $\alpha \in G(\boldsymbol{Q}(\zeta)/\boldsymbol{Q})$ and $\xi \in \Delta_b$, we put $y_m(\alpha) = Y_m(\xi)$ if $\zeta^\alpha = \zeta^r$ and $h_b(\xi) = \begin{pmatrix} r & 0 \\ 0 & 1 \end{pmatrix}$. By (4.10.5), we have

$$y_m(\alpha\beta) = y_m(\alpha)^\beta y_m(\beta) \qquad (\alpha, \beta \in G(\boldsymbol{Q}(\zeta)/\boldsymbol{Q})) .$$

Therefore, by Weil [23, Th. 3], we find, for each m, a projective variety W_m', defined over $\boldsymbol{Q}$, and a biregular isomorphism z_m of W_m' to W_m, defined over $\boldsymbol{Q}(\zeta)$, such that $y_m(\alpha) = z_m^\alpha \circ z_m^{-1}$. Put $\varphi_m' = z_0^{-1} \circ \varphi_m \circ z_m$ and $f_m' = z_m^{-1} \circ f_m \circ z_0$. Since $\varphi_m^\alpha \circ y_m(\alpha) = y_0(\alpha) \circ \varphi_m$ and $y_m(\alpha) \circ f_m = f_m^\alpha \circ y_0(\alpha)$ for every $\alpha \in G(\boldsymbol{Q}(\zeta)/\boldsymbol{Q})$, we have $(\varphi_m')^\alpha = \varphi_m'$ and $(f_m')^\alpha = f_m'$, so that φ_m' and f_m' are defined over $\boldsymbol{Q}$. Thus we get $\{V', W_m', \varphi_m', f_m'\}$, z_0 and z_m satisfying (7.4.1–4). The uniqueness follows from [23, Th. 3] or can be proved in a straightforward way.

7.5. Let us take the system $\{V', W_m', \varphi_m', f_m'\}$ of 7.4 in place of $\{V, W_m, \varphi_m, f_m\}$ and denote it again by $\{V, W_m, \varphi_m, f_m\}$. This amounts to assuming that

(7.5.1) $\{V,\ W_m,\ \varphi_m,\ f_m\}$ *is defined over* $\boldsymbol{Q}$.

(7.5.2) $Y_0(\xi) = D_V$ *and* $Y_m(\xi) = D_{W_m}$ *if* $\xi \in \Delta_b$ *and* $h_b(\xi) = \begin{pmatrix} r & 0 \\ 0 & 1 \end{pmatrix}$ *with* $r \in \boldsymbol{Z}/b_0\boldsymbol{Z}$, *where* D_U *is the diagonal on* $U \times U$ (cf. 2.6).

Then, with $\Gamma = \Gamma_b$,

(7.5.3) $X_m(\Gamma\xi\Gamma)$ *is defined over* $\boldsymbol{Q}$ *if* $\xi \in \Delta_b$ *and* $h_b(\xi) = \begin{pmatrix} a & 0 \\ 0 & d \end{pmatrix}$.

In fact, applying 4.11 to $X_m(\Gamma\alpha\Gamma)$, $Y_m(\psi)$ and $Y_m(\chi)$ with $\psi = \chi$ and $h_b(\psi) = \begin{pmatrix} r & 0 \\ 0 & 1 \end{pmatrix}$, we get $X_m(\Gamma\alpha\Gamma)^{\rho r} = X_m(\Gamma\alpha\Gamma)$ on account of (7.5.2), which proves (7.5.3). By the same reasoning, we get

(7.5.4) $Y_m(\eta)$ *is defined over* $\boldsymbol{Q}$, *if* $\eta \in \Delta_b$, *and* $h_b(\eta) = \begin{pmatrix} a & 0 \\ 0 & d \end{pmatrix}$.

Let the notation and the assumption be the same as in 5.14. If an algebro-geometric object X is rational over $\boldsymbol{Q}$, then $\mathfrak{p}(X)$ is the same as the reduction of X modulo (p). As in § 6, we write it $\tilde{X}$. Let $\alpha \in \mathfrak{o}$ be such that $\det(\alpha) = p$ and $h_b(\alpha) = \begin{pmatrix} 1 & 0 \\ 0 & p \end{pmatrix}$. (For the existence of such an element, see 7.1). Then we have $h_b(\alpha') = \begin{pmatrix} p & 0 \\ 0 & 1 \end{pmatrix}$, so that the formula of 5.14 is now written in the form

(7.5.5) $$\tilde{X}_m(\Gamma\alpha\Gamma) = \Pi_{mp} + \Pi_{mp}^* \circ \tilde{Y}_m(\alpha) .$$

Here we write Π_{mp} and Π_{mp}^* instead of Π_{mp} and Π_{mp}^* for an obvious reason.

By 1.8, there exists an element ε of Γ_1 such that $h_b(\varepsilon) = \begin{pmatrix} 0 & 1 \\ -1 & 0 \end{pmatrix}$. We have then $\varepsilon\alpha \equiv \alpha'\varepsilon \bmod b$, so that $Y_m(\varepsilon)^\rho \circ Y_m(\alpha) = Y_m(\varepsilon)$, where ρ is an automorphism of $\boldsymbol{Q}(\zeta)$ such that $\zeta^\rho = \zeta^p$. Hence we get $\tilde{Y}_m(\varepsilon)^\rho \circ \tilde{Y}_m(\alpha) = \tilde{Y}_m(\varepsilon)$. Here we have to consider reduction modulo $\mathfrak{p}$ for a prime factor $\mathfrak{p}$ of p in $\boldsymbol{Q}(\zeta)$.

By (2.5.5), we have

(7.5.6) $$\Pi_{mp}^{*} \circ \tilde{Y}_m(\alpha) = \tilde{Y}_m(\varepsilon)^{-1} \circ \Pi_{mp}^{*} \circ \tilde{Y}_m(\varepsilon) .$$

Let δ be an element of Γ_1 defined as in 1.11 and 5.16. Then $h_{\mathfrak{b}}(p\delta) = \begin{pmatrix} 1 & 0 \\ 0 & p^2 \end{pmatrix}$. From (7.5.5) and (7.5.6) we obtain

(7.5.7) $$F_n\big(\tilde{X}_m(\Gamma\alpha\Gamma),\, p\tilde{X}_m(\Gamma p\delta\Gamma)\big) = (\Pi_{mp})^n + \tilde{Y}_m(\varepsilon)^{-1} \circ (\Pi_{mp}^{*})^n \circ \tilde{Y}_m(\varepsilon) ,$$

where F_n is the polynomial defined by (1.16.1).

7.6. Let r and η_r be as in 7.3. Take an element ξ_r of $\Delta_{\mathfrak{b}}$ such that $h_{\mathfrak{b}}(\xi_r)$ $= \begin{pmatrix} 1 & 0 \\ 0 & r \end{pmatrix}$. By (7.5.2), (7.5.4), (4.10.5), and (4.10.6), we have

(7.6.1) $$Y_m(r) = Y_m(\xi_r) = Y_m(\eta_r) = X_m(\Gamma_{\mathfrak{b}}\eta_r\Gamma_{\mathfrak{b}}) .$$

We see that $r \to Y_m(r)$ is a homomorphism of $\mathrm{GL}_1(\mathbf{Z}/b_0\mathbf{Z})$ to a group of biregular automorphisms of W_m, defined over $\mathbf{Q}$. The kernel is $\{Y_m(\pm 1)\}$ or $\{Y_m(1)\}$ according as $m = 0$ or $m > 0$.

7.7. Main Theorem III. *Let $\{V,\ W_m,\ \varphi_m,\ f_m\}$ be the fibre system of abelian varieties constructed in 3.2–4 for an integral two sided $\mathfrak{o}$-ideal $\mathfrak{b}$. Suppose that $\mathfrak{b}$ is prime to the discriminant $d(\Phi)$, and the conditions (7.5.1) and (7.5.2) are satisfied. (The existence of models satisfying these conditions is ensured by 7.4.) Let p be a prime number, α an element of $\mathfrak{o}$ such that $\det(\alpha) = p$, $h_{\mathfrak{b}}(\alpha) = \begin{pmatrix} 1 & 0 \\ 0 & p \end{pmatrix}$, and δ an element of Γ_1 such that $p\delta \equiv \alpha^2 \bmod \mathfrak{b}$. Let σ be a character of $\mathrm{GL}_1(\mathbf{Z}/b_0\mathbf{Z})$. Suppose that $\sigma(-1) = 1$ if $m = 0$. Then, for almost all p, and for $m \geq 0$, one has*

$$L(u;\, p(W_m),\, \mathbf{Z}/p\mathbf{Z},\, \sigma)$$
$$= \prod_{j=0}^{2m} [(1 - p^j u)(1 - p^{j+1}u)]^{-e \cdot a(m,2j,0)}$$
$$\times \prod_{i=0}^{4m} \prod_{\nu=0}^{i} \det[1 - (\Gamma_{\mathfrak{b}}\alpha\Gamma_{\mathfrak{b}})_{\nu+2,\sigma}^{*}\, p^{(i-\nu)/2}u + (\Gamma_{\mathfrak{b}}p\delta\Gamma_{\mathfrak{b}})_{\nu+2,\sigma}^{*}\, p^{i-\nu+1}u^2]^{(-1)^i a(m,i,\nu)} .$$

Here $p(W_m)$ is the reduction of W_m modulo p; $L(u;\, p(W_m), \mathbf{Z}/p\mathbf{Z}, \sigma)$ is defined with respect to a homomorphism $r \to Y_m(r)$ of $\mathrm{GL}_1(\mathbf{Z}/b_0\mathbf{Z})$ to a group of birational automorphisms of W_m; $e = 1$ or 0 according as σ is the identity character or not; $(\Gamma_{\mathfrak{b}}\xi\Gamma_{\mathfrak{b}})_{k,\sigma}^{*}$ is defined in 7.3.

Proof. Let η_r be as in 7.3, and ε be as in 7.5. Then we have $\varepsilon^{-1}\eta_r\varepsilon \equiv \eta_r^{-1}$ mod $\mathfrak{b}$, so that $Y_m(\varepsilon^{-1}) \circ Y_m(\eta_r) \circ Y_m(\varepsilon) = Y_m(\eta_r)^{-1}$. It follows that

$$\tilde{Y}_m(r)^{-1} \circ \tilde{Y}_m(\varepsilon)^{-1} \circ (\Pi_{mp}^{*})^n \circ \tilde{Y}_m(\varepsilon) = \tilde{Y}_m(\varepsilon)^{-1} \circ [(\Pi_{mp}^{*})^n \circ \tilde{Y}_m(r)] \circ \tilde{Y}_m(\varepsilon) .$$

Therefore, by 6.3, 6.4, and (7.5.7), we have

$$I_0\big(Y_m(r)^{-1} \circ F_n(X_m(\Gamma\alpha\Gamma),\, pX_m(\Gamma p\delta\Gamma))\big) = 2 \cdot I_0\big(\tilde{Y}_m(r)^{-1} \circ \Pi_{mp}^n\big)$$

with $\Gamma = \Gamma_b$. Hence, by 6.2 and 7.2, we get

$$(d/du) \log L\big(u; \, p(W_m), \, \mathbf{Z}/p\mathbf{Z}, \, \sigma\big)$$
$$= t^{-1} \sum_{n=1}^{\infty} u^{n-1} \sum_{(r,\,b_0)=1} \sigma(r) \cdot I_0\big(\tilde{Y}_m(r)^{-1} \circ \Pi_{mp}^n \big)$$
$$= (2t)^{-1} \sum_{n=1}^{\infty} u^{n-1} \sum_{(r,\,b_0)=1} \sigma(r) \cdot I_0\big[Y_m(r)^{-1} \circ F_n\big(X_m(\Gamma\alpha\Gamma), \, pX_m(\Gamma p\delta\Gamma)\big)\big]$$
$$= \sum_{j=0}^{2m} a(m, 2j, 0)P_j + \sum_{i=0}^{4m} \sum_{\nu=0}^{i} (-1)^{i+1} a(m, i, \nu) Q_{i\nu},$$

where t = the order of $\mathrm{GL}_1(\mathbf{Z}/b_0\mathbf{Z})$,

$$P_j = \sum_{n=1}^{\infty} u^{n-1} p^{nj} t^{-1} \sum_r \sigma(r) F_n\big(d(\Gamma\alpha\Gamma), \, p \cdot d(\Gamma p\delta\Gamma)\big),$$
$$Q_{i\nu} = \sum_{n=1}^{\infty} u^{n-1} p^{n(i-\nu)/2} t^{-1} \sum_r \sigma(r) \cdot \mathrm{tr}\big[(\Gamma \eta_r \Gamma)_{\nu+2} F_n\big((\Gamma\alpha\Gamma)_{\nu+2}, \, p(\Gamma p\delta\Gamma)_{\nu+2}\big)\big].$$

By (1.18.1), we get

$$P_j = \begin{cases} 0 & \text{if} \quad \sigma \neq 1, \\ -(d/du) \log \big[(1 - p^j u)(1 - p^{j+1}u)\big] & \text{if} \quad \sigma = 1. \end{cases}$$

By (7.3.2) and (1.16.2),

$$Q_{i\nu} = -(d/du) \log \big[\det \big(1 - (\Gamma\alpha\,\Gamma)^*_{\nu+2,\sigma} p^{(i-\nu)/2} u + (\Gamma p\delta\Gamma)^*_{\nu+2,\sigma} \, p^{i-\nu+1}u^2\big)\big].$$

From this we get our assertion.

7.8. Corollary. *Let the notation be as in 7.7. Then, for almost all p and for $m \geq 0$, one has*

$$Z(u; \, p(W_m), \, \mathbf{Z}/p\mathbf{Z})$$
$$= \prod_{j=0}^{2m} \big[(1 - p^j u)(1 - p^{j+1}u)\big]^{-a(m,2j,0)}$$
$$\times \prod_{i=0}^{4m} \prod_{\nu=0}^{i} \det \big[1 - (\Gamma_b \alpha \Gamma_b)_{\nu+2} p^{(i-\nu)/2} u + (\Gamma_b p\delta\Gamma_b)_{\nu+2} p^{i-\nu+1} u^2\big]^{(-1)^i a(m,i,\nu)}.$$

Furthermore, let $W_m/\{\pm 1\}$ be the quotient variety of W_m by $\{Y_m(1), Y_m(-1)\}$. Then, for almost all p and for $m \geq 0$, one has

$$Z(u; \, p(W_m/\{\pm 1\}), \, \mathbf{Z}/p\mathbf{Z})$$
$$= \prod_{j=0}^{2m} \big[(1 - p^j u)(1 - p^{j+1}u)\big]^{-a(m,2j,0)}$$
$$\times \prod_{i=0}^{4m} \prod_{\substack{0 \leq \nu \leq i \\ \nu:\,\mathrm{even}}} \det \big[1 - (\Gamma_b \alpha \Gamma_b)_{\nu+2} p^{(i-\nu)/2} u + (\Gamma_b p\delta\Gamma_b)_{\nu+2} p^{i-\nu+1} u^2\big]^{(-1)^i a(m,i,\nu)},$$

where the only difference from the above formula is the restriction of ν to even integers.

Proof. The first formula can be obtained by taking the product of the $L(u; \, p(W_m), \, \mathbf{Z}/p\mathbf{Z}, \, \sigma)$ for all characters σ of $\mathrm{GL}_1(\mathbf{Z}/b_0\mathbf{Z})$. As for the second, we take the product for all the characters σ such that $\sigma(-1) = 1$. We notice that

$$\sum_{\sigma}' \mathfrak{S}_k^*(\Gamma_{\mathfrak{b}}, \sigma) = \begin{cases} \mathfrak{S}_k(\Gamma_{\mathfrak{b}}) & \text{if } k \text{ is even,} \\ \{0\} & \text{if } k \text{ is odd,} \end{cases}$$

where $\sum_{\sigma}'$ means the direct sum taken over all the characters σ such that $\sigma(-1) = 1$. Hence we get the second formula.

7.9. To get the global zeta function of W_m over $\boldsymbol{Q}$, define $D(s; \mathfrak{b}, k)$ by

$$D(s; \mathfrak{b}, k) = \sum (\Gamma_{\mathfrak{t}} \xi \Gamma_{\mathfrak{b}})_k \det (\xi)^{-s},$$

where the summation is taken over all the distinct cosets $\Gamma_{\mathfrak{t}} \xi \Gamma_{\mathfrak{b}}$ with $\xi \in \Delta_{\mathfrak{b}}^*$ (cf. 7.1). For each prime number p, let α_p be an element of $\Delta_{\mathfrak{b}}^*$ such that $\det (\alpha_p) = p$, and δ_p be such that $p\delta_p \equiv \alpha_p^2 \bmod \mathfrak{b}$. Then $D(s; \mathfrak{b}, k)$ can be expressed as an Euler product:

$$D(s; \mathfrak{b}, k) = \prod_{p \mid d(\Phi),\, p \nmid b_0} [1 - (\Gamma_{\mathfrak{b}} \alpha_p \Gamma_{\mathfrak{b}})_k p^{-s}]^{-1}$$
$$\times \prod_{p \nmid b_0 d(\Phi)} [1 - (\Gamma_{\mathfrak{b}} \alpha_p \Gamma_{\mathfrak{b}})_k p^{-s} + (\Gamma_{\mathfrak{b}} p \delta \Gamma_{\mathfrak{b}})_k p^{1-2s}]^{-1}.$$

Moreover $D(s; \mathfrak{b}, k)$ can be continued holomorphically to the whole s-plane and satisfies a functional equation [14, 15].

The global zeta function of W_m over $\boldsymbol{Q}$ is defined by

$$Z(s;\ W_m,\ \boldsymbol{Q}) = \prod_p' Z(p^{-s};\ p(W_m),\ \boldsymbol{Z}/p\boldsymbol{Z}),$$

the product being taken over all "good" p's.

7.10. MAIN THEOREM IV. *Let the notation and assumption be the same as in 7.7, 7.8, and 7.9. Let $\zeta(s)$ be Riemann's zeta function. Then*

$$Z(s;\ W_m,\ \boldsymbol{Q}) = \prod_{p \in \Lambda^*} f(p^{-s}) \cdot \prod_{j=0}^{2m} [\zeta(s-j)\zeta(s-j-1)]^{a(m, 2j, 0)}$$
$$\times \prod_{i=0}^{4m} \prod_{\nu=0}^{i} \det \left[D\left(s - \frac{i-\nu}{2};\ \mathfrak{b},\ \nu + 2 \right) \right]^{(-1)^{i+1} a(m, i, \nu)}.$$

Here Λ^* is a finite set of prime numbers, and $f(p^{-s})$ is a rational function of p^{-s}, both depending on Φ, m, $\mathfrak{b}$.

One can get similarly the global zeta function of $W_m/\{\pm 1\}$ over $\boldsymbol{Q}$. As for the estimate for the characteristic roots of $(\Gamma_{\mathfrak{b}} \alpha_p \Gamma_{\mathfrak{b}})_k$, one gets an assertion similar to 6.12. See also 6.14.

7.11. *Remark.* The results of §§ 6, 7, specialized to the case $m = 0$, are more general than [14] and Konno [7]. In fact, these previous works treated only the case where the level ideal $\mathfrak{b}$ is prime to the discriminant, while the present investigation includes the assertions 6.8 and 6.11 in which $\mathfrak{b}$ can be arbitrary.

A result of this kind was first obtained by Eichler [4] for the curves uniformized by certain elliptic modular functions, and this was generalized by [11].

Rangachari [9] treated L-functions with characters in this case. It is easy to reformulate these results in the form of our 6.8 and 6.11 with $m = 0$. One should also notice that Igusa [26] succeeded in shutting up exceptional primes in the smallest possible range. As for the problem of extending the results to the case $m > 0$, see Introduction and Footnote 4.

University of Tokyo
Princeton University

References

1. A. Borel and A. Haefliger, *La classe d'homologie fondamentales d'un espace analytique*, Bull. Soc. Math. France, 89 (1961), 461-513.
2. W. L. Chow, *On equivalence classes of cycles in an algebraic variety*, Ann. of Math. 64 (1956), 450-479.
3. M. Eichler, *Allgemeine Kongruenzklasseneinteilungen der Ideale einfacher Algebren über algebraischen Zahlkörpern und ihre L-Reihen*, J. Reine Angew. Math. 179 (1938), 227-251.
4. ———, *Quaternäre quadratische Formen und die Riemannsche Vermutung für die Kongruenzzetafunktion*, Arch. Math. 5 (1954), 355-366.
5. S. Koizumi, *On specialization of the Albanese and Picard varieties*, Mem. Coll. Sci. Univ. of Kyoto, Ser. A, 32 (1960), 371-382.
6. ——— and G. Shimura, *On specializations of abelian varieties*, Sci. Papers Coll. Gen. Ed. Univ. of Tokyo, 9 (1959), 187-211.
7. S. Konno, *On Artin's L-functions of the algebraic curves uniformized by certain automorphic functions*, J. Math. Soc. Japan, 15 (1963), 89-100.
8. M. Kuga, Fibre varieties over a symmetric space whose fibres are abelian varieties, lecture note, University of Chicago, 1963-64.
9. S. S. Rangachari, *Modulare Korrespondenzen und L-Reihen*, J. Reine Angew. Math. 205 (1961), 119-155.
10. G. Shimura, *Reduction of algebraic varieties with respect to a discrete valuation of the basic field*, Amer. J. Math. 77 (1955), 134-176.
11. ———, *Correspondances modulaires et les fonctions ζ de courbes algébriques*, J. Math. Soc. Japan, 10 (1958), 1-28.
12. ———, *On the theory of automorphic functions*, Ann. of Math. 70 (1959), 101-144.
13. ———, *Sur les intégrales attachées aux formes automorphes*, J. Math. Soc. Japan, 11 (1959), 291-311.
14. ———, *On the zeta-functions of the algebraic curves uniformized by certain automorphic functions*, J. Math. Soc. Japan, 13 (1961), 275-331.
15. ———, *On Dirichlet series and abelian varieties attached to automorphic forms*, Ann. of Math. 76 (1962), 237-294.
16. ———, *On analytic families of polarized abelian varieties and automorphic functions*, Ann. of Math. 78 (1963), 149-192.
17. ———, *On the field of definition for a field of automorphic functions*, I, II, Ann. of Math. 80 (1964), 160-189, 81 (1965), 124-165.
18. ——— and Y. Taniyama, Complex multiplication of abelian varieties and its applications to number theory, Publ. Math. Soc. Japan, No. 6, 1961.
19. A. Weil, Foundations of algebraic geometry, Amer. Math. Soc. Coll. Publ. vol. 29, revised and enlarged edition, 1962.
20. ———, Sur les courbes algébriques et les variétés qui s'en déduisent, Herman, Paris, 1948.

21. A. WEIL, *Variétés abéliennes et courbes algébriques*, Hermann, Paris, 1948.
22. ———, *Number of solutions of equations in finite fields*, Bull. Amer. Math. Soc., 55 (1949), 497-508.
23. ———, *The field of definition of a variety*, Ammer. J. Math. 78 (1956), 509-524.
24. M. EICHLER, *Quadratische Formen und Modulfunktionen*, Acta Arith. 4 (1958), 217-239.
25. H. SHIMIZU, *On zeta functions of quaternion algebras*, Ann. of Math. 81 (1965), 166-193.
26. J. IGUSA, *Kroneckerian model of fields of elliptic modular functions*, Amer. J. Math. 81 (1959), 561-577.

(Received March 23, 1965)

A reciprocity law in non-solvable extensions

Journal für die reine und angewandte Mathematik, 221 (1966), 209-220

The purpose of this note is to give some examples of non-solvable extensions of the rational number field, for which the law of decomposition of primes is given explicitly. Our result can be obtained, by a certain routine consideration and a simple numerical calculation, almost directly from the known result on the Hasse zeta function of an algebraic curve uniformized by modular functions. Therefore it will be possible to get many more examples of the same kind by the same method. However, I have not tried to get the most extensive result, but only intended to give an explicit and comprehensible statement, which suggests doubtlessly the existence of a general reciprocity law for non-abelian or non-solvable extensions of number fields.

Notation. As usual, Z, Q and C will stand for the ring of rational integers, the rational number field, and the complex number field, respectively. If A is a ring with an identity element, $GL_n(A)$ denotes the group of invertible matrices of size n with entries in A, and $SL_n(A)$ the subgroup of $GL_n(A)$ consisting of the elements of determinant 1. The identity element of $GL_n(A)$ will be denoted by 1_n. For a normal extension K of a field F, $G(K/F)$ denotes the Galois group of K over F.

1. Statement of the result

Theorem[1]). *For each prime number l such that $7 \leq l \leq 97$, there exists a normal extension K_l of Q with the following properties:*

(i) *There is an isomorphism $\sigma \to R(\sigma)$ of the Galois group $G(K_l/Q)$ onto $GL_2(Z/lZ)$.*

(ii) *The subfield of K_l corresponding to the subgroup $R^{-1}(SL_2(Z/lZ))$ of $G(K_l/Q)$ is the cyclotomic field $Q(e^{2\pi i/l})$.*

(iii) *Every prime number p, other than 11 and l, is unramified in K_l.*

(iv) *Let p be a prime number, other than 11 and l. Let $\mathfrak{p}$ be a prime ideal in K_l dividing p, and $\sigma_{\mathfrak{p}}$ a Frobenius automorphism of K_l over Q for $\mathfrak{p}$. Then one has*

$$\det[X \cdot 1_2 - R(\sigma_{\mathfrak{p}})] \equiv X^2 - c_p X + p \mod (l),$$

where X is an indeterminate, and c_p is an integer determined by

$$\sum_{m=1}^{\infty} c_m X^m = X \cdot \prod_{n=1}^{\infty} (1 - X^n)^2 (1 - X^{11n})^2.$$

[1]) A somewhat weaker form of the result for $l = 7$ was noted in [17, § 15]. Cf. also [16, § 6. 3].

If we define a Dirichlet series $D(s)$ by

$$D(s) = \sum_{m=1}^{\infty} c_m m^{-s},$$

then it converges absolutely for $\mathrm{Re}(s) > 3/2$, and has an Euler product:

$$D(s) = (1 - 11^{-s})^{-1} \cdot \prod_{p \neq 11} (1 - c_p p^{-s} + p^{1-2s})^{-1}.$$

Moreover, $D(s)$ is holomorphically continued on the whole complex plane and satisfies a functional equation

$$D^*(s) = D^*(2-s) \quad \text{with} \quad D^*(s) = \Gamma(s)(2\pi)^{-s} 11^{s/2} D(s).$$

To be brief, the reciprocity law for the extension K_l/Q can be described in terms of the coefficients of a Dirichlet series with a functional equation and an Euler product.

The assertion (iv) can not give a necessary and sufficient condition for a prime to be divisible by a prime ideal in K_l of absolute degree one. Such a condition can be obtained for certain subfields of K_l. Define five subfields of K_l as follows by giving the corresponding subgroups of $GL_2(Z/lZ)$:

subfield of K_l: corresponding subgroup of $GL_2(Z/lZ)$.

$$K'_l : \{\pm 1\}$$

$$M_l : \left\{ \begin{pmatrix} 1 & 0 \\ c & d \end{pmatrix} \,\middle|\, c, d \in Z/lZ, \ (d, l) = 1 \right\}$$

$$M'_l : \left\{ \pm \begin{pmatrix} 1 & 0 \\ c & d \end{pmatrix} \,\middle|\, c, d \in Z/lZ, \ (d, l) = 1 \right\}$$

$$H_l : \left\{ \begin{pmatrix} a & 0 \\ 0 & a \end{pmatrix} \,\middle|\, a \in Z/lZ, \ (a, l) = 1 \right\}$$

$$F_l : \left\{ \begin{pmatrix} a & 0 \\ c & d \end{pmatrix} \,\middle|\, a, c, d \in Z/lZ, \ (a, l) = (d, l) = 1 \right\}.$$

Then K_l (resp. K'_l, H_l) is the smallest normal extension of Q containing M_l (resp. M'_l, F_l). From the above theorem we get easily

Corollary 1. *In each one of the fields M_l, M'_l and F_l, a prime p, other than 11 and l, is divisible by a prime ideal of absolute degree one, if and only if each corresponding one of the following conditions is satisfied:*

$$M_l : \ 1 - c_p + p \equiv 0 \ \bmod (l),$$

$$M'_l : \ 1 - \varepsilon c_p + p \equiv 0 \ \bmod (l) \ \text{with} \ \varepsilon = 1 \ \text{or} \ -1,$$

$$F_l : \ X^2 - c_p X + p \ \text{is reducible in } Z/lZ.$$

As for the fields M'_l (resp. F_l), we can give *explicitly* an irreducible polynomial $A_l(X)$ (resp. $U_l(X)$) of degree $(l^2 - 1)/2$ (resp. $l + 1$), with coefficients in Q, such that $M'_l = Q(\xi)$ (resp. $F_l = Q(\eta)$) for a root ξ (resp. η) of $A_l(X)$ (resp. $U_l(X)$). Then the condition given in the above corollary is a necessary and sufficient condition for that $A_l(X) \equiv 0$ (resp. $U_l(X) \equiv 0$) $\bmod (p)$ has a root in Z/pZ, naturally with a finite number of exceptional p.

2. The Artin L-functions of K_l[2])

Let us identify $G(K_l/Q)$ with $GL_2(Z/lZ)$ through R. Let Ξ_l be the set of all linear combinations of the simple (complex valued) characters of $GL_2(Z/lZ)$ with integral coefficients. The rank of Ξ_l over Z is $l^2 - 1$. To every $\chi \in \Xi_l$, we can attach an L-function $L(s; K_l/Q; \chi)$ by

$$(d/ds) \log L(s; K_l/Q; \chi) = - \sum_p \sum_{m=1}^{\infty} \chi^*(p^m) \, p^{-ms} \log p,$$

$$\chi^*(p^m) = [T_\mathfrak{p} : 1]^{-1} \sum_{\gamma \in T_\mathfrak{p}} \chi(\sigma_\mathfrak{p}^m \gamma),$$

where $\mathfrak{p}$ is a prime ideal in K_l dividing p, $T_\mathfrak{p}$ is the innertia group for $\mathfrak{p}$, and $\sigma_\mathfrak{p}$ is a Frobenius automorphism of K_l over Q for $\mathfrak{p}$. If p is different from 11 and l,

$$\chi^*(p^m) = \chi(\sigma_\mathfrak{p}^m).$$

Let Ξ_l' be the subset of Ξ_l consisting of the functions χ which can be written in the form

$$(2.1) \qquad \chi(\sigma) = \xi(tr(\sigma),\ \det(\sigma)) \qquad\qquad (\sigma \in GL_2(Z/lZ))$$

with a function ξ on $(Z/lZ) \times GL_1(Z/lZ)$. Obviously Ξ_l' is a submodule of Ξ_l of rank $l^2 - l$. Now take the logarithmic derivative of $D(s)$:

$$(d/ds) \log D(s) = - \sum_p \sum_{m=1}^{\infty} d(p^m) \, p^{-ms} \log p.$$

Here the $d(p^m)$ are integers obtained from the following relation:

$$\sum_{m=1}^{\infty} d(p^m) \, X^m = \begin{cases} (c_p X - 2p X^2)/(1 - c_p X + p X^2) & (p \neq 11), \\ X/(1 - X) & (p = 11). \end{cases}$$

If p is other than 11 and l, then, from our theorem, we see easily that $tr(\sigma_\mathfrak{p}^m) \equiv d(p^m)$, $\det(\sigma_\mathfrak{p}^m) \equiv p^m \bmod (l)$. Hence we get

Corollary 2. *Let χ be an element of Ξ_l', defined by* (2.1) *with a function ξ on* $(Z/lZ) \times GL_1(Z/lZ)$. *Then, for every prime p other than* 11 *and l, the p-factor of the L-function $L(s; K_l/Q; \chi)$ is determined by $\chi^*(p^m) = \xi(d(p^m), p^m)$.*

In other words, we can compute, up to the factors involving the primes 11 and l, the Artin L-function $L(s; K_l/Q; \chi)$ for every χ of Ξ_l' by applying a simple operation to $(d/ds) \log D(s)$.

The same principle is applicable to the extension $K_l/Q(e^{2\pi i/l})$. In this case, the Galois group is $SL_2(Z/lZ)$, which has exactly $l + 4$ simple characters. Let Φ_l be the set of all linear combinations of simple characters of $SL_2(Z/lZ)$ with integral coefficients, and Φ_l' the subset of Φ_l consisting of all the φ such that $\varphi(\sigma) = \eta(tr(\sigma))$ with a function η on Z/lZ. Then Φ_l' is a submodule of Φ_l of rank l. By the same argument as above, we get, for every φ in Φ_l' with η,

$$L(s; K_l/Q(e^{2\pi i/l}); \varphi) = \prod_\mathfrak{p} L_\mathfrak{p}(s; K_l/Q(e^{2\pi i/l}); \varphi),$$

$$(d/ds) \log L_\mathfrak{p}(s; K_l/Q(e^{2\pi i/l}); \varphi) = - \sum_{m=1}^{\infty} \eta(d(N(\mathfrak{p})^m)) N(\mathfrak{p})^{-ms} \log N(\mathfrak{p}) \quad (\mathfrak{p} \nmid 11, l).$$

---

[2]) For the formulation of the result in this section, I owe much to A. Weil.

It follows that

$$(d/ds) \log \Big[\prod_{\mathfrak{p} \,\nmid\, 11, l} L_{\mathfrak{p}}(s;\, K_l/Q(e^{2\pi i/l});\, \varphi) \Big] = -(l-1) \sum \eta(d(p^n)) p^{-ns} \log p,$$

where the sum is taken over all the prime powers $p^n > 1$ such that $p^n \equiv 1 \bmod (l)$ and $p \neq 11$.

3. Division points of an elliptic curve [3]

Let E be an elliptic curve, i. e., an abelian variety of dimension one, defined over Q. E has a structure of an additive group. For a prime number l and a natural number n, put

$$g(l^n) = g(l^n, E) = \{ t \in E \mid l^n t = 0 \},$$

$$g(l^\infty) = g(l^\infty, E) = \bigcup_{n=1}^{\infty} g(l^n, E).$$

It is well-known that

$$g(l^n) \cong Z/l^n Z \oplus Z/l^n Z,$$

$$g(l^\infty) \cong Q_l/Z_l \oplus Q_l/Z_l,$$

where Q_l denotes the l-adic number field and Z_l the ring of l-adic integers. Let $K(l^n)$ resp. $K(l^\infty)$ be the field generated by the coordinates of the points in $g(l^n)$ resp. in $g(l^\infty)$. It can be easily seen that $K(l^n)$ resp. $K(l^\infty)$ is a finite resp. an infinite normal extension of Q. Taking a basis of $g(l^n)$ resp. $g(l^\infty)$, we get a representation R_n resp. R_∞ of the Galois group $G(K(l^n)/Q)$ resp. $G(K(l^\infty)/Q)$ by matrices in $GL_2(Z/l^n Z)$ resp. $GL_2(Z_l)$. We may assume that

$$R_n(\sigma') \equiv R_\infty(\sigma) \bmod (l^n)$$

if σ' is the restriction of an element σ of $G(K(l^\infty)/Q)$ to $K(l^n)$.

Now let us consider the equation for E modulo a prime number p [4]. For all except a finite number of p, we get an elliptic curve E_p defined over the finite field Z/pZ. For a point x on E_p, let x^p denote the point on E_p whose coordinates are the p-th powers of the coordinates of x. The mapping $\pi_p \colon x \to x^p$ is an endomorphism of E_p. On the curve E_p, define $g(l^n, E_p)$ and $g(l^\infty, E_p)$ in the same way as above. These modules have the same structure as $g(l^n, E)$ resp. $g(l^\infty, E)$, provided that $p \neq l$. Since π_p gives an endomorphism of $g(l^\infty, E_p)$, we can represent π_p by a matrix $S(\pi_p)$ in $GL_2(Z_l)$. It is a classical result that

$$(3.1) \qquad \det [X \cdot 1_2 - S(\pi_p)] = X^2 - a_p X + p,$$

where a_p is determined by

$$(3.2) \quad 1 - a_p + p = \text{the number of points on } E_p \text{ with coordinates in } Z/pZ \,[5].$$

Let $\mathfrak{P}$ be a prime divisor of p in $K(l^\infty)$, and $\mathfrak{p}$ the restriction of $\mathfrak{P}$ to $K(l^n)$. Let σ be a Frobenius automorphism for $\mathfrak{P}$. The restriction σ' of σ to $K(l^n)$ is of course a Frobenius automorphism for $\mathfrak{p}$. Now reduction modulo $\mathfrak{P}$ defines an isomorphism of $g(l^\infty, E)$ onto $g(l^\infty, E_p)$, provided that $p \neq l$ [18, § 11. 1, Prop. 13]. Moreover, from the definition of Frobenius automorphism, we see easily

$$t^\sigma \bmod \mathfrak{P} = \pi_p(t \bmod \mathfrak{P}) \qquad\qquad (t \in g(l^\infty, E)).$$

---

[3] The consideration in this section can be generalized to any abelian variety defined over an algebraic number field. Cf., for example, ([18], pp. 149—151).

[4] For a general theory of reduction modulo p, we refer to [2], [14] and ([18]. Ch. III).

[5] For a general tratment of these results, see Weil [20], [21].

Therefore, choosing suitable bases of $g(l^\infty, E)$ and $g(l^\infty, E_p)$, we get $R_\infty(\sigma) = S(\pi_p)$, so that

$$\det [X \cdot 1_2 - R_\infty(\sigma)] = X^2 - a_p X + p,$$

$$\det [X \cdot 1_2 - R_n(\sigma')] \equiv X^2 - a_p X + p \bmod (l^n).$$

This implies especially that σ is uniquely determined by $\mathfrak{P}$. Therefore $\mathfrak{P}$ is unramified if $p \neq l$ and reduction modulo $\mathfrak{P}$ works well. On the other hand, by means of the symbol $e_{X,a}(s, t)$ introduced in [21, No. 75], we can easily prove that $K(l^n)$ contains a primitive l^n-th root of unity ζ, and

$$\zeta^\tau = \zeta^{\det (R_n(\tau))}$$

for every $\tau \in G(K(l^n)/Q)$. Hence, if R_1 is surjective, the assertions of our theorem are true with the present $K(l)$, a_p and R_1, instead of K_l, c_p and R, for all except a finite number of p, instead of "all p other than 11 and l". Even if R_1 is not surjective, this gives a good deal of information on the arithmetic of the extension $K(l)$. It is also noteworthy that the law of decomposition of primes in the fields $K(l^n)$ for *distinct* l's can be described in terms of the *same* a_p.

It is therefore important to get a_p more explicitly as a function of p. Deuring [3] proved that if E has complex multiplication, the a_p are related to Hecke's Grössen-characters. In this case, as the classical theory of complex multiplication tells us, the fields $K(l^n)$ generate abelian extensions over an imaginary quadratic field. The result of Deuring can be extended to the case of higher dimensional abelian varieties with sufficiently many complex multiplication. Namely, the Hasse zeta function of such an abelian variety can be expressed as a product of L-functions with Grössen-characters (Taniyama [19], Weil [22]; cf. also [18], § 18). However, with such an abelian variety, we always come across abelian extensions [18]. To get non-abelian or non-solvable extensions, we have to consider an abelian variety with no or few multiplications[6]. In fact our theorem concerns a certain elliptic curve without complex multiplication. But before going into detail of description of the curve, let us make a supplementary remark on $S(\pi_p)$ with $a_p^2 - 4p \equiv 0 \bmod (l)$, and discuss a problem of finding a defining equation for our extensions.

4. Frobenius automorphism with the discriminant divisible by l

If $a_p^2 - 4p \equiv 0 \bmod (l)$, $R_1(\sigma')$ is conjugate either to $\begin{pmatrix} b & 0 \\ 0 & b \end{pmatrix}$ or to $\begin{pmatrix} b & 1 \\ 0 & b \end{pmatrix}$ with an element b of Z/lZ. In general, one can not distinguish one from the other by only a_p. However, we have at least the following weak criterion.

Lemma 1. *If* $a_p^2 - 4p = ld$ *with an integer d which is not divisible by l, then $R_1(\sigma')$ is conjugate to a matrix of the form* $\begin{pmatrix} b & 1 \\ 0 & b \end{pmatrix}$.

Proof. Our assumption implies that $Q_l(\pi_p)$ is a ramified extension of Q_l of degree 2, and $Z_l[\pi_p]$ is the ring of all integers in $Q_l(\pi_p)$. It follows that any two $Z_l[\pi_p]$-modules, which are free Z_l-modules of rank 2, are isomorphic. Therefore, if V and W are two matrices with coefficients in Z_l, with the same characteristic polynomial $X^2 - a_p X + p$, then there exists an element T of $GL_2(Z_l)$ such that $TVT^{-1} = W$. In particular, we

---

[6]) One should notice that $K(l^n)$ is always a solvable extension of $K(l)$ for every n. Therefore the un-solvability may occur only if the first step $K(l)/Q$ is unsolvable.

find an element T of $GL_2(Z_l)$ such that $T S(\pi_p) T^{-1} = \begin{pmatrix} 0 & p \\ -1 & a_p \end{pmatrix}$. If $R_1(\sigma') = \begin{pmatrix} b & 0 \\ 0 & b \end{pmatrix}$, then we get $\begin{pmatrix} 0 & p \\ -1 & a_p \end{pmatrix} \equiv \begin{pmatrix} b & 0 \\ 0 & b \end{pmatrix}$ mod (l), which is a contradiction. This proves our lemma.

5. Defining equation

It is well-known that every elliptic curve E, defined over Q, is birationally equivalent over Q to a curve

$$(5.1) \qquad Y^2 = 4X^3 - \gamma X - \gamma',$$

with rational numbers γ and γ'. The number $j = \gamma^3/(\gamma^3 - 27\gamma'^2)$ is called the *invariant* of E. Hereafter we assume that E has the form (5.1) and $\gamma, \gamma' \in Q$. We take the point at infinity as the origin of E. Let h be the function on E taking the X-coordinate, i. e., $h(x, y) = x$ for a point (x, y) on E. Then h is finite on E except at the origin, and $h(u) = h(v)$ if and only if $u = \pm v$. Suppose that $G(K(l)/Q)$ is isomorphic to $GL_2(Z/lZ)$. Let t be a point on E of order l. It can easily be verified that $Q(h(t))$ is a subfield of $K(l)$ corresponding to a subgroup of $GL_2(Z/lZ)$ which is conjugate to

$$(5.2) \qquad \left\{ \pm \begin{pmatrix} 1 & 0 \\ c & d \end{pmatrix} \,\middle|\, c, d \in Z/lZ, (d, l) = 1 \right\}.$$

Now, if l is odd, there exist two polynomials A_l and B_l with rational coefficients such that

$$h(lv) = B_l(h(v))/A_l(h(v))^2 \qquad\qquad (v \in E),$$
$$\deg(A_l) = (l^2 - 1)/2, \quad \deg(B_l) = l^2.$$

This is only the multiplication formula of Weierstrass' function, and can be found in any text book of elliptic functions, for example, [6, p. 184]. The polynomials A_l and B_l depend on γ and γ'; so let us write $A_l(X) = A_l(X; \gamma, \gamma')$. By the property of h, we know that $h(t)$ is a root of the equation $A_l(X) = 0$. Since $\deg(A_l) = (l^2 - 1)/2$, this is irreducible over Q. Hence all its roots generate the subfield of $K(l)$ corresponding to the subgroup $\{\pm 1\}$ of $GL_2(Z/lZ)$.

Let $E_0 \colon Y^2 = 4X^3 - \gamma_0 X - \gamma_0'$ be another elliptic curve, with γ_0 and γ_0' in Q, and with the same invariant as E. Then there exists an isomorphism λ of E to E_0. Let $K_0(l)$ and h_0 be defined for E_0 correspondingly. For simplicity, assume that E has no automorphism other than ± 1. Then it can be easily proved that, for a suitable α in Q, one has $\lambda(x, y) = (\alpha x, \alpha^{3/2} y)$ for $(x, y) \in E$, and $\gamma_0 = \gamma \alpha^2$, $\gamma_0' = \gamma' \alpha^3$. Therefore λ is defined over $Q(\alpha^{1/2})$, and $h_0(\lambda v) = \alpha h(v)$ for $v \in E$. Hence if we adjoin $\alpha^{1/2}$ to $K(l)$ and $K_0(l)$, we get the same field. But we can not have $K(l) = K_0(l)$ in general. However, we have clearly $Q(h_0(\lambda t)) = Q(h(t))$, so that we may take $A_l(X; \gamma_0, \gamma_0') = 0$ as a defining equation for the field $Q(h(t))$. For example, for a given invariant j, other than 0 and 1, define δ by $j = \delta/(\delta - 27)$. Then $Q(h(t))$ can be generated by a root of

$$A_l(X; \delta, \delta) = 0.$$

To get an equation for the subfield of $K(l)$ corresponding to

$$(5.3) \qquad \left\{ \begin{pmatrix} a & 0 \\ c & d \end{pmatrix} \,\middle|\, a, c, d \in Z/lZ, (a, l) = (d, l) = 1 \right\},$$

take an elliptic curve E' isomorphic to E/Zt with a point t of order l. Let j' be the invariant of E'. It can be easily proved that $Q(j')$ is the subfield of $K(l)$ corresponding to the subgroup (5. 2). Now it is known that there is an equation $U_l(j, j') = 0$, of degree $l + 1$ and with rational coefficients, which is called the transformation equation of level l (cf. for example [6], pp. 342—349). In the present case, j is a rational number, so that $U_l(j, X) = 0$ is an irreducible equation over Q, satisfied by j'.

6. Curves uniformized by modular functions

For every positive integer q, put

$$\Gamma_0(q) = \left\{ \begin{pmatrix} a & b \\ c & d \end{pmatrix} \in SL_2(Z) \,\Big|\, c \equiv 0 \bmod (q) \right\}.$$

Then $\Gamma_0(q)$ is a properly discontinuous group operating on the upper half plane

$$H = \{z \in C \mid \mathrm{Im}\,(z) > 0\}.$$

Let H^* be the set-theoretical union of H and the rational points on the real line, including the point at infinity. We can define a structure of compact Riemann surface on the quotient space $H^*/\Gamma_0(q)$ in a natural manner. Therefore, we can find a non-singular projective curve V_q, isomorphic to $H^*/\Gamma_0(q)$. Let φ be the natural projection of H^* to V_q. For a prime number p, not dividing q, let T_p be the subset of $V_q \times V_q$ consisting of the points of the form $\varphi(z) \times \varphi(pz)$ for all $z \in H^*$. It can be easily shown that T_p is an algebraic curve on $V_q \times V_q$.

Now the modular functions belonging to the group $\Gamma_0(q)$ form a field, which is generated by $J(z)$ and $J(qz)$ over C, where $J(z)$ is the classical modular function of level 1. Therefore we can take the curce V_q in such a way that V_q is defined over Q, and the function field of V_q over Q (not over C) is exactly $Q(J(z), J(qz))$. With this choice of V_q, it turns out that the T_p are all defined over Q.

Let us consider reduction modulo p of V_q and T_p. For all except a finite number of p, we get a non-singular curve $(V_q)_p$ defined over Z/pZ. On this curve, let Π_p resp. $'\Pi_p$ be the locus of $x \times x^p$ resp. $x^p \times x$, where x^p denotes, as in § 3, the point whose coordinates are the p-th powers of those of x. Now, for all except a finite number of p, we have

$$(6. 1) \qquad (T_p)_p = \Pi_p + '\Pi_p \cdot (Y_p)_p$$

on $(V_q)_p \times (V_q)_p$. Here $(\ )_p$ denotes the variety reduced modulo p, and Y_p is a certain birational automorphism of V_q. This was first proved by Eichler [5], and later generalized by [15]; the latter includes also the result of the same kind for the principal congruence subgroups.

Our theorem can be derived from (6. 1) in the special case $q = 11$. In this case V_q is an elliptic curve, so that the discussion of § 3 is applicable. There is only one differential form of the first kind on V_{11}, up to constant factors. This corresponds to a cusp form on H:

$$(6. 2) \qquad f(z) = \left(\Delta(z)\,\Delta(11z)\right)^{1/12}$$

$$= x \cdot \prod_{n=1}^{\infty} (1 - x^n)^2 (1 - x^{11n})^2 \qquad (x = e^{2\pi i z})$$

$$= \sum_{m=1}^{\infty} c_m x^m,$$

where $\Delta(z) = x \cdot \prod_{n=1}^{\infty} (1 - x^n)^{24}$. Let ω be a differential form on V_{11} of the first kind. It is an elementary fact in the theory of algebraic curves that if T_p transforms ω to $\mu\omega$

28*

with a complex number μ, then the representation of T_p in $g(l^\infty, V_{11})$ has the characteristic roots μ and $\bar{\mu}$. By virtue of Hecke's theory [7], we know that T_p transforms f to $c_p f$. Moreover Y_p is the identity mapping in the present case. Therefore, from (6. 1) it follows that c_p is equal to the number a_p defined by (3. 1) or (3. 2) with V_{11} as the elliptic curve E of § 3 [7]). In other words, the a_p for the elliptic curve V_{11} are determined as the coefficients c_p of (6. 2), for all except a finite number of p. But Igusa [9] proved that this is true for all $p \neq 11$, by showing that the reduction process works well for $p \neq 11$.

Let K_l denote the field $K(l)$ defined for V_{11}. The proof of our theorem will be completed if we show the surjectivity of the isomorphism of $G(K_l/Q)$ to $GL_2(Z/lZ)$ for every l such that $7 \leq l \leq 97$. This will be done in the following § 7.

The Euler product for $D(s) = \sum\limits_{m=1}^{\infty} c_m m^{-s}$ is a consequence of a general result of Hecke [7], [8]. The functional equation can be obtained by combining $f(-1/11z) = -11z^2 f(z)$ with $\Gamma(s)(2\pi)^{-s} D(s) = \int\limits_0^\infty f(iy) y^{s-1} dy$. It will be worth while mentioning that the c_m are related to the number of representations of an integer by a quadratic form [8, § 11, Beispiel 2].

7. Determination of Galois group

Let G_l denote the image of $G(K_l/Q)$ into $GL_2(Z/lZ)$, and let $S_l = G_l \cap SL_2(Z/lZ)$. Our discussion shows at least that G_l contains an element with the characteristic polynomial $X^2 - c_p X + p$, for every prime number $p \neq 11, l$. In particular this implies that $G_l \cdot SL_2(Z/lZ) = GL_2(Z/lZ)$. Therefore, to prove $G_l = GL_2(Z/lZ)$, it is sufficient to show $S_l = SL_2(Z/lZ)$. In Burnside [1, pp. 436—451], all the subgroups of $SL_2(Z/lZ)/\{\pm 1\}$ are determined (the result due to Gierster, cf. also [6], pp. 465—475). We can derive from this result the following three lemmas[8]).

Lemma 2. *Let S be a proper subgroup of $SL_2(Z/lZ)/\{\pm 1\}$. Then either S is isomorphic to a subgroup of the alternating group of degree 5, or the order of S is a divisor of at least one of the numbers $l - 1$, $l + 1$, $l(l-1)/2$, 24.*

Lemma 3. *If $l \geq 11$, no proper subgroup of $SL_2(Z/lZ)$ can contain elements with all the possible types of characteristic polynomial with the constant term 1.*

Lemma 4. *If $l \geq 5$, no proper subgroup of $GL_2(Z/lZ)$ can contain elements with all the possible types of characteristic polynomial.*

Let us now consider, for example, the case $l = 59$. From the table at the end of the paper, we get $c_{827} = -52$. Hence G_{59} contains an element α whose characteristic polynomial is

$$X^2 + 52X + 827 \equiv X^2 - 7X + 1 \mod (59).$$

By a simple calculation, we can verify easily that α is of order 29. Applying the same to $p = 1889$, we find an element β of order 12 in S_l such that $\beta^6 = -1$. By Lemma 2, we have $S_l = SL_2(Z/lZ)$, so that $G_l = GL_2(Z/lZ)$ for $l = 59$. We can treat all the primes l in our theorem in the same manner, except $l = 7$. If $l = 7$, we find $c_{29} = 0$,

---

[7]) For detail, see [5], [15].

[8]) For these I owe much to H. Nagao, who communicated me the proofs of Lemmas 3, 4. Actually, Lemma 2 is sufficient for our purpose, and Lemmas 3 and 4 are not necessary. However, it will not be useless to mention these facts here.

so that S_7 contains an element α, satisfying $\alpha^2 + 1 = 0$, which is of order 4. Further we have $c_{23} = -1$, so that $c_{23}^2 - 4 \cdot 23 = -7 \cdot 13$. By Lemma 1 of § 4, this shows that G_7 contains an element conjugate to $\begin{pmatrix} 4 & 1 \\ 0 & 4 \end{pmatrix}$. From these facts we can conclude that S_7 contains -1, and the order of $S_7/\{\pm 1\}$ is divisible by 14. By lemma 2, this proves $S_7 = SL_2(Z/7Z)$, hence $G_7 = GL_2(Z/7Z)$. Instead of Lemma 1, we can make use of Lemma 4. Of course the prime 97 is not the limit. One may get more l's with the property $G_l = GL_2(Z/lZ)$, even within the effect of our table of c_p. The result tempts us to conjecture that $G_l = GL_2(Z/lZ)$ for infinitely many l, or even for almost all l.

By Fricke [6, pp. 404—407], we know that the elliptic curve V_{11} has the invariant $-2^6 \cdot 3^{-3} \cdot 11^{-5} \cdot 31^3$ (cf. also [10], II, pp. 401—444). Therefore, as indicated in § 5, we can write explicitly defining equations for the fields M_l' and F_l of § 1, without any exact information about the equation for V_{11}. However, we can actually write explicitly an equation for V_{11}. In fact, it can be verified that the curve $\sigma^2 = \tau(\tau^3 - 20\tau^2 + 56\tau - 44)$ obtained in [6], p. 406, (13) gives a model for the function field $Q(J(z), J(11z))$. By a simple calculation, we can transform it rationally over Q, to

$$(7.1) \qquad Y^2 = 4X^3 - (4 \cdot 31/3) X - (41 \cdot 61/27).$$

Hence there is no ambiguity about the fields M_l and K_l. The last equation can be found also in ([10], II, p. 444).

In a certain case we can prove that the Galois group of $K(l^\infty)$ over Q is isomorphic to $GL_2(Z_l)$ by means of

Lemma 5. *Let H be a closed subgroup of $GL_2(Z_l)$. Suppose that:* (i) *l is odd;* (ii) *the natural map $Z_l \to Z_l/lZ_l$ gives a surjective map of H to $GL_2(Z/lZ)$;* (iii) *H contains an element* $Y \equiv 1 + l \cdot \begin{pmatrix} r & 0 \\ 0 & s \end{pmatrix} \bmod (l^2)$ *with elements r and s of Z_l such that $r^2 - s^2 \not\equiv 0 \bmod (l)$. Then $H = GL_2(Z_l)$.*

Proof. If $n = l^{\nu-1}$ and $\nu > 0$, we have $Y^n \equiv 1 + l^\nu \cdot \begin{pmatrix} r & 0 \\ 0 & s \end{pmatrix} \bmod (l^{\nu+1})$. For every $B \in GL_2(Z_l)$, take an element U of H such that $U \equiv B \bmod (l)$. Then $UY^nU^{-1} \equiv 1 + l^\nu B \begin{pmatrix} r & 0 \\ 0 & s \end{pmatrix} B^{-1} \bmod (l^{\nu+1})$. Under the assumption (iii), we can easily verify that the matrices of the form $B \begin{pmatrix} r & 0 \\ 0 & s \end{pmatrix} B^{-1}$ span (over Z) the whole matrix ring modulo (l). Hence, for every $V \in GL_2(Z_l)$ such that $V \equiv 1 \bmod (l^\nu)$, we can find a product W of several elements of the form UY^nU^{-1} with $U \in H$ so that $V \equiv W \bmod (l^{\nu+1})$. Since H is closed, this fact combined with (ii) shows $H = GL_2(Z_l)$.

Let us now consider the extension $K(l^\infty)$ obtained from the curve V_{11} (which is equivalent to (7.1)). Let G_l' be the image of $G(K(l^\infty)/Q)$ under R_∞. By the consideration of § 3, G_l' contains an element of $GL_2(Z_l)$ whose characteristic polynomial is $X^2 - c_p X + p$. Let us take, for example, l to be 7. For $p = 5$, we have $c_p = 1$, so that

$$X^2 - c_p X + p \equiv (X - 2)(X + 1) \bmod (7).$$

Hence $X^2 - c_p X + p$ has two distinct roots a and b in Z_7, such that $a - b$ is a unit of Z_7. Therefore, choosing a suitable basis, we may assume that G_7' contains an element $U = \begin{pmatrix} a & 0 \\ 0 & b \end{pmatrix}$. (If $a \equiv b \bmod (7)$, this reasoning is false.) By a simple computation, we have

$$a^6 \equiv 1 + 7 \cdot 5, \quad b^6 \equiv 1 + 7 \bmod (7^2).$$

Applying Lemma 5 to G_7' with U^6 as Y, we get $R_\infty[G(K(7^\infty)/Q)] = GL_2(Z_7)$. As a simple consequence, we notice that $G(K(7^n)/Q)$ is isomorphic to $GL_2(Z/7^nZ)$ for every positive integer n. The verification for larger l's is left to the voluntary reader.

8. Concluding remarks

If the curve V_q is of genus > 1, we have to consider the Jacobian variety W_q of V_q. Then the reciprocity law for the fields generated by the coordinates of the points of finite order on W_q can be described in terms of the eigen-values of Hecke operators[9]. We can prove an analogous result for algebraic curves uniformized by automorphic functions belonging to an indefinite quaternion algebra [16]. The eigen-values of Hecke operators can be obtained by the trace-formula of Eichler and Selberg. In any case, the determination of Hasse zeta function of an algebraic curve includes, as a natural consequence, a reciprocity law of certain algebraic extensions, though we have no characterization, other than the properties such as given in our theorem, of those extensions, except for the case of complex multiplication (of dimension ≥ 1). The subject is closely connected with the theory of automorphic forms with respect to an arithmetically defined discontinuous group. In the investigations [5], [15], [16], only cusp forms of weight 2 came into the problem. Now it can be shown [12] that the automorphic forms of higher weight are also connected with the zeta function of an algebraic variety, the discontinuous group being a unit group in an indefinite division quaternion algebra. In this case, M. Kuga [11] has obtained an interesting result; namely, the eigen-values of Hecke operators for such forms are again related to the decomposition of primes in the number fields of our type.

Needless to say we should aim at putting all these results into one unified theory. Even though we are, at present, far from the completion of the task, it is quite certain that here is a vast fertile plain in number theory, little of which has been brought under cultivation.

Table of c_p for $p < 2000$; $\displaystyle\sum_{m=1}^{\infty} c_m x^m = x \cdot \prod_{n=1}^{\infty} (1 - x^n)^2 (1 - x^{11n})^2.$ [10]

p	c_p	p	c_p	p	c_p	p	c_p	p	c_p	p	c_p
2	—2	239	—30	563	4	887	—22	1259	—25	1619	—20
3	—1	241	—8	569	0	907	—12	1277	—47	1621	22
5	1	251	—23	571	—28	911	12	1279	—15	1627	78
7	—2	257	—2	577	33	919	10	1283	—36	1637	33
11	1	263	14	587	28	929	—30	1289	0	1657	—2
13	4	269	10	593	44	937	8	1291	—8	1663	4
17	—2	271	—28	599	40	941	42	1297	48	1667	48
19	0	277	—2	601	2	947	—27	1301	27	1669	50
23	—1	281	—18	607	—22	953	34	1303	39	1693	—6
29	0	283	4	613	—16	967	—32	1307	28	1697	—42
31	7	293	24	617	18	971	47	1319	—30	1699	40
37	3	307	8	619	—25	977	—27	1321	47	1709	—45
41	—8	311	12	631	7	983	39	1327	68	1721	—3

[9] In certain cases, the Jacobian variety W_q turns out to be simple and of dimension > 1 (cf. [4], [13]). It will be interesting to determine the Galois groups of the fields analogous to $K(l)$ in these cases.

[10] I wish to acknowledge my gratitude to H. F. Trotter for making the table by an electronic computer.

p	c_p	p	c_p	p	c_p	p	c_p	p	c_p	p	c_p
43	−6	313	−1	641	−33	991	−8	1361	12	1723	−46
47	8	317	13	643	29	997	38	1367	−72	1733	−6
53	−6	331	7	647	−7	1009	−10	1373	39	1741	17
59	5	337	−22	653	−41	1013	39	1381	−68	1747	−57
61	12	347	28	659	10	1019	−10	1399	60	1753	34
67	−7	349	30	661	37	1021	22	1409	−15	1759	−40
71	−3	353	−21	673	14	1031	32	1423	29	1777	8
73	4	359	−20	677	−42	1033	−16	1427	−12	1783	59
79	−10	367	−17	683	−16	1039	5	1429	−70	1787	−57
83	−6	373	−26	691	17	1049	−55	1433	54	1789	10
89	15	379	−5	701	2	1051	2	1439	0	1801	52
97	−7	383	−1	709	−25	1061	−13	1447	28	1811	12
101	2	389	−15	719	15	1063	44	1451	52	1823	−56
103	−16	397	−2	727	3	1069	−20	1453	−71	1831	−43
107	18	401	2	733	−36	1087	8	1459	−20	1847	−52
109	10	409	−30	739	50	1091	−58	1471	22	1861	62
113	9	419	20	743	4	1093	−51	1481	32	1867	28
127	8	421	22	751	−23	1097	−42	1483	49	1871	−3
131	−18	431	−18	757	−22	1103	−51	1487	58	1873	−6
137	−7	433	−11	761	12	1109	−30	1489	−15	1877	18
139	10	439	40	769	20	1117	48	1493	−36	1879	−35
149	−10	443	−11	773	−6	1123	24	1499	55	1889	70
151	2	449	35	787	−32	1129	50	1511	37	1901	77
157	−7	457	−12	797	53	1151	2	1523	−41	1907	−52
163	4	461	12	809	0	1153	−31	1531	32	1913	−36
167	−12	463	−11	811	−38	1163	34	1543	−36	1931	−18
173	−6	467	−27	821	22	1171	−3	1549	−15	1933	54
179	−15	479	20	823	39	1181	−18	1553	−56	1949	−40
181	7	487	23	827	−52	1187	−12	1559	−60	1951	−23
191	17	491	−8	829	25	1193	−21	1567	−52	1973	79
193	4	499	20	839	−5	1201	2	1571	−28	1979	30
197	−2	503	−26	853	14	1213	−41	1579	−30	1987	−22
199	0	509	15	857	8	1217	−42	1583	34	1993	−66
211	12	521	−3	859	−15	1223	14	1597	−32	1997	−72
223	19	523	−16	863	24	1229	60	1601	2	1999	−20
227	18	541	−8	877	−12	1231	−18	1607	33		
229	15	547	8	881	−43	1237	18	1609	−10		
233	24	557	−2	883	4	1249	40	1613	−6		

References

[1] *W. Burnside*, Theory of groups of finite order, Cambridge, 2nd ed. 1911.

[2] *M. Deuring*, Reduktion algebraischer Funktionenkörper nach Primdivisoren des Konstantenkörpers, Math. Zeitschr. **47** (1942), 643—654.

[3] *M. Deuring*, Die Zetafunktion einer algebraischen Kurve vom Geschlechte Eins. I, II, III, IV, Nachr. Akad. Wiss. Göttingen (**1953**) 85—94, (**1955**) 13—42, (**1956**) 37—76, (**1957**) 55—80.

[4] *K. Doi*, On the jacobian varieties of the fields of elliptic modular functions, Osaka Math. J. **15** (1963), 249—256.

[5] *M. Eichler*, Quaternäre quadratische Formen und die Riemannsche Vermutung für die Kongruenzzetafunktion, Arch. Math. **5** (1954) 355—366.

[6] *R. Fricke*, Die elliptischen Funktionen und ihre Anwendungen. II, Leipzig und Berlin 1922.

[7] *E. Hecke*, Über Modulfunktionen und die Dirichletschen Reihen mit Eulerscher Produktentwicklung. I, II, Math. Ann. **114** (1937), 1—28, 316—351.

[8] *E. Hecke*, Analytische Arithmetik der positiven quadratischen Formen, Dansk. Vidensk. Selsk. Math-fys. Meddel. **17**, 12 (Kobenhavn, 1940).

[9] *J. Igusa*, Kroneckerian model of fields of elliptic modular functions, Amer. J. Math. **81** (1959), 561—577.

[10] *F. Klein* und *R. Fricke*, Vorlesungen über die Theorie der Modulfunktionen. I, II, Leipzig 1890—92.

[11] *M. Kuga*, Fibre varieties over a symmetric space whose fibres are abelian varieties, Lecture note, University of Chicago 1964, to appear.

[12] *M. Kuga* and *G. Shimura*, On the zeta function of a fibre variety whose fibres are abelian varieties, to appear.

[13] *T. Matsui*, On the endomorphism algebra of jacobian varieties attached to the fields of elliptic modular functions, Osaka J. Math. **1** (1964), 25—31.

[14] *G. Shimura*, Reduction of algebraic varieties with respect to a discrete valuation of the basic field, Amer. J. Math. **77** (1955), 134—176.

[15] *G. Shimura*, Correspondances modulaires et les fonctions ζ de courbes algébriques, J. Math. Soc. Japan **10** (1958), 1—28.

[16] *G. Shimura*, On the zeta-funtions of the algebraic curves uniformized by certain automorphic functions, J. Math. Japan **13** (1961), 275—331.

[17] *G. Shimura*, Automorphic functions and number theory. II (in Japanese), Sûgaku **13** (1961), 65—80.

[18] *G. Shimura* and *Y. Taniyama*, Complex multiplication of abelian varieties and its applications to number theory, Publ. Math. Soc. Japan, No. 6, 1961.

[19] *Y. Taniyama*, *L*-functions of number fields and zeta functions of abelian varieties, J. Math. Soc. Japan **9** (1957), 330—366.

[20] *A. Weil*, Sur les courbes algébriques et les variétés qui s'en déduisent, Paris 1948.

[21] *A. Weil*, Variétés abéliennes et courbes algébriques, Paris 1948.

[22] *A. Weil*, Jacobi sums as „Grössencharaktere", Trans. Amer. Math. Soc. **73** (1952), 487—495.

Eingegangen 9. März 1965

Moduli and fibre systems of abelian varieties

Annals of Mathematics, 83 (1966), 294-338

Let A be an abelian variety of dimension n, defined over a subfield of the complex number field C, and $\mathcal{C}$ a polarization of A. The Riemann form (or the cohomology class) of a basic polar divisor in $\mathcal{C}$ determines uniquely an alternating matrix of the form $E = \begin{pmatrix} 0 & -e \\ e & 0 \end{pmatrix}$, with a diagonal matrix e whose diagonal elements $e_1, \cdots, e_n$ are positive integers such that $e_1 = 1$, $e_{i+1} \equiv 0 \bmod (e_i)$. We call $\mathcal{P} = (A, \mathcal{C})$ a polarized abelian variety of type (e). A natural question concerning the moduli of polarized abelian varieties is to construct, for each given e, a variety V_e, defined over the rational number field Q, and to assign a point $\mathfrak{v}_e(\mathcal{P})$ of V_e to every polarized abelian variety $\mathcal{P}$ of type (e), so that the following conditions (1–4) are satisfied.

(1) $\mathfrak{v}_e(\mathcal{P}) = \mathfrak{v}_e(\mathcal{P}')$ *if and only if $\mathcal{P}$ is isomorphic to $\mathcal{P}'$.*

(2) *If $\mathcal{P}'$ is a specialization of $\mathcal{P}$ over a field k, then $\big(\mathfrak{v}_e(\mathcal{P}'), \mathcal{P}'\big)$ is a specialization of $\big(\mathfrak{v}_e(\mathcal{P}), \mathcal{P}\big)$ over k.*

(3) $Q\big(\mathfrak{v}_e(\mathcal{P})\big)$ *is the field of moduli of $\mathcal{P}$.*

(4) *Every point of V_e is of the form $\mathfrak{v}_e(\mathcal{P})$.*

If $\mathcal{P} = (A, \mathcal{C})$ is as above, then A is isomorphic to a complex torus $C_n/D_e(z)$ with a lattice $D_e(z)$ generated by the column vectors of an $(n \times 2n)$-matrix $\omega_e(z) = (z\ e)$, for a point z of the Siegel space $\mathfrak{S}_n$ of degree n. Let $P_e = \begin{pmatrix} 1 & 0 \\ 0 & e \end{pmatrix}$, $\Gamma_e = \mathrm{Sp}(n, R) \cap \{P_e^{-1}\mathrm{SL}(2n, Z)P_e\}$. Then one may require

(5) *There exists a holomorphic mapping φ of $\mathfrak{S}_n$ onto V_e which induces a biregular isomorphism of $\mathfrak{S}_n/\Gamma_e$ to V_e and such that $\mathfrak{v}_e(\mathcal{P}) = \varphi(z)$ if z corresponds to $\mathcal{P}$ in the above sense.*

Furthermore, take all the points $t_1, \cdots, t_s$ on A of order N for a positive integer N, and consider the structure $\mathfrak{Q} = (A, \mathcal{C}; t_1, \cdots, t_s)$. Let

$$\Gamma_e^N = \{X \in \Gamma_e \mid P_e X P_e^{-1} \equiv 1 \bmod (N)\}.$$

Then the quotient $\mathfrak{S}_n/\Gamma_e^N$ is in one-to-one correspondence with all the isomorphism-classes of such structures. Therefore one can ask the existence of a variety V_e^N, defined over a certain algebraic number field k_N, and an assignment $\mathfrak{v}_e^N$ which assigns to $\mathfrak{Q}$ a point $\mathfrak{v}_e^N(\mathfrak{Q})$ of V_e^N, with the properties analogous to (1-5). Moreover, for a sufficiently large N, one can expect to construct a

* Research supported by NSF grant GP-3454.

fibre variety, defined over k_N, of which the base is V_e^N, and the fibres are such structures $\mathfrak{A}$ corresponding nicely to the points of $\mathfrak{S}_n$.

The purpose of this paper is to answer these questions affirmatively. We shall do this actually in a more general framework. Taking a division algebra L over $\boldsymbol{Q}$ with a positive involution ρ, and a representation Φ of L by complex matrices, we consider a structure $\mathfrak{A} = (A, C, \theta; t_1, \cdots, t_s)$, formed by a polarized abelian variety (A, C), an isomorphism θ of L into $\mathrm{End}_Q(A)$, and points $t_1, \cdots, t_s$ of A of finite order, satisfying the following conditions (6, 7).

(6) *There exists a holomorphic mapping ξ of C^n onto A which induces an isomorphism of a complex torus C^n/D to A such that $\xi(\Phi(\alpha)x) = \theta(\alpha)\Phi(x)$ for every $\alpha \in \theta^{-1}(\theta(L) \cap \mathrm{End}(A))$ and $x \in C^n$.*

(7) *If $*$ denotes the involution of $\mathrm{End}_Q(A)$ determined by C, then $\theta(\alpha)^* = \theta(\alpha^\rho)$ for every $\alpha \in L$.*

Let L^m be a left L-module of rank m, where $m = 2n/[L:Q]$, and let $L_R^m = L^m \otimes_Q R$. If (6) and (7) are satisfied, then there exist an R-linear isomorphism $\mathfrak{y}$ of L_R^m onto C^n and an L-valued anti-hermitian form T on $L^m \times L^m$ such that:

(8) $\mathfrak{y}(\alpha u) = \Phi(\alpha)\mathfrak{y}(u)$ *for every* $\alpha \in L$ *and* $u \in L_R^m$.

(9) *If $\mathfrak{E}$ is the Riemann form on C^n/D corresponding to a basic polar divisor in C, then $\mathfrak{E}(\mathfrak{y}(x), \mathfrak{y}(x')) = \mathrm{tr}(T(x, x'))$ for $(x, x') \in L^m \times L^m$, where tr is the reduced trace of L to $\boldsymbol{Q}$.*

Let $\mathfrak{y}^{-1}(D) = \mathfrak{M}$, and let $v_1, \cdots, v_s$ be elements of L^m such that $\xi(\mathfrak{y}(v_i)) = t_i$. We put

$$\Omega = (L, \Phi, \rho; T, \mathfrak{M}; v_1, \cdots, v_s),$$

and call $\mathfrak{A}$ a PEL-structure of type Ω. (Here P stands for polarization, E for endomorphism, and L for level.) The collection of data, Ω, will be called a PEL-type.

We shall show that there exists an algebraic number field k_Ω of finite degree, which is characterized by the property

(10) *Let $\mathfrak{A}$ be any PEL-structure of type Ω, and σ an automorphism of C. Then $\mathfrak{A}^\sigma$ is of type Ω if and only if σ leaves every element of k_Ω invariant.* Then we shall obtain a variety V, defined over k_Ω, and assign a point $\mathfrak{v}(\mathfrak{A})$ of V to every PEL-structure $\mathfrak{A}$ of type Ω, so that the conditions analogous to (1-4) are satisfied. Such a variety V is isomorphic to the quotient of a bounded symmetric domain $\mathcal{H}$ by a discontinuous group

$$\Gamma = \{\alpha \in \mathrm{GL}_m(L) \mid T(x\alpha, y\alpha) = T(x, y), \mathfrak{M}\alpha = \mathfrak{M}, v_i\alpha \equiv v_i \bmod \mathfrak{M} \, (i=1, \cdots, s)\}.$$

If Γ has no element of finite order other than the identity element, we can construct a fibre variety of which the base is V, and the fibres are PEL-structures of type Ω, everything being rational over k_Ω, as expected. If $L = Q$ and the v_j

generate $N^{-1}\mathfrak{M}$, then we have $k_\Omega = Q^{(2\pi i/N)} = k_N$. For a more precise statement, we refer the reader to 5.1, 5.3, 5.4, 5.5, 6.2, 6.7 in the text. These theorems generalize and sharpen the results of the same kind in the previous papers [14, Exp. 20], [17], [18], [19]. For the investigations in the same or closer direction, we mention the works of Baily [1], [2], Igusa [4], [5], Kuga [9], Matsusaka [11], Mumford [12], and Siegel [22]. We have been informed that Matsusaka has obtained a similar result for the maximal families of abelian varieties, i.e., $(V_e, \mathfrak{d}_e)$ in our notation; and we have had a look at a sketch of the proof. The method seems different from ours.

To get our result, we first construct a fibre system $\mathfrak{F}$ over C by means of theta-functions and the projective embedding of $\mathcal{H}/\Gamma$ due to Baily-Borel [3]. Here we need the classical transformation formula of theta functions, and a holomorphic property of the Chow points of projective varieties belonging to an analytic family. No analysis of Fourier coefficients of theta functions or automorphic forms is necessary. Then we prove a universal mapping property of $\mathfrak{F}$ from a deformation-theoretical view-point (4.1). This property enables us, first, to define the number field k_Ω; and, secondly, to apply Weil's criterion to lower the field of rationality to k_Ω. Up to this point, the results are obtained by assuming that Γ has no element of finite order other than the identity element. Finally, taking the quotient with respect to a finite group, we get the desired result for $(V, \mathfrak{d})$ in the general case.

Notation and Terminology. We shall mean by C^n (resp. R^n, Z^n) the set of all n-dimensional *column vectors* with components in C (resp. R, Z). However, for an algebra L with positive involution, L^m (resp. L_R^m) will mean the set of all m-dimensional *row vectors* with components in L (resp. L_R). For an associative ring S with identity element, we denote by $M_r(S)$ and $\mathrm{GL}_r(S)$ the ring of all matrices of size r with coefficients in S and the group of all invertible elements in $M_r(S)$, respectively. For instance, $M_n(C)$ acts on C^n by the left multiplication, and $\mathrm{GL}_m(L)$ on L^m by the right multiplication. A diagonal matrix with diagonal elements $e_1, \cdots, e_n$ will be denoted by $\mathrm{diag}\,[e_1, \cdots, e_n]$.

As for the language of algebraic geometry, we shall mainly follow Weil [23], [24]. A variety will mean an abstract variety in the sense of [23, Ch. VII]. (As a matter of fact, we shall lose nothing of importance by assuming every variety to be a Zariski open subset of a projective variety.) The divisor of a rational or meromorphic function f will be denoted by $\mathrm{div}(f)$. By a *discrete place* $\mathfrak{p}$ of a field k, we shall understand a place $\mathfrak{p}$ such that $\{a \in k \mid \mathfrak{p}(a) \neq \infty\}$ is either a discrete valuation ring of rank one, or equal to k. In the latter case, $\mathfrak{p}$ is an isomorphism of k. If X is an algebro-geometric object, rational over k, we denote by $\mathfrak{p}(X)$ the reduction of X modulo $\mathfrak{p}$. For a general treatment of reduction modulo $\mathfrak{p}$, we refer to [16], [21, Ch. III].

1. Generalities on a fibre system of abelian varieties

1.1. We call $\{V, W, h, f\}$ a *fibre system of abelian varieties*, defined over a field k, if the following conditions are satisfied.

(1.1.1) *V and W are non-singular varieties, h is a morphism of W into V, and f is a morphism of V into W, all defined over k.*

(1.1.2) *$h \circ f$ is the idendity mapping of V.*

(1.1.3) *If Z is the graph of h, then Z and $W \times u$ intersect properly on $W \times V$ for every $u \in V$, and $Z \cdot (W \times u) = A_u \times u$ with a variety A_u.*

(1.1.4) *A_u is an abelian variety with the origin $f(u)$.*

We shall often write A_u as $h^{-1}(u)$. We notice that the abelian variety A_u is defined over $k(u)$. If x is a generic point of V over k, then $(A_u, f(u))$ is a specialization of $(A_x, f(x))$ over k. By Koizumi [7, Th. 3], the graph of the law of addition on A_x is specialized to the graph of the law of addition on A_u. In other words, if $\mathfrak{p}$ is a discrete place of $k(x)$ such that $\mathfrak{p}(x) = u$ and $\mathfrak{p}(a) = a$ for every $a \in k$, then A_x is without defect for $\mathfrak{p}$ in the sense of [21, § 11.1] and $\mathfrak{p}(A_x) = A_u$.

With a divisor Y on W, rational over k, we call $\{V, W, h, f, Y\}$ a *fibre system of polarized abelian varieties*, defined over k, if the following condition is satisfied.

(1.1.5) *For every $u \in V$, Y and A_u intersect properly on W, and $Y \cdot A_u$ defines a polarization of A_u.*

1.2. PROPOSITION. *Let $\{V, W, h, f\}$ be a fibre system of abelian varieties, defined over a field k, and Y a divisor on W, rational over an extension K of k, satisfying (1.1.5). Let $\bar{k}$ be the algebraic closure of k. Then there exists a divisor Z of W, rational over $\bar{k}$, such that, for every $u \in V$, Z and A_u intersect properly on W, and $Z \cdot A_u$ defines the same polarization as $Y \cdot A_u$ on A_u. Moreover, suppose that, for a generic point x of V over k, the polarization of A_x containing $Y \cdot A_x$ is defined over $k(x)$. Then Z can be chosen so as to be rational over k.*

PROOF. To prove the first assertion, we may assume that K is a finitely generated extension of $\bar{k}$, and $\dim_k K = 1$, since the assertion in the general case can be easily derived from a successive application of this particular case. Let Λ be the set of all $\bar{k}$-valued discrete places $\mathfrak{p}$ of K such that $\mathfrak{p}(a) = a$ for every $a \in \bar{k}$. We are going to prove the following assertion (P_r) by induction on r.

(P$_r$) *For every Zariski K-closed subset U of V, of dimension r, there exist a finite number of non-zero elements $t_1, \cdots, t_m$ of K such that, if $\mathfrak{p} \in \Lambda$ and $\mathfrak{p}(t_i) \neq 0$ for every i, then $\mathfrak{p}(Y)$ intersects properly with A_u for every $u \in \mathfrak{p}(U)$.*

It is sufficient to prove the case where U has a generic point v over K. By [21, § 12.3, Prop. 19], there exist a finite number of non-zero elements $s_1, \cdots, s_\alpha$ of $K(v)$ such that, if $\mathfrak{q}$ is a discrete place of $K(v)$ and $\mathfrak{q}(s_i) \neq 0$ for every i, then $\mathfrak{q}(Y \cap A_v) = \mathfrak{q}(Y) \cap \mathfrak{q}(A_v)$. Put

$$U_i = \{w \in U \mid (v, s_i) \to (w, 0) \text{ ref. } K\} \qquad (i = 1, \cdots, \alpha).$$

Then U_i is a Zariski K-closed subset of U of dimension less than r. Let D be the projective line, and T_i the locus of (v, s_i) on $V \times D$ over K. Let φ be the projection map of $V \times D$ to V. Then $U_i = \varphi[T_i \cap (U \times 0)]$. By [21, § 12.3, Prop. 19], there exist a finite number of non-zero elements $s_1', \cdots, s_\beta'$ of K such that if $\mathfrak{p} \in \Lambda$ and $\mathfrak{p}(s_j') \neq 0$ for every j, then

$$\mathfrak{p}(T_i) \cap \big(\mathfrak{p}(U) \times 0\big) = \mathfrak{p}\big(T_i \cap (U \times 0)\big)$$

for every i. For such a $\mathfrak{p}$, we have $\mathfrak{p}(U_i) = \{w \in \mathfrak{p}(U) \mid (v, s_i) \to (w, 0) \text{ ref. } \mathfrak{p}\}$. Now assuming (P$_0$), $\cdots$, (P$_{r-1}$) to be true, we find a finite number of non-zero elements $t_1, \cdots, t_\gamma$ of K such that if $\mathfrak{p} \in \Lambda$ and $\mathfrak{p}(t_l) \neq 0$ for every l, then $\mathfrak{p}(Y)$ intersects properly with A_u for every $u \in \bigcup_{i=1}^{\alpha} \mathfrak{p}(U_i)$. Let $\mathfrak{p} \in \Lambda$ and $\mathfrak{p}(s_j') \neq 0$ for every j and $\mathfrak{p}(t_l) \neq 0$ for every l, and let $u \in \mathfrak{p}(U) - \bigcup_{i=1}^{\alpha} \mathfrak{p}(U_i)$. Since $u \in \mathfrak{p}(U)$, we can find, by [16, Prop. 26], a discrete place $\mathfrak{q}$ of $K(v)$ such that $\mathfrak{q} = \mathfrak{p}$ on K and $\mathfrak{q}(v) = u$. Since $u \notin \bigcup_{i=1}^{\alpha} \mathfrak{p}(U_i)$, we have $\mathfrak{q}(s_i) \neq 0$ for every i, so that $\mathfrak{q}(Y \cap A_v) = \mathfrak{q}(Y) \cap \mathfrak{q}(A_v) = \mathfrak{p}(Y) \cap A_u$. Since reduction modulo $\mathfrak{q}$ does not augment the dimension, this implies that $\mathfrak{p}(Y)$ intersects properly with A_u. This completes the proof of (P$_r$) by induction.

Now take r to be $\dim(V)$ and U to be V. Let $t_1, \cdots, t_m$ be as in (P$_r$). There exists clearly a place $\mathfrak{p} \in \Lambda$ such that $\mathfrak{p}(t_i) \neq 0$ for every i. Put $Z = \mathfrak{p}(Y)$ with such a $\mathfrak{p}$. Then Z is rational over $\bar{k}$, and Z intersects properly with A_u for every $u \in V$. Let x be a generic point of V over K, and M the algebraic closure of $k(x)$. Since K and M are linearly disjoint over $\bar{k}$, Z is a specialization of Y over M, so that $Z \cdot A_x$ is a specialization of $Y \cdot A_x$ over M. Therefore $Z \cdot A_x$ is algebraically equivalent to $Y \cdot A_x$ on A_x. For every $u \in V$, we have a specialization

$$(A_x, Z \cdot A_x, Y \cdot A_x) \to (A_u, Z \cdot A_u, Y \cdot A_u)$$

over K. Hence $Z \cdot A_u$ is algebraically equivalent to $Y \cdot A_u$ on A_u. (This can be shown, for example, by means of [21, p. 96, Prop. 14].)

Suppose that the polarization $\mathcal{C}$ of A_x containing the divisor $Y \cdot A_x$ is defined over $k(x)$. Then, for every automorphism σ of $\bar{k}$ over k, $Z^\sigma \cdot A_x$ is contained in $\mathcal{C}$. In fact, extend σ to an automorphism τ of $\bar{k}(x)$ so that $\tau = \sigma$ on $\bar{k}$, $x^\tau = x$. Then $Z^\sigma \cdot A_x = (Z \cdot A_x)^\tau \in \mathcal{C}^\tau = \mathcal{C}$. Let $Z' = p^e \sum Z^\sigma$ with all the distinct conjugates Z^σ of Z over k, and with a suitably high power p^e of the characteristic p. Then Z' is rational over k, Z' intersects properly with A_u for every $u \in V$, and $Z' \cdot A_x \in \mathcal{C}$, so that $mZ' \cdot A_x$ is algebraically equivalent to $m'Y \cdot A_x$ with positive integers m and m'. Taking again a specialization over $x \to u$ ref. K, we find that $Z' \cdot A_u$ defines the same polarization as $Y \cdot A_u$ on A_u. This completes the proof.

1.3. PROPOSITION. *Let $\{V, W, h, f\}$ and $\{V', W', h', f'\}$ be fibre systems of abelian varieties, and r a morphism of V into V', all defined over k. Define a subset W_0 of $W \times W'$ and mappings h_0 and f_0 by*

$$W_0 = \{(x, x') \in W \times W' \mid r(h(x)) = h'(x')\},$$
$$h_0(x, x') = h(x) \qquad\qquad ((x, x') \in W_0),$$
$$f_0(u) = f(u) \times f'(r(u)) \qquad\qquad (u \in V).$$

Then $\{V, W_0, h_0, f_0\}$ is a fibre system of abelian varieties, defined over k, and $h_0^{-1}(u) = h^{-1}(u) \times h'^{-1}(r(u))$ for every $u \in V$.

PROOF. Put $A_u = h^{-1}(u)$, $A'_u = h'^{-1}(r(u))$ for $u \in V$. It is clear that $W_0 = \bigcup_{u \in V}(A_u \times A'_u)$. Let z be a generic point of V over k, and (x, y) a generic point of $A_z \times A'_z$ over $k(z)$. Let $u \in V$. Since (u, A_u, A'_u) is a specialization of (z, A_z, A'_z) over k, we see that W_0 is the locus of (x, y) over k. Let T be the locus of (x, y, z) over k, i.e., the graph of h_0. By Weil [23, VIII$_3$, Th. 6], we have $(W \times W' \times z) \cdot T = A_z \times A'_z \times z$. We see easily that $A_u \times A'_u \times u$ is the only component of $(W \times W' \times u) \cap T$. Hence, specializing z to u, we get $(W \times W' \times u) \cdot T = A_u \times A'_u \times u$. By the converse part of [23, VI$_2$, Th. 6], every point of $A_u \times A'_u \times u$ is simple on T. Since T and W_0 are biregularly isomorphic, this shows that W_0 is non-singular. Using the same reasoning about the intersection product, we get easily $(W_0 \times u) \cdot T = A_u \times A'_u \times u$, the intersection product being taken on $W_0 \times V$. This proves our proposition.

1.4. PROPOSITION. *The notation being as in 1.3, put $A_u = h^{-1}(u)$ and $A'_u = h'^{-1}(r(u))$ for $u \in V$. Let S be a subvariety of $W \times W'$, defined over k. Then the following three conditions are equivalent.*

(1.4.1) There exist a generic point x of W over k and a homomorphism λ of $A_{h(x)}$ into $A'_{h(x)}$, defined over $k(h(x))$, such that S is the locus of $x \times \lambda x$ over k.

(1.4.2) $S \subset W_0$. For every $u \in V$, S and $A_u \times A'_u$ intersect properly on W_0, and $S \cdot (A_u \times A'_u)$ is the graph of a homomorphism of A_u into A'_u.

(1.4.3) S is the graph of a morphism of W into W' whose restriction to

A_u is a homomorphism of A_u into A'_u for every $u \in V$.

PROOF. Let $d = \dim(V)$, $n = \dim(A_u)$, $n' = \dim(A'_u)$. Then $\dim(W) = d + n$, $\dim(W_0) = d + n + n'$. Let y be a generic point of V over k, and x a generic point of A_y over $k(y)$. Assume that (1.4.2) is satisfied. Then $\dim(S) = d + n$, and $S \cdot (A_y \times A'_y)$ is the graph of a homomorphism λ of A_y into A'_y. We observe that λ is defined over $k(y)$, and $x \times \lambda x \in S$. Since x is generic on W over k, we have $\dim_k(x, \lambda x) \geqq d + n$, so that S is the locus of $x \times \lambda x$ over k, which implies (1.4.1). The implication (1.4.3) $\Rightarrow$ (1.4.1) is obvious. Conversely, suppose that (1.4.1) is satisfied. Put $y = h(x)$. Let $u \in V$ and

$$(z, w) \in S \cap (A_u \times A'_u).$$

Then (u, z, w) is a specialization of $(y, x, \lambda x)$ over k. Let $\mathfrak{p}$ be a discrete place of $k(x)$ such that $\mathfrak{p}(y, x, \lambda x) = (u, z, w)$ and $\mathfrak{p}(a) = a$ for every $a \in k$. By [21, § 11.1, Prop. 12], $\mathfrak{p}(\lambda)$ is a well-defined homomorphism of A_u into A'_u. Put $\mu = \mathfrak{p}(\lambda)$. Then $w = \mu z$, and reduction modulo $\mathfrak{p}$ sends the graph G_λ of λ to the graph G_μ of μ. Since μ is defined over an algebraic extension of $k(u)$, we get $\dim_{k(u)}(z, w) = \dim_{k(u)}(z) \leqq \dim(A_u) = n$. Here the equality holds if (z, w) is generic on G_μ over the algebraic closure of $k(u)$. Since λ is defined over $k(y)$, and $k(y) \subset k(x)$, we have $\dim(S) = \dim_k(x, \lambda x) = \dim_k(x) = d + n$. It follows that S and $A_u \times A'_u$ intersect properly on W_0. By [16, Th. 17], $S \cdot (A_u \times A'_u) = \mathfrak{p}(S \cdot (A_y \times A'_y)) = \mathfrak{p}(G_\lambda) = G_\mu$, hence (1.4.2) is satisfied. This shows also that, if (z, w) is a specialization of $(x, \lambda x)$ over k, then w is uniquely determined by z and $w = \mu z$. Hence the rational mapping of W to W', with S as its graph, is defined everywhere on W, and its restriction to A_u is a homomorphism into A'_u. This implies (1.4.3) and completes our proof.

1.5. Let L be an algebra over Q with identity element. By a PEL-*structure*, having L as endomorphism algebra, we mean a structure $\mathfrak{Q} = (A, \mathcal{C}, \theta; t_1, \cdots, t_s)$ formed by an abelian variety A, a polarization $\mathcal{C}$ of A, an isomorphism θ of L into $\mathrm{End}_Q(A)$, which maps the identity element of L to the identity element of $\mathrm{End}_Q(A)$, and points $t_1, \cdots, t_s$ on A of finite order. We note that a permutation of $(t_1, \cdots, t_s)$ may define a distinct PEL-structure. Let $\mathfrak{Q}' = (A', \mathcal{C}', \theta'; t'_1, \cdots, t'_s)$ be another PEL-structure with the same L and s. An isomorphism (resp. isogeny) λ of A to A' is called an *isomorphism* (resp. *isogeny*) of $\mathfrak{Q}$ to $\mathfrak{Q}'$ if λ^{-1} sends $\mathcal{C}'$ to $\mathcal{C}$, $\lambda\theta(a) = \theta'(a)\lambda$ for all $a \in L$, and $\lambda t_i = t'_i$ for every i. We say that $\mathfrak{Q}$ is defined over a field k, if k is a field of rationality for the abelian variety A, at least one divisor of $\mathcal{C}$, the elements of $\theta(L) \cap \mathrm{End}(A)$, and the points t_i. This being so, for an isomorphism σ of k into a field k', we get a PEL-structure $\mathfrak{Q}^\sigma = (A^\sigma, \mathcal{C}^\sigma, \theta^\sigma; t_1^\sigma, \cdots, t_s^\sigma)$, with the same L, defined over k'. In [19, II, 1.4], we proved that there exists a field

k_0 with the following property.

(1.5.1) *Two automorphisms σ and τ of the universal domain coincide on k_0 if and only if $\mathfrak{Q}^\sigma$ and $\mathfrak{Q}^\tau$ are isomorphic.*

If the universal domain is of characteristic 0, k_0 is uniquely determined by this property. We call then k_0 *the field of moduli of* $\mathfrak{Q}$.

Let $\mathfrak{o}$ be an order in L, i.e., a subring of L, containing Z, which is a free Z-submodule of L of maximal rank. Let $\{V, W, h, f, Y\}$ be as in 1.1. We call

$$\mathfrak{F} = \{V,\ W,\ h,\ f,\ Y,\ S(a)\ (a \in \mathfrak{o}),\ f_1,\ \cdots,\ f_s\}$$

a *fibre system of* PEL-*structures*, with L as endomorphism algebra, if the following conditions are satisfied.

(1.5.2) *For every $a \in \mathfrak{o}$, $S(a)$ is the graph of a morphism of W into W whose restriction to A_u is an endomorphism $\theta'_u(a)$ of A_u for every $u \in V$.*

(1.5.3) *For every $u \in V$, θ'_u is an isomorphism of $\mathfrak{o}$ into $\mathrm{End}(A_u)$, which maps the identity to the identity.*

(1.5.4) $f_1,\ \cdots,\ f_s$ *are morphisms of V into W, and the $f_i(u)$ are points on A_u of finite order, for every $u \in V$.*

Then we can define naturally a PEL-structure, for every $u \in V$,

$$\mathfrak{Q}_u = \left(A_u,\ \mathcal{C}_u,\ \theta_u;\ f_1(u),\ \cdots,\ f_s(u)\right),$$

where $\mathcal{C}_u$ is the polarization of A_u containing $Y \cdot A_u$, and θ_u is an isomorphism of L into $\mathrm{End}_Q(A_u)$, which is a unique extension of θ'_u. Each $\mathfrak{Q}_u$ is called a *fibre* of $\mathfrak{F}$. We call k a *field of rationality* (or *definition*) for $\mathfrak{F}$ if k is a field of rationality for $V, W, h, f, Y, S(a)$ $(a \in \mathfrak{o})$, $f_1, \cdots, f_s$. Then, for every $u \in V$, $\mathfrak{Q}_u$ is defined over $k(u)$.

Let $\{b_1, \cdots, b_r\}$ be a basis of $\mathfrak{o}$ over Z, and let k_0 be a field of rationality for V, W, h, f and $S(b_1), \cdots, S(b_r)$. Let x be a generic point of W over k_0, and $y = h(x)$. Then the $\theta_y(b_i)$ are all defined over $k_0(y)$. Hence $\theta_y(a)$ is defined over $k_0(y)$ for every $a \in \mathfrak{o}$. Let $S'(a)$ be the locus of $x \times \theta_y(a)x$ over k_0. By 1.4, we see that $S(a) = S'(a)$, so that $S(a)$ is defined over k_0 for every $a \in \mathfrak{o}$. This shows especially that there exists a field of rationality for $\mathfrak{F}$ which is finitely generated over the prime field.

1.6. Let z be a generic point of V over a field k of rationality for $\mathfrak{F}$. Then Y_z defines an involution $\lambda \to \lambda^*$ of $\mathrm{End}_Q(A_z)$. Now *suppose that L has an involution ρ such that $\theta_z(a)^* = \theta_z(a^\rho)$ for $a \in L$. Then for every $u \in V$, the involution of $\mathrm{End}_Q(A_u)$ defined by Y_u induces in $\theta_u(L)$ an involution $\theta_u(a) \to \theta_u(a^\rho)$.* In fact, taking an l-adic coordinate system on A_z, we have $E_l(Y_z)M_l\big(\theta_z(a^\rho)\big) = {}^t M_l\big(\theta_z(a)\big)E_l(Y_z)$, the notation E_l and M_l being as in Weil [24]. Since $\left(A_u, Y_u, \theta_u(a)\ (a \in \mathfrak{o})\right)$ is a specialization of $\left(A_z, Y_z, \theta_z(a)\ (a \in \mathfrak{o})\right)$

over k, we find, with respect to a suitable l-adic coordinate system on A_u, $E_l(Y_u)M_l(\theta_u(a^\rho)) = {}^t M_l(\theta_u(a))E_l(Y_u)$, by virtue of [21, § 11.1, Prop. 14], which proves the assertion.

1.7. PROPOSITION. *Let L be an algebra over $\mathbf{Q}$, and $\mathfrak{o}$, $\mathfrak{o}'$ be two orders in L. Let $\mathfrak{F} = \{V, W, h, f, Y, S(a) \, (a \in \mathfrak{o}), f_1, \cdots, f_s\}$ and*

$$\mathfrak{F}' = \{V', W', h', f', Y', S'(a) \, (a \in \mathfrak{o}'), f'_1, \cdots, f'_s\}$$

be two systems of PEL*-structures with L as endomorphism algebra, defined over k. Let $\mathfrak{A}_u(u \in V) \, (\text{resp. } \mathfrak{A}'_v(v \in V'))$ be the fibres of $\mathfrak{F}$ (resp. $\mathfrak{F}'$). Suppose that there exist a morphism r of V into V', defined over k, a generic point z of V over k, and an isomorphism λ of $\mathfrak{A}_z$ to $\mathfrak{A}'_{r(z)}$, defined over $k(z)$. Then there exists a morphism R of W to W', defined over k, whose restriction λ_u to A_u is an isomorphism of $\mathfrak{A}_u$ to $\mathfrak{A}'_{r(u)}$ for every $u \in V$, and $\lambda_z = \lambda$. Moreover, if r is a biregular morphism of V onto V', then R is a biregular morphism of W onto W'.*

PROOF. Let x be a generic point of A_z over $k(z)$, and R the locus of $x \times \lambda x$ over k on $W \times W'$. By 1.4, R defines a morphism of W to W', whose restriction λ_u to A_u is a homomorphism of A_u to $A'_{r(u)}$. Since λ is an *isomorphism* and λ_u is a specialization of λ (cf. the proof of 1.4), λ_u should be an *isomorphism* [21, § 11.1, Prop. 12]. By our assumption, λ is an isomorphism of $\mathfrak{A}_z$ to $\mathfrak{A}'_{r(z)}$. Specializing z to u, one can easily verify that λ_u is an isomorphism of $\mathfrak{A}_u$ to $\mathfrak{A}'_{r(u)}$. If r is biregular and surjective, we get the last assertion by applying our result to r^{-1} and λ^{-1}.

We call a couple (R, r) of surjective biregular morphisms an *isomorphism* (*automorphism* if $\mathfrak{F} = \mathfrak{F}'$) of $\mathfrak{F}$ to $\mathfrak{F}'$, if they are in the above situation.

1.8. We are going to consider the quotient of a fibre system of PEL-structures by a finite group of automorphisms. First we recall some elementary results concerning the quotient of a variety by a finite group. Let V be a Zariski open subset of a projective variety V^*, and G a finite group of biregular automorphisms of V. Let k be a field of rationality for V, V^* and the elements of G. Then there exists a variety U and a morphism j of V onto U, both defined over k, satisfying the following conditions.

(1.8.1) $j(x) = j(y)$ *if and only if $x = g(y)$ for some $g \in G$.*

(1.8.2) *If j' is a rational mapping of V into a variety U' satisfying $j' \circ g = j'$ for every $g \in G$, then there exists a rational mapping F of U to U' such that $j' = F \circ j$, and F is defined at $j(a)$ whenever j' is defined at a.*

Such U and j are uniquely determined up to biregular isomorphisms over k. They can be obtained as follows (cf. Serre [15, Ch. III, no. 12] and [21, § 4.3]). Let x be a generic point of V over k, and y the Chow point of a 0-dimensional cycle $\sum_{g\in G} g(x)$. Let U_1^* be the locus of y over k, and U^* the normalization of U_1 with respect to k. Then there is a natural rational mapping j of V^* to U^*. It can be easily seen that j is defined everywhere on V, and $j(V)$ is a Zariski open subset of U^*. Putting $U = j(V)$, we get (U, j) satisfying the above conditions. This construction shows that one may assume

(1.8.3) *U is a Zariski open subset of a projective variety U^*, which is normal and defined over k.*

We call U *the quotient of V by G and j the natural morphism of V to U*. We notice that if a is simple on V, and the $g(a)$ for $g \in G$ are all distinct, then $j(a)$ is simple on U.

1.9. Let $\mathfrak{F} = \{V, W, h, f, Y, S(a)\ (a \in \mathfrak{o}), f_1, \cdots, f_s\}$ and $\mathfrak{A}_u\ (u \in V)$ be as in 1.5, and k a field of rationality for $\mathfrak{F}$. Let G be a finite group of automorphisms of $\mathfrak{F}$ (in the sense of 1.7), defined over k. Each element g of G is a couple (R_g, r_g) formed by biregular automorphisms R_g and r_g of W and V, respectively. Suppose that *no element r_g, with g other than the identity element, leaves any point of V fixed.* Then a similar assertion is true for the R_g. Suppose further that V and W are Zariski open subsets of projective varieties. Let V' (resp. W') be the quotient of V (resp. W) by $\{r_g\}$ (resp. $\{R_g\}$), and j (resp. J) the natural morphism of V to V' (resp. W to W'). Then *one can define a fibre system of* PEL-*structures*,

$$\mathfrak{F}' = \{V', W', h', f', Y', S'(a)\ (a \in \mathfrak{o}), f_1', \cdots, f_s'\},$$

with fibres

$$\mathfrak{A}_u' = \big(A_u', C_u', 0_u'; f_1'(u), \cdots, f_s'(u)\big) \qquad (u \in V'),$$

so that the restriction of J to A_u is an isomorphism of $\mathfrak{A}_u$ to $\mathfrak{A}_{j(u)}'$ for every $u \in V$. $\mathfrak{F}'$ can be taken so as to be defined over k.

To prove this, let us put, for simplicity, $g(w) = R_g(w)$ for $w \in W$ and $g(v) = r_g(v)$ for $v \in V$. Since G is fixed-point-free, V' and W' are non-singular. In view of (1.8.2), we find a morphism h' of W' to V' and a morphism f' of V' to W' such that $h' \circ J = j \circ h$ and $f' \circ j = J \circ f$. Let x be a generic point of W over k, and let T be the locus of $h(x) \times J(x)$ on $V \times W'$ over k. Then

$$T \cap \big(V \times J(y)\big) = \bigcup_{g \in G}\big(g(h(y)) \times J(y)\big)$$

for every $y \in W$, hence T and $V \times J(y)$ intersect properly on $V \times W'$. In particular, $T \cdot (V \times J(x)) = \sum_{g \in G} g(h(x)) \times J(x)$. Specializing x to y, we get $T \cdot (V \times J(y)) = \sum_{g \in G} g(h(y)) \times J(y)$. By the converse part of [23, VI$_2$, Th. 6], we see that T is a non-singular variety. Let F be a morphism of W to T such that $F(x) = h(x) \times J(x)$. It is easy to see that F is one-to-one. Since T is non-singular, it follows that F is biregular. In particular, if we put $J(A_u) = A_{j(u)}'$ for $u \in V$, then $F(A_u) = u \times A_{j(u)}'$. This shows that $A_{j(u)}'$ has a structure of abelian variety with the origin $f'(j(u))$, and J induces an isomorphism of A_u to $A'_{j(u)}$. Put $z = h(x)$. Since A_z is a locus of x over $k(z)$, $A'_{j(z)}$ is the locus of $J(x)$ over $k(z)$. Since $k(h(x))$ and $k(J(x))$ are linearly disjoint over $k(j(z))$, $A'_{j(z)}$ is the locus of $J(x)$ over $k(j(z))$. Let Z be the graph of h'. Then

$$Z \cdot \big(W' \times j(z)\big) = A'_{j(z)} \times j(z)\,,$$

and $Z \cap (W' \times j(u)) = A'_{j(u)} \times j(u)$ for every $u \in V$. Since F is biregular and (u, A_u) is a specialization of (z, A_z) over k, we see that $(j(u), A'_{j(u)})$ is a specialization of $(j(z), A'_{j(z)})$ over k. Hence $Z \cdot (W' \times j(u)) = A'_{j(u)} \times j(u)$ for every $u \in V$. This proves that $\{V', W', h', f'\}$ is a fibre system of abelian varieties.

Let λ_u be the isomorphism of A_u to $A'_{j(u)}$ induced by J. Define $\theta'_{j(z)}$ by $\theta'_{j(z)}(a) = \lambda_z \theta_z(a)\lambda_z^{-1}$ $(a \in \mathfrak{o})$. Then we see that $\theta'_{j(z)}(a) = \lambda_{g(z)}\theta_{g(z)}(a)\lambda_{g(z)}^{-1}$ for every $g \in G$. It follows that $\theta'_{j(z)}(a)$ is invariant under every automorphism of $k(z)$ over $k(j(z))$. Hence $\theta'_{j(z)}(a)$ is defined over $k(j(z))$. Let $S'(a)$ be the locus of $J(x) \times \theta'_{j(z)}(a)J(x)$ over k. By 1.4, $S'(a) \cdot (A'_v \times A'_v)$ is the graph of an endomorphism $\theta'_v(a)$ of A'_v for every $v \in V'$. Specializing z to u, we get $\theta'_{j(u)}(a)\lambda_u = \lambda_u \theta_u(a)$ $(a \in \mathfrak{o})$. This shows especially that θ'_v is an isomorphism of $\mathfrak{o}$ into $\mathrm{End}(A'_v)$.

Since $F(Y) = \bigcup_{u \in V}\big(u \times \lambda_u(Y_u)\big)$, we have

$$F(Y) \cap (V \times A'_{j(u)}) = \bigcup_{g \in G}\big(g(u) \times \lambda_{g(u)}(Y_{g(u)})\big)\,,$$

so that $\{F(Y) \cdot (V \times A'_{j(u)})\}_{V \times W'} = \sum_{g \in G}m_{g,u}\big(g(u) \times \lambda_{g(u)}(Y_{g(u)})\big)$ with positive integers $m_{g,u}$. Put $Y' = \mathrm{pr}_{W'}(F(Y))$. By [23, VIII$_3$, Th. 8], Y' and $A'_{j(u)}$ intersect properly on W', and $Y' \cdot A'_{j(u)} = \sum m_{g,u} \cdot \lambda_{g(u)}(Y_{g(u)})$. Since $\lambda_{g(u)}^{-1}\lambda_u$ is an isomorphism of $\mathfrak{Q}_u$ to $\mathfrak{Q}_{g(u)}$, this shows that $Y' \cdot A'_{j(u)}$ defines a polarization on $A'_{j(u)}$, which is sent to $\mathcal{C}_u$, by λ_u^{-1}.

Finally, in view of (1.8.2), we obtain morphisms f'_i of V' into W', defined over k, such that $f'_i \circ j = J \circ f_i$. Then $\lambda_u\big(f_i(u)\big) = J\big(f_i(u)\big) = f'_i\big(j(u)\big)$ for every $u \in V$. This completes the proof of our assertion.

2. Maximal families of polarized abelian varieties[1]

2.1. *Riemann forms and theta functions.*[2] Let D be a lattice in C^n. By a *Riemann form* on the complex torus C^n/D, we understand an R-valued R-bilinear form $\mathfrak{E}(x, y)$ on $C^n \times C^n$ satisfying the following conditions:

(2.1.1) *The values $\mathfrak{E}(u, v)$ for $(u, v) \in D \times D$ are integers.*

(2.1.2) $\mathfrak{E}(u, v) = -\mathfrak{E}(v, u)$.

(2.1.3) $\mathfrak{E}(u, \sqrt{-1}\, v)$ *is symmetric in* (u, v) *and* $\mathfrak{E}(u, \sqrt{-1}\, u) \geq 0$.

We say that $\mathfrak{E}$ is *non-degenerate* if it is a non-degenerate bilinear form in

---

[1] A large part of this section is of expository nature. For an exposition of the same kind, cf. [14, No. 20].

[2] A detailed account of the subject is given in Weil [26]. In this book, a Riemann form is defined to be a *hermitian* form whose imaginary part satisfies (2.1.1-3).

the usual sense. Let ω be an $n \times 2n$ matrix whose columns generate D over Z, and E a matrix of size $2n$ such that

$$(2.1.4) \qquad \mathfrak{E}(\omega x, \omega y) = {}^t x E y \qquad ((x, y) \in R^{2n} \times R^{2n}).$$

Then the conditions (2.1.1-3) with a non-degenerate $\mathfrak{E}$ are equivalent to the following (2.1.5-8).

(2.1.5) *The coefficients of E are integers.*

(2.1.6) ${}^t E = -E$.

(2.1.7) $\omega \cdot {}^t E^{-1} \cdot {}^t \omega = 0$.

(2.1.8) $\sqrt{-1}\, \omega \cdot {}^t E^{-1} \cdot {}^t \bar{\omega}$ *is a positive definite hermitian matrix.*

We call a holomorphic function f on C^n a *theta function* on C^n relative to D, if

$$(2.1.9) \qquad f(u + d) = f(u) \cdot \exp\left[2\pi\sqrt{-1}\,(L_d(u) + c_d)\right] \qquad (u \in C^n, d \in D),$$

where L_d is a C-valued C-linear form on C^n, and c_d is a complex number, both depending on d. We see easily that $L_{d+d'} = L_d + L_{d'}$. Hence there exists a C-valued R-bilinear form $F(u, v)$ on $C^n \times C^n$ such that $L_d(v) = F(d, v)$. Put

$$(2.1.10) \qquad \mathfrak{E}(u, v) = F(u, v) - F(v, u).$$

Then it can be verified that $\mathfrak{E}$ is a Riemann form on C^n/D. We call $\mathfrak{E}$ *the Riemann form determined by f.* Let X be a positive divisor on C^n/D. Then there exists a theta function f on C^n relative to D such that $\mathrm{div}(f) = X$. The Riemann form determined by f will be called also *the Riemann form determined by X.* This Riemann form depends on only X and is independent of the choice of f. Every Riemann form is determined by a positive divisor.

A complex torus C^n/D has a structure of abelian variety if and only if it has a non-degenerate Riemann form. Let f and $\mathfrak{E}$ be as in (2.1.9-10). Define E by (2.1.4). Then the theta functions satisfying (2.1.9) with fixed multiplicators $\exp\left[2\pi\sqrt{-1}\,(L_d(u) + c_d)\right]\,(d \in D)$ form a vector space $\mathfrak{L}$ over C of dimension $\det(E)^{1/2}$. Let $\{\vartheta_0, \cdots, \vartheta_N\}$ be a basis of $\mathfrak{L}$ over C, with $N = \det(E)^{1/2} - 1$. If the elementary divisors of E are all ≥ 3, then

$$\mathrm{rank}\begin{pmatrix} \vartheta_0(u) & \partial\vartheta_0/\partial u_1 & \cdots & \partial\vartheta_0/\partial u_n \\ \cdots & \cdots & \cdots & \cdots \\ \vartheta_N(u) & \partial\vartheta_N/\partial u_1 & \cdots & \partial\vartheta_N/\partial u_n \end{pmatrix} = n + 1$$

everywhere on C^n, where $u_1, \cdots, u_n$ are components of the vector u. Let Θ be the mapping of C^n into the projective space P^N of dimension N, defined by $\Theta(u) = (\vartheta_0(u), \cdots, \vartheta_N(u))\,(u \in C^n)$. Then Θ induces a biregular isomorphism of C^n/D onto an abelian variety A in P^N. Moreover, under the mapping Θ, the set of divisors $\{\mathrm{div}(\vartheta) \mid \vartheta \in \mathfrak{L}\}$ on C^n/D is mapped onto the complete linear system on A formed by all the hyperplane sections of A.

2.2. *Riemann form as a cohomology class.* Let V be a complex manifold, and X a divisor on V. We can attach to X a 2-cohomology class $a(X)$ as follows. One can find a covering $\{U_\lambda\}$ of V (which is locally finite and simple in the sense of [26, p. 86]), so that, for each λ, $D \cap U_\lambda = \mathrm{div}(f_\lambda)$ with a meromorphic function f_λ in U_λ. Then there exists, for each λ, a differential form η_λ of degree 1 in U_λ such that $\eta_\mu - \eta_\lambda = (2\pi i)^{-1} d \log(f_\lambda^{-1} f_\mu)$ in $U_\lambda \cap U_\mu$. Furthermore there exists a differential form ε of degree 2 on V such that $\varepsilon = d\eta_\lambda$ on U_λ for every λ. We denote by $a(X)$ the cohomology class of the form ε, and call it *the cohomology class of X* (cf. [26, Ch. V]).

Now let us consider the case where V is a complex torus C^n/D. Let $g_1, \cdots, g_{2n}$ be a basis of D over Z, and $x_1(u), \cdots, x_{2n}(u)$ the R-valued R-linear forms on C^n defined by $u = \sum_{i=1}^{2n} x_i(u) g_i$ $(u \in C^n)$. Then every invariant differential form α of degree p in C^n/D can be expressed in the form

$$\alpha = \sum a_{(i)} dx_{i_1} \wedge \cdots \wedge dx_{i_p}$$

with $a_{(i)}$ in C. Let X be a positive divisor on C^n/D, and $\mathfrak{E}$ the Riemann form determined by X. Then the cohomology class $a(X)$ of X can be represented by the 2-form

$$\sum_{i<j} \mathfrak{E}(g_i, g_j) dx_i \wedge dx_j \qquad \text{(cf. [26, p. 112, Prop. 4])} .$$

2.3. *Paramodular groups.* Let $e_1, \cdots, e_n$ be n positive integers such that $e_1 = 1$, and e_{i+1} is divisible by e_i. Put

$$e = \mathrm{diag}[e_1, \cdots, e_n], \qquad E = \begin{pmatrix} 0 & -e \\ e & 0 \end{pmatrix} .$$

Let $\mathfrak{S}_n$ be the Siegel space of degree n, i.e., the set of all complex matrices z of degree n such that $\mathrm{Im}(z)$ is positive definite. One can easily verify that a *period matrix* $\omega = (u\ v)$, with u and v in $M_n(C)$, satisfies (2.1.7-8) for the present E, if and only if u and v are invertible, and $ev^{-1}u \in \mathfrak{S}_n$. For every $z \in \mathfrak{S}_n$, put

$$(2.3.1) \qquad\qquad \omega_e(z) = (z\ e) \qquad\qquad ((n \times 2n)\text{-matrix}) ,$$

$$\Gamma'_e = \{T \in \mathrm{GL}_{2n}(Z) \mid {}^t T E T = E\} ,$$

$$(2.3.2) \qquad \Gamma_e = \{P_e^{-1} \cdot {}^t T P_e \mid T \in \Gamma'_e\} , \qquad P_e = \begin{pmatrix} 1 & 0 \\ 0 & e \end{pmatrix} ,$$

Then $\omega_e(z)$ satisfies (2.1.7-8), and Γ_e is a discrete subgroup of $\mathrm{Sp}(n, R)$. In particular, $\Gamma_1 = \Gamma'_1 = \mathrm{Sp}(n, Z)$. Let T be an element of $\mathrm{GL}_{2n}(R)$ such that ${}^t T E T = rE$ with a positive number r. Put $P_e^{-1} \cdot {}^t T^{-1} P_e = \begin{pmatrix} a & b \\ c & d \end{pmatrix}$ with a, b, c, d in $M_n(R)$, and define the operation of T on $\mathfrak{S}_n$ by

$$T(z) = (az + b)(cz + d)^{-1} \qquad (z \in \mathfrak{S}_n).$$

Furthermore, put

$$J(T, z) = {}^t(cz + d)^{-1}.$$

If $T(z) = z'$, we have

$$P_e^{-1} \cdot {}^t T^{-1} P_e \binom{z}{1} = \binom{z'}{1}(cz + d),$$

so that $J(T, z)\omega_e(z) = \omega_e(z')T$.

2.4. *Maximal families of polarized abelian varieties.* The notation being as in 2.3, let $D_e(z)$ be the lattice in C^n generated by the columns of $\omega_e(z)$. Then $C^n/D_e(z)$ has a structure of abelian variety, and the form $\mathfrak{E}$ defined by

$$\mathfrak{E}\big(\omega_e(z)x,\ \omega_e(z)y\big) = {}^t xEy,$$

is a non-degenerate Riemann form on $C^n/D_e(z)$. Let μ be a positive integer, and let g be an element of R^n such that $\mu eg \in Z^n$. Put, for $u \in C^n$ and $z \in \mathfrak{S}_n$,

$$\vartheta_{\mu,g}(u, z) = \sum_{a \in Z^n} \exp\left[2\mu\pi i\left(\frac{1}{2}\cdot{}^t(a + g)z(a + g) + {}^t(a + g)u\right)\right].$$

One can easily verify that, for $x \in R^{2n}$ and $b \in Z^{2n}$,

$$\vartheta_{\mu,g}\big(\omega_e(z)(x + b),\ z\big) = \vartheta_{\mu,g}\big(\omega_e(z)x\big)\cdot\exp\left[2\pi i\left({}^t bHx + \frac{1}{2}\,{}^t bH_0 b\right)\right],$$

where

$$H = -\mu\binom{z\ \ e}{0\ \ 0}, \qquad H_0 = -\mu\binom{z\ \ 0}{0\ \ 0}.$$

Hence $H - {}^t H = \mu E$. This means that $\vartheta_{\mu,g}(u, z)$ is a theta function relative to $D_e(z)$, and $\mu\mathfrak{E}$ is the Riemann form determined by $\vartheta_{\mu,g}$. Moreover, the $\vartheta_{\mu,g}$, for all g such that $\mu eg = {}^t(c_1, \cdots, c_n)$, $0 \leqq c_i < \mu e_i$, form a basis of the vector space of all theta functions with a system of multiplicators

$$\{\exp[2\pi i({}^t bHx + 2^{-1}\cdot{}^t bH_0 b)]\}_{b \in Z^{2n}}.$$

Put $N = \mu^n\det(e) - 1$, and denote by $\vartheta_0(u, z), \cdots, \vartheta_N(u, z)$ these $\vartheta_{\mu,g}(u, z)$ with a fixed order. Further choose and fix any positive integer $\mu \geqq 3$, and define a holomorphic mapping Θ_e of $C^n \times \mathfrak{S}_n$ into the projective space P^N of dimension N, by

$$\Theta_e(u, z) = \big(\vartheta_0(u, z), \cdots, \vartheta_N(u, z)\big).$$

By virtue of the general principle stated in 2.1, we see that, for each $z \in \mathfrak{S}_n$, the mapping $u \to \Theta_e(u, z)$ gives an isomorphism of $C^n/D_e(z)$ to an abelian variety in P^N, which we denote by $A_e(z)$. Let $\mathcal{C}_e(z)$ be the polarization of $A_e(z)$ determined by hyperplane sections. Put

$$\mathscr{P}_e(z) = \big(A_e(z),\, C_e(z)\big).$$

Thus we get a family $\Sigma_e = \{\mathscr{P}_e(z) \mid z \in \mathfrak{S}_n\}$ of polarized abelian varieties, which we call a *maximal family of polarized abelian varieties with elementary divisors* $e_1, \cdots, e_n$. This family has the property

$$(2.4.1) \qquad \deg\big(A_e(z)\big) = \mu^n \det(e) n! \qquad\qquad \textit{for every } z \in \mathfrak{S}_n.$$

In fact, by Nishi [13], if X is a hyperplane section of $A_e(z)$, the dimension of the complete linear system determined by X is $\deg(X^n)/n!$, and $\deg(X^n)$ is, in the present situation, the degree of the projective variety $A_e(z)$. This proves (2.4.1). This formula is known classically as Poincaré's theorem on common zeros of theta functions.

Let λ be an isogeny of $\mathscr{P}_e(z)$ to $\mathscr{P}_e(z')$. Then λ determines an element Y of $\mathrm{GL}_n(C)$ and an element T of $M_{2n}(Z)$ such that

$$Y\omega_e(z) = \omega_e(z')T, \qquad {}^tTET = rE$$

with a positive integer r. If λ is an isomorphism, then $T \in \Gamma'_e$, and one can verify easily that $T(z) = z'$ and $Y = J(T, z)$. Conversely, if $T(z) = z'$ with an element T of Γ'_e, the linear transformation $u \to J(T, z)u$ in C^n gives an isomorphism of $\mathscr{P}_e(z)$ to $\mathscr{P}_e(z')$ (cf. 2.3). Thus $\mathscr{P}_e(z)$ and $\mathscr{P}_e(z')$ are isomorphic if and only if $z' = T(z)$ with an element T of Γ'_e. Now the following proposition is fundamental for our construction of our fibre systems.

2.5. Proposition. *There exists a positive integer ν_0, depending on only e and μ, such that, if $T \in \Gamma'_e$ and $T \equiv 1_{2n} \bmod(\nu_0)$, then*

$$\Theta_e\big(J(T, z)u,\, T(z)\big) = \Theta_e(u, z) \qquad\qquad \textit{for } (u, z) \in C^n \times \mathfrak{S}_n.$$

This follows easily from a general transformation formula of theta functions due to Krazer-Prym [8, pp. 117–122]. For our purpose, the strongest result is not necessary; even the specialized formula [8, p. 121, (L_1)] is more than enough. We can take any positive multiple of $2 \cdot (\mu e_n)^3$ as ν_0. Another exact formula is given in Siegel [22, Satz 9, 10]. It is desirable to obtain a simpler proof by means of a general Poisson-formula of Weil [27].

3. The family Σ_Ω of PEL-structures

3.1. PEL-*type*. Let L be a division algebra over Q with a positive involution ρ, and Φ a representation of L in a complex vector space C^n, such that the direct sum of Φ and its complex conjugate is equivalent to a rational representation of L. Then $2n$ is divisible by $[L : Q]$. Put

$$m = 2n/[L : Q].$$

Let L^m be the left L-module of all m-dimensional row vectors with components in L, and T a ρ-anti-hermitian form on L^m, i.e., an L-valued Q-bilinear form on $L^m \times L^m$ satisfying

$$T(ax, by) = aT(x, y)b^\rho, \quad T(x, y)^\rho = -T(y, x) \qquad (a, b \in L; \, x, y \in L^m) .$$

Suppose that T is non-degenerate, i.e., $T(x, L^m) = 0$ only if $x = 0$. We put $L_R = L \otimes_Q R$, $L_R^m = L^m \otimes_Q R$, and denote by $\mathrm{tr}(a)$ the reduced trace of an element a of L_R to R. Then T can be extended uniquely to an L_R-valued R-bilinear form on $L_R^m \times L_R^m$, which we denote again by T. Let $\mathfrak{M}$ be a free Z-submodule of L^m of rank $2n$. We assume that

$$(3.1.1) \qquad \mathrm{tr}\big(T(\mathfrak{M}, \mathfrak{M})\big) = Z .$$

For any given T, a suitable rational multiple of T satisfies this condition. Let $v_1, \cdots, v_s$ be elements of L^m. Now we put

$$\Omega = (L, \Phi, \rho; \, T, \mathfrak{M}; \, v_1, \cdots, v_s) ,$$

and call Ω a PEL-*type*. Two PEL-types $\Omega = (L, \Phi, \rho; \, T, \mathfrak{M}; \, v_1, \cdots, v_s)$ and $\Omega' = (L', \Phi', \rho'; \, T', \mathfrak{M}'; \, v_1', \cdots, v_s')$ are said to be *equivalent*, if $L = L'$, $\rho = \rho'$, $s = s'$, Φ and Φ' are equivalent representations of L, and there exists an L-linear automorphism α of L^m such that

$$T'(x\alpha, y\alpha) = T(x, y), \quad \mathfrak{M}' = \mathfrak{M}\alpha, \quad v_i' \equiv v_i\alpha \mod \mathfrak{M}' \qquad \text{for every } i.$$

3.2. A PEL-structure $\mathfrak{Q} = (A, \mathcal{C}, \theta; \, t_1, \cdots, t_s)$ is said to be of type $\Omega = (L, \Phi, \rho; \, T, \mathfrak{M}; \, v_1, \cdots, v_s)$ if there exist a complex torus C^n/D and a commutative diagram

$$
\begin{array}{ccccccccc}
0 & \longrightarrow & \mathfrak{M} & \longrightarrow & L_R^m & \longrightarrow & L_R^m/\mathfrak{M} & \longrightarrow & 0 \\
& & \downarrow{\mathfrak{h}} & & \downarrow{\mathfrak{h}} & & \downarrow & & \\
0 & \longrightarrow & D & \longrightarrow & C^n & \overset{\xi}{\longrightarrow} & A & \longrightarrow & 0
\end{array}
\qquad
\begin{array}{l}
\text{(exact)} \\
\\
\text{(exact)}
\end{array}
$$

satisfying the following conditions (3.2.1-4).

(3.2.1) *ξ gives a biregular isomorphism of C^n/D to A, and $\xi\big(\Phi(a)u\big) = \theta(a)\xi(u)$ for $a \in L$, $u \in C^n$, provided that $\theta(a) \in \mathrm{End}(A)$.*

(3.2.2) *$\mathfrak{h}$ is an R-linear isomorphism of L_R^m to C^n satisfying $\mathfrak{h}(\mathfrak{M}) = D$ and $\mathfrak{h}(ax) = \Phi(a)\mathfrak{h}(x)$ $(a \in L, \, x \in L_R^m)$.*

(3.2.3) *$\mathcal{C}$ contains a divisor which determines a Riemann form $\mathfrak{E}$ on C^n/D such that $\mathfrak{E}\big(\mathfrak{h}(x), \mathfrak{h}(x')\big) = \mathrm{tr}\big(T(x, x')\big)$ $(x, x' \in L_R^m)$.*

$$(3.2.4) \qquad t_i = \xi\big(\mathfrak{h}(v_i)\big) \qquad (i = 1, \cdots, s) .$$

From (3.2.3) it follows that

$$(3.2.5) \qquad \mathfrak{E}\big(\Phi(a)u, v\big) = \mathfrak{E}\big(u, \Phi(a^\rho)v\big) \qquad (a \in L; \, u, v \in C^n) .$$

In other words, C determines an involution of $\mathrm{End}_Q(A)$ whose restriction to $\theta(L)$ is the map $\theta(a) \to \theta(a^\rho)$. We say that $\mathscr{P} = (A, C, \theta)$ *is of type* (L, Φ, ρ), if this condition on C and ρ is satisfied, and there exists a holomorphic mapping ξ of a complex torus C^n/D to A satisfying (3.2.1). This being so, one can find T and $\mathfrak{M}$ so that $(A, C, \theta; 0)$ is of type $(L, \Phi, \rho; T, \mathfrak{M}; 0)$ (cf. [18], [19, I, 1.5, 1.9]). Put

$$(3.2.6) \qquad \mathfrak{o} = \{a \in L \mid a\mathfrak{M} \subset \mathfrak{M}\}.$$

Then $\mathfrak{o}$ is an order in L, and $\theta(\mathfrak{o}) = \theta(L) \cap \mathrm{End}(A)$.

We notice that, if $\mathfrak{Q}$ is of type Ω and of type

$$\Omega' = (L, \Phi', \rho; T', \mathfrak{M}'; v_1', \cdots, v_s'),$$

then Ω and Ω' are equivalent.

If ρ is an involution of the second kind (so that L is an algebra of (Type IV) of [18, Prop. 1]), one should impose a certain condition [18, p. 160, (25)] on the *signatures* of T, in order to insure the existence of a PEL-structure of type Ω. But there is no need, for our later use, to describe the condition explicitly.

3.3. Define an algebraic group $\mathrm{U}(T)$, a Lie group $\mathrm{U}_R(T)$, discrete subgroups $\Gamma(T, \mathfrak{M})$ and $\Gamma(T, \mathfrak{N}/\mathfrak{M})$ of $\mathrm{U}_R(T)$ by

$$\mathrm{U}(T) = \{\alpha \in \mathrm{GL}_m(L) \mid T(x\alpha, y\alpha) = T(x, y)\},$$
$$\mathrm{U}_R(T) = \{\alpha \in \mathrm{GL}_m(L_R) \mid T(x\alpha, y\alpha) = T(x, y)\},$$
$$\Gamma(T, \mathfrak{M}) = \{\alpha \in \mathrm{U}(T) \mid \mathfrak{M}\alpha = \mathfrak{M}\},$$
$$\mathfrak{N} = \mathfrak{M} + \sum_{i=1}^{s} \mathfrak{o}v_i,$$
$$\Gamma(T, \mathfrak{N}/\mathfrak{M}) = \{\alpha \in \Gamma(T, \mathfrak{M}) \mid \mathfrak{N}(1 - \alpha) \subset \mathfrak{M}\}.$$

Let $\mathfrak{H}$ be a bounded symmetric domain isomorphic to the quotient of $\mathrm{U}_R(T)$ by a maximal compact subgroup. In [18] and [19], we have constructed a family $\Sigma_\Omega = \{\mathfrak{Q}_z \mid z \in \mathfrak{H}\}$ of PEL-structures

$$(3.3.1) \qquad \mathfrak{Q}_z = \left(A_z, C_z, \theta_z; t_1(z), \cdots, t_s(z)\right)$$

of type Ω parametrized by the point z of $\mathfrak{H}$. We are going to recall some properties of this family.

In [18, 2.6, p. 163 (cf. also 3.2)], we obtained holomorphic mappings $z \to \mathfrak{x}_i(z, T)$ $(i = 1, \cdots, m)$ of $\mathfrak{H}$ into C^n, which generate the *periods* of A_z as follows. For $b = (b_1, \cdots, b_m) \in L_R^m$ with $b_i \in L_R$ and for $z \in \mathfrak{H}$, put

$$(3.3.2) \qquad \mathfrak{y}(b, z) = \sum_{i=1}^{m} \Phi(b_i)\mathfrak{x}_i(z, T).$$

Then $b \to \mathfrak{y}(b, z)$ gives an R-linear isomorphism of L_R^m to C^n. Let D_z be the image of $\mathfrak{M}$ by this mapping. On the complex torus C^n/D_z, one gets a non-degenerate Riemann form $\mathfrak{E}_z$ such that

$$(3.3.3) \qquad \mathfrak{E}_z\big(\mathfrak{y}(b, z), \mathfrak{y}(b', z)\big) = \mathrm{tr}\big(T(b, b')\big) \qquad (b, b' \in L_R^m) .$$

Since $\mathrm{tr}\big(T(b, b')\big)$ is a non-degenerate alternating form in (b, b'), we can find a basis $\{w_1, \cdots, w_{2n}\}$ of $\mathfrak{M}$ over Z so that

$$E = \big(\mathrm{tr}(T(w_i, w_j))\big)_{i, j=1, \cdots, 2n} = \begin{pmatrix} 0 & -e \\ e & 0 \end{pmatrix} ,$$

$$e = \mathrm{diag}[e_1, \cdots, e_n], \ e_{i+1} \equiv 0 \bmod (e_i) ,$$

with positive integers $e_1, \cdots, e_n$. In view of the assumption (3.1.1), we have $e_1 = 1$. With such a basis $\{w_i\}$ of $\mathfrak{M}$ over Z, let $\mathfrak{u}(z)$ resp. $\mathfrak{v}(z)$ be matrices of size n whose columns are $\mathfrak{y}(w_1, z), \cdots, \mathfrak{y}(w_n, z)$, resp. $\mathfrak{y}(w_{n+1}, z), \cdots, \mathfrak{y}(w_{2n}, z)$. Put

$$(3.3.4) \qquad \mathfrak{s}(z) = e\mathfrak{v}(z)^{-1}\mathfrak{u}(z) \qquad (z \in \mathfrak{H}) .$$

As remarked in 2.3, $\mathfrak{s}(z)$ is a point of the Siegel space $\mathfrak{S}_n$. The C-linear map $u \to e\mathfrak{v}(z)^{-1}u \ (u \in C^n)$ defines clearly an isomorphism of C^n/D_z to $C^n/D_e(\mathfrak{s}(z))$. With the notation of 2.4, put

$$(3.3.5) \qquad A_z = A_e(\mathfrak{s}(z)) , \qquad \mathcal{C}_z = \mathcal{C}_e(\mathfrak{s}(z)) ,$$

$$(3.3.6) \qquad \Xi(u, z) = \Theta_e\big(e\mathfrak{v}(z)^{-1}u, \mathfrak{s}(z)\big) \qquad (u \in C^n, z \in \mathfrak{H}) .$$

Then Ξ is a holomorphic mapping of $C^n \times \mathfrak{H}$ into P^N, and, for each $z \in \mathfrak{H}$, $u \to \Xi(u, z)$ is an isomorphism of C^n/D_z to A_z. For every $a \in \mathfrak{o}$, $\Phi(a)$ defines an endomorphism of C^n/D_z, which corresponds to the endomorphism $\theta_z(a)$ of A_z:

$$(3.3.7) \qquad \theta_z(a)\Xi(u, z) = \Xi\big(\Phi(a)u, z\big) \qquad (a \in \mathfrak{o}, u \in C^n, z \in \mathfrak{H}) .$$

Put

$$(3.3.8) \qquad \xi(x, z) = \Xi\big(\mathfrak{y}(x, z), z\big) \qquad (x \in L_R^m, z \in \mathfrak{H}) .$$

For each $z \in \mathfrak{H}$, the map $x \to \xi(x, z)$ gives an isomorphism of $L_R^m/\mathfrak{M}$ to A_z. Then the points $t_i(z)$ on A_z are defined by

$$(3.3.9) \qquad t_i(z) = \xi(v_i, z) \qquad (z \in \mathfrak{H}; \ i = 1, \cdots, s) .$$

This finishes the description of $\mathfrak{Q}_z = (A_z, \mathcal{C}_z, \theta_z; \{t_i(z)\})$. We put also

$$(3.3.10) \qquad \mathcal{P}_z = (A_z, \mathcal{C}_z, \theta_z) .$$

3.4. Let α be an element of $\mathrm{GL}_m(L)$ such that $\mathfrak{M}\alpha \subset \mathfrak{M}$ and $T(x\alpha, y\alpha) = \delta T(x, y)$ with a positive integer δ. By [19, II, 4.2], there exists an element $\Lambda(\alpha, z)$ of $\mathrm{GL}_n(C)$ such that

$$(3.4.1) \qquad \Lambda(\alpha, z)\mathfrak{y}(x, z) = \mathfrak{y}\big(x\alpha, \alpha^{-1}(z)\big) \qquad (x \in L_R^m, z \in \mathfrak{H}) .$$

From $\Lambda(\alpha, z)$ we obtain an isogeny $\iota(\alpha, z)$ of $\mathcal{P}_z$ to $\mathcal{P}_w$, with $w = \alpha^{-1}(z)$, such that

$$(3.4.2) \qquad \iota(\alpha, z)\Xi(u, z) = \Xi\big(\Lambda(\alpha, z)u, \alpha^{-1}(z)\big) \qquad\qquad (u \in \mathbf{C}^n, z \in \mathfrak{H}),$$

$$(3.4.3) \qquad \iota(\alpha, z)\xi(x, z) = \xi\big(x\alpha, \alpha^{-1}(z)\big) \qquad\qquad (x \in L_R^m, z \in \mathfrak{H}).$$

The isogeny $\iota(\alpha, z)$ is an isomorphism of $\mathcal{P}_z$ to $\mathcal{P}_w$ if $\alpha \in \Gamma(T, \mathfrak{M})$. It is an isomorphism of $\mathfrak{Q}_z$ to $\mathfrak{Q}_w$ if and only if $\alpha \in \Gamma(T, \mathfrak{N}/\mathfrak{M})$. Moreover, *every isomorphism of $\mathfrak{Q}_z$ to another member of Σ_Ω is of the form $\iota(\alpha, z)$ with* $\alpha \in \Gamma(T, \mathfrak{N}/\mathfrak{M})$. Two members $\mathfrak{Q}_z$ and $\mathfrak{Q}_w$ are isomorphic if and only if $z = \alpha(w)$ for some $\alpha \in \Gamma(T, \mathfrak{N}/\mathfrak{M})$. For the proof of these results, see [18, 2.8], [19, II, 4.2, 4.4].

From our construction, it is clear that every $\mathfrak{Q}_z$ is of type Ω. Conversely, *every* PEL-*structure $\mathfrak{Q} = \big(A, \mathcal{C}, \theta; \{t_i\}\big)$ *of type Ω is isomorphic to a member of Σ_Ω.* In fact, by [18, Th. 1], there exists an isomorphism η of $(A, \mathcal{C}, \theta)$ to $\mathcal{P}_z$ for some $z \in \mathfrak{H}$. Take an element v_i' so that $\eta(t_i) = \xi(v_i', z)$. Then $\mathfrak{Q}$ is of type $(L, \Phi, \rho; T, \mathfrak{M}; \{v_i'\})$. As remarked at the end of 3.2, there exists an element α of $\mathrm{GL}_m(L)$ such that $T(x\alpha, y\alpha) = T(x, y)$, $\mathfrak{M}\alpha = \mathfrak{M}$ and $v_i \equiv v_i'\alpha \bmod \mathfrak{M}$. Put $\alpha^{-1}(z) = w$, $\varepsilon = \iota(\alpha, z)\circ\eta$. Then ε is an isomorphism of $(A, \mathcal{C}, \theta)$ to $\mathcal{P}_w$, and $\varepsilon(t_i) = \xi\big(v_i'\alpha, \alpha^{-1}(z)\big) = \xi(v_i, w) = t_i(w)$. Hence ε gives an isomorphism of $\mathfrak{Q}$ to $\mathfrak{Q}_w$, q.e.d. (Cf. also 3.16 below.)

3.5. PROPOSITION. *There exists a positive integer ν_0 such that $\iota(\alpha, z)$ is the identity mapping of A_z if $\alpha \in \Gamma(T, \nu_0^{-1}\mathfrak{M}/\mathfrak{M})$.*

PROOF. For $\alpha \in \Gamma(T, \mathfrak{M})$, define an element $B = (b_{ij})$ of $\mathrm{GL}_{2n}(\mathbf{Z})$ by $w_j\alpha = \sum_{i=1}^{2n} w_i b_{ij}(j = 1, \cdots, 2n)$. Then B belongs to the group Γ_z' defined by (2.3.2). From (3.4.1) we obtain

$$\Lambda(\alpha, z)\big(\mathfrak{u}(z) \quad \mathfrak{v}(z)\big) = \big(\mathfrak{u}(\alpha^{-1}(z)) \quad \mathfrak{v}(\alpha^{-1}(z))\big)B,$$

so that

$$[e\mathfrak{v}(\alpha^{-1}(z))^{-1}\Lambda(\alpha, z)\mathfrak{v}(z)e^{-1}]\omega_e\big(\mathfrak{s}(z)\big) = \omega_e\big(\mathfrak{s}(\alpha^{-1}(z))\big)B.$$

Hence $\Lambda(\alpha, z) = \mathfrak{v}(\alpha^{-1}(z))e^{-1}J(B, \mathfrak{s}(z))e\mathfrak{v}(z)^{-1}$ and $\mathfrak{s}(\alpha^{-1}(z)) = B(\mathfrak{s}(z))$ with the notation of 2.3. If $\alpha \in \Gamma(T, \nu_0^{-1}\mathfrak{M}/\mathfrak{M})$, then $B \in \Gamma_z'$, $\beta \equiv 1_{2n} \bmod (\nu_0)$. Therefore, by 2.5, for a suitable ν_0, we have

$$\begin{aligned}
\Xi\big(\Lambda(\alpha, z)u, \alpha^{-1}(z)\big) &= \Xi\big(\mathfrak{v}(\alpha^{-1}(z))e^{-1}J(B, \mathfrak{s}(z))e\mathfrak{v}(z)^{-1}u, \alpha^{-1}(z)\big) \\
&= \Theta_e\big(J(B, \mathfrak{s}(z))e\mathfrak{v}(z)^{-1}u, B(\mathfrak{s}(z))\big) \\
&= \Theta_e\big(e\mathfrak{v}(z)^{-1}u, \mathfrak{s}(z)\big) = \Xi(u, z),
\end{aligned}$$

which shows that $\iota(\alpha, z)$ is the identity mapping.

3.6. By Baily-Borel [3], there exists an embedding of $\mathfrak{H}/\Gamma(T, \mathfrak{N}/\mathfrak{M})$ into a projective space P^M. We denote by φ the mapping of $\mathfrak{H}$ into P^M which induces the embedding of $\mathfrak{H}/\Gamma(T, \mathfrak{N}/\mathfrak{M})$. We put $V = \varphi(\mathfrak{H})$ and denote by V^* the Zariski closure of V. Then the result of [3] asserts that

(3.6.1) *V is Zariski open in V^*.*

(3.6.2) *V^* is projectively normal.*

(3.6.3) *Every meromorphic function on $\mathfrak{K}$, invariant under $\Gamma(T, \mathfrak{N}/\mathfrak{M})$, may be regarded as a rational function on V, provided that either $\dim(\mathfrak{K}) > 1$, or $\mathfrak{K}/\Gamma(T, \mathfrak{N}/\mathfrak{M})$ is compact.*

3.7. We are going to construct a fibre system of PEL-structures of which the base is V, and each fibre is isomorphic to $\mathfrak{Q}_z$. For that purpose, we impose the following conditions on Ω.

(3.7.1) *Either $\dim(\mathfrak{K}) > 1$, or $\mathfrak{K}/\Gamma(T, \mathfrak{N}/\mathfrak{M})$ is compact.*

This is *not* satisfied if and only if $\Gamma(T, \mathfrak{N}/\mathfrak{M})$ is commensurable with a conjugate of $SL_2(Z)$. The condition (3.7.1) does not exclude the case where $\mathfrak{K}$ consists of only one point (cf. [18, p. 163]).

(3.7.2) *$\Gamma(T, \mathfrak{N}/\mathfrak{M})$ has no element of finite order other than the identity element.*

This implies especially

(3.7.3) *V is a non-singular variety.*

(3.7.4) *For every $z \in \mathfrak{K}$, $\mathfrak{Q}_z$ has no automorphism other than the identity mapping.*

This follows from the assertions of 3.4 (cf. [19, II, 5.8]). From (3.6.3) we obtain (under the assumption (3.7.1))

(3.7.5) *If $\mathfrak{g}$ is a holomorphic mapping of $\mathfrak{K}$ into a projective space P^κ, invariant under $\Gamma(T, \mathfrak{N}/\mathfrak{M})$, then there exists a morphism g of V into P^κ such that $g(\varphi(z)) = \mathfrak{g}(z)$.*

3.8. PROPOSITION. *Under the assumptions (3.7.1-2), there exists a fibre system of PEL-structures*

$$\mathfrak{F} = \{V, W, h, f, Y, S(a) \ (a \in \mathfrak{o}), f_1, \cdots, f_s\},$$

with fibres $\mathfrak{Q}_u = (A_u, \mathcal{C}_u, \theta_u; f_1(u), \cdots, f_s(u)) \ (u \in V)$, in the sense of 1.5, such that:

(3.8.1) *V is a projective embedding of $\mathfrak{K}/\Gamma(T, \mathfrak{N}/\mathfrak{M})$ by a biregular mapping φ as stated in 3.6.*

(3.8.2) *For every $z \in \mathfrak{K}$ and $u = \varphi(z)$, $\mathfrak{Q}_u$ is isomorphic to $\mathfrak{Q}_z$.*

(3.8.3) *W is a Zariski open subset of a projective variety.*

The proof will be given in the following paragraphs 3.9-3.13.

3.9. Let $\mathfrak{c}(z)$ (resp. $\mathfrak{d}_a(z)$) denote the Chow point of A_z (resp. the Chow point of the graph of $\theta_z(a)$), a being any element of $\mathfrak{o}$. By [14, Exposé 19, Th. 3], $\mathfrak{c}(z)$ and the $\mathfrak{d}_a(z)$ are holomorphic in z. In fact, as for $\mathfrak{c}(z)$, the condition of [14, Exposé 19, Th. 3] is clearly satisfied, in view of (2.4.1) and our construction of Θ_e. As for $\mathfrak{d}_a$, one has to verify that the degree of the graph of $\theta_z(a)$ is independent of z. But this can be done easily by following the argument of [17, pp. 118-121].

3.10. To prove 3.8, first we assume that $\mathfrak{N} \supset \nu_0^{-1}\mathfrak{M}$ for a positive integer ν_0 with the property of 3.5. Then if $\gamma \in \Gamma(T, \mathfrak{N}/\mathfrak{M})$, $\iota(\gamma, z)$ is the identity mapping of A_z, so that $\mathfrak{Q}_z = \mathfrak{Q}_{\gamma(z)}$. Put $\mathfrak{f}(z) = \xi(0, z)$. Then $\mathfrak{f}(z)$, $t_i(z)$, $\mathfrak{c}(z)$, $\mathfrak{d}_a(z)$ are $\Gamma(T, \mathfrak{N}/\mathfrak{M})$-invariant holomorphic mappings of $\mathfrak{H}$ into projective spaces. By (3.7.5), there exist morphisms f', f'_i, c, d_a of V into projective spaces such that $\mathfrak{f}(z) = f'(\varphi(z))$, $t_i(z) = f'_i(\varphi(z))$, $\mathfrak{c}(z) = c(\varphi(z))$, $\mathfrak{d}_a(z) = d_a(\varphi(z))$. Let k be a field of rationality for V and these morphisms f', f'_i, c, d_a. For every $u \in V$, we take a point z of $\mathfrak{H}$ so that $\varphi(z) = u$, and put $A'_u = A_z$, $\theta'_u(a) = \theta_z(a)$. This does not depend on the choice of z. Since $k(c(u), f'(u)) \subset k(u)$, we see that the abelian variety A'_u, with the origin $f'(u)$, is defined over $k(u)$. Let y be a generic point of V over k, and x a generic point of A'_y over $k(y)$. Let W be the locus of (y, x) over k in $V \times P^N$. Since $(u, c(u), A'_u)$ is a specialization of $(y, c(y), A'_y)$, we see easily that $W = \bigcup_{u \in V}(u \times A'_u)$. By [23, VIII$_3$, Th. 6], we have $(y \times P^N) \cdot W = y \times A'_y$. For every $u \in V$, $(u \times P^N) \cap W = u \times A'_u$, hence $u \times P^N$ and W intersect properly on $V \times P^N$. Since (u, A'_u) is a specialization of (y, A'_y) over k, we get, by [16, Th. 23], $(u \times P^N) \cdot W = u \times A'_u$. By the converse part of [23, VI$_2$, Th. 6], every point of $u \times A'_u$ is simple on W. This proves that W is a non-singular variety. Let h be the natural projection of W to V, and let $f(u) = u \times f'(u)$. If Z is the graph of h, i.e., the locus of (y, y, x) over k on $V \times W$, then one can prove, specializing $Z \cdot (y \times W) = y \times y \times A'_y$, that $Z \cdot (u \times W) = u \times u \times A'_u$ on $V \times W$ for every $u \in V$. Therefore $\{V, W, h, f\}$ is a fibre system of abelian varieties, and $h^{-1}(u) = u \times A'_u$. We put $A_u = u \times A'_u$ and denote by $\theta_u(a)$ the element of $\mathrm{End}_Q(A_u)$ corresponding to $\theta'_u(a)$ by an obvious isomorphism of A'_u to A_u.

3.11. Let H be a hyperplane in P^N, rational over k. Then $H \cdot A'_u$ is well-defined for every $u \in V$, since $A'_u = A_z$ for point $z \in \mathfrak{H}$ with $u = \varphi(z)$, and A_z is a projective embedding of a complex torus by theta-functions. Put $Y_u = u \times (H \cdot A'_u)$ and $Y = (V \times H) \cdot W$, the latter intersection product being taken in $V \times P^N$. By [23, VIII$_4$, Th. 10], we have

$$Y_u = \{(u \times A'_u) \cdot (V \times H)\}_{V \times P^N} = \{(u \times A'_u) \cdot Y\}_W .$$

Hence Y satisfies (1.1.5) for the present fibre system.

3.12. The points x and y being as in 3.10, let $\bar{x} = y \times x \in A_y$, and let $S(a)$ be the locus of $\bar{x} \times \theta_y(a)\bar{x}$ over k for each $a \in \mathfrak{o}$. Since $d_a(y)$ is rational over $k(y)$, $\theta_y(a)$ is defined over $k(y)$, hence we can apply 1.4 to the present case. Since $(u, d_a(u))$ is a specialization of $(y, d_a(y))$ over k, we see that $S(a) \cdot (A_u \times A_u)$

is the graph of $\theta_u(a)$. Thus we have obtained, under the assumption $\nu_0^{-1}\mathfrak{M} \subset \mathfrak{N}$, a fibre system of PEL-structures

$$\mathfrak{F} = \{V,\ W,\ h, f,\ Y,\ S(a)\ (a \in \mathfrak{o}),\ f_1,\ \cdots,\ f_s\}$$

with fibres $\mathfrak{Q}_u = (A_u,\ \mathcal{C}_u,\ \theta_u;\ \{f_i(u)\})$, where $f_i(u) = u \times f_i'(u)$. From our construction, it is clear that $\mathfrak{Q}_u$ is isomorphic to $\mathfrak{Q}_z$ if $z = \varphi(u)$.

3.13. Let $v_1^*,\ \cdots,\ v_t^*$ be elements of L^m, and let $\mathfrak{N}^* = \mathfrak{M} + \sum_{i=1}^{t} \mathfrak{o}v_i^*$, $\Omega^* = (L,\ \Phi,\ \rho;\ T,\ \mathfrak{M};\ v_1^*,\ \cdots,\ v_t^*)$. Assume that (3.7.2) is satisfied with $\mathfrak{N}^*$ in place of $\mathfrak{N}$. Take a suitably large multiple μ of ν_0 so that $\mu^{-1}\mathfrak{M} \supset \mathfrak{N}^*$, and put $\mathfrak{N} = \mu^{-1}\mathfrak{M} = \sum_{i=1}^{s} \mathfrak{o}v_i$, $\Omega = (L,\ \Phi,\ \rho;\ T,\ \mathfrak{M};\ v_1,\ \cdots,\ v_s)$. We choose the v_i so that $v_i = v_i^*$ for $i = 1,\ \cdots, t$. (We need no linear independence of the v_i.) We see that $\Gamma(T, \mathfrak{N}/\mathfrak{M})$ is a normal subgroup of $\Gamma(T, \mathfrak{N}^*/\mathfrak{M})$. Put $G = \Gamma(T, \mathfrak{N}^*/\mathfrak{M})/\Gamma(T, \mathfrak{N}/\mathfrak{M})$. By our choice of $\mathfrak{N}$, we can apply the result of 3.10-12 to the present Ω. Let $\mathfrak{F}$ and $\mathfrak{Q}_u$ be defined as in 3.12, with these Ω and $\mathfrak{N}$.

Let $\gamma \in \Gamma(T, \mathfrak{N}^*/\mathfrak{M})$. Then $z \to \gamma(z)$ defines a biregular automorphism of V, which depends only on the class of γ modulo $\Gamma(T, \mathfrak{N}/\mathfrak{M})$. We denote this biregular automorphism of V by r_g with the element g of G represented by γ. Restricting the f_i to the first t members $f_1,\ \cdots,\ f_t$, we obtain

$$\mathfrak{Q}_u' = (A_u,\ \mathcal{C}_u,\ \theta_u;\ f_1(u),\ \cdots,\ f_t(u))$$

from $\mathfrak{Q}_u$, and the fibre system $\mathfrak{F}'$ with fibres $\mathfrak{Q}_u'$. The only distinction between $\mathfrak{F}$ and $\mathfrak{F}'$ is the difference $\{f_{t+1},\ \cdots,\ f_s\}$. We see that $\iota(\gamma, z)$ (with γ in $\Gamma(T, \mathfrak{N}^*/\mathfrak{M})$) is an isomorphism of $\mathscr{P}_z$ to $\mathscr{P}_{\gamma^{-1}(z)}$ which maps $\xi(v_i^*, z)$ to $\xi(v_i^*, \gamma^{-1}(z))$. Hence there is an isomorphism of $\mathfrak{Q}_u'$ to $\mathfrak{Q}_{r_g(u)}'$ for every $u \in V$, which is a unique isomorphism of $\mathfrak{Q}_u'$ to $\mathfrak{Q}_{r_g(u)}'$, in view of (3.7.4), since $\Gamma(T, \mathfrak{N}^*/\mathfrak{M})$ satisfies (3.7.2). Let k be a field of rationality for $\mathfrak{F}$ and all the r_g, and y a generic point of V over k. Let λ_g be the unique isomorphism of $\mathfrak{Q}_y'$ to $\mathfrak{Q}_{r_g(y)}'$. From the uniqueness, it follows that λ_g is defined over $k(y)$, since $\mathfrak{Q}_y'$ and $\mathfrak{Q}_{r_g(y)}'$ are defined over $k(y)$. By 1.7, one can find a biregular automorphism R_g of $\mathfrak{F}'$, which induces an isomorphism of $\mathfrak{Q}_u'$ to $\mathfrak{Q}_{r_g(u)}'$, for every $u \in V$. From the uniqueness of λ_g, one can derive $R_{gg'} = R_g R_{g'}$ ($g, g' \in G$). By virtue of the result of 1.9, taking the quotient of $\mathfrak{F}'$ by G, we obtain a fibre system $\mathfrak{F}^*$ with fibres $\mathfrak{Q}_u^*$. Then it is obvious that this $\mathfrak{F}^*$ has the required property of 3.8 for Ω^* and $\mathfrak{N}^*$. This completes the proof of 3.8.

3.14. We can define the fibre variety W as complex manifold also in the following way. First define a law of multiplication on $L_R^m \times \mathrm{U}_R(T)$ by

$$(x, \alpha)(y, \beta) = (x\beta + y, \alpha\beta) \qquad (x, y \in L_R^m;\ \alpha, \beta \in \mathrm{U}_R(T)).$$

Then $L_R^m \times \mathrm{U}_R(T)$ becomes a group with respect to this law. We let every element of $L_R^m \times \mathrm{U}_R(T)$ operate on $L_R^m \times \mathcal{H}$ by the rule

$$(x, z)(y, \beta) = (x\beta + y,\ \beta^{-1}(z)) \qquad\qquad (x, y \in L_R^m;\ z \in \mathcal{H};\ \beta \in \mathrm{U}_R(T)).$$

We see that $(x, z) \to (\mathfrak{y}(x, z), z)$ is a real analytic isomorphism of $L_R^m \times \mathcal{H}$ to $C^n \times \mathcal{H}$, and $\mathfrak{y}((x, z)(0, \beta)) = \Lambda(\beta, z)\mathfrak{y}(x, z)$. We transfer the complex structure of $C^n \times \mathcal{H}$ to $L_R^m \times \mathcal{H}$ by this isomorphism. Then the operation of every element of $L_R^m \times \Gamma(T, \mathfrak{M})$ on $L_R^m \times \mathcal{H}$ is complex analytic [19, II, 5.2]. We see that $\mathfrak{M} \times \Gamma(T, \mathfrak{N}/\mathfrak{M})$ is a discrete subgroup of $L_R^m \times \mathrm{U}_R(T)$, and hence

$$(L_R^m \times \mathcal{H})/(\mathfrak{M} \times \Gamma(T, \mathfrak{N}/\mathfrak{M}))$$

is a complex manifold, provided that (3.7.2) is satisfied. Now assuming $\mathfrak{N} \supset \nu_0^{-1}\mathfrak{M}$, define a mapping ψ of $L_R^m \times \mathcal{H}$ to $V \times P^N$ by

$$(3.14.1) \qquad\qquad \psi(x, z) = \varphi(z) \times \xi(x, z) \qquad\qquad (x \in L_R^m, z \in \mathcal{H}).$$

In the general case where $\mathfrak{N} \supset \nu_0^{-1}\mathfrak{M}$ does not necessarily hold, take $\mathfrak{N}_1$ so that $\mathfrak{N}_1 \supset \mathfrak{N} + \nu_0^{-1}\mathfrak{M}$ and $\Gamma(T, \mathfrak{N}_1/\mathfrak{M})$ is a normal subgroup of $\Gamma(T, \mathfrak{N}/\mathfrak{M})$. Then we obtain a mapping ψ_1 of $L_R^m \times \mathcal{H}$ to $V \times P^N$ defined by (3.14.1) with this $\mathfrak{N}_1$. Let W (resp. W_1) be the fibre variety defined with respect to $\mathfrak{N}$ (resp. $\mathfrak{N}_1$). Then there is a natural projection p of W_1 to W, for a reason similar to the discussion in 3.13. Define ψ by $\psi(x, z) = p(\psi_1(x, z))$. Then, in any case, we see that ψ gives a biregular isomorphism of the complex manifold

$$(L_R^m \times \mathcal{H})/(\mathfrak{M} \times \Gamma(T, \mathfrak{N}/\mathfrak{M}))$$

onto W. This follows easily from the fact that the correspondence is one-to-one, and the jacobian matrix of ψ is of maximal rank everywhere on $L_R^m \times \mathcal{H}$. This gives another proof of non-singularity of W. In the present investigation, however, we shall make no use of this expression of W as a quotient of $L_R^m \times \mathcal{H}$.

3.15. From the discussion of 3.4, we see that our fibre system $\mathfrak{F}$ of 3.8 has the following property:

(3.15.1) *Every* PEL-*structure of type Ω is isomorphic to one and only one fibre $\mathfrak{Q}_u$ of $\mathfrak{F}$.*

Let k be a field of rationality for $\mathfrak{F}$, finitely generated over Q. Let K_u be the field of moduli of $\mathcal{P}_u$. Then

$$(3.15.2) \qquad\qquad k(u) = k \cdot K_u \qquad\qquad for\ every\ u \in V.$$

In fact the inclusion $k(u) \supset k \cdot K_u$ is obvious, since $\mathfrak{Q}_u$ is defined over $k(u)$. Let σ be an automorphism of C over $k \cdot K_u$. Then $\mathfrak{Q}_{u^\sigma} = (\mathfrak{Q}_u)^\sigma$ is isomorphic to $\mathfrak{Q}_u$ in view of the property (1.5.1) of the field of moduli. By (3.15.1), we have $u^\sigma = u$, so that σ is the identity mapping on $k(u)$. This proves $k(u) \subset k \cdot K_u$, hence (3.15.2).

3.16. Let $\mathfrak{A}$ be an arbitrary PEL-structure of type Ω. Let C^n/D and $\mathfrak{y}$ be as in 3.2. Let d_j be the element of L^m whose j^{th} component is 1, and all the other components are 0. Then $d_1, \cdots, d_m$ form a basis of L^m over L. As remarked in 3.4, $\mathfrak{A}$ is isomorphic to a member $\mathfrak{A}_z$ of Σ_Ω. The discussion of [18, pp. 157-163] shows that the point z can be obtained from the $\mathfrak{y}(d_i)$ explicitly. In fact putting $\mathfrak{x}_i = \mathfrak{y}(d_i)$ for $i = 1, \cdots, m$, there is a well-defined mapping

$$(\mathfrak{x}_1, \cdots, \mathfrak{x}_m) \to z \in \mathcal{H} ,$$

which depends on only Ω. (With the notation of [18, pp. 157-163], this mapping is obtained in the following steps:

$$(\mathfrak{x}_1, \cdots, \mathfrak{x}_m) \to (X_1, \cdots, X_g) \to (X_1 \bar{W}_1^{-1}, \cdots, X_g \bar{W}_g^{-1}) \to z = (z_1, \cdots, z_g) .)$$

We notice that z is holomorphic in $(\mathfrak{x}_1, \cdots, \mathfrak{x}_m)$. Now, z being determined by the $\mathfrak{x}_i$, there exists an element B of $\mathrm{GL}_n(C)$ (Λ of [18, pp. 161-162]), commuting with the $\Phi(a)$ ($a \in L$), such that $B\mathfrak{y}(x) = \mathfrak{y}(x, z)$ for $x \in L_R^m$. Then B induces an isomorphism of C^n/D to C^n/D_z with D_z defined in 3.3, which gives an isomorphism of $\mathfrak{A}$ to $\mathfrak{A}_z$.

4. Deformation of PEL-structures

4.1. PROPOSITION. *Let $\Omega = \big(L, \Phi, \rho; T, \mathfrak{M}; \{v_i\}\big)$ be a PEL-type, and $\mathfrak{o}$ an order in L. Let $\{V, W, h, f, Y, S(a)\ (a \in \mathfrak{o}), f_1, \cdots, f_s\}$ be an arbitrary fibre system of PEL-structures with L as endomorphism algebra, with fibres $\mathfrak{A}_u = \big(A_u, C_u, \theta_u; \{f_i(u)\}\big)\ (u \in V)$, which is not necessarily the one constructed in § 3. Suppose that $\mathfrak{A}_u$ is of type Ω for at least one point u of V. Then $\mathfrak{A}_v$ is of type Ω for all $v \in V$. Moreover, there exist a neighbourhood $\mathfrak{N}$ of u on V and a continuous map $t \to z(t)$ of $\mathfrak{N}$ into $\mathcal{H}$ such that $\mathfrak{A}_t$ is isomorphic to $\mathfrak{A}_{z(t)}$ for every $t \in \mathfrak{N}$, where $\mathfrak{A}_z$ is a member of $\Sigma_\Omega = \{\mathfrak{A}_z \mid z \in \mathcal{H}\}$ constructed in 3.3.*

The following paragraphs 4.2-7 are devoted to the proof, which is rather straightforward and with no trick.

4.2. Let us first recall a result of Kodaira-Spencer [6] concerning the deformation of complex tori. Let $\mathfrak{S}$ be the set of all complex matrices s of size n such that $\det(\mathrm{Im}(s)) \neq 0$, and let $\omega(s)$ be a matrix with n rows and $2n$ columns defined by

$$\omega(s) = (s \ 1_n) \qquad\qquad (s \in \mathfrak{S}) .$$

Let $D(s)$ be the lattice in C^n generated by the column vectors of $\omega(s)$. Furthermore let G denote a group of analytic automorphisms of $C^n \times \mathfrak{S}$ consisting of the mappings

$$(z, s) \rightarrow (z + \omega(s)y, s) \qquad\qquad (z \in C^n, s \in S)$$

for all $y \in Z^{2n}$. Then the quotient $\mathfrak{B} = (C^n \times S)/G$ is a complex manifold, and there is an obvious projection $\pi : \mathfrak{B} \rightarrow S$ such that $\pi^{-1}(s)$ is a complex torus $C^n/D(s)$. Put $B_s = \pi^{-1}(s) = C^n/D(s)$ for every $s \in S$.

The notation being as in 4.1, let $n = \dim(A_u)$, and $X = R^{2n}/Z^{2n}$. In view of our definition in 1.1, for every $u \in V$, there exist an open neighborhood $\mathfrak{N}$ of u on V and a diffeomorphism g of $X \times \mathfrak{N}$ to $h^{-1}(\mathfrak{N})$ such that, for every $t \in \mathfrak{N}$, g is a diffeomorphism of $X \times t$ to A_t.

Let ι be a complex analytic isomorphism of A_u to B_{s_0} with a point s_0 of S. By [6, pp. 408-411], we may choose such an $\mathfrak{N}$ so that there exist a differential map $\sigma : \mathfrak{N} \rightarrow S$ and a differentiable map $\tau : h^{-1}(\mathfrak{N}) \rightarrow \mathfrak{B}$ with the property that for each $t \in \mathfrak{N}$, τ maps A_t biregularly onto $B_{\sigma(t)}$, $\sigma(u) = s_0$, and the restriction of h to A_u coincides with ι.

Let J be a map of $R^{2n} \times S$ to $C^n \times S$ such that

$$(4.2.1) \qquad J(x, s) = \big(\omega(s)x, s\big) \qquad\qquad (x \in R^{2n}, s \in S),$$

and let j be a diffeomorphism of $X \times S$ to $\mathfrak{B}$ which is induced from J in an obvious way. Then we have a commutative diagram:

$$
\begin{array}{ccccccc}
 & & C^n \times S & \xleftarrow{\quad J \quad} & R^{2n} \times S & & \\
 & & \downarrow{\scriptstyle (G)} & & \downarrow{\scriptstyle (Z^{2n})} & & \\
X \times \mathfrak{N} \xrightarrow{\ g\ } h^{-1}(\mathfrak{N}) \xrightarrow{\ \tau\ } & & \mathfrak{B} & \xleftarrow{\ j\ } & X \times S & & \\
\downarrow \qquad\qquad \downarrow{\scriptstyle h} & & \downarrow{\scriptstyle \pi} & & \downarrow & & \\
\mathfrak{N} \xrightarrow{\ \mathrm{id.}\ } \mathfrak{N} \xrightarrow{\ \sigma\ } & & S & \xleftarrow{\ \mathrm{id.}\ } & S & &
\end{array}
$$

The manifolds X, A_t and B_s have a natural structure of additive group. The map j, restricted to each fibre, preserves the law of addition, but g and τ may not have the same property. We are going to replace g and τ by the ones with such a property.

Let f be as in 4.1. We remind the reader that, for every $v \in V$, $f(v)$ is the origin of A_v. Restrict f to $\mathfrak{N}$ and define a map $p: \mathfrak{N} \rightarrow X$ and a map $q_t: X \rightarrow X$ by

$$g^{-1} \circ f(t) = \big(p(t), t\big) \qquad\qquad (t \in \mathfrak{N}),$$
$$j^{-1} \circ \tau \circ g(x, t) = \big(q_t(x), \sigma(t)\big) \qquad\qquad (x \in X, t \in \mathfrak{N}).$$

Then p is differentiable, and q_t is a diffeomorphism of X to itself, for every $t \in \mathfrak{N}$. Now define a map $H: X \times \mathfrak{N} \rightarrow \mathfrak{B}$ and a map $F: X \times \mathfrak{N} \rightarrow X \times \mathfrak{N}$ by

$$H(x, t) = j\big(x, \sigma(t)\big),$$
$$F(x, t) = \big(q_t(x) - q_t(p(t)), t\big) \qquad\qquad (x \in X, t \in \mathfrak{N}).$$

Then H is differentiable, and F is a diffeomorphism of $X \times \mathfrak{N}$ to itself. Put $g_0 = g \circ F^{-1}$ and $\tau_0 = H \circ F \circ g^{-1}$. Then g_0 is a diffeomorphism of $X \times \mathfrak{N}$ to $h^{-1}(\mathfrak{N})$, and τ_0 is a differentiable map of $h^{-1}(\mathfrak{N})$ to $\mathfrak{B}$. We have now

$$(4.2.2) \qquad j^{-1} \circ \tau_0 \circ g_0(x, t) = j^{-1} \circ H(x, t) = \big(x, \sigma(t)\big) .$$

If $w = g(x, t) \in A_t$, then

$$\begin{aligned}
\tau_0(w) &= H \circ F(x, t) = j\big(q_t(x) - q_t(p(t)), \sigma(t)\big) \\
&= j\big(q_t(x), \sigma(t)\big) - j\big(q_t(p(t)), \sigma(t)\big) \\
&= \tau \circ g(x, t) - \tau \circ g\big(p(t), t\big) \\
&= \tau(w) - \tau\big(f(t)\big) ,
\end{aligned}$$

the difference being with respect to the law of addition on $B_{\sigma(t)}$. This implies that, on each fibre A_t, τ_0 induces a biregular isomorphism of A_t to $B_{\sigma(t)}$, which maps the origin $f(t)$ of A_t to the origin $j(0, \sigma(t))$ of $B_{\sigma(t)}$. Therefore, for every $t \in \mathfrak{N}$, $B_{\sigma(t)}$ has a structure of abelian variety, and τ_0 gives an isomorphism of A_t to $B_{\sigma(t)}$, both being understood as abelian varieties. Moreover, if we start with a mapping $\iota : A_u \to B_{s_0}$ which maps origin to origin, then $\tau_0(w) = \tau(w) = \iota(w)$ for $w \in A_u$.

4.3. Let us now take account of the polarization $\mathcal{C}_u$ of A_u. As indicated in 2.2, the divisor $Y \cap h^{-1}(\mathfrak{N})$ on $h^{-1}(\mathfrak{N})$ defines a 2-form ε on $h^{-1}(\mathfrak{N})$ whose cohomology class is $a(Y \cap h^{-1}(\mathfrak{N}))$. Then it is easy to verify that, for every $t \in \mathfrak{N}$, ε defines a form ε_t on A_t whose cohomology class is exactly $a(Y \cdot A_t)$. In view of our assumption (1.1.5), this can be checked in a straightforward way. Let $(x_1, \cdots, x_{2n})$ be the coordinates in R^{2n} with respect to the ordinary basis of Z^{2n}, and $(t_1, \cdots, t_r)$ be a coordinate system on $\mathfrak{N}$. Put

$$\varepsilon \circ g_0 = \sum_{\alpha, \beta} e_{\alpha\beta}(x, t) dx_\alpha \wedge dx_\beta + \varepsilon',$$

where ε' means the sum of all terms involving dt_i. If g_t is the restriction of g_0 to $X \times t$, then $\varepsilon_t \circ g_t = \sum_{\alpha, \beta} e_{\alpha\beta}(x, t) dx_\alpha \wedge dx_\beta$. Let $\{c_{\lambda\mu} \mid \lambda, \mu = 1, \cdots, 2n\}$ be a basis of 2-cycles on X such that

$$\int_{c_{\lambda\mu}} dx_\alpha \wedge dx_\beta = \delta_{\lambda\alpha} \cdot \delta_{\mu\beta}, \qquad c_{\lambda\mu} = -c_{\mu\lambda} ,$$

where $\delta_{\lambda\alpha}$ is Kronecker's symbol. Put

$$E(t) = \big(e_{\alpha\beta}(t)\big) , \qquad e_{\alpha\beta}(t) = \int_{c_{\alpha\beta}} \varepsilon_t \circ g_t .$$

Then $\varepsilon_t \circ g_t$ is cohomologous to $\sum_{\alpha, \beta} e_{\alpha\beta}(t) dx_\alpha \wedge dx_\beta$, and as remarked in 2.2, the divisor $Y \cdot A_t$ of A_t determines a Riemann form represented by the alternating matrix $E(t)$ with respect to the coordinates $x_i \circ g_t^{-1}$ of A_t. Therefore, the $e_{\alpha\beta}(t)$ are integers. On the other hand, $e_{\alpha\beta}(x, t)$ is differentiable in (x, t), so that

$e_{\alpha\beta}(t)$ is continuous in t. It follows that the $e_{\alpha\beta}(t)$ are constants on $\mathfrak{N}$. We put

$$E_0 = (e_{\alpha\beta}), \qquad e_{\alpha\beta} = e_{\alpha\beta}(t) \,.$$

Changing the coordinate system in R^{2n} by a suitable element of $\mathrm{GL}_{2n}(Z)$, we may assume that

$$(4.3.1) \qquad E_0 = \mu E, \qquad E = \begin{pmatrix} 0 & -e \\ e & 0 \end{pmatrix}, \qquad e = \mathrm{diag}[e_1, \cdots, e_n] \,,$$

$$e_1 = 1, \qquad e_{i+1} \equiv 0 \mod (e_i) \,,$$

with positive integers $\mu, e_1, \cdots, e_n$.

4.4. In the next place, we consider endomorphisms $\theta_t(a)$ $(a \in \mathfrak{o})$ of A_t. For every $a \in \mathfrak{o}$ and for every $t \in \mathfrak{N}$, there exists an endomorphism $\psi_t(a)$ of X such that

$$\theta_t(a)g_0(x, t) = g_0(\psi_t(a)x, t) \qquad\qquad (x \in X,\, t \in \mathfrak{N}) \,.$$

Since $S(a)$ is a graph of a morphism, the map

$$X \times \mathfrak{N} \ni (x, t) \to g_0(x, t) \times g_0(\psi_t(a)x, t) \in S(a)$$

is continuous. Hence $\psi_t(a)x$ is continuous in t for every fixed x. Since $\psi_t(a)$ is an endomorphism of $X = R^{2n}/Z^{2n}$, this implies that $\psi_t(a)$ is independent of t for every $a \in \mathfrak{o}$.

Let $\Psi(a)$ denote the R-linear endomorphism of R^{2n} which gives $\psi_t(a)$ on X. Then $a \to \Psi(a)$ is an isomorphism of $\mathfrak{o}$ into $M_{2n}(Z)$. Extend this to an isomorphism of L into $M_{2n}(R)$, and denote it again by Ψ. In 4.2, we have seen that, for every $t \in \mathfrak{N}$, the restriction of τ_0 to A_t is an isomorphism to $B_{\sigma(t)} = C^n/D(\sigma(t))$. Let $\Phi_t(a)$ denote the C-linear endomorphism of C^n corresponding to $\theta_t(a)$. In view of (4.2.2), for every $a \in \mathfrak{o}$, we have

$$\begin{aligned} j(\Psi(a)x, \sigma(t)) &= \tau_0 \circ g_0(\Psi(a)x, t) = \tau_0[\theta_t(a)g_0(x, t)] \\ &= \Phi_t(a)[\tau_0 \circ g_0(x, t)] = \Phi_t(a)j(x, \sigma(t)) \qquad \mod D(\sigma(t)) \,, \end{aligned}$$

so that with matrix notation,

$$\Phi_t(a)\omega(\sigma(t)) = \omega(\sigma(t))\Psi(a) \,.$$

Since $\mathfrak{A}_u$ is of type Ω, there exists a non-singular matrix R such that $\Phi_u(a) = R\Phi(a)R^{-1}$ $(a \in L)$. Put

$$P_t = \begin{pmatrix} \alpha_t & \beta_t \\ \gamma_t & \delta_t \end{pmatrix} = \begin{pmatrix} \omega(\sigma(t)) \\ \overline{\omega(\sigma(t))} \end{pmatrix}\begin{pmatrix} \omega(\sigma(u)) \\ \overline{\omega(\sigma(u))} \end{pmatrix}^{-1}\begin{pmatrix} R & 0 \\ 0 & \bar{R} \end{pmatrix} \,.$$

Then we have

$$\begin{pmatrix} \Phi_t(a) & 0 \\ 0 & \overline{\Phi_t(a)} \end{pmatrix} P_t = P_t \begin{pmatrix} \Phi(a) & 0 \\ 0 & \overline{\Phi(a)} \end{pmatrix} \,,$$

and hence $\Phi_t(a)\alpha_t = \alpha_t\Phi(a)$. Since $\alpha_u = R$, and α_t is continuous in t, we see that α_t is non-singular in a neighborhood $\mathfrak{N}'$ of u, contained in $\mathfrak{N}$. For simplicity, we write this $\mathfrak{N}'$ again as $\mathfrak{N}$. Put

$$\omega_t = \alpha_t^{-1}\omega\big(\sigma(t)\big) , \qquad D_t = \alpha_t^{-1}D_{\sigma(t)} ,$$
$$\tau_t(x) = \alpha_t^{-1}\tau_0(x) \qquad (\mathrm{mod}\ D_t) \qquad\qquad (t \in \mathfrak{N},\ x \in A_t) .$$

Then $\tau_t : A_t \to C^n/D_t$ is an isomorphism, and

$$(4.4.1) \qquad\qquad \Phi(a)\omega_t = \omega_t\Psi(a) , \qquad \tau_t\big(\theta_t(a)x\big) = \Phi(a)\tau_t(x) \qquad\qquad (x \in A_t) .$$

4.5. Since $\mathfrak{Q}_u$ is of type Ω, there is an R-linear isomorphism $\kappa : L_R^m \to R^{2n}$ such that

$$(4.5.1) \qquad\qquad\qquad \kappa(ax) = \Psi(a)\kappa(x) \qquad\qquad\qquad (a \in L,\ x \in L_R^m) ,$$
$$(4.5.2) \qquad\qquad\qquad \kappa(\mathfrak{M}) = Z^{2n} ,$$
$$(4.5.3) \qquad\qquad\qquad g_0\big(\kappa(v_i),\ u\big) = f_i(u) \qquad\qquad\qquad (i = 1, \cdots, s) ,$$
$$(4.5.4) \qquad\qquad\qquad \mathrm{tr}\big(T(x,\ y)\big) = {}^t\kappa(x)E\kappa(y) \qquad\qquad\qquad (x,\ y \in L_R^m) .$$

Since the f_i are continuous mappings of V to W such that $h(f_i(t)) = t$, we can find continuous maps $p_i : \mathfrak{N} \to L_R^m$ so that

$$f_i(t) = g_0\big(\kappa(p_i(t)),\ t\big) \qquad\qquad\qquad (t \in \mathfrak{N};\ i = 1, \cdots, s) .$$

The p_i should be constant on $\mathfrak{N}$, since the $f_i(t)$ are points of finite order on A_t. In view of (4.5.3), we may put $p_i(t) = v_i$, hence

$$(4.5.5) \qquad\quad \tau_t\big(f_i(t)\big) = \alpha_t^{-1}[\tau_0 \circ g_0\big(\kappa(v_i),\ t\big)] = \alpha_t^{-1}\cdot j\big(\kappa(v_i),\ \sigma(t)\big)$$
$$= \omega_t\cdot\kappa(v_i) \qquad \mathrm{mod}\ D_t \qquad\qquad (t \in \mathfrak{N};\ i = 1, \cdots, s) .$$

Let $\{d_1, \cdots, d_m\}$ be the basis of L^m over L defined in 3.16, and let $y_i(t) = \omega_t\kappa(d_i)$ $(t \in \mathfrak{N};\ i = 1, \cdots, m)$. For every $(a_1, \cdots, a_m) \in L_R^m$, we have, by (4.4.1) and (4.5.1),

$$(4.5.6) \qquad\qquad \sum_{i=1}^m \Phi(a_i)y_i(t) = \omega_t\kappa\big((a_1, \cdots, a_m)\big) \qquad\qquad (t \in \mathfrak{N}) .$$

We see that, for every $t \in \mathfrak{N}$, the map

$$(4.5.7) \qquad\qquad\qquad (a_1, \cdots, a_m) \to \sum_{i=1}^m \Phi(a_i)y_i(t)$$

gives an isomorphism of $L_R^m/\mathfrak{M}$ to C^n/D_t. Let $\mathfrak{E}_t$ be the Riemann form on C^n/D_t determined by the divisor $\tau_t(Y\cdot A_t)$. In view of the result of 4.3 and (4.2.2), $\mathfrak{E}_t$ is given by

$$\mathfrak{E}_t(\omega_t x,\ \omega_t y) = {}^t x E_0 y \qquad\qquad\qquad (x,\ y \in R^{2n}) ,$$

E_0 being as in (4.3.1). By (4.5.4) and (4.5.6), we have

$$\mu\cdot\mathrm{tr}\big(T(a,\ b)\big) = \mathfrak{E}_t\big(\sum_{i=1}^m \Phi(a_i)y_i(t),\ \sum_{i=1}^m \Phi(b_i)y_i(t)\big)$$

for $a = (a_1, \cdots, a_m) \in L_R^m$, $b = (b_1, \cdots, b_m) \in L_R^m$. Therefore, the map (4.5.7) plays the role of the map $\mathfrak{y}$ in the definition 3.2; and τ_i^{-1}, the role of ξ. Thus we have proved that $\mathfrak{A}_t$ is of type Ω for all $t \in \mathfrak{N}$.

4.6. Since $\mathfrak{N}$ contains a generic point of V over a field of rationality for $\mathfrak{F}$, we see, by virtue of 1.6, that for every $x \in V$, Y_x determines an involution of $\mathrm{End}_Q(A_x)$ whose restriction to $\theta_x(L)$ is the map $\theta_x(a) \to \theta_x(a^\rho)$ $(a \in L)$. Therefore $\mathfrak{A}_x$ is of type $\Omega' = (L, \Phi', \rho; T', \mathfrak{M}'; \{v_i'\})$ with certain Φ', T', $\mathfrak{M}'$, $\{v_i'\}$ (cf. [18], [19, I, 1.5, 1.9], [19, II, 3.4]). Our discussion of 4.2-4.5 shows that $\mathfrak{A}_y$ is of type Ω' for every y in a neighborhood of x. Let V_1 be the set of all points t of V such that $\mathfrak{A}_t$ is of type Ω. Then V_1 is an open subset of V containing u, and the complement $V - V_1$ is open for the same reason. Since V is connected, we have $V = V_1$, which proves our first assertion of 4.1.

4.7. Coming back to the stage of 4.5, we regard the vectors $y_1(t), \cdots, y_m(t)$ as the vectors $\mathfrak{y}(d_1), \cdots, \mathfrak{y}(d_m)$ of 3.16. As shown in 3.16, they determine a point z of $\mathcal{H}$ in a well-defined way, so that $\mathfrak{A}_t$ is isomorphic to $\mathfrak{A}_z$, and z depends continuously on the $y_i(t)$ $(= \mathfrak{y}(d_i))$. Put $z = z(t)$ for $t \in \mathfrak{N}$. Since the $y_i(t)$ are continuous in t, the map $t \to z(t)$ is continuous. This proves the last assertion of 4.1.

5. Main theorems concerning fibre systems

5.1. THEOREM. *For every* PEL-*type* $\Omega = (L, \Phi, \rho; T, \mathfrak{M}; v_1, \cdots, v_s)$, *for which* (3.7.1) *is satisfied, there exists an algebraic number field* k_Ω *of finite degree with the following properties.*

(5.1.1) *Let* $\mathfrak{A}$ *be a* PEL-*structure of type* Ω, *and* σ *an automorphism of* C. *Then* $\mathfrak{A}^\sigma$ *is of type* Ω *if and only if* σ *is the identity mapping on* k_Ω.

(5.1.2) *If* $\mathfrak{A}$ *is of type* Ω, *the field of moduli of* $\mathfrak{A}$ *contains* k_Ω.

(5.1.3) *Let* K' *be the field generated over* Q *by* $\mathrm{Tr}\,(\Phi(\alpha))$ *for all* α *in the center of* L. *Then* $K' \subset k_\Omega$.
Moreover the field k_Ω *can be uniquely determined by the property* (5.1.1).

PROOF. We can find a positive integer q so that $\mathfrak{N} \subset q^{-1}\mathfrak{M}$ and $\Gamma(T, q^{-1}\mathfrak{M}/\mathfrak{M})$ has no element of finite order. Take a set of elements $\{v_1^*, \cdots, v_t^*\}$ of L^m such that $q^{-1}\mathfrak{M} = \sum_{i=1}^t \mathfrak{o}v_i^*$, $v_j^* = v_j$ for $j = 1, \cdots, s$. Let Ω^* be a PEL-type obtained from Ω by replacing $\{v_1, \cdots, v_s\}$ with $\{v_1^*, \cdots, v_t^*\}$. Let us consider the fibre system of PEL-structures

$$\mathfrak{F}^* = \{V, W, h, f, Y, S(a)\,(a \in \mathfrak{o}), f_1^*, \cdots, f_t^*\}$$

with fibres $\mathfrak{A}_u^* = (A_u, C_u, \theta_u; f_i^*(u))$ of type Ω^*, constructed in 3.8. Taking only the first s morphisms $f_1^*, \cdots, f_s^*$, one obtains a fibre system $\mathfrak{F}$ from $\mathfrak{F}^*$, whose fibres $\mathfrak{A}_u$ are of type Ω. We see that every PEL-structure $\mathfrak{A}$ of type Ω is isomorphic to a fibre $\mathfrak{A}_u$ for some $u \in V$. (The point u may not be uniquely determined.) Let σ be an automorphism of C, and let $u \in V$. Then $(\mathcal{Q}_u)^\sigma$ is of type Ω' for a PEL-type $\Omega' = (L, \Phi^\sigma, \rho; T', \mathfrak{M}': \{v'_i\})$ with certain T', $\mathfrak{M}'$ and $\{v'_i\}$ [19, I, 1.5]. By 4.1, Ω' does not depend on the choice of u. Therefore we shall write $\Omega^\sigma = \Omega'$. If τ is another automorphism of C, we have $\Omega^{\sigma\tau} = (\Omega^\sigma)^\tau$. Let K_u be the field of moduli of $\mathcal{Q}_u$. Then K_u is finitely generated over Q. Let k be the algebraic closure of Q in K_u, and let $t = (t_1, \cdots, t_\lambda)$ be a set of elements such that $K_u = k(t)$. Let X be the locus of t over k in an affine space, and X_0 the set of all generic points of X over k. Let σ and τ be automorphisms of C over k. If $t^\sigma = t^\tau$, then $\sigma = \tau$ on K_u, so that $\mathcal{Q}^\sigma$ is isomorphic to $\mathcal{Q}^\tau$. Hence $\Omega^\sigma = \Omega^\tau$. Therefore, if $s = t^\sigma$, we may write Ω^σ as Ω_s. The notation Ω_s can be defined for every $s \in X_0$, since we can always find an automorphism of C over k which sends t to s. We have especially $\Omega = \Omega_t$. Now we observe that there are only countably many inequivalent PEL-types with a given L, ρ, and n. For simplicity, let us identify equivalent PEL-types. By Lemma 5.2 below, we can find two independent generic points s and r of X over k such that $\Omega_r = \Omega_s$. Let x and y be arbitrary independent generic points of X over k. There is an automorphism σ of C over k such that $x = r^\sigma$ and $y = s^\sigma$. Then $\Omega_x = (\Omega_r)^\sigma = (\Omega_s)^\sigma = \Omega_y$. Let $z \in X_0$. Take a generic point x of X over $k(t, z)$. By virtue of what has just been proved, we have $\Omega_x = \Omega_t$ and $\Omega_z = \Omega_x$. This implies that $\Omega_z = \Omega_t = \Omega$ for ever $z \in X_0$. In other words, for every automorphism σ of C, if σ is the identity mapping on k, then $\Omega^\sigma = \Omega$. Recall that k is a finite algebraic extension of Q. Take a finite normal extension k' of Q, containing k; and let $G(k'/Q)$ denote the Galois group of k' over Q. For every $\sigma \in G(k'/Q)$, take an automorphism τ of C which coincides with σ on k' and put $\Omega^\sigma = \Omega^\tau$. In view of the property of k established above, Ω^σ depends actually on only σ and is independent of the choice of τ. Let

$$H = \{\sigma \in G(k'/Q) \mid \Omega^\sigma = \Omega\} \, ,$$

and let k_Ω be the subfield of k' corresponding to H. Then it is clear that k_Ω has the property (5.1.1). By our construction of k_Ω, we have $k_\Omega \subset k \subset K_u$, which proves (5.1.2). If an automorphism σ of C is the identity mapping on k_Ω, then $\Omega^\sigma = \Omega$, hence Φ^σ is equivalent to Φ. It follows that σ is the identity mapping on K'. This proves (5.1.3). Finally it is obvious that the property (5.1.1) characterizes k_Ω.

5.2. LEMMA. *Let k be a subfield of C with countably many elements, and X a variety defined over k. Let X_0 be the set of all generic points of X over k. Suppose that X_0 is a disjoint union of countably many subsets X_1, X_2, $\cdots$. Then at least one X_j contains at least two independent generic points of X over k.*

PROOF. We may assume that the X_i are non-empty. For each i, take an element r_i of X_i. Since the field $k(r_1, r_2, \cdots)$ has countably many elements, there exists a generic point s of X over $k(r_1, r_2, \cdots)$. Since X_0 is the union of the X_i, the point s is contained in some X_j. Then r_j and s are independent generic points of X over k, contained in X_j.

5.3. THEOREM. *Let* $\Omega = (L, \Phi, \rho; T, \mathfrak{M}; v_1, \cdots, v_s)$ *be a* PEL-*type, and let*

$$\mathfrak{o} = \{a \in L \mid a\mathfrak{M} \subset \mathfrak{M}\}, \qquad \mathfrak{N} = \mathfrak{M} + \sum_{i=1}^{s} \mathfrak{o}v_i \ .$$

Let $\Gamma(T, \mathfrak{N}/\mathfrak{M})$ *and* $\mathfrak{H}$ *be defined as in* § 3. *Suppose that* (3.7.1) *and* (3.7.2) *are satisfied. Let* k_Ω *be as in* 5.1. *Then there exists a fibre system of* PEL-*structures*

$$\mathfrak{F} = \{V, W, h, Y, S(a)\,(a \in \mathfrak{o}), f_1, \cdots, f_s\} \ ,$$

with fibres $\mathfrak{Q}_u = \left(A_u, \mathcal{C}_u, \theta_u; \{f_i(u)\}\right)(u \in V)$, *which has the following properties.*

(5.3.0) $\mathfrak{F}$ *is defined over* k_Ω.

(5.3.1) $\mathfrak{Q}_u$ *is of type* Ω *for every* $u \in V$.

(5.3.2) *For every* PEL-*structure* $\mathfrak{Q}$ *of type* Ω, *there exists one and only one point* u *of* V *such that* $\mathfrak{Q}$ *is isomorphic to* $\mathfrak{Q}_u$.

(5.3.3) $k_\Omega(u)$ *is the field of moduli of* $\mathfrak{Q}$ *for every* $u \in V$.

(5.3.4) V *and* W *are Zariski open subsets of projective varieties.*

(5.3.5) *There is a holomorphic mapping* φ *of* $\mathfrak{H}$ *onto* V *which induces a biregular isomorphism of* $\mathfrak{H}/\Gamma(T, \mathfrak{N}/\mathfrak{M})$ *onto* V *and such that, if* $u = \varphi(z)$ *with* $z \in \mathfrak{H}$ *then* $\mathfrak{Q}_u$ *is isomorphic to* $\mathfrak{Q}_z$ *defined in* 3.3.

5.4. THEOREM. *There exists only one fibre system (resp. fibre system defined over* k_Ω) *with the property* (5.3.2), *up to isomorphisms (resp. isomorphisms defined over* k_Ω) *in the sense of* 1.7.

5.5. THEOREM. *Let the notation and the assumption be the same as in* 5.3. *Let* $\mathfrak{F}' = \{V', W', h', f', Y', S'(a)\,(a \in \mathfrak{o}), f_1', \cdots, f_s'\}$ *be a fibre system of* PEL-*structures, with fibres* $\mathfrak{Q}_v'\,(v \in V')$, *and let* k *be a field of definition for* $\mathfrak{F}'$. *Suppose that at least one fibre* $\mathfrak{Q}_v'$ *is of type* Ω. *Then* $k_\Omega \subset k$, *and there exist a morphism* r *of* V' *to* V, *and a morphism* R *of* W' *to* W, *both defined over* k, *such that the restriction of* R *to* $\mathfrak{Q}_v'$ *gives an isomorphism of* $\mathfrak{Q}_v'$ *to* $\mathfrak{Q}_{r(v)}$.

Proofs will be given in the following 5.6–5.8.

5.6. Let $\mathfrak{F}$ be a fibre system obtained in 3.8, and let $\mathfrak{F}'$ be as in 5.5. By 4.1, every fibre $\mathfrak{Q}_v'$ is of type Ω. By (3.15.1), for every $v \in V'$, there exists one and only one point $r(v)$ of V such that $\mathfrak{Q}_v'$ is isomorphic to $\mathfrak{Q}_{r(v)}$. In this way we get a mapping $r\colon V' \to V$. By 4.1, for every $v \in V'$, there exists a neighborhood $\mathfrak{N}$ of v on V' and a continuous map $t \to z(t)$ of $\mathfrak{N}$ into $\mathfrak{H}$ such that $\mathfrak{A}'_t$ is isomorphic to $\mathfrak{A}_{z(t)}$ defined in 3.3. Since $\mathfrak{A}_{\varphi(z(t))}$ is isomorphic to $\mathfrak{A}_{z(t)}$, we have $r(t) = \varphi(z(t))$. This implies that r is continuous.

Let k be a field of rationality for $\mathfrak{F}$ and $\mathfrak{F}'$, which is finitely generated over $\mathbf{Q}$. Let K'_v denote the field of moduli of $\mathfrak{A}'_v$. Since $\mathfrak{A}'_v$ is defined over $k(v)$, we have $K'_v \subset k(v)$. K'_v is also the field of moduli of $\mathfrak{A}_{r(v)}$, since $\mathfrak{A}_{r(v)}$ is isomorphic to $\mathfrak{A}'_v$. Hence, by (3.15.2), $k(r(v)) = k \cdot K'_v \subset k(v)$ for every $v \in V'$.

Let y be a generic point of V' over k, and let Z^* be the locus of $y \times r(y)$ over k on $V' \times V$. Let w be another generic point of V' over k. Then there exists an isomorphism τ of $k(y)$ to $k(w)$ over k such that $y^\tau = w$. Then $\mathfrak{A}'_w = (\mathfrak{A}'_y)^\tau \cong (\mathfrak{A}_{r(y)})^\tau = \mathfrak{A}_{r(y)^\tau}$, hence $r(w) = r(y)^\tau$. It follows that $w \times r(w)$ is a generic specialization of $y \times r(y)$ over k. Therefore Z^* does not depend on the choice of y. Let $Z = \{(t, r(t)) \mid t \in V'\}$, and let $(t, r(t)) \in Z$. We can find a sequence $\{y_i\}$ of generic points of V' over k which converges to t in the sense of usual topology on the complex manifold V'. Since r is continuous, $\{(y_i, r(y_i))\}$ converges to $(t, r(t))$. Since $(y_i, r(y_i)) \in Z^*$ and Z^* is closed, we have $(t, r(t)) \in Z^*$, so that $Z \subset Z^*$. Now if (u, v) is a generic point of Z^* over k, then u is generic on V' over k, so that $v = r(u)$ in view of the above discussion, hence $(u, v) \in Z$. The set of all generic points of Z^* over k is dense in Z^* in the usual topology. Since Z is the graph of a continuous map, it follows that $Z = Z^*$. Since V' is non-singular, this shows that r is a morphism of V' into V.

Now there exists an isomorphism λ of $\mathfrak{A}'_y$ to $\mathfrak{A}_{r(y)}$. Since $\mathfrak{A}_{r(y)}$ has no automorphisms other than the identity mapping (cf. 3.7.4), λ is uniquely determined. Since $\mathfrak{A}_y$ and $\mathfrak{A}_{r(y)}$ are defined over $k(y)(\supset k(r(y)))$, the uniqueness of λ implies that λ is defined over $k(y)$. By 1.7, there exists a morphism R of W' into W whose restriction λ_v to $\mathfrak{A}'_v$ is an isomorphism of $\mathfrak{A}'_v$ to $\mathfrak{A}_{r(v)}$ for every $v \in V'$, and $\lambda_y = \lambda$.

5.7. $\mathfrak{F}$ being as the one in 3.8, let σ be an automorphism of C over k_Ω. We put

$$\mathfrak{F}^\sigma = \{V^\sigma, W^\sigma, h^\sigma, f^\sigma, Y^\sigma, S(a)^\sigma \, (a \in \mathfrak{o}); f_1^\sigma, \cdots, f_s^\sigma\}$$

and take $\mathfrak{F}^\sigma$ as the fibre system $\mathfrak{F}'$ in the discussion of 5.6. Then we find a morphism r of V^σ into V and a morphism R of W^σ into W whose restriction to $\mathfrak{A}'_v (v \in V^\sigma)$ is an isomorphism of $\mathfrak{A}'_v$ to $\mathfrak{A}_{r(v)}$. Now let $\mathfrak{A}$ be an arbitrary PEL-structure of type Ω. Then $\mathfrak{A}^{\sigma^{-1}}$ is of type Ω, so that $\mathfrak{A}^{\sigma^{-1}}$ is isomorphic to $\mathfrak{A}_u$ for exactly one point u of V. Then $\mathfrak{A}$ is isomorphic to $\mathfrak{A}'_v$ with $v = u^\sigma \in V^\sigma$. It is clear that v is uniquely determined by $\mathfrak{A}$. From this fact it follows that r is a one-to-one surjective mapping of V^σ to V. Since V is non-singular, this implies that r is biregular. By 1.7, R is a biregular morphism of W^σ onto W.

Let us write R^{-1} and r^{-1} as R_σ and r_σ. If τ is another automorphism of C over k_Ω, in view of (3.7.4) and (3.15.1), it is clear that

$$R_{\sigma\tau} = R_\sigma^\tau \circ R_\tau, \, r_{\sigma\tau} = r_\sigma^\tau \circ r_\tau \, .$$

Hence, by Weil [25], there exist varieties V_0 and W_0, defined over k_Ω, a biregular morphism r_0 of V_0 to V, and a biregular morphism R_0 of W_0 to W, such that $r_\sigma = r_0^\sigma \circ r_0^{-1}$ and $R_\sigma = R_0^\sigma \circ R_0^{-1}$. We may assume that V_0 and W_0 are Zariski open subsets of projective varieties. Put $h_0 = r_0^{-1} \circ h \circ R_0$, $f_0 = R_0^{-1} \circ f \circ r_0$. Then $\{V_0, W_0, h_0, f_0\}$ is clearly a fibre system of abelian varieties. We have

$$(5.7.1) \qquad h_0^\sigma = (r_0^\sigma)^{-1} \circ h^\sigma \circ R_0^\sigma = (r_0^\sigma)^{-1} \circ r_\sigma \circ h \circ R_\sigma^{-1} \circ R_0^\sigma = r_0^{-1} \circ h \circ R_0 = h_0 \, ,$$

so that h_0 is defined over k_Ω. It is shown similarly that f_0 is defined over k_Ω. Moreover, transferring Y, $S(a)$, f_i to the objects Y_0, $S_0(a)$, f_{0i} defined on $\{V_0, W_0, h_0, f_0\}$, we can define a fibre system of PEL-structures

$$\mathfrak{F}_0 = \{V_0, W_0, h_0, f_0, Y_0, S_0(a) \, (a \in \mathfrak{o}), f_{01}, \cdots, f_{0s}\} \, ,$$

so that (R_0, r_0) is an isomorphism of $\mathfrak{F}_0$ to $\mathfrak{F}$. By means of an argument similar to (5.7.1), we see easily that the $S_0(a)$ and the f_{0i} are defined over k_Ω. As for the divisor Y_0, the matter is not so simple. Let k be a field of rationality for $\mathfrak{F}$ and R_0, and let x be a generic point of V_0 over k. Let σ be an automorphism of C over $k_\Omega(x)$. Put $z = r_0(x)$. Then $z^\sigma = r_0^\sigma(x)$. Let λ be the restriction of R_0 to $h_0^{-1}(x)$, and $\mathcal{C}_{0x}$ the polarization of $h_0^{-1}(x)$ defined by Y_{0x}. Then λ is an isomorphism of the polarized abelian variety $(h_0^{-1}(x), \mathcal{C}_{0x})$ to $(A_z, \mathcal{C}_z)$, and λ^σ is an isomorphism of $(h_0^{-1}(x), \mathcal{C}_{0x}^\sigma)$ to $(A_z^\sigma, \mathcal{C}_z^\sigma)$, where $Y_z \in \mathcal{C}_z$. Since $R_\sigma = R_0^\sigma \circ R_0^{-1}$ gives an isomorphism of $(A_z, \mathcal{C}_z)$ to $(A_z^\sigma, \mathcal{C}_z^\sigma)$, and since λ^σ is the restriction of R_0^σ to $h_0^{-1}(x)$, we see that $\mathcal{C}_{0x}^\sigma = \mathcal{C}_{0x}$. Now we can find a divisor X in $\mathcal{C}_{0x}$, rational over an algebraic extension of $k_\Omega(x)$, since $h_0^{-1}(x)$ is defined over $k_\Omega(x)$. Every conjugate of X over $k_\Omega(x)$ should be contained in $\mathcal{C}_{0x}$ in view of the equality $\mathcal{C}_{0x}^\sigma = \mathcal{C}_{0x}$. Taking their sum, we obtain a divisor in $\mathcal{C}_{0x}$, rational over $k_\Omega(x)$. Hence, by virtue of 1.2, we can replace Y_0 by a divisor which is rational over k_Ω, without changing the polarization of each fibre. Since $\mathfrak{F}_0$ is isomorphic to $\mathfrak{F}$, this proves the existence of a fibre system satisfying $(5.3.0, 1, 2, 4, 5)$. With such a fibre system, we can take k_Ω in place of the field k of (3.15.2). Then we get (5.3.3) in view of (5.1.2). This completes the proof of 5.3.

5.8. As to 5.5, we obtain the inclusion $k_\Omega \subset k$ easily from the characterization (5.1.1) of k_Ω and 4.1. Now, in the discussion of 5.6, we can take $\mathfrak{F}$ to be the fibre system obtained in 5.3. Then the result of 5.6 proves 5.5. Furthermore, if $\mathfrak{F}'$ satisfies (5.3.2), then r is one-to-one and surjective, hence biregular. By (1.7), (R, r) is an isomorphism of $\mathfrak{F}'$ to $\mathfrak{F}$. This proves 5.4.

5.9. REMARK. The notation being as in 5.3, let x be a generic point of V over k_Ω. Then k_Ω is the algebraic closure of Q in the field of moduli of $\mathfrak{Q}_x$. Therefore k_Ω coincides with the field k in [19, II, Th. 4.5].

5.10. REMARK. We may repeat the remark of 3.14 for the fibre system $\mathfrak{F}$ of 5.3, since it is isomorphic to the one in 3.8. Namely there exists a holomorphic mapping ψ of $L_R^m \times \mathfrak{H}$ to W, which gives rise to a (canonical) biregular isomorphism of $(L_R^m \times \mathfrak{H})/(\mathfrak{M} \times \Gamma(T, \mathfrak{N}/\mathfrak{M}))$ to W.

5.11. REMARK. For a given Ω, it may happen that $\mathfrak{H}$ consists of only one point [18, p. 163, Remarks 1, 2]. In this case there is only one $\mathfrak{Q}$ of type Ω up to isomorphisms. Then k_Ω is the field of moduli of $\mathfrak{Q}$. The underlying abelian variety is a product of abelian varieties belonging to certain CM-types [18, Prop. 14, 15].

5.12. *Fibre systems of elliptic curves.* In the above theorems, we have excluded the case where $\dim(\mathfrak{H}) = 1$ and $\mathfrak{H}/\Gamma(T, \mathfrak{N}/\mathfrak{M})$ is not compact. In this case, one can verify that every member of Σ_Ω is a product of elliptic curves, and $\Gamma(T, \mathfrak{N}/\mathfrak{M})$ is conjugate to a group which is commensurable to $\mathrm{SL}_2(Z)$. Therefore, the investigation of such a family reduces essentially to the investigation of the family of elliptic curves, which we shall now treat briefly.

Let $\Omega = (L, \Phi, \rho; T, \mathfrak{M}; v_1, \cdots, v_s)$ be as follows: $L = Q$, $\Phi(a) = a$, $a^\rho = a$ for $a \in Q$, $\mathfrak{M} = Z^2 \subset Q^2$, $T((x, y), (x', y')) = x'y - xy'$; $v_1, \cdots, v_s \in Q^2$. Then $\mathfrak{H} = \{z \in C \mid \mathrm{Im}(z) > 0\}$, $\Gamma(T, \mathfrak{M}) = \mathrm{SL}_2(Z)$. The action of every element $\gamma = \begin{pmatrix} a & b \\ c & d \end{pmatrix}$ of $\mathrm{SL}_2(Z)$ on $\mathfrak{H}$ is defined by $\gamma(z) = (az + b)/(cz + d)$. Now we consider Weierstrass functions $\wp(u; \omega_1, \omega_2)$, $\wp'(u; \omega_1, \omega_2)$, $g_2(\omega_1, \omega_2)$, $g_3(\omega_1, \omega_2)$ ($u \in C$; $\mathrm{Im}(\omega_1/\omega_2) > 0$) satisfying

$$(5.12.1) \qquad \wp'^2 = 4\wp^3 - g_2\wp - g_3 .$$

Put, for $z \in \mathfrak{H}$,

$$d(z) = e^{\pi i z/6} \prod_{n=1}^{\infty} (1 - e^{2n\pi i z})^2 ,$$
$$r(z) = d(z)^{-4} g_2(z, 1), \qquad\qquad s(z) = d(z)^{-6} g_3(z, 1) ,$$
$$p(u, z) = d(z)^{-2}\wp(u; z, 1) , \qquad q(u, z) = d(z)^{-3}\wp'(u; z, 1) ,$$
$$\Gamma_N = \Gamma(T, N^{-1}\mathfrak{M}/\mathfrak{M}) = \{\alpha \in \mathrm{SL}_2(Z) \mid \alpha \equiv 1 \quad \mathrm{mod}\,(N)\} .$$

Then r, s are modular functions of level 12, and p, q are invariant under the transformation $(u, z) \to ((cz + d)^{-1}u, \gamma(z))$ for every $\gamma = \begin{pmatrix} a & b \\ c & d \end{pmatrix} \in \Gamma_{12}$. Let $D_z = Zz + Z$, and let A_z be an elliptic curve in projective plane P^2 defined by

$$X_0 X_2^2 = 4X_1^3 - r(z)X_0^2 X_1 - s(z)X_0^3 .$$

This is isomorphic to C/D_z, and to (5.12.1) with $z = \omega_1/\omega_2$, through the map $(\wp(u), \wp'(u)) \to (1, p(u, z), q(u, z))$. We take the point $(0, 0, 1)$ as the origin of A_z.

Let N be a positive multiple of 12, and φ be a holomorphic mapping of $\mathcal{H}$ into a projective space P^M which induces a biregular map of $\mathcal{H}/\Gamma_N$ onto a (non-complete) curve V. Since $A_{\gamma(z)} = A_z$ for $\gamma \in \Gamma_N$, we put $A'_u = A_z$ for $u = \varphi(z)$ with $z \in \mathcal{H}$. Put, for $v = (a, b) \in N^{-1}\mathbf{Z}^2$,

$$\mathfrak{f}(v, z) = \begin{cases} (0, 0, 1) & \text{if } v \in \mathbf{Z}^2, \\ \big((1,\, p((aZ + b)/N, z),\, q((az + b)/N, z)\big) & \text{if } v \notin \mathbf{Z}^2. \end{cases}$$

Then $\mathfrak{f}(v, z)$ is a point on A_z of order at most N, and the coordinates of $\mathfrak{f}(v, z)$ are modular functions of level N. Hence there exists a morphism $u \to f(v, u)$ of V into P^2 such that $f(v, \varphi(z)) = \mathfrak{f}(v, z)$. We find similarly a morphism c of V into P^2 such that $c(\varphi(z)) = (1, r(z), s(z))$. Put $W = \bigcup_{u \in V}(u \times A'_u)$. Since A'_u is defined over $\mathbf{Q}(c(u), f(0, u))$, we can repeat the proof of 3.10. Then we get a fibre system of PEL-structures of type Ω with the v_i such that $N^{-1}\mathbf{Z}^2 = \sum_i \mathbf{Z}v_i$. We may take the locus of $u \times f(0, u)$ as the divisor Y. Taking a quotient with respect to a finite group, as discussed in 3.13, we can remove the assumption that N is divisible by 12. Thus the assertion of 3.8 is true in the case $L = \mathbf{Q}, m = 2$, under the assumption (3.7.2). Then the assertions of Theorems 5.1, 5.3, 5.4, 5.5 can be verified in this case without any modification of the proofs, assuming (3.7.2) for the last three theorems.

6. Moduli of PEL-structures

6.1. Let $\mathfrak{Q} = (A, \mathcal{C}, \theta; t_1, \cdots, t_s)$ be a PEL-structures defined over a field k, and $\mathfrak{p}$ a discrete place of k. Suppose that A is without defect for $\mathfrak{p}$ in the sense of [21, §11,1]. Let A' be the abelian variety obtained from A by reduction modulo $\mathfrak{p}$. Then we get naturally a PEL-structure $\mathfrak{Q}' = (A', \mathcal{C}', \theta'; t'_1, \cdots, t'_s)$ defined over the residue field $\mathfrak{p}(k)$. Here $\theta'(a) = \mathfrak{p}(\theta(a))$, $t'_i = \mathfrak{p}(t_i)$, and $\mathcal{C}'$ is the polarization of A' containing the divisor $\mathfrak{p}(X)$ for a divisor X in $\mathcal{C}$ rational over k. Since X is non-degenerate, $\mathfrak{p}(X)$ is non-degenerate [21, p. 96, Prop. 14]. We call $\mathfrak{Q}'$ *the reduction of $\mathfrak{Q}$ modulo* $\mathfrak{p}$, and denote it by $\mathfrak{p}(\mathfrak{Q})$. When $\mathfrak{p}$ is an extension of a place $\mathfrak{p}'$ of a subfield k' of k, we call $\mathfrak{Q}'$ a *specialization of $\mathfrak{Q}$ over* $\mathfrak{p}'$. In particular if $\mathfrak{p}'(c) = c$ for every $c \in k'$, we call $\mathfrak{Q}'$ a *specialization of $\mathfrak{Q}$ over k.*

The divisor X being as above, take a positive integer g so that gX is ample on A and $g \cdot \mathfrak{p}(X)$ is ample on A'. Let $\mathfrak{L}$ (resp. $\mathfrak{L}'$) the set of functions, rational over k (resp. $\mathfrak{p}(k)$), whose divisors are $g(X)$ (resp. $g \cdot \mathfrak{p}(X)$). By Nishi [13], $\mathfrak{L}$ and $\mathfrak{L}'$ have the same dimension. By [21, pp. 86–87], we find a basis of $\mathfrak{L}$ over k, whose reduction modulo $\mathfrak{p}$ forms a basis of $\mathfrak{L}'$ over $\mathfrak{p}(k)$. Embedding $\mathfrak{Q}$ and $\mathfrak{Q}'$ into projective spaces by means of these bases, we obtain PEL-structures $\mathfrak{Q}_1$

741

and $\mathfrak{Q}'_1$ in projective spaces such that $\mathfrak{p}(\mathfrak{Q}_1) = \mathfrak{Q}'_1$, and $\mathfrak{Q}$ (resp. $\mathfrak{Q}'$) is isomorphic to $\mathfrak{Q}_1$ (resp. $\mathfrak{Q}'_1$). In this way we can always replace the specialization $\mathfrak{Q} \to \mathfrak{Q}'$ by a specialization of a PEL-structure defined in a projective space.

Let $\mathfrak{F}$ be a fibre system, defined over a field k, with fibres $\mathfrak{Q}_u$ ($u \in V$) as given in 1.5. Let u and v be points on V such that u is a specialization of v over k. Then it is clear that $(u, \mathfrak{Q}_u)$ is a specialization of $(v, \mathfrak{Q}_v)$ over k, i.e., $\mathfrak{p}(v) = u$ and $\mathfrak{p}(\mathfrak{Q}_v) = \mathfrak{Q}_u$ for a discrete place $\mathfrak{p}$ of $k(v)$ such that $\mathfrak{p}(c) = c$ ($c \in k$).

6.2. THEOREM. *Let $\Omega = (L, \Phi, \rho; T, \mathfrak{M}; v_1, \cdots, v_s)$ be a PEL-type. Suppose that either (3.7.1) is satisfied or $L = \mathbf{Q}$. Let k_Ω be the algebraic number field defined in 5.1. Then one can define a variety V, and assign, to every PEL-structure $\mathfrak{Q}$ of type Ω, exactly one point $\mathfrak{v}(\mathfrak{Q})$ of V so that the following conditions are satisfied.*

(6.2.1) V is defined over k_Ω, and everywhere normal.

(6.2.2) $\mathfrak{v}(\mathfrak{Q}) = \mathfrak{v}(\mathfrak{Q}')$ if and only if $\mathfrak{Q}$ is isomorphic to $\mathfrak{Q}'$.

(6.2.3) Let $\mathfrak{Q} = (A, C, \theta; t_1, \cdots, t_s)$ be a PEL-structure of type Ω, defined over a field K, and $\mathfrak{p}$ a C-valued discrete place of K. Suppose that A is without defect for $\mathfrak{p}$, and $\mathfrak{p}(c) = c$ for every $c \in k_\Omega$. Then $\mathfrak{p}(\mathfrak{Q})$ is of type Ω, $\mathfrak{v}(\mathfrak{Q})$ is rational over K, and $\mathfrak{p}(\mathfrak{v}(\mathfrak{Q})) = \mathfrak{v}(\mathfrak{p}(\mathfrak{Q}))$.

(6.2.4) $k_\Omega(\mathfrak{v}(\mathfrak{Q}))$ is the field of moduli of $\mathfrak{Q}$.

(6.2.5) There exists a holomorphic mapping φ of $\mathcal{H}$ onto V, which induces a biregular isomorphism of $\mathcal{H}/\Gamma(T, \mathfrak{N}/\mathfrak{M})$ onto V, and such that $\mathfrak{v}(\mathfrak{Q}_z) = \varphi(z)$ for a member $\mathfrak{Q}_z$ of Σ_Ω defined in 3.3.

(6.2.6) V is a Zariski open subset of a projective variety.

We shall prove this in the following 6.3–6.5.

6.3. Let $\mathfrak{o} = \{a \in L \mid a\mathfrak{M} \subset \mathfrak{M}\}$, and $\mathfrak{N} = \mathfrak{M} + \sum_{i=1}^{s} \mathfrak{o}v_i$. We can find a positive integer q so that $\mathfrak{N} \subset q^{-1}\mathfrak{M}$, and $\Gamma(T, q^{-1}\mathfrak{M}/\mathfrak{M})$ has no element of finite order other than the identity element. Let $\mathfrak{N}^* = q^{-1}\mathfrak{M} = \sum_{i=1}^{r} \mathfrak{o}v_i^*$, and $\Omega^* = (L, \Phi, \rho; T, \mathfrak{M}; v_1, \cdots, v_s, v_1^*, \cdots, v_r^*)$. Let $\mathfrak{F}^* = \{V^*, W, h, f, Y, S(a)$ $(a \in \mathfrak{o}), f_1, \cdots, f_s, f_1^*, \cdots, f_r^*\}$ be a fibre system of PEL-structures of type Ω^*, with the properties of 5.3. We denote the fibres by

$$\mathfrak{Q}_u^* = \big(A_u, C_u, \theta_u; f_1(u), \cdots, f_s(u), f_1^*(u), \cdots, f_r^*(u)\big) \qquad (u \in V^*)$$

and put $\mathfrak{Q}_u = (A_u, C_u, \theta_u; f_1(u), \cdots, f_s(u))$ $(u \in V^*)$. Then the $\mathfrak{Q}_u$ are all of type Ω, and every PEL-structure $\mathfrak{Q}$ of type Ω is isomorphic to $\mathfrak{Q}_u$ for some $u \in V^*$. Let φ^* be a holomorphic mapping of $\mathcal{H}$ onto V^* which induces an isomorphism of $\mathcal{H}/\Gamma(T, \mathfrak{N}^*/\mathfrak{M})$ to V^* as stated in (5.3.5). Let Δ be the set of all elements of $\Gamma(T, \mathfrak{N}/\mathfrak{M})$ which give the identity mapping on $\mathcal{H}$. Put $G = \Gamma(T, \mathfrak{N}/\mathfrak{M})/\Gamma(T, \mathfrak{N}^*/\mathfrak{M})\Delta$. Let $\gamma \in \Gamma(T, \mathfrak{N}/\mathfrak{M})$. Then $z \to \gamma(z)$ defines a biregular automorphism of V^*, which depends on only the class of γ modulo $\Gamma(T, \mathfrak{N}^*/\mathfrak{M})\Delta$.

In this way G may be regarded as a finite group of biregular automorphisms of V^*. From the characterization (5.1.1), we see easily that $k_\Omega \subset k_{\Omega^*}$. For simplicity, put $k = k_{\Omega^*}$. From [19, II, 1.11] and 5.9, we see that k is a finite Galois extension of k_Ω. Let x be a generic point of V^* over k. Let $g \in G$ and $g(x) = y$. Then there exists an isomorphism λ of $\mathfrak{A}_x$ to $\mathfrak{A}_y$ (corresponding to $\iota(\gamma, z)$ if $\varphi^*(z) = x$ and g is the class of γ^{-1}). Put $f_i' = \lambda f_i^*(x) \, (i = 1, \cdots, r)$. Then λ is an isomorphism of $\mathfrak{A}_x^*$ to $\big(A_y, \mathcal{C}_y, \theta_y; \{f_j(y), f_i'\}\big)$. Since $\mathfrak{N}^* = q^{-1}\mathfrak{M} = \sum_{i=1}^{r} \mathfrak{o}v_i^*$, we have

$$\sum_{i=1}^{r} \theta_u(\mathfrak{o})f_i^*(u) = \{t \in A_u \mid qt = 0\} \qquad\qquad (u \in V^*).$$

Hence

$$\begin{aligned}
\sum_{i=1}^{r} \theta_y(\mathfrak{o})f_i' &= \lambda[\{t \in A_x \mid qt = 0\}] = \{t \in A_y \mid qt = 0\} \\
&= \sum_{i=1}^{r} \theta_y(\mathfrak{o})f_i^*(y) \, .
\end{aligned}$$

It follows from this that $(A_y, \mathcal{C}_y, \theta_y; \{f_j(y), f_i'\})$ has the same field of moduli as $\mathfrak{A}_y^*$. Hence $\mathfrak{A}_x^*$ and $\mathfrak{A}_y^*$ have the same field of moduli. We have therefore $k(x) = k(y)$ in view of (5.3.3). This shows that the biregular automorphism g of V^* is defined over k. Let V be a quotient of V^* by G, and π the natural map $V^* \to V$, both defined over k. As remarked in (1.8.3), we may assume that V is normal and is a Zariski open subset of a projective variety. Put $\varphi = \pi \circ \varphi^*$. Then it is clear that φ is holomorphic and induces a biregular isomorphism of $\mathcal{H}/\Gamma(T, \mathfrak{N}/\mathfrak{M})$ onto V. Moreover, if $u = \varphi^*(z)$ with $z \in \mathcal{H}$, $\mathfrak{A}_u$ is isomorphic to the member $\mathfrak{A}_z$ of Σ_Ω (cf. 3.3). Hence, for two points u and v of V^*, $\mathfrak{A}_u$ is isomorphic to $\mathfrak{A}_v$ if and only if $\pi(u) = \pi(v)$.

6.4. Let $\mathfrak{A}$ be an arbitrary PEL-structure of type Ω. Then $\mathfrak{A}$ is isomorphic to $\mathfrak{A}_u$ for some $u \in V^*$. Put $\mathfrak{v}(\mathfrak{A}) = \pi(u)$. By the last remark of 6.3, we see that $\mathfrak{v}(\mathfrak{A})$ is uniquely determined by only the isomorphism-class of $\mathfrak{A}$. It is clear that $\mathfrak{v}(\mathfrak{A}^\sigma) = \mathfrak{v}(\mathfrak{A})^\sigma$ for every automorphism σ of $\boldsymbol{C}$ over k. This implies especially

(6.4.1) $k\big(\mathfrak{v}(\mathfrak{A})\big)$ *is the composite of* k *and the field of moduli of* $\mathfrak{A}$. It follows that $\mathfrak{v}(\mathfrak{A})$ is rational over every field of definition for $\mathfrak{A}$ containing k. Now we want to prove

(6.4.2) *Let* $\mathfrak{A}$ *and* $\mathfrak{p}$ *be as in* (6.2.3). *Suppose that* $k \subset K$, *and* $\mathfrak{p}(\alpha) = \alpha$ *for every* $\alpha \in k$. *Then* $\mathfrak{p}(\mathfrak{A})$ *is of type* Ω *and* $\mathfrak{p}\big(\mathfrak{v}(\mathfrak{A})\big) = \mathfrak{v}(\mathfrak{p}(\mathfrak{A}))$.

In view of the consideration in 6.1, we may assume that $\mathfrak{A}$ is defined in a projective space, and the polarization of $\mathfrak{A}$ is determined by hyperplane sections. We can find points $t_1^*, \cdots, t_r^*$ on A so that $\mathfrak{A}^* = (A, \mathcal{C}, \theta; t_1, \cdots, t_s, t_1^*, \cdots, t_r^*)$ is of type Ω^*. Let $K' = K(t_1^*, \cdots, t_r^*)$. Extend $\mathfrak{p}$ to a discrete place of K', and denote it again by $\mathfrak{p}$. Take a divisor X in $\mathcal{C}$, rational over K, and a basis $\{\varepsilon_1, \cdots, \varepsilon_\lambda\}$

of $\mathfrak{o}$ over Z. Let b, c and d_i be the Chow points of A, X and the graph of $\theta(\varepsilon_i)$, respectively. Let e be the origin of A, and let

$$a = (b,\, c,\, d_1,\, \cdots,\, d_\lambda,\, e,\, t_1,\, \cdots,\, t_s,\, t_1^*,\, \cdots,\, t_r^*)\,.$$

Then $k(a) \subset K'$, and $\mathfrak{Q}^*$ is defined over $k(a)$. Let $x = (a, \mathfrak{v}(\mathfrak{Q}))$ and $x' = \mathfrak{p}(x)$. Then x' is a specialization of x over k. If $\dim_k x = \dim_k x'$, there exists an isomorphism τ of $k(x)$ onto $k(x')$ over k such that $x^\tau = x'$. Since $\mathfrak{p}(a) = a^\tau$, we have $\mathfrak{p}(\mathfrak{Q}) = \mathfrak{Q}^\tau$, so that $\mathfrak{p}(\mathfrak{Q})$ is of type Ω, and $\mathfrak{p}(\mathfrak{v}(\mathfrak{Q})) = \mathfrak{v}(\mathfrak{Q})^\tau = \mathfrak{v}(\mathfrak{Q}^\tau) = \mathfrak{v}(\mathfrak{p}(\mathfrak{Q}))$. Let us therefore assume that $\dim_k x > \dim_k x'$. Then x' is a non-generic specialization of x over k. We can find a generic specialization y of x over k and a finitely generated extension k' of k such that x' is a specialization of y over k', and $k'(y)$ is a regular extension of k' of dimension one. (For a proof of this fact, see, for example, [23, first edition, p. 276, Prop. 7 and its proof].) Let U be the locus of y over k'. We can find a complete non-singular curve $\tilde{U}_1$, defined over k', with a generic point z over k' and a birational morphism $\psi \colon \tilde{U}_1 \to U$, defined over k', such that $\psi(z) = y$. Let σ be an isomorphism of $k(x)$ to $k(y)$ over k such that $x^\sigma = y$. Put

$$(\mathfrak{Q}^*)^\sigma = \big(A^\sigma,\, C^\sigma,\, \theta^\sigma;\, \{t_i^\sigma,\, t_j^{*\sigma}\}\big)\,, \qquad \mathfrak{p}(\mathfrak{Q}^*) = \big(A',\, C',\, \theta';\, \{t_i',\, t_j^{*\prime}\}\big)\,.$$

Since x' is a specialization of y over k', there exists a discrete place $\mathfrak{q}$ of $k'(y)$ such that $\mathfrak{q}(y) = x'$ and $\mathfrak{q}(\alpha) = \alpha$ for every $\alpha \in k'$. Since $x = (a, \mathfrak{v}(\mathfrak{Q}))$, we have

$$(6.4.3) \qquad\qquad \mathfrak{q}(a^\sigma) = \mathfrak{p}(a)\,, \qquad \mathfrak{q}\big(\mathfrak{v}(\mathfrak{Q})^\sigma\big) = \mathfrak{p}\big(\mathfrak{v}(\mathfrak{Q})\big)\,.$$

From our definition of a, we obtain $\mathfrak{q}\big((\mathfrak{Q}^*)^\sigma\big) = \mathfrak{p}(\mathfrak{Q}^*)$ and $\mathfrak{q}(\mathfrak{Q}^\sigma) = \mathfrak{p}(\mathfrak{Q})$. Since $\mathfrak{Q}^*$ is defined over $k(a)$, we see that $(\mathfrak{Q}^*)^\sigma$ is defined over $k'(y) = k'(z)$. Let w be a generic point of A^σ over $k'(z)$, and let $\tilde{W}_1$ be the locus of $w \times z$ over k' in $P^N \times \tilde{U}_1$, where P^N is the ambient space for A^σ. Then by [23, VIII$_3$, Th. 6],

$$(6.4.4) \qquad\qquad \tilde{W}_1 \cdot (P^N \times z) = A^\sigma \times z \qquad\qquad \text{on } P^N \times \tilde{U}_1.$$

Put $\mathfrak{q}(z) = u$. Then $\psi(u) = \mathfrak{q}(y) = x'$. Since $\tilde{W}_1$ is not contained in $P^N \times u$, $\tilde{W}_1$ and $P^N \times u$ intersect properly on $P^N \times \tilde{U}_1$. Applying $\mathfrak{q}$ to (6.4.4), we get $\tilde{W}_1 \cdot (P^N \times u) = A' \times u$. Let $\tilde{U}_2$ be the set of all points ξ of $\tilde{U}_1$ such that

$$\tilde{W}_1 \cdot (P^N \times \xi) = A'_\xi \times \xi$$

with an abelian variety A'_ξ. Since $\tilde{U}_1$ is a curve, we see easily that $\tilde{U}_2$ is a Zariski k'-open subset of $\tilde{U}_1$, containing u, and $A'_z = A^\sigma$, $A'_u = A'$. Take a hyperplane H in P^N, rational over k', which does not contain A'. Let $\tilde{U}$ be the set of all points ξ of $\tilde{U}_2$ such that H does not contain A'_ξ. Then $\tilde{U}$ is a Zariski k'-open subset of $\tilde{U}_2$, which contains u, since $A' = A'_u$. Let $\tilde{W} = \tilde{W}_1 \cap (P^N \times \tilde{U})$. Then we have again $\tilde{W} \cdot (P^N \times \xi) = A'_\xi \times \xi$ on $P^N \times \tilde{U}$ for every $\xi \in \tilde{U}$. By the converse part of [23, VI$_2$, Th. 6], every point of $A'_\xi \times \xi$ is simple on $\tilde{W}$. Hence

$\widetilde{W}$ is non-singular. Let $\tilde{h}$ be the projection map of $\widetilde{W}$ to $\widetilde{U}$, and Z the graph of $\tilde{h}$, i.e., the locus of $w \times z \times z$ over k'. Then $Z \cap (\widetilde{W} \times \xi) = A'_\xi \times \xi \times \xi$ for every $\xi \in \widetilde{U}$, and $Z \cdot (\widetilde{W} \times z) = A'_z \times z \times z$ on $\widetilde{W} \times \widetilde{U}$. Specializing z to ξ, we find $Z \cdot (\widetilde{W} \times \xi) = A'_\xi \times \xi \times \xi$ on $\widetilde{W} \times \widetilde{U}$ for every $\xi \in \widetilde{U}$. Let $\tilde{f}, \tilde{f}_1, \cdots, \tilde{f}_s, \tilde{f}_1^*, \cdots, \tilde{f}_r^*$ be the rational maps of $\widetilde{U}$ to $\widetilde{W}$, defined over k', such that $\tilde{f}(z) = e^\sigma \times z, \tilde{f}_i(z) = t_i^\sigma \times z$, $\tilde{f}_j^*(z) = t_j^{*\sigma}$. Since $\widetilde{U}$ is a non-singular curve, these rational maps are defined at every point ξ of $\widetilde{U}$ and take values in $A'_\xi \times \xi$. Then our discussion shows that $\{\widetilde{U}, \widetilde{W}, \tilde{h}, \tilde{f}\}$ is a fibre system of abelian varieties. Put $\widetilde{A}_\xi = \tilde{h}^{-1}(\xi) = A'_\xi \times \xi$ for every $\xi \in \widetilde{U}$. Put $\widetilde{Y} = \widetilde{W} \cdot (H \times \widetilde{U})$. By the same reasoning as in 3.11, we can verify that $\widetilde{Y} \cdot \widetilde{A}_\xi = (H \cdot A'_\xi) \times \xi$ for every $\xi \in \widetilde{U}$. Let $\tilde{\theta}_z(\alpha)$ (with $\alpha \in \mathfrak{o}$) be the endomorphism of $\widetilde{A}_z$ corresponding to $\theta^\sigma(\alpha)$ through an obvious isomorphism of $A^\sigma = A'_z$ to $\widetilde{A}_z = A'_z \times z$. We see that $\tilde{\theta}_z(\alpha)$ is defined over $k'(z)$. Let $\widetilde{S}(\alpha)$ be the locus of $w \times \tilde{\theta}_z(\alpha)w$ on $\widetilde{W} \times \widetilde{W}$. By virtue of 1.4, we obtain thus a fibre system of PEL-structures

$$\widetilde{\widetilde{\mathfrak{F}}} = \{\widetilde{U}, \widetilde{W}, \tilde{h}, \tilde{f}, \widetilde{S}(\alpha) \, (\alpha \in \mathfrak{o}), \tilde{f}_i, \tilde{f}_j^*\} \, .$$

Let $\widetilde{\widetilde{\mathfrak{Q}}}_\xi$ denote the fibre of $\widetilde{\widetilde{\mathfrak{F}}}$ standing on ξ. Then $\widetilde{\widetilde{\mathfrak{Q}}}_z$ is isomorphic to $(\mathfrak{Q}^*)^\sigma$. Since $\mathfrak{q}(z) = u$, we see that $\widetilde{\widetilde{\mathfrak{Q}}}_u = \mathfrak{q}(\widetilde{\widetilde{\mathfrak{Q}}}_z)$ is isomorphic to $\mathfrak{q}((\mathfrak{Q}^*)^\sigma) = \mathfrak{p}(\mathfrak{Q}^*)$.

Now, by 5.5, there exists a morphism j of $\widetilde{U}$ to the variety V^* of 6.3, defined over k', such that $\widetilde{\widetilde{\mathfrak{Q}}}_\xi$ is isomorphic to $\mathfrak{Q}_{j(\xi)}^*$. Then $\mathfrak{p}(\mathfrak{Q}^*)$ resp. $(\mathfrak{Q}^*)^\sigma$ is isomorphic to $\mathfrak{Q}_{j(u)}^*$ resp. $\mathfrak{Q}_{j(z)}^*$. Disregarding the f_j^*, we see that $\mathfrak{p}(\mathfrak{Q})$ resp. $\mathfrak{Q}^\sigma$ is isomorphic to $\mathfrak{Q}_{j(u)}$ resp. $\mathfrak{Q}_{j(z)}$. Hence $\mathfrak{p}(\mathfrak{Q})$ is of type Ω, and $\mathfrak{v}(\mathfrak{p}(\mathfrak{Q})) = \pi(j(u))$, $\mathfrak{v}(\mathfrak{Q}^\sigma) = \pi(j(z))$. Hence, by (6.4.3), we have $\mathfrak{p}(\mathfrak{v}(\mathfrak{Q})) = \mathfrak{q}(\mathfrak{v}(\mathfrak{Q}^\sigma)) = \mathfrak{q}(\pi(j(z))) = \pi(j(\mathfrak{q}(z))) = \pi(j(u)) = \mathfrak{v}(\mathfrak{p}(\mathfrak{Q}))$, which completes the proof of (6.4.2).

6.5. Let V and k be as in 6.3, and let σ be an automorphism of C over k_Ω. For a PEL-structure $\mathfrak{Q}$ of type Ω, we define a point $\mathfrak{v}^\sigma(\mathfrak{Q})$ of V^σ by $\mathfrak{v}^\sigma(\mathfrak{Q}) = \mathfrak{v}(\mathfrak{Q}^{\sigma^{-1}})^\sigma$. Let τ be another automorphism of C over k_Ω such that $\tau = \sigma$ on k. By a remark at the beginning of 6.4, we have $\mathfrak{v}(\mathfrak{Q}^{\tau^{-1}}) = \mathfrak{v}((\mathfrak{Q}^{\sigma^{-1}})^{\sigma\tau^{-1}}) = \mathfrak{v}(\mathfrak{Q}^{\sigma^{-1}})^{\sigma\tau^{-1}}$, hence $\mathfrak{v}^\tau(\mathfrak{Q}) = \mathfrak{v}^\sigma(\mathfrak{Q})$. In other words, $\mathfrak{v}^\sigma$ depends on only the effect of σ on k. We can define a one-to-one map r_σ of V^σ onto V by means of the relation $r_\sigma(\mathfrak{v}^\sigma(\mathfrak{Q})) = \mathfrak{v}(\mathfrak{Q})$. If K is the field of moduli of $\mathfrak{Q}$, then $K^{\sigma^{-1}}$ is the field of moduli of $\mathfrak{Q}^{\sigma^{-1}}$, so that, by (6.4.1), $K^{\sigma^{-1}} \cdot k = k(\mathfrak{v}(\mathfrak{Q}^{\sigma^{-1}}))$, hence $K \cdot k = k(\mathfrak{v}^\sigma(\mathfrak{Q}))$. It follows from this and (6.4.1) that $k(r_\sigma(v)) = k(v)$ for every $v \in V^\sigma$. We are going to prove that r_σ is a biregular morphism of V^σ onto V. To prove this, take a generic point x of $V^{*\sigma}$ over k, and let $\pi^\sigma(x) = y$. Let η be an arbitrary point of V^σ, and let (η, ζ) be a specialization of $(y, r_\sigma(y))$ over k. Extend this to a specialization $(x, y, r_\sigma(y)) \to (\xi, \eta, \zeta)$ over k. Since V is the quotient of V^* by G, we see that ξ is actually a point of $V^{*\sigma}$, and $\pi^\sigma(\xi) = \eta$. (For the moment, we do not know whether or not ζ is actually a point of V, since V may not be complete.) Let $\mathfrak{Q}^*$ and $\mathfrak{Q}^{*\prime}$ be the fibres of the fibre system $\mathfrak{F}^{*\sigma}$, standing on x and ξ, respectively. Let $\mathfrak{Q}$ resp. $\mathfrak{Q}'$ be the PEL-structure obtained from $\mathfrak{Q}^*$ resp. $\mathfrak{Q}^{*\prime}$ by removing the points corresponding to the $f_i^{*\sigma}$. Then they are of type Ω. We can find a discrete place $\mathfrak{p}$ of $k(x)$ such that $\mathfrak{p}(x, y, r_\sigma(y)) = (\xi, \eta, \zeta)$ and $\mathfrak{p}(\alpha) = \alpha$ for every $\alpha \in k$. Then $\mathfrak{p}(\mathfrak{Q}) = \mathfrak{Q}'$. Put $x^{\sigma^{-1}} = a$, $\xi^{\sigma^{-1}} = b$. Then a, b are points of V^*. Let $\mathfrak{Q}_a$ and $\mathfrak{Q}_b$ be the fibres defined in 6.3. Then $(\mathfrak{Q}_a)^\sigma = \mathfrak{Q}$, $(\mathfrak{Q}_b)^\sigma = \mathfrak{Q}'$, so that $\mathfrak{v}^\sigma(\mathfrak{Q}) = \mathfrak{v}(\mathfrak{Q}_a)^\sigma = \pi(a)^\sigma = \pi^\sigma(x) = y$, $\mathfrak{v}^\sigma(\mathfrak{Q}') = \mathfrak{v}(\mathfrak{Q}_b)^\sigma = \pi(b)^\sigma = \pi^\sigma(\xi) = \eta$. Hence $\mathfrak{v}(\mathfrak{Q}) = r_\sigma(\mathfrak{v}^\sigma(\mathfrak{Q})) = r_\sigma(y)$ and $\mathfrak{v}(\mathfrak{Q}') = r_\sigma(\mathfrak{v}^\sigma(\mathfrak{Q}')) = r_\sigma(\eta)$. By (6.4.2) we have $\zeta = \mathfrak{p}(r_\sigma(y)) = \mathfrak{p}(\mathfrak{v}(\mathfrak{Q})) = \mathfrak{v}(\mathfrak{p}(\mathfrak{Q})) = \mathfrak{v}(\mathfrak{Q}') = r_\sigma(\eta)$. Therefore ζ is uniquely determined by η and equals $r_\sigma(\eta)$. Since V^σ is normal, it follows from this that r_σ is a morphism of V^σ to V, defined over k. Moreover, since r_σ is one-to-one and V is normal, r_σ is biregular.

Let $G(k/k_\Omega)$ denote the Galois group of k over k_Ω. As remarked above, we can understand that r_σ is determined for an element σ of $G(k/k_\Omega)$. By virtue of the relation $r_\sigma(\mathfrak{v}^\sigma(\mathfrak{Q})) = \mathfrak{v}(\mathfrak{Q})$, we obtain easily $r_{\sigma\tau} = r_\tau \circ (r_\sigma)^\tau$ for any two elements σ and τ of $G(k/k_\Omega)$. By [25], we obtain a variety V_0, defined over k_Ω, and a biregular morphism r of V to V_0, defined over k, such that $r_\sigma = r^{-1} \circ r^\sigma$. We may assume that V_0 is a Zariski open subset of a projective variety. Put $\varphi_0 = r \circ \varphi$, $\mathfrak{v}_0(\mathfrak{Q}) = r(\mathfrak{v}(\mathfrak{Q}))$ for every $\mathfrak{Q}$ of type Ω. We are going to show that these V_0, $\mathfrak{v}_0$, φ_0 have the required properties of 6.2. By our construction, (6.2.1, 2, 5, 6) are all satisfied by $(V_0, \mathfrak{v}_0, \varphi_0)$.

Let τ be an automorphism of C over k_Ω. Then we have

$$\mathfrak{v}_0(\mathfrak{Q})^\tau = r(\mathfrak{v}(\mathfrak{Q}))^\tau = r^\tau(\mathfrak{v}^\tau(\mathfrak{Q}^\tau)) = r \circ r_\tau(\mathfrak{v}^\tau(\mathfrak{Q}^\tau)) = r(\mathfrak{v}(\mathfrak{Q}^\tau)) = \mathfrak{v}_0(\mathfrak{Q}^\tau) \ .$$

Since $\mathfrak{v}_0(\mathfrak{Q})$ depends on only the isomorphism-class of $\mathfrak{Q}$, this, combined with (5.1.2), proves (6.2.4) with $\mathfrak{v}_0$ in place of $\mathfrak{v}$.

Let $\mathfrak{Q}$, K and $\mathfrak{p}$ be as in (6.2.3). Since $\mathfrak{Q}$ is defined over K, K contains the field of moduli of $\mathfrak{Q}$, so that $\mathfrak{v}_0(\mathfrak{Q})$ is rational over K. By the consideration in 6.1, we may assume that $\mathfrak{Q}$ is defined in a projective space. As in the proof of (6.4.2), we take a collection d of the Chow points of (a finite number of) objects which determine $\mathfrak{Q}$ completely. Put $e = (d, \mathfrak{v}_0(\mathfrak{Q}))$, $e' = \mathfrak{p}(e)$. Then e' is a specialization of e over k_Ω. We can find a generic specialization g of e over k_Ω, which has e' as a specialization over k. Let σ be an isomorphism of $k_\Omega(e)$ to $k_\Omega(g)$ over k_Ω such that $e^\sigma = g$. Since $\mathfrak{Q}$ is defined over $k_\Omega(e)$, $\mathfrak{Q}^\sigma$ is meaningful. Let $\mathfrak{q}$ be a discrete place of $k(g)$ such that $\mathfrak{q}(g) = e'$ and $\mathfrak{q}(c) = c$ for every $c \in k$. Since $g = (d^\sigma, \mathfrak{v}_0(\mathfrak{Q})^\sigma)$, we have $\mathfrak{q}(d^\sigma) = \mathfrak{p}(d)$ and $\mathfrak{q}(\mathfrak{v}_0(\mathfrak{Q})^\sigma) = \mathfrak{p}(\mathfrak{v}_0(\mathfrak{Q}))$. Since σ is the identity mapping on k_Ω, $\mathfrak{Q}^\sigma$ is of type Ω. By (6.4.2), $\mathfrak{q}(\mathfrak{Q}^\sigma)$ is of type Ω, and $\mathfrak{v}(\mathfrak{q}(\mathfrak{Q}^\sigma)) = \mathfrak{q}(\mathfrak{v}(\mathfrak{Q}^\sigma))$. Since $\mathfrak{q}(d^\sigma) = \mathfrak{p}(d)$, we have $\mathfrak{q}(\mathfrak{Q}^\sigma) = \mathfrak{p}(\mathfrak{Q})$. Therefore $\mathfrak{p}(\mathfrak{Q})$

is of type Ω, and $\mathfrak{v}_0(\mathfrak{p}(\mathfrak{Q})) = \mathfrak{v}_0(\mathfrak{q}(\mathfrak{Q}^\sigma)) = r(\mathfrak{v}(\mathfrak{q}(\mathfrak{Q}^\sigma))) = r(\mathfrak{q}(\mathfrak{v}(\mathfrak{Q}^\sigma))) = \mathfrak{q}(r(\mathfrak{v}(\mathfrak{Q}^\sigma))) = \mathfrak{q}(\mathfrak{v}_0(\mathfrak{Q}^\sigma)) = \mathfrak{q}(\mathfrak{v}_0(\mathfrak{Q})^\sigma) = \mathfrak{p}(\mathfrak{v}_0(\mathfrak{Q}))$. This proves that $(V_0, \mathfrak{v}_0)$ satisfies (6.2.3) and completes the proof of 6.2.

6.6. REMARK. The property (6.2.3) includes, as a special case,

(6.6.1) $\mathfrak{v}(\mathfrak{Q}^\sigma) = \mathfrak{v}(\mathfrak{Q})^\sigma$ *for every automorphism σ of C over k_Ω.*
This, together with (5.1.2), implies (6.2.4).

6.7. THEOREM. *The notation and assumption being as in 6.2, let V' be a variety defined over an extension k' of k_Ω, and $\mathfrak{v}'$ an "assignment" which assigns a point $\mathfrak{B}'(\mathfrak{Q})$ of V' to every* PEL*-structure $\mathfrak{Q}$ of type Ω, satisfying the following conditions.*

(6.7.1) *If $\mathfrak{Q}_1$ and $\mathfrak{Q}_2$ are isomorphic, $\mathfrak{v}'(\mathfrak{Q}_1) = \mathfrak{v}'(\mathfrak{Q}_2)$.*

(6.7.2) *Let $\mathfrak{Q} = (A, C, \theta; t_1, \cdots, t_s)$ be a* PEL*-structure of type Ω defined over a field K containing k', and $\mathfrak{p}$ a C-valued discrete place of K. Suppose that A is without defect for $\mathfrak{p}$, and $\mathfrak{p}(c) = c$ for every $c \in k'$. Then $\mathfrak{v}'(\mathfrak{Q})$ is rational over K, and $\mathfrak{p}(\mathfrak{v}'(\mathfrak{Q})) = \mathfrak{v}'(\mathfrak{p}(\mathfrak{Q}))$.*

Then there exists a morphism j of V into V', defined over k', such that $\mathfrak{v}' = j \circ \mathfrak{v}$. Moreover, suppose that V' is normal over k', the converse of (6.7.1) is true, and every point of V' is of the form $\mathfrak{v}'(\mathfrak{Q})$. Then j is a biregular isomorphism of V onto V'.

PROOF. In view of (6.7.1), we can define a mapping j of V onto V' such that $j(\mathfrak{v}(\mathfrak{Q})) = \mathfrak{v}'(\mathfrak{Q})$. As remarked in 6.6, (6.7.2) implies (6.6.1) with k' in place of k_Ω. Hence, by (6.7.1), we have

(6.7.3) $k'(\mathfrak{v}'(\mathfrak{Q})) \subset k' \cdot$ (the field of moduli of $\mathfrak{Q}$) $= k'(\mathfrak{v}(\mathfrak{Q}))$,
so that $k'(j(w)) \subset k'(w)$ for every $w \in V$. Let V^*, $\mathfrak{F}^*$, $\mathfrak{Q}_u$ and k be as in 6.3. Our proof of 6.3–6.5 shows that there exists a morphism t of V^* onto V, defined over k, such that $\mathfrak{v}(\mathfrak{Q}_u) = t(u)$ for every $u \in V^*$. Put $k'' = k' \cdot k$. Let z be a generic point of V^* over k'', and let $x = t(z)$. Let y be an arbitrary point of V, and let (y, η, ζ) be a specialization of $(x, j(x), z)$ over k''. Since V may be regarded as a quotient of V^* by a finite group, ζ is actually a point of V^*. Since $\mathfrak{Q}_z$ is defined over $k(z)$, we have $k_\Omega(x) \subset k(z)$. We can find a discrete place $\mathfrak{p}$ of $k''(z)$ such that

$$\mathfrak{p}(x, j(x), z) = (y, \eta, \zeta)$$

and $\mathfrak{p}(c) = c$ for every $c \in k''$. Then we have $\mathfrak{p}(\mathfrak{Q}_z) = \mathfrak{Q}_\zeta$, so that $y = \mathfrak{p}(x) = \mathfrak{p}(\mathfrak{v}(\mathfrak{Q}_z)) = \mathfrak{v}(\mathfrak{p}(\mathfrak{Q}_z)) = \mathfrak{v}(\mathfrak{Q}_\zeta)$ and $\eta = \mathfrak{p}(j(x)) = \mathfrak{p}(\mathfrak{v}'(\mathfrak{Q}_z)) = \mathfrak{v}'(\mathfrak{p}(\mathfrak{Q}_z)) = \mathfrak{v}'(\mathfrak{Q}_\zeta) = j(y)$. Therefore η is uniquely determined by y and equal to $j(y)$. Since V is normal and $k'(x) \supset k'(j(x))$, this shows that j is a morphism of V into V', defined over k'.

If the converse of (6.7.1) is true, then j is injective. If further V' is normal, then j is biregular. This proves the last assertion.

6.8. REMARK. In view of 6.7, the couple $(V, \mathfrak{b})$ of 6.2 is uniquely determined, up to biregular isomorphisms over k_Ω, by the properties (6.2.1–3) and the property that every point of V is of the form $\mathfrak{b}(\mathfrak{Q})$. The last condition is insured by (6.2.5). It is also easy to see that $(V, \mathfrak{b})$ can be characterized by (6.2.1, 4, 5).

7. The number field k_Ω and concluding remarks.

7.1. All the algebras L over $\boldsymbol{Q}$ with positive involution are classified into the following four types.

(*Type* I). Totally real algebraic number field F.

(*Type* II). Quaternion algebra L over F such that the simple components of $L \otimes_Q R$ are all isomorphic to $M_2(\boldsymbol{R})$.

(*Type* III). Quaternion algebra L over F such that the simple components of $L \otimes_Q R$ are all isomorphic to the division ring of real quaternions.

(*Type* IV). Normal simple algebra over a totally imaginary quadratic extension K of F, with an involution of the second kind.

This result is due to Albert. For a simple proof and the possibility for positive involutions, see [18, § 1].

Let L be an algebra of the above types, and let $\Omega = (L, \Phi, \rho; T, \mathfrak{M}; v_1, \cdots, v_s)$ be a PEL-type. We remind the reader that the direct sum of Φ and its complex conjugate should be equivalent to a rational representation. For the first three types, this means that Φ is equivalent to a multiple of a reduced representation of L with respect to $\boldsymbol{Q}$. For this and the description of Φ in the case of (*Type* IV), we refer to [18, 2.1].

Let K' be the field generated over $\boldsymbol{Q}$ by the $\operatorname{tr}(\Phi(\alpha))$ for all α in the center of L. Then we obtain

(*Type* I, II, III). $K' = \boldsymbol{Q}$,

(*Type* IV). $K' = \begin{cases} \boldsymbol{Q} \text{ if } \Phi \text{ is equivalent to its complex conjugate,} \\ \text{a totally imaginary quadratic extension of a} \\ \text{totally real algebraic number field } F'', \text{ otherwise.} \end{cases}$

For (*Type* I, II, III), this is obvious. For an algebra L of (*Type* IV), a proof of this fact is given in [19, I, 4.3] in the case $L = K$. The proof applies to the general case.

Let k_Ω be the field determined in 5.1. By (5.1.3), $K' \subset k_\Omega$. In many cases we verified that k_Ω is an abelian extension of K'. One may conjecture that this is true in general, except for the case where $\dim(\mathfrak{K}) = 0$ and L is of

(*Type* III), i.e., the case discussed in [18, Prop. 15]. We shall now summarize the known facts and make a few comments.

7.2. Let F be a totally real algebraic number field, and $\mathfrak{g}$ the ring of integers in F. If $L = F$, we obtain [19, II, 7.2]:

(7.2.1) *If* $\mathfrak{g}\mathfrak{M} = \mathfrak{M}$ *and* $\mathfrak{M} + \sum_{i=1}^{s} \mathfrak{g} v_i = \mathfrak{a}^{-1}\mathfrak{M}$ *for an integral ideal* $\mathfrak{a}$ *of* $\mathfrak{g}$, *then* $k_\Omega = \boldsymbol{Q}(e^{2\pi i/b})$, *with a positive integer* b *such that* $\mathfrak{a} \cap \boldsymbol{Z} = b\boldsymbol{Z}$.

Let L be an algebra of (*Type* II) with F as center, and $\mathfrak{o}$ a maximal order in L. Then [19, II, 6.8] gives

(7.2.2.) *If* $\mathfrak{o}\mathfrak{M} = \mathfrak{M}$ *and* $\mathfrak{M} + \sum_{i=1}^{s} \mathfrak{o} v_i = \mathfrak{a}^{-1}\mathfrak{M}$ *for an integral two-sided* $\mathfrak{o}$*-ideal* $\mathfrak{a}$, *then* $k_\Omega = \boldsymbol{Q}(e^{2\pi i/b})$ *with a positive integer* b *such that* $\mathfrak{a} \cap \boldsymbol{Z} = b\boldsymbol{Z}$.

In both cases, if $\mathfrak{a} = \mathfrak{g}$ or $\mathfrak{a} = \mathfrak{o}$, we get $k_\Omega = \boldsymbol{Q}$ (cf. also [19, I, §3].). It is easy to derive, from these results, that k_Ω is abelian over $\boldsymbol{Q}$, without the assumptions of (7.2.1) or (7.2.2), so far as L is of (*Type* I or II). This can be done by combining Remark 5.9 with [19, II, 1.12].

The field k_Ω is not necessarily the *smallest* field of definition for the field of automorphic functions with respect to $\boldsymbol{Q}$. As shown in [19, II, 6.10–11, 7.3], $\boldsymbol{Q}$ can be taken as a field of definition for the function field, under the assumptions of (7.2.1) or (7.2.2); in the latter case, one has to assume further that $\mathfrak{a}$ is prime to the discriminant of L over F. In these cases, one can construct a fibre system of abelian varieties, defined over $\boldsymbol{Q}$, by means of an argument similar to [10, 7.4]. But we shall not discuss the problem here.

7.3. Let L be an algebra of (*Type* IV), and K the center of L. Let $\mathfrak{r}$ be the ring of integers in K. In [19] we have shown, in the case $\dim(\mathcal{H}) > 0$,

(7.3.1) *If* $L = K$, $\mathfrak{r}\mathfrak{M} = \mathfrak{M}$ *and* $\mathfrak{c}^{-1}\mathfrak{M} = \mathfrak{M} + \sum_{i=1}^{s} \mathfrak{r} v_i$ *with an integral ideal* $\mathfrak{c}$ *in* F, *then* k_Ω *is an abelian extension of* K'.

Moreover, the law of reciprocity in the extension k_Ω/K' can be described in terms of $(T, \mathfrak{M}, \{v_i\})$. One may notice that the result is true also in the case $\dim(\mathcal{H}) = 0$, as asserted by the theory of complex multiplication of abelian varieties [21]. Now it seems possible to generalize, without great difficulty, these results to the case where L is not necessarily commutative. This will be done first by investigating the classes and genera of a unitary group of T; and, secondly, by following the argument of [19]. As for the first point, one needs a (strong) approximation theorem in the (special) unitary group, in order to determine the classes in a genus. Such a theorem has been proved by Kneser in a more general case (unpublished). As for the second point, the proof in [19] will be greatly simplified by the results in the present paper. In fact, the varieties $\mathcal{V}(T, \mathfrak{M})$ and $\mathcal{V}(T, \mathfrak{M}, \{v_i\})$ of [19] have been obtained in a certain constructive way; they are characterized birationally over k_Ω, but not biregularly. We may replace them by the varieties V (or even by fibre systems) given in 5.3 and 6.2, which are characterized biregularly over k_Ω. This replacement will also improve the formulation of the results.

7.4. By introducing a certain real analytic structure into the members of the family Σ_Ω, we get a subfamily which is parametrized again by a certain bounded symmetric domain. In this case we can not apply the argument of the proof of 5.1, in order to find an algebraic number field like k_Ω. However, one is still able to handle the family number-theoretically, by choosing sufficiently many special members. We hope to discuss this problem in a subsequent paper.

Kuga and Satake have investigated, in great generality, a family of abelian varieties parametrized by the points of a bounded symmetric domain $\mathcal{H}$ modulo an arithmetically defined discontinuous group Γ. All the possible types of symmetric domains (and probably all the possible types of discontinuous groups) have been determined by Satake. On the other hand, Kuga proved that if $\mathcal{H}/\Gamma$ is compact, one can construct a fibre system of abelian varieties $\{V, W, h, f\}$ in the sense of 1.1 with $V = \mathcal{H}/\Gamma$ and a projective variety W, over the complex number field. We notice that our proof of the construction of a fibre system in 3.8 is obviously applicable to such a general case, regardless of the compactness of $\mathcal{H}/\Gamma$. In fact, to get a fibre system *over the complex number field*, we do not need any algebro-geometrical characterization of the members.

7.5. Theorems 5.3 and 6.2 show the existence of meromorphic functions on $\mathcal{H}$ whose values at z generate the field of moduli of $\mathfrak{Q}_z$. Now the abelian varieties belonging to a CM-type in the sense of [21] occur as a member $\mathfrak{Q}_z$ with a fixed point z of a *modular correspondence* on $\mathcal{H}$. Then a class-field can be generated by the values of those meromorphic functions at z. For example, if one takes L to be a totally real algebraic number field F and m to be 2, then $\mathcal{H}$ is a product of the usual upper half planes, and the functions are the so-called Hilbert modular functions. It is easy to determine special members in this particular case, and formulate the result of [21] from this view-point. However, it is not trivial, and actually a complicated task to determine special members in the general case. As some examples indicate (cf. for example, [19, I, § 4], [20, § 4]), there are some interesting features in this problem, and it will be worth while working it out. One may also notice an implicit connection between this question and the characterization, if any, of a family of abelian varieties which are not determined only by a PEL-type.

PRINCETON UNIVERSITY

REFERENCES

1. W. L. BAILY, JR., *On the theory of θ-functions, the moduli of abelian varieties, and the moduli of curves*, Ann. of Math., 75 (1962), 342-381.
2. ———, *On the moduli of abelian varieties with multiplications*, J. Math. Soc. Japan, 15 (1963), 367-386.
3. ———, and A. BOREL, *On the compactification of arithmetically defined quotients of bounded symmetric domains*, Bull. Amer. Math. Soc., 70 (1964), 588-593.
4. J. IGUSA, *Arithmetic variety of moduli for genus two*, Ann. of Math., 72 (1960), 612-649.
5. ———, *On the graded ring of theta constants*, Amer. J. Math., 86 (1964), 219-246.
6. K. KODAIRA and D. C. SPENCER, *On deformations of complex analytic structures: I-II*, Ann. of Math., 67 (1958), 328-466.
7. S. KOIZUMI, *On specialization of the Albanese and Picard varieties*, Mem. Coll. Sci. Univ. of Kyoto, Ser. A. 32 (1960), 371-382.
8. A. KRAZER und F. PRYM, Neue Grundlagen einer Theorie der Allgemeinen Thetafunktionen, Teubner, Leipzig, 1892.
9. M. KUGA, *Fibre varieties over a symmetric space whose fibres are abelian varieties*, lecture note, 1963-64 Univ. of Chicago, (1965).
10. ———, and G. SHIMURA, *On the zeta function of a fibre variety whose fibres are abelian varieties*, Ann. of Math., 82 (1965), 478-539.
11. T. MATSUSAKA, *Polarized varieties, fields of moduli and generalized Kummer varieties of polarized abelian varieties*, Amer. J. Math., 80 (1958) 45-82.
12. D. MUMFORD, " Projective invariants of projective structures and applications ", Proc. Int. Cong. Math., 1962, Stockholm, (1963), 526-530.
13. M. NISHI, *Some results on abelian varieties*, Natural Science Report, Ochanomizu Univ., 9 (1958), 1-12.
14. Séminaire H. CARTAN, 1957/1958, Fonctions automorphes, Paris, 1958.
15. J. -P. SERRE, Groupes algébriques et corps de classes, Hermann, Paris, 1959.
16. G. SHIMURA, *Reduction of algebraic varieties with respect to a discrete valuation of the basic field*, Amer. J. Math., 77 (1955), 134-176.
17. ———, *On the theory of automorphic functions*, Ann. of Math. 70 (1959), 101-144.
18. ———, *On analytic families of polarized abelian varieties and automorphic functions*, Ann. of Math. 78 (1963), 149-192.
19. ———, *On the field of definition for a field of automorphic functions: I. II*, Ann. of Math., 80 (1964), 160-189, 81 (1965), 124-165.
20. ———, *On the class-fields obtained by complex multiplication of abelian varieties*, Osaka Math. J., 14 (1962), 33-44.
21. ———, and Y. TANIYAMA, Complex multiplication of abelian varieties and its applications to number theory, Publ. Math. Soc. Japan, No. 6, Tokyo, 1961.
22. C. L. SIEGEL, *Moduln Abelscher Funktionen*, Nach. Akad. Wiss. Göttingen, 1963, No. 25, 365-427.
23. A. WEIL, Foundations of Algebraic Geometry, IInd edition, 1962.
24. ———, Variétés abéliennes et courbes algébriques, Hermann, Paris, 1948.
25. ———, *The field of definition of a variety*, Amer. J. Math., 78 (1956), 509-524.
26. ———, Introduction à l'étude des variétés Kählériennes, Hermann, Paris, 1958.
27. ———, *Sur certains groupes d'opérateurs unitaires*, Acta Math., 111 (1964), 143-211.

(Received April 12, 1965)

On the field of definition for a field of automorphic functions: III

Annals of Mathematics, 83 (1966), 377-385

The purpose of this paper is to complement the main theorems of the previous papers with the same title, referred to hereafter as [FI] and [FII], and to simplify the proofs. In [FI, II], it was shown that a certain abelian extension k of a certain algebraic number field K' can be determined by a family $\Sigma(\Omega)$ of abelian varieties belonging to a given type Ω of structures. The structures consist of endomorphism algebra, polarization and points of finite order, of an abelian variety. The case of structures without points of finite order was treated in [FI]. We observed that k is unramified over K', and the law of reciprocity for the extension k/K' can be described in terms of the inter-relation between Ω and its "transform". One of the aims of [FII] was to extend this result to the general case involving points of finite order and ramified abelian extensions. However, we determined only the ideal group in K' corresponding to k, but did not give an explicit reciprocity law in such a general case. This will be done in the present paper. We shall also reformulate the results in a more comprehensible manner, and give a simplified proof by means of the characterization of the field k provided by [2, 5.1]. This characterization enables us to dispense with artificial considerations which were necessary in the original proofs. We devote one section to clarify a point concerning the existence of a special member of the family $\Sigma(\Omega)$, of which the previous proof in [FI] was not with complete clarity.

List of symbols. They were introduced in [FI, II], occasionally with a somewhat general meaning.

F : an algebraic number field of finite degree, assumed to be totally real in §§ 2, 3.

$g = [F : Q]$.

K : a quadratic extension of F, assumed to be totally imaginary in §§ 2, 3.

ρ : the non-trivial automorphism of K over F.

V : a vector space over K of finite dimension.

m : the dimension of V over K.

$\mathfrak{g}$: the ring of integers in F.

$\mathfrak{r}$: the ring of integers in K.

$\{\tau_1, \cdots, \tau_g\}$:   a set of isomorphisms of K into C whose restrictions to F are distinct.

Φ :   a representation of K by complex matrices of size mg such that $\mathrm{tr}(\Phi(a)) = \sum_{\nu=1}^{g}(r_\nu a^{\tau_\nu} + s_\nu a^{\rho\tau_\nu})$ with non-negative integers r_ν and s_ν satisfying $r_\nu + s_\nu = m$ for every ν.

K' :   the field generated over $\boldsymbol{Q}$ by the numbers $\mathrm{tr}(\Phi(a))$ for all $a \in K$.

T :   a ρ-anti-hermitian form on V satisfying the condition [FI, (5.1.1)].

1. The equivalence of frames with respect to a unitary group

1.1. Throughout this section we shall use the same notation and terminology as in [FII, § 8] and [3], except for the definition of genus of $\mathfrak{c}$-frames, or oriented $\mathfrak{c}$-frames, which will be modified slightly.

Let S be a non-degenerate ρ-hermitian form on V. We define the group $G(S)$ of similitudes of S, the unitary group $U(S)$ of S and their localizations $G_\mathfrak{p}(S)$, $U_\mathfrak{p}(S)$ for every prime ideal $\mathfrak{p}$ of F [FII, 8.5].

Suppose that m is even. Let $\mathfrak{c}$ be an integral ideal in F, and Λ a genus of $\mathfrak{r}$-lattices in V with respect to $G(S)$. Let J denote the set of all ideals $\mathfrak{a}$ in K such that $N_{K/F}(\mathfrak{a}) = \mathfrak{g}$. Define $D(\Lambda; \mathfrak{c})$ and $f(a)$ as in [FII, 8.6] and put

$$B(\Lambda; \mathfrak{c}) = J \times D(\Lambda; \mathfrak{c}) ,$$
$$B_0(\Lambda; \mathfrak{c}) = \{(\mathfrak{r}a, f(a)) \in J \times D(\Lambda; \mathfrak{c}) \mid a \in K, aa^\rho = 1\} .$$

We regard these sets as multiplicative groups in an obvious way.

Let $\mathfrak{L}^* = (\mathfrak{L}; x_1, \cdots, x_m)$ and $\mathfrak{M}^* = (\mathfrak{M}; y_1, \cdots, y_m)$ be $\mathfrak{c}$-frames [FII, 8.7]. We assume that $\mathfrak{L}$ and $\mathfrak{M}$ belong to Λ. Now we modify the definition of genus of $\mathfrak{L}^*$ as follows. The $\mathfrak{c}$-frames $\mathfrak{L}^*$ and $\mathfrak{M}^*$ are said to belong to the same *genus*, if, for every prime ideal $\mathfrak{p}$ in F, there exists an element $\alpha_\mathfrak{p}$ of $G_\mathfrak{p}(S)$ such that $\mu(\alpha_\mathfrak{p})$ is a unit of $\mathfrak{g}_\mathfrak{p}$, $\mathfrak{L}_\mathfrak{p}\alpha_\mathfrak{p} = \mathfrak{M}_\mathfrak{p}$ and $x_i\alpha_\mathfrak{p} \equiv y_i \bmod \mathfrak{M}_\mathfrak{p}$ for every i. The only difference from the original definition of [FII, 8.7] is the additional condition on $\mu(\alpha_\mathfrak{p})$.

By [3, 5.21, (ii)], we have $[\mathfrak{L}/\mathfrak{M}] \in J$. Denote by $[\mathfrak{L}^*/\mathfrak{M}^*]$ the element of $B(\Lambda; \mathfrak{c})/B_0(\Lambda; \mathfrak{c})$ represented by $([\mathfrak{L}/\mathfrak{M}], d(\mathfrak{L}^*/\mathfrak{M}^*))$, where $d(\mathfrak{L}^*/\mathfrak{M}^*)$ is the element of $D(\Lambda; \mathfrak{c})$ defined in [FII, 8.7]. We can easily verify that

$$[\mathfrak{L}^*/\mathfrak{M}^*][\mathfrak{M}^*/\mathfrak{N}^*] = [\mathfrak{L}^*/\mathfrak{N}^*]$$

for three $\mathfrak{c}$-frames $\mathfrak{L}^*$, $\mathfrak{M}^*$, $\mathfrak{N}^*$ belonging to the same genus.

We say that $\mathfrak{L}^*$ and $\mathfrak{M}^*$ are *equivalent* if there exists an element α of $U(S)$ such that $\mathfrak{L}\alpha = \mathfrak{M}$ and $x_i\alpha \equiv y_i \bmod \mathfrak{M}$ for every i. Now [FII, 8.8] asserts that, when S is indefinite,

(1.1.1) $\mathfrak{L}^*$ *and* $\mathfrak{M}^*$ *are equivalent if and only if* $[\mathfrak{L}^*/\mathfrak{M}^*] = 1$.

1.2. Let us now consider the case of odd m; put $m = 2l + 1$. For an integral ideal c in F, let $B(c)$ denote the set of all pairs (a, b) formed by an ideal a in K and an element b of F such that $aa^\rho = rb$ and a is prime to c. We consider $B(c)$ as a group by the law of multiplication $(a, b)(a', b') = (aa', bb')$, and define a subgroup $B_0(c)$ of $B(c)$ by

$$B_0(c) = \{(ru, uu^\rho) \mid u \in K, u \neq 0, u - 1 \in c_\mathfrak{p} \text{ for every } \mathfrak{p} \mid c\} \,.$$

(If we restrict (a, b) to the pairs with totally positive b, the factor group $B(c)/B_0(c)$ is the same as the group $\mathfrak{C}(K; c)$ introduced in [5, 16.1].)

Let $\mathfrak{L}^* = (\mathfrak{L}; a; x_1, \cdots, x_m)$, $\mathfrak{M}^* = (\mathfrak{M}; b; y_1, \cdots, y_m)$ be oriented c-frames [FII, 8.17]. We say that $\mathfrak{L}^*$ and $\mathfrak{M}^*$ belong to the same *genus*, if, for every prime ideal $\mathfrak{p}$ in F, there exists an element $\alpha_\mathfrak{p}$ of $G_\mathfrak{p}(S)$ such that $a^{-1}b\mu(\alpha_\mathfrak{p})$ is a unit of $g_\mathfrak{p}$, $\mathfrak{L}_\mathfrak{p}\alpha_\mathfrak{p} = \mathfrak{M}_\mathfrak{p}$, and $x_i\alpha_\mathfrak{p} \equiv y_i \bmod \mathfrak{M}_\mathfrak{p}$ for every i. Here again the difference from the original definition is the condition on $\mu(\alpha_\mathfrak{p})$. Now, by the approximation theorem in number fields, we can find an element v of K so that $v^{-1}\mu(\alpha_\mathfrak{p})^{-l} \det(\alpha_\mathfrak{p}) \equiv 1 \bmod c_\mathfrak{p}$ for every prime ideal $\mathfrak{p}$ dividing c. Then we have $vr_\mathfrak{p} = (a^{-1}b)^l \det(\alpha_\mathfrak{p})r_\mathfrak{p} = (a^{-1}b)^l[\mathfrak{L}/\mathfrak{M}]_\mathfrak{p}$ for every such $\mathfrak{p}$, hence $v^{-1}(a^{-1}b)^l[\mathfrak{L}/\mathfrak{M}]$ is an ideal in K prime to c. Since $N_{K/F}([\mathfrak{L}/\mathfrak{M}])_\mathfrak{p} = \mu(\alpha_\mathfrak{p})^m g_\mathfrak{p} = (ab^{-1})^m g_\mathfrak{p}$ for any $\mathfrak{p}$, we have $N_{K/F}([\mathfrak{L}/\mathfrak{M}]) = (ab^{-1})^m g$. Therefore we see that

$$(1.2.1) \qquad\qquad \left(v^{-1}(a^{-1}b)^l[\mathfrak{L}/\mathfrak{M}], (vv^\rho)^{-1}ab^{-1}\right)$$

is an element of $B(c)$. Furthermore, the change of v amounts to the multiplication by an element of $B_0(c)$. Hence the class of (1.2.1) modulo $B_0(c)$ is uniquely determined by only $\mathfrak{L}^*$ and $\mathfrak{M}^*$. We denote by $[\mathfrak{L}^*/\mathfrak{M}^*]$ this element of $B(c)/B_0(c)$ represented by (1.2.1). This symbol is obviously multiplicative:

$$[\mathfrak{L}^*/\mathfrak{M}^*][\mathfrak{M}^*/\mathfrak{N}^*] = [\mathfrak{L}^*/\mathfrak{N}^*] \,.$$

We say that $\mathfrak{L}^*$ and $\mathfrak{M}^*$ are *equivalent* if there exists an element α of $G(S)$ such that $\mathfrak{L}\alpha = \mathfrak{M}$, $\mu(\alpha)b = a$, $x_i\alpha \equiv y_i \bmod \mathfrak{M}$ for every i. Then, assuming S to be indefinite, and $m > 1$, we obtain

(1.2.2) $\mathfrak{L}^*$ *and* $\mathfrak{M}^*$ *are equivalent if and only if* $[\mathfrak{L}^*/\mathfrak{M}^*] = 1$.

The *only if* part can be seen by putting $v = \mu(\alpha)^{-l} \det(\alpha)$. Conversely, if $[\mathfrak{L}^*/\mathfrak{M}^*] = 1$, the notation $\alpha_\mathfrak{p}$ and v being as above, there exists an element u of K such that $u - 1 \in c_\mathfrak{p}$ for every $\mathfrak{p} \mid c$ and $v^{-1}(a^{-1}b)^l[\mathfrak{L}/\mathfrak{M}] = ru$, $(vv^\rho)^{-1}(ab^{-1}) = uu^\rho$. Then $(a^{-1}b)^l[\mathfrak{L}/\mathfrak{M}] = ruv$, $ab^{-1} = (uv)(uv)^\rho$, and $(uv)^{-1}\mu(\alpha_\mathfrak{p})^{-l} \det(\alpha_\mathfrak{p}) \equiv 1 \bmod c_\mathfrak{p}$ for every prime factor $\mathfrak{p}$ of c. Therefore from [FII, 8.18] we get the *if* part of (1.2.2).

2. A supplementary consideration about the existence of a special member in the family of abelian varieties

2.1. In [FI, 5.7], we obtained an abelian variety, belonging to a CM-type

$(M; \{\alpha_i\})$, with a Riemann form (in the strict sense) defined by $\mathrm{Tr}_{M/\mathbf{Q}}(exy^\rho)$ for an element e of M. In order to insure that $\mathrm{Tr}_{M/\mathbf{Q}}(exy^\rho)$ defines actually a Riemann form, one has to verify that $\mathrm{Im}\,(e^{\alpha_i}) > 0$ for every i. Since this point has not been clearly stated in [FI], we supply here a proof of this fact. More precisely, we show that, if T satisfies [FI, (5.1.1)], then the assertion of [FI, 4.10] is still true under the additional condition (v) $\mathrm{Im}\,(e^{\alpha_i}) > 0$, for $1 \leq i \leq mg$. To prove this, we first generalize [FI, 4.12] in the following form.

2.2. LEMMA. *Let R be an algebraic number field of finite degree, and P a finite algebraic extension of R. Let $x_1, \cdots, x_n, z$ be $n+1$ independent variables, and $\psi(z, x_1, \cdots, x_n)$ an irreducible polynomial in $P[z, x_1, \cdots, x_n]$. Let $R(x_1, \cdots, x_n, y_1^{(t)}, \cdots, y_s^{(t)})$, for $t = 1, \cdots, w$, be algebraic extensions of $R(x_1, \cdots, x_n)$. Suppose that R is algebraically closed in $R(x_1, \cdots, x_n, y_1^{(t)}, \cdots, y_s^{(t)})$ for every t. Then there exist rational numbers $a_1, \cdots, a_n$ and algebraic numbers $b_1^{(t)}, \cdots, b_s^{(t)}$ such that:*

(i) $(a_1, \cdots, a_n, b_1^{(1)}, \cdots, b_s^{(1)}, \cdots, b_1^{(w)}, \cdots, b_s^{(w)})$ *is a specialization of* $(x_1, \cdots, x_n, y_1^{(1)}, \cdots, y_s^{(1)}, \cdots, y_1^{(w)}, \cdots, y_s^{(w)})$ *over P;*

(ii) $\psi(z, a_1, \cdots, a_n)$ *is irreducible over P;*

(iii) $R(b_1^{(t)}, \cdots, b_s^{(t)})$ *and P are linearly disjoint over R, for every t.*

This can be proved by an obvious modification of the original proof of [FI, 4.12].

2.3. Now let us consider the symbols v_{pkl} defined in [FI, 4.13]. The v_{pkl} can be determined by the choice of r_ν roots $u_{\nu 1}, \cdots, u_{\nu r_\nu}$ among the roots of $\psi_\nu(z)$ for each ν. Put $w = \prod_{\nu=1}^{g} m!/[r_\nu! \cdot (m - r_\nu)!]$. Then we observe that there are exactly w ways of defining the set of elements $\{v_{pkl}\}$, depending on such choices. Let $v^{(t)} = \{v_{pkl}^{(t)}\}$, for each t in the range $1 \leq t \leq w$, denote the set of elements defined by each fixed choice. Then the content of [FI, 4.14] can be understood as follows: *For every t, K' is algebraically closed in $\mathbf{Q}(x_{ij}^p, v^{(t)})$, the field generated by the elements x_{ij}^p and $v_{pkl}^{(t)}$, for all p, i, j, k, l and fixed t.*

By means of 2.2, we find g matrices $B_p = (b_{ij}^p)(p = 1, \cdots, g)$ with rational entries and algebraic numbers $c_{pkl}^{(t)}$ such that:

(i) $(b_{ij}^p, c_{pkl}^{(t)})$ is a specialization of $(x_{ij}^p, v_{pkl}^{(t)})$ over P, where both sets are the sets containing the elements with all p, i, j, k, l, t;

(ii) $\psi_1(z, B_1, \cdots, B_g)$ is irreducible over P;

(iii) For every t, P and $K'(c_{pkl}^{(t)})$ ($=$ the field generated by the $c_{pkl}^{(t)}$ with fixed t and all p, k, l) are linearly disjoint over K'.

This is a slight modification of the first consideration in [FI, 4.15]. Let $u_{\nu 1}, \cdots, u_{\nu m}$ be the roots of $\psi_\nu(z)$ in S_0 (in an arbitrary order), and let $(y_{11}, \cdots, y_{1m}, \cdots, y_{g1}, \cdots, y_{gm})$ be a specialization of $(u_{11}, \cdots, u_{1m}, \cdots, u_{g1}, \cdots, u_{gm})$

over the specialization (i). Then repeat the argument of [FI, 4.15] with $y_1 = y_{11}$, except for the following point: we define a set of isomorphisms $\{\alpha_{\nu i}^{(t)}\}$ of M so that the images of y_1 under the $\alpha_{\nu i}^{(t)}$ for $i = 1, \cdots, r_\nu$ are exactly the specializations of the r_ν roots of $\psi_\nu(z)$ which have been adopted for the definition of $v^{(t)}$. In this way, we can define, for each t, a CM-type $\left(M; \{\alpha_{\nu i}^{(t)}\}\right)$ and an element e, satisfying [FI, 4.10, (i-iv)]. The element e is common for all t. We notice that these w CM-types are exactly all the CM-types, with the same M, satisfying [FI, (4.6.1, 2)].

2.4. Let f be a K-linear isomorphism of M to V such that $T(f(x), f(y)) = \mathrm{Tr}_{M/K}(exy^\rho)$, which is assured by [FI, 4.10, (ii)] for the present M and e. For each ν, let $\beta_{\nu 1}, \cdots, \beta_{\nu m}$ be all the extensions of τ_ν to M. Then there exists a complex matrix Z_ν such that $Z_\nu T^{\tau_\nu} \cdot {}^t\bar{Z}_\nu$ equals the diagonal matrix with diagonal elements $e^{\beta_{\nu 1}}, \cdots, e^{\beta_{\nu m}}$. In view of the assumption [FI, (5.1.1)], we may arrange $\beta_{\nu i}$ so that $\mathrm{Im}\,(e^{\beta_{\nu i}}) > 0$ for $1 \leq i \leq r_\nu$ and $\mathrm{Im}\,(e^{\beta_{\nu i}}) < 0$ for $r_\nu < i \leq m$. Now define $\alpha_{\nu i}$ by

$$\alpha_{\nu i} = \begin{cases} \beta_{\nu i} & \text{for} \quad 1 \leq i \leq r_\nu\,, \\ \rho\beta_{\nu i} & \text{for} \quad r_\nu < i \leq m\,. \end{cases}$$

Then $(M; \{\alpha_{\nu i}\})$ satisfies [FI, (4.6.1, 2)]. Therefore, by the remark at the end of 2.3, $\{\alpha_{\nu i}\}$ must coincide with $\{\alpha_{\nu i}^{(t)}\}$, as a whole, for some t. Since $\mathrm{Im}\,(e^{\alpha_{\nu i}}) > 0$ for every ν and i, this completes our proof.

3. An explicit reciprocity law

3.1. PROPOSITION. *Let $\Omega = (L, \Phi, \rho; T, \mathfrak{M}; x_1, \cdots, x_s)$ be a PEL-type in the sense of [2, 3.1]. Suppose that either [2, (3.7.1)] is satisfied or $L = \mathbf{Q}$. Then, for every automorphism σ of C, there is a PEL-type Ω^σ which is characterized, up to equivalence, by the following condition.*

(3.1.1) If $\mathfrak{A}$ is a PEL-structure of type Ω, then $\mathfrak{A}^\sigma$ is of type Ω^σ.

Moreover, let $k(\Omega)$ be the algebraic number field determined by [2, 5.1], and $V(\Omega)$ be a variety with the properties of [2, 6.2]. When [2, (3.7.2)] is satisfied, let $\mathfrak{F}(\Omega)$ be a fibre system of PEL-structures with the properties of [2, 5.3]. Then:

(3.1.2) Ω^σ is equivalent to Ω^τ in the sense of [2, 3.2] if and only if $\sigma = \tau$ on $k(\Omega)$.

(3.1.3) $k(\Omega^\sigma) = k(\Omega)^\sigma$.

(3.1.4) $V(\Omega)^\sigma$ is biregularly isomorphic to $V(\Omega^\sigma)$ over $k(\Omega^\sigma)$.

(3.1.5) There is an isomorphism of $\mathfrak{F}(\Omega)^\sigma$ to $\mathfrak{F}(\Omega^\sigma)$, defined over $k(\Omega^\sigma)$, in the sense of [2, 1.7].

PROOF. The first assertion is included in the proof of [2, 5.1]. We see that $\Omega^{\sigma\tau} = (\Omega^{\sigma})^{\tau}$ for two automorphisms σ and τ of C. Then (3.1.2) follows from [2, (5.1.1)], and (3.1.3) follows easily from (3.1.2). The assertions (3.1.4) and (3.1.5) are easy consequences of the uniqueness of $V(\Omega)$ and $\mathfrak{F}(\Omega)$ [2, 6.8 and 5.4].

More strongly than (3.1.4, 5), one can find a *canonical* isomorphism of $V(\Omega)^{\sigma}$ to $V(\Omega^{\sigma})$ (resp. $\mathfrak{F}(\Omega)^{\sigma}$ to $\mathfrak{F}(\Omega^{\sigma})$), whose description will be left to the reader.

3.2. *Remark.* Let $\mathcal{H}$ be a bounded symmetric domain and Γ a discontinuous group operating on $\mathcal{H}$ such that $\mathcal{H}/\Gamma$ is isomorphic to a Zariski open subset V of a projective variety. Then one may conjecture that, for every automorphism σ of C, V^{σ} is isomorphic to $\mathcal{H}'/\Gamma'$ with a bounded symmetric domain $\mathcal{H}'$ and a discontinuous group Γ', presumably with $\mathcal{H}' = \mathcal{H}$. The above assertion (3.1.4) shows that this is true for a fairly wide class of arithmetical discontinuous groups. If Ω is as in 3.1, one has $\Omega^{\sigma} = (L, \Phi^{\sigma}, \rho; T', \mathfrak{M}'; x_1', \cdots, x_s')$ with certain $T', \mathfrak{M}', x_1', \cdots, x_s'$ (cf. [FI, § 1] and [FII, § 3]).

3.3. Let the notation be as in the list of symbols. We assume $\sum_{\nu=1}^{g} r_{\nu}s_{\nu} > 0$. Let $\mathfrak{c}$ be an integral ideal in F, $\mathfrak{M}$ an $\mathfrak{r}$-lattice in V, and $x_1, \cdots, x_m$ elements of V such that

$$\mathfrak{c}^{-1}\mathfrak{M} = \mathfrak{M} + \mathfrak{r}x_1 + \cdots + \mathfrak{r}x_m \ .$$

Now we consider a PEL-type

(3.3.1) $\Omega = (K, \Phi, \rho; T, \mathfrak{M}; x_1, \cdots, x_m) \ .$

If σ is an automorphism of C over K', we have

$$\Omega^{\sigma} = (K, \Phi, \rho; cT, \mathfrak{N}; y_1, \cdots, y_m)$$

with a totally positive element c of F, an $\mathfrak{r}$-lattice $\mathfrak{N}$ in V, and elements $y_1, \cdots, y_m$ such that $\mathfrak{c}^{-1}\mathfrak{N} = \mathfrak{N} + \mathfrak{r}y_1 + \cdots + \mathfrak{r}y_m$. If m is even, we can put $c = 1$. These facts were shown in [FII, 9.5].

As in [FII, 9.2], we take an element ζ of K such that $\zeta^{\rho} = -\zeta$, and consider $\mathfrak{c}$-frames, oriented $\mathfrak{c}$-frames and the symbol $[\mathfrak{L}^*/\mathfrak{M}^*]$ with respect to the hermitian form ζT. Put

$$\mathfrak{M}^* = \begin{cases} (\mathfrak{M}; x_1, \cdots, x_m) & \text{if } m \text{ is even}, \\ (\mathfrak{M}; 1; x_1, \cdots, x_m) & \text{if } m \text{ is odd}, \end{cases}$$

$$\mathfrak{N}^* = \begin{cases} (\mathfrak{N}; y_1, \cdots, y_m) & \text{if } m \text{ is even}, \\ (\mathfrak{N}; c; y_1, \cdots, y_m) & \text{if } m \text{ is odd}. \end{cases}$$

From [FII, 3.3, 3.4] we see that $\mathfrak{c}$-frames or oriented $\mathfrak{c}$-frames $\mathfrak{M}^*$ and $\mathfrak{N}^*$ belong to the same genus in the sense of § 1. We notice also that there is no conflict between the equivalence of PEL-types [2, 3.2] and the equivalence of $\mathfrak{c}$-frames or oriented $\mathfrak{c}$-frames (§ 1). We put

$$[\Omega^\sigma/\Omega] = [\mathfrak{N}^*/\mathfrak{M}^*] \ .$$

This is an element of $B(\Lambda; \mathfrak{c})/B_0(\Lambda; \mathfrak{c})$ or $B(\mathfrak{c})/B_0(\mathfrak{c})$ according as m is even or odd, where Λ is the genus of $\mathfrak{M}$ and $\mathfrak{N}$ with respect to $G(\zeta T)$.

3.4. Define isomorphisms $\sigma_1, \cdots, \sigma_{g'}$ of K' into C and integers v_λ as in [FI, 4.4, 4.7]. Put, for every ideal $\mathfrak{a}$ in K',

$$\psi(\mathfrak{a}) = \begin{cases} \mathfrak{r} & \text{if } m \text{ is even and } K' = \boldsymbol{Q} \ , \\[2ex] \prod_{\lambda=1}^{g'} \left(\dfrac{\mathfrak{a}^{\sigma_\lambda}}{\mathfrak{a}^{\sigma_\lambda \rho}} \right)^{v_\lambda} & \text{if } m \text{ is even and } K' \neq \boldsymbol{Q} \ , \\[2ex] \prod_{\lambda=1}^{g'} \mathfrak{a}^{\sigma_\lambda} \cdot \left(\dfrac{\mathfrak{a}^{\sigma_\lambda}}{\mathfrak{a}^{\sigma_\lambda \rho}} \right)^{v_\lambda} & \text{if } m \text{ is odd} \ . \end{cases}$$

Then $\psi(\mathfrak{a})$ is an ideal in K [FI, 4.5]. Let $\mathfrak{d}$ denote the relative discriminant of K with respect to F. Let $H_\mathfrak{c}$ be the set of ideals $\mathfrak{a}$ in K' defined by

$$H_\mathfrak{c} = \{\mathfrak{a} \mid \text{prime to } N(\mathfrak{c}\mathfrak{d}), \ \left(\psi(\mathfrak{a})^{-1}, \left(N(\mathfrak{a})^{m/2}, N(\mathfrak{a})\right)\right) \in B_0(\Lambda; \mathfrak{c})\} \quad \text{if } m \text{ is even} \ ,$$
$$H_\mathfrak{c} = \{\mathfrak{a} \mid \text{prime to } N(\mathfrak{c}), \ \left(\psi(\mathfrak{a}), N(\mathfrak{a})\right) \in B_0(\mathfrak{c})\} \quad \text{if } m \text{ is odd} \ .$$

In the case of even m, $(N(\mathfrak{a})^{m/2}, N(\mathfrak{a}))$ means the element of $D(\Lambda; \mathfrak{c})$ whose components are the cosets containing this pair. Then, in either case, $H_\mathfrak{c}$ is an ideal group in K' (in the sense of class-field theory), and coincides with the one defined by [FII, (9.2.2), (9.2.3)]. If m is even, $H_\mathfrak{c}$ depends on both Λ and $\mathfrak{c}$, while if m is odd, $H_\mathfrak{c}$ does not depend on the genus of $\mathfrak{M}$.

3.5. THEOREM. *Let Ω be defined by (3.3.1), and let $k(\Omega)$ be the algebraic number field determined by [2, 5.1]. Then $k(\Omega)$ is a class-field over K', corresponding to the ideal group $H_\mathfrak{c}$ in K'. Moreover, if $\sigma = \left(\dfrac{k(\Omega)/K'}{\mathfrak{a}} \right)$ with an ideal $\mathfrak{a}$ in K', then*

$$[\Omega^\sigma/\Omega] = \text{the class of } \left(\psi(\mathfrak{a}), \left(N(\mathfrak{a})^{m/2}, N(\mathfrak{a})\right)^{-1}\right) \bmod B_0(\Lambda; \mathfrak{c}) \text{ if } m \text{ is even} \ ,$$
$$[\Omega^\sigma/\Omega] = \text{the class of } \left(\psi(\mathfrak{a}), N(\mathfrak{a})\right) \bmod B_0(\mathfrak{c}) \text{ if } m \text{ is odd} \ .$$

PROOF. Let $k_\mathfrak{c}$ denote the class-field over K' corresponding to the ideal-group $H_\mathfrak{c}$. Let us take M, N, $\{\alpha_i\}$ and e as in [FI, 4.10] with $P = k_\mathfrak{c} \cdot k(\Omega)$, and with an additional condition that $\mathrm{Im}\,(e^{\alpha_i}) > 0$ for every i, as assured in § 2. Then we obtain a polarized abelian variety $\mathscr{P} = (A, \mathcal{C}, \theta)$ belonging to the CM-type $(M; \{\alpha_i\})$ [FI, 5.7]. As shown in the same place, if we denote by θ_1 the restriction of θ to K, then $(A, \mathcal{C}, \theta_1)$ is of type $(K, \Phi, \rho; T, \mathfrak{M})$. Hence we can find points $t_1, \cdots, t_s$ on A so that $\mathscr{Q} = (A, \mathcal{C}, \theta; t_1, \cdots, t_s)$ is a PEL-structure of type Ω. Let $(M'; \{\beta_j\})$ be the dual of $(M; \{\alpha_i\})$ (see [FI, 4.6] and [5, 8.3]). Then $\mathscr{Q}$ has the property stated in [FII, 9.7]; the structure $\mathscr{Q}_y$ in [FII, 9.7] is just a projective embedding of $\mathscr{Q}$. Therefore we may identify $\mathscr{Q}$ with $\mathscr{Q}_y$.

Let f denote $N(\mathfrak{c}\mathfrak{d})$ or $N(\mathfrak{c})$ according as m is even or odd. Let $\mathfrak{a}$ be an ideal in K', prime to f, and let $\sigma = \left(\frac{k_c/K'}{\mathfrak{a}}\right)$. Since we have chosen $k_c \cdot k(\Omega)$ as the field P in [FI, 4.10], M' and k_c are linearly disjoint over K'. Therefore we can extend σ to an automorphism τ of C over M'. Let $\Omega^\tau = (K, \Phi, \rho; cT, \mathfrak{N}; \{y_i\})$, with $c = 1$ if m is even. Then $\mathfrak{A}^\tau$ is of type Ω^τ by virtue of (3.1.1). Let k and b be as in [FII, 9.7], and R a Galois extension of K containing k, $k(\Omega)$ and k_c. We can find a prime ideal $\mathfrak{q}$ of M', not dividing bf, and a prime factor $\mathfrak{Q}$ of $\mathfrak{q}$ in R such that τ induces on R a Frobenius automorphism for $\mathfrak{Q}/\mathfrak{q}$. Put $\mathfrak{h} = N_{M'/K'}(\mathfrak{q})$. Then we have

$$\left(\frac{k_c/K'}{\mathfrak{a}}\right) = \left(\frac{k_c/K'}{\mathfrak{h}}\right),$$

so that $\mathfrak{a}^{-1}\mathfrak{h} \in H_c$. By [FII, 9.7] and [FI, 4.7], there exists an element γ of $G(\zeta T)$ such that

$$(3.5.1) \qquad \mathfrak{N} \supset \mathfrak{M}\gamma,$$

$$(3.5.2) \qquad [\mathfrak{N}/\mathfrak{M}\gamma] = \begin{cases} N(\mathfrak{h})^{m/2}\psi(\mathfrak{h}) & \text{if } m \text{ is even}, \\ N(\mathfrak{h})^{(m-1)/2}\psi(\mathfrak{h}) & \text{if } m \text{ is odd}, \end{cases}$$

$$(3.5.3) \qquad \mu(\gamma)c = N(\mathfrak{h}) \qquad\qquad (c = 1 \text{ if } m \text{ is even}),$$

$$(3.5.4) \qquad x_i\gamma \equiv y_i \bmod \mathfrak{N}.$$

Case of even m. If $\mathfrak{p}$ is a prime ideal dividing $c\mathfrak{b}$, then $\mathfrak{p}$ is prime to $N(\mathfrak{h})$, so that (3.5.1, 2) imply $\mathfrak{N}_\mathfrak{p} = \mathfrak{M}_\mathfrak{p}\gamma$. Therefore $d(\mathfrak{M}^*/\mathfrak{N}^*)$ can be represented by $(\det(\gamma), N(\mathfrak{h}))$. Put $a = \det(\gamma)^{-1}N(\mathfrak{h})^{m/2}$. Then $[\mathfrak{N}/\mathfrak{M}] = a \cdot \psi(\mathfrak{h})$ and $d(\mathfrak{N}^*/\mathfrak{M}^*) = (aN(\mathfrak{h})^{-m/2}, N(\mathfrak{h})^{-1})$. Hence $[\Omega^\tau/\Omega] = [\mathfrak{N}^*/\mathfrak{M}^*] = $ the class of $(\psi(\mathfrak{h}), (N(\mathfrak{h})^{m/2}, N(\mathfrak{h}))^{-1}) \bmod B_0(\Lambda; c)$. Since $\mathfrak{a}^{-1}\mathfrak{h} \in H_c$, we get

$$(3.5.5) \quad [\Omega^\tau/\Omega] = \text{the class of } (\psi(\mathfrak{a}), (N(\mathfrak{a})^{m/2}, N(\mathfrak{a}))^{-1}) \bmod B_0(\Lambda; c).$$

Therefore we see that

$\tau = $ identity on $k(\Omega)M'$

$\Leftrightarrow \Omega^\tau = \Omega$ (3.1.2)

$\Leftrightarrow \mathfrak{M}^*$ and $\mathfrak{N}^*$ are equivalent

$\Leftrightarrow (\psi(\mathfrak{a}), (N(\mathfrak{a})^{m/2}, N(\mathfrak{a}))^{-1}) \in B_0(\Lambda; c)$ (1.1.1)

$\Leftrightarrow \mathfrak{a} \in H_c$ (definition of H_c)

$\Leftrightarrow \tau = $ identity on $k_c M'$.

This proves $k(\Omega)M' = k_c M'$. Since M' and $k(\Omega)k_c$ are linearly disjoint, we obtain $k(\Omega) = k_c$; and (3.5.5) is exactly the desired "law of reciprocity" in our theorem.

Case of odd m. If $\mathfrak{p}$ is a prime ideal dividing c, then $\mathfrak{p}$ is prime to $N(\mathfrak{h})$, so that (3.5.1, 2) imply $\mathfrak{N}_\mathfrak{p} = \mathfrak{M}_\mathfrak{p}\gamma$. Therefore, in the definition of $[\mathfrak{N}^*/\mathfrak{M}^*]$ in 1.2, we can take a, b, $\alpha_\mathfrak{p}$ and v to be c, 1, γ^{-1} and $\mu(\gamma)^l \det(\gamma)^{-1}$, respectively. Then

$v^{-1}(a^{-1}b)'[\mathfrak{N}/\mathfrak{M}] = \psi(\mathfrak{y})$, and $(vv^{\rho})^{-1}ab^{-1} = \mu(\gamma)c = N(\mathfrak{y})$. Hence we have

$$[\Omega^{\tau}/\Omega] = [\mathfrak{N}^*/\mathfrak{M}^*] = \text{the class of } \big(\psi(\mathfrak{y}),\, N(\mathfrak{y})\big) \bmod B_0(c)$$
$$= \text{the class of } \big(\psi(a),\, N(a)\big) \bmod B_0(c)\ .$$

Repeating the same argument as in the case of even m, with (1.2.2) instead of (1.1.1), we obtain the required result.

3.6. *Remark.* As mentioned in [2, 5.9], the field $k(\Omega)$ is the algebraic closure of Q in the field of moduli of a "generic member" of the family $\Sigma(\Omega)$ of PEL-structures of type Ω. In the previous formulation of [FI, 5.10] and [FII, 9.4], we defined the number field $k(\Omega)$ in this way.

The main difference of the present proof from the original ones in [FI, II] lies in the following point. In [FI, II], one had to worry about the possibility that $\mathfrak{A}^{\tau}$ might belong to some other family which is different from $\Sigma(\Omega^{\tau})$. For this reason, we needed troublesome considerations such as [FI, 2.5–9, 5.4], the first part of the proof of [FI, 5.10] involving the symbols $\nu(T, \mathfrak{M})$, $e(T, \mathfrak{M})$, $W(\mathfrak{N})$, etc., and the corresponding assertions with points of finite order [FII, p. 160, footnote 9]. In the new proof, we could cast off all these things by means of Proposition 3.1, which is a direct consequence of the theory of [2]. It is also possible, if one wishes, to express the reciprocity law in terms of biregular equivalence of moduli varieties or fibre systems, in view of (3.1.4, 5).

One should notice that two varieties $V(\Omega)$ and $V(\Omega')$ in the sense of (3.1) may be biregularly isomorphic over C, even if Ω and Ω' are not equivalent. For an investigation of this question, we refer the reader to [4].

PRINCETON UNIVERSITY

REFERENCES

1. G. SHIMURA, *On the field of definition for a field of automorphic functions*: I, II, Ann. of Math., 80 (1964), 160-189; 81 (1965), 124-165, referred to as [FI], [FII].
2. ———, *Moduli and fibre systems of abelian varieties*, Ann. of Math., 83 (1966) 294-338.
3. ———, *Arithmetic of unitary groups*, Ann. of Math., 79 (1964). 369-409.
4. ———, *Class-fields and automorphic functions*, Ann. of Math., 80 (1964), 444-463.
5. ———, and Y. TANIYAMA, Complex multiplication of abelian varieties and its applications to number theory, Publ. Math. Soc. Japan, No. 6, 1961.

(Received June 23, 1965)

Moduli of abelian varieties and number theory

Proceedings of the Symposium on Pure Mathematics, 9,
Algebraic groups and discontinuous subgroups, 1965,
American Mathematical Society, (1966), 312-332

There are many arithmetically defined discontinuous groups Γ operating on a bounded symmetric domain S such that S/Γ parametrizes a family of abelian varieties. For such a Γ and S, one can naturally consider a fibre variety of which the base space is S/Γ and the fibres are the abelian varieties of the family. The cohomological or analytical aspects of the theory will be discussed in the lectures of Kuga and Satake (cf. also the articles by Baily and Mumford). Therefore my talk will be confined to the algebro-geometric and number-theoretical aspects. I shall consider a family $\Sigma_\Omega = \{\mathcal{Q}_z | z \in S\}$ of abelian varieties $\mathcal{Q}_z$ with a prescribed type Ω of structures. Here Ω describes the type of polarization, endomorphism-ring and points of finite order on an abelian variety. My main purpose is to provide a brief account of principal results concerning the following three problems.

(i) The existence of a "nice moduli-variety" and a "nice fibre variety" for the family Σ_Ω, defined over a well-defined algebraic number field k_Ω.

(ii) The description of k_Ω as a class-field.

(iii) The Hasse zeta-function of the fibre variety in a special case (collaboration with Kuga).

All the results will be stated without proofs. For the detailed proofs of (i), (ii) and (iii), see [12], [11] and [4], respectively. I devote (partly by the chairman's request) the first three sections to an exposition of basic notions and some elementary results on abelian varieties (mostly over C), Riemann forms, and the maximal families of abelian varieties.

1. **Abelian varieties** (cf. [5], [16], [17]). A projective variety A defined over a field k of characteristic $p \geq 0$ is an *abelian variety* if there exist morphisms (of algebraic varieties) $f: A \times A \to A$ and $g: A \to A$ which define a group structure on A by $f(x, y) = x + y, g(x) = -x$. Additive notation is used since any such group structure on a projective variety can be shown to be commutative. The neutral element is accordingly denoted by 0. If f and g are defined over k, then we say that the abelian variety A *is defined over* k. This is so if and only if the neutral element is rational over k.

Let A and B be two abelian varieties defined over k. By a *homomorphism* of A into B, or an *endomorphism* when $A = B$, we shall always understand a morphism λ of A into B, satisfying $\lambda(x + y) = \lambda(x) + \lambda(y)$; if λ is birational, we call it an *isomorphism*, or an *automorphism* when $A = B$. Suppose that A and B have the

312

761

same dimension. Then a homomorphism λ of A into B is surjective if and only if the kernel of λ is finite. Such a λ is called an *isogeny* of A to B. If there exists an isogeny of A to B, A and B are said to be *isogenous*.

We denote by $\mathrm{Hom}(A, B)$ the module of all homomorphisms of A into B, and put $\mathrm{End}(A) = \mathrm{Hom}(A, A)$, $\mathrm{End}_Q(A) = \mathrm{End}(A) \otimes_Z Q$. Then $\mathrm{Hom}(A, B)$ is a finitely generated free Z-module, and $\mathrm{End}_Q(A)$ with a natural structure of ring is a semisimple algebra over Q of finite rank. Moreover it can be shown that $\mathrm{End}_Q(A)$ has a positive involution. Here by an *involution* of an algebra L over Q, we mean a Q-linear map $\rho: L \to L$ such that $(xy)^\rho = y^\rho x^\rho$ and $(x^\rho)^\rho = x$; ρ is said to be *positive* if $\mathrm{Tr}_{L/Q}(xx^\rho) > 0$ for every $x \in L$, $\neq 0$, where $\mathrm{Tr}_{L/Q}$ denotes the reduced trace from L to Q. We shall discuss in §4 an explicit way of obtaining a positive involution of $\mathrm{End}_Q(A)$ in the case of characteristic 0.

An abelian variety A is said to be *simple* if A and $\{0\}$ are the only abelian subvarieties of A. Every abelian variety is isogenous to a product of simple abelian varieties. An abelian variety A is simple if and only if $\mathrm{End}_Q(A)$ is a division algebra.

We shall now consider the case where the universal domain is C. Every abelian variety defined over a subfield of C, viewed as a complex manifold, is isomorphic to a complex torus. But the converse is not necessarily true. To describe the condition, let C^n/D be a complex torus, with a lattice D in C^n (i.e., a discrete subgroup of C^n of rank $2n$). An R-valued R-bilinear form $E(x, y)$ on C^n is called a *Riemann form* on C^n/D if it satisfies the following three conditions.

(1.1) *The value $E(x, y)$ is an integer for every $(x, y) \in D \times D$.*

(1.2) $E(x, y) = -E(y, x)$.

(1.3) *The R-bilinear form $E(x, \sqrt{(-1)}y)$ in (x, y) is symmetric and positive definite.*

Then one knows that a complex torus has a structure of abelian variety if and only if there exists a Riemann form on it.

Let A be an abelian variety of dimension n defined over a subfield of C, and η an isomorphism of A to a complex torus C^n/D. The pair $(C^n/D, \eta)$ is called an *analytic coordinate system* of A. Take a basis $\{g_1, \cdots, g_{2n}\}$ of D over Z and define real coordinate functions $x_i: C^n \to R$ by $u = \sum_{i=1}^{2n} x_i(u)g_i$ for $u \in C^n$. Then, given a Riemann form E on C^n/D, there exists a divisor X of A such that

$$\sum_{i<j} E(g_i, g_j) \, dx_i \wedge dx_j$$

represents the cohomology class of $\eta(X)$. (A *divisor* of an algebraic variety V is an element of the free Z-module formally generated by all the subvarieties of V of codimension one.) In this case we say that X *determines* E (with respect to η). If two divisors X and X' on A determine Riemann forms E and E' respectively, then X is algebraically equivalent to X' if and only if $E = E'$. (If the reader does not know what algebraic equivalence means, he can adopt this as a definition of algebraic equivalence.)

Let A and A' be two abelian varieties with analytic coordinate-systems $(C^n/D, \eta)$ and $(C^m/D', \eta')$ respectively. For every homomorphism $\lambda: A \to A'$ there exists a C-linear mapping $\Lambda : C^n \to C^m$ such that $\Lambda(D) \subset D'$ with the relation $\eta'\lambda = \Lambda\eta$, and conversely every such linear mapping $\Lambda : C^n \to C^m$ corresponds to a homomorphism of A into A'. Assume especially $A = A'$, $(C^n/D, \eta) = (C^m/D', \eta')$. Then the mapping $\lambda \mapsto \Lambda$ is uniquely extended to a representation $\Phi : \mathrm{End}_Q(A) \to \mathrm{End}(C^n)$, which is called the *analytic representation* of $\mathrm{End}_Q(A)$. Further we observe that $Q \cdot D$ is a vector space of dimension $2n$ over Q, and Λ gives an endomorphism of $Q \cdot D$. Therefore, with respect to a basis of $Q \cdot D$ over Q, we get a representation Ψ of $\mathrm{End}_Q(A)$ by matrices of size $2n$ with entries in Q. We call Ψ the *rational representation* of $\mathrm{End}_Q(A)$. Since a basis of $Q \cdot D$ over Q is a basis of C^n over R, it can be easily shown that Ψ is equivalent to the direct sum of the analytic representation Φ and its complex conjugate $\bar{\Phi}$.

2. **Polarized abelian varieties.** Let A be an abelian variety defined over a subfield of C with an analytic coordinate system $(C^n/D, \eta)$. A *polarization* of A is a set $\mathscr{C}$ of divisors of A satisfying the following three conditions.

(2.1) *Every X in $\mathscr{C}$ determines a Riemann form on C^n/D.*

(2.2) *If X and X' are in $\mathscr{C}$, then there exist positive integers m and m' such that mX is algebraically equivalent to (i.e. homologous to) $m'X'$.*

(2.3) *$\mathscr{C}$ is a maximal set of divisors satisfying* (2.1, 2.2). (If the universal domain is not necessarily of characteristic 0, one can define polarization by replacing (2.1) with (2.1') *$\mathscr{C}$ contains an ample divisor.*) Now a *polarized abelian variety* is a pair $(A, \mathscr{C})$ formed by an abelian variety A and a polarization $\mathscr{C}$ of A. One can find a divisor Y in $\mathscr{C}$ such that every divisor in $\mathscr{C}$ is algebraically equivalent to a multiple hY with a positive integer h. We call Y a *basic divisor* in $\mathscr{C}$.

An isomorphism (resp. isogeny) λ of A to A' is called an *isomorphism* (resp. *isogeny*) of $(A, \mathscr{C})$ to $(A', \mathscr{C}')$ if $\lambda^{-1}(X') \in \mathscr{C}$ for every $X' \in \mathscr{C}'$. If a Riemann form E on a complex torus C^n/D is given, then C^n/D and E determine a polarized abelian variety $\mathscr{P}$ up to isomorphism. Let E' be a Riemann form on another complex torus C^n/D' of the same dimension, and let $\mathscr{P}'$ be a polarized abelian variety determined by C^n/D' and E'. Then an element Λ of $\mathrm{End}(C^n)$ gives an isogeny of $\mathscr{P}$ to $\mathscr{P}'$ if and only if $\Lambda(D) \subset D'$ and $E'(\Lambda x, \Lambda y) = c \cdot E(x, y)$ with a positive rational number c. Now we recall a classical

LEMMA 1 (FROBENIUS). *Let B be an invertible alternating matrix of size $2n$ with entries in Z. Then there exists an element U of $\mathrm{GL}_{2n}(Z)$ such that*

$$
{}^t\!UBU = \begin{bmatrix} 0 & -e \\ e & 0 \end{bmatrix}, \qquad e = \begin{bmatrix} e_1 & & \\ & \ddots & \\ & & e_n \end{bmatrix},
$$

where the e_i are positive integers satisfying $e_{i+1} \equiv 0 \bmod(e_i)$.

A and $(C^n/D, \eta)$ being as above, let X be a divisor in a polarization $\mathscr{C}$ of A, and let E be the Riemann form determined by X. In view of (1.1) and (1.2), E can be represented by an alternating matrix B with entries in Z, with respect to a basis of D over Z. Applying Lemma 1 to B, we get n integers $e_1, \cdots, e_n$, which may be called *the elementary divisors* of X. Then X is a basic divisor in $\mathscr{C}$ if and only if $e_1 = 1$. Thus we are led to the following definition. Let e be as in Lemma 1 with positive integers $e_1, \cdots, e_n$ such that $e_1 = 1$, $e_{i+1} \equiv 0 \bmod(e_i)$, and

$$B = \begin{bmatrix} 0 & -e \\ e & 0 \end{bmatrix}.$$

Define an alternating form $B(x, y)$ on $\mathbf{R}^{2n}$ by $B(x, y) = {}^t x B y$, regarding the elements of $\mathbf{R}^{2n}$ as column vectors. A polarized abelian variety $(A, \mathscr{C})$ is said to be of type (e) if there exist a lattice D in $\mathbf{C}^n$ and a commutative diagram

$$(2.4)\qquad \begin{array}{ccccccccc} 0 & \longrightarrow & \mathbf{Z}^{2n} & \longrightarrow & \mathbf{R}^{2n} & \longrightarrow & \mathbf{R}^{2n}/\mathbf{Z}^{2n} & \longrightarrow & 0 \\ & & \downarrow & & f\downarrow & & \downarrow & & \\ 0 & \longrightarrow & D & \longrightarrow & \mathbf{C}^n & \overset{\xi}{\longrightarrow} & A & \longrightarrow & 0 \end{array}$$

satisfying the following three conditions:

(2.5) ξ *gives a holomorphic isomorphism of* $\mathbf{C}^n/D$ *to* A.

(2.6) f *is an* $\mathbf{R}$-*linear isomorphism, and* $f(\mathbf{Z}^{2n}) = D$.

(2.7) $\mathscr{C}$ *has a basic divisor* X *which determines a Riemann form* $E(x, y)$ *on* $\mathbf{C}^n/D$ *such that* $E(f(x), f(y)) = B(x, y)$ $((x, y) \in \mathbf{R}^{2n} \times \mathbf{R}^{2n})$.

From the above discussion we see that every polarized abelian variety is of type (e) for some unique (e).

3. **Maximal families of abelian varieties.** Let $(A, \mathscr{C})$ be a polarized abelian variety of type (e). Let $\{d_1, \cdots, d_{2n}\}$ be the standard basis of $\mathbf{Z}^{2n}$ over $\mathbf{Z}$. Regard the elements of $\mathbf{C}^n$ as column vectors. Let f be as in the diagram (2.4). Define an $n \times 2n$ matrix w by $w = (f(d_1) \cdots f(d_{2n}))$, and write $w = (u \ v)$ with square matrices u and v of size n. Put $z = ev^{-1}u$. By a simple calculation we can verify that ${}^t z = z$ and $\mathrm{Im}(z) > 0$ (positive definite). We put

$$S_n = \{ z \in M_n(\mathbf{C}) \mid {}^t z = z,\ \mathrm{Im}(z) > 0 \},$$

and call S_n the *Siegel space of degree* n. If we define an element g of $\mathrm{End}(\mathbf{C}^n)$ by $g(x) = ev^{-1}x$ for $x \in \mathbf{C}^n$, and take $\{\xi \circ g^{-1}, g \circ f, g(D), E(g^{-1}(x), g^{-1}(y))\}$ in place of $\{\xi, f, D, E\}$, then we have a diagram similar to (2.4), and $(g \circ f(d_1) \cdots g \circ f(d_{2n})) = (z \ e)$. In other words, if $(A, \mathscr{C})$ is of type (e), then we can choose the diagram (2.4) so that $(f(d_1) \cdots f(d_{2n})) = (z \ e)$ with a point z of S_n.

Conversely, for any $z \in S_n$, define $f_z : \mathbf{R}^{2n} \to \mathbf{C}^n$ by $f_z(a) = f(a, z) = (e \ z)a$ for $a \in \mathbf{R}^{2n}$ and let $D_z = f_z(\mathbf{Z}^{2n})$, $E_z(x, y) = B(f_z^{-1}(x), f_z^{-1}(y))$. Then E_z is a Riemann form on $\mathbf{C}^n/D_z$, so that we get an abelian variety A_z isomorphic to $\mathbf{C}^n/D_z$. Let $\mathscr{C}_z$ be the polarization of A_z determined by E_z, and let $\mathscr{P}_z = (A_z, \mathscr{C}_z)$. Thus we obtain a family $\Sigma_e = \{\mathscr{P}_z \mid z \in S_n\}$ of polarized abelian varieties of type (e). The above discussion shows that every polarized abelian variety of type (e) is isomorphic to a member of Σ_e. In this sense Σ_e may be called a maximal family of polarized abelian varieties. (For the moment we do not consider any specific projective embeddings of A_z.)

To determine the isomorphism classes of $\mathscr{P}_z$, let

$$P_e = \begin{bmatrix} 1 & 0 \\ 0 & e \end{bmatrix},$$

$$\Gamma'_e = \{T \in \mathrm{GL}_{2n}(\mathbf{Z}) \,|\, {}^t TBT = B\}, \qquad \Gamma_e = \{P_e^{-1} \cdot {}^t TP_e \,|\, T \in \Gamma'_e\}.$$

Then Γ_e is a discrete subgroup of $\mathrm{Sp}(n, \mathbf{R})$ commensurable with $\mathrm{Sp}(n, \mathbf{Z})$. If $e = 1_n$, we have $\Gamma_e = \Gamma'_e = \mathrm{Sp}(n, \mathbf{Z})$. For an element T of Γ'_e, let

$$P_e^{-1} \cdot {}^t T^{-1} P_e = \begin{bmatrix} a & b \\ c & d \end{bmatrix}$$

with $a, b, c, d \in M_n(\mathbf{R})$, and define the action of T on S_n by $T(z) = (az + b)(cz + d)^{-1}$ for $z \in S_n$. Now two members $\mathscr{P}_z$ and $\mathscr{P}_w$ of Σ_e, with z and w in S_n, are isomorphic if and only if there exists an element T of Γ'_e such that $T(z) = w$. Hence S_n/Γ'_e is in one-to-one correspondence with all the isomorphism classes of polarized abelian varieties of type (e).

4. Families of abelian varieties with a prescribed type of polarization and endomorphisms. If A is an abelian variety defined over a subfield of C, every polarization $\mathscr{C}$ of A determines a positive involution ρ of $\mathrm{End}_Q(A)$ as follows. Take a complex torus C^n/D isomorphic to A. Then $\mathscr{C}$ determines a Riemann form E on C^n/D up to rational factors. If $\phi(\lambda)$ denotes the element of $\mathrm{End}(C^n)$ corresponding to an element λ of $\mathrm{End}_Q(A)$, then, for every $\lambda \in \mathrm{End}_Q(A)$, there exists an element λ^ρ of $\mathrm{End}_Q(A)$ such that

$$E(\phi(\lambda)x, y) = E(x, \phi(\lambda^\rho)y) \qquad ((x, y) \in C^n \times C^n).$$

It can be easily shown that ρ is a positive involution of $\mathrm{End}_Q(A)$. Therefore, writing $\mathscr{L}$ for $\mathrm{End}_Q(A)$, we see that any polarized abelian variety $(A, \mathscr{C})$ determines a triple $(\mathscr{L}, \phi, \rho)$ formed by a semisimple algebra $\mathscr{L}$ over Q, a representation ϕ of $\mathscr{L}$ by complex matrices, and a positive involution ρ of $\mathscr{L}$.

This observation motivates the following problem. Let L be a semisimple algebra over Q with a positive involution ρ, and Φ a representation of L by complex matrices. Determine all families of polarized abelian varieties $(A, \mathscr{C})$ such that: (i) $\mathrm{End}_Q(A)$ contains L, (ii) $\mathscr{C}$ gives the involution ρ on L, and (iii) Φ is equivalent to the restriction (to L) of the analytic representation of $\mathrm{End}_Q(A)$. Since L or $\mathrm{End}_Q(A)$ may have many automorphisms and we shall deal with many distinct A's, it is convenient to take L outside $\mathrm{End}_Q(A)$, and specify an isomorphism of L into $\mathrm{End}_Q(A)$. To be more precise, we say that a triple $\mathscr{P} = (A, \mathscr{C}, \theta)$ is a polarized abelian variety of type $\{L, \Phi, \rho\}$, or more briefly, $\mathscr{P}$ belongs to $\{L, \Phi, \rho\}$, if the following three conditions are satisfied:

(4.1) *A is an abelian variety defined over a subfield of C.*

(4.2) *θ is an isomorphism of L into $\mathrm{End}_Q(A)$; if ψ is the analytic representation of $\mathrm{End}_Q(A)$, then $\psi \circ \theta$ is equivalent to Φ.*

(4.3) $\mathscr{C}$ *is a polarization of* A; *and the involution of* $\mathrm{End}_Q(A)$ *determined by* $\mathscr{C}$ *coincides on* $\theta(L)$ *with the involution* $\theta(a) \mapsto \theta(a^\rho)$.

In the following treatment, we always assume

(4.4) Φ *maps the identity element of* L *to the identity matrix*,

(4.5) L *is a division algebra*,

though a more general case is worth while considering. If n is the degree of Φ, (4.4) implies $\dim(A) = n$.

The representation Φ cannot be arbitrary in order to ensure the existence of a polarized abelian variety of type $\{L, \Phi, \rho\}$. In fact, if $(A, \mathscr{C}, \theta)$ belongs to $\{L, \Phi, \rho\}$, then the rational representation of $\mathrm{End}_Q(A)$ induces a rational representation of L of degree $2n$, equivalent to the sum of Φ and $\bar{\Phi}$, where $n = \dim(A)$. Therefore $\Phi + \bar{\Phi}$ must contain all the absolutely irreducible representations of L with the same multiplicity. In view of (4.5), $[L:Q]$ must divide $2n$.

Now we shall determine the polarized abelian varieties of type $\{L, \Phi, \rho\}$. Let $(A, \mathscr{C}, \theta)$ belong to $\{L, \Phi, \rho\}$. Take a complex torus C^n/D isomorphic to A. By the action of $\Phi(L)$, $Q \cdot D$ can be considered as an L-module, whose rank is obviously $2n/[L:Q]$. Let us take and fix a (left) L-module V of rank m, where $m = 2n/[L:Q]$. Then there is an L-isomorphism $f: V \to Q \cdot D$. Put $\mathfrak{M} = f^{-1}(D)$. Then $\mathfrak{M}$ is a free Z-module of rank $2n$ in V. Let X be a divisor in $\mathscr{C}$, and E the Riemann form on C^n/D determined by X. Put $B(x, y) = E(f(x), f(y))$ for $(x, y) \in V \times V$. Then $B: V \times V \to Q$ is a Q-bilinear form satisfying

$$(4.6) \qquad B(x, y) = -B(y, x), \quad B(ax, y) = B(x, a^\rho y) \qquad (x, y \in V; a \in L).$$

If X is a basic divisor of $\mathscr{C}$, we have $B(\mathfrak{M}, \mathfrak{M}) = Z$.

LEMMA 2. *Let* L *be a division algebra over* Q *with an involution* ρ, V *a left* L-*module, and* $B: V \times V \to Q$ *a* Q-*bilinear form satisfying* (4.6). *Then there exists a* Q-*bilinear map* $T: V \times V \to L$ *such that*

$$(4.7) \qquad\qquad\qquad B(x, y) = \mathrm{Tr}_{L/Q}(T(x, y)),$$

$$(4.8) \quad T(x, y)^\rho = -T(y, x), \quad T(ax, by) = a \cdot T(x, y) \cdot b^\rho \qquad (x, y \in V; a, b \in L)$$

A Q-bilinear map $T: V \times V \to L$ is called an L-*valued* ρ-*antihermitian form* on V if it satisfies (4.8). We apply this lemma to the above Q-bilinear form B and obtain an L-valued ρ-antihermitian form T on V such that

$$\mathrm{Tr}_{L/Q}(T(x, y)) = E(f(x), f(y)).$$

Thus from $(A, \mathscr{C}, \theta)$, we obtain an L-isomorphism f of V to $Q \cdot D$, a lattice $\mathfrak{M}$ in V, and a ρ-antihermitian form T on V. As D is a lattice in C^n, f can be extended to an R-linear isomorphism of $V_R = V \otimes_Q R$ to C^n. Then f maps V_R/D isomorphically to A (through the isomorphism of C^n/D to A), and V/D corresponds to the set of all points of finite order on A. Therefore, if we take points $t_1, \cdots, t_s$ on A of finite order, there exist points $x_i \in V$ such that $f(x_i) \bmod D$ represents x_i.

This observation leads us to the following formulation. We call a collection of objects $\Omega = (L, \Phi, \rho; V, T, \mathfrak{M}; x_1, \cdots, x_s)$ a PEL-*type* if the objects are as follows:

L: a division algebra over $\mathbf{Q}$.

ρ: a positive involution of L.

Φ: a representation of L into $M_n(\mathbf{C})$ such that $\Phi + \bar{\Phi}$ is equivalent to a rational representation.

V: a left L-module of rank m, where $m = 2n/[L:\mathbf{Q}]$.

T: a nondegenerate L-valued ρ-antihermitian form on V.

$\mathfrak{M}$: a free $\mathbf{Z}$-submodule of V of rank $2n$.

x_i: elements of V.

The consideration of the points t_i and the x_i is motivated by the need of treating congruence subgroups of discontinuous groups. We say that $\mathcal{Q} = (A, \mathcal{C}, \theta; t_1, \cdots, t_s)$ is a PEL-*structure* if

A: an abelian variety defined over a subfield of $\mathbf{C}$.

$\mathcal{C}$: a polarization of A.

θ: an isomorphism of L into $\mathrm{End}_\mathbf{Q}(A)$.

t_i: points of finite order on A.

Now we say that $\mathcal{Q}$ is of type Ω if there exists a commutative diagram

$$(4.9) \quad \begin{array}{ccccccccc} 0 & \longrightarrow & \mathfrak{M} & \longrightarrow & V_{\mathbf{R}} & \longrightarrow & V_{\mathbf{R}}/\mathfrak{M} & \longrightarrow & 0 \\ & & \downarrow & & {\scriptstyle f}\downarrow & & \downarrow & & \\ 0 & \longrightarrow & D & \longrightarrow & C^n & \overset{\xi}{\longrightarrow} & A & \longrightarrow & 0 \end{array} \qquad (V_{\mathbf{R}} = V \otimes_\mathbf{Q} \mathbf{R})$$

such that the following conditions (4.10–4.14) are satisfied.

(4.10) ξ *gives a holomorphic isomorphism of* C^n/D *to* A.

(4.11) f *is an* $\mathbf{R}$-*linear isomorphism, and* $f(\mathfrak{M}) = D$.

(4.12) $f(\alpha x) = \Phi(\alpha) f(x)$, *and* $\Phi(\alpha)$ *defines* $\theta(\alpha)$ *for every* $\alpha \in L$.

(4.13) $\mathcal{C}$ *contains a basic divisor* X *which determines a Riemann form* E *on* C^n/D *such that* $E(f(x), f(y)) = \mathrm{Tr}_{L/\mathbf{Q}}(T(x, y))$ *for* $(x, y) \in V \times V$.

(4.14) $t_i = \xi(f(x_i))$ *for every* i.

Since $\mathrm{Tr}_{L/\mathbf{Q}}(T(x, y))$ corresponds to a *basic* divisor, we have to assume

$$(4.15) \qquad\qquad \mathrm{Tr}_{L/\mathbf{Q}}(T(\mathfrak{M}, \mathfrak{M})) = \mathbf{Z}.$$

(If this is not satisfied, we can replace T by a suitable rational multiple of it satisfying this condition.)

Let $\mathcal{Q} = (A, \mathcal{C}, \theta, ; t_1, \cdots, t_s)$ and $\mathcal{Q}' = (A', \mathcal{C}', \theta'; t'_1, \cdots, t'_s)$ be two PEL-structures. An isomorphism $\lambda: (A, \mathcal{C}) \to (A', \mathcal{C}')$ is called an *isomorphism* of $\mathcal{Q}$ to $\mathcal{Q}'$ if $\lambda\theta(\alpha) = \theta'(\alpha)\lambda$ for every $\alpha \in L$ and $\lambda(t_i) = t'_i$ for every i.

We say that Ω is *equivalent* to another PEL-type $\Omega' = (L', \Phi', \rho'; V', T', \mathfrak{M}'; x'_1, \cdots, x'_{s'})$ if $L = L'$, $\rho = \rho'$, $V = V'$, $s = s'$, Φ and Φ' are equivalent as representations of L, and there exists an L-linear automorphism μ of V such that $T'(x\mu, y\mu) = T(x, y)$, $\mathfrak{M}\mu = \mathfrak{M}'$, $x_i\mu \equiv x'_i \bmod \mathfrak{M}'$ for every i. Let $\mathcal{Q}$ be of type Ω. Then $\mathcal{Q}$ is of type Ω' if and only if Ω is equivalent to Ω'. A PEL-type Ω is said to be *admissible* if there exists at least one PEL-structure of type Ω.

THEOREM 3. *For every admissible PEL-structure* $\Omega = (L, \Phi, \rho; V, T, \mathfrak{M}; x_1, \cdots, x_s)$, *there exists a bounded symmetric domain S and an analytic map $f: V_{\mathbf{R}} \times S \to \mathbf{C}^n$ holomorphic in the variable $z \in S$ with the following properties:*

(1) *If we put* $f_z(x) = f(x, z)$ *for* $x \in V_{\mathbf{R}}$ *and* $z \in S$, *then for every* $z \in S$, f_z *is an* $\mathbf{R}$*-linear isomorphism of* $V_{\mathbf{R}}$ *to* $\mathbf{C}^n$. *Further put* $D_z = f_z(\mathfrak{M})$, $A_z = \mathbf{C}^n/D_z$, $E_z(u, v)$ $= \mathrm{Tr}_{L/\mathbf{Q}}(T(f_z^{-1}(u), f_z^{-1}(v)))\,((u, v) \in \mathbf{C}^n \times \mathbf{C}^n)$, $t_i(z) = f_z(x_i) \bmod D_z$, $\theta_z(a) = \Phi(a)$ $(a \in L)$. *Then* E_z *defines a polarization* $\mathscr{C}_z$ *on* A_z, *and*

$$\mathscr{Q}_z = (A_z, \mathscr{C}_z, \theta_z; t_1(z), \cdots, t_s(z))$$

is a PEL-structure of type Ω.

(2) *Every PEL-structure of type* Ω *is isomorphic to* $\mathscr{Q}_z$ *for some* $z \in S$.

Thus we obtain a maximal family $\Sigma_\Omega = \{\mathscr{Q}_z | z \in S\}$ of PEL-structures of type Ω. Let G be an algebraic group defined over $\mathbf{Q}$ such that $G_{\mathbf{Q}}$ can be identified with the group of all L-linear automorphisms α of V satisfying $T(x\alpha, y\alpha) = T(x, y)$. Then the bounded symmetric domain S in the above theorem can be obtained as the quotient of $G_{\mathbf{R}}$ by a maximal compact subgroup. Let

$$\Gamma = \{\alpha \in G_{\mathbf{Q}} | \mathfrak{M}\alpha = \mathfrak{M}, x_i\alpha \equiv x_i \bmod \mathfrak{M}\ (i = 1, \cdots, s)\}.$$

Then Γ is commensurable with $G_{\mathbf{Z}}$ and acts naturally on S.

THEOREM 4. *Two members* $\mathscr{Q}_z$ *and* $\mathscr{Q}_w$ *of* Σ_Ω *are isomorphic if and only if* $z = \gamma(w)$ *for some* $\gamma \in \Gamma$.

In other words $\Gamma \backslash S$ is in one-to-one correspondence with all the isomorphism-classes of PEL-structures of type Ω.

5. **Classification of** L, Φ, ρ, S, **and the admissibility of** Ω. Let F be the set of elements x in the center of L such that $x^\rho = x$. Then it can be shown that F is a totally real algebraic number field. Let g denote the degree of F over $\mathbf{Q}$. According to Albert, all the division algebras L with positive involution are classified into the following four types:

(Type I) $L = F$.

(Type II) L is a totally indefinite quaternion algebra over F, i.e., L has F as its center, and $L \otimes_{\mathbf{Q}} R = M_2(\mathbf{R}) \times \cdots \times M_2(\mathbf{R})$ (g copies).

(Type III) L is a totally definite quaternion algebra over F, i.e., L has F as its center, and $L \otimes_{\mathbf{Q}} R = K \times \cdots \times K$ (g copies), where K means the division ring of real quaternions.

(Type IV) The center K of L is a totally imaginary quadratic extension of F.

If $L = F$, ρ should be the identity mapping. If L is of (Type III), the standard quaternion conjugate is the only positive involution of L. If L is of (Type II, IV) and L is not commutative, L has infinitely many positive involutions. As for (Type IV), ρ must induce the nontrivial automorphism of K over F.

Now the direct sum of Φ and $\bar{\Phi}$ is equivalent to a rational representation if and only if Φ satisfies the following condition:

(5.1) (Type I, II, III) Φ *is a multiple of a reduced representation of* L *over* Q.

(Type IV) *Let* $\tau_1, \cdots, \tau_g$ *be* g *isomorphisms of* K *into* C, *such that* $\tau_1, \cdots, \tau_g$, $\rho\tau_1, \cdots, \rho\tau_g$ *form the set of all isomorphisms of* K *into* C, *and let* $[L:K] = q^2$. *Let* qr_v, *resp.* qs_v, *be the multiplicity of* τ_v, *resp.* $\rho\tau_v$, *in the restriction of* Φ *to* K. *Then* $r_v + s_v = mq$ *for* $v = 1, \cdots, g$, *where* $m = 2n/[L:Q]$ (cf. *definition of* PEL-*type*).

The PEL-type Ω is admissible if L is of (Type I, II, III) and the conditions (4.15) and (5.1) are satisfied. When L is of (Type IV), Ω is admissible if and only if the following condition is satisfied besides (4.15) and (5.1).

(5.2) *Let* ϕ_v *be a homomorphism of* $M_m(L)$ *into* $M_{mq}(C)$ *such that* $\phi_v(a) = a^{\tau_v}1_{mq}$ *for* $a \in K$ *and*

$$\phi_v({}^tU^\rho) = {}^t\overline{\phi_v(U)} \quad for \quad U \in M_m(L).$$

Then, for each v, *the complex hermitian matrix* $\sqrt{(-1)}\phi_v(T_0)^{-1}$ *has* r_v *positive eigenvalues and* s_v *negative eigenvalues, where* T_0 *is an element of* $M_m(L)$ *which represents* T *with respect to a basis of* V *over* L.

The bounded symmetric domain S can be described as follows according to the type of L.

$$\text{(Type I)} \qquad S = S_{m/2} \times \cdots \times S_{m/2},$$

$$\text{(Type II)} \qquad S = S_m \times \cdots \times S_m,$$

$$\text{(Type III)} \qquad S = S'_m \times \cdots \times S'_m,$$

$$\text{(Type IV)} \qquad S = S_{r_1, s_1} \times \cdots \times S_{r_g, s_g}.$$

Here the number of copies is g in each case; S_r is the Siegel space of degree r (see §3); S'_m is the space all complex *alternating* matrices z of size m such that $1 - {}^t\bar{z}z$ is positive hermitian; $S_{r,s}$ is the space of all complex matrices z with r rows and s columns such that $1 - {}^t\bar{z}z$ is positive hermitian. If either $r = 0$ or $s = 0$, we understand by $S_{r,s}$ a space consisting of only one point. Therefore, if $\Sigma^g_{v=1}r_vs_v = 0$, S consists of only one point, so that Σ_Ω has only one member. For (Type III) with $m = 1$, we have also a family with the only member. In all these families with single member, the abelian variety in question has sufficiently many complex multiplications.

6. **The moduli-variety for the family** Σ_Ω. Let V be a variety in a projective space P^N, defined by a system of equations $F_i(X_0, \cdots, X_N) = 0$ ($i \in I$). Let σ be an automorphism of C. Then we denote by V^σ the variety defined by $F_i^\sigma(X_0, \cdots, X_N) = 0$ ($i \in I$), where F_i^σ is the polynomial whose coefficients are transforms of coefficients of F_i under σ. Let L and $\mathcal{Q} = (A, \mathcal{C}, \theta; t_1, \cdots, t_s)$ be as before. Then we can define $\mathcal{Q}^\sigma = (A^\sigma, \mathcal{C}^\sigma, \theta^\sigma; t_1^\sigma, \cdots, t_s^\sigma)$ as follows: $\mathcal{C}^\sigma$ is the polarization of A^σ containing a divisor X^σ for a divisor X in $\mathcal{C}$; $\theta^\sigma(a) = \theta(a)^\sigma$ for $a \in L$.

THEOREM 5. *For every PEL-structure $\mathscr{Q} = (A, \mathscr{C}, \theta; t_1, \cdots, t_s)$, there exists a unique subfield k_0 of C with the following property: σ and τ being two automorphisms of C, $\mathscr{Q}^\sigma$ is isomorphic to $\mathscr{Q}^\tau$ if and only if $\sigma = \tau$ on k_0.*

We call k_0 *the field of moduli* of $\mathscr{Q}$. For example, let $L = \mathscr{Q}$, $t_1 = \cdots = t_s = 0$, and $E = C/(Z + Z\tau)$ with $\mathrm{Im}(\tau) > 0$, so that E is an elliptic curve. Then the field of moduli of $(E, \mathscr{C}, \theta; 0, \cdots, 0)$, with obvious $\mathscr{C}$ and θ, is $Q(j(\tau))$ for the value $j(\tau)$ of the classical modular function j.

We say that a PEL-type Ω is *abnormal* if (i) $\dim(S) = 1$, (ii) $\Gamma\backslash S$ is not compact, and (iii) $L \neq Q$. Ω is said to be *normal* if at least one of these conditions is not satisfied.

THEOREM 6. *Let Ω be an admissible and normal PEL-type. For every automorphism σ of C, there exists a PEL-type Ω^σ which is determined, up to equivalence, by the following property: if $\mathscr{Q}$ is of type Ω, then $\mathscr{Q}^\sigma$ is of type Ω^σ. (Ω^σ depends only on Ω and σ, and is independent of $\mathscr{Q}$.)*

THEOREM 7. *For every admissible and normal PEL-type Ω, there exists a unique algebraic number field k_Ω of finite degree with the following properties:*

(1) An automorphism σ of C is the identity mapping on k_Ω if and only if Ω^σ is equivalent to Ω. (This property characterizes k_Ω.)

(2) If $\mathscr{Q}$ is of type Ω, then k_Ω is contained in the field of moduli of $\mathscr{Q}$.

(3) Let $\Omega = (L, \Phi, \rho; V, T, \mathfrak{M}; x_1, \cdots, x_s)$, and let K_Φ be the field generated over Q by $\mathrm{tr}\, \Phi(\alpha)$ for all α in the center of L. Then $K_\Phi \subset k_\Omega$.

Now by a result of Baily and Borel (cf. [1] and Baily's talk), one can embed $\Gamma\backslash S$ onto a Zariski open subset of a projective variety. ($\Gamma\backslash S$ may or may not be compact.) Our first main theorem asserts that there exists a nice model for $\Gamma\backslash S$.

MAIN THEOREM I. *For every admissible and normal PEL-type Ω, there exists a couple (V, v) and ψ with the following properties:*

(1) V is a Zariski open subset of a projective variety.

(2) V is defined over k_Ω.

(3) v is an "assignment" which assigns a point $v(\mathscr{Q})$ of V to every PEL-structure $\mathscr{Q}$ of type Ω; and $\mathscr{Q}$ is isomorphic to $\mathscr{Q}'$ if and only if $v(\mathscr{Q}) = v(\mathscr{Q}')$.

(4) The coordinates of $v(\mathscr{Q})$ generate over k_Ω the field of moduli of $\mathscr{Q}$.

(5) ψ is a holomorphic mapping of S into a projective space which induces a biregular morphism of $\Gamma\backslash S$ onto V, and such that $\psi(z) = v(\mathscr{Q}_z)$ for every $z \in S$, where $\mathscr{Q}_z$ is a member of the family Σ_Ω defined in Theorem 3.

(6) Let $\mathscr{Q}$ and $\mathscr{Q}'$ be of type Ω, and $\mathfrak{p}$ a C-valued place of a field of definition for $\mathscr{Q}$ such that $\mathfrak{p}(x) = x$ for $x \in k_\Omega$ and the reduction of $\mathscr{Q}$ modulo $\mathfrak{p}$ is $\mathscr{Q}'$. Then $\mathfrak{p}(v(\mathscr{Q})) = v(\mathscr{Q}')$.

We call (V, v) a *moduli-variety* for PEL-structures of type Ω, understanding that $v(\mathscr{Q})$ is the "modulus" of $\mathscr{Q}$. (V, v) is uniquely determined, up to biregular isomorphisms over k_Ω, by the above properties. Let us set $V = V(\Omega)$.

THEOREM 8. *Let Ω be an admissible and normal PEL-type, and σ an automorphism of C. Then $k(\Omega^\sigma) = k(\Omega)^\sigma$, and $V(\Omega^\sigma)$ is biregularly isomorphic to $V(\Omega)^\sigma$ over $k(\Omega)^\sigma$.*

In general, one can make the following conjecture. Let S be a bounded symmetric domain, Γ an arithmetic discontinuous group operating on S, and σ an automorphism of C. Then (i) $\Gamma\backslash S$ has a projective embedding V defined over an algebraic number field; (ii) the transform of V under σ is biregularly isomorphic to $\Gamma'\backslash S$ with another arithmetic discontinuous group Γ'.

The above results show that this conjecture is true in the case of Γ and S obtained from an admissible and normal PEL-type.

7. The number field k_Ω as a class-field.

PROPOSITION 9. *Let K_Φ be the field defined in (3) of Theorem 7. Then (i) $K_\Phi = Q$ if Φ is equivalent to $\bar\Phi$; (ii) K_Φ is a totally imaginary quadratic extension of a totally real algebraic number field if Φ is not equivalent to $\bar\Phi$.*

Therefore $K_\Phi = Q$ if L is of (Type I, II, III), and both cases $K_\Phi = Q$ and $K_\Phi \neq Q$ may occur if L is of (Type IV).

MAIN THEOREM II. *The field k_Ω is an abelian extension of K_Φ in the following cases:*
(1) *L is of (Type I) or of (Type II);*
(2) *L is of (Type IV) and commutative.*

One can make the following general conjecture: If Ω is admissible and normal, then k_Ω is an abelian extension of K_Φ except for the case where $\dim(S) = 0$ and L is of (Type III).

THEOREM 10. *Let $\mathfrak{o}$ be an order in L defined by $\mathfrak{o} = \{a \in L | a\mathfrak{M} \subset \mathfrak{M}\}$.*
(1) *Suppose that L is of (Type I or II), $\mathfrak{o}$ is a maximal order in L, and $N^{-1}\mathfrak{M} = \mathfrak{M} + \sum_{i=1}^{s} \mathfrak{o}x_i$ with a positive integer N. (The last equality means that the x_i generate the points of order N. If $x_1 = \cdots = x_s = 0$, $N = 1$.) Then $k_\Omega = Q(e^{2\pi i/N})$.*
(2) *Suppose that L is of (Type I) (hence $L = F$) and $x_1 = \cdots = x_s = 0$. Let c be the smallest positive integer such that $c^{-1}\mathfrak{o}$ contains the maximal order in F. Let H be the set of all rational integers μ, prime to c, which occur as the multiplier of an F-linear similitude α of T, i.e., $T(x\alpha, y\alpha) = \mu T(x, y)$, such that $\mathfrak{M}\alpha \subset \mathfrak{M}$. Then k_Ω is the subfield of $Q(\zeta)$, consisting of all the elements of $Q(\zeta)$ invariant under the automorphisms $\zeta \to \zeta^\mu$ for all $\mu \in H$, where $\zeta = e^{2\pi i/c}$.*

If L is of (Type IV) and $\dim(S) = 0$, the description of k_Ω as a class-field over K_Φ is exactly the theory of complex multiplication [15]. One can get a similar result also in the case $\dim(S) > 0$. To explain this, we have to introduce the notion of class and genus of lattices. Let F denote, for a while, an arbitrary algebraic number field of finite degree, and K a quadratic extension of F. Let V be a vector space over K of dimension m, and T a nondegenerate K-valued antihermitian form on V. We denote by G the unitary group of T. Let $\mathfrak{o}_F$ resp. $\mathfrak{o}_K$ be the ring of integers in F resp. K. Let $\mathfrak{M}$ be an $\mathfrak{o}_K$-lattice in V. For every prime ideal $\mathfrak{p}$ in F, let $F_\mathfrak{p}$ denote the completion of F with respect to $\mathfrak{p}$, and $K_\mathfrak{p} = K \otimes_F F_\mathfrak{p}$, $V_\mathfrak{p} = V \otimes_F F_\mathfrak{p}$. Then $V_\mathfrak{p}$ is a $K_\mathfrak{p}$-module, and T is uniquely extended to a $K_\mathfrak{p}$-valued antihermitian form on $V_\mathfrak{p}$. So we can define the unitary group $G_\mathfrak{p}$ on $V_\mathfrak{p}$. By a *class* of $\mathfrak{o}_K$-lattices in V with respect to G, we understand a maximal set of $\mathfrak{o}_K$-lattices in V whose members are transformed to each other by elements of G. By a *genus* of $\mathfrak{o}_K$-lattices in V with respect to G, we understand a maximal set Λ of $\mathfrak{o}_K$-lattices with the following properties: If $\mathfrak{M}$ and $\mathfrak{N}$ are members of Λ, then, for every prime ideal $\mathfrak{p}$ in F, there exists an element $\alpha_\mathfrak{p}$ of $G_\mathfrak{p}$ such that $\mathfrak{M}_\mathfrak{p}\alpha_\mathfrak{p} = \mathfrak{N}_\mathfrak{p}$. A genus Λ consists of a finite number of classes. Let J denote the group of ideals $\mathfrak{a}$ in K such that $N_{K/F}(\mathfrak{a}) = \mathfrak{o}_F$, and J_0 the group of principal ideals $a\mathfrak{o}_K$ such that $N_{K/F}(a) = 1$. Then the number of classes in Λ equals $2^{e(\Lambda)} \cdot [J:J_0]$, where $e(\Lambda)$ is a nonnegative integer depending on Λ. If T is indefinite and $\dim_K V$ is even, there exists a genus Λ such that $e(\Lambda) = 0$. Let Λ be such a genus and $\mathfrak{M}$ an $\mathfrak{o}_K$-lattice in Λ. Then the map $\Lambda \ni \mathfrak{N} \mapsto [\mathfrak{M}/\mathfrak{N}]$ gives a one-to-one correspondence between the classes in Λ and J/J_0. Here $[\mathfrak{M}/\mathfrak{N}]$ means a fractional ideal in K generated over $\mathfrak{o}_K$ by $\det(\alpha)$ for all K-linear automorphisms α of V such that $\mathfrak{M}\alpha \subset \mathfrak{N}$. We call such a genus *nice*. (For details, see [**10**].)

Let us come back to a PEL-type, and assume that F is totally real, and K is totally imaginary. We take L to be K.

MAIN THEOREM III. *Let* $\Omega = (K, \Phi, \rho; V, T, \mathfrak{M}; 0, \cdots, 0)$ *be an admissible PEL-type. Suppose that* T *is indefinite,* $K_\Phi \neq Q$, $\dim_K V$ *is even, and the genus of* $\mathfrak{M}$ *is nice. Let* $\mathfrak{a}$ *be an ideal in* K_Φ, *and let* σ *be an automorphism of* C *such that* σ *coincides with the Frobenius automorphism* $((k_\Omega/K_\Phi)/\mathfrak{a})$ *on* k_Ω. *Then we have*

$$\Omega^\sigma = (K, \Phi, \rho; V, T, \mathfrak{N}; 0, \cdots, 0)$$

with a lattice $\mathfrak{N}$ *belonging to the same genus as* $\mathfrak{M}$, *and*

$$[\mathfrak{M}/\mathfrak{N}] = \prod_{i=1}^{h} (\mathfrak{a}^{\sigma_i}/\mathfrak{a}^{\bar{\sigma}_i})^{v_i} \bmod(J_0),$$

where $\{\sigma_1, \cdots, \sigma_h, \bar{\sigma}_1, \cdots, \bar{\sigma}_h\}$ *is the set of all isomorphisms of* K_Φ *into* C, *and* $v_1, \cdots, v_h$ *are certain integers determined only by* K *and* Φ.

Since the class of $\mathfrak{N}$ (and hence Ω^σ) can be completely determined by $[\mathfrak{M}/\mathfrak{N}] \bmod(J_0)$, this theorem affords an "explicit reciprocity-law" for the abelian extension k_Ω/K_Φ. One can prove an analogous and somewhat complicated relation for odd m, and also in a more general case with nonzero $x_1, \cdots, x_s$.

8. **Fibre systems of abelian varieties.** We say that $\{V, W, h, f\}$ is a *fibre system of abelian varieties* defined over a field k, if the following conditions are satisfied:

(8.1) V and W are nonsingular varieties defined over k.

(8.2) h is a morphism of W into V, f is a morphism of V into W, both defined over k, such that $h \circ f$ is the identity mapping of V.

(8.3) If $Z \subset W \times V$ is the graph of h, then for every $u \in V$, Z and $W \times u$ intersect properly on $W \times V$, and $Z \cdot (W \times u) = A_u \times u$ with an abelian variety A_u whose neutral element is $f(u)$.

MAIN THEOREM IV. *Let* $\Omega = (L, \Phi, \rho; V, T, \mathfrak{M}; x_1, \cdots, x_s)$ *be an admissible and normal PEL-type, and let* $\mathfrak{o} = \{a \in L \,|\, a\mathfrak{M} \subset \mathfrak{M}\}$. *Suppose that* Γ *has no element of finite order other than the identity element. Then there exists a system*

$$\mathfrak{F} = \{V, W, h, f, Y, \Theta(a)(a \in \mathfrak{o}), f_1, \cdots, f_s\}$$

with the following properties.

(1) $\{V, W, h, f\}$ *is a fibre system of abelian varieties defined over* k_Ω.

(2) *With a suitably defined* v, (V, v) *is a moduli-variety for PEL-structures of type* Ω *(cf. Main Theorem I).*

(3) Y *is a divisor on* W, *rational over* k_Ω, *such that* $Y \cdot A_u$ *defines a polarization* $\mathscr{C}_u$ *on* A_u *for every* $u \in V$.

(4) *For every* $a \in \mathfrak{o}$, $\Theta(a)$ *is a morphism of* W *to* W *defined over* k_Ω *such that the restriction of* $\Theta(a)$ *to* A_u *is an endomorphism of* A_u *for every* $u \in V$; *denote it by* $\theta_u(a)$.

(5) *The* $f_i : V \to W$ $(i = 1, \cdots, s)$ *are sections which are morphisms, defined over* k_Ω.

(6) *For each* $u \in V$, $\mathscr{Q}_u = (A_u, \mathscr{C}_u, \theta_u; f_1(u), \cdots, f_s(u))$ *is a PEL-structure of type* Ω.

(7) *Let* $\Sigma_\Omega = \{\mathscr{Q}_z \,|\, z \in S\}$ *and* $\psi : S \to V$ *be as in Main Theorem I. Then* $\mathscr{Q}_{\psi(z)}$ *is isomorphic to* $\mathscr{Q}_z$.

9. Families of abelian varieties characterized by a certain nonholomorphic structure.

Let F be as before a totally real algebraic number field of degree g, and B a quaternion algebra over F. Let h be the number of archimedean primes of F for which B is unramified. Then

$$B \otimes_Q R = M_2(R) \times \cdots \times M_2(R) \times K \times \cdots \times K,$$

taking the $M_2(R)$, h times, and the K, $g - h$ times, (where K is the division ring of real quaternions). Let m be a positive integer and let G be an algebraic group defined over Q, which can be identified with the group

$$\{X \in \mathrm{GL}_m(B) \,|\, {}^t X^\iota \cdot X = 1_m\}$$

where ι denotes the main involution of B (ι is not a positive involution unless $h = 0$). Then the quotient S of G_R by a maximal compact subgroup is the product of h copies of the Siegel space of degree m. Let $\Gamma = G_Z$. Now this group Γ does not occur in our theory of PEL-structure, except when $m = 1$ or $g = h$. (If $g = h$, G is exactly the group attached to a PEL-type with an algebra $L = B$, which is of (Type II). If $m = 1$, Γ is commensurable with the group obtained from an algebra of (Type IV).) However, we can still construct a family Σ' of abelian varieties parametrized by $\Gamma\backslash S$, which is actually a subfamily of Σ_Ω of (Type IV). The members of Σ' can be characterized by the possession of a certain non-holomorphic endomorphism, or a certain rational 2-cohomology class, which is not of $(1, 1)$-type. It should be mentioned that there are infinitely many distinct such families attached to the same S and Γ. We can also find a projective variety, isomorphic to $\Gamma\backslash S$, defined over an algebraic number field of finite degree. This field is again of *abelian* nature. The same type of discussion can be also made for the unitary group of an ι-antihermitian form over B. A full detail of all these results will be given in [**14**].

10. **Hecke operators.** Let L be an indefinite division quaternion algebra over Q, which is, by definition, a division algebra over Q such that $L \otimes_Q R = M_2(R)$. Let $\mathfrak{o}$ be a maximal order in L. (It should be noted that if $\mathfrak{o}_1$ and $\mathfrak{o}_2$ are two maximal orders in L, there exists an element a of L such that $a\mathfrak{o}_1 a^{-1} = \mathfrak{o}_2$ and $\det(a) > 0$.) Regarding L as a subring of $M_2(R)$, let

$$\Gamma_b = \{\gamma \in \mathfrak{o} \mid \det(\gamma) = 1, \gamma \equiv 1 \bmod b\mathfrak{o}\}$$

for every positive integer b. We denote by H the complex upper half plane $\{z \in C \mid \mathrm{Im}(z) > 0\}$, and by $\mathrm{GL}_2^+(R)$ the group of elements in $M_2(R)$ with positive determinant. For

$$\alpha = \begin{bmatrix} a & b \\ c & d \end{bmatrix} \in M_2(R),$$

we put

$$\alpha' = \begin{bmatrix} d & -b \\ -c & a \end{bmatrix}.$$

Further for

$$\alpha = \begin{bmatrix} a & b \\ c & d \end{bmatrix} \in \mathrm{GL}_2^+(R)$$

and $z \in H$, we put $\alpha(z) = (az + b)/(cz + d)$ and $j(\alpha, z) = cz + d$. Then Γ_b, regarded as a discrete subgroup of $\mathrm{SL}_2(R)$, gives a properly discontinuous group of transformations on H with compact quotient $\Gamma_b\backslash H$. Moreover, $\Gamma_b\backslash H$ is a special case of $\Gamma\backslash S$ considered in §§4–8. In fact, for every $z \in H$, define a lattice D_z in C^2 by

$$D_z = \mathfrak{o} \cdot \begin{bmatrix} z \\ 1 \end{bmatrix},$$

elements of $\mathfrak{o}$ being considered as real 2×2 matrices. Then the complex torus C^2/D_z has a structure of abelian variety. Let $A_z = C^2/D_z$. For a suitable $v \in L$ such that v^2 is a negative integer, we can define a Riemann form E on C^2/D_z by

$$E\left(x\begin{bmatrix} z \\ 1 \end{bmatrix}, y\begin{bmatrix} z \\ 1 \end{bmatrix}\right) = \mathrm{Tr}_{L/Q}(vxy') \qquad (x, y \in M_2(R)).$$

Let $\mathscr{C}_z$ be the polarization of A_z determined by E. For every $\alpha \in L$, the linear transformation of C^2 given by the matrix α defines an element of $\mathrm{End}_Q(A_z)$, which we denote by $\theta_z(\alpha)$. Let t_z be the point on A_z corresponding to

$$b^{-1}\begin{bmatrix} z \\ 1 \end{bmatrix}.$$

Put $\mathscr{Q}_z = (A_z, \mathscr{C}_z, \theta_z; t_z)$. In this way we get a family $\Sigma = \{\mathscr{Q}_z | z \in H\}$, which is a special case of Σ_Ω considered in §4. Two members $\mathscr{Q}_z$ and $\mathscr{Q}_w$ are isomorphic if and only if $z = \gamma(w)$ for some $\gamma \in \Gamma_b$ (cf. Theorem 4).

From now on we always assume that $b > 2$. Then Γ_b has no elements of finite order other than the identity. Applying our Main Theorem IV to the present case, we get a fibre variety $W \to V$, of which the base V is biregularly isomorphic to $\Gamma_b \backslash H$, and each fibre is isomorphic to A_z. We can take the field of rationality to be $Q(\zeta)$ with $\zeta = e^{2\pi i/b}$. For every nonnegative integer m, one can construct, over $Q(\zeta)$, a fibre system of abelian varieties $W_m \to V$, whose fibre standing on $u \in V$ is the product of m copies of A_u, where A_u is a fibre of W on u. The purpose of the remaining part of this lecture is to determine the zeta-function of W_m in the sense of Hasse–Weil.

First we have to introduce the Hecke ring in L. Let

$$\Delta_b = \{\alpha \in \mathfrak{o} | \det(\alpha) > 0, (\det(\alpha), b) = 1\}.$$

Then, for every $\alpha \in \Delta_b$, one has $\Gamma_b \alpha \Gamma_b = \bigcup_i \Gamma_b \alpha_i$ for a finite number of elements α_i. Let $R(\Gamma_b, \Delta_b)$ denote the free Z-module consisting of all the formal finite sums $\sum_\lambda c_\lambda \Gamma_b \alpha_\lambda \Gamma_b$ with $c_\lambda \in Z$, $\alpha_\lambda \in \Delta_b$. The module $R(\Gamma_b, \Delta_b)$ is called the Hecke ring associated with Γ_b and Δ_b, when a law of multiplication is defined as follows: If $\Gamma_b \alpha \Gamma_b = \bigcup_i \Gamma_b \alpha_i$ and $\Gamma_b \beta \Gamma_b = \bigcup_j \Gamma_b \beta_j$ are disjoint unions, then

$$(\Gamma_b \alpha \Gamma_b) \cdot (\Gamma_b \beta \Gamma_b) = \sum \mu_\xi \cdot \Gamma_b \xi \Gamma_b,$$

where the summation is taken over all the distinct double cosets $\Gamma_b \xi \Gamma_b \subset \Gamma_b \alpha \Gamma_b \beta \Gamma_b$, and $\mu_\xi = $ the number of (i, j) such that $\Gamma_b \alpha_i \beta_j = \Gamma_b \xi$. (It can be shown that this number μ_ξ depends only on $\Gamma_b \xi \Gamma_b, \Gamma_b \alpha \Gamma_b, \Gamma_b \beta \Gamma_b$; it does not depend on the choice of representatives.) This law of multiplication is associative. In the special case $b = 1$, one can show that $R(\Gamma_1, \Delta_1)$ is commutative.

For a positive integer k, let $S_k(\Gamma_b)$ denote the set of all holomorphic functions f on H such that $f(\gamma(z))j(\gamma, z)^{-k} = f(z)$ for every $\gamma \in \Gamma_b$. An element of $S_k(\Gamma_b)$ is called a *holomorphic automorphic form of weight k* with respect to Γ_b. If $\alpha \in \Delta_b$, we can define the action $(\Gamma_b \alpha \Gamma_b)_k$ of $\Gamma_b \alpha \Gamma_b$ on $S_k(\Gamma_b)$ in the following way. Let $\Gamma_b \alpha \Gamma_b = \bigcup_{i=1}^d \Gamma_b \alpha_i$ be a disjoint union. For every $f \in S_k(\Gamma_b)$, we put

$$f|(\Gamma_b \alpha \Gamma_b)_k = \det(\alpha)^{k-1} \sum_{i=1}^d f(\alpha_i(z))j(\alpha_i, z)^{-k}.$$

Then $\Gamma_b \alpha \Gamma_b \to (\Gamma_b \alpha \Gamma_b)_k$ defines a representation of the ring $R(\Gamma_b, \Delta_b)$ in the complex vector space $S_k(\Gamma_b)$. The linear transformations $(\Gamma_b \alpha \Gamma_b)_k$ may be called the *Hecke operators*.

For our purpose we need also another type of Hecke operators. Let G_b denote the group of invertible elements of o/bo, and ρ a representation of G_b in a complex vector space U of finite dimension. Let $S_{k,\rho}(\Gamma_1)$ denote the set of all holomorphic maps $f : H \to U$ such that $f(\gamma(z))j(\gamma, z)^{-k} = \rho(\gamma)f(z)$ for every $\gamma \in \Gamma_1$. For $\Gamma_1\alpha\Gamma_1 = \bigcup_{i=1}^{d}\Gamma_1\alpha_i$ with $\alpha \in \Delta_b$, and for $f \in S_{k,\rho}(\Gamma_1)$, we define

$$f|(\Gamma_1\alpha\Gamma_1)_{k,\rho} = \det(\alpha)^{k-1} \sum_{i=1}^{d} \rho(\alpha_i^{-1})f(\alpha_i(z))j(\alpha_i, z)^{-k}.$$

Then $\Gamma_1\alpha\Gamma_1 \to (\Gamma_1\alpha\Gamma_1)_{k,\rho}$ defines a representation of $R(\Gamma_1, \Delta_b)$ in $S_{k,\rho}(\Gamma_1)$.

LEMMA 11. *Let ρ be a regular representation of G_b, $\chi(\alpha) = \operatorname{tr}\rho(\alpha)$, and let $\Gamma_1 = \bigcup_{i=1}^{N}\Gamma_b\gamma_i$, $N = [\Gamma_1 : \Gamma_b]$. Then, for $\xi \in \Delta_b$,*

$$\operatorname{tr}[(\Gamma_1\xi\Gamma_1)_{k,\rho}] = N^{-1} \sum_{i=1}^{N} \chi(\xi'\gamma_i) \cdot \operatorname{tr}[(\Gamma_b\gamma_i^{-1}\Gamma_b)_k \cdot (\Gamma_b\xi\Gamma_b)_k].$$

Now we define a Dirichlet series $D(s; k, b, \rho)$ by

$$D(s; k, b, \rho) = \sum (\Gamma_1\alpha\Gamma_1)_{k,\rho} \det(\alpha)^{-s},$$

where the summation is taken over all the distinct double cosets $\Gamma_1\alpha\Gamma_1$ with α in Δ_b. Then $D(s; k, b, \rho)$ can be expressed as an Euler product:

$$D(s; k, b, \rho) = \prod_{p|d_0, p\nmid b} [1 - (\Gamma_1\alpha_p\Gamma_1)_{k,\rho}p^{-s}]^{-1}$$

$$\times \prod_{p\nmid bd_0} [1 - (\Gamma_1\alpha_p\Gamma_1)_{k,\rho}p^{-s} + (\Gamma_1 p\Gamma_1)_{k,\rho}p^{1-2s}]^{-1},$$

where d_0 is the discriminant of L, and α_p is an element of o such that $\det(\alpha_p) = p$. Moreover, $D(s; k, b, \rho)$ can be continued holomorphically to the whole s-plane, and satisfies a functional equation [7], [8].

11. **Algebraic correspondences on W_m and their congruence relations.** First we observe that the fibre variety W_m, as a real analytic manifold, can be constructed as follows (cf. Kuga's lecture). Let L_R^m denote the product of m copies of $L_R = M_2(R)$. The product $\operatorname{GL}_2^+(R) \times L_R^m$ forms a group with respect to the law of multiplication

$$(\alpha, u)(\beta, v) = (\alpha\beta, v\beta' + u) \qquad (\alpha, \beta \in \operatorname{GL}_2^+(R); u, v \in L_R^m).$$

We let $\operatorname{GL}_2^+(R) \times L_R^m$ act on $H \times L_R^m$ by

$$(\alpha, x)(z, y) = (\alpha(z), y\alpha' + x) \qquad (\alpha \in \operatorname{GL}_2^+(R); x, y \in L_R^m, z \in H).$$

(Recall that $\alpha \to \alpha'$ is the main involution of $M_2(R)$. We regard L_R^m as a left and right L_R-module.) Let o^m denote the product of m copies of o. Then $\Gamma_b \times o^m$ is a discrete subgroup of $\operatorname{GL}_2^+(R) \times L_R^m$, and W_m can be obtained as the quotient:

$W_m = (\Gamma_b \times \mathfrak{o}^m)\backslash(H \times L_R^m)$. We may write: $W_0 = V = \Gamma_b \backslash H$. Then we have naturally a commutative diagram:

$$
\begin{array}{ccc}
H \times L_R^m & \xrightarrow{\phi_m} & W_m \\
\downarrow & & \downarrow \\
H & \xrightarrow{\phi_0} & V
\end{array}
$$

For an element α of Δ_b, put $\alpha^* = (\alpha, 0)$ $(\in GL_2^+(R) \times L_R^m)$, and

$$X_m = \{\phi_m(s) \times \phi_m(\alpha^* s) | s \in H \times L_R^m\} \qquad (m \geqq 0).$$

It can be proved that X_m is an algebraic subvariety of W_m defined over $Q(\zeta)$. Furthermore X_m is determined only by $\Gamma_b \alpha \Gamma_b$, and independent of the choice of the representative α. We write therefore $X_m = X_m(\Gamma_b \alpha \Gamma_b)$. If $m = 0$, we have

$$V \times V \supset X_0(\Gamma_b \alpha \Gamma_b) = \{\phi_0(z) \times \phi_0(\alpha(z)) | z \in H\}.$$

Still α being an element of Δ_b, let σ be an automorphism of $Q(\zeta)$ such that $\zeta^\sigma = \zeta^{\det(\alpha)}$. Then one can find a biregular morphism $Y_m(\alpha): W_m \to W_m^\sigma$ with the following properties:

$$(11.1) \qquad
\begin{array}{ccc}
W_m & \xrightarrow{Y_m(\alpha)} & W_m^\sigma \\
\downarrow & & \downarrow \\
V & \xrightarrow{Y_0(\alpha)} & V
\end{array}
\qquad \text{is commutative.}$$

(11.2) $Y_m(\alpha)$ depends only on the class of α mod $b\mathfrak{o}$.

(11.3) If β is another element of Δ_b, then $Y_m(\beta\alpha) = Y_m(\beta)^\sigma \circ Y_m(\alpha)$, where σ is as above.

(11.4) If $\alpha \in \Gamma_1$, then $Y_m(\alpha) = X_m(\Gamma_b \alpha \Gamma_b)$.

Let p be a prime number, and $\mathfrak{p}$ a prime ideal in $Q(\zeta)$ dividing p. We shall denote by $\mathfrak{p}(X)$ or $\tilde{X}$ the reduction modulo $\mathfrak{p}$ of an algebro-geometric object X. Now, for almost all $\mathfrak{p}$, we have a "nice reduction" $\mathfrak{p}(W_m) \to \mathfrak{p}(V)$, which forms a fibre system of abelian varieties defined over the residue field modulo $\mathfrak{p}$.

LEMMA 12. *Let p be a prime number which does not divide b, and α an element of $\mathfrak{o}$ such that $\det(\alpha) = p$. Then there exists an element δ of Γ_1 such that $\alpha^2 \equiv p\delta$ mod $b\mathfrak{o}$.*

THEOREM 13. *Let p, α and δ be as in Lemma 12, and let $\mathfrak{p}$ be a prime ideal in $Q(\zeta)$ dividing p. Then, for almost all $\mathfrak{p}$, we have*

$$\tilde{X}_m(\Gamma_b \alpha \Gamma_b) = {}^t\tilde{Y}_m(\alpha') \circ \Pi + \Pi^* \circ \tilde{Y}_m(\alpha),$$

$$p \cdot \tilde{X}_m(\Gamma_b p\delta \Gamma_b) = [{}^t\tilde{Y}_m(\alpha') \circ \Pi] \circ [\Pi^* \circ \tilde{Y}_m(\alpha)],$$

where Π is the Frobenius correspondence on $\tilde{W}_m \times \tilde{W}_m^p$, i.e., the locus of $v \times v^p$ with $v \in \tilde{W}_m$, and Π^* is the locus of $v^p \times p \cdot v$ on $\tilde{W}_m^p \times \tilde{W}_m$ with $v \in \tilde{W}_m$. (Here $p \cdot v = v + \cdots + v$ (p times) on the fibre abelian variety containing v.)

12. Calculation of the zeta-function of W_m.

Let $I(X)$ denote the intersection number of an algebraic correspondence X with the diagonal.

LEMMA 14. (Kuga [3]). *Let $d(\Gamma_b \alpha \Gamma_b)$ denote the number of right cosets in $\Gamma_b \alpha \Gamma_b$. Define integers $a(m, i, v)$ by*

$$a(0, 0, 0) = 1,$$

$$a(m, i, v) = \binom{2m}{(i + v)/2}\binom{2m}{(i - v)/2} - \binom{2m}{(i + v)/2 + 1}\binom{2m}{(i - v)/2 - 1}$$

$$\text{if } \quad i \equiv v \ \mathrm{mod}(2),$$

$$= 0 \quad \text{if } \quad i \not\equiv v \ \mathrm{mod}(2),$$

where the expressions in parentheses mean the binomial coefficient. Then, for every $m \geq 0$,

$$2^{-1} I(X_m(\Gamma_b \alpha \Gamma_b)) = \sum_{j=0}^{2m} a(m, 2j, 0) \det(\alpha)^j \, d(\Gamma_b \alpha \Gamma_b)$$

$$+ \sum_{i=0}^{4m} \sum_{v=0}^{i} (-1)^{i+1} a(m, i, v) \det(\alpha)^{(i-v)/2} \, \mathrm{Re}[\mathrm{tr}(\Gamma_b \alpha \Gamma_b)].$$

The method of proof of this "trace-formula" will be indicated in Kuga's talk. To make our later calculation smooth, we introduce the following notation. Let x, y, u be indeterminates. Define polynomials $F_n(x, y)$ (with coefficients in Z) by

$$-\frac{d}{du} \log(1 - xu + yu^2) = \sum_{n=1}^{\infty} F_n(x, y) u^{n-1}.$$

If $x = z + w$ and $y = zw$, then $F_n(x, y) = z^n + w^n$. If X and Y are commuting matrices, then

$$\frac{d}{du} \log[\det(1 - Xu + Yu^2)^{-1}] = \sum_{n=1}^{\infty} \mathrm{tr}[F_n(X, Y)] u^{n-1}.$$

From Theorem 13 we obtain easily

LEMMA 15. $F_n(\tilde{X}_m(\Gamma_b \alpha \Gamma_b), p\tilde{X}_m(\Gamma_b p \delta \Gamma_b)) = ({}^t\tilde{Y}_m(\alpha') \circ \Pi)^n + (\Pi^* \circ \tilde{Y}_m(\alpha))^n.$

LEMMA 16. *Let $\Pi_{\mathfrak{p}}^{(n)}$ be the locus of $v \times v^{p^n}$ on $\tilde{W} \times \tilde{W}^{p^n}$ with $v \in \tilde{W}$, and $\Pi_{\mathfrak{p}}^{*(n)}$ the locus of $v^{p^n} \times (p^n \cdot v)$ on $\tilde{W}^{p^n} \times \tilde{W}$ with $v \in \tilde{W}$. Let β be an element of $\mathfrak{o}$ such that $\det(\beta) = p^n$. Then*

$$I({}^t\tilde{Y}_m(\beta) \circ \Pi_{\mathfrak{p}}^{(n)}) = I(\Pi_{\mathfrak{p}}^{*(n)} \circ \tilde{Y}(\beta)).$$

Now the zeta-function $Z_{\mathfrak{p}}(u)$ of $\mathfrak{p}(W_m)$ over the residue field modulo $\mathfrak{p}$ is defined by

$$\frac{d}{du} \log Z_{\mathfrak{p}}(u) = \sum_{n=1}^{\infty} I(\Pi_{\mathfrak{p}}^{(fn)}) u^{n-1}; \qquad N(\mathfrak{p}) = p^f.$$

If $\mathfrak{q}$ is another prime ideal in $Q(\zeta)$ dividing p, then it is easily seen that $Z_{\mathfrak{q}} = Z_{\mathfrak{p}}$.

Let χ, N and $\{\gamma_i\}$ be as in Lemma 11, and let $\phi(b) = [G_b : \Gamma_1/\Gamma_b]$. Then

$$(Z) \qquad \frac{d}{du} \log\left[\prod_{\mathfrak{p}|p} Z_\mathfrak{p}(u^f)\right] = \phi(b) \sum_{n=1}^{\infty} I(\Pi_\mathfrak{p}^{(fn)})u^{(n-1)f} \cdot u^{f-1}$$

$$= N^{-1} \sum_{n=1}^{\infty} u^{n-1} \sum_{\sigma}{}' \chi(\sigma)I({}^t\tilde{Y}_m(\sigma) \circ \Pi_\mathfrak{p}^{(n)}).$$

Here we write $Y_m(\sigma) = Y_m(\alpha)$ if $\alpha \bmod b\mathfrak{o}$ represents σ, and $\sum_{\sigma}'$ is the summation taken over all $\sigma \in G_b$ such that $\det(\alpha) = p^n$. We see easily

$$(Z) = N^{-1} \sum_{n=1}^{\infty} u^{n-1} \sum_{j=1}^{N} \chi(\alpha'^n\gamma_j)I({}^t\tilde{Y}_m(\alpha'^n\gamma_j) \circ \Pi_\mathfrak{p}^{(n)})$$

with an element α of $\mathfrak{o}$ such that $\det(\alpha) = p$. Similarly

$$(Z) = N^{-1} \sum_{n=1}^{\infty} u^{n-1} \sum_{j=1}^{N} \chi(\gamma_j'\alpha^n)I({}^t\tilde{Y}_m(\gamma_j'\alpha^n) \circ \Pi_\mathfrak{p}^{(n)})$$

$$= N^{-1} \sum_{n=1}^{\infty} u^{n-1} \sum_{j=1}^{N} \chi(\gamma_j'\alpha^n)I(\Pi_\mathfrak{p}^{*(n)} \circ \tilde{Y}_m(\gamma_j'\alpha^n)) \quad \text{(Lemma 16)}.$$

By (11.3) we have $\Pi_\mathfrak{p}^{*(n)} \circ \tilde{Y}_m(\gamma_j'\alpha^n) = {}^t\tilde{Y}_m(\gamma_j) \circ \Pi_\mathfrak{p}^{*(n)} \circ \tilde{Y}_m(\alpha^n)$, hence

$$\frac{d}{du} \log\left[\prod_{\mathfrak{p}|p} Z_\mathfrak{p}(u^f)\right] = (2N)^{-1} \sum_{n=1}^{\infty} u^{n-1} \sum_{j=1}^{N} \chi(\alpha'^n\gamma_j)I({}^t\tilde{Y}_m(\gamma_j) \circ U_n),$$

where $U_n = {}^t\tilde{Y}_m(\alpha'^n) \circ \Pi_\mathfrak{p}^{(n)} + \Pi_\mathfrak{p}^{*(n)} \circ \tilde{Y}_m(\alpha^n)$. By Lemma 15 and the property (11.4), we have $U_n = F_n(\tilde{X}_m(\Gamma_b\alpha\Gamma_b), p\tilde{X}_m(\Gamma_b p\delta\Gamma_b))$. Applying Lemma 14 to this correspondence, we find

$$\frac{d}{du} \log\left[\prod_{\mathfrak{p}|p} Z_\mathfrak{p}(u^f)\right] = \sum_{j=0}^{2m} a(m, 2j, 0)P_j + \sum_{i=0}^{4m} \sum_{v=0}^{i} (-1)^{i+1}a(m, i, v)Q_{iv}$$

with

$$P_j = N^{-1} \sum_{n=1}^{\infty} u^{n-1} \sum_{j=1}^{N} \chi(\alpha'^n\gamma_j)p^{ni}F_n(d(\Gamma_b\alpha\Gamma_b), p \cdot d(\Gamma_b p\delta\Gamma_b)),$$

$$Q_{iv} = N^{-1} \sum_{n=1}^{\infty} u^{n-1} \sum_{j=1}^{N} \chi(\alpha'^n\gamma_j)p^{n(i-v)/2}$$

$$\cdot \operatorname{Re}\{\operatorname{tr}[(\Gamma_b\gamma_j\Gamma_b)_{v+2} \cdot F_n((\Gamma_b\alpha\Gamma_b)_{v+2}, p \cdot (\Gamma_b p\delta\Gamma_b)_{v+2})]\}.$$

By Lemma 11, we get

$$\prod_{\mathfrak{p}|p} Z_\mathfrak{p}(u^f) = \prod_{j=0}^{m} \left[\prod_{\mathfrak{p}|p}(1 - N(\mathfrak{p})^j u^f)(1 - N(\mathfrak{p})^{j+1}u^f)\right]^{-a(m,2j,0)}$$

$$\cdot \prod_{i=0}^{4m} \prod_{v=0}^{i} \det[1 - (\Gamma_1\alpha\Gamma_1)_{v+2,\rho}p^{(i-v)/2}u$$

$$+ (\Gamma_1 p\Gamma_1)_{v+2,\rho}p^{i-v+1}u^2]^{(-1)^i a(m,i,v)}.$$

Now we define the global zeta function $Z(s; W_m, Q(\zeta))$ by

$$Z(s; W_m, Q(\zeta)) = \prod_{\mathfrak{p}} Z_{\mathfrak{p}}(N(\mathfrak{p})^{-s}),$$

the product being taken over all prime ideals $\mathfrak{p}$ in $Q(\zeta)$ with nice reduction of W_m. The above calculation proves

MAIN THEOREM V. *Let* $Z(s; Q(\zeta))$ *denote the Dedekind zeta function of* $Q(\zeta)$. *Then, up to a finite number of* $\mathfrak{p}$*-factors,* $Z(s; W_m, Q(\zeta))$ *is equal to the product*

$$\prod_{j=0}^{m} [Z(s - j; Q(\zeta))Z(s - j - 1; Q(\zeta))]^{a(m,2j,0)}$$

$$\times \prod_{i=0}^{4m} \prod_{v=0}^{i} \det[D(s - (i - v)/2; b, v + 2, \rho)]^{(-1)^{i+1}a(m,i,v)}.$$

Here ρ *is the regular representation of* G_b.

COROLLARY. *Let* α *be an element of* $\mathfrak{o}$ *such that* $\det(\alpha) = p$. *If Weil's conjecture on* $Z_{\mathfrak{p}}(u)$ *is true, then, for every representation* ψ *of* G_b *and for every integer* $k > 2$, *the absolute value of the characteristic roots of* $(\Gamma_1 \alpha \Gamma_1)_{k,\psi}$ *and* $(\Gamma_b \alpha \Gamma_b)_k$ *does not exceed* $2p^{(k-1)/2}$, *except for a finite number of exceptional* p's.

Since Weil's conjecture is true for curves, the absolute value of the characteristic roots of $(\Gamma_1 \alpha \Gamma_1)_{2,\psi}$ and $(\Gamma_b \alpha \Gamma_b)_2$ do not exceed $2 \cdot p^{\frac{1}{2}}$ for almost all p.

THEOREM 17. *Suppose that* b *is prime to the discriminant of* L. *Then* W_m *has a model defined over* Q, *and the zeta-function* $Z_{\mathfrak{p}}(u)$ *of* $W_m \bmod(p)$ *over the prime field is given by*

$$Z_{\mathfrak{p}}(u) = \prod_{j=0}^{2m} [(1 - p^j u)(1 - p^{j+1}u)]^{-a(m,2j,0)}$$

$$\times \prod_{i=0}^{4m} \prod_{v=0}^{i} \det[1 - (\Gamma_b \alpha \Gamma_b)_{v+2} p^{(i-v)/2}u + (\Gamma_b p \delta \Gamma_b)_{v+2} p^{i-v+1}u^2]^{(-1)^i a(m,i,v)}.$$

In view of the results of Eichler [2] and Shimizu [6], we know that there are some linear relations between the Dirichlet series $D(s; b, k, \rho)$ and Hecke's Dirichlet series attached to cusp forms with respect to congruence subgroups of $SL_2(Z)$. Therefore, from the above corollary, one can derive an estimate for the Fourier coefficients of certain cusp forms, assuming Weil's conjecture to be true. For example, let

$$f(z) = [\Delta(z)\Delta(2z)\Delta(3z)\Delta(6z)]^{\frac{1}{2}} = \sum_{n=1}^{\infty} a_n e^{2\pi i n z},$$

where $\Delta(z) = q \cdot \prod_{n=1}^{\infty}(1 - q^n)^{24}$, $q = e^{2\pi i z}$. Then $f(z)$ is a modular form of weight $k = 4$ and of level 6. One can show that $\sum_{n=1}^{\infty} a_n n^{-s}$ is among our $D(s; k, b, \rho)$ for the quaternion algebra of discriminant 6, with $k = 4, b = 1$, $\rho = $ identity representation. Hence, Weil's conjecture together with the above result implies $|a_p| < 2p^{\frac{1}{2}}$ for almost all p.

By a simple observation about the fixed points of $X_0(\Gamma_b \xi \Gamma_b)$, together with the trace-formula of Eichler and Selberg, we can obtain the following "asymptotic estimate" for the eigenvalues. Let p be a prime number not dividing b and the discriminant of L, and let α be an element of $\mathfrak{o}$ such that $\det(\alpha) = p$. Let $r = \dim S_k(\Gamma_b)$ and let $\lambda_1, \cdots, \lambda_r$ be the eigenvalues of $(\Gamma_b \alpha \Gamma_b)_k$. Suppose that $k \geq 3$ and $b > 2$. Then

$$r^{-1} \sum_{i=1}^{r} |\lambda_i|^2 \leq (1 + p)p^{k-2}[1 + A_k(1 + (p + 1)/(g - 1))].$$

Here g is the genus of $\Gamma_b \backslash H$, and A_k is a positive constant which depends only on k; $A_k = 1$ if k is odd, and $A_k \leq 1/3$ if k is even. Let us now take a sequence of positive integers $b_1, b_2, \cdots$ which tends to infinity such that all the b_ν are prime to p, and consider a sequence of groups $\Gamma_{b_1}, \Gamma_{b_2}, \cdots$. Let $v_p(b_\nu)$ denote

$$r^{-1} \sum_{i=1}^{r} |\lambda_i|^2$$

defined for $\Gamma_{b_\nu} \alpha \Gamma_{b_\nu}$. Then, for a fixed p, we have

$$\limsup_{\nu \to \infty} v_p(b_\nu) \leq (4/3) \cdot (1 + p)p^{k-2} \qquad \text{if } k \text{ is even,}$$
$$\leq 2(1 + p)p^{k-2} \qquad \text{if } k \text{ is odd.}$$

This is neither stronger nor weaker than the conjecture $|\lambda_i| \leq 2p^{(k-1)/2}$.

References

1. W. L. Baily, Jr. and A. Borel, *On the compactification of arithmetically defined quotients of bounded symmetric domains*, Bull. Amer. Math. Soc. **70** (1964), 588–593.

2. M. Eichler, *Quadratische Formen und Modulfunktionen*, Acta Arith. **4** (1958), 217–239.

3. M. Kuga, *Fibre varieties over a symmetric space whose fibres are abelian varieties*, Lecture Notes, University of Chicago, Chicago, Illinois, 1963–1964.

4. M. Kuga and G. Shimura, *On the zeta-function of a fibre variety whose fibres are abelian varieties*, Ann. of Math. (2) **82** (1965), 478–539.

5. S. Lang, *Abelian varieties*, Interscience, New York, 1959.

6. H. Shimizu, *On zeta-functions of quaternion algebras*, Ann. of Math. (2) **81** (1965), 166–193.

7. G. Shimura, *On the zeta-functions of the algebraic curves uniformized by certain automorphic functions*, J. Math. Soc. Japan **13** (1961), 275–331.

8. ———, *On Dirichlet series and abelian varieties attached to automorphic forms*, Ann. of Math. (2) **76** (1962), 237–294.

9. ———, *On analytic families of polarized abelian varieties and automorphic functions*, Ann. of Math. (2) **78** (1963), 149–192.

10. ———, *Arithmetic of unitary groups*, Ann. of Math. (2) **79** (1964), 369–409.

11. ———, *On the field of definition for a field of automorphic functions*, Ann. of Math. (2) (a) I, **80** (1964), 160–189; (b) II, ibid. **81** (1965), 124–165; (c) III, ibid. (to appear).

12. ———, *Moduli and fibre systems of abelian varieties*, Ann. of Math. (to appear).

13. ———, *Class-fields and automorphic functions*, Ann. of Math. (2) **80** (1964), 444–463.

14. ———, *Discontinuous groups and abelian varieties*, (to appear).

15. G. Shimura and Y. Taniyama, *Complex multiplication of abelian varieties and its applications to number theory*, Publ. Math. Soc. Japan, No. 6, Math. Soc. Japan, Tokyo, 1961.

16. A. Weil, *Variétés abéliennes et courbes algébriques*, Hermann, Paris, 1948.

17. ———, *Introduction à l'étude des variétés kählériennes*, Hermann, Paris, 1958.

Notes I

Numbers in brackets set in boldface such as [**79a**] mean items in the list of articles; those in roman such as [17] and [94b] are references originally cited in the article to which notes are given. Some of the corrections were already made in the original articles. For example, a list of corrections to [**67b**] was given at the end of [**67c**], but they are superseded by the corrections made in these notes.

54. A note on the normalization-theorem of an integral domain

p. 8, line 3: Strictly speaking, Lemma 5 does not apply here, since K is an algebraic number field in that lemma. However, we easily see that Lemma 5 is valid for the field of quotients K of a Dedekind domain J, if we understand that "the prime divisors" of that lemma are those corresponding to the prime ideals of J. This justifies the reasoning of page 8.

55. Reduction of algebraic varieties with respect to a discrete valuation of the basic field

p. 136, line 3: For "$\mathfrak{K}$" read "$\mathfrak{R}$".

p. 146, Proposition 17: It is stated that the proposition can be proved by the same argument as in [7, Chapter V, Section 3, Proposition 19]. The first part that asserts $\partial \overline{F}/\partial Z(\eta, \zeta) \neq 0$ can be proved, indeed, in the same manner, but the remaining part requires some new ideas. A complete proof is given in [**59d**, p. 206, Theorem and Lemma 1].

p. 150, last line: For "Obviously, (η)" read "Obviously, (ξ)".

p. 151, first line: For "(τ_{ij}, τ_i)" read "$(\delta_{ij}, \varepsilon_i)$".

p. 155, Corollary to Theorem 10: This corollary should read as follows: Let V be a variety defined over k and $\mathfrak{V}$ a component of $\overline{V}$. If $\mathfrak{V}$ is simple on V, then $\mu(V, \mathfrak{V}) = 1$ and $[\mathfrak{V} : \kappa]_i = 1$.

p. 160, footnote 10': For "$1 \leq i \leq \tau$" read "$1 \leq i \leq r$".

p. 162, line 14 from the bottom: For "$\mathfrak{g}_\alpha$" read "$\mathfrak{F}_\alpha$".

Section 6: Better formulations of the results of this part can be given by changing the meaning of "for almost all" as follows: *We say that a statement S holds for almost all* $\mathfrak{p}_\lambda$ *if there exist a finite number of elements* $a_1, \ldots, a_t$ *of K such that S holds whenever* $a_i \notin \mathfrak{p}_\lambda$ *for* $1 \leq i \leq t$. Then we can easily verify that Propositions 29 and 30, Lemma 3, and Theorem 26 are true with this new terminology, with no change of the methods of proof. This fact is noted in [**61a**, §12.2], which is reproduced in [**98**, §12.2].

58a. Correspondances modulaires et les fonctions zeta de courbes algébriques

p. 4, line 19: Insert "X" after "coordonnées".

58b. Modules des variétés abéliennes polarisées et fonctions modulaires

Exposé I. Lemma 3 was originally attributed to Matsusaka, which turned out to be wrong; see notes to [**59b**].

The field of moduli (le corps du module) of (A, C) is defined in Matsusaka [1, Section 4] in a somewhat different way, but it can easily be seen that it coincides with ours. However, the property of (W, h) in that paper, that is, [1, Theorem 8, (iii)] corresponding to (W3) at the end of the present article is stated without the isomorphism σ. But that property combined with our Theorem 2 gives (W3). A more direct and self-contained proof of the existence of (W, h) satisfying (W1), (W2), (W3) is given in [**61a**, p. 30, Proposition 16; p. 35, Theorem 3], which is reproduced in [**98**, p. 28, Proposition 16; p. 32, Theorem 3].

Exposé II. The following fact is employed in the proof of Lemmas 6 and 7:

If $G_0, \ldots, G_r$ *are relatively prime polynomials with complex coefficients and* $G_0 \neq 0$, *then there exist r complex numbers* h_λ *such that* G_0 *and* $\sum_{\lambda=1}^{r} h_\lambda G_\lambda$ *are relatively prime.*

Here is a simple proof. Clearly we may assume that G_0 is not a constant. For each non-constant factor g of G_0 put $V_g = \{(h_\lambda) \in \mathbf{C}^r | g \text{ divides } \sum_{\lambda=1}^{r} h_\lambda G_\lambda\}$. Suppose G_0 and $\sum_{\lambda=1}^{r} h_\lambda G_\lambda$ are not relatively prime for every $(h_\lambda) \in \mathbf{C}^r$; then $\mathbf{C}^r = \bigcup_g V_g$. Since each V_g is a vector space, we have $\mathbf{C}^r = V_g$ for some g, which means that g divides G_λ for $1 \leq \lambda \leq r$, a contradiction.

We apply this to (ii) and (iii). Take (h_λ) so that $\sum_{\lambda=1}^{r} h_\lambda G_\lambda(u, y)$ and $G_0(u, y)$ are relatively prime. From (ii) we see that $G_0(u, y)$ divides "$\sum_{\lambda=1}^{r} h_\lambda \cdot G_\lambda(u, y) \sum_{\nu=1}^{p} \alpha'_\nu M_\nu(u)$, and hence $G_0(u, y)$ divides $\sum_{\nu=1}^{p} \alpha'_\nu M_\nu(u)$, as stated there. Similarly from (iii) we see that $G_0(u, v_1)$ divides $H_0(u, v_1)$ as stated there.

The proof of (iv) is as follows: Put $\psi_k(u, s) = f_k(u, s)/f_0(u, s)$ for $1 \le k \le N$. Since $L(t')$ is transversal to the tangent space to $V(s_0)$ at p_1, we can find $N - n$ polynomials $H_\lambda(X_1, \ldots, X_N)$ $(1 \le \lambda \le N - n)$ such that $H_\lambda\big(\psi_1(u, s), \ldots, \psi_N(u, s)\big) = 0$ and

$$(1) \qquad \det \begin{bmatrix} \partial H_1/\partial X_1 & \cdots & \partial H_{N-n}/\partial X_1 & t'_{11} & \cdots & t'_{n1} \\ \cdots & \cdots & \cdots & \cdots & \cdots & \cdots \\ \partial H_1/\partial X_N & \cdots & \partial H_{N-n}/\partial X_N & t'_{1N} & \cdots & t'_{nN} \end{bmatrix} \ne 0$$

at the affine point $(X_\lambda) = \big(\psi_\lambda(u^1, s_0)\big)$. Since $f_k = f_0\psi_k$, we have

$$(2) \qquad \sum_{k=1}^{N} \partial H_\lambda/\partial X_k(\partial f_k/\partial u_j - \psi_k \partial f_0/\partial u_j) = f_0 \sum_{k=1}^{N} \partial H_\lambda/\partial X_k \cdot \partial \psi_k/\partial u_j = 0.$$

Subtracting ψ_k times the first column of (S1) from the column including f_k for each $k > 0$, we find that

$$(3) \qquad \operatorname{rank} \begin{bmatrix} \partial f_1/\partial u_1 - \psi_1 \partial f_0/\partial u_1 & \cdots & \partial f_N/\partial u_1 - \psi_N \partial f_0/\partial u_1 \\ \cdots & \cdots & \cdots \\ \partial f_1/\partial u_n - \psi_1 \partial f_0/\partial u_n & \cdots & \partial f_N/\partial u_n - \psi_N \partial f_0/\partial u_n \end{bmatrix} = n.$$

Multiply the matrix of (1) on the right of the matrix of (3). Then, by (2) we obtain $\operatorname{rank}[0 \quad A] = n$ with $A_{ji} = \sum_{k=1}^{N} t'_{ik}(\partial f_k/\partial u_j - \psi_k \partial f_0/\partial u_j)(u^1, s_0)$. Since $p_1 \in L(t')$, we have $\sum_{k=1}^{N} t'_{ik} f_k(u^1, s_0) = -t'_{i0} f_0(u^1, s_0)$, and hence

$$\sum_{k=1}^{N} t'_{ik}(\psi_k \partial f_0/\partial u_j)(u^1, s_0) = -t'_{i0}(\partial f_0/\partial u_j)(u^1, s_0).$$

Thus $A_{ji} = \sum_{k=1}^{N} \big(t'_{ik}\partial f_k/\partial u_j + t'_{i0}\partial f_0/\partial u_j\big)(u^1, s_0) = (\partial \varphi_i/\partial u_j)(t', u^1, s_0)$, and so we obtain (iv).

Exposé III. In the proof of Theorem 4 we employed the existence of elements $a_1, \ldots, a_t$ in connection with the irreducibility of a specialization of $\mathfrak{F}_\delta(z_0)$. Such a_i are guaranteed by [**55**, Proposition 30]. Strictly speaking, we have to modify the terminology; see notes to [**55**, Section 6]. See also [**61a**, §12.2, Proposition 17], which is reproduced in [**98**, §12.2, Proposition 17].

59a. Fonctions automorphes et correspondances modulaires

p. 336, line 13 from the bottom: For "$M(\tau_\nu)$" read "$M(\alpha_\nu)$".

59b. On the theory of automorphic functions

p. 102, line 2 from the bottom: Insert "$\{(f) + X \mid f \in L(X)\}$ has no fixed component and" after "if".

p. 103, Lemma 3: This is due to Nishi, as noted in [**58b**, Exposé I]. In the spring of 1958, in Paris, Weil told me that the result was communicated to him by Matsusaka. Therefore, in that exposition I originally attributed it to Matsusaka. Later the involvement of Nishi was mentioned, and by the time I was preparing the present article in Princeton, the result was known to most experts. However, I was not sure of a proper attribution, which is why Lemma 3 is stated with no name attached. I believe that the result is completely due to Nishi, and also that Weil was not responsible for the misattribution.

p. 109, line 9: Insert "of K'" after "K'_1".

p. 110, line 16: Insert "of" before "moduli".

p. 111, line 17: For "on $\mathfrak{r}'$" read "to $\mathfrak{r}'$".

p. 116, line 4 from the bottom: For "hyerplane" read "hyperplane".

p. 117, lines 17, 18: These two lines should read: "and a positive integer g such that

$$\det(U) = \pm 1 \text{ and } {}^tUEU = g \begin{bmatrix} 0 & -\delta \\ \delta & 0 \end{bmatrix}."$$

p. 118, line 2: The second Λ should read Λ_δ.

p. 121, line 16: Insert "be" after "$\Psi(z)$".

p. 123, line 6: Insert "over $\mathbf{Q}$" after "Ψ".

p. 127, line 5 from the bottom: For Proposition 2" read "Proposition 12".

p. 129, line 8: For "$\alpha \in \mathfrak{a}$" read $\alpha \in \mathfrak{o}$".

p. 130, line 14 from the bottom: Insert "positive" after "X is".

p. 131, line 2: Insert "positive non-degenerate" before "divisor".

p. 139, line 7: For "dy_2" read "dx_2".

p. 143, last line: Insert "nonzero" before "elements".

59c. Sur les intégrales attachées aux formes automorphes

Section 1: Though all the statements are correct, we have to take $\mathbf{C}$ as the basic field instead of $\mathbf{R}$, since the symbols are later employed in that case.

p. 300, line 2 from the bottom: For "$\sum_{m=1}^{e_\mu-1}$" read "$\sum_{m=0}^{e_\mu-1}$".

p. 302, line 5 from the bottom: For "Z" read "$\mathbf{Z}$".

p. 305, line 13: Insert "et de ξ" after "$\{\rho_\mu\}$".

p. 308, line 15: For "$2h(m)$" read "$\leq 2h(m)$".

p. 308, line 16: For "C" read "$\mathbf{C}$".

p. 308, line 15 from the bottom: For "$h(m)$" read "$\leq h(m)$".

p. 309, lines 9, 10: For "le produit ... «Grössencharakteren»" read "la fonction L attachée à un «Grössencharakter»".

59d. On specializations of abelian varieties

p. 191, line 18 from the bottom: Insert "Suppose that f is defined along $\widetilde{V}$." after "$F(x, y) = f(x)f(y)^{-1}$."

p. 196, lines 4, 5 from the bottom: The functions f_i/f_0 are defined and finite at $\bar{a}$ for the following reason. Since $(f_i/f_0) = (f_i) + X - ((f_0) + X)$ and $(f_0) + X \succ 0$, we have $(f_i/f_0)_\infty \prec (f_0) + X$, and so $\mathfrak{p}((f_i/f_0)_\infty) \prec (\widetilde{f_0}) + \widetilde{X}$; thus $\bar{a} \notin \mathfrak{p}((f_i/f_0)_\infty)$. Now, in general, if φ is a function on a $\mathfrak{p}$-simple $\mathfrak{p}$-variety V and ξ is a point on $\widetilde{V}$ simple on V, then φ is defined and finite at ξ if and only if $\xi \notin \mathfrak{p}((\varphi)_\infty)$. Applying this fact to f_i/f_0 and $\bar{a}$, we obtain the desired result.

p. 209, line 10 from the bottom; p. 210, line 14 from the bottom: For "Theorem 1" read "the theorem on page 206".

61b. On the zeta functions of the algebraic curves uniformized by certain automorphic functions

p. 276, line 6: For "lead" read "leads".

p. 279, line 16: The congruence "$b \equiv N(\beta)$ mod. $\mathfrak{a}$" should be understood in the sense of multiplicative congruence.

p. 281, line 17: The numbers $p + 1$ and 1 should be interchanged.

p. 286, (14): This equality should read $g^*(u) = N(\alpha)^{\kappa/2-1} \sum_\nu f^*(\beta_\nu u)$ with $\beta_\nu = N(\alpha)^{-1/2}\alpha_\nu$.

p. 286, line 7 from the bottom: For "$\kappa - 1$" read "$\kappa/2 - 1$".

p. 286, line 5 from the bottom: For "$\kappa + 1$" read "$\kappa/2 + 1$".

p. 293, line 17: For "$\mathfrak{T}_\kappa(p, \mathfrak{o})$" read "$\mathfrak{T}_\kappa(p, \mathfrak{o})^{-1}$".

p. 304, line 21: For "A" read "$\varphi(A)$".

p. 304, line 8 from the bottom: For "P_M" read "P^M".

p. 316, line 6 from the bottom: For "$\mathbf{p}(X)$" read "$\mathbf{p}_1(X)$".

p. 319, line 18 from the bottom: For "$\mathfrak{g}_0 = \{0\}$" read "$\widetilde{\mathfrak{g}}_0 = \{0\}$".

p. 319, line 17 from the bottom: For "$\mathfrak{g}_i$ is " read "$\widetilde{\mathfrak{g}}_i$ is ".

p. 319, line 16 from the bottom: For "3]" read "13]".

62a. On Dirichlet series and abelian varieties attached to automorphic forms

p. 245, line 18: For "$\mathfrak{k}$" read "$\mathfrak{K}$".

p. 257, line 6: For "$\mathrm{Im}\big(\alpha_i^r(z_r)^{n_r}$" read "$\mathrm{Im}\big(\alpha_i^r(z_r)\big)^{n_r}$".

p. 259, line 19: For "c_p" read "c_n".

p. 259, line 6 from the bottom: For "σ" read "$\sigma_{\mathfrak{p}}$".

p. 263, line 2: The left-hand side of the equality should read "$\varphi_{\mathfrak{p}}(x_{\mathfrak{p}})$".

p. 266, line 16: For "$N_{D/)}$" read "$N_{D/\mathbf{Q}}$".

p. 273, line 13 from the bottom: For "operator" read "operators".

p. 273, line 8 from the bottom: For "$N_{N/Q}$" read "$N_{D/Q}$".

p. 274, line 16 from the bottom: For "replaing" read "replacing".

p. 277, line 7 from the bottom: For "$\mathrm{tr}(X)$" read "$e\big(\mathrm{tr}(X)\big)$".

p. 287, line 3 from the bottom: For "ω" read "w".

p. 292, line 6 from the bottom: For " properties " read " property ".

p. 293, line 7 from the bottom: For " Abgebren " read " Algebren ".

62b. On the class-fields obtained by complex multiplication of abelian varieties

p. 40, line 1: For "$\tau_1\sigma$" read "$\tau_1\rho$".

p. 40, line 19: For "F_0" read "F_0^γ".

p. 44, line 12: Insert " of " after " composite ".

63b. On analytic families of polarized abelian varieties and automorphic functions

p. 151, line 12 from the bottom: Insert " over the center" after " a matric algebra".

p. 153, footnote 3: For the proof of this fact, see Theorem 1 below.

p. 153, line 12: For " poitive " read " positive ".

p. 159, (21): A simple proof of the equivalence (13) $\Longleftrightarrow$ (21) can be given as follows. Given $U = (u_{ij}) \in M_m(L_{\mathbf{R}})$, put $\mathfrak{x}'_i = \sum_{j=1}^m \Phi(u_{ij})\mathfrak{x}_j$ for $1 \le i \le m$ and denote by X'_ν the matrix of (17) defined with $\mathfrak{x}'_i$ in place of $\mathfrak{x}_i$. Then $X'_\nu = X_\nu \cdot {}^t\omega_\nu(U)$ for $1 \le \nu \le g$. Suppose now $\det(X_\nu) \ne 0$ for every ν and $\sum_{j=1}^m \Phi(a_j)\mathfrak{x}_j = 0$ with $a_j \in L_{\mathbf{R}}$. Define $U = (u_{ij})$ with $u_{ij} = a_j$ for every (i, j). Then $\mathfrak{x}'_i = 0$, so that $X_\nu \cdot {}^t\omega_\nu(U) = 0$; thus $\omega_\nu(U) = 0$ for every ν, which means that $a_1 = \cdots = a_m = 0$. Since $2n = m[L : \mathbf{Q}]$, we obtain (13). This proves (21) $\Longrightarrow$ (13). Now we can easily find the $\mathfrak{x}_i$ satisfying (21). Suppose (13) is satisfied by some $\mathfrak{x}'_i$ in place of $\mathfrak{x}_i$, that is, $\mathbf{C}^n = \sum_{i=1}^m \Phi(L_{\mathbf{R}})\mathfrak{x}'_i$. Then $\mathfrak{x}_i = \sum_{j=1}^m \Phi(u_{ij})\mathfrak{x}'_j$ with some $u_{ij} \in L_{\mathbf{R}}$, and so $X_\nu = X'_\nu \cdot {}^t\omega_\nu(U)$. Since $\det(X_\nu) \ne 0$, we have $\det(X'_\nu) \ne 0$, which proves that (13) $\Longrightarrow$ (21).

p. 165, line 17: Insert a comma between "z'" and "T'".

p. 192, Reference 4: For " 1938" read " 1958".

The classification of simple algebras with involutions over number fields were essentially done by Albert, as mentioned on pp. 152–153. In the case of involutions of the first kind, the algebra must be a matrix algebra over a field or a division quaternion algebra. Therefore the question of positive involutions reduces to Propositions 1, 2, and 3. As for involutions of the second kind, we have the following two theorems. However, we do not really need these results in various investigations, since we can always start with the setting that the algebra in question has a positive involution of the second kind, as we did in this article.

Theorem 1. *Let L be a central simple algebra over F_0 of (Type IV) as in Proposition 1 with an involution σ whose restriction to F_0 generates $\mathrm{Gal}(F_0/F)$. Then L has a positive involution that coincides with σ on F_0.*

PROOF. Let $[F : \mathbf{Q}] = g$ and $[L : F_0] = q^2$; then $L_{\mathbf{R}} = M_q(\mathbf{C})^g$ as in §2.1. Denote by the same letter σ the $\mathbf{R}$-linear extension of σ to $L_{\mathbf{R}}$. The center of $L_{\mathbf{R}}$ is $\mathbf{C}^g$, and for $c = (c_\nu)_{\nu=1}^g \in \mathbf{C}^g$ we have $c^\sigma = (\overline{c}_\nu)_{\nu=1}^g$. Therefore, for $x = (x_\nu)_{\nu=1}^g \in M_q(\mathbf{C})^g$ we have $x^\sigma = (a_\nu \cdot {}^t\overline{x}_\nu a_\nu^{-1})_{\nu=1}^g$ with $a_\nu \in GL_q(\mathbf{C})$. Since $(x^\sigma)^\sigma = x$, we see that ${}^t\overline{a}_\nu = e_\nu a_\nu$ with $e_\nu \in \mathbf{C}$ such that $e_\nu\overline{e}_\nu = 1$. Take $b_\nu \in \mathbf{C}^\times$ so that $e_\nu = b_\nu/\overline{b}_\nu$. Changing a_ν for $b_\nu a_\nu$, we may assume that ${}^t\overline{a}_\nu = a_\nu$. Put $W = \{w \in M_q(\mathbf{C}) | {}^t\overline{w} = w\}$, $Y = \{y \in L | y^\sigma = y\}$, and $Y_{\mathbf{R}} = Y \otimes_{\mathbf{Q}} \mathbf{R}$. Then $Y_{\mathbf{R}}$ can be identified with $\prod_{\nu=1}^g Wa_\nu^{-1}$. Let W_+ be the set of all positive definite elements of W. Then $\prod_{\nu=1}^g W_+ a_\nu^{-1}$ is a nonempty open subset of $Y_{\mathbf{R}}$. Since Y is dense in $Y_{\mathbf{R}}$, we can find an element h of $Y \cap \prod_{\nu=1}^g W_+ a_\nu^{-1}$. Put $x^\tau = hx^\sigma h^{-1}$ for $x \in L_{\mathbf{R}}$. Clearly τ gives an involution of L. For $x = (x_\nu)$ as above we have $x^\tau = (k_\nu \cdot {}^t\overline{x}_\nu k_\nu^{-1})$ with $k_\nu = h_\nu a_\nu$. Since $k_\nu \in W_+$ for every ν, we see that τ is a positive involution.

Theorem 2. *Let K be a quadratic extension of an algebraic number field F of finite degree, π the generator of $\mathrm{Gal}(K/F)$, and A a central simple algebra over K. Then*

A has an involution σ such that $\sigma = \pi$ on K if and only if $\mathrm{inv}_q(A) + \mathrm{inv}_{\pi(q)}(A) = 0$ for every (archimedean or nonarchimedean) prime q of K and $\mathrm{inv}_q(A) = 0$ for every prime q of K such that $\pi(q) = q$, where $\mathrm{inv}_q(A)$, an element of $\mathbf{Q}/\mathbf{Z}$, is the invariant associated with A and q in the standard way.

We first prove three lemmas, in which K is a field of any characteristic.

Lemma 1. *Let A and B be central simple algebras over K, and let $C = A \otimes_K B$. Suppose that C and A have involutions σ and τ respectively, which coincide on K. Then B has an involution that coincides with σ on K.*

PROOF. If $B = M_n(D)$ with a division algebra D over K, then C can be identified with $M_n(A) \otimes_K D$. The map $x \mapsto {}^t x^\tau$ is an involution of $M_n(A)$ that coincides with σ on K. Once we have an involution of D that coincides with σ on K, then we can define an involution of $B = M_n(D)$ with the same property. Thus it is sufficient to prove the case where B is a division algebra. Now $x \mapsto x^{\tau\sigma}$ gives a K-linear ring-injection of A into C, so that there exists an element r of $C^\times$ such that $x^{\tau\sigma} = r^{-1}xr$ for all $x \in A$. Substituting x^τ for x, we find that $x^\tau = rx^\sigma r^{-1}$ for all $x \in A$ and $A = rA^\sigma r^{-1}$. Taking the commutor, we obtain $B = rB^\sigma r^{-1}$. Therefore we can define an anti-automorphism α of B by $y^\alpha = ry^\sigma r^{-1}$ for $y \in B$. Put $s = r^\sigma r^{-1}$. Then, for $x \in A$ we have $x = (x^\tau)^\tau = r(rx^\sigma r^{-1})^\sigma r^{-1} = s^{-1}xs$, and hence $s \in B$. Similarly $(y^\alpha)^\alpha = s^{-1}ys$ for all $y \in B$. If $s = -1$, then α is an involution of B that coincides with σ on K. Suppose $s \neq -1$; put $t = r + r^\sigma$. Then $1 + s \in B^\times$ as B is a division algebra; thus $t = (1+s)r \in C^\times$. Put $y^\beta = ty^\sigma t^{-1}$ for $y \in B$. Then $y^\beta = (1+s)y^\alpha(1+s)^{-1} \in B$, and $(y^\beta)^\beta = y$. Thus β is an involution of B with the desired property.

Lemma 2. *If B is a central simple algebra over K, and $M_n(B)$ has an involution σ, then B has an involution that coincides with σ on K.*

PROOF. Take $A = M_n(K)$ and put $x^\tau = {}^t(x_{ij}^\sigma)$ for $x = (x_{ij}) \in A$ with $x_{ij} \in K$. Applying Lemma 1 to this situation, we obtain our lemma, since $M_n(B) = M_n(K) \otimes_K B$.

Lemma 3. *Let C be a quaternion algebra over K with an involution σ of the second kind and ι the canonical involution of C. Put $F = \{x \in K \,|\, x^\sigma = x\}$ and $D = \{x \in C \,|\, x^\sigma = x^\iota\}$. Then D is a quaternion algebra over F and $C = D \otimes_F K$.*

PROOF. We first prove that $\sigma\iota = \iota\sigma$. Let $x \in C, \notin K$. Then x^ι is the unique element of C such that both xx^ι and $x + x^\iota$ belong to K. Now $xx^{\sigma\iota\sigma} = (x^{\sigma\iota}x^\sigma)^\sigma \in K$ and $x + x^{\sigma\iota\sigma} = (x^\sigma + x^{\sigma\iota})^\sigma \in K$, and hence $x^{\sigma\iota\sigma} = x^\iota$, which proves that $x^{\sigma\iota} = x^{\iota\sigma}$ if $x \notin K$. Since the last equality is trivially true for $x \in K$, we obtain $\sigma\iota = \iota\sigma$. Once we know this, our lemma is a special case of [**67a**, Proposition 1.2].

PROOF OF THEOREM 2. For archimedean or nonarchimedean primes p of F and q of K denote by F_p and K_q their completions at p and q. First suppose that q is nonarchimedean and $\pi(q) \neq q$; put $r = \pi(q)$. Extend π to an isomorphism of K_q to K_r and denote it again by π. Suppose A has an involution σ that coincides with π on K. Then the map $x \otimes y \mapsto x^\sigma \otimes y^\pi$ for $x \in A$ and $y \in K_q$ gives an anti-isomorphism of A_q to A_r. Applying this anti-isomorphism to the canonical expression of A_q as a cyclic algebra, we easily see that $\mathrm{inv}_r(A) = -\mathrm{inv}_q(A)$. Next suppose $\pi(q) = q$; then π can be extended to an automorphism of K_q, and the map $x \otimes y \mapsto x^\sigma \otimes y^\pi$ for $x \in A$ and $y \in K_q$ gives an involution of A_q, so that $\mathrm{inv}_q(A) = -\mathrm{inv}_q(A)$ for the same reason. Suppose $\mathrm{inv}_q(A) \neq 0$; then $\mathrm{inv}_q(A)$ is represented by $1/2$, so that $A_q = M_n(B)$ with a division quaternion algebra B over K_q. By Lemma 2, B has an involution that coincides with π on K_q. By Lemma 3, $B = D \otimes_{F_p} K_q$ with a quaternion algebra D over a quadratic subfield F_p of K_q. Since D splits over every quadratic extension of F_p, B cannot be a division algebra, a contradiction. Thus $\mathrm{inv}_q(A) = 0$. Next suppose q is archimedean. If q is imaginary, then $\mathrm{inv}_q(A) = 0$. Suppose q is real and put $r = \pi(q)$; then r is real, and again we have an an anti-isomorphism of A_q to A_r, and clearly $\mathrm{inv}_r(A) = -\mathrm{inv}_q(A)$.

Conversely, to prove the existence of σ under the stated conditions on $\mathrm{inv}_q(A)$, denote by S the set of all archimedean and nonarchimedean primes q of K such that $\mathrm{inv}_q(A) \neq 0$; put $[A : K] = n^2$. By the so-called Grunwald's lemma proved by Wang (Ann. of Math. 51 (1950), 471–484, Corollary 2), there exists a cyclic extension W of F such that $[W : F] = n$ and $[W_\mathfrak{p} : F_p]\mathrm{inv}_q(A) = 0$ for every $q \in S$, where $p = q \cap F$ and $\mathfrak{p}$ is any prime of W dividing p. We may assume that W is linearly disjoint with K over F. (Indeed, take a prime q of K not contained in S such that $[K_q : F_p] = 2$; by Wang's result, we can take W so that $W_\mathfrak{p} = F_p$.) Put $Z = WK$. The condition on $[W_\mathfrak{p} : F_p]$ implies that A splits over Z, so that A can be given as a cyclic algebra (Z, μ, c), where μ is a generator of $\mathrm{Gal}(Z/K)$ and $c \in K^\times$. Take a cyclic algebra $B = (W, \mu, cc^\pi)$, where we identify $\mathrm{Gal}(Z/K)$ with $\mathrm{Gal}(W/F)$. Given a prime q of K, let $\mathfrak{q}$ resp. $\mathfrak{r}$ be a prime of Z dividing q resp. r, where $r = \pi(q)$; put $\mathfrak{p} = \mathfrak{q} \cap W$ and $p = q \cap F$. By a well-known principle, A_q (resp. A_r) belongs to the same Brauer class as $(Z_\mathfrak{q}, \mu^m, c)$ (resp. $(Z_\mathfrak{r}, \mu^m, c)$), where $m = n/[Z_\mathfrak{q} : K_q]$, and similarly $B_p = B \otimes_F F_p$ belongs to the same Brauer class as $(W_\mathfrak{p}, \mu^{m'}, cc^\pi)$ with $m' = n/[W_\mathfrak{p} : F_p]$. Suppose $q \neq r$; then π can be extended to an isomorphism of K_r to K_q, and then to that of $(Z_\mathfrak{r}, \mu^m, c)$ to $(Z_\mathfrak{q}, \mu^m, c^\pi)$. From this fact we can derive that $\mathrm{inv}_r(A)$ is the invariant of $(Z_\mathfrak{q}, \mu^m, c^\pi)$. Now $(Z_\mathfrak{q}, \mu^m, cc^\pi)$ belongs to the same Brauer class as $(Z_\mathfrak{q}, \mu^m, c) \otimes_{K_q} (Z_\mathfrak{q}, \mu^m, c^\pi)$, whose invariant is $\mathrm{inv}_q(A) + \mathrm{inv}_r(A)$, which is 0 by our assumption. Therefore $cc^\pi \in N_{Z_\mathfrak{q}/K_q}(Z_\mathfrak{q}^\times) = N_{W_\mathfrak{p}/F_p}(W_\mathfrak{p}^\times)$. If $\pi(q) = q$, then $\mathrm{inv}_q(A) = 0$, so that $c \in N_{Z_\mathfrak{q}/K_q}(Z_\mathfrak{q}^\times)$, and hence $cc^\pi \in N_{Z_\mathfrak{q}/F_p}(Z_\mathfrak{q}^\times) \subset N_{W_\mathfrak{p}/F_p}(W_\mathfrak{p}^\times)$. In either case B splits over F_p for every p, so that B is a matrix algebra over F. Thus $cc^\pi \in N_{W/F}(d)$ with $d \in W^\times$. Put $A = \sum_{i=0}^{n-1} Zu^i$ with an element u such that $uzu^{-1} = z^\mu$ for $z \in Z$ and $u^n = c$. Define an F-linear map $\sigma : A \to A$ by $w^\sigma = w$ for $w \in W, \sigma = \pi$ on K, and $u^\sigma = du^{-1}$. Then we can easily verify that σ gives an involution of A that coincides with π on K. This completes the proof of Theorem 2.

The above proof of Theorem 2, as well as the above three lemmas, was taken from my letter to David Mumford, dated Feb. 25, 1969, written in reply to his letter I received on Feb. 24, in which he asked whether I knew a necessary and sufficient condition in terms of $\mathrm{inv}_q(A)$ for the existence of an involution of the second kind.

63c. On the cohomology groups attached to certain vector valued differential forms on the product of the upper half planes

p. 434, line 3 from the bottom of the text: This expression for ω should be disregarded, since it is incorrect and unnecessary. The wrong formula is caused by the confusion of a form on G with that on the product of upper half planes. The correct argument is as follows. By Proposition 4.4, if we change the complex structure of H_{l_j} as described, then the form on G corresponding to ω is of the form $e_{0\cdots0} \otimes g^\circ$ with the function g° on G corresponding to an element g of $\mathcal{F}(\Gamma, m_1 + 2, \ldots, m_N + 2, \sigma)$. Substituting $\bar{z}_{l_j}$ for z_{l_j} in $g(z)$, we obtain the element f of line 2 from the bottom of the text.

63d. On modular correpondences for $Sp(n, \mathbf{Z})$ and their congruence relations

Theorem 4: Insert "and $\mathcal{K}_N = \mathcal{K}'\mathbf{C}$" after "over $\mathbf{Q}$".

63e. On the fields of definition for fields of automorphic functions

A 120-page volume of mimeographed notes titled "Proceedings of the 1963 Number Theory Conference" was made and sent to various institutions. The conference was supported by National Science Foundation. Apparently my present article caught the attention of some Russian mathematicians.

Theorem 3 is indeed true with no restriction on $[F : \mathbf{Q}]$, as proved in [**64c**, Theorem 1].

See also notes to [**64e**].

64a. Arithmetic of unitary groups

p. 369, line 3 from the bottom: For "F" read "K".

p. 393, line 9: For "F^*" read "K^*".

p. 408, line 16 from the bottom: Insert "L" before " in V".

64b. On the field of definition for a field of automorphic functions

p. 173, (4.4.2) : For "$1 \le \lambda \le m$" read "$1 \le \lambda \le g'$".

p. 177, line 9: The quantities v_{pkl} depend on the choice of the order of the quantities u_{vi}. Lemma 4.14 holds for any fixed choice. See also §2.3 of [**66c**].

p. 178, line 22: For "y_1" (two places) read "y_{11}".

p. 179, lines 16, 17: For "$\mu(T)$" read "$\mu(S)$".

64c. Class-fields and automorphic functions

p. 444, last line: For F read $F_{\mathfrak{p}}$.

p. 453, lines 13 from the bottom: Move "by means of 1.12" to the end of the next line.

p. 457, line 8 from the bottom: The first entry of the matrix is a_i.

p. 459, line 5 from the bottom: For "$\xi^{-1}(\mathfrak{o})$" read "$\tau\xi^{-1}(\mathfrak{o})$".

p. 460, line 7 from the bottom: After "K" insert " (defined with any choice of $\{\tau_1, \ldots, \tau_g\}$ as in §4.4)".

p. 462, line 18: To be precise, for every $\gamma \in \Gamma(\mathfrak{o})$, there exists an element γ' of $\Gamma(\mathfrak{o}')$ such that $\alpha^{-1}\gamma\alpha$ has the same action as γ' on $\mathfrak{H}'$. Therefore $a_i^{-1}h_j(\gamma)a_i = \pm h_i(\gamma')$, and the sign $\pm$ depends on i. Fix one i; since $-1 \in \Gamma(\mathfrak{o}) \cap \Gamma(\mathfrak{o}')$, we can determine $\gamma \mapsto \gamma'$ by $a_i^{-1}h_j(\gamma)a_i = h_i(\gamma')$ for that particular i. Then $\gamma \mapsto \gamma'$ can be extended to an automorphism ω of D over $\mathbf{Q}$. By (iii), $x^\omega = b^{-1}xb$ for every $x \in D$ with $b \in D^\times$. Then $a_i^{-1}h_j(x)a_i = h_i(b^{-1}xb)$. Taking x from F, we find that $i = j$. Thus $a_i^{-1}h_i(x)a_i = h_i(b^{-1}xb)$ for every $x \in D$. For another index k we have similarly $a_k^{-1}h_k(x)a_k = h_k(d^{-1}xd)$ for every $x \in D$ with $d \in D^\times$. Then $d^{-1}\gamma d = \pm b^{-1}\gamma b$ for every $\gamma \in \Gamma(\mathfrak{o})$, so that $d^{-1}\gamma^2 d = b^{-1}\gamma^2 b$ for every $\gamma \in \Gamma(\mathfrak{o})$. Since all such γ^2 generate D over F, we have $bd^{-1} \in F$, and $a_i^{-1}h_i(x)a_i = h_i(b^{-1}xb)$ for every i and every $x \in D$. After this we can repeat our argument.

64d. On purely transcendental fields of automorphic functions of several variables

p. 1, line 9 of the text: Insert "other than $\overline{S}$" after "$\mathbf{Q}$".

p. 9, line 5: For " anp " read " and ".

p. 12, line 15: For the second "W" read "M".

64e. The zeta function of an algebraic variety and automorphic functions

This is the text of my lectures at the Summer Institute on Algebraic Geometry held at Woods Hole. Mimeographed notes were distributed to the participants.

At a party in the fall of 1964 given by a member of the Institute for Advanced Study, Jean-Pierre Serre came to me and said that my results on the zeta function of modular curves were not of much value, as they didn't apply to an arbitrary elliptic curve over $\mathbf{Q}$. I responded by saying that I believed that such a curve should always be a quotient of the jacobian variety of a modular curve. Serre mentioned this to André Weil who was not at the party. A few days later, I met Weil at the Institute, and he asked me whether I really made that statement. I said: "Yes, I did; don't you think it plausible?" Then he said something to the effect that "Since both sets are countable, I don't see any reason against it, but at the same time I don't see any reason for the statement." (These are taken from my letter of September 16, 1986 to Freydoon Shahidi, with minor rhetorical changes. I record these conversations here in connection with the present article [**64e**], since Serre also participated in the Summer Institute, and what he said was his reaction to my lectures and the notes.)

After a few months, in response to a question of Weil of which the exact nature I don't remember, I told Weil the following: If f is an elliptic modular cusp form of weight 2 with **Q**-rational Fourier coefficients which is a Hecke eigenform, then we can prove: (i) $f(z)dz$, viewed as the differential form on the curve $\Gamma_0(N)\backslash H^*$, defines a **Q**-rational elliptic curve E, which is a quotient of the jacobian variety of that modular curve; (ii) the zeta function of E is the Mellin transform of f. These results were eventually generalized to the case of an arbitrary Hecke eigenform of weight 2; see [**71d**, Theorems 7.14, 7.15] and [**73d**]; see also [**71d**, Proposition 7.19]. In 1963–64 I had these results at least when the factor of the jacobian variety of a modular curve is one-dimensional, but published them with proofs only in [**71d**]. Weil acknowledged my result (i) on the curve E associated to f in his paper [1967a], but he did not mention (ii).

In any case, the conjecture means the correspondence $f \mapsto E$ is reversible. However, I always thought that the direction $f \mapsto E$ was far more fundamental than its opposite, a point few people seem to have understood. See also notes to [**89a**].

Though my idea on the modularity of **Q**-rational elliptic curves was acknowledged by Weil only in 1979 and 1986, there are other instances of my oral communications that were acknowledged sooner, if not consistently. Here are two examples.

At the Boulder conference of 1963 (see notes to [**63e**]) I met Brian Birch, who told me his ideas on the significance of $Z_E(1)$, where Z_E is the zeta function of a **Q**-rational elliptic curve E. As he was not familiar with the curves uniformized by modular functions, I explained to him the results of Eichler and myself. Furthermore, since he was naturally interested in the question of vanishing or nonvanishing of $Z_E(1)$, I told him the following three facts:

(1) If E is given as $\Gamma_0(N)\backslash H^*$, then $Z_E(1)$ is a constant times $\int_0^\infty f(iy)dy$ with the cusp form f in question, and therefore $Z_E(1)$ is practically a period. If $N = 11$ for example, the explicit form of f tells that $f(iy)$ is always positive, so that $Z_E(1) \neq 0$, and the same holds in some other cases.

(2) If g is the twist of f by a Dirichlet character, then g is a cusp form of higher level.

(3) In particular, if the character is quadratic, then the Mellin transform of g gives Z_D for the twist D of E. From the functional equation of Z_D we can obtain examples of D such that $Z_D(1) = 0$.

These, if easy, were all new to him. He acknowledged these in his papers in the early 1960's. I think I told him only about the cases in which $\Gamma_0(N)\backslash H$ is of genus 1. I think I knew that D was a factor of the Jacobian at the higher level, but probably I didn't tell him about it beyond (2) and (3). These facts were of course in my mind when I formed the idea of my conjecture. There was one more factor; see [**96b**].

Another concerns what I told Atiyah and Bott at the Summer Institute of 1964 where the present article [**64e**] was presented. In the introduction of "Notes on the Lefschetz fixed point theorem for elliptic complexes," Harvard University, Fall 1964, they wrote: "Our main formula also generalizes a result of Eichler on algebraic curves which was brought to our attention by Shimura during the recent conference at Woods Hole on algebraic geometry. In fact, this work resulted precisely from our attempt to prove Shimura's conjectures in this direction."

Besides, their article in Bull. Amer. Math. Soc. 72 (1966), 245–250 contains the following sentence: "The first of these (which means Theorem 2 in that article) was conjectured to us by Shimura and was proved by Eichler for dimension one." See also [01a] (not reproduced in these volumes). I had that conjectural formula for some time, and at the conference I told it to Atiyah and Bott. Immediately they and also many other participants became interested in it, and started to work on the question excitedly.

In fact, their statements about Eichler's work were not completely accurate for the following reason. They proved a theorem concerning a holomorphic map of a variety (or a vector bundle) into itself, but I was considering more generally the case of an algebraic correspondence. Besides, Eichler's result applies to an algebraic correspondence of a curve, and so "a result of Eichler" should have been "a special case of a result of Eichler." The paper of Eichler was not mentioned.

65a. On the field of definition for a field of automorphic functions: II

p. 128, line 13: Insert "This implies especially that f_σ is defined over k_1" before "Therefore".

p. 135, lines 12, 13 from the bottom: Delete "with the notation ... of $w(z)$".

p. 136, line 2: Insert " and $\delta \in \mathbf{Z}$" after "$\mathfrak{M}\alpha \subset \mathfrak{M}$".

p. 159, line 3 from the bottom: For "auch" read "such".

p. 161, line 18 from the bottom: The first "$\mathfrak{K}$" should read "$\mathfrak{K}_\mathfrak{c}$" .

p. 161, line 12 from the bottom: For "$p|\mathfrak{c}$" read "$\mathfrak{p}|\mathfrak{c}$".

p. 163, line 4 from the bottom: For "$\xi^{-1}(\mathfrak{o})$" read "$\mathfrak{r} \cdot \xi^{-1}(\mathfrak{o})$".

65b. On the zeta function of a fibre variety whose fibres are abelian varieties

p. 484, line 18: For "$\Gamma_1\alpha_j$" read "$\Gamma_1\alpha_i$".

p. 486, lines 2, 3 from the bottom: For "ξ" read "$\det(\xi)$".

p. 487, line 9 from the top and line 6 from the bottom: For "$\xi \equiv 1$" read "$\det(\xi) \equiv 1$".

p. 502, lines 6, 7: For "η'" read "ξ'".

p. 528, lines 5, 6: For "$\Pi_\mathfrak{p}^{nf}$" read "$\Pi_{\mathfrak{p}^\mathfrak{r}}^{nf}$".

p. 530, line 1: For "fnnction" read "function".

66b. Moduli and fibre systems of abelian varieties

p. 295, line 10: For "$\Phi(x)$" read "$\xi(x)$".

p. 302, line 11 from the bottom: Insert " nonsingular" at the beginning of the line.

p. 308, Proposition 2.5: For this result Krazer-Prym [8] is quoted. A simpler and self-contained proof is given in [**76a**, pp. 676–679], and also in [**98**, Section 28]; the latter is more detailed than the former. The transformation formulas in a more general case of $Sp(n, F)$ with an arbitrary totally real F are given in [**85a**, Propositions 1.1 and 1.2]. This accomplishes the desire expressed on lines 8–9 from the bottom of p. 308, as the proof employs Weil's metaplectic groups. The case of an arbitrary number field is treated in [**93b**] with the same idea.

p. 323, line 10: For "$\Omega^\sigma = \Omega^\tau$" read "$\Omega^\sigma$ is equivalent to Ω^τ".

p. 324, line 12: Insert "$f,$" after "$h,$".

p. 325, line 18 from the bottom: Insert " defined over k" at the end of the line.

p. 325, line 14 from the bottom: Insert " defined over k" at the end of the line.

p. 325, line 12 from the bottom: Insert " Notice that r and R are uniquely determined." at the end of the line.

p. 326, the first three lines: To be more precise, we first observe that r^σ and R^σ are uniquely determined for σ. If V is defined over k, then, as shown in §5.6, r^σ and R^σ are defined over kk^σ.

p. 328, line 7 from the bottom: For "k" read "k'".

p. 334, line 9: For "$\mathfrak{B}'$" read "$\mathfrak{v}'$".

66c. On the field of definition for a field of automorphic functions: III

p. 381, line 6: For "3.2" read "3.1".

66d. Moduli of abelian varieties and number theory

p. 331, line 17: For " Z_p" read "Z_p".

p. 331, line 6 from the bottom: The exponent $\frac{1}{2}$ should read $1/12$.

p. 332, line 14: For "b_v" read "b_v".

MIX
Papier aus verantwortungsvollen Quellen
Paper from responsible sources
FSC® C105338

If you have any concerns about our products,
you can contact us on
ProductSafety@springernature.com

In case Publisher is established outside the EU,
the EU authorized representative is:
Springer Nature Customer Service Center GmbH
Europaplatz 3, 69115 Heidelberg, Germany

Printed by Libri Plureos GmbH
in Hamburg, Germany